RECIPROCALS OF BASIC FUNCTIONS

18. $\displaystyle\int \frac{1}{1 \pm \sin u}\,du = \tan u \mp \sec u + C$

19. $\displaystyle\int \frac{1}{1 \pm \cos u}\,du = -\cot u \pm \csc u + C$

20. $\displaystyle\int \frac{1}{1 \pm \tan u}\,du = \tfrac{1}{2}(u \pm \ln|\cos u \pm \sin u|) + C$

21. $\displaystyle\int \frac{1}{\sin u \cos u}\,du = \ln|\tan u| + C$

22. $\displaystyle\int \frac{1}{1 \pm \cot u}\,du$ [illegible]

23. $\displaystyle\int \frac{1}{1 \pm \sec u}\,du$ [illegible]

24. $\displaystyle\int \frac{1}{1 \pm \csc u}\,du = u$ [illegible]

25. $\displaystyle\int \frac{1}{1 \pm e^u}\,du = u - \ln(1 \pm e^u) + C$

POWERS OF TRIGONOMETRIC FUNCTIONS

26. $\displaystyle\int \sin^2 u\,du = \tfrac{1}{2}u - \tfrac{1}{4}\sin 2u + C$

27. $\displaystyle\int \cos^2 u\,du = \tfrac{1}{2}u + \tfrac{1}{4}\sin 2u + C$

28. $\displaystyle\int \tan^2 u\,du = \tan u - u + C$

29. $\displaystyle\int \sin^n u\,du = -\frac{1}{n}\sin^{n-1} u \cos u + \frac{n-1}{n}\int \sin^{n-2} u\,du$

30. $\displaystyle\int \cos^n u\,du = \frac{1}{n}\cos^{n-1} u \sin u + \frac{n-1}{n}\int \cos^{n-2} u\,du$

31. $\displaystyle\int \tan^n u\,du = \frac{1}{n-1}\tan^{n-1} u - \int \tan^{n-2} u\,du$

32. $\displaystyle\int \cot^2 u\,du = -\cot u - u + C$

33. $\displaystyle\int \sec^2 u\,du = \tan u + C$

34. $\displaystyle\int \csc^2 u\,du = -\cot u + C$

35. $\displaystyle\int \cot^n u\,du = -\frac{1}{n-1}\cot^{n-1} u - \int \cot^{n-2} u\,du$

36. $\displaystyle\int \sec^n u\,du = \frac{1}{n-1}\sec^{n-2} u \tan u + \frac{n-2}{n-1}\int \sec^{n-2} u\,du$

37. $\displaystyle\int \csc^n u\,du = -\frac{1}{n-1}\csc^{n-2} u \cot u + \frac{n-2}{n-1}\int \csc^{n-2} u\,du$

PRODUCTS OF TRIGONOMETRIC FUNCTIONS

38. $\displaystyle\int \sin mu \sin nu\,du = -\frac{\sin(m+n)u}{2(m+n)} + \frac{\sin(m-n)u}{2(m-n)} + C$

39. $\displaystyle\int \cos mu \cos nu\,du = \frac{\sin(m+n)u}{2(m+n)} + \frac{\sin(m-n)u}{2(m-n)} + C$

40. $\displaystyle\int \sin mu \cos nu\,du = -\frac{\cos(m+n)u}{2(m+n)} - \frac{\cos(m-n)u}{2(m-n)} + C$

41. $\displaystyle\int \sin^m u \cos^n u\,du = -\frac{\sin^{m-1} u \cos^{n+1} u}{m+n} + \frac{m-1}{m+n}\int \sin^{m-2} u \cos^n u\,du$

$$= \frac{\sin^{m+1} u \cos^{n-1} u}{m+n} + \frac{n-1}{m+n}\int \sin^m u \cos^{n-2} u\,du$$

PRODUCTS OF TRIGONOMETRIC AND EXPONENTIAL FUNCTIONS

42. $\displaystyle\int e^{au}\sin bu\,du = \frac{e^{au}}{a^2+b^2}(a \sin bu - b \cos bu) + C$

43. $\displaystyle\int e^{au}\cos bu\,du = \frac{e^{au}}{a^2+b^2}(a \cos bu + b \sin bu) + C$

POWERS OF u MULTIPLYING OR DIVIDING BASIC FUNCTIONS

44. $\displaystyle\int u \sin u\,du = \sin u - u \cos u + C$

45. $\displaystyle\int u \cos u\,du = \cos u + u \sin u + C$

46. $\displaystyle\int u^2 \sin u\,du = 2u \sin u + (2 - u^2)\cos u + C$

47. $\displaystyle\int u^2 \cos u\,du = 2u \cos u + (u^2 - 2)\sin u + C$

48. $\displaystyle\int u^n \sin u\,du = -u^n \cos u + n\int u^{n-1}\cos u\,du$

49. $\displaystyle\int u^n \cos u\,du = u^n \sin u - n\int u^{n-1}\sin u\,du$

50. $\displaystyle\int u^n \ln u\,du = \frac{u^{n+1}}{(n+1)^2}[(n+1)\ln u - 1] + C$

51. $\displaystyle\int ue^u\,du = e^u(u-1) + C$

52. $\displaystyle\int u^n e^u\,du = u^n e^u - n\int u^{n-1}e^u\,du$

53. $\displaystyle\int u^n a^u\,du = \frac{u^n a^u}{\ln a} - \frac{n}{\ln a}\int u^{n-1}a^u\,du + C$

54. $\displaystyle\int \frac{e^u\,du}{u^n} = -\frac{e^u}{(n-1)u^{n-1}} + \frac{1}{n-1}\int \frac{e^u\,du}{u^{n-1}}$

55. $\displaystyle\int \frac{a^u\,du}{u^n} = -\frac{a^u}{(n-1)u^{n-1}} + \frac{\ln a}{n-1}\int \frac{a^u\,du}{u^{n-1}}$

56. $\displaystyle\int \frac{du}{u \ln u} = \ln|\ln u| + C$

POLYNOMIALS MULTIPLYING BASIC FUNCTIONS

57. $\displaystyle\int p(u)e^{au}\,du = \frac{1}{a}p(u)e^{au} - \frac{1}{a^2}p'(u)e^{au} + \frac{1}{a^3}p''(u)e^{au} - \cdots$ [signs alternate: $+ - + - \cdots$]

58. $\displaystyle\int p(u)\sin au\,du = -\frac{1}{a}p(u)\cos au + \frac{1}{a^2}p'(u)\sin au + \frac{1}{a^3}p''(u)\cos au - \cdots$ [signs alternate in pairs after first term: $+ + - - + + - - \cdots$]

59. $\displaystyle\int p(u)\cos au\,du = \frac{1}{a}p(u)\sin au + \frac{1}{a^2}p'(u)\cos au - \frac{1}{a^3}p''(u)\sin au - \cdots$ [signs alternate in pairs: $+ + - - + + - - \cdots$]

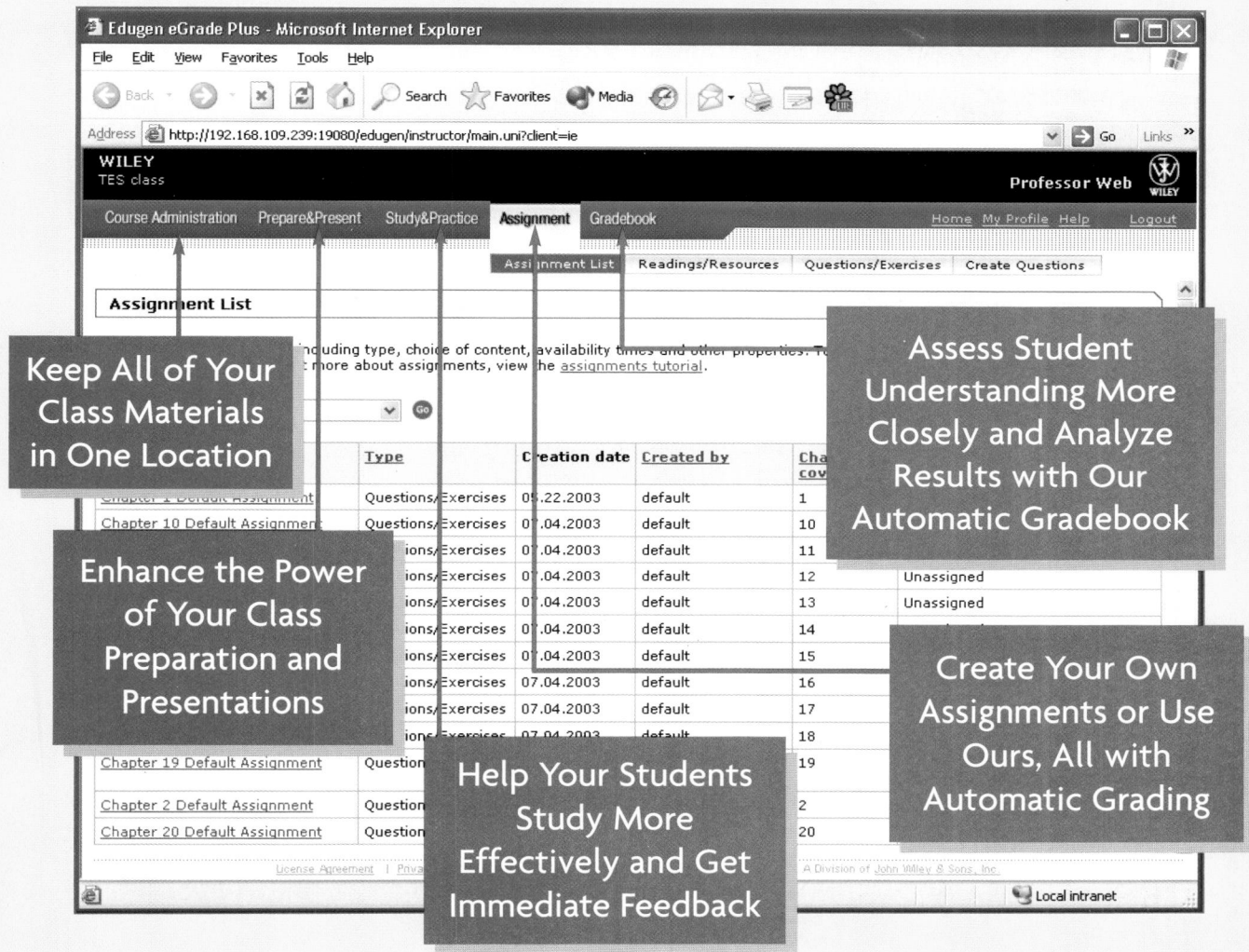

Students, eGrade Plus Allows You to:

Study More Effectively

Get Immediate Feedback When You Practice on Your Own

eGrade Plus problems link directly to relevant sections of the **electronic book content,** so that you can review the text while you study and complete homework online. Additional resources include **hyperlinks to the Student Study Guide, the Student Solutions Manual, calculus explorations,** and **Calculus Solutions, powered by JustAsk!**

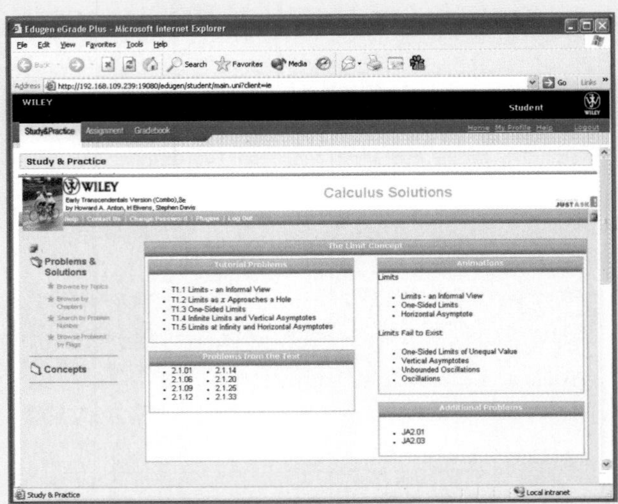

Complete Assignments / Get Help with Problem Solving

An **Assignment** area keeps all your assigned work in one location, making it easy for you to stay "on task." In addition, many homework problems contain a **link** to the relevant section of the **multimedia book,** providing you with a text explanation to help you conquer problem-solving obstacles as they arise. You will have access to a variety of resources for building your confidence and understanding.

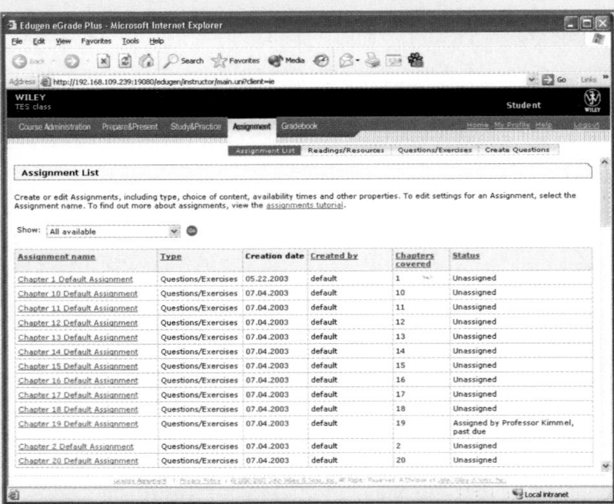

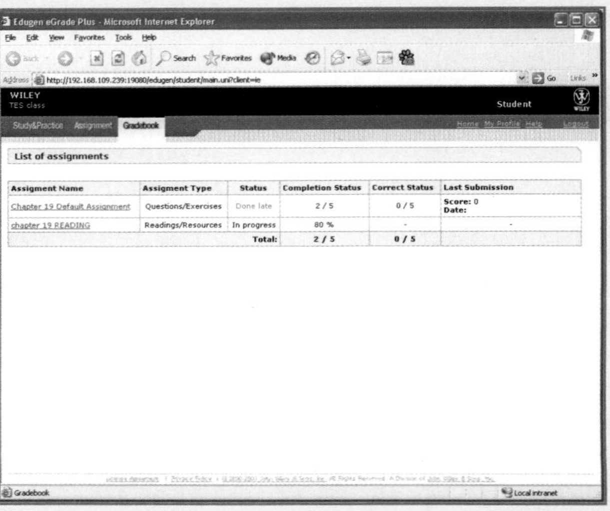

Keep Track of How You're Doing

A **Personal Gradebook** allows you to view your results from past assignments at any time.

Calculus

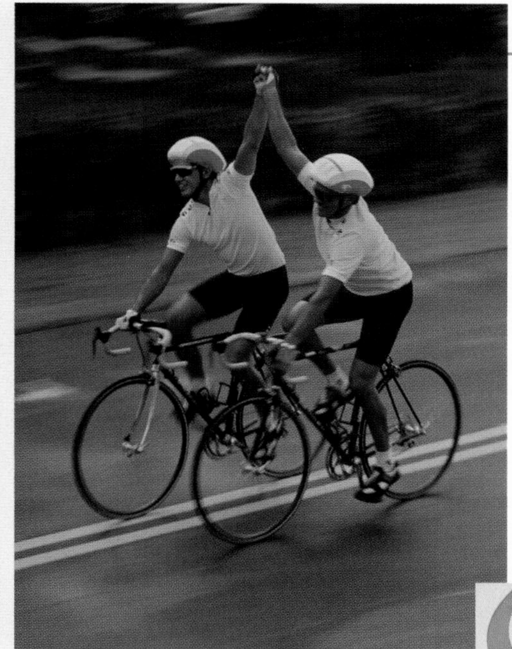

e i g h t h   e d i t i o n

Calculus

HOWARD ANTON

■ Drexel University

IRL BIVENS

■ Davidson College

STEPHEN DAVIS

■ Davidson College

WILEY

JOHN WILEY & SONS, INC.

Associate Publisher: Laurie Rosatone
Freelance Developmental Editor: Anne Scanlan-Rohrer
Senior Marketing Manager: Angela Battle
Associate Editor: Jennifer Battista
Editorial Assistants: Danielle Amico/Kelly Boyle
Senior Production Editor: Ken Santor
Senior Designer: Karin Kincheloe
Cover Design: David Levy
Cover Photo: © Arthur Tilley/Taxi Getty Images
Text Design: Nancy Field
Photo Editor: Hilary Newman/Ellinor Wagoner
Illustration Editor: Sigmund Malinowski
Illustration Studio: Techsetters, Inc.

This book was set in Times Roman by Techsetters, Inc., and printed and bound by Von Hoffmann Press. The cover was printed by Von Hoffmann Press.

This book is printed on acid-free paper. ♾

The paper in this book was manufactured by a mill whose forest management programs include sustained yield harvesting of its timberlands. Sustained yield harvesting principles ensure that the numbers of trees cut each year does not exceed the amount of new growth.

ISBN 0-471-48273-0

Printed in the United States of America

10 9 8 7 6 5 4 3 2 1

ABOUT HOWARD ANTON

Howard Anton obtained his B.A. from Lehigh University, his M.A. from the University of Illinois, and his Ph.D. from the Polytechnic University of Brooklyn, all in mathematics. In the early 1960s he worked for Burroughs Corporation and Avco Corporation at Cape Canaveral, Florida, where he was involved with the manned space program. In 1968 he joined the Mathematics Department at Drexel University, where he taught full time until 1983. Since that time he has been an adjunct professor at Drexel and has devoted the majority of his time to textbook writing and activities for mathematical associations. Dr. Anton was president of the EPADEL Section of the Mathematical Association of America (MAA), served on the board of Governors of that organization, and guided the creation of the Student Chapters of the MAA. He has published numerous research papers in functional analysis, approximation theory, and topology, as well as pedagogical papers. He is best known for his textbooks in mathematics, which are among the most widely used in the world. There are currently more than one hundred versions of his books, including translations into Spanish, Arabic, Portuguese, Italian, Indonesian, French, Japanese, Chinese, Hebrew, and German. For relaxation, Dr. Anton enjoys traveling and photography.

ABOUT IRL BIVENS

Irl C. Bivens, recipient of the George Polya Award and the Merten M. Hasse Prize for Expository Writing in Mathematics, received his A.B. from Pfeiffer College and his Ph.D. from the University of North Carolina at Chapel Hill, both in mathematics. Since 1982, he has taught at Davidson College, where he currently holds the position of professor of mathematics. A typical academic year sees him teaching courses in calculus, topology, and geometry. Dr. Bivens also enjoys mathematical history, and his annual History of Mathematics seminar is a perennial favorite with Davidson mathematics majors. He has published numerous articles on undergraduate mathematics, as well as research papers in his specialty, differential geometry. He is currently a member of the editorial board for the MAA Problem Book series and is a reviewer for *Mathematical Reviews*. When he is not pursuing mathematics, Professor Bivens enjoys juggling, swimming, walking, and spending time with his son Robert.

ABOUT STEPHEN DAVIS

Stephen L. Davis received his B.A. from Lindenwood College and his Ph.D. from Rutgers University in mathematics. Having previously taught at Rutgers University and Ohio State University, Dr. Davis came to Davidson College in 1981, where he is currently a professor of mathematics. He regularly teaches calculus, linear algebra, abstract algebra, and computer science. A sabbatical in 1995–1996 took him to Swarthmore College as a visiting associate professor. Professor Davis has published numerous articles on calculus reform and testing, as well as research papers on finite group theory, his specialty. Professor Davis has held several offices in the Southeastern section of the MAA, including chair and secretary-treasurer. He is currently a faculty consultant for the Educational Testing Service Advanced Placement Calculus Test, a board member of the North Carolina Association of Advanced Placement Mathematics Teachers, and is actively involved in nurturing mathematically talented high school students through leadership in the Charlotte Mathematics Club. He was formerly North Carolina state director for the MAA. For relaxation, he plays basketball, juggles, and travels. Professor Davis and his wife Elisabeth have three children, Laura, Anne, and James, all former calculus students.

To
My Wife Pat
My Children: Brian, David, and Lauren

In Memory of
My Mother Shirley
My Father Benjamin
My Esteemed Colleague Albert Herr
My Benefactor Stephen Girard (1750–1831)
—HA

To
My Son Robert
—IB

To
My Wife Elisabeth
My Children: Laura, Anne, and James
—SD

PREFACE

ABOUT THIS EDITION

A major focus of this edition was to ***increase student comprehension*** through judicious streamlining of the exposition; the creation of new problem types, particularly the Quick Check and Focus on Concepts exercises; and revision of many examples to add more steps and reformat them for clarity.

Multiple Versions For greater flexibility, there are two versions of this text—*late transcendentals* and *early transcendentals*. The late transcendentals version covers logarithmic, exponential, and inverse trigonometric functions *after* all of the basic material on differentiation and integration has been developed; in the early transcendentals version, logarithmic, exponential, and inverse trigonometric functions are discussed earlier. Both versions of this text are available in two volumes, a single variable volume and a multivariable volume.

Technology This edition provides many examples and exercises for instructors who want to use graphing calculators, computer algebra systems, or other programs. However, these are implemented in a way that allows the text to be used in courses where technology is used extensively, moderately, or not at all. To provide a sound foundation for the technology material, we have included a section entitled *Graphing Functions Using Calculators* and *Computer Algebra Systems* (Section 1.2). New ***Technology Mastery*** comments direct students to useful "just in time" applications of technology. Exercises that require technology are marked with icons for easy identification.

Internet This text is supplemented by a Web site:

www.wiley.com/college/anton

Streamlined Exposition Every page, every explanation, and every example in the text was examined critically and the exposition was streamlined, where needed, to get students right to the heart of concepts. Many examples have been revised to make them clearer and more inviting. In addition, Appendices A, B, C, and D from the seventh edition were moved to the companion Web site. *Expanding the Calculus Horizon* modules are now posted on the text's Web site, and students are still directed to the modules by a preview paragraph and Web link at the end of appropriate chapters in the text.

NEW FEATURES IN THE EIGHTH EDITION

New and Updated Exercises

- New ***Quick Check Exercises*** starting each section's exercise set contain a basic set of 4–8 exercises designed to cover key skills and concepts in the section. Students can use these as a concise way of testing their knowledge of each section. Answers to the Quick Check exercises appear at the end of each section.
- New ***Focus on Concepts*** throughout each exercise set highlight exercises of a conceptual nature.
- ***Review Exercises*** replace the Supplementary Exercises at the end of each chapter. A selection of these exercises can be used to review important concepts within the chapter or to construct a chapter test. In addition, exercise sets were revised and expanded to include more variety and better pairings between odd and even exercises.

Margin Notes General margin comments call attention to ideas in the text or provide further insights. These general comments and the ***Technology Mastery*** comments replace the *For the Reader* comments from previous editions.

Derivative Notation The notation for the definition of the derivative has been brought into alignment with that used in standard calculus texts.

Analysis of Functions The traditional "curve sketching" is part of the Analysis of Functions (Sections 4.1–4.3). Section 4.3 has been revised to strike a better balance between the methods of calculus and the use of technology in the graphing of functions. The section on rectilinear motion has been moved to the end of the chapter to facilitate the transition from the discussion of graphing to the topic of maxima and minima for functions. This allows applied max/min problems to be covered earlier in the chapter.

Techniques of Differentiation The Techniques of Differentiation (Section 3.3) now concentrates on basic rules: derivatives of a constant and powers of x, the constant multiple rule, and the sum/difference rules. The product and quotient rules have been moved to a section of their own (3.4).

OTHER FEATURES

Flexibility This edition has a built-in flexibility that is designed to serve a broad spectrum of calculus philosophies—from traditional to reform. Technology can be emphasized or not, and the order of many topics can be permuted freely to accommodate the instructor's specific needs.

Trigonometry Review Deficiencies in trigonometry plague many students, so we have included a substantial trigonometry review in Appendix A.

Historical Notes The biographies and historical notes have been a hallmark of this text from its first edition and have been maintained. All of the biographical materials have been distilled from standard sources with the goal of capturing the personalities of the great mathematicians and bringing them to life for the students.

Graded Exercise Sets Section Exercise Sets are "graded" to begin with routine problems and progress gradually toward problems of greater difficulty.

Rigor The challenge of writing a good calculus book is to strike the right balance between rigor and clarity. Our goal is to present precise mathematics to the fullest extent possible in an introductory treatment. Where clarity and rigor conflict, we choose clarity; however, we believe it to be important that the student understand the difference between a careful proof and an informal argument, so we have tried to make it clear to the reader when the arguments being presented are informal or motivational. Theory involving ϵ-δ arguments appear in separate sections so that they can be covered or not, as preferred by the instructor.

Mathematical Level This text is written at a mathematical level that will prepare students for a wide variety of careers that require a sound mathematics background, including engineering, the various sciences, and business.

Computer Graphics This edition makes extensive use of modern computer graphics to clarify concepts and to develop the student's ability to visualize mathematical objects, particularly those in 3-space. For those students who are working with graphing technology, there are many exercises that are designed to develop the student's ability to generate and analyze mathematical curves and surfaces.

Applicability of Calculus One of the primary goals of this edition is to link calculus to the real world and the student's own experience. This theme is carried through in the examples, exercises, and modules. Applications given in the exercises have been chosen to provide the student a sense of how calculus can be applied.

Early Differential Equations Basic ideas about differential equations, initial-value problems, direction fields, and integral curves are introduced concurrently with integration and then revisited in more detail in Chapter 9.

Early Parametric Option In keeping with the current trend of discussing parametric equations early, parametric curves are introduced in Section 1.7 and then revisited in Chapter 11, where calculus-related matters are discussed. Instructors who prefer the traditional late discussion of parametric equations will have no problem deferring the material in Section 1.7 until the discussion of analytic geometry in Chapter 11.

Principles of Integral Evaluation The traditional Techniques of Integration is entitled "Principles of Integral Evaluation" to reflect its more modern approach to the material. The chapter emphasizes general methods and the role of technology rather than specific tricks for evaluating complicated or obscure integrals.

Appendix on Polynomial Equations Because many calculus students are weak in solving polynomial equations, we have included an appendix (Appendix B) that reviews the Factor Theorem, the Remainder Theorem, and procedures for finding rational roots.

Rule of Four The "rule of four" refers to presenting concepts from the verbal, algebraic, visual, and numerical points of view. In keeping with current pedagogical philosophy, we used this approach whenever appropriate.

SUPPLEMENTS

SUPPLEMENTS FOR THE STUDENT

Student Solutions Manual, Neil Wigley
The Student Solutions Manual provides students with detailed solutions to odd-numbered exercises from the text.
Single Variable ISBN: 0-471-67210-6
Multivariable ISBN: 0-471-67212-2

Student Study Guide, Brian Camp
The Student Study Guide contains key ideas and study suggestions, as well as sample tests for each section and chapter of the text.
Single Variable ISBN: 0-471-67211-4
Multivariable ISBN: 0-471-67213-0

SUPPLEMENTS FOR THE INSTRUCTOR

SUPPLEMENTS FOR THE INSTRUCTOR CAN BE OBTAINED BY SENDING A REQUEST ON YOUR INSTITUTIONAL LETTERHEAD TO MATHEMATICS MARKETING MANAGER, JOHN WILEY & SONS, INC., 111 RIVER STREET, HOBOKEN, NJ 07030, OR BY CONTACTING YOUR LOCAL WILEY REPRESENTATIVE.

Instructor's Manual, Irl Bivens and Stephen Davis
The Instructor's Manual provides suggested time allocations and teaching plans for each section in the text. Most of the teaching plans contain a bulleted list of key points to emphasize. The discussion of each section concludes with a sample homework assignment.
ISBN: 0-471-67207-6

Instructor's Solutions Manual, Neil Wigley
The Instructor's Solutions Manual contains detailed solutions to all exercises in the text.
Single Variable ISBN: 0-471-67208-4
Multivariable ISBN: 0-471-72429-7

Test Bank, Henry Smith
The Test Bank contains a variety of questions and answers for every section in the text.
ISBN: 0-471-67209-2

FOR THE STUDENT AND THE INSTRUCTOR

Web Horizon Modules
Selected chapters end with references to Web modules called *Expanding the Calculus Horizon*. As the name implies, these modules are intended to take the student a step beyond the traditional calculus text. The modules, all of which are optional, can be assigned either as individual or group projects and can be used by instructors to tailor the calculus course to meet their specific needs and teaching philosophies. For example, there are modules that touch on iteration and dynamical systems, equations of motion, application of integration to railroad design, collision of comets with Earth, and hurricane modeling. These can be found on the Web site,

www.wiley.com/college/anton

OTHER RESOURCES

eGrade Plus is a powerful online tool that provides instructors and students with an integrated suite of teaching and learning resources in one easy-to-use Web site. eGrade Plus is organized around the essential activities you and your students perform in class:

For Instructors

- **Prepare & Present:** Create class presentations using a wealth of Wiley-provided resources—such as an online version of the textbook, PowerPoint slides, and interactive simulations—making your preparation time more efficient. You may easily adapt, customize, and add to this content to meet the needs of your course.
- **Create Assignments:** Automate the assigning and grading of homework or quizzes by using Wiley-provided question banks, or by writing your own. Student results will be automatically graded and recorded in your gradebook. eGrade Plus can link homework problems to the relevant section of the online text, providing students with context-sensitive help.
- **Track Student Progress:** Keep track of your students' progress via an instructor's gradebook, which allows you to analyze individual and overall class results to determine their progress and level of understanding.
- **Administer Your Course:** eGrade Plus can easily be integrated with another course management system, gradebook, or other resources you are using in your class, providing you with the flexibility to build your course, your way.

For Students
Wiley's eGrade Plus provides immediate feedback on student assignments and a wealth of support materials. This powerful study tool will help your students develop their conceptual understanding of the class material and increase their ability to solve problems.

- **A "Study and Practice"** area links directly to text content, allowing students to review the text while they study and complete homework assignments. This package includes the following:

    - **Calculus Solutions powered by JustAsk!(TM)** include problems that correlate to chapter materials, interactive tutorials, detailed solutions and answers, and solution guidelines.

  - **Calculus Explorations** comprise a series of interactive Java applets that allow students to explore the geometric significance of many major concepts of Calculus 1.
  - **Algebra & Trigonometry Refresher** is a self-paced, guided review of key algebra and trigonometry topics that are essential for mastering calculus.
  - **Student Solutions Manual** contains detailed solutions to selected problems in the text.
  - **Student Study Guide** offers study hints and tips, key ideas and concepts, and sample quizzes and tests.
  - **Calculus WebQuizzes** provide opportunity for student self-assessment.

- **An "Assignment"** area keeps all the work you want your students to complete in one location, making it easy for them to stay "on task." Students will have access to a variety of interactive problem-solving tools, as well as other resources for building their confidence and understanding. In addition, many homework problems contain a link to the relevant section of the multimedia book, providing students with context-sensitive help that allows them to conquer problem-solving obstacles as they arise.
- **A Personal Gradebook** for each student will allow students to view their results from past assignments at any time.

Please visit **www.wiley.com/college/anton**, or view our online demo at **www.wiley.com/college/egradeplus**. Here you will find additional information about the features and benefits of eGrade Plus, how to request a "test drive" of eGrade Plus for this title, and how to adopt it for class use.

The Faculty Resource Network The *Faculty Resource Network* is a peer-to-peer network of academic faculty dedicated to the effective use of technology in the classroom. This group can help you apply innovative classroom techniques, implement specific software packages, and tailor the technology experience to the specific needs of each individual class. Ask your Wiley representative for more details.

ACKNOWLEDGMENTS

It has been our good fortune to have the advice and guidance of many talented people whose knowledge and skills have enhanced this book in many ways. For their valuable help we thank:

REVIEWERS AND CONTRIBUTORS TO THE EIGHTH EDITION

Gregory Adams, *Bucknell University*
Bill Allen, *Reedley College–Clovis Center*
Jerry Allison, *Black Hawk College*
Stella Ashford, *Southern University and A&M College*
Christopher Barker, *San Joaquin Delta College*
David Bradley, *University of Maine*
Paul Britt, *Louisiana State University*
Andrew Bulleri, *Howard Community College*
Miriam Castroconde, *Irvine Valley College*
Neena Chopra, *The Pennsylvania State University*
Gaemus Collins, *University of California, San Diego*
Danielle Cross, *Northern Essex Community College*
Stephan DeLong, *Tidewater Community College–Virginia Beach Campus*
Ryness Doherty, *Community College of Denver*
T. J. Duda, *Columbus State Community College*
Peter Embalabala, *Lincoln Land Community College*
Laurene Fausett, *Georgia Southern University*
Richard Hall, *Cochise College*
Noal Harbertson, *California State University, Fresno*
Donald Hartig, *California Polytechnic State University*
Konrad Heuvers, *Michigan Technological University*
John Johnson, *George Fox University*
Grant Karamyan, *University of California, Los Angeles*
Cecilia Knoll, *Florida Institute of Technology*
Carole King Krueger, *The University of Texas at Arlington*
Richard Lane, *University of Montana*
James Martin, *Wake Technical Community College*
Vania Mascioni, *Ball State University*
Tamra Mason, *Albuquerque TVI Community College*
Roy Mathias, *The College of William & Mary*
John Michaels, *SUNY Brockport*
Darrell Minor, *Columbus State Community College*
Darren Narayan, *Rochester Institute of Technology*
Efton Park, *Texas Christian University*
Joanne Peeples, *El Paso Community College*
Richard Ponticelli, *North Shore Community College*
Holly Puterbaugh, *University of Vermont*
Robert Rock, *Daniel Webster College*
John Saccoman, *Seton Hall University*
Paul Seeburger, *Monroe Community College*
Charlotte Simmons, *University of Central Oklahoma*
Bryan Stewart, *Tarrant County College–Southeast Campus*
Bradley Stoll, *The Harker School*
Eleanor Storey, *Front Range Community College*
Richard Swanson, *Montana State University*
Helen Tyler, *Manhattan College*
Paramanathan Varatharajah, *North Carolina A&T State University*
David Voss, *Western Illinois University*
Jim Voss, *Front Range Community College*
Richard Watkins, *Tidewater Community College*
Jane West, *Trident Technical College*
Janine Wittwer, *Williams College*
Richard Zang, *University of New Hampshire*
Diane Zych, *Erie Community College–North Campus*

REVIEWERS AND CONTRIBUTORS TO EARLIER EDITIONS

Edith Ainsworth, *University of Alabama*
Loren Argabright, *Drexel University*
David Armacost, *Amherst College*
Dan Arndt, *University of Texas at Dallas*
Ajay Arora, *McMaster University*
Mary Lane Baggett, *University of Mississippi*
John Bailey, *Clark State Community College*
Robert C. Banash, *St. Ambrose University*
William H. Barker, *Bowdoin College*
George R. Barnes, *University of Louisville*
Scott E. Barnett, *Wayne State University*
Larry Bates, *University of Calgary*
John P. Beckwith, *Michigan Technological University*
Joan E. Bell, *Northeastern Oklahoma State University*
Harry N. Bixler, *Baruch College, CUNY*
Kbenesh Blayneh, *Florida A&M University*
Marilyn Blockus, *San Jose State University*
Ray Boersma, *Front Range Community College*
Barbara Bohannon, *Hofstra University*
David Bolen, *Virginia Military Institute*
Daniel Bonar, *Denison University*
George W. Booth, *Brooklyn College*

Phyllis Boutilier, *Michigan Technological University*
Linda Bridge, *Long Beach City College*
Mark Bridger, *Northeastern University*
Judith Broadwin, *Jericho High School*
John Brothers, *Indiana University*
Stephen L. Brown, *Olivet Nazarene University*
Virginia Buchanan, *Hiram College*
Robert C. Bucker, *Western Kentucky University*
Robert Bumcrot, *Hofstra University*
Christopher Butler, *Case Western Reserve University*
Carlos E. Caballero, *Winthrop University*
Cheryl Cantwell, *Seminole Community College*
James Caristi, *Valparaiso University*
Judith Carter, *North Shore Community College*
Stan R. Chadick, *Northwestern State University*
Hongwei Chen, *Christopher Newport University*
Chris Christensen, *Northern Kentucky University*
Robert D. Cismowski, *San Bernardino Valley College*
Patricia Clark, *Rochester Institute of Technology*
Hannah Clavner, *Drexel University*
Ted Clinkenbeard, *Des Moines Area Community College*
David Clydesdale, *Sauk Valley Community College*
David Cohen, *University of California, Los Angeles*
Michael Cohen, *Hofstra University*
Pasquale Condo, *University of Lowell*
Robert Conley, *Precision Visuals*
Mary Ann Connors, *U.S. Military Academy at West Point*
Cecil J. Coone, *State Technical Institute at Memphis*
Norman Cornish, *University of Detroit*
Fielden Cox, *Centennial College*
Terrance Cremeans, *Oakland Community College*
Gary Crown, *Wichita State University*
Lawrence Cusick, *California State University–Fresno*
Michael Dagg, *Numerical Solutions, Inc.*
Art Davis, *San Jose State University*
A. L. Deal, *Virginia Military Institute*
Charles Denlinger, *Millersville University*
William H. Dent, *Maryville College*
Blaise DeSesa, *Allentown College of St. Francis de Sales*
Blaise DeSesa, *Drexel University*
Debbie A. Desrochers, *Napa Valley College*
Dennis DeTurck, *University of Pennsylvania*
Jacqueline Dewar, *Loyola Marymount University*
Preston Dinkins, *Southern University*
Gloria S. Dion, *Educational Testing Service*
Irving Drooyan, *Los Angeles Pierce College*
Tom Drouet, *East Los Angeles College*
Clyde Dubbs, *New Mexico Institute of Mining and Technology*
Della Duncan, *California State University–Fresno*
Ken Dunn, *Dalhousie University*
Sheldon Dyck, *Waterloo Maple Software*
Hugh B. Easler, *College of William and Mary*
Scott Eckert, *Cuyamaca College*
Joseph M. Egar, *Cleveland State University*
Judith Elkins, *Sweet Briar College*
Brett Elliott, *Southeastern Oklahoma State University*
William D. Emerson, *Metropolitan State College*
Garret J. Etgen, *University of Houston*
Benny Evans, *Oklahoma State University*
Philip Farmer, *Diablo Valley College*
Victor Feser, *University of Maryland*
Iris Brann Fetta, *Clemson University*
James H. Fife, *Educational Testing Service*
Sally E. Fischbeck, *Rochester Institute of Technology*
Dorothy M. Fitzgerald, *Golden West College*
Barbara Flajnik, *Virginia Military Institute*
Daniel Flath, *University of South Alabama*
Ernesto Franco, *California State University–Fresno*
Nicholas E. Frangos, *Hofstra University*
Katherine Franklin, *Los Angeles Pierce College*
Marc Frantz, *Indiana University–Purdue University at Indianapolis*
Michael Frantz, *University of La Verne*
Susan L. Friedman, *Bernard M. Baruch College, CUNY*
William R. Fuller, *Purdue University*
Beverly Fusfield
Daniel B. Gallup, *Pasadena City College*
Bradley E. Garner, *Boise State University*
Carrie Garner
Susan Gerstein
Mahmood Ghamsary, *Long Beach City College*
Rob Gilchrist, *U.S. Air Force Academy*
G. S. Gill, *Brigham Young University*
Michael Gilpin, *Michigan Technological University*
Kaplana Godbole, *Michigan Technological Institute*
S. B. Gokhale, *Western Illinois University*
Morton Goldberg, *Broome Community College*
Mardechai Goodman, *Rosary College*
Sid Graham, *Michigan Technological University*
Bob Grant, *Mesa Community College*
Raymond Greenwell, *Hofstra University*
Dixie Griffin, Jr., *Louisiana Tech University*
Gary Grimes, *Mt. Hood Community College*
David Gross, *University of Connecticut*
Jane Grossman, *University of Lowell*
Michael Grossman, *University of Lowell*
Dennis Hadah, *Saddleback Community College*
Diane Hagglund, *Waterloo Maple Software*
Douglas W. Hall, *Michigan State University*
Nancy A. Harrington, *University of Lowell*
Kent Harris, *Western Illinois University*
Karl Havlak, *Angelo State University*
J. Derrick Head, *University of Minnesota–Morris*
Jim Hefferson, *St. Michael College*
Albert Herr, *Drexel University*
Peter Herron, *Suffolk County Community College*
Warland R. Hersey, *North Shore Community College*
Konrad J. Heuvers, *Michigan Technological University*
Dean Hickerson
Robert Higgins, *Quantics Corporation*
Rebecca Hill, *Rochester Institute of Technology*
Tommie Ann Hill-Natter, *Prairie View A&M University*
Holly Hirst, *Appalachian State University*
Edwin Hoefer, *Rochester Institute of Technology*
Louis F. Hoelzle, *Bucks County Community College*
Robert Homolka, *Kansas State University–Salina*
Henry Horton, *University of West Florida*
Joe Howe, *St. Charles County Community College*
Shirley Huffman, *Southwest Missouri State University*
Hugh E. Huntley, *University of Michigan*
Fatenah Issa, *Loyola University of Chicago*
Gary S. Itzkowitz, *Rowan University*
Emmett Johnson, *Grambling State University*
Jerry Johnson, *University of Nevada–Reno*
John M. Johnson, *George Fox College*
Wells R. Johnson, *Bowdoin College*
Kenneth Kalmanson, *Montclair State University*
Herbert Kasube, *Bradley University*
Phil Kavanagh, *Mesa State College*
David Keller, *Kirkwood Community College*
Maureen Kelley, *Northern Essex Community College*
Dan Kemp, *South Dakota State University*
Harvey B. Keynes, *University of Minnesota*
Lynn Kiaer, *Rose-Hulman Institute of Technology*
Vesna Kilibarda, *Indiana University Northwest*
Cecilia Knoll, *Florida Institute of Technology*
Holly A. Kresch, *Diablo Valley College*
Richard Krikorian, *Westchester Community College*
John Kubicek, *Southwest Missouri State University*
Paul Kumpel, *SUNY, Stony Brook*
Theodore Lai, *Hudson County Community College*
Fat C. Lam, *Gallaudet University*
Leo Lampone, *Quantics Corporation*
James F. Lanahan, *University of Detroit–Mercy*
Bruce Landman, *University of North Carolina at Greensboro*
Jeuel LaTorre, *Clemson University*
Kuen Hung Lee, *Los Angeles Trade–Technology College*
Marshall J. Leitman, *Case Western Reserve University*
Benjamin Levy, *Lexington H.S., Lexington, Mass.*
Darryl A. Linde, *Northeastern Oklahoma State University*
Phil Locke, *University of Maine, Orono*
Leland E. Long, *Muscatine Community College*
John Lucas, *University of Wisconsin–Oshkosh*
Stanley M. Lukawecki, *Clemson University*
Phoebe Lutz, *Delta College*
Nicholas Macri, *Temple University*
Michael Magill, *Purdue University*

Ernest Manfred, *U.S. Coast Guard Academy*
Melvin J. Maron, *University of Louisville*
Mauricio Marroquin, *Los Angeles Valley College*
Thomas W. Mason, *Florida A&M University*
Majid Masso, *Brookdale Community College*
Larry Matthews, *Concordia College*
Thomas McElligott, *University of Lowell*
Phillip McGill, *Illinois Central College*
Judith McKinney, *California State Polytechnic University, Pomona*
Joseph Meier, *Millersville University*
Robert Meitz, *Arizona State University*
Laurie Haskell Messina, *University of Oklahoma*
Aileen Michaels, *Hofstra University*
Janet S. Milton, *Radford University*
Robert Mitchell, *Rowan College of New Jersey*
Marilyn Molloy, *Our Lady of the Lake University*
Ron Moore, *Ryerson Polytechnical Institute*
Barbara Moses, *Bowling Green State University*
Eric Murphy, *U.S. Air Force Academy*
David Nash, *VP Research, Autofacts, Inc.*
Doug Nelson, *Central Oregon Community College*
Lawrence J. Newberry, *Glendale College*
Kylene Norman, *Clark State Community College*
Roxie Novak, *Radford University*
Richard Nowakowski, *Dalhousie University*
Stanley Ocken, *City College–CUNY*
Ralph Okojie, *Elizabeth City State University*
Ann Ostberg
Judith Palagallo, *The University of Akron*
Donald Passman, *University of Wisconsin*
David Patterson, *West Texas A&M*
Walter M. Patterson, *Lander University*
Steven E. Pav, *Alfred University*
Edward Peifer, *Ulster County Community College*
Gary L. Peterson, *James Madison University*
Lefkios Petevis, *Kirkwood Community College*
Robert Phillips, *University of South Carolina at Aiken*
Mark A. Pinsky, *Northeastern University*
Catherine H. Pirri, *Northern Essex Community College*
Thomas W. Polaski, *Winthrop University*
Father Bernard Portz, *Creighton University*
Irwin Pressman, *Carleton University*
Douglas Quinney, *University of Keele*
David Randall, *Oakland Community College*
B. David Redman, Jr., *Delta College*
Irmgard Redman, *Delta College*
Richard Remzowski, *Broome Community College*
Guanshen Ren, *College of Saint Scholastica*
William H. Richardson, *Wichita State University*
John Rickert, *Rose-Hulman Institute of Technology*
David Robbins, *Trinity College*
Lila F. Roberts, *Georgia Southern University*
David Rollins, *University of Central Florida*
Naomi Rose, *Mercer County Community College*
Sharon Ross, *DeKalb College*
David Ryeburn, *Simon Fraser University*
David Sandell, *U.S. Coast Guard Academy*
Avinash Sathaye, *University of Kentucky*
Ned W. Schillow, *Lehigh County Community College*
Dennis Schneider, *Knox College*
George W. Schultz, *St. Petersburg Junior College*
Dan Seth, *Morehead State University*
Richard B. Shad, *Florida Community College–Jacksonville*
George Shapiro, *Brooklyn College*
Parashu R. Sharma, *Grambling State University*
Michael D. Shaw, *Florida Institute of Technology*
Donald R. Sherbert, *University of Illinois*
Howard Sherwood, *University of Central Florida*
Mary Margaret Shoaf-Grubbs, *College of New Rochelle*
Bhagat Singh, *University of Wisconsin Centers*
Ann Sitomer, *Portland Community College*
Martha Sklar, *Los Angeles City College*
Henry Smith, *Southeastern Louisiana University*
Jeanne Smith, *Saddleback Community College*
John L. Smith, *Rancho Santiago Community College*
Wolfe Snow, *Brooklyn College*
Ian Spatz, *Brooklyn College*
Jean Springer, *Mount Royal College*
Rajalakshmi Sriram, *Okaloosa-Walton Community College*
Norton Starr, *Amherst College*
Mark Stevenson, *Oakland Community College*
Gary S. Stoudt, *University of Indiana of Pennsylvania*
John A. Suvak, *Memorial University of Newfoundland*
P. Narayana Swamy, *Southern Illinois University*
Richard B. Thompson, *The University of Arizona*
Skip Thompson, *Radford University*
Josef S. Torok, *Rochester Institute of Technology*
William F. Trench, *Trinity University*
Walter W. Turner, *Western Michigan University*
Thomas Vanden Eynden, *Thomas More College*
Paul Vesce, *University of Missouri–Kansas City*
Richard C. Vile, *Eastern Michigan University*
David Voss, *Western Illinois University*
Ronald Wagoner, *California State University–Fresno*
Shirley Wakin, *University of New Haven*
James E. Ward, *Bowdoin College*
James Warner, *Precision Visuals*
Peter Waterman, *Northern Illinois University*
Evelyn Weinstock, *Glassboro State College*
Bruce R. Wenner, *University of Missouri–Kansas City*
Candice A. Weston, *University of Lowell*
Bruce F. White, *Lander University*
Neil Wigley, *University of Windsor*
Ted Wilcox, *Rochester Institute of Technology*
Gary L. Wood, *Azusa Pacific University*
Yihren Wu, *Hofstra University*
Richard Yuskaitis, *Precision Visuals*
Michael Zeidler, *Milwaukee Area Technical College*
Michael L. Zwilling, *Mount Union College*

The following people read the eighth edition at various stages for mathematical and pedagogical accuracy and/or assisted with the critically important job of preparing answers to exercises:

Elka Block, *Twin Prime Editorial*
Dean Hickerson, *University of California, Davis*
Thomas Polaski, *Winthrop University*
Frank Purcell, *Twin Prime Editorial*
David Ryeburn, *Simon Fraser University*

CONTENTS

chapter one

FUNCTIONS

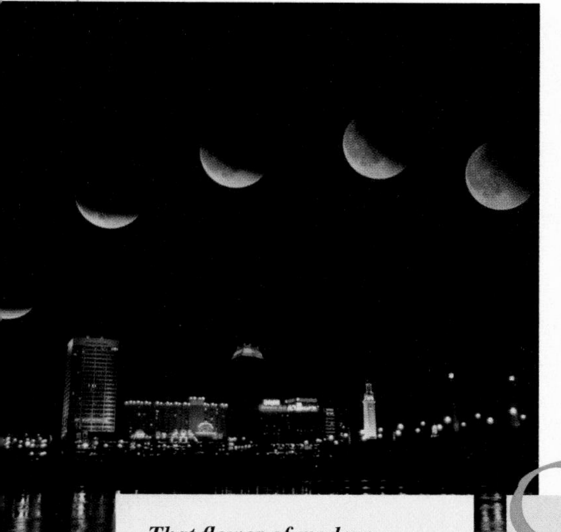

That flower of modern mathematical thought—the notion of a function.
—Thomas J. McCormack
Science Essayist and Translator

One of the important themes in calculus is the analysis of relationships between physical or mathematical quantities. Such relationships can be described in terms of graphs, formulas, numerical data, or words. In this chapter we will develop the concept of a "function," which is the basic idea that underlies almost all mathematical and physical relationships, regardless of the form in which they are expressed. We will study properties of some of the most basic functions that occur in calculus. For readers who would like to pursue applications in more depth, we will provide an optional section that introduces the use of regression curves to model real-world data. We will conclude the chapter with a discussion of curves in the plane that are most conveniently described using a pair of functions. (This material on "parametric curves" may be postponed until later if desired.)

Photo: ***The development of calculus in the 17th and 18th centuries was motivated by the need to understand physical phenomena such as the tides, the phases of the moon, the nature of light, and gravity.***

1.1 FUNCTIONS

In this section we will define and develop the concept of a "function," which is the basic mathematical object that scientists and mathematicians use to describe relationships between variable quantities. Functions play a central role in calculus and its applications.

DEFINITION OF A FUNCTION

Many scientific laws and engineering principles describe how one quantity depends on another. This idea was formalized in 1673 by Leibniz who coined the term *function* to indicate the dependence of one quantity on another, as described in the following definition.

1.1.1 DEFINITION. If a variable y depends on a variable x in such a way that each value of x determines exactly one value of y, then we say that ***y is a function of x***.

Four common methods for representing functions are:

- Numerically by tables
- Algebraically by formulas
- Geometrically by graphs
- Verbally

Table 1.1.1
INDIANAPOLIS 500
QUALIFYING SPEEDS

YEAR t	SPEED S (mi/h)
1987	215.390
1988	219.198
1989	223.885
1990	225.301
1991	224.113
1992	232.482
1993	223.967
1994	228.011
1995	231.604
1996	233.100
1997	218.263
1998	223.503
1999	225.179
2000	223.471
2001	226.037
2002	231.342
2003	231.725
2004	222.024

The method of representation often depends on how the function arises. For example:

- Table 1.1.1 shows the top qualifying speed S for the Indianapolis 500 auto race as a function of the year t. There is exactly one value of S for each value of t.
- Figure 1.1.1 is a graphical record of an earthquake recorded on a seismograph. The graph describes the deflection D of the seismograph needle as a function of the time T elapsed since the wave left the earthquake's epicenter. There is exactly one value of D for each value of T.
- Some of the most familiar functions arise from formulas; for example, the formula $C = 2\pi r$ expresses the circumference C of a circle as a function of its radius r. There is exactly one value of C for each value of r.
- Sometimes functions are described in words. For example, Isaac Newton's Law of Universal Gravitation is often stated as follows: The gravitational force of attraction between two bodies in the Universe is directly proportional to the product of their masses and inversely proportional to the square of the distance between them. This is the verbal description of the formula
$$F = G\frac{m_1 m_2}{r^2}$$
in which F is the force of attraction, m_1 and m_2 are the masses, r is the distance between them, and G is a constant. If the masses are constant, then the verbal description defines F as a function of r. There is exactly one value of F for each value of r.

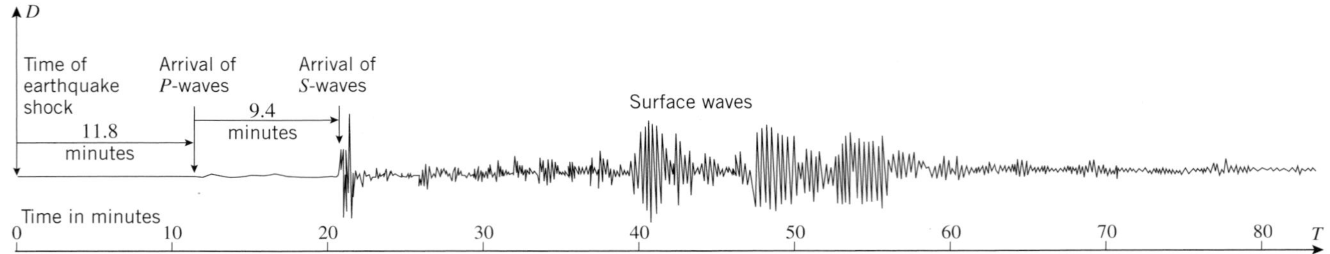

Figure 1.1.1

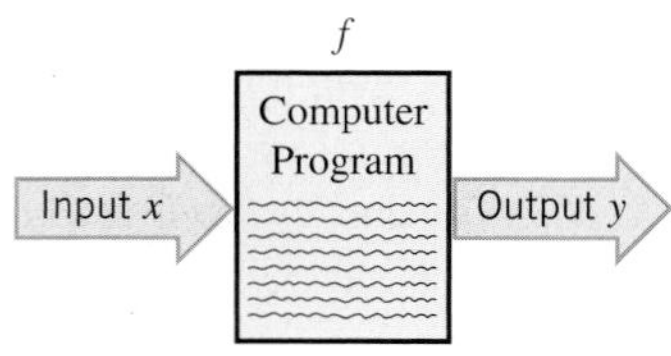

Figure 1.1.2

In the mid-eighteenth century the Swiss mathematician Leonhard Euler (pronounced "oiler") conceived the idea of denoting functions by letters of the alphabet, thereby making it possible to refer to functions without stating specific formulas, graphs, or tables. To understand Euler's idea, think of a function as a computer program that takes an *input* x, operates on it in some way, and produces exactly one *output* y. The computer program is an object in its own right, so we can give it a name, say f. Thus, the function f (the computer program) associates a unique output y with each input x (Figure 1.1.2). This suggests the following definition.

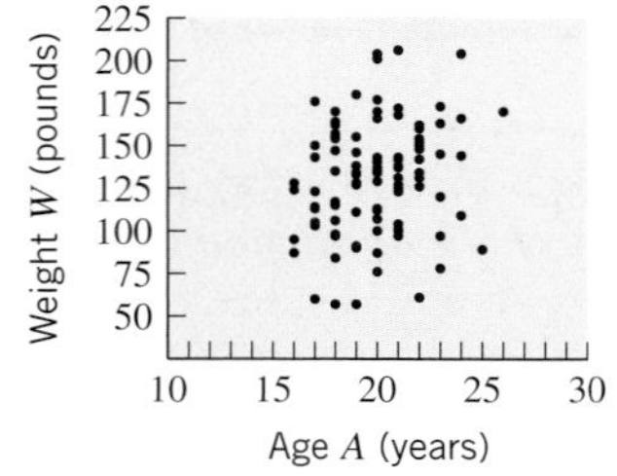

Figure 1.1.3

> **1.1.2 DEFINITION.** A ***function*** f is a rule that associates a unique output with each input. If the input is denoted by x, then the output is denoted by $f(x)$ (read "f of x").

In this definition the term *unique* means "exactly one." Thus, a function cannot assign two different outputs to the same input. For example, Figure 1.1.3 shows a plot of weight versus age for a random sample of 100 college students. This plot does *not* describe

W as a function of A because there are some values of A with more than one corresponding value of W. This is to be expected, since two people with the same age can have different weights.

INDEPENDENT AND DEPENDENT VARIABLES

For a given input x, the output of a function f is called the ***value*** of f at x or the ***image*** of x under f. Sometimes we will want to denote the output by a single letter, say y, and write

$$y = f(x)$$

This equation expresses y as a function of x; the variable x is called the ***independent variable*** (or ***argument***) of f, and the variable y is called the ***dependent variable*** of f. This terminology is intended to suggest that x is free to vary, but that once x has a specific value a corresponding value of y is determined. For now we will only consider functions in which the independent and dependent variables are real numbers, in which case we say that f is a ***real-valued function of a real variable***. Later, we will consider other kinds of functions.

Table 1.1.2

x	0	1	2	3
y	3	4	−1	6

▶ **Example 1** Table 1.1.2 describes a functional relationship $y = f(x)$ for which

$f(0) = 3$ — f associates $y = 3$ with $x = 0$.

$f(1) = 4$ — f associates $y = 4$ with $x = 1$.

$f(2) = -1$ — f associates $y = -1$ with $x = 2$.

$f(3) = 6$ — f associates $y = 6$ with $x = 3$. ◀

▶ **Example 2** The equation

$$y = 3x^2 - 4x + 2$$

has the form $y = f(x)$ in which the function f is given by the formula

$$f(x) = 3x^2 - 4x + 2$$

Leonhard Euler (1707–1783) Euler was probably the most prolific mathematician who ever lived. It has been said that "Euler wrote mathematics as effortlessly as most men breathe." He was born in Basel, Switzerland, and was the son of a Protestant minister who had himself studied mathematics. Euler's genius developed early. He attended the University of Basel, where by age 16 he obtained both a Bachelor of Arts degree and a Master's degree in philosophy. While at Basel, Euler had the good fortune to be tutored one day a week in mathematics by a distinguished mathematician, Johann Bernoulli. At the urging of his father, Euler then began to study theology. The lure of mathematics was too great, however, and by age 18 Euler had begun to do mathematical research. Nevertheless, the influence of his father and his theological studies remained, and throughout his life Euler was a deeply religious, unaffected person. At various times Euler taught at St. Petersburg Academy of Sciences (in Russia), the University of Basel, and the Berlin Academy of Sciences. Euler's energy and capacity for work were virtually boundless. His collected works form more than 100 quarto-sized volumes and it is believed that much of his work has been lost. What is particularly astonishing is that Euler was blind for the last 17 years of his life, and this was one of his most productive periods! Euler's flawless memory was phenomenal. Early in his life he memorized the entire *Aeneid* by Virgil, and at age 70 he could not only recite the entire work but could also state the first and last sentence on each page of the book from which he memorized the work. His ability to solve problems in his head was beyond belief. He worked out in his head major problems of lunar motion that baffled Isaac Newton and once did a complicated calculation in his head to settle an argument between two students whose computations differed in the fiftieth decimal place.

Following the development of calculus by Leibniz and Newton, results in mathematics developed rapidly in a disorganized way. Euler's genius gave coherence to the mathematical landscape. He was the first mathematician to bring the full power of calculus to bear on problems from physics. He made major contributions to virtually every branch of mathematics as well as to the theory of optics, planetary motion, electricity, magnetism, and general mechanics.

For each input x, the corresponding output y is obtained by substituting x in this formula. For example,

$$f(0) = 3(0)^2 - 4(0) + 2 = 2$$
f associates $y = 2$ with $x = 0$.

$$f(-1.7) = 3(-1.7)^2 - 4(-1.7) + 2 = 17.47$$
f associates $y = 17.47$ with $x = -1.7$.

$$f(\sqrt{2}) = 3(\sqrt{2})^2 - 4\sqrt{2} + 2 = 8 - 4\sqrt{2}$$
f associates $y = 8 - 4\sqrt{2}$ with $x = \sqrt{2}$. ◀

GRAPHS OF FUNCTIONS

If f is a real-valued function of a real variable, then the ***graph*** of f in the xy-plane is defined to be the graph of the equation $y = f(x)$. For example, the graph of the function $f(x) = x$ is the graph of the equation $y = x$, shown in Figure 1.1.4. That figure also shows the graphs of some other basic functions that may already be familiar to you. In the next section we will discuss techniques for graphing functions using graphing technology.

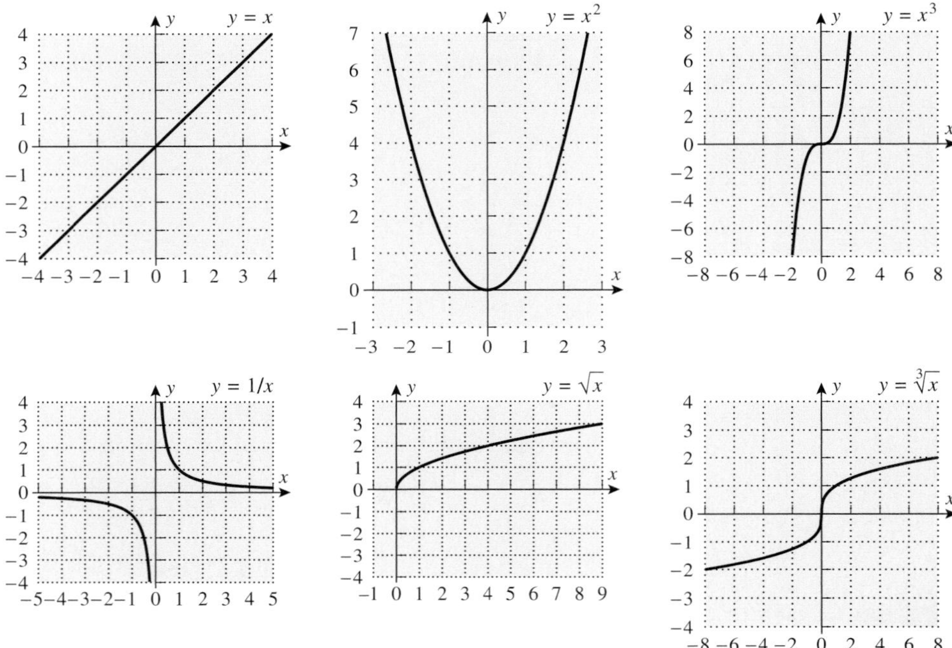

Figure 1.1.4

Since $\sqrt{x}$ is imaginary for negative values of x, there are no points on the graph of $y = \sqrt{x}$ in the region where $x < 0$.

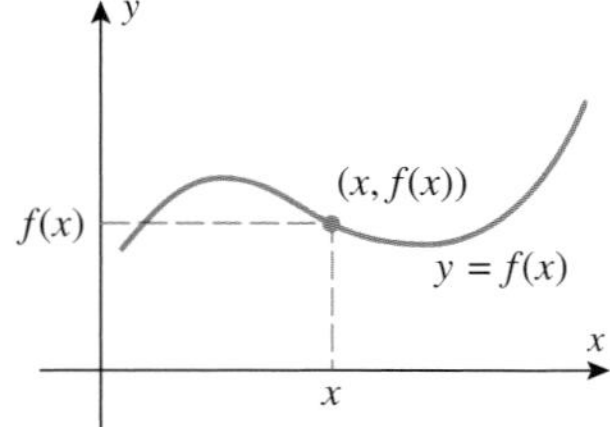

Figure 1.1.5 The y-coordinate of a point on the graph of $y = f(x)$ is the value of f at the corresponding x-coordinate.

Graphs can provide valuable visual information about a function. For example, since the graph of a function f in the xy-plane is the graph of the equation $y = f(x)$, the points on the graph of f are of the form $(x, f(x))$; that is, *the y-coordinate of a point on the graph of f is the value of f at the corresponding x-coordinate* (Figure 1.1.5). The values of x for which $f(x) = 0$ are the x-coordinates of the points where the graph of f intersects the x-axis (Figure 1.1.6). These values are called the ***zeros*** of f, the ***roots*** of $f(x) = 0$, or the ***x-intercepts*** of $y = f(x)$.

THE VERTICAL LINE TEST

Not every curve in the xy-plane is the graph of a function. For example, consider the curve in Figure 1.1.7, which is cut at two distinct points, (a, b) and (a, c), by a vertical line. This curve cannot be the graph of $y = f(x)$ for any function f; otherwise, we would have

$$f(a) = b \quad \text{and} \quad f(a) = c$$

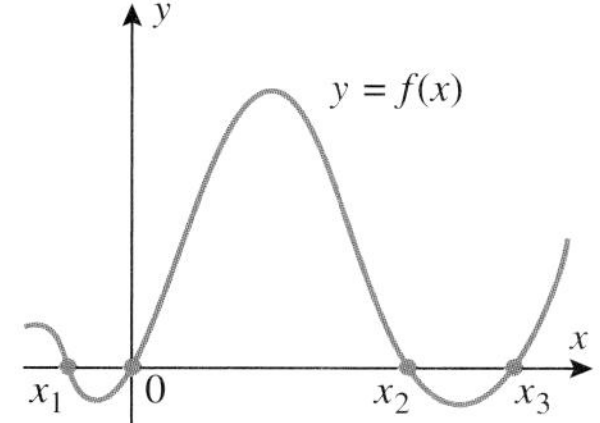

Figure 1.1.6 f has zeros at x_1, 0, x_2, and x_3.

which is impossible, since f cannot assign two different values to a. Thus, there is no function f whose graph is the given curve. This illustrates the following general result, which we will call the ***vertical line test***.

> **1.1.3 THE VERTICAL LINE TEST.** *A curve in the xy-plane is the graph of some function f if and only if no vertical line intersects the curve more than once.*

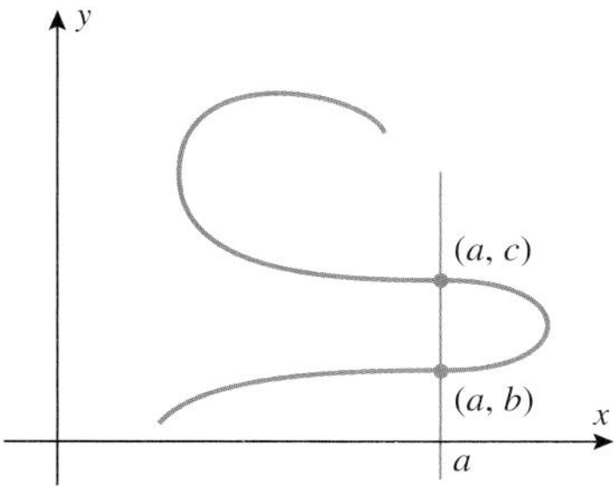

Figure 1.1.7 This curve cannot be the graph of a function.

► **Example 3** The graph of the equation

$$x^2 + y^2 = 25 \tag{1}$$

is a circle of radius 5 centered at the origin and hence there are vertical lines that cut the graph more than once. This can also be seen algebraically by solving Equation (1) for y in terms of x:

$$y = \pm\sqrt{25 - x^2}$$

This equation does not define y as a function of x because the right side is "multiple valued" in the sense that values of x in the interval $(-5, 5)$ produce two corresponding values of y. For example, if $x = 4$, then $y = \pm 3$, and hence $(4, 3)$ and $(4, -3)$ are two points on the circle that lie on the same vertical line (Figure 1.1.8). However, we can regard the circle as the union of the two semicircles

$$y = \sqrt{25 - x^2} \quad \text{and} \quad y = -\sqrt{25 - x^2}$$

each of which defines y as a function of x (Figure 1.1.9). ◄

See Web Appendix G for a review of circles.

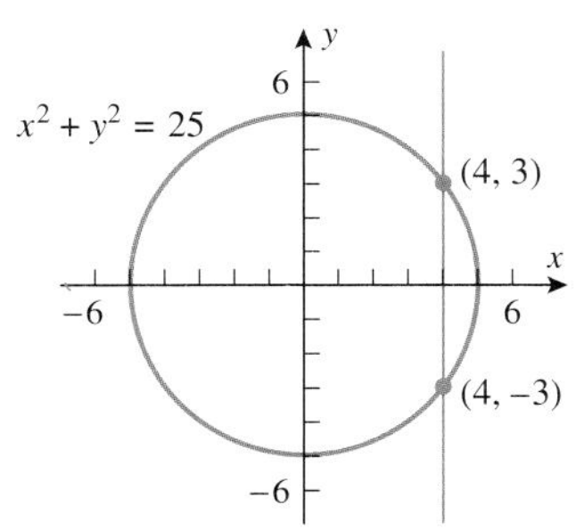

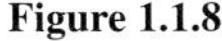

Figure 1.1.8

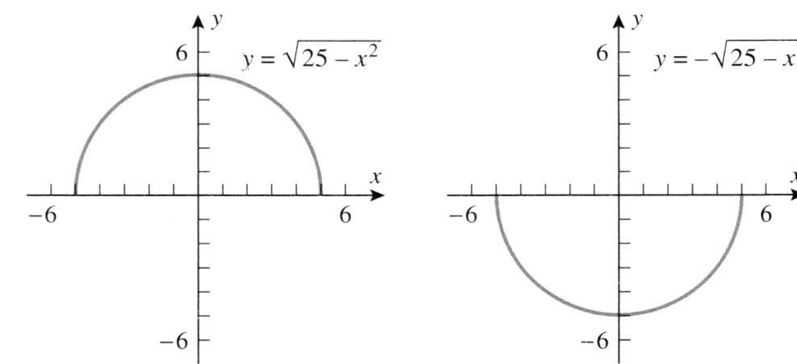

Figure 1.1.9 The union of these semicircles is the full circle.

THE ABSOLUTE VALUE FUNCTION

Recall that the ***absolute value*** or ***magnitude*** of a real number x is defined by

$$|x| = \begin{cases} x, & x \ge 0 \\ -x, & x < 0 \end{cases}$$

Symbols such as $+x$ and $-x$ are deceptive, since it is tempting to conclude that $+x$ is positive and $-x$ is negative. However, this need not be so, since x itself can be positive or negative. For example, if x is negative, say $x = -3$, then $-x = 3$ is positive and $+x = -3$ is negative.

The effect of taking the absolute value of a number is to strip away the minus sign if the number is negative and to leave the number unchanged if it is nonnegative. Thus,

$$|5| = 5, \quad \left|-\tfrac{4}{7}\right| = \tfrac{4}{7}, \quad |0| = 0$$

A more detailed discussion of the properties of absolute value is given in Web Appendix E. However, for convenience we provide the following summary of its algebraic properties.

1.1.4 PROPERTIES OF ABSOLUTE VALUE. *If a and b are real numbers, then*

(*a*) $|-a| = |a|$	A number and its negative have the same absolute value.
(*b*) $|ab| = |a|\,|b|$	The absolute value of a product is the product of the absolute values.
(*c*) $|a/b| = |a|/|b|,\ b \neq 0$	The absolute value of a ratio is the ratio of the absolute values.
(*d*) $|a+b| \leq |a| + |b|$	The ***triangle inequality***

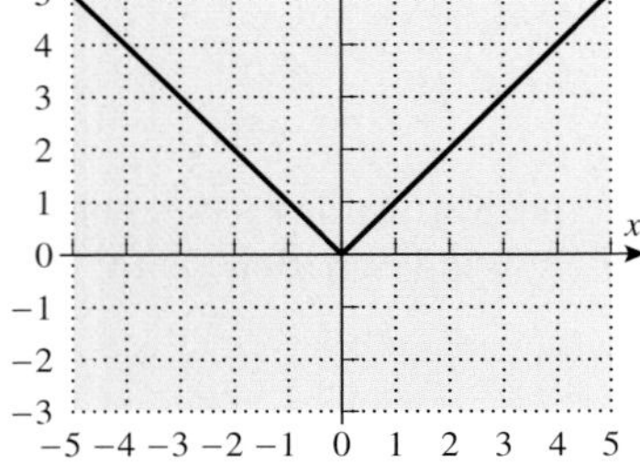

Figure 1.1.10

The graph of the function $f(x) = |x|$ can be obtained by graphing the two parts of the equation

$$y = \begin{cases} x, & x \geq 0 \\ -x, & x < 0 \end{cases}$$

separately. For $x \geq 0$, the graph of $y = x$ is a ray of slope 1 with its endpoint at the origin, and for $x < 0$, the graph of $y = -x$ is a ray of slope -1 with its endpoint at the origin. Combining the two parts produces the V-shaped graph in Figure 1.1.10.

Absolute values have important relationships to square roots. To see why this is so, recall from algebra that every positive real number x has two square roots, one positive and one negative. By definition, the symbol $\sqrt{x}$ denotes the *positive* square root of x. To denote the negative square root you must write $-\sqrt{x}$. For example, the positive square root of 9 is $\sqrt{9} = 3$, and the negative square root is $-\sqrt{9} = -3$. (Do not make the mistake of writing $\sqrt{9} = \pm 3$.)

Care must be exercised in simplifying expressions of the form $\sqrt{x^2}$, since it is *not* always true that $\sqrt{x^2} = x$. This equation is correct if x is nonnegative, but it is false for negative x. For example, if $x = -4$, then

$$\sqrt{x^2} = \sqrt{(-4)^2} = \sqrt{16} = 4 \neq x$$

TECHNOLOGY MASTERY

Verify (2) by using a graphing utility to show that the equations $y = \sqrt{x^2}$ and $y = |x|$ have the same graph.

A statement that is correct for all real values of x is

$$\sqrt{x^2} = |x| \tag{2}$$

FUNCTIONS DEFINED PIECEWISE

The absolute value function $f(x) = |x|$ is an example of a function that is defined ***piecewise*** in the sense that the formula for f changes, depending on the value of x.

► **Example 4** Sketch the graph of the function defined piecewise by the formula

$$f(x) = \begin{cases} 0, & x \leq -1 \\ \sqrt{1-x^2}, & -1 < x < 1 \\ x, & x \geq 1 \end{cases}$$

Solution. The formula for f changes at the points $x = -1$ and $x = 1$. (We call these the ***breakpoints*** for the formula.) A good procedure for graphing functions defined piecewise is to graph the function separately over the open intervals determined by the breakpoints, and then graph f at the breakpoints themselves. For the function f in this example the graph is the horizontal ray $y = 0$ on the interval $(-\infty, -1)$, it is the semicircle $y = \sqrt{1-x^2}$ on the interval $(-1, 1)$, and it is the ray $y = x$ on the interval $(1, +\infty)$. The formula for f specifies that the equation $y = 0$ applies at the breakpoint -1 [so $y = f(-1) = 0$], and it specifies that the equation $y = x$ applies at the breakpoint 1 [so $y = f(1) = 1$]. The graph of f is shown in Figure 1.1.11. ◄

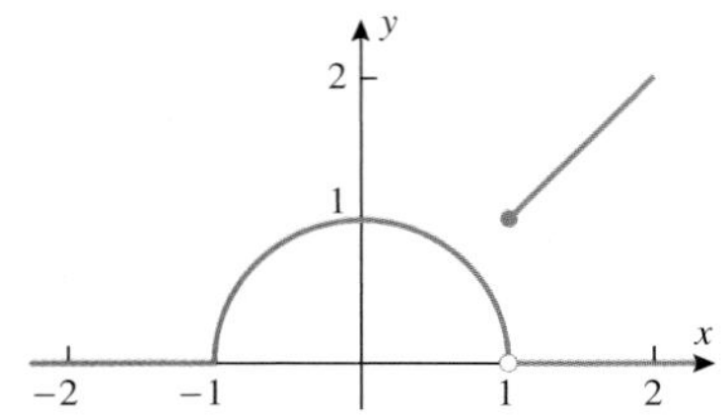

Figure 1.1.11

In Figure 1.1.11 the solid dot and open circle at the breakpoint $x = 1$ serve to emphasize that the point on the graph lies on the ray and not the semicircle. There is no ambiguity at the breakpoint $x = -1$ because the two parts of the graph join together continuously there.

▶ **Example 5** Increasing the speed at which air moves over a person's skin increases the rate of moisture evaporation and makes the person feel cooler. (This is why we fan ourselves in hot weather.) The ***wind chill index*** is the temperature at a wind speed of 4 mi/h that would produce the same sensation on exposed skin as the current temperature and wind speed combination. An empirical formula (i.e., a formula based on experimental data) for the wind chill index W at 32°F for a wind speed of v mi/h is

$$W = \begin{cases} 32, & 0 \le v \le 3 \\ 55.628 - 22.07v^{0.16}, & 3 < v \end{cases}$$

A computer-generated graph of $W(v)$ is shown in Figure 1.1.12. ◀

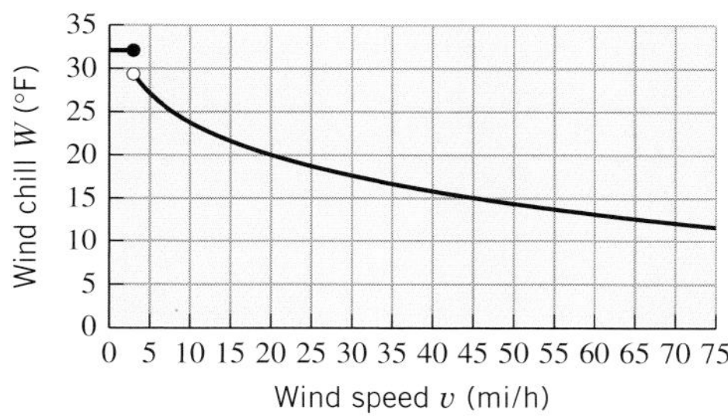

Figure 1.1.12 Wind chill versus wind speed at 32°F

DOMAIN AND RANGE

If x and y are related by the equation $y = f(x)$, then the set of all allowable inputs (x-values) is called the ***domain*** of f, and the set of outputs (y-values) that result when x varies over the domain is called the ***range*** of f. For example, if f is the function defined by the table in Example 1, then the domain is the set $\{0, 1, 2, 3\}$ and the range is the set $\{3, 4, -1, 6\}$.

Sometimes physical or geometric considerations impose restrictions on the allowable inputs of a function. For example, if y denotes the area of a square of side x, then these variables are related by the equation $y = x^2$. Although this equation produces a unique value of y for every real number x, the fact that lengths must be nonnegative imposes the requirement that $x \ge 0$.

One might argue that a physical square cannot have a side of length zero. However, it is often convenient mathematically to allow zero lengths, and we will do so throughout this text.

When a function is defined by a mathematical formula, the formula itself may impose restrictions on the allowable inputs. For example, if $y = 1/x$, then $x = 0$ is not an allowable input since division by zero is undefined, and if $y = \sqrt{x}$, then negative values of x are not allowable inputs because they produce imaginary values for y and we have agreed to consider only real-valued functions of a real variable. In general, we make the following definition.

1.1.5 DEFINITION. If a real-valued function of a real variable is defined by a formula, and if no domain is stated explicitly, then it is to be understood that the domain consists of all real numbers for which the formula yields a real value. This is called the ***natural domain*** of the function.

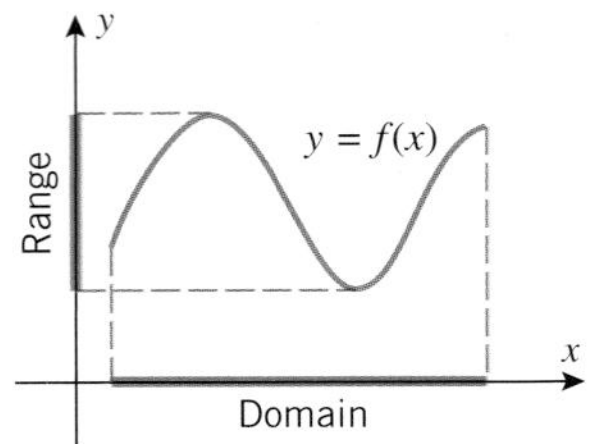

Figure 1.1.13 The projection of $y = f(x)$ on the x-axis is the set of allowable x-values for f, and the projection on the y-axis is the set of corresponding y-values.

The domain and range of a function f can be pictured by projecting the graph of $y = f(x)$ onto the coordinate axes as shown in Figure 1.1.13.

► **Example 6** Find the natural domain of

(a) $f(x) = x^3$ (b) $f(x) = 1/[(x-1)(x-3)]$
(c) $f(x) = \tan x$ (d) $f(x) = \sqrt{x^2 - 5x + 6}$

Solution (a). The function f has real values for all real x, so its natural domain is the interval $(-\infty, +\infty)$.

Solution (b). The function f has real values for all real x, except $x = 1$ and $x = 3$, where divisions by zero occur. Thus, the natural domain is

$$\{x : x \neq 1 \text{ and } x \neq 3\} = (-\infty, 1) \cup (1, 3) \cup (3, +\infty)$$

Solution (c). Since $f(x) = \tan x = \sin x / \cos x$, the function f has real values except where $\cos x = 0$, and this occurs when x is an odd integer multiple of $\pi/2$. Thus, the natural domain consists of all real numbers except

$$x = \pm\frac{\pi}{2}, \pm\frac{3\pi}{2}, \pm\frac{5\pi}{2}, \ldots$$

For a review of trigonometry see Appendix A.

Solution (d). The function f has real values, except when the expression inside the radical is negative. Thus the natural domain consists of all real numbers x such that

$$x^2 - 5x + 6 = (x-3)(x-2) \geq 0$$

This inequality is satisfied if $x \leq 2$ or $x \geq 3$ (verify), so the natural domain of f is

$$(-\infty, 2] \cup [3, +\infty) \blacktriangleleft$$

In some cases we will include the domain explicitly when defining a function. For example, if $f(x) = x^2$ is the area of a square of side x, then we can write

$$f(x) = x^2, \quad x \geq 0$$

to indicate that we take the domain of f to be the set of nonnegative real numbers (Figure 1.1.14).

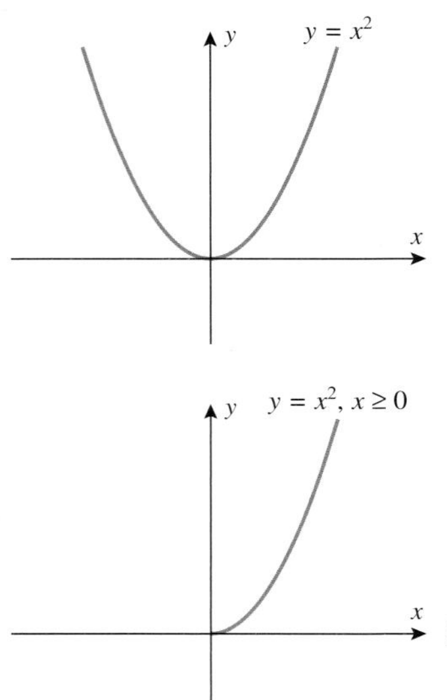

Figure 1.1.14

THE EFFECT OF ALGEBRAIC OPERATIONS ON THE DOMAIN

Algebraic expressions are frequently simplified by canceling common factors in the numerator and denominator. However, care must be exercised when simplifying formulas for functions in this way, since this process can alter the domain.

► **Example 7** The natural domain of the function

$$f(x) = \frac{x^2 - 4}{x - 2} \tag{3}$$

consists of all real x except $x = 2$. However, if we factor the numerator and then cancel the common factor in the numerator and denominator, we obtain

$$f(x) = \frac{(x-2)(x+2)}{x-2} = x + 2 \tag{4}$$

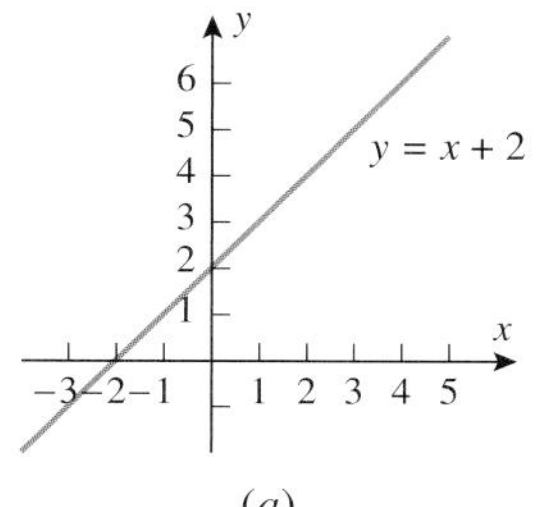

(a)

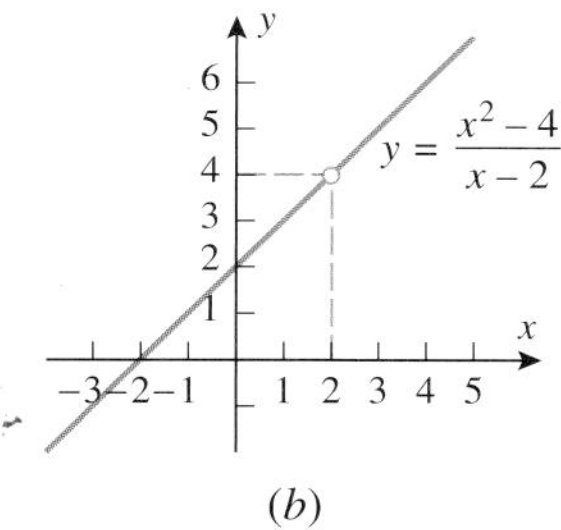

(b)

Figure 1.1.15

Since the right side of (4) has a value of $f(2) = 4$, whereas $f(2)$ was undefined in (3), the algebraic simplification has changed the function. Geometrically, the graph of (4) is the line in Figure 1.1.15*a*, whereas the graph of (3) is the same line but with a hole at $x = 2$, since the function is undefined there (Figure 1.1.15*b*). In short, the geometric effect of the algebraic cancellation is to eliminate the hole in the original graph. ◀

Sometimes alterations to the domain of a function that result from algebraic simplification are irrelevant to the problem at hand and can be ignored. However, if the domain must be preserved, then one must impose the restrictions on the simplified function explicitly. For example, if we wanted to preserve the domain of the function in Example 7, then we would have to express the simplified form of the function as

$$f(x) = x + 2, \quad x \neq 2$$

▶ **Example 8** Find the domain and range of

(a) $f(x) = 2 + \sqrt{x-1}$ (b) $f(x) = (x+1)/(x-1)$

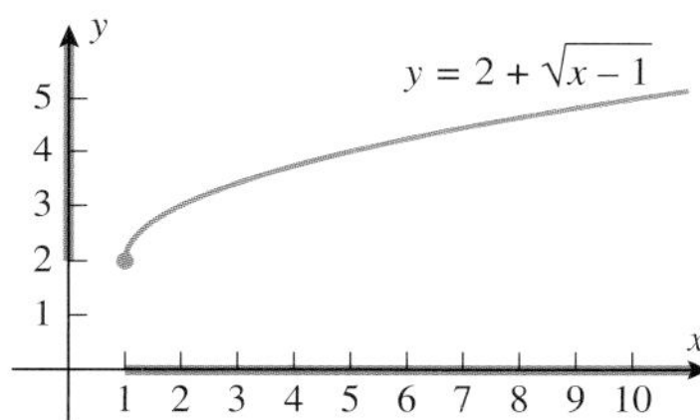

Figure 1.1.16

Solution (a). Since no domain is stated explicitly, the domain of f is the natural domain $[1, +\infty)$. As x varies over the interval $[1, +\infty)$, the value of $\sqrt{x-1}$ varies over the interval $[0, +\infty)$, so the value of $f(x) = 2 + \sqrt{x-1}$ varies over the interval $[2, +\infty)$, which is the range of f. The domain and range are highlighted in green on the x- and y-axes in Figure 1.1.16.

Solution (b). The given function f is defined for all real x, except $x = 1$, so the natural domain of f is

$$\{x : x \neq 1\} = (-\infty, 1) \cup (1, +\infty)$$

To determine the range it will be convenient to introduce a dependent variable

$$y = \frac{x+1}{x-1} \tag{5}$$

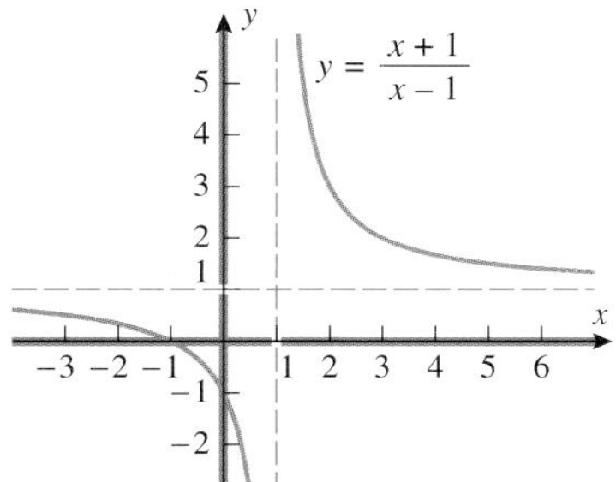

Figure 1.1.17

Although the set of possible y-values is not immediately evident from this equation, the graph of (5), which is shown in Figure 1.1.17, suggests that the range of f consists of all y, except $y = 1$. To see that this is so, we solve (5) for x in terms of y:

$$\begin{aligned}(x-1)y &= x+1\\ xy - y &= x+1\\ xy - x &= y+1\\ x(y-1) &= y+1\\ x &= \frac{y+1}{y-1}\end{aligned}$$

It is now evident from the right side of this equation that $y = 1$ is not in the range; otherwise we would have a division by zero. No other values of y are excluded by this equation, so the range of the function f is $\{y : y \neq 1\} = (-\infty, 1) \cup (1, +\infty)$, which agrees with the result obtained graphically. ◀

DOMAIN AND RANGE IN APPLIED PROBLEMS

In applications, physical considerations often impose restrictions on the domain and range of a function.

► **Example 9** An open box is to be made from a 16-inch by 30-inch piece of cardboard by cutting out squares of equal size from the four corners and bending up the sides (Figure 1.1.18*a*).

(a) Let V be the volume of the box that results when the squares have sides of length x. Find a formula for V as a function of x.

(b) Find the domain of V.

(c) Use the graph of V given in Figure 1.1.18*c* to estimate the range of V.

(d) Describe in words what the graph tells you about the volume.

Solution (a). As shown in Figure 1.1.18*b*, the resulting box has dimensions $16 - 2x$ by $30 - 2x$ by x, so the volume $V(x)$ is given by

$$V(x) = (16 - 2x)(30 - 2x)x = 480x - 92x^2 + 4x^3$$

Solution (b). The domain is the set of x-values and the range is the set of V-values. Because x is a length, it must be nonnegative, and because we cannot cut out squares whose sides are more than 8 in long (why?), the x-values in the domain must satisfy

$$0 \le x \le 8$$

Solution (c). From the graph of V versus x in Figure 1.1.18*c* we estimate that the V-values in the range satisfy

$$0 \le V \le 725$$

Note that this is an approximation. Later we will show how to find the range exactly.

Solution (d). The graph tells us that the box of maximum volume occurs for a value of x that is between 3 and 4 and that the maximum volume is approximately 725 in^3. Moreover, the volume decreases toward zero as x gets closer to 0 or 8. ◄

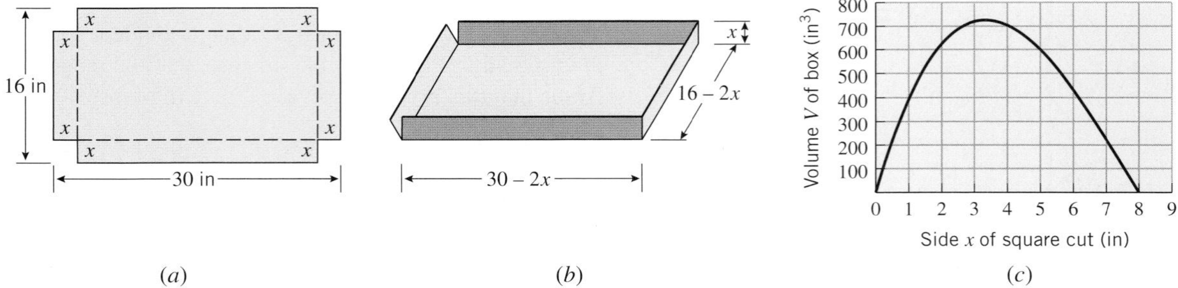

Figure 1.1.18

In applications involving time, formulas for functions are often expressed in terms of a variable t whose starting value is taken to be $t = 0$.

► **Example 10** At 8:05 A.M. a car is clocked at 100 ft/s by a radar detector that is positioned at the edge of a straight highway. Assuming that the car maintains a constant speed between 8:05 A.M. and 8:06 A.M., find a function $D(t)$ that expresses the distance traveled by the car during that time interval as a function of the time t.

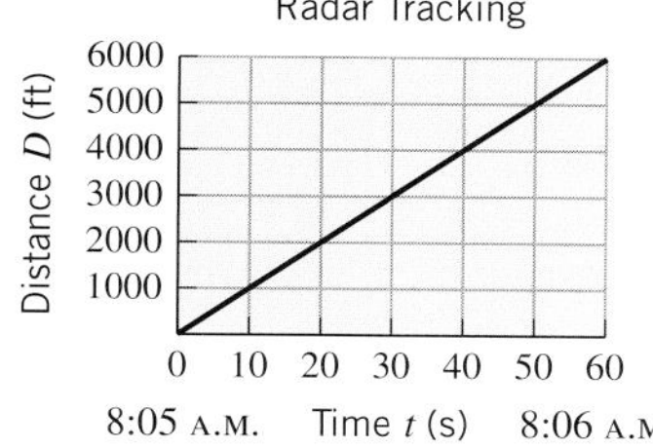

Figure 1.1.19

Solution. It would be clumsy to use clock time for the variable t, so let us agree to measure the elapsed time in seconds, starting with $t = 0$ at 8:05 A.M. and ending with $t = 60$ at 8:06 A.M. At each instant, the distance traveled (in ft) is equal to the speed of the car (in ft/s) multiplied by the elapsed time (in s). Thus,

$$D(t) = 100t, \quad 0 \leq t \leq 60$$

The graph of D versus t is shown in Figure 1.1.19. ◀

ISSUES OF SCALE AND UNITS

In geometric problems where you want to preserve the "true" shape of a graph, you must use units of equal length on both axes. For example, if you graph a circle in a coordinate system in which 1 unit in the y-direction is smaller than 1 unit in the x-direction, then the circle will be squashed vertically into an elliptical shape (Figure 1.1.20). You must also use units of equal length when you want to apply the distance formula

$$d = \sqrt{(x_2 - x_1)^2 + (y_2 - y_1)^2}$$

to calculate the distance between two points (x_1, y_1) and (x_2, y_2) in the xy-plane.

However, sometimes it is inconvenient or impossible to display a graph using units of equal length. For example, consider the equation

$$y = x^2$$

If we want to show the portion of the graph over the interval $-3 \leq x \leq 3$, then there is no problem using units of equal length, since y only varies from 0 to 9 over that interval. However, if we want to show the portion of the graph over the interval $-10 \leq x \leq 10$, then there is a problem keeping the units equal in length, since the value of y varies between 0 and 100. In this case the only reasonable way to show all of the graph that occurs over the interval $-10 \leq x \leq 10$ is to compress the unit of length along the y-axis, as illustrated in Figure 1.1.21.

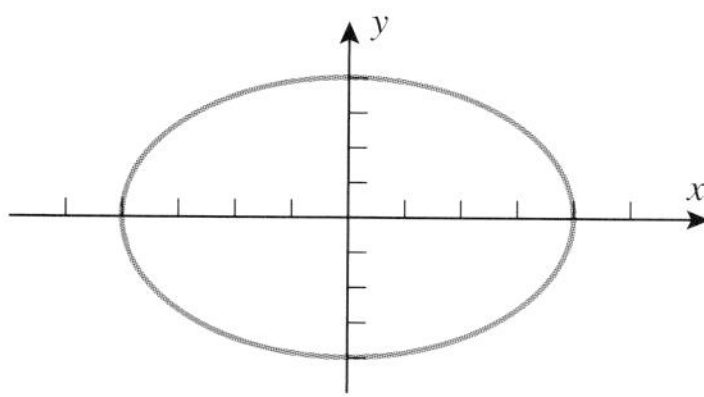

The circle is squashed because 1 unit on the y-axis has a smaller length than 1 unit on the x-axis.

Figure 1.1.20

In applications where the variables on the two axes have unrelated units (say, centimeters on the y-axis and seconds on the x-axis), then nothing is gained by requiring the units to have equal lengths; choose the lengths to make the graph as clear as possible.

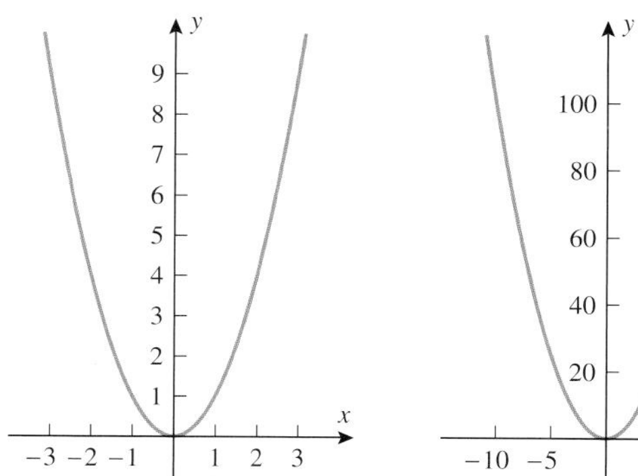

Figure 1.1.21

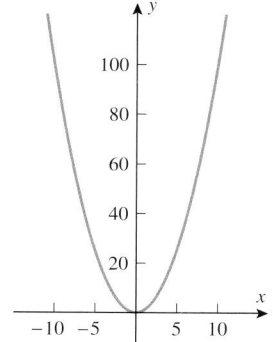

✔QUICK CHECK EXERCISES 1.1 *(See page 16 for answers.)*

1. Let $f(x) = \sqrt{x+1} + 4$.
   (a) The natural domain of f is ________.
   (b) $f(3) =$ ________
   (c) $f(t^2 - 1) =$ ________
   (d) $f(x) = 7$ if $x =$ ________
   (e) The range of f is ________.

2. Line segments in an xy-plane form "letters" as depicted.

   (a) If the y-axis is parallel to the letter I, which of the letters represent the graph of $y = f(x)$ for some function f?

(b) If the y-axis is perpendicular to the letter I, which of the letters represent the graph of $y = f(x)$ for some function f?

3. Use the accompanying graph of $y = f(x)$ to complete each part.
(a) The domain of f is ________.
(b) The range of f is ________.
(c) $f(-3) =$ ________
(d) $f\left(\frac{1}{2}\right) =$ ________
(e) The solutions to $f(x) = -\frac{3}{2}$ are $x =$ ________ and $x =$ ________.

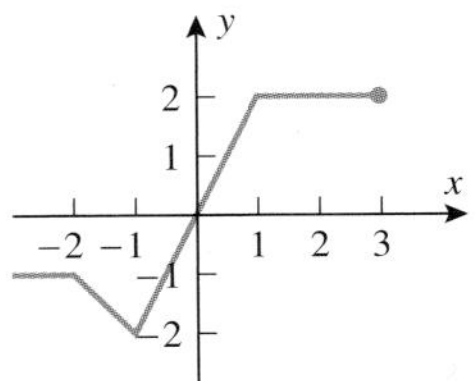

4. The accompanying table gives a 5-day forecast of high and low temperatures in degrees Fahrenheit (°F).

	MON	TUE	WED	THURS	FRI
HIGH	75	71	65	70	73
LOW	52	56	48	50	52

(a) Suppose that x and y denote the respective high and low temperature predictions for each of the 5 days. Is y a function of x? If so, give the domain and range of this function.
(b) Suppose that x and y denote the respective low and high temperature predictions for each of the 5 days. Is y a function of x? If so, give the domain and range of this function.

5. Let l, w, and A denote the length, width, and area of a rectangle, respectively, and suppose that the width of the rectangle is half the length.
(a) If l is expressed as a function of w, then $l =$ ________.
(b) If A is expressed as a function of l, then $A =$ ________.
(c) If w is expressed as a function of A, then $w =$ ________.

EXERCISE SET 1.1 Graphing Utility

1. Use the accompanying graph to answer the following questions, making reasonable approximations where needed.
(a) For what values of x is $y = 1$?
(b) For what values of x is $y = 3$?
(c) For what values of y is $x = 3$?
(d) For what values of x is $y \leq 0$?
(e) What are the maximum and minimum values of y and for what values of x do they occur?

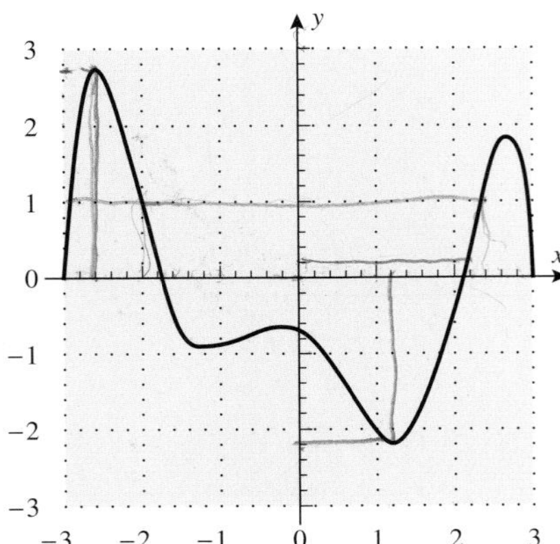

Figure Ex-1

2. Use the accompanying table to answer the questions posed in Exercise 1.

x	−2	−1	0	2	3	4	5	6
y	5	1	−2	7	−1	1	0	9

Table Ex-2

3. In each part of the accompanying figure, determine whether the graph defines y as a function of x.

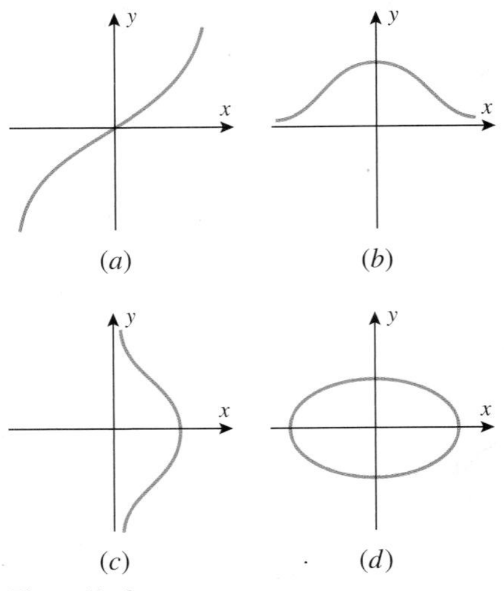

Figure Ex-3

4. In each part, compare the natural domains of f and g.

(a) $f(x) = \dfrac{x^2 + x}{x + 1}$; $g(x) = x$

(b) $f(x) = \dfrac{x\sqrt{x} + \sqrt{x}}{x + 1}$; $g(x) = \sqrt{x}$

FOCUS ON CONCEPTS

5. Use the cigarette consumption graph in the accompanying figure to answer the following questions, making reasonable approximations where needed.

(a) When did the annual cigarette consumption reach 3000 per adult for the first time?

(b) When did the annual cigarette consumption per adult reach its peak, and what was the peak value?

(c) Can you tell from the graph how many cigarettes were consumed in a given year? If not, what additional information would you need to make that determination?

(d) What factors are likely to cause a sharp increase in annual cigarette consumption per adult?

(e) What factors are likely to cause a sharp decline in annual cigarette consumption per adult?

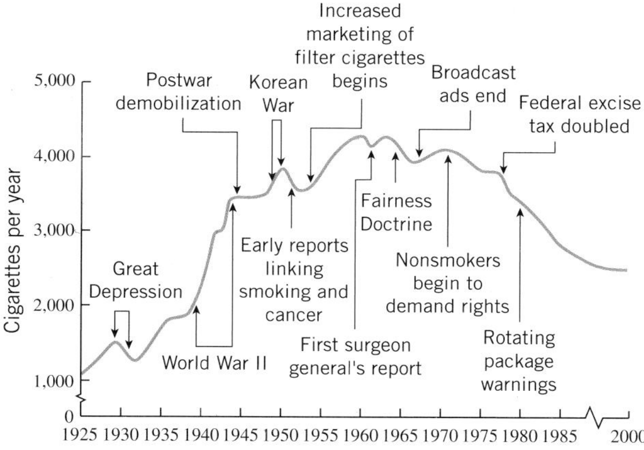

Source: U.S. Department of Health and Human Services.

Figure Ex-5

6. Use the cigarette consumption graph in Exercise 5 to answer the following questions, making reasonable approximations where needed.

(a) When did the annual cigarette consumption fall to 3000 per adult?

(b) Between the year of the first surgeon general's report and the year 1970, when was the annual cigarette consumption per adult at a minumum?

(c) Which was greater, the rate of growth of per capita cigarette consumption during World War II, or the rate of growth between the end of World War II and the beginning of the Korean War?

(d) Does it appear that the per capita cigarette consumption will eventually fall to pre–World War II levels?

7. The accompanying graph shows the median income in U.S. households (adjusted for inflation) between 1985 and 2001. Use the graph to answer the following questions, making reasonable approximations where needed.

(a) When was the median income at its maximum value, and what was the median income when that occurred?

(b) When was the median income at its minimum value, and what was the median income when that occurred?

(c) The median income was declining during the 2-year period between 1999 and 2001. Was it declining more rapidly during the first year or the second year of that period? Explain your reasoning.

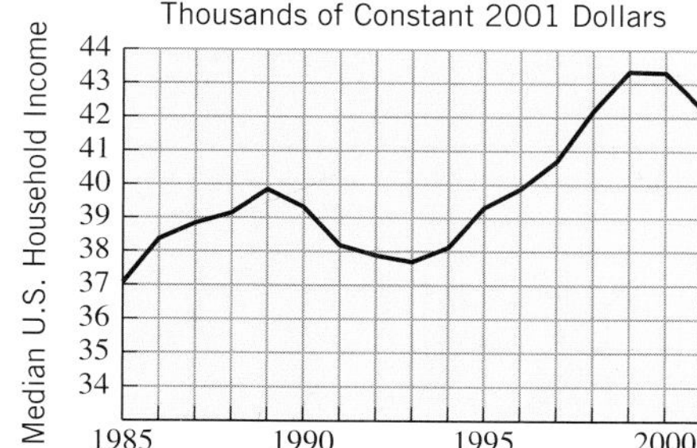

Source: U.S. Census Bureau, July 2003

Figure Ex-7

8. Use the median income graph in Exercise 7 to answer the following questions, making reasonable approximations where needed.

(a) What was the average yearly growth of median income between 1993 and 1999?

(b) The median income was increasing during the 6-year period between 1993 and 1999. Was it increasing more rapidly during the first 3 years or the second 3 years of that period? Explain your reasoning.

(c) Consider the statement: "After years of decline, median income this year was finally higher than that of last year." In what year would this statement have been correct?

9. Find $f(0)$, $f(2)$, $f(-2)$, $f(3)$, $f(\sqrt{2})$, and $f(3t)$.

(a) $f(x) = 3x^2 - 2$ (b) $f(x) = \begin{cases} \dfrac{1}{x}, & x > 3 \\ 2x, & x \le 3 \end{cases}$

10. Find $g(3)$, $g(-1)$, $g(\pi)$, $g(-1.1)$, and $g(t^2 - 1)$.

(a) $g(x) = \dfrac{x+1}{x-1}$ (b) $g(x) = \begin{cases} \sqrt{x+1}, & x \ge 1 \\ 3, & x < 1 \end{cases}$

11–14 Find the natural domain of the function algebraically, and confirm that your result is consistent with the graph produced by your graphing utility. [*Note:* Set your graphing utility to radian mode when graphing trigonometric functions.]

11. (a) $f(x) = \dfrac{1}{x-3}$ (b) $g(x) = \sqrt{x^2 - 3}$

(c) $G(x) = \sqrt{x^2 - 2x + 5}$ (d) $f(x) = \dfrac{x}{|x|}$

(e) $h(x) = \dfrac{1}{1 - \sin x}$

12. (a) $f(x) = \dfrac{1}{5x+7}$ (b) $h(x) = \sqrt{x - 3x^2}$
(c) $G(x) = \sqrt{\dfrac{x^2-4}{x-4}}$ (d) $f(x) = \dfrac{x^2-1}{x+1}$
(e) $h(x) = \dfrac{3}{2 - \cos x}$

13. (a) $f(x) = \sqrt{3-x}$ (b) $g(x) = \sqrt{4-x^2}$
(c) $h(x) = 3 + \sqrt{x}$ (d) $G(x) = x^3 + 2$
(e) $H(x) = 3\sin x$

14. (a) $f(x) = \sqrt{3x-2}$ (b) $g(x) = \sqrt{9-4x^2}$
(c) $h(x) = \dfrac{1}{3+\sqrt{x}}$ (d) $G(x) = \dfrac{3}{x}$
(e) $H(x) = \sin^2\sqrt{x}$

FOCUS ON CONCEPTS

15. (a) If you had a device that could record the Earth's population continuously, would you expect the graph of population versus time to be a continuous (unbroken) curve? Explain what might cause breaks in the curve.
(b) Suppose that a hospital patient receives an injection of an antibiotic every 8 hours and that between injections the concentration C of the antibiotic in the bloodstream decreases as the antibiotic is absorbed by the tissues. What might the graph of C versus the elapsed time t look like?

16. (a) If you had a device that could record the temperature of a room continuously over a 24-hour period, would you expect the graph of temperature versus time to be a continuous (unbroken) curve? Explain your reasoning.
(b) If you had a computer that could track the number of boxes of cereal on the shelf of a market continuously over a 1-week period, would you expect the graph of the number of boxes on the shelf versus time to be a continuous (unbroken) curve? Explain your reasoning.

17. A boat is bobbing up and down on some gentle waves. Suddenly it gets hit by a large wave and sinks. Sketch a rough graph of the height of the boat above the ocean floor as a function of time.

18. A cup of hot coffee sits on a table. You pour in some cool milk and let it sit for an hour. Sketch a rough graph of the temperature of the coffee as a function of time.

19. Use the equation $y = x^2 - 6x + 8$ to answer the following questions.
(a) For what values of x is $y = 0$?
(b) For what values of x is $y = -10$?
(c) For what values of x is $y \geq 0$?
(d) Does y have a minimum value? A maximum value? If so, find them.

20. Use the equation $y = 1 + \sqrt{x}$ to answer the following questions.
(a) For what values of x is $y = 4$?
(b) For what values of x is $y = 0$?
(c) For what values of x is $y \geq 6$?
(d) Does y have a minimum value? A maximum value? If so, find them.

21. As shown in the accompanying figure, a pendulum of constant length L makes an angle θ with its vertical position. Express the height h as a function of the angle θ.

22. Express the length L of a chord of a circle with radius 10 cm as a function of the central angle θ (see the accompanying figure).

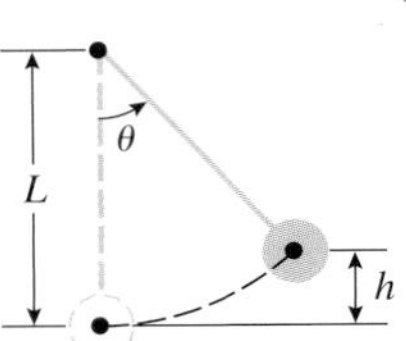

Figure Ex-21

Figure Ex-22

23–24 Express the function in piecewise form without using absolute values. [*Suggestion:* It may help to generate the graph of the function.]

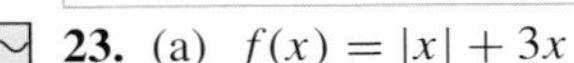

23. (a) $f(x) = |x| + 3x + 1$ (b) $g(x) = |x| + |x-1|$

24. (a) $f(x) = 3 + |2x - 5|$ (b) $g(x) = 3|x-2| - |x+1|$

25. As shown in the accompanying figure, an open box is to be constructed from a rectangular sheet of metal, 8 in by 15 in, by cutting out squares with sides of length x from each corner and bending up the sides.
(a) Express the volume V as a function of x.
(b) Find the domain of V.
(c) Plot the graph of the function V obtained in part (a) and estimate the range of this function.
(d) In words, describe how the volume V varies with x, and discuss how one might construct boxes of maximum volume.

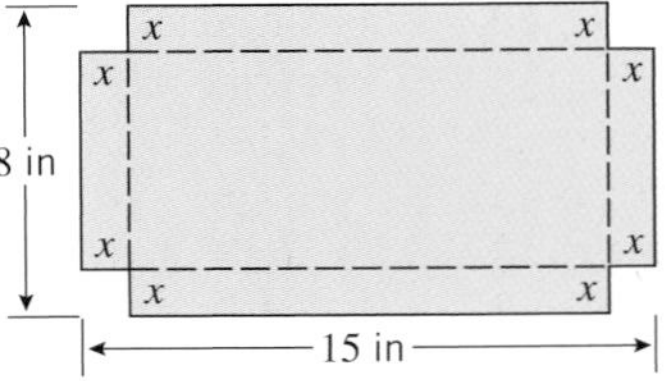

Figure Ex-25

26. Repeat Exercise 25 assuming the box is constructed in the same fashion from a 6-inch-square sheet of metal.

27. A construction company has adjoined a 1000-ft^2 rectangular enclosure to its office building. Three sides of the enclosure are fenced in. The side of the building adjacent to the enclosure is 100 ft long and a portion of this side is used as the fourth side of the enclosure. Let x and y be the dimensions of the enclosure, where x is measured parallel to the building, and let L be the length of fencing required for those dimensions.

(a) Find a formula for L in terms of x and y.
(b) Find a formula that expresses L as a function of x alone.
(c) What is the domain of the function in part (b)?
(d) Plot the function in part (b) and estimate the dimensions of the enclosure that minimize the amount of fencing required.

28. As shown in the accompanying figure, a camera is mounted at a point 3000 ft from the base of a rocket launching pad. The rocket rises vertically when launched, and the camera's elevation angle is continually adjusted to follow the bottom of the rocket.

(a) Express the height x as a function of the elevation angle θ.
(b) Find the domain of the function in part (a).
(c) Plot the graph of the function in part (a) and use it to estimate the height of the rocket when the elevation angle is $\pi/4 \approx 0.7854$ radian. Compare this estimate to the exact height. [*Suggestion:* If you are using a graphing calculator, the trace and zoom features will be helpful here.]

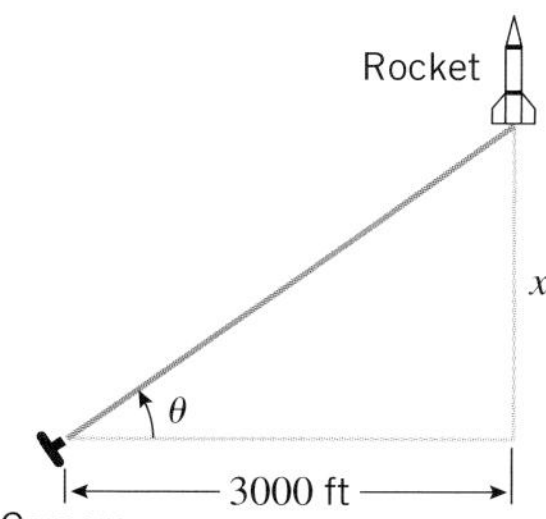

Figure Ex-28

29. A soup company wants to manufacture a can in the shape of a right circular cylinder that will hold 500 cm^3 of liquid. The material for the top and bottom costs 0.02 cent/cm^2, and the material for the sides costs 0.01 cent/cm^2.

(a) Estimate the radius r and the height h of the can that costs the least to manufacture. [*Suggestion:* Express the cost C in terms of r.]
(b) Suppose that the tops and bottoms of radius r are punched out from square sheets with sides of length $2r$ and the scraps are waste. If you allow for the cost of the waste, would you expect the can of least cost to be taller or shorter than the one in part (a)? Explain.
(c) Estimate the radius, height, and cost of the can in part (b), and determine whether your conjecture was correct.

30. The designer of a sports facility wants to put a quarter-mile (1320 ft) running track around a football field, oriented as in the accompanying figure. The football field is 360 ft long (including the end zones) and 160 ft wide. The track consists of two straightaways and two semicircles, with the straightaways extending at least the length of the football field.

(a) Show that it is possible to construct a quarter-mile track around the football field. [*Suggestion:* Find the shortest track that can be constructed around the field.]
(b) Let L be the length of a straightaway (in feet), and let x be the distance (in feet) between a sideline of the football field and a straightaway. Make a graph of L versus x.
(c) Use the graph to estimate the value of x that produces the shortest straightaways, and then find this value of x exactly.
(d) Use the graph to estimate the length of the longest possible straightaways, and then find that length exactly.

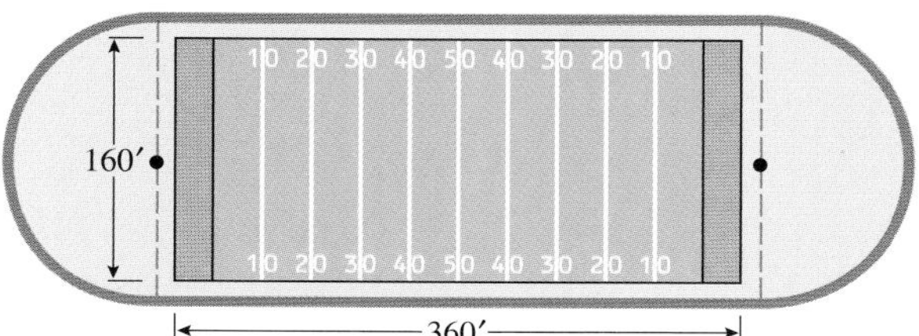

Figure Ex-30

31–32 (i) Explain why the function f has one or more holes in its graph, and state the x-values at which those holes occur. (ii) Find a function g whose graph is identical to that of f, but without the holes.

31. $f(x) = \dfrac{(x+2)(x^2-1)}{(x+2)(x-1)}$ **32.** $f(x) = \dfrac{x^2+|x|}{|x|}$

33. In 2001 the National Weather Service introduced a new wind chill temperature (WCT) index. For a given outside temperature T and wind speed v, the wind chill temperature index is the equivalent temperature that exposed skin would feel with a wind speed of v mi/h. Based on a more accurate model of cooling due to wind, the new formula is

$$\text{WCT} = \begin{cases} T, & 0 \le v \le 3 \\ 35.74 + 0.6215T - 35.75v^{0.16} + 0.4275Tv^{0.16}, & 3 < v \end{cases}$$

where T is the temperature in °F, v is the wind speed in mi/h, and WCT is the equivalent temperature in °F. Find the WCT to the nearest degree if $T = 25$°F and
(a) $v = 3$ mi/h (b) $v = 15$ mi/h (c) $v = 46$ mi/h.

Source: Adapted from UMAP Module 658, *Windchill*, W. Bosch and L. Cobb, COMAP, Arlington, MA.

34–36 Use the formula for the wind chill temperature index described in Exercise 33.

34. Find the air temperature to the nearest degree if the WCT is reported as -60°F with a wind speed of 48 mi/h.

35. Find the air temperature to the nearest degree if the WCT is reported as $-10°$F with a wind speed of 48 mi/h.

36. Find the wind speed to the nearest mile per hour if the WCT is reported as $5°$F with an air temperature of $20°$F.

QUICK CHECK ANSWERS 1.1

1. (a) $[-1, +\infty)$ (b) 6 (c) $|t| + 4$ (d) 8 (e) $[4, +\infty)$ **2.** (a) M (b) I **3.** (a) $(-\infty, 3]$ (b) $[-2, 2]$ (c) -1 (d) 1 (e) $-\frac{3}{4}$; $-\frac{3}{2}$ **4.** (a) yes; domain: $\{65, 70, 71, 73, 75\}$; range: $\{48, 50, 52, 56\}$ (b) no **5.** (a) $l = 2w$ (b) $A = l^2/2$ (c) $w = \sqrt{A/2}$

1.2 GRAPHING FUNCTIONS USING CALCULATORS AND COMPUTER ALGEBRA SYSTEMS

In this section we will discuss issues that relate to generating graphs of equations and functions with graphing utilities (graphing calculators and computers). Because graphing utilities vary widely, it is difficult to make general statements about them. Therefore, at various places in this section we will ask you to refer to the documentation for your own graphing utility for specific details about the way it operates.

GRAPHING CALCULATORS AND COMPUTER ALGEBRA SYSTEMS

The development of new technology has significantly changed how and where mathematicians, engineers, and scientists perform their work, as well as their approach to problem solving. Among the most significant of these developments are programs called ***Computer Algebra Systems*** (abbreviated CAS), the most common being *Mathematica*, *Maple*, and *Derive*.* Computer algebra systems not only have graphing capabilities, but, as their name suggests, they can perform many of the symbolic computations that occur in algebra, calculus, and branches of higher mathematics. For example, it is a trivial task for a CAS to perform the factorization

$$x^6 + 23x^5 + 147x^4 - 139x^3 - 3464x^2 - 2112x + 23040 = (x+5)(x-3)^2(x+8)^3$$

or the exact numerical computation

$$\left(\frac{63456}{3177295} - \frac{43907}{22854377}\right)^3 = \frac{2251912457164208291259320230122866923}{382895955819369204449565945369203764688375}$$

Technology has also made it possible to generate graphs of equations and functions in seconds that in the past might have taken hours. Figure 1.2.1 shows the graphs of the

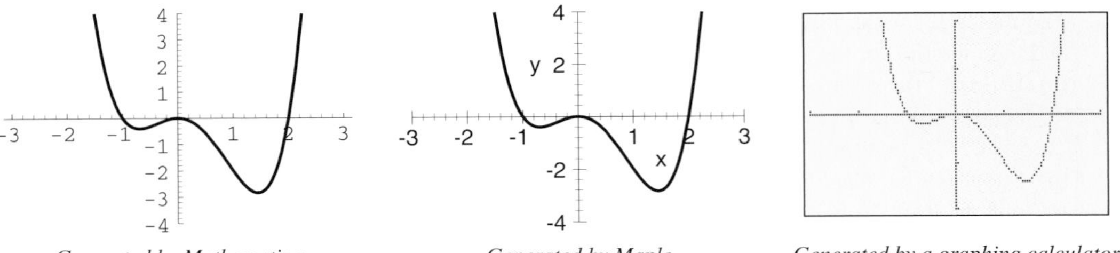

Generated by Mathematica *Generated by Maple* *Generated by a graphing calculator*

Figure 1.2.1

---

* *Mathematica* is a product of Wolfram Research, Inc.; *Maple* is a product of Waterloo Maple Software, Inc.; and *Derive* is a product of Soft Warehouse, Inc.

function $f(x) = x^4 - x^3 - 2x^2$ produced with various graphing utilities; the first two were generated with the CAS programs, *Mathematica* and *Maple*, and the third with a graphing calculator. Graphing calculators produce coarser graphs than most computer programs but have the advantage of being compact and portable.

VIEWING WINDOWS

Graphing utilities can only show a portion of the xy-plane in the viewing screen, so the first step in graphing an equation is to determine which rectangular portion of the xy-plane you want to display. This region is called the ***viewing window*** (or ***viewing rectangle***). For example, in Figure 1.2.1 the viewing window extends over the interval $[-3, 3]$ in the x-direction and over the interval $[-4, 4]$ in the y-direction, so we denote the viewing window by $[-3, 3] \times [-4, 4]$ (read "$[-3, 3]$ by $[-4, 4]$"). In general, if the viewing window is $[a, b] \times [c, d]$, then the window extends between $x = a$ and $x = b$ in the x-direction and between $y = c$ and $y = d$ in the y-direction. We will call $[a, b]$ the ***x-interval*** for the window and $[c, d]$ the ***y-interval*** for the window (Figure 1.2.2).

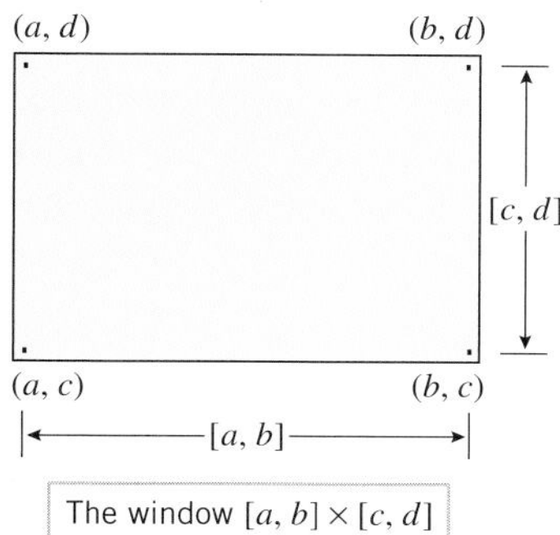

Figure 1.2.2

Different graphing utilities designate viewing windows in different ways. For example, the first two graphs in Figure 1.2.1 were produced by the commands

```
Plot[x^4 - x^3 -2*x^2, {x, -3, 3}, PlotRange->{-4, 4}]
```
(*Mathematica*)

```
plot(x^4 - x^3 -2*x^2, x = -3..3, y = -4..4);
```
(*Maple*)

and the last graph was produced on a graphing calculator by pressing the GRAPH button after setting the values for the variables that determine the x-interval and y-interval to be

$$x\text{Min} = -3, \quad x\text{Max} = 3, \quad y\text{Min} = -4, \quad y\text{Max} = 4$$

TECHNOLOGY MASTERY

Use your graphing utility to generate the graph of the function

$$f(x) = x^4 - x^3 - 2x^2$$

in the window $[-3, 3] \times [-4, 4]$.

TICK MARKS AND GRID LINES

To help locate points in a viewing window, graphing utilities provide methods for drawing ***tick marks*** (also called ***scale marks***). With computer programs such as *Mathematica* and *Maple*, there are specific commands for designating the spacing between tick marks, but if the user does not specify the spacing, then the programs make certain *default* choices. For example, in the first two parts of Figure 1.2.1, the tick marks shown were the default choices.

On some graphing calculators the spacing between tick marks is determined by two ***scale variables*** (also called ***scale factors***), which we will denote by

$$x\text{Scl} \quad \text{and} \quad y\text{Scl}$$

(The notation varies among calculators.) These variables specify the spacing between the tick marks in the x- and y-directions, respectively. For example, in the third part of Figure 1.2.1 the window and tick marks were designated by the settings

$$\begin{aligned} x\text{Min} &= -3 & x\text{Max} &= 3 \\ y\text{Min} &= -4 & y\text{Max} &= 4 \\ x\text{Scl} &= 1 & y\text{Scl} &= 1 \end{aligned}$$

Most graphing utilities allow for variations in the design and positioning of tick marks. For example, Figure 1.2.3 shows two variations of the graphs in Figure 1.2.1; the first was generated on a computer using an option for placing the ticks and numbers on the edges of a box, and the second was generated on a graphing calculator using an option for drawing grid lines to simulate graph paper.

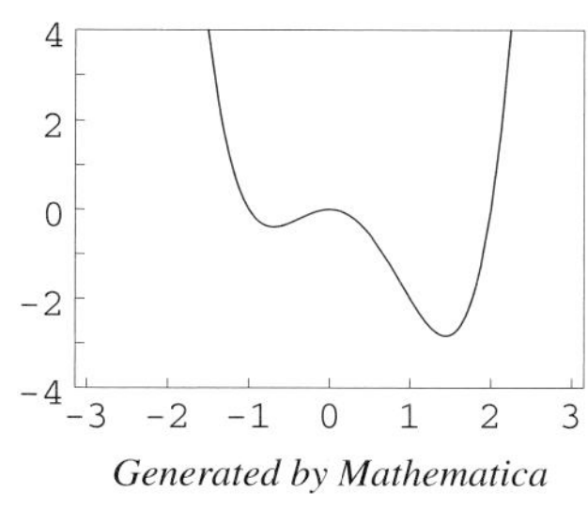

Generated by Mathematica

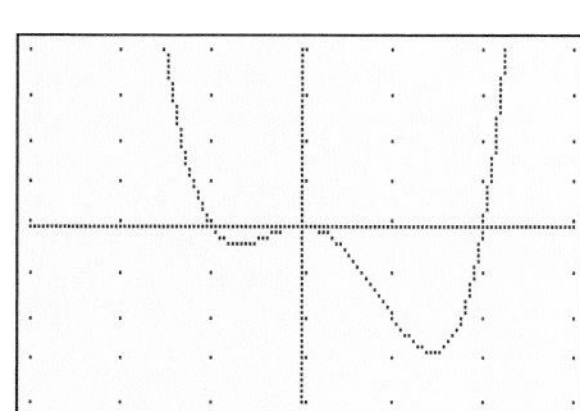

Generated by a graphing calculator

Figure 1.2.3

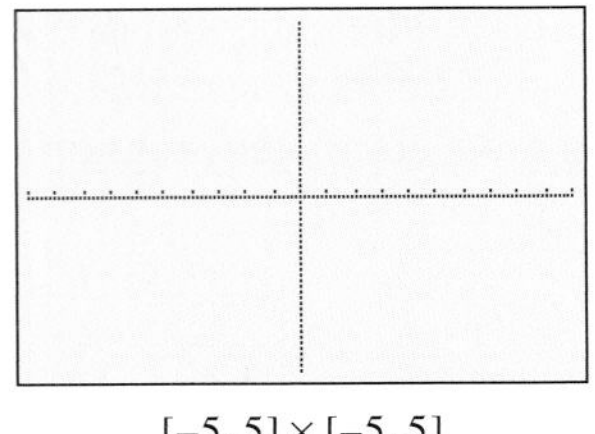

$[-5, 5] \times [-5, 5]$
$x\text{Scl} = 0.5$, $y\text{Scl} = 10$

Figure 1.2.4

► **Example 1** Figure 1.2.4 shows the window $[-5, 5] \times [-5, 5]$ with the tick marks spaced 0.5 unit apart in the x-direction and 10 units apart in the y-direction. No tick marks are actually visible in the y-direction because the tick mark at the origin is covered by the x-axis, and all other tick marks in that direction fall outside of the viewing window. ◄

► **Example 2** Figure 1.2.5 shows the window $[-10, 10] \times [-10, 10]$ with the tick marks spaced 0.1 unit apart in the x- and y-directions. In this case the tick marks are so close together that they create thick lines on the coordinate axes. When this occurs you will usually want to increase the scale factors to reduce the number of tick marks to make them legible. ◄

TECHNOLOGY MASTERY

Graphing calculators provide a way of clearing all settings and returning them to *default values*. For example, on one calculator the default window is $[-10, 10] \times [-10, 10]$ and the default scale factors are $x\text{Scl} = 1$ and $y\text{Scl} = 1$. Read your documentation to determine the default values for your calculator and how to restore the default settings. If you are using a CAS, read your documentation to determine the commands for specifying the spacing between tick marks.

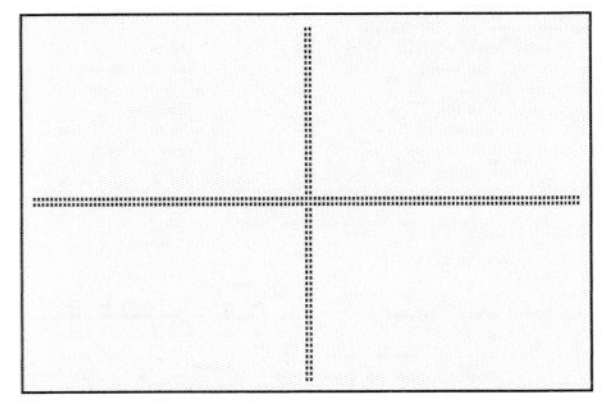

$[-10, 10] \times [-10, 10]$
$x\text{Scl} = 0.1$, $y\text{Scl} = 0.1$

Figure 1.2.5

CHOOSING A VIEWING WINDOW

When the graph of a function extends indefinitely in some direction, no single viewing window can show it all. In such cases the choice of the viewing window can affect one's perception of how the graph looks. For example, Figure 1.2.6 shows a computer-generated graph of $y = 9 - x^2$, and Figure 1.2.7 shows four views of this graph generated on a calculator.

- In part (a) the graph falls completely outside of the window, so the window is blank (except for the ticks and axes).
- In part (b) the graph is broken into two pieces because it passes in and out of the window.
- In part (c) the graph appears to be a straight line because we have zoomed in on a very small segment of the curve.
- In part (d) we have a more revealing picture of the graph shape because the window encompasses the high point on the graph and the intersections with the x-axis.

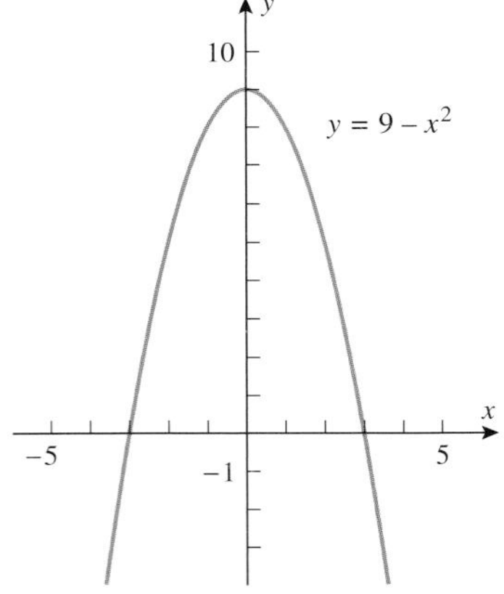

Figure 1.2.6

The following example illustrates how the domain and range of a function can be used to find a good viewing window when the graph of the function does not extend indefinitely in both the x- and y-directions.

► **Example 3** Use the domain and range of the function $f(x) = \sqrt{12 - 3x^2}$ to determine a viewing window that contains the entire graph.

Solution. The natural domain of f is $[-2, 2]$ and the range is $[0, \sqrt{12}]$ (verify), so the entire graph will be contained in the viewing window $[-2, 2] \times [0, \sqrt{12}]$. For clarity, it is desirable to use a slightly larger window to avoid having the graph too close to the edges of the screen. For example, taking the viewing window to be $[-3, 3] \times [-1, 4]$ yields the graph in Figure 1.2.8. ◄

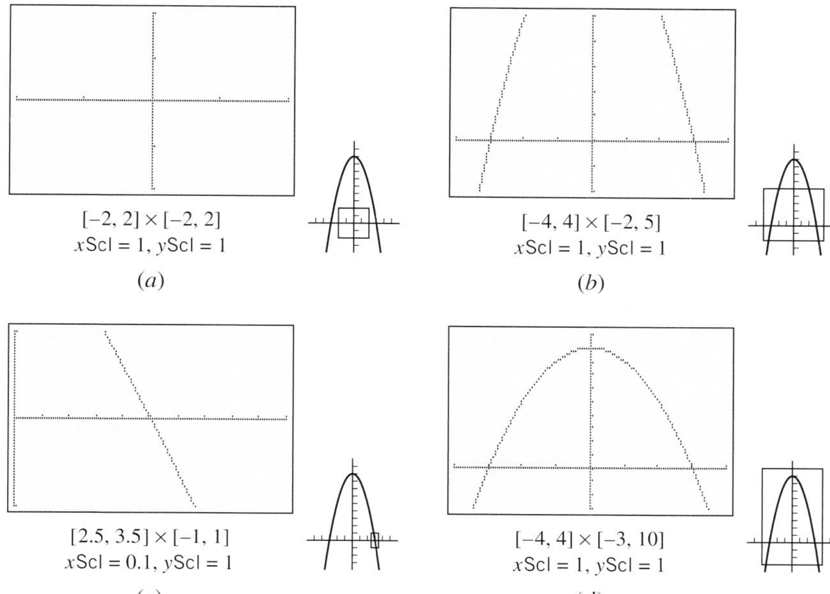

Figure 1.2.7 Four views of $y = 9 - x^2$

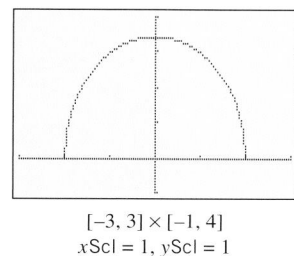

[−3, 3] × [−1, 4]
xScl = 1, yScl = 1

Figure 1.2.8

Sometimes it will be impossible to find a single window that shows all important features of a graph, in which case you will need to decide what is most important for the problem at hand and choose the window appropriately.

▶ **Example 4** Graph the equation $y = x^3 - 12x^2 + 18$ in the following windows and discuss the advantages and disadvantages of each window.

(a) $[-10, 10] \times [-10, 10]$ with $x\text{Scl} = 1$, $y\text{Scl} = 1$

(b) $[-20, 20] \times [-20, 20]$ with $x\text{Scl} = 1$, $y\text{Scl} = 1$

(c) $[-20, 20] \times [-300, 20]$ with $x\text{Scl} = 1$, $y\text{Scl} = 20$

(d) $[-5, 15] \times [-300, 20]$ with $x\text{Scl} = 1$, $y\text{Scl} = 20$

(e) $[1, 2] \times [-1, 1]$ with $x\text{Scl} = 0.1$, $y\text{Scl} = 0.1$

Solution (a). The window in Figure 1.2.9a has chopped off the portion of the graph that intersects the y-axis, and it shows only two of three possible real roots for the given cubic polynomial. To remedy these problems we need to widen the window in both the x- and y-directions.

Solution (b). The window in Figure 1.2.9b shows the intersection of the graph with the y-axis and the three real roots, but it has chopped off the portion of the graph between the two positive roots. Moreover, the ticks in the y-direction are nearly illegible because they are so close together. We need to extend the window in the negative y-direction and increase yScl. We do not know how far to extend the window, so some experimentation will be required to obtain what we want.

Solution (c). The window in Figure 1.2.9c shows all of the main features of the graph. However, we have some wasted space in the x-direction. We can improve the picture by shortening the window in the x-direction appropriately.

Solution (d). The window in Figure 1.2.9d shows all of the main features of the graph without a lot of wasted space. However, the window does not provide a clear view of the roots. To get a closer view of the roots we must forget about showing all of the main features of the graph and choose windows that zoom in on the roots themselves.

Solution (e). The window in Figure 1.2.9e displays very little of the graph, but it clearly shows that the root in the interval [1, 2] is approximately 1.3. ◄

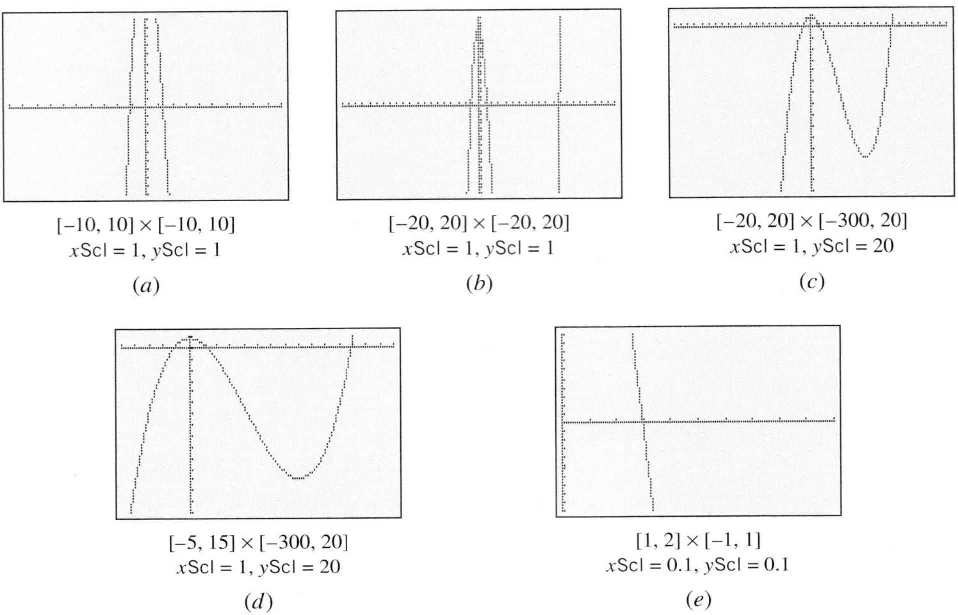

Figure 1.2.9

TECHNOLOGY MASTERY

Sometimes you will want to determine the viewing window by choosing the x-interval and allowing the graphing utility to determine a y-interval that encompasses the maximum and minimum values of the function over the x-interval. Most graphing utilities provide some method for doing this, so read your documentation to determine how to use this feature. Allowing the graphing utility to determine the y-interval of the window takes some of the guesswork out of problems like that in part (b) of the preceding example.

ZOOMING

The process of enlarging or reducing the size of a viewing window is called ***zooming***. If you reduce the size of the window, you see *less* of the graph as a whole, but more detail of the part shown; this is called ***zooming in***. In contrast, if you enlarge the size of the window, you see *more* of the graph as a whole, but less detail of the part shown; this is called ***zooming out***. Most graphing calculators provide menu items for zooming in or zooming out by fixed factors. For example, on one calculator the amount of enlargement or reduction is controlled by setting values for two ***zoom factors***, denoted by xFact and yFact. If

$$x\text{Fact} = 10 \quad \text{and} \quad y\text{Fact} = 5$$

then each time a zoom command is executed the viewing window is enlarged or reduced by a factor of 10 in the x-direction and a factor of 5 in the y-direction. With computer programs such as *Mathematica* and *Maple*, zooming is controlled by adjusting the x-interval and y-interval directly; however, there are ways to automate this by programming.

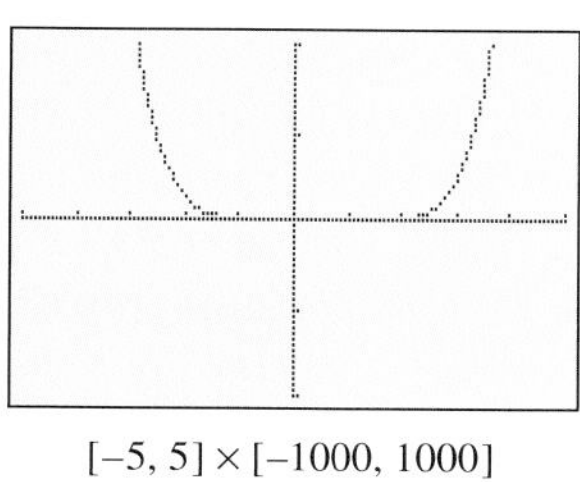

$[-5, 5] \times [-1000, 1000]$
xScl = 1, yScl = 500
(a)

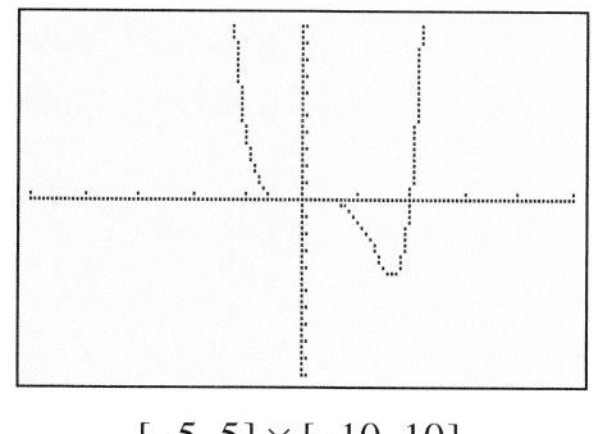

$[-5, 5] \times [-10, 10]$
xScl = 1, yScl = 1
(b)

Figure 1.2.10

COMPRESSION

Enlarging the viewing window for a graph has the geometric effect of compressing the graph, since more of the graph is packed into the calculator screen. If the compression is sufficiently great, then some of the detail in the graph may be lost. Thus, the choice of the viewing window frequently depends on whether you want to see more of the graph or more of the detail. Figure 1.2.10 shows two views of the equation

$$y = x^5(x - 2)$$

In part (*a*) of the figure the y-interval is very large, resulting in a vertical compression that obscures the detail in the vicinity of the x-axis. In part (*b*) the y-interval is smaller, and consequently we see more of the detail in the vicinity of the x-axis but less of the graph in the y-direction.

► **Example 5** The function $f(x) = x + 0.01\sin(50\pi x)$ is the sum of $f_1(x) = x$, whose graph is the line $y = x$, and $f_2(x) = 0.01\sin(50\pi x)$, whose graph is a sinusoidal curve with amplitude 0.01 and period $2\pi/50\pi = 0.04$. This suggests that the graph of $f(x)$ will follow the general path of the line $y = x$ but will have bumps resulting from the contributions of the sinusoidal oscillations, as shown in part (*c*) of Figure 1.2.11. Generate the four graphs shown in Figure 1.2.11 and explain why the oscillations are visible only in part (*c*).

Solution. To generate the four graphs, you first need to put your utility in radian mode.* Because the windows in successive parts of the example are decreasing in size by a factor of 10, calculator users can generate successive graphs by using the zoom feature with the zoom factors set to 10 in both the x- and y-directions.

(a) In Figure 1.2.11*a* the graph appears to be a straight line because the vertical compression has hidden the small sinusoidal oscillations (their amplitude is only 0.01).

(b) In Figure 1.2.11*b* small bumps begin to appear on the line because there is less vertical compression.

(c) In Figure 1.2.11*c* the oscillations have become clear because the vertical scale is more in keeping with the amplitude of the oscillations.

(d) In Figure 1.2.11*d* the graph appears to be a straight line because we have zoomed in on such a small portion of the curve. ◄

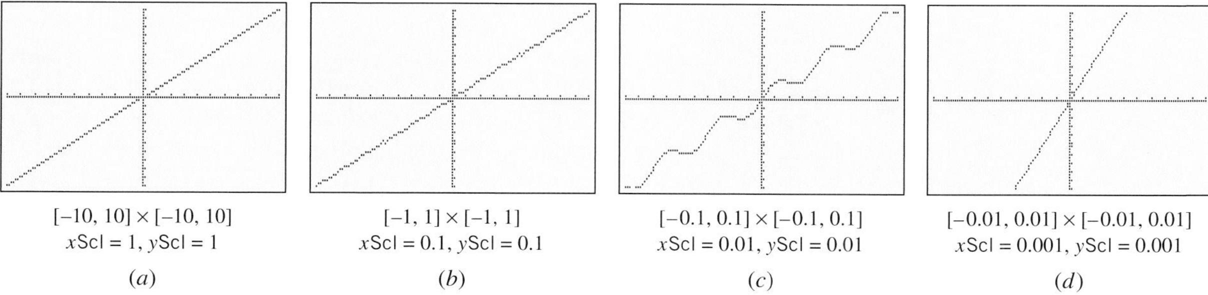

$[-10, 10] \times [-10, 10]$ xScl = 1, yScl = 1 (a)	$[-1, 1] \times [-1, 1]$ xScl = 0.1, yScl = 0.1 (b)	$[-0.1, 0.1] \times [-0.1, 0.1]$ xScl = 0.01, yScl = 0.01 (c)	$[-0.01, 0.01] \times [-0.01, 0.01]$ xScl = 0.001, yScl = 0.001 (d)

Figure 1.2.11

* In this text we follow the convention that angles are measured in radians unless degree measure is specified.

ASPECT RATIO DISTORTION

Figure 1.2.12a shows a circle of radius 5 and two perpendicular lines graphed in the window $[-10, 10] \times [-10, 10]$ with $x\text{Scl} = 1$ and $y\text{Scl} = 1$. However, the circle is distorted and the lines do not appear perpendicular because the calculator has not used the same length for 1 unit on the x-axis and 1 unit on the y-axis. (Compare the spacing between the ticks on the axes.) This is called ***aspect ratio distortion***. Many calculators provide a menu item for automatically correcting the distortion by adjusting the viewing window appropriately. For example, one calculator makes this correction to the viewing window $[-10, 10] \times [-10, 10]$ by changing it to

$$[-16.9970674487, 16.9970674487] \times [-10, 10]$$

(Figure 1.2.12b). With computer programs such as *Mathematica* and *Maple*, aspect ratio distortion is controlled with adjustments to the physical dimensions of the viewing window on the computer screen, rather than altering the x- and y-intervals of the viewing window.

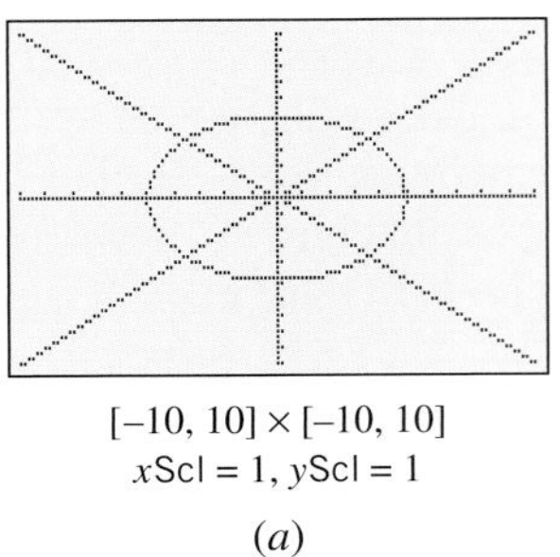

$[-10, 10] \times [-10, 10]$
$x\text{Scl} = 1, y\text{Scl} = 1$
(a)

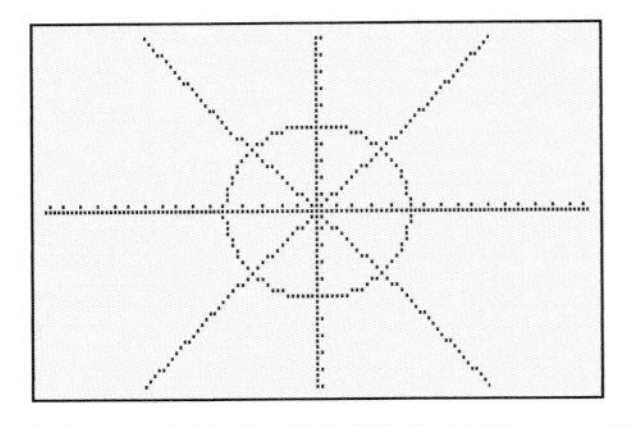

$[-16.9970674487, 16.9970674487] \times [-10, 10]$
$x\text{Scl} = 1, y\text{Scl} = 1$
(b)

Figure 1.2.12

SAMPLING ERROR

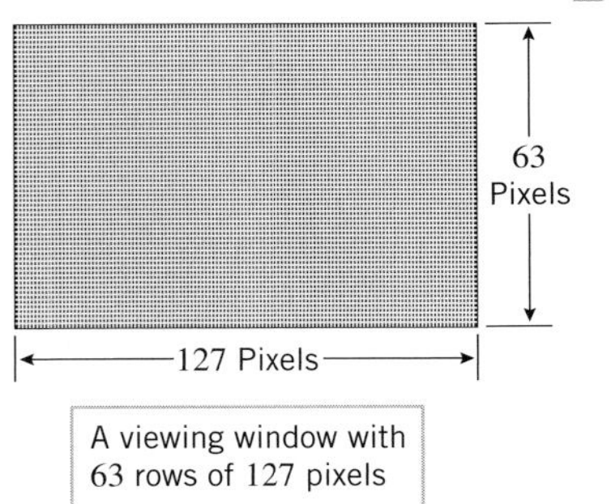

Figure 1.2.13

TECHNOLOGY MASTERY

If you are using a graphing calculator, read the documentation to determine its resolution.

The viewing window of a graphing utility is composed of a rectangular grid of small rectangular blocks called ***pixels***. For black-and-white displays each pixel has two states—an activated (or dark) state and a deactivated (or light state). A graph is formed by activating appropriate pixels to produce the curve shape. In one popular calculator the grid of pixels consists of 63 rows of 127 pixels each (Figure 1.2.13), in which case we say that the screen has a ***resolution*** of 127×63 (pixels in each row $\times$ number of rows). A typical resolution for a computer screen is 1024×768. The greater the resolution, the smoother the graphs tend to appear on the screen.

The procedure that a graphing utility uses to generate a graph is similar to plotting points by hand: When an equation is entered and a window is chosen, the utility *selects* the x-coordinates of certain pixels (the choice of which depends on the window being used) and *computes* the corresponding y-coordinates. It then activates the pixels whose coordinates most closely match those of the calculated points and uses a built-in algorithm to activate additional intermediate pixels to create the curve shape. This process is not perfect, and it is possible that a particular window will produce a false impression about the graph shape because important characteristics of the graph occur between the computed points. This is called ***sampling error***. For example, Figure 1.2.14 shows the graph of $y = \cos(10\pi x)$ produced by a popular calculator in four different windows. (Your calculator may produce different results.) The graph in part (a) has the correct shape, but the other three do not because of sampling error:

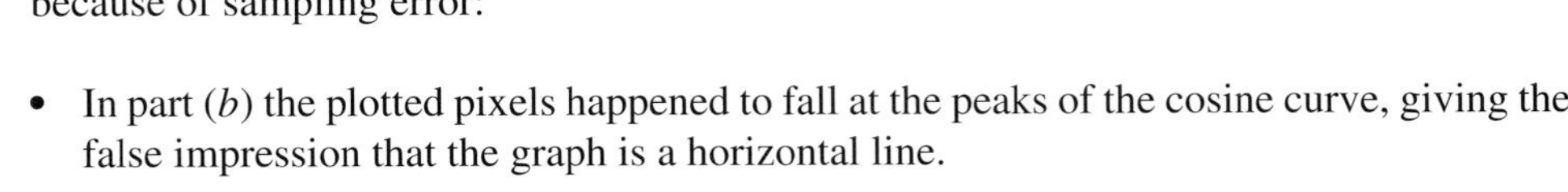

- In part (b) the plotted pixels happened to fall at the peaks of the cosine curve, giving the false impression that the graph is a horizontal line.
- In part (c) the plotted pixels fell at successively higher points along the graph.
- In part (d) the plotted points fell in some regular pattern that created yet another misleading impression of the graph shape.

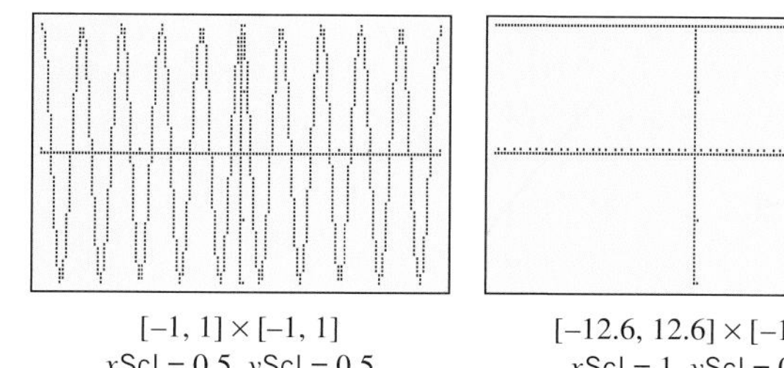
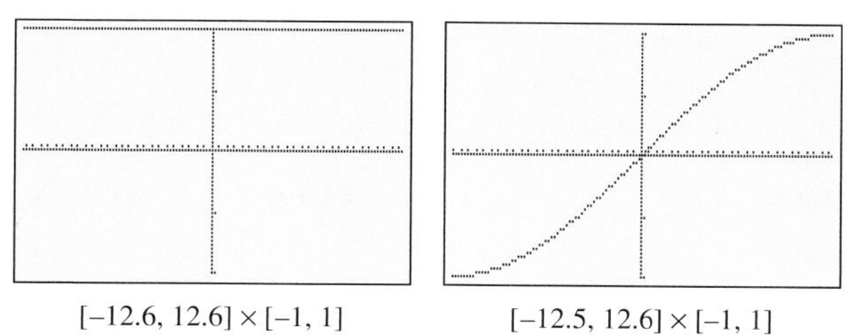
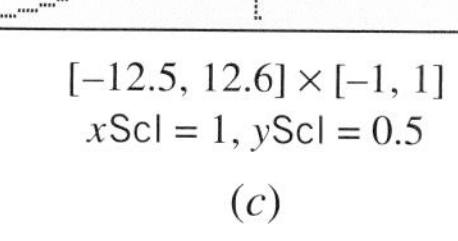
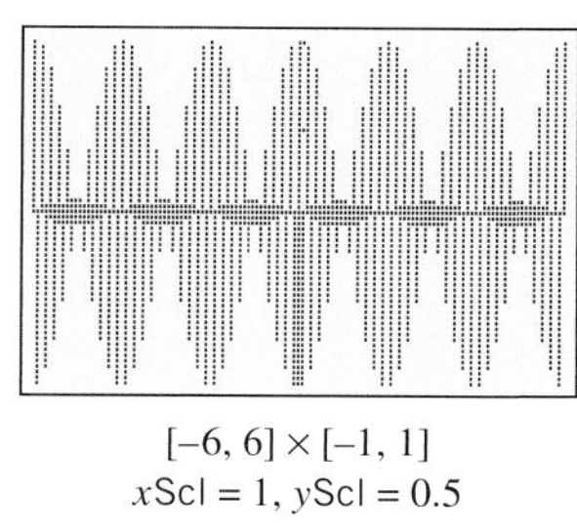

[−1, 1] × [−1, 1]
xScl = 0.5, yScl = 0.5
(*a*)

[−12.6, 12.6] × [−1, 1]
xScl = 1, yScl = 0.5
(*b*)

[−12.5, 12.6] × [−1, 1]
xScl = 1, yScl = 0.5
(*c*)

[−6, 6] × [−1, 1]
xScl = 1, yScl = 0.5
(*d*)

Figure 1.2.14

> For trigonometric graphs with rapid oscillations, Figure 1.2.14 suggests that restricting the x-interval to a few periods is likely to produce a more accurate representation about the graph shape.

FALSE GAPS

Sometimes graphs that are continuous appear to have gaps when they are generated on a calculator. These ***false gaps*** typically occur where the graph rises so rapidly that vertical space is opened up between successive pixels.

▶ **Example 6** Figure 1.2.15 shows the graph of the semicircle $y = \sqrt{9 - x^2}$ in two viewing windows. Although this semicircle has x-intercepts at the points $x = \pm 3$, part (*a*) of the figure shows false gaps at those points because there are no pixels with x-coordinates ± 3 in the window selected. In part (*b*) no gaps occur because there are pixels with x-coordinates $x = \pm 3$ in the window being used. ◀

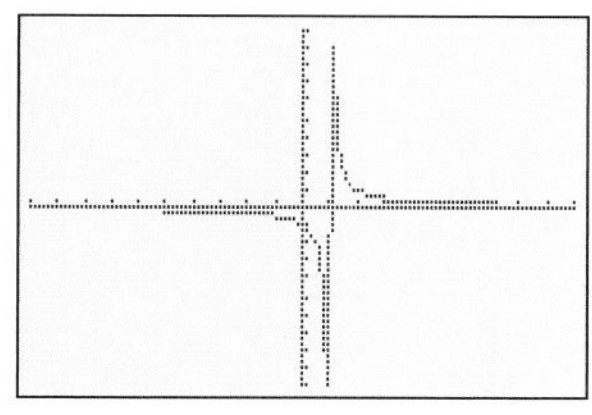

[−10, 10] × [−10, 10]
xScl = 1, yScl = 1

$y = 1/(x - 1)$ with false line segments

(*a*)

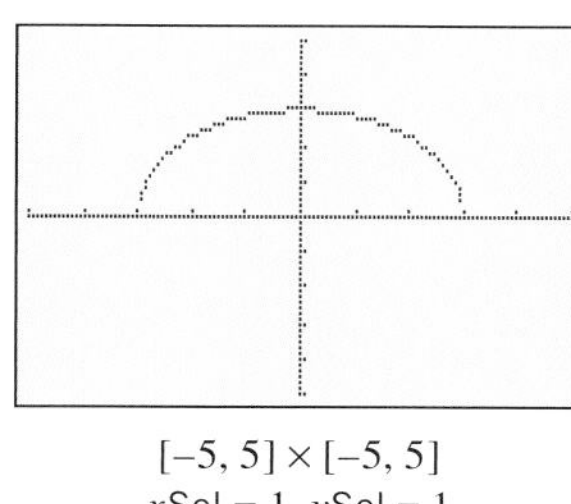

[−5, 5] × [−5, 5]
xScl = 1, yScl = 1

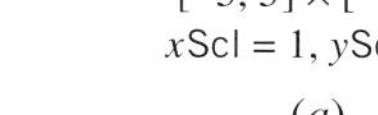

(*a*)

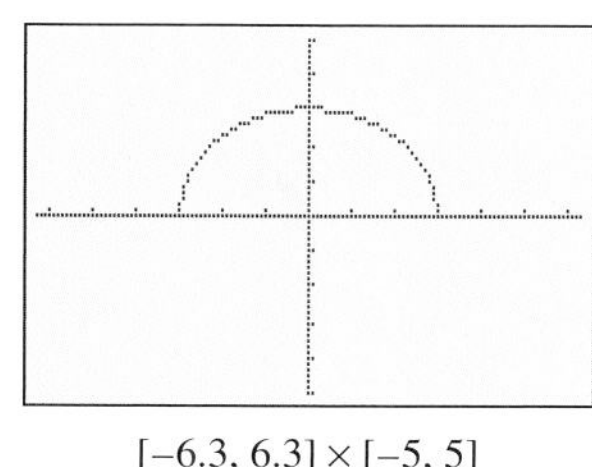

[−6.3, 6.3] × [−5, 5]
xScl = 1, yScl = 1

(*b*)

Figure 1.2.15

FALSE LINE SEGMENTS

In addition to creating false gaps in continuous graphs, calculators can err in the opposite direction by placing ***false line segments*** in the gaps of discontinuous curves.

▶ **Example 7** Figure 1.2.16*a* shows the graph of $y = 1/(x - 1)$ in the default window on a calculator. Although the graph appears to contain vertical line segments near $x = 1$, they should not be there. There is actually a gap in the curve at $x = 1$, since a division by zero occurs at that point (Figure 1.2.16*b*). ◀

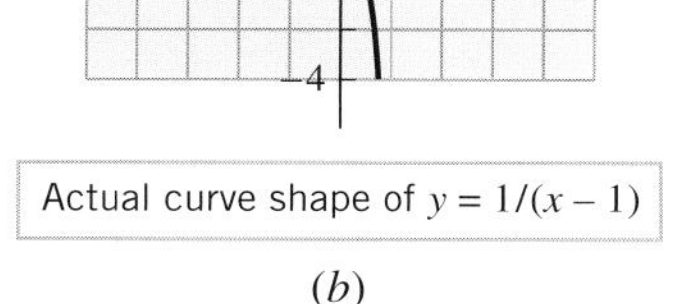

Actual curve shape of $y = 1/(x - 1)$

(*b*)

Figure 1.2.16

ERRORS OF OMISSION

Most graphing utilities use logarithms to evaluate functions with fractional exponents such as $f(x) = x^{2/3} = \sqrt[3]{x^2}$. However, because logarithms are only defined for positive numbers, many (but not all) graphing utilities will omit portions of the graphs of functions with fractional exponents. For example, one calculator graphs $y = x^{2/3}$ as in Figure 1.2.17*a*, whereas the actual graph is as in Figure 1.2.17*b*. (See the discussion preceding Exercise 29 for a way of circumventing this problem.)

TECHNOLOGY MASTERY

Determine whether your graphing utility produces the graph of the equation $y = x^{2/3}$ for both positive and negative values of x.

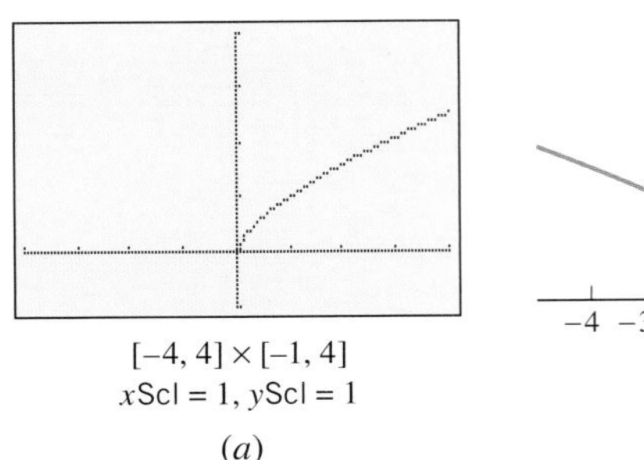

$[-4, 4] \times [-1, 4]$
$x\text{Scl} = 1$, $y\text{Scl} = 1$
(*a*)

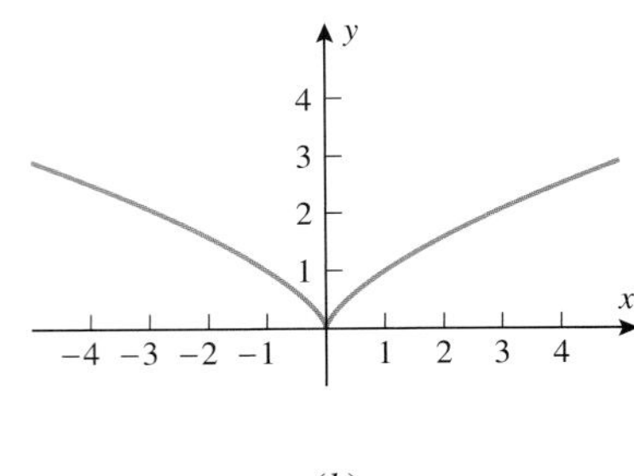

(*b*)

Figure 1.2.17

WHAT IS THE TRUE SHAPE OF A GRAPH?

Although graphing utilities are powerful tools for generating graphs quickly, they can produce misleading graphs as a result of compression, sampling error, false gaps, and false line segments. In short, *graphing utilities can suggest graph shapes, but they cannot establish them with certainty*. Thus, the more you know about the functions you are graphing, the easier it will be to choose good viewing windows, and the better you will be able to judge the reasonableness of the results produced by your graphing utility.

MORE INFORMATION ON GRAPHING AND CALCULATING UTILITIES

The main source of information about your graphing utility is its own documentation, and from time to time we will suggest that you refer to that documentation to learn some particular technique.

✔QUICK CHECK EXERCISES 1.2 *(See page 26 for answers.)*

1. Use a graphing utility to graph the equation

$$y = 0.4x^3 + \sin(3^x)$$

in the given windows, and discuss the advantages of each window.

(a) $[-6, 6] \times [-30, 50]$ (b) $[-4, 4] \times [-25, 25]$
(c) $[1.2, 1.4] \times [0, 0.2]$ (d) $[5.99, 6.01] \times [84, 88]$

2. Use the domain and range of f to find a viewing window that displays the entire graph for $f(x) = 2 - \sqrt{1 - 25x^2}$.

3. Explain how the graph of $y = |x|$ would look if drawn in the viewing window $[-0.01, 0.01] \times [-10, 10]$ on a square display.

4. Explain how the graph of $y = |x|$ would look if drawn in the viewing window $[-10, 10] \times [-0.01, 0.01]$ on a square display.

EXERCISE SET 1.2 Graphing Utility

1–4 Use a graphing utility to generate the graph of f in the given viewing windows, and specify the window that you think gives the best view of the graph.

1. $f(x) = x^4 - x^2$
(a) $[-50, 50] \times [-50, 50]$ (b) $[-5, 5] \times [-5, 5]$
(c) $[-2, 2] \times [-2, 2]$ (d) $[-2, 2] \times [-1, 1]$
(e) $[-1.5, 1.5] \times [-0.5, 0.5]$

2. $f(x) = x^5 - x^3$
(a) $[-50, 50] \times [-50, 50]$ (b) $[-5, 5] \times [-5, 5]$
(c) $[-2, 2] \times [-2, 2]$ (d) $[-2, 2] \times [-1, 1]$
(e) $[-1.5, 1.5] \times [-0.5, 0.5]$

3. $f(x) = x^2 + 12$
(a) $[-1, 1] \times [13, 15]$ (b) $[-2, 2] \times [11, 15]$
(c) $[-4, 4] \times [10, 28]$ (d) A window of your choice

4. $f(x) = -12 - x^2$
(a) $[-1, 1] \times [-15, -13]$ (b) $[-2, 2] \times [-15, -11]$
(c) $[-4, 4] \times [-28, -10]$ (d) A window of your choice

5–6 Use the domain and range of f to determine a viewing window that contains the entire graph, and generate the graph in that window.

5. $f(x) = \sqrt{16 - 2x^2}$ 6. $f(x) = \sqrt{3 - 2x - x^2}$

FOCUS ON CONCEPTS

7. Graph the function $f(x) = x^3 - 15x^2 - 3x + 45$ using the stated windows and tick spacing, and discuss the advantages and disadvantages of each window.
(a) $[-10, 10] \times [-10, 10]$ with $x\text{Scl} = 1$ and $y\text{Scl} = 1$
(b) $[-20, 20] \times [-20, 20]$ with $x\text{Scl} = 1$ and $y\text{Scl} = 1$
(c) $[-5, 20] \times [-500, 50]$ with $x\text{Scl} = 5$ and $y\text{Scl} = 50$
(d) $[-2, -1] \times [-1, 1]$ with $x\text{Scl} = 0.1$ and $y\text{Scl} = 0.1$
(e) $[9, 11] \times [-486, -484]$ with $x\text{Scl} = 0.1$ and $y\text{Scl} = 0.1$

8. Graph the function $f(x) = -x^3 - 12x^2 + 4x + 48$ using the stated windows and tick spacing, and discuss the advantages and disadvantages of each window.
(a) $[-10, 10] \times [-10, 10]$ with $x\text{Scl} = 1$ and $y\text{Scl} = 1$
(b) $[-20, 20] \times [-20, 20]$ with $x\text{Scl} = 1$ and $y\text{Scl} = 1$
(c) $[-16, 4] \times [-250, 50]$ with $x\text{Scl} = 2$ and $y\text{Scl} = 25$
(d) $[-3, -1] \times [-1, 1]$ with $x\text{Scl} = 0.1$ and $y\text{Scl} = 0.1$
(e) $[-9, -7] \times [-241, -239]$ with $x\text{Scl} = 0.1$ and $y\text{Scl} = 0.1$

9–16 Generate the graph of f in a viewing window that you think is appropriate.

9. $f(x) = x^2 - 9x - 36$ 10. $f(x) = \dfrac{x + 7}{x - 9}$

11. $f(x) = 2\cos(80x)$ 12. $f(x) = 12\sin(x/80)$

13. $f(x) = 300 - 10x^2 + 0.01x^3$

14. $f(x) = x(30 - 2x)(25 - 2x)$

15. $f(x) = x^2 + \dfrac{1}{x}$ 16. $f(x) = \sqrt{11x - 18}$

17–18 Generate the graph of f and determine whether your graphs contain false line segments. Sketch the actual graph and see if you can make the false line segments disappear by changing the viewing window.

17. $f(x) = \dfrac{x}{x^2 - 1}$ 18. $f(x) = \dfrac{x^2}{4 - x^2}$

19. The graph of the equation $x^2 + y^2 = 16$ is a circle of radius 4 centered at the origin.
(a) Find a function whose graph is the upper semicircle and graph it.
(b) Find a function whose graph is the lower semicircle and graph it.
(c) Graph the upper and lower semicircles together. If the combined graphs do not appear circular, see if you can adjust the viewing window to eliminate the aspect ratio distortion.
(d) Graph the portion of the circle in the first quadrant.
(e) Is there a function whose graph is the right half of the circle? Explain.

20. In each part, graph the equation by solving for y in terms of x and graphing the resulting functions together.
(a) $x^2/4 + y^2/9 = 1$ (b) $y^2 - x^2 = 1$

21. Read the documentation for your graphing utility to determine how to graph functions involving absolute values, and graph the given equation.
(a) $y = |x|$ (b) $y = |x - 1|$
(c) $y = |x| - 1$ (d) $y = |\sin x|$
(e) $y = \sin |x|$ (f) $y = |x| - |x + 1|$

22. Based on your knowledge of the absolute value function, sketch the graph of $f(x) = |x|/x$. Check your result using a graphing utility.

FOCUS ON CONCEPTS

23. Make a conjecture about the relationship between the graph of $y = f(x)$ and the graph of $y = |f(x)|$; check your conjecture with some specific functions.

24. Make a conjecture about the relationship between the graph of $y = f(x)$ and the graph of $y = f(|x|)$; check your conjecture with some specific functions.

25. (a) Based on your knowledge of the absolute value function, sketch the graph of $y = |x - a|$, where a is a constant. Check your result using a graphing utility and some specific values of a.
(b) Sketch the graph of $y = |x - 1| + |x - 2|$; check your result with a graphing utility.

26. How are the graphs of $y = |x|$ and $y = \sqrt{x^2}$ related? Check your answer with a graphing utility.

27–28 Most graphing utilities provide some way of graphing functions that are defined piecewise; read the documentation for your graphing utility to find out how to do this. However, if your goal is just to find the general shape of the graph, you can graph each portion of the function separately and combine the pieces with a hand-drawn sketch. Use this method in these exercises.

27. Draw the graph of

$$f(x) = \begin{cases} \sqrt[3]{x-2}, & x \le 2 \\ x^3 - 2x - 4, & x > 2 \end{cases}$$

28. Draw the graph of

$$f(x) = \begin{cases} x^3 - x^2, & x \le 1 \\ \dfrac{1}{1-x}, & 1 < x < 4 \\ x^2 \cos\sqrt{x}, & 4 \le x \end{cases}$$

29–30 We noted in the text that for functions involving fractional exponents (or radicals), graphing utilities sometimes omit portions of the graph. If $f(x) = x^{p/q}$, where p/q is a positive fraction in *lowest terms*, then you can circumvent this problem as follows:

- If p is even and q is odd, then graph $g(x) = |x|^{p/q}$ instead of $f(x)$.
- If p is odd and q is odd, then graph $g(x) = (|x|/x)|x|^{p/q}$ instead of $f(x)$.

We will explain why this works in the exercises of the next section.

29. (a) Generate the graphs of $f(x) = x^{2/5}$ and $g(x) = |x|^{2/5}$, and determine whether your graphing utility missed part of the graph of f.
(b) Generate the graphs of the functions $f(x) = x^{1/5}$ and $g(x) = (|x|/x)|x|^{1/5}$, and determine whether your graphing utility missed part of the graph of f.
(c) Generate a graph of the function $f(x) = (x-1)^{4/5}$ that shows all of its important features.
(d) Generate a graph of the function $f(x) = (x+1)^{3/4}$ that shows all of its important features.

30. The graphs of $y = (x^2-4)^{2/3}$ and $y = [(x^2-4)^2]^{1/3}$ should be the same. Does your graphing utility produce the same graph for both equations? If not, what do you think is happening?

31. In each part, graph the function for various values of c, and write a paragraph or two that describes how changes in c affect the graph in each case.
(a) $y = cx^2$ (b) $y = x^2 + cx$ (c) $y = x^2 + x + c$

32. The graph of an equation of the form $y^2 = x(x-a)(x-b)$ (where $0 < a < b$) is called a ***bipartite cubic***. The accompanying figure shows a typical graph of this type.
(a) Graph the bipartite cubic $y^2 = x(x-1)(x-2)$ by solving for y in terms of x and graphing the two resulting functions.
(b) Find the x-intercepts of the bipartite cubic

$$y^2 = x(x-a)(x-b)$$

and make a conjecture about how changes in the values of a and b would affect the graph. Test your conjecture by graphing the bipartite cubic for various values of a and b.

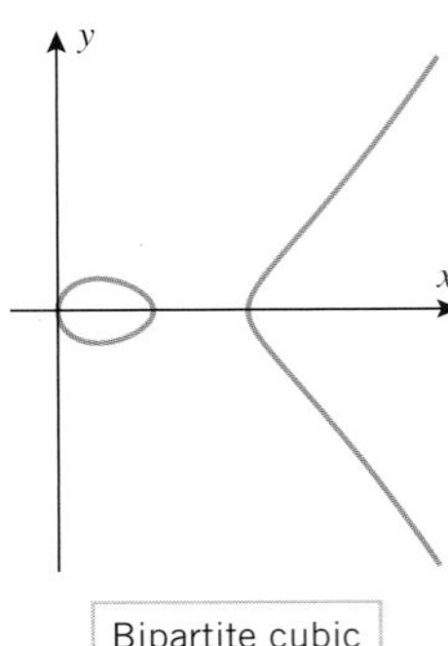

Figure Ex-32

33. Based on your knowledge of the graphs of $y = x$ and $y = \sin x$, make a sketch of the graph of $y = x \sin x$. Check your conclusion using a graphing utility.

34. What do you think the graph of $y = \sin(1/x)$ looks like? Test your conclusion using a graphing utility. [*Suggestion:* Examine the graph on a succession of smaller and smaller intervals centered at $x = 0$.]

✔QUICK CHECK ANSWERS 1.2

1. (a)

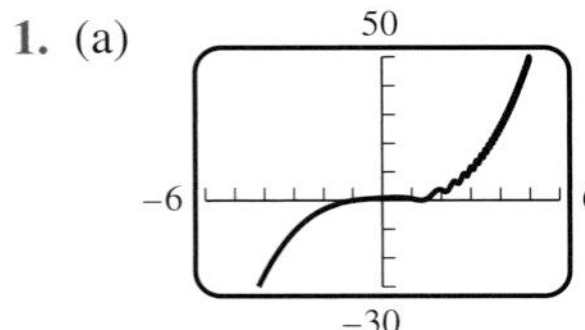

The graph is similar to a cubic, with some variation near the origin.

(b)

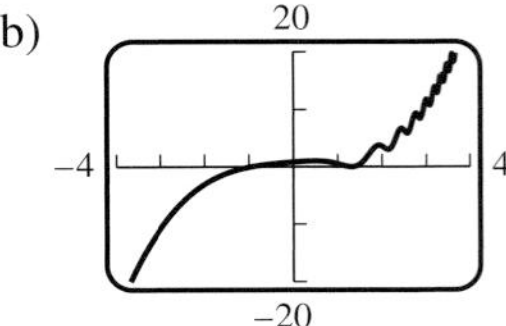

The zero near $x = -1$ is more evident and there is a possible zero just right of $x = 1$.

(c)

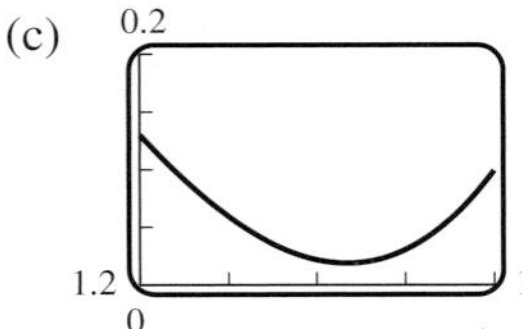

The function does not have a zero to the right of $x = 1$.

(d)

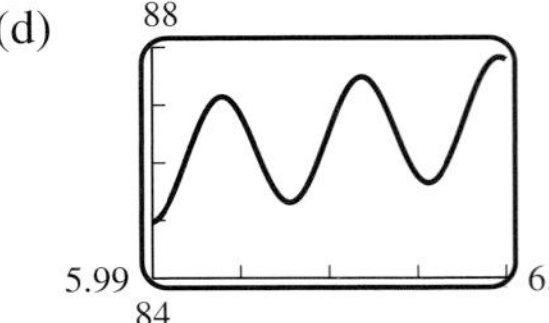

The term $\sin(3^x)$ produces rapid oscillations in the graph as x increases. The $0.4x^3$ term causes the graph to rise steeply for larger x-values, hiding this oscillation in part (a).

2. $[-0.2, 0.2] \times [1, 2]$ **3.** The graph is indistinguishable from the x-axis.
4. The graph is indistinguishable from the nonnegative y-axis.

1.3 NEW FUNCTIONS FROM OLD

Just as numbers can be added, subtracted, multiplied, and divided to produce other numbers, so functions can be added, subtracted, multiplied, and divided to produce other functions. In this section we will discuss these operations and some others that have no analogs in ordinary arithmetic.

ARITHMETIC OPERATIONS ON FUNCTIONS

Two functions, f and g, can be added, subtracted, multiplied, and divided in a natural way to form new functions $f+g$, $f-g$, fg, and f/g. For example, $f+g$ is defined by the formula

$$(f+g)(x) = f(x) + g(x) \tag{1}$$

which states that for each input the value of $f+g$ is obtained by adding the values of f and g. For example, if

$$f(x) = x \quad \text{and} \quad g(x) = x^2$$

then

$$(f+g)(x) = f(x) + g(x) = x + x^2$$

Equation (1) provides a formula for $f+g$ but does not say anything about the domain of $f+g$. However, for the right side of this equation to be defined, x must lie in the domains of both f and g, so we define the domain of $f+g$ to be the intersection of these two domains. More generally, we make the following definition.

1.3.1 DEFINITION. Given functions f and g, we define

$$(f+g)(x) = f(x) + g(x)$$
$$(f-g)(x) = f(x) - g(x)$$
$$(fg)(x) = f(x)g(x)$$
$$(f/g)(x) = f(x)/g(x)$$

For the functions $f+g$, $f-g$, and fg we define the domain to be the intersection of the domains of f and g, and for the function f/g we define the domain to be the intersection of the domains of f and g but with the points where $g(x) = 0$ excluded (to avoid division by zero).

If f is a constant function, that is $f(x) = c$ for all x, then the product of f and g is cg, so multiplying a function by a constant is a special case of multiplying two functions.

▶ **Example 1** Let

$$f(x) = 1 + \sqrt{x-2} \quad \text{and} \quad g(x) = x - 3$$

Find the domains and formulas for the functions $f+g$, $f-g$, fg, f/g, and $7f$.

Solution. First, we will find the formulas and then the domains. The formulas are

$$(f+g)(x) = f(x) + g(x) = (1 + \sqrt{x-2}) + (x-3) = x - 2 + \sqrt{x-2} \tag{2}$$

$$(f-g)(x) = f(x) - g(x) = (1 + \sqrt{x-2}) - (x-3) = 4 - x + \sqrt{x-2} \tag{3}$$

$$(fg)(x) = f(x)g(x) = (1 + \sqrt{x-2})(x-3) \tag{4}$$

$$(f/g)(x) = f(x)/g(x) = \frac{1 + \sqrt{x-2}}{x-3} \tag{5}$$

$$(7f)(x) = 7f(x) = 7 + 7\sqrt{x-2} \tag{6}$$

The domains of f and g are $[2, +\infty)$ and $(-\infty, +\infty)$, respectively (their natural domains). Thus, it follows from Definition 1.3.1 that the domains of $f+g$, $f-g$, and fg are the intersection of these two domains, namely,

$$[2, +\infty) \cap (-\infty, +\infty) = [2, +\infty) \tag{7}$$

Moreover, since $g(x) = 0$ if $x = 3$, the domain of f/g is (7) with $x = 3$ removed, namely,

$$[2, 3) \cup (3, +\infty)$$

Finally, the domain of $7f$ is the same as the domain of f. ◀

It turned out in the last example that the domains of the functions $f+g$, $f-g$, fg, and f/g were the natural domains resulting from the formulas obtained for these functions. This will not always be the case, and here is an example.

▶ **Example 2** Show that if $f(x) = \sqrt{x}$, $g(x) = \sqrt{x}$, and $h(x) = x$, then the domain of fg is not the same as the natural domain of h.

Solution. The natural domain of $h(x) = x$ is $(-\infty, +\infty)$. Note that

$$(fg)(x) = \sqrt{x}\sqrt{x} = x = h(x)$$

on the domain of fg. The domains of both f and g are $[0, +\infty)$, so the domain of fg is

$$[0, +\infty) \cap [0, +\infty) = [0, +\infty)$$

by Definition 1.3.1. Since the domains of fg and h are different, it would be misleading to write $(fg)(x) = x$ without including the restriction that this formula holds only for $x \geq 0$. ◀

COMPOSITION OF FUNCTIONS

We now consider an operation on functions, called *composition*, which has no direct analog in ordinary arithmetic. Informally stated, the operation of composition is performed by substituting some function for the independent variable of another function. For example, suppose that

$$f(x) = x^2 \quad \text{and} \quad g(x) = x + 1$$

If we substitute $g(x)$ for x in the formula for f, we obtain a new function

$$f(g(x)) = (g(x))^2 = (x+1)^2$$

which we denote by $f \circ g$. Thus,

$$(f \circ g)(x) = f(g(x)) = (g(x))^2 = (x+1)^2$$

In general, we make the following definition.

Although the domain of $f \circ g$ may seem complicated at first glance, it makes sense intuitively: To compute $f(g(x))$ one needs x in the domain of g to compute $g(x)$, then one needs $g(x)$ in the domain of f to compute $f(g(x))$.

1.3.2 DEFINITION. Given functions f and g, the ***composition*** of f with g, denoted by $f \circ g$, is the function defined by

$$(f \circ g)(x) = f(g(x))$$

The domain of $f \circ g$ is defined to consist of all x in the domain of g for which $g(x)$ is in the domain of f.

▶ **Example 3** Let $f(x) = x^2 + 3$ and $g(x) = \sqrt{x}$. Find

(a) $(f \circ g)(x)$ (b) $(g \circ f)(x)$

Solution (a). The formula for $f(g(x))$ is

$$f(g(x)) = [g(x)]^2 + 3 = (\sqrt{x}\,)^2 + 3 = x + 3$$

Since the domain of g is $[0, +\infty)$ and the domain of f is $(-\infty, +\infty)$, the domain of $f \circ g$ consists of all x in $[0, +\infty)$ such that $g(x) = \sqrt{x}$ lies in $(-\infty, +\infty)$; thus, the domain of $f \circ g$ is $[0, +\infty)$. Therefore,

$$(f \circ g)(x) = x + 3, \quad x \geq 0$$

Solution (b). The formula for $g(f(x))$ is

$$g(f(x)) = \sqrt{f(x)} = \sqrt{x^2 + 3}$$

Since the domain of f is $(-\infty, +\infty)$ and the domain of g is $[0, +\infty)$, the domain of $g \circ f$ consists of all x in $(-\infty, +\infty)$ such that $f(x) = x^2 + 3$ lies in $[0, +\infty)$. Thus, the domain of $g \circ f$ is $(-\infty, +\infty)$. Therefore,

$$(g \circ f)(x) = \sqrt{x^2 + 3}$$

There is no need to indicate that the domain is $(-\infty, +\infty)$, since this is the natural domain of $\sqrt{x^2 + 3}$. ◀

Note that the functions $f \circ g$ and $g \circ f$ in Example 3 are not the same. Thus, the order in which functions are composed can (and usually will) make a difference in the end result.

Compositions can also be defined for three or more functions; for example, $(f \circ g \circ h)(x)$ is computed as

$$(f \circ g \circ h)(x) = f(g(h(x)))$$

In other words, first find $h(x)$, then find $g(h(x))$, and then find $f(g(h(x)))$.

▶ **Example 4** Find $(f \circ g \circ h)(x)$ if

$$f(x) = \sqrt{x}, \quad g(x) = 1/x, \quad h(x) = x^3$$

Solution.

$$(f \circ g \circ h)(x) = f(g(h(x))) = f(g(x^3)) = f(1/x^3) = \sqrt{1/x^3} = 1/x^{3/2}$$ ◀

EXPRESSING A FUNCTION AS A COMPOSITION

Many problems in mathematics are attacked by "decomposing" functions into compositions of simpler functions. For example, consider the function h given by

$$h(x) = (x + 1)^2$$

To evaluate $h(x)$ for a given value of x, we would first compute $x + 1$ and then square the result. These two operations are performed by the functions

$$g(x) = x + 1 \quad \text{and} \quad f(x) = x^2$$

We can express h in terms of f and g by writing

$$h(x) = (x + 1)^2 = [g(x)]^2 = f(g(x))$$

so we have succeeded in expressing h as the composition $h = f \circ g$.

The thought process in this example suggests a general procedure for decomposing a function h into a composition $h = f \circ g$:

- Think about how you would evaluate $h(x)$ for a specific value of x, trying to break the evaluation into two steps performed in succession.
- The first operation in the evaluation will determine a function g and the second a function f.
- The formula for h can then be written as $h(x) = f(g(x))$.

For descriptive purposes, we will refer to g as the "inside function" and f as the "outside function" in the expression $f(g(x))$. The inside function performs the first operation and the outside function performs the second.

▶ **Example 5** Express $h(x) = (x - 4)^5$ as a composition of two functions.

Solution. To evaluate $h(x)$ for a given value of x we would first compute $x - 4$ and then raise the result to the fifth power. Therefore, the inside function (first operation) is

$$g(x) = x - 4$$

and the outside function (second operation) is

$$f(x) = x^5$$

so $h(x) = f(g(x))$. As a check,

$$f(g(x)) = [g(x)]^5 = (x - 4)^5 = h(x) \blacktriangleleft$$

▶ **Example 6** Express $\sin(x^3)$ as a composition of two functions.

Solution. To evaluate $\sin(x^3)$, we would first compute x^3 and then take the sine, so $g(x) = x^3$ is the inside function and $f(x) = \sin x$ the outside function. Therefore,

$$\sin(x^3) = f(g(x)) \quad \boxed{g(x) = x^3 \text{ and } f(x) = \sin x} \blacktriangleleft$$

Table 1.3.1 gives some more examples of decomposing functions into compositions.

Table 1.3.1

FUNCTION	$g(x)$ INSIDE	$f(x)$ OUTSIDE	COMPOSITION
$(x^2+1)^{10}$	x^2+1	x^{10}	$(x^2+1)^{10} = f(g(x))$
$\sin^3 x$	$\sin x$	x^3	$\sin^3 x = f(g(x))$
$\tan(x^5)$	x^5	$\tan x$	$\tan(x^5) = f(g(x))$
$\sqrt{4-3x}$	$4-3x$	$\sqrt{x}$	$\sqrt{4-3x} = f(g(x))$
$8+\sqrt{x}$	$\sqrt{x}$	$8+x$	$8+\sqrt{x} = f(g(x))$
$\dfrac{1}{x+1}$	$x+1$	$\dfrac{1}{x}$	$\dfrac{1}{x+1} = f(g(x))$

There is always more than one way to express a function as a composition. For example, here are two ways to express $(x^2+1)^{10}$ as a composition that differ from that in Table 1.3.1:

$$(x^2+1)^{10} = [(x^2+1)^2]^5 = f(g(x)) \qquad g(x) = (x^2+1)^2 \text{ and } f(x) = x^5$$

$$(x^2+1)^{10} = [(x^2+1)^3]^{10/3} = f(g(x)) \qquad g(x) = (x^2+1)^3 \text{ and } f(x) = x^{10/3}$$

NEW FUNCTIONS FROM OLD

The remainder of this section will be devoted to considering the geometric effect of performing basic operations on functions. This will enable us to use known graphs of functions to visualize or sketch graphs of related functions. For example, Figure 1.3.1 shows the graphs of yearly new car sales $N(t)$ and used car sales $U(t)$ over a certain time period. Those graphs can be used to construct the graph of the total car sales $T(t) = N(t) + U(t)$ by adding the values of $N(t)$ and $U(t)$ for each value of t. In general, the graph of $y = f(x) + g(x)$ can be constructed from the graphs of $y = f(x)$ and $y = g(x)$ by adding corresponding y-values for each x.

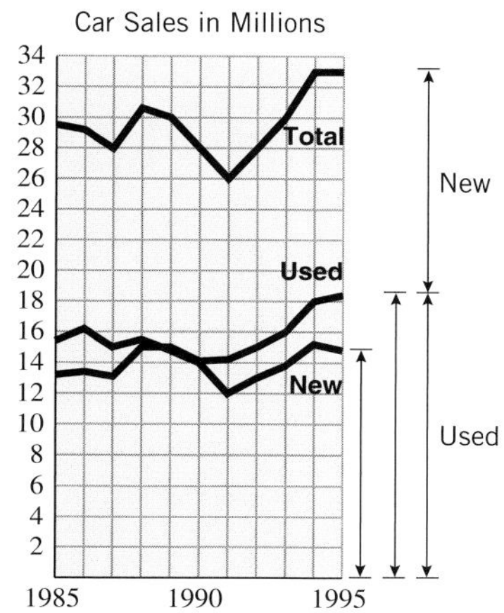

Source: NADA.

Figure 1.3.1

► **Example 7** Referring to Figure 1.1.4 for the graphs of $y = \sqrt{x}$ and $y = 1/x$, make a sketch that shows the general shape of the graph of $y = \sqrt{x} + 1/x$ for $x \geq 0$.

Solution. To add the corresponding y-values of $y = \sqrt{x}$ and $y = 1/x$ graphically, just imagine them to be "stacked" on top of one another. This yields the sketch in Figure 1.3.2. ◄

Use the technique in Example 7 to sketch the graph of the function

$$\sqrt{x} - \frac{1}{x}$$

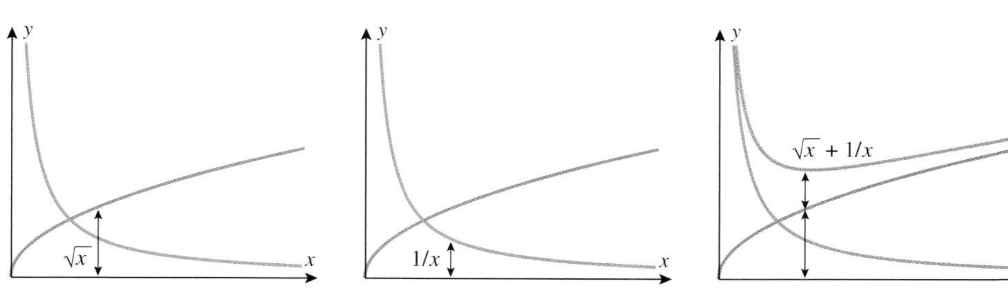

Figure 1.3.2 Add the y-coordinates of $\sqrt{x}$ and $1/x$ to obtain the y-coordinate of $\sqrt{x} + 1/x$.

TRANSLATIONS

Table 1.3.2 illustrates the geometric effect on the graph of $y = f(x)$ of adding or subtracting a *positive* constant c to f or to its independent variable x. For example, the first result in the table illustrates that adding a positive constant c to a function f adds c to each y-coordinate of its graph, thereby shifting the graph of f up by c units. Similarly, subtracting c from f shifts the graph down by c units. On the other hand, if a positive constant c is added to x, then the value of $y = f(x+c)$ at $x - c$ is $f(x)$; and since the point $x - c$ is c units to the left of x on the x-axis, the graph of $y = f(x+c)$ must be the graph of $y = f(x)$ shifted left by c units. Similarly, subtracting c from x shifts the graph of $y = f(x)$ right by c units.

Before proceeding to the next examples, it will be helpful to review the graphs in Figures 1.1.4 and 1.1.10.

Table 1.3.2

OPERATION ON $y = f(x)$	Add a positive constant c to $f(x)$	Subtract a positive constant c from $f(x)$	Add a positive constant c to x	Subtract a positive constant c from x
NEW EQUATION	$y = f(x) + c$	$y = f(x) - c$	$y = f(x + c)$	$y = f(x - c)$
GEOMETRIC EFFECT	Translates the graph of $y = f(x)$ up c units	Translates the graph of $y = f(x)$ down c units	Translates the graph of $y = f(x)$ left c units	Translates the graph of $y = f(x)$ right c units
EXAMPLE	$y = x^2 + 2$; $y = x^2$; 2	$y = x^2$; $y = x^2 - 2$; −2	$y = (x + 2)^2$; $y = x^2$; −2	$y = (x - 2)^2$; $y = x^2$; 2

▶ **Example 8** Sketch the graph of

$$\text{(a)}\ y = \sqrt{x - 3} \qquad \text{(b)}\ y = \sqrt{x + 3}$$

Solution. The graph of the equation $y = \sqrt{x - 3}$ can be obtained by translating the graph of $y = \sqrt{x}$ right 3 units, and the graph of $y = \sqrt{x + 3}$ can be obtained by translating the graph of $y = \sqrt{x}$ left 3 units (Figure 1.3.3). ◀

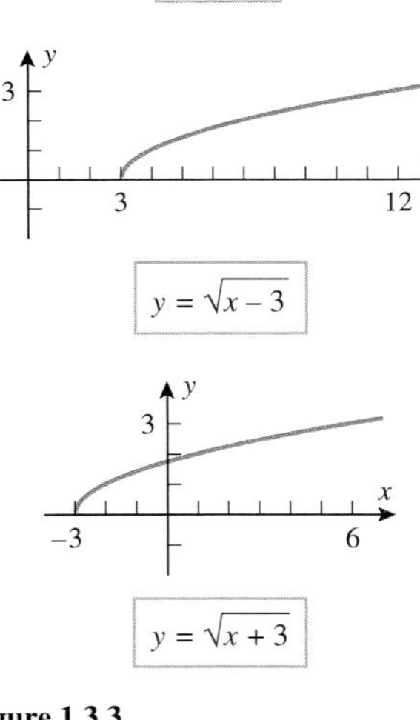

Figure 1.3.3

▶ **Example 9** Sketch the graph of $y = |x - 3| + 2$.

Solution. The graph can be obtained by two translations: first translate the graph of $y = |x|$ right 3 units to obtain the graph of $y = |x - 3|$, then translate this graph up 2 units to obtain the graph of $y = |x - 3| + 2$ (Figure 1.3.4). If desired, the same result can be obtained by performing the translations in the opposite order: first translate the graph of $|x|$ up 2 units to obtain the graph of $y = |x| + 2$, then translate this graph right 3 units to obtain the graph of $y = |x - 3| + 2$. ◀

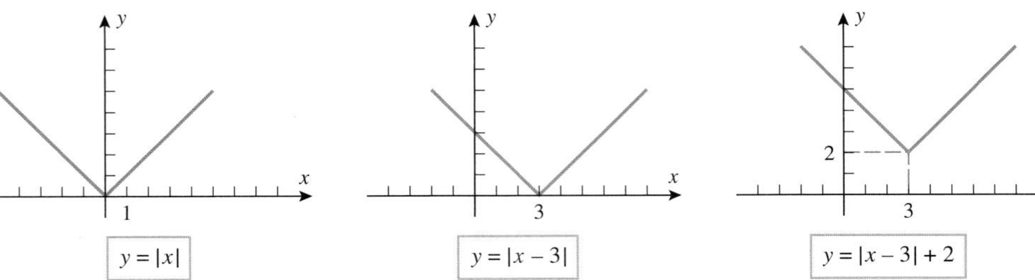

Figure 1.3.4

▶ **Example 10** Sketch the graph of $y = x^2 - 4x + 5$.

Solution. Completing the square on the first two terms yields

$$y = (x^2 - 4x + 4) - 4 + 5 = (x - 2)^2 + 1$$

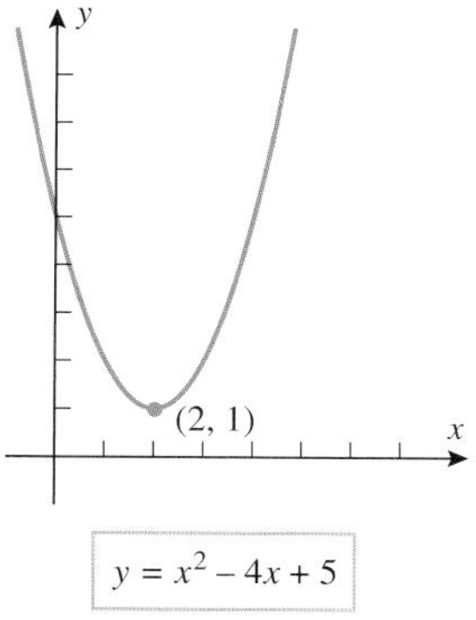

Figure 1.3.5

(see Web Appendix G for a review of this technique). In this form we see that the graph can be obtained by translating the graph of $y = x^2$ right 2 units because of the $x - 2$, and up 1 unit because of the $+1$ (Figure 1.3.5). ◀

REFLECTIONS

The graph of $y = f(-x)$ is the reflection of the graph of $y = f(x)$ about the y-axis because the point (x, y) on the graph of $f(x)$ is replaced by $(-x, y)$. Similarly, the graph of $y = -f(x)$ is the reflection of the graph of $y = f(x)$ about the x-axis because the point (x, y) on the graph of $f(x)$ is replaced by $(x, -y)$ [the equation $y = -f(x)$ is equivalent to $-y = f(x)$]. This is summarized in Table 1.3.3.

Table 1.3.3

OPERATION ON $y = f(x)$	Replace x by $-x$	Multiply $f(x)$ by -1
NEW EQUATION	$y = f(-x)$	$y = -f(x)$
GEOMETRIC EFFECT	Reflects the graph of $y = f(x)$ about the y-axis	Reflects the graph of $y = f(x)$ about the x-axis
EXAMPLE	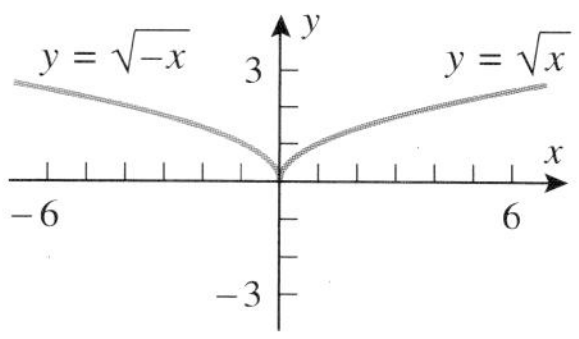	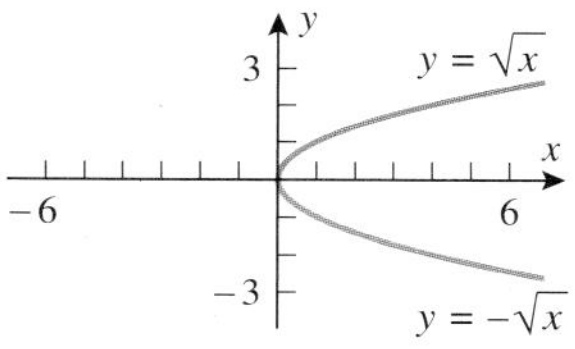

▶ **Example 11** Sketch the graph of $y = \sqrt[3]{2 - x}$.

Solution. The graph can be obtained by a reflection and a translation: first reflect the graph of $y = \sqrt[3]{x}$ about the y-axis to obtain the graph of $y = \sqrt[3]{-x}$, then translate this graph right 2 units to obtain the graph of the equation $y = \sqrt[3]{-(x - 2)} = \sqrt[3]{2 - x}$ (Figure 1.3.6). ◀

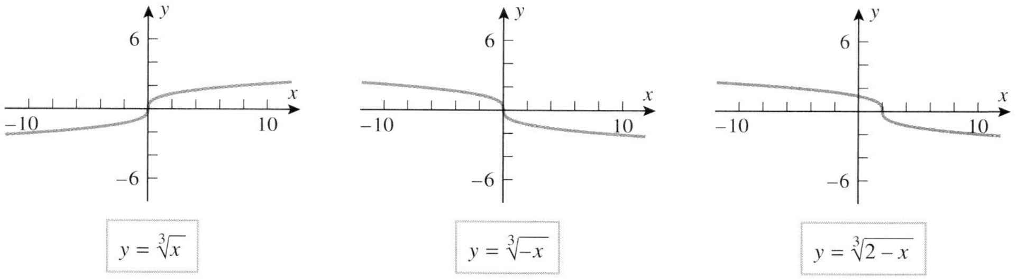

Figure 1.3.6

▶ **Example 12** Sketch the graph of $y = 4 - |x - 2|$.

Solution. The graph can be obtained by a reflection and two translations: first translate the graph of $y = |x|$ right 2 units to obtain the graph of $y = |x - 2|$; then reflect this graph about the x-axis to obtain the graph of $y = -|x - 2|$; and then translate this graph up 4 units to obtain the graph of the equation $y = -|x - 2| + 4 = 4 - |x - 2|$ (Figure 1.3.7). ◀

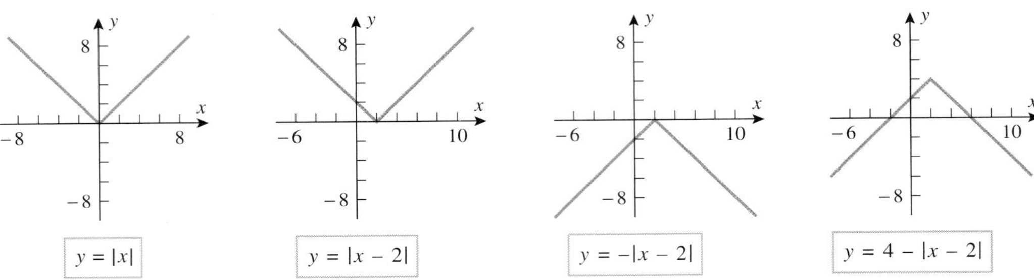

Figure 1.3.7

STRETCHES AND COMPRESSIONS

Multiplying $f(x)$ by a *positive* constant c has the geometric effect of stretching the graph of $y = f(x)$ in the y-direction by a factor of c if $c > 1$ and compressing it in the y-direction by a factor of $1/c$ if $0 < c < 1$. For example, multiplying $f(x)$ by 2 doubles each y-coordinate, thereby stretching the graph vertically by a factor of 2, and multiplying by $\frac{1}{2}$ cuts each y-coordinate in half, thereby compressing the graph vertically by a factor of 2. Similarly, multiplying x by a *positive* constant c has the geometric effect of compressing the graph of $y = f(x)$ by a factor of c in the x-direction if $c > 1$ and stretching it by a factor of $1/c$ if $0 < c < 1$. [If this seems backwards to you, then think of it this way: The value of $2x$ changes twice as fast as x, so a point moving along the x-axis from the origin will only have to move half as far for $y = f(2x)$ to have the same value as $y = f(x)$, thereby creating a horizontal compression of the graph.] All of this is summarized in Table 1.3.4.

> Describe the geometric effect of multiplying a function f by a *negative* constant in terms of reflection and stretching or compressing. What is the geometric effect of multiplying the independent variable of a function f by a *negative* constant?

Table 1.3.4

OPERATION ON $y = f(x)$	Multiply $f(x)$ by c $(c > 1)$	Multiply $f(x)$ by c $(0 < c < 1)$	Multiply x by c $(c > 1)$	Multiply x by c $(0 < c < 1)$
NEW EQUATION	$y = cf(x)$	$y = cf(x)$	$y = f(cx)$	$y = f(cx)$
GEOMETRIC EFFECT	Stretches the graph of $y = f(x)$ vertically by a factor of c	Compresses the graph of $y = f(x)$ vertically by a factor of $1/c$	Compresses the graph of $y = f(x)$ horizontally by a factor of c	Stretches the graph of $y = f(x)$ horizontally by a factor of $1/c$
EXAMPLE	$y = 2\cos x$; $y = \cos x$	$y = \cos x$; $y = \frac{1}{2}\cos x$	$y = \cos x$; $y = \cos 2x$	$y = \cos \frac{1}{2}x$; $y = \cos x$

SYMMETRY

Figure 1.3.8 illustrates three types of symmetries: ***symmetry about the x-axis***, ***symmetry about the y-axis***, and ***symmetry about the origin***. As illustrated in the figure, a curve is symmetric about the x-axis if for each point (x, y) on the graph the point $(x, -y)$ is also on the graph, and it is symmetric about the y-axis if for each point (x, y) on the graph the point $(-x, y)$ is also on the graph. A curve is symmetric about the origin if for each point (x, y) on the graph, the point $(-x, -y)$ is also on the graph. (Equivalently, a graph is symmetric about the origin if rotating the graph 180° about the origin leaves it unchanged.) This suggests the following symmetry tests.

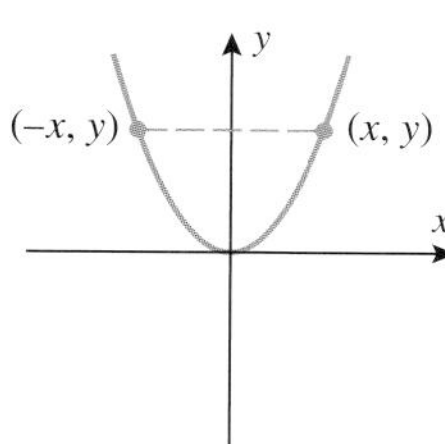

Symmetric about the x-axis

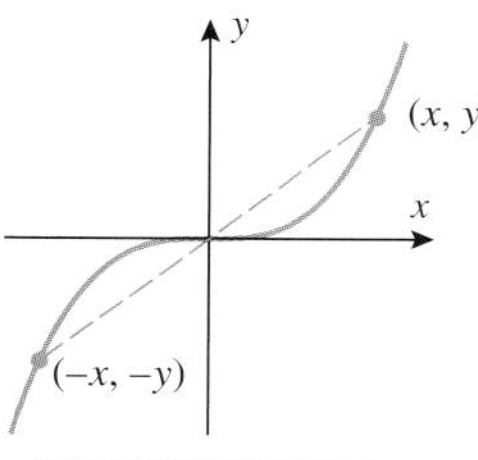

Symmetric about the y-axis

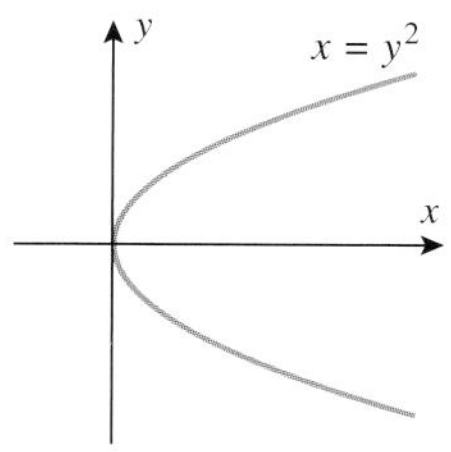

Symmetric about the origin

Figure 1.3.8

1.3.3 THEOREM (*Symmetry Tests*).

(a) *A plane curve is symmetric about the y-axis if and only if replacing x by $-x$ in its equation produces an equivalent equation.*

(b) *A plane curve is symmetric about the x-axis if and only if replacing y by $-y$ in its equation produces an equivalent equation.*

(c) *A plane curve is symmetric about the origin if and only if replacing both x by $-x$ and y by $-y$ in its equation produces an equivalent equation.*

▶ **Example 13** Use Theorem 1.3.3 to identify symmetries in the graph of $x = y^2$.

Solution. Replacing y by $-y$ yields $x = (-y)^2$, which simplifies to the original equation $x = y^2$. Thus, the graph is symmetric about the x-axis. The graph is not symmetric about the y-axis because replacing x by $-x$ yields $-x = y^2$, which is not equivalent to the original equation $x = y^2$. Similarly, the graph is not symmetric about the origin because replacing x by $-x$ and y by $-y$ yields $-x = (-y)^2$, which simplifies to $-x = y^2$, and this is again not equivalent to the original equation. These results are consistent with the graph of $x = y^2$ shown in Figure 1.3.9. ◀

y

$x = y^2$

x

Figure 1.3.9

EVEN AND ODD FUNCTIONS

A function f is said to be an ***even function*** if

$$f(-x) = f(x) \tag{8}$$

and is said to be an ***odd function*** if

$$f(-x) = -f(x) \tag{9}$$

Geometrically, the graphs of even functions are symmetric about the y-axis because replacing x by $-x$ in the equation $y = f(x)$ yields $y = f(-x)$, which is equivalent to the original equation $y = f(x)$ by (8) (see Figure 1.3.10). Similarly, it follows from (9) that graphs of odd functions are symmetric about the origin (see Figure 1.3.11). Some examples of even functions are x^2, x^4, x^6, and $\cos x$; and some examples of odd functions are x^3, x^5, x^7, and $\sin x$.

Explain why the graph of a nonzero function cannot be symmetric about the x-axis.

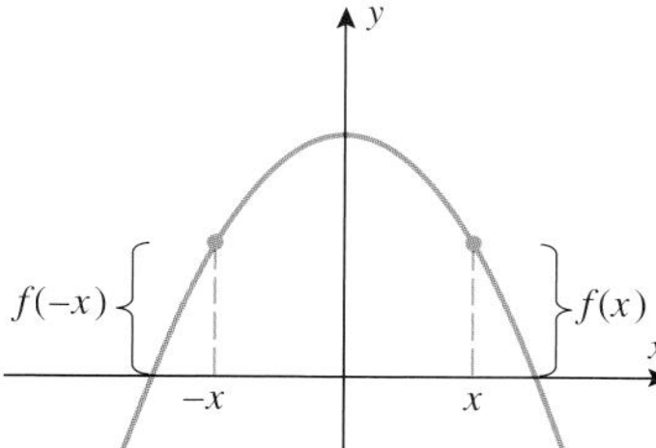

Figure 1.3.10 This is the graph of an even function since $f(-x) = f(x)$.

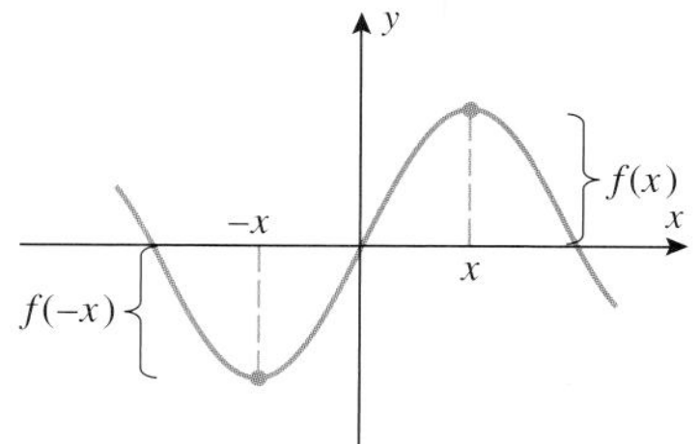

Figure 1.3.11 This is the graph of an odd function since $f(-x) = -f(x)$.

✔ QUICK CHECK EXERCISES 1.3

(See page 39 for answers.)

1. Let $f(x) = 3\sqrt{x} - 2$ and $g(x) = |x|$. In each part, give the formula for the function and state the corresponding domain.
   (a) $f + g$: ________ Domain: ________
   (b) $f - g$: ________ Domain: ________
   (c) fg: ________ Domain: ________
   (d) f/g: ________ Domain: ________

2. Let $f(x) = 2 - x^2$ and $g(x) = \sqrt{x}$. In each part, give the formula for the composition and state the corresponding domain.
   (a) $f \circ g$: ________ Domain: ________
   (b) $g \circ f$: ________ Domain: ________

3. The graph of $y = 1 + (x - 2)^2$ may be obtained by shifting the graph of $y = x^2$ ________ (left/right) by ________ unit(s) and then shifting this new graph ________ (up/down) by ________ unit(s).

4. Let

$$f(x) = \begin{cases} |x+1|, & -2 \le x \le 0 \\ |x-1|, & 0 < x \le 2 \end{cases}$$

   (a) The letter of the alphabet that most resembles the graph of f is ________.
   (b) Is f an even function?

EXERCISE SET 1.3

FOCUS ON CONCEPTS

1. The graph of a function f is shown in the accompanying figure. Sketch the graphs of the following equations.
   (a) $y = f(x) - 1$ (b) $y = f(x - 1)$
   (c) $y = \frac{1}{2}f(x)$ (d) $y = f\left(-\frac{1}{2}x\right)$

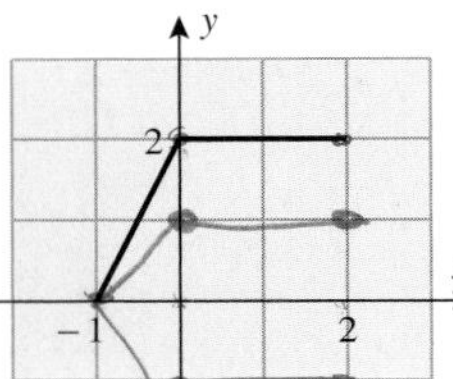

Figure Ex-1

2. Use the graph in Exercise 1 to sketch the graphs of the following equations.
   (a) $y = -f(-x)$ (b) $y = f(2 - x)$
   (c) $y = 1 - f(2 - x)$ (d) $y = \frac{1}{2}f(2x)$

3. The graph of a function f is shown in the accompanying figure. Sketch the graphs of the following equations.
   (a) $y = f(x + 1)$ (b) $y = f(2x)$
   (c) $y = |f(x)|$ (d) $y = 1 - |f(x)|$

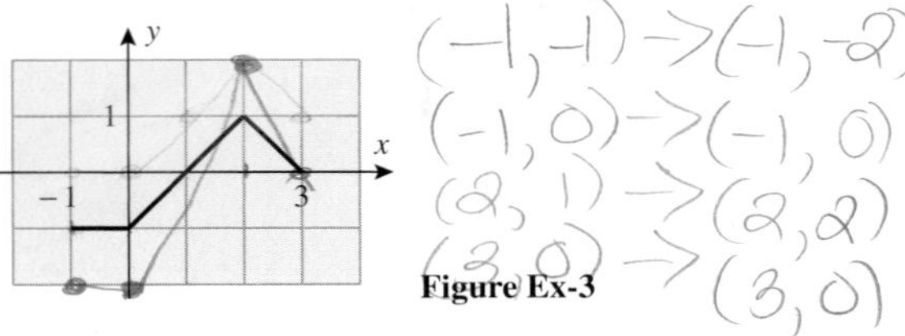

Figure Ex-3

4. Use the graph in Exercise 3 to sketch the graph of the equation $y = f(|x|)$.

5–10 Sketch the graph of the equation by translating, reflecting, compressing, and stretching the graph of $y = x^2$ appropriately, and then use a graphing utility to confirm that your sketch is correct.

5. $y = -2(x + 1)^2 - 3$
6. $y = \frac{1}{2}(x - 3)^2 + 2$
7. $y = x^2 + 6x$
8. $y = x^2 + 6x - 10$
9. $y = 1 + 2x - x^2$
10. $y = \frac{1}{2}(x^2 - 2x + 3)$

11–14 Sketch the graph of the equation by translating, reflecting, compressing, and stretching the graph of $y = \sqrt{x}$ appropriately, and then use a graphing utility to confirm that your sketch is correct.

11. $y = 3 - \sqrt{x + 1}$
12. $y = 1 + \sqrt{x - 4}$
13. $y = \frac{1}{2}\sqrt{x} + 1$
14. $y = -\sqrt{3x}$

15–18 Sketch the graph of the equation by translating, reflecting, compressing, and stretching the graph of $y = 1/x$ appropriately, and then use a graphing utility to confirm that your sketch is correct.

15. $y = \dfrac{1}{x - 3}$
16. $y = \dfrac{1}{1 - x}$
17. $y = 2 - \dfrac{1}{x + 1}$
18. $y = \dfrac{x - 1}{x}$

19–22 Sketch the graph of the equation by translating, reflecting, compressing, and stretching the graph of $y = |x|$ appropriately, and then use a graphing utility to confirm that your sketch is correct.

19. $y = |x + 2| - 2$
20. $y = 1 - |x - 3|$

21. $y = |2x - 1| + 1$

22. $y = \sqrt{x^2 - 4x + 4}$

23–26 Sketch the graph of the equation by translating, reflecting, compressing, and stretching the graph of $y = \sqrt[3]{x}$ appropriately, and then use a graphing utility to confirm that your sketch is correct.

23. $y = 1 - 2\sqrt[3]{x}$

24. $y = \sqrt[3]{x - 2} - 3$

25. $y = 2 + \sqrt[3]{x + 1}$

26. $y + \sqrt[3]{x - 2} = 0$

27. (a) Sketch the graph of $y = x + |x|$ by adding the corresponding y-coordinates on the graphs of $y = x$ and $y = |x|$.
(b) Express the equation $y = x + |x|$ in piecewise form with no absolute values, and confirm that the graph you obtained in part (a) is consistent with this equation.

28. Sketch the graph of $y = x + (1/x)$ by adding corresponding y-coordinates on the graphs of $y = x$ and $y = 1/x$. Use a graphing utility to confirm that your sketch is correct.

29–30 Find formulas for $f + g$, $f - g$, fg, and f/g, and state the domains of the functions.

29. $f(x) = 2\sqrt{x - 1}$, $g(x) = \sqrt{x - 1}$

30. $f(x) = \dfrac{x}{1 + x^2}$, $g(x) = \dfrac{1}{x}$

31. Let $f(x) = \sqrt{x}$ and $g(x) = x^3 + 1$. Find
(a) $f(g(2))$ (b) $g(f(4))$
(c) $f(f(16))$ (d) $g(g(0))$.

32. Let $g(x) = \pi - x^2$ and $h(x) = \cos x$. Find
(a) $g(h(0))$ (b) $h(g(\sqrt{\pi/2}))$
(c) $g(g(1))$ (d) $h(h(\pi/2))$.

33. Let $f(x) = x^2 + 1$. Find
(a) $f(t^2)$ (b) $f(t + 2)$ (c) $f(x + 2)$
(d) $f\left(\dfrac{1}{x}\right)$ (e) $f(x + h)$ (f) $f(-x)$
(g) $f(\sqrt{x})$ (h) $f(3x)$.

34. Let $g(x) = \sqrt{x}$. Find
(a) $g(5s + 2)$ (b) $g(\sqrt{x} + 2)$ (c) $3g(5x)$
(d) $\dfrac{1}{g(x)}$ (e) $g(g(x))$ (f) $(g(x))^2 - g(x^2)$
(g) $g(1/\sqrt{x})$ (h) $g((x - 1)^2)$.

35–38 Find formulas for $f \circ g$ and $g \circ f$, and state the domains of the compositions.

35. $f(x) = x^2$, $g(x) = \sqrt{1 - x}$

36. $f(x) = \sqrt{x - 3}$, $g(x) = \sqrt{x^2 + 3}$

37. $f(x) = \dfrac{1 + x}{1 - x}$, $g(x) = \dfrac{x}{1 - x}$

38. $f(x) = \dfrac{x}{1 + x^2}$, $g(x) = \dfrac{1}{x}$

39–40 Find a formula for $f \circ g \circ h$.

39. $f(x) = x^2 + 1$, $g(x) = \dfrac{1}{x}$, $h(x) = x^3$

40. $f(x) = \dfrac{1}{1 + x}$, $g(x) = \sqrt[3]{x}$, $h(x) = \dfrac{1}{x^3}$

41–44 Express f as a composition of two functions; that is, find g and h such that $f = g \circ h$. [*Note:* Each exercise has more than one solution.]

41. (a) $f(x) = \sqrt{x + 2}$ (b) $f(x) = |x^2 - 3x + 5|$

42. (a) $f(x) = x^2 + 1$ (b) $f(x) = \dfrac{1}{x - 3}$

43. (a) $f(x) = \sin^2 x$ (b) $f(x) = \dfrac{3}{5 + \cos x}$

44. (a) $f(x) = 3\sin(x^2)$ (b) $f(x) = 3\sin^2 x + 4\sin x$

45–46 Express F as a composition of three functions; that is, find f, g, and h such that $F = f \circ g \circ h$. [*Note:* Each exercise has more than one solution.]

45. (a) $F(x) = \left(1 + \sin(x^2)\right)^3$ (b) $F(x) = \sqrt{1 - \sqrt[3]{x}}$

46. (a) $F(x) = \dfrac{1}{1 - x^2}$ (b) $F(x) = |5 + 2x|$

FOCUS ON CONCEPTS

47. Use the data in the accompanying table to make a plot of $y = f(g(x))$.

x	–3	–2	–1	0	1	2	3
$f(x)$	–4	–3	–2	–1	0	1	2
$g(x)$	–1	0	1	2	3	–2	–3

Table Ex-47

48. Find the domain of $g \circ f$ for the functions f and g in Exercise 47.

49. Sketch the graph of $y = f(g(x))$ for the functions graphed in the accompanying figure.

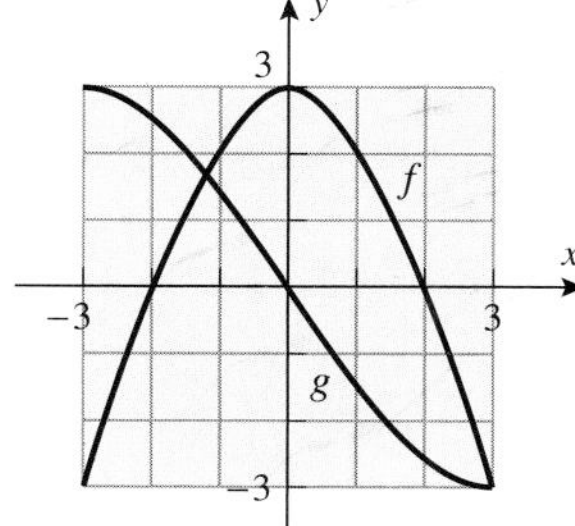

Figure Ex-49

50. Sketch the graph of $y = g(f(x))$ for the functions graphed in Exercise 49.

51. Use the graphs of f and g in Exercise 49 to estimate the solutions of the equations $f(g(x)) = 0$ and $g(f(x)) = 0$.

52. Use the table given in Exercise 47 to solve the equations $f(g(x)) = 0$ and $g(f(x)) = 0$.

53–56 Find

$$\frac{f(w) - f(x)}{w - x} \quad \text{and} \quad \frac{f(x+h) - f(x)}{h}$$

Simplify as much as possible.

53. $f(x) = 3x^2 - 5$

54. $f(x) = x^2 + 6x$

55. $f(x) = 1/x$

56. $f(x) = 1/x^2$

57. Classify the functions whose values are given in the accompanying table as even, odd, or neither.

x	−3	−2	−1	0	1	2	3
$f(x)$	5	3	2	3	1	−3	5
$g(x)$	4	1	−2	0	2	−1	−4
$h(x)$	2	−5	8	−2	8	−5	2

Table Ex-57

58. Complete the accompanying table so that the graph of $y = f(x)$ is symmetric about
(a) the y-axis (b) the origin.

x	−3	−2	−1	0	1	2	3
$f(x)$	1		−1	0		−5	

Table Ex-58

59. The accompanying figure shows a portion of a graph. Complete the graph so that the entire graph is symmetric about
(a) the x-axis (b) the y-axis (c) the origin.

60. The accompanying figure shows a portion of the graph of a function f. Complete the graph assuming that
(a) f is an even function (b) f is an odd function.

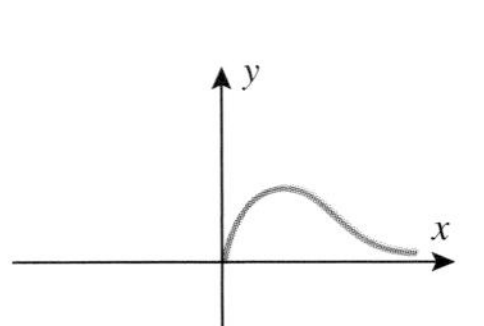

Figure Ex-59

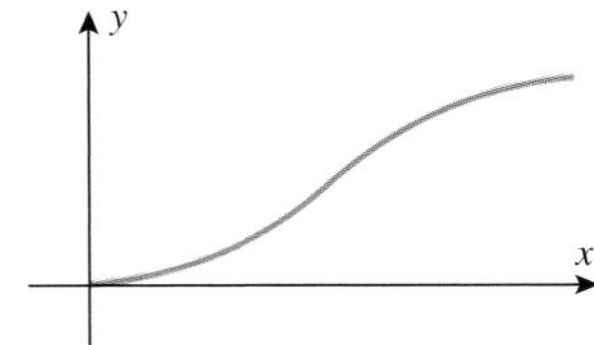

Figure Ex-60

61. Classify the functions graphed in the accompanying figure as even, odd, or neither.

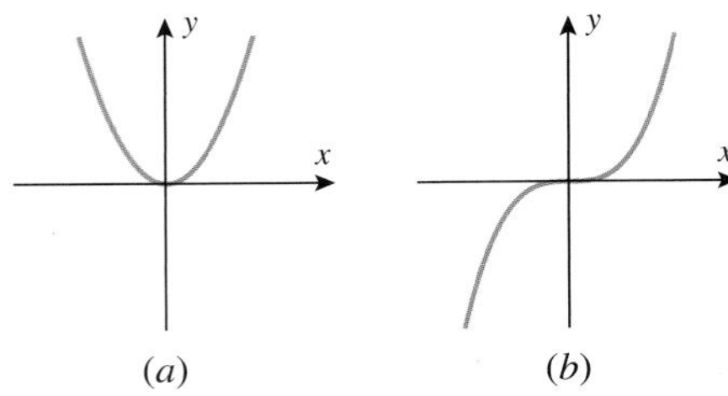

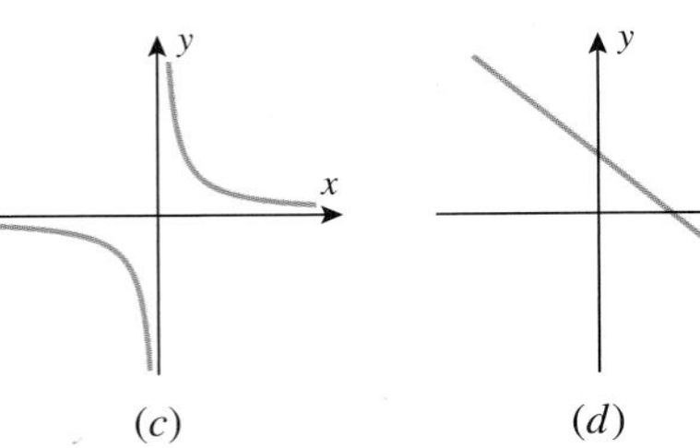

Figure Ex-61

62. In each part of the accompanying figure determine whether the graph is symmetric about the x-axis, the y-axis, the origin, or none of the preceding.

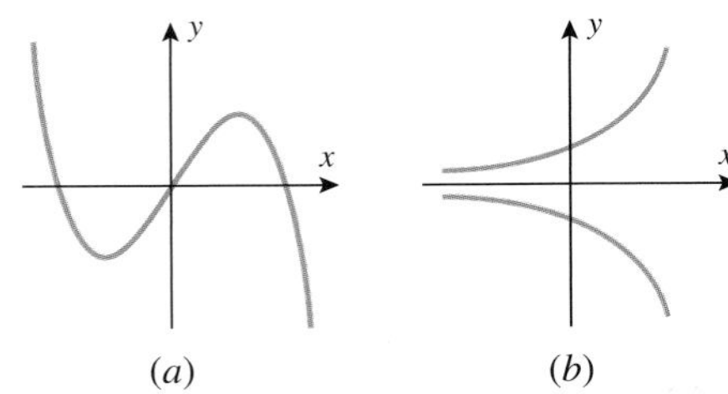

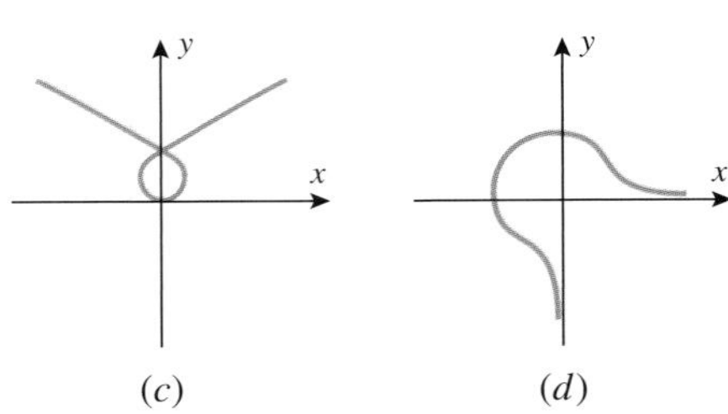

Figure Ex-62

63. In each part, classify the function as even, odd, or neither.
(a) $f(x) = x^2$ (b) $f(x) = x^3$
(c) $f(x) = |x|$ (d) $f(x) = x + 1$
(e) $f(x) = \dfrac{x^5 - x}{1 + x^2}$ (f) $f(x) = 2$

64. Suppose that the function f has domain all real numbers. Determine whether each function can be classified as even or odd. Explain.
(a) $g(x) = \dfrac{f(x) + f(-x)}{2}$ (b) $h(x) = \dfrac{f(x) - f(-x)}{2}$

65. Suppose that the function f has domain all real numbers. Show that f can be written as the sum of an even function and an odd function. [*Hint:* See Exercise 64.]

66–67 Use Theorem 1.3.3 to determine whether the graph has symmetries about the x-axis, the y-axis, or the origin.

66. (a) $x = 5y^2 + 9$ (b) $x^2 - 2y^2 = 3$
(c) $xy = 5$

67. (a) $x^4 = 2y^3 + y$ (b) $y = \dfrac{x}{3 + x^2}$
(c) $y^2 = |x| - 5$

68–69 (i) Use a graphing utility to graph the equation in the first quadrant. [*Note:* To do this you will have to solve the equation for y in terms of x.] (ii) Use symmetry to make a hand-drawn sketch of the entire graph. (iii) Confirm your work by generating the graph of the equation in the remaining three quadrants.

68. $9x^2 + 4y^2 = 36$ **69.** $4x^2 + 16y^2 = 16$

70. The graph of the equation $x^{2/3} + y^{2/3} = 1$, which is shown in the accompanying figure, is called a ***four-cusped hypocycloid***.
(a) Use Theorem 1.3.3 to confirm that this graph is symmetric about the x-axis, the y-axis, and the origin.
(b) Find a function f whose graph in the first quadrant coincides with the four-cusped hypocycloid, and use a graphing utility to confirm your work.
(c) Repeat part (b) for the remaining three quadrants.

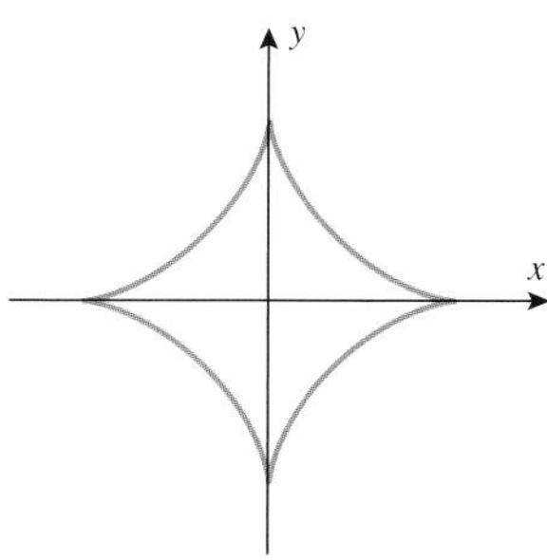

Four-cusped hypocycloid

Figure Ex-70

71. The equation $y = |f(x)|$ can be written as

$$y = \begin{cases} f(x), & f(x) \geq 0 \\ -f(x), & f(x) < 0 \end{cases}$$

which shows that the graph of $y = |f(x)|$ can be obtained from the graph of $y = f(x)$ by retaining the portion that lies on or above the x-axis and reflecting about the x-axis the portion that lies below the x-axis. Use this method to obtain the graph of $y = |2x - 3|$ from the graph of $y = 2x - 3$.

72–73 Use the method described in Exercise 71.

72. Sketch the graph of $y = |1 - x^2|$.

73. Sketch the graph of
(a) $f(x) = |\cos x|$ (b) $f(x) = \cos x + |\cos x|$.

74. The ***greatest integer function***, $\lfloor x \rfloor$, is defined to be the greatest integer that is less than or equal to x. For example, $\lfloor 2.7 \rfloor = 2$, $\lfloor -2.3 \rfloor = -3$, and $\lfloor 4 \rfloor = 4$. Sketch the graph of $y = f(x)$.
(a) $f(x) = \lfloor x \rfloor$ (b) $f(x) = \lfloor x^2 \rfloor$
(c) $f(x) = \lfloor x \rfloor^2$ (d) $f(x) = \lfloor \sin x \rfloor$

75. Is it ever true that $f \circ g = g \circ f$ if f and g are nonconstant functions? If not, prove it; if so, give some examples for which it is true.

76. In the discussion preceding Exercise 29 of Section 1.2, we gave a procedure for generating a complete graph of $f(x) = x^{p/q}$ in which we suggested graphing the function $g(x) = |x|^{p/q}$ instead of $f(x)$ when p is even and q is odd and graphing $g(x) = (|x|/x)|x|^{p/q}$ if p is odd and q is odd. Show that in both cases $f(x) = g(x)$ if $x > 0$ or $x < 0$. [*Hint:* Show that $f(x)$ is an even function if p is even and q is odd and is an odd function if p is odd and q is odd.]

✔ QUICK CHECK ANSWERS 1.3

1. (a) $(f + g)(x) = 3\sqrt{x} - 2 + x$; $x \geq 0$ (b) $(f - g)(x) = 3\sqrt{x} - 2 - x$; $x \geq 0$ (c) $(fg)(x) = 3x^{3/2} - 2x$; $x \geq 0$ (d) $(f/g)(x) = \dfrac{3\sqrt{x} - 2}{x}$; $x > 0$ **2.** (a) $(f \circ g)(x) = 2 - x$; $x \geq 0$ (b) $(g \circ f)(x) = \sqrt{2 - x^2}$; $-\sqrt{2} \leq x \leq \sqrt{2}$
3. right; 2; up; 1 **4.** (a) W (b) yes

1.4 FAMILIES OF FUNCTIONS

Functions are often grouped into families according to the form of their defining formulas or other common characteristics. In this section we will discuss some of the most basic families of functions.

FAMILIES OF CURVES

The graph of a constant function $f(x) = c$ is the graph of the equation $y = c$, which is the horizontal line shown in Figure 1.4.1*a*. If we vary c, then we obtain a set or ***family*** of horizontal lines such as those in Figure 1.4.1*b*.

Constants that are varied to produce families of curves are called ***parameters***. For example, recall that an equation of the form $y = mx + b$ represents a line of slope m and y-intercept b. If we keep b fixed and treat m as a parameter, then we obtain a family of lines whose members all have y-intercept b (Figure 1.4.2*a*), and if we keep m fixed and treat b as a parameter, we obtain a family of parallel lines whose members all have slope m (Figure 1.4.2*b*).

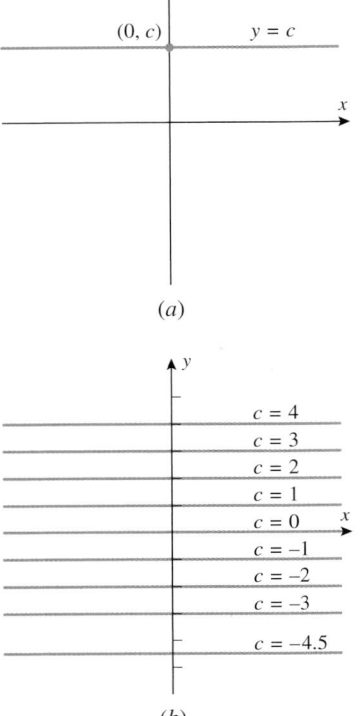

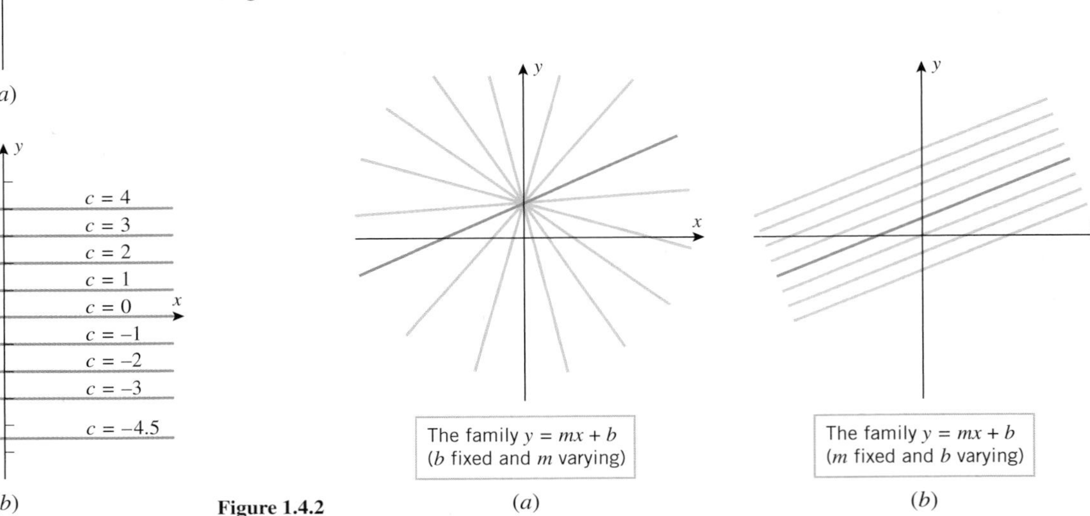

(*a*) (*b*)

Figure 1.4.1

Figure 1.4.2

THE FAMILY $y = x^n$

A function of the form $f(x) = x^p$, where p is constant, is called a ***power function***. For the moment, let us consider the case where p is a positive integer, say $p = n$. The graphs of the curves $y = x^n$ for $n = 1, 2, 3, 4$, and 5 are shown in Figure 1.4.3. The first graph is the line with slope 1 that passes through the origin, and the second is a parabola that opens up and has its vertex at the origin (see Web Appendix G).

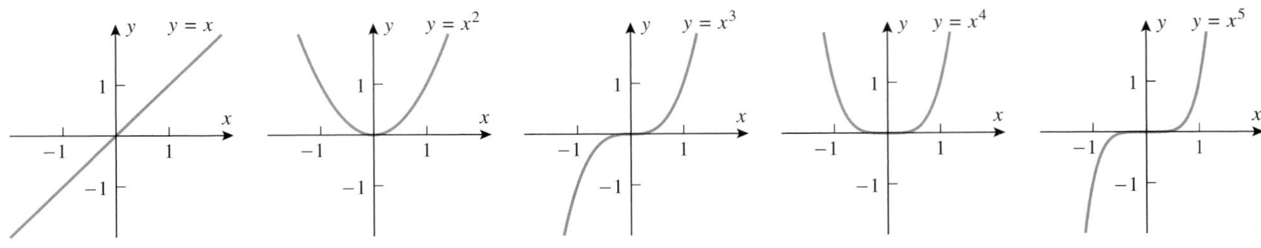

Figure 1.4.3

For $n \geq 2$ the shape of the curve $y = x^n$ depends on whether n is even or odd (Figure 1.4.4):

- For even values of n, the functions $f(x) = x^n$ are even, so their graphs are symmetric about the y-axis. The graphs all have the general shape of the parabola $y = x^2$ (though they are not actually parabolas if $n > 2$), and each graph passes through the points $(-1, 1)$ and $(1, 1)$. As n increases, the graphs become flatter over the interval $-1 < x < 1$ and steeper over the intervals $x > 1$ and $x < -1$.
- For odd values of n, the functions $f(x) = x^n$ are odd, so their graphs are symmetric about the origin. The graphs all have the general shape of the curve $y = x^3$, and each graph passes through the points $(1, 1)$ and $(-1, -1)$. As n increases, the graphs become flatter over the interval $-1 < x < 1$ and steeper over the intervals $x > 1$ and $x < -1$.

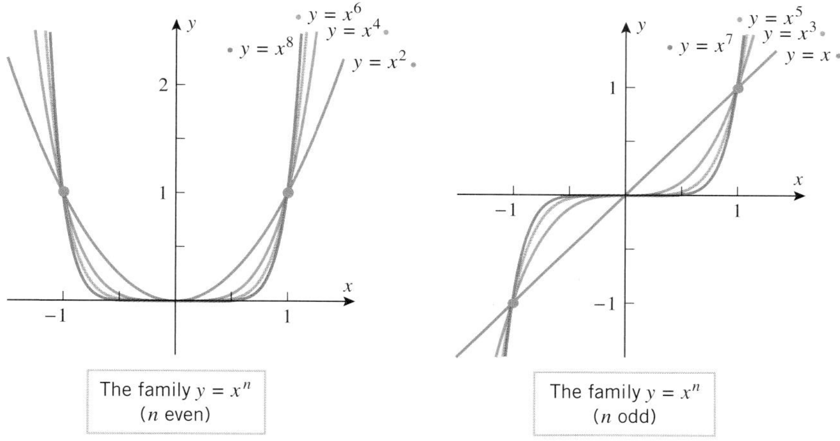

Figure 1.4.4

> The flattening and steepening effects can be understood by considering what happens when a number x is raised to higher and higher powers: If $-1 < x < 1$, then the absolute value of x^n *decreases* as n increases, thereby causing the graphs to become flatter on this interval as n increases (try raising $\frac{1}{2}$ or $-\frac{1}{2}$ to higher and higher powers). On the other hand, if $x > 1$ or $x < -1$, then the absolute value of x^n *increases* as n increases, thereby causing the graphs to become steeper on these intervals as n increases (try raising 2 or -2 to higher and higher powers).

THE FAMILY $y = x^{-n}$

If p is a negative integer, say $p = -n$, then the power functions $f(x) = x^p$ have the form $f(x) = x^{-n} = 1/x^n$. Figure 1.4.5 shows the graphs of $y = 1/x$ and $y = 1/x^2$. The graph of $y = 1/x$ is called an ***equilateral hyperbola*** (for reasons to be discussed later).

As illustrated in Figure 1.4.5, the shape of the curve $y = 1/x^n$ depends on whether n is even or odd:

- For even values of n, the functions $f(x) = 1/x^n$ are even, so their graphs are symmetric about the y-axis. The graphs all have the general shape of the curve $y = 1/x^2$, and each graph passes through the points $(-1, 1)$ and $(1, 1)$. As n increases, the graphs become steeper over the intervals $-1 < x < 0$ and $0 < x < 1$ and become flatter over the intervals $x > 1$ and $x < -1$.

By considering the value of $1/x^n$ for a fixed x as n increases, explain why the graphs become flatter or steeper as described here for increasing values of n.

- For odd values of n, the functions $f(x) = 1/x^n$ are odd, so their graphs are symmetric about the origin. The graphs all have the general shape of the curve $y = 1/x$, and each graph passes through the points (1, 1) and (−1, −1). As n increases, the graphs become steeper over the intervals $-1 < x < 0$ and $0 < x < 1$ and become flatter over the intervals $x > 1$ and $x < -1$.
- For both even and odd values of n the graph $y = 1/x^n$ has a break at the origin (called a ***discontinuity***), which occurs because division by zero is undefined.

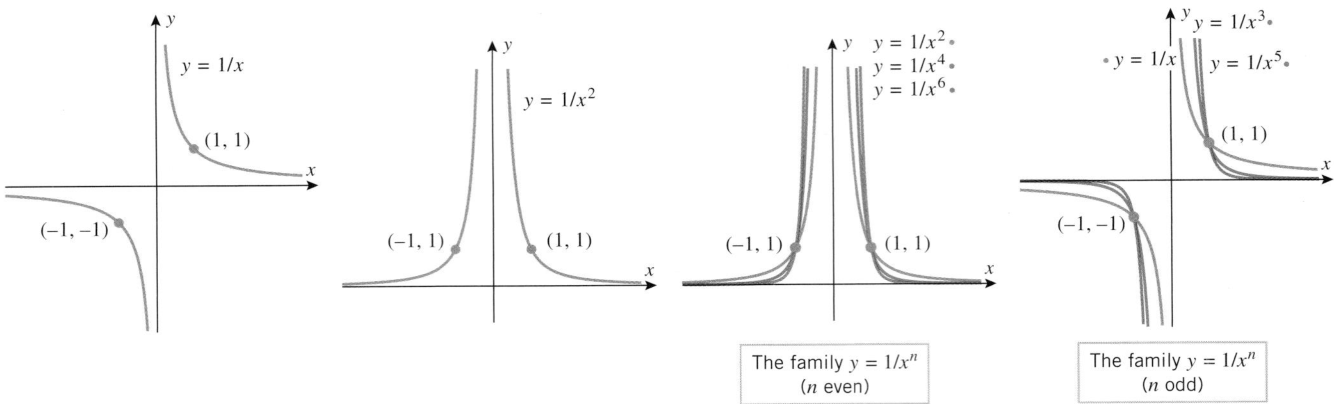

Figure 1.4.5

INVERSE PROPORTIONS

Recall that a variable y is said to be ***inversely proportional to a variable x*** if there is a positive constant k, called the ***constant of proportionality***, such that

$$y = \frac{k}{x} \tag{1}$$

Since k is assumed to be positive, the graph of (1) has the same shape as $y = 1/x$ but is compressed or stretched in the y-direction. Also, it should be evident from (1) that doubling x multiplies y by $\frac{1}{2}$, tripling x multiplies y by $\frac{1}{3}$, and so forth.

Equation (1) can be expressed as $xy = k$, which tells us that the product of inversely proportional variables is a positive constant. This is a useful form for identifying inverse proportionality in experimental data.

Table 1.4.1
EXPERIMENTAL DATA

x	0.8	1	2.5	4	6.25	10
y	6.25	5	2	1.25	0.8	0.5

▶ **Example 1** Table 1.4.1 shows some experimental data.

(a) Explain why the data suggest that y is inversely proportional to x.

(b) Express y as a function of x.

(c) Graph your function and the data together for $x > 0$.

Solution. For every data point we have $xy = 5$, so y is inversely proportional to x and $y = 5/x$. The graph of this equation with the data points is shown in Figure 1.4.6. ◀

Inverse proportions arise in various laws of physics. For example, ***Boyle's law*** in physics states that *if a fixed amount of an ideal gas is held at a constant temperature, then the product* *of the pressure P exerted by the gas and the volume V that it occupies is constant*; that is,

$$PV = k$$

This implies that the variables P and V are inversely proportional to one another. Figure 1.4.7 shows a typical graph of volume versus pressure under the conditions of Boyle's law. Note how doubling the pressure reduces the volume to half as much, as expected.

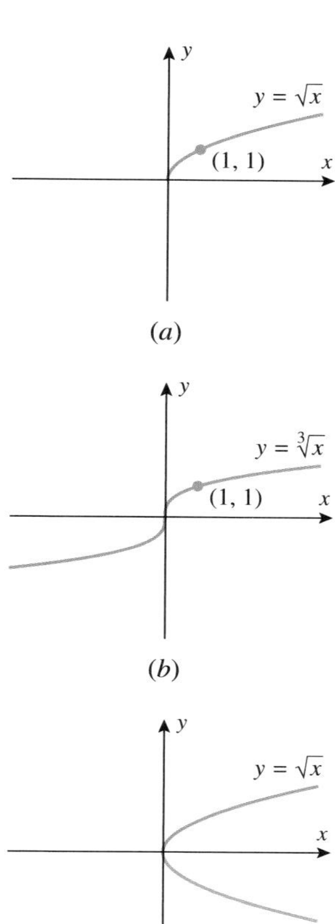

Figure 1.4.8

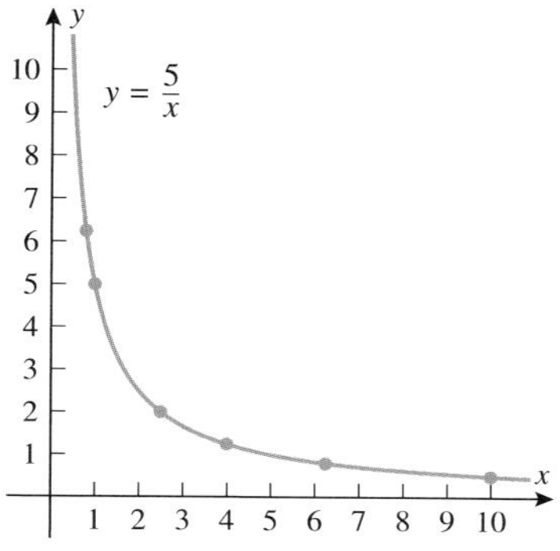

Figure 1.4.6

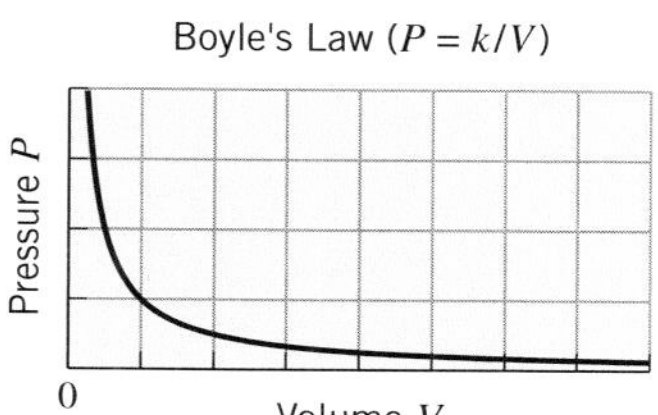

Figure 1.4.7

POWER FUNCTIONS WITH NONINTEGER EXPONENTS

If $p = 1/n$, where n is a positive integer, then the power functions $f(x) = x^p$ have the form $f(x) = x^{1/n} = \sqrt[n]{x}$. In particular, if $n = 2$, then $f(x) = \sqrt{x}$, and if $n = 3$, then $f(x) = \sqrt[3]{x}$. The graphs of these functions are shown in parts (*a*) and (*b*) of Figure 1.4.8.

Since every real number has a real cube root, the domain of the function $f(x) = \sqrt[3]{x}$ is $(-\infty, +\infty)$, and hence the graph of $y = \sqrt[3]{x}$ extends over the entire x-axis. In contrast, the graph of $y = \sqrt{x}$ extends only over the interval $[0, +\infty)$ because $\sqrt{x}$ is imaginary for negative x. As illustrated in Figure 1.4.8*c*, the graphs of $y = \sqrt{x}$ and $y = -\sqrt{x}$ form the upper and lower halves of the parabola $x = y^2$. In general, the graph of $y = \sqrt[n]{x}$ extends over the entire x-axis if n is odd, but extends only over the interval $[0, +\infty)$ if n is even.

Power functions can have other fractional exponents. Some examples are

$$f(x) = x^{2/3}, \quad f(x) = \sqrt[5]{x^3}, \quad f(x) = x^{-7/8} \tag{2}$$

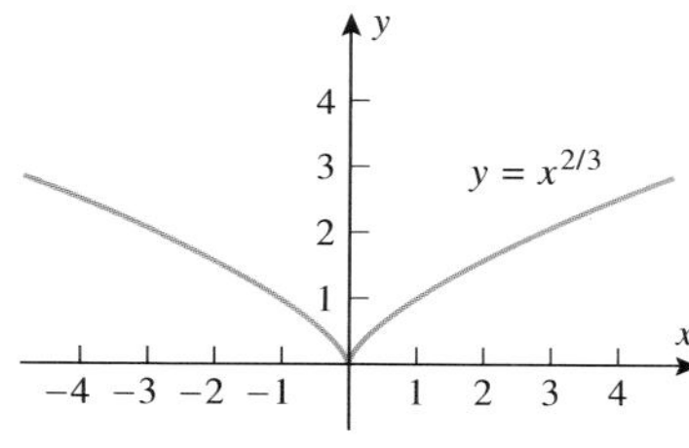

Figure 1.4.9

The graph of $f(x) = x^{2/3}$ shown in Figure 1.4.9 was discussed in Section 1.2. We will discuss expressions involving irrational exponents later.

TECHNOLOGY MASTERY

Read the note preceding Exercise 29 of Section 1.2, and use a graphing utility to generate graphs of $f(x) = \sqrt[5]{x^3}$ and $f(x) = x^{-7/8}$ that show all of their significant features.

POLYNOMIALS

A ***polynomial in x*** is a function that is expressible as a sum of finitely many terms of the form cx^n, where c is a constant and n is a nonnegative integer. Some examples of polynomials are

$$2x + 1, \quad 3x^2 + 5x - \sqrt{2}, \quad x^3, \quad 4\ (= 4x^0), \quad 5x^7 - x^4 + 3$$

The function $(x^2 - 4)^3$ is also a polynomial because it can be expanded by the binomial formula (see the inside front cover) and expressed as a sum of terms of the form cx^n:

$$(x^2 - 4)^3 = (x^2)^3 - 3(x^2)^2(4) + 3(x^2)(4^2) - (4^3) = x^6 - 12x^4 + 48x^2 - 64 \tag{3}$$

A more detailed review of polynomials appears in Appendix B.

A general polynomial can be written in either of the following forms, depending on whether one wants the powers of x in ascending or descending order:

$$c_0 + c_1x + c_2x^2 + \cdots + c_nx^n$$

$$c_nx^n + c_{n-1}x^{n-1} + \cdots + c_1x + c_0$$

The constants $c_0, c_1, \ldots, c_n$ are called the ***coefficients*** of the polynomial. When a polynomial is expressed in one of these forms, the highest power of x that occurs with a nonzero coefficient is called the ***degree*** of the polynomial. Nonzero constant polynomials are considered to have degree 0, since we can write $c = cx^0$. Polynomials of degree 1, 2, 3, 4, and 5 are described as ***linear***, ***quadratic***, ***cubic***, ***quartic***, and ***quintic***, respectively. For example,

$3 + 5x$	$x^2 - 3x + 1$	$2x^3 - 7$
Has degree 1 (linear)	Has degree 2 (quadratic)	Has degree 3 (cubic)
$8x^4 - 9x^3 + 5x - 3$	$\sqrt{3} + x^3 + x^5$	$(x^2 - 4)^3$
Has degree 4 (quartic)	Has degree 5 (quintic)	Has degree 6 [see (3)]

The natural domain of a polynomial in x is $(-\infty, +\infty)$, since the only operations involved are multiplication and addition; the range depends on the particular polynomial. We already know that the graphs of polynomials of degree 0 and 1 are lines and that the graphs of polynomials of degree 2 are parabolas. Figure 1.4.10 shows the graphs of some typical polynomials of higher degree. Later, we will discuss polynomial graphs in detail, but for now it suffices to observe that graphs of polynomials are very well behaved in the sense that they have no discontinuities or sharp corners. As illustrated in Figure 1.4.10, the graphs of polynomials wander up and down for awhile in a roller-coaster fashion, but eventually that behavior stops and the graphs steadily rise or fall indefinitely as one travels along the curve in either the positive or negative direction. We will see later that the number of peaks and valleys is less than the degree of the polynomial.

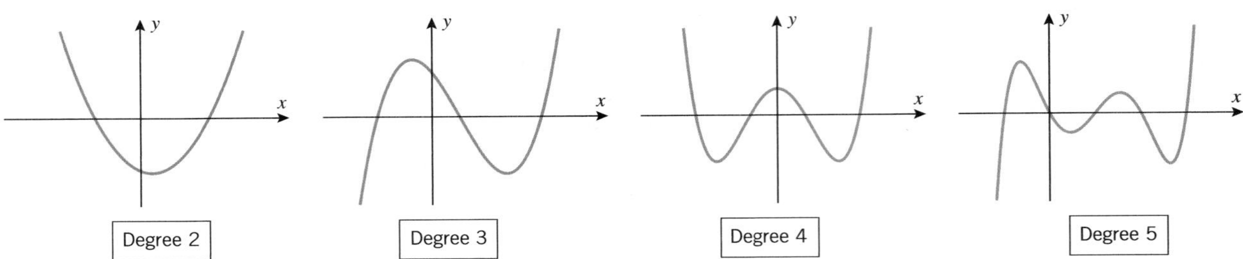

Figure 1.4.10

RATIONAL FUNCTIONS

A function that can be expressed as a ratio of two polynomials is called a ***rational function***. If $P(x)$ and $Q(x)$ are polynomials, then the domain of the rational function

$$f(x) = \frac{P(x)}{Q(x)}$$

consists of all values of x such that $Q(x) \neq 0$. For example, the domain of the rational function

$$f(x) = \frac{x^2 + 2x}{x^2 - 1}$$

consists of all values of x, except $x = 1$ and $x = -1$. Its graph is shown in Figure 1.4.11 along with the graphs of two other typical rational functions.

The graphs of rational functions with nonconstant denominators differ from the graphs of polynomials in some essential ways:

- Unlike polynomials whose graphs are continuous (unbroken) curves, the graphs of rational functions have discontinuities at the points where the denominator is zero.
- Unlike polynomials, rational functions may have numbers at which they are not defined. Near such points, many rational functions have graphs that closely approximate a vertical line, called a ***vertical asymptote***. These are represented by the dashed vertical lines in Figure 1.4.11.
- Unlike the graphs of nonconstant polynomials, which eventually rise or fall indefinitely, the graphs of many rational functions eventually get closer and closer to some horizontal line, called a ***horizontal asymptote***, as one traverses the curve in either the positive or negative direction. The horizontal asymptotes are represented by the dashed horizontal lines in the first two parts of Figure 1.4.11. In the third part of the figure the x-axis is a horizontal asymptote.

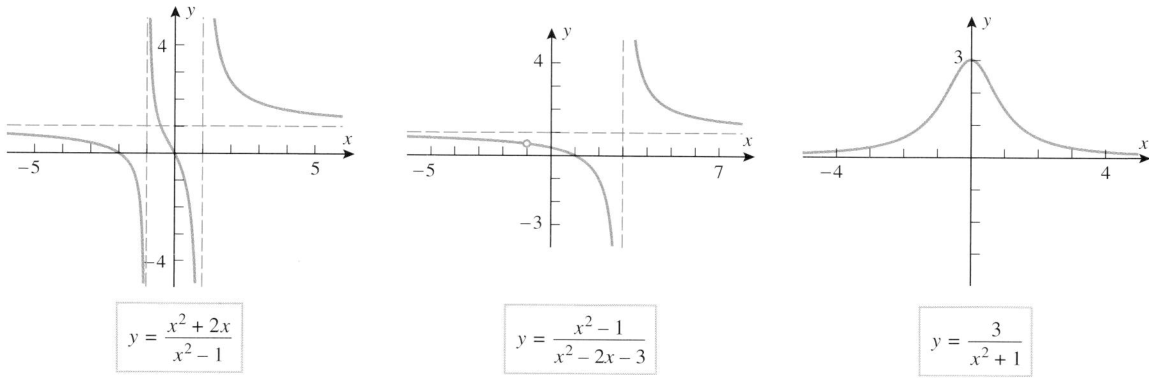

Figure 1.4.11

ALGEBRAIC FUNCTIONS

Functions that can be constructed from polynomials by applying finitely many algebraic operations (addition, subtraction, multiplication, division, and root extraction) are called ***algebraic functions***. Some examples are

$$f(x) = \sqrt{x^2 - 4}, \quad f(x) = 3\sqrt[3]{x}(2 + x), \quad f(x) = x^{2/3}(x + 2)^2$$

As illustrated in Figure 1.4.12, the graphs of algebraic functions vary widely, so it is difficult to make general statements about them. Later in this text we will develop general calculus methods for analyzing such functions.

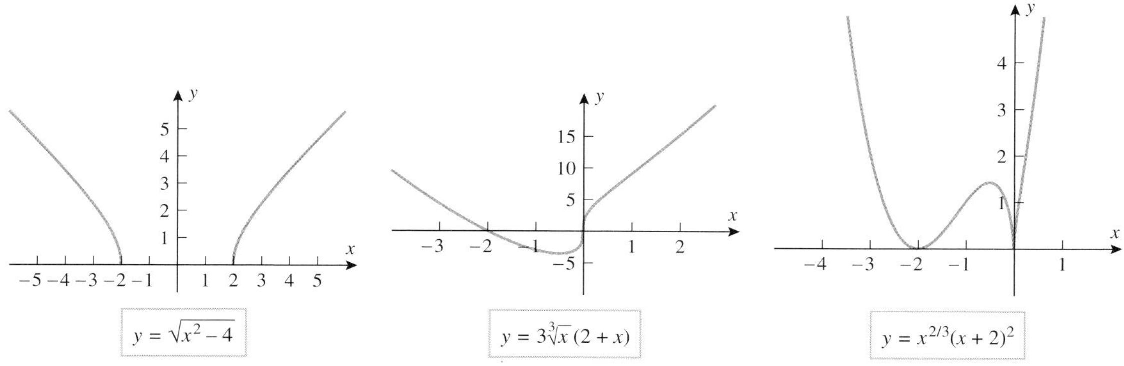

Figure 1.4.12

In this text we will assume that the independent variable of a trigonometric function is in radians unless otherwise stated. A review of trigonometric functions can be found in Appendix A.

THE FAMILIES $y = A \sin Bx$ AND $y = A \cos Bx$

Many important applications lead to trigonometric functions of the form

$$f(x) = A\sin(Bx - C) \quad \text{and} \quad g(x) = A\cos(Bx - C) \tag{4}$$

where A, B, and C are nonzero constants. The graphs of such functions can be obtained by stretching, compressing, translating, and reflecting the graphs of $y = \sin x$ and $y = \cos x$ appropriately. To see why this is so, let us start with the case where $C = 0$ and consider how the graphs of the equations

$$y = A\sin Bx \quad \text{and} \quad y = A\cos Bx$$

relate to the graphs of $y = \sin x$ and $y = \cos x$. If A and B are positive, then the effect of the constant A is to stretch or compress the graphs of $y = \sin x$ and $y = \cos x$ vertically and the effect of the constant B is to compress or stretch the graphs of $\sin x$ and $\cos x$ horizontally. For example, the graph of $y = 2\sin 4x$ can be obtained by stretching the graph of $y = \sin x$ vertically by a factor of 2 and compressing it horizontally by a factor of 4. (Recall from Section 1.3 that the multiplier of x *stretches* when it is less than 1 and *compresses* when it is greater than 1.) Thus, as shown in Figure 1.4.13, the graph of $y = 2\sin 4x$ varies between -2 and 2, and repeats every $2\pi/4 = \pi/2$ units.

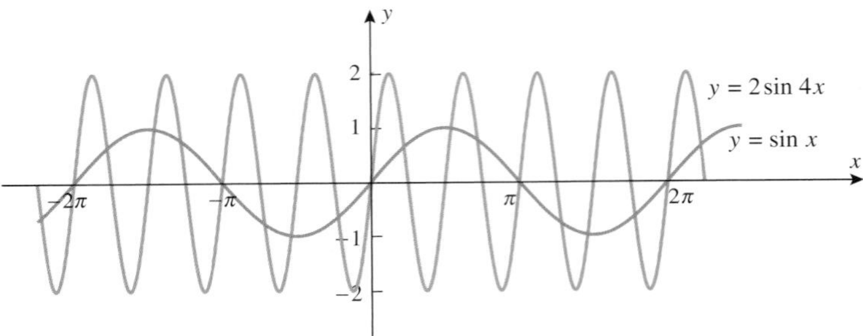

Figure 1.4.13

In general, if A and B are positive numbers, then the graphs of

$$y = A\sin Bx \quad \text{and} \quad y = A\cos Bx$$

oscillate between $-A$ and A and repeat every $2\pi/B$ units, so we say that these functions have ***amplitude*** A and ***period*** $2\pi/B$. In addition, we define the ***frequency*** of these functions to be the reciprocal of the period, that is, the frequency is $B/2\pi$. If A or B is negative, then these constants cause reflections of the graphs about the axes as well as compressing or stretching them; and in this case the amplitude, period, and frequency are given by

$$\text{amplitude} = |A|, \quad \text{period} = \frac{2\pi}{|B|}, \quad \text{frequency} = \frac{|B|}{2\pi}$$

▶ **Example 2** Make sketches of the following graphs that show the period and amplitude.

(a) $y = 3\sin 2\pi x$ (b) $y = -3\cos 0.5x$ (c) $y = 1 + \sin x$

Solution (a). The equation is of the form $y = A\sin Bx$ with $A = 3$ and $B = 2\pi$, so the graph has the shape of a sine function, but it has an amplitude of $A = 3$ and a period of $2\pi/B = 2\pi/2\pi = 1$ (Figure 1.4.14a).

Solution (b). The equation is of the form $y = A\cos Bx$ with $A = -3$ and $B = 0.5$, so the graph has the shape of a cosine curve that has been reflected about the x-axis (because $A = -3$ is negative), but with amplitude $|A| = 3$ and period $2\pi/B = 2\pi/0.5 = 4\pi$ (Figure 1.4.14*b*).

Solution (c). The graph has the shape of a sine curve that has been translated up 1 unit (Figure 1.4.14*c*). ◀

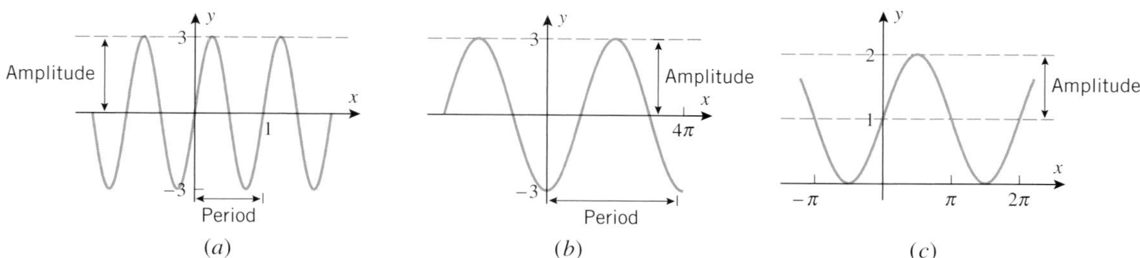

Figure 1.4.14

THE FAMILIES $y = A\sin(Bx - C)$ AND $y = A\cos(Bx - C)$

To investigate the graphs of the more general families

$$y = A\sin(Bx - C) \quad \text{and} \quad y = A\cos(Bx - C)$$

it will be helpful to rewrite these equations as

$$y = A\sin\left[B\left(x - \frac{C}{B}\right)\right] \quad \text{and} \quad y = A\cos\left[B\left(x - \frac{C}{B}\right)\right]$$

In this form we see that the graphs of these equations can be obtained by translating the graphs of $y = A\sin Bx$ and $y = A\cos Bx$ to the left or right, depending on the sign of C/B. For example, if $C/B > 0$, then the graph of

$$y = A\sin[B(x - C/B)] = A\sin(Bx - C)$$

can be obtained by translating the graph of $y = A\sin Bx$ to the right by C/B units (Figure 1.4.15). If $C/B < 0$, the graph of $y = A\sin(Bx - C)$ is obtained by translating the graph of $y = A\sin Bx$ to the left by $|C/B|$ units.

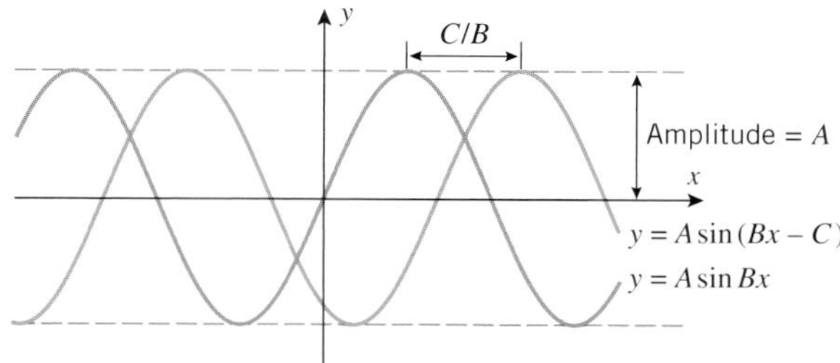

Figure 1.4.15

▶ **Example 3** Find the amplitude and period of

$$y = 3\cos\left(2x + \frac{\pi}{2}\right)$$

and determine how the graph of $y = 3\cos 2x$ should be translated to produce the graph of this equation. Confirm your results by graphing the equation on a calculator or computer.

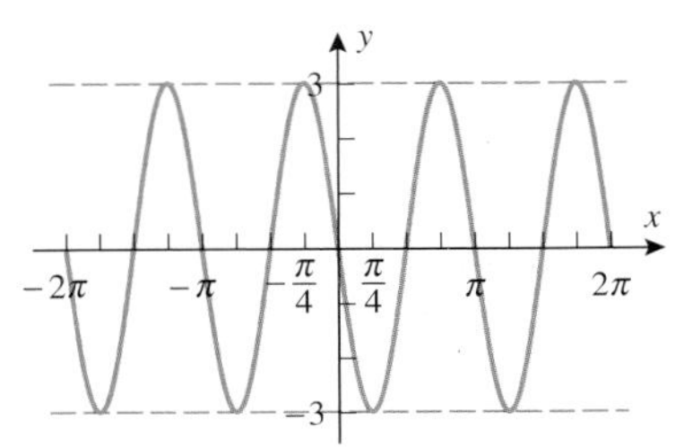

Figure 1.4.16

Solution. The equation can be rewritten as

$$y = 3\cos\left[2x - \left(-\frac{\pi}{2}\right)\right] = 3\cos\left[2\left(x - \left(-\frac{\pi}{4}\right)\right)\right]$$

which is of the form

$$y = A\cos\left[B\left(x - \frac{C}{B}\right)\right]$$

with $A = 3$, $B = 2$, and $C/B = -\pi/4$. It follows that the amplitude is $A = 3$, the period is $2\pi/B = \pi$, and the graph is obtained by translating the graph of $y = 3\cos 2x$ left by $|C/B| = \pi/4$ units (Figure 1.4.16). ◀

✔QUICK CHECK EXERCISES 1.4 *(See page 51 for answers.)*

1. Consider the family of functions $y = x^n$, where n is an integer. The graphs of $y = x^n$ are symmetric with respect to the y-axis if n is ________. These graphs are symmetric with respect to the origin if n is ________. The y-axis is a vertical asymptote for these graphs if n is ________.
2. What is the natural domain of a polynomial?
3. Consider the family of functions $y = x^{1/n}$, where n is a nonzero integer. Find the natural domain of these functions if n is
   (a) positive and even (b) positive and odd
   (c) negative and even (d) negative and odd.
4. Classify each equation as a polynomial, rational, algebraic, or not an algebraic function.
   (a) $y = \sqrt{x} + 2$ (b) $y = \sqrt{3}x^4 - x + 1$
   (c) $y = 5x^3 + \cos 4x$ (d) $y = \dfrac{x^2+5}{2x-7}$
   (e) $y = 3x^2 + 4x^{-2}$
5. The graph of $y = A\sin Bx$ has amplitude ________ and is periodic with period ________.

EXERCISE SET 1.4 Graphing Utility

1. (a) Find an equation for the family of lines whose members have slope $m = 3$.
   (b) Find an equation for the member of the family that passes through $(-1, 3)$.
   (c) Sketch some members of the family, and label them with their equations. Include the line in part (b).
2. Find an equation for the family of lines whose members are perpendicular to those in Exercise 1.
3. (a) Find an equation for the family of lines with y-intercept $b = 2$.
   (b) Find an equation for the member of the family whose angle of inclination is $135°$.
   (c) Sketch some members of the family, and label them with their equations. Include the line in part (b).
4. Find an equation for
   (a) the family of lines that pass through the origin
   (b) the family of lines with x-intercept $a = 1$
   (c) the family of lines that pass through the point $(1, -2)$
   (d) the family of lines parallel to $2x + 4y = 1$.
5. Find an equation for the family of lines tangent to the circle with center at the origin and radius 3.
6. Find an equation for the family of lines that pass through the intersection of $5x - 3y + 11 = 0$ and $2x - 9y + 7 = 0$.
7. The U.S. Internal Revenue Service uses a 10-year linear depreciation schedule to determine the value of various business items. This means that an item is assumed to have a value of zero at the end of the tenth year and that at intermediate times the value is a linear function of the elapsed time. Sketch some typical depreciation lines, and explain the practical significance of the y-intercepts.
8. Find all lines through $(6, -1)$ for which the product of the x- and y-intercepts is 3.

FOCUS ON CONCEPTS

9–10 State a geometric property common to all lines in the family, and sketch five of the lines.

9. (a) The family $y = -x + b$
   (b) The family $y = mx - 1$
   (c) The family $y = m(x + 4) + 2$
   (d) The family $x - ky = 1$

10. (a) The family $y = b$
(b) The family $Ax + 2y + 1 = 0$
(c) The family $2x + By + 1 = 0$
(d) The family $y - 1 = m(x + 1)$

11. In each part, match the equation with one of the accompanying graphs.
(a) $y = \sqrt[5]{x}$ (b) $y = 2x^5$
(c) $y = -1/x^8$ (d) $y = \sqrt{x^2 - 1}$
(e) $y = \sqrt[4]{x} - 2$ (f) $y = -\sqrt[5]{x^2}$

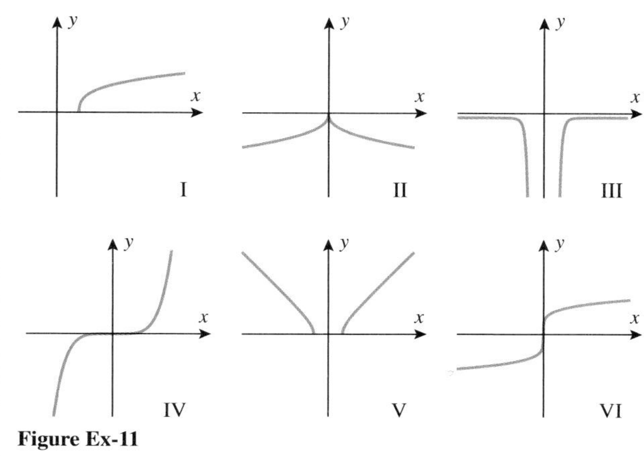

Figure Ex-11

12. The accompanying table gives approximate values of three functions: one of the form kx^2, one of the form kx^{-3}, and one of the form $kx^{3/2}$. Identify which is which, and estimate k in each case.

x	0.25	0.37	2.1	4.0	5.8	6.2	7.9	9.3
$f(x)$	640	197	1.08	0.156	0.0513	0.0420	0.0203	0.0124
$g(x)$	0.0312	0.0684	2.20	8.00	16.8	19.2	31.2	43.2
$h(x)$	0.250	0.450	6.09	16.0	27.9	30.9	44.4	56.7

Table Ex-12

13–14 Sketch the graph of the equation for $n = 1, 3,$ and 5 in one coordinate system and for $n = 2, 4,$ and 6 in another coordinate system. Check your work with a graphing utility.

13. (a) $y = -x^n$ (b) $y = 2x^{-n}$ (c) $y = (x - 1)^{1/n}$

14. (a) $y = 2x^n$ (b) $y = -x^{-n}$
(c) $y = -3(x + 2)^{1/n}$

15. (a) Sketch the graph of $y = ax^2$ for $a = \pm 1, \pm 2,$ and ± 3 in a single coordinate system.
(b) Sketch the graph of $y = x^2 + b$ for $b = \pm 1, \pm 2,$ and ± 3 in a single coordinate system.
(c) Sketch some typical members of the family of curves $y = ax^2 + b$.

16. (a) Sketch the graph of $y = a\sqrt{x}$ for $a = \pm 1, \pm 2,$ and ± 3 in a single coordinate system.
(b) Sketch the graph of $y = \sqrt{x} + b$ for $b = \pm 1, \pm 2,$ and ± 3 in a single coordinate system.
(c) Sketch some typical members of the family of curves $y = a\sqrt{x} + b$.

17–20 Sketch the graph of the equation by making appropriate transformations to the graph of a basic power function. Check your work with a graphing utility.

17. (a) $y = 2(x + 1)^2$ (b) $y = -3(x - 2)^3$
(c) $y = \dfrac{-3}{(x + 1)^2}$ (d) $y = \dfrac{1}{(x - 3)^5}$

18. (a) $y = 1 - \sqrt{x + 2}$ (b) $y = 1 - \sqrt[3]{x + 2}$
(c) $y = \dfrac{5}{(1 - x)^3}$ (d) $y = \dfrac{2}{(4 + x)^4}$

19. (a) $y = \sqrt[3]{x + 1}$ (b) $y = 1 - \sqrt{x - 2}$
(c) $y = (x - 1)^5 + 2$ (d) $y = \dfrac{x + 1}{x}$

20. (a) $y = 1 + \dfrac{1}{x - 2}$ (b) $y = \dfrac{1}{1 + 2x + x^2}$
(c) $y = -\dfrac{2}{x^7}$ (d) $y = x^2 + 2x$

21. Sketch the graph of $y = x^2 + 2x$ by completing the square and making appropriate transformations to the graph of $y = x^2$.

22. (a) Use the graph of $y = \sqrt{x}$ to help sketch the graph of $y = \sqrt{|x|}$.
(b) Use the graph of $y = \sqrt[3]{x}$ to help sketch the graph of $y = \sqrt[3]{|x|}$.

23. As discussed in this section, Boyle's law states that at a constant temperature the pressure P exerted by a gas is related to the volume V by the equation $PV = k$.
(a) Find the appropriate units for the constant k if pressure (which is force per unit area) is in newtons per square meter (N/m^2) and volume is in cubic meters (m^3).
(b) Find k if the gas exerts a pressure of 20,000 N/m^2 when the volume is 1 liter (0.001 m^3).
(c) Make a table that shows the pressures for volumes of 0.25, 0.5, 1.0, 1.5, and 2.0 liters.
(d) Make a graph of P versus V.

24. A manufacturer of cardboard drink containers wants to construct a closed rectangular container that has a square base and will hold $\frac{1}{10}$ liter (100 cm^3). Estimate the dimension of the container that will require the least amount of material for its manufacture.

25–26 A variable y is said to be ***inversely proportional to the square of a variable*** x if y is related to x by an equation of the form $y = k/x^2$, where k is a nonzero constant, called the ***constant of proportionality***. This terminology is used in these exercises.

25. According to ***Coulomb's law***, the force F of attraction between positive and negative point charges is inversely proportional to the square of the distance x between them.
(a) Assuming that the force of attraction between two point charges is 0.0005 newton when the distance between them is 0.3 meter, find the constant of proportionality (with proper units).
(b) Find the force of attraction between the point charges when they are 3 meters apart.
(c) Make a graph of force versus distance for the two charges.
(d) What happens to the force as the particles get closer and closer together? What happens as they get farther and farther apart?

26. It follows from Newton's Law of Universal Gravitation that the weight W of an object (relative to the Earth) is inversely proportional to the square of the distance x between the object and the center of the Earth, that is, $W = C/x^2$.
(a) Assuming that a weather satellite weighs 2000 pounds on the surface of the Earth and that the Earth is a sphere of radius 4000 miles, find the constant C.
(b) Find the weight of the satellite when it is 1000 miles above the surface of the Earth.
(c) Make a graph of the satellite's weight versus its distance from the center of the Earth.
(d) Is there any distance from the center of the Earth at which the weight of the satellite is zero? Explain your reasoning.

FOCUS ON CONCEPTS

27. In each part, match the equation with one of the accompanying graphs, and give the equations for the horizontal and vertical asymptotes.

(a) $y = \dfrac{x^2}{x^2 - x - 2}$ (b) $y = \dfrac{x - 1}{x^2 - x - 6}$

(c) $y = \dfrac{2x^4}{x^4 + 1}$ (d) $y = \dfrac{4}{(x + 2)^2}$

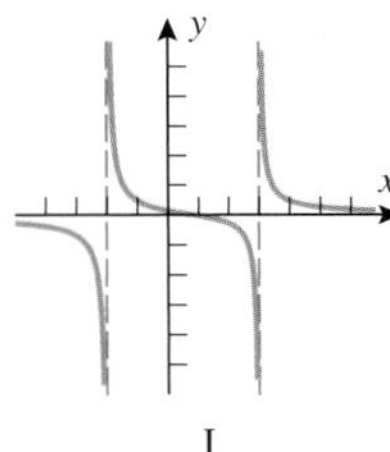

I

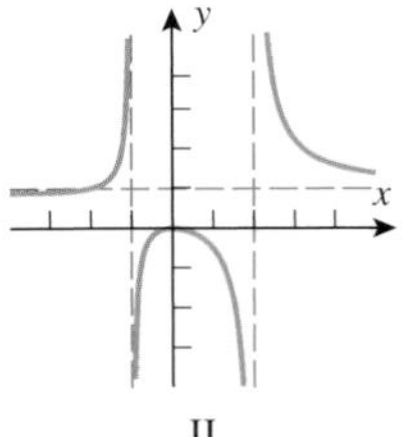

II

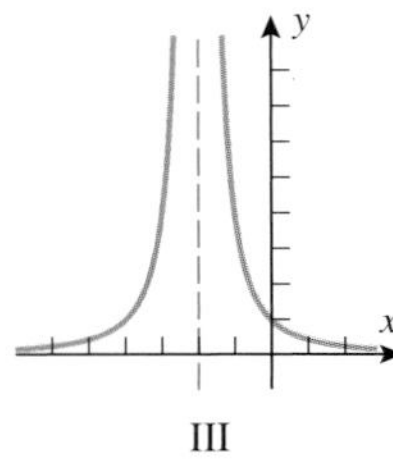

III

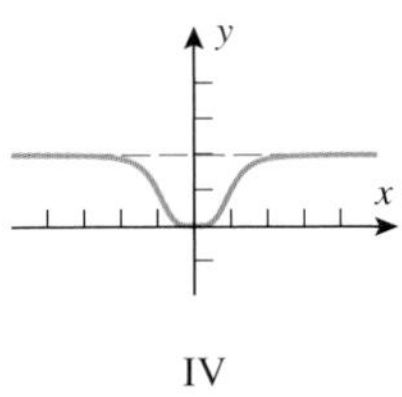

IV

Figure Ex-27

28. Find an equation of the form $y = k/(x^2 + bx + c)$ whose graph is a reasonable match to that in the accompanying figure. Check your work with a graphing utility.

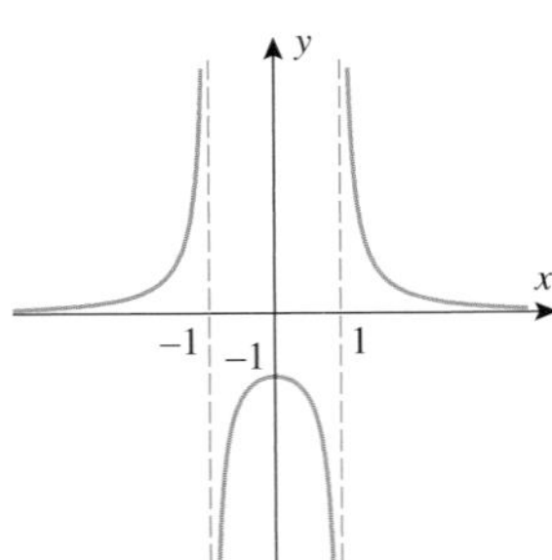

Figure Ex-28

29–30 Find an equation of the form $y = D + A \sin Bx$ or $y = D + A \cos Bx$ for each graph.

29.

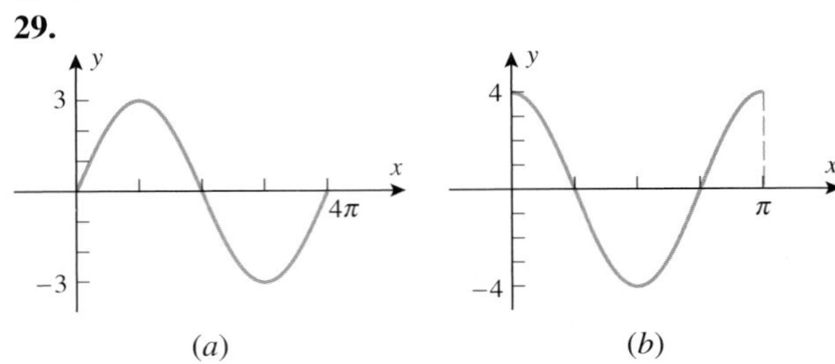

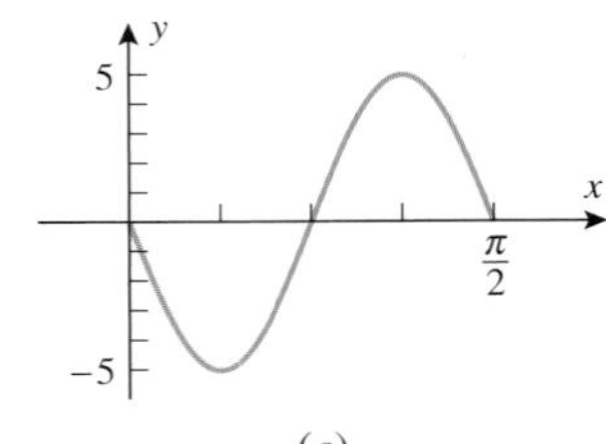

Figure Ex-29

30.

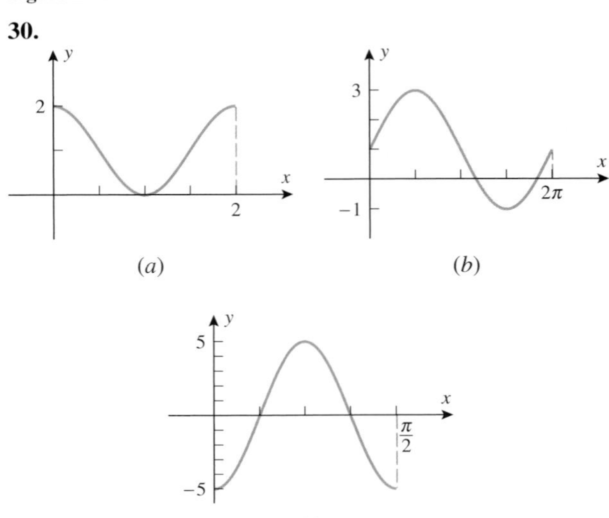

Figure Ex-30

31. In each part, find an equation for the graph that has the form $y = y_0 + A\sin(Bx - C)$.

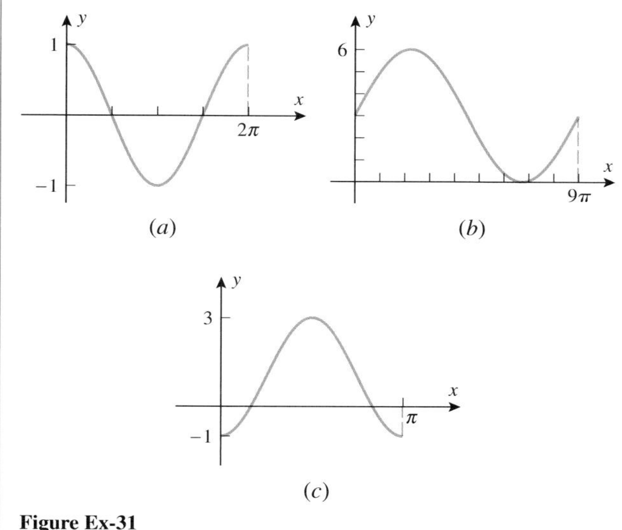

Figure Ex-31

32. In the United States, a standard electrical outlet supplies sinusoidal electrical current with a maximum voltage of $V = 120\sqrt{2}$ volts (V) at a frequency of 60 hertz (Hz). Write an equation that expresses V as a function of the time t, assuming that $V = 0$ if $t = 0$. [*Note:* 1 Hz = 1 cycle per second.]

33–34 Find the amplitude and period, and sketch at least two periods of the graph by hand. Check your work with a graphing utility.

33. (a) $y = 3\sin 4x$ (b) $y = -2\cos \pi x$
(c) $y = 2 + \cos\left(\dfrac{x}{2}\right)$

34. (a) $y = -1 - 4\sin 2x$ (b) $y = \frac{1}{2}\cos(3x - \pi)$
(c) $y = -4\sin\left(\dfrac{x}{3} + 2\pi\right)$

35. Equations of the form

$$x = A_1 \sin \omega t + A_2 \cos \omega t$$

arise in the study of vibrations and other periodic motion.
(a) Use the trigonometric identity for $\sin(\alpha + \beta)$ to show that this equation can be expressed in the form

$$x = A\sin(\omega t + \theta)$$

where $A^2 = A_1^2 + A_2^2$ and $\tan\theta = A_2/A_1$.
(b) Express the equation

$$x = \sqrt{2}\sin 2\pi t + \sqrt{6}\cos 2\pi t$$

in the form $x = A\sin(\omega t + \theta)$, and use a graphing utility to confirm that both equations have the same graph.

36. Determine the number of solutions of $x = 2\sin x$, and use a graphing or calculating utility to estimate them.

✔ QUICK CHECK ANSWERS 1.4

1. even; odd; negative 2. $(-\infty, +\infty)$ 3. (a) $[0, +\infty)$ (b) $(-\infty, +\infty)$ (c) $(0, +\infty)$ (d) $(-\infty, 0) \cup (0, +\infty)$ 4. (a) algebraic (b) polynomial (c) not algebraic (d) rational (e) rational 5. $|A|$; $2\pi/|B|$

1.5 INVERSE FUNCTIONS

In everyday language the term "inversion" conveys the idea of a reversal. For example, in meteorology a temperature inversion is a reversal in the usual temperature properties of air layers, and in music a melodic inversion reverses an ascending interval to the corresponding descending interval. In mathematics the term ***inverse*** *is used to describe functions that reverse one another in the sense that each undoes the effect of the other. In this section we discuss this fundamental mathematical idea.*

INVERSE FUNCTIONS

The idea of solving an equation $y = f(x)$ for x as a function of y, say $x = g(y)$, is one of the most important ideas in mathematics. Sometimes, solving an equation is a simple process; for example, using basic algebra the equation

$$y = x^3 + 1 \qquad \boxed{y = f(x)}$$

can be solved for x as a function of y:

$$x = \sqrt[3]{y - 1} \qquad \boxed{x = g(y)}$$

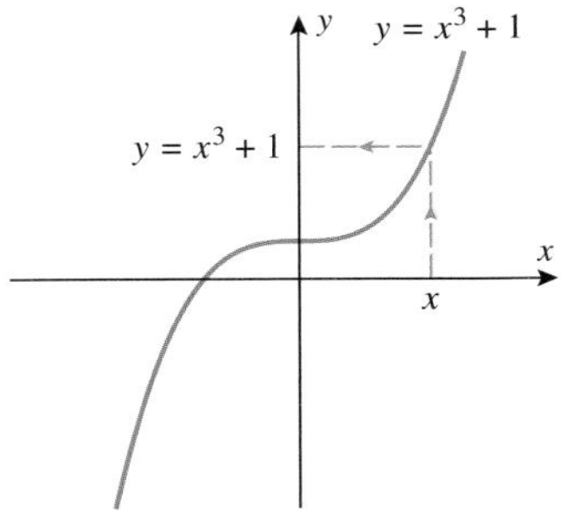

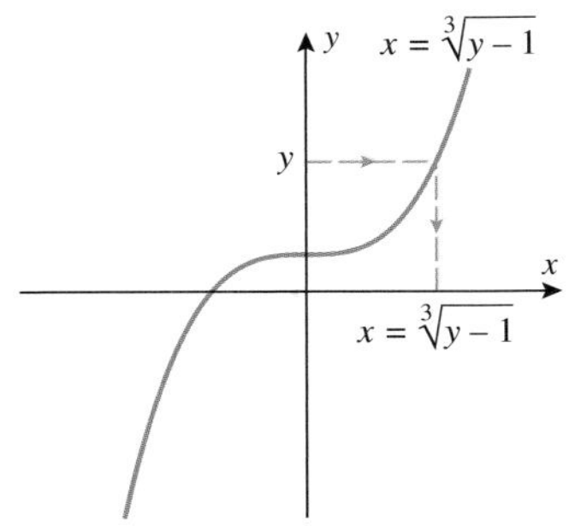

Figure 1.5.1

The first equation is better for computing y if x is known, and the second is better for computing x if y is known (Figure 1.5.1).

Our primary interest in this section is to identify relationships that may exist between the functions f and g when an equation $y = f(x)$ is expressed as $x = g(y)$, or conversely. For example, consider the functions $f(x) = x^3 + 1$ and $g(y) = \sqrt[3]{y-1}$ discussed above. When these functions are composed in either order they cancel out the effect of one another in the sense that

$$\begin{aligned} g(f(x)) &= \sqrt[3]{f(x)-1} = \sqrt[3]{(x^3+1)-1} = x \\ f(g(y)) &= [g(y)]^3 + 1 = (\sqrt[3]{y-1})^3 + 1 = y \end{aligned} \tag{1}$$

Pairs of functions with these two properties are so important that there is some terminology for them.

1.5.1 DEFINITION. If the functions f and g satisfy the two conditions

$$\begin{aligned} g(f(x)) &= x \text{ for every } x \text{ in the domain of } f \\ f(g(y)) &= y \text{ for every } y \text{ in the domain of } g \end{aligned}$$

then we say that ***f is an inverse of g*** and ***g is an inverse of f*** or that ***f and g are inverse functions***.

WARNING

If f is a function, then the -1 in the symbol f^{-1} always denotes an inverse and *never* an exponent. That is,

$$f^{-1}(x) \text{ never means } \frac{1}{f(x)}$$

It can be shown (Exercise 34) that if a function f has an inverse, then that inverse is unique. Thus, if a function f has an inverse, then we are entitled to talk about "the" inverse of f, in which case we denote it by the symbol f^{-1}.

▶ **Example 1** The computations in (1) show that $g(y) = \sqrt[3]{y-1}$ is the inverse of $f(x) = x^3 + 1$. Thus, we can express g in inverse notation as

$$f^{-1}(y) = \sqrt[3]{y-1}$$

and we can express the equations in Definition 1.5.1 as

$$\begin{aligned} f^{-1}(f(x)) &= x \quad \text{for every } x \text{ in the domain of } f \\ f(f^{-1}(y)) &= y \quad \text{for every } y \text{ in the domain of } f^{-1} \end{aligned} \tag{2}$$

We will call these the ***cancellation equations*** for f and f^{-1}. ◀

CHANGING THE INDEPENDENT VARIABLE

The formulas in (2) use x as the independent variable for f and y as the independent variable for f^{-1}. Although it is often convenient to use different independent variables for f and f^{-1}, there will be occasions on which it is desirable to use the same independent variable for both. For example, if we want to graph the functions f and f^{-1} together in the same xy-coordinate system, then we would want to use x as the independent variable and y as the dependent variable for both functions. Thus, to graph the functions $f(x) = x^3 + 1$ and $f^{-1}(y) = \sqrt[3]{y-1}$ of Example 1 in the same xy-coordinate system, we would change the independent variable y to x, use y as the dependent variable for both functions, and graph the equations

$$y = x^3 + 1 \quad \text{and} \quad y = \sqrt[3]{x-1}$$

We will talk more about graphs of inverse functions later in this section, but for reference we give the following reformulation of the cancellation equations in (2) using x as the independent variable for both f and f^{-1}:

$$\begin{aligned} f^{-1}(f(x)) &= x \quad \text{for every } x \text{ in the domain of } f \\ f(f^{-1}(x)) &= x \quad \text{for every } x \text{ in the domain of } f^{-1} \end{aligned} \tag{3}$$

► **Example 2** Confirm each of the following.

(a) The inverse of $f(x) = 2x$ is $f^{-1}(x) = \frac{1}{2}x$.

(b) The inverse of $f(x) = x^3$ is $f^{-1}(x) = x^{1/3}$.

Solution (a).

$$f^{-1}(f(x)) = f^{-1}(2x) = \tfrac{1}{2}(2x) = x$$
$$f(f^{-1}(x)) = f\left(\tfrac{1}{2}x\right) = 2\left(\tfrac{1}{2}x\right) = x$$

Solution (b).

$$f^{-1}(f(x)) = f^{-1}(x^3) = \left(x^3\right)^{1/3} = x$$
$$f(f^{-1}(x)) = f(x^{1/3}) = \left(x^{1/3}\right)^3 = x \quad ◄$$

The results in Example 2 should make sense to you intuitively, since the operations of multiplying by 2 and multiplying by $\frac{1}{2}$ in either order cancel the effect of one another, as do the operations of cubing and taking a cube root.

► **Example 3** Given that the function f has an inverse and that $f(3) = 5$, find $f^{-1}(5)$.

Solution. Apply f^{-1} to both sides of the equation $f(3) = 5$ to obtain

$$f^{-1}(f(3)) = f^{-1}(5)$$

and now apply the first equation in (3) to conclude that $f^{-1}(5) = 3$. ◄

In general, if f has an inverse and $f(a) = b$, then the procedure in Example 3 shows that $a = f^{-1}(b)$; that is, f^{-1} maps each output of f back into the corresponding input (Figure 1.5.2).

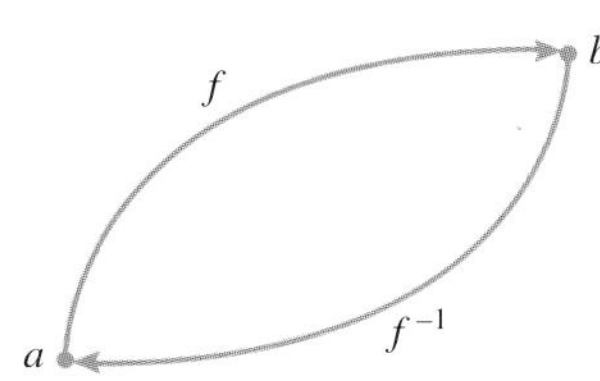

Figure 1.5.2 If f maps a to b, then f^{-1} maps b back to a.

DOMAIN AND RANGE OF INVERSE FUNCTIONS

The equations in (3) imply the following relationships between the domains and ranges of f and f^{-1}:

$$\begin{aligned} \text{domain of } f^{-1} &= \text{range of } f \\ \text{range of } f^{-1} &= \text{domain of } f \end{aligned} \tag{4}$$

One way to show that two sets are the same is to show that each is a subset of the other. Thus we can establish the first equality in (4) by showing that the domain of f^{-1} is a subset of the range of f and that the range of f is a subset of the domain of f^{-1}. We do this as follows: The first equation in (3) implies that f^{-1} is defined at $f(x)$ for all values of x in the domain of f, and this implies that the range of f is a subset of the domain of f^{-1}. Conversely, if x is in the domain of f^{-1}, then the second equation in (3) implies that x is in the range of f because it is the image of $f^{-1}(x)$. Thus, the domain of f^{-1} is a subset of the range of f. We leave the proof of the second equation in (4) as an exercise.

A METHOD FOR FINDING INVERSE FUNCTIONS

At the beginning of this section we observed that solving $y = f(x) = x^3 + 1$ for x as a function of y produces $x = f^{-1}(y) = \sqrt[3]{y-1}$. The following theorem shows that this is not accidental.

1.5.2 THEOREM. *If an equation $y = f(x)$ can be solved for x as a function of y, say $x = g(y)$, then f has an inverse and that inverse is $g(y) = f^{-1}(y)$.*

PROOF. Substituting $y = f(x)$ into $x = g(y)$ yields $x = g(f(x))$, which confirms the first equation in Definition 1.5.1, and substituting $x = g(y)$ into $y = f(x)$ yields $y = f(g(y))$, which confirms the second equation in Definition 1.5.1. ■

Theorem 1.5.2 provides us with the following procedure for finding the inverse of a function.

A Procedure for Finding the Inverse of a Function f

Step 1. Write down the equation $y = f(x)$.

Step 2. If possible, solve this equation for x as a function of y.

Step 3. The resulting equation will be $x = f^{-1}(y)$, which provides a formula for f^{-1} with y as the independent variable.

Step 4. If y is acceptable as the independent variable for the inverse function, then you are done, but if you want to have x as the independent variable, then you need to interchange x and y in the equation $x = f^{-1}(y)$ to obtain $y = f^{-1}(x)$.

An alternative way to obtain a formula for $f^{-1}(x)$ with x as the independent variable is to reverse the roles of x and y at the outset and solve the equation $x = f(y)$ for y as a function of x.

▶ **Example 4** Find a formula for the inverse of $f(x) = \sqrt{3x-2}$ with x as the independent variable, and state the domain of f^{-1}.

Solution. Following the procedure stated above, we first write

$$y = \sqrt{3x-2}$$

Then we solve this equation for x as a function of y:

$$y^2 = 3x - 2$$
$$x = \tfrac{1}{3}(y^2 + 2)$$

Since we want x to be the independent variable, we reverse x and y in the last equation to produce the formula

$$f^{-1}(x) = \tfrac{1}{3}(x^2 + 2) \tag{5}$$

We know from (4) that the domain of f^{-1} is the range of f. In general, this need not be the same as the natural domain of the fomula for f^{-1}. Indeed, in this example the natural domain of (5) is $(-\infty, +\infty)$, whereas the range of $f(x) = \sqrt{3x-2}$ is $[0, +\infty)$. Thus, if we want to make the domain of f^{-1} clear, we must express it explicitly by rewriting (5) as

$$f^{-1}(x) = \tfrac{1}{3}(x^2 + 2), \quad x \geq 0 \blacktriangleleft$$

EXISTENCE OF INVERSE FUNCTIONS

The procedure we gave above for finding the inverse of a function f was based on solving the equation $y = f(x)$ for x as a function of y. This procedure can fail for two reasons—the function f may not have an inverse, or it may have an inverse but the equation $y = f(x)$ cannot be solved explicitly for x as a function of y. Thus, it is important to establish conditions that ensure the existence of an inverse, even if it cannot be found explicitly.

If a function f has an inverse, then it must assign distinct outputs to distinct inputs. For example, the function $f(x) = x^2$ cannot have an inverse because it assigns the same value to $x = 2$ and $x = -2$, namely, $f(2) = f(-2) = 4$. Thus, if $f(x) = x^2$ were to have an inverse, then the equation $f(2) = 4$ would imply that $f^{-1}(4) = 2$, and the equation $f(-2) = 4$ would imply that $f^{-1}(4) = -2$. But this is impossible because $f^{-1}(4)$ cannot have two different values. Another way to see that $f(x) = x^2$ has no inverse is to attempt to find the inverse by solving the equation $y = x^2$ for x as a function of y. We run into trouble immediately because the resulting equation $x = \pm\sqrt{y}$ does not express x as a *single* function of y.

A function that assigns distinct outputs to distinct inputs is said to be ***one-to-one*** or ***invertible***, so we know from the preceding discussion that if a function f has an inverse, then it must be one-to-one. The converse is also true, thereby establishing the following theorem.

1.5.3 THEOREM. *A function has an inverse if and only if it is one-to-one.*

Stated algebraically, a function f is one-to-one if and only if $f(x_1) \neq f(x_2)$ whenever $x_1 \neq x_2$; stated geometrically, a function f is one-to-one if and only if the graph of $y = f(x)$ is cut at most once by any horizontal line (Figure 1.5.3). The latter statement together with Theorem 1.5.3 provides the following geometric test for determining whether a function has an inverse.

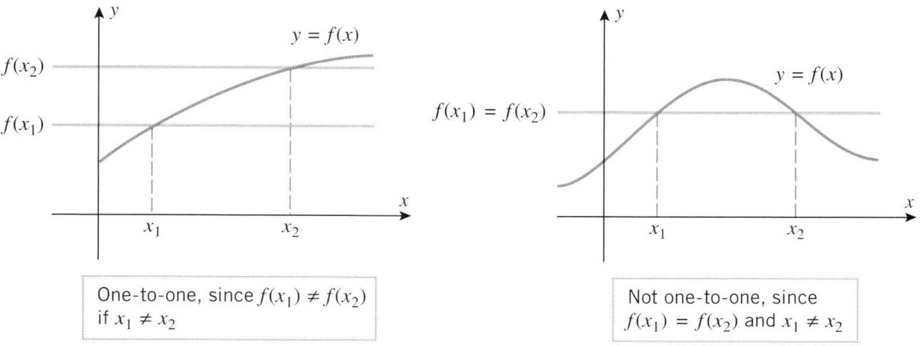

Figure 1.5.3

1.5.4 THEOREM (*The Horizontal Line Test*). *A function has an inverse function if and only if its graph is cut at most once by any horizontal line.*

▶ **Example 5** Use the horizontal line test to show that $f(x) = x^2$ has no inverse but that $f(x) = x^3$ does.

Solution. Figure 1.5.4 shows a horizontal line that cuts the graph of $y = x^2$ more than once, so $f(x) = x^2$ is not invertible. Figure 1.5.5 shows that the graph of $y = x^3$ is cut at most once by any horizontal line, so $f(x) = x^3$ is invertible. [Recall from Example 2 that the inverse of $f(x) = x^3$ is $f^{-1}(x) = x^{1/3}$.] ◀

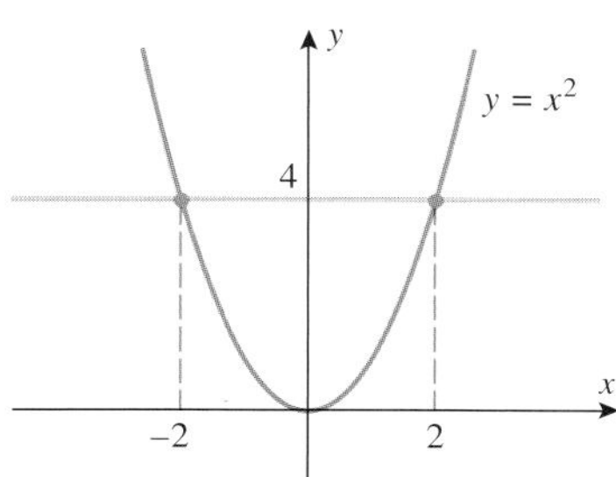

Figure 1.5.4

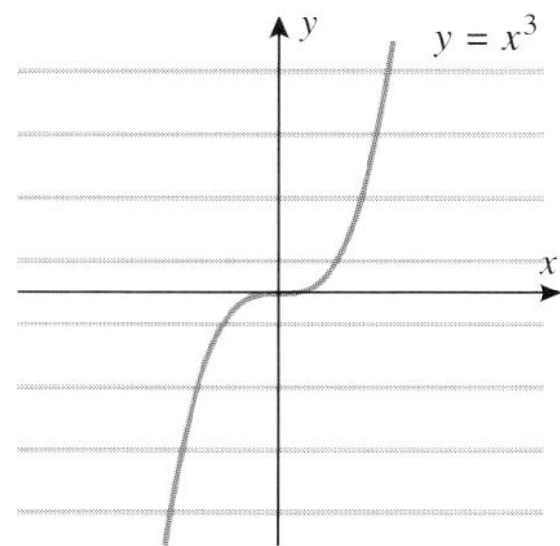

Figure 1.5.5

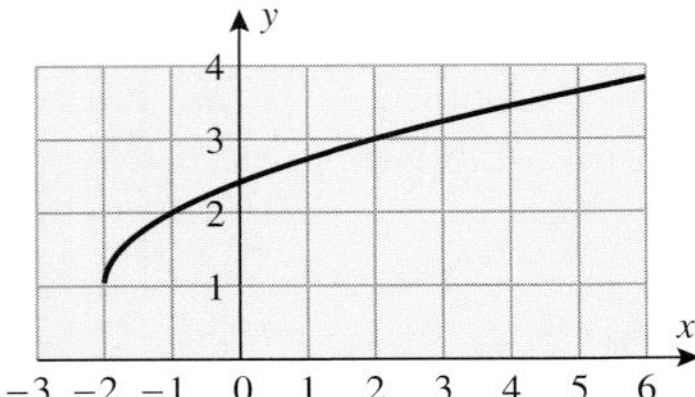

Figure 1.5.6

▶ **Example 6** Explain why the function f that is graphed in Figure 1.5.6 has an inverse, and find $f^{-1}(3)$.

Solution. The function f has an inverse since its graph passes the horizontal line test. To evaluate $f^{-1}(3)$, we view $f^{-1}(3)$ as that number x for which $f(x) = 3$. From the graph we see that $f(2) = 3$, so $f^{-1}(3) = 2$. ◀

The function $f(x) = x^3$ in Figure 1.5.5 is an example of an increasing function. Give an example of a decreasing function and compute its inverse.

INCREASING OR DECREASING FUNCTIONS ARE INVERTIBLE

A function whose graph is always rising as it is traversed from left to right is said to be an ***increasing function***, and a function whose graph is always falling as it is traversed from left to right is said to be a ***decreasing function***. If x_1 and x_2 are points in the domain of a function f, then f is increasing if

$$f(x_1) < f(x_2) \quad \text{whenever } x_1 < x_2$$

and f is decreasing if

$$f(x_1) > f(x_2) \quad \text{whenever } x_1 < x_2$$

(Figure 1.5.7). It is evident geometrically that increasing and decreasing functions pass the horizontal line test and hence are invertible.

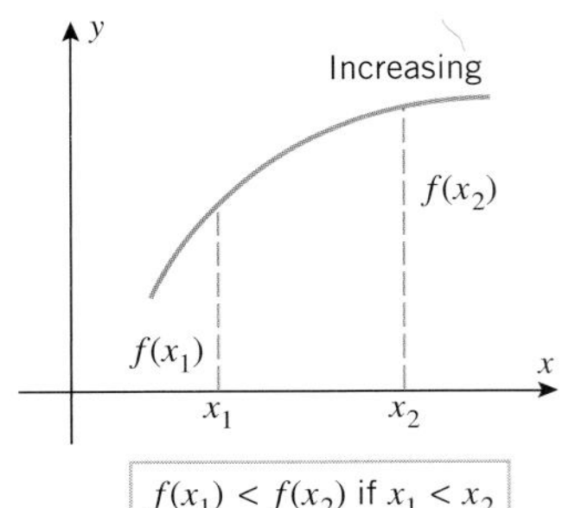

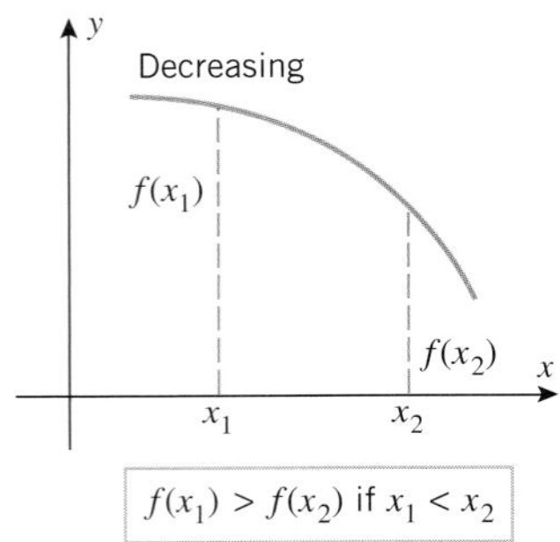

Figure 1.5.7

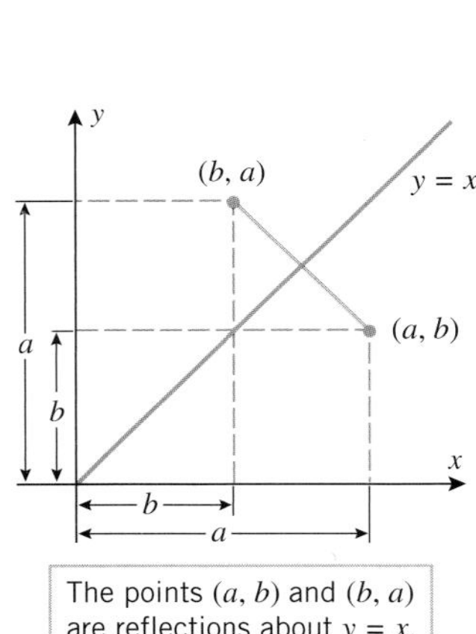

Figure 1.5.8

GRAPHS OF INVERSE FUNCTIONS

Our next objective is to explore the relationship between the graphs of f and f^{-1}. For this purpose, it will be desirable to use x as the independent variable for both functions so we can compare the graphs of $y = f(x)$ and $y = f^{-1}(x)$.

If (a, b) is a point on the graph $y = f(x)$, then $b = f(a)$. This is equivalent to the statement that $a = f^{-1}(b)$, which means that (b, a) is a point on the graph of $y = f^{-1}(x)$. In short, reversing the coordinates of a point on the graph of f produces a point on the graph of f^{-1}. Similarly, reversing the coordinates of a point on the graph of f^{-1} produces a point on the graph of f (verify). However, the geometric effect of reversing the coordinates of a point is to reflect that point about the line $y = x$ (Figure 1.5.8), and hence the graphs of $y = f(x)$ and $y = f^{-1}(x)$ are reflections of one another about this line (Figure 1.5.9). In summary, we have the following result.

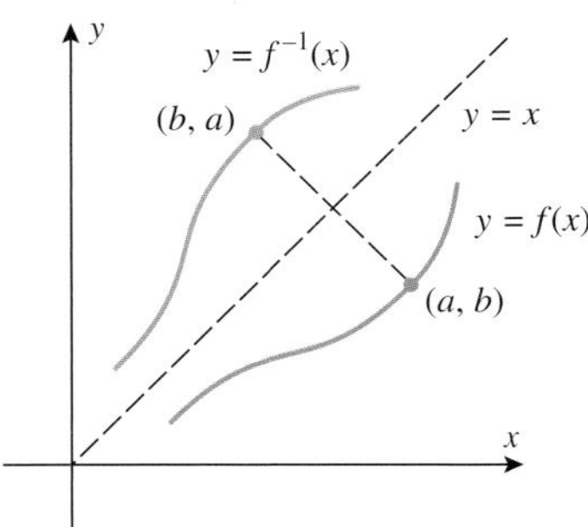

Figure 1.5.9

1.5.5 THEOREM. *If f has an inverse, then the graphs of $y = f(x)$ and $y = f^{-1}(x)$ are reflections of one another about the line $y = x$; that is, each graph is the mirror image of the other with respect to that line.*

▶ **Example 7** Figure 1.5.10 shows the graphs of the inverse functions discussed in Examples 2 and 4. ◀

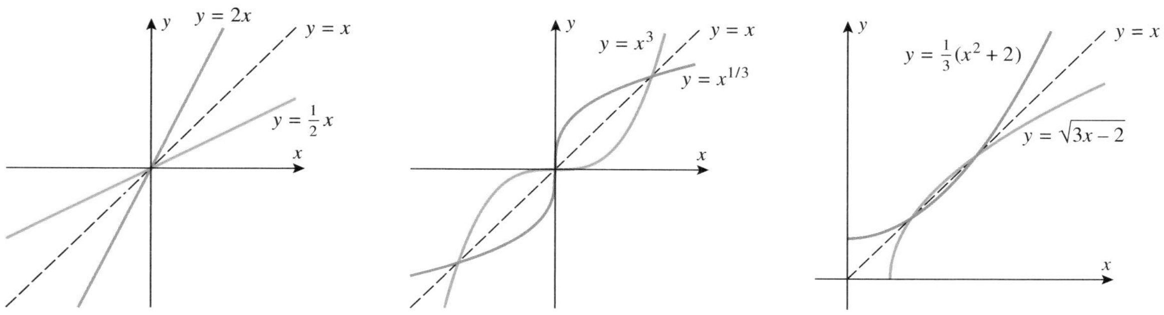

Figure 1.5.10

RESTRICTING DOMAINS FOR INVERTIBILITY

If a function g is obtained from a function f by placing restrictions on the domain of f, then g is called a ***restriction of f***. Thus, for example, the function

$$g(x) = x^3, \quad x \geq 0$$

is a restriction of the function $f(x) = x^3$. More precisely, it is called the restriction of x^3 to the interval $[0, +\infty)$.

Sometimes it is possible to create an invertible function from a function that is not invertible by restricting the domain appropriately. For example, we showed earlier that $f(x) = x^2$ is not invertible. However, consider the restricted functions

$$f_1(x) = x^2, \quad x \geq 0 \quad \text{and} \quad f_2(x) = x^2, \quad x \leq 0$$

the union of whose graphs is the complete graph of $f(x) = x^2$ (Figure 1.5.11). These restricted functions are each one-to-one (hence invertible), since their graphs pass the horizontal line test. As illustrated in Figure 1.5.12, their inverses are

$$f_1^{-1}(x) = \sqrt{x} \quad \text{and} \quad f_2^{-1}(x) = -\sqrt{x}$$

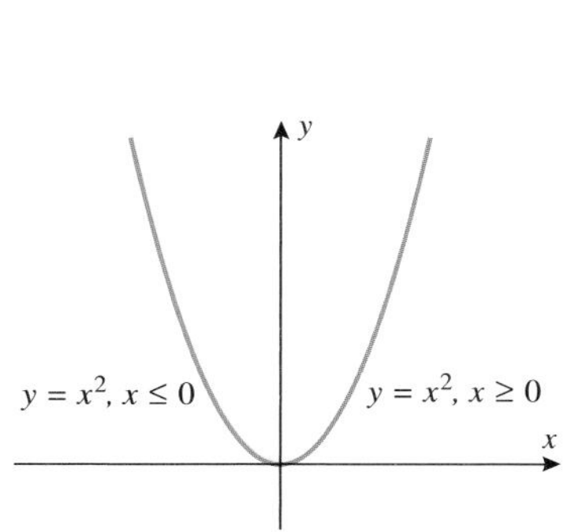

Figure 1.5.11

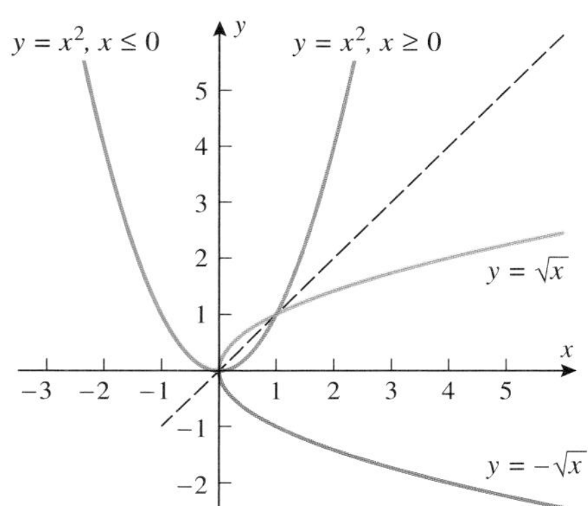

Figure 1.5.12

✔ QUICK CHECK EXERCISES 1.5

(See page 59 for answers.)

1. In each part, determine whether the function f is one-to-one.
   (a) $f(t)$ is the number of people in line at a movie theater at time t.
   (b) $f(x)$ is the measured high temperature (rounded to the nearest °F) in a city on the xth day of the year.
   (c) $f(v)$ is the weight of v cubic inches of lead.

2. A student enters a number on a calculator, doubles it, adds 8 to the result, divides the sum by 2, subtracts 3 from the quotient, and then cubes the difference. If the resulting number is x, then ________ was the student's original number.

3. If $(3, -2)$ is a point on the graph of an odd invertible function f, then ________ and ________ are points on the graph of f^{-1}.

EXERCISE SET 1.5 Graphing Utility

1. In (a)–(d), determine whether f and g are inverse functions.
   (a) $f(x) = 4x,\ g(x) = \frac{1}{4}x$
   (b) $f(x) = 3x + 1,\ g(x) = 3x - 1$
   (c) $f(x) = \sqrt[3]{x-2},\ g(x) = x^3 + 2$
   (d) $f(x) = x^4,\ g(x) = \sqrt[4]{x}$

2. Check your answers to Exercise 1 with a graphing utility by determining whether the graphs of f and g are reflections of one another about the line $y = x$.

3. In each part, use the horizontal line test to determine whether the function f is one-to-one.
   (a) $f(x) = 3x + 2$ (b) $f(x) = \sqrt{x-1}$
   (c) $f(x) = |x|$ (d) $f(x) = x^3$
   (e) $f(x) = x^2 - 2x + 2$ (f) $f(x) = \sin x$

4. In each part, generate the graph of the function f with a graphing utility, and determine whether f is one-to-one.
   (a) $f(x) = x^3 - 3x + 2$ (b) $f(x) = x^3 - 3x^2 + 3x - 1$

FOCUS ON CONCEPTS

5. In each part, determine whether the function f defined by the table is one-to-one.

(a)

x	1	2	3	4	5	6
$f(x)$	−2	−1	0	1	2	3

(b)

x	1	2	3	4	5	6
$f(x)$	4	−7	6	−3	1	4

6. A face of a broken clock lies in the xy-plane with the center of the clock at the origin and 3:00 in the direction of the positive x-axis. When the clock broke, the tip of the hour hand stopped on the graph of $y = f(x)$, where f is a function that satisfies $f(0) = 0$.
   (a) Are there any times of the day that cannot appear in such a configuration? Explain.
   (b) How does your answer to part (a) change if f must be an invertible function?
   (c) How do your answers to parts (a) and (b) change if it was the tip of the minute hand that stopped on the graph of f?

7. (a) The accompanying figure shows the graph of a function f over its domain $-8 \le x \le 8$. Explain why f has an inverse, and use the graph to find $f^{-1}(2)$, $f^{-1}(-1)$, and $f^{-1}(0)$.
   (b) Find the domain and range of f^{-1}.
   (c) Sketch the graph of f^{-1}.

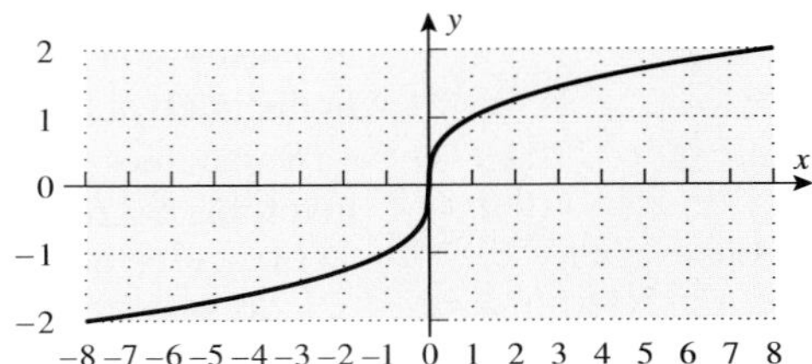

Figure Ex-7

8. (a) Explain why the function f graphed in the accompanying figure has no inverse function on its domain $-3 \le x \le 4$.
   (b) Subdivide the domain into three adjacent intervals on each of which the function f has an inverse.

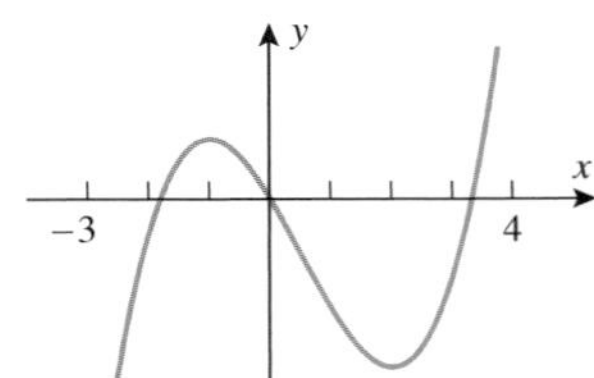

Figure Ex-8

9–17 Find a formula for $f^{-1}(x)$.

9. $f(x) = 7x - 6$
10. $f(x) = \dfrac{x+1}{x-1}$
11. $f(x) = 3x^3 - 5$
12. $f(x) = \sqrt[5]{4x+2}$
13. $f(x) = \sqrt[3]{2x-1}$

14. $f(x) = 5/(x^2 + 1), \quad x \geq 0$

15. $f(x) = 3/x^2, \quad x < 0$ **16.** $f(x) = \begin{cases} 2x, & x \leq 0 \\ x^2, & x > 0 \end{cases}$

17. $f(x) = \begin{cases} 5/2 - x, & x < 2 \\ 1/x, & x \geq 2 \end{cases}$

18. Find a formula for $p^{-1}(x)$, given that

$$p(x) = x^3 - 3x^2 + 3x - 1$$

19–23 Find a formula for $f^{-1}(x)$, and state the domain of the function f^{-1}.

19. $f(x) = (x + 2)^4, \quad x \geq 0$

20. $f(x) = \sqrt{x + 3}$ **21.** $f(x) = -\sqrt{3 - 2x}$

22. $f(x) = 3x^2 + 5x - 2, \quad x \geq 0$

23. $f(x) = x - 5x^2, \quad x \geq 1$

FOCUS ON CONCEPTS

24. The formula $F = \frac{9}{5}C + 32$, where $C \geq -273.15$ expresses the Fahrenheit temperature F as a function of the Celsius temperature C.
(a) Find a formula for the inverse function.
(b) In words, what does the inverse function tell you?
(c) Find the domain and range of the inverse function.

25. (a) One meter is about 6.214×10^{-4} miles. Find a formula $y = f(x)$ that expresses a length y in meters as a function of the same length x in miles.
(b) Find a formula for the inverse of f.
(c) Describe what the formula $x = f^{-1}(y)$ tells you in practical terms.

26. Let $f(x) = x^2$, $x > 1$, and $g(x) = \sqrt{x}$.
(a) Show that $f(g(x)) = x$, $x > 1$, and $g(f(x)) = x$, $x > 1$.
(b) Show that f and g are *not* inverses by showing that the graphs of $y = f(x)$ and $y = g(x)$ are not reflections of one another about $y = x$.
(c) Do parts (a) and (b) contradict one another? Explain.

27. (a) Show that $f(x) = (3 - x)/(1 - x)$ is its own inverse.
(b) What does the result in part (a) tell you about the graph of f?

28. Let $f(x) = ax^2 + bx + c$, $a > 0$. Find f^{-1} if the domain of f is restricted to
(a) $x \geq -b/(2a)$ (b) $x \leq -b/(2a)$.

29. Let $f(x) = 2x^3 + 5x + 3$. Find x if $f^{-1}(x) = 1$.

30. Let $f(x) = \dfrac{x^3}{x^2 + 1}$. Find x if $f^{-1}(x) = 2$.

31. Prove that if $a^2 + bc \neq 0$, then the graph of

$$f(x) = \frac{ax + b}{cx - a}$$

is symmetric about the line $y = x$.

32. (a) Prove: If f and g are one-to-one, then so is the composition $f \circ g$.
(b) Prove: If f and g are one-to-one, then

$$(f \circ g)^{-1} = g^{-1} \circ f^{-1}$$

33. Sketch the graph of a function that is one-to-one on $(-\infty, +\infty)$, yet not increasing on $(-\infty, +\infty)$ and not decreasing on $(-\infty, +\infty)$.

34. Prove: A one-to-one function f cannot have two different inverses.

✔ QUICK CHECK ANSWERS 1.5

1. (a) not one-to-one (b) not one-to-one (c) one-to-one 2. $\sqrt[3]{x} - 1$ 3. $(-2, 3)$; $(2, -3)$

1.6 MATHEMATICAL MODELS

In this section we will discuss the idea of "mathematical modeling," which is the procedure that mathematicians and scientists use to describe physical phenomena mathematically. In particular, we will explain the method of "least squares," which is a mathematical technique used to obtain functions that best describe a set of observed data.

MATHEMATICAL MODELS

A ***mathematical model*** of a physical law is a description of that law in the language of mathematics. Such models make it possible to use mathematical methods to deduce results about the physical world that are not evident or have never been observed. For example, the possibility of placing a satellite in orbit around the Earth was deduced mathematically from Issac Newton's model of mechanics nearly 200 years before the launching of *Sputnik*, and Albert Einstein (1879–1955) gave a relativistic model of mechanics in 1915 that explained a precession (position shift) in the perihelion of the planet Mercury that was not confirmed by physical measurement until 1967.

In a typical modeling situation, a scientist wants to obtain a mathematical relationship between two variables x and y using a set of n ordered pairs of measurements

$$(x_1, y_1),\ (x_2, y_2),\ (x_3, y_3), \ldots,\ (x_n, y_n) \tag{1}$$

that relate corresponding values of the variables. We distinguish between two types of physical phenomena—***deterministic phenomena*** in which each value of x determines one value of y, and ***probabilistic phenomena*** in which the value of y associated with a specific x is not uniquely determined. For example, if y is the amount that a force x stretches a certain spring, then each value of x determines a unique y, so this is a deterministic phenomenon. In contrast, if y is the weight of a person whose height is x, then y is not uniquely determined by x, since people with the same height can have different weights. Nevertheless, there is a "correlation" between weight and height that makes it more likely for a taller person to weigh more, so this is a probabilistic phenomenon. We will be concerned with deterministic phenomena only.

In a deterministic model the variable y is a function of the variable x, and the goal is to find a formula $y = f(x)$ that best describes the data. One way to model a set of deterministic data is to look for a function f whose graph passes through all of the data points; this is called an ***interpolating function***. Although interpolating functions are appropriate in certain situations, they do not adequately account for measurement errors in data. For example, suppose that the relationship between x and y is known to be linear but accuracy limitations in the measuring devices and random variations in experimental conditions produce the data plotted in Figure 1.6.1*a*. One could use a computer program to find a polynomial of degree 10 whose graph passes through all of the data points (Figure 1.6.1*b*). However, such a polynomial model does not successfully convey the underlying linear relationship. A better approach is to look for a linear equation $y = mx + b$ whose graph more accurately describes the linear relationship, even if its graph does not pass through all (or any) of the data points (Figure 1.6.1*c*).

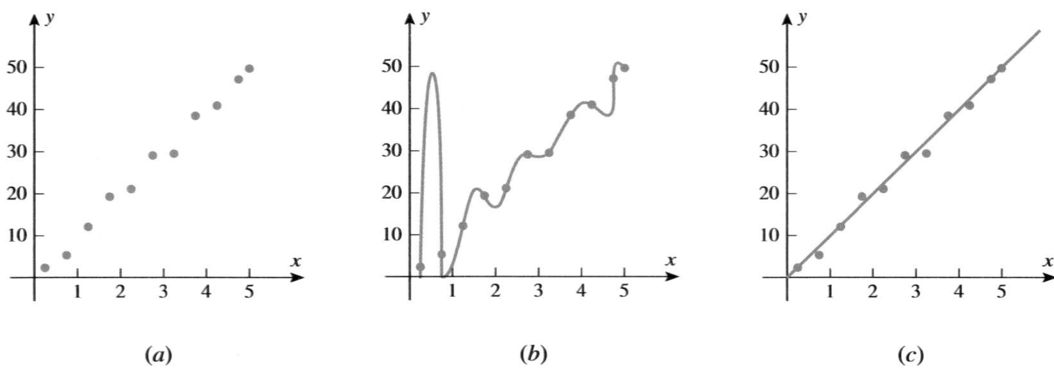

Figure 1.6.1

LINEAR MODELS

The most important methods for finding linear models are based on the following idea: For any proposed linear model $y = mx + b$, draw a vertical connector from each data point

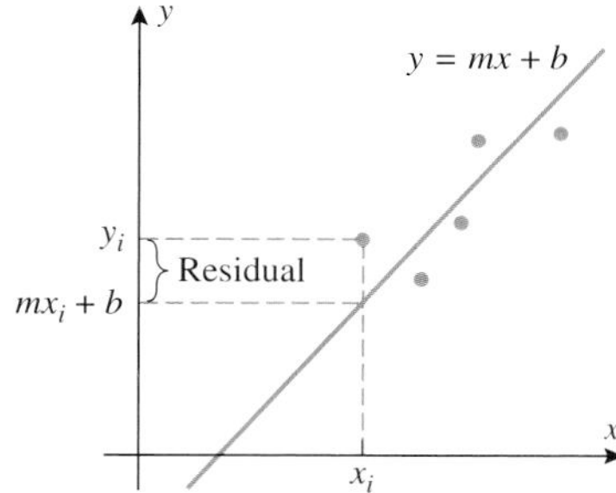

Figure 1.6.2

Most graphing calculators, CAS programs, and spreadsheet programs provide methods for finding regression lines. You will need access to one of these technologies for the examples in this section.

(x_i, y_i) to the line, and consider the differences $y_i - y$ (Figure 1.6.2). These differences, which are called ***residuals***, may be viewed as "errors" that result when the line is used to model the data. Points above the line have positive errors, points below the line have negative errors, and points on the line have no error.

One way to choose a linear model is to look for a line $y = mx + b$ in which the sum of the residuals is zero, the logic being that this makes the positive and negative errors balance out. However, one can find examples where this procedure produces unacceptably poor models, so for reasons that we cannot discuss here the most common method for finding a linear model is to look for a line $y = mx + b$ in which the *sum of the squares* of the residuals is as small as possible. This is called the ***least squares line of best fit*** or the ***regression line***.

It is possible to compute a regression line, even in cases where the data have no apparent linear pattern. Thus, it is important to have some quantitative method of determining whether a linear model is appropriate for the data. The most common measure of linearity in data is called the ***correlation coefficient***. Following convention, we denote the correlation coefficient by the letter r. Although a detailed discussion of correlation coefficients is beyond the scope of this text, here are some of the basic facts:

- The values of r are in the interval $-1 \le r \le 1$, where r has the same sign as the slope of the regression line.
- If r is equal to 1 or -1, then the data points all lie on a line, so a linear model is a perfect fit for the data.
- If $r = 0$, then the data points exhibit no linear tendency, so a linear model is inappropriate for the data.

The closer r is to 1 or -1, the more tightly the data points hug the regression line and the more appropriate the regression line is as a model; the closer r is to 0, the more scattered the points and the less appropriate the regression line is as a model (Figure 1.6.3).

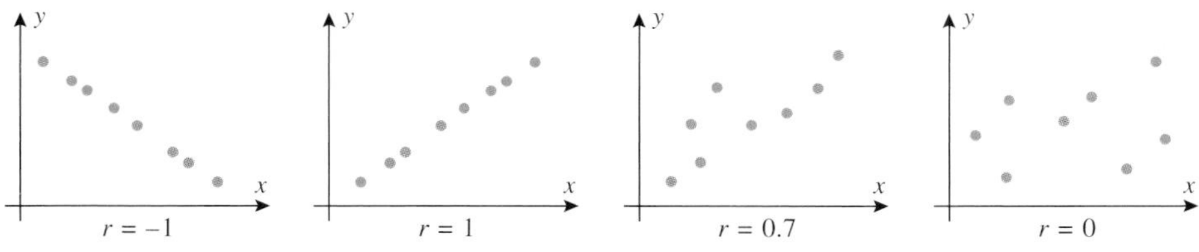

Figure 1.6.3

Roughly stated, the value of r^2 is a measure of the percentage of data points that fall in a "tight linear band." Thus $r = 0.5$ means that 25% of the points fall in a tight linear band, and $r = 0.9$ means that 81% of the points fall in a tight linear band. (A precise explanation of what is meant by a "tight linear band" requires ideas from statistics.)

Table 1.6.1

TEMPERATURE T (°C)	PRESSURE p (atm)
0	2.54
50	3.06
100	3.46
150	4.00
200	4.41

▶ **Example 1** Table 1.6.1 gives a set of data points relating the pressure p in atmospheres (atm) and the temperature T (in °C) of a fixed quantity of carbon dioxide in a closed cylinder. The associated plot in Figure 1.6.4a suggests that there is a linear relationship between the pressure and the temperature.

(a) Use your calculating utility to find the least squares line for the data. If your utility can produce the correlation coefficient, then find it.

(b) Use the model obtained in part (a) to predict the pressure when the temperature is 250°C.

(c) Use the model obtained in part (a) to predict a temperature at which the pressure of the gas will be zero.

Solution (a). The least squares line is given by $p = 0.00936T + 2.558$ (Figure 1.6.4*b*) with correlation coefficient $r = 0.998979$.

Solution (b). If $T = 250$, then $p = (0.00936)(250) + 2.558 = 4.898$ (atm).

Solution (c). Solving the equation $0 = p = 0.00936T + 2.558$ yields $T \approx -273.291°\text{C}$. ◀

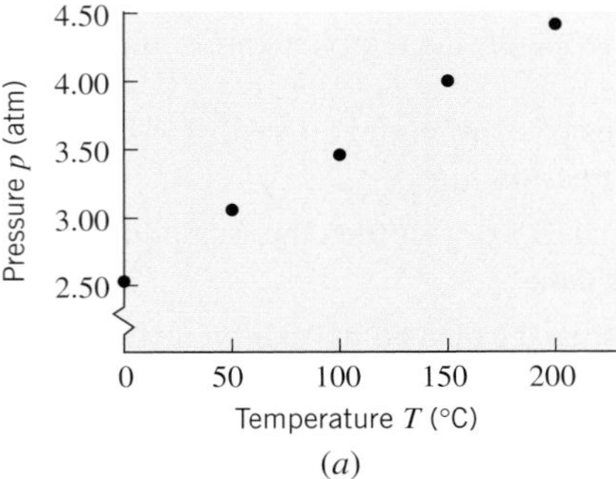

(*a*)

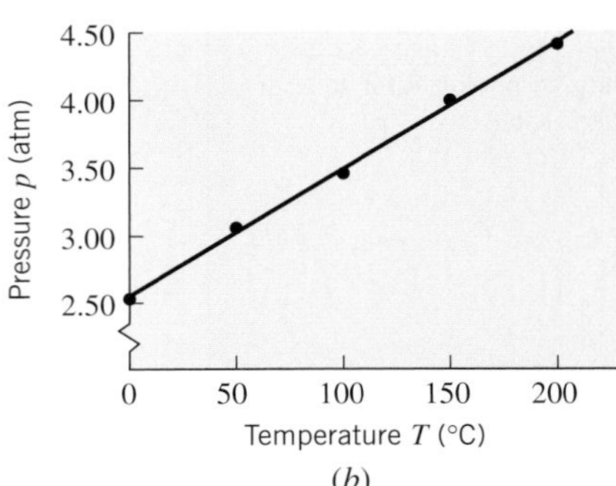

(*b*)

Figure 1.6.4

It is not always convenient (or necessary) to obtain the least squares line for a linear phenomenon in order to create a model. In some cases, more elementary methods suffice. Here is an example.

▶ **Example 2** Figure 1.6.5*a* shows a graph of temperature versus altitude that was transmitted by the *Magellan* spacecraft when it entered the atmosphere of Venus in October 1991. The graph strongly suggests that there is a linear relationship between temperature and altitude for altitudes between 35 and 60 km.

(a) Use the graph transmitted by the *Magellan* spacecraft to find a linear model of temperature versus altitude in the Venusian atmosphere that is valid for altitudes between 35 and 60 km.

(b) Use the model obtained in part (a) to estimate the temperature at the surface of Venus, and discuss the assumptions you are making in obtaining the estimate.

Solution (a). Let T be the temperature in kelvins and h the altitude in kilometers. We will first estimate the slope m of the linear portion of the graph, then estimate the coordinates of a data point (h_1, T_1) on that portion of the graph, and then use the point-slope form of a line

$$T - T_1 = m(h - h_1) \tag{2}$$

The graph nearly passes through the point (60, 250), so we will take $h_1 \approx 60$ and $T_1 \approx 250$. In Figure 1.6.5*b* we have sketched a line that closely approximates the linear portion of the data. Using the intersections of that line with the edges of the grid box, we estimate the slope to be

$$m \approx \frac{100-490}{78-30} = -\frac{390}{48} = -8.125 \text{ K/km}$$

Substituting our estimates of h_1, T_1, and m into (2) yields the equation

$$T - 250 = -8.125(h - 60)$$

or equivalently,

$$T = -8.125h + 737.5 \tag{3}$$

The method of Example 2 is crude, at best, since it relies on extracting rough estimates of numerical data from a graph. Nevertheless, the final result is quite good, since information from NASA places the surface temperature of Venus at about 737 K (hot enough to melt lead).

Solution (b). The *Magellan* spacecraft stopped transmitting data at an altitude of approximately 35 km, so we cannot be certain that the linear model applies at lower altitudes. However, if we *assume* that the model is valid at all lower altitudes, then we can approximate the temperature at the surface of Venus by setting $h = 0$ in (3). We obtain $T \approx 737.5$ K. ◀

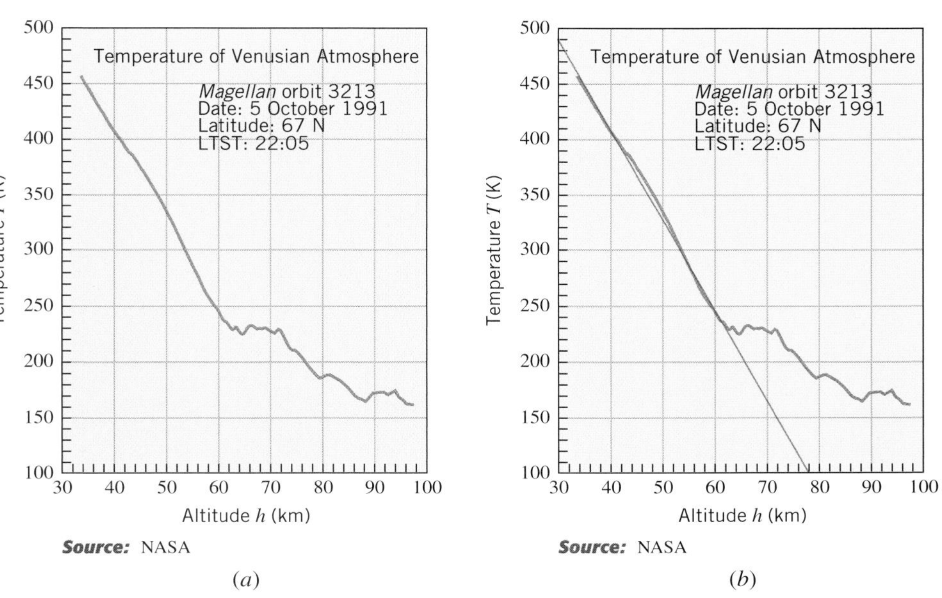

Figure 1.6.5

QUADRATIC AND TRIGONOMETRIC FUNCTIONS AS MODELS

Although linear models are simple, they are not always appropriate. It may be, for example, that a quadratic model $y = ax^2 + bx + c$ is suggested by the graphical form of the data or by some known law. Most calculators, CAS programs, and spreadsheets can compute a ***quadratic regression curve*** for a set of (x, y) data pairs that minimizes the sum of the squares of the residuals.

▶ **Example 3** To study the equations of motion of a falling body, a student in a physics laboratory collects the data in Table 1.6.2 showing the height of a body at various times over a 0.15-s time interval. If air resistance is neglected and if the acceleration due to gravity is assumed constant, then known principles of physics state that the height h should be a quadratic function of time t. This is consistent with Figure 1.6.6*a*, in which the plotted data suggest the shape of an inverted parabola.

Table 1.6.2

TIME t (s)	HEIGHT h (cm)
0.008333	98.4
0.025	96.9
0.04167	95.1
0.05833	92.9
0.075	90.8
0.09167	88.1
0.10833	85.3
0.125	82.1
0.14167	78.6
0.15833	74.9

(a) Determine the quadratic regression curve for the data in Table 1.6.2.

(b) According to the model obtained in part (a), when will the object strike the ground?

Solution (a). Using the quadratic regression routine on a calculator, we find that the quadratic curve that best fits the data in Table 1.6.2 has equation

$$h = 99.02 - 73.21t - 499.13t^2$$

Figure 1.6.6b shows the data points and the graph of this quadratic function on the same set of axes. It appears that we have excellent agreement between our curve and the data.

Solution (b). Solving the equation $0 = h = 99.02 - 73.21t - 499.13t^2$, we find that the object will strike the ground at $t \approx 0.38$ s. ◀

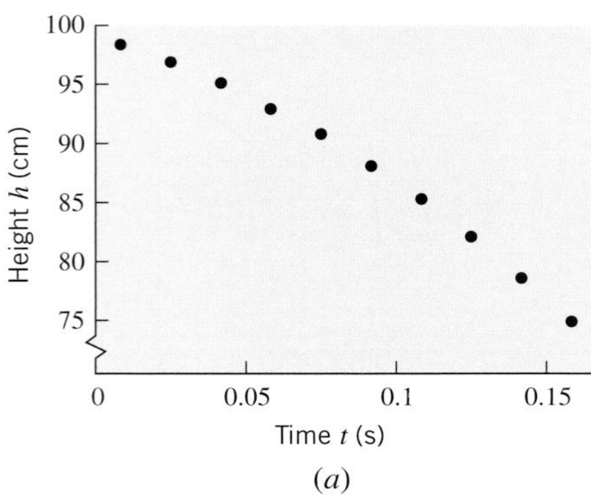

(a)

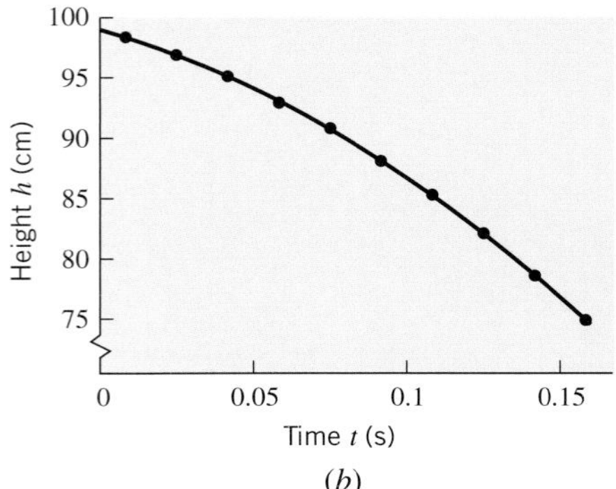

(b)

Figure 1.6.6

The trigonometric functions $y = A\sin(Bx - C)$ and $y = A\cos(Bx - C)$ are particularly useful for modeling periodic phenomena.

▶ **Example 4** Figure 1.6.7a shows a table and a plot of temperature data recorded over a 24-hour period in the city of Philadelphia.* Find a function that models the data, and graph your function and data together.

Solution. The pattern of the data suggests that the relationship between the temperature T and the time t can be modeled by a sinusoidal function that has been translated both horizontally and vertically, so we will look for an equation of the form

$$T = D + A\sin[Bt - C] = D + A\sin\left[B\left(t - \frac{C}{B}\right)\right] \tag{4}$$

Since the highest temperature is 95°F and the lowest temperature is 75°F, we take $2A = 20$ or $A = 10$. The midpoint between the high and low is 85°F, so we have a vertical shift of $D = 85$. The period seems to be about 24, so $2\pi/B = 24$ or $B = \pi/12$. The horizontal shift appears to be about 10 (verify), so $C/B = 10$. Substituting these values in (4) yields the equation

$$T = 85 + 10\sin\left[\frac{\pi}{12}(t - 10)\right] \tag{5}$$

whose graph is shown in Figure 1.6.7b. ◀

*This example is based on the article "Everybody Talks About It!—Weather Investigations," by Gloria S. Dion and Iris Brann Fetta, *The Mathematics Teacher*, Vol. 89, No. 2, February 1996, pp. 160–165.

PHILADELPHIA TEMPERATURES
FROM 1:00 A.M. TO 12:00 MIDNIGHT ON 27 AUGUST 1993
(t = HOURS AFTER MIDNIGHT AND T = DEGREES FAHRENHEIT)

	A.M.		P.M.	
	t	T	t	T
1:00	1	78°	13	91°
2:00	2	77°	14	93°
3:00	3	77°	15	94°
4:00	4	76°	16	95°
5:00	5	76°	17	93°
6:00	6	75°	18	92°
7:00	7	75°	19	89°
8:00	8	77°	20	86°
9:00	9	79°	21	84°
10:00	10	83°	22	83°
11:00	11	87°	23	81°
12:00	12	90°	24	79°

Source: *Philadelphia Inquirer*, 28 August 1993.

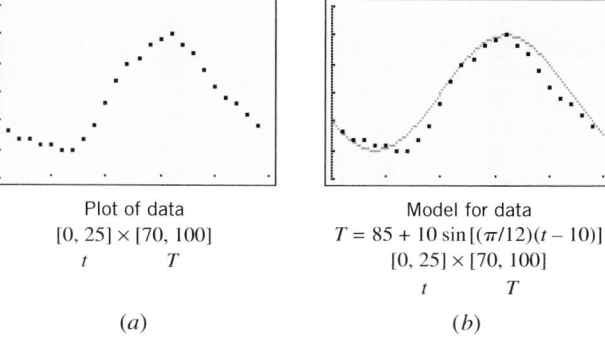

Plot of data
$[0, 25] \times [70, 100]$
t T

(*a*)

Model for data
$T = 85 + 10\sin[(\pi/12)(t - 10)]$
$[0, 25] \times [70, 100]$
t T

(*b*)

Figure 1.6.7

TECHNOLOGY MASTERY

Note that Equation (5) in this example was *not* obtained by minimizing the sum of the squares of the residuals; rather, we used the calculator's graphing capability to see that the proposed model gave a *reasonable* fit to the data. Using a computer program, we can show that the curve of the form $T = D + A\sin(Bt - C)$ that minimizes the sum of the squares of the residuals for the data in Figure 1.6.7 is

$$y = 84.2037 + 9.5964\sin(0.2849t - 2.9300)$$

Use your technology utility to compare the graph of this best-fit curve to the graph of (5).

✔ QUICK CHECK EXERCISES 1.6 *(See page 69 for answers.)*

1. In each part, state an appropriate model (linear, quadratic, trigonometric) for a collection of data points (x, y) for which the given situation applies.
   (a) y is the distance traveled at time x of an automobile accelerating at a constant rate: ________.
   (b) y is the speed at time x of an automobile accelerating at a constant rate: ________.
   (c) y is the stride of a man x inches tall: ________.
   (d) y is the fraction of the moon illuminated x days after the beginning of the year: ________.

2. For the data set $\{(-1, 0), (0, 0), (1, 2), (3, 3)\}$ and the line $y = x + 1$ the sum of the squares of the residuals is ________.

3. Suppose that a data set has a linear model with correlation coefficient 0.5 and that a second data set has a linear model with correlation coefficient -0.75. Which model appears to be more appropriate for its corresponding data set?

4. Consider a data set consisting of the two points $(0, -1)$ and $(0, 1)$. Let L denote the line $y = mx + b$.
   (a) The sum of squares of the residuals from the two points to L is ________.
   (b) If L minimizes the sum of the squares of the residuals from the two points, then $b =$ ________.

EXERCISE SET 1.6 Graphing Utility

1. One of the lines in the accompanying figure is the regression line. Which one is it?

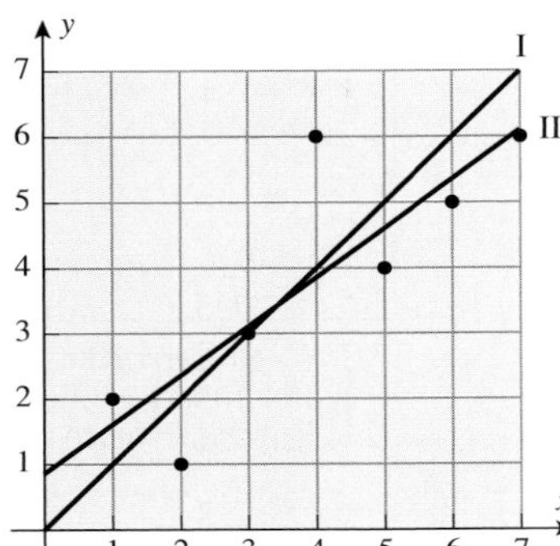

Figure Ex-1

2. In each part, determine whether a model (linear, quadratic, or trigonometric) reasonably describes the plot of data in the accompanying figure.

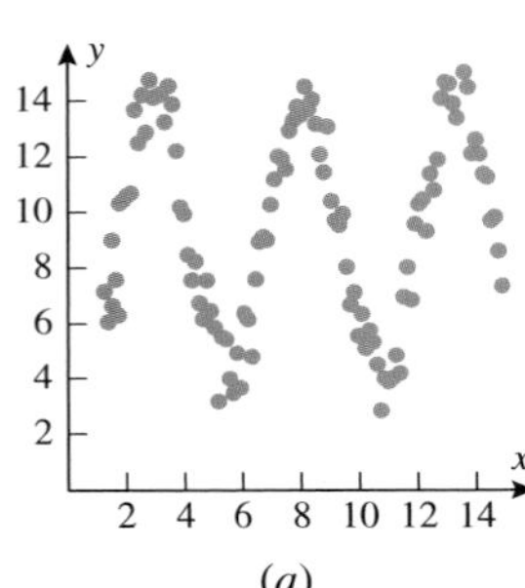

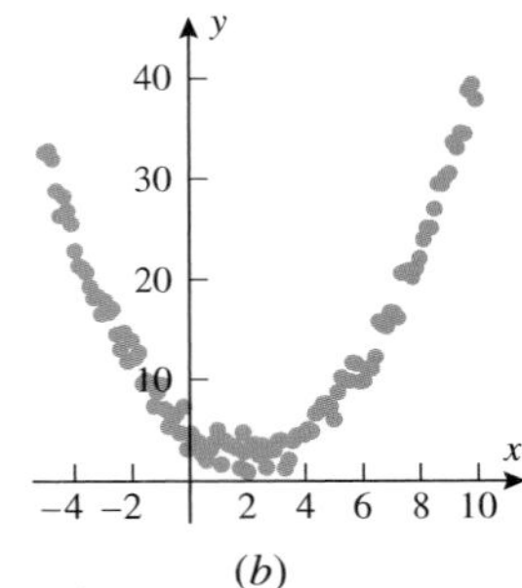

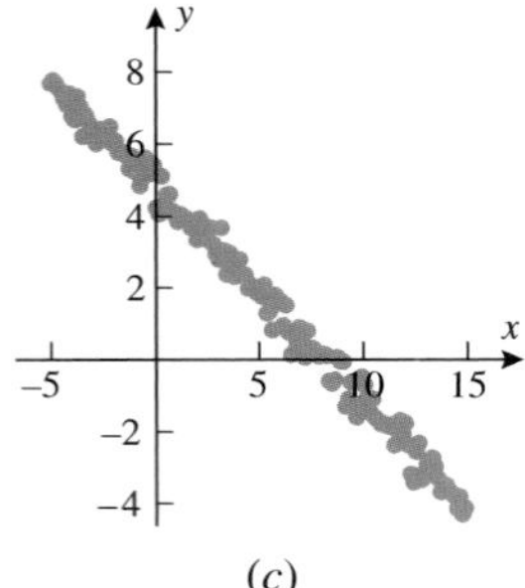

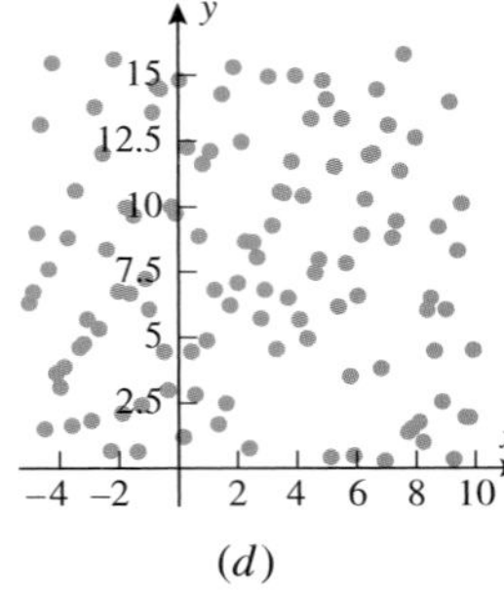

Figure Ex-2

3. Table 1.1.1 provides data for the top qualifying speeds at the Indianapolis 500 from 1987 to 2004. Find the least squares line for these data. What is the correlation coefficient? Sketch the least squares line on a plot of the data points.

4. A 25-liter container holds 150 g of O_2. The pressure p of the gas is measured at various temperatures T (see the accompanying table).
   (a) Determine the least squares line for the data given in the table.
   (b) Use the model obtained in part (a) to estimate the pressure of the gas at a temperature of $-50°$C.

TEMPERATURE T (°C)	PRESSURE p (atm)
0	4.18
50	4.96
100	5.74
150	6.49
200	7.26

Table Ex-4

5. A 20-liter container holds 100 g of N_2. The pressure p of this gas is measured at various temperatures T (see the accompanying table).
   (a) Find the least squares line for this collection of data points. If your calculating utility can produce the correlation coefficient, then find it.
   (b) Use the model obtained in part (a) to predict the pressure of the gas at a temperature of $-50°$C.
   (c) Use the model obtained in part (a) to predict a temperature at which the pressure of the gas will be zero.

TEMPERATURE T (°C)	PRESSURE p (atm)
0	3.99
25	4.34
50	4.70
75	5.08
100	5.45

Table Ex-5

6. A 40-liter container holds 20 g of H_2. The pressure p of this gas is measured at various temperatures T (see the accompanying table).
   (a) Find the least squares line for this collection of data points. If your calculating utility can produce the correlation coefficient, then find it.
   (b) Use the model obtained in part (a) to predict a temperature at which the pressure of the gas will be zero.
   (c) At approximately what temperature of the gas will a 10°C increase in temperature result in a 5% increase in pressure?

TEMPERATURE T (°C)	PRESSURE p (atm)
0	5.55
30	6.13
60	6.75
90	7.35
120	7.98

Table Ex-6

7. The ***resistivity*** of a metal is a measure of the extent to which a wire made from the metal will resist the flow of electrical current. (The actual *resistance* of the wire will depend on both the resistivity of the metal and the dimensions of the wire.) A common unit for resistivity is the ohm-meter ($\Omega \cdot m$). Experiments show that lowering the temperature of a metal also lowers its resistivity. The accompanying table gives the resistivity of copper at various temperatures.
(a) Find the least squares line for this collection of data points.
(b) Using the model obtained in part (a), at what temperature will copper have a resistivity of zero?

TEMPERATURE (°C)	RESISTIVITY ($10^{-8}\Omega \cdot m$)
–100	0.82
–50	1.19
0	1.54
50	1.91
100	2.27
150	2.63

Table Ex-7

8. The accompanying table gives the resistivity of tungsten at various temperatures.
(a) Find the least squares line for this collection of data points.
(b) Using the model obtained in part (a), at what temperature will tungsten have a resistivity of zero?

TEMPERATURE (°C)	RESISTIVITY ($10^{-8}\Omega \cdot m$)
–100	2.43
–50	3.61
0	4.78
50	5.96
100	7.16
150	8.32

Table Ex-8

9. The accompanying table gives the number of inches that a spring is stretched by various attached weights.
(a) Use linear regression to express the amount of stretch of the spring as a function of the weight attached.
(b) Use the model obtained in part (a) to determine the weight required to stretch the spring 8 in.

WEIGHT (lb)	STRETCH (in)
0	0
2	0.99
4	2.01
6	2.99
8	4.00
10	5.03
12	6.01

Table Ex-9

10. The accompanying table gives the number of inches that a spring is stretched by various attached weights.
(a) Use linear regression to express the amount of stretch of the spring as a function of the weight attached.
(b) Suppose that the spring has been stretched a certain amount by a weight and that adding another 5 lb to the weight doubles the stretch of the spring. Use the model obtained in part (a) to determine the original amount that the spring was stretched.

WEIGHT (lb)	STRETCH (in)
0	0
1	0.73
2	1.50
3	2.24
4	3.02
5	3.77

Table Ex-10

11. The accompanying table provides the heights and rebounds per minute for players on the 2002–2003 Davidson College women's basketball team who played more than 100 minutes during the season.
(a) Find the least squares line for these data. If your calculating utility can produce the correlation coefficient, then find it.
(b) Sketch the least squares line on a plot of the data points.
(c) Is the least squares line a good model for these data? Explain.

HEIGHT	REBOUNDS PER MINUTE
6′0″	0.132
6′0″	0.252
5′6″	0.126
5′10″	0.139
6′2″	0.227
6′3″	0.299
6′0″	0.170
5′5″	0.071
6′2″	0.222

Table Ex-11

12. The accompanying table provides the heights and weights for players on the 2002–2003 Davidson College men's basketball team.
(a) Find the least squares line for these data. If your calculating utility can produce the correlation coefficient, then find it.
(b) Sketch the least squares line on a plot of the data points.
(c) Use this model to predict the weight of the team's new 7-ft recruit.

HEIGHT	WEIGHT (lb)	HEIGHT	WEIGHT (lb)
6′2″	185	6′5″	210
6′6″	215	6′4″	185
6′1″	175	6′5″	190
6′1″	180	6′3″	180
6′6″	210	6′8″	235
5′10″	175	6′8″	215
6′9″	210	6′10″	235
6′1″	180		

Table Ex-12

FOCUS ON CONCEPTS

13. A data set consists of the three points (0, 0), (1, 0), and (1, 1).

(a) For this data set and the line $y = x + b$, express the sum of the squares of the residuals as a function of the parameter b.

(b) Using your answer in part (a), find the line of slope 1 for which the sum of the squares of the residuals is a minimum.

14. A data set consists of two distinct points with the same x-coordinate. Explain why there are infinitely many linear regression lines for this data set. All the regression lines have something in common. What is it?

15. A data set consists of three distinct points, exactly two of which have the same x-coordinate. Explain how to use your answer in Exercise 14 to find the linear regression line for this data set.

16. A data set consists of four points that are the vertices of a trapezoid with two sides parallel to the y-axis. Explain how to use your answer in Exercise 14 to find the linear regression line for this data set.

17. (**The Age of the Universe**) In the early 1900s the astronomer Edwin P. Hubble (1889–1953) noted an unexpected relationship between the radial velocity of a galaxy and its distance d from any reference point (Earth, for example). That relationship, now known as ***Hubble's law***, states that the galaxies are receding with a velocity v that is directly proportional to the distance d. This is usually expressed as $v = Hd$, where H (the constant of proportionality) is called ***Hubble's constant***. When applying this formula it is usual to express v in kilometers per second (km/s) and d in millions of light-years (Mly), in which case H has units of km/s/Mly. The accompanying figure shows an original plot and trend line of the velocity-distance relationship obtained by Hubble and a collaborator Milton L. Humason (1891–1972).

(a) Use the trend line in the figure to estimate Hubble's constant.

(b) An estimate of the age of the universe can be obtained by assuming that the galaxies move with constant velocity v, in which case v and d are related by $d = vt$. Assuming that the Universe began with a "big bang" that initiated its expansion, show that the Universe is roughly 1.5×10^{10} years old. [Use the conversion 1 Mly $\approx$ 9.048×10^{18} km and take $H = 20$ km/s/Mly, which is in keeping with current estimates that place H between 15 and 27 km/s/Mly. (Note that the current estimates are significantly less than that resulting from Hubble's data.)]

(c) In a more realistic model of the Universe, the velocity v would decrease with time. What effect would that have on your estimate in part (b)?

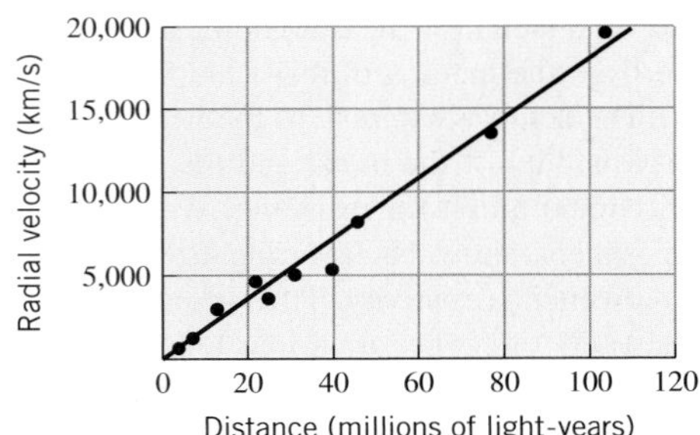

Figure Ex-17

18. The accompanying table gives data for the U.S. population at 10-year intervals from 1790 to 1850. Use quadratic regression to model the U.S. population as a function of time since 1790. What does your model predict as the population of the United States in the year 2000? How accurate is this prediction?

U.S. POPULATION

YEAR t	POPULATION P (millions)
1790	3.9
1800	5.3
1810	7.2
1820	9.6
1830	12
1840	17
1850	23

Source: *The World Almanac.* **Table Ex-18**

19. The accompanying table gives the minutes of daylight predicted for Davidson, North Carolina, in 10-day increments during the year 2000. Find a function that models the data in this table, and graph your function on a plot of the data.

DAY	DAYLIGHT (min)	DAY	DAYLIGHT (min)
10	716	190	986
20	727	200	975
30	744	210	961
40	762	220	944
50	783	230	926
60	804	240	905
70	826	250	883
80	848	260	861
90	872	270	839
100	894	280	817
110	915	290	795
120	935	300	774
130	954	310	755
140	969	320	738
150	982	330	723
160	990	340	712
170	993	350	706
180	992	360	706

Table Ex-19

20. The accompanying table gives the fraction of the Moon that is illuminated (as seen from Earth) at midnight (Eastern Standard Time) in 2-day intervals for the first 60 days of 1999. Find a function that models the data in this table, and graph your function on a plot of the data.

DAY	ILLUMINATION	DAY	ILLUMINATION
2	1	32	1
4	0.94	34	0.93
6	0.81	36	0.79
8	0.63	38	0.62
10	0.44	40	0.43
12	0.26	42	0.25
14	0.12	44	0.10
16	0.02	46	0.01
18	0	48	0.01
20	0.07	50	0.11
22	0.22	52	0.29
24	0.43	54	0.51
26	0.66	56	0.73
28	0.85	58	0.90
30	0.97	60	0.99

Table Ex-20

21. The accompanying table provides data about the relationship between distance d traveled in meters and elapsed time t in seconds for an object dropped near the Earth's surface. Plot time versus distance and make a guess at a "square-root function" that provides a reasonable model for t in terms of d. Use a graphing utility to confirm the reasonableness of your guess.

d (meters)	0	2.5	5	10	15	20	25
t (seconds)	0	0.7	1.0	1.4	1.7	2	2.3

Table Ex-21

22. (a) The accompanying table provides data on five moons of the planet Saturn. In this table r is the *orbital radius* (the average distance between the moon and Saturn) and t is the time in days required for the moon to complete one orbit around Saturn. For each data pair calculate $tr^{-3/2}$, and use your results to find a reasonable model for r as a function of t.

(b) Use the model obtained in part (a) to estimate the orbital radius of the moon Enceladus, given that its orbit time is $t \approx 1.370$ days.

(c) Use the model obtained in part (a) to estimate the orbit time of the moon Tethys, given that its orbital radius is $r \approx 2.9467 \times 10^5$ km.

MOON	RADIUS (100,000 km)	ORBIT TIME (days)
1980S28	1.3767	0.602
1980S27	1.3935	0.613
1980S26	1.4170	0.629
1980S3	1.5142	0.694
1980S1	1.5147	0.695

Table Ex-22

✔QUICK CHECK ANSWERS 1.6

1. (a) quadratic (b) linear (c) linear (d) trigonometric **2.** 2 **3.** the model with correlation coefficient -0.75
4. (a) $2b^2 + 2$ (b) 0

1.7 PARAMETRIC EQUATIONS

Thus far, our study of graphs has focused on graphs of functions. However, because such graphs must pass the vertical line test, this limitation precludes curves with self-intersections or even such basic curves as circles. In this section we will study an alternative method for describing curves algebraically that is not subject to the severe restriction of the vertical line test.

This material is placed here to provide an early parametric option. However, it can be deferred until Chapter 11, if preferred.

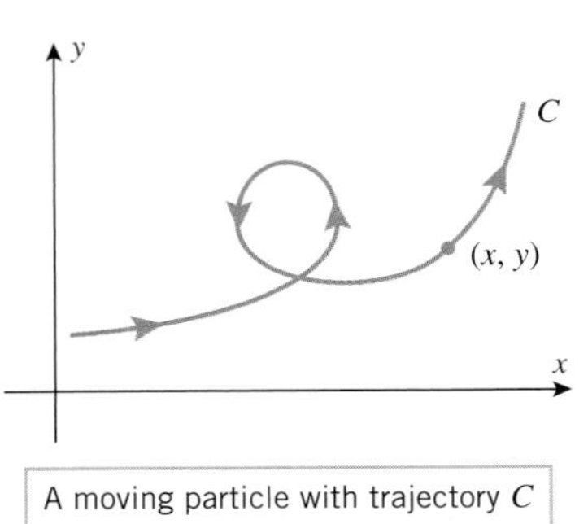

A moving particle with trajectory C

Figure 1.7.1

PARAMETRIC EQUATIONS

Suppose that a particle moves along a curve C in the xy-plane in such a way that its x- and y-coordinates, as functions of time, are

$$x = f(t), \quad y = g(t)$$

We call these the ***parametric equations*** of motion for the particle and refer to C as the ***trajectory*** of the particle or the ***graph*** of the equations (Figure 1.7.1). The variable t is called the ***parameter*** for the equations.

▶ **Example 1** Sketch the trajectory over the time interval $0 \le t \le 10$ of the particle whose parametric equations of motion are

$$x = t - 3\sin t, \quad y = 4 - 3\cos t \tag{1}$$

Solution. One way to sketch the trajectory is to choose a representative succession of times, plot the (x, y) coordinates of points on the trajectory at those times, and connect the points with a smooth curve. The trajectory in Figure 1.7.2 was obtained in this way from the data in Table 1.7.1 in which the approximate coordinates of the particle are given at time increments of 1 unit. Observe that there is no t-axis in the picture; the values of t appear only as labels on the plotted points, and even these are usually omitted unless it is important to emphasize the locations of the particle at specific times. ◀

TECHNOLOGY MASTERY

Read the documentation for your graphing utility to learn how to graph parametric equations, and then generate the trajectory in Example 1. Explore the behavior of the particle beyond time $t = 10$.

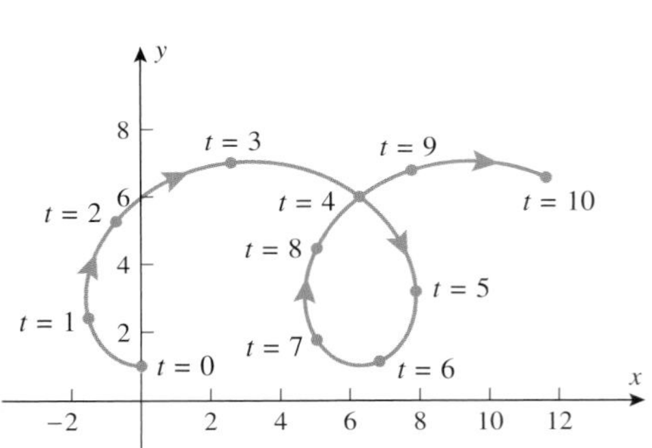

Figure 1.7.2

Table 1.7.1

t	x	y
0	0.0	1.0
1	−1.5	2.4
2	−0.7	5.2
3	2.6	7.0
4	6.3	6.0
5	7.9	3.1
6	6.8	1.1
7	5.0	1.7
8	5.0	4.4
9	7.8	6.7
10	11.6	6.5

Although parametric equations commonly arise in problems of motion with time as the parameter, they arise in other contexts as well. Thus, unless the problem dictates that the parameter t in the equations

$$x = f(t), \quad y = g(t)$$

represents time, it should be viewed simply as an independent variable that varies over some interval of real numbers. (In fact, there is no need to use the letter t for the parameter; any letter not reserved for another purpose can be used.) If no restrictions on the parameter are stated explicitly or implied by the equations, then it is understood that it varies from $-\infty$ to $+\infty$. To indicate that a parameter t is restricted to an interval $[a, b]$, we will write

$$x = f(t), \quad y = g(t) \qquad (a \le t \le b)$$

▶ **Example 2** Find the graph of the parametric equations

$$x = \cos t, \quad y = \sin t \qquad (0 \le t \le 2\pi) \tag{2}$$

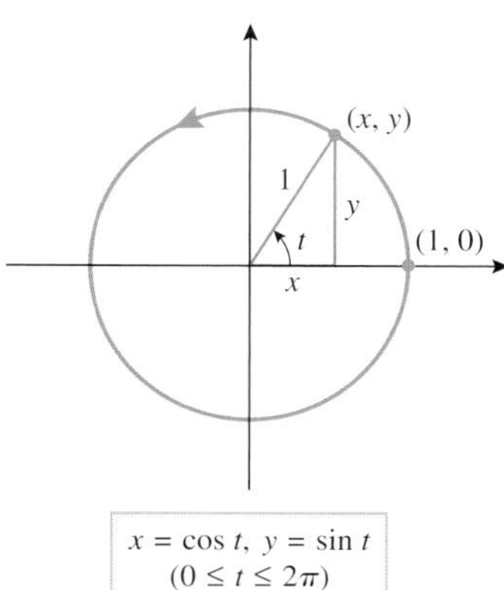

Figure 1.7.3

Solution. One way to find the graph is to eliminate the parameter t by noting that

$$x^2 + y^2 = \sin^2 t + \cos^2 t = 1$$

Thus, the graph is contained in the unit circle $x^2 + y^2 = 1$. Geometrically, the parameter t can be interpreted as the angle swept out by the radial line from the origin to the point $(x, y) = (\cos t, \sin t)$ on the unit circle (Figure 1.7.3). As t increases from 0 to 2π, the point traces the circle counterclockwise, starting at $(1, 0)$ when $t = 0$ and completing one full revolution when $t = 2\pi$. One can obtain different portions of the circle by varying the interval over which the parameter varies. For example,

$$x = \cos t, \quad y = \sin t \qquad (0 \le t \le \pi) \tag{3}$$

represents just the upper semicircle in Figure 1.7.3. ◀

ORIENTATION

The direction in which the graph of a pair of parametric equations is traced as the parameter increases is called the ***direction of increasing parameter*** or sometimes the ***orientation*** imposed on the curve by the equations. Thus, we make a distinction between a ***curve***, which is a set of points, and a ***parametric curve***, which is a curve with an orientation imposed on it by a set of parametric equations. For example, we saw in Example 2 that the circle represented parametrically by (2) is traced counterclockwise as t increases and hence has *counterclockwise orientation*. As shown in Figures 1.7.2 and 1.7.3, the orientation of a parametric curve can be indicated by arrowheads.

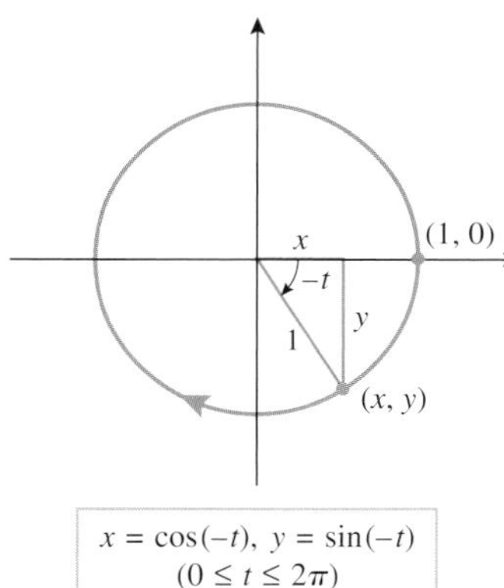

Figure 1.7.4

To obtain parametric equations for the unit circle with *clockwise orientation*, we can replace t by $-t$ in (2) and use the identities $\cos(-t) = \cos t$ and $\sin(-t) = -\sin t$. This yields

$$x = \cos t, \quad y = -\sin t \qquad (0 \le t \le 2\pi)$$

Here, the circle is traced clockwise by a point that starts at $(1, 0)$ when $t = 0$ and completes one full revolution when $t = 2\pi$ (Figure 1.7.4).

TECHNOLOGY MASTERY

When parametric equations are graphed using a calculator, the orientation can often be determined by watching the direction in which the graph is traced on the screen. However, many computers graph so fast that it is often hard to discern the orientation. See if you can use your graphing utility to confirm that (3) has a counterclockwise orientation.

▶ **Example 3** Graph the parametric curve

$$x = 2t - 3, \quad y = 6t - 7$$

by eliminating the parameter, and indicate the orientation on the graph.

Solution. To eliminate the parameter we will solve the first equation for t as a function of x, and then substitute this expression for t into the second equation:

$$t = \left(\tfrac{1}{2}\right)(x + 3)$$

$$y = 6\left(\tfrac{1}{2}\right)(x + 3) - 7$$

$$y = 3x + 2$$

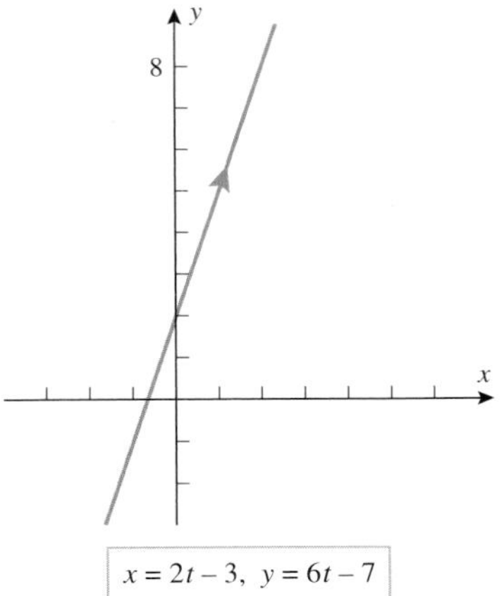

Figure 1.7.5

Thus, the graph is a line of slope 3 and y-intercept 2. To find the orientation we must look to the original equations; the direction of increasing t can be deduced by observing that x increases as t increases *or* by observing that y increases as t increases. Either piece of information tells us that the line is traced left to right as shown in Figure 1.7.5. ◀

Not all parametric equations produce curves with definite orientations; if the equations are badly behaved, then the point tracing the curve may leap around sporadically or move back and forth, failing to determine a definite direction. For example, if

$$x = \sin t, \quad y = \sin^2 t$$

then the point (x, y) moves along the parabola $y = x^2$. However, the value of x varies periodically between -1 and 1, so the point (x, y) moves periodically back and forth along the parabola between the points $(-1, 1)$ and $(1, 1)$ (as shown in Figure 1.7.6). Later in the text we will discuss restrictions that eliminate such erratic behavior, but for now we will just avoid such complications.

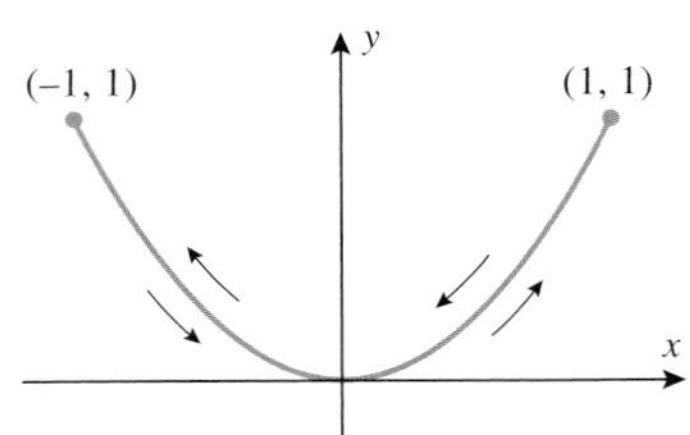

Figure 1.7.6

EXPRESSING ORDINARY FUNCTIONS PARAMETRICALLY

An equation $y = f(x)$ can be expressed in parametric form by introducing the parameter $t = x$; this yields the parametric equations $x = t$, $y = f(t)$. For example, the portion of the curve $y = \cos x$ over the interval $[-2\pi, 2\pi]$ can be expressed parametrically as

$$x = t, \quad y = \cos t \qquad (-2\pi \le t \le 2\pi)$$

(Figure 1.7.7).

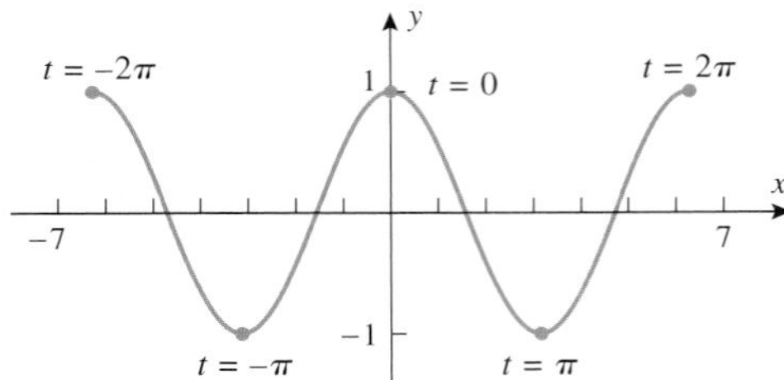

Figure 1.7.7

GENERATING PARAMETRIC CURVES WITH GRAPHING UTILITIES

Many graphing utilities allow you to graph equations of the form $y = f(x)$ but not equations of the form $x = g(y)$. Sometimes you will be able to rewrite $x = g(y)$ in the form $y = f(x)$; however, if this is inconvenient or impossible, then you can graph $x = g(y)$ by introducing a parameter $t = y$ and expressing the equation in the parametric form $x = g(t)$, $y = t$. (You may have to experiment with various intervals for t to produce a complete graph.)

▶ **Example 4** Use a graphing utility to graph the equation $x = 3y^5 - 5y^3 + 1$.

Solution. If we let $t = y$ be the parameter, then the equation can be written in parametric form as

$$x = 3t^5 - 5t^3 + 1, \quad y = t$$

Figure 1.7.8 shows the graph of these equations for $-1.5 \le t \le 1.5$. ◀

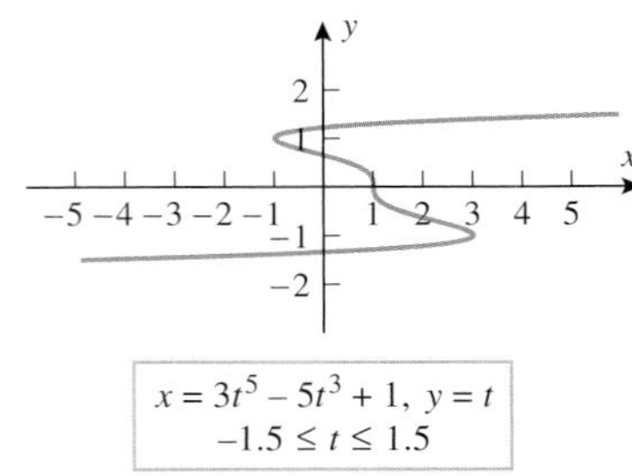

Figure 1.7.8

Some parametric curves are so complex that it is virtually impossible to visualize them without using some kind of graphing utility. Figure 1.7.9 shows three such curves.

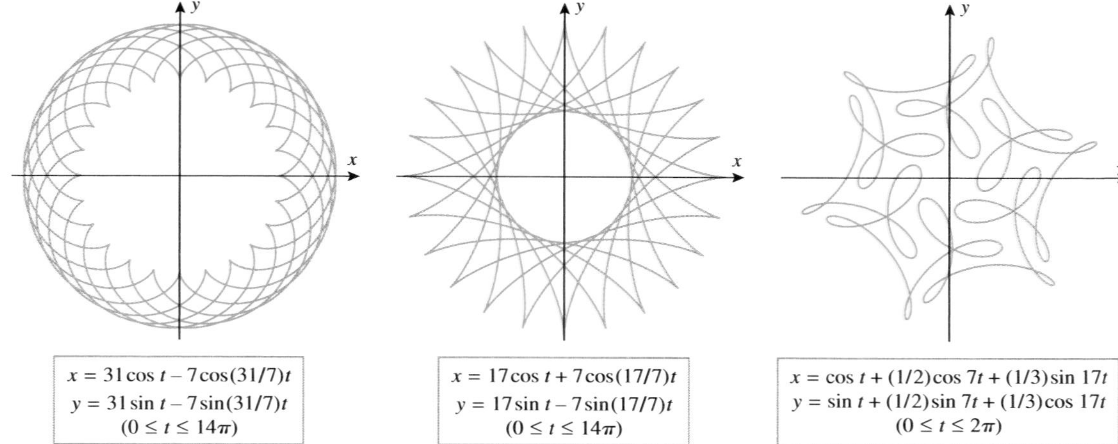

Figure 1.7.9

TECHNOLOGY MASTERY

Try your hand at using a graphing utility to generate some parametric curves that you think are interesting or beautiful.

GRAPHING INVERSE FUNCTIONS WITH GRAPHING UTILITIES

Most graphing utilities cannot graph inverse functions directly. However, there is a way of graphing inverse functions by expressing the graphs parametrically. To see how this can be done, suppose that we are interested in graphing the inverse of a one-to-one function f. We know that the equation $y = f(x)$ can be expressed parametrically as

$$x = t, \quad y = f(t) \tag{4}$$

and we know that the graph of f^{-1} can be obtained by interchanging x and y, since this reflects the graph of f about the line $y = x$. Thus, from (4) the graph of f^{-1} can be represented parametrically as

$$x = f(t), \quad y = t \tag{5}$$

For example, Figure 1.7.10 shows the graph of $f(x) = x^5 + x + 1$ and its inverse generated with a graphing utility. The graph of f was generated from the parametric equations

$$x = t, \quad y = t^5 + t + 1$$

and the graph of f^{-1} was generated from the parametric equations

$$x = t^5 + t + 1, \quad y = t$$

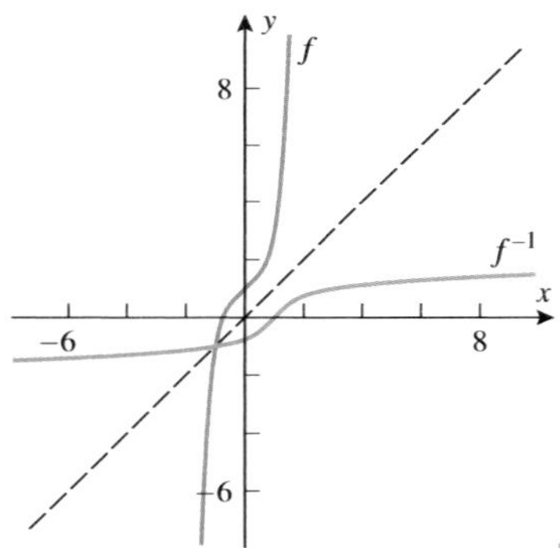

Figure 1.7.10

TRANSLATION

If a parametric curve C is given by the equations $x = f(t)$, $y = g(t)$, then adding a constant to $f(t)$ translates the curve C in the x-direction, and adding a constant to $g(t)$ translates it in the y-direction. Thus, a circle of radius r, centered at (x_0, y_0) can be represented parametrically as

$$x = x_0 + r\cos t, \quad y = y_0 + r\sin t \qquad (0 \le t \le 2\pi) \tag{6}$$

(Figure 1.7.11). If desired, we can eliminate the parameter from these equations by noting that

$$(x - x_0)^2 + (y - y_0)^2 = (r\cos t)^2 + (r\sin t)^2 = r^2$$

Thus, we have obtained the familiar equation in rectangular coordinates for a circle of radius r, centered at (x_0, y_0):

$$(x - x_0)^2 + (y - y_0)^2 = r^2 \tag{7}$$

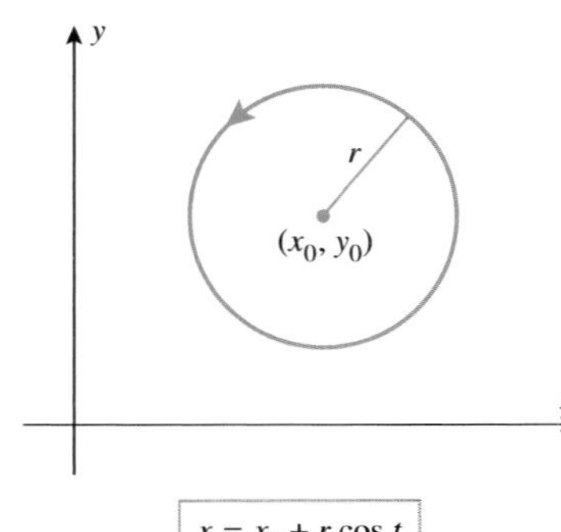

Figure 1.7.11

SCALING

If a parametric curve C is given by the equations $x = f(t)$, $y = g(t)$, then multiplying $f(t)$ by a constant stretches or compresses C in the x-direction, and multiplying $g(t)$ by a

TECHNOLOGY MASTERY

Use the parametric capability of your graphing utility to generate a circle of radius 5 that is centered at $(3, -2)$.

constant stretches or compresses C in the y-direction. For example, we would expect the parametric equations

$$x = 3\cos t, \quad y = 2\sin t \qquad (0 \le t \le 2\pi)$$

to represent an ellipse, centered at the origin, since the graph of these equations results from stretching the unit circle

$$x = \cos t, \quad y = \sin t \qquad (0 \le t \le 2\pi)$$

by a factor of 3 in the x-direction and a factor of 2 in the y-direction. In general, if a and b are positive constants, then the parametric equations

$$x = a\cos t, \quad y = b\sin t \qquad (0 \le t \le 2\pi) \tag{8}$$

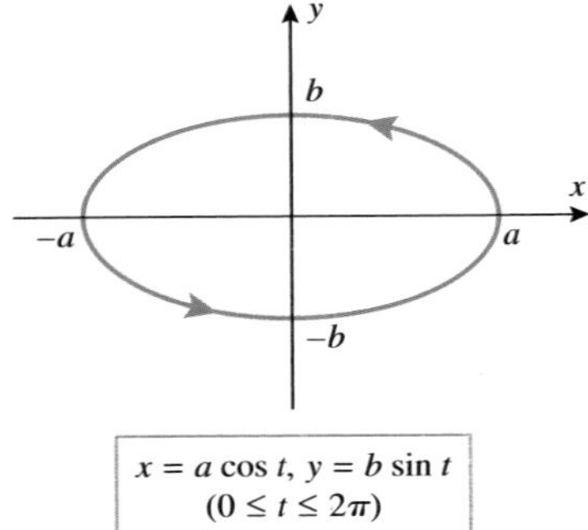

Figure 1.7.12

represent an ellipse, centered at the origin, and extending between $-a$ and a on the x-axis and between $-b$ and b on the y-axis (Figure 1.7.12). The numbers a and b are called the ***semiaxes*** of the ellipse. If desired, we can eliminate the parameter t in (8) and rewrite the equations in rectangular coordinates as

$$\frac{x^2}{a^2} + \frac{y^2}{b^2} = 1 \tag{9}$$

TECHNOLOGY MASTERY

Use the parametric capability of your graphing utility to generate an ellipse that is centered at the origin and that extends between -4 and 4 in the x-direction and between -3 and 3 in the y-direction. Generate an ellipse with the same dimensions, but translated so that its center is at the point $(2, 3)$.

LISSAJOUS CURVES

In the mid-1850s the French physicist Jules Antoine Lissajous (1822–1880) became interested in parametric equations of the form

$$x = \sin at, \quad y = \sin bt \tag{10}$$

in the course of studying vibrations that combine two perpendicular sinusoidal motions. The first equation in (10) describes a sinusoidal oscillation in the x-direction with frequency $a/2\pi$, and the second describes a sinusoidal oscillation in the y-direction with frequency $b/2\pi$. If a/b is a rational number, then the combined effect of the oscillations is a periodic motion along a path called a ***Lissajous curve***. Figure 1.7.13 shows some typical Lissajous curves.

TECHNOLOGY MASTERY

Generate some Lissajous curves on your graphing utility, and also see if you can figure out when each of the curves in Figure 1.7.13 begins to repeat.

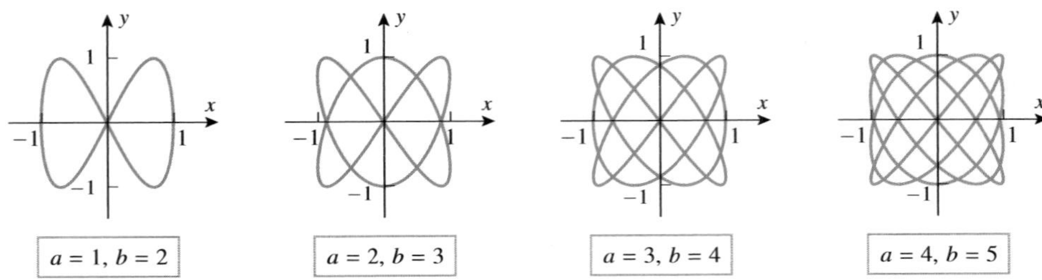

Figure 1.7.13

CYCLOIDS

If a wheel rolls in a straight line along a flat road, then a point on the rim of the wheel will trace a curve called a ***cycloid*** (Figure 1.7.14). This curve has a fascinating history, which we will discuss shortly; but first we will show how to obtain parametric equations for it. For this purpose, let us assume that the wheel has radius a and rolls along the positive x-axis of a rectangular coordinate system. Let $P(x, y)$ be the point on the rim that traces the cycloid, and assume that P is initially at the origin. We will take as our parameter the angle θ that is swept out by the radial line to P as the wheel rolls (Figure 1.7.14). It is standard here to regard θ to be positive, even though it is generated by a clockwise rotation.

The motion of P is a combination of the movement of the wheel's center parallel to the x-axis and the rotation of P around the center. As the radial line sweeps out an angle θ, the point P traverses an arc of length $a\theta$, and the wheel moves a distance $a\theta$ along the x-axis (why?). Thus, as suggested by Figure 1.7.15, the center moves to the point $(a\theta, a)$, and the coordinates of $P(x, y)$ are

$$x = a\theta - a\sin\theta, \quad y = a - a\cos\theta \tag{11}$$

These are the equations of the cycloid in terms of the parameter θ.

TECHNOLOGY MASTERY

Use your graphing utility to generate two "arches" of the cycloid produced by a point on the rim of a wheel of radius 1.

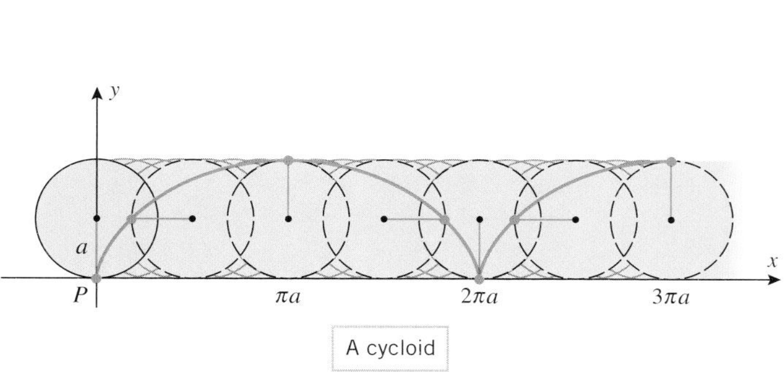

Figure 1.7.14

Figure 1.7.15

THE ROLE OF THE CYCLOID IN MATHEMATICS HISTORY

The cycloid is of interest because it provides the solution to two famous mathematical problems—the ***brachistochrone problem*** (from Greek words meaning "shortest time") and the ***tautochrone problem*** (from Greek words meaning "equal time"). The brachistochrone problem is to determine the shape of a wire along which a bead might slide from a point P to another point Q, not directly below, in the *shortest time*. The tautochrone problem is to find the shape of a wire from P to Q such that two beads started at any points on the wire between P and Q reach Q in the same amount of time. The solution to both problems turns out to be an inverted cycloid (Figure 1.7.16).

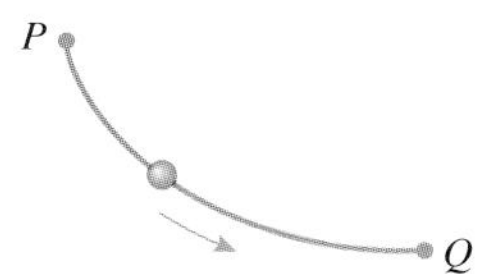

Figure 1.7.16

In June of 1696, Johann Bernoulli posed the brachistochrone problem in the form of a challenge to other mathematicians. At first, one might conjecture that the wire should form a straight line, since that shape results in the shortest distance from P to Q. However, the inverted cycloid allows the bead to fall more rapidly at first, building up sufficient initial speed to reach Q in the shortest time, even though it travels a longer distance. The problem was solved by Newton and Leibniz as well as by Johann Bernoulli and his older brother Jakob; it was formulated and solved *incorrectly* years earlier by Galileo, who thought the answer was a circular arc.

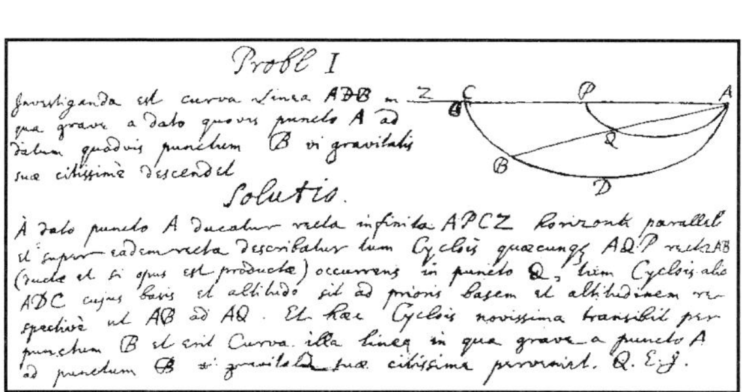

Probl I

Investiganda est curva linea ADB in qua grave a dato quovis puncto A ad datum quodvis punctum B vi gravitatis suae citissime descendet.

Solutio.

A dato puncto A ducatur recta infinita APCZ horizonti parallela et super eadem recta describatur tum Cyclois quaecunque AQP rectae AB (ductae et si opus est productae) occurrens in puncto Q, tum Cyclois alia ADC cujus basis et altitudo sit ad prioris basem et altitudinem respective ut AB ad AQ. Et haec Cyclois novissima transibit per punctum B et erit Curva illa linea in qua grave a puncto A ad punctum B vi gravitatis suae citissime perveniet. Q.E.I.

Newton's solution of the brachistochrone problem in his own handwriting

Johann (left) **and Jakob** (right) **Bernoulli** Members of an amazing Swiss family that included several generations of outstanding mathematicians and scientists. Nikolaus Bernoulli (1623–1708), a druggist, fled from Antwerp to escape religious persecution and ultimately settled in Basel, Switzerland. There he had three sons, Jakob I (also called Jacques or James), Nikolaus, and Johann I (also called Jean or John). The Roman numerals are used to distinguish family members with identical names (see the family tree below). Following Newton and Leibniz, the Bernoulli brothers, Jakob I and Johann I, are considered by some to be the two most important founders of calculus. Jakob I was self-taught in mathematics. His father wanted him to study for the ministry, but he turned to mathematics and in 1686 became a professor at the University of Basel. When he started working in mathematics, he knew nothing of Newton's and Leibniz' work. He eventually became familiar with Newton's results, but because so little of Leibniz' work was published, Jakob duplicated many of Leibniz' results.

Jakob's younger brother Johann I was urged to enter into business by his father. Instead, he turned to medicine and studied mathematics under the guidance of his older brother. He eventually became a mathematics professor at Gröningen in Holland, and then, when Jakob died in 1705, Johann succeeded him as mathematics professor at Basel. Throughout their lives, Jakob I and Johann I had a mutual passion for criticizing each other's work, which frequently erupted into ugly confrontations. Leibniz tried to mediate the disputes, but Jakob, who resented Leibniz' superior intellect, accused him of siding with Johann, and thus Leibniz became entangled in the arguments. The brothers often worked on common problems that they posed as challenges to one another. Johann, interested in gaining fame, often used unscrupulous means to make himself appear the originator of his brother's results; Jakob occasionally retaliated. Thus, it is often difficult to determine who deserves credit for many results. However, both men made major contributions to the development of calculus. In addition to his work on calculus, Jakob helped establish fundamental principles in probability, including the Law of Large Numbers, which is a cornerstone of modern probability theory.

Among the other members of the Bernoulli family, Daniel, son of Johann I, is the most famous. He was a professor of mathematics at St. Petersburg Academy in Russia and subsequently a professor of anatomy and then physics at Basel. He did work in calculus and probability, but is best known for his work in physics. A basic law of fluid flow, called Bernoulli's principle, is named in his honor. He won the annual prize of the French Academy 10 times for work on vibrating strings, tides of the sea, and kinetic theory of gases.

Johann II succeeded his father as professor of mathematics at Basel. His research was on the theory of heat and sound. Nikolaus I was a mathematician and law scholar who worked on probability and series. On the recommendation of Leibniz, he was appointed professor of mathematics at Padua and then went to Basel as a professor of logic and then law. Nikolaus II was professor of jurisprudence in Switzerland and then professor of mathematics at St. Petersburg Academy. Johann III was a professor of mathematics and astronomy in Berlin and Jakob II succeeded his uncle Daniel as professor of mathematics at St. Petersburg Academy in Russia. Truly an incredible family!

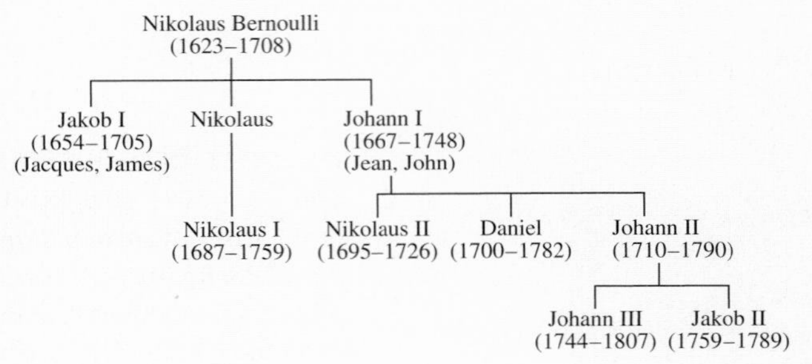

✔ QUICK CHECK EXERCISES 1.7 (See page 80 for answers.)

1. Find parametric equations for a circle of radius 2, centered at (3, 5).
2. Find parametric equations for the ellipse
$$\frac{x^2}{a^2} + \frac{y^2}{b^2} = 1$$
3. The graph of the curve described by the parametric equations $x = 4t - 1$, $y = 3t + 2$ is a straight line with slope ________ and y-intercept ________.
4. Suppose that a parametric curve C is given by the equations $x = f(t)$, $y = g(t)$ for $0 \le t \le 1$. Find parametric equations for C that reverse the direction the curve is traced as the parameter increases from 0 to 1.
5. If $f(x)$ is a one-to-one function, then parametric equations for the curve $y = f^{-1}(x)$ are given by $x =$ ________ and $y = t$.

EXERCISE SET 1.7 Graphing Utility

1. (a) By eliminating the parameter, sketch the trajectory over the time interval $0 \le t \le 5$ of the particle whose parametric equations of motion are
$$x = t - 1, \quad y = t + 1$$
(b) Indicate the direction of motion on your sketch.
(c) Make a table of x- and y-coordinates of the particle at times $t = 0, 1, 2, 3, 4, 5$.
(d) Mark the position of the particle on the curve at the times in part (c), and label those positions with the values of t.

2. (a) By eliminating the parameter, sketch the trajectory over the time interval $0 \le t \le 1$ of the particle whose parametric equations of motion are

$$x = \cos(\pi t), \quad y = \sin(\pi t)$$

(b) Indicate the direction of motion on your sketch.
(c) Make a table of x- and y-coordinates of the particle at times $t = 0, 0.25, 0.5, 0.75, 1$.
(d) Mark the position of the particle on the curve at the times in part (c), and label those positions with the values of t.

3–12 Sketch the curve by eliminating the parameter, and indicate the direction of increasing t.

3. $x = 3t - 4,\ y = 6t + 2$
4. $x = t - 3,\ y = 3t - 7 \quad (0 \le t \le 3)$
5. $x = 2\cos t,\ y = 5\sin t \quad (0 \le t \le 2\pi)$
6. $x = \sqrt{t},\ y = 2t + 4$
7. $x = 3 + 2\cos t,\ y = 2 + 4\sin t \quad (0 \le t \le 2\pi)$
8. $x = \sec t,\ y = \tan t \quad (\pi \le t < 3\pi/2)$
9. $x = \cos 2t,\ y = \sin t \quad (-\pi/2 \le t \le \pi/2)$
10. $x = 4t + 3,\ y = 16t^2 - 9$
11. $x = 2\sin^2 t,\ y = 3\cos^2 t \quad (0 \le t \le \pi/2)$
12. $x = \sec^2 t,\ y = \tan^2 t \quad (-\pi/2 < t < \pi/2)$

13–18 Find parametric equations for the curve, and check your work by generating the curve with a graphing utility.

13. A circle of radius 5, centered at the origin, oriented clockwise.
14. The portion of the circle $x^2 + y^2 = 1$ that lies in the third quadrant, oriented counterclockwise.
15. A vertical line intersecting the x-axis at $x = 2$, oriented upward.
16. The ellipse $x^2/4 + y^2/9 = 1$, oriented counterclockwise.
17. The portion of the parabola $x = y^2$ joining $(1, -1)$ and $(1, 1)$, oriented down to up.
18. The circle of radius 4, centered at $(1, -3)$, oriented counterclockwise.
19. (a) Use a graphing utility to generate the trajectory of a particle whose equations of motion over the time interval $0 \le t \le 5$ are

$$x = 6t - \tfrac{1}{2}t^3, \quad y = 1 + \tfrac{1}{2}t^2$$

(b) Make a table of x- and y-coordinates of the particle at times $t = 0, 1, 2, 3, 4, 5$.
(c) At what times is the particle on the y-axis?
(d) During what time interval is $y < 5$?
(e) At what time does the x-coordinate of the particle reach a maximum?

20. (a) Use a graphing utility to generate the trajectory of a paper airplane whose equations of motion for $t \ge 0$ are

$$x = t - 2\sin t, \quad y = 3 - 2\cos t$$

(b) Assuming that the plane flies in a room in which the floor is at $y = 0$, explain why the plane will not crash into the floor. [For simplicity, ignore the physical size of the plane by treating it as a particle.]
(c) How high must the ceiling be to ensure that the plane does not touch or crash into it?

21–22 Graph the equation using a graphing utility.

21. (a) $x = y^2 + 2y + 1$
(b) $x = \sin y,\ -2\pi \le y \le 2\pi$
22. (a) $x = y + 2y^3 - y^5$
(b) $x = \tan y,\ -\pi/2 < y < \pi/2$
23. (a) By eliminating the parameter, show that the equations

$$x = x_0 + (x_1 - x_0)t, \quad y = y_0 + (y_1 - y_0)t$$

represent the line passing through the points (x_0, y_0) and (x_1, y_1).
(b) Show that if $0 \le t \le 1$, then the equations in part (a) represent the line segment joining (x_0, y_0) and (x_1, y_1), oriented in the direction from (x_0, y_0) to (x_1, y_1).
(c) Use the result in part (b) to find parametric equations for the line segment joining the points $(1, -2)$ and $(2, 4)$, oriented in the direction from $(1, -2)$ to $(2, 4)$.
(d) Use the result in part (b) to find parametric equations for the line segment in part (c), but oriented in the direction from $(2, 4)$ to $(1, -2)$.

24. Use the result in Exercise 23 to find
(a) parametric equations for the line segment joining the points $(-3, -4)$ and $(-5, 1)$, oriented from $(-3, -4)$ to $(-5, 1)$
(b) parametric equations for the line segment traced from $(0, b)$ to $(a, 0)$, oriented from $(0, b)$ to $(a, 0)$.

25. (a) Suppose that the line segment from the point $P(x_0, y_0)$ to $Q(x_1, y_1)$ is represented parametrically by

$$x = x_0 + (x_1 - x_0)t, \quad y = y_0 + (y_1 - y_0)t \quad (0 \le t \le 1)$$

and that $R(x, y)$ is the point on the line segment corresponding to a specified value of t (see the accompanying figure). Show that $t = r/q$, where r is the distance from P to R and q is the distance from P to Q.
(b) What value of t produces the midpoint between points P and Q?
(c) What value of t produces the point that is three-fourths of the way from P to Q?

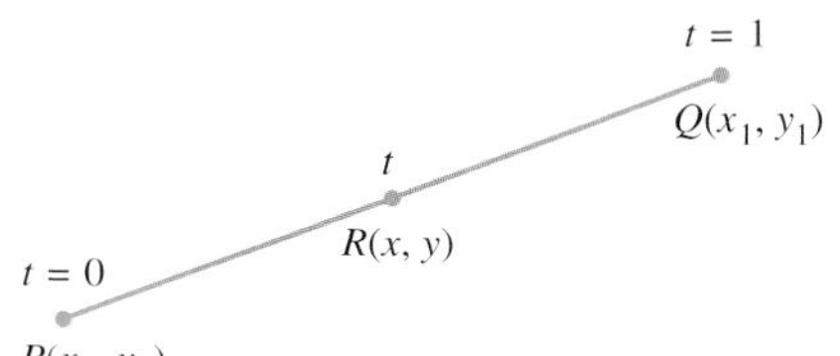

Figure Ex-25

26. Find parametric equations for the line segment joining $P(2, -1)$ and $Q(3, 1)$, and use the result in Exercise 25 to find
(a) the midpoint between P and Q
(b) the point that is one-fourth of the way from P to Q
(c) the point that is three-fourths of the way from P to Q.

FOCUS ON CONCEPTS

27. In each part, match the parametric equation with one of the curves labeled (I)–(VI), and explain your reasoning.
(a) $x = \sqrt{t},\ y = \sin 3t$ (b) $x = 2\cos t,\ y = 3\sin t$
(c) $x = t\cos t,\ y = t\sin t$
(d) $x = \dfrac{3t}{1+t^3},\ y = \dfrac{3t^2}{1+t^3}$
(e) $x = \dfrac{t^3}{1+t^2},\ y = \dfrac{2t^2}{1+t^2}$
(f) $x = \frac{1}{2}\cos t,\ y = \sin 2t$

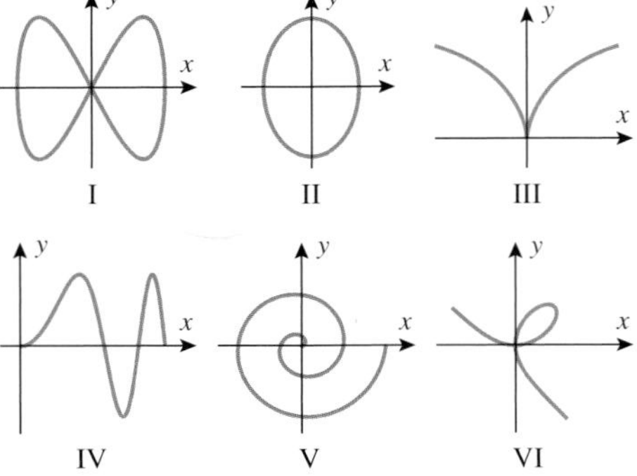

Figure Ex-27

28. Use a graphing utility to generate the curves in Exercise 27, and in each case identify the orientation.

29. Explain why the parametric curve

$$x = t^2, \quad y = t^4 \quad (-1 \le t \le 1)$$

does not have a definite orientation.

30. (a) In parts (a) and (b) of Exercise 23 we obtained parametric equations for a line segment in which the parameter varied from $t = 0$ to $t = 1$. Sometimes it is desirable to have parametric equations for a line segment in which the parameter varies over some other interval, say $t_0 \le t \le t_1$. Use the ideas in Exercise 23 to show that the line segment joining the points (x_0, y_0) and (x_1, y_1) can be represented parametrically as

$$x = x_0 + (x_1 - x_0)\frac{t - t_0}{t_1 - t_0}, \quad y = y_0 + (y_1 - y_0)\frac{t - t_0}{t_1 - t_0} \quad (t_0 \le t \le t_1)$$

(b) Which way is the line segment oriented?
(c) Find parametric equations for the line segment traced from $(3, -1)$ to $(1, 4)$ as t varies from 1 to 2, and check your result with a graphing utility.

31. (a) By eliminating the parameter, show that if a and c are not both zero, then the graph of the parametric equations

$$x = at + b, \quad y = ct + d \quad (t_0 \le t \le t_1)$$

is a line segment.
(b) Sketch the parametric curve

$$x = 2t - 1, \quad y = t + 1 \quad (1 \le t \le 2)$$

and indicate its orientation.

32. (a) What can you say about the line in Exercise 31 if a or c (but not both) is zero?
(b) What do the equations represent if a and c are both zero?

33–36 Use a graphing utility and parametric equations to display the graphs of f and f^{-1} on the same screen.

33. $f(x) = x^3 + 0.2x - 1, \quad -1 \le x \le 2$

34. $f(x) = \sqrt{x^2 + 2} + x, \quad -5 \le x \le 5$

35. $f(x) = \cos(\cos 0.5x), \quad 0 \le x \le 3$

36. $f(x) = x + \sin x, \quad 0 \le x \le 6$

37. Parametric curves can be defined piecewise by using different formulas for different values of the parameter. Sketch the curve that is represented piecewise by the parametric equations

$$\begin{cases} x = 2t, & y = 4t^2 \quad (0 \le t \le \frac{1}{2}) \\ x = 2 - 2t, & y = 2t \quad (\frac{1}{2} \le t \le 1) \end{cases}$$

38. Find parametric equations for the rectangle in the accompanying figure, assuming that the rectangle is traced counterclockwise as t varies from 0 to 1, starting at $(\frac{1}{2}, \frac{1}{2})$ when $t = 0$. [*Hint:* Represent the rectangle piecewise, letting t vary from 0 to $\frac{1}{4}$ for the first edge, from $\frac{1}{4}$ to $\frac{1}{2}$ for the second edge, and so forth.]

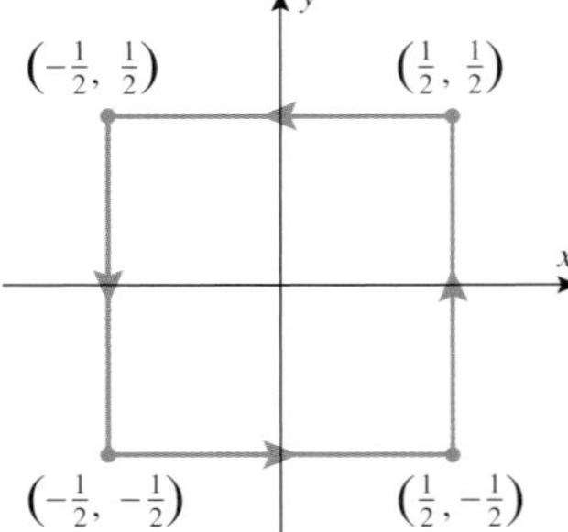

Figure Ex-38

39. (a) Find parametric equations for the ellipse that is centered at the origin and has intercepts $(4, 0)$, $(-4, 0)$, $(0, 3)$, and $(0, -3)$.
(b) Find parametric equations for the ellipse that results by translating the ellipse in part (a) so that its center is at $(-1, 2)$.
(c) Confirm your results in parts (a) and (b) using a graphing utility.

40. We will show later in the text that if a projectile is fired from ground level with an initial speed of v_0 meters per second at an angle α with the horizontal, and if air resistance is neglected, then its position after t seconds, relative to the coordinate system in the accompanying figure is

$$x = (v_0 \cos \alpha)t, \quad y = (v_0 \sin \alpha)t - \tfrac{1}{2}gt^2$$

where $g \approx 9.8 \text{ m/s}^2$.

(a) By eliminating the parameter, show that the trajectory is a parabola.

(b) Sketch the trajectory if $\alpha = 30°$ and $v_0 = 1000$ m/s.

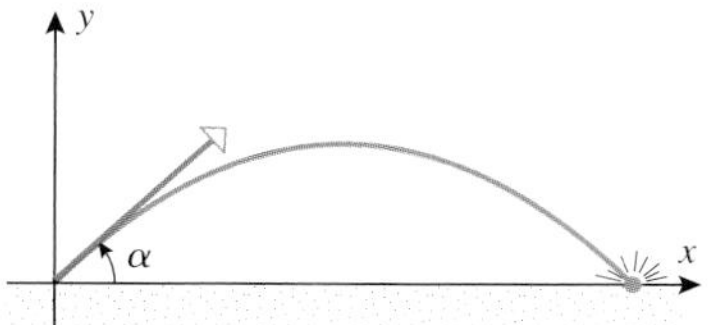

Figure Ex-40

41. A shell is fired from a cannon at an angle of $\alpha = 45°$ with an initial speed of $v_0 = 800$ m/s.

(a) Find parametric equations for the shell's trajectory relative to the coordinate system in Figure Ex-40.

(b) How high does the shell rise?

(c) How far does the shell travel horizontally?

42. A robot arm, designed to buff flat surfaces on an automobile, consists of two attached rods, one that moves back and forth horizontally, and a second, with the buffing pad at the end, that moves up and down (see Figure Ex-42).

(a) Suppose that the horizontal arm of the robot moves so that the x-coordinate of the buffer's center at time t is $x = 25 \sin \pi t$ and the vertical arm moves so that the y-coordinate of the buffer's center at time t is $y = 12.5 \sin \pi t$. Graph the trajectory of the center of the buffing pad.

(b) Suppose that the x- and y-coordinates in part (a) are $x = 25 \sin \pi a t$ and $y = 12.5 \sin \pi b t$, where the constants a and b can be controlled by programming the robot arm. Graph the trajectory of the center of the pad if $a = 4$ and $b = 5$.

(c) Investigate the trajectories that result in part (b) for various choices of a and b.

43. Describe the family of curves described by the parametric equations

$$x = a \cos t + h, \quad y = b \sin t + k \qquad (0 \le t \le 2\pi)$$

if

(a) h and k are fixed but a and b can vary

(b) a and b are fixed but h and k can vary

(c) $a = 1$ and $b = 1$, but h and k vary so that $h = k + 1$.

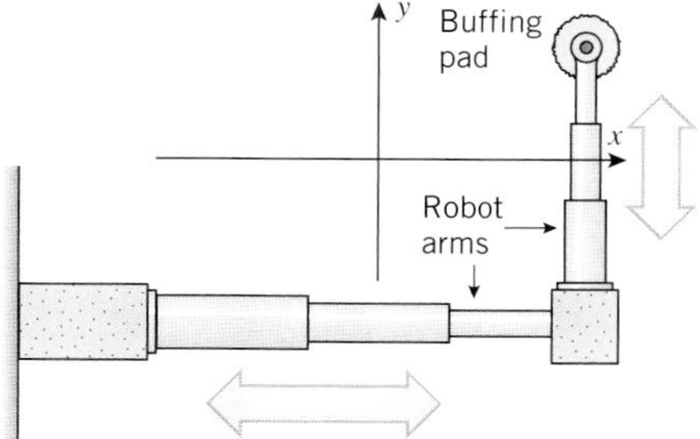

Figure Ex-42

44. A ***hypocycloid*** is a curve traced by a point P on the circumference of a circle that rolls inside a larger fixed circle. Suppose that the fixed circle has radius a, the rolling circle has radius b, and the fixed circle is centered at the origin. Let ϕ be the angle shown in the accompanying figure, and assume that the point P is at $(a, 0)$ when $\phi = 0$. Show that the hypocycloid generated is given by the parametric equations

$$x = (a - b)\cos \phi + b \cos\left(\frac{a - b}{b}\phi\right)$$

$$y = (a - b)\sin \phi - b \sin\left(\frac{a - b}{b}\phi\right)$$

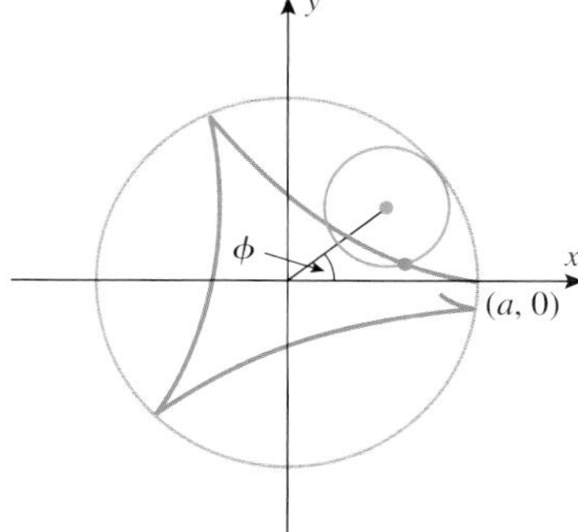

Figure Ex-44

45. If $b = \frac{1}{4}a$ in Exercise 44, then the resulting curve is called a four-cusped hypocycloid.

(a) Sketch this curve.

(b) Show that the curve is given by the parametric equations

$$x = a \cos^3 \phi, \quad y = a \sin^3 \phi$$

(c) Show that the curve is given by the equation

$$x^{2/3} + y^{2/3} = a^{2/3}$$

in rectangular coordinates.

46. (a) Use a graphing utility to study how the curves in the family

$$x = 2a \cos^2 t, \quad y = 2a \cos t \sin t \qquad (-2\pi < t < 2\pi)$$

change as a varies from 0 to 5.

(b) Confirm your conclusion algebraically.

(c) Write a brief paragraph that describes your findings.

✔ QUICK CHECK ANSWERS 1.7

1. $x = 3 + 2\cos t,\ y = 5 + 2\sin t\ (0 \le t \le 2\pi)$ **2.** $x = a\cos t,\ y = b\sin t\ (0 \le t \le 2\pi)$ **3.** $\frac{3}{4}$; 2.75
4. $x = f(1-t),\ y = g(1-t)$ **5.** $f(t)$

CHAPTER REVIEW EXERCISES

Graphing Utility

1. Sketch the graph of the function

$$f(x) = \begin{cases} -1, & x \le -5 \\ \sqrt{25 - x^2}, & -5 < x < 5 \\ x - 5, & x \ge 5 \end{cases}$$

2. Use the graphs of the functions f and g in the accompanying figure to solve the following problems.
(a) Find the values of $f(-2)$ and $g(3)$.
(b) For what values of x is $f(x) = g(x)$?
(c) For what values of x is $f(x) < 2$?
(d) What are the domain and range of f?
(e) What are the domain and range of g?
(f) Find the zeros of f and g.

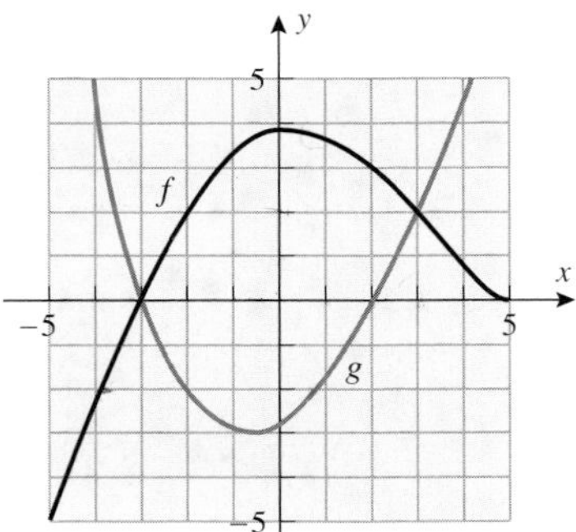

Figure Ex-2

3. A glass filled with water that has a temperature of 40°F is placed in a room in which the temperature is a constant 70°F. Sketch a rough graph that reasonably describes the temperature of the water in the glass as a function of the elapsed time.

4. You want to paint the top of a circular table. Find a formula that expresses the amount of paint required as a function of the radius, and discuss all of the assumptions you have made in finding the formula.

5. A rectangular storage container with an open top and a square base has a volume of 8 cubic meters. Material for the base costs \$5 per square meter and material for the sides \$2 per square meter.
(a) Find a formula that expresses the total cost of materials as a function of the length of a side of the base.
(b) What is the domain of the cost function obtained in part (a)?

6. A ball of radius 3 inches is coated uniformly with plastic.
(a) Express the volume of the plastic as a function of its thickness.
(b) What is the domain of the volume function obtained in part (a)?

7. A box with a closed top is to be made from a 6-ft by 10-ft piece of cardboard by cutting out four squares of equal size (see the accompanying figure), folding along the dashed lines, and tucking the two extra flaps inside.
(a) Find a formula that expresses the volume of the box as a function of the length of the sides of the cut-out squares.
(b) Find an inequality that specifies the domain of the function in part (a).
(c) Use the graph of the volume function to estimate the dimensions of the box of largest volume.

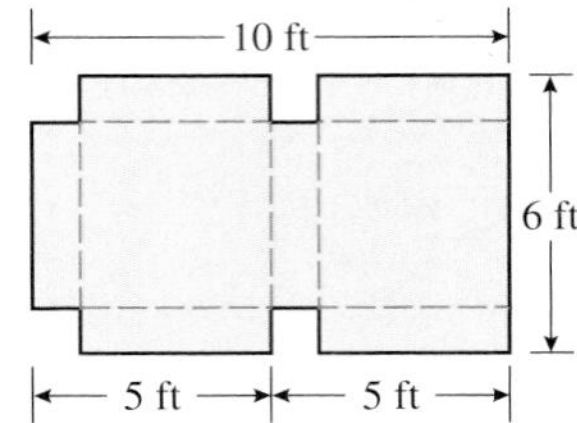

Figure Ex-7

8. Let C denote the graph of $y = 1/x,\ x > 0$.
(a) Express the distance between the point $P(1, 0)$ and a point Q on C as a function of the x-coordinate of Q.
(b) What is the domain of the distance function obtained in part (a)?
(c) Use the graph of the distance function obtained in part (a) to estimate the point Q on C that is closest to the point P.

9. Sketch the graph of the equation $x^2 - 4y^2 = 0$.

10. Generate the graph of $f(x) = x^4 - 24x^3 - 25x^2$ in two different viewing windows, each of which illustrates a different property of f. Identify each viewing window and a characteristic of the graph of f that is illustrated well in the window.

11. Complete the following table.

x	−4	−3	−2	−1	0	1	2	3	4
$f(x)$	0	−1	2	1	3	−2	−3	4	−4
$g(x)$	3	2	1	−3	−1	−4	4	−2	0
$(f \circ g)(x)$									
$(g \circ f)(x)$									

12. Let $f(x) = -x^2$ and $g(x) = 1/\sqrt{x}$. Find formulas for $f \circ g$ and $g \circ f$ and state the domain of each composition.

13. Given that $f(x) = x^2 + 1$ and $g(x) = 3x + 2$, find all values of x such that $f(g(x)) = g(f(x))$.

14. Let $f(x) = (2x - 1)/(x + 1)$ and $g(x) = 1/(x - 1)$.
(a) Find $f(g(x))$.
(b) Is the natural domain of the function $h(x) = (3 - x)/x$ the same as the domain of $f \circ g$? Explain.

15. Given that

$$f(x) = \frac{x}{x-1}, \quad g(x) = \frac{1}{x}, \quad h(x) = x^2 - 1$$

find a formula for $f \circ g \circ h$ and state the domain of this composition.

16. Given that $f(x) = 2x + 1$ and $h(x) = 2x^2 + 4x + 1$, find a function g such that $f(g(x)) = h(x)$.

17. In each part, classify the function as even, odd, or neither.
(a) $x^2 \sin x$ (b) $\sin^2 x$ (c) $x + x^2$ (d) $\sin x \tan x$

18. (a) Write an equation for the graph that is obtained by reflecting the graph of $y = |x - 1|$ about the y-axis, then stretching that graph vertically by a factor of 2, then translating that graph down 3 units, and then reflecting that graph about the x-axis.
(b) Sketch the original graph and the final graph.

19. In each part, describe the family of curves.
(a) $(x - a)^2 + (y - a^2)^2 = 1$
(b) $y = a + (x - 2a)^2$

20. Find an equation for a parabola that passes through the points $(2, 0)$, $(8, 18)$, and $(-8, 18)$.

21. Suppose that the expected low temperature in Anchorage, Alaska (in °F), is modeled by the equation

$$T = 50 \sin \frac{2\pi}{365}(t - 101) + 25$$

where t is in days and $t = 0$ corresponds to January 1.
(a) Sketch the graph of T versus t for $0 \le t \le 365$.
(b) Use the model to predict when the coldest day of the year will occur.
(c) Based on this model, how many days during the year would you expect the temperature to be below 0°F?

22. The accompanying figure shows a model for the tide variation in an inlet to San Francisco Bay during a 24-hour period. Find an equation of the form $y = y_0 + y_1 \sin(at + b)$ for the model, assuming that $t = 0$ corresponds to midnight.

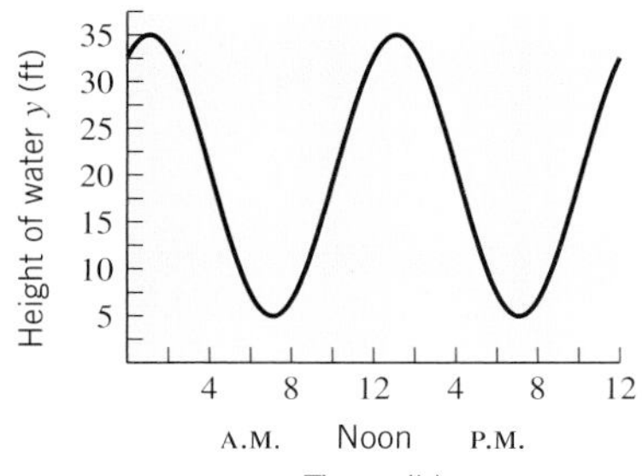

Figure Ex-22

23. The accompanying figure shows the graphs of the equations $y = 1 + 2\sin x$ and $y = 2\sin(x/2) + 2\cos(x/2)$ for $-2\pi \le x \le 2\pi$. Without the aid of a calculator, label each curve by its equation, and find the coordinates of the points A, B, C, and D. Explain your reasoning.

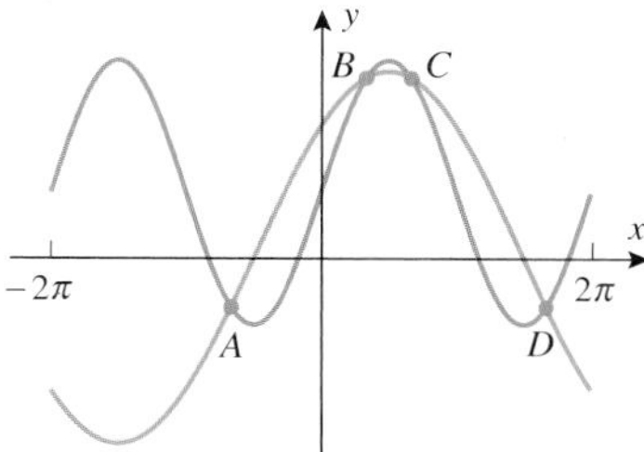

Figure Ex-23

24. The electrical resistance R in ohms (Ω) for a pure metal wire is related to its temperature T in °C by the formula

$$R = R_0(1 + kT)$$

in which R_0 and k are positive constants.
(a) Make a hand-drawn sketch of the graph of R versus T, and explain the geometric significance of R_0 and k for your graph.
(b) In theory, the resistance R of a pure metal wire drops to zero when the temperature reaches absolute zero ($T = -273$°C). What information does this give you about k?
(c) A tungsten bulb filament has a resistance of 1.1 Ω at a temperature of 20°C. What information does this give you about R_0 for the filament?
(d) At what temperature will the tungsten filament have a resistance of 1.5 Ω?

25. (a) State conditions under which two functions, f and g, will be inverses, and give several examples of such functions.
(b) In words, what is the relationship between the graphs of $y = f(x)$ and $y = g(x)$ when f and g are inverse functions?
(c) What is the relationship between the domains and ranges of inverse functions f and g?
(d) What condition must be satisfied for a function f to have an inverse? Give some examples of functions that do not have inverses.

26. In each part, find $f^{-1}(x)$ if the inverse exists.
(a) $f(x) = 8x^3 - 1$ (b) $f(x) = x^2 - 2x + 1$
(c) $f(x) = 3/(x + 1)$ (d) $f(x) = (x + 2)/(x - 1)$

27. An important problem addressed by calculus is that of finding a good linear approximation to the function $f(x)$ near a particular x-value. One possible approach (not the best) is to sample values of the function near the specified x-value, find the least squares line for this sample, and translate the least squares line so that it passes through the point on the graph of $y = f(x)$ corresponding to the given x-value. Let $f(x) = x^2 \sin x$.

(a) Make a table of $(x, f(x))$ values for $x = 1.9, 1.92, 1.94, \ldots, 2.1$.

(b) Find a least squares line for the data in part (a).

(c) Find the equation of the line passing through the point $(2, f(2))$ and parallel to the least squares line.

(d) Using a graphing utility with a graphing window containing $(2, f(2))$, graph $y = f(x)$ and the line you found in part (c). How do the graphs compare as you zoom closer to the point $(2, f(2))$?

[*Note:* The best linear approximation to $y = x^2 \sin x$ near $x = 2$ is given by $y \approx 1.9726x - 0.308015$. Later, we will see how to use the tools of calculus to find this answer.]

28. An extension of the linear approximation problem is finding a good polynomial approximation to the function $f(x)$ near a particular x-value. One possible approach (not the best) is to sample values of the function near the specified x-value, apply polynomial regression to this sample, and translate the regression curve so that it passes through the point on the graph of $y = f(x)$ corresponding to the given x-value. Let $f(x) = \cos x$.

(a) Make a table of $(x, f(x))$ values for $x = -0.1, -0.08, -0.06, \ldots, 0.1$.

(b) Use quadratic regression to model the data in part (a) with a quadratic polynomial.

(c) Translate your quadratic modeling function from part (b) to obtain a quadratic function that passes through the point $(0, f(0))$.

(d) Using a graphing utility with a graphing window containing $(0, f(0))$, graph $y = f(x)$ and the polynomial you found in part (c). How do the graphs compare as you zoom closer to the point $(0, f(0))$?

[*Note:* The best quadratic approximation to $y = \cos x$ near $x = 0$ is given by $y \approx -0.5x^2 + 1$. In Chapter 10, we will see how to use the tools of calculus to find this answer.]

29. Table Ex-29 gives the water level (in meters above the mean low-water mark) at a Cape Hatteras, North Carolina, fishing pier, recorded in 2-hour increments starting from midnight, July 1, 1999. Why should we expect that a trigonometric function should fit these data? Find a function that models the data, and graph your function on a plot of the data.

30. A professor wishes to use midterm grades as a predictor of final grades in a small seminar that he teaches once a year. The midterm grades and final grades for last year's seminar are listed in the Table Ex-30.

(a) Find the linear regression model that expresses the final grade in terms of the midterm grade.

(b) Suppose that a student in this year's seminar earned a midterm grade of 88. Use the model obtained in part (a) to predict the student's final grade in the seminar.

HOUR	WATER LEVEL (m)	HOUR	WATER LEVEL (m)
0	0.526	36	0.534
2	0.157	38	0.192
4	0.161	40	0.141
6	0.486	42	0.426
8	0.779	44	0.849
10	0.740	46	1.032
12	0.412	48	0.765
14	0.141	50	0.281
16	0.260	52	0.042
18	0.633	54	0.157
20	1.015	56	0.587
22	1.021	58	0.777
24	0.670	60	0.620
26	0.231	62	0.241
28	0.128	64	0.045
30	0.345	66	0.195
32	0.697	68	0.613
34	0.821	70	0.945

Table Ex-29

MIDTERM GRADE	FINAL GRADE
78	78
94	91
78	76
84	82
95	92
96	93
77	75

Table Ex-30

31. Find parametric equations for the portion of the circle $x^2 + y^2 = 2$ that lies outside the first quadrant, oriented clockwise. Check your work by generating the curve with a graphing utility.

32. (a) Suppose that the equations $x = f(t)$, $y = g(t)$ describe a curve C as t increases from 0 to 1. Find parametric equations that describe the same curve C but traced in the opposite direction as t increases from 0 to 1.

(b) Check your work using the parametric graphing feature of a graphing utility by generating the line segment between $(1, 2)$ and $(4, 0)$ in both possible directions as t increases from 0 to 1.

33. Sketch the curve described by the parametric equations

$$x = t\cos(2\pi t), \quad y = t\sin(2\pi t)$$

and check your result with a graphing utility.

EXPANDING THE CALCULUS HORIZON

Ecologists use mathematical models based on the process of iteration to predict the growth and decline of animal populations. To learn more about this process and its applications, and to build upon the mathematics learned in this chapter, go to

www.wiley.com/college/anton

LIMITS AND CONTINUITY

A limit is a peculiar and fundamental conception, the use of which in proving the propositions of Higher Geometry cannot be superseded by any combination of other hypotheses and definitions.
—William Whewell
Science Philosopher

The development of calculus in the seventeenth century by Newton and Leibniz provided scientists with their first real understanding of what is meant by an "instantaneous rate of change" such as velocity and acceleration. Once the idea was understood conceptually, efficient computational methods followed, and science took a quantum leap forward. The fundamental building block on which rates of change rest is the concept of a "limit," an idea that is so important that all other calculus concepts are now based on it.

In this chapter we will develop the concept of a limit in stages, proceeding from an informal, intuitive notion to a precise mathematical definition. We will also develop theorems and procedures for calculating limits, and we will conclude the chapter by using the limits to study "continuous" curves.

Photo: *Air resistance prevents the velocity of a skydiver from increasing indefinitely. The velocity approaches a limit, called the "terminal velocity".*

2.1 LIMITS (AN INTUITIVE APPROACH)

The concept of a "limit" is the fundamental building block on which all calculus concepts are based. In this section we will study limits informally, with the goal of developing an intuitive feel for the basic ideas. In the next three sections we will focus on computational methods and precise definitions.

Many of the ideas of calculus originated with the following two geometric problems:

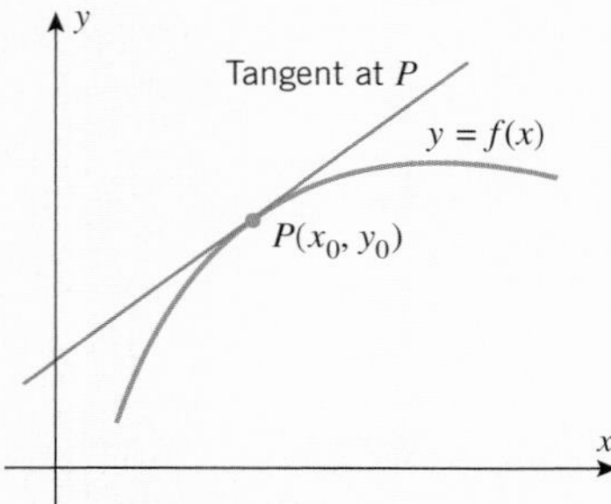

Figure 2.1.1

THE TANGENT LINE PROBLEM. Given a function f and a point $P(x_0, y_0)$ on its graph, find an equation of the line that is tangent to the graph at P (Figure 2.1.1).

THE AREA PROBLEM. Given a function f, find the area between the graph of f and an interval $[a, b]$ on the x-axis (Figure 2.1.2).

Traditionally, that portion of calculus arising from the tangent line problem is called ***differential calculus*** and that arising from the area problem is called ***integral calculus***.

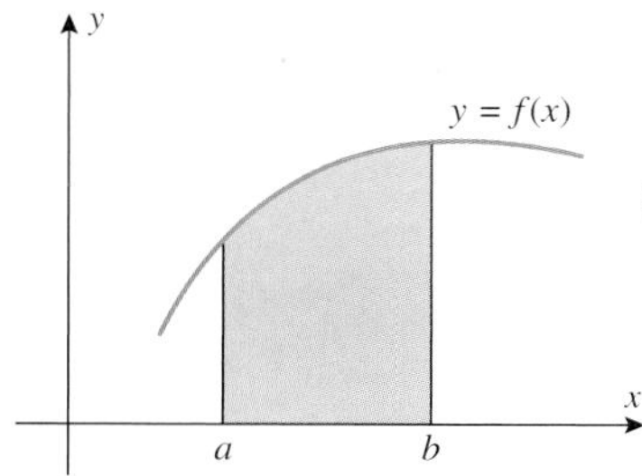

Figure 2.1.2

However, we will see later that the tangent line and area problems are so closely related that the distinction between differential and integral calculus is somewhat artificial.

TANGENT LINES AND LIMITS

In plane geometry, a line is called *tangent* to a circle if it meets the circle at precisely one point (Figure 2.1.3*a*). However, this definition is not appropriate for more general curves. For example, in Figure 2.1.3*b*, the line meets the curve exactly once but is obviously not what we would regard to be a tangent line; and in Figure 2.1.3*c*, the line appears to be tangent to the curve, yet it intersects the curve more than once.

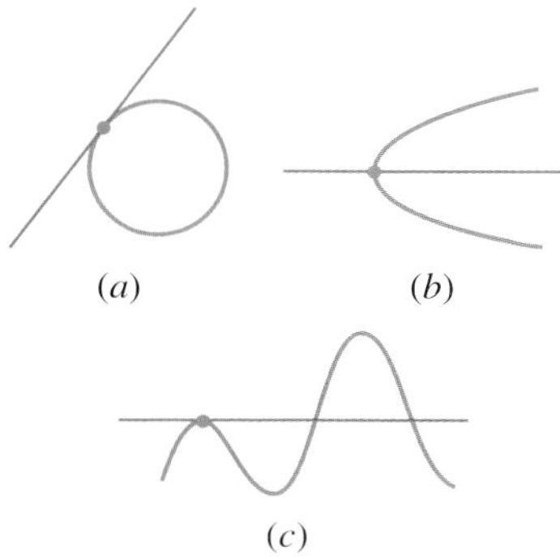

Figure 2.1.3

To obtain a definition of a tangent line that applies to curves other than circles, we must view tangent lines another way. For this purpose, suppose that we are interested in the tangent line at a point P on a curve in the xy-plane and that Q is any point that lies on the curve and is different from P. The line through P and Q is called a ***secant line*** for the curve at P. Intuition suggests that if we move the point Q along the curve toward P, then the secant line will rotate toward a *limiting position*. The line in this limiting position is what we will consider to be the ***tangent line*** at P (Figure 2.1.4*a*). As suggested by Figure 2.1.4*b*, this new concept of a tangent line coincides with the traditional concept when applied to circles.

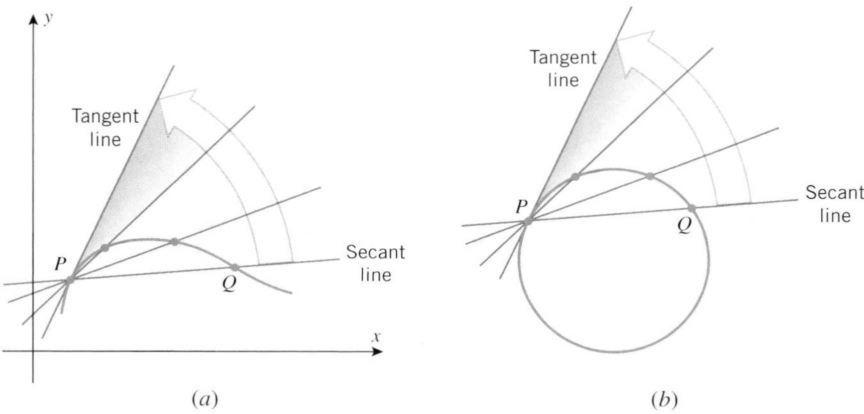

Figure 2.1.4

▶ **Example 1** Find an equation for the tangent line to the parabola $y = x^2$ at the point $P(1, 1)$.

Solution. If we can find the slope $m_{\tan}$ of the tangent line at P, then we can use the point P and the point-slope formula for a line (Web Appendix F) to write the equation of the tangent line as

$$y - 1 = m_{\tan}(x - 1) \tag{1}$$

To find the slope $m_{\tan}$, consider the secant line through P and a point $Q(x, x^2)$ on the parabola that is distinct from P. The slope $m_{\sec}$ of this secant line is

$$m_{\sec} = \frac{x^2 - 1}{x - 1} \tag{2}$$

Why are we requiring that P and Q be distinct?

Figure 2.1.4*a* suggests that if we now let Q move along the parabola, getting closer and closer to P, then the limiting position of the secant line through P and Q will coincide with that of the tangent line at P. This in turn suggests that the value of $m_{\sec}$ will get closer and closer to the value of $m_{\tan}$ as P moves toward Q along the curve. However, to say that $Q(x, x^2)$ gets closer and closer to $P(1, 1)$ is algebraically equivalent to saying that x gets

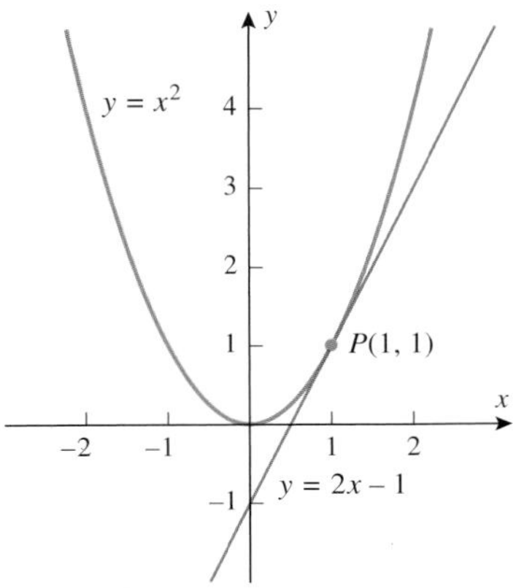

Figure 2.1.5

closer and closer to 1. Thus, the problem of finding $m_{\tan}$ reduces to finding the "limiting value" of $m_{\sec}$ in Formula (2) as x gets closer and closer to 1 (but with $x \neq 1$ to ensure that P and Q remain distinct).

Since $x \neq 1$, we can rewrite (2) as

$$m_{\sec} = \frac{x^2 - 1}{x - 1} = \frac{(x - 1)(x + 1)}{(x - 1)} = x + 1$$

from which it is evident that $m_{\sec}$ gets closer and closer to 2 as x gets closer and closer to 1. Thus, $m_{\tan} = 2$ and (1) implies that the equation of the tangent line is

$$y - 1 = 2(x - 1) \quad \text{or equivalently} \quad y = 2x - 1$$

Figure 2.1.5 shows the graph of $y = x^2$ and this tangent line. ◀

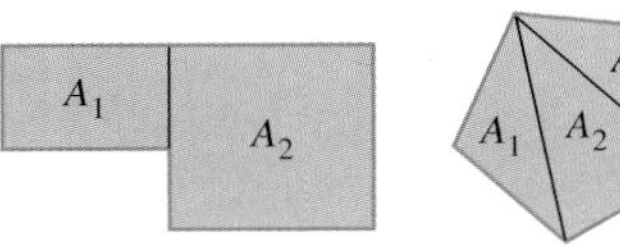

Figure 2.1.6

AREAS AND LIMITS

Just as the general notion of a tangent line leads to the concept of *limit*, so does the general notion of area. For plane regions with straight-line boundaries, areas can often be calculated by subdividing the region into rectangles or triangles and adding the areas of the constituent parts (Figure 2.1.6). However, for regions with curved boundaries, such as that in Figure 2.1.7*a*, a more general approach is needed. One such approach is to begin by approximating the area of the region by inscribing a number of rectangles of equal width under the curve and adding the areas of these rectangles (Figure 2.1.7*b*). Intuition suggests that if we repeat that approximation process using more and more rectangles, then the rectangles will tend to fill in the gaps under the curve, and the approximations will get closer and closer to the exact area under the curve (Figure 2.1.7*c*). This suggests that we can define the area under the curve to be the limiting value of these approximations. This idea will be considered in detail later, but the point to note here is that once again the concept of a limit comes into play.

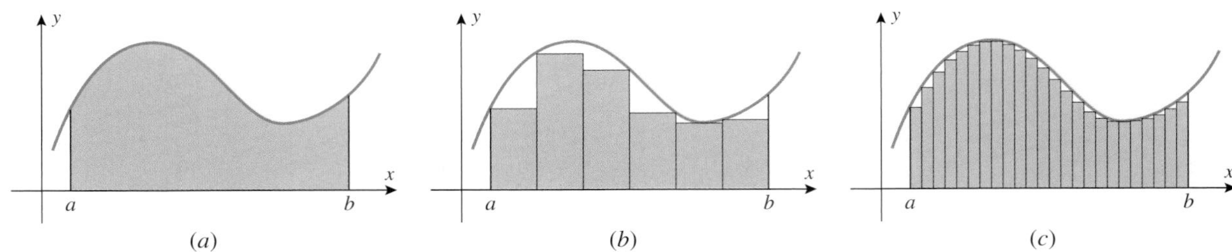

Figure 2.1.7

DECIMALS AND LIMITS

Limits also arise in the familiar context of decimals. For example, the decimal expansion of the fraction $\frac{1}{3}$ is

$$\frac{1}{3} = 0.33333\ldots \tag{3}$$

in which the dots indicate that the digit 3 repeats indefinitely. Although you may not have thought about decimals in this way, we can write (3) as

$$\frac{1}{3} = 0.33333\ldots = 0.3 + 0.03 + 0.003 + 0.0003 + 0.00003 + \cdots \tag{4}$$

which is a sum with "infinitely many" terms. As we will discuss in more detail later, we interpret (4) to mean that the succession of finite sums

$$0.3,\ \ 0.3 + 0.03,\ \ 0.3 + 0.03 + 0.003,\ \ 0.3 + 0.03 + 0.003 + 0.0003, \ldots$$

gets closer and closer to a limiting value of $\frac{1}{3}$ as more and more terms are included. Thus, limits even occur in the familiar context of decimal representations of real numbers.

LIMITS

Now that we have seen how limits arise in various ways, let us focus on the limit concept itself.

The most basic use of limits is to describe how a function behaves as the independent variable approaches a given value. For example, let us examine the behavior of the function

$$f(x) = x^2 - x + 1$$

for x-values closer and closer to 2. It is evident from the graph and table in Figure 2.1.8 that the values of $f(x)$ get closer and closer to 3 as values of x are selected closer and closer to 2 on either the left or the right side of 2. We describe this by saying that the "limit of $x^2 - x + 1$ is 3 as x approaches 2 from either side," and we write

$$\lim_{x\to 2} (x^2 - x + 1) = 3 \tag{5}$$

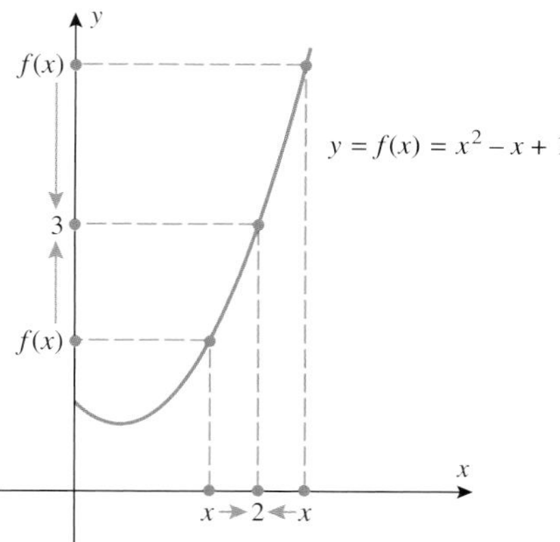

x	1.0	1.5	1.9	1.95	1.99	1.995	1.999	2	2.001	2.005	2.01	2.05	2.1	2.5	3.0
$f(x)$	1.000000	1.750000	2.710000	2.852500	2.970100	2.985025	2.997001		3.003001	3.015025	3.030100	3.152500	3.310000	4.750000	7.000000

Left side → ← Right side

Figure 2.1.8

This leads us to the following general idea.

> **2.1.1 LIMITS (AN INFORMAL VIEW).** If the values of $f(x)$ can be made as close as we like to L by taking values of x sufficiently close to a (but not equal to a), then we write
>
> $$\lim_{x\to a} f(x) = L \tag{6}$$
>
> which is read "the limit of $f(x)$ as x approaches a is L" or "$f(x)$ approaches L as x approaches a." The expression in (6) can also be written as
>
> $$f(x) \to L \quad \text{as} \quad x \to a \tag{7}$$

Since x is required to be different from a in (6), the value of f at a, or even whether f is defined at a, has no bearing on the limit L. The limit describes the behavior of f *close to* a but not *at* a.

▶ **Example 2** Use numerical evidence to make a conjecture about the value of

$$\lim_{x \to 1} \frac{x - 1}{\sqrt{x} - 1} \tag{8}$$

Solution. Although the function

$$f(x) = \frac{x - 1}{\sqrt{x} - 1} \tag{9}$$

is undefined at $x = 1$, this has no bearing on the limit. Table 2.1.1 shows sample x-values approaching 1 from the left side and from the right side. In both cases the corresponding values of $f(x)$, calculated to six decimal places, appear to get closer and closer to 2, and hence we conjecture that

$$\lim_{x \to 1} \frac{x - 1}{\sqrt{x} - 1} = 2$$

This is consistent with the graph of f shown in Figure 2.1.9. In the next section we will show how to obtain this result algebraically. ◀

TECHNOLOGY MASTERY

Use a graphing utility to generate the graph of the equation $y = f(x)$ for the function in (9). Find a window containing $x = 1$ in which all values of $f(x)$ are within 0.5 of $y = 2$ and one in which all values of $f(x)$ are within 0.1 of $y = 2$.

Table 2.1.1

x	0.99	0.999	0.9999	0.99999	0	1.00001	1.0001	1.001	1.01
$f(x)$	1.994987	1.999500	1.999950	1.999995		2.000005	2.000050	2.000500	2.004988

Left side → ← Right side

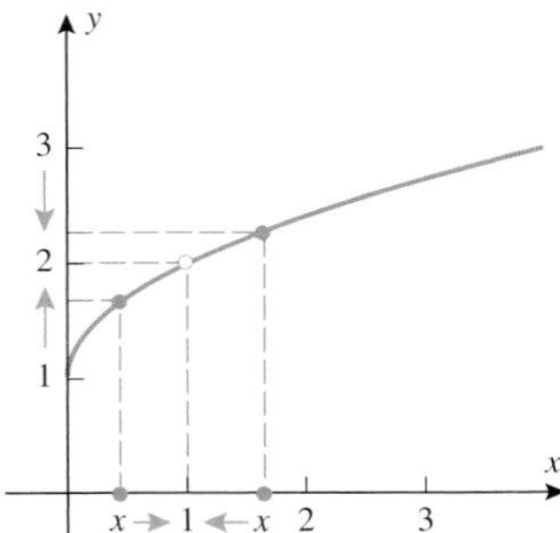

Figure 2.1.9

▶ **Example 3** Use numerical evidence to make a conjecture about the value of

$$\lim_{x \to 0} \frac{\sin x}{x} \tag{10}$$

Solution. With the help of a calculating utility set in radian mode, we obtain Table 2.1.2. The data in the table suggest that

$$\lim_{x \to 0} \frac{\sin x}{x} = 1 \tag{11}$$

The result is consistent with the graph of $f(x) = (\sin x)/x$ shown in Figure 2.1.10. Later in this chapter we will give a geometric argument to prove that our conjecture is correct. ◀

Table 2.1.2

x (RADIANS)	$y = \frac{\sin x}{x}$
±1.0	0.84147
±0.9	0.87036
±0.8	0.89670
±0.7	0.92031
±0.6	0.94107
±0.5	0.95885
±0.4	0.97355
±0.3	0.98507
±0.2	0.99335
±0.1	0.99833
±0.01	0.99998

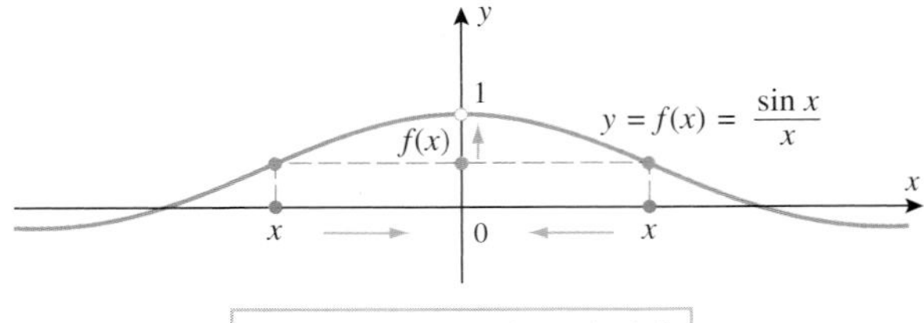

Figure 2.1.10

SAMPLING PITFALLS

Numerical evidence can sometimes lead to incorrect conclusions about limits because of roundoff error or because the sample values chosen do not reveal the true limiting behavior. For example, one might *incorrectly* conclude from Table 2.1.3 that

$$\lim_{x \to 0} \sin\left(\frac{\pi}{x}\right)$$

Use numerical evidence to determine whether the limit in (11) changes if x is measured in degrees.

is zero. The fact that this is not correct is evidenced by the graph of f in Figure 2.1.11. The graph reveals that the values of f oscillate between -1 and 1 with increasing rapidity as $x \to 0$ and hence do not approach a limit. The data in the table deceived us because the x-values selected all happened to be x-intercepts for $f(x)$. This points out the need for having alternative methods for corroborating limits conjectured from numerical evidence.

Table 2.1.3

x	$\dfrac{\pi}{x}$	$f(x) = \sin\left(\dfrac{\pi}{x}\right)$
$x = \pm 1$	$\pm\pi$	$\sin(\pm\pi) = 0$
$x = \pm 0.1$	$\pm 10\pi$	$\sin(\pm 10\pi) = 0$
$x = \pm 0.01$	$\pm 100\pi$	$\sin(\pm 100\pi) = 0$
$x = \pm 0.001$	$\pm 1000\pi$	$\sin(\pm 1000\pi) = 0$
$x = \pm 0.0001$	$\pm 10{,}000\pi$	$\sin(\pm 10{,}000\pi) = 0$
$\vdots$	$\vdots$	$\vdots$

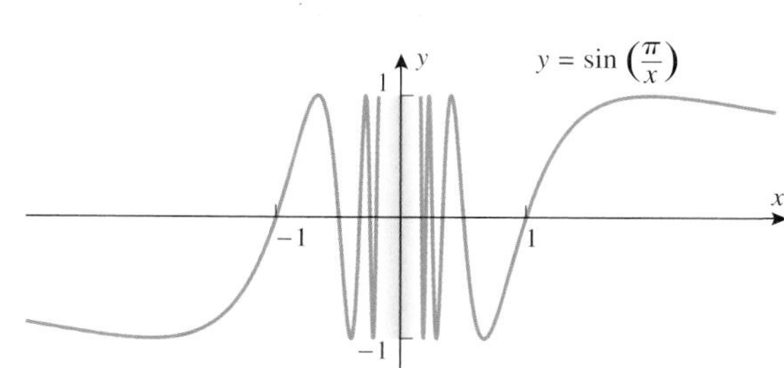

Figure 2.1.11

ONE-SIDED LIMITS

The limit in (6) is called a ***two-sided limit*** because it requires the values of $f(x)$ to get closer and closer to L as values of x are taken from *either* side of $x = a$. However, some functions exhibit different behaviors on the two sides of an x-value a, in which case it is necessary to distinguish whether values of x near a are on the left side or on the right side of a for purposes of investigating limiting behavior. For example, consider the function

$$f(x) = \frac{|x|}{x} = \begin{cases} 1, & x > 0 \\ -1, & x < 0 \end{cases} \tag{12}$$

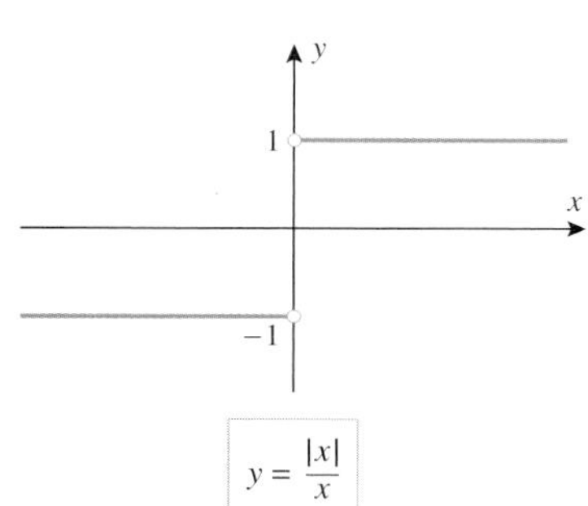

Figure 2.1.12

which is graphed in Figure 2.1.12. As x approaches 0 from the *right*, the values of $f(x)$ approach a limit of 1 [in fact, the values of $f(x)$ are exactly 1 for all such x], and similarly, as x approaches 0 from the *left*, the values of $f(x)$ approach a limit of -1. We denote these limits by writing

$$\lim_{x\to 0^+} \frac{|x|}{x} = 1 \quad \text{and} \quad \lim_{x\to 0^-} \frac{|x|}{x} = -1 \tag{13}$$

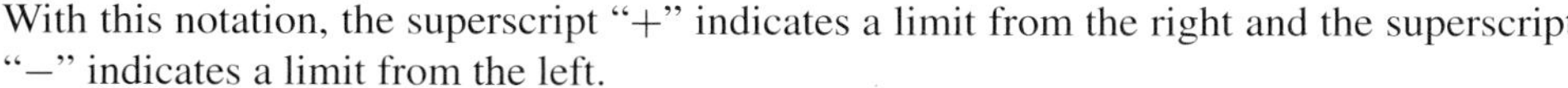

With this notation, the superscript "+" indicates a limit from the right and the superscript "−" indicates a limit from the left.

This leads to the general idea of a ***one-sided limit***.

As with two-sided limits, the one-sided limits in (14) and (15) can also be written as

$$f(x) \to L \text{ as } x \to a^+$$

and

$$f(x) \to L \text{ as } x \to a^-$$

respectively.

2.1.2 ONE-SIDED LIMITS (AN INFORMAL VIEW). If the values of $f(x)$ can be made as close as we like to L by taking values of x sufficiently close to a (but greater than a), then we write

$$\lim_{x\to a^+} f(x) = L \tag{14}$$

and if the values of $f(x)$ can be made as close as we like to L by taking values of x sufficiently close to a (but less than a), then we write

$$\lim_{x\to a^-} f(x) = L \tag{15}$$

Expression (14) is read "the limit of $f(x)$ as x approaches a from the right is L" or "$f(x)$ approaches L as x approaches a from the right." Similarly, expression (15) is read "the limit of $f(x)$ as x approaches a from the left is L" or "$f(x)$ approaches L as x approaches a from the left."

THE RELATIONSHIP BETWEEN ONE-SIDED LIMITS AND TWO-SIDED LIMITS

In general, there is no guarantee that a function f will have a two-sided limit at a given point a; that is, the values of $f(x)$ may not get closer and closer to any *single* real number L as $x \to a$. In this case we say that

$$\lim_{x \to a} f(x) \quad \textit{does not exist}$$

(and similarly for one-sided limits).

In order for the two-sided limit of a function $f(x)$ to exist at a point a, the values of $f(x)$ must approach some real number L as x approaches a, and this number must be the same regardless of whether x approaches a from the left or the right. This suggests the following result, which we state without formal proof.

> **2.1.3 THE RELATIONSHIP BETWEEN ONE-SIDED AND TWO-SIDED LIMITS.** The two-sided limit of a function $f(x)$ exists at a if and only if both of the one-sided limits exist at a and have the same value; that is,
>
> $$\lim_{x \to a} f(x) = L \quad \text{if and only if} \quad \lim_{x \to a^-} f(x) = L = \lim_{x \to a^+} f(x)$$

► **Example 4** Explain why

$$\lim_{x \to 0} \frac{|x|}{x}$$

does not exist.

Solution. As x approaches 0, the values of $f(x) = |x|/x$ approach -1 from the left and approach 1 from the right [see (13)]. Thus, the one-sided limits at 0 are not the same. ◄

► **Example 5** For the functions in Figure 2.1.13, find the one-sided and two-sided limits at $x = a$ if they exist.

Solution. The functions in all three figures have the same one-sided limits as $x \to a$, since the functions are identical, except at $x = a$. These limits are

$$\lim_{x \to a^+} f(x) = 3 \quad \text{and} \quad \lim_{x \to a^-} f(x) = 1$$

In all three cases the two-sided limit does not exist as $x \to a$ because the one-sided limits are not equal. ◄

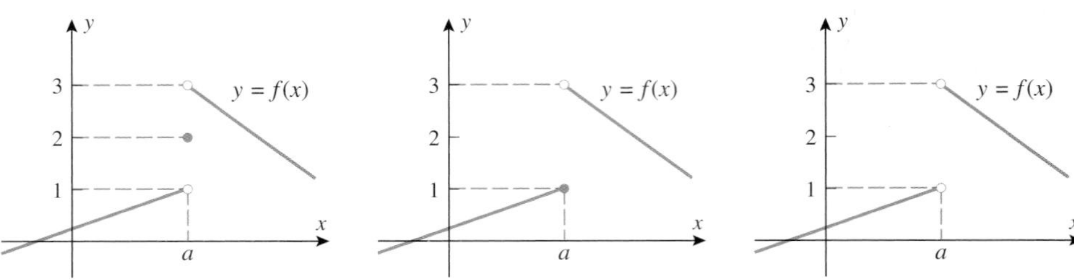

Figure 2.1.13

► **Example 6** For the functions in Figure 2.1.14, find the one-sided and two-sided limits at $x = a$ if they exist.

Solution. As in the preceding example, the value of f at $x = a$ has no bearing on the limits as $x \to a$, so in all three cases we have

$$\lim_{x \to a^+} f(x) = 2 \quad \text{and} \quad \lim_{x \to a^-} f(x) = 2$$

Since the one-sided limits are equal, the two-sided limit exists and

$$\lim_{x \to a} f(x) = 2 \blacktriangleleft$$

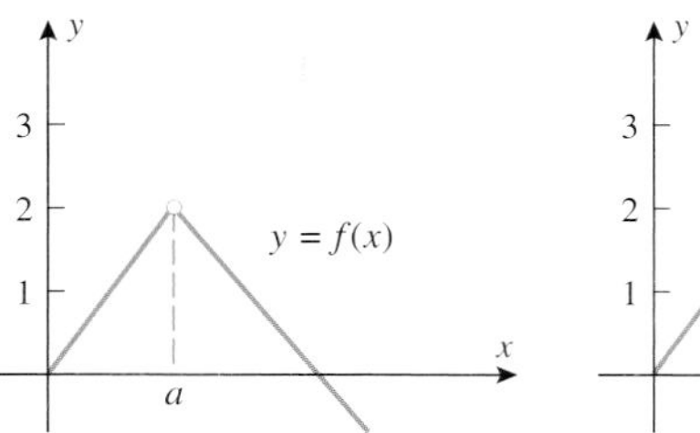

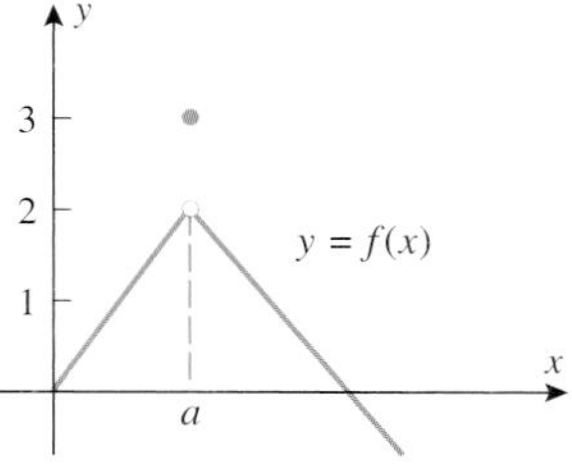

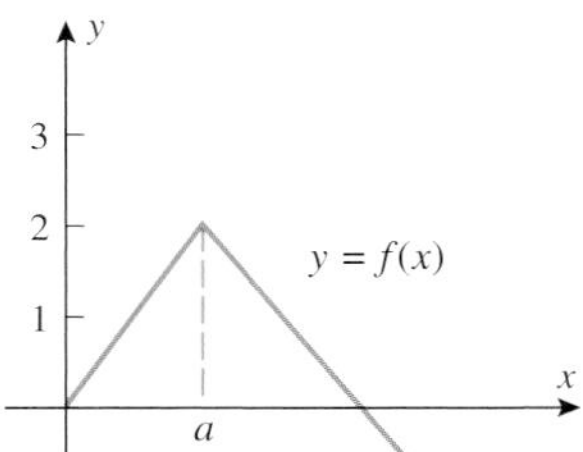

Figure 2.1.14

INFINITE LIMITS

We observed in Section 1.4 that certain rational functions have points at which they are not defined and near which the graphs closely approximate a vertical line called a vertical asymptote. For example, consider the behavior of $f(x) = 1/x$ for values of x near 0. It is evident from the table and graph in Figure 2.1.15 that as x-values are taken closer and closer to 0 from the right, the values of $1/x$ are positive and increase indefinitely. That is, for any specified number M, values of $1/x$ will exceed M for positive values of x taken near enough to 0. In this case, $f(x) = 1/x$ is said to ***increase without bound*** as x approaches 0 from the right. Similarly, x-values taken closer and closer to 0 from the left produce values of $1/x$ that are negative and decrease indefinitely, so we say that $f(x) = 1/x$ ***decreases without bound*** as x approaches 0 from the left. Since values of $1/x$ do not approach a real number as x approaches 0 either from the right or from the left, neither one-sided limit exists at 0. However, we describe the particular way that these limits fail to exist by writing

$$\lim_{x \to 0^+} \frac{1}{x} = +\infty \quad \text{and} \quad \lim_{x \to 0^-} \frac{1}{x} = -\infty$$

The symbols $+\infty$ and $-\infty$ here are *not* real numbers; they simply describe particular ways in which the limits fail to exist. Do not make the mistake of manipulating these symbols using rules of algebra. For example, it is *incorrect* to write $(+\infty) - (+\infty) = 0$.

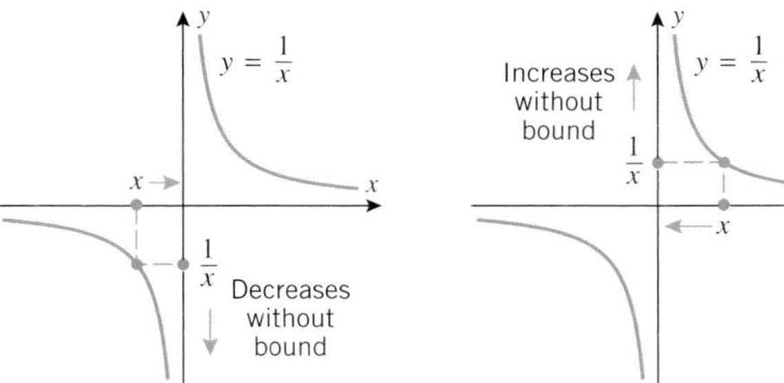

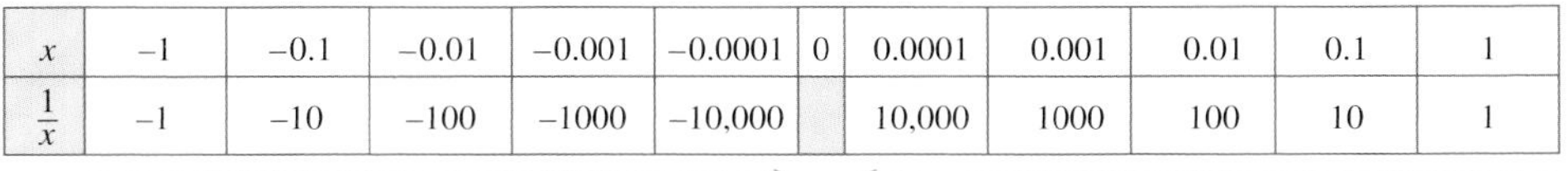

x	−1	−0.1	−0.01	−0.001	−0.0001	0	0.0001	0.001	0.01	0.1	1
$\frac{1}{x}$	−1	−10	−100	−1000	−10,000		10,000	1000	100	10	1

Left side → ← Right side

Figure 2.1.15

2.1.4 INFINITE LIMITS (AN INFORMAL VIEW). The expressions

$$\lim_{x \to a^-} f(x) = +\infty \quad \text{and} \quad \lim_{x \to a^+} f(x) = +\infty$$

denote that $f(x)$ increases without bound as x approaches a from the left and from the right, respectively. If both are true, then we write

$$\lim_{x \to a} f(x) = +\infty$$

Similarly, the expressions

$$\lim_{x \to a^-} f(x) = -\infty \quad \text{and} \quad \lim_{x \to a^+} f(x) = -\infty$$

denote that $f(x)$ decreases without bound as x approaches a from the left and from the right, respectively. If both are true, then we write

$$\lim_{x \to a} f(x) = -\infty$$

▶ **Example 7** For the functions in Figure 2.1.16, describe the limits at $x = a$ in appropriate limit notation.

Solution (a). In Figure 2.1.16a, the function increases without bound as x approaches a from the right and decreases without bound as x approaches a from the left. Thus,

$$\lim_{x \to a^+} \frac{1}{x-a} = +\infty \quad \text{and} \quad \lim_{x \to a^-} \frac{1}{x-a} = -\infty$$

Solution (b). In Figure 2.1.16b, the function increases without bound as x approaches a from both the left and right. Thus,

$$\lim_{x \to a} \frac{1}{(x-a)^2} = \lim_{x \to a^+} \frac{1}{(x-a)^2} = \lim_{x \to a^-} \frac{1}{(x-a)^2} = +\infty$$

Solution (c). In Figure 2.1.16c, the function decreases without bound as x approaches a from the right and increases without bound as x approaches a from the left. Thus,

$$\lim_{x \to a^+} \frac{-1}{x-a} = -\infty \quad \text{and} \quad \lim_{x \to a^-} \frac{-1}{x-a} = +\infty$$

Solution (d). In Figure 2.1.16d, the function decreases without bound as x approaches a from both the left and right. Thus,

$$\lim_{x \to a} \frac{-1}{(x-a)^2} = \lim_{x \to a^+} \frac{-1}{(x-a)^2} = \lim_{x \to a^-} \frac{-1}{(x-a)^2} = -\infty \ ◀$$

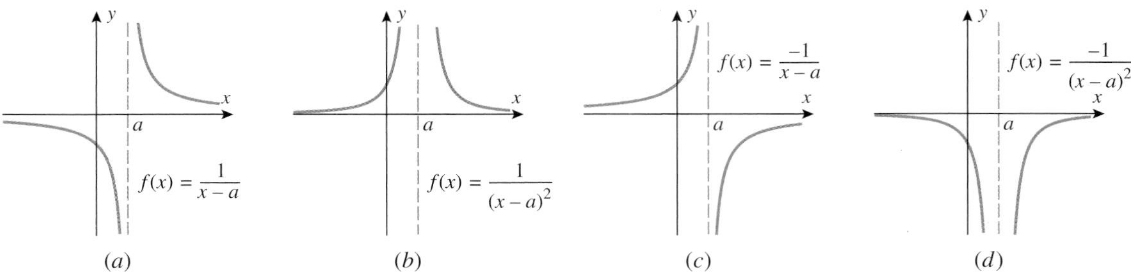

Figure 2.1.16

VERTICAL ASYMPTOTES

Figure 2.1.17 illustrates geometrically what happens when any of the following situations occur:

$$\lim_{x \to a^-} f(x) = +\infty, \quad \lim_{x \to a^+} f(x) = +\infty, \quad \lim_{x \to a^-} f(x) = -\infty, \quad \lim_{x \to a^+} f(x) = -\infty$$

In each case the graph of $y = f(x)$ either rises or falls without bound, squeezing closer and closer to the vertical line $x = a$ as x approaches a from the side indicated in the limit. The line $x = a$ is called a ***vertical asymptote*** of the curve $y = f(x)$ (from the Greek word *asymptotos*, meaning "nonintersecting").

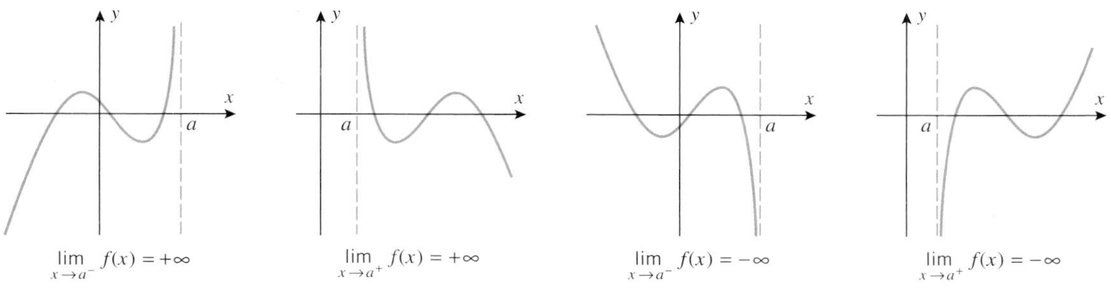

Figure 2.1.17

For the function in (16), find expressions for the left- and right-hand limits at each asymptote.

► **Example 8** Referring to Figure 1.4.11 we see that $x = -1$ and $x = 1$ are vertical asymptotes of the graph of

$$f(x) = \frac{x^2 + 2x}{x^2 - 1} \quad ◄ \tag{16}$$

✔ QUICK CHECK EXERCISES 2.1 *(See page 96 for answers.)*

1. Suppose that a function f has the property that for all real numbers x, the distance between $f(x)$ and 3 is at most $|x|$. From this we can conclude that $f(x) \to$ ______ as $x \to$ ______.

2. Use the accompanying graph of $y = f(x)$ $(-\infty < x < 3)$ to determine the limits.

(a) $\lim_{x \to 0} f(x) =$ ______

(b) $\lim_{x \to 2^-} f(x) =$ ______

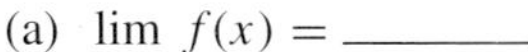

(c) $\lim_{x \to 2^+} f(x) =$ ______

(d) $\lim_{x \to 3^-} f(x) =$ ______

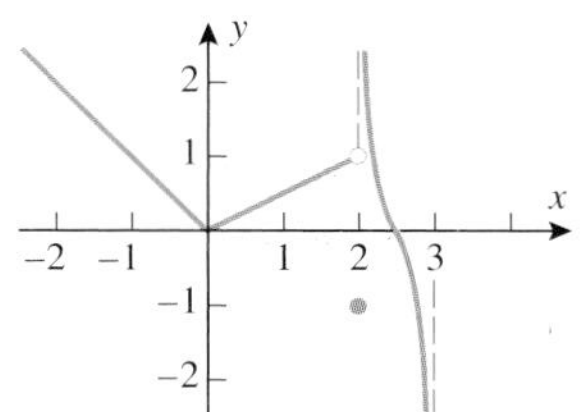

Figure Ex-2

3. The slope of the secant line through $P(2, 4)$ and $Q(x, x^2)$ on the parabola $y = x^2$ is $m_{\text{sec}} = x + 2$. It follows that the slope of the tangent line to this parabola at the point P is ______.

EXERCISE SET 2.1 Graphing Utility [C] CAS

1–6 In these exercises, make reasonable assumptions about the graph of the indicated function outside of the region depicted.

1. For the function F graphed in the accompanying figure (on the next page), find

(a) $\lim_{x \to -2^-} F(x)$ (b) $\lim_{x \to -2^+} F(x)$

(c) $\lim_{x \to -2} F(x)$ (d) $F(-2)$.

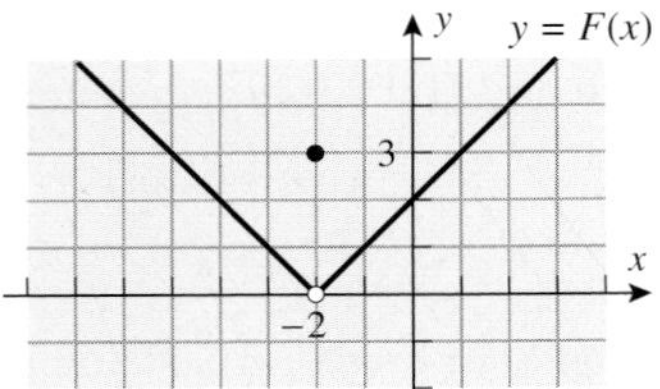

Figure Ex-1

2. For the function ϕ graphed in the accompanying figure, find
(a) $\lim_{x\to 4^-} \phi(x)$ (b) $\lim_{x\to 4^+} \phi(x)$
(c) $\lim_{x\to 4} \phi(x)$ (d) $\phi(4)$.

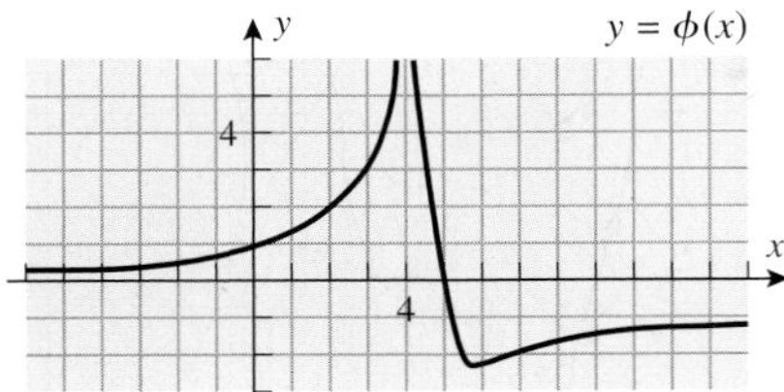

Figure Ex-2

3. For the function f graphed in the accompanying figure, find
(a) $\lim_{x\to 3^-} f(x)$ (b) $\lim_{x\to 3^+} f(x)$
(c) $\lim_{x\to 3} f(x)$ (d) $f(3)$.

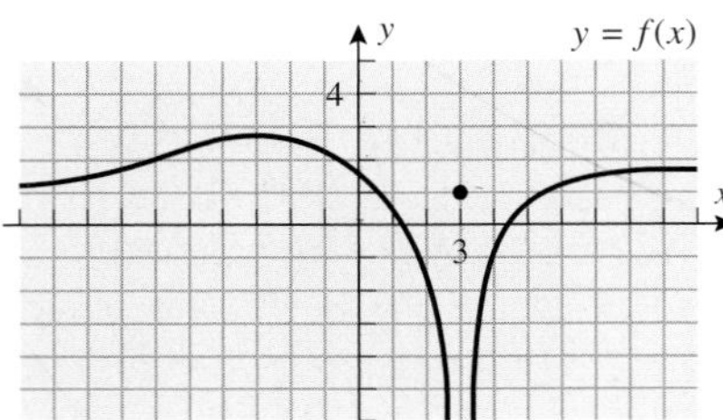

Figure Ex-3

4. For the function f graphed in the accompanying figure, find
(a) $\lim_{x\to 0^-} f(x)$ (b) $\lim_{x\to 0^+} f(x)$
(c) $\lim_{x\to 0} f(x)$ (d) $f(0)$.

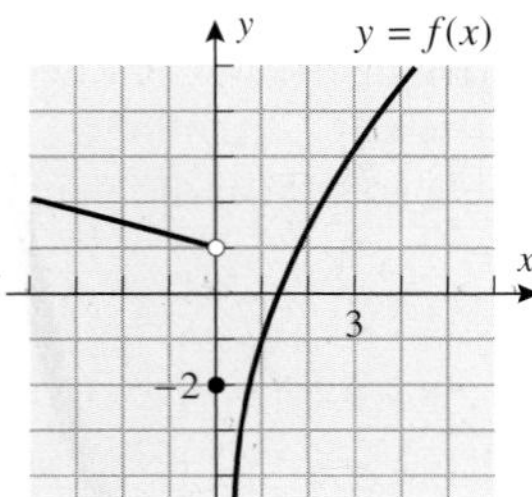

Figure Ex-4

5. Consider the function g graphed in the accompanying figure. For what values of x_0, $-7 \le x_0 \le 4$, does $\lim_{x\to x_0} g(x)$ exist?

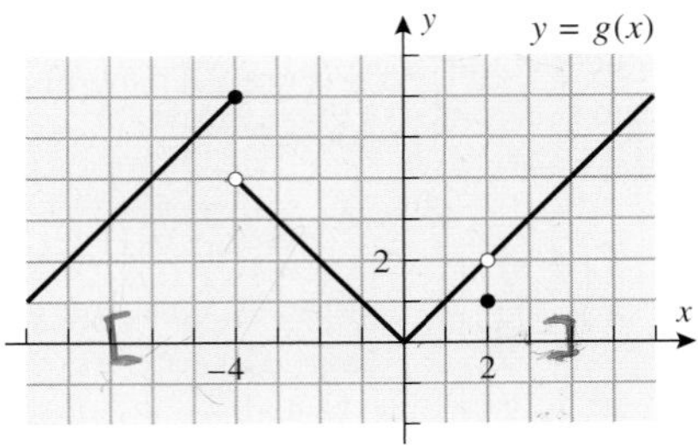

Figure Ex-5

6. Consider the function f graphed in the accompanying figure. For what values of x_0, $-9 \le x_0 \le 4$, does $\lim_{x\to x_0} f(x)$ exist?

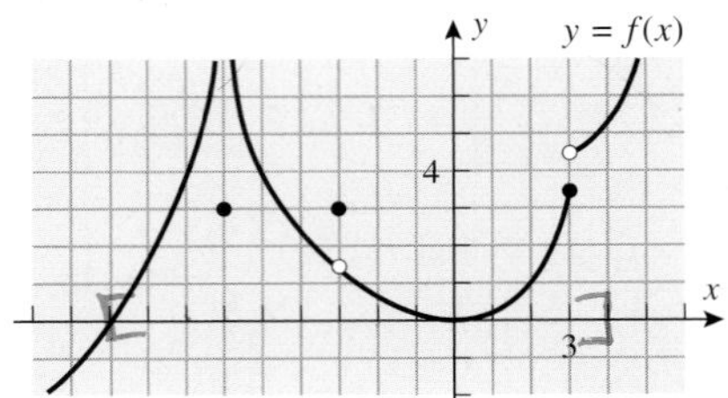

Figure Ex-6

FOCUS ON CONCEPTS

7–12 Sketch a possible graph for a function f with the specified properties. (Many different solutions are possible.)

7. (i) the domain of f is $[-1, 1]$
(ii) $f(-1) = f(0) = f(1) = 0$
(iii) $\lim_{x\to -1^+} f(x) = \lim_{x\to 0} f(x) = \lim_{x\to 1^-} f(x) = 1$

8. (i) the domain of f is $[-2, 1]$
(ii) $f(-2) = f(0) = f(1) = 0$
(iii) $\lim_{x\to -2^+} f(x) = 2$, $\lim_{x\to 0} f(x) = 0$, and $\lim_{x\to 1^-} f(x) = 1$

9. (i) the domain of f is $(-\infty, 0]$
(ii) $f(-2) = f(0) = 1$
(iii) $\lim_{x\to -2} f(x) = +\infty$

10. (i) the domain of f is $(0, +\infty)$
(ii) $f(1) = 0$
(iii) the y-axis is a vertical asymptote for the graph of f
(iv) $f(x) < 0$ if $0 < x < 1$

11. (i) $f(-3) = f(0) = f(2) = 0$
(ii) $\lim_{x\to -2^-} f(x) = +\infty$ and $\lim_{x\to -2^+} f(x) = -\infty$
(iii) $\lim_{x\to 1} f(x) = +\infty$

12. (i) $f(-1) = 0$, $f(0) = 1$, $f(1) = 0$
(ii) $\lim_{x\to -1^-} f(x) = 0$ and $\lim_{x\to -1^+} f(x) = +\infty$
(iii) $\lim_{x\to 1^-} f(x) = 1$ and $\lim_{x\to 1^+} f(x) = +\infty$

13–16 (i) Make a guess at the limit (if it exists) by evaluating the function at the specified x-values. (ii) Confirm your conclusions about the limit by graphing the function over an appropriate interval. (iii) If you have a CAS, then use it to find the limit. [*Note:* For the trigonometric functions, be sure to put your calculating and graphing utilities in radian mode.]

C **13.** (a) $\lim_{x\to 1} \frac{x-1}{x^3-1}$; $x = 2, 1.5, 1.1, 1.01, 1.001, 0, 0.5, 0.9, 0.99, 0.999$

(b) $\lim_{x\to 1^+} \frac{x+1}{x^3-1}$; $x = 2, 1.5, 1.1, 1.01, 1.001, 1.0001$

(c) $\lim_{x\to 1^-} \frac{x+1}{x^3-1}$; $x = 0, 0.5, 0.9, 0.99, 0.999, 0.9999$

C **14.** (a) $\lim_{x\to 0} \frac{\sqrt{x+1}-1}{x}$; $x = \pm 0.25, \pm 0.1, \pm 0.001, \pm 0.0001$

(b) $\lim_{x\to 0^+} \frac{\sqrt{x+1}+1}{x}$; $x = 0.25, 0.1, 0.001, 0.0001$

(c) $\lim_{x\to 0^-} \frac{\sqrt{x+1}+1}{x}$; $x = -0.25, -0.1, -0.001, -0.0001$

C **15.** (a) $\lim_{x\to 0} \frac{\sin 3x}{x}$; $x = \pm 0.25, \pm 0.1, \pm 0.001, \pm 0.0001$

(b) $\lim_{x\to -1} \frac{\cos x}{x+1}$; $x = 0, -0.5, -0.9, -0.99, -0.999, -1.5, -1.1, -1.01, -1.001$

C **16.** (a) $\lim_{x\to -1} \frac{\tan(x+1)}{x+1}$; $x = 0, -0.5, -0.9, -0.99, -0.999, -1.5, -1.1, -1.01, -1.001$

(b) $\lim_{x\to 0} \frac{\sin(5x)}{\sin(2x)}$; $x = \pm 0.25, \pm 0.1, \pm 0.001, \pm 0.0001$

17–20 Modify the argument of Example 1 to find the equation of the tangent line to the specified graph at the point given.

17. The graph of $y = x^2$ at $(-1, 1)$.

18. The graph of $y = x^2$ at $(0, 0)$.

19. The graph of $y = x^4$ at $(1, 1)$.

20. The graph of $y = x^4$ at $(-1, 1)$.

FOCUS ON CONCEPTS

21. (a) Let

$$f(x) = (1 + x^2)^{1.1/x^2}$$

Graph f in the window

$$[-1, 1] \times [2.5, 3.5]$$

and use the calculator's trace feature to make a conjecture about the limit of $f(x)$ as $x \to 0$.

(b) Graph f in the window

$$[-0.001, 0.001] \times [2.5, 3.5]$$

and use the calculator's trace feature to make a conjecture about the limit of $f(x)$ as $x \to 0$.

(c) Graph f in the window

$$[-0.000001, 0.000001] \times [2.5, 3.5]$$

and use the calculator's trace feature to make a conjecture about the limit of $f(x)$ as $x \to 0$.

(d) Later we will be able to show that

$$\lim_{x\to 0} (1 + x^2)^{1.1/x^2} \approx 3.00416602$$

What flaw do your graphs reveal about using numerical evidence (as revealed by the graphs you obtained) to make conjectures about limits?

Roundoff error is one source of inaccuracy in calculator and computer computations. Another source of error, called ***catastrophic subtraction***, occurs when two nearly equal numbers are subtracted, and the result is used as part of another calculation. For example, by hand calculation we have

$$(0.123456789012345 - 0.123456789012344) \times 10^{15} = 1$$

However, a calculator that can only store 14 decimal digits produces a value of 0 for this computation, since the numbers being subtracted are identical in the first 14 digits. Catastrophic subtraction can sometimes be avoided by rearranging formulas algebraically, but your best defense is to be aware that it can occur. Watch out for it in the next exercise.

C **22.** (a) Let

$$f(x) = \frac{x - \sin x}{x^3}$$

Make a conjecture about the limit of f as $x \to 0^+$ by evaluating $f(x)$ at $x = 0.1, 0.01, 0.001, 0.0001$.

(b) Evaluate the function $f(x)$ at $x = 0.000001$, 0.0000001, 0.00000001, 0.000000001, 0.0000000001, and make another conjecture.

(c) What flaw does this reveal about using numerical evidence to make conjectures about limits?

(d) If you have a CAS, use it to show that the exact value of the limit is $\frac{1}{6}$.

23. (a) The accompanying figure (next page) shows two different views of the graph of the function in Exercise 22, as generated by *Mathematica*. What is happening?

(b) Use your graphing utility to generate the graphs, and see whether the same problem occurs.

(c) Would you expect a similar problem to occur in the vicinity of $x = 0$ for the function

$$f(x) = \frac{1 - \cos x}{x}?$$

See if it does.

24. In the special theory of relativity the mass m of a moving object is a function $m = m(v)$ of the object's speed v. The accompanying figure (next page), in which c denotes the speed of light, displays some of the qualitative features of this function.

(a) What is the physical interpretation of m_0?

(b) What is $\lim_{v\to c^-} m(v)$? What is the physical significance of this limit?

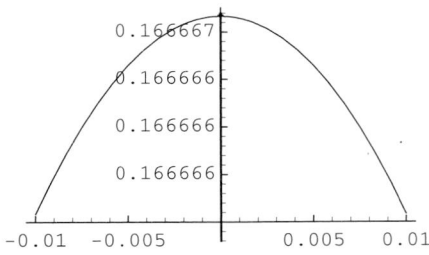

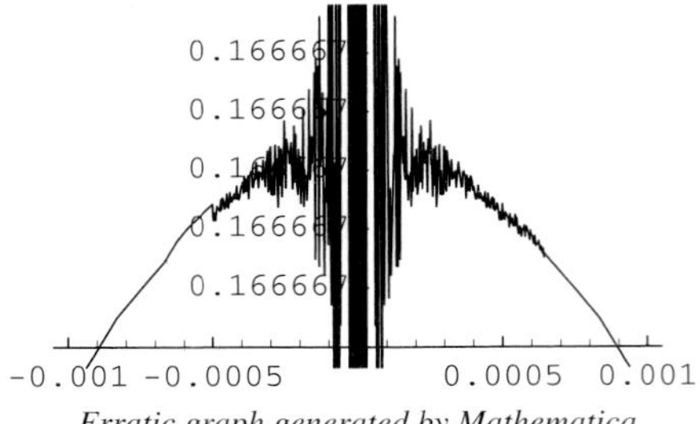

Erratic graph generated by Mathematica

Figure Ex-23

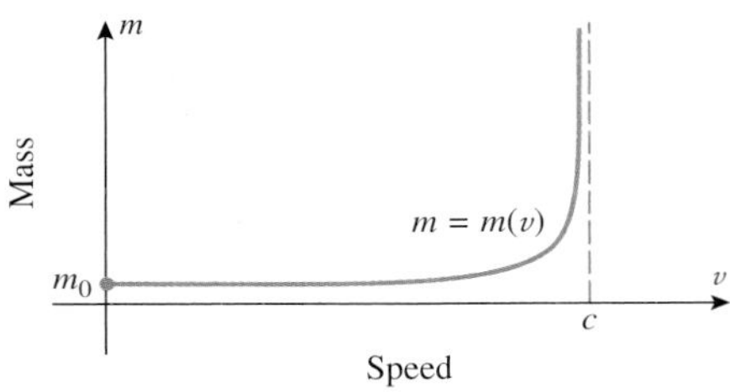

Figure Ex-24

25. In the special theory of relativity the length l of a narrow rod moving longitudinally is a function $l = l(v)$ of the rod's speed v. The accompanying figure, in which c denotes the speed of light, displays some of the qualitative features of this function.

(a) What is the physical interpretation of l_0?

(b) What is $\lim_{v \to c^-} l(v)$? What is the physical significance of this limit?

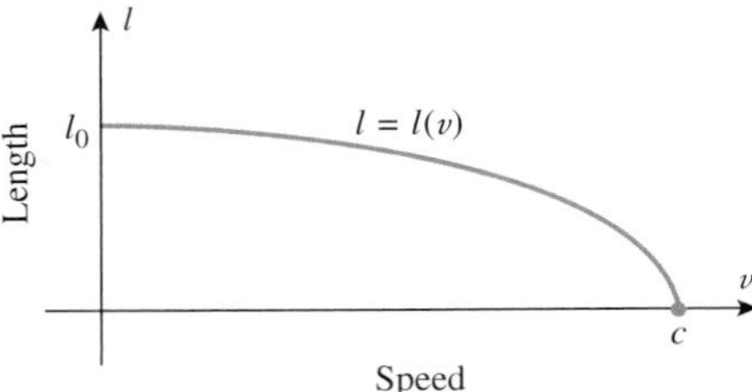

Figure Ex-25

✔QUICK CHECK ANSWERS 2.1

1. 3; 0 **2.** (a) 0 (b) 1 (c) $+\infty$ (d) $-\infty$ **3.** 4

2.2 COMPUTING LIMITS

In this section we will discuss algebraic techniques for computing limits of many functions. We base these results on the informal development of the limit concept discussed in the preceding section. A more formal derivation of these results is possible after Section 2.4.

SOME BASIC LIMITS

Our strategy for finding limits algebraically has two parts:

- First we will obtain the limits of some simple functions.
- Then we will develop a repertoire of theorems that will enable us to use the limits of those simple functions as building blocks for finding limits of more complicated functions.

We start with the following basic result, which is illustrated in Figure 2.2.1.

2.2.1 THEOREM. *Let a and k be real numbers.*

(*a*) $\lim_{x \to a} k = k$ (*b*) $\lim_{x \to a} x = a$ (*c*) $\lim_{x \to 0^-} \frac{1}{x} = -\infty$ (*d*) $\lim_{x \to 0^+} \frac{1}{x} = +\infty$

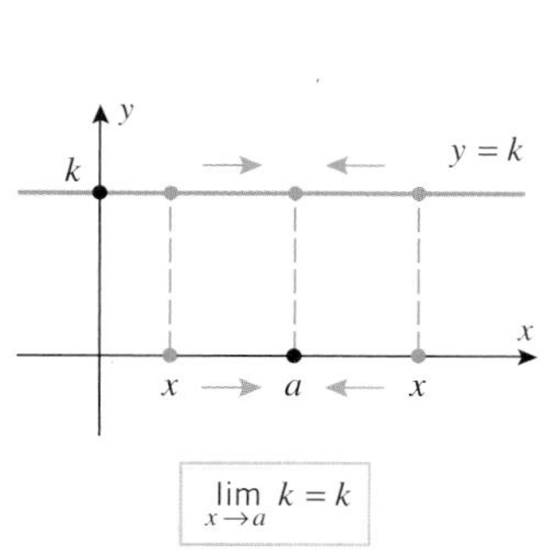

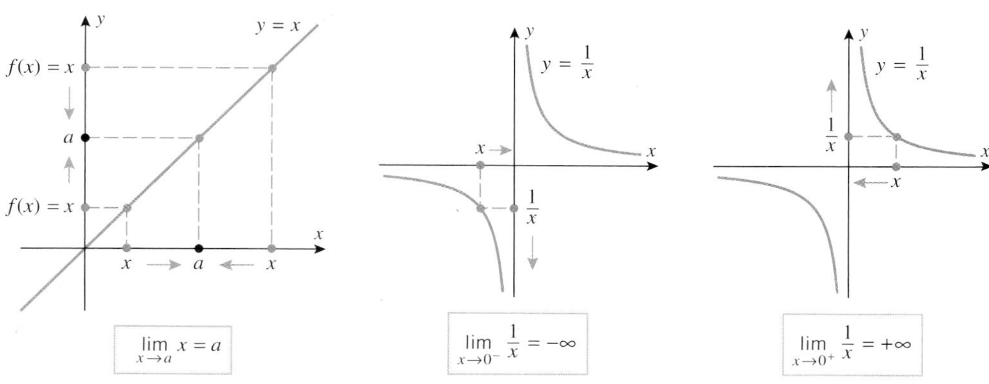

Figure 2.2.1

► **Example 1** If $f(x) = k$ is a constant function, then the values of $f(x)$ remain fixed at k as x varies, which explains why $f(x) \to k$ as $x \to a$ for all values of a. For example,

$$\lim_{x \to -25} 3 = 3, \qquad \lim_{x \to 0} 3 = 3, \qquad \lim_{x \to \pi} 3 = 3 \blacktriangleleft$$

► **Example 2** If $f(x) = x$, then as $x \to a$ it must also be true that $f(x) \to a$. For example,

$$\lim_{x \to 0} x = 0, \qquad \lim_{x \to -2} x = -2, \qquad \lim_{x \to \pi} x = \pi \blacktriangleleft$$

Do not confuse the algebraic size of a number with its closeness to zero. For positive numbers, the smaller the number the closer it is to zero, but for negative numbers, the larger the number the closer it is to zero. For example, -2 is larger than -4, but it is closer to zero.

► **Example 3** You should know from your experience with fractions that for a fixed nonzero numerator, the closer the denominator is to zero, the larger the absolute value of the fraction. This fact and the data in Table 2.2.1 suggest why $1/x \to +\infty$ as $x \to 0^+$ and why $1/x \to -\infty$ as $x \to 0^-$. ◄

Table 2.2.1

	VALUES						CONCLUSION
x	−1	−0.1	−0.01	−0.001	−0.0001	···	As $x \to 0^-$ the value of $1/x$
$1/x$	−1	−10	−100	−1000	−10,000	···	decreases without bound.
x	1	0.1	0.01	0.001	0.0001	···	As $x \to 0^+$ the value of $1/x$
$1/x$	1	10	100	1000	10,000	···	increases without bound.

The following theorem, parts of which are proved in Appendix C, will be our basic tool for finding limits algebraically.

2.2.2 THEOREM. *Let a be a real number, and suppose that*

$$\lim_{x \to a} f(x) = L_1 \quad \textit{and} \quad \lim_{x \to a} g(x) = L_2$$

That is, the limits exist and have values L_1 and L_2, respectively. Then:

(*a*) $\lim_{x \to a} [f(x) + g(x)] = \lim_{x \to a} f(x) + \lim_{x \to a} g(x) = L_1 + L_2$

(*b*) $\lim_{x \to a} [f(x) - g(x)] = \lim_{x \to a} f(x) - \lim_{x \to a} g(x) = L_1 - L_2$

(*c*) $\lim_{x \to a} [f(x)g(x)] = \left(\lim_{x \to a} f(x)\right)\left(\lim_{x \to a} g(x)\right) = L_1 L_2$

(*d*) $\lim_{x \to a} \dfrac{f(x)}{g(x)} = \dfrac{\lim_{x \to a} f(x)}{\lim_{x \to a} g(x)} = \dfrac{L_1}{L_2}$, *provided* $L_2 \neq 0$

(*e*) $\lim_{x \to a} \sqrt[n]{f(x)} = \sqrt[n]{\lim_{x \to a} f(x)} = \sqrt[n]{L_1}$, *provided* $L_1 > 0$ *if n is even.*

Moreover, these statements are also true for the one-sided limits as $x \to a^-$ *or as* $x \to a^+$.

This theorem can be stated informally as follows:

(*a*) *The limit of a sum is the sum of the limits.*
(*b*) *The limit of a difference is the difference of the limits.*
(*c*) *The limit of a product is the product of the limits.*
(*d*) *The limit of a quotient is the quotient of the limits, provided the limit of the denominator is not zero.*
(*e*) *The limit of an nth root is the nth root of the limit.*

For the special case of part (*c*) in which $f(x) = k$ is a constant function, we have

$$\lim_{x \to a} (kg(x)) = \lim_{x \to a} k \cdot \lim_{x \to a} g(x) = k \lim_{x \to a} g(x) \tag{1}$$

and similarly for one-sided limits. This result can be rephrased as:

A constant factor can be moved through a limit symbol.

Although parts (*a*) and (*c*) of Theorem 2.2.2 are stated for two functions, the results hold for any finite number of functions. Moreover, the various parts of the theorem can be used in combination to reformulate expressions involving limits.

► **Example 4**

$$\lim_{x\to a}[f(x) - g(x) + 2h(x)] = \lim_{x\to a} f(x) - \lim_{x\to a} g(x) + 2\lim_{x\to a} h(x)$$

$$\lim_{x\to a}[f(x)g(x)h(x)] = \left(\lim_{x\to a} f(x)\right)\left(\lim_{x\to a} g(x)\right)\left(\lim_{x\to a} h(x)\right)$$

$$\lim_{x\to a}[f(x)]^3 = \left(\lim_{x\to a} f(x)\right)^3$$

Take $g(x) = h(x) = f(x)$ in the last equation.

$$\lim_{x\to a}[f(x)]^n = \left(\lim_{x\to a} f(x)\right)^n$$

The extension of Theorem 2.2.2(*c*) in which there are n factors, each of which is $f(x)$

$$\lim_{x\to a} x^n = \left(\lim_{x\to a} x\right)^n = a^n$$

Apply the previous result with $f(x) = x$. ◄

LIMITS OF POLYNOMIALS AND RATIONAL FUNCTIONS AS $x \to a$

► **Example 5** Find $\lim_{x\to 5}(x^2 - 4x + 3)$.

Solution.

$$\begin{aligned}\lim_{x\to 5}(x^2 - 4x + 3) &= \lim_{x\to 5} x^2 - \lim_{x\to 5} 4x + \lim_{x\to 5} 3 && \text{Theorem 2.2.2(}a\text{), (}b\text{)}\\ &= \lim_{x\to 5} x^2 - 4\lim_{x\to 5} x + \lim_{x\to 5} 3 && \text{A constant can be moved through a limit symbol.}\\ &= 5^2 - 4(5) + 3 && \text{The last part of Example 4}\\ &= 8\end{aligned}$$

◄

Observe that in Example 5 the limit of the polynomial $p(x) = x^2 - 4x + 3$ as $x \to 5$ turned out to be the same as $p(5)$. This is not accidental: The next result shows that, in general, the limit of a polynomial $p(x)$ as $x \to a$ is the same as the value of the polynomial at a. Knowing this fact allows us to reduce the computation of limits of polynomials to simply evaluating the polynomial at the appropriate point.

2.2.3 THEOREM. *For any polynomial*

$$p(x) = c_0 + c_1x + \cdots + c_nx^n$$

and any real number a,

$$\lim_{x\to a} p(x) = c_0 + c_1a + \cdots + c_na^n = p(a)$$

PROOF.

$$\begin{aligned}\lim_{x\to a} p(x) &= \lim_{x\to a}\left(c_0 + c_1x + \cdots + c_nx^n\right)\\ &= \lim_{x\to a} c_0 + \lim_{x\to a} c_1x + \cdots + \lim_{x\to a} c_nx^n\\ &= \lim_{x\to a} c_0 + c_1\lim_{x\to a} x + \cdots + c_n\lim_{x\to a} x^n\\ &= c_0 + c_1a + \cdots + c_na^n = p(a)\end{aligned}$$

■

▶ Example 6 Find $\lim_{x \to 1} (x^7 - 2x^5 + 1)^{35}$.

Solution. The function involved is a polynomial (why?), so the limit can be obtained by evaluating this polynomial at $x = 1$. This yields

$$\lim_{x \to 1} (x^7 - 2x^5 + 1)^{35} = 0 \blacktriangleleft$$

Recall that a rational function is a ratio of two polynomials. The following example illustrates how Theorems 2.2.2(*d*) and 2.2.3 can sometimes be used in combination to compute limits of rational functions.

▶ Example 7 Find $\lim_{x \to 2} \dfrac{5x^3 + 4}{x - 3}$.

Solution.

$$\lim_{x \to 2} \frac{5x^3 + 4}{x - 3} = \frac{\lim_{x \to 2} (5x^3 + 4)}{\lim_{x \to 2} (x - 3)} \quad \boxed{\text{Theorem 2.2.2}(d)}$$

$$= \frac{5 \cdot 2^3 + 4}{2 - 3} = -44 \quad \boxed{\text{Theorem 2.2.3}} \blacktriangleleft$$

The method used in the last example will not work for rational functions in which the limit of the denominator is zero because Theorem 2.2.2(*d*) is not applicable. There are two cases of this type to be considered—the case where the limit of the denominator is zero and the limit of the numerator is not, and the case where the limits of the numerator and denominator are both zero. If the limit of the denominator is zero but the limit of the numerator is not, then one can prove that the limit of the rational function does not exist and that one of the following situations occurs:

- The limit may be $-\infty$.
- The limit may be $+\infty$.
- The limit may be $-\infty$ from one side and $+\infty$ from the other.

Figure 2.2.2 illustrates these three possibilities graphically for rational functions of the form $1/(x - a)$, $1/(x - a)^2$, and $-1/(x - a)^2$.

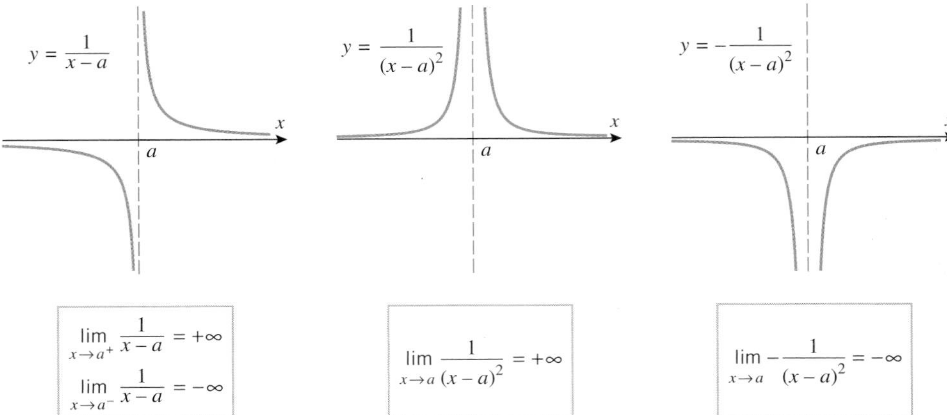

Figure 2.2.2

▶ Example 8 Find

(a) $\lim_{x\to 4^+} \dfrac{2-x}{(x-4)(x+2)}$ (b) $\lim_{x\to 4^-} \dfrac{2-x}{(x-4)(x+2)}$ (c) $\lim_{x\to 4} \dfrac{2-x}{(x-4)(x+2)}$

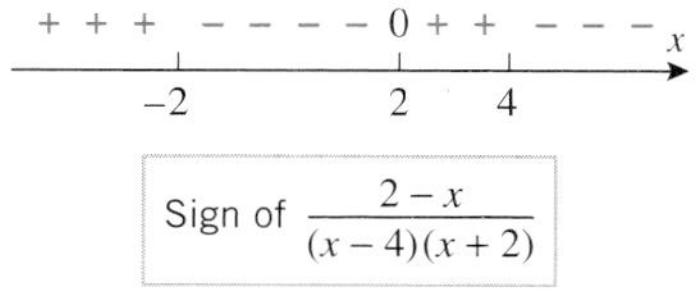

+ + + − − − − 0 + + − − − x

−2 2 4

Sign of $\dfrac{2-x}{(x-4)(x+2)}$

Figure 2.2.3

Solution. In all three parts the limit of the numerator is −2, and the limit of the denominator is 0, so the limit of the ratio does not exist. To be more specific than this, we need to analyze the sign of the ratio. The sign of the ratio, which is given in Figure 2.2.3, is determined by the signs of $2-x$, $x-4$, and $x+2$. (The method of test points, discussed in Web Appendix D, provides a way of finding the sign of the ratio here.) It follows from this figure that as x approaches 4 from the right, the ratio is always negative; and as x approaches 4 from the left, the ratio is eventually positive. Thus,

$$\lim_{x\to 4^+} \frac{2-x}{(x-4)(x+2)} = -\infty \quad\text{and}\quad \lim_{x\to 4^-} \frac{2-x}{(x-4)(x+2)} = +\infty$$

Because the one-sided limits have opposite signs, all we can say about the two-sided limit is that it does not exist. ◀

In the case where $p(x)/q(x)$ is a rational function for which $p(a)=0$ and $q(a)=0$, the numerator and denominator must have one or more common factors of $x-a$. In this case the limit of $p(x)/q(x)$ as $x\to a$ can be found by canceling all common factors of $x-a$ and using one of the methods already considered to find the limit of the simplified function. Here are some examples.

In Example 9, the simplified function $x+2$ is defined at $x=2$ but the original function is not. However, this has no effect on the limit as x *approaches* 2 since the two functions are identical if $x \neq 2$.

▶ Example 9 Find $\lim_{x\to 2} \dfrac{x^2-4}{x-2}$.

Solution. Since 2 is a zero of both the numerator and the denominator, they share a common factor of $x-2$. The limit can be obtained as follows:

$$\lim_{x\to 2} \frac{x^2-4}{x-2} = \lim_{x\to 2} \frac{(x-2)(x+2)}{x-2} = \lim_{x\to 2}(x+2) = 4 \ ◀$$

▶ Example 10 Find

(a) $\lim_{x\to 3} \dfrac{x^2-6x+9}{x-3}$ (b) $\lim_{x\to -4} \dfrac{2x+8}{x^2+x-12}$ (c) $\lim_{x\to 5} \dfrac{x^2-3x-10}{x^2-10x+25}$

Solution (a). The numerator and the denominator both have a zero at $x=3$, so there is a common factor of $x-3$. Then

$$\lim_{x\to 3} \frac{x^2-6x+9}{x-3} = \lim_{x\to 3} \frac{(x-3)^2}{x-3} = \lim_{x\to 3}(x-3) = 0$$

Solution (b). The numerator and the denominator both have a zero at $x=-4$, so there is a common factor of $x-(-4) = x+4$. Then

$$\lim_{x\to -4} \frac{2x+8}{x^2+x-12} = \lim_{x\to -4} \frac{2(x+4)}{(x+4)(x-3)} = \lim_{x\to -4} \frac{2}{x-3} = -\frac{2}{7}$$

Solution (c). The numerator and the denominator both have a zero at $x=5$, so there is a common factor of $x-5$. Then

$$\lim_{x\to 5} \frac{x^2-3x-10}{x^2-10x+25} = \lim_{x\to 5} \frac{(x-5)(x+2)}{(x-5)(x-5)} = \lim_{x\to 5} \frac{x+2}{x-5}$$

However,

$$\lim_{x \to 5}(x+2) = 7 \neq 0 \quad \text{and} \quad \lim_{x \to 5}(x-5) = 0$$

so

$$\lim_{x \to 5} \frac{x^2 - 3x - 10}{x^2 - 10x + 25} = \lim_{x \to 5} \frac{x+2}{x-5}$$

does not exist. More precisely, the sign analysis in Figure 2.2.4 implies that

$$\lim_{x \to 5^+} \frac{x^2 - 3x - 10}{x^2 - 10x + 25} = \lim_{x \to 5^+} \frac{x+2}{x-5} = +\infty$$

and

$$\lim_{x \to 5^-} \frac{x^2 - 3x - 10}{x^2 - 10x + 25} = \lim_{x \to 5^-} \frac{x+2}{x-5} = -\infty \blacktriangleleft$$

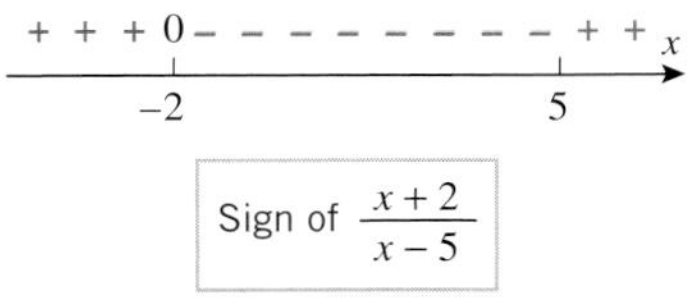

+ + + 0 − − − − − − − − − + + x

−2 5

Sign of $\frac{x+2}{x-5}$

Figure 2.2.4

A quotient $f(x)/g(x)$ in which the numerator and denominator both have a limit of zero as $x \to a$ is called an ***indeterminate form of type* $0/0$**. The problem with such limits is that it is difficult to tell by inspection whether the limit exists, and, if so, its value. Informally stated, this is because there are two conflicting influences at work: The value of $f(x)/g(x)$ would tend to zero as $f(x)$ approached zero if $g(x)$ were to remain at some fixed nonzero value, whereas the value of this ratio would tend to increase or decrease without bound as $g(x)$ approached zero if $f(x)$ were to remain at some fixed nonzero value. But with both $f(x)$ and $g(x)$ approaching zero, the behavior of the ratio depends on precisely how these conflicting tendencies offset one another for the particular f and g.

Discuss the logical errors in the following statement: "An indeterminate form of type $0/0$ must have a limit of zero because zero divided by anything is zero."

Sometimes, limits of indeterminate forms of type $0/0$ can be found by algebraic simplification, as in the last two examples, but frequently this will not work and other methods must be used. We will study such methods in later sections.

The following theorem summarizes our observations about limits of rational functions.

2.2.4 THEOREM. *Let*

$$f(x) = \frac{p(x)}{q(x)}$$

be a rational function, and let a be any real number.

(a) If $q(a) \neq 0$, then $\lim_{x \to a} f(x) = f(a)$.

(b) If $q(a) = 0$ but $p(a) \neq 0$, then $\lim_{x \to a} f(x)$ does not exist.

LIMITS INVOLVING RADICALS

▶ **Example 11** Find $\lim_{x \to 1} \dfrac{x-1}{\sqrt{x}-1}$.

Solution. In Example 2 of Section 2.1 we used numerical evidence to conjecture that this limit is 2. Here we will confirm this algebraically. Since this limit is an indeterminate form of type $0/0$, we will need to devise some strategy for making the limit (if it exists) evident. One such strategy is to rationalize the denominator of the function. This yields

$$\frac{x-1}{\sqrt{x}-1} = \frac{(x-1)(\sqrt{x}+1)}{(\sqrt{x}-1)(\sqrt{x}+1)} = \frac{(x-1)(\sqrt{x}+1)}{x-1} = \sqrt{x}+1 \quad (x \neq 1)$$

Therefore,

$$\lim_{x \to 1} \frac{x-1}{\sqrt{x}-1} = \lim_{x \to 1} (\sqrt{x}+1) = 2 \blacktriangleleft$$

LIMITS OF PIECEWISE-DEFINED FUNCTIONS

For functions that are defined piecewise, a two-sided limit at a point where the formula changes is best obtained by first finding the one-sided limits at that point.

▶ Example 12 Let

$$f(x) = \begin{cases} 1/(x+2), & x < -2 \\ x^2 - 5, & -2 < x \le 3 \\ \sqrt{x+13}, & x > 3 \end{cases}$$

Find

$$\text{(a)} \lim_{x \to -2} f(x) \qquad \text{(b)} \lim_{x \to 0} f(x) \qquad \text{(c)} \lim_{x \to 3} f(x)$$

Solution (a). We will determine the stated two-sided limit by first considering the corresponding one-sided limits. For each one-sided limit, we must use that part of the formula that is applicable on the interval over which x varies. For example, as x approaches -2 from the left, the applicable part of the formula is

$$f(x) = \frac{1}{x+2}$$

and as x approaches -2 from the right, the applicable part of the formula near -2 is

$$f(x) = x^2 - 5$$

Thus,

$$\lim_{x \to -2^-} f(x) = \lim_{x \to -2^-} \frac{1}{x+2} = -\infty$$

$$\lim_{x \to -2^+} f(x) = \lim_{x \to -2^+} (x^2 - 5) = (-2)^2 - 5 = -1$$

from which it follows that $\lim_{x \to -2} f(x)$ does not exist.

Solution (b). The applicable part of the formula is $f(x) = x^2 - 5$ on both sides of 0, so there is no need to consider one-sided limits here. We see directly that

$$\lim_{x \to 0} f(x) = \lim_{x \to 0} (x^2 - 5) = 0^2 - 5 = -5$$

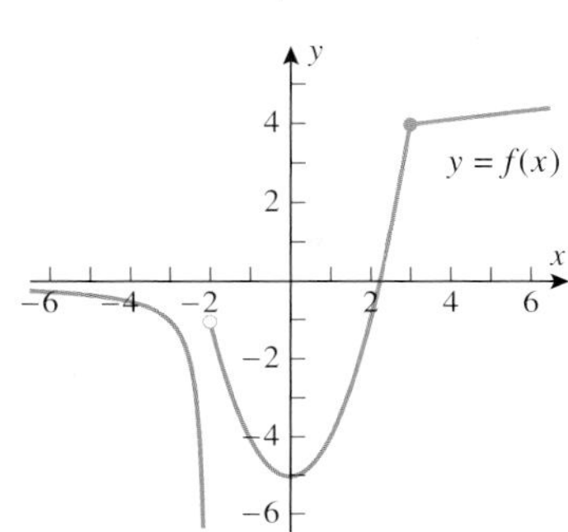

Figure 2.2.5

Solution (c). Using the applicable parts of the formula for $f(x)$, we obtain

$$\lim_{x \to 3^-} f(x) = \lim_{x \to 3^-} (x^2 - 5) = 3^2 - 5 = 4$$

$$\lim_{x \to 3^+} f(x) = \lim_{x \to 3^+} \sqrt{x+13} = \sqrt{\lim_{x \to 3^+} (x+13)} = \sqrt{3+13} = 4$$

Since the one-sided limits are equal, we have

$$\lim_{x \to 3} f(x) = 4$$

We note that the limit calculations in parts (a), (b), and (c) are consistent with the graph of f shown in Figure 2.2.5. ◀

✔ QUICK CHECK EXERCISES 2.2 (See page 105 for answers.)

1. In each part, find the limit by inspection.

(a) $\lim_{x \to 8} 7 =$ ______ (b) $\lim_{y \to 3^+} 12y =$ ______

(c) $\lim_{x \to 0^-} \dfrac{x}{|x|} =$ ______ (d) $\lim_{w \to 5} \dfrac{w}{|w|} =$ ______

(e) $\lim_{z \to 1^-} \dfrac{1}{1-z} =$ ______

2. Given that $\lim_{x \to a} f(x) = 1$ and $\lim_{x \to a} g(x) = 2$, find the limits that exist.

(a) $\lim_{x \to a} [3f(x) + 2g(x)] =$ ______

(b) $\lim_{x \to a} \dfrac{2f(x)+1}{1 - f(x)g(x)} =$ ______

(c) $\lim_{x \to a} \dfrac{\sqrt{f(x)+3}}{g(x)} =$ ______

(d) $\lim_{x \to a} \dfrac{f(x) - g(x)}{4 - [g(x)]^2} =$ ______

3. Find the limits.

(a) $\lim_{x \to -1} (x^3 + x^2 + x)^{101} =$ ______

(b) $\lim_{x \to 2^-} \dfrac{(x-1)(x-2)}{x+1} =$ ______

(c) $\lim_{x \to -1^+} \dfrac{(x-1)(x-2)}{x+1} =$ ______

(d) $\lim_{x \to 4} \dfrac{x^2 - 16}{x - 4} =$ ______

4. Let

$$f(x) = \begin{cases} x+1, & x \le 1 \\ x-1, & x > 1 \end{cases}$$

Find the limits that exist.

(a) $\lim_{x \to 1^-} f(x) =$ ______

(b) $\lim_{x \to 1^+} f(x) =$ ______

(c) $\lim_{x \to 1} f(x) =$ ______

EXERCISE SET 2.2

FOCUS ON CONCEPTS

1. Given that

$$\lim_{x \to a} f(x) = 2, \quad \lim_{x \to a} g(x) = -4, \quad \lim_{x \to a} h(x) = 0$$

find the limits that exist. If the limit does not exist, explain why.

(a) $\lim_{x \to a} [f(x) + 2g(x)]$

(b) $\lim_{x \to a} [h(x) - 3g(x) + 1]$

(c) $\lim_{x \to a} [f(x)g(x)]$

(d) $\lim_{x \to a} [g(x)]^2$

(e) $\lim_{x \to a} \sqrt[3]{6 + f(x)}$

(f) $\lim_{x \to a} \dfrac{2}{g(x)}$

(g) $\lim_{x \to a} \dfrac{3f(x) - 8g(x)}{h(x)}$

(h) $\lim_{x \to a} \dfrac{7g(x)}{2f(x) + g(x)}$

2. Use the graphs of f and g in the accompanying figure to find the limits that exist. If the limit does not exist, explain why.

(a) $\lim_{x \to 2} [f(x) + g(x)]$ (b) $\lim_{x \to 0} [f(x) + g(x)]$

(c) $\lim_{x \to 0^+} [f(x) + g(x)]$ (d) $\lim_{x \to 0^-} [f(x) + g(x)]$

(e) $\lim_{x \to 2} \dfrac{f(x)}{1 + g(x)}$ (f) $\lim_{x \to 2} \dfrac{1 + g(x)}{f(x)}$

(g) $\lim_{x \to 0^+} \sqrt{f(x)}$ (h) $\lim_{x \to 0^-} \sqrt{f(x)}$

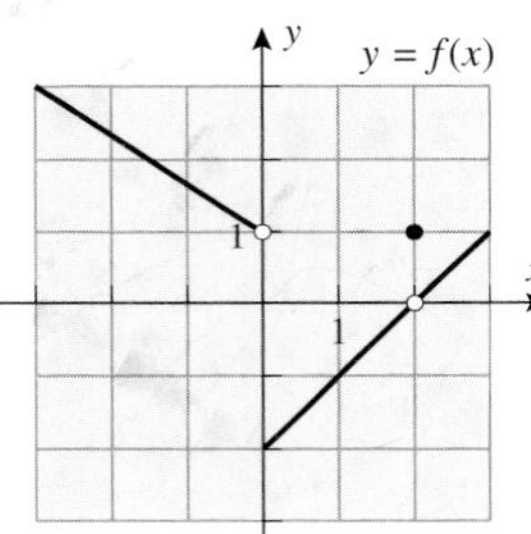

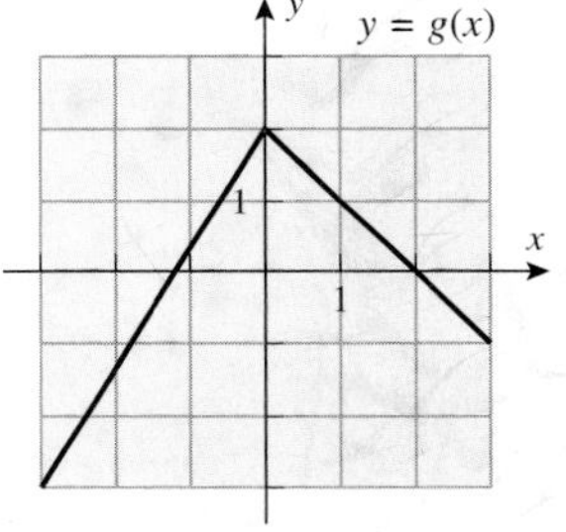

Figure Ex-2

3–30 Find the limits.

3. $\lim_{x \to 2} x(x-1)(x+1)$

4. $\lim_{x \to 3} (x^3 - 3x^2 + 9x)$

5. $\lim_{x \to 3} \dfrac{x^2 - 2x}{x + 1}$

6. $\lim_{x \to 0} \dfrac{6x - 9}{x^3 - 12x + 3}$

7. $\lim_{x \to 1^+} \dfrac{x^4 - 1}{x - 1}$

8. $\lim_{t \to -2} \dfrac{t^3 + 8}{t + 2}$

9. $\lim_{x \to -1} \dfrac{x^2 + 6x + 5}{x^2 - 3x - 4}$

10. $\lim_{x \to 2} \dfrac{x^2 - 4x + 4}{x^2 + x - 6}$

11. $\lim_{x \to -1} \dfrac{2x^2 + x - 1}{x + 1}$

12. $\lim_{x \to 1} \dfrac{3x^2 - x - 2}{2x^2 + x - 3}$

13. $\lim_{t \to 2} \dfrac{t^3 + 3t^2 - 12t + 4}{t^3 - 4t}$

14. $\lim_{t \to 1} \dfrac{t^3 + t^2 - 5t + 3}{t^3 - 3t + 2}$

15. $\lim_{x \to 3^+} \dfrac{x}{x - 3}$

16. $\lim_{x \to 3^-} \dfrac{x}{x - 3}$

17. $\lim_{x \to 3} \dfrac{x}{x - 3}$

18. $\lim_{x \to 2^+} \dfrac{x}{x^2 - 4}$

19. $\lim_{x \to 2^-} \frac{x}{x^2 - 4}$

20. $\lim_{x \to 2} \frac{x}{x^2 - 4}$

21. $\lim_{y \to 6^+} \frac{y + 6}{y^2 - 36}$

22. $\lim_{y \to 6^-} \frac{y + 6}{y^2 - 36}$

23. $\lim_{y \to 6} \frac{y + 6}{y^2 - 36}$

24. $\lim_{x \to 4^+} \frac{3 - x}{x^2 - 2x - 8}$

25. $\lim_{x \to 4^-} \frac{3 - x}{x^2 - 2x - 8}$

26. $\lim_{x \to 4} \frac{3 - x}{x^2 - 2x - 8}$

27. $\lim_{x \to 2^+} \frac{1}{|2 - x|}$

28. $\lim_{x \to 3^-} \frac{1}{|x - 3|}$

29. $\lim_{x \to 9} \frac{x - 9}{\sqrt{x} - 3}$

30. $\lim_{y \to 4} \frac{4 - y}{2 - \sqrt{y}}$

31. Let

$$f(x) = \begin{cases} x - 1, & x \leq 3 \\ 3x - 7, & x > 3 \end{cases}$$

Find

(a) $\lim_{x \to 3^-} f(x)$ (b) $\lim_{x \to 3^+} f(x)$ (c) $\lim_{x \to 3} f(x)$.

32. Let

$$g(t) = \begin{cases} t^2, & t \geq 0 \\ t - 2, & t < 0 \end{cases}$$

Find

(a) $\lim_{t \to 0^-} g(t)$ (b) $\lim_{t \to 0^+} g(t)$ (c) $\lim_{t \to 0} g(t)$.

FOCUS ON CONCEPTS

33. Let

$$f(x) = \frac{x^3 - 1}{x - 1}$$

(a) Find $\lim_{x \to 1} f(x)$.
(b) Sketch the graph of $y = f(x)$.

34. Let

$$f(x) = \begin{cases} \dfrac{x^2 - 9}{x + 3}, & x \neq -3 \\ k, & x = -3 \end{cases}$$

(a) Find k so that $f(-3) = \lim_{x \to -3} f(x)$.
(b) With k assigned the value $\lim_{x \to -3} f(x)$, show that $f(x)$ can be expressed as a polynomial.

35. (a) Explain why the following calculation is incorrect.

$$\lim_{x \to 0^+} \left(\frac{1}{x} - \frac{1}{x^2} \right) = \lim_{x \to 0^+} \frac{1}{x} - \lim_{x \to 0^+} \frac{1}{x^2}$$
$$= +\infty - (+\infty) = 0$$

(b) Show that $\lim_{x \to 0^+} \left(\frac{1}{x} - \frac{1}{x^2} \right) = -\infty$.

36. Find $\lim_{x \to 0^-} \left(\frac{1}{x} + \frac{1}{x^2} \right)$.

37–38 First rationalize the numerator and then find the limit.

37. $\lim_{x \to 0} \frac{\sqrt{x + 4} - 2}{x}$

38. $\lim_{x \to 0} \frac{\sqrt{x^2 + 4} - 2}{x}$

39. Let $p(x)$ and $q(x)$ be polynomials, and suppose $q(x_0) = 0$. Discuss the behavior of the graph of $y = p(x)/q(x)$ in the vicinity of $x = x_0$. Give examples to support your conclusions.

40. Let

$$f(x) = \frac{(a + b)x + (a - b)|x|}{2x}$$

Assuming that a and b are constants, find

(a) $\lim_{x \to 0^-} f(x)$ (b) $\lim_{x \to 0^+} f(x)$

(c) all values of a and b such that $\lim_{x \to 0} f(x)$ exists.

✔ QUICK CHECK ANSWERS 2.2

1. (a) 7 (b) 36 (c) −1 (d) 1 (e) $+\infty$ **2.** (a) 7 (b) −3 (c) 1 (d) does not exist **3.** (a) −1 (b) 0 (c) $+\infty$ (d) 8 **4.** (a) 2 (b) 0 (c) does not exist

2.3 LIMITS AT INFINITY; END BEHAVIOR OF A FUNCTION

Up to now we have been concerned with limits that describe the behavior of a function $f(x)$ as x approaches some real number a. In this section we will be concerned with the behavior of $f(x)$ as x increases or decreases without bound.

LIMITS AT INFINITY AND HORIZONTAL ASYMPTOTES

If the values of a variable x increase without bound, then we write $x \to +\infty$, and if the values of x decrease without bound, then we write $x \to -\infty$. The behavior of a function $f(x)$ as x increases without bound or decreases without bound is sometimes called the ***end***

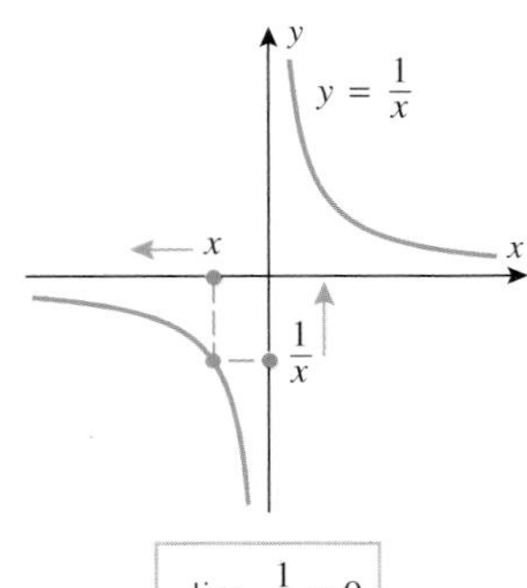

$\lim_{x\to-\infty} \frac{1}{x} = 0$

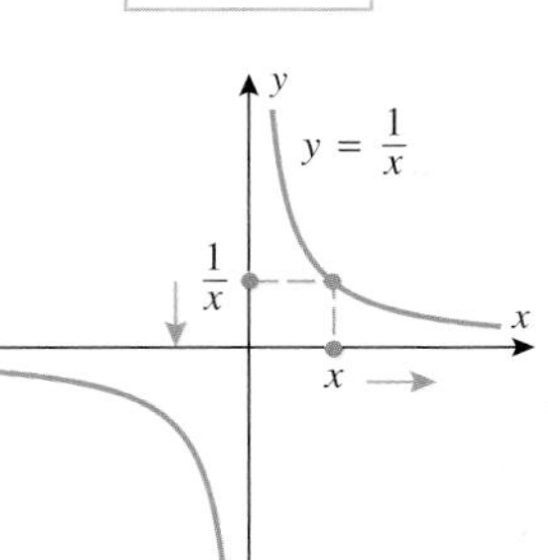

$\lim_{x\to+\infty} \frac{1}{x} = 0$

Figure 2.3.1

behavior of the function. For example,

$$\lim_{x\to-\infty} \frac{1}{x} = 0 \quad \text{and} \quad \lim_{x\to+\infty} \frac{1}{x} = 0 \tag{1–2}$$

are illustrated numerically in Table 2.3.1 and geometrically in Figure 2.3.1.

Table 2.3.1

	VALUES						CONCLUSION
x	−1	−10	−100	−1000	−10,000	···	As $x \to -\infty$ the value of $1/x$ increases toward zero.
$1/x$	−1	−0.1	−0.01	−0.001	−0.0001	···	
x	1	10	100	1000	10,000	···	As $x \to +\infty$ the value of $1/x$ decreases toward zero.
$1/x$	1	0.1	0.01	0.001	0.0001	···	

In general, we will use the following notation.

2.3.1 LIMITS AT INFINITY (AN INFORMAL VIEW). If the values of $f(x)$ eventually get as close as we like to a number L as x increases without bound, then we write

$$\lim_{x\to+\infty} f(x) = L \quad \text{or} \quad f(x) \to L \text{ as } x \to +\infty \tag{3}$$

Similarly, if the values of $f(x)$ eventually get as close as we like to a number L as x decreases without bound, then we write

$$\lim_{x\to-\infty} f(x) = L \quad \text{or} \quad f(x) \to L \text{ as } x \to -\infty \tag{4}$$

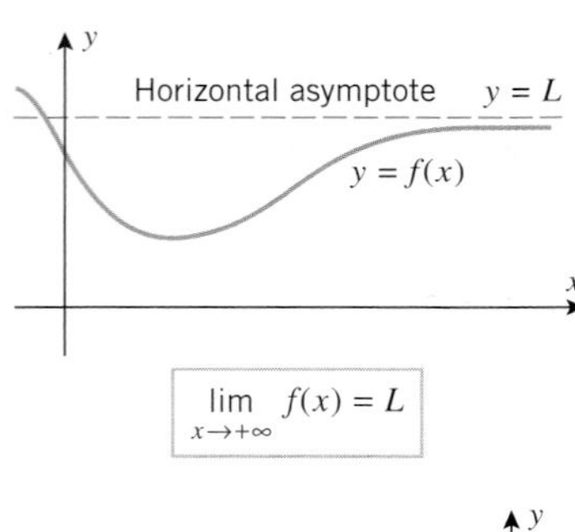

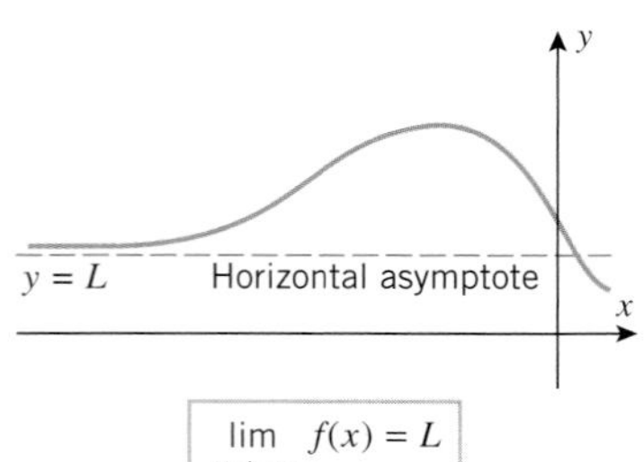

Figure 2.3.2

Figure 2.3.2 illustrates the end behavior of a function f when

$$\lim_{x\to+\infty} f(x) = L \quad \text{or} \quad \lim_{x\to-\infty} f(x) = L$$

In the first case the graph of f eventually squeezes as close as we like to the line $y = L$ as x increases without bound, and in the second case it eventually squeezes as close as we like to the line $y = L$ as x decreases without bound. If either limit holds, we call the line $y = L$ a ***horizontal asymptote*** for the graph of f.

► **Example 1** It follows from (1) and (2) that $y = 0$ is a horizontal asymptote for the graph of $f(x) = 1/x$ in both the positive and negative directions. This is consistent with the graph of $y = 1/x$ shown in Figure 2.3.1. ◄

LIMIT LAWS FOR LIMITS AT INFINITY

It can be shown that the limit laws in Theorem 2.2.2 carry over without change to limits at $+\infty$ and $-\infty$. Moreover, it follows by the same argument used in Section 2.2 that if n is a positive integer, then

$$\lim_{x\to+\infty} (f(x))^n = \left(\lim_{x\to+\infty} f(x)\right)^n \qquad \lim_{x\to-\infty} (f(x))^n = \left(\lim_{x\to-\infty} f(x)\right)^n \tag{5–6}$$

provided the indicated limit of $f(x)$ exists. It also follows that constants can be moved through the limit symbols for limits at infinity:

$$\lim_{x\to+\infty} kf(x) = k \lim_{x\to+\infty} f(x) \qquad \lim_{x\to-\infty} kf(x) = k \lim_{x\to-\infty} f(x) \tag{7–8}$$

provided the indicated limit of $f(x)$ exists.

Finally, if $f(x) = k$ is a constant function, then the values of f do not change as $x \to +\infty$ or as $x \to -\infty$, so

$$\lim_{x\to+\infty} k = k \qquad \lim_{x\to-\infty} k = k \tag{9–10}$$

▶ Example 2 It follows from (1), (2), (5), and (6) that if n is a positive integer, then

$$\lim_{x\to+\infty} \frac{1}{x^n} = \left(\lim_{x\to+\infty} \frac{1}{x}\right)^n = 0 \quad \text{and} \quad \lim_{x\to-\infty} \frac{1}{x^n} = \left(\lim_{x\to-\infty} \frac{1}{x}\right)^n = 0 \ ◀$$

INFINITE LIMITS AT INFINITY

Limits at infinity, like limits at a real number a, can fail to exist for various reasons. One such possibility is that the values of $f(x)$ increase or decrease without bound as $x \to +\infty$ or as $x \to -\infty$. We will use the following notation to describe this situation.

> **2.3.2 INFINITE LIMITS AT INFINITY (AN INFORMAL VIEW).** If the values of $f(x)$ increase without bound as $x \to +\infty$ or as $x \to -\infty$, then we write
>
> $$\lim_{x\to+\infty} f(x) = +\infty \quad \text{or} \quad \lim_{x\to-\infty} f(x) = +\infty$$
>
> as appropriate; and if the values of $f(x)$ decrease without bound as $x \to +\infty$ or as $x \to -\infty$, then we write
>
> $$\lim_{x\to+\infty} f(x) = -\infty \quad \text{or} \quad \lim_{x\to-\infty} f(x) = -\infty$$
>
> as appropriate.

LIMITS OF x^n AS $x \to \pm\infty$

Figure 2.3.3 illustrates the end behavior of the polynomials x^n for $n = 1, 2, 3$, and 4. These are special cases of the following general results:

$$\lim_{x\to+\infty} x^n = +\infty, \quad n = 1, 2, 3, \ldots \qquad \lim_{x\to-\infty} x^n = \begin{cases} -\infty, & n = 1, 3, 5, \ldots \\ +\infty, & n = 2, 4, 6, \ldots \end{cases} \tag{11–12}$$

Multiplying x^n by a positive real number does not affect limits (11) and (12), but multiplying by a negative real number reverses the sign.

▶ Example 3

$$\lim_{x\to+\infty} 2x^5 = +\infty, \qquad \lim_{x\to-\infty} 2x^5 = -\infty$$

$$\lim_{x\to+\infty} -7x^6 = -\infty, \qquad \lim_{x\to-\infty} -7x^6 = -\infty \ ◀$$

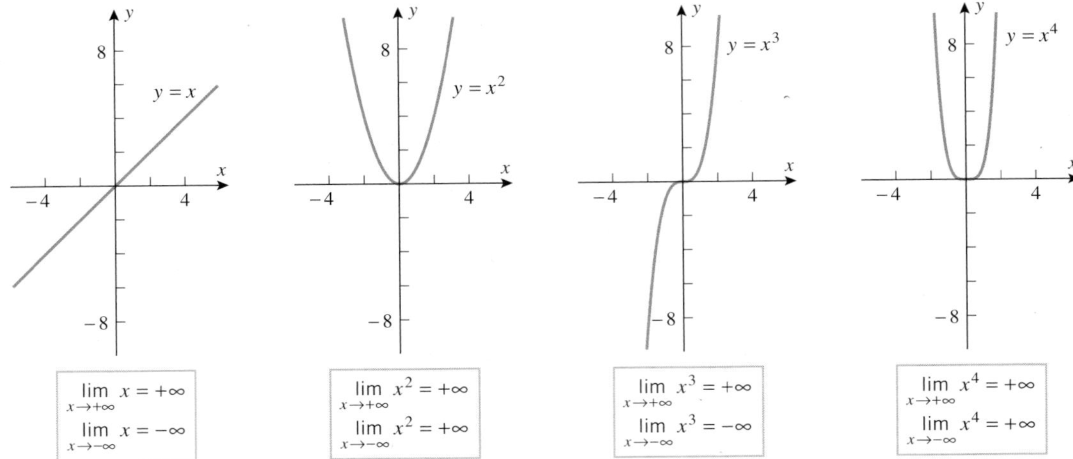

Figure 2.3.3

LIMITS OF POLYNOMIALS AS $x \to \pm\infty$

There is a useful principle about polynomials which, expressed informally, states that:

The end behavior of a polynomial matches the end behavior of its highest degree term.

More precisely, if $c_n \neq 0$, then

$$\lim_{x \to -\infty} (c_0 + c_1 x + \cdots + c_n x^n) = \lim_{x \to -\infty} c_n x^n \tag{13}$$

$$\lim_{x \to +\infty} (c_0 + c_1 x + \cdots + c_n x^n) = \lim_{x \to +\infty} c_n x^n \tag{14}$$

We can motivate these results by factoring out the highest power of x from the polynomial and examining the limit of the factored expression. Thus,

$$c_0 + c_1 x + \cdots + c_n x^n = x^n \left(\frac{c_0}{x^n} + \frac{c_1}{x^{n-1}} + \cdots + c_n \right)$$

As $x \to -\infty$ or $x \to +\infty$, it follows from (1) and (2) that all of the terms with positive powers of x in the denominator approach 0, so (13) and (14) are certainly plausible.

▶ Example 4

$$\lim_{x \to -\infty} (7x^5 - 4x^3 + 2x - 9) = \lim_{x \to -\infty} 7x^5 = -\infty$$

$$\lim_{x \to -\infty} (-4x^8 + 17x^3 - 5x + 1) = \lim_{x \to -\infty} -4x^8 = -\infty \blacktriangleleft$$

LIMITS OF RATIONAL FUNCTIONS AS $x \to \pm\infty$

One technique for determining the end behavior of a rational function is to divide each term in the numerator and denominator by the highest power of x that occurs in the denominator, after which the limiting behavior can be determined using results we have already established. Here are some examples.

Example 5 Find $\lim_{x\to+\infty} \dfrac{3x+5}{6x-8}$.

Solution. Divide each term in the numerator and denominator by the highest power of x that occurs in the denominator, namely, $x^1 = x$. We obtain

$$\lim_{x\to+\infty} \frac{3x+5}{6x-8} = \lim_{x\to+\infty} \frac{3+\dfrac{5}{x}}{6-\dfrac{8}{x}}$$ Divide each term by x.

$$= \frac{\lim\limits_{x\to+\infty}\left(3+\dfrac{5}{x}\right)}{\lim\limits_{x\to+\infty}\left(6-\dfrac{8}{x}\right)}$$ Limit of a quotient is the quotient of the limits.

$$= \frac{\lim\limits_{x\to+\infty} 3 + \lim\limits_{x\to+\infty} \dfrac{5}{x}}{\lim\limits_{x\to+\infty} 6 - \lim\limits_{x\to+\infty} \dfrac{8}{x}}$$ Limit of a sum is the sum of the limits.

$$= \frac{3+5\lim\limits_{x\to+\infty}\dfrac{1}{x}}{6-8\lim\limits_{x\to+\infty}\dfrac{1}{x}} = \frac{3+0}{6+0} = \frac{1}{2}$$ A constant can be moved through a limit symbol: Formulas (2) and (9). ◀

Example 6 Find $\lim_{x\to-\infty} \dfrac{4x^2-x}{2x^3-5}$.

Solution. Divide each term in the numerator and denominator by the highest power of x that occurs in the denominator, namely, x^3. We obtain

$$\lim_{x\to-\infty} \frac{4x^2-x}{2x^3-5} = \lim_{x\to-\infty} \frac{\dfrac{4}{x}-\dfrac{1}{x^2}}{2-\dfrac{5}{x^3}}$$ Divide each term by x^3.

$$= \frac{\lim\limits_{x\to-\infty}\left(\dfrac{4}{x}-\dfrac{1}{x^2}\right)}{\lim\limits_{x\to-\infty}\left(2-\dfrac{5}{x^3}\right)}$$ Limit of a quotient is the quotient of the limits.

$$= \frac{\lim\limits_{x\to-\infty}\dfrac{4}{x} - \lim\limits_{x\to-\infty}\dfrac{1}{x^2}}{\lim\limits_{x\to-\infty} 2 - \lim\limits_{x\to-\infty}\dfrac{5}{x^3}}$$ Limit of a difference is the difference of the limits.

$$= \frac{4\lim\limits_{x\to-\infty}\dfrac{1}{x} - \lim\limits_{x\to-\infty}\dfrac{1}{x^2}}{2-5\lim\limits_{x\to-\infty}\dfrac{1}{x^3}} = \frac{0-0}{2-0} = 0$$ A constant can be moved through a limit symbol: Formula (10) and Example 2. ◀

▶ **Example 7** Find $\lim_{x\to+\infty} \dfrac{5x^3 - 2x^2 + 1}{1 - 3x}$.

Solution. Divide each term in the numerator and denominator by the highest power of x that occurs in the denominator, namely, $x^1 = x$. We obtain

$$\lim_{x\to+\infty} \frac{5x^3 - 2x^2 + 1}{1 - 3x} = \lim_{x\to+\infty} \frac{5x^2 - 2x + \dfrac{1}{x}}{\dfrac{1}{x} - 3} \tag{15}$$

In this case we cannot argue that the limit of the quotient is the quotient of the limits because the limit of the numerator does not exist. However, we have

$$\lim_{x\to+\infty} (5x^2 - 2x) = +\infty, \quad \lim_{x\to+\infty} \frac{1}{x} = 0, \quad \lim_{x\to+\infty} \left(\frac{1}{x} - 3\right) = -3$$

Thus, the numerator on the right side of (15) approaches $+\infty$ and the denominator has a finite *negative* limit. We conclude from this that the quotient approaches $-\infty$; that is,

$$\lim_{x\to+\infty} \frac{5x^3 - 2x^2 + 1}{1 - 3x} = \lim_{x\to+\infty} \frac{5x^2 - 2x + \dfrac{1}{x}}{\dfrac{1}{x} - 3} = -\infty \quad ◀$$

A QUICK METHOD FOR FINDING LIMITS OF RATIONAL FUNCTIONS AS $x \to +\infty$ OR $x \to -\infty$

Since the end behavior of a polynomial matches the end behavior of its highest degree term, one can reasonably conclude that:

The end behavior of a rational function matches the end behavior of the quotient of the highest degree term in the numerator divided by the highest degree term in the denominator.

▶ **Example 8** Use the preceding observation to compute the limits in Examples 5, 6, and 7.

Solution.

$$\lim_{x\to+\infty} \frac{3x + 5}{6x - 8} = \lim_{x\to+\infty} \frac{3x}{6x} = \lim_{x\to+\infty} \frac{1}{2} = \frac{1}{2}$$

$$\lim_{x\to-\infty} \frac{4x^2 - x}{2x^3 - 5} = \lim_{x\to-\infty} \frac{4x^2}{2x^3} = \lim_{x\to-\infty} \frac{2}{x} = 0$$

$$\lim_{x\to+\infty} \frac{5x^3 - 2x^2 + 1}{1 - 3x} = \lim_{x\to+\infty} \frac{5x^3}{(-3x)} = \lim_{x\to+\infty} \left(-\frac{5}{3}x^2\right) = -\infty \quad ◀$$

LIMITS INVOLVING RADICALS

▶ **Example 9** Find $\lim_{x\to+\infty} \sqrt[3]{\dfrac{3x + 5}{6x - 8}}$.

Solution.

$$\lim_{x\to+\infty} \sqrt[3]{\frac{3x + 5}{6x - 8}} = \sqrt[3]{\lim_{x\to+\infty} \frac{3x + 5}{6x - 8}} = \sqrt[3]{\frac{1}{2}} \quad ◀$$

The limit of an nth root is the nth root of the limit.

▶ **Example 10** Find

$$\text{(a)} \lim_{x\to+\infty} \frac{\sqrt{x^2+2}}{3x-6} \qquad \text{(b)} \lim_{x\to-\infty} \frac{\sqrt{x^2+2}}{3x-6}$$

In both parts it would be helpful to manipulate the function so that the powers of x are transformed to powers of $1/x$. This can be achieved in both cases by dividing the numerator and denominator by $|x|$ and using the fact that $\sqrt{x^2} = |x|$.

Solution (a). As $x \to +\infty$, the values of x under consideration are positive, so we can replace $|x|$ by x where helpful. We obtain

$$\lim_{x\to+\infty} \frac{\sqrt{x^2+2}}{3x-6} = \lim_{x\to+\infty} \frac{\dfrac{\sqrt{x^2+2}}{|x|}}{\dfrac{3x-6}{|x|}} = \lim_{x\to+\infty} \frac{\dfrac{\sqrt{x^2+2}}{\sqrt{x^2}}}{\dfrac{3x-6}{x}}$$

$$= \lim_{x\to+\infty} \frac{\sqrt{1+\dfrac{2}{x^2}}}{3-\dfrac{6}{x}} = \frac{\lim\limits_{x\to+\infty} \sqrt{1+\dfrac{2}{x^2}}}{\lim\limits_{x\to+\infty} \left(3-\dfrac{6}{x}\right)}$$

$$= \frac{\sqrt{\lim\limits_{x\to+\infty} \left(1+\dfrac{2}{x^2}\right)}}{\lim\limits_{x\to+\infty} \left(3-\dfrac{6}{x}\right)} = \frac{\sqrt{\left(\lim\limits_{x\to+\infty} 1\right) + \left(2 \lim\limits_{x\to+\infty} \dfrac{1}{x^2}\right)}}{\left(\lim\limits_{x\to+\infty} 3\right) - \left(6 \lim\limits_{x\to+\infty} \dfrac{1}{x}\right)}$$

$$= \frac{\sqrt{1+(2\cdot 0)}}{3-(6\cdot 0)} = \frac{1}{3}$$

TECHNOLOGY MASTERY

It follows from Example 10 that the function

$$f(x) = \frac{\sqrt{x^2+2}}{3x-6}$$

has an asymptote of $y = \frac{1}{3}$ in the positive direction and an asymptote of $y = -\frac{1}{3}$ in the negative direction. Confirm this using a graphing utility.

Solution (b). As $x \to -\infty$, the values of x under consideration are negative, so we can replace $|x|$ by $-x$ where helpful. We obtain

$$\lim_{x\to-\infty} \frac{\sqrt{x^2+2}}{3x-6} = \lim_{x\to-\infty} \frac{\dfrac{\sqrt{x^2+2}}{|x|}}{\dfrac{3x-6}{|x|}} = \lim_{x\to-\infty} \frac{\dfrac{\sqrt{x^2+2}}{\sqrt{x^2}}}{\dfrac{3x-6}{(-x)}}$$

$$= \lim_{x\to-\infty} \frac{\sqrt{1+\dfrac{2}{x^2}}}{-3+\dfrac{6}{x}} = -\frac{1}{3} \quad ◀$$

▶ **Example 11** Find

$$\text{(a)} \lim_{x\to+\infty} (\sqrt{x^6+5} - x^3) \qquad \text{(b)} \lim_{x\to+\infty} (\sqrt{x^6+5x^3} - x^3)$$

We noted in Section 2.1 that the standard rules of algebra do not apply to the symbols $+\infty$ and $-\infty$. Part (b) of Example 11 illustrates this: The terms $\sqrt{x^6+5x^3}$ and x^3 both approach $+\infty$ as $x \to +\infty$, but their difference does not approach 0.

Solution. Graphs of the functions $f(x) = \sqrt{x^6+5} - x^3$, and $g(x) = \sqrt{x^6+5x^3} - x^3$ for $x \geq 0$, are shown in Figure 2.3.4. From the graphs we might conjecture that the requested limits are 0 and $\frac{5}{2}$, respectively. To confirm this, we treat each function as a fraction with a

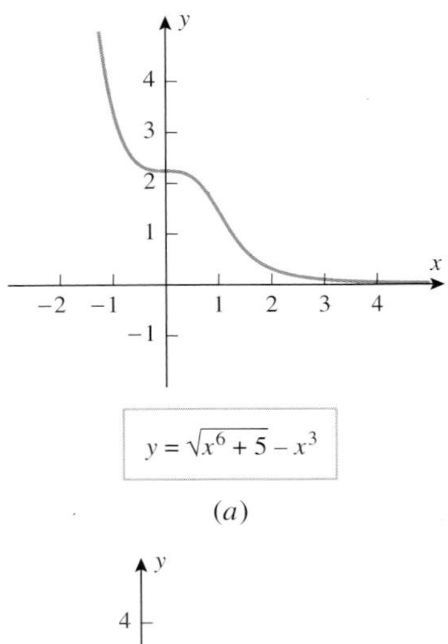

(a)

$y = \sqrt{x^6+5x^3} - x^3,\ x \geq 0$

(b)

Figure 2.3.4

denominator of 1 and rationalize the numerator.

$$\lim_{x\to+\infty}(\sqrt{x^6+5}-x^3) = \lim_{x\to+\infty}(\sqrt{x^6+5}-x^3)\left(\frac{\sqrt{x^6+5}+x^3}{\sqrt{x^6+5}+x^3}\right)$$

$$= \lim_{x\to+\infty}\frac{(x^6+5)-x^6}{\sqrt{x^6+5}+x^3} = \lim_{x\to+\infty}\frac{5}{\sqrt{x^6+5}+x^3}$$

$$= \lim_{x\to+\infty}\frac{\dfrac{5}{x^3}}{\sqrt{1+\dfrac{5}{x^6}}+1} \qquad \boxed{\sqrt{x^6} = x^3 \text{ for } x > 0}$$

$$= \frac{0}{\sqrt{1+0}+1} = 0$$

$$\lim_{x\to+\infty}(\sqrt{x^6+5x^3}-x^3) = \lim_{x\to+\infty}(\sqrt{x^6+5x^3}-x^3)\left(\frac{\sqrt{x^6+5x^3}+x^3}{\sqrt{x^6+5x^3}+x^3}\right)$$

$$= \lim_{x\to+\infty}\frac{(x^6+5x^3)-x^6}{\sqrt{x^6+5x^3}+x^3} = \lim_{x\to+\infty}\frac{5x^3}{\sqrt{x^6+5x^3}+x^3}$$

$$= \lim_{x\to+\infty}\frac{5}{\sqrt{1+\dfrac{5}{x^3}}+1} \qquad \boxed{\sqrt{x^6} = x^3 \text{ for } x > 0}$$

$$= \frac{5}{\sqrt{1+0}+1} = \frac{5}{2} \blacktriangleleft$$

END BEHAVIOR OF TRIGONOMETRIC FUNCTIONS

Consider the function $f(x) = \sin x$ that is graphed in Figure 2.3.5. For this function the limits as $x \to +\infty$ and as $x \to -\infty$ fail to exist not because $f(x)$ increases or decreases without bound, but rather because the values vary between -1 and 1 without approaching some specific real number. In general, the trigonometric functions fail to have limits as $x \to +\infty$ and as $x \to -\infty$ because of periodicity. There is no specific notation to denote this kind of behavior.

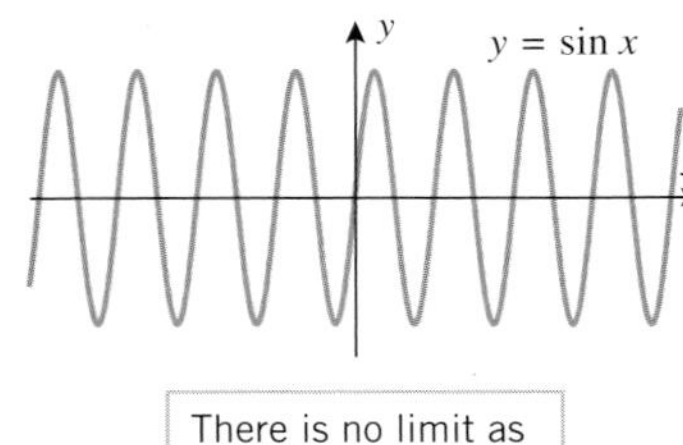

There is no limit as $x \to +\infty$ or $x \to -\infty$.

Figure 2.3.5

✔ QUICK CHECK EXERCISES 2.3

(See page 115 for answers.)

1. Find the limits.
   (a) $\lim_{x\to-\infty}(-3) =$ ______
   (b) $\lim_{h\to+\infty}(-2h) =$ ______
   (c) $\lim_{y\to-\infty}\dfrac{y}{|y|} =$ ______
   (d) $\lim_{z\to-\infty}(3-z) =$ ______
   (e) $\lim_{h\to-\infty}\left(5-\dfrac{1}{h}\right) =$ ______

2. Find the limits that exist.
   (a) $\lim_{x\to-\infty}\dfrac{2x^2+x}{4x^2-3} =$ ______
   (b) $\lim_{x\to+\infty}\dfrac{1}{2+\sin x} =$ ______

3. Given that

$$\lim_{x\to+\infty} f(x) = 2 \quad \text{and} \quad \lim_{x\to+\infty} g(x) = -3$$

find the limits that exist.

(a) $\lim_{x\to+\infty} [3f(x) - g(x)] =$ ______

(b) $\lim_{x\to+\infty} \dfrac{f(x)}{g(x)} =$ ______

(c) $\lim_{x\to+\infty} \dfrac{2f(x) + 3g(x)}{3f(x) + 2g(x)} =$ ______

(d) $\lim_{x\to+\infty} \sqrt{10 - f(x)g(x)} =$ ______

4. Consider the graphs of $y = 1/(x+1)$, $y = x/(x+1)$, and $y = x^2/(x+1)$. Which of these graphs, if any, has a horizontal asymptote?

EXERCISE SET 2.3 Graphing Utility

FOCUS ON CONCEPTS

1–4 In these exercises, make reasonable assumptions about the end behavior of the indicated function.

1. For the function g graphed in the accompanying figure, find
(a) $\lim_{x\to-\infty} g(x)$ (b) $\lim_{x\to+\infty} g(x)$.

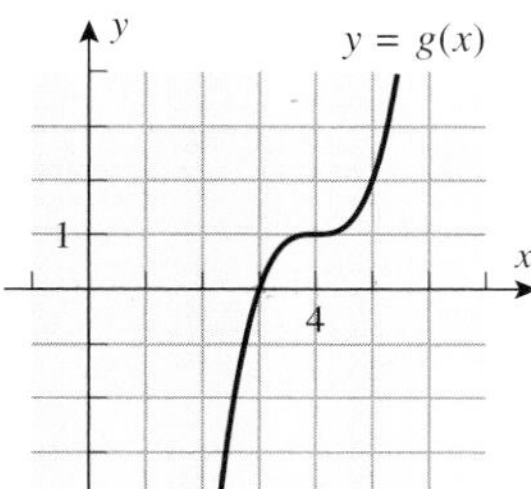

Figure Ex-1

2. For the function ϕ graphed in the accompanying figure, find
(a) $\lim_{x\to-\infty} \phi(x)$ (b) $\lim_{x\to+\infty} \phi(x)$.

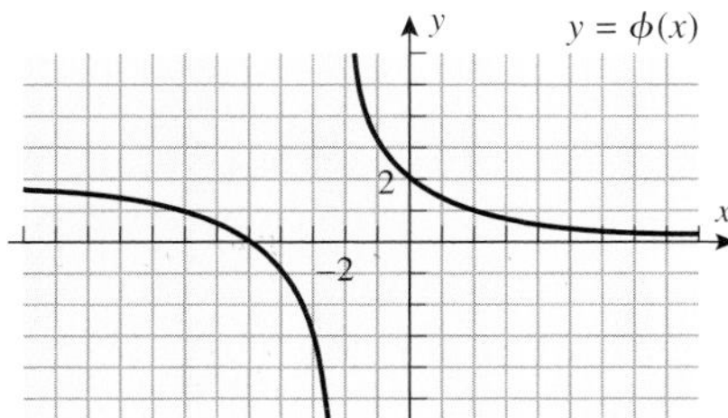

Figure Ex-2

3. For the function ϕ graphed in the accompanying figure, find
(a) $\lim_{x\to-\infty} \phi(x)$ (b) $\lim_{x\to+\infty} \phi(x)$.

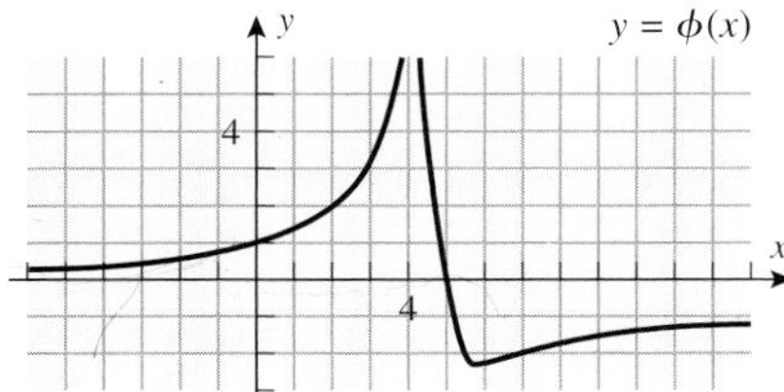

Figure Ex-3

4. For the function G graphed in the accompanying figure, find
(a) $\lim_{x\to-\infty} G(x)$ (b) $\lim_{x\to+\infty} G(x)$.

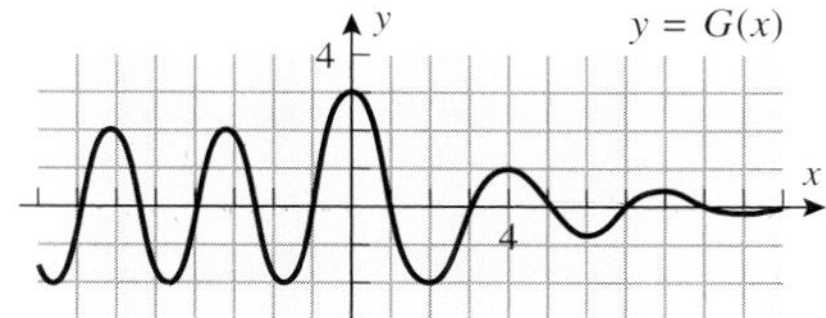

Figure Ex-4

5. Given that

$$\lim_{x\to+\infty} f(x) = 3, \quad \lim_{x\to+\infty} g(x) = -5, \quad \lim_{x\to+\infty} h(x) = 0$$

find the limits that exist. If the limit does not exist, explain why.

(a) $\lim_{x\to+\infty} [f(x) + 3g(x)]$

(b) $\lim_{x\to+\infty} [h(x) - 4g(x) + 1]$

(c) $\lim_{x\to+\infty} [f(x)g(x)]$ (d) $\lim_{x\to+\infty} [g(x)]^2$

(e) $\lim_{x\to+\infty} \sqrt[3]{5 + f(x)}$ (f) $\lim_{x\to+\infty} \dfrac{3}{g(x)}$

(g) $\lim_{x\to+\infty} \dfrac{3h(x) + 4}{x^2}$ (h) $\lim_{x\to+\infty} \dfrac{6f(x)}{5f(x) + 3g(x)}$

6. Given that

$$\lim_{x\to-\infty} f(x) = 7 \quad \text{and} \quad \lim_{x\to-\infty} g(x) = -6$$

find the limits that exist. If the limit does not exist, explain why.

(a) $\lim_{x\to-\infty} [2f(x) - g(x)]$ (b) $\lim_{x\to-\infty} [6f(x) + 7g(x)]$

(c) $\lim_{x\to-\infty} [x^2 + g(x)]$ (d) $\lim_{x\to-\infty} [x^2 g(x)]$

(e) $\lim_{x\to-\infty} \sqrt[3]{f(x)g(x)}$ (f) $\lim_{x\to-\infty} \dfrac{g(x)}{f(x)}$

(g) $\lim_{x\to-\infty} \left[f(x) + \dfrac{g(x)}{x}\right]$ (h) $\lim_{x\to-\infty} \dfrac{xf(x)}{(2x+3)g(x)}$

7–28 Find the limits.

7. $\lim_{x\to+\infty} (1 + 2x - 3x^5)$

8. $\lim_{x\to+\infty} (2x^3 - 100x + 5)$

9. $\lim_{x\to+\infty}\sqrt{x}$

10. $\lim_{x\to-\infty}\sqrt{5-x}$

11. $\lim_{x\to+\infty}\dfrac{3x+1}{2x-5}$

12. $\lim_{x\to+\infty}\dfrac{5x^2-4x}{2x^2+3}$

13. $\lim_{y\to-\infty}\dfrac{3}{y+4}$

14. $\lim_{x\to+\infty}\dfrac{1}{x-12}$

15. $\lim_{x\to-\infty}\dfrac{x-2}{x^2+2x+1}$

16. $\lim_{x\to+\infty}\dfrac{5x^2+7}{3x^2-x}$

17. $\lim_{x\to+\infty}\sqrt[3]{\dfrac{2+3x-5x^2}{1+8x^2}}$

18. $\lim_{s\to+\infty}\sqrt[3]{\dfrac{3s^7-4s^5}{2s^7+1}}$

19. $\lim_{x\to-\infty}\dfrac{\sqrt{5x^2-2}}{x+3}$

20. $\lim_{x\to+\infty}\dfrac{\sqrt{5x^2-2}}{x+3}$

21. $\lim_{y\to-\infty}\dfrac{2-y}{\sqrt{7+6y^2}}$

22. $\lim_{y\to+\infty}\dfrac{2-y}{\sqrt{7+6y^2}}$

23. $\lim_{x\to-\infty}\dfrac{\sqrt{3x^4+x}}{x^2-8}$

24. $\lim_{x\to+\infty}\dfrac{\sqrt{3x^4+x}}{x^2-8}$

25. $\lim_{x\to+\infty}\dfrac{7-6x^5}{x+3}$

26. $\lim_{t\to-\infty}\dfrac{5-2t^3}{t^2+1}$

27. $\lim_{t\to+\infty}\dfrac{6-t^3}{7t^3+3}$

28. $\lim_{x\to-\infty}\dfrac{x+4x^3}{1-x^2+7x^3}$

FOCUS ON CONCEPTS

29. Assume that a particle is accelerated by a constant force. The two curves $v=n(t)$ and $v=e(t)$ in the accompanying figure provide velocity versus time curves for the particle as predicted by classical physics and by the special theory of relativity, respectively. The parameter c represents the speed of light. Using the language of limits, describe the differences in the long-term predictions of the two theories.

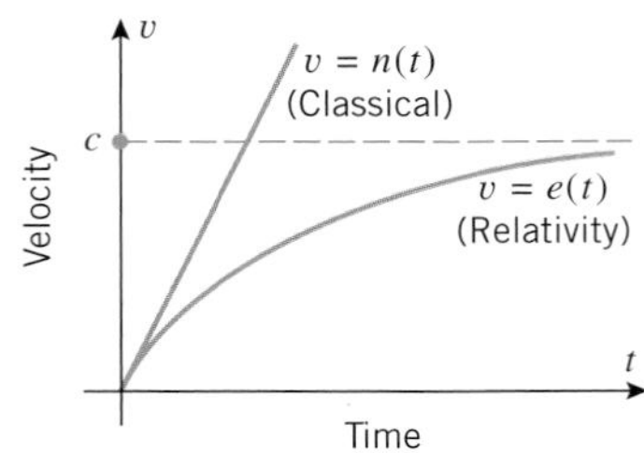

Figure Ex-29

30. Let $T=f(t)$ denote the temperature of a baked potato t minutes after it has been removed from a hot oven. The accompanying figure shows the temperature versus time curve for the potato, where r is the temperature of the room.

(a) What is the physical significance of $\lim_{t\to0^+} f(t)$?

(b) What is the physical significance of $\lim_{t\to+\infty} f(t)$?

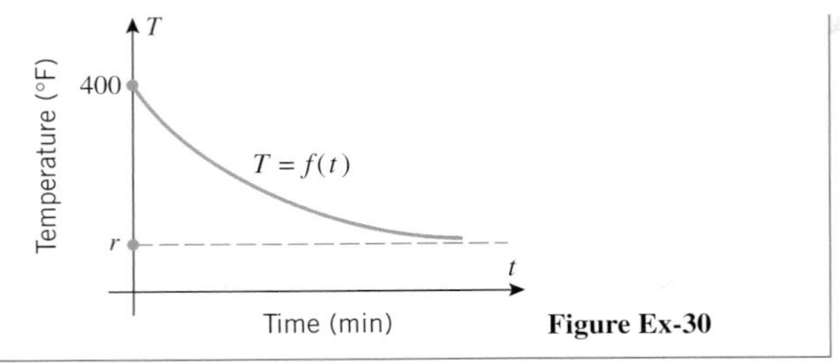

Figure Ex-30

31. Let

$$f(x)=\begin{cases}2x^2+5, & x<0\\[4pt] \dfrac{3-5x^3}{1+4x+x^3}, & x\geq 0\end{cases}$$

Find

(a) $\lim_{x\to-\infty} f(x)$ (b) $\lim_{x\to+\infty} f(x)$.

32. Let

$$g(t)=\begin{cases}\dfrac{2+3t}{5t^2+6}, & t<1{,}000{,}000\\[4pt] \dfrac{\sqrt{36t^2-100}}{5-t}, & t>1{,}000{,}000\end{cases}$$

Find

(a) $\lim_{t\to-\infty} g(t)$ (b) $\lim_{t\to+\infty} g(t)$.

33–36 Find the limits.

33. $\lim_{x\to+\infty}(\sqrt{x^2+3}-x)$

34. $\lim_{x\to+\infty}(\sqrt{x^2-3x}-x)$

35. $\lim_{x\to+\infty}(\sqrt{x^2+ax}-x)$

36. $\lim_{x\to+\infty}(\sqrt{x^2+ax}-\sqrt{x^2+bx})$

37. Discuss the limits of $p(x)=(1-x)^n$ as $x\to+\infty$ and $x\to-\infty$ for positive integer values of n.

38. Let $p(x)=(1-x)^n$ and $q(x)=(1-x)^m$. Discuss the limits of $p(x)/q(x)$ as $x\to+\infty$ and $x\to-\infty$ for positive integer values of m and n.

39. Let $p(x)$ be a polynomial of degree n. Discuss the limits of $p(x)/x^m$ as $x\to+\infty$ and $x\to-\infty$ for positive integer values of m.

40. In each part, find examples of polynomials $p(x)$ and $q(x)$ that satisfy the stated condition and such that $p(x)\to+\infty$ and $q(x)\to+\infty$ as $x\to+\infty$.

(a) $\lim_{x\to+\infty}\dfrac{p(x)}{q(x)}=1$ (b) $\lim_{x\to+\infty}\dfrac{p(x)}{q(x)}=0$

(c) $\lim_{x\to+\infty}\dfrac{p(x)}{q(x)}=+\infty$ (d) $\lim_{x\to+\infty}[p(x)-q(x)]=3$

41. Assuming that m and n are positive integers, find

$$\lim_{x\to-\infty}\frac{2+3x^n}{1-x^m}$$

[*Hint:* Your answer will depend on whether $m<n$, $m=n$, or $m>n$.]

42. Find

$$\lim_{x\to+\infty}\frac{c_0+c_1x+\cdots+c_nx^n}{d_0+d_1x+\cdots+d_mx^m}$$

where $c_n \neq 0$ and $d_m \neq 0$. [*Hint:* Your answer will depend on whether $m < n$, $m = n$, or $m > n$.]

FOCUS ON CONCEPTS

43. Suppose that $f(x)$ denotes a function such that

$$\lim_{t \to 0} f\left(\frac{1}{t}\right) = L$$

What can be said about the limits

$$\lim_{x \to +\infty} f(x) \quad \text{and} \quad \lim_{x \to -\infty} f(x)?$$

44. (a) Suppose that $f(x)$ denotes a function such that

$$\lim_{t \to +\infty} f(t) = L$$

What can be said about the limit

$$\lim_{x \to 0^+} f\left(\frac{1}{x}\right)?$$

(b) Suppose that $f(x)$ denotes a function such that

$$\lim_{t \to -\infty} f(t) = L$$

What can be said about the limit

$$\lim_{x \to 0^-} f\left(\frac{1}{x}\right)?$$

45–50 The notion of an asymptote can be extended to include curves as well as lines. Specifically, we say that curves $y = f(x)$ and $y = g(x)$ are ***asymptotic as*** $x \to +\infty$ provided

$$\lim_{x \to +\infty} [f(x) - g(x)] = 0$$

and are ***asymptotic as*** $x \to -\infty$ provided

$$\lim_{x \to -\infty} [f(x) - g(x)] = 0$$

Informally stated, two curves are asymptotic as $x \to +\infty$ provided they remain as close together as we like for sufficiently large values of x. Similarly, two curves are asymptotic as $x \to -\infty$ provided they remain as close together as we like for negative numbers x of sufficiently large magnitude. For example, if

$$f(x) = x^2 + \frac{2}{x-1} \quad \text{and} \quad g(x) = x^2$$

then $y = f(x)$ is asymptotic to $y = g(x)$ as $x \to +\infty$ and as $x \to -\infty$ since

$$\lim_{x \to +\infty} [f(x) - g(x)] = \lim_{x \to +\infty} \frac{2}{x-1} = 0$$

$$\lim_{x \to -\infty} [f(x) - g(x)] = \lim_{x \to -\infty} \frac{2}{x-1} = 0$$

This asymptotic behavior is illustrated in the following figure, which also shows the vertical asymptote of $f(x)$ at $x = 1$.

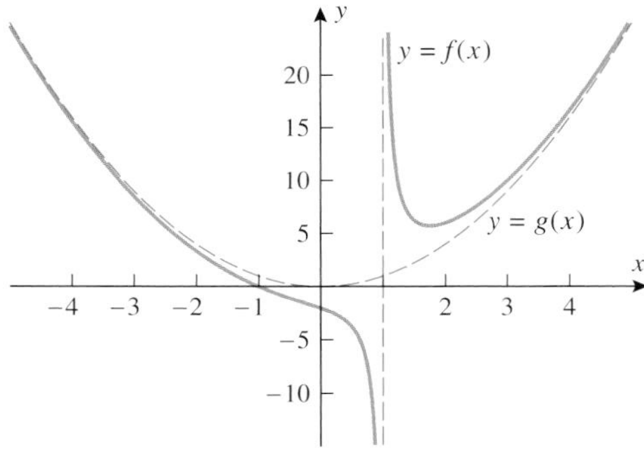

In these exercises, determine a simpler function $g(x)$ such that $y = f(x)$ is asymptotic to $y = g(x)$ as $x \to +\infty$ or $x \to -\infty$. Use a graphing utility to generate the graphs of $y = f(x)$ and $y = g(x)$ and identify all vertical asymptotes.

45. $f(x) = \dfrac{x^2 - 2}{x - 2}$

46. $f(x) = \dfrac{x^3 - x + 3}{x}$

47. $f(x) = \dfrac{-x^3 + 3x^2 + x - 1}{x - 3}$

48. $f(x) = \dfrac{x^5 - x^3 + 3}{x^2 - 1}$

49. $f(x) = \sin x + \dfrac{1}{x - 1}$

50. $f(x) = \sqrt{\dfrac{x^3 - x^2 + 2}{x - 1}}$

✔ QUICK CHECK ANSWERS 2.3

1. (a) -3 (b) $-\infty$ (c) -1 (d) $+\infty$ (e) 5 2. (a) $\frac{1}{2}$ (b) does not exist 3. (a) 9 (b) $-\frac{2}{3}$ (c) does not exist (d) 4
4. The graphs of $y = 1/(x + 1)$ and $y = x/(x + 1)$ have horizontal asymptotes.

2.4 LIMITS (DISCUSSED MORE RIGOROUSLY)

In the previous sections of this chapter we focused on the discovery of values of limits, either by sampling selected x-values or by applying limit theorems that were stated without proof. Our main goal in this section is to define the notion of a limit precisely, thereby making it possible to establish limits with certainty and to prove theorems about them. This will also provide us with a deeper understanding of some of the more subtle properties of functions.

MOTIVATION FOR THE DEFINITION OF A TWO-SIDED LIMIT

The statement $\lim_{x\to a} f(x) = L$ can be interpreted informally to mean that we can make the value of $f(x)$ as close as we like to the real number L by making the value of x sufficiently close to a. It is our goal to make the informal phrases "as close as we like to L" and "sufficiently close to a" mathematically precise.

To do this, consider the function f graphed in Figure 2.4.1a for which $f(x)\to L$ as $x\to a$. For visual simplicity we have drawn the graph of f to be increasing on an open interval containing a, and we have intentionally placed a hole in the graph at $x = a$ to emphasize that f need not be defined at $x = a$ to have a limit there.

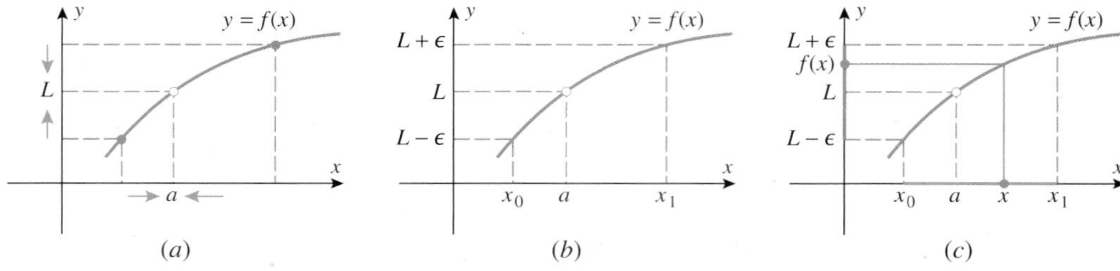

Figure 2.4.1

Next, let us choose any positive number ϵ and ask how close x must be to a in order for the values of $f(x)$ to be within ϵ units of L. We can answer this geometrically by drawing horizontal lines from the points $L+\epsilon$ and $L-\epsilon$ on the y-axis until they meet the curve $y = f(x)$, and then drawing vertical lines from those points on the curve to the x-axis (Figure 2.4.1b). As indicated in the figure, let x_0 and x_1 be the points where those vertical lines intersect the x-axis.

Now imagine that x gets closer and closer to a (from either side). Eventually, x will lie inside the interval (x_0, x_1), which is marked in green in Figure 2.4.1c; and when this happens, the value of $f(x)$ will fall between $L-\epsilon$ and $L+\epsilon$, marked in red in the figure. Thus, we conclude:

> *If $f(x)\to L$ as $x\to a$, then for any positive number ϵ, we can find an open interval (x_0, x_1) on the x-axis that contains a and has the property that for each x in that interval (except possibly for $x = a$), the value of $f(x)$ is between $L-\epsilon$ and $L+\epsilon$.*

What is important about this result is that it holds no matter how small we make ϵ. However, making ϵ smaller and smaller forces $f(x)$ *closer and closer* to L—which is precisely the concept we were trying to capture mathematically.

Observe that in Figure 2.4.1c the interval (x_0, x_1) extends farther on the right side of a than on the left side. However, for many purposes it is preferable to have an interval that extends the same distance on both sides of a. For this purpose, let us choose any

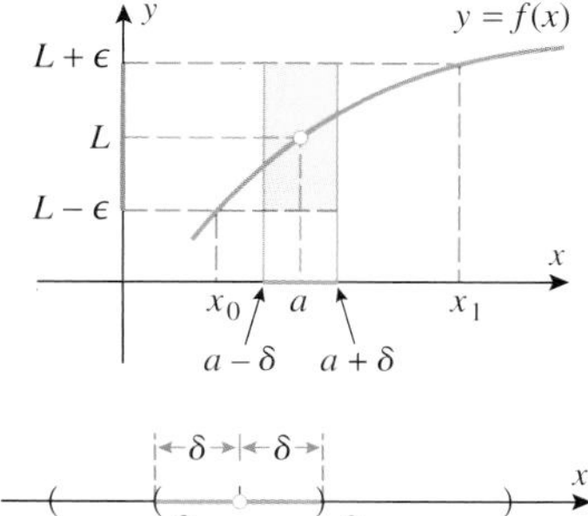

Figure 2.4.2

positive number δ that is smaller than both $x_1 - a$ and $a - x_0$, and consider the interval $(a - \delta, a + \delta)$. This interval extends the same distance δ on both sides of a and lies inside of the interval (x_0, x_1) (Figure 2.4.2). Moreover, the condition $L - \epsilon < f(x) < L + \epsilon$ holds for every x in this interval (except possibly $x = a$), since this condition holds on the larger interval (x_0, x_1).

Since the condition $L - \epsilon < f(x) < L + \epsilon$ can be expressed as

$$|f(x) - L| < \epsilon$$

and the condition that x lies in the interval $(a - \delta, a + \delta)$, but $x \neq a$, can be expressed as

$$0 < |x - a| < \delta$$

we are led to the following precise definition of a two-sided limit.

The definitions of one-sided limits require minor adjustments to Definition 2.4.1. For example, for a limit from the right we need only assume that $f(x)$ is defined on an interval (a, b) extending to the right of a and that the ϵ condition is met for x in an interval $a < x < a + \delta$ extending to the right of a. A similar adjustment must be made for a limit from the left.

2.4.1 LIMIT DEFINITION. Let $f(x)$ be defined for all x in some open interval containing the number a, with the possible exception that $f(x)$ need not be defined at a. We will write

$$\lim_{x \to a} f(x) = L$$

if given any number $\epsilon > 0$ we can find a number $\delta > 0$ such that

$$|f(x) - L| < \epsilon \quad \text{if} \quad 0 < |x - a| < \delta$$

This definition, which is attributed to the German mathematician Karl Weierstrass and is commonly called the "epsilon-delta" definition of a two-sided limit, makes the transition from an informal concept of a limit to a precise definition. Specifically, the informal phrase "as close as we like to L" is given quantitative meaning by our ability to choose the positive number ϵ arbitrarily, and the phrase "sufficiently close to a" is quantified by the positive number δ.

In the preceding sections we illustrated various numerical and graphical methods for *guessing* at limits. Now that we have a precise definition to work with, we can actually confirm the validity of those guesses with mathematical proof. Here is a typical example of such a proof.

Karl Weierstrass (1815–1897) Weierstrass, the son of a customs officer, was born in Ostenfelde, Germany. As a youth Weierstrass showed outstanding skills in languages and mathematics. However, at the urging of his dominant father, Weierstrass entered the law and commerce program at the University of Bonn. To the chagrin of his family, the rugged and congenial young man concentrated instead on fencing and beer drinking. Four years later he returned home without a degree. In 1839 Weierstrass entered the Academy of Münster to study for a career in secondary education, and he met and studied under an excellent mathematician named Christof Gudermann. Gudermann's ideas greatly influenced the work of Weierstrass. After receiving his teaching certificate, Weierstrass spent the next 15 years in secondary education teaching German, geography, and mathematics. In addition, he taught handwriting to small children. During this period much of Weierstrass's mathematical work was ignored because he was a secondary schoolteacher and not a college professor. Then, in 1854, he published a paper of major importance that created a sensation in the mathematics world and catapulted him to international fame overnight. He was immediately given an honorary Doctorate at the University of Königsberg and began a new career in college teaching at the University of Berlin in 1856. In 1859 the strain of his mathematical research caused a temporary nervous breakdown and led to spells of dizziness that plagued him for the rest of his life. Weierstrass was a brilliant teacher and his classes overflowed with multitudes of auditors. In spite of his fame, he never lost his early beer-drinking congeniality and was always in the company of students, both ordinary and brilliant. Weierstrass was acknowledged as the leading mathematical analyst in the world. He and his students opened the door to the modern school of mathematical analysis.

▶ **Example 1** Use Definition 2.4.1 to prove that $\lim_{x\to 2}(3x-5)=1$.

Solution. We must show that given any positive number ϵ, we can find a positive number δ such that

$$|\underbrace{(3x-5)}_{f(x)} - \underbrace{1}_{L}| < \epsilon \quad \text{if} \quad 0 < |x - \underbrace{2}_{a}| < \delta \tag{1}$$

There are two things to do. First, we must *discover* a value of δ for which this statement holds, and then we must *prove* that the statement holds for that δ. For the discovery part we begin by simplifying (1) and writing it as

$$|3x-6| < \epsilon \quad \text{if} \quad 0 < |x-2| < \delta$$

Next we will rewrite this statement in a form that will facilitate the discovery of an appropriate δ:

$$\begin{aligned} 3|x-2| &< \epsilon \quad \text{if} \quad 0 < |x-2| < \delta \\ |x-2| &< \epsilon/3 \quad \text{if} \quad 0 < |x-2| < \delta \end{aligned} \tag{2}$$

It should be self-evident that this last statement holds if $\delta = \epsilon/3$, which completes the discovery portion of our work. Now we need to prove that (1) holds for this choice of δ. However, statement (1) is equivalent to (2), and (2) holds with $\delta = \epsilon/3$, so (1) also holds with $\delta = \epsilon/3$. This proves that $\lim_{x\to 2}(3x-5)=1$. ◀

Example 1 illustrates the general form of a limit proof: We *assume* that we are given a positive number ϵ, and we try to *prove* that we can find a positive number δ such that

$$|f(x)-L| < \epsilon \quad \text{if} \quad 0 < |x-a| < \delta \tag{3}$$

This is done by first discovering δ, and then proving that the discovered δ works. Since the argument has to be general enough to work for all positive values of ϵ, the quantity δ has to be expressed as a function of ϵ. In Example 1 we found the function $\delta = \epsilon/3$ by some simple algebra; however, most limit proofs require a little more algebraic and logical ingenuity. Thus, if you find our ensuing discussion of "ϵ-δ" proofs challenging, do not become discouraged; the concepts and techniques are intrinsically difficult. In fact, a precise understanding of limits evaded the finest mathematical minds for more than 150 years after the basic concepts of calculus were discovered.

▶ **Example 2** Prove that $\lim_{x\to 0^+}\sqrt{x}=0$.

Solution. Note that the domain of $\sqrt{x}$ is $0 \le x$, so it is valid to discuss the limit as $x \to 0^+$. We must show that given $\epsilon > 0$, there exists a $\delta > 0$ such that

$$|\sqrt{x}-0| < \epsilon \quad \text{if} \quad 0 < x < 0 + \delta$$

or more simply,

$$\sqrt{x} < \epsilon \quad \text{if} \quad 0 < x < \delta \tag{4}$$

But, by squaring both sides of the inequality $\sqrt{x} < \epsilon$, we can rewrite (4) as

$$x < \epsilon^2 \quad \text{if} \quad 0 < x < \delta \tag{5}$$

It should be self-evident that (5) is true if $\delta = \epsilon^2$; and since (5) is a reformulation of (4), we have shown that (4) holds with $\delta = \epsilon^2$. This proves that $\lim_{x\to 0^+}\sqrt{x}=0$. ◀

In Example 2 the limit from the left and the two-sided limit do not exist at $x = 0$ because the domain of $\sqrt{x}$ includes no numbers to the left of 0.

THE VALUE OF δ IS NOT UNIQUE

In preparation for our next example, we note that the value of δ in Definition 2.4.1 is not unique; once we have found a value of δ that fulfills the requirements of the definition, then any *smaller* positive number δ_1 will also fulfill those requirements. That is, if it is true that

$$|f(x) - L| < \epsilon \quad \text{if} \quad 0 < |x - a| < \delta$$

then it will also be true that

$$|f(x) - L| < \epsilon \quad \text{if} \quad 0 < |x - a| < \delta_1$$

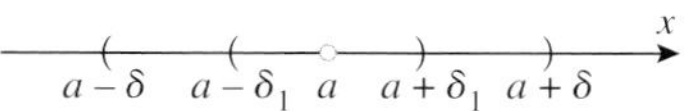

Figure 2.4.3

This is because $\{x : 0 < |x - a| < \delta_1\}$ is a subset of $\{x : 0 < |x - a| < \delta\}$ (Figure 2.4.3), and hence if $|f(x) - L| < \epsilon$ is satisfied for all x in the larger set, then it will automatically be satisfied for all x in the subset. Thus, in Example 1, where we used $\delta = \epsilon/3$, we could have used any smaller value of δ such as $\delta = \epsilon/4$, $\delta = \epsilon/5$, or $\delta = \epsilon/6$.

▶ **Example 3** Prove that $\lim_{x \to 3} x^2 = 9$.

Solution. We must show that given any positive number ϵ, we can find a positive number δ such that

$$|x^2 - 9| < \epsilon \quad \text{if} \quad 0 < |x - 3| < \delta \tag{6}$$

Because $|x - 3|$ occurs on the right side of this "if statement," it will be helpful to factor the left side to introduce a factor of $|x - 3|$. This yields the following alternative form of (6):

$$|x + 3||x - 3| < \epsilon \quad \text{if} \quad 0 < |x - 3| < \delta \tag{7}$$

> If you are wondering how we knew to make the restriction $\delta \le 1$, as opposed to $\delta \le 5$ or $\delta \le \frac{1}{2}$, for example, the answer is that 1 is merely a convenient choice—any restriction of the form $\delta \le c$ would work equally well.

We wish to bound the factor $|x + 3|$. If we knew, for example, that $\delta \le 1$, then we would have $-1 < x - 3 < 1$, so $5 < x + 3 < 7$, and consequently $|x + 3| < 7$. Thus, if $\delta \le 1$ and $0 < |x - 3| < \delta$, then

$$|x + 3||x - 3| < 7\delta$$

It follows that (7) will be satisfied for any positive δ such that $\delta \le 1$ and $7\delta < \epsilon$. We can achieve this by taking δ to be the minimum of the numbers 1 and $\epsilon/7$, which is sometimes written as $\delta = \min(1, \epsilon/7)$. This proves that $\lim_{x \to 3} x^2 = 9$. ◀

LIMITS AS $x \to \pm\infty$

In Section 2.3 we discussed the limits

$$\lim_{x \to +\infty} f(x) = L \quad \text{and} \quad \lim_{x \to -\infty} f(x) = L$$

from an intuitive point of view. The first limit can be interpreted to mean that we can make the value of $f(x)$ as close as we like to L by taking x sufficiently large, and the second can be interpreted to mean that we can make the value of $f(x)$ as close as we like to L by taking x sufficiently far to the left of 0. These ideas are captured in the following definitions and are illustrated in Figure 2.4.4.

> **2.4.2 DEFINITION.** Let $f(x)$ be defined for all x in some infinite open interval extending in the positive x-direction. We will write
>
> $$\lim_{x \to +\infty} f(x) = L$$
>
> if given any number $\epsilon > 0$, there corresponds a positive number N such that
>
> $$|f(x) - L| < \epsilon \quad \text{if} \quad x > N$$

2.4.3 DEFINITION. Let $f(x)$ be defined for all x in some infinite open interval extending in the negative x-direction. We will write

$$\lim_{x \to -\infty} f(x) = L$$

if given any number $\epsilon > 0$, there corresponds a negative number N such that

$$|f(x) - L| < \epsilon \quad \text{if} \quad x < N$$

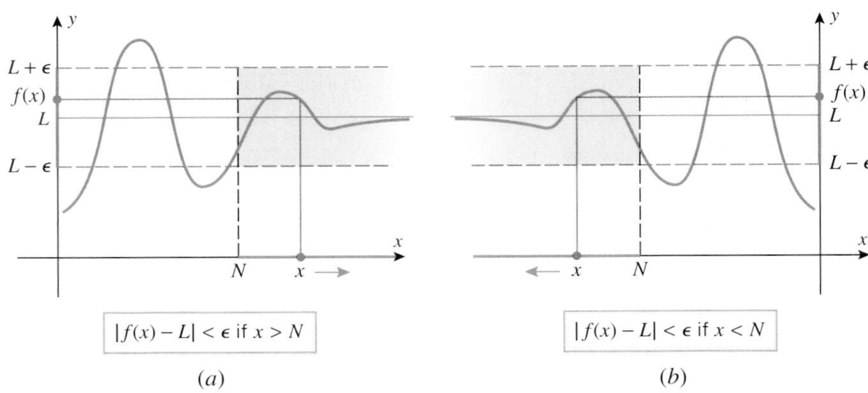

Figure 2.4.4

To see how these definitions relate to our informal concepts of these limits, suppose that $f(x) \to L$ as $x \to +\infty$, and for a given ϵ let N be the positive number described in Definition 2.4.2. If x is allowed to increase indefinitely, then eventually x will lie in the interval $(N, +\infty)$, which is marked in green in Figure 2.4.4*a*; when this happens, the value of $f(x)$ will fall between $L - \epsilon$ and $L + \epsilon$, marked in red in the figure. Since this is true for all positive values of ϵ (no matter how small), we can force the values of $f(x)$ as close as we like to L by making N sufficiently large. This agrees with our informal concept of this limit. Similarly, Figure 2.4.4*b* illustrates Definition 2.4.3.

▶ **Example 4** Prove that $\lim_{x \to +\infty} \dfrac{1}{x} = 0$.

Solution. Applying Definition 2.4.2 with $f(x) = 1/x$ and $L = 0$, we must show that given $\epsilon > 0$, we can find a number $N > 0$ such that

$$\left|\frac{1}{x} - 0\right| < \epsilon \quad \text{if} \quad x > N \tag{8}$$

Because $x \to +\infty$ we can assume that $x > 0$. Thus, we can eliminate the absolute values in this statement and rewrite it as

$$\frac{1}{x} < \epsilon \quad \text{if} \quad x > N$$

or, on taking reciprocals,

$$x > \frac{1}{\epsilon} \quad \text{if} \quad x > N \tag{9}$$

It is self-evident that $N = 1/\epsilon$ satisfies this requirement, and since (9) and (8) are equivalent for $x > 0$, the proof is complete. ◀

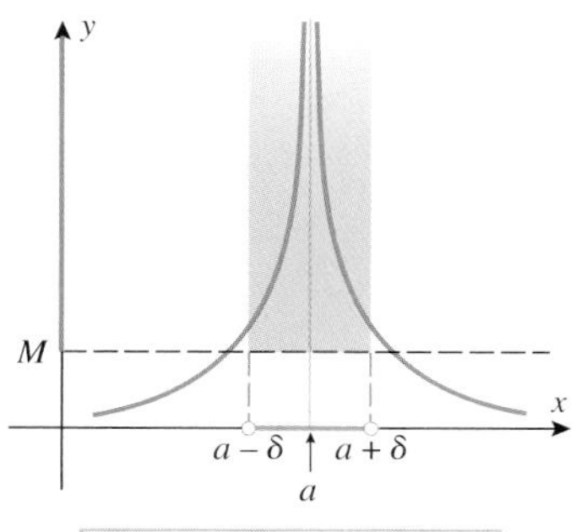

(a)

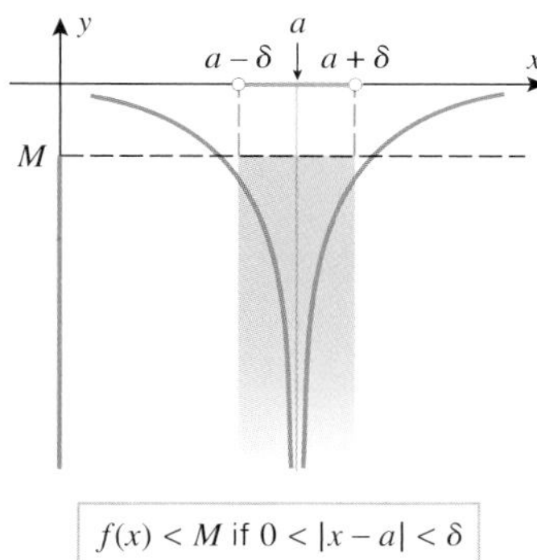

(b)

Figure 2.4.5

INFINITE LIMITS

In Section 2.1 we discussed limits of the following type from an intuitive viewpoint:

$$\lim_{x \to a} f(x) = +\infty, \qquad \lim_{x \to a} f(x) = -\infty \tag{10}$$

$$\lim_{x \to a^+} f(x) = +\infty, \qquad \lim_{x \to a^+} f(x) = -\infty \tag{11}$$

$$\lim_{x \to a^-} f(x) = +\infty, \qquad \lim_{x \to a^-} f(x) = -\infty \tag{12}$$

Recall that each of these expressions describes a particular way in which the limit fails to exist. The $+\infty$ indicates that the limit fails to exist because $f(x)$ increases without bound, and the $-\infty$ indicates that the limit fails to exist because $f(x)$ decreases without bound. These ideas are captured more precisely in the following definitions and are illustrated in Figure 2.4.5.

2.4.4 DEFINITION. Let $f(x)$ be defined for all x in some open interval containing a, except that $f(x)$ need not be defined at a. We will write

$$\lim_{x \to a} f(x) = +\infty$$

if given any positive number M, we can find a number $\delta > 0$ such that $f(x)$ satisfies

$$f(x) > M \quad \text{if} \quad 0 < |x - a| < \delta$$

2.4.5 DEFINITION. Let $f(x)$ be defined for all x in some open interval containing a, except that $f(x)$ need not be defined at a. We will write

$$\lim_{x \to a} f(x) = -\infty$$

if given any negative number M, we can find a number $\delta > 0$ such that $f(x)$ satisfies

$$f(x) < M \quad \text{if} \quad 0 < |x - a| < \delta$$

How would you define these limits?

$\lim_{x \to a^+} f(x) = +\infty \qquad \lim_{x \to a^+} f(x) = -\infty$

$\lim_{x \to a^-} f(x) = +\infty \qquad \lim_{x \to a^-} f(x) = -\infty$

$\lim_{x \to +\infty} f(x) = +\infty \qquad \lim_{x \to +\infty} f(x) = -\infty$

$\lim_{x \to -\infty} f(x) = +\infty \qquad \lim_{x \to -\infty} f(x) = -\infty$

To see how these definitions relate to our informal concepts of these limits, suppose that $f(x) \to +\infty$ as $x \to a$, and for a given M let δ be the corresponding positive number described in Definition 2.4.4. Next, imagine that x gets closer and closer to a (from either side). Eventually, x will lie in the interval $(a - \delta, a + \delta)$, which is marked in green in Figure 2.4.5*a*; when this happens the value of $f(x)$ will be greater than M, marked in red in the figure. Since this is true for any positive value of M (no matter how large), we can force the values of $f(x)$ to be as large as we like by making x sufficiently close to a. This agrees with our informal concept of this limit. Similarly, Figure 2.4.5*b* illustrates Definition 2.4.5.

► **Example 5** Prove that $\lim_{x \to 0} \dfrac{1}{x^2} = +\infty$.

Solution. Applying Definition 2.4.4 with $f(x) = 1/x^2$ and $a = 0$, we must show that given a number $M > 0$, we can find a number $\delta > 0$ such that

$$\frac{1}{x^2} > M \quad \text{if} \quad 0 < |x - 0| < \delta \tag{13}$$

or, on taking reciprocals and simplifying,

$$x^2 < \frac{1}{M} \quad \text{if} \quad 0 < |x| < \delta \tag{14}$$

But $x^2 < 1/M$ if $|x| < 1/\sqrt{M}$, so that $\delta = 1/\sqrt{M}$ satisfies (14). Since (13) is equivalent to (14), the proof is complete. ◄

✔QUICK CHECK EXERCISES 2.4 (See page 125 for answers.)

1. The definition of a two-sided limit states: $\lim_{x \to a} f(x) = L$ if given any number ________ there is a number ________ such that $|f(x) - L| < \epsilon$ if ________.

2. Find the largest open interval centered at $x = 1$ such that for each x in the interval, $f(x)$ is within ϵ units of $f(1) = 5$.
(a) $f(x) = x + 4,\ \epsilon = 0.01$
(b) $f(x) = 5x,\ \epsilon = 0.02$
(c) $f(x) = 3x^2 + 2,\ \epsilon = 0.1212$

3. Suppose that ϵ is any positive number. Find the largest value of δ such that $|5x - 10| < \epsilon$ if $0 < |x - 2| < \delta$.

4. The definition of limit at $+\infty$ states: $\lim_{x \to +\infty} f(x) = L$ if given any number ________ there is a positive number ________ such that $|f(x) - L| < \epsilon$ if ________.

5. Find the smallest positive number N such that for each $x > N$, the value of $f(x) = 1/\sqrt{x}$ is within ϵ units of 0.
(a) $\epsilon = 0.1$ (b) $\epsilon = 0.01$ (c) $\epsilon = 0.001$

EXERCISE SET 2.4 Graphing Utility

1. (a) Find the largest open interval, centered at the origin on the x-axis, such that for each x in the interval the value of the function $f(x) = x + 2$ is within 0.1 unit of the number $f(0) = 2$.
(b) Find the largest open interval, centered at $x = 3$, such that for each x in the interval the value of the function $f(x) = 4x - 5$ is within 0.01 unit of the number $f(3) = 7$.
(c) Find the largest open interval, centered at $x = 4$, such that for each x in the interval the value of the function $f(x) = x^2$ is within 0.001 unit of the number $f(4) = 16$.

2. In each part, find the largest open interval, centered at $x = 0$, such that for each x in the interval the value of $f(x) = 2x + 3$ is within ϵ units of the number $f(0) = 3$.
(a) $\epsilon = 0.1$ (b) $\epsilon = 0.01$ (c) $\epsilon = 0.0012$

3. (a) Find the values of x_0 and x_1 in the accompanying figure.
(b) Find a positive number δ such that $|\sqrt{x} - 2| < 0.05$ if $0 < |x - 4| < \delta$.

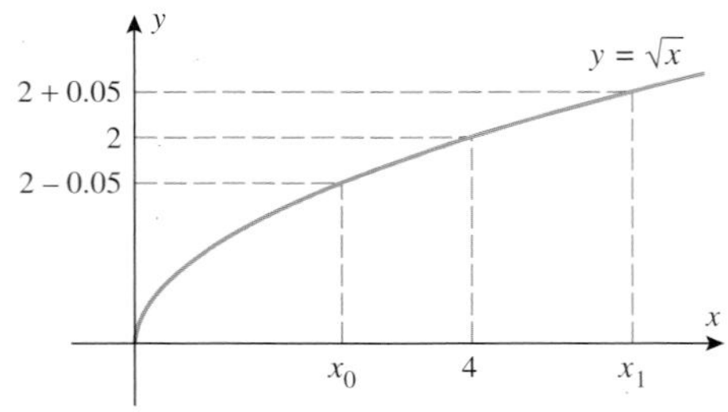

Figure Ex-3

4. (a) Find the values of x_0 and x_1 in the accompanying figure.
(b) Find a positive number δ such that $|(1/x) - 1| < 0.1$ if $0 < |x - 1| < \delta$.

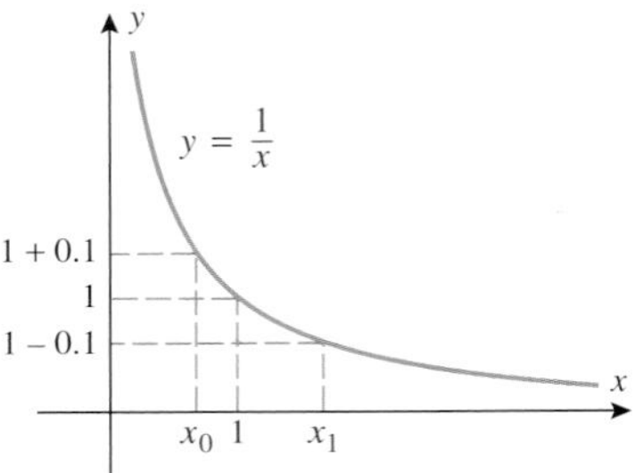

Figure Ex-4

5. Generate the graph of $f(x) = x^3 - 4x + 5$ with a graphing utility, and use the graph to find a number δ such that $|f(x) - 2| < 0.05$ if $0 < |x - 1| < \delta$. [*Hint:* Show that the inequality $|f(x) - 2| < 0.05$ can be rewritten as $1.95 < x^3 - 4x + 5 < 2.05$, and estimate the values of x for which $x^3 - 4x + 5 = 1.95$ and $x^3 - 4x + 5 = 2.05$.]

6. Use the method of Exercise 5 to find a number δ such that $|\sqrt{5x + 1} - 4| < 0.5$ if $0 < |x - 3| < \delta$.

7. Let $f(x) = x + \sqrt{x}$ with $L = \lim_{x \to 1} f(x)$ and let $\epsilon = 0.2$. Use a graphing utility and its trace feature to find a positive number δ such that $|f(x) - L| < \epsilon$ if $0 < |x - 1| < \delta$.

8. Let $f(x) = (\sin 2x)/x$ and use a graphing utility to conjecture the value of $L = \lim_{x \to 0} f(x)$. Then let $\epsilon = 0.1$ and use the graphing utility and its trace feature to find a positive number δ such that $|f(x) - L| < \epsilon$ if $0 < |x| < \delta$.

9–16 A positive number ϵ and the limit L of a function f at a are given. Find a number δ such that $|f(x) - L| < \epsilon$ if $0 < |x - a| < \delta$.

9. $\lim_{x \to 4} 2x = 8;\ \epsilon = 0.1$

10. $\lim_{x \to 3} (5x - 2) = 13;\ \epsilon = 0.01$

11. $\lim_{x \to 3} \dfrac{x^2 - 9}{x - 3} = 6;\ \epsilon = 0.05$

12. $\lim_{x \to -1/2} \dfrac{4x^2 - 1}{2x + 1} = -2;\ \epsilon = 0.05$

13. $\lim_{x \to 2} x^3 = 8;\ \epsilon = 0.001$

14. $\lim_{x \to 4} \sqrt{x} = 2;\ \epsilon = 0.001$

15. $\lim_{x \to 5} \frac{1}{x} = \frac{1}{5}$; $\epsilon = 0.05$

16. $\lim_{x \to 0} |x| = 0$; $\epsilon = 0.05$

FOCUS ON CONCEPTS

17. Suppose that $f(x)$ is a function and that for any given $\epsilon > 0$, the condition $0 < |x - 4| < \epsilon/5$ guarantees that $|f(x) - 3| < \epsilon$.
(a) What limit is described by this statement?
(b) Find a value of δ such that $0 < |x - 4| < \delta$ guarantees that $|10f(x) - 30| < 0.005$.

18. Suppose that $f(x)$ and $g(x)$ are functions and that for any given $\epsilon > 0$ the following conditions hold:
(i) If $0 < |x - 3| < \epsilon^2$, then $|f(x) - 7| < \epsilon$.
(ii) If $0 < |x - 3| < \min(1/2, \epsilon/8)$, then $|g(x) - 5| < \epsilon$.
(a) What limits are described by these two statements?
(b) Find a value of δ such that $0 < |x - 3| < \delta$ guarantees that $|3f(x) - 21| < 0.03$.

19. Suppose that $f(x)$ is the function in Exercise 17. Find a value of δ such that $0 < |x - 4| < \delta$ guarantees that $|10f(x) + 2x - 38| < 0.01$.
Hint: By the triangle inequality,

$$|(10f(x) - 30) + (2x - 8)| \le |10f(x) - 30| + |2x - 8|$$

20. Suppose that $f(x)$ and $g(x)$ are the functions in Exercise 18. Find a value of δ such that $0 < |x - 3| < \delta$ guarantees that $|3f(x) + g(x) - 26| < 0.06$.
Hint: By the triangle inequality,

$$|(3f(x) - 21) + (g(x) - 5)| \le |3f(x) - 21| + |g(x) - 5|$$

21–26 Use Definition 2.4.1 to prove that the stated limit is correct.

21. $\lim_{x \to 5} 3x = 15$

22. $\lim_{x \to -1} (7x + 5) = -2$

23. $\lim_{x \to 0} \frac{2x^2 + x}{x} = 1$

24. $\lim_{x \to -3} \frac{x^2 - 9}{x + 3} = -6$

25. $\lim_{x \to 1} f(x) = 3$, where $f(x) = \begin{cases} x + 2, & x \ne 1 \\ 10, & x = 1 \end{cases}$

26. $\lim_{x \to 2} f(x) = 5$, where $f(x) = \begin{cases} 9 - 2x, & x \ne 2 \\ 49, & x = 2 \end{cases}$

FOCUS ON CONCEPTS

27. (a) Show that

$$|(3x^2 + 2x - 20) - 300| = |3x + 32| \cdot |x - 10|$$

(b) Find an upper bound for $|3x + 32|$ if x satisfies $|x - 10| < 1$.
(c) Fill in the blanks to complete a proof that

$$\lim_{x \to 10} [3x^2 + 2x - 20] = 300$$

Suppose that $\epsilon > 0$. Set $\delta = \min(1, \underline{\hspace{2cm}})$ and assume that $0 < |x - 10| < \delta$. Then

$$\begin{aligned} |(3x^2 + 2x - 20) - 300| &= |3x + 32| \cdot |x - 10| \\ &< \underline{\hspace{2cm}} \cdot |x - 10| \\ &< \underline{\hspace{2cm}} \cdot \underline{\hspace{2cm}} \\ &= \epsilon \end{aligned}$$

28. (a) Show that

$$\left| \frac{28}{3x + 1} - 4 \right| = \left| \frac{12}{3x + 1} \right| \cdot |x - 2|$$

(b) Is $|12/(3x + 1)|$ bounded if $|x - 2| < 4$? If not, explain; if so, give a bound.
(c) Is $|12/(3x + 1)|$ bounded if $|x - 2| < 1$? If not, explain; if so, give a bound.
(d) Fill in the blanks to complete a proof that

$$\lim_{x \to 2} \left[\frac{28}{3x + 1} \right] = 4$$

Suppose that $\epsilon > 0$. Set $\delta = \min(1, \underline{\hspace{2cm}})$ and assume that $0 < |x - 2| < \delta$. Then

$$\begin{aligned} \left| \frac{28}{3x + 1} - 4 \right| &= \left| \frac{12}{3x + 1} \right| \cdot |x - 2| \\ &< \underline{\hspace{2cm}} \cdot |x - 2| \\ &< \underline{\hspace{2cm}} \cdot \underline{\hspace{2cm}} \\ &= \epsilon \end{aligned}$$

29–34 Use Definition 2.4.1 to prove that the stated limit is correct.

29. $\lim_{x \to 1} 2x^2 = 2$

30. $\lim_{x \to 3} (x^2 + x) = 12$

31. $\lim_{x \to -2} \frac{1}{x + 1} = -1$

32. $\lim_{x \to 1/2} \frac{2x + 3}{x} = 8$

33. $\lim_{x \to 4} \sqrt{x} = 2$

34. $\lim_{x \to 2} f(x) = 5$, where $f(x) = \begin{cases} 9 - 2x, & x < 2 \\ 3x - 1, & x > 2 \end{cases}$

35. (a) Find the smallest positive number N such that for each x in the interval $(N, +\infty)$, the value of the function $f(x) = 1/x^2$ is within 0.1 unit of $L = 0$.
(b) Find the smallest positive number N such that for each x in the interval $(N, +\infty)$, the value of $f(x) = x/(x + 1)$ is within 0.01 unit of $L = 1$.
(c) Find the largest negative number N such that for each x in the interval $(-\infty, N)$, the value of the function $f(x) = 1/x^3$ is within 0.001 unit of $L = 0$.
(d) Find the largest negative number N such that for each x in the interval $(-\infty, N)$, the value of the function $f(x) = x/(x + 1)$ is within 0.01 unit of $L = 1$.

36. In each part, find the smallest positive value of N such that for each x in the interval $(N, +\infty)$, the function $f(x) = 1/x^3$ is within ϵ units of the number $L = 0$.
(a) $\epsilon = 0.1$ (b) $\epsilon = 0.01$ (c) $\epsilon = 0.001$

37. (a) Find the values of x_1 and x_2 in the accompanying figure.
(b) Find a positive number N such that

$$\left|\frac{x^2}{1+x^2} - 1\right| < \epsilon$$

for $x > N$.
(c) Find a negative number N such that

$$\left|\frac{x^2}{1+x^2} - 1\right| < \epsilon$$

for $x < N$.

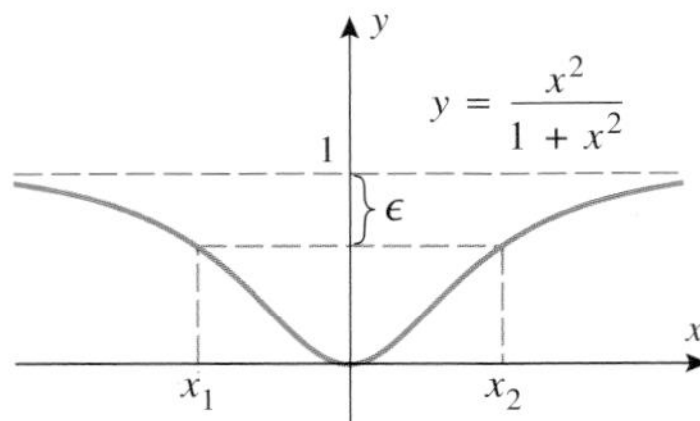

Figure Ex-37

38. (a) Find the values of x_1 and x_2 in the accompanying figure.
(b) Find a positive number N such that

$$\left|\frac{1}{\sqrt[3]{x}} - 0\right| = \left|\frac{1}{\sqrt[3]{x}}\right| < \epsilon$$

for $x > N$.
(c) Find a negative number N such that

$$\left|\frac{1}{\sqrt[3]{x}} - 0\right| = \left|\frac{1}{\sqrt[3]{x}}\right| < \epsilon$$

for $x < N$.

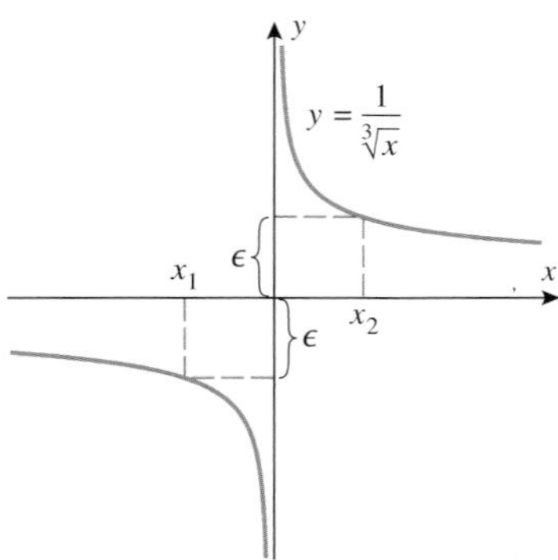

Figure Ex-38

39–42 A positive number ϵ and the limit L of a function f at $+\infty$ are given. Find a positive number N such that $|f(x) - L| < \epsilon$ if $x > N$.

39. $\lim_{x\to+\infty} \frac{1}{x^2} = 0;\ \epsilon = 0.01$

40. $\lim_{x\to+\infty} \frac{1}{x+2} = 0;\ \epsilon = 0.005$

41. $\lim_{x\to+\infty} \frac{x}{x+1} = 1;\ \epsilon = 0.001$

42. $\lim_{x\to+\infty} \frac{4x-1}{2x+5} = 2;\ \epsilon = 0.1$

43–46 A positive number ϵ and the limit L of a function f at $-\infty$ are given. Find a negative number N such that $|f(x) - L| < \epsilon$ if $x < N$.

43. $\lim_{x\to-\infty} \frac{1}{x+2} = 0;\ \epsilon = 0.005$

44. $\lim_{x\to-\infty} \frac{1}{x^2} = 0;\ \epsilon = 0.01$

45. $\lim_{x\to-\infty} \frac{4x-1}{2x+5} = 2;\ \epsilon = 0.1$

46. $\lim_{x\to-\infty} \frac{x}{x+1} = 1;\ \epsilon = 0.001$

47–52 Use Definition 2.4.2 or 2.4.3 to prove that the stated limit is correct.

47. $\lim_{x\to+\infty} \frac{1}{x^2} = 0$

48. $\lim_{x\to+\infty} \frac{1}{x+2} = 0$

49. $\lim_{x\to-\infty} \frac{4x-1}{2x+5} = 2$

50. $\lim_{x\to-\infty} \frac{x}{x+1} = 1$

51. $\lim_{x\to+\infty} \frac{2\sqrt{x}}{\sqrt{x}-1} = 2$

52. $\lim_{x\to-\infty} \frac{\sqrt[3]{x}}{\sqrt[3]{x}+2} = 1$

53. (a) Find the largest open interval, centered at the origin on the x-axis, such that for each x in the interval, other than the center, the values of $f(x) = 1/x^2$ are greater than 100.
(b) Find the largest open interval, centered at $x = 1$, such that for each x in the interval, other than the center, the values of the function $f(x) = 1/|x-1|$ are greater than 1000.
(c) Find the largest open interval, centered at $x = 3$, such that for each x in the interval, other than the center, the values of the function $f(x) = -1/(x-3)^2$ are less than -1000.
(d) Find the largest open interval, centered at the origin on the x-axis, such that for each x in the interval, other than the center, the values of $f(x) = -1/x^4$ are less than $-10{,}000$.

54. In each part, find the largest open interval centered at $x = 1$, such that for each x in the interval, other than the center, the value of $f(x) = 1/(x-1)^2$ is greater than M.
(a) $M = 10$ (b) $M = 1000$ (c) $M = 100{,}000$

55–60 Use Definition 2.4.4 or 2.4.5 to prove that the stated limit is correct.

55. $\lim_{x\to 3} \frac{1}{(x-3)^2} = +\infty$

56. $\lim_{x\to 3} \frac{-1}{(x-3)^2} = -\infty$

57. $\lim_{x\to 0} \frac{1}{|x|} = +\infty$

58. $\lim_{x\to 1} \frac{1}{|x-1|} = +\infty$

59. $\lim_{x\to 0} \left(-\frac{1}{x^4}\right) = -\infty$

60. $\lim_{x\to 0} \frac{1}{x^4} = +\infty$

61–66 Use the marginal note on page 117 to prove that the stated limit is correct.

61. $\lim_{x \to 2^+} (x+1) = 3$ **62.** $\lim_{x \to 1^-} (3x+2) = 5$

63. $\lim_{x \to 4^+} \sqrt{x-4} = 0$ **64.** $\lim_{x \to 0^-} \sqrt{-x} = 0$

65. $\lim_{x \to 2^+} f(x) = 2$, where $f(x) = \begin{cases} x, & x > 2 \\ 3x, & x \le 2 \end{cases}$

66. $\lim_{x \to 2^-} f(x) = 6$, where $f(x) = \begin{cases} x, & x > 2 \\ 3x, & x \le 2 \end{cases}$

67–70 Write out the definition for the corresponding limit in the marginal note on page 121, and use your definition to prove that the stated limit is correct.

67. (a) $\lim_{x \to 1^+} \dfrac{1}{1-x} = -\infty$ (b) $\lim_{x \to 1^-} \dfrac{1}{1-x} = +\infty$

68. (a) $\lim_{x \to 0^+} \dfrac{1}{x} = +\infty$ (b) $\lim_{x \to 0^-} \dfrac{1}{x} = -\infty$

69. (a) $\lim_{x \to +\infty} (x+1) = +\infty$ (b) $\lim_{x \to -\infty} (x+1) = -\infty$

70. (a) $\lim_{x \to +\infty} (x^2-3) = +\infty$ (b) $\lim_{x \to -\infty} (x^3+5) = -\infty$

71. Prove the result in Example 3 under the assumption that $\delta \le 2$ rather than $\delta \le 1$.

72. (a) In Definition 2.4.1 there is a condition requiring that $f(x)$ be defined for all x in some open interval containing a, except possibly at a itself. What is the purpose of this requirement?
(b) Why is $\lim_{x \to 0} \sqrt{x} = 0$ an incorrect statement?
(c) Is $\lim_{x \to 0.01} \sqrt{x} = 0.1$ a correct statement?

73. According to Ohm's law, when a voltage of V volts is applied across a resistor with a resistance of R ohms, a current of $I = V/R$ amperes flows through the resistor.
(a) How much current flows if a voltage of 3.0 volts is applied across a resistance of 7.5 ohms?
(b) If the resistance varies by ± 0.1 ohm, and the voltage remains constant at 3.0 volts, what is the resulting range of values for the current?
(c) If temperature variations cause the resistance to vary by $\pm\delta$ from its value of 7.5 ohms, and the voltage remains constant at 3.0 volts, what is the resulting range of values for the current?
(d) If the current is not allowed to vary by more than $\epsilon = \pm 0.001$ ampere at a voltage of 3.0 volts, what variation of $\pm\delta$ from the value of 7.5 ohms is allowable?
(e) Certain alloys become ***superconductors*** as their temperature approaches absolute zero ($-273°$C), meaning that their resistance approaches zero. If the voltage remains constant, what happens to the current in a superconductor as $R \to 0^+$?

✔QUICK CHECK ANSWERS 2.4

1. $\epsilon > 0$; $\delta > 0$; $0 < |x-a| < \delta$ **2.** (a) (0.99, 1.01) (b) (0.996, 1.004) (c) (0.98, 1.02) **3.** $\delta = \epsilon/5$ **4.** $\epsilon > 0$; N; $x > N$
5. (a) $N = 100$ (b) $N = 10{,}000$ (c) $N = 1{,}000{,}000$

2.5 CONTINUITY

A thrown baseball cannot vanish at some point and reappear someplace else to continue its motion. Thus, we perceive the path of the ball as an unbroken curve. In this section, we translate "unbroken curve" into a precise mathematical formulation called continuity, and develop some fundamental properties of continuous curves.

DEFINITION OF CONTINUITY

Intuitively, the graph of a function can be described as a "continuous curve" if it has no breaks or holes. To make this idea more precise we need to understand what properties of a function can cause breaks or holes. Referring to Figure 2.5.1, we see that the graph of a function has a break or hole if any of the following conditions occur:

- The function f is undefined at c (Figure 2.5.1a).
- The limit of $f(x)$ does not exist as x approaches c (Figures 2.5.1b, 2.5.1c).
- The value of the function and the value of the limit at c are different (Figure 2.5.1d).

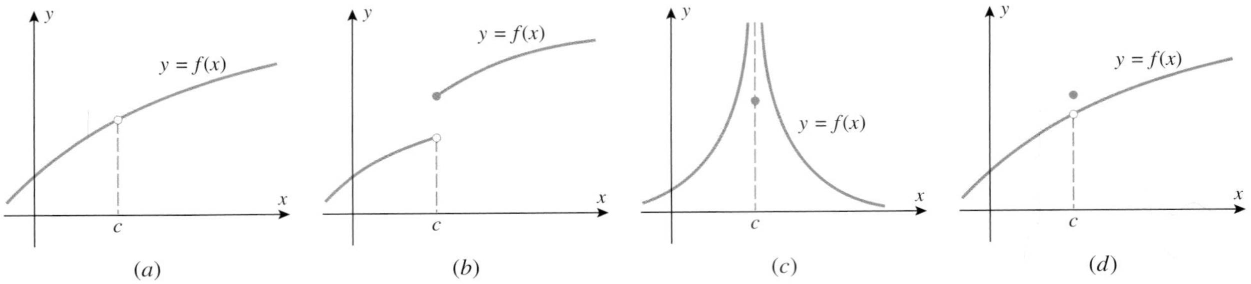

Figure 2.5.1

This suggests the following definition.

> The third condition in Definition 2.5.1 actually implies the first two, since it is tacitly understood in the statement
>
> $$\lim_{x \to c} f(x) = f(c)$$
>
> that the limit exists and the function is defined at c. Thus, when we want to establish continuity at c our usual procedure will be to verify the third condition only.

2.5.1 DEFINITION. A function f is said to be ***continuous at*** $\boldsymbol{x = c}$ provided the following conditions are satisfied:

1. $f(c)$ is defined.
2. $\lim_{x \to c} f(x)$ exists.
3. $\lim_{x \to c} f(x) = f(c)$.

If one or more of the conditions of this definition fails to hold, then we will say that f has a ***discontinuity at*** $\boldsymbol{x = c}$. Each function drawn in Figure 2.5.1 illustrates a discontinuity at $x = c$. In Figure 2.5.1*a*, the function is not defined at c, violating the first condition of Definition 2.5.1. In Figures 2.5.1*b* and 2.5.1*c*, $\lim_{x \to c} f(x)$ does not exist, violating the second condition of Definition 2.5.1. In Figure 2.5.1*d*, the function is defined at c and $\lim_{x \to c} f(x)$ exists, but these two values are not equal, violating the third condition of Definition 2.5.1.

▶ **Example 1** Determine whether the following functions are continuous at $x = 2$.

$$f(x) = \frac{x^2 - 4}{x - 2}, \qquad g(x) = \begin{cases} \dfrac{x^2 - 4}{x - 2}, & x \neq 2 \\ 3, & x = 2, \end{cases} \qquad h(x) = \begin{cases} \dfrac{x^2 - 4}{x - 2}, & x \neq 2 \\ 4, & x = 2 \end{cases}$$

Solution. In each case we must determine whether the limit of the function as $x \to 2$ is the same as the value of the function at $x = 2$. In all three cases the functions are identical, except at $x = 2$, and hence all three have the same limit at $x = 2$, namely

$$\lim_{x \to 2} f(x) = \lim_{x \to 2} g(x) = \lim_{x \to 2} h(x) = \lim_{x \to 2} \frac{x^2 - 4}{x - 2} = \lim_{x \to 2} (x + 2) = 4$$

The function f is undefined at $x = 2$, and hence is not continuous at $x = 2$ (Figure 2.5.2*a*). The function g is defined at $x = 2$, but its value there is $g(2) = 3$, which is not the same as the limit as x approaches 2; hence, g is also not continuous at $x = 2$ (Figure 2.5.2*b*). The value of the function h at $x = 2$ is $h(2) = 4$, which is the same as the limit as x approaches 2; hence, h is continuous at $x = 2$ (Figure 2.5.2*c*). (Note that the function h could have been written more simply as $h(x) = x + 2$, but we wrote it in piecewise form to emphasize its relationship to f and g.) ◀

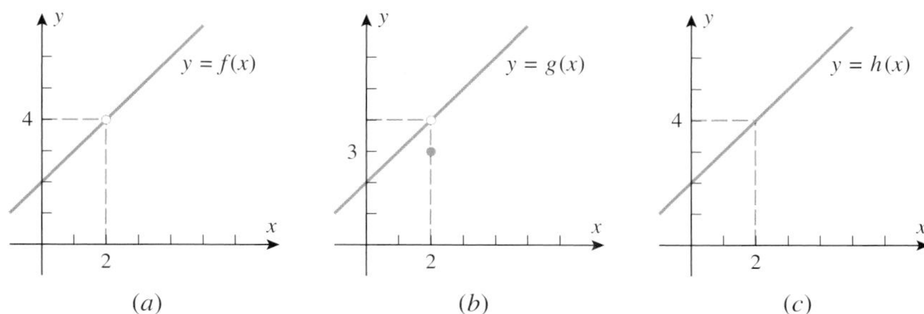

Figure 2.5.2

CONTINUITY IN APPLICATIONS

In applications, discontinuities often signal the occurrence of important physical events. For example, Figure 2.5.3*a* is a graph of voltage versus time for an underground cable that is accidentally cut by a work crew at time $t = t_0$ (the voltage drops to zero when the line is cut). Figure 2.5.3*b* shows the graph of inventory versus time for a company that restocks its warehouse to y_1 units when the inventory falls to y_0 units. The discontinuities occur at those times when restocking occurs.

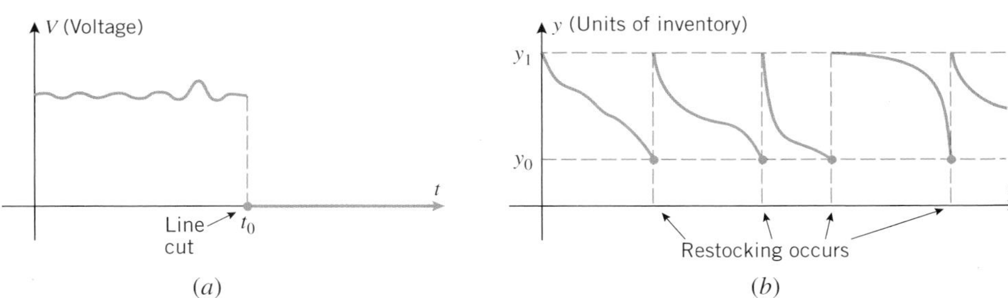

Figure 2.5.3

CONTINUITY ON AN INTERVAL

If a function f is continuous at each number in an open interval (a, b), then we say that f is ***continuous on*** **(*a*, *b*)**. This definition applies to infinite open intervals of the form $(a, +\infty)$, $(-\infty, b)$, and $(-\infty, +\infty)$. In the case where f is continuous on $(-\infty, +\infty)$, we will say that f is ***continuous everywhere***.

Because Definition 2.5.1 involves a two-sided limit, that definition does not generally apply at the endpoints of a closed interval $[a, b]$ or at the endpoint of an interval of the form $[a, b)$, $(a, b]$, $(-\infty, b]$, or $[a, +\infty)$. To remedy this problem, we will agree that a function is continuous at an endpoint of an interval if its value at the endpoint is equal to the appropriate one-sided limit at that endpoint. For example, the function graphed in Figure 2.5.4 is continuous at the right endpoint of the interval $[a, b]$ because

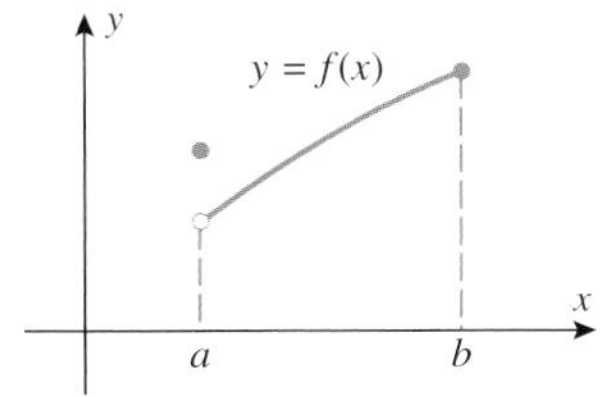

Figure 2.5.4

$$\lim_{x \to b^-} f(x) = f(b)$$

but it is not continuous at the left endpoint because

$$\lim_{x \to a^+} f(x) \neq f(a)$$

In general, we will say a function f is ***continuous from the left*** at c if

$$\lim_{x \to c^-} f(x) = f(c)$$

and is ***continuous from the right*** at c if

$$\lim_{x\to c^+} f(x) = f(c)$$

Using this terminology we define continuity on a closed interval as follows.

> Modify Definition 2.5.2 appropriately so that it applies to intervals of the form $[a, +\infty)$, $(-\infty, b]$, $(a, b]$, and $[a, b)$.

2.5.2 DEFINITION. A function f is said to be ***continuous on a closed interval [a, b]*** if the following conditions are satisfied:

1. f is continuous on (a, b).
2. f is continuous from the right at a.
3. f is continuous from the left at b.

▸ **Example 2** What can you say about the continuity of the function $f(x) = \sqrt{9 - x^2}$?

Solution. Because the natural domain of this function is the closed interval $[-3, 3]$, we will need to investigate the continuity of f on the open interval $(-3, 3)$ and at the two endpoints. If c is any point in the interval $(-3, 3)$, then it follows from Theorem 2.2.2(*e*) that

$$\lim_{x\to c} f(x) = \lim_{x\to c} \sqrt{9 - x^2} = \sqrt{\lim_{x\to c}(9 - x^2)} = \sqrt{9 - c^2} = f(c)$$

which proves f is continuous at each point in the interval $(-3, 3)$. The function f is also continuous at the endpoints since

$$\lim_{x\to 3^-} f(x) = \lim_{x\to 3^-} \sqrt{9 - x^2} = \sqrt{\lim_{x\to 3^-}(9 - x^2)} = 0 = f(3)$$

$$\lim_{x\to -3^+} f(x) = \lim_{x\to -3^+} \sqrt{9 - x^2} = \sqrt{\lim_{x\to -3^+}(9 - x^2)} = 0 = f(-3)$$

Thus, f is continuous on the closed interval $[-3, 3]$. ◂

SOME PROPERTIES OF CONTINUOUS FUNCTIONS

The following theorem, which is a consequence of Theorem 2.2.2, will enable us to reach conclusions about the continuity of functions that are obtained by adding, subtracting, multiplying, and dividing continuous functions.

2.5.3 THEOREM. *If the functions f and g are continuous at c, then*

(*a*) *$f + g$ is continuous at c.*

(*b*) *$f - g$ is continuous at c.*

(*c*) *fg is continuous at c.*

(*d*) *f/g is continuous at c if $g(c) \neq 0$ and has a discontinuity at c if $g(c) = 0$.*

We will prove part (*d*). The remaining proofs are similar and will be left to the exercises.

PROOF. First, consider the case where $g(c) = 0$. In this case $f(c)/g(c)$ is undefined, so the function f/g has a discontinuity at c.

Next, consider the case where $g(c) \neq 0$. To prove that f/g is continuous at c, we must show that

$$\lim_{x\to c} \frac{f(x)}{g(x)} = \frac{f(c)}{g(c)} \tag{1}$$

Since f and g are continuous at c,

$$\lim_{x \to c} f(x) = f(c) \quad \text{and} \quad \lim_{x \to c} g(x) = g(c)$$

Thus, by Theorem 2.2.2(d)

$$\lim_{x \to c} \frac{f(x)}{g(x)} = \frac{\lim_{x \to c} f(x)}{\lim_{x \to c} g(x)} = \frac{f(c)}{g(c)}$$

which proves (1). ■

CONTINUITY OF POLYNOMIALS AND RATIONAL FUNCTIONS

The general procedure for showing that a function is continuous everywhere is to show that it is continuous at an *arbitrary* point. For example, we know from Theorem 2.2.3 that if $p(x)$ is a polynomial and a is *any* real number, then

$$\lim_{x \to a} p(x) = p(a)$$

This shows that polynomials are continuous everywhere. Moreover, since rational functions are ratios of polynomials, it follows from part (d) of Theorem 2.5.3 that rational functions are continuous at points other than the zeros of the denominator, and at these zeros they have discontinuities. Thus, we have the following result.

2.5.4 THEOREM.

(a) *A polynomial is continuous everywhere.*

(b) *A rational function is continuous at every point where the denominator is nonzero, and has discontinuities at the points where the denominator is zero.*

TECHNOLOGY MASTERY

If you use a graphing utility to generate the graph of the equation in Example 3, there is a good chance you will see the discontinuity at $x = 2$ but not at $x = 3$. Try it, and explain what you think is happening.

► **Example 3** For what values of x is there a discontinuity in the graph of

$$y = \frac{x^2 - 9}{x^2 - 5x + 6}?$$

Solution. The function being graphed is a rational function, and hence is continuous at every number where the denominator is nonzero. Solving the equation

$$x^2 - 5x + 6 = 0$$

yields discontinuities at $x = 2$ and at $x = 3$. ◄

► **Example 4** Show that $|x|$ is continuous everywhere (Figure 1.1.10).

Solution. We can write $|x|$ as

$$|x| = \begin{cases} x & \text{if} \quad x > 0 \\ 0 & \text{if} \quad x = 0 \\ -x & \text{if} \quad x < 0 \end{cases}$$

so $|x|$ is the same as the polynomial x on the interval $(0, +\infty)$ and is the same as the polynomial $-x$ on the interval $(-\infty, 0)$. But polynomials are continuous everywhere, so $x = 0$ is the only possible discontinuity for $|x|$. Since $|0| = 0$, to prove the continuity at $x = 0$ we must show that

$$\lim_{x \to 0} |x| = 0 \qquad (2)$$

Because the piecewise formula for $|x|$ changes at 0, it will be helpful to consider the one-sided limits at 0 rather than the two-sided limit. We obtain

$$\lim_{x\to 0^+} |x| = \lim_{x\to 0^+} x = 0 \quad \text{and} \quad \lim_{x\to 0^-} |x| = \lim_{x\to 0^-} (-x) = 0$$

Thus, (2) holds and $|x|$ is continuous at $x = 0$. ◄

CONTINUITY OF COMPOSITIONS

The following theorem, whose proof is given in Appendix C, will be useful for calculating limits of compositions of functions.

In words, Theorem 2.5.5 states that a limit symbol can be moved through a function sign provided that the limit of the expression inside the function sign exists and the function is continuous at this limit.

2.5.5 THEOREM. *If $\lim_{x\to c} g(x) = L$ and if the function f is continuous at L, then $\lim_{x\to c} f(g(x)) = f(L)$. That is,*

$$\lim_{x\to c} f(g(x)) = f\left(\lim_{x\to c} g(x)\right)$$

This equality remains valid if $\lim_{x\to c}$ is replaced everywhere by one of $\lim_{x\to c^+}$, $\lim_{x\to c^-}$, $\lim_{x\to +\infty}$, or $\lim_{x\to -\infty}$.

In the special case of this theorem where $f(x) = |x|$, the fact that $|x|$ is continuous everywhere allows us to write

$$\lim_{x\to c} |g(x)| = \left|\lim_{x\to c} g(x)\right| \tag{3}$$

provided $\lim_{x\to c} g(x)$ exists. Thus, for example,

$$\lim_{x\to 3} |5 - x^2| = \left|\lim_{x\to 3} (5 - x^2)\right| = |-4| = 4$$

The following theorem is concerned with the continuity of compositions of functions; the first part deals with continuity at a specific number and the second with continuity everywhere.

2.5.6 THEOREM.

(a) *If the function g is continuous at c, and the function f is continuous at $g(c)$, then the composition $f \circ g$ is continuous at c.*

(b) *If the function g is continuous everywhere and the function f is continuous everywhere, then the composition $f \circ g$ is continuous everywhere.*

PROOF. We will prove part (*a*) only; the proof of part (*b*) can be obtained by applying part (*a*) at an arbitrary number c. To prove that $f \circ g$ is continuous at c, we must show that the value of $f \circ g$ and the value of its limit are the same at $x = c$. But this is so, since we can write

$$\lim_{x\to c} (f \circ g)(x) = \lim_{x\to c} f(g(x)) = f\left(\lim_{x\to c} g(x)\right) = f(g(c)) = (f \circ g)(c) \qquad ■$$

Theorem 2.5.5 g is continuous at c.

Can the absolute value of a function that is not continuous everywhere be continuous everywhere? Justify your answer.

We know from Example 4 that the function $|x|$ is continuous everywhere. Thus, if $g(x)$ is continuous at c, then by part (*a*) of Theorem 2.5.6, the function $|g(x)|$ must also be

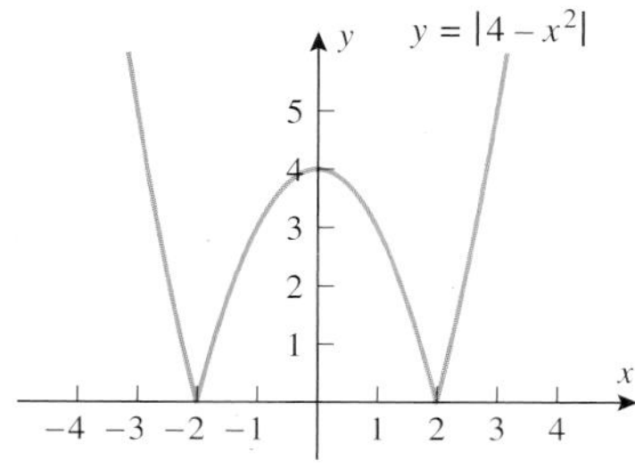

Figure 2.5.5

continuous at c; and, more generally, if $g(x)$ is continuous everywhere, then so is $|g(x)|$. Stated informally:

The absolute value of a continuous function is continuous.

For example, the polynomial $g(x) = 4 - x^2$ is continuous everywhere, so we can conclude that the function $|4 - x^2|$ is also continuous everywhere (Figure 2.5.5).

CONTINUITY OF INVERSE FUNCTIONS

Since the graphs of a one-to-one function f and its inverse f^{-1} are reflections of one another about the line $y = x$, it is clear geometrically that if the graph of f has no breaks or holes in it, then neither does the graph of f^{-1}. This, and the fact that the range of f is the domain of f^{-1}, suggests the following result, which we state without formal proof.

To paraphrase Theorem 2.5.7, *the inverse of a continuous function is continuous.*

2.5.7 THEOREM. *If f is a one-to-one function that is continuous at each point of its domain, then f^{-1} is continuous at each point of its domain; that is f^{-1} is continuous at each point in the range of f.*

► **Example 5** Is the inverse of $f(x) = x^3 + x$ continuous at each point in $(-\infty, +\infty)$?

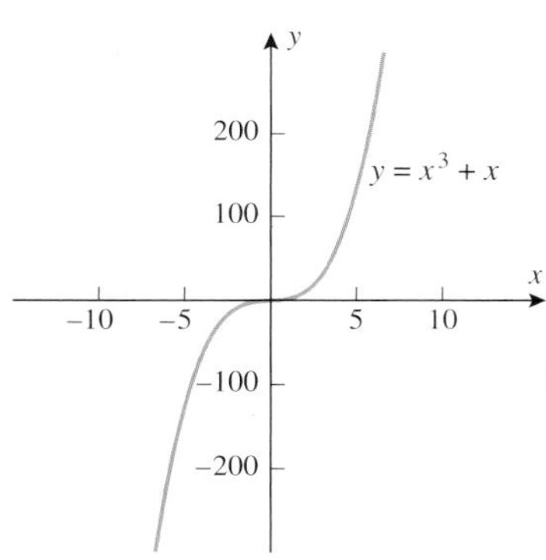

Figure 2.5.6

Solution. Note that f is increasing for all x (why?), so f is one-to-one and has an inverse. Also notice that f is a polynomial, and so is continuous everywhere. From Figure 2.5.6 we infer that the range of f is $(-\infty, +\infty)$. Although a formula for $f^{-1}(x)$ cannot be found easily, we can use Theorem 2.5.7 to conclude that f^{-1} is continous on $(-\infty, +\infty)$. ◄

THE INTERMEDIATE-VALUE THEOREM

Figure 2.5.7 shows the graph of a function that is continuous on the closed interval $[a, b]$. The figure suggests that if we draw any horizontal line $y = k$, where k is between $f(a)$ and $f(b)$, then that line will cross the curve $y = f(x)$ at least once over the interval $[a, b]$. Stated in numerical terms, if f is continuous on $[a, b]$, then the function f must take on every value k between $f(a)$ and $f(b)$ at least once as x varies from a to b. For example, the polynomial $p(x) = x^5 - x + 3$ has a value of 3 at $x = 1$ and a value of 33 at $x = 2$. Thus, it follows from the continuity of p that the equation $x^5 - x + 3 = k$ has at least one solution in the interval $[1, 2]$ for every value of k between 3 and 33. This idea is stated more precisely in the following theorem.

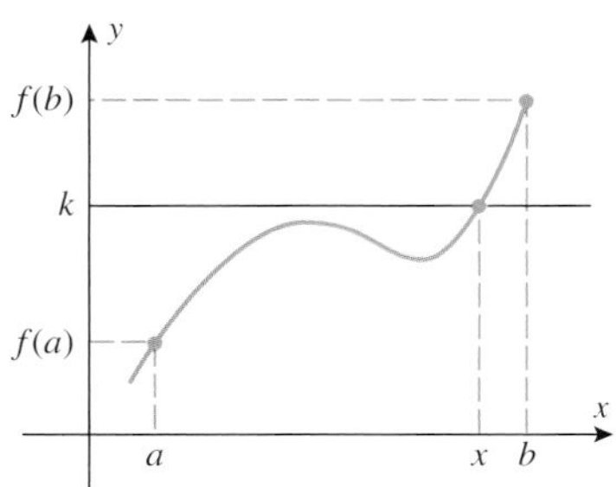

Figure 2.5.7

2.5.8 THEOREM (*Intermediate-Value Theorem*). *If f is continuous on a closed interval $[a, b]$ and k is any number between $f(a)$ and $f(b)$, inclusive, then there is at least one number x in the interval $[a, b]$ such that $f(x) = k$.*

Although this theorem is intuitively obvious, its proof depends on a mathematically precise development of the real number system, which is beyond the scope of this text.

APPROXIMATING ROOTS USING THE INTERMEDIATE-VALUE THEOREM

A variety of problems can be reduced to solving an equation $f(x) = 0$ for its roots. Sometimes it is possible to solve for the roots exactly using algebra, but often this is not possible and one must settle for decimal approximations of the roots. One procedure for approximating roots is based on the following consequence of the Intermediate-Value Theorem.

2.5.9 THEOREM. *If f is continuous on $[a, b]$, and if $f(a)$ and $f(b)$ are nonzero and have opposite signs, then there is at least one solution of the equation $f(x) = 0$ in the interval (a, b).*

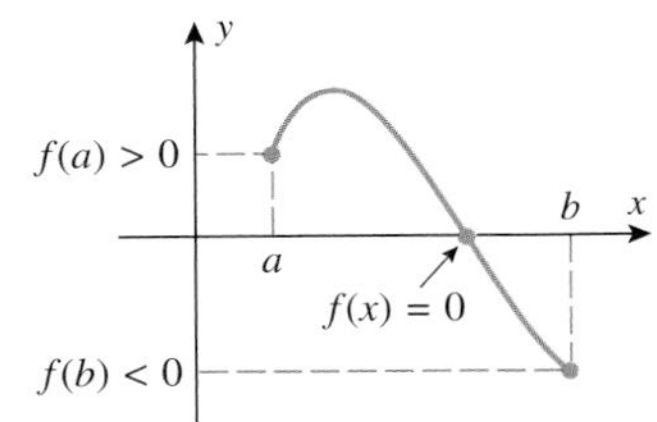

Figure 2.5.8

This result, which is illustrated in Figure 2.5.8, can be proved as follows.

PROOF. Since $f(a)$ and $f(b)$ have opposite signs, 0 is between $f(a)$ and $f(b)$. Thus, by the Intermediate-Value Theorem there is at least one number x in the interval $[a, b]$ such that $f(x) = 0$. However, $f(a)$ and $f(b)$ are nonzero, so x must lie in the interval (a, b), which completes the proof. ■

Before we illustrate how this theorem can be used to approximate roots, it will be helpful to discuss some standard terminology for describing errors in approximations. If x is an approximation to a quantity x_0, then we call

$$\epsilon = |x - x_0|$$

the ***absolute error*** or (less precisely) the ***error*** in the approximation. The terminology in Table 2.5.1 is used to describe the size of such errors.

Table 2.5.1

ERROR	DESCRIPTION
$\|x - x_0\| \le 0.1$	x approximates x_0 with an error of at most 0.1.
$\|x - x_0\| \le 0.01$	x approximates x_0 with an error of at most 0.01.
$\|x - x_0\| \le 0.001$	x approximates x_0 with an error of at most 0.001.
$\|x - x_0\| \le 0.0001$	x approximates x_0 with an error of at most 0.0001.
$\|x - x_0\| \le 0.5$	x approximates x_0 to the nearest integer.
$\|x - x_0\| \le 0.05$	x approximates x_0 to 1 decimal place (i.e., to the nearest tenth).
$\|x - x_0\| \le 0.005$	x approximates x_0 to 2 decimal places (i.e., to the nearest hundredth).
$\|x - x_0\| \le 0.0005$	x approximates x_0 to 3 decimal places (i.e., to the nearest thousandth).

▶ **Example 6** The equation

$$x^3 - x - 1 = 0$$

cannot be solved algebraically very easily because the left side has no simple factors. However, if we graph $p(x) = x^3 - x - 1$ with a graphing utility (Figure 2.5.9), then we are led to conjecture that there is one real root and that this root lies inside the interval $[1, 2]$. The existence of a root in this interval is also confirmed by Theorem 2.5.9, since $p(1) = -1$ and $p(2) = 5$ have opposite signs. Approximate this root to two decimal-place accuracy.

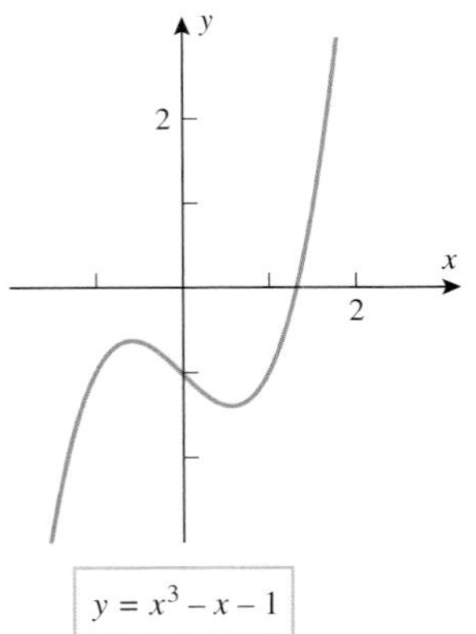

Figure 2.5.9

Solution. Our objective is to approximate the unknown root x_0 with an error of at most 0.005. It follows that if we can find an interval of length 0.01 that contains the root, then the midpoint of that interval will approximate the root with an error of at most $0.01/2 = 0.005$, which will achieve the desired accuracy.

We know that the root x_0 lies in the interval $[1, 2]$. However, this interval has length 1, which is too large. We can pinpoint the location of the root more precisely by dividing the interval $[1, 2]$ into 10 equal parts and evaluating p at the points of subdivision using a calculating utility (Table 2.5.2). In this table $p(1.3)$ and $p(1.4)$ have opposite signs, so we know that the root lies in the interval $[1.3, 1.4]$. This interval has length 0.1, which is still too large, so we repeat the process by dividing the interval $[1.3, 1.4]$ into 10 parts and evaluating p at the points of subdivision; this yields Table 2.5.3, which tells us that the root is inside the interval $[1.32, 1.33]$ (Figure 2.5.10). Since this interval has length 0.01, its midpoint 1.325 will approximate the root with an error of at most 0.005. Thus, $x_0 \approx 1.325$ to two decimal-place accuracy. ◄

Table 2.5.2

x	1	1.1	1.2	1.3	1.4	1.5	1.6	1.7	1.8	1.9	2
$p(x)$	−1	−0.77	−0.47	−0.10	0.34	0.88	1.50	2.21	3.03	3.96	5

Figure 2.5.10

Table 2.5.3

x	1.3	1.31	1.32	1.33	1.34	1.35	1.36	1.37	1.38	1.39	1.4
$p(x)$	−0.103	−0.062	−0.020	0.023	0.066	0.110	0.155	0.201	0.248	0.296	0.344

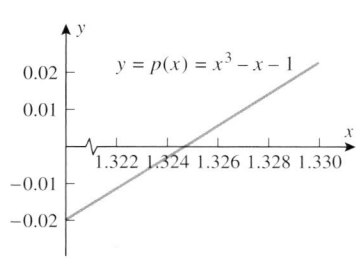

APPROXIMATING ROOTS BY ZOOMING WITH A GRAPHING UTILITY

The method illustrated in Example 6 can also be implemented with a graphing utility as follows.

Approximating a Root by Zooming

Step 1. Figure 2.5.11a shows the graph of p in the window $[-5, 5] \times [-5, 5]$ with $x\text{Scl} = 1$ and $y\text{Scl} = 1$. That graph places the root between $x = 1$ and $x = 2$.

Step 2. Since we know that the root lies between $x = 1$ and $x = 2$, we will zoom in by regraphing p over an x-interval that extends between these values and in which $x\text{Scl} = 0.1$. The y-interval and $y\text{Scl}$ are not critical, as long as the y-interval extends above and below the x-axis. Figure 2.5.11b shows the graph of p in the window $[1, 2] \times [-1, 1]$ with $x\text{Scl} = 0.1$ and $y\text{Scl} = 0.1$. That graph places the root between $x = 1.3$ and $x = 1.4$.

Step 3. Since we know that the root lies between $x = 1.3$ and $x = 1.4$, we will zoom in again by regraphing p over an x-interval that extends between these values and in which $x\text{Scl} = 0.01$. Figure 2.5.11c shows the graph of the function p in the window $[1.3, 1.4] \times [-0.1, 0.1]$ with $x\text{Scl} = 0.01$ and $y\text{Scl} = 0.01$. That graph places the root between $x = 1.32$ and $x = 1.33$.

Step 4. Since the interval in Step 3 has length 0.01, its midpoint 1.325 approximates the root with an error of at most 0.005, so $x_0 \approx 1.325$ to two decimal-place accuracy.

[−5, 5] × [−5, 5]
xScl = 1, yScl = 1
(a)

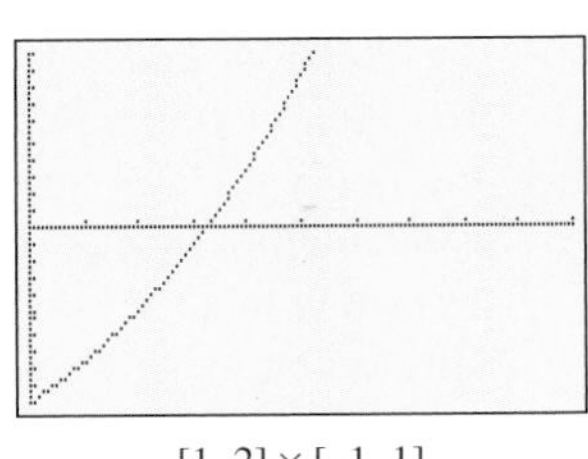

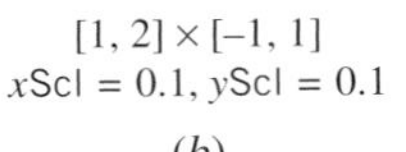

[1, 2] × [−1, 1]
xScl = 0.1, yScl = 0.1
(b)

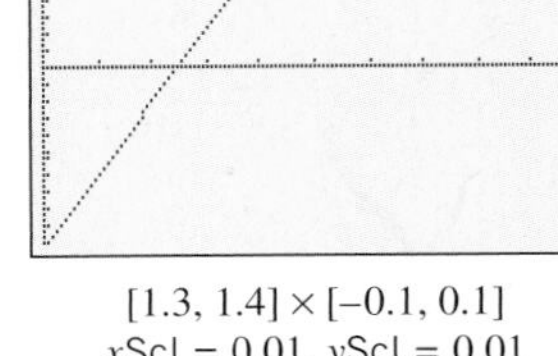

[1.3, 1.4] × [−0.1, 0.1]
xScl = 0.01, yScl = 0.01
(c)

Figure 2.5.11

TECHNOLOGY MASTERY

Use a graphing or calculating utility to show that the root x_0 in Example 6 can be approximated as $x_0 \approx 1.3245$ to three decimal-place accuracy.

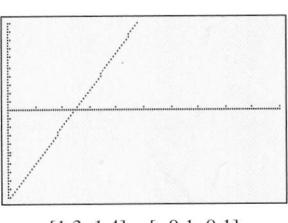

To say that x approximates x_0 to n decimal places does *not* mean that the first n decimal places of x and x_0 will be the same when the numbers are rounded to n decimal places. For example, $x = 1.084$ approximates $x_0 = 1.087$ to two decimal places because $|x - x_0| = 0.003\ (< 0.005)$. However, if we round these values to two decimal places, then we obtain $x \approx 1.08$ and $x_0 \approx 1.09$. Thus, if you approximate a number to n decimal places, then you should display that approximation to at least $n + 1$ decimal places to preserve the accuracy.

✔QUICK CHECK EXERCISES 2.5 (See page 137 for answers.)

1. A function f is continuous at $x = c$ if $f(c)$ is defined, $\lim_{x \to c} f(x)$ exists, and ________.

2. Consider the functions

$$f(x) = \begin{cases} 1, & x \neq 4 \\ -1, & x = 4 \end{cases} \quad \text{and} \quad g(x) = \begin{cases} 4x - 10, & x \neq 4 \\ -6, & x = 4 \end{cases}$$

In each part, is the given function continuous at $x = 4$?
(a) $f(x)$ (b) $g(x)$ (c) $-g(x)$ (d) $|f(x)|$
(e) $f(x)g(x)$ (f) $g(f(x))$ (g) $g(x) - 6f(x)$

3. For what values of x, if any, is the function

$$f(x) = \frac{x^2 - 16}{x^2 - 5x + 4}$$

discontinuous?

4. Suppose that a function f is continuous everywhere and that $f(-2) = 3$, $f(-1) = -1$, $f(0) = -4$, $f(1) = 1$, and $f(2) = 5$. Does the Intermediate-Value Theorem guarantee that f has a root on the following intervals?
(a) $[-2, -1]$ (b) $[-1, 0]$ (c) $[-1, 1]$ (d) $[0, 2]$

EXERCISE SET 2.5 Graphing Utility

FOCUS ON CONCEPTS

1–4 Let f be the function whose graph is shown. On which of the following intervals, if any, is f continuous?
(a) $[1, 3]$ (b) $(1, 3)$ (c) $[1, 2]$
(d) $(1, 2)$ (e) $[2, 3]$ (f) $(2, 3)$
For each interval on which f is not continuous, indicate which conditions for the continuity of f do not hold.

1. **2.**

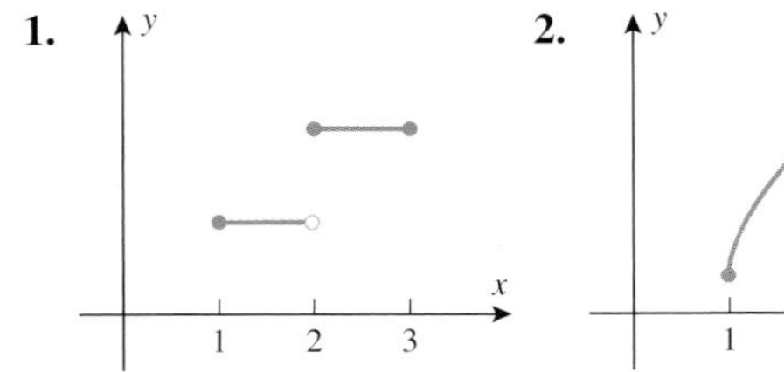

3.

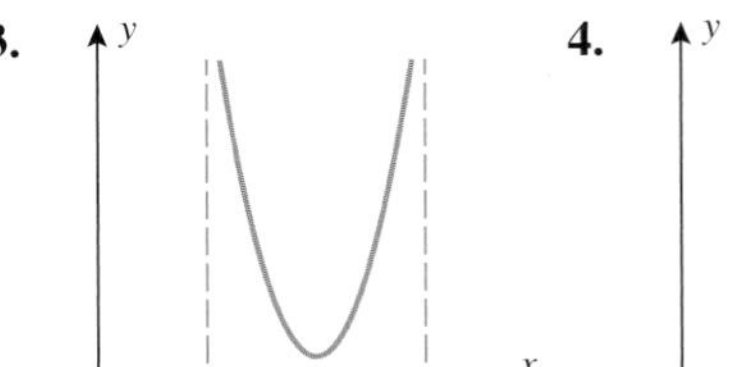

4.

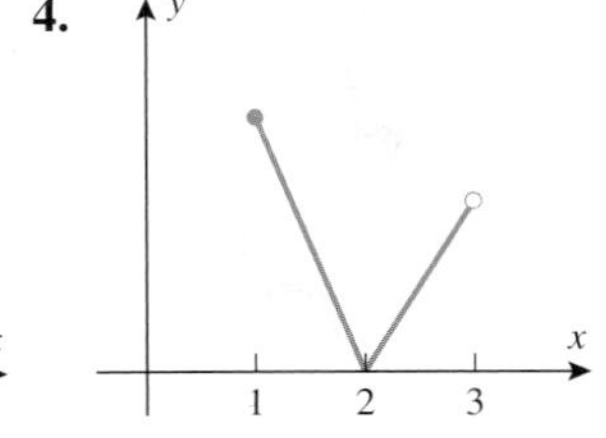

5. Suppose that f and g are continuous functions such that $f(2) = 1$ and $\lim_{x \to 2} [f(x) + 4g(x)] = 13$. Find
(a) $g(2)$ (b) $\lim_{x \to 2} g(x)$.

6. Suppose that f and g are continuous functions such that $\lim_{x \to 3} g(x) = 5$ and $f(3) = -2$. Find $\lim_{x \to 3} [f(x)/g(x)]$.

7. In each part sketch the graph of a function f that satisfies the stated conditions.
   (a) f is continuous everywhere except at $x = 3$, at which point it is continuous from the right.
   (b) f has a two-sided limit at $x = 3$, but it is not continuous at $x = 3$.
   (c) f is not continuous at $x = 3$, but if its value at $x = 3$ is changed from $f(3) = 1$ to $f(3) = 0$, it becomes continuous at $x = 3$.
   (d) f is continuous on the interval $[0, 3)$ and is defined on the closed interval $[0, 3]$; but f is not continuous on the interval $[0, 3]$.

8. Find formulas for some functions that are continuous on the intervals $(-\infty, 0)$ and $(0, +\infty)$, but are not continuous on the interval $(-\infty, +\infty)$.

9. A student parking lot at a university charges \$2.00 for the first half hour (or any part) and \$1.00 for each subsequent half hour (or any part) up to a daily maximum of \$10.00.
   (a) Sketch a graph of cost as a function of the time parked.
   (b) Discuss the significance of the discontinuities in the graph to a student who parks there.

10. In each part determine whether the function is continuous or not, and explain your reasoning.
   (a) The Earth's population as a function of time.
   (b) Your exact height as a function of time.
   (c) The cost of a taxi ride in your city as a function of the distance traveled.
   (d) The volume of a melting ice cube as a function of time.

11–22 Find values of x, if any, at which f is not continuous.

11. $f(x) = 5x^4 - 3x + 7$

12. $f(x) = \sqrt[3]{x - 8}$

13. $f(x) = \dfrac{x + 2}{x^2 + 4}$

14. $f(x) = \dfrac{x + 2}{x^2 - 4}$

15. $f(x) = \dfrac{x}{2x^2 + x}$

16. $f(x) = \dfrac{2x + 1}{4x^2 + 4x + 5}$

17. $f(x) = \dfrac{3}{x} + \dfrac{x - 1}{x^2 - 1}$

18. $f(x) = \dfrac{5}{x} + \dfrac{2x}{x + 4}$

19. $f(x) = \dfrac{x^2 + 6x + 9}{|x| + 3}$

20. $f(x) = \left|4 - \dfrac{8}{x^4 + x}\right|$

21. $f(x) = \begin{cases} 2x + 3, & x \le 4 \\ 7 + \dfrac{16}{x}, & x > 4 \end{cases}$

22. $f(x) = \begin{cases} \dfrac{3}{x - 1}, & x \ne 1 \\ 3, & x = 1 \end{cases}$

23–24 Find a value of the constant k, if possible, that will make the function continuous everywhere.

23. (a) $f(x) = \begin{cases} 7x - 2, & x \le 1 \\ kx^2, & x > 1 \end{cases}$
   (b) $f(x) = \begin{cases} kx^2, & x \le 2 \\ 2x + k, & x > 2 \end{cases}$

24. (a) $f(x) = \begin{cases} 9 - x^2, & x \ge -3 \\ k/x^2, & x < -3 \end{cases}$
   (b) $f(x) = \begin{cases} 9 - x^2, & x \ge 0 \\ k/x^2, & x < 0 \end{cases}$

25. Find values of the constants k and m, if possible, that will make the function f continuous everywhere.
$$f(x) = \begin{cases} x^2 + 5, & x > 2 \\ m(x + 1) + k, & -1 < x \le 2 \\ 2x^3 + x + 7, & x \le -1 \end{cases}$$

26. On which of the following intervals is
$$f(x) = \frac{1}{\sqrt{x - 2}}$$
continuous?
(a) $[2, +\infty)$ (b) $(-\infty, +\infty)$ (c) $(2, +\infty)$ (d) $[1, 2)$

27–30 A function f is said to have a ***removable discontinuity*** at $x = c$ if $\lim_{x \to c} f(x)$ exists but f is not continuous at $x = c$, either because f is not defined at c or because the definition for $f(c)$ differs from the value of the limit. This terminology will be needed in these exercises.

27. (a) Sketch the graph of a function with a removable discontinuity at $x = c$ for which $f(c)$ is undefined.
   (b) Sketch the graph of a function with a removable discontinuity at $x = c$ for which $f(c)$ is defined.

28. (a) The terminology *removable discontinuity* is appropriate because a removable discontinuity of a function f at $x = c$ can be "removed" by redefining the value of f appropriately at $x = c$. What value for $f(c)$ removes the discontinuity?
   (b) Show that the following functions have removable discontinuities at $x = 1$, and sketch their graphs.
$$f(x) = \frac{x^2 - 1}{x - 1} \quad \text{and} \quad g(x) = \begin{cases} 1, & x > 1 \\ 0, & x = 1 \\ 1, & x < 1 \end{cases}$$
   (c) What values should be assigned to $f(1)$ and $g(1)$ to remove the discontinuities?

29–30 Find the values of x (if any) at which f is not continuous, and determine whether each such value is a removable discontinuity.

29. (a) $f(x) = \dfrac{|x|}{x}$ (b) $f(x) = \dfrac{x^2 + 3x}{x + 3}$
   (c) $f(x) = \dfrac{x - 2}{|x| - 2}$

30. (a) $f(x) = \dfrac{x^2 - 4}{x^3 - 8}$ (b) $f(x) = \begin{cases} 2x - 3, & x \le 2 \\ x^2, & x > 2 \end{cases}$
   (c) $f(x) = \begin{cases} 3x^2 + 5, & x \ne 1 \\ 6, & x = 1 \end{cases}$

31. (a) Use a graphing utility to generate the graph of the function $f(x) = (x+3)/(2x^2+5x-3)$, and then use the graph to make a conjecture about the number and locations of all discontinuities.
(b) Check your conjecture by factoring the denominator.

32. (a) Use a graphing utility to generate the graph of the function $f(x) = x/(x^3 - x + 2)$, and then use the graph to make a conjecture about the number and locations of all discontinuities.
(b) Use the Intermediate-Value Theorem to approximate the locations of all discontinuities to two decimal places.

33. Prove that $f(x) = x^{3/5}$ is continuous everywhere, carefully justifying each step.

34. Prove that $f(x) = 1/\sqrt{x^4 + 7x^2 + 1}$ is continuous everywhere, carefully justifying each step.

35. Let f and g be discontinuous at c. Give examples to show that
(a) $f + g$ can be continuous or discontinuous at c
(b) fg can be continuous or discontinuous at c.

36. Prove part (b) of Theorem 2.5.4.

37. Prove:
(a) part (a) of Theorem 2.5.3
(b) part (b) of Theorem 2.5.3
(c) part (c) of Theorem 2.5.3.

38. Prove: If f and g are continuous on $[a, b]$, and $f(a) > g(a)$, $f(b) < g(b)$, then there is at least one solution of the equation $f(x) = g(x)$ in (a, b). [*Hint:* Consider $f(x) - g(x)$.]

FOCUS ON CONCEPTS

39. Give an example of a function f that is defined on a closed interval, and whose values at the endpoints have opposite signs, but for which the equation $f(x) = 0$ has no solution in the interval.

40. Let f be the function whose graph is shown in Exercise 2. For each interval, determine (i) whether the hypothesis of the Intermediate-Value Theorem is satisfied, and (ii) whether the conclusion of the Intermediate-Value Theorem is satisfied.
(a) $[1, 2]$ (b) $[2, 3]$ (c) $[1, 3]$

41. Use the Intermediate-Value Theorem to show that there is a right circular cylinder of height h and radius less than r whose volume is equal to that of a right circular cone of height h and radius r.

42. Use the Intermediate-Value Theorem to show that there is a square with a diagonal length that is between r and $2r$ and an area that is half the area of a circle of radius r.

43. Show that the equation $x^3 + x^2 - 2x = 1$ has at least one solution in the interval $[-1, 1]$.

44. Prove: If $p(x)$ is a polynomial of odd degree, then the equation $p(x) = 0$ has at least one real solution.

45. The accompanying figure shows the graph of the equation $y = x^4 + x - 1$. Use the method of Example 6 to approximate the x-intercepts with an error of at most 0.05.

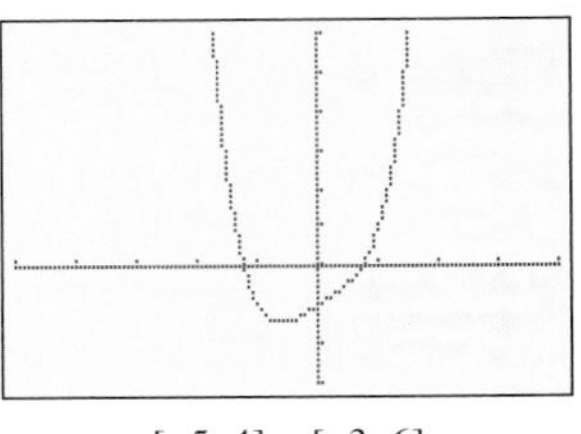

$[-5, 4] \times [-3, 6]$
xScl = 1, yScl = 1

Figure Ex-45

46. Use a graphing utility to solve the problem in Exercise 45 by zooming.

47. The accompanying figure shows the graph of the equation $y = 5 - x - x^4$. Use the method of Example 6 to approximate the roots of the equation $5 - x - x^4 = 0$ to two decimal-place accuracy.

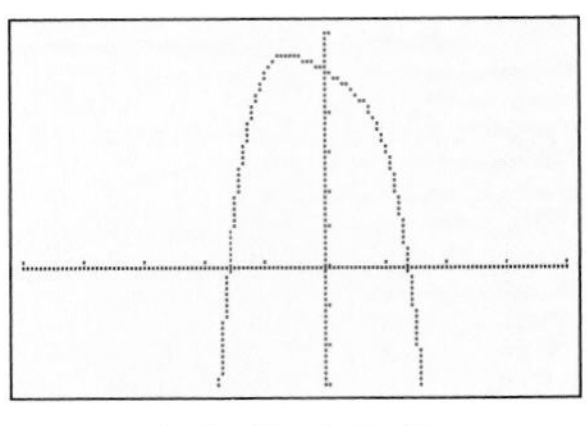

$[-5, 4] \times [-3, 6]$
xScl = 1, yScl = 1

Figure Ex-47

48. Use a graphing utility to solve the problem in Exercise 47 by zooming.

49. Use the fact that $\sqrt{5}$ is a solution of $x^2 - 5 = 0$ to approximate $\sqrt{5}$ with an error of at most 0.005.

50. Prove that if a and b are positive, then the equation

$$\frac{a}{x-1} + \frac{b}{x-3} = 0$$

has at least one solution in the interval $(1, 3)$.

51. A sphere of unknown radius x consists of a spherical core and a coating that is 1 cm thick (see the accompanying figure). Given that the volume of the coating and the volume of the core are the same, approximate the radius of the sphere to three decimal-place accuracy.

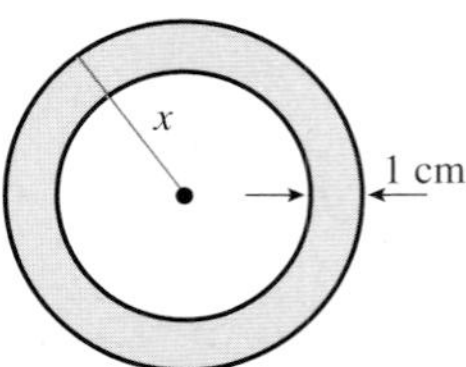

Figure Ex-51

52. A monk begins walking up a mountain road at 12:00 noon and reaches the top at 12:00 midnight. He meditates and rests until 12:00 noon the next day, at which time he begins walking down the same road, reaching the bottom at 12:00

midnight. Show that there is at least one point on the road that he reaches at the same time of day on the way up as on the way down.

53. Let f be defined at c. Prove that f is continuous at c if, given $\epsilon > 0$, there exists a $\delta > 0$ such that $|f(x) - f(c)| < \epsilon$ if $|x - c| < \delta$.

54. Suppose that f is continuous on the interval $[0, 1]$ and that $0 \leq f(x) \leq 1$ for all x in this interval.
(a) Sketch the graph of $y = x$ together with a possible graph for f over the interval $[0, 1]$.
(b) Use the Intermediate-Value Theorem to help prove that there is at least one number c in the interval $[0, 1]$ such that $f(c) = c$.

55. Let $f(x) = x^6 + 3x + 5$, $x \geq 0$. Show that f has an inverse function and that f^{-1} is continuous on $[5, +\infty)$.

56. Suppose that f is an invertible function, $f(0) = 0$, f is continuous at 0, and $\lim_{x\to 0} f(x)/x$ exists. Given that $L = \lim_{x\to 0} f(x)/x$, show

$$\lim_{x\to 0} \frac{x}{f^{-1}(x)} = L$$

[*Hint:* Apply Theorem 2.5.5 to the composition $h \circ g$ where

$$h(x) = \begin{cases} f(x)/x, & x \neq 0 \\ L, & x = 0 \end{cases}$$

and $g(x) = f^{-1}(x)$.]

✔ QUICK CHECK ANSWERS 2.5

1. $\lim_{x\to c} f(x) = f(c)$ **2.** (a) no (b) no (c) no (d) yes (e) yes (f) no (g) yes **3.** $x = 1, 4$
4. (a) yes (b) no (c) yes (d) yes

2.6 CONTINUITY OF TRIGONOMETRIC FUNCTIONS

In this section we will discuss the continuity properties of trigonometric functions. We will also discuss some important limits involving such functions.

CONTINUITY OF TRIGONOMETRIC FUNCTIONS

Recall from trigonometry that the graphs of $\sin x$ and $\cos x$ are drawn as continuous curves. We will not formally prove that these functions are continuous, but we can motivate this fact by letting c be a fixed angle in radian measure and x a variable angle in radian measure. If, as illustrated in Figure 2.6.1, the angle x approaches the angle c, then the point $P(\cos x, \sin x)$ moves along the unit circle toward $Q(\cos c, \sin c)$, and the coordinates of P approach the corresponding coordinates of Q. This implies that

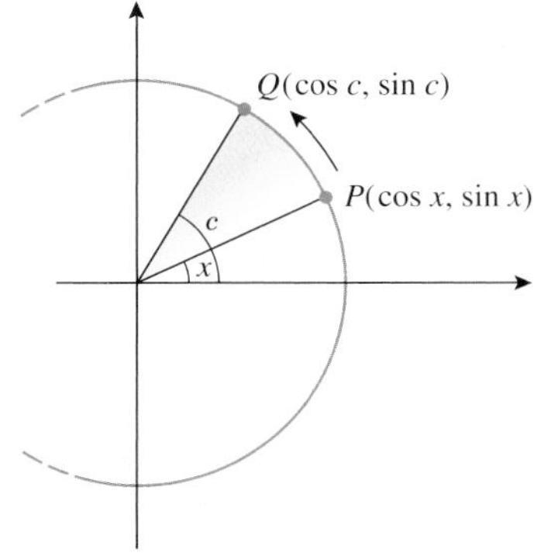

Figure 2.6.1

$$\lim_{x\to c} \sin x = \sin c \quad \text{and} \quad \lim_{x\to c} \cos x = \cos c \tag{1}$$

Thus, $\sin x$ and $\cos x$ are continuous at the arbitrary point c; that is, these functions are continuous everywhere.

The formulas in (1) can be used to find limits of the remaining trigonometric functions by expressing them in terms of $\sin x$ and $\cos x$; for example, if $\cos c \neq 0$, then

$$\lim_{x\to c} \tan x = \lim_{x\to c} \frac{\sin x}{\cos x} = \frac{\sin c}{\cos c} = \tan c$$

Thus, we are led to the following theorem.

Theorem 2.6.1 implies that the six basic trigonometric functions are continuous on their domains. In particular, $\sin x$ and $\cos x$ are continuous everywhere.

2.6.1 THEOREM. *If c is any number in the natural domain of the stated trigonometric function, then*

$$\lim_{x\to c} \sin x = \sin c \qquad \lim_{x\to c} \cos x = \cos c \qquad \lim_{x\to c} \tan x = \tan c$$

$$\lim_{x\to c} \csc x = \csc c \qquad \lim_{x\to c} \sec x = \sec c \qquad \lim_{x\to c} \cot x = \cot c$$

▶ **Example 1** Find the limit

$$\lim_{x\to 1} \cos\left(\frac{x^2-1}{x-1}\right)$$

Solution. Since the cosine function is continuous everywhere, it follows from Theorem 2.5.5 that

$$\lim_{x\to 1} \cos(g(x)) = \cos\left(\lim_{x\to 1} g(x)\right)$$

provided $\lim_{x\to 1} g(x)$ exists. Thus,

$$\lim_{x\to 1} \cos\left(\frac{x^2-1}{x-1}\right) = \lim_{x\to 1} \cos(x+1) = \cos\left(\lim_{x\to 1}(x+1)\right) = \cos 2 \blacktriangleleft$$

OBTAINING LIMITS BY SQUEEZING

In Section 2.1 we used numerical evidence to conjecture that

$$\lim_{x\to 0} \frac{\sin x}{x} = 1 \tag{2}$$

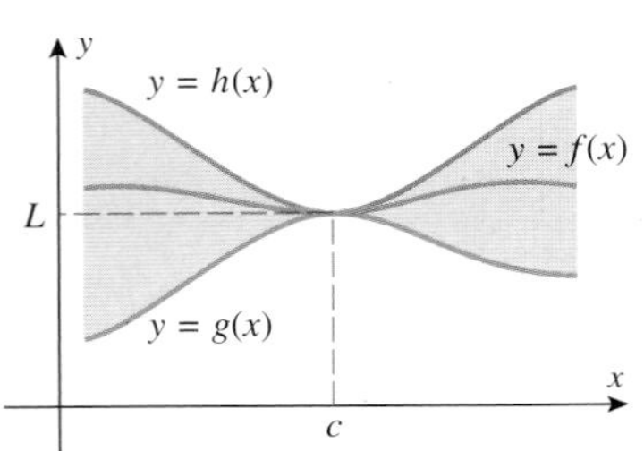

Figure 2.6.2

However, this limit is not easy to establish with certainty. The limit is an indeterminate form of type 0/0, and there is no simple algebraic manipulation that one can perform to obtain the limit. Later in the text we will develop general methods for finding limits of indeterminate forms, but in this particular case we can use a technique called *squeezing*.

The method of squeezing is used to prove that $f(x)\to L$ as $x\to c$ by "trapping" or "squeezing" f between two functions, g and h, whose limits as $x\to c$ are known with *certainty* to be L. As illustrated in Figure 2.6.2, this forces f to have a limit of L as well. This is the idea behind the following theorem, which we state without proof.

The Squeezing Theorem also holds for one-sided limits and limits at $+\infty$ and $-\infty$. How do you think the hypotheses would change in those cases?

2.6.2 THEOREM (*The Squeezing Theorem*). *Let f, g, and h be functions satisfying*

$$g(x) \le f(x) \le h(x)$$

for all x in some open interval containing the number c, with the possible exception that the inequalities need not hold at c. If g and h have the same limit as x approaches c, say

$$\lim_{x\to c} g(x) = \lim_{x\to c} h(x) = L$$

then f also has this limit as x approaches c, that is,

$$\lim_{x\to c} f(x) = L$$

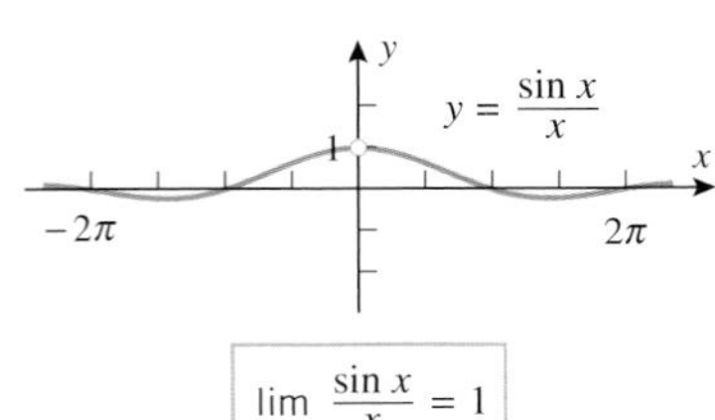

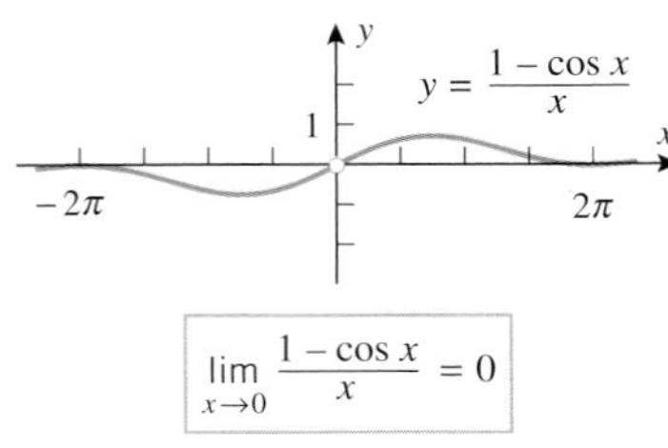

Figure 2.6.3

To illustrate how the Squeezing Theorem works, we will prove the following result, which is illustrated in Figure 2.6.3.

2.6.3 THEOREM.

$$(a)\quad \lim_{x\to 0} \frac{\sin x}{x} = 1 \qquad (b)\quad \lim_{x\to 0} \frac{1-\cos x}{x} = 0$$

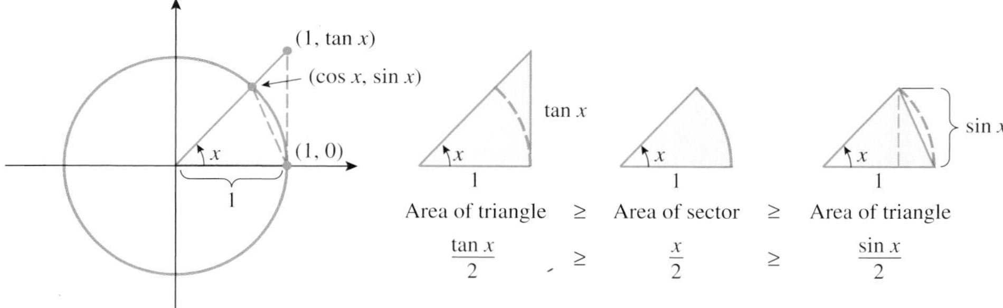

Figure 2.6.4

PROOF (*a*). In this proof we will interpret x as an angle in radian measure, and we will assume to start that $0 < x < \pi/2$. As illustrated in Figure 2.6.4, the area of a sector with central angle x and radius 1 lies between the areas of two triangles, one with area $\frac{1}{2}\tan x$ and the other with area $\frac{1}{2}\sin x$. Since the sector has area $\frac{1}{2}x$ (see marginal note), it follows that

$$\frac{1}{2}\tan x \geq \frac{1}{2}x \geq \frac{1}{2}\sin x$$

Multiplying through by $2/(\sin x)$ and using the fact that $\sin x > 0$ for $0 < x < \pi/2$, we obtain

$$\frac{1}{\cos x} \geq \frac{x}{\sin x} \geq 1$$

Next, taking reciprocals reverses the inequalities, so we obtain

$$\cos x \leq \frac{\sin x}{x} \leq 1 \tag{3}$$

which squeezes the function $(\sin x)/x$ between the functions $\cos x$ and 1. Although we derived these inequalities by assuming that $0 < x < \pi/2$, they also hold for $-\pi/2 < x < 0$ [since replacing x by $-x$ and using the identities $\sin(-x) = -\sin x$, and $\cos(-x) = \cos x$ leaves (3) unchanged]. Finally, since

$$\lim_{x\to 0} \cos x = 1 \quad \text{and} \quad \lim_{x\to 0} 1 = 1$$

the Squeezing Theorem implies that

$$\lim_{x\to 0} \frac{\sin x}{x} = 1$$

Recall that the area A of a sector of radius r and central angle θ is

$$A = \frac{1}{2}r^2\theta$$

This can be derived from the relationship

$$\frac{A}{\pi r^2} = \frac{\theta}{2\pi}$$

which states that the area of the sector is to the area of the circle as the central angle of the sector is to the central angle of the circle.

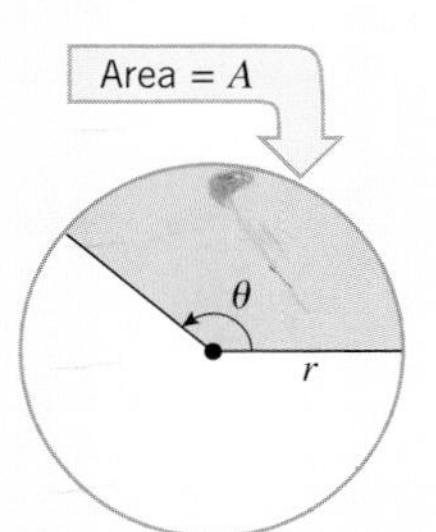

PROOF (*b*). For this proof we will use the limit in part (*a*), the continuity of the sine function, and the trigonometric identity $\sin^2 x = 1 - \cos^2 x$. We obtain

$$\lim_{x\to 0} \frac{1-\cos x}{x} = \lim_{x\to 0} \left[\frac{1-\cos x}{x} \cdot \frac{1+\cos x}{1+\cos x}\right] = \lim_{x\to 0} \frac{\sin^2 x}{(1+\cos x)x}$$

$$= \left(\lim_{x\to 0} \frac{\sin x}{x}\right)\left(\lim_{x\to 0} \frac{\sin x}{1+\cos x}\right) = (1)\left(\frac{0}{1+1}\right) = 0 \quad \blacksquare$$

▶ **Example 2** Find

$$\text{(a)} \lim_{x\to 0} \frac{\tan x}{x} \qquad \text{(b)} \lim_{\theta\to 0} \frac{\sin 2\theta}{\theta} \qquad \text{(c)} \lim_{x\to 0} \frac{\sin 3x}{\sin 5x}$$

***Solution** (a).*

$$\lim_{x\to 0} \frac{\tan x}{x} = \lim_{x\to 0} \left(\frac{\sin x}{x} \cdot \frac{1}{\cos x}\right) = \left(\lim_{x\to 0} \frac{\sin x}{x}\right)\left(\lim_{x\to 0} \frac{1}{\cos x}\right) = (1)(1) = 1$$

Solution (b). The trick is to multiply and divide by 2, which will make the denominator the same as the argument of the sine function [just as in Theorem 2.6.3(*a*)]:

$$\lim_{\theta\to 0} \frac{\sin 2\theta}{\theta} = \lim_{\theta\to 0} 2 \cdot \frac{\sin 2\theta}{2\theta} = 2 \lim_{\theta\to 0} \frac{\sin 2\theta}{2\theta}$$

Now make the substitution $x = 2\theta$, and use the fact that $x \to 0$ as $\theta \to 0$. This yields

$$\lim_{\theta\to 0} \frac{\sin 2\theta}{\theta} = 2 \lim_{\theta\to 0} \frac{\sin 2\theta}{2\theta} = 2 \lim_{x\to 0} \frac{\sin x}{x} = 2(1) = 2$$

TECHNOLOGY MASTERY

Use a graphing utility to confirm the limits in Example 2, and if you have a CAS, use it to obtain the limits.

Solution (c).

$$\lim_{x\to 0} \frac{\sin 3x}{\sin 5x} = \lim_{x\to 0} \frac{\dfrac{\sin 3x}{x}}{\dfrac{\sin 5x}{x}} = \lim_{x\to 0} \frac{3 \cdot \dfrac{\sin 3x}{3x}}{5 \cdot \dfrac{\sin 5x}{5x}} = \frac{3 \cdot 1}{5 \cdot 1} = \frac{3}{5} \blacktriangleleft$$

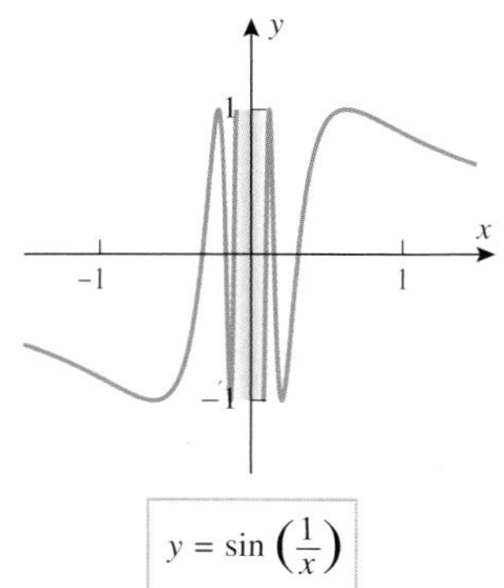

Figure 2.6.5

► **Example 3** Discuss the limits

$$\text{(a)} \lim_{x\to 0} \sin\left(\frac{1}{x}\right) \qquad \text{(b)} \lim_{x\to 0} x \sin\left(\frac{1}{x}\right)$$

Solution (a). Let us view $1/x$ as an angle in radian measure. As $x \to 0^+$, the angle $1/x$ approaches $+\infty$, so the values of $\sin(1/x)$ keep oscillating between -1 and 1 without approaching a limit. Similarly, as $x \to 0^-$, the angle $1/x$ approaches $-\infty$, so again the values of $\sin(1/x)$ keep oscillating between -1 and 1 without approaching a limit. These conclusions are consistent with the graph shown in Figure 2.6.5. Note that the oscillations become more and more rapid as $x \to 0$ because $1/x$ increases (or decreases) more and more rapidly as x approaches 0.

Solution (b). Since

$$-1 \le \sin\left(\frac{1}{x}\right) \le 1$$

it follows that if $x \ne 0$, then

$$-|x| \le x \sin\left(\frac{1}{x}\right) \le |x| \tag{4}$$

Confirm (4) by considering the cases $x > 0$ and $x < 0$ separately.

Since $|x| \to 0$ as $x \to 0$, the inequalities in (4) and the Squeezing Theorem imply that

$$\lim_{x\to 0} x \sin\left(\frac{1}{x}\right) = 0$$

This is consistent with the graph shown in Figure 2.6.6. ◄

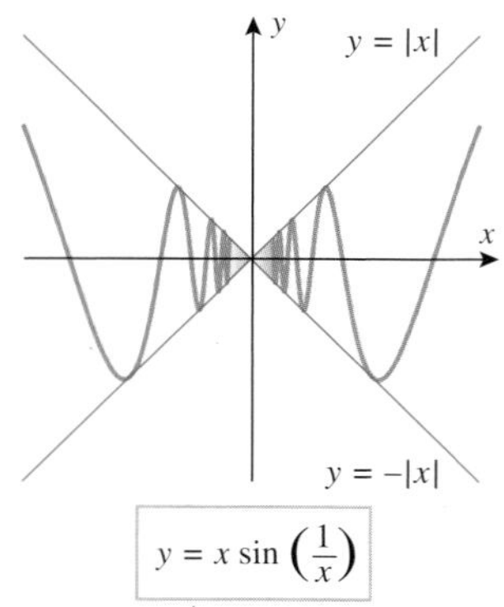

Figure 2.6.6

It follows from part (b) of this example that the function

$$f(x) = \begin{cases} x\sin(1/x), & x \ne 0 \\ 0, & x = 0 \end{cases}$$

is continuous at $x = 0$, since the value of the function and the value of the limit are the same at 0. This shows that the behavior of a function can be very complex in the vicinity of $x = c$, even though the function is continuous at c.

✔QUICK CHECK EXERCISES 2.6 *(See page 143 for answers.)*

1. In each part, is the given function continuous on the interval $[0, \pi/2)$?
(a) $\sin x$ (b) $\cos x$ (c) $\tan x$
(d) $\csc x$ (e) $\sec x$ (f) $\cot x$

2. Evaluate.
(a) $\lim_{x \to 0} \frac{\sin x}{x}$ (b) $\lim_{x \to 0} \frac{1 - \cos x}{x}$

3. Let $f(x) = x^2 \cos(\pi/x)$, $g(x) = -x^2$, and $h(x) = x^2$.
(a) Explain why the functions g and h, together with the Squeezing Theorem, can be used to find $\lim_{x \to 0} f(x)$ but not $\lim_{x \to 1} f(x)$.
(b) Evaluate $\lim_{x \to 0} f(x)$ and $\lim_{x \to 1} f(x)$.

4. Evaluate.
(a) $\lim_{x \to 0} \sin\left(\frac{x^2 - 3x}{x}\right)$ (b) $\lim_{x \to 3^+} \sin\left(\frac{x}{x^2 - 3x}\right)$
(c) $\lim_{x \to 0} \frac{\sin(x^2 - 3x)}{x}$ (d) $\lim_{x \to 3} \frac{\sin(x^2 - 3x)}{x}$

EXERCISE SET 2.6 Graphing Utility

1–8 Find the discontinuities, if any.

1. $f(x) = \sin(x^2 - 2)$
2. $f(x) = \cos\left(\frac{x}{x - \pi}\right)$
3. $f(x) = |\cot x|$
4. $f(x) = \sec x$
5. $f(x) = \csc x$
6. $f(x) = \frac{1}{1 + \sin^2 x}$
7. $f(x) = \frac{1}{1 - 2\sin x}$
8. $f(x) = \sqrt{2 + \tan^2 x}$

9. Use Theorem 2.5.6 to show that the following functions are continuous everywhere by expressing them as compositions of simpler functions that are known to be continuous.
(a) $\sin(x^3 + 7x + 1)$ (b) $|\sin x|$
(c) $\cos^3(x + 1)$ (d) $\sqrt{3 + \sin 2x}$
(e) $\sin(\sin x)$ (f) $\cos^5 x - 2\cos^3 x + 1$

10. (a) Prove that if $g(x)$ is continuous everywhere, then so are $\sin(g(x))$, $\cos(g(x))$, $g(\sin(x))$, and $g(\cos(x))$.
(b) Illustrate the result in part (a) with some of your own choices for g.

11–32 Find the limits.

11. $\lim_{x \to +\infty} \cos\left(\frac{1}{x}\right)$
12. $\lim_{x \to +\infty} \sin\left(\frac{\pi x}{2 - 3x}\right)$
13. $\lim_{\theta \to 0} \frac{\sin 3\theta}{\theta}$
14. $\lim_{h \to 0} \frac{\sin h}{2h}$
15. $\lim_{\theta \to 0^+} \frac{\sin \theta}{\theta^2}$
16. $\lim_{\theta \to 0} \frac{\sin^2 \theta}{\theta}$
17. $\lim_{x \to 0} \frac{\tan 7x}{\sin 3x}$
18. $\lim_{x \to 0} \frac{\sin 6x}{\sin 8x}$
19. $\lim_{x \to 0^+} \frac{\sin x}{5\sqrt{x}}$
20. $\lim_{x \to 0} \frac{\sin^2 x}{3x^2}$
21. $\lim_{x \to 0} \frac{\sin x^2}{x}$
22. $\lim_{h \to 0} \frac{\sin h}{1 - \cos h}$
23. $\lim_{t \to 0} \frac{t^2}{1 - \cos^2 t}$
24. $\lim_{x \to 0} \frac{x}{\cos\left(\frac{1}{2}\pi - x\right)}$
25. $\lim_{\theta \to 0} \frac{\theta^2}{1 - \cos \theta}$
26. $\lim_{h \to 0} \frac{1 - \cos 3h}{\cos^2 5h - 1}$
27. $\lim_{x \to 0^+} \sin\left(\frac{1}{x}\right)$
28. $\lim_{x \to 0} \frac{x^2 - 3\sin x}{x}$
29. $\lim_{x \to 0} \frac{2 - \cos 3x - \cos 4x}{x}$
30. $\lim_{x \to 0} \frac{\tan 3x^2 + \sin^2 5x}{x^2}$
31. $\lim_{x \to 0} \frac{\tan ax}{\sin bx}$, $(a \neq 0, b \neq 0)$
32. $\lim_{x \to 0} \frac{\sin^2(kx)}{x^2}$, $k \neq 0$

33–36 (a) Construct a table to estimate the limit by evaluating the function near the limiting value. (b) Find the exact value of the limit.

33. $\lim_{x \to 5} \frac{\sin(x - 5)}{x^2 - 25}$
34. $\lim_{x \to 2} \frac{\sin(2x - 4)}{x^2 - 4}$
35. $\lim_{x \to -2} \frac{\sin(x^2 + 3x + 2)}{x + 2}$
36. $\lim_{x \to -1} \frac{\sin(x^2 + 3x + 2)}{x^3 + 1}$

FOCUS ON CONCEPTS

37. In Example 3 we used the Squeezing Theorem to prove that

$$\lim_{x \to 0} x \sin\left(\frac{1}{x}\right) = 0$$

Why couldn't we have obtained the same result by writing

$$\lim_{x \to 0} x \sin\left(\frac{1}{x}\right) = \lim_{x \to 0} x \cdot \lim_{x \to 0} \sin\left(\frac{1}{x}\right) = 0 \cdot \lim_{x \to 0} \sin\left(\frac{1}{x}\right) = 0?$$

38. Find a value for the constant k that makes

$$f(x) = \begin{cases} \dfrac{\sin 3x}{x}, & x \neq 0 \\ k, & x = 0 \end{cases}$$

continuous at $x = 0$.

39. Find a nonzero value for the constant k that makes

$$f(x) = \begin{cases} \dfrac{\tan kx}{x}, & x < 0 \\ 3x + 2k^2, & x \geq 0 \end{cases}$$

continuous at $x = 0$.

40. Is

$$f(x) = \begin{cases} \dfrac{\sin x}{|x|}, & x \neq 0 \\ 1, & x = 0 \end{cases}$$

continuous at $x = 0$? Explain.

41. In parts (a)–(c), find the limit by making the indicated substitution.

(a) $\displaystyle\lim_{x\to+\infty} x \sin \frac{1}{x}$; $t = \dfrac{1}{x}$

(b) $\displaystyle\lim_{x\to-\infty} x \left(1 - \cos \frac{1}{x}\right)$; $t = \dfrac{1}{x}$

(c) $\displaystyle\lim_{x\to\pi} \frac{\pi - x}{\sin x}$; $t = \pi - x$

42. Find $\displaystyle\lim_{x\to 2} \frac{\cos(\pi/x)}{x-2}$. $\left[\textit{Hint:} \text{ Let } t = \dfrac{\pi}{2} - \dfrac{\pi}{x}.\right]$

43. Find $\displaystyle\lim_{x\to 1} \frac{\sin(\pi x)}{x-1}$.

44. Find $\displaystyle\lim_{x\to\pi/4} \frac{\tan x - 1}{x - \pi/4}$.

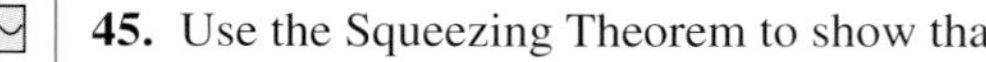

FOCUS ON CONCEPTS

45. Use the Squeezing Theorem to show that

$$\lim_{x\to 0} x \cos \frac{50\pi}{x} = 0$$

and illustrate the principle involved by using a graphing utility to graph the equations $y = |x|$, $y = -|x|$, and $y = x \cos(50\pi/x)$ on the same screen in the window $[-1, 1] \times [-1, 1]$.

46. Use the Squeezing Theorem to show that

$$\lim_{x\to 0} x^2 \sin\left(\frac{50\pi}{\sqrt[3]{x}}\right) = 0$$

and illustrate the principle involved by using a graphing utility to graph the equations $y = x^2$, $y = -x^2$, and $y = x^2 \sin(50\pi/\sqrt[3]{x}\,)$ on the same screen in the window $[-0.5, 0.5] \times [-0.25, 0.25]$.

47. Sketch the graphs of the curves $y = 1 - x^2$, $y = \cos x$, and $y = f(x)$, where f is a function that satisfies the inequalities

$$1 - x^2 \leq f(x) \leq \cos x$$

for all x in the interval $(-\pi/2, \pi/2)$. What can you say about the limit of $f(x)$ as $x \to 0$? Explain.

48. Sketch the graphs of the curves $y = 1/x$, $y = -1/x$, and $y = f(x)$, where f is a function that satisfies the inequalities

$$-\frac{1}{x} \leq f(x) \leq \frac{1}{x}$$

for all x in the interval $[1, +\infty)$. What can you say about the limit of $f(x)$ as $x \to +\infty$? Explain your reasoning.

49. Find formulas for functions g and h such that $g(x) \to 0$ and $h(x) \to 0$ as $x \to +\infty$ and such that

$$g(x) \leq \frac{\sin x}{x} \leq h(x)$$

for positive values of x. What can you say about the limit

$$\lim_{x\to+\infty} \frac{\sin x}{x}?$$

Explain your reasoning.

50. Draw pictures analogous to Figure 2.6.2 that illustrate the Squeezing Theorem for limits of the forms $\lim_{x\to+\infty} f(x)$ and $\lim_{x\to-\infty} f(x)$.

51–52 Recall that unless stated otherwise the variable x in trigonometric functions such as $\sin x$ and $\cos x$ is assumed to be in radian measure. The limits in Theorem 2.6.3 are based on that assumption. These exercises explore what happens to those limits if degree measure is used for x.

51. (a) Show that if x is in degrees, then

$$\lim_{x\to 0} \frac{\sin x}{x} = \frac{\pi}{180}$$

(b) Confirm that the limit in part (a) is consistent with the results produced by your calculating utility by setting the utility to degree measure and calculating $(\sin x)/x$ for some values of x that get closer and closer to 0.

52. What is the limit of $(1 - \cos x)/x$ as $x \to 0$ if x is in degrees?

53. It follows from part (*a*) of Theorem 2.6.3 that if θ is small (near zero) and measured in radians, then one should expect the approximation

$$\sin\theta \approx \theta$$

to be good.

(a) Find $\sin 10°$ using a calculating utility.

(b) Estimate $\sin 10°$ using the approximation above.

54. (a) Use the approximation of $\sin\theta$ that is given in Exercise 53 together with the identity $\cos 2\alpha = 1 - 2\sin^2\alpha$ with $\alpha = \theta/2$ to show that if θ is small (near zero) and measured in radians, then one should expect the approximation

$$\cos\theta \approx 1 - \tfrac{1}{2}\theta^2$$

to be good.

(b) Find $\cos 10°$ using a calculating utility.

(c) Estimate $\cos 10°$ using the approximation above.

55. It follows from part (a) of Example 2 that if θ is small (near zero) and measured in radians, then one should expect the approximation

$$\tan\theta \approx \theta$$

to be good.

(a) Find tan 5° using a calculating utility.
(b) Find tan 5° using the approximation above.

56. Referring to the accompanying figure, suppose that the angle of elevation of the top of a building, as measured from a point L feet from its base, is found to be α degrees.
(a) Use the relationship $h = L \tan \alpha$ to calculate the height of a building for which $L = 500$ ft and $\alpha = 6°$.
(b) Show that if L is large compared to the building height h, then one should expect good results in approximating h by $h \approx \pi L\alpha/180$.
(c) Use the result in part (b) to approximate the building height h in part (a).

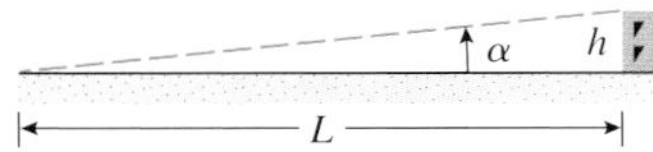

Figure Ex-56

57. (a) Use the Intermediate-Value Theorem to show that the equation $x = \cos x$ has at least one solution in the interval $[0, \pi/2]$.
(b) Show graphically that there is exactly one solution in the interval.
(c) Approximate the solution to three decimal places.

58. (a) Use the Intermediate-Value Theorem to show that the equation $x + \sin x = 1$ has at least one solution in the interval $[0, \pi/6]$.
(b) Show graphically that there is exactly one solution in the interval.
(c) Approximate the solution to three decimal places.

59. In the study of falling objects near the surface of the Earth, the ***acceleration g due to gravity*** is commonly taken to be a constant 9.8 m/s^2. However, the elliptical shape of the Earth and other factors cause variations in this value that depend on latitude. The following formula, known as the World Geodetic System 1984 (WGS 84) Ellipsoidal Gravity Formula, is used to predict the value of g at a latitude of ϕ degrees (either north or south of the equator):

$$g = 9.7803253359\frac{1 + 0.0019318526461 \sin^2 \phi}{\sqrt{1 - 0.0066943799901 \sin^2 \phi}} \text{ m/s}^2$$

(a) Use a graphing utility to graph the curve $y = g(\phi)$ for $0° \le \phi \le 90°$. What do the values of g at $\phi = 0°$ and at $\phi = 90°$ tell you about the WGS 84 ellipsoid model for the Earth?
(b) Show that $g = 9.8$ m/s^2 somewhere between latitudes of 38° and 39°.

60. Let

$$f(x) = \begin{cases} 1 & \text{if } x \text{ is a rational number} \\ 0 & \text{if } x \text{ is an irrational number} \end{cases}$$

(a) Make a conjecture about the limit of $f(x)$ as $x \to 0$.
(b) Make a conjecture about the limit of $xf(x)$ as $x \to 0$.
(c) Prove your conjectures.

✔ QUICK CHECK ANSWERS 2.6

1. (a) yes (b) yes (c) yes (d) no (e) yes (f) no **2.** (a) 1 (b) 0
3. (a) Since $-1 \le \cos(\pi/x) \le 1$ for $x \ne 0$, it follows that $-x^2 \le f(x) \le x^2$. Note that $\lim_{x\to 0} g(x) = 0 = \lim_{x\to 0} h(x)$, so the Squeezing Theorem can be used for $\lim_{x\to 0} f(x)$. However, $\lim_{x\to 1} g(x) = -1$ and $\lim_{x\to 0} h(x) = 1 \ne -1$, so the Squeezing Theorem cannot be used for $\lim_{x\to 1} f(x)$. (b) 0; −1 **4.** (a) sin(−3) (b) does not exist (c) −3 (d) 0

CHAPTER REVIEW EXERCISES

Graphing Utility

1. For the function f graphed in the accompanying figure, find the limit if it exists.
(a) $\lim_{x\to 1} f(x)$ (b) $\lim_{x\to 2} f(x)$ (c) $\lim_{x\to 3} f(x)$
(d) $\lim_{x\to 4} f(x)$ (e) $\lim_{x\to +\infty} f(x)$ (f) $\lim_{x\to -\infty} f(x)$
(g) $\lim_{x\to 3^+} f(x)$ (h) $\lim_{x\to 3^-} f(x)$ (i) $\lim_{x\to 0} f(x)$

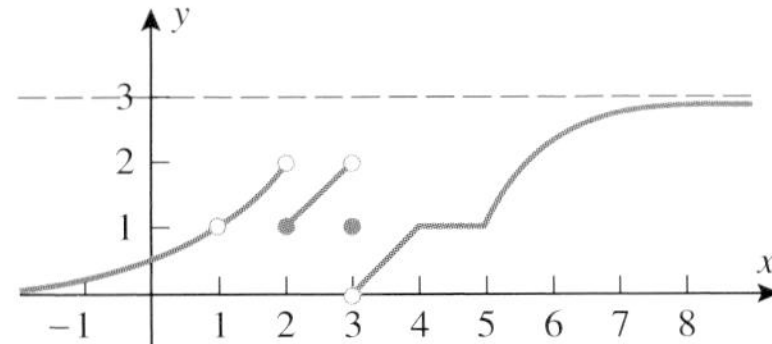

Figure Ex-1

2. In each part, evaluate the function for the stated values of x, and make a conjecture about the value of the limit. Confirm your conjecture by finding the limit algebraically.
(a) $f(x) = \dfrac{x-2}{x^2-4}$; $\lim_{x\to 2^+} f(x)$; $x = 2.5, 2.1, 2.01, 2.001, 2.0001, 2.00001$
(b) $f(x) = \dfrac{\tan 4x}{x}$; $\lim_{x\to 0} f(x)$; $x = \pm 1.0, \pm 0.1, \pm 0.01, \pm 0.001, \pm 0.0001, \pm 0.00001$

3. (a) Approximate the value for the limit

$$\lim_{x\to 0} \frac{3^x - 2^x}{x}$$

to three decimal places by constructing an appropriate table of values.
(b) Confirm your approximation using graphical evidence.

4. Approximate

$$\lim_{x\to 3}\frac{2^x-8}{x-3}$$

both by looking at a graph and by calculating values for some appropriate choices of x.

5–10 Find the limits.

5. $\lim_{x\to -1}\frac{x^3-x^2}{x-1}$ **6.** $\lim_{x\to 1}\frac{x^3-x^2}{x-1}$

7. $\lim_{x\to -3}\frac{3x+9}{x^2+4x+3}$ **8.** $\lim_{x\to 2^-}\frac{x+2}{x-2}$

9. $\lim_{x\to +\infty}\frac{(2x-1)^5}{(3x^2+2x-7)(x^3-9x)}$

10. $\lim_{x\to 0}\frac{\sqrt{x^2+4}-2}{x^2}$

11. In each part, find the horizontal asymptotes, if any.
(a) $y=\frac{2x-7}{x^2-4x}$ (b) $y=\frac{x^3-x^2+10}{3x^2-4x}$
(c) $y=\frac{2x^2-6}{x^2+5x}$

12. In each part, find $\lim_{x\to a} f(x)$, if it exists, where a is replaced by $0, 5^+, -5^-, -5, 5, -\infty$, and $+\infty$.
(a) $f(x)=\sqrt{5-x}$
(b) $f(x)=\begin{cases}(x-5)/|x-5|, & x\neq 5\\ 0, & x=5\end{cases}$

13–17 Find the limits.

13. $\lim_{x\to 0}\frac{\sin 3x}{\tan 3x}$ **14.** $\lim_{x\to 0}\frac{x\sin x}{1-\cos x}$

15. $\lim_{x\to 0}\frac{3x-\sin(kx)}{x}$, $k\neq 0$

16. $\lim_{\theta\to 0}\tan\left(\frac{1-\cos\theta}{\theta}\right)$ **17.** $\lim_{x\to -1}\frac{\sin(x+1)}{x^2-1}$

18. (a) Write a paragraph or two that describes how the limit of a function can fail to exist at $x=a$, and accompany your description with some specific examples.
(b) Write a paragraph or two that describes how the limit of a function can fail to exist as $x\to+\infty$ or $x\to-\infty$, and accompany your description with some specific examples.
(c) Write a paragraph or two that describes how a function can fail to be continuous at $x=a$, and accompany your description with some specific examples.

19. (a) Find a formula for a rational function that has a vertical asymptote at $x=1$ and a horizontal asymptote at $y=2$.
(b) Check your work by using a graphing utility to graph the function.

20. Paraphrase the ϵ-δ definition for $\lim_{x\to a} f(x)=L$ in terms of a graphing utility viewing window centered at the point (a, L).

21. Suppose that $f(x)$ is a function and that for any given $\epsilon>0$, the condition $0<|x-2|<\frac{3}{4}\epsilon$ guarantees that $|f(x)-5|<\epsilon$.
(a) What limit is described by this statement?
(b) Find a value of δ such that $0<|x-2|<\delta$ guarantees that $|8f(x)-40|<0.048$.

22. The limit

$$\lim_{x\to 0}\frac{\sin x}{x}=1$$

ensures that there is a number δ such that

$$\left|\frac{\sin x}{x}-1\right|<0.001$$

if $0<|x|<\delta$. Estimate the largest such δ.

23. In each part, a positive number ϵ and the limit L of a function f at a are given. Find a number δ such that $|f(x)-L|<\epsilon$ if $0<|x-a|<\delta$.
(a) $\lim_{x\to 2}(4x-7)=1;\ \epsilon=0.01$
(b) $\lim_{x\to 3/2}\frac{4x^2-9}{2x-3}=6;\ \epsilon=0.05$
(c) $\lim_{x\to 4}x^2=16;\ \epsilon=0.001$

24. Use Definition 2.4.1 to prove the stated limits are correct.
(a) $\lim_{x\to 2}(4x-7)=1$ (b) $\lim_{x\to 3/2}\frac{4x^2-9}{2x-3}=6$

25. Suppose that f is continuous at x_0 and that $f(x_0)>0$. Give either an ϵ-δ proof or a convincing verbal argument to show that there must be an open interval containing x_0 on which $f(x)>0$.

26. (a) Let

$$f(x)=\frac{\sin x-\sin 1}{x-1}$$

Approximate $\lim_{x\to 1} f(x)$ by graphing f and calculating values for some appropriate choices of x.
(b) Use the identity

$$\sin\alpha-\sin\beta=2\sin\frac{\alpha-\beta}{2}\cos\frac{\alpha+\beta}{2}$$

to find the exact value of $\lim_{x\to 1} f(x)$.

27. Find values of x, if any, at which the given function is not continuous.
(a) $f(x)=\frac{x}{x^2-1}$ (b) $f(x)=|x^3-2x^2|$
(c) $f(x)=\frac{x+3}{|x^2+3x|}$

28. Determine where f is continuous.
(a) $f(x)=\frac{x}{|x|-3}$ (b) $f(x)=\cos\left(\frac{1}{x}\right)$
(c) $f(x)=\frac{2x-1}{2x^2+3x-2}$

29. Suppose that

$$f(x)=\begin{cases}-x^4+3, & x\le 2\\ x^2+9, & x>2\end{cases}$$

Is f continuous everywhere? Justify your conclusion.

30. One dictionary describes a continuous function as "one whose value at each point is closely approached by its values at neighboring points."

(a) How would you explain the meaning of the terms "neighboring points" and "closely approached" to a nonmathematician?

(b) Write a paragraph that explains why the dictionary definition is consistent with Definition 2.5.1.

31. Show that the conclusion of the Intermediate-Value Theorem may be false if f is not continuous on the interval $[a, b]$.

32. Suppose that f is continuous on the interval $[0, 1]$, that $f(0) = 2$, and that f has no zeros in the interval. Prove that $f(x) > 0$ for all x in $[0, 1]$.

33. Show that the equation $x^4 + 5x^3 + 5x - 1 = 0$ has at least two real solutions in the interval $[-6, 2]$.

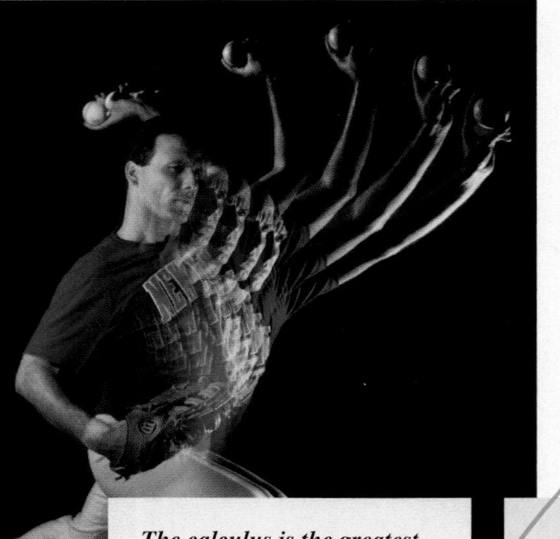

THE DERIVATIVE

The calculus is the greatest aid we have to the appreciation of physical truth in the broadest sense of the word.

—William Fogg Osgood
Mathematician

Many real-world phenomena involve changing quantities—the speed of a rocket, the inflation of currency, the number of bacteria in a culture, the shock intensity of an earthquake, the voltage of an electrical signal, and so forth. In this chapter we will develop the concept of a "derivative," which is the mathematical tool for studying the rate at which one quantity changes relative to another. The study of rates of change is closely related to the geometric concept of a tangent line to a curve, so we will also be discussing the general definition of a tangent line and methods for finding its slope and equation. Finally we will show how derivatives can be used to approximate nonlinear functions by linear functions.

Photo: ***One of the crowning achievements of calculus is its ability to capture continuous motion mathematically, allowing that motion to be analyzed instant-by-instant.***

3.1 TANGENT LINES, VELOCITY, AND GENERAL RATES OF CHANGE

In this section we will discuss three ideas: tangent lines to curves, the velocity of an object moving along a line, and the rate at which one variable changes relative to another. Our goal is to show how these seemingly unrelated ideas are, in actuality, closely linked.

TANGENT LINES

In Example 1 of Section 2.1, we showed how the notion of a limit could be used to find an equation of a tangent line to a curve. At that stage in the text we did not have precise definitions of tangent lines and limits to work with, so the argument was intuitive and informal. However, now that limits have been defined precisely, we are in a position to give a mathematical definition of the tangent line to a curve $y = f(x)$ at a point $P(x_0, f(x_0))$ on the curve. As illustrated in Figure 3.1.1, consider a point $Q(x, f(x))$ on the curve that is distinct from P, and compute the slope m_{PQ} of the secant line through P and Q:

$$m_{PQ} = \frac{f(x) - f(x_0)}{x - x_0}$$

If we let x approach x_0, then the point Q will move along the curve and approach the point P. If the secant line through P and Q approaches a limiting position as $x \to x_0$, then we will regard that position to be the position of the tangent line at P. Stated another way, if the slope m_{PQ} of the secant line through P and Q approaches a limit as $x \to x_0$, then we regard that limit to be the slope $m_{\tan}$ of the tangent line at P. Thus, we make the following definition.

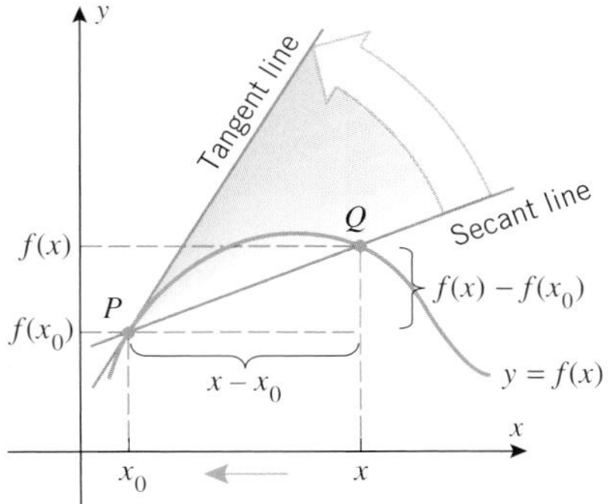

Figure 3.1.1

3.1.1 DEFINITION. Suppose that x_0 is in the domain of the function f. The ***tangent line*** to the curve $y = f(x)$ at the point $P(x_0, f(x_0))$ is the line with equation

$$y - f(x_0) = m_{\tan}(x - x_0)$$

where

$$m_{\tan} = \lim_{x \to x_0} \frac{f(x) - f(x_0)}{x - x_0} \tag{1}$$

provided the limit exists. For simplicity, we will also call this the tangent line to $y = f(x)$ at x_0.

▶ **Example 1** Use Definition 3.1.1 to find an equation for the tangent line to the parabola $y = x^2$ at the point $P(1, 1)$, and confirm the result agrees with that obtained in Example 1 of Section 2.1.

Solution. Applying Formula (1) with $f(x) = x^2$ and $x_0 = 1$, we have

$$\begin{aligned} m_{\tan} &= \lim_{x \to 1} \frac{f(x) - f(1)}{x - 1} \\ &= \lim_{x \to 1} \frac{x^2 - 1}{x - 1} \\ &= \lim_{x \to 1} \frac{(x - 1)(x + 1)}{x - 1} = \lim_{x \to 1}(x + 1) = 2 \end{aligned}$$

Thus, the tangent line to $y = x^2$ at $(1, 1)$ has equation

$$y - 1 = 2(x - 1) \quad \text{or equivalently} \quad y = 2x - 1$$

which agrees with Example 1 of Section 2.1. ◀

There is an alternative way of expressing Formula (1) that is commonly used. If we let h denote the difference

$$h = x - x_0$$

then the statement that $x \to x_0$ is equivalent to the statement $h \to 0$, so we can rewrite (1) in terms of x_0 and h as

$$m_{\tan} = \lim_{h \to 0} \frac{f(x_0 + h) - f(x_0)}{h} \tag{2}$$

Although this formula looks different from (1), it is really just an alternative way of expressing the slope of the tangent line as a limit of slopes of secant lines (Figure 3.1.2).

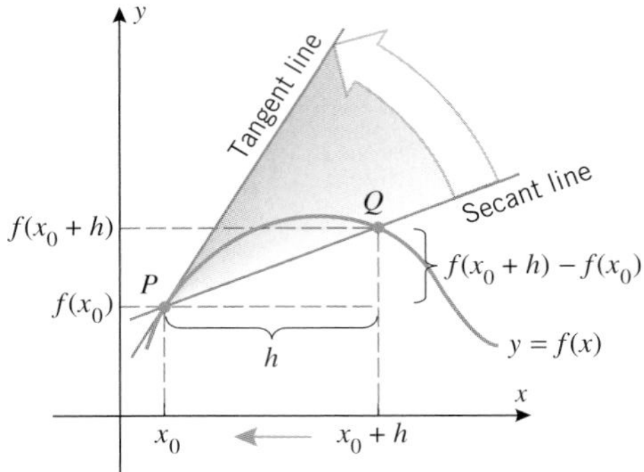

Figure 3.1.2

▶ Example 2 Compute the slope in Example 1 using Formula (2).

Formulas (1) and (2) for $m_{\tan}$ usually lead to indeterminate forms of type 0/0, so you will generally need to perform algebraic simplifications or use other methods to determine limits of such indeterminate forms.

Solution. Applying Formula (2) with $f(x) = x^2$ and $x_0 = 1$, we obtain

$$\begin{aligned} m_{\tan} &= \lim_{h\to 0} \frac{f(1+h) - f(1)}{h} \\ &= \lim_{h\to 0} \frac{(1+h)^2 - 1^2}{h} \\ &= \lim_{h\to 0} \frac{1 + 2h + h^2 - 1}{h} = \lim_{h\to 0}(2+h) = 2 \end{aligned}$$

which agrees with the slope found in Example 1. ◀

▶ Example 3 Find an equation for the tangent line to the curve $y = 2/x$ at the point (2, 1) on this curve.

Solution. First, we will find the slope of the tangent line by applying Formula (2) with $f(x) = 2/x$ and $x_0 = 2$. This yields

$$\begin{aligned} m_{\tan} &= \lim_{h\to 0} \frac{f(2+h) - f(2)}{h} \\ &= \lim_{h\to 0} \frac{\dfrac{2}{2+h} - 1}{h} = \lim_{h\to 0} \frac{\left(\dfrac{2-(2+h)}{2+h}\right)}{h} \\ &= \lim_{h\to 0} \frac{-h}{h(2+h)} = -\lim_{h\to 0} \frac{1}{2+h} \\ &= -\frac{1}{2} \end{aligned}$$

Thus, an equation of the tangent line at (2, 1) is

$$y - 1 = -\tfrac{1}{2}(x-2) \quad \text{or equivalently} \quad y = -\tfrac{1}{2}x + 2$$

(see Figure 3.1.3). ◀

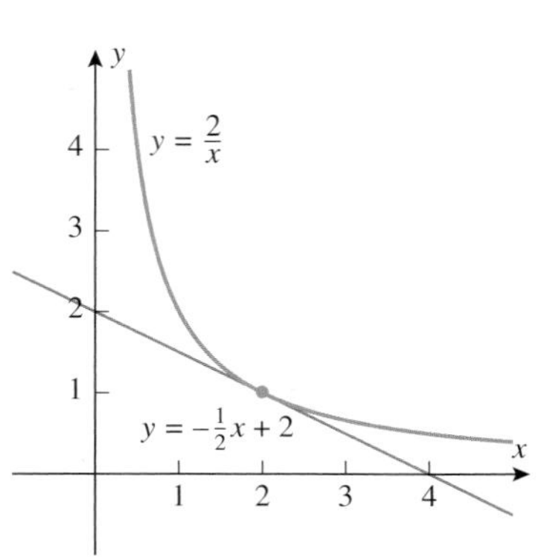

Figure 3.1.3

► **Example 4** Find the slopes of the tangent lines to the curve $y = \sqrt{x}$ at $x_0 = 1$, $x_0 = 4$, and $x_0 = 9$.

Solution. We could compute each of these slopes separately, but it will be more efficient to find the slope for a general value of x_0 and then substitute the specific numerical values. Proceeding in this way we obtain

$$
\begin{aligned}
m_{\tan} &= \lim_{h \to 0} \frac{f(x_0 + h) - f(x_0)}{h} \\
&= \lim_{h \to 0} \frac{\sqrt{x_0 + h} - \sqrt{x_0}}{h} \\
&= \lim_{h \to 0} \frac{\sqrt{x_0 + h} - \sqrt{x_0}}{h} \cdot \frac{\sqrt{x_0 + h} + \sqrt{x_0}}{\sqrt{x_0 + h} + \sqrt{x_0}} \\
&= \lim_{h \to 0} \frac{x_0 + h - x_0}{h(\sqrt{x_0 + h} + \sqrt{x_0})} \\
&= \lim_{h \to 0} \frac{h}{h(\sqrt{x_0 + h} + \sqrt{x_0})} \\
&= \lim_{h \to 0} \frac{1}{\sqrt{x_0 + h} + \sqrt{x_0}} = \frac{1}{2\sqrt{x_0}}
\end{aligned}
$$

Rationalize the numerator to help eliminate the indeterminate form of the limit.

The slopes at $x_0 = 1, 4$, and 9 can now be obtained by substituting these values into our general formula for $m_{\tan}$. Thus,

$$\text{slope at } x_0 = 1 \text{ is } \frac{1}{2\sqrt{1}} = \frac{1}{2}, \quad \text{slope at } x_0 = 4 \text{ is } \frac{1}{2\sqrt{4}} = \frac{1}{4}$$

$$\text{slope at } x_0 = 9 \text{ is } \frac{1}{2\sqrt{9}} = \frac{1}{6}$$

(see Figure 3.1.4). ◄

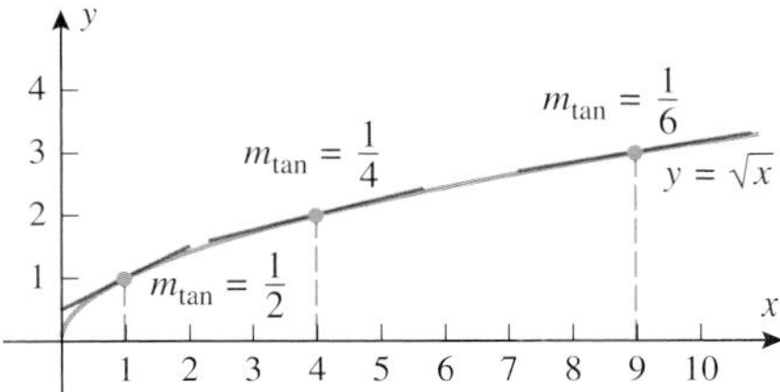

Figure 3.1.4

RECTILINEAR MOTION

One of the important themes in calculus is the study of motion. To describe the motion of an object completely, one must specify its ***speed*** (how fast it is going) and the direction in which it is moving. The speed and the direction of motion together comprise what is called the ***velocity*** of the object. For example, knowing that the speed of an aircraft is 500 mi/h tells us how fast it is going, but not which way it is moving. In contrast, knowing that the velocity of the aircraft is 500 mi/h *due south* pins down the speed and the direction of motion.

Later, we will study the motion of objects that move along curves in two- or three-dimensional space, but for now we will only consider motion along a line; this is called ***rectilinear motion***. Some examples are a piston moving up and down in a cylinder, a race car moving along a straight track, an object dropped from the top of a building and falling straight down, a ball thrown straight up and then falling down along the same line, and so forth.

For computational purposes, we will assume that a particle in rectilinear motion moves along a coordinate line, such as an x-axis or a y-axis. In general discussions where we need not be specific, we will call the coordinate line the s-axis. A graphical description of rectilinear motion along an s-axis can be obtained by making a plot of the s-coordinate of the particle versus the elapsed time t from starting time $t = 0$. This is called the ***position versus time curve*** for the particle. Figure 3.1.5 shows two typical position versus time curves. The first is for a car that starts at the origin and moves only in the positive direction of the s-axis. In this case s increases as t increases. The second is for a ball that is thrown straight up in the positive direction of an s-axis from some initial height s_0, and then falls straight down in the negative direction. In this case s increases as the ball moves up and decreases as it moves down.

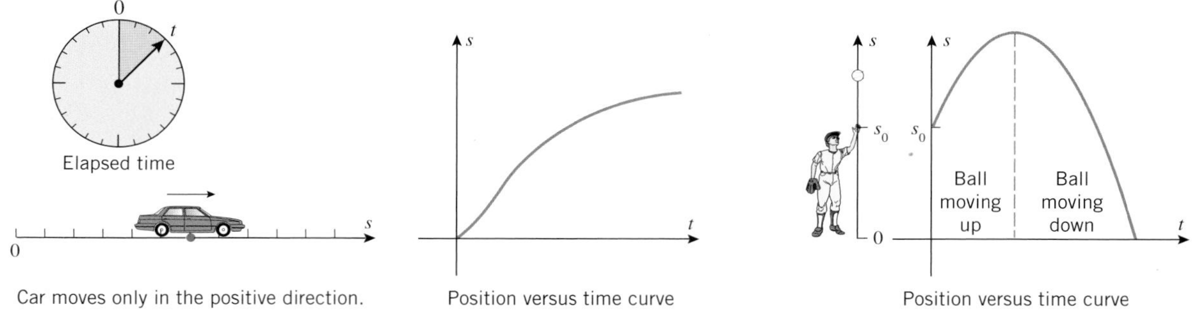

Car moves only in the positive direction. Position versus time curve Position versus time curve

Figure 3.1.5

If a particle in rectilinear motion moves along an s-axis so that its position coordinate function of the elapsed time t is

$$s = f(t) \tag{3}$$

then f is called the ***position function of the particle***; the graph of (3) is the position versus time curve.

DISPLACEMENT AND AVERAGE VELOCITY

The key to describing the velocity of a particle in rectilinear motion is the notion of "displacement." If $[t_0, t_0 + h]$ is a given time interval, then we define the ***displacement*** or ***change in position*** of the particle over this time interval to be the difference between its final and initial position coordinates:

$$\text{displacement over the interval } [t_0, t_0 + h] = f(t_0 + h) - f(t_0) \tag{4}$$

The displacement is positive if the final position is in the positive direction relative to the initial position, negative if the final position is in the negative direction relative to the initial position, and zero if the final position coincides with the initial position (Figure 3.1.6).

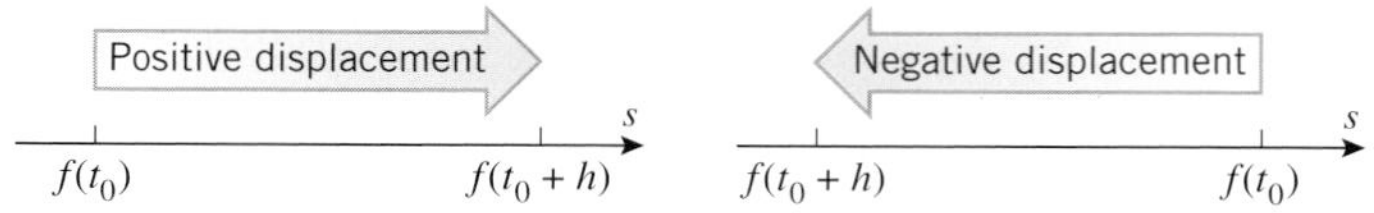

Positive displacement — s: $f(t_0)$, $f(t_0 + h)$ | Negative displacement — s: $f(t_0 + h)$, $f(t_0)$

Figure 3.1.6

If, over a given time interval, a particle in rectilinear motion moves only in the positive direction, then the displacement over that time interval is the same as the distance traveled. However, if the particle can move in either direction over that time interval, then the displacement and distance traveled may differ. For example, if the particle moves 100 units in the positive direction and then 100 units in the negative direction, the distance traveled is 200 units but the displacement is zero (since the final position coincides with the initial position).

For a particle in rectilinear motion it is important to distinguish between its *speed* (how fast it is moving) and its *velocity* (how fast and in what direction). This is done by using negative velocity for motion in the negative direction and positive velocity for motion in the positive direction. Thus, a particle with a velocity of -2 m/s has a speed of 2 m/s and is moving in the negative direction, and a particle with a velocity of 2 m/s has a speed of 2 m/s and is moving in the positive direction.

Suppose that a particle in rectilinear motion along an s-axis has position function $s = f(t)$. The ***average velocity*** of the particle over a time interval $[t_0, t_0 + h]$, $h > 0$, is defined to be

Show that (5) is also correct for a time interval $[t_0 + h, t_0]$, $h < 0$.

$$v_{\text{ave}} = \frac{\text{displacement}}{\text{time elapsed}} = \frac{f(t_0 + h) - f(t_0)}{h} \tag{5}$$

Average speed is defined using the distance the particle travels, as opposed to its displacement. For example, if a particle moves 5 m in the positive direction and then retreats 2 m in the negative direction, its displacement is 3 m but it has traveled a distance of $5 + 2 = 7$ m. We define *average speed* to be the ratio of distance traveled to time elapsed:

$$\text{speed}_{\text{ave}} = \frac{\text{distance traveled}}{\text{time elapsed}}$$

In the case where the particle moves only in the positive direction, its displacement and distance traveled over any time interval are the same. In this case, average speed and average velocity are also the same.

▶ **Example 5** Suppose that $s = f(t) = 1 + 3t - 2t^2$ is the position function of a particle, where s is in meters and t is in seconds. Find the displacements and average velocities of the particle over the time intervals (a) $[0, 1]$ and (b) $[1, 3]$.

Solution (a). Applying (4) with $t_0 = 0$ and $h = 1$, we see that the displacement is

$$f(t_0 + h) - f(t_0) = f(1) - f(0) = 2 - 1 = 1 \text{ m}$$

It follows from (5) that the average velocity over the interval $[0, 1]$ is $1/1 = 1$ m/s.

Solution (b). Applying (4) with $t_0 = 1$ and $h = 2$, we see that the displacement is

$$f(t_0 + h) - f(t_0) = f(3) - f(1) = -8 - 2 = -10 \text{ m}$$

It follows from (5) that the average velocity over the interval $[1, 3]$ is $-10/2 = -5$ m/s. ◀

INSTANTANEOUS VELOCITY

For a particle in rectilinear motion, average velocity describes its behavior over an *interval* of time. We are interested in the particle's "instantaneous velocity," which describes its behavior at a specific *instant* in time. Formula (5) is not directly applicable for computing instantaneous velocity because the "time elapsed" at a specific instant is zero, so (5) is undefined. One way to circumvent this problem is to compute average velocities for small time intervals between $t = t_0$ and $t = t_0 + h$. These average velocities may be viewed as approximations to the "instantaneous velocity" of the particle at time t_0. If these average velocities have a limit as h approaches zero, then we can take that limit to be the *instantaneous velocity* of the particle at time t_0. Here is an example.

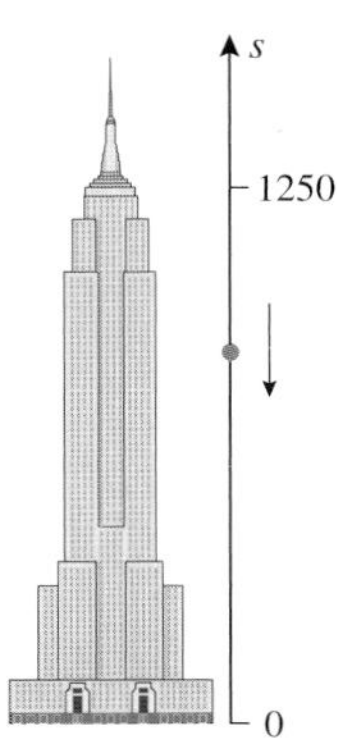

Figure 3.1.7

Note the negative values for the velocities in Example 6. This is consistent with the fact that the object is moving in the negative direction along the s-axis.

Table 3.1.1

TIME INTERVAL	AVERAGE VELOCITY (ft/s)
$5.0 \le t \le 6.0$	−176
$5.0 \le t \le 5.1$	−161.6
$5.0 \le t \le 5.01$	−160.16
$5.0 \le t \le 5.001$	−160.016
$5.0 \le t \le 5.0001$	−160.0016

▶ **Example 6** Suppose that an object is released from rest (i.e., its initial velocity is zero) from the Empire State Building from a height of 1250 ft above street level. It is shown in physics that with appropriate simplifying assumptions the object's height s (in feet) above the street level, t seconds after its release, can be modeled by the position function

$$s = f(t) = 1250 - 16t^2$$

(see Figure 3.1.7). Verify that the object has not reached the ground at $t = 5$ s, and find its instantaneous velocity at that time.

Solution. We first note that $f(5) = 1250 - 400 = 850$ ft, so the object is still falling 5 s after it is released. As a first approximation to the object's instantaneous velocity at time $t = 5$ s, let us compute the average velocity over the time interval from $t = 5$ to $t = 6$. It follows from (5) with $t_0 = 5$ and $h = 1$ that

$$v_{\text{ave}} = \frac{f(t_0 + h) - f(t_0)}{h} = \frac{f(6) - f(5)}{1} = \frac{674 - 850}{1} = -176 \text{ ft/s}$$

To improve on this initial approximation we will compute the average velocity over a succession of smaller and smaller time intervals. We leave it to you to verify the results in Table 3.1.1. The average velocities in this table appear to be approaching a limit of -160 ft/s, providing strong evidence that the instantaneous velocity at time $t = 5$ s is -160 ft/s. To confirm this analytically, we start by computing the object's average velocity over a general time interval between $t = 5$ and $t = 5 + h$ using Formula (5):

$$v_{\text{ave}} = \frac{f(5 + h) - f(5)}{h} = \frac{[1250 - 16(5 + h)^2] - 850}{h}$$

The object's instantaneous velocity at time $t = 5$ is calculated as a limit as $h \to 0$:

$$\begin{aligned}
\text{instantaneous velocity} &= \lim_{h \to 0} \frac{[1250 - 16(5 + h)^2] - 850}{h} \\
&= \lim_{h \to 0} \frac{400 - 16(25 + 10h + h^2)}{h} \\
&= \lim_{h \to 0} \frac{-16(10h + h^2)}{h} = \lim_{h \to 0} -16(10 + h) = -160
\end{aligned}$$

This confirms our numerical conjecture that the instantaneous velocity after 5 s is -160 ft/s. ◀

Consider a particle in rectilinear motion with position function $s = f(t)$. Motivated by Example 6, we define the instantaneous velocity v_{inst} of the particle at time t_0 to be the limit as $h \to 0$ of its average velocities v_{ave} over time intervals between $t = t_0$ and $t = t_0 + h$. Thus, from (5) we obtain

$$v_{\text{inst}} = \lim_{h \to 0} \frac{f(t_0 + h) - f(t_0)}{h} \tag{6}$$

Geometrically, the average velocity v_{ave} between $t = t_0$ and $t = t_0 + h$ is the slope of the secant line through points $P(t_0, f(t_0))$ and $Q(t_0 + h, f(t_0 + h))$ on the position versus time curve, and the instantaneous velocity v_{inst} at time t_0 is the slope of the tangent line to the position versus time curve at the point $P(t_0, f(t_0))$ (Figure 3.1.8).

The instantaneous velocity is positive when the particle is moving in the positive direction, is negative when the particle is moving in the negative direction, and is zero when the particle is momentarily stopped. We define the ***instantaneous speed*** at time t_0 to be the absolute value of the velocity:

$$\text{speed}_{\text{inst}} = |v_{\text{inst}}| = \left| \lim_{h\to 0} \frac{f(t_0+h) - f(t_0)}{h} \right| \tag{7}$$

This tells us how fast the particle is moving at time $t = t_0$ but not its direction. Thus, in Example 6, the object's speed at time $t = 5$ s is $|-160| = 160$ ft/s, whereas its instantaneous velocity at that instant is -160 ft/s.

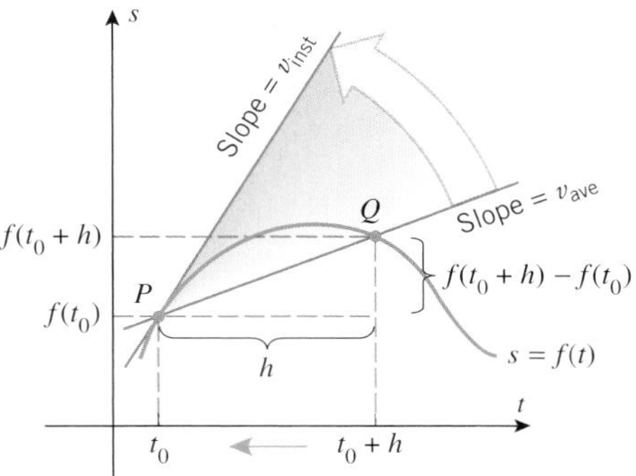

Figure 3.1.8

SLOPES AND RATES OF CHANGE

Velocity can be viewed as *rate of change*—the rate of change of position with respect to time. Rates of change occur in other applications as well. For example:

- A microbiologist might be interested in the rate at which the number of bacteria in a colony changes with time.
- An engineer might be interested in the rate at which the length of a metal rod changes with temperature.
- An economist might be interested in the rate at which production cost changes with the quantity of a product that is manufactured.
- A medical researcher might be interested in the rate at which the radius of an artery changes with the concentration of alcohol in the bloodstream.

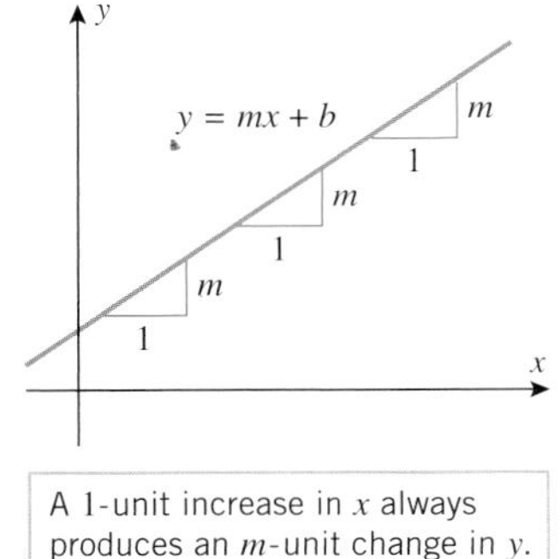

A 1-unit increase in x always produces an m-unit change in y.

Figure 3.1.9

Our next objective is to define precisely what is meant by the "rate of change of y with respect to x" when y is a function of x. In the case where y is a linear function of x, say $y = mx + b$, the slope m is the natural measure of the rate of change of y with respect to x. As illustrated in Figure 3.1.9, each 1-unit increase in x anywhere along the line produces an m-unit change in y, so we see that y changes at a constant rate with respect to x along the line and that m measures this rate of change.

▶ **Example 7** Find the rate of change of y with respect to x if

$$\text{(a)}\ \ y = 2x - 1 \qquad \text{(b)}\ \ y = -5x + 1$$

Solution. In part (a) the rate of change of y with respect to x is $m = 2$, so each 1-unit increase in x produces a 2-unit increase in y. In part (b) the rate of change of y with respect to x is $m = -5$, so each 1-unit increase in x produces a 5-unit decrease in y. ◀

In applied problems, changing the units of measurement can change the slope of a line, so it is essential to include the units when calculating the slope and describing rates of change. The following example illustrates this.

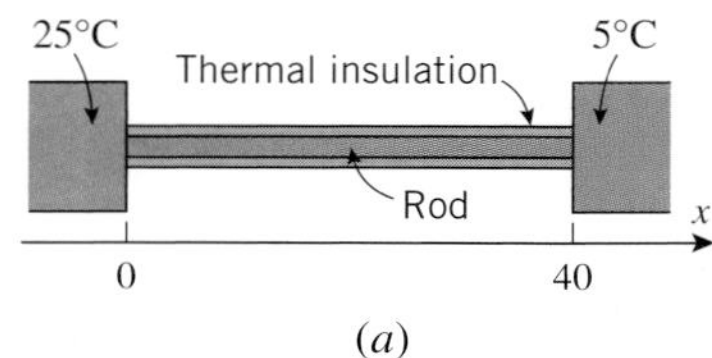

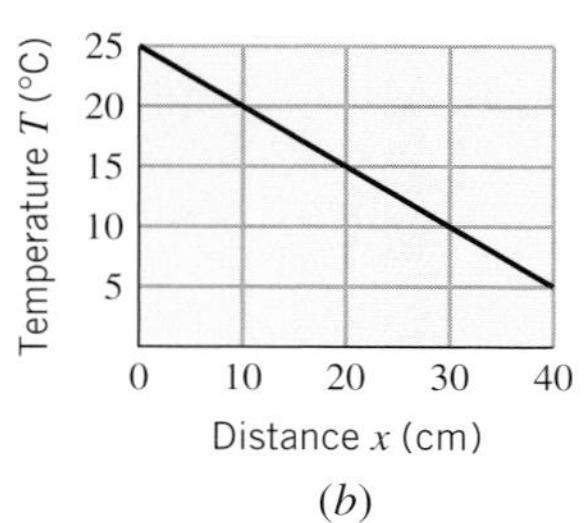

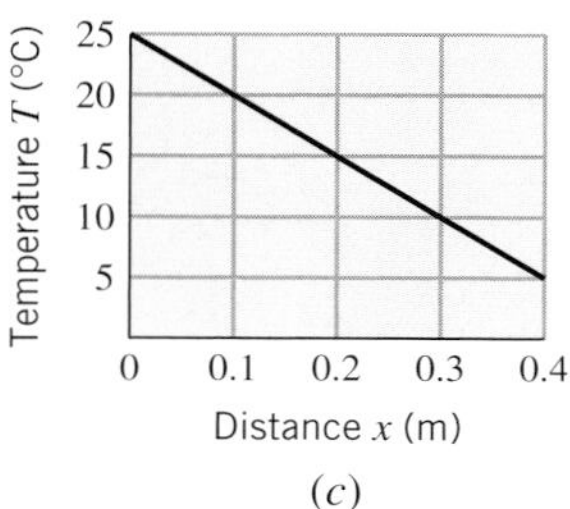

Figure 3.1.10

► **Example 8** Suppose that a uniform rod of length 40 cm (= 0.4 m) is thermally insulated around the lateral surface and that the exposed ends of the rod are held at constant temperatures of 25°C and 5°C, respectively (Figure 3.1.10*a*). It is shown in physics that under appropriate conditions the graph of the temperature T versus the distance x from the left-hand end of the rod will be a straight line. Parts (*b*) and (*c*) of Figure 3.1.10 show two such graphs: one in which x is measured in centimeters and one in which it is measured in meters. The slopes in the two cases are

$$m = \frac{5 - 25}{40 - 0} = \frac{-20}{40} = -0.5 \tag{8}$$

$$m = \frac{5 - 25}{0.4 - 0} = \frac{-20}{0.4} = -50 \tag{9}$$

The slope in (8) implies that the temperature *decreases* at a rate of 0.5°C per centimeter of distance from the left end of the rod, and the slope in (9) implies that the temperature decreases at a rate of 50°C per meter of distance from the left end of the rod. The two statements are equivalent physically, even though the slopes differ. ◄

Although the rate of change of y with respect to x is constant along a nonvertical line $y = mx + b$, this is not true for a general curve $y = f(x)$. For example, in Figure 3.1.11 the change in y that results from a 1-unit increase in x tends to have greater magnitude in regions where the curve rises or falls rapidly than in regions where it rises or falls slowly. As with velocity, we will distinguish between the average rate of change over an interval and the instantaneous rate of change at a specific point.

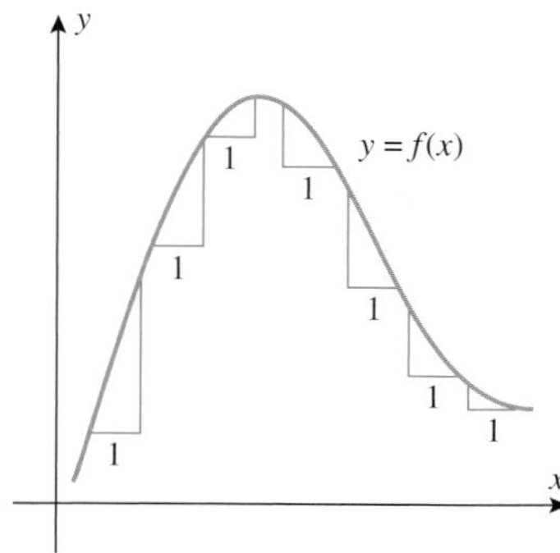

Figure 3.1.11

If $y = f(x)$, then we define the ***average rate of change of y with respect to x over the interval*** $[x_0, x_1]$ to be

$$r_{\text{ave}} = \frac{f(x_1) - f(x_0)}{x_1 - x_0} \tag{10}$$

and we define the ***instantaneous rate of change of y with respect to x at*** x_0 to be

$$r_{\text{inst}} = \lim_{x_1 \to x_0} \frac{f(x_1) - f(x_0)}{x_1 - x_0} \tag{11}$$

Geometrically, the average rate of change of y with respect to x over the interval $[x_0, x_1]$ is the slope of the secant line through the points $P(x_0, f(x_0))$ and $Q(x_1, f(x_1))$ (Figure 3.1.12), and the instantaneous rate of change of y with respect to x at x_0 is the slope of the tangent line at the point $P(x_0, f(x_0))$ (since it is the limit of the slopes of the secant lines through P).

If desired, we can let $h = x_1 - x_0$, and rewrite (10) and (11) as

$$r_{\text{ave}} = \frac{f(x_0 + h) - f(x_0)}{h} \tag{12}$$

$$r_{\text{inst}} = \lim_{h \to 0} \frac{f(x_0 + h) - f(x_0)}{h} \tag{13}$$

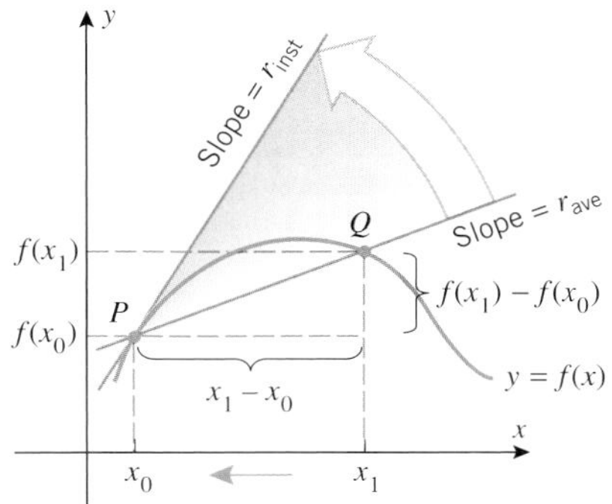

Figure 3.1.12

► **Example 9** Let $y = x^2 + 1$.

(a) Find the average rate of change of y with respect to x over the interval $[3, 5]$.

(b) Find the instantaneous rate of change of y with respect to x when $x = -4$.

Solution (a). We will apply Formula (10) with $f(x) = x^2 + 1$, $x_0 = 3$, and $x_1 = 5$. This yields

$$r_{ave} = \frac{f(x_1) - f(x_0)}{x_1 - x_0} = \frac{f(5) - f(3)}{5 - 3} = \frac{26 - 10}{2} = 8$$

Thus, y increases an average of 8 units per unit increase in x over the interval $[3, 5]$.

Solution (b). We will apply Formula (11) with $f(x) = x^2 + 1$ and $x_0 = -4$. This yields

$$r_{inst} = \lim_{x_1 \to x_0} \frac{f(x_1) - f(x_0)}{x_1 - x_0} = \lim_{x_1 \to -4} \frac{f(x_1) - f(-4)}{x_1 - (-4)} = \lim_{x_1 \to -4} \frac{(x_1^2 + 1) - 17}{x_1 + 4}$$

$$= \lim_{x_1 \to -4} \frac{x_1^2 - 16}{x_1 + 4} = \lim_{x_1 \to -4} \frac{(x_1 + 4)(x_1 - 4)}{x_1 + 4} = \lim_{x_1 \to -4} (x_1 - 4) = -8$$

Thus, a small increase in x from $x = -4$ will produce approximately an 8-fold decrease in y. ◄

Perform the calculations in Example 9 using Formulas (12) and (13).

RATES OF CHANGE IN APPLICATIONS

In applied problems, average and instantaneous rates of change must be accompanied by appropriate units. In general, the units for a rate of change of y with respect to x are obtained by "dividing" the units of y by the units of x and then simplifying according to the standard rules of algebra. Here are some examples:

- If y is in degrees Fahrenheit (°F) and x is in inches (in), then a rate of change of y with respect to x has units of degrees Fahrenheit per inch (°F/in).
- If y is in feet per second (ft/s) and x is in seconds (s), then a rate of change of y with respect to x has units of feet per second per second (ft/s/s), which would usually be written as ft/s^2.
- If y is in newton-meters (N·m) and x is in meters (m), then a rate of change of y with respect to x has units of newtons (N), since N·m/m = N.
- If y is in foot-pounds (ft·lb) and x is in hours (h), then a rate of change of y with respect to x has units of foot-pounds per hour (ft·lb/h).

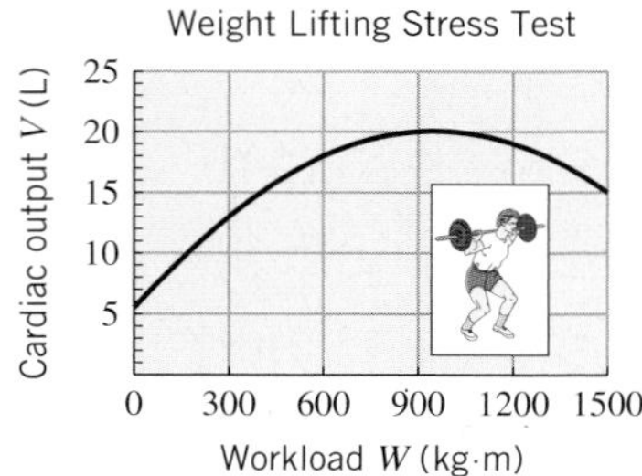

Figure 3.1.13

▶ **Example 10** The limiting factor in athletic endurance is cardiac output, that is, the volume of blood that the heart can pump per unit of time during an athletic competition. Figure 3.1.13 shows a stress-test graph of cardiac output V in liters (L) of blood versus workload W in kilogram-meters (kg·m) for 1 minute of weight lifting. This graph illustrates the known medical fact that cardiac output increases with the workload, but after reaching a peak value begins to decrease.

(a) Use the secant line shown in Figure 3.1.14*a* to estimate the average rate of change of cardiac output with respect to workload as the workload increases from 300 to 1200 kg·m.

(b) Use the line segment shown in Figure 3.1.14*b* to estimate the instantaneous rate of change of cardiac output with respect to workload at the point where the workload is 300 kg·m.

Solution (a). Using the estimated points (300, 13) and (1200, 19) to find the slope of the secant line, we obtain

$$r_{\text{ave}} \approx \frac{19 - 13}{1200 - 300} \approx 0.0067 \frac{\text{L}}{\text{kg·m}}$$

This means that on average a 1-unit increase in workload produced a 0.0067-L increase in cardiac output over the interval.

Solution (b). We estimate the slope of the cardiac output curve at $W = 300$ by sketching a line that appears to meet the curve at $W = 300$ with slope equal to that of the curve (Figure 3.1.14*b*). Estimating points (0, 7) and (900, 25) on this line, we obtain

$$r_{\text{inst}} \approx \frac{25 - 7}{900 - 0} = 0.02 \frac{\text{L}}{\text{kg·m}}$$ ◀

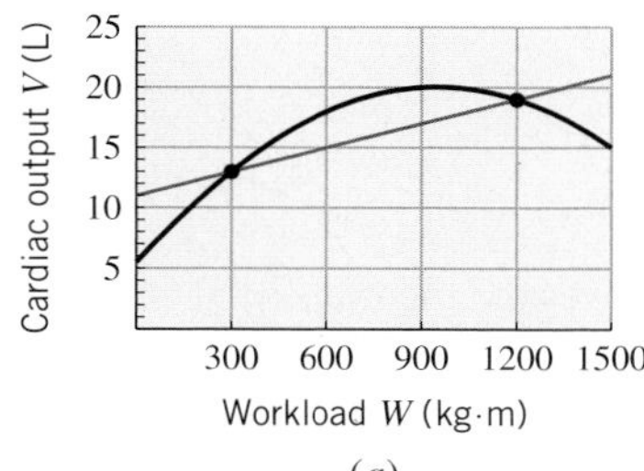

(*a*)

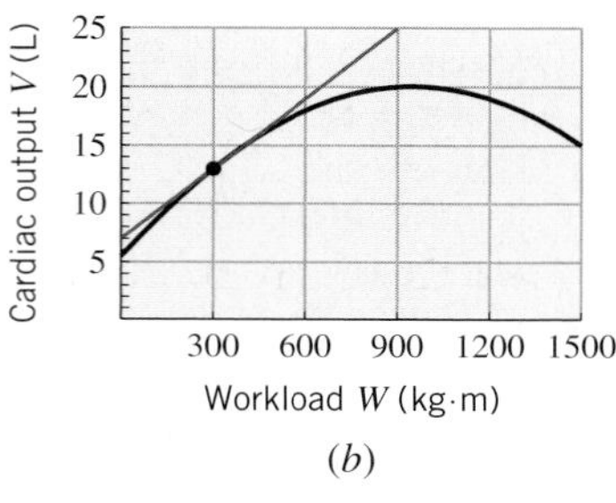

(*b*)

Figure 3.1.14

✔ QUICK CHECK EXERCISES 3.1 (See page 159 for answers.)

1. The slope $m_{\tan}$ of the tangent line to the curve $y = f(x)$ at the point $P(x_0, f(x_0))$ is given by

$$m_{\tan} = \lim_{x \to x_0} \underline{\qquad} = \lim_{h \to 0} \underline{\qquad}$$

2. The tangent line to the curve $y = (x - 1)^2$ at the point $(-1, 4)$ has equation $4x + y = 0$. Thus, the value of the limit

$$\lim_{x \to -1} \frac{x^2 - 2x - 3}{x + 1}$$

is ________.

3. A particle is moving along an s-axis, where s is in feet. During the first 5 seconds of motion, the position of the particle is given by

$$s = 10 - (3 - t)^2, \quad 0 \le t \le 5$$

Use this position function to complete each part.
(a) Initially, the particle moves a distance of ________ ft in the (positive/negative) ________ direction; then it reverses direction, traveling a distance of ________ ft during the remainder of the 5-second period.
(b) The average velocity of the particle over the 5-second period is ________.
(c) The average speed of the particle over the 5-second period is ________.

4. Let $s = f(t)$ be the equation of a position versus time curve for a particle in rectilinear motion, where s is in meters and t is in seconds. Assume that $s = -1$ when $t = 2$ and that the instantaneous velocity of the particle at this instant is 3 m/s. The equation of the tangent line to the position versus time curve at time $t = 2$ is ________.

5. Suppose that $y = x^2 + x$.
(a) The average rate of change of y with respect to x over the interval $2 \le x \le 5$ is ________.
(b) The instantaneous rate of change of y with respect to x at $x = 0$ is ________.

EXERCISE SET 3.1

FOCUS ON CONCEPTS

1. The accompanying figure shows the position versus time curve for an elevator that moves upward a distance of 60 m and then discharges its passengers.
(a) Estimate the instantaneous velocity of the elevator at $t = 10$ s.
(b) Sketch a velocity versus time curve for the motion of the elevator for $0 \le t \le 20$.

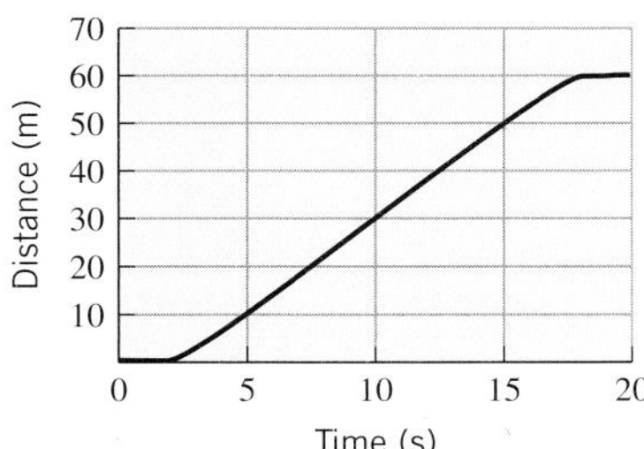

Figure Ex-1

2. The accompanying figure shows the position versus time curve for an automobile over a period of time of 10 s. Use the line segments shown in the figure to estimate the instantaneous velocity of the automobile at time $t = 4$ s and again at time $t = 8$ s.

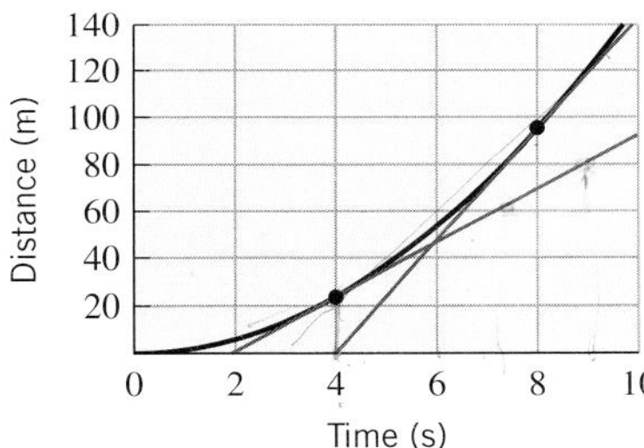

Figure Ex-2

3. A sky diver falls vertically downward from an airplane. The accompanying figure shows the graph of the distance s fallen by the sky diver versus the time t since leaping from the plane.
(a) Use the line segment in the accompanying graph to estimate the instantaneous speed of the sky diver at time $t = 5$ s.
(b) Estimate the instantaneous speed of the sky diver at time $t = 17.5$ s. What appears to be happening to the speed of the sky diver over time?

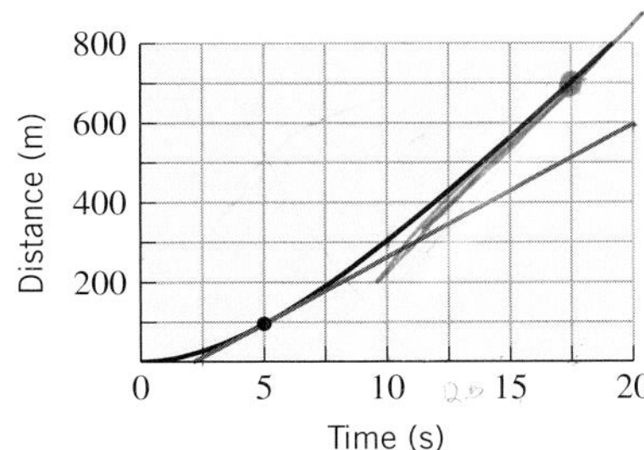

Figure Ex-3

4. The accompanying figure shows the position versus time curve for a certain particle moving along a straight line. Estimate each of the following from the graph:
(a) the average velocity over the interval $0 \le t \le 3$
(b) the values of t at which the instantaneous velocity is zero
(c) the values of t at which the instantaneous velocity is either a maximum or a minimum
(d) the instantaneous velocity when $t = 3$ s.

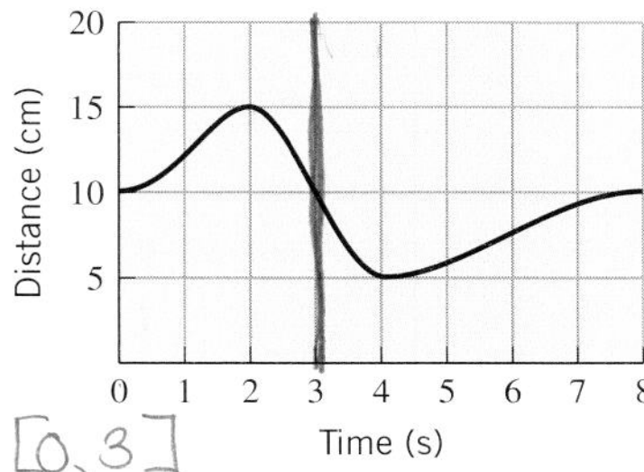

Figure Ex-4

5. The accompanying figure shows the position versus time curve for a certain particle moving on a straight line.
(a) Is the particle moving faster at time t_0 or time t_2? Explain.
(b) The portion of the curve near the origin is horizontal. What does this tell us about the initial velocity of the particle?
(c) Is the particle speeding up or slowing down in the interval $[t_0, t_1]$? Explain.
(d) Is the particle speeding up or slowing down in the interval $[t_1, t_2]$? Explain.

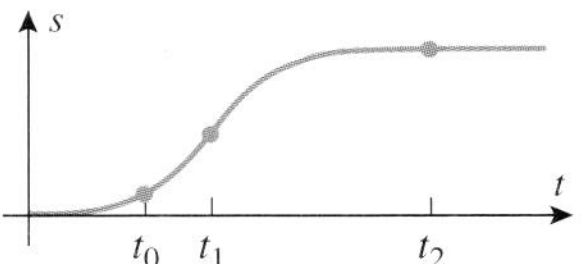

Figure Ex-5

6. An automobile, initially at rest, begins to move along a straight track. The velocity increases steadily until suddenly the driver sees a concrete barrier in the road and applies the brakes sharply at time t_0. The car decelerates rapidly, but it is too late—the car crashes into the barrier at time t_1 and instantaneously comes to rest. Sketch a position versus time curve that might represent the motion of the car.

7. If a particle moves at constant velocity, what can you say about its position versus time curve?

8. The accompanying figure (next page) shows the position versus time curves of four different particles moving on a straight line. For each particle, determine whether its instantaneous velocity is increasing or decreasing with time.

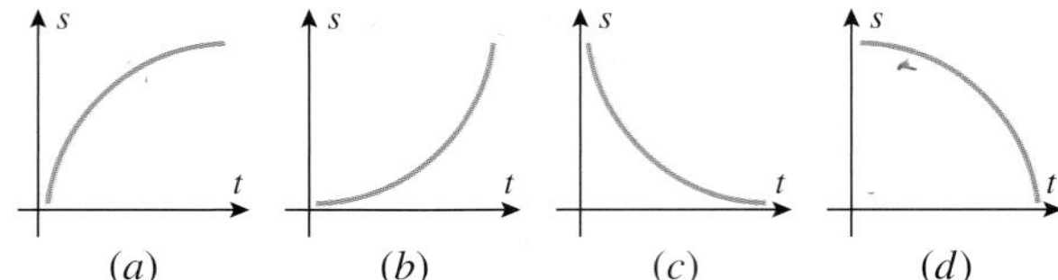

Figure Ex-8

9–12 A function $y = f(x)$ and values of x_0 and x_1 are given.

(a) Find the average rate of change of y with respect to x over the interval $[x_0, x_1]$.

(b) Find the instantaneous rate of change of y with respect to x at the specified value of x_0.

(c) Find the instantaneous rate of change of y with respect to x at an arbitrary value of x_0.

(d) The average rate of change in part (a) is the slope of a certain secant line, and the instantaneous rate of change in part (b) is the slope of a certain tangent line. Sketch the graph of $y = f(x)$ together with those two lines.

9. $y = 2x^2;\ x_0 = 0,\ x_1 = 1$

10. $y = x^3;\ x_0 = 1,\ x_1 = 2$

11. $y = 1/x;\ x_0 = 2,\ x_1 = 3$

12. $y = 1/x^2;\ x_0 = 1,\ x_1 = 2$

13–16 A function $y = f(x)$ and an x-value x_0 are given.

(a) Find a formula for the slope of the tangent line to the graph of f at a general point $x = x_0$.

(b) Use the formula obtained in part (a) to find the slope of the tangent line for the given value of x_0.

13. $f(x) = x^2 - 1;\ x_0 = -1$

14. $f(x) = x^2 + 3x + 2;\ x_0 = 2$

15. $f(x) = \sqrt{x};\ x_0 = 1$

16. $f(x) = 1/\sqrt{x};\ x_0 = 4$

FOCUS ON CONCEPTS

17. Suppose that the outside temperature versus time curve over a 24-hour period is as shown in the accompanying figure.

(a) Estimate the maximum temperature and the time at which it occurs.

(b) The temperature rise is fairly linear from 8 A.M. to 2 P.M. Estimate the rate at which the temperature is increasing during this time period.

(c) Estimate the time at which the temperature is decreasing most rapidly. Estimate the instantaneous rate of change of temperature with respect to time at this instant.

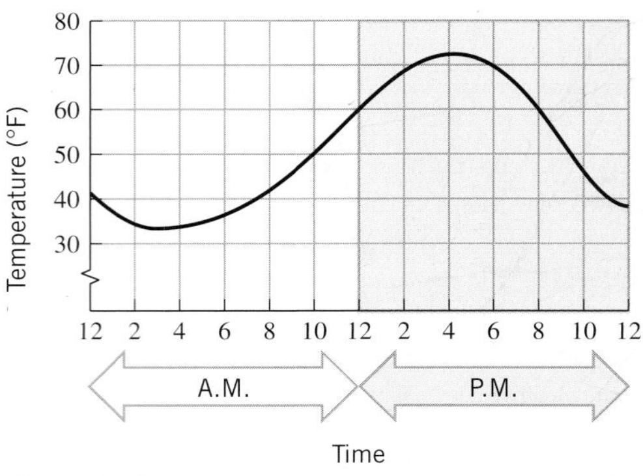

Figure Ex-17

18. The accompanying figure shows the graph of the pressure p in atmospheres (atm) versus the volume V in liters (L) of 1 mole of an ideal gas at a constant temperature of 300 K (kelvins). Use the line segments shown in the figure to estimate the rate of change of pressure with respect to volume at the points where $V = 10$ L and $V = 25$ L.

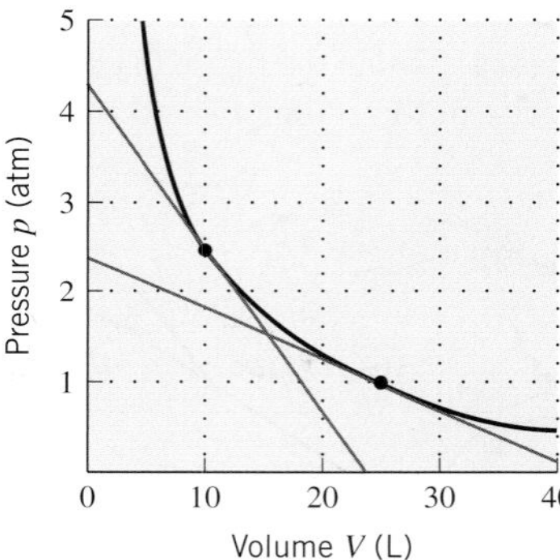

Figure Ex-18

19. The accompanying figure shows the graph of the height h in centimeters versus the age t in years of an individual from birth to age 20.

(a) When is the growth rate greatest?

(b) Estimate the growth rate at age 5.

(c) At approximately what age between 10 and 20 is the growth rate greatest? Estimate the growth rate at this age.

(d) Draw a rough graph of the growth rate versus age.

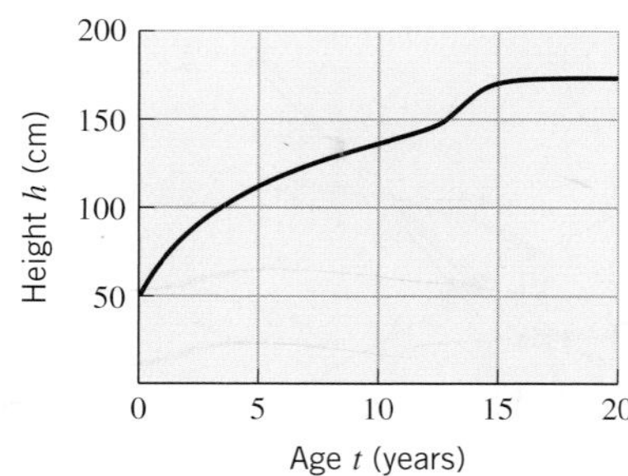

Figure Ex-19

20. A rock is dropped from a height of 576 ft and falls toward Earth in a straight line. In t seconds the rock drops a distance of $s = 16t^2$ ft.
(a) How many seconds after release does the rock hit the ground?
(b) What is the average velocity of the rock during the time it is falling?
(c) What is the average velocity of the rock for the first 3 s?
(d) What is the instantaneous velocity of the rock when it hits the ground?

21. During the first 40 s of a rocket flight, the rocket is propelled straight up so that in t seconds it reaches a height of $s = t^3/\sqrt{10}$ ft.
(a) How high does the rocket travel in 40 s?
(b) What is the average velocity of the rocket during the first 40 s?
(c) What is the average velocity of the rocket during the first 135 ft of its flight?
(d) What is the instantaneous velocity of the rocket at the end of 40 s?

22. An automobile is driven down a straight highway such that after $0 \le t \le 12$ seconds it is $s = 4.5t^2$ feet from its initial position.
(a) Find the average velocity of the car over the interval [0, 12].
(b) Find the instantaneous velocity of the car at $t = 6$.

23. A particle moves in the positive direction along a straight line so that after t minutes its distance is $s = 6t^4$ feet from the origin.
(a) Find the average velocity of the particle over the interval [2, 4].
(b) Find the instantaneous velocity at $t = 2$.

✔ QUICK CHECK ANSWERS 3.1

1. $\dfrac{f(x) - f(x_0)}{x - x_0}$; $\dfrac{f(x_0 + h) - f(x_0)}{h}$ **2.** -4 **3.** (a) 9; positive; 4 (b) 1 ft/s (c) $\dfrac{13}{5}$ ft/s **4.** $s = 3t - 7$
5. (a) $r_{\text{ave}} = 8$ (b) $r_{\text{inst}} = 1$

3.2 THE DERIVATIVE FUNCTION

In this section we will discuss the concept of a "derivative," which is the primary mathematical tool that is used to calculate and study rates of change.

DEFINITION OF THE DERIVATIVE FUNCTION

In the last section we showed that if the limit

$$\lim_{h \to 0} \frac{f(x_0 + h) - f(x_0)}{h}$$

exists, then it can be interpreted either as the slope of the tangent line to the curve $y = f(x)$ at $x = x_0$ or as the instantaneous rate of change of y with respect to x at $x = x_0$ [see Formulas (2) and (13) of that section]. This limit is so important that it has a special notation:

$$f'(x_0) = \lim_{h \to 0} \frac{f(x_0 + h) - f(x_0)}{h} \tag{1}$$

You can think of f' (read "f prime") as a function whose input is x_0 and whose output is the number $f'(x_0)$ that represents either the slope of the tangent line to $y = f(x)$ at $x = x_0$ or the instantaneous rate of change of y with respect to x at $x = x_0$. To emphasize this function point of view, we will replace x_0 by x in (1) and make the following definition.

The expression

$$\frac{f(x+h) - f(x)}{h}$$

that appears in (2) is commonly called the "difference quotient."

3.2.1 DEFINITION. The function f' defined by the formula

$$f'(x) = \lim_{h \to 0} \frac{f(x+h) - f(x)}{h} \tag{2}$$

is called the ***derivative of f with respect to x***. The domain of f' consists of all x in the domain of f for which the limit exists.

The term "derivative" is used because the function f' is *derived* from the function f by a limiting process.

▶ **Example 1** Find the derivative with respect to x of $f(x) = x^2 + 1$, and use it to find the equation of the tangent line to $y = x^2 + 1$ at $x = 2$.

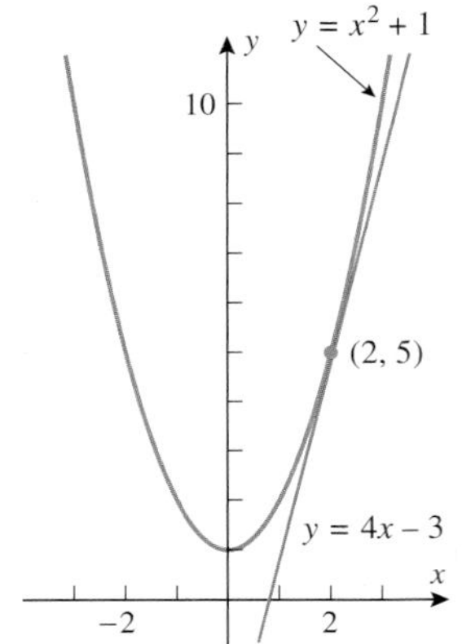

Figure 3.2.1

Solution. It follows from (2) that

$$\begin{aligned} f'(x) &= \lim_{h \to 0} \frac{f(x+h) - f(x)}{h} = \lim_{h \to 0} \frac{[(x+h)^2 + 1] - [x^2 + 1]}{h} \\ &= \lim_{h \to 0} \frac{x^2 + 2xh + h^2 + 1 - x^2 - 1}{h} = \lim_{h \to 0} \frac{2xh + h^2}{h} \\ &= \lim_{h \to 0} (2x + h) = 2x \end{aligned}$$

Thus, the slope of the tangent line to $y = x^2 + 1$ at $x = 2$ is $f'(2) = 4$. Since $y = 5$ if $x = 2$, the point-slope form of the tangent line is

$$y - 5 = 4(x - 2)$$

which we can rewrite in slope-intercept form as $y = 4x - 3$ (Figure 3.2.1). ◀

You can think of f' as a "slope-producing function" in the sense that the value of $f'(x)$ at $x = x_0$ is the slope of the tangent line to the graph of f at $x = x_0$. This aspect of the derivative is illustrated in Figure 3.2.2, which shows the graphs of $f(x) = x^2 + 1$ and its derivative $f'(x) = 2x$ (obtained in Example 1). The figure illustrates that the values of $f'(x) = 2x$ at $x = -2$, 0, and 2 correspond to the slopes of the tangent lines to the graph of $f(x) = x^2 + 1$ at those values of x.

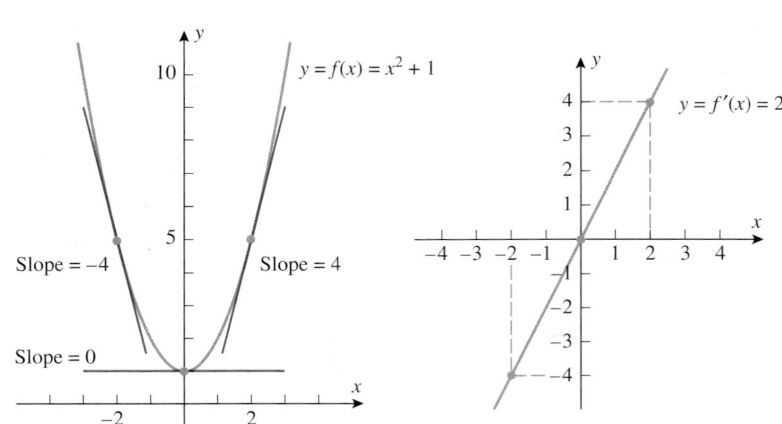

Figure 3.2.2

In general, if $f'(x)$ is defined at $x = x_0$, then the point-slope form of the equation of the tangent line to the graph of $y = f(x)$ at the point $(x_0, f(x_0))$ is given by

$$y - f(x_0) = f'(x_0)(x - x_0)$$

or, equivalently,

$$y = f(x_0) + f'(x_0)(x - x_0) \tag{3}$$

▶ Example 2

(a) Find the derivative with respect to x of $f(x) = x^3 - x$.

(b) Graph f and f' together, and discuss the relationship between the two graphs.

Solution (a).

$$\begin{aligned} f'(x) &= \lim_{h\to 0} \frac{f(x+h) - f(x)}{h} \\ &= \lim_{h\to 0} \frac{[(x+h)^3 - (x+h)] - [x^3 - x]}{h} \\ &= \lim_{h\to 0} \frac{[x^3 + 3x^2h + 3xh^2 + h^3 - x - h] - [x^3 - x]}{h} \\ &= \lim_{h\to 0} \frac{3x^2h + 3xh^2 + h^3 - h}{h} \\ &= \lim_{h\to 0} [3x^2 + 3xh + h^2 - 1] = 3x^2 - 1 \end{aligned}$$

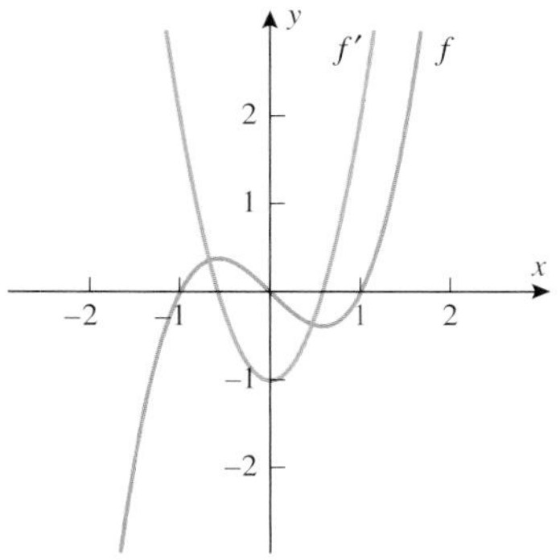

Figure 3.2.3

Solution (b). Since $f'(x)$ can be interpreted as the slope of the tangent line to the graph of $y = f(x)$ at x, it follows that $f'(x)$ is positive where the tangent line has positive slope, is negative where the tangent line has negative slope, and is zero where the tangent line is horizontal. We leave it for you to verify that this is consistent with the graphs of $f(x) = x^3 - x$ and $f'(x) = 3x^2 - 1$ shown in Figure 3.2.3. ◀

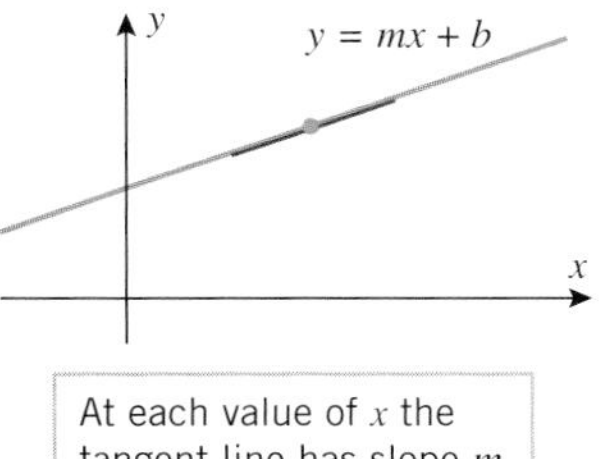
y
y = mx + b
x
At each value of x the tangent line has slope m.

Figure 3.2.4

▶ Example 3 At each value of x, the tangent line to a line $y = mx + b$ coincides with the line itself (Figure 3.2.4), and hence all tangent lines have slope m. This suggests geometrically that if $f(x) = mx + b$, then $f'(x) = m$ for all x. This is confirmed by the following computations:

$$\begin{aligned} f'(x) &= \lim_{h\to 0} \frac{f(x+h) - f(x)}{h} \\ &= \lim_{h\to 0} \frac{[m(x+h) + b] - [mx + b]}{h} \\ &= \lim_{h\to 0} \frac{mh}{h} = \lim_{h\to 0} m = m \end{aligned}$$ ◀

The result in Example 3 is consistent with our earlier observation that the rate of change of y with respect to x along a line $y = mx + b$ is constant and that constant is m.

▶ Example 4

(a) Find the derivative with respect to x of $f(x) = \sqrt{x}$.

(b) Find the slope of the tangent line to $y = \sqrt{x}$ at $x = 9$.

(c) Find the limits of $f'(x)$ as $x \to 0^+$ and as $x \to +\infty$, and explain what those limits say about the graph of f.

Solution (a). Recall from Example 4 of Section 3.1 that the slope of the tangent line to $y = \sqrt{x}$ at $x = x_0$ is given by $m_{\tan} = 1/(2\sqrt{x_0}\,)$. Thus, $f'(x) = 1/(2\sqrt{x}\,)$.

Solution (b). The slope of the tangent line at $x = 9$ is $f'(9)$. From part (a), this slope is $f'(9) = 1/(2\sqrt{9}\,) = \frac{1}{6}$.

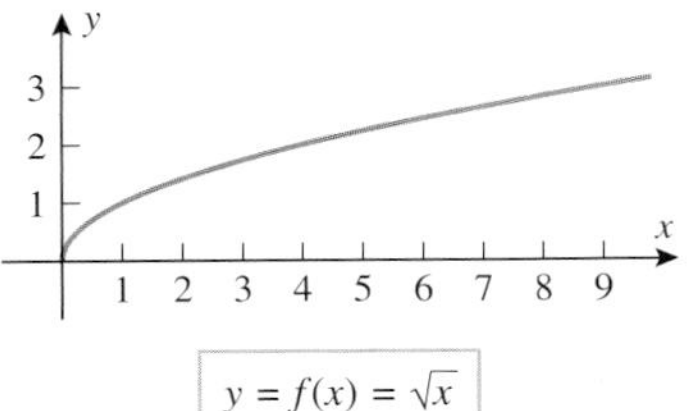

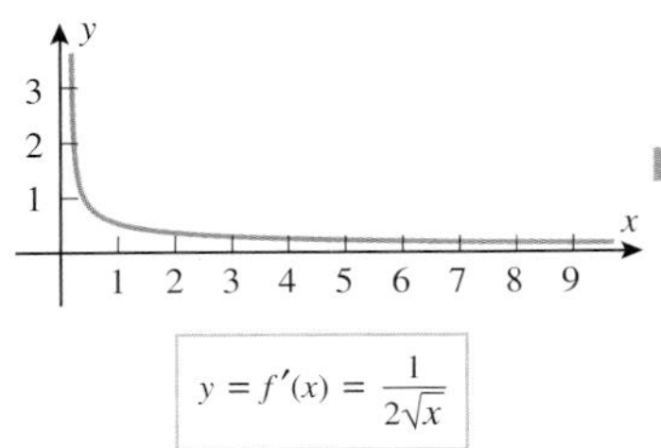

Figure 3.2.5

Solution (c). The graphs of $f(x) = \sqrt{x}$ and $f'(x) = 1/(2\sqrt{x})$ are shown in Figure 3.2.5. Observe that $f'(x) > 0$ if $x > 0$, which means that all tangent lines to the graph of $y = \sqrt{x}$ have positive slope at all points in this interval. Since

$$\lim_{x \to 0^+} \frac{1}{2\sqrt{x}} = +\infty \quad \text{and} \quad \lim_{x \to +\infty} \frac{1}{2\sqrt{x}} = 0$$

the graph becomes more and more vertical as $x \to 0^+$ and more and more horizontal as $x \to +\infty$. ◀

USING DERIVATIVES TO COMPUTE INSTANTANEOUS VELOCITY

It follows from Formula (6) of Section 3.1 (with t replacing t_0) that if $s = f(t)$ is the position function of a particle in rectilinear motion, then the instantaneous velocity at an arbitrary time t is given by

$$v_{\text{inst}} = \lim_{h \to 0} \frac{f(t+h) - f(t)}{h}$$

Since the right side of this equation is the derivative of the function f (with t rather than x as the independent variable), it follows that if $f(t)$ is the position function of a particle in rectilinear motion, then the function

$$v(t) = f'(t) = \lim_{h \to 0} \frac{f(t+h) - f(t)}{h} \tag{4}$$

represents the instantaneous velocity of the particle at time t. Accordingly, we call (4) the ***instantaneous velocity function*** or, more simply, the ***velocity function*** of the particle.

▶ **Example 5** Recall from Example 6 of Section 3.1 that, under appropriate assumptions, the position function for an object dropped from the Empire State Building from 1250 ft above street level can be modeled by the position function $s = f(t) = 1250 - 16t^2$. Here, $f(t)$ is measured in feet above street level and t is measured in seconds after the object is released.

(a) Find the velocity function of the object.

(b) Find the time interval over which the velocity function is valid.

(c) What is the velocity of the object when it hits the ground?

Solution (a). It follows from (4) that the velocity function is

$$\begin{aligned} v(t) &= \lim_{h \to 0} \frac{f(t+h) - f(t)}{h} = \lim_{h \to 0} \frac{[1250 - 16(t+h)^2] - [1250 - 16t^2]}{h} \\ &= \lim_{h \to 0} \frac{-16[t^2 + 2th + h^2 - t^2]}{h} = -16\left(\lim_{h \to 0} \frac{2th + h^2}{h}\right) \\ &= -16 \cdot \lim_{h \to 0}(2t + h) = -32t \end{aligned}$$

where the units of velocity are feet per second.

Solution (b). The velocity function in part (a) is valid from the time the object is released ($t = 0$) until the time t_1 that it hits the ground, that is, when

$$1250 - 16t_1^2 = 0 \quad \text{or equivalently} \quad 16t_1^2 = 1250$$

Solving for the positive value of t_1 tells us the velocity function is valid until time

$$t_1 = \sqrt{\frac{1250}{16}} \approx 8.8 \text{ s}$$

Solution (c). To find the velocity of the object when it hits the ground, we substitute the value of t_1 obtained in part (b) into the velocity function $v(t) = -32t$. This yields

$$v(t_1) = -32t_1 = -32\sqrt{\frac{1250}{16}} \approx -282.8 \text{ ft/s} \blacktriangleleft$$

DIFFERENTIABILITY

It is possible that the limit that defines the derivative of a function f may not exist at certain points in the domain of f. At such points the derivative is undefined. To account for this possibility we make the following definition.

3.2.2 DEFINITION. A function f is said to be ***differentiable at*** x_0 if the limit

$$f'(x_0) = \lim_{h \to 0} \frac{f(x_0 + h) - f(x_0)}{h} \tag{5}$$

exists. If f is differentiable at each point of the open interval (a, b), then we say that it is ***differentiable on (a, b)***, and similarly for open intervals of the form $(a, +\infty)$, $(-\infty, b)$, and $(-\infty, +\infty)$. In the last case we say that f is ***differentiable everywhere***.

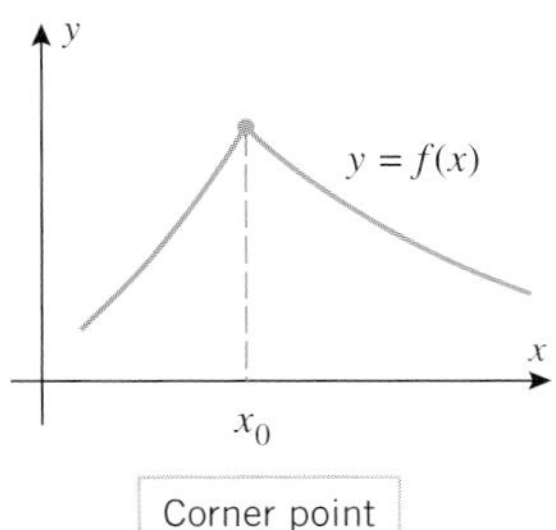

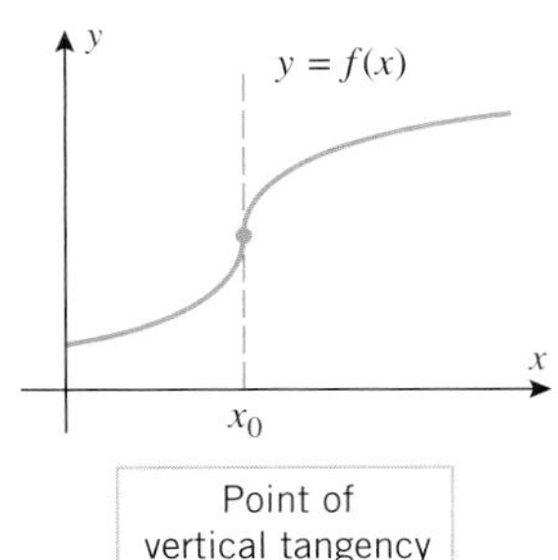

Figure 3.2.6

Geometrically, a function f is differentiable at x_0 if the graph of f has a tangent line at x_0. Thus, f is not differentiable at any point x_0 where the secant lines from $P(x_0, f(x_0))$ to points $Q(x, f(x))$ distinct from P do not approach a unique *nonvertical* limiting position as $x \to x_0$. Figure 3.2.6 illustrates two common ways in which a function that is continuous at x_0 can fail to be differentiable at x_0. These can be described informally as

- corner points
- points of vertical tangency

At a corner point, the slopes of the secant lines have different limits from the left and from the right, and hence the *two-sided* limit that defines the derivative does not exist (Figure 3.2.7). At a point of vertical tangency the slopes of the secant lines approach $+\infty$ or $-\infty$ from the left and from the right (Figure 3.2.8), so again the limit that defines the derivative does not exist.

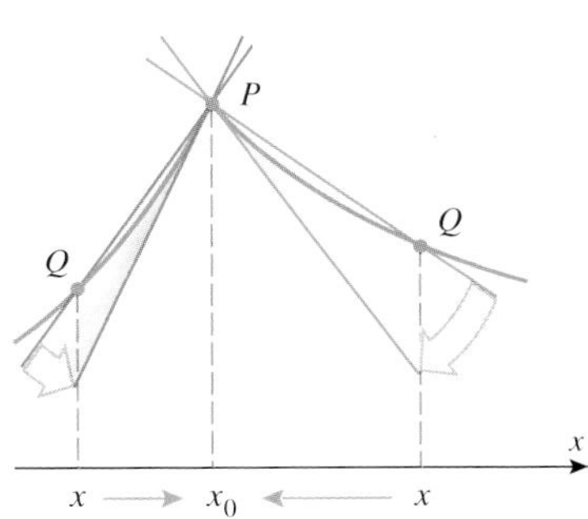

Figure 3.2.7

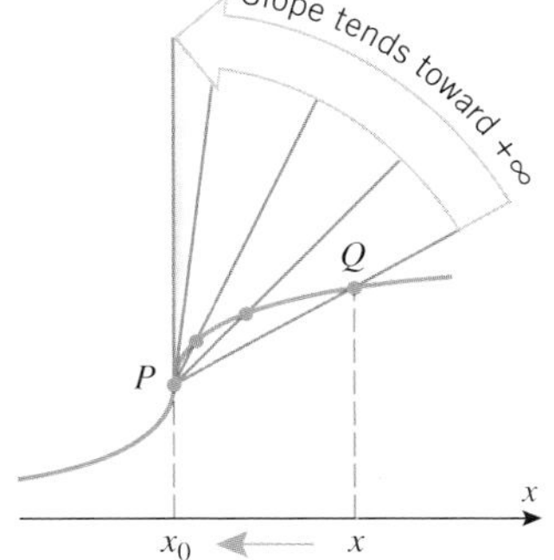

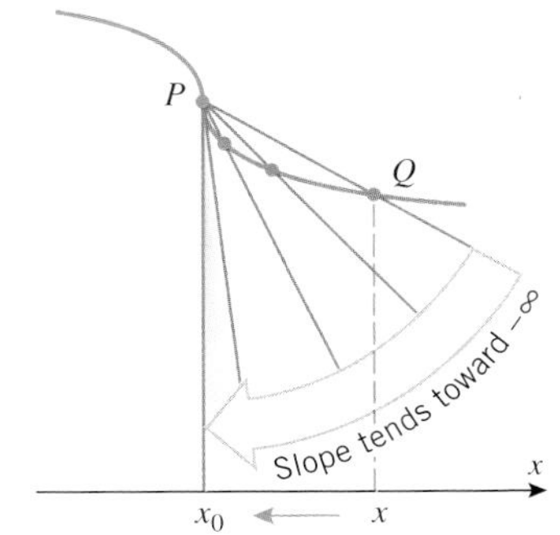

Figure 3.2.8

There are other less obvious circumstances under which a function may fail to be differentiable. (See Exercise 45, for example.)

As illustrated in Figure 3.2.9, differentiability at x_0 can also be described informally in terms of the behavior of the graph of f under increasingly stronger magnification at the point $P(x_0, f(x_0))$. If f is differentiable at x_0, then under sufficiently strong magnification at P the graph looks like a nonvertical line (the tangent line); if a corner point occurs at x_0, then no matter how great the magnification at P the corner persists and the graph never looks like a nonvertical line; and if vertical tangency occurs at x_0, then the graph of f looks like a vertical line under sufficiently strong magnification at P.

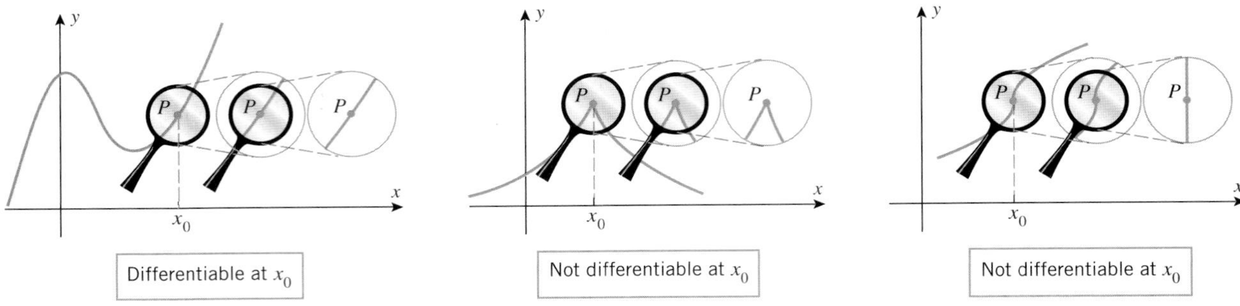

Figure 3.2.9

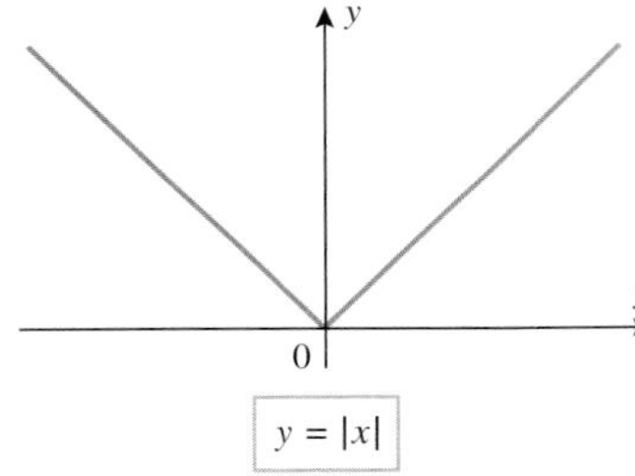

Figure 3.2.10

► **Example 6** The graph of $y = |x|$ in Figure 3.2.10 has a corner at $x = 0$, which implies that $f(x) = |x|$ is not differentiable at $x = 0$.

(a) Prove that $f(x) = |x|$ is not differentiable at $x = 0$ by showing that the limit in Definition 3.2.2 does not exist at $x = 0$.

(b) Find a formula for $f'(x)$.

Solution (a). From Formula (5) with $x_0 = 0$, the value of $f'(0)$, if it were to exist, would be given by

$$f'(0) = \lim_{h\to 0} \frac{f(0+h) - f(0)}{h} = \lim_{h\to 0} \frac{f(h) - f(0)}{h} = \lim_{h\to 0} \frac{|h| - |0|}{h} = \lim_{h\to 0} \frac{|h|}{h} \tag{6}$$

But

$$\frac{|h|}{h} = \begin{cases} 1, & h > 0 \\ -1, & h < 0 \end{cases}$$

so that

$$\lim_{h\to 0^-} \frac{|h|}{h} = -1 \quad \text{and} \quad \lim_{h\to 0^+} \frac{|h|}{h} = 1$$

Since these one-sided limits are not equal, the two-sided limit in (5) does not exist, and hence f is not differentiable at $x = 0$.

Solution (b). A formula for the derivative of $f(x) = |x|$ can be obtained by writing $|x|$ in piecewise form and treating the cases $x > 0$ and $x < 0$ separately. If $x > 0$, then $f(x) = x$ and $f'(x) = 1$; if $x < 0$, then $f(x) = -x$ and $f'(x) = -1$. Thus,

$$f'(x) = \begin{cases} 1, & x > 0 \\ -1, & x < 0 \end{cases}$$

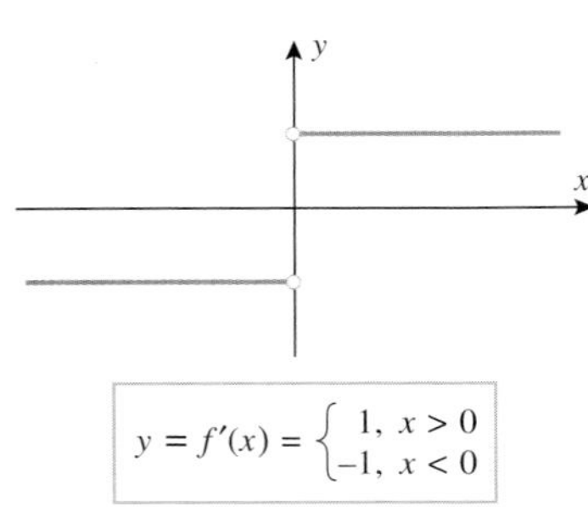

Figure 3.2.11

The graph of f' is shown in Figure 3.2.11. Observe that f' is not continuous at $x = 0$, so this example shows that a function that is continuous everywhere may have a derivative that fails to be continuous everywhere. ◄

THE RELATIONSHIP BETWEEN DIFFERENTIABILITY AND CONTINUITY

We already know that functions are not differentiable at corner points and points of vertical tangency. The next theorem shows that functions are not differentiable at points of discontinuity. We will do this by proving that if f is differentiable at a point, then it must be continuous at that point.

A theorem that says "If statement A is true, then statement B is true" is equivalent to the theorem that says "If statement B is not true, then statement A is not true." The two theorems are called ***contrapositive forms*** of one another. Thus, Theorem 3.2.3 can be rewritten in contrapositive form as "If a function f is not continuous at x_0, then f is not differentiable at x_0."

3.2.3 THEOREM. *If a function f is differentiable at x_0, then f is continuous at x_0.*

PROOF. We are given that f is differentiable at x_0, so it follows from (5) that $f'(x_0)$ exists and is given by

$$f'(x_0) = \lim_{h \to 0} \left[\frac{f(x_0 + h) - f(x_0)}{h} \right] \tag{7}$$

To show that f is continuous at x_0, we must show that $\lim_{x \to x_0} f(x) = f(x_0)$ or, equivalently,

$$\lim_{x \to x_0} [f(x) - f(x_0)] = 0$$

Expressing this in terms of the variable $h = x - x_0$, we must prove that

$$\lim_{h \to 0} [f(x_0 + h) - f(x_0)] = 0$$

However, this can be proved using (7) as follows:

$$\begin{aligned} \lim_{h \to 0} [f(x_0 + h) - f(x_0)] &= \lim_{h \to 0} \left[\frac{f(x_0 + h) - f(x_0)}{h} \cdot h \right] \\ &= \lim_{h \to 0} \left[\frac{f(x_0 + h) - f(x_0)}{h} \right] \cdot \lim_{h \to 0} h \\ &= f'(x_0) \cdot 0 = 0 \end{aligned}$$

■

WARNING

The converse of Theorem 3.2.3 is false; that is, *a function may be continuous at a point but not differentiable at that point.* This occurs, for example, at corner points of continuous functions. For instance, $f(x) = |x|$ is continuous at $x = 0$ but not differentiable there (Example 6).

The relationship between continuity and differentiability was of great historical significance in the development of calculus. In the early nineteenth century mathematicians believed that if a continuous function had many points of nondifferentiability, these points, like the tips of a sawblade, would have to be separated from one another and joined by smooth curve segments (Figure 3.2.12). This misconception was corrected by a series of discoveries beginning in 1834. In that year a Bohemian priest, philosopher, and mathematician named Bernhard Bolzano discovered a procedure for constructing a continuous function that is not differentiable at any point. Later, in 1860, the great German mathematician Karl Weierstrass (biography on p. 117) produced the first formula for such a function. The graphs of such functions are impossible to draw; it is as if the corners are so numerous that any segment of the curve, when suitably enlarged, reveals more corners. The discovery of these functions was important in that it made mathematicians distrustful of their geometric intuition and more reliant on precise mathematical proof. Recently, such functions have started to play a fundamental role in the study of geometric objects called ***fractals***. Fractals have revealed an order to natural phenomena that were previously dismissed as random and chaotic.

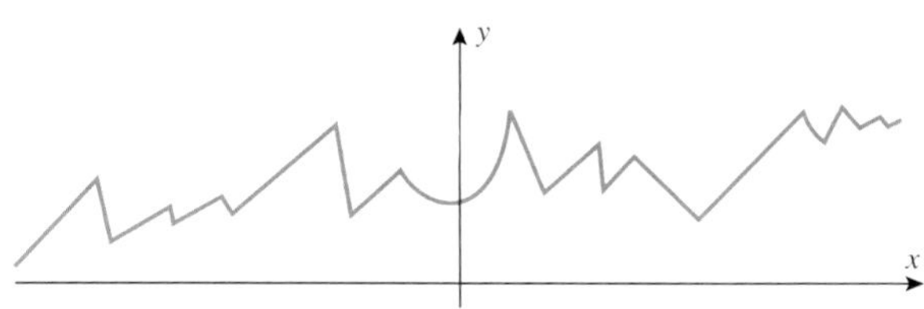

Figure 3.2.12

DERIVATIVES AT THE ENDPOINTS OF AN INTERVAL

If a function f is defined on a closed interval $[a, b]$ but not outside that interval, then f' is not defined at the endpoints of the interval because derivatives are two-sided limits. To deal with this we define ***left-hand derivatives*** and ***right-hand derivatives*** by

$$f'_-(x) = \lim_{h\to 0^-} \frac{f(x+h) - f(x)}{h} \quad \text{and} \quad f'_+(x) = \lim_{h\to 0^+} \frac{f(x+h) - f(x)}{h}$$

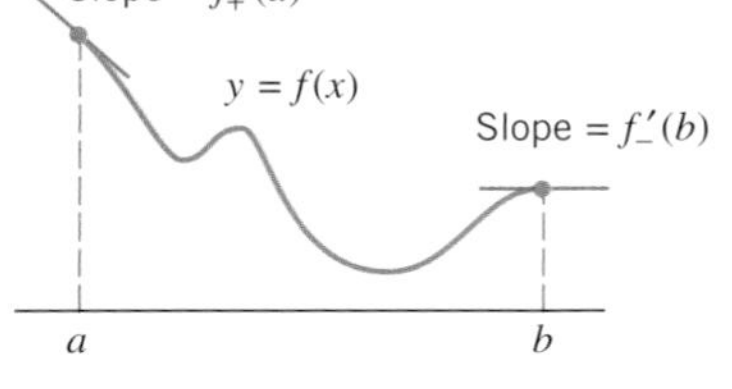

Figure 3.2.13

respectively. These are called ***one-sided derivatives***. Geometrically, $f'_-(x)$ is the limit of the slopes of the secant lines as x is approached from the left and $f'_+(x)$ is the limit of the slopes of the secant lines as x is approached from the right. For a closed interval $[a, b]$, we will understand the derivative at the left endpoint to be $f'_+(a)$ and at the right endpoint to be $f'_-(b)$ (Figure 3.2.13).

In general, we will say that f is ***differentiable*** on an interval of the form $[a, b]$, $[a, +\infty)$, $(-\infty, b]$, $[a, b)$, or $(a, b]$ if it is differentiable at all points inside the interval and the appropriate one-sided derivative exists at each included endpoint.

It can be proved that a function f is continuous from the left at those points where the left-hand derivative exists and is continuous from the right at those points where the right-hand derivative exists.

OTHER DERIVATIVE NOTATIONS

The process of finding a derivative is called ***differentiation***. You can think of differentiation as an *operation* on functions that associates a function f' with a function f. When the independent variable is x, the differentiation operation is also commonly denoted by

$$f'(x) = \frac{d}{dx}[f(x)] \quad \text{or} \quad f'(x) = D_x[f(x)]$$

In the case where there is a dependent variable $y = f(x)$, the derivative is also commonly denoted by

$$f'(x) = y'(x) \quad \text{or} \quad f'(x) = \frac{dy}{dx}$$

Later, the symbols dy and dx will be given specific meanings. However, for the time being do not regard dy/dx as a ratio, but rather as a single symbol denoting the derivative.

With the above notations, the value of the derivative at a point x_0 can be expressed as

$$f'(x_0) = \frac{d}{dx}[f(x)]\bigg|_{x=x_0}, \quad f'(x_0) = D_x[f(x)]\big|_{x=x_0}, \quad f'(x_0) = y'(x_0), \quad f'(x_0) = \frac{dy}{dx}\bigg|_{x=x_0}$$

If a variable w changes from some initial value w_0 to some final value w_1, then the final value minus the initial value is called an ***increment*** in w and is denoted by

$$\Delta w = w_1 - w_0 \tag{8}$$

Increments can be positive or negative, depending on whether the final value is larger or smaller than the initial value. The increment symbol in (8) should not be interpreted as a product; rather, Δw should be regarded as a single symbol representing the change in the value of w.

Bernhard Bolzano (1781–1848) Bolzano, the son of an art dealer, was born in Prague, Bohemia (Czech Republic). He was educated at the University of Prague, and eventually won enough mathematical fame to be recommended for a mathematics chair there. However, Bolzano became an ordained Roman Catholic priest, and in 1805 he was appointed to a chair of Philosophy at the University of Prague. Bolzano was a man of great human compassion; he spoke out for educational reform, he voiced the right of individual conscience over government demands, and he lectured on the absurdity of war and militarism. His views so disenchanted Emperor Franz I of Austria that the emperor pressed the Archbishop of Prague to have Bolzano recant his statements. Bolzano refused and was then forced to retire in 1824 on a small pension. Bolzano's main contribution to mathematics was philosophical. His work helped convince mathematicians that sound mathematics must ultimately rest on rigorous proof rather than intuition. In addition to his work in mathematics, Bolzano investigated problems concerning space, force, and wave propagation.

It is common to regard the variable h in the derivative formula

$$f'(x) = \lim_{h \to 0} \frac{f(x+h) - f(x)}{h} \tag{9}$$

as an increment Δx in x and write (9) as

$$f'(x) = \lim_{\Delta x \to 0} \frac{f(x+\Delta x) - f(x)}{\Delta x} \tag{10}$$

Moreover, if $y = f(x)$, then the numerator in (10) can be regarded as the increment

$$\Delta y = f(x+\Delta x) - f(x) \tag{11}$$

in which case

$$\frac{dy}{dx} = \lim_{\Delta x \to 0} \frac{\Delta y}{\Delta x} = \lim_{\Delta x \to 0} \frac{f(x+\Delta x) - f(x)}{\Delta x} \tag{12}$$

The geometric interpretations of Δx and Δy are shown in Figure 3.2.14.

Sometimes it is desirable to express derivatives in a form that does not use increments at all. For example, if we let $w = x + h$ in Formula (9), then $h = w - x$ and $w \to x$ as $h \to 0$, so we can rewrite that formula as

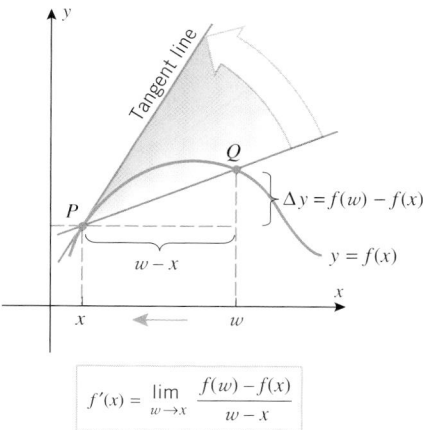

$$f'(x) = \lim_{w \to x} \frac{f(w) - f(x)}{w - x} \tag{13}$$

(Compare Figures 3.2.14 and 3.2.15.)

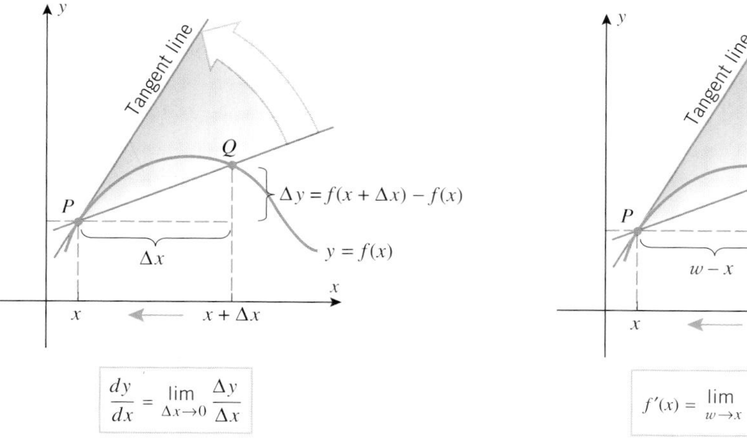

Figure 3.2.14

Figure 3.2.15

When letters other than x and y are used for the independent and dependent variables, then the derivative notations must be adjusted accordingly. Thus, for example, if $s = f(t)$ is the position function for a particle in rectilinear motion, then the velocity function $v(t)$ in (4) can be expressed as

$$v(t) = \frac{ds}{dt} = \lim_{\Delta t \to 0} \frac{\Delta s}{\Delta t} = \lim_{\Delta t \to 0} \frac{f(t+\Delta t) - f(t)}{\Delta t} \tag{14}$$

✔ QUICK CHECK EXERCISES 3.2

(See page 171 for answers.)

1. (a) The function $f'(x)$ is defined by the formula

$$f'(x) = \lim_{h \to 0} \_\_\_\_\_\_$$

   (b) Suppose that $f(2+h) = 3h \cos h - 1$ for all h. The value of $f'(2)$ is ______.

2. A point on the graph of $y = \sqrt{x}$ at which the tangent line has slope equal to the y-coordinate of the point is ______.

3. Suppose that the line $2x + 3y = 5$ is tangent to the graph of $y = f(x)$ at $x = 1$. The value of $f(1)$ is ______ and the value of $f'(1)$ is ______.

4. Which theorem guarantees us that if

$$\lim_{h \to 0} \frac{f(x_0+h) - f(x_0)}{h}$$

exists, then $\lim_{x \to x_0} f(x) = f(x_0)$?

EXERCISE SET 3.2  Graphing Utility

FOCUS ON CONCEPTS

1. Use the graph of $y = f(x)$ in the accompanying figure to estimate the value of $f'(1)$, $f'(3)$, $f'(5)$, and $f'(6)$.

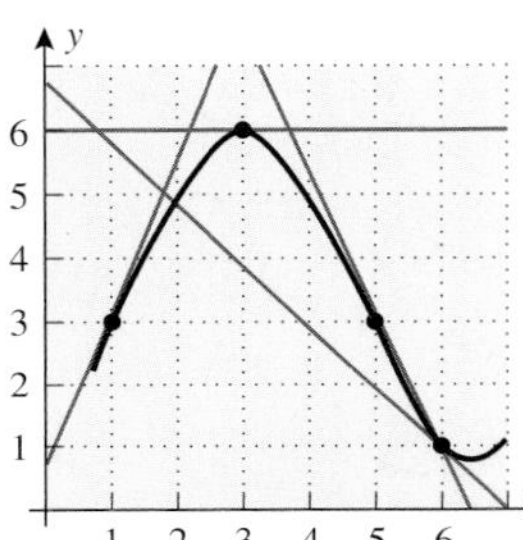

Figure Ex-1

2. For the function graphed in the accompanying figure, arrange the numbers 0, $f'(-3)$, $f'(0)$, $f'(2)$, and $f'(4)$ in increasing order.

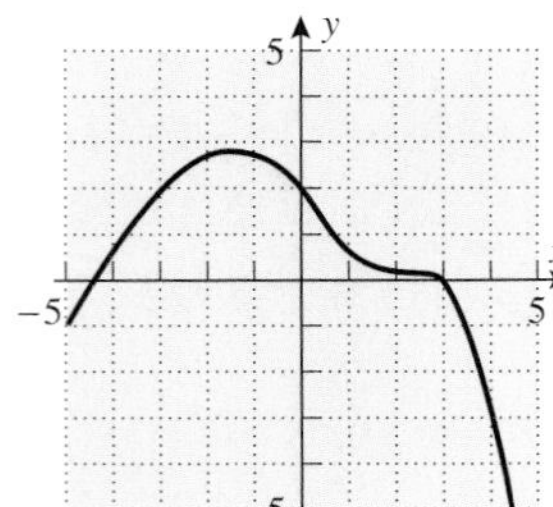

Figure Ex-2

3. (a) If you are given an equation for the tangent line at the point $(a, f(a))$ on a curve $y = f(x)$, how would you go about finding $f'(a)$?
   (b) Given that the tangent line to the graph of $y = f(x)$ at the point $(2, 5)$ has the equation $y = 3x - 1$, find $f'(2)$.
   (c) For the function $y = f(x)$ in part (b), what is the instantaneous rate of change of y with respect to x at $x = 2$?

4. Given that the tangent line to $y = f(x)$ at the point $(1, 2)$ passes through the point $(-1, -1)$, find $f'(1)$.

5. Sketch the graph of a function f for which $f(0) = -1$, $f'(0) = 0$, $f'(x) < 0$ if $x < 0$, and $f'(x) > 0$ if $x > 0$.

6. Sketch the graph of a function f for which $f(0) = 0$, $f'(0) = 0$, and $f'(x) > 0$ if $x < 0$ or $x > 0$.

7. Given that $f(3) = -1$ and $f'(3) = 5$, find an equation for the tangent line to the graph of $y = f(x)$ at $x = 3$.

8. Given that $f(-2) = 3$ and $f'(-2) = -4$, find an equation for the tangent line to the graph of $y = f(x)$ at $x = -2$.

9–14 Use Definition 3.2.1 to find $f'(x)$, and then find the tangent line to the graph of $y = f(x)$ at $x = a$.

9. $f(x) = 2x^2;\ a = 1$
10. $f(x) = 1/x^2;\ a = -1$
11. $f(x) = x^3;\ a = 0$
12. $f(x) = 2x^3 + 1;\ a = -1$
13. $f(x) = \sqrt{x+1};\ a = 8$
14. $f(x) = \sqrt{2x+1};\ a = 4$

15–20 Use Formula (12) to find dy/dx.

15. $y = \dfrac{1}{x}$
16. $y = \dfrac{1}{x+1}$
17. $y = x^2 - x$
18. $y = x^4$
19. $y = \dfrac{1}{\sqrt{x}}$
20. $y = \dfrac{1}{\sqrt{x-1}}$

21–22 Use Definition 3.2.1 (with appropriate change in notation) to obtain the derivative requested.

21. Find $f'(t)$ if $f(t) = 4t^2 + t$.
22. Find dV/dr if $V = \frac{4}{3}\pi r^3$.

FOCUS ON CONCEPTS

23. Match the graphs of the functions shown in (a)–(f) with the graphs of their derivatives in (A)–(F).

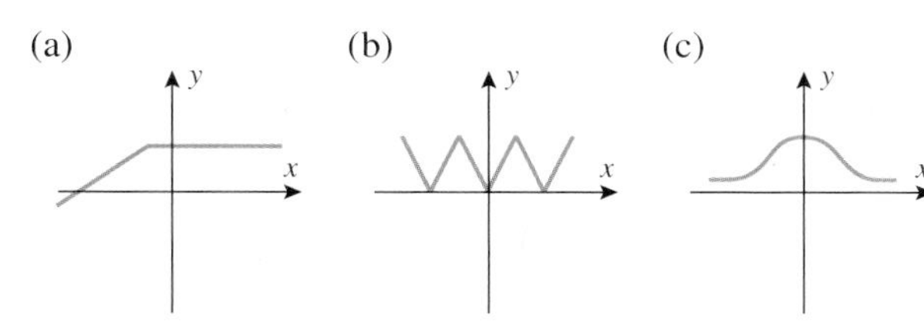

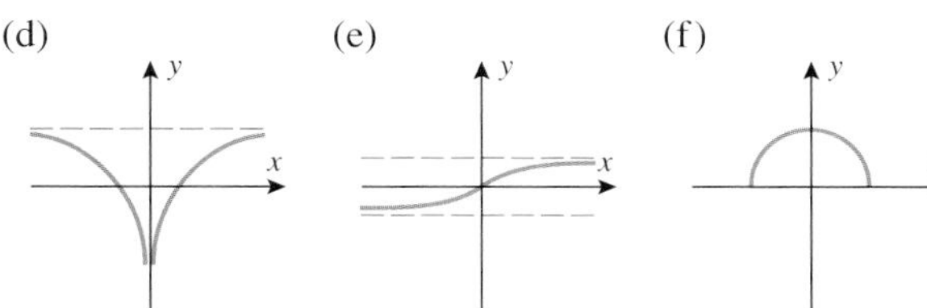

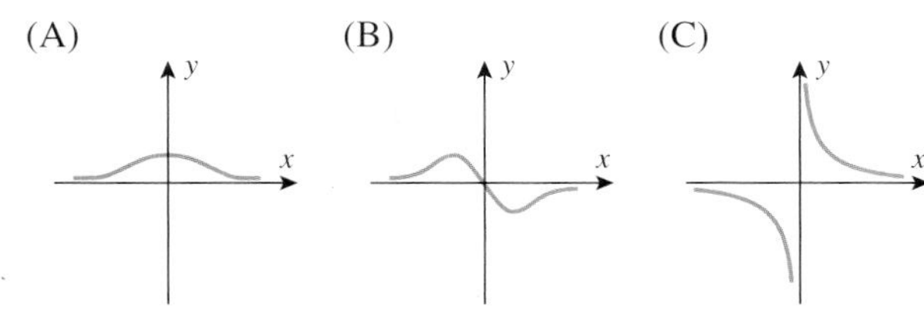

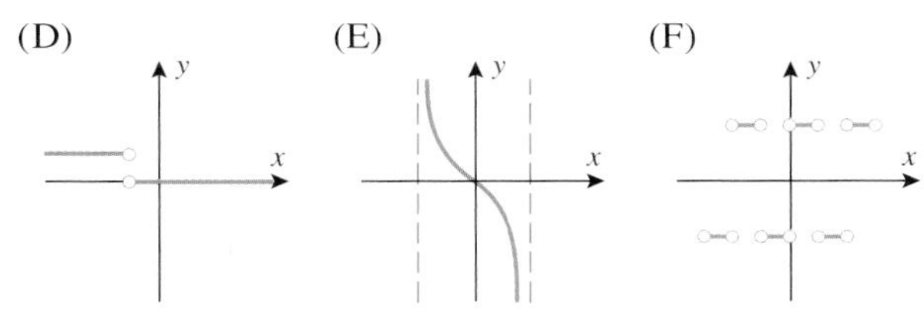

24. Let $f(x) = \sqrt{1 - x^2}$. Use a geometric argument to find $f'(\sqrt{2}/2)$.

25–26 Sketch the graph of the derivative of the function whose graph is shown.

25. (a) (b) (c)

26. (a) (b) (c)

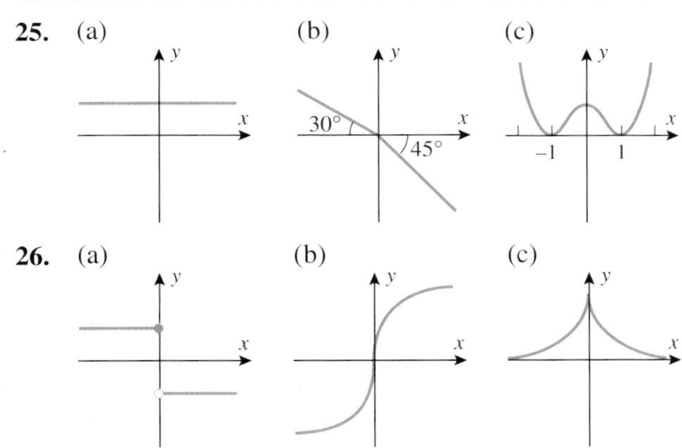

27–28 The given limit represents $f'(a)$ for some function f and some number a. Find $f(x)$ and a in each case.

27. (a) $\lim_{\Delta x \to 0} \frac{\sqrt{1+\Delta x} - 1}{\Delta x}$ (b) $\lim_{x_1 \to 3} \frac{x_1^2 - 9}{x_1 - 3}$

28. (a) $\lim_{h \to 0} \frac{\cos(\pi + h) + 1}{h}$ (b) $\lim_{x \to 1} \frac{x^7 - 1}{x - 1}$

29. Find $dy/dx|_{x=1}$, given that $y = 1 - x^2$.

30. Find $dy/dx|_{x=-2}$, given that $y = (x+2)/x$.

31. Find an equation for the line that is tangent to the curve $y = x^3 - 2x + 1$ at the point $(0, 1)$, and use a graphing utility to graph the curve and its tangent line on the same screen.

32. Use a graphing utility to graph the following on the same screen: the curve $y = x^2/4$, the tangent line to this curve at $x = 1$, and the secant line joining the points $(0, 0)$ and $(2, 1)$ on this curve.

33. Let $f(x) = 2^x$. Estimate $f'(1)$ by

(a) using a graphing utility to zoom in at an appropriate point until the graph looks like a straight line, and then estimating the slope

(b) using a calculating utility to estimate the limit in Formula (13) by making a table of values for a succession of values of w approaching 1.

34. Let $f(x) = \sin x$. Estimate $f'(\pi/4)$ by

(a) using a graphing utility to zoom in at an appropriate point until the graph looks like a straight line, and then estimating the slope

(b) using a calculating utility to estimate the limit in Formula (13) by making a table of values for a succession of values of w approaching $\pi/4$.

FOCUS ON CONCEPTS

35. Suppose that the cost of drilling x feet for an oil well is $C = f(x)$ dollars.

(a) What are the units of $f'(x)$?

(b) In practical terms, what does $f'(x)$ mean in this case?

(c) What can you say about the sign of $f'(x)$?

(d) Estimate the cost of drilling an additional foot, starting at a depth of 300 ft, given that $f'(300) = 1000$.

36. A paint manufacturing company estimates that it can sell $g = f(p)$ gallons of paint at a price of p dollars.

(a) What are the units of dg/dp?

(b) In practical terms, what does dg/dp mean in this case?

(c) What can you say about the sign of dg/dp?

(d) Given that $dg/dp|_{p=10} = -100$, what can you say about the effect of increasing the price from \$10 per gallon to \$11 per gallon?

37. It is a fact that when a flexible rope is wrapped around a rough cylinder, a small force of magnitude F_0 at one end can resist a large force of magnitude F at the other end. The size of F depends on the angle θ through which the rope is wrapped around the cylinder (see the accompanying figure on the next page). The figure shows the graph of F (in pounds) versus θ (in radians), where F is the magnitude of the force that can be resisted by a force with magnitude $F_0 = 10$ lb for a certain rope and cylinder.

(a) Estimate the values of F and $dF/d\theta$ when the angle $\theta = 10$ radians.

(b) It can be shown that the force F satisfies the equation $dF/d\theta = \mu F$, where the constant μ is called the ***coefficient of friction***. Use the results in part (a) to estimate the value of μ.

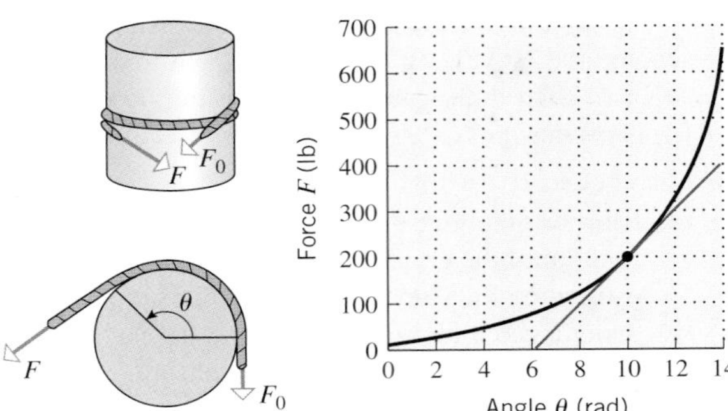

Figure Ex-37

38. The accompanying figure shows the velocity versus time curve for a rocket in outer space where the only significant force on the rocket is from its engines. It can be shown that the mass $M(t)$ (in slugs) of the rocket at time t seconds satisfies the equation

$$M(t) = \frac{T}{dv/dt}$$

where T is the thrust (in lb) of the rocket's engines and v is the velocity (in ft/s) of the rocket. The thrust of the first stage of a *Saturn V* rocket is $T = 7{,}680{,}982$ lb. Use this value of T and the line segment in the figure to estimate the mass of the rocket at time $t = 100$.

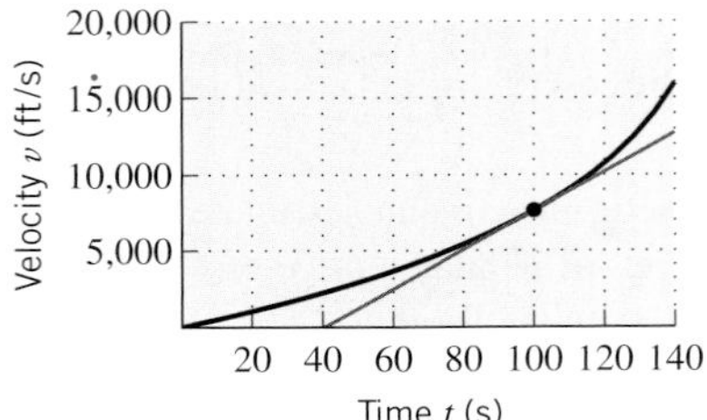

Figure Ex-38

39. According to ***Newton's Law of Cooling***, the rate of change of an object's temperature is proportional to the difference between the temperature of the object and that of the surrounding medium. The accompanying figure shows the graph of the temperature T (in degrees Fahrenheit) versus time t (in minutes) for a cup of coffee, initially with a temperature of 200° F, that is allowed to cool in a room with a constant temperature of 75° F.

(a) Estimate T and dT/dt when $t = 10$ min.

(b) Newton's Law of Cooling can be expressed as

$$\frac{dT}{dt} = k(T - T_0)$$

where k is the constant of proportionality and T_0 is the temperature (assumed constant) of the surrounding medium. Use the results in part (a) to estimate the value of k.

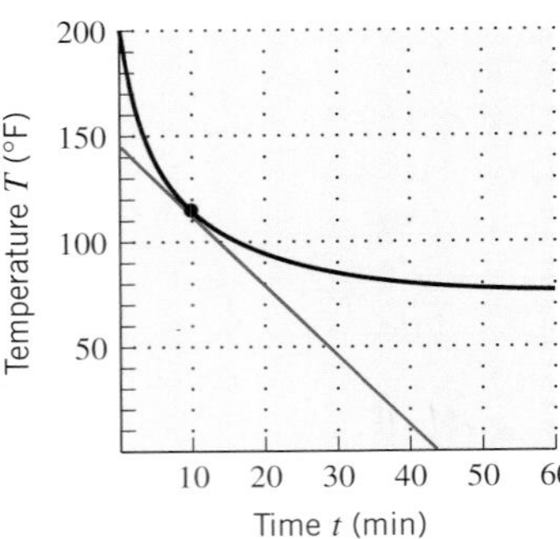

Figure Ex-39

40. Write a paragraph that explains what it means for a function to be differentiable. Include some examples of functions that are not differentiable, and explain the relationship between differentiability and continuity.

41. Show that $f(x) = \sqrt[3]{x}$ is continuous at $x = 0$ but not differentiable at $x = 0$. Sketch the graph of f.

42. Show that $f(x) = \sqrt[3]{(x-2)^2}$ is continuous at $x = 2$ but not differentiable at $x = 2$. Sketch the graph of f.

43. Show that

$$f(x) = \begin{cases} x^2 + 1, & x \le 1 \\ 2x, & x > 1 \end{cases}$$

is continuous and differentiable at $x = 1$. Sketch the graph of f.

44. Show that

$$f(x) = \begin{cases} x^2 + 2, & x \le 1 \\ x + 2, & x > 1 \end{cases}$$

is continuous but not differentiable at $x = 1$. Sketch the graph of f.

45. Show that

$$f(x) = \begin{cases} x \sin(1/x), & x \ne 0 \\ 0, & x = 0 \end{cases}$$

is continuous but not differentiable at $x = 0$. Sketch the graph of f near $x = 0$. (See Figure 2.6.6 and the remark following Example 3 in Section 2.6.)

46. Show that

$$f(x) = \begin{cases} x^2 \sin(1/x), & x \ne 0 \\ 0, & x = 0 \end{cases}$$

is continuous and differentiable at $x = 0$. Sketch the graph of f near $x = 0$.

FOCUS ON CONCEPTS

47. Suppose that a function f is differentiable at x_0 and that $f'(x_0) > 0$. Prove that there exists an open interval containing x_0 such that if x_1 and x_2 are any two points in this interval with $x_1 < x_0 < x_2$, then $f(x_1) < f(x_0) < f(x_2)$.

48. Suppose that a function f is differentiable at x_0 and define $g(x) = f(mx + b)$, where m and b are constants. Prove that if x_1 is a point at which $mx_1 + b = x_0$, then $g(x)$ is differentiable at x_1 and $g'(x_1) = mf'(x_0)$.

49. Suppose that a function f is differentiable at $x = 0$ with $f(0) = f'(0) = 0$, and let $y = mx$, $m \neq 0$, denote any line of nonzero slope through the origin.
(a) Prove that there exists an open interval containing 0 such that for all x in this interval $|f(x)| < \left|\frac{1}{2}mx\right|$. [*Hint:* Let $\epsilon = \frac{1}{2}|m|$ and apply Definition 2.4.1 to (5) with $x_0 = 0$.]
(b) Conclude from part (a) and the triangle inequality that there exists an open interval containing 0 such that $|f(x)| < |f(x) - mx|$ for all x in this interval.
(c) Explain why the result obtained in part (b) may be interpreted to mean that the tangent line to the graph of f at the origin is the best *linear* approximation to f at that point.

50. Suppose that f is differentiable at x_0. Modify the argument of Exercise 49 to prove that the tangent line to the graph of f at the point $P(x_0, f(x_0))$ provides the best linear approximation to f at P. [*Hint:* Suppose that $y = f(x_0) + m(x - x_0)$ is any line through $P(x_0, f(x_0))$ with slope $m \neq f'(x_0)$. Apply Definition 2.4.1 to (5) with $x = x_0 + h$ and $\epsilon = \frac{1}{2}|f'(x_0) - m|$.]

✔ QUICK CHECK ANSWERS 3.2

1. (a) $\dfrac{f(x+h) - f(x)}{h}$ (b) 3 **2.** $(1/2, \sqrt{2}/2)$ **3.** 1; $-\dfrac{2}{3}$
4. Theorem 3.2.3: If f is differentiable at x_0, then f is continuous at x_0.

3.3 TECHNIQUES OF DIFFERENTIATION

In the last section we defined the derivative of a function f as a limit, and we used that limit to calculate a few simple derivatives. In this section we will develop some important theorems that will enable us to calculate derivatives more efficiently.

DERIVATIVE OF A CONSTANT

The simplest kind of function is a constant function $f(x) = c$. Since the graph of f is a horizontal line of slope 0, the tangent line to the graph of f has slope 0 for every x; and hence we can see geometrically that $f'(x) = 0$ (Figure 3.3.1). We can also see this algebraically since

$$f'(x) = \lim_{h \to 0} \frac{f(x+h) - f(x)}{h} = \lim_{h \to 0} \frac{c - c}{h} = \lim_{h \to 0} 0 = 0$$

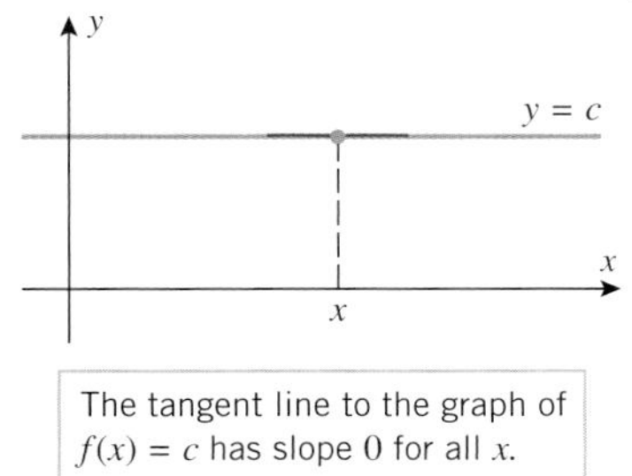

The tangent line to the graph of $f(x) = c$ has slope 0 for all x.

Figure 3.3.1

Thus, we have established the following result.

3.3.1 THEOREM. *The derivative of a constant function is 0; that is, if c is any real number, then*

$$\frac{d}{dx}[c] = 0 \tag{1}$$

▶ Example 1

$$\frac{d}{dx}[1] = 0, \quad \frac{d}{dx}[-3] = 0, \quad \frac{d}{dx}[\pi] = 0, \quad \frac{d}{dx}[-\sqrt{2}] = 0 \ ◀$$

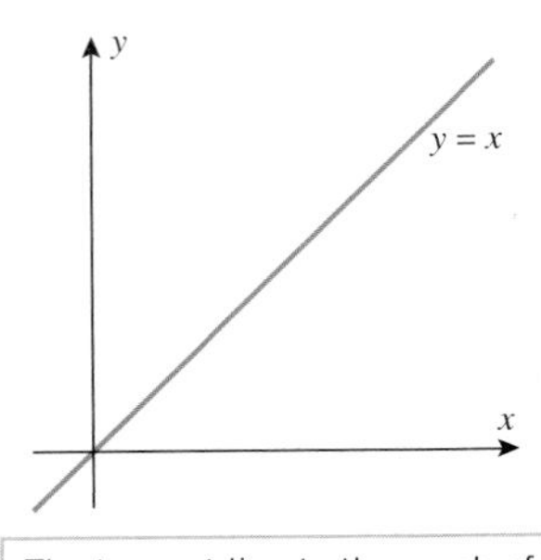

The tangent line to the graph of $f(x) = x$ has slope 1 for all x.

Figure 3.3.2

DERIVATIVES OF INTEGER POWERS OF x

The simplest power function is $f(x) = x$. Since the graph of f is a line of slope 1, it follows from Example 3 of Section 3.2 that $f'(x) = 1$ for all x (Figure 3.3.2). In other words,

$$\frac{d}{dx}[x] = 1 \tag{2}$$

This formula is a special case of the following more general result.

> **3.3.2 THEOREM (*The Power Rule*).** *If n is a positive integer, then*
>
> $$\frac{d}{dx}[x^n] = nx^{n-1} \tag{3}$$

Verify that Formula (2) is the special case of (3) in which $n = 1$.

PROOF. Let $f(x) = x^n$. Thus, from the definition of a derivative and the binomial theorem for expanding the expression $(x+h)^n$, we obtain

$$\begin{aligned}\frac{d}{dx}[x^n] = f'(x) &= \lim_{h\to 0}\frac{f(x+h)-f(x)}{h} = \lim_{h\to 0}\frac{(x+h)^n - x^n}{h}\\ &= \lim_{h\to 0}\frac{\left[x^n + nx^{n-1}h + \frac{n(n-1)}{2!}x^{n-2}h^2 + \cdots + nxh^{n-1} + h^n\right] - x^n}{h}\\ &= \lim_{h\to 0}\frac{nx^{n-1}h + \frac{n(n-1)}{2!}x^{n-2}h^2 + \cdots + nxh^{n-1} + h^n}{h}\\ &= \lim_{h\to 0}\left[nx^{n-1} + \frac{n(n-1)}{2!}x^{n-2}h + \cdots + nxh^{n-2} + h^{n-1}\right]\\ &= nx^{n-1} + 0 + \cdots + 0 + 0\\ &= nx^{n-1}\end{aligned}$$

■

In words, *the derivative of x raised to a positive integer power is the product of the integer exponent and x raised to the next lower integer power.*

▶ Example 2

$$\frac{d}{dx}[x^2] = 2x, \quad \frac{d}{dx}[x^3] = 3x^2, \quad \frac{d}{dx}[x^4] = 4x^3, \quad \frac{d}{dt}[t^{12}] = 12t^{11}$$ ◀

Although the power rule in Formula (3) applies only to *positive* integer powers of x, we will eventually show that the same formula holds for any real exponent. As a first step in this direction, we will show that the formula holds for all integer powers of x.

> **3.3.3 THEOREM (*Extended Power Rule*).** *If n is any integer, then*
>
> $$\frac{d}{dx}[x^n] = nx^{n-1} \tag{4}$$

PROOF. The result has already been established in the case where $n > 0$. If $n < 0$, then let $m = -n$ so that

$$x^n = x^{-m} = \frac{1}{x^m}$$

Then

$$\begin{aligned}
\frac{d}{dx}[x^n] = \frac{d}{dx}\left[\frac{1}{x^m}\right] &= \lim_{h\to 0}\frac{\dfrac{1}{(x+h)^m} - \dfrac{1}{x^m}}{h} = \lim_{h\to 0}\frac{x^m - (x+h)^m}{(x+h)^m x^m h} \\
&= -\lim_{h\to 0}\frac{(x+h)^m - x^m}{h}\cdot\lim_{h\to 0}\frac{1}{(x+h)^m x^m} \\
&= \underbrace{-mx^{m-1}}\cdot\frac{1}{x^{2m}} = -mx^{-m-1} \\
&= nx^{n-1}
\end{aligned}$$

By the proof of Theorem 3.3.2.

which proves (4). In the case $n = 0$, Formula (4) reduces to

$$\frac{d}{dx}[1] = 0\cdot x^{-1} = 0$$

which is correct by Theorem 3.3.1. ■

► Example 3

$$\frac{d}{dx}[x^{-9}] = -9x^{-9-1} = -9x^{-10}$$

$$\frac{d}{dx}\left[\frac{1}{x}\right] = \frac{d}{dx}[x^{-1}] = (-1)x^{-1-1} = -x^{-2} = -\frac{1}{x^2}$$

$$\frac{d}{dw}\left[\frac{1}{w^{100}}\right] = \frac{d}{dw}[w^{-100}] = -100w^{-101} = -\frac{100}{w^{101}}$$ ◄

Looking back, we can see that Formula (4) is also valid for $n = \frac{1}{2}$, since we showed in Example 4 of Section 3.2 that

$$\frac{d}{dx}[\sqrt{x}] = \frac{1}{2\sqrt{x}} \tag{5}$$

which can be expressed as

$$\frac{d}{dx}[x^{1/2}] = \frac{1}{2}x^{-1/2}$$

■ DERIVATIVE OF A CONSTANT TIMES A FUNCTION

Formula (6) can also be expressed in function notation as

$$(cf)' = cf'$$

3.3.4 THEOREM (*Constant Multiple Rule*). *If f is differentiable at x and c is any real number, then cf is also differentiable at x and*

$$\frac{d}{dx}[cf(x)] = c\frac{d}{dx}[f(x)] \tag{6}$$

PROOF.

$$\begin{aligned}\frac{d}{dx}[cf(x)] &= \lim_{h\to 0}\frac{cf(x+h)-cf(x)}{h}\\ &= \lim_{h\to 0} c\left[\frac{f(x+h)-f(x)}{h}\right]\\ &= c\lim_{h\to 0}\frac{f(x+h)-f(x)}{h}\\ &= c\frac{d}{dx}[f(x)]\end{aligned}$$

A constant factor can be moved through a limit sign.

■

In words, *a constant factor can be moved through a derivative sign.*

► **Example 4**

$$\frac{d}{dx}[4x^8] = 4\frac{d}{dx}[x^8] = 4[8x^7] = 32x^7$$

$$\frac{d}{dx}[-x^{12}] = (-1)\frac{d}{dx}[x^{12}] = -12x^{11}$$

$$\frac{d}{dx}\left[\frac{\pi}{x}\right] = \pi\frac{d}{dx}[x^{-1}] = \pi(-x^{-2}) = -\frac{\pi}{x^2}$$ ◄

DERIVATIVES OF SUMS AND DIFFERENCES

Formulas (7) and (8) can also be expressed as

$$(f+g)' = f' + g'$$
$$(f-g)' = f' - g'$$

3.3.5 THEOREM (*Sum and Difference Rules*). *If f and g are differentiable at x, then so are $f+g$ and $f-g$ and*

$$\frac{d}{dx}[f(x)+g(x)] = \frac{d}{dx}[f(x)] + \frac{d}{dx}[g(x)] \tag{7}$$

$$\frac{d}{dx}[f(x)-g(x)] = \frac{d}{dx}[f(x)] - \frac{d}{dx}[g(x)] \tag{8}$$

PROOF. Formula (7) can be proved as follows:

$$\begin{aligned}\frac{d}{dx}[f(x)+g(x)] &= \lim_{h\to 0}\frac{[f(x+h)+g(x+h)]-[f(x)+g(x)]}{h}\\ &= \lim_{h\to 0}\frac{[f(x+h)-f(x)]+[g(x+h)-g(x)]}{h}\\ &= \lim_{h\to 0}\frac{f(x+h)-f(x)}{h} + \lim_{h\to 0}\frac{g(x+h)-g(x)}{h}\\ &= \frac{d}{dx}[f(x)] + \frac{d}{dx}[g(x)]\end{aligned}$$

The limit of a sum is the sum of the limits.

Formula (8) can be proved in a similar manner or, alternatively, by writing $f(x)-g(x)$ as $f(x)+(-1)g(x)$ and then applying Formulas (6) and (7). ■

In words, *the derivative of a sum equals the sum of the derivatives*, and *the derivative of a difference equals the difference of the derivatives.*

▶ Example 5

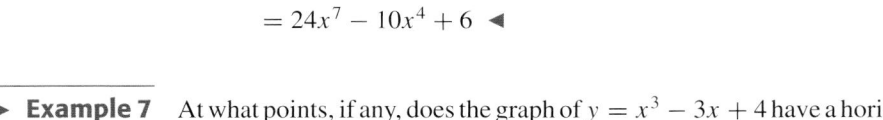

$$\frac{d}{dx}[2x^6 + x^{-9}] = \frac{d}{dx}[2x^6] + \frac{d}{dx}[x^{-9}] = 12x^5 + (-9)x^{-10} = 12x^5 - 9x^{-10}$$

$$\frac{d}{dx}[1 - 2\sqrt{x}\,] = \frac{d}{dx}[1] - \frac{d}{dx}[2\sqrt{x}\,] = 0 - 2\left(\frac{1}{2\sqrt{x}}\right) = -\frac{1}{\sqrt{x}}$$

See Formula (5). ◀

Although Formulas (7) and (8) are stated for sums and differences of two functions, they can be extended to any finite number of functions. For example, by grouping and applying Formula (7) twice we obtain

$$(f + g + h)' = [(f + g) + h]' = (f + g)' + h' = f' + g' + h'$$

As illustrated in the following example, the constant multiple rule together with the extended versions of the sum and difference rules can be used to differentiate any polynomial.

▶ Example 6 Find dy/dx if $y = 3x^8 - 2x^5 + 6x + 1$.

Solution.

$$\begin{aligned}\frac{dy}{dx} &= \frac{d}{dx}[3x^8 - 2x^5 + 6x + 1] \\ &= \frac{d}{dx}[3x^8] - \frac{d}{dx}[2x^5] + \frac{d}{dx}[6x] + \frac{d}{dx}[1] \\ &= 24x^7 - 10x^4 + 6\end{aligned}$$ ◀

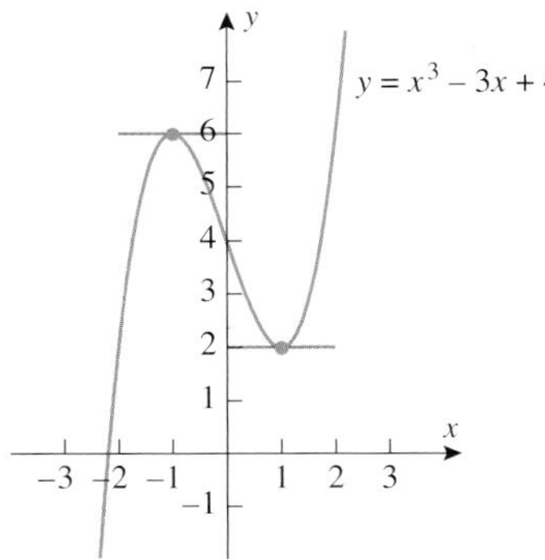

Figure 3.3.3

▶ Example 7 At what points, if any, does the graph of $y = x^3 - 3x + 4$ have a horizontal tangent line?

Solution. Horizontal tangent lines have slope zero, so we must find those values of x for which $y'(x) = 0$. Differentiating yields

$$y'(x) = \frac{d}{dx}[x^3 - 3x + 4] = 3x^2 - 3$$

Thus, horizontal tangent lines occur at those values of x for which $3x^2 - 3 = 0$, that is, if $x = -1$ or $x = 1$. The corresponding points on the curve $y = x^3 - 3x + 4$ are $(-1, 6)$ and $(1, 2)$ (see Figure 3.3.3). ◀

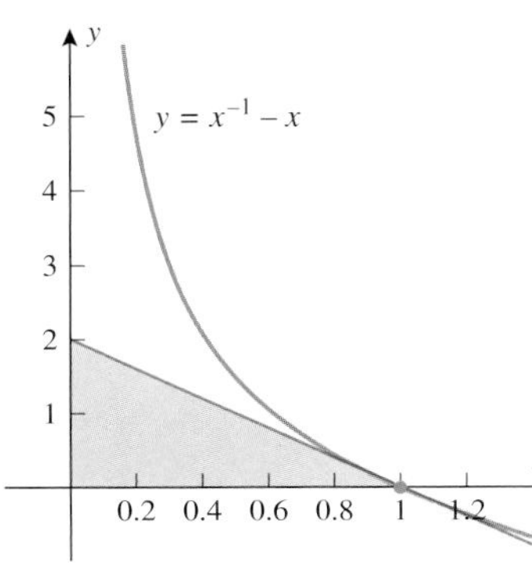

Figure 3.3.4

▶ Example 8 Find the area of the triangle formed by the coordinate axes and the tangent line to the curve $y = x^{-1} - x$ at the point $(1, 0)$.

Solution. Since the derivative of y with respect to x is

$$y'(x) = \frac{d}{dx}[x^{-1} - x] = \frac{d}{dx}[x^{-1}] - \frac{d}{dx}[x] = -x^{-2} - 1$$

the slope of the tangent line at the point $(1, 0)$ is $y'(1) = -2$. Thus, the equation of the tangent line at this point is

$$y - 0 = -2(x - 1) \quad \text{or equivalently} \quad y = -2x + 2$$

Since the y-intercept of this line is 2, the area of the triangle formed by the coordinate axes and the tangent line is 1 square unit (Figure 3.3.4). ◀

HIGHER DERIVATIVES

The derivative f' of a function f is itself a function and hence may have a derivative of its own. If f' is differentiable, then its derivative is denoted by f'' and is called the ***second derivative*** of f. As long as we have differentiability, we can continue the process of differentiating to obtain third, fourth, fifth, and even higher derivatives of f. These successive derivatives are denoted by

$$f', \quad f'' = (f')', \quad f''' = (f'')', \quad f^{(4)} = (f''')', \quad f^{(5)} = (f^{(4)})', \ldots$$

If $y = f(x)$, then successive derivatives can also be denoted by

$$y', \quad y'', \quad y''', \quad y^{(4)}, \quad y^{(5)}, \ldots,$$

Other common notations are

$$y' = \frac{dy}{dx} = \frac{d}{dx}[f(x)]$$

$$y'' = \frac{d^2y}{dx^2} = \frac{d}{dx}\left[\frac{d}{dx}[f(x)]\right] = \frac{d^2}{dx^2}[f(x)]$$

$$y''' = \frac{d^3y}{dx^3} = \frac{d}{dx}\left[\frac{d^2}{dx^2}[f(x)]\right] = \frac{d^3}{dx^3}[f(x)]$$

$$\vdots \qquad\qquad\qquad\qquad \vdots$$

These are called, in succession, the *first derivative*, the *second derivative*, the *third derivative*, and so forth. The number of times that f is differentiated is called the ***order*** of the derivative. A general nth order derivative can be denoted by

$$\frac{d^ny}{dx^n} = f^{(n)}(x) = \frac{d^n}{dx^n}[f(x)] \tag{9}$$

and the value of a general nth order derivative at a specific point $x = x_0$ can be denoted by

$$\left.\frac{d^ny}{dx^n}\right|_{x=x_0} = f^{(n)}(x_0) = \left.\frac{d^n}{dx^n}[f(x)]\right|_{x=x_0} \tag{10}$$

▶ **Example 9** If $f(x) = 3x^4 - 2x^3 + x^2 - 4x + 2$, then

$$f'(x) = 12x^3 - 6x^2 + 2x - 4$$
$$f''(x) = 36x^2 - 12x + 2$$
$$f'''(x) = 72x - 12$$
$$f^{(4)}(x) = 72$$
$$f^{(5)}(x) = 0$$
$$\vdots$$
$$f^{(n)}(x) = 0 \quad (n \geq 5)$$ ◀

We will discuss the significance of second derivatives and those of higher order in later sections.

✔ QUICK CHECK EXERCISES 3.3

(See page 179 for answers.)

1. In each part, determine $f'(x)$.
   (a) $f(x) = -x^2 + x + \dfrac{3}{x^2}$ (b) $f(x) = \dfrac{x^2+1}{x}$
   (c) $f(x) = (x^2+1)(x^2-1)$ (d) $f(x) = \sqrt{3x}$
2. The equation of the tangent line to the curve $y = x^2 + x + 1$ at $x = 1$ is ______.
3. The line $y = 9x - 5$ is tangent to the curve $y = x^3 + 3x^2$ at the point ______.
4. The x- and y-intercepts of the tangent line to the graph of $y = 1/x$ at the point $(2, \frac{1}{2})$ are $x =$ ______ and $y =$ ______.
5. Find all solutions of the equation $f''(x) = f'(x)$ given that $f(x) = 3x^3 - 3x^2 + x + 1$.

EXERCISE SET 3.3 Graphing Utility

1–8 Find dy/dx.

1. $y = 4x^7$
2. $y = -3x^{12}$
3. $y = 3x^8 + 2x + 1$
4. $y = \frac{1}{2}(x^4 + 7)$
5. $y = \pi^3$
6. $y = \sqrt{2}x + (1/\sqrt{2})$
7. $y = -\frac{1}{3}(x^7 + 2x - 9)$
8. $y = \dfrac{x^2+1}{5}$

9–16 Find $f'(x)$.

9. $f(x) = x^{-3} + \dfrac{1}{x^7}$
10. $f(x) = \sqrt{x} + \dfrac{1}{x}$
11. $f(x) = -3x^{-8} + 2\sqrt{x}$
12. $f(x) = 7x^{-6} - 5\sqrt{x}$
13. $f(x) = (3x^2 + 1)^2$
14. $f(x) = (x^5 + 2x)^2$
15. $f(x) = ax^3 + bx^2 + cx + d$ $(a, b, c, d$ constant)
16. $f(x) = \dfrac{1}{a}\left(x^2 + \dfrac{1}{b}x + c\right)$ $(a, b, c$ constant)

17–18 Find $y'(1)$.

17. $y = 5x^2 - 3x + 1$
18. $y = \dfrac{x^{3/2}+2}{x}$

19–20 Find dx/dt.

19. $x = t^2 - t$
20. $x = \dfrac{t^2+1}{3t}$

21–24 Find $dy/dx|_{x=1}$.

21. $y = 1 + x + x^2 + x^3 + x^4 + x^5$
22. $y = \dfrac{1 + x + x^2 + x^3 + x^4 + x^5 + x^6}{x^3}$
23. $y = (1 - x)(1 + x)(1 + x^2)(1 + x^4)$
24. $y = x^{24} + 2x^{12} + 3x^8 + 4x^6$

25–26 Approximate $f'(1)$ by considering the difference quotient
$$\frac{f(1+h) - f(1)}{h}$$
for values of h near 0, and then find the exact value of $f'(1)$ by differentiating.

25. $f(x) = x^3 - 3x + 1$
26. $f(x) = \dfrac{1}{x^2}$

27–28 Use a graphing utility to estimate the value of $f'(1)$ by zooming in on the graph of f, and then compare your estimate to the exact value obtained by differentiating.

27. $f(x) = \dfrac{x^2+1}{x}$
28. $f(x) = \dfrac{x + 2x^{3/2}}{\sqrt{x}}$

29–32 Find the indicated derivative.

29. $\dfrac{d}{dt}[16t^2]$
30. $\dfrac{dC}{dr}$, where $C = 2\pi r$
31. $V'(r)$, where $V = \pi r^3$
32. $\dfrac{d}{d\alpha}[2\alpha^{-1} + \alpha]$
33. A spherical balloon is being inflated.
    (a) Find a general formula for the instantaneous rate of change of the volume V with respect to the radius r, given that $V = \frac{4}{3}\pi r^3$.
    (b) Find the rate of change of V with respect to r at the instant when the radius is $r = 5$.
34. Find $\dfrac{d}{d\lambda}\left[\dfrac{\lambda\lambda_0 + \lambda^6}{2 - \lambda_0}\right]$ (λ_0 is constant).
35. Find an equation of the tangent line to the graph of $y = f(x)$ at $x = -3$ if $f(-3) = 2$ and $f'(-3) = 5$.
36. Find an equation of the tangent line to the graph of $y = f(x)$ at $x = 2$ if $f(2) = -2$ and $f'(2) = -1$.

37–38 Find d^2y/dx^2.

37. (a) $y = 7x^3 - 5x^2 + x$ (b) $y = 12x^2 - 2x + 3$
(c) $y = \dfrac{x+1}{x}$ (d) $y = (5x^2 - 3)(7x^3 + x)$

38. (a) $y = 4x^7 - 5x^3 + 2x$ (b) $y = 3x + 2$
(c) $y = \dfrac{3x-2}{5x}$ (d) $y = (x^3 - 5)(2x + 3)$

39–40 Find y'''.

39. (a) $y = x^{-5} + x^5$ (b) $y = 1/x$
(c) $y = ax^3 + bx + c$ (a, b, c constant)

40. (a) $y = 5x^2 - 4x + 7$ (b) $y = 3x^{-2} + 4x^{-1} + x$
(c) $y = ax^4 + bx^2 + c$ (a, b, c constant)

41. Find
(a) $f'''(2)$, where $f(x) = 3x^2 - 2$
(b) $\left.\dfrac{d^2y}{dx^2}\right|_{x=1}$, where $y = 6x^5 - 4x^2$
(c) $\left.\dfrac{d^4}{dx^4}[x^{-3}]\right|_{x=1}$.

42. Find
(a) $y'''(0)$, where $y = 4x^4 + 2x^3 + 3$
(b) $\left.\dfrac{d^4y}{dx^4}\right|_{x=1}$, where $y = \dfrac{6}{x^4}$.

43. Show that $y = x^3 + 3x + 1$ satisfies $y''' + xy'' - 2y' = 0$.

44. Show that if $x \neq 0$, then $y = 1/x$ satisfies the equation $x^3y'' + x^2y' - xy = 0$.

45–46 Use a graphing utility to make rough estimates of the locations of all horizontal tangent lines, and then find their exact locations by differentiating.

45. $y = \frac{1}{3}x^3 - \frac{3}{2}x^2 + 2x$

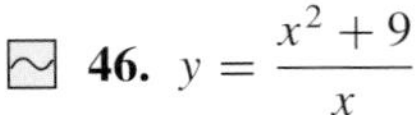

46. $y = \dfrac{x^2+9}{x}$

FOCUS ON CONCEPTS

47. Find a function $y = ax^2 + bx + c$ whose graph has an x-intercept of 1, a y-intercept of -2, and a tangent line with a slope of -1 at the y-intercept.

48. Find k if the curve $y = x^2 + k$ is tangent to the line $y = 2x$.

49. Find the x-coordinate of the point on the graph of $y = x^2$ where the tangent line is parallel to the secant line that cuts the curve at $x = -1$ and $x = 2$.

50. Find the x-coordinate of the point on the graph of $y = \sqrt{x}$ where the tangent line is parallel to the secant line that cuts the curve at $x = 1$ and $x = 4$.

51. Find the coordinates of all points on the graph of $y = 1 - x^2$ at which the tangent line passes through the point $(2, 0)$.

52. Show that any two tangent lines to the parabola $y = ax^2$, $a \neq 0$, intersect at a point that is on the vertical line halfway between the points of tangency.

53. Suppose that L is the tangent line at $x = x_0$ to the graph of the cubic equation $y = ax^3 + bx$. Find the x-coordinate of the point where L intersects the graph a second time.

54. Show that the segment of the tangent line to the graph of $y = 1/x$ that is cut off by the coordinate axes is bisected by the point of tangency.

55. Show that the triangle that is formed by any tangent line to the graph of $y = 1/x$, $x > 0$, and the coordinate axes has an area of 2 square units.

56. Find conditions on a, b, c, and d so that the graph of the polynomial $f(x) = ax^3 + bx^2 + cx + d$ has
(a) exactly two horizontal tangents
(b) exactly one horizontal tangent
(c) no horizontal tangents.

57. Newton's Law of Universal Gravitation states that the magnitude F of the force exerted by a point with mass M on a point with mass m is

$$F = \frac{GmM}{r^2}$$

where G is a constant and r is the distance between the bodies. Assuming that the points are moving, find a formula for the instantaneous rate of change of F with respect to r.

58. In the temperature range between 0°C and 700°C the resistance R [in ohms (Ω)] of a certain platinum resistance thermometer is given by

$$R = 10 + 0.04124T - 1.779 \times 10^{-5}T^2$$

where T is the temperature in degrees Celsius. Where in the interval from 0°C to 700°C is the resistance of the thermometer most sensitive and least sensitive to temperature changes? [*Hint:* Consider the size of dR/dT in the interval $0 \le T \le 700$.]

59–60 Use a graphing utility to make rough estimates of the intervals on which $f'(x) > 0$, and then find those intervals exactly by differentiating.

59. $f(x) = x - \dfrac{1}{x}$ **60.** $f(x) = x^3 - 3x$

61–64 You are asked in these exercises to determine whether a piecewise-defined function f is differentiable at a value $x = x_0$, where f is defined by different formulas on different sides of x_0. You may use the following result, which is a consequence of the Mean-Value Theorem (discussed in Section 4.7). ***Theorem.*** *Let f be continuous at x_0 and suppose that $\lim_{x\to x_0} f'(x)$ exists. Then f is differentiable at x_0, and $f'(x_0) = \lim_{x\to x_0} f'(x)$.*

61. Show that

$$f(x) = \begin{cases} x^2 + x + 1, & x \le 1 \\ 3x, & x > 1 \end{cases}$$

is continuous at $x = 1$. Determine whether f is differentiable at $x = 1$. If so, find the value of the derivative there. Sketch the graph of f.

62. Let
$$f(x) = \begin{cases} x^2 - 16x, & x < 9 \\ \sqrt{x}, & x \geq 9 \end{cases}$$
Is f continuous at $x = 9$? Determine whether f is differentiable at $x = 9$. If so, find the value of the derivative there.

63. Let
$$f(x) = \begin{cases} x^2, & x \leq 1 \\ \sqrt{x}, & x > 1 \end{cases}$$
Determine whether f is differentiable at $x = 1$. If so, find the value of the derivative there.

64. Let
$$f(x) = \begin{cases} x^3 + \frac{1}{16}, & x < \frac{1}{2} \\ \frac{3}{4}x^2, & x \geq \frac{1}{2} \end{cases}$$
Determine whether f is differentiable at $x = \frac{1}{2}$. If so, find the value of the derivative there.

65. Find all points where f fails to be differentiable. Justify your answer.
(a) $f(x) = |3x - 2|$ (b) $f(x) = |x^2 - 4|$

66. In each part compute f', f'', f''' and then state the formula for $f^{(n)}$.
(a) $f(x) = 1/x$ (b) $f(x) = 1/x^2$
[*Hint:* The expression $(-1)^n$ has a value of 1 if n is even and -1 if n is odd. Use this expression in your answer.]

67. (a) Prove:
$$\frac{d^2}{dx^2}[cf(x)] = c\frac{d^2}{dx^2}[f(x)]$$
$$\frac{d^2}{dx^2}[f(x) + g(x)] = \frac{d^2}{dx^2}[f(x)] + \frac{d^2}{dx^2}[g(x)]$$
(b) Do the results in part (a) generalize to nth derivatives? Justify your answer.

68. Let $f(x) = x^8 - 2x + 3$; find
$$\lim_{w \to 2} \frac{f'(w) - f'(2)}{w - 2}$$

69. (a) Find $f^{(n)}(x)$ if $f(x) = x^n$, $n = 1, 2, 3, \ldots$
(b) Find $f^{(n)}(x)$ if $f(x) = x^k$ and $n > k$, where k is a positive integer.
(c) Find $f^{(n)}(x)$ if
$$f(x) = a_0 + a_1x + a_2x^2 + \cdots + a_nx^n$$

70. (a) Prove: If $f''(x)$ exists for each x in (a, b), then both f and f' are continuous on (a, b).
(b) What can be said about the continuity of f and its derivatives if $f^{(n)}(x)$ exists for each x in (a, b)?

71. Let $f(x) = (mx + b)^n$, where m and b are constants and n is an integer. Use the result of Exercise 48 in Section 3.2 to prove that $f'(x) = nm(mx + b)^{n-1}$.

72–73 Verify the result of Exercise 71 for $f(x)$.

72. $f(x) = (2x + 3)^2$ **73.** $f(x) = (3x - 1)^3$

74–77 Use the result of Exercise 71 to compute the derivative of the given function $f(x)$.

74. $f(x) = \dfrac{1}{x - 1}$ **75.** $f(x) = \dfrac{3}{(2x + 1)^2}$

76. $f(x) = \dfrac{x}{x + 1}$ **77.** $f(x) = \dfrac{2x^2 + 4x + 3}{x^2 + 2x + 1}$

✔ QUICK CHECK ANSWERS 3.3

1. (a) $f'(x) = -2x + 1 - \dfrac{6}{x^3}$ (b) $f'(x) = 1 - \dfrac{1}{x^2}$ (c) $f'(x) = 4x^3$ (d) $f'(x) = \dfrac{1}{2}\sqrt{\dfrac{3}{x}}$ **2.** $y = 3x$ **3.** $(1, 4)$ **4.** 4; 1 **5.** $\dfrac{1}{3}, \dfrac{7}{3}$

3.4 THE PRODUCT AND QUOTIENT RULES

In this section we will develop techniques for differentiating products and quotients of functions whose derivatives are known.

DERIVATIVE OF A PRODUCT

You might be tempted to conjecture that the derivative of a product of two functions is the product of their derivatives. However, a simple example will show this to be false. Consider the functions
$$f(x) = x \quad \text{and} \quad g(x) = x^2$$

The product of their derivatives is

$$f'(x)g'(x) = (1)(2x) = 2x$$

but their product is $h(x) = f(x)g(x) = x^3$, so the derivative of the product is

$$h'(x) = 3x^2$$

Thus, the derivative of the product is not equal to the product of the derivatives. The correct relationship, which is credited to Leibniz, is given by the following theorem.

Formula (1) can also be expressed as

$$(f \cdot g)' = f \cdot g' + g \cdot f'$$

3.4.1 THEOREM (*The Product Rule*). *If f and g are differentiable at x, then so is the product $f \cdot g$, and*

$$\frac{d}{dx}[f(x)g(x)] = f(x)\frac{d}{dx}[g(x)] + g(x)\frac{d}{dx}[f(x)] \tag{1}$$

PROOF. Whereas the proofs of the derivative rules in the last section were straightforward applications of the derivative definition, a key step in this proof involves adding and subtracting the quantity $f(x+h)g(x)$ to the numerator in the derivative definition. This yields

$$\begin{aligned}
\frac{d}{dx}[f(x)g(x)] &= \lim_{h\to 0} \frac{f(x+h)\cdot g(x+h) - f(x)\cdot g(x)}{h} \\
&= \lim_{h\to 0} \frac{f(x+h)g(x+h) - f(x+h)g(x) + f(x+h)g(x) - f(x)g(x)}{h} \\
&= \lim_{h\to 0} \left[f(x+h)\cdot \frac{g(x+h)-g(x)}{h} + g(x)\cdot \frac{f(x+h)-f(x)}{h} \right] \\
&= \lim_{h\to 0} f(x+h)\cdot \lim_{h\to 0} \frac{g(x+h)-g(x)}{h} + \lim_{h\to 0} g(x)\cdot \lim_{h\to 0} \frac{f(x+h)-f(x)}{h} \\
&= \left[\lim_{h\to 0} f(x+h)\right]\frac{d}{dx}[g(x)] + \left[\lim_{h\to 0} g(x)\right]\frac{d}{dx}[f(x)] \\
&= f(x)\frac{d}{dx}[g(x)] + g(x)\frac{d}{dx}[f(x)]
\end{aligned}$$

[*Note:* In the last step $f(x+h) \to f(x)$ as $h \to 0$ because f is continuous at x by Theorem 3.2.3. Also, $g(x) \to g(x)$ as $h \to 0$ because $g(x)$ does not involve h and hence is treated as constant for the limit.] ■

In words, *the derivative of a product of two functions is the first function times the derivative of the second plus the second function times the derivative of the first.*

▶ **Example 1** Find dy/dx if $y = (4x^2 - 1)(7x^3 + x)$.

Solution. There are two methods that can be used to find dy/dx. We can either use the product rule or we can multiply out the factors in y and then differentiate. We will give both methods.

Method 1. (*Using the Product Rule*)

$$\begin{aligned}\frac{dy}{dx} &= \frac{d}{dx}[(4x^2-1)(7x^3+x)] \\ &= (4x^2-1)\frac{d}{dx}[7x^3+x] + (7x^3+x)\frac{d}{dx}[4x^2-1] \\ &= (4x^2-1)(21x^2+1) + (7x^3+x)(8x) = 140x^4 - 9x^2 - 1\end{aligned}$$

Method 2. (*Multiplying First*)

$$y = (4x^2-1)(7x^3+x) = 28x^5 - 3x^3 - x$$

Thus,

$$\frac{dy}{dx} = \frac{d}{dx}[28x^5 - 3x^3 - x] = 140x^4 - 9x^2 - 1$$

which agrees with the result obtained using the product rule. ◀

▶ **Example 2** Find ds/dt if $s = (1+t)\sqrt{t}$.

Solution. Applying the product rule yields

$$\begin{aligned}\frac{ds}{dt} &= \frac{d}{dt}[(1+t)\sqrt{t}] \\ &= (1+t)\frac{d}{dt}[\sqrt{t}] + \sqrt{t}\frac{d}{dt}[1+t] \\ &= \frac{1+t}{2\sqrt{t}} + \sqrt{t} = \frac{1+3t}{2\sqrt{t}}\end{aligned}$$ ◀

DERIVATIVE OF A QUOTIENT

Just as the derivative of a product is not generally the product of the derivatives, so the derivative of a quotient is not generally the quotient of the derivatives. The correct relationship is given by the following theorem.

3.4.2 THEOREM (*The Quotient Rule*). *If f and g are both differentiable at x and if $g(x) \neq 0$, then f/g is differentiable at x and*

$$\frac{d}{dx}\left[\frac{f(x)}{g(x)}\right] = \frac{g(x)\dfrac{d}{dx}[f(x)] - f(x)\dfrac{d}{dx}[g(x)]}{[g(x)]^2} \tag{2}$$

Formula (2) can also be expressed as

$$\left(\frac{f}{g}\right)' = \frac{g \cdot f' - f \cdot g'}{g^2}$$

PROOF.

$$\frac{d}{dx}\left[\frac{f(x)}{g(x)}\right] = \lim_{h\to 0}\frac{\dfrac{f(x+h)}{g(x+h)} - \dfrac{f(x)}{g(x)}}{h} = \lim_{h\to 0}\frac{f(x+h)\cdot g(x) - f(x)\cdot g(x+h)}{h\cdot g(x)\cdot g(x+h)}$$

Adding and subtracting $f(x) \cdot g(x)$ in the numerator yields

$$\frac{d}{dx}\left[\frac{f(x)}{g(x)}\right] = \lim_{h\to 0} \frac{f(x+h)\cdot g(x) - f(x)\cdot g(x) - f(x)\cdot g(x+h) + f(x)\cdot g(x)}{h\cdot g(x)\cdot g(x+h)}$$

$$= \lim_{h\to 0} \frac{\left[g(x)\cdot \dfrac{f(x+h)-f(x)}{h}\right] - \left[f(x)\cdot \dfrac{g(x+h)-g(x)}{h}\right]}{g(x)\cdot g(x+h)}$$

$$= \frac{\lim\limits_{h\to 0} g(x)\cdot \lim\limits_{h\to 0}\dfrac{f(x+h)-f(x)}{h} - \lim\limits_{h\to 0} f(x)\cdot \lim\limits_{h\to 0}\dfrac{g(x+h)-g(x)}{h}}{\lim\limits_{h\to 0} g(x)\cdot \lim\limits_{h\to 0} g(x+h)}$$

$$= \frac{\left[\lim\limits_{h\to 0} g(x)\right]\cdot \dfrac{d}{dx}[f(x)] - \left[\lim\limits_{h\to 0} f(x)\right]\cdot \dfrac{d}{dx}[g(x)]}{\lim\limits_{h\to 0} g(x)\cdot \lim\limits_{h\to 0} g(x+h)}$$

$$= \frac{g(x)\dfrac{d}{dx}[f(x)] - f(x)\dfrac{d}{dx}[g(x)]}{[g(x)]^2}$$

[See the note at the end of the proof of Theorem 3.4.1 for an explanation of the last step.] ■

In words, *the derivative of a quotient of two functions is the denominator times the derivative of the numerator minus the numerator times the derivative of the denominator, all divided by the denominator squared.*

Sometimes it is better to simplify a function than to apply the quotient rule blindly. For example, it is easier to differentiate

$$f(x) = \frac{x^{3/2}+x}{\sqrt{x}}$$

by rewriting it as

$$f(x) = x + \sqrt{x}$$

as opposed to using the quotient rule.

▶ Example 3 Find $y'(x)$ for $y = \dfrac{x^3+2x^2-1}{x+5}$.

Solution. Applying the quotient rule yields

$$\frac{dy}{dx} = \frac{d}{dx}\left[\frac{x^3+2x^2-1}{x+5}\right] = \frac{(x+5)\dfrac{d}{dx}[x^3+2x^2-1] - (x^3+2x^2-1)\dfrac{d}{dx}[x+5]}{(x+5)^2}$$

$$= \frac{(x+5)(3x^2+4x) - (x^3+2x^2-1)(1)}{(x+5)^2}$$

$$= \frac{(3x^3+19x^2+20x) - (x^3+2x^2-1)}{(x+5)^2}$$

$$= \frac{2x^3+17x^2+20x+1}{(x+5)^2} \blacktriangleleft$$

▶ Example 4 Let $f(x) = \dfrac{x^2-1}{x^4+1}$.

(a) Graph $y = f(x)$, and use your graph to make rough estimates of the locations of all horizontal tangent lines.

(b) By differentiating, find the exact locations of the horizontal tangent lines.

Solution (a). In Figure 3.4.1 we have shown the graph of the equation $y = f(x)$ in the window $[-2.5, 2.5] \times [-1, 1]$. This graph suggests that horizontal tangent lines occur at $x = 0$, $x \approx 1.5$, and $x \approx -1.5$.

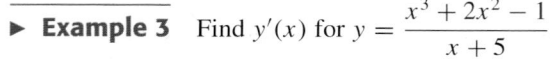

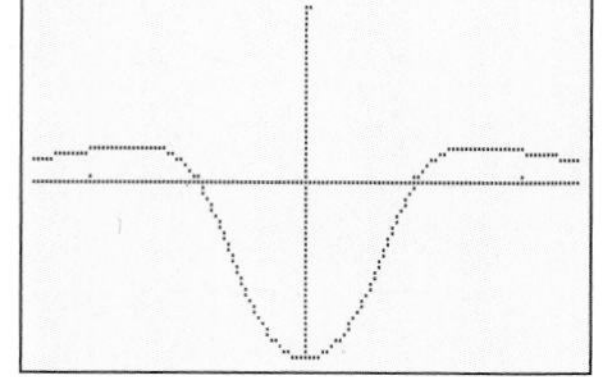

$[-2.5, 2.5] \times [-1, 1]$
$x\text{Scl} = 1, y\text{Scl} = 1$

$$y = \frac{x^2-1}{x^4+1}$$

Figure 3.4.1

Solution (b). To find the exact locations of the horizontal tangent lines, we must find the points where $dy/dx = 0$. We start by finding dy/dx:

$$\frac{dy}{dx} = \frac{d}{dx}\left[\frac{x^2-1}{x^4+1}\right] = \frac{(x^4+1)\frac{d}{dx}[x^2-1] - (x^2-1)\frac{d}{dx}[x^4+1]}{(x^4+1)^2}$$

$$= \frac{(x^4+1)(2x)-(x^2-1)(4x^3)}{(x^4+1)^2}$$

The differentiation is complete. The rest is simplification.

$$= \frac{-2x^5+4x^3+2x}{(x^4+1)^2} = -\frac{2x(x^4-2x^2-1)}{(x^4+1)^2}$$

Now we will set $dy/dx = 0$ and solve for x. We obtain

$$-\frac{2x(x^4-2x^2-1)}{(x^4+1)^2} = 0$$

The solutions of this equation are the values of x for which the numerator is 0, that is,

$$2x(x^4-2x^2-1) = 0$$

The first factor yields the solution $x = 0$. Other solutions can be found by solving the equation

$$x^4-2x^2-1 = 0$$

This can be treated as a quadratic equation in x^2 and solved by the quadratic formula. This yields

$$x^2 = \frac{2 \pm \sqrt{8}}{2} = 1 \pm \sqrt{2}$$

The minus sign yields imaginary values of x, which we ignore since they are not relevant to the problem. The plus sign yields the solutions

$$x = \pm\sqrt{1+\sqrt{2}}$$

In summary, horizontal tangent lines occur at

$$x = 0, \quad x = \sqrt{1+\sqrt{2}} \approx 1.55, \quad \text{and} \quad x = -\sqrt{1+\sqrt{2}} \approx -1.55$$

which is consistent with the rough estimates that we obtained graphically in part (a). ◀

Derive the following rule for differentiating a reciprocal:

$$\left(\frac{1}{g}\right)' = -\frac{g'}{g^2}$$

Use it to find the derivative of

$$f(x) = \frac{1}{x^2+1}$$

SUMMARY OF DIFFERENTIATION RULES

The following table summarizes the differentiation rules that we have encountered thus far.

$\frac{d}{dx}[c] = 0$	$(f+g)' = f' + g'$	$(f \cdot g)' = f \cdot g' + g \cdot f'$	$\left(\frac{1}{g}\right)' = -\frac{g'}{g^2}$
$(cf)' = cf'$	$(f-g)' = f' - g'$	$\left(\frac{f}{g}\right)' = \frac{g \cdot f' - f \cdot g'}{g^2}$	$\frac{d}{dx}[x^n] = nx^{n-1}$ (n an integer)

✔QUICK CHECK EXERCISES 3.4 *(See page 185 for answers.)*

1. (a) $\frac{d}{dx}\left[\frac{x+1}{x-1}\right] =$ ______
   (b) $\frac{d}{dx}[(4-x^3)(x^2+x-1)] =$ ______
   (c) $\frac{d}{dx}\left[\frac{1}{x^3-x^2}\right] =$ ______

2. Find $F'(1)$ given that $f(1) = -1$, $f'(1) = 2$, $g(1) = 3$, and $g'(1) = -1$.
   (a) $F(x) = 2f(x) - 3g(x)$
   (b) $F(x) = [f(x)]^2$
   (c) $F(x) = f(x)g(x)$
   (d) $F(x) = f(x)/g(x)$

3. The equation of the tangent line to the graph of
$$y=\frac{3x+4}{x^2+1}$$
at $x=2$ is ______.

4. Assuming that f is twice differentiable, find a formula for the second derivative of $g(x)=[f(x)]^2$ in terms of f, f', and f''.

EXERCISE SET 3.4

Graphing Utility

1–4 Compute the derivative of the given function $f(x)$ by (a) multiplying and then differentiating and (b) using the product rule. Verify that (a) and (b) yield the same result.

1. $f(x)=(x+1)(2x-1)$ **2.** $f(x)=(3x^2-1)(x^2+2)$

3. $f(x)=(x^2+1)(x^2-1)$

4. $f(x)=(x+1)(x^2-x+1)$

5–10 Find $f'(x)$.

5. $f(x)=(3x^2+6)\left(2x-\frac{1}{4}\right)$

6. $f(x)=(2-x-3x^3)(7+x^5)$

7. $f(x)=(x^3+7x^2-8)(2x^{-3}+x^{-4})$

8. $f(x)=\left(\frac{1}{x}+\frac{1}{x^2}\right)(3x^3+27)$

9. $f(x)=(x-2)(x^2+2x+4)$

10. $f(x)=(x^2+x)(x^2-x)$

11–16 Find $dy/dx|_{x=1}$.

11. $y=\frac{1}{5x-3}$ **12.** $y=\frac{3}{\sqrt{x}+2}$

13. $y=\frac{2x-1}{x+3}$ **14.** $y=\frac{4x+1}{x^2-5}$

15. $y=\left(\frac{3x+2}{x}\right)(x^{-5}+1)$ **16.** $y=(2x^7-x^2)\left(\frac{x-1}{x+1}\right)$

17–18 Use a graphing utility to estimate the value of $f'(1)$ by zooming in on the graph of f, and then compare your estimate to the exact value obtained by differentiating.

17. $f(x)=\frac{x}{x^2+1}$ **18.** $f(x)=\frac{x^2-1}{x^2+1}$

19. Find $g'(4)$ given that $f(4)=3$ and $f'(4)=-5$.
(a) $g(x)=\sqrt{x}f(x)$ (b) $g(x)=\frac{f(x)}{x}$

20. Find $g'(3)$ given that $f(3)=-2$ and $f'(3)=4$.
(a) $g(x)=3x^2-5f(x)$ (b) $g(x)=\frac{2x+1}{f(x)}$

21. Find $F'(2)$ given that $f(2)=-1$, $f'(2)=4$, $g(2)=1$, and $g'(2)=-5$.
(a) $F(x)=5f(x)+2g(x)$ (b) $F(x)=f(x)-3g(x)$
(c) $F(x)=f(x)g(x)$ (d) $F(x)=f(x)/g(x)$

22. Find $F'(\pi)$ given that $f(\pi)=10$, $f'(\pi)=-1$, $g(\pi)=-3$, and $g'(\pi)=2$.
(a) $F(x)=6f(x)-5g(x)$ (b) $F(x)=x(f(x)+g(x))$
(c) $F(x)=2f(x)g(x)$ (d) $F(x)=\frac{f(x)}{4+g(x)}$

23–28 Find all values of x at which the tangent line to the given curve satisfies the stated property.

23. $y=\frac{x^2-1}{x+2}$; horizontal **24.** $y=\frac{x^2+1}{x-1}$; horizontal

25. $y=\frac{x^2+1}{x+1}$; parallel to the line $y=x$

26. $y=\frac{x+3}{x+2}$; perpendicular to the line $y=x$

27. $y=\frac{1}{x+4}$; passes through the origin

28. $y=\frac{2x+5}{x+2}$; y-intercept 2

FOCUS ON CONCEPTS

29. (a) Define what it should mean to say that two curves intersect at right angles.
(b) Prove that the curves $y=1/x$ and $y=1/(2-x)$ intersect at right angles.

30. Find all values of a such that the curves $y=a/(x-1)$ and $y=x^2-2x+1$ intersect at right angles.

31. Find a general formula for $F''(x)$ if $F(x)=xf(x)$ and f and f' are differentiable at x.

32. Suppose that the function f is differentiable everywhere and $F(x)=xf(x)$.
(a) Express $F'''(x)$ in terms of x and derivatives of f.
(b) For $n\geq 2$, conjecture a formula for $F^{(n)}(x)$.

33. Apply the product rule (Theorem 3.4.1) twice to show that if f, g, and h are differentiable functions, then $f\cdot g\cdot h$ is differentiable, and
$$(f\cdot g\cdot h)'=f'\cdot g\cdot h+f\cdot g'\cdot h+f\cdot g\cdot h'$$

34. Based on the result in Exercise 33, make a conjecture about a formula for differentiating a product of n functions.

35. Use the formula in Exercise 33 to find
(a) $\frac{d}{dx}\left[(2x+1)\left(1+\frac{1}{x}\right)(x^{-3}+7)\right]$
(b) $\frac{d}{dx}[(x^7+2x-3)^3]$.

36. Use the formula you obtained in Exercise 34 to find
(a) $\frac{d}{dx}[x^{-5}(x^2+2x)(4-3x)(2x^9+1)]$
(b) $\frac{d}{dx}[(x^2+1)^{50}]$.

37. Given that f is a differentiable function and n is a positive integer, use the result of Exercise 34 to determine a formula for the derivative of $g(x) = (f(x))^n$.

38. Use the result of Exercise 37 to find the derivative of $g(x) = (x^2-1)^{10}$.

39. Use the quotient rule (Theorem 3.4.2) to derive the formula for the derivative of $f(x) = x^{-n}$, where n is a positive integer.

40. Assuming that $h(x) = f(x)/g(x)$ is differentiable, use the product rule to derive Formula (2). [*Hint:* Differentiate both sides of the equation $h(x) \cdot g(x) = f(x)$ and solve for $h'(x)$.]

✔ QUICK CHECK ANSWERS 3.4

1. (a) $-\frac{2}{(x-1)^2}$ (b) $(4-x^3)(2x+1)+(x^2+x-1)(-3x^2)$ (c) $\frac{2x-3x^2}{(x^3-x^2)^2}$ 2. (a) 7 (b) −4 (c) 7 (d) $\frac{5}{9}$
3. $y = -x + 4$ 4. $2f'^2 + 2ff''$

3.5 DERIVATIVES OF TRIGONOMETRIC FUNCTIONS

The main objective of this section is to obtain formulas for the derivatives of the six basic trigonometric functions. If needed, you will find a review of trigonometric functions in Appendix A.

We will assume in this section that the variable x in the trigonometric functions $\sin x$, $\cos x$, $\tan x$, $\cot x$, $\sec x$, and $\csc x$ is measured in radians. Also, we will need the limits in Theorem 2.6.3, but restated as follows using h rather than x as the variable:

$$\lim_{h\to 0} \frac{\sin h}{h} = 1 \quad \text{and} \quad \lim_{h\to 0} \frac{1-\cos h}{h} = 0 \tag{1–2}$$

Let us start with the problem of differentiating $f(x) = \sin x$. Using the definition of the derivative we obtain

$$\begin{aligned}
f'(x) &= \lim_{h\to 0} \frac{f(x+h)-f(x)}{h} \\
&= \lim_{h\to 0} \frac{\sin(x+h)-\sin x}{h} \\
&= \lim_{h\to 0} \frac{\sin x\cos h + \cos x \sin h - \sin x}{h} && \text{By the addition formula for sine} \\
&= \lim_{h\to 0} \left[\sin x\left(\frac{\cos h - 1}{h}\right) + \cos x\left(\frac{\sin h}{h}\right)\right] \\
&= \lim_{h\to 0} \left[\cos x\left(\frac{\sin h}{h}\right) - \sin x\left(\frac{1-\cos h}{h}\right)\right] && \text{Algebraic reorganization} \\
&= \lim_{h\to 0} \cos x \cdot \lim_{h\to 0} \frac{\sin h}{h} - \lim_{h\to 0} \sin x \cdot \lim_{h\to 0} \frac{1-\cos h}{h} \\
&= \left(\lim_{h\to 0} \cos x\right)(1) - \left(\lim_{h\to 0} \sin x\right)(0) && \text{Formulas (1) and (2)} \\
&= \lim_{h\to 0} \cos x = \cos x && \text{$\cos x$ does not involve the variable h and hence is treated as a constant in the limit computation.}
\end{aligned}$$

Recall that the derivative formulas for the trigonometric functions are only valid if x is in radians. See Exercise 47 for an explanation of how these formulas change when x is measured in degrees.

Thus, we have shown that

$$\frac{d}{dx}[\sin x] = \cos x \tag{3}$$

In the exercises we will ask you to use the same method to derive the following formula for the derivative of $\cos x$:

$$\frac{d}{dx}[\cos x] = -\sin x \tag{4}$$

▶ **Example 1** Find dy/dx if $y = x \sin x$.

Solution. Using Formula (3) and the product rule we obtain

$$\begin{aligned}\frac{dy}{dx} &= \frac{d}{dx}[x \sin x] \\ &= x\frac{d}{dx}[\sin x] + \sin x \frac{d}{dx}[x] \\ &= x \cos x + \sin x \quad \blacktriangleleft\end{aligned}$$

▶ **Example 2** Find dy/dx if $y = \dfrac{\sin x}{1 + \cos x}$.

Solution. Using the quotient rule together with Formulas (3) and (4) we obtain

$$\begin{aligned}\frac{dy}{dx} &= \frac{(1+\cos x)\cdot \frac{d}{dx}[\sin x] - \sin x \cdot \frac{d}{dx}[1+\cos x]}{(1+\cos x)^2} \\ &= \frac{(1+\cos x)(\cos x) - (\sin x)(-\sin x)}{(1+\cos x)^2} \\ &= \frac{\cos x + \cos^2 x + \sin^2 x}{(1+\cos x)^2} = \frac{\cos x + 1}{(1+\cos x)^2} = \frac{1}{1+\cos x} \quad \blacktriangleleft\end{aligned}$$

The derivative formulas for the trigonometric functions should be memorized. An aid for memorizing these is given in Exercise 48.

The derivatives of the remaining trigonometric functions are

$$\frac{d}{dx}[\tan x] = \sec^2 x \qquad \frac{d}{dx}[\sec x] = \sec x \tan x \tag{5–6}$$

$$\frac{d}{dx}[\cot x] = -\csc^2 x \qquad \frac{d}{dx}[\csc x] = -\csc x \cot x \tag{7–8}$$

These can all be obtained using the definition of the derivative, but it is easier to use Formulas (3) and (4) and apply the quotient rule to the relationships

$$\tan x = \frac{\sin x}{\cos x}, \quad \cot x = \frac{\cos x}{\sin x}, \quad \sec x = \frac{1}{\cos x}, \quad \csc x = \frac{1}{\sin x}$$

For example,

$$\begin{aligned}\frac{d}{dx}[\tan x] &= \frac{d}{dx}\left[\frac{\sin x}{\cos x}\right] = \frac{\cos x \cdot \frac{d}{dx}[\sin x] - \sin x \cdot \frac{d}{dx}[\cos x]}{\cos^2 x} \\ &= \frac{\cos x \cdot \cos x - \sin x \cdot (-\sin x)}{\cos^2 x} = \frac{\cos^2 x + \sin^2 x}{\cos^2 x} = \frac{1}{\cos^2 x} = \sec^2 x\end{aligned}$$

When finding the value of a derivative at a specific point $x = x_0$, it is important to substitute x_0 *after* the derivative is obtained. Thus, in Example 3 we made the substitution $x = \pi/4$ after f'' was calculated. What would have happened had we *incorrectly* substituted $x = \pi/4$ into $f'(x)$ before calculating f''?

▶ **Example 3** Find $f''(\pi/4)$ if $f(x) = \sec x$.

$$f'(x) = \sec x \tan x$$

$$\begin{aligned} f''(x) &= \sec x \cdot \frac{d}{dx}[\tan x] + \tan x \cdot \frac{d}{dx}[\sec x] \\ &= \sec x \cdot \sec^2 x + \tan x \cdot \sec x \tan x \\ &= \sec^3 x + \sec x \tan^2 x \end{aligned}$$

Thus,

$$\begin{aligned} f''(\pi/4) &= \sec^3(\pi/4) + \sec(\pi/4)\tan^2(\pi/4) \\ &= (\sqrt{2})^3 + (\sqrt{2})(1)^2 = 3\sqrt{2} \end{aligned}$$ ◀

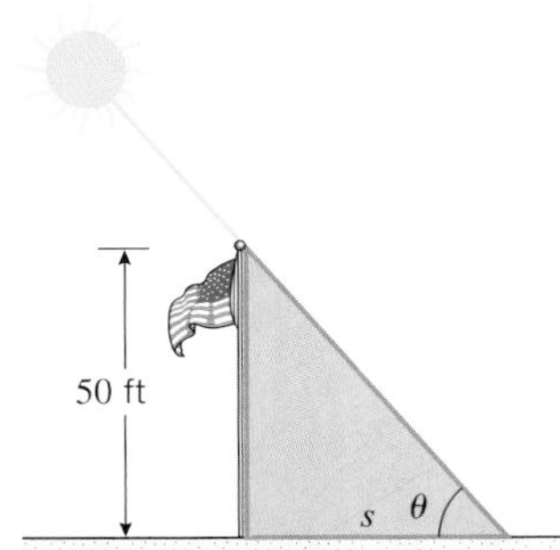

Figure 3.5.1

▶ **Example 4** On a sunny day, a 50-ft flagpole casts a shadow that changes with the angle of elevation of the Sun. Let s be the length of the shadow and θ the angle of elevation of the Sun (Figure 3.5.1). Find the rate at which the length of the shadow is changing with respect to θ when $\theta = 45°$. Express your answer in units of feet/degree.

Solution. The variables s and θ are related by $\tan\theta = 50/s$ or, equivalently,

$$s = 50 \cot\theta \tag{9}$$

If θ is measured in radians, then Formula (7) is applicable, which yields

$$\frac{ds}{d\theta} = -50 \csc^2\theta$$

which is the rate of change of shadow length with respect to the elevation angle θ in units of feet/radian. When $\theta = 45°$ (or equivalently $\theta = \pi/4$ radians), we obtain

$$\left.\frac{ds}{d\theta}\right|_{\theta=\pi/4} = -50\csc^2(\pi/4) = -100 \text{ feet/radian}$$

Converting radians (rad) to degrees (deg) yields

$$-100 \frac{\text{ft}}{\text{rad}} \cdot \frac{\pi}{180} \frac{\text{rad}}{\text{deg}} = -\frac{5}{9}\pi \frac{\text{ft}}{\text{deg}} \approx -1.75 \text{ ft/deg}$$

Thus, when $\theta = 45°$, the shadow length is decreasing (because of the minus sign) at an approximate rate of 1.75 ft/deg increase in the angle of elevation. ◀

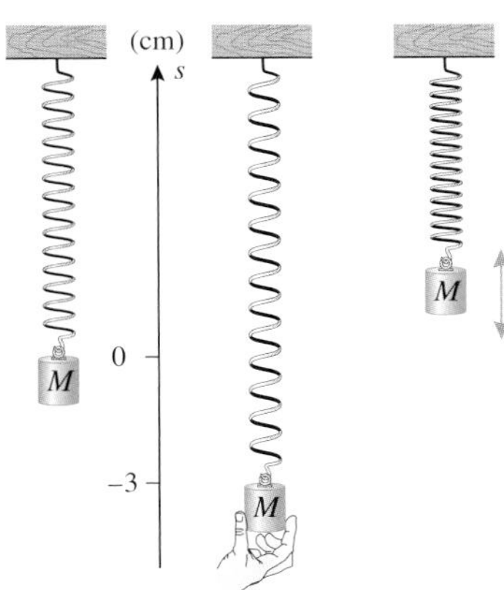

Figure 3.5.2

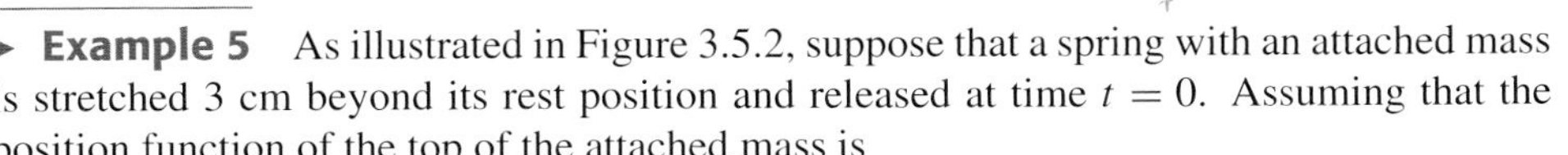

▶ **Example 5** As illustrated in Figure 3.5.2, suppose that a spring with an attached mass is stretched 3 cm beyond its rest position and released at time $t = 0$. Assuming that the position function of the top of the attached mass is

$$s = -3\cos t \tag{10}$$

where s is in centimeters and t is in seconds, find the velocity function and discuss the motion of the attached mass.

Solution. The velocity function is

$$v = \frac{ds}{dt} = \frac{d}{dt}[-3\cos t] = 3\sin t$$

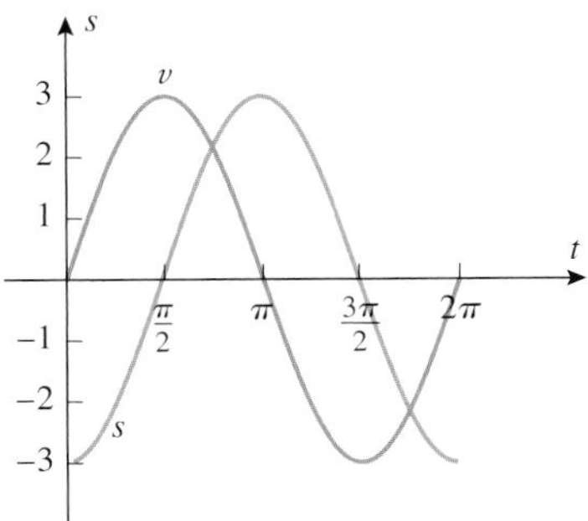

Figure 3.5.3

Figure 3.5.3 shows the graphs of the position and velocity functions. The position function tells us that the top of the mass oscillates between a low point of $s = -3$ and a high point of $s = 3$ with one complete oscillation occuring every 2π seconds [the period of (10)]. The

> The top of the mass has its maximum speed when it passes through its rest position. Why? What is that maximum speed?

top of the mass is moving up (the positive s-direction) when v is positive, is moving down when v is negative, and is at a high or low point when $v = 0$. Thus, for example, the top of the mass moves up from time $t = 0$ to time $t = \pi$, at which time it reaches the high point $s = 3$ and then moves down until time $t = 2\pi$, at which time it reaches the low point of $s = -3$. The motion then repeats periodically. ◀

✔ QUICK CHECK EXERCISES 3.5 *(See page 190 for answers.)*

1. Find dy/dx.
   (a) $y = \sin x$ (b) $y = \cos x$
   (c) $y = \tan x$ (d) $y = \sec x$
2. Find $f'(x)$ and $f'(\pi/3)$ if $f(x) = \sin x \cos x$.
3. Use a derivative to evaluate each limit.
   (a) $\lim\limits_{h \to 0} \dfrac{\sin\left(\frac{\pi}{2} + h\right) - 1}{h}$ (b) $\lim\limits_{h \to 0} \dfrac{\csc(x + h) - \csc x}{h}$

EXERCISE SET 3.5  Graphing Utility

1–18 Find $f'(x)$.

1. $f(x) = 4\cos x + 2\sin x$
2. $f(x) = \dfrac{5}{x^2} + \sin x$
3. $f(x) = -4x^2 \cos x$
4. $f(x) = 2\sin^2 x$
5. $f(x) = \dfrac{5 - \cos x}{5 + \sin x}$
6. $f(x) = \dfrac{\sin x}{x^2 + \sin x}$
7. $f(x) = \sec x - \sqrt{2}\tan x$
8. $f(x) = (x^2 + 1)\sec x$
9. $f(x) = 4\csc x - \cot x$
10. $f(x) = \cos x - x\csc x$
11. $f(x) = \sec x \tan x$
12. $f(x) = \csc x \cot x$
13. $f(x) = \dfrac{\cot x}{1 + \csc x}$
14. $f(x) = \dfrac{\sec x}{1 + \tan x}$
15. $f(x) = \sin^2 x + \cos^2 x$
16. $f(x) = \sec^2 x - \tan^2 x$
17. $f(x) = \dfrac{\sin x \sec x}{1 + x\tan x}$
18. $f(x) = \dfrac{(x^2 + 1)\cot x}{3 - \cos x \csc x}$

19–24 Find d^2y/dx^2.

19. $y = x\cos x$
20. $y = \csc x$
21. $y = x\sin x - 3\cos x$
22. $y = x^2\cos x + 4\sin x$
23. $y = \sin x \cos x$
24. $y = \tan x$
25. Find the equation of the line tangent to the graph of $\tan x$ at
   (a) $x = 0$ (b) $x = \pi/4$ (c) $x = -\pi/4$.
26. Find the equation of the line tangent to the graph of $\sin x$ at
   (a) $x = 0$ (b) $x = \pi$ (c) $x = \pi/4$.
27. (a) Show that $y = x\sin x$ is a solution to $y'' + y = 2\cos x$.
   (b) Show that $y = x\sin x$ is a solution of the equation $y^{(4)} + y'' = -2\cos x$.
28. (a) Show that $y = \cos x$ and $y = \sin x$ are solutions of the equation $y'' + y = 0$.
   (b) Show that $y = A\sin x + B\cos x$ is a solution of the equation $y'' + y = 0$ for all constants A and B.
29. Find all values in the interval $[-2\pi, 2\pi]$ at which the graph of f has a horizontal tangent line.
   (a) $f(x) = \sin x$ (b) $f(x) = x + \cos x$
   (c) $f(x) = \tan x$ (d) $f(x) = \sec x$
30. (a) Use a graphing utility to make rough estimates of the values in the interval $[0, 2\pi]$ at which the graph of $y = \sin x \cos x$ has a horizontal tangent line.
   (b) Find the exact locations of the points where the graph has a horizontal tangent line.
31. A 10-ft ladder leans against a wall at an angle θ with the horizontal, as shown in the accompanying figure. The top of the ladder is x feet above the ground. If the bottom of the ladder is pushed toward the wall, find the rate at which x changes with respect to θ when $\theta = 60°$. Express the answer in units of feet/degree.

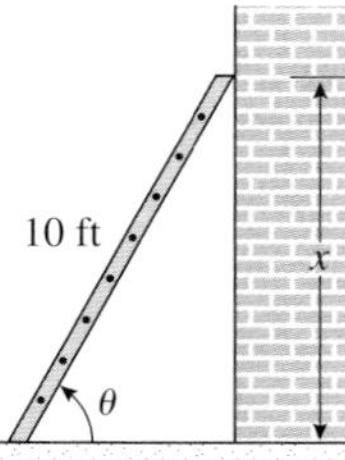

Figure Ex-31

32. An airplane is flying on a horizontal path at a height of 3800 ft, as shown in the accompanying figure. At what rate is the distance s between the airplane and the fixed point P changing with respect to θ when $\theta = 30°$? Express the answer in units of feet/degree.

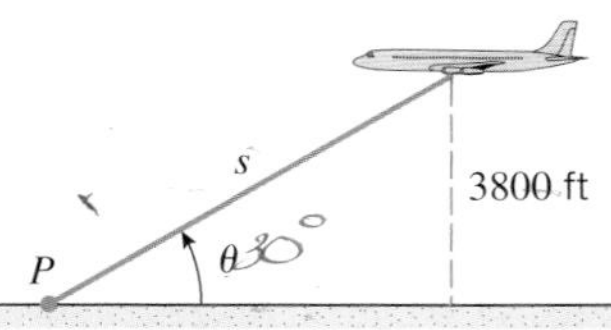

Figure Ex-32

soh cah toa
sin 30° = 3800/s
θ = C/B
s=30

33. A searchlight is trained on the side of a tall building. As the light rotates, the spot it illuminates moves up and down the side of the building. That is, the distance D between ground level and the illuminated spot on the side of the building is a function of the angle θ formed by the light beam and the horizontal (see the accompanying figure). If the searchlight is located 50 m from the building, find the rate at which D is changing with respect to θ when $\theta = 45°$. Express your answer in units of meters/degree.

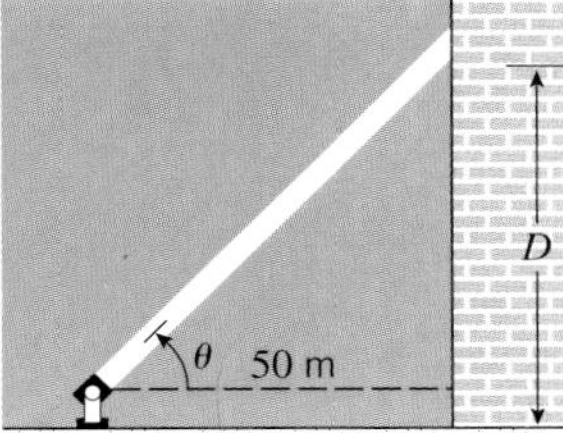

Figure Ex-33

34. An Earth-observing satellite can see only a portion of the Earth's surface. The satellite has horizon sensors that can detect the angle θ shown in the accompanying figure. Let r be the radius of the Earth (assumed spherical) and h the distance of the satellite from the Earth's surface.
(a) Show that $h = r(\csc\theta - 1)$.
(b) Using $r = 6378$ km, find the rate at which h is changing with respect to θ when $\theta = 30°$. Express the answer in units of kilometers/degree.

Source: Adapted from *Space Mathematics*, NASA, 1985.

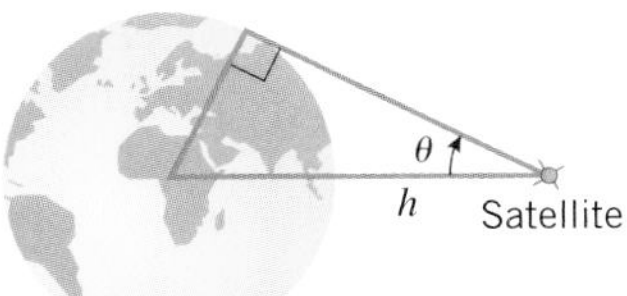

Figure Ex-34

35–36 Make a conjecture about the derivative by calculating the first few derivatives and observing the resulting pattern.

35. (a) $\dfrac{d^{87}}{dx^{87}}[\sin x]$ (b) $\dfrac{d^{100}}{dx^{100}}[\cos x]$

36. $\dfrac{d^{17}}{dx^{17}}[x \sin x]$

37. Let $f(x) = \cos x$. Find all positive integers n for which $f^{(n)}(x) = \sin x$.

38. Let $f(x) = \sin x$. Find all positive integers n for which $f^{(n)}(x) = \sin x$.

FOCUS ON CONCEPTS

39. In each part, determine where f is differentiable.
(a) $f(x) = \sin x$ (b) $f(x) = \cos x$
(c) $f(x) = \tan x$ (d) $f(x) = \cot x$
(e) $f(x) = \sec x$ (f) $f(x) = \csc x$
(g) $f(x) = \dfrac{1}{1 + \cos x}$ (h) $f(x) = \dfrac{1}{\sin x \cos x}$
(i) $f(x) = \dfrac{\cos x}{2 - \sin x}$

40. (a) Derive Formula (4) using the definition of a derivative.
(b) Use Formulas (3) and (4) to obtain (7).
(c) Use Formula (4) to obtain (6).
(d) Use Formula (3) to obtain (8).

41. Use Formula (1), the alternative form for the definition of derivative given in Formula (13) of Section 3.2, that is,
$$f'(x) = \lim_{w \to x} \frac{f(w) - f(x)}{w - x}$$
and the difference identity
$$\sin\alpha - \sin\beta = 2\sin\left(\frac{\alpha - \beta}{2}\right)\cos\left(\frac{\alpha + \beta}{2}\right)$$
to show that $\dfrac{d}{dx}[\sin x] = \cos x$.

42. Follow the directions of Exercise 41 using the difference identity
$$\cos\alpha - \cos\beta = -2\sin\left(\frac{\alpha - \beta}{2}\right)\sin\left(\frac{\alpha + \beta}{2}\right)$$
to show that $\dfrac{d}{dx}[\cos x] = -\sin x$.

43. (a) Show that $\displaystyle\lim_{h \to 0} \frac{\tan h}{h} = 1$.
(b) Use the result in part (a) to help derive the formula for the derivative of $\tan x$ directly from the definition of a derivative.

44. Without using any trigonometric identities, find
$$\lim_{x \to 0} \frac{\tan(x + y) - \tan y}{x}$$
[*Hint:* Relate the given limit to the definition of the derivative of an appropriate function of y.]

45. Show that if k is a constant, then
$$\frac{d}{dx}[\cos kx] = -k \sin kx$$

46. A spring with an attached mass is stretched 4 cm below its rest position and released at time $t = 0$. After release, the mass moves along a vertical axis, for which we take the positive direction to be up and the origin to match the position of the top of the mass at its rest position. Assume that the position function of the top of the mass is $s = -4\cos \pi t$, where s is in centimeters and t is in seconds after the release of the mass. Use the result of Exercise 45 to find the velocity function for the mass. What is the velocity of the mass on its initial pass through the rest position?

47. The derivative formulas for $\sin x$, $\cos x$, $\tan x$, $\cot x$, $\sec x$, and $\csc x$ were obtained under the assumption that x is measured in radians. Using the results of Exercises 51 and 52 of Section 2.6, prove that if x is measured in degrees, then

(a) $\frac{d}{dx}[\sin x] = \frac{\pi}{180}\cos x$

(b) $\frac{d}{dx}[\cos x] = -\frac{\pi}{180}\sin x$

48. Let us agree to call the functions $\cos x$, $\cot x$, and $\csc x$ the *cofunctions* of $\sin x$, $\tan x$, and $\sec x$, respectively. Convince yourself that the derivative of any cofunction can be obtained from the derivative of the corresponding function by introducing a minus sign and replacing each function in the derivative by its cofunction. Memorize the derivatives of $\sin x$, $\tan x$, and $\sec x$ and then use the above observation to deduce the derivatives of the cofunctions.

✔QUICK CHECK ANSWERS 3.5

1. (a) $\cos x$ (b) $-\sin x$ (c) $\sec^2 x$ (d) $\sec x \tan x$ **2.** $f'(x) = \cos^2 x - \sin^2 x$, $f'(\pi/3) = -\frac{1}{2}$

3. (a) $\left.\frac{d}{dx}[\sin x]\right|_{x=\pi/2} = 0$ (b) $\frac{d}{dx}[\csc x] = -\csc x \cot x$

3.6 THE CHAIN RULE

In this section we will derive a formula that expresses the derivative of a composition $f \circ g$ in terms of the derivatives of f and g. This formula will enable us to differentiate complicated functions using known derivatives of simpler functions.

DERIVATIVES OF COMPOSITIONS

Consider the differentiation problem

$$\frac{d}{dx}[(x^2+1)^{100}] \tag{1}$$

Two ways we can perform this differentiation with the tools available so far are to expand the function $(x^2+1)^{100}$ by the binomial formula and differentiate term by term or to apply the derivative definition. Either way the computations are prohibitive, so we need to find a new approach. Let $h(x) = (x^2+1)^{100}$. Our strategy will be to write h as a composition of simpler functions that we can differentiate easily and then express (1) in terms of the derivatives of the simpler functions. For example, let us express $h(x)$ as the composition

$$h(x) = (f \circ g)(x) = f(g(x)) \quad \text{where} \quad g(x) = x^2+1, \quad f(x) = x^{100}$$

Since we know that

$$g'(x) = 2x \quad \text{and} \quad f'(x) = 100x^{99}$$

what we need is some way to express $h'(x)$ in terms of these known derivatives. The key to doing this is to introduce the dependent variables

$$y = (x^2+1)^{100} \quad \text{and} \quad u = g(x) = x^2+1$$

from which it follows

$$y = u^{100}$$

Thus, we want to use the known derivatives

$$\frac{dy}{du} = 100u^{99} \quad \text{and} \quad \frac{du}{dx} = 2x \tag{2}$$

to find the unknown derivative

$$\frac{dy}{dx} = \frac{d}{dx}[(x^2+1)^{100}] \tag{3}$$

To do this, let us think of the derivatives in (2) and (3) as rates of change. Thus, we are interested in using the known rates of change dy/du and du/dx to find the unknown rate

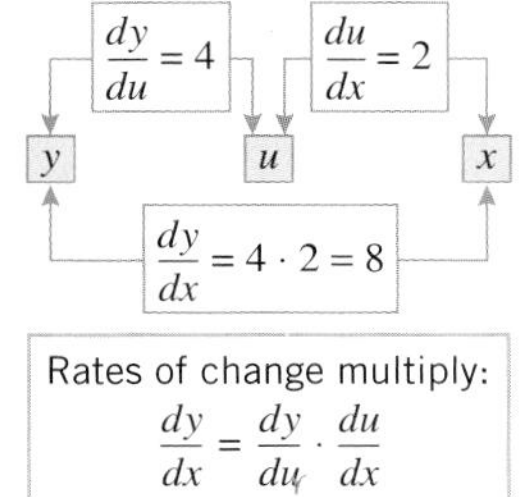

Figure 3.6.1

of change dy/dx. But intuition suggests that rates of change multiply. For example, if y changes 4 times as fast as u and u changes 2 times as fast as x, then y changes $4 \times 2 = 8$ times as fast as x (Figure 3.6.1). This suggests that

$$\frac{dy}{dx} = \frac{dy}{du} \cdot \frac{du}{dx} = 100u^{99} \cdot 2x = 100(x^2+1)^{99} \cdot 2x = 200x(x^2+1)^{99}$$

Note that this is a more efficient process for performing the differentiation in (1) than resorting to the tedious binomial expansion.

The following theorem, the proof of which is given in Appendix C, formalizes the preceding ideas.

> The name "chain rule" is appropriate because the desired derivative is obtained by a two-link "chain" of simpler derivatives.

3.6.1 THEOREM (*The Chain Rule*). *If g is differentiable at x and f is differentiable at $g(x)$, then the composition $f \circ g$ is differentiable at x. Moreover, if*

$$y = f(g(x)) \quad \text{and} \quad u = g(x)$$

then $y = f(u)$ and

$$\frac{dy}{dx} = \frac{dy}{du} \cdot \frac{du}{dx} \tag{4}$$

> Formula (4) is easy to remember because the left side is exactly what results if we "cancel" the du's on the right side. This "canceling" device provides a good way of deducing the correct form of the chain rule when different variables are used. For example, if w is a function of x and x is a function of t, then the chain rule takes the form
>
> $$\frac{dw}{dt} = \frac{dw}{dx} \cdot \frac{dx}{dt}$$

▶ **Example 1** Find dy/dx if $y = \cos(x^3)$.

Solution. Let $u = x^3$ and express y as $y = \cos u$. Applying Formula (4) yields

$$\begin{aligned}\frac{dy}{dx} &= \frac{dy}{du} \cdot \frac{du}{dx} \\ &= \frac{d}{du}[\cos u] \cdot \frac{d}{dx}[x^3] \\ &= (-\sin u) \cdot (3x^2) \\ &= (-\sin(x^3)) \cdot (3x^2) = -3x^2 \sin(x^3) \blacktriangleleft\end{aligned}$$

▶ **Example 2** Find dw/dt if $w = \tan x$ and $x = 4t^3 + t$.

Solution. In this case the chain rule computations take the form

$$\begin{aligned}\frac{dw}{dt} &= \frac{dw}{dx} \cdot \frac{dx}{dt} \\ &= \frac{d}{dx}[\tan x] \cdot \frac{d}{dt}[4t^3 + t] \\ &= (\sec^2 x) \cdot (12t^2 + 1) \\ &= (\sec^2(4t^3 + t)) \cdot (12t^2 + 1) = (12t^2 + 1)\sec^2(4t^3 + t) \blacktriangleleft\end{aligned}$$

AN ALTERNATIVE VERSION OF THE CHAIN RULE

Formula (4) for the chain rule can be unwieldy in some problems because it involves so many variables. As you become more comfortable with the chain rule, you may want to dispense with writing out the dependent variables by expressing (4) in the form

> Confirm that (5) is an alternative version of (4) by letting $y = f(g(x))$ and $u = g(x)$.

$$\frac{d}{dx}[f(g(x))] = (f \circ g)'(x) = f'(g(x))g'(x) \tag{5}$$

A convenient way to remember this formula is to call f the "outside function" and g the "inside function" in the composition $f(g(x))$ and then express (5) in words as:

The derivative of $f(g(x))$ is the derivative of the outside function evaluated at the inside function times the derivative of the inside function.

$$\frac{d}{dx}[f(g(x))] = \underbrace{f'(g(x))}\cdot\underbrace{g'(x)}$$

Derivative of the outside function evaluated at the inside function

Derivative of the inside function

▶ **Example 3** (*Example 1 revisited*) Find $h'(x)$ if $h(x) = \cos(x^3)$.

Solution. We can think of h as a composition $f(g(x))$ in which $g(x) = x^3$ is the inside function and $f(x) = \cos x$ is the outside function. Thus, Formula (5) yields

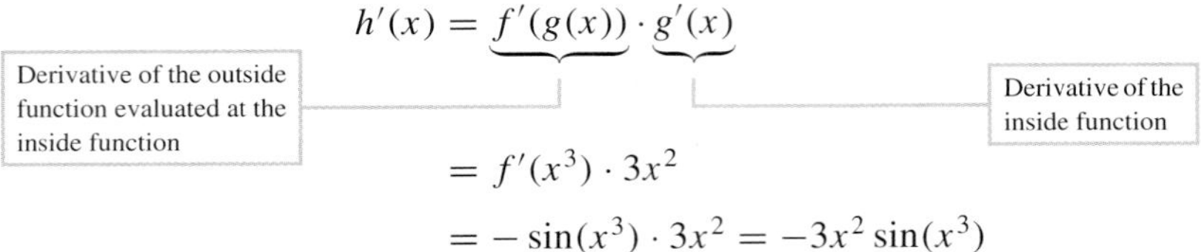

$$\begin{aligned} h'(x) &= \underbrace{f'(g(x))}\cdot\underbrace{g'(x)} \\ &= f'(x^3)\cdot 3x^2 \\ &= -\sin(x^3)\cdot 3x^2 = -3x^2\sin(x^3) \end{aligned}$$

which agrees with the result obtained in Example 1. ◀

▶ **Example 4**

$$\frac{d}{dx}[\tan^2 x] = \frac{d}{dx}[(\tan x)^2] = \underbrace{(2\tan x)}\cdot\underbrace{(\sec^2 x)} = 2\tan x\sec^2 x$$

Derivative of the outside function evaluated at the inside function

Derivative of the inside function

$$\frac{d}{dx}[\sqrt{x^2+1}] = \underbrace{\frac{1}{2\sqrt{x^2+1}}}\cdot\underbrace{2x} = \frac{x}{\sqrt{x^2+1}}$$

See Formula (5) of Section 3.3. ◀

Derivative of the outside function evaluated at the inside function

Derivative of the inside function

GENERALIZED DERIVATIVE FORMULAS

There is a useful third variation of the chain rule that strikes a middle ground between Formulas (4) and (5). If we let $u = g(x)$ in (5), then we can rewrite that formula as

$$\frac{d}{dx}[f(u)] = f'(u)\frac{du}{dx} \tag{6}$$

This result, called the ***generalized derivative formula*** for f, provides a way of using the derivative of $f(x)$ to produce the derivative of $f(u)$, where u is a function of x. Table 3.6.1 gives some examples of this formula.

Table 3.6.1

GENERALIZED DERIVATIVE FORMULAS	
$\frac{d}{dx}[u^n] = nu^{n-1}\frac{du}{dx}$ (n an integer)	$\frac{d}{dx}[\sqrt{u}] = \frac{1}{2\sqrt{u}}\frac{du}{dx}$
$\frac{d}{dx}[\sin u] = \cos u\frac{du}{dx}$	$\frac{d}{dx}[\cos u] = -\sin u\frac{du}{dx}$
$\frac{d}{dx}[\tan u] = \sec^2 u\frac{du}{dx}$	$\frac{d}{dx}[\cot u] = -\csc^2 u\frac{du}{dx}$
$\frac{d}{dx}[\sec u] = \sec u \tan u\frac{du}{dx}$	$\frac{d}{dx}[\csc u] = -\csc u \cot u\frac{du}{dx}$

▶ **Example 5** Find

(a) $\frac{d}{dx}[\sin(2x)]$ (b) $\frac{d}{dx}[\tan(x^2+1)]$ (c) $\frac{d}{dx}[\sqrt{x^3+\csc x}]$

(d) $\frac{d}{dx}[(1+x^5\cot x)^{-8}]$ (e) $\frac{d}{dx}\left[\frac{1}{x^3+2x-3}\right]$

Solution (a). Taking $u = 2x$ in the generalized derivative formula for $\sin u$ yields

$$\frac{d}{dx}[\sin(2x)] = \frac{d}{dx}[\sin u] = \cos u\frac{du}{dx} = \cos 2x \cdot \frac{d}{dx}[2x] = \cos 2x \cdot 2 = 2\cos 2x$$

Solution (b). Taking $u = x^2+1$ in the generalized derivative formula for $\tan u$ yields

$$\begin{aligned}\frac{d}{dx}[\tan(x^2+1)] &= \frac{d}{dx}[\tan u] = \sec^2 u\frac{du}{dx}\\ &= \sec^2(x^2+1)\cdot\frac{d}{dx}[x^2+1] = \sec^2(x^2+1)\cdot 2x\\ &= 2x\sec^2(x^2+1)\end{aligned}$$

Solution (c). Taking $u = x^3+\csc x$ in the generalized derivative formula for $\sqrt{u}$ yields

$$\begin{aligned}\frac{d}{dx}[\sqrt{x^3+\csc x}] &= \frac{d}{dx}[\sqrt{u}] = \frac{1}{2\sqrt{u}}\frac{du}{dx} = \frac{1}{2\sqrt{x^3+\csc x}}\cdot\frac{d}{dx}[x^3+\csc x]\\ &= \frac{1}{2\sqrt{x^3+\csc x}}\cdot(3x^2-\csc x\cot x) = \frac{3x^2-\csc x\cot x}{2\sqrt{x^3+\csc x}}\end{aligned}$$

Solution (d). Taking $u = 1+x^5\cot x$ in the generalized derivative formula for u^{-8} yields

$$\begin{aligned}\frac{d}{dx}[(1+x^5\cot x)^{-8}] &= \frac{d}{dx}[u^{-8}] = -8u^{-9}\frac{du}{dx}\\ &= -8(1+x^5\cot x)^{-9}\cdot\frac{d}{dx}[1+x^5\cot x]\\ &= -8(1+x^5\cot x)^{-9}\cdot(x^5(-\csc^2 x)+5x^4\cot x)\\ &= (8x^5\csc^2 x-40x^4\cot x)(1+x^5\cot x)^{-9}\end{aligned}$$

Solution (e). Taking $u = x^3 + 2x - 3$ in the generalized derivative formula for u^{-1} yields

$$\begin{aligned}\frac{d}{dx}\left[\frac{1}{x^3+2x-3}\right] &= \frac{d}{dx}[(x^3+2x-3)^{-1}] = \frac{d}{dx}[u^{-1}] \\ &= -u^{-2}\frac{du}{dx} = -(x^3+2x-3)^{-2}\frac{d}{dx}[x^3+2x-3] \\ &= -(x^3+2x-3)^{-2}(3x^2+2) = -\frac{3x^2+2}{(x^3+2x-3)^2} \quad \blacktriangleleft\end{aligned}$$

Sometimes you will have to make adjustments in notation or apply the chain rule more than once to calculate a derivative.

▶ Example 6 Find

(a) $\dfrac{d}{dx}[\sin(\sqrt{1+\cos x})]$ (b) $\dfrac{d\mu}{dt}$ if $\mu = \sec\sqrt{\omega t}$ (ω constant)

Solution (a). Taking $u = \sqrt{1+\cos x}$ in the generalized derivative formula for $\sin u$ yields

$$\begin{aligned}\frac{d}{dx}[\sin(\sqrt{1+\cos x})] &= \frac{d}{dx}[\sin u] = \cos u\frac{du}{dx} \\ &= \cos(\sqrt{1+\cos x}) \cdot \frac{d}{dx}[\sqrt{1+\cos x}] \\ &= \cos(\sqrt{1+\cos x}) \cdot \frac{-\sin x}{2\sqrt{1+\cos x}} \\ &= -\frac{\sin x\cos(\sqrt{1+\cos x})}{2\sqrt{1+\cos x}}\end{aligned}$$

We used the generalized derivative formula for $\sqrt{u}$ with $u = 1 + \cos x$.

Solution (b).

$$\begin{aligned}\frac{d\mu}{dt} = \frac{d}{dt}[\sec\sqrt{\omega t}] &= \sec\sqrt{\omega t}\tan\sqrt{\omega t}\frac{d}{dt}[\sqrt{\omega t}] \\ &= \sec\sqrt{\omega t}\tan\sqrt{\omega t}\frac{\omega}{2\sqrt{\omega t}} \quad \blacktriangleleft\end{aligned}$$

We used the generalized derivative formula for $\sec u$ with $u = \sqrt{\omega t}$.

We used the generalized derivative formula for $\sqrt{u}$ with $u = \omega t$.

DIFFERENTIATING USING COMPUTER ALGEBRA SYSTEMS

Even with the chain rule and other differentiation rules, some derivative computations can be tedious to perform. For complicated derivatives, engineers and scientists often use computer algebra systems such as *Mathematica*, *Maple*, or *Derive*. For example, although we have all the mathematical tools to compute

$$\frac{d}{dx}\left[\frac{(x^2+1)^{10}\sin^3(\sqrt{x})}{\sqrt{1+\csc x}}\right] \tag{7}$$

by hand, the computation is sufficiently involved that it may be more efficient (and less error-prone) to use a computer algebra system.

TECHNOLOGY MASTERY

If you have a CAS, use it to perform the differentiation in (7).

✔ QUICK CHECK EXERCISES 3.6

(See page 198 for answers.)

1. The chain rule states that the derivative of the composition of two functions is the derivative of the ______ function evaluated at the ______ function times the derivative of the ______ function.
2. If y is a differentiable function of u, and u is a differentiable function of x, then
$$\frac{dy}{dx} = \underline{\qquad} \cdot \underline{\qquad}$$
3. Find dy/dx.
   (a) $y = (x^2 + 5)^{10}$ (b) $y = \sqrt{1 + 6x}$
4. Find dy/dx.
   (a) $y = \sin(3x + 2)$ (b) $y = (x^2 \tan x)^4$
5. Suppose that $f(2) = 3$, $f'(2) = 4$, $g(3) = 6$, and $g'(3) = -5$. Evaluate
   (a) $h'(2)$, where $h(x) = g(f(x))$
   (b) $k'(3)$, where $k(x) = f\left(\frac{1}{3}g(x)\right)$.

EXERCISE SET 3.6

Graphing Utility C CAS

1. Given that $f'(0) = 2$, $g(0) = 0$, and $g'(0) = 3$, find $(f \circ g)'(0)$.
2. Given that $f'(9) = 5$, $g(2) = 9$, and $g'(2) = -3$, find $(f \circ g)'(2)$.
3. Let $f(x) = x^5$ and $g(x) = 2x - 3$.
   (a) Find $(f \circ g)(x)$ and $(f \circ g)'(x)$.
   (b) Find $(g \circ f)(x)$ and $(g \circ f)'(x)$.
4. Let $f(x) = 5\sqrt{x}$ and $g(x) = 4 + \cos x$.
   (a) Find $(f \circ g)(x)$ and $(f \circ g)'(x)$.
   (b) Find $(g \circ f)(x)$ and $(g \circ f)'(x)$.

FOCUS ON CONCEPTS

5. Given the following table of values, find the indicated derivatives in parts (a) and (b).

x	$f(x)$	$f'(x)$	$g(x)$	$g'(x)$
3	5	–2	5	7
5	3	–1	12	4

   (a) $F'(3)$, where $F(x) = f(g(x))$
   (b) $G'(3)$, where $G(x) = g(f(x))$
6. Given the following table of values, find the indicated derivatives in parts (a) and (b).

x	$f(x)$	$f'(x)$	$g(x)$	$g'(x)$
–1	2	3	2	–3
2	0	4	1	–5

   (a) $F'(-1)$, where $F(x) = f(g(x))$
   (b) $G'(-1)$, where $G(x) = g(f(x))$

7–26 Find $f'(x)$.

7. $f(x) = (x^3 + 2x)^{37}$
8. $f(x) = (3x^2 + 2x - 1)^6$
9. $f(x) = \left(x^3 - \dfrac{7}{x}\right)^{-2}$
10. $f(x) = \dfrac{1}{(x^5 - x + 1)^9}$
11. $f(x) = \dfrac{4}{(3x^2 - 2x + 1)^3}$
12. $f(x) = \sqrt{x^3 - 2x + 5}$
13. $f(x) = \sqrt{4 + \sqrt{3x}}$
14. $f(x) = \sqrt[4]{x} \quad (= \sqrt{\sqrt{x}})$
15. $f(x) = \sin\left(\dfrac{1}{x^2}\right)$
16. $f(x) = \tan\sqrt{x}$
17. $f(x) = 4\cos^5 x$
18. $f(x) = 4x + 5\sin^4 x$
19. $f(x) = \cos^2(3\sqrt{x})$
20. $f(x) = \tan^4(x^3)$
21. $f(x) = 2\sec^2(x^7)$
22. $f(x) = \cos^3\left(\dfrac{x}{x + 1}\right)$
23. $f(x) = \sqrt{\cos(5x)}$
24. $f(x) = \sqrt{3x - \sin^2(4x)}$
25. $f(x) = [x + \csc(x^3 + 3)]^{-3}$
26. $f(x) = [x^4 - \sec(4x^2 - 2)]^{-4}$

27–40 Find dy/dx.

27. $y = x^3 \sin^2(5x)$
28. $y = \sqrt{x}\tan^3(\sqrt{x})$
29. $y = x^5 \sec(1/x)$
30. $y = \dfrac{\sin x}{\sec(3x + 1)}$
31. $y = \cos(\cos x)$
32. $y = \sin(\tan 3x)$
33. $y = \cos^3(\sin 2x)$
34. $y = \dfrac{1 + \csc(x^2)}{1 - \cot(x^2)}$
35. $y = (5x + 8)^7 \left(1 - \sqrt{x}\right)^6$
36. $y = (x^2 + x)^5 \sin^8 x$
37. $y = \left(\dfrac{x - 5}{2x + 1}\right)^3$
38. $y = \left(\dfrac{1 + x^2}{1 - x^2}\right)^{17}$
39. $y = \dfrac{(2x + 3)^3}{(4x^2 - 1)^8}$
40. $y = [1 + \sin^3(x^5)]^{12}$

41–42 Use a CAS to find dy/dx.

C 41. $y = [x \sin 2x + \tan^4(x^7)]^5$

C 42. $y = \tan^4\left(2 + \dfrac{(7 - x)\sqrt{3x^2 + 5}}{x^3 + \sin x}\right)$

43–50 Find an equation for the tangent line to the graph at the specified value of x.

43. $y = x\cos 3x,\ x = \pi$

44. $y = \sin(1 + x^3),\ x = -3$

45. $y = \sec^3\left(\frac{\pi}{2} - x\right),\ x = -\frac{\pi}{2}$

46. $y = \left(x - \frac{1}{x}\right)^3,\ x = 2$

47. $y = \tan(4x^2),\ x = \sqrt{\pi}$

48. $y = 3\cot^4 x,\ x = \frac{\pi}{4}$

49. $y = x^2\sqrt{5 - x^2},\ x = 1$

50. $y = \frac{x}{\sqrt{1 - x^2}},\ x = 0$

51–54 Find d^2y/dx^2.

51. $y = x\cos(5x) - \sin^2 x$

52. $y = \sin(3x^2)$

53. $y = \frac{1 + x}{1 - x}$

54. $y = x\tan\left(\frac{1}{x}\right)$

55–58 Find the indicated derivative.

55. $y = \cot^3(\pi - \theta)$; find $\frac{dy}{d\theta}$.

56. $\lambda = \left(\frac{au + b}{cu + d}\right)^6$; find $\frac{d\lambda}{du}$ $(a, b, c, d$ constants).

57. $\frac{d}{d\omega}[a\cos^2 \pi\omega + b\sin^2 \pi\omega]$ $(a, b$ constants)

58. $x = \csc^2\left(\frac{\pi}{3} - y\right)$; find $\frac{dx}{dy}$.

59. (a) Use a graphing utility to obtain the graph of the function $f(x) = x\sqrt{4 - x^2}$.
(b) Use the graph in part (a) to make a rough sketch of the graph of f'.
(c) Find $f'(x)$, and then check your work in part (b) by using the graphing utility to obtain the graph of f'.
(d) Find the equation of the tangent line to the graph of f at $x = 1$, and graph f and the tangent line together.

60. (a) Use a graphing utility to obtain the graph of the function $f(x) = \sin x^2 \cos x$ over the interval $[-\pi/2, \pi/2]$.
(b) Use the graph in part (a) to make a rough sketch of the graph of f' over the interval.
(c) Find $f'(x)$, and then check your work in part (b) by using the graphing utility to obtain the graph of f' over the interval.
(d) Find the equation of the tangent line to the graph of f at $x = 1$, and graph f and the tangent line together over the interval.

61. If an object suspended from a spring is displaced vertically from its equilibrium position by a small amount and released, and if the air resistance and the mass of the spring are ignored, then the resulting oscillation of the object is called ***simple harmonic motion***. Under appropriate conditions the displacement y from equilibrium in terms of time t is given by

$$y = A\cos\omega t$$

where A is the initial displacement at time $t = 0$, and ω is a constant that depends on the mass of the object and the stiffness of the spring (see the accompanying figure). The constant $|A|$ is called the ***amplitude*** of the motion and ω the ***angular frequency***.
(a) Show that

$$\frac{d^2y}{dt^2} = -\omega^2 y$$

(b) The ***period*** T is the time required to make one complete oscillation. Show that $T = 2\pi/\omega$.
(c) The ***frequency*** f of the vibration is the number of oscillations per unit time. Find f in terms of the period T.
(d) Find the amplitude, period, and frequency of an object that is executing simple harmonic motion given by $y = 0.6\cos 15t$, where t is in seconds and y is in centimeters.

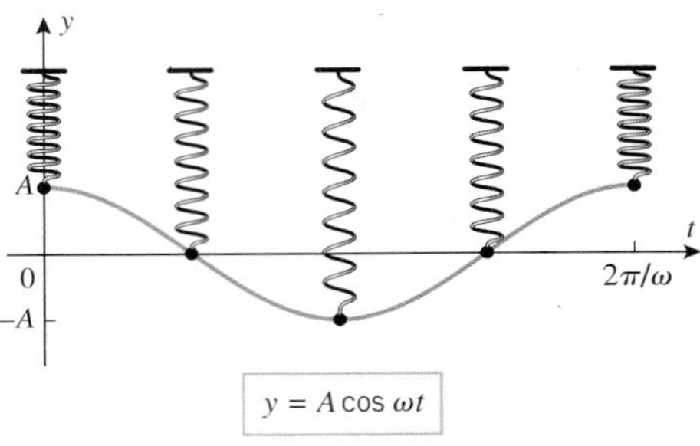

Figure Ex-61

62. Find the value of the constant A so that $y = A\sin 3t$ satisfies the equation

$$\frac{d^2y}{dt^2} + 2y = 4\sin 3t$$

FOCUS ON CONCEPTS

63. Consider the function f graphed in the accompanying figure. Evaluate

$$\frac{d}{dx}\left[\sqrt{x + f(x)}\right]\Big|_{x=-1}$$

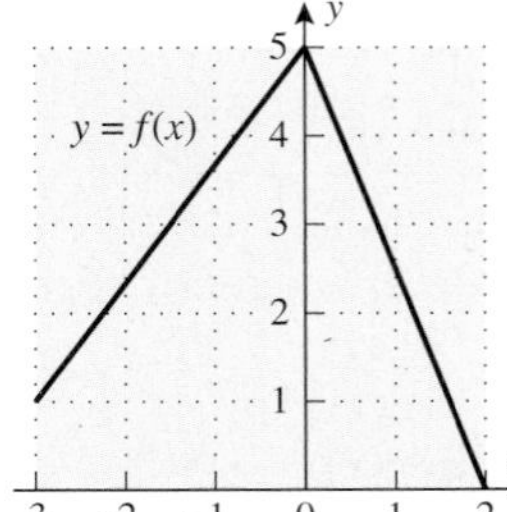

Figure Ex-63

64. Using the function f in Exercise 63, evaluate

$$\frac{d}{dx}[f(2\sin x)]\Big|_{x=\pi/6}$$

65. The accompanying figure shows the graph of atmospheric pressure p (lb/in^2) versus the altitude h (mi) above sea level.
(a) From the graph and the tangent line at $h = 2$ shown on the graph, estimate the values of p and dp/dh at an altitude of 2 mi.
(b) If the altitude of a space vehicle is increasing at the rate of 0.3 mi/s at the instant when it is 2 mi above sea level, how fast is the pressure changing with time at this instant?

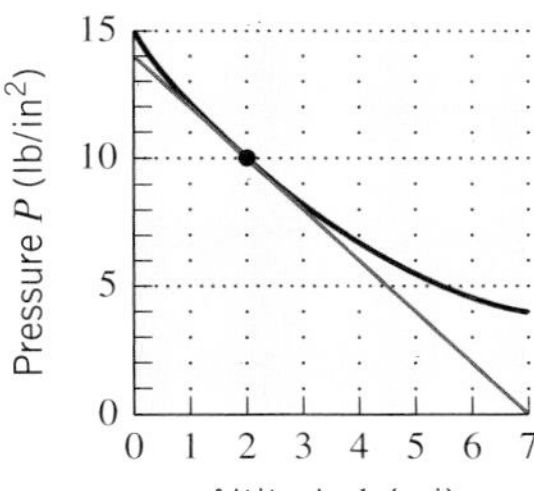

Figure Ex-65

66. The force F (in pounds) acting at an angle θ with the horizontal that is needed to drag a crate weighing W pounds along a horizontal surface at a constant velocity is given by

$$F = \frac{\mu W}{\cos\theta + \mu\sin\theta}$$

where μ is a constant called the ***coefficient of sliding friction*** between the crate and the surface (see the accompanying figure). Suppose that the crate weighs 150 lb and that $\mu = 0.3$.
(a) Find $dF/d\theta$ when $\theta = 30°$. Express the answer in units of pounds/degree.
(b) Find dF/dt when $\theta = 30°$ if θ is decreasing at the rate of 0.5°/s at this instant.

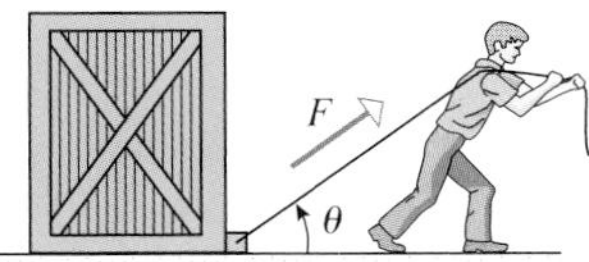

Figure Ex-66

67. Recall that

$$\frac{d}{dx}(|x|) = \begin{cases} 1, & x > 0 \\ -1, & x < 0 \end{cases}$$

Use this result and the chain rule to find

$$\frac{d}{dx}(|\sin x|)$$

for nonzero x in the interval $(-\pi, \pi)$.

68. Use the derivative formula for $\sin x$ and the identity

$$\cos x = \sin\left(\frac{\pi}{2} - x\right)$$

to obtain the derivative formula for $\cos x$.

69. Let

$$f(x) = \begin{cases} x\sin\dfrac{1}{x}, & x \neq 0 \\ 0, & x = 0 \end{cases}$$

(a) Show that f is continuous at $x = 0$.
(b) Use Definition 3.2.1 to show that $f'(0)$ does not exist.
(c) Find $f'(x)$ for $x \neq 0$.
(d) Determine whether $\lim_{x\to 0} f'(x)$ exists.

70. Let

$$f(x) = \begin{cases} x^2\sin\dfrac{1}{x}, & x \neq 0 \\ 0, & x = 0 \end{cases}$$

(a) Show that f is continuous at $x = 0$.
(b) Use Definition 3.2.1 to find $f'(0)$.
(c) Find $f'(x)$ for $x \neq 0$.
(d) Show that f' is not continuous at $x = 0$.

71. Given the following table of values, find the indicated derivatives in parts (a) and (b).

x	$f(x)$	$f'(x)$
2	1	7
8	5	–3

(a) $g'(2)$, where $g(x) = [f(x)]^3$
(b) $h'(2)$, where $h(x) = f(x^3)$

72. Given that $f'(x) = \sqrt{3x+4}$ and $g(x) = x^2 - 1$, find $F'(x)$ if $F(x) = f(g(x))$.

73. Given that $f'(x) = \dfrac{x}{x^2+1}$ and $g(x) = \sqrt{3x-1}$, find $F'(x)$ if $F(x) = f(g(x))$.

74. Find $f'(x^2)$ if $\dfrac{d}{dx}[f(x^2)] = x^2$.

75. Find $\dfrac{d}{dx}[f(x)]$ if $\dfrac{d}{dx}[f(3x)] = 6x$.

76. Recall that a function f is ***even*** if $f(-x) = f(x)$ and ***odd*** if $f(-x) = -f(x)$, for all x in the domain of f. Assuming that f is differentiable, prove:
(a) f' is odd if f is even
(b) f' is even if f is odd.

77. Draw some pictures to illustrate the results in Exercise 76, and write a paragraph that gives an informal explanation of why the results are true.

78. Let $y = f_1(u)$, $u = f_2(v)$, $v = f_3(w)$, and $w = f_4(x)$. Express dy/dx in terms of dy/du, dw/dx, du/dv, and dv/dw.

79. Find a formula for

$$\frac{d}{dx}[f(g(h(x)))]$$

✔ QUICK CHECK ANSWERS 3.6

1. outside; inside; inside 2. $\dfrac{dy}{du} \cdot \dfrac{du}{dx}$ 3. (a) $10(x^2+5)^9 \cdot 2x = 20x(x^2+5)^9$ (b) $\dfrac{1}{2\sqrt{1+6x}} \cdot 6 = \dfrac{3}{\sqrt{1+6x}}$

4. (a) $3\cos(3x+2)$ (b) $4(x^2 \tan x)^3(2x \tan x + x^2 \sec^2 x)$ 5. (a) $g'(f(2))f'(2) = -20$ (b) $f'\left(\frac{1}{3}g(3)\right) \cdot \frac{1}{3}g'(3) = -\frac{20}{3}$

3.7 IMPLICIT DIFFERENTIATION

Up to now we have been concerned with differentiating functions that are given by equations of the form $y = f(x)$. In this section we will consider methods for differentiating functions for which it is inconvenient or impossible to express them in this form.

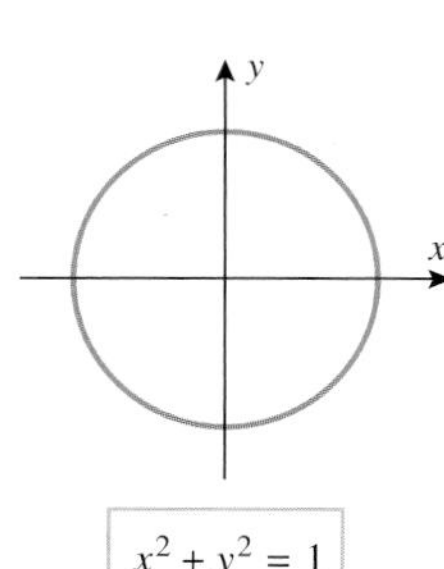

$x^2 + y^2 = 1$

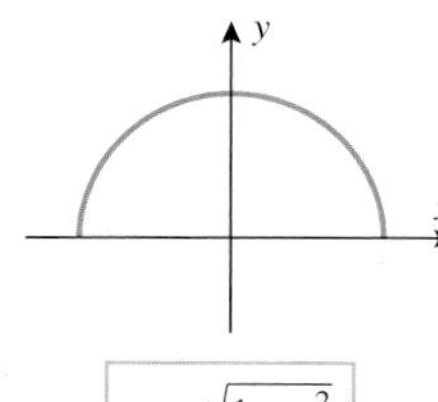

$y = \sqrt{1-x^2}$

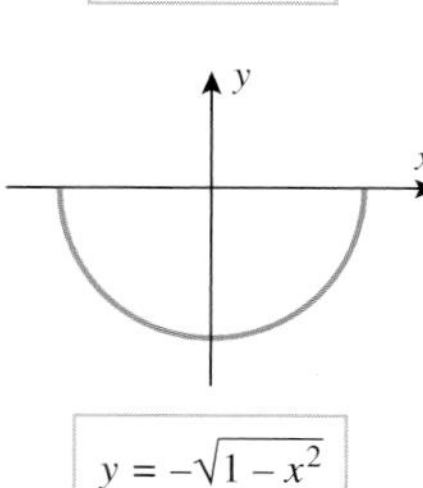

$y = -\sqrt{1-x^2}$

Figure 3.7.1

■ FUNCTIONS DEFINED EXPLICITLY AND IMPLICITLY

An equation of the form $y = f(x)$ is said to define y ***explicitly*** as a function of x because the variable y appears alone on one side of the equation. However, sometimes functions are defined by equations in which y is not alone on one side; for example, the equation

$$yx + y + 1 = x \tag{1}$$

is not of the form $y = f(x)$, but it still defines y as a function of x since it can be rewritten as

$$y = \frac{x-1}{x+1}$$

Thus, we say that (1) defines y ***implicitly*** as a function of x, the function being

$$f(x) = \frac{x-1}{x+1}$$

An equation in x and y can implicitly define more than one function of x. This can occur when the graph of the equation fails the vertical line test, so it is not the graph of a function of x. For example, if we solve the equation of the circle

$$x^2 + y^2 = 1 \tag{2}$$

for y in terms of x, we obtain $y = \pm\sqrt{1-x^2}$, so we have found two functions that are defined implicitly by (2), namely,

$$f_1(x) = \sqrt{1-x^2} \quad \text{and} \quad f_2(x) = -\sqrt{1-x^2} \tag{3}$$

The graphs of these functions are the upper and lower semicircles of the circle $x^2 + y^2 = 1$ (Figure 3.7.1). This leads us to the following definition.

3.7.1 DEFINITION. We will say that a given equation in x and y defines the function f ***implicitly*** if the graph of $y = f(x)$ coincides with a portion of the graph of the equation.

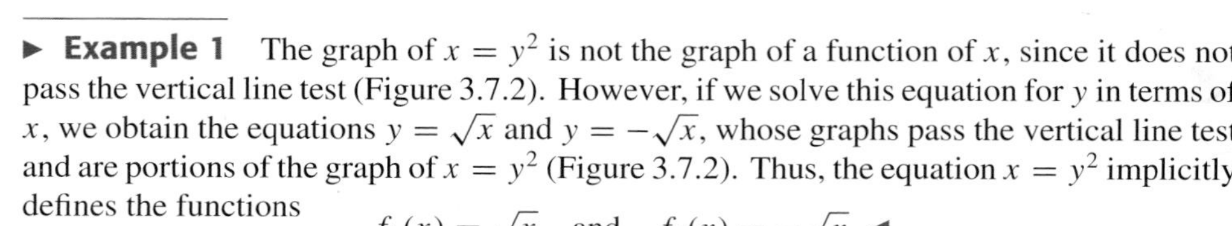

► **Example 1** The graph of $x = y^2$ is not the graph of a function of x, since it does not pass the vertical line test (Figure 3.7.2). However, if we solve this equation for y in terms of x, we obtain the equations $y = \sqrt{x}$ and $y = -\sqrt{x}$, whose graphs pass the vertical line test and are portions of the graph of $x = y^2$ (Figure 3.7.2). Thus, the equation $x = y^2$ implicitly defines the functions

$$f_1(x) = \sqrt{x} \quad \text{and} \quad f_2(x) = -\sqrt{x} \blacktriangleleft$$

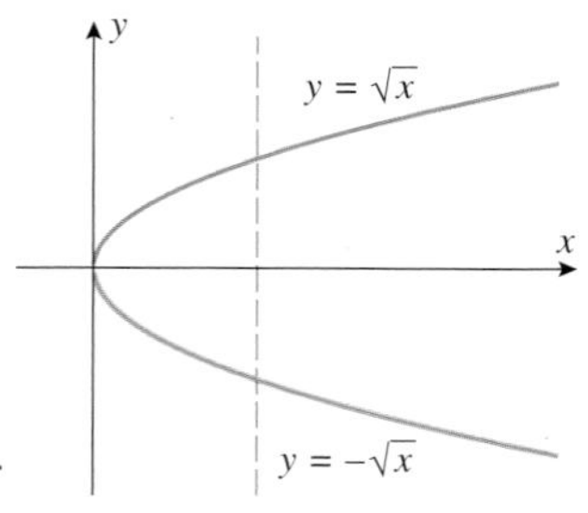

Figure 3.7.2 The graph of $x = y^2$ does not pass the vertical line test, but the graphs of $y = \sqrt{x}$ and $y = -\sqrt{x}$ do.

Although it was a trivial matter in the last example to solve the equation $x = y^2$ for y in terms of x, it is difficult or impossible to do this for some equations. For example, the equation

$$x^3 + y^3 = 3xy \tag{4}$$

can be solved for y in terms of x, but the resulting formulas are too complicated to be practical. Other equations, such as $\sin(xy) = y$, cannot be solved for y by any elementary method. Thus, even though an equation may define one or more functions of x, it may not be possible or practical to find explicit formulas for those functions.

Fortunately, CAS programs, such as *Mathematica* and *Maple*, have "implicit plotting" capabilities that can graph equations such as (4). The graph of this equation, which is called the ***Folium of Descartes***, is shown in Figure 3.7.3*a*. Parts (*b*) and (*c*) of the figure show the graphs (in blue) of two functions that are defined implicitly by (4).

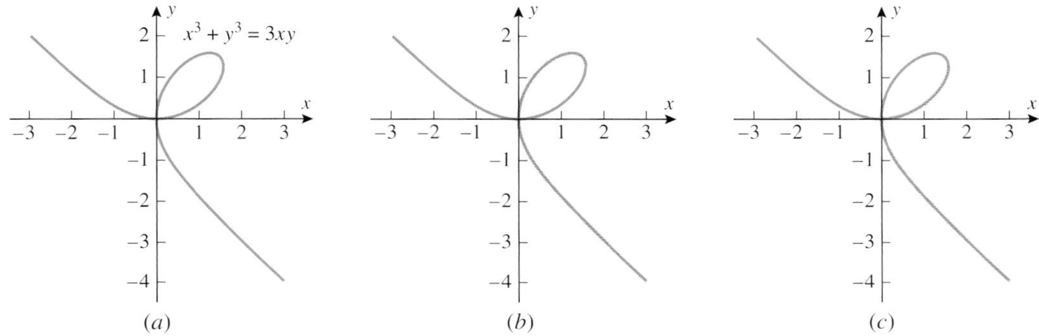

Figure 3.7.3

IMPLICIT DIFFERENTIATION

In general, it is not necessary to solve an equation for y in terms of x in order to differentiate the functions defined implicitly by the equation. To illustrate this, let us consider the simple equation

$$xy = 1 \tag{5}$$

One way to find dy/dx is to rewrite this equation as

$$y = \frac{1}{x} \tag{6}$$

from which it follows that

$$\frac{dy}{dx} = -\frac{1}{x^2} \tag{7}$$

However, there is another way to obtain this derivative. We can differentiate both sides of (5) *before* solving for y in terms of x, treating y as a (temporarily unspecified) differentiable

René Descartes (1596–1650) Descartes, a French aristocrat, was the son of a government official. He graduated from the University of Poitiers with a law degree at age 20. After a brief probe into the pleasures of Paris he became a military engineer, first for the Dutch Prince of Nassau and then for the German Duke of Bavaria. It was during his service as a soldier that Descartes began to pursue mathematics seriously and develop his analytic geometry. After the wars, he returned to Paris where he stalked the city as an eccentric, wearing a sword in his belt and a plumed hat. He lived in leisure, seldom arose before 11 A.M., and dabbled in the study of human physiology, philosophy, glaciers, meteors, and rainbows. He eventually moved to Holland, where he published his *Discourse on the Method*, and finally to Sweden where he died while serving as tutor to Queen Christina. Descartes is regarded as a genius of the first magnitude. In addition to major contributions in mathematics and philosophy he is considered, along with William Harvey, to be a founder of modern physiology.

function of x. With this approach we obtain

$$\frac{d}{dx}[xy] = \frac{d}{dx}[1]$$

$$x\frac{d}{dx}[y] + y\frac{d}{dx}[x] = 0$$

$$x\frac{dy}{dx} + y = 0$$

$$\frac{dy}{dx} = -\frac{y}{x}$$

If we now substitute (6) into the last expression, we obtain

$$\frac{dy}{dx} = -\frac{1}{x^2}$$

which agrees with Equation (7). This method of obtaining derivatives is called ***implicit differentiation***.

► **Example 2** Use implicit differentiation to find dy/dx if $5y^2 + \sin y = x^2$.

$$\frac{d}{dx}[5y^2 + \sin y] = \frac{d}{dx}[x^2]$$

$$5\frac{d}{dx}[y^2] + \frac{d}{dx}[\sin y] = 2x$$

$$5\left(2y\frac{dy}{dx}\right) + (\cos y)\frac{dy}{dx} = 2x$$

The chain rule was used here because y is a function of x.

$$10y\frac{dy}{dx} + (\cos y)\frac{dy}{dx} = 2x$$

Solving for dy/dx we obtain

$$\frac{dy}{dx} = \frac{2x}{10y + \cos y} \tag{8}$$

Note that this formula involves both x and y. In order to obtain a formula for dy/dx that involves x alone, we would have to solve the original equation for y in terms of x and then substitute in (8). However, it is impossible to do this, so we are forced to leave the formula for dy/dx in terms of x and y. ◄

► **Example 3** Use implicit differentiation to find d^2y/dx^2 if $4x^2 - 2y^2 = 9$.

Solution. Differentiating both sides of $4x^2 - 2y^2 = 9$ implicitly yields

$$8x - 4y\frac{dy}{dx} = 0$$

from which we obtain

$$\frac{dy}{dx} = \frac{2x}{y} \tag{9}$$

Differentiating both sides of (9) implicitly yields

$$\frac{d^2y}{dx^2} = \frac{(y)(2) - (2x)(dy/dx)}{y^2} \tag{10}$$

Substituting (9) into (10) and simplifying using the original equation, we obtain

$$\frac{d^2y}{dx^2} = \frac{2y - 2x(2x/y)}{y^2} = \frac{2y^2 - 4x^2}{y^3} = -\frac{9}{y^3}$$ ◄

In Examples 2 and 3, the resulting formulas for dy/dx involved both x and y. Although it is usually more desirable to have the formula for dy/dx expressed in terms of x alone, having the formula in terms of x and y is not an impediment to finding slopes and equations of tangent lines provided the x- and y-coordinates of the point of tangency are known. This is illustrated in the following example.

▶ **Example 4** Find the slopes of the tangent lines to the curve $y^2 - x + 1 = 0$ at the points $(2, -1)$ and $(2, 1)$.

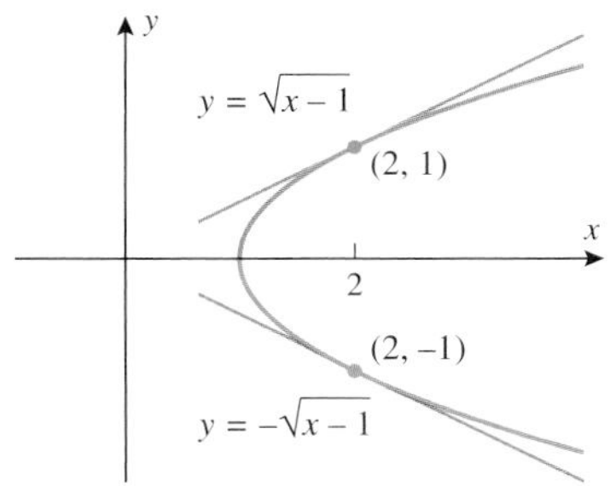

Figure 3.7.4

Solution. We could proceed by solving the equation for y in terms of x, and then evaluating the derivative of $y = \sqrt{x-1}$ at $(2, 1)$ and the derivative of $y = -\sqrt{x-1}$ at $(2, -1)$ (Figure 3.7.4). However, implicit differentiation is more efficient since it can be used for the slopes of *both* tangent lines. Differentiating implicitly yields

$$\frac{d}{dx}[y^2 - x + 1] = \frac{d}{dx}[0]$$

$$\frac{d}{dx}[y^2] - \frac{d}{dx}[x] + \frac{d}{dx}[1] = \frac{d}{dx}[0]$$

$$2y\frac{dy}{dx} - 1 = 0$$

$$\frac{dy}{dx} = \frac{1}{2y}$$

At $(2, -1)$ we have $y = -1$, and at $(2, 1)$ we have $y = 1$, so the slopes of the tangent lines to the curve at those points are

$$\left.\frac{dy}{dx}\right|_{\substack{x=2\\y=-1}} = -\frac{1}{2} \quad \text{and} \quad \left.\frac{dy}{dx}\right|_{\substack{x=2\\y=1}} = \frac{1}{2}$$ ◀

▶ **Example 5**

(a) Use implicit differentiation to find dy/dx for the Folium of Descartes $x^3 + y^3 = 3xy$.

(b) Find an equation for the tangent line to the Folium of Descartes at the point $\left(\frac{3}{2}, \frac{3}{2}\right)$.

(c) At what point(s) in the first quadrant is the tangent line to the Folium of Descartes horizontal?

Solution (a). Differentiating both sides of the given equation implicitly yields

$$\frac{d}{dx}[x^3 + y^3] = \frac{d}{dx}[3xy]$$

$$3x^2 + 3y^2\frac{dy}{dx} = 3x\frac{dy}{dx} + 3y$$

$$x^2 + y^2\frac{dy}{dx} = x\frac{dy}{dx} + y$$

$$(y^2 - x)\frac{dy}{dx} = y - x^2$$

$$\frac{dy}{dx} = \frac{y - x^2}{y^2 - x} \tag{11}$$

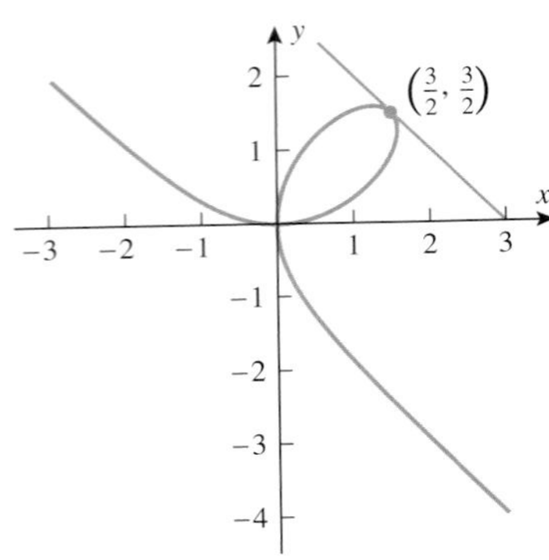

Figure 3.7.5

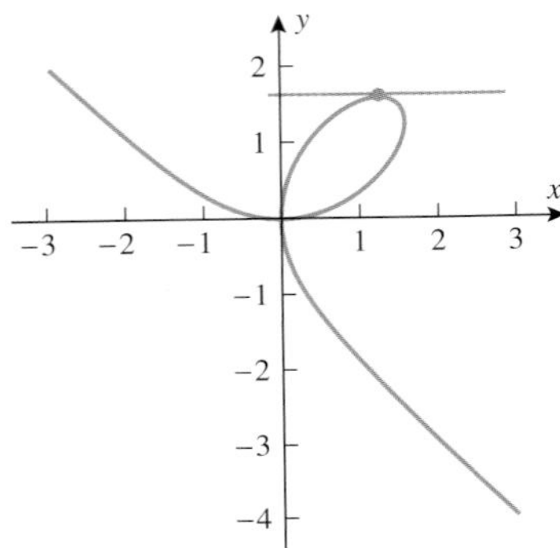

Figure 3.7.6

Solution (b). At the point $(\frac{3}{2}, \frac{3}{2})$, we have $x = \frac{3}{2}$ and $y = \frac{3}{2}$, so from (11) the slope $m_{\tan}$ of the tangent line at this point is

$$m_{\tan} = \left.\frac{dy}{dx}\right|_{\substack{x=3/2 \\ y=3/2}} = \frac{(3/2) - (3/2)^2}{(3/2)^2 - (3/2)} = -1$$

Thus, the equation of the tangent line at the point $(\frac{3}{2}, \frac{3}{2})$ is

$$y - \tfrac{3}{2} = -1\left(x - \tfrac{3}{2}\right) \quad \text{or} \quad x + y = 3$$

which is consistent with Figure 3.7.5.

Solution (c). The tangent line is horizontal at the points where $dy/dx = 0$, and from (11) this occurs only where $y - x^2 = 0$ or

$$y = x^2 \tag{12}$$

Substituting this expression for y in the equation $x^3 + y^3 = 3xy$ for the curve yields

$$x^3 + (x^2)^3 = 3x^3$$
$$x^6 - 2x^3 = 0$$
$$x^3(x^3 - 2) = 0$$

whose solutions are $x = 0$ and $x = 2^{1/3}$. From (12), the solutions $x = 0$ and $x = 2^{1/3}$ yield the points $(0, 0)$ and $(2^{1/3}, 2^{2/3})$, respectively. Of these two, only $(2^{1/3}, 2^{2/3})$ is in the first quadrant. Substituting $x = 2^{1/3}$, $y = 2^{2/3}$ into (11) yields

$$\left.\frac{dy}{dx}\right|_{\substack{x=2^{1/3} \\ y=2^{2/3}}} = \frac{0}{2^{4/3} - 2^{2/3}} = 0$$

We conclude that $(2^{1/3}, 2^{2/3}) \approx (1.26, 1.59)$ is the only point on the Folium of Descartes in the first quadrant at which the tangent line is horizontal (Figure 3.7.6). ◀

DIFFERENTIABILITY OF FUNCTIONS DEFINED IMPLICITLY

When differentiating implicitly, it is assumed that y represents a differentiable function of x. If this is not so, then the resulting calculations may be nonsense. For example, if we differentiate the equation

$$x^2 + y^2 + 1 = 0 \tag{13}$$

we obtain

$$2x + 2y\frac{dy}{dx} = 0 \quad \text{or} \quad \frac{dy}{dx} = -\frac{x}{y}$$

However, this derivative is meaningless because there are no real values of x and y that satisfy (13) (why?). This tells us that (13) has no real graph and hence certainly does not define any functions implicitly.

The nonsensical conclusion of these computations conveys the importance of knowing whether an equation in x and y that is to be differentiated implicitly actually defines some differentiable function of x implicitly. Unfortunately, this can be a difficult problem, so we will leave the discussion of such matters for more advanced courses in analysis.

> Formula (11) cannot be evaluated at (0, 0) and hence provides no information about the nature of the Folium of Descartes at the origin. Based on the graphs in Figure 3.7.3, what can you say about the differentiability of the implicitly defined functions graphed in blue in parts (*b*) and (*c*) of the figure?

DERIVATIVES OF RATIONAL POWERS OF *x*

In Theorem 3.3.3 and the discussion immediately following it, we showed that the formula

$$\frac{d}{dx}[x^n] = nx^{n-1} \tag{14}$$

holds for integer values of n and for $n = \frac{1}{2}$. We will now use implicit differentiation to show that this formula holds for any rational exponent. More precisely, we will show that if r is a rational number, then

$$\frac{d}{dx}[x^r] = rx^{r-1} \tag{15}$$

wherever x^r and x^{r-1} are defined. For now, we will assume without proof that x^r is differentiable; the justification for this will be considered later.

Let $y = x^r$. Since r is a rational number, it can be expressed as a ratio of integers $r = m/n$. Thus, $y = x^r = x^{m/n}$ can be written as

$$y^n = x^m \quad \text{so that} \quad \frac{d}{dx}[y^n] = \frac{d}{dx}[x^m]$$

By differentiating implicitly with respect to x and using (14), we obtain

$$ny^{n-1}\frac{dy}{dx} = mx^{m-1} \tag{16}$$

But

$$y^{n-1} = [x^{m/n}]^{n-1} = x^{m-(m/n)}$$

Thus, (16) can be written as

$$nx^{m-(m/n)}\frac{dy}{dx} = mx^{m-1}$$

so that

$$\frac{dy}{dx} = \frac{m}{n}x^{(m/n)-1} = rx^{r-1}$$

which establishes (15).

▶ **Example 6** From (15)

$$\frac{d}{dx}[x^{4/5}] = \frac{4}{5}x^{(4/5)-1} = \frac{4}{5}x^{-1/5}$$

$$\frac{d}{dx}[x^{-7/8}] = -\frac{7}{8}x^{(-7/8)-1} = -\frac{7}{8}x^{-15/8}$$

$$\frac{d}{dx}[\sqrt[3]{x}] = \frac{d}{dx}[x^{1/3}] = \frac{1}{3}x^{-2/3} = \frac{1}{3\sqrt[3]{x^2}}$$ ◀

If u is a differentiable function of x, and r is a rational number, then the chain rule yields the following generalization of (15):

$$\frac{d}{dx}[u^r] = ru^{r-1} \cdot \frac{du}{dx} \tag{17}$$

▶ **Example 7**

$$\frac{d}{dx}[x^2 - x + 2]^{3/4} = \frac{3}{4}(x^2 - x + 2)^{-1/4} \cdot \frac{d}{dx}[x^2 - x + 2]$$

$$= \frac{3}{4}(x^2 - x + 2)^{-1/4}(2x - 1)$$

$$\frac{d}{dx}[(\sec \pi x)^{-4/5}] = -\frac{4}{5}(\sec \pi x)^{-9/5} \cdot \frac{d}{dx}[\sec \pi x]$$

$$= -\frac{4}{5}(\sec \pi x)^{-9/5} \cdot \sec \pi x \tan \pi x \cdot \pi$$

$$= -\frac{4\pi}{5}(\sec \pi x)^{-4/5} \tan \pi x$$ ◀

✔ QUICK CHECK EXERCISES 3.7

(See page 206 for answers.)

1. Find dy/dx.
   (a) $y = x^{2/3}$ (b) $y = \dfrac{1}{\sqrt[3]{x^4}}$
   (c) $y = \sqrt[3]{4x-1}$ (d) $y = (\tan^3 x)^{4/5}$
2. The equation $xy^3 = 1$ defines y implicitly as a function of x.
   (a) Differentiating the equation implicitly yields $dy/dx =$ ______.
   (b) Confirm that the result in part (a) agrees with the result obtained by solving the given equation for y as a function of x and then differentiating.
3. Find dy/dx by implicit differentiation.
   (a) $x^2 - y^3 = xy$ (b) $\sin x \cos y = \tan y$
4. The equation of the tangent line to the graph of $x + y + xy = 3$ at the point $(1, 1)$ is ______.
5. Use implicit differentiation to find d^2y/dx^2 for $\sin y = x$.

EXERCISE SET 3.7

C CAS

1–8 Find dy/dx.

1. $y = \sqrt[3]{2x-5}$
2. $y = \sqrt[3]{2 + \tan(x^2)}$
3. $y = \left(\dfrac{x+1}{x-2}\right)^{2/3}$
4. $y = \sqrt{\dfrac{x^2+1}{x^2-5}}$
5. $y = x^3(5x^2+1)^{-2/3}$
6. $y = \dfrac{\sqrt[3]{2x-1}}{x}$
7. $y = [\sin(3/x)]^{5/2}$
8. $y = [\cos(x^3)]^{-1/2}$

9–10 (a) Find dy/dx by differentiating implicitly. (b) Solve the equation for y as a function of x, and find dy/dx from that equation. (c) Confirm that the two results are consistent by expressing the derivative in part (a) as a function of x alone.

9. $x + xy - 2x^3 = 2$
10. $\sqrt{y} - \sin x = 2$

11–20 Find dy/dx by implicit differentiation.

11. $x^2 + y^2 = 100$
12. $x^3 + y^3 = 3xy^2$
13. $x^2y + 3xy^3 - x = 3$
14. $x^3y^2 - 5x^2y + x = 1$
15. $\dfrac{1}{\sqrt{x}} + \dfrac{1}{\sqrt{y}} = 1$
16. $x^2 = \dfrac{x+y}{x-y}$
17. $\sin(x^2y^2) = x$
18. $\cos(xy^2) = y$
19. $\tan^3(xy^2 + y) = x$
20. $\dfrac{xy^3}{1 + \sec y} = 1 + y^4$

21–26 Find d^2y/dx^2 by implicit differentiation.

21. $2x^2 - 3y^2 = 4$
22. $x^3 + y^3 = 1$
23. $x^3y^3 - 4 = 0$
24. $xy + y^2 = 2$
25. $y + \sin y = x$
26. $x \cos y = y$

27–28 Find the slope of the tangent line to the curve at the given points in two ways: first by solving for y in terms of x and differentiating and then by implicit differentiation.

27. $x^2 + y^2 = 1$; $(1/2, \sqrt{3}/2)$, $(1/2, -\sqrt{3}/2)$
28. $y^2 - x + 1 = 0$; $(10, 3)$, $(10, -3)$

29–32 Use implicit differentiation to find the slope of the tangent line to the curve at the specified point, and check that your answer is consistent with the accompanying graph.

29. $x^4 + y^4 = 16$; $(1, \sqrt[4]{15})$ [***Lamé's special quartic***]
30. $y^3 + yx^2 + x^2 - 3y^2 = 0$; $(0, 3)$ [***trisectrix***]
31. $2(x^2 + y^2)^2 = 25(x^2 - y^2)$; $(3, 1)$ [***lemniscate***]
32. $x^{2/3} + y^{2/3} = 4$; $(-1, 3\sqrt{3})$ [***four-cusped hypocycloid***]

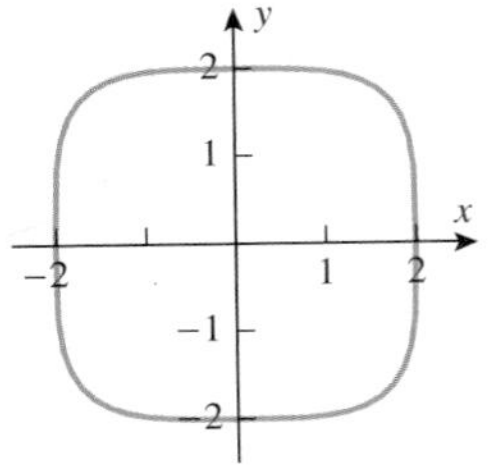

Figure Ex-29

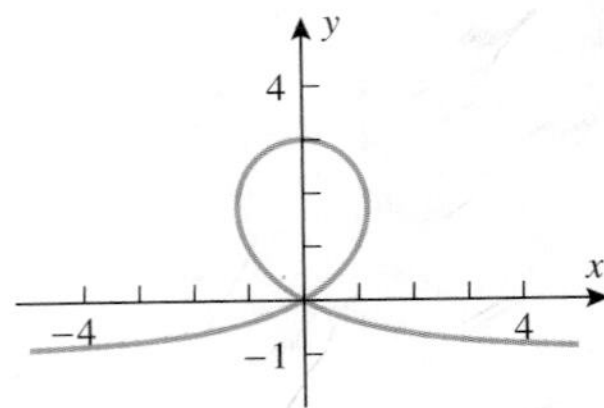

Figure Ex-30

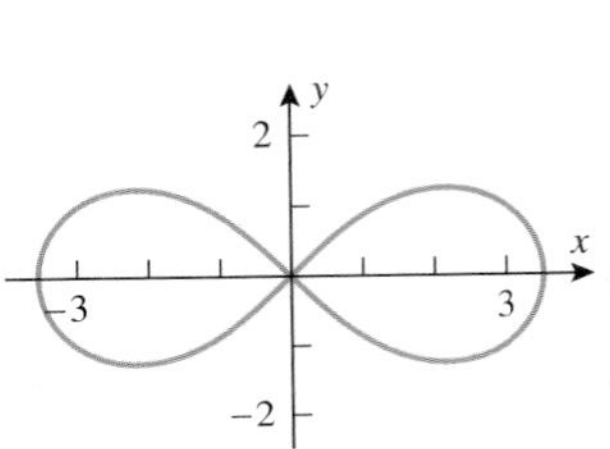

Figure Ex-31

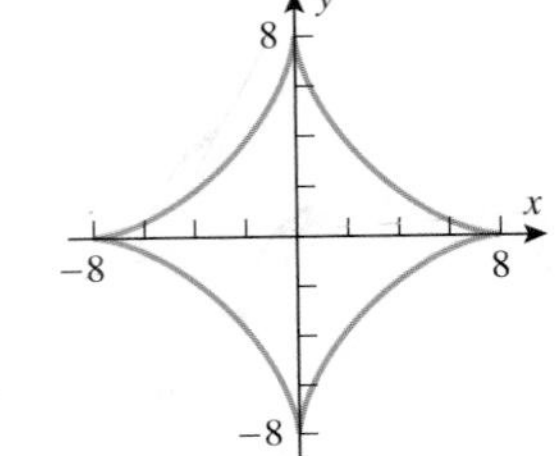

Figure Ex-32

33–36 Use implicit differentiation to find the specified derivative.

33. $a^4 - t^4 = 6a^2t$; da/dt **34.** $\sqrt{u} + \sqrt{v} = 5$; du/dv

35. $a^2\omega^2 + b^2\lambda^2 = 1$ (a, b constants); $d\omega/d\lambda$

36. $y = \sin x$; dx/dy

C **37.** (a) Use the implicit plotting capability of a CAS to graph the equation $y^4 + y^2 = x(x-1)$.
(b) Use implicit differentiation to help explain why the graph in part (a) has no horizontal tangent lines.
(c) Solve the equation $y^4 + y^2 = x(x-1)$ for x in terms of y and explain why the graph in part (a) consists of two parabolas.

C **38.** Curves with equations of the form $y^2 = x(x-a)(x-b)$, where $a < b$, are called ***bipartite cubics***.
(a) Use the implicit plotting capability of a CAS to graph the bipartite cubic $y^2 = x(x-1)(x-2)$.
(b) At what points does the curve in part (a) have a horizontal tangent line?
(c) Solve the equation in part (a) for y in terms of x, and use the result to explain why the graph consists of two separate parts (i.e., is *bipartite*).
(d) Graph the equation in part (a) without using the implicit plotting capability of the CAS.

39–40 These exercises deal with the rotated ellipse C whose equation is $x^2 - xy + y^2 = 4$.

C **39.** (a) Use the implicit plotting capability of a CAS to graph C.
(b) Use the graph to estimate the x-coordinates of all points at which the graph has a horizontal tangent line.
(c) Find the exact values for the x-coordinates in part (b).

C **40.** (a) Use the implicit plotting capability of a CAS to graph C.
(b) Use the graph to estimate the x-coordinates of all points at which the graph has a vertical tangent line.
(c) Find the exact values for the x-coordinates in part (b).

FOCUS ON CONCEPTS

41–42 These exercises deal with the rotated ellipse C whose equation is $x^2 - xy + y^2 = 4$.

41. Show that the line $y = x$ intersects C at two points P and Q and that the tangent lines to C at P and Q are parallel.

42. Prove that if $P(a, b)$ is a point on C, then so is $Q(-a, -b)$ and that the tangent lines to C through P and Q are parallel.

43. Find the values of a and b for the curve $x^2y + ay^2 = b$ if the point $(1, 1)$ is on its graph and the tangent line at $(1, 1)$ has the equation $4x + 3y = 7$.

44. At what point(s) is the tangent line to the curve $y^3 = 2x^2$ perpendicular to the line $x + 2y - 2 = 0$?

C **45.** (a) Use the implicit plotting capability of a CAS to graph the curve C whose equation is $x^3 - 2xy + y^3 = 0$.
(b) Use the graph in part (a) to estimate the x-coordinates of a point in the first quadrant that is on C and at which the tangent line to C is parallel to the x-axis.
(c) Find the exact value of the x-coordinate in part (b).

C **46.** (a) Use the implicit plotting capability of a CAS to graph the curve C whose equation is $x^3 - 2xy + y^3 = 0$.
(b) Use the graph to guess the coordinates of a point in the first quadrant that is on C and at which the tangent line to C is parallel to the line $y = -x$.
(c) Use implicit differentiation to verify your conjecture in part (b).

FOCUS ON CONCEPTS

47. (a) Use analytic geometry to find equations for two lines through the origin that are tangent to the circle $x^2 - 4x + y^2 + 3 = 0$.
(b) Use implicit differentiation to find the requisite lines in part (a).

48. Find equations for two lines through the origin that are tangent to the ellipse $2x^2 - 4x + y^2 + 1 = 0$.

49. Let a, b, and c be constants with $b \neq 0$, and let r denote a nonzero rational number. Use implicit differentiation to show that the tangent line to the curve $ax^r + by^r = c$ at (x_0, y_0) is $ax_0^{r-1}x + by_0^{r-1}y = c$.

50. Prove that for every nonzero rational number r, the tangent line to the graph of $x^r + y^r = 2$ at the point $(1, 1)$ has slope -1.

51. Find dy/dx if

$$2y^3t + t^3y = 1 \quad \text{and} \quad \frac{dt}{dx} = \frac{1}{\cos t}$$

52. (a) Show that $f(x) = x^{4/3}$ is differentiable at 0, but not twice differentiable at 0.
(b) Show that $f(x) = x^{7/3}$ is twice differentiable at 0, but not three times differentiable at 0.
(c) Find an exponent k such that $f(x) = x^k$ is $(n-1)$ times differentiable at 0, but not n times differentiable at 0.

53–54 Find all rational values of r such that $y = x^r$ satisfies the given equation.

53. $3x^2y'' + 4xy' - 2y = 0$ **54.** $16x^2y'' + 24xy' + y = 0$

55–56 Two curves are said to be ***orthogonal*** if their tangent lines are perpendicular at each point of intersection, and two families of curves are said to be ***orthogonal trajectories*** of one another if each member of one family is orthogonal to each member of the other family. This terminology is used in these exercises.

55. The accompanying figure (next page) shows some typical members of the families of circles $x^2 + (y-c)^2 = c^2$ (black curves) and $(x-k)^2 + y^2 = k^2$ (gray curves). Show that these families are orthogonal trajectories of one another. [*Hint:* For the tangent lines to be perpendicular at a point of intersection, the slopes of those tangent lines must be negative reciprocals of one another.]

56. The accompanying figure shows some typical members of the families of hyperbolas $xy = c$ (black curves) and $x^2 - y^2 = k$ (gray curves), where $c \neq 0$ and $k \neq 0$. Use the hint in Exercise 55 to show that these families are orthogonal trajectories of one another.

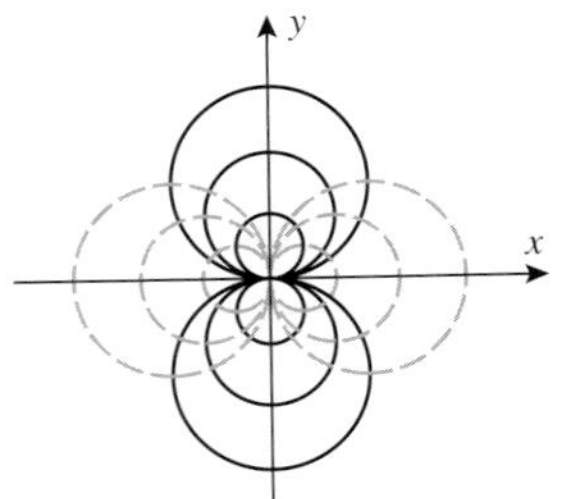

Figure Ex-55

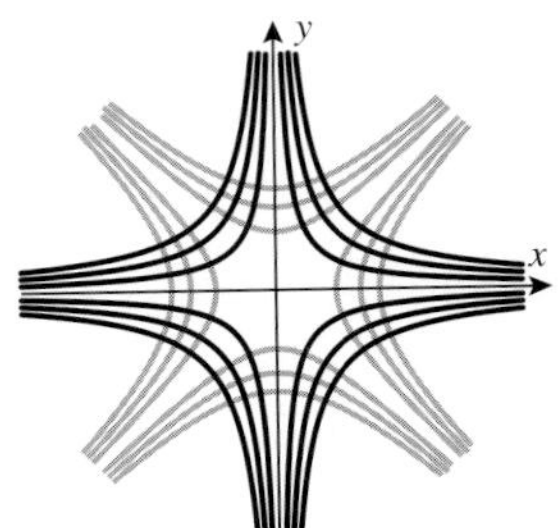

Figure Ex-56

✔ QUICK CHECK ANSWERS 3.7

1. (a) $\dfrac{dy}{dx} = \dfrac{2}{3}x^{-1/3}$ (b) $\dfrac{dy}{dx} = -\dfrac{4}{3}x^{-7/3}$ (c) $\dfrac{dy}{dx} = \dfrac{4}{3}(4x-1)^{-2/3}$ (d) $\dfrac{dy}{dx} = \dfrac{12}{5}(\tan x)^{7/5}\sec^2 x$

2. (a) $\dfrac{dy}{dx} = -\dfrac{y}{3x}$ (b) $\dfrac{dy}{dx} = -\dfrac{1}{3}x^{-4/3} = -\dfrac{x^{-1/3}}{3x} = -\dfrac{y}{3x}$ **3.** (a) $\dfrac{dy}{dx} = \dfrac{2x-y}{x+3y^2}$ (b) $\dfrac{dy}{dx} = \dfrac{\cos x \cos y}{\sec^2 y + \sin x \sin y}$

4. $y = 2 - x$ **5.** $\dfrac{d^2y}{dx^2} = \sec^2 y \tan y$

3.8 RELATED RATES

In this section we will study related rates problems. In such problems one tries to find the rate at which some quantity is changing by relating the quantity to other quantities whose rates of change are known.

DIFFERENTIATING EQUATIONS TO RELATE RATES

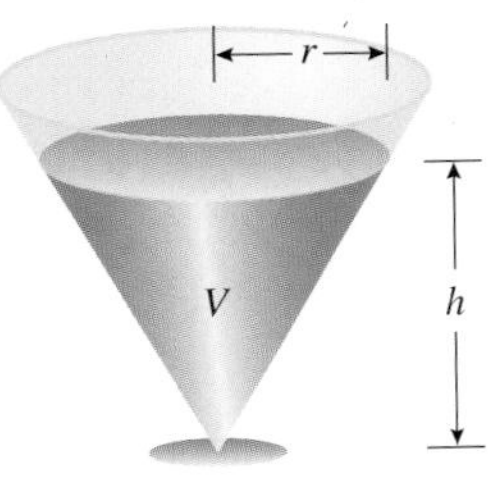

Figure 3.8.1 shows a liquid draining through a conical filter. As the liquid drains, its volume V, height h, and radius r are functions of the elapsed time t, and at each instant these variables are related by the equation

$$V = \frac{\pi}{3}r^2h$$

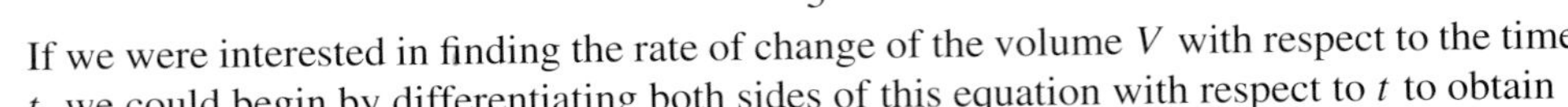

If we were interested in finding the rate of change of the volume V with respect to the time t, we could begin by differentiating both sides of this equation with respect to t to obtain

$$\frac{dV}{dt} = \frac{\pi}{3}\left[r^2\frac{dh}{dt} + h\left(2r\frac{dr}{dt}\right)\right] = \frac{\pi}{3}\left(r^2\frac{dh}{dt} + 2rh\frac{dr}{dt}\right)$$

Thus, to find dV/dt at a specific time t from this equation we would need to have values for r, h, dh/dt, and dr/dt at that time. This is called a ***related rates problem*** because the goal is to find an unknown rate of change by *relating* it to other variables whose values and whose rates of change at time t are known or can be found in some way. Let us begin with a simple example.

Figure 3.8.1

► **Example 1** Suppose that x and y are differentiable functions of t and are related by the equation $y = x^3$. Find dy/dt at time $t = 1$ if $x = 2$ and $dx/dt = 4$ at time $t = 1$.

Solution. Using the chain rule to differentiate both sides of the equation $y = x^3$ with respect to t yields

$$\frac{dy}{dt} = \frac{d}{dt}[x^3] = 3x^2\frac{dx}{dt}$$

Thus, the value of dy/dt at time $t = 1$ is

$$\left.\frac{dy}{dt}\right|_{t=1} = 3(2)^2 \left.\frac{dx}{dt}\right|_{t=1} = 12 \cdot 4 = 48 \blacktriangleleft$$

▶ **Example 2** Assume that oil spilled from a ruptured tanker spreads in a circular pattern whose radius increases at a constant rate of 2 ft/s. How fast is the area of the spill increasing when the radius of the spill is 60 ft?

Solution. Let

t = number of seconds elapsed from the time of the spill
r = radius of the spill in feet after t seconds
A = area of the spill in square feet after t seconds

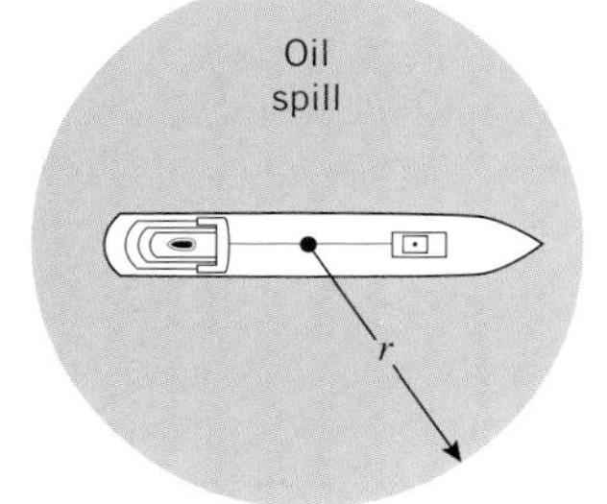

Figure 3.8.2

(Figure 3.8.2). We know the rate at which the radius is increasing, and we want to find the rate at which the area is increasing at the instant when $r = 60$; that is, we want to find

$$\left.\frac{dA}{dt}\right|_{r=60} \quad \text{given that} \quad \frac{dr}{dt} = 2 \text{ ft/s}$$

This suggests that we look for an equation relating A and r that we can differentiate with respect to t to produce a relationship between dA/dt and dr/dt. But A is the area of a circle of radius r, so

$$A = \pi r^2 \tag{1}$$

Differentiating both sides of (1) with respect to t yields

$$\frac{dA}{dt} = 2\pi r \frac{dr}{dt} \tag{2}$$

Thus, when $r = 60$ the area of the spill is increasing at the rate of

$$\left.\frac{dA}{dt}\right|_{r=60} = 2\pi(60)(2) = 240\pi \text{ ft}^2/\text{s} \approx 754 \text{ ft}^2/\text{s} \blacktriangleleft$$

With some minor variations, the method used in Example 2 can be used to solve a variety of related rates problems. We can break the method down into five steps.

A Strategy for Solving Related Rates Problems

Step 1. Assign letters to all quantities that vary with time and any others that seem relevant to the problem. Give a definition for each letter.

Step 2. Identify the rates of change that are known and the rate of change that is to be found. Interpret each rate as a derivative.

Step 3. Find an equation that relates the variables whose rates of change were identified in Step 2. To do this, it will often be helpful to draw an appropriately labeled figure that illustrates the relationship.

Step 4. Differentiate both sides of the equation obtained in Step 3 with respect to time to produce a relationship between the known rates of change and the unknown rate of change.

Step 5. *After* completing Step 4, substitute all known values for the rates of change and the variables, and then solve for the unknown rate of change.

WARNING

We have italicized the word "After" in Step 5 because it is a common error to substitute numerical values before performing the differentiation. For instance, in Example 2 had we substituted the known value of $r = 60$ in (1) before differentiating, we would have obtained $dA/dt = 0$, which is obviously incorrect.

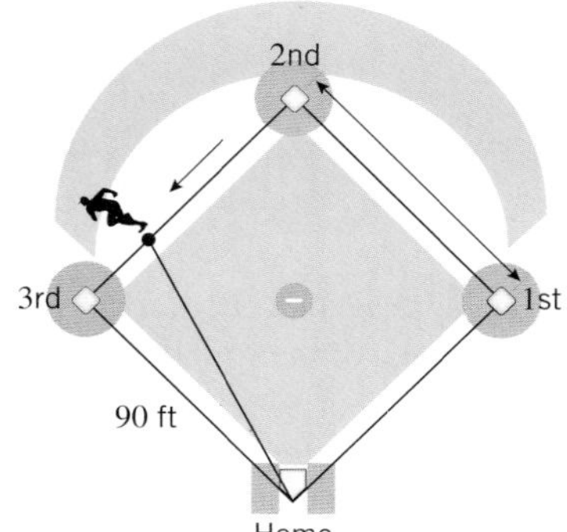

Figure 3.8.3

Example 3 A baseball diamond is a square whose sides are 90 ft long (Figure 3.8.3). Suppose that a player running from second base to third base has a speed of 30 ft/s at the instant when he or she is 20 ft from third base. At what rate is the player's distance from home plate changing at that instant?

Solution. We are given a constant speed with which the player is approaching third base, and we want to find the rate of change of the distance between the player and home plate at a particular instant. Thus, let

t = number of seconds since the player left second base
x = distance in feet from the player to third base
y = distance in feet from the player to home plate

Thus, we want to find

$$\left.\frac{dy}{dt}\right|_{x=20} \quad \text{given that} \quad \left.\frac{dx}{dt}\right|_{x=20} = -30 \text{ ft/s}$$

The quantity

$$\left.\frac{dx}{dt}\right|_{x=20}$$

is negative because x is decreasing with respect to t.

As suggested by Figure 3.8.4, an equation relating the variables x and y can be obtained using the Theorem of Pythagoras:

$$x^2 + 90^2 = y^2 \tag{3}$$

Differentiating both sides of this equation with respect to t yields

$$2x\frac{dx}{dt} = 2y\frac{dy}{dt}$$

from which we obtain

$$\frac{dy}{dt} = \frac{x}{y}\frac{dx}{dt} \tag{4}$$

When $x = 20$, it follows from (3) that

$$y = \sqrt{20^2 + 90^2} = \sqrt{8500} = 10\sqrt{85}$$

so that (4) yields

$$\left.\frac{dy}{dt}\right|_{x=20} = \frac{20}{10\sqrt{85}}(-30) = -\frac{60}{\sqrt{85}} \approx -6.51 \text{ ft/s}$$

The negative sign in the answer tells us that y is decreasing, which makes sense physically from Figure 3.8.4. ◀

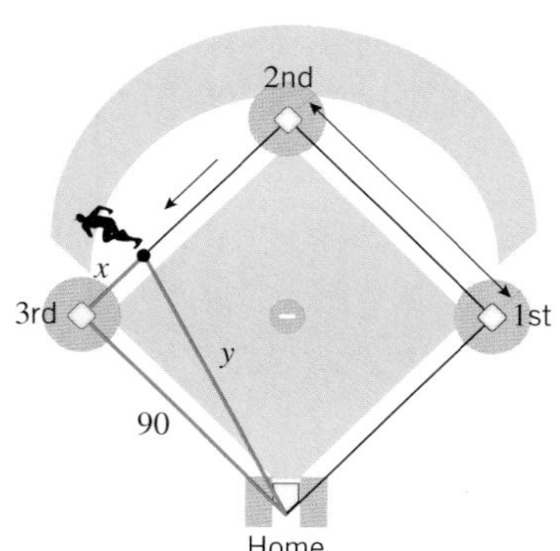

Figure 3.8.4

Example 4 In Figure 3.8.5 we have shown a camera mounted at a point 3000 ft from the base of a rocket launching pad. If the rocket is rising vertically at 880 ft/s when it is 4000 ft above the launching pad, how fast must the camera elevation angle change at that instant to keep the camera aimed at the rocket?

Solution. Let

t = number of seconds elapsed from the time of launch
ϕ = camera elevation angle in radians after t seconds
h = height of the rocket in feet after t seconds

(Figure 3.8.6). At each instant the rate at which the camera elevation angle must change is $d\phi/dt$, and the rate at which the rocket is rising is dh/dt. We want to find

$$\left.\frac{d\phi}{dt}\right|_{h=4000} \quad \text{given that} \quad \left.\frac{dh}{dt}\right|_{h=4000} = 880 \text{ ft/s}$$

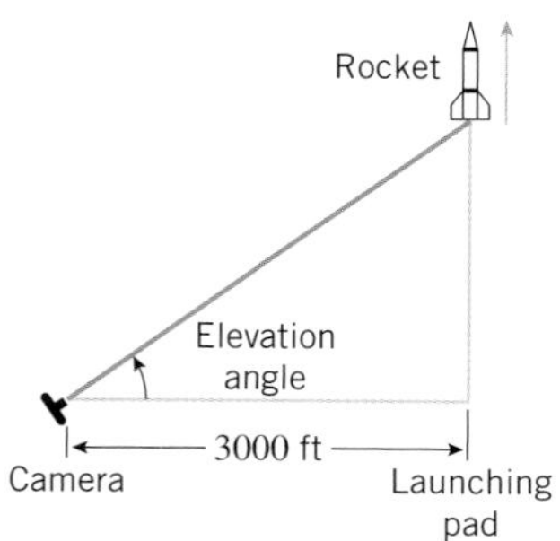

Figure 3.8.5

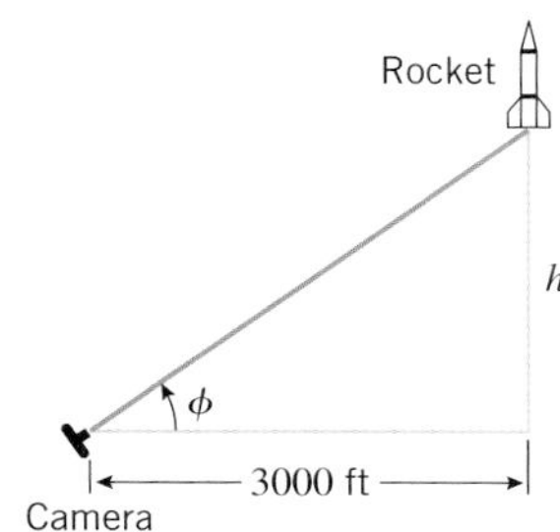

Figure 3.8.6

From Figure 3.8.6 we see that

$$\tan\phi = \frac{h}{3000} \tag{5}$$

Differentiating both sides of (5) with respect to t yields

$$(\sec^2\phi)\frac{d\phi}{dt} = \frac{1}{3000}\frac{dh}{dt} \tag{6}$$

Evaluating when $h = 4000$, it follows that

$$(\sec\phi)\Big|_{h=4000} = \frac{5000}{3000} = \frac{5}{3}$$

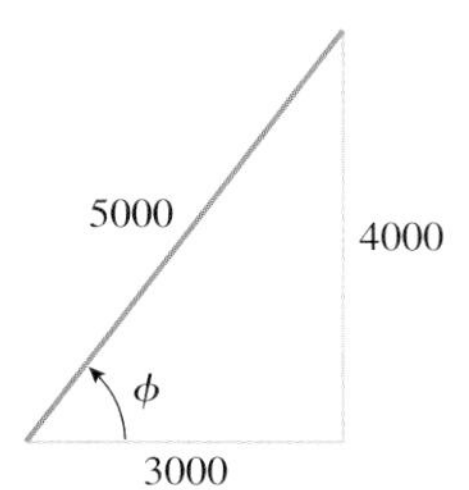

Figure 3.8.7

(see Figure 3.8.7), so that from (6)

$$\left(\frac{5}{3}\right)^2 \frac{d\phi}{dt}\bigg|_{h=4000} = \frac{1}{3000}\cdot 880 = \frac{22}{75}$$

$$\frac{d\phi}{dt}\bigg|_{h=4000} = \frac{22}{75}\cdot\frac{9}{25} = \frac{66}{625} \approx 0.11 \text{ rad/s} \approx 6.05 \text{ deg/s} \blacktriangleleft$$

▶ **Example 5** Suppose that liquid is to be cleared of sediment by allowing it to drain through a conical filter that is 16 cm high and has a radius of 4 cm at the top (Figure 3.8.8). Suppose also that the liquid is forced out of the cone at a constant rate of 2 cm^3/min.

(a) Do you think that the depth of the liquid will decrease at a constant rate? Give a verbal argument that justifies your conclusion.

(b) Find a formula that expresses the rate at which the depth of the liquid is changing in terms of the depth, and use that formula to determine whether your conclusion in part (a) is correct.

(c) At what rate is the depth of the liquid changing at the instant when the liquid in the cone is 8 cm deep?

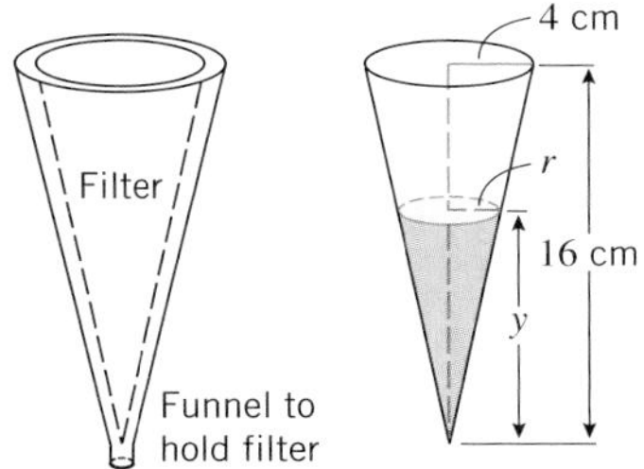

Figure 3.8.8

Solution (a). For the volume of liquid to decrease by a *fixed amount*, it requires a greater decrease in depth when the cone is close to empty than when it is almost full (Figure 3.8.9). This suggests that for the volume to decrease at a constant rate, the depth must decrease at an increasing rate.

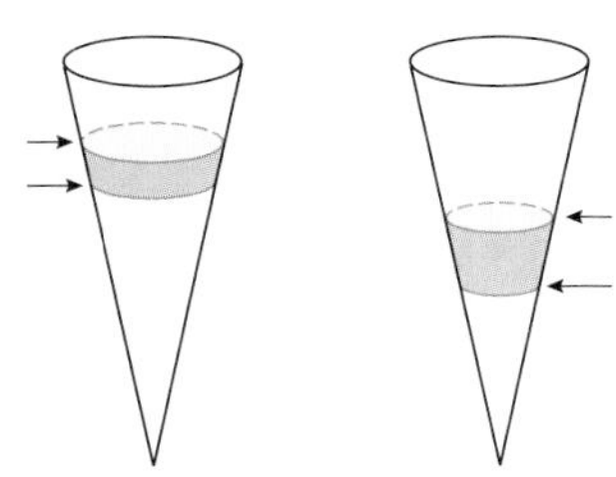

The same volume has drained, but the change in height is greater near the bottom than near the top.

Figure 3.8.9

Solution (b). Let

t = time elapsed from the initial observation (min)
V = volume of liquid in the cone at time t (cm^3)
y = depth of the liquid in the cone at time t (cm)
r = radius of the liquid surface at time t (cm)

(Figure 3.8.8). At each instant the rate at which the volume of liquid is changing is dV/dt, and the rate at which the depth is changing is dy/dt. We want to express dy/dt in terms of y given that dV/dt has a constant value of $dV/dt = -2$. (We must use a minus sign here because V *decreases* as t increases.)

From the formula for the volume of a cone, the volume V, the radius r, and the depth y are related by

$$V = \tfrac{1}{3}\pi r^2 y \tag{7}$$

If we differentiate both sides of (7) with respect to t, the right side will involve the quantity dr/dt. Since we have no direct information about dr/dt, it is desirable to eliminate r from

(7) before differentiating. This can be done using similar triangles. From Figure 3.8.8 we see that

$$\frac{r}{y} = \frac{4}{16} \quad \text{or} \quad r = \frac{1}{4}y$$

Substituting this expression in (7) gives

$$V = \frac{\pi}{48}y^3 \tag{8}$$

Differentiating both sides of (8) with respect to t we obtain

$$\frac{dV}{dt} = \frac{\pi}{48}\left(3y^2\frac{dy}{dt}\right)$$

or

$$\frac{dy}{dt} = \frac{16}{\pi y^2}\frac{dV}{dt} = \frac{16}{\pi y^2}(-2) = -\frac{32}{\pi y^2} \tag{9}$$

which expresses dy/dt in terms of y. The minus sign tells us that y is decreasing with time, and

$$\left|\frac{dy}{dt}\right| = \frac{32}{\pi y^2}$$

tells us how fast y is decreasing. From this formula we see that $|dy/dt|$ increases as y decreases, which confirms our conjecture in part (a) that the depth of the liquid decreases more quickly as the liquid drains through the filter.

Solution (c). The rate at which the depth is changing when the depth is 8 cm can be obtained from (9) with $y = 8$:

$$\left.\frac{dy}{dt}\right|_{y=8} = -\frac{32}{\pi(8^2)} = -\frac{1}{2\pi} \approx -0.16 \text{ cm/min} \blacktriangleleft$$

✔ QUICK CHECK EXERCISES 3.8 *(See page 213 for answers.)*

1. If $A = x^2$ and $\dfrac{dx}{dt} = 3$, find $\left.\dfrac{dA}{dt}\right|_{x=10}$.

2. If $A = x^2$ and $\dfrac{dA}{dt} = 3$, find $\left.\dfrac{dx}{dt}\right|_{x=10}$.

3. A 10-foot ladder stands on a horizontal floor and leans against a vertical wall. Use x to denote the distance along the floor from the wall to the foot of the ladder, and use y to denote the distance along the wall from the floor to the top of the ladder. If the foot of the ladder is dragged away from the wall, find an equation that relates rates of change of x and y with respect to time.

4. Suppose that a block of ice in the shape of a right circular cylinder melts so that it retains its cylindrical shape. Find an equation that relates the rates of change of the volume (V), height (h), and radius (r) of the block of ice.

EXERCISE SET 3.8

1–4 Both x and y denote functions of t that are related by the given equation. Use this equation and the given derivative information to find the specified derivative.

1. Equation: $y = 3x + 5$.
(a) Given that $dx/dt = 2$, find dy/dt when $x = 1$.
(b) Given that $dy/dt = -1$, find dx/dt when $x = 0$.

2. Equation: $x + 4y = 3$.
(a) Given that $dx/dt = 1$, find dy/dt when $x = 2$.
(b) Given that $dy/dt = 4$, find dx/dt when $x = 3$.

3. Equation: $4x^2 + 9y^2 = 1$.
(a) Given that $dx/dt = 3$, find dy/dt when $(x, y) = \left(\frac{1}{2\sqrt{2}}, \frac{1}{3\sqrt{2}}\right)$.
(b) Given that $dy/dt = 8$, find dx/dt when $(x, y) = \left(\frac{1}{3}, -\frac{\sqrt{5}}{9}\right)$.

4. Equation: $x^2 + y^2 = 2x + 4y$.
(a) Given that $dx/dt = -5$, find dy/dt when $(x, y) = (3, 1)$.
(b) Given that $dy/dt = 6$, find dx/dt when $(x, y) = (1 + \sqrt{2}, 2 + \sqrt{3})$.

FOCUS ON CONCEPTS

5. Let A be the area of a square whose sides have length x, and assume that x varies with the time t.
(a) Draw a picture of the square with the labels A and x placed appropriately.
(b) Write an equation that relates A and x.
(c) Use the equation in part (b) to find an equation that relates dA/dt and dx/dt.
(d) At a certain instant the sides are 3 ft long and increasing at a rate of 2 ft/min. How fast is the area increasing at that instant?

6. Let A be the area of a circle of radius r, and assume that r increases with the time t.
(a) Draw a picture of the circle with the labels A and r placed appropriately.
(b) Write an equation that relates A and r.
(c) Use the equation in part (b) to find an equation that relates dA/dt and dr/dt.
(d) At a certain instant the radius is 5 cm and increasing at the rate of 2 cm/s. How fast is the area increasing at that instant?

7. Let V be the volume of a cylinder having height h and radius r, and assume that h and r vary with time.
(a) How are dV/dt, dh/dt, and dr/dt related?
(b) At a certain instant, the height is 6 in and increasing at 1 in/s, while the radius is 10 in and decreasing at 1 in/s. How fast is the volume changing at that instant? Is the volume increasing or decreasing at that instant?

8. Let l be the length of a diagonal of a rectangle whose sides have lengths x and y, and assume that x and y vary with time.
(a) How are dl/dt, dx/dt, and dy/dt related?
(b) If x increases at a constant rate of $\frac{1}{2}$ ft/s and y decreases at a constant rate of $\frac{1}{4}$ ft/s, how fast is the size of the diagonal changing when $x = 3$ ft and $y = 4$ ft? Is the diagonal increasing or decreasing at that instant?

9. Let θ (in radians) be an acute angle in a right triangle, and let x and y, respectively, be the lengths of the sides adjacent to and opposite θ. Suppose also that x and y vary with time.
(a) How are $d\theta/dt$, dx/dt, and dy/dt related?
(b) At a certain instant, $x = 2$ units and is increasing at 1 unit/s, while $y = 2$ units and is decreasing at $\frac{1}{4}$ unit/s. How fast is θ changing at that instant? Is θ increasing or decreasing at that instant?

10. Suppose that $z = x^3y^2$, where both x and y are changing with time. At a certain instant when $x = 1$ and $y = 2$, x is decreasing at the rate of 2 units/s, and y is increasing at the rate of 3 units/s. How fast is z changing at this instant? Is z increasing or decreasing?

11. The minute hand of a certain clock is 4 in long. Starting from the moment when the hand is pointing straight up, how fast is the area of the sector that is swept out by the hand increasing at any instant during the next revolution of the hand?

12. A stone dropped into a still pond sends out a circular ripple whose radius increases at a constant rate of 3 ft/s. How rapidly is the area enclosed by the ripple increasing at the end of 10 s?

13. Oil spilled from a ruptured tanker spreads in a circle whose area increases at a constant rate of 6 mi^2/h. How fast is the radius of the spill increasing when the area is 9 mi^2?

14. A spherical balloon is inflated so that its volume is increasing at the rate of 3 ft^3/min. How fast is the diameter of the balloon increasing when the radius is 1 ft?

15. A spherical balloon is to be deflated so that its radius decreases at a constant rate of 15 cm/min. At what rate must air be removed when the radius is 9 cm?

16. A 17-ft ladder is leaning against a wall. If the bottom of the ladder is pulled along the ground away from the wall at a constant rate of 5 ft/s, how fast will the top of the ladder be moving down the wall when it is 8 ft above the ground?

17. A 13-ft ladder is leaning against a wall. If the top of the ladder slips down the wall at a rate of 2 ft/s, how fast will the foot be moving away from the wall when the top is 5 ft above the ground?

18. A 10-ft plank is leaning against a wall. If at a certain instant the bottom of the plank is 2 ft from the wall and is being pushed toward the wall at the rate of 6 in/s, how fast is the acute angle that the plank makes with the ground increasing?

19. A softball diamond is a square whose sides are 60 ft long. Suppose that a player running from first to second base has a speed of 25 ft/s at the instant when she is 10 ft from second base. At what rate is the player's distance from home plate changing at that instant?

20. A rocket, rising vertically, is tracked by a radar station that is on the ground 5 mi from the launchpad. How fast is the rocket rising when it is 4 mi high and its distance from the radar station is increasing at a rate of 2000 mi/h?

21. For the camera and rocket shown in Figure 3.8.5, at what rate is the camera-to-rocket distance changing when the rocket is 4000 ft up and rising vertically at 880 ft/s?

22. For the camera and rocket shown in Figure 3.8.5, at what rate is the rocket rising when the elevation angle is $\pi/4$ radians and increasing at a rate of 0.2 radian/s?

23. A satellite is in an elliptical orbit around the Earth. Its distance r (in miles) from the center of the Earth is given by

$$r = \frac{4995}{1 + 0.12\cos\theta}$$

where θ is the angle measured from the point on the orbit nearest the Earth's surface (see the accompanying figure on the next page).

(a) Find the altitude of the satellite at ***perigee*** (the point nearest the surface of the Earth) and at ***apogee*** (the point farthest from the surface of the Earth). Use 3960 mi as the radius of the Earth.

(b) At the instant when θ is $120°$, the angle θ is increasing at the rate of $2.7°/\text{min}$. Find the altitude of the satellite and the rate at which the altitude is changing at this instant. Express the rate in units of mi/min.

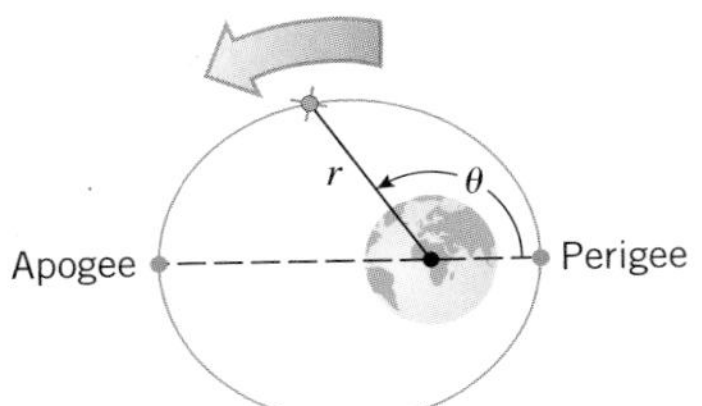

Figure Ex-23

24. An aircraft is flying horizontally at a constant height of 4000 ft above a fixed observation point (see the accompanying figure). At a certain instant the angle of elevation θ is $30°$ and decreasing, and the speed of the aircraft is 300 mi/h.

(a) How fast is θ decreasing at this instant? Express the result in units of degrees/s.

(b) How fast is the distance between the aircraft and the observation point changing at this instant? Express the result in units of ft/s. Use 1 mi = 5280 ft.

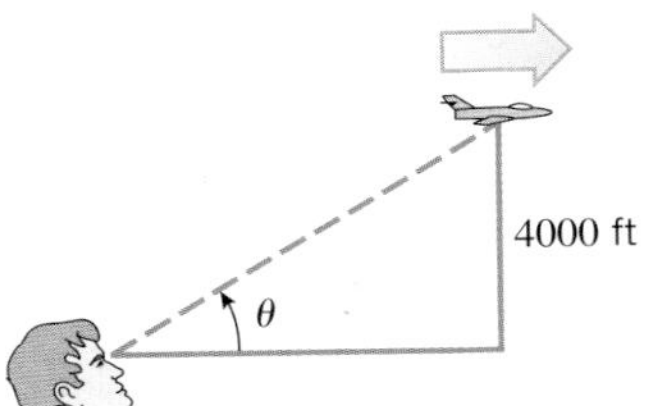

Figure Ex-24

25. A conical water tank with vertex down has a radius of 10 ft at the top and is 24 ft high. If water flows into the tank at a rate of $20\ \text{ft}^3/\text{min}$, how fast is the depth of the water increasing when the water is 16 ft deep?

26. Grain pouring from a chute at the rate of $8\ \text{ft}^3/\text{min}$ forms a conical pile whose height is always twice its radius. How fast is the height of the pile increasing at the instant when the pile is 6 ft high?

27. Sand pouring from a chute forms a conical pile whose height is always equal to the diameter. If the height increases at a constant rate of 5 ft/min, at what rate is sand pouring from the chute when the pile is 10 ft high?

28. Wheat is poured through a chute at the rate of $10\ \text{ft}^3/\text{min}$, and falls in a conical pile whose bottom radius is always half the altitude. How fast will the circumference of the base be increasing when the pile is 8 ft high?

29. An aircraft is climbing at a $30°$ angle to the horizontal. How fast is the aircraft gaining altitude if its speed is 500 mi/h?

30. A boat is pulled into a dock by means of a rope attached to a pulley on the dock (see the accompanying figure). The rope is attached to the bow of the boat at a point 10 ft below the pulley. If the rope is pulled through the pulley at a rate of 20 ft/min, at what rate will the boat be approaching the dock when 125 ft of rope is out?

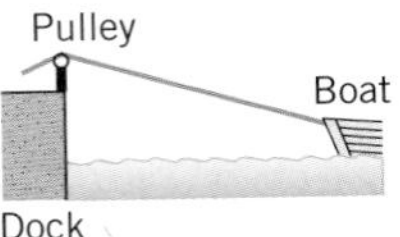

Figure Ex-30

31. For the boat in Exercise 30, how fast must the rope be pulled if we want the boat to approach the dock at a rate of 12 ft/min at the instant when 125 ft of rope is out?

32. A man 6 ft tall is walking at the rate of 3 ft/s toward a streetlight 18 ft high (see the accompanying figure).

(a) At what rate is his shadow length changing?

(b) How fast is the tip of his shadow moving?

Figure Ex-32

33. A beacon that makes one revolution every 10 s is located on a ship anchored 4 kilometers from a straight shoreline. How fast is the beam moving along the shoreline when it makes an angle of $45°$ with the shore?

34. An aircraft is flying at a constant altitude with a constant speed of 600 mi/h. An antiaircraft missile is fired on a straight line perpendicular to the flight path of the aircraft so that it will hit the aircraft at a point P (see the accompanying figure). At the instant the aircraft is 2 mi from the impact point P the missile is 4 mi from P and flying at 1200 mi/h. At that instant, how rapidly is the distance between missile and aircraft decreasing?

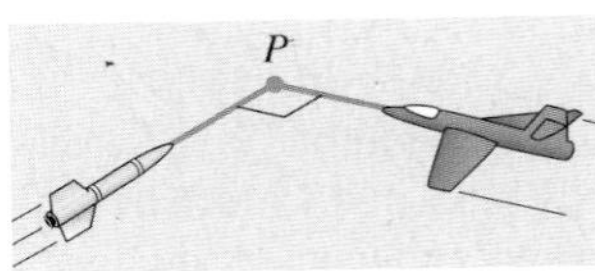

Figure Ex-34

35. Solve Exercise 34 under the assumption that the angle between the flight paths is $120°$ instead of the assumption that the paths are perpendicular. [*Hint:* Use the law of cosines.]

36. A police helicopter is flying due north at 100 mi/h and at a constant altitude of $\frac{1}{2}$ mi. Below, a car is traveling west on a highway at 75 mi/h. At the moment the helicopter crosses over the highway the car is 2 mi east of the helicopter.

(a) How fast is the distance between the car and helicopter changing at the moment the helicopter crosses the highway?

(b) Is the distance between the car and helicopter increasing or decreasing at that moment?

37. A particle is moving along the curve whose equation is

$$\frac{xy^3}{1+y^2} = \frac{8}{5}$$

Assume that the x-coordinate is increasing at the rate of 6 units/s when the particle is at the point $(1, 2)$.

(a) At what rate is the y-coordinate of the point changing at that instant?

(b) Is the particle rising or falling at that instant?

38. A point P is moving along the curve whose equation is $y = \sqrt{x^3 + 17}$. When P is at $(2, 5)$, y is increasing at the rate of 2 units/s. How fast is x changing?

39. A point P is moving along the line whose equation is $y = 2x$. How fast is the distance between P and the point $(3, 0)$ changing at the instant when P is at $(3, 6)$ if x is decreasing at the rate of 2 units/s at that instant?

40. A point P is moving along the curve whose equation is $y = \sqrt{x}$. Suppose that x is increasing at the rate of 4 units/s when $x = 3$.

(a) How fast is the distance between P and the point $(2, 0)$ changing at this instant?

(b) How fast is the angle of inclination of the line segment from P to $(2, 0)$ changing at this instant?

41. A particle is moving along the curve $y = x/(x^2 + 1)$. Find all values of x at which the rate of change of x with respect to time is three times that of y. [Assume that dx/dt is never zero.]

42. A particle is moving along the curve $16x^2 + 9y^2 = 144$. Find all points (x, y) at which the rates of change of x and y with respect to time are equal. [Assume that dx/dt and dy/dt are never both zero at the same point.]

43. The ***thin lens equation*** in physics is

$$\frac{1}{s} + \frac{1}{S} = \frac{1}{f}$$

where s is the object distance from the lens, S is the image distance from the lens, and f is the focal length of the lens. Suppose that a certain lens has a focal length of 6 cm and that an object is moving toward the lens at the rate of 2 cm/s. How fast is the image distance changing at the instant when the object is 10 cm from the lens? Is the image moving away from the lens or toward the lens?

44. Water is stored in a cone-shaped reservoir (vertex down). Assuming the water evaporates at a rate proportional to the surface area exposed to the air, show that the depth of the water will decrease at a constant rate that does not depend on the dimensions of the reservoir.

45. A meteor enters the Earth's atmosphere and burns up at a rate that, at each instant, is proportional to its surface area. Assuming that the meteor is always spherical, show that the radius decreases at a constant rate.

46. On a certain clock the minute hand is 4 in long and the hour hand is 3 in long. How fast is the distance between the tips of the hands changing at 9 o'clock?

47. Coffee is poured at a uniform rate of 20 cm^3/s into a cup whose inside is shaped like a truncated cone (see the accompanying figure). If the upper and lower radii of the cup are 4 cm and 2 cm and the height of the cup is 6 cm, how fast will the coffee level be rising when the coffee is halfway up? [*Hint:* Extend the cup downward to form a cone.]

Figure Ex-47

✔QUICK CHECK ANSWERS 3.8

1. 60 2. $\frac{3}{20}$ 3. $x\frac{dx}{dt} + y\frac{dy}{dt} = 0$ 4. $\frac{dV}{dt} = 2\pi rh\frac{dr}{dt} + \pi r^2\frac{dh}{dt}$

3.9 LOCAL LINEAR APPROXIMATION; DIFFERENTIALS

In this section we will show how derivatives can be used to approximate nonlinear functions by linear functions. Also, up to now we have been interpreting dy/dx as a single entity representing the derivative. In this section we will define the quantities dx and dy themselves, thereby allowing us to interpret dy/dx as an actual ratio.

Recall from Section 3.2 that if a function f is differentiable at x_0, then a sufficiently magnified portion of the graph of f centered at the point $P(x_0, f(x_0))$ takes on the appearance

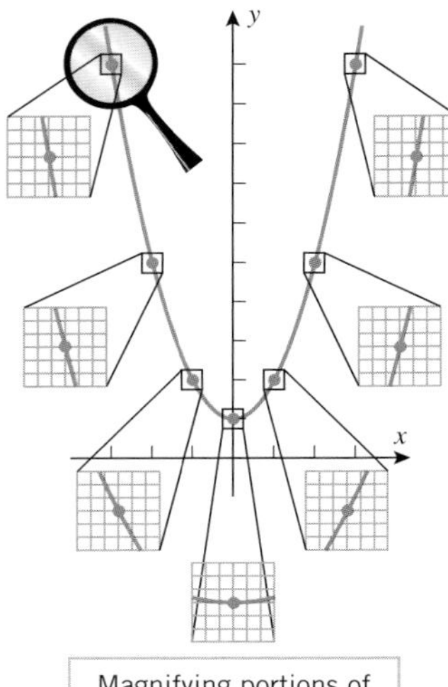

Magnifying portions of the graph of $y = x^2 + 1$

Figure 3.9.1

of a straight line segment. Figure 3.9.1 illustrates this at several points on the graph of $y = x^2 + 1$. For this reason, a function that is differentiable at x_0 is sometimes said to be ***locally linear*** at x_0.

The line that best approximates the graph of f in the vicinity of $P(x_0, f(x_0))$ is the tangent line to the graph of f at x_0, given by the equation

$$y = f(x_0) + f'(x_0)(x - x_0)$$

[see Formula (3) of Section 3.2]. Thus, for values of x near x_0 we can approximate values of $f(x)$ by

$$f(x) \approx f(x_0) + f'(x_0)(x - x_0) \tag{1}$$

This is called the ***local linear approximation*** of f at x_0. This formula can also be expressed in terms of the increment $\Delta x = x - x_0$ as

$$f(x_0 + \Delta x) \approx f(x_0) + f'(x_0)\Delta x \tag{2}$$

▶ Example 1

(a) Find the local linear approximation of $f(x) = \sqrt{x}$ at $x_0 = 1$.

(b) Use the local linear approximation obtained in part (a) to approximate $\sqrt{1.1}$, and compare your approximation to the result produced directly by a calculating utility.

Solution (a). Since $f'(x) = 1/(2\sqrt{x})$, it follows from (1) that the local linear approximation of $\sqrt{x}$ at a point x_0 is

$$\sqrt{x} \approx \sqrt{x_0} + \frac{1}{2\sqrt{x_0}}(x - x_0)$$

Thus, the local linear approximation at $x_0 = 1$ is

$$\sqrt{x} \approx 1 + \tfrac{1}{2}(x - 1) \tag{3}$$

The graphs of $y = \sqrt{x}$ and the local linear approximation $y = 1 + \frac{1}{2}(x - 1)$ are shown in Figure 3.9.2.

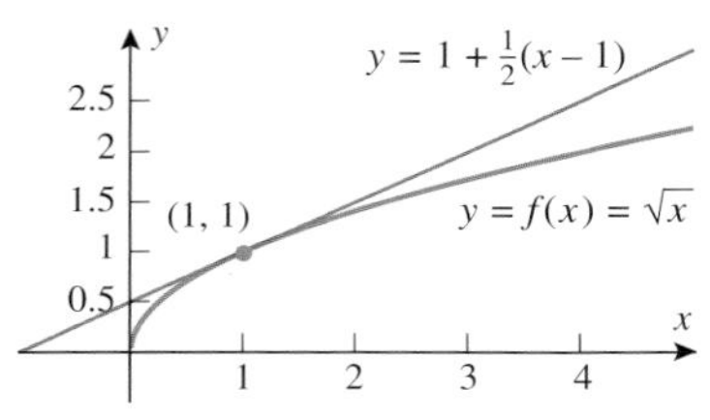

Figure 3.9.2

Solution (b). Applying (3) with $x = 1.1$ yields

$$\sqrt{1.1} \approx 1 + \tfrac{1}{2}(1.1 - 1) = 1.05$$

Since the tangent line $y = 1 + \frac{1}{2}(x - 1)$ in Figure 3.9.2 lies above the graph of $f(x) = \sqrt{x}$, we would expect this approximation to be slightly too large. This expectation is confirmed by the calculator approximation $\sqrt{1.1} \approx 1.04881$. ◀

Examples 1 and 2 illustrate important ideas and are not meant to suggest that you should use local linear approximations for computations that your calculating utility can perform. The main application of local linear approximation is in modeling problems where it is useful to replace complicated functions by simpler ones.

▶ Example 2

(a) Find the local linear approximation of $f(x) = \sin x$ at $x_0 = 0$.

(b) Use the local linear approximation obtained in part (a) to approximate $\sin 2°$, and compare your approximation to the result produced directly by your calculating device.

Solution (a). Since $f'(x) = \cos x$, it follows from (1) that the local linear approximation of $\sin x$ at a point x_0 is

$$\sin x \approx \sin x_0 + (\cos x_0)(x - x_0)$$

Thus, the local linear approximation at $x_0 = 0$ is

$$\sin x \approx \sin 0 + (\cos 0)(x - 0)$$

which simplifies to

$$\sin x \approx x \tag{4}$$

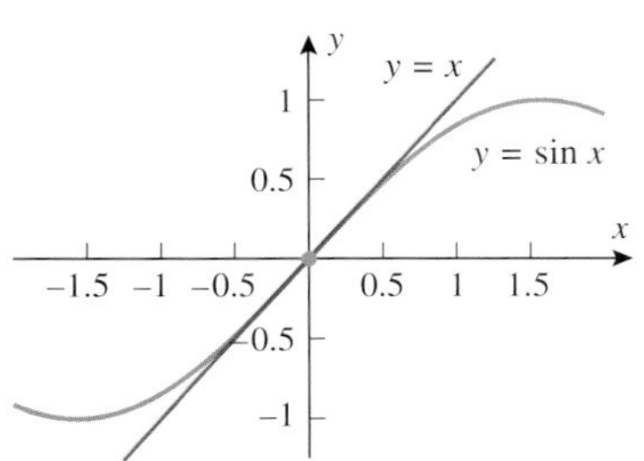

Figure 3.9.3

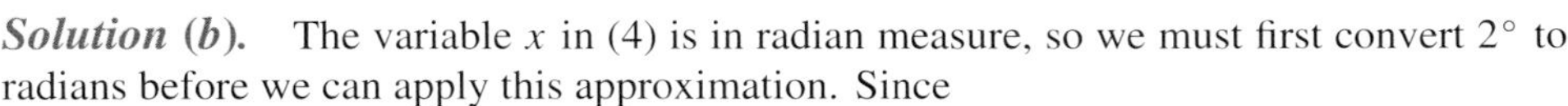

Solution (b). The variable x in (4) is in radian measure, so we must first convert 2° to radians before we can apply this approximation. Since

$$2^\circ = 2\left(\frac{\pi}{180}\right) = \frac{\pi}{90} \approx 0.0349066 \text{ radian}$$

it follows from (4) that $\sin 2^\circ \approx 0.0349066$. Comparing the two graphs in Figure 3.9.3, we would expect this approximation to be slightly larger than the exact value. The calculator approximation $\sin 2^\circ \approx 0.0348995$ shows that this is indeed the case. ◄

ERROR IN LOCAL LINEAR APPROXIMATIONS

As a general rule, the accuracy of the local linear approximation to $f(x)$ at x_0 will deteriorate as x gets progressively farther from x_0. To illustrate this for the approximation $\sin x \approx x$ in Example 2, let us graph the function

$$E(x) = |\sin x - x|$$

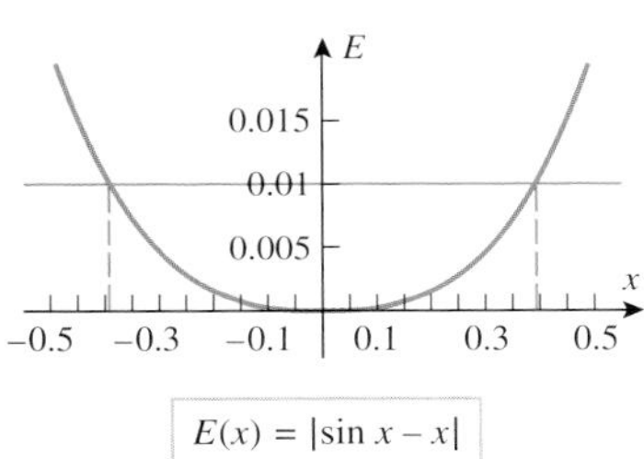

Figure 3.9.4

which is the absolute value of the error in the approximation (Figure 3.9.4).

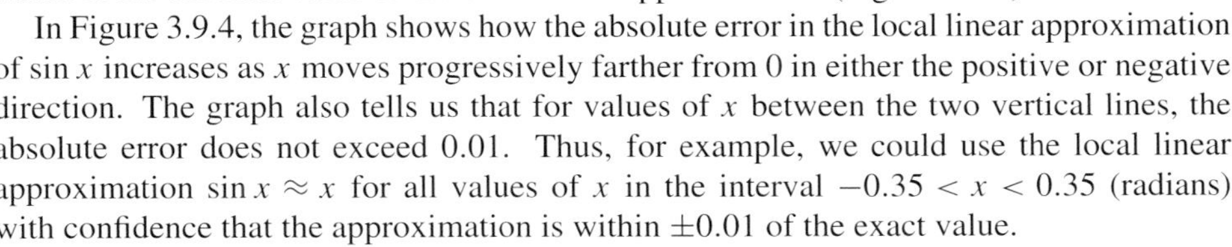

In Figure 3.9.4, the graph shows how the absolute error in the local linear approximation of $\sin x$ increases as x moves progressively farther from 0 in either the positive or negative direction. The graph also tells us that for values of x between the two vertical lines, the absolute error does not exceed 0.01. Thus, for example, we could use the local linear approximation $\sin x \approx x$ for all values of x in the interval $-0.35 < x < 0.35$ (radians) with confidence that the approximation is within ± 0.01 of the exact value.

DIFFERENTIALS

Newton and Leibniz each used a different notation when they published their discoveries of calculus, thereby creating a notational divide between Britain and the European continent that lasted for more than 50 years. The ***Leibniz notation*** dy/dx eventually prevailed because it suggests correct formulas in a natural way, the chain rule

$$\frac{dy}{dx} = \frac{dy}{du} \cdot \frac{du}{dx}$$

being a good example.

Up to now we have interpreted dy/dx as a single entity representing the derivative of y with respect to x; the symbols "dy" and "dx," which are called ***differentials***, have had no meanings attached to them. Our next goal is to define these symbols in such a way that dy/dx can be treated as an actual ratio. To do this, assume that f is differentiable at a point x, *define* dx to be an independent variable that can have any real value, and *define* dy by the formula

$$dy = f'(x)\,dx \tag{5}$$

If $dx \neq 0$, then we can divide both sides of (5) by dx to obtain

$$\frac{dy}{dx} = f'(x) \tag{6}$$

Thus, we have achieved our goal of defining dy and dx so their ratio is $f'(x)$. Formula (5) is said to express (6) in ***differential form***.

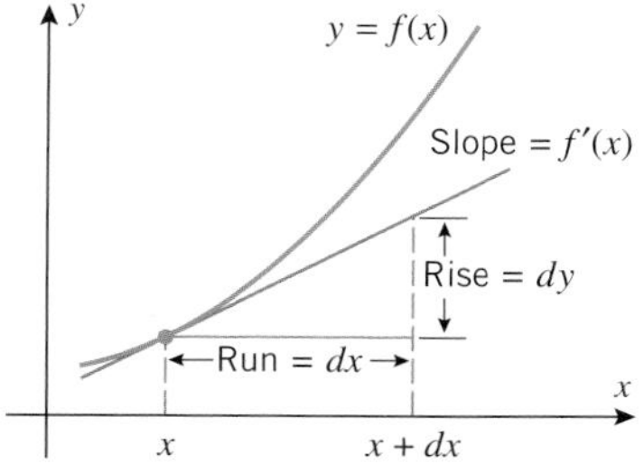

Figure 3.9.5

To interpret (5) geometrically, note that $f'(x)$ is the slope of the tangent line to the graph of f at x. The differentials dy and dx can be viewed as a corresponding rise and run of this tangent line (Figure 3.9.5).

▶ **Example 3** Express the derivative with respect to x of $y = x^2$ in differential form, and discuss the relationship between dy and dx at $x = 1$.

Solution. The derivative of y with respect to x is $dy/dx = 2x$, which can be expressed in differential form as

$$dy = 2x\,dx$$

When $x = 1$ this becomes

$$dy = 2\,dx$$

This tells us that if we travel along the tangent line to the curve $y = x^2$ at $x = 1$, then a change of dx units in x produces a change of $2\,dx$ units in y. Thus, for example, a run of $dx = 2$ units produces a rise of $dy = 4$ units along the tangent line (Figure 3.9.6). ◀

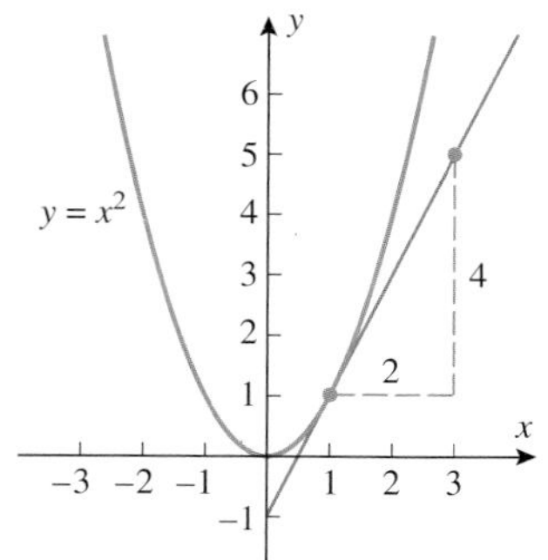

Figure 3.9.6

It is important to understand the distinction between the increment Δy and the differential dy. To see the difference, let us assign the independent variables dx and Δx the same value, so $dx = \Delta x$. Then Δy represents the change in y that occurs when we start at x and travel *along the curve* $y = f(x)$ until we have moved Δx $(= dx)$ units in the x-direction, while dy represents the change in y that occurs if we start at x and travel *along the tangent line* until we have moved dx $(= \Delta x)$ units in the x-direction (Figure 3.9.7).

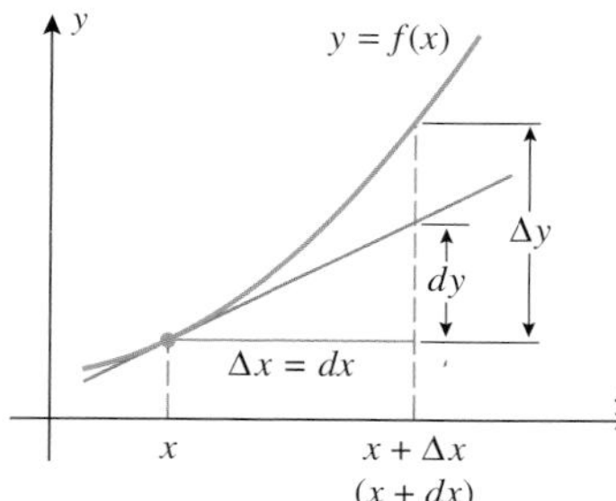

Figure 3.9.7

▶ **Example 4** Let $y = \sqrt{x}$. Find dy and Δy at $x = 4$ with $dx = \Delta x = 3$. Then make a sketch of $y = \sqrt{x}$, showing dy and Δy in the picture.

Solution. With $f(x) = \sqrt{x}$ we obtain

$$\Delta y = f(x + \Delta x) - f(x) = \sqrt{x + \Delta x} - \sqrt{x} = \sqrt{7} - \sqrt{4} \approx 0.65$$

If $y = \sqrt{x}$, then

$$\frac{dy}{dx} = \frac{1}{2\sqrt{x}}, \quad \text{so} \quad dy = \frac{1}{2\sqrt{x}}\,dx = \frac{1}{2\sqrt{4}}(3) = \frac{3}{4} = 0.75$$

Figure 3.9.8 shows the curve $y = \sqrt{x}$ together with dy and Δy. ◀

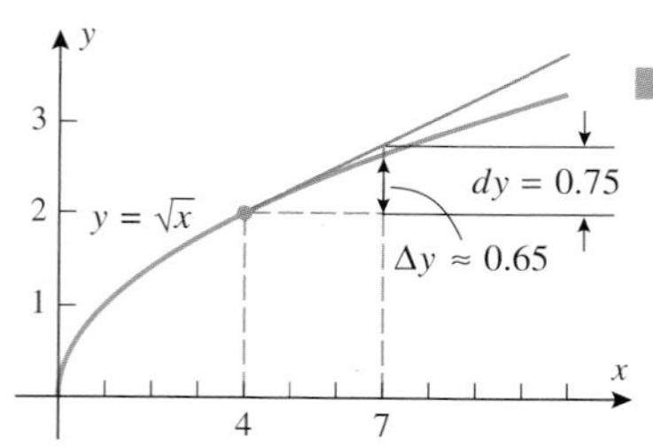

Figure 3.9.8

LOCAL LINEAR APPROXIMATION FROM THE DIFFERENTIAL POINT OF VIEW

Although Δy and dy are generally different, the differential dy will nonetheless be a good approximation of Δy provided $dx = \Delta x$ is close to 0. To see this, recall from Section 3.2 that

$$f'(x) = \lim_{\Delta x \to 0} \frac{\Delta y}{\Delta x}$$

It follows that if Δx is close to 0, then we will have $f'(x) \approx \Delta y/\Delta x$ or, equivalently,

$$\Delta y \approx f'(x)\Delta x$$

If we agree to let $dx = \Delta x$, then we can rewrite this as

$$\Delta y \approx f'(x)\,dx = dy \tag{7}$$

In words, this states that for values of dx near zero the differential dy closely approximates the increment Δy (Figure 3.9.7). But this is to be expected since the graph of the tangent line at x is the local linear approximation of the graph of f.

ERROR PROPAGATION IN APPLICATIONS

In applications, small errors invariably occur in measured quantities. When these quantities are used in computations, those errors produce errors in the computed quantities. This is called ***error propagation***. Our goal is to show how to use local linear approximation and differentials to estimate errors in computed quantities from estimates of errors in the measured quantities. For this purpose, suppose that

x_0 is the exact value of the quantity being measured
$y_0 = f(x_0)$ is the exact value of the quantity being computed
x is the measured value of x_0
$y = f(x)$ is the computed value of y

We define

$dx\ (= \Delta x) = x - x_0$ to be the ***measurement error*** of x
$\Delta y = f(x) - f(x_0)$ to be the ***propagated error*** of y

Note that measurement error is positive if the measured value is greater than the exact value and is negative if it is less than the exact value. The sign of the propagated error conveys similar information.

It follows from (7) with x_0 replacing x that the propagated error Δy can be approximated by

$$\Delta y \approx dy = f'(x_0)\,dx \tag{8}$$

Unfortunately, there is a practical difficulty in applying this formula since the value of x_0 is unknown. (Keep in mind that only the measured value x is known to the researcher.) This being the case, it is standard practice in research to use the measured value x in place of x_0 in (8) and use the approximation

$$\Delta y \approx dy = f'(x)\,dx \tag{9}$$

for the propagated error.

Explain why an error estimate of at most $\pm\frac{1}{32}$ inch is reasonable for a ruler that is calibrated in sixteenths of an inch.

▶ **Example 5** Suppose that the side of a square is measured with a ruler to be 10 inches with a measurement error of at most $\pm\frac{1}{32}$ in. Estimate the error in the computed area of the square.

Solution. Let x denote the exact length of a side and y the exact area so that $y = x^2$. It follows from (9) with $f(x) = x^2$ that if dx is the measurement error, then the propagated error Δy can be approximated as

$$\Delta y \approx dy = 2x\,dx$$

Substituting the measured value $x = 10$ into this equation yields

$$dy = 20\,dx \tag{10}$$

But to say that the measurement error is at most $\pm\frac{1}{32}$ means that

$$-\frac{1}{32} \le dx \le \frac{1}{32}$$

Multiplying these inequalities through by 20 and applying (10) yields

$$20\left(-\tfrac{1}{32}\right) \le dy \le 20\left(\tfrac{1}{32}\right) \quad \text{or equivalently} \quad -\tfrac{5}{8} \le dy \le \tfrac{5}{8}$$

Thus, the propagated error in the area is estimated to be within $\pm\frac{5}{8}$ in^2. ◀

If the true value of a quantity is q and a measurement or calculation produces an error Δq, then $\Delta q/q$ is called the ***relative error*** in the measurement or calculation; when expressed as a percentage, $\Delta q/q$ is called the ***percentage error***. As a practical matter, the true value q is usually unknown, so that the measured or calculated value of q is used instead; and the relative error is approximated by dq/q.

▶ **Example 6** The radius of a sphere is measured with a percentage error within $\pm 0.04\%$. Estimate the percentage error in the calculated volume of the sphere.

Solution. The volume V of a sphere is $V = \frac{4}{3}\pi r^3$, so

$$\frac{dV}{dr} = 4\pi r^2$$

Formula (11) tells us that, as a rule of thumb, the percentage error in the computed volume of a sphere is approximately 3 times the percentage error in the measured value of its radius. As a rule of thumb, how is the percentage error in the computed area of a square related to the percentage error in the measured value of a side?

from which it follows that $dV = 4\pi r^2\,dr$. Thus, the relative error in V is approximately

$$\frac{dV}{V} = \frac{4\pi r^2\,dr}{\frac{4}{3}\pi r^3} = 3\frac{dr}{r} \tag{11}$$

We are given that the relative error in the measured value of r is $\pm 0.04\%$, which means that

$$-0.0004 \le \frac{dr}{r} \le 0.0004$$

Multiplying these inequalities through by 3 and applying (11) yields

$$3(-0.0004) \le \frac{dV}{V} \le 3(0.0004) \quad \text{or equivalently} \quad -0.0012 \le \frac{dV}{V} \le 0.0012$$

Thus, we estimate the percentage error in the calculated value of V to be within $\pm 0.12\%$. ◀

MORE NOTATION; DIFFERENTIAL FORMULAS

The symbol df is another common notation for the differential of a function $y = f(x)$. For example, if $f(x) = \sin x$, then we can write $df = \cos x\,dx$. We can also view the symbol "d" as an *operator* that acts on a function to produce the corresponding differential. For example, $d[x^2] = 2x\,dx$, $d[\sin x] = \cos x\,dx$, and so on. All of the general rules of differentiation then have corresponding differential versions:

DERIVATIVE FORMULA	DIFFERENTIAL FORMULA
$\frac{d}{dx}[c] = 0$	$d[c] = 0$
$\frac{d}{dx}[cf] = c\frac{df}{dx}$	$d[cf] = c\,df$
$\frac{d}{dx}[f+g] = \frac{df}{dx} + \frac{dg}{dx}$	$d[f+g] = df + dg$
$\frac{d}{dx}[fg] = f\frac{dg}{dx} + g\frac{df}{dx}$	$d[fg] = f\,dg + g\,df$
$\frac{d}{dx}\left[\frac{f}{g}\right] = \frac{g\frac{df}{dx} - f\frac{dg}{dx}}{g^2}$	$d\left[\frac{f}{g}\right] = \frac{g\,df - f\,dg}{g^2}$

For example,

$$\begin{aligned} d[x^2 \sin x] &= (x^2 \cos x + 2x \sin x)\,dx \\ &= x^2(\cos x\,dx) + (2x\,dx)\sin x \\ &= x^2 d[\sin x] + (\sin x)\,d[x^2] \end{aligned}$$

illustrates the differential version of the product rule.

✔QUICK CHECK EXERCISES 3.9 *(See page 221 for answers.)*

1. The local linear approximation of f at x_0 uses the ________ line to the graph of $y = f(x)$ at $x = x_0$ to approximate values of ________ for values of x near ________.

2. Find an equation for the local linear approximation to $y = 5 - x^2$ at $x_0 = 2$.

3. Let $y = 5 - x^2$. Find dy and Δy at $x = 2$ with $dx = \Delta x = 0.1$.

4. The intensity of light from a light source is a function $I = f(x)$ of the distance x from the light source. Suppose that a small gemstone is measured to be 10 m from a light source, $f(10) = 0.2\ \text{W/m}^2$, and $f'(10) = -0.04\ \text{W/m}^3$. If the distance $x = 10$ m was obtained with a measurement error within ± 0.05 m, estimate the percentage error in the calculated intensity of the light on the gemstone.

EXERCISE SET 3.9

1. (a) Use Formula (1) to obtain the local linear approximation of x^3 at $x_0 = 1$.
   (b) Use Formula (2) to rewrite the approximation obtained in part (a) in terms of Δx.
   (c) Use the result obtained in part (a) to approximate $(1.02)^3$, and confirm that the formula obtained in part (b) produces the same result.

2. (a) Use Formula (1) to obtain the local linear approximation of $1/x$ at $x_0 = 2$.
   (b) Use Formula (2) to rewrite the approximation obtained in part (a) in terms of Δx.
   (c) Use the result obtained in part (a) to approximate $1/2.05$, and confirm that the formula obtained in part (b) produces the same result.

FOCUS ON CONCEPTS

3. (a) Find the local linear approximation of the function $f(x) = \sqrt{1+x}$ at $x_0 = 0$, and use it to approximate $\sqrt{0.9}$ and $\sqrt{1.1}$.
   (b) Graph f and its tangent line at x_0 together, and use the graphs to illustrate the relationship between the exact values and the approximations of $\sqrt{0.9}$ and $\sqrt{1.1}$.

4. (a) Find the local linear approximation of the function $f(x) = 1/\sqrt{x}$ at $x_0 = 4$, and use it to approximate $1/\sqrt{3.9}$ and $1/\sqrt{4.1}$.
   (b) Graph f and its tangent line at x_0 together, and use the graphs to illustrate the relationship between the exact values and the approximations of $1/\sqrt{3.9}$ and $1/\sqrt{4.1}$.

5–8 Confirm that the stated formula is the local linear approximation at $x_0 = 0$.

5. $(1+x)^{15} \approx 1 + 15x$

6. $\dfrac{1}{\sqrt{1-x}} \approx 1 + \dfrac{1}{2}x$

7. $\tan x \approx x$

8. $\dfrac{1}{1+x} \approx 1 - x$

9–12 Confirm that the stated formula is the local linear approximation of f at $x_0 = 1$, where $\Delta x = x - 1$.

9. $f(x) = x^4;\ (1+\Delta x)^4 \approx 1 + 4\Delta x$

10. $f(x) = \sqrt{x};\ \sqrt{1+\Delta x} \approx 1 + \frac{1}{2}\Delta x$

11. $f(x) = \dfrac{1}{2+x};\ \dfrac{1}{3+\Delta x} \approx \dfrac{1}{3} - \dfrac{1}{9}\Delta x$

12. $f(x) = (4+x)^3;\ (5+\Delta x)^3 \approx 125 + 75\Delta x$

13–16 Confirm that the formula is the local linear approximation at $x_0 = 0$, and use a graphing utility to estimate an interval of x-values on which the error is at most ± 0.1.

13. $\sqrt{x+3} \approx \sqrt{3} + \dfrac{1}{2\sqrt{3}}x$

14. $\dfrac{1}{\sqrt{9-x}} \approx \dfrac{1}{3} + \dfrac{1}{54}x$

15. $\tan 2x \approx 2x$

16. $\dfrac{1}{(1+2x)^5} \approx 1 - 10x$

17. (a) Use the local linear approximation of $\sin x$ at $x_0 = 0$ obtained in Example 2 to approximate $\sin 1^\circ$, and compare the approximation to the result produced directly by your calculating device.
    (b) How would you choose x_0 to approximate $\sin 44^\circ$?
    (c) Approximate $\sin 44^\circ$; compare the approximation to the result produced directly by your calculating device.

18. (a) Use the local linear approximation of $\tan x$ at $x_0 = 0$ to approximate $\tan 2^\circ$, and compare the approximation to the result produced directly by your calculating device.
    (b) How would you choose x_0 to approximate $\tan 61^\circ$?
    (c) Approximate $\tan 61^\circ$; compare the approximation to the result produced directly by your calculating device.

19–27 Use an appropriate local linear approximation to estimate the value of the given quantity.

19. $(3.02)^4$
20. $(1.97)^3$
21. $\sqrt{65}$
22. $\sqrt{24}$
23. $\sqrt{80.9}$
24. $\sqrt{36.03}$
25. $\sin 0.1$
26. $\tan 0.2$
27. $\cos 31^\circ$

FOCUS ON CONCEPTS

28. The approximation $(1+x)^k \approx 1+kx$ is commonly used by engineers for quick calculations.
(a) Derive this result, and use it to make a rough estimate of $(1.001)^{37}$.
(b) Compare your estimate to that produced directly by your calculating device.
(c) Show that this formula produces a very bad estimate of $(1.1)^{37}$, and explain why.

29. (a) Let $y = x^2$. Find dy and Δy at $x = 2$ with $dx = \Delta x = 1$.
(b) Sketch the graph of $y = x^2$, showing dy and Δy in the picture.

30. (a) Let $y = x^3$. Find dy and Δy at $x = 1$ with $dx = \Delta x = 1$.
(b) Sketch the graph of $y = x^3$, showing dy and Δy in the picture.

31. (a) Let $y = 1/x$. Find dy and Δy at $x = 1$ with $dx = \Delta x = -0.5$.
(b) Sketch the graph of $y = 1/x$, showing dy and Δy in the picture.

32. (a) Let $y = \sqrt{x}$. Find dy and Δy at $x = 9$ with $dx = \Delta x = -1$.
(b) Sketch the graph of $y = \sqrt{x}$, showing dy and Δy in the picture.

33–36 Find formulas for dy and Δy.

33. $y = x^3$ **34.** $y = 8x - 4$
35. $y = x^2 - 2x + 1$ **36.** $y = \sin x$

37–40 Find the differential dy.

37. (a) $y = 4x^3 - 7x^2$ (b) $y = x\cos x$
38. (a) $y = 1/x$ (b) $y = 5\tan x$
39. (a) $y = x\sqrt{1-x}$ (b) $y = (1+x)^{-17}$
40. (a) $y = \dfrac{1}{x^3 - 1}$ (b) $y = \dfrac{1-x^3}{2-x}$

41–44 Use the differential dy to approximate Δy when x changes as indicated.

41. $y = \sqrt{3x-2}$; from $x = 2$ to $x = 2.03$
42. $y = \sqrt{x^2+8}$; from $x = 1$ to $x = 0.97$
43. $y = \dfrac{x}{x^2+1}$; from $x = 2$ to $x = 1.96$
44. $y = x\sqrt{8x+1}$; from $x = 3$ to $x = 3.05$

45. The side of a square is measured to be 10 ft, with a possible error of ± 0.1 ft.
(a) Use differentials to estimate the error in the calculated area.
(b) Estimate the percentage errors in the side and the area.

46. The side of a cube is measured to be 25 cm, with a possible error of ± 1 cm.
(a) Use differentials to estimate the error in the calculated volume.
(b) Estimate the percentage errors in the side and volume.

47. The hypotenuse of a right triangle is known to be 10 in exactly, and one of the acute angles is measured to be $30°$, with a possible error of $\pm 1°$.
(a) Use differentials to estimate the errors in the sides opposite and adjacent to the measured angle.
(b) Estimate the percentage errors in the sides.

48. One side of a right triangle is known to be 25 cm exactly. The angle opposite to this side is measured to be $60°$, with a possible error of $\pm 0.5°$.
(a) Use differentials to estimate the errors in the adjacent side and the hypotenuse.
(b) Estimate the percentage errors in the adjacent side and hypotenuse.

49. The electrical resistance R of a certain wire is given by $R = k/r^2$, where k is a constant and r is the radius of the wire. Assuming that the radius r has a possible error of $\pm 5\%$, use differentials to estimate the percentage error in R. (Assume k is exact.)

50. A 12-foot ladder leaning against a wall makes an angle θ with the floor. If the top of the ladder is h feet up the wall, express h in terms of θ and then use dh to estimate the change in h if θ changes from $60°$ to $59°$.

51. The area of a right triangle with a hypotenuse of H is calculated using the formula $A = \frac{1}{4}H^2 \sin 2\theta$, where θ is one of the acute angles. Use differentials to approximate the error in calculating A if $H = 4$ cm (exactly) and θ is measured to be $30°$, with a possible error of $\pm 15'$.

52. The side of a square is measured with a possible percentage error of $\pm 1\%$. Use differentials to estimate the percentage error in the area.

53. The side of a cube is measured with a possible percentage error of $\pm 2\%$. Use differentials to estimate the percentage error in the volume.

54. The volume of a sphere is to be computed from a measured value of its radius. Estimate the maximum permissible percentage error in the measurement if the percentage error in the volume must be kept within $\pm 3\%$. ($V = \frac{4}{3}\pi r^3$ is the volume of a sphere of radius r.)

55. The area of a circle is to be computed from a measured value of its diameter. Estimate the maximum permissible percentage error in the measurement if the percentage error in the area must be kept within $\pm 1\%$.

56. A steel cube with 1-in sides is coated with 0.01 in of copper. Use differentials to estimate the volume of copper in the coating. [*Hint:* Let ΔV be the change in the volume of the cube.]

57. A metal rod 15 cm long and 5 cm in diameter is to be covered (except for the ends) with insulation that is 0.1 cm thick. Use differentials to estimate the volume of insulation. [*Hint:* Let ΔV be the change in volume of the rod.]

58. The time required for one complete oscillation of a pendulum is called its ***period***. If L is the length of the pendulum and the oscillation is small, then the period is given by $P = 2\pi\sqrt{L/g}$, where g is the constant acceleration due to gravity. Use differentials to show that the percentage error in P is approximately half the percentage error in L.

59. If the temperature T of a metal rod of length L is changed by an amount ΔT, then the length will change by the amount $\Delta L = \alpha L \Delta T$, where α is called the ***coefficient of linear expansion***. For moderate changes in temperature α is taken as constant.

(a) Suppose that a rod 40 cm long at 20°C is found to be 40.006 cm long when the temperature is raised to 30°C. Find α.

(b) If an aluminum pole is 180 cm long at 15°C, how long is the pole if the temperature is raised to 40°C? [Take $\alpha = 2.3 \times 10^{-5}/°\text{C}$.]

60. If the temperature T of a solid or liquid of volume V is changed by an amount ΔT, then the volume will change by the amount $\Delta V = \beta V \Delta T$, where β is called the ***coefficient of volume expansion***. For moderate changes in temperature β is taken as constant. Suppose that a tank truck loads 4000 gallons of ethyl alcohol at a temperature of 35°C and delivers its load sometime later at a temperature of 15°C. Using $\beta = 7.5 \times 10^{-4}/°\text{C}$ for ethyl alcohol, find the number of gallons delivered.

✔ QUICK CHECK ANSWERS 3.9

1. tangent; $f(x)$; x_0 2. $y = 1 + (-4)(x-2)$ or $y = -4x + 9$ 3. $dy = -0.4$, $\Delta y = -0.41$ 4. within $\pm 1\%$

CHAPTER REVIEW EXERCISES  

1. Explain the difference between average and instantaneous rates of change, and discuss how they are calculated.

2. Complete each part for the function $y = \frac{1}{2}x^2$.

(a) Find the average rate of change of y with respect to x over the interval $[3, 4]$.

(b) Find the instantaneous rate of change of y with respect to x at $x = 3$.

(c) Find the instantaneous rate of change of y with respect to x at a general x-value.

(d) Sketch the graph of $y = \frac{1}{2}x^2$ together with the secant line whose slope is given by the result in part (a), and indicate graphically the slope of the tangent line that corresponds to the result in part (b).

3. Complete each part for the function $f(x) = x^2 + 1$.

(a) Find the slope of the tangent line to the graph of f at a general x-value.

(b) Find the slope of the tangent line to the graph of f at $x = 2$.

4. A car is traveling on a straight road that is 120 mi long. For the first 100 mi the car travels at an average velocity of 50 mi/h. Show that no matter how fast the car travels for the final 20 mi it cannot bring the average velocity up to 60 mi/h for the entire trip.

5. At time $t = 0$ a car moves into the passing lane to pass a slow-moving truck. The average velocity of the car from $t = 1$ to $t = 1 + h$ is

$$v_{\text{ave}} = \frac{3(h+1)^{2.5} + 580h - 3}{10h}$$

Estimate the instantaneous velocity of the car at $t = 1$, where time is in seconds and distance is in feet.

6. A sky diver jumps from an airplane. Suppose that the distance she falls during the first t seconds before her parachute opens is $s(t) = 976((0.835)^t - 1) + 176t$, where s is in feet. Graph s versus t for $0 \le t \le 20$, and use your graph to estimate the instantaneous velocity at $t = 15$.

7. A particle moves on a line away from its initial position so that after t hours it is $s = 3t^2 + t$ miles from its initial position.

(a) Find the average velocity of the particle over the interval $[1, 3]$.

(b) Find the instantaneous velocity at $t = 1$.

8. State the definition of a derivative, and give two interpretations of it.

9. Use the definition of a derivative to find dy/dx, and check your answer by calculating the derivative using appropriate derivative formulas.

(a) $y = \sqrt{9 - 4x}$ (b) $y = \dfrac{x}{x+1}$

10. Suppose that $f(x) = \begin{cases} x^2 - 1, & x \le 1 \\ k(x-1), & x > 1. \end{cases}$

For what values of k is f

(a) continuous? (b) differentiable?

11. The accompanying figure (next page) shows the graph of $y = f'(x)$ for an unspecified function f.

(a) For what values of x does the curve $y = f(x)$ have a horizontal tangent line?

(b) Over what intervals does the curve $y = f(x)$ have tangent lines with positive slope?

(c) Over what intervals does the curve $y = f(x)$ have tangent lines with negative slope?

(d) Given that $g(x) = f(x)\sin x$, find $g''(0)$.

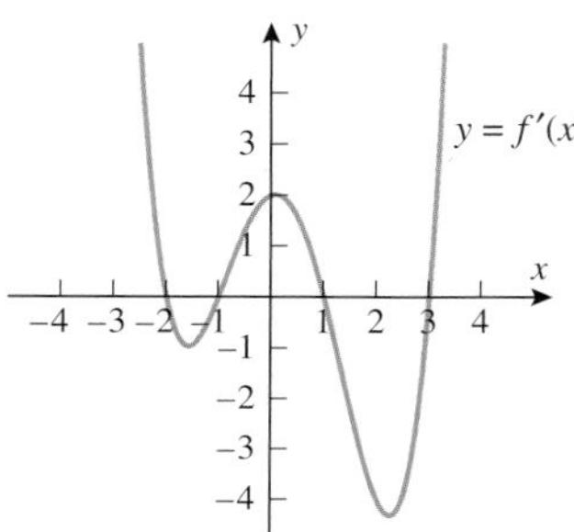

Figure Ex-11

12. Sketch the graph of a function f for which $f(0) = 1$, $f'(0) = 0$, $f'(x) > 0$ if $x < 0$, and $f'(x) < 0$ if $x > 0$.

13. According to the U.S. Bureau of the Census, the estimated and projected midyear world population, N, in billions for the years 1950, 1975, 2000, 2025, and 2050 was 2.555, 4.088, 6.080, 7.841, and 9.104, respectively. Although the increase in population is not a continuous function of the time t, we can apply the ideas in this section if we are willing to approximate the graph of N versus t by a continuous curve, as shown in the accompanying figure.

(a) Use the tangent line at $t = 2000$ shown in the figure to approximate the value of dN/dt there. Interpret your result as a rate of change.

(b) The instantaneous ***growth rate*** is defined as

$$\frac{dN/dt}{N}$$

Use your answer to part (a) to approximate the instantaneous growth rate at the start of the year 2000. Express the result as a percentage and include the proper units.

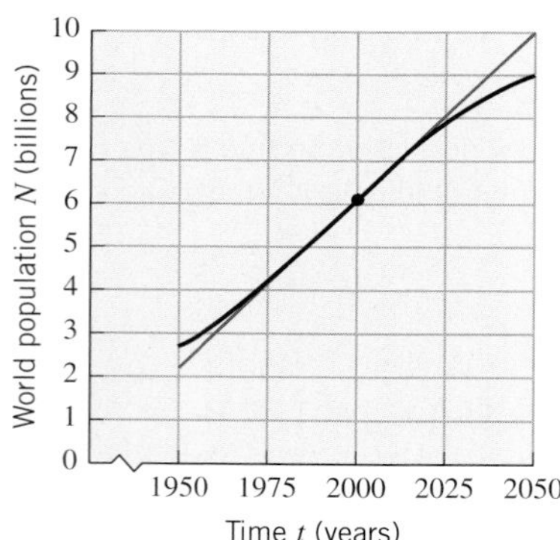

Figure Ex-13

14. Use a graphing utility to graph the function

$$f(x) = |x^4 - x - 1| - x$$

and estimate the values of x where the derivative of this function does not exist.

15–18 (a) Use a CAS to find $f'(x)$ via Definition 3.2.1; (b) check the result by finding the derivative by hand; (c) use the CAS to find $f''(x)$.

15. $f(x) = x^2 \sin x$ **16.** $f(x) = \sqrt{x} + \cos^2 x$

17. $f(x) = \dfrac{2x^2 - x + 5}{3x + 2}$ **18.** $f(x) = \dfrac{\tan x}{1 + x^2}$

19. The amount of water in a tank t minutes after it has started to drain is given by $W = 100(t - 15)^2$ gal.

(a) At what rate is the water running out at the end of 5 min?

(b) What is the average rate at which the water flows out during the first 5 min?

20. Use the formula $V = l^3$ for the volume of a cube of side l to find

(a) the average rate at which the volume of a cube changes with l as l increases from $l = 2$ to $l = 4$

(b) the instantaneous rate at which the volume of a cube changes with l when $l = 5$.

21–22 Zoom in on the graph of f on an interval containing $x = x_0$ until the graph looks like a straight line. Estimate the slope of this line and then check your answer by finding the exact value of $f'(x_0)$.

21. (a) $f(x) = x^2 - 1,\ x_0 = 1.8$

(b) $f(x) = \dfrac{x^2}{x - 2},\ x_0 = 3.5$

22. (a) $f(x) = x^3 - x^2 + 1,\ x_0 = 2.3$

(b) $f(x) = \dfrac{x}{x^2 + 1},\ x_0 = -0.5$

23. Suppose that a function f is differentiable at $x = 1$ and

$$\lim_{h \to 0} \frac{f(1 + h)}{h} = 5$$

Find $f(1)$ and $f'(1)$.

24. Suppose that f has the property $f(x + y) = f(x)f(y)$ for all values of x and y and that $f(0) = f'(0) = 1$. Show that f is differentiable and $f'(x) = f(x)$. [*Hint:* Start by expressing $f'(x)$ as a limit.]

25. Find the equations of all lines through the origin that are tangent to the curve $y = x^3 - 9x^2 - 16x$.

26. Find all values of x for which the tangent line to the curve $y = 2x^3 - x^2$ is perpendicular to the line $x + 4y = 10$.

27. Let $f(x) = x^2$. Show that for any distinct values of a and b, the slope of the tangent line to $y = f(x)$ at $x = \frac{1}{2}(a + b)$ is equal to the slope of the secant line through the points (a, a^2) and (b, b^2). Draw a picture to illustrate this result.

28. In each part, evaluate the expression given that $f(1) = 1$, $g(1) = -2$, $f'(1) = 3$, and $g'(1) = -1$.

(a) $\left.\dfrac{d}{dx}[f(x)g(x)]\right|_{x=1}$ (b) $\left.\dfrac{d}{dx}\left[\dfrac{f(x)}{g(x)}\right]\right|_{x=1}$

(c) $\left.\dfrac{d}{dx}\left[\sqrt{f(x)}\right]\right|_{x=1}$ (d) $\dfrac{d}{dx}[f(1)g'(1)]$

29–30 Find $f'(x)$.

29. (a) $f(x) = x^8 - 3\sqrt{x} + 5x^{-3}$
(b) $f(x) = (2x + 1)^{101}(5x^2 - 7)$
(c) $f(x) = \sqrt{3x + 1}(x - 1)^2$
(d) $f(x) = \left(\dfrac{3x + 1}{x^2}\right)^3$

30. (a) $f(x) = \sin x + 2\cos^3 x$
(b) $f(x) = (1 + \sec x)(x^2 - \tan x)$
(c) $f(x) = \cot\left(\dfrac{\csc 2x}{x^3 + 5}\right)$
(d) $f(x) = \dfrac{1}{2x + \sin^3 x}$

31–32 Find the values of x at which the curve $y = f(x)$ has a horizontal tangent line.

31. $f(x) = (2x + 7)^6(x - 2)^5$ **32.** $f(x) = \dfrac{(x - 3)^4}{x^2 + 2x}$

33. Find all lines that are simultaneously tangent to the graph of $y = x^2 + 1$ and to the graph of $y = -x^2 - 1$.

34. (a) Let n denote an even positive integer. Generalize the result of Exercise 33 by finding all lines that are simultaneously tangent to the graph of $y = x^n + n - 1$ and to the graph of $y = -x^n - n + 1$.
(b) Let n denote an odd positive integer. Are there any lines that are simultaneously tangent to the graph of $y = x^n + n - 1$ and to the graph of $y = -x^n - n + 1$? Explain.

35. Find all values of x for which the line that is tangent to $y = 3x - \tan x$ is parallel to the line $y - x = 2$.

36. Approximate the values of x at which the tangent line to the graph of $y = x^3 - \sin x$ is horizontal.

37. Suppose that $f(x) = M\sin x + N\cos x$ for some constants M and N. If $f(\pi/4) = 3$ and $f'(\pi/4) = 1$, find an equation for the tangent line to $y = f(x)$ at $x = 3\pi/4$.

38. Suppose that $f(x) = M\tan x + N\sec x$ for some constants M and N. If $f(\pi/4) = 2$ and $f'(\pi/4) = 0$, find an equation for the tangent line to $y = f(x)$ at $x = 0$.

39. Suppose that $f'(x) = 2x \cdot f(x)$ and $f(2) = 5$.
(a) Find $g'(\pi/3)$ if $g(x) = f(\sec x)$.
(b) Find $h'(2)$ if $h(x) = [f(x)/(x - 1)]^4$.

40–42 Find dy/dx.

40. $y = \sqrt[4]{6x - 5}$ **41.** $y = \sqrt[3]{x^2 + x}$ **42.** $y = \dfrac{(3 - 2x)^{4/3}}{x^2}$

43–44 (a) Find dy/dx using the method of implicit differentiation. (b) Solve the equation for y as a function of x, and find dy/dx from that equation. (c) Confirm that the two results are consistent by expressing the derivative in part (a) as a function of x alone.

43. $x^3 + xy - 2x = 1$ **44.** $xy = x - y$

45–48 Find dy/dx by implicit differentiation.

45. $\dfrac{1}{y} + \dfrac{1}{x} = 1$ **46.** $x^3 - y^3 = 6xy$

47. $\sec(xy) = y$ **48.** $x^2 = \dfrac{\cot y}{1 + \csc y}$

49. If $3x^2 - 4y^2 = 7$, find d^2y/dx^2 by implicit differentiation.

50. Under certain conditions, the period T of a clock pendulum (i.e., the time required for one back-and-forth movement) is given in terms of its length L by $T = 2\pi\sqrt{L/g}$, where g is the constant acceleration due to gravity.
(a) Assuming that the length of a clock pendulum can vary (say, due to temperature changes), find the rate of change of the period T with respect to the length L.
(b) If L is in meters (m) and T is in seconds (s), what are the units for the rate of change in part (a)?
(c) If a pendulum clock is running slow, should the length of the pendulum be increased or decreased to correct the problem?
(d) The constant g generally decreases with altitude. If you move a pendulum clock from sea level to a higher elevation, will it run faster or slower?
(e) Assuming the length of the pendulum to be constant, find the rate of change of the period T with respect to g.
(f) Assuming that T is in seconds (s) and g is in meters per second squared (m/s^2), find the units for the rate of change in part (e).

51. An oil slick on a lake is surrounded by a floating circular containment boom. As the boom is pulled in, the circular containment area shrinks. If the boom is pulled in at the rate of 5 m/min, at what rate is the containment area shrinking when the containment area has a diameter of 100 m?

52. The hypotenuse of a right triangle is growing at a constant rate of a centimeters per second and one leg is decreasing at a constant rate of b centimeters per second. How fast is the acute angle between the hypotenuse and the other leg changing at the instant when both legs are 1 cm?

53. In each part, use the given information to find Δx, Δy, and dy.
(a) $y = 1/(x - 1)$; x decreases from 2 to 1.5.
(b) $y = \tan x$; x increases from $-\pi/4$ to 0.
(c) $y = \sqrt{25 - x^2}$; x increases from 0 to 3.

54. Use an appropriate local linear approximation to estimate the value of $\cot 46°$, and compare your answer to the value obtained with a calculating device.

55. The base of the Great Pyramid at Giza is a square that is 230 m on each side.
(a) As illustrated in the accompanying figure, suppose that an archaeologist standing at the center of a side measures the angle of elevation of the apex to be $\phi = 51°$ with an error of $\pm 0.5°$. What can the archaeologist reasonably say about the height of the pyramid?

(b) Use differentials to estimate the allowable error in the elevation angle that will ensure that the error in calculating the height is at most ± 5 m.

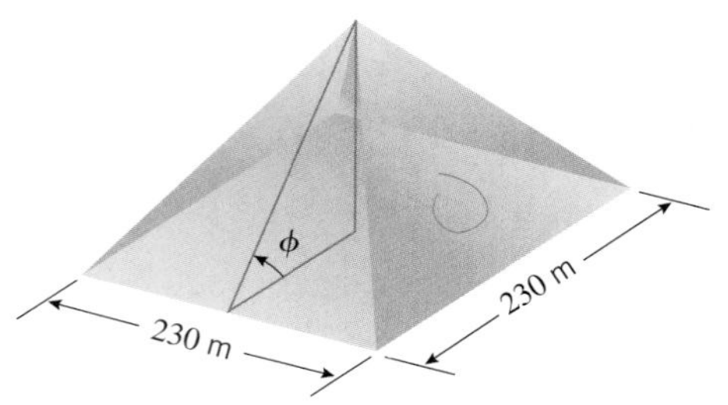

Figure Ex-55

EXPANDING THE CALCULUS HORIZON

Wouldn't it be great if you had a robot to perform your basic but boring errands and chores, so you could concentrate on your studies? How easy would it be to design such a robot? To learn more about the mathematics of robotics, and to build upon the mathematics learned in this chapter, go to

www.wiley.com/college/anton

c h a p t e r f o u r

THE DERIVATIVE IN GRAPHING AND APPLICATIONS

In the fall of 1972 President Nixon announced that the rate of increase of inflation was decreasing. This was the first time a sitting president used the third derivative to advance his case for reelection.

—Hugo Rossi
Mathematician

In this chapter we will study various applications of the derivative. For example, we will use methods of calculus to analyze functions and their graphs. In the process, we will show how calculus and graphing utilities, working together, can provide most of the important information about the behavior of functions. Another important application of the derivative will be in the solution of optimization problems. For example, if time is the main consideration in a problem, we might be interested in finding the quickest way to perform a task, and if cost is the main consideration, we might be interested in finding the least expensive way to perform a task. Mathematically, optimization problems can be reduced to finding the largest or smallest value of a function on some interval, and determining where the largest or smallest value occurs. Using the derivative, we will develop the mathematical tools necessary for solving such problems. We will also use the derivative to study the motion of a particle moving along a line, and we will show how the derivative can help us to approximate solutions of equations.

Photo: *Derivatives can help to find the most cost-effective location for an offshore oil-drilling rig.*

4.1 ANALYSIS OF FUNCTIONS I: INCREASE, DECREASE, AND CONCAVITY

Although graphing utilities are useful for determining the general shape of a graph, many problems require more precision than graphing utilities are capable of producing. The purpose of this section is to develop mathematical tools that can be used to determine the exact shape of a graph and the precise locations of its key features.

INCREASING AND DECREASING FUNCTIONS

The terms *increasing*, *decreasing*, and *constant* are used to describe the behavior of a function over an interval as we travel left to right along its graph. For example, the function graphed in Figure 4.1.1 can be described as increasing on the interval $(-\infty, 0]$, decreasing on the interval $[0, 2]$, increasing again on the interval $[2, 4]$, and constant on the interval $[4, +\infty)$.

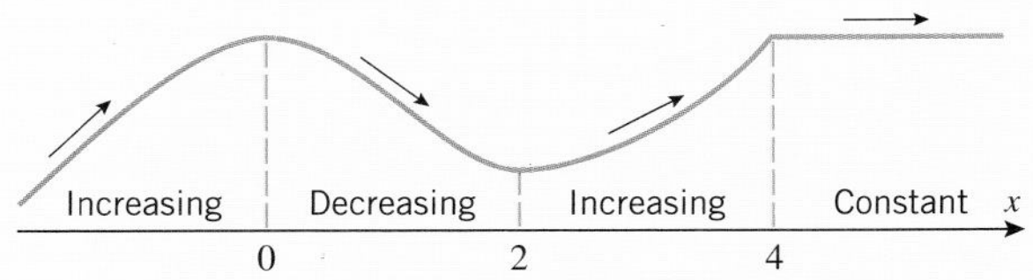

Figure 4.1.1

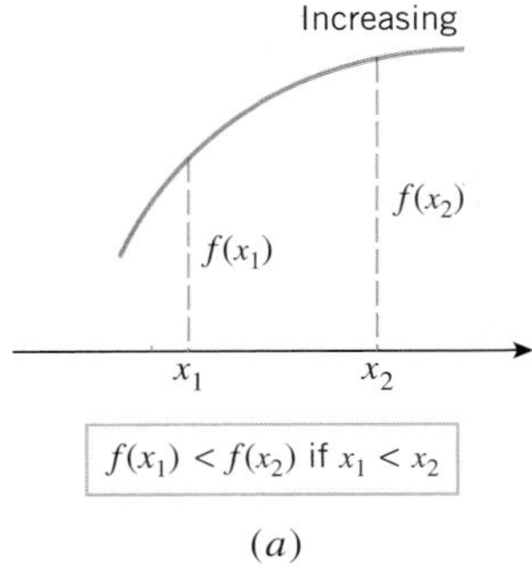

(a)

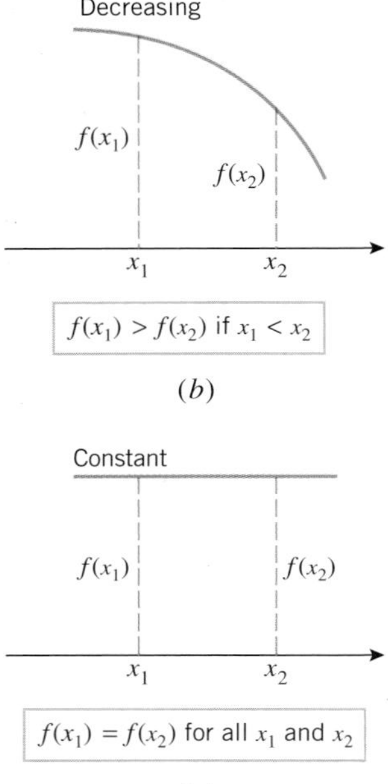

(c)

Figure 4.1.2

The following definition, which is illustrated in Figure 4.1.2, expresses these intuitive ideas precisely.

> **4.1.1 DEFINITION.** Let f be defined on an interval, and let x_1 and x_2 denote points in that interval.
>
> (a) f is ***increasing*** on the interval if $f(x_1) < f(x_2)$ whenever $x_1 < x_2$.
>
> (b) f is ***decreasing*** on the interval if $f(x_1) > f(x_2)$ whenever $x_1 < x_2$.
>
> (c) f is ***constant*** on the interval if $f(x_1) = f(x_2)$ for all points x_1 and x_2.

Figure 4.1.3 suggests that a differentiable function f is increasing on any interval where each tangent line to its graph has positive slope, is decreasing on any interval where each tangent line to its graph has negative slope, and is constant on any interval where each tangent line to its graph has zero slope. This intuitive observation suggests the following important theorem that will be proved in Section 4.7.

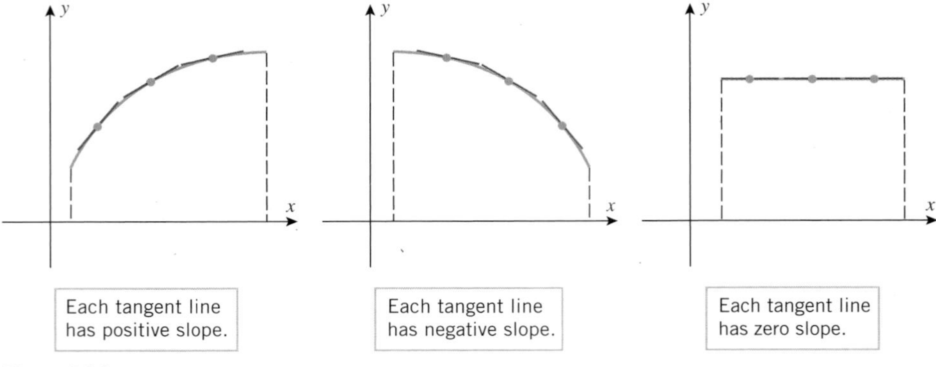

Figure 4.1.3

> **4.1.2 THEOREM.** *Let f be a function that is continuous on a closed interval $[a, b]$ and differentiable on the open interval (a, b).*
>
> (a) *If $f'(x) > 0$ for every value of x in (a, b), then f is increasing on $[a, b]$.*
>
> (b) *If $f'(x) < 0$ for every value of x in (a, b), then f is decreasing on $[a, b]$.*
>
> (c) *If $f'(x) = 0$ for every value of x in (a, b), then f is constant on $[a, b]$.*

Observe that the derivative conditions in Theorem 4.1.2 are only required to hold *inside* the interval $[a, b]$, even though the conclusions apply to the entire interval.

Although stated for closed intervals, Theorem 4.1.2 is applicable on any interval I on which f is continuous. For example, if f is continuous on $[a, +\infty)$ and $f'(x) > 0$ on $(a, +\infty)$, then f is increasing on $[a, +\infty)$; and if f is continuous on $(-\infty, +\infty)$ and $f'(x) < 0$ on $(-\infty, +\infty)$, then f is decreasing on $(-\infty, +\infty)$.

► **Example 1** Find the intervals on which $f(x) = x^2 - 4x + 3$ is increasing and the intervals on which it is decreasing.

Solution. The graph of f in Figure 4.1.4 suggests that f is decreasing for $x \le 2$ and increasing for $x \ge 2$. To confirm this, we analyze the sign of f'. The derivative of f is

$$f'(x) = 2x - 4 = 2(x - 2)$$

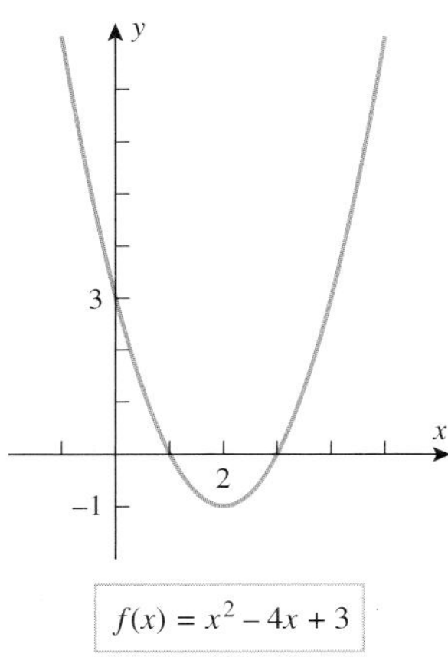

Figure 4.1.4

It follows that

$$f'(x) < 0 \quad \text{if} \quad x < 2$$
$$f'(x) > 0 \quad \text{if} \quad 2 < x$$

Since f is continuous everywhere, it follows from the comment after Theorem 4.1.2 that

$$f \text{ is decreasing on } (-\infty, 2]$$
$$f \text{ is increasing on } [2, +\infty)$$

These conclusions are consistent with the graph of f in Figure 4.1.4. ◀

▶ **Example 2** Find the intervals on which $f(x) = x^3$ is increasing and the intervals on which it is decreasing.

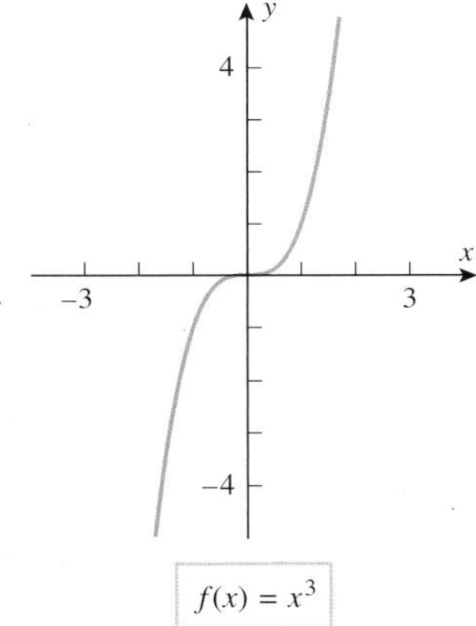

Figure 4.1.5

Solution. The graph of f in Figure 4.1.5 suggests that f is increasing over the entire x-axis. To confirm this, we differentiate f to obtain $f'(x) = 3x^2$. Thus,

$$f'(x) > 0 \quad \text{if} \quad x < 0$$
$$f'(x) > 0 \quad \text{if} \quad 0 < x$$

Since f is continuous everywhere,

$$f \text{ is increasing on } (-\infty, 0]$$
$$f \text{ is increasing on } [0, +\infty)$$

Since f is increasing on the "abutting" intervals $(-\infty, 0]$ and $[0, +\infty)$, it follows that f is increasing on their union $(-\infty, +\infty)$ (see Exercise 51). ◀

▶ **Example 3**

(a) Use the graph of $f(x) = 3x^4 + 4x^3 - 12x^2 + 2$ in Figure 4.1.6 to make a conjecture about the intervals on which f is increasing or decreasing.

(b) Use Theorem 4.1.2 to determine whether your conjecture is correct.

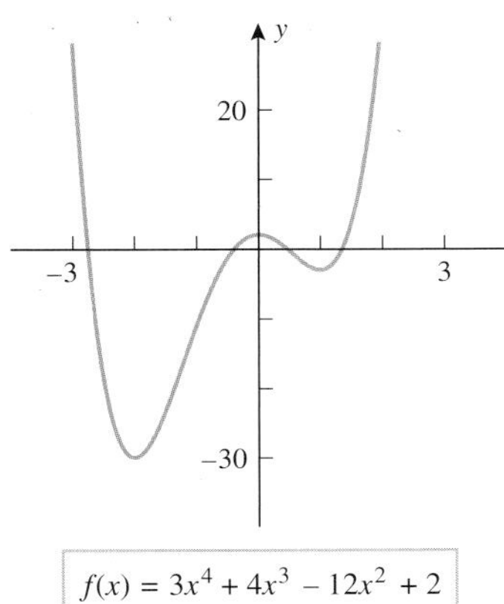

Figure 4.1.6

Solution (a). The graph suggests that the function f is decreasing if $x \le -2$, increasing if $-2 \le x \le 0$, decreasing if $0 \le x \le 1$, and increasing if $x \ge 1$.

Solution (b). Differentiating f we obtain

$$f'(x) = 12x^3 + 12x^2 - 24x = 12x(x^2 + x - 2) = 12x(x + 2)(x - 1)$$

The sign analysis of f' in Table 4.1.1 can be obtained using the method of test points discussed in Web Appendix D. The conclusions in Table 4.1.1 confirm the conjecture in part (a). ◀

Table 4.1.1

INTERVAL	$(12x)(x+2)(x-1)$	$f'(x)$	CONCLUSION
$x < -2$	(−) (−) (−)	−	f is decreasing on $(-\infty, -2]$
$-2 < x < 0$	(−) (+) (−)	+	f is increasing on $[-2, 0]$
$0 < x < 1$	(+) (+) (−)	−	f is decreasing on $[0, 1]$
$1 < x$	(+) (+) (+)	+	f is increasing on $[1, +\infty)$

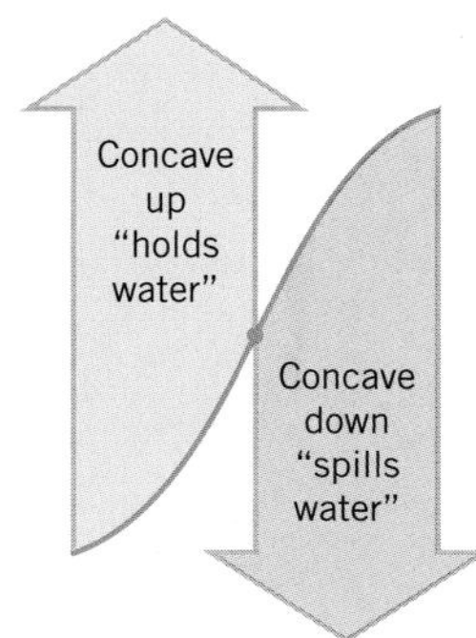

Figure 4.1.7

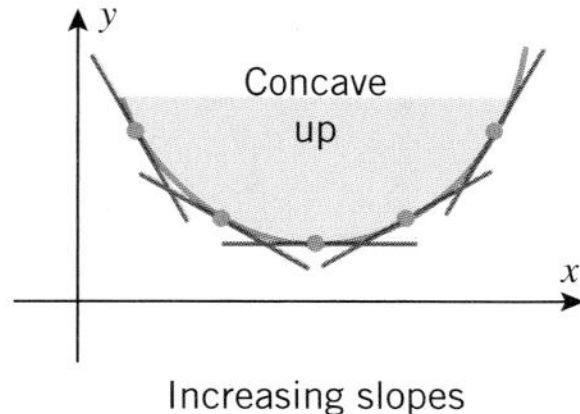

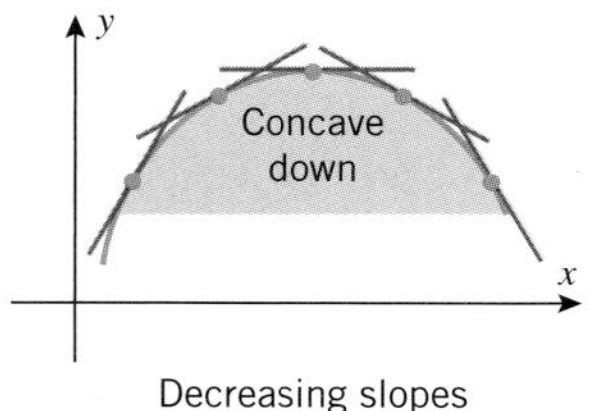

Figure 4.1.8

CONCAVITY

Although the sign of the derivative of f reveals where the graph of f is increasing or decreasing, it does not reveal the direction of *curvature*. For example, the graph is increasing on both sides of the point in Figure 4.1.7, but on the left side it has an upward curvature ("holds water") and on the right side it has a downward curvature ("spills water"). On intervals where the graph of f has upward curvature we say that f is *concave up*, and on intervals where the graph has downward curvature we say that f is *concave down*.

Figure 4.1.8 suggests two ways to characterize the concavity of a differentiable function f on an open interval:

- f is concave up on an open interval if its tangent lines have increasing slopes on that interval and is concave down if they have decreasing slopes.
- f is concave up on an open interval if its graph lies above its tangent lines on that interval and is concave down if it lies below its tangent lines.

Our formal definition for "concave up" and "concave down" corresponds to the first of these characterizations.

> **4.1.3 DEFINITION.** If f is differentiable on an open interval I, then f is said to be ***concave up*** on I if f' is increasing on I, and f is said to be ***concave down*** on I if f' is decreasing on I.

Since the slopes of the tangent lines to the graph of a differentiable function f are the values of its derivative f', it follows from Theorem 4.1.2 (applied to f' rather than f) that f' will be increasing on intervals where f'' is positive and that f' will be decreasing on intervals where f'' is negative. Thus, we have the following theorem.

> **4.1.4 THEOREM.** *Let f be twice differentiable on an open interval I.*
>
> *(a) If $f''(x) > 0$ for every value of x in I, then f is concave up on I.*
>
> *(b) If $f''(x) < 0$ for every value of x in I, then f is concave down on I.*

► **Example 4** Figure 4.1.4 suggests that the function $f(x) = x^2 - 4x + 3$ is concave up on the interval $(-\infty, +\infty)$. This is consistent with Theorem 4.1.4, since $f'(x) = 2x - 4$ and $f''(x) = 2$, so

$$f''(x) > 0 \quad \text{on the interval } (-\infty, +\infty)$$

Also, Figure 4.1.5 suggests that $f(x) = x^3$ is concave down on the interval $(-\infty, 0)$ and concave up on the interval $(0, +\infty)$. This agrees with Theorem 4.1.4, since $f'(x) = 3x^2$ and $f''(x) = 6x$, so

$$f''(x) < 0 \quad \text{if } x < 0 \quad \text{and} \quad f''(x) > 0 \quad \text{if } x > 0$$ ◄

INFLECTION POINTS

We see from Example 4 and Figure 4.1.5 that the graph of $f(x) = x^3$ changes from concave down to concave up at $x = 0$. Points where a curve changes from concave up to concave down or vice versa are of special interest, so there is some terminology associated with them.

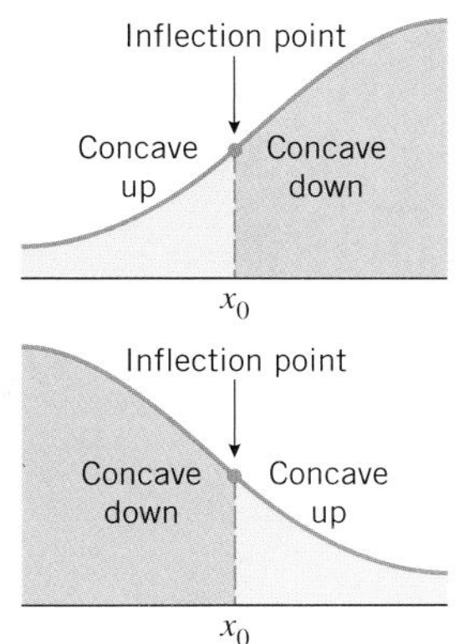

Figure 4.1.9

> **4.1.5 DEFINITION.** If f is continuous on an open interval containing a value x_0, and if f changes the direction of its concavity at the point $(x_0, f(x_0))$, then we say that f has an ***inflection point at*** $\boldsymbol{x_0}$, and we call the point $(x_0, f(x_0))$ on the graph of f an ***inflection point*** of f (Figure 4.1.9).

▸ **Example 5** Figure 4.1.10 shows the graph of the function $f(x) = x^3 - 3x^2 + 1$. Use the first and second derivatives of f to determine the intervals on which f is increasing, decreasing, concave up, and concave down. Locate all inflection points and confirm that your conclusions are consistent with the graph.

Solution. Calculating the first two derivatives of f we obtain

$$f'(x) = 3x^2 - 6x = 3x(x - 2)$$

$$f''(x) = 6x - 6 = 6(x - 1)$$

The sign analysis of these derivatives is shown in the following tables:

INTERVAL	$(3x)(x-2)$	$f'(x)$	CONCLUSION
$x < 0$	$(-)(-)$	$+$	f is increasing on $(-\infty, 0]$
$0 < x < 2$	$(+)(-)$	$-$	f is decreasing on $[0, 2]$
$x > 2$	$(+)(+)$	$+$	f is increasing on $[2, +\infty)$

INTERVAL	$6(x-1)$	$f''(x)$	CONCLUSION
$x < 1$	$(-)$	$-$	f is concave down on $(-\infty, 1)$
$x > 1$	$(+)$	$+$	f is concave up on $(1, +\infty)$

2

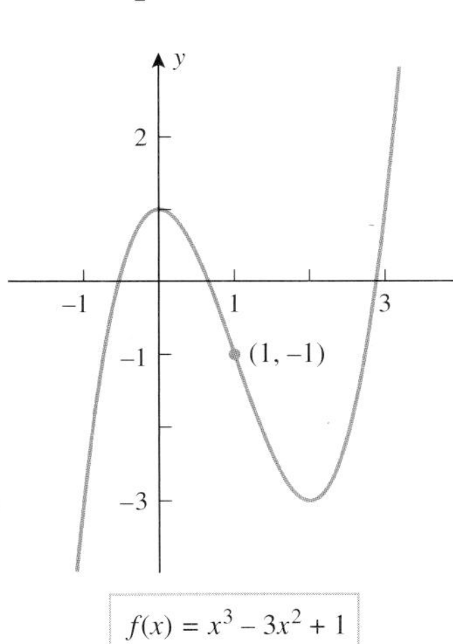

Figure 4.1.10

The second table shows that there is an inflection point at $x = 1$, since f changes from concave down to concave up at that point. The inflection point is $(1, f(1)) = (1, -1)$. All of these conclusions are consistent with the graph of f. ◂

One can correctly guess from Figure 4.1.10 that the function $f(x) = x^3 - 3x^2 + 1$ has an inflection point at $x = 1$ without actually computing derivatives. However, sometimes changes in concavity are so subtle that calculus is essential to confirm their existence and identify their location.

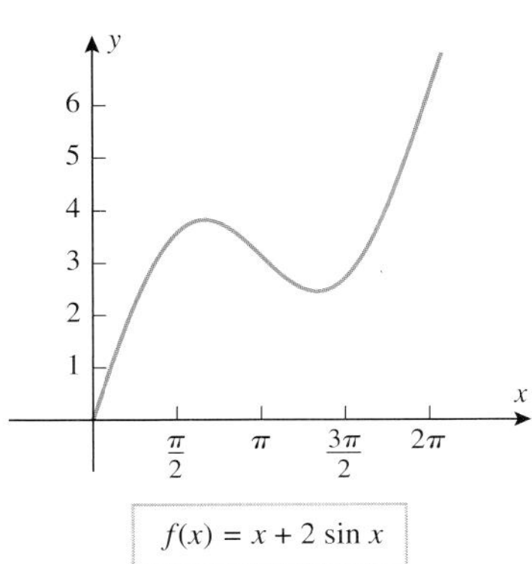

Figure 4.1.11

▸ **Example 6** Figure 4.1.11 shows the graph of the function $f(x) = x + 2\sin x$ over the interval $[0, 2\pi]$. Use the first and second derivatives of f to determine where f is increasing, decreasing, concave up, and concave down. Locate all inflection points and confirm that your conclusions are consistent with the graph.

Solution. Calculating the first two derivatives of f we obtain

$$f'(x) = 1 + 2\cos x$$

$$f''(x) = -2\sin x$$

Since f' is a continuous function, it changes sign on the interval $(0, 2\pi)$ only at points where $f'(x) = 0$ (why?). These values are solutions of the equation

$$1 + 2\cos x = 0 \quad \text{or equivalently} \quad \cos x = -\tfrac{1}{2}$$

There are two solutions of this equation in the interval $(0, 2\pi)$, namely, $x = 2\pi/3$ and $x = 4\pi/3$ (verify). Similarly, f'' is a continuous function, so its sign changes in the interval $(0, 2\pi)$ will occur only at values of x for which $f''(x) = 0$. These values are solutions of the equation

$$-2\sin x = 0$$

There is one solution of this equation in the interval $(0, 2\pi)$, namely, $x = \pi$. With the help of these "sign transition points" we obtain the sign analysis shown in the following tables:

INTERVAL	$f'(x) = 1 + 2\cos x$	CONCLUSION
$0 < x < 2\pi/3$	+	f is increasing on $[0, 2\pi/3]$
$2\pi/3 < x < 4\pi/3$	–	f is decreasing on $[2\pi/3, 4\pi/3]$
$4\pi/3 < x < 2\pi$	+	f is increasing on $[4\pi/3, 2\pi]$

Two ways to determine the signs in the tables in Example 6 are the method of test points and using the unit circle definition of sine and cosine. See Web Appendix D and Appendix A.

INTERVAL	$f''(x) = -2\sin x$	CONCLUSION
$0 < x < \pi$	–	f is concave down on $(0, \pi)$
$\pi < x < 2\pi$	+	f is concave up on $(\pi, 2\pi)$

The second table shows that there is an inflection point at $x = \pi$, since f changes from concave down to concave up at that point. All of these conclusions are consistent with the graph of f. ◀

In the preceding examples the inflection points of f occurred wherever $f''(x) = 0$. However, this is not always the case. Here is a specific example.

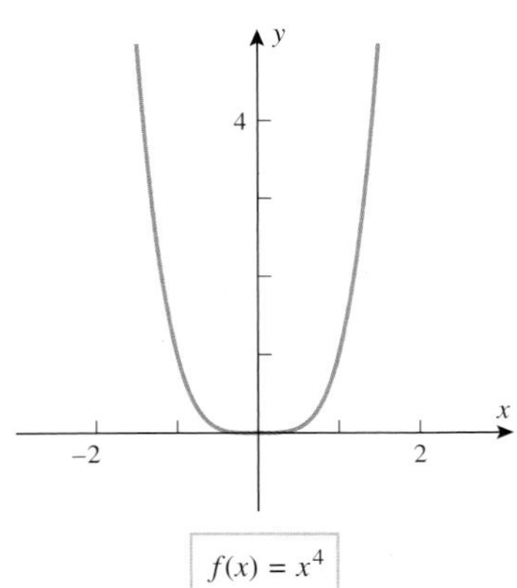

Figure 4.1.12

► **Example 7** Find the inflection points, if any, of $f(x) = x^4$.

Solution. Calculating the first two derivatives of f we obtain

$$f'(x) = 4x^3$$
$$f''(x) = 12x^2$$

Since $f''(x)$ is positive for $x < 0$ and for $x > 0$, the function f is concave up on the interval $(-\infty, 0)$ and on the interval $(0, +\infty)$. Thus, there is no change in concavity and hence no inflection point at $x = 0$, even though $f''(0) = 0$ (Figure 4.1.12). ◀

We will see later that if a function f has an inflection point at $x = x_0$ and $f''(x_0)$ exists, then $f''(x_0) = 0$. Also, we will see in Section 4.3 that an inflection point may also occur where $f''(x)$ is not defined.

INFLECTION POINTS IN APPLICATIONS

Inflection points of a function f are those points on the graph of $y = f(x)$ where the slopes of the tangent lines change from increasing to decreasing or vice versa (Figure 4.1.13). Since the slope of the tangent line at a point on the graph of $y = f(x)$ can be interpreted as the rate of change of y with respect to x at that point, we can interpret inflection points in the following way:

Give an argument to show that the function $f(x) = x^4$ graphed in Figure 4.1.12 is concave up on the interval $(-\infty, +\infty)$.

Inflection points mark the places on the curve $y = f(x)$ where the rate of change of y with respect to x changes from increasing to decreasing, or vice versa.

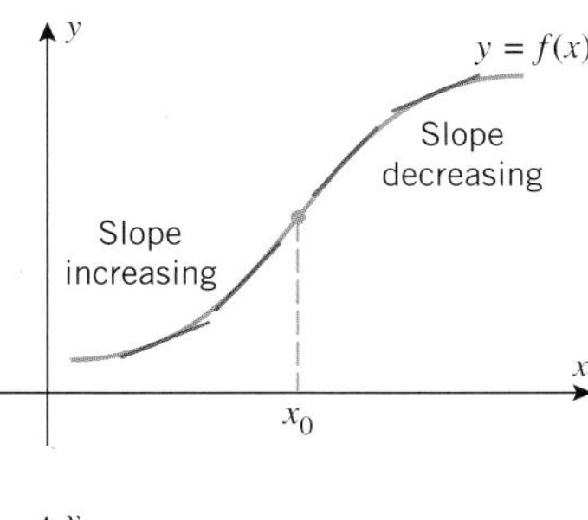

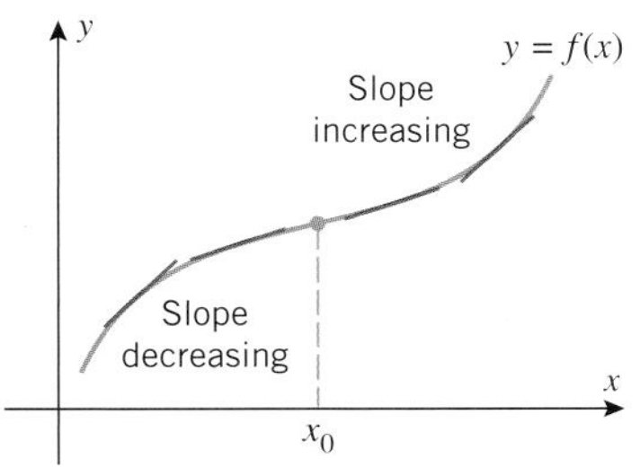

Figure 4.1.13

This is a subtle idea, since we are dealing with a change in a rate of change. It can help with your understanding of this idea to realize that inflection points may have interpretations in more familiar contexts. For example, consider the statement "Oil prices rose sharply during the first half of the year but have since begun to level off." If the price of oil is plotted as a function of time of year, this statement suggests the existence of an inflection point on the graph near the end of June. (Why?) To give a more visual example, consider the flask shown in Figure 4.1.14. Suppose that water is added to the flask so that the volume increases at a constant rate with respect to the time t, and let us examine the rate at which the water level y rises with respect to t. Initially, the level y will rise at a slow rate because of the wide base. However, as the diameter of the flask narrows, the rate at which the level y rises will increase until the level is at the narrow point in the neck. From that point on the rate at which the level rises will decrease as the diameter gets wider and wider. Thus, the narrow point in the neck is the point at which the rate of change of y with respect to t changes from increasing to decreasing.

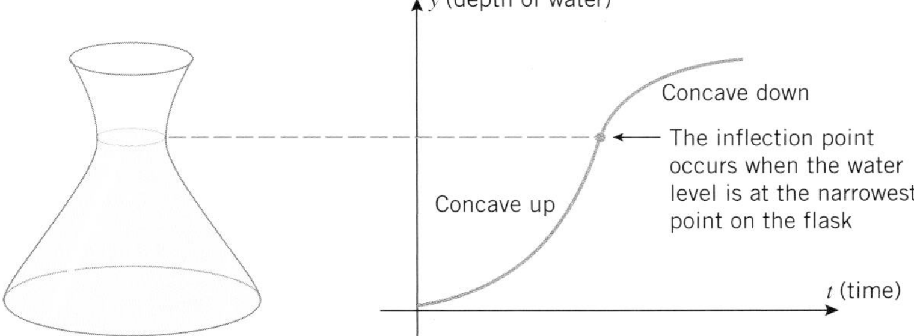

Figure 4.1.14

✔ QUICK CHECK EXERCISES 4.1

(See page 234 for answers.)

1. (a) A function f is increasing on (a, b) if ________ whenever $a < x_1 < x_2 < b$.
(b) A function f is decreasing on (a, b) if ________ whenever $a < x_1 < x_2 < b$.
(c) A function f is concave up on (a, b) if f' is ________ on (a, b).
(d) If $f''(a)$ exists and f has an inflection point at $x = a$, then $f''(a)$ ________.

2. Let $f(x) = 0.1(x^3 - 3x^2 - 9x)$.
(a) Solutions to $f'(x) = 0$ are $x =$ ________.
(b) The function f is increasing on the interval(s) ________.
(c) The function f is concave down on the interval(s) ________.
(d) ________ is an inflection point on the graph of f.

3. If $f(x)$ has the derivative $f'(x) = x(x-4)^2$ then:
(a) The function f is increasing on the interval(s) ________.
(b) The function f is concave up on the interval(s) ________.
(c) The function f is concave down on the interval(s) ________.

4. Consider the statement "The rise in the cost of living slowed during the first half of the year." If we graph the cost of living versus time for the first half of the year, how does the graph reflect this statement?

EXERCISE SET 4.1

 Graphing Utility C CAS

FOCUS ON CONCEPTS

1. In each part, sketch the graph of a function f with the stated properties, and discuss the signs of f' and f''.
(a) The function f is concave up and increasing on the interval $(-\infty, +\infty)$.
(b) The function f is concave down and increasing on the interval $(-\infty, +\infty)$.
(c) The function f is concave up and decreasing on the interval $(-\infty, +\infty)$.
(d) The function f is concave down and decreasing on the interval $(-\infty, +\infty)$.

2. In each part, sketch the graph of a function f with the stated properties.
(a) f is increasing on $(-\infty, +\infty)$, has an inflection point at the origin, and is concave up on $(0, +\infty)$.

(b) f is increasing on $(-\infty, +\infty)$, has an inflection point at the origin, and is concave down on $(0, +\infty)$.

(c) f is decreasing on $(-\infty, +\infty)$, has an inflection point at the origin, and is concave up on $(0, +\infty)$.

(d) f is decreasing on $(-\infty, +\infty)$, has an inflection point at the origin, and is concave down on $(0, +\infty)$.

3. Use the graph of the equation $y = f(x)$ in the accompanying figure to find the signs of dy/dx and d^2y/dx^2 at the points A, B, and C.

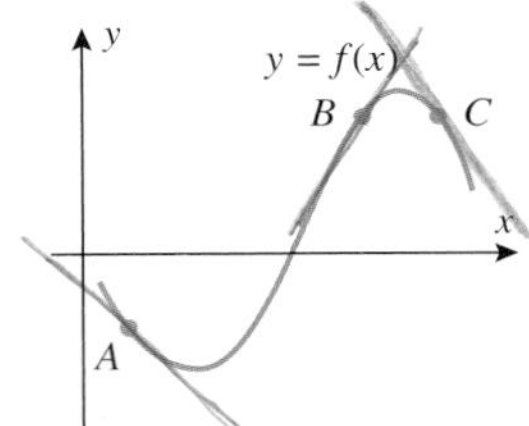

Figure Ex-3

4. Use the graph of the equation $y = f'(x)$ in the accompanying figure to find the signs of dy/dx and d^2y/dx^2 at the points A, B, and C.

5. Use the graph of $y = f''(x)$ in the accompanying figure to determine the x-coordinates of all inflection points of f. Explain your reasoning.

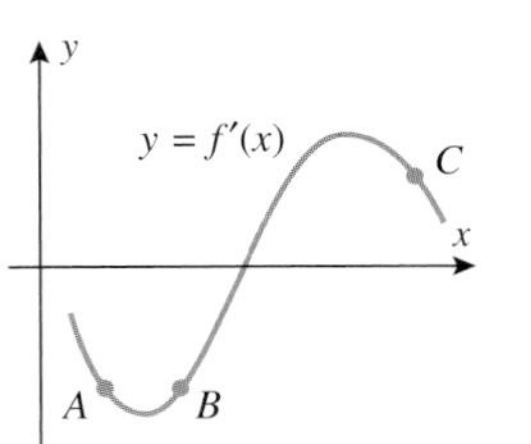

Figure Ex-4

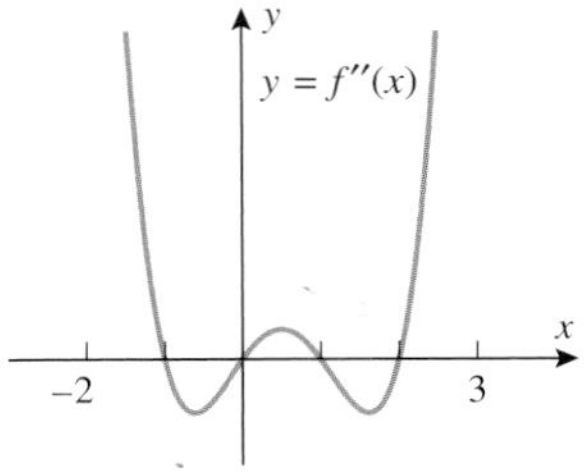

Figure Ex-5

6. Use the graph of $y = f'(x)$ in the accompanying figure to replace the question mark with $<$, $=$, or $>$, as appropriate. Explain your reasoning.

(a) $f(0)$? $f(1)$ (b) $f(1)$? $f(2)$ (c) $f'(0)$? 0

(d) $f'(1)$? 0 (e) $f''(0)$? 0 (f) $f''(2)$? 0

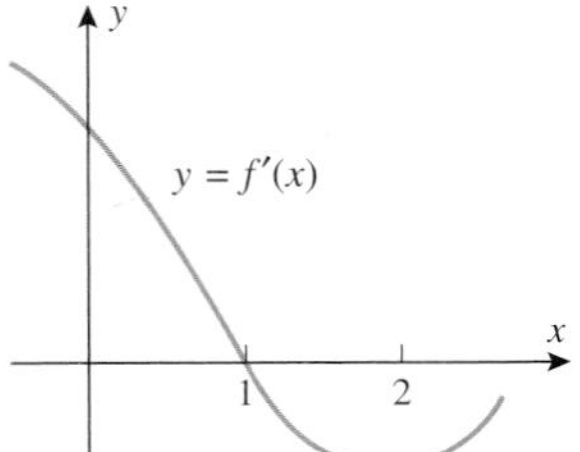

Figure Ex-6

7. In each part, use the graph of $y = f(x)$ in the accompanying figure to find the requested information.

(a) Find the intervals on which f is increasing.

(b) Find the intervals on which f is decreasing.

(c) Find the open intervals on which f is concave up.

(d) Find the open intervals on which f is concave down.

(e) Find all values of x at which f has an inflection point.

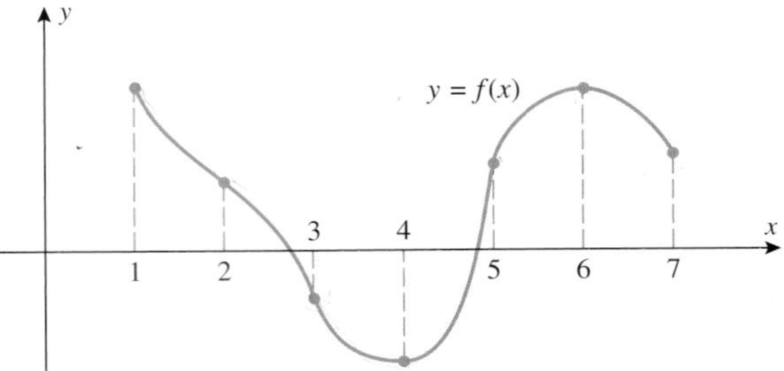

Figure Ex-7

8. Use the graph in Exercise 7 to make a table that shows the signs of f' and f'' over the intervals (1, 2), (2, 3), (3, 4), (4, 5), (5, 6), and (6, 7).

9–10 A sign chart is presented for the first and second derivatives of a function f. Assuming that f is continuous everywhere, find: (a) the intervals on which f is increasing, (b) the intervals on which f is decreasing, (c) the open intervals on which f is concave up, (d) the open intervals on which f is concave down, and (e) the x-coordinates of all inflection points.

9.

INTERVAL	SIGN OF $f'(x)$	SIGN OF $f''(x)$
$x < 1$	$-$	$+$
$1 < x < 2$	$+$	$+$
$2 < x < 3$	$+$	$-$
$3 < x < 4$	$-$	$-$
$4 < x$	$-$	$+$

10.

INTERVAL	SIGN OF $f'(x)$	SIGN OF $f''(x)$
$x < 1$	$+$	$+$
$1 < x < 3$	$+$	$-$
$3 < x$	$+$	$+$

11–22 Find: (a) the intervals on which f is increasing, (b) the intervals on which f is decreasing, (c) the open intervals on which f is concave up, (d) the open intervals on which f is concave down, and (e) the x-coordinates of all inflection points.

11. $f(x) = x^2 - 3x + 8$

12. $f(x) = 5 - 4x - x^2$

13. $f(x) = (2x + 1)^3$

14. $f(x) = 5 + 12x - x^3$

15. $f(x) = 3x^4 - 4x^3$

16. $f(x) = x^4 - 5x^3 + 9x^2$

17. $f(x) = \dfrac{x - 2}{(x^2 - x + 1)^2}$

18. $f(x) = \dfrac{x}{x^2 + 2}$

19. $f(x) = \sqrt[3]{x^2 + x + 1}$

20. $f(x) = x^{4/3} - x^{1/3}$

21. $f(x) = (x^{2/3} - 1)^2$

22. $f(x) = x^{2/3} - x$

23–28 Analyze the trigonometric function f over the specified interval, stating where f is increasing, decreasing, concave up, and concave down, and stating the x-coordinates of all inflection points. Confirm that your results are consistent with the graph of f generated with a graphing utility.

23. $f(x) = \sin x - \cos x$; $[-\pi, \pi]$

24. $f(x) = \sec x \tan x$; $(-\pi/2, \pi/2)$

25. $f(x) = 1 - \tan(x/2)$; $(-\pi, \pi)$

26. $f(x) = 2x + \cot x$; $(0, \pi)$

27. $f(x) = (\sin x + \cos x)^2$; $[-\pi, \pi]$

28. $f(x) = \sin^2 2x$; $[0, \pi]$

FOCUS ON CONCEPTS

29. In each part sketch a continuous curve $y = f(x)$ with the stated properties.
(a) $f(2) = 4$, $f'(2) = 0$, $f''(x) > 0$ for all x
(b) $f(2) = 4$, $f'(2) = 0$, $f''(x) < 0$ for $x < 2$, $f''(x) > 0$ for $x > 2$
(c) $f(2) = 4$, $f''(x) < 0$ for $x \neq 2$ and $\lim_{x\to 2^+} f'(x) = +\infty$, $\lim_{x\to 2^-} f'(x) = -\infty$

30. In each part sketch a continuous curve $y = f(x)$ with the stated properties.
(a) $f(2) = 4$, $f'(2) = 0$, $f''(x) < 0$ for all x
(b) $f(2) = 4$, $f'(2) = 0$, $f''(x) > 0$ for $x < 2$, $f''(x) < 0$ for $x > 2$
(c) $f(2) = 4$, $f''(x) > 0$ for $x \neq 2$ and $\lim_{x\to 2^+} f'(x) = -\infty$, $\lim_{x\to 2^-} f'(x) = +\infty$

31. Suppose that $g(x)$ is a function that is defined and differentiable for all real numbers x and that $g(x)$ has the following properties:
(i) $g(0) = 2$ and $g'(0) = -\frac{2}{3}$.
(ii) $g(4) = 3$ and $g'(4) = 3$.
(iii) $g(x)$ is concave up for $x < 4$ and concave down for $x > 4$.
(iv) $g(x) \geq -10$ for all x.

(a) How many zeros does g have?
(b) How many zeros does g' have?
(c) Exactly one of the following limits is possible:

$$\lim_{x\to+\infty} g'(x) = -5, \quad \lim_{x\to+\infty} g'(x) = 0, \quad \lim_{x\to+\infty} g'(x) = 5$$

Identify which of these results is possible and draw a rough sketch of the graph of such a function $g(x)$. Explain why the other two results are impossible.

32. In each part, assume that a is a constant and find the inflection points, if any.
(a) $f(x) = (x - a)^3$ (b) $f(x) = (x - a)^4$

33. Given that a is a constant and n is a positive integer, what can you say about the existence of inflection points of the function $f(x) = (x - a)^n$? Justify your answer.

34. Suppose that f is an increasing function on $[a, b]$ and that x_0 is in (a, b). Use the definition of $f'(x_0)$ to prove that if f is differentiable at x_0, then $f'(x_0) \geq 0$.

35–38 If f is increasing on an interval $[0, b)$, then it follows from Definition 4.1.1 that $f(0) < f(x)$ for each x in the interval $(0, b)$. Use this result in these exercises.

35. Show that $\sqrt[3]{1+x} < 1 + \frac{1}{3}x$ if $x > 0$, and confirm the inequality with a graphing utility. [*Hint:* Show that the function $f(x) = 1 + \frac{1}{3}x - \sqrt[3]{1+x}$ is increasing on $[0, +\infty)$.]

36. Show that $x < \tan x$ if $0 < x < \pi/2$, and confirm the inequality with a graphing utility. [*Hint:* Show that the function $f(x) = \tan x - x$ is increasing on $[0, \pi/2)$.]

37. Use a graphing utility to make a conjecture about the relative sizes of x and $\sin x$ for $x \geq 0$, and prove your conjecture.

38. Use a graphing utility to make a conjecture about the relative sizes of $1 - x^2/2$ and $\cos x$ for $x \geq 0$, and prove your conjecture. [*Hint:* Use the result of Exercise 37.]

39–40 Use a graphing utility to generate the graphs of f' and f'' over the stated interval; then use those graphs to estimate the x-coordinates of the inflection points of f, the intervals on which f is concave up or down, and the intervals on which f is increasing or decreasing. Check your estimates by graphing f.

39. $f(x) = x^4 - 24x^2 + 12x$, $-5 \leq x \leq 5$

40. $f(x) = \dfrac{1}{1+x^2}$, $-5 \leq x \leq 5$

41–42 Use a CAS to find f'' and to approximate the x-coordinates of the inflection points to six decimal places. Confirm that your answer is consistent with the graph of f.

C **41.** $f(x) = \dfrac{10x - 3}{3x^2 - 5x + 8}$

C **42.** $f(x) = \dfrac{x^3 - 8x + 7}{\sqrt{x^2+1}}$

43. Use Definition 4.1.1 to prove that $f(x) = x^2$ is increasing on $[0, +\infty)$.

44. Use Definition 4.1.1 to prove that $f(x) = 1/x$ is decreasing on $(0, +\infty)$.

FOCUS ON CONCEPTS

45–46 Determine whether the statements are true or false. If a statement is false, find functions for which the statement fails to hold.

45. (a) If f and g are increasing on an interval, then so is $f + g$.
(b) If f and g are increasing on an interval, then so is $f \cdot g$.

46. (a) If f and g are concave up on an interval, then so is $f + g$.
(b) If f and g are concave up on an interval, then so is $f \cdot g$.

47. In each part, find functions f and g that are increasing on $(-\infty, +\infty)$ and for which $f - g$ has the stated property.
(a) $f - g$ is decreasing on $(-\infty, +\infty)$.
(b) $f - g$ is constant on $(-\infty, +\infty)$.
(c) $f - g$ is increasing on $(-\infty, +\infty)$.

48. In each part, find functions f and g that are positive and increasing on $(-\infty, +\infty)$ and for which f/g has the stated property.
(a) f/g is decreasing on $(-\infty, +\infty)$.
(b) f/g is constant on $(-\infty, +\infty)$.
(c) f/g is increasing on $(-\infty, +\infty)$.

49. (a) Prove that a general cubic polynomial

$$f(x) = ax^3 + bx^2 + cx + d \quad (a \neq 0)$$

has exactly one inflection point.
(b) Prove that if a cubic polynomial has three x-intercepts, then the inflection point occurs at the average value of the intercepts.
(c) Use the result in part (b) to find the inflection point of the cubic polynomial $f(x) = x^3 - 3x^2 + 2x$, and check your result by using f'' to determine where f is concave up and concave down.

50. From Exercise 49, the polynomial $f(x) = x^3 + bx^2 + 1$ has one inflection point. Use a graphing utility to reach a conclusion about the effect of the constant b on the location of the inflection point. Use f'' to explain what you have observed graphically.

51. Use Definition 4.1.1 to prove:
(a) If f is increasing on the intervals $(a, c]$ and $[c, b)$, then f is increasing on (a, b).
(b) If f is decreasing on the intervals $(a, c]$ and $[c, b)$, then f is decreasing on (a, b).

52. Use part (a) of Exercise 51 to show that $f(x) = x + \sin x$ is increasing on the interval $(-\infty, +\infty)$.

53. Use part (b) of Exercise 51 to show that $f(x) = \cos x - x$ is decreasing on the interval $(-\infty, +\infty)$.

54. Let $y = 1/(1 + x^2)$. Find the values of x for which y is increasing most rapidly or decreasing most rapidly.

FOCUS ON CONCEPTS

55–58 Suppose that water is flowing at a constant rate into the container shown. Make a rough sketch of the graph of the water level y versus the time t. Make sure that your sketch conveys where the graph is concave up and concave down, and label the y-coordinates of the inflection points.

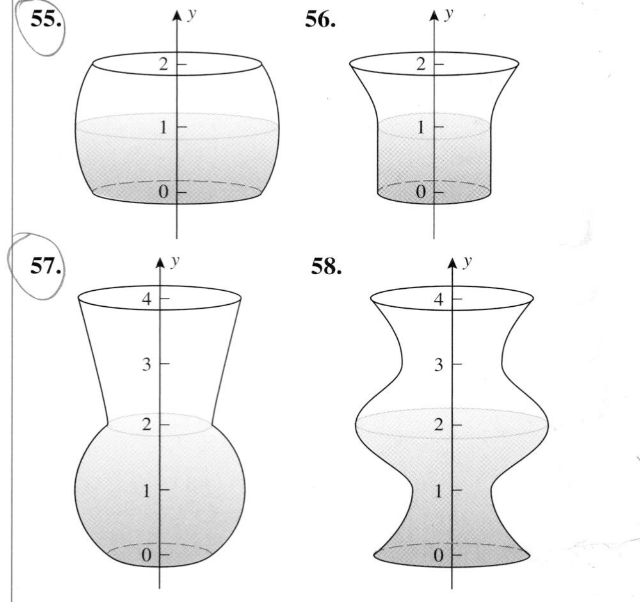

✔ QUICK CHECK ANSWERS 4.1

1. (a) $f(x_1) < f(x_2)$ (b) $f(x_1) > f(x_2)$ (c) increasing (d) $= 0$ 2. (a) $-1, 3$ (b) $(-\infty, -1]$ and $[3, +\infty)$ (c) $(-\infty, 1)$ (d) $(1, -1.1)$ 3. (a) $(0, +\infty)$ (b) $(-\infty, 4/3), (4, +\infty)$ (c) $(4/3, 4)$ 4. The graph is increasing and concave down.

4.2 ANALYSIS OF FUNCTIONS II: RELATIVE EXTREMA; GRAPHING POLYNOMIALS

In this section we will develop methods for finding the high and low points on the graph of a function and we will discuss procedures for analyzing the graphs of polynomials.

RELATIVE MAXIMA AND MINIMA

If we imagine the graph of a function f to be a two-dimensional mountain range with hills and valleys, then the tops of the hills are called "relative maxima," and the bottoms of the valleys are called "relative minima" (Figure 4.2.1). The relative maxima are the high points

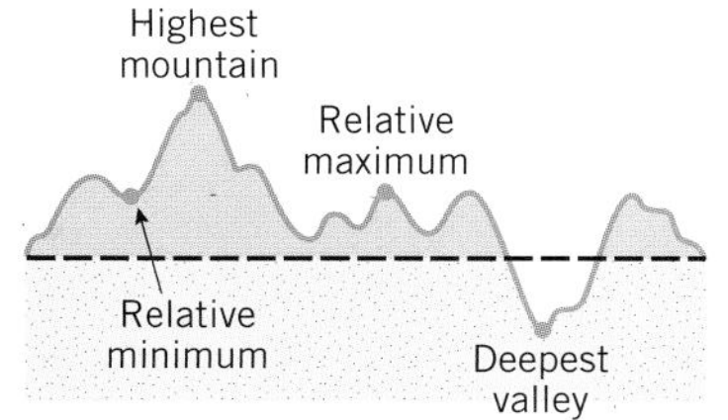

Figure 4.2.1

in their *immediate vicinity*, and the relative minima are the low points. A relative maximum need not be the highest point in the entire mountain range, and a relative minimum need not be the lowest point—they are just high and low points *relative* to the nearby terrain. These ideas are captured in the following definition.

4.2.1 DEFINITION. A function f is said to have a ***relative maximum*** at x_0 if there is an open interval containing x_0 on which $f(x_0)$ is the largest value, that is, $f(x_0) \geq f(x)$ for all x in the interval. Similarly, f is said to have a ***relative minimum*** at x_0 if there is an open interval containing x_0 on which $f(x_0)$ is the smallest value, that is, $f(x_0) \leq f(x)$ for all x in the interval. If f has either a relative maximum or a relative minimum at x_0, then f is said to have a ***relative extremum*** at x_0.

▶ Example 1 One can see from Figure 4.2.2 that:

- $f(x) = x^2$ has a relative minimum at $x = 0$ but no relative maxima.
- $f(x) = x^3$ has no relative extrema.
- $f(x) = x^3 - 3x + 3$ has a relative maximum at $x = -1$ and a relative minimum at $x = 1$.
- $f(x) = \cos x$ has relative maxima at all even multiples of π and relative minima at all odd multiples of π. ◀

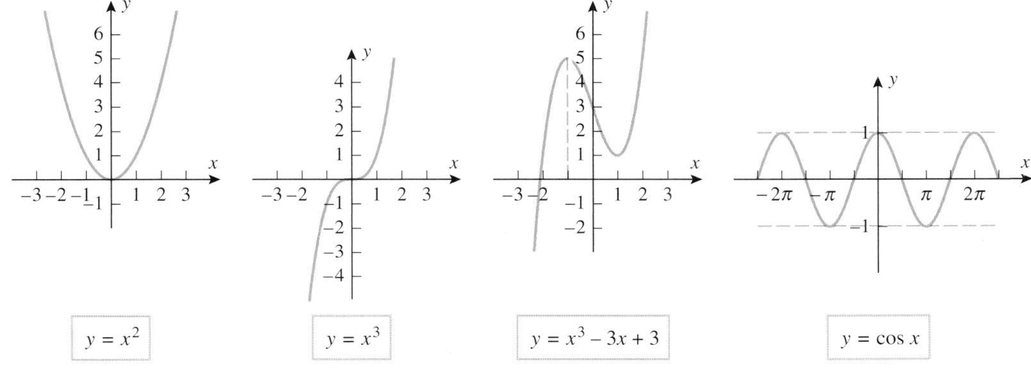

Figure 4.2.2

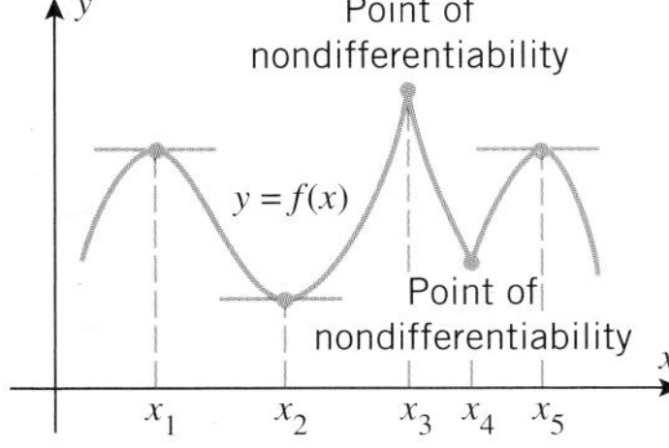

Figure 4.2.3 The points x_1, x_2, x_3, x_4, and x_5 are critical points. Of these, x_1, x_2, and x_5 are stationary points.

The relative extrema for the four functions in Example 1 occur at points where the graphs of the functions have horizontal tangent lines. Figure 4.2.3 illustrates that a relative extremum can also occur at a point where a function is not differentiable. In general, we define a ***critical point*** for a function f to be a point in the domain of f at which either the graph of f has a horizontal tangent line or f is not differentiable. To distinguish between the two types of critical points we call x a ***stationary point*** of f if $f'(x) = 0$. The following theorem, which is proved in Appendix C, states that the critical points for a function form a complete set of candidates for relative extrema on the interior of the domain of the function.

4.2.2 THEOREM. *Suppose that f is a function defined on an open interval containing the point x_0. If f has a relative extremum at $x = x_0$, then $x = x_0$ is a critical point of f; that is, either $f'(x_0) = 0$ or f is not differentiable at x_0.*

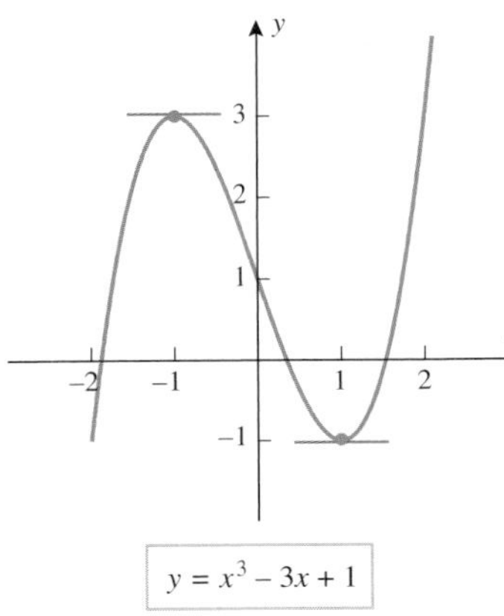

Figure 4.2.4

What is the maximum number of critical points that a polynomial of degree n can have? Why?

► **Example 2** Find all critical points of $f(x) = x^3 - 3x + 1$.

Solution. The function f, being a polynomial, is differentiable everywhere, so its critical points are all stationary points. To find these points we must solve the equation $f'(x) = 0$. Since

$$f'(x) = 3x^2 - 3 = 3(x + 1)(x - 1)$$

we conclude that the critical points occur at $x = -1$ and $x = 1$. This is consistent with the graph of f in Figure 4.2.4. ◄

► **Example 3** Find all critical points of $f(x) = 3x^{5/3} - 15x^{2/3}$.

Solution. The function f is continuous everywhere and its derivative is

$$f'(x) = 5x^{2/3} - 10x^{-1/3} = 5x^{-1/3}(x - 2) = \frac{5(x - 2)}{x^{1/3}}$$

We see from this that $f'(x) = 0$ if $x = 2$ and $f'(x)$ is undefined if $x = 0$. Thus $x = 0$ and $x = 2$ are critical points and $x = 2$ is a stationary point. This is consistent with the graph of f shown in Figure 4.2.5. ◄

TECHNOLOGY MASTERY

As discussed on pp. 23 and 24, your graphing utility may have trouble producing portions of the graph in Figure 4.2.5 because of the fractional exponents. Use the note preceding Exercise 29 of Section 1.2 to help generate the graph shown in the figure.

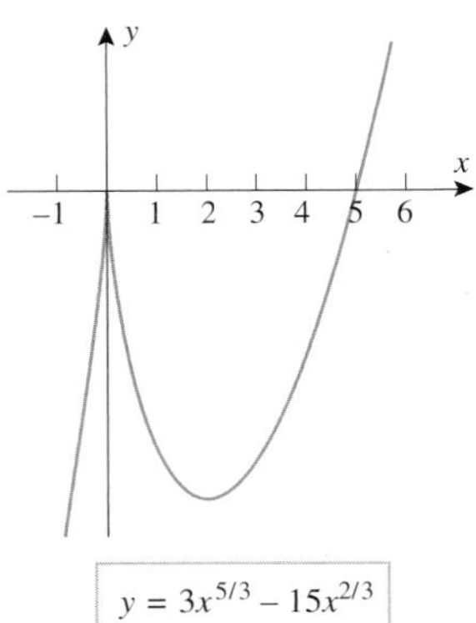

Figure 4.2.5

FIRST DERIVATIVE TEST

Theorem 4.2.2 asserts that the relative extrema must occur at critical points, but it does *not* say that a relative extremum occurs at *every* critical point. For example, for the eight critical points in Figure 4.2.6, relative extrema occur at each x_0 in the top row but not at any x_0 in the bottom row. Moreover, at the critical points in the first row the derivatives have opposite signs on the two sides of x_0, whereas at the critical points in the second row the signs of the derivatives are the same on both sides. This suggests that:

A function f has a relative extremum at those critical points where f' changes sign.

We can actually take this a step further. At the two relative maxima in Figure 4.2.6 the derivative is positive on the left side and negative on the right side, and at the two relative minima the derivative is negative on the left side and positive on the right side. All of this is summarized more precisely in the following theorem.

Informally stated, parts (*a*) and (*b*) of Theorem 4.2.3 tell us that for a continuous function, relative maxima occur at critical points where the derivative changes from + to − and relative minima where it changes from − to +.

4.2.3 THEOREM (*First Derivative Test*). *Suppose that f is continuous at a critical point x_0.*

(*a*) *If $f'(x) > 0$ on an open interval extending left from x_0 and $f'(x) < 0$ on an open interval extending right from x_0, then f has a relative maximum at x_0.*

(*b*) *If $f'(x) < 0$ on an open interval extending left from x_0 and $f'(x) > 0$ on an open interval extending right from x_0, then f has a relative minimum at x_0.*

(*c*) *If $f'(x)$ has the same sign on an open interval extending left from x_0 as it does on an open interval extending right from x_0, then f does not have a relative extremum at x_0.*

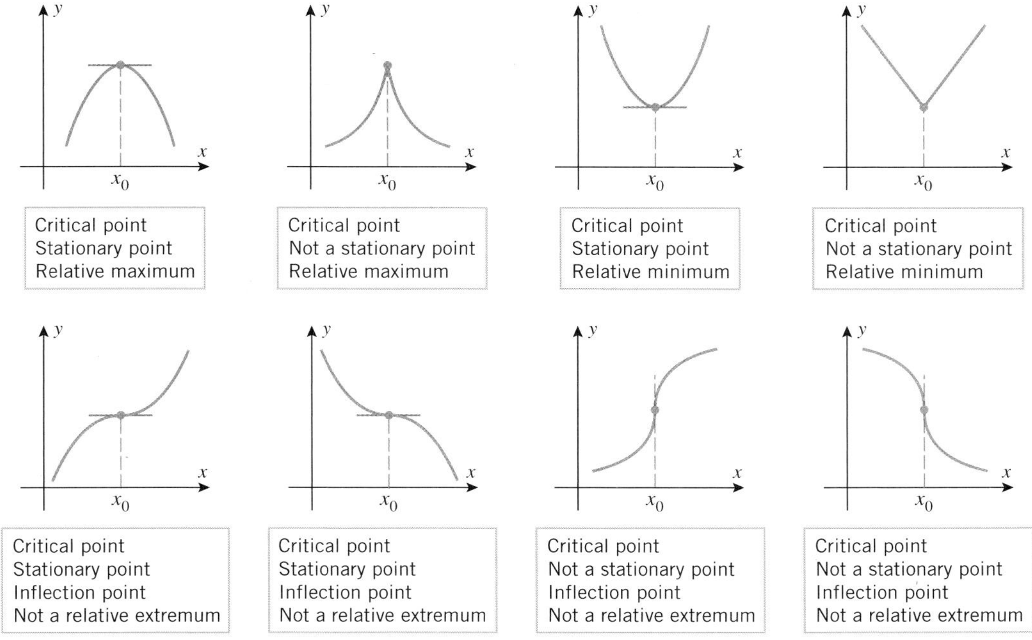

Figure 4.2.6

PROOF. We will prove part (a) and leave parts (b) and (c) as exercises. We are assuming that $f'(x) > 0$ on the interval (a, x_0) and that $f'(x) < 0$ on the interval (x_0, b), and we want to show that

$$f(x_0) \geq f(x)$$

for all x in the interval (a, b). However, the two hypotheses, together with Theorem 4.1.2 and its associated marginal note imply that f is increasing on the interval $(a, x_0]$ and decreasing on the interval $[x_0, b)$. Thus, $f(x_0) \geq f(x)$ for all x in (a, b) with equality only at x_0. ■

Table 4.2.1

INTERVAL	$5(x-2)/x^{1/3}$	$f'(x)$
$x < 0$	$(-)/(-)$	+
$0 < x < 2$	$(-)/(+)$	–
$x > 2$	$(+)/(+)$	+

► **Example 4** We showed in Example 3 that the function $f(x) = 3x^{5/3} - 15x^{2/3}$ has critical points at $x = 0$ and $x = 2$. Figure 4.2.5 suggests that f has a relative maximum at $x = 0$ and a relative minimum at $x = 2$. Confirm this using the first derivative test.

Solution. We showed in Example 3 that

$$f'(x) = \frac{5(x-2)}{x^{1/3}}$$

A sign analysis of this derivative is shown in Table 4.2.1. The sign of f' changes from $+$ to $-$ at $x = 0$, so there is a relative maximum at that point. The sign changes from $-$ to $+$ at $x = 2$, so there is a relative minimum at that point. ◄

$f'' < 0$
Concave down

$f'' > 0$
Concave up

Relative maximum

Relative minimum

Figure 4.2.7

SECOND DERIVATIVE TEST

There is another test for relative extrema that is often easier to apply than the first derivative test. It is based on the geometric observation that a function f has a relative maximum at a stationary point if the graph of f is concave down on an open interval containing that point, and it has a relative minimum if it is concave up (Figure 4.2.7).

4.2.4 THEOREM (*Second Derivative Test*). *Suppose that f is twice differentiable at the point x_0.*

(a) If $f'(x_0) = 0$ and $f''(x_0) > 0$, then f has a relative minimum at x_0.

(b) If $f'(x_0) = 0$ and $f''(x_0) < 0$, then f has a relative maximum at x_0.

(c) If $f'(x_0) = 0$ and $f''(x_0) = 0$, then the test is inconclusive; that is, f may have a relative maximum, a relative minimum, or neither at x_0.

We will prove parts (a) and (c) and leave part (b) as an exercise.

PROOF (a). We are given that $f'(x_0) = 0$ and $f''(x_0) > 0$, and we want to show that f has a relative minimum at x_0. Expressing $f''(x_0)$ as a limit and using the two given conditions we obtain

$$f''(x_0) = \lim_{x \to x_0} \frac{f'(x) - f'(x_0)}{x - x_0} = \lim_{x \to x_0} \frac{f'(x)}{x - x_0} > 0$$

This implies that for x sufficiently close to but different from x_0 we have

$$\frac{f'(x)}{x - x_0} > 0 \tag{1}$$

Thus, there is an open interval extending left from x_0 and an open interval extending right from x_0 on which (1) holds. On the open interval extending left the denominator in (1) is negative, so $f'(x) < 0$, and on the open interval extending right the denominator is positive, so $f'(x) > 0$. It now follows from part (b) of the first derivative test (Theorem 4.2.3) that f has a relative minimum at x_0.

PROOF (c). To prove this part of the theorem we need only provide functions for which $f'(x_0) = 0$ and $f''(x_0) = 0$ at some point x_0, but with one having a relative minimum at x_0, one having a relative maximum at x_0, and one having neither at x_0. We leave it as an exercise for you to show that three such functions are $f(x) = x^4$ (relative minimum at $x = 0$), $f(x) = -x^4$ (relative maximum at $x = 0$), and $f(x) = x^3$ (neither a relative maximum nor a relative minimum at x_0). ■

▶ **Example 5** Find the relative extrema of $f(x) = 3x^5 - 5x^3$.

Solution. We have

$$f'(x) = 15x^4 - 15x^2 = 15x^2(x^2 - 1) = 15x^2(x + 1)(x - 1)$$
$$f''(x) = 60x^3 - 30x = 30x(2x^2 - 1)$$

Solving $f'(x) = 0$ yields the stationary points $x = 0$, $x = -1$, and $x = 1$. As shown in the following table, we can conclude from the second derivative test that f has a relative maximum at $x = -1$ and a relative minimum at $x = 1$.

STATIONARY POINT	$30x(2x^2 - 1)$	$f''(x)$	SECOND DERIVATIVE TEST
$x = -1$	-30	$-$	f has a relative maximum
$x = 0$	0	0	Inconclusive
$x = 1$	30	$+$	f has a relative minimum

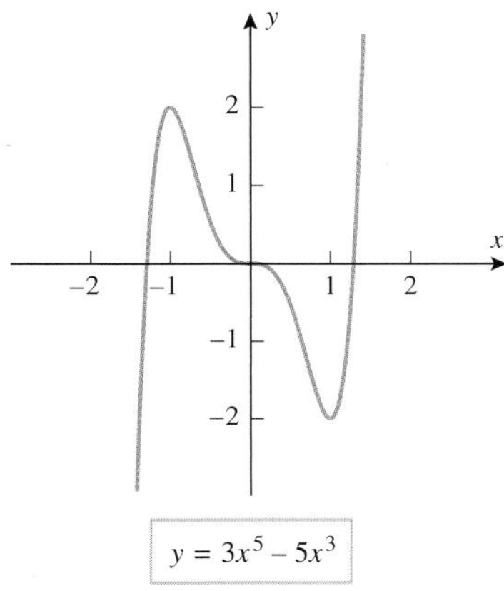

Figure 4.2.8

The test is inconclusive at $x = 0$, so we will try the first derivative test at that point. A sign analysis of f' is given in the following table:

INTERVAL	$15x^2(x+1)(x-1)$	$f'(x)$
$-1 < x < 0$	$(+)(+)(-)$	$-$
$0 < x < 1$	$(+)(+)(-)$	$-$

Since there is no sign change in f' at $x = 0$, there is neither a relative maximum nor a relative minimum at that point. All of this is consistent with the graph of f shown in Figure 4.2.8. ◀

GEOMETRIC IMPLICATIONS OF MULTIPLICITY

Our final goal in this section is to outline a general procedure that can be used to analyze and graph polynomials. To do so, it will be helpful to understand how the graph of a polynomial behaves in the vicinity of its roots. For example, it would be nice to know what property of the polynomial in Example 5 produced the inflection point and horizontal tangent at the root $x = 0$.

Recall that a root $x = r$ of a polynomial $p(x)$ has ***multiplicity m*** if $(x - r)^m$ divides $p(x)$ but $(x - r)^{m+1}$ does not. A root of multiplicity 1 is called a ***simple root***. Figure 4.2.9 and the following theorem show that the behavior of a polynomial in the vicinity of a real root is determined by the multiplicity of that root (we omit the proof).

4.2.5 THE GEOMETRIC IMPLICATIONS OF MULTIPLICITY. *Suppose that $p(x)$ is a polynomial with a root of multiplicity m at $x = r$.*

(a) *If m is even, then the graph of $y = p(x)$ is tangent to the x-axis at $x = r$, does not cross the x-axis there, and does not have an inflection point there.*

(b) *If m is odd and greater than 1, then the graph is tangent to the x-axis at $x = r$, crosses the x-axis there, and also has an inflection point there.*

(c) *If $m = 1$ (so that the root is simple), then the graph is not tangent to the x-axis at $x = r$, crosses the x-axis there, and may or may not have an inflection point there.*

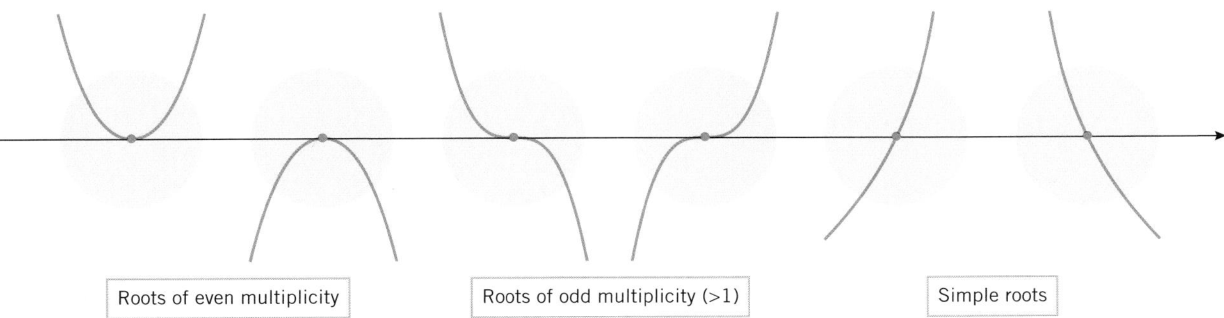

Figure 4.2.9

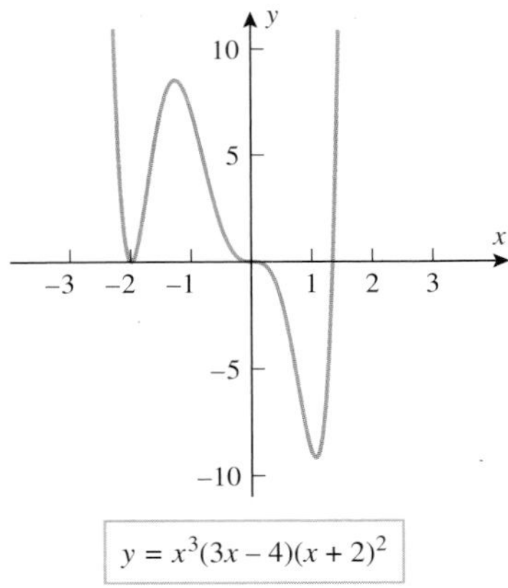

$y = x^3(3x-4)(x+2)^2$

Figure 4.2.10

▶ **Example 6** Make a conjecture about the behavior of the graph of

$$y = x^3(3x-4)(x+2)^2$$

in the vicinity of its x-intercepts, and test your conjecture by generating the graph.

Solution. The x-intercepts occur at $x = 0$, $x = \frac{4}{3}$, and $x = -2$. The root $x = 0$ has multiplicity 3, which is odd, so at that point the graph should be tangent to the x-axis, cross the x-axis, and have an inflection point there. The root $x = -2$ has multiplicity 2, which is even, so the graph should be tangent to but not cross the x-axis there. The root $x = \frac{4}{3}$ is simple, so at that point the curve should cross the x-axis without being tangent to it. All of this is consistent with the graph in Figure 4.2.10. ◀

ANALYSIS OF POLYNOMIALS

Historically, the term "curve sketching" meant using calculus to help draw the graph of a function by hand—the graph was the goal. Since graphs can now be produced with great precision using calculators and computers, the purpose of curve sketching has changed. Today, we typically start with a graph produced by a calculator or computer, then use curve sketching to identify important features of the graph that the calculator or computer might have missed. Thus, the goal of curve sketching is no longer the graph itself, but rather the information it reveals about the function.

Polynomials are among the simplest functions to graph and analyze. Their significant features are symmetry, intercepts, relative extrema, inflection points, and the behavior as $x \to +\infty$ and as $x \to -\infty$. Figure 4.2.11 shows the graphs of four polynomials in x. The graphs in Figure 4.2.11 have properties that are common to all polynomials:

> For each of the graphs in Figure 4.2.11, count the number of x-intercepts, relative extrema, and inflection points, and confirm that your count is consistent with the degree of the polynomial.

- The natural domain of a polynomial is $(-\infty, +\infty)$.
- Polynomials are continuous everywhere.
- Polynomials are differentiable everywhere, so their graphs have no corners or vertical tangent lines.
- The graph of a nonconstant polynomial eventually increases or decreases without bound as $x \to +\infty$ and as $x \to -\infty$. This is because the limit of a nonconstant polynomial as $x \to +\infty$ or as $x \to -\infty$ is $\pm\infty$, depending on the sign of the term of highest degree and whether the polynomial has even or odd degree [see Formulas (13) and (14) of Section 2.3 and the related discussion].
- The graph of a polynomial of degree n (> 2) has at most n x-intercepts, at most $n-1$ relative extrema, and at most $n-2$ inflection points. This is because the x-intercepts, relative extrema, and inflection points of a polynomial $p(x)$ are among the real solutions of the equations $p(x) = 0$, $p'(x) = 0$, and $p''(x) = 0$, and the polynomials in these equations have degree n, $n-1$, and $n-2$, respectively. Thus, for example, the graph of a quadratic polynomial has at most two x-intercepts, one relative extremum, and no inflection points; and the graph of a cubic polynomial has at most three x-intercepts, two relative extrema, and one inflection point.

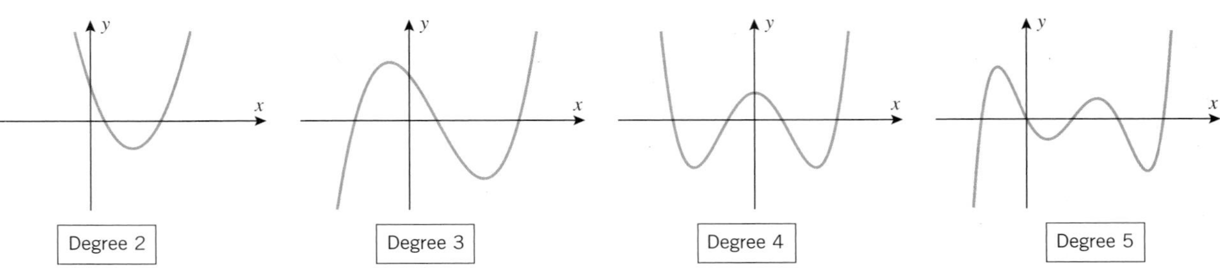

Figure 4.2.11

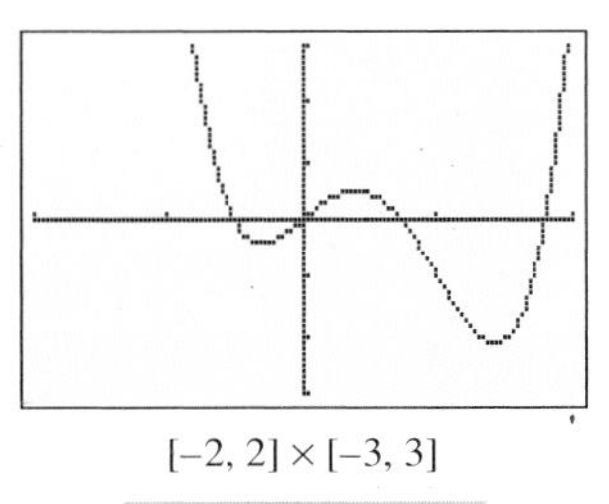

$[-2, 2] \times [-3, 3]$

$y = 3x^4 - 6x^3 + 2x$

Figure 4.2.12

▶ **Example 7** Figure 4.2.12 shows the graph of

$$y = 3x^4 - 6x^3 + 2x$$

produced on a graphing calculator. Confirm that the graph is not missing any significant features.

Solution. We can be confident that the graph shows all significant features of the polynomial because the polynomial has degree 4 and we can account for four roots, three relative extrema, and two inflection points. Moreover, the graph suggests the correct behavior as $x \to +\infty$ and as $x \to -\infty$, since

$$\lim_{x \to +\infty} (3x^4 - 6x^3 + 2x) = \lim_{x \to +\infty} 3x^4 = +\infty$$

$$\lim_{x \to -\infty} (3x^4 - 6x^3 + 2x) = \lim_{x \to -\infty} 3x^4 = +\infty \blacktriangleleft$$

▶ **Example 8** Sketch the graph of the equation

$$y = x^3 - 3x + 2$$

and identify the locations of the intercepts, relative extrema, and inflection points.

Solution. The following analysis will produce the information needed to sketch the graph:

A review of polynomial factoring is given in Appendix B.

- *x-intercepts:* Factoring the polynomial yields

$$x^3 - 3x + 2 = (x + 2)(x - 1)^2$$

  which tells us that the x-intercepts are $x = -2$ and $x = 1$.
- *y-intercept:* Setting $x = 0$ yields $y = 2$.
- *End behavior:* We have

$$\lim_{x \to +\infty} (x^3 - 3x + 2) = \lim_{x \to +\infty} x^3 = +\infty$$

$$\lim_{x \to -\infty} (x^3 - 3x + 2) = \lim_{x \to -\infty} x^3 = -\infty$$

  so the graph increases without bound as $x \to +\infty$ and decreases without bound as $x \to -\infty$.
- *Derivatives:*

$$\frac{dy}{dx} = 3x^2 - 3 = 3(x - 1)(x + 1)$$

$$\frac{d^2y}{dx^2} = 6x$$

- *Increase, decrease, relative extrema, inflection points:* Figure 4.2.13 gives a sign analysis of the first and second derivatives and indicates its geometric significance. There are stationary points at $x = -1$ and $x = 1$. Since the sign of dy/dx changes from $+$ to $-$ at $x = -1$, there is a relative maximum there, and since it changes from $-$ to $+$ at $x = 1$, there is a relative minimum there. The sign of d^2y/dx^2 changes from $-$ to $+$ at $x = 0$, so there is an inflection point there.
- *Final sketch:* Figure 4.2.14 shows the final sketch with the coordinates of the intercepts, relative extrema, and inflection point labeled. ◀

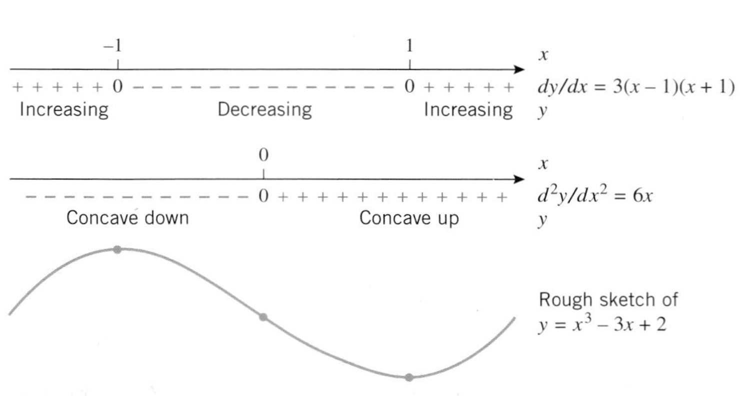

Figure 4.2.13

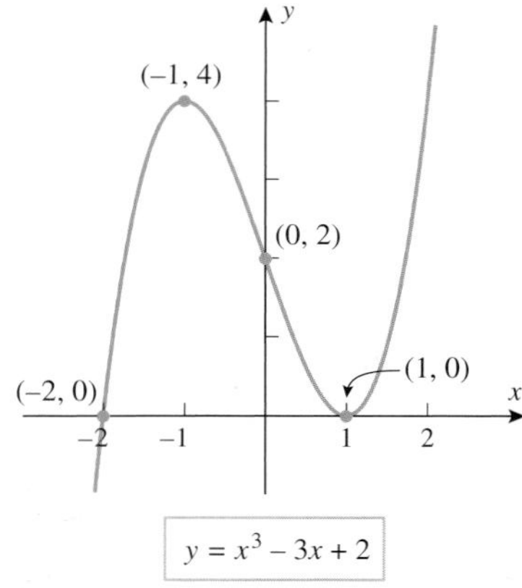

Figure 4.2.14

✔QUICK CHECK EXERCISES 4.2 *(See page 244 for answers.)*

1. A function f has a relative maximum at x_0 if there is an open interval containing x_0 on which $f(x)$ is ________ $f(x_0)$ for every x in the interval.

2. Suppose that f is defined everywhere and $x = 2, 3, 5, 7$ are critical points for f. If $f'(x)$ is positive on the intervals $(-\infty, 2)$ and $(5, 7)$, and if $f'(x)$ is negative on the intervals $(2, 3)$, $(3, 5)$, and $(7, +\infty)$, then f has relative maxima at $x =$ ________ and f has relative minima at $x =$ ________.

3. Suppose that f is defined everywhere and $x = -2$ and $x = 1$ are critical points for f. If $f''(x) = 2x + 1$, then f has a relative ________ at $x = -2$ and f has a relative ________ at $x = 1$.

4. Let $f(x) = (x^2 - 4)^2$. Then $f'(x) = 4x(x^2 - 4)$ and $f''(x) = 4(3x^2 - 4)$. Identify the locations of the (a) relative maxima, (b) relative minima, and (c) inflection points on the graph of f.

EXERCISE SET 4.2 Graphing Utility C CAS

FOCUS ON CONCEPTS

1. In each part, sketch the graph of a continuous function f with the stated properties.
   (a) f is concave up on the interval $(-\infty, +\infty)$ and has exactly one relative extremum.
   (b) f is concave up on the interval $(-\infty, +\infty)$ and has no relative extrema.
   (c) The function f has exactly two relative extrema on the interval $(-\infty, +\infty)$, and $f(x) \to +\infty$ as $x \to +\infty$.
   (d) The function f has exactly two relative extrema on the interval $(-\infty, +\infty)$, and $f(x) \to -\infty$ as $x \to +\infty$.

2. In each part, sketch the graph of a continuous function f with the stated properties.
   (a) f has exactly one relative extremum on $(-\infty, +\infty)$, and $f(x) \to 0$ as $x \to +\infty$ and as $x \to -\infty$.
   (b) f has exactly two relative extrema on $(-\infty, +\infty)$, and $f(x) \to 0$ as $x \to +\infty$ and as $x \to -\infty$.
   (c) f has exactly one inflection point and one relative extremum on $(-\infty, +\infty)$.
   (d) f has infinitely many relative extrema, and $f(x) \to 0$ as $x \to +\infty$ and as $x \to -\infty$.

3. (a) Use both the first and second derivative tests to show that $f(x) = 3x^2 - 6x + 1$ has a relative minimum at $x = 1$.
   (b) Use both the first and second derivative tests to show that $f(x) = x^3 - 3x + 3$ has a relative minimum at $x = 1$ and a relative maximum at $x = -1$.

4. (a) Use both the first and second derivative tests to show that $f(x) = \sin^2 x$ has a relative minimum at $x = 0$.
   (b) Use both the first and second derivative tests to show that $g(x) = \tan^2 x$ has a relative minimum at $x = 0$.
   (c) Give an informal verbal argument to explain without calculus why the functions in parts (a) and (b) have relative minima at $x = 0$.

5. (a) Show that both of the functions $f(x) = (x - 1)^4$ and $g(x) = x^3 - 3x^2 + 3x - 2$ have stationary points at $x = 1$.
   (b) What does the second derivative test tell you about the nature of these stationary points?
   (c) What does the first derivative test tell you about the nature of these stationary points?

6. (a) Show that $f(x) = 1 - x^5$ and $g(x) = 3x^4 - 8x^3$ both have stationary points at $x = 0$.
(b) What does the second derivative test tell you about the nature of these stationary points?
(c) What does the first derivative test tell you about the nature of these stationary points?

7–14 Locate the critical points and identify which critical points are stationary points.

7. $f(x) = 4x^4 - 16x^2 + 17$ **8.** $f(x) = 3x^4 + 12x$

9. $f(x) = \dfrac{x+1}{x^2+3}$ **10.** $f(x) = \dfrac{x^2}{x^3+8}$

11. $f(x) = \sqrt[3]{x^2-25}$ **12.** $f(x) = x^2(x-1)^{2/3}$

13. $f(x) = |\sin x|$ **14.** $f(x) = \sin |x|$

FOCUS ON CONCEPTS

15–18 Use the graph of f' shown in the figure to estimate all values of x at which f has (a) relative minima, (b) relative maxima, and (c) inflection points.

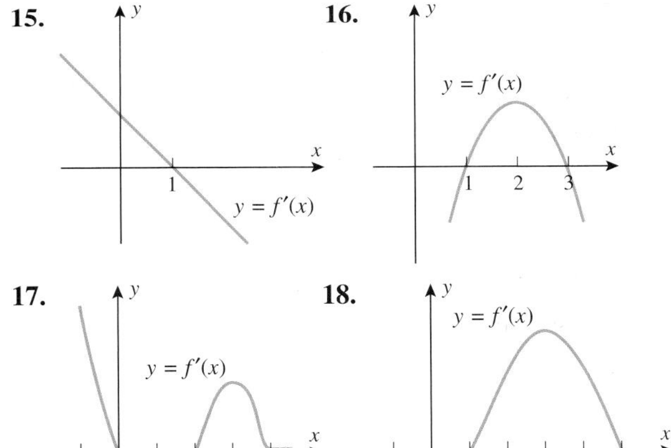

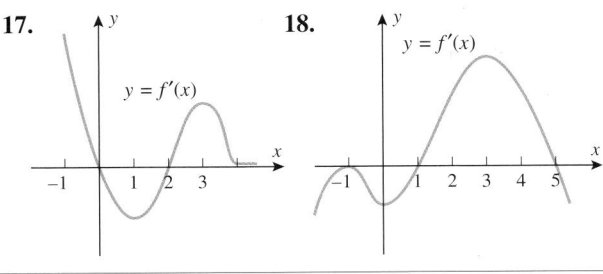

19–22 Use the given derivative to find all critical points of f, and at each critical point determine whether a relative maximum, relative minimum, or neither occurs. Assume in each case that f is continuous everywhere.

19. $f'(x) = x^2(x^3 - 5)$ **20.** $f'(x) = 4x^3 - 9x$

21. $f'(x) = \dfrac{2-3x}{\sqrt[3]{x+2}}$ **22.** $f'(x) = \dfrac{x^2-7}{\sqrt[3]{x^2+4}}$

23–26 Find the relative extrema using both first and second derivative tests.

23. $f(x) = 1 + 8x - 3x^2$

24. $f(x) = x^4 - 12x^3$

25. $f(x) = \sin 2x,\ 0 < x < \pi$

26. $f(x) = x + \sin 2x,\ 0 < x < \pi$

27–36 Use any method to find the relative extrema of the function f.

27. $f(x) = x^4 - 4x^3 + 4x^2$ **28.** $f(x) = x(x-4)^3$

29. $f(x) = x^3(x+1)^2$ **30.** $f(x) = x^2(x+1)^3$

31. $f(x) = 2x + 3x^{2/3}$ **32.** $f(x) = 2x + 3x^{1/3}$

33. $f(x) = \dfrac{x+3}{x-2}$ **34.** $f(x) = \dfrac{x^2}{x^4+16}$

35. $f(x) = |3x - x^2|$ **36.** $f(x) = |1 + \sqrt[3]{x}|$

37–46 Give a graph of the polynomial and label the coordinates of the intercepts, stationary points, and inflection points. Check your work with a graphing utility.

37. $p(x) = x^2 - 3x - 4$ **38.** $p(x) = 1 + 8x - x^2$

39. $p(x) = 2x^3 - 3x^2 - 36x + 5$

40. $p(x) = 2 - x + 2x^2 - x^3$

41. $p(x) = (x+1)^2(2x - x^2)$

42. $p(x) = x^4 - 6x^2 + 5$

43. $p(x) = x^4 - 2x^3 + 2x - 1$ **44.** $p(x) = 4x^3 - 9x^4$

45. $p(x) = x(x^2-1)^2$ **46.** $p(x) = x(x^2-1)^3$

47. In each part: (i) Make a conjecture about the behavior of the graph in the vicinity of its x-intercepts. (ii) Make a rough sketch of the graph based on your conjecture and the limits of the polynomial as $x \to +\infty$ and as $x \to -\infty$. (iii) Compare your sketch to the graph generated with a graphing utility.
(a) $y = x(x-1)(x+1)$ (b) $y = x^2(x-1)^2(x+1)^2$
(c) $y = x^2(x-1)^2(x+1)^3$ (d) $y = x(x-1)^5(x+1)^4$

48. Sketch the graph of $y = (x-a)^m(x-b)^n$ for the stated values of m and n, assuming that $a < b$ (six graphs in total).
(a) $m = 1,\ n = 1, 2, 3$ (b) $m = 2,\ n = 2, 3$
(c) $m = 3,\ n = 3$

49–52 Find the relative extrema in the interval $0 < x < 2\pi$, and confirm that your results are consistent with the graph of f generated by a graphing utility.

49. $f(x) = |\sin 2x|$ **50.** $f(x) = \sqrt{3}x + 2\sin x$

51. $f(x) = \cos^2 x$ **52.** $f(x) = \dfrac{\sin x}{2 - \cos x}$

53–54 Use a graphing utility to generate the graphs of f' and f'' over the stated interval, and then use those graphs to estimate the x-coordinates of the relative extrema of f. Check that your estimates are consistent with the graph of f.

53. $f(x) = x^4 - 24x^2 + 12x,\quad -5 \le x \le 5$

54. $f(x) = \sin \frac{1}{2}x \cos x,\quad -\pi/2 \le x \le \pi/2$

55–58 Use a CAS to graph f' and f'', and then use those graphs to estimate the x-coordinates of the relative extrema of f. Check that your estimates are consistent with the graph of f.

C **55.** $f(x) = \dfrac{10x^3 - 3}{3x^2 - 5x + 8}$ C **56.** $f(x) = \dfrac{x^3 - x^2}{x^2 + 1}$

C **57.** $f(x) = \sqrt{x^4 + \cos^2 x}$ C **58.** $f(x) = \dfrac{x^3 - 8x + 7}{\sqrt{x^2 + 1}}$

59. In each part, find k so that f has a relative extremum at the point where $x = 3$.

(a) $f(x) = x^2 + \dfrac{k}{x}$ (b) $f(x) = \dfrac{x}{x^2 + k}$

C **60.** (a) Use a CAS to graph the function

$$f(x) = \frac{x^4 + 1}{x^2 + 1}$$

and use the graph to estimate the x-coordinates of the relative extrema.

(b) Find the exact x-coordinates by using the CAS to solve the equation $f'(x) = 0$.

FOCUS ON CONCEPTS

61. The two graphs in the accompanying figure depict a function $r(x)$ and its derivative $r'(x)$.

(a) Approximate the coordinates of each inflection point on the graph of $y = r(x)$.

(b) Suppose that $f(x)$ is a function that is continuous everywhere and whose *derivative* satisfies

$$f'(x) = (x^2 - 4) \cdot r(x)$$

What are the critical points for $f(x)$? At each critical point, identify whether $f(x)$ has a (relative) maximum, minimum, or neither a maximum or minimum. Approximate $f''(1)$.

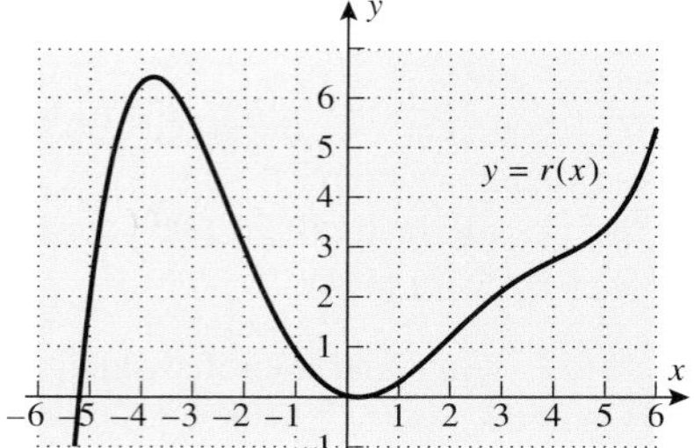

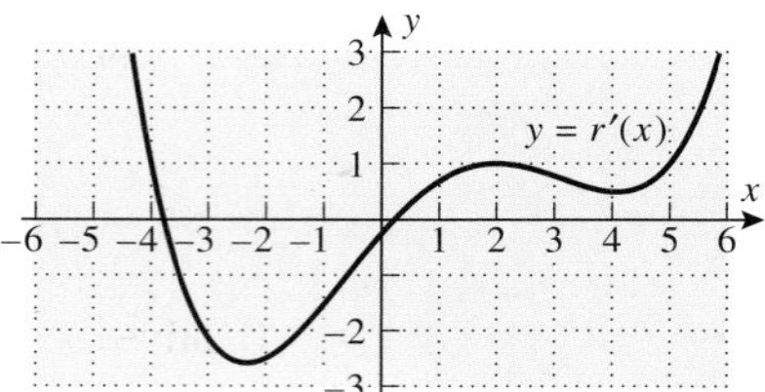

Figure Ex-61

62. With $r(x)$ as provided in Exercise 61, let $g(x)$ be a function that is continuous everywhere such that $g'(x) = x - r(x)$. For which values of x does $g(x)$ have an inflection point?

63. Find values of a, b, c, and d so that the function

$$f(x) = ax^3 + bx^2 + cx + d$$

has a relative minimum at $(0, 0)$ and a relative maximum at $(1, 1)$.

64. Let h and g have relative maxima at x_0. Prove or disprove:

(a) $h + g$ has a relative maximum at x_0

(b) $h - g$ has a relative maximum at x_0.

65. Sketch some curves that show that the three parts of the first derivative test (Theorem 4.2.3) can be false without the assumption that f is continuous at x_0.

✔ QUICK CHECK ANSWERS 4.2

1. less than or equal to **2.** 2, 7; 5 **3.** maximum; minimum **4.** (a) $(0, 16)$ (b) $(-2, 0)$ and $(2, 0)$ (c) $(-2/\sqrt{3}, 64/9)$ and $(2/\sqrt{3}, 64/9)$

4.3 MORE ON CURVE SKETCHING: RATIONAL FUNCTIONS; CURVES WITH CUSPS AND VERTICAL TANGENT LINES; USING TECHNOLOGY

In this section we will discuss procedures for graphing rational functions and other kinds of curves. We will also discuss the interplay between calculus and technology in curve sketching.

PROPERTIES OF GRAPHS

In many problems, the properties of interest in the graph of a function are:

- symmetries
- periodicity
- x-intercepts
- y-intercepts
- relative extrema
- concavity
- intervals of increase and decrease
- inflection points
- asymptotes
- behavior as $x \to +\infty$ or as $x \to -\infty$

Some of these properties may not be relevant in certain cases; for example, asymptotes are characteristic of rational functions but not of polynomials, and periodicity is characteristic of trigonometric functions but not of polynomial or rational functions. Thus, when analyzing the graph of a function f, it helps to know something about the general properties of the family to which it belongs.

In a given problem you will usually have a definite objective for your analysis of a graph. For example, you may be interested in showing all of the important characteristics of the function, you may only be interested in the behavior of the graph as $x \to +\infty$ or as $x \to -\infty$, or you may be interested in some specific feature such as a particular inflection point. Thus, your objectives in the problem will dictate those characteristics on which you want to focus.

GRAPHING RATIONAL FUNCTIONS

Recall that a rational function is a function of the form $f(x) = P(x)/Q(x)$ in which $P(x)$ and $Q(x)$ are polynomials. Graphs of rational functions are more complicated than those of polynomials because of the possibility of asymptotes and discontinuities (see Figure 1.4.11, for example). If $P(x)$ and $Q(x)$ have no common factors, then the information obtained in the following steps will usually be sufficient to obtain an accurate sketch of the graph of a rational function.

How to Sketch the Graph of a Rational Function $f(x) = P(x)/Q(x)$, Where $P(x)$ and $Q(x)$ Have No Common Factors

Step 1. Determine whether there is symmetry about the y-axis or the origin.

Step 2. Find the x- and y-intercepts.

Step 3. Find the values of x for which $Q(x) = 0$. The graph has a vertical asymptote at each such value.

Step 4. Determine the end behavior of the graph by computing the limits of $f(x)$ as $x \to +\infty$ and as $x \to -\infty$. If either limit has a finite value L, then the line $y = L$ is a horizontal asymptote.

Step 5. The only places where $f(x)$ can change sign are at the x-intercepts or the vertical asymptotes. Mark the points on the x-axis at which these occur and calculate a sample value of $f(x)$ in each of the open intervals determined by these points. This will tell you whether $f(x)$ is positive or negative over that interval.

Step 6. Use $f'(x)$ and $f''(x)$ to determine the intervals where $f(x)$ is increasing, decreasing, concave up, and concave down. Determine the locations of all stationary points, relative extrema, and inflection points.

▶ **Example 1** Sketch a graph of the equation

$$y = \frac{2x^2 - 8}{x^2 - 16}$$

and identify the locations of the intercepts, relative extrema, inflection points, and asymptotes.

Solution. The numerator and denominator have no common factors, so we will use the procedure just outlined.

- *Symmetries:* Replacing x by $-x$ does not change the equation, so the graph is symmetric about the y-axis.
- *x-intercepts:* Setting $y = 0$ yields the x-intercepts $x = -2$ and $x = 2$.
- *y-intercept:* Setting $x = 0$ yields the y-intercept $y = \frac{1}{2}$.
- *Vertical asymptotes:* Setting $x^2 - 16 = 0$ yields the solutions $x = -4$ and $x = 4$. Since neither solution is a root of $2x^2 - 8$, the graph has vertical asymptotes at $x = -4$ and $x = 4$.
- *Sign of y:* The set of points where x-intercepts or vertical asymptotes occur is $\{-4, -2, 2, 4\}$. These points divide the x-axis into the open intervals

$$(-\infty, -4), \quad (-4, -2), \quad (-2, 2), \quad (2, 4), \quad (4, +\infty)$$

  We can find the sign of y on each interval by choosing an arbitrary test point in the interval and evaluating $y = f(x)$ at the test point (Table 4.3.1). This analysis is summarized on the first line of Figure 4.3.1*a*.
- *Horizontal asymptotes:* The limits

$$\lim_{x \to +\infty} \frac{2x^2 - 8}{x^2 - 16} = \lim_{x \to +\infty} \frac{2 - (8/x^2)}{1 - (16/x^2)} = 2$$
$$\lim_{x \to -\infty} \frac{2x^2 - 8}{x^2 - 16} = \lim_{x \to -\infty} \frac{2 - (8/x^2)}{1 - (16/x^2)} = 2$$

  yield the horizontal asymptote $y = 2$.
- *Derivatives:*

$$\frac{dy}{dx} = \frac{(x^2 - 16)(4x) - (2x^2 - 8)(2x)}{(x^2 - 16)^2} = -\frac{48x}{(x^2 - 16)^2}$$
$$\frac{d^2y}{dx^2} = \frac{48(16 + 3x^2)}{(x^2 - 16)^3} \quad \text{(verify)}$$

Conclusions:

- The sign analysis of y in Figure 4.3.1*a* reveals the behavior of the graph in the vicinity of the vertical asymptotes $x = -4$ and $x = 4$: The graph increases without bound as

If we cancel the common factor x in the numerator and denominator of

$$y = \frac{x(2x^2 - 8)}{x(x^2 - 16)}$$

we obtain the equation graphed in Example 1. Is there any difference between the graph of this equation and that obtained in Example 1? In general, what is the effect on the graph of a rational function if common factors in the numerator and denominator are canceled?

$x \to -4^-$ and decreases without bound as $x \to -4^+$; and the graph decreases without bound as $x \to 4^-$ and increases without bound as $x \to 4^+$ (Figure 4.3.1*b*).

- The sign analysis of dy/dx in Figure 4.3.1*a* shows that there is a relative maximum at the stationary point $x = 0$. There are no relative minima.
- The sign analysis of d^2y/dx^2 in Figure 4.3.1*a* shows that the graph is concave up to the left of $x = -4$, is concave down between $x = -4$ and $x = 4$, and is concave up to the right of $x = 4$. There are no inflection points.

The graph is shown in Figure 4.3.1*c*. ◀

Table 4.3.1

INTERVAL	TEST POINT	$y = \dfrac{2x^2 - 8}{x^2 - 16}$	SIGN OF y
$(-\infty, -4)$	$x = -5$	$y = 14/3$	+
$(-4, -2)$	$x = -3$	$y = -10/7$	−
$(-2, 2)$	$x = 0$	$y = 1/2$	+
$(2, 4)$	$x = 3$	$y = -10/7$	−
$(4, +\infty)$	$x = 5$	$y = 14/3$	+

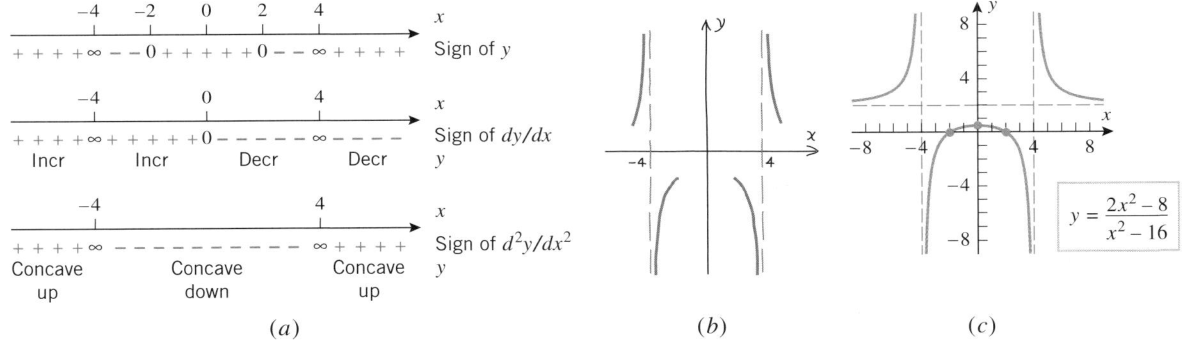

Figure 4.3.1

▶ **Example 2** Sketch a graph of

$$y = \frac{x^2 - 1}{x^3}$$

and identify the locations of all asymptotes, intercepts, relative extrema, and inflection points.

Solution.

- *Symmetries:* Replacing x by $-x$ and y by $-y$ yields an equation that simplifies to the original equation, so the graph is symmetric about the origin.
- *x-intercepts:* Setting $y = 0$ yields the x-intercepts $x = -1$ and $x = 1$.
- *y-intercept:* Setting $x = 0$ leads to a division by zero, so that there is no y-intercept.
- *Vertical asymptotes:* Setting $x^3 = 0$ yields the solution $x = 0$. This is not a root of $x^2 - 1$, so $x = 0$ is a vertical asymptote.

Table 4.3.2

INTERVAL	TEST POINT	$y = \dfrac{x^2-1}{x^3}$	SIGN OF y
$(-\infty, -1)$	-2	$-\frac{3}{8}$	$-$
$(-1, 0)$	$-\frac{1}{2}$	6	$+$
$(0, 1)$	$\frac{1}{2}$	-6	$-$
$(1, +\infty)$	2	$\frac{3}{8}$	$+$

- *Sign of y:* The set of points where x-intercepts or vertical asymptotes occur is $\{-1, 0, 1\}$. These points divide the x-axis into the open intervals

$$(-\infty, -1), \quad (-1, 0), \quad (0, 1), \quad (1, +\infty)$$

Table 4.3.2 uses the method of test points to produce the sign of y on each of these intervals.

- *Horizontal asymptotes:* The limits

$$\lim_{x\to+\infty} \frac{x^2-1}{x^3} = \lim_{x\to+\infty}\left(\frac{1}{x} - \frac{1}{x^3}\right) = 0$$

$$\lim_{x\to-\infty} \frac{x^2-1}{x^3} = \lim_{x\to-\infty}\left(\frac{1}{x} - \frac{1}{x^3}\right) = 0$$

yield the horizontal asymptote $y = 0$.

- *Derivatives:*

$$\frac{dy}{dx} = \frac{x^3(2x) - (x^2-1)(3x^2)}{(x^3)^2} = \frac{3-x^2}{x^4} = \frac{(\sqrt{3}+x)(\sqrt{3}-x)}{x^4}$$

$$\frac{d^2y}{dx^2} = \frac{x^4(-2x) - (3-x^2)(4x^3)}{(x^4)^2} = \frac{2(x^2-6)}{x^5} = \frac{2(x-\sqrt{6})(x+\sqrt{6})}{x^5}$$

Conclusions:

- The sign analysis of y in Figure 4.3.2*a* reveals the behavior of the graph in the vicinity of the vertical asymptote $x = 0$: The graph increases without bound as $x \to 0^-$ and decreases without bound as $x \to 0^+$ (Figure 4.3.2*b*).
- The sign analysis of dy/dx in Figure 4.3.2*a* shows that there is a relative minimum at $x = -\sqrt{3}$ and a relative maximum at $x = \sqrt{3}$.
- The sign analysis of d^2y/dx^2 in Figure 4.3.2*a* shows that the graph changes concavity at the vertical asymptote $x = 0$ and that there are inflection points at $x = -\sqrt{6}$ and $x = \sqrt{6}$.

The graph is shown in Figure 4.3.2*c*. To produce a slightly more accurate sketch, we used a graphing utility to help plot the relative extrema and inflection points. You should confirm that the approximate coordinates of the inflection points are $(-2.45, -0.34)$ and $(2.45, 0.34)$ and that the approximate coordinates of the relative minimum and relative maximum are $(-1.73, -0.38)$ and $(1.73, 0.38)$, respectively. ◀

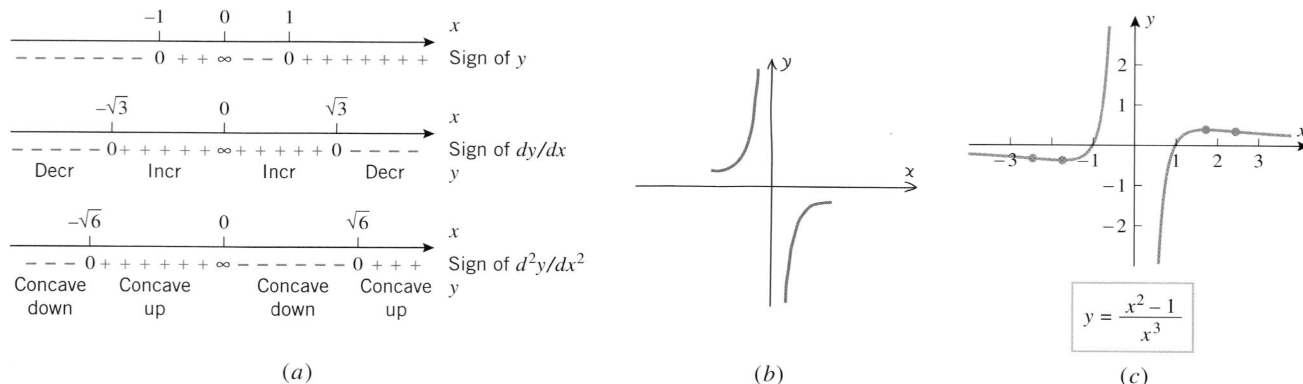

Figure 4.3.2

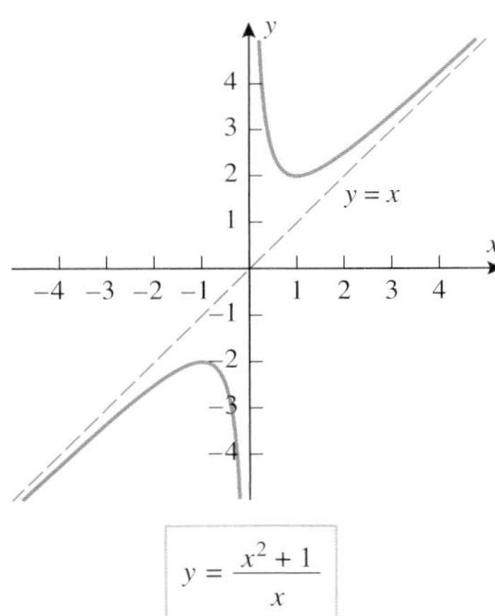

Figure 4.3.3

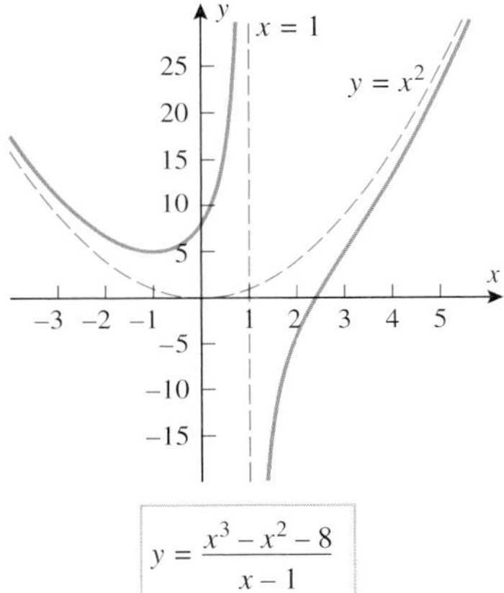

Figure 4.3.4

RATIONAL FUNCTIONS WITH OBLIQUE OR CURVILINEAR ASYMPTOTES

In the rational functions of Examples 1 and 2, the degree of the numerator did not exceed the degree of the denominator, and the asymptotes were either vertical or horizontal. If the numerator of a rational function has greater degree than the denominator, then other kinds of "asymptotes" are possible. For example, consider the rational functions

$$f(x) = \frac{x^2+1}{x} \quad \text{and} \quad g(x) = \frac{x^3-x^2-8}{x-1} \tag{1}$$

By division we can rewrite these as

$$f(x) = x + \frac{1}{x} \quad \text{and} \quad g(x) = x^2 - \frac{8}{x-1}$$

Since the second terms both approach 0 as $x \to +\infty$ or as $x \to -\infty$, it follows that

$$(f(x) - x) \to 0 \quad \text{as } x \to +\infty \text{ or as } x \to -\infty$$
$$(g(x) - x^2) \to 0 \quad \text{as } x \to +\infty \text{ or as } x \to -\infty$$

Geometrically, this means that the graph of $y = f(x)$ eventually gets closer and closer to the line $y = x$ as $x \to +\infty$ or as $x \to -\infty$. The line $y = x$ is called an ***oblique*** or ***slant asymptote*** of f. Similarly the graph of $y = g(x)$ eventually gets closer and closer to the parabola $y = x^2$ as $x \to +\infty$ or as $x \to -\infty$. The parabola is called a ***curvilinear asymptote*** of g. The graphs of the functions in (1) are shown in Figures 4.3.3 and 4.3.4.

In general, if $f(x) = P(x)/Q(x)$ is a rational function, then we can find quotient and remainder polynomials $q(x)$ and $r(x)$ such that

$$f(x) = q(x) + \frac{r(x)}{Q(x)}$$

and the degree of $r(x)$ is less than the degree of $Q(x)$. Then $r(x)/Q(x) \to 0$ as $x \to +\infty$ and as $x \to -\infty$, so $y = q(x)$ is an asymptote of f. This asymptote will be an oblique line if the degree of $P(x)$ is one greater than the degree of $Q(x)$, and it will be curvilinear if the degree of $P(x)$ exceeds that of $Q(x)$ by two or more. Problems involving these kinds of asymptotes are given in the exercises (Exercises 17 and 18).

GRAPHS WITH VERTICAL TANGENTS AND CUSPS

Figure 4.3.5 shows four curve elements that are commonly found in graphs of functions that involve radicals or fractional exponents. In all four cases, the function is not differentiable at x_0 because the secant line through $(x_0, f(x_0))$ and $(x, f(x))$ approaches a vertical position as x approaches x_0 from either side. Thus, in each case, the curve has a vertical tangent line at $(x_0, f(x_0))$. In parts (a) and (b) of the figure, there is an inflection point at x_0 because there is a change in concavity at that point. In parts (c) and (d), where $f'(x)$ approaches $+\infty$ from one side of x_0 and $-\infty$ from the other side, we say that the graph has a ***cusp*** at x_0.

Note that inflection points can be distinguished from cusps at vertical tangents by the limiting behavior of f'. An inflection point occurs at x_0 if the one-sided limits of f' at x_0 have the same sign, and a cusp occurs if they have opposite signs.

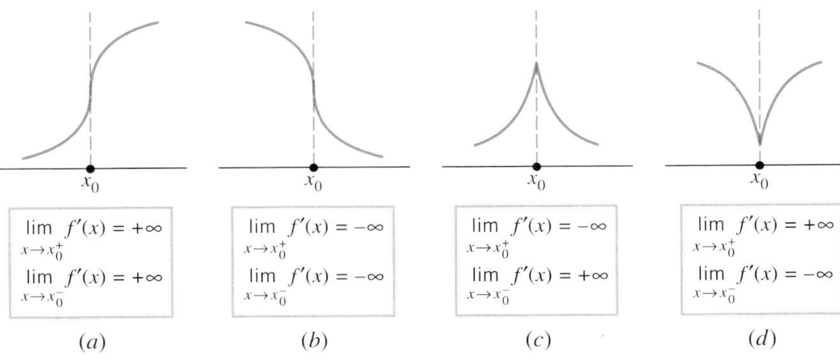

Figure 4.3.5

▶ **Example 3** Sketch the graph of $y = (x - 4)^{2/3}$.

- *Symmetries:* There are no symmetries about the coordinate axes or the origin (verify). However, the graph of $y = (x - 4)^{2/3}$ is symmetric about the line $x = 4$, since it is a translation (four units to the right) of the graph of $y = x^{2/3}$, which is symmetric about the y-axis.
- *x-intercepts:* Setting $y = 0$ yields the x-intercept $x = 4$.
- *y-intercepts:* Setting $x = 0$ yields the y-intercept $y = \sqrt[3]{16} \approx 2.5$.
- *Vertical asymptotes:* None, since $f(x) = (x - 4)^{2/3}$ is continuous everywhere.
- *Horizontal asymptotes:* None, since

$$\lim_{x\to+\infty} (x - 4)^{2/3} = +\infty \quad \text{and} \quad \lim_{x\to-\infty} (x - 4)^{2/3} = +\infty$$

- *Derivatives:*

$$\frac{dy}{dx} = f'(x) = \frac{2}{3}(x - 4)^{-1/3} = \frac{2}{3(x - 4)^{1/3}}$$

$$\frac{d^2y}{dx^2} = f''(x) = -\frac{2}{9}(x - 4)^{-4/3} = -\frac{2}{9(x - 4)^{4/3}}$$

- *Vertical tangent lines:* There is a vertical tangent line and cusp at $x = 4$ of the type in Figure 4.3.5*d* since $f(x) = (x - 4)^{2/3}$ is continuous at $x = 4$ and

$$\lim_{x\to4^+} f'(x) = \lim_{x\to4^+} \frac{2}{3(x - 4)^{1/3}} = +\infty$$

$$\lim_{x\to4^-} f'(x) = \lim_{x\to4^-} \frac{2}{3(x - 4)^{1/3}} = -\infty$$

Conclusions:

- The function $f(x) = (x - 4)^{2/3} = ((x - 4)^{1/3})^2$ is nonnegative for all x. There is a zero for f at $x = 4$.
- There is a critical point at $x = 4$, since f is not differentiable there. We saw above that a cusp occurs at this point. The sign analysis of dy/dx in Figure 4.3.6*a* and the first derivative test show that there is a relative minimum at this cusp, since $f'(x) < 0$ if $x < 4$ and $f'(x) > 0$ if $x > 4$.
- The sign analysis of d^2y/dx^2 in Figure 4.3.6*a* shows that the graph is concave down on both sides of the cusp.

The graph is shown in Figure 4.3.6*b*. ◀

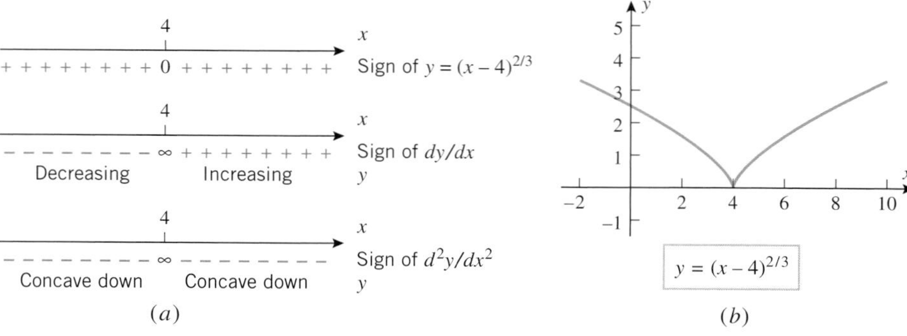

Figure 4.3.6

GRAPHING USING CALCULUS AND TECHNOLOGY TOGETHER

Thus far in this chapter we have used calculus to produce graphs of functions; the graph was the end result. Now we will work in the reverse direction by *starting* with a graph produced by a graphing utility. Our goal will be to use the tools of calculus to determine the exact locations of relative extrema, inflection points, and other features suggested by that graph and to determine whether the graph may be missing some important features that we would like to see.

▶ **Example 4** Use a graphing utility to generate the graph of $f(x) = 6x^{1/3} + 3x^{4/3}$, and discuss what it tells you about relative extrema, inflection points, asymptotes, and end behavior. Use calculus to find the exact locations of all key features of the graph.

Solution. Figure 4.3.7*b* shows a graph of f produced by a graphing utility. The graph suggests that there is an x-intercept at $x = 0$, a relative minimum between $x = -1$ and $x = 0$, no horizontal or vertical asymptotes, a vertical tangent at $x = 0$, and an inflection point at $x = 0$. For a more precise analysis of this information we need to consider the derivatives

$$f'(x) = 2x^{-2/3} + 4x^{1/3} = 2x^{-2/3}(1 + 2x) = \frac{2(2x+1)}{x^{2/3}}$$

$$f''(x) = -\frac{4}{3}x^{-5/3} + \frac{4}{3}x^{-2/3} = \frac{4}{3}x^{-5/3}(-1 + x) = \frac{4(x-1)}{3x^{5/3}}$$

- *x-intercepts:* Setting $f(x) = 6x^{1/3} + 3x^{4/3} = 3x^{1/3}(2 + x) = 0$ yields the x-intercepts $x = 0$ and $x = -2$.
- *y-intercept:* Setting $x = 0$ yields the y-intercept $y = 0$.
- *Vertical asymptotes:* None, since $f(x) = 6x^{1/3} + 3x^{4/3}$ is continuous everywhere.
- *Horizontal asymptotes:* None, since

$$\lim_{x\to+\infty} (6x^{1/3} + 3x^{4/3}) = \lim_{x\to+\infty} 3x^{1/3}(2 + x) = +\infty$$

$$\lim_{x\to-\infty} (6x^{1/3} + 3x^{4/3}) = \lim_{x\to-\infty} 3x^{1/3}(2 + x) = +\infty$$

- *Relative extrema:* From the formula for $f'(x)$ we see that there is a stationary point at $x = -\frac{1}{2}$ and a critical point at $x = 0$ at which f is not differentiable. The sign analysis of f' in Figure 4.3.7*a* and the first derivative test show that there is a relative minimum at the stationary point $x = -\frac{1}{2}$ (verify) and that there is no relative extremum at the critical point $x = 0$ (verify).
- *Inflection points:* The sign analysis of f'' in Figure 4.3.7*a* shows that there are inflection points at $x = 0$ and at $x = 1$ (verify). Note that the inflection point at $x = 1$ is so subtle that it is not evident from the graph in Figure 4.3.7*b*.
- *Vertical tangent lines:* There is a vertical tangent line and inflection point of the type in Figure 4.3.5*a* at $x = 0$ since f is continuous at that point, there is a change in concavity there, and

$$\lim_{x\to 0^+} f'(x) = \lim_{x\to 0^+} \frac{2(2x+1)}{x^{2/3}} = +\infty$$

$$\lim_{x\to 0^-} f'(x) = \lim_{x\to 0^-} \frac{2(2x+1)}{x^{2/3}} = +\infty \quad ◀$$

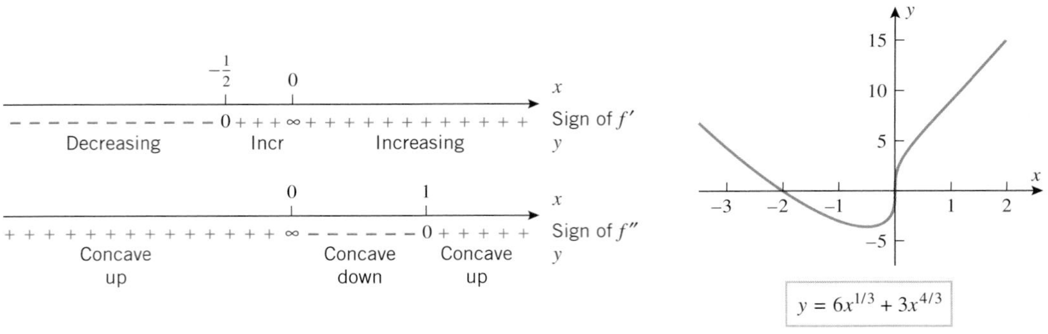

x: $-\frac{1}{2}$, 0

Sign of f': – – – – – – – – – – – – 0 + + + ∞ + + + + + + + + + + + + +

y: Decreasing | Incr | Increasing

x: 0, 1

Sign of f'': + + + + + + + + + + + + + + + + ∞ – – – – – – 0 + + + + +

y: Concave up | Concave down | Concave up

(a)

$y = 6x^{1/3} + 3x^{4/3}$

(b)

Figure 4.3.7

✔QUICK CHECK EXERCISES 4.3 (See page 254 for answers.)

1. Let $f(x) = \dfrac{3(x+1)(x-3)}{(x+2)(x-4)}$. Given that

$$f'(x) = \frac{-30(x-1)}{(x+2)^2(x-4)^2}, \qquad f''(x) = \frac{90(x^2-2x+4)}{(x+2)^3(x-4)^3}$$

determine the following properties of the graph of f.
(a) The x- and y-intercepts are ______.
(b) The vertical asymptotes are ______.
(c) The horizontal asymptote is ______.
(d) The graph is above the x-axis on the intervals ______.
(e) The graph is increasing on the intervals ______.
(f) The graph is concave up on the intervals ______.
(g) The relative maximum point on the graph is ______.

2. Let $f(x) = \dfrac{x^2-4}{x^{8/3}}$. Given that

$$f'(x) = \frac{-2(x^2-16)}{3x^{11/3}}, \qquad f''(x) = \frac{2(5x^2-176)}{9x^{14/3}}$$

determine the following properties of the graph of f.
(a) The x-intercepts are ______.
(b) The vertical asymptote is ______.
(c) The horizontal asymptote is ______.
(d) The graph is above the x-axis on the intervals ______.
(e) The graph is increasing on the intervals ______.
(f) The graph is concave up on the intervals ______.
(g) Inflection points occur at $x =$ ______.

EXERCISE SET 4.3 Graphing Utility

1–14 Give a graph of the rational function and label the coordinates of the stationary points and inflection points. Show the horizontal and vertical asymptotes and label them with their equations. Label point(s), if any, where the graph crosses a horizontal asymptote. Check your work with a graphing utility.

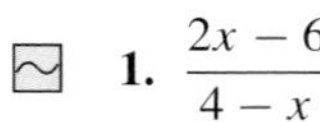

1. $\dfrac{2x-6}{4-x}$ **2.** $\dfrac{8}{x^2-4}$ **3.** $\dfrac{x}{x^2-4}$

4. $\dfrac{x^2}{x^2-4}$ **5.** $\dfrac{x^2}{x^2+4}$ **6.** $\dfrac{(x^2-1)^2}{x^4+1}$

7. $\dfrac{x^3+1}{x^3-1}$ **8.** $2 - \dfrac{1}{3x^2+x^3}$

9. $\dfrac{4}{x^2} - \dfrac{2}{x} + 3$ **10.** $\dfrac{3(x+1)^2}{(x-1)^2}$

11. $\dfrac{(3x+1)^2}{(x-1)^2}$ **12.** $3 + \dfrac{x+1}{(x-1)^4}$

13. $\dfrac{x^2+x}{1-x^2}$ **14.** $\dfrac{x^2}{1-x^3}$

15. In each part, make a rough sketch of the graph using asymptotes and appropriate limits but no derivatives. Compare your sketch to that generated with a graphing utility.
(a) $y = \dfrac{3x^2-8}{x^2-4}$ (b) $y = \dfrac{x^2+2x}{x^2-1}$
(c) $y = \dfrac{2x-x^2}{x^2+x-2}$ (d) $y = \dfrac{x^2}{x^2-x-2}$

16. (a) Sketch the graph of

$$y = \frac{1}{(x-a)(x-b)}$$

assuming that $a \neq b$.

(b) Prove that if $a \neq b$, then the function

$$f(x) = \frac{1}{(x-a)(x-b)}$$

is symmetric about the line $x = (a+b)/2$.

17. Show that $y = x + 3$ is an oblique asymptote of the graph of $f(x) = x^2/(x-3)$. Sketch the graph of $y = f(x)$ showing this asymptotic behavior.

18. Show that $y = 3 - x^2$ is a curvilinear asymptote of the graph of $f(x) = (2 + 3x - x^3)/x$. Sketch the graph of $y = f(x)$ showing this asymptotic behavior.

19–24 Sketch a graph of the rational function and label the coordinates of the stationary points and inflection points. Show the horizontal, vertical, oblique, and curvilinear asymptotes and label them with their equations. Label point(s), if any, where the graph crosses an asymptote. Check your work with a graphing utility.

19. $x^2 - \dfrac{1}{x}$ **20.** $\dfrac{x^2-2}{x}$ **21.** $\dfrac{(x-2)^3}{x^2}$

22. $x - \dfrac{1}{x} - \dfrac{1}{x^2}$ **23.** $\dfrac{x^3-4x-8}{x+2}$ **24.** $\dfrac{x^5}{x^2+1}$

FOCUS ON CONCEPTS

25. In each part, match the function with graphs I–VI without using a graphing utility, and then use a graphing utility to generate the graphs.
(a) $x^{1/3}$ (b) $x^{1/4}$ (c) $x^{1/5}$
(d) $x^{2/5}$ (e) $x^{4/3}$ (f) $x^{-1/3}$

I
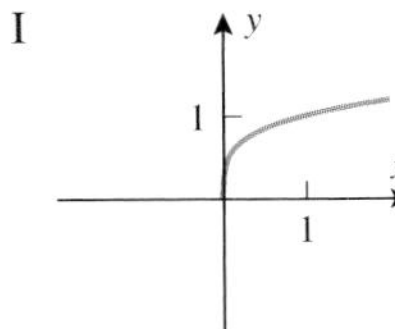

II
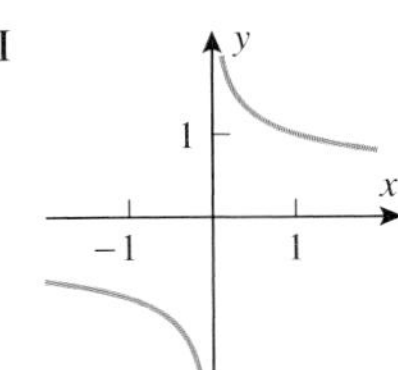

III
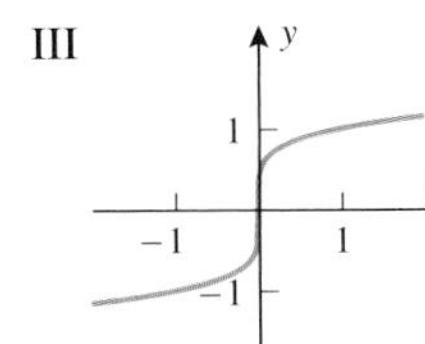

IV
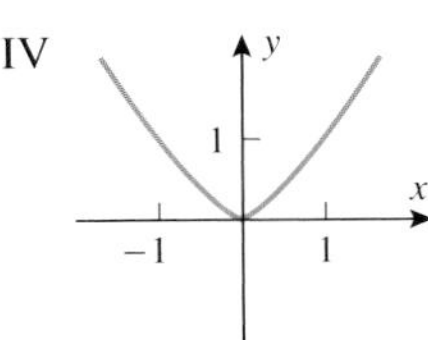

V
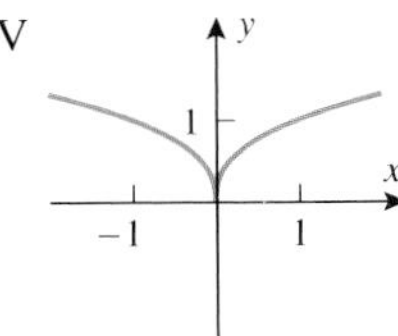

VI
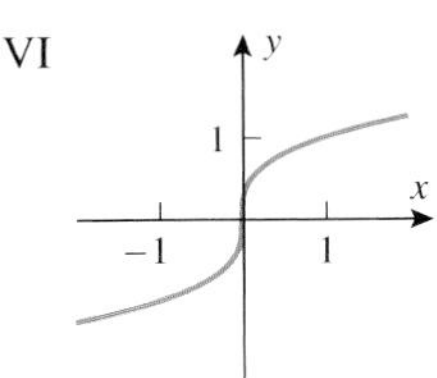

Figure Ex-25

26. Sketch the general shape of the graph of $y = x^{1/n}$, and then explain in words what happens to the shape of the graph as n increases if
(a) n is a positive even integer
(b) n is a positive odd integer.

27–34 Give a graph of the function and identify the locations of all critical points and inflection points. Check your work with a graphing utility.

27. $\sqrt{4x^2-1}$ **28.** $\sqrt[3]{x^2-4}$

29. $2x + 3x^{2/3}$ **30.** $2x^2 - 3x^{4/3}$

31. $4x^{1/3} - x^{4/3}$ **32.** $5x^{2/3} + x^{5/3}$

33. $\dfrac{8+x}{2+\sqrt[3]{x}}$ **34.** $\dfrac{8(\sqrt{x}-1)}{x}$

35–40 Give a graph of the function and identify the locations of all relative extrema and inflection points. Check your work with a graphing utility.

35. $x + \sin x$ **36.** $x - \tan x$

37. $\sqrt{3}\cos x + \sin x$ **38.** $\sin x + \cos x$

39. $\sin^2 x - \cos x, \quad -\pi \le x \le 3\pi$

40. $\sqrt{\tan x}, \quad 0 \le x < \pi/2$

FOCUS ON CONCEPTS

41. The accompanying figure shows the graph of the *derivative* of a function h that is defined and continuous on the interval $(-\infty, +\infty)$. Assume that the graph of h' has a vertical asymptote at $x = 3$ and that

$$h'(x) \to 0^+ \text{ as } x \to -\infty$$
$$h'(x) \to -\infty \text{ as } x \to +\infty$$

(a) What are the critical points for $h(x)$?
(b) Identify the intervals on which $h(x)$ is increasing.
(c) Identify the x-coordinates of relative extrema for $h(x)$ and classify each as a relative maximum or relative minimum.
(d) Estimate the x-coordinates of inflection points for $h(x)$.

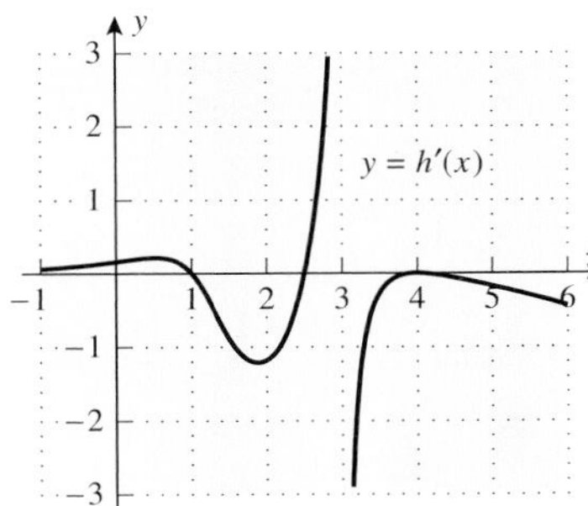

Figure Ex-41

42. Let $f(x) = (1 - 2x)h(x)$, where $h(x)$ is as given in Exercise 41. Suppose that $x = 5$ is a critical point for $f(x)$.
(a) Estimate $h(5)$.
(b) Use the second derivative test to determine whether $f(x)$ has a relative maximum or a relative minimum at $x = 5$.

43. A rectangular plot of land is to be fenced off so that the area enclosed will be 400 ft^2. Let L be the length of fencing needed and x the length of one side of the rectangle. Show that $L = 2x + 800/x$ for $x > 0$, and sketch the graph of L versus x for $x > 0$.

44. A box with a square base and open top is to be made from sheet metal so that its volume is 500 in^3. Let S be the area of the surface of the box and x the length of a side of the square base. Show that $S = x^2 + 2000/x$ for $x > 0$, and sketch the graph of S versus x for $x > 0$.

45. The accompanying figure shows a computer-generated graph of the polynomial $y = 0.1x^5(x - 1)$ using a viewing window of $[-2, 2.5] \times [-1, 5]$. Show that the choice of the vertical scale caused the computer to miss important features of the graph. Find the features that were missed and make your own sketch of the graph that shows the missing features.

46. The accompanying figure shows a computer-generated graph of the polynomial $y = 0.1x^5(x + 1)^2$ using a viewing window of $[-2, 1.5] \times [-0.2, 0.2]$. Show that the choice of the vertical scale caused the computer to miss important features of the graph. Find the features that were missed and make your own sketch of the graph that shows the missing features.

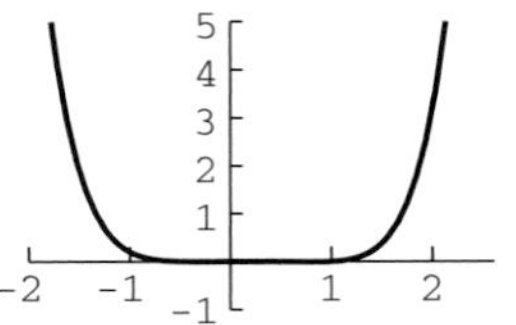

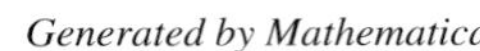
Generated by Mathematica

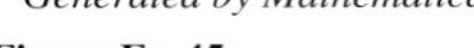
Figure Ex-45

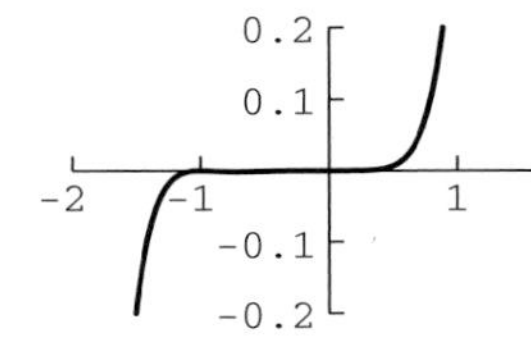

Generated by Mathematica

Figure Ex-46

✔ QUICK CHECK ANSWERS 4.3

1. (a) $(-1, 0)$, $(3, 0)$, $\left(0, \frac{9}{8}\right)$ (b) $x = -2$ and $x = 4$ (c) $y = 3$ (d) $(-\infty, -2)$, $(-1, 3)$, and $(4, +\infty)$ (e) $(-\infty, -2)$ and $(-2, 1]$ (f) $(-\infty, -2)$ and $(4, +\infty)$ (g) $\left(1, \frac{4}{3}\right)$ 2. (a) $(-2, 0)$, $(2, 0)$ (b) $x = 0$ (c) $y = 0$ (d) $(-\infty, -2)$ and $(2, +\infty)$ (e) $(-\infty, -4]$ and $(0, 4]$ (f) $(-\infty, -4\sqrt{11/5})$ and $(4\sqrt{11/5}, +\infty)$ (g) $\pm 4\sqrt{11/5} \approx \pm 5.93$

4.4 ABSOLUTE MAXIMA AND MINIMA

At the beginning of Section 4.2 we observed that if the graph of a function f is viewed as a two-dimensional mountain range (Figure 4.2.1), then the relative maxima and minima correspond to the tops of the hills and the bottoms of the valleys; that is, they are the high and low points in their immediate vicinity. In this section we will be concerned with the more encompassing problem of finding the highest and lowest points over the entire mountain range, that is, we will be looking for the top of the highest hill and the bottom of the deepest valley. In mathematical terms, we will be looking for the largest and smallest values of a function over an interval.

■ ABSOLUTE EXTREMA

We will begin with some terminology for describing the largest and smallest values of a function on an interval.

4.4.1 DEFINITION. Let I be an interval in the domain of a function f. We say that f has an ***absolute maximum*** at a point x_0 in I if $f(x) \le f(x_0)$ for all x in I, and we say that f has an ***absolute minimum*** at x_0 if $f(x_0) \le f(x)$ for all x in I. We say that f has an ***absolute extremum*** at x_0 if it has either an absolute maximum or an absolute minimum at that point.

If f has an absolute maximum at the point x_0 on an interval I, then $f(x_0)$ is the largest value of f on I, and if f has an absolute minimum at x_0, then $f(x_0)$ is the smallest value of f on I. In general, there is no guarantee that a function will actually have an absolute maximum or minimum on a given interval (Figure 4.4.1).

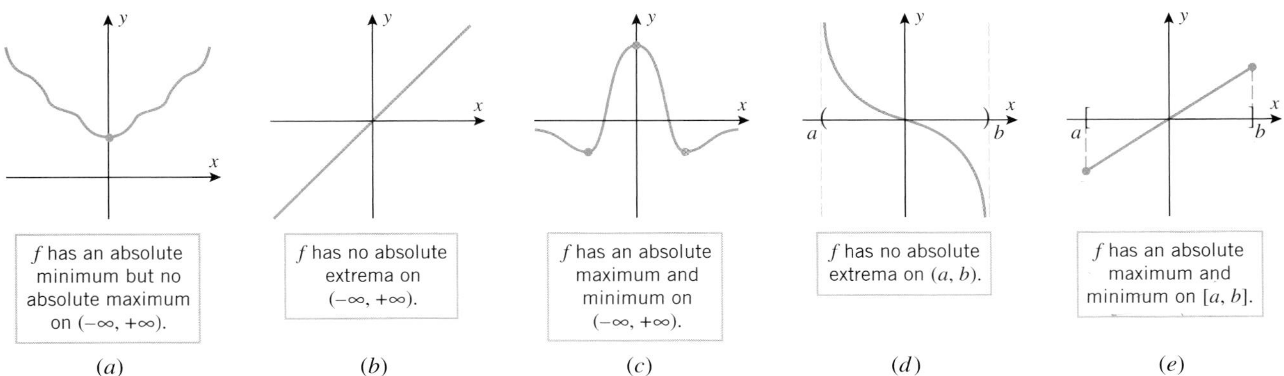

f has an absolute minimum but no absolute maximum on $(-\infty, +\infty)$.
f has no absolute extrema on $(-\infty, +\infty)$.
f has an absolute maximum and minimum on $(-\infty, +\infty)$.
f has no absolute extrema on (a, b).
f has an absolute maximum and minimum on $[a, b]$.
(a)
(b)
(c)
(d)
(e)

Figure 4.4.1

THE EXTREME VALUE THEOREM

Parts (a)–(d) of Figure 4.4.1 show that a continuous function may or may not have absolute maxima or minima on an infinite interval or on a finite open interval. However, the following theorem shows that a continuous function must have both an absolute maximum and an absolute minimum on every *finite closed* interval [see part (e) of Figure 4.4.1].

The hypotheses in the Extreme-Value Theorem are essential. That is, if either the interval is not closed or f is not continuous on the interval, then f need not have absolute extrema on the interval (Exercises 4–6).

4.4.2 THEOREM (*Extreme-Value Theorem*). *If a function f is continuous on a finite closed interval $[a, b]$ then f has both an absolute maximum and an absolute minimum on $[a, b]$.*

Although the proof of this theorem is too difficult to include here, you should be able to convince yourself of its validity with a little experimentation—try graphing various continuous functions over the interval $[0, 1]$, and convince yourself that there is no way to avoid having a highest and lowest point on a graph. As a physical analogy, if you imagine the graph to be a roller coaster track starting at $x = 0$ and ending at $x = 1$, the roller coaster will have to pass through a highest point and a lowest point during the trip.

The Extreme-Value Theorem is an example of what mathematicians call an ***existence theorem***. Such theorems state conditions under which certain objects exist, in this case absolute extrema. However, knowing that an object exists and finding it are two separate things. We will now address methods for determining the locations of absolute extrema under the conditions of the Extreme-Value Theorem.

If f is continuous on the finite closed interval $[a, b]$, then the absolute extrema of f occur either at the endpoints of the interval or inside on the open interval (a, b). If the absolute extrema happen to fall inside, then the following theorem tells us that they must occur at critical points of f.

Theorem 4.4.3 is also valid on infinite open intervals, that is, intervals of the form $(-\infty, +\infty)$, $(a, +\infty)$, and $(-\infty, b)$.

4.4.3 THEOREM. *If f has an absolute extremum on an open interval (a, b), then it must occur at a critical point of f.*

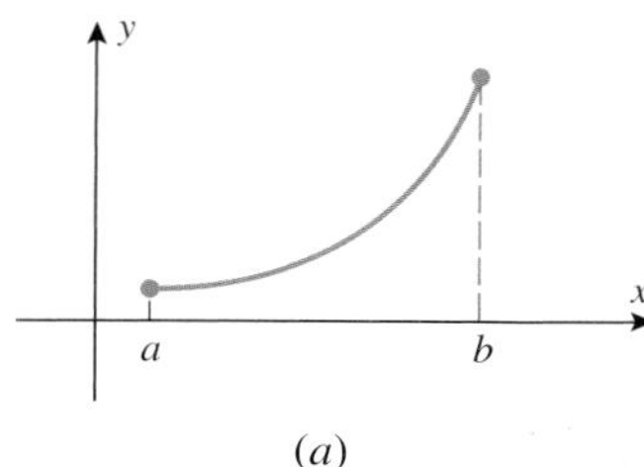

(a)

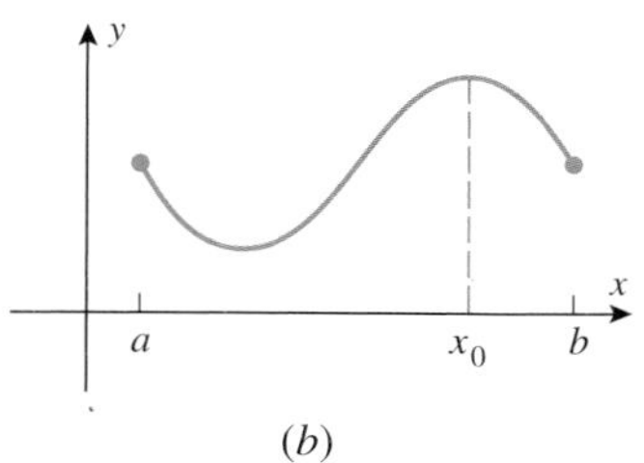

(b)

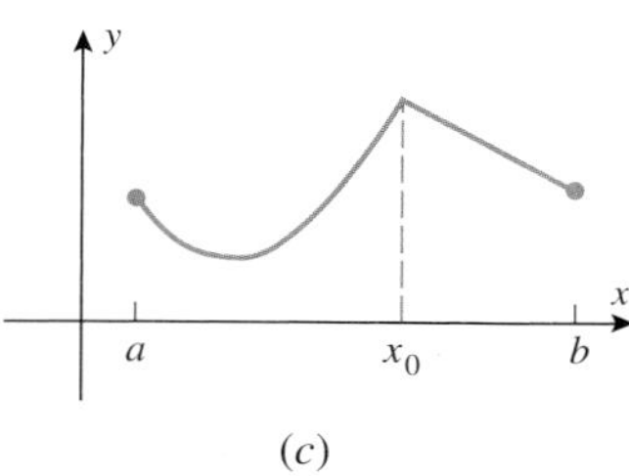

(c)

Figure 4.4.2 In part (a) the absolute maximum occurs at an endpoint of $[a, b]$, in part (b) it occurs at a stationary point in (a, b), and in part (c) it occurs at a critical point in (a, b) where f is not differentiable.

PROOF. If f has an absolute maximum on (a, b) at x_0, then $f(x_0)$ is also a relative maximum for f; for if $f(x_0)$ is the largest value of f on all (a, b), then $f(x_0)$ is certainly the largest value for f in the immediate vicinity of x_0. Thus, x_0 is a critical point of f by Theorem 4.2.2. The proof for absolute minima is similar. ■

It follows from this theorem that if f is continuous on the finite closed interval $[a, b]$, then the absolute extrema occur either at the endpoints of the interval or at critical points inside the interval (Figure 4.4.2). Thus, we can use the following procedure to find the absolute extrema of a continuous function on a finite closed interval $[a, b]$.

***A Procedure for Finding the Absolute Extrema of a Continuous Function f on a Finite Closed Interval* [a, b]**

Step 1. Find the critical points of f in (a, b).

Step 2. Evaluate f at all the critical points and at the endpoints a and b.

Step 3. The largest of the values in Step 2 is the absolute maximum value of f on $[a, b]$ and the smallest value is the absolute minimum.

▶ **Example 1** Find the absolute maximum and minimum values of the function $f(x) = 2x^3 - 15x^2 + 36x$ on the interval $[1, 5]$, and determine where these values occur.

Solution. Since f is continuous and differentiable everywhere, the absolute extrema must occur either at endpoints of the interval or at solutions to the equation $f'(x) = 0$ in the open interval $(1, 5)$. The equation $f'(x) = 0$ can be written as

$$6x^2 - 30x + 36 = 6(x^2 - 5x + 6) = 6(x - 2)(x - 3) = 0$$

Thus, there are stationary points at $x = 2$ and at $x = 3$. Evaluating f at the endpoints, at $x = 2$ and at $x = 3$ yields

$$\begin{aligned} f(1) &= 2(1)^3 - 15(1)^2 + 36(1) = 23 \\ f(2) &= 2(2)^3 - 15(2)^2 + 36(2) = 28 \\ f(3) &= 2(3)^3 - 15(3)^2 + 36(3) = 27 \\ f(5) &= 2(5)^3 - 15(5)^2 + 36(5) = 55 \end{aligned}$$

from which we conclude that the absolute minimum of f on $[1, 5]$ is 23, occurring at $x = 1$, and the absolute maximum of f on $[1, 5]$ is 55, occurring at $x = 5$. This is consistent with the graph of f in Figure 4.4.3. ◀

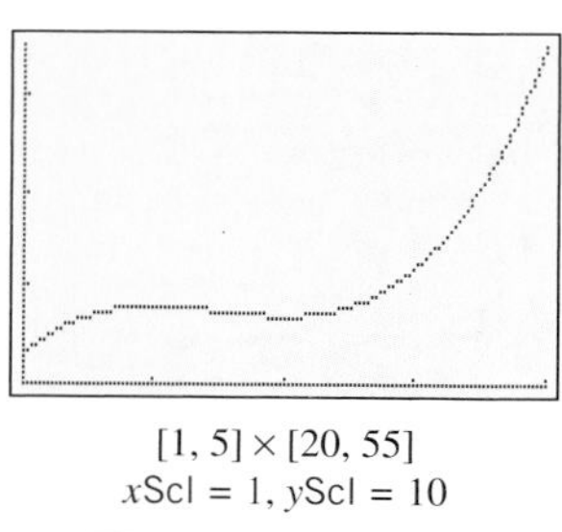

$[1, 5] \times [20, 55]$
$x\text{Scl} = 1, y\text{Scl} = 10$

$y = 2x^3 - 15x^2 + 36x$

Figure 4.4.3

▶ **Example 2** Find the absolute extrema of $f(x) = 6x^{4/3} - 3x^{1/3}$ on the interval $[-1, 1]$, and determine where these values occur.

Solution. Note that f is continuous everywhere and therefore the Extreme-Value Theorem guarantees that f has a maximum and a minimum value in the interval $[-1, 1]$. Differentiating, we obtain

$$f'(x) = 8x^{1/3} - x^{-2/3} = x^{-2/3}(8x - 1) = \frac{8x - 1}{x^{2/3}}$$

Thus, $f'(x) = 0$ at $x = \frac{1}{8}$, and $f'(x)$ is undefined at $x = 0$. Evaluating f at these critical points and endpoints yields Table 4.4.1, from which we conclude that an absolute minimum value of $-\frac{9}{8}$ occurs at $x = \frac{1}{8}$, and an absolute maximum value of 9 occurs at $x = -1$. ◀

Table 4.4.1

x	-1	0	$\frac{1}{8}$	1
$f(x)$	9	0	$-\frac{9}{8}$	3

ABSOLUTE EXTREMA ON INFINITE INTERVALS

We observed earlier that a continuous function may or may not have absolute extrema on an infinite interval (see Figure 4.4.1). However, certain conclusions about the existence of absolute extrema of a continuous function f on $(-\infty, +\infty)$ can be drawn from the behavior of $f(x)$ as $x \to -\infty$ and as $x \to +\infty$ (Table 4.4.2).

Table 4.4.2

LIMITS	$\lim_{x\to-\infty} f(x) = +\infty$ $\lim_{x\to+\infty} f(x) = +\infty$	$\lim_{x\to-\infty} f(x) = -\infty$ $\lim_{x\to+\infty} f(x) = -\infty$	$\lim_{x\to-\infty} f(x) = -\infty$ $\lim_{x\to+\infty} f(x) = +\infty$	$\lim_{x\to-\infty} f(x) = +\infty$ $\lim_{x\to+\infty} f(x) = -\infty$
CONCLUSION IF *f* IS CONTINUOUS EVERYWHERE	f has an absolute minimum but no absolute maximum on $(-\infty, +\infty)$.	f has an absolute maximum but no absolute minimum on $(-\infty, +\infty)$.	f has neither an absolute maximum nor an absolute minimum on $(-\infty, +\infty)$.	f has neither an absolute maximum nor an absolute minimum on $(-\infty, +\infty)$.
GRAPH				

▶ **Example 3** What can you say about the existence of absolute extrema on $(-\infty, +\infty)$ for polynomials?

Solution. If $p(x)$ is a polynomial of odd degree, then

$$\lim_{x\to+\infty} p(x) \quad \text{and} \quad \lim_{x\to-\infty} p(x) \tag{1}$$

have opposite signs (one is $+\infty$ and the other is $-\infty$), so there are no absolute extrema. On the other hand, if $p(x)$ has even degree, then the limits in (1) have the same sign (both $+\infty$ or both $-\infty$). If the leading coefficient is positive, then both limits are $+\infty$, and there is an absolute minimum but no absolute maximum; if the leading coefficient is negative, then both limits are $-\infty$, and there is an absolute maximum but no absolute minimum. ◀

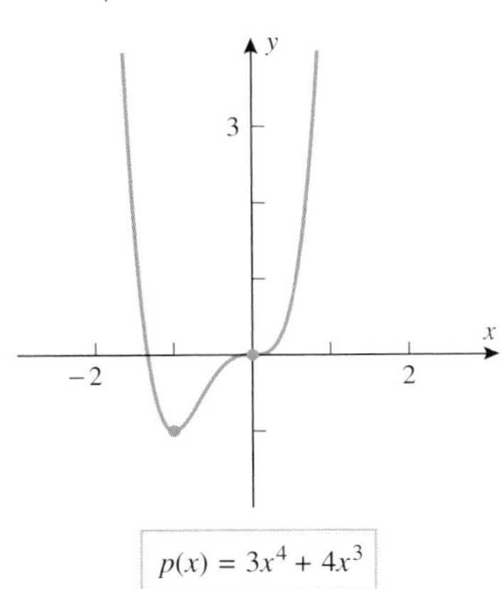

$p(x) = 3x^4 + 4x^3$

Figure 4.4.4

▶ **Example 4** Determine by inspection whether $p(x) = 3x^4 + 4x^3$ has any absolute extrema. If so, find them and state where they occur.

Solution. Since $p(x)$ has even degree and the leading coefficient is positive, $p(x) \to +\infty$ as $x \to \pm\infty$. Thus, there is an absolute minimum but no absolute maximum. From Theorem 4.4.3 [applied to the interval $(-\infty, +\infty)$], the absolute minimum must occur at a critical point of p. Since p is differentiable everywhere, we can find all critical points by solving the equation $p'(x) = 0$. This equation is

$$12x^3 + 12x^2 = 12x^2(x + 1) = 0$$

from which we conclude that the critical points are $x = 0$ and $x = -1$. Evaluating p at these critical points yields

$$p(0) = 0 \quad \text{and} \quad p(-1) = -1$$

Therefore, p has an absolute minimum of -1 at $x = -1$ (Figure 4.4.4). ◀

ABSOLUTE EXTREMA ON OPEN INTERVALS

We know that a continuous function may or may not have absolute extrema on an open interval. However, certain conclusions about the existence of absolute extrema of a continuous function f on a finite open interval (a, b) can be drawn from the behavior of $f(x)$ as $x \to a^+$ and as $x \to b^-$ (Table 4.4.3). Similar conclusions can be drawn for intervals of the form $(-\infty, b)$ or $(a, +\infty)$.

Table 4.4.3

LIMITS	$\lim_{x\to a^+} f(x) = +\infty$ $\lim_{x\to b^-} f(x) = +\infty$	$\lim_{x\to a^+} f(x) = -\infty$ $\lim_{x\to b^-} f(x) = -\infty$	$\lim_{x\to a^+} f(x) = -\infty$ $\lim_{x\to b^-} f(x) = +\infty$	$\lim_{x\to a^+} f(x) = +\infty$ $\lim_{x\to b^-} f(x) = -\infty$
CONCLUSION IF f IS CONTINUOUS ON (a, b)	f has an absolute minimum but no absolute maximum on (a, b).	f has an absolute maximum but no absolute minimum on (a, b).	f has neither an absolute maximum nor an absolute minimum on (a, b).	f has neither an absolute maximum nor an absolute minimum on (a, b).
GRAPH				

▶ **Example 5** Determine whether the function

$$f(x) = \frac{1}{x^2 - x}$$

has any absolute extrema on the interval $(0, 1)$. If so, find them and state where they occur.

Solution. Since f is continuous on the interval $(0, 1)$ and

$$\lim_{x\to 0^+} f(x) = \lim_{x\to 0^+} \frac{1}{x^2 - x} = \lim_{x\to 0^+} \frac{1}{x(x-1)} = -\infty$$

$$\lim_{x\to 1^-} f(x) = \lim_{x\to 1^-} \frac{1}{x^2 - x} = \lim_{x\to 1^-} \frac{1}{x(x-1)} = -\infty$$

the function f has an absolute maximum but no absolute minimum on the interval $(0, 1)$. By Theorem 4.4.3 the absolute maximum must occur at a critical point of f in the interval $(0, 1)$. We have

$$f'(x) = -\frac{2x - 1}{(x^2 - x)^2}$$

so the only solution of the equation $f'(x) = 0$ is $x = \frac{1}{2}$. Although f is not differentiable at $x = 0$ or at $x = 1$, these values are doubly disqualified since they are neither in the domain of f nor in the interval $(0, 1)$. Thus, the absolute maximum occurs at $x = \frac{1}{2}$, and this absolute maximum is

$$f\left(\tfrac{1}{2}\right) = \frac{1}{\left(\frac{1}{2}\right)^2 - \frac{1}{2}} = -4$$

(Figure 4.4.5). ◀

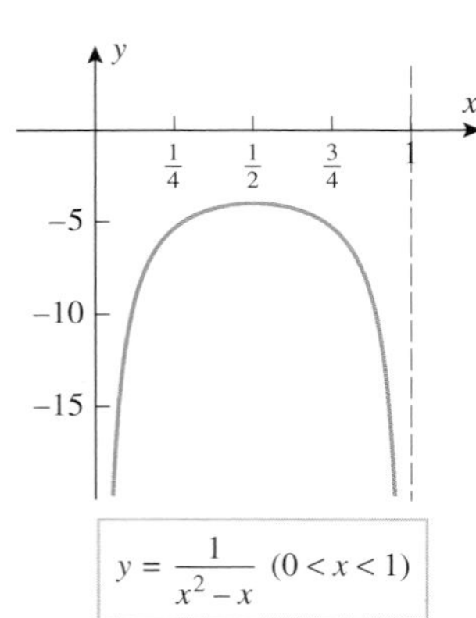

Figure 4.4.5

ABSOLUTE EXTREMA OF FUNCTIONS WITH ONE RELATIVE EXTREMUM

If a continuous function has only one relative extremum on a finite or infinite interval I, then that relative extremum must of necessity also be an absolute extremum. To understand why this is so, suppose that f has a relative maximum at x_0 in an interval I, and there are no other relative extrema of f on I. If $f(x_0)$ is *not* the absolute maximum of f on I, then the graph of f has to make an upward turn somewhere on I to rise above $f(x_0)$. However, this cannot happen because in the process of making an upward turn it would produce a second relative extremum on I (Figure 4.4.6). Thus, $f(x_0)$ must be the absolute maximum as well as a relative maximum. This idea is captured in the following theorem, which we state without proof.

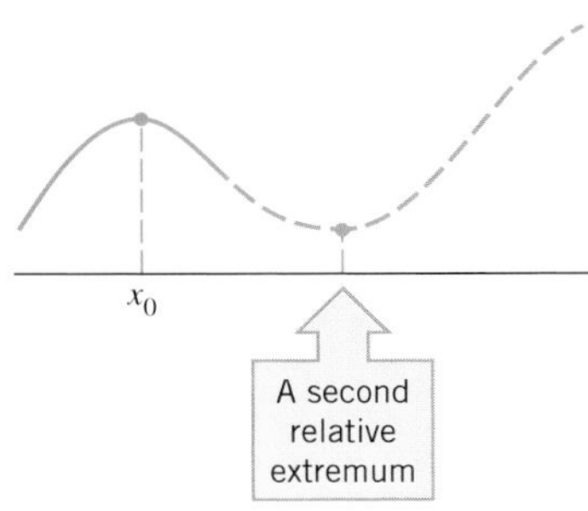

Figure 4.4.6

4.4.4 THEOREM. *Suppose that f is continuous and has exactly one relative extremum on an interval I, say at x_0.*

(a) If f has a relative minimum at x_0, then $f(x_0)$ is the absolute minimum of f on I.

(b) If f has a relative maximum at x_0, then $f(x_0)$ is the absolute maximum of f on I.

This theorem is often helpful in situations where other methods are difficult or tedious to apply.

Example 6 Find the absolute extrema, if any, of the function $f(x) = x^3 - 3x^2 + 4$ on the interval $(0, +\infty)$.

Solution. We have $\lim_{x\to+\infty} f(x) = +\infty$ (verify), so f does not have an absolute maximum on the interval $(0, +\infty)$. However, continuity of f and the fact that $\lim_{x\to 0^+} f(x) = f(0) = 4$ is finite allow for the possibility that f has an absolute minimum on $(0, +\infty)$. If so, it would have to occur at a critical point of f, so we consider the derivative

$$f'(x) = 3x^2 - 6x = 3x(x - 2)$$

We see that $x = 0$ and $x = 2$ are the critical points of f. Of these only $x = 2$ is in the interval $(0, +\infty)$, so this is the only point at which an absolute minimum could occur. To see whether an absolute minimum actually does occur at this point, we can apply part (a) of Theorem 4.4.4. Since

$$f''(x) = 6x - 6$$

we have $f''(2) = 6 > 0$, so a relative minimum occurs at $x = 2$ by the second derivative test. Thus, $f(x)$ has an absolute minimum at $x = 2$, and this absolute minimum is $f(2) = 0$ (Figure 4.4.7). ◀

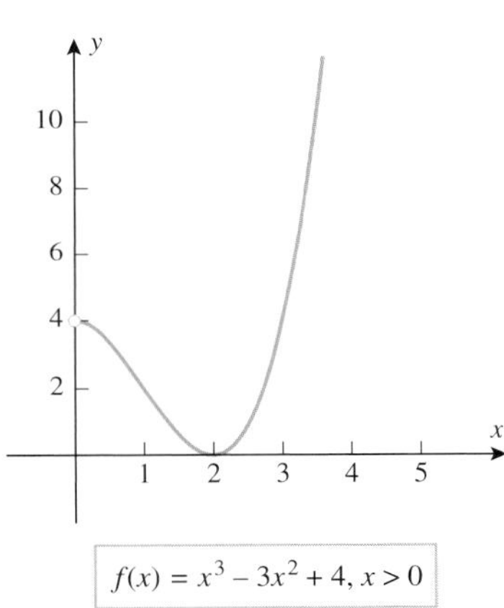

Figure 4.4.7

Does the function in Example 6 have an absolute minimum on the interval $(-\infty, +\infty)$?

✔QUICK CHECK EXERCISES 4.4 *(See page 262 for answers.)*

1. Use the accompanying graph to find the x-coordinates of the relative extrema and absolute extrema of f on $[0, 6]$.

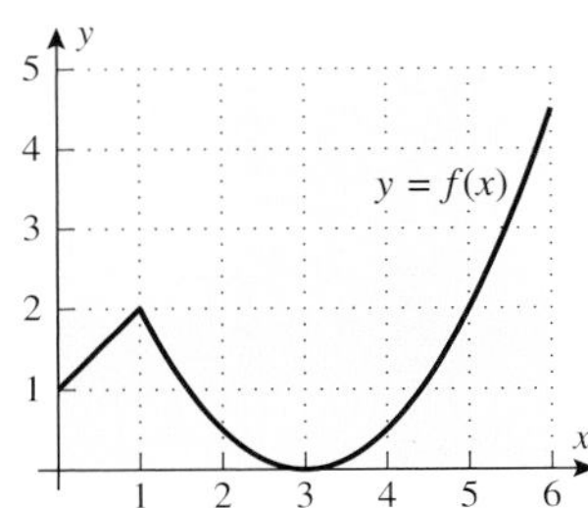

Figure Ex-1

2. True or false:
   (a) If a function f is continuous on $[a, b]$, then f has an absolute maximum on $[a, b]$.
   (b) If a function f is continuous on (a, b), then f has an absolute minimum on (a, b).
   (c) If a function f has an absolute minimum value on (a, b), then there is a critical point of f in (a, b).
   (d) If a function f is continuous on $[a, b]$ and f has no relative extreme values in (a, b), then the absolute maximum value of f exists and occurs either at $x = a$ or at $x = b$.

3. Suppose that a function f is continuous on $[-4, 4]$ and has critical points at $x = -3, 0, 2$. Use the accompanying table to determine the absolute maximum and absolute minimum values, if any, for f on the indicated intervals.
(a) $[1, 4]$ (b) $[-2, 2]$ (c) $[-4, 4]$ (d) $(-4, 4)$

x	−4	−3	−2	−1	0	1	2	3	4
$f(x)$	2224	−1333	0	1603	2096	2293	2400	2717	6064

4. Let $f(x) = x^3 - 3x^2 - 9x + 25$. Use the derivative $f'(x) = 3(x + 1)(x - 3)$ to determine the absolute maximum and absolute minimum values, if any, for f on each of the given intervals.
(a) $[0, 4]$ (b) $[-2, 4]$ (c) $[-4, 2]$
(d) $[-5, 10]$ (e) $(-5, 4)$

EXERCISE SET 4.4 Graphing Utility C CAS

FOCUS ON CONCEPTS

1–2 Use the graph to find x-coordinates of the relative extrema and absolute extrema of f on $[0, 7]$.

1.

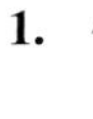

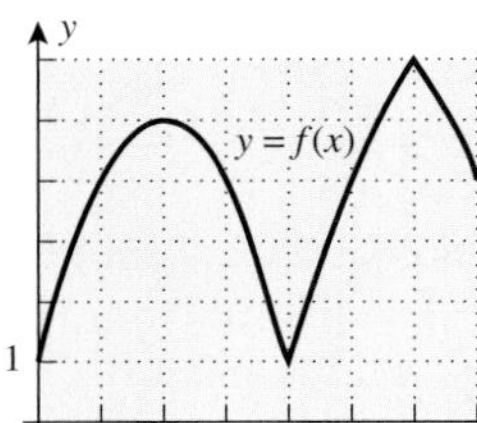

2.

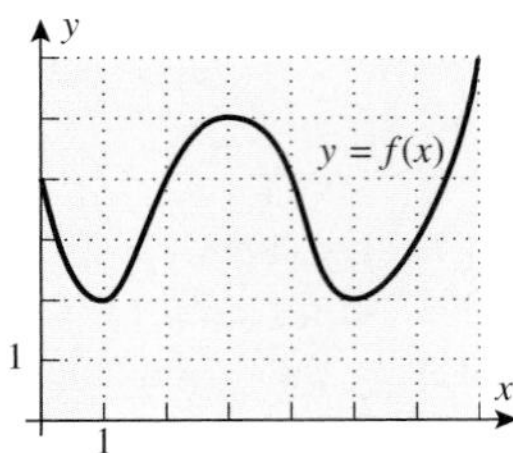

3. In each part, sketch the graph of a continuous function f with the stated properties on the interval $[0, 10]$.
(a) f has an absolute minimum at $x = 0$ and an absolute maximum at $x = 10$.
(b) f has an absolute minimum at $x = 2$ and an absolute maximum at $x = 7$.
(c) f has relative minima at $x = 1$ and $x = 8$, has relative maxima at $x = 3$ and $x = 7$, has an absolute minimum at $x = 5$, and has an absolute maximum at $x = 10$.

4. In each part, sketch the graph of a continuous function f with the stated properties on the interval $(-\infty, +\infty)$.
(a) f has no relative extrema or absolute extrema.
(b) f has an absolute minimum at $x = 0$ but no absolute maximum.
(c) f has an absolute maximum at $x = -5$ and an absolute minimum at $x = 5$.

5. Let
$$f(x) = \begin{cases} \dfrac{1}{1-x}, & 0 \le x < 1 \\ 0, & x = 1 \end{cases}$$
Explain why f has a minimum value but no maximum value on the closed interval $[0, 1]$.

6. Let
$$f(x) = \begin{cases} x, & 0 < x < 1 \\ \frac{1}{2}, & x = 0, 1 \end{cases}$$
Explain why f has neither a minimum value nor a maximum value on the closed interval $[0, 1]$.

7–16 Find the absolute maximum and minimum values of f on the given closed interval, and state where those values occur.

7. $f(x) = 4x^2 - 12x + 10$; $[1, 2]$
8. $f(x) = 8x - x^2$; $[0, 6]$
9. $f(x) = (x - 2)^3$; $[1, 4]$
10. $f(x) = 2x^3 + 3x^2 - 12x$; $[-3, 2]$
11. $f(x) = \dfrac{3x}{\sqrt{4x^2 + 1}}$; $[-1, 1]$
12. $f(x) = (x^2 + x)^{2/3}$; $[-2, 3]$
13. $f(x) = x - 2\sin x$; $[-\pi/4, \pi/2]$
14. $f(x) = \sin x - \cos x$; $[0, \pi]$
15. $f(x) = 1 + |9 - x^2|$; $[-5, 1]$
16. $f(x) = |6 - 4x|$; $[-3, 3]$

17–24 Find the absolute maximum and minimum values of f, if any, on the given interval, and state where those values occur.

17. $f(x) = x^2 - x - 2$; $(-\infty, +\infty)$
18. $f(x) = 3 - 4x - 2x^2$; $(-\infty, +\infty)$
19. $f(x) = 4x^3 - 3x^4$; $(-\infty, +\infty)$
20. $f(x) = x^4 + 4x$; $(-\infty, +\infty)$
21. $f(x) = 2x^3 - 6x + 2$; $(-\infty, +\infty)$
22. $f(x) = x^3 - 9x + 1$; $(-\infty, +\infty)$
23. $f(x) = \dfrac{x^2 + 1}{x + 1}$; $(-5, -1)$
24. $f(x) = \dfrac{x - 2}{x + 1}$; $(-1, 5]$

25–34 Use a graphing utility to estimate the absolute maximum and minimum values of f, if any, on the stated interval, and then use calculus methods to find the exact values.

25. $f(x) = (x^2 - 2x)^2$; $(-\infty, +\infty)$

26. $f(x) = (x-1)^2(x+2)^2$; $(-\infty, +\infty)$

27. $f(x) = x^{2/3}(20-x)$; $[-1, 20]$

28. $f(x) = \dfrac{x}{x^2+2}$; $[-1, 4]$

29. $f(x) = 1 + \dfrac{1}{x}$; $(0, +\infty)$

30. $f(x) = \dfrac{2x^2-3x+3}{x^2-2x+2}$; $[1, +\infty)$

31. $f(x) = \dfrac{2-\cos x}{\sin x}$; $[\pi/4, 3\pi/4]$

32. $f(x) = \sin^2 x + \cos x$; $[-\pi, \pi]$

33. $f(x) = \sin(\cos x)$; $[0, 2\pi]$

34. $f(x) = \cos(\sin x)$; $[0, \pi]$

35. Find the absolute maximum and minimum values of

$$f(x) = \begin{cases} 4x-2, & x < 1 \\ (x-2)(x-3), & x \geq 1 \end{cases}$$

on $\left[\frac{1}{2}, \frac{7}{2}\right]$.

36. Let $f(x) = x^2 + px + q$. Find the values of p and q such that $f(1) = 3$ is an extreme value of f on $[0, 2]$. Is this value a maximum or minimum?

37–38 If f is a periodic function, then the locations of all absolute extrema on the interval $(-\infty, +\infty)$ can be obtained by finding the locations of the absolute extrema for one period and using the periodicity to locate the rest. Use this idea in these exercises to find the absolute maximum and minimum values of the function, and state the x-values at which they occur.

37. $f(x) = 2\cos x + \cos 2x$

38. $f(x) = 3\cos\dfrac{x}{3} + 2\cos\dfrac{x}{2}$

39–40 One way of proving that $f(x) \leq g(x)$ for all x in a given interval is to show that $0 \leq g(x) - f(x)$ for all x in the interval; and one way of proving the latter inequality is to show that the absolute minimum value of $g(x) - f(x)$ on the interval is nonnegative. Use this idea to prove the inequalities in these exercises.

39. Prove that $\sin x \leq x$ for all x in the interval $[0, 2\pi]$.

40. Prove that $\cos x \geq 1 - (x^2/2)$ for all x in the interval $[0, 2\pi]$.

41. What is the smallest possible slope for a tangent to the graph of the equation $y = x^3 - 3x^2 + 5x$?

42. (a) Show that $f(x) = \sec x + \csc x$ has a minimum value but no maximum value on the interval $(0, \pi/2)$.
(b) Find the minimum value in part (a).

C 43. Show that the absolute minimum value of

$$f(x) = x^2 + \frac{x^2}{(8-x)^2}, \quad x > 8$$

occurs at $x = 10$ by using a CAS to find $f'(x)$ and to solve the equation $f'(x) = 0$.

C 44. The vertical displacement $f(t)$ of a cork bobbing up and down on the ocean's surface may be modelled by the function

$$f(t) = A\cos t + B\sin t$$

where $A > 0$ and $B > 0$. Use a CAS to find the maximum and minimum values of $f(t)$ in terms of A and B.

45. It can be proved that if f is differentiable on (a, b) and L is a line that does not intersect the curve $y = f(x)$ over an interval (a, b), then the points at which the curve is closest to or farthest from the line L, if any, occur at points where the tangent line to the curve is parallel to L (see the accompanying figure). Use this result to find the points on the graph of $y = -x^2$, $-1 \leq x \leq 1.5$, that are closest to and farthest from the line $y = 2 - x$.

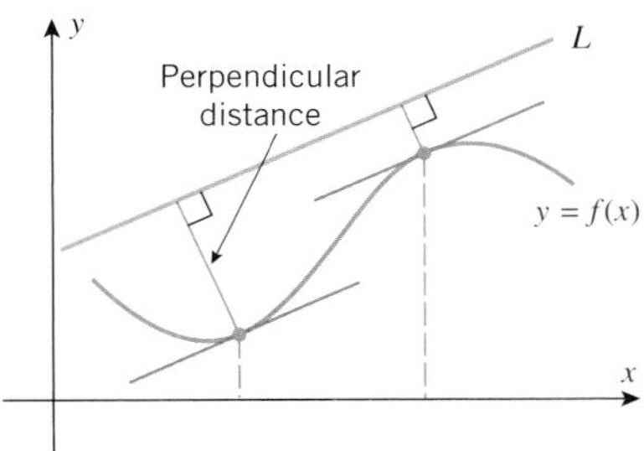

Figure Ex-45

46. Use the idea discussed in Exercise 45 to find the coordinates of all points on the graph of $y = x^3$, $-1 \leq x \leq 1$, closest to and farthest from the line $y = \frac{4}{3}x - 1$.

47. Suppose that the equations of motion of a paper airplane during the first 12 seconds of flight are

$$x = t - 2\sin t, \quad y = 2 - 2\cos t \quad (0 \leq t \leq 12)$$

What are the highest and lowest points in the trajectory, and when is the airplane at those points?

48. The accompanying figure shows the path of a fly whose equations of motion are

$$x = \frac{\cos t}{2 + \sin t}, \quad y = 3 + \sin(2t) - 2\sin^2 t \quad (0 \leq t \leq 2\pi)$$

(a) How high and low does it fly?
(b) How far left and right of the origin does it fly?

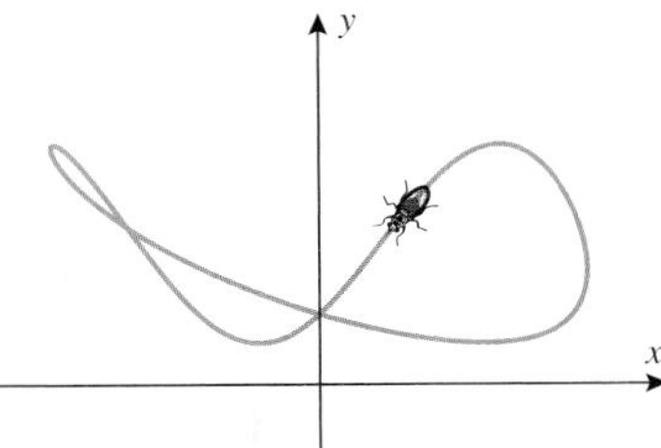

Figure Ex-48

49. Let $f(x) = ax^2 + bx + c$, where $a > 0$. Prove that $f(x) \geq 0$ for all x if and only if $b^2 - 4ac \leq 0$. [*Hint:* Find the minimum of $f(x)$.]

50. Prove Theorem 4.4.3 in the case where the extreme value is a minimum.

✔QUICK CHECK ANSWERS 4.4

1. There is a relative minimum at $x = 3$, a relative maximum at $x = 1$, an absolute minimum at $x = 3$, and an absolute maximum at $x = 6$. **2.** (a) true (b) false (c) true (d) true **3.** (a) max, 6064; min, 2293 (b) max, 2400; min, 0 (c) max, 6064; min, -1333 (d) no max; min, -1333 **4.** (a) max, $f(0) = 25$; min, $f(3) = -2$ (b) max, $f(-1) = 30$; min, $f(3) = -2$ (c) max, $f(-1) = 30$; min, $f(-4) = -51$ (d) max, $f(10) = 635$; min, $f(-5) = -130$ (e) max, $f(-1) = 30$; no min

4.5 APPLIED MAXIMUM AND MIMIMUM PROBLEMS

In this section we will show how the methods discussed in the last section can be used to solve various applied optimization problems.

CLASSIFICATION OF OPTIMIZATION PROBLEMS

The applied optimization problems that we will consider in this section fall into the following two categories:

- Problems that reduce to maximizing or minimizing a continuous function over a finite closed interval.
- Problems that reduce to maximizing or minimizing a continuous function over an infinite interval or a finite interval that is not closed.

For problems of the first type the Extreme-Value Theorem (4.4.2) guarantees that the problem has a solution, and we know that the solution can be obtained by examining the values of the function at the critical points and at the endpoints. However, for problems of the second type there may or may not be a solution. If the function is continuous and has exactly one relative extremum of the appropriate type on the interval, then Theorem 4.4.4 guarantees the existence of a solution and provides a method for finding it. In cases where this theorem is not applicable some ingenuity may be required to solve the problem.

PROBLEMS INVOLVING FINITE CLOSED INTERVALS

In his *On a Method for the Evaluation of Maxima and Minima*, the seventeenth century French mathematician Pierre de Fermat solved an optimization problem very similar to the one posed in our first example. Fermat's work on such optimization problems prompted the French mathematician Laplace to proclaim Fermat the "true inventor of the differential calculus." Although this honor must still reside with Newton and Leibniz, it is the case that Fermat developed procedures that anticipated parts of differential calculus.

► **Example 1** A garden is to be laid out in a rectangular area and protected by a chicken wire fence. What is the largest possible area of the garden if only 100 running feet of chicken wire is available for the fence?

Solution. Let

x = length of the rectangle (ft)
y = width of the rectangle (ft)
A = area of the rectangle (ft^2)

Then

$$A = xy \tag{1}$$

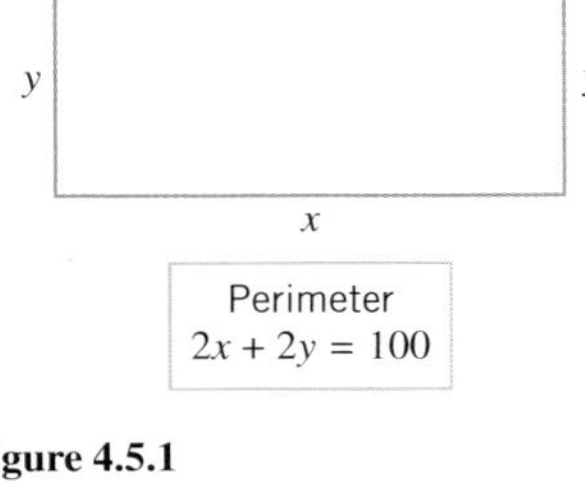

Figure 4.5.1

Since the perimeter of the rectangle is 100 ft, the variables x and y are related by the equation

$$2x + 2y = 100 \quad \text{or} \quad y = 50 - x \tag{2}$$

(See Figure 4.5.1.) Substituting (2) in (1) yields

$$A = x(50 - x) = 50x - x^2 \tag{3}$$

Because x represents a length it cannot be negative, and because the two sides of length x cannot have a combined length exceeding the total perimeter of 100 ft, the variable x must satisfy

$$0 \le x \le 50 \tag{4}$$

Thus, we have reduced the problem to that of finding the value (or values) of x in $[0, 50]$, for which A is maximum. Since A is a polynomial in x, it is continuous on $[0, 50]$, and so the maximum must occur at an endpoint of this interval or at a critical point.

From (3) we obtain

$$\frac{dA}{dx} = 50 - 2x$$

Setting $dA/dx = 0$ we obtain

$$50 - 2x = 0$$

or $x = 25$. Thus, the maximum occurs at one of the values

$$x = 0, \quad x = 25, \quad x = 50$$

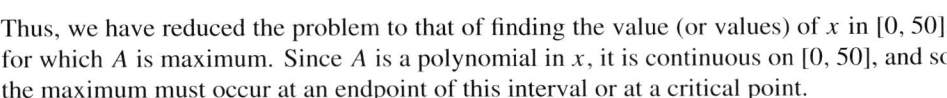

Pierre de Fermat (1601–1665) Fermat, the son of a successful French leather merchant, was a lawyer who practiced mathematics as a hobby. He received a Bachelor of Civil Laws degree from the University of Orleans in 1631 and subsequently held various government positions, including a post as councillor to the Toulouse parliament. Although he was apparently financially successful, confidential documents of that time suggest that his performance in office and as a lawyer was poor, perhaps because he devoted so much time to mathematics. Throughout his life, Fermat fought all efforts to have his mathematical results published. He had the unfortunate habit of scribbling his work in the margins of books and often sent his results to friends without keeping copies for himself. As a result, he never received credit for many major achievements until his name was raised from obscurity in the mid-nineteenth century. It is now known that Fermat, simultaneously and independently of Descartes, developed analytic geometry. Unfortunately, Descartes and Fermat argued bitterly over various problems so that there was never any real cooperation between these two great geniuses.

Fermat solved many fundamental calculus problems. He obtained the first procedure for differentiating polynomials, and solved many important maximization, minimization, area, and tangent problems. His work served to inspire Isaac Newton. Fermat is best known for his work in number theory, the study of properties of and relationships between whole numbers. He was the first mathematician to make substantial contributions to this field after the ancient Greek mathematician Diophantus. Unfortunately, none of Fermat's contemporaries appreciated his work in this area, a fact that eventually pushed Fermat into isolation and obscurity in later life. In addition to his work in calculus and number theory, Fermat was one of the founders of probability theory and made major contributions to the theory of optics. Outside mathematics, Fermat was a classical scholar of some note, was fluent in French, Italian, Spanish, Latin, and Greek, and he composed a considerable amount of Latin poetry.

One of the great mysteries of mathematics is shrouded in Fermat's work in number theory. In the margin of a book by Diophantus, Fermat scribbled that for integer values of n greater than 2, the equation $x^n + y^n = z^n$ has no nonzero integer solutions for x, y, and z. He stated, "I have discovered a truly marvelous proof of this, which however the margin is not large enough to contain." This result, which became known as "Fermat's last theorem," appeared to be true, but its proof evaded the greatest mathematical geniuses for 300 years until Professor Andrew Wiles of Princeton University presented a proof in June 1993 in a dramatic series of three lectures that drew international media attention (see *New York Times*, June 27, 1993). As it turned out, that proof had a serious gap that Wiles and Richard Taylor fixed and published in 1995. A prize of 100,000 German marks was offered in 1908 for the solution, but it is worthless today because of inflation.

Table 4.5.1

x	0	25	50
A	0	625	0

Substituting these values in (3) yields Table 4.5.1, which tells us that the maximum area of 625 ft^2 occurs at $x = 25$, which is consistent with the graph of (3) in Figure 4.5.2. From (2) the corresponding value of y is 25, so the rectangle of perimeter 100 ft with greatest area is a square with sides of length 25 ft. ◀

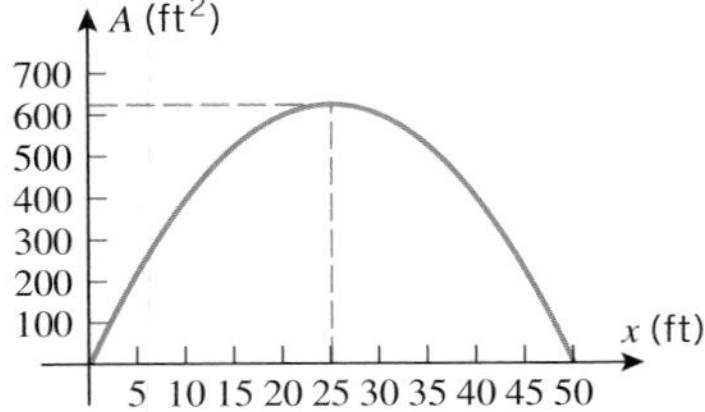

Figure 4.5.2

Example 1 illustrates the following five-step procedure that can be used for solving many applied maximum and minimum problems.

A Procedure for Solving Applied Maximum and Minimum Problems

Step 1. Draw an appropriate figure and label the quantities relevant to the problem.

Step 2. Find a formula for the quantity to be maximized or minimized.

Step 3. Using the conditions stated in the problem to eliminate variables, express the quantity to be maximized or minimized as a function of one variable.

Step 4. Find the interval of possible values for this variable from the physical restrictions in the problem.

Step 5. If applicable, use the techniques of the preceding section to obtain the maximum or minimum.

In Example 1 we included $x = 0$ and $x = 50$ as possible values of x, even though these correspond to rectangles with two sides of length zero. If we view this as a purely mathematical problem, then there is nothing wrong with this. However, if we view this as an applied problem in which the rectangle will be formed from physical material, then these values should be excluded.

▶ **Example 2** An open box is to be made from a 16-inch by 30-inch piece of cardboard by cutting out squares of equal size from the four corners and bending up the sides (Figure 4.5.3). What size should the squares be to obtain a box with the largest volume?

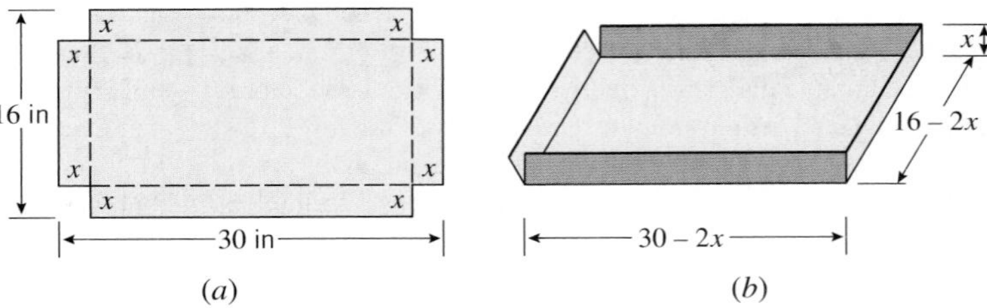

Figure 4.5.3

Solution. For emphasis, we explicitly list the steps of the five-step problem-solving procedure given above as an outline for the solution of this problem. (In later examples we will follow these guidelines without listing the steps.)

- *Step 1:* Figure 4.5.3*a* illustrates the cardboard piece with squares removed from its corners. Let

$$x = \text{length (in inches) of the sides of the squares to be cut out}$$
$$V = \text{volume (in cubic inches) of the resulting box}$$

- *Step 2:* Because we are removing a square of side x from each corner, the resulting box will have dimensions $16 - 2x$ by $30 - 2x$ by x (Figure 4.5.3*b*). Since the volume of a box is the product of its dimensions, we have

$$V = (16 - 2x)(30 - 2x)x = 480x - 92x^2 + 4x^3 \tag{5}$$

- *Step 3:* Note that our expression for volume is already in terms of the single variable x.
- *Step 4:* The variable x in (5) is subject to certain restrictions. Because x represents a length, it cannot be negative, and because the width of the cardboard is 16 inches, we cannot cut out squares whose sides are more than 8 inches long. Thus, the variable x in (5) must satisfy

$$0 \le x \le 8$$

and hence we have reduced our problem to finding the value (or values) of x in the interval $[0, 8]$ for which (5) is a maximum.

Table 4.5.2

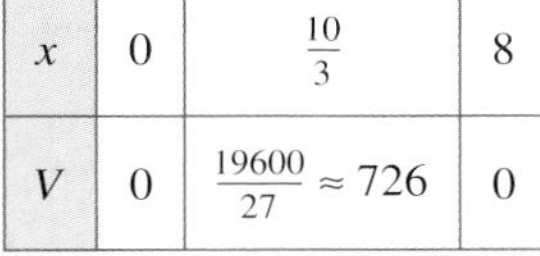

x	0	$\frac{10}{3}$	8
V	0	$\frac{19600}{27} \approx 726$	0

- *Step 5:* From (5) we obtain

$$\begin{aligned}\frac{dV}{dx} &= 480 - 184x + 12x^2 = 4(120 - 46x + 3x^2)\\ &= 4(x - 12)(3x - 10)\end{aligned}$$

Setting $dV/dx = 0$ yields

$$x = \tfrac{10}{3} \quad \text{and} \quad x = 12$$

Since $x = 12$ falls outside the interval $[0, 8]$, the maximum value of V occurs either at the critical point $x = \frac{10}{3}$ or at the endpoints $x = 0$, $x = 8$. Substituting these values into (5) yields Table 4.5.2, which tells us that the greatest possible volume $V = \frac{19600}{27}\text{ in}^3 \approx 726\text{ in}^3$ occurs when we cut out squares whose sides have length $\frac{10}{3}$ inches. This is consistent with the graph of (5) shown in Figure 4.5.4. ◀

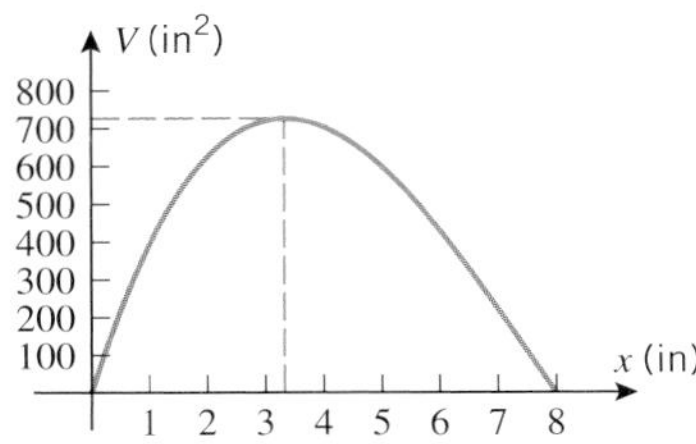

Figure 4.5.4

▶ **Example 3** Figure 4.5.5 shows an offshore oil well located at a point W that is 5 km from the closest point A on a straight shoreline. Oil is to be piped from W to a shore point B that is 8 km from A by piping it on a straight line under water from W to some shore point P between A and B and then on to B via pipe along the shoreline. If the cost of laying pipe is \$1,000,000/km under water and \$500,000/km over land, where should the point P be located to minimize the cost of laying the pipe?

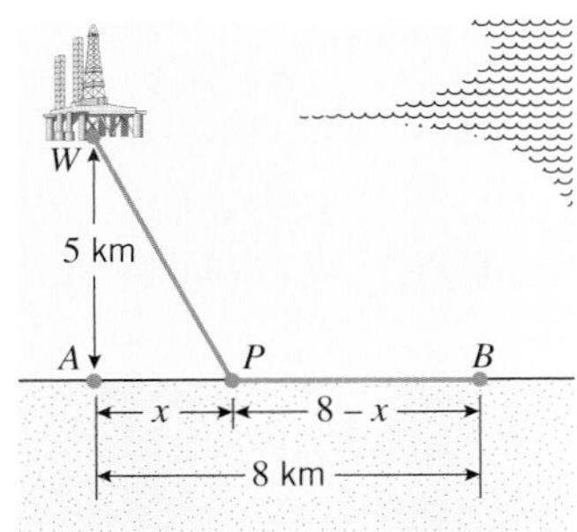

Figure 4.5.5

Solution. Let

$$x = \text{distance (in kilometers) between } A \text{ and } P$$
$$c = \text{cost (in millions of dollars) for the entire pipeline}$$

From Figure 4.5.5 the length of pipe under water is the distance between W and P. By the Theorem of Pythagoras that length is

$$\sqrt{x^2 + 25} \tag{6}$$

Also from Figure 4.5.5, the length of pipe over land is the distance between P and B, which is

$$8 - x \tag{7}$$

From (6) and (7) it follows that the total cost c (in millions of dollars) for the pipeline is

$$c = 1(\sqrt{x^2 + 25}) + \tfrac{1}{2}(8 - x) = \sqrt{x^2 + 25} + \tfrac{1}{2}(8 - x) \tag{8}$$

Because the distance between A and B is 8 km, the distance x between A and P must satisfy

$$0 \le x \le 8$$

We have thus reduced our problem to finding the value (or values) of x in the interval $[0, 8]$ for which c is a minimum. Since c is a continuous function of x on the closed interval $[0, 8]$, we can use the methods developed in the preceding section to find the minimum.

From (8) we obtain

$$\frac{dc}{dx} = \frac{x}{\sqrt{x^2 + 25}} - \frac{1}{2}$$

Setting $dc/dx = 0$ and solving for x yields

$$\frac{x}{\sqrt{x^2+25}} = \frac{1}{2} \tag{9}$$

$$x^2 = \frac{1}{4}(x^2+25)$$

$$x = \pm\frac{5}{\sqrt{3}}$$

TECHNOLOGY MASTERY

If you have a CAS, use it to check all of the computations in Example 3. Specifically, differentiate c with respect to x, solve the equation $dc/dx = 0$, and perform all of the numerical calculations.

The number $-5/\sqrt{3}$ is not a solution of (9) and must be discarded, leaving $x = 5/\sqrt{3}$ as the only critical point. Since this point lies in the interval [0, 8], the minimum must occur at one of the values

$$x = 0, \quad x = 5/\sqrt{3}, \quad x = 8$$

Substituting these values into (8) yields Table 4.5.3, which tells us that the least possible cost of the pipeline (to the nearest dollar) is $c = \$8{,}330{,}127$, and this occurs when the point P is located at a distance of $5/\sqrt{3} \approx 2.89$ km from A. ◄

Table 4.5.3

x	0	$\frac{5}{\sqrt{3}}$	8
c	9	$\frac{10}{\sqrt{3}} + \left(4 - \frac{5}{2\sqrt{3}}\right) \approx 8.330127$	$\sqrt{89} \approx 9.433981$

► **Example 4** Find the radius and height of the right circular cylinder of largest volume that can be inscribed in a right circular cone with radius 6 inches and height 10 inches (Figure 4.5.6*a*).

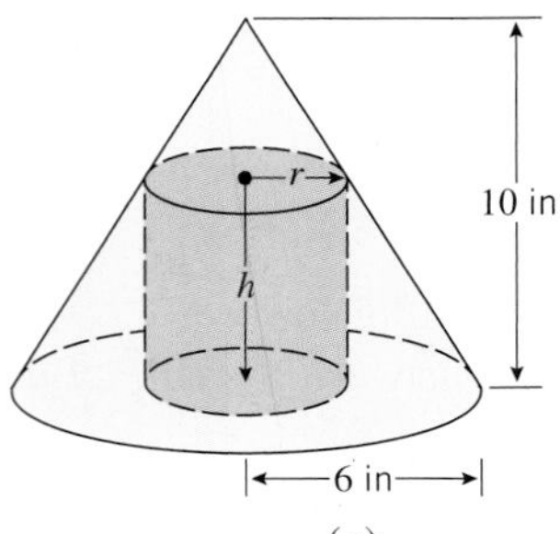

(*a*)

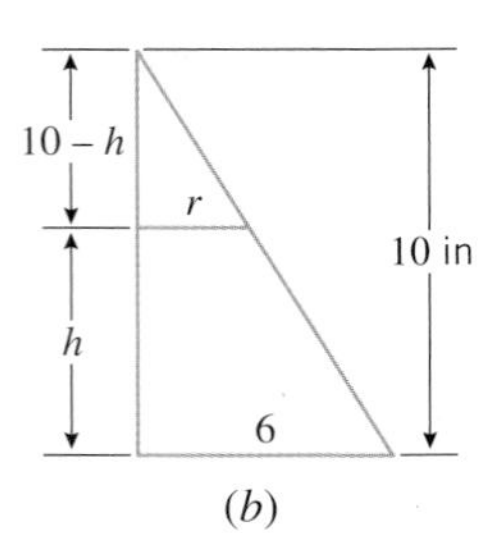

(*b*)

Figure 4.5.6

Solution. Let

r = radius (in inches) of the cylinder
h = height (in inches) of the cylinder
V = volume (in cubic inches) of the cylinder

The formula for the volume of the inscribed cylinder is

$$V = \pi r^2 h \tag{10}$$

To eliminate one of the variables in (10) we need a relationship between r and h. Using similar triangles (Figure 4.5.6*b*) we obtain

$$\frac{10-h}{r} = \frac{10}{6} \quad \text{or} \quad h = 10 - \frac{5}{3}r \tag{11}$$

Substituting (11) into (10) we obtain

$$V = \pi r^2\left(10 - \frac{5}{3}r\right) = 10\pi r^2 - \frac{5}{3}\pi r^3 \tag{12}$$

which expresses V in terms of r alone. Because r represents a radius it cannot be negative, and because the radius of the inscribed cylinder cannot exceed the radius of the cone, the variable r must satisfy

$$0 \le r \le 6$$

Thus, we have reduced the problem to that of finding the value (or values) of r in [0, 6] for which (12) is a maximum. Since V is a continuous function of r on [0, 6], the methods developed in the preceding section apply.

From (12) we obtain

$$\frac{dV}{dr} = 20\pi r - 5\pi r^2 = 5\pi r(4 - r)$$

Setting $dV/dr = 0$ gives

$$5\pi r(4 - r) = 0$$

so $r = 0$ and $r = 4$ are critical points. Since these lie in the interval $[0, 6]$, the maximum must occur at one of the values

$$r = 0, \quad r = 4, \quad r = 6$$

Table 4.5.4

r	0	4	6
V	0	$\frac{160}{3}\pi$	0

Substituting these values into (12) yields Table 4.5.4, which tells us that the maximum volume $V = \frac{160}{3}\pi \approx 168$ in^3 occurs when the inscribed cylinder has radius 4 in. When $r = 4$ it follows from (11) that $h = \frac{10}{3}$. Thus, the inscribed cylinder of largest volume has radius $r = 4$ in and height $h = \frac{10}{3}$ in. ◀

PROBLEMS INVOLVING INTERVALS THAT ARE NOT BOTH FINITE AND CLOSED

▶ **Example 5** A closed cylindrical can is to hold 1 liter (1000 cm^3) of liquid. How should we choose the height and radius to minimize the amount of material needed to manufacture the can?

Solution. Let

h = height (in cm) of the can

r = radius (in cm) of the can

S = surface area (in cm^2) of the can

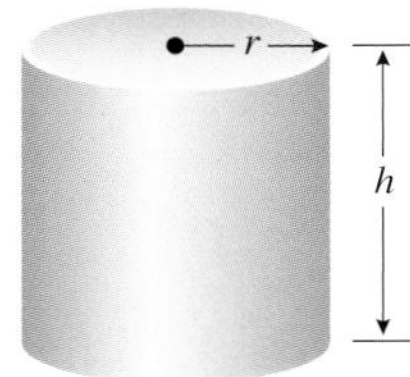

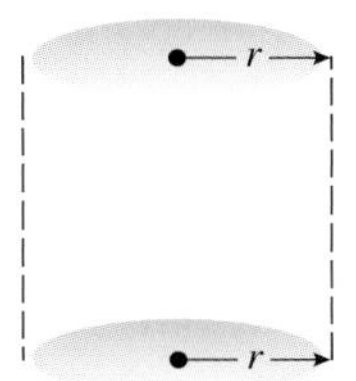

Area $2\pi r^2$

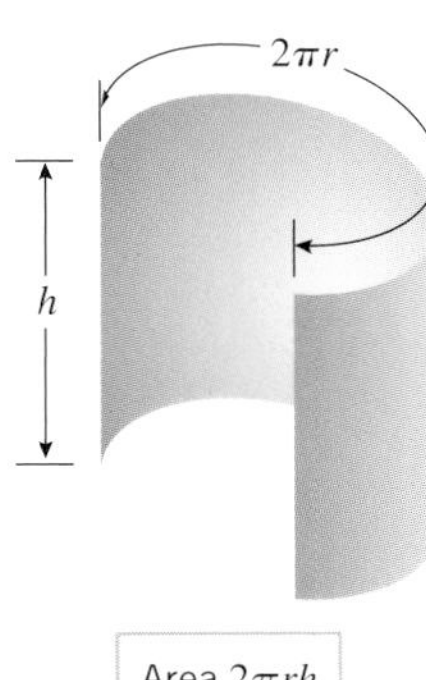

Area $2\pi rh$

Figure 4.5.7

Assuming there is no waste or overlap, the amount of material needed for manufacture will be the same as the surface area of the can. Since the can consists of two circular disks of radius r and a rectangular sheet with dimensions h by $2\pi r$ (Figure 4.5.7), the surface area will be

$$S = 2\pi r^2 + 2\pi rh \tag{13}$$

Since S depends on two variables, r and h, we will look for some condition in the problem that will allow us to express one of these variables in terms of the other. For this purpose, observe that the volume of the can is 1000 cm^3, so it follows from the formula $V = \pi r^2 h$ for the volume of a cylinder that

$$1000 = \pi r^2 h \quad \text{or} \quad h = \frac{1000}{\pi r^2} \tag{14–15}$$

Substituting (15) in (13) yields

$$S = 2\pi r^2 + \frac{2000}{r} \tag{16}$$

Thus, we have reduced the problem to finding a value of r in the interval $(0, +\infty)$ for which S is minimum. Since S is a continuous function of r on the interval $(0, +\infty)$ and

$$\lim_{r\to 0^+}\left(2\pi r^2 + \frac{2000}{r}\right) = +\infty \quad \text{and} \quad \lim_{r\to +\infty}\left(2\pi r^2 + \frac{2000}{r}\right) = +\infty$$

the analysis in Table 4.4.3 implies that S does have a minimum on the interval $(0, +\infty)$. Since this minimum must occur at a critical point, we calculate

$$\frac{dS}{dr} = 4\pi r - \frac{2000}{r^2} \tag{17}$$

Setting $dS/dr = 0$ gives

$$r = \frac{10}{\sqrt[3]{2\pi}} \approx 5.4 \tag{18}$$

Since (18) is the only critical point in the interval $(0, +\infty)$, this value of r yields the minimum value of S. From (15) the value of h corresponding to this r is

$$h = \frac{1000}{\pi(10/\sqrt[3]{2\pi})^2} = \frac{20}{\sqrt[3]{2\pi}} = 2r$$

It is not an accident here that the minimum occurs when the height of the can is equal to the diameter of its base (Exercise 27).

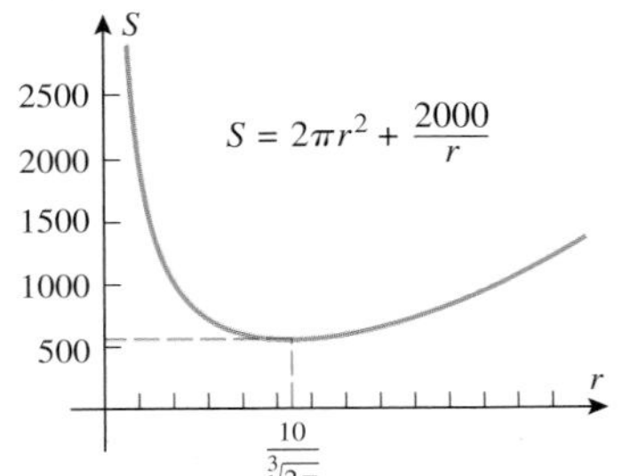

Figure 4.5.8

Second Solution. The conclusion that a minimum occurs at the value of r in (18) can be deduced from Theorem 4.4.4 and the second derivative test by noting that

$$\frac{d^2S}{dr^2} = 4\pi + \frac{4000}{r^3}$$

is positive if $r > 0$ and hence is positive if $r = 10/\sqrt[3]{2\pi}$. This implies that a relative minimum, and therefore a minimum, occurs at the critical point $r = 10/\sqrt[3]{2\pi}$.

Third Solution. An alternative justification that the critical point $r = 10/\sqrt[3]{2\pi}$ corresponds to a minimum for S is to view the graph of S versus r (Figure 4.5.8). ◀

In Example 5, the surface area S has no absolute maximum, since S increases without bound as the radius r approaches 0 (Figure 4.5.8). Thus, had we asked for the dimensions of the can requiring the *maximum* amount of material for its manufacture, there would have been no solution to the problem. Optimization problems with no solution are sometimes called *ill posed.*

▶ **Example 6** Find a point on the curve $y = x^2$ that is closest to the point $(18, 0)$.

Solution. The distance L between $(18, 0)$ and an arbitrary point (x, y) on the curve $y = x^2$ (Figure 4.5.9) is given by

$$L = \sqrt{(x-18)^2 + (y-0)^2}$$

Since (x, y) lies on the curve, x and y satisfy $y = x^2$; thus,

$$L = \sqrt{(x-18)^2 + x^4} \tag{19}$$

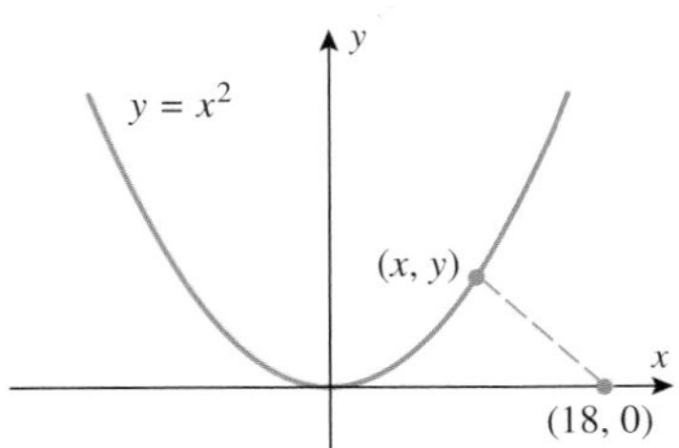

Figure 4.5.9

Because there are no restrictions on x, the problem reduces to finding a value of x in $(-\infty, +\infty)$ for which (19) is a minimum. The distance L and the square of the distance L^2 are minimized at the same value (see Exercise 62). Thus, the minimum value of L in (19) and the minimum value of

$$S = L^2 = (x-18)^2 + x^4 \tag{20}$$

occur at the same x-value.

From (20),

$$\frac{dS}{dx} = 2(x-18) + 4x^3 = 4x^3 + 2x - 36 \tag{21}$$

so the critical points satisfy $4x^3 + 2x - 36 = 0$ or, equivalently,

$$2x^3 + x - 18 = 0 \tag{22}$$

To solve for x we will begin by checking the divisors of -18 to see whether the polynomial on the left side has any integer roots (see Appendix B). These divisors are $\pm 1, \pm 2, \pm 3, \pm 6, \pm 9$, and ± 18. A check of these values shows that $x = 2$ is a root, so $x - 2$ is a factor of the polynomial. After dividing the polynomial by this factor we can rewrite (22) as

$$(x-2)(2x^2 + 4x + 9) = 0$$

Thus, the remaining solutions of (22) satisfy the quadratic equation

$$2x^2 + 4x + 9 = 0$$

But this equation has no real solutions (using the quadratic formula), so $x = 2$ is the only critical point of S. To determine the nature of this critical point we will use the second derivative test. From (21),

$$\frac{d^2S}{dx^2} = 12x^2 + 2, \quad \text{so} \quad \left.\frac{d^2S}{dx^2}\right|_{x=2} = 50 > 0$$

which shows that a relative minimum occurs at $x = 2$. Since $x = 2$ yields the only relative extremum for L, it follows from Theorem 4.4.4 that an absolute minimum value of L also occurs at $x = 2$. Thus, the point on the curve $y = x^2$ closest to (18, 0) is

$$(x, y) = (x, x^2) = (2, 4) \blacktriangleleft$$

AN APPLICATION TO ECONOMICS

Three functions of importance to an economist or a manufacturer are

$C(x)$ = total cost of producing x units of a product during some time period

$R(x)$ = total revenue from selling x units of the product during the time period

$P(x)$ = total profit obtained by selling x units of the product during the time period

These are called, respectively, the ***cost function***, ***revenue function***, and ***profit function***. If all units produced are sold, then these are related by

$$\underset{\text{[profit]} \;=\; \text{[revenue]} - \text{[cost]}}{P(x) = R(x) - C(x)} \tag{23}$$

The total cost $C(x)$ of producing x units can be expressed as a sum

$$C(x) = a + M(x) \tag{24}$$

where a is a constant, called ***overhead***, and $M(x)$ is a function representing ***manufacturing cost***. The overhead, which includes such fixed costs as rent and insurance, does not depend on x; it must be paid even if nothing is produced. On the other hand, the manufacturing cost $M(x)$, which includes such items as cost of materials and labor, depends on the number of items manufactured. It is shown in economics that with suitable simplifying assumptions, $M(x)$ can be expressed in the form

$$M(x) = bx + cx^2$$

where b and c are constants. Substituting this in (24) yields

$$C(x) = a + bx + cx^2 \tag{25}$$

If a manufacturing firm can sell all the items it produces for p dollars apiece, then its total revenue $R(x)$ (in dollars) will be

$$R(x) = px \tag{26}$$

and its total profit $P(x)$ (in dollars) will be

$$P(x) = [\text{total revenue}] - [\text{total cost}] = R(x) - C(x) = px - C(x)$$

Thus, if the cost function is given by (25),

$$P(x) = px - (a + bx + cx^2) \tag{27}$$

Depending on such factors as number of employees, amount of machinery available, economic conditions, and competition, there will be some upper limit l on the number of items a manufacturer is capable of producing and selling. Thus, during a fixed time period the variable x in (27) will satisfy

$$0 \le x \le l$$

By determining the value or values of x in $[0, l]$ that maximize (27), the firm can determine how many units of its product must be manufactured and sold to yield the greatest profit. This is illustrated in the following numerical example.

▶ **Example 7** A liquid form of penicillin manufactured by a pharmaceutical firm is sold in bulk at a price of \$200 per unit. If the total production cost (in dollars) for x units is

$$C(x) = 500{,}000 + 80x + 0.003x^2$$

and if the production capacity of the firm is at most 30,000 units in a specified time, how many units of penicillin must be manufactured and sold in that time to maximize the profit?

Solution. Since the total revenue for selling x units is $R(x) = 200x$, the profit $P(x)$ on x units will be

$$P(x) = R(x) - C(x) = 200x - (500{,}000 + 80x + 0.003x^2) \qquad (28)$$

Since the production capacity is at most 30,000 units, x must lie in the interval $[0, 30{,}000]$. From (28)

$$\frac{dP}{dx} = 200 - (80 + 0.006x) = 120 - 0.006x$$

Setting $dP/dx = 0$ gives

$$120 - 0.006x = 0 \quad \text{or} \quad x = 20{,}000$$

Since this critical point lies in the interval $[0, 30{,}000]$, the maximum profit must occur at one of the values

$$x = 0, \quad x = 20{,}000, \quad \text{or} \quad x = 30{,}000$$

Substituting these values in (28) yields Table 4.5.5, which tells us that the maximum profit $P = \$700{,}000$ occurs when $x = 20{,}000$ units are manufactured and sold in the specified time. ◀

Table 4.5.5

x	0	20,000	30,000
$P(x)$	−500,000	700,000	400,000

MARGINAL ANALYSIS

Economists call $P'(x)$, $R'(x)$, and $C'(x)$ the ***marginal profit***, ***marginal revenue***, and ***marginal cost***, respectively; and they interpret these quantities as the *additional* profit, revenue, and cost that result from producing and selling one additional unit of the product when the production and sales levels are at x units. These interpretations follow from the local linear approximations of the profit, revenue, and cost functions. For example, it follows from Formula (2) of Section 3.9 that when the production and sales levels are at x units the local linear approximation of the profit function is

$$P(x + \Delta x) \approx P(x) + P'(x)\Delta x$$

Thus, if $\Delta x = 1$ (one additional unit produced and sold), this formula implies

$$P(x + 1) \approx P(x) + P'(x)$$

and hence the *additional* profit that results from producing and selling one additional unit can be approximated as

$$P(x + 1) - P(x) \approx P'(x)$$

A BASIC PRINCIPLE OF ECONOMICS

It follows from (23) that $P'(x) = 0$ has the same solution as $C'(x) = R'(x)$, and this implies that the maximum profit must occur where the marginal revenue is equal to the marginal cost; that is:

The maximum profit occurs where the cost of manufacturing and selling an additional unit of a product is approximately equal to the revenue generated by the additional unit.

In Example 7, the maximum profit occurs when $x = 20{,}000$ units. Note that

$$C(20{,}001) - C(20{,}000) = \$200.003 \quad \text{and} \quad R(20{,}001) - R(20{,}000) = \$200$$

which is consistent with this basic economic principle.

✔QUICK CHECK EXERCISES 4.5

(See page 275 for answers.)

1. A positive number and its reciprocal are added together. The smallest possible value of this sum is ________.
2. If the sum of two positive numbers is 10, then the largest their product could be is ________.
3. If $x + 2y = 2$, then the smallest possible value of $x^2 + y^2$ is ________.
4. A rectangle is inscribed in the triangle whose vertices are the origin, the point $(0, 2)$, and the point $(1, 0)$. If each side of the rectangle is either parallel to, or coincident with, one of the coordinate axes, then the largest possible area for the rectangle is ________.

EXERCISE SET 4.5

1. Find a number in the closed interval $\left[\frac{1}{2}, \frac{3}{2}\right]$ such that the sum of the number and its reciprocal is
   (a) as small as possible
   (b) as large as possible.
2. How should two nonnegative numbers be chosen so that their sum is 1 and the sum of their squares is
   (a) as large as possible
   (b) as small as possible?
3. A rectangular field is to be bounded by a fence on three sides and by a straight stream on the fourth side. Find the dimensions of the field with maximum area that can be enclosed using 1000 ft of fence.
4. A field has boundary a right triangle with hypotenuse along a straight stream. A fence bounds the other two sides of the field. Find the dimensions of the field with maximum area that can be enclosed using 1000 ft of fence.
5. A rectangular plot of land is to be fenced in using two kinds of fencing. Two opposite sides will use heavy-duty fencing selling for \$3 a foot, while the remaining two sides will use standard fencing selling for \$2 a foot. What are the dimensions of the rectangular plot of greatest area that can be fenced in at a cost of \$6000?
6. A rectangle is to be inscribed in a right triangle having sides of length 6 in, 8 in, and 10 in. Find the dimensions of the rectangle with greatest area assuming the rectangle is positioned as in Figure Ex-6.
7. Solve the problem in Exercise 6 assuming the rectangle is positioned as in Figure Ex-7.

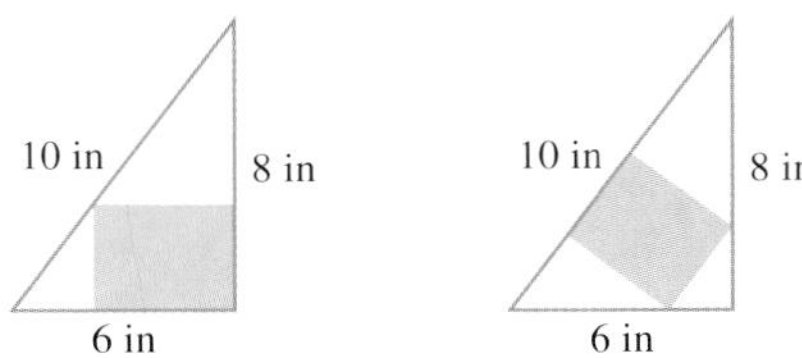

Figure Ex-6 **Figure Ex-7**

8. A rectangle has its two lower corners on the x-axis and its two upper corners on the curve $y = 16 - x^2$. For all such rectangles, what are the dimensions of the one with largest area?
9. Find the dimensions of the rectangle with maximum area that can be inscribed in a circle of radius 10.
10. What is the largest possible area of a region in the plane that is contained in both the rectangle with corners at $(\pm 8, \pm 10)$ and a square whose sides are parallel to the coordinate axes and whose lower left corner is on the line $y = -4x$?
11. A rectangular area of 3200 ft^2 is to be fenced off. Two opposite sides will use fencing costing \$1 per foot and the

remaining sides will use fencing costing \$2 per foot. Find the dimensions of the rectangle of least cost.

12. Show that among all rectangles with perimeter p, the square has the maximum area.

13. Show that among all rectangles with area A, the square has the minimum perimeter.

14. A wire of length 12 in can be bent into a circle, bent into a square, or cut into two pieces to make both a circle and a square. How much wire should be used for the circle if the total area enclosed by the figure(s) is to be
(a) a maximum (b) a minimum?

15. A field in the shape of an isosceles triangle is to be bounded by a fence on the two equal sides of the triangle, and by a straight stream on the third side. Find the dimensions of the field of largest area that can be enclosed by 300 yards of fence.

16. A church window consisting of a rectangle topped by a semicircle is to have a perimeter p. Find the radius of the semicircle if the area of the window is to be maximum.

17. A box with a square base is taller than it is wide. In order to send the box through the U.S. mail, the height of the box and the perimeter of the base can sum to no more than 108 in. What is the maximum volume for such a box?

18. A box with a square base is wider than it is tall. In order to send the box through the U.S. mail, the width of the box and the perimeter of one of the (nonsquare) sides of the box can sum to no more than 108 in. What is the maximum volume for such a box?

19. An open box is to be made from a 3-ft by 8-ft rectangular piece of sheet metal by cutting out squares of equal size from the four corners and bending up the sides. Find the maximum volume that the box can have.

20. A closed rectangular container with a square base is to have a volume of 2250 in^3. The material for the top and bottom of the container will cost \$2 per in^2, and the material for the sides will cost \$3 per in^2. Find the dimensions of the container of least cost.

21. A closed rectangular container with a square base is to have a volume of 2000 cm^3. It costs twice as much per square centimeter for the top and bottom as it does for the sides. Find the dimensions of the container of least cost.

22. A container with square base, vertical sides, and open top is to be made from 1000 ft^2 of material. Find the dimensions of the container with greatest volume.

23. A rectangular container with two square sides and an open top is to have a volume of V cubic units. Find the dimensions of the container with minimum surface area.

24. Find the dimensions of the right circular cylinder of largest volume that can be inscribed in a sphere of radius R.

25. Find the dimensions of the right circular cylinder of greatest surface area that can be inscribed in a sphere of radius R.

26. Show that the right circular cylinder of greatest volume that can be inscribed in a right circular cone has volume that is $\frac{4}{9}$ the volume of the cone (Figure Ex-26).

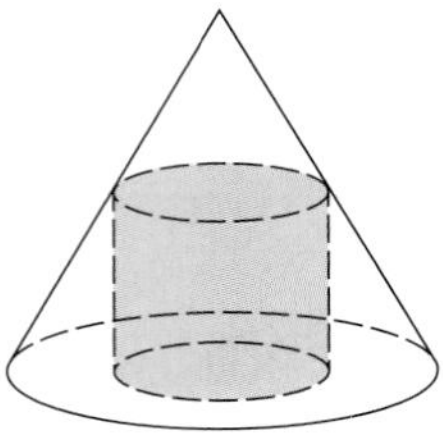

Figure Ex-26

27. A closed, cylindrical can is to have a volume of V cubic units. Show that the can of minimum surface area is achieved when the height is equal to the diameter of the base.

28. A closed cylindrical can is to have a surface area of S square units. Show that the can of maximum volume is achieved when the height is equal to the diameter of the base.

29. A cylindrical can, open at the top, is to hold 500 cm^3 of liquid. Find the height and radius that minimize the amount of material needed to manufacture the can.

30. A soup can in the shape of a right circular cylinder of radius r and height h is to have a prescribed volume V. The top and bottom are cut from squares as shown in Figure Ex-30. If the shaded corners are wasted, but there is no other waste, find the ratio r/h for the can requiring the least material (including waste).

31. A box-shaped wire frame consists of two identical wire squares whose vertices are connected by four straight wires of equal length (Figure Ex-31). If the frame is to be made from a wire of length L, what should the dimensions be to obtain a box of greatest volume?

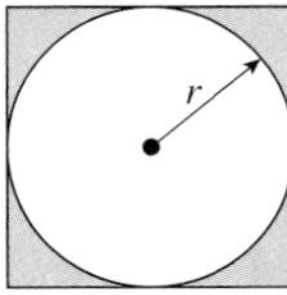

Figure Ex-30

Figure Ex-31

32. Suppose that the sum of the surface areas of a sphere and a cube is a constant.
(a) Show that the sum of their volumes is smallest when the diameter of the sphere is equal to the length of an edge of the cube.
(b) When will the sum of their volumes be greatest?

33. Find the height and radius of the cone of slant height L whose volume is as large as possible.

34. A cone is made from a circular sheet of radius R by cutting out a sector and gluing the cut edges of the remaining piece together (Figure Ex-34). What is the maximum volume attainable for the cone?

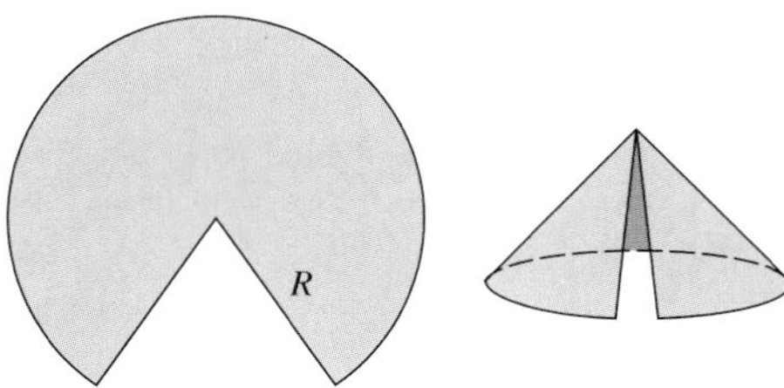

Figure Ex-34

35. A cone-shaped paper drinking cup is to hold 10 cm^3 of water. Find the height and radius of the cup that will require the least amount of paper.

36. Find the dimensions of the isosceles triangle of least area that can be circumscribed about a circle of radius R.

37. Find the height and radius of the right circular cone with least volume that can be circumscribed about a sphere of radius R.

38. A trapezoid is inscribed in a semicircle of radius 2 so that one side is along the diameter (Figure Ex-38). Find the maximum possible area for the trapezoid. [*Hint:* Express the area of the trapezoid in terms of θ.]

39. A drainage channel is to be made so that its cross section is a trapezoid with equally sloping sides (Figure Ex-39). If the sides and bottom all have a length of 5 ft, how should the angle θ $(0 \le \theta \le \pi/2)$ be chosen to yield the greatest cross-sectional area of the channel?

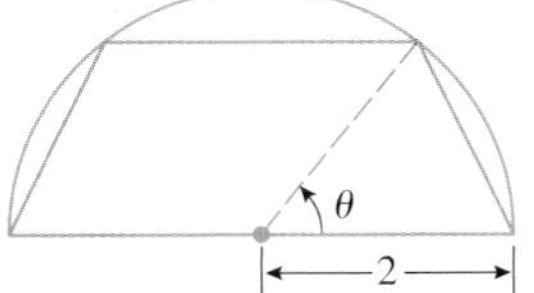

Figure Ex-38

Figure Ex-39

40. A lamp is suspended above the center of a round table of radius r. How high above the table should the lamp be placed to achieve maximum illumination at the edge of the table? [Assume that the illumination I is directly proportional to the cosine of the angle of incidence ϕ of the light rays and inversely proportional to the square of the distance l from the light source (Figure Ex-40).]

41. A plank is used to reach over a fence 8 ft high to support a wall that is 1 ft behind the fence (Figure Ex-41). What is the length of the shortest plank that can be used? [*Hint:* Express the length of the plank in terms of the angle θ shown in the figure.]

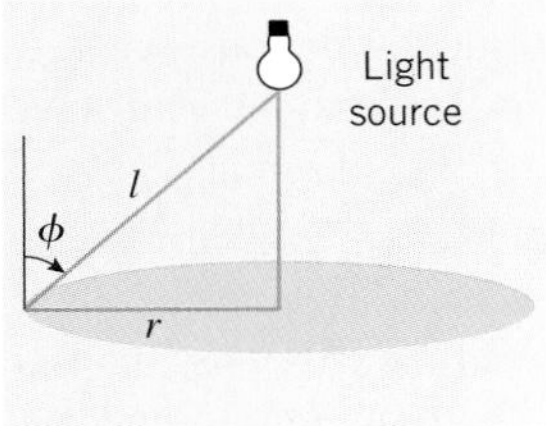

Figure Ex-40

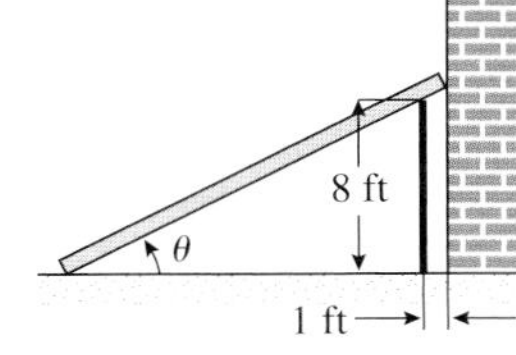

Figure Ex-41

42. A commercial cattle ranch currently allows 20 steers per acre of grazing land; on the average its steers weigh 2000 lb at market. Estimates by the Agriculture Department indicate that the average market weight per steer will be reduced by 50 lb for each additional steer added per acre of grazing land. How many steers per acre should be allowed in order for the ranch to get the largest possible total market weight for its cattle?

43. (a) A chemical manufacturer sells sulfuric acid in bulk at a price of \$100 per unit. If the daily total production cost in dollars for x units is

$$C(x) = 100{,}000 + 50x + 0.0025x^2$$

and if the daily production capacity is at most 7000 units, how many units of sulfuric acid must be manufactured and sold daily to maximize the profit?

(b) Would it benefit the manufacturer to expand the daily production capacity?

(c) Use marginal analysis to approximate the effect on profit if daily production could be increased from 7000 to 7001 units.

44. A firm determines that x units of its product can be sold daily at p dollars per unit, where

$$x = 1000 - p$$

The cost of producing x units per day is

$$C(x) = 3000 + 20x$$

(a) Find the revenue function $R(x)$.

(b) Find the profit function $P(x)$.

(c) Assuming that the production capacity is at most 500 units per day, determine how many units the company must produce and sell each day to maximize the profit.

(d) Find the maximum profit.

(e) What price per unit must be charged to obtain the maximum profit?

45. In a certain chemical manufacturing process, the daily weight y of defective chemical output depends on the total weight x of all output according to the empirical formula

$$y = 0.01x + 0.00003x^2$$

where x and y are in pounds. If the profit is \$100 per pound of nondefective chemical produced and the loss is \$20 per pound of defective chemical produced, how many pounds of chemical should be produced daily to maximize the total daily profit?

46. An independent truck driver charges a client \$15 for each hour of driving, plus the cost of fuel. At highway speeds of v miles per hour, the trucker's rig gets $10 - 0.07v$ miles per gallon of diesel fuel. If diesel fuel costs \$1.50 per gallon, what speed v will minimize the cost to the client?

47. Two particles, A and B, are in motion in the xy-plane. Their coordinates at each instant of time t $(t \ge 0)$ are given by $x_A = t$, $y_A = 2t$, $x_B = 1 - t$, and $y_B = t$. Find the minimum distance between A and B.

48. Follow the directions of Exercise 47, with $x_A = t$, $y_A = t^2$, $x_B = 2t$, and $y_B = 2$.

49. Prove that $(1, 0)$ is the closest point on $x^2 + y^2 = 1$ to $(2, 0)$.

50. Find all points on the curve $y = \sqrt{x}$ for $0 \le x \le 3$ that are closest to, and at the greatest distance from, the point $(2, 0)$.

FOCUS ON CONCEPTS

51. Suppose that $f(x) = mx + b$ is a linear function of x and that Q is any point in the xy-plane.
(a) Without using calculus, explain how to find the point on the graph of f closest to Q.
(b) Use the derivative to verify your answer in part (a).

52. Let C denote a circle in the xy-plane with center P, and let Q denote any point in the plane distinct from P.
(a) Without using calculus, explain how to find the points on C that are closest to, and at a greatest distance from, the point Q.
(b) Use the derivative to verify your answer in part (a).

53. (a) Find all points P that lie on the rotated ellipse $x^2 - xy + y^2 = 4$ where the tangent line is perpendicular to the line through P and the origin.
(b) Give a geometric explanation for why the points found in part (a) are the points on the ellipse that are closest to, or at the greatest distance from, the origin.

54. Using the derivative, explain why the points found in Exercise 53(a) are the points on the ellipse that are closest to, or at the greatest distance from, the origin.

55. Find the coordinates of the point P on the curve

$$y = \frac{1}{x^2} \quad (x > 0)$$

where the segment of the tangent line at P that is cut off by the coordinate axes has its shortest length.

56. Find the x-coordinate of the point P on the parabola

$$y = 1 - x^2 \quad (0 < x \le 1)$$

where the triangle that is enclosed by the tangent line at P and the coordinate axes has the smallest area.

57. Where on the curve $y = (1 + x^2)^{-1}$ does the tangent line have the greatest slope?

58. A man is floating in a rowboat 1 mile from the (straight) shoreline of a large lake. A town is located on the shoreline 1 mile from the point on the shoreline closest to the man. As suggested in Figure Ex-58, he intends to row in a straight line to some point P on the shoreline and then walk the remaining distance to the town. To what point should he row in order to reach his destination in the least time if
(a) he can walk 5 mi/h and row 3 mi/h
(b) he can walk 5 mi/h and row 4 mi/h?

59. A pipe of negligible diameter is to be carried horizontally around a corner from a hallway 8 ft wide into a hallway 4 ft wide (Figure Ex-59). What is the maximum length that the pipe can have?

Source: An interesting discussion of this problem in the case where the diameter of the pipe is not neglected is given by Norman Miller in the *American Mathematical Monthly*, Vol. 56, 1949, pp. 177–179.

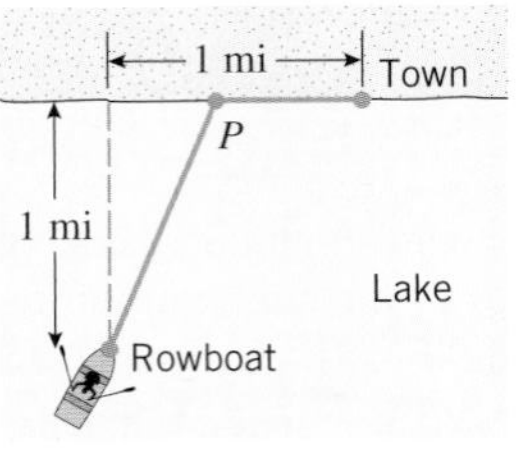

Figure Ex-58

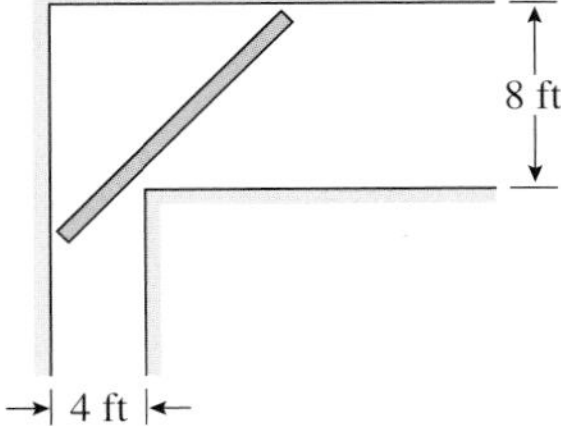

Figure Ex-59

60. If an unknown physical quantity x is measured n times, the measurements $x_1, x_2, \ldots, x_n$ often vary because of uncontrollable factors such as temperature, atmospheric pressure, and so forth. Thus, a scientist is often faced with the problem of using n different observed measurements to obtain an estimate $\bar{x}$ of an unknown quantity x. One method for making such an estimate is based on the ***least squares principle***, which states that the estimate $\bar{x}$ should be chosen to minimize

$$s = (x_1 - \bar{x})^2 + (x_2 - \bar{x})^2 + \cdots + (x_n - \bar{x})^2$$

which is the sum of the squares of the deviations between the estimate $\bar{x}$ and the measured values. Show that the estimate resulting from the least squares principle is

$$\bar{x} = \frac{1}{n}(x_1 + x_2 + \cdots + x_n)$$

that is, $\bar{x}$ is the arithmetic average of the observed values.

61. Suppose that the intensity of a point light source is directly proportional to the strength of the source and inversely proportional to the square of the distance from the source. Two point light sources with strengths of S and $8S$ are separated by a distance of 90 cm. Where on the line segment between the two sources is the total intensity a minimum?

62. Prove: If $f(x) \ge 0$ on an interval I and if $f(x)$ has a maximum value on I at x_0, then $\sqrt{f(x)}$ also has a maximum value at x_0. Similarly for minimum values. [*Hint:* Use the fact that $\sqrt{x}$ is an increasing function on the interval $[0, +\infty)$.]

63. Given points $A(2, 1)$ and $B(5, 4)$, find the point P in the interval $[2, 5]$ on the x-axis that maximizes angle APB.

64. The lower edge of a painting, 10 ft in height, is 2 ft above an observer's eye level. Assuming that the best view is obtained when the angle subtended at the observer's eye by the painting is maximum, how far from the wall should the observer stand?

65. ***Fermat's principle*** (biography on p. 263) in optics states that light traveling from one point to another follows that path for which the total travel time is minimum. In a uniform medium, the paths of "minimum time" and "shortest distance" turn out to be the same, so that light, if unobstructed, travels along a straight line. Assume that we have a light source, a flat mirror, and an observer in a uniform medium. If a light ray leaves the source, bounces off the mirror, and travels on to the observer, then its path will consist of two line segments, as shown in Figure Ex-65. According to Fermat's principle, the path will be such that the total travel time t is minimum or, since the medium is uniform, the path will be such that the total distance traveled from A to P to B is as small as possible. Assuming the minimum occurs when $dt/dx = 0$, show that the light ray will strike the mirror at the point P where the "angle of incidence" θ_1 equals the "angle of reflection" θ_2.

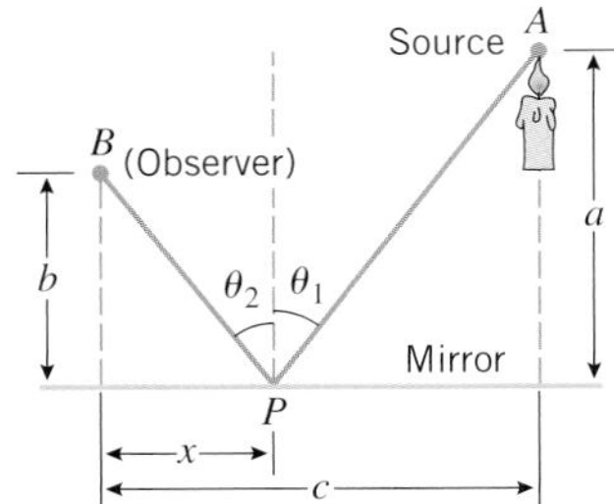

Figure Ex-65

66. Fermat's principle (Exercise 65) also explains why light rays traveling between air and water undergo bending (refraction). Imagine that we have two uniform media (such as air and water) and a light ray traveling from a source A in one medium to an observer B in the other medium (Figure Ex-66). It is known that light travels at a constant speed in a uniform medium, but more slowly in a dense medium (such as water) than in a thin medium (such as air). Consequently, the path of shortest time from A to B is not necessarily a straight line, but rather some broken line path A to P to B allowing the light to take greatest advantage of its higher speed through the thin medium. ***Snell's law of refraction*** states that the path of the light ray will be such that

$$\frac{\sin\theta_1}{v_1} = \frac{\sin\theta_2}{v_2}$$

where v_1 is the speed of light in the first medium, v_2 is the speed of light in the second medium, and θ_1 and θ_2 are the angles shown in Figure Ex-66. Show that this follows from the assumption that the path of minimum time occurs when $dt/dx = 0$.

67. A farmer wants to walk at a constant rate from her barn to a straight river, fill her pail, and carry it to her house in the least time.
(a) Explain how this problem relates to Fermat's principle and the light-reflection problem in Exercise 65.
(b) Use the result of Exercise 65 to describe geometrically the best path for the farmer to take.
(c) Use part (b) to determine where the farmer should fill her pail if her house and barn are located as in Figure Ex-67.

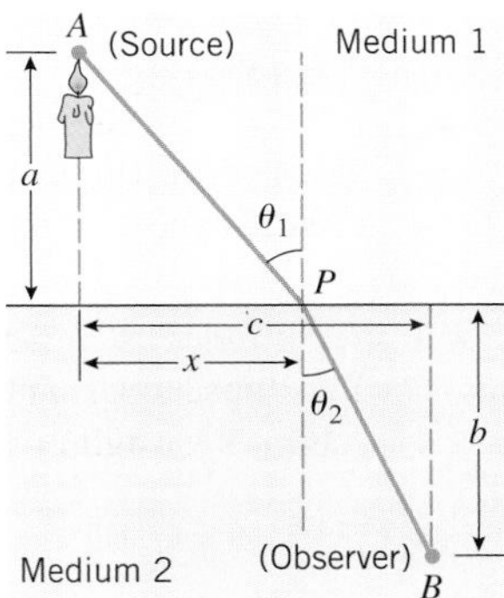

Figure Ex-66

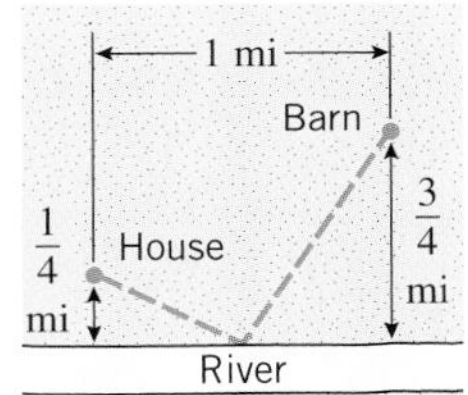

Figure Ex-67

✔ QUICK CHECK ANSWERS 4.5

1. 2 **2.** 25 **3.** $\frac{4}{5}$ **4.** $\frac{1}{2}$

Willebrord van Roijen Snell (1591–1626) Dutch mathematician. Snell, who succeeded his father to the post of Professor of Mathematics at the University of Leiden in 1613, is most famous for the result of light refraction that bears his name. Although this phenomenon was studied as far back as the ancient Greek astronomer Ptolemy, until Snell's work the relationship was incorrectly thought to be $\theta_1/v_1 = \theta_2/v_2$. Snell's law was published by Descartes in 1638 without giving proper credit to Snell. Snell also discovered a method for determining distances by triangulation that founded the modern technique of mapmaking.

4.6 NEWTON'S METHOD

In Section 2.5 we showed how to approximate the roots of an equation $f(x) = 0$ by using the Intermediate-Value Theorem and also by zooming in on the x-intercepts of $y = f(x)$ with a graphing utility. In this section we will study a technique, called "Newton's Method," that is usually more efficient than either of those methods. Newton's Method is the technique used by many commercial and scientific computer programs for finding roots.

NEWTON'S METHOD

In beginning algebra one learns that the solution of a first-degree equation $ax + b = 0$ is given by the formula $x = -b/a$, and the solutions of a second-degree equation

$$ax^2 + bx + c = 0$$

are given by the quadratic formula. Formulas also exist for the solutions of all third- and fourth-degree equations, although they are too complicated to be of practical use. In 1826 it was shown by the Norwegian mathematician Niels Henrik Abel that it is impossible to construct a similar formula for the solutions of a *general* fifth-degree equation or higher. Thus, for a *specific* fifth-degree polynomial equation such as

$$x^5 - 9x^4 + 2x^3 - 5x^2 + 17x - 8 = 0$$

it may be difficult or impossible to find exact values for all of the solutions. Similar difficulties occur for nonpolynomial equations such as

$$x - \cos x = 0$$

For such equations the solutions are generally approximated in some way, often by the method we will now discuss.

Niels Henrik Abel (1802–1829) Norwegian mathematician. Abel was the son of a poor Lutheran minister and a remarkably beautiful mother from whom he inherited strikingly good looks. In his brief life of 26 years Abel lived in virtual poverty and suffered a succession of adversities, yet he managed to prove major results that altered the mathematical landscape forever. At the age of thirteen he was sent away from home to a school whose better days had long passed. By a stroke of luck the school had just hired a teacher named Bernt Michael Holmboe, who quickly discovered that Abel had extraordinary mathematical ability. Together, they studied the calculus texts of Euler and works of Newton and the later French mathematicians. By the time he graduated, Abel was familar with most of the great mathematical literature. In 1820 his father died, leaving the family in dire financial straits. Abel was able to enter the University of Christiania in Oslo only because he was granted a free room and several professors supported him directly from their salaries. The University had no advanced courses in mathematics, so Abel took a preliminary degree in 1822 and then continued to study mathematics on his own. In 1824 he published at his own expense the proof that it is impossible to solve the general fifth-degree polynomial equation algebraically. With the hope that this landmark paper would lead to his recognition and acceptance by the European mathematical community, Abel sent the paper to the great German mathematician Gauss, who casually declared it to be a "monstrosity" and tossed it aside. However, in 1826 Abel's paper on the fifth-degree equation and other work was published in the first issue of a new journal, founded by his friend, Leopold Crelle. In the summer of 1826 he completed a landmark work on transcendental functions, which he submitted to the French Academy of Sciences. He hoped to establish himself as a major mathematician, for many young mathematicians had gained quick distinction by having their work accepted by the Academy. However, Abel waited in vain because the paper was either ignored or misplaced by one of the referees, and it did not surface again until two years after his death. That paper was later described by one major mathematician as "...the most important mathematical discovery that has been made in our century...." After submitting his paper, Abel returned to Norway, ill with tuberculosis and in heavy debt. While eking out a meager living as a tutor, he continued to produce great work and his fame spread. Soon great efforts were being made to secure a suitable mathematical position for him. Fearing that his great work had been lost by the Academy, he mailed a proof of the main results to Crelle in January of 1829. In April he suffered a violent hemorrhage and died. Two days later Crelle wrote to inform him that an appointment had been secured for him in Berlin and his days of poverty were over! Abel's great paper was finally published by the Academy twelve years after his death.

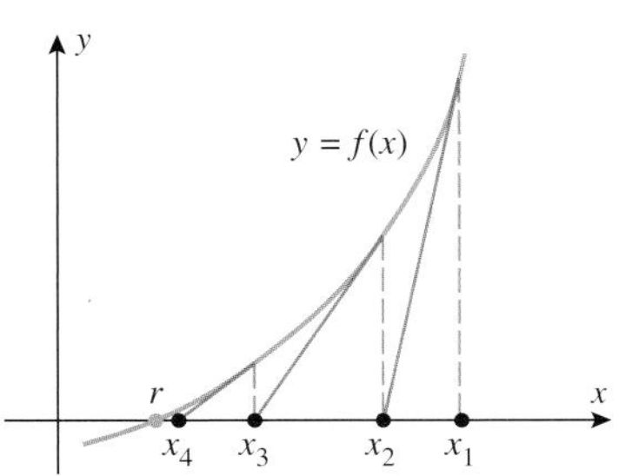

Figure 4.6.1

Suppose that we are trying to find a root r of the equation $f(x) = 0$, and suppose that by some method we are able to obtain an initial rough estimate, x_1, of r, say by generating the graph of $y = f(x)$ with a graphing utility and examining the x-intercept. If $f(x_1) = 0$, then $r = x_1$. If $f(x_1) \neq 0$, then we consider an easier problem, that of finding a root to a linear equation. The best linear approximation to $y = f(x)$ near $x = x_1$ is given by the tangent line to the graph of f at x_1, so it might be reasonable to expect that the x-intercept to this tangent line provides an improved approximation to r. Call this intercept x_2 (Figure 4.6.1). We can now treat x_2 in the same way we did x_1. If $f(x_2) = 0$, then $r = x_2$. If $f(x_2) \neq 0$, then construct the tangent line to the graph of f at x_2, and take x_3 to be the x-intercept of this tangent line. Continuing in this way we can generate a succession of values $x_1, x_2, x_3, x_4, \ldots$ that will usually approach r. This procedure for approximating r is called ***Newton's Method***.

To implement Newton's Method analytically, we must derive a formula that will tell us how to calculate each improved approximation from the preceding approximation. For this purpose, we note that the point-slope form of the tangent line to $y = f(x)$ at the initial approximation x_1 is

$$y - f(x_1) = f'(x_1)(x - x_1) \tag{1}$$

If $f'(x_1) \neq 0$, then this line is not parallel to the x-axis and consequently it crosses the x-axis at some point $(x_2, 0)$. Substituting the coordinates of this point in (1) yields

$$-f(x_1) = f'(x_1)(x_2 - x_1)$$

Solving for x_2 we obtain

$$x_2 = x_1 - \frac{f(x_1)}{f'(x_1)} \tag{2}$$

The next approximation can be obtained more easily. If we view x_2 as the starting approximation and x_3 the new approximation, we can simply apply (2) with x_2 in place of x_1 and x_3 in place of x_2. This yields

$$x_3 = x_2 - \frac{f(x_2)}{f'(x_2)} \tag{3}$$

provided $f'(x_2) \neq 0$. In general, if x_n is the nth approximation, then it is evident from the pattern in (2) and (3) that the improved approximation x_{n+1} is given by

Newton's Method

$$x_{n+1} = x_n - \frac{f(x_n)}{f'(x_n)}, \qquad n = 1, 2, 3, \ldots \tag{4}$$

▶ Example 1 Use Newton's Method to approximate the real solutions of

$$x^3 - x - 1 = 0$$

Solution. Let $f(x) = x^3 - x - 1$, so $f'(x) = 3x^2 - 1$ and (4) becomes

$$x_{n+1} = x_n - \frac{x_n^3 - x_n - 1}{3x_n^2 - 1} \tag{5}$$

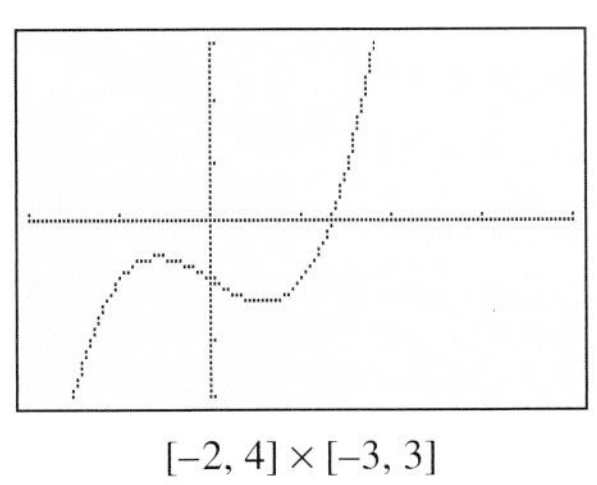

$[-2, 4] \times [-3, 3]$
xScl = 1, yScl = 1

$y = x^3 - x - 1$

Figure 4.6.2

From the graph of f in Figure 4.6.2, we see that the given equation has only one real solution. This solution lies between 1 and 2 because $f(1) = -1 < 0$ and $f(2) = 5 > 0$. We will use $x_1 = 1.5$ as our first approximation ($x_1 = 1$ or $x_1 = 2$ would also be reasonable choices).

Letting $n = 1$ in (5) and substituting $x_1 = 1.5$ yields

$$x_2 = 1.5 - \frac{(1.5)^3 - 1.5 - 1}{3(1.5)^2 - 1} \approx 1.34782609 \tag{6}$$

(We used a calculator that displays nine digits.) Next, we let $n = 2$ in (5) and substitute x_2 to obtain

$$x_3 = x_2 - \frac{x_2^3 - x_2 - 1}{3x_2^2 - 1} \approx 1.32520040 \tag{7}$$

TECHNOLOGY MASTERY

Many calculators and computer programs calculate internally with more digits than they display. Where possible, you should use stored calculated values rather than values displayed from earlier calculations. Thus, in Example 1 the value of x_2 used in (7) should be the stored value, not the value in (6).

If we continue this process until two identical approximations are generated in succession, we obtain

$$\begin{aligned} x_1 &= 1.5 \\ x_2 &\approx 1.34782609 \\ x_3 &\approx 1.32520040 \\ x_4 &\approx 1.32471817 \\ x_5 &\approx 1.32471796 \\ x_6 &\approx 1.32471796 \end{aligned}$$

At this stage there is no need to continue further because we have reached the display accuracy limit of our calculator, and all subsequent approximations that the calculator generates will likely be the same. Thus, the solution is approximately $x \approx 1.32471796$. ◀

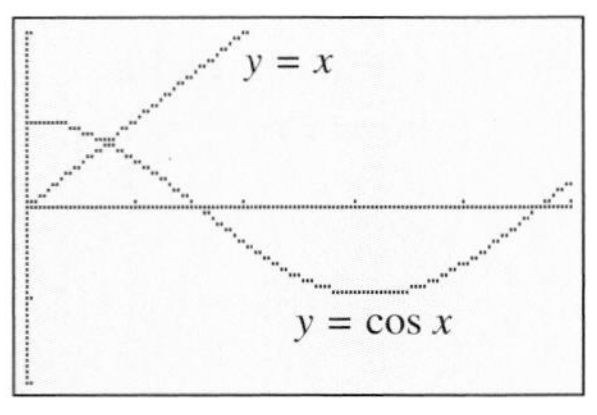

$[0, 5] \times [-2, 2]$
xScl = 1, yScl = 1

Figure 4.6.3

▶ **Example 2** It is evident from Figure 4.6.3 that if x is in radians, then the equation

$$\cos x = x$$

has a solution between 0 and 1. Use Newton's Method to approximate it.

Solution. Rewrite the equation as

$$x - \cos x = 0$$

and apply (4) with $f(x) = x - \cos x$. Since $f'(x) = 1 + \sin x$, (4) becomes

$$x_{n+1} = x_n - \frac{x_n - \cos x_n}{1 + \sin x_n} \tag{8}$$

From Figure 4.6.3, the solution seems closer to $x = 1$ than $x = 0$, so we will use $x_1 = 1$ (radian) as our initial approximation. Letting $n = 1$ in (8) and substituting $x_1 = 1$ yields

$$x_2 = 1 - \frac{1 - \cos 1}{1 + \sin 1} \approx 0.750363868$$

Next, letting $n = 2$ in (8) and substituting this value of x_2 yields

$$x_3 = x_2 - \frac{x_2 - \cos x_2}{1 + \sin x_2} \approx 0.739112891$$

If we continue this process until two identical approximations are generated in succession, we obtain

$$\begin{aligned} x_1 &= 1 \\ x_2 &\approx 0.750363868 \\ x_3 &\approx 0.739112891 \\ x_4 &\approx 0.739085133 \\ x_5 &\approx 0.739085133 \end{aligned}$$

Thus, to the accuracy limit of our calculator, the solution of the equation $\cos x = x$ is $x \approx 0.739085133$. ◀

SOME DIFFICULTIES WITH NEWTON'S METHOD

When Newton's Method works, the approximations usually converge toward the solution with dramatic speed. However, there are situations in which the method fails. For example, if $f'(x_n) = 0$ for some n, then (4) involves a division by zero, making it impossible to

x_3 cannot be generated.

Figure 4.6.4

generate x_{n+1}. However, this is to be expected because the tangent line to $y = f(x)$ is parallel to the x-axis where $f'(x_n) = 0$, and hence this tangent line does not cross the x-axis to generate the next approximation (Figure 4.6.4).

Newton's Method can fail for other reasons as well; sometimes it may overlook the root you are trying to find and converge to a different root, and sometimes it may fail to converge altogether. For example, consider the equation

$$x^{1/3} = 0$$

which has $x = 0$ as its only solution, and try to approximate this solution by Newton's Method with a starting value of $x_0 = 1$. Letting $f(x) = x^{1/3}$, Formula (4) becomes

$$x_{n+1} = x_n - \frac{(x_n)^{1/3}}{\frac{1}{3}(x_n)^{-2/3}} = x_n - 3x_n = -2x_n$$

Beginning with $x_1 = 1$, the successive values generated by this formula are

$$x_1 = 1, \quad x_2 = -2, \quad x_3 = 4, \quad x_4 = -8, \ldots$$

which obviously do not converge to $x = 0$. Figure 4.6.5 illustrates what is happening geometrically in this situation.

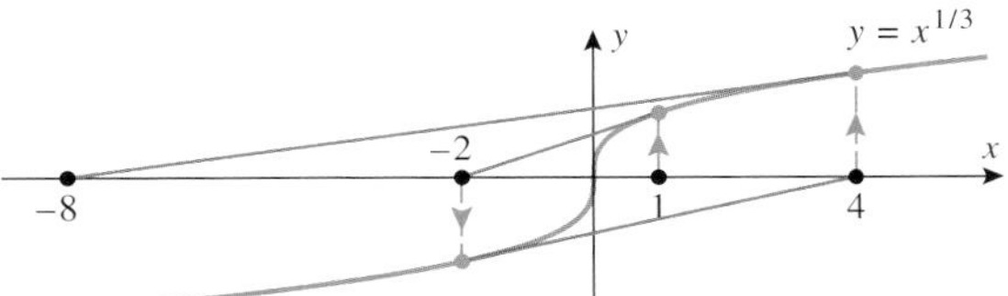

Figure 4.6.5

To learn more about the conditions under which Newton's Method converges and for a discussion of error questions, you should consult a book on numerical analysis. For a more in-depth discussion of Newton's Method and its relationship to contemporary studies of chaos and fractals, you may want to read the article, "Newton's Method and Fractal Patterns," by Phillip Straffin, which appears in *Applications of Calculus*, MAA Notes, Vol. 3, No. 29, 1993, published by the Mathematical Association of America.

✔ QUICK CHECK EXERCISES 4.6 *(See page 281 for answers.)*

1. Use the accompanying graph to estimate x_2 and x_3 if Newton's Method is applied to the equation $y = f(x)$ with $x_1 = 8$.

2. Suppose that $f(1) = 2$ and $f'(1) = 4$. If Newton's Method is applied to $y = f(x)$ with $x_1 = 1$, then $x_2 =$ ______.

3. Suppose we are given that $f(0) = 3$ and that $x_2 = 3$ when Newton's Method is applied to $y = f(x)$ with $x_1 = 0$. Then $f'(0) =$ ______.

4. If Newton's Method is applied to $y = x^5 - 2$ with $x_1 = 1$, then $x_2 =$ ______.

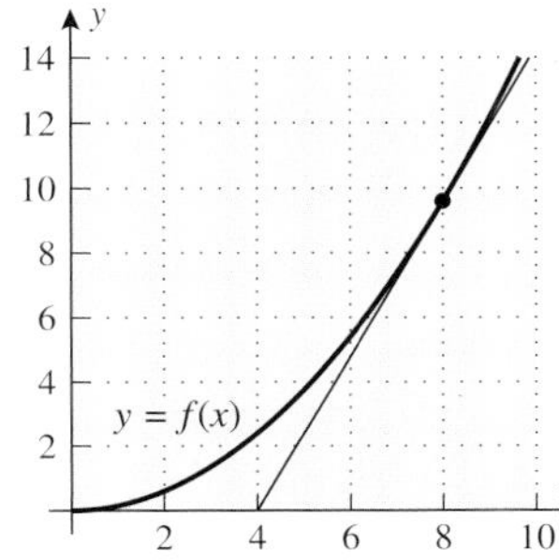

Figure Ex-1

EXERCISE SET 4.6 Graphing Utility

In this exercise set express your answer with as many decimal digits as your calculating utility can display, but use the procedure in the Technology Mastery on p. 278.

1. Approximate $\sqrt{2}$ by applying Newton's Method to the equation $x^2 - 2 = 0$.

2. Approximate $\sqrt{5}$ by applying Newton's Method to the equation $x^2 - 5 = 0$.

3. Approximate $\sqrt[3]{6}$ by applying Newton's Method to the equation $x^3 - 6 = 0$.

4. To what equation would you apply Newton's Method to approximate the nth root of a?

5–8 The given equation has one real solution. Approximate it by Newton's Method.

5. $x^3 - 2x - 2 = 0$

6. $x^3 + x - 1 = 0$

7. $x^5 + x^4 - 5 = 0$

8. $x^5 - 3x + 3 = 0$

9–14 Use a graphing utility to determine how many solutions the equation has, and then use Newton's Method to approximate the solution that satisfies the stated condition.

9. $x^4 + x^2 - 4 = 0;\ x < 0$

10. $x^5 - 5x^3 - 2 = 0;\ x > 0$

11. $2\cos x = x;\ x > 0$

12. $\sin x = x^2;\ x > 0$

13. $x - \tan x = 0;\ \pi/2 < x < 3\pi/2$

14. $1 + x^2 \sin x = 0;\ \pi/2 < x < 3\pi/2$

15–18 Use a graphing utility to determine the number of times the curves intersect; and then apply Newton's Method, where needed, to approximate the x-coordinates of all intersections.

15. $y = x^3$ and $y = 1 - x$

16. $y = \sin x$ and $y = x^3 - 2x^2 + 1$

17. $y = x^2$ and $y = \sqrt{2x + 1}$

18. $y = \frac{1}{8}x^3 - 1$ and $y = \cos x - 2$

19. The ***mechanic's rule*** for approximating square roots states that $\sqrt{a} \approx x_{n+1}$, where

$$x_{n+1} = \frac{1}{2}\left(x_n + \frac{a}{x_n}\right), \quad n = 1, 2, 3, \ldots$$

and x_1 is any positive approximation to $\sqrt{a}$.

(a) Apply Newton's Method to

$$f(x) = x^2 - a$$

to derive the mechanic's rule.

(b) Use the mechanic's rule to approximate $\sqrt{10}$.

20. Many calculators compute reciprocals using the approximation $1/a \approx x_{n+1}$, where

$$x_{n+1} = x_n(2 - ax_n), \quad n = 1, 2, 3, \ldots$$

and x_1 is an initial approximation to $1/a$. This formula makes it possible to perform divisions using multiplications and subtractions, which is a faster procedure than dividing directly.

(a) Apply Newton's Method to

$$f(x) = \frac{1}{x} - a$$

to derive this approximation.

(b) Use the formula to approximate $\frac{1}{17}$.

21. Use Newton's Method to approximate the absolute minimum of $f(x) = \frac{1}{4}x^4 + x^2 - 5x$.

22. Use Newton's Method to approximate the absolute maximum of $f(x) = x \sin x$ on the interval $[0, \pi]$.

23. Use Newton's Method to approximate the coordinates of the point on the parabola $y = x^2$ that is closest to the point $(1, 0)$.

24. Use Newton's Method to approximate the dimensions of the rectangle of largest area that can be inscribed under the curve $y = \cos x$ for $0 \le x \le \pi/2$, as shown in the accompanying figure.

25. (a) Show that on a circle of radius r, the central angle θ that subtends an arc whose length is 1.5 times the length L of its chord satisfies the equation $\theta = 3\sin(\theta/2)$ (Figure Ex-25).

(b) Use Newton's Method to approximate θ.

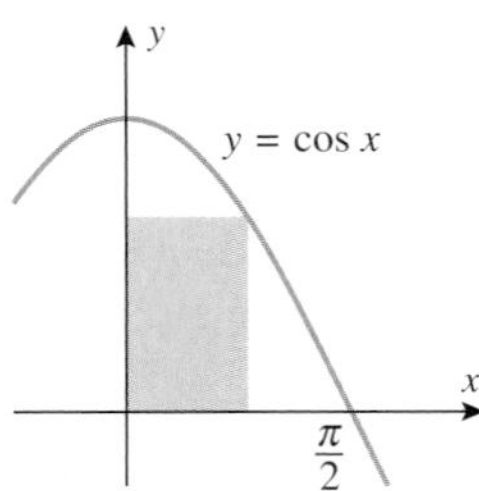

Figure Ex-24

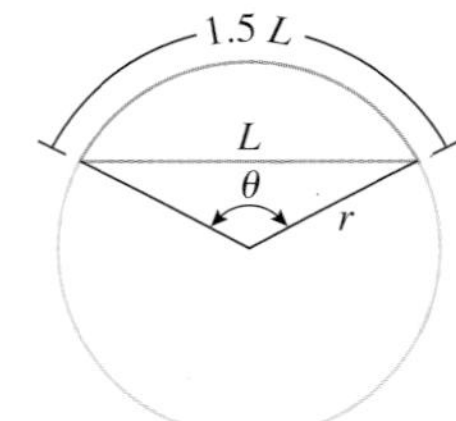

Figure Ex-25

26. A ***segment*** of a circle is the region enclosed by an arc and its chord (Figure Ex-26). If r is the radius of the circle and θ the angle subtended at the center of the circle, then it can be shown that the area A of the segment is $A = \frac{1}{2}r^2(\theta - \sin\theta)$, where θ is in radians. Find the value of θ for which the area of the segment is one-fourth the area of the circle. Give θ to the nearest degree.

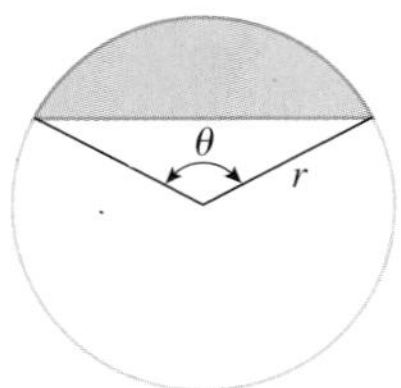

Figure Ex-26

27–28 Use Newton's Method to approximate all real values of y satisfying the given equation for the indicated value of x.

27. $xy^4 + x^3y = 1;\ x = 1$ **28.** $xy - \cos\left(\frac{1}{2}xy\right) = 0;\ x = 2$

29. An ***annuity*** is a sequence of equal payments that are paid or received at regular time intervals. For example, you may want to deposit equal amounts at the end of each year into an interest-bearing account for the purpose of accumulating a lump sum at some future time. If, at the end of each year, interest of $i \times 100\%$ on the account balance for that year is added to the account, then the account is said to pay $i \times 100\%$ interest, ***compounded annually***. It can be shown that if payments of Q dollars are deposited at the end of each year into an account that pays $i \times 100\%$ compounded annually, then at the time when the nth payment and the accrued interest for the past year are deposited, the amount $S(n)$ in the account is given by the formula

$$S(n) = \frac{Q}{i}[(1+i)^n - 1]$$

Suppose that you can invest \$5000 in an interest-bearing account at the end of each year, and your objective is to have \$250,000 on the 25th payment. Approximately what annual compound interest rate must the account pay for you to achieve your goal? [*Hint:* Show that the interest rate i satisfies the equation $50i = (1+i)^{25} - 1$, and solve it using Newton's Method.]

FOCUS ON CONCEPTS

30. (a) Use a graphing utility to generate the graph of

$$f(x) = \frac{x}{x^2+1}$$

and use it to explain what happens if you apply Newton's Method with a starting value of $x_1 = 2$. Check your conclusion by computing x_2, x_3, x_4, and x_5.

(b) Use the graph generated in part (a) to explain what happens if you apply Newton's Method with a starting value of $x_1 = 0.5$. Check your conclusion by computing x_2, x_3, x_4, and x_5.

31. (a) Apply Newton's Method to $f(x) = x^2 + 1$ with a starting value of $x_1 = 0.5$, and determine if the values of $x_2, \ldots, x_{10}$ appear to converge.

(b) Explain what is happening.

32. In each part, explain what happens if you apply Newton's Method to a function f when the given condition is satisfied for some value of n.

(a) $f(x_n) = 0$ (b) $x_{n+1} = x_n$

(c) $x_{n+2} = x_n \neq x_{n+1}$

33. Suppose that f is a function whose derivative is continuous everywhere. Assume that there exists a real number c such that when Newton's Method is applied to f, the inequality

$$|x_n - c| < \frac{1}{n}$$

is satisfied for all values of $n = 1, 2, 3, \ldots$.

(a) Explain why

$$|x_{n+1} - x_n| < \frac{2}{n}$$

for all values of $n = 1, 2, 3, \ldots$.

(b) Show that there exists a positive constant M such that

$$|f(x_n)| \leq M|x_{n+1} - x_n| < \frac{2M}{n}$$

for all values of $n = 1, 2, 3, \ldots$.

(c) Prove that if $f(c) \neq 0$, then there exists a positive integer N such that

$$\frac{|f(c)|}{2} < |f(x_n)|$$

if $n > N$. [*Hint:* Argue that $f(x) \to f(c)$ as $x \to c$ and then apply Definition 2.4.1 with $\epsilon = \frac{1}{2}|f(c)|$.]

(d) What can you conclude from parts (b) and (c)?

34. What are the important elements in the argument suggested by Exercise 33? Can you extend this argument to a wider collection of functions?

✔ QUICK CHECK ANSWERS 4.6

1. $x_2 \approx 4$, $x_3 \approx 2$ **2.** $\frac{1}{2}$ **3.** -1 **4.** 1.2

4.7 ROLLE'S THEOREM; MEAN-VALUE THEOREM

In this section we will discuss a result called the Mean-Value Theorem. This theorem has so many important consequences that it is regarded as one of the major principles in calculus.

ROLLE'S THEOREM

We will begin with a special case of the Mean-Value Theorem, called Rolle's Theorem, in honor of the mathematician Michel Rolle. This theorem states the geometrically obvious

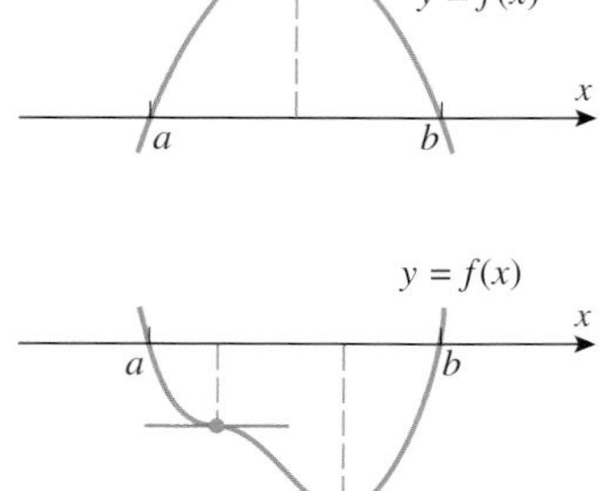

Figure 4.7.1

fact that if the graph of a differentiable function intersects the x-axis at two places, a and b, then somewhere between a and b there must be at least one place where the tangent line is horizontal (Figure 4.7.1). The precise statement of the theorem is as follows.

4.7.1 THEOREM (*Rolle's Theorem*). *Let f be continuous on the closed interval $[a, b]$ and differentiable on the open interval (a, b). If*

$$f(a) = 0 \quad \text{and} \quad f(b) = 0$$

then there is at least one point c in the interval (a, b) such that $f'(c) = 0$.

PROOF. We will divide the proof into three cases: the case where $f(x) = 0$ for all x in (a, b), the case where $f(x) > 0$ at some point in (a, b), and the case where $f(x) < 0$ at some point in (a, b).

CASE 1. If $f(x) = 0$ for all x in (a, b), then $f'(c) = 0$ at every point c in (a, b) because f is a constant function on that interval.

CASE 2. Assume that $f(x) > 0$ at some point in (a, b). Since f is continuous on $[a, b]$, it follows from the Extreme-Value Theorem (4.4.2) that f has an absolute maximum on $[a, b]$. The absolute maximum value cannot occur at an endpoint of $[a, b]$ because $f(a) = f(b) = 0$, and we have assumed that $f(x) > 0$ at some point in (a, b). Thus, the absolute maximum must occur at some point c in (a, b). It follows from Theorem 4.4.3 that c is a critical point of f, and since f is differentiable on (a, b), this critical point must be a stationary point; that is, $f'(c) = 0$.

CASE 3. Assume that $f(x) < 0$ at some point in (a, b). The proof of this case is similar to Case 2 and will be omitted. ■

▶ **Example 1** Find the two x-intercepts of the function $f(x) = x^2 - 5x + 4$ and confirm that $f'(c) = 0$ at some point c between those intercepts.

Solution. The function f can be factored as

$$x^2 - 5x + 4 = (x - 1)(x - 4)$$

Michel Rolle (1652–1719) French mathematician. Rolle, the son of a shopkeeper, received only an elementary education. He married early and as a young man struggled hard to support his family on the meager wages of a transcriber for notaries and attorneys. In spite of his financial problems and minimal education, Rolle studied algebra and Diophantine analysis (a branch of number theory) on his own. Rolle's fortune changed dramatically in 1682 when he published an elegant solution of a difficult, unsolved problem in Diophantine analysis. The public recognition of his achievement led to a patronage under minister Louvois, a job as an elementary mathematics teacher, and eventually to a short-term administrative post in the Ministry of War. In 1685 he joined the Académie des Sciences in a low-level position for which he received no regular salary until 1699. He stayed at the Académie until he died of apoplexy in 1719.

While Rolle's forte was always Diophantine analysis, his most important work was a book on the algebra of equations, called *Traité d'algèbre*, published in 1690. In that book Rolle firmly established the notation $\sqrt[n]{a}$ [earlier written as $\sqrt{ⓝ\, a}$] for the nth root of a, and proved a polynomial version of the theorem that today bears his name. (Rolle's Theorem was named by Giusto Bellavitis in 1846.) Ironically, Rolle was one of the most vocal early antagonists of calculus. He strove intently to demonstrate that it gave erroneous results and was based on unsound reasoning. He quarreled so vigorously on the subject that the Académie des Sciences was forced to intervene on several occasions. Among his several achievements, Rolle helped advance the currently accepted size order for negative numbers. Descartes, for example, viewed -2 as smaller than -5. Rolle preceded most of his contemporaries by adopting the current convention in 1691.

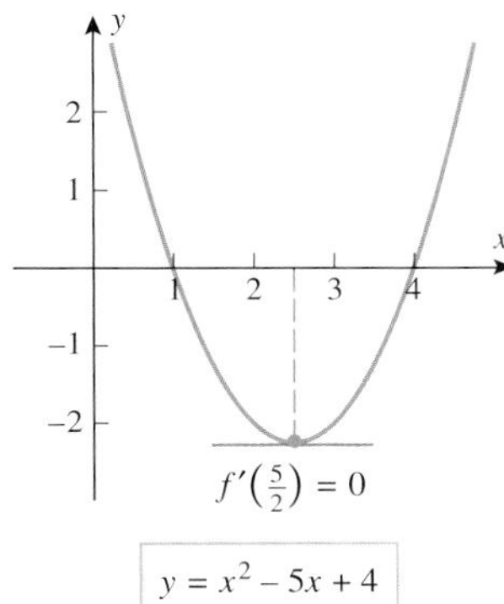

Figure 4.7.2

so the x-intercepts are $x = 1$ and $x = 4$. Since the polynomial f is continuous and differentiable everywhere, the hypotheses of Rolle's Theorem are satisfied on the interval $[1, 4]$. Thus, we are guaranteed the existence of at least one point c in the interval $(1, 4)$ such that $f'(c) = 0$. Differentiating f yields

$$f'(x) = 2x - 5$$

Solving the equation $f'(x) = 0$ yields $x = \frac{5}{2}$, so $c = \frac{5}{2}$ is a point in the interval $(1, 4)$ at which $f'(c) = 0$ (Figure 4.7.2). ◀

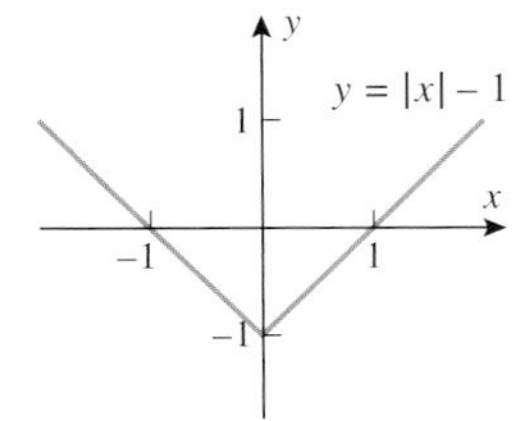

Figure 4.7.3

▶ **Example 2** The differentiability requirement in Rolle's Theorem is critical. If f fails to be differentiable at even one place in the interval (a, b), then the conclusion of the theorem may not hold. For example, the function $f(x) = |x| - 1$ graphed in Figure 4.7.3 has roots at $x = -1$ and $x = 1$, yet there is no horizontal tangent to the graph of f over the interval $(-1, 1)$. ◀

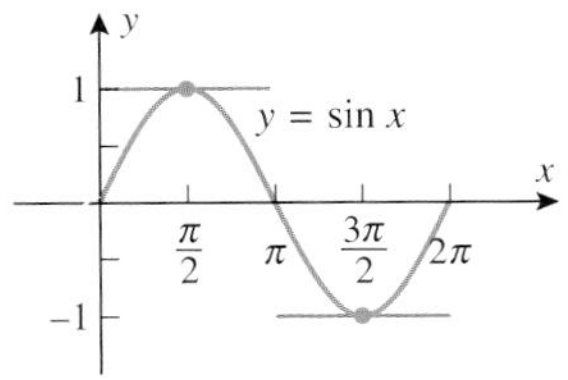

Figure 4.7.4

▶ **Example 3** If f satisfies the conditions of Rolle's Theorem on $[a, b]$, then the theorem guarantees the existence of *at least* one point c in (a, b) at which $f'(c) = 0$. There may, however, be more than one such c. For example, the function $f(x) = \sin x$ is continuous and differentiable everywhere, so the hypotheses of Rolle's Theorem are satisfied on the interval $[0, 2\pi]$ whose endpoints are roots of f. As indicated in Figure 4.7.4, there are two points in the interval $[0, 2\pi]$ at which the graph of f has a horizontal tangent, $c_1 = \pi/2$ and $c_2 = 3\pi/2$. ◀

In Examples 1 and 3 we were able to find exact values of c because the equation $f'(x) = 0$ was easy to solve. However, in the applications of Rolle's Theorem it is usually the *existence of* c that is important and not its actual value.

THE MEAN-VALUE THEOREM

Rolle's Theorem is a special case of a more general result, called the ***Mean-Value Theorem***. Geometrically, this theorem states that between any two points $A(a, f(a))$ and $B(b, f(b))$ on the graph of a differentiable function f, there is at least one place where the tangent line to the graph is parallel to the secant line joining A and B (Figure 4.7.5).

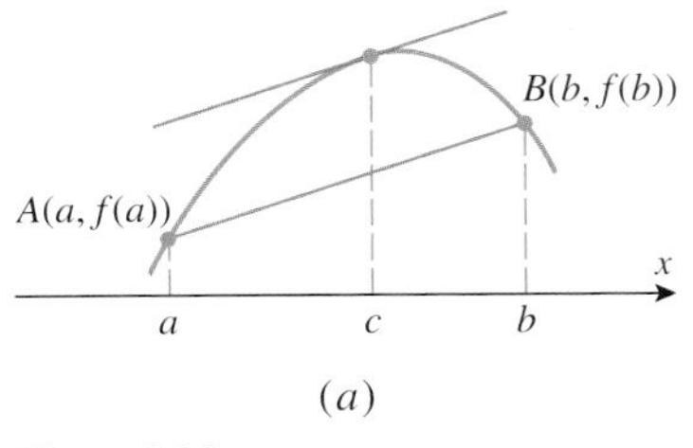

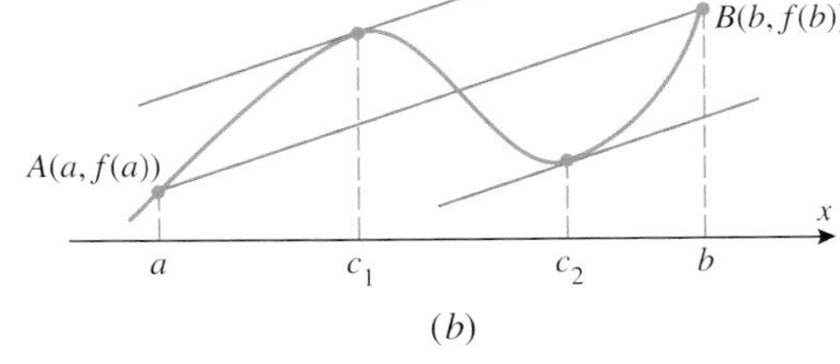

Figure 4.7.5

Note that the slope of the secant line joining $A(a, f(a))$ and $B(b, f(b))$ is

$$\frac{f(b) - f(a)}{b - a}$$

and that the slope of the tangent line at c in Figure 4.7.5*a* is $f'(c)$. Similarly, in Figure 4.7.5*b* the slopes of the tangent lines at c_1 and c_2 are $f'(c_1)$ and $f'(c_2)$, respectively. Since nonvertical parallel lines have the same slope, the Mean-Value Theorem can be stated precisely as follows.

4.7.2 THEOREM (*Mean-Value Theorem*). *Let f be continuous on the closed interval $[a, b]$ and differentiable on the open interval (a, b). Then there is at least one point c in (a, b) such that*

$$f'(c) = \frac{f(b) - f(a)}{b - a} \tag{1}$$

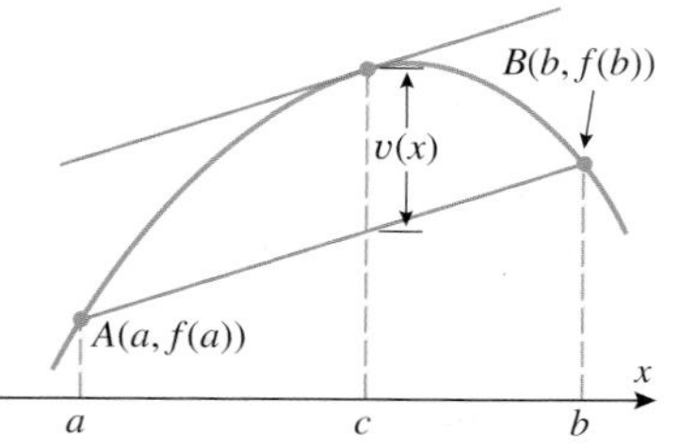

The tangent line is parallel to the secant line where the vertical distance $v(x)$ between the secant line and the graph of f is maximum.

Figure 4.7.6

MOTIVATION FOR THE PROOF OF THEOREM 4.7.2. Figure 4.7.6 suggests that (1) will hold (i.e., the tangent line will be parallel to the secant line) at a point c where the vertical distance between the curve and the secant line is maximum. Thus, to prove the Mean-Value Theorem it is natural to begin by looking for a formula for the vertical distance $v(x)$ between the curve $y = f(x)$ and the secant line joining $(a, f(a))$ and $(b, f(b))$.

PROOF OF THEOREM 4.7.2. Since the two-point form of the equation of the secant line joining $(a, f(a))$ and $(b, f(b))$ is

$$y - f(a) = \frac{f(b) - f(a)}{b - a}(x - a)$$

or, equivalently,

$$y = \frac{f(b) - f(a)}{b - a}(x - a) + f(a)$$

the difference $v(x)$ between the height of the graph of f and the height of the secant line is

$$v(x) = f(x) - \left[\frac{f(b) - f(a)}{b - a}(x - a) + f(a)\right] \tag{2}$$

Since $f(x)$ is continuous on $[a, b]$ and differentiable on (a, b), so is $v(x)$. Moreover,

$$v(a) = 0 \quad \text{and} \quad v(b) = 0$$

so that $v(x)$ satisfies the hypotheses of Rolle's Theorem on the interval $[a, b]$. Thus, there is a point c in (a, b) such that $v'(c) = 0$. But from Equation (2)

$$v'(x) = f'(x) - \frac{f(b) - f(a)}{b - a}$$

so

$$v'(c) = f'(c) - \frac{f(b) - f(a)}{b - a}$$

Since $v'(c) = 0$, we have

$$f'(c) = \frac{f(b) - f(a)}{b - a}$$ ■

▶ **Example 4** Show that the function $f(x) = \frac{1}{4}x^3 + 1$ satisfies the hypotheses of the Mean-Value Theorem over the interval $[0, 2]$, and find all values of c in the interval $(0, 2)$ at which the tangent line to the graph of f is parallel to the secant line joining the points $(0, f(0))$ and $(2, f(2))$.

Solution. The function f is continuous and differentiable everywhere because it is a polynomial. In particular, f is continuous on $[0, 2]$ and differentiable on $(0, 2)$, so the hypotheses of the Mean-Value Theorem are satisfied with $a = 0$ and $b = 2$. But

$$f(a) = f(0) = 1, \qquad f(b) = f(2) = 3$$

$$f'(x) = \frac{3x^2}{4}, \qquad f'(c) = \frac{3c^2}{4}$$

so in this case Equation (1) becomes

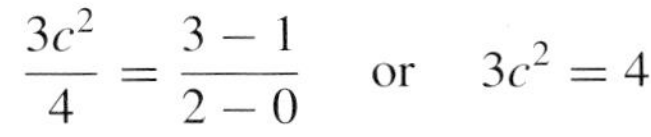

$$\frac{3c^2}{4} = \frac{3-1}{2-0} \quad \text{or} \quad 3c^2 = 4$$

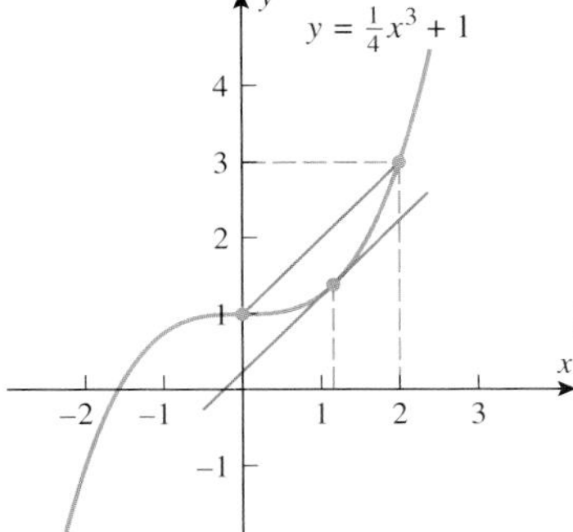

Figure 4.7.7

which has the two solutions $c = \pm 2/\sqrt{3} \approx \pm 1.15$. However, only the positive solution lies in the interval $(0, 2)$; this value of c is consistent with Figure 4.7.7. ◀

VELOCITY INTERPRETATION OF THE MEAN-VALUE THEOREM

There is a nice interpretation of the Mean-Value Theorem in the situation where $x = f(t)$ is the position versus time curve for a car moving along a straight road. In this case, the right side of (1) is the average velocity of the car over the time interval from $a \le t \le b$, and the left side is the instantaneous velocity at time $t = c$. Thus, the Mean-Value Theorem implies that at least once during the time interval the instantaneous velocity must equal the average velocity. This agrees with our real-world experience—if the average velocity for a trip is 40 mi/h, then sometime during the trip the speedometer has to read 40 mi/h.

▶ **Example 5** You are driving on a straight highway on which the speed limit is 55 mi/h. At 8:05 A.M. a police car clocks your velocity at 50 mi/h and at 8:10 A.M. a second police car posted 5 mi down the road clocks your velocity at 55 mi/h. Explain why the police have a right to charge you with a speeding violation.

Solution. You traveled 5 mi in 5 min $\left(= \frac{1}{12}\text{ h}\right)$, so your average velocity was 60 mi/h. Therefore, the Mean-Value Theorem guarantees the police that your instantaneous velocity was 60 mi/h at least once over the 5-mi section of highway. ◀

CONSEQUENCES OF THE MEAN-VALUE THEOREM

We stated at the beginning of this section that the Mean-Value Theorem is the starting point for many important results in calculus. As an example of this, we will use it to prove Theorem 4.1.2, which was one of our fundamental tools for analyzing graphs of functions.

> **4.1.2 THEOREM (*Revisited*).** *Let f be a function that is continuous on a closed interval $[a, b]$ and differentiable on the open interval (a, b).*
>
> (*a*) *If $f'(x) > 0$ for every value of x in (a, b), then f is increasing on $[a, b]$.*
>
> (*b*) *If $f'(x) < 0$ for every value of x in (a, b), then f is decreasing on $[a, b]$.*
>
> (*c*) *If $f'(x) = 0$ for every value of x in (a, b), then f is constant on $[a, b]$.*

PROOF (*a*). Suppose that x_1 and x_2 are points in $[a, b]$ such that $x_1 < x_2$. We must show that $f(x_1) < f(x_2)$. Because the hypotheses of the Mean-Value Theorem are satisfied on the entire interval $[a, b]$, they are satisfied on the subinterval $[x_1, x_2]$. Thus, there is some point c in the open interval (x_1, x_2) such that

$$f'(c) = \frac{f(x_2) - f(x_1)}{x_2 - x_1}$$

or, equivalently,

$$f(x_2) - f(x_1) = f'(c)(x_2 - x_1) \tag{3}$$

Since c is in the open interval (x_1, x_2), it follows that $a < c < b$; thus, $f'(c) > 0$. However, $x_2 - x_1 > 0$ since we assumed that $x_1 < x_2$. It follows from (3) that $f(x_2) - f(x_1) > 0$ or, equivalently, $f(x_1) < f(x_2)$, which is what we were to prove. The proofs of parts (*b*) and (*c*) are similar and are left as exercises. ■

THE CONSTANT DIFFERENCE THEOREM

We know from our earliest study of derivatives that the derivative of a constant is zero. Part (*c*) of Theorem 4.1.2 is the converse of that result; that is, a function whose derivative is zero on an interval must be constant on that interval. If we apply this to the difference of two functions, we obtain the following useful theorem.

> **4.7.3 THEOREM (*Constant Difference Theorem*).** *If f and g are differentiable on an interval I, and if $f'(x) = g'(x)$ for all x in I, then $f - g$ is constant on I; that is, there is a constant k such that $f(x) - g(x) = k$ or, equivalently,*
>
> $$f(x) = g(x) + k$$
>
> *for all x in I.*

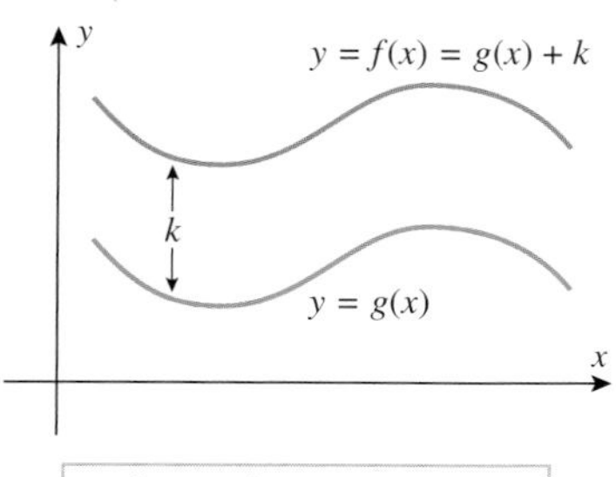

If $f'(x) = g'(x)$ on an interval, then the graphs of f and g are vertical translations of one another.

Figure 4.7.8

PROOF. Let x_1 and x_2 be any points in I such that $x_1 < x_2$. Since the functions f and g are differentiable on I, they are continuous on I. Since $[x_1, x_2]$ is a subinterval of I, it follows that f and g are continuous on $[x_1, x_2]$ and differentiable on (x_1, x_2). Moreover, it follows from the basic properties of derivatives and continuity that the same is true of the function

$$F(x) = f(x) - g(x)$$

Since $F'(x) = f'(x) - g'(x) = 0$, it follows from part (*c*) of Theorem 4.1.2 that $F(x) = f(x) - g(x)$ is constant on the interval $[x_1, x_2]$. This means that $f(x) - g(x)$ has the same value at any two points x_1 and x_2 in I, and this implies that $f - g$ is constant on I. ■

Geometrically, the Constant Difference Theorem tells us that if f and g have the same derivative on an interval, then the graphs of f and g are vertical translations of one another over that interval (Figure 4.7.8).

✔QUICK CHECK EXERCISES 4.7 (See page 289 for answers.)

1. Let $f(x) = x^2 - x$.
   (a) An interval on which f satisfies the hypotheses of Rolle's Theorem is ______.
   (b) Find all values of c that satisfy the conclusion of Rolle's Theorem for the function f on the interval in part (a).

2. Use the accompanying graph of f to find an interval $[a, b]$ on which Rolle's Theorem applies, and find all values of c in that interval that satisfy the conclusion of the theorem.

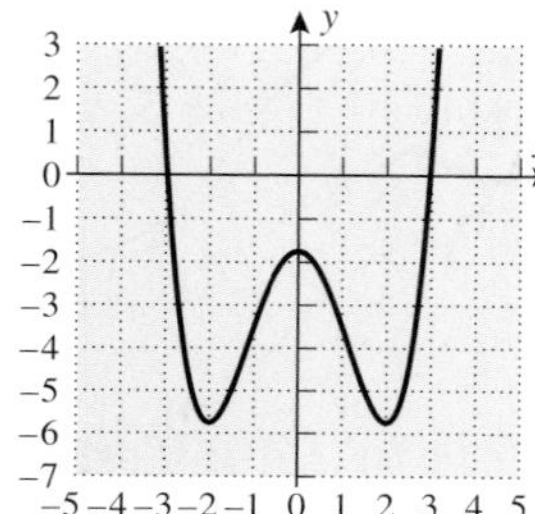

Figure Ex-2

3. Let $f(x) = x^2 - x$.
   (a) Find a point b such that the slope of the secant line through $(0, 0)$ and $(b, f(b))$ is 1.
   (b) Find all values of c that satisfy the conclusion of the Mean-Value Theorem for the function f on the interval $[0, b]$, where b is the point found in part (a).

4. Use the graph of f in the accompanying figure to estimate all values of c that satisfy the conclusion of the Mean-Value Theorem on the interval
   (a) $[0, 8]$ (b) $[0, 4]$.

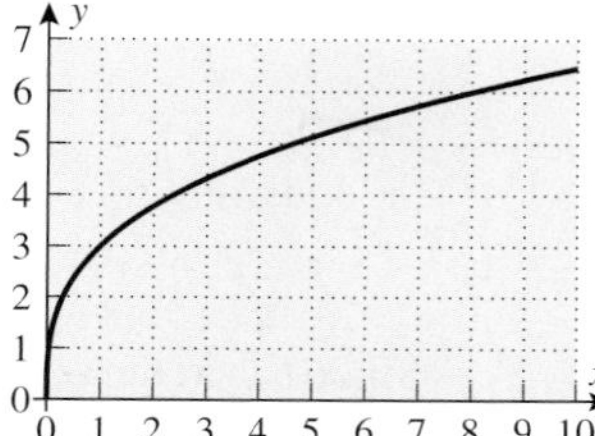

Figure Ex-4

5. Find a function f such that the graph of f contains the point $(1, 5)$ and such that for every value of x_0 the tangent line to the graph of f at x_0 is parallel to the tangent line to the graph of $y = x^2$ at x_0.

EXERCISE SET 4.7 Graphing Utility

1–6 Verify that the hypotheses of Rolle's Theorem are satisfied on the given interval, and find all values of c in that interval that satisfy the conclusion of the theorem.

1. $f(x) = x^2 - 8x + 15;\ [3, 5]$
2. $f(x) = x^3 - 3x^2 + 2x;\ [0, 2]$
3. $f(x) = \cos x;\ [\pi/2, 3\pi/2]$
4. $f(x) = (x^2 - 1)/(x - 2);\ [-1, 1]$
5. $f(x) = \frac{1}{2}x - \sqrt{x};\ [0, 4]$
6. $f(x) = \dfrac{1}{x^2} - \dfrac{4}{3x} + \dfrac{1}{3};\ [1, 3]$

7–12 Verify that the hypotheses of the Mean-Value Theorem are satisfied on the given interval, and find all values of c in that interval that satisfy the conclusion of the theorem.

7. $f(x) = x^2 - x;\ [-3, 5]$
8. $f(x) = x^3 + x - 4;\ [-1, 2]$
9. $f(x) = \sqrt{x + 1};\ [0, 3]$
10. $f(x) = x - \dfrac{1}{x};\ [3, 4]$
11. $f(x) = \sqrt{25 - x^2};\ [-5, 3]$
12. $f(x) = \dfrac{1}{x - 1};\ [2, 5]$

13. (a) Find an interval $[a, b]$ on which
$$f(x) = x^4 + x^3 - x^2 + x - 2$$
satisfies the hypotheses of Rolle's Theorem.
(b) Generate the graph of $f'(x)$, and use it to make rough estimates of all values of c in the interval obtained in part (a) that satisfy the conclusion of Rolle's Theorem.
(c) Use Newton's Method to improve on the rough estimates obtained in part (b).

14. Let $f(x) = x^3 - 4x$.
(a) Find the equation of the secant line through the points $(-2, f(-2))$ and $(1, f(1))$.
(b) Show that there is only one point c in the interval $(-2, 1)$ that satisfies the conclusion of the Mean-Value Theorem for the secant line in part (a).
(c) Find the equation of the tangent line to the graph of f at the point $(c, f(c))$.
(d) Use a graphing utility to generate the secant line in part (a) and the tangent line in part (c) in the same coordinate system, and confirm visually that the two lines seem parallel.

FOCUS ON CONCEPTS

15. Let $f(x) = \tan x$.
(a) Show that there is no point c in the interval $(0, \pi)$ such that $f'(c) = 0$, even though $f(0) = f(\pi) = 0$.
(b) Explain why the result in part (a) does not violate Rolle's Theorem.

16. Let $f(x) = x^{2/3}$, $a = -1$, and $b = 8$.
(a) Show that there is no point c in (a, b) such that
$$f'(c) = \frac{f(b) - f(a)}{b - a}$$
(b) Explain why the result in part (a) does not violate the Mean-Value Theorem.

17. (a) Show that if f is differentiable on $(-\infty, +\infty)$, and if $y = f(x)$ and $y = f'(x)$ are graphed in the same coordinate system, then between any two x-intercepts of f there is at least one x-intercept of f'.
(b) Give some examples that illustrate this.

18. Review Formulas (10) and (11) in Section 3.1 and use the Mean-Value Theorem to show that if f is differentiable on $(-\infty, +\infty)$, then for any interval $[x_0, x_1]$ there is at least one point in (x_0, x_1) where the instantaneous rate of change of y with respect to x is equal to the average rate of change over the interval.

19–21 Use the result of Exercise 18 in these exercises.

19. An automobile travels 4 mi along a straight road in 5 min. Show that the speedometer reads exactly 48 mi/h at least once during the trip.

20. At 11 A.M. on a certain morning the outside temperature was 76°F. At 11 P.M. that evening it had dropped to 52°F.
(a) Show that at some instant during this period the temperature was decreasing at the rate of 2°F/h.
(b) Suppose that you know that the temperature reached a high of 88°F sometime between 11 A.M. and 11 P.M. Show that at some instant during this period the temperature was decreasing at a rate greater than 3°F/h.

21. Suppose that two runners in a 100-m dash finish in a tie. Show that they had the same velocity at least once during the race.

22. Use the fact that
$$\frac{d}{dx}(3x^4 + x^2 - 4x) = 12x^3 + 2x - 4$$
to show that the equation $12x^3 + 2x - 4 = 0$ has at least one solution in the interval $(0, 1)$.

23. (a) Use the Constant Difference Theorem (4.7.3) to show that if $f'(x) = g'(x)$ for all x in the interval $(-\infty, +\infty)$, and if f and g have the same value at some point x_0, then $f(x) = g(x)$ for all x in $(-\infty, +\infty)$.
(b) Use the result in part (a) to confirm the trigonometric identity $\sin^2 x + \cos^2 x = 1$.

24. (a) Use the Constant Difference Theorem (4.7.3) to show that if $f'(x) = g'(x)$ for all x in $(-\infty, +\infty)$, and if $f(x_0) - g(x_0) = c$ at some point x_0, then

$$f(x) - g(x) = c$$

for all x in $(-\infty, +\infty)$.
(b) Use the result in part (a) to show that the function

$$h(x) = (x-1)^3 - (x^2+3)(x-3)$$

is constant for all x in $(-\infty, +\infty)$, and find the constant.
(c) Check the result in part (b) by multiplying out and simplifying the formula for $h(x)$.

FOCUS ON CONCEPTS

25. (a) Use the Mean-Value Theorem to show that if f is differentiable on an interval I, and if $|f'(x)| \leq M$ for all values of x in I, then

$$|f(x) - f(y)| \leq M|x - y|$$

for all values of x and y in I.
(b) Use the result in part (a) to show that

$$|\sin x - \sin y| \leq |x - y|$$

for all real values of x and y.

26. (a) Use the Mean-Value Theorem to show that if f is differentiable on an open interval I, and if $|f'(x)| \geq M$ for all values of x in I, then

$$|f(x) - f(y)| \geq M|x - y|$$

for all values of x and y in I.
(b) Use the result in part (a) to show that

$$|\tan x - \tan y| \geq |x - y|$$

for all values of x and y in the interval $(-\pi/2, \pi/2)$.
(c) Use the result in part (b) to show that

$$|\tan x + \tan y| \geq |x + y|$$

for all values of x and y in the interval $(-\pi/2, \pi/2)$.

27. (a) Use the Mean-Value Theorem to show that

$$\sqrt{y} - \sqrt{x} < \frac{y - x}{2\sqrt{x}}$$

if $0 < x < y$.
(b) Use the result in part (a) to show that if $0 < x < y$, then $\sqrt{xy} < \frac{1}{2}(x + y)$.

28. Show that if f is differentiable on an open interval I and $f'(x) \neq 0$ on I, the equation $f(x) = 0$ can have at most one real root in I.

29. Use the result in Exercise 28 to show the following:
(a) The equation $x^3 + 4x - 1 = 0$ has exactly one real root.
(b) If $b^2 - 3ac < 0$ and if $a \neq 0$, then the equation

$$ax^3 + bx^2 + cx + d = 0$$

has exactly one real root.

30. Use the inequality $\sqrt{3} < 1.8$ to prove that

$$1.7 < \sqrt{3} < 1.75$$

[*Hint:* Let $f(x) = \sqrt{x}$, $a = 3$, and $b = 4$ in the Mean-Value Theorem.]

31. Use the Mean-Value Theorem to prove that

$$x - \frac{x^3}{6} < \sin x < x \quad (x > 0)$$

32. Show that if f and g are functions for which

$$f'(x) = g(x) \quad \text{and} \quad g'(x) = f(x)$$

for all x, then $f^2(x) - g^2(x)$ is a constant.

33. (a) Show that if f and g are functions for which

$$f'(x) = g(x) \quad \text{and} \quad g'(x) = -f(x)$$

for all x, then $f^2(x) + g^2(x)$ is a constant.
(b) Give an example of functions f and g with this property.

FOCUS ON CONCEPTS

34. Let f and g be continuous on $[a, b]$ and differentiable on (a, b). Prove: If $f(a) = g(a)$ and $f(b) = g(b)$, then there is a point c in (a, b) such that $f'(c) = g'(c)$.

35. Illustrate the result in Exercise 34 by drawing an appropriate picture.

36. (a) Prove that if $f''(x) > 0$ for all x in (a, b), then $f'(x) = 0$ at most once in (a, b).
(b) Give a geometric interpretation of the result in (a).

37. (a) Prove part (*b*) of Theorem 4.1.2.
(b) Prove part (*c*) of Theorem 4.1.2.

38. Use the Mean-Value Theorem to prove the following result: Let f be continuous at x_0 and suppose that $\lim_{x \to x_0} f'(x)$ exists. Then f is differentiable at x_0, and

$$f'(x_0) = \lim_{x \to x_0} f'(x)$$

[*Hint:* The derivative $f'(x_0)$ is given by

$$f'(x_0) = \lim_{x \to x_0} \frac{f(x) - f(x_0)}{x - x_0}$$

provided this limit exists.]

FOCUS ON CONCEPTS

39. Let

$$f(x) = \begin{cases} 3x^2, & x \leq 1 \\ ax + b, & x > 1 \end{cases}$$

Find the values of a and b so that f will be differentiable at $x = 1$.

40. (a) Let
$$f(x) = \begin{cases} x^2, & x \le 0 \\ x^2 + 1, & x > 0 \end{cases}$$
Show that
$$\lim_{x \to 0^-} f'(x) = \lim_{x \to 0^+} f'(x)$$
but that $f'(0)$ does not exist.

(b) Let
$$f(x) = \begin{cases} x^2, & x \le 0 \\ x^3, & x > 0 \end{cases}$$
Show that $f'(0)$ exists but $f''(0)$ does not.

41. Use the Mean-Value Theorem to prove the following result: The graph of a function f has a point of vertical tangency at $(x_0, f(x_0))$ if f is continuous at x_0 and $f'(x)$ approaches either $+\infty$ or $-\infty$ as $x \to x_0^+$ and as $x \to x_0^-$.

✔ QUICK CHECK ANSWERS 4.7

1. (a) $[0, 1]$ (b) $c = \frac{1}{2}$ 2. $[-3, 3]$; $c = -2, 0, 2$ 3. (a) $b = 2$ (b) $c = 1$ 4. (a) 1.5 (b) 0.8 5. $f(x) = x^2 + 4$

4.8 RECTILINEAR MOTION

In this section we will continue the study of rectilinear motion that we began in Section 3.1. We will define the notion of "acceleration" mathematically, and we will show how the tools of calculus developed earlier in this chapter can be used to analyze rectilinear motion in more depth.

REVIEW OF TERMINOLOGY

Recall from Section 3.1 that a particle that can move in either direction along a coordinate line is said to be in ***rectilinear motion***. The line might be an x-axis, a y-axis, or a coordinate line inclined at some angle. In general discussions we will designate the coordinate line as the s-axis. We will assume that units are chosen for measuring distance and time and that we begin observing the motion of the particle at time $t = 0$. As the particle moves along the s-axis, its coordinate s will be some function of time, say $s = s(t)$. We call $s(t)$ the ***position function*** of the particle,* and we call the graph of s versus t the ***position versus time curve***. If the coordinate of a particle at time t_1 is $s(t_1)$ and the coordinate at a later time t_2 is $s(t_2)$, then $s(t_2) - s(t_1)$ is called the ***displacement*** of the particle over the time interval $[t_1, t_2]$. The displacement describes the change in position of the particle.

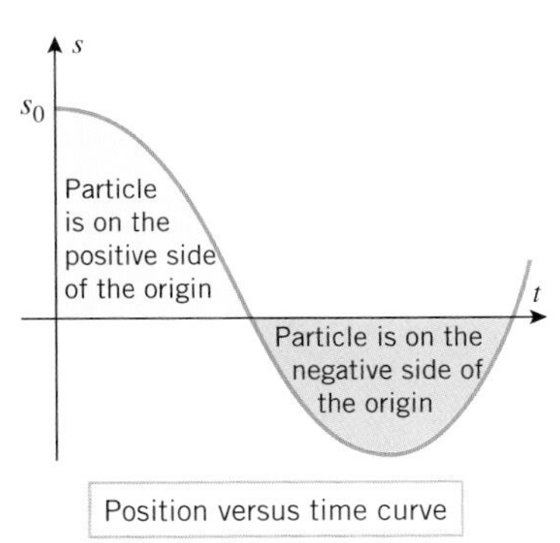

Figure 4.8.1

Figure 4.8.1 shows a typical position versus time curve for a particle in rectilinear motion. We can tell from that graph that the coordinate of the particle at time $t = 0$ is s_0, and we can tell from the sign of s when the particle is on the negative or the positive side of the origin as it moves along the coordinate line.

▶ **Example 1** Figure 4.8.2*a* shows the position versus time curve for a particle moving along an s-axis. In words, describe how the position of the particle changes with time.

Solution. The particle is at $s = -3$ at time $t = 0$. It moves in the positive direction until time $t = 4$, since s is increasing. At time $t = 4$ the particle is at position $s = 3$. At that time it turns around and travels in the negative direction until time $t = 7$, since s is decreasing.

* In writing $s = s(t)$, rather than the more familiar $s = f(t)$, we are using the letter s both as the dependent variable and the name of the function. This is common practice in engineering and physics.

At time $t = 7$ the particle is at position $s = -1$, and it remains stationary thereafter, since s is constant for $t > 7$. This is illustrated schematically in Figure 4.8.2*b*. ◄

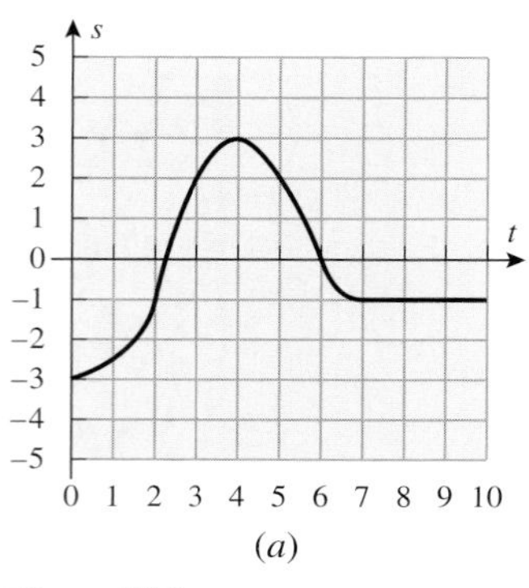

(*a*)

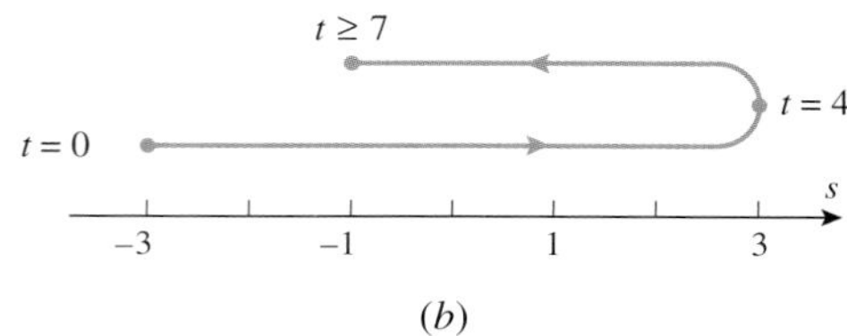

(*b*)

Figure 4.8.2

We should more properly call $v(t)$ the ***instantaneous velocity function*** to distinguish instantaneous velocity from average velocity. However, we will follow the standard practice of referring to it as the "velocity function," leaving it understood that it describes instantaneous velocity.

VELOCITY AND SPEED

Recall from Formulas (6) and (7) of Section 3.1 and Formula (4) of Section 3.2 that the instantaneous velocity of a particle in rectilinear motion is the derivative of the position function and the instantaneous speed is the absolute value of the instantaneous velocity. Thus, if a particle in rectilinear motion has position function $s(t)$, then we define its ***velocity function*** $v(t)$ to be

$$v(t) = s'(t) = \frac{ds}{dt} \tag{1}$$

and we define its ***speed function*** to be

$$|v(t)| = |s'(t)| = \left|\frac{ds}{dt}\right| \tag{2}$$

The sign of the velocity tells which way the particle is moving—a positive value for $v(t)$ means that s is increasing with time, so the particle is moving in the positive direction, and a negative value for $v(t)$ means that s is decreasing with time, so the particle is moving in the negative direction. If $v(t) = 0$, then the particle has momentarily stopped. The speed function, which is always nonnegative, tells us how fast the particle is moving but not its direction of motion.

► **Example 2** Let $s(t) = t^3 - 6t^2$ be the position function of a particle moving along an s-axis, where s is in meters and t is in seconds. Find the velocity and speed functions, and show the graphs of position, velocity, and speed versus time.

Solution. From (1) and (2), the velocity and speed functions are given by

$$v(t) = \frac{ds}{dt} = 3t^2 - 12t \quad \text{and} \quad |v(t)| = |3t^2 - 12t|$$

The graphs of position, velocity, and speed versus time are shown in Figure 4.8.3. Observe that velocity and speed both have units of meters per second (m/s), since s is in meters (m) and time is in seconds (s). ◄

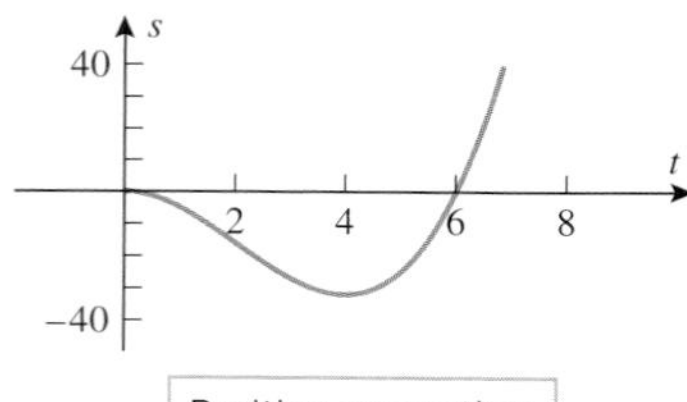

Position versus time

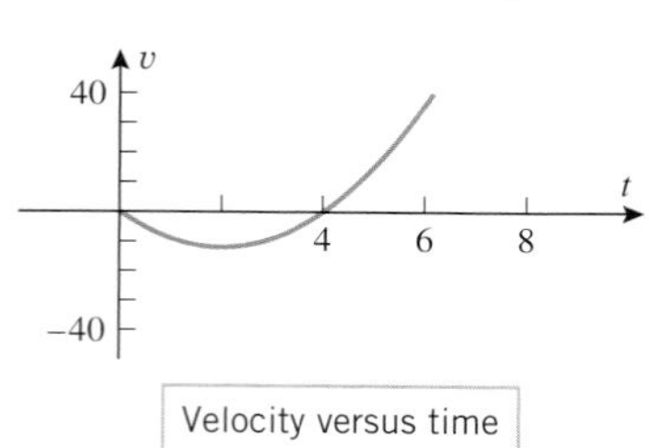

Velocity versus time

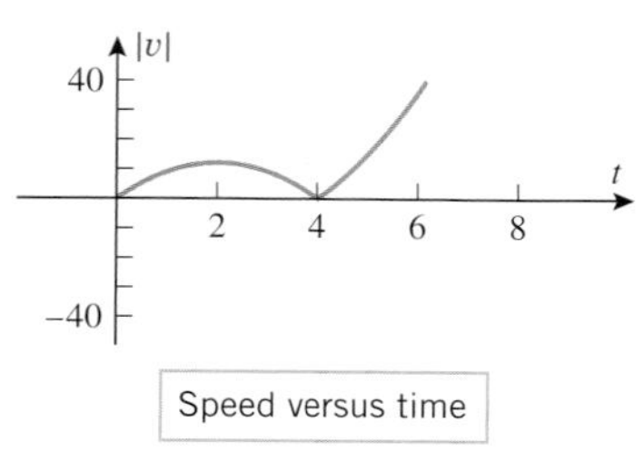

Speed versus time

Figure 4.8.3

The graphs in Figure 4.8.3 provide a wealth of visual information about the motion of the particle. For example, the position versus time curve tells us that the particle is on the negative side of the origin for $0 < t < 6$, is on the positive side of the origin for $t > 6$, and is at the origin at times $t = 0$ and $t = 6$. The velocity versus time curve tells us that the particle is moving in the negative direction if $0 < t < 4$, is moving in the positive direction if $t > 4$, and is momentarily stopped at times $t = 0$ and $t = 4$ (the velocity is zero at those times). The speed versus time curve tells us that the speed of the particle is increasing for $0 < t < 2$, decreasing for $2 < t < 4$, and increasing again for $t > 4$.

ACCELERATION

In rectilinear motion, the rate at which the instantaneous velocity of a particle changes with time is called its ***instantaneous acceleration***. Thus, if a particle in rectilinear motion has velocity function $v(t)$, then we define its ***acceleration function*** to be

$$a(t) = v'(t) = \frac{dv}{dt} \tag{3}$$

Alternatively, we can use the fact that $v(t) = s'(t)$ to express the acceleration function in terms of the position function as

$$a(t) = s''(t) = \frac{d^2s}{dt^2} \tag{4}$$

▶ **Example 3** Let $s(t) = t^3 - 6t^2$ be the position function of a particle moving along an s-axis, where s is in meters and t is in seconds. Find the acceleration function $a(t)$, and show the graph of acceleration versus time.

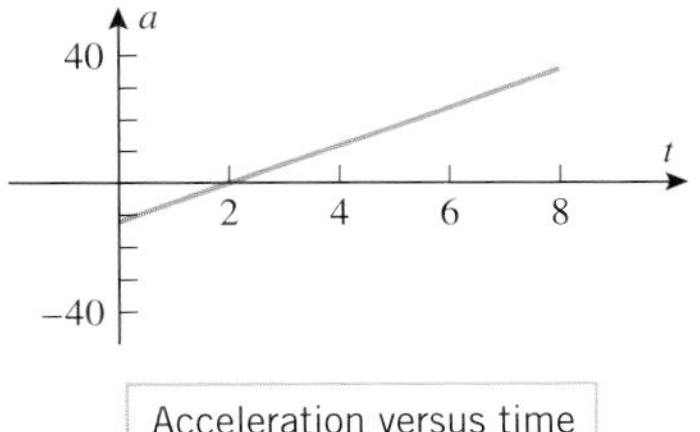

Figure 4.8.4

Solution. From Example 2, the velocity function of the particle is $v(t) = 3t^2 - 12t$, so the acceleration function is

$$a(t) = \frac{dv}{dt} = 6t - 12$$

and the acceleration versus time curve is the line shown in Figure 4.8.4. Note that in this example the acceleration has units of m/s^2, since v is in meters per second (m/s) and time is in seconds (s). ◀

SPEEDING UP AND SLOWING DOWN

We will say that a particle in rectilinear motion is ***speeding up*** when its speed is increasing and is ***slowing down*** when its speed is decreasing. In everyday language an object that is speeding up is said to be "accelerating" and an object that is slowing down is said to be "decelerating"; thus, one might expect that a particle in rectilinear motion will be speeding up when its acceleration is positive and slowing down when it is negative. Although this is true for a particle moving in the positive direction, it is *not* true for a particle moving in the negative direction—a particle with negative velocity is speeding up when its acceleration is negative and slowing down when its acceleration is positive. This is because a positive acceleration implies an increasing velocity, and increasing a negative velocity decreases its absolute value; similarly, a negative acceleration implies a decreasing velocity, and decreasing a negative velocity increases its absolute value.

The preceding informal discussion can be summarized as follows (Exercise 33):

If $a(t) = 0$ over a certain time interval, what does this tell you about the motion of the particle during that time?

INTERPRETING THE SIGN OF ACCELERATION. *A particle in rectilinear motion is speeding up when its velocity and acceleration have the same sign and slowing down when they have opposite signs.*

▶ **Example 4** In Examples 2 and 3 we found the velocity versus time curve and the acceleration versus time curve for a particle with position function $s(t) = t^3 - 6t^2$. Use those curves to determine when the particle is speeding up and slowing down, and confirm that your results are consistent with the speed versus time curve obtained in Example 2.

Solution. Over the time interval $0 < t < 2$ the velocity and acceleration are negative, so the particle is speeding up. This is consistent with the speed versus time curve, since the speed is increasing over this time interval. Over the time interval $2 < t < 4$ the velocity is negative and the acceleration is positive, so the particle is slowing down. This is also consistent with the speed versus time curve, since the speed is decreasing over this time interval. Finally, on the time interval $t > 4$ the velocity and acceleration are positive, so the particle is speeding up, which again is consistent with the speed versus time curve. ◀

ANALYZING THE POSITION VERSUS TIME CURVE

The position versus time curve contains all of the significant information about the position and velocity of a particle in rectilinear motion:

- If $s(t) > 0$, the particle is on the positive side of the s-axis.
- If $s(t) < 0$, the particle is on the negative side of the s-axis.
- The slope of the curve at any time is equal to the instantaneous velocity at that time.
- Where the curve has positive slope, the velocity is positive and the particle is moving in the positive direction.
- Where the curve has negative slope, the velocity is negative and the particle is moving in the negative direction.
- Where the slope of the curve is zero, the velocity is zero, and the particle is momentarily stopped.

Information about the acceleration of a particle in rectilinear motion can also be deduced from the position versus time curve by examining its concavity. For example, we know that the position versus time curve will be concave up on intervals where $s''(t) > 0$ and will be concave down on intervals where $s''(t) < 0$. But we know from (4) that $s''(t)$ is the acceleration, so that on intervals where the position versus time curve is concave up the particle has a positive acceleration, and on intervals where it is concave down the particle has a negative acceleration.

Table 4.8.1 summarizes our observations about the position versus time curve.

▶ **Example 5** Use the position versus time curve in Figure 4.8.2 to determine when the particle in Example 1 is speeding up and slowing down.

Solution. From $t = 0$ to $t = 2$, the acceleration and velocity are positive, so the particle is speeding up. From $t = 2$ to $t = 4$, the acceleration is negative and the velocity is positive, so the particle is slowing down. At $t = 4$, the velocity is zero, so the particle has momentarily stopped. From $t = 4$ to $t = 6$, the acceleration is negative and the velocity is negative, so the particle is speeding up. From $t = 6$ to $t = 7$, the acceleration is positive and the velocity is negative, so the particle is slowing down. Thereafter, the velocity is zero, so the particle has stopped. ◀

Table 4.8.1

POSITION VERSUS TIME CURVE	CHARACTERISTICS OF THE CURVE AT $t = t_0$	BEHAVIOR OF THE PARTICLE AT TIME $t = t_0$
	• $s(t_0) > 0$ • Curve has positive slope. • Curve is concave down.	• Particle is on the positive side of the origin. • Particle is moving in the positive direction. • Velocity is decreasing. • Particle is slowing down.
	• $s(t_0) > 0$ • Curve has negative slope. • Curve is concave down.	• Particle is on the positive side of the origin. • Particle is moving in the negative direction. • Velocity is decreasing. • Particle is speeding up.
	• $s(t_0) < 0$ • Curve has negative slope. • Curve is concave up.	• Particle is on the negative side of the origin. • Particle is moving in the negative direction. • Velocity is increasing. • Particle is slowing down.
	• $s(t_0) > 0$ • Curve has zero slope. • Curve is concave down.	• Particle is on the positive side of the origin. • Particle is momentarily stopped. • Velocity is decreasing.

▶ **Example 6** Suppose that the position function of a particle moving on a coordinate line is given by $s(t) = 2t^3 - 21t^2 + 60t + 3$. Analyze the motion of the particle for $t \geq 0$.

Solution. The velocity and acceleration functions are

$$v(t) = s'(t) = 6t^2 - 42t + 60 = 6(t-2)(t-5)$$
$$a(t) = v'(t) = 12t - 42 = 12\left(t - \tfrac{7}{2}\right)$$

- *Direction of motion:* The sign analysis of the velocity function in Figure 4.8.5 shows that the particle is moving in the positive direction over the time interval $0 \leq t < 2$, stops momentarily at time $t = 2$, moves in the negative direction over the time interval $2 < t < 5$, stops momentarily at time $t = 5$, and then moves in the positive direction thereafter.

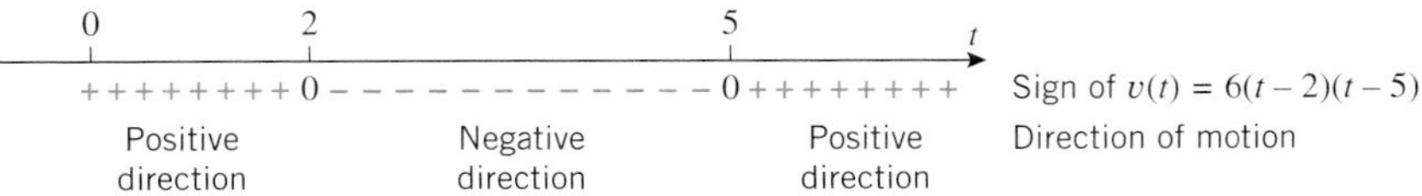

Figure 4.8.5

- *Change in speed:* A comparison of the signs of the velocity and acceleration functions is shown in Figure 4.8.6. Since the particle is speeding up when the signs are the same and is slowing down when they are opposite, we see that the particle is slowing down over the time interval $0 \le t < 2$ and stops momentarily at time $t = 2$. It is then speeding up over the time interval $2 < t < \frac{7}{2}$. At time $t = \frac{7}{2}$ the instantaneous acceleration is zero, so the particle is neither speeding up nor slowing down. It is then slowing down over the time interval $\frac{7}{2} < t < 5$ and stops momentarily at time $t = 5$. Thereafter, it is speeding up.

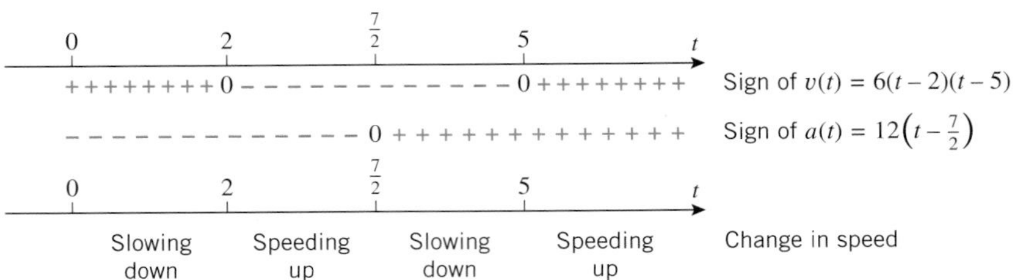

Figure 4.8.6

Conclusions: The diagram in Figure 4.8.7 summarizes the above information schematically. The curved line is descriptive only; the actual path is back and forth on the coordinate line. The coordinates of the particle at times $t = 0$, $t = 2$, $t = \frac{7}{2}$, and $t = 5$ were computed from $s(t)$. Segments in red indicate that the particle is speeding up and segments in blue indicate that it is slowing down. ◀

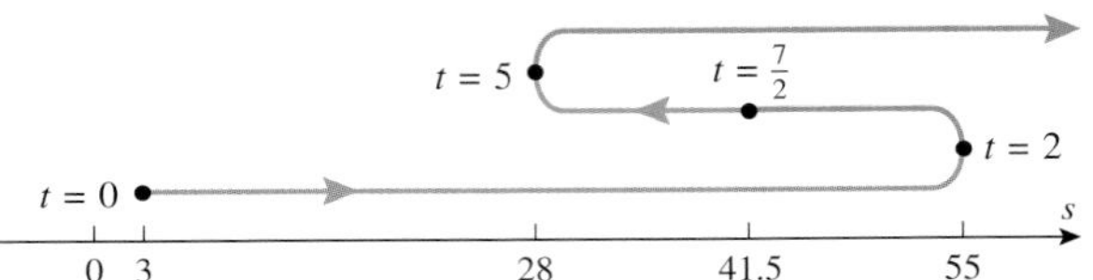

Figure 4.8.7

✔QUICK CHECK EXERCISES 4.8 *(See page 297 for answers.)*

1. For a particle in rectilinear motion, the velocity and position functions $v(t)$ and $s(t)$ are related by the equation ________, and the acceleration and velocity functions $a(t)$ and $v(t)$ are related by the equation ________.
2. Suppose that a particle moving along the s-axis has position function $s(t) = 7t - 2t^2$. At time $t = 3$, the particle's position is ________, its velocity is ________, its speed is ________, and its acceleration is ________.
3. A particle in rectilinear motion is speeding up if the signs of its velocity and acceleration are ________, and it is slowing down if these signs are ________.
4. Suppose that a particle moving along the s-axis has position function $s(t) = t^4 - 24t^2$ over the time interval $t \ge 0$. The particle slows down over the time interval(s) ________.

EXERCISE SET 4.8 Graphing Utility

FOCUS ON CONCEPTS

1. The graphs of three position functions are shown in the accompanying figure. In each case determine the signs of the velocity and acceleration, and then determine whether the particle is speeding up or slowing down.

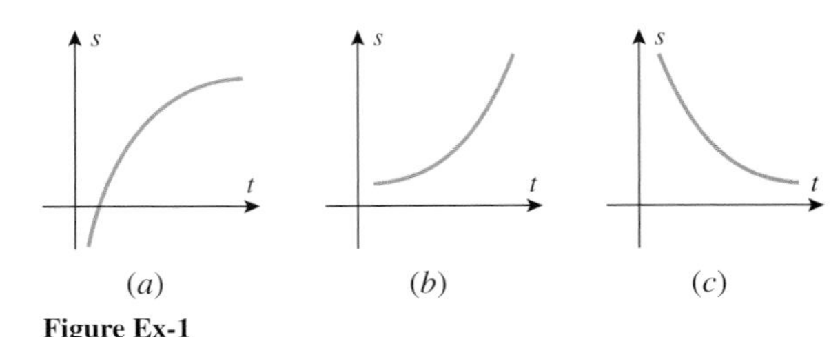

Figure Ex-1

2. The graphs of three velocity functions are shown in the accompanying figure. In each case determine the sign of the acceleration, and then determine whether the particle is speeding up or slowing down.

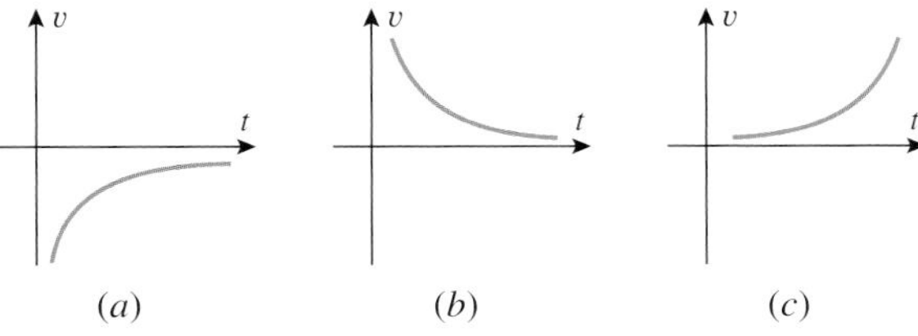

Figure Ex-2

3. The position function of a particle moving on a horizontal x-axis is shown in the accompanying figure.
(a) Is the particle moving left or right at time t_0?
(b) Is the acceleration positive or negative at time t_0?
(c) Is the particle speeding up or slowing down at time t_0?
(d) Is the particle speeding up or slowing down at time t_1?

Figure Ex-3

4. For the graphs in the accompanying figure, match the position functions with their corresponding velocity functions.

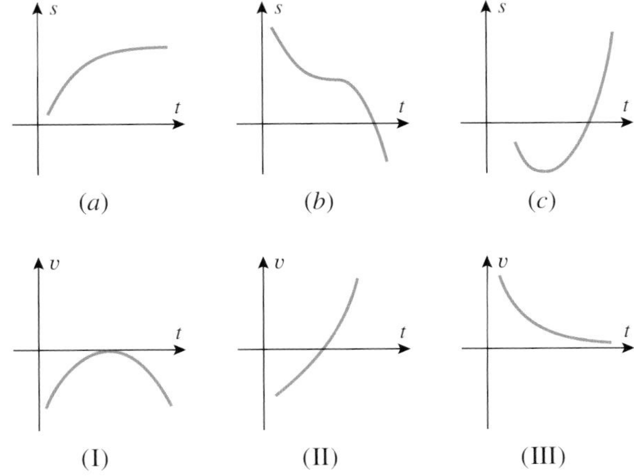

Figure Ex-4

5. Sketch a reasonable graph of s versus t for a mouse that is trapped in a narrow corridor (an s-axis with the positive direction to the right) and scurries back and forth as follows. It runs right with a constant speed of 1.2 m/s for awhile, then gradually slows down to 0.6 m/s, then quickly speeds up to 2.0 m/s, then gradually slows to a stop but immediately reverses direction and quickly speeds up to 1.2 m/s.

6. The accompanying figure shows the graph of s versus t for an ant that moves along a narrow vertical pipe (an s-axis with the positive direction up).
(a) When, if ever, is the ant above the origin?
(b) When, if ever, does the ant have velocity zero?
(c) When, if ever, is the ant moving down the pipe?

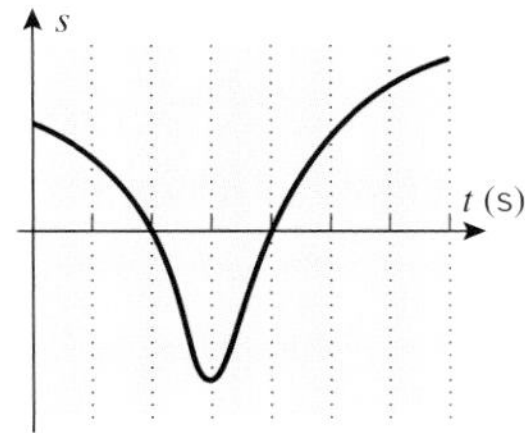

Figure Ex-6

7. The accompanying figure shows the graph of velocity versus time for a particle moving along a coordinate line. Make a rough sketch of the graphs of speed versus time and acceleration versus time.

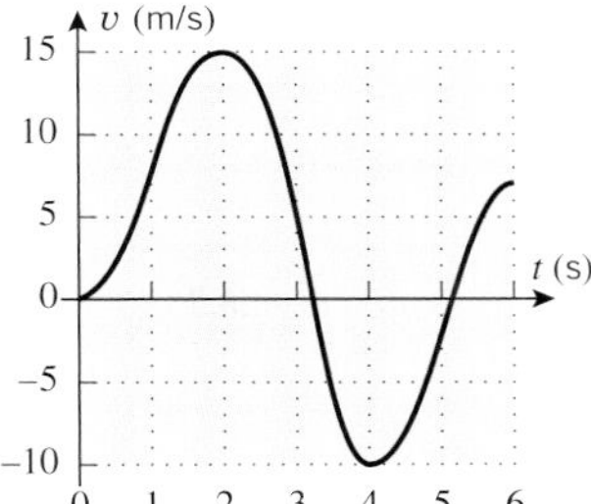

Figure Ex-7

8. The accompanying figure shows the position versus time graph for an elevator that ascends 40 m from one stop to the next.
(a) Estimate the velocity when the elevator is halfway up to the top.
(b) Sketch rough graphs of the velocity versus time curve and the acceleration versus time curve.

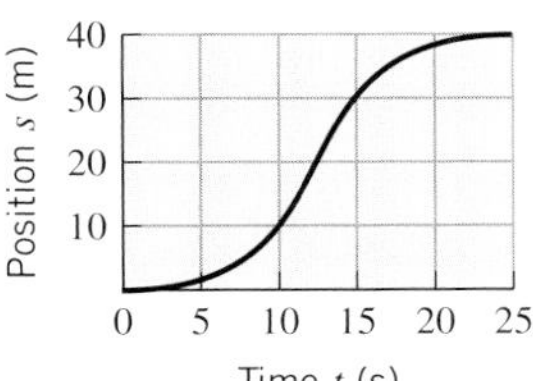

Figure Ex-8

9. The accompanying figure (next page) shows the velocity versus time graph for a test run on a Pontiac Grand Prix GTP. Using this graph, estimate
(a) the acceleration at 60 mi/h (in ft/s^2)
(b) the time at which the maximum acceleration occurs.

Source: Data from *Car and Driver Magazine*, July 2003.

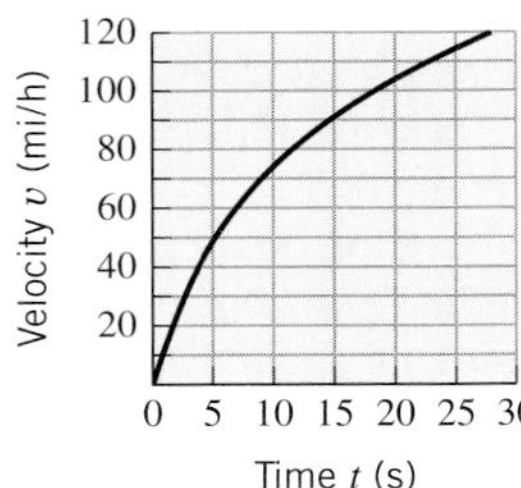

Figure Ex-9

10. The accompanying figure shows the velocity versus time graph for a test run on a Chevrolet Malibu. Using this graph, estimate
(a) the acceleration at 60 mi/h (in ft/s^2)
(b) the time at which the maximum acceleration occurs.

Source: Data from *Car and Driver Magazine*, November 2003.

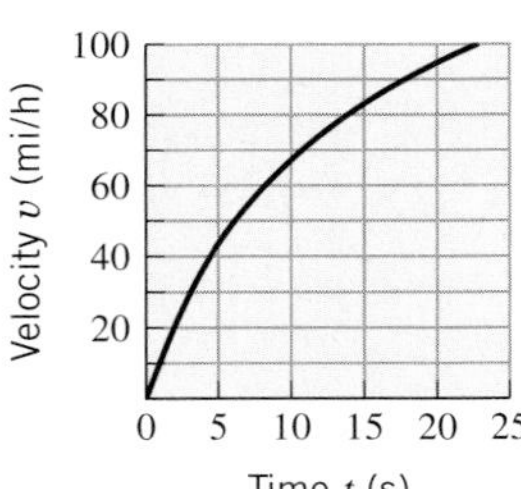

Figure Ex-10

11–12 The function $s(t)$ describes the position of a particle moving along a coordinate line, where s is in meters and t is in seconds.
(a) Make a table showing the position, velocity, and acceleration to two decimal places at times $t = 1, 2, 3, 4, 5$.
(b) At each of the times in part (a), determine whether the particle is stopped; if it is not, state its direction of motion.
(c) At each of the times in part (a), determine whether the particle is speeding up, slowing down, or neither.

11. $s(t) = \sin \dfrac{\pi t}{4}$ **12.** $f(t) = 2\cos\left(\dfrac{\pi}{3}t - \dfrac{2\pi}{3}\right)$

13–16 The function $s(t)$ describes the position of a particle moving along a coordinate line, where s is in feet and t is in seconds.
(a) Find the velocity and acceleration functions.
(b) Find the position, velocity, speed, and acceleration at time $t = 1$.
(c) At what times is the particle stopped?
(d) When is the particle speeding up? Slowing down?
(e) Find the total distance traveled by the particle from time $t = 0$ to time $t = 5$.

13. $s(t) = t^3 - 3t^2, \quad t \geq 0$

14. $s(t) = t^4 - 4t^2 + 4, \quad t \geq 0$

15. $s(t) = 9 - 9\cos(\pi t/3), \quad 0 \leq t \leq 5$

16. $s(t) = \dfrac{t}{t^2 + 4}, \quad t \geq 0$

17. Let $s(t) = t/(t^2 + 5)$ be the position function of a particle moving along a coordinate line, where s is in meters and t is in seconds. Use a graphing utility to generate the graphs of $s(t)$, $v(t)$, and $a(t)$ for $t \geq 0$, and use those graphs where needed.
(a) Use the appropriate graph to make a rough estimate of the time at which the particle first reverses the direction of its motion; and then find the time exactly.
(b) Find the exact position of the particle when it first reverses the direction of its motion.
(c) Use the appropriate graphs to make a rough estimate of the time intervals on which the particle is speeding up and on which it is slowing down; and then find those time intervals exactly.

18. Let $s(t) = 4t^2/(2t^4 + 3)$ be the position function of a particle moving along a coordinate line, where s is in meters and t is in seconds. Use a graphing utility to generate the graphs of $s(t)$, $v(t)$, and $a(t)$ for $t \geq 0$, and use those graphs where needed.
(a) Use the appropriate graph to make a rough estimate of the time at which the particle first reverses the direction of its motion; and then find the time exactly.
(b) Find the exact position of the particle when it first reverses the direction of its motion.
(c) Use the appropriate graphs to make a rough estimate of the time intervals on which the particle is speeding up and on which it is slowing down; and then find those time intervals exactly.

19–24 A position function of a particle moving along a coordinate line is given. Use the method of Example 6 to analyze the motion of the particle for $t \geq 0$, and give a schematic picture of the motion (as in Figure 4.8.7).

19. $s = -4t + 3$ **20.** $s = 5t^2 - 20t$

21. $s = t^3 - 9t^2 + 24t$ **22.** $s = t + \dfrac{25}{t+2}$

23. $s = \begin{cases} \cos t, & 0 \leq t < 2\pi \\ 1, & t \geq 2\pi \end{cases}$

24. $s = \begin{cases} 2t(t-2)^2, & 0 \leq t < 3 \\ 13 - 7(t-4)^2, & t \geq 3 \end{cases}$

25. Let $s(t) = 5t^2 - 22t$ be the position function of a particle moving along a coordinate line, where s is in feet and t is in seconds.
(a) Find the maximum speed of the particle during the time interval $1 \leq t \leq 3$.
(b) When, during the time interval $1 \leq t \leq 3$, is the particle farthest from the origin? What is its position at that instant?

26. Let $s = 100/(t^2 + 12)$ be the position function of a particle moving along a coordinate line, where s is in feet and t is in seconds. Find the maximum speed of the particle for $t \geq 0$, and find the direction of motion of the particle when it has its maximum speed.

27–28 A position function of a particle moving along a coordinate line is provided. (a) Evaluate s and v when $a = 0$. (b) Evaluate s and a when $v = 0$.

27. $s = \sin 2t,\ 0 \leq t \leq \pi/2$ **28.** $s = t^3 - 6t^2 + 1$

29. Let $s = \sqrt{2t^2 + 1}$ be the position function of a particle moving along a coordinate line.
(a) Use a graphing utility to generate the graph of v versus t, and make a conjecture about the velocity of the particle as $t \to +\infty$.
(b) Check your conjecture by finding $\lim_{t \to +\infty} v$.

30. (a) Use the chain rule to show that for a particle in rectilinear motion $a = v(dv/ds)$.
(b) Let $s = \sqrt{3t + 7},\ t \geq 0$. Find a formula for v in terms of s and use the equation in part (a) to find the acceleration when $s = 5$.

31. Suppose that the position functions of two particles, P_1 and P_2, in motion along the same line are

$$s_1 = \tfrac{1}{2}t^2 - t + 3 \quad \text{and} \quad s_2 = -\tfrac{1}{4}t^2 + t + 1$$

respectively, for $t \geq 0$.
(a) Prove that P_1 and P_2 do not collide.
(b) How close do P_1 and P_2 get to one another?
(c) During what intervals of time are they moving in opposite directions?

32. Let $s_A = 15t^2 + 10t + 20$ and $s_B = 5t^2 + 40t,\ t \geq 0$, be the position functions of cars A and B that are moving along parallel straight lanes of a highway.
(a) How far is car A ahead of car B when $t = 0$?
(b) At what instants of time are the cars next to one another?
(c) At what instant of time do they have the same velocity? Which car is ahead at this instant?

33. Prove that a particle is speeding up if the velocity and acceleration have the same sign, and slowing down if they have opposite signs. [*Hint:* Let $r(t) = |v(t)|$ and find $r'(t)$ using the chain rule.]

✔ QUICK CHECK ANSWERS 4.8

1. $v(t) = s'(t)$; $a(t) = v'(t)$ 2. 3; -5; 5; -4 3. the same; opposite 4. $2 < t < 2\sqrt{3}$

CHAPTER REVIEW EXERCISES  Graphing Utility C CAS

1. (a) If $x_1 < x_2$, what relationship must hold between $f(x_1)$ and $f(x_2)$ if f is increasing on an interval containing x_1 and x_2? Decreasing? Constant?
(b) What condition on f' ensures that f is increasing on an interval $[a, b]$? Decreasing? Constant?

2. (a) What condition on f' ensures that f is concave up on an open interval I? Concave down?
(b) What condition on f'' ensures that f is concave up on an open interval I? Concave down?
(c) In words, what is an inflection point of f?

3–8 Find: (a) the intervals on which f is increasing, (b) the intervals on which f is decreasing, (c) the open intervals on which f is concave up, (d) the open intervals on which f is concave down, and (e) the x-coordinates of all inflection points.

3. $f(x) = x^2 - 5x + 6$ **4.** $f(x) = x^4 - 8x^2 + 16$

5. $f(x) = \dfrac{x^2}{x^2 + 2}$ **6.** $f(x) = \sqrt[3]{x + 2}$

7. $f(x) = x^{1/3}(x + 4)$ **8.** $f(x) = x^{4/3} - x^{1/3}$

9–12 Analyze the trigonometric function f over the specified interval, stating where f is increasing, decreasing, concave up, and concave down, and stating the x-coordinates of all inflection points. Confirm that your results are consistent with the graph of f generated with a graphing utility.

9. $f(x) = \cos x$; $[0, 2\pi]$

10. $f(x) = \tan x$; $(-\pi/2, \pi/2)$

11. $f(x) = \sin x \cos x$; $[0, \pi]$

12. $f(x) = \cos^2 x - 2 \sin x$; $[0, 2\pi]$

13. In each part, sketch a continuous curve $y = f(x)$ with the stated properties.
(a) $f(2) = 4,\ f'(2) = 1,\ f''(x) < 0$ for $x < 2$, $f''(x) > 0$ for $x > 2$
(b) $f(2) = 4,\ f''(x) > 0$ for $x < 2,\ f''(x) < 0$ for $x > 2$, $\lim_{x \to 2^-} f'(x) = +\infty,\ \lim_{x \to 2^+} f'(x) = +\infty$
(c) $f(2) = 4,\ f''(x) < 0$ for $x \neq 2,\ \lim_{x \to 2^-} f'(x) = 1$, $\lim_{x \to 2^+} f'(x) = -1$

14. In parts (a)–(d), the graph of a polynomial with degree at most 6 is given. Find equations for polynomials that produce graphs with these shapes, and check your answers with a graphing utility.

(a)
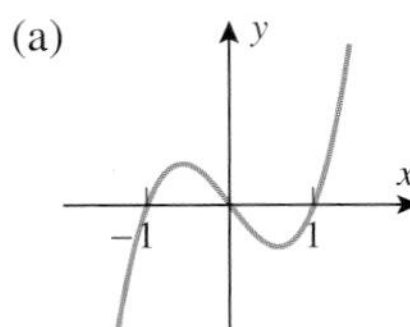
(b)
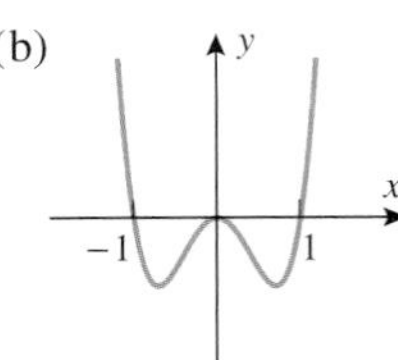
(c)
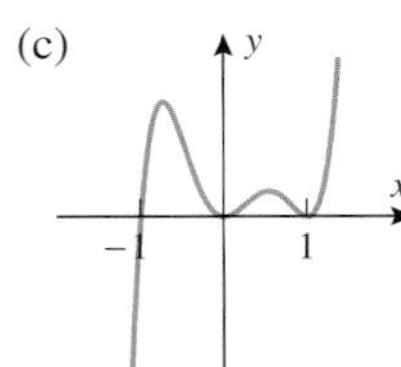
(d)
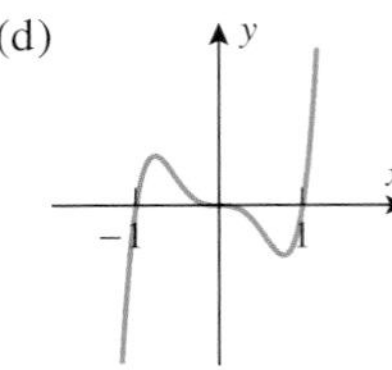

15. For a general quadratic polynomial

$$f(x) = ax^2 + bx + c \quad (a \neq 0)$$

find conditions on a, b, and c to ensure that f is always increasing or always decreasing on $[0, +\infty)$.

16. For the general cubic polynomial

$$f(x) = ax^3 + bx^2 + cx + d \quad (a \neq 0)$$

find conditions on a, b, c, and d to ensure that f is always increasing or always decreasing on $(-\infty, +\infty)$.

17. (a) Where on the graph of $y = f(x)$ would you expect y to be increasing or decreasing most rapidly with respect to x?
(b) In words, what is a relative extremum?
(c) State a procedure for determining where the relative extrema of f occur.

18. Determine whether the statement is true or false. If it is false, give an example for which the statement fails.
(a) If f has a relative maximum at x_0, then $f(x_0)$ is the largest value that $f(x)$ can have.
(b) If the largest value for f on the interval (a, b) is at x_0, then f has a relative maximum at x_0.
(c) A function f has a relative extremum at each of its critical points.

19. (a) According to the first derivative test, what conditions ensure that f has a relative maximum at x_0? A relative minimum?
(b) According to the second derivative test, what conditions ensure that f has a relative maximum at x_0? A relative minimum?

20–22 Locate the critical points and identify which critical points correspond to stationary points.

20. (a) $f(x) = x^3 + 3x^2 - 9x + 1$
(b) $f(x) = x^4 - 6x^2 - 3$

21. (a) $f(x) = \dfrac{x}{x^2 + 2}$ (b) $f(x) = \dfrac{x^2 - 3}{x^2 + 1}$

22. (a) $f(x) = x^{1/3}(x - 4)$ (b) $f(x) = x^{4/3} - 6x^{1/3}$

23. In each part, find all critical points, and use the first derivative test to classify them as relative maxima, relative minima, or neither.
(a) $f(x) = x^{1/3}(x - 7)^2$
(b) $f(x) = 2\sin x - \cos 2x, \quad 0 \le x \le 2\pi$
(c) $f(x) = 3x - (x - 1)^{3/2}$

24. In each part, find all critical points, and use the second derivative test (where possible) to classify them as relative maxima, relative minima, or neither.
(a) $f(x) = x^{-1/2} + \frac{1}{9}x^{1/2}$
(b) $f(x) = x^2 + 8/x$
(c) $f(x) = \sin^2 x - \cos x, \quad 0 \le x \le 2\pi$

25–32 Give a graph of f, and identify the limits as $x \to \pm\infty$, as well as locations of all relative extrema, inflection points, and asymptotes (as appropriate).

25. $f(x) = x^4 - 3x^3 + 3x^2 + 1$

26. $f(x) = x^5 - 4x^4 + 4x^3$

27. $f(x) = \tan(x^2 + 1)$ **28.** $f(x) = x - \cos x$

29. $f(x) = \dfrac{x^2}{x^2 + 2x + 5}$ **30.** $f(x) = \dfrac{25 - 9x^2}{x^3}$

31. $f(x) = \begin{cases} \frac{1}{2}x^2, & x \le 0 \\ -x^2, & x > 0 \end{cases}$

32. $f(x) = (1 + x)^{2/3}(3 - x)^{1/3}$

33–38 Use any method to find the relative extrema of the function f.

33. $f(x) = x^3 + 5x - 2$ **34.** $f(x) = x^4 - 2x^2 + 7$

35. $f(x) = x^{4/5}$ **36.** $f(x) = 2x + x^{2/3}$

37. $f(x) = \dfrac{x^2}{x^2 + 1}$ **38.** $f(x) = \dfrac{x}{x + 2}$

39–40 When using a graphing utility, important features of a graph may be missed if the viewing window is not chosen appropriately. This is illustrated in Exercises 39 and 40.

39. (a) Generate the graph of $f(x) = \frac{1}{3}x^3 - \frac{1}{400}x$ over the interval $[-5, 5]$, and make a conjecture about the locations and nature of all critical points.
(b) Find the exact locations of all the critical points, and classify them as relative maxima, relative minima, or neither.
(c) Confirm the results in part (b) by graphing f over an appropriate interval.

40. (a) Generate the graph of
$$f(x) = \tfrac{1}{5}x^5 - \tfrac{7}{8}x^4 + \tfrac{1}{3}x^3 + \tfrac{7}{2}x^2 - 6x$$
over the interval $[-5, 5]$, and make a conjecture about the locations and nature of all critical points.
(b) Find the exact locations of all the critical points, and classify them as relative maxima, relative minima, or neither.
(c) Confirm the results in part (b) by graphing portions of f over appropriate intervals. [*Note:* It will not be possible to find a single window in which all of the critical points are discernible.]

41. (a) Use a graphing utility to generate the graphs of $y = x$ and $y = (x^3 - 8)/(x^2 + 1)$ together over the interval $[-5, 5]$, and make a conjecture about the relationship between the two graphs.
(b) Confirm your conjecture in part (a).

42. Use implicit differentiation to show that a function defined implicitly by $\sin x + \cos y = 2y$ has a critical point whenever $\cos x = 0$. Then use either the first or second derivative test to classify these critical points as relative maxima or minima.

43. Let
$$f(x) = \frac{2x^3 + x^2 - 15x + 7}{(2x - 1)(3x^2 + x - 1)}$$
Graph $y = f(x)$, and find the equations of all horizontal and vertical asymptotes. Explain why there is no vertical asymptote at $x = \frac{1}{2}$, even though the denominator of f is zero at that point.

C 44. Let
$$f(x) = \frac{x^5 - x^4 - 3x^3 + 2x + 4}{x^7 - 2x^6 - 3x^5 + 6x^4 + 4x - 8}$$
(a) Use a CAS to factor the numerator and denominator of f, and use the results to determine the locations of all vertical asymptotes.
(b) Confirm that your answer is consistent with the graph of f.

45. (a) What inequality must $f(x)$ satisfy for the function f to have an absolute maximum on an interval I at x_0?
(b) What inequality must $f(x)$ satisfy for f to have an absolute minimum on I at x_0?
(c) What is the difference between an absolute extremum and a relative extremum?

46. According to the Extreme-Value Theorem, what conditions on a function f and an interval I guarantee that f will have both an absolute maximum and an absolute minimum on I?

47. In each part, determine whether the statement is true or false, and justify your answer.
(a) If f is differentiable on the open interval (a, b), and if f has an absolute extremum on that interval, then it must occur at a stationary point of f.
(b) If f is continuous on the open interval (a, b), and if f has an absolute extremum on that interval, then it must occur at a stationary point of f.

48–50 In each part, find the absolute minimum m and the absolute maximum M of f on the given interval (if they exist), and state where the absolute extrema occur.

48. (a) $f(x) = 1/x;\ [-2, -1]$
(b) $f(x) = x^3 - x^4;\ [-1, \frac{3}{2}]$
(c) $f(x) = x - \tan x;\ [-\pi/4, \pi/4]$

49. (a) $f(x) = x^2 - 3x - 1;\ (-\infty, +\infty)$
(b) $f(x) = x^3 - 3x - 2;\ (-\infty, +\infty)$
(c) $f(x) = -|x^2 - 2x|;\ (-\infty, +\infty)$

50. (a) $f(x) = 2x^5 - 5x^4 + 7;\ (-1, 3)$
(b) $f(x) = (3 - x)/(2 - x);\ (0, 2)$
(c) $f(x) = 2x/(x^2 + 3);\ (0, 2]$
(d) $f(x) = x^2(x - 2)^{1/3};\ (0, 3]$

51. In each part, use a graphing utility to estimate the absolute maximum and minimum values of f, if any, on the stated interval, and then use calculus methods to find the exact values.
(a) $f(x) = (x^2 - 1)^2;\ (-\infty, +\infty)$
(b) $f(x) = x/(x^2 + 1);\ [0, +\infty)$
(c) $f(x) = 2\sec x - \tan x;\ [0, \pi/4]$

52. Prove that $\tan x > x$ for all x in $(0, \pi/2)$.

C 53. Let
$$f(x) = \frac{x^3 + 2}{x^4 + 1}$$
(a) Generate the graph of $y = f(x)$, and use the graph to make rough estimates of the coordinates of the absolute extrema.
(b) Use a CAS to solve the equation $f'(x) = 0$ and then use it to make more accurate approximations of the coordinates in part (a).

54. A church window consists of a blue semicircular section surmounting a clear rectangular section as shown in the accompanying figure. The blue glass lets through half as much light per unit area as the clear glass. Find the radius r of the window that admits the most light if the perimeter of the entire window is to be P feet.

55. Find the dimensions of the rectangle of maximum area that can be inscribed inside the ellipse $(x/4)^2 + (y/3)^2 = 1$ (see the accompanying figure).

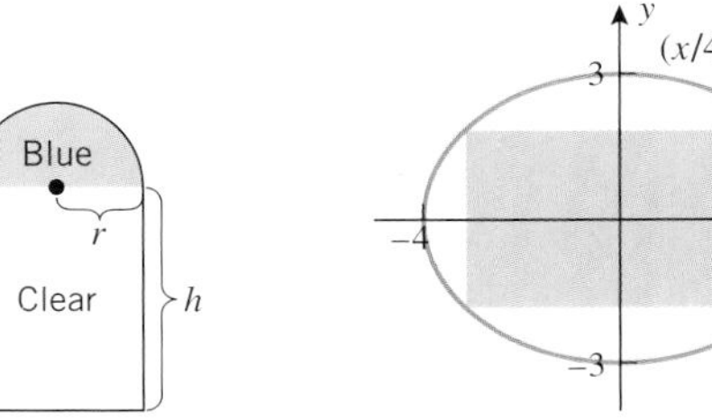

Figure Ex-54 Figure Ex-55

C 56. As shown in the accompanying figure (next page), suppose that a boat enters the river at the point $(1, 0)$ and maintains a heading toward the origin. As a result of the strong current, the boat follows the path

$$y = \frac{x^{10/3} - 1}{2x^{2/3}}$$

where x and y are in miles.

(a) Graph the path taken by the boat.

(b) Can the boat reach the origin? If not, discuss its fate and find how close it comes to the origin.

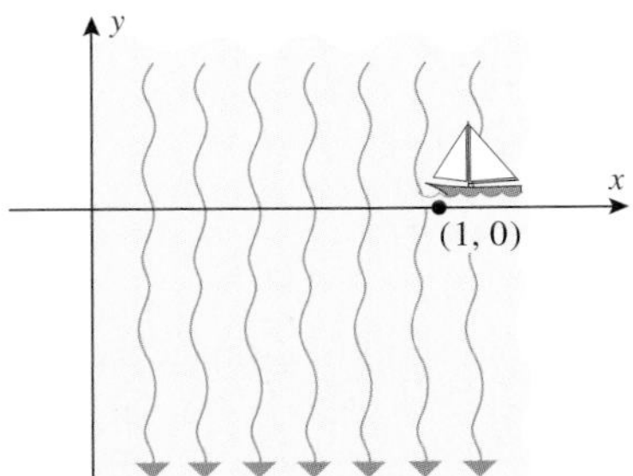

Figure Ex-56

57. A sheet of cardboard 12 in square is used to make an open box by cutting squares of equal size from the four corners and folding up the sides. What size squares should be cut to obtain a box with largest possible volume?

58. Draw an appropriate picture, and describe the basic idea of Newton's Method without using any formulas.

59. Use Newton's Method to approximate all three solutions of $x^3 - 4x + 1 = 0$.

60. Use Newton's Method to approximate the smallest positive solution of $\sin x + \cos x = 0$.

61. Use a graphing utility to determine the number of times the curve $y = x^3$ intersects the curve $y = (x/2) - 1$. Then apply Newton's Method to approximate the x-coordinates of all intersections.

62. According to ***Kepler's law***, the planets in our solar system move in elliptical orbits around the Sun. If a planet's closest approach to the Sun occurs at time $t = 0$, then the distance r from the center of the planet to the center of the Sun at some later time t can be determined from the equation

$$r = a(1 - e\cos\phi)$$

where a is the average distance between centers, e is a positive constant that measures the "flatness" of the elliptical orbit, and ϕ is the solution of *Kepler's equation*

$$\frac{2\pi t}{T} = \phi - e\sin\phi$$

in which T is the time it takes for one complete orbit of the planet. Estimate the distance from the Earth to the Sun when $t = 90$ days. [First find ϕ from Kepler's equation, and then use this value of ϕ to find the distance. Use $a = 150 \times 10^6$ km, $e = 0.0167$, and $T = 365$ days.]

63. Using the formulas in Exercise 62, find the distance from the planet Mars to the Sun when $t = 1$ year. For Mars use $a = 228 \times 10^6$ km, $e = 0.0934$, and $T = 1.88$ years.

64. Suppose that f is continuous on the closed interval $[a, b]$ and differentiable on the open interval (a, b), and suppose that $f(a) = f(b)$. Is it true or false that f must have at least one stationary point in (a, b)? Justify your answer.

65. In each part, determine whether all of the hypotheses of Rolle's Theorem are satisfied on the stated interval. If not, state which hypotheses fail; if so, find all values of c guaranteed in the conclusion of the theorem.

(a) $f(x) = \sqrt{4 - x^2}$ on $[-2, 2]$

(b) $f(x) = x^{2/3} - 1$ on $[-1, 1]$

(c) $f(x) = \sin(x^2)$ on $[0, \sqrt{\pi}\,]$

66. In each part, determine whether all of the hypotheses of the Mean-Value Theorem are satisfied on the stated interval. If not, state which hypotheses fail; if so, find all values of c guaranteed in the conclusion of the theorem.

(a) $f(x) = |x - 1|$ on $[-2, 2]$

(b) $f(x) = \dfrac{x + 1}{x - 1}$ on $[2, 3]$

(c) $f(x) = \begin{cases} 3 - x^2 & \text{if } x \le 1 \\ 2/x & \text{if } x > 1 \end{cases}$ on $[0, 2]$

67. Use the fact that

$$\frac{d}{dx}(x^6 - 2x^2 + x) = 6x^5 - 4x + 1$$

to show that the equation $6x^5 - 4x + 1 = 0$ has at least one solution in the interval $(0, 1)$.

68. Let $g(x) = x^3 - 4x + 6$. Find $f(x)$ so that $f'(x) = g'(x)$ and $f(1) = 2$.

69. (a) Can an object in rectilinear motion reverse direction if its acceleration is constant? Justify your answer using a velocity versus time curve.

(b) Can an object in rectilinear motion have increasing speed and decreasing acceleration? Justify your answer using a velocity versus time curve.

70. Suppose that the position function of a particle in rectilinear motion is given by the formula $s(t) = t/(2t^2 + 8)$ for $t \ge 0$.

(a) Use a graphing utility to generate the position, velocity, and acceleration versus time curves.

(b) Use the appropriate graph to make a rough estimate of the time when the particle reverses direction, and then find that time exactly.

(c) Find the position, velocity, and acceleration at the instant when the particle reverses direction.

(d) Use the appropriate graphs to make rough estimates of the time intervals on which the particle is speeding up and the time intervals on which it is slowing down, and then find those time intervals exactly.

(e) When does the particle have its maximum and minimum velocities?

C **71.** Suppose that the position function of a particle in rectilinear motion is given by the formula

$$s(t) = \frac{t^2 + 1}{t^4 + 1}, \quad t \ge 0$$

(a) Use a CAS to find simplified formulas for the velocity function $v(t)$ and the acceleration function $a(t)$.
(b) Graph the position, velocity, and acceleration versus time curves.
(c) Use the appropriate graph to make a rough estimate of the time at which the particle is farthest from the origin and its distance from the origin at that time.
(d) Use the appropriate graph to make a rough estimate of the time interval during which the particle is moving in the positive direction.
(e) Use the appropriate graphs to make rough estimates of the time intervals during which the particle is speeding up and the time intervals during which it is slowing down.
(f) Use the appropriate graph to make a rough estimate of the maximum speed of the particle and the time at which the maximum speed occurs.

72. Is it true or false that a particle in rectilinear motion is speeding up when its velocity is increasing and slowing down when its velocity is decreasing? Justify your answer.

Blammo the Human Cannonball will be fired from a cannon to land in a small net at the opposite end of the circus area. As Blammo's manager, your job is to do the mathematical calculations that will allow Blammo to perform his death-defying act safely. The methods that you will use are from the field of ballistics. To learn more about ballistics, and to build upon the mathematics learned in this chapter, go to

www.wiley.com/college/anton

chapter five

INTEGRATION

Give me a place to stand and I will move the earth.

—Archimedes
Ancient Greek Scientist and Mathematician

In this chapter we will begin with an overview of the problem of finding areas—we will discuss what the term "area" means, and we will outline two approaches to defining and calculating areas. Following this overview, we will discuss the Fundamental Theorem of Calculus, which is the theorem that relates the problems of finding tangent lines and areas, and we will discuss techniques for calculating areas. Finally, we will use the ideas in this chapter to continue our study of rectilinear motion and to examine some consequences of the chain rule in integral calculus.

Photo: *If the velocity of a dragster is known over a certain time interval, it is possible to find the distance it travels during that time interval using techniques studied in this chapter.*

5.1 AN OVERVIEW OF THE AREA PROBLEM

In this introductory section we will consider the problem of calculating areas of plane regions with curvilinear boundaries. All of the results in this section will be reexamined in more detail later in this chapter, so our purpose here is simply to introduce the fundamental concepts.

THE AREA PROBLEM

Formulas for the areas of polygons, such as squares, rectangles, triangles, and trapezoids, were well known in many early civilizations. However, the problem of finding formulas for regions with curved boundaries (a circle being the simplest example) caused difficulties for early mathematicians.

The first real progress in dealing with the general area problem was made by the Greek mathematician Archimedes, who obtained areas of regions bounded by circular arcs, parabolas, spirals, and various other curves using an ingenious procedure that was later called the *method of exhaustion*. The method, when applied to a circle, consists of inscribing a succession of regular polygons in the circle and allowing the number of sides to increase indefinitely (Figure 5.1.1). As the number of sides increases, the polygons tend to "exhaust" the region inside the circle, and the areas of the polygons become better and better approximations of the exact area of the circle.

To see how this works numerically, let $A(n)$ denote the area of a regular n-sided polygon inscribed in a circle of radius 1. Table 5.1.1 shows the values of $A(n)$ for various choices of n. Note that for large values of n the area $A(n)$ appears to be close to π (square units), as one would expect. This suggests that for a circle of radius 1, the method of exhaustion

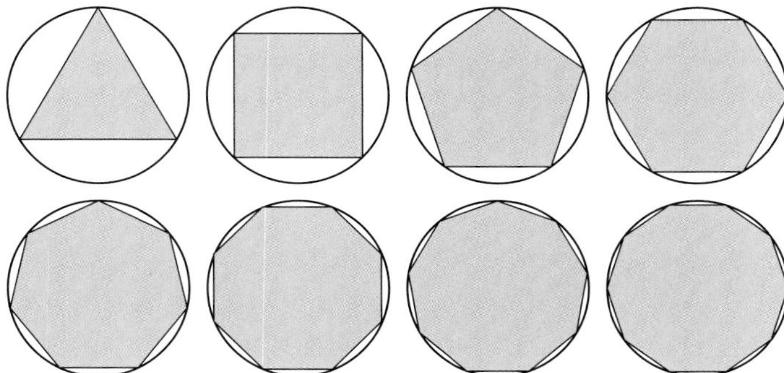

Figure 5.1.1

Archimedes (287 B.C.–212 B.C.) Greek mathematician and scientist. Born in Syracuse, Sicily, Archimedes was the son of the astronomer Pheidias and possibly related to Heiron II, king of Syracuse. Most of the facts about his life come from the Roman biographer, Plutarch, who inserted a few tantalizing pages about him in the massive biography of the Roman soldier, Marcellus. In the words of one writer, "the account of Archimedes is slipped like a tissue-thin shaving of ham in a bull-choking sandwich."

Archimedes ranks with Newton and Gauss as one of the three greatest mathematicians who ever lived, and he is certainly the greatest mathematician of antiquity. His mathematical work is so modern in spirit and technique that it is barely distinguishable from that of a seventeenth-century mathematician, yet it was all done without benefit of algebra or a convenient number system. Among his mathematical achievements, Archimedes developed a general method (exhaustion) for finding areas and volumes, and he used the method to find areas bounded by parabolas and spirals and to find volumes of cylinders, paraboloids, and segments of spheres. He gave a procedure for approximating π and bounded its value between $3\frac{10}{71}$ and $3\frac{1}{7}$. In spite of the limitations of the Greek numbering system, he devised methods for finding square roots and invented a method based on the Greek myriad (10,000) for representing numbers as large as 1 followed by 80 million billion zeros.

Of all his mathematical work, Archimedes was most proud of his discovery of a method for finding the volume of a sphere—he showed that the volume of a sphere is two-thirds the volume of the smallest cylinder that can contain it. At his request, the figure of a sphere and cylinder was engraved on his tombstone.

In addition to mathematics, Archimedes worked extensively in mechanics and hydrostatics. Nearly every schoolchild knows Archimedes as the absent-minded scientist who, on realizing that a floating object displaces its weight of liquid, leaped from his bath and ran naked through the streets of Syracuse shouting, "Eureka, Eureka!"—(meaning, "I have found it!"). Archimedes actually created the discipline of hydrostatics and used it to find equilibrium positions for various floating bodies. He laid down the fundamental postulates of mechanics, discovered the laws of levers, and calculated centers of gravity for various flat surfaces and solids. In the excitement of discovering the mathematical laws of the lever, he is said to have declared, "Give me a place to stand and I will move the earth."

Although Archimedes was apparently more interested in pure mathematics than its applications, he was an engineering genius. During the second Punic war, when Syracuse was attacked by the Roman fleet under the command of Marcellus, it was reported by Plutarch that Archimedes' military inventions held the fleet at bay for three years. He invented super catapults that showered the Romans with rocks weighing a quarter ton or more, and fearsome mechanical devices with iron "beaks and claws" that reached over the city walls, grasped the ships, and spun them against the rocks. After the first repulse, Marcellus called Archimedes a "geometrical Briareus (a hundred-armed mythological monster) who uses our ships like cups to ladle water from the sea."

Eventually the Roman army was victorious and contrary to Marcellus' specific orders the 75-year-old Archimedes was killed by a Roman soldier. According to one report of the incident, the soldier cast a shadow across the sand in which Archimedes was working on a mathematical problem. When the annoyed Archimedes yelled, "Don't disturb my circles," the soldier flew into a rage and cut the old man down.

Although there is no known likeness or statue of this great man, nine works of Archimedes have survived to the present day. Especially important is his treatise, *The Method of Mechanical Theorems*, which was part of a palimpsest found in Constantinople in 1906. In this treatise Archimedes explains how he made some of his discoveries, using reasoning that anticipated ideas of the integral calculus. Thought to be lost, the Archimedes palimpsest later resurfaced in 1998, when it was purchased by an anonymous private collector for two million dollars.

Table 5.1.1

n	$A(n)$
100	3.13952597647
200	3.14107590781
300	3.14136298250
400	3.14146346236
500	3.14150997084
1000	3.14157198278
2000	3.14158748588
3000	3.14159035683
4000	3.14159136166
5000	3.14159182676
10000	3.14159244688

is equivalent to an equation of the form

$$\lim_{n\to\infty} A(n) = \pi$$

Since Greek mathematicians were suspicious of the concept of "infinity," they avoided its use in mathematical arguments. As a result, computation of area using the method of exhaustion was a very cumbersome procedure. It remained for Newton and Leibniz to obtain a general method for finding areas that explicitly used the notion of a limit. We will discuss their method in the context of the following problem.

5.1.1 THE AREA PROBLEM. Given a function f that is continuous and nonnegative on an interval $[a, b]$, find the area between the graph of f and the interval $[a, b]$ on the x-axis (Figure 5.1.2).

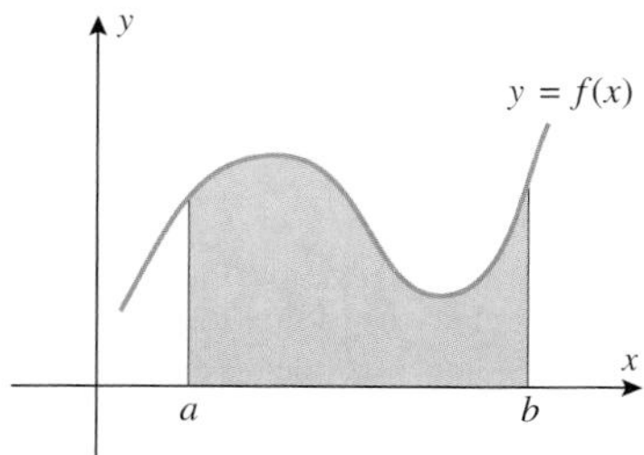

Figure 5.1.2

Logically speaking, we cannot really talk about computing areas without a precise mathematical definition of the term "area." Later in this chapter we will give such a definition, but for now we will treat the concept intuitively.

THE RECTANGLE METHOD FOR FINDING AREAS

One approach to the area problem is to use Archimedes' method of exhaustion in the following way:

- Divide the interval $[a, b]$ into n equal subintervals, and over each subinterval construct a rectangle that extends from the x-axis to any point on the curve $y = f(x)$ that is above the subinterval; the particular point does not matter—it can be above the center, above an endpoint, or above any other point in the subinterval. In Figure 5.1.3 it is above the center.
- For each n, the total area of the rectangles can be viewed as an *approximation* to the exact area under the curve over the interval $[a, b]$. Moreover, it is evident intuitively that as n increases these approximations will get better and better and will approach the exact area as a limit (Figure 5.1.4). That is, if A denotes the exact area under the curve and A_n denotes the approximation to A using n rectangles, then

$$A = \lim_{n\to+\infty} A_n$$

We will call this the ***rectangle method*** for computing A.

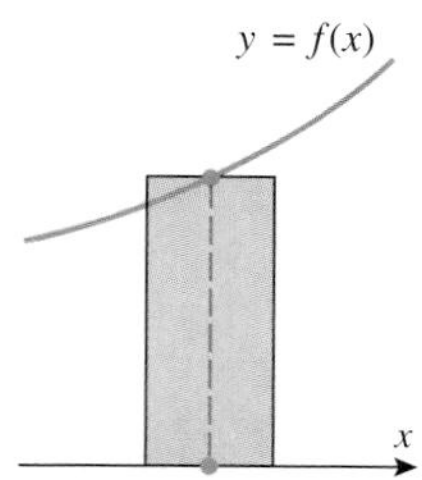

Figure 5.1.3

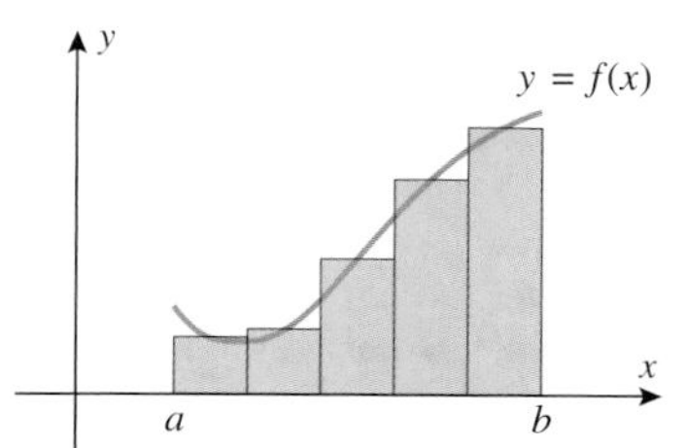

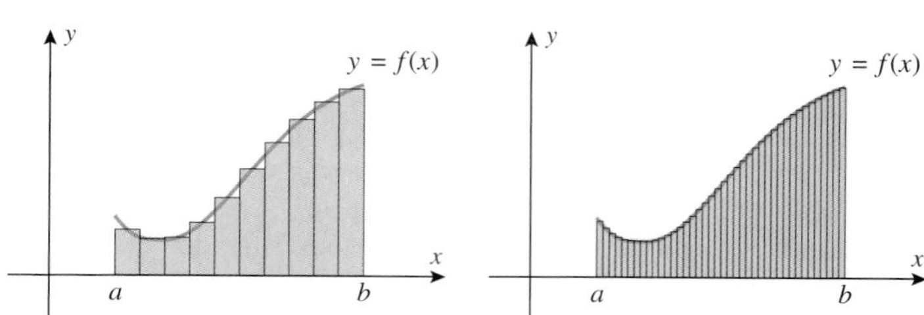

Figure 5.1.4

To illustrate this idea, we will use the rectangle method to approximate the area under the curve $y = x^2$ over the interval $[0, 1]$ (Figure 5.1.5). We will begin by dividing the interval $[0, 1]$ into n equal subintervals, from which it follows that each subinterval has length $1/n$; the endpoints of the subintervals occur at

$$0, \frac{1}{n}, \frac{2}{n}, \frac{3}{n}, \ldots, \frac{n-1}{n}, 1$$

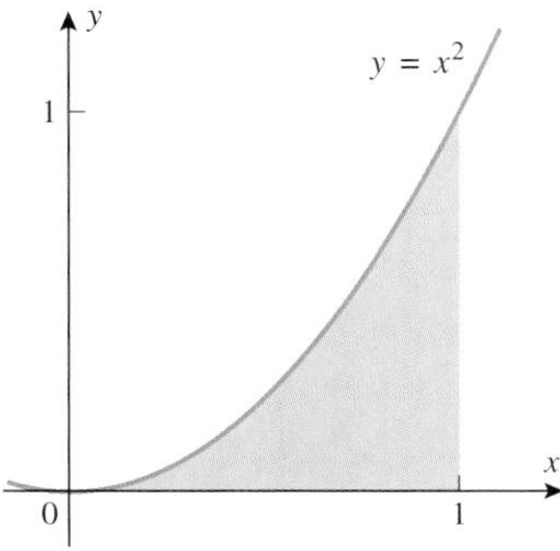

Figure 5.1.5

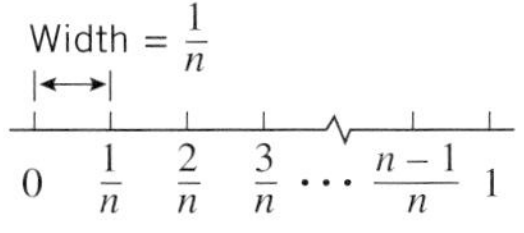

Figure 5.1.6

(Figure 5.1.6). We want to construct a rectangle over each of these subintervals whose height is the value of the function $f(x) = x^2$ at some point in the subinterval. To be specific, let us use the right endpoints, in which case the heights of our rectangles will be

$$\left(\frac{1}{n}\right)^2, \left(\frac{2}{n}\right)^2, \left(\frac{3}{n}\right)^2, \ldots, 1^2$$

and since each rectangle has a base of width $1/n$, the total area A_n of the n rectangles will be

$$A_n = \left[\left(\frac{1}{n}\right)^2 + \left(\frac{2}{n}\right)^2 + \left(\frac{3}{n}\right)^2 + \cdots + 1^2\right]\left(\frac{1}{n}\right) \tag{1}$$

For example, if $n = 4$, then the total area of the four approximating rectangles would be

$$A_4 = \left[\left(\tfrac{1}{4}\right)^2 + \left(\tfrac{2}{4}\right)^2 + \left(\tfrac{3}{4}\right)^2 + 1^2\right]\left(\tfrac{1}{4}\right) = \tfrac{15}{32} = 0.46875$$

Table 5.1.2 shows the result of evaluating (1) on a computer for some increasingly large values of n. These computations suggest that the exact area is close to $\frac{1}{3}$. Later in this chapter we will prove that this area is exactly $\frac{1}{3}$ by showing that

$$\lim_{n\to\infty} A_n = \tfrac{1}{3}$$

TECHNOLOGY MASTERY

Use a calculating utility to compute the value of A_{10} in Table 5.1.2. Some calculating utilities have special commands for computing sums such as that in (1) for any specified value of n. If your utility has this feature, use it to compute A_{100} as well.

Table 5.1.2

n	4	10	100	1000	10,000	100,000
A_n	0.468750	0.385000	0.338350	0.333834	0.333383	0.333338

THE ANTIDERIVATIVE METHOD FOR FINDING AREAS

Although the rectangle method is appealing intuitively, the limits that result can only be evaluated in certain cases. For this reason, progress on the area problem remained at a rudimentary level until the latter part of the seventeenth century when Isaac Newton and Gottfried Leibniz independently discovered a fundamental relationship between areas and derivatives. Briefly stated, they showed that if f is a nonnegative continuous function on the interval $[a, b]$, and if $A(x)$ denotes the area under the graph of f over the interval $[a, x]$, where x is any point in the interval $[a, b]$ (Figure 5.1.7), then

$$A'(x) = f(x) \tag{2}$$

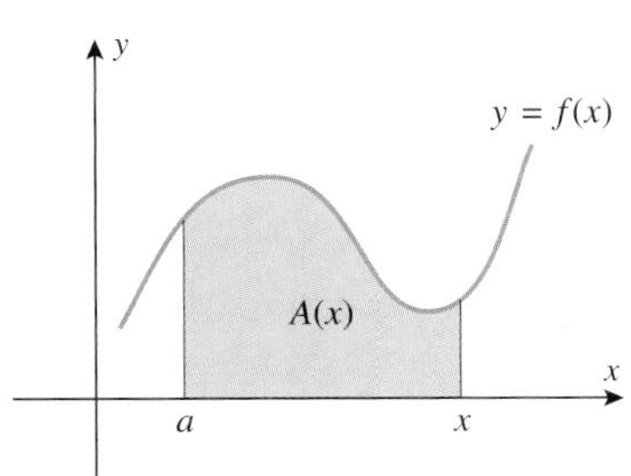

Figure 5.1.7

The following example confirms Formula (2) in some cases where a formula for $A(x)$ can be found using elementary geometry.

▶ **Example 1** For each of the functions f, find the area $A(x)$ between the graph of f and the interval $[a, x] = [-1, x]$, and find the derivative $A'(x)$ of this area function.

(a) $f(x) = 2$ (b) $f(x) = x + 1$ (c) $f(x) = 2x + 3$

Solution (a). From Figure 5.1.8a we see that

$$A(x) = 2(x - (-1)) = 2(x + 1) = 2x + 2$$

is the area of a rectangle of height 2 and base $x + 1$. For this area function,

$$A'(x) = 2 = f(x)$$

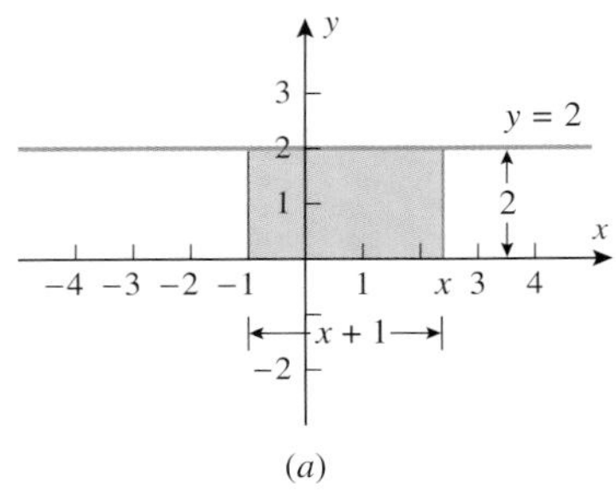

(a)

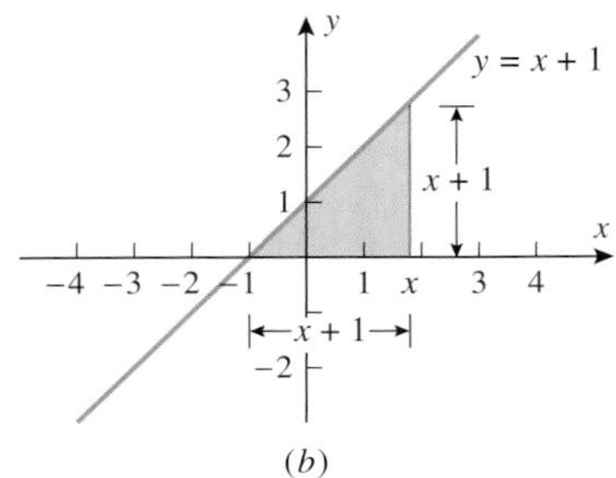

(b)

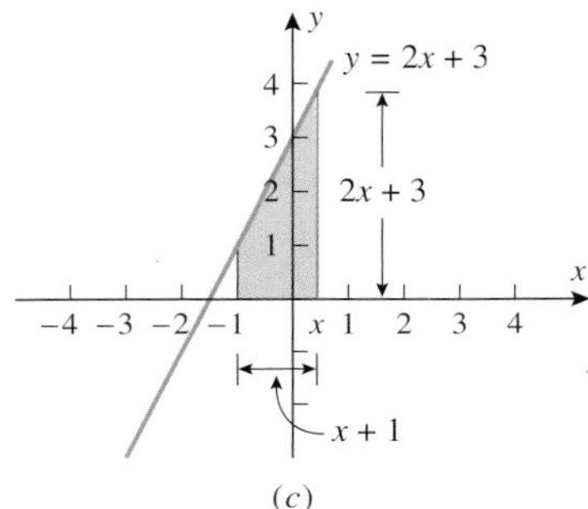

(c)

Figure 5.1.8

Solution (b). From Figure 5.1.8*b* we see that

$$A(x) = \frac{1}{2}(x+1)(x+1) = \frac{x^2}{2} + x + \frac{1}{2}$$

is the area of an isosceles right triangle with base and height equal to $x + 1$. For this area function,

$$A'(x) = x + 1 = f(x)$$

Solution (c). Recall that the formula for the area of a trapezoid is $A = \frac{1}{2}(b + b')h$, where b and b' denote the lengths of the parallel sides of the trapezoid, and the altitude h denotes the distance between the parallel sides. From Figure 5.1.8*c* we see that

$$A(x) = \tfrac{1}{2}((2x+3)+1)(x-(-1)) = x^2 + 3x + 2$$

is the area of a trapezoid with parallel sides of lengths 1 and $2x + 3$ and with altitude $x - (-1) = x + 1$. For this area function,

$$A'(x) = 2x + 3 = f(x) \blacktriangleleft$$

Formula (2) is important because it relates the area function A and the region-bounding function f. Although a formula for $A(x)$ may be difficult to obtain directly, its derivative, $f(x)$, is given. If a formula for $A(x)$ can be recovered from the given formula for $A'(x)$, then the area under the graph of f over the interval $[a, b]$ can be obtained by computing $A(b)$.

The process of finding a function from its derivative is called ***antidifferentiation***, and a procedure for finding areas via antidifferentiation is called the ***antiderivative method***. To illustrate this method, let us revisit the problem of finding the area in Figure 5.1.5.

▶ **Example 2** Use the antiderivative method to find the area under the graph of $y = x^2$ over the interval $[0, 1]$.

Solution. Let x be any point in the interval $[0, 1]$, and let $A(x)$ denote the area under the graph of $f(x) = x^2$ over the interval $[0, x]$. It follows from (2) that

$$A'(x) = x^2 \tag{3}$$

To find $A(x)$ we must look for a function whose derivative is x^2. By guessing, we see that one such function is $\frac{1}{3}x^3$ so by Theorem 4.7.3

$$A(x) = \tfrac{1}{3}x^3 + C \tag{4}$$

for some real constant C. We can determine the specific value for C by considering the case where $x = 0$. In this case (4) implies that

$$A(0) = C \tag{5}$$

But if $x = 0$, then the interval $[0, x]$ reduces to a single point. If we agree that the area above a single point should be taken as zero, then $A(0) = 0$ and (5) implies that $C = 0$. Thus, it follows from (4) that

$$A(x) = \tfrac{1}{3}x^3$$

is the area function we are seeking. This implies that the area under the graph of $y = x^2$ over the interval $[0, 1]$ is

$$A(1) = \tfrac{1}{3}(1^3) = \tfrac{1}{3}$$

This is consistent with the result that we previously obtained numerically. ◀

As Example 2 illustrates, antidifferentiation is essentially a guessing process in which one tries to "undo" a differentiation. One of the objectives in this chapter is to develop efficient antidifferentiation procedures.

THE RECTANGLE METHOD AND THE ANTIDERIVATIVE METHOD COMPARED

The rectangle method and the antiderivative method provide two very different approaches to the area problem, each of which is important. The antiderivative method is usually the more efficient way to *compute* areas, but it is the rectangle method that is used to formally *define* the notion of area, thereby allowing us to prove mathematical results about areas. The underlying idea of the rectangle approach is also important because it can be adapted readily to such diverse problems as finding the volume of a solid, the length of a curve, the mass of an object, and the work done in pumping water out of a tank, to name a few.

✔QUICK CHECK EXERCISES 5.1

(See page 308 for answers.)

1. Let R denote the region below the graph of $f(x) = \sqrt{1 - x^2}$ and above the interval $[-1, 1]$.
   (a) Use a geometric argument to find the area of R.
   (b) What estimate results if the area of R is approximated by the total area within the rectangles of the accompanying figure?

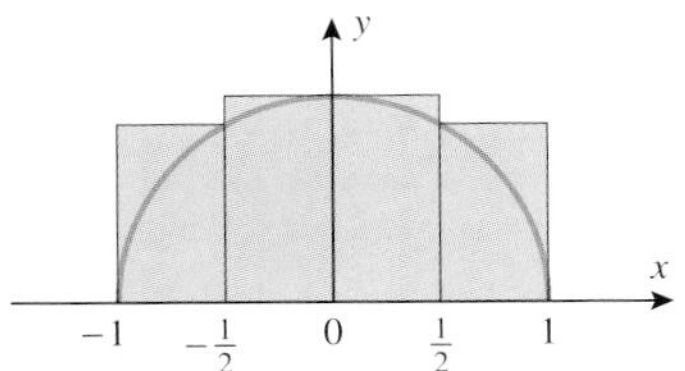

Figure Ex-1

2. Suppose that the rectangle method with n rectangles is applied to the graph of a function f plotted over an interval $[a, b]$. If, for every positive integer n, the approximation $A(n) = 2 + (2/n)$ results, the area between the graph of f and the interval $[a, b]$ is ________.
3. The area under the graph of $y = x^2$ over the interval $[0, 3]$ is ________.
4. Find a formula for the area $A(x)$ between the graph of the function $f(x) = x$ and the interval $[0, x]$, and verify that $A'(x) = f(x)$.
5. The area under the graph of $y = f(x)$ over the interval $[0, x]$ is $A(x) = x + \sin x$. It follows that $f(x) =$ ________.

EXERCISE SET 5.1

1–8 Estimate the area between the graph of the function f and the interval $[a, b]$. Use an approximation scheme with n rectangles similar to our treatment of $f(x) = x^2$ in this section. If your calculating utility will perform automatic summations, estimate the specified area using $n = 10, 50$, and 100 rectangles. Otherwise, estimate this area using $n = 2, 5$, and 10 rectangles.

1. $f(x) = \sqrt{x};\ [a, b] = [0, 1]$
2. $f(x) = \dfrac{1}{x+1};\ [a, b] = [0, 1]$
3. $f(x) = \sin x;\ [a, b] = [0, \pi]$
4. $f(x) = \cos x;\ [a, b] = [0, \pi/2]$
5. $f(x) = \dfrac{1}{x};\ [a, b] = [1, 2]$
6. $f(x) = \cos x;\ [a, b] = [-\pi/2, \pi/2]$
7. $f(x) = \sqrt{1 - x^2};\ [a, b] = [0, 1]$
8. $f(x) = \sqrt{1 - x^2};\ [a, b] = [-1, 1]$

9–14 Use simple area formulas from geometry to find the area function $A(x)$ that gives the area between the graph of the specified function f and the interval $[a, x]$. Confirm that $A'(x) = f(x)$ in every case.

9. $f(x) = 3;\ [a, x] = [1, x]$
10. $f(x) = 5;\ [a, x] = [2, x]$
11. $f(x) = 2x + 2;\ [a, x] = [0, x]$
12. $f(x) = 3x - 3;\ [a, x] = [1, x]$
13. $f(x) = 2x + 2;\ [a, x] = [1, x]$
14. $f(x) = 3x - 3;\ [a, x] = [2, x]$
15. How do the area functions in Exercises 11 and 13 compare? Explain.
16. Let $f(x)$ denote a *linear function* that is nonnegative on the interval $[a, b]$. For each value of x in $[a, b]$, define $A(x)$ to be the area between the graph of f and the interval $[a, x]$.
    (a) Prove that $A(x) = \frac{1}{2}[f(a) + f(x)](x - a)$.
    (b) Use part (a) to verify that $A'(x) = f(x)$.

FOCUS ON CONCEPTS

17. Explain how to use the formula for $A(x)$ found in the solution to Example 2 to determine the area between the graph of $y = x^2$ and the interval $[3, 6]$.
18. Modify the solution of Example 2 to find the area under the graph of $y = x^2$ and over the interval $[-3, 9]$.

19–20 The area $A(x)$ under the graph of f and over the interval $[a, x]$ is given. Find the function f and the value of a.

19. $A(x) = x^2 - 4$ **20.** $A(x) = x^2 - x$

21. Let A denote the area between the graph of $f(x) = \sqrt{x}$ and the interval $[0, 1]$, and let B denote the area between the graph of $f(x) = x^2$ and the interval $[0, 1]$. Explain geometrically why $A + B = 1$.

22. Let A denote the area between the graph of $f(x) = 1/x$ and the interval $[1, 2]$, and let B denote the area between the graph of f and the interval $\left[\frac{1}{2}, 1\right]$. Explain geometrically why $A = B$.

✔ QUICK CHECK ANSWERS 5.1

1. (a) $\frac{\pi}{2}$ (b) $1 + \frac{\sqrt{3}}{2}$ 2. 2 3. 9 4. $A(x) = \frac{x^2}{2}$; $A'(x) = \frac{2x}{2} = x = f(x)$ 5. $1 + \cos x$

5.2 THE INDEFINITE INTEGRAL

In the last section we saw how antidifferentiation could be used to find exact areas. In this section we will develop some fundamental results about antidifferentiation.

ANTIDERIVATIVES

5.2.1 DEFINITION. A function F is called an ***antiderivative*** of a function f on a given interval I if $F'(x) = f(x)$ for all x in the interval.

For example, the function $F(x) = \frac{1}{3}x^3$ is an antiderivative of $f(x) = x^2$ on the interval $(-\infty, +\infty)$ because for each x in this interval

$$F'(x) = \frac{d}{dx}\left[\tfrac{1}{3}x^3\right] = x^2 = f(x)$$

However, $F(x) = \frac{1}{3}x^3$ is not the only antiderivative of f on this interval. If we add any constant C to $\frac{1}{3}x^3$, then the function $G(x) = \frac{1}{3}x^3 + C$ is also an antiderivative of f on $(-\infty, +\infty)$, since

$$G'(x) = \frac{d}{dx}\left[\tfrac{1}{3}x^3 + C\right] = x^2 + 0 = f(x)$$

In general, once any single antiderivative is known, other antiderivatives can be obtained by adding constants to the known antiderivative. Thus,

$$\tfrac{1}{3}x^3, \quad \tfrac{1}{3}x^3 + 2, \quad \tfrac{1}{3}x^3 - 5, \quad \tfrac{1}{3}x^3 + \sqrt{2}$$

are all antiderivatives of $f(x) = x^2$.

It is reasonable to ask if there are antiderivatives of a function f that cannot be obtained by adding some constant to a known antiderivative F. The answer is *no*—once a single antiderivative of f on an interval I is known, all other antiderivatives on that interval are obtainable by adding constants to the known antiderivative. This is so because Theorem 4.7.3 tells us that if two functions are differentiable on an open interval I such that their derivatives are equal on I, then the functions differ by a constant on I. The following theorem summarizes these observations.

> **5.2.2 THEOREM.** *If $F(x)$ is any antiderivative of $f(x)$ on an interval I, then for any constant C the function $F(x)+C$ is also an antiderivative on that interval. Moreover, each antiderivative of $f(x)$ on the interval I can be expressed in the form $F(x)+C$ by choosing the constant C appropriately.*

THE INDEFINITE INTEGRAL

Extract from the manuscript of Leibniz dated October 29, 1675 in which the integral sign first appeared.

The process of finding antiderivatives is called ***antidifferentiation*** or ***integration***. Thus, if

$$\frac{d}{dx}[F(x)] = f(x) \tag{1}$$

then ***integrating*** (or ***antidifferentiating***) the function $f(x)$ produces an antiderivative of the form $F(x)+C$. To emphasize this process, Equation (1) is recast using ***integral notation***,

$$\int f(x)\,dx = F(x) + C \tag{2}$$

where C is understood to represent an arbitrary constant. It is important to note that (1) and (2) are just different notations to express the same fact. For example,

$$\int x^2\,dx = \tfrac{1}{3}x^3 + C \quad \text{is equivalent to} \quad \frac{d}{dx}\left[\tfrac{1}{3}x^3\right] = x^2$$

Note that if we differentiate an antiderivative of $f(x)$, we obtain $f(x)$ back again. Thus,

$$\frac{d}{dx}\left[\int f(x)\,dx\right] = f(x) \tag{3}$$

The expression $\int f(x)\,dx$ is called an ***indefinite integral***. The adjective "indefinite" emphasizes that the result of antidifferentiation is a "generic" function, described only up to a constant term. The "elongated s" that appears on the left side of (2) is called an ***integral sign***,* the function $f(x)$ is called the ***integrand***, and the constant C is called the ***constant of integration***. Equation (2) should be read as:

The integral sign and the differential serve as delimiters enclosing the integrand. In particular, we do **not** write $\int dx\, f(x)$ when we intend $\int f(x)\,dx$.

The integral of $f(x)$ with respect to x is equal to $F(x)$ plus a constant.

The differential symbol, dx, in the differentiation and antidifferentiation operations

$$\frac{d}{dx}[\ ] \quad \text{and} \quad \int [\ ]\,dx$$

serves to identify the independent variable. If an independent variable other than x is used, say t, then the notation must be adjusted appropriately. Thus,

$$\frac{d}{dt}[F(t)] = f(t) \quad \text{and} \quad \int f(t)\,dt = F(t) + C$$

are equivalent statements. Here are some examples of derivative formulas and their equivalent integration formulas:

---

*This notation was devised by Leibniz. In his early papers Leibniz used the notation "omn." (an abbreviation for the Latin word "omnes") to denote integration. Then on October 29, 1675 he wrote, "It will be useful to write $\int$ for omn., thus $\int l$ for omn. l . . . ." Two or three weeks later he refined the notation further and wrote $\int [\ ]\,dx$ rather than $\int$ alone. This notation is so useful and so powerful that its development by Leibniz must be regarded as a major milestone in the history of mathematics and science.

DERIVATIVE FORMULA	EQUIVALENT INTEGRATION FORMULA
$\frac{d}{dx}[x^3] = 3x^2$	$\int 3x^2\,dx = x^3 + C$
$\frac{d}{dx}[\sqrt{x}] = \frac{1}{2\sqrt{x}}$	$\int \frac{1}{2\sqrt{x}}\,dx = \sqrt{x} + C$
$\frac{d}{dt}[\tan t] = \sec^2 t$	$\int \sec^2 t\,dt = \tan t + C$
$\frac{d}{du}[u^{3/2}] = \frac{3}{2}u^{1/2}$	$\int \frac{3}{2}u^{1/2}\,du = u^{3/2} + C$

For simplicity, the dx is sometimes absorbed into the integrand. For example,

$$\int 1\,dx \quad \text{can be written as} \quad \int dx$$

$$\int \frac{1}{x^2}\,dx \quad \text{can be written as} \quad \int \frac{dx}{x^2}$$

INTEGRATION FORMULAS

Integration is essentially educated guesswork—given the derivative f of a function F, one tries to guess what the function F is. However, many basic integration formulas can be obtained directly from their companion differentiation formulas. Some of the most important are given in Table 5.2.1.

Table 5.2.1

DIFFERENTIATION FORMULA	INTEGRATION FORMULA
1. $\frac{d}{dx}[x] = 1$	$\int dx = x + C$
2. $\frac{d}{dx}\left[\frac{x^{r+1}}{r+1}\right] = x^r \quad (r \neq -1)$	$\int x^r\,dx = \frac{x^{r+1}}{r+1} + C \quad (r \neq -1)$
3. $\frac{d}{dx}[\sin x] = \cos x$	$\int \cos x\,dx = \sin x + C$
4. $\frac{d}{dx}[-\cos x] = \sin x$	$\int \sin x\,dx = -\cos x + C$
5. $\frac{d}{dx}[\tan x] = \sec^2 x$	$\int \sec^2 x\,dx = \tan x + C$
6. $\frac{d}{dx}[-\cot x] = \csc^2 x$	$\int \csc^2 x\,dx = -\cot x + C$
7. $\frac{d}{dx}[\sec x] = \sec x \tan x$	$\int \sec x \tan x\,dx = \sec x + C$
8. $\frac{d}{dx}[-\csc x] = \csc x \cot x$	$\int \csc x \cot x\,dx = -\csc x + C$

► **Example 1** The second integration formula in Table 5.2.1 will be easier to remember if you express it in words:

To integrate a power of x (other than -1), add 1 to the exponent and divide by the new exponent.

Here are some examples:

$$\int x^2\,dx = \frac{x^3}{3} + C \qquad \boxed{r = 2}$$

$$\int x^3\,dx = \frac{x^4}{4} + C \qquad \boxed{r = 3}$$

$$\int \frac{1}{x^5}\,dx = \int x^{-5}\,dx = \frac{x^{-5+1}}{-5+1} + C = -\frac{1}{4x^4} + C \qquad \boxed{r = -5}$$

$$\int \sqrt{x}\,dx = \int x^{\frac{1}{2}}\,dx = \frac{x^{\frac{1}{2}+1}}{\frac{1}{2}+1} + C = \tfrac{2}{3}x^{\frac{3}{2}} + C = \tfrac{2}{3}(\sqrt{x})^3 + C \qquad \boxed{r = \tfrac{1}{2}}$$ ◀

PROPERTIES OF THE INDEFINITE INTEGRAL

Our first properties of antiderivatives follow directly from the simple constant factor, sum, and difference rules for derivatives.

5.2.3 THEOREM. *Suppose that $F(x)$ and $G(x)$ are antiderivatives of $f(x)$ and $g(x)$, respectively, and that c is a constant. Then:*

(*a*) *A constant factor can be moved through an integral sign; that is,*

$$\int cf(x)\,dx = cF(x) + C$$

(*b*) *An antiderivative of a sum is the sum of the antiderivatives; that is,*

$$\int [f(x) + g(x)]\,dx = F(x) + G(x) + C$$

(*c*) *An antiderivative of a difference is the difference of the antiderivatives; that is,*

$$\int [f(x) - g(x)]\,dx = F(x) - G(x) + C$$

PROOF. In general, to establish the validity of an equation of the form

$$\int h(x)\,dx = H(x) + C$$

one must show that

$$\frac{d}{dx}[H(x)] = h(x)$$

We are given that $F(x)$ and $G(x)$ are antiderivatives of $f(x)$ and $g(x)$, respectively, so we know that

$$\frac{d}{dx}[F(x)] = f(x) \quad \text{and} \quad \frac{d}{dx}[G(x)] = g(x)$$

Thus,

$$\frac{d}{dx}[cF(x)] = c\frac{d}{dx}[F(x)] = cf(x)$$

$$\frac{d}{dx}[F(x) + G(x)] = \frac{d}{dx}[F(x)] + \frac{d}{dx}[G(x)] = f(x) + g(x)$$

$$\frac{d}{dx}[F(x) - G(x)] = \frac{d}{dx}[F(x)] - \frac{d}{dx}[G(x)] = f(x) - g(x)$$

which proves the three statements of the theorem. ■

The statements in Theorem 5.2.3 can be summarized by the following formulas:

$$\int cf(x)\,dx = c\int f(x)\,dx \tag{4}$$

$$\int [f(x)+g(x)]\,dx = \int f(x)\,dx + \int g(x)\,dx \tag{5}$$

$$\int [f(x)-g(x)]\,dx = \int f(x)\,dx - \int g(x)\,dx \tag{6}$$

However, these equations must be applied carefully to avoid errors and unnecessary complexities arising from the constants of integration. For example, if you were to use (4) to integrate $0x$ by incorrectly equating $\int 0x\,dx$ with $0\int x\,dx = 0$, then you will have erroneously lost a constant of integration (where?). If you use (4) to integrate $2x$ by writing

$$\int 2x\,dx = 2\int x\,dx = 2\left(\frac{x^2}{2}+C\right) = x^2+2C$$

then you will have an unnecessarily complicated form of the arbitrary constant. Similarly, if you use (5) to integrate $1+x$ by writing

$$\int (1+x)\,dx = \int 1\,dx + \int x\,dx = (x+C_1)+\left(\frac{x^2}{2}+C_2\right) = x+\frac{x^2}{2}+C_1+C_2$$

then you will have two arbitrary constants when one will suffice. These three kinds of problems are caused by introducing constants of integration too soon and can be avoided by inserting the constant of integration in the final result, rather than in intermediate computations.

► **Example 2** Evaluate

$$\text{(a)}\ \int 4\cos x\,dx \qquad \text{(b)}\ \int (x+x^2)\,dx$$

Solution (a). Since $F(x)=\sin x$ is an antiderivative for $f(x)=\cos x$ (Table 5.2.1), we obtain

$$\int 4\cos x\,dx \underset{(4)}{=} 4\int \cos x\,dx = 4\sin x + C$$

Solution (b). From Table 5.2.1 we obtain

$$\int (x+x^2)\,dx \underset{(5)}{=} \int x\,dx + \int x^2\,dx = \frac{x^2}{2}+\frac{x^3}{3}+C$$ ◄

Parts (b) and (c) of Theorem 5.2.3 can be extended to more than two functions, which in combination with part (a) results in the following general formula:

$$\int [c_1 f_1(x) + c_2 f_2(x) + \cdots + c_n f_n(x)]\,dx$$
$$= c_1\int f_1(x)\,dx + c_2\int f_2(x)\,dx + \cdots + c_n\int f_n(x)\,dx \tag{7}$$

▶ Example 3

$$\int (3x^6 - 2x^2 + 7x + 1)\,dx = 3\int x^6\,dx - 2\int x^2\,dx + 7\int x\,dx + \int 1\,dx$$

$$= \frac{3x^7}{7} - \frac{2x^3}{3} + \frac{7x^2}{2} + x + C \blacktriangleleft$$

Sometimes it is useful to rewrite an integrand in a different form before performing the integration.

▶ Example 4 Evaluate

$$\text{(a)} \int \frac{\cos x}{\sin^2 x}\,dx \qquad \text{(b)} \int \frac{t^2 - 2t^4}{t^4}\,dt$$

Solution (a).

$$\int \frac{\cos x}{\sin^2 x}\,dx = \int \frac{1}{\sin x}\,\frac{\cos x}{\sin x}\,dx = \int \csc x \cot x\,dx = -\csc x + C$$

Formula 8 in Table 5.2.1

Solution (b).

$$\int \frac{t^2 - 2t^4}{t^4}\,dt = \int \left(\frac{1}{t^2} - 2\right) dt = \int (t^{-2} - 2)\,dt$$

$$= \frac{t^{-1}}{-1} - 2t + C = -\frac{1}{t} - 2t + C \blacktriangleleft$$

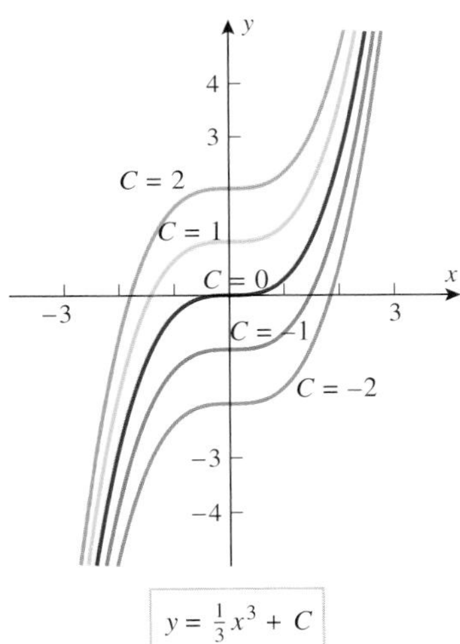

Figure 5.2.1

INTEGRAL CURVES

Graphs of antiderivatives of a function f are called ***integral curves*** of f. We know from Theorem 5.2.2 that if $y = F(x)$ is any integral curve of $f(x)$, then all other integral curves are vertical translations of this curve, since they have equations of the form $y = F(x) + C$. For example, $y = \frac{1}{3}x^3$ is one integral curve for $f(x) = x^2$, so all the other integral curves have equations of the form $y = \frac{1}{3}x^3 + C$; conversely, the graph of any equation of this form is an integral curve (Figure 5.2.1).

In many problems one is interested in finding a function whose derivative satisfies specified conditions. The following example illustrates a geometric problem of this type.

▶ Example 5 Suppose that a point moves along some unknown curve $y = f(x)$ in the xy-plane in such a way that at each point (x, y) on the curve, the tangent line has slope x^2. Find an equation for the curve given that it passes through the point $(2, 1)$.

Solution. We know that $dy/dx = x^2$, so

$$y = \int x^2\,dx = \tfrac{1}{3}x^3 + C$$

Since the curve passes through $(2, 1)$, a specific value for C can be found by using the fact that $y = 1$ if $x = 2$. Substituting these values in the above equation yields

$$1 = \tfrac{1}{3}(2^3) + C \quad \text{or} \quad C = -\tfrac{5}{3}$$

In Example 5, the requirement that the graph of f pass through the point $(2, 1)$ selects the single integral curve $y = \frac{1}{3}x^3 - \frac{5}{3}$ from the family of curves $y = \frac{1}{3}x^3 + C$ (Figure 5.2.2).

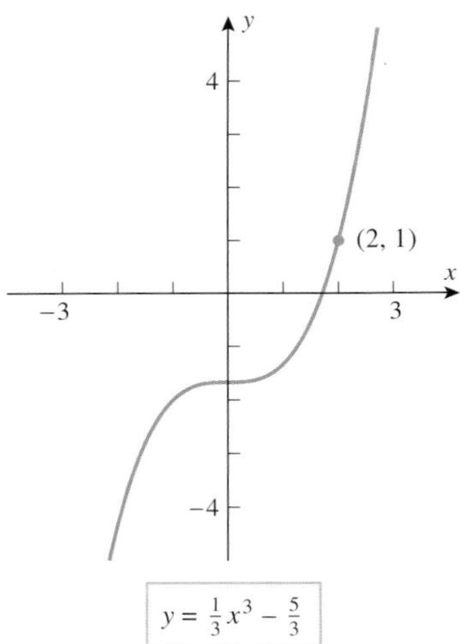

Figure 5.2.2

so an equation of the curve is

$$y = \tfrac{1}{3}x^3 - \tfrac{5}{3}$$

(Figure 5.2.2). ◀

INTEGRATION FROM THE VIEWPOINT OF DIFFERENTIAL EQUATIONS

We will now consider another way of looking at integration that will be useful in our later work. Suppose that $f(x)$ is a known function and we are interested in finding a function $F(x)$ such that $y = F(x)$ satisfies the equation

$$\frac{dy}{dx} = f(x) \tag{8}$$

The solutions of this equation are the antiderivatives of $f(x)$, and we know that these can be obtained by integrating $f(x)$. For example, the solutions of the equation

$$\frac{dy}{dx} = x^2 \tag{9}$$

are

$$y = \int x^2\,dx = \frac{x^3}{3} + C$$

Equation (8) is called a ***differential equation*** because it involves a derivative of an unknown function. Differential equations are different from the kinds of equations we have encountered so far in that the unknown is a *function* and not a *number* as in an equation such as $x^2 + 5x - 6 = 0$.

Sometimes we will not be interested in finding all of the solutions of (8), but rather we will want only the solution whose integral curve passes through a specified point (x_0, y_0). For example, in Example 5 we solved (9) for the integral curve that passed through the point $(2, 1)$.

For simplicity, it is common in the study of differential equations to denote a solution of $dy/dx = f(x)$ as $y(x)$ rather than $F(x)$, as earlier. With this notation, the problem of finding a function $y(x)$ whose derivative is $f(x)$ and whose integral curve passes through the point (x_0, y_0) is expressed as

$$\frac{dy}{dx} = f(x), \quad y(x_0) = y_0 \tag{10}$$

This is called an ***initial-value problem***, and the requirement that $y(x_0) = y_0$ is called the ***initial condition*** for the problem.

▶ **Example 6** Solve the initial-value problem

$$\frac{dy}{dx} = \cos x, \quad y(0) = 1$$

Solution. The solution of the differential equation is

$$y = \int \cos x\,dx = \sin x + C \tag{11}$$

The initial condition $y(0) = 1$ implies that $y = 1$ if $x = 0$; substituting these values in (11) yields

$$1 = \sin(0) + C \quad \text{or} \quad C = 1$$

Thus, the solution of the initial-value problem is $y = \sin x + 1$. ◀

SLOPE FIELDS

If we interpret dy/dx as the slope of a tangent line, then at a point (x, y) on an integral curve of the equation $dy/dx = f(x)$, the slope of the tangent line is $f(x)$. What is interesting about this is that the slopes of the tangent lines to the integral curves can be obtained without actually solving the differential equation. For example, if

$$\frac{dy}{dx} = \sqrt{x^2 + 1}$$

then we know without solving the equation that at the point where $x = 1$ the tangent line to an integral curve has slope $\sqrt{1^2 + 1} = \sqrt{2}$; and more generally, at a point where $x = a$, the tangent line to an integral curve has slope $\sqrt{a^2 + 1}$.

A geometric description of the integral curves of a differential equation $dy/dx = f(x)$ can be obtained by choosing a rectangular grid of points in the xy-plane, calculating the slopes of the tangent lines to the integral curves at the gridpoints, and drawing small portions of the tangent lines through those points. The resulting picture, which is called a ***slope field*** or ***direction field*** for the equation, shows the "direction" of the integral curves at the gridpoints. With sufficiently many gridpoints it is often possible to visualize the integral curves themselves; for example, Figure 5.2.3a shows a slope field for the differential equation $dy/dx = x^2$, and Figure 5.2.3b shows that same field with the integral curves imposed on it—the more gridpoints that are used, the more completely the slope field reveals the shape of the integral curves. However, the amount of computation can be considerable, so computers are usually used when slope fields with many gridpoints are needed.

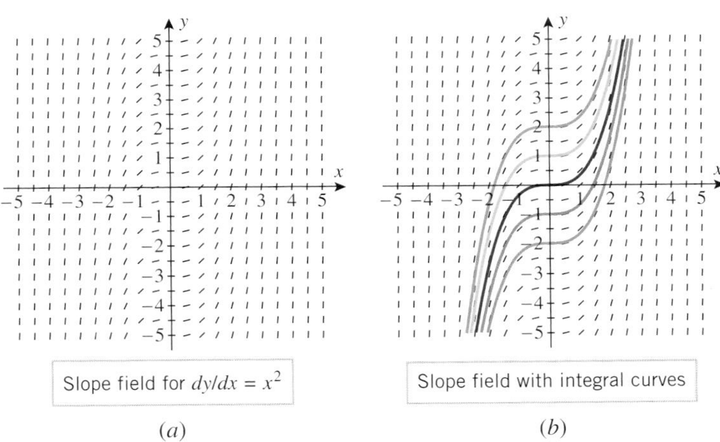

Figure 5.2.3

✔QUICK CHECK EXERCISES 5.2 *(See page 318 for answers.)*

1. A function F is an antiderivative of a function f on an interval I if ________ for all x in I.

2. Write an equivalent integration formula for each given derivative formula.

   (a) $\dfrac{d}{dx}[\sqrt{x}] = \dfrac{1}{2\sqrt{x}}$ (b) $\dfrac{d}{dx}[\sin x] = \cos x$

3. Evaluate the integrals.

   (a) $\displaystyle\int [x^3 + x + 5]\,dx$ (b) $\displaystyle\int [\sec^2 x - \csc x \cot x]\,dx$

4. The graph of $y = x^2 + x$ is an integral curve for the function $f(x) =$ ________. If G is a function whose graph is also an integral curve for f, and if $G(1) = 5$, then $G(x) =$ ________.

5. A slope field for the differential equation

$$\frac{dy}{dx} = \frac{2x}{x^2 - 4}$$

has a line segment with slope ________ through the point $(0, 5)$ and has a line segment with slope ________ through the point $(-4, 1)$.

EXERCISE SET 5.2 Graphing Utility C CAS

1. In each part, confirm that the formula is correct, and state a corresponding integration formula.

(a) $\dfrac{d}{dx}[\sqrt{1+x^2}] = \dfrac{x}{\sqrt{1+x^2}}$

(b) $\dfrac{d}{dx}\left[\frac{1}{3}\sin(1+x^3)\right] = x^2\cos(1+x^3)$

2. In each part, confirm that the stated formula is correct by differentiating.

(a) $\int x\sin x\,dx = \sin x - x\cos x + C$

(b) $\int \dfrac{dx}{(1-x^2)^{3/2}} = \dfrac{x}{\sqrt{1-x^2}} + C$

FOCUS ON CONCEPTS

3. What is a *constant of integration*? Why does an answer to an integration problem involve a constant of integration?

4. What is an *integral curve* of a function f? How are two integral curves of a function f related?

5–8 Find the derivative and state a corresponding integration formula.

5. $\dfrac{d}{dx}[\sqrt{x^3+5}]$ **6.** $\dfrac{d}{dx}\left[\dfrac{x}{x^2+3}\right]$

7. $\dfrac{d}{dx}[\sin(2\sqrt{x})]$ **8.** $\dfrac{d}{dx}[\sin x - x\cos x]$

9–10 Evaluate the integral by rewriting the integrand appropriately, if required, and applying the power rule (Formula 2 in Table 5.2.1).

9. (a) $\int x^8\,dx$ (b) $\int x^{5/7}\,dx$ (c) $\int x^3\sqrt{x}\,dx$

10. (a) $\int \sqrt[3]{x^2}\,dx$ (b) $\int \dfrac{1}{x^6}\,dx$ (c) $\int x^{-7/8}\,dx$

11–14 Evaluate each integral by applying Theorem 5.2.3 and Formula 2 in Table 5.2.1 appropriately.

11. $\int \left[5x + \dfrac{2}{3x^5}\right]dx$ **12.** $\int \left[x^{-1/2} - 3x^{7/5} + \frac{1}{9}\right]dx$

13. $\int [x^{-3} - 3x^{1/4} + 8x^2]\,dx$

14. $\int \left[\dfrac{10}{y^{3/4}} - \sqrt[3]{y} + \dfrac{4}{\sqrt{y}}\right]dy$

15–30 Evaluate the integral and check your answer by differentiating.

15. $\int x(1+x^3)\,dx$ **16.** $\int (2+y^2)^2\,dy$

17. $\int x^{1/3}(2-x)^2\,dx$ **18.** $\int (1+x^2)(2-x)\,dx$

19. $\int \dfrac{x^5+2x^2-1}{x^4}\,dx$ **20.** $\int \dfrac{1-2t^3}{t^3}\,dt$

21. $\int [3\sin x - 2\sec^2 x]\,dx$ **22.** $\int [\csc^2 t - \sec t\tan t]\,dt$

23. $\int \sec x(\sec x + \tan x)\,dx$ **24.** $\int \csc x(\sin x + \cot x)\,dx$

25. $\int \dfrac{\sec\theta}{\cos\theta}\,d\theta$ **26.** $\int \dfrac{dy}{\csc y}$

27. $\int \dfrac{\sin x}{\cos^2 x}\,dx$ **28.** $\int \left[\phi + \dfrac{2}{\sin^2\phi}\right]d\phi$

29. $\int [1+\sin^2\theta\csc\theta]\,d\theta$ **30.** $\int \dfrac{\sec x + \cos x}{2\cos x}\,dx$

31. Evaluate the integral

$$\int \frac{1}{1+\sin x}\,dx$$

by multiplying the numerator and denominator by an appropriate expression.

32. Use the double-angle formula $\cos 2x = 2\cos^2 x - 1$ to evaluate the integral

$$\int \frac{1}{1+\cos 2x}\,dx$$

33. Use a graphing utility to generate some representative integral curves of the function $f(x) = 5x^4 - \sec^2 x$ over the interval $(-\pi/2, \pi/2)$.

34. Use a graphing utility to generate some representative integral curves of the function $f(x) = (x^2-1)/x^2$ over the interval $(0, 5)$.

35. Suppose that a point moves along a curve $y = f(x)$ in the xy-plane in such a way that at each point (x, y) on the curve the tangent line has slope $-\sin x$. Find an equation for the curve, given that it passes through the point $(0, 2)$.

36. Suppose that a point moves along a curve $y = f(x)$ in the xy-plane in such a way that at each point (x, y) on the curve the tangent line has slope $(x+1)^2$. Find an equation for the curve, given that it passes through the point $(-2, 8)$.

37–38 Solve the initial-value problems.

37. (a) $\dfrac{dy}{dx} = \sqrt[3]{x},\ y(1) = 2$

(b) $\dfrac{dy}{dt} = \sin t + 1,\ y\left(\dfrac{\pi}{3}\right) = \dfrac{1}{2}$

(c) $\dfrac{dy}{dx} = \dfrac{x+1}{\sqrt{x}},\ y(1) = 0$

38. (a) $\dfrac{dy}{dx} = \dfrac{1}{(2x)^3}, \; y(1) = 0$

(b) $\dfrac{dy}{dt} = \sec^2 t - \sin t, \; y\left(\dfrac{\pi}{4}\right) = 1$

(c) $\dfrac{dy}{dx} = x^2\sqrt{x^3}, \; y(0) = 0$

39. Find the general form of a function whose second derivative is $\sqrt{x}$. [*Hint:* Solve the equation $f''(x) = \sqrt{x}$ for $f(x)$ by integrating both sides twice.]

40. Find a function f such that $f''(x) = x + \cos x$ and such that $f(0) = 1$ and $f'(0) = 2$. [*Hint:* Integrate both sides of the equation twice.]

41–43 Find an equation of the curve that satisfies the given conditions.

41. At each point (x, y) on the curve the slope is $2x + 1$; the curve passes through the point $(-3, 0)$.

42. At each point (x, y) on the curve the slope equals the square of the distance between the point and the y-axis; the point $(-1, 2)$ is on the curve.

43. At each point (x, y) on the curve, y satisfies the condition $d^2y/dx^2 = 6x$; the line $y = 5 - 3x$ is tangent to the curve at the point where $x = 1$.

C **44.** In each part, use a CAS to solve the initial-value problem.

(a) $\dfrac{dy}{dx} = x^2 \cos 3x, \; y(\pi/2) = -1$

(b) $\dfrac{dy}{dx} = \dfrac{x^3}{(4 + x^2)^{3/2}}, \; y(0) = -2$

45. (a) Use a graphing utility to generate a slope field for the differential equation $dy/dx = x$ in the region $-5 \le x \le 5$ and $-5 \le y \le 5$.

(b) Graph some representative integral curves of the function $f(x) = x$.

(c) Find an equation for the integral curve that passes through the point $(2, 1)$.

46. (a) Use a graphing utility to generate a slope field for the differential equation $dy/dx = \sqrt{x}$ in the region $0 \le x \le 10$ and $-5 \le y \le 5$.

(b) Graph some representative integral curves of the function $f(x) = \sqrt{x}$ for $x > 0$.

(c) Find an equation for the integral curve that passes through the point $(4, 10/3)$.

47–50 The given slope field figure corresponds to one of the differential equations below. Identify the differential equation that matches the figure, and sketch solution curves through the highlighted points.

(a) $\dfrac{dy}{dx} = 2$ (b) $\dfrac{dy}{dx} = -x$

(c) $\dfrac{dy}{dx} = x^2 - 4$ (d) $\dfrac{dy}{dx} = \sin x$

47. **48.**

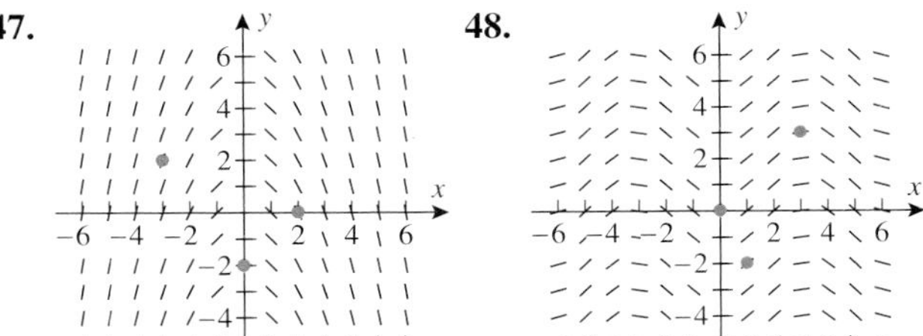

49. **50.**

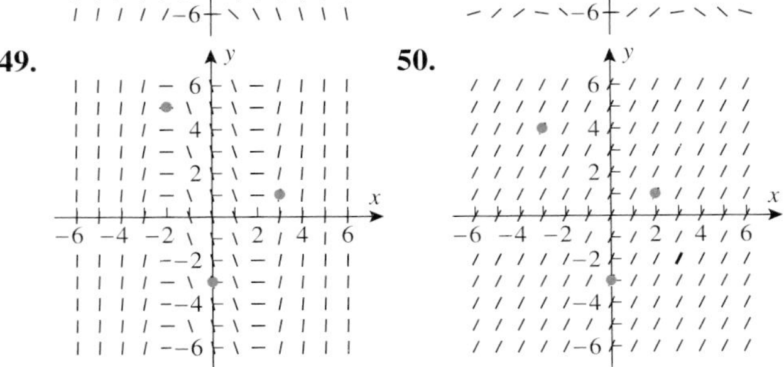

FOCUS ON CONCEPTS

51. (a) Show that

$$F(x) = \tfrac{1}{6}(3x + 4)^2 \quad \text{and} \quad G(x) = \tfrac{3}{2}x^2 + 4x$$

differ by a constant by showing that they are antiderivatives of the same function.

(b) Find the constant C such that $F(x) - G(x) = C$ by evaluating the functions $F(x)$ and $G(x)$ at a particular value of x.

(c) Check your answer in part (b) by simplifying the expression $F(x) - G(x)$ algebraically.

52. Follow the directions of Exercise 51 using

$$F(x) = \frac{3x^2 + 5}{4x^2} \quad \text{and} \quad G(x) = \frac{5}{4x^2}$$

53. Let F and G be the functions defined by

$$F(x) = \frac{x^2 + 3x}{x} \quad \text{and} \quad G(x) = \begin{cases} x + 3, & x > 0 \\ x, & x < 0 \end{cases}$$

(a) Show that F and G have the same derivative.

(b) Show that $G(x) \ne F(x) + C$ for any constant C.

(c) Do parts (a) and (b) violate Theorem 5.2.2? Explain.

54. Follow the directions of Exercise 53 using

$$F(x) = \frac{x \sin x}{x} \quad \text{and} \quad G(x) = \begin{cases} 2 + \sin x, & x > 0 \\ -1 + \sin x, & x < 0 \end{cases}$$

55–56 Use a trigonometric identity to evaluate the integral.

55. $\displaystyle\int \tan^2 x \, dx$ **56.** $\displaystyle\int \cot^2 x \, dx$

57–58 Use the identities $\cos 2\theta = 1 - 2\sin^2\theta = 2\cos^2\theta - 1$ to help evaluate the integrals.

57. $\displaystyle\int \sin^2(x/2) \, dx$ **58.** $\displaystyle\int \cos^2(x/2) \, dx$

59. The speed of sound in air at 0°C (or 273 K on the Kelvin scale) is 1087 ft/s, but the speed v increases as the temperature T rises. Experimentation has shown that the rate of change of v with respect to T is

$$\frac{dv}{dT} = \frac{1087}{2\sqrt{273}}T^{-1/2}$$

where v is in feet per second and T is in kelvins (K). Find a formula that expresses v as a function of T.

60. Suppose that a uniform metal rod 50 cm long is insulated laterally, and the temperatures at the exposed ends are maintained at 25°C and 85°C, respectively. Assume that an x-axis is chosen as in the accompanying figure and that the temperature $T(x)$ satisfies the equation

$$\frac{d^2T}{dx^2} = 0$$

Find $T(x)$ for $0 \leq x \leq 50$.

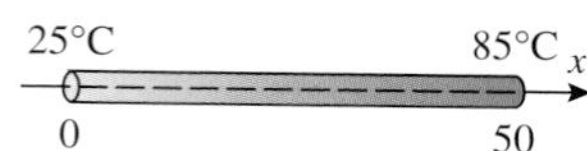

Figure Ex-60

✔ QUICK CHECK ANSWERS 5.2

1. $F'(x) = f(x)$ 2. (a) $\int \frac{1}{2\sqrt{x}}\,dx = \sqrt{x} + C$ (b) $\int \cos x\,dx = \sin x + C$
3. (a) $\frac{1}{4}x^4 + \frac{1}{2}x^2 + 5x + C$ (b) $\tan x + \csc x + C$ 4. $2x + 1$; $x^2 + x + 3$ 5. 0; $-\frac{2}{3}$

5.3 INTEGRATION BY SUBSTITUTION

*In this section we will study a technique, called **substitution**, that can often be used to transform complicated integration problems into simpler ones.*

■ u-SUBSTITUTION

The method of substitution can be motivated by examining the chain rule from the viewpoint of antidifferentiation. For this purpose, suppose that F is an antiderivative of f and that g is a differentiable function. The chain rule implies that the derivative of $F(g(x))$ can be expressed as

$$\frac{d}{dx}[F(g(x))] = F'(g(x))g'(x)$$

which we can write in integral form as

$$\int F'(g(x))g'(x)\,dx = F(g(x)) + C \tag{1}$$

or since F is an antiderivative of f,

$$\int f(g(x))g'(x)\,dx = F(g(x)) + C \tag{2}$$

For our purposes it will be useful to let $u = g(x)$ and to write $du/dx = g'(x)$ in the differential form $du = g'(x)\,dx$. With this notation (2) can be expressed as

$$\int f(u)\,du = F(u) + C \tag{3}$$

The process of evaluating an integral of form (2) by converting it into form (3) with the substitution

$$u = g(x) \quad \text{and} \quad du = g'(x)\,dx$$

is called the ***method of u-substitution***. Here our emphasis is *not* on the interpretation of the expression $du = g'(x)\,dx$. Rather, the differential notation serves primarily as a useful "bookkeeping" device for the method of u-substitution. The following example illustrates how the method works.

Example 1 Evaluate $\int (x^2+1)^{50} \cdot 2x\,dx$.

Solution. If we let $u = x^2 + 1$, then $du/dx = 2x$, which implies that $du = 2x\,dx$. Thus, the given integral can be written as

$$\int (x^2+1)^{50} \cdot 2x\,dx = \int u^{50}\,du = \frac{u^{51}}{51} + C = \frac{(x^2+1)^{51}}{51} + C$$

It is important to realize that in the method of u-substitution you have control over the choice of u, but once you make that choice you have no control over the resulting expression for du. Thus, in the last example we *chose* $u = x^2 + 1$ but $du = 2x\,dx$ was *computed*. Fortunately, our choice of u, combined with the computed du, worked out perfectly to produce an integral involving u that was easy to evaluate. However, in general, the method of u-substitution will fail if the chosen u and the computed du cannot be used to produce an integrand in which no expressions involving x remain, or if you cannot evaluate the resulting integral. Thus, for example, the substitution $u = x^2$, $du = 2x\,dx$ will not work for the integral

$$\int 2x \sin x^4\,dx$$

because this substitution results in the integral

$$\int \sin u^2\,du$$

which still cannot be evaluated in terms of familiar functions.

In general, there are no hard and fast rules for choosing u, and in some problems no choice of u will work. In such cases other methods need to be used, some of which will be discussed later. Making appropriate choices for u will come with experience, but you may find the following *guidelines*, combined with a mastery of the basic integrals in Table 5.2.1, helpful.

Guidelines for u-Substitution

Step 1. Look for some composition $f(g(x))$ within the integrand for which the substitution

$$u = g(x), \quad du = g'(x)\,dx$$

produces an integral that is expressed entirely in terms of u and du. This may or may not be possible.

Step 2. If you are successful in Step 1, then try to evaluate the resulting integral in terms of u. Again, this may or may not be possible.

Step 3. If you are successful in Step 2, then replace u by $g(x)$ to express your final answer in terms of x.

EASY TO RECOGNIZE SUBSTITUTIONS

The easiest substitutions occur when the integrand is the derivative of a known function, except for a constant added to or subtracted from the independent variable.

▶ Example 2

$$\int \sin(x+9)\,dx = \int \sin u\,du = -\cos u + C = -\cos(x+9) + C$$

$u = x + 9$
$du = 1 \cdot dx = dx$

$$\int (x-8)^{23}\,dx = \int u^{23}\,du = \frac{u^{24}}{24} + C = \frac{(x-8)^{24}}{24} + C \blacktriangleleft$$

$u = x - 8$
$du = 1 \cdot dx = dx$

Another easy u-substitution occurs when the integrand is the derivative of a known function, except for a constant that multiplies or divides the independent variable. The following example illustrates two ways to evaluate such integrals.

▶ Example 3 Evaluate $\int \cos 5x\,dx$.

Solution.

$$\int \cos 5x\,dx = \int (\cos u) \cdot \frac{1}{5}\,du = \frac{1}{5}\int \cos u\,du = \frac{1}{5}\sin u + C = \frac{1}{5}\sin 5x + C$$

$u = 5x$
$du = 5\,dx$ or $dx = \frac{1}{5}\,du$

Alternative Solution. There is a variation of the preceding method that some people prefer. The substitution $u = 5x$ requires $du = 5\,dx$. If there were a factor of 5 in the integrand, then we could group the 5 and dx together to form the du required by the substitution. Since there is no factor of 5, we will insert one and compensate by putting a factor of $\frac{1}{5}$ in front of the integral. The computations are as follows:

$$\int \cos 5x\,dx = \frac{1}{5}\int \cos 5x \cdot 5\,dx = \frac{1}{5}\int \cos u\,du = \frac{1}{5}\sin u + C = \frac{1}{5}\sin 5x + C \blacktriangleleft$$

$u = 5x$
$du = 5\,dx$

More generally, if the integrand is a composition of the form $f(ax+b)$, where $f(x)$ is an easy to integrate function, then the substitution $u = ax + b$, $du = a\,dx$ will work.

▶ Example 4

$$\int \frac{dx}{\left(\frac{1}{3}x - 8\right)^5} = \int \frac{3\,du}{u^5} = 3\int u^{-5}\,du = -\frac{3}{4}u^{-4} + C = -\frac{3}{4}\left(\frac{1}{3}x - 8\right)^{-4} + C \blacktriangleleft$$

$u = \frac{1}{3}x - 8$
$du = \frac{1}{3}\,dx$ or $dx = 3\,du$

With the help of Theorem 5.2.3, a complicated integral can sometimes be computed by expressing it as a sum of simpler integrals.

▶ Example 5

$$\int \left(\frac{1}{x^2} + \sec^2 \pi x\right) dx = \int \frac{dx}{x^2} + \int \sec^2 \pi x \, dx$$

$$= -\frac{1}{x} + \int \sec^2 \pi x \, dx$$

$$= -\frac{1}{x} + \frac{1}{\pi}\int \sec^2 u \, du$$

$u = \pi x$
$du = \pi \, dx$ or $dx = \frac{1}{\pi} du$

$$= -\frac{1}{x} + \frac{1}{\pi}\tan u + C = -\frac{1}{x} + \frac{1}{\pi}\tan \pi x + C \blacktriangleleft$$

The next three examples illustrate a substitution $u = g(x)$ where $g(x)$ is a nonlinear function.

▶ Example 6 Evaluate $\int \sin^2 x \cos x \, dx$.

Solution. If we let $u = \sin x$, then

$$\frac{du}{dx} = \cos x, \quad \text{so} \quad du = \cos x \, dx$$

Thus,

$$\int \sin^2 x \cos x \, dx = \int u^2 \, du = \frac{u^3}{3} + C = \frac{\sin^3 x}{3} + C \blacktriangleleft$$

▶ Example 7 Evaluate $\int \frac{\cos \sqrt{x}}{\sqrt{x}} dx$.

Solution. If we let $u = \sqrt{x}$, then

$$\frac{du}{dx} = \frac{1}{2\sqrt{x}}, \quad \text{so} \quad du = \frac{1}{2\sqrt{x}} dx \quad \text{or} \quad 2\,du = \frac{1}{\sqrt{x}} dx$$

Thus,

$$\int \frac{\cos \sqrt{x}}{\sqrt{x}} dx = \int 2\cos u \, du = 2\int \cos u \, du = 2\sin u + C = 2\sin \sqrt{x} + C \blacktriangleleft$$

▶ Example 8 Evaluate $\int t^4 \sqrt[3]{3 - 5t^5} \, dt$.

Solution.

$$\int t^4 \sqrt[3]{3 - 5t^5} \, dt = -\frac{1}{25}\int \sqrt[3]{u} \, du = -\frac{1}{25}\int u^{1/3} \, du$$

$u = 3 - 5t^5$
$du = -25t^4 \, dt$ or $-\frac{1}{25} du = t^4 \, dt$

$$= -\frac{1}{25}\frac{u^{4/3}}{4/3} + C = -\frac{3}{100}(3 - 5t^5)^{4/3} + C \blacktriangleleft$$

LESS APPARENT SUBSTITUTIONS

The method of substitution is relatively straightforward, provided the integrand contains an easily recognized composition $f(g(x))$ and the remainder of the integrand is a constant multiple of $g'(x)$. If this is not the case, the method may still apply but can require more computation.

▶ Example 9 Evaluate $\int x^2\sqrt{x-1}\,dx$.

Solution. The composition $\sqrt{x-1}$ suggests the substitution

$$u = x - 1 \quad \text{so that} \quad du = dx \tag{4}$$

From the first equality in (4)

$$x^2 = (u+1)^2 = u^2 + 2u + 1$$

so that

$$\begin{aligned}\int x^2\sqrt{x-1}\,dx &= \int (u^2+2u+1)\sqrt{u}\,du = \int (u^{5/2}+2u^{3/2}+u^{1/2})\,du\\ &= \tfrac{2}{7}u^{7/2} + \tfrac{4}{5}u^{5/2} + \tfrac{2}{3}u^{3/2} + C\\ &= \tfrac{2}{7}(x-1)^{7/2} + \tfrac{4}{5}(x-1)^{5/2} + \tfrac{2}{3}(x-1)^{3/2} + C \ \blacktriangleleft\end{aligned}$$

▶ Example 10 Evaluate $\int \cos^3 x\,dx$.

Solution. The only compositions in the integrand that suggest themselves are

$$\cos^3 x = (\cos x)^3 \quad \text{and} \quad \cos^2 x = (\cos x)^2$$

However, neither the substitution $u = \cos x$ nor the substitution $u = \cos^2 x$ work (verify). In this case, an appropriate substitution is not suggested by the composition contained in the integrand. On the other hand, note from Equation (2) that the derivative $g'(x)$ appears as a factor in the integrand. This suggests that we write

$$\int \cos^3 x\,dx = \int \cos^2 x \cos x\,dx$$

and solve the equation $du = \cos x\,dx$ for $u = \sin x$. Since $\sin^2 x + \cos^2 x = 1$, we then have

$$\begin{aligned}\int \cos^3 x\,dx &= \int \cos^2 x\cos x\,dx = \int (1-\sin^2 x)\cos x\,dx = \int (1-u^2)\,du\\ &= u - \frac{u^3}{3} + C = \sin x - \frac{1}{3}\sin^3 x + C \ \blacktriangleleft\end{aligned}$$

TECHNOLOGY MASTERY

If you have a CAS, use it to calculate the integrals in the examples in this section. If your CAS produces an answer that is different from the one in the text, then confirm algebraically that the two answers agree. Also, explore the effect of using the CAS to simplify the expressions it produces for the integrals.

INTEGRATION USING COMPUTER ALGEBRA SYSTEMS

The advent of computer algebra systems has made it possible to evaluate many kinds of integrals that would be laborious to evaluate by hand. For example, *Derive*, running on a handheld calculator, evaluated the integral

$$\int \frac{5x^2}{(1+x)^{1/3}}\,dx = \frac{3(x+1)^{2/3}(5x^2-6x+9)}{8} + C$$

in about a second. The computer algebra system *Mathematica*, running on a personal computer, required even less time to evaluate this same integral. However, just as one would not want to rely on a calculator to compute $2+2$, so one would not want to use a CAS to integrate a simple function such as $f(x)=x^2$. Thus, even if you have a CAS, you will want to develop a reasonable level of competence in evaluating basic integrals. Moreover, the mathematical techniques that we will introduce for evaluating basic integrals are precisely the techniques that computer algebra systems use to evaluate more complicated integrals.

✔ QUICK CHECK EXERCISES 5.3

(See page 324 for answers.)

1. Indicate the u-substitution.

(a) $\int 3x^2(1+x^3)^{25}\,dx = \int u^{25}\,du$ if $u =$ ______ and $du =$ ______.

(b) $\int 2x\sin x^2\,dx = \int \sin u\,du$ if $u =$ ______ and $du =$ ______.

(c) $\int \dfrac{18x}{\sqrt{1+9x^2}}\,dx = \int \dfrac{1}{\sqrt{u}}\,du$ if $u =$ ______ and $du =$ ______.

2. Evaluate the integrals in Quick Check Exercise 1.

3. Supply the missing integrand corresponding to the indicated u-substitution.

(a) $\int 5(5x-3)^{-1/3}\,dx = \int$ ______ du; $u=5x-3$

(b) $\int (3-\tan x)\sec^2 x\,dx = \int$ ______ du; $u=3-\tan x$

(c) $\int \dfrac{\sqrt[3]{8+\sqrt{x}}}{\sqrt{x}}\,dx = \int$ ______ du; $u=8+\sqrt{x}$

4. Evaluate the integrals in Quick Check Exercise 3.

EXERCISE SET 5.3

Graphing Utility | C CAS

1–4 Evaluate the integrals using the indicated substitutions.

1. (a) $\int 2x(x^2+1)^{23}\,dx$; $u=x^2+1$

(b) $\int \cos^3 x\sin x\,dx$; $u=\cos x$

(c) $\int \dfrac{1}{\sqrt{x}}\sin\sqrt{x}\,dx$; $u=\sqrt{x}$

(d) $\int \dfrac{3x\,dx}{\sqrt{4x^2+5}}$; $u=4x^2+5$

2. (a) $\int \sec^2(4x+1)\,dx$; $u=4x+1$

(b) $\int y\sqrt{1+2y^2}\,dy$; $u=1+2y^2$

(c) $\int \sqrt{\sin\pi\theta}\cos\pi\theta\,d\theta$; $u=\sin\pi\theta$

(d) $\int (2x+7)(x^2+7x+3)^{4/5}\,dx$; $u=x^2+7x+3$

3. (a) $\int \cot x\csc^2 x\,dx$; $u=\cot x$

(b) $\int (1+\sin t)^9\cos t\,dt$; $u=1+\sin t$

(c) $\int \cos 2x\,dx$; $u=2x$ (d) $\int x\sec^2 x^2\,dx$; $u=x^2$

4. (a) $\int x^2\sqrt{1+x}\,dx$; $u=1+x$

(b) $\int [\csc(\sin x)]^2\cos x\,dx$; $u=\sin x$

(c) $\int \sin(x-\pi)\,dx$; $u=x-\pi$

(d) $\int \dfrac{5x^4}{(x^5+1)^2}\,dx$; $u=x^5+1$

FOCUS ON CONCEPTS

5. Explain the connection between the chain rule for differentiation and the method of u-substitution for integration.

6. Explain how the substitution $u=ax+b$ helps to perform an integration in which the integrand is $f(ax+b)$, where $f(x)$ is an easy to integrate function.

7–32 Evaluate the integrals using appropriate substitutions.

7. $\int (4x-3)^9\,dx$

8. $\int x^3\sqrt{5+x^4}\,dx$

9. $\int \sin 7x\,dx$

10. $\int \cos\dfrac{x}{3}\,dx$

11. $\int \sec 4x\tan 4x\,dx$

12. $\int \sec^2 5x\,dx$

13. $\int t\sqrt{7t^2+12}\,dt$

14. $\int \dfrac{x}{\sqrt{4-5x^2}}\,dx$

15. $\int \dfrac{6}{(1-2x)^3}\,dx$

16. $\int \dfrac{x^2+1}{\sqrt{x^3+3x}}\,dx$

17. $\displaystyle\int \frac{x^3}{(5x^4+2)^3}\,dx$

18. $\displaystyle\int \frac{\sin(1/x)}{3x^2}\,dx$

19. $\displaystyle\int \frac{\sin(5/x)}{x^2}\,dx$

20. $\displaystyle\int \frac{\sec^2(\sqrt{x})}{\sqrt{x}}\,dx$

21. $\displaystyle\int \cos^4 3t \sin 3t\,dt$

22. $\displaystyle\int \cos 2t \sin^5 2t\,dt$

23. $\displaystyle\int x\sec^2(x^2)\,dx$

24. $\displaystyle\int \frac{\cos 4\theta}{(1+2\sin 4\theta)^4}\,d\theta$

25. $\displaystyle\int \cos 4\theta\sqrt{2-\sin 4\theta}\,d\theta$

26. $\displaystyle\int \tan^3 5x \sec^2 5x\,dx$

27. $\displaystyle\int \sec^3 2x \tan 2x\,dx$

28. $\displaystyle\int [\sin(\sin\theta)]\cos\theta\,d\theta$

29. $\displaystyle\int \frac{y}{\sqrt{2y+1}}\,dx$

30. $\displaystyle\int x\sqrt{4-x}\,dx$

31. $\displaystyle\int \sin^3 2\theta\,d\theta$

32. $\displaystyle\int \sec^4 3\theta\,d\theta$ [*Hint:* Apply a trigonometric identity.]

33–35 Evaluate the integrals assuming that n is a positive integer and $b \neq 0$.

33. $\displaystyle\int (a+bx)^n\,dx$

34. $\displaystyle\int \sqrt[n]{a+bx}\,dx$

35. $\displaystyle\int \sin^n(a+bx)\cos(a+bx)\,dx$

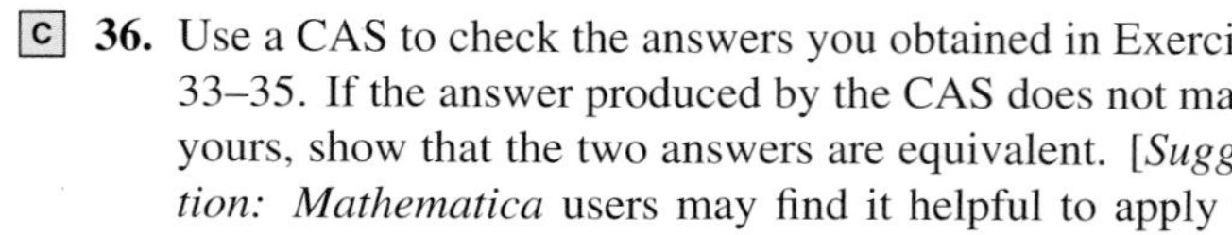

36. Use a CAS to check the answers you obtained in Exercises 33–35. If the answer produced by the CAS does not match yours, show that the two answers are equivalent. [*Suggestion: Mathematica* users may find it helpful to apply the Simplify command to the answer.]

FOCUS ON CONCEPTS

37. (a) Evaluate the integral $\int \sin x\cos x\,dx$ by two methods: first by letting $u=\sin x$, and then by letting $u=\cos x$.
(b) Explain why the two apparently different answers obtained in part (a) are really equivalent.

38. (a) Evaluate the integral $\int (5x-1)^2\,dx$ by two methods: first square and integrate, then let $u=5x-1$.
(b) Explain why the two apparently different answers obtained in part (a) are really equivalent.

39–40 Solve the initial-value problems.

39. $\dfrac{dy}{dx}=\sqrt{5x+1},\ y(3)=-2$

40. $\dfrac{dy}{dx}=2+\sin 3x,\ y(\pi/3)=0$

41. (a) Evaluate $\int [x/\sqrt{x^2+1}]\,dx$.
(b) Use a graphing utility to generate some typical integral curves of $f(x)=x/\sqrt{x^2+1}$ over the interval $(-5,5)$.

42. (a) Evaluate $\int 2x\sin(25-x^2)\,dx$.
(b) Use a graphing utility to generate some typical integral curves of $f(x)=2x\sin(25-x^2)$ over the interval $(-5,5)$.

43. Find a function f such that the slope of the tangent line at a point (x,y) on the curve $y=f(x)$ is $\sqrt{3x+1}$, and the curve passes through the point $(0,1)$.

44. A population of minnows in a lake is estimated to be 100,000 at the beginning of the year 2005. Suppose that t years after the beginning of 2005 the rate of growth of the population $p(t)$ (in thousands) is given by $p'(t)=(3+0.12t)^{3/2}$. Estimate the projected population at the beginning of the year 2010.

✔ QUICK CHECK ANSWERS 5.3

1. (a) $1+x^3$; $3x^2\,dx$ (b) x^2; $2x\,dx$ (c) $1+9x^2$; $18x\,dx$ 2. (a) $\frac{1}{26}(1+x^3)^{26}+C$ (b) $-\cos x^2+C$ (c) $2\sqrt{1+9x^2}+C$
3. (a) $u^{-1/3}$ (b) $-u$ (c) $2\sqrt[3]{u}$ 4. (a) $\frac{3}{2}(5x-3)^{2/3}+C$ (b) $-\frac{1}{2}(3-\tan x)^2+C$ (c) $\frac{3}{2}(8+\sqrt{x})^{4/3}+C$

5.4 THE DEFINITION OF AREA AS A LIMIT; SIGMA NOTATION

Our main goal in this section is to use the rectangle method to give a precise mathematical definition of the "area under a curve." To simplify our computations, we will begin by discussing a useful notation for expressing lengthy sums in a compact form.

SIGMA NOTATION

The notation we will discuss is called ***sigma notation*** or ***summation notation*** because it uses the uppercase Greek letter Σ (sigma) to denote various kinds of sums. To illustrate how this notation works, consider the sum

$$1^2+2^2+3^2+4^2+5^2$$

in which each term is of the form k^2, where k is one of the integers from 1 to 5. In sigma notation this sum can be written as

$$\sum_{k=1}^{5} k^2$$

which is read "the summation of k^2, where k runs from 1 to 5." The notation tells us to form the sum of the terms that result when we substitute successive integers for k in the expression k^2, starting with $k = 1$ and ending with $k = 5$.

More generally, if $f(k)$ is a function of k, and if m and n are integers such that $m \leq n$, then

$$\sum_{k=m}^{n} f(k) \tag{1}$$

denotes the sum of the terms that result when we substitute successive integers for k, starting with $k = m$ and ending with $k = n$ (Figure 5.4.1).

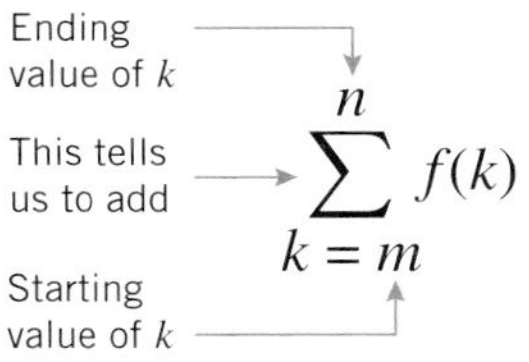

Ending value of k

This tells us to add

$$\sum_{k=m}^{n} f(k)$$

Starting value of k

Figure 5.4.1

▶ Example 1

$$\sum_{k=4}^{8} k^3 = 4^3 + 5^3 + 6^3 + 7^3 + 8^3$$

$$\sum_{k=1}^{5} 2k = 2 \cdot 1 + 2 \cdot 2 + 2 \cdot 3 + 2 \cdot 4 + 2 \cdot 5 = 2 + 4 + 6 + 8 + 10$$

$$\sum_{k=0}^{5} (2k + 1) = 1 + 3 + 5 + 7 + 9 + 11$$

$$\sum_{k=0}^{5} (-1)^k (2k + 1) = 1 - 3 + 5 - 7 + 9 - 11$$

$$\sum_{k=-3}^{1} k^3 = (-3)^3 + (-2)^3 + (-1)^3 + 0^3 + 1^3 = -27 - 8 - 1 + 0 + 1$$

$$\sum_{k=1}^{3} k \sin\left(\frac{k\pi}{5}\right) = \sin\frac{\pi}{5} + 2\sin\frac{2\pi}{5} + 3\sin\frac{3\pi}{5} \blacktriangleleft$$

The numbers m and n in (1) are called, respectively, the ***lower*** and ***upper limits of summation***; and the letter k is called the ***index of summation***. It is not essential to use k as the index of summation; any letter not reserved for another purpose will do. For example,

$$\sum_{i=1}^{6} \frac{1}{i}, \quad \sum_{j=1}^{6} \frac{1}{j}, \quad \text{and} \quad \sum_{n=1}^{6} \frac{1}{n}$$

all denote the sum

$$1 + \frac{1}{2} + \frac{1}{3} + \frac{1}{4} + \frac{1}{5} + \frac{1}{6}$$

If the upper and lower limits of summation are the same, then the "sum" in (1) reduces to a single term. For example,

$$\sum_{k=2}^{2} k^3 = 2^3 \quad \text{and} \quad \sum_{i=1}^{1} \frac{1}{i+2} = \frac{1}{1+2} = \frac{1}{3}$$

In the sums

$$\sum_{i=1}^{5} 2 \quad \text{and} \quad \sum_{j=0}^{2} x^3$$

the expression to the right of the Σ sign does not involve the index of summation. In such cases, we take all the terms in the sum to be the same, with one term for each allowable value of the summation index. Thus,

$$\sum_{i=1}^{5} 2 = 2+2+2+2+2 \quad \text{and} \quad \sum_{j=0}^{2} x^3 = x^3+x^3+x^3$$

CHANGING THE LIMITS OF SUMMATION

A sum can be written in more than one way using sigma notation with different limits of summation and correspondingly different summands. For example,

$$\sum_{i=1}^{5} 2i = 2+4+6+8+10 = \sum_{j=0}^{4}(2j+2) = \sum_{k=3}^{7}(2k-4)$$

On occasion we will want to change the sigma notation for a given sum to a sigma notation with different limits of summation.

► **Example 2** Express

$$\sum_{k=3}^{7} 5^{k-2}$$

in sigma notation so that the lower limit of summation is 0 rather than 3.

Solution.

$$\begin{aligned}\sum_{k=3}^{7} 5^{k-2} &= 5^1+5^2+5^3+5^4+5^5\\ &= 5^{0+1}+5^{1+1}+5^{2+1}+5^{3+1}+5^{4+1}\\ &= \sum_{j=0}^{4} 5^{j+1} = \sum_{k=0}^{4} 5^{k+1}\end{aligned}$$ ◄

PROPERTIES OF SUMS

When stating general properties of sums it is often convenient to use a subscripted letter such as a_k in place of the function notation $f(k)$. For example,

$$\sum_{k=1}^{5} a_k = a_1+a_2+a_3+a_4+a_5 = \sum_{j=1}^{5} a_j = \sum_{k=-1}^{3} a_{k+2}$$

$$\sum_{k=1}^{n} a_k = a_1+a_2+\cdots+a_n = \sum_{j=1}^{n} a_j = \sum_{k=-1}^{n-2} a_{k+2}$$

Our first properties provide some basic rules for manipulating sums.

5.4.1 THEOREM.

(a) $\displaystyle\sum_{k=1}^{n} ca_k = c\sum_{k=1}^{n} a_k$ *(if c does not depend on k)*

(b) $\displaystyle\sum_{k=1}^{n}(a_k+b_k) = \sum_{k=1}^{n} a_k + \sum_{k=1}^{n} b_k$

(c) $\displaystyle\sum_{k=1}^{n}(a_k-b_k) = \sum_{k=1}^{n} a_k - \sum_{k=1}^{n} b_k$

We will prove parts (*a*) and (*b*) and leave part (*c*) as an exercise.

PROOF (*a*).

$$\sum_{k=1}^{n} ca_k = ca_1 + ca_2 + \cdots + ca_n = c(a_1 + a_2 + \cdots + a_n) = c\sum_{k=1}^{n} a_k$$

PROOF (*b*).

$$\sum_{k=1}^{n}(a_k + b_k) = (a_1 + b_1) + (a_2 + b_2) + \cdots + (a_n + b_n)$$

$$= (a_1 + a_2 + \cdots + a_n) + (b_1 + b_2 + \cdots + b_n) = \sum_{k=1}^{n} a_k + \sum_{k=1}^{n} b_k \quad \blacksquare$$

Restating Theorem 5.4.1 in words:

(*a*) *A constant factor can be moved through a sigma sign.*

(*b*) *Sigma distributes across sums.*

(*c*) *Sigma distributes across differences.*

SUMMATION FORMULAS

The following theorem lists some useful formulas for sums of powers of integers. The derivations of these formulas are given in Appendix C.

TECHNOLOGY MASTERY

If you have access to a CAS, it will provide a method for finding closed forms such as those in Theorem 5.4.2. Use your CAS to confirm the formulas in that theorem, and then find closed forms for

$$\sum_{k=1}^{n} k^4 \quad \text{and} \quad \sum_{k=1}^{n} k^5$$

5.4.2 THEOREM.

(*a*) $\displaystyle\sum_{k=1}^{n} k = 1 + 2 + \cdots + n = \frac{n(n+1)}{2}$

(*b*) $\displaystyle\sum_{k=1}^{n} k^2 = 1^2 + 2^2 + \cdots + n^2 = \frac{n(n+1)(2n+1)}{6}$

(*c*) $\displaystyle\sum_{k=1}^{n} k^3 = 1^3 + 2^3 + \cdots + n^3 = \left[\frac{n(n+1)}{2}\right]^2$

▶ **Example 3** Evaluate $\displaystyle\sum_{k=1}^{30} k(k+1)$.

Solution.

$$\sum_{k=1}^{30} k(k+1) = \sum_{k=1}^{30}(k^2 + k) = \sum_{k=1}^{30} k^2 + \sum_{k=1}^{30} k$$

$$= \frac{30(31)(61)}{6} + \frac{30(31)}{2} = 9920 \qquad \text{Theorem 5.4.2}(a), (b) \quad ◀$$

TECHNOLOGY MASTERY

Many calculating utilities provide some way of evaluating sums expressed in sigma notation. If your utility has this capability, use it to confirm that the result in Example 3 is correct.

In formulas such as

$$\sum_{k=1}^{n} k = \frac{n(n+1)}{2} \quad \text{or} \quad 1 + 2 + \cdots + n = \frac{n(n+1)}{2}$$

the left side of the equality is said to express the sum in ***open form*** and the right side is said to express it in ***closed form***. The open form indicates the summands and the closed form is an explicit formula for the sum.

▶ **Example 4** Express $\sum_{k=1}^{n}(3+k)^2$ in closed form.

Solution.

$$\begin{aligned}\sum_{k=1}^{n}(3+k)^2 &= 4^2+5^2+\cdots+(3+n)^2\\ &= [1^2+2^2+3^3+4^2+5^2+\cdots+(3+n)^2]-[1^2+2^2+3^2]\\ &= \left(\sum_{k=1}^{3+n}k^2\right)-14\\ &= \frac{(3+n)(4+n)(7+2n)}{6}-14=\frac{1}{6}(73n+21n^2+2n^3)\end{aligned}$$ ◀

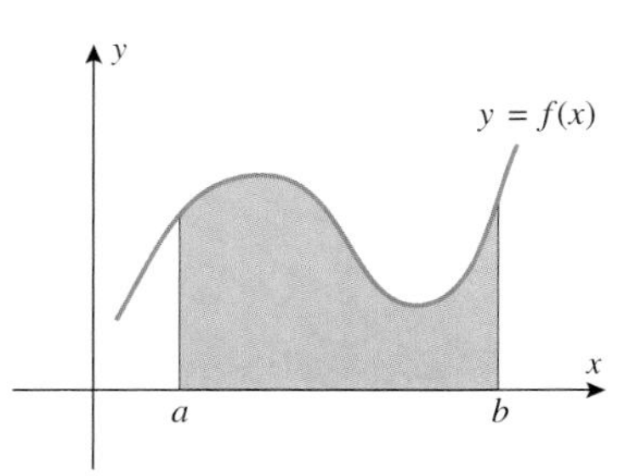

Figure 5.4.2

A DEFINITION OF AREA

We now turn to the problem of giving a precise definition of what is meant by the "area under a curve." Specifically, suppose that the function f is continuous and nonnegative on the interval $[a, b]$, and let R denote the region bounded below by the x-axis, bounded on the sides by the vertical lines $x = a$ and $x = b$, and bounded above by the curve $y = f(x)$ (Figure 5.4.2). Using the rectangle method of Section 5.1, we can motivate a definition for the area of R as follows:

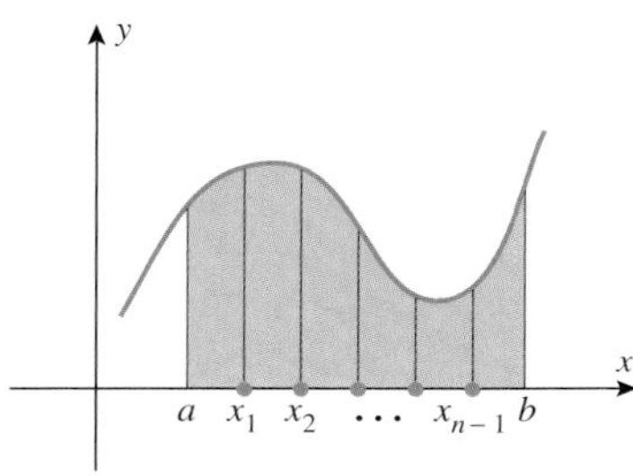

Figure 5.4.3

- Divide the interval $[a, b]$ into n equal subintervals by inserting $n - 1$ equally spaced points between a and b, and denote those points by

$$x_1, x_2, \ldots, x_{n-1}$$

(Figure 5.4.3). Each of these subintervals has width $(b - a)/n$, which is customarily denoted by

$$\Delta x = \frac{b-a}{n}$$

- Over each subinterval construct a rectangle whose height is the value of f at an arbitrarily selected point in the subinterval. Thus, if

$$x_1^*, x_2^*, \ldots, x_n^*$$

denote the points selected in the subintervals, then the rectangles will have heights $f(x_1^*), f(x_2^*), \ldots, f(x_n^*)$ and areas

$$f(x_1^*)\Delta x, \quad f(x_2^*)\Delta x, \ldots, \quad f(x_n^*)\Delta x$$

(Figure 5.4.4).

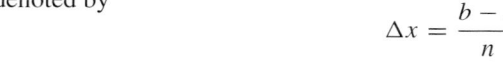

- The union of the rectangles forms a region R_n whose area can be regarded as an approximation to the area A of the region R; that is,

$$A = \text{area}(R) \approx \text{area}(R_n) = f(x_1^*)\Delta x + f(x_2^*)\Delta x + \cdots + f(x_n^*)\Delta x$$

(Figure 5.4.5). This can be expressed more compactly in sigma notation as

$$A \approx \sum_{k=1}^{n} f(x_k^*)\Delta x$$

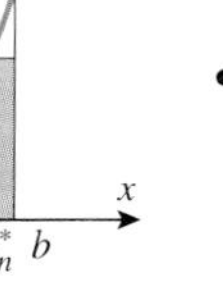

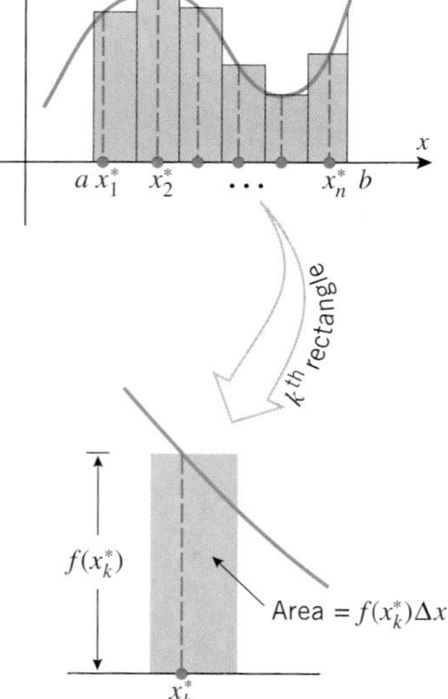

Figure 5.4.4

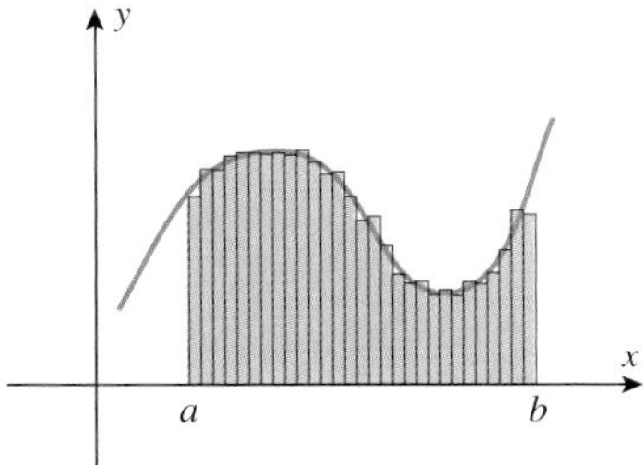

Figure 5.4.5 area(R_n) ≈ area(R)

The limit in (2) is interpreted to mean that given any number $\epsilon > 0$ the inequality

$$\left| A - \sum_{k=1}^{n} f(x_k^*)\Delta x \right| < \epsilon$$

holds when n is sufficiently large, no matter how the points x_k^* are selected.

- Repeat the process using more and more subdivisions, and define the area of R to be the "limit" of the areas of the approximating regions R_n as n increases without bound. That is, we define the area A as

$$A = \lim_{n \to +\infty} \sum_{k=1}^{n} f(x_k^*)\Delta x$$

In summary, we make the following definition.

5.4.3 DEFINITION (*Area Under a Curve*). If the function f is continuous on $[a, b]$ and if $f(x) \geq 0$ for all x in $[a, b]$, then the ***area*** under the curve $y = f(x)$ over the interval $[a, b]$ is defined by

$$A = \lim_{n \to +\infty} \sum_{k=1}^{n} f(x_k^*)\Delta x \qquad (2)$$

There is a difference in interpretation between $\lim_{n\to+\infty}$ and $\lim_{x\to+\infty}$, where n represents a positive integer and x represents a real number. Later we will study limits of the type $\lim_{n\to+\infty}$ in detail, but for now suffice it to say that the computational techniques we have used for limits of type $\lim_{x\to+\infty}$ will also work for $\lim_{n\to+\infty}$.

The values of $x_1^*, x_2^*, \ldots, x_n^*$ in (2) can be chosen arbitrarily, so it is conceivable that different choices of these values might produce different values of A. Were this to happen, then Definition 5.4.3 would not be an acceptable definition of area. Fortunately, this does not happen; it is proved in advanced courses that if f is continuous (as we have assumed), then the same value of A results no matter how the x_k^* are chosen. In practice they are chosen in some systematic fashion, some common choices being

- the left endpoint of each subinterval
- the right endpoint of each subinterval
- the midpoint of each subinterval

To be more specific, suppose that the interval $[a, b]$ is divided into n equal parts of length $\Delta x = (b-a)/n$ by the points $x_1, x_2, \ldots, x_{n-1}$, and let $x_0 = a$ and $x_n = b$ (Figure 5.4.6). Then,

$$x_k = a + k\Delta x \quad \text{for } k = 0, 1, 2, \ldots, n$$

Thus, the left endpoint, right endpoint, and midpoint choices for $x_1^*, x_2^*, \ldots, x_n^*$ are given by

$$x_k^* = x_{k-1} = a + (k-1)\Delta x \quad \text{Left endpoint} \qquad (3)$$

$$x_k^* = x_k = a + k\Delta x \quad \text{Right endpoint} \qquad (4)$$

$$x_k^* = \tfrac{1}{2}(x_{k-1} + x_k) = a + \left(k - \tfrac{1}{2}\right)\Delta x \quad \text{Midpoint} \qquad (5)$$

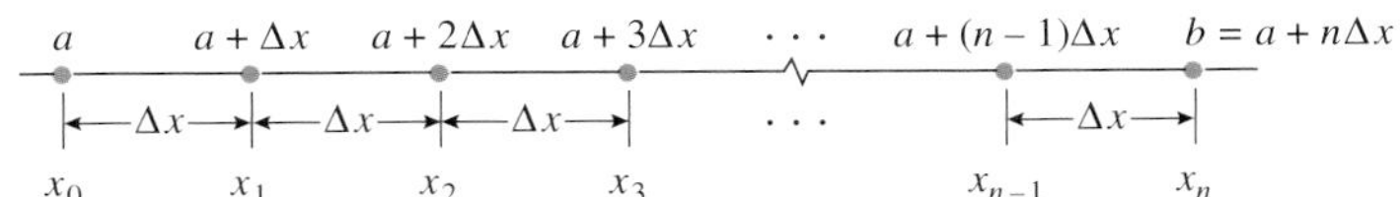

Figure 5.4.6

When applicable, the antiderivative method will be the method of choice for finding exact areas. However, the following examples will help to reinforce the ideas that we have just discussed.

▶ **Example 5** Use Definition 5.4.3 with x_k^* as the right endpoint of each subinterval to find the area between the graph of $f(x) = x^2$ and the interval $[0, 1]$.

Solution. The length of each subinterval is

$$\Delta x = \frac{b-a}{n} = \frac{1-0}{n} = \frac{1}{n}$$

so it follows from (4) that

$$x_k^* = a + k\Delta x = \frac{k}{n}$$

Thus,

$$\begin{aligned}\sum_{k=1}^{n} f(x_k^*)\Delta x = \sum_{k=1}^{n}(x_k^*)^2\Delta x &= \sum_{k=1}^{n}\left(\frac{k}{n}\right)^2\frac{1}{n} = \frac{1}{n^3}\sum_{k=1}^{n}k^2\\ &= \frac{1}{n^3}\left[\frac{n(n+1)(2n+1)}{6}\right] \qquad \text{Part } (b) \text{ of Theorem 5.4.2}\\ &= \frac{1}{3} + \frac{1}{2n} + \frac{1}{6n^2}\end{aligned}$$

from which it follows that

$$A = \lim_{n\to+\infty}\sum_{k=1}^{n} f(x_k^*)\Delta x = \lim_{n\to+\infty}\left(\frac{1}{3} + \frac{1}{2n} + \frac{1}{6n^2}\right) = \frac{1}{3}$$

Observe that this is consistent with the results in Table 5.1.2 and the related discussion in Section 5.1. ◀

In the solution to Example 5 we made use of one of the "closed form" summation formulas from Theorem 5.4.2. The next result collects some consequences of Theorem 5.4.2 that can facilitate computations of area using Definition 5.4.3.

5.4.4 THEOREM.

(*a*) $\displaystyle\lim_{n\to+\infty}\frac{1}{n}\sum_{k=1}^{n}1 = 1$ (*b*) $\displaystyle\lim_{n\to+\infty}\frac{1}{n^2}\sum_{k=1}^{n}k = \frac{1}{2}$

(*c*) $\displaystyle\lim_{n\to+\infty}\frac{1}{n^3}\sum_{k=1}^{n}k^2 = \frac{1}{3}$ (*d*) $\displaystyle\lim_{n\to+\infty}\frac{1}{n^4}\sum_{k=1}^{n}k^3 = \frac{1}{4}$

The proof of Theorem 5.4.4 is left as an exercise for the reader.

▶ **Example 6** Use Definition 5.4.3 with x_k^* as the midpoint of each subinterval to find the area under the parabola $y = f(x) = 9 - x^2$ and over the interval $[0, 3]$.

Solution. Each subinterval has length

$$\Delta x = \frac{b-a}{n} = \frac{3-0}{n} = \frac{3}{n}$$

so it follows from (5) that

$$x_k^* = a + \left(k - \frac{1}{2}\right)\Delta x = \left(k - \frac{1}{2}\right)\left(\frac{3}{n}\right)$$

Thus,

$$\begin{aligned} f(x_k^*)\Delta x = [9 - (x_k^*)^2]\Delta x &= \left[9 - \left(k - \frac{1}{2}\right)^2\left(\frac{3}{n}\right)^2\right]\left(\frac{3}{n}\right) \\ &= \left[9 - \left(k^2 - k + \frac{1}{4}\right)\left(\frac{9}{n^2}\right)\right]\left(\frac{3}{n}\right) \\ &= \frac{27}{n} - \frac{27}{n^3}k^2 + \frac{27}{n^3}k - \frac{27}{4n^3} \end{aligned}$$

from which it follows that

$$\begin{aligned} A &= \lim_{n\to+\infty}\sum_{k=1}^{n} f(x_k^*)\Delta x \\ &= \lim_{n\to+\infty}\sum_{k=1}^{n}\left(\frac{27}{n} - \frac{27}{n^3}k^2 + \frac{27}{n^3}k - \frac{27}{4n^3}\right) \\ &= \lim_{n\to+\infty} 27\left[\frac{1}{n}\sum_{k=1}^{n}1 - \frac{1}{n^3}\sum_{k=1}^{n}k^2 + \frac{1}{n}\left(\frac{1}{n^2}\sum_{k=1}^{n}k\right) - \frac{1}{4n^2}\left(\frac{1}{n}\sum_{k=1}^{n}1\right)\right] \\ &= 27\left[1 - \frac{1}{3} + 0\cdot\frac{1}{2} - 0\cdot 1\right] = 18 \qquad \text{Theorem 5.4.4} \end{aligned}$$

◀

NET SIGNED AREA

In Definition 5.4.3 we assumed that f is continuous and nonnegative on the interval $[a, b]$. If f is continuous and attains both positive and negative values on $[a, b]$, then the limit

$$\lim_{n\to+\infty}\sum_{k=1}^{n} f(x_k^*)\Delta x \tag{6}$$

no longer represents the area between the curve $y = f(x)$ and the interval $[a, b]$ on the x-axis; rather, it represents a difference of areas—the area of the region that is above the interval $[a, b]$ and below the curve $y = f(x)$ minus the area of the region that is below the interval $[a, b]$ and above the curve $y = f(x)$. We call this the ***net signed area*** between the graph of $y = f(x)$ and the interval $[a, b]$. For example, in Figure 5.4.7a, the net signed area between the curve $y = f(x)$ and the interval $[a, b]$ is

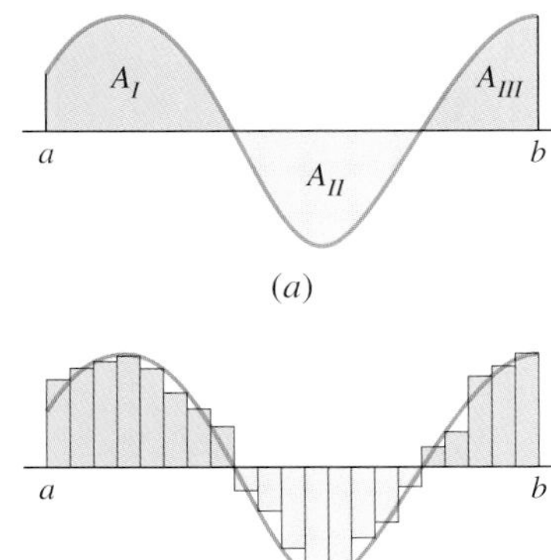

Figure 5.4.7

$$(A_I + A_{III}) - A_{II} = \big[\text{area above } [a, b]\big] - \big[\text{area below } [a, b]\big]$$

To explain why the limit in (6) represents this net signed area, let us subdivide the interval $[a, b]$ in Figure 5.4.7a into n equal subintervals and examine the terms in the sum

$$\sum_{k=1}^{n} f(x_k^*)\Delta x \tag{7}$$

If $f(x_k^*)$ is positive, then the product $f(x_k^*)\Delta x$ represents the area of the rectangle with height $f(x_k^*)$ and base Δx (the pink rectangles in Figure 5.4.7b). However, if $f(x_k^*)$ is negative, then the product $f(x_k^*)\Delta x$ is the *negative* of the area of the rectangle with height $|f(x_k^*)|$ and base Δx (the green rectangles in Figure 5.4.7b). Thus, (7) represents the total area of the pink rectangles minus the total area of the green rectangles. As n increases, the pink rectangles fill out the regions with areas A_I and A_{III} and the green rectangles fill out the region with area A_{II}, which explains why the limit in (6) represents the signed area between $y = f(x)$ and the interval $[a, b]$. We formalize this in the following definition.

As with Definition 5.4.3, it can be proved that the limit in (8) always exists and that the same value of A results no matter how the points in the subintervals are chosen.

5.4.5 DEFINITION (*Net Signed Area*). If the function f is continuous on $[a, b]$, then the ***net signed area*** A between $y = f(x)$ and the interval $[a, b]$ is defined by

$$A = \lim_{n \to +\infty} \sum_{k=1}^{n} f(x_k^*)\Delta x \tag{8}$$

▶ **Example 7** Use Definition 5.4.5 with x_k^* as the left endpoint of each subinterval to find the net signed area between the graph of $y = f(x) = x - 1$ and the interval $[0, 2]$.

Solution. Each subinterval has length

$$\Delta x = \frac{b-a}{n} = \frac{2-0}{n} = \frac{2}{n}$$

so it follows from (3) that

$$x_k^* = a + (k-1)\Delta x = (k-1)\left(\frac{2}{n}\right)$$

Thus,

$$\sum_{k=1}^{n} f(x_k^*)\Delta x = \sum_{k=1}^{n} (x_k^* - 1)\Delta x = \sum_{k=1}^{n} \left[(k-1)\left(\frac{2}{n}\right) - 1\right]\left(\frac{2}{n}\right)$$

$$= \sum_{k=1}^{n} \left[\left(\frac{4}{n^2}\right)k - \frac{4}{n^2} - \frac{2}{n}\right]$$

from which it follows that

$$A = \lim_{n \to +\infty} \sum_{k=1}^{n} f(x_k^*)\Delta x = \lim_{n \to +\infty} \left[4\left(\frac{1}{n^2}\sum_{k=1}^{n} k\right) - \frac{4}{n}\left(\frac{1}{n}\sum_{k=1}^{n} 1\right) - 2\left(\frac{1}{n}\sum_{k=1}^{n} 1\right)\right]$$

$$= 4\left(\frac{1}{2}\right) - 0 \cdot 1 - 2 \cdot 1 = 0 \qquad \text{Theorem 5.4.4}$$

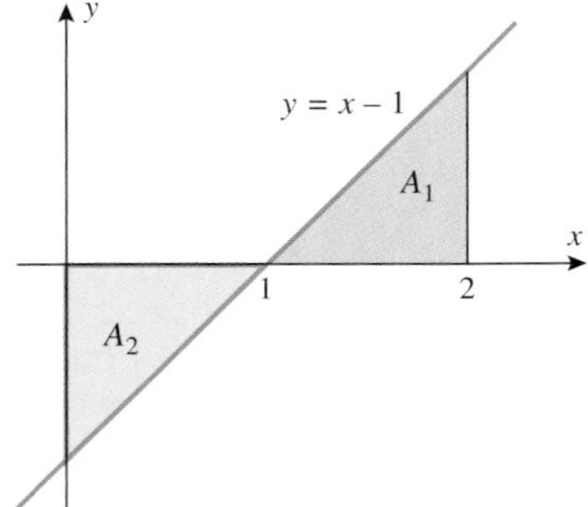

Figure 5.4.8

Since the net signed area is zero, the area A_1 below the graph of f and above the interval $[0, 2]$ must equal the area A_2 above the graph of f and below the interval $[0, 2]$. This conclusion agrees with the graph of f shown in Figure 5.4.8. ◀

NUMERICAL APPROXIMATIONS OF AREA

Although the antiderivative method discussed in Section 5.1 (and to be studied in more detail later) is generally more efficient than Definition 5.4.3 for calculating exact areas, this definition is useful for *approximating* areas. If follows from this definition that if n is large, then

$$\sum_{k=1}^{n} f(x_k^*)\Delta x = \Delta x \sum_{k=1}^{n} f(x_k^*) = \Delta x[f(x_1^*) + f(x_2^*) + \cdots + f(x_n^*)] \tag{9}$$

will be a good approximation to the area A. If one of Formulas (3), (4), or (5) is used to choose the x_k^* in (9), then the result is called the ***left endpoint approximation***, the ***right endpoint approximation***, or the ***midpoint approximation***, respectively (Figure 5.4.9).

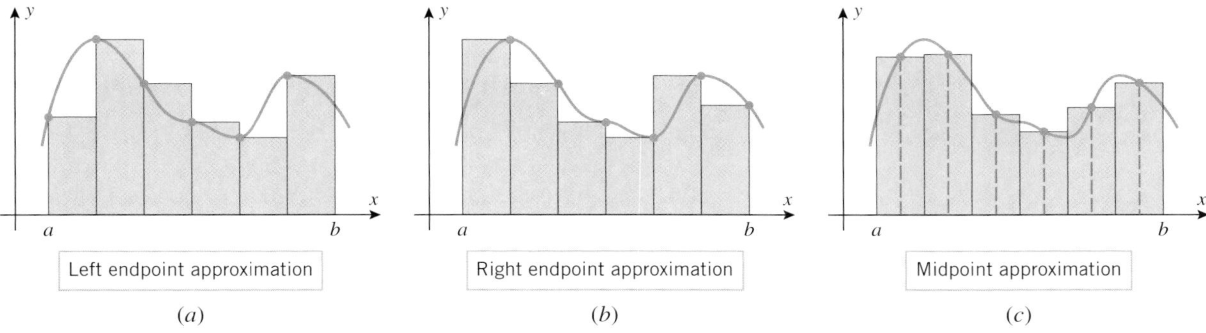

Figure 5.4.9

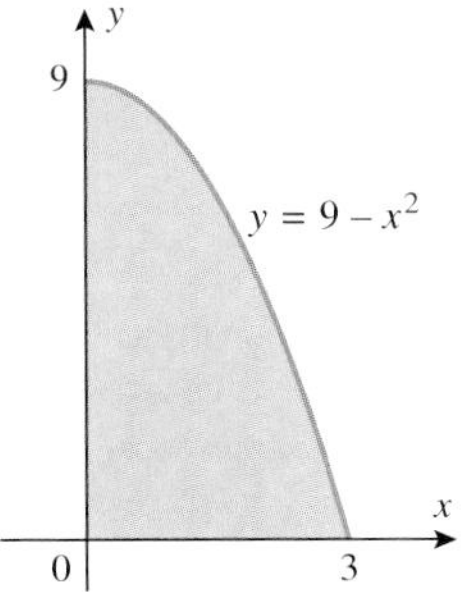

Figure 5.4.10

▶ **Example 8** Find the left endpoint, right endpoint, and midpoint approximations of the area under the curve $y = 9 - x^2$ over the interval $[0, 3]$ with $n = 10$, $n = 20$, and $n = 50$ (Figure 5.4.10). Compare the accuracies of these three methods.

Solution. Details of the computations for the case $n = 10$ are shown to six decimal places in Table 5.4.1 and the results of all the computations are given in Table 5.4.2. We showed in Example 6 that the exact area is 18 (i.e., 18 square units), so in this case the midpoint approximation is more accurate than the endpoint approximations. This is also evident geometrically from Figure 5.4.11. You can also see from the figure that in this case the left endpoint approximation overestimates the area and the right endpoint approximation underestimates it. Later in the text we will investigate the error that results when an area is approximated by the midpoint rule. ◀

Table 5.4.1

$n = 10,\ \Delta x = (b-a)/n = (3-0)/10 = 0.3$

	LEFT ENDPOINT APPROXIMATION		RIGHT ENDPOINT APPROXIMATION		MIDPOINT APPROXIMATION	
k	x_k^*	$9-(x_k^*)^2$	x_k^*	$9-(x_k^*)^2$	x_k^*	$9-(x_k^*)^2$
1	0.0	9.000000	0.3	8.910000	0.15	8.977500
2	0.3	8.910000	0.6	8.640000	0.45	8.797500
3	0.6	8.640000	0.9	8.190000	0.75	8.437500
4	0.9	8.190000	1.2	7.560000	1.05	7.897500
5	1.2	7.560000	1.5	6.750000	1.35	7.177500
6	1.5	6.750000	1.8	5.760000	1.65	6.277500
7	1.8	5.760000	2.1	4.590000	1.95	5.197500
8	2.1	4.590000	2.4	3.240000	2.25	3.937500
9	2.4	3.240000	2.7	1.710000	2.55	2.497500
10	2.7	1.710000	3.0	0.000000	2.85	0.877500
		64.350000		55.350000		60.075000
$\Delta x \sum_{k=1}^{n} f(x_k^*)$		(0.3)(64.350000) = 19.305000		(0.3)(55.350000) = 16.605000		(0.3)(60.075000) = 18.022500

Table 5.4.2

n	LEFT ENDPOINT APPROXIMATION	RIGHT ENDPOINT APPROXIMATION	MIDPOINT APPROXIMATION
10	19.305000	16.605000	18.022500
20	18.663750	17.313750	18.005625
50	18.268200	17.728200	18.000900

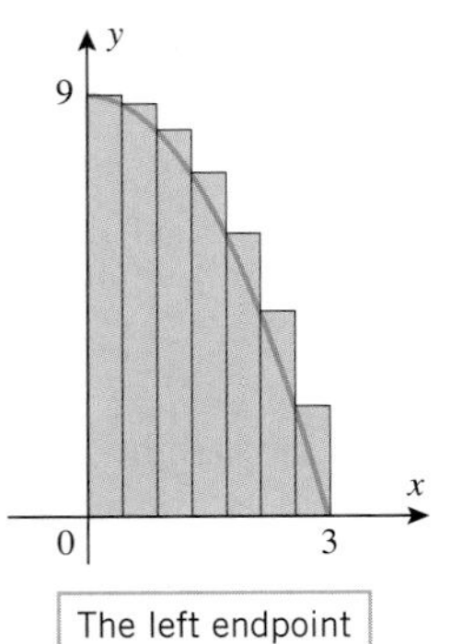

The right endpoint approximation underestimates the area.

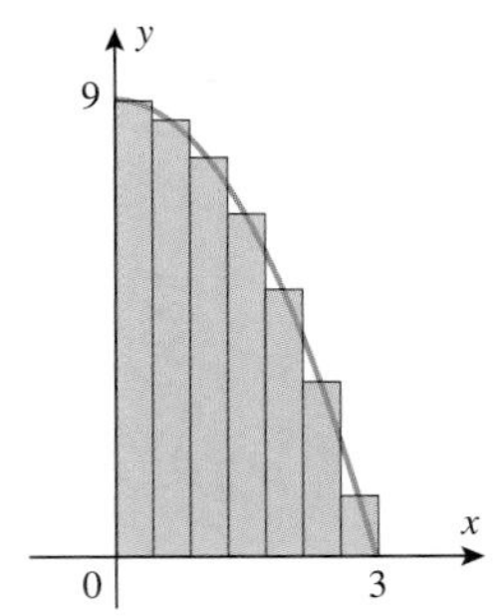

Figure 5.4.11

✔ QUICK CHECK EXERCISES 5.4 *(See page 337 for answers.)*

1. (a) Write the sum in two ways:

$$\frac{1}{2}+\frac{1}{4}+\frac{1}{6}+\frac{1}{8}=\sum_{k=1}^{4}\_\_\_\_\_\_=\sum_{j=0}^{3}\_\_\_\_\_\_$$

(b) Express the sum $10+10^2+10^3+10^4+10^5$ using sigma notation.

2. Express the sums in closed form.

(a) $\sum_{k=1}^{n} k$ (b) $\sum_{k=1}^{n}(6k+1)$ (c) $\sum_{k=1}^{n} k^2$

3. Divide the interval $[1, 3]$ into $n = 4$ subintervals of equal length.

(a) Each subintervals has width ______.

(b) The left endpoints of the subintervals are ______.

(c) The midpoints of the subintervals are ______.

(d) The right endpoints of the subintervals are ______.

4. Find the left endpoint approximation for the area between the curve $y = x^2$ and the interval $[1, 3]$ using $n = 4$ equal subdivisions of the interval.

5. The right endpoint approximation for the net signed area between $y = f(x)$ and an interval $[a, b]$ is given by

$$\sum_{k=1}^{n}\frac{6k}{n^2}$$

Find the exact value of this net signed area.

EXERCISE SET 5.4 [C] CAS

1. Evaluate.

(a) $\sum_{k=1}^{3} k^3$ (b) $\sum_{j=2}^{6}(3j-1)$ (c) $\sum_{i=-4}^{1}(i^2-i)$

(d) $\sum_{n=0}^{5} 1$ (e) $\sum_{k=0}^{4}(-2)^k$ (f) $\sum_{n=1}^{6}\sin n\pi$

2. Evaluate.

(a) $\sum_{k=1}^{4} k\sin\frac{k\pi}{2}$ (b) $\sum_{j=0}^{5}(-1)^j$ (c) $\sum_{i=7}^{20}\pi^2$

(d) $\sum_{m=3}^{5} 2^{m+1}$ (e) $\sum_{n=1}^{6}\sqrt{n}$ (f) $\sum_{k=0}^{10}\cos k\pi$

3–8 Write each expression in sigma notation but do not evaluate.

3. $1 + 2 + 3 + \cdots + 10$

4. $3 \cdot 1 + 3 \cdot 2 + 3 \cdot 3 + \cdots + 3 \cdot 20$

5. $2 + 4 + 6 + 8 + \cdots + 20$ **6.** $1 + 3 + 5 + 7 + \cdots + 15$

7. $1 - 3 + 5 - 7 + 9 - 11$ **8.** $1 - \frac{1}{2} + \frac{1}{3} - \frac{1}{4} + \frac{1}{5}$

9. (a) Express the sum of the even integers from 2 to 100 in sigma notation.
(b) Express the sum of the odd integers from 1 to 99 in sigma notation.

10. Express in sigma notation.
(a) $a_1 - a_2 + a_3 - a_4 + a_5$
(b) $-b_0 + b_1 - b_2 + b_3 - b_4 + b_5$
(c) $a_0 + a_1x + a_2x^2 + \cdots + a_nx^n$
(d) $a^5 + a^4b + a^3b^2 + a^2b^3 + ab^4 + b^5$

11–16 Use Theorem 5.4.2 to evaluate the sums. Check your answers using the summation feature of a calculating utility.

11. $\sum_{k=1}^{100} k$ **12.** $\sum_{k=1}^{100} (7k + 1)$ **13.** $\sum_{k=1}^{20} k^2$

14. $\sum_{k=4}^{20} k^2$ **15.** $\sum_{k=1}^{30} k(k-2)(k+2)$

16. $\sum_{k=1}^{6} (k - k^3)$

17–20 Express the sums in closed form.

17. $\sum_{k=1}^{n} \frac{3k}{n}$ **18.** $\sum_{k=1}^{n-1} \frac{k^2}{n}$ **19.** $\sum_{k=1}^{n-1} \frac{k^3}{n^2}$

20. $\sum_{k=1}^{n} \left(\frac{5}{n} - \frac{2k}{n}\right)$

C **21.** For each of the sums that you obtained in Exercises 17–20, use a CAS to check your answer. If the answer produced by the CAS does not match your own, show that the two answers are equivalent.

22. Solve the equation $\sum_{k=1}^{n} k = 465$.

23–26 Express the function of n in closed form and then find the limit.

23. $\lim_{n\to+\infty} \frac{1 + 2 + 3 + \cdots + n}{n^2}$

24. $\lim_{n\to+\infty} \frac{1^2 + 2^2 + 3^2 + \cdots + n^2}{n^3}$

25. $\lim_{n\to+\infty} \sum_{k=1}^{n} \frac{5k}{n^2}$ **26.** $\lim_{n\to+\infty} \sum_{k=1}^{n-1} \frac{2k^2}{n^3}$

27. Express $1 + 2 + 2^2 + 2^3 + 2^4 + 2^5$ in sigma notation with
(a) $j = 0$ as the lower limit of summation
(b) $j = 1$ as the lower limit of summation
(c) $j = 2$ as the lower limit of summation.

28. Express

$$\sum_{k=5}^{9} k2^{k+4}$$

in sigma notation with
(a) $k = 1$ as the lower limit of summation
(b) $k = 13$ as the upper limit of summation.

FOCUS ON CONCEPTS

29. (a) Write the first three and final two summands in the sum

$$\sum_{k=1}^{n} \left(2 + k \cdot \frac{3}{n}\right)^4 \frac{3}{n}$$

Explain why this sum gives the right endpoint approximation for the area under the curve $y = x^4$ over the interval $[2, 5]$.
(b) Show that a change in the index range of the sum in part (a) can produce the left endpoint approximation for the area under the curve $y = x^4$ over the interval $[2, 5]$.

30. For a function f that is continuous on $[a, b]$, Definition 5.4.5 says that the net signed area A between $y = f(x)$ and the interval $[a, b]$ is

$$A = \lim_{n\to+\infty} \sum_{k=1}^{n} f(x_k^*)\Delta x$$

Give geometric interpretations for the symbols n, x_k^*, and Δx. Explain how to interpret the limit in this definition.

31–34 Divide the specified interval into $n = 4$ subintervals of equal length and then compute

$$\sum_{k=1}^{4} f(x_k^*)\Delta x$$

with x_k^* as (a) the left endpoint of each subinterval, (b) the midpoint of each subinterval, and (c) the right endpoint of each subinterval. Illustrate each part with a graph of f that includes the rectangles whose areas are represented in the sum.

31. $f(x) = 3x + 1$; $[2, 6]$ **32.** $f(x) = 1/x$; $[1, 9]$

33. $f(x) = \cos x$; $[0, \pi]$ **34.** $f(x) = 2x - x^2$; $[-1, 3]$

35–38 Use a calculating utility with summation capabilities or a CAS to obtain an approximate value for the area between the curve $y = f(x)$ and the specified interval with $n = 10$, 20, and 50 subintervals using the (a) left endpoint, (b) midpoint, and (c) right endpoint approximations.

C **35.** $f(x) = 1/x$; $[1, 2]$ C **36.** $f(x) = 1/x^2$; $[1, 3]$

C 37. $f(x)=\sqrt{x};\ [0,4]$ C 38. $f(x)=\sin x;\ [0,\pi/2]$

39–44 Use Definition 5.4.3 with x_k^* as the *right* endpoint of each subinterval to find the area under the curve $y=f(x)$ over the specified interval.

39. $f(x)=x/2;\ [1,4]$
40. $f(x)=5-x;\ [0,5]$
41. $f(x)=9-x^2;\ [0,3]$
42. $f(x)=4-\frac{1}{4}x^2;\ [0,3]$
43. $f(x)=x^3;\ [2,6]$
44. $f(x)=1-x^3;\ [-3,-1]$

45–48 Use Definition 5.4.3 with x_k^* as the *left* endpoint of each subinterval to find the area under the curve $y=f(x)$ over the specified interval.

45. $f(x)=x/2;\ [1,4]$
46. $f(x)=5-x;\ [0,5]$
47. $f(x)=9-x^2;\ [0,3]$
48. $f(x)=4-\frac{1}{4}x^2;\ [0,3]$

49–52 Use Definition 5.4.3 with x_k^* as the *midpoint* of each subinterval to find the area under the curve $y=f(x)$ over the specified interval.

49. $f(x)=2x;\ [0,4]$
50. $f(x)=6-x;\ [1,5]$
51. $f(x)=x^2;\ [0,1]$
52. $f(x)=x^2;\ [-1,1]$

53–56 Use Definition 5.4.5 with x_k^* as the *right* endpoint of each subinterval to find the net signed area between the curve $y=f(x)$ and the specified interval.

53. $f(x)=x;\ [-1,1]$. Verify your answer with a simple geometric argument.
54. $f(x)=x;\ [-1,2]$. Verify your answer with a simple geometric argument.
55. $f(x)=x^2-1;\ [0,2]$
56. $f(x)=x^3;\ [-1,1]$
57. Use Definition 5.4.3 with x_k^* as the left endpoint of each subinterval to find the area under the graph of $y=mx$ and over the interval $[a,b]$, where $m>0$ and $a\geq 0$.
58. Use Definition 5.4.5 with x_k^* as the right endpoint of each subinterval to find the net signed area between the graph of $y=mx$ and the interval $[a,b]$.
59. (a) Show that the area under the graph of $y=x^3$ and over the interval $[0,b]$ is $b^4/4$.
    (b) Find a formula for the area under $y=x^3$ over the interval $[a,b]$, where $a\geq 0$.
60. Find the area between the graph of $y=\sqrt{x}$ and the interval $[0,1]$. [*Hint:* Use the result of Exercise 21 of Section 5.1.]
61. An artist wants to create a rough triangular design using uniform square tiles glued edge to edge. She places n tiles in a row to form the base of the triangle and then makes each successive row two tiles shorter than the preceding row. Find a formula for the number of tiles used in the design. [*Hint:* Your answer will depend on whether n is even or odd.]
62. An artist wants to create a sculpture by gluing together uniform spheres. She creates a rough rectangular base that has 50 spheres along one edge and 30 spheres along the other. She then creates successive layers by gluing spheres in the grooves of the preceding layer. How many spheres will there be in the sculpture?

63–66 Consider the sum

$$\sum_{k=1}^{4}[(k+1)^3-k^3]=[5^3-4^3]+[4^3-3^3]+[3^3-2^3]+[2^3-1^3]=5^3-1^3=124$$

For convenience, the terms are listed in reverse order. Note how cancellation allows the entire sum to collapse like a telescope. A sum is said to ***telescope*** when part of each term cancels part of an adjacent term, leaving only portions of the first and last terms uncanceled. Evaluate the telescoping sums in these exercises.

63. $\displaystyle\sum_{k=5}^{17}(3^k-3^{k-1})$
64. $\displaystyle\sum_{k=1}^{50}\left(\frac{1}{k}-\frac{1}{k+1}\right)$
65. $\displaystyle\sum_{k=2}^{20}\left(\frac{1}{k^2}-\frac{1}{(k-1)^2}\right)$
66. $\displaystyle\sum_{k=1}^{100}(2^{k+1}-2^k)$
67. (a) Show that
$$\frac{1}{1\cdot 3}+\frac{1}{3\cdot 5}+\cdots+\frac{1}{(2n-1)(2n+1)}=\frac{n}{2n+1}$$
$$\left[\textit{Hint: }\frac{1}{(2n-1)(2n+1)}=\frac{1}{2}\left(\frac{1}{2n-1}-\frac{1}{2n+1}\right).\right]$$
    (b) Use the result in part (a) to find
$$\lim_{n\to+\infty}\sum_{k=1}^{n}\frac{1}{(2k-1)(2k+1)}$$
68. (a) Show that
$$\frac{1}{1\cdot 2}+\frac{1}{2\cdot 3}+\frac{1}{3\cdot 4}+\cdots+\frac{1}{n(n+1)}=\frac{n}{n+1}$$
$$\left[\textit{Hint: }\frac{1}{n(n+1)}=\frac{1}{n}-\frac{1}{n+1}.\right]$$
    (b) Use the result in part (a) to find
$$\lim_{n\to+\infty}\sum_{k=1}^{n}\frac{1}{k(k+1)}$$
69. Let $\bar{x}$ denote the arithmetic average of the n numbers $x_1,x_2,\ldots,x_n$. Use Theorem 5.4.1 to prove that
$$\sum_{i=1}^{n}(x_i-\bar{x})=0$$
70. Let
$$S=\sum_{k=0}^{n}ar^k$$
Show that $S-rS=a-ar^{n+1}$ and hence that
$$\sum_{k=0}^{n}ar^k=\frac{a-ar^{n+1}}{1-r}\quad(r\neq 1)$$
(A sum of this form is called a ***geometric sum***.)

71. In each part, rewrite the sum, if necessary, so that the lower limit is 0, and then use the formula derived in Exercise 70 to evaluate the sum. Check your answers using the summation feature of a calculating utility.

(a) $\sum_{k=1}^{20} 3^k$ (b) $\sum_{k=5}^{30} 2^k$ (c) $\sum_{k=0}^{100} (-1)^{k+1} \frac{1}{2^k}$

C **72.** In each part, make a conjecture about the limit by using a CAS to evaluate the sum for $n = 10, 20$, and 50. Check your conjecture by using the formula in Exercise 70 to express the sum in closed form, and then find the limit exactly.

(a) $\lim_{n \to +\infty} \sum_{k=0}^{n} \frac{1}{2^k}$ (b) $\lim_{n \to +\infty} \sum_{k=1}^{n} \left(\frac{3}{4}\right)^k$

73. By writing out the sums, determine whether the following are valid identities.

(a) $\int \left[\sum_{i=1}^{n} f_i(x)\right] dx = \sum_{i=1}^{n} \left[\int f_i(x)\, dx\right]$

(b) $\frac{d}{dx}\left[\sum_{i=1}^{n} f_i(x)\right] = \sum_{i=1}^{n} \left[\frac{d}{dx}[f_i(x)]\right]$

74. Which of the following are valid identities?

(a) $\sum_{i=1}^{n} a_i b_i = \sum_{i=1}^{n} a_i \sum_{i=1}^{n} b_i$ (b) $\sum_{i=1}^{n} \frac{a_i}{b_i} = \sum_{i=1}^{n} a_i \Big/ \sum_{i=1}^{n} b_i$

(c) $\sum_{i=1}^{n} a_i^2 = \left(\sum_{i=1}^{n} a_i\right)^2$

75. Prove part (*c*) of Theorem 5.4.1.

76. Prove Theorem 5.4.4.

✔ QUICK CHECK ANSWERS 5.4

1. (a) $\frac{1}{2k}$; $\frac{1}{2(j+1)}$ (b) $\sum_{k=1}^{5} 10^k$ **2.** (a) $\frac{n(n+1)}{2}$ (b) $3n(n+1)+n$ (c) $\frac{n(n+1)(2n+1)}{6}$ **3.** (a) 0.5 (b) 1, 1.5, 2, 2.5 (c) 1.25, 1.75, 2.25, 2.75 (d) 1.5, 2, 2.5, 3 **4.** 6.75 **5.** $\lim_{n \to +\infty} \frac{3n^2 + 4n}{n^2} = 3$

5.5 THE DEFINITE INTEGRAL

In this section we will introduce the concept of a "definite integral," which will link the concept of area to other important concepts such as length, volume, density, probability, and work.

RIEMANN SUMS AND THE DEFINITE INTEGRAL

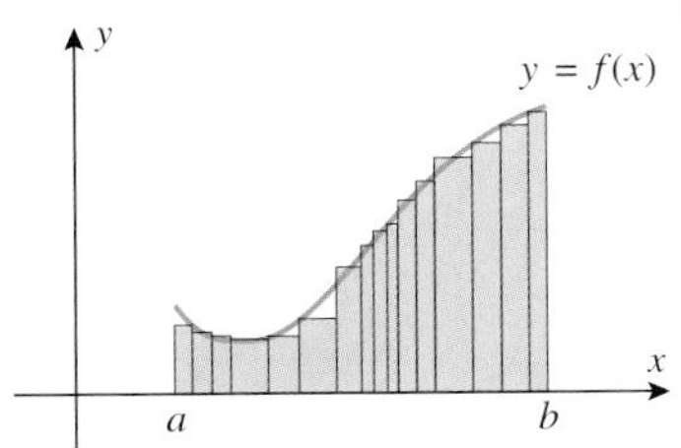

Figure 5.5.1

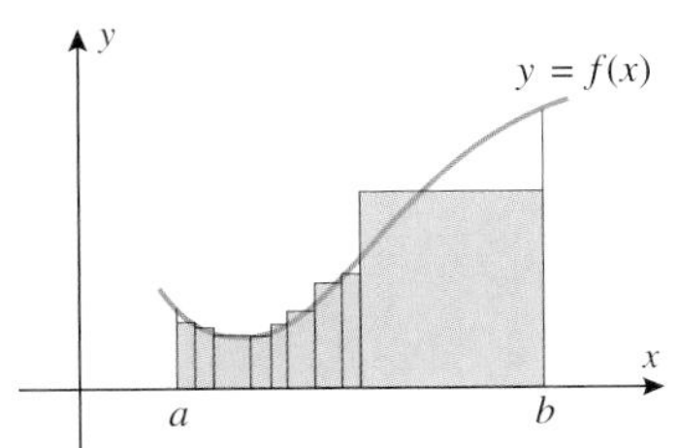

Figure 5.5.2

In our definition of net signed area (Definition 5.4.5), we assumed that for each positive number n, the interval $[a, b]$ was subdivided into n subintervals of equal length to create bases for the approximating rectangles. For some functions it may be more convenient to use rectangles with different widths (see Exercise 39); however, if we are to "exhaust" an area with rectangles of different widths, then it is important that successive subdivisions be constructed in such a way that the widths of the rectangles approach zero as n increases (Figure 5.5.1). Thus, we must preclude the kind of situation that occurs in Figure 5.5.2 in which the right half of the interval is never subdivided. If this kind of subdivision were allowed, the error in the approximation would not approach zero as n increased.

A ***partition*** of the interval $[a, b]$ is a collection of points

$$a = x_0 < x_1 < x_2 < \cdots < x_{n-1} < x_n = b$$

that divides $[a, b]$ into n subintervals of lengths

$$\Delta x_1 = x_1 - x_0, \quad \Delta x_2 = x_2 - x_1, \quad \Delta x_3 = x_3 - x_2, \ldots, \quad \Delta x_n = x_n - x_{n-1}$$

The partition is said to be ***regular*** provided the subintervals all have the same length

$$\Delta x_k = \Delta x = \frac{b-a}{n}$$

For a regular partition, the widths of the approximating rectangles approach zero as n is made large. Since this need not be the case for a general partition, we need some way to measure the "size" of these widths. One approach is to let $\max \Delta x_k$ denote the largest of the subinterval widths. The magnitude $\max \Delta x_k$ is called the ***mesh size*** of the partition. For example, Figure 5.5.3 shows a partition of the interval $[0, 6]$ into four subintervals with a mesh size of 2.

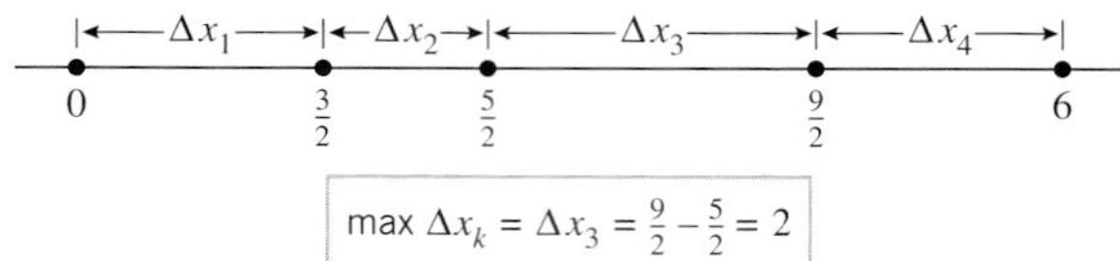

Figure 5.5.3

If we are to generalize Definition 5.4.5 so that it allows for unequal subinterval widths, we must replace the constant length Δx by the variable length Δx_k. When this is done the sum

$$\sum_{k=1}^{n} f(x_k^*)\Delta x \quad \text{is replaced by} \quad \sum_{k=1}^{n} f(x_k^*)\Delta x_k$$

Some writers use the symbol $\|\Delta\|$ rather than $\max \Delta x_k$ for the mesh size of the partition, in which case $\max \Delta x_k \to 0$ would be replaced by $\|\Delta\| \to 0$.

We also need to replace the expression $n \to +\infty$ by an expression that guarantees us that the lengths of all subintervals approach zero. We will use the expression $\max \Delta x_k \to 0$ for this purpose. Based on our intuitive concept of area, we would then expect the net signed area A between the graph of f and the interval $[a, b]$ to satisfy the equation

$$A = \lim_{\max \Delta x_k \to 0} \sum_{k=1}^{n} f(x_k^*)\Delta x_k$$

(We will see in a moment that this is the case.) The limit that appears in this expression is one of the fundamental concepts of integral calculus and forms the basis for the following definition.

5.5.1 DEFINITION. A function f is said to be ***integrable*** on a finite closed interval $[a, b]$ if the limit

$$\lim_{\max \Delta x_k \to 0} \sum_{k=1}^{n} f(x_k^*)\Delta x_k$$

exists and does not depend on the choice of partitions or on the choice of the points x_k^* in the subintervals. When this is the case we denote the limit by the symbol

$$\int_a^b f(x)\,dx = \lim_{\max \Delta x_k \to 0} \sum_{k=1}^{n} f(x_k^*)\Delta x_k$$

which is called the ***definite integral*** of f from a to b. The numbers a and b are called the ***lower limit of integration*** and the ***upper limit of integration***, respectively, and $f(x)$ is called the ***integrand***.

The notation used for the definite integral deserves some comment. Historically, the expression "$f(x)\,dx$" was interpreted to be the "infinitesimal area" of a rectangle with height $f(x)$ and "infinitesimal" width dx. By "summing" these infinitesimal areas, the entire area under the curve was obtained. The integral symbol "$\int$" is an "elongated s" that was used to indicate this summation. For us, the integral symbol "$\int$" and the symbol

"dx" can serve as reminders that the definite integral is actually a limit of a *summation* as $\Delta x_k \to 0$. The sum that appears in Definition 5.5.1 is called a ***Riemann sum***, and the definite integral is sometimes called the ***Riemann integral*** in honor of the German mathematician Bernhard Riemann who formulated many of the basic concepts of integral calculus. (The reason for the similarity in notation between the definite integral and the indefinite integral will become clear in the next section, where we will establish a link between the two types of "integration.")

The limit that appears in Definition 5.5.1 is somewhat different from the kinds of limits discussed in Chapter 2. Loosely phrased, the expression

$$\lim_{\max \Delta x_k \to 0} \sum_{k=1}^{n} f(x_k^*)\Delta x_k = L$$

is intended to convey the idea that we can force the Riemann sums to be as close as we please to L, regardless of how the values of x_k^* are chosen, by making the mesh size of the partition sufficiently small. While it is possible to give a more formal definition of this limit, we will simply rely on intuitive arguments when applying Definition 5.5.1.

Although a function need not be continuous on an interval to be integrable on that interval (Exercise 35), we will be interested primarily in definite integrals of continuous functions. The following theorem, which we will state without proof, says that if a function is continuous on a finite closed interval, then it is integrable on that interval, and its definite integral is the net signed area between the graph of the function and the interval.

5.5.2 THEOREM. *If a function f is continuous on an interval $[a, b]$, then f is integrable on $[a, b]$, and the net signed area A between the graph of f and the interval $[a, b]$ is*

$$A = \int_a^b f(x)\,dx \qquad (1)$$

Georg Friedrich Bernhard Riemann (1826–1866) German mathematician. Bernhard Riemann, as he is commonly known, was the son of a Protestant minister. He received his elementary education from his father and showed brilliance in arithmetic at an early age. In 1846 he enrolled at Göttingen University to study theology and philology, but he soon transferred to mathematics. He studied physics under W. E. Weber and mathematics under Carl Friedrich Gauss, whom some people consider to be the greatest mathematician who ever lived. In 1851 Riemann received his Ph.D. under Gauss, after which he remained at Göttingen to teach. In 1862, one month after his marriage, Riemann suffered an attack of pleuritis, and for the remainder of his life was an extremely sick man. He finally succumbed to tuberculosis in 1866 at age 39.

An interesting story surrounds Riemann's work in geometry. For his introductory lecture prior to becoming an associate professor, Riemann submitted three possible topics to Gauss. Gauss surprised Riemann by choosing the topic Riemann liked the least, the foundations of geometry. The lecture was like a scene from a movie. The old and failing Gauss, a giant in his day, watching intently as his brilliant and youthful protégé skillfully pieced together portions of the old man's own work into a complete and beautiful system. Gauss is said to have gasped with delight as the lecture neared its end, and on the way home he marveled at his student's brilliance. Gauss died shortly thereafter. The results presented by Riemann that day eventually evolved into a fundamental tool that Einstein used some 50 years later to develop relativity theory.

In addition to his work in geometry, Riemann made major contributions to the theory of complex functions and mathematical physics. The notion of the definite integral, as it is presented in most basic calculus courses, is due to him. Riemann's early death was a great loss to mathematics, for his mathematical work was brilliant and of fundamental importance.

Formula (1) follows from the integrability of f, since the integrability allows us to use any partitions to evaluate the integral. In particular, if we use *regular* partitions of $[a, b]$, then

$$\Delta x_k = \Delta x = \frac{b-a}{n}$$

for all values of k. This implies that $\max \Delta x_k = (b-a)/n$, from which it follows that $\max \Delta x_k \to 0$ if and only if $n \to +\infty$. Thus,

$$\int_a^b f(x)\,dx = \lim_{\max \Delta x_k \to 0} \sum_{k=1}^{n} f(x_k^*)\Delta x_k = \lim_{n \to +\infty} \sum_{k=1}^{n} f(x_k^*)\Delta x = A$$

In the simplest cases, definite integrals of continuous functions can be calculated using formulas from plane geometry to compute signed areas.

▶ **Example 1** Sketch the region whose area is represented by the definite integral, and evaluate the integral using an appropriate formula from geometry.

$$\text{(a)}\ \int_1^4 2\,dx \qquad \text{(b)}\ \int_{-1}^2 (x+2)\,dx \qquad \text{(c)}\ \int_0^1 \sqrt{1-x^2}\,dx$$

Solution (a). The graph of the integrand is the horizontal line $y = 2$, so the region is a rectangle of height 2 extending over the interval from 1 to 4 (Figure 5.5.4*a*). Thus,

$$\int_1^4 2\,dx = (\text{area of rectangle}) = 2(3) = 6$$

Solution (b). The graph of the integrand is the line $y = x + 2$, so the region is a trapezoid whose base extends from $x = -1$ to $x = 2$ (Figure 5.5.4*b*). Thus,

$$\int_{-1}^2 (x+2)\,dx = (\text{area of trapezoid}) = \frac{1}{2}(1+4)(3) = \frac{15}{2}$$

Solution (c). The graph of $y = \sqrt{1-x^2}$ is the upper semicircle of radius 1, centered at the origin, so the region is the right quarter-circle extending from $x = 0$ to $x = 1$ (Figure 5.5.4*c*). Thus,

$$\int_0^1 \sqrt{1-x^2}\,dx = (\text{area of quarter-circle}) = \frac{1}{4}\pi(1^2) = \frac{\pi}{4} \blacktriangleleft$$

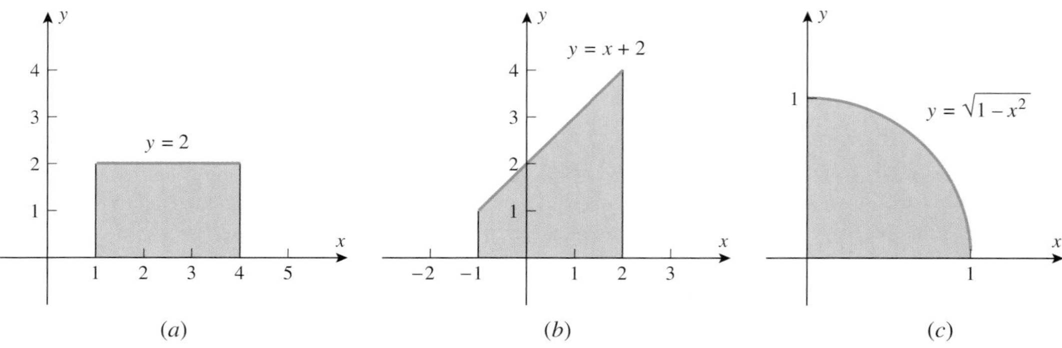

Figure 5.5.4

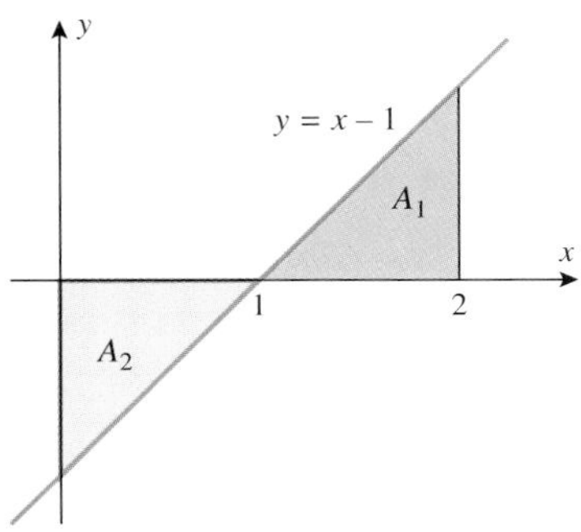

Figure 5.5.5

▶ **Example 2** Evaluate

$$\text{(a)} \int_0^2 (x-1)\,dx \qquad \text{(b)} \int_0^1 (x-1)\,dx$$

Solution. The graph of $y = x - 1$ is shown in Figure 5.5.5, and we leave it for you to verify that the shaded triangular regions both have area $\frac{1}{2}$. Over the interval $[0, 2]$ the net signed area is $A_1 - A_2 = \frac{1}{2} - \frac{1}{2} = 0$, and over the interval $[0, 1]$ the net signed area is $-A_2 = -\frac{1}{2}$. Thus,

$$\int_0^2 (x-1)\,dx = 0 \quad \text{and} \quad \int_0^1 (x-1)\,dx = -\tfrac{1}{2}$$

(Recall that in Example 7 of Section 5.4, we used Definition 5.4.5 to show that the net signed area between the graph of $y = x - 1$ and the interval $[0, 2]$ is 0.) ◀

PROPERTIES OF THE DEFINITE INTEGRAL

It is assumed in Definition 5.5.1 that $[a, b]$ is a finite closed interval with $a < b$, and hence the upper limit of integration in the definite integral is greater than the lower limit of integration. However, it will be convenient to extend this definition to allow for cases in which the upper and lower limits of integration are equal or the lower limit of integration is greater than the upper limit of integration. For this purpose we make the following special definitions.

5.5.3 DEFINITION.

(a) If a is in the domain of f, we define

$$\int_a^a f(x)\,dx = 0$$

(b) If f is integrable on $[a, b]$, then we define

$$\int_b^a f(x)\,dx = -\int_a^b f(x)\,dx$$

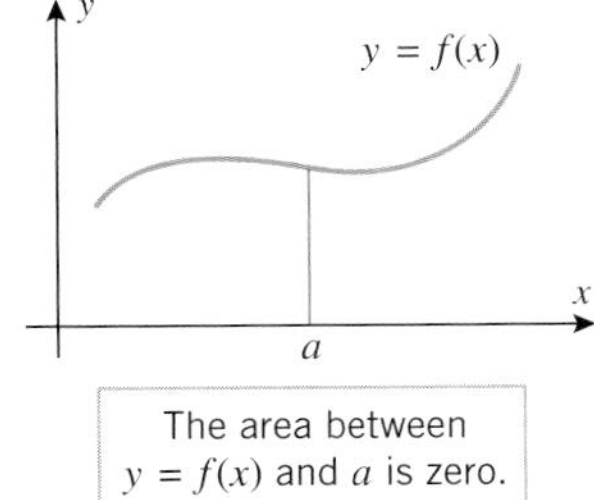

The area between $y = f(x)$ and a is zero.

Figure 5.5.6

Part (a) of this definition is consistent with the intuitive idea that the area between a point on the x-axis and a curve $y = f(x)$ should be zero (Figure 5.5.6). Part (b) of the definition is simply a useful convention; it states that interchanging the limits of integration reverses the sign of the integral.

▶ **Example 3**

(a) $\displaystyle\int_1^1 x^2\,dx = 0$

(b) $\displaystyle\int_1^0 \sqrt{1-x^2}\,dx = -\int_0^1 \sqrt{1-x^2}\,dx = -\frac{\pi}{4}$ ◀

Example 1(c)

Because definite integrals are defined as limits, they inherit many of the properties of limits. For example, we know that constants can be moved through limit signs and that the limit of a sum or difference is the sum or difference of the limits. Thus, you should not be surprised by the following theorem, which we state without formal proof.

5.5.4 THEOREM. *If f and g are integrable on $[a, b]$ and if c is a constant, then cf, $f + g$, and $f - g$ are integrable on $[a, b]$ and*

(*a*) $$\int_a^b cf(x)\,dx = c\int_a^b f(x)\,dx$$

(*b*) $$\int_a^b [f(x) + g(x)]\,dx = \int_a^b f(x)\,dx + \int_a^b g(x)\,dx$$

(*c*) $$\int_a^b [f(x) - g(x)]\,dx = \int_a^b f(x)\,dx - \int_a^b g(x)\,dx$$

Part (*b*) of this theorem can be extended to more than two functions. More precisely,

$$\int_a^b [f_1(x) + f_2(x) + \cdots + f_n(x)]\,dx = \int_a^b f_1(x)\,dx + \int_a^b f_2(x)\,dx + \cdots + \int_a^b f_n(x)\,dx$$

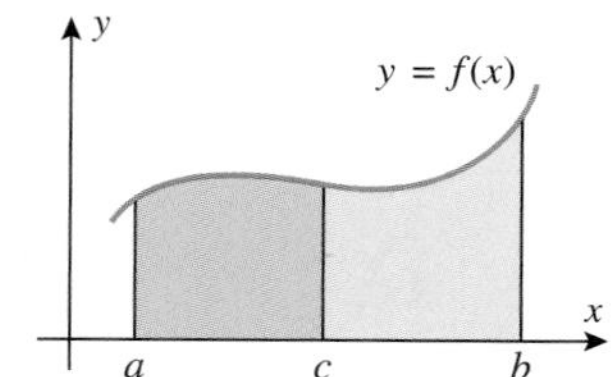

Figure 5.5.7

Some properties of definite integrals can be motivated by interpreting the integral as an area. For example, if f is continuous and nonnegative on the interval $[a, b]$, and if c is a point between a and b, then the area under $y = f(x)$ over the interval $[a, b]$ can be split into two parts and expressed as the area under the graph from a to c plus the area under the graph from c to b (Figure 5.5.7), that is,

$$\int_a^b f(x)\,dx = \int_a^c f(x)\,dx + \int_c^b f(x)\,dx$$

This is a special case of the following theorem about definite integrals, which we state without proof.

5.5.5 THEOREM. *If f is integrable on a closed interval containing the three points a, b, and c, then*

$$\int_a^b f(x)\,dx = \int_a^c f(x)\,dx + \int_c^b f(x)\,dx$$

no matter how the points are ordered.

The following theorem, which we state without formal proof, can also be motivated by interpreting definite integrals as areas.

Part (*b*) of Theorem 5.5.6 states that the sense (direction) of the inequality $f(x) \geq g(x)$ is unchanged if one integrates both sides. Moreover, if $b > a$, then both parts of the theorem remain true if $\geq$ is replaced by $\leq$, $>$, or $<$ throughout.

5.5.6 THEOREM.

(*a*) *If f is integrable on $[a, b]$ and $f(x) \geq 0$ for all x in $[a, b]$, then*

$$\int_a^b f(x)\,dx \geq 0$$

(*b*) *If f and g are integrable on $[a, b]$ and $f(x) \geq g(x)$ for all x in $[a, b]$, then*

$$\int_a^b f(x)\,dx \geq \int_a^b g(x)\,dx$$

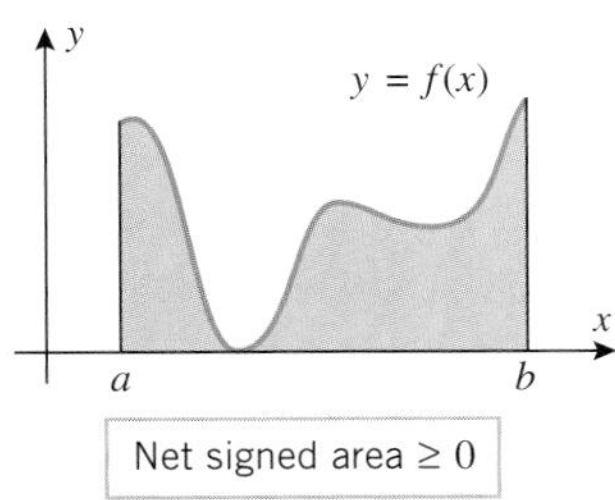

Figure 5.5.8

Geometrically, part (*a*) of this theorem states the obvious fact that if f is nonnegative on $[a, b]$, then the net signed area between the graph of f and the interval $[a, b]$ is also nonnegative (Figure 5.5.8). Part (*b*) has its simplest interpretation when f and g are nonnegative on $[a, b]$, in which case the theorem states that if the graph of f does not go below the graph of g, then the area under the graph of f is at least as large as the area under the graph of g (Figure 5.5.9).

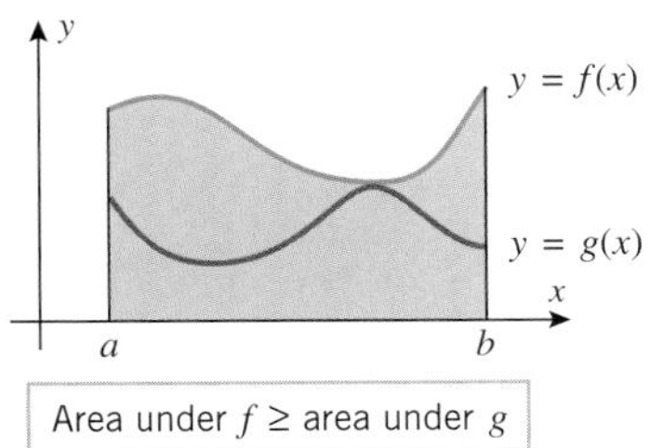

Figure 5.5.9

▸ **Example 4** Evaluate

$$\int_0^1 (5 - 3\sqrt{1 - x^2}\,)\,dx$$

Solution. From parts (*a*) and (*c*) of Theorem 5.5.4 we can write

$$\int_0^1 (5 - 3\sqrt{1 - x^2}\,)\,dx = \int_0^1 5\,dx - \int_0^1 3\sqrt{1 - x^2}\,dx = \int_0^1 5\,dx - 3\int_0^1 \sqrt{1 - x^2}\,dx$$

The first integral in this difference can be interpreted as the area of a rectangle of height 5 and base 1, so its value is 5, and from Example 1 the value of the second integral is $\pi/4$. Thus,

$$\int_0^1 (5 - 3\sqrt{1 - x^2}\,)\,dx = 5 - 3\left(\frac{\pi}{4}\right) = 5 - \frac{3\pi}{4}$$ ◂

DISCONTINUITIES AND INTEGRABILITY

The problem of determining when functions with discontinuities are integrable is quite complex and beyond the scope of this text. However, there are a few basic results about integrability that are important to know; we begin with a definition.

5.5.7 DEFINITION. A function f that is defined on an interval I is said to be ***bounded*** on I if there is a positive number M such that

$$-M \leq f(x) \leq M$$

for all x in the interval I. Geometrically, this means that the graph of f over the interval I lies between the lines $y = -M$ and $y = M$.

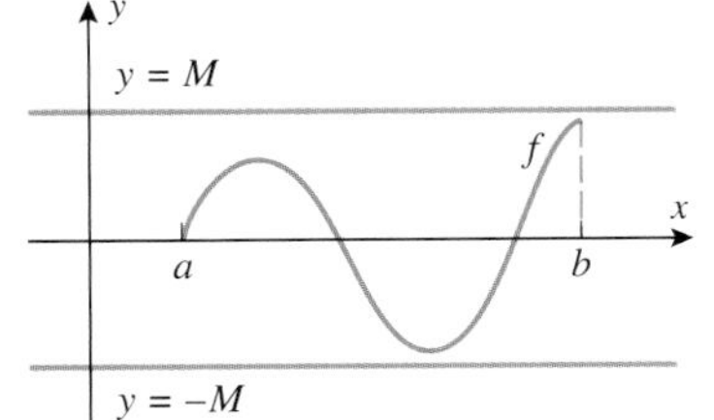

Figure 5.5.10

For example, a continuous function f is bounded on *every* finite closed interval because the Extreme-Value Theorem (4.4.2) implies that f has an absolute maximum and an absolute minimum on the interval; hence, its graph will lie between the line $y = -M$ and $y = M$, provided we make M large enough (Figure 5.5.10). In contrast, a function that has a vertical

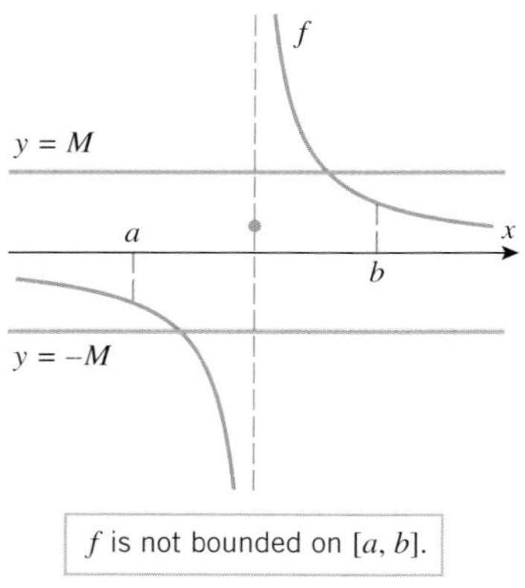

Figure 5.5.11

asymptote inside of an interval is not bounded on that interval because its graph over the interval cannot be made to lie between the lines $y = -M$ and $y = M$, no matter how large we make the value of M (Figure 5.5.11).

The following theorem, which we state without proof, provides some facts about integrability for functions with discontinuities.

5.5.8 THEOREM. *Let f be a function that is defined on the finite closed interval $[a, b]$.*

(a) *If f has finitely many discontinuities in $[a, b]$ but is bounded on $[a, b]$, then f is integrable on $[a, b]$.*

(b) *If f is not bounded on $[a, b]$, then f is not integrable on $[a, b]$.*

QUICK CHECK EXERCISES 5.5

(See page 347 for answers.)

1. In each part, use the partition of [2, 7] in the accompanying figure.

2 3 4.5 6.5 7

(a) What is n, the number of subintervals in this partition?
(b) $x_0 =$ _____; $x_1 =$ _____; $x_2 =$ _____;
$x_3 =$ _____; $x_4 =$ _____
(c) $\Delta x_1 =$ _____; $\Delta x_2 =$ _____; $\Delta x_3 =$ _____;
$\Delta x_4 =$ _____
(d) The mesh of this partition is _____.

2. Let $f(x) = 4x - 12$. Use the partition of [2, 7] in Quick Check Exercise 1 and the choices $x_1^* = 2$, $x_2^* = 4$, $x_3^* = 5$, and $x_4^* = 7$ to evaluate the Riemann sum

$$\sum_{k=1}^{4} f(x_k^*)\Delta x_k$$

3. (a) Sketch the region whose signed area is represented by

$$\int_2^7 [4x - 12]\,dx$$

(b) Use appropriate formulas from geometry to evaluate the integral.

4. Suppose that $g(x)$ is a function for which

$$\int_{-2}^{1} g(x)\,dx = 5 \quad \text{and} \quad \int_{1}^{2} g(x)\,dx = -2$$

Use this information and appropriate formulas from geometry to evaluate the definite integrals.

(a) $\int_1^2 5g(x)\,dx$ (b) $\int_{-2}^{2} g(x)\,dx$

(c) $\int_{-2}^{1} [4 - 3g(x)]\,dx$ (d) $\int_{-2}^{2} [g(x) + \sqrt{4 - x^2}]\,dx$

EXERCISE SET 5.5

1–4 Find the value of

(a) $\sum_{k=1}^{n} f(x_k^*)\,\Delta x_k$ (b) $\max \Delta x_k$.

1. $f(x) = x + 1$; $a = 0, b = 4$; $n = 3$;
$\Delta x_1 = 1, \Delta x_2 = 1, \Delta x_3 = 2$;
$x_1^* = \frac{1}{3}, x_2^* = \frac{3}{2}, x_3^* = 3$

2. $f(x) = \cos x$; $a = 0, b = 2\pi$; $n = 4$;
$\Delta x_1 = \pi/2, \Delta x_2 = 3\pi/4, \Delta x_3 = \pi/2, \Delta x_4 = \pi/4$;
$x_1^* = \pi/4, x_2^* = \pi, x_3^* = 3\pi/2, x_4^* = 7\pi/4$

3. $f(x) = 4 - x^2$; $a = -3, b = 4$; $n = 4$;
$\Delta x_1 = 1, \Delta x_2 = 2, \Delta x_3 = 1, \Delta x_4 = 3$;
$x_1^* = -\frac{5}{2}, x_2^* = -1, x_3^* = \frac{1}{4}, x_4^* = 3$

4. $f(x) = x^3$; $a = -3, b = 3$; $n = 4$;
$\Delta x_1 = 2, \Delta x_2 = 1, \Delta x_3 = 1, \Delta x_4 = 2$;
$x_1^* = -2, x_2^* = 0, x_3^* = 0, x_4^* = 2$

5–8 Use the given values of a and b to express the following limits as integrals. (Do not evaluate the integrals.)

5. $\lim_{\max \Delta x_k \to 0} \sum_{k=1}^{n} (x_k^*)^2 \Delta x_k$; $a = -1, b = 2$

6. $\lim_{\max \Delta x_k \to 0} \sum_{k=1}^{n} (x_k^*)^3 \Delta x_k;\ a = 1, b = 2$

7. $\lim_{\max \Delta x_k \to 0} \sum_{k=1}^{n} 4x_k^*(1 - 3x_k^*) \Delta x_k;\ a = -3, b = 3$

8. $\lim_{\max \Delta x_k \to 0} \sum_{k=1}^{n} (\sin^2 x_k^*) \Delta x_k;\ a = 0, b = \pi/2$

9–10 Use Definition 5.5.1 to express the integrals as limits of Riemann sums. (Do not evaluate the integrals.)

9. (a) $\int_1^2 2x\,dx$ (b) $\int_0^1 \frac{x}{x+1}\,dx$

10. (a) $\int_1^2 \sqrt{x}\,dx$ (b) $\int_{-\pi/2}^{\pi/2} (1 + \cos x)\,dx$

11–14 Sketch the region whose signed area is represented by the definite integral, and evaluate the integral using an appropriate formula from geometry, where needed.

11. (a) $\int_0^3 x\,dx$ (b) $\int_{-2}^{-1} x\,dx$

(c) $\int_{-1}^{4} x\,dx$ (d) $\int_{-5}^{5} x\,dx$

12. (a) $\int_0^2 \left(1 - \frac{1}{2}x\right) dx$ (b) $\int_{-1}^{1} \left(1 - \frac{1}{2}x\right) dx$

(c) $\int_2^3 \left(1 - \frac{1}{2}x\right) dx$ (d) $\int_0^3 \left(1 - \frac{1}{2}x\right) dx$

13. (a) $\int_0^5 2\,dx$ (b) $\int_0^{\pi} \cos x\,dx$

(c) $\int_{-1}^{2} |2x - 3|\,dx$ (d) $\int_{-1}^{1} \sqrt{1 - x^2}\,dx$

14. (a) $\int_{-10}^{-5} 6\,dx$ (b) $\int_{-\pi/3}^{\pi/3} \sin x\,dx$

(c) $\int_0^3 |x - 2|\,dx$ (d) $\int_0^2 \sqrt{4 - x^2}\,dx$

15. In each part, evaluate the integral, given that

$$f(x) = \begin{cases} |x - 2|, & x \geq 0 \\ x + 2, & x < 0 \end{cases}$$

(a) $\int_{-2}^{0} f(x)\,dx$ (b) $\int_{-2}^{2} f(x)\,dx$

(c) $\int_0^6 f(x)\,dx$ (d) $\int_{-4}^{6} f(x)\,dx$

16. In each part, evaluate the integral, given that

$$f(x) = \begin{cases} 2x, & x \leq 1 \\ 2, & x > 1 \end{cases}$$

(a) $\int_0^1 f(x)\,dx$ (b) $\int_{-1}^{1} f(x)\,dx$

(c) $\int_1^{10} f(x)\,dx$ (d) $\int_{1/2}^{5} f(x)\,dx$

FOCUS ON CONCEPTS

17–18 Use the areas shown in the figure to find

(a) $\int_a^b f(x)\,dx$ (b) $\int_b^c f(x)\,dx$

(c) $\int_a^c f(x)\,dx$ (d) $\int_a^d f(x)\,dx$.

17.

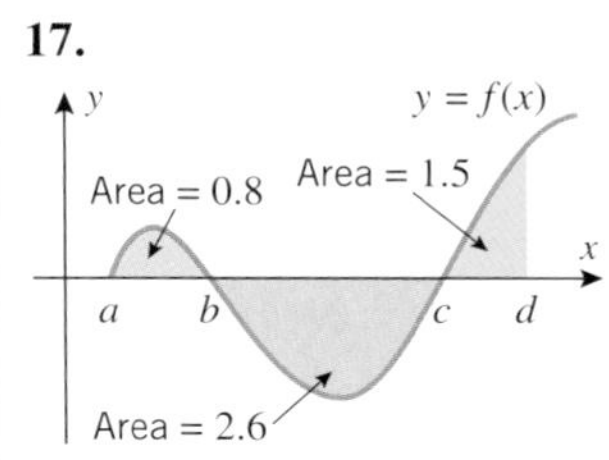

18.

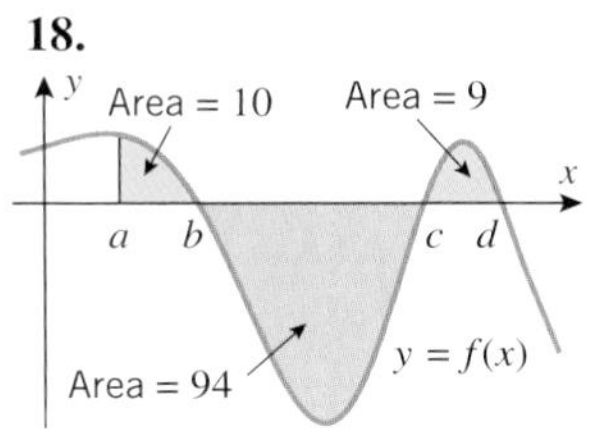

19. Find $\int_{-1}^{2} [f(x) + 2g(x)]\,dx$ if

$$\int_{-1}^{2} f(x)\,dx = 5 \quad \text{and} \quad \int_{-1}^{2} g(x)\,dx = -3$$

20. Find $\int_1^4 [3f(x) - g(x)]\,dx$ if

$$\int_1^4 f(x)\,dx = 2 \quad \text{and} \quad \int_1^4 g(x)\,dx = 10$$

21. Find $\int_1^5 f(x)\,dx$ if

$$\int_0^1 f(x)\,dx = -2 \quad \text{and} \quad \int_0^5 f(x)\,dx = 1$$

22. Find $\int_3^{-2} f(x)\,dx$ if

$$\int_{-2}^{1} f(x)\,dx = 2 \quad \text{and} \quad \int_1^3 f(x)\,dx = -6$$

23–26 Use Theorem 5.5.4 and appropriate formulas from geometry to evaluate the integrals.

23. $\int_{-1}^{3} (4 - 5x)\,dx$ **24.** $\int_{-2}^{2} (1 - 3|x|)\,dx$

25. $\int_0^1 (x + 2\sqrt{1 - x^2})\,dx$ **26.** $\int_{-3}^{0} (2 + \sqrt{9 - x^2})\,dx$

27–28 Use Theorem 5.5.6 to determine whether the value of the integral is positive or negative.

27. (a) $\int_2^3 \frac{\sqrt{x}}{1 - x}\,dx$ (b) $\int_0^4 \frac{x^2}{3 - \cos x}\,dx$

28. (a) $\int_{-3}^{-1} \frac{x^4}{\sqrt{3 - x}}\,dx$ (b) $\int_{-2}^{2} \frac{x^3 - 9}{|x| + 1}\,dx$

29–30 Evaluate the integrals by completing the square and applying appropriate formulas from geometry.

29. $\int_0^{10} \sqrt{10x - x^2}\,dx$ **30.** $\int_0^3 \sqrt{6x - x^2}\,dx$

31–32 Evaluate the limit by expressing it as a definite integral over the interval $[a, b]$ and applying appropriate formulas from geometry.

31. $\lim_{\max \Delta x_k \to 0} \sum_{k=1}^{n} (3x_k^* + 1)\Delta x_k;\ a = 0, b = 1$

32. $\lim_{\max \Delta x_k \to 0} \sum_{k=1}^{n} \sqrt{4 - (x_k^*)^2}\,\Delta x_k;\ a = -2, b = 2$

FOCUS ON CONCEPTS

33. Let $f(x) = C$ be a constant function.

(a) Use a formula from geometry to show that

$$\int_a^b f(x)\,dx = C(b - a)$$

(b) Show that any Riemann sum for $f(x)$ over $[a, b]$ evaluates to $C(b - a)$. Use Definition 5.5.1 to show that

$$\int_a^b f(x)\,dx = C(b - a)$$

34. In each part, use Theorems 5.5.2 and 5.5.8 to determine whether the function f is integrable on the interval $[-1, 1]$.

(a) $f(x) = \cos x$

(b) $f(x) = \begin{cases} x/|x|, & x \neq 0 \\ 0, & x = 0 \end{cases}$

(c) $f(x) = \begin{cases} 1/x^2, & x \neq 0 \\ 0, & x = 0 \end{cases}$

(d) $f(x) = \begin{cases} \sin 1/x, & x \neq 0 \\ 0, & x = 0 \end{cases}$

35. Define a function f on $[0, 1]$ by

$$f(x) = \begin{cases} 1, & 0 < x \leq 1 \\ 0, & x = 0 \end{cases}$$

Use Definition 5.5.1 to show that

$$\int_0^1 f(x)\,dx = 1$$

36. It can be shown that every interval contains both rational and irrational numbers. Accepting this to be so, do you believe that the function

$$f(x) = \begin{cases} 1 & \text{if } x \text{ is rational} \\ 0 & \text{if } x \text{ is irrational} \end{cases}$$

is integrable on a closed interval $[a, b]$? Explain your reasoning.

37. Find the largest and smallest values that the Riemann sum

$$\sum_{k=1}^{3} f(x_k^*)\Delta x_k$$

can have on the interval $[0, \pi]$ for the function $f(x) = \sin x$ and $\Delta x_1 = \pi/4$, $\Delta x_2 = 7\pi/12$, and $\Delta x_3 = \pi/6$.

38. Find the largest and smallest values that the Riemann sum

$$\sum_{k=1}^{3} f(x_k^*)\Delta x_k$$

can have on the interval $[0, 4]$ if $f(x) = x^2 - 3x + 4$ and $\Delta x_1 = 1$, $\Delta x_2 = 2$, and $\Delta x_3 = 1$.

39. The function $f(x) = \sqrt{x}$ is continuous on $[0, 4]$ and therefore integrable on this interval. Evaluate

$$\int_0^4 \sqrt{x}\,dx$$

by using Definition 5.5.1. Use subintervals of unequal length given by the partition

$$0 < 4(1)^2/n^2 < 4(2)^2/n^2 < \cdots < 4(n-1)^2/n^2 < 4$$

and let x_k^* be the right endpoint of the kth subinterval.

FOCUS ON CONCEPTS

40. Suppose that f is defined on the interval $[a, b]$ and that $f(x) = 0$ for $a < x \leq b$. Use Definition 5.5.1 to prove that

$$\int_a^b f(x)\,dx = 0$$

41. Suppose that g is a continuous function on the interval $[a, b]$ and that f is a function defined on $[a, b]$ with $f(x) = g(x)$ for $a < x \leq b$. Prove that

$$\int_a^b f(x)\,dx = \int_a^b g(x)\,dx$$

[*Hint:* Write

$$\int_a^b f(x)\,dx = \int_a^b [(f(x) - g(x)) + g(x)]\,dx$$

and use the result of Exercise 40 along with Theorem 5.5.4(*b*).]

42. Define the function f by

$$f(x) = \begin{cases} \dfrac{1}{x}, & x \neq 0 \\ 0, & x = 0 \end{cases}$$

It follows from Theorem 5.5.8(*b*) that f is not integrable on the interval $[0, 1]$. Prove this to be the case by applying Definition 5.5.1. [*Hint:* Argue that no matter how small the mesh size is for a partition of $[0, 1]$, there will always be a choice of x_1^* that will make the Riemann sum in Definition 5.5.1 as large as we like.]

✔ QUICK CHECK ANSWERS 5.5

1. (a) $n = 4$ (b) 2, 3, 4.5, 6.5, 7 (c) 1, 1.5, 2, 0.5 (d) 2 **2.** 26 **3.** (a) See Figure QCA-3 (b) 30
4. (a) -10 (b) 3 (c) -3 (d) $3 + 2\pi$

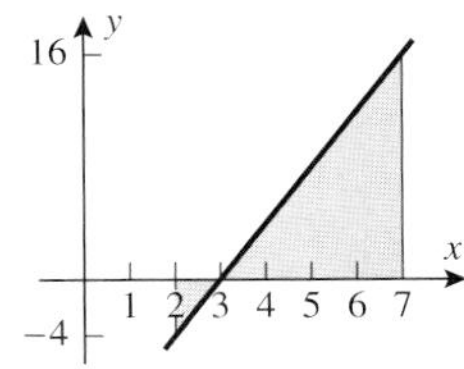

Figure QCA-3

5.6 THE FUNDAMENTAL THEOREM OF CALCULUS

In this section we will establish two basic relationships between definite and indefinite integrals that together constitute a result called the "Fundamental Theorem of Calculus." One part of this theorem will relate the rectangle and antiderivative methods for calculating areas, and the second part will provide a powerful method for evaluating definite integrals using antiderivatives.

THE FUNDAMENTAL THEOREM OF CALCULUS

As in earlier sections, let us begin by assuming that f is nonnegative and continuous on an interval $[a, b]$, in which case the area A under the graph of f over the interval $[a, b]$ is represented by the definite integral

$$A = \int_a^b f(x)\,dx \tag{1}$$

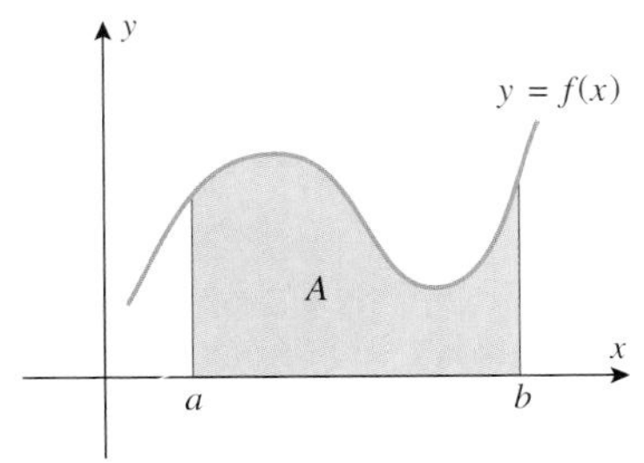

Figure 5.6.1

(Figure 5.6.1).

Recall that our discussion of the antiderivative method in Section 5.1 suggested that if $A(x)$ is the area under the graph of f from a to x (Figure 5.6.2), then

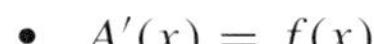

- $A'(x) = f(x)$
- $A(a) = 0$ — The area under the curve from a to a is the area above the single point a, and hence is zero.
- $A(b) = A$ — The area under the curve from a to b is A.

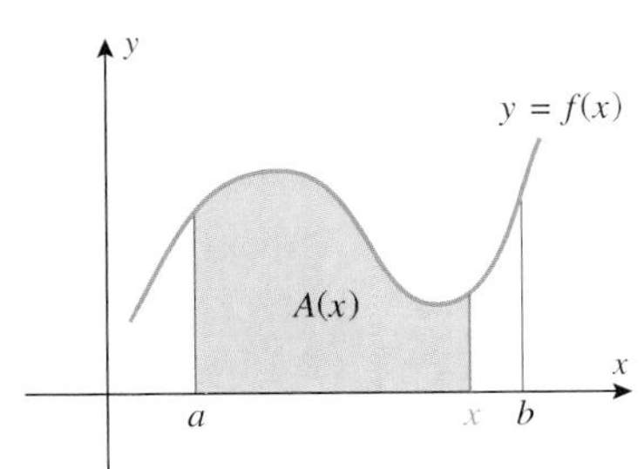

Figure 5.6.2

The formula $A'(x) = f(x)$ states that $A(x)$ is an antiderivative of $f(x)$, which implies that every other antiderivative of $f(x)$ on $[a, b]$ can be obtained by adding a constant to $A(x)$. Accordingly, let

$$F(x) = A(x) + C$$

be any antiderivative of $f(x)$, and consider what happens when we subtract $F(a)$ from $F(b)$:

$$F(b) - F(a) = [A(b) + C] - [A(a) + C] = A(b) - A(a) = A - 0 = A$$

Hence (1) can be expressed as

$$\int_a^b f(x)\,dx = F(b) - F(a)$$

In words, this equation states:

The definite integral can be evaluated by finding any antiderivative of the integrand and then subtracting the value of this antiderivative at the lower limit of integration from its value at the upper limit of integration.

Although our evidence for this result assumed that f is nonnegative on $[a, b]$, this assumption is not essential.

5.6.1 THEOREM (*The Fundamental Theorem of Calculus, Part 1*). *If f is continuous on $[a, b]$ and F is any antiderivative of f on $[a, b]$, then*

$$\int_a^b f(x)\,dx = F(b) - F(a) \tag{2}$$

PROOF. Let $x_1, x_2, \ldots, x_{n-1}$ be any points in $[a, b]$ such that

$$a < x_1 < x_2 < \cdots < x_{n-1} < b$$

These values divide $[a, b]$ into n subintervals

$$[a, x_1], [x_1, x_2], \ldots, [x_{n-1}, b] \tag{3}$$

whose lengths, as usual, we denote by

$$\Delta x_1, \Delta x_2, \ldots, \Delta x_n$$

By hypothesis, $F'(x) = f(x)$ for all x in $[a, b]$, so F satisfies the hypotheses of the Mean-Value Theorem (4.7.2) on each subinterval in (3). Hence, we can find points $x_1^*, x_2^*, \ldots, x_n^*$ in the respective subintervals in (3) such that

$$\begin{aligned}
F(x_1) - F(a) &= F'(x_1^*)(x_1 - a) = f(x_1^*)\Delta x_1\\
F(x_2) - F(x_1) &= F'(x_2^*)(x_2 - x_1) = f(x_2^*)\Delta x_2\\
F(x_3) - F(x_2) &= F'(x_3^*)(x_3 - x_2) = f(x_3^*)\Delta x_3\\
&\vdots\\
F(b) - F(x_{n-1}) &= F'(x_n^*)(b - x_{n-1}) = f(x_n^*)\Delta x_n
\end{aligned}$$

Adding the preceding equations yields

$$F(b) - F(a) = \sum_{k=1}^{n} f(x_k^*)\Delta x_k \tag{4}$$

Let us now increase n in such a way that $\max \Delta x_k \to 0$. Since f is assumed to be continuous, the right side of (4) approaches $\int_a^b f(x)\,dx$ by Theorem 5.5.2 and Definition 5.5.1. However, the left side of (4) is independent of n; that is, the left side of (4) remains constant as n increases. Thus,

$$F(b) - F(a) = \lim_{\max \Delta x_k \to 0} \sum_{k=1}^{n} f(x_k^*)\Delta x_k = \int_a^b f(x)\,dx \qquad \blacksquare$$

It is standard to denote the difference $F(b) - F(a)$ as

$$F(x)\Big]_a^b = F(b) - F(a) \quad \text{or} \quad \Big[F(x)\Big]_a^b = F(b) - F(a)$$

For example, using the first of these notations we can express (2) as

$$\int_a^b f(x)\,dx = F(x)\Big]_a^b \tag{5}$$

We will sometimes write

$$F(x)\Big]_{x=a}^b = F(b) - F(a)$$

when it is important to emphasize that a and b are values for the variable x.

The integral in Example 1 represents the area of a certain trapezoid. Sketch the trapezoid, and find its area using geometry.

▶ **Example 1** Evaluate $\int_1^2 x\,dx$.

Solution. The function $F(x) = \frac{1}{2}x^2$ is an antiderivative of $f(x) = x$; thus, from (2)

$$\int_1^2 x\,dx = \frac{1}{2}x^2\bigg]_1^2 = \frac{1}{2}(2)^2 - \frac{1}{2}(1)^2 = 2 - \frac{1}{2} = \frac{3}{2} \blacktriangleleft$$

▶ **Example 2** In Example 6 of Section 5.4 we used the definition of area to show that the area under the graph of $y = 9 - x^2$ over the interval $[0, 3]$ is 18 (square units). We can now solve that problem much more easily using the Fundamental Theorem of Calculus:

$$A = \int_0^3 (9 - x^2)\,dx = \left[9x - \frac{x^3}{3}\right]_0^3 = \left(27 - \frac{27}{3}\right) - 0 = 18 \blacktriangleleft$$

▶ **Example 3**

(a) Find the area under the curve $y = \cos x$ over the interval $[0, \pi/2]$ (Figure 5.6.3).

(b) Make a conjecture about the value of the integral

$$\int_0^\pi \cos x\,dx$$

and confirm your conjecture using the Fundamental Theorem of Calculus.

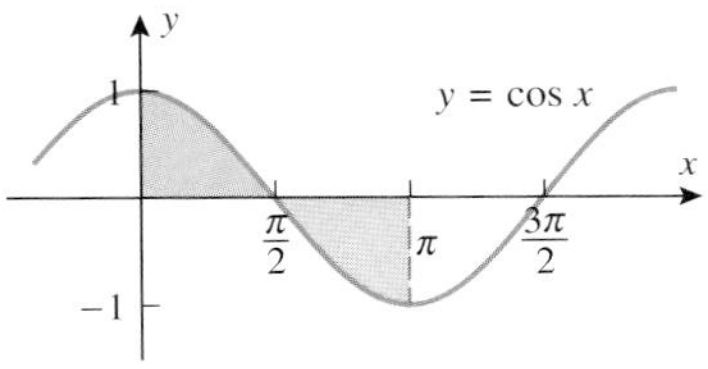

Figure 5.6.3

Solution (a). Since $\cos x \geq 0$ over the interval $[0, \pi/2]$, the area A under the curve is

$$A = \int_0^{\pi/2} \cos x\,dx = \sin x\bigg]_0^{\pi/2} = \sin\frac{\pi}{2} - \sin 0 = 1$$

Solution (b). The given integral can be interpreted as the signed area between the graph of $y = \cos x$ and the interval $[0, \pi]$. The graph in Figure 5.6.3 suggests that over the interval $[0, \pi]$ the portion of area above the x-axis is the same as the portion of area below the x-axis, so we conjecture that the signed area is zero; this implies that the value of the integral is zero. This is confirmed by the computations

$$\int_0^\pi \cos x\,dx = \sin x\bigg]_0^\pi = \sin\pi - \sin 0 = 0 \blacktriangleleft$$

THE RELATIONSHIP BETWEEN DEFINITE AND INDEFINITE INTEGRALS

Observe that in the preceding examples we did not include a constant of integration in the antiderivatives. In general, when applying the Fundamental Theorem of Calculus there is no need to include a constant of integration because it will drop out anyway. To see that this is so, let F be any antiderivative of the integrand on $[a, b]$, and let C be any constant; then

$$\int_a^b f(x)\,dx = \big[F(x) + C\big]_a^b = [F(b) + C] - [F(a) + C] = F(b) - F(a)$$

Thus, for purposes of evaluating a definite integral we can omit the constant of integration in

$$\int_a^b f(x)\,dx = \big[F(x) + C\big]_a^b$$

and express (5) as

$$\int_a^b f(x)\,dx = \int f(x)\,dx\Big]_a^b \tag{6}$$

which relates the definite and indefinite integrals.

▶ Example 4

$$\int_1^9 \sqrt{x}\,dx = \int x^{1/2}\,dx\Big]_1^9 = \frac{2}{3}x^{3/2}\Big]_1^9 = \frac{2}{3}(27-1) = \frac{52}{3} \blacktriangleleft$$

▶ Example 5 Table 5.2.1 will be helpful for the following computations.

$$\int_4^9 x^2\sqrt{x}\,dx = \int_4^9 x^{5/2}\,dx = \frac{2}{7}x^{7/2}\Big]_4^9 = \frac{2}{7}(2187-128) = \frac{4118}{7} = 588\frac{2}{7}$$

$$\int_0^{\pi/2} \frac{\sin x}{5}\,dx = -\frac{\cos x}{5}\Big]_0^{\pi/2} = -\frac{1}{5}\left[\cos\left(\frac{\pi}{2}\right) - \cos 0\right] = -\frac{1}{5}[0-1] = \frac{1}{5}$$

$$\int_0^{\pi/3} \sec^2 x\,dx = \tan x\Big]_0^{\pi/3} = \tan\left(\frac{\pi}{3}\right) - \tan 0 = \sqrt{3} - 0 = \sqrt{3}$$

$$\int_{-\pi/4}^{\pi/4} \sec x \tan x\,dx = \sec x\Big]_{-\pi/4}^{\pi/4} = \sec\left(\frac{\pi}{4}\right) - \sec\left(-\frac{\pi}{4}\right) = \sqrt{2} - \sqrt{2} = 0 \blacktriangleleft$$

TECHNOLOGY MASTERY

If you have a CAS, read the documentation on evaluating definite integrals and then check the results in Example 5.

WARNING

The requirements in the Fundamental Theorem of Calculus that f be continuous on $[a, b]$ and that F be an antiderivative for f over the entire interval $[a, b]$ are important to keep in mind. Disregarding these assumptions will likely lead to incorrect results. For example, the function $f(x) = 1/x^2$ fails on two counts to be continuous at $x = 0$: $f(x)$ is not defined at $x = 0$ and $\lim_{x\to 0} f(x)$ does not exist. Thus, the Fundamental Theorem of Calculus should not be used to integrate f on any interval that contains $x = 0$. However, if we ignore this and blindly apply Formula (2) over the interval $[-1, 1]$, we might *incorrectly* compute $\int_{-1}^{1}(1/x^2)\,dx$ by evaluating an antiderivative, $-1/x$, at the endpoints, arriving at the answer

$$-\frac{1}{x}\Big]_{-1}^{1} = -[1-(-1)] = -2$$

But $f(x) = 1/x^2$ is a nonnegative function, so clearly a negative value for the definite integral is impossible.

The Fundamental Theorem of Calculus can be applied without modification to definite integrals in which the lower limit of integration is greater than or equal to the upper limit of integration.

▶ Example 6

$$\int_1^1 x^2\,dx = \frac{x^3}{3}\Big]_1^1 = \frac{1}{3} - \frac{1}{3} = 0$$

$$\int_4^0 x\,dx = \frac{x^2}{2}\Big]_4^0 = \frac{0}{2} - \frac{16}{2} = -8$$

The latter result is consistent with the result that would be obtained by first reversing the limits of integration in accordance with Definition 5.5.3(b):

$$\int_4^0 x\,dx = -\int_0^4 x\,dx = -\frac{x^2}{2}\Big]_0^4 = -\left[\frac{16}{2} - \frac{0}{2}\right] = -8 \blacktriangleleft$$

To integrate a continuous function that is defined piecewise on an interval $[a, b]$, split this interval into subintervals at the breakpoints of the function, and integrate separately over each subinterval in accordance with Theorem 5.5.5.

▶ **Example 7** Evaluate $\int_0^6 f(x)\,dx$ if

$$f(x) = \begin{cases} x^2, & x < 2 \\ 3x - 2, & x \geq 2 \end{cases}$$

Solution. From Theorem 5.5.5

$$\begin{aligned} \int_0^6 f(x)\,dx &= \int_0^2 f(x)\,dx + \int_2^6 f(x)\,dx \\ &= \int_0^2 x^2\,dx + \int_2^6 (3x - 2)\,dx \\ &= \frac{x^3}{3}\Big]_0^2 + \left[\frac{3x^2}{2} - 2x\right]_2^6 = \left(\frac{8}{3} - 0\right) + (42 - 2) = \frac{128}{3} \blacktriangleleft \end{aligned}$$

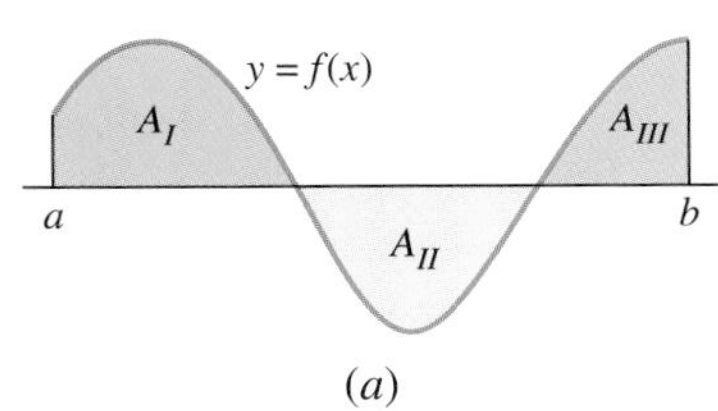

(a)

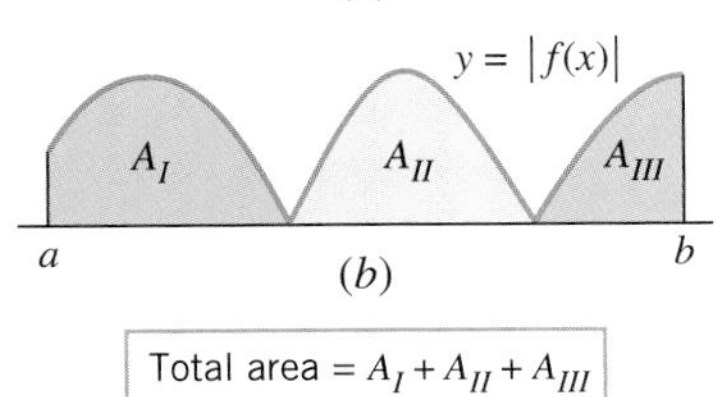

(b)

Total area $= A_I + A_{II} + A_{III}$

Figure 5.6.4

If f is a continuous function on the interval $[a, b]$, then we define the ***total area*** between the curve $y = f(x)$ and the interval $[a, b]$ to be

$$\text{total area} = \int_a^b |f(x)|\,dx \tag{7}$$

(Figure 5.6.4). To compute total area using Formula (7), begin by dividing the interval of integration into subintervals on which $f(x)$ does not change sign. On the subintervals for which $0 \leq f(x)$ replace $|f(x)|$ by $f(x)$, and on the subintervals for which $f(x) \leq 0$ replace $|f(x)|$ by $-f(x)$. Adding the resulting integrals then yields the total area.

▶ **Example 8** Find the total area between the curve $y = 1 - x^2$ and the x-axis over the interval $[0, 2]$ (Figure 5.6.5).

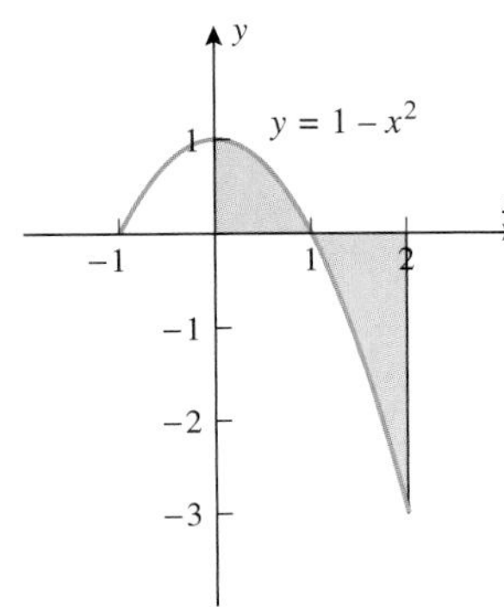

Figure 5.6.5

Solution. The area A is given by

$$\begin{aligned} A = \int_0^2 |1 - x^2|\,dx &= \int_0^1 (1 - x^2)\,dx + \int_1^2 -(1 - x^2)\,dx \\ &= \left[x - \frac{x^3}{3}\right]_0^1 - \left[x - \frac{x^3}{3}\right]_1^2 \\ &= \frac{2}{3} - \left(-\frac{4}{3}\right) = 2 \blacktriangleleft \end{aligned}$$

DUMMY VARIABLES

To evaluate a definite integral using the Fundamental Theorem of Calculus, one needs to be able to find an antiderivative of the integrand; thus, it is important to know what kinds of functions have antiderivatives. It is our next objective to show that all continuous functions have antiderivatives, but to do this we will need some preliminary results.

Formula (6) shows that there is a close relationship between the integrals

$$\int_a^b f(x)\,dx \quad \text{and} \quad \int f(x)\,dx$$

However, the definite and indefinite integrals differ in some important ways. For one thing, the two integrals are different kinds of objects—the definite integral is a *number* (the net signed area between the graph of $y = f(x)$ and the interval $[a, b]$), whereas the indefinite integral is a *function*, or more accurately a set of functions [the antiderivatives of $f(x)$]. However, the two types of integrals also differ in the role played by the variable of integration. In an indefinite integral, the variable of integration is "passed through" to the antiderivative in the sense that integrating a function of x produces a function of x, integrating a function of t produces a function of t, and so forth. For example,

$$\int x^2\,dx = \frac{x^3}{3} + C \quad \text{and} \quad \int t^2\,dt = \frac{t^3}{3} + C$$

In contrast, the variable of integration in a definite integral is not passed through to the end result, since the end result is a number. Thus, integrating a function of x over an interval and integrating the same function of t over the same interval of integration produce the same value for the integral. For example,

$$\int_1^3 x^2\,dx = \frac{x^3}{3}\bigg]_{x=1}^3 = \frac{27}{3} - \frac{1}{3} = \frac{26}{3} \quad \text{and} \quad \int_1^3 t^2\,dt = \frac{t^3}{3}\bigg]_{t=1}^3 = \frac{27}{3} - \frac{1}{3} = \frac{26}{3}$$

However, this latter result should not be surprising, since the area under the graph of the curve $y = f(x)$ over an interval $[a, b]$ on the x-axis is the same as the area under the graph of the curve $y = f(t)$ over the interval $[a, b]$ on the t-axis (Figure 5.6.6).

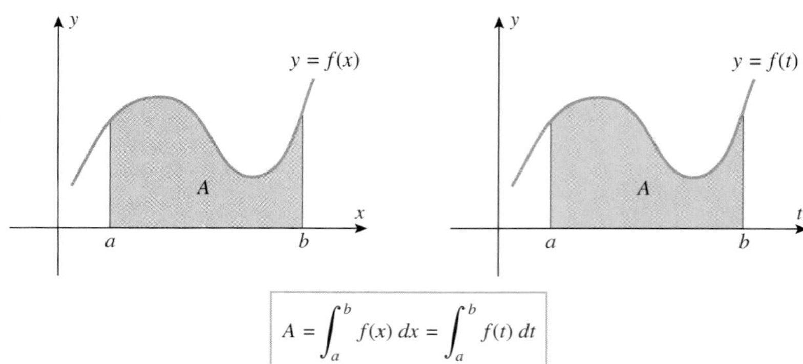

Figure 5.6.6

Because the variable of integration in a definite integral plays no role in the end result, it is often referred to as a ***dummy variable***. In summary:

> *Whenever you find it convenient to change the letter used for the variable of integration in a definite integral, you can do so without changing the value of the integral.*

THE MEAN-VALUE THEOREM FOR INTEGRALS

To reach our goal of showing that continuous functions have antiderivatives, we will need to develop a basic property of definite integrals, known as the *Mean-Value Theorem for Integrals.* In the next chapter we will use this theorem to extend the familiar idea of "average value" so that it applies to continuous functions, but here we will need it as a tool for developing other results.

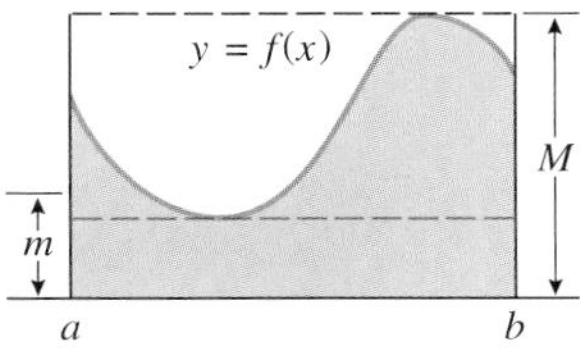

Let f be a continuous nonnegative function on $[a, b]$, and let m and M be the minimum and maximum values of $f(x)$ on this interval. Consider the rectangles of heights m and M over the interval $[a, b]$ (Figure 5.6.7). It is clear geometrically from this figure that the area

Figure 5.6.7

$$A = \int_a^b f(x)\,dx$$

under $y = f(x)$ is at least as large as the area of the rectangle of height m and no larger than the area of the rectangle of height M. It seems reasonable, therefore, that there is a rectangle over the interval $[a, b]$ of some appropriate height $f(x^*)$ between m and M whose area is precisely A; that is,

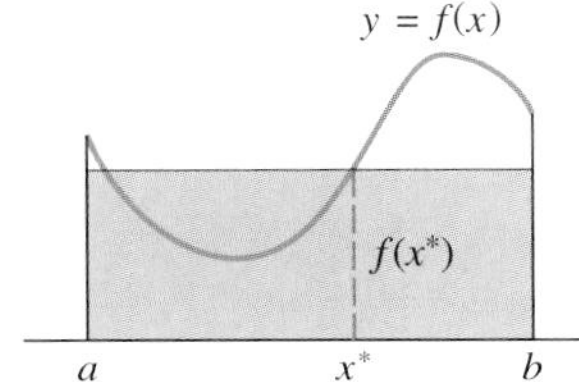

Figure 5.6.8

$$\int_a^b f(x)\,dx = f(x^*)(b-a)$$

(Figure 5.6.8). This is a special case of the following result.

5.6.2 THEOREM (*The Mean-Value Theorem for Integrals*). *If f is continuous on a closed interval $[a, b]$, then there is at least one point x^* in $[a, b]$ such that*

$$\int_a^b f(x)\,dx = f(x^*)(b-a) \tag{8}$$

PROOF. By the Extreme-Value Theorem (4.4.2), f assumes a maximum value M and a minimum value m on $[a, b]$. Thus, for all x in $[a, b]$,

$$m \le f(x) \le M$$

and from Theorem 5.5.6(b)

$$\int_a^b m\,dx \le \int_a^b f(x)\,dx \le \int_a^b M\,dx$$

or

$$m(b-a) \le \int_a^b f(x)\,dx \le M(b-a) \tag{9}$$

or

$$m \le \frac{1}{b-a}\int_a^b f(x)\,dx \le M$$

This implies that

$$\frac{1}{b-a}\int_a^b f(x)\,dx \tag{10}$$

is a number between m and M, and since $f(x)$ assumes the values m and M on $[a, b]$, it follows from the Intermediate-Value Theorem (2.5.8) that $f(x)$ must assume the value (10) at some x^* in $[a, b]$; that is,

$$\frac{1}{b-a}\int_a^b f(x)\,dx = f(x^*) \quad \text{or} \quad \int_a^b f(x)\,dx = f(x^*)(b-a) \qquad \blacksquare$$

▶ **Example 9** Since $f(x) = x^2$ is continuous on the interval $[1, 4]$, the Mean-Value Theorem for Integrals guarantees that there is a point x^* in $[1, 4]$ such that

$$\int_1^4 x^2\,dx = f(x^*)(4-1) = (x^*)^2(4-1) = 3(x^*)^2$$

But

$$\int_1^4 x^2\,dx = \frac{x^3}{3}\bigg]_1^4 = 21$$

so that

$$3(x^*)^2 = 21 \quad \text{or} \quad (x^*)^2 = 7 \quad \text{or} \quad x^* = \pm\sqrt{7}$$

Thus, $x^* = \sqrt{7} \approx 2.65$ is the point in the interval $[1, 4]$ whose existence is guaranteed by the Mean-Value Theorem for Integrals. ◀

PART 2 OF THE FUNDAMENTAL THEOREM OF CALCULUS

In Section 5.1 we suggested that if f is continuous and nonnegative on $[a, b]$, and if $A(x)$ is the area under the graph of $y = f(x)$ over the interval $[a, x]$ (Figure 5.6.2), then $A'(x) = f(x)$. But $A(x)$ can be expressed as the definite integral

$$A(x) = \int_a^x f(t)\,dt$$

(where we have used t rather than x as the variable of integration to avoid confusion with the x that appears as the upper limit of integration). Thus, the relationship $A'(x) = f(x)$ can be expressed as

$$\frac{d}{dx}\left[\int_a^x f(t)\,dt\right] = f(x)$$

This is a special case of the following more general result, which applies even if f has negative values.

5.6.3 THEOREM (*The Fundamental Theorem of Calculus, Part 2*). *If f is continuous on an interval I, then f has an antiderivative on I. In particular if a is any point in I, then the function F defined by*

$$F(x) = \int_a^x f(t)\,dt$$

is an antiderivative of f on I; that is, $F'(x) = f(x)$ for each x in I, or in an alternative notation

$$\frac{d}{dx}\left[\int_a^x f(t)\,dt\right] = f(x) \tag{11}$$

PROOF. We will show first that $F(x)$ is defined at each x in the interval I. If $x > a$ and x is in the interval I, then Theorem 5.5.2 applied to the interval $[a, x]$ and the continuity of f on I ensure that $F(x)$ is defined; and if x is in the interval I and $x \le a$, then Definition 5.5.3 combined with Theorem 5.5.2 ensures that $F(x)$ is defined. Thus, $F(x)$ is defined for all x in I.

Next we will show that $F'(x) = f(x)$ for each x in the interval I. If x is not an endpoint of I, then it follows from the definition of a derivative that

$$
\begin{aligned}
F'(x) &= \lim_{h \to 0} \frac{F(x+h) - F(x)}{h} \\
&= \lim_{h \to 0} \frac{1}{h}\left[\int_a^{x+h} f(t)\,dt - \int_a^x f(t)\,dt\right] \\
&= \lim_{h \to 0} \frac{1}{h}\left[\int_a^{x+h} f(t)\,dt + \int_x^a f(t)\,dt\right] \\
&= \lim_{h \to 0} \frac{1}{h}\int_x^{x+h} f(t)\,dt \qquad \text{Theorem 5.5.5} \qquad (12)
\end{aligned}
$$

Applying the Mean-Value Theorem for Integrals (Theorem 5.6.2) to the integral in (12) we obtain

$$\frac{1}{h}\int_x^{x+h} f(t)\,dt = \frac{1}{h}[f(t^*) \cdot h] = f(t^*) \qquad (13)$$

where t^* is some number between x and $x+h$. Because t^* is trapped between x and $x+h$, it follows that $t^* \to x$ as $h \to 0$. Thus, the continuity of f at x implies that $f(t^*) \to f(x)$ as $h \to 0$. Therefore, it follows from (12) and (13) that

$$F'(x) = \lim_{h \to 0}\left(\frac{1}{h}\int_x^{x+h} f(t)\,dt\right) = \lim_{h \to 0} f(t^*) = f(x)$$

If x is an endpoint of the interval I, then the two-sided limits in the proof must be replaced by the appropriate one-sided limits, but otherwise the arguments are identical. ■

In words, Formula (11) states:

If a definite integral has a variable upper limit of integration, a constant lower limit of integration, and a continuous integrand, then the derivative of the integral with respect to its upper limit is equal to the integrand evaluated at the upper limit.

► **Example 10** Find

$$\frac{d}{dx}\left[\int_1^x t^3\,dt\right]$$

by applying Part 2 of the Fundamental Theorem of Calculus, and then confirm the result by performing the integration and then differentiating.

Solution. The integrand is a continuous function, so from (11)

$$\frac{d}{dx}\left[\int_1^x t^3\,dt\right] = x^3$$

Alternatively, evaluating the integral and then differentiating yields

$$\int_1^x t^3\,dt = \frac{t^4}{4}\bigg]_{t=1}^x = \frac{x^4}{4} - \frac{1}{4}, \qquad \frac{d}{dx}\left[\frac{x^4}{4} - \frac{1}{4}\right] = x^3$$

so the two methods for differentiating the integral agree. ◄

▶ **Example 11** Since

$$f(x) = \frac{\sin x}{x}$$

is continuous on any interval that does not contain the origin, it follows from (11) that on the interval $(0, +\infty)$ we have

$$\frac{d}{dx}\left[\int_1^x \frac{\sin t}{t}\,dt\right] = \frac{\sin x}{x}$$

Unlike the preceding example, there is no way to evaluate the integral in terms of familiar functions, so Formula (11) provides the only simple method for finding the derivative. ◀

DIFFERENTIATION AND INTEGRATION ARE INVERSE PROCESSES

The two parts of the Fundamental Theorem of Calculus, when taken together, tell us that differentiation and integration are inverse processes in the sense that each undoes the effect of the other. To see why this is so, note that Part 1 of the Fundamental Theorem of Calculus (5.6.1) implies that

$$\int_a^x f'(t)\,dt = f(x) - f(a)$$

which tells us that if the value of $f(a)$ is known, then the function f can be recovered from its derivative f' by integrating. Conversely, Part 2 of the Fundamental Theorem of Calculus (5.6.3) states that

$$\frac{d}{dx}\left[\int_a^x f(t)\,dt\right] = f(x)$$

which tells us that the function f can be recovered from its integral by differentiating. Thus, differentiation and integration can be viewed as inverse processes.

It is common to treat parts 1 and 2 of the Fundamental Theorem of Calculus as a single theorem, and refer to it simply as the *Fundamental Theorem of Calculus*. This theorem ranks as one of the greatest discoveries in the history of science, and its formulation by Newton and Leibniz is generally regarded to be the "discovery of calculus."

INTEGRATING RATES OF CHANGE

The Fundamental Theorem of Calculus

$$\int_a^b f(x)\,dx = F(b) - F(a) \tag{14}$$

has a useful interpretation that can be seen by rewriting it in a slightly different form. Since F is an antiderivative of f on the interval $[a, b]$, we can use the relationship $F'(x) = f(x)$ to rewrite (14) as

$$\int_a^b F'(x)\,dx = F(b) - F(a) \tag{15}$$

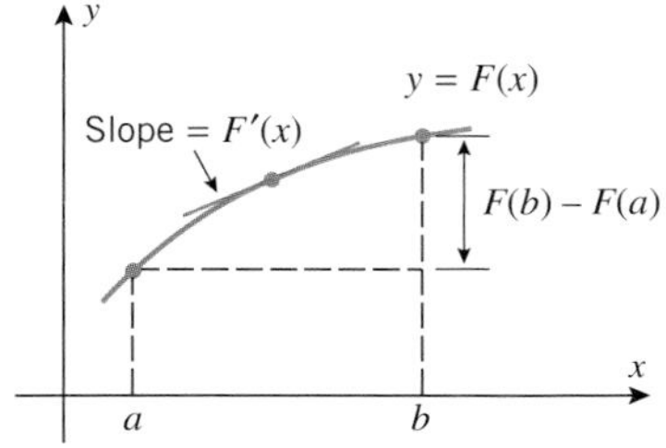

Integrating the slope of $y = F(x)$ over the interval $[a, b]$ produces the change $F(b) - F(a)$ in the value of $F(x)$.

Figure 5.6.9

In this formula we can view $F'(x)$ as the rate of change of $F(x)$ with respect to x, and we can view $F(b) - F(a)$ as the *change* in the value of $F(x)$ as x increases from a to b (Figure 5.6.9). Thus, we have the following useful principle.

> **5.6.4** INTEGRATING A RATE OF CHANGE. Integrating the rate of change of $F(x)$ with respect to x over an interval $[a, b]$ produces the change in the value of $F(x)$ that occurs as x increases from a to b.

Here are some examples of this idea:

- If $P(t)$ is a population (e.g., plants, animals, or people) at time t, then $P'(t)$ is the rate at which the population is changing at time t, and
$$\int_{t_1}^{t_2} P'(t)\,dt = P(t_2) - P(t_1)$$
is the change in the population between times t_1 and t_2.
- If $A(t)$ is the area of an oil spill at time t, then $A'(t)$ is the rate at which the area of the spill is changing at time t, and
$$\int_{t_1}^{t_2} A'(t)\,dt = A(t_2) - A(t_1)$$
is the change in the area of the spill between times t_1 and t_2.
- If $P'(x)$ is the marginal profit that results from producing and selling x units of a product (see p. 270), then
$$\int_{x_1}^{x_2} P'(x)\,dx = P(x_2) - P(x_1)$$
is the change in the profit that results when the production level increases from x_1 units to x_2 units.

✔QUICK CHECK EXERCISES 5.6 (See page 360 for answers.)

1. (a) If $F(x)$ is an antiderivative for $f(x)$, then
$$\int_a^b f(x)\,dx = \_\_\_\_\_\_$$
(b) $\int_a^b F'(x)\,dx = \_\_\_\_\_\_$

(c) $\dfrac{d}{dx}\left[\int_a^x f(t)\,dt\right] = \_\_\_\_\_\_$

2. (a) $\int_1^4 (x^2 - 2x)\,dx = \_\_\_\_\_\_$

(b) $\int_{\pi/6}^{\pi/4} \cos x\,dx = \_\_\_\_\_\_$

3. The total area between the graph of $y = 2x + 2$ and the interval $[-4, 2]$ is ______.

4. For the function $f(x) = 1/x^3$ and the interval $[1, 4]$, the point x^* guaranteed by the Mean-Value Theorem for Integrals is ______.

5. The area of an oil spill is increasing at a rate of $25t$ ft^2/s t seconds after the spill. Between times $t = 2$ and $t = 4$ the area of the spill increases by ______.

EXERCISE SET 5.6 ⌇ Graphing Utility [C] CAS

1. In each part, use a definite integral to find the area of the region, and check your answer using an appropriate formula from geometry.

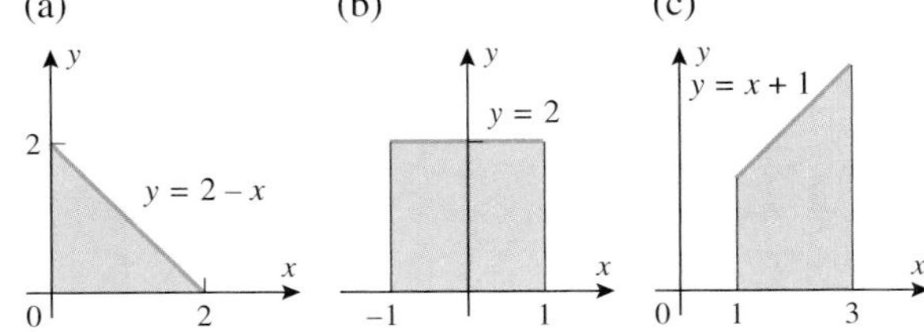

2. In each part, use a definite integral to find the area under the curve $y = f(x)$ over the stated interval, and check your answer using an appropriate formula from geometry.
(a) $f(x) = x$; $[0, 5]$
(b) $f(x) = 5$; $[3, 9]$
(c) $f(x) = x + 3$; $[-1, 2]$

3–6 Find the area under the curve $y = f(x)$ over the stated interval.

3. $f(x) = x^3$; $[2, 3]$

4. $f(x) = x^4$; $[-1, 1]$

5. $f(x) = 3\sqrt{x}$; $[1, 4]$

6. $f(x) = x^{-2/3}$; $[1, 27]$

7–19 Evaluate the integrals using Part 1 of the Fundamental Theorem of Calculus.

7. $\int_{-2}^{1} (x^2 - 6x + 12)\, dx$
8. $\int_{-1}^{2} 4x(1 - x^2)\, dx$
9. $\int_{1}^{4} \frac{4}{x^2}\, dx$
10. $\int_{1}^{2} \frac{1}{x^6}\, dx$
11. $\int_{4}^{9} 2x\sqrt{x}\, dx$
12. $\int_{1}^{4} \frac{1}{x\sqrt{x}}\, dx$
13. $\int_{-\pi/2}^{\pi/2} \sin\theta\, d\theta$
14. $\int_{0}^{\pi/4} \sec^2\theta\, d\theta$
15. $\int_{-\pi/4}^{\pi/4} \cos x\, dx$
16. $\int_{0}^{\pi/3} (2x - \sec x \tan x)\, dx$
17. $\int_{1}^{4} \left(\frac{1}{\sqrt{t}} - 3\sqrt{t}\right) dt$
18. $\int_{4}^{9} (4y^{-1/2} + 2y^{1/2} + y^{-5/2})\, dy$
19. $\int_{\pi/6}^{\pi/2} \left(x + \frac{2}{\sin^2 x}\right) dx$

C 20. Use a CAS to evaluate the integral

$$\int_{a}^{4a} (a^{1/2} - x^{1/2})\, dx$$

and check the answer by hand.

21–22 Use Theorem 5.5.5 to evaluate the given integrals.

21. (a) $\int_{-1}^{1} |2x - 1|\, dx$ (b) $\int_{0}^{3\pi/4} |\cos x|\, dx$

22. (a) $\int_{-1}^{2} \sqrt{2 + |x|}\, dx$ (b) $\int_{0}^{\pi/2} \left|\frac{1}{2} - \cos x\right|\, dx$

C 23. (a) CAS programs provide methods for entering functions that are defined piecewise. Check your documentation to see how this is done, and then use the CAS to evaluate

$$\int_{0}^{2} f(x)\, dx, \quad \text{where} \quad f(x) = \begin{cases} x, & x \le 1 \\ x^2, & x > 1 \end{cases}$$

Use Theorem 5.5.5 to check the answer by hand.

(b) Find a formula for an antiderivative F of f on the interval $[0, 4]$ and verify that

$$\int_{0}^{2} f(x)\, dx = F(2) - F(0)$$

C 24. (a) Use a CAS to evaluate

$$\int_{0}^{4} f(x)\, dx, \quad \text{where} \quad f(x) = \begin{cases} \sqrt{x}, & 0 \le x < 1 \\ 1/x^2, & x \ge 1 \end{cases}$$

Use Theorem 5.5.5 to check the answer by hand.

(b) Find a formula for an antiderivative F of f on the interval $[0, 4]$ and verify that

$$\int_{0}^{4} f(x)\, dx = F(4) - F(0)$$

25–28 Use a calculating utility to find the midpoint approximation of the integral using $n = 20$ subintervals, and then find the exact value of the integral using Part 1 of the Fundamental Theorem of Calculus.

25. $\int_{1}^{3} \frac{1}{x^2}\, dx$
26. $\int_{0}^{\pi/2} \sin x\, dx$
27. $\int_{-1}^{1} \sec^2 x\, dx$
28. $\int_{1}^{3} \frac{1}{x^2}\, dx$

29. Find the area under the curve $y = x^2 + 1$ over the interval $[0, 3]$. Make a sketch of the region.

30. Find the area that is above the x-axis but below the curve $y = x - x^2$. Make a sketch of the region.

31. Find the area under the curve $y = 3\sin x$ over the interval $[0, 2\pi/3]$. Sketch the region.

32. Find the area below the interval $[-2, -1]$ but above the curve $y = x^3$. Make a sketch of the region.

33–36 Sketch the curve and find the total area between the curve and the given interval on the x-axis.

33. $y = x^2 - x$; $[0, 2]$
34. $y = \sin x$; $[0, 3\pi/2]$
35. $y = 2\sqrt{x+1} - 3$; $[0, 3]$
36. $y = \frac{x^2 - 1}{x^2}$; $[\frac{1}{2}, 2]$

37. A student wants to find the area enclosed by the graphs of $y = \cos x$, $y = 0$, $x = 0$, and $x = 0.8$.
(a) Show that the exact area is $\sin 0.8$.
(b) A student uses a calculator to approximate the result in part (a) to three decimal places and obtains an incorrect answer of 0.014. What was the student's error? Find the correct approximation.

FOCUS ON CONCEPTS

38. (a) Use a graphing utility to generate the graph of

$$f(x) = \frac{1}{100}(x + 2)(x + 1)(x - 3)(x - 5)$$

and use the graph to make a conjecture about the sign of the integral

$$\int_{-2}^{5} f(x)\, dx$$

(b) Check your conjecture by evaluating the integral.

39. (a) Let f be an odd function; that is, $f(-x) = -f(x)$. Invent a theorem that makes a statement about the value of an integral of the form

$$\int_{-a}^{a} f(x)\, dx$$

(b) Confirm that your theorem works for the integrals

$$\int_{-1}^{1} x^3\, dx \quad \text{and} \quad \int_{-\pi/2}^{\pi/2} \sin x\, dx$$

(c) Let f be an even function; that is, $f(-x) = f(x)$. Invent a theorem that makes a statement about the relationship between the integrals

$$\int_{-a}^{a} f(x)\,dx \quad \text{and} \quad \int_{0}^{a} f(x)\,dx$$

(d) Confirm that your theorem works for the integrals

$$\int_{-1}^{1} x^2\,dx \quad \text{and} \quad \int_{-\pi/2}^{\pi/2} \cos x\,dx$$

C **40.** Use the theorem you invented in Exercise 39(a) to evaluate the integral

$$\int_{-5}^{5} \frac{x^7 - x^5 + x}{x^4 + x^2 + 7}\,dx$$

and check your answer with a CAS.

41. Define $F(x)$ by

$$F(x) = \int_{1}^{x} (3t^2 - 3)\,dt$$

(a) Use Part 2 of the Fundamental Theorem of Calculus to find $F'(x)$.
(b) Check the result in part (a) by first integrating and then differentiating.

42. Define $F(x)$ by

$$F(x) = \int_{\pi/4}^{x} \cos 2t\,dt$$

(a) Use Part 2 of the Fundamental Theorem of Calculus to find $F'(x)$.
(b) Check the result in part (a) by first integrating and then differentiating.

43–46 Use Part 2 of the Fundamental Theorem of Calculus to find the derivatives.

43. (a) $\dfrac{d}{dx}\displaystyle\int_{1}^{x} \sin(t^2)\,dt$ (b) $\dfrac{d}{dx}\displaystyle\int_{1}^{x} \sqrt{1-\cos t}\,dt$

44. (a) $\dfrac{d}{dx}\displaystyle\int_{0}^{x} \frac{dt}{1+\sqrt{t}}$ (b) $\dfrac{d}{dx}\displaystyle\int_{2}^{x} \frac{dt}{t^2+3t-4}$

45. $\dfrac{d}{dx}\displaystyle\int_{x}^{0} t\sec t\,dt$ [*Hint:* Use Definition 5.5.3(b).]

46. $\dfrac{d}{du}\displaystyle\int_{0}^{u} |x|\,dx$

47. Let $F(x) = \displaystyle\int_{4}^{x} \sqrt{t^2+9}\,dt$. Find

(a) $F(4)$ (b) $F'(4)$ (c) $F''(4)$.

48. Let $F(x) = \displaystyle\int_{0}^{x} \frac{\cos t}{t^2+3t+5}\,dt$. Find

(a) $F(0)$ (b) $F'(0)$ (c) $F''(0)$.

FOCUS ON CONCEPTS

49. Let $F(x) = \displaystyle\int_{0}^{x} \frac{t-3}{t^2+7}\,dt$ for $-\infty < x < +\infty$.

(a) Find the value of x where F attains its minimum value.
(b) Find intervals over which F is only increasing or only decreasing.
(c) Find open intervals over which F is only concave up or only concave down.

C **50.** Use the plotting and numerical integration commands of a CAS to generate the graph of the function F in Exercise 49 over the interval $-20 \le x \le 20$, and confirm that the graph is consistent with the results obtained in that exercise.

51. (a) Over what open interval does the formula

$$F(x) = \int_{1}^{x} \frac{dt}{t}$$

represent an antiderivative of $f(x) = 1/x$?
(b) Find a point where the graph of F crosses the x-axis.

52. (a) Over what open interval does the formula

$$F(x) = \int_{1}^{x} \frac{1}{t^2-9}\,dt$$

represent an antiderivative of

$$f(x) = \frac{1}{x^2-9}?$$

(b) Find a point where the graph of F crosses the x-axis.

53–54 Find all values of x^* in the stated interval that satisfy Equation (8) in the Mean-Value Theorem for Integrals (5.6.2), and explain what these numbers represent.

53. (a) $f(x) = \sqrt{x}$; $[0, 3]$
(b) $f(x) = x^2 + x$; $[-12, 0]$

54. (a) $f(x) = \sin x$; $[-\pi, \pi]$ (b) $f(x) = 1/x^2$; $[1, 3]$

55–56 It was shown in the proof of the Mean-Value Theorem for Integrals (5.6.2) that if f is continuous on $[a, b]$, and if $m \le f(x) \le M$ on $[a, b]$, then

$$m(b-a) \le \int_{a}^{b} f(x)\,dx \le M(b-a)$$

[see (9)]. These inequalities make it possible to obtain bounds on the size of a definite integral from bounds on the size of its integrand. This is illustrated in Exercises 55 and 56.

55. Find the maximum and minimum values of $\sqrt{x^3+2}$ for $0 \le x \le 3$, and use these values to find bounds on the value of the integral

$$\int_{0}^{3} \sqrt{x^3+2}\,dx$$

56. Find values of m and M such that $m \le x\sin x \le M$ for $0 \le x \le \pi$, and use these values to find bounds on the value of the integral

$$\int_{0}^{\pi} x\sin x\,dx$$

57. Prove:

(a) $\left[cF(x)\right]_a^b = c\left[F(x)\right]_a^b$

(b) $\left[F(x) + G(x)\right]_a^b = F(x)\Big]_a^b + G(x)\Big]_a^b$

(c) $\left[F(x) - G(x)\right]_a^b = F(x)\Big]_a^b - G(x)\Big]_a^b$.

FOCUS ON CONCEPTS

58. Explain why the Fundamental Theorem of Calculus may be applied without modification to definite integrals in which the lower limit of integration is greater than or equal to the upper limit of integration.

59. (a) If $h'(t)$ is the rate of change of a child's height measured in inches per year, what does the integral $\int_0^{10} h'(t)\,dt$ represent, and what are its units?
(b) If $r'(t)$ is the rate of change of the radius of a spherical balloon measured in centimeters per second, what does the integral $\int_1^2 r'(t)\,dt$ represent, and what are its units?
(c) If $H(t)$ is the rate of change of the speed of sound with respect to temperature measured in ft/s per °F, what does the integral $\int_{32}^{100} H(t)\,dt$ represent, and what are its units?
(d) If $v(t)$ is the velocity of a particle in rectilinear motion, measured in cm/h, what does the integral $\int_{t_1}^{t_2} v(t)\,dt$ represent, and what are its units?

60. (a) Suppose that sludge is emptied into a river at the rate of $V(t)$ gallons per minute, starting at time $t = 0$. Write an integral that represents the total volume of sludge that is emptied into the river during the first hour.
(b) Suppose that the tangent line to a curve $y = f(x)$ has slope $m(x)$ at x. What does the integral $\int_{x_1}^{x_2} m(x)\,dx$ represent?

61. (a) Suppose that a reservoir supplies water to an industrial park at a constant rate of $r = 4$ gallons per minute (gal/min) between 8:30 A.M. and 9:00 A.M. How much water does the reservoir supply during that time period?
(b) Suppose that one of the industrial plants increases its water consumption between 9:00 A.M. and 10:00 A.M. and that the rate at which the reservoir supplies water increases linearly, as shown in the accompanying figure. How much water does the reservoir supply during that 1-hour time period?
(c) Suppose that from 10:00 A.M. to 12 noon the rate at which the reservoir supplies water is given by the formula $r(t) = 10 + \sqrt{t}$ gal/min, where t is the time (in minutes) since 10:00 A.M. How much water does the reservoir supply during that 2-hour time period?

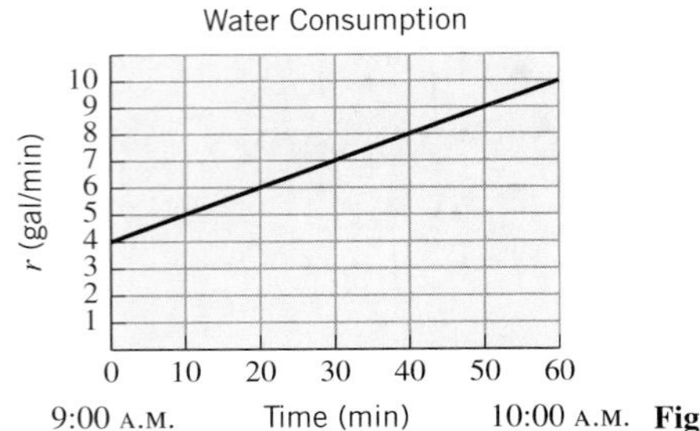

Figure Ex-61

62. A traffic engineer monitors the rate at which cars enter the main highway during the afternoon rush hour. From her data she estimates that between 4:30 P.M. and 5:30 P.M. the rate $R(t)$ at which cars enter the highway is given by the formula $R(t) = 100(1 - 0.0001t^2)$ cars per minute, where t is the time (in minutes) since 4:30 P.M.
(a) When does the peak traffic flow into the highway occur?
(b) Estimate the number of cars that enter the highway during the rush hour.

63–64 Evaluate each limit by interpreting it as a Riemann sum in which the given interval is divided into n subintervals of equal width.

63. $\displaystyle\lim_{n\to+\infty} \sum_{k=1}^{n} \frac{\pi}{4n} \sec^2\left(\frac{\pi k}{4n}\right);\ \left[0, \frac{\pi}{4}\right]$

64. $\displaystyle\lim_{n\to+\infty} \sum_{k=1}^{n} \frac{n}{(n+k)^2};\ [1, 2]$

65. Prove the Mean-Value Theorem for Integrals (Theorem 5.6.2) by applying the Mean-Value Theorem (4.7.2) to an antiderivative F for f.

✔ QUICK CHECK ANSWERS 5.6

1. (a) $F(b) - F(a)$ (b) $F(b) - F(a)$ (c) $f(x)$ 2. (a) 6 (b) $\dfrac{\sqrt{2}-1}{2}$ 3. 18 4. $\sqrt[3]{32/5}$ 5. 150 ft^2

5.7 RECTILINEAR MOTION REVISITED USING INTEGRATION

In Section 4.8 we used the derivative to define the notions of instantaneous velocity and acceleration for a particle in rectilinear motion. In this section we will resume the study of such motion using the tools of integration.

FINDING POSITION AND VELOCITY BY INTEGRATION

Recall from Formulas (1) and (3) of Section 4.8 that if a particle in rectilinear motion has position function $s(t)$, then its instantaneous velocity and acceleration are given by the formulas

$$v(t) = s'(t) \quad \text{and} \quad a(t) = v'(t)$$

It follows from these formulas that $s(t)$ is an antiderivative of $v(t)$ and $v(t)$ is an antiderivative of $a(t)$; that is,

$$s(t) = \int v(t)\,dt \qquad \text{and} \qquad v(t) = \int a(t)\,dt \tag{1–2}$$

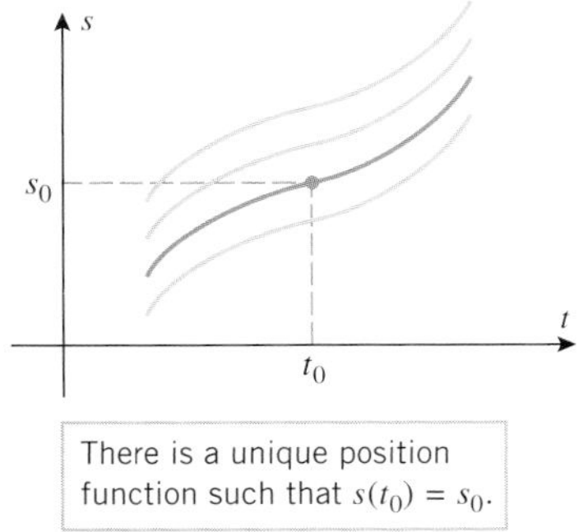

There is a unique position function such that $s(t_0) = s_0$.

Figure 5.7.1

By Formula (1), if we know the velocity function $v(t)$ of a particle in rectilinear motion, then by integrating $v(t)$ we can produce a family of position functions with that velocity function. If, in addition, we know the position s_0 of the particle at any time t_0, then we have sufficient information to find the constant of integration and determine a unique position function (Figure 5.7.1). Similarly, if we know the acceleration function $a(t)$ of the particle, then by integrating $a(t)$ we can produce a family of velocity functions with that acceleration function. If, in addition, we know the velocity v_0 of the particle at any time t_0, then we have sufficient information to find the constant of integration and determine a unique velocity function (Figure 5.7.2).

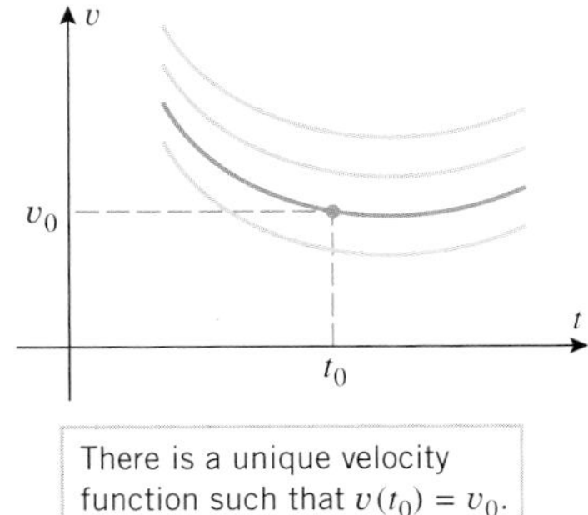

There is a unique velocity function such that $v(t_0) = v_0$.

Figure 5.7.2

▶ **Example 1** Suppose that a particle moves with velocity $v(t) = \cos \pi t$ along a coordinate line. Assuming that the particle has coordinate $s = 4$ at time $t = 0$, find its position function.

Solution. The position function is

$$s(t) = \int v(t)\,dt = \int \cos \pi t\,dt = \frac{1}{\pi}\sin \pi t + C$$

Since $s = 4$ when $t = 0$, it follows that

$$4 = s(0) = \frac{1}{\pi}\sin 0 + C = C$$

Thus,

$$s(t) = \frac{1}{\pi}\sin \pi t + 4 \;◀$$

COMPUTING DISPLACEMENT AND DISTANCE TRAVELED BY INTEGRATION

Recall that the displacement over a time interval of a particle in rectilinear motion is its final coordinate minus its initial coordinate. Thus, if the position function of the particle is $s(t)$, then its displacement over the time interval $[t_0, t_1]$ is $s(t_1) - s(t_0)$. This can be written in integral form as

Interpret Formula (3) as a special case of Formula (15) in Section 5.6.

$$\begin{bmatrix}\text{displacement}\\ \text{over the time}\\ \text{interval } [t_0, t_1]\end{bmatrix} = \int_{t_0}^{t_1} v(t)\,dt = \int_{t_0}^{t_1} s'(t)\,dt = s(t_1) - s(t_0) \tag{3}$$

In contrast, to find the distance traveled by the particle over the time interval $[t_0, t_1]$ (distance traveled in the positive direction plus the distance traveled in the negative direction), we must integrate the absolute value of the velocity function; that is,

$$\begin{bmatrix}\text{distance traveled}\\ \text{during time}\\ \text{interval } [t_0, t_1]\end{bmatrix} = \int_{t_0}^{t_1} |v(t)|\,dt \tag{4}$$

Since the absolute value of velocity is speed, Formulas (3) and (4) can be summarized informally as follows:

Integrating velocity over a time interval produces displacement, and integrating speed over a time interval produces distance traveled.

► **Example 2** Suppose that a particle moves on a coordinate line so that its velocity at time t is $v(t) = t^2 - 2t$ m/s.

(a) Find the displacement of the particle during the time interval $0 \le t \le 3$.

(b) Find the distance traveled by the particle during the time interval $0 \le t \le 3$.

In physical problems it is important to associate correct units with definite integrals. In general, the units for

$$\int_a^b f(x)\,dx$$

are units of $f(x)$ times units of x, since the integral is the limit of Riemann sums, each of whose terms has these units. For example, if $v(t)$ is in meters per second (m/s) and t is in seconds (s), then

$$\int_a^b v(t)\,dt$$

is in meters since

$$(\text{m/s}) \times \text{s} = \text{m}$$

Solution (a). From (3) the displacement is

$$\int_0^3 v(t)\,dt = \int_0^3 (t^2 - 2t)\,dt = \left[\frac{t^3}{3} - t^2\right]_0^3 = 0$$

Thus, the particle is at the same position at time $t = 3$ as at $t = 0$.

Solution (b). The velocity can be written as $v(t) = t^2 - 2t = t(t-2)$, from which we see that $v(t) \le 0$ for $0 \le t \le 2$ and $v(t) \ge 0$ for $2 \le t \le 3$. Thus, it follows from (4) that the distance traveled is

$$\begin{aligned}\int_0^3 |v(t)|\,dt &= \int_0^2 -v(t)\,dt + \int_2^3 v(t)\,dt\\ &= \int_0^2 -(t^2 - 2t)\,dt + \int_2^3 (t^2 - 2t)\,dt\\ &= -\left[\frac{t^3}{3} - t^2\right]_0^2 + \left[\frac{t^3}{3} - t^2\right]_2^3 = \frac{4}{3} + \frac{4}{3} = \frac{8}{3}\ \text{m}\end{aligned}$$ ◄

ANALYZING THE VELOCITY VERSUS TIME CURVE

In Section 4.8 we showed how to use the position versus time curve to obtain information about the behavior of a particle in rectilinear motion (Table 4.8.1). Similarly, there is valuable information that can be obtained from the velocity versus time curve. For example, the integral in (3) can be interpreted geometrically as the *net signed area* between the graph of $v(t)$ and the interval $[t_0, t_1]$, and the integral in (4) can be interpreted as the *total area* between the graph of $v(t)$ and the interval $[t_0, t_1]$. Thus we have the following result.

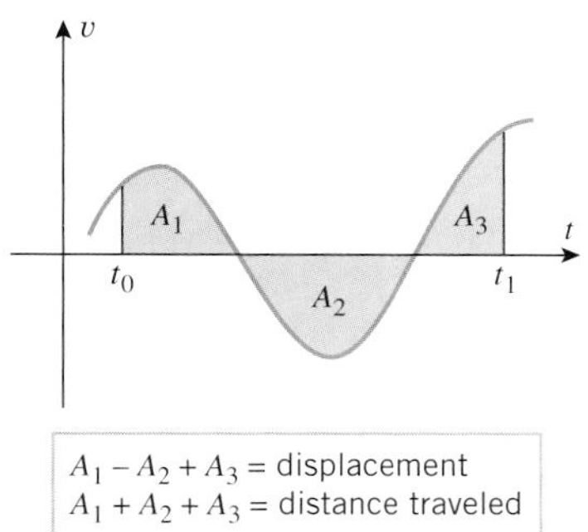

Figure 5.7.3

5.7.1 FINDING DISPLACEMENT AND DISTANCE TRAVELED FROM THE VELOCITY VERSUS TIME CURVE. For a particle in rectilinear motion, the net signed area between the velocity versus time curve and the interval $[t_0, t_1]$ on the t-axis represents the displacement of the particle over that time interval, and the total area between the velocity versus time curve and the interval $[t_0, t_1]$ on the t-axis represents the distance traveled by the particle over that time interval (Figure 5.7.3).

▶ **Example 3** Figure 5.7.4 shows three velocity versus time curves for a particle in rectilinear motion along a horizontal line with the positive direction to the right. In each case find the displacement and the distance traveled over the time interval $0 \leq t \leq 4$, and explain what that information tells you about the motion of the particle.

Solution (a). In part (*a*) of the figure the area and the net signed area over the interval are both 2. Thus, at the end of the time period the particle is 2 units to the right of its starting point and has traveled a distance of 2 units.

Solution (b). In part (*b*) of the figure the net signed area is -2, and the total area is 2. Thus, at the end of the time period the particle is 2 units to the left of its starting point and has traveled a distance of 2 units.

Solution (c). In part (*c*) of the figure the net signed area is 0, and the total area is 2. Thus, at the end of the time period the particle is back at its starting point and has traveled a distance of 2 units. More specifically, it traveled 1 unit to the right over the time interval $0 \leq t \leq 1$ and then 1 unit to the left over the time interval $1 \leq t \leq 2$ (why?). ◀

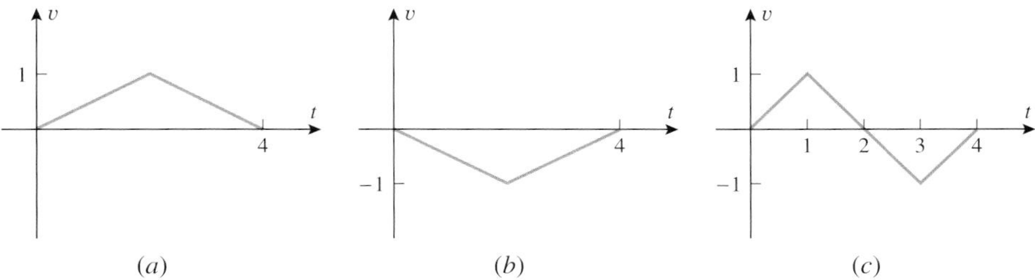

Figure 5.7.4

UNIFORMLY ACCELERATED MOTION

One of the most important cases of rectilinear motion occurs when a particle has constant acceleration. We call this ***uniformly accelerated motion***.

We will show that if a particle moves with constant acceleration along an s-axis, and if the position and velocity of the particle are known at some point in time, say when $t = 0$, then it is possible to derive formulas for the position $s(t)$ and the velocity $v(t)$ at any time t. To see how this can be done, suppose that the particle has constant acceleration

$$a(t) = a \tag{5}$$

and

$$s = s_0 \quad \text{when} \quad t = 0 \tag{6}$$

$$v = v_0 \quad \text{when} \quad t = 0 \tag{7}$$

where s_0 and v_0 are known. We call (6) and (7) the ***initial conditions*** for the motion.

With (5) as a starting point, we can integrate $a(t)$ to obtain $v(t)$, and we can integrate $v(t)$ to obtain $s(t)$, using an initial condition in each case to determine the constant of integration. The computations are as follows:

$$v(t) = \int a(t)\,dt = \int a\,dt = at + C_1 \tag{8}$$

To determine the constant of integration C_1 we apply initial condition (7) to this equation to obtain

$$v_0 = v(0) = a \cdot 0 + C_1 = C_1$$

Substituting this in (8) and putting the constant term first yields

$$v(t) = v_0 + at$$

Since v_0 is constant, it follows that

$$s(t) = \int v(t)\,dt = \int (v_0 + at)\,dt = v_0 t + \tfrac{1}{2}at^2 + C_2 \tag{9}$$

To determine the constant C_2 we apply initial condition (6) to this equation to obtain

$$s_0 = s(0) = v_0 \cdot 0 + \tfrac{1}{2}a \cdot 0 + C_2 = C_2$$

Substituting this in (9) and putting the constant term first yields

$$s(t) = s_0 + v_0 t + \tfrac{1}{2}at^2$$

In summary, we have the following result.

How can you tell from the graph of the velocity versus time curve whether a particle moving along a line has uniformly accelerated motion?

5.7.2 UNIFORMLY ACCELERATED MOTION. If a particle moves with constant acceleration a along an s-axis, and if the position and velocity at time $t = 0$ are s_0 and v_0, respectively, then the position and velocity functions of the particle are

$$s(t) = s_0 + v_0 t + \tfrac{1}{2}at^2 \tag{10}$$

$$v(t) = v_0 + at \tag{11}$$

▶ **Example 4** Suppose that an intergalactic spacecraft uses a sail and the "solar wind" to produce a constant acceleration of 0.032 m/s^2. Assuming that the spacecraft has a velocity of 10,000 m/s when the sail is first raised, how far will the spacecraft travel in 1 hour, and what will its velocity be at the end of this hour?

Solution. In this problem the choice of a coordinate axis is at our discretion, so we will choose it to make the computations as simple as possible. Accordingly, let us introduce an s-axis whose positive direction is in the direction of motion, and let us take the origin to coincide with the position of the spacecraft at the time $t = 0$ when the sail is raised. Thus, the Formulas (10) and (11) for uniformly accelerated motion apply with

$$s_0 = s(0) = 0, \quad v_0 = v(0) = 10{,}000, \quad \text{and} \quad a = 0.032$$

Since 1 hour corresponds to $t = 3600$ s, it follows from (10) that in 1 hour the spacecraft travels a distance of

$$s(3600) = 10{,}000(3600) + \tfrac{1}{2}(0.032)(3600)^2 \approx 36{,}200{,}000 \text{ m}$$

and it follows from (11) that after 1 hour its velocity is

$$v(3600) = 10{,}000 + (0.032)(3600) \approx 10{,}100 \text{ m/s} \blacktriangleleft$$

► **Example 5** A bus has stopped to pick up riders, and a woman is running at a constant velocity of 5 m/s to catch it. When she is 11 m behind the front door the bus pulls away with a constant acceleration of 1 m/s^2. From that point in time, how long will it take for the woman to reach the front door of the bus if she keeps running with a velocity of 5 m/s?

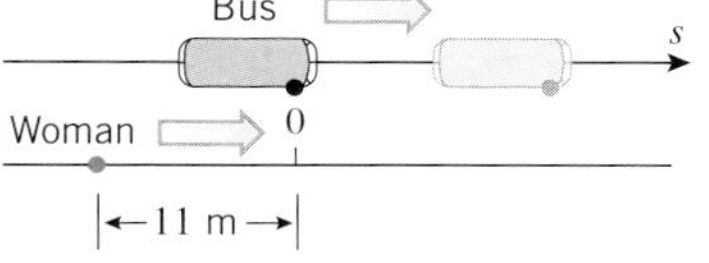

Figure 5.7.5

Solution. As shown in Figure 5.7.5, choose the s-axis so that the bus and the woman are moving in the positive direction, and the front door of the bus is at the origin at the time $t = 0$ when the bus begins to pull away. To catch the bus at some later time t, the woman will have to cover a distance $s_w(t)$ that is equal to 11 m plus the distance $s_b(t)$ traveled by the bus; that is, the woman will catch the bus when

$$s_w(t) = s_b(t) + 11 \tag{12}$$

Since the woman has a constant velocity of 5 m/s, the distance she travels in t seconds is $s_w(t) = 5t$. Thus, (12) can be written as

$$s_b(t) = 5t - 11 \tag{13}$$

Since the bus has a constant acceleration of $a = 1$ m/s^2, and since $s_0 = v_0 = 0$ at time $t = 0$ (why?), it follows from (10) that

$$s_b(t) = \tfrac{1}{2}t^2$$

Substituting this equation into (13) and reorganizing the terms yields the quadratic equation

$$\tfrac{1}{2}t^2 - 5t + 11 = 0 \quad \text{or} \quad t^2 - 10t + 22 = 0$$

Solving this equation for t using the quadratic formula yields two solutions:

$$t = 5 - \sqrt{3} \approx 3.3 \quad \text{and} \quad t = 5 + \sqrt{3} \approx 6.7$$

(verify). Thus, the woman can reach the door at two different times, $t = 3.3$ s and $t = 6.7$ s. The reason that there are two solutions can be explained as follows: When the woman first reaches the door, she is running faster than the bus and can run past it if the driver does not see her. However, as the bus speeds up, it eventually catches up to her, and she has another chance to flag it down. ◄

FREE-FALL MODEL

Motion that occurs when an object near the Earth is imparted some initial velocity (up or down) and thereafter moves along a vertical line is called ***free-fall motion***. In modeling free-fall motion we assume that the only force acting on the object is the Earth's gravity and that the object stays sufficiently close to the Earth that the gravitational force is constant. In particular, air resistance and the gravitational pull of other celestial bodies are neglected.

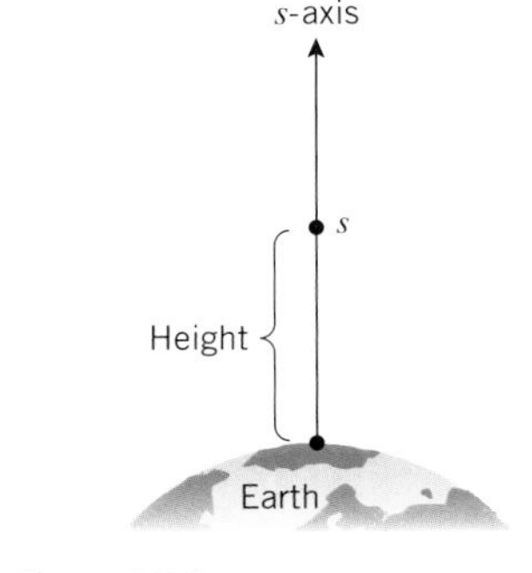

Figure 5.7.6

In our model we will ignore the physical size of the object by treating it as a particle, and we will assume that it moves along an s-axis whose origin is at the surface of the Earth and whose positive direction is up. With this convention, the s-coordinate of the particle is the height of the particle above the surface of the Earth (Figure 5.7.6).

It is a fact of physics that a particle with free-fall motion has constant acceleration. The magnitude of this constant, denoted by the letter g, is called the ***acceleration due to gravity*** and is approximately 9.8 m/s^2 or 32 ft/s^2, depending on whether distance is measured in meters or feet.*

*Strictly speaking, the constant g varies with the latitude and the distance from the Earth's center. However, for motion at a fixed latitude and near the surface of the Earth, the assumption of a constant g is satisfactory for many applications.

Recall that a particle is speeding up when its velocity and acceleration have the same sign and is slowing down when they have opposite signs. Thus, because we have chosen the positive direction to be up, it follows that the acceleration $a(t)$ of a particle in free fall is negative for all values of t. To see that this is so, observe that an upward-moving particle (positive velocity) is slowing down, so its acceleration must be negative; and a downward-moving particle (negative velocity) is speeding up, so its acceleration must also be negative. Thus, we conclude that

$$a(t) = -g \tag{14}$$

It now follows from this and Formulas (10) and (11) for uniformly accelerated motion that the position and velocity functions for a particle in free-fall motion are

How would Formulas (14), (15), and (16) change if we choose the direction of the positive s-axis to be down?

$$s(t) = s_0 + v_0 t - \tfrac{1}{2}gt^2 \tag{15}$$

$$v(t) = v_0 - gt \tag{16}$$

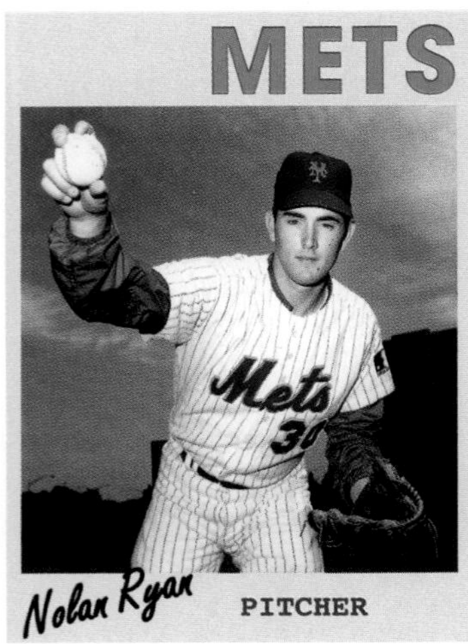

Nolan Ryan's rookie baseball card

► **Example 6** Nolan Ryan, one of the fastest baseball pitchers of all time, was capable of throwing a baseball 150 ft/s (over 102 mi/h). During his career, he had the opportunity to pitch in the Houston Astrodome, home to the Houston Astros Baseball Team from 1965 to 1999. The Astrodome was an indoor stadium with a ceiling 208 ft high. Could Nolan Ryan have hit the ceiling of the Astrodome if he were capable of giving a baseball an upward velocity of 100 ft/s from a height of 7 ft?

Solution. Since distance is in feet, we take $g = 32$ ft/s^2. Initially, we have $s_0 = 7$ ft and $v_0 = 100$ ft/s, so from (15) and (16) we have

$$s(t) = 7 + 100t - 16t^2$$
$$v(t) = 100 - 32t$$

The ball will rise until $v(t) = 0$, that is, until $100 - 32t = 0$. Solving this equation we see that the ball is at its maximum height at time $t = \frac{25}{8}$. To find the height of the ball at this instant we substitute this value of t into the position function to obtain

$$s\left(\tfrac{25}{8}\right) = 7 + 100\left(\tfrac{25}{8}\right) - 16\left(\tfrac{25}{8}\right)^2 = 163.25 \text{ ft}$$

which is roughly 45 ft short of hitting the ceiling. ◄

In Example 6 the ball is moving up when the velocity is positive and is moving down when the velocity is negative, so it makes sense physically that the velocity is zero when the ball reaches its peak.

► **Example 7** A penny is released from rest near the top of the Empire State Building at a point that is 1250 ft above the ground (Figure 5.7.7). Assuming that the free-fall model applies, how long does it take for the penny to hit the ground, and what is its speed at the time of impact?

Solution. Since distance is in feet, we take $g = 32$ ft/s^2. Initially, we have $s_0 = 1250$ and $v_0 = 0$, so from (15)

$$s(t) = 1250 - 16t^2 \tag{17}$$

Impact occurs when $s(t) = 0$. Solving this equation for t, we obtain

$$1250 - 16t^2 = 0$$
$$t^2 = \frac{1250}{16} = \frac{625}{8}$$
$$t = \pm\frac{25}{\sqrt{8}} \approx \pm 8.8 \text{ s}$$

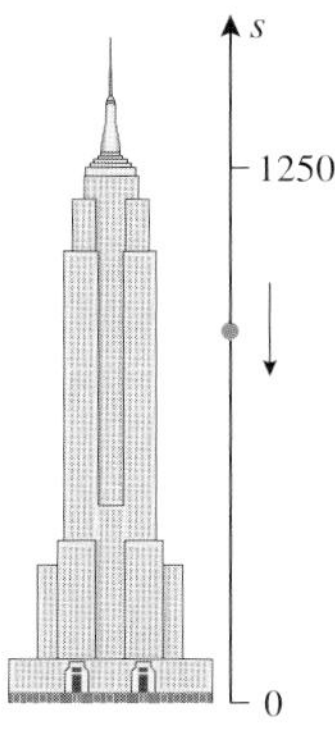

Figure 5.7.7

Since $t \geq 0$, we can discard the negative solution and conclude that it takes $25/\sqrt{8} \approx 8.8$ s for the penny to hit the ground. To obtain the velocity at the time of impact, we substitute $t = 25/\sqrt{8}$, $v_0 = 0$, and $g = 32$ in (16) to obtain

$$v\left(\frac{25}{\sqrt{8}}\right) = 0 - 32\left(\frac{25}{\sqrt{8}}\right) = -200\sqrt{2} \approx -282.8 \text{ ft/s}$$

Thus, the speed at the time of impact is

$$\left|v\left(\frac{25}{\sqrt{8}}\right)\right| = 200\sqrt{2} \approx 282.8 \text{ ft/s}$$

which is more than 192 mi/h. ◀

✔QUICK CHECK EXERCISES 5.7 *(See page 370 for answers.)*

1. Suppose that a particle is moving along an s-axis with velocity $v(t) = 2t + 1$. If at time $t = 0$ the particle is at position $s = 2$, the position function of the particle is $s(t) =$ ________.
2. Let $v(t)$ denote the velocity function of a particle that is moving along an s-axis with constant acceleration $a = -2$. If $v(1) = 4$, then $v(t) =$ ________.
3. Let $v(t)$ denote the velocity function of a particle in rectilinear motion. Suppose that $v(0) = -1$, $v(3) = 2$, and the velocity versus time curve is a straight line. The displacement of the particle between times $t = 0$ and $t = 3$ is ________, and the distance traveled by the particle over this period of time is ________.
4. Based on the free-fall model, from what height must a coin be dropped so that it strikes the ground with speed 48 ft/s?

EXERCISE SET 5.7 Graphing Utility C CAS

FOCUS ON CONCEPTS

1. In each part, the velocity versus time curve is given for a particle moving along a line. Use the curve to find the displacement and the distance traveled by the particle over the time interval $0 \leq t \leq 3$.

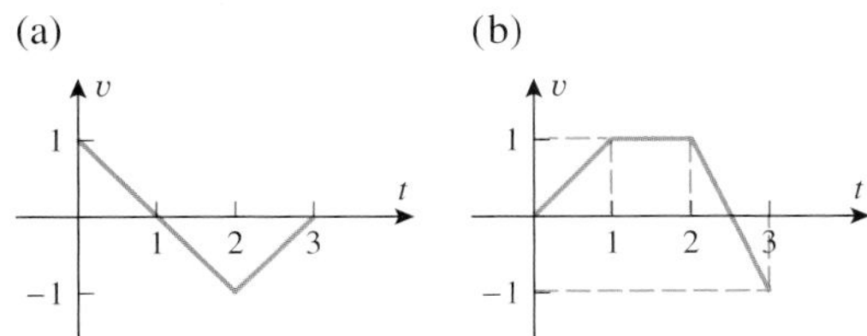

2. Sketch a velocity versus time curve for a particle that travels a distance of 5 units along a coordinate line during the time interval $0 \leq t \leq 10$ and has a displacement of 0 units.
3. The accompanying figure shows the acceleration versus time curve for a particle moving along a coordinate line. If the initial velocity of the particle is 20 m/s, estimate
   (a) the velocity at time $t = 4$ s
   (b) the velocity at time $t = 6$ s.

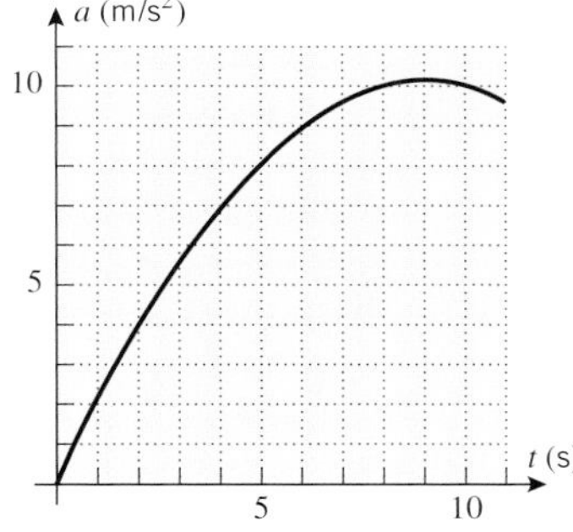

Figure Ex-3

4. The accompanying figure (next page) shows the velocity versus time curve over the time interval $1 \leq t \leq 5$ for a particle moving along a horizontal coordinate line.
   (a) What can you say about the sign of the acceleration over the time interval?
   (b) When is the particle speeding up? Slowing down?
   (c) What can you say about the location of the particle at time $t = 5$ relative to its location at time $t = 1$? Explain your reasoning.

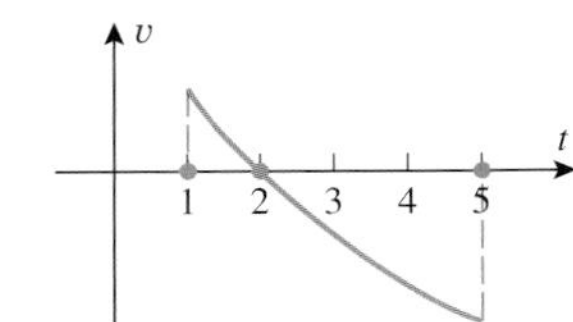

Figure Ex-4

5–8 A particle moves along an s-axis. Use the given information to find the position function of the particle.

5. (a) $v(t) = 3t^2 - 2t$; $s(0) = 1$
(b) $a(t) = 3\sin 3t$; $v(0) = 3$; $s(0) = 3$

6. (a) $v(t) = 1 + \sin t$; $s(0) = -3$
(b) $a(t) = t^2 - 3t + 1$; $v(0) = 0$; $s(0) = 0$

7. (a) $v(t) = 3t + 1$; $s(2) = 4$
(b) $a(t) = 2t^{-3}$; $v(1) = 0$; $s(1) = 2$

8. (a) $v(t) = t^{2/3}$; $s(8) = 0$
(b) $a(t) = \sqrt{t}$; $v(4) = 1$; $s(4) = -5$

9–12 A particle moves with a velocity of $v(t)$ m/s along an s-axis. Find the displacement and the distance traveled by the particle during the given time interval.

9. (a) $v(t) = \sin t$; $0 \le t \le \pi/2$
(b) $v(t) = \cos t$; $\pi/2 \le t \le 2\pi$

10. (a) $v(t) = 3t - 2$; $0 \le t \le 2$
(b) $v(t) = |1 - 2t|$; $0 \le t \le 2$

11. (a) $v(t) = t^3 - 3t^2 + 2t$; $0 \le t \le 3$
(b) $v(t) = \sqrt{t} - 2$; $0 \le t \le 3$

12. (a) $v(t) = t - \sqrt{t}$; $0 \le t \le 4$
(b) $v(t) = \dfrac{1}{\sqrt{t+1}}$; $0 \le t \le 3$

13–16 A particle moves with acceleration $a(t)$ m/s^2 along an s-axis and has velocity v_0 m/s at time $t = 0$. Find the displacement and the distance traveled by the particle during the given time interval.

13. $a(t) = 3$; $v_0 = -1$; $0 \le t \le 2$

14. $a(t) = t - 2$; $v_0 = 0$; $1 \le t \le 5$

15. $a(t) = 1/\sqrt{3t+1}$; $v_0 = \frac{4}{3}$; $1 \le t \le 5$

16. $a(t) = \sin t$; $v_0 = 1$; $\pi/4 \le t \le \pi/2$

17. In each part, use the given information to find the position, velocity, speed, and acceleration at time $t = 1$.
(a) $v = \sin \frac{1}{2}\pi t$; $s = 0$ when $t = 0$
(b) $a = -3t$; $s = 1$ and $v = 0$ when $t = 0$

18. In each part, use the given information to find the position, velocity, speed, and acceleration at time $t = 1$.
(a) $v = \cos \frac{1}{3}\pi t$; $s = 0$ when $t = \frac{3}{2}$
(b) $a = 5t - t^3$; $s = 1$ and $v = -1/2$ when $t = 0$

19. Suppose that a particle moves along a line so that its velocity v at time t is given by

$$v(t) = \begin{cases} 5t, & 0 \le t < 1 \\ 6\sqrt{t} - 1, & 1 \le t \end{cases}$$

where t is in seconds and v is in centimeters per second (cm/s). Estimate the time(s) at which the particle is 4 cm from its starting position.

20. Suppose that a particle moves along a line so that its velocity v at time t is given by

$$v(t) = -3t^2 + 7t, \quad t \ge 0$$

where t is in seconds and v is in centimeters per second (cm/s). Estimate the time(s) at which the particle is 2 cm from its starting position.

21. Suppose that the velocity function of a particle moving along an s-axis is $v(t) = 20t^2 - 110t + 120$ ft/s and that the particle is at the origin at time $t = 0$. Use a graphing utility to generate the graphs of $s(t)$, $v(t)$, and $a(t)$ for the first 6 s of motion.

22. Suppose that the acceleration function of a particle moving along an s-axis is $a(t) = 4t - 30$ m/s^2 and that the position and velocity at time $t = 0$ are $s_0 = -5$ m and $v_0 = 3$ m/s. Use a graphing utility to generate the graphs of $s(t)$, $v(t)$, and $a(t)$ for the first 25 s of motion.

23–24 For the given velocity function $v(t)$:
(a) Generate the velocity versus time curve, and use it to make a conjecture about the sign of the displacement over the given time interval.
(b) Use a CAS to find the displacement.

C **23.** $v(t) = 0.5 - t\sin t$; $0 \le t \le 5$

C **24.** $v(t) = 0.5 - t\cos \pi t$; $0 \le t \le 1$

25. Suppose that at time $t = 0$ a particle is at the origin of an x-axis and has a velocity of $v_0 = 25$ cm/s. For the first 4 s thereafter it has no acceleration, and then it is acted on by a retarding force that produces a constant negative acceleration of $a = -10$ cm/s^2.
(a) Sketch the acceleration versus time curve over the interval $0 \le t \le 12$.
(b) Sketch the velocity versus time curve over the time interval $0 \le t \le 12$.
(c) Find the x-coordinate of the particle at times $t = 8$ s and $t = 12$ s.
(d) What is the maximum x-coordinate of the particle over the time interval $0 \le t \le 12$?

26. Formulas (10) and (11) for uniformly accelerated motion can be rearranged in various useful ways. For simplicity, let $s = s(t)$ and $v = v(t)$, and derive the following variations of those formulas.
(a) $a = \dfrac{v^2 - v_0^2}{2(s - s_0)}$ (b) $t = \dfrac{2(s - s_0)}{v_0 + v}$
(c) $s = s_0 + vt - \frac{1}{2}at^2$ [Note how this differs from (10).]

27–34 In these exercises assume that the object is moving with uniformly accelerated motion in the positive direction of a coordinate line, and apply Formulas (10) and (11) or those from Exercise 26, as appropriate. In some of these problems you will need the fact that 88 ft/s = 60 mi/h.

27. (a) An automobile traveling on a straight road decelerates uniformly from 55 mi/h to 40 mi/h in 10 s. Find its acceleration in ft/s^2.
(b) A bicycle rider traveling on a straight path accelerates uniformly from rest to 30 km/h in 1 min. Find his acceleration in km/s^2.

28. A car traveling 60 mi/h along a straight road decelerates at a constant rate of 11 ft/s^2.
(a) How long will it take until the speed is 45 mi/h?
(b) How far will the car travel before coming to a stop?

29. Spotting a police car, you hit the brakes on your new Porsche to reduce your speed from 90 mi/h to 60 mi/h at a constant rate over a distance of 200 ft.
(a) Find the acceleration in ft/s^2.
(b) How long does it take for you to reduce your speed to 55 mi/h?
(c) At the acceleration obtained in part (a), how long would it take for you to bring your Porsche to a complete stop from 90 mi/h?

30. A particle moving along a straight line is accelerating at a constant rate of 5 m/s^2. Find the initial velocity if the particle moves 60 m in the first 4 s.

31. A motorcycle, starting from rest, speeds up with a constant acceleration of 2.6 m/s^2. After it has traveled 120 m, it slows down with a constant acceleration of -1.5 m/s^2 until it attains a speed of 12 m/s. What is the distance traveled by the motorcycle at that point?

32. A sprinter in a 100-m race explodes out of the starting block with an acceleration of 4.0 m/s^2, which she sustains for 2.0 s. Her acceleration then drops to zero for the rest of race.
(a) What is her time for the race?
(b) Make a graph of her distance from the starting block versus time.

33. A car that has stopped at a toll booth leaves the booth with a constant acceleration of 4 ft/s^2. At the time the car leaves the booth it is 2500 ft behind a truck traveling with a constant velocity of 50 ft/s. How long will it take for the car to catch the truck, and how far will the car be from the toll booth at that time?

34. In the final sprint of a rowing race the challenger is rowing at a constant speed of 12 m/s. At the point where the leader is 100 m from the finish line and the challenger is 15 m behind, the leader is rowing at 8 m/s but starts accelerating at a constant 0.5 m/s^2. Who wins?

35–44 Assume that a free-fall model applies. Solve these exercises by applying Formulas (15) and (16) or, if appropriate, use those from Exercise 26 with $a = -g$. In these exercises take $g = 32$ ft/s^2 or $g = 9.8$ m/s^2, depending on the units.

35. A projectile is launched vertically upward from ground level with an initial velocity of 112 ft/s.
(a) Find the velocity at $t = 3$ s and $t = 5$ s.
(b) How high will the projectile rise?
(c) Find the speed of the projectile when it hits the ground.

36. A projectile fired downward from a height of 112 ft reaches the ground in 2 s. What is its initial velocity?

37. A projectile is fired vertically upward from ground level with an initial velocity of 16 ft/s.
(a) How long will it take for the projectile to hit the ground?
(b) How long will the projectile be moving upward?

38. In 1939, Joe Sprinz of the San Francisco Seals Baseball Club attempted to catch a ball dropped from a blimp at a height of 800 ft (for the purpose of breaking the record for catching a ball dropped from the greatest height set the preceding year by members of the Cleveland Indians).
(a) How long does it take for a ball to drop 800 ft?
(b) What is the velocity of a ball in miles per hour after an 800-ft drop (88 ft/s = 60 mi/h)?

[*Note:* As a practical matter, it is unrealistic to ignore wind resistance in this problem; however, even with the slowing effect of wind resistance, the impact of the ball slammed Sprinz's glove hand into his face, fractured his upper jaw in 12 places, broke five teeth, and knocked him unconscious. He dropped the ball!]

39. A projectile is launched upward from ground level with an initial speed of 60 m/s.
(a) How long does it take for the projectile to reach its highest point?
(b) How high does the projectile go?
(c) How long does it take for the projectile to drop back to the ground from its highest point?
(d) What is the speed of the projectile when it hits the ground?

40. (a) Use the results in Exercise 39 to make a conjecture about the relationship between the initial and final speeds of a projectile that is launched upward from ground level and returns to ground level.
(b) Prove your conjecture.

41. A projectile is fired vertically upward with an initial velocity of 49 m/s from a tower 150 m high.
(a) How long will it take for the projectile to reach its maximum height?
(b) What is the maximum height?
(c) How long will it take for the projectile to pass its starting point on the way down?

(d) What is the velocity when it passes the starting point on the way down?
(e) How long will it take for the projectile to hit the ground?
(f) What will be its speed at impact?

42. A man drops a stone from a bridge. What is the height of the bridge if
(a) the stone hits the water 4 s later
(b) the sound of the splash reaches the man 4 s later? [Take 1080 ft/s as the speed of sound.]

43. In Example 6, how fast would Nolan Ryan have to throw a ball upward from a height of 7 feet in order to hit the ceiling of the Astrodome?

44. A rock thrown downward with an unknown initial velocity from a height of 1000 ft reaches the ground in 5 s. Find the velocity of the rock when it hits the ground.

✔ QUICK CHECK ANSWERS 5.7

1. $t^2 + t + 2$ **2.** $6 - 2t$ **3.** $\frac{3}{2}; \frac{5}{2}$ **4.** 36 ft

5.8 EVALUATING DEFINITE INTEGRALS BY SUBSTITUTION

In this section we will discuss two methods for evaluating definite integrals in which a substitution is required.

TWO METHODS FOR MAKING SUBSTITUTIONS IN DEFINITE INTEGRALS

Recall from Section 5.3 that indefinite integrals of the form

$$\int f(g(x))g'(x)\,dx$$

can sometimes be evaluated by making the u-substitution

$$u = g(x), \quad du = g'(x)\,dx \tag{1}$$

which converts the integral to the form

$$\int f(u)\,du$$

To apply this method to a definite integral of the form

$$\int_a^b f(g(x))g'(x)\,dx$$

we need to account for the effect that the substitution has on the x-limits of integration. There are two ways of doing this.

Method 1.

First evaluate the indefinite integral

$$\int f(g(x))g'(x)\,dx$$

by substitution, and then use the relationship

$$\int_a^b f(g(x))g'(x)\,dx = \left[\int f(g(x))g'(x)\,dx\right]_a^b$$

to evaluate the definite integral. This procedure does not require any modification of the x-limits of integration.

Method 2.

Make the substitution (1) directly in the definite integral, and then use the relationship $u = g(x)$ to replace the x-limits, $x = a$ and $x = b$, by corresponding u-limits, $u = g(a)$ and $u = g(b)$. This produces a new definite integral

$$\int_{g(a)}^{g(b)} f(u)\,du$$

that is expressed entirely in terms of u.

▸ **Example 1** Use the two methods above to evaluate $\int_0^2 x(x^2+1)^3\,dx$.

Solution by Method 1. If we let

$$u = x^2 + 1 \quad \text{so that} \quad du = 2x\,dx \tag{2}$$

then we obtain

$$\int x(x^2+1)^3\,dx = \frac{1}{2}\int u^3\,du = \frac{u^4}{8} + C = \frac{(x^2+1)^4}{8} + C$$

Thus,

$$\int_0^2 x(x^2+1)^3\,dx = \left[\int x(x^2+1)^3\,dx\right]_{x=0}^{2}$$

$$= \frac{(x^2+1)^4}{8}\Bigg]_{x=0}^{2} = \frac{625}{8} - \frac{1}{8} = 78$$

Solution by Method 2. If we make the substitution $u = x^2 + 1$ in (2), then

$$\begin{aligned} &\text{if} \quad x = 0, \quad u = 1 \\ &\text{if} \quad x = 2, \quad u = 5 \end{aligned}$$

Thus,

$$\int_0^2 x(x^2+1)^3\,dx = \frac{1}{2}\int_1^5 u^3\,du$$

$$= \frac{u^4}{8}\Bigg]_{u=1}^{5} = \frac{625}{8} - \frac{1}{8} = 78$$

which agrees with the result obtained by Method 1. ◂

The following theorem states precise conditions under which Method 2 can be used.

5.8.1 THEOREM. *If g' is continuous on $[a, b]$ and f is continuous on an interval containing the values of $g(x)$ for $a \le x \le b$, then*

$$\int_a^b f(g(x))g'(x)\,dx = \int_{g(a)}^{g(b)} f(u)\,du$$

PROOF. Since f is continuous on an interval containing the values of $g(x)$ for $a \le x \le b$, it follows that f has an antiderivative F on that interval. If we let $u = g(x)$, then the chain rule implies that

$$\frac{d}{dx}F(g(x)) = \frac{d}{dx}F(u) = \frac{dF}{du}\frac{du}{dx} = f(u)\frac{du}{dx} = f(g(x))g'(x)$$

for each x in $[a, b]$. Thus, $F(g(x))$ is an antiderivative of $f(g(x))g'(x)$ on $[a, b]$. Therefore, by Part 1 of the Fundamental Theorem of Calculus (Theorem 5.6.1)

$$\int_a^b f(g(x))g'(x)\,dx = F(g(x))\Big]_a^b = F(g(b)) - F(g(a)) = \int_{g(a)}^{g(b)} f(u)\,du \quad \blacksquare$$

The choice of methods for evaluating definite integrals by substitution is generally a matter of taste, but in the following examples we will use the second method, since the idea is new.

▶ **Example 2** Evaluate

$$\text{(a)} \int_0^{\pi/8} \sin^5 2x \cos 2x\,dx \qquad \text{(b)} \int_2^5 (2x-5)(x-3)^9\,dx$$

Solution (a). Let

$$u = \sin 2x \quad \text{so that} \quad du = 2\cos 2x\,dx \quad \left(\text{or } \tfrac{1}{2}\,du = \cos 2x\,dx\right)$$

With this substitution,

$$\text{if} \quad x = 0, \quad u = \sin(0) = 0$$
$$\text{if} \quad x = \pi/8, \quad u = \sin(\pi/4) = 1/\sqrt{2}$$

so

$$\int_0^{\pi/8} \sin^5 2x \cos 2x\,dx = \frac{1}{2}\int_0^{1/\sqrt{2}} u^5\,du$$

$$= \frac{1}{2}\cdot\frac{u^6}{6}\Big]_{u=0}^{1/\sqrt{2}} = \frac{1}{2}\left[\frac{1}{6(\sqrt{2})^6} - 0\right] = \frac{1}{96}$$

Solution (b). Let

$$u = x - 3 \quad \text{so that} \quad du = dx$$

This leaves a factor of $2x + 5$ unresolved in the integrand. However,

$$x = u + 3, \quad \text{so} \quad 2x - 5 = 2(u+3) - 5 = 2u + 1$$

With this substitution,

$$\text{if} \quad x = 2, \quad u = 2 - 3 = -1$$
$$\text{if} \quad x = 5, \quad u = 5 - 3 = 2$$

so

$$\int_2^5 (2x-5)(x-3)^9\,dx = \int_{-1}^{2} (2u+1)u^9\,du = \int_{-1}^{2} (2u^{10} + u^9)\,du$$

$$= \left[\frac{2u^{11}}{11} + \frac{u^{10}}{10}\right]_{u=-1}^{2} = \left(\frac{2^{12}}{11} + \frac{2^{10}}{10}\right) - \left(-\frac{2}{11} + \frac{1}{10}\right)$$

$$= \frac{52{,}233}{110} \approx 474.8 \blacktriangleleft$$

▶ Example 3 Evaluate $\displaystyle\int_1^3 \frac{\cos(\pi/x)}{x^2}\,dx$.

Solution. Let

$$u = \frac{\pi}{x} \quad \text{so that} \quad du = -\frac{\pi}{x^2}\,dx = -\pi \cdot \frac{1}{x^2}\,dx \quad \text{or} \quad -\frac{1}{\pi}\,du = \frac{1}{x^2}\,dx$$

The u-substitution in Example 3 produces an integral in which the upper u-limit is smaller than the lower u-limit. Use Definition 5.5.3(b) to convert this integral to one whose lower limit is smaller than the upper limit and verify that it produces an integral with the same value as that in the example.

With this substitution,

$$\text{if} \quad x = 1, \quad u = \pi$$
$$\text{if} \quad x = 3, \quad u = \pi/3$$

Thus,

$$\int_1^3 \frac{\cos(\pi/x)}{x^2}\,dx = -\frac{1}{\pi}\int_{\pi}^{\pi/3} \cos u\,du$$
$$= -\frac{1}{\pi}\sin u\bigg]_{u=\pi}^{\pi/3} = -\frac{1}{\pi}(\sin(\pi/3) - \sin\pi)$$
$$= -\frac{\sqrt{3}}{2\pi} \approx -0.2757 \blacktriangleleft$$

✔QUICK CHECK EXERCISES 5.8 (See page 375 for answers.)

1. Assume that g' is continuous on $[a, b]$ and that f is continuous on an interval containing the values of $g(x)$ for $a \le x \le b$. If F is an antiderivative for f, then

$$\int_a^b f(g(x))g'(x)\,dx = \_\_\_\_\_\_\_\_$$

2. In each part, use the substitution to replace the given integral with an integral involving the variable u. (Do not evaluate the integral.)
   (a) $\displaystyle\int_0^2 3x^2(1+x^3)^3\,dx;\ u = 1 + x^3$
   (b) $\displaystyle\int_0^2 \frac{x}{\sqrt{5-x^2}}\,dx;\ u = 5 - x^2$

3. Evaluate the integral by making an appropriate substitution.
   (a) $\displaystyle\int_{-\pi}^0 \sin(3x - \pi)\,dx = \_\_\_\_\_\_\_\_$
   (b) $\displaystyle\int_0^{\pi/2} \sqrt[3]{\sin x}\cos x\,dx = \_\_\_\_\_\_\_\_$

EXERCISE SET 5.8 Graphing Utility C CAS

1–2 Express the integral in terms of the variable u, but do not evaluate it.

1. (a) $\displaystyle\int_1^3 (2x-1)^3\,dx;\ u = 2x - 1$
   (b) $\displaystyle\int_0^4 3x\sqrt{25-x^2}\,dx;\ u = 25 - x^2$
   (c) $\displaystyle\int_{-1/2}^{1/2} \cos(\pi\theta)\,d\theta;\ u = \pi\theta$
   (d) $\displaystyle\int_0^1 (x+2)(x+1)^5\,dx;\ u = x + 1$

2. (a) $\displaystyle\int_{-1}^4 (5-2x)^8\,dx;\ u = 5 - 2x$
   (b) $\displaystyle\int_{-\pi/3}^{2\pi/3} \frac{\sin x}{\sqrt{2+\cos x}}\,dx;\ u = 2 + \cos x$
   (c) $\displaystyle\int_0^{\pi/4} \tan^2 x\sec^2 x\,dx;\ u = \tan x$
   (d) $\displaystyle\int_0^1 x^3\sqrt{x^2+3}\,dx;\ u = x^2 + 3$

3–12 Evaluate the definite integral two ways: first by a u-substitution in the definite integral and then by a u-substitution in the corresponding indefinite integral.

3. $\int_0^1 (2x+1)^3\,dx$

4. $\int_1^2 (4x-2)^3\,dx$

5. $\int_0^1 (2x-1)^3\,dx$

6. $\int_1^2 (4-3x)^8\,dx$

7. $\int_0^8 x\sqrt{1+x}\,dx$

8. $\int_{-3}^0 x\sqrt{1-x}\,dx$

9. $\int_0^{\pi/2} 4\sin(x/2)\,dx$

10. $\int_0^{\pi/6} 2\cos 3x\,dx$

11. $\int_{-2}^{-1} \frac{x}{(x^2+2)^3}\,dx$

12. $\int_{1-\pi}^{1+\pi} \sec^2\left(\frac{1}{4}x-\frac{1}{4}\right)dx$

13–16 Evaluate the definite integral by expressing it in terms of u and evaluating the resulting integral using a formula from geometry.

13. $\int_{-5/3}^{5/3} \sqrt{25-9x^2}\,dx;\ u=3x$

14. $\int_0^2 x\sqrt{16-x^4}\,dx;\ u=x^2$

15. $\int_{\pi/3}^{\pi/2} \sin\theta\sqrt{1-4\cos^2\theta}\,d\theta;\ u=2\cos\theta$

16. $\int_{-3}^1 \sqrt{3-2x-x^2}\,dx;\ u=x+1$

17. Find the area under the curve $y=\sin\pi x$ over the interval $[0,1]$.

18. Find the area under the curve $y=3\cos 2x$ over the interval $[0,\pi/8]$.

19. Find the area under the curve $y=9/(x+2)^2$ over the interval $[-1,1]$.

20. Find the area under the curve $y=1/(3x+1)^2$ over the interval $[0,1]$.

21–34 Evaluate the integrals by any method.

21. $\int_1^5 \frac{dx}{\sqrt{2x-1}}$

22. $\int_1^2 \sqrt{5x-1}\,dx$

23. $\int_{-1}^1 \frac{x^2\,dx}{\sqrt{x^3+9}}$

24. $\int_{\pi/2}^{\pi} 6\sin x(\cos x+1)^5\,dx$

25. $\int_1^3 \frac{x+2}{\sqrt{x^2+4x+7}}\,dx$

26. $\int_1^2 \frac{dx}{x^2-6x+9}$

27. $\int_0^{\pi/4} 4\sin x\cos x\,dx$

28. $\int_0^{\pi/4} \sqrt{\tan x}\sec^2 x\,dx$

29. $\int_0^{\sqrt{\pi}} 5x\cos(x^2)\,dx$

30. $\int_{\pi^2}^{4\pi^2} \frac{1}{\sqrt{x}}\sin\sqrt{x}\,dx$

31. $\int_{\pi/12}^{\pi/9} \sec^2 3\theta\,d\theta$

32. $\int_{\pi/6}^{\pi/3} \csc^2 2\theta\,d\theta$

33. $\int_0^1 \frac{y^2\,dy}{\sqrt{4-3y}}$

34. $\int_{-1}^4 \frac{x\,dx}{\sqrt{5+x}}$

C 35. (a) Use a CAS to find the exact value of the integral

$$\int_0^{\pi/6} \sin^4 x\cos^3 x\,dx$$

(b) Confirm the exact value by hand calculation. [*Hint:* Use the identity $\cos^2 x = 1-\sin^2 x$.]

C 36. (a) Use a CAS to find the exact value of the integral

$$\int_{-\pi/4}^{\pi/4} \tan^4 x\,dx$$

(b) Confirm the exact value by hand calculation. [*Hint:* Use the identity $1+\tan^2 x=\sec^2 x$.]

37. (a) Find $\int_0^1 f(3x+1)\,dx$ if $\int_1^4 f(x)\,dx=5$.
(b) Find $\int_0^3 f(3x)\,dx$ if $\int_0^9 f(x)\,dx=5$.
(c) Find $\int_{-2}^0 xf(x^2)\,dx$ if $\int_0^4 f(x)\,dx=1$.

38. Given that m and n are positive integers, show that

$$\int_0^1 x^m(1-x)^n\,dx=\int_0^1 x^n(1-x)^m\,dx$$

by making a substitution. Do not attempt to evaluate the integrals.

39. Given that n is a positive integer, show that

$$\int_0^{\pi/2} \sin^n x\,dx=\int_0^{\pi/2}\cos^n x\,dx$$

by using a trigonometric identity and making a substitution. Do not attempt to evaluate the integrals.

40. Given that n is a positive integer, evaluate the integral

$$\int_0^1 x(1-x)^n\,dx$$

C 41. (a) Find the limit

$$\lim_{n\to+\infty}\sum_{k=1}^{n}\frac{\sin(k\pi/n)}{n}$$

by evaluating an appropriate definite integral over the interval $[0,1]$.
(b) Check your answer to part (a) by evaluating the limit directly with a CAS.

FOCUS ON CONCEPTS

42. Show that if f and g are continuous functions, then

$$\int_0^t f(t-x)g(x)\,dx=\int_0^t f(x)g(t-x)\,dx$$

43. (a) Let

$$I=\int_0^a \frac{f(x)}{f(x)+f(a-x)}\,dx$$

Show that $I=a/2$.

[*Hint:* Let $u=a-x$, and then note the difference between the resulting integrand and 1.]

(b) Use the result of part (a) to find

$$\int_0^3 \frac{\sqrt{x}}{\sqrt{x}+\sqrt{3-x}}\,dx$$

(c) Use the result of part (a) to find

$$\int_0^{\pi/2} \frac{\sin x}{\sin x+\cos x}\,dx$$

44. Let

$$I = \int_{-1}^{1} \frac{1}{1+x^2}\,dx$$

Show that the substitution $x = 1/u$ results in

$$I = -\int_{-1}^{1} \frac{1}{1+u^2}\,du = -I$$

so $2I = 0$, which implies that $I = 0$. However, this is impossible since the integrand of the given integral is positive over the interval of integration. Where is the error?

45. (a) Prove that if f is an odd function, then

$$\int_{-a}^{a} f(x)\,dx = 0$$

and give a geometric explanation of this result. [*Hint:* One way to prove that a quantity q is zero is to show that $q = -q$.]

(b) Prove that if f is an even function, then

$$\int_{-a}^{a} f(x)\,dx = 2\int_0^a f(x)\,dx$$

and give a geometric explanation of this result. [*Hint:* Split the interval of integration from $-a$ to a into two parts at 0.]

46. Evaluate

(a) $\int_{-1}^{1} x\sqrt{\cos(x^2)}\,dx$

(b) $\int_0^{\pi} \sin^8 x \cos^5 x\,dx$.

[*Hint:* Use the substitution $u = x - (\pi/2)$.]

✔ QUICK CHECK ANSWERS 5.8

1. $F(g(b)) - F(g(a))$ **2.** (a) $\int_1^9 u^3\,du$ (b) $\int_1^5 \frac{1}{2\sqrt{u}}\,du$ **3.** (a) $\frac{2}{3}$ (b) $\frac{3}{4}$

CHAPTER REVIEW EXERCISES

Graphing Utility C CAS

1. Write a paragraph that describes the *rectangle method* for defining the area under a curve $y = f(x)$ over an interval $[a, b]$.

2. (a) Devise a procedure for finding upper and lower estimates of the area (in cm^2) of the shaded region in the accompanying figure.
(b) Use your procedure to find upper and lower estimates of the area.
(c) Improve on the estimates you obtained in part (b).

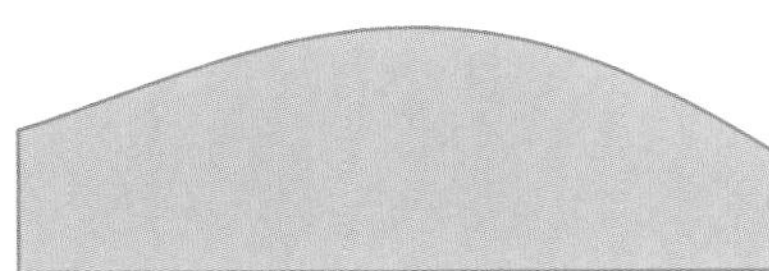

Figure Ex-2

3–6 Evaluate the integrals.

3. $\int \left[\frac{1}{2x^3} + 4\sqrt{x}\right] dx$

4. $\int [u^3 - 2u + 7]\,du$

5. $\int [4\sin x + 2\cos x]\,dx$

6. $\int \sec x(\tan x + \cos x)\,dx$

7. Solve the initial-value problems.

(a) $\frac{dy}{dx} = \frac{1-x}{\sqrt{x}},\ y(1) = 0$

(b) $\frac{dy}{dx} = \cos x - 5x,\ y(0) = 1$

8. The accompanying figure shows the slope field for a differential equation $dy/dx = f(x)$. Which of the following functions is most likely to be $f(x)$?

$$\sqrt{x},\quad \sin x,\quad x^4,\quad x$$

Explain your reasoning.

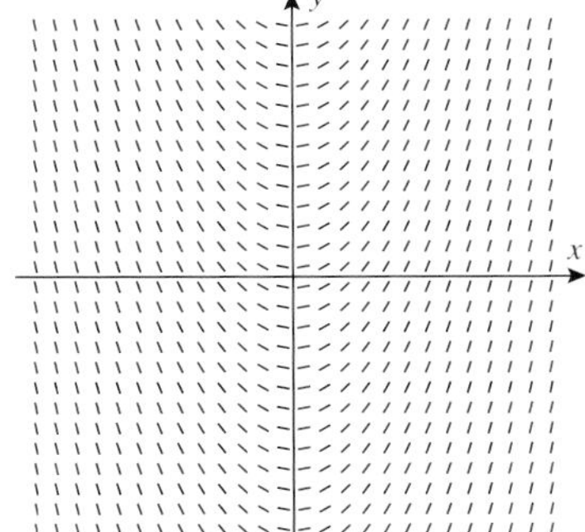

Figure Ex-8

9. (a) Show that the substitutions $u = \sec x$ and $u = \tan x$ produce different values for the integral

$$\int \sec^2 x \tan x\, dx$$

(b) Explain why both are correct.

10. Use the two substitutions in Exercise 9 to evaluate the definite integral

$$\int_0^{\pi/4} \sec^2 x \tan x\, dx$$

and confirm that they produce the same result.

11. Evaluate the integral

$$\int \frac{x^7}{\sqrt{x^4+2}}\, dx$$

by making the substitution $u = x^4 + 2$.

12. Evaluate the integral

$$\int \sqrt{1 + x^{-2/3}}\, dx$$

by making the substitution $u = 1 + x^{2/3}$.

13–16 Evaluate the integrals by hand, and check your answers with a CAS if you have one.

13. $\displaystyle\int \frac{\cos 3x}{\sqrt{5 + 2\sin 3x}}\, dx$ **14.** $\displaystyle\int \frac{\sqrt{3+\sqrt{x}}}{\sqrt{x}}\, dx$

15. $\displaystyle\int \frac{x^2}{(ax^3+b)^2}\, dx$ **16.** $\displaystyle\int x \sec^2(ax^2)\, dx$

17. In each part, confirm the stated equality.
(a) $1 \cdot 2 + 2 \cdot 3 + \cdots + n(n+1) = \frac{1}{3}n(n+1)(n+2)$

(b) $\displaystyle\lim_{n\to+\infty} \sum_{k=1}^{n-1} \left(\frac{9}{n} - \frac{k}{n^2}\right) = \frac{17}{2}$

(c) $\displaystyle\sum_{i=1}^{3} \left(\sum_{j=1}^{2} (i+j)\right) = 21$

18. Express

$$\sum_{k=4}^{18} k(k-3)$$

in sigma notation with
(a) $k = 0$ as the lower limit of summation
(b) $k = 5$ as the lower limit of summation.

19. The accompanying figure shows a square that is n units by n units that has been subdivided into a 1-unit square and $n - 1$ "L-shaped" regions. Use this figure to show that the sum of the first n consecutive positive odd integers is n^2.

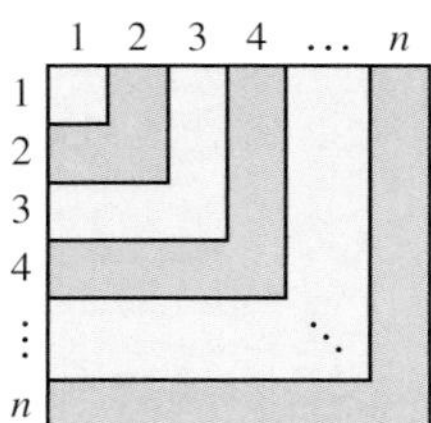

Figure Ex-19

20. Derive the result of Exercise 19 by writing

$$1 + 3 + 5 + \cdots + (2n-1) = \sum_{k=1}^{n} (2k-1)$$

21. The accompanying figure shows five points on the graph of an unknown function f. Devise a strategy for using the known points to approximate the area A under the graph of $y = f(x)$ over the interval $[1, 5]$. Describe your strategy, and use it to approximate A.

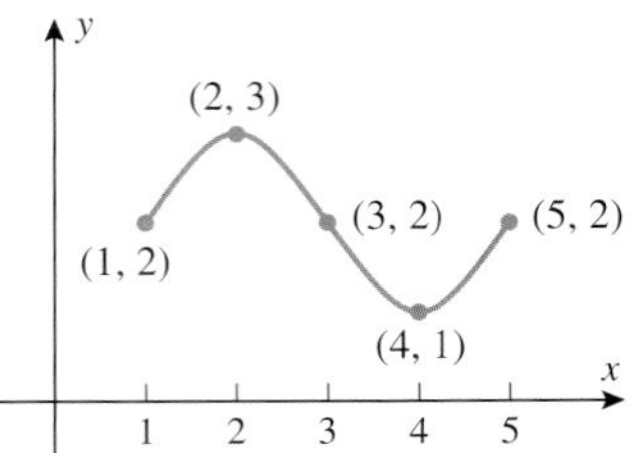

Figure Ex-21

22. Find the left endpoint, right endpoint, and midpoint approximations of the area under the curve $y = 3x/(x+1)$ over the interval $[0, 5]$ using $n = 5$ subintervals.

23. Find the area under the graph of $f(x) = 4x - x^2$ over the interval $[0, 4]$ using Definition 5.4.3 with x_k^* as the *right* endpoint of each subinterval.

24. Find the area under the graph of $f(x) = 5x - x^2$ over the interval $[0, 5]$ using Definition 5.4.3 with x_k^* as the *left* endpoint of each subinterval.

25–27 Use a calculating utility to find the left endpoint, right endpoint, and midpoint approximations to the area under the curve $y = f(x)$ over the stated interval using $n = 10$ subintervals.

25. $y = 1/x$; $[1, 2]$ **26.** $y = \tan x$; $[0, 1]$

27. $y = \sin x$; $[0, \pi]$

28. The *definite integral* of f over the interval $[a, b]$ is defined as the limit

$$\int_a^b f(x)\, dx = \lim_{\max \Delta x_k \to 0} \sum_{k=1}^{n} f(x_k^*)\Delta x_k$$

Explain what the various symbols on the right side of this equation mean.

29. Suppose that

$$\int_0^1 f(x)\, dx = \tfrac{1}{2}, \quad \int_1^2 f(x)\, dx = \tfrac{1}{4},$$

$$\int_0^3 f(x)\, dx = -1, \quad \int_0^1 g(x)\, dx = 2$$

In each part, use this information to evaluate the given integral, if possible. If there is not enough information to evaluate the integral, then say so.

(a) $\int_0^2 f(x)\,dx$ (b) $\int_1^3 f(x)\,dx$ (c) $\int_2^3 5f(x)\,dx$

(d) $\int_1^0 g(x)\,dx$ (e) $\int_0^1 g(2x)\,dx$ (f) $\int_0^1 [g(x)]^2\,dx$

30. In each part, use the information in Exercise 29 to evaluate the given integral. If there is not enough information to evaluate the integral, then say so.

(a) $\int_0^1 [f(x)+g(x)]\,dx$ (b) $\int_0^1 f(x)g(x)\,dx$

(c) $\int_0^1 \frac{f(x)}{g(x)}\,dx$ (d) $\int_0^1 [4g(x)-3f(x)]\,dx$

31. In each part, evaluate the integral. Where appropriate, you may use a geometric formula.

(a) $\int_{-1}^1 (1+\sqrt{1-x^2})\,dx$

(b) $\int_0^3 (x\sqrt{x^2+1}-\sqrt{9-x^2})\,dx$

(c) $\int_0^1 x\sqrt{1-x^4}\,dx$

32. Evaluate the integral $\int_0^1 |2x-1|\,dx$, and sketch the region whose area it represents.

33. One of the numbers π, $\pi/2$, $35\pi/128$, $1-\pi$ is the correct value of the integral

$$\int_0^\pi \sin^8 x\,dx$$

Use the accompanying graph of $y=\sin^8 x$ and a logical process of elimination to find the correct value. [Do not attempt to evaluate the integral.]

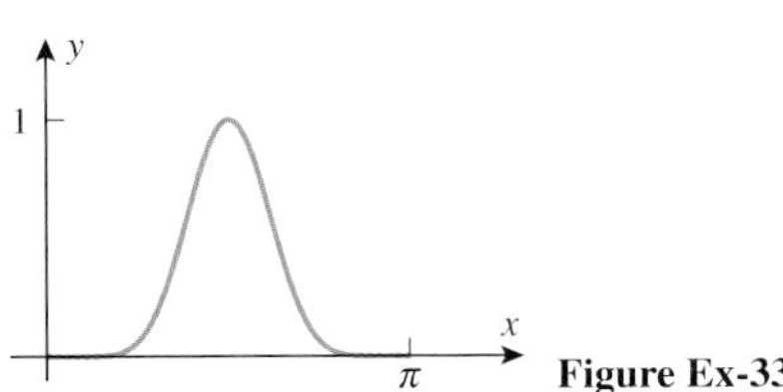

Figure Ex-33

34. In each part, find the limit by interpreting it as a limit of Riemann sums in which the interval [0, 1] is divided into n subintervals of equal length.

(a) $\lim_{n\to+\infty} \frac{\sqrt{1}+\sqrt{2}+\sqrt{3}+\cdots+\sqrt{n}}{n^{3/2}}$

(b) $\lim_{n\to+\infty} \frac{1^4+2^4+3^4+\cdots+n^4}{n^5}$

35. (a) Express the equation

$$\int_a^b [f_1(x)+f_2(x)+\cdots+f_n(x)]\,dx$$
$$=\int_a^b f_1(x)\,dx+\int_a^b f_2(x)\,dx+\cdots+\int_a^b f_n(x)\,dx$$

in sigma notation.

(b) If $c_1, c_2, \ldots, c_n$ are constants and $f_1, f_2, \ldots, f_n$ are integrable functions on $[a, b]$, do you think it is always true that

$$\int_a^b \left(\sum_{k=1}^n c_k f_k(x)\right) dx = \sum_{k=1}^n \left[c_k \int_a^b f_k(x)\,dx\right]?$$

Explain your reasoning.

36. Find a formula (defined piecewise) for the upper boundary of the trapezoid shown in the accompanying figure, and then integrate that function to derive the formula for the area of the trapezoid given on the inside front cover of this text.

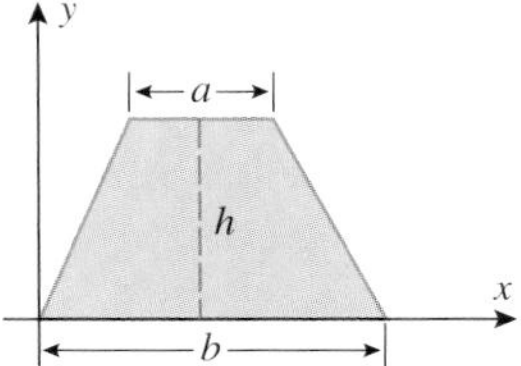

Figure Ex-36

37–38 Find the area under the curve $y=f(x)$ over the stated interval.

37. $f(x)=\sqrt{x}$; $[1, 9]$

38. $f(x)=x^{-3/5}$; $[1, 4]$

39–46 Evaluate the integrals using the Fundamental Theorem of Calculus and (if necessary) properties of the definite integral.

39. $\int_{-3}^0 (x^2-4x+7)\,dx$

40. $\int_{-1}^2 x(1+x^3)\,dx$

41. $\int_1^3 \frac{1}{x^2}\,dx$

42. $\int_1^8 (5x^{2/3}-4x^{-2})\,dx$

43. $\int_0^1 (x-\sec x\tan x)\,dx$

44. $\int_1^4 \left(\frac{3}{\sqrt{t}}-5\sqrt{t}-t^{-3/2}\right) dt$

45. $\int_0^2 |2x-3|\,dx$

46. $\int_0^{\pi/2} \left|\tfrac{1}{2}-\sin x\right|\,dx$

47. Find the area that is above the x-axis but below the curve $y=(1-x)(x-2)$. Make a sketch of the region.

C **48.** Use a CAS to find the area of the region in the first quadrant that lies below the curve $y=x+x^2-x^3$ and above the x-axis.

49. Define $F(x)$ by

$$F(x)=\int_1^x (t^3+1)\,dt$$

(a) Use Part 2 of the Fundamental Theorem of Calculus to find $F'(x)$.

(b) Check the result in part (a) by first integrating and then differentiating.

50. Define $F(x)$ by

$$F(x)=\int_4^x \frac{1}{\sqrt{t}}\,dt$$

(a) Use Part 2 of the Fundamental Theorem of Calculus to find $F'(x)$.
(b) Check the result in part (a) by first integrating and then differentiating.

51–54 Use Part 2 of the Fundamental Theorem of Calculus to find the derivatives.

51. $\frac{d}{dx}\left[\int_0^x \frac{1}{t^4+5}\,dt\right]$ **52.** $\frac{d}{dx}\left[\int_0^x \frac{t}{\cos t^2}\,dt\right]$

53. $\frac{d}{dx}\left[\int_0^x |t-1|\,dt\right]$ **54.** $\frac{d}{dx}\left[\int_\pi^x \cos\sqrt{t}\,dt\right]$

55. State the two parts of the Fundamental Theorem of Calculus, and explain what is meant by the statement "Differentiation and integration are inverse processes."

C **56.** Let $F(x) = \int_0^x \frac{t^2-3}{t^4+7}\,dt$.
(a) Find the intervals on which F is increasing and those on which F is decreasing.
(b) Find the open intervals on which F is concave up and those on which F is concave down.
(c) Find the x-values, if any, at which the function F has absolute extrema.
(d) Use a CAS to graph F, and confirm that the results in parts (a), (b), and (c) are consistent with the graph.

57. Prove that the function

$$F(x) = \int_0^x \frac{1}{1+t^2}\,dt + \int_0^{1/x} \frac{1}{1+t^2}\,dt$$

is constant on the interval $(0, +\infty)$.

58. What is the natural domain of the function

$$F(x) = \int_1^x \frac{1}{t^2-9}\,dt?$$

Explain your reasoning.

59. In each part, determine the values of x for which $F(x)$ is positive, negative, or zero without performing the integration; explain your reasoning.
(a) $F(x) = \int_1^x \frac{t^4}{t^2+3}\,dt$ (b) $F(x) = \int_{-1}^x \sqrt{4-t^2}\,dt$

C **60.** Use a CAS to approximate the largest and smallest values of the integral

$$\int_{-1}^x \frac{t}{\sqrt{2+t^3}}\,dt$$

for $1 \le x \le 3$.

61. Find all values of x^* in the stated interval that are guaranteed to exist by the Mean-Value Theorem for Integrals, and explain what these numbers represent.
(a) $f(x) = \sqrt{x}$; $[0, 3]$ (b) $f(x) = 2x - x^2$; $[0, 2]$

62. A 10-gram tumor is discovered in a laboratory rat on March 1. The tumor is growing at a rate of $r(t) = t/7$ grams per week, where t denotes the number of weeks since March 1. What will be the mass of the tumor on June 7?

63. Derive the formulas for the position and velocity functions of a particle that moves with uniformly accelerated motion along a coordinate line.

64. The velocity of a particle moving along an s-axis is measured at 5-s intervals for 40 s, and the velocity function is modeled by a smooth curve. (The curve and the data points are shown in the accompanying figure.) Use this model in each part.
(a) Does the particle have constant acceleration? Explain your reasoning.
(b) Is there any 15-s time interval during which the acceleration is constant? Explain your reasoning.
(c) Estimate the distance traveled by the particle from time $t = 0$ to time $t = 40$.
(d) Estimate the average velocity of the particle over the 40-s time period.
(e) Is the particle ever slowing down during the 40-s time period? Explain your reasoning.
(f) Is there sufficient information for you to determine the s-coordinate of the particle at time $t = 10$? If so, find it. If not, explain what additional information you need.

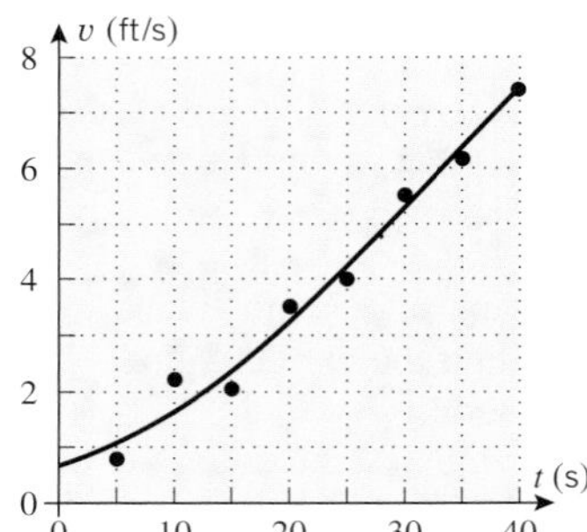

Figure Ex-64

65–68 A particle moves along an s-axis. Use the given information to find the position function of the particle.

65. $v(t) = t^3 - 2t^2 + 1$; $s(0) = 1$

66. $a(t) = 4\cos 2t$; $v(0) = -1$, $s(0) = -3$

67. $v(t) = 2t - 3$; $s(1) = 5$

68. $a(t) = \cos t - 2t$; $v(0) = 0$, $s(0) = 0$

69–72 A particle moves with a velocity of $v(t)$ m/s along an s-axis. Find the displacement and the distance traveled by the particle during the given time interval.

69. $v(t) = 2t - 4$; $0 \le t \le 6$

70. $v(t) = |t-3|$; $0 \le t \le 5$

71. $v(t) = \frac{1}{2} - \frac{1}{t^2}$; $1 \le t \le 3$

72. $v(t) = \frac{3}{\sqrt{t}}$; $4 \le t \le 9$

73–74 A particle moves with acceleration $a(t)$ m/s^2 along an s-axis and has velocity v_0 m/s at time $t = 0$. Find the displacement and the distance traveled by the particle during the given time interval.

73. $a(t) = -2;\ v_0 = 3;\ 1 \le t \le 4$

74. $a(t) = \dfrac{1}{\sqrt{5t+1}};\ v_0 = 2;\ 0 \le t \le 3$

75–77 Sketch the curve and find the total area between the curve and the given interval on the x-axis.

75. $y = x^2 - 1;\ [0, 3]$

76. $y = \sqrt{x+1} - 1;\ [-1, 1]$

77. $y = x^3 - 4x^2 + 3x;\ [0, 3]$

78. Suppose that the velocity function of a particle moving along an s-axis is $v(t) = 20t^2 - 100t + 50$ ft/s and that the particle is at the origin at time $t = 0$. Use a graphing utility to generate the graphs of $s(t)$, $v(t)$, and $a(t)$ for the first 6 s of motion.

79. A car traveling 60 mi/h (= 88 ft/s) along a straight road decelerates at a constant rate of 10 ft/s^2.
(a) How long will it take until the speed is 45 mi/h?
(b) How far will the car travel before coming to a stop?

80. A particle moving along a straight line is accelerating at a constant rate of 3 m/s^2. Find the initial velocity if the particle moves 40 m in the first 4 s.

81. A ball is thrown vertically upwards from a height of s_0 ft with an initial velocity of v_0 ft/s. If the ball is caught at height s_0, determine its average speed through the air using the free-fall model.

82. A rock, dropped from an unknown height, strikes the ground with a speed of 24 m/s. Find the height from which the rock was dropped.

83–87 Evaluate the integrals by making an appropriate substitution.

83. $\displaystyle\int_0^1 (2x+1)^4\,dx$

84. $\displaystyle\int_{-5}^0 x\sqrt{4-x}\,dx$

85. $\displaystyle\int_0^1 \frac{dx}{\sqrt{3x+1}}$

86. $\displaystyle\int_0^{\sqrt{\pi}} x\sin x^2\,dx$

87. $\displaystyle\int_0^1 \sin^2(\pi x)\cos(\pi x)\,dx$

88. Solve the initial-value problems.
(a) $\dfrac{dy}{dx} = \sqrt[3]{x},\ y(1) = 2$
(b) $\dfrac{dy}{dx} = x\sin x^2;\ y(\sqrt{\pi/2}) = 1$

89. Find a function f and a number a such that

$$2 + \int_a^x f(t)\,dt = \frac{8}{x+3}$$

APPLICATIONS OF THE DEFINITE INTEGRAL IN GEOMETRY, SCIENCE, AND ENGINEERING

The experimental verification of a theory concerning any natural phenomenon generally rests on the result of an integration.

—J.W. Mellor
Chemist

In the last chapter we introduced the definite integral as the limit of Riemann sums in the context of finding areas. However, Riemann sums and definite integrals have applications that extend far beyond the area problem. In this chapter we will show how Riemann sums and definite integrals arise in such problems as finding the volume and surface area of a solid, finding the length of a plane curve, calculating the work done by a force, and finding the pressure and force exerted by a fluid on a submerged object.

Although these problems are diverse, the required calculations can all be approached by the same procedure that we used to find areas—breaking the required calculation into "small parts," making an approximation for each part, adding the approximations from the parts to produce a Riemann sum that approximates the entire quantity to be calculated, and then taking the limit of the Riemann sums to produce an exact result.

Photo: *Calculus is essential for the computations required to land an astronaut on the moon.*

6.1 AREA BETWEEN TWO CURVES

In the last chapter we showed how to find the area between a curve $y = f(x)$ and an interval on the x-axis. Here we will show how to find the area between two curves.

A REVIEW OF RIEMANN SUMS

Before we consider the problem of finding the area between two curves it will be helpful to review the basic principle that underlies the calculation of area as a definite integral. Recall that if f is continuous and nonnegative on $[a, b]$, then the definite integral for the area A under $y = f(x)$ over the interval $[a, b]$ is obtained in four steps (Figure 6.1.1):

- Divide the interval $[a, b]$ into n subintervals, and use those subintervals to divide the region under the curve $y = f(x)$ into n strips.
- Assuming that the width of the kth strip is Δx_k, approximate the area of that strip by the area $f(x_k^*)\Delta x_k$ of a rectangle of width Δx_k and height $f(x_k^*)$, where x_k^* is a point in the kth subinterval.

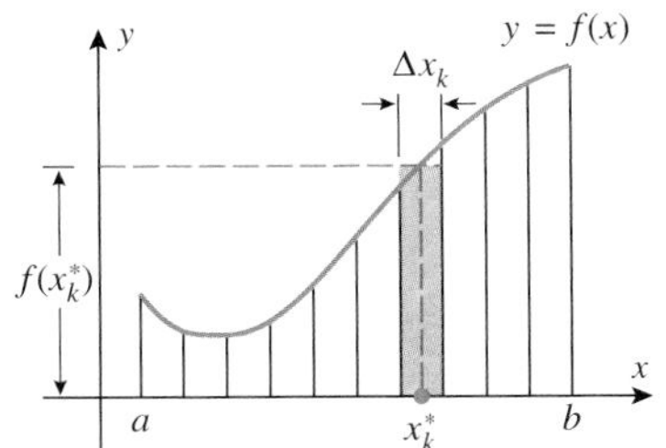

Figure 6.1.1

- Add the approximate areas of the strips to approximate the entire area A by the Riemann sum:

$$A \approx \sum_{k=1}^{n} f(x_k^*)\Delta x_k$$

- Take the limit of the Riemann sums as the number of subintervals increases and their widths approach zero. This causes the error in the approximations to approach zero and produces the following definite integral for the exact area A:

$$A = \lim_{\max \Delta x_k \to 0} \sum_{k=1}^{n} f(x_k^*)\Delta x_k = \int_a^b f(x)\,dx$$

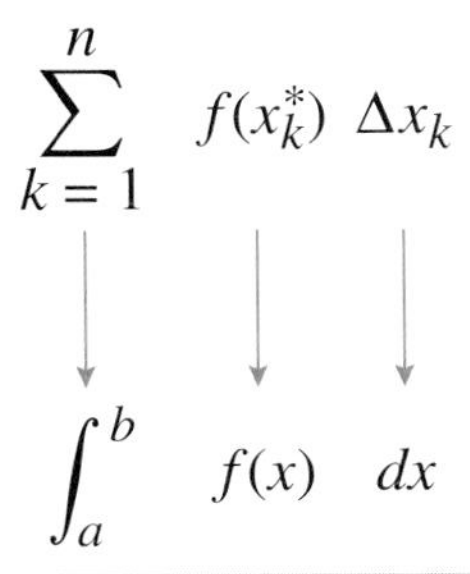

Effect of the limit process on the Riemann sum.

Figure 6.1.2

Figure 6.1.2 illustrates the effect that the limit process has on the various parts of the Riemann sum:

- The quantity x_k^* in the Riemann sum becomes the variable x in the definite integral.
- The interval width Δx_k in the Riemann sum becomes the dx in the definite integral.
- The interval $[a, b]$, which is the union of the subintervals with widths $\Delta x_1, \Delta x_2, \ldots,$ Δx_n, does not appear explicitly in the Riemann sum but is represented by the upper and lower limits of integration in the definite integral.

AREA BETWEEN $y = f(x)$ AND $y = g(x)$

We will now consider the following extension of the area problem.

> **6.1.1 FIRST AREA PROBLEM.** Suppose that f and g are continuous functions on an interval $[a, b]$ and
>
> $$f(x) \geq g(x) \quad \text{for} \quad a \leq x \leq b$$
>
> [This means that the curve $y = f(x)$ lies above the curve $y = g(x)$ and that the two can touch but not cross.] Find the area A of the region bounded above by $y = f(x)$, below by $y = g(x)$, and on the sides by the lines $x = a$ and $x = b$ (Figure 6.1.3a).

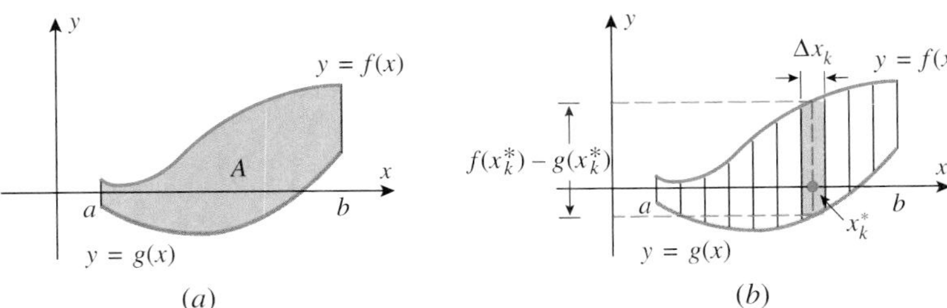

Figure 6.1.3

To solve this problem we divide the interval $[a, b]$ into n subintervals, which has the effect of subdividing the region into n strips (Figure 6.1.3b). If we assume that the width of the kth strip is Δx_k, then the area of the strip can be approximated by the area of a rectangle of width Δx_k and height $f(x_k^*) - g(x_k^*)$, where x_k^* is a point in the kth subinterval. Adding these approximations yields the following Riemann sum that approximates the area A:

$$A \approx \sum_{k=1}^{n} [f(x_k^*) - g(x_k^*)]\Delta x_k$$

Taking the limit as n increases and the widths of the subintervals approach zero yields the following definite integral for the area A between the curves:

$$A = \lim_{\max \Delta x_k \to 0} \sum_{k=1}^{n} [f(x_k^*) - g(x_k^*)]\Delta x_k = \int_a^b [f(x) - g(x)]\,dx$$

In summary, we have the following result.

6.1.2 AREA FORMULA. If f and g are continuous functions on the interval $[a, b]$, and if $f(x) \geq g(x)$ for all x in $[a, b]$, then the area of the region bounded above by $y = f(x)$, below by $y = g(x)$, on the left by the line $x = a$, and on the right by the line $x = b$ is

$$A = \int_a^b [f(x) - g(x)]\,dx \tag{1}$$

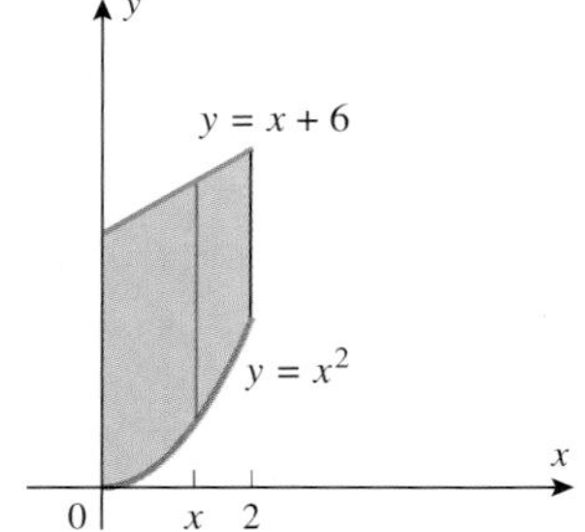

Figure 6.1.4

► Example 1 Find the area of the region bounded above by $y = x + 6$, bounded below by $y = x^2$, and bounded on the sides by the lines $x = 0$ and $x = 2$.

Solution. The region and a cross section are shown in Figure 6.1.4. The cross section extends from $g(x) = x^2$ on the bottom to $f(x) = x + 6$ on the top. If the cross section is moved through the region, then its leftmost position will be $x = 0$ and its rightmost position will be $x = 2$. Thus, from (1)

$$A = \int_0^2 [(x + 6) - x^2]\,dx = \left[\frac{x^2}{2} + 6x - \frac{x^3}{3}\right]_0^2 = \frac{34}{3} - 0 = \frac{34}{3} \blacktriangleleft$$

What does the integral in (1) represent if the graphs of f and g cross one another over the interval $[a, b]$? How would you find the area between the curves in this case?

It is possible that the upper and lower boundaries of a region may intersect at one or both endpoints, in which case the sides of the region will be points, rather than vertical line segments (Figure 6.1.5). When that occurs you will have to determine the points of intersection to obtain the limits of integration.

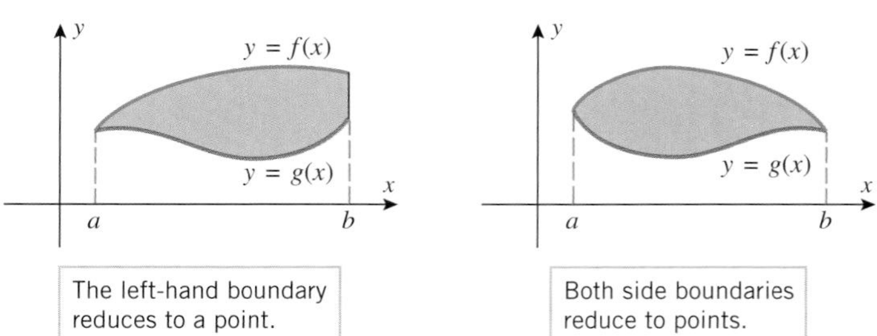

Figure 6.1.5

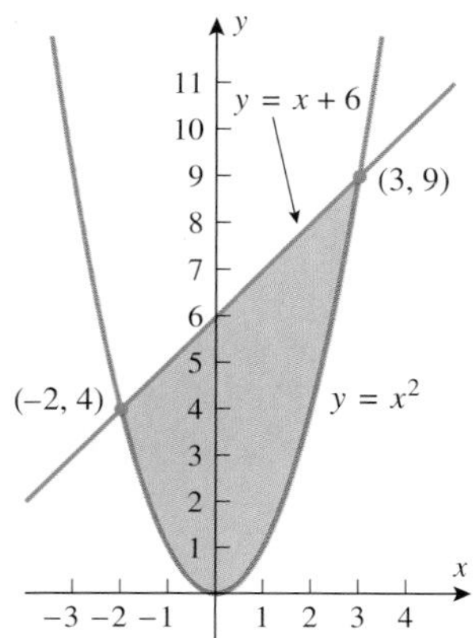

Figure 6.1.6

► Example 2 Find the area of the region that is enclosed between the curves $y = x^2$ and $y = x + 6$.

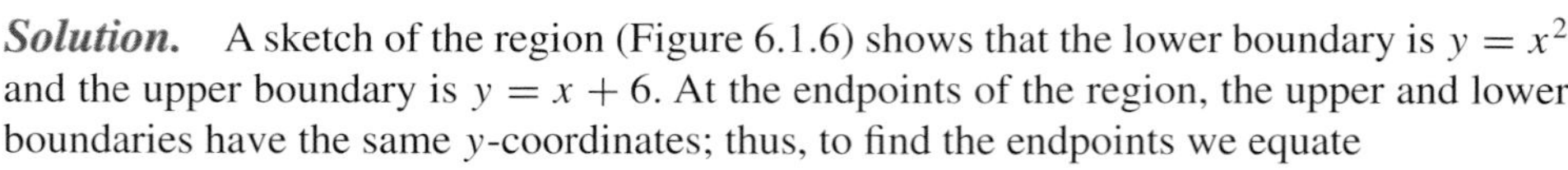

Solution. A sketch of the region (Figure 6.1.6) shows that the lower boundary is $y = x^2$ and the upper boundary is $y = x + 6$. At the endpoints of the region, the upper and lower boundaries have the same y-coordinates; thus, to find the endpoints we equate

$$y = x^2 \quad \text{and} \quad y = x + 6 \tag{2}$$

This yields

$$x^2 = x + 6 \quad \text{or} \quad x^2 - x - 6 = 0 \quad \text{or} \quad (x + 2)(x - 3) = 0$$

from which we obtain

$$x = -2 \quad \text{and} \quad x = 3$$

Although the y-coordinates of the endpoints are not essential to our solution, they may be obtained from (2) by substituting $x = -2$ and $x = 3$ in either equation. This yields $y = 4$ and $y = 9$, so the upper and lower boundaries intersect at $(-2, 4)$ and $(3, 9)$.

From (1) with $f(x) = x + 6$, $g(x) = x^2$, $a = -2$, and $b = 3$, we obtain the area

$$A = \int_{-2}^{3} [(x+6) - x^2]\, dx = \left[\frac{x^2}{2} + 6x - \frac{x^3}{3}\right]_{-2}^{3} = \frac{27}{2} - \left(-\frac{22}{3}\right) = \frac{125}{6} \blacktriangleleft$$

In the case where f and g are *nonnegative* on the interval $[a, b]$, the formula

$$A = \int_a^b [f(x) - g(x)]\, dx = \int_a^b f(x)\, dx - \int_a^b g(x)\, dx$$

states that the area A between the curves can be obtained by subtracting the area under $y = g(x)$ from the area under $y = f(x)$ (Figure 6.1.7).

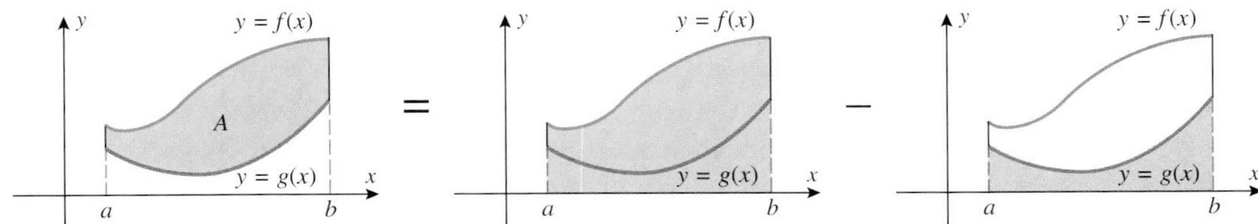

Figure 6.1.7

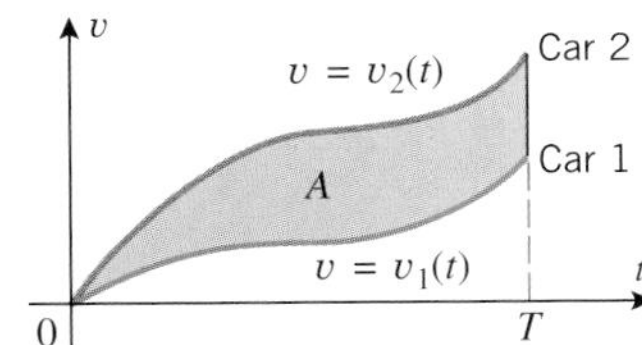

Figure 6.1.8

► **Example 3** Figure 6.1.8 shows velocity versus time curves for two race cars that move along a straight track, starting from rest at the same time. What does the area A between the curves over the interval $0 \le t \le T$ represent?

Solution. From (1)

$$A = \int_0^T [v_2(t) - v_1(t)]\, dt = \int_0^T v_2(t)\, dt - \int_0^T v_1(t)\, dt$$

Since $0 \le v_1(t) \le v_2(t)$ on $[0, T]$, it follows from Formula (4) of Section 5.7 that the integral of v_2 is the distance traveled by car 2 during the time interval, and the integral of v_1 is the distance traveled by car 1. Thus, A is the distance by which car 2 is ahead of car 1 at time T. ◄

Some regions may require careful thought to determine the integrand and limits of integration in (1). Here is a systematic procedure that you can follow to set up this formula.

It is not necessary to make an extremely accurate sketch in Step 1; the only purpose of the sketch is to determine which curve is the upper boundary and which is the lower boundary.

Finding the Limits of Integration for the Area Between Two Curves

Step 1. Sketch the region and then draw a vertical line segment through the region at an arbitrary point x on the x-axis, connecting the top and bottom boundaries (Figure 6.1.9*a*).

Step 2. The y-coordinate of the top endpoint of the line segment sketched in Step 1 will be $f(x)$, the bottom one $g(x)$, and the length of the line segment will be $f(x) - g(x)$. This is the integrand in (1).

Step 3. To determine the limits of integration, imagine moving the line segment left and then right. The leftmost position at which the line segment intersects the region is $x = a$ and the rightmost is $x = b$ (Figures 6.1.9*b* and 6.1.9*c*).

(a) (b) (c)

Figure 6.1.9

There is a useful way of thinking about this procedure:

If you view the vertical line segment as the "cross section" of the region at the point x, then Formula (1) states that the area between the curves is obtained by integrating the length of the cross section over the interval $[a, b]$.

It is possible for the upper or lower boundary of a region to consist of two or more different curves, in which case it will be convenient to subdivide the region into smaller pieces in order to apply Formula (1). This is illustrated in the next example.

▶ **Example 4** Find the area of the region enclosed by $x = y^2$ and $y = x - 2$.

Solution. To determine the appropriate boundaries of the region, we need to know where the curves $x = y^2$ and $y = x - 2$ intersect. In Example 2 we found intersections by equating the expressions for y. Here it is easier to rewrite the latter equation as $x = y + 2$ and equate the expressions for x, namely,

$$x = y^2 \quad \text{and} \quad x = y + 2 \tag{3}$$

This yields

$$y^2 = y + 2 \quad \text{or} \quad y^2 - y - 2 = 0 \quad \text{or} \quad (y + 1)(y - 2) = 0$$

from which we obtain $y = -1$, $y = 2$. Substituting these values in either equation in (3) we see that the corresponding x-values are $x = 1$ and $x = 4$, respectively, so the points of intersection are $(1, -1)$ and $(4, 2)$ (Figure 6.1.10*a*).

To apply Formula (1), the equations of the boundaries must be written so that y is expressed explicitly as a function of x. The upper boundary can be written as $y = \sqrt{x}$ (rewrite $x = y^2$ as $y = \pm\sqrt{x}$ and choose the $+$ for the upper portion of the curve). The lower boundary consists of two parts:

$$y = -\sqrt{x} \quad \text{for} \quad 0 \le x \le 1 \qquad \text{and} \qquad y = x - 2 \quad \text{for} \quad 1 \le x \le 4$$

(Figure 6.1.10*b*). Because of this change in the formula for the lower boundary, it is necessary to divide the region into two parts and find the area of each part separately.

From (1) with $f(x) = \sqrt{x}$, $g(x) = -\sqrt{x}$, $a = 0$, and $b = 1$, we obtain

$$A_1 = \int_0^1 [\sqrt{x} - (-\sqrt{x})]\,dx = 2\int_0^1 \sqrt{x}\,dx = 2\left[\frac{2}{3}x^{3/2}\right]_0^1 = \frac{4}{3} - 0 = \frac{4}{3}$$

From (1) with $f(x) = \sqrt{x}$, $g(x) = x - 2$, $a = 1$, and $b = 4$, we obtain

$$A_2 = \int_1^4 [\sqrt{x} - (x - 2)]\,dx = \int_1^4 (\sqrt{x} - x + 2)\,dx$$

$$= \left[\frac{2}{3}x^{3/2} - \frac{1}{2}x^2 + 2x\right]_1^4 = \left(\frac{16}{3} - 8 + 8\right) - \left(\frac{2}{3} - \frac{1}{2} + 2\right) = \frac{19}{6}$$

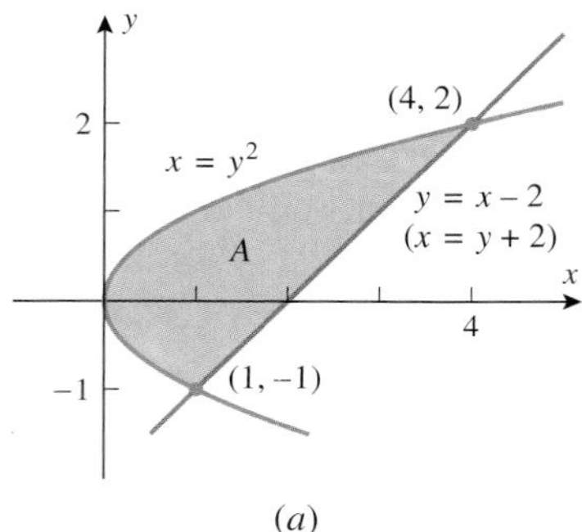

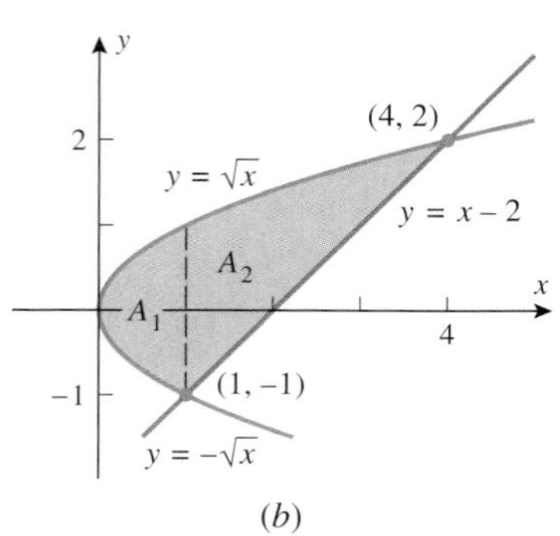

Figure 6.1.10

Thus, the area of the entire region is

$$A = A_1 + A_2 = \frac{4}{3} + \frac{19}{6} = \frac{9}{2} \blacktriangleleft$$

REVERSING THE ROLES OF *x* AND *y*

Sometimes it is possible to avoid splitting a region into parts by integrating with respect to y rather than x. We will now show how this can be done.

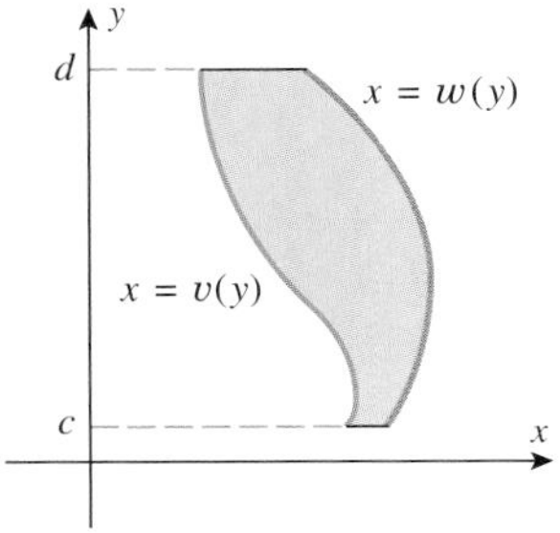

Figure 6.1.11

> **6.1.3 SECOND AREA PROBLEM.** Suppose that w and v are continuous functions of y on an interval $[c, d]$ and that
>
> $$w(y) \geq v(y) \quad \text{for} \quad c \leq y \leq d$$
>
> [This means that the curve $x = w(y)$ lies to the right of the curve $x = v(y)$ and that the two can touch but not cross.] Find the area A of the region bounded on the left by $x = v(y)$, on the right by $x = w(y)$, and above and below by the lines $y = d$ and $y = c$ (Figure 6.1.11).

Proceeding as in the derivation of (1), but with the roles of x and y reversed, leads to the following analog of 6.1.2.

> **6.1.4 AREA FORMULA.** If w and v are continuous functions and if $w(y) \geq v(y)$ for all y in $[c, d]$, then the area of the region bounded on the left by $x = v(y)$, on the right by $x = w(y)$, below by $y = c$, and above by $y = d$ is
>
> $$A = \int_c^d [w(y) - v(y)]\, dy \tag{4}$$

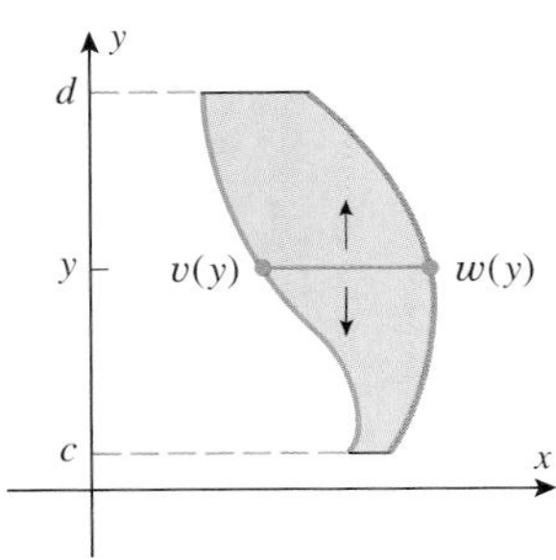

Figure 6.1.12

The guiding principle in applying this formula is the same as with (1): The integrand in (4) can be viewed as the length of the horizontal cross section at an arbitrary point y on the y-axis, in which case Formula (4) states that the area can be obtained by integrating the length of the horizontal cross section over the interval $[c, d]$ on the y-axis (Figure 6.1.12).

In Example 4, we split the region into two parts to facilitate integrating with respect to x. In the next example we will see that splitting this region can be avoided if we integrate with respect to y.

The choice between Formulas (1) and (4) is usually dictated by the shape of the region and which formula requires the least amount of splitting. However, sometimes one might choose the formula that requires more splitting because it is easier to evaluate the resulting integrals.

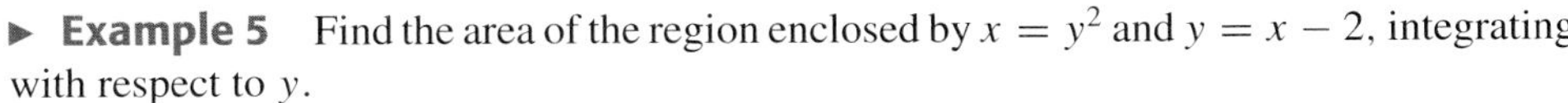

▶ **Example 5** Find the area of the region enclosed by $x = y^2$ and $y = x - 2$, integrating with respect to y.

Solution. As indicated in Figure 6.1.10*a* the left boundary is $x = y^2$, the right boundary is $y = x - 2$, and the region extends over the interval $-1 \leq y \leq 2$. However, to apply (4) the equations for the boundaries must be written so that x is expressed explicitly as a function of y. Thus, we rewrite $y = x - 2$ as $x = y + 2$. It now follows from (4) that

$$A = \int_{-1}^{2} [(y + 2) - y^2]\, dy = \left[\frac{y^2}{2} + 2y - \frac{y^3}{3}\right]_{-1}^{2} = \frac{9}{2}$$

which agrees with the result obtained in Example 4. ◀

✔ QUICK CHECK EXERCISES 6.1

(See page 387 for answers.)

1. An integral expression for the area of the region between the curves $y = 20 - 3x^2$ and $y = 3\sqrt{x}$ and bounded on the sides by $x = 0$ and $x = 2$ is ______. The value of this integral is ______.
2. An integral expression for the area of the parallelogram bounded by $y = 2x + 8$, $y = 2x - 3$, $x = -1$, and $x = 5$ is ______. The value of this integral is ______.
3. (a) The points of intersection for the circle $x^2 + y^2 = 4$ and the line $y = x + 2$ are ______ and ______.
   (b) Expressed as a definite integral with respect to x, ______ gives the area of the region inside the circle $x^2 + y^2 = 4$ and above the line $y = x + 2$.
   (c) Expressed as a definite integral with respect to y, ______ gives the area of the region described in part (b).
   (d) Applying appropriate formulas from geometry, the area of the region described in part (b) is ______.
4. The area of the region enclosed by the curves $y = x^2$ and $y = \sqrt[3]{x}$ is ______.

EXERCISE SET 6.1 Graphing Utility C CAS

1–4 Find the area of the shaded region.

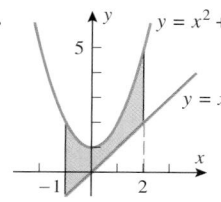

1. 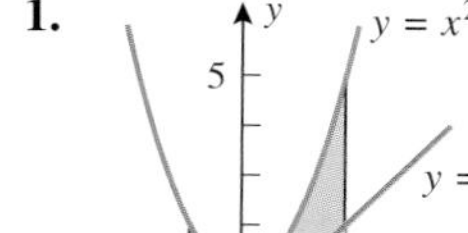

2. 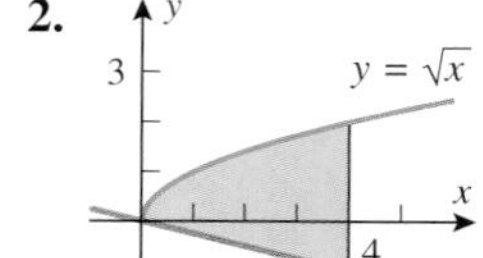

3. 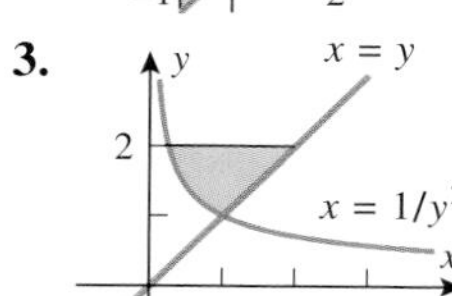

4. 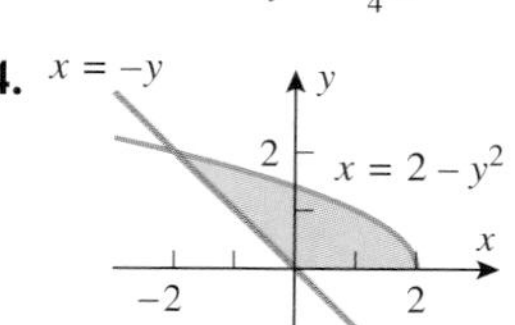

5. Find the area of the region enclosed by the curves $y = x^2$ and $y = 4x$ by integrating
   (a) with respect to x (b) with respect to y.
6. Find the area of the region enclosed by the curves $y^2 = 4x$ and $y = 2x - 4$ by integrating
   (a) with respect to x (b) with respect to y.

7–14 Sketch the region enclosed by the curves and find its area.

7. $y = x^2$, $y = \sqrt{x}$, $x = \frac{1}{4}$, $x = 1$
8. $y = x^3 - 4x$, $y = 0$, $x = 0$, $x = 2$
9. $y = \cos 2x$, $y = 0$, $x = \pi/4$, $x = \pi/2$
10. $y = \sec^2 x$, $y = 2$, $x = -\pi/4$, $x = \pi/4$
11. $x = \sin y$, $x = 0$, $y = \pi/4$, $y = 3\pi/4$
12. $x^2 = y$, $x = y - 2$
13. $y = 2 + |x - 1|$, $y = -\frac{1}{5}x + 7$
14. $y = x$, $y = 4x$, $y = -x + 2$

15–20 Use a graphing utility, where helpful, to find the area of the region enclosed by the curves.

15. $y = x^3 - 4x^2 + 3x$, $y = 0$
16. $y = x^3 - 2x^2$, $y = 2x^2 - 3x$
17. $y = \sin x$, $y = \cos x$, $x = 0$, $x = 2\pi$
18. $y = x^3 - 4x$, $y = 0$
19. $x = y^3 - y$, $x = 0$
20. $x = y^3 - 4y^2 + 3y$, $x = y^2 - y$
21. Use a CAS to find the area enclosed by $y = 3 - 2x$ and $y = x^6 + 2x^5 - 3x^4 + x^2$.
22. Use a CAS to find the exact area enclosed by the curves $y = x^5 - 2x^3 - 3x$ and $y = x^3$.
23. Find a horizontal line $y = k$ that divides the area between $y = x^2$ and $y = 9$ into two equal parts.
24. Find a vertical line $x = k$ that divides the area enclosed by $x = \sqrt{y}$, $x = 2$, and $y = 0$ into two equal parts.
25. (a) Find the area of the region enclosed by the parabola $y = 2x - x^2$ and the x-axis.
    (b) Find the value of m so that the line $y = mx$ divides the region in part (a) into two regions of equal area.
26. Find the area between the curve $y = \sin x$ and the line segment joining the points $(0, 0)$ and $(5\pi/6, 1/2)$ on the curve.

27–28 Use Newton's Method (Section 4.6), where needed, to approximate the x-coordinates of the intersections of the curves to at least four decimal places, and then use those approximations to approximate the area of the region.

27. The region that lies below the curve $y = \sin x$ and above the line $y = 0.2x$, where $x \geq 0$.
28. The region enclosed by the graphs of $y = x^2$ and $y = \cos x$.
29. Find the area of the region that is enclosed by the curves $y = x^2 - 1$ and $y = 2\sin x$.

C **30.** Referring to the accompanying figure, use a CAS to estimate the value of k so that the areas of the shaded regions are equal.

Source: This exercise is based on Problem A1 that was posed in the Fifty-Fourth Annual William Lowell Putnam Mathematical Competition.

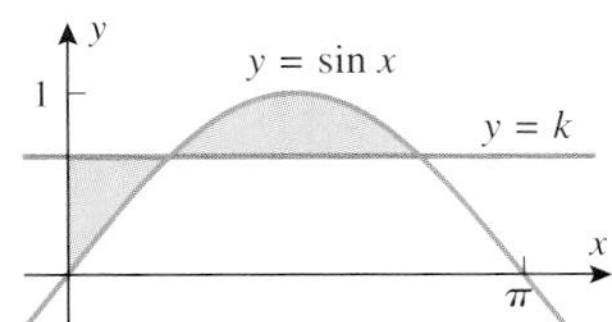

Figure Ex-30

FOCUS ON CONCEPTS

31. Two racers in adjacent lanes move with velocity functions $v_1(t)$ m/s and $v_2(t)$ m/s, respectively. Suppose that the racers are even at time $t = 60$ s. Interpret the value of the integral

$$\int_0^{60} [v_2(t) - v_1(t)]\, dt$$

in this context.

32. The accompanying figure shows acceleration versus time curves for two cars that move along a straight track, accelerating from rest at the starting line. What does the area A between the curves over the interval $0 \le t \le T$ represent? Justify your answer.

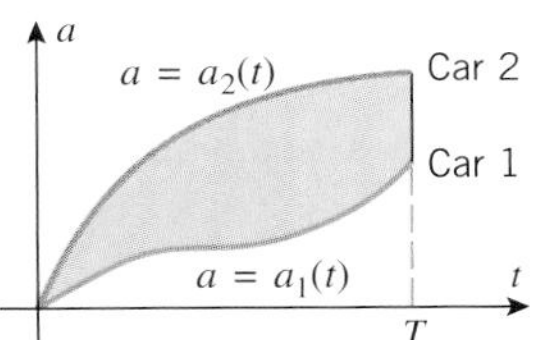

Figure Ex-32

33. Suppose that f and g are integrable on $[a, b]$, but neither $f(x) \ge g(x)$ nor $g(x) \ge f(x)$ holds for all x in $[a, b]$ [i.e., the curves $y = f(x)$ and $y = g(x)$ are intertwined].

(a) What is the geometric significance of the integral

$$\int_a^b [f(x) - g(x)]\, dx?$$

(b) What is the geometric significance of the integral

$$\int_a^b |f(x) - g(x)|\, dx?$$

34. Let $A(n)$ be the area in the first quadrant enclosed by the curves $y = \sqrt[n]{x}$ and $y = x$.

(a) By considering how the graph of $y = \sqrt[n]{x}$ changes as n increases, make a conjecture about the limit of $A(n)$ as $n \to +\infty$.

(b) Confirm your conjecture by calculating the limit.

35. Find the area of the region enclosed between the curve $x^{1/2} + y^{1/2} = a^{1/2}$ and the coordinate axes.

36. Show that the area of the ellipse in the accompanying figure is πab. [*Hint:* Use a formula from geometry.]

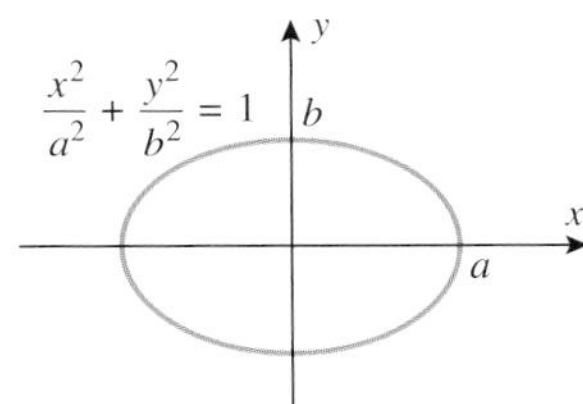

Figure Ex-36

37. A rectangle with edges parallel to the coordinate axes has one vertex at the origin and the diagonally opposite vertex on the curve $y = kx^m$ at the point where $x = b$ ($b > 0$, $k > 0$, and $m \ge 0$). Show that the fraction of the area of the rectangle that lies between the curve and the x-axis depends on m but not on k or b.

✔ QUICK CHECK ANSWERS 6.1

1. $\int_0^2 [(20 - 3x^2) - 3\sqrt{x}]\, dx$; $32 - 4\sqrt{2}$ 2. $\int_{-1}^5 [(2x + 8) - (2x - 3)]\, dx$; 66

3. (a) $(-2, 0)$; $(0, 2)$ (b) $\int_{-2}^0 [\sqrt{4 - x^2} - (x + 2)]\, dx$ (c) $\int_0^2 [(y - 2) + \sqrt{4 - y^2}]\, dy$ (d) $\pi - 2$ 4. $\frac{5}{12}$

6.2 VOLUMES BY SLICING; DISKS AND WASHERS

In the last section we showed that the area of a plane region bounded by two curves can be obtained by integrating the length of a general cross section over an appropriate interval. In this section we will see that the same basic principle can be used to find volumes of certain three-dimensional solids.

VOLUMES BY SLICING

Recall that the underlying principle for finding the area of a plane region is to divide the region into thin strips, approximate the area of each strip by the area of a rectangle, add the approximations to form a Riemann sum, and take the limit of the Riemann sums to produce an integral for the area. Under appropriate conditions, the same strategy can be used to find the volume of a solid. The idea is to divide the solid into thin slabs, approximate the volume of each slab, add the approximations to form a Riemann sum, and take the limit of the Riemann sums to produce an integral for the volume (Figure 6.2.1).

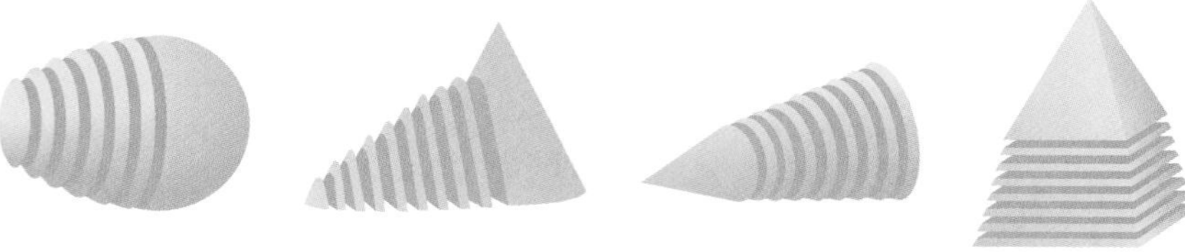

Figure 6.2.1

What makes this method work is the fact that a *thin* slab has cross sections that do not vary much in size or shape, which, as we will see, makes its volume easy to approximate (Figure 6.2.2). Moreover, the thinner the slab, the less variation in its cross sections and the better the approximation. Thus, once we approximate the volumes of the slabs, we can set up a Riemann sum whose limit is the volume of the entire solid. We will give the details shortly, but first we need to discuss how to find the volume of a solid whose cross sections do not vary in size and shape (i.e., are congruent).

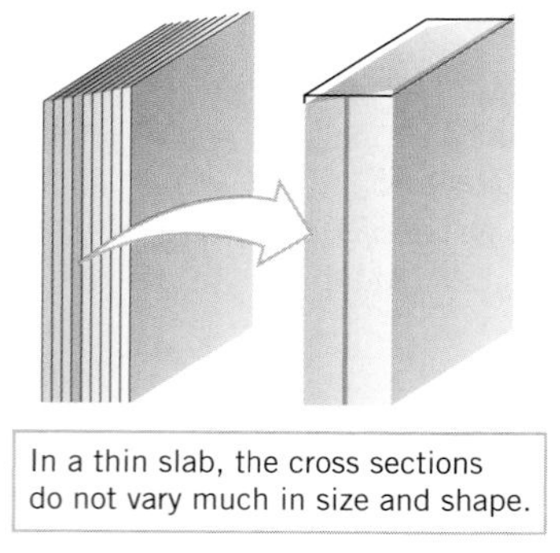

In a thin slab, the cross sections do not vary much in size and shape.

Figure 6.2.2

One of the simplest examples of a solid with congruent cross sections is a right circular cylinder of radius r, since all cross sections taken perpendicular to the central axis are circular regions of radius r. The volume V of a right circular cylinder of radius r and height h can be expressed in terms of the height and the area of a cross section as

$$V = \pi r^2 h = [\text{area of a cross section}] \times [\text{height}] \tag{1}$$

This is a special case of a more general volume formula that applies to solids called right cylinders. A ***right cylinder*** is a solid that is generated when a plane region is translated along a line or ***axis*** that is perpendicular to the region (Figure 6.2.3).

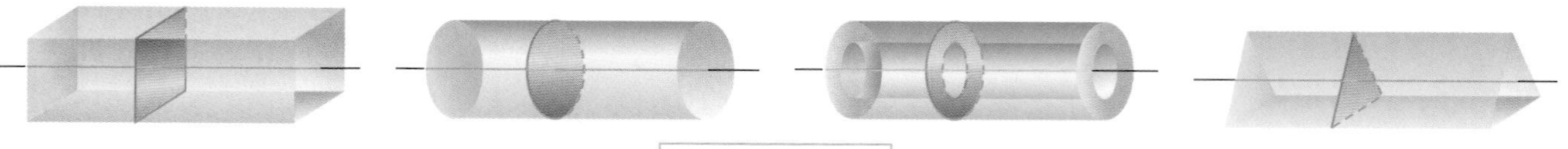

Some right cylinders

Figure 6.2.3

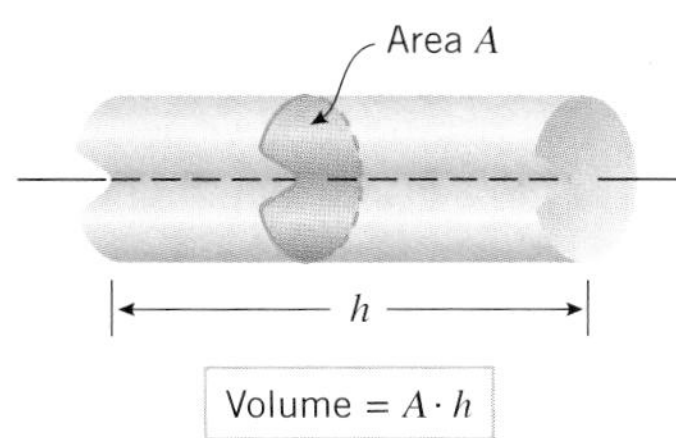

Figure 6.2.4

If a right cylinder is generated by translating a region of area A through a distance h, then h is called the ***height*** (or sometimes the ***width***) of the cylinder, and the volume V of the cylinder is defined to be

$$V = A \cdot h = [\text{area of a cross section}] \times [\text{height}] \tag{2}$$

(Figure 6.2.4). Note that this is consistent with Formula (1) for the volume of a right *circular* cylinder.

We now have all of the tools required to solve the following problem.

> **6.2.1 PROBLEM.** Let S be a solid that extends along the x-axis and is bounded on the left and right, respectively, by the planes that are perpendicular to the x-axis at $x = a$ and $x = b$ (Figure 6.2.5). Find the volume V of the solid, assuming that its cross-sectional area $A(x)$ is known at each x in the interval $[a, b]$.

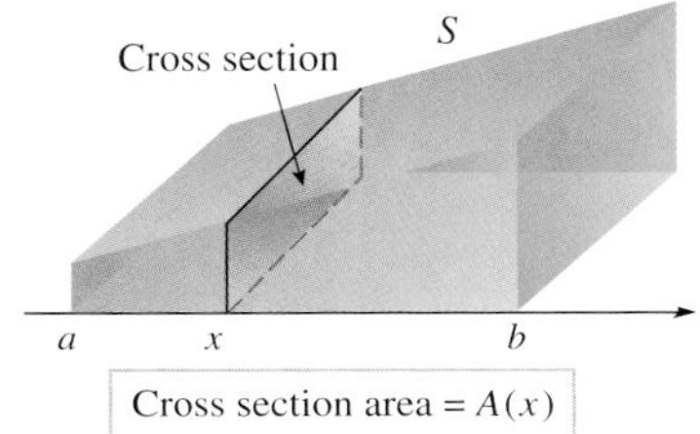

Figure 6.2.5

To solve this problem we begin by dividing the interval $[a, b]$ into n subintervals, thereby dividing the solid into n slabs as shown in the left part of Figure 6.2.6. If we assume that the width of the kth subinterval is Δx_k, then the volume of the kth slab can be approximated by the volume $A(x_k^*)\Delta x_k$ of a right cylinder of width (height) Δx_k and cross-sectional area $A(x_k^*)$, where x_k^* is a point in the kth subinterval (see the right part of Figure 6.2.6).

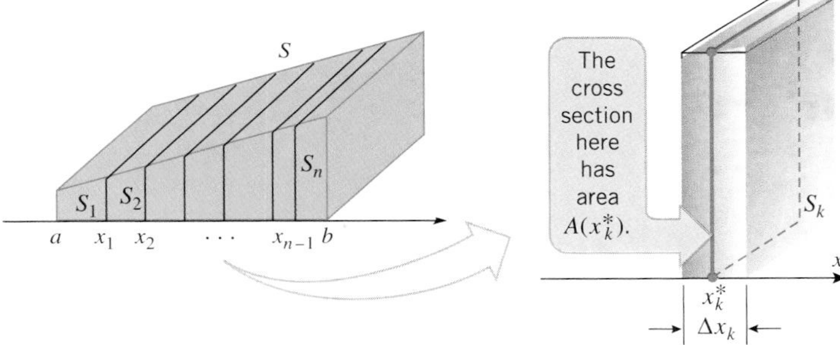

Figure 6.2.6

Adding these approximations yields the following Riemann sum that approximates the volume V:

$$V \approx \sum_{k=1}^{n} A(x_k^*)\Delta x_k$$

Taking the limit as n increases and the widths of the subintervals approach zero yields the definite integral

$$V = \lim_{\max \Delta x_k \to 0} \sum_{k=1}^{n} A(x_k^*)\Delta x_k = \int_a^b A(x)\,dx$$

In summary, we have the following result.

> **6.2.2 VOLUME FORMULA.** Let S be a solid bounded by two parallel planes perpendicular to the x-axis at $x = a$ and $x = b$. If, for each x in $[a, b]$, the cross-sectional area of S perpendicular to the x-axis is $A(x)$, then the volume of the solid is
>
> $$V = \int_a^b A(x)\,dx \tag{3}$$
>
> provided $A(x)$ is integrable.

There is a similar result for cross sections perpendicular to the y-axis.

6.2.3 VOLUME FORMULA. Let S be a solid bounded by two parallel planes perpendicular to the y-axis at $y = c$ and $y = d$. If, for each y in $[c, d]$, the cross-sectional area of S perpendicular to the y-axis is $A(y)$, then the volume of the solid is

$$V = \int_c^d A(y)\, dy \tag{4}$$

provided $A(y)$ is integrable.

In words, these formulas state:

The volume of a solid can be obtained by integrating the cross-sectional area from one end of the solid to the other.

▶ **Example 1** Derive the formula for the volume of a right pyramid whose altitude is h and whose base is a square with sides of length a.

Solution. As illustrated in Figure 6.2.7a, we introduce a rectangular coordinate system in which the y-axis passes through the apex and is perpendicular to the base, and the x-axis passes through the base and is parallel to a side of the base.

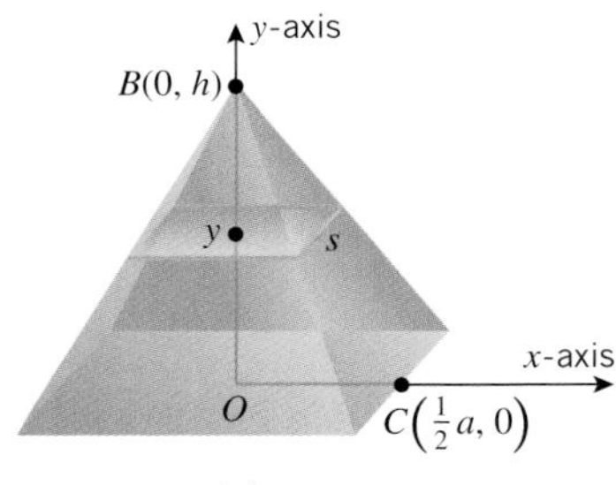

(a)

(b)

Figure 6.2.7

At any y in the interval $[0, h]$ on the y-axis, the cross section perpendicular to the y-axis is a square. If s denotes the length of a side of this square, then by similar triangles (Figure 6.2.7b)

$$\frac{\frac{1}{2}s}{\frac{1}{2}a} = \frac{h-y}{h} \quad \text{or} \quad s = \frac{a}{h}(h-y)$$

Thus, the area $A(y)$ of the cross section at y is

$$A(y) = s^2 = \frac{a^2}{h^2}(h-y)^2$$

and by (4) the volume is

$$V = \int_0^h A(y)\, dy = \int_0^h \frac{a^2}{h^2}(h-y)^2\, dy = \frac{a^2}{h^2}\int_0^h (h-y)^2\, dy$$

$$= \frac{a^2}{h^2}\left[-\frac{1}{3}(h-y)^3\right]_{y=0}^{h} = \frac{a^2}{h^2}\left[0 + \frac{1}{3}h^3\right] = \frac{1}{3}a^2 h$$

That is, the volume is $\frac{1}{3}$ of the area of the base times the altitude. ◀

SOLIDS OF REVOLUTION

A ***solid of revolution*** is a solid that is generated by revolving a plane region about a line that lies in the same plane as the region; the line is called the ***axis of revolution***. Many familiar solids are of this type (Figure 6.2.8).

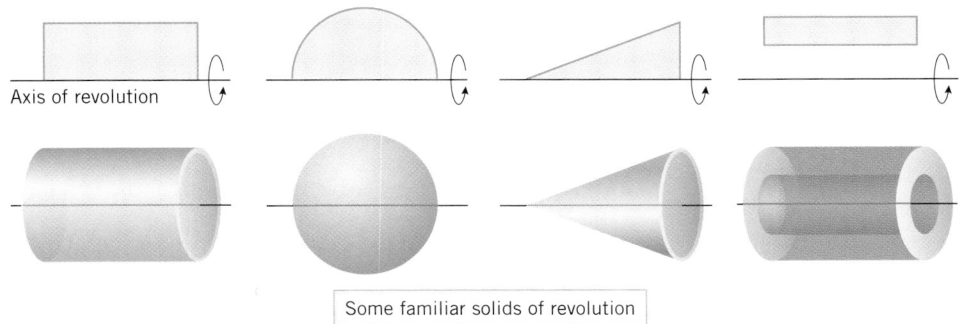

Figure 6.2.8

VOLUMES BY DISKS PERPENDICULAR TO THE x-AXIS

We will be interested in the following general problem.

> **6.2.4 PROBLEM.** Let f be continuous and nonnegative on $[a, b]$, and let R be the region that is bounded above by $y = f(x)$, below by the x-axis, and on the sides by the lines $x = a$ and $x = b$ (Figure 6.2.9a). Find the volume of the solid of revolution that is generated by revolving the region R about the x-axis.

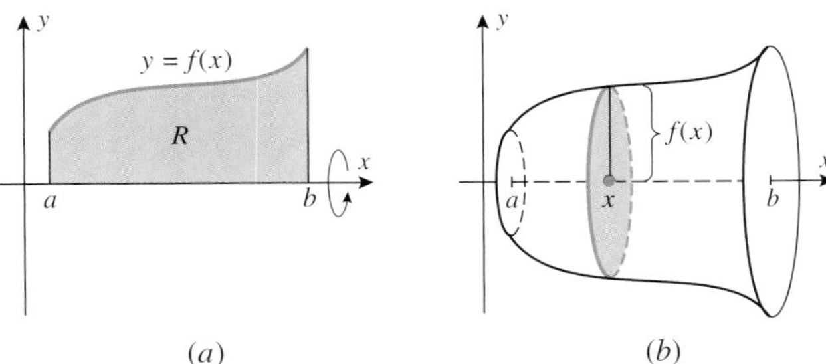

Figure 6.2.9

We can solve this problem by slicing. For this purpose, observe that the cross section of the solid taken perpendicular to the x-axis at the point x is a circular disk of radius $f(x)$ (Figure 6.2.9b). The area of this region is

$$A(x) = \pi[f(x)]^2$$

Thus, from (3) the volume of the solid is

$$V = \int_a^b \pi[f(x)]^2\,dx \tag{5}$$

Because the cross sections are disk shaped, the application of this formula is called the ***method of disks***.

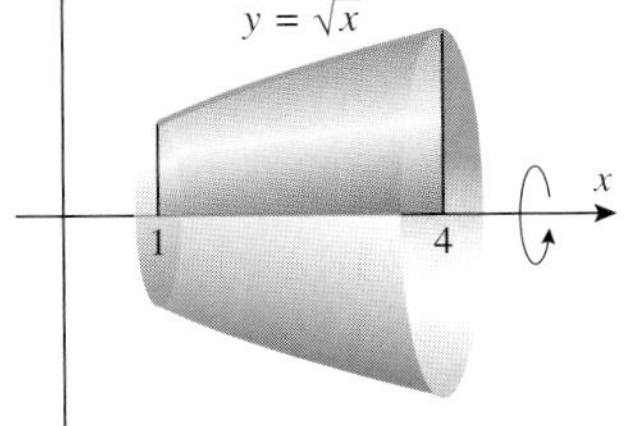

Figure 6.2.10

▶ **Example 2** Find the volume of the solid that is obtained when the region under the curve $y = \sqrt{x}$ over the interval $[1, 4]$ is revolved about the x-axis (Figure 6.2.10).

Solution. From (5), the volume is

$$V = \int_a^b \pi[f(x)]^2\,dx = \int_1^4 \pi x\,dx = \left.\frac{\pi x^2}{2}\right]_1^4 = 8\pi - \frac{\pi}{2} = \frac{15\pi}{2} \blacktriangleleft$$

▶ **Example 3** Derive the formula for the volume of a sphere of radius r.

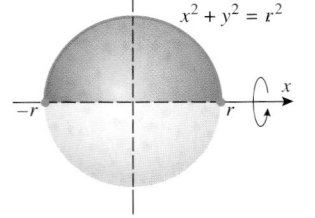

Solution. As indicated in Figure 6.2.11, a sphere of radius r can be generated by revolving the upper semicircular disk enclosed between the x-axis and

$$x^2 + y^2 = r^2$$

about the x-axis. Since the upper half of this circle is the graph of $y = f(x) = \sqrt{r^2 - x^2}$, it follows from (5) that the volume of the sphere is

$$V = \int_a^b \pi[f(x)]^2\,dx = \int_{-r}^{r} \pi(r^2 - x^2)\,dx = \pi\left[r^2x - \frac{x^3}{3}\right]_{-r}^{r} = \frac{4}{3}\pi r^3 \blacktriangleleft$$

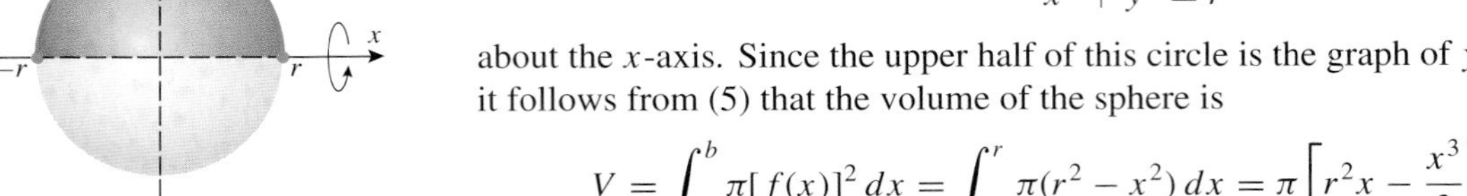

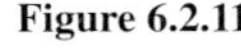

Figure 6.2.11

VOLUMES BY WASHERS PERPENDICULAR TO THE *x*-AXIS

Not all solids of revolution have solid interiors; some have holes or channels that create interior surfaces, as in the last part of Figure 6.2.8. Thus, we will be interested in problems of the following type.

> **6.2.5 PROBLEM.** Let f and g be continuous and nonnegative on $[a, b]$, and suppose that $f(x) \geq g(x)$ for all x in the interval $[a, b]$. Let R be the region that is bounded above by $y = f(x)$, below by $y = g(x)$, and on the sides by the lines $x = a$ and $x = b$ (Figure 6.2.12*a*). Find the volume of the solid of revolution that is generated by revolving the region R about the x-axis.

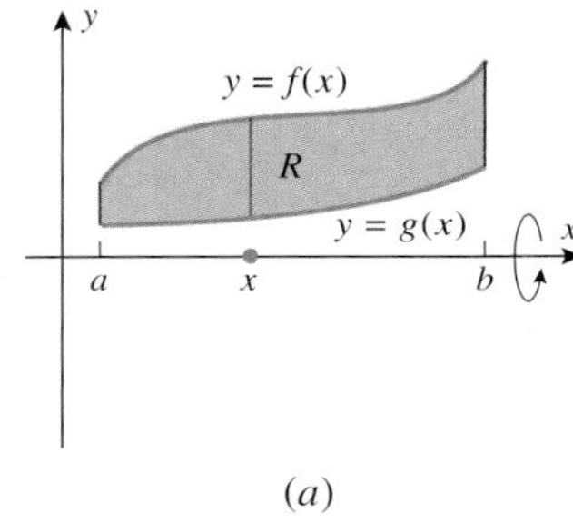

(*a*)

(*b*)

Figure 6.2.12

We can solve this problem by slicing. For this purpose, observe that the cross section of the solid taken perpendicular to the x-axis at the point x is the annular or "washer-shaped" region with inner radius $g(x)$ and outer radius $f(x)$ (Figure 6.2.12*b*); hence its area is

$$A(x) = \pi[f(x)]^2 - \pi[g(x)]^2 = \pi([f(x)]^2 - [g(x)]^2)$$

Thus, from (3) the volume of the solid is

$$V = \int_a^b \pi([f(x)]^2 - [g(x)]^2)\,dx \qquad (6)$$

Because the cross sections are washer shaped, the application of this formula is called the ***method of washers***.

▶ **Example 4** Find the volume of the solid generated when the region between the graphs of the equations $f(x) = \frac{1}{2} + x^2$ and $g(x) = x$ over the interval $[0, 2]$ is revolved about the x-axis (Figure 6.2.13).

Solution. From (6) the volume is

$$V = \int_a^b \pi([f(x)]^2 - [g(x)]^2)\,dx = \int_0^2 \pi\left(\left[\tfrac{1}{2} + x^2\right]^2 - x^2\right)dx$$

$$= \int_0^2 \pi\left(\frac{1}{4} + x^4\right)dx = \pi\left[\frac{x}{4} + \frac{x^5}{5}\right]_0^2 = \frac{69\pi}{10} \blacktriangleleft$$

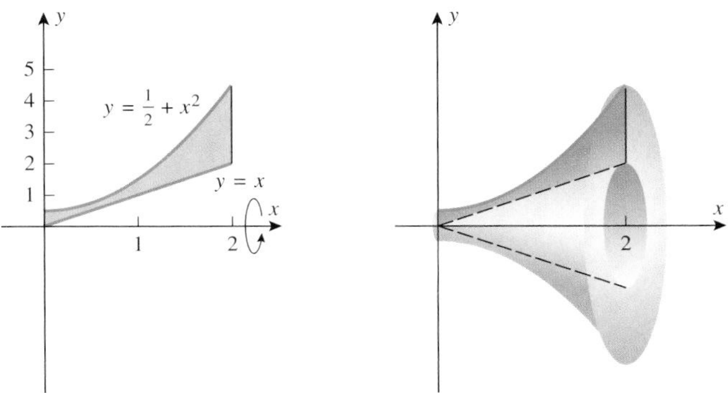

Figure 6.2.13

VOLUMES BY DISKS AND WASHERS PERPENDICULAR TO THE y-AXIS

The methods of disks and washers have analogs for regions that are revolved about the y-axis (Figures 6.2.14 and 6.2.15). Using the method of slicing and Formula (4), you should have no trouble deducing the following formulas for the volumes of the solids in the figures.

$$V = \int_c^d \pi[u(y)]^2\,dy \qquad V = \int_c^d \pi([w(y)]^2 - [v(y)]^2)\,dy \tag{7–8}$$

Disks　　Washers

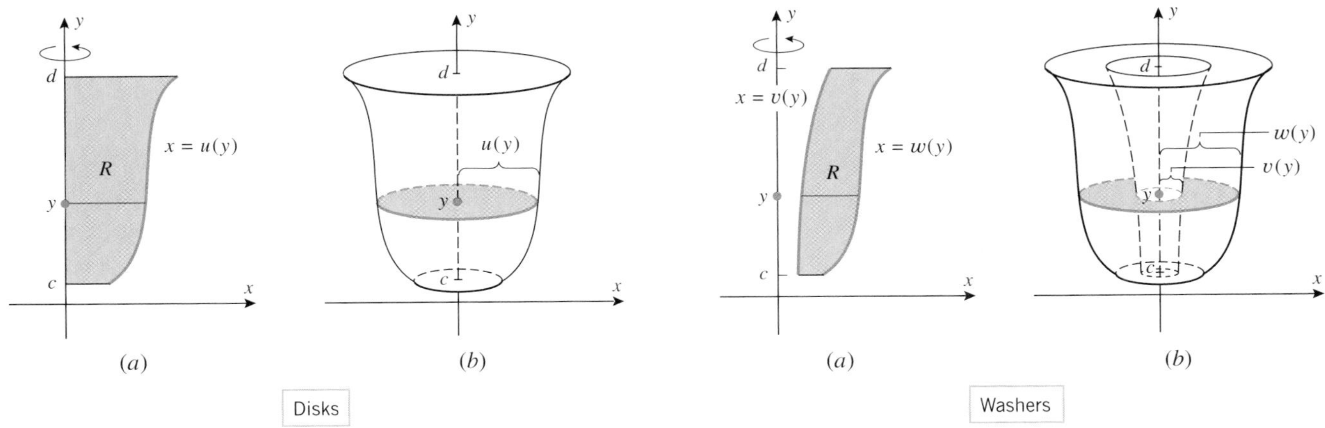

Figure 6.2.14　　Figure 6.2.15

▶ **Example 5** Find the volume of the solid generated when the region enclosed by $y = \sqrt{x}$, $y = 2$, and $x = 0$ is revolved about the y-axis (Figure 6.2.16).

Solution. The cross sections taken perpendicular to the y-axis are disks, so we will apply (7). But first we must rewrite $y = \sqrt{x}$ as $x = y^2$. Thus, from (7) with $u(y) = y^2$, the volume is

$$V = \int_c^d \pi[u(y)]^2\,dy = \int_0^2 \pi y^4\,dy = \left.\frac{\pi y^5}{5}\right]_0^2 = \frac{32\pi}{5} \quad \blacktriangleleft$$

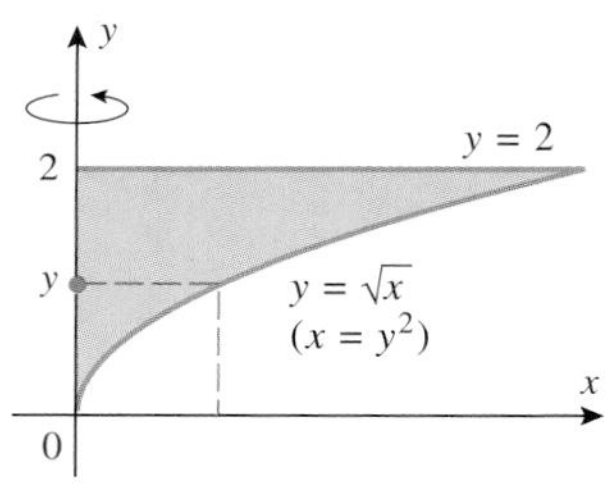

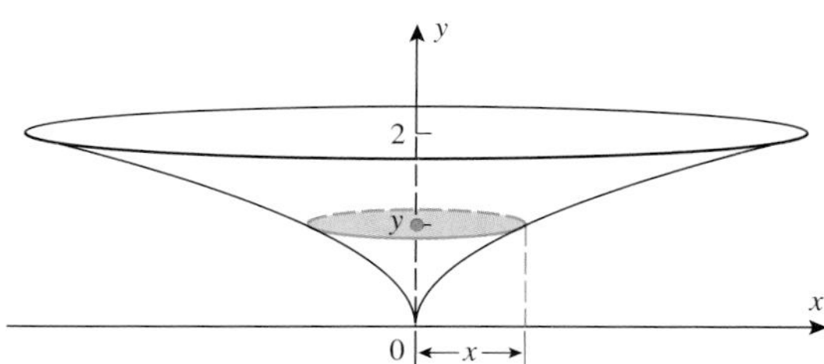

Figure 6.2.16

✔QUICK CHECK EXERCISES 6.2 (See page 397 for answers.)

1. A solid S extends along the x-axis from $x = 1$ to $x = 3$. For x between 1 and 3, the cross-sectional area of S perpendicular to the x-axis is $3x^2$. An integral expression for the volume of S is ________. The value of this integral is ________.

2. A solid S is generated by revolving the region between the x-axis and the curve $y = \sqrt{\sin x}$ $(0 \le x \le \pi)$ about the x-axis.
   (a) For x between 0 and π, the cross-sectional area of S perpendicular to the x-axis at x is $A(x) =$ ________.
   (b) An integral expression for the volume of S is ________.
   (c) The value of the integral in part (b) is ________.

3. A solid S is generated by revolving the region enclosed by the line $y = 2x + 1$ and the curve $y = x^2 + 1$ about the x-axis.
   (a) For x between ________ and ________, the cross-sectional area of S perpendicular to the x-axis at x is $A(x) =$ ________.
   (b) An integral expression for the volume of S is ________.
   (c) The value of the integral in part (b) is ________.

4. A solid S is generated by revolving the region enclosed by the line $y = x + 1$ and the curve $y = x^2 + 1$ about the y-axis.
   (a) For y between ________ and ________, the cross-sectional area of S perpendicular to the y-axis at y is $A(y) =$ ________.
   (b) An integral expression for the volume of S is ________.
   (c) The value of the integral in part (b) is ________.

EXERCISE SET 6.2  CAS

1–4 Find the volume of the solid that results when the shaded region is revolved about the indicated axis.

1.

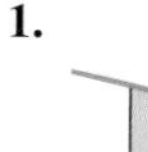

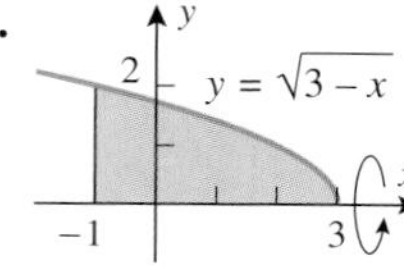

2.

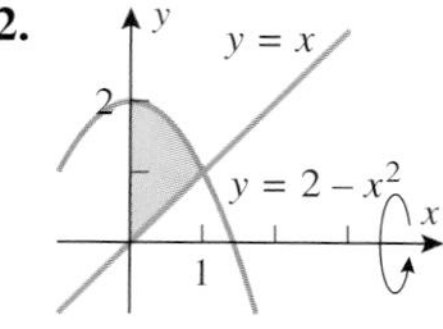

3.

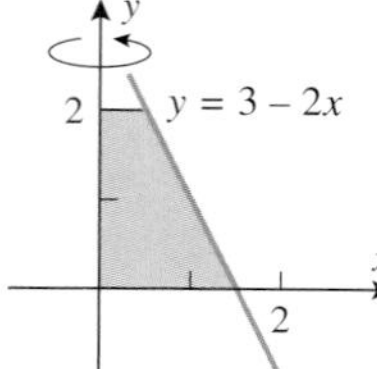

4.

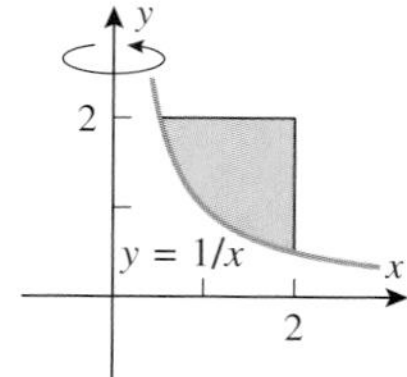

5. Find the volume of the solid whose base is the region bounded between the curve $y = x^2$ and the x-axis from $x = 0$ to $x = 2$ and whose cross sections taken perpendicular to the x-axis are squares.

6. Find the volume of the solid whose base is the region bounded between the curve $y = \sec x$ and the x-axis from $x = \pi/4$ to $x = \pi/3$ and whose cross sections taken perpendicular to the x-axis are squares.

7–12 Find the volume of the solid that results when the region enclosed by the given curves is revolved about the x-axis.

7. $y = \sqrt{\cos x}$, $x = \pi/4$, $x = \pi/2$, $y = 0$

8. $y = x^2$, $y = x^3$

9. $y = \sqrt{25 - x^2}$, $y = 3$

10. $y = 9 - x^2$, $y = 0$

11. $x = \sqrt{y}$, $x = y/4$

12. $y = \sin x$, $y = \cos x$, $x = 0$, $x = \pi/4$.
[*Hint:* Use the identity $\cos 2x = \cos^2 x - \sin^2 x$.]

13. Find the volume of the solid whose base is the region bounded between the curve $y = x^3$ and the y-axis from $y = 0$ to $y = 1$ and whose cross sections taken perpendicular to the y-axis are squares.

14. Find the volume of the solid whose base is the region enclosed between the curve $x = 1 - y^2$ and the y-axis and whose cross sections taken perpendicular to the y-axis are squares.

15–20 Find the volume of the solid that results when the region enclosed by the given curves is revolved about the y-axis.

15. $x = \sqrt{1+y},\ x = 0,\ y = 3$

16. $y = x^2 - 1,\ x = 2,\ y = 0$

17. $x = \csc y,\ y = \pi/4,\ y = 3\pi/4,\ x = 0$

18. $y = x^2,\ x = y^2$

19. $x = y^2,\ x = y + 2$

20. $x = 1 - y^2,\ x = 2 + y^2,\ y = -1,\ y = 1$

21. Find the volume of the solid that results when the region above the x-axis and below the ellipse

$$\frac{x^2}{a^2} + \frac{y^2}{b^2} = 1 \quad (a > 0, b > 0)$$

is revolved about the x-axis.

22. Let V be the volume of the solid that results when the region enclosed by $y = 1/x$, $y = 0$, $x = 2$, and $x = b$ $(0 < b < 2)$ is revolved about the x-axis. Find the value of b for which $V = 3$.

23. Find the volume of the solid generated when the region enclosed by $y = \sqrt{x+1}$, $y = \sqrt{2x}$, and $y = 0$ is revolved about the x-axis. [*Hint:* Split the solid into two parts.]

24. Find the volume of the solid generated when the region enclosed by $y = \sqrt{x}$, $y = 6 - x$, and $y = 0$ is revolved about the x-axis. [*Hint:* Split the solid into two parts.]

FOCUS ON CONCEPTS

25. Suppose that f is a continuous function on $[a, b]$, and let R be the region between the curve $y = f(x)$ and the line $y = k$ from $x = a$ to $x = b$. Using the method of disks, derive with explanation a formula for the volume of a solid generated by revolving R about the line $y = k$. State and explain additional assumptions, if any, that you need about f for your formula.

26. Suppose that v and w are continuous functions on $[c, d]$, and let R be the region between the curves $x = v(y)$ and $x = w(y)$ from $y = c$ to $y = d$. Using the method of washers, derive with explanation a formula for the volume of a solid generated by revolving R about the line $x = k$. State and explain additional assumptions, if any, that you need about v and w for your formula.

27. Consider the solid generated by revolving the shaded region in Exercise 1 about the line $y = 2$.
(a) Make a conjecture as to which is larger: the volume of this solid or the volume of the solid in Exercise 1. Explain the basis of your conjecture.
(b) Check your conjecture by calculating this volume and comparing it to the volume obtained in Exercise 1.

28. Consider the solid generated by revolving the shaded region in Exercise 4 about the line $x = 2.5$.
(a) Make a conjecture as to which is larger: the volume of this solid or the volume of the solid in Exercise 4. Explain the basis of your conjecture.
(b) Check your conjecture by calculating this volume and comparing it to the volume obtained in Exercise 4.

29. Find the volume of the solid that results when the region enclosed by $y = \sqrt{x}$, $y = 0$, and $x = 9$ is revolved about the line $x = 9$.

30. Find the volume of the solid that results when the region in Exercise 29 is revolved about the line $y = 3$.

31. Find the volume of the solid that results when the region enclosed by $x = y^2$ and $x = y$ is revolved about the line $y = -1$.

32. Find the volume of the solid that results when the region in Exercise 31 is revolved about the line $x = -1$.

33. A nose cone for a space reentry vehicle is designed so that a cross section, taken x ft from the tip and perpendicular to the axis of symmetry, is a circle of radius $\frac{1}{4}x^2$ ft. Find the volume of the nose cone given that its length is 20 ft.

34. A certain solid is 1 ft high, and a horizontal cross section taken x ft above the bottom of the solid is an annulus of inner radius x^2 ft and outer radius $\sqrt{x}$ ft. Find the volume of the solid.

35. Find the volume of the solid whose base is the region bounded between the curves $y = x$ and $y = x^2$, and whose cross sections perpendicular to the x-axis are squares.

36. The base of a certain solid is the region enclosed by $y = \sqrt{x}$, $y = 0$, and $x = 4$. Every cross section perpendicular to the x-axis is a semicircle with its diameter across the base. Find the volume of the solid.

37. In parts (a)–(c) find the volume of the solid whose base is enclosed by the circle $x^2 + y^2 = 1$ and whose cross sections taken perpendicular to the x-axis are
(a) semicircles (b) squares

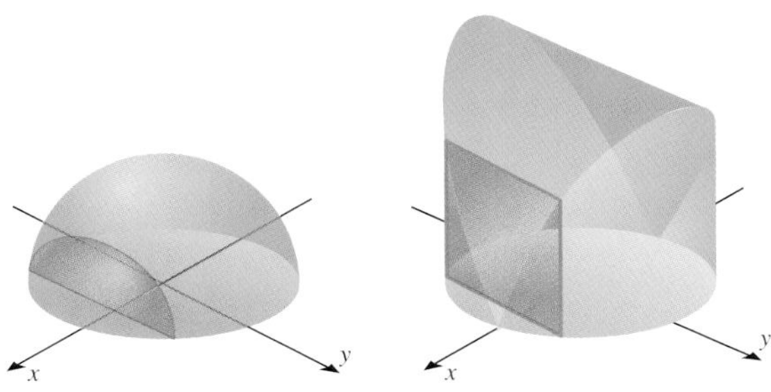

(c) equilateral triangles.

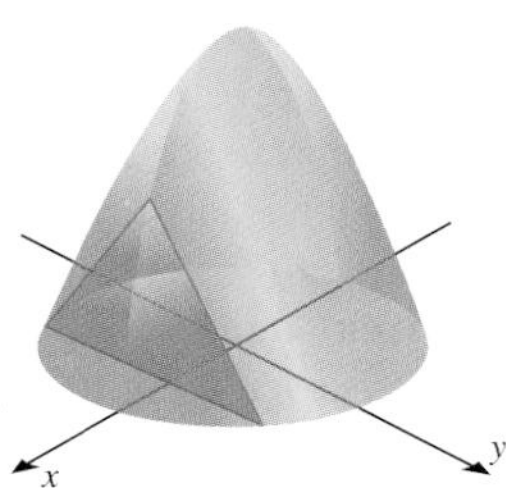

38. As shown in the accompanying figure, a cathedral dome is designed with three semicircular supports of radius r so that each horizontal cross section is a regular hexagon. Show that the volume of the dome is $r^3\sqrt{3}$.

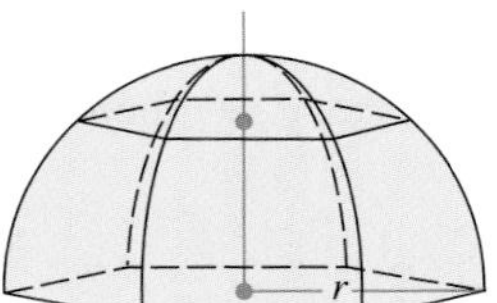

Figure Ex-38

39–40 Use a CAS to estimate the volume of the solid that results when the region enclosed by the curves is revolved about the stated axis.

C **39.** $y = \sin^8 x,\ \ y = 2x/\pi,\ \ x = 0,\ \ x = \pi/2;\ x\text{-axis}$

C **40.** $y = \pi^2 \sin x \cos^3 x,\ \ y = 4x^2,\ \ x = 0,\ \ x = \pi/4;\ x\text{-axis}$

41. The accompanying figure shows a ***spherical cap*** of radius ρ and height h cut from a sphere of radius r. Show that the volume V of the spherical cap can be expressed as
(a) $V = \frac{1}{3}\pi h^2(3r - h)$ (b) $V = \frac{1}{6}\pi h(3\rho^2 + h^2)$.

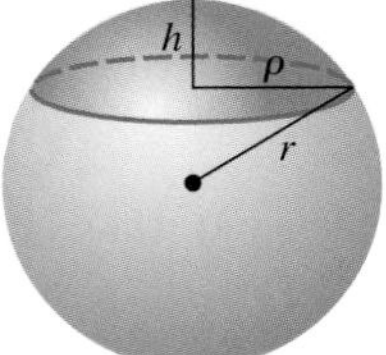

Figure Ex-41

42. If fluid enters a hemispherical bowl with a radius of 10 ft at a rate of $\frac{1}{2}$ ft^3/min, how fast will the fluid be rising when the depth is 5 ft? [*Hint:* See Exercise 41.]

43. The accompanying figure shows the dimensions of a small lightbulb at 10 equally spaced points.

(a) Use formulas from geometry to make a rough estimate of the volume enclosed by the glass portion of the bulb.

(b) Use the average of left and right endpoint approximations to approximate the volume.

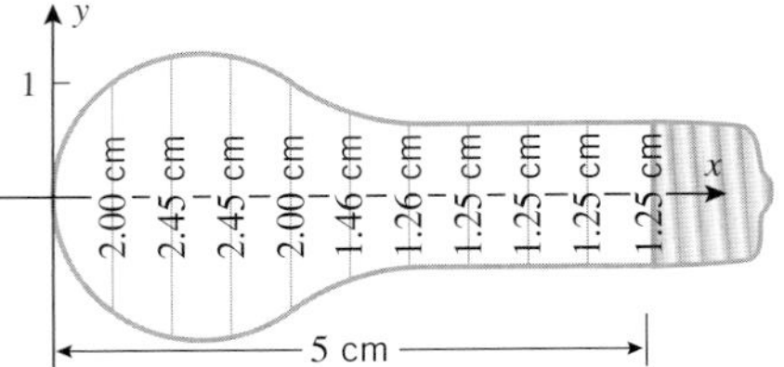

Figure Ex-43

44. Use the result in Exercise 41 to find the volume of the solid that remains when a hole of radius $r/2$ is drilled through the center of a sphere of radius r, and then check your answer by integrating.

45. As shown in the accompanying figure, a cocktail glass with a bowl shaped like a hemisphere of diameter 8 cm contains a cherry with a diameter of 2 cm. If the glass is filled to a depth of h cm, what is the volume of liquid it contains? [*Hint:* First consider the case where the cherry is partially submerged, then the case where it is totally submerged.]

Figure Ex-45

46. Find the volume of the torus that results when the region enclosed by the circle of radius r with center at $(h, 0)$, $h > r$, is revolved about the y-axis. [*Hint:* Use an appropriate formula from plane geometry to help evaluate the definite integral.]

47. A wedge is cut from a right circular cylinder of radius r by two planes, one perpendicular to the axis of the cylinder and the other making an angle θ with the first. Find the volume of the wedge by slicing perpendicular to the y-axis as shown in the accompanying figure.

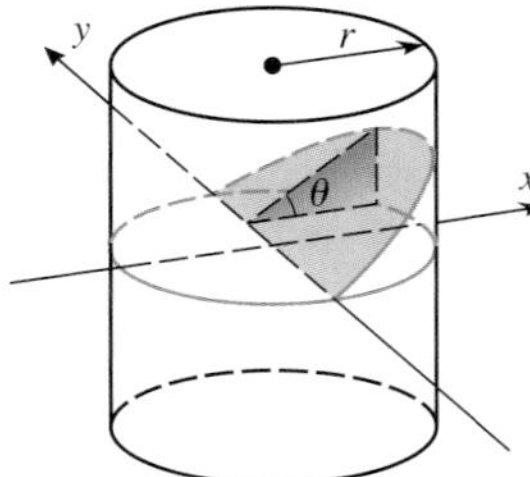

Figure Ex-47

48. Find the volume of the wedge described in Exercise 47 by slicing perpendicular to the x-axis.

49. Two right circular cylinders of radius r have axes that intersect at right angles. Find the volume of the solid common to the two cylinders. [*Hint:* One-eighth of the solid is sketched in the accompanying figure.]

50. In 1635 Bonaventura Cavalieri, a student of Galileo, stated the following result, called ***Cavalieri's principle***: *If two solids have the same height, and if the areas of their cross sections taken parallel to and at equal distances from their bases are always equal, then the solids have the same volume.* Use this result to find the volume of the oblique cylinder in the accompanying figure.

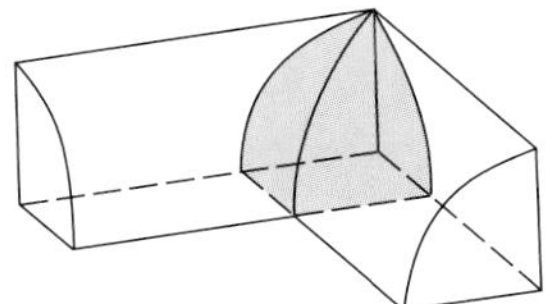

Figure Ex-49

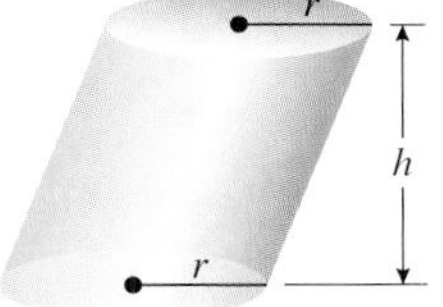

Figure Ex-50

✔ QUICK CHECK ANSWERS 6.2

1. $\int_1^3 3x^2\,dx$; 26 **2.** (a) $\pi \sin x$ (b) $\int_0^{\pi} \pi \sin x\,dx$ (c) 2π **3.** (a) 0; 2; $\pi[(2x+1)^2-(x^2+1)^2] = \pi[-x^4+2x^2+4x]$ (b) $\int_0^2 \pi[-x^4+2x^2+4x]\,dx$ (c) $\frac{104}{15}\pi$ **4.** (a) 1; 2; $\pi[(y-1)-(y-1)^2] = \pi[-y^2+3y-2]$ (b) $\int_1^2 \pi[-y^2+3y-2]\,dy$ (c) $\frac{\pi}{6}$

6.3 VOLUMES BY CYLINDRICAL SHELLS

The methods for computing volumes that have been discussed so far depend on our ability to compute the cross-sectional area of the solid and to integrate that area across the solid. In this section we will develop another method for finding volumes that may be applicable when the cross-sectional area cannot be found or the integration is too difficult.

CYLINDRICAL SHELLS

In this section we will be interested in the following problem.

> **6.3.1 PROBLEM.** Let f be continuous and nonnegative on $[a, b]$ $(0 \le a < b)$, and let R be the region that is bounded above by $y = f(x)$, below by the x-axis, and on the sides by the lines $x = a$ and $x = b$. Find the volume V of the solid of revolution S that is generated by revolving the region R about the y-axis (Figure 6.3.1).

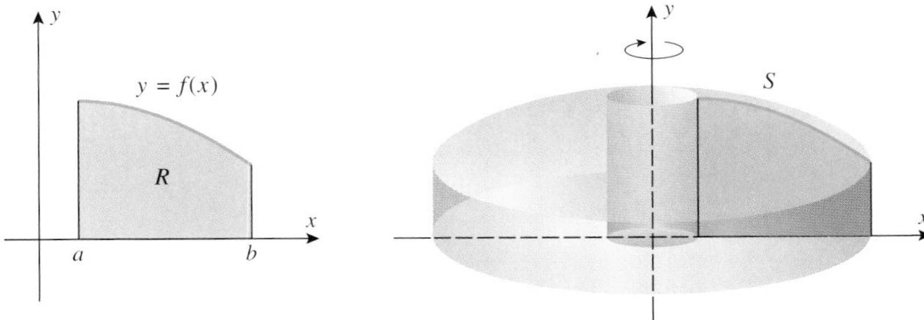

Figure 6.3.1

Sometimes problems of the above type can be solved by the method of disks or washers perpendicular to the y-axis, but when that method is not applicable or the resulting integral is difficult, the *method of cylindrical shells*, which we will discuss here, will often work.

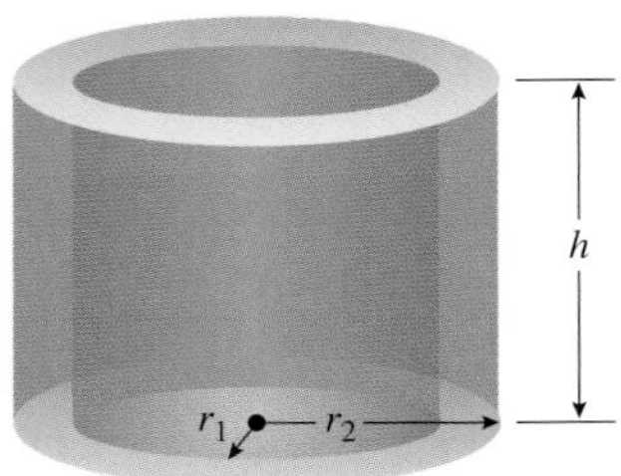

Figure 6.3.2

A ***cylindrical shell*** is a solid enclosed by two concentric right circular cylinders (Figure 6.3.2). The volume V of a cylindrical shell with inner radius r_1, outer radius r_2, and height h can be written as

$$\begin{aligned} V &= [\text{area of cross section}] \cdot [\text{height}] \\ &= (\pi r_2^2 - \pi r_1^2)h \\ &= \pi(r_2 + r_1)(r_2 - r_1)h \\ &= 2\pi \cdot \left[\tfrac{1}{2}(r_1 + r_2)\right] \cdot h \cdot (r_2 - r_1) \end{aligned}$$

But $\frac{1}{2}(r_1 + r_2)$ is the average radius of the shell and $r_2 - r_1$ is its thickness, so

$$V = 2\pi \cdot [\text{average radius}] \cdot [\text{height}] \cdot [\text{thickness}] \tag{1}$$

We will now show how this formula can be used to solve Problem 6.3.1. The underlying idea is to divide the interval $[a, b]$ into n subintervals, thereby subdividing the region R into n strips, $R_1, R_2, \ldots, R_n$ (Figure 6.3.3a). When the region R is revolved about the y-axis, these strips generate "tube-like" solids $S_1, S_2, \ldots, S_n$ that are nested one inside the other and together comprise the entire solid S (Figure 6.3.3b). Thus, the volume V of the solid can be obtained by adding together the volumes of the tubes; that is,

$$V = V(S_1) + V(S_2) + \cdots + V(S_n)$$

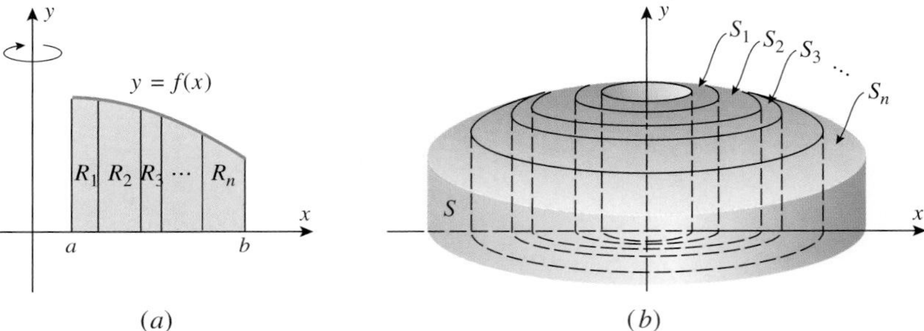

Figure 6.3.3

As a rule, the tubes will have curved upper surfaces, so there will be no simple formulas for their volumes. However, if the strips are thin, then we can approximate each strip by a rectangle (Figure 6.3.4a). These rectangles, when revolved about the y-axis, will produce cylindrical shells whose volumes closely approximate the volumes of the tubes generated by the original strips (Figure 6.3.4b). We will show that by adding the volumes of the cylindrical shells we can obtain a Riemann sum that approximates the volume V, and by taking the limit of the Riemann sums we can obtain an integral for the exact volume V.

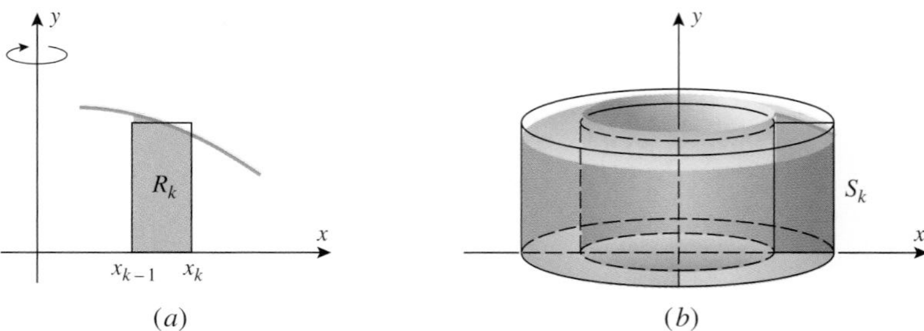

Figure 6.3.4

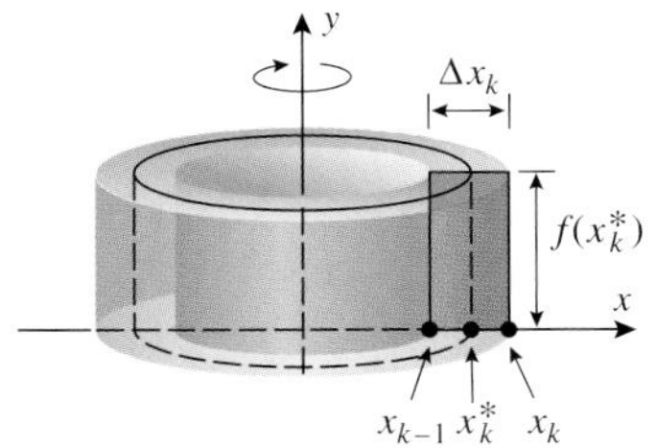

Figure 6.3.5

To implement this idea, suppose that the kth strip extends from x_{k-1} to x_k and that the width of this strip is

$$\Delta x_k = x_k - x_{k-1}$$

If we let x_k^* be the *midpoint* of the interval $[x_{k-1}, x_k]$, and if we construct a rectangle of height $f(x_k^*)$ over the interval, then revolving this rectangle about the y-axis produces a cylindrical shell of average radius x_k^*, height $f(x_k^*)$, and thickness Δx_k (Figure 6.3.5). From (1), the volume V_k of this cylindrical shell is

$$V_k = 2\pi x_k^* f(x_k^*)\Delta x_k$$

Adding the volumes of the n cylindrical shells yields the following Riemann sum that approximates the volume V:

$$V \approx \sum_{k=1}^{n} 2\pi x_k^* f(x_k^*)\Delta x_k$$

Taking the limit as n increases and the widths of the subintervals approach zero yields the definite integral

$$V = \lim_{\max \Delta x_k \to 0} \sum_{k=1}^{n} 2\pi x_k^* f(x_k^*)\Delta x_k = \int_a^b 2\pi x f(x)\, dx$$

In summary, we have the following result.

6.3.2 VOLUME BY CYLINDRICAL SHELLS ABOUT THE y-AXIS. Let f be continuous and nonnegative on $[a, b]$ $(0 \le a < b)$, and let R be the region that is bounded above by $y = f(x)$, below by the x-axis, and on the sides by the lines $x = a$ and $x = b$. Then the volume V of the solid of revolution that is generated by revolving the region R about the y-axis is given by

$$V = \int_a^b 2\pi x f(x)\, dx \qquad (2)$$

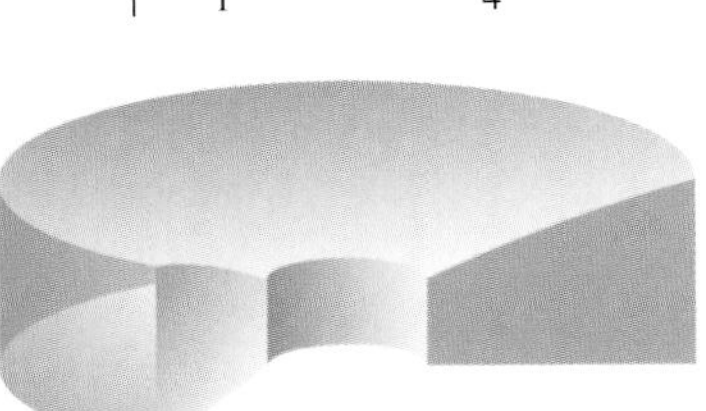

Cutaway view of the solid

Figure 6.3.6

▶ **Example 1** Use cylindrical shells to find the volume of the solid generated when the region enclosed between $y = \sqrt{x}$, $x = 1$, $x = 4$, and the x-axis is revolved about the y-axis (Figure 6.3.6).

Solution. Since $f(x) = \sqrt{x}$, $a = 1$, and $b = 4$, Formula (2) yields

$$V = \int_1^4 2\pi x\sqrt{x}\, dx = 2\pi \int_1^4 x^{3/2}\, dx = \left[2\pi \cdot \frac{2}{5}x^{5/2}\right]_1^4 = \frac{4\pi}{5}[32 - 1] = \frac{124\pi}{5} \quad ◀$$

VARIATIONS OF THE METHOD OF CYLINDRICAL SHELLS

The method of cylindrical shells is applicable in a variety of situations that do not fit the conditions required by Formula (2). For example, the region may be enclosed between two curves, or the axis of revolution may be some line other than the y-axis. However, rather than develop a separate formula for every possible situation, we will give a general way of thinking about the method of cylindrical shells that can be adapted to each new situation as it arises.

For this purpose, we will need to reexamine the integrand in Formula (2): At each x in the interval $[a, b]$, the vertical line segment from the x-axis to the curve $y = f(x)$ can be viewed as the cross section of the region R at x (Figure 6.3.7a). When the region R is revolved about the y-axis, the cross section at x sweeps out the *surface* of a right circular cylinder of height $f(x)$ and radius x (Figure 6.3.7*b*). The area of this surface is

$$2\pi x f(x)$$

(Figure 6.3.7*c*), which is the integrand in (2). Thus, Formula (2) can be viewed informally in the following way.

6.3.3 AN INFORMAL VIEWPOINT ABOUT CYLINDRICAL SHELLS. The volume V of a solid of revolution that is generated by revolving a region R about an axis can be obtained by integrating the area of the surface generated by an arbitrary cross section of R taken parallel to the axis of revolution.

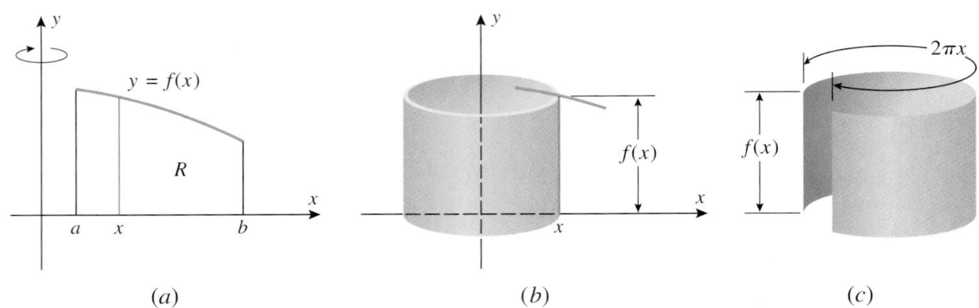

Figure 6.3.7

The following examples illustrate how to apply this result in situations where Formula (2) is not applicable.

▶ **Example 2** Use cylindrical shells to find the volume of the solid generated when the region R in the first quadrant enclosed between $y = x$ and $y = x^2$ is revolved about the y-axis (Figure 6.3.8).

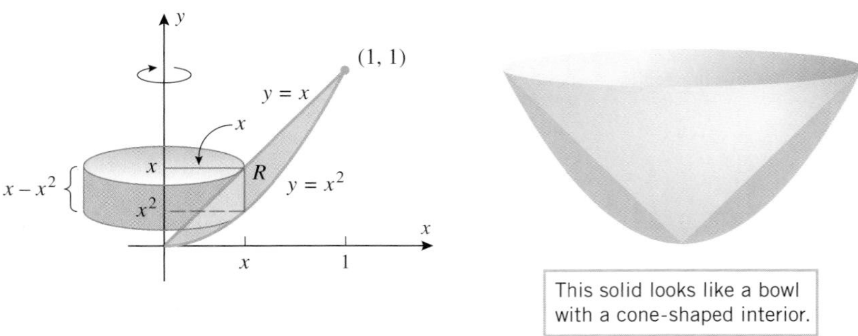

Figure 6.3.8

Solution. As illustrated in the figure, at each x in $[0, 1]$ the cross section of R parallel to the y-axis generates a cylindrical surface of height $x - x^2$ and radius x. Since the area of this surface is

$$2\pi x(x - x^2)$$

the volume of the solid is

$$V = \int_0^1 2\pi x(x - x^2)\,dx = 2\pi \int_0^1 (x^2 - x^3)\,dx$$
$$= 2\pi \left[\frac{x^3}{3} - \frac{x^4}{4}\right]_0^1 = 2\pi\left[\frac{1}{3} - \frac{1}{4}\right] = \frac{\pi}{6} \blacktriangleleft$$

▶ **Example 3** Use cylindrical shells to find the volume of the solid generated when the region R under $y = x^2$ over the interval $[0, 2]$ is revolved about the line $y = -1$ (Figure 6.3.9).

Solution. As illustrated in the figure, at each y in the interval $0 \le y \le 4$, the cross section of R parallel to the x-axis generates a cylindrical surface of height $2 - \sqrt{y}$ and radius $y + 1$. Since the area of this surface is

$$2\pi(y + 1)(2 - \sqrt{y})$$

it follows that the volume of the solid is

The volumes in Examples 2 and 3 can also be obtained by the method of washers. Confirm that the volumes produced by that method agree with those obtained by cylindrical shells.

$$\int_0^4 2\pi(y + 1)(2 - \sqrt{y})\,dy = 2\pi \int_0^4 (2y - y^{3/2} + 2 - y^{1/2})\,dy$$
$$= 2\pi\left[y^2 - \frac{2}{5}y^{5/2} + 2y - \frac{2}{3}y^{3/2}\right]_0^4 = \frac{176\pi}{15} \blacktriangleleft$$

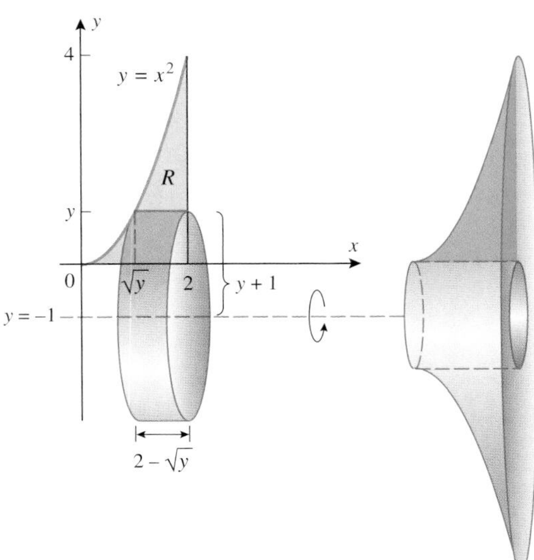

Figure 6.3.9

✔ QUICK CHECK EXERCISES 6.3 (See page 403 for answers.)

1. Let R be the region between the x-axis and the curve $y = 1 + \sqrt{x}$ for $1 \le x \le 4$.
   (a) For x between 1 and 4, the area of the cylindrical surface generated by revolving the vertical cross section of R at x about the y-axis is ______.
   (b) Using cylindrical shells, an integral expression for the volume of the solid generated by revolving R about the y-axis is ______.
   (c) The value of the integral in part (b) is ______.

2. Let R be the region described in Quick Check Exercise 1.
   (a) For x between 1 and 4, the area of the cylindrical surface generated by revolving the vertical cross section of R at x about the line $x = 5$ is ________.
   (b) Using cylindrical shells, an integral expression for the volume of the solid generated by revolving R about the line $x = 5$ is ________.
   (c) The value of the integral in part (b) is ________.

3. A solid S is generated by revolving the region enclosed by the curves $x = (y - 2)^2$ and $x = 4$ about the x-axis. Using cylindrical shells, an integral expression for the volume of S is ________. The value of this integral is ________.

EXERCISE SET 6.3 [C] CAS

1–4 Use cylindrical shells to find the volume of the solid generated when the shaded region is revolved about the indicated axis.

1.
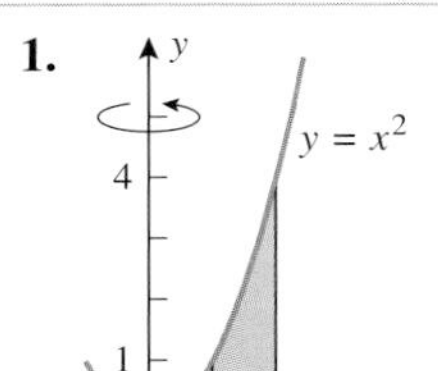

2.
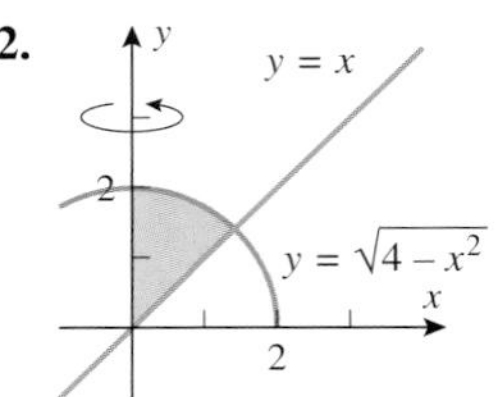

3.
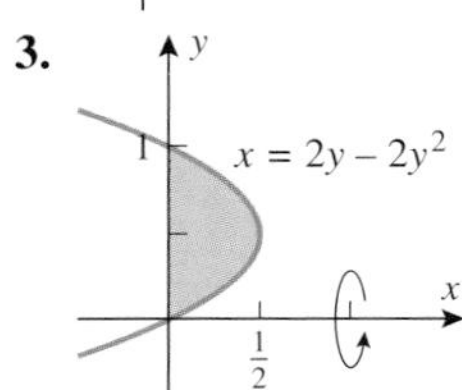

4.
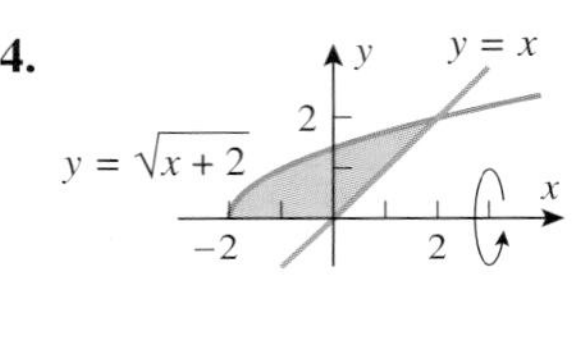

5–10 Use cylindrical shells to find the volume of the solid generated when the region enclosed by the given curves is revolved about the y-axis.

5. $y = x^3,\ x = 1,\ y = 0$
6. $y = \sqrt{x},\ x = 4,\ x = 9,\ y = 0$
7. $y = 1/x,\ y = 0,\ x = 1,\ x = 3$
8. $y = \cos(x^2),\ x = 0,\ x = \frac{1}{2}\sqrt{\pi},\ y = 0$
9. $y = 2x - 1,\ y = -2x + 3,\ x = 2$
10. $y = 2x - x^2,\ y = 0$

11–14 Use cylindrical shells to find the volume of the solid generated when the region enclosed by the given curves is revolved about the x-axis.

11. $y^2 = x,\ y = 1,\ x = 0$
12. $x = 2y,\ y = 2,\ y = 3,\ x = 0$
13. $y = x^2,\ x = 1,\ y = 0$
14. $xy = 4,\ x + y = 5$

[C] 15. Use a CAS to find the volume of the solid generated when the region enclosed by $y = \sin x$ and $y = 0$ for $0 \le x \le \pi$ is revolved about the y-axis.

[C] 16. Use a CAS to find the volume of the solid generated when the region enclosed by $y = \cos x$, $y = 0$, and $x = 0$ for $0 \le x \le \pi/2$ is revolved about the y-axis.

[C] 17. Consider the region to the right of the y-axis, to the left of the vertical line $x = k$ $(0 < k < \pi)$, and between the curve $y = \sin x$ and the x-axis. Use a CAS to estimate the value of k so that the solid generated by revolving the region about the y-axis has a volume of 8 cubic units.

FOCUS ON CONCEPTS

18. Let R_1 and R_2 be regions of the form shown in the accompanying figure. Use cylindrical shells to find a formula for the volume of the solid that results when
    (a) region R_1 is revolved about the y-axis.
    (b) region R_2 is revolved about the x-axis.

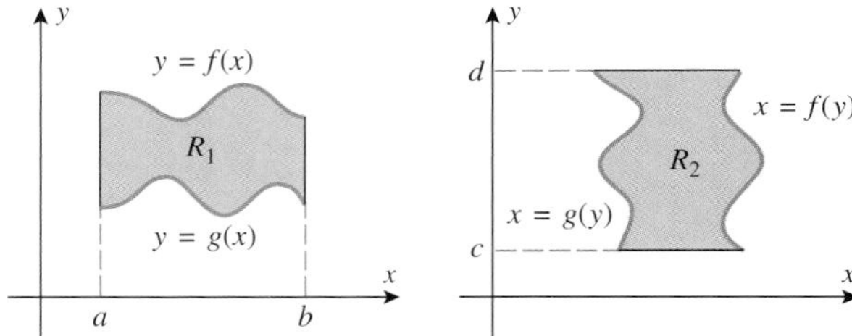

Figure Ex-18

19. (a) Use cylindrical shells to find the volume of the solid that is generated when the region under the curve
$$y = x^3 - 3x^2 + 2x$$
over $[0, 1]$ is revolved about the y-axis.
    (b) For this problem, is the method of cylindrical shells easier or harder than the method of slicing discussed in the last section? Explain.

20. Let f be continuous and nonnegative on $[a, b]$, and let R be the region that is enclosed by $y = f(x)$ and $y = 0$ for $a \le x \le b$. Using the method of cylindrical shells, derive with explanation a formula for the volume of the solid generated by revolving R about the line $x = k$, where $k \le a$.

21. Use cylindrical shells to find the volume of the solid that is generated when the region that is enclosed by $y = 1/x^3$, $x = 1$, $x = 2$, $y = 0$ is revolved about the line $x = -1$.

22. Use cylindrical shells to find the volume of the solid that is generated when the region that is enclosed by $y = x^3$, $y = 1$, $x = 0$ is revolved about the line $y = 1$.

23. Use cylindrical shells to find the volume of the cone generated when the triangle with vertices $(0, 0)$, $(0, r)$, $(h, 0)$, where $r > 0$ and $h > 0$, is revolved about the x-axis.

24. The region enclosed between the curve $y^2 = kx$ and the line $x = \frac{1}{4}k$ is revolved about the line $x = \frac{1}{2}k$. Use cylindrical shells to find the volume of the resulting solid. (Assume $k > 0$.)

25. As shown in the accompanying figure, a cylindrical hole is drilled all the way through the center of a sphere. Show that the volume of the remaining solid depends only on the length L of the hole, not on the size of the sphere.

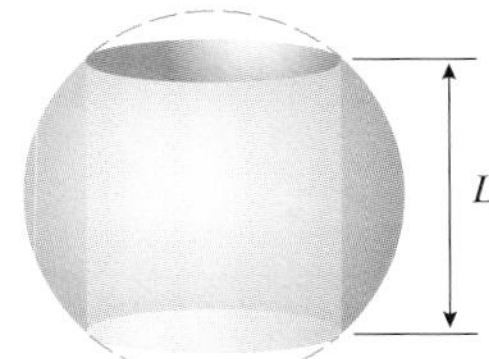

Figure Ex-25

26. Use cylindrical shells to find the volume of the torus obtained by revolving the circle $x^2 + y^2 = a^2$ about the line $x = b$, where $b > a > 0$. [*Hint:* It may help in the integration to think of an integral as an area.]

27. Let V_x and V_y be the volumes of the solids that result when the region enclosed by $y = 1/x$, $y = 0$, $x = \frac{1}{2}$, and $x = b$ $(b > \frac{1}{2})$ is revolved about the x-axis and y-axis, respectively. Is there a value of b for which $V_x = V_y$?

✔ QUICK CHECK ANSWERS 6.3

1. (a) $2\pi x(1 + \sqrt{x})$ (b) $\int_1^4 2\pi x(1 + \sqrt{x})\,dx$ (c) 39.8π **2.** (a) $2\pi(5 - x)(1 + \sqrt{x})$ (b) $\int_1^4 2\pi(5 - x)(1 + \sqrt{x})\,dx$ (c) $\frac{553}{15}\pi$
3. $\int_0^4 2\pi y[4 - (y - 2)^2]\,dy$; $\frac{128}{3}\pi$

6.4 LENGTH OF A PLANE CURVE

In this section we will use the tools of calculus to study the problem of finding the length of a plane curve.

ARC LENGTH

Our first objective is to define what we mean by the ***length*** (also called the ***arc length***) of a plane curve $y = f(x)$ over an interval $[a, b]$ (Figure 6.4.1). Once that is done we will be able to focus on the problem of computing arc lengths. To avoid some complications that would otherwise occur, we will impose the requirement that f' be continuous on $[a, b]$, in which case we will say that $y = f(x)$ is a ***smooth curve*** on $[a, b]$ or that f is a ***smooth function*** on $[a, b]$. Thus, we will be concerned with the following problem.

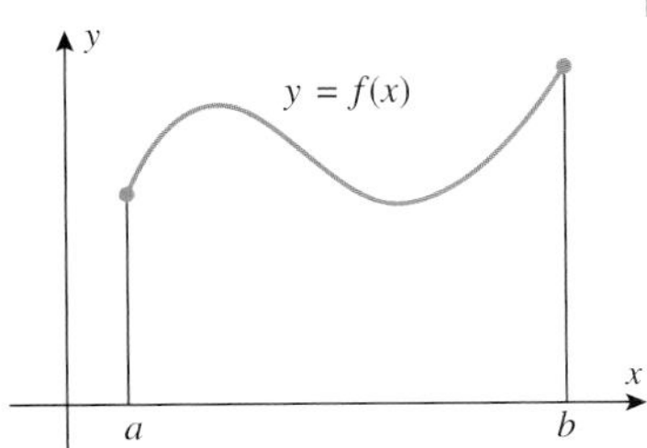

Figure 6.4.1

Intuitively, you might think of the arc length of a curve as the number obtained by aligning a piece of string with the curve and then measuring the length of the string after it is straightened out.

> **6.4.1 ARC LENGTH PROBLEM.** Suppose that $y = f(x)$ is a smooth curve on the interval $[a, b]$. Define and find a formula for the arc length L of the curve $y = f(x)$ over the interval $[a, b]$.

To define the arc length of a curve we start by breaking the curve into small segments. Then we approximate the curve segments by line segments and add the lengths of the line segments to form a Riemann sum. Figure 6.4.2 illustrates how such line segments tend to become better and better approximations to a curve as the number of segments increases. As the number of segments increases the corresponding Riemann sums approach a definite integral whose value we will take to be the arc length L of the curve.

To implement our idea for solving Problem 6.4.1, divide the interval $[a, b]$ into n subintervals by inserting points $x_1, x_2, \ldots, x_{n-1}$ between $a = x_0$ and $b = x_n$. As shown in the

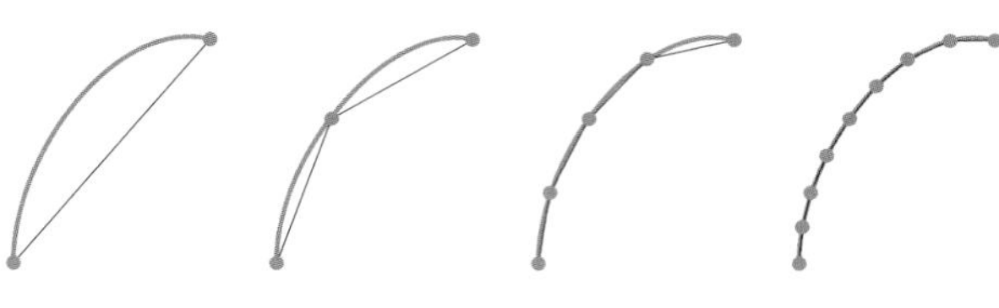

Figure 6.4.2

top part of Figure 6.4.3, let $P_0, P_1, \ldots, P_n$ be the points on the curve with x-coordinates $a = x_0, x_1, x_2, \ldots, x_{n-1}, b = x_n$ and join these points with straight line segments. These line segments form a ***polygonal path*** that we can regard as an approximation to the curve $y = f(x)$. As indicated in the bottom part of Figure 6.4.3, the length L_k of the kth line segment in the polygonal path is

$$L_k = \sqrt{(\Delta x_k)^2 + (\Delta y_k)^2} = \sqrt{(\Delta x_k)^2 + [f(x_k) - f(x_{k-1})]^2} \tag{1}$$

If we now add the lengths of these line segments, we obtain the following approximation to the length L of the curve

$$L \approx \sum_{k=1}^{n} L_k = \sum_{k=1}^{n} \sqrt{(\Delta x_k)^2 + [f(x_k) - f(x_{k-1})]^2} \tag{2}$$

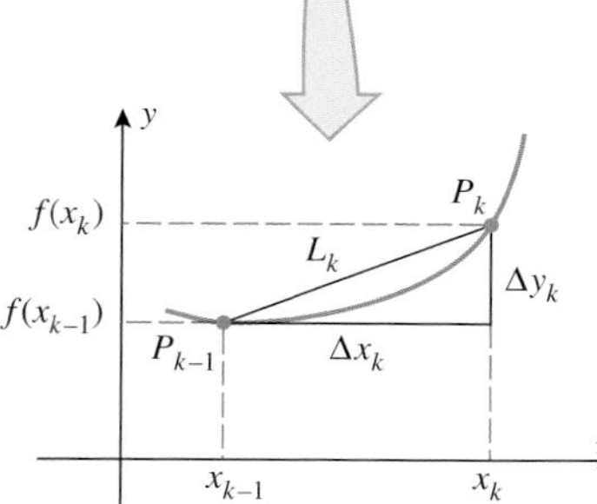

Figure 6.4.3

To put this in the form of a Riemann sum we will apply the Mean-Value Theorem (4.7.2). This theorem implies that there is a point x_k^* between x_{k-1} and x_k such that

$$\frac{f(x_k) - f(x_{k-1})}{x_k - x_{k-1}} = f'(x_k^*) \quad \text{or} \quad f(x_k) - f(x_{k-1}) = f'(x_k^*)\Delta x_k$$

and hence we can rewrite (2) as

$$L \approx \sum_{k=1}^{n} \sqrt{1 + [f'(x_k^*)]^2}\, \Delta x_k$$

Explain why the approximation in (2) cannot be greater than L.

Thus, taking the limit as n increases and the widths of the subintervals approach zero yields the following integral that defines the arc length L:

$$L = \lim_{\max \Delta x_k \to 0} \sum_{k=1}^{n} \sqrt{1 + [f'(x_k^*)]^2}\, \Delta x_k = \int_a^b \sqrt{1 + [f'(x)]^2}\, dx$$

In summary, we have the following definition:

6.4.2 DEFINITION. If $y = f(x)$ is a smooth curve on the interval $[a, b]$, then the arc length L of this curve over $[a, b]$ is defined as

$$L = \int_a^b \sqrt{1 + [f'(x)]^2}\, dx \tag{3}$$

This result provides both a definition and a formula for computing arc lengths. Where convenient, (3) can also be expressed as

$$L = \int_a^b \sqrt{1 + [f'(x)]^2}\, dx = \int_a^b \sqrt{1 + \left(\frac{dy}{dx}\right)^2}\, dx \tag{4}$$

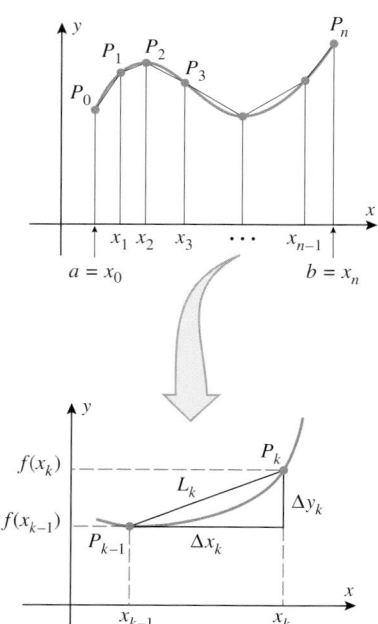

Moreover, for a curve expressed in the form $x = g(y)$, where g' is continuous on $[c, d]$, the arc length L from $y = c$ to $y = d$ can be expressed as

$$L = \int_c^d \sqrt{1 + [g'(y)]^2}\, dy = \int_c^d \sqrt{1 + \left(\frac{dx}{dy}\right)^2}\, dy \tag{5}$$

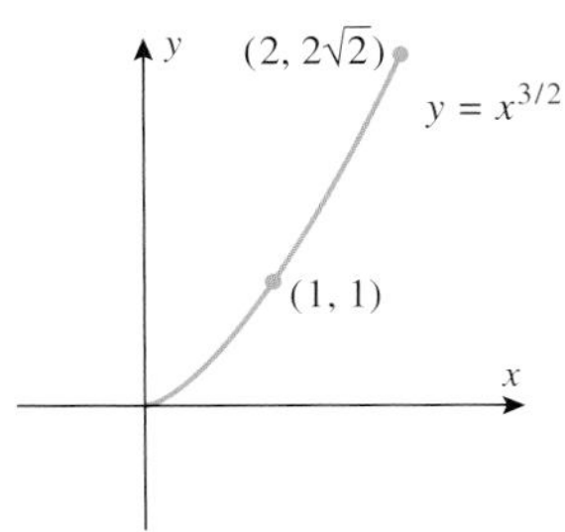

Figure 6.4.4

▶ **Example 1** Find the arc length of the curve $y = x^{3/2}$ from $(1, 1)$ to $(2, 2\sqrt{2})$ (Figure 6.4.4) in two ways: (a) using Formula (4) and (b) using Formula (5).

Solution (a).

$$\frac{dy}{dx} = \frac{3}{2}x^{1/2}$$

and since the curve extends from $x = 1$ to $x = 2$, it follows from (4) that

$$L = \int_1^2 \sqrt{1 + \tfrac{9}{4}x}\, dx$$

To evaluate this integral we make the u-substitution

$$u = 1 + \tfrac{9}{4}x, \quad du = \tfrac{9}{4}\, dx$$

and then change the x-limits of integration ($x = 1, x = 2$) to the corresponding u-limits $\left(u = \frac{13}{4}, u = \frac{22}{4}\right)$:

$$\begin{aligned} L = \frac{4}{9}\int_{13/4}^{22/4} u^{1/2}\, du = \frac{8}{27}u^{3/2}\Bigg]_{13/4}^{22/4} &= \frac{8}{27}\left[\left(\frac{22}{4}\right)^{3/2} - \left(\frac{13}{4}\right)^{3/2}\right] \\ &= \frac{22\sqrt{22} - 13\sqrt{13}}{27} \approx 2.09 \end{aligned}$$

Solution (b). To apply Formula (5) we must first rewrite the equation $y = x^{3/2}$ so that x is expressed as a function of y. This yields $x = y^{2/3}$ and

$$\frac{dx}{dy} = \frac{2}{3}y^{-1/3}$$

Since the curve extends from $y = 1$ to $y = 2\sqrt{2}$, it follows from (5) that

$$L = \int_1^{2\sqrt{2}} \sqrt{1 + \tfrac{4}{9}y^{-2/3}}\, dy = \frac{1}{3}\int_1^{2\sqrt{2}} y^{-1/3}\sqrt{9y^{2/3} + 4}\, dy$$

To evaluate this integral we make the u-substitution

$$u = 9y^{2/3} + 4, \quad du = 6y^{-1/3}\, dy$$

and change the y-limits of integration ($y = 1, y = 2\sqrt{2}$) to the corresponding u-limits ($u = 13, u = 22$). This gives

$$L = \frac{1}{18}\int_{13}^{22} u^{1/2}\, du = \frac{1}{27}u^{3/2}\Bigg]_{13}^{22} = \frac{1}{27}[(22)^{3/2} - (13)^{3/2}] = \frac{22\sqrt{22} - 13\sqrt{13}}{27}$$

This result agrees with that in part (a); however, the integration here is more tedious. In problems where there is a choice between using (4) or (5), it is often the case that one of the formulas leads to a simpler integral than the other. ◀

The arc from the point $(1, 1)$ to the point $(2, 2\sqrt{2})$ in Figure 6.4.4 is nearly a straight line, so the arc length should be only slightly larger than the straight-line distance between these points. Show that this is so.

FINDING ARC LENGTH BY NUMERICAL METHODS

In Chapter 8 we will develop some techniques of integration that will enable us to find exact values of more integrals encountered in arc length calculations; however, generally speaking, most such integrals are impossible to evaluate in terms of elementary functions. In these cases one usually approximates the integral using a numerical method such as the midpoint rule discussed in Section 5.4.

TECHNOLOGY MASTERY

If your calculating utility has a numerical integration capability, use it to confirm that the arc length L in Example 2 is approximately $L \approx 3.8202$.

► Example 2 From (4), the arc length of $y = \sin x$ from $x = 0$ to $x = \pi$ is given by the integral

$$L = \int_0^{\pi} \sqrt{1 + (\cos x)^2}\, dx$$

This integral cannot be evaluated in terms of elementary functions; however, using a calculating utility with a numerical integration capability yields the approximation $L \approx 3.8202$. ◄

ARC LENGTH OF PARAMETRIC CURVES

The following result provides a formula for finding the arc length of a curve from parametric equations for the curve. Its derivation is similar to that of Formula (3) and will be omitted.

Formulas (4) and (5) can be viewed as special cases of (6). For example, Formula (4) can be obtained from (6) by writing $y = f(x)$ parametrically as

$$x = t, \quad y = f(t)$$

and Formula (5) can be obtained by writing $x = g(y)$ parametrically as

$$x = g(t), \quad y = t$$

(see Exercise 14).

6.4.3 ARC LENGTH FORMULA FOR PARAMETRIC CURVES. If no segment of the curve represented by the parametric equations

$$x = x(t), \quad y = y(t) \qquad (a \le t \le b)$$

is traced more than once as t increases from a to b, and if dx/dt and dy/dt are continuous functions for $a \le t \le b$, then the arc length L of the curve is given by

$$L = \int_a^b \sqrt{\left(\frac{dx}{dt}\right)^2 + \left(\frac{dy}{dt}\right)^2}\, dt \tag{6}$$

► Example 3 Use (6) to find the circumference of a circle of radius a from the parametric equations

$$x = a\cos t, \quad y = a\sin t \qquad (0 \le t \le 2\pi)$$

Solution.

$$L = \int_0^{2\pi} \sqrt{\left(\frac{dx}{dt}\right)^2 + \left(\frac{dy}{dt}\right)^2}\, dt = \int_0^{2\pi} \sqrt{(-a\sin t)^2 + (a\cos t)^2}\, dt$$

$$= \int_0^{2\pi} a\, dt = at\Big]_0^{2\pi} = 2\pi a \;◄$$

✔QUICK CHECK EXERCISES 6.4 (See page 408 for answers.)

1. A function f is smooth on $[a, b]$ if f' is ______ on $[a, b]$.
2. If a function f is smooth on $[a, b]$, then the length of the curve $y = f(x)$ over $[a, b]$ is ______.
3. The distance between points $(1, 0)$ and $(\pi, 1)$ is ______.
4. Let L be the length of the curve $y = x^2$ from $(0, 0)$ to $(2, 4)$.
   (a) Integrating with respect to x, an integral expression for L is ______.
   (b) Integrating with respect to y, an integral expression for L is ______.

EXERCISE SET 6.4 Graphing Utility [C] CAS

1. Use the Theorem of Pythagoras to find the length of the line segment $y = 2x$ from $(1, 2)$ to $(2, 4)$, and confirm that the value is consistent with the length computed using
(a) Formula (4) (b) Formula (5).

2. Use the Theorem of Pythagoras to find the length of the line segment $x = t$, $y = 5t$ $(0 \le t \le 1)$, and confirm that the value is consistent with the length computed using Formula (6).

3–8 Find the exact arc length of the curve over the stated interval.

3. $y = 3x^{3/2} - 1$ from $x = 0$ to $x = 1$
4. $x = \frac{1}{3}(y^2 + 2)^{3/2}$ from $y = 0$ to $y = 1$
5. $y = x^{2/3}$ from $x = 1$ to $x = 8$
6. $y = (x^6 + 8)/(16x^2)$ from $x = 2$ to $x = 3$
7. $24xy = y^4 + 48$ from $y = 2$ to $y = 4$
8. $x = \frac{1}{8}y^4 + \frac{1}{4}y^{-2}$ from $y = 1$ to $y = 4$

9–12 Find the exact arc length of the parametric curve without eliminating the parameter.

9. $x = \frac{1}{3}t^3$, $y = \frac{1}{2}t^2$ $(0 \le t \le 1)$
10. $x = (1 + t)^2$, $y = (1 + t)^3$ $(0 \le t \le 1)$
11. $x = \cos 2t$, $y = \sin 2t$ $(0 \le t \le \pi/2)$
12. $x = \cos t + t \sin t$, $y = \sin t - t \cos t$ $(0 \le t \le \pi)$

FOCUS ON CONCEPTS

13. Consider the curve $y = x^{2/3}$.
(a) Sketch the portion of the curve between $x = -1$ and $x = 8$.
(b) Explain why Formula (4) cannot be used to find the arc length of the curve sketched in part (a).
(c) Find the arc length of the curve sketched in part (a).

14. Derive Formulas (4) and (5) from Formula (6) by choosing appropriate parametrizations of the curves.

15. Consider the curve segments $y = x^2$ from $x = \frac{1}{2}$ to $x = 2$ and $y = \sqrt{x}$ from $x = \frac{1}{4}$ to $x = 4$.
(a) Graph the two curve segments and use your graphs to explain why the lengths of these two curve segments should be equal.
(b) Set up integrals that give the arc lengths of the curve segments by integrating with respect to x. Demonstrate a substitution that verifies that these two integrals are equal.
(c) Set up integrals that give the arc lengths of the curve segments by integrating with respect to y.
(d) Approximate the arc length of each curve segment using Formula (2) with $n = 10$ equal subintervals.
(e) Which of the two approximations in part (d) is more accurate? Explain.
(f) Use the midpoint approximation with $n = 10$ subintervals to approximate each arc length integral in part (b).
(g) Use a calculating utility with numerical integration capabilities to approximate the arc length integrals in part (b) to four decimal places.

16. Follow the directions of Exercise 15 for the curve segments $y = x^{8/3}$ from $x = 10^{-3}$ to $x = 1$ and $y = x^{3/8}$ from $x = 10^{-8}$ to $x = 1$.

17. Follow the directions of Exercise 15 for the curve segment $y = 1 + 1/x$ from $x = 1$ to $x = 3$ and for the curve segment $y = 1/(x - 1)$ from $x = 4/3$ to $x = 2$.

18. Let $y = f(x)$ be a smooth curve on the closed interval $[a, b]$. Prove that if m and M are nonnegative numbers such that $m \le |f'(x)| \le M$ for all x in $[a, b]$, then the arc length L of $y = f(x)$ over the interval $[a, b]$ satisfies the inequalities

$$(b - a)\sqrt{1 + m^2} \le L \le (b - a)\sqrt{1 + M^2}$$

19. Use the result of Exercise 18 to show that the arc length L of $y = \sec x$ over the interval $0 \le x \le \pi/3$ satisfies

$$\frac{\pi}{3} \le L \le \frac{\pi}{3}\sqrt{13}$$

[C] 20. A basketball player makes a successful shot from the free throw line. Suppose that the path of the ball from the moment of release to the moment it enters the hoop is described by

$$y = 2.15 + 2.09x - 0.41x^2, \quad 0 \le x \le 4.6$$

where x is the horizontal distance (in meters) from the point of release, and y is the vertical distance (in meters) above the floor. Use a CAS or a scientific calculator with numerical integration capabilities to approximate the distance the ball travels from the moment it is released to the moment it enters the hoop. Round your answer to two decimal places.

[C] 21. Find a positive value of k (to two decimal places) such that the curve $y = k \sin x$ has an arc length of $L = 5$ units over the interval from $x = 0$ to $x = \pi$. [*Hint:* Find an integral for the arc length L in terms of k, and then use a CAS or a scientific calculator with a numeric integration capability to find integer values of k at which the values of $L - 5$ have opposite signs. Complete the solution by using the Intermediate-Value Theorem (2.5.8) to approximate the value of k to two decimal places.]

[C] 22. As shown in the accompanying figure (next page), a horizontal beam with dimensions 2 in × 6 in × 16 ft is fixed at both ends and is subjected to a uniformly distributed load of 120 lb/ft. As a result of the load, the centerline of the beam undergoes a deflection that is described by

$$y = -1.67 \times 10^{-8}(x^4 - 2Lx^3 + L^2x^2)$$

($0 \le x \le 192$), where $L = 192$ inches is the length of the unloaded beam, x is the horizontal distance along the beam measured in inches from the left end, and y is the deflection of the centerline in inches.

(a) Graph y versus x for $0 \le x \le 192$.

(b) Find the maximum deflection of the centerline.

(c) Use a CAS or a calculator with a numerical integration capability to find the length of the centerline of the loaded beam. Round your answer to two decimal places.

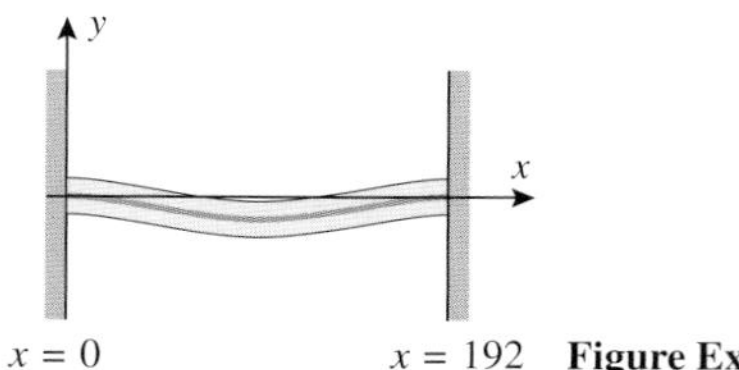

Figure Ex-22

C **23.** A golfer makes a successful chip shot to the green. Suppose that the path of the ball from the moment it is struck to the moment it hits the green is described by

$$y = 12.54x - 0.41x^2$$

where x is the horizontal distance (in yards) from the point where the ball is struck, and y is the vertical distance (in yards) above the fairway. Use a CAS or a calculating utility with a numerical integration capability to find the distance the ball travels from the moment it is struck to the moment it hits the green. Assume that the fairway and green are at the same level and round your answer to two decimal places.

C **24.** (a) Recall from Section 1.7 that a cycloid is the path traced by a point on the rim of a wheel that rolls along a line (Figure 1.7.14). Use the parametric equations in Formula (11) of that section to show that the length L of one arch of a cycloid is given by the integral

$$L = a\int_0^{2\pi} \sqrt{2(1 - \cos\theta)}\, d\theta$$

(b) Use a CAS to show that L is eight times the radius of the wheel (see the accompanying figure).

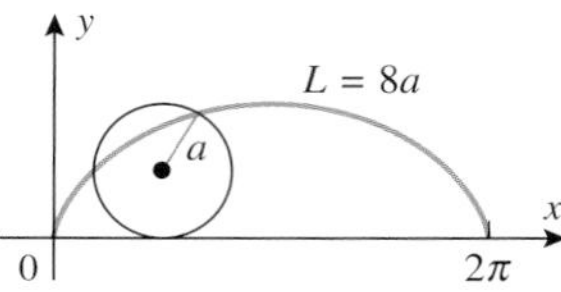

Figure Ex-24

25. It was stated in Exercise 45 of Section 1.7 that the curve given parametrically by the equations

$$x = a\cos^3\phi, \quad y = a\sin^3\phi$$

is called a *four-cusped hypocycloid* (also called an ***astroid***).

(a) Use a graphing utility to generate the graph in the case where $a = 1$, so that it is traced exactly once.

(b) Find the exact arc length of the curve in part (a).

26. Show that the total arc length of the ellipse $x = a\cos t$, $y = b\sin t$, $0 \le t \le 2\pi$ for $a > b > 0$ is given by

$$4a\int_0^{\pi/2} \sqrt{1 - k^2\cos^2 t}\, dt$$

where $k = \sqrt{a^2 - b^2}/a$.

C **27.** (a) Show that the total arc length of the ellipse

$$x = 2\cos t, \quad y = \sin t \quad (0 \le t \le 2\pi)$$

is given by

$$4\int_0^{\pi/2} \sqrt{1 + 3\sin^2 t}\, dt$$

(b) Use a CAS or a scientific calculator with numerical integration capabilities to approximate the arc length in part (a). Round your answer to two decimal places.

(c) Suppose that the parametric equations in part (a) describe the path of a particle moving in the xy-plane, where t is time in seconds and x and y are in centimeters. Use a CAS or a scientific calculator with numerical integration capabilities to approximate the distance traveled by the particle from $t = 1.5$ s to $t = 4.8$ s. Round your answer to two decimal places.

✔ QUICK CHECK ANSWERS 6.4

1. continuous 2. $\int_a^b \sqrt{1 + [f'(x)]^2}\, dx$ 3. $\sqrt{(\pi - 1)^2 + 1}$ 4. (a) $\int_0^2 \sqrt{1 + 4x^2}\, dx$ (b) $\int_0^4 \sqrt{1 + \frac{1}{4y}}\, dy$

6.5 AREA OF A SURFACE OF REVOLUTION

In this section we will consider the problem of finding the area of a surface that is generated by revolving a plane curve about a line.

SURFACE AREA

A ***surface of revolution*** is a surface that is generated by revolving a plane curve about an axis that lies in the same plane as the curve. For example, the surface of a sphere can be generated by revolving a semicircle about its diameter, and the lateral surface of a right circular cylinder can be generated by revolving a line segment about an axis that is parallel to it (Figure 6.5.1).

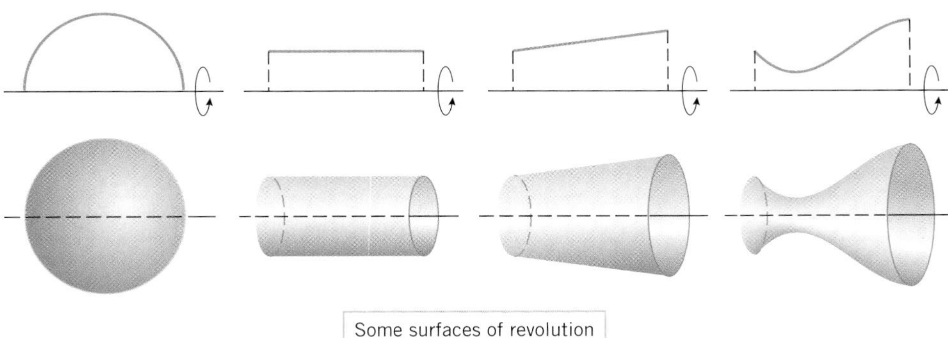

Figure 6.5.1

In this section we will be concerned with the following problem.

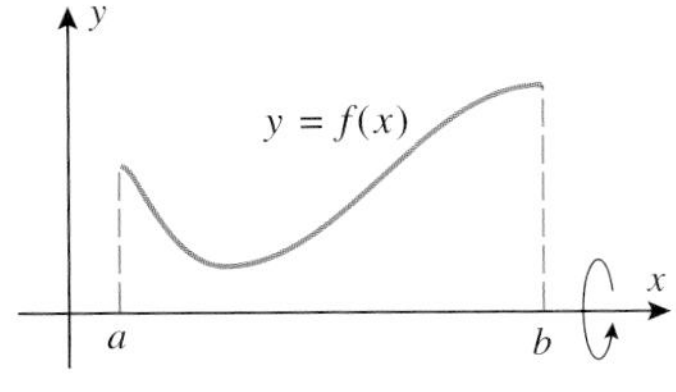

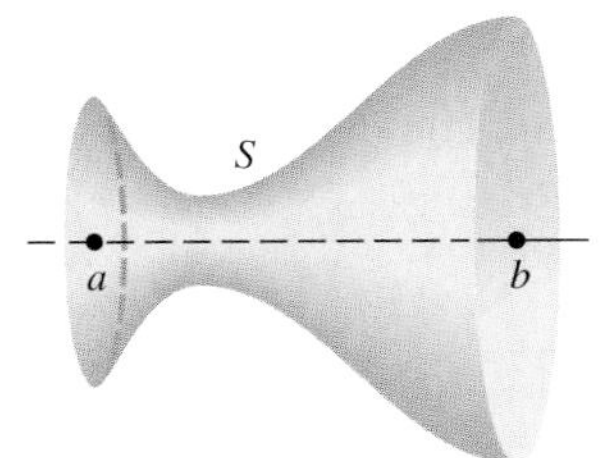

Figure 6.5.2

6.5.1 SURFACE AREA PROBLEM. Suppose that f is a smooth, nonnegative function on $[a, b]$ and that a surface of revolution is generated by revolving the portion of the curve $y = f(x)$ between $x = a$ and $x = b$ about the x-axis (Figure 6.5.2). Define what is meant by the *area* S of the surface, and find a formula for computing it.

To motivate an appropriate definition for the area S of a surface of revolution, we will decompose the surface into small sections whose areas can be approximated by elementary formulas, add the approximations of the areas of the sections to form a Riemann sum that approximates S, and then take the limit of the Riemann sums to obtain an integral for the exact value of S.

To implement this idea, divide the interval $[a, b]$ into n subintervals by inserting points $x_1, x_2, \ldots, x_{n-1}$ between $a = x_0$ and $b = x_n$. As illustrated in Figure 6.5.3a, the corresponding points on the graph of f define a polygonal path that approximates the curve $y = f(x)$ over the interval $[a, b]$. When this polygonal path is revolved about the x-axis, it generates a surface consisting of n parts, each of which is a frustum of a right circular cone (Figure 6.5.3b). Thus, the area of each part of the approximating surface can be obtained from the formula

$$S = \pi(r_1 + r_2)l \tag{1}$$

for the lateral area S of a frustum of slant height l and base radii r_1 and r_2 (Figure 6.5.4). As suggested by Figure 6.5.5, the kth frustum has radii $f(x_{k-1})$ and $f(x_k)$ and height Δx_k.

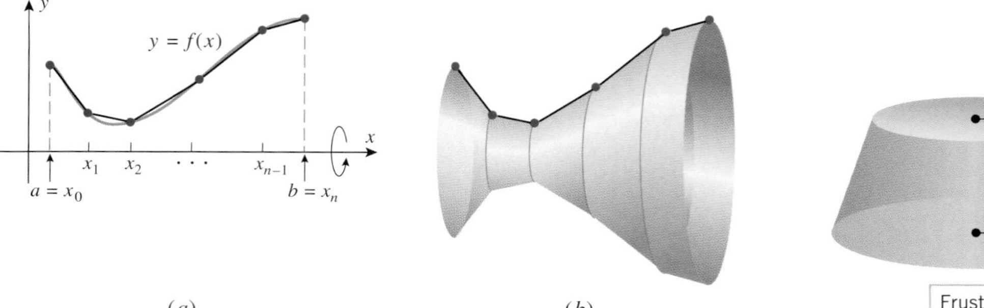

Figure 6.5.3

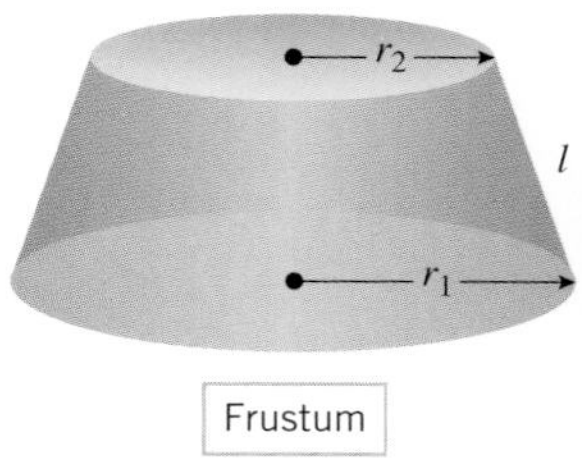

Figure 6.5.4

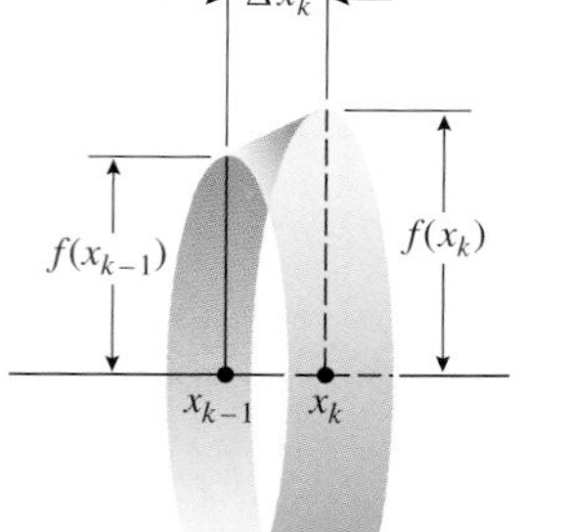

Figure 6.5.5

Its slant height is the length L_k of the kth line segment in the polygonal path, which from Formula (1) of Section 6.4 is

$$L_k = \sqrt{(\Delta x_k)^2 + [f(x_k) - f(x_{k-1})]^2}$$

Thus, the lateral area S_k of the kth frustum is

$$S_k = \pi[f(x_{k-1}) + f(x_k)]\sqrt{(\Delta x_k)^2 + [f(x_k) - f(x_{k-1})]^2}$$

If we add these areas, we obtain the following approximation to the area S of the entire surface:

$$S \approx \sum_{k=1}^{n} \pi[f(x_{k-1}) + f(x_k)]\sqrt{(\Delta x_k)^2 + [f(x_k) - f(x_{k-1})]^2} \tag{2}$$

To put this in the form of a Riemann sum we will apply the Mean-Value Theorem (4.7.2). This theorem implies that there is a point x_k^* between x_{k-1} and x_k such that

$$\frac{f(x_k) - f(x_{k-1})}{x_k - x_{k-1}} = f'(x_k^*) \quad \text{or} \quad f(x_k) - f(x_{k-1}) = f'(x_k^*)\Delta x_k$$

and hence we can rewrite (2) as

$$S \approx \sum_{k=1}^{n} \pi[f(x_{k-1}) + f(x_k)]\sqrt{1 + [f'(x_k^*)]^2}\, \Delta x_k \tag{3}$$

However, this is not yet a Riemann sum because it involves the variables x_{k-1} and x_k. To eliminate these variables from the expression, observe that the average value of the numbers $f(x_{k-1})$ and $f(x_k)$ lies between these numbers, so the continuity of f and the Intermediate-Value Theorem (2.5.8) imply that there is a point x_k^{**} between x_{k-1} and x_k such that

$$\tfrac{1}{2}[f(x_{k-1}) + f(x_k)] = f(x_k^{**})$$

Thus, (2) can be expressed as

$$S \approx \sum_{k=1}^{n} 2\pi f(x_k^{**})\sqrt{1 + [f'(x_k^*)]^2}\, \Delta x_k$$

Although this expression is close to a Riemann sum in form, it is not a true Riemann sum because it involves two variables x_k^* and x_k^{**}, rather than x_k^* alone. However, it is proved in advanced calculus courses that this has no effect on the limit because of the continuity of f. Thus, we can assume that $x_k^{**} = x_k^*$ when taking the limit, and this suggests that S can be defined as

$$S = \lim_{\max \Delta x_k \to 0} \sum_{k=1}^{n} 2\pi f(x_k^{**})\sqrt{1 + [f'(x_k^*)]^2}\, \Delta x_k = \int_a^b 2\pi f(x)\sqrt{1 + [f'(x)]^2}\, dx$$

In summary, we have the following definition.

6.5.2 DEFINITION. If f is a smooth, nonnegative function on $[a, b]$, then the surface area S of the surface of revolution that is generated by revolving the portion of the curve $y = f(x)$ between $x = a$ and $x = b$ about the x-axis is defined as

$$S = \int_a^b 2\pi f(x)\sqrt{1 + [f'(x)]^2}\, dx$$

This result provides both a definition and a formula for computing surface areas. Where convenient, this formula can also be expressed as

$$S = \int_a^b 2\pi f(x)\sqrt{1 + [f'(x)]^2}\, dx = \int_a^b 2\pi y\sqrt{1 + \left(\frac{dy}{dx}\right)^2}\, dx \tag{4}$$

Moreover, if g is nonnegative and $x = g(y)$ is a smooth curve on the interval $[c, d]$, then the area of the surface that is generated by revolving the portion of a curve $x = g(y)$ between $y = c$ and $y = d$ about the y-axis can be expressed as

$$S = \int_c^d 2\pi g(y)\sqrt{1 + [g'(y)]^2}\, dy = \int_c^d 2\pi x\sqrt{1 + \left(\frac{dx}{dy}\right)^2}\, dy \tag{5}$$

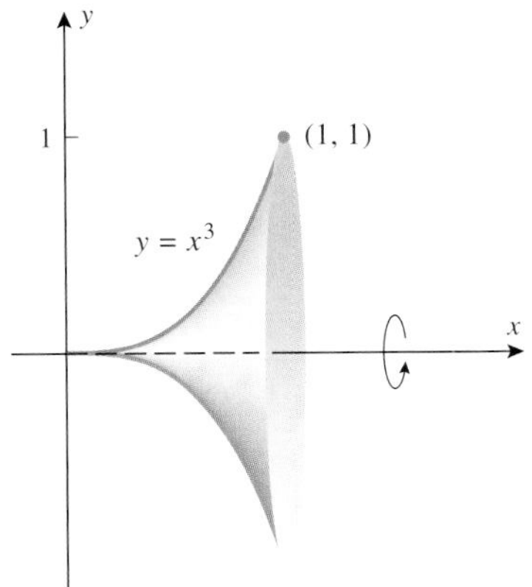

Figure 6.5.6

▶ **Example 1** Find the area of the surface that is generated by revolving the portion of the curve $y = x^3$ between $x = 0$ and $x = 1$ about the x-axis (Figure 6.5.6).

Solution. Since $y = x^3$, we have $dy/dx = 3x^2$, and hence from (4) the surface area S is

$$\begin{aligned}
S &= \int_0^1 2\pi y\sqrt{1 + \left(\frac{dy}{dx}\right)^2}\, dx \\
&= \int_0^1 2\pi x^3\sqrt{1 + (3x^2)^2}\, dx \\
&= 2\pi \int_0^1 x^3(1 + 9x^4)^{1/2}\, dx \\
&= \frac{2\pi}{36} \int_1^{10} u^{1/2}\, du \qquad \begin{bmatrix} u = 1 + 9x^4 \\ du = 36x^3\, dx \end{bmatrix} \\
&= \frac{2\pi}{36} \cdot \frac{2}{3} u^{3/2}\bigg]_{u=1}^{10} = \frac{\pi}{27}(10^{3/2} - 1) \approx 3.56 \quad ◀
\end{aligned}$$

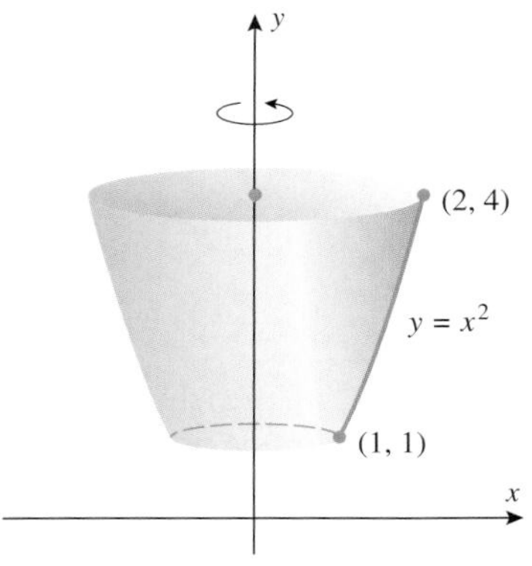

Figure 6.5.7

▶ **Example 2** Find the area of the surface that is generated by revolving the portion of the curve $y = x^2$ between $x = 1$ and $x = 2$ about the y-axis (Figure 6.5.7).

Solution. Because the curve is revolved about the y-axis we will apply Formula (5). Toward this end, we rewrite $y = x^2$ as $x = \sqrt{y}$ and observe that the y-values corresponding to $x = 1$ and $x = 2$ are $y = 1$ and $y = 4$. Since $x = \sqrt{y}$, we have $dx/dy = 1/(2\sqrt{y})$, and hence from (5) the surface area S is

$$\begin{aligned} S &= \int_1^4 2\pi x \sqrt{1 + \left(\frac{dx}{dy}\right)^2}\, dy \\ &= \int_1^4 2\pi\sqrt{y}\sqrt{1 + \left(\frac{1}{2\sqrt{y}}\right)^2}\, dy \\ &= \pi \int_1^4 \sqrt{4y+1}\, dy \\ &= \frac{\pi}{4}\int_5^{17} u^{1/2}\, du \qquad \boxed{\begin{array}{l} u = 4y+1 \\ du = 4\,dy \end{array}} \\ &= \frac{\pi}{4}\cdot\frac{2}{3}u^{3/2}\Big]_{u=5}^{17} = \frac{\pi}{6}(17^{3/2} - 5^{3/2}) \approx 30.85 \end{aligned}$$

◀

✔ QUICK CHECK EXERCISES 6.5 (See page 414 for answers.)

1. If f is a smooth, nonnegative function on $[a, b]$, then the surface area S of the surface of revolution generated by revolving the portion of the curve $y = f(x)$ between $x = a$ and $x = b$ about the x-axis is ______.
2. The lateral area of the frustum with slant height $\sqrt{10}$ and base radii $r_1 = 1$ and $r_2 = 2$ is ______.
3. An integral expression for the area of the surface generated by rotating the line segment joining (3, 1) and (6, 2) about the x-axis is ______. The value of this integral is ______.
4. An integral expression for the area of the surface generated by rotating the line segment joining (3, 1) and (6, 2) about the y-axis is ______. The value of this integral is ______.

EXERCISE SET 6.5 [C] CAS

1–4 Find the area of the surface generated by revolving the given curve about the x-axis.

1. $y = 7x,\ 0 \le x \le 1$
2. $y = \sqrt{x},\ 1 \le x \le 4$
3. $y = \sqrt{4 - x^2},\ -1 \le x \le 1$
4. $x = \sqrt[3]{y},\ 1 \le y \le 8$

5–8 Find the area of the surface generated by revolving the given curve about the y-axis.

5. $x = 9y + 1,\ 0 \le y \le 2$
6. $x = y^3,\ 0 \le y \le 1$
7. $x = \sqrt{9 - y^2},\ -2 \le y \le 2$
8. $x = 2\sqrt{1 - y},\ -1 \le y \le 0$

9–12 Use a CAS to find the exact area of the surface generated by revolving the curve about the stated axis.

[C] 9. $y = \sqrt{x} - \frac{1}{3}x^{3/2},\ 1 \le x \le 3$; x-axis

[C] 10. $y = \frac{1}{3}x^3 + \frac{1}{4}x^{-1},\ 1 \le x \le 2$; x-axis

[C] 11. $8xy^2 = 2y^6 + 1,\ 1 \le y \le 2$; y-axis

[C] 12. $x = \sqrt{16 - y},\ 0 \le y \le 15$; y-axis

13–14 Use a CAS or a calculating utility with numerical integration capabilities to approximate the area of the surface generated by revolving the curve about the stated axis. Round your answer to two decimal places.

C **13.** $y = x \sin x,\ 0 \le x \le \pi$; x-axis

C **14.** $x = \tan y,\ 0 \le y \le \pi/4$; y-axis

15–16 Approximate the area of the surface using Formula (2) with $n = 20$ subintervals of equal width. Round your answer to two decimal places.

15. The surface of Exercise 13.

16. The surface of Exercise 14.

17. Use Formula (4) to show that the lateral area S of a right circular cone with height h and base radius r is

$$S = \pi r\sqrt{r^2 + h^2}$$

18. Show that the area of the surface of a sphere of radius r is $4\pi r^2$. [*Hint:* Revolve the semicircle $y = \sqrt{r^2 - x^2}$ about the x-axis.]

FOCUS ON CONCEPTS

19. (a) The figure in Exercise 41 of Section 6.2 shows a spherical cap of height h cut from a sphere of radius r. Show that the surface area S of the cap is $S = 2\pi rh$. [*Hint:* Revolve an appropriate portion of the circle $x^2 + y^2 = r^2$ about the y-axis.]

(b) The portion of a sphere that is cut by two parallel planes is called a ***zone***. Use the result in part (a) to show that the surface area of a zone depends on the radius of the sphere and the distance between the planes, but not on the location of the zone.

20. (a) If a cone of slant height l and base radius r is cut along a lateral edge and laid flat, then as shown in the accompanying figure it becomes a sector of a circle of radius l. Use the formula $A = \frac{1}{2}l^2\theta$ for the area of a sector with radius l and central angle θ (in radians) to show that the lateral surface area of the cone is πrl.

(b) Use the result in part (a) to obtain Formula (1) for the lateral surface area of a frustum.

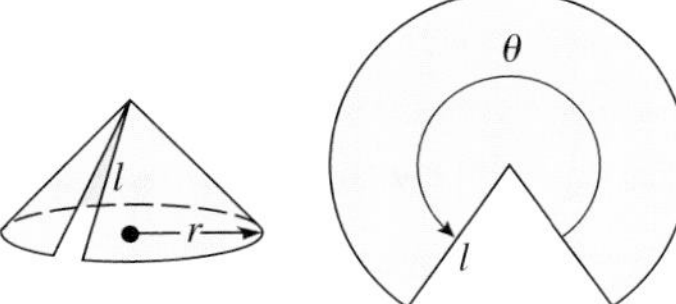

Figure Ex-20

21. Assume that $y = f(x)$ is a smooth curve on the interval $[a, b]$ and assume that $f(x) \ge 0$ for $a \le x \le b$. Derive a formula for the surface area generated when the curve $y = f(x)$, $a \le x \le b$, is revolved about the line $y = -k$ $(k > 0)$.

22. Let $y = f(x)$ be a smooth curve on the interval $[a, b]$ and assume that $f(x) \ge 0$ for $a \le x \le b$. By the Extreme-Value Theorem (4.4.2), the function f has a maximum value K and a minimum value k on $[a, b]$. Prove: If L is the arc length of the curve $y = f(x)$ between $x = a$ and $x = b$, and if S is the area of the surface that is generated by revolving this curve about the x-axis, then

$$2\pi kL \le S \le 2\pi KL$$

23. Use the results of Exercise 22 above and Exercise 19 in Section 6.4 to show that the area S of the surface generated by revolving the curve $y = \sec x$, $0 \le x \le \pi/3$, about the x-axis satisfies

$$\frac{2\pi^2}{3} \le S \le \frac{4\pi^2}{3}\sqrt{13}$$

24. Let $y = f(x)$ be a smooth curve on $[a, b]$ and assume that $f(x) \ge 0$ for $a \le x \le b$. Let A be the area under the curve $y = f(x)$ between $x = a$ and $x = b$, and let S be the area of the surface obtained when this section of curve is revolved about the x-axis.

(a) Prove that $2\pi A \le S$.

(b) For what functions f is $2\pi A = S$?

25–26 For these exercises, divide the interval $[a, b]$ into n subintervals by inserting points $t_1, t_2, \ldots, t_{n-1}$ between $a = t_0$ and $b = t_n$, and assume that $x'(t)$ and $y'(t)$ are continuous functions and that no segment of the curve

$$x = x(t), \quad y = y(t) \quad (a \le t \le b)$$

is traced more than once.

25. Let S be the area of the surface generated by revolving the curve $x = x(t)$, $y = y(t)$ $(a \le t \le b)$ about the x-axis. Explain how S can be approximated by

$$S \approx \sum_{k=1}^{n} (\pi[y(t_{k-1}) + y(t_k)] \times \sqrt{[x(t_k) - x(t_{k-1})]^2 + [y(t_k) - y(t_{k-1})]^2}\,)$$

Using results from advanced calculus, it can be shown that as $\max \Delta t_k \to 0$, this sum converges to

$$S = \int_a^b 2\pi y(t)\sqrt{[x'(t)]^2 + [y'(t)]^2}\,dt \qquad \text{(A)}$$

26. Let S be the area of the surface generated by revolving the curve $x = x(t)$, $y = y(t)$ $(a \le t \le b)$ about the y-axis. Explain how S can be approximated by

$$S \approx \sum_{k=1}^{n} (\pi[x(t_{k-1}) + x(t_k)] \times \sqrt{[x(t_k) - x(t_{k-1})]^2 + [y(t_k) - y(t_{k-1})]^2}\,)$$

Using results from advanced calculus, it can be shown that as $\max \Delta t_k \to 0$, this sum converges to

$$S = \int_a^b 2\pi x(t)\sqrt{[x'(t)]^2 + [y'(t)]^2}\,dt \qquad \text{(B)}$$

27–33 Use Formulas (A) and (B) from Exercises 25 and 26.

27. Find the area of the surface generated by revolving the parametric curve $x = t^2$, $y = 2t$ $(0 \le t \le 4)$ about the x-axis.

C **28.** Use a CAS to find the area of the surface generated by revolving the parametric curve

$$x = \cos^2 t, \quad y = 5 \sin t \qquad (0 \le t \le \pi/2)$$

about the x-axis.

29. Find the area of the surface generated by revolving the parametric curve $x = t$, $y = 2t^2$ $(0 \le t \le 1)$ about the y-axis.

30. Find the area of the surface generated by revolving the parametric curve $x = \cos^2 t$, $y = \sin^2 t$ $(0 \le t \le \pi/2)$ about the y-axis.

31. By revolving the semicircle

$$x = r \cos t, \quad y = r \sin t \qquad (0 \le t \le \pi)$$

about the x-axis, show that the surface area of a sphere of radius r is $4\pi r^2$.

32. The equations

$$x = a\phi - a \sin \phi, \quad y = a - a \cos \phi \qquad (0 \le \phi \le 2\pi)$$

represent one arch of a cycloid. Show that the surface area generated by revolving this curve about the x-axis is $S = 64\pi a^2/3$. [*Hint:* Use the identities

$$\sin^2 \frac{\phi}{2} = \frac{1 - \cos \phi}{2} \quad \text{and} \quad \sin^3 \phi = (1 - \cos^2 \phi) \sin \phi$$

to help with the integration.]

33. Derive Formulas (4) and (5) from Formulas (A) and (B) in Exercises 25 and 26 by choosing appropriate parametrizations for the curves $y = f(x)$ and $x = g(y)$.

✔ QUICK CHECK ANSWERS 6.5

1. $\int_a^b 2\pi f(x)\sqrt{1 + [f'(x)]^2}\, dx$ 2. $3\sqrt{10}\pi$ 3. $\int_3^6 (2\pi)\left(\frac{x}{3}\right)\sqrt{\frac{10}{9}}\, dx = \int_3^6 \frac{2\sqrt{10}\pi}{9} x\, dx$; $3\sqrt{10}\pi$
4. $\int_1^2 (2\pi)(3y)\sqrt{10}\, dy$; $9\sqrt{10}\pi$

6.6 AVERAGE VALUE OF A FUNCTION AND ITS APPLICATIONS

In this section we will define the notion of the "average value" of a function, and we will give various applications of this idea.

AVERAGE VALUE OF A CONTINUOUS FUNCTION

In scientific work, numerical information is often summarized by an *average value* or *mean value* of the observed data. There are various kinds of averages, but the most common is the ***arithmetic mean*** or ***arithmetic average***, which is formed by adding the data and dividing by the number of data points. Thus, the arithmetic average $\overline{a}$ of n numbers $a_1, a_2, \ldots, a_n$ is

$$\overline{a} = \frac{1}{n}(a_1 + a_2 + \cdots + a_n) = \frac{1}{n}\sum_{k=1}^{n} a_k$$

In the case where the a_k's are values of a function f, say,

$$a_1 = f(x_1),\ a_2 = f(x_2),\ \ldots,\ a_n = f(x_n)$$

then the arithmetic average $\overline{a}$ of these function values is

$$\overline{a} = \frac{1}{n}\sum_{k=1}^{n} f(x_k)$$

We will now show how to extend this concept so that we can compute not only the arithmetic average of finitely many function values but an average of *all* values of $f(x)$ as

x varies over a closed interval $[a, b]$. For this purpose recall the Mean-Value Theorem for Integrals (5.6.2), which states that if f is continuous on the interval $[a, b]$, then there is at least one point x^* in this interval such that

$$\int_a^b f(x)\,dx = f(x^*)(b-a)$$

The quantity

$$f(x^*) = \frac{1}{b-a}\int_a^b f(x)\,dx$$

will be our candidate for the average value of f over the interval $[a, b]$. To explain what motivates this, divide the interval $[a, b]$ into n subintervals of equal length

$$\Delta x = \frac{b-a}{n} \tag{1}$$

and choose arbitrary points $x_1^*, x_2^*, \ldots, x_n^*$ in successive subintervals. Then the arithmetic average of the values $f(x_1^*), f(x_2^*), \ldots, f(x_n^*)$ is

$$\text{ave} = \frac{1}{n}[f(x_1^*) + f(x_2^*) + \cdots + f(x_n^*)]$$

or from (1)

$$\text{ave} = \frac{1}{b-a}[f(x_1^*)\Delta x + f(x_2^*)\Delta x + \cdots + f(x_n^*)\Delta x] = \frac{1}{b-a}\sum_{k=1}^{n} f(x_k^*)\Delta x$$

Taking the limit as $n \to +\infty$ yields

$$\lim_{n\to+\infty}\frac{1}{b-a}\sum_{k=1}^{n} f(x_k^*)\Delta x = \frac{1}{b-a}\int_a^b f(x)\,dx$$

Since this equation describes what happens when we compute the average of "more and more" values of $f(x)$, we are led to the following definition.

Note that the Mean-Value Theorem for Integrals, when expressed in form (2), ensures that there is always at least one point x^* in $[a, b]$ at which the value of f is equal to the average value of f over the interval.

6.6.1 DEFINITION. If f is continuous on $[a, b]$, then the ***average value*** (or ***mean value***) of f on $[a, b]$ is defined to be

$$f_{\text{ave}} = \frac{1}{b-a}\int_a^b f(x)\,dx \tag{2}$$

When f is nonnegative on $[a, b]$, the quantity f_{ave} has a simple geometric interpretation, which can be seen by writing (2) as

$$f_{\text{ave}} \cdot (b-a) = \int_a^b f(x)\,dx$$

The left side of this equation is the area of a rectangle with a height of f_{ave} and base of length $b - a$, and the right side is the area under $y = f(x)$ over $[a, b]$. Thus, f_{ave} is the height of a rectangle constructed over the interval $[a, b]$, whose area is the same as the area under the graph of f over that interval (Figure 6.6.1).

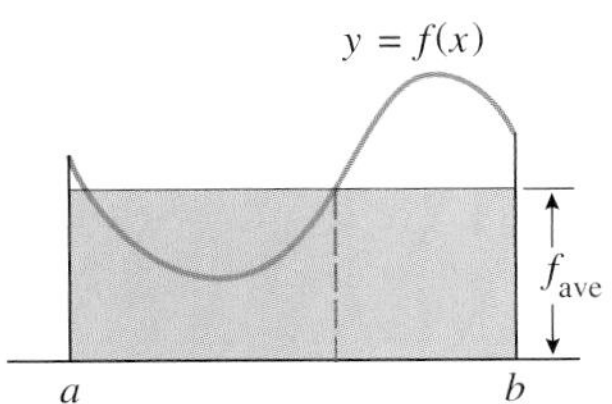

Figure 6.6.1

▶ **Example 1** Find the average value of the function $f(x) = \sqrt{x}$ over the interval $[1, 4]$, and find all points in the interval at which the value of f is the same as the average.

Solution.

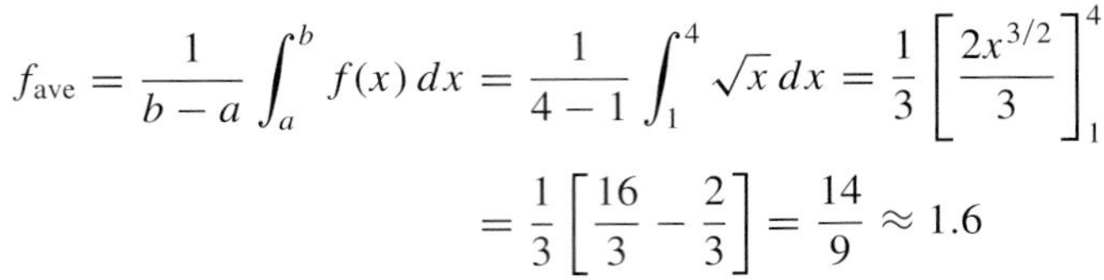

$$f_{\text{ave}} = \frac{1}{b-a}\int_a^b f(x)\,dx = \frac{1}{4-1}\int_1^4 \sqrt{x}\,dx = \frac{1}{3}\left[\frac{2x^{3/2}}{3}\right]_1^4$$

$$= \frac{1}{3}\left[\frac{16}{3} - \frac{2}{3}\right] = \frac{14}{9} \approx 1.6$$

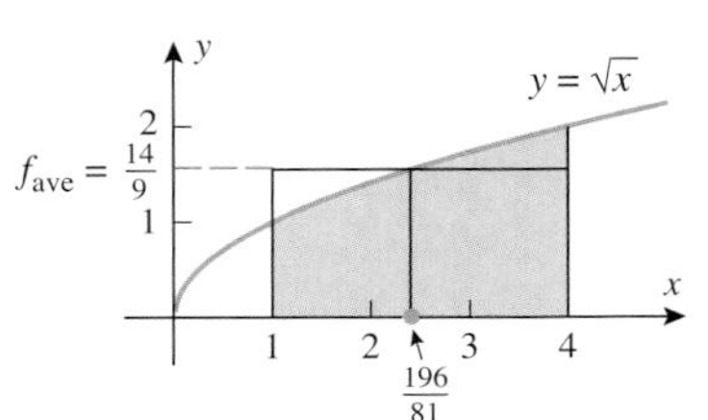

Figure 6.6.2

The x-values at which $f(x) = \sqrt{x}$ is the same as this average satisfy $\sqrt{x} = 14/9$, from which we obtain $x = 196/81 \approx 2.4$ (Figure 6.6.2). ◀

AVERAGE VELOCITY REVISITED

Consider a particle in rectilinear motion. In Section 3.1 we defined the average velocity of the particle over a time interval to be its displacement over the time interval divided by the time elapsed. Thus, if the particle has position function $s(t)$, then its average velocity v_{ave} over a time interval $[t_0, t_1]$ is

$$v_{\text{ave}} = \frac{s(t_1) - s(t_0)}{t_1 - t_0}$$

However, the displacement $s(t_1) - s(t_0)$ is the integral of velocity over the time interval $[t_0, t_1]$ [Formula (3) of Section 5.7]. Thus, we can express v_{ave} as

$$v_{\text{ave}} = \frac{1}{t_1 - t_0}\int_{t_0}^{t_1} v(t)\,dt \tag{3}$$

which, by Definition 6.6.1, is the average value of the velocity function over the time interval $[t_0, t_1]$. Thus, we have shown that:

For a particle in rectilinear motion, the average value of the velocity function over a time interval is the same as the average velocity of the particle over that interval; that is, the average value of the velocity function is equal to the displacement of the particle divided by the time elapsed.

Since velocity functions are generally continuous, it follows from the marginal note associated with Definition 6.6.1 that a particle's average velocity over a time interval matches the particle's velocity at some time in the interval.

▶ **Example 2** Show that if a body released from rest (initial velocity zero) is in free fall, then its average velocity over a time interval $[0, T]$ during its fall is its velocity at time $t = T/2$.

The result of Example 2 can be generalized to show that the average velocity of a particle in uniformly accelerated motion during a time interval $[a, b]$ is the velocity at time $t = (a + b)/2$. (See Exercise 14.)

Solution. It follows from Formula (16) of Section 5.7 with $v_0 = 0$ that the velocity function of the body is $v(t) = -gt$. Thus, its average velocity over a time interval $[0, T]$ is

$$\begin{aligned} v_{\text{ave}} &= \frac{1}{T-0}\int_0^T v(t)\,dt \\ &= \frac{1}{T}\int_0^T -gt\,dt \\ &= -\frac{g}{T}\left[\frac{1}{2}t^2\right]_0^T = -g\cdot\frac{T}{2} = v\left(\frac{T}{2}\right) \end{aligned}$$ ◀

✔QUICK CHECK EXERCISES 6.6

(See page 418 for answers.)

1. The arithmetic average of n numbers, $a_1, a_2, \ldots, a_n$ is ______.

2. If f is continuous on $[a, b]$, then the average value of f on $[a, b]$ is ______.

3. If f is continuous on $[a, b]$, then the ______ Theorem for Integrals guarantees that for at least one point x^* in $[a, b]$, $f(x^*)$ equals the average value of f on $[a, b]$.

4. The average value of $f(x) = 4x^3$ on $[1, 3]$ is ______.

EXERCISE SET 6.6 [C] CAS

1. (a) Find f_{ave} of $f(x) = 2x$ over $[0, 4]$.
   (b) Find a point x^* in $[0, 4]$ such that $f(x^*) = f_{\text{ave}}$.
   (c) Sketch a graph of $f(x) = 2x$ over $[0, 4]$, and construct a rectangle over the interval whose area is the same as the area under the graph of f over the interval.

2. (a) Find f_{ave} of $f(x) = x^2$ over $[0, 2]$.
   (b) Find a point x^* in $[0, 2]$ such that $f(x^*) = f_{\text{ave}}$.
   (c) Sketch a graph of $f(x) = x^2$ over $[0, 2]$, and construct a rectangle over the interval whose area is the same as the area under the graph of f over the interval.

3–8 Find the average value of the function over the given interval.

3. $f(x) = 3x;\ [1, 3]$
4. $f(x) = \sqrt[3]{x};\ [-1, 8]$
5. $f(x) = \sin x;\ [0, \pi]$
6. $f(x) = \sec x \tan x;\ [0, \pi/3]$
7. $f(x) = \dfrac{x}{(5x^2+1)^2};\ [0, 2]$
8. $f(x) = \sec^2 \pi x;\ \left[-\frac{1}{4}, \frac{1}{4}\right]$

FOCUS ON CONCEPTS

9. Let $f(x) = 3x^2$.
   (a) Find the arithmetic average of the values $f(0.4)$, $f(0.8)$, $f(1.2)$, $f(1.6)$, and $f(2.0)$.
   (b) Find the arithmetic average of the values $f(0.1)$, $f(0.2)$, $f(0.3), \ldots, f(2.0)$.
   (c) Find the average value of f on $[0, 2]$.
   (d) Explain why the answer to part (c) is less than the answers to parts (a) and (b).

10. Let $f(x) = 1 + \dfrac{1}{x^2}$.
    (a) Find the arithmetic average of the values $f\left(\frac{6}{5}\right)$, $f\left(\frac{7}{5}\right)$, $f\left(\frac{8}{5}\right)$, $f\left(\frac{9}{5}\right)$, and $f(2)$.
    (b) Find the arithmetic average of the values $f(1.1)$, $f(1.2)$, $f(1.3), \ldots, f(2)$.
    (c) Find the average value of f on $[1, 2]$.
    (d) Explain why the answer to part (c) is greater than the answers to parts (a) and (b).

11. In each part, the velocity versus time curve is given for a particle moving along a line. Use the curve to find the average velocity of the particle over the time interval $0 \le t \le 3$.

(a)

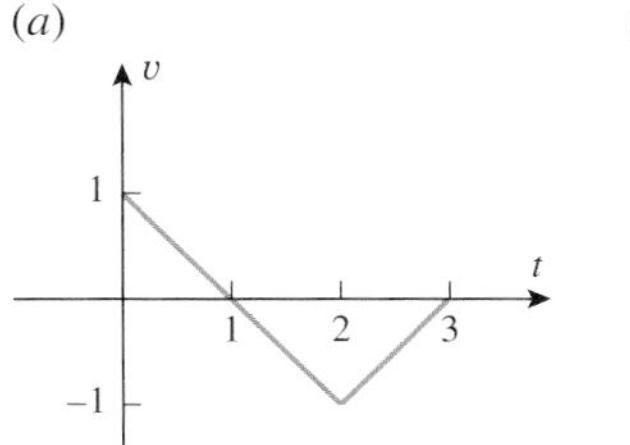

(b)

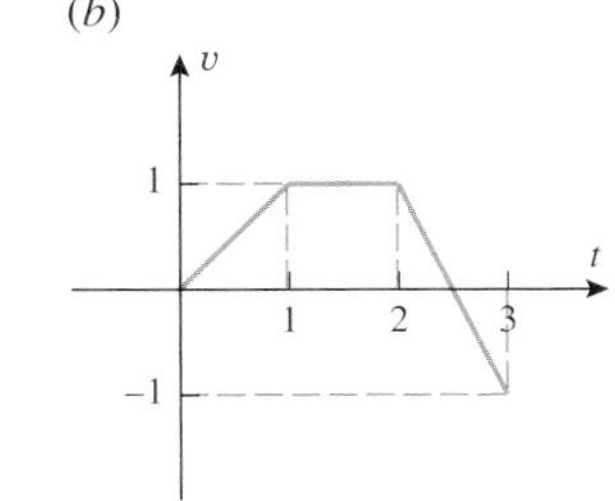

12. Suppose that a particle moving along a line starts from rest and has an average velocity of 2 ft/s over the time interval $0 \le t \le 5$. Sketch a velocity versus time curve for the particle assuming that the particle is also at rest at time $t = 5$. Explain how your curve satisfies the required properties.

13. Suppose that f is a linear function. Using the graph of f, explain why the average value of f on $[a, b]$ is
$$f\left(\frac{a+b}{2}\right)$$

14. Suppose that a particle moves along a coordinate line with constant acceleration. Show that the average velocity of the particle during a time interval $[a, b]$ matches the velocity of the particle at the midpoint of the interval.

15. (a) Suppose that the velocity function of a particle moving along a coordinate line is $v(t) = 3t^3 + 2$. Find the average velocity of the particle over the time interval $1 \le t \le 4$ by integrating.
    (b) Suppose that the position function of a particle moving along a coordinate line is $s(t) = 6t^2 + t$. Find the average velocity of the particle over the time interval $1 \le t \le 4$ algebraically.

16. (a) Suppose that the acceleration function of a particle moving along a coordinate line is $a(t) = t + 1$. Find the average acceleration of the particle over the time interval $0 \le t \le 5$ by integrating.
    (b) Suppose that the velocity function of a particle moving along a coordinate line is $v(t) = \cos t$. Find the average acceleration of the particle over the time interval $0 \le t \le \pi/4$ algebraically.

17. Water is run at a constant rate of 1 ft^3/min to fill a cylindrical tank of radius 3 ft and height 5 ft. Assuming that the tank is initially empty, make a conjecture about the average weight of the water in the tank over the time period required to fill it, and then check your conjecture by integrating. [Take the weight density of water to be 62.4 lb/ft^3.]

18. (a) The temperature of a 10-m-long metal bar is 15°C at one end and 30°C at the other end. Assuming that the temperature increases linearly from the cooler end to the hotter end, what is the average temperature of the bar?
(b) Explain why there must be a point on the bar where the temperature is the same as the average, and find it.

19. A traffic engineer monitors the rate at which cars enter the main highway during the afternoon rush hour. From her data she estimates that between 4:30 P.M. and 5:30 P.M. the rate $R(t)$ at which cars enter the highway is given by the formula $R(t) = 100(1 - 0.0001t^2)$ cars per minute, where t is the time (in minutes) since 4:30 P.M. Find the average rate, in cars per minute, at which cars enter the highway during the first half-hour of rush hour.

20. Suppose that the value of a yacht in dollars after t years of use is $V(t) = 275{,}000\sqrt{\frac{20}{t+20}}$. What is the average value of the yacht over its first 10 years of use?

21. (a) The accompanying table shows the fraction of the Moon that is illuminated (as seen from Earth) at midnight (Eastern Standard Time) for the first week of 2005. Find the average fraction of the Moon illuminated during the first week of 2005.

Source: Data from the U.S Naval Observatory Astronomical Applications Department.

(b) The function $f(x) = 0.5 + 0.5\sin(0.213x + 2.481)$ models data for illumination of the Moon for the first 60 days of 2005. Find the average value of this illumination function over the interval $[0, 7]$.

DAY	1	2	3	4	5	6	7
ILLUMINATION	0.74	0.65	0.56	0.45	0.35	0.25	0.16

Table Ex-21

C **22.** The function J_0 defined by

$$J_0(x) = \frac{1}{\pi}\int_0^{\pi} \cos(x \sin t)\,dt$$

is called the ***Bessel function of order zero***.
(a) Find a function f and an interval $[a, b]$ for which $J_0(1)$ is the average value of f over $[a, b]$.
(b) Estimate $J_0(1)$.
(c) Use a CAS to graph the equation $y = J_0(x)$ over the interval $0 \le x \le 8$.
(d) Estimate the smallest positive zero of J_0.

23. Find a positive value of k such that the average value of $f(x) = \sqrt{3x}$ over the interval $[0, k]$ is 6.

24. Find a positive value of k such that the average value of $f(x) = 1/(kx^2)$ over the interval $[k, 1]$ is 2.

25. Electricity is supplied to homes in the form of ***alternating current***, which means that the voltage has a sinusoidal waveform described by an equation of the form

$$V = V_p \sin(2\pi f t)$$

(see the accompanying figure). In this equation, V_p is called the ***peak voltage*** or ***amplitude*** of the current, f is called its ***frequency***, and $1/f$ is called its ***period***. The voltages V and V_p are measured in volts (V), the time t is measured in seconds (s), and the frequency is measured in hertz (Hz). (1 Hz = 1 cycle per second; a ***cycle*** is the electrical term for one period of the waveform.) Most alternating-current voltmeters read what is called the ***rms*** or ***root-mean-square*** value of V. By definition, this is the square root of the average value of V^2 over one period.
(a) Show that

$$V_{\text{rms}} = \frac{V_p}{\sqrt{2}}$$

[*Hint:* Compute the average over the cycle from $t = 0$ to $t = 1/f$, and use the identity $\sin^2\theta = \frac{1}{2}(1 - \cos 2\theta)$ to help evaluate the integral.]
(b) In the United States, electrical outlets supply alternating current with an rms voltage of 120 V at a frequency of 60 Hz. What is the peak voltage at such an outlet?

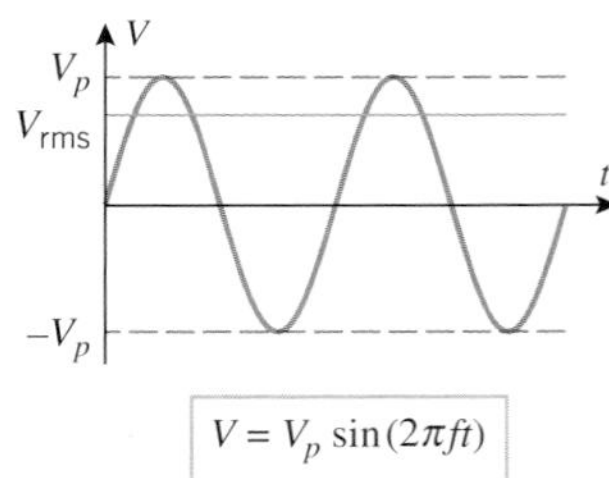

Figure Ex-25

26. Suppose that a tumor grows at the rate of $r(t) = kt$ grams per week for some positive constant k, where t is the number of weeks since the tumor appeared. When, during the second 26 weeks of growth, is the mass of the tumor the same as its average mass during that period?

✔ QUICK CHECK ANSWERS 6.6

1. $\frac{1}{n}\sum_{k=1}^{n} a_k$ 2. $\frac{1}{b-a}\int_a^b f(x)\,dx$ 3. Mean-Value 4. 40

6.7 WORK

In this section we will use the integration tools developed in the preceding chapter to study some of the basic principles of "work," which is one of the fundamental concepts in physics and engineering.

THE ROLE OF WORK IN PHYSICS AND ENGINEERING

In this section we will be concerned with two related concepts, *work* and *energy*. To put these ideas in a familiar setting, when you push a stalled car for a certain distance you are performing work, and the effect of your work is to make the car move. The energy of motion caused by the work is called the *kinetic energy* of the car. The exact connection between work and kinetic energy is governed by a principle of physics called the *work–energy relationship*. Although we will touch on this idea in this section, a detailed study of the relationship between work and energy will be left for courses in physics and engineering. Our primary goal here will be to explain the role of integration in the study of work.

WORK DONE BY A CONSTANT FORCE APPLIED IN THE DIRECTION OF MOTION

When a stalled car is pushed, the speed that the car attains depends on the force F with which it is pushed and the distance d over which that force is applied (Figure 6.7.1). Thus, force and distance are the ingredients of work in the following definition.

6.7.1 DEFINITION. If a constant force of magnitude F is applied in the direction of motion of an object, and if that object moves a distance d, then we define the ***work*** W performed by the force on the object to be

$$W = F \cdot d \tag{1}$$

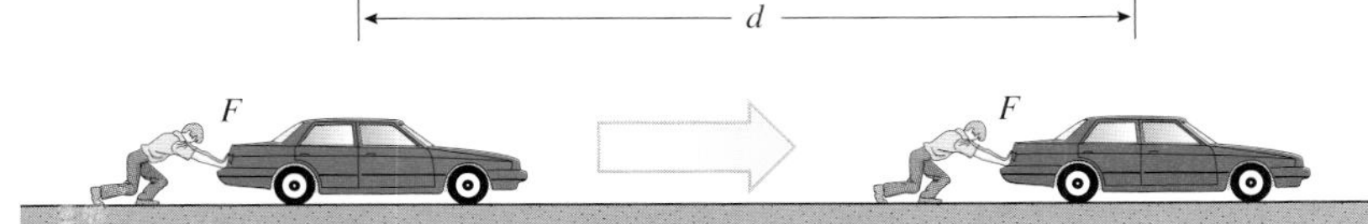

Figure 6.7.1

If you push against an immovable object, such as a brick wall, you may tire yourself out, but you will not perform any work. Why?

Common units for measuring force are newtons (N) in the International System of Units (SI), dynes (dyn) in the centimeter-gram-second (CGS) system, and pounds (lb) in the British Engineering (BE) system. One newton is the force required to give a mass of 1 kg an acceleration of 1 m/s^2, one dyne is the force required to give a mass of 1 g an acceleration of 1 cm/s^2, and one pound of force is the force required to give a mass of 1 slug an acceleration of 1 ft/s^2.

It follows from Definition 6.7.1 that work has units of force times distance. The most common units of work are newton-meters (N·m), dyne-centimeters (dyn·cm), and foot-pounds (ft·lb). As indicated in Table 6.7.1, one newton-meter is also called a ***joule*** (J), and one dyne-centimeter is also called an ***erg***. One foot-pound is approximately 1.36 J.

▶ **Example 1** An object moves 5 ft along a line while subjected to a constant force of 100 lb in its direction of motion. The work done is

$$W = F \cdot d = 100 \cdot 5 = 500 \text{ ft·lb}$$

Table 6.7.1

SYSTEM	FORCE ×	DISTANCE =	WORK
SI	newton (N)	meter (m)	joule (J)
CGS	dyne (dyn)	centimeter (cm)	erg
BE	pound (lb)	foot (ft)	foot-pound (ft·lb)
CONVERSION FACTORS:			
$1\text{ N} = 10^5\text{ dyn} \approx 0.225\text{ lb}$		$1\text{ lb} \approx 4.45\text{ N}$	
$1\text{ J} = 10^7\text{ erg} \approx 0.738\text{ ft·lb}$		$1\text{ ft·lb} \approx 1.36\text{ J} = 1.36 \times 10^7\text{ erg}$	

An object moves 25 m along a line while subjected to a constant force of 4 N in its direction of motion. The work done is

$$W = F \cdot d = 4 \cdot 25 = 100\text{ N·m} = 100\text{ J} \blacktriangleleft$$

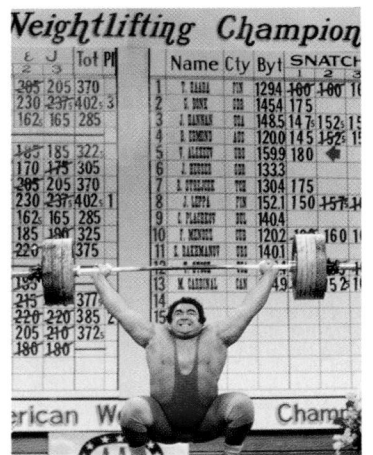

Vasili Alexeev lifting a record-breaking 562 lb in the 1976 Olympics

► **Example 2** In the 1976 Olympics, Vasili Alexeev astounded the world by lifting a record-breaking 562 lb from the floor to above his head (about 2 m). Equally astounding was the feat of strongman Paul Anderson, who in 1957 braced himself on the floor and used his back to lift 6270 lb of lead and automobile parts a distance of 1 cm. Who did more work?

Solution. To lift an object one must apply sufficient force to overcome the gravitational force that the Earth exerts on that object. The force that the Earth exerts on an object is that object's weight; thus, in performing their feats, Alexeev applied a force of 562 lb over a distance of 2 m and Anderson applied a force of 6270 lb over a distance of 1 cm. Pounds are units in the BE system, meters are units in SI, and centimeters are units in the CGS system. We will need to decide on the measurement system we want to use and be consistent. Let us agree to use SI and express the work of the two men in joules. Using the conversion factor in Table 6.7.1 we obtain

$$562\text{ lb} \approx 562\text{ lb} \times 4.45\text{ N/lb} \approx 2500\text{ N}$$
$$6270\text{ lb} \approx 6270\text{ lb} \times 4.45\text{ N/lb} \approx 27{,}900\text{ N}$$

Using these values and the fact that $1\text{ cm} = 0.01\text{ m}$ we obtain

$$\text{Alexeev's work} = (2500\text{ N}) \times (2\text{ m}) = 5000\text{ J}$$
$$\text{Anderson's work} = (27{,}900\text{ N}) \times (0.01\text{ m}) = 279\text{ J}$$

Therefore, even though Anderson's lift required a tremendous upward force, it was applied over such a short distance that Alexeev did more work. ◄

WORK DONE BY A VARIABLE FORCE APPLIED IN THE DIRECTION OF MOTION

Many important problems are concerned with finding the work done by a *variable* force that is applied in the direction of motion. For example, Figure 6.7.2*a* shows a spring in its natural state (neither compressed nor stretched). If we want to pull the block horizontally (Figure 6.7.2*b*), then we would have to apply more and more force to the block to overcome the increasing force of the stretching spring. Thus, our next objective is to define what is meant by the work performed by a variable force and to find a formula for computing it. This will require calculus.

Natural position

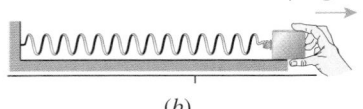

(*a*)

Force must be exerted to stretch spring

(*b*)

Figure 6.7.2

6.7.2 PROBLEM. Suppose that an object moves in the positive direction along a coordinate line while subjected to a variable force $F(x)$ that is applied in the direction of motion. Define what is meant by the *work* W performed by the force on the object as the object moves from $x = a$ to $x = b$, and find a formula for computing the work.

The basic idea for solving this problem is to break up the interval $[a, b]$ into subintervals that are sufficiently small that the force does not vary much on each subinterval. This will allow us to treat the force as constant on each subinterval and to approximate the work on each subinterval using Formula (1). By adding the approximations to the work on the subintervals, we will obtain a Riemann sum that approximates the work W over the entire interval, and by taking the limit of the Riemann sums we will obtain an integral for W.

To implement this idea, divide the interval $[a, b]$ into n subintervals by inserting points $x_1, x_2, \ldots, x_{n-1}$ between $a = x_0$ and $b = x_n$. We can use Formula (1) to approximate the work W_k done in the kth subinterval by choosing any point x_k^* in this interval and regarding the force to have a constant value $F(x_k^*)$ throughout the interval. Since the width of the kth subinterval is $x_k - x_{k-1} = \Delta x_k$, this yields the approximation

$$W_k \approx F(x_k^*)\Delta x_k$$

Adding these approximations yields the following Riemann sum that approximates the work W done over the entire interval:

$$W \approx \sum_{k=1}^{n} F(x_k^*)\Delta x_k$$

Taking the limit as n increases and the widths of the subintervals approach zero yields the definite integral

$$W = \lim_{\max \Delta x_k \to 0} \sum_{k=1}^{n} F(x_k^*)\Delta x_k = \int_a^b F(x)\,dx$$

In summary, we have the following result.

6.7.3 DEFINITION. Suppose that an object moves in the positive direction along a coordinate line over the interval $[a, b]$ while subjected to a variable force $F(x)$ that is applied in the direction of motion. Then we define the ***work*** W performed by the force on the object to be

$$W = \int_a^b F(x)\,dx \tag{2}$$

Hooke's law [Robert Hooke (1635–1703), English physicist] states that under appropriate conditions a spring that is stretched x units beyond its natural length pulls back with a force

$$F(x) = kx$$

where k is a constant (called the ***spring constant*** or ***spring stiffness***). The value of k depends on such factors as the thickness of the spring and the material used in its composition. Since $k = F(x)/x$, the constant k has units of force per unit length.

► **Example 3** A spring exerts a force of 5 N when stretched 1 m beyond its natural length.

(a) Find the spring constant k.

(b) How much work is required to stretch the spring 1.8 m beyond its natural length?

Solution (a). From Hooke's law,

$$F(x) = kx$$

From the data, $F(x) = 5$ N when $x = 1$ m, so $5 = k \cdot 1$. Thus, the spring constant is $k = 5$ newtons per meter (N/m). This means that the force $F(x)$ required to stretch the spring x meters is

$$F(x) = 5x \tag{3}$$

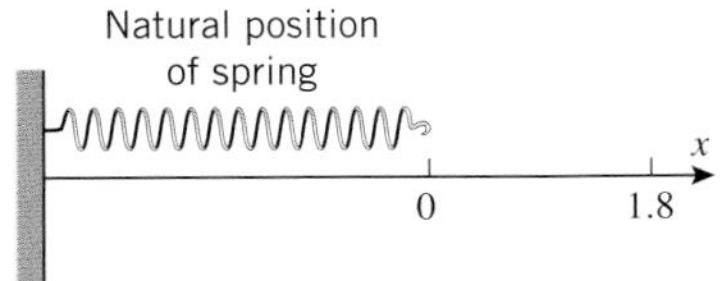

Figure 6.7.3

Solution (b). Place the spring along a coordinate line as shown in Figure 6.7.3. We want to find the work W required to stretch the spring over the interval from $x = 0$ to $x = 1.8$. From (2) and (3) the work W required is

$$W = \int_a^b F(x)\,dx = \int_0^{1.8} 5x\,dx = \frac{5x^2}{2}\bigg]_0^{1.8} = 8.1 \text{ J} \blacktriangleleft$$

▶ **Example 4** An astronaut's *weight* (or more precisely, *Earth weight*) is the force exerted on the astronaut by the Earth's gravity. As the astronaut moves upward into space, the gravitational pull of the Earth decreases, and hence so does his or her weight. We will show later in the text that if the Earth is assumed to be a sphere of radius 4000 mi, then an astronaut who weighs 150 lb on Earth will have a weight of

$$w(x) = \frac{2{,}400{,}000{,}000}{x^2} \text{ lb}, \quad x \geq 4000$$

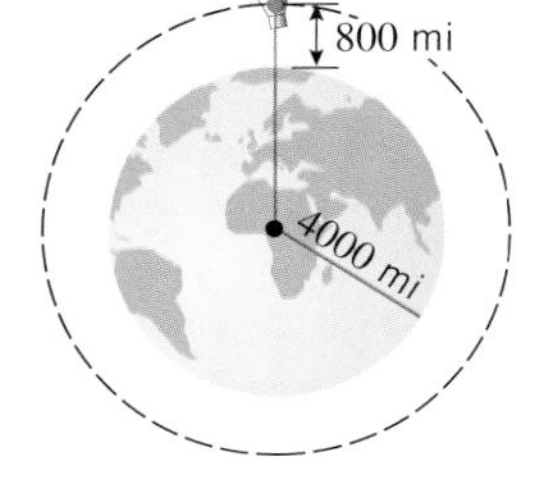

Figure 6.7.4

at a distance of x mi from the Earth's center. Use this formula to determine the work in foot-pounds required to lift the astronaut to a point that is 800 mi above the surface of the Earth (Figure 6.7.4).

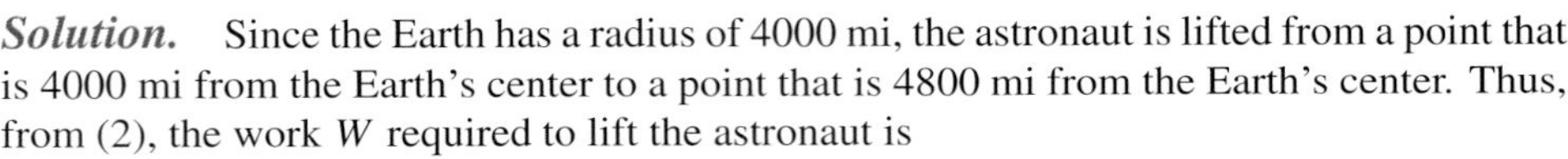
Solution. Since the Earth has a radius of 4000 mi, the astronaut is lifted from a point that is 4000 mi from the Earth's center to a point that is 4800 mi from the Earth's center. Thus, from (2), the work W required to lift the astronaut is

$$\begin{aligned}
W &= \int_{4000}^{4800} \frac{2{,}400{,}000{,}000}{x^2}\,dx \\
&= -\frac{2{,}400{,}000{,}000}{x}\bigg]_{4000}^{4800} \\
&= -500{,}000 + 600{,}000 \\
&= 100{,}000 \text{ mile-pounds} \\
&= (100{,}000 \text{ mi}\cdot\text{lb}) \times (5280 \text{ ft/mi}) \\
&= 5.28 \times 10^8 \text{ ft}\cdot\text{lb} \blacktriangleleft
\end{aligned}$$

CALCULATING WORK FROM BASIC PRINCIPLES

Some problems cannot be solved by mechanically substituting into formulas, and one must return to basic principles to obtain solutions. This is illustrated in the next example.

▶ **Example 5** A conical water tank of radius 10 ft and height 30 ft is filled with water to a depth of 15 ft (Figure 6.7.5*a*). How much work is required to pump all of the water out through a hole in the top of the tank?

Solution. Our strategy will be to divide the water into thin layers, approximate the work required to move each layer to the top of the tank, add the approximations for the layers to obtain a Riemann sum that approximates the total work, and then take the limit of the Riemann sums to produce an integral for the total work.

To implement this idea, introduce an x-axis as shown in Figure 6.7.5a, and divide the water into n layers with Δx_k denoting the thickness of the kth layer. This division induces a partition of the interval [15, 30] into n subintervals. Although the upper and lower surfaces of the kth layer are at different distances from the top, the difference will be small if the layer is thin, and we can reasonably assume that the entire layer is concentrated at a single point x_k^* (Figure 6.7.5a). Thus, the work W_k required to move the kth layer to the top of the tank is approximately

$$W_k \approx F_k x_k^* \tag{4}$$

where F_k is the force required to lift the kth layer. But the force required to lift the kth layer is the force needed to overcome gravity, and this is the same as the weight of the layer. If the layer is very thin, we can approximate the volume of the kth layer with the volume of a cylinder of height Δx_k and radius r_k, where (by similar triangles)

$$\frac{r_k}{x_k^*} = \frac{10}{30} = \frac{1}{3}$$

or, equivalently, $r_k = x_k^*/3$ (Figure 6.7.5b). Therefore, the volume of the kth layer of water is approximately

$$\pi r_k^2 \Delta x_k = \pi (x_k^*/3)^2 \Delta x_k = \frac{\pi}{9}(x_k^*)^2 \Delta x_k$$

Since the weight density of water is 62.4 lb/ft^3, it follows that

$$F_k \approx \frac{62.4\pi}{9}(x_k^*)^2 \Delta x_k$$

Thus, from (4)

$$W_k \approx \left(\frac{62.4\pi}{9}(x_k^*)^2 \Delta x_k\right) x_k^* = \frac{62.4\pi}{9}(x_k^*)^3 \Delta x_k$$

and hence the work W required to move all n layers has the approximation

$$W = \sum_{k=1}^{n} W_k \approx \sum_{k=1}^{n} \frac{62.4\pi}{9}(x_k^*)^3 \Delta x_k$$

To find the *exact* value of the work we take the limit as $\max \Delta x_k \to 0$. This yields

$$\begin{aligned} W &= \lim_{\max \Delta x_k \to 0} \sum_{k=1}^{n} \frac{62.4\pi}{9}(x_k^*)^3 \Delta x_k = \int_{15}^{30} \frac{62.4\pi}{9} x^3\,dx \\ &= \frac{62.4\pi}{9}\left(\frac{x^4}{4}\right)\Bigg]_{15}^{30} = 1{,}316{,}250\pi \approx 4{,}135{,}000 \text{ ft·lb} \end{aligned}$$ ◀

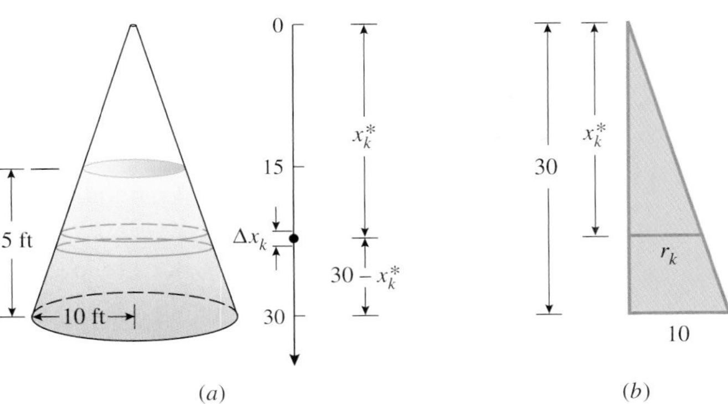

Figure 6.7.5

THE WORK–ENERGY RELATIONSHIP

When you see an object in motion, you can be certain that somehow work has been expended to create that motion. For example, when you drop a stone from a building, the stone gathers speed because the force of the Earth's gravity is performing work on it, and when a hockey player strikes a puck with a hockey stick, the work performed on the puck during the brief period of contact with the stick creates the enormous speed of the puck across the ice. However, experience shows that the speed obtained by an object depends not only on the amount of work done, but also on the mass of the object. For example, the work required to throw a 5-oz baseball 50 mi/h would accelerate a 10-lb bowling ball to less than 9 mi/h.

Using the method of substitution for definite integrals, we will derive a simple equation that relates the work done on an object to the object's mass and velocity. Furthermore, this equation will allow us to motivate an appropriate definition for the "energy of motion" of an object. As in Definition 6.7.3, we will assume that an object moves in the positive direction along a coordinate line over the interval $[a, b]$ while subjected to a force $F(x)$ that is applied in the direction of motion. We let $x = x(t)$, $v = v(t) = x'(t)$, and $v'(t)$ denote the respective position, velocity, and acceleration of the object at time t. It follows from Newton's Second Law of Motion that

$$F(x(t)) = mv'(t)$$

where m is the mass of the object. Assume that

$$x(t_0) = a \quad \text{and} \quad x(t_1) = b$$

with

$$v(t_0) = v_i \quad \text{and} \quad v(t_1) = v_f$$

the initial and final velocities of the object, respectively. Then

$$\begin{aligned} W &= \int_a^b F(x)\,dx = \int_{x(t_0)}^{x(t_1)} F(x)\,dx \\ &= \int_{t_0}^{t_1} F(x(t))x'(t)\,dt && \text{By Theorem 5.8.1 with } x = x(t),\ dx = x'(t)\,dt \\ &= \int_{t_0}^{t_1} mv'(t)v(t)\,dt = \int_{t_0}^{t_1} mv(t)v'(t)\,dt \\ &= \int_{v(t_0)}^{v(t_1)} mv\,dv && \text{By Theorem 5.8.1 with } v = v(t),\ dv = v'(t)\,dt \\ &= \int_{v_i}^{v_f} mv\,dv = \tfrac{1}{2}mv^2\Big|_{v_i}^{v_f} = \tfrac{1}{2}mv_f^2 - \tfrac{1}{2}mv_i^2 \end{aligned}$$

We see from the equation

$$W = \tfrac{1}{2}mv_f^2 - \tfrac{1}{2}mv_i^2 \tag{5}$$

that the work done on the object is equal to the change in the quantity $\frac{1}{2}mv^2$ from its initial value to its final value. We will refer to Equation (5) as the ***work–energy relationship***. If we define the "energy of motion" or ***kinetic energy*** of our object to be given by

$$K = \tfrac{1}{2}mv^2 \tag{6}$$

then Equation (5) tells us that the work done on an object is equal to the *change* in the object's kinetic energy. Loosely speaking, we may think of work done on an object as being "transformed" into kinetic energy of the object. The units of kinetic energy are the same as the units of work. For example, in SI kinetic energy is measured in joules (J).

▶ **Example 6** A space probe of mass $m = 5.00 \times 10^4$ kg travels in deep space subjected only to the force of its own engine. Starting at a time when the speed of the probe is $v = 1.10 \times 10^4$ m/s, the engine is fired continuously over a distance of 2.50×10^6 m with a constant force of 4.00×10^5 N in the direction of motion. What is the final speed of the probe?

Solution. Since the force applied by the engine is constant and in the direction of motion, the work W expended by the engine on the probe is

$$W = \text{force} \times \text{distance} = (4.00 \times 10^5 \text{ N}) \times (2.50 \times 10^6 \text{ m}) = 1.00 \times 10^{12} \text{ J}$$

From (5), the final kinetic energy $K_f = \frac{1}{2}mv_f^2$ of the probe can be expressed in terms of the work W and the initial kinetic energy $K_i = \frac{1}{2}mv_i^2$ as

$$K_f = W + K_i$$

Thus, from the known mass and initial speed we have

$$K_f = (1.00 \times 10^{12} \text{ J}) + \tfrac{1}{2}(5.00 \times 10^4 \text{ kg})(1.10 \times 10^4 \text{ m/s})^2 = 4.025 \times 10^{12} \text{ J}$$

The final kinetic energy is $K_f = \frac{1}{2}mv_f^2$, so the final speed of the probe is

$$v_f = \sqrt{\frac{2K_f}{m}} = \sqrt{\frac{2(4.025 \times 10^{12})}{5.00 \times 10^4}} \approx 1.27 \times 10^4 \text{ m/s} \blacktriangleleft$$

✔QUICK CHECK EXERCISES 6.7 (See page 427 for answers.)

1. If a constant force of 5 lb moves an object 10 ft, then the work done by the force on the object is ________.
2. A newton-meter is also called a ________. A dyne-centimeter is also called an ________.
3. Suppose that an object moves in the positive direction along a coordinate line over the interval $[a, b]$. The work performed on the object by a variable force $F(x)$ applied in the direction of motion is $W =$ ________.
4. A force $F(x) = 10 - 2x$ N applied in the positive x-direction moves an object 3 m from $x = 2$ to $x = 5$. The work done by the force on the object is ________.

EXERCISE SET 6.7

1. Find the work done when
   (a) a constant force of 30 lb in the positive x-direction moves an object from $x = -2$ to $x = 5$ ft
   (b) a variable force of $F(x) = 1/x^2$ lb in the positive x-direction moves an object from $x = 1$ to $x = 6$ ft.

2. A variable force $F(x)$ in the positive x-direction is graphed in the accompanying figure. Find the work done by the force on a particle that moves from $x = 0$ to $x = 5$.

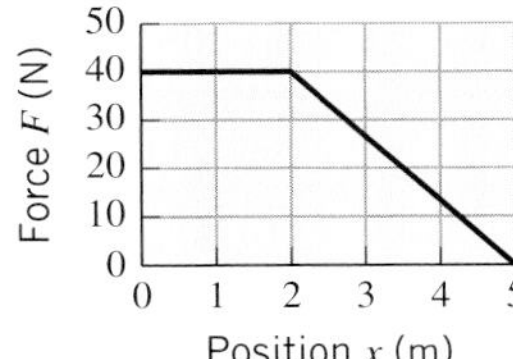

Figure Ex-2

FOCUS ON CONCEPTS

3. For the variable force $F(x)$ in Exercise 2, consider the distance d for which the work done by the force on the particle when the particle moves from $x = 0$ to $x = d$ is half the work done when the particle moves from $x = 0$ to $x = 5$. By inspecting the graph of F, is d more or less than 2.5? Explain, and then find the exact value of d.
4. Suppose that a variable force $F(x)$ is applied in the positive x-direction so that an object moves from $x = a$ to $x = b$. Relate the work done by the force on the object and the average value of F over $[a, b]$, and illustrate this relationship graphically.
5. In which scenario do you perform more work: by raising a cup of coffee from a table to your mouth, or by holding a calculus textbook at shoulder level for 5 minutes? Explain.

6. A constant force of 40 N in the positive x-direction is applied to a particle whose velocity versus time curve is shown in the accompanying figure. Find the work done by the force on the particle from time $t = 0$ to $t = 15$.

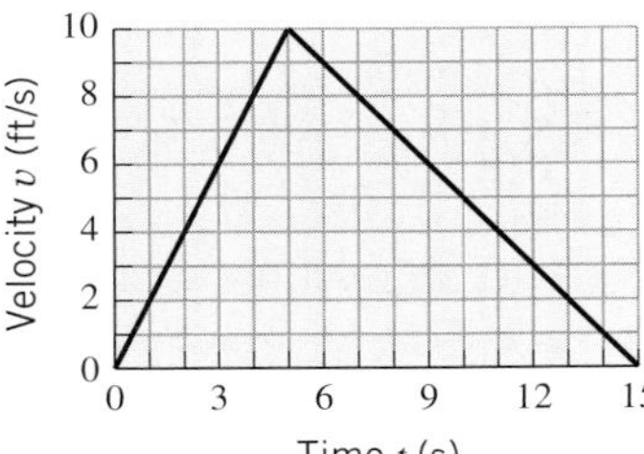

Figure Ex-6

7. A constant force of 10 lb in the positive x-direction is applied to a particle whose velocity versus time curve is shown in the accompanying figure. Find the work done by the force on the particle from time $t = 0$ to $t = 5$.

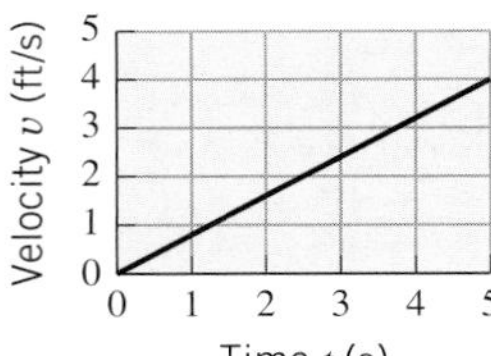

Figure Ex-7

8. A spring whose natural length is 15 cm exerts a force of 45 N when stretched to a length of 20 cm.
(a) Find the spring constant (in newtons/meter).
(b) Find the work that is done in stretching the spring 3 cm beyond its natural length.
(c) Find the work done in stretching the spring from a length of 20 cm to a length of 25 cm.

9. A spring exerts a force of 100 N when it is stretched 0.2 m beyond its natural length. How much work is required to stretch the spring 0.8 m beyond its natural length?

10. Assume that a force of 6 N is required to compress a spring from a natural length of 4 m to a length of $3\frac{1}{2}$ m. Find the work required to compress the spring from its natural length to a length of 2 m. (Hooke's law applies to compression as well as extension.)

11. Assume that 10 ft·lb of work is required to stretch a spring 1 ft beyond its natural length. What is the spring constant?

12. A cylindrical tank of radius 5 ft and height 9 ft is two-thirds filled with water. Find the work required to pump all the water over the upper rim.

13. Solve Exercise 12 assuming that the tank is two-thirds filled with a liquid that weighs ρ lb/ft^3.

14. A cone-shaped water reservoir is 20 ft in diameter across the top and 15 ft deep. If the reservoir is filled to a depth of 10 ft, how much work is required to pump all the water to the top of the reservoir?

15. The vat shown in the accompanying figure contains water to a depth of 2 m. Find the work required to pump all the water to the top of the vat. [Use 9810 N/m^3 as the weight density of water.]

16. The cylindrical tank shown in the accompanying figure is filled with a liquid weighing 50 lb/ft^3. Find the work required to pump all the liquid to a level 1 ft above the top of the tank.

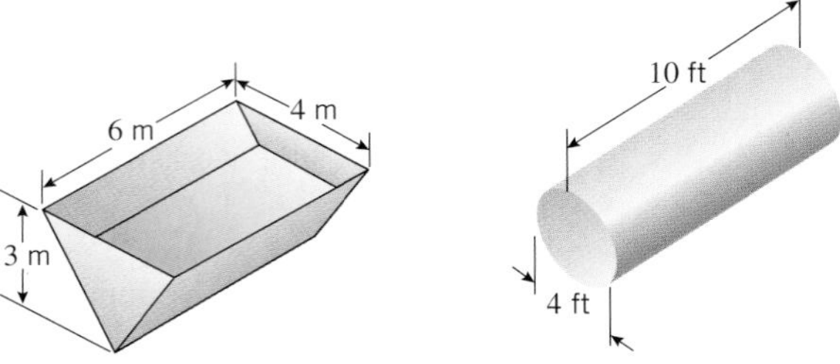

Figure Ex-15 **Figure Ex-16**

17. A swimming pool is built in the shape of a rectangular parallelepiped 10 ft deep, 15 ft wide, and 20 ft long.
(a) If the pool is filled to 1 ft below the top, how much work is required to pump all the water into a drain at the top edge of the pool?
(b) A one-horsepower motor can do 550 ft·lb of work per second. What size motor is required to empty the pool in 1 hour?

18. How much work is required to fill the swimming pool in Exercise 17 to 1 ft below the top if the water is pumped in through an opening located at the bottom of the pool?

19. A 100-ft length of steel chain weighing 15 lb/ft is dangling from a pulley. How much work is required to wind the chain onto the pulley?

20. A 3-lb bucket containing 20 lb of water is hanging at the end of a 20-ft rope that weighs 4 oz/ft. The other end of the rope is attached to a pulley. How much work is required to wind the length of rope onto the pulley, assuming that the rope is wound onto the pulley at a rate of 2 ft/s and that as the bucket is being lifted, water leaks from the bucket at a rate of 0.5 lb/s?

21. A rocket weighing 3 tons is filled with 40 tons of liquid fuel. In the initial part of the flight, fuel is burned off at a constant rate of 2 tons per 1000 ft of vertical height. How much work is done in lifting the rocket to 3000 ft?

22. It follows from Coulomb's law in physics that two like electrostatic charges repel each other with a force inversely proportional to the square of the distance between them. Suppose that two charges A and B repel with a force of k newtons when they are positioned at points $A(-a, 0)$ and $B(a, 0)$, where a is measured in meters. Find the work W required to move charge A along the x-axis to the origin if charge B remains stationary.

23. It is a law of physics that the gravitational force exerted by the Earth on an object above the Earth's surface varies inversely as the square of its distance from the Earth's center. Thus, an object's weight $w(x)$ is related to its distance x from the Earth's center by a formula of the form

$$w(x) = \frac{k}{x^2}$$

where k is a constant of proportionality that depends on the mass of the object.

(a) Use this fact and the assumption that the Earth is a sphere of radius 4000 mi to obtain the formula for $w(x)$ in Example 4.

(b) Find a formula for the weight $w(x)$ of a satellite that is x mi from the Earth's surface if its weight on Earth is 6000 lb.

(c) How much work is required to lift the satellite from the surface of the Earth to an orbital position that is 1000 mi high?

24. (a) The formula $w(x) = k/x^2$ in Exercise 23 is applicable to all celestial bodies. Assuming that the Moon is a sphere of radius 1080 mi, find the force that the Moon exerts on an astronaut who is x mi from the surface of the Moon if her weight on the Moon's surface is 20 lb.

(b) How much work is required to lift the astronaut to a point that is 10.8 mi above the Moon's surface?

25. The world's first commercial high-speed magnetic levitation (MAGLEV) train, a 30-km double-track project connecting Shanghai, China, to Pudong International Airport, began full revenue service in 2003. Suppose that a MAGLEV train has a mass $m = 4.00 \times 10^5$ kg and that starting at a time when the train has a speed of 20 m/s the engine applies a force of 6.40×10^5 N in the direction of motion over a distance of 3.00×10^3 m. Use the work–energy relationship (5) to find the final speed of the train.

26. Assume that a Mars probe of mass $m = 2.00 \times 10^3$ kg is subjected only to the force of its own engine. Starting at a time when the speed of the probe is $v = 1.00 \times 10^4$ m/s, the engine is fired continuously over a distance of 2.00×10^5 m with a constant force of 2.00×10^5 N in the direction of motion. Use the work–energy relationship (5) to find the final speed of the probe.

27. On August 10, 1972 a meteorite with an estimated mass of 4×10^6 kg and an estimated speed of 15 km/s skipped across the atmosphere above the western United States and Canada but fortunately did not hit the Earth.

(a) Assuming that the meteorite had hit the Earth with a speed of 15 km/s, what would have been its change in kinetic energy in joules (J)?

(b) Express the energy as a multiple of the explosive energy of 1 megaton of TNT, which is 4.2×10^{15} J.

(c) The energy associated with the Hiroshima atomic bomb was 13 kilotons of TNT. To how many such bombs would the meteorite impact have been equivalent?

✔ QUICK CHECK ANSWERS 6.7

1. 50 ft-lb 2. joule; erg 3. $\int_a^b F(x)\,dx$ 4. 9 J

6.8 FLUID PRESSURE AND FORCE

In this section we will use the integration tools developed in the preceding chapter to study the pressures and forces exerted by fluids on submerged objects.

WHAT IS A FLUID?

A ***fluid*** is a substance that flows to conform to the boundaries of any container in which it is placed. Fluids include *liquids*, such as water, oil, and mercury, as well as *gases*, such as helium, oxygen, and air. The study of fluids falls into two categories: *fluid statics* (the study of fluids at rest) and *fluid dynamics* (the study of fluids in motion). In this section we will be concerned only with fluid statics; toward the end of this text we will investigate problems in fluid dynamics.

THE CONCEPT OF PRESSURE

The effect that a force has on an object depends on how that force is spread over the surface of the object. For example, when you walk on soft snow with boots, the weight of your body crushes the snow and you sink into it. However, if you put on a pair of snowshoes to spread the weight of your body over a greater surface area, then the weight of your body has less of a crushing effect on the snow. The concept that accounts for both the magnitude of a force and the area over which it is applied is called "pressure".

6.8.1 DEFINITION. If a force of magnitude F is applied to a surface of area A, then we define the ***pressure*** P exerted by the force on the surface to be

$$P = \frac{F}{A} \tag{1}$$

It follows from this definition that pressure has units of force per unit area. The most common units of pressure are newtons per square meter (N/m^2) in SI and pounds per square inch (lb/in^2) or pounds per square foot (lb/ft^2) in the BE system. As indicated in Table 6.8.1, one newton per square meter is called a *pascal* (Pa). A pressure of 1 Pa is quite small (1 Pa $= 1.45 \times 10^{-4}$ lb/in^2), so in countries using SI, tire pressure gauges are usually calibrated in kilopascals (kPa), which is 1000 pascals.

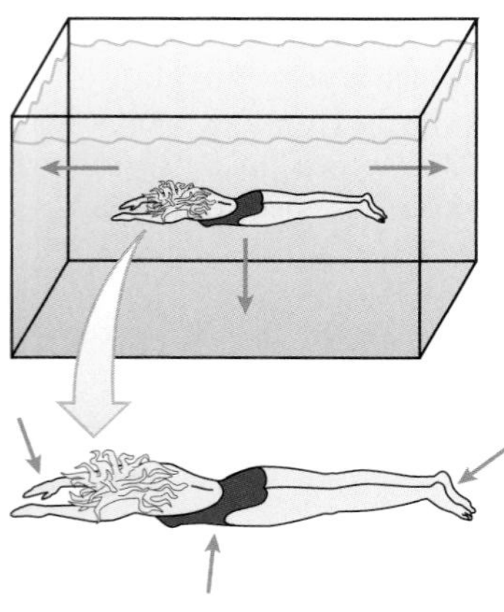

Fluid forces always act perpendicular to the surface of a submerged object.

Figure 6.8.1

Table 6.8.1

SYSTEM	FORCE ÷	AREA =	PRESSURE
SI	newton (N)	square meter (m^2)	pascal (Pa)
BE	pound (lb)	square foot (ft^2)	lb/ft^2
BE	pound (lb)	square inch (in^2)	lb/in^2 (psi)

CONVERSION FACTORS:
1 Pa $\approx 1.45 \times 10^{-4}$ lb/in^2 $\approx 2.09 \times 10^{-2}$ lb/ft^2
1 lb/in^2 $\approx 6.89 \times 10^3$ Pa   1 lb/ft^2 ≈ 47.9 Pa

In this section we will be interested in pressures and forces on objects submerged in fluids. Pressures themselves have no directional characteristics, but the forces that they create always act perpendicular to the face of the submerged object. Thus, in Figure 6.8.1 the water pressure creates horizontal forces on the sides of the tank, vertical forces on the bottom of the tank, and forces that vary in direction, so as to be perpendicular to the different parts of the swimmer's body.

Blaise Pascal (1623–1662) French mathematician and scientist. Pascal's mother died when he was three years old and his father, a highly educated magistrate, personally provided the boy's early education. Although Pascal showed an inclination for science and mathematics, his father refused to tutor him in those subjects until he mastered Latin and Greek. Pascal's sister and primary biographer claimed that he independently discovered the first thirty-two propositions of Euclid without ever reading a book on geometry. (However, it is generally agreed that the story is apocryphal.) Nevertheless, the precocious Pascal published a highly respected essay on conic sections by the time he was sixteen years old. Descartes, who read the essay, thought it so brilliant that he could not believe that it was written by such a young man. By age 18 his health began to fail and until his death he was in frequent pain. However, his creativity was unimpaired.

Pascal's contributions to physics include the discovery that air pressure decreases with altitude and the principle of fluid pressure that bears his name. However, the originality of his work is questioned by some historians. Pascal made major contributions to a branch of mathematics called "projective geometry," and he helped to develop probability theory through a series of letters with Fermat.

In 1646, Pascal's health problems resulted in a deep emotional crisis that led him to become increasingly concerned with religious matters. Although born a Catholic, he converted to a religious doctrine called Jansenism and spent most of his final years writing on religion and philosophy.

► **Example 1** Referring to Figure 6.8.1, suppose that the back of the swimmer's hand has a surface area of 8.4×10^{-3} m^2 and that the pressure acting on it is 5.1×10^4 Pa (a realistic value near the bottom of a deep diving pool). Find the force that acts on the swimmer's hand.

Solution. From (1), the force F is

$$F = PA = (5.1 \times 10^4 \text{ N/m}^2)(8.4 \times 10^{-3} \text{ m}^2) \approx 4.3 \times 10^2 \text{ N}$$

This is quite a large force (nearly 100 lb in the BE system). ◄

FLUID DENSITY

Scuba divers know that the pressure and forces on their bodies increase with the depth they dive. This is caused by the weight of the water and air above—the deeper the diver goes, the greater the weight above and hence the greater the pressure and force exerted on the diver.

To calculate pressures and forces on submerged objects, we need to know something about the characteristics of the fluids in which they are submerged. For simplicity, we will assume that the fluids under consideration are *homogeneous*, by which we mean that any two samples of the fluid with the same volume have the same mass. It follows from this assumption that the mass per unit volume is a constant δ that depends on the physical characteristics of the fluid but not on the size or location of the sample; we call

$$\delta = \frac{m}{V} \tag{2}$$

the ***mass density*** of the fluid. Sometimes it is more convenient to work with weight per unit volume than with mass per unit volume. Thus, we define the ***weight density*** ρ of a fluid to be

$$\rho = \frac{w}{V} \tag{3}$$

where w is the weight of a fluid sample of volume V. Thus, if the weight density of a fluid is known, then the weight w of a fluid sample of volume V can be computed from the formula $w = \rho V$. Table 6.8.2 shows some typical weight densities.

Table 6.8.2

WEIGHT DENSITIES

SI	N/m^3
Machine oil	4,708
Gasoline	6,602
Fresh water	9,810
Seawater	10,045
Mercury	133,416
BE SYSTEM	**lb/ft^3**
Machine oil	30.0
Gasoline	42.0
Fresh water	62.4
Seawater	64.0
Mercury	849.0

All densities are affected by variations in temperature and pressure. Weight densities are also affected by variations in g.

FLUID PRESSURE

To calculate fluid pressures and forces we will need to make use of an experimental observation. Suppose that a flat surface of area A is submerged in a homogeneous fluid of weight density ρ such that the entire surface lies between depths h_1 and h_2, where $h_1 \le h_2$ (Figure 6.8.2). Experiments show that on both sides of the surface, the fluid exerts a force that is perpendicular to the surface and whose magnitude F satisfies the inequalities

$$\rho h_1 A \le F \le \rho h_2 A \tag{4}$$

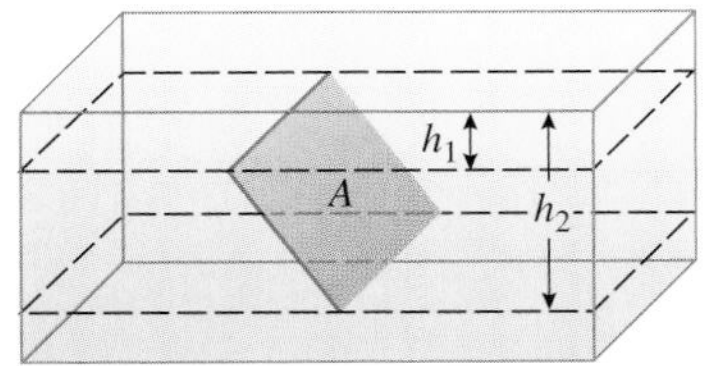

Figure 6.8.2

Thus, it follows from (1) that the pressure $P = F/A$ on a given side of the surface satisfies the inequalities

$$\rho h_1 \le P \le \rho h_2 \tag{5}$$

Note that it is now a straightforward matter to calculate fluid force and pressure on a flat surface that is submerged *horizontally* at depth h, for then $h = h_1 = h_2$ and inequalities (4) and (5) become the *equalities*

$$F = \rho h A \tag{6}$$

and

$$P = \rho h \tag{7}$$

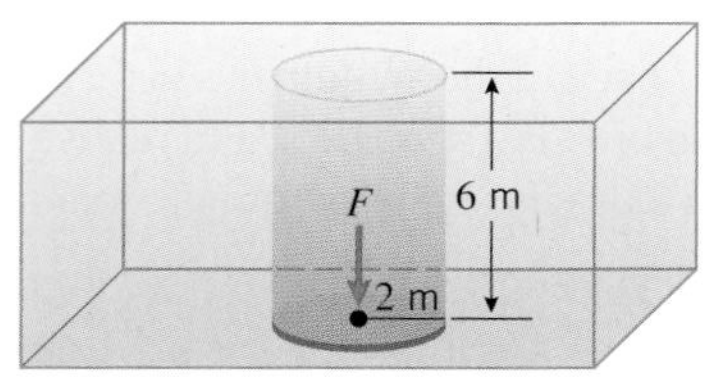

Figure 6.8.3

▶ **Example 2** Find the fluid pressure and force on the top of a flat circular plate of radius 2 m that is submerged horizontally in water at a depth of 6 m (Figure 6.8.3).

Solution. Since the weight density of water is $\rho = 9810\ \text{N/m}^3$, it follows from (7) that the fluid pressure is

$$P = \rho h = (9810)(6) = 58{,}860\ \text{Pa}$$

and it follows from (6) that the fluid force is

$$F = \rho h A = \rho h(\pi r^2) = (9810)(6)(4\pi) = 235{,}440\pi \approx 739{,}700\ \text{N} \ ◀$$

FLUID FORCE ON A VERTICAL SURFACE

It was easy to calculate the fluid force on the horizontal plate in Example 2 because each point on the plate was at the same depth. The problem of finding the fluid force on a vertical surface is more complicated because the depth, and hence the pressure, is not constant over the surface. To find the fluid force on a vertical surface we will need calculus.

> **6.8.2 PROBLEM.** Suppose that a flat surface is immersed vertically in a fluid of weight density ρ and that the submerged portion of the surface extends from $x = a$ to $x = b$ along an x-axis whose positive direction is down (Figure 6.8.4a). For $a \le x \le b$, suppose that $w(x)$ is the width of the surface and that $h(x)$ is the depth of the point x. Define what is meant by the *fluid force F* on the surface, and find a formula for computing it.

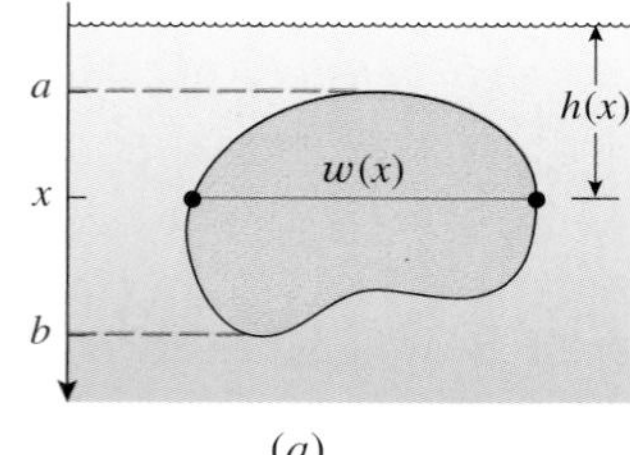

(a)

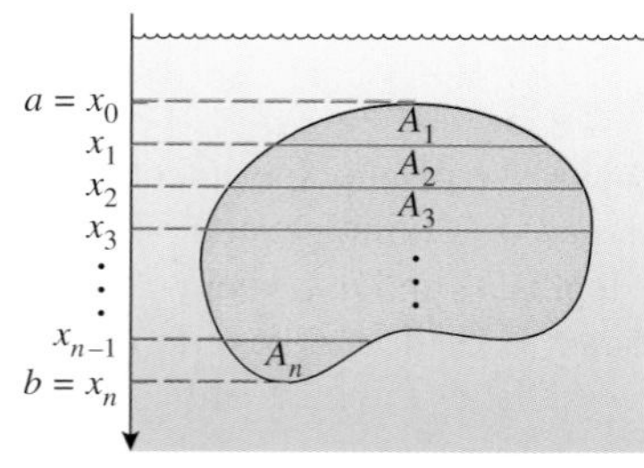

(b)

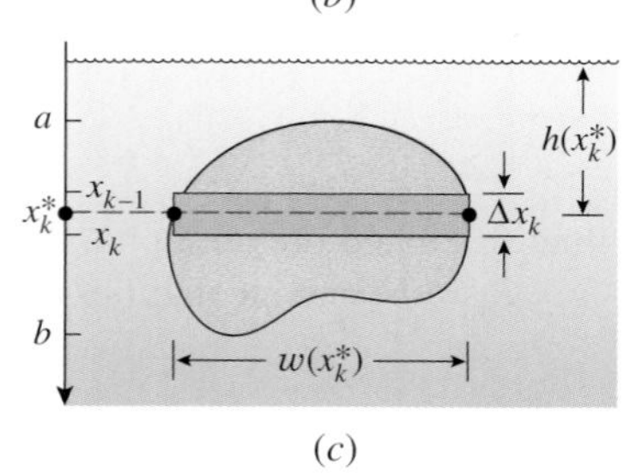

(c)

Figure 6.8.4

The basic idea for solving this problem is to divide the surface into horizontal strips whose areas may be approximated by areas of rectangles. These area approximations, along with inequalities (4), will allow us to create a Riemann sum that approximates the total force on the surface. By taking a limit of Riemann sums we will then obtain an integral for F.

To implement this idea, we divide the interval $[a, b]$ into n subintervals by inserting the points $x_1, x_2, \ldots, x_{n-1}$ between $a = x_0$ and $b = x_n$. This has the effect of dividing the surface into n strips of area A_k, $k = 1, 2, \ldots, n$ (Figure 6.8.4b). It follows from (4) that the force F_k on the kth strip satisfies the inequalities

$$\rho h(x_{k-1})A_k \le F_k \le \rho h(x_k)A_k$$

or, equivalently,

$$h(x_{k-1}) \le \frac{F_k}{\rho A_k} \le h(x_k)$$

Since the depth function $h(x)$ increases linearly, there must exist a point x_k^* between x_{k-1} and x_k such that

$$h(x_k^*) = \frac{F_k}{\rho A_k}$$

or, equivalently,

$$F_k = \rho h(x_k^*)A_k$$

We now approximate the area A_k of the kth strip of the surface by the area of a rectangle of width $w(x_k^*)$ and height $\Delta x_k = x_k - x_{k-1}$ (Figure 6.8.4c). It follows that F_k may be approximated as

$$F_k = \rho h(x_k^*)A_k \approx \rho h(x_k^*) \cdot \underbrace{w(x_k^*)\Delta x_k}_{\text{Area of rectangle}}$$

Adding these approximations yields the following Riemann sum that approximates the total force F on the surface:

$$F = \sum_{k=1}^{n} F_k \approx \sum_{k=1}^{n} \rho h(x_k^*) w(x_k^*) \Delta x_k$$

Taking the limit as n increases and the widths of the subintervals approach zero yields the definite integral

$$F = \lim_{\max \Delta x_k \to 0} \sum_{k=1}^{n} \rho h(x_k^*) w(x_k^*) \Delta x_k = \int_a^b \rho h(x) w(x)\, dx$$

In summary, we have the following result.

6.8.3 DEFINITION. Suppose that a flat surface is immersed vertically in a fluid of weight density ρ and that the submerged portion of the surface extends from $x = a$ to $x = b$ along an x-axis whose positive direction is down (Figure 6.8.4a). For $a \le x \le b$, suppose that $w(x)$ is the width of the surface and that $h(x)$ is the depth of the point x. Then we define the ***fluid force*** F on the surface to be

$$F = \int_a^b \rho h(x) w(x)\, dx \tag{8}$$

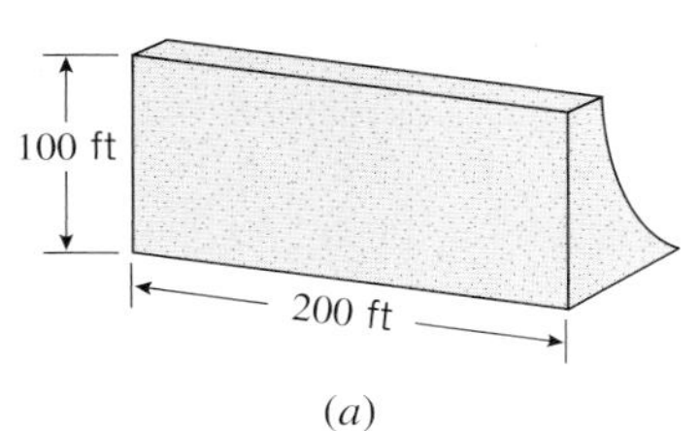

(a)

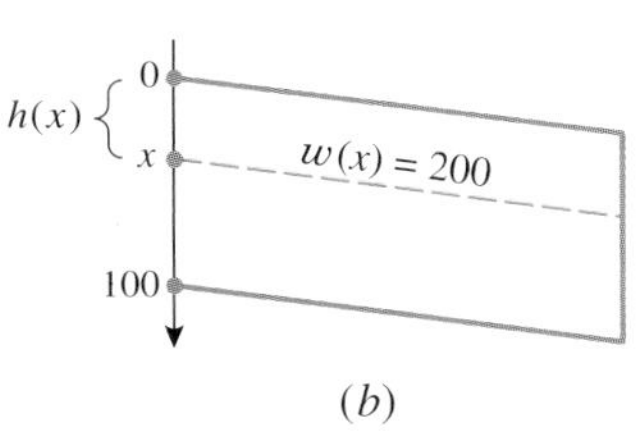

(b)

Figure 6.8.5

▶ Example 3 The face of a dam is a vertical rectangle of height 100 ft and width 200 ft (Figure 6.8.5a). Find the total fluid force exerted on the face when the water surface is level with the top of the dam.

Solution. Introduce an x-axis with its origin at the water surface as shown in Figure 6.8.5b. At a point x on this axis, the width of the dam in feet is $w(x) = 200$ and the depth in feet is $h(x) = x$. Thus, from (8) with $\rho = 62.4$ lb/ft^3 (the weight density of water) we obtain as the total force on the face

$$F = \int_0^{100} (62.4)(x)(200)\, dx = 12{,}480 \int_0^{100} x\, dx$$

$$= 12{,}480 \left. \frac{x^2}{2} \right]_0^{100} = 62{,}400{,}000 \text{ lb} \blacktriangleleft$$

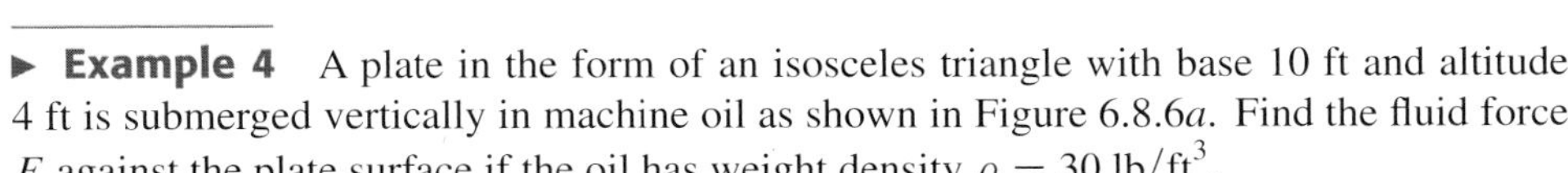

▶ Example 4 A plate in the form of an isosceles triangle with base 10 ft and altitude 4 ft is submerged vertically in machine oil as shown in Figure 6.8.6a. Find the fluid force F against the plate surface if the oil has weight density $\rho = 30$ lb/ft^3.

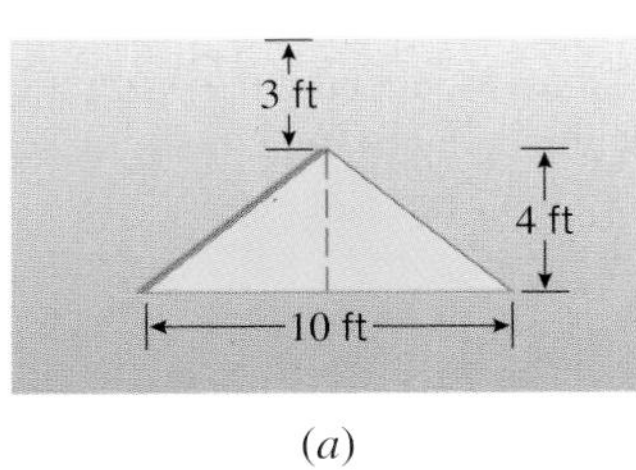

(a)

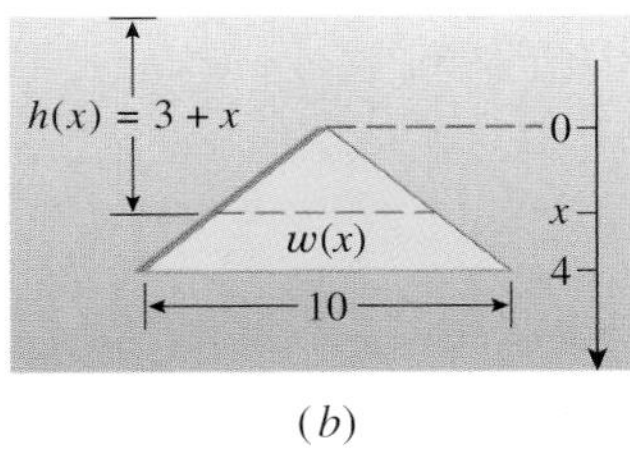

(b)

Figure 6.8.6

Solution. Introduce an x-axis as shown in Figure 6.8.6b. By similar triangles, the width of the plate, in feet, at a depth of $h(x) = (3 + x)$ ft satisfies

$$\frac{w(x)}{10} = \frac{x}{4}, \quad \text{so} \quad w(x) = \frac{5}{2}x$$

Thus, it follows from (8) that the force on the plate is

$$F = \int_a^b \rho h(x) w(x)\, dx = \int_0^4 (30)(3 + x)\left(\frac{5}{2}x\right) dx$$

$$= 75 \int_0^4 (3x + x^2)\, dx = 75 \left[\frac{3x^2}{2} + \frac{x^3}{3} \right]_0^4 = 3400 \text{ lb} \blacktriangleleft$$

✔QUICK CHECK EXERCISES 6.8 *(See page 433 for answers.)*

1. The pressure unit equivalent to a newton per square meter (N/m^2) is called a ________. The pressure unit psi stands for ________.

2. Given that the weight density of water is 9810 N/m^3, the fluid pressure on a rectangular 2 m × 3 m flat plate submerged horizontally in water at a depth of 10 m is ________. The fluid force on the plate is ________.

3. Suppose that a flat surface is immersed vertically in a fluid of weight density ρ and that the submerged portion of the surface extends from $x = a$ to $x = b$ along an x-axis whose positive direction is down. If, for $a \le x \le b$, the surface has width $w(x)$ and depth $h(x)$, then the fluid force on the surface is $F =$ ________.

4. A rectangular plate 2 m wide and 3 m high is submerged vertically in water so that the top of the plate is 5 m below the water surface. An integral expression for the force of the water on the plate surface is $F =$ ________. The value of this integral is ________.

EXERCISE SET 6.8

In this exercise set, refer to Table 6.8.2 for weight densities of fluids, where needed.

1. A flat rectangular plate is submerged horizontally in water.
   (a) Find the force (in lb) and the pressure (in lb/ft^2) on the top surface of the plate if its area is 100 ft^2 and the surface is at a depth of 5 ft.
   (b) Find the force (in N) and the pressure (in Pa) on the top surface of the plate if its area is 25 m^2 and the surface is at a depth of 10 m.

2. (a) Find the force (in N) on the deck of a sunken ship if its area is 160 m^2 and the pressure acting on it is 6.0×10^5 Pa.
   (b) Find the force (in lb) on a diver's face mask if its area is 60 in^2 and the pressure acting on it is 100 lb/in^2.

3–8 The flat surfaces shown are submerged vertically in water. Find the fluid force against each surface.

3.
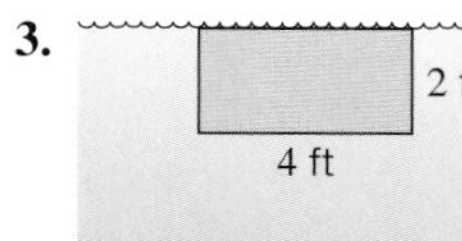

4.
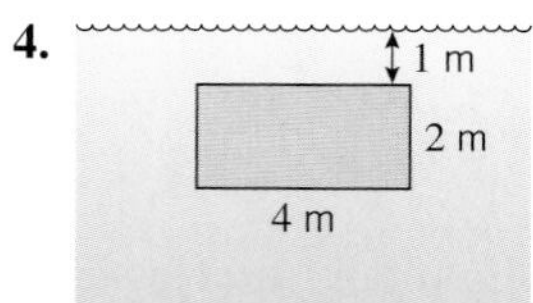

5.
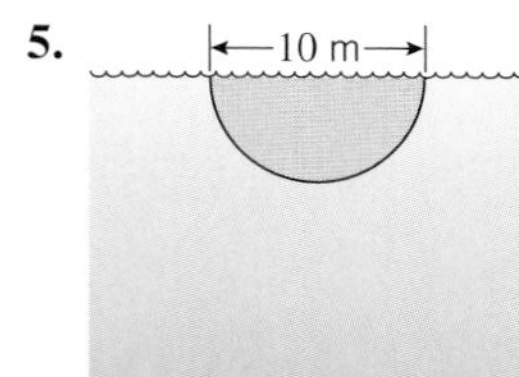

6.
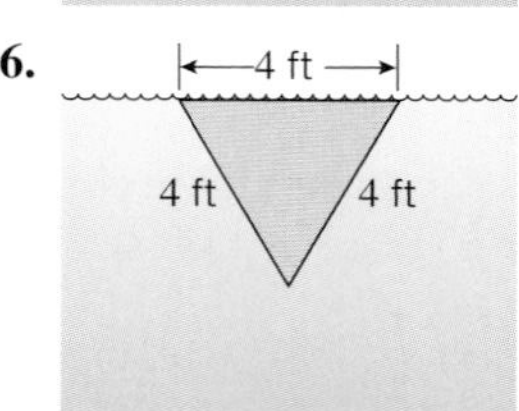

7.
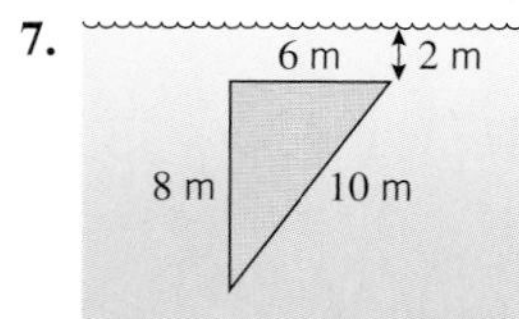

8.
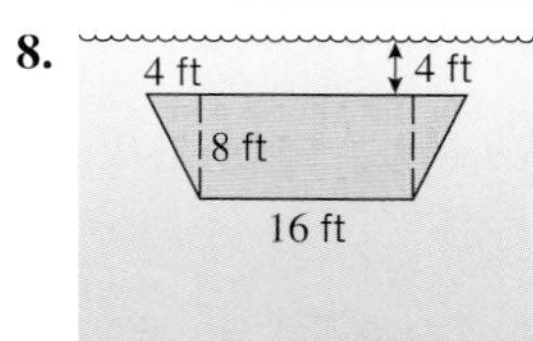

9. Suppose that a flat surface is immersed vertically in a fluid of weight density ρ. If ρ is doubled, is the force on the plate also doubled? Explain your reasoning.

10. An oil tank is shaped like a right circular cylinder of diameter 4 ft. Find the total fluid force against one end when the axis is horizontal and the tank is half filled with oil of weight density 50 lb/ft^3.

11. A square plate of side a feet is dipped in a liquid of weight density ρ lb/ft^3. Find the fluid force on the plate if a vertex is at the surface and a diagonal is perpendicular to the surface.

12–15 Formula (8) gives the fluid force on a flat surface immersed vertically in a fluid. More generally, if a flat surface is immersed so that it makes an angle of $0 \le \theta < \pi/2$ with the vertical, then the fluid force on the surface is given by

$$F = \int_a^b \rho h(x) w(x) \sec\theta \, dx$$

Use this formula in these exercises.

12. Derive the formula given above for the fluid force on a flat surface immersed at an angle in a fluid.

13. The accompanying figure shows a rectangular swimming pool whose bottom is an inclined plane. Find the fluid force on the bottom when the pool is filled to the top.

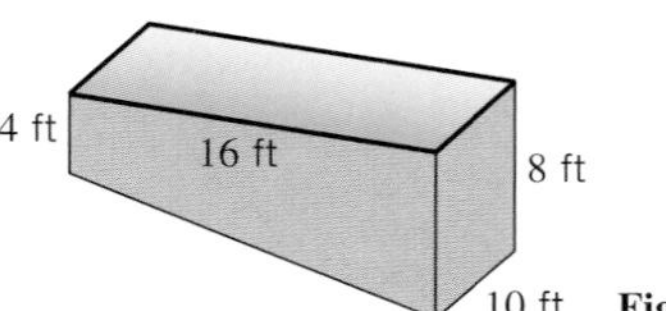

Figure Ex-13

14. By how many feet should the water in the pool of Exercise 13 be lowered in order for the force on the bottom to be reduced by a factor of $\frac{1}{2}$?

15. The accompanying figure shows a dam whose face is an inclined rectangle. Find the fluid force on the face when the water is level with the top of this dam.

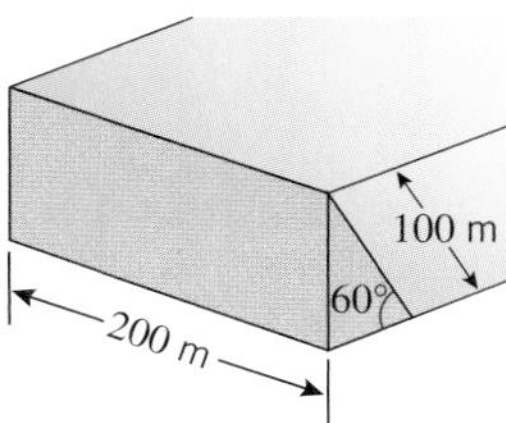

Figure Ex-15

16. An observation window on a submarine is a square with 2-ft sides. Using ρ_0 for the weight density of seawater, find the fluid force on the window when the submarine has descended so that the window is vertical and its top is at a depth of h feet.

FOCUS ON CONCEPTS

17. (a) Show: If the submarine in Exercise 16 descends vertically at a constant rate, then the fluid force on the window increases at a constant rate.
(b) At what rate is the force on the window increasing if the submarine is descending vertically at 20 ft/min?

18. (a) Let $D = D_a$ denote a disk of radius a submerged in a fluid of weight density ρ such that the center of D is h units below the surface of the fluid. For each value of r in the interval $(0, a]$, let D_r denote the disk of radius r that is concentric with D. Select a side of the disk D and define $P(r)$ to be the fluid pressure on the chosen side of D_r. Use (5) to prove that

$$\lim_{r \to 0^+} P(r) = \rho h$$

(b) Explain why the result in part (a) may be interpreted to mean that *fluid pressure at a given depth is the same in all directions*. (This statement is one version of a result known as ***Pascal's Principle***.)

✔QUICK CHECK ANSWERS 6.8

1. pascal; pounds per square inch 2. 98,100 Pa; 588,600 N 3. $\int_a^b \rho h(x) w(x)\, dx$ 4. $\int_0^3 9810[(5+x)2]\, dx$; 382,590 N

CHAPTER REVIEW EXERCISES

1. Describe the method of slicing for finding volumes, and use that method to derive an integral formula for finding volumes by the method of disks.

2. State an integral formula for finding a volume by the method of cylindrical shells, and use Riemann sums to derive the formula.

3. State an integral formula for finding the arc length of a smooth curve $y = f(x)$ over an interval $[a, b]$, and use Riemann sums to derive the formula.

4. State an integral formula for the work W done by a variable force $F(x)$ applied in the direction of motion to an object moving from $x = a$ to $x = b$, and use Riemann sums to derive the formula.

5. State an integral formula for the fluid force F exerted on a vertical flat surface immersed in a fluid of weight density ρ, and use Riemann sums to derive the formula.

6. Let R be the region in the first quadrant enclosed by $y = x^2$, $y = 2 + x$, and $x = 0$. In each part, set up, but *do not evaluate*, an integral or a sum of integrals that will solve the problem.
(a) Find the area of R by integrating with respect to x.
(b) Find the area of R by integrating with respect to y.
(c) Find the volume of the solid generated by revolving R about the x-axis by integrating with respect to x.
(d) Find the volume of the solid generated by revolving R about the x-axis by integrating with respect to y.
(e) Find the volume of the solid generated by revolving R about the y-axis by integrating with respect to x.
(f) Find the volume of the solid generated by revolving R about the y-axis by integrating with respect to y.
(g) Find the volume of the solid generated by revolving R about the line $y = -3$ by integrating with respect to x.
(h) Find the volume of the solid generated by revolving R about the line $x = 5$ by integrating with respect to x.

7. (a) Set up a sum of definite integrals that represents the total shaded area between the curves $y = f(x)$ and $y = g(x)$ in the accompanying figure.
(b) Find the total area enclosed between $y = x^3$ and $y = x$ over the interval $[-1, 2]$.

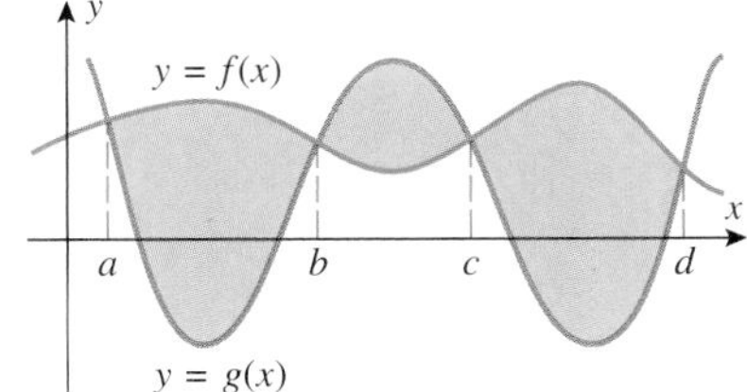

Figure Ex-7

8. The accompanying figure shows velocity versus time curves for two cars that move along a straight track, accelerating from rest at a common starting line.
(a) How far apart are the cars after 60 seconds?
(b) How far apart are the cars after T seconds, where $0 \le T \le 60$?

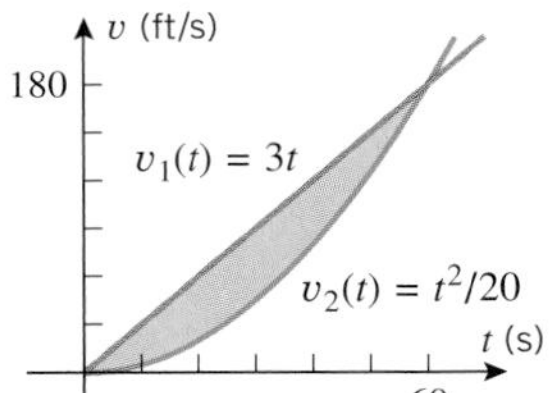

Figure Ex-8

9. Let R be the region enclosed by the curves $y = x^2 + 4$, $y = x^3$, and the y-axis. Find and evaluate a definite integral that represents the volume of the solid generated by revolving R about the x-axis.

10. A football has the shape of the solid generated by revolving the region bounded between the x-axis and the parabola $y = 4R(x^2 - \frac{1}{4}L^2)/L^2$ about the x-axis. Find its volume.

11. Find the arc length in the second quadrant of the curve $x^{2/3} + y^{2/3} = 4$ from $x = -8$ to $x = -1$.

12. Let C be the curve $y = x^3$ between $x = 1$ and $x = 3$. In each part, set up, but *do not evaluate*, an integral that solves the problem.
(a) Find the arc length of C by integrating with respect to x.
(b) Find the arc length of C by integrating with respect to y.

13. Find the area of the surface generated by revolving the curve $y = \sqrt{25 - x}$, $9 \le x \le 16$, about the x-axis.

14. Let C be the curve $27x - y^3 = 0$ between $y = 0$ and $y = 2$. In each part, set up, but *do not evaluate*, an integral or a sum of integrals that solves the problem.
(a) Find the area of the surface generated by revolving C about the x-axis by integrating with respect to x.
(b) Find the area of the surface generated by revolving C about the y-axis by integrating with respect to y.
(c) Find the area of the surface generated by revolving C about the line $y = -2$ by integrating with respect to y.

15. Use the graph of f shown in the accompanying figure to find the average value of f on the interval $[0, 10]$.

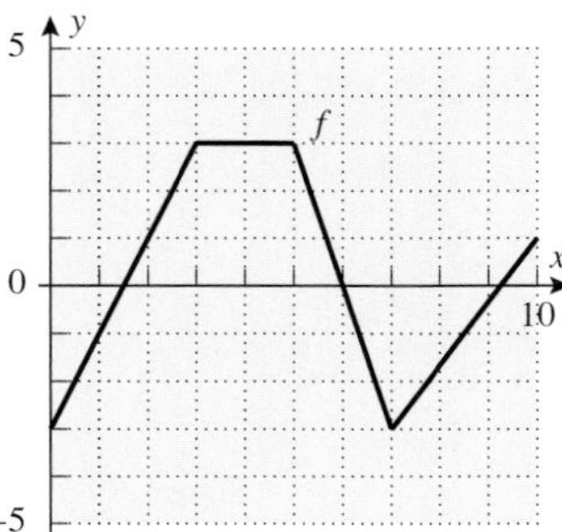

Figure Ex-15

16. Find the average value of $f(x) = x^2 + \dfrac{1}{x^2}$ over the interval $[\frac{1}{2}, 2]$.

17. Consider the solid generated by revolving the region enclosed by $y = \sec x$, $x = 0$, $x = \pi/3$, and $y = 0$ about the x-axis. Find the average value of the area of a cross section of this solid taken perpendicular to the x-axis.

18. Suppose that f is a smooth function on $[a, b]$. Show that the average rate of change of f over $[a, b]$ is the same as the average value of f' over $[a, b]$.

19. (a) A spring exerts a force of 0.5 N when stretched 0.25 m beyond its natural length. Assuming that Hooke's law applies, how much work was performed in stretching the spring to this length?
(b) How far beyond its natural length can the spring be stretched with 25 J of work?

20. A boat is anchored so that the anchor is 150 ft below the surface of the water. In the water, the anchor weighs 2000 lb and the chain weighs 30 lb/ft. How much work is required to raise the anchor to the surface?

21. In each part, set up, but *do not evaluate*, an integral that solves the problem.
(a) Find the fluid force exerted on a side of a box that has a 3-m-square base and is filled to a depth of 1 m with a liquid of weight density ρ N/m^3.
(b) Find the fluid force exerted by a liquid of weight density ρ lb/ft^3 on a face of the vertical plate shown in part (*a*) of the accompanying figure.
(c) Find the fluid force exerted on the parabolic dam in part (*b*) of the accompanying figure by water that extends to the top of the dam.

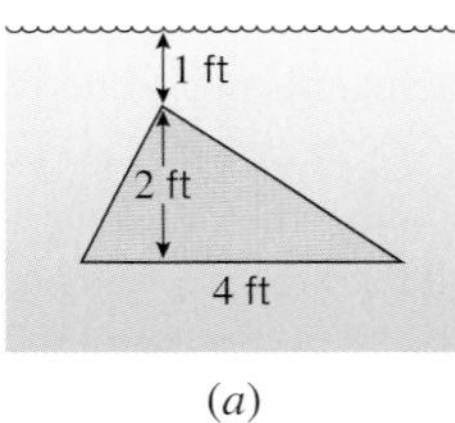

(*a*)

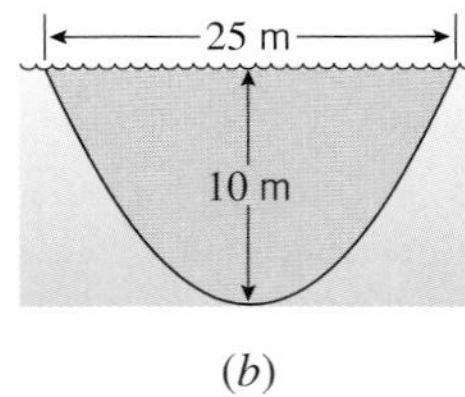

(*b*)

Figure Ex-21

EXPONENTIAL, LOGARITHMIC, AND INVERSE TRIGONOMETRIC FUNCTIONS

Who has not been amazed to learn that the function $y = e^x$, like a phoenix rising again from its own ashes, is its own derivative?
—**François le Lionnais**
Science Writer and Chess Master

We begin this chapter with a review of exponential and logarithmic functions. These functions have important applications, from modeling population growth and the spread of disease, to the measurement of the magnitude of an earthquake or the perceived loudness of a sound. Logarithmic and exponential functions are best understood within the context of inverse functions and we will derive an important relationship between the derivative of a function and the derivative of its inverse. This connection will allow us to compute derivative formulas for logarithmic and exponential functions, along with their associated integration formulas. Later in the chapter we will exploit this connection again, to find the derivatives of inverse trigonometric functions, together with some related integration formulas. Along the way, we will discuss L'Hôpital's rule, a powerful tool for evaluating limits. We conclude the chapter with a study of some important combinations of exponential functions known as "hyperbolic functions."

Photo: *The growth and decline of animal populations and natural resources can be modeled using basic functions studied in this chapter.*

7.1 EXPONENTIAL AND LOGARITHMIC FUNCTIONS

When logarithms were introduced in the seventeenth century as a computational tool, they provided scientists of that period computing power that was previously unimaginable. Although computers and calculators have replaced logarithm tables for numerical calculations, the logarithmic functions have wide-ranging applications in mathematics and science. In this section we will review some properties of exponents and logarithms and then develop results about exponential and logarithmic functions.

IRRATIONAL EXPONENTS

Recall from algebra that if b is a nonzero real number, then nonzero *integer* powers of b are defined by

$$b^n = \underbrace{b \times b \times \cdots \times b}_{n \text{ factors}} \quad \text{and} \quad b^{-n} = \frac{1}{b^n}$$

and if $n = 0$, then $b^0 = 1$. Also, if p/q is a positive *rational* number expressed in lowest terms, then

$$b^{p/q} = \sqrt[q]{b^p} = (\sqrt[q]{b})^p \quad \text{and} \quad b^{-p/q} = \frac{1}{b^{p/q}}$$

If b is negative, then some fractional powers of b will have imaginary values—the quantity $(-2)^{1/2} = \sqrt{-2}$, for example. To avoid this complication, we will assume throughout this section that $b > 0$, even if it is not stated explicitly.

There are various methods for defining *irrational* powers such as

$$2^{\pi}, \quad 3^{\sqrt{2}}, \quad \pi^{-\sqrt{7}}$$

One approach is to define irrational powers of b via successive approximations using rational powers of b. For example, to define 2^{π} consider the decimal representation of π:

$$3.1415926\ldots$$

From this decimal we can form a sequence of rational numbers that gets closer and closer to π, namely,

$$3.1, \quad 3.14, \quad 3.141, \quad 3.1415, \quad 3.14159$$

and from these we can form a sequence of *rational* powers of 2:

$$2^{3.1}, \quad 2^{3.14}, \quad 2^{3.141}, \quad 2^{3.1415}, \quad 2^{3.14159}$$

Since the exponents of the terms in this sequence get successively closer to π, it seems plausible that the terms themselves will get successively closer to some number. It is that number that we *define* to be 2^{π}. This is illustrated in Table 7.1.1, which we generated using a calculator. The table suggests that to four decimal places the value of 2^{π} is

$$2^{\pi} \approx 8.8250 \tag{1}$$

Table 7.1.1

x	2^x
3	8.000000
3.1	8.574188
3.14	8.815241
3.141	8.821353
3.1415	8.824411
3.14159	8.824962
3.141592	8.824974
3.1415926	8.824977

TECHNOLOGY MASTERY

Use a calculating utility to verify the results in Table 7.1.1, and then verify (1) by using the utility to compute 2^{π} directly.

With this notion for irrational powers, we remark without proof that the following familiar laws of exponents hold for all real values of p and q:

$$b^p b^q = b^{p+q}, \quad \frac{b^p}{b^q} = b^{p-q}, \quad (b^p)^q = b^{pq}$$

THE FAMILY OF EXPONENTIAL FUNCTIONS

A function of the form $f(x) = b^x$, where $b > 0$, is called an ***exponential function with base b***. Some examples are

$$f(x) = 2^x, \quad f(x) = \left(\tfrac{1}{2}\right)^x, \quad f(x) = \pi^x$$

Note that an exponential function has a constant base and variable exponent. Thus, functions such as $f(x) = x^2$ and $f(x) = x^{\pi}$ would *not* be classified as exponential functions, since they have a variable base and a constant exponent.

Figure 7.1.1 illustrates that the graph of $y = b^x$ has one of three general forms, depending on the value of b. Notice in this figure that the value of b^x increases as x increases if $b > 1$, it decreases as x increases if $0 < b < 1$, and it is constant if $b = 1$. The graphs all pass through the point $(0, 1)$ because $b^0 = 1$.

If $b > 1$, then as you traverse the graph of $y = b^x$ from left to right the values of b^x increase indefinitely, whereas if you traverse the graph from right to left, the values of b^x decrease toward zero but never reach zero. Similarly, if $0 < b < 1$, then as you traverse the graph from left to right the values of b^x decrease toward zero but never reach zero, whereas if you traverse the graph from right to left the values of b^x increase indefinitely.

Some typical members of the family of exponential functions are graphed in Figure 7.1.2. This figure illustrates that the graph of $y = (1/b)^x$ is the reflection of the graph of $y = b^x$ about the y-axis. This is because replacing x by $-x$ in the equation $y = b^x$ yields

$$y = b^{-x} = (1/b)^x$$

The figure also conveys that the larger the base $b > 1$, the more rapidly $f(x) = b^x$ increases for $x > 0$.

Since it is not our objective in this section to develop properties of exponential functions in rigorous mathematical detail, we will simply observe without proof that the following

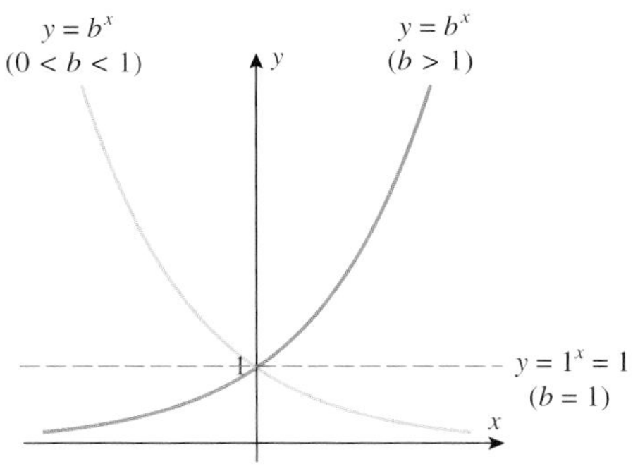

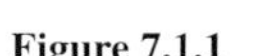

Figure 7.1.1

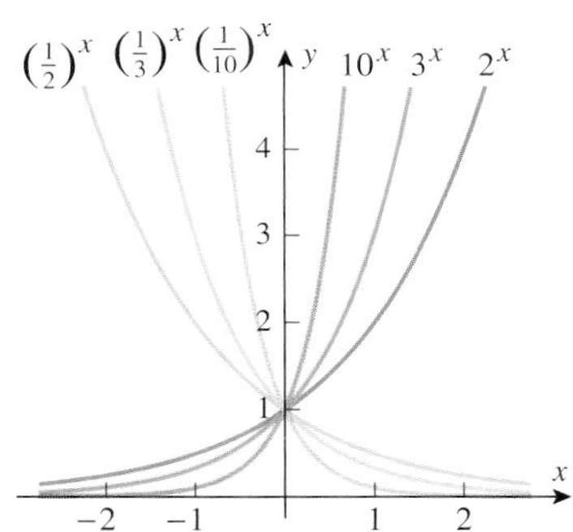

Figure 7.1.2 The family $y = b^x\,(b > 0)$

properties of exponential functions are consistent with the graphs shown in Figures 7.1.1 and 7.1.2.

> **7.1.1 THEOREM.** *If $b > 0$ and $b \neq 1$, then:*
>
> (*a*) *The function $f(x) = b^x$ is defined for all real values of x, so its natural domain is $(-\infty, +\infty)$.*
>
> (*b*) *The function $f(x) = b^x$ is continuous on the interval $(-\infty, +\infty)$, and its range is $(0, +\infty)$.*

▶ **Example 1** Sketch the graph of the function $f(x) = 1 - 2^x$ and find its domain and range.

Solution. Start with a graph of $y = 2^x$. Reflect this graph across the x-axis to obtain the graph of $y = -2^x$, then translate that graph upward by 1 unit to obtain the graph of $y = 1 - 2^x$ (Figure 7.1.3). The dashed line in the third part of Figure 7.1.3 is a horizontal asymptote for the graph. You should be able to see from the graph that the domain of f is $(-\infty, +\infty)$ and the range is $(-\infty, 1)$. ◀

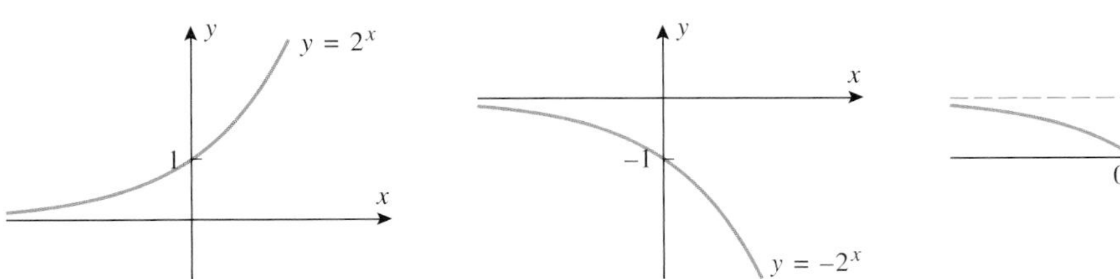

Figure 7.1.3

THE NATURAL EXPONENTIAL FUNCTION

Among all possible bases for exponential functions there is one particular base that plays a special role in calculus. That base, denoted by the letter e, is a certain irrational number whose value to six decimal places is

$$e \approx 2.718282 \tag{2}$$

The use of the letter e is in honor of the Swiss mathematician Leonhard Euler (biography on p. 3) who is credited with recognizing the mathematical importance of this constant.

This base is important in calculus because, as we will prove later, $b = e$ is the only base for which the slope of the tangent line to the curve $y = b^x$ at any point P on the curve is equal

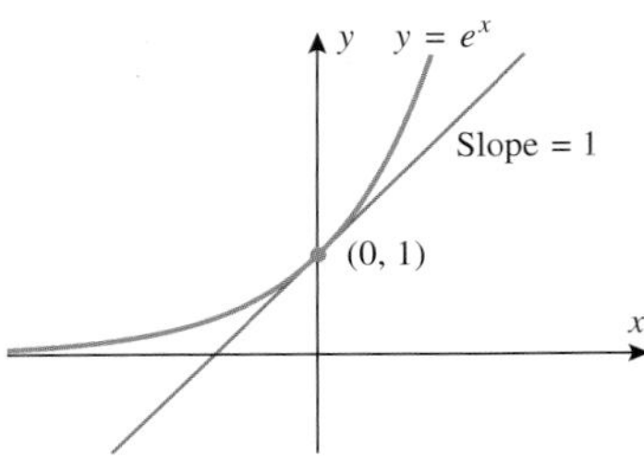

Figure 7.1.4 The tangent line to the graph of $y = e^x$ at (0, 1) has slope 1.

to the y-coordinate at P. Thus, for example, the tangent line to $y = e^x$ at (0, 1) has slope 1 (Figure 7.1.4).

The function $f(x) = e^x$ is called the ***natural exponential function***. Since the number e is between 2 and 3, the graph of $y = e^x$ fits between the graphs of $y = 2^x$ and $y = 3^x$ as shown in Figure 7.1.5. To simplify typography, the natural exponential function is sometimes written as $\exp(x)$ in which case the relationship $e^{x_1+x_2} = e^{x_1}e^{x_2}$ would be expressed as

$$\exp(x_1 + x_2) = \exp(x_1)\exp(x_2)$$

TECHNOLOGY MASTERY

Your technology utility should have keys or commands for approximating e and for graphing the natural exponential function. Read your documentation on how to do this and use your utility to confirm (2) and to generate the graphs in Figure 7.1.5.

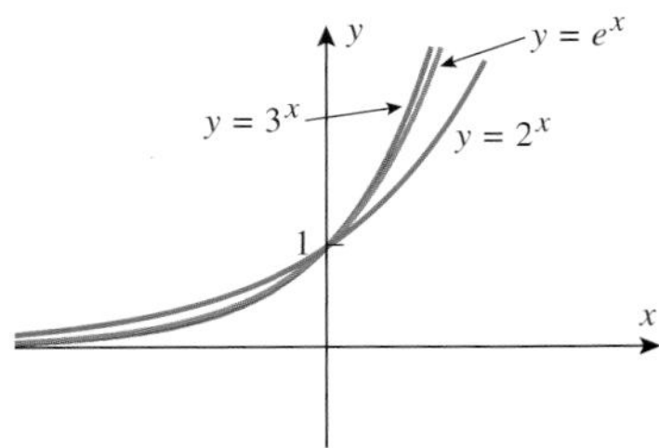

Figure 7.1.5

The constant e also arises in the context of the graph of the equation

$$y = \left(1 + \frac{1}{x}\right)^x$$

As shown in Figure 7.1.6 and Table 7.1.2, $y = e$ is a horizontal asymptote of this graph.

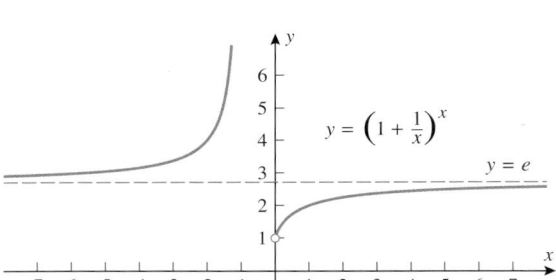

Figure 7.1.6

Table 7.1.2

THE VALUES OF $(1 + 1/x)^x$ APPROACH e AS $x \to +\infty$

x	$1 + \frac{1}{x}$	$\left(1 + \frac{1}{x}\right)^x$
1	2	≈ 2.000000
10	1.1	2.593742
100	1.01	2.704814
1000	1.001	2.716924
10,000	1.0001	2.718146
100,000	1.00001	2.718268
1,000,000	1.000001	2.718280

This is expressed by the limits

$$\lim_{x \to +\infty} \left(1 + \frac{1}{x}\right)^x = e \quad \text{and} \quad \lim_{x \to -\infty} \left(1 + \frac{1}{x}\right)^x = e \tag{3–4}$$

These limits can be derived from the limit

$$\lim_{x \to 0} (1 + x)^{1/x} = e \tag{5}$$

which is sometimes taken as the definition of e.

LOGARITHMIC FUNCTIONS

Recall from algebra that a logarithm is an exponent. More precisely, if $b > 0$ and $b \neq 1$, then for a positive value of x the expression

$$\log_b x$$

(read "the logarithm to the base b of x") denotes that exponent to which b must be raised to produce x. Thus, for example,

$$\log_{10} 100 = 2, \quad \log_{10}(1/1000) = -3, \quad \log_2 16 = 4, \quad \log_b 1 = 0, \quad \log_b b = 1$$

$10^2 = 100$ | $10^{-3} = 1/1000$ | $2^4 = 16$ | $b^0 = 1$ | $b^1 = b$

Logarithms with base 10 are called ***common logarithms*** and are often written without explicit reference to the base. Thus, the symbol $\log x$ generally denotes $\log_{10} x$.

We call the function $f(x) = \log_b x$ the ***logarithmic function with base b***.

Logarithmic functions can also be viewed as inverses of exponential functions. To see why this is so, observe from Figure 7.1.1 that if $b > 0$ and $b \neq 1$, then the graph of $f(x) = b^x$ passes the horizontal line test, so b^x has an inverse. We can find a formula for this inverse with x as the independent variable by solving the equation

$$x = b^y$$

for y as a function of x. But this equation states that y is the logarithm to the base b of x, so it can be rewritten as

$$y = \log_b x$$

Thus, we have established the following result.

7.1.2 THEOREM. *If $b > 0$ and $b \neq 1$, then b^x and $\log_b x$ are inverse functions.*

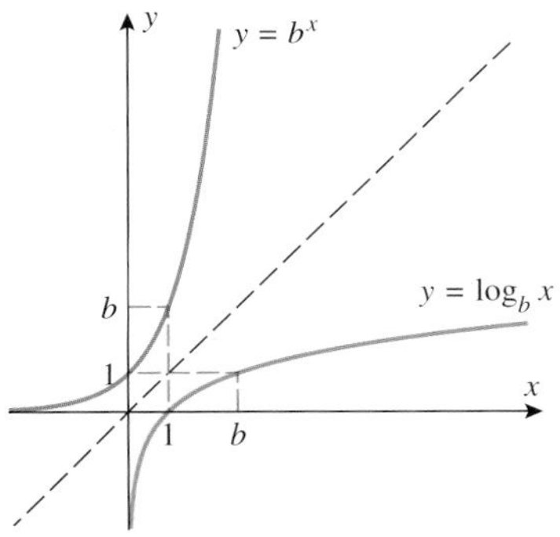

Figure 7.1.7 The functions b^x and $\log_b x$ are inverses.

It follows from this theorem that the graphs of $y = b^x$ and $y = \log_b x$ are reflections of one another about the line $y = x$ (see Figure 7.1.7 for the case where $b > 1$). Figure 7.1.8 shows the graphs of $y = \log_b x$ for various values of b. Observe that they all pass through the point $(1, 0)$.

The most important logarithms in applications are those with base e. These are called ***natural logarithms*** because the function $\log_e x$ is the inverse of the natural exponential function e^x. It is standard to denote the natural logarithm of x by $\ln x$ (read "ell en of x"), rather than $\log_e x$. For example,

$$\ln 1 = 0, \quad \ln e = 1, \quad \ln 1/e = -1, \quad \ln(e^2) = 2$$

Since $e^0 = 1$ | Since $e^1 = e$ | Since $e^{-1} = 1/e$ | Since $e^2 = e^2$

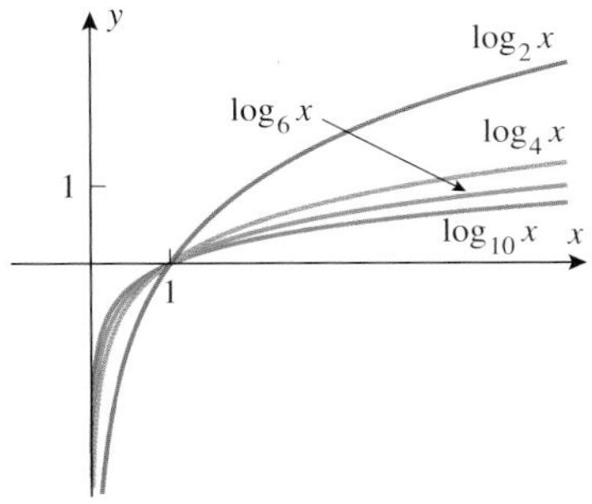

Figure 7.1.8 The family $y = \log_b x$ $(b > 1)$

In general,

$$y = \ln x \quad \text{if and only if} \quad x = e^y$$

As shown in Table 7.1.3, the inverse relationship between b^x and $\log_b x$ produces a correspondence between some basic properties of those functions.

TECHNOLOGY MASTERY

Use your graphing utility to generate the graphs of $y = \ln x$ and $y = \log x$.

Table 7.1.3
CORRESPONDENCE BETWEEN PROPERTIES OF LOGARITHMIC AND EXPONENTIAL FUNCTIONS

PROPERTY OF b^x	PROPERTY OF $\log_b x$
$b^0 = 1$	$\log_b 1 = 0$
$b^1 = b$	$\log_b b = 1$
Range is $(0, +\infty)$	Domain is $(0, +\infty)$
Domain is $(-\infty, +\infty)$	Range is $(-\infty, +\infty)$

It also follows from the cancellation properties of inverse functions [see Formula (3) in Section 1.5] that

$$\begin{aligned} \log_b(b^x) &= x \quad \text{for all real values of } x \\ b^{\log_b x} &= x \quad \text{for } x > 0 \end{aligned} \tag{6}$$

In the special case where $b = e$, these equations become

$$\begin{aligned} \ln(e^x) &= x \quad \text{for all real values of } x \\ e^{\ln x} &= x \quad \text{for } x > 0 \end{aligned} \tag{7}$$

In words, the functions b^x and $\log_b x$ cancel out the effect of one another when composed in either order; for example,

$$\log 10^x = x, \quad 10^{\log x} = x, \quad \ln e^x = x, \quad e^{\ln x} = x, \quad \ln e^5 = 5, \quad e^{\ln \pi} = \pi$$

SOLVING EQUATIONS INVOLVING EXPONENTIALS AND LOGARITHMS

You should be familiar with the following properties of logarithms from your earlier studies.

7.1.3 THEOREM (*Algebraic Properties of Logarithms*). *If $b > 0$, $b \neq 1$, $a > 0$, $c > 0$, and r is any real number, then:*

(*a*) $\log_b(ac) = \log_b a + \log_b c$ — Product property

(*b*) $\log_b(a/c) = \log_b a - \log_b c$ — Quotient property

(*c*) $\log_b(a^r) = r \log_b a$ — Power property

(*d*) $\log_b(1/c) = -\log_b c$ — Reciprocal property

Expressions of the form $\log_b(u + v)$ and $\log_b(u - v)$ have no useful simplifications. In particular,

$$\log_b(u + v) \neq \log_b(u) + \log_b(v)$$
$$\log_b(u - v) \neq \log_b(u) - \log_b(v)$$

These properties are often used to expand a single logarithm into sums, differences, and multiples of other logarithms and, conversely, to condense sums, differences, and multiples of logarithms into a single logarithm. For example,

$$\log \frac{xy^5}{\sqrt{z}} = \log xy^5 - \log \sqrt{z} = \log x + \log y^5 - \log z^{1/2} = \log x + 5 \log y - \tfrac{1}{2} \log z$$

$$5 \log 2 + \log 3 - \log 8 = \log 32 + \log 3 - \log 8 = \log \frac{32 \cdot 3}{8} = \log 12$$

$$\tfrac{1}{3} \ln x - \ln(x^2 - 1) + 2 \ln(x + 3) = \ln x^{1/3} - \ln(x^2 - 1) + \ln(x + 3)^2 = \ln \frac{\sqrt[3]{x}(x + 3)^2}{x^2 - 1}$$

The inverse relationship between logarithmic and exponential functions provides the following useful result for solving equations involving natural exponentials and logarithms:

$$y = e^x \text{ is equivalent to } x = \ln y \text{ if } y > 0 \text{ and } x \text{ is any real number} \tag{8}$$

More generally, if $b > 0$ and $b \neq 1$, then

$$y = b^x \text{ is equivalent to } x = \log_b y \text{ if } y > 0 \text{ and } x \text{ is any real number} \tag{9}$$

An equation of the form $\log_b x = k$ can be solved for x by rewriting it in the exponential form $x = b^k$, and an equation of the form $b^x = k$ can be solved by rewriting it in the logarithm form $x = \log_b k$. Alternatively, the equation $b^x = k$ can be solved by taking *any* logarithm of both sides (but usually log or ln) and applying part (*c*) of Theorem 7.1.3. These ideas are illustrated in the following example.

▸ **Example 2** Find x such that

$$\text{(a) } \log x = \sqrt{2} \qquad \text{(b) } \ln(x+1) = 5 \qquad \text{(c) } 5^x = 7$$

Solution (a). Converting the equation to exponential form yields

$$x = 10^{\sqrt{2}} \approx 25.95$$

Solution (b). Converting the equation to exponential form yields

$$x + 1 = e^5 \quad \text{or} \quad x = e^5 - 1 \approx 147.41$$

Solution (c). Taking the natural logarithm of both sides and using the power property of logarithms yields

$$x \ln 5 = \ln 7 \quad \text{or} \quad x = \frac{\ln 7}{\ln 5} \approx 1.21 \blacktriangleleft$$

▸ **Example 3** A satellite that requires 7 watts of power to operate at full capacity is equipped with a radioisotope power supply whose power output P in watts is given by the equation

$$P = 75e^{-t/125}$$

where t is the time in days that the supply is used. How long can the satellite operate at full capacity?

Solution. The power P will fall to 7 watts when

$$7 = 75e^{-t/125}$$

The solution for t is as follows:

$$\begin{aligned}
7/75 &= e^{-t/125} \\
\ln(7/75) &= \ln(e^{-t/125}) \\
\ln(7/75) &= -t/125 \\
t &= -125 \ln(7/75) \approx 296.4
\end{aligned}$$

so the satellite can operate at full capacity for about 296 days. ◂

Here is a more complicated example.

▸ **Example 4** Solve $\dfrac{e^x - e^{-x}}{2} = 1$ for x.

Solution. Multiplying both sides of the given equation by 2 yields

$$e^x - e^{-x} = 2$$

or equivalently,

$$e^x - \frac{1}{e^x} = 2$$

Multiplying through by e^x yields

$$e^{2x} - 1 = 2e^x \quad \text{or} \quad e^{2x} - 2e^x - 1 = 0$$

This is really a quadratic equation in disguise, as can be seen by rewriting it in the form

$$(e^x)^2 - 2e^x - 1 = 0$$

and letting $u = e^x$ to obtain

$$u^2 - 2u - 1 = 0$$

Solving for u by the quadratic formula yields

$$u = \frac{2 \pm \sqrt{4+4}}{2} = \frac{2 \pm \sqrt{8}}{2} = 1 \pm \sqrt{2}$$

or, since $u = e^x$,

$$e^x = 1 \pm \sqrt{2}$$

But e^x cannot be negative, so we discard the negative value $1 - \sqrt{2}$; thus,

$$\begin{aligned} e^x &= 1 + \sqrt{2} \\ \ln e^x &= \ln(1 + \sqrt{2}) \\ x &= \ln(1 + \sqrt{2}) \approx 0.881 \end{aligned}$$ ◄

CHANGE OF BASE FORMULA FOR LOGARITHMS

Scientific calculators generally provide keys for evaluating common logarithms and natural logarithms but have no keys for evaluating logarithms with other bases. However, this is not a serious deficiency because it is possible to express a logarithm with any base in terms of logarithms with any other base (see Exercise 40). For example, the following formula expresses a logarithm with base b in terms of natural logarithms:

$$\log_b x = \frac{\ln x}{\ln b} \tag{10}$$

We can derive this result by letting $y = \log_b x$, from which it follows that $b^y = x$. Taking the natural logarithm of both sides of this equation we obtain $y \ln b = \ln x$, from which (10) follows.

▶ **Example 5** Use a calculating utility to evaluate $\log_2 5$ by expressing this logarithm in terms of natural logarithms.

Solution. From (10) we obtain

$$\log_2 5 = \frac{\ln 5}{\ln 2} \approx 2.321928$$ ◄

LOGARITHMIC SCALES IN SCIENCE AND ENGINEERING

Logarithms are used in science and engineering to deal with quantities whose units vary over an excessively wide range of values. For example, the "loudness" of a sound can be measured by its ***intensity*** I (in watts per square meter), which is related to the energy transmitted by the sound wave—the greater the intensity, the greater the transmitted energy, and the louder the sound is perceived by the human ear. However, intensity units are unwieldy because they vary over an enormous range. For example, a sound at the threshold of human hearing has an intensity of about 10^{-12} W/m^2, a close whisper has an intensity that is about 100 times the hearing threshold, and a jet engine at 50 meters has an intensity that is about $10{,}000{,}000{,}000{,}000 = 10^{13}$ times the hearing threshold. To see how logarithms can be used to reduce this wide spread, observe that if

$$y = \log x$$

then increasing x by a *factor* of 10 *adds* 1 unit to y since

$$\log 10x = \log 10 + \log x = 1 + y$$

Table 7.1.4

β (dB)	I/I_0
0	$10^0 = 1$
10	$10^1 = 10$
20	$10^2 = 100$
30	$10^3 = 1{,}000$
40	$10^4 = 10{,}000$
50	$10^5 = 100{,}000$
⋮	⋮
120	$10^{12} = 1{,}000{,}000{,}000{,}000$

Physicists and engineers take advantage of this property by measuring loudness in terms of the ***sound level*** β, which is defined by

$$\beta = 10\log(I/I_0)$$

where $I_0 = 10^{-12}$ W/m^2 is a reference intensity close to the threshold of human hearing. The units of β are ***decibels*** (dB), named in honor of the telephone inventor Alexander Graham Bell. With this scale of measurement, *multiplying* the intensity I by a factor of 10 *adds* 10 dB to the sound level β (verify). This results in a more tractable scale than intensity for measuring sound loudness (Table 7.1.4). Some other familiar logarithmic scales are the ***Richter scale*** used to measure earthquake intensity and the **pH *scale*** used to measure acidity in chemistry, both of which are discussed in the exercises.

Peter Townsend of The Who sustained permanent hearing reduction due to the high decibel level of his band's music.

▸ **Example 6** In 1976 the rock group The Who set a record for the loudest concert: 120 dB. By comparison, a jackhammer positioned at the same spot as The Who would have produced a sound level of 92 dB. What is the ratio of the sound intensity of The Who to the sound intensity of a jackhammer?

Solution. Let I_1 and β_1 (= 120 dB) denote the intensity and sound level of The Who, and let I_2 and β_2 (= 92 dB) denote the intensity and sound level of the jackhammer. Then

$$\begin{aligned}
I_1/I_2 &= (I_1/I_0)/(I_2/I_0) \\
\log(I_1/I_2) &= \log(I_1/I_0) - \log(I_2/I_0) \\
10\log(I_1/I_2) &= 10\log(I_1/I_0) - 10\log(I_2/I_0) \\
10\log(I_1/I_2) &= \beta_1 - \beta_2 = 120 - 92 = 28 \\
\log(I_1/I_2) &= 2.8
\end{aligned}$$

Thus, $I_1/I_2 = 10^{2.8} \approx 631$, which tells us that the sound intensity of The Who was 630 times greater than a jackhammer! ◂

EXPONENTIAL AND LOGARITHMIC GROWTH

Table 7.1.5

x	e^x	$\ln x$
1	2.72	0.00
2	7.39	0.69
3	20.09	1.10
4	54.60	1.39
5	148.41	1.61
6	403.43	1.79
7	1096.63	1.95
8	2980.96	2.08
9	8103.08	2.20
10	22026.47	2.30
100	2.69×10^{43}	4.61
1000	1.97×10^{434}	6.91

The growth patterns of e^x and $\ln x$ illustrated in Table 7.1.5 are worth noting. Both functions increase as x increases, but they increase in dramatically different ways—the value of e^x increases extremely rapidly and that of $\ln x$ increases extremely slowly. For example, at $x = 10$ the value of e^x is over 22,000, but at $x = 1000$ the value of $\ln x$ has not even reached 7.

Table 7.1.5 strongly suggests that $f(x) = e^x$ increases without bound, which is consistent with the fact that the range of this function is $(0, +\infty)$. Indeed, if we choose any positive number M, then we will have $e^x = M$ when $x = \ln M$, and since the values of e^x increase as x increases, we will have

$$e^x > M \quad \text{if} \quad x > \ln M$$

(Figure 7.1.9). It is not clear from Table 7.1.5 whether $\ln x$ increases without bound as x increases because the values grow so slowly, but we know this to be so since the range of this function is $(-\infty, +\infty)$. To see this algebraically, let M be any positive number. We will have $\ln x = M$ when $x = e^M$, and since the values of $\ln x$ increase as x increases, we will have

$$\ln x > M \quad \text{if} \quad x > e^M$$

(Figure 7.1.10).

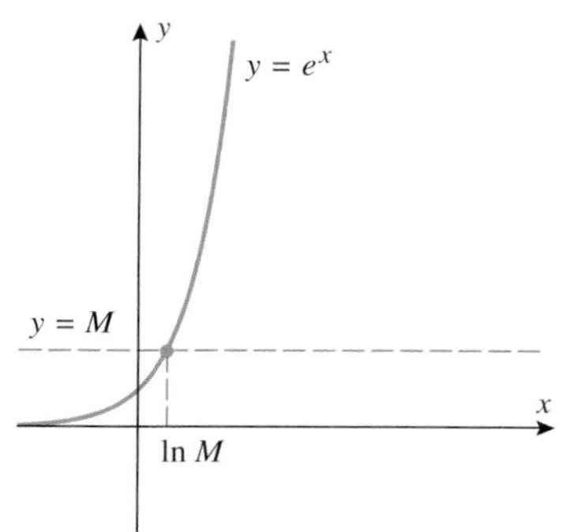

Figure 7.1.9 The value of $y = e^x$ will exceed an arbitrary positive value of M when $x > \ln M$.

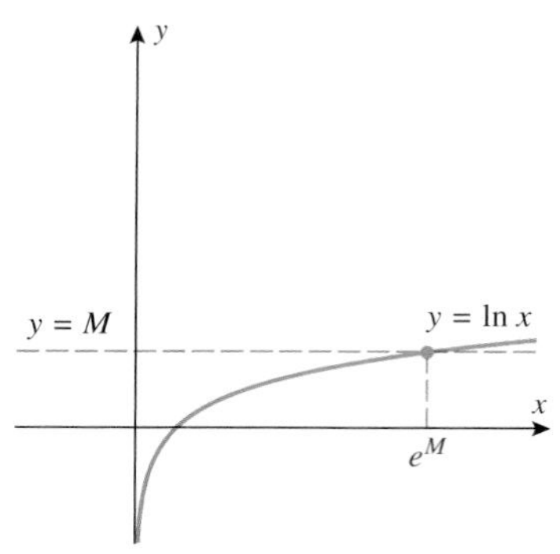

Figure 7.1.10 The value of $y = \ln x$ will exceed an arbitrary positive value of M when $x > e^M$.

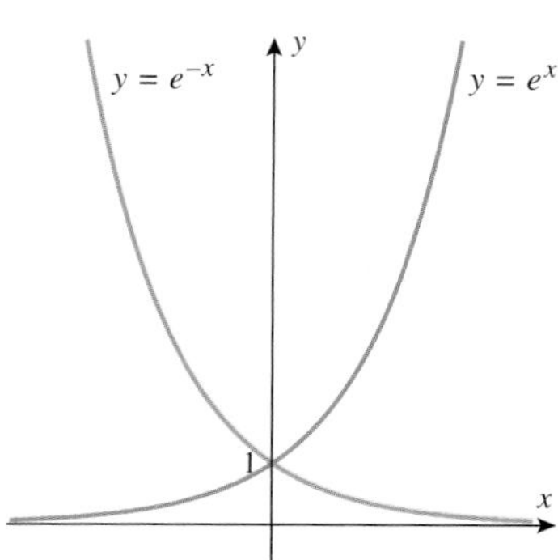

Figure 7.1.11

In summary,

$$\lim_{x\to+\infty} e^x = +\infty \qquad \lim_{x\to+\infty} \ln x = +\infty \tag{11–12}$$

The following limits can be deduced numerically by constructing appropriate tables of values (verify):

$$\lim_{x\to-\infty} e^x = 0 \qquad \lim_{x\to 0^+} \ln x = -\infty \tag{13–14}$$

The following limits can be deduced numerically, but they can be seen more readily by noting that the graph of $y = e^{-x}$ is the reflection about the y-axis of the graph of $y = e^x$ (Figure 7.1.11):

$$\lim_{x\to+\infty} e^{-x} = 0 \qquad \lim_{x\to-\infty} e^{-x} = +\infty \tag{15–16}$$

✔ QUICK CHECK EXERCISES 7.1 *(See page 446 for answers.)*

1. The function $y = \left(\frac{1}{2}\right)^x$ has domain ________ and range ________.

2. The function $y = \ln(1 - x)$ has domain ________ and range ________.

3. Express as a power of 4:
(a) 1 (b) 2 (c) $\frac{1}{16}$ (d) $\sqrt{8}$ (e) 5.

4. Solve each equation for x.
(a) $e^x = \frac{1}{2}$ (b) $10^{3x} = 1{,}000{,}000$
(c) $7e^{3x} = 56$

5. Solve each equation for x.
(a) $\ln x = 3$ (b) $\log(x - 1) = 2$
(c) $2\log x - \log(x + 1) = \log 4 - \log 3$

EXERCISE SET 7.1 Graphing Utility

1–2 Simplify the expression without using a calculating utility.

1. (a) $-8^{2/3}$ (b) $(-8)^{2/3}$ (c) $8^{-2/3}$
2. (a) 2^{-4} (b) $4^{1.5}$ (c) $9^{-0.5}$

3–4 Use a calculating utility to approximate the expression. Round your answer to four decimal places.

3. (a) $2^{1.57}$ (b) $5^{-2.1}$
4. (a) $\sqrt[5]{24}$ (b) $\sqrt[8]{0.6}$

5–6 Find the exact value of the expression without using a calculating utility.

5. (a) $\log_2 16$ (b) $\log_2\left(\frac{1}{32}\right)$
(c) $\log_4 4$ (d) $\log_9 3$
6. (a) $\log_{10}(0.001)$ (b) $\log_{10}(10^4)$
(c) $\ln(e^3)$ (d) $\ln(\sqrt{e})$

7–8 Use a calculating utility to approximate the expression. Round your answer to four decimal places.

7. (a) $\log 23.2$ (b) $\ln 0.74$
8. (a) $\log 0.3$ (b) $\ln \pi$

9–10 Use the logarithm properties in Theorem 7.1.3 to rewrite the expression in terms of r, s, and t, where $r = \ln a$, $s = \ln b$, and $t = \ln c$.

9. (a) $\ln a^2\sqrt{bc}$ (b) $\ln \dfrac{b}{a^3c}$

10. (a) $\ln \dfrac{\sqrt[3]{c}}{ab}$ (b) $\ln \sqrt{\dfrac{ab^3}{c^2}}$

11–12 Expand the logarithm in terms of sums, differences, and multiples of simpler logarithms.

11. (a) $\log(10x\sqrt{x-3})$ (b) $\ln \dfrac{x^2 \sin^3 x}{\sqrt{x^2+1}}$

12. (a) $\log \dfrac{\sqrt[3]{x+2}}{\cos 5x}$ (b) $\ln \sqrt{\dfrac{x^2+1}{x^3+5}}$

13–15 Rewrite the expression as a single logarithm.

13. $4\log 2 - \log 3 + \log 16$

14. $\frac{1}{2}\log x - 3\log(\sin 2x) + 2$

15. $2\ln(x+1) + \frac{1}{3}\ln x - \ln(\cos x)$

16–25 Solve for x without using a calculating utility.

16. $\log_{10}(1+x) = 3$

17. $\log_{10}(\sqrt{x}) = -1$

18. $\ln(x^2) = 4$

19. $\ln(1/x) = -2$

20. $\log_3(3^x) = 7$

21. $\log_5(5^{2x}) = 8$

22. $\log_{10} x^2 + \log_{10} x = 30$

23. $\log_{10} x^{3/2} - \log_{10}\sqrt{x} = 5$

24. $\ln 4x - 3\ln(x^2) = \ln 2$

25. $\ln(1/x) + \ln(2x^3) = \ln 3$

26–31 Solve for x without using a calculating utility. Use the natural logarithm anywhere that logarithms are needed.

26. $3^x = 2$

27. $5^{-2x} = 3$

28. $3e^{-2x} = 5$

29. $2e^{3x} = 7$

30. $e^x - 2xe^x = 0$

31. $xe^{-x} + 2e^{-x} = 0$

32–33 Rewrite the given equation as a quadratic equation in u, where $u = e^x$; then solve for x.

32. $e^{2x} - e^x = 6$

33. $e^{-2x} - 3e^{-x} = -2$

FOCUS ON CONCEPTS

34–36 Sketch the graph of the equation without using a graphing utility.

34. (a) $y = 1 + \ln(x-2)$ (b) $y = 3 + e^{x-2}$

35. (a) $y = \left(\frac{1}{2}\right)^{x-1} - 1$ (b) $y = \ln|x|$

36. (a) $y = 1 - e^{-x+1}$ (b) $y = 3\ln\sqrt[3]{x-1}$

37. Use a calculating utility and the change of base formula (10) to find the values of $\log_2 7.35$ and $\log_5 0.6$, rounded to four decimal places.

38–39 Graph the functions on the same screen of a graphing utility. [Use the change of base formula (10), where needed.]

38. $\ln x,\ e^x,\ \log x,\ 10^x$

39. $\log_2 x,\ \ln x,\ \log_5 x,\ \log x$

40. (a) Derive the general change of base formula

$$\log_b x = \frac{\log_a x}{\log_a b}$$

(b) Use the result in part (a) to find the exact value of $(\log_2 81)(\log_3 32)$ without using a calculating utility.

41. Use a graphing utility to estimate the two points of intersection of the graphs of $y = 1.3^x$ and $y = \log_{1.3} x$.

42. The United States public debt D, in billions of dollars, has been modeled as $D = 0.051517(1.1306727)^x$, where x is the number of years since 1900. Based on this model, when did the debt first reach one trillion dollars?

FOCUS ON CONCEPTS

43. (a) Is the curve in the accompanying figure the graph of an exponential function? Explain your reasoning.
(b) Find the equation of an exponential function that passes through the point (4, 2).
(c) Find the equation of an exponential function that passes through the point $(2, \frac{1}{4})$.
(d) Use a graphing utility to generate the graph of an exponential function that passes through the point (2, 5).

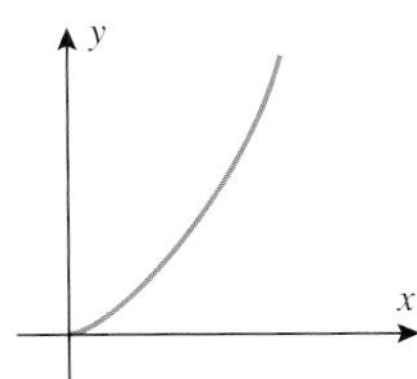

Figure Ex-43

44. (a) Make a conjecture about the general shape of the graph of $y = \log(\log x)$, and sketch the graph of this equation and $y = \log x$ in the same coordinate system.
(b) Check your work in part (a) with a graphing utility.

45. Find the fallacy in the following "proof" that $\frac{1}{8} > \frac{1}{4}$. Multiply both sides of the inequality $3 > 2$ by $\log \frac{1}{2}$ to get

$$3\log\tfrac{1}{2} > 2\log\tfrac{1}{2}$$
$$\log\left(\tfrac{1}{2}\right)^3 > \log\left(\tfrac{1}{2}\right)^2$$
$$\log\tfrac{1}{8} > \log\tfrac{1}{4}$$
$$\tfrac{1}{8} > \tfrac{1}{4}$$

46. Prove the four algebraic properties of logarithms in Theorem 7.1.3.

47. If equipment in the satellite of Example 3 requires 15 watts to operate correctly, what is the operational lifetime of the power supply?

48. The equation $Q = 12e^{-0.055t}$ gives the mass Q in grams of radioactive potassium-42 that will remain from some initial quantity after t hours of radioactive decay.
(a) How many grams were there initially?
(b) How many grams remain after 4 hours?
(c) How long will it take to reduce the amount of radioactive potassium-42 to half of the initial amount?

49. The acidity of a substance is measured by its pH value, which is defined by the formula

$$\text{pH} = -\log[H^+]$$

where the symbol $[H^+]$ denotes the concentration of hydrogen ions measured in moles per liter. Distilled water has a pH of 7; a substance is called *acidic* if it has pH < 7 and *basic* if it has pH > 7. Find the pH of each of the following substances and state whether it is acidic or basic.

	SUBSTANCE	$[H^+]$
(a)	Arterial blood	3.9×10^{-8} mol/L
(b)	Tomatoes	6.3×10^{-5} mol/L
(c)	Milk	4.0×10^{-7} mol/L
(d)	Coffee	1.2×10^{-6} mol/L

50. Use the definition of pH in Exercise 49 to find $[H^+]$ in a solution having a pH equal to
(a) 2.44 (b) 8.06.

51. The perceived loudness β of a sound in decibels (dB) is related to its intensity I in watts per square meter (W/m^2) by the equation

$$\beta = 10\log(I/I_0)$$

where $I_0 = 10^{-12}$ W/m^2. Damage to the average ear occurs at 90 dB or greater. Find the decibel level of each of the following sounds and state whether it will cause ear damage.

	SOUND	I
(a)	Jet aircraft (from 50 ft)	1.0×10^2 W/m^2
(b)	Amplified rock music	1.0 W/m^2
(c)	Garbage disposal	1.0×10^{-4} W/m^2
(d)	TV (mid volume from 10 ft)	3.2×10^{-5} W/m^2

52–54 Use the definition of the decibel level of a sound (see Exercise 51).

52. If one sound is three times as intense as another, how much greater is its decibel level?

53. According to one source, the noise inside a moving automobile is about 70 dB, whereas an electric blender generates 93 dB. Find the ratio of the intensity of the noise of the blender to that of the automobile.

54. Suppose that the intensity level of an echo is $\frac{2}{3}$ the intensity level of the original sound. If each echo results in another echo, how many echoes will be heard from a 120-dB sound given that the average human ear can hear a sound as low as 10 dB?

55. On the ***Richter scale***, the magnitude M of an earthquake is related to the released energy E in joules (J) by the equation

$$\log E = 4.4 + 1.5M$$

(a) Find the energy E of the 1906 San Francisco earthquake that registered $M = 8.2$ on the Richter scale.
(b) If the released energy of one earthquake is 10 times that of another, how much greater is its magnitude on the Richter scale?

56. Suppose that the magnitudes of two earthquakes differ by 1 on the Richter scale. Find the ratio of the released energy of the larger earthquake to that of the smaller earthquake. [*Note:* See Exercise 55 for terminology.]

✔QUICK CHECK ANSWERS 7.1

1. $(-\infty, +\infty)$; $(0, +\infty)$ **2.** $(-\infty, 1)$; $(-\infty, +\infty)$ **3.** (a) 4^0 (b) $4^{1/2}$ (c) 4^{-2} (d) $4^{3/4}$ (e) $4^{\log_4 5}$
4. (a) $\ln\frac{1}{2} = -\ln 2$ (b) 2 (c) $\ln 2$ **5.** (a) e^3 (b) 101 (c) 2

7.2 DERIVATIVES AND INTEGRALS INVOLVING LOGARITHMIC FUNCTIONS

In this section we will obtain derivative formulas for logarithmic functions, and we will explain why the natural logarithm function is preferred over logarithms with other bases in calculus. The derivative formulas which we derive will allow us to find and use corresponding integral formulas.

DERIVATIVES OF LOGARITHMIC FUNCTIONS

We begin by establishing that $f(x) = \ln x$ is differentiable for $x > 0$ by using the derivative definition to find its derivative. To obtain this derivative, we need the fact that $\ln x$ is continuous for $x > 0$. Since e^x is continuous by Theorem 7.1.1(b), we know that $\ln x$ is continuous for $x > 0$ by Theorem 2.5.7. We will also need the limit

$$\lim_{v \to 0}(1 + v)^{1/v} = e \tag{1}$$

that was given in Formula (5) of Section 7.1 (with x rather than v as the variable). Using the definition of a derivative, we obtain

$$\begin{aligned}
\frac{d}{dx}[\ln x] &= \lim_{h\to 0}\frac{\ln(x+h) - \ln x}{h} \\
&= \lim_{h\to 0}\frac{1}{h}\ln\left(\frac{x+h}{x}\right) && \text{The quotient property of logarithms in Theorem 7.1.3} \\
&= \lim_{h\to 0}\frac{1}{h}\ln\left(1+\frac{h}{x}\right) \\
&= \lim_{v\to 0}\frac{1}{vx}\ln(1+v) && \text{Let } v = h/x \text{ and note that } v\to 0 \text{ if and only if } h\to 0. \\
&= \frac{1}{x}\lim_{v\to 0}\frac{1}{v}\ln(1+v) && x \text{ is fixed in this limit computation, so } 1/x \text{ can be moved through the limit sign.} \\
&= \frac{1}{x}\lim_{v\to 0}\ln(1+v)^{1/v} && \text{The power property of logarithms in Theorem 7.1.3} \\
&= \frac{1}{x}\ln\left[\lim_{v\to 0}(1+v)^{1/v}\right] && \ln x \text{ is continuous on } (0,+\infty) \text{ so we can move the limit through the function symbol.} \\
&= \frac{1}{x}\ln e \\
&= \frac{1}{x} && \text{Since } \ln e = 1
\end{aligned}$$

Thus,

$$\frac{d}{dx}[\ln x] = \frac{1}{x}, \quad x > 0 \tag{2}$$

A derivative formula for the general logarithmic function $\log_b x$ can be obtained from (2) by using Formula (10) of Section 7.1 to write

$$\frac{d}{dx}[\log_b x] = \frac{d}{dx}\left[\frac{\ln x}{\ln b}\right] = \frac{1}{\ln b}\frac{d}{dx}[\ln x]$$

It follows from this that

$$\frac{d}{dx}[\log_b x] = \frac{1}{x\ln b}, \quad x > 0 \tag{3}$$

Note that, among all possible bases, the base $b = e$ produces the simplest formula for the derivative of $\log_b x$. This is one of the reasons why the natural logarithm function is preferred over other logarithms in calculus.

▶ Example 1

(a) Figure 7.2.1 shows the graph of $y = \ln x$ and its tangent lines at the points $x = \frac{1}{2}, 1, 3,$ and 5. Find the slopes of those tangent lines.

(b) Does the graph of $y = \ln x$ have any horizontal tangent lines? Use the derivative of $\ln x$ to justify your answer.

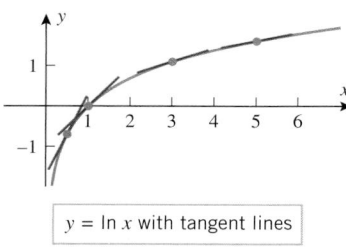

$y = \ln x$ with tangent lines

Figure 7.2.1

Solution (a). From (2), the slopes of the tangent lines at the points $x = \frac{1}{2}, 1, 3,$ and 5 are $1/x = 2, 1, \frac{1}{3},$ and $\frac{1}{5}$, respectively, which is consistent with Figure 7.2.1.

Solution (b). It does not appear from the graph of $y = \ln x$ that there are any horizontal tangent lines. This is confirmed by the fact that $dy/dx = 1/x$ is not equal to zero for any real value of x. ◀

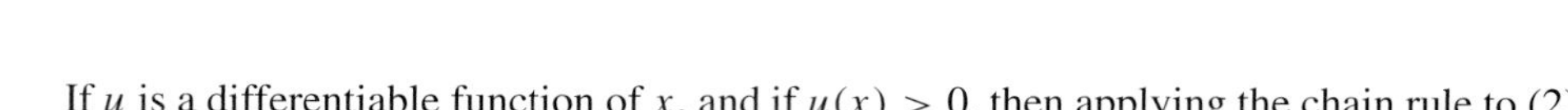

If u is a differentiable function of x, and if $u(x) > 0$, then applying the chain rule to (2) and (3) produces the following generalized derivative formulas:

$$\frac{d}{dx}[\ln u] = \frac{1}{u} \cdot \frac{du}{dx} \quad \text{and} \quad \frac{d}{dx}[\log_b u] = \frac{1}{u \ln b} \cdot \frac{du}{dx} \tag{4–5}$$

▶ Example 2 Find $\dfrac{d}{dx}[\ln(x^2 + 1)]$.

Solution. Using (4) with $u = x^2 + 1$ we obtain

$$\frac{d}{dx}[\ln(x^2 + 1)] = \frac{1}{x^2 + 1} \cdot \frac{d}{dx}[x^2 + 1] = \frac{1}{x^2 + 1} \cdot 2x = \frac{2x}{x^2 + 1} \quad ◀$$

When possible, the properties of logarithms in Theorem 7.1.3 should be used to convert products, quotients, and exponents into sums, differences, and constant multiples *before* differentiating a function involving logarithms.

▶ Example 3

$$\begin{aligned} \frac{d}{dx}\left[\ln\left(\frac{x^2 \sin x}{\sqrt{1 + x}}\right)\right] &= \frac{d}{dx}\left[2 \ln x + \ln(\sin x) - \frac{1}{2}\ln(1 + x)\right] \\ &= \frac{2}{x} + \frac{\cos x}{\sin x} - \frac{1}{2(1 + x)} \\ &= \frac{2}{x} + \cot x - \frac{1}{2 + 2x} \quad ◀ \end{aligned}$$

Figure 7.2.2 shows the graph of $f(x) = \ln |x|$. This function is important because it "extends" the domain of the natural logarithm function in the sense that the values of $\ln |x|$ and $\ln x$ are the same for $x > 0$, but $\ln |x|$ is defined for all nonzero values of x, and $\ln x$ is only defined for positive values of x.

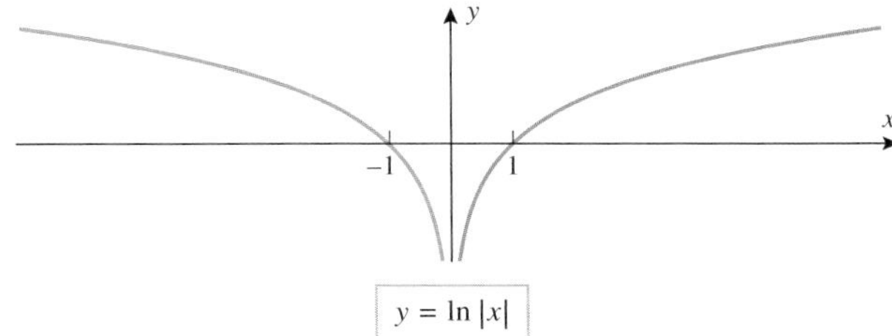

Figure 7.2.2

The derivative of $\ln|x|$ for $x \neq 0$ can be obtained by considering the cases $x > 0$ and $x < 0$ separately:

Case $x > 0$. In this case $|x| = x$, so

$$\frac{d}{dx}[\ln|x|] = \frac{d}{dx}[\ln x] = \frac{1}{x}$$

Case $x < 0$. In this case $|x| = -x$, so it follows from (4) that

$$\frac{d}{dx}[\ln|x|] = \frac{d}{dx}[\ln(-x)] = \frac{1}{(-x)} \cdot \frac{d}{dx}[-x] = \frac{1}{x}$$

Since the same formula results in both cases, we have shown that

$$\frac{d}{dx}[\ln|x|] = \frac{1}{x} \quad \text{if } x \neq 0 \tag{6}$$

▶ **Example 4** From (6) and the chain rule,

$$\frac{d}{dx}[\ln|\sin x|] = \frac{1}{\sin x} \cdot \frac{d}{dx}[\sin x] = \frac{\cos x}{\sin x} = \cot x \ ◀$$

LOGARITHMIC DIFFERENTIATION

We now consider a technique called ***logarithmic differentiation*** that is useful for differentiating functions that are composed of products, quotients, and powers.

▶ **Example 5** The derivative of

$$y = \frac{x^2\sqrt[3]{7x-14}}{(1+x^2)^4} \tag{7}$$

is messy to calculate directly. However, if we first take the natural logarithm of both sides and then use its properties, we can write

$$\ln y = 2\ln x + \tfrac{1}{3}\ln(7x-14) - 4\ln(1+x^2)$$

Differentiating both sides with respect to x yields

$$\frac{1}{y}\frac{dy}{dx} = \frac{2}{x} + \frac{7/3}{7x-14} - \frac{8x}{1+x^2}$$

Thus, on solving for dy/dx and using (7) we obtain

$$\frac{dy}{dx} = \frac{x^2\sqrt[3]{7x-14}}{(1+x^2)^4}\left[\frac{2}{x} + \frac{1}{3x-6} - \frac{8x}{1+x^2}\right] \ ◀$$

Since $\ln y$ is only defined for $y > 0$, the computations in Example 5 are only valid for $x > 2$ (verify). However, because the derivative of $\ln y$ is the same as the derivative of $\ln |y|$, and because $\ln |y|$ is defined for $y < 0$ as well as $y > 0$, it follows that the formula obtained for dy/dx is valid for $x < 2$ as well as $x > 2$. In general, whenever a derivative dy/dx is obtained by logarithmic differentiation, the resulting derivative formula will be valid for all values of x for which $y \neq 0$. It may be valid at those points as well, but it is not guaranteed.

INTEGRALS INVOLVING ln *x*

Formula (2) states that the function $\ln x$ is an antiderivative of $1/x$ on the interval $(0, +\infty)$, whereas Formula (6) states that the function $\ln |x|$ is an antiderivative of $1/x$ on each of the intervals $(-\infty, 0)$ and $(0, +\infty)$. Thus we have the companion integration formula to (6),

$$\int \frac{1}{u}\, du = \ln |u| + C \tag{8}$$

with the implicit understanding that this formula is applicable only across an interval that does not contain 0.

► **Example 6** Applying Formula (8),

$$\int_1^e \frac{1}{x}\, dx = \ln |x| \Big]_1^e = \ln |e| - \ln |1| = 1 - 0 = 1$$

$$\int_{-e}^{-1} \frac{1}{x}\, dx = \ln |x| \Big]_{-e}^{-1} = \ln |-1| - \ln |-e| = 0 - 1 = -1 \blacktriangleleft$$

► **Example 7** Evaluate $\displaystyle\int \frac{3x^2}{x^3+5}\, dx$.

Solution. Make the substitution

$$u = x^3 + 5, \quad du = 3x^2\, dx$$

so that

$$\int \frac{3x^2}{x^3+5}\, dx = \int \frac{1}{u}\, du = \ln |u| + C = \ln |x^3 + 5| + C \blacktriangleleft$$

Formula (8)

► **Example 8** Evaluate $\displaystyle\int \tan x\, dx$.

Solution.

$$\int \tan x\, dx = \int \frac{\sin x}{\cos x}\, dx = -\int \frac{1}{u}\, du = -\ln |u| + C = -\ln |\cos x| + C \blacktriangleleft$$

$u = \cos x$
$du = -\sin x\, dx$

The last two examples illustrate an important point: any integral of the form

$$\int \frac{g'(x)}{g(x)}\, dx$$

(where the numerator of the integrand is the derivative of the denominator) can be evaluated by the u-substitution $u = g(x)$, $du = g'(x)\, dx$, since this substitution yields

$$\int \frac{g'(x)}{g(x)}\, dx = \int \frac{du}{u} = \ln |u| + C = \ln |g(x)| + C$$

DERIVATIVES OF IRRATIONAL POWERS OF *x*

We know from Formula (15) of Section 3.7 that the differentiation formula

$$\frac{d}{dx}[x^r] = rx^{r-1} \tag{9}$$

holds for rational values of r. We will now use logarithmic differentiation to show that this formula holds if r is *any* real number (rational or irrational). In our computations we will assume that x^r is a differentiable function and that the familiar laws of exponents hold for real exponents.

Let $y = x^r$, where r is a real number. The derivative dy/dx can be obtained by logarithmic differentiation as follows:

$$\ln|y| = \ln|x^r| = r\ln|x|$$

$$\frac{d}{dx}[\ln|y|] = \frac{d}{dx}[r\ln|x|]$$

$$\frac{1}{y}\frac{dy}{dx} = \frac{r}{x}$$

$$\frac{dy}{dx} = \frac{r}{x}y = \frac{r}{x}x^r = rx^{r-1}$$

Example 9

$$\frac{d}{dx}[x^\pi] = \pi x^{\pi-1}, \quad \frac{d}{dx}[x^{\sqrt{2}}] = \sqrt{2}x^{\sqrt{2}-1}, \quad \frac{d}{dx}[x^{-e}] = -ex^{-e-1} \blacktriangleleft$$

✔QUICK CHECK EXERCISES 7.2 *(See page 453 for answers.)*

1. The equation of the tangent line to the graph of $y = \ln x$ at $x = e^2$ is ________.
2. Find dy/dx.
   (a) $y = x^{\sqrt{3}}$ (b) $y = \ln 3x$
   (c) $y = \ln\sqrt{x}$ (d) $y = \log(1/|x|)$
3. Use logarithmic differentiation to find the derivative of
$$f(x) = \frac{\sqrt{x+1}}{\sqrt[3]{x-1}}$$
4. $\lim_{h\to 0}\frac{\ln(1+h)}{h} =$ ________
5. $\int_2^5 \frac{1}{t}\,dt =$ ________

EXERCISE SET 7.2

1–26 Find dy/dx.

1. $y = \ln 5x$
2. $y = \ln\frac{x}{3}$
3. $y = \ln|1+x|$
4. $y = \ln(2+\sqrt{x})$
5. $y = \ln|x^2-1|$
6. $y = \ln|x^3-7x^2-3|$
7. $y = \ln\left(\frac{x}{1+x^2}\right)$
8. $y = \ln\left|\frac{1+x}{1-x}\right|$
9. $y = \ln x^2$
10. $y = (\ln x)^3$
11. $y = \sqrt{\ln x}$
12. $y = \ln\sqrt{x}$
13. $y = x\ln x$
14. $y = x^3\ln x$
15. $y = x^2\log_2(3-2x)$
16. $y = x[\log_2(x^2-2x)]^3$
17. $y = \frac{x^2}{1+\log x}$
18. $y = \frac{\log x}{1+\log x}$
19. $y = \ln(\ln x)$
20. $y = \ln(\ln(\ln x))$
21. $y = \ln(\tan x)$
22. $y = \ln(\cos x)$
23. $y = \cos(\ln x)$
24. $y = \sin^2(\ln x)$
25. $y = \log(\sin^2 x)$
26. $y = \log(1-\sin^2 x)$

27–30 Use the method of Example 3 to help perform the indicated differentiation.

27. $\dfrac{d}{dx}[\ln((x-1)^3(x^2+1)^4)]$

28. $\dfrac{d}{dx}[\ln((\cos^2 x)\sqrt{1+x^4})]$

29. $\dfrac{d}{dx}\left[\ln \dfrac{\cos x}{\sqrt{4-3x^2}}\right]$ **30.** $\dfrac{d}{dx}\left[\ln \sqrt{\dfrac{x-1}{x+1}}\right]$

31–34 Find dy/dx using logarithmic differentiation.

31. $y = x\sqrt[3]{1+x^2}$ **32.** $y = \sqrt[5]{\dfrac{x-1}{x+1}}$

33. $y = \dfrac{(x^2-8)^{1/3}\sqrt{x^3+1}}{x^6-7x+5}$ **34.** $y = \dfrac{\sin x \cos x \tan^3 x}{\sqrt{x}}$

35. Find $f'(x)$ if $f(x) = x^e$. **36.** Find $\dfrac{dy}{dx}$ if $y = \dfrac{1}{x^{\sqrt{10}}}$.

37. Find
(a) $\dfrac{d}{dx}[\log_x e]$ (b) $\dfrac{d}{dx}[\log_x 2]$.

38. Find
(a) $\dfrac{d}{dx}[\log_{(1/x)} e]$ (b) $\dfrac{d}{dx}[\log_{(\ln x)} e]$.

39–42 Find the equation of the tangent line to the graph of $y = f(x)$ at $x = x_0$.

39. $f(x) = \ln x;\ x_0 = e^{-1}$ **40.** $f(x) = \log x;\ x_0 = 10$

41. $f(x) = \ln(-x);\ x_0 = -e$ **42.** $f(x) = \ln|x|;\ x_0 = -2$

FOCUS ON CONCEPTS

43. Find the equation of a line through the origin that is tangent to the graph of $y = \ln x$.

44. Explain why the y-intercept of a tangent line to the curve $y = \ln x$ must be 1 unit less than the y-coordinate of the point of tangency.

45. Find a formula for the area $A(w)$ of the triangle bounded by the tangent line to the graph of $y = \ln x$ at $P(w, \ln w)$, the horizontal line through P, and the y-axis.

46. Find a formula for the area $A(w)$ of the triangle bounded by the tangent line to the graph of $y = \ln x^2$ at $P(w, \ln w^2)$, the horizontal line through P, and the y-axis.

47. Verify that $y = \ln(x+e)$ satisfies $dy/dx = e^{-y}$, with $y = 1$ when $x = 0$.

48. Verify that $y = -\ln(e^2 - x)$ satisfies $dy/dx = e^y$, with $y = -2$ when $x = 0$.

49. Find a function f such that $y = f(x)$ satisfies $dy/dx = e^{-y}$, with $y = 0$ when $x = 0$.

50. Find a function f such that $y = f(x)$ satisfies $dy/dx = e^y$, with $y = -\ln 2$ when $x = 0$.

51–52 Find the limit by interpreting the expression as an appropriate derivative.

51. (a) $\displaystyle\lim_{\Delta x \to 0} \frac{\ln(e^2+\Delta x) - 2}{\Delta x}$ (b) $\displaystyle\lim_{w \to 1} \frac{\ln w}{w-1}$

52. (a) $\displaystyle\lim_{x \to 0} \frac{\ln(\cos x)}{x}$ (b) $\displaystyle\lim_{h \to 0} \frac{(1+h)^{\sqrt{2}} - 1}{h}$

53–54 Evaluate the integral and check your answer by differentiating.

53. $\displaystyle\int \left[\frac{2}{x} + 3\sin x\right] dx$ **54.** $\displaystyle\int \left[\frac{1}{2t} + 2t\right] dt$

55–56 Evaluate the integrals using the indicated substitutions.

55. $\displaystyle\int \frac{dx}{x\ln x};\ u = \ln x$

56. $\displaystyle\int \frac{\sin 3\theta}{1+\cos 3\theta}\,d\theta;\ u = 1+\cos 3\theta$

57–58 Evaluate the integrals using appropriate substitutions.

57. $\displaystyle\int \frac{x^4}{1+x^5}\,dx$ **58.** $\displaystyle\int \frac{dx}{2x}$

59–60 Evaluate each integral by first modifying the form of the integrand and then making an appropriate substitution, if needed.

59. $\displaystyle\int \frac{t+1}{t}\,dt$ **60.** $\displaystyle\int \cot x\,dx$

61–62 Evaluate the integrals.

61. $\displaystyle\int_0^2 \frac{3x}{1+x^2}\,dx$ **62.** $\displaystyle\int_{1/2}^1 \frac{1}{2x}\,dx$

63. Evaluate the definite integral by making the indicated u-substitution.

$$\int_e^{e^2} \frac{\ln x}{x}\,dx;\quad u = \ln x$$

64. Evaluate the definite integral by expressing it in terms of u and evaluating the resulting integral using a formula from geometry.

$$\int_{e^{-3}}^{e^3} \frac{\sqrt{9-(\ln x)^2}}{x}\,dx;\quad u = \ln x$$

65–66 Evaluate the integrals by any method.

65. $\displaystyle\int_0^e \frac{dx}{2x+e}$

66. $\displaystyle\int_0^{\pi/3} \frac{\sin x}{1+\cos x}\,dx$

67. Solve the initial-value problem.

$$\frac{dy}{dt} = \frac{1}{t}, \quad y(-1) = 5$$

✔ QUICK CHECK ANSWERS 7.2

1. $y = \dfrac{x}{e^2} + 1$ **2.** (a) $\dfrac{dy}{dx} = \sqrt{3}x^{\sqrt{3}-1}$ (b) $\dfrac{dy}{dx} = \dfrac{1}{x}$ (c) $\dfrac{dy}{dx} = \dfrac{1}{2x}$ (d) $\dfrac{dy}{dx} = -\dfrac{1}{x \ln 10}$ **3.** $\dfrac{\sqrt{x+1}}{\sqrt[3]{x-1}}\left[\dfrac{1}{2(x+1)} - \dfrac{1}{3(x-1)}\right]$

4. 1 **5.** $\ln\left(\dfrac{5}{2}\right)$

7.3 DERIVATIVES OF INVERSE FUNCTIONS; DERIVATIVES AND INTEGRALS INVOLVING EXPONENTIAL FUNCTIONS

In this section we will establish conditions under which the inverse of a one-to-one differentiable function is differentiable, and we will show how the derivative of a one-to-one function can be used to obtain the derivative of its inverse. This will enable us to obtain derivative formulas for exponential functions from the derivative formulas for logarithmic functions. We will then find the corresponding integration formulas for exponential functions.

See Section 1.5 for a review of one-to-one functions and inverse functions.

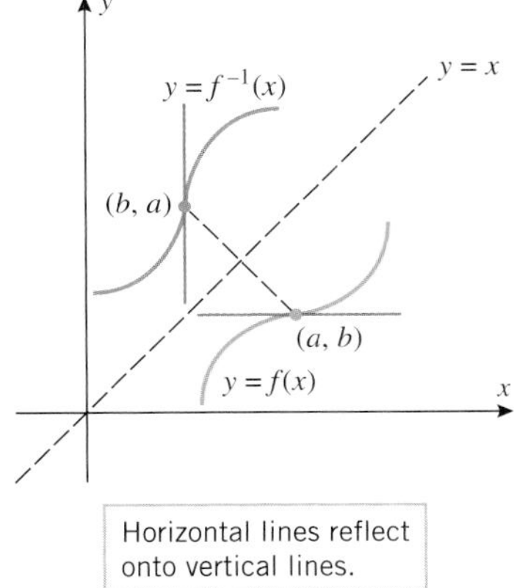
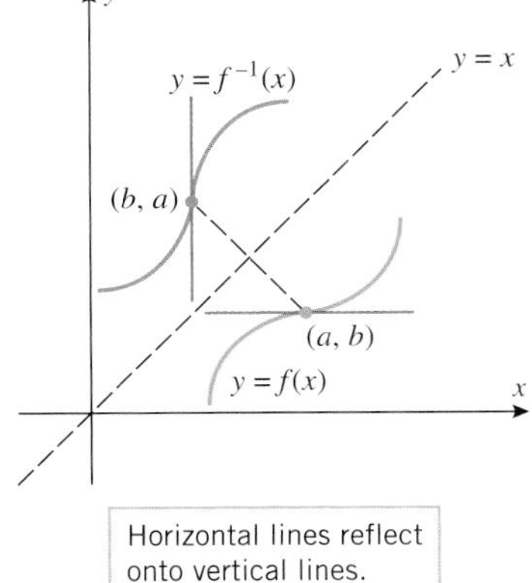

Figure 7.3.1

DIFFERENTIABILITY OF INVERSE FUNCTIONS

Our first goal in this section is to establish conditions under which the inverse of a one-to-one differentiable function f will be differentiable. Geometrically, a function is differentiable at those points where its graph has a nonvertical tangent line, and since the graph of $y = f^{-1}(x)$ is the reflection of the graph of $y = f(x)$ about the line $y = x$, it follows that the points where f^{-1} is not differentiable are reflections of the points where the graph of f has a horizontal tangent line (Figure 7.3.1). Stated algebraically, f^{-1} will fail to be differentiable at a point (b, a) on its graph if $f'(a) = 0$.

Assuming that f is differentiable at the point (a, b) and that $f'(a) \neq 0$, let us now try to find a relationship between the slope of the tangent line at the point (a, b) on the graph of f and the slope of the tangent line at the point (b, a) on the graph of f^{-1}. We know that the equation of the tangent line to the graph of f at the point (a, b) is

$$y - b = f'(a)(x - a)$$

An equation for the reflection of this line about the line $y = x$ can be obtained by interchanging x and y, so it follows that the tangent line to the graph of f^{-1} at the point (b, a) is

$$x - b = f'(a)(y - a)$$

which we can rewrite as

$$y - a = \frac{1}{f'(a)}(x - b)$$

This equation tells us that the slope $(f^{-1})'(b)$ of the tangent line to the graph of f^{-1} at the point (b, a) is

$$(f^{-1})'(b) = \frac{1}{f'(a)} \quad \text{or equivalently} \quad (f^{-1})'(b) = \frac{1}{f'(f^{-1}(b))}$$

(Figure 7.3.2). In summary, we have the following result.

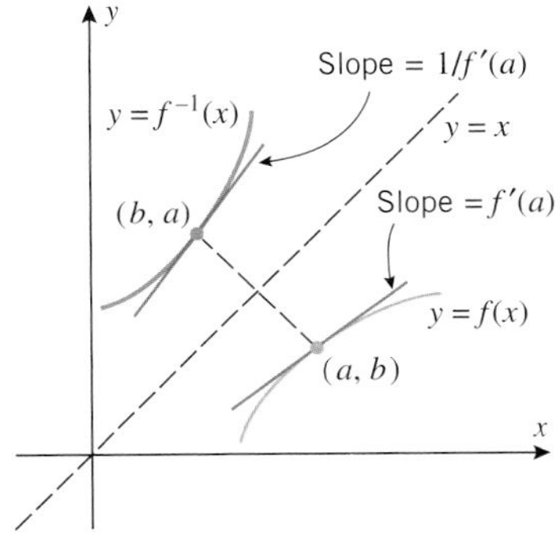

Figure 7.3.2

7.3.1 THEOREM (*Differentiability of Inverse Functions*). *Suppose that the domain of a function f is an open interval I and that f is differentiable and one-to-one on this interval. Then f^{-1} is differentiable at any point x in the range of f at which $f'(f^{-1}(x)) \neq 0$, and its derivative is*

$$\frac{d}{dx}[f^{-1}(x)] = \frac{1}{f'(f^{-1}(x))} \tag{1}$$

Formula (1) can be expressed in a less forbidding form by introducing the dependent variable $y = f^{-1}(x)$ and rewriting this equation as $x = f(y)$. The two sides of (1) can then be expressed as

$$\frac{d}{dx}[f^{-1}(x)] = \frac{dy}{dx} \quad \text{and} \quad \frac{1}{f'(f^{-1}(x))} = \frac{1}{f'(y)} = \frac{1}{dx/dy}$$

Thus, we obtain the following alternative version of Formula (1):

$$\frac{dy}{dx} = \frac{1}{dx/dy} \tag{2}$$

► **Example 1** Confirm Formula (2) for the function $f(x) = x^3 + 1$.

Solution. We showed in Section 1.5 that if we solve $y = f(x) = x^3 + 1$ for x in terms of y, then we obtain $x = f^{-1}(y) = \sqrt[3]{y-1}$. Thus,

$$\frac{dy}{dx} = \frac{d}{dx}[x^3+1] = 3x^2 \quad \text{and} \quad \frac{dx}{dy} = \frac{d}{dy}[\sqrt[3]{y-1}] = \frac{d}{dy}[(y-1)^{1/3}] = \frac{1}{3}(y-1)^{-2/3}$$

If we now substitute $x = \sqrt[3]{y-1}$ in the expression for dy/dx, we obtain

$$\frac{dy}{dx} = 3(\sqrt[3]{y-1})^2 = 3(y-1)^{2/3} = \frac{1}{dx/dy}$$

which confirms the validity of (2) in this case. ◄

INCREASING OR DECREASING FUNCTIONS ARE ONE-TO-ONE

In Example 1 we were able to find an explicit formula for f^{-1} by solving the equation $y = f(x)$ for x in terms of y. However, the real importance of Formulas (1) and (2) is in the case where we know that f is a one-to-one differentiable function but we cannot solve $y = f(x)$ for x in terms of y. In that case, we can use Formula (1) or (2) to obtain properties of f^{-1} without having an explicit formula for this inverse. Our next objective is to develop a theorem that will enable us do this.

If the graph of a function f is always increasing or always decreasing over the domain of f, then a horizontal line will cut the graph of f in at most one point (Figure 7.3.3), so f must have an inverse function. In Theorem 4.1.2 we saw that f must be increasing on any interval on which $f'(x) > 0$ and must be decreasing on any interval on which $f'(x) < 0$. Combining this with Theorem 7.3.1 we obtain the following result.

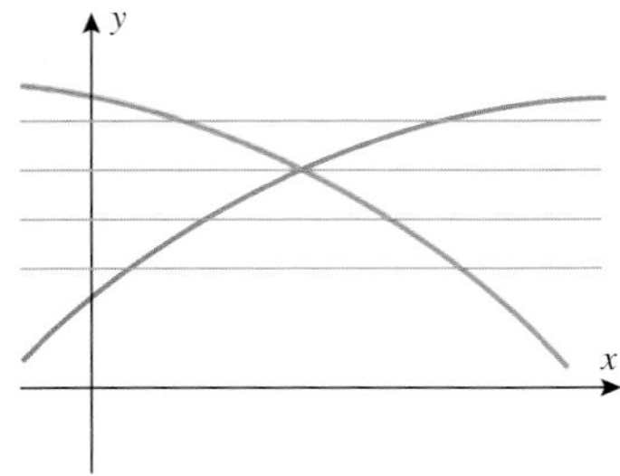

Figure 7.3.3 The graph of an increasing function or a decreasing function is cut at most once by any horizontal line.

7.3.2 THEOREM. *Suppose that the domain of a function f is an open interval I on which $f'(x) > 0$ or on which $f'(x) < 0$. Then f is one-to-one, $f^{-1}(x)$ is differentiable at all values of x in the range of f, and the derivative of $f^{-1}(x)$ is given by Formula (1).*

▶ **Example 2** Consider the function $f(x) = x^5 + x + 1$.

(a) Show that f is one-to-one on the interval $(-\infty, +\infty)$.
(b) Show that f^{-1} is differentiable on the interval $(-\infty, +\infty)$.
(c) Find a formula for the derivative of f^{-1} using Formula (2).
(d) Find a formula for the derivative of f^{-1} using implicit differentiation.

Solution (a). Since

$$f'(x) = 5x^4 + 1 > 0$$

for all real values of x, it follows from Theorem 7.3.2 that f is one-to-one on the interval $(-\infty, +\infty)$.

Solution (b). Since f is a polynomial of odd degree, its range is $(-\infty, +\infty)$. (Why?) Thus, it follows from Theorem 7.3.2 that f^{-1} is differentiable on the interval $(-\infty, +\infty)$.

Solution (c). If we let $y = f^{-1}(x)$, then

$$x = f(y) = y^5 + y + 1 \tag{3}$$

from which it follows that $dx/dy = 5y^4 + 1$. Then, from Formula (2),

$$\frac{dy}{dx} = \frac{1}{dx/dy} = \frac{1}{5y^4 + 1} \tag{4}$$

Since we are unable to solve (3) for y in terms of x, we must leave (4) in terms of y.

Solution (d). Differentiating (3) implicitly with respect to x yields

$$\frac{d}{dx}[x] = \frac{d}{dx}[y^5 + y + 1]$$

$$1 = (5y^4 + 1)\frac{dy}{dx}$$

$$\frac{dy}{dx} = \frac{1}{5y^4 + 1}$$

which agrees with (4). ◀

In general, once it is established that f^{-1} is differentiable, one has the option of calculating the derivative of f^{-1} using Formula (1) or (2), or by differentiating implicitly, as in Example 2.

DERIVATIVES OF EXPONENTIAL FUNCTIONS

Our next objective is to show that the general exponential function b^x $(b > 0, b \neq 1)$ is differentiable everywhere and to find its derivative. To do this, we will use the fact that b^x is the inverse of the function $f(x) = \log_b x$. If $b > 1$ then $\ln b > 0$, so

$$f'(x) = \frac{d}{dx}[\log_b x] = \frac{1}{x \ln b} > 0 \quad \text{for all } x \text{ in the interval } (0, +\infty)$$

It now follows from Theorem 7.3.2 that for $b > 1$, $f^{-1}(x) = b^x$ is differentiable for all x in the range of $f(x) = \log_b x$. But we know from Table 7.1.3 that the range of $\log_b x$ is $(-\infty, +\infty)$, so we have established that b^x is differentiable everywhere. This result is also true for the case $0 < b < 1$ [Exercise 6(d)].

To obtain a derivative formula for b^x we rewrite $y = b^x$ as

$$x = \log_b y$$

and differentiate implicitly using Formula (5) of Section 7.2 to obtain

$$1 = \frac{1}{y \ln b} \cdot \frac{dy}{dx}$$

Solving for dy/dx and replacing y by b^x we have

$$\frac{dy}{dx} = y \ln b = b^x \ln b$$

Thus, we have shown that

$$\frac{d}{dx}[b^x] = b^x \ln b \tag{5}$$

In the special case where $b = e$ we have $\ln e = 1$, so that (5) becomes

$$\frac{d}{dx}[e^x] = e^x \tag{6}$$

In Section 7.1 we stated that $b = e$ is the only base for which the slope of the tangent line to the curve $y = b^x$ at any point P on the curve is the y-coordinate at P (see pages 437–438). Verify this statement.

Moreover, if u is a differentiable function of x, then it follows from (5) and (6) that

$$\frac{d}{dx}[b^u] = b^u \ln b \cdot \frac{du}{dx} \quad \text{and} \quad \frac{d}{dx}[e^u] = e^u \cdot \frac{du}{dx} \tag{7–8}$$

It is important to distinguish between differentiating an exponential function b^x (variable exponent and constant base) and a power function x^b (variable base and constant exponent). For example, compare the derivative

$$\frac{d}{dx}[x^2] = 2x$$

to the derivative of 2^x in Example 3.

▶ Example 3 The following computations use Formulas (7) and (8).

$$\frac{d}{dx}[2^x] = 2^x \ln 2$$

$$\frac{d}{dx}[e^{-2x}] = e^{-2x} \cdot \frac{d}{dx}[-2x] = -2e^{-2x}$$

$$\frac{d}{dx}[e^{x^3}] = e^{x^3} \cdot \frac{d}{dx}[x^3] = 3x^2 e^{x^3}$$

$$\frac{d}{dx}[e^{\cos x}] = e^{\cos x} \cdot \frac{d}{dx}[\cos x] = -(\sin x)e^{\cos x} \quad ◀$$

Functions of the form $f(x) = u^v$ in which u and v are *nonconstant* functions of x are neither exponential functions nor power functions. Functions of this form can be differentiated using logarithmic differentiation.

▶ Example 4 Use logarithmic differentiation to find $\frac{d}{dx}[(x^2+1)^{\sin x}]$.

Solution. Setting $y = (x^2+1)^{\sin x}$ we have

$$\ln y = \ln[(x^2+1)^{\sin x}] = (\sin x)\ln(x^2+1)$$

Differentiating both sides with respect to x yields

$$\begin{aligned} \frac{1}{y}\frac{dy}{dx} &= \frac{d}{dx}[(\sin x)\ln(x^2+1)] \\ &= (\sin x)\frac{1}{x^2+1}(2x) + (\cos x)\ln(x^2+1) \end{aligned}$$

Thus,

$$\frac{dy}{dx} = y\left[\frac{2x \sin x}{x^2+1} + (\cos x)\ln(x^2+1)\right]$$
$$= (x^2+1)^{\sin x}\left[\frac{2x \sin x}{x^2+1} + (\cos x)\ln(x^2+1)\right] \blacktriangleleft$$

INTEGRALS INVOLVING EXPONENTIAL FUNCTIONS

Associated with derivatives (7) and (8) are the companion integration formulas

$$\int b^u\,du = \frac{b^u}{\ln b} + C \quad \text{and} \quad \int e^u\,du = e^u + C \tag{9–10}$$

▶ Example 5

$$\int 2^x\,dx = \frac{2^x}{\ln 2} + C \blacktriangleleft$$

▶ Example 6 Evaluate $\int e^{5x}\,dx$.

Solution. Let $u = 5x$ so that $du = 5\,dx$ or $dx = \frac{1}{5}\,du$, which yields

$$\int e^{5x}\,dx = \frac{1}{5}\int e^u\,du = \frac{1}{5}e^u + C = \frac{1}{5}e^{5x} + C \blacktriangleleft$$

▶ Example 7 The following computations use Formula (10).

$$\int e^{-x}\,dx = -\int e^u\,du = -e^u + C = -e^{-x} + C$$

$u = -x$
$du = -dx$

$$\int x^2 e^{x^3}\,dx = \frac{1}{3}\int e^u\,du = \frac{1}{3}e^u + C = \frac{1}{3}e^{x^3} + C$$

$u = x^3$
$du = 3x^2\,dx$

$$\int \frac{e^{\sqrt{x}}}{\sqrt{x}}\,dx = 2\int e^u\,du = 2e^u + C = 2e^{\sqrt{x}} + C \blacktriangleleft$$

$u = \sqrt{x}$
$du = \frac{1}{2\sqrt{x}}\,dx$

▶ Example 8 Evaluate $\int_0^{\ln 3} e^x(1+e^x)^{1/2}\,dx$.

Solution. Make the u-substitution

$$u = 1 + e^x, \quad du = e^x\,dx$$

and change the x-limits of integration ($x = 0$, $x = \ln 3$) to the u-limits

$$u = 1 + e^0 = 2, \quad u = 1 + e^{\ln 3} = 1 + 3 = 4$$

This yields

$$\int_0^{\ln 3} e^x(1+e^x)^{1/2}\,dx = \int_2^4 u^{1/2}\,du = \frac{2}{3}u^{3/2}\Big]_2^4 = \frac{2}{3}[4^{3/2} - 2^{3/2}] = \frac{16 - 4\sqrt{2}}{3} \blacktriangleleft$$

✔QUICK CHECK EXERCISES 7.3 (See page 460 for answers.)

1. Suppose that a one-to-one function f has tangent line $y = 5x + 3$ at the point $(1, 8)$. Evaluate $(f^{-1})'(8)$.
2. In each case, from the given derivative, determine whether the function f is invertible.
   (a) $f'(x) = x^2 + 1$ (b) $f'(x) = x^2 - 1$
   (c) $f'(x) = \sin x$ (d) $f'(x) = \frac{1}{2} - e^{x^2}$
3. Evaluate the derivative.
   (a) $\frac{d}{dx}[e^x]$ (b) $\frac{d}{dx}[7^x]$
   (c) $\frac{d}{dx}[\cos(e^x + 1)]$ (d) $\frac{d}{dx}[e^{3x-2}]$
4. $\int_0^2 e^{-3x}\,dx =$ ______

EXERCISE SET 7.3 Graphing Utility

FOCUS ON CONCEPTS

1. Let $f(x) = x^5 + x^3 + x$.
   (a) Show that f is one-to-one and confirm that $f(1) = 3$.
   (b) Find $(f^{-1})'(3)$.
2. Let $f(x) = x^3 + 2e^x$.
   (a) Show that f is one-to-one and confirm that $f(0) = 2$.
   (b) Find $(f^{-1})'(2)$.

3–4 Find $(f^{-1})'(x)$ using Formula (1), and check your answer by differentiating f^{-1} directly.

3. $f(x) = 2/(x + 3)$ 4. $f(x) = \ln(2x + 1)$

5–6 Determine whether the function f is one-to-one by examining the sign of $f'(x)$.

5. (a) $f(x) = x^2 + 8x + 1$
   (b) $f(x) = 2x^5 + x^3 + 3x + 2$
   (c) $f(x) = 2x + \sin x$
   (d) $f(x) = \left(\frac{1}{2}\right)^x$
6. (a) $f(x) = x^3 + 3x^2 - 8$
   (b) $f(x) = x^5 + 8x^3 + 2x - 1$
   (c) $f(x) = \dfrac{x}{x+1}$
   (d) $f(x) = \log_b x,\quad 0 < b < 1$

7–10 Find the derivative of f^{-1} by using Formula (2), and check your result by differentiating implicitly.

7. $f(x) = 5x^3 + x - 7$ 8. $f(x) = 1/x^2,\quad x > 0$
9. $f(x) = 2x^5 + x^3 + 1$
10. $f(x) = 5x - \sin 2x,\quad -\frac{\pi}{4} < x < \frac{\pi}{4}$

11–22 Find dy/dx.

11. $y = e^{7x}$ 12. $y = e^{-5x^2}$
13. $y = x^3 e^x$ 14. $y = e^{1/x}$
15. $y = \dfrac{e^x - e^{-x}}{e^x + e^{-x}}$ 16. $y = \sin(e^x)$
17. $y = e^{x \tan x}$ 18. $y = \dfrac{e^x}{\ln x}$
19. $y = e^{(x - e^{3x})}$ 20. $y = \exp(\sqrt{1 + 5x^3})$
21. $y = \ln(1 - xe^{-x})$ 22. $y = \ln(\cos e^x)$

23–26 Find $f'(x)$ by Formula (7) and then by logarithmic differentiation.

23. $f(x) = 2^x$ 24. $f(x) = 3^{-x}$
25. $f(x) = \pi^{\sin x}$ 26. $f(x) = \pi^{x \tan x}$

27–30 Find dy/dx using the method of logarithmic differentiation.

27. $y = (x^3 - 2x)^{\ln x}$ 28. $y = x^{\sin x}$
29. $y = (\ln x)^{\tan x}$ 30. $y = (x^2 + 3)^{\ln x}$
31. Find $f'(x)$ if $f(x) = x^e$.
32. (a) Explain why Formula (5) cannot be used to find $(d/dx)[x^x]$.
    (b) Find this derivative by logarithmic differentiation.

FOCUS ON CONCEPTS

33. (a) Show that $f(x) = x^3 - 3x^2 + 2x$ is not one-to-one on $(-\infty, +\infty)$.
    (b) Find the largest value of k such that f is one-to-one on the interval $(-k, k)$.
34. (a) Show that the function $f(x) = x^4 - 2x^3$ is not one-to-one on $(-\infty, +\infty)$.
    (b) Find the smallest value of k such that f is one-to-one on the interval $[k, +\infty)$.
35. Let $f(x) = x^4 + x^3 + 1,\ 0 \le x \le 2$.
    (a) Show that f is one-to-one.
    (b) Let $g(x) = f^{-1}(x)$ and define $F(x) = f(2g(x))$. Find an equation for the tangent line to $y = F(x)$ at $x = 3$.
36. Let $f(x) = \dfrac{\exp(4 - x^2)}{x},\ x > 0$.
    (a) Show that f is one-to-one.
    (b) Let $g(x) = f^{-1}(x)$ and define $F(x) = f([g(x)]^2)$. Find $F'\left(\frac{1}{2}\right)$.

37. Let $f(x) = e^{kx}$ and $g(x) = e^{-kx}$. Find
(a) $f^{(n)}(x)$ (b) $g^{(n)}(x)$.

38. Find dy/dt if $y = e^{-\lambda t}(A \sin \omega t + B \cos \omega t)$, where A, B, λ, and ω are constants.

39. Find $f'(x)$ if
$$f(x) = \frac{1}{\sqrt{2\pi}\sigma} \exp\left[-\frac{1}{2}\left(\frac{x-\mu}{\sigma}\right)^2\right]$$
where μ and σ are constants and $\sigma \neq 0$.

40. Show that for any constants A and k, the function $y = Ae^{kt}$ satisfies the equation $dy/dt = ky$.

41. Show that for any constants A and B, the function
$$y = Ae^{2x} + Be^{-4x}$$
satisfies the equation
$$y'' + 2y' - 8y = 0$$

42. Show that
(a) $y = xe^{-x}$ satisfies the equation $xy' = (1-x)y$
(b) $y = xe^{-x^2/2}$ satisfies the equation $xy' = (1-x^2)y$.

43. Show that the rate of change of $y = 100e^{-0.2x}$ with respect to x is proportional to y.

44. Show that the rate of change of $y = 3^{(5x+1)}4^{(-x/2)}$ with respect to x is proportional to y.

45. Show that
$$y = \frac{60}{5+7e^{-t}} \quad \text{satisfies} \quad \frac{dy}{dt} = r\left(1 - \frac{y}{K}\right)y$$
for some constants r and K, and determine the values of these constants.

46. Suppose that the population of oxygen-dependent bacteria in a pond is modeled by the equation
$$P(t) = \frac{60}{5+7e^{-t}}$$
where $P(t)$ is the population (in billions) t days after an initial observation at time $t = 0$.
(a) Use a graphing utility to graph the function $P(t)$.
(b) In words, explain what happens to the population over time. Check your conclusion by finding $\lim_{t\to+\infty} P(t)$.
(c) In words, what happens to the *rate* of population growth over time? Check your conclusion by graphing $P'(t)$.

FOCUS ON CONCEPTS

47–48 Find the limit by interpreting the expression as an appropriate derivative.

47. $\lim_{h\to 0} \frac{10^h - 1}{h}$ **48.** $\lim_{\Delta x\to 0} \frac{(2+\Delta x)^{(2+\Delta x)} - 4}{\Delta x}$

49–50 Evaluate the integral and check your answer by differentiating.

49. $\int \left[\frac{2}{x} + 3e^x\right] dx$ **50.** $\int \left[\frac{1}{2t} - \sqrt{2}e^t\right] dt$

51–52 Evaluate the integrals using the indicated substitutions.

51. $\int e^{-5x}\,dx;\ u = -5x$ **52.** $\int \frac{e^x}{1+e^x}\,dx;\ u = 1 + e^x$

53–60 Evaluate the integrals using appropriate substitutions.

53. $\int e^{2x}\,dx$ **54.** $\int e^{-x/2}\,dx$ **55.** $\int e^{\sin x}\cos x\,dx$

56. $\int x^3 e^{x^4}\,dx$ **57.** $\int x^2 e^{-2x^3}\,dx$ **58.** $\int \frac{e^x + e^{-x}}{e^x - e^{-x}}\,dx$

59. $\int \frac{dx}{e^x}$ **60.** $\int \sqrt{e^x}\,dx$

61–62 Evaluate each integral by first modifying the form of the integrand and then making an appropriate substitution, if needed.

61. $\int [\ln(e^x) + \ln(e^{-x})]\,dx$ **62.** $\int e^{2\ln x}\,dx$

63–64 Evaluate the integrals.

63. $\int_{\ln 2}^{3} 5e^x\,dx$ **64.** $\int_0^1 (e^x - x)\,dx$

65. Evaluate the definite integral by making the indicated u-substitution.
$$\int_0^1 e^{2x-1}\,dx; \quad u = 2x - 1$$

66–68 Evaluate the integrals by any method.

66. $\int_0^{\ln 5} e^x(3 - 4e^x)\,dx$ **67.** $\int_{-\ln 3}^{\ln 3} \frac{e^x}{e^x + 4}\,dx$

68. $\int_1^{\sqrt{2}} xe^{-x^2}\,dx$

69. Suppose that at time $t = 0$ there are 750 bacteria in a growth medium and the bacteria population $y(t)$ grows at the rate $y'(t) = 802.137e^{1.528t}$ bacteria per hour. How many bacteria will there be in 12 hours?

70. Suppose that a particle moving along a coordinate line has velocity $v(t) = 25 + 10e^{-0.05t}$ ft/s.
(a) What is the distance traveled by the particle from time $t = 0$ to time $t = 10$?
(b) Does the term $10e^{-0.05t}$ have much effect on the distance traveled by the particle over that time interval? Explain your reasoning.

71. Find a positive value of k such that the area under the graph of $y = e^{2x}$ over the interval $[0, k]$ is 3 square units.

72. Solve the initial-value problem.
$$\frac{dy}{dt} = -e^{2t}, \quad y(0) = 6$$

✔ QUICK CHECK ANSWERS 7.3

1. $\frac{1}{5}$ 2. (a) yes (b) no (c) no (d) yes 3. (a) e^x (b) $7^x \ln 7$ (c) $-e^x \sin(e^x + 1)$ (d) $3e^{3x-2}$ 4. $-\frac{1}{3}e^{-6} + \frac{1}{3}$

7.4 GRAPHS AND APPLICATIONS INVOLVING LOGARITHMIC AND EXPONENTIAL FUNCTIONS

In this section we will apply the techniques developed in Chapter 4 to graphing functions involving logarithmic or exponential functions. We will also look at applications of differentiation and integration in some contexts that involve logarithmic or exponential functions.

SOME PROPERTIES OF e^x AND $\ln x$

In Figure 7.4.1 we present computer-generated graphs of $y = e^x$ and $y = \ln x$. Since $f(x) = e^x$ and $g(x) = \ln x$ are inverses, their graphs are reflections of each other about the line $y = x$. Table 7.4.1 summarizes some important properties of e^x and $\ln x$.

Figure 7.4.1 The functions e^x and $\ln x$ are inverses.

Table 7.4.1

PROPERTIES OF e^x	PROPERTIES OF $\ln x$
$e^x > 0$ for all x	$\ln x > 0$ if $x > 1$ $\ln x < 0$ if $0 < x < 1$ $\ln x = 0$ if $x = 1$
e^x is increasing on $(-\infty, +\infty)$	$\ln x$ is increasing on $(0, +\infty)$
The graph of e^x is concave up on $(-\infty, +\infty)$	The graph of $\ln x$ is concave down on $(0, +\infty)$

We can verify that $y = e^x$ is increasing and its graph is concave up from its first and second derivatives. For all x in $(-\infty, +\infty)$ we have

$$\frac{d}{dx}[e^x] = e^x > 0 \quad \text{and} \quad \frac{d^2}{dx^2}[e^x] = \frac{d}{dx}[e^x] = e^x > 0$$

The first of these inequalities demonstrates that e^x is increasing on $(-\infty, +\infty)$, and the second inequality shows that the graph of $y = e^x$ is concave up on $(-\infty, +\infty)$.

Similarly, for all x in $(0, +\infty)$ we have

$$\frac{d}{dx}[\ln x] = \frac{1}{x} > 0 \quad \text{and} \quad \frac{d^2}{dx^2}[\ln x] = \frac{d}{dx}\left[\frac{1}{x}\right] = -\frac{1}{x^2} < 0$$

The first of these inequalities demonstrates that $\ln x$ is increasing on $(0, +\infty)$, and the second inequality shows that the graph of $y = \ln x$ is concave down on $(0, +\infty)$.

GRAPHING EXPONENTIAL AND LOGARITHMIC FUNCTIONS

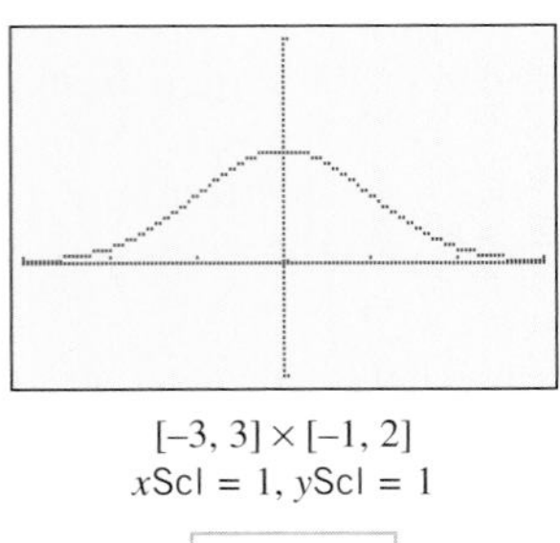

$[-3, 3] \times [-1, 2]$
xScl = 1, yScl = 1

$y = e^{-x^2/2}$

Figure 7.4.2

► **Example 1** Generate or sketch a graph of $y = e^{-x^2/2}$ and identify the exact locations of all relative extrema and inflection points.

Solution. Figure 7.4.2 shows a graph of $y = e^{-x^2/2}$ produced by a graphing utility in the window $[-3, 3] \times [-1, 2]$. This figure suggests that the graph is symmetric about the y-axis and has a relative maximum at $x = 0$, a horizontal asymptote $y = 0$, and two inflection points. The following analysis will confirm these features and identify their exact locations.

- *Symmetries:* Replacing x by $-x$ does not change the equation, so the graph is symmetric about the y-axis.
- *x-intercepts:* Setting $y = 0$ leads to the equation $e^{-x^2/2} = 0$, which has no solutions since all powers of e have positive values. Thus, there are no x-intercepts.
- *y-intercepts:* Setting $x = 0$ yields the y-intercept $y = 1$.
- *Vertical asymptotes:* There are no vertical asymptotes since $e^{-x^2/2}$ is continuous on $(-\infty, +\infty)$.
- *Horizontal asymptotes:* The x-axis ($y = 0$) is a horizontal asymptote since it follows from Formula (13) in Section 7.1 that

$$\lim_{x\to-\infty} e^{-x^2/2} = \lim_{x\to+\infty} e^{-x^2/2} = 0$$

- *Derivatives:*

$$\frac{dy}{dx} = e^{-x^2/2}\frac{d}{dx}\left[-\frac{x^2}{2}\right] = -xe^{-x^2/2}$$

$$\frac{d^2y}{dx^2} = -x\frac{d}{dx}[e^{-x^2/2}] + e^{-x^2/2}\frac{d}{dx}[-x]$$

$$= x^2e^{-x^2/2} - e^{-x^2/2} = (x^2-1)e^{-x^2/2}$$

Conclusions:

- The sign analysis of y in Figure 7.4.3 is based on the fact that $e^{-x^2/2} > 0$ for all x. This shows that the graph is always above the x-axis.
- The sign analysis of dy/dx in Figure 7.4.3 is based on the fact that $dy/dx = -xe^{-x^2/2}$ has the same sign as $-x$. This analysis and the first derivative test show that there is a stationary point at $x = 0$ at which there is a relative maximum. The value of y at the relative maximum is $y = e^0 = 1$.
- Since $d^2y/dx^2 = (x^2-1)e^{-x^2/2}$, d^2y/dx^2 has the same sign as $x^2 - 1$. The sign analysis of d^2y/dx^2 in Figure 7.4.3 shows that there are inflection points at $x = -1$ and $x = 1$. The graph changes from concave up to concave down at $x = -1$ and from concave down to concave up at $x = 1$. The coordinates of the inflection points are $(-1, e^{-1/2}) \approx (-1, 0.61)$ and $(1, e^{-1/2}) \approx (1, 0.61)$. ◀

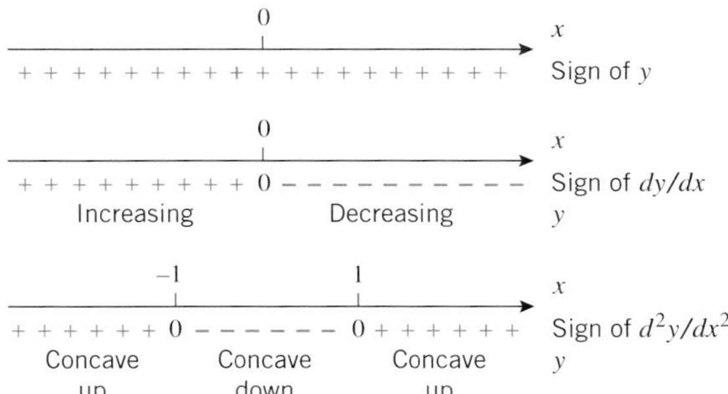

Figure 7.4.3

▶ **Example 2** Use a graphing utility to generate the graph of $f(x) = (\ln x)/x$, and discuss what it tells you about relative extrema, inflection points, asymptotes, and end behavior. Use calculus to find the locations of all key features of the graph.

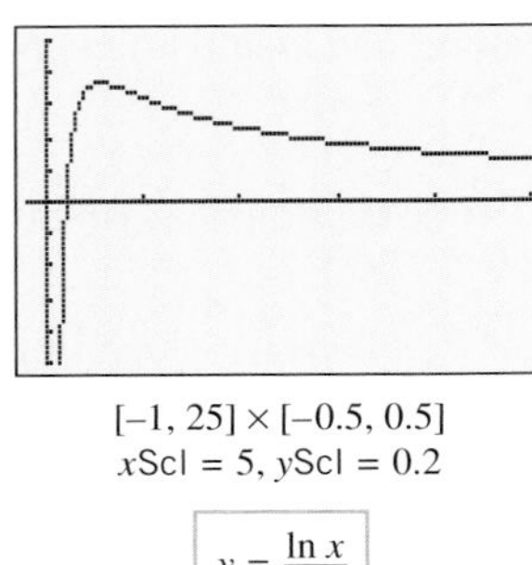

Figure 7.4.4

Solution. Figure 7.4.4 shows a graph of f produced by a graphing utility. The graph suggests that there is an x-intercept near $x = 1$, a relative maximum somewhere between $x = 0$ and $x = 5$, an inflection point near $x = 5$, a vertical asymptote at $x = 0$, and possibly a horizontal asymptote $y = 0$. For a more precise analysis of this information we need to consider the derivatives

$$f'(x) = \frac{x\left(\frac{1}{x}\right) - (\ln x)(1)}{x^2} = \frac{1 - \ln x}{x^2}$$

$$f''(x) = \frac{x^2\left(-\frac{1}{x}\right) - (1 - \ln x)(2x)}{x^4} = \frac{2x\ln x - 3x}{x^4} = \frac{2\ln x - 3}{x^3}$$

- *Relative extrema:* Solving $f'(x) = 0$ yields the stationary point $x = e$ (verify). Since
$$f''(e) = \frac{2-3}{e^3} = -\frac{1}{e^3} < 0$$
there is a relative maximum at $x = e \approx 2.7$ by the second derivative test.
- *Inflection points:* Since $f(x) = (\ln x)/x$ is only defined for positive values of x, the second derivative $f''(x)$ has the same sign as $2\ln x - 3$. We leave it for you to use the inequalities $(2\ln x - 3) < 0$ and $(2\ln x - 3) > 0$ to show that $f''(x) < 0$ if $x < e^{3/2}$ and $f''(x) > 0$ if $x > e^{3/2}$. Thus, there is an inflection point at $x = e^{3/2} \approx 4.5$.
- *Vertical asymptotes:* Since
$$\lim_{x\to 0^+} \frac{1}{x} = +\infty \quad \text{and} \quad \lim_{x\to 0^+} \ln x = -\infty$$
it follows that the values of
$$f(x) = \frac{\ln x}{x} = \frac{1}{x}(\ln x)$$
will decrease without bound as $x \to 0^+$, so
$$\lim_{x\to 0^+} \frac{\ln x}{x} = -\infty$$
and the graph has a vertical asymptote $x = 0$.
- *Horizontal asymptotes:* Note that $(\ln x)/x > 0$ for $x > 1$. Also, $f'(x) < 0$ for $x > e$ so $f(x)$ is decreasing and positive for $x > e$ (verify). We will develop a technique in Section 7.5 that will allow us to conclude that
$$\lim_{x\to +\infty} \frac{\ln x}{x} = 0$$
so that $y = 0$ is a horizontal asymptote.
- *Intercepts:* Setting $f(x) = 0$ yields $(\ln x)/x = 0$. The only real solution of this equation is $x = 1$, so there is an x-intercept at this point. ◀

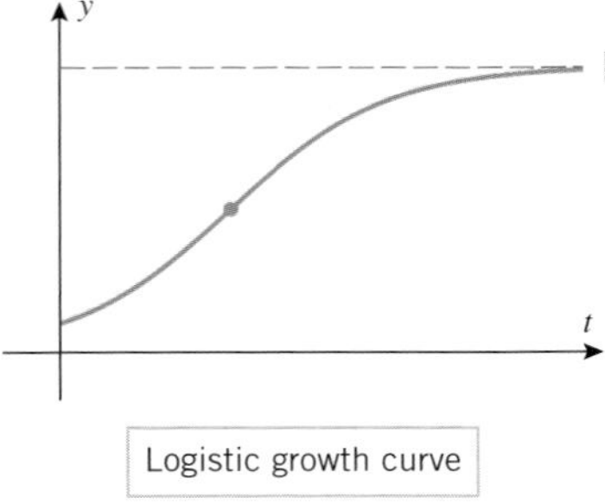

Figure 7.4.5

LOGISTIC CURVES

When a population grows in an environment in which space or food is limited, the graph of population versus time is typically an S-shaped curve of the form shown in Figure 7.4.5. The scenario described by this curve is a population that grows slowly at first and then more and more rapidly as the number of individuals producing offspring increases. However, at a certain point in time (where the inflection point occurs) the environmental factors begin to show their effect, and the growth rate begins a steady decline. Over an extended period of time the population approaches a limiting value that represents the upper limit on the number of individuals that the available space or food can sustain. Population growth curves of this type are called ***logistic growth curves***.

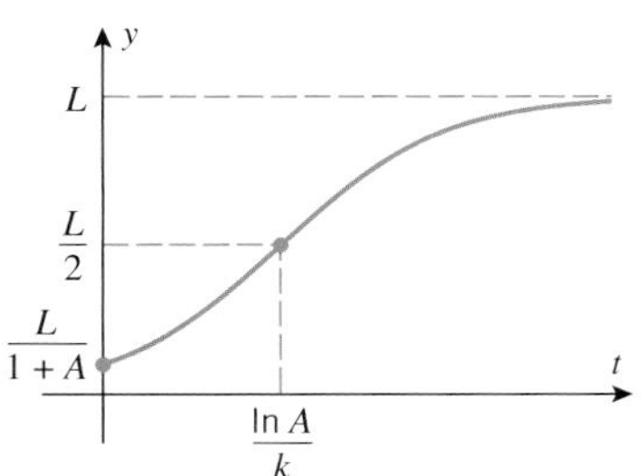

Figure 7.4.6

► **Example 3** We will show in a later chapter that logistic growth curves arise from equations of the form

$$y = \frac{L}{1 + Ae^{-kt}} \tag{1}$$

where y is the population at time t ($t \geq 0$) and A, k, and L are positive constants. Show that Figure 7.4.6 correctly describes the graph of this equation when $A > 1$.

Solution. We leave it for you to confirm that at time $t = 0$ the value of y is

$$y = \frac{L}{1 + A} < \frac{L}{2}$$

and that for $t \geq 0$ the population y satisfies

$$\frac{L}{1 + A} \leq y < L$$

This is consistent with the graph in Figure 7.4.6. The horizontal asymptote at $y = L$ is confirmed by the limit

$$\lim_{t \to +\infty} \frac{L}{1 + Ae^{-kt}} = \frac{L}{1 + 0} = L$$

Physically, L represents the upper limit on the size of the population.

To investigate intervals of increase or decrease, concavity, and inflection points, we need the first and second derivatives of y with respect to t. We leave it for you to confirm that

$$\frac{dy}{dt} = \frac{k}{L} y(L - y) \tag{2}$$

$$\frac{d^2y}{dt^2} = \frac{k^2}{L^2} y(L - y)(L - 2y) \tag{3}$$

Since $k > 0$, $y > 0$, and $L - y > 0$, it follows from (2) that $dy/dt > 0$ for all t. Thus, y is always increasing, which is consistent with Figure 7.4.6.

Since $y > 0$ and $L - y > 0$, it follows from (3) that

$$\frac{d^2y}{dt^2} > 0 \quad \text{if} \quad L - 2y > 0$$

$$\frac{d^2y}{dt^2} < 0 \quad \text{if} \quad L - 2y < 0$$

Thus, the graph of y versus t is concave up if $y < L/2$, concave down if $y > L/2$, and has an inflection point where $y = L/2$, all of which is consistent with Figure 7.4.6.

Finally, we leave it for you to solve the equation

$$\frac{L}{2} = \frac{L}{1 + Ae^{-kt}}$$

for t to show that the inflection point occurs at

$$t = \frac{1}{k} \ln A = \frac{\ln A}{k} \tag{4}$$ ◄

NEWTON'S LAW OF COOLING

► **Example 4** A glass of lemonade with a temperature of 40°F is left to sit in a room whose temperature is a constant 70°F. Using a principle of physics called ***Newton's Law of Cooling***, one can show that if the temperature of the lemonade reaches 52°F in 1 hour, then the temperature T of the lemonade as a function of the elapsed time t is modeled by the equation

$$T = 70 - 30e^{-0.5t}$$

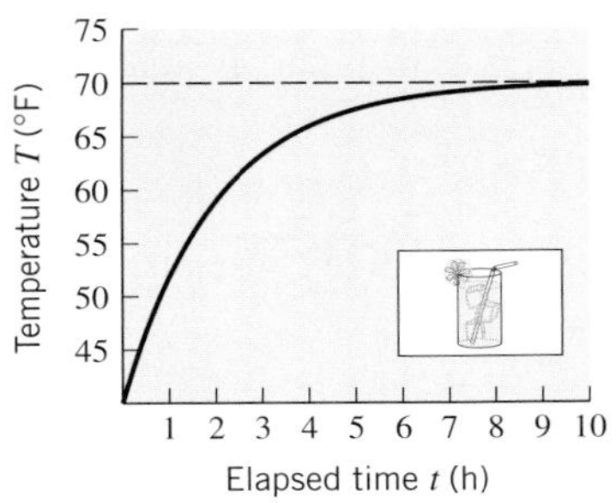

Figure 7.4.7

where T is in degrees Fahrenheit and t is in hours. The graph of this equation, shown in Figure 7.4.7, conforms to our everyday experience that the temperature of the lemonade gradually approaches the temperature of the room.

(a) In words, what happens to the *rate* of temperature rise over time?

(b) Use a derivative to confirm your conclusion in part (a).

(c) Find the average temperature T_{ave} of the lemonade over the first 5 hours.

Solution (a). The rate of change of temperature with respect to time is the slope of the curve $T = 70 - 30e^{-0.5t}$. As t increases, the curve rises to a horizontal asymptote, so the slope of the curve decreases to zero. Thus, the temperature rises at an ever-decreasing rate.

Solution (b). The rate of change in temperature with respect to time is

$$\frac{dT}{dt} = \frac{d}{dt}[70 - 30e^{-0.5t}] = -30(-0.5e^{-0.5t}) = 15e^{-0.5t}$$

As t increases, this derivative decreases to zero, which confirms the conclusion in part (a).

Solution (c). From Definition 6.6.1 the average value of T over the time interval $[0, 5]$ is

$$T_{\text{ave}} = \frac{1}{5}\int_0^5 (70 - 30e^{-0.5t})\,dt \qquad (5)$$

To evaluate this integral, we make the substitution

$$u = -0.5t \quad \text{so that} \quad du = -0.5\,dt \quad [\text{or } dt = -2\,du]$$

With this substitution,

$$\begin{aligned} &\text{if} \quad t = 0, \quad u = 0 \\ &\text{if} \quad t = 5, \quad u = (-0.5)5 = -2.5 \end{aligned}$$

Thus, (5) can be expressed as

$$\begin{aligned} T_{\text{ave}} &= \frac{1}{5}\int_0^{-2.5} (70 - 30e^u)(-2)\,du = -\frac{2}{5}\int_0^{-2.5} (70 - 30e^u)\,du \\ &= -\frac{2}{5}\Big[70u - 30e^u\Big]_{u=0}^{-2.5} = -\frac{2}{5}\big[(-175 - 30e^{-2.5}) - (-30)\big] \\ &= 58 + 12e^{-2.5} \approx 59^\circ\text{F} \end{aligned}$$ ◄

✔QUICK CHECK EXERCISES 7.4 (See page 467 for answers.)

1. Consider a function f whose *derivative* is given by $f'(x) = (x - 4)^2 e^{-x/2}$.
   (a) The function f is increasing on the interval(s) ________.
   (b) The function f is concave up on the interval(s) ________.
   (c) The function f is concave down on the interval(s) ________.

2. Let $f(x) = x^2(2\ln x - 3)$. Given that
   $$f'(x) = 4x(\ln x - 1), \qquad f''(x) = 4\ln x$$
   determine the following properties of the graph of f.
   (a) The graph is increasing on the interval ________.
   (b) The graph is concave down on the interval ________.

3. Let $f(x) = (x - 2)^2 e^{x/2}$. Given that
   $$f'(x) = \tfrac{1}{2}(x^2 - 4)e^{x/2}, \qquad f''(x) = \tfrac{1}{4}(x^2 + 4x - 4)e^{x/2}$$
   determine the following properties of the graph of f.
   (a) The graph is above the x-axis on the intervals ________.
   (b) The graph is increasing on the intervals ________.
   (c) The graph is concave up on the intervals ________.
   (d) The relative minimum point on the graph is ________.
   (e) The relative maximum point on the graph is ________.
   (f) Inflection points occur at $x =$ ________.

EXERCISE SET 7.4 Graphing Utility C CAS

1–4 Use the given derivative to find all critical points of f, and at each critical point determine whether a relative maximum, relative minimum, or neither occurs. Assume in each case that f is continuous everywhere.

1. $f'(x) = xe^{1-x^2}$

2. $f'(x) = x^4(e^x - 3)$

3. $f'(x) = \ln\left(\dfrac{2}{1+x^2}\right)$

4. $f'(x) = e^{2x} - 5e^x + 6$

5–8 Use a graphing utility to estimate the absolute maximum and minimum values of f, if any, on the stated interval, and then use calculus methods to find the exact values.

5. $f(x) = x^3e^{-2x}$; $[1, 4]$

6. $f(x) = \dfrac{\ln(2x)}{x}$; $[1, e]$

7. $f(x) = 5\ln(x^2+1) - 3x$; $[0, 4]$

8. $f(x) = (x^2-1)e^x$; $[-2, 2]$

9–18 We will develop techniques in Section 7.5 to verify that

$$\lim_{x\to+\infty} \frac{e^x}{x} = +\infty, \quad \lim_{x\to+\infty} \frac{x}{e^x} = 0, \quad \lim_{x\to-\infty} xe^x = 0$$

In these exercises: (a) Use these results, as necessary, to find the limits of $f(x)$ as $x \to +\infty$ and as $x \to -\infty$. (b) Sketch a graph of $f(x)$ and identify all relative extrema, inflection points, and asymptotes (as appropriate). Check your work with a graphing utility.

9. $f(x) = xe^x$

10. $f(x) = xe^{-x}$

11. $f(x) = x^2e^{-2x}$

12. $f(x) = x^2e^{2x}$

13. $f(x) = x^2e^{-x^2}$

14. $f(x) = e^{-1/x^2}$

15. $f(x) = \dfrac{e^x}{1-x}$

16. $f(x) = x^{2/3}e^x$

17. $f(x) = x^2e^{1-x}$

18. $f(x) = x^3e^{x-1}$

19–24 We will develop techniques in Section 7.5 to verify that

$$\lim_{x\to+\infty} \frac{\ln x}{x^r} = 0, \quad \lim_{x\to+\infty} \frac{x^r}{\ln x} = +\infty, \quad \lim_{x\to 0^+} x^r \ln x = 0$$

for any positive real number r. In these exercises: (a) Use these results, as necessary, to find the limits of $f(x)$ as $x \to +\infty$ and as $x \to 0^+$. (b) Sketch a graph of $f(x)$ and identify all relative extrema, inflection points, and asymptotes (as appropriate). Check your work with a graphing utility.

19. $f(x) = x\ln x$

20. $f(x) = x^2\ln x$

21. $f(x) = x^2\ln(2x)$

22. $f(x) = \ln(x^2+1)$

23. $f(x) = x^{2/3}\ln x$

24. $f(x) = x^{-1/3}\ln x$

FOCUS ON CONCEPTS

25. Consider the family of curves $y = xe^{-bx}$ $(b > 0)$.
(a) Use a graphing utility to generate some members of this family.
(b) Discuss the effect of varying b on the shape of the graph, and discuss the locations of the relative extrema and inflection points.

26. Consider the family of curves $y = e^{-bx^2}$ $(b > 0)$.
(a) Use a graphing utility to generate some members of this family.
(b) Discuss the effect of varying b on the shape of the graph, and discuss the locations of the relative extrema and inflection points.

27. (a) Determine whether the following limits exist, and if so, find them:

$$\lim_{x\to+\infty} e^x\cos x, \quad \lim_{x\to-\infty} e^x\cos x$$

(b) Sketch the graphs of the equations $y = e^x$, $y = -e^x$, and $y = e^x\cos x$ in the same coordinate system, and label any points of intersection.
(c) Use a graphing utility to generate some members of the family $y = e^{ax}\cos bx$ $(a > 0$ and $b > 0)$, and discuss the effect of varying a and b on the shape of the curve.

28. Consider the family of curves $y = x^ne^{-x^2/n}$, where n is a positive integer.
(a) Use a graphing utility to generate some members of this family.
(b) Discuss the effect of varying n on the shape of the graph, and discuss the locations of the relative extrema and inflection points.

29. Suppose that a population y grows according to the logistic model given by Formula (1).
(a) At what rate is y increasing at time $t = 0$?
(b) In words, describe how the rate of growth of y varies with time.
(c) At what time is the population growing most rapidly?

30. Suppose that the number of individuals at time t in a certain wildlife population is given by

$$N(t) = \frac{340}{1 + 9(0.77)^t}, \quad t \ge 0$$

where t is in years. Use a graphing utility to estimate the time at which the size of the population is increasing most rapidly.

31. Suppose that the spread of a flu virus on a college campus is modeled by the function

$$y(t) = \frac{1000}{1 + 999e^{-0.9t}}$$

where $y(t)$ is the number of infected students at time t (in days, starting with $t = 0$). Use a graphing utility to estimate the day on which the virus is spreading most rapidly.

32. Assuming that A, k, and L are positive constants, verify that the graph of $y = L/(1 + Ae^{-kt})$ has an inflection point at $(\frac{1}{k}\ln A, \frac{1}{2}L)$.

33. Suppose that the number of bacteria in a culture at time t is given by $N = 5000(25 + te^{-t/20})$.
(a) Find the largest and smallest number of bacteria in the culture during the time interval $0 \le t \le 100$.
(b) At what time during the time interval in part (a) is the number of bacteria decreasing most rapidly?

34. The concentration $C(t)$ of a drug in the bloodstream t hours after it has been injected is commonly modeled by an equation of the form

$$C(t) = \frac{K(e^{-bt} - e^{-at})}{a - b}$$

where $K > 0$ and $a > b > 0$.
(a) At what time does the maximum concentration occur?
(b) Let $K = 1$ for simplicity, and use a graphing utility to check your result in part (a) by graphing $C(t)$ for various values of a and b.

35. The equilibrium constant k of a balanced chemical reaction changes with the absolute temperature T according to the law

$$k = k_0 \exp\left(-\frac{q(T - T_0)}{2T_0T}\right)$$

where k_0, q, and T_0 are constants. Find the rate of change of k with respect to T.

36. Let $s(t) = t/e^t$ be the position function of a particle moving along a coordinate line, where s is in meters and t is in seconds. Use a graphing utility to generate the graphs of $s(t)$, $v(t)$, and $a(t)$ for $t \ge 0$, and use those graphs where needed.
(a) Use the appropriate graph to make a rough estimate of the time at which the particle first reverses the direction of its motion; and then find the time exactly.
(b) Find the exact position of the particle when it first reverses the direction of its motion.
(c) Use the appropriate graphs to make a rough estimate of the time intervals on which the particle is speeding up and on which it is slowing down; and then find those time intervals exactly.

37–38 Find the area under the curve $y = f(x)$ over the stated interval.

37. $f(x) = e^{2x}$; $[0, \ln 2]$ **38.** $f(x) = \dfrac{1}{x}$; $[1, 5]$

39–40 Sketch the region enclosed by the curves and find its area.

39. $y = e^x$, $y = e^{2x}$, $x = 0$, $x = \ln 2$

40. $x = 1/y$, $x = 0$, $y = 1$, $y = e$

41–42 Sketch the curve and find the total area between the curve and the given interval on the x-axis.

41. $y = e^x - 1$; $[-1, 1]$ **42.** $y = \dfrac{x-2}{x}$; $[1, 3]$

43–45 Find the average value of the function over the given interval.

43. $f(x) = 1/x$; $[1, e]$ **44.** $f(x) = e^x$; $[-1, \ln 5]$

45. $f(x) = e^{-2x}$; $[0, 4]$

46. Suppose that the value of a yacht in dollars after t years of use is $V(t) = 275{,}000e^{-0.17t}$. What is the average value of the yacht over its first 10 years of use?

47–48 For the given velocity function $v(t)$:
(a) Generate the velocity versus time curve, and use it to make a conjecture about the sign of the displacement over the given time interval.
(b) Use a CAS to find the displacement.

C **47.** $v(t) = 0.5 - te^{-t}$; $0 \le t \le 5$

C **48.** $v(t) = t\ln(t + 0.1)$; $0 \le t \le 1$

49–50 Use a graphing utility to determine the number of times the curves intersect; and then apply Newton's Method, where needed, to approximate the x-coordinates of all intersections.

49. $y = 1$ and $y = e^x \sin x$; $0 < x < \pi$

50. $y = e^{-x}$ and $y = \ln x$

51. For the function

$$f(x) = \frac{e^{-x}}{1 + x^2}$$

use Newton's Method to approximate the x-coordinates of the inflection points to two decimal places.

52. (a) Show that $e^x \ge 1 + x$ if $x \ge 0$.
(b) Show that $e^x \ge 1 + x + \frac{1}{2}x^2$ if $x \ge 0$.
(c) Confirm the inequalities in parts (a) and (b) with a graphing utility.

53–54 Find the volume of the solid that results when the region enclosed by the given curves is revolved about the x-axis.

53. $y = e^x$, $y = 0$, $x = 0$, $x = \ln 3$

54. $y = e^{-2x}$, $y = 0$, $x = 0$, $x = 1$

55–56 Use cylindrical shells to find the volume of the solid generated when the region enclosed by the given curves is revolved about the y-axis.

55. $y = \dfrac{1}{x^2+1}, \; x = 0, \; x = 1, \; y = 0$

56. $y = e^{x^2}, \; x = 1, \; x = \sqrt{3}, \; y = 0$

57–58 Find the exact arc length of the parametric curve without eliminating the parameter.

57. $x = e^t \cos t, \; y = e^t \sin t \quad (0 \le t \le \pi/2)$

58. $x = e^t(\sin t + \cos t), \; y = e^t(\cos t - \sin t) \quad (1 \le t \le 4)$

59–60 Express the exact arc length of the curve over the given interval as an integral that has been simplified to eliminate the radical, and then evaluate the integral using a CAS.

C **59.** $y = \ln(\sec x)$ from $x = 0$ to $x = \pi/4$

C **60.** $y = \ln(\sin x)$ from $x = \pi/4$ to $x = \pi/2$

61–62 Use a CAS or a calculating utility with numerical integration capabilities to approximate the area of the surface generated by revolving the curve about the stated axis. Round your answer to two decimal places.

C **61.** $y = e^x, \; 0 \le x \le 1$; x-axis

C **62.** $y = e^x, \; 1 \le y \le e$; y-axis

C **63.** Use a CAS to find the area of the surface generated by revolving the parametric curve $x = e^t \cos t, \; y = e^t \sin t$ $(0 \le t \le \pi/2)$ about the x-axis.

✔ QUICK CHECK ANSWERS 7.4

1. (a) $(-\infty, +\infty)$ (b) $(4, 8)$ (c) $(-\infty, 4), (8, +\infty)$ **2.** (a) $(e, +\infty)$ (b) $(0, 1)$ **3.** (a) $(-\infty, 2)$ and $(2, +\infty)$ (b) $(-\infty, -2]$ and $[2, +\infty)$ (c) $(-\infty, -2-2\sqrt{2})$ and $(-2+2\sqrt{2}, +\infty)$ (d) $(2, 0)$ (e) $(-2, 16e^{-1}) \approx (-2, 5.89)$ (f) $-2 \pm 2\sqrt{2}$

7.5 L'HÔPITAL'S RULE; INDETERMINATE FORMS

In this section we will discuss a general method for using derivatives to find limits. This method will enable us to establish limits with certainty that earlier in the text we were only able to conjecture using numerical or graphical evidence. The method that we will discuss in this section is an extremely powerful tool that is used internally by many computer programs to calculate limits of various types.

INDETERMINATE FORMS OF TYPE 0/0

Recall that a limit of the form

$$\lim_{x \to a} \frac{f(x)}{g(x)} \tag{1}$$

in which $f(x) \to 0$ and $g(x) \to 0$ as $x \to a$ is called an ***indeterminate form of type* 0/0**. Some examples encountered earlier in the text are

$$\lim_{x \to 1} \frac{x^2 - 1}{x - 1} = 2, \quad \lim_{x \to 0} \frac{\sin x}{x} = 1, \quad \lim_{x \to 0} \frac{1 - \cos x}{x} = 0$$

The first limit was obtained algebraically by factoring the numerator and canceling the common factor of $x - 1$, and the second two limits were obtained using geometric methods. However, there are many indeterminate forms for which neither algebraic nor geometric methods will produce the limit, so we need to develop a more general method.

To motivate such a method, suppose that (1) is an indeterminate form of type 0/0 in which f' and g' are continuous at $x = a$ and $g'(a) \ne 0$. Since f and g can be closely approximated by their local linear approximations near a, it is reasonable to expect that

$$\lim_{x \to a} \frac{f(x)}{g(x)} = \lim_{x \to a} \frac{f(a) + f'(a)(x - a)}{g(a) + g'(a)(x - a)} \tag{2}$$

Since we are assuming that f' and g' are continuous at $x = a$, we have

$$\lim_{x \to a} f'(x) = f'(a) \quad \text{and} \quad \lim_{x \to a} g'(x) = g'(a)$$

and since the differentiability of f and g at $x = a$ implies the continuity of f and g at $x = a$, we have

$$f(a) = \lim_{x \to a} f(x) = 0 \quad \text{and} \quad g(a) = \lim_{x \to a} g(x) = 0$$

Thus, we can rewrite (2) as

$$\lim_{x \to a} \frac{f(x)}{g(x)} = \lim_{x \to a} \frac{f'(a)(x-a)}{g'(a)(x-a)} = \lim_{x \to a} \frac{f'(a)}{g'(a)} = \lim_{x \to a} \frac{f'(x)}{g'(x)} \tag{3}$$

This result, called ***L'Hôpital's rule***, converts the given indeterminate form into a limit involving derivatives that is often easier to evaluate.

Although we motivated (3) by assuming that f and g have continuous derivatives at $x = a$ and that $g'(a) \neq 0$, the result is true under less stringent conditions and is also valid for one-sided limits and limits at $+\infty$ and $-\infty$. The proof of the following precise statement of L'Hôpital's rule is omitted.

7.5.1 THEOREM (*L'Hôpital's Rule for Form* 0/0). *Suppose that f and g are differentiable functions on an open interval containing $x = a$, except possibly at $x = a$, and that*

$$\lim_{x \to a} f(x) = 0 \quad \textit{and} \quad \lim_{x \to a} g(x) = 0$$

If $\lim_{x \to a} [f'(x)/g'(x)]$ exists, or if this limit is $+\infty$ or $-\infty$, then

$$\lim_{x \to a} \frac{f(x)}{g(x)} = \lim_{x \to a} \frac{f'(x)}{g'(x)}$$

Moreover, this statement is also true in the case of a limit as $x \to a^-$, $x \to a^+$, $x \to -\infty$, or as $x \to +\infty$.

WARNING

Note that in L'Hôpital's rule the numerator and denominator are differentiated individually. This is *not* the same as differentiating $f(x)/g(x)$.

In the examples that follow we will apply L'Hôpital's rule using the following three-step process:

Applying L'Hôpital's Rule

Step 1. Check that the limit of $f(x)/g(x)$ is an indeterminate form of type 0/0.

Step 2. Differentiate f and g separately.

Step 3. Find the limit of $f'(x)/g'(x)$. If this limit is finite, $+\infty$, or $-\infty$, then it is equal to the limit of $f(x)/g(x)$.

► **Example 1** Find the limit

$$\lim_{x \to 2} \frac{x^2 - 4}{x - 2}$$

using L'Hôpital's rule, and check the result by factoring.

Solution. The numerator and denominator have a limit of 0, so the limit is an indeterminate form of type 0/0. Applying L'Hôpital's rule yields

$$\lim_{x\to 2}\frac{x^2-4}{x-2}=\lim_{x\to 2}\frac{\dfrac{d}{dx}[x^2-4]}{\dfrac{d}{dx}[x-2]}=\lim_{x\to 2}\frac{2x}{1}=4$$

This agrees with the computation

$$\lim_{x\to 2}\frac{x^2-4}{x-2}=\lim_{x\to 2}\frac{(x-2)(x+2)}{x-2}=\lim_{x\to 2}(x+2)=4 \blacktriangleleft$$

The limit in Example 1 can be interpreted as the limit form of a certain derivative. Use that derivative to evaluate the limit.

▶ **Example 2** In each part confirm that the limit is an indeterminate form of type 0/0, and evaluate it using L'Hôpital's rule.

(a) $\lim_{x\to 0}\frac{\sin 2x}{x}$ (b) $\lim_{x\to \pi/2}\frac{1-\sin x}{\cos x}$ (c) $\lim_{x\to 0}\frac{e^x-1}{x^3}$

(d) $\lim_{x\to 0^-}\frac{\tan x}{x^2}$ (e) $\lim_{x\to 0}\frac{1-\cos x}{x^2}$ (f) $\lim_{x\to +\infty}\frac{x^{-4/3}}{\sin(1/x)}$

WARNING

Applying L'Hôpital's rule to limits that are not indeterminate forms can produce incorrect results. For example, the computation

$$\lim_{x\to 0}\frac{x+6}{x+2}=\lim_{x\to 0}\frac{\dfrac{d}{dx}[x+6]}{\dfrac{d}{dx}[x+2]}=\lim_{x\to 0}\frac{1}{1}=1$$

is ***not valid***, since the limit is not an indeterminate form. The correct result is

$$\lim_{x\to 0}\frac{x+6}{x+2}=\frac{0+6}{0+2}=3$$

Solution (a). The numerator and denominator have a limit of 0, so the limit is an indeterminate form of type 0/0. Applying L'Hôpital's rule yields

$$\lim_{x\to 0}\frac{\sin 2x}{x}=\lim_{x\to 0}\frac{\dfrac{d}{dx}[\sin 2x]}{\dfrac{d}{dx}[x]}=\lim_{x\to 0}\frac{2\cos 2x}{1}=2$$

Observe that this result agrees with that obtained by substitution in Example 2(b) of Section 2.6.

Solution (b). The numerator and denominator have a limit of 0, so the limit is an indeterminate form of type 0/0. Applying L'Hôpital's rule yields

$$\lim_{x\to \pi/2}\frac{1-\sin x}{\cos x}=\lim_{x\to \pi/2}\frac{\dfrac{d}{dx}[1-\sin x]}{\dfrac{d}{dx}[\cos x]}=\lim_{x\to \pi/2}\frac{-\cos x}{-\sin x}=\frac{0}{-1}=0$$

Solution (c). The numerator and denominator have a limit of 0, so the limit is an indeterminate form of type 0/0. Applying L'Hôpital's rule yields

$$\lim_{x\to 0}\frac{e^x-1}{x^3}=\lim_{x\to 0}\frac{\dfrac{d}{dx}[e^x-1]}{\dfrac{d}{dx}[x^3]}=\lim_{x\to 0}\frac{e^x}{3x^2}=+\infty$$

Guillaume François Antoine de L'Hôpital (1661–1704) French mathematician. L'Hôpital, born to parents of the French high nobility, held the title of Marquis de Sainte-Mesme Comte d'Autrement. He showed mathematical talent quite early and at age 15 solved a difficult problem about cycloids posed by Pascal. As a young man he served briefly as a cavalry officer, but resigned because of nearsightedness. In his own time he gained fame as the author of the first textbook ever published on differential calculus, *L'Analyse des Infiniment Petits pour l'Intelligence des Lignes Courbes* (1696). L'Hôpital's rule appeared for the first time in that book. Actually, L'Hôpital's rule and most of the material in the calculus text were due to John Bernoulli, who was L'Hôpital's teacher. L'Hôpital dropped his plans for a book on integral calculus when Leibniz informed him that he intended to write such a text. L'Hôpital was apparently generous and personable, and his many contacts with major mathematicians provided the vehicle for disseminating major discoveries in calculus throughout Europe.

Solution (d). The numerator and denominator have a limit of 0, so the limit is an indeterminate form of type 0/0. Applying L'Hôpital's rule yields

$$\lim_{x\to 0^-} \frac{\tan x}{x^2} = \lim_{x\to 0^-} \frac{\sec^2 x}{2x} = -\infty$$

Solution (e). The numerator and denominator have a limit of 0, so the limit is an indeterminate form of type 0/0. Applying L'Hôpital's rule yields

$$\lim_{x\to 0} \frac{1-\cos x}{x^2} = \lim_{x\to 0} \frac{\sin x}{2x}$$

Since the new limit is another indeterminate form of type 0/0, we apply L'Hôpital's rule again:

$$\lim_{x\to 0} \frac{1-\cos x}{x^2} = \lim_{x\to 0} \frac{\sin x}{2x} = \lim_{x\to 0} \frac{\cos x}{2} = \frac{1}{2}$$

Solution (f). The numerator and denominator have a limit of 0, so the limit is an indeterminate form of type 0/0. Applying L'Hôpital's rule yields

$$\lim_{x\to +\infty} \frac{x^{-4/3}}{\sin(1/x)} = \lim_{x\to +\infty} \frac{-\frac{4}{3}x^{-7/3}}{(-1/x^2)\cos(1/x)} = \lim_{x\to +\infty} \frac{\frac{4}{3}x^{-1/3}}{\cos(1/x)} = \frac{0}{1} = 0 \blacktriangleleft$$

INDETERMINATE FORMS OF TYPE ∞/∞

When we want to indicate that the limit (or a one-sided limit) of a function is $+\infty$ or $-\infty$ without being specific about the sign, we will say that the limit is ∞. For example,

$$\lim_{x\to a^+} f(x) = \infty \quad \text{means} \quad \lim_{x\to a^+} f(x) = +\infty \quad \text{or} \quad \lim_{x\to a^+} f(x) = -\infty$$

$$\lim_{x\to +\infty} f(x) = \infty \quad \text{means} \quad \lim_{x\to +\infty} f(x) = +\infty \quad \text{or} \quad \lim_{x\to +\infty} f(x) = -\infty$$

$$\lim_{x\to a} f(x) = \infty \quad \text{means} \quad \lim_{x\to a^+} f(x) = \pm\infty \quad \text{and} \quad \lim_{x\to a^-} f(x) = \pm\infty$$

The limit of a ratio, $f(x)/g(x)$, in which the numerator has limit ∞ and the denominator has limit ∞ is called an ***indeterminate form of type*** ∞/∞. The following version of L'Hôpital's rule, which we state without proof, can often be used to evaluate limits of this type.

7.5.2 THEOREM (*L'Hôpital's Rule for Form* ∞/∞). *Suppose that f and g are differentiable functions on an open interval containing $x = a$, except possibly at $x = a$, and that*

$$\lim_{x\to a} f(x) = \infty \quad \textit{and} \quad \lim_{x\to a} g(x) = \infty$$

If $\lim_{x\to a} [f'(x)/g'(x)]$ exists, or if this limit is $+\infty$ or $-\infty$, then

$$\lim_{x\to a} \frac{f(x)}{g(x)} = \lim_{x\to a} \frac{f'(x)}{g'(x)}$$

Moreover, this statement is also true in the case of a limit as $x \to a^-$, $x \to a^+$, $x \to -\infty$, or as $x \to +\infty$.

▶ **Example 3** In each part confirm that the limit is an indeterminate form of type ∞/∞ and apply L'Hôpital's rule.

$$\text{(a)} \lim_{x\to +\infty} \frac{x}{e^x} \qquad \text{(b)} \lim_{x\to 0^+} \frac{\ln x}{\csc x}$$

Solution (a). The numerator and denominator both have a limit of $+\infty$, so we have an indeterminate form of type ∞/∞. Applying L'Hôpital's rule yields

$$\lim_{x\to+\infty}\frac{x}{e^x} = \lim_{x\to+\infty}\frac{1}{e^x} = 0$$

Solution (b). The numerator has a limit of $-\infty$ and the denominator has a limit of $+\infty$, so we have an indeterminate form of type ∞/∞. Applying L'Hôpital's rule yields

$$\lim_{x\to 0^+}\frac{\ln x}{\csc x} = \lim_{x\to 0^+}\frac{1/x}{-\csc x\cot x} \tag{4}$$

This last limit is again an indeterminate form of type ∞/∞. Moreover, any additional applications of L'Hôpital's rule will yield powers of $1/x$ in the numerator and expressions involving $\csc x$ and $\cot x$ in the denominator; thus, repeated application of L'Hôpital's rule simply produces new indeterminate forms. We must try something else. The last limit in (4) can be rewritten as

$$\lim_{x\to 0^+}\left(-\frac{\sin x}{x}\tan x\right) = -\lim_{x\to 0^+}\frac{\sin x}{x}\cdot\lim_{x\to 0^+}\tan x = -(1)(0) = 0$$

Thus,

$$\lim_{x\to 0^+}\frac{\ln x}{\csc x} = 0 \blacktriangleleft$$

ANALYZING THE GROWTH OF EXPONENTIAL FUNCTIONS USING L'HÔPITAL'S RULE

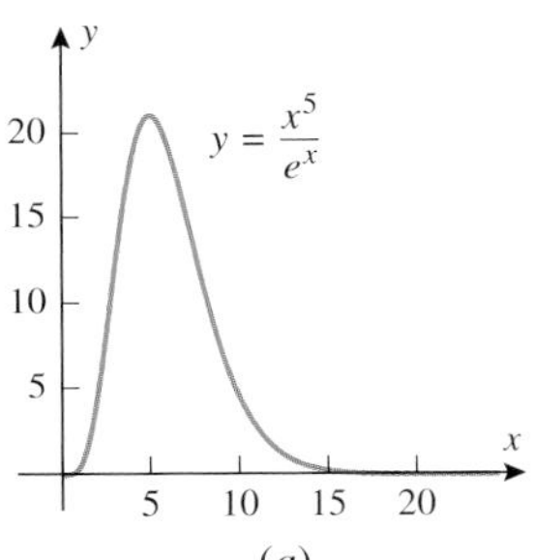

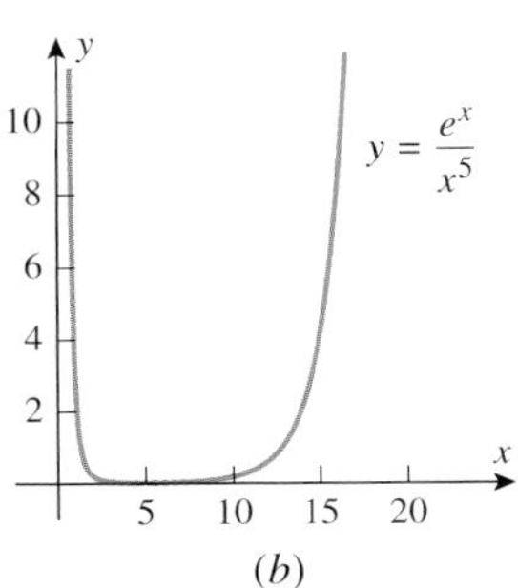

Figure 7.5.1

If n is any positive integer, then $x^n \to +\infty$ as $x \to +\infty$. Such integer powers of x are sometimes used as "measuring sticks" to describe how rapidly other functions grow. For example, we know that $e^x \to +\infty$ as $x \to +\infty$ and that the growth of e^x is very rapid (Table 7.1.5); however, the growth of x^n is also rapid when n is a high power, so it is reasonable to ask whether high powers of x grow more or less rapidly than e^x. One way to investigate this is to examine the behavior of the ratio x^n/e^x as $x \to +\infty$. For example, Figure 7.5.1a shows the graph of $y = x^5/e^x$. This graph suggests that $x^5/e^x \to 0$ as $x \to +\infty$, and this implies that the growth of the function e^x is sufficiently rapid that its values eventually overtake those of x^5 and force the ratio toward zero. Stated informally, "e^x eventually grows more rapidly than x^5." The same conclusion could have been reached by putting e^x on top and examining the behavior of e^x/x^5 as $x \to +\infty$ (Figure 7.5.1b). In this case the values of e^x eventually overtake those of x^5 and force the ratio toward $+\infty$. More generally, we can use L'Hôpital's rule to show that e^x *eventually grows more rapidly than any positive integer power of* x, that is,

$$\lim_{x\to+\infty}\frac{x^n}{e^x} = 0 \quad\text{and}\quad \lim_{x\to+\infty}\frac{e^x}{x^n} = +\infty \tag{5–6}$$

Both limits are indeterminate forms of type ∞/∞ that can be evaluated using L'Hôpital's rule. For example, to establish (5), we will need to apply L'Hôpital's rule n times. For this purpose, observe that successive differentiations of x^n reduce the exponent by 1 each time, thus producing a constant for the nth derivative. For example, the successive derivatives of x^3 are $3x^2$, $6x$, and 6. In general, the nth derivative of x^n is $n(n-1)(n-2)\cdots 1 = n!$ (verify).* Thus, applying L'Hôpital's rule n times to (5) yields

$$\lim_{x\to+\infty}\frac{x^n}{e^x} = \lim_{x\to+\infty}\frac{n!}{e^x} = 0$$

Limit (6) can be established similarly.

---

*Recall that for $n \geq 1$ the expression n!, read ***n-factorial***, denotes the product of the first n positive integers.

INDETERMINATE FORMS OF TYPE $0 \cdot \infty$

Thus far we have discussed indeterminate forms of type $0/0$ and ∞/∞. However, these are not the only possibilities; in general, the limit of an expression that has one of the forms

$$\frac{f(x)}{g(x)}, \quad f(x)\cdot g(x), \quad f(x)^{g(x)}, \quad f(x)-g(x), \quad f(x)+g(x)$$

is called an *indeterminate form* if the limits of $f(x)$ and $g(x)$ individually exert conflicting influences on the limit of the entire expression. For example, the limit

$$\lim_{x\to 0^+} x\ln x$$

is an ***indeterminate form of type*** $\mathbf{0\cdot\infty}$ because the limit of the first factor is 0, the limit of the second factor is $-\infty$, and these two limits exert conflicting influences on the product. On the other hand, the limit

$$\lim_{x\to +\infty} [\sqrt{x}(1-x^2)]$$

is not an indeterminate form because the first factor has a limit of $+\infty$, the second factor has a limit of $-\infty$, and these influences work together to produce a limit of $-\infty$ for the product.

Indeterminate forms of type $0\cdot\infty$ can sometimes be evaluated by rewriting the product as a ratio, and then applying L'Hôpital's rule for indeterminate forms of type $0/0$ or ∞/∞.

WARNING

It is tempting to argue that an indeterminate form of type $0\cdot\infty$ has value 0 since "zero times anything is zero." However, this is fallacious since $0\cdot\infty$ is not a product of numbers, but rather a statement about limits. For example, here are two indeterminate forms of type $0\cdot\infty$ whose limits are *not* zero:

$$\lim_{x\to 0}\left(x\cdot\frac{1}{x}\right)=\lim_{x\to 0} 1 = 1$$

$$\lim_{x\to 0^+}\left(\sqrt{x}\cdot\frac{1}{x}\right)=\lim_{x\to 0^+}\left(\frac{1}{\sqrt{x}}\right) = +\infty$$

▶ Example 4 Evaluate

(a) $\displaystyle\lim_{x\to 0^+} x\ln x$ (b) $\displaystyle\lim_{x\to\pi/4} (1-\tan x)\sec 2x$

Solution (a). The factor x has a limit of 0 and the factor $\ln x$ has a limit of $-\infty$, so the stated problem is an indeterminate form of type $0\cdot\infty$. There are two possible approaches: we can rewrite the limit as

$$\lim_{x\to 0^+}\frac{\ln x}{1/x} \quad\text{or}\quad \lim_{x\to 0^+}\frac{x}{1/\ln x}$$

the first being an indeterminate form of type ∞/∞ and the second an indeterminate form of type $0/0$. However, the first form is the preferred initial choice because the derivative of $1/x$ is less complicated than the derivative of $1/\ln x$. That choice yields

$$\lim_{x\to 0^+} x\ln x = \lim_{x\to 0^+}\frac{\ln x}{1/x} = \lim_{x\to 0^+}\frac{1/x}{-1/x^2} = \lim_{x\to 0^+}(-x) = 0$$

Solution (b). The stated problem is an indeterminate form of type $0\cdot\infty$. We will convert it to an indeterminate form of type $0/0$:

$$\lim_{x\to\pi/4}(1-\tan x)\sec 2x = \lim_{x\to\pi/4}\frac{1-\tan x}{1/\sec 2x} = \lim_{x\to\pi/4}\frac{1-\tan x}{\cos 2x}$$

$$= \lim_{x\to\pi/4}\frac{-\sec^2 x}{-2\sin 2x} = \frac{-2}{-2} = 1 \; ◀$$

INDETERMINATE FORMS OF TYPE $\infty - \infty$

A limit problem that leads to one of the expressions

$$(+\infty)-(+\infty), \quad (-\infty)-(-\infty),$$
$$(+\infty)+(-\infty), \quad (-\infty)+(+\infty)$$

is called an ***indeterminate form of type*** $\boldsymbol{\infty-\infty}$. Such limits are indeterminate because the two terms exert conflicting influences on the expression: one pushes it in the positive direction and the other pushes it in the negative direction. However, limit problems that lead to one of the expressions

$$(+\infty) + (+\infty), \quad (+\infty) - (-\infty),$$
$$(-\infty) + (-\infty), \quad (-\infty) - (+\infty)$$

are not indeterminate, since the two terms work together (those on the top produce a limit of $+\infty$ and those on the bottom produce a limit of $-\infty$).

Indeterminate forms of type $\infty - \infty$ can sometimes be evaluated by combining the terms and manipulating the result to produce an indeterminate form of type $0/0$ or ∞/∞.

▶ **Example 5** Evaluate $\lim_{x \to 0^+} \left(\frac{1}{x} - \frac{1}{\sin x} \right)$.

Solution. Both terms have a limit of $+\infty$, so the stated problem is an indeterminate form of type $\infty - \infty$. Combining the two terms yields

$$\lim_{x \to 0^+} \left(\frac{1}{x} - \frac{1}{\sin x} \right) = \lim_{x \to 0^+} \frac{\sin x - x}{x \sin x}$$

which is an indeterminate form of type $0/0$. Applying L'Hôpital's rule twice yields

$$\begin{aligned} \lim_{x \to 0^+} \frac{\sin x - x}{x \sin x} &= \lim_{x \to 0^+} \frac{\cos x - 1}{\sin x + x \cos x} \\ &= \lim_{x \to 0^+} \frac{-\sin x}{\cos x + \cos x - x \sin x} = \frac{0}{2} = 0 \end{aligned}$$ ◀

INDETERMINATE FORMS OF TYPE 0^0, ∞^0, 1^∞

Limits of the form

$$\lim f(x)^{g(x)}$$

can give rise to ***indeterminate forms of the types*** $\mathbf{0^0}$, $\mathbf{\infty^0}$, ***and*** $\mathbf{1^\infty}$. (The interpretations of these symbols should be clear.) For example, the limit

$$\lim_{x \to 0^+} (1 + x)^{1/x}$$

whose value we know to be e [see Formula (5) of Section 7.1] is an indeterminate form of type 1^∞. It is indeterminate because the expressions $1 + x$ and $1/x$ exert two conflicting influences: the first approaches 1, which drives the expression toward 1, and the second approaches $+\infty$, which drives the expression toward $+\infty$.

Indeterminate forms of types 0^0, ∞^0, and 1^∞ can sometimes be evaluated by first introducing a dependent variable

$$y = f(x)^{g(x)}$$

and then computing the limit of $\ln y$. Since

$$\ln y = \ln[f(x)^{g(x)}] = g(x) \cdot \ln[f(x)]$$

the limit of $\ln y$ will be an indeterminate form of type $0 \cdot \infty$ (verify), which can be evaluated by methods we have already studied. Once the limit of $\ln y$ is known, it is a straightforward matter to determine the limit of $y = f(x)^{g(x)}$, as we will illustrate in the next example.

▶ **Example 6** Show that $\lim_{x \to 0} (1 + x)^{1/x} = e$.

Solution. As discussed above, we begin by introducing a dependent variable

$$y = (1 + x)^{1/x}$$

and taking the natural logarithm of both sides:

$$\ln y = \ln(1+x)^{1/x} = \frac{1}{x}\ln(1+x) = \frac{\ln(1+x)}{x}$$

Thus,

$$\lim_{x\to 0} \ln y = \lim_{x\to 0} \frac{\ln(1+x)}{x}$$

which is an indeterminate form of type 0/0, so by L'Hôpital's rule

$$\lim_{x\to 0} \ln y = \lim_{x\to 0} \frac{\ln(1+x)}{x} = \lim_{x\to 0} \frac{1/(1+x)}{1} = 1$$

Since we have shown that $\ln y \to 1$ as $x \to 0$, the continuity of the exponential function implies that $e^{\ln y} \to e^1$ as $x \to 0$, and this implies that $y \to e$ as $x \to 0$. Thus,

$$\lim_{x\to 0} (1+x)^{1/x} = e \quad ◄$$

✔QUICK CHECK EXERCISES 7.5 *(See page 476 for answers.)*

1. In each part, does L'Hôpital's rule apply to the given limit?
   (a) $\lim_{x\to 1} \frac{2x-2}{x^3+x-2}$ (b) $\lim_{x\to 0} \frac{\cos x}{x}$
   (c) $\lim_{x\to 0} \frac{e^{2x}-1}{\tan x}$
2. Evaluate each of the limits in Quick Check Exercise 1.
3. Using L'Hôpital's rule,
   $$\lim_{x\to+\infty} \frac{e^x}{500x^2} = \_\_\_\_\_\_$$

EXERCISE SET 7.5 Graphing Utility C CAS

1–2 Evaluate the given limit without using L'Hôpital's rule, and then check that your answer is correct using L'Hôpital's rule.

1. (a) $\lim_{x\to 2} \frac{x^2-4}{x^2+2x-8}$ (b) $\lim_{x\to+\infty} \frac{2x-5}{3x+7}$
2. (a) $\lim_{x\to 0} \frac{\sin x}{\tan x}$ (b) $\lim_{x\to 1} \frac{x^2-1}{x^3-1}$

3–4 For the given functions $f(x)$ and $g(x)$, and specified point a: (a) Verify that $\lim_{x\to a} f(x)/g(x)$ is an indeterminate form of type 0/0. (b) Find the local linear approximations $T_f(x)$ and $T_g(x)$ to the functions $f(x)$ and $g(x)$, respectively, at $x = a$. (c) Without using L'Hôpital's rule, verify that

$$\lim_{x\to a} \frac{f(x)}{g(x)} = \lim_{x\to a} \frac{T_f(x)}{T_g(x)}$$

3. $f(x) = x^2 - 1,\ g(x) = x^2 - x - 2,\ a = -1$
4. $f(x) = \cos x,\ g(x) = \cot x,\ a = \pi/2$

5–34 Find the limit.

5. $\lim_{x\to 0} \frac{e^x-1}{\sin x}$
6. $\lim_{x\to 0} \frac{\sin 2x}{\sin 5x}$
7. $\lim_{\theta\to 0} \frac{\tan\theta}{\theta}$
8. $\lim_{t\to 0} \frac{te^t}{1-e^t}$
9. $\lim_{x\to\pi^+} \frac{\sin x}{x-\pi}$
10. $\lim_{x\to 0^+} \frac{\sin x}{x^2}$
11. $\lim_{x\to+\infty} \frac{\ln x}{x}$
12. $\lim_{x\to+\infty} \frac{e^{3x}}{x^2}$
13. $\lim_{x\to 0^+} \frac{\cot x}{\ln x}$
14. $\lim_{x\to 0^+} \frac{1-\ln x}{e^{1/x}}$
15. $\lim_{x\to+\infty} \frac{x^{100}}{e^x}$
16. $\lim_{x\to 0^+} \frac{\ln(\sin x)}{\ln(\tan x)}$
17. $\lim_{x\to+\infty} xe^{-x}$
18. $\lim_{x\to\pi^-} (x-\pi)\tan\frac{1}{2}x$
19. $\lim_{x\to+\infty} x\sin\frac{\pi}{x}$
20. $\lim_{x\to 0^+} \tan x \ln x$
21. $\lim_{x\to\pi/2^-} \sec 3x \cos 5x$
22. $\lim_{x\to\pi} (x-\pi)\cot x$
23. $\lim_{x\to+\infty} (1-3/x)^x$
24. $\lim_{x\to 0} (1+2x)^{-3/x}$
25. $\lim_{x\to 0} (e^x+x)^{1/x}$
26. $\lim_{x\to+\infty} (1+a/x)^{bx}$
27. $\lim_{x\to 1} (2-x)^{\tan[(\pi/2)x]}$
28. $\lim_{x\to+\infty} [\cos(2/x)]^{x^2}$
29. $\lim_{x\to 0} (\csc x - 1/x)$
30. $\lim_{x\to 0} \left(\frac{1}{x^2} - \frac{\cos 3x}{x^2}\right)$
31. $\lim_{x\to+\infty} (\sqrt{x^2+x} - x)$
32. $\lim_{x\to 0} \left(\frac{1}{x} - \frac{1}{e^x-1}\right)$
33. $\lim_{x\to+\infty} [x - \ln(x^2+1)]$
34. $\lim_{x\to+\infty} [\ln x - \ln(1+x)]$
35. C Use a CAS to check the answers you obtained in Exercises 29–34.

36. Show that for any positive integer n

(a) $\lim_{x\to+\infty} \dfrac{\ln x}{x^n} = 0$ (b) $\lim_{x\to+\infty} \dfrac{x^n}{\ln x} = +\infty.$

FOCUS ON CONCEPTS

37. (a) Find the error in the following calculation:

$$\lim_{x\to 1} \frac{x^3 - x^2 + x - 1}{x^3 - x^2} = \lim_{x\to 1} \frac{3x^2 - 2x + 1}{3x^2 - 2x} = \lim_{x\to 1} \frac{6x - 2}{6x - 2} = 1$$

(b) Find the correct limit.

38. (a) Find the error in the following calculation:

$$\lim_{x\to 2} \frac{e^{3x^2-12x+12}}{x^4 - 16} = \lim_{x\to 2} \frac{(6x - 12)e^{3x^2-12x+12}}{4x^3} = 0$$

(b) Find the correct limit.

39–42 Make a conjecture about the limit by graphing the function involved with a graphing utility; then check your conjecture using L'Hôpital's rule.

39. $\lim_{x\to+\infty} \dfrac{\ln(\ln x)}{\sqrt{x}}$ **40.** $\lim_{x\to 0^+} x^x$

41. $\lim_{x\to 0^+} (\sin x)^{3/\ln x}$ **42.** $\lim_{x\to(\pi/2)^-} \dfrac{4\tan x}{1 + \sec x}$

43–46 Make a conjecture about the equations of horizontal asymptotes, if any, by graphing the equation with a graphing utility; then check your answer using L'Hôpital's rule.

43. $y = \ln x - e^x$ **44.** $y = x - \ln(1 + 2e^x)$

45. $y = (\ln x)^{1/x}$ **46.** $y = \left(\dfrac{x+1}{x+2}\right)^x$

47. Limits of the type

$$0/\infty, \quad \infty/0, \quad 0^\infty, \quad \infty\cdot\infty, \quad +\infty + (+\infty),$$
$$+\infty - (-\infty), \quad -\infty + (-\infty), \quad -\infty - (+\infty)$$

are *not* indeterminate forms. Find the following limits by inspection.

(a) $\lim_{x\to 0^+} \dfrac{x}{\ln x}$ (b) $\lim_{x\to+\infty} \dfrac{x^3}{e^{-x}}$

(c) $\lim_{x\to(\pi/2)^-} (\cos x)^{\tan x}$ (d) $\lim_{x\to 0^+} (\ln x)\cot x$

(e) $\lim_{x\to 0^+} \left(\dfrac{1}{x} - \ln x\right)$ (f) $\lim_{x\to-\infty} (x + x^3)$

48. There is a myth that circulates among beginning calculus students which states that all indeterminate forms of types 0^0, ∞^0, and 1^∞ have value 1 because "anything to the zero power is 1" and "1 to any power is 1." The fallacy is that 0^0, ∞^0, and 1^∞ are not powers of numbers, but rather descriptions of limits. The following examples, which were suggested by Prof. Jack Staib of Drexel University, show that such indeterminate forms can have any positive real value:

(a) $\lim_{x\to 0^+} [x^{(\ln a)/(1+\ln x)}] = a$ (form 0^0)

(b) $\lim_{x\to+\infty} [x^{(\ln a)/(1+\ln x)}] = a$ (form ∞^0)

(c) $\lim_{x\to 0} [(x + 1)^{(\ln a)/x}] = a$ (form 1^∞).

Verify these results.

49–52 Verify that L'Hôpital's rule is of no help in finding the limit; then find the limit, if it exists, by some other method.

49. $\lim_{x\to+\infty} \dfrac{x + \sin 2x}{x}$ **50.** $\lim_{x\to+\infty} \dfrac{2x - \sin x}{3x + \sin x}$

51. $\lim_{x\to+\infty} \dfrac{x(2 + \sin 2x)}{x + 1}$ **52.** $\lim_{x\to+\infty} \dfrac{x(2 + \sin x)}{x^2 + 1}$

53. The accompanying schematic diagram represents an electrical circuit consisting of an electromotive force that produces a voltage V, a resistor with resistance R, and an inductor with inductance L. It is shown in electrical circuit theory that if the voltage is first applied at time $t = 0$, then the current I flowing through the circuit at time t is given by

$$I = \frac{V}{R}(1 - e^{-Rt/L})$$

What is the effect on the current at a fixed time t if the resistance approaches 0 (i.e., $R \to 0^+$)?

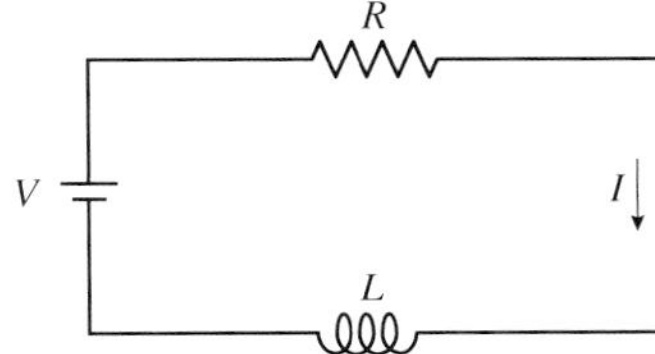

Figure Ex-53

54. (a) Show that $\lim_{x\to\pi/2} (\pi/2 - x)\tan x = 1.$

(b) Show that

$$\lim_{x\to\pi/2} \left(\frac{1}{\pi/2 - x} - \tan x\right) = 0$$

(c) It follows from part (b) that the approximation

$$\tan x \approx \frac{1}{\pi/2 - x}$$

should be good for values of x near $\pi/2$. Use a calculator to find $\tan x$ and $1/(\pi/2 - x)$ for $x = 1.57$; compare the results.

55. (a) Use a CAS to show that if k is a positive constant, then

$$\lim_{x\to+\infty} x(k^{1/x} - 1) = \ln k$$

(b) Confirm this result using L'Hôpital's rule. [*Hint:* Express the limit in terms of $t = 1/x$.]

(c) If n is a positive integer, then it follows from part (a) with $x = n$ that the approximation

$$n(\sqrt[n]{k} - 1) \approx \ln k$$

should be good when n is large. Use this result and the square root key on a calculator to approximate the values of $\ln 0.3$ and $\ln 2$ with $n = 1024$, then compare the values obtained with values of the logarithms generated directly from the calculator. [*Hint:* The nth roots for which n is a power of 2 can be obtained as successive square roots.]

56. Find all values of k and l such that

$$\lim_{x\to 0}\frac{k+\cos lx}{x^2}=-4$$

FOCUS ON CONCEPTS

57. Let $f(x)=x^2\sin(1/x)$.

(a) Are the limits $\lim_{x\to 0^+} f(x)$ and $\lim_{x\to 0^-} f(x)$ indeterminate forms?

(b) Use a graphing utility to generate the graph of f, and use the graph to make conjectures about the limits in part (a).

(c) Use the Squeezing Theorem (2.6.2) to confirm that your conjectures in part (b) are correct.

58. (a) Explain why L'Hôpital's rule does not apply to the problem

$$\lim_{x\to 0}\frac{x^2\sin(1/x)}{\sin x}$$

(b) Find the limit.

59. Find $\lim_{x\to 0^+}\dfrac{x\sin(1/x)}{\sin x}$ if it exists.

60. Suppose that functions f and g are differentiable at $x=a$ and that $f(a)=g(a)=0$. If $g'(a)\neq 0$, show that

$$\lim_{x\to a}\frac{f(x)}{g(x)}=\frac{f'(a)}{g'(a)}$$

without using L'Hôpital's rule. [*Hint:* divide the numerator and denominator of $f(x)/g(x)$ by $x-a$ and use the definitions for $f'(a)$ and $g'(a)$.]

✔ QUICK CHECK ANSWERS 7.5

1. (a) yes (b) no (c) yes 2. (a) $\frac{1}{2}$ (b) does not exist (c) 2 3. $+\infty$

7.6 LOGARITHMIC FUNCTIONS FROM THE INTEGRAL POINT OF VIEW

In Section 7.1 we defined the natural logarithm function $\ln x$ *to be the inverse of* e^x*. In this section we will define* $\ln x$ *as a particular definite integral in which* x *is the upper limit of integration. This new approach will allow us to prove rigorously that* $\ln x$ *and* e^x *are differentiable and will provide us with an alternative treatment of irrational exponents. Furthermore, we will see that it can be convenient to define other functions as definite integrals with variable upper limits of integration. In particular, we will show that such functions often arise in the solution of certain initial-value problems.*

THE CONNECTION BETWEEN NATURAL LOGARITHMS AND INTEGRALS

The connection between natural logarithms and integrals was made in the middle of the seventeenth century in the course of investigating areas under the curve $y=1/t$. The problem being considered was to find values of $t_1, t_2, t_3, \ldots, t_n, \ldots$ for which the areas $A_1, A_2, A_3, \ldots, A_n, \ldots$ in Figure 7.6.1a would be equal. Through the combined work of Isaac Newton, the Belgian Jesuit priest Gregory of St. Vincent (1584–1667), and Gregory's student Alfons A. de Sarasa (1618–1667), it was shown that by taking the points to be

$$t_1=e,\quad t_2=e^2,\quad t_3=e^3,\ldots,\quad t_n=e^n,\ldots$$

each of the areas would be 1 (Figure 7.6.1b). Thus, in modern integral notation

$$\int_1^{e^n}\frac{1}{t}\,dt=n$$

which can be expressed as

$$\int_1^{e^n}\frac{1}{t}\,dt=\ln(e^n)$$

By comparing the upper limit of the integral and the expression inside the logarithm, it is a natural leap to the more general result

$$\int_1^{x}\frac{1}{t}\,dt=\ln x$$

which today we take as the formal definition of the natural logarithm.

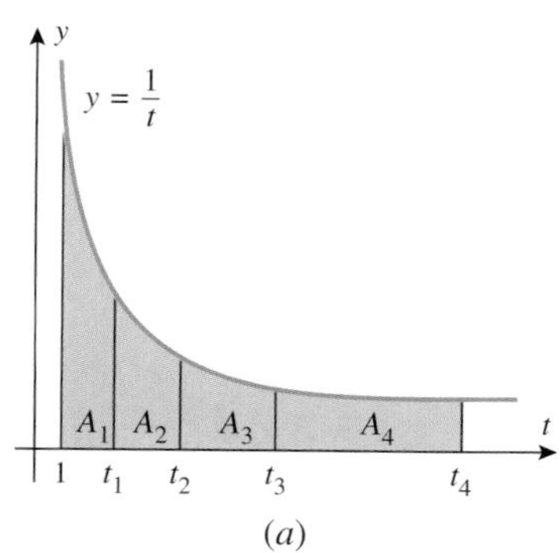

(a)

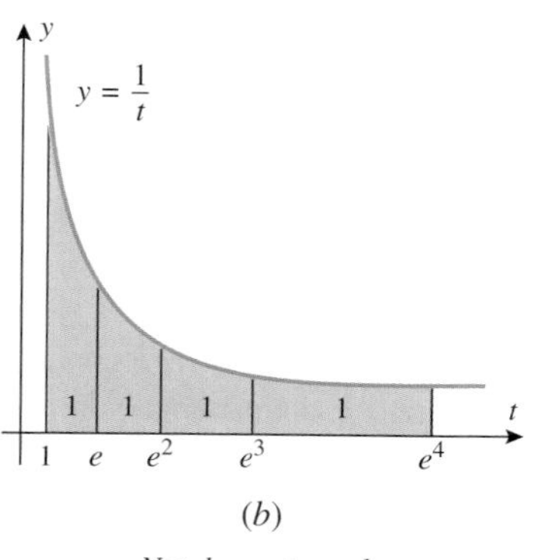

(b)

Not drawn to scale

Figure 7.6.1

Review Theorem 5.5.8 and then explain why x is required to be positive in Definition 7.6.1.

7.6.1 DEFINITION. The ***natural logarithm*** of x is denoted by $\ln x$ and is defined by the integral

$$\ln x = \int_1^x \frac{1}{t}\,dt, \quad x > 0 \tag{1}$$

None of the properties of $\ln x$ obtained in this section should be new, but now, for the first time, we give them a sound mathematical footing.

Our strategy for putting the study of logarithmic and exponential functions on a sound mathematical footing is to use (1) as a starting point and then define e^x as the inverse of $\ln x$. This is the exact opposite of our previous approach in which we defined $\ln x$ to be the inverse of e^x. However, whereas previously we had to *assume* that e^x is continuous, the continuity of e^x will now follow from our definitions as a *theorem*. Our first challenge is to demonstrate that the properties of $\ln x$ resulting from Definition 7.6.1 are consistent with those obtained earlier. To start, observe that Part 2 of the Fundamental Theorem of Calculus (Theorem 5.6.3) implies that $\ln x$ is differentiable and

$$\frac{d}{dx}[\ln x] = \frac{d}{dx}\left[\int_1^x \frac{1}{t}\,dt\right] = \frac{1}{x} \quad (x > 0) \tag{2}$$

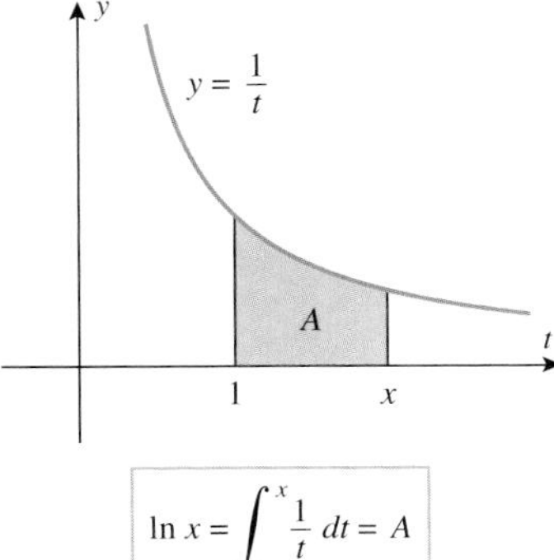

$\ln x = \int_1^x \frac{1}{t}\,dt = A$

(*a*)

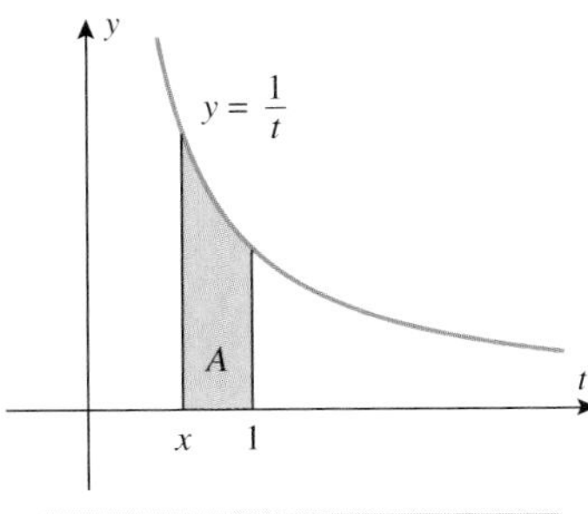

$\ln x = \int_1^x \frac{1}{t}\,dt = -\int_x^1 \frac{1}{t}\,dt = -A$

(*b*)

Figure 7.6.2

This is consistent with the derivative formula for $\ln x$ that we obtained previously. Moreover, because differentiability implies continuity, it follows that $\ln x$ is a continuous function on the interval $(0, +\infty)$.

Other properties of $\ln x$ can be obtained by interpreting the integral in (1) geometrically: In the case where $x > 1$, this integral represents the area under the curve $y = 1/t$ from $t = 1$ to $t = x$ (Figure 7.6.2*a*); in the case where $0 < x < 1$, the integral represents the negative of the area under the curve $y = 1/t$ from $t = x$ to $t = 1$ (Figure 7.6.2*b*); and in the case where $x = 1$, the integral has value 0 because its upper and lower limits of integration are the same. These geometric observations imply that

$$\begin{aligned} \ln x > 0 &\quad \text{if} \quad x > 1 \\ \ln x < 0 &\quad \text{if} \quad 0 < x < 1 \\ \ln x = 0 &\quad \text{if} \quad x = 1 \end{aligned}$$

Also, since $1/x$ is positive for $x > 0$, it follows from (2) that $\ln x$ is an increasing function on the interval $(0, +\infty)$. This is all consistent with the graph of $\ln x$ in Figure 7.6.3.

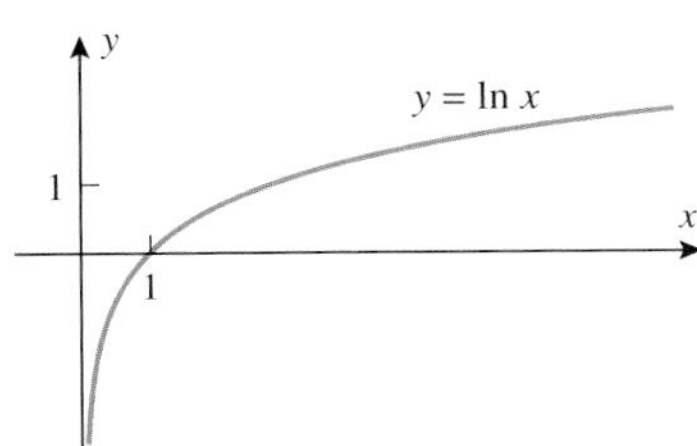

Figure 7.6.3

ALGEBRAIC PROPERTIES OF ln *x*

We can use (1) to show that Definition 7.6.1 produces the standard algebraic properties of logarithms.

7.6.2 THEOREM. *For any positive numbers a and c and any rational number r:*

(*a*) $\ln ac = \ln a + \ln c$ (*b*) $\ln \frac{1}{c} = -\ln c$

(*c*) $\ln \frac{a}{c} = \ln a - \ln c$ (*d*) $\ln a^r = r \ln a$

PROOF (*a*). Treating a as a constant, consider the function $f(x) = \ln(ax)$. Then

$$f'(x) = \frac{1}{ax} \cdot \frac{d}{dx}(ax) = \frac{1}{ax} \cdot a = \frac{1}{x}$$

Thus, $\ln ax$ and $\ln x$ have the same derivative on $(0, +\infty)$, so these functions must differ by a constant on this interval. That is, there is a constant k such that

$$\ln ax - \ln x = k \tag{3}$$

on $(0, +\infty)$. Substituting $x = 1$ into this equation we conclude that $\ln a = k$ (verify). Thus, (3) can be written as

$$\ln ax - \ln x = \ln a$$

Setting $x = c$ establishes that

$$\ln ac - \ln c = \ln a \quad \text{or} \quad \ln ac = \ln a + \ln c$$

PROOFS (*b*) AND (*c*). Part (*b*) follows immediately from part (*a*) by substituting $1/c$ for a (verify). Then

$$\ln \frac{a}{c} = \ln \left(a \cdot \frac{1}{c}\right) = \ln a + \ln \frac{1}{c} = \ln a - \ln c$$

PROOF (*d*). First, we will argue that part (*d*) is satisfied if r is any nonnegative integer. If $r = 1$, then (*d*) is clearly satisfied; if $r = 0$, then (*d*) follows from the fact that $\ln 1 = 0$. Suppose that we know (*d*) is satisfied for r equal to some integer n. It then follows from part (*a*) that

$$\ln a^{n+1} = \ln[a \cdot a^n] = \ln a + \ln a^n = \ln a + n \ln a = (n+1)\ln a$$

> How is the proof of Theorem 7.6.2(*d*) for the case where r is a nonnegative integer analogous to a row of falling dominos? (This "domino" argument uses an informal version of a property of the integers known as the *principle of mathematical induction.*)

That is, if (*d*) is valid for r equal to some integer n, then it is also valid for $r = n + 1$. However, since we know (*d*) is satisfied if $r = 1$, it follows that (*d*) is valid for $r = 2$. But this implies that (*d*) is satisfied for $r = 3$, which in turn implies that (*d*) is valid for $r = 4$, and so forth. We conclude that (*d*) is satisfied if r is any nonnegative integer.

Next, suppose that $r = -m$ is a negative integer. Then

$$\begin{aligned} \ln a^r = \ln a^{-m} = \ln \frac{1}{a^m} &= -\ln a^m && \text{By part } (b) \\ &= -m \ln a && \text{Part } (d) \text{ is valid for positive powers.} \\ &= r \ln a \end{aligned}$$

which shows that (*d*) is valid for any negative integer r. Combining this result with our previous conclusion that (*d*) is satisfied for a nonnegative integer r shows that (*d*) is valid if r is *any* integer.

Finally, suppose that $r = m/n$ is any rational number, where $m \neq 0$ and $n \neq 0$ are integers. Then

$$\begin{aligned} \ln a^r = \frac{n \ln a^r}{n} &= \frac{\ln[(a^r)^n]}{n} && \text{Part } (d) \text{ is valid for integer powers.} \\ &= \frac{\ln a^{rn}}{n} && \text{Property of exponents} \\ &= \frac{\ln a^m}{n} && \text{Definition of } r \\ &= \frac{m \ln a}{n} && \text{Part } (d) \text{ is valid for integer powers.} \\ &= \frac{m}{n} \ln a = r \ln a \end{aligned}$$

which shows that (*d*) is valid for any rational number r. ■

APPROXIMATING ln *x* NUMERICALLY

For specific values of x, the value of $\ln x$ can be approximated numerically by approximating the definite integral in (1), say by using the midpoint approximation that was discussed in Section 5.4.

Example 1 Approximate ln 2 using the midpoint approximation with $n = 10$.

Solution. From (1), the exact value of ln 2 is represented by the integral

$$\ln 2 = \int_1^2 \frac{1}{t}\,dt$$

The midpoint rule is given in Formulas (5) and (9) of Section 5.4. Expressed in terms of t, the latter formula is

$$\int_a^b f(t)\,dt \approx \Delta t \sum_{k=1}^{n} f(t_k^*)$$

where Δt is the common width of the subintervals and $t_1^*, t_2^*, \ldots, t_n^*$ are the midpoints. In this case we have 10 subintervals, so $\Delta t = (2-1)/10 = 0.1$. The computations to six decimal places are shown in Table 7.6.1. By comparison, a calculator set to display six decimal places gives $\ln 2 \approx 0.693147$, so the magnitude of the error in the midpoint approximation is about 0.000311. Greater accuracy in the midpoint approximation can be obtained by increasing n. For example, the midpoint approximation with $n = 100$ yields $\ln 2 \approx 0.693144$, which is correct to five decimal places. ◄

Table 7.6.1

$n = 10$
$\Delta t = (b-a)/n = (2-1)/10 = 0.1$

k	t_k^*	$1/t_k^*$
1	1.05	0.952381
2	1.15	0.869565
3	1.25	0.800000
4	1.35	0.740741
5	1.45	0.689655
6	1.55	0.645161
7	1.65	0.606061
8	1.75	0.571429
9	1.85	0.540541
10	1.95	0.512821
		6.928355

$$\Delta t \sum_{k=1}^{n} f(t_k^*) \approx (0.1)(6.928355) \approx 0.692836$$

DOMAIN, RANGE, AND END BEHAVIOR OF ln *x*

7.6.3 THEOREM.

(*a*) *The domain of* $\ln x$ *is* $(0, +\infty)$.

(*b*) $\lim_{x\to 0^+} \ln x = -\infty$ *and* $\lim_{x\to +\infty} \ln x = +\infty$

(*c*) *The range of* $\ln x$ *is* $(-\infty, +\infty)$.

PROOF (*a*) AND (*b*). We have already shown that $\ln x$ is defined and increasing on the interval $(0, +\infty)$. To prove that $\ln x \to +\infty$ as $x \to +\infty$, we must show that given any number $M > 0$, the value of $\ln x$ exceeds M for sufficiently large values of x. To do this, let N be any integer. If $x > 2^N$, then

$$\ln x > \ln 2^N = N \ln 2 \tag{4}$$

by Theorem 7.6.2(*d*). Since

$$\ln 2 = \int_1^2 \frac{1}{t}\,dt > 0$$

it follows that $N \ln 2$ can be made arbitrarily large by choosing N sufficiently large. In particular, we can choose N so that $N \ln 2 > M$. It now follows from (4) that if $x > 2^N$, then $\ln x > M$, and this proves that

$$\lim_{x\to +\infty} \ln x = +\infty$$

Furthermore, by observing that $v = 1/x \to +\infty$ as $x \to 0^+$, we can use the preceding limit and Theorem 7.6.2(*b*) to conclude that

$$\lim_{x\to 0^+} \ln x = \lim_{v\to +\infty} \ln \frac{1}{v} = \lim_{v\to +\infty} (-\ln v) = -\infty$$

PROOF (*c*). It follows from part (*a*), the continuity of $\ln x$, and the Intermediate-Value Theorem (2.5.8) that $\ln x$ assumes every real value as x varies over the interval $(0, +\infty)$ (why?). ■

DEFINITION OF e^x

In Section 7.1 we defined $\ln x$ to be the inverse of the natural exponential function e^x. Now that we have a formal definition of $\ln x$ in terms of an integral, we will define the natural exponential function to be the inverse of $\ln x$.

Since $\ln x$ is increasing and continuous on $(0, +\infty)$ with range $(-\infty, +\infty)$, there is exactly one (positive) solution to the equation $\ln x = 1$. We *define* e to be the unique solution to $\ln x = 1$, so

$$\ln e = 1 \tag{5}$$

Furthermore, if x is any real number, there is a unique positive solution y to $\ln y = x$, so for irrational values of x we *define* e^x to be this solution. That is, when x is irrational, e^x is defined by

$$\ln e^x = x \tag{6}$$

Note that for rational values of x, we also have $\ln e^x = x \ln e = x$ from Theorem 7.6.2(d). Moreover, it follows immediately that $e^{\ln x} = x$ for any $x > 0$. Thus, (6) defines the exponential function for all real values of x as the inverse of the natural logarithm function.

7.6.4 DEFINITION. The inverse of the natural logarithm function $\ln x$ is denoted by e^x and is called the ***natural exponential function***.

We can now establish the differentiability of e^x and confirm that

$$\frac{d}{dx}[e^x] = e^x$$

7.6.5 THEOREM. *The natural exponential function e^x is differentiable, and hence continuous, on $(-\infty, +\infty)$, and its derivative is*

$$\frac{d}{dx}[e^x] = e^x$$

PROOF. Because $\ln x$ is differentiable and

$$\frac{d}{dx}[\ln x] = \frac{1}{x} > 0$$

for all x in $(0, +\infty)$, it follows from Theorem 7.3.1, with $f(x) = \ln x$ and $f^{-1}(x) = e^x$, that e^x is differentiable on $(-\infty, +\infty)$ and its derivative is

$$\frac{d}{dx}\underbrace{[e^x]}_{f^{-1}(x)} = \underbrace{\frac{1}{1/e^x}}_{f'(f^{-1}(x))} = e^x \qquad \blacksquare$$

IRRATIONAL EXPONENTS

Recall from Theorem 7.6.2(d) that if $a > 0$ and r is a rational number, then $\ln a^r = r \ln a$. Then $a^r = e^{\ln a^r} = e^{r \ln a}$ for any positive value of a and any rational number r. But the expression $e^{r \ln a}$ makes sense for *any* real number r, whether rational or irrational, so it is a good candidate to give meaning to a^r for any real number r.

Use Definition 7.6.6 to prove that if $a > 0$ and r is a real number, then $\ln a^r = r \ln a$.

7.6.6 DEFINITION. If $a > 0$ and r is a real number, a^r is defined by

$$a^r = e^{r \ln a} \tag{7}$$

With this definition it can be shown that the standard algebraic properties of exponents, such as

$$a^p a^q = a^{p+q}, \quad \frac{a^p}{a^q} = a^{p-q}, \quad (a^p)^q = a^{pq}, \quad (a^p)(b^p) = (ab)^p$$

hold for any real values of a, b, p, and q, where a and b are positive. In addition, using (7) for a real exponent r, we can define the power function x^r whose domain consists of all positive real numbers and, for a positive base b, we can define the ***base b exponential function b^x*** whose domain consists of all real numbers.

7.6.7 THEOREM.

(*a*) *For any real number r, the power function x^r is differentiable on $(0, +\infty)$ and its derivative is*

$$\frac{d}{dx}[x^r] = rx^{r-1}$$

(*b*) *For $b > 0$ and $b \neq 1$, the base b exponential function b^x is differentiable on $(-\infty, +\infty)$ and its derivative is*

$$\frac{d}{dx}[b^x] = b^x \ln b$$

PROOF. The differentiability of $x^r = e^{r \ln x}$ and $b^x = e^{x \ln b}$ on their domains follows from the differentiability of $\ln x$ on $(0, +\infty)$ and of e^x on $(-\infty, +\infty)$:

$$\frac{d}{dx}[x^r] = \frac{d}{dx}[e^{r \ln x}] = e^{r \ln x} \cdot \frac{d}{dx}[r \ln x] = x^r \cdot \frac{r}{x} = rx^{r-1}$$

$$\frac{d}{dx}[b^x] = \frac{d}{dx}[e^{x \ln b}] = e^{x \ln b} \cdot \frac{d}{dx}[x \ln b] = b^x \ln b \qquad \blacksquare$$

We expressed e as the value of a limit in Formulas (3) and (4) of Section 7.1 and in Formula (1) of Section 7.2. We now have the mathematical tools necessary to prove the existence of these limits.

7.6.8 THEOREM.

(*a*) $\lim_{x \to 0} (1 + x)^{1/x} = e$ (*b*) $\lim_{x \to +\infty} \left(1 + \frac{1}{x}\right)^x = e$ (*c*) $\lim_{x \to -\infty} \left(1 + \frac{1}{x}\right)^x = e$

PROOF. We will prove part (*a*); the proofs of parts (*b*) and (*c*) follow from this limit and are left as exercises. We first observe that

$$\left.\frac{d}{dx}[\ln(x+1)]\right|_{x=0} = \left.\frac{1}{x+1} \cdot 1\right|_{x=0} = 1$$

However, using the definition of the derivative, we obtain

$$1 = \left.\frac{d}{dx}[\ln(x+1)]\right|_{x=0} = \lim_{h \to 0} \frac{\ln(0+h+1) - \ln(0+1)}{h}$$

$$= \lim_{h \to 0} \left[\frac{1}{h} \cdot \ln(1+h)\right]$$

or, equivalently,

$$\lim_{x \to 0} \frac{1}{x} \cdot \ln(1+x) = 1 \tag{8}$$

Now

$$\begin{aligned}
\lim_{x\to 0}(1+x)^{1/x} &= \lim_{x\to 0} e^{\frac{1}{x}\cdot\ln(1+x)} && \text{Definition 7.6.6} \\
&= e^{[\lim_{x\to 0}\frac{1}{x}\cdot\ln(1+x)]} && \text{Theorem 2.5.5} \\
&= e^1 && \text{Equation (8)} \\
&= e
\end{aligned}$$

■

GENERAL LOGARITHMS

We note that for $b > 0$ and $b \neq 1$, the function b^x is one-to-one, and so has an inverse function. Using the definition of b^x, we can solve $y = b^x$ for x as a function of y:

$$\begin{aligned}
y &= b^x = e^{x\ln b} \\
\ln y &= \ln(e^{x\ln b}) = x\ln b \\
\frac{\ln y}{\ln b} &= x
\end{aligned}$$

Thus, the inverse function for b^x is $(\ln x)/(\ln b)$.

7.6.9 DEFINITION. For $b > 0$ and $b \neq 1$, the ***base b logarithm*** function, denoted $\log_b x$, is defined by

$$\log_b x = \frac{\ln x}{\ln b} \tag{9}$$

It follows immediately from this definition that $\log_b x$ is the inverse function for b^x and satisfies the properties in Table 7.1.3. Furthermore, $\log_b x$ is differentiable, and hence continuous, on $(0, +\infty)$, and its derivative is

$$\frac{d}{dx}[\log_b x] = \frac{1}{x\ln b}$$

As a final note of consistency, we observe that $\log_e x = \ln x$.

FUNCTIONS DEFINED BY INTEGRALS

The functions we have dealt with thus far in this text are called ***elementary functions***; they include polynomial, rational, power, exponential, logarithmic, and trigonometric functions, and all other functions that can be obtained from these by addition, subtraction, multiplication, division, root extraction, and composition.

However, there are many important functions that do not fall into this category. Such functions occur in many ways, but they commonly arise in the course of solving initial-value problems of the form

$$\frac{dy}{dx} = f(x), \quad y(x_0) = y_0 \tag{10}$$

Recall from Example 6 of Section 5.2 and the discussion preceding it that the basic method for solving (10) is to integrate $f(x)$, and then use the initial condition to determine the constant of integration. It can be proved that if f is continuous, then (10) has a unique solution and that this procedure produces it. However, there is another approach: Instead of solving each initial-value problem individually, we can find a general formula for the solution of (10), and then apply that formula to solve specific problems. We will now show that

$$y(x) = y_0 + \int_{x_0}^{x} f(t)\,dt \tag{11}$$

is a formula for the solution of (10). To confirm this we must show that $dy/dx = f(x)$ and that $y(x_0) = y_0$. The computations are as follows:

$$\frac{dy}{dx} = \frac{d}{dx}\left[y_0 + \int_{x_0}^{x} f(t)\,dt\right] = 0 + f(x) = f(x)$$

$$y(x_0) = y_0 + \int_{x_0}^{x_0} f(t)\,dt = y_0 + 0 = y_0$$

▶ **Example 2** In Example 6 of Section 5.2 we showed that the solution of the initial-value problem

$$\frac{dy}{dx} = \cos x, \quad y(0) = 1$$

is $y(x) = 1 + \sin x$. This initial-value problem can also be solved by applying Formula (11) with $f(x) = \cos x$, $x_0 = 0$, and $y_0 = 1$. This yields

$$y(x) = 1 + \int_0^x \cos t\,dt = 1 + \left[\sin t\right]_{t=0}^{x} = 1 + \sin x \quad ◀$$

In the last example we were able to perform the integration in Formula (11) and express the solution of the initial-value problem as an elementary function. However, sometimes this will not be possible, in which case the solution of the initial-value problem must be left in terms of an "unevaluated" integral. For example, from (11), the solution of the initial-value problem

$$\frac{dy}{dx} = e^{-x^2}, \quad y(0) = 1$$

is

$$y(x) = 1 + \int_0^x e^{-t^2}\,dt$$

However, it can be shown that there is no way to express the integral in this solution as an elementary function. Thus, we have encountered a *new* function, which we regard to be *defined* by the integral. A close relative of this function, known as the ***error function***, plays an important role in probability and statistics; it is denoted by $\text{erf}(x)$ and is defined as

$$\text{erf}(x) = \frac{2}{\sqrt{\pi}}\int_0^x e^{-t^2}\,dt \tag{12}$$

Indeed, many of the most important functions in science and engineering are defined as integrals that have special names and notations associated with them. For example, the functions defined by

$$S(x) = \int_0^x \sin\left(\frac{\pi t^2}{2}\right) dt \quad \text{and} \quad C(x) = \int_0^x \cos\left(\frac{\pi t^2}{2}\right) dt \tag{13–14}$$

are called the ***Fresnel sine and cosine functions***, respectively, in honor of the French physicist Augustin Fresnel (1788–1827), who first encountered them in his study of diffraction of light waves.

EVALUATING AND GRAPHING FUNCTIONS DEFINED BY INTEGRALS

The following values of $S(1)$ and $C(1)$ were produced by a CAS that has a built-in algorithm for approximating definite integrals:

$$S(1) = \int_0^1 \sin\left(\frac{\pi t^2}{2}\right) dt \approx 0.438259, \qquad C(1) = \int_0^1 \cos\left(\frac{\pi t^2}{2}\right) dt \approx 0.779893$$

To generate graphs of functions defined by integrals, computer programs choose a set of x-values in the domain, approximate the integral for each of those values, and then plot the resulting points. Thus, there is a lot of computation involved in generating such graphs, since each plotted point requires the approximation of an integral. The graphs of the Fresnel functions in Figure 7.6.4 were generated in this way using a CAS.

Although it required a considerable amount of computation to generate the graphs of the Fresnel functions, the derivatives of $S(x)$ and $C(x)$ are easy to obtain using Part 2 of the Fundamental Theorem of Calculus (5.6.3); they are

$$S'(x) = \sin\left(\frac{\pi x^2}{2}\right) \quad \text{and} \quad C'(x) = \cos\left(\frac{\pi x^2}{2}\right) \tag{15–16}$$

These derivatives can be used to determine the locations of the relative extrema and inflection points and to investigate other properties of $S(x)$ and $C(x)$.

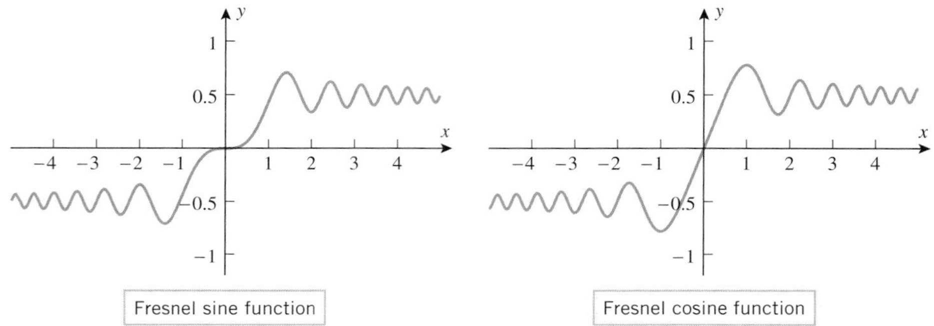

Figure 7.6.4

INTEGRALS WITH FUNCTIONS AS LIMITS OF INTEGRATION

Various applications can lead to integrals in which at least one of the limits of integration is a function of x. Some examples are

$$\int_x^1 \sqrt{\sin t}\, dt, \quad \int_{x^2}^{\sin x} \sqrt{t^3+1}\, dt, \quad \int_{\ln x}^{\pi} \frac{dt}{t^7-8}$$

We will complete this section by showing how to differentiate integrals of the form

$$\int_a^{g(x)} f(t)\, dt \tag{17}$$

where a is constant. Derivatives of other kinds of integrals with functions as limits of integration will be discussed in the exercises.

To differentiate (17) we can view the integral as a composition $F(g(x))$, where

$$F(x) = \int_a^x f(t)\, dt$$

If we now apply the chain rule, we obtain

$$\frac{d}{dx}\left[\int_a^{g(x)} f(t)\, dt\right] = \frac{d}{dx}[F(g(x))] = F'(g(x))g'(x) = f(g(x))g'(x)$$

Theorem 5.6.3

Thus,

$$\frac{d}{dx}\left[\int_a^{g(x)} f(t)\,dt\right] = f(g(x))g'(x) \tag{18}$$

In words:

To differentiate an integral with a constant lower limit and a function as the upper limit, substitute the upper limit into the integrand, and multiply by the derivative of the upper limit.

▶ Example 3

$$\frac{d}{dx}\left[\int_1^{\sin x} (1-t^2)\,dt\right] = (1-\sin^2 x)\cos x = \cos^3 x \quad ◀$$

✔QUICK CHECK EXERCISES 7.6 *(See page 488 for answers.)*

1. (a) $\int_1^{1/e} \frac{1}{t}\,dt =$ ______

(b) If $\ln a = 2$ and $\ln b = -3$, then $\int_1^{ab^2} \frac{1}{t}\,dt =$ ______.

2. Estimate ln 2 using Definition 7.6.1 and
(a) a left endpoint approximation with $n = 2$
(b) a right endpoint approximation with $n = 2$.

3. $\pi^{1/(\ln \pi)} =$ ______

4. A solution to the initial-value problem

$$\frac{dy}{dx} = \sin x, \quad y(0) = \pi$$

is $y =$ ______.

5. $\frac{d}{dx}\left[\int_0^{e^{-x}} \frac{1}{1+t^4}\,dt\right] =$ ______

EXERCISE SET 7.6 Graphing Utility C CAS

1. Sketch the curve $y = 1/t$, and shade a region under the curve whose area is
(a) ln 2 (b) $-\ln 0.5$ (c) 2.

2. Sketch the curve $y = 1/t$, and shade two different regions under the curve whose areas are ln 1.5.

3. Given that $\ln a = 2$ and $\ln c = 5$, find
(a) $\int_1^{ac} \frac{1}{t}\,dt$ (b) $\int_1^{1/c} \frac{1}{t}\,dt$
(c) $\int_1^{a/c} \frac{1}{t}\,dt$ (d) $\int_1^{a^3} \frac{1}{t}\,dt$.

4. Given that $\ln a = 9$, find
(a) $\int_1^{\sqrt{a}} \frac{1}{t}\,dt$ (b) $\int_1^{2a} \frac{1}{t}\,dt$
(c) $\int_1^{2/a} \frac{1}{t}\,dt$ (d) $\int_2^{a} \frac{1}{t}\,dt$.

5. Approximate ln 5 using the midpoint rule with $n = 10$, and estimate the magnitude of the error by comparing your answer to that produced directly by a calculating utility.

6. Approximate ln 3 using the midpoint rule with $n = 20$, and estimate the magnitude of the error by comparing your answer to that produced directly by a calculating utility.

7. Simplify the expression and state the values of x for which your simplification is valid.
(a) $e^{-\ln x}$ (b) $e^{\ln x^2}$
(c) $\ln\left(e^{-x^2}\right)$ (d) $\ln(1/e^x)$
(e) $\exp(3\ln x)$ (f) $\ln(xe^x)$
(g) $\ln\left(e^{x-\sqrt[3]{x}}\right)$ (h) $e^{x-\ln x}$

8. (a) Let $f(x) = e^{-2x}$. Find the simplest exact value of the function $f(\ln 3)$.
(b) Let $f(x) = e^x + 3e^{-x}$. Find the simplest exact value of the function $f(\ln 2)$.

9–10 Express the given quantity as a power of e.

9. (a) 3^π (b) $2^{\sqrt{2}}$

10. (a) π^{-x} (b) $x^{2x}, \quad x > 0$

11–12 Find the limits by making appropriate substitutions in the limits given in Theorem 7.6.8.

11. (a) $\lim_{x\to+\infty}\left(1+\frac{1}{2x}\right)^x$ (b) $\lim_{x\to 0}(1+2x)^{1/x}$

12. (a) $\lim_{x\to+\infty}\left(1+\frac{3}{x}\right)^x$ (b) $\lim_{x\to 0}(1+x)^{1/(3x)}$

13–14 Find $g'(x)$ using Part 2 of the Fundamental Theorem of Calculus, and check your answer by evaluating the integral and then differentiating.

13. $g(x)=\int_1^x (t^2-t)\,dt$ **14.** $g(x)=\int_\pi^x (1-\cos t)\,dt$

15–16 Find the derivative using Formula (18), and check your answer by evaluating the integral and then differentiating the result.

15. (a) $\frac{d}{dx}\int_1^{x^3}\frac{1}{t}\,dt$ (b) $\frac{d}{dx}\int_1^{\ln x} e^t\,dt$

16. (a) $\frac{d}{dx}\int_{-1}^{x^2}\sqrt{t+1}\,dt$ (b) $\frac{d}{dx}\int_\pi^{1/x}\sin t\,dt$

17. Let $F(x)=\int_0^x \frac{\sin t}{t^2+1}\,dt$. Find
(a) $F(0)$ (b) $F'(0)$ (c) $F''(0)$.

18. Let $F(x)=\int_2^x \sqrt{3t^2+1}\,dt$. Find
(a) $F(2)$ (b) $F'(2)$ (c) $F''(2)$.

C **19.** (a) Use Formula (18) to find

$$\frac{d}{dx}\int_1^{x^2} t\sqrt{1+t}\,dt$$

(b) Use a CAS to evaluate the integral and differentiate the resulting function.
(c) Use the simplification command of the CAS, if necessary, to confirm that the answers in parts (a) and (b) are the same.

20. Show that
(a) $\frac{d}{dx}\left[\int_x^a f(t)\,dt\right]=-f(x)$
(b) $\frac{d}{dx}\left[\int_{g(x)}^a f(t)\,dt\right]=-f(g(x))g'(x)$.

21–22 Use the results in Exercise 20 to find the derivative.

21. (a) $\frac{d}{dx}\int_x^\pi \cos(t^3)\,dt$ (b) $\frac{d}{dx}\int_{\tan x}^3 \frac{t^2}{1+t^2}\,dt$

22. (a) $\frac{d}{dx}\int_x^0 \frac{1}{(t^2+1)^2}\,dt$ (b) $\frac{d}{dx}\int_{1/x}^\pi \cos^3 t\,dt$

23. Find

$$\frac{d}{dx}\left[\int_{3x}^{x^2}\frac{t-1}{t^2+1}\,dt\right]$$

by writing

$$\int_{3x}^{x^2}\frac{t-1}{t^2+1}\,dt=\int_{3x}^{0}\frac{t-1}{t^2+1}\,dt+\int_{0}^{x^2}\frac{t-1}{t^2+1}\,dt$$

24. Use Exercise 20(b) and the idea in Exercise 23 to show that

$$\frac{d}{dx}\int_{h(x)}^{g(x)} f(t)\,dt=f(g(x))g'(x)-f(h(x))h'(x)$$

25. Use the result obtained in Exercise 24 to perform the following differentiations:
(a) $\frac{d}{dx}\int_{x^2}^{x^3}\sin^2 t\,dt$ (b) $\frac{d}{dx}\int_{-x}^{x}\frac{1}{1+t}\,dt$.

26. Prove that the function

$$F(x)=\int_x^{5x}\frac{1}{t}\,dt$$

is constant on the interval $(0,+\infty)$ by using Exercise 24 to find $F'(x)$. What is that constant?

FOCUS ON CONCEPTS

27. Let $F(x)=\int_0^x f(t)\,dt$, where f is the function whose graph is shown in the accompanying figure.
(a) Find $F(0)$, $F(3)$, $F(5)$, $F(7)$, and $F(10)$.
(b) On what subintervals of the interval $[0, 10]$ is F increasing? Decreasing?
(c) Where does F have its maximum value? Its minimum value?
(d) Sketch the graph of F.

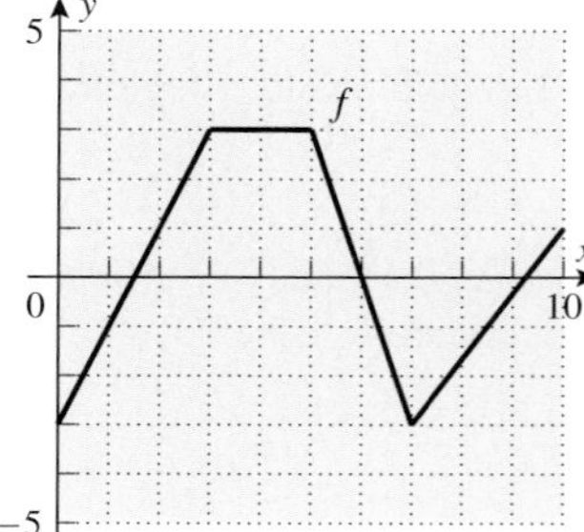

Figure Ex-27

28. Determine the inflection point(s) for the graph of F in Exercise 27.

29–30 Express $F(x)$ in a piecewise form that does not involve an integral.

29. $F(x)=\int_{-1}^x |t|\,dt$

30. $F(x)=\int_0^x f(t)\,dt$, where $f(x)=\begin{cases} x, & 0\le x\le 2\\ 2, & x>2\end{cases}$

31–34 Use Formula (11) to solve the initial-value problem.

31. $\dfrac{dy}{dx} = \dfrac{2x^2+1}{x}, \; y(1) = 2$ **32.** $\dfrac{dy}{dx} = \dfrac{x+1}{\sqrt{x}}, \; y(1) = 0$

33. $\dfrac{dy}{dx} = \sec^2 x - \sin x, \; y(\pi/4) = 1$

34. $\dfrac{dy}{dx} = \dfrac{1}{x \ln x}, \; y(e) = 1$

35. Suppose that at time $t = 0$ there are P_0 individuals who have disease X, and suppose that a certain model for the spread of the disease predicts that the disease will spread at the rate of $r(t)$ individuals per day. Write a formula for the number of individuals who will have disease X after x days.

36. Suppose that $v(t)$ is the velocity function of a particle moving along an s-axis. Write a formula for the coordinate of the particle at time T if the particle is at s_1 at time $t = 1$.

FOCUS ON CONCEPTS

37. The accompanying figure shows the graphs of $y = f(x)$ and $y = \int_0^x f(t)\,dt$. Determine which graph is which, and explain your reasoning.

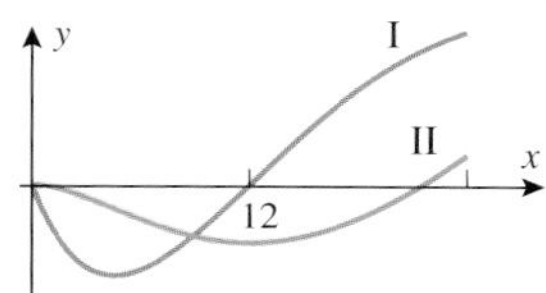

Figure Ex-37

38. (a) Make a conjecture about the value of the limit

$$\lim_{k \to 0} \int_1^b t^{k-1}\,dt \quad (b > 0)$$

(b) Check your conjecture by evaluating the integral and finding the limit. [*Hint:* Interpret the limit as the definition of the derivative of an exponential function.]

39. Let $F(x) = \int_0^x f(t)\,dt$, where f is the function graphed in the accompanying figure.
(a) Where do the relative minima of F occur?
(b) Where do the relative maxima of F occur?
(c) Where does the absolute maximum of F on the interval $[0, 5]$ occur?
(d) Where does the absolute minimum of F on the interval $[0, 5]$ occur?
(e) Where is F concave up? Concave down?
(f) Sketch the graph of F.

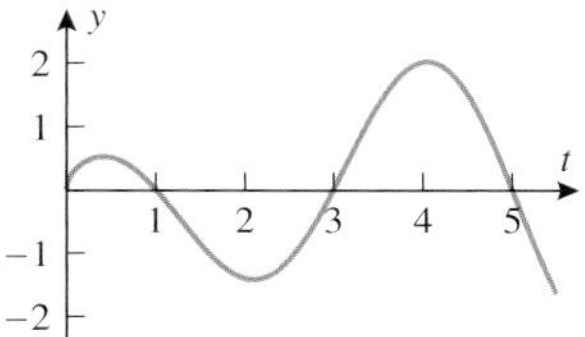

Figure Ex-39

C **40.** CAS programs have commands for working with most of the important nonelementary functions. Check your CAS documentation for information about the error function $\operatorname{erf}(x)$ [see Formula (12)], and then complete the following.
(a) Generate the graph of $\operatorname{erf}(x)$.
(b) Use the graph to make a conjecture about the existence and location of any relative maxima and minima of $\operatorname{erf}(x)$.
(c) Check your conjecture in part (b) using the derivative of $\operatorname{erf}(x)$.
(d) Use the graph to make a conjecture about the existence and location of any inflection points of $\operatorname{erf}(x)$.
(e) Check your conjecture in part (d) using the second derivative of $\operatorname{erf}(x)$.
(f) Use the graph to make a conjecture about the existence of horizontal asymptotes of $\operatorname{erf}(x)$.
(g) Check your conjecture in part (f) by using the CAS to find the limits of $\operatorname{erf}(x)$ as $x \to \pm\infty$.

41. The Fresnel sine and cosine functions $S(x)$ and $C(x)$ were defined in Formulas (13) and (14) and graphed in Figure 7.6.4. Their derivatives were given in Formulas (15) and (16).
(a) At what points does $C(x)$ have relative minima? Relative maxima?
(b) Where do the inflection points of $C(x)$ occur?
(c) Confirm that your answers in parts (a) and (b) are consistent with the graph of $C(x)$.

42. Find the limit

$$\lim_{h \to 0} \frac{1}{h} \int_x^{x+h} \ln t\,dt$$

43. Find a function f and a number a such that

$$4 + \int_a^x f(t)\,dt = e^{2x}$$

FOCUS ON CONCEPTS

44. (a) Give a geometric argument to show that

$$\frac{1}{x+1} < \int_x^{x+1} \frac{1}{t}\,dt < \frac{1}{x}, \quad x > 0$$

(b) Use the result in part (a) to prove that

$$\frac{1}{x+1} < \ln\left(1 + \frac{1}{x}\right) < \frac{1}{x}, \quad x > 0$$

(c) Use the result in part (b) to prove that

$$e^{x/(x+1)} < \left(1 + \frac{1}{x}\right)^x < e, \quad x > 0$$

and hence that

$$\lim_{x \to +\infty} \left(1 + \frac{1}{x}\right)^x = e$$

(d) Use the result in part (b) to prove that

$$\left(1 + \frac{1}{x}\right)^x < e < \left(1 + \frac{1}{x}\right)^{x+1}, \quad x > 0$$

45. Use a graphing utility to generate the graph of
$$y = \left(1 + \frac{1}{x}\right)^{x+1} - \left(1 + \frac{1}{x}\right)^{x}$$
in the window $[0, 100] \times [0, 0.2]$, and use that graph and part (d) of Exercise 44 to make a rough estimate of the error in the approximation
$$e \approx \left(1 + \frac{1}{50}\right)^{50}$$

46. Prove: If f is continuous on an open interval I and a is any point in I, then
$$F(x) = \int_a^x f(t)\,dt$$
is continuous on I.

✔QUICK CHECK ANSWERS 7.6

1. (a) -1 (b) -4 **2.** (a) $\frac{5}{6}$ (b) $\frac{7}{12}$ **3.** e **4.** $-\cos x + \pi + 1$ **5.** $-\frac{e^{-x}}{1 + e^{-4x}}$

7.7 DERIVATIVES AND INTEGRALS INVOLVING INVERSE TRIGONOMETRIC FUNCTIONS

A common problem in trigonometry is to find an angle x using a known value of $\sin x$, $\cos x$, *or some other trigonometric function. Problems of this type involve the computation of inverse trigonometric functions. In this section we will study these functions from the viewpoint of general inverse functions, with the goal of developing derivative formulas for the inverse trigonometric functions. We will also derive some related integration formulas that involve inverse trigonometric functions.*

INVERSE TRIGONOMETRIC FUNCTIONS

The six basic trigonometric functions do not have inverses because their graphs repeat periodically and hence do not pass the horizontal line test. To circumvent this problem we will restrict the domains of the trigonometric functions to produce one-to-one functions and then define the "inverse trigonometric functions" to be the inverses of these restricted functions. The top part of Figure 7.7.1 shows geometrically how these restrictions are made for $\sin x$, $\cos x$, $\tan x$, and $\sec x$, and the bottom part of the figure shows the graphs of the corresponding inverse functions

$$\sin^{-1} x, \quad \cos^{-1} x, \quad \tan^{-1} x, \quad \sec^{-1} x$$

(also denoted by $\arcsin x$, $\arccos x$, $\arctan x$, and $\operatorname{arcsec} x$). Inverses of $\cot x$ and $\csc x$ are of lesser importance and will be considered in the exercises.

If you have trouble visualizing the correspondence between the top and bottom parts of Figure 7.7.1, keep in mind that a reflection about $y = x$ converts vertical lines into horizontal lines, and vice versa, and converts x-intercepts into y-intercepts, and vice versa.

The following formal definitions summarize the preceding discussion.

7.7.1 DEFINITION. The ***inverse sine function***, denoted by $\sin^{-1}$, is defined to be the inverse of the restricted sine function
$$\sin x, \quad -\pi/2 \le x \le \pi/2$$

7.7.2 DEFINITION. The ***inverse cosine function***, denoted by $\cos^{-1}$, is defined to be the inverse of the restricted cosine function
$$\cos x, \quad 0 \le x \le \pi$$

The notations $\sin^{-1} x$, $\cos^{-1} x$, ... are reserved exclusively for the inverse trigonometric functions and are not used for reciprocals of the trigonometric functions. If we wanted to express the reciprocal $1/\sin x$ using an exponent, we would write $(\sin x)^{-1}$ and *never* $\sin^{-1} x$.

7.7.3 DEFINITION. The ***inverse tangent function***, denoted by $\tan^{-1}$, is defined to be the inverse of the restricted tangent function
$$\tan x, \quad -\pi/2 < x < \pi/2$$

> **7.7.4 DEFINITION.*** The ***inverse secant function***, denoted by $\sec^{-1}$, is defined to be the inverse of the restricted secant function
>
> $$\sec x, \quad 0 \le x \le \pi \text{ with } x \ne \pi/2$$

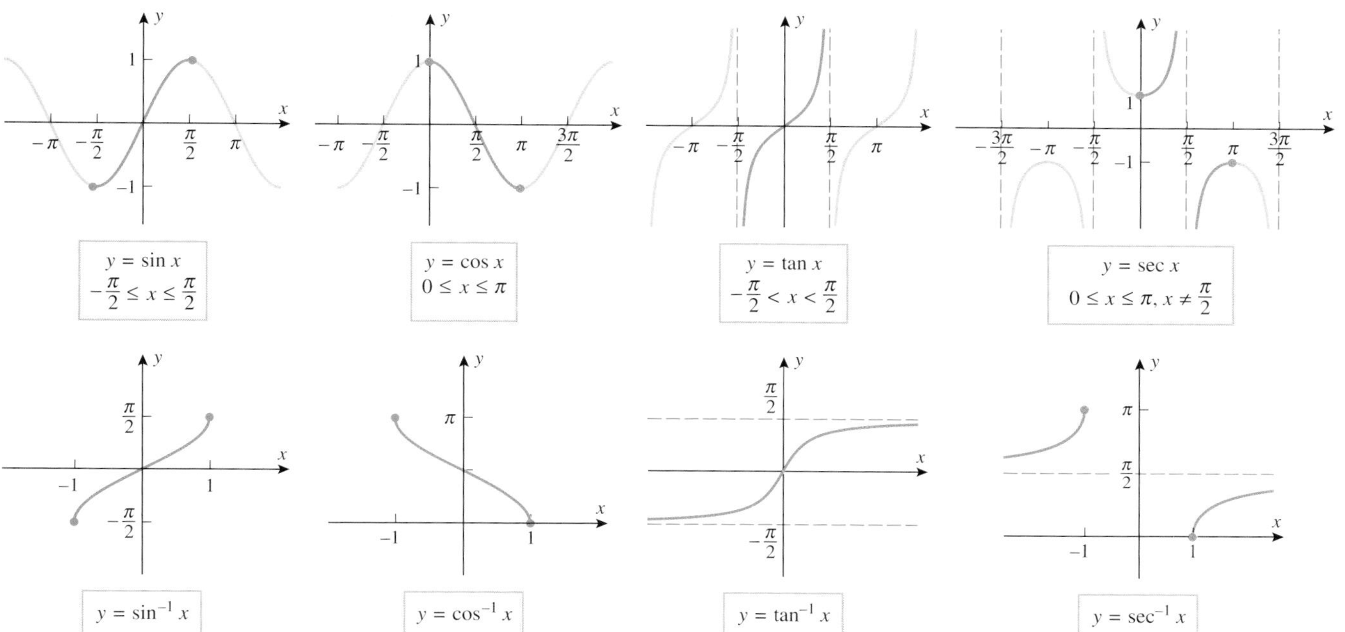

Figure 7.7.1

Table 7.7.1 summarizes the basic properties of the inverse trigonometric functions we have considered. You should confirm that the domains and ranges listed in this table are consistent with the graphs shown in Figure 7.7.1.

Table 7.7.1

FUNCTION	DOMAIN	RANGE	BASIC RELATIONSHIPS
$\sin^{-1}$	$[-1, 1]$	$[-\pi/2, \pi/2]$	$\sin^{-1}(\sin x) = x$ if $-\pi/2 \le x \le \pi/2$ $\sin(\sin^{-1} x) = x$ if $-1 \le x \le 1$
$\cos^{-1}$	$[-1, 1]$	$[0, \pi]$	$\cos^{-1}(\cos x) = x$ if $0 \le x \le \pi$ $\cos(\cos^{-1} x) = x$ if $-1 \le x \le 1$
$\tan^{-1}$	$(-\infty, +\infty)$	$(-\pi/2, \pi/2)$	$\tan^{-1}(\tan x) = x$ if $-\pi/2 < x < \pi/2$ $\tan(\tan^{-1} x) = x$ if $-\infty < x < +\infty$
$\sec^{-1}$	$(-\infty, -1] \cup [1, +\infty)$	$[0, \pi/2) \cup (\pi/2, \pi]$	$\sec^{-1}(\sec x) = x$ if $0 \le x \le \pi, x \ne \pi/2$ $\sec(\sec^{-1} x) = x$ if $\lvert x\rvert \ge 1$

*There is no universal agreement on the definition of $\sec^{-1} x$, and some mathematicians prefer to restrict the domain of $\sec x$ so that $0 \le x < \pi/2$ or $\pi \le x < 3\pi/2$, which was the definition used in some earlier editions of this text. Each definition has advantages and disadvantages, but we will use the current definition to conform with the conventions used by the CAS programs *Mathematica*, *Maple*, and *Derive*.

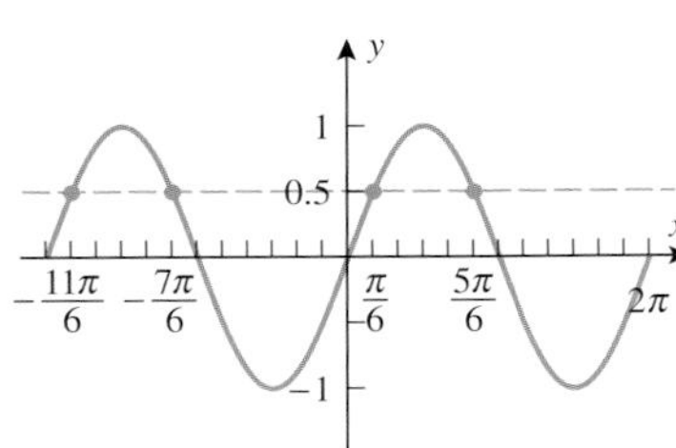

Figure 7.7.2

EVALUATING INVERSE TRIGONOMETRIC FUNCTIONS

A common problem in trigonometry is to find an angle whose sine is known. For example, you might want to find an angle x in radian measure such that

$$\sin x = \tfrac{1}{2} \tag{1}$$

and, more generally, for a given value of y in the interval $-1 \le y \le 1$ you might want to solve the equation

$$\sin x = y \tag{2}$$

Because $\sin x$ repeats periodically, this equation has infinitely many solutions for x; however, if we solve this equation as

$$x = \sin^{-1} y$$

then we isolate the specific solution that lies in the interval $[-\pi/2, \pi/2]$, since this is the range of the inverse sine. For example, Figure 7.7.2 shows four solutions of Equation (1), namely, $-11\pi/6$, $-7\pi/6$, $\pi/6$, and $5\pi/6$. Of these, $\pi/6$ is the solution in the interval $[-\pi/2, \pi/2]$, so

$$\sin^{-1}\left(\tfrac{1}{2}\right) = \pi/6 \tag{3}$$

TECHNOLOGY MASTERY

Refer to the documentation for your calculating utility to determine how to calculate inverse sines, inverse cosines, and inverse tangents; and then confirm Equation (3) numerically by showing that

$$\sin^{-1}(0.5) \approx 0.523598775598\ldots \approx \pi/6$$

In general, if we view $x = \sin^{-1} y$ as an angle in radian measure whose sine is y, then the restriction $-\pi/2 \le x \le \pi/2$ imposes the geometric requirement that the angle x in standard position terminate in either the first or fourth quadrant or on an axis adjacent to those quadrants.

▶ **Example 1** Find exact values of

$$\text{(a) } \sin^{-1}(1/\sqrt{2}) \qquad \text{(b) } \sin^{-1}(-1)$$

by inspection, and confirm your results numerically using a calculating utility.

Solution (a). Because $\sin^{-1}(1/\sqrt{2}) > 0$, we can view $x = \sin^{-1}(1/\sqrt{2})$ as that angle in the first quadrant such that $\sin\theta = 1/\sqrt{2}$. Thus, $\sin^{-1}(1/\sqrt{2}) = \pi/4$. You can confirm this with your calculating utility by showing that $\sin^{-1}(1/\sqrt{2}) \approx 0.785 \approx \pi/4$.

Solution (b). Because $\sin^{-1}(-1) < 0$, we can view $x = \sin^{-1}(-1)$ as an angle in the fourth quadrant (or an adjacent axis) such that $\sin x = -1$. Thus, $\sin^{-1}(-1) = -\pi/2$. You can confirm this with your calculating utility by showing that $\sin^{-1}(-1) \approx -1.57 \approx -\pi/2$. ◀

If $x = \cos^{-1} y$ is viewed as an angle in radian measure whose cosine is y, in what possible quadrants can x lie? Answer the same question for

$$x = \tan^{-1} y \quad \text{and} \quad x = \sec^{-1} y$$

TECHNOLOGY MASTERY

Most calculators do not provide a direct method for calculating inverse secants. In such situations the identity

$$\sec^{-1} x = \cos^{-1}(1/x) \tag{4}$$

is useful (Exercise 16). Use this formula to show that

$$\sec^{-1}(2.25) \approx 1.11 \quad \text{and} \quad \sec^{-1}(-2.25) \approx 2.03$$

If you have a calculating utility (such as a CAS) that can find $\sec^{-1} x$ directly, use it to check these values.

IDENTITIES FOR INVERSE TRIGONOMETRIC FUNCTIONS

If we interpret $\sin^{-1} x$ as an angle in radian measure whose sine is x, and if that angle is *nonnegative*, then we can represent $\sin^{-1} x$ geometrically as an angle in a right triangle in which the hypotenuse has length 1 and the side opposite to the angle $\sin^{-1} x$ has length x (Figure 7.7.3*a*). By the Theorem of Pythagoras the side adjacent to the angle $\sin^{-1} x$ has length $\sqrt{1-x^2}$. Moreover, the third angle in Figure 7.7.3*a* is $\cos^{-1} x$, since the cosine of that angle is x (Figure 7.7.3*b*). This triangle motivates a number of useful identities involving inverse trigonometric functions that are valid for $-1 \le x \le 1$; for example,

$$\sin^{-1} x + \cos^{-1} x = \frac{\pi}{2} \tag{5}$$

$$\cos(\sin^{-1} x) = \sqrt{1-x^2} \tag{6}$$

$$\sin(\cos^{-1} x) = \sqrt{1-x^2} \tag{7}$$

$$\tan(\sin^{-1} x) = \frac{x}{\sqrt{1-x^2}} \tag{8}$$

There is little to be gained by memorizing these identities. What is important is the mastery of the *method* used to obtain them.

In a similar manner, $\tan^{-1} x$ and $\sec^{-1} x$ can be represented as angles in the right triangles shown in Figures 7.7.3*c* and 7.7.3*d* (verify). Those triangles reveal additional useful identities; for example,

$$\sec(\tan^{-1} x) = \sqrt{1+x^2} \tag{9}$$

$$\sin(\sec^{-1} x) = \frac{\sqrt{x^2-1}}{x} \quad (x \ge 1) \tag{10}$$

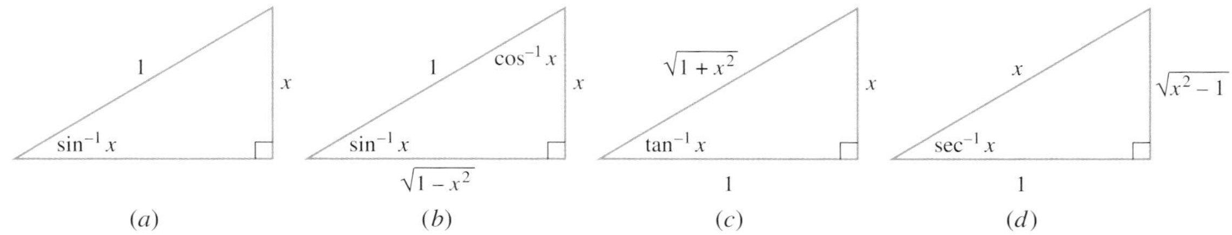

Figure 7.7.3

The triangle technique does not always produce the most general form of an identity. For example, in Exercise 90 we will ask you to derive the following extension of Formula (10) that is valid for $x \le -1$ as well as $x \ge 1$:

$$\sin(\sec^{-1} x) = \frac{\sqrt{x^2-1}}{|x|} \quad (|x| \ge 1) \tag{11}$$

Referring to Figure 7.7.1, observe that the inverse sine and inverse tangent are odd functions; that is,

$$\sin^{-1}(-x) = -\sin^{-1}(x) \quad \text{and} \quad \tan^{-1}(-x) = -\tan^{-1}(x) \tag{12–13}$$

▶ **Example 2** Figure 7.7.4 shows a computer-generated graph of $y = \sin^{-1}(\sin x)$. One might think that this graph should be the line $y = x$, since $\sin^{-1}(\sin x) = x$. Why isn't it?

Solution. The relationship $\sin^{-1}(\sin x) = x$ is valid on the interval $-\pi/2 \le x \le \pi/2$, so we can say with certainty that the graphs of $y = \sin^{-1}(\sin x)$ and $y = x$ coincide on this interval (which is confirmed by Figure 7.7.4). However, outside of this interval the relationship $\sin^{-1}(\sin x) = x$ does not hold. For example, if the quantity x lies in the interval $\pi/2 \le x \le 3\pi/2$, then the quantity $x - \pi$ lies in the interval $-\pi/2 \le x \le \pi/2$, so

$$\sin^{-1}[\sin(x-\pi)] = x - \pi$$

Thus, by using the identity $\sin(x-\pi) = -\sin x$ and the fact that $\sin^{-1}$ is an odd function, we can express $\sin^{-1}(\sin x)$ as

$$\sin^{-1}(\sin x) = \sin^{-1}[-\sin(x-\pi)] = -\sin^{-1}[\sin(x-\pi)] = -(x-\pi)$$

This shows that on the interval $\pi/2 \le x \le 3\pi/2$ the graph of $y = \sin^{-1}(\sin x)$ coincides with the line $y = -(x - \pi)$, which has slope -1 and an x-intercept at $x = \pi$. This agrees with Figure 7.7.4. ◀

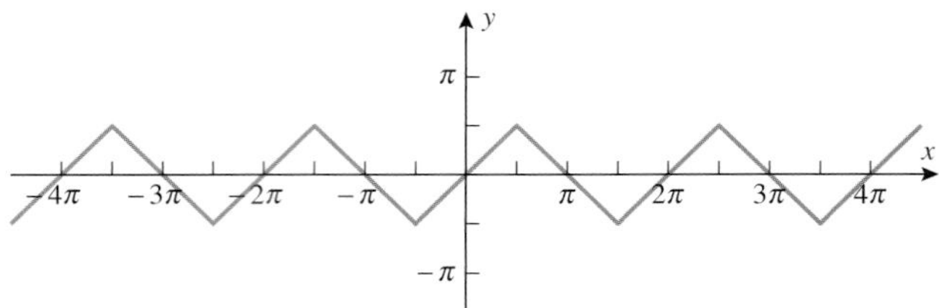

Figure 7.7.4

DERIVATIVES OF THE INVERSE TRIGONOMETRIC FUNCTIONS

Let us begin by investigating the differentiability of the function $\sin^{-1} x$. If we let $f(x) = \sin x$ $(-\pi/2 \le x \le \pi/2)$, then it follows from Theorem 7.3.1 that $f^{-1}(x) = \sin^{-1} x$ will be differentiable at any point x where $\cos(\sin^{-1} x) \neq 0$. This is equivalent to the condition

$$\sin^{-1} x \neq -\frac{\pi}{2} \quad \text{and} \quad \sin^{-1} x \neq \frac{\pi}{2}$$

so it follows that $\sin^{-1} x$ is differentiable on the interval $(-1, 1)$.

Observe that $\sin^{-1} x$ is only differentiable on the interval $(-1, 1)$, even though its domain is $[-1, 1]$. This is because the graph of $y = \sin x$ has horizontal tangent lines at the points $(\pi/2, 1)$ and $(-\pi/2, -1)$, so the graph of $y = \sin^{-1} x$ has vertical tangent lines at $x = \pm 1$.

A derivative formula for $\sin^{-1} x$ on $(-1, 1)$ can be obtained by using Formula (1) or (2) in Section 7.3 or by differentiating implicitly. We will use the latter method. Rewriting the equation $y = \sin^{-1} x$ as $x = \sin y$ and differentiating implicitly with respect to x, we obtain

$$\frac{d}{dx}[x] = \frac{d}{dx}[\sin y]$$

$$1 = \cos y \cdot \frac{dy}{dx}$$

$$\frac{dy}{dx} = \frac{1}{\cos y} = \frac{1}{\cos(\sin^{-1} x)}$$

At this point we have succeeded in obtaining the derivative; however, this derivative formula can be simplified using Formula (6). This yields

$$\frac{dy}{dx} = \frac{1}{\sqrt{1 - x^2}}$$

Thus, we have shown that

$$\frac{d}{dx}[\sin^{-1} x] = \frac{1}{\sqrt{1 - x^2}} \qquad (-1 < x < 1)$$

More generally, if u is a differentiable function of x, then the chain rule produces the following generalized version of this formula:

$$\frac{d}{dx}[\sin^{-1} u] = \frac{1}{\sqrt{1 - u^2}} \frac{du}{dx} \qquad (-1 < u < 1)$$

The method used to derive this formula can be used to obtain generalized derivative formulas for the remaining inverse trigonometric functions. The following is a complete list of these formulas, each of which is valid on the natural domain of the function that multiplies du/dx.

$$\frac{d}{dx}[\sin^{-1} u] = \frac{1}{\sqrt{1 - u^2}} \frac{du}{dx} \qquad \frac{d}{dx}[\cos^{-1} u] = -\frac{1}{\sqrt{1 - u^2}} \frac{du}{dx} \tag{14–15}$$

$$\frac{d}{dx}[\tan^{-1} u] = \frac{1}{1+u^2}\frac{du}{dx} \qquad \frac{d}{dx}[\cot^{-1} u] = -\frac{1}{1+u^2}\frac{du}{dx} \tag{16–17}$$

$$\frac{d}{dx}[\sec^{-1} u] = \frac{1}{|u|\sqrt{u^2-1}}\frac{du}{dx} \qquad \frac{d}{dx}[\csc^{-1} u] = -\frac{1}{|u|\sqrt{u^2-1}}\frac{du}{dx} \tag{18–19}$$

▶ **Example 3** Find dy/dx if

(a) $y = \sin^{-1}(x^3)$ (b) $y = \sec^{-1}(e^x)$

Solution (a). From (14)

$$\frac{dy}{dx} = \frac{1}{\sqrt{1-(x^3)^2}}(3x^2) = \frac{3x^2}{\sqrt{1-x^6}}$$

Solution (b). From (18)

$$\frac{dy}{dx} = \frac{1}{e^x\sqrt{(e^x)^2-1}}(e^x) = \frac{1}{\sqrt{e^{2x}-1}} \quad ◀$$

INTEGRATION FORMULAS

Differentiation formulas (14)–(19) yield useful integration formulas. Those most commonly needed are

$$\int \frac{du}{\sqrt{1-u^2}} = \sin^{-1} u + C \tag{20}$$

$$\int \frac{du}{1+u^2} = \tan^{-1} u + C \tag{21}$$

See Exercise 39 for a discussion of Formula (22).

$$\int \frac{du}{u\sqrt{u^2-1}} = \sec^{-1} |u| + C \tag{22}$$

▶ **Example 4** Evaluate $\displaystyle\int \frac{dx}{1+3x^2}$.

Solution. Substituting

$$u = \sqrt{3}x, \quad du = \sqrt{3}\,dx$$

yields

$$\int \frac{dx}{1+3x^2} = \frac{1}{\sqrt{3}}\int \frac{du}{1+u^2} = \frac{1}{\sqrt{3}}\tan^{-1} u + C = \frac{1}{\sqrt{3}}\tan^{-1}(\sqrt{3}x) + C \quad ◀$$

▶ **Example 5** Evaluate $\displaystyle\int \frac{e^x}{\sqrt{1-e^{2x}}}\,dx$.

Solution. Substituting

$$u = e^x, \quad du = e^x\,dx$$

yields

$$\int \frac{e^x}{\sqrt{1-e^{2x}}}\,dx = \int \frac{du}{\sqrt{1-u^2}} = \sin^{-1} u + C = \sin^{-1}(e^x) + C \quad ◀$$

Example 6 Evaluate $\int \frac{dx}{a^2 + x^2}\, dx$, where $a \neq 0$ is a constant.

Solution. Some simple algebra and an appropriate u-substitution will allow us to use (21).

$$\int \frac{dx}{a^2 + x^2} = \frac{1}{a}\int \frac{dx/a}{1 + (x/a)^2} = \frac{1}{a}\int \frac{du}{1 + u^2} = \frac{1}{a}\tan^{-1} u + C = \frac{1}{a}\tan^{-1}\frac{x}{a} + C \blacktriangleleft$$

$u = x/a$
$du = dx/a$

The method of Example 6 leads to the following generalizations of (20), (21), and (22) for $a > 0$:

$$\int \frac{du}{a^2 + u^2} = \frac{1}{a}\tan^{-1}\frac{u}{a} + C \tag{23}$$

$$\int \frac{du}{\sqrt{a^2 - u^2}} = \sin^{-1}\frac{u}{a} + C \tag{24}$$

$$\int \frac{du}{u\sqrt{u^2 - a^2}} = \frac{1}{a}\sec^{-1}\left|\frac{u}{a}\right| + C \tag{25}$$

Example 7 Evaluate $\int \frac{dx}{\sqrt{2 - x^2}}$.

Solution. Applying (24) with $u = x$ and $a = \sqrt{2}$ yields

$$\int \frac{dx}{\sqrt{2 - x^2}} = \sin^{-1}\frac{x}{\sqrt{2}} + C \blacktriangleleft$$

✔ QUICK CHECK EXERCISES 7.7 *(See page 498 for answers.)*

1. In each part, determine the exact value without using a calculating utility.
(a) $\sin^{-1}(-1) =$ ______
(b) $\tan^{-1}(1) =$ ______
(c) $\sin^{-1}\left(\frac{1}{2}\sqrt{3}\right) =$ ______
(d) $\cos^{-1}\left(\frac{1}{2}\right) =$ ______
(e) $\sec^{-1}(-2) =$ ______

2. In each part, determine the exact value without using a calculating utility.
(a) $\sin^{-1}(\sin \pi/7) =$ ______
(b) $\sin^{-1}(\sin 5\pi/7) =$ ______
(c) $\tan^{-1}(\tan 13\pi/6) =$ ______
(d) $\cos^{-1}(\cos 12\pi/7) =$ ______

3. $\frac{d}{dx}[\sin^{-1}(2x)] =$ ______

4. $\int_{-1/2}^{1/2} \frac{1}{\sqrt{1 - x^2}}\, dx =$ ______

EXERCISE SET 7.7 Graphing Utility

1. Given that $\theta = \tan^{-1}\left(\frac{4}{3}\right)$, find the exact values of $\sin\theta$, $\cos\theta$, $\cot\theta$, $\sec\theta$, and $\csc\theta$.

2. Given that $\theta = \sec^{-1} 2.6$, find the exact values of $\sin\theta$, $\cos\theta$, $\tan\theta$, $\cot\theta$, and $\csc\theta$.

3. For which values of x is it true that
(a) $\cos^{-1}(\cos x) = x$ (b) $\cos(\cos^{-1} x) = x$
(c) $\tan^{-1}(\tan x) = x$ (d) $\tan(\tan^{-1} x) = x$?

4–5 Find the exact value of the given quantity.

4. $\sec\left[\sin^{-1}\left(-\frac{3}{4}\right)\right]$ 5. $\sin\left[2\cos^{-1}\left(\frac{3}{5}\right)\right]$

6–7 Complete the identities using the triangle method (Figure 7.7.3).

6. (a) $\sin(\cos^{-1} x) = ?$ (b) $\tan(\cos^{-1} x) = ?$
(c) $\csc(\tan^{-1} x) = ?$ (d) $\sin(\tan^{-1} x) = ?$

7. (a) $\cos(\tan^{-1} x) = ?$ (b) $\tan(\cos^{-1} x) = ?$
(c) $\sin(\sec^{-1} x) = ?$ (d) $\cot(\sec^{-1} x) = ?$

8. (a) Use a calculating utility set to radian measure to make tables of values of $y = \sin^{-1} x$ and $y = \cos^{-1} x$ for $x = -1, -0.8, -0.6, \ldots, 0, 0.2, \ldots, 1$. Round your answers to two decimal places.
(b) Plot the points obtained in part (a), and use the points to sketch the graphs of $y = \sin^{-1} x$ and $y = \cos^{-1} x$. Confirm that your sketches agree with those in Figure 7.7.1.
(c) Use your graphing utility to graph $y = \sin^{-1} x$ and $y = \cos^{-1} x$; confirm that the graphs agree with those in Figure 7.7.1.

9. In each part, sketch the graph and check your work with a graphing utility.
(a) $y = \sin^{-1} 2x$ (b) $y = \tan^{-1} \frac{1}{2}x$

10–12 Use a calculating utility to approximate the solution of each equation. Where radians are used, express your answer to four decimal places, and where degrees are used, express it to the nearest tenth of a degree. [*Note:* In each part, the solution is not in the range of the relevant inverse trigonometric function.]

10. (a) $\sin x = 0.37,\ \pi/2 < x < \pi$
(b) $\sin \theta = -0.61,\ 180° < \theta < 270°$

11. (a) $\cos x = -0.85,\ \pi < x < 3\pi/2$
(b) $\cos \theta = 0.23,\ -90° < \theta < 0°$

12. (a) $\tan x = 3.16,\ -\pi < x < -\pi/2$
(b) $\tan \theta = -0.45,\ 90° < \theta < 180°$

FOCUS ON CONCEPTS

13. (a) Use a calculating utility to evaluate $\sin^{-1}(\sin^{-1} 0.25)$ and $\sin^{-1}(\sin^{-1} 0.9)$, and explain what you think is happening in the second calculation.
(b) For what values of x in the interval $-1 \le x \le 1$ will your calculating utility produce a real value for the function $\sin^{-1}(\sin^{-1} x)$?

14. A soccer player kicks a ball with an initial speed of 14 m/s at an angle θ with the horizontal (see the accompanying figure). The ball lands 18 m down the field. If air resistance is neglected, then the ball will have a parabolic trajectory and the horizontal range R will be given by

$$R = \frac{v^2}{g}\sin 2\theta$$

where v is the initial speed of the ball and g is the acceleration due to gravity. Using $g = 9.8\ \text{m/s}^2$, approximate two values of θ, to the nearest degree, at which the ball could have been kicked. Which angle results in the shorter time of flight? Why?

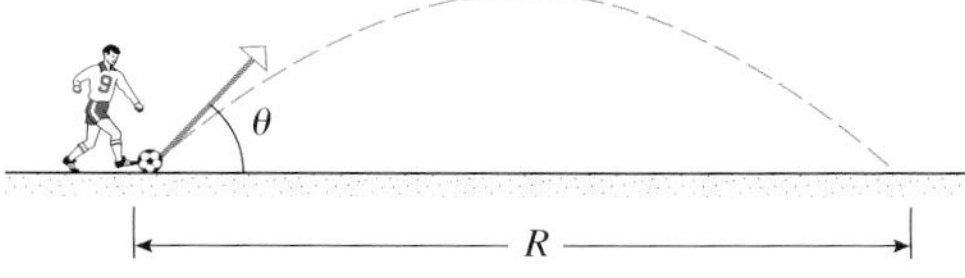

Figure Ex-14

15–16 The function $\cot^{-1} x$ is defined to be the inverse of the restricted cotangent function

$$\cot x,\quad 0 < x < \pi$$

and the function $\csc^{-1} x$ is defined to be the inverse of the restricted cosecant function

$$\csc x,\quad -\pi/2 < x < \pi/2,\quad x \ne 0$$

Use these definitions in these and in all subsequent exercises that involve these functions.

15. (a) Sketch the graphs of $\cot^{-1} x$ and $\csc^{-1} x$.
(b) Find the domain and range of $\cot^{-1} x$ and $\csc^{-1} x$.

16. Show that

(a) $\cot^{-1} x = \begin{cases} \tan^{-1}(1/x), & \text{if } x > 0 \\ \pi + \tan^{-1}(1/x), & \text{if } x < 0 \end{cases}$

(b) $\sec^{-1} x = \cos^{-1} \dfrac{1}{x}$, if $|x| \ge 1$

(c) $\csc^{-1} x = \sin^{-1} \dfrac{1}{x}$, if $|x| \ge 1$.

17–32 Find dy/dx.

17. $y = \sin^{-1}(3x)$
18. $y = \cos^{-1}\left(\dfrac{x+1}{2}\right)$
19. $y = \sin^{-1}(1/x)$
20. $y = \cos^{-1}(\cos x)$
21. $y = \tan^{-1}(x^3)$
22. $y = \sec^{-1}(x^5)$
23. $y = (\tan x)^{-1}$
24. $y = \dfrac{1}{\tan^{-1} x}$
25. $y = e^x \sec^{-1} x$
26. $y = \ln(\cos^{-1} x)$
27. $y = \sin^{-1} x + \cos^{-1} x$
28. $y = x^2(\sin^{-1} x)^3$
29. $y = \sec^{-1} x + \csc^{-1} x$
30. $y = \csc^{-1}(e^x)$
31. $y = \cot^{-1}(\sqrt{x})$
32. $y = \sqrt{\cot^{-1} x}$

33–34 Find dy/dx by implicit differentiation.

33. $x^3 + x \tan^{-1} y = e^y$
34. $\sin^{-1}(xy) = \cos^{-1}(x - y)$

35. Differentiate $x = \tan y$ implicitly and use identity (9) to find the derivative of $y = \tan^{-1} x$.

36. (a) Use Theorem 7.3.1 to prove that

$$\frac{d}{dx}[\cot^{-1} x]\bigg|_{x=0} = -1$$

(b) Use part (a) above, part (a) of Exercise 16, and the chain rule to show that

$$\frac{d}{dx}[\cot^{-1} x] = -\frac{1}{1+x^2}$$

for $-\infty < x < +\infty$.

(c) Conclude from part (b) that

$$\frac{d}{dx}[\cot^{-1} u] = -\frac{1}{1+u^2}\frac{du}{dx}$$

for $-\infty < u < +\infty$.

37. Use identity (5) and Formula (14) to obtain the derivative of $y = \cos^{-1} x$.

38. Use the identity in part (b) of Exercise 16 and the result of Exercise 37 to obtain the derivative of $y = \sec^{-1} x$.

39. Use the result of Exercise 38 to verify Formula (22).

40. (a) Use part (c) of Exercise 16 and the chain rule to show that

$$\frac{d}{dx}[\csc^{-1} x] = -\frac{1}{|x|\sqrt{x^2-1}}$$

for $1 < |x|$.

(b) Conclude from part (a) that

$$\frac{d}{dx}[\csc^{-1} u] = -\frac{1}{|u|\sqrt{u^2-1}}\frac{du}{dx}$$

for $1 < |u|$.

41–42 Evaluate the integral and check your answer by differentiating.

41. $\displaystyle\int \left[\frac{1}{2\sqrt{1-x^2}} - \frac{3}{1+x^2}\right] dx$

42. $\displaystyle\int \left[\frac{4}{x\sqrt{x^2-1}} + \frac{1+x+x^3}{1+x^2}\right] dx$

43–60 Evaluate the integral.

43. $\displaystyle\int \frac{dx}{\sqrt{1-4x^2}}$

44. $\displaystyle\int \frac{dx}{1+16x^2}$

45. $\displaystyle\int \frac{e^x}{1+e^{2x}}\, dx$

46. $\displaystyle\int \frac{t}{t^4+1}\, dt$

47. $\displaystyle\int \frac{\sec^2 x\, dx}{\sqrt{1-\tan^2 x}}$

48. $\displaystyle\int \frac{\sin\theta}{\cos^2\theta+1}\, d\theta$

49. $\displaystyle\int_0^{1/\sqrt{2}} \frac{dx}{\sqrt{1-x^2}}$

50. $\displaystyle\int_{-1}^{1} \frac{dx}{1+x^2}$

51. $\displaystyle\int_{\sqrt{2}}^{2} \frac{dx}{x\sqrt{x^2-1}}$

52. $\displaystyle\int_{-\sqrt{2}}^{-2/\sqrt{3}} \frac{dx}{x\sqrt{x^2-1}}$

53. $\displaystyle\int_1^{\sqrt{3}} \frac{\sqrt{\tan^{-1} x}}{1+x^2}\, dx$

54. $\displaystyle\int_1^{\sqrt{e}} \frac{dx}{x\sqrt{1-(\ln x)^2}}$

55. $\displaystyle\int_1^{3} \frac{dx}{\sqrt{x}\,(x+1)}$

56. $\displaystyle\int_{\ln 2}^{\ln(2/\sqrt{3})} \frac{e^{-x}\, dx}{\sqrt{1-e^{-2x}}}$

57. $\displaystyle\int_0^{1} \frac{x}{\sqrt{4-3x^4}}\, dx$

58. $\displaystyle\int_1^{2} \frac{1}{\sqrt{x}\sqrt{4-x}}\, dx$

59. $\displaystyle\int_0^{1/\sqrt{3}} \frac{1}{1+9x^2}\, dx$

60. $\displaystyle\int_1^{\sqrt{2}} \frac{x}{3+x^4}\, dx$

61–62 Use Formulas (23), (24), and (25) to help evaluate the integrals.

61. (a) $\displaystyle\int \frac{dx}{\sqrt{9-x^2}}$ (b) $\displaystyle\int \frac{dx}{5+x^2}$ (c) $\displaystyle\int \frac{dx}{x\sqrt{x^2-\pi}}$

62. (a) $\displaystyle\int \frac{e^x}{4+e^{2x}}\, dx$ (b) $\displaystyle\int \frac{dx}{\sqrt{9-4x^2}}$ (c) $\displaystyle\int \frac{dy}{y\sqrt{5y^2-3}}$

63. Most scientific calculators have keys for the values of only $\sin^{-1} x$, $\cos^{-1} x$, and $\tan^{-1} x$. The formulas in Exercise 16 show how a calculator can be used to obtain values of $\cot^{-1} x$, $\sec^{-1} x$, and $\csc^{-1} x$. Use these formulas and a calculator to find numerical values for each of the following inverse trigonometric functions. Express your answers in degrees, rounded to the nearest tenth of a degree.
(a) $\cot^{-1} 0.7$ (b) $\sec^{-1} 1.2$ (c) $\csc^{-1} 2.3$

64. An Earth-observing satellite has horizon sensors that can measure the angle θ shown in the accompanying figure. Let R be the radius of the Earth (assumed spherical) and h the distance between the satellite and the Earth's surface.
(a) Show that $\sin\theta = \dfrac{R}{R+h}$.
(b) Find θ, to the nearest degree, for a satellite that is 10,000 km from the Earth's surface (use $R = 6378$ km).

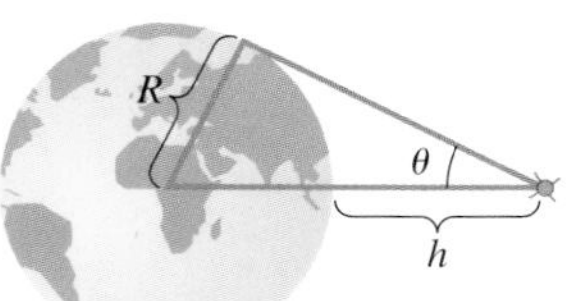

Figure Ex-64

65. The number of hours of daylight on a given day at a given point on the Earth's surface depends on the latitude λ of the point, the angle γ through which the Earth has moved in its orbital plane during the time period from the vernal equinox (March 21), and the angle of inclination ϕ of the Earth's axis of rotation measured from ecliptic north ($\phi \approx 23.45°$). The number of hours of daylight h can be approximated by the formula

$$h = \begin{cases} 24, & D \ge 1 \\ 12 + \frac{2}{15}\sin^{-1} D, & |D| < 1 \\ 0, & D \le -1 \end{cases}$$

where

$$D = \frac{\sin\phi \sin\gamma \tan\lambda}{\sqrt{1-\sin^2\phi \sin^2\gamma}}$$

and $\sin^{-1} D$ is in degree measure. Given that Fairbanks, Alaska, is located at a latitude of $\lambda = 65°$ N and also that

$\gamma = 90°$ on June 20 and $\gamma = 270°$ on December 20, approximate
(a) the maximum number of daylight hours at Fairbanks to one decimal place
(b) the minimum number of daylight hours at Fairbanks to one decimal place.

Source: This problem was adapted from *TEAM, A Path to Applied Mathematics*, The Mathematical Association of America, Washington, D.C., 1985.

66. The ***law of cosines*** states that

$$c^2 = a^2 + b^2 - 2ab\cos\theta$$

where a, b, and c are the lengths of the sides of a triangle and θ is the angle formed by sides a and b. Find θ, to the nearest degree, for the triangle with $a = 2$, $b = 3$, and $c = 4$.

67. An airplane is flying at a constant height of 3000 ft above water at a speed of 400 ft/s. The pilot is to release a survival package so that it lands in the water at a sighted point P. If air resistance is neglected, then the package will follow a parabolic trajectory whose equation relative to the coordinate system in the accompanying figure is

$$y = 3000 - \frac{g}{2v^2}x^2$$

where g is the acceleration due to gravity and v is the speed of the airplane. Using $g = 32$ ft/s^2, find the "line of sight" angle θ, to the nearest degree, that will result in the package hitting the target point.

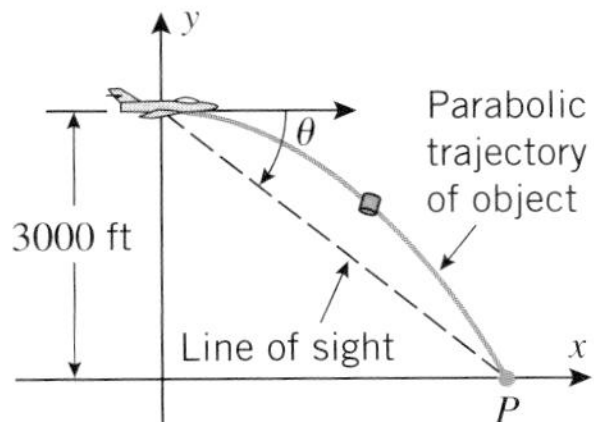

Figure Ex-67

68. A camera is positioned x feet from the base of a missile launching pad (see the accompanying figure). If a missile of length a feet is launched vertically, show that when the base of the missile is b feet above the camera lens, the angle θ subtended at the lens by the missile is

$$\theta = \cot^{-1}\frac{x}{a+b} - \cot^{-1}\frac{x}{b}$$

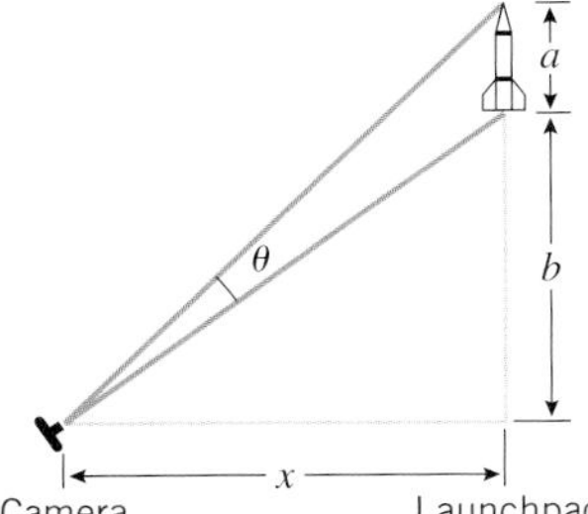

Figure Ex-68

69–70 Sketch the region enclosed by the curves and find its area.

69. $y = \dfrac{2}{1+x^2}$, $y = |x|$

70. $y = \dfrac{1}{\sqrt{1-x^2}}$, $y = 2$

71. Estimate the value of k $(0 < k < 1)$ so that the region enclosed by $y = 1/\sqrt{1-x^2}$, $y = x$, $x = 0$, and $x = k$ has an area of 1 square unit.

72. Estimate the area of the region in the first quadrant enclosed by $y = \sin 2x$ and $y = \sin^{-1} x$.

73. Find the volume of the solid that results when the region enclosed by the curves $y = \dfrac{1}{\sqrt{4+x^2}}$, $x = -2$, $x = 2$, and $y = 0$ is revolved about the x-axis.

C **74.** Use a CAS to estimate the volume of the solid that results when the region enclosed by the curves $y = x\sqrt{\tan^{-1} x}$ and $y = x$ is revolved about the x-axis.

C **75.** Consider the region enclosed by $y = \sin^{-1} x$, $y = 0$, and $x = 1$. Find the volume of the solid generated by revolving the region about the x-axis using
(a) disks (b) cylindrical shells.

76. (a) Find the volume V of the solid generated when the region bounded by $y = 1/(1+x^4)$, $y = 0$, $x = 1$, and $x = b$ $(b > 1)$ is revolved about the y-axis.
(b) Find $\lim_{b\to+\infty} V$.

77–78 Find the average value of the function over the given interval.

77. $f(x) = \dfrac{1}{1+x^2}$; $[1, \sqrt{3}]$

78. $f(x) = \dfrac{1}{\sqrt{1-x^2}}$; $\left[-\frac{1}{2}, 0\right]$

79. Use the Mean-Value Theorem to prove that

$$\frac{x}{1+x^2} < \tan^{-1} x < x \quad (x > 0)$$

80. Find a positive value of k such that the average value of $f(x) = 1/(k^2 + x^2)$ over the interval $[-k, k]$ is π.

81–83 Solve the initial-value problems.

81. $\dfrac{dy}{dt} = \dfrac{3}{\sqrt{1-t^2}}$, $y\left(\dfrac{\sqrt{3}}{2}\right) = 0$

82. $\dfrac{dy}{dx} = \dfrac{x^2-1}{x^2+1}$, $y(1) = \dfrac{\pi}{2}$

83. $\dfrac{dy}{dt} = \dfrac{1}{25+9t^2}$, $y\left(-\dfrac{5}{3}\right) = \dfrac{\pi}{30}$

84. Evaluate the limit by interpreting it as a Riemann sum in which the interval $[0, 1]$ is divided into n subintervals of equal length:

$$\lim_{n\to+\infty} \sum_{k=1}^{n} \frac{n}{n^2+k^2}$$

85. Prove:
(a) $\sin^{-1}(-x) = -\sin^{-1} x$
(b) $\tan^{-1}(-x) = -\tan^{-1} x$.

86. Prove:
(a) $\cos^{-1}(-x) = \pi - \cos^{-1} x$
(b) $\sec^{-1}(-x) = \pi - \sec^{-1} x$.

87. Prove:
(a) $\sin^{-1} x = \tan^{-1} \dfrac{x}{\sqrt{1-x^2}} \quad (|x| < 1)$

(b) $\cos^{-1} x = \dfrac{\pi}{2} - \tan^{-1} \dfrac{x}{\sqrt{1-x^2}} \quad (|x| < 1)$.

88. Prove:

$$\tan^{-1} x + \tan^{-1} y = \tan^{-1}\left(\frac{x+y}{1-xy}\right)$$

provided $-\pi/2 < \tan^{-1} x + \tan^{-1} y < \pi/2$. [*Hint:* Use an identity for $\tan(\alpha + \beta)$.]

89. Use the result in Exercise 88 to show that
(a) $\tan^{-1} \frac{1}{2} + \tan^{-1} \frac{1}{3} = \pi/4$
(b) $2\tan^{-1} \frac{1}{3} + \tan^{-1} \frac{1}{7} = \pi/4$.

90. Use identities (4) and (7) to obtain identity (11).

✔ QUICK CHECK ANSWERS 7.7

1. (a) $-\pi/2$ (b) $\pi/4$ (c) $\pi/3$ (d) $\pi/3$ (e) $2\pi/3$ **2.** (a) $\pi/7$ (b) $2\pi/7$ (c) $\pi/6$ (d) $2\pi/7$ **3.** $\dfrac{2}{\sqrt{1-4x^2}}$ **4.** $\pi/3$

7.8 HYPERBOLIC FUNCTIONS AND HANGING CABLES

In this section we will study certain combinations of e^x and e^{-x}, called "hyperbolic functions." These functions, which arise in various engineering applications, have many properties in common with the trigonometric functions. This similarity is somewhat surprising, since there is little on the surface to suggest that there should be any relationship between exponential and trigonometric functions. This is because the relationship occurs within the context of complex numbers, a topic which we will leave for more advanced courses.

DEFINITIONS OF HYPERBOLIC FUNCTIONS

To introduce the hyperbolic functions, observe that the function e^x can be expressed in the following way as the sum of an even function and an odd function:

$$e^x = \underbrace{\frac{e^x + e^{-x}}{2}}_{\text{Even}} + \underbrace{\frac{e^x - e^{-x}}{2}}_{\text{Odd}}$$

These functions are sufficiently important that there are names and notation associated with them: the odd function is called the *hyperbolic sine* of x and the even function is called the *hyperbolic cosine* of x. They are denoted by

$$\sinh x = \frac{e^x - e^{-x}}{2} \quad \text{and} \quad \cosh x = \frac{e^x + e^{-x}}{2}$$

where sinh is pronounced "cinch" and cosh rhymes with "gosh." From these two building blocks we can create four more functions to produce the following set of six ***hyperbolic functions***.

The terms "tanh," "sech," and "csch" are pronounced "tanch," "seech," and "coseech," respectively.

7.8.1 DEFINITION.

Hyperbolic sine $\sinh x = \dfrac{e^x - e^{-x}}{2}$

Hyperbolic cosine $\cosh x = \dfrac{e^x + e^{-x}}{2}$

Hyperbolic tangent $\tanh x = \dfrac{\sinh x}{\cosh x} = \dfrac{e^x - e^{-x}}{e^x + e^{-x}}$

Hyperbolic cotangent $\coth x = \dfrac{\cosh x}{\sinh x} = \dfrac{e^x + e^{-x}}{e^x - e^{-x}}$

Hyperbolic secant $\operatorname{sech} x = \dfrac{1}{\cosh x} = \dfrac{2}{e^x + e^{-x}}$

Hyperbolic cosecant $\operatorname{csch} x = \dfrac{1}{\sinh x} = \dfrac{2}{e^x - e^{-x}}$

TECHNOLOGY MASTERY

Computer algebra systems have built-in capabilities for evaluating hyperbolic functions directly, but some calculators do not. However, if you need to evaluate a hyperbolic function on a calculator, you can do so by expressing it in terms of exponential functions, as in Example 1.

▸ Example 1

$$\sinh 0 = \frac{e^0 - e^{-0}}{2} = \frac{1-1}{2} = 0$$

$$\cosh 0 = \frac{e^0 + e^{-0}}{2} = \frac{1+1}{2} = 1$$

$$\sinh 2 = \frac{e^2 - e^{-2}}{2} \approx 3.6269 \blacktriangleleft$$

GRAPHS OF THE HYPERBOLIC FUNCTIONS

The graphs of the hyperbolic functions, which are shown in Figure 7.8.1, can be generated with a graphing utility, but it is worthwhile to observe that the general shape of the graph of $y = \cosh x$ can be obtained by sketching the graphs of $y = \frac{1}{2}e^x$ and $y = \frac{1}{2}e^{-x}$ separately and adding the corresponding y-coordinates [see part (a) of the figure]. Similarly, the general shape of the graph of $y = \sinh x$ can be obtained by sketching the graphs of $y = \frac{1}{2}e^x$ and $y = -\frac{1}{2}e^{-x}$ separately and adding corresponding y-coordinates [see part (b) of the figure].

Observe that $\sinh x$ has a domain of $(-\infty, +\infty)$ and a range of $(-\infty, +\infty)$, whereas $\cosh x$ has a domain of $(-\infty, +\infty)$ and a range of $[1, +\infty)$. Observe also that $y = \frac{1}{2}e^x$ and $y = \frac{1}{2}e^{-x}$ are ***curvilinear asymptotes*** for $y = \cosh x$ in the sense that the graph of $y = \cosh x$ gets closer and closer to the graph of $y = \frac{1}{2}e^x$ as $x \to +\infty$ and gets closer and closer to the graph of $y = \frac{1}{2}e^{-x}$ as $x \to -\infty$. (See Section 4.3.) Similarly, $y = \frac{1}{2}e^x$ is a curvilinear asymptote for $y = \sinh x$ as $x \to +\infty$ and $y = -\frac{1}{2}e^{-x}$ is a curvilinear asymptote as $x \to -\infty$. Other properties of the hyperbolic functions are explored in the exercises.

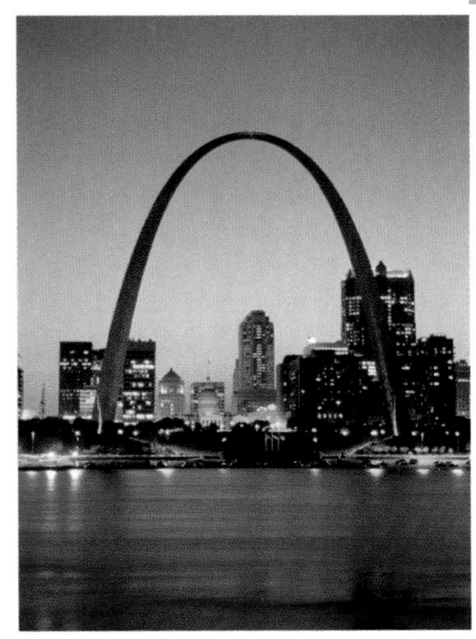

The design of the Gateway Arch near St. Louis is based on an inverted hyperbolic cosine curve (Exercise 73).

HANGING CABLES AND OTHER APPLICATIONS

Hyperbolic functions arise in vibratory motions inside elastic solids and more generally in many problems where mechanical energy is gradually absorbed by a surrounding medium. They also occur when a homogeneous, flexible cable is suspended between two points, as with a telephone line hanging between two poles. Such a cable forms a curve, called a ***catenary*** (from the Latin *catena*, meaning "chain"). If, as in Figure 7.8.2, a coordinate

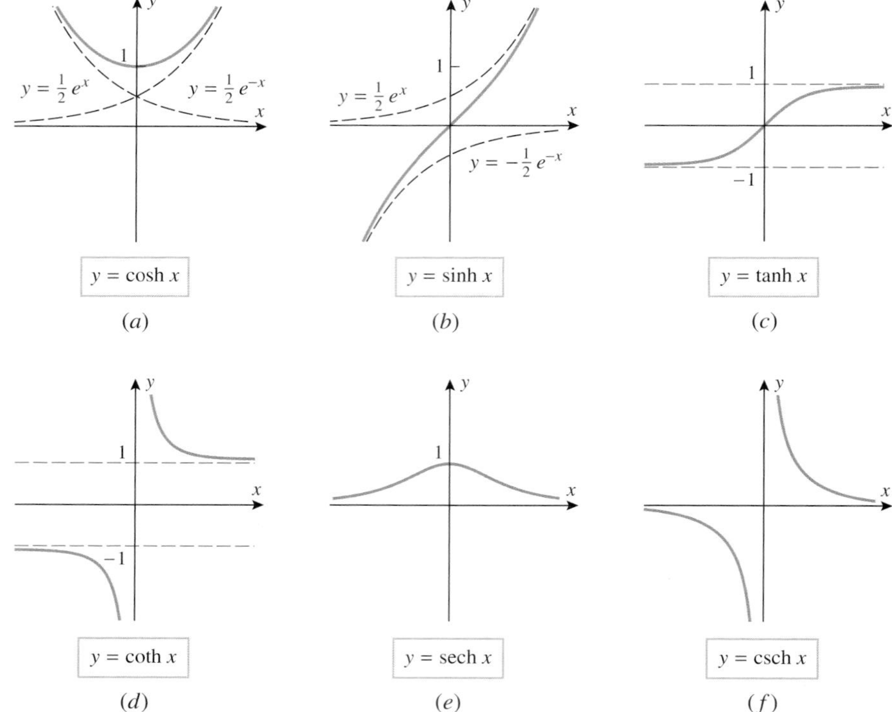

Figure 7.8.1

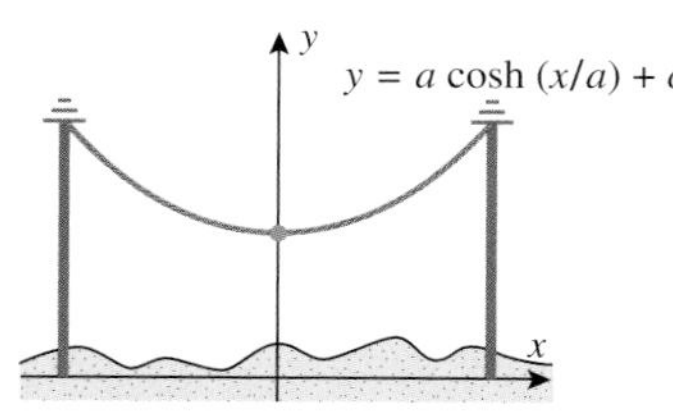

Figure 7.8.2

system is introduced so that the low point of the cable lies on the y-axis, then it can be shown using principles of physics that the cable has an equation of the form

$$y = a\cosh\left(\frac{x}{a}\right) + c$$

HYPERBOLIC IDENTITIES

The hyperbolic functions satisfy various identities that are similar to identities for trigonometric functions. The most fundamental of these is

$$\cosh^2 x - \sinh^2 x = 1 \tag{1}$$

which can be proved by writing

$$\begin{aligned}\cosh^2 x - \sinh^2 x &= (\cosh x + \sinh x)(\cosh x - \sinh x)\\ &= \left(\frac{e^x + e^{-x}}{2} + \frac{e^x - e^{-x}}{2}\right)\left(\frac{e^x + e^{-x}}{2} - \frac{e^x - e^{-x}}{2}\right)\\ &= e^x \cdot e^{-x} = 1\end{aligned}$$

Other hyperbolic identities can be derived in a similar manner or, alternatively, by performing algebraic operations on known identities. For example, if we divide (1) by $\cosh^2 x$, we obtain

$$1 - \tanh^2 x = \operatorname{sech}^2 x$$

and if we divide (1) by $\sinh^2 x$, we obtain

$$\coth^2 x - 1 = \operatorname{csch}^2 x$$

The following theorem summarizes some of the more useful hyperbolic identities. The proofs of those not already obtained are left as exercises.

7.8.2 THEOREM.

$\cosh x + \sinh x = e^x$	$\sinh(x + y) = \sinh x \cosh y + \cosh x \sinh y$
$\cosh x - \sinh x = e^{-x}$	$\cosh(x + y) = \cosh x \cosh y + \sinh x \sinh y$
$\cosh^2 x - \sinh^2 x = 1$	$\sinh(x - y) = \sinh x \cosh y - \cosh x \sinh y$
$1 - \tanh^2 x = \text{sech}^2 x$	$\cosh(x - y) = \cosh x \cosh y - \sinh x \sinh y$
$\coth^2 x - 1 = \text{csch}^2 x$	$\sinh 2x = 2 \sinh x \cosh x$
$\cosh(-x) = \cosh x$	$\cosh 2x = \cosh^2 x + \sinh^2 x$
$\sinh(-x) = -\sinh x$	$\cosh 2x = 2\sinh^2 x + 1 = 2\cosh^2 x - 1$

WHY THEY ARE CALLED HYPERBOLIC FUNCTIONS

Recall that the parametric equations

$$x = \cos t, \quad y = \sin t \qquad (0 \le t \le 2\pi)$$

represent the unit circle $x^2 + y^2 = 1$ (Figure 7.8.3*a*), as may be seen by writing

$$x^2 + y^2 = \cos^2 t + \sin^2 t = 1$$

If $0 \le t \le 2\pi$, then the parameter t can be interpreted as the angle in radians from the positive x-axis to the point $(\cos t, \sin t)$ or, alternatively, as twice the shaded area of the sector in Figure 7.8.3*a* (verify). Analogously, the parametric equations

$$x = \cosh t, \quad y = \sinh t \qquad (-\infty < t < +\infty)$$

represent a portion of the curve $x^2 - y^2 = 1$, as may be seen by writing

$$x^2 - y^2 = \cosh^2 t - \sinh^2 t = 1$$

and observing that $x = \cosh t > 0$. This curve, which is shown in Figure 7.8.3*b*, is the right half of a larger curve called the ***unit hyperbola***; this is the reason why the functions in this section are called *hyperbolic* functions. It can be shown that if $t \ge 0$, then the parameter t can be interpreted as twice the shaded area in Figure 7.8.3*b*. (We omit the details.)

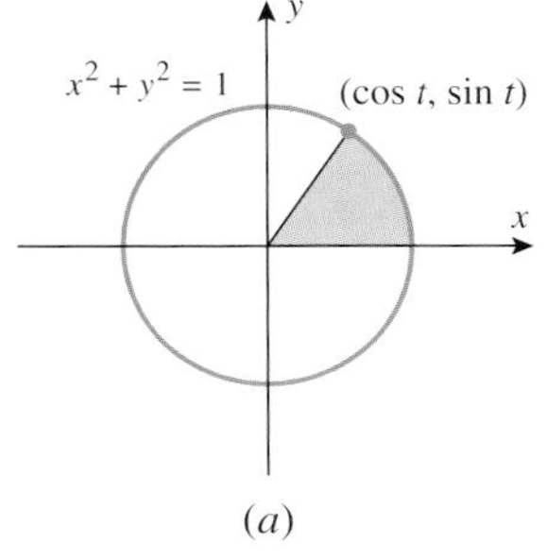

(*a*)

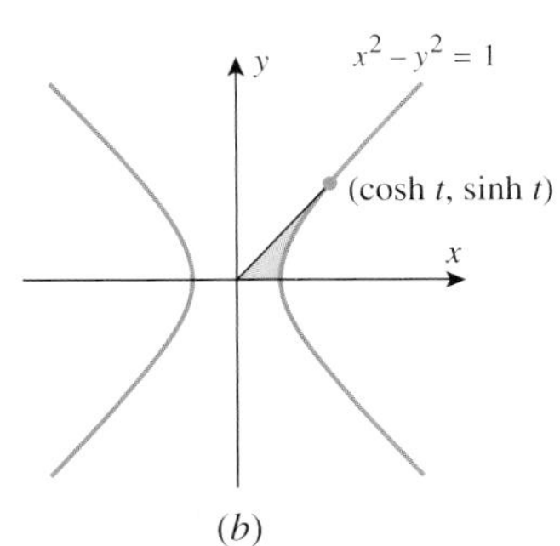

(*b*)

Figure 7.8.3

DERIVATIVE AND INTEGRAL FORMULAS

Derivative formulas for $\sinh x$ and $\cosh x$ can be obtained by expressing these functions in terms of e^x and e^{-x}:

$$\frac{d}{dx}[\sinh x] = \frac{d}{dx}\left[\frac{e^x - e^{-x}}{2}\right] = \frac{e^x + e^{-x}}{2} = \cosh x$$

$$\frac{d}{dx}[\cosh x] = \frac{d}{dx}\left[\frac{e^x + e^{-x}}{2}\right] = \frac{e^x - e^{-x}}{2} = \sinh x$$

Derivatives of the remaining hyperbolic functions can be obtained by expressing them in terms of sinh and cosh and applying appropriate identities. For example,

$$\frac{d}{dx}[\tanh x] = \frac{d}{dx}\left[\frac{\sinh x}{\cosh x}\right] = \frac{\cosh x \dfrac{d}{dx}[\sinh x] - \sinh x \dfrac{d}{dx}[\cosh x]}{\cosh^2 x}$$

$$= \frac{\cosh^2 x - \sinh^2 x}{\cosh^2 x} = \frac{1}{\cosh^2 x} = \text{sech}^2 x$$

The following theorem provides a complete list of the generalized derivative formulas and corresponding integration formulas for the hyperbolic functions.

7.8.3 THEOREM.

$$\frac{d}{dx}[\sinh u] = \cosh u \frac{du}{dx} \qquad \int \cosh u\, du = \sinh u + C$$

$$\frac{d}{dx}[\cosh u] = \sinh u \frac{du}{dx} \qquad \int \sinh u\, du = \cosh u + C$$

$$\frac{d}{dx}[\tanh u] = \operatorname{sech}^2 u \frac{du}{dx} \qquad \int \operatorname{sech}^2 u\, du = \tanh u + C$$

$$\frac{d}{dx}[\coth u] = -\operatorname{csch}^2 u \frac{du}{dx} \qquad \int \operatorname{csch}^2 u\, du = -\coth u + C$$

$$\frac{d}{dx}[\operatorname{sech} u] = -\operatorname{sech} u \tanh u \frac{du}{dx} \qquad \int \operatorname{sech} u \tanh u\, du = -\operatorname{sech} u + C$$

$$\frac{d}{dx}[\operatorname{csch} u] = -\operatorname{csch} u \coth u \frac{du}{dx} \qquad \int \operatorname{csch} u \coth u\, du = -\operatorname{csch} u + C$$

▶ **Example 2**

$$\frac{d}{dx}[\cosh(x^3)] = \sinh(x^3) \cdot \frac{d}{dx}[x^3] = 3x^2 \sinh(x^3)$$

$$\frac{d}{dx}[\ln(\tanh x)] = \frac{1}{\tanh x} \cdot \frac{d}{dx}[\tanh x] = \frac{\operatorname{sech}^2 x}{\tanh x}$$ ◀

▶ **Example 3**

$$\int \sinh^5 x \cosh x\, dx = \tfrac{1}{6}\sinh^6 x + C$$

$u = \sinh x$
$du = \cosh x\, dx$

$$\int \tanh x\, dx = \int \frac{\sinh x}{\cosh x}\, dx$$
$$= \ln|\cosh x| + C$$
$$= \ln(\cosh x) + C$$

$u = \cosh x$
$du = \sinh x\, dx$

We were justified in dropping the absolute value signs since $\cosh x > 0$ for all x. ◀

▶ **Example 4** A 100-ft wire is attached at its ends to the tops of two 50-ft poles that are positioned 90 ft apart (Figure 7.8.4). How high above the ground is the middle of the wire?

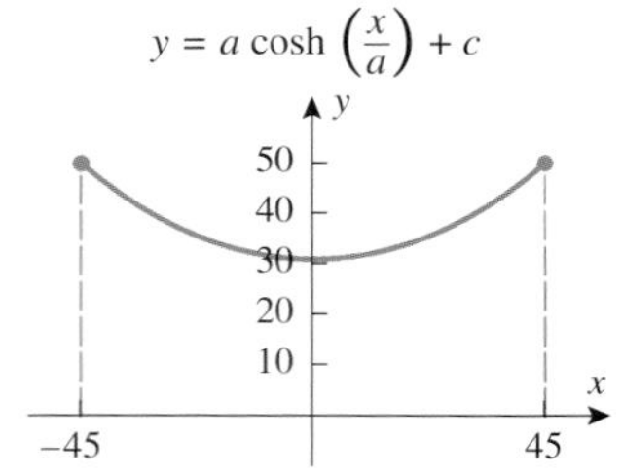

Figure 7.8.4

Solution. From above, the wire forms a catenary curve with equation

$$y = a\cosh\left(\frac{x}{a}\right) + c$$

where the origin is on the ground midway between the poles. Using Formula (4) of Section 6.4 for the length of the catenary, we have

$$100 = \int_{-45}^{45} \sqrt{1 + \left(\frac{dy}{dx}\right)^2}\, dx$$
$$= 2\int_{0}^{45} \sqrt{1 + \left(\frac{dy}{dx}\right)^2}\, dx$$

By symmetry about the y-axis

$$= 2\int_{0}^{45} \sqrt{1 + \sinh^2\left(\frac{x}{a}\right)}\, dx$$

$$= 2\int_0^{45} \cosh\left(\frac{x}{a}\right) dx \qquad \text{By (1) and the fact that } \cosh x > 0$$

$$= 2a \sinh\left(\frac{x}{a}\right)\Big]_0^{45} = 2a \sinh\left(\frac{45}{a}\right)$$

Using a calculating utility's numeric solver to solve

$$100 = 2a \sinh\left(\frac{45}{a}\right)$$

for a gives $a \approx 56.01$. Then

$$50 = y(45) = 56.01 \cosh\left(\frac{45}{56.01}\right) + c \approx 75.08 + c$$

so $c \approx -25.08$. Thus, the middle of the wire is $y(0) \approx 56.01 - 25.08 = 30.93$ ft above the ground. ◀

INVERSES OF HYPERBOLIC FUNCTIONS

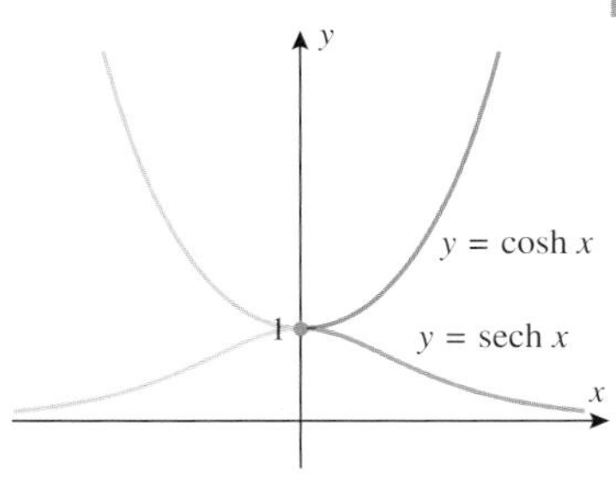

With the restriction that $x \geq 0$, the curves $y = \cosh x$ and $y = \operatorname{sech} x$ pass the horizontal line test.

Figure 7.8.5

Referring to Figure 7.8.1, it is evident that the graphs of $\sinh x$, $\tanh x$, $\coth x$, and $\operatorname{csch} x$ pass the horizontal line test, but the graphs of $\cosh x$ and $\operatorname{sech} x$ do not. In the latter case restricting x to be nonnegative makes the functions invertible (Figure 7.8.5). The graphs of the six inverse hyperbolic functions in Figure 7.8.6 were obtained by reflecting the graphs of the hyperbolic functions (with the appropriate restrictions) about the line $y = x$.

Table 7.8.1 summarizes the basic properties of the inverse hyperbolic functions. You should confirm that the domains and ranges listed in this table agree with the graphs in Figure 7.8.6.

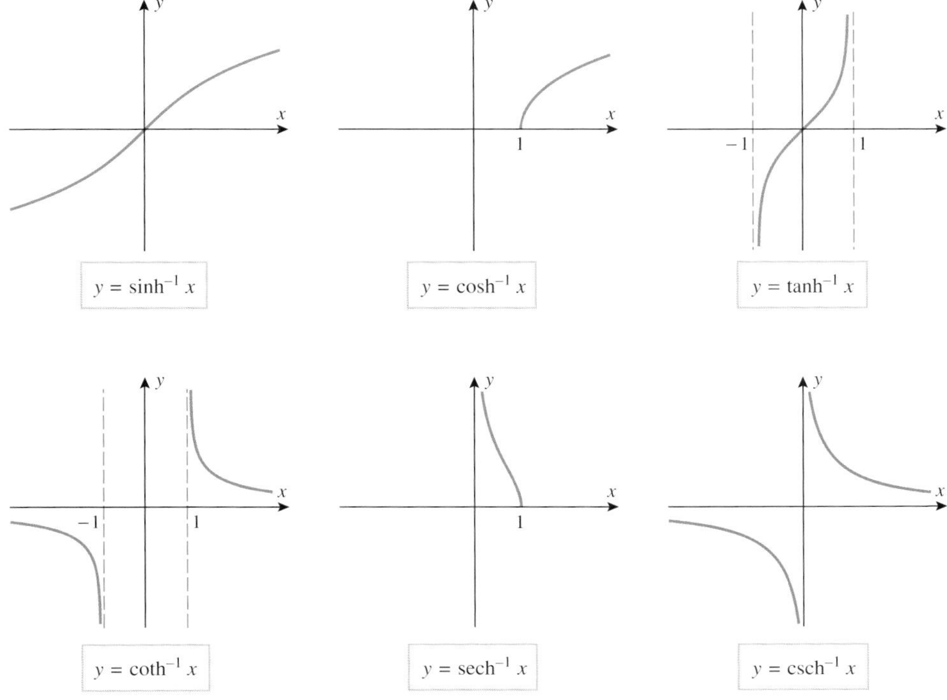

Figure 7.8.6

Table 7.8.1

FUNCTION	DOMAIN	RANGE	BASIC RELATIONSHIPS
$\sinh^{-1} x$	$(-\infty, +\infty)$	$(-\infty, +\infty)$	$\sinh^{-1}(\sinh x) = x$ if $-\infty < x < +\infty$ $\sinh(\sinh^{-1} x) = x$ if $-\infty < x < +\infty$
$\cosh^{-1} x$	$[1, +\infty)$	$[0, +\infty)$	$\cosh^{-1}(\cosh x) = x$ if $x \geq 0$ $\cosh(\cosh^{-1} x) = x$ if $x \geq 1$
$\tanh^{-1} x$	$(-1, 1)$	$(-\infty, +\infty)$	$\tanh^{-1}(\tanh x) = x$ if $-\infty < x < +\infty$ $\tanh(\tanh^{-1} x) = x$ if $-1 < x < 1$
$\coth^{-1} x$	$(-\infty, -1) \cup (1, +\infty)$	$(-\infty, 0) \cup (0, +\infty)$	$\coth^{-1}(\coth x) = x$ if $x < 0$ or $x > 0$ $\coth(\coth^{-1} x) = x$ if $x < -1$ or $x > 1$
$\text{sech}^{-1} x$	$(0, 1]$	$[0, +\infty)$	$\text{sech}^{-1}(\text{sech}\, x) = x$ if $x \geq 0$ $\text{sech}(\text{sech}^{-1} x) = x$ if $0 < x \leq 1$
$\text{csch}^{-1} x$	$(-\infty, 0) \cup (0, +\infty)$	$(-\infty, 0) \cup (0, +\infty)$	$\text{csch}^{-1}(\text{csch}\, x) = x$ if $x < 0$ or $x > 0$ $\text{csch}(\text{csch}^{-1} x) = x$ if $x < 0$ or $x > 0$

LOGARITHMIC FORMS OF INVERSE HYPERBOLIC FUNCTIONS

Because the hyperbolic functions are expressible in terms of e^x, it should not be surprising that the inverse hyperbolic functions are expressible in terms of natural logarithms; the next theorem shows that this is so.

7.8.4 THEOREM. *The following relationships hold for all x in the domains of the stated inverse hyperbolic functions*:

$$\sinh^{-1} x = \ln(x + \sqrt{x^2 + 1}) \qquad \cosh^{-1} x = \ln(x + \sqrt{x^2 - 1})$$

$$\tanh^{-1} x = \frac{1}{2} \ln\left(\frac{1 + x}{1 - x}\right) \qquad \coth^{-1} x = \frac{1}{2} \ln\left(\frac{x + 1}{x - 1}\right)$$

$$\text{sech}^{-1} x = \ln\left(\frac{1 + \sqrt{1 - x^2}}{x}\right) \qquad \text{csch}^{-1} x = \ln\left(\frac{1}{x} + \frac{\sqrt{1 + x^2}}{|x|}\right)$$

We will show how to derive the first formula in this theorem, and leave the rest as exercises. The basic idea is to write the equation $x = \sinh y$ in terms of exponential functions and solve this equation for y as a function of x. This will produce the equation $y = \sinh^{-1} x$ with $\sinh^{-1} x$ expressed in terms of natural logarithms. Expressing $x = \sinh y$ in terms of exponentials yields

$$x = \sinh y = \frac{e^y - e^{-y}}{2}$$

which can be rewritten as

$$e^y - 2x - e^{-y} = 0$$

Multiplying this equation through by e^y we obtain

$$e^{2y} - 2xe^y - 1 = 0$$

and applying the quadratic formula yields

$$e^y = \frac{2x \pm \sqrt{4x^2+4}}{2} = x \pm \sqrt{x^2+1}$$

Since $e^y > 0$, the solution involving the minus sign is extraneous and must be discarded. Thus,

$$e^y = x + \sqrt{x^2+1}$$

Taking natural logarithms yields

$$y = \ln(x + \sqrt{x^2+1}) \quad \text{or} \quad \sinh^{-1} x = \ln(x + \sqrt{x^2+1})$$

▶ Example 5

$$\sinh^{-1} 1 = \ln(1 + \sqrt{1^2+1}) = \ln(1+\sqrt{2}) \approx 0.8814$$

$$\tanh^{-1}\left(\frac{1}{2}\right) = \frac{1}{2}\ln\left(\frac{1+\frac{1}{2}}{1-\frac{1}{2}}\right) = \frac{1}{2}\ln 3 \approx 0.5493 \quad ◀$$

DERIVATIVES AND INTEGRALS INVOLVING INVERSE HYPERBOLIC FUNCTIONS

Theorem 7.3.1 can be used to establish the differentiability of the inverse hyperbolic functions (we omit the details), and formulas for the derivatives can be obtained from Theorem 7.8.4. For example,

> Show that the derivative of the function $\sinh^{-1} x$ can also be obtained by letting $y = \sinh^{-1} x$ and then differentiating $x = \sinh y$ implicitly.

$$\frac{d}{dx}[\sinh^{-1} x] = \frac{d}{dx}[\ln(x + \sqrt{x^2+1})] = \frac{1}{x+\sqrt{x^2+1}}\left(1 + \frac{x}{\sqrt{x^2+1}}\right)$$

$$= \frac{\sqrt{x^2+1}+x}{(x+\sqrt{x^2+1})(\sqrt{x^2+1})} = \frac{1}{\sqrt{x^2+1}}$$

This computation leads to two integral formulas, a formula that involves $\sinh^{-1} x$ and an equivalent formula that involves logarithms:

$$\int \frac{dx}{\sqrt{x^2+1}} = \sinh^{-1} x + C = \ln(x + \sqrt{x^2+1}) + C$$

The following two theorems list the generalized derivative formulas and corresponding integration formulas for the inverse hyperbolic functions. Some of the proofs appear as exercises.

7.8.5 THEOREM.

$$\frac{d}{dx}(\sinh^{-1} u) = \frac{1}{\sqrt{1+u^2}}\frac{du}{dx} \qquad \frac{d}{dx}(\coth^{-1} u) = \frac{1}{1-u^2}\frac{du}{dx}, \quad |u| > 1$$

$$\frac{d}{dx}(\cosh^{-1} u) = \frac{1}{\sqrt{u^2-1}}\frac{du}{dx}, \quad u > 1 \qquad \frac{d}{dx}(\operatorname{sech}^{-1} u) = -\frac{1}{u\sqrt{1-u^2}}\frac{du}{dx}, \quad 0 < u < 1$$

$$\frac{d}{dx}(\tanh^{-1} u) = \frac{1}{1-u^2}\frac{du}{dx}, \quad |u| < 1 \qquad \frac{d}{dx}(\operatorname{csch}^{-1} u) = -\frac{1}{|u|\sqrt{1+u^2}}\frac{du}{dx}, \quad u \neq 0$$

7.8.6 THEOREM. *If $a > 0$, then*

$$\int \frac{du}{\sqrt{a^2+u^2}} = \sinh^{-1}\left(\frac{u}{a}\right) + C \quad or \quad \ln(u + \sqrt{u^2+a^2}) + C$$

$$\int \frac{du}{\sqrt{u^2-a^2}} = \cosh^{-1}\left(\frac{u}{a}\right) + C \quad or \quad \ln(u + \sqrt{u^2-a^2}) + C, \quad u > a$$

$$\int \frac{du}{a^2-u^2} = \begin{cases} \dfrac{1}{a}\tanh^{-1}\left(\dfrac{u}{a}\right) + C, & |u| < a \\ \dfrac{1}{a}\coth^{-1}\left(\dfrac{u}{a}\right) + C, & |u| > a \end{cases} \quad or \quad \frac{1}{2a}\ln\left|\frac{a+u}{a-u}\right| + C, \quad |u| \neq a$$

$$\int \frac{du}{u\sqrt{a^2-u^2}} = -\frac{1}{a}\operatorname{sech}^{-1}\left|\frac{u}{a}\right| + C \quad or \quad -\frac{1}{a}\ln\left(\frac{a+\sqrt{a^2-u^2}}{|u|}\right) + C, \quad 0 < |u| < a$$

$$\int \frac{du}{u\sqrt{a^2+u^2}} = -\frac{1}{a}\operatorname{csch}^{-1}\left|\frac{u}{a}\right| + C \quad or \quad -\frac{1}{a}\ln\left(\frac{a+\sqrt{a^2+u^2}}{|u|}\right) + C, \quad u \neq 0$$

▶ **Example 6** Evaluate $\displaystyle\int \frac{dx}{\sqrt{4x^2-9}}$, $x > \frac{3}{2}$.

Solution. Let $u = 2x$. Thus, $du = 2\,dx$ and

$$\int \frac{dx}{\sqrt{4x^2-9}} = \frac{1}{2}\int \frac{2\,dx}{\sqrt{4x^2-9}} = \frac{1}{2}\int \frac{du}{\sqrt{u^2-3^2}}$$
$$= \frac{1}{2}\cosh^{-1}\left(\frac{u}{3}\right) + C = \frac{1}{2}\cosh^{-1}\left(\frac{2x}{3}\right) + C$$

Alternatively, we can use the logarithmic equivalent of $\cosh^{-1}(2x/3)$,

$$\cosh^{-1}\left(\frac{2x}{3}\right) = \ln(2x + \sqrt{4x^2-9}) - \ln 3$$

(verify), and express the answer as

$$\int \frac{dx}{\sqrt{4x^2-9}} = \frac{1}{2}\ln(2x + \sqrt{4x^2-9}) + C \quad ◀$$

✔QUICK CHECK EXERCISES 7.8 *(See page 509 for answers.)*

1. $\cosh x =$ ______ $\sinh x =$ ______
   $\tanh x =$ ______

2. Complete the table.

	$\cosh x$	$\sinh x$	$\tanh x$	$\coth x$	$\operatorname{sech} x$	$\operatorname{csch} x$
DOMAIN						
RANGE						

3. The parametric equations

$$x = \cosh t, \quad y = \sinh t \quad (-\infty < t < +\infty)$$

represent the right half of the curve called a ______. Eliminating the parameter, the equation of this curve is ______.

4. $\dfrac{d}{dx}[\cosh x] =$ ______ $\dfrac{d}{dx}[\sinh x] =$ ______
   $\dfrac{d}{dx}[\tanh x] =$ ______

5. $\int \cosh x\,dx =$ ______ $\int \sinh x\,dx =$ ______

$\int \tanh x\,dx =$ ______

6. $\frac{d}{dx}[\cosh^{-1} x] =$ ______ $\frac{d}{dx}[\sinh^{-1} x] =$ ______

$\frac{d}{dx}[\tanh^{-1} x] =$ ______

EXERCISE SET 7.8

Graphing Utility   C CAS

1–2 Approximate the expression to four decimal places.

1. (a) $\sinh 3$ (b) $\cosh(-2)$ (c) $\tanh(\ln 4)$
(d) $\sinh^{-1}(-2)$ (e) $\cosh^{-1} 3$ (f) $\tanh^{-1} \frac{3}{4}$

2. (a) $\operatorname{csch}(-1)$ (b) $\operatorname{sech}(\ln 2)$ (c) $\coth 1$
(d) $\operatorname{sech}^{-1} \frac{1}{2}$ (e) $\coth^{-1} 3$ (f) $\operatorname{csch}^{-1}(-\sqrt{3})$

3. Find the exact numerical value of each expression.
(a) $\sinh(\ln 3)$ (b) $\cosh(-\ln 2)$
(c) $\tanh(2\ln 5)$ (d) $\sinh(-3\ln 2)$

4. In each part, rewrite the expression as a ratio of polynomials.
(a) $\cosh(\ln x)$ (b) $\sinh(\ln x)$
(c) $\tanh(2\ln x)$ (d) $\cosh(-\ln x)$

5. In each part, a value for one of the hyperbolic functions is given at an unspecified positive number x_0. Use appropriate identities to find the exact values of the remaining five hyperbolic functions at x_0.
(a) $\sinh x_0 = 2$ (b) $\cosh x_0 = \frac{5}{4}$ (c) $\tanh x_0 = \frac{4}{5}$

6. Obtain the derivative formulas for $\operatorname{csch} x$, $\operatorname{sech} x$, and $\coth x$ from the derivative formulas for $\sinh x$, $\cosh x$, and $\tanh x$.

7. Find the derivatives of $\sinh^{-1} x$, $\cosh^{-1} x$, and $\tanh^{-1} x$ by differentiating the equations $x = \sinh y$, $x = \cosh y$, and $x = \tanh y$ implicitly.

C **8.** Use a CAS to find the derivatives of $\sinh^{-1} x$, $\cosh^{-1} x$, $\tanh^{-1} x$, $\coth^{-1} x$, $\operatorname{sech}^{-1} x$, and $\operatorname{csch}^{-1} x$, and confirm that your answers are consistent with those in Theorem 7.8.5.

9–28 Find dy/dx.

9. $y = \sinh(4x - 8)$
10. $y = \cosh(x^4)$
11. $y = \coth(\ln x)$
12. $y = \ln(\tanh 2x)$
13. $y = \operatorname{csch}(1/x)$
14. $y = \operatorname{sech}(e^{2x})$
15. $y = \sqrt{4x + \cosh^2(5x)}$
16. $y = \sinh^3(2x)$
17. $y = x^3 \tanh^2(\sqrt{x})$
18. $y = \sinh(\cos 3x)$
19. $y = \sinh^{-1}\left(\frac{1}{3}x\right)$
20. $y = \sinh^{-1}(1/x)$
21. $y = \ln(\cosh^{-1} x)$
22. $y = \cosh^{-1}(\sinh^{-1} x)$
23. $y = \dfrac{1}{\tanh^{-1} x}$
24. $y = (\coth^{-1} x)^2$
25. $y = \cosh^{-1}(\cosh x)$
26. $y = \sinh^{-1}(\tanh x)$
27. $y = e^x \operatorname{sech}^{-1} \sqrt{x}$
28. $y = (1 + x \operatorname{csch}^{-1} x)^{10}$

C **29.** Use a CAS to find the derivatives in Example 2. If the answers produced by the CAS do not match those in the text, then use appropriate identities to show that the answers are equivalent.

C **30.** For each of the derivatives you obtained in Exercises 9–28, use a CAS to check your answer. If the answer produced by the CAS does not match your own, show that the two answers are equivalent.

31–46 Evaluate the integrals.

31. $\int \sinh^6 x \cosh x\,dx$
32. $\int \cosh(2x - 3)\,dx$
33. $\int \sqrt{\tanh x}\,\operatorname{sech}^2 x\,dx$
34. $\int \operatorname{csch}^2(3x)\,dx$
35. $\int \tanh x\,dx$
36. $\int \coth^2 x \operatorname{csch}^2 x\,dx$
37. $\int_{\ln 2}^{\ln 3} \tanh x \operatorname{sech}^3 x\,dx$
38. $\int_0^{\ln 3} \frac{e^x - e^{-x}}{e^x + e^{-x}}\,dx$
39. $\int \frac{dx}{\sqrt{1 + 9x^2}}$
40. $\int \frac{dx}{\sqrt{x^2 - 2}}$ $(x > \sqrt{2})$
41. $\int \frac{dx}{\sqrt{1 - e^{2x}}}$ $(x < 0)$
42. $\int \frac{\sin\theta\,d\theta}{\sqrt{1 + \cos^2\theta}}$
43. $\int \frac{dx}{x\sqrt{1 + 4x^2}}$
44. $\int \frac{dx}{\sqrt{9x^2 - 25}}$ $(x > 5/3)$
45. $\int_0^{1/2} \frac{dx}{1 - x^2}$
46. $\int_0^{\sqrt{3}} \frac{dt}{\sqrt{t^2 + 1}}$

C **47.** For each of the integrals you evaluated in Exercises 31–46, use a CAS to check your answer. If the answer produced by the CAS does not match your own, show that the two answers are equivalent.

48. Use a graphing utility to generate the graphs of $\sinh x$, $\cosh x$, and $\tanh x$ by expressing these functions in terms of e^x and e^{-x}. If your graphing utility can graph the hyperbolic functions directly, then generate the graphs that way as well.

49. Find the area enclosed by $y = \sinh 2x$, $y = 0$, and $x = \ln 3$.

50. Find the volume of the solid that is generated when the region enclosed by $y = \operatorname{sech} x$, $y = 0$, $x = 0$, and $x = \ln 2$ is revolved about the x-axis.

51. Find the volume of the solid that is generated when the region enclosed by $y = \cosh 2x$, $y = \sinh 2x$, $x = 0$, and $x = 5$ is revolved about the x-axis.

52. Approximate the positive value of the constant a such that the area enclosed by $y = \cosh ax$, $y = 0$, $x = 0$, and $x = 1$ is 2 square units. Express your answer to at least five decimal places.

53. Find the arc length of the catenary $y = \cosh x$ between $x = 0$ and $x = \ln 2$.

54. Find the arc length of the catenary $y = a\cosh(x/a)$ between $x = 0$ and $x = x_1$ $(x_1 > 0)$.

55. Find the limits, and confirm that they are consistent with the graphs in Figures 7.8.1 and 7.8.6.

(a) $\lim_{x\to+\infty} \sinh x$ (b) $\lim_{x\to-\infty} \sinh x$

(c) $\lim_{x\to+\infty} \tanh x$ (d) $\lim_{x\to-\infty} \tanh x$

(e) $\lim_{x\to+\infty} \sinh^{-1} x$ (f) $\lim_{x\to 1^-} \tanh^{-1} x$

FOCUS ON CONCEPTS

56. Explain how to obtain the asymptotes for $y = \tanh x$ from the curvilinear asymptotes for $y = \cosh x$ and $y = \sinh x$.

57. Prove that $\sinh x$ is an odd function of x and that $\cosh x$ is an even function of x, and check that this is consistent with the graphs in Figure 7.8.1.

58–59 Prove the identities.

58. (a) $\cosh x + \sinh x = e^x$

(b) $\cosh x - \sinh x = e^{-x}$

(c) $\sinh(x + y) = \sinh x \cosh y + \cosh x \sinh y$

(d) $\sinh 2x = 2\sinh x \cosh x$

(e) $\cosh(x + y) = \cosh x \cosh y + \sinh x \sinh y$

(f) $\cosh 2x = \cosh^2 x + \sinh^2 x$

(g) $\cosh 2x = 2\sinh^2 x + 1$

(h) $\cosh 2x = 2\cosh^2 x - 1$

59. (a) $1 - \tanh^2 x = \operatorname{sech}^2 x$

(b) $\tanh(x + y) = \dfrac{\tanh x + \tanh y}{1 + \tanh x \tanh y}$

(c) $\tanh 2x = \dfrac{2\tanh x}{1 + \tanh^2 x}$

60. Prove:

(a) $\cosh^{-1} x = \ln(x + \sqrt{x^2 - 1}), \quad x \ge 1$

(b) $\tanh^{-1} x = \dfrac{1}{2}\ln\left(\dfrac{1 + x}{1 - x}\right), \quad -1 < x < 1.$

61. Use Exercise 60 to obtain the derivative formulas for $\cosh^{-1} x$ and $\tanh^{-1} x$.

62. Prove:

$$\operatorname{sech}^{-1} x = \cosh^{-1}(1/x), \quad 0 < x \le 1$$

$$\coth^{-1} x = \tanh^{-1}(1/x), \quad |x| > 1$$

$$\operatorname{csch}^{-1} x = \sinh^{-1}(1/x), \quad x \ne 0$$

63. Use Exercise 62 to express the integral

$$\int \frac{du}{1 - u^2}$$

entirely in terms of $\tanh^{-1}$.

64. Show that

(a) $\dfrac{d}{dx}[\operatorname{sech}^{-1}|x|] = -\dfrac{1}{x\sqrt{1 - x^2}}$

(b) $\dfrac{d}{dx}[\operatorname{csch}^{-1}|x|] = -\dfrac{1}{x\sqrt{1 + x^2}}$.

65. In each part, find the limit.

(a) $\lim_{x\to+\infty} (\cosh^{-1} x - \ln x)$ (b) $\lim_{x\to+\infty} \dfrac{\cosh x}{e^x}$

66. Use the first and second derivatives to show that the graph of $y = \tanh^{-1} x$ is always increasing and has an inflection point at the origin.

67. The integration formulas for $1/\sqrt{u^2 - a^2}$ in Theorem 7.8.6 are valid for $u > a$. Show that the following formula is valid for $u < -a$:

$$\int \frac{du}{\sqrt{u^2 - a^2}} = -\cosh^{-1}\left(-\frac{u}{a}\right) + C \quad \text{or} \quad \ln\left|u + \sqrt{u^2 - a^2}\right| + C$$

68. Show that $(\sinh x + \cosh x)^n = \sinh nx + \cosh nx$.

69. Show that

$$\int_{-a}^{a} e^{tx}\,dx = \frac{2\sinh at}{t}$$

70. A cable is suspended between two poles as shown in Figure 7.8.2. Assume that the equation of the curve formed by the cable is $y = a\cosh(x/a)$, where a is a positive constant. Suppose that the x-coordinates of the points of support are $x = -b$ and $x = b$, where $b > 0$.

(a) Show that the length L of the cable is given by

$$L = 2a \sinh \frac{b}{a}$$

(b) Show that the sag S (the vertical distance between the highest and lowest points on the cable) is given by

$$S = a\cosh \frac{b}{a} - a$$

71–72 These exercises refer to the hanging cable described in Exercise 70.

71. Assuming that the poles are 400 ft apart and the sag in the cable is 30 ft, approximate the length of the cable by approximating a. Express your final answer to the nearest tenth of a foot. [*Hint:* First let $u = 200/a$.]

72. Assuming that the cable is 120 ft long and the poles are 100 ft apart, approximate the sag in the cable by approximating a. Express your final answer to the nearest tenth of a foot. [*Hint:* First let $u = 50/a$.]

73. The design of the Gateway Arch in St. Louis, Missouri, by architect Eero Saarinan was implemented using equations provided by Dr. Hannskarl Badel. The equation used for the centerline of the arch was

$$y = 693.8597 - 68.7672 \cosh(0.0100333x) \text{ ft}$$

for x between -299.2239 and 299.2239.

(a) Use a graphing utility to graph the centerline of the arch.

(b) Find the length of the centerline to four decimal places.

(c) For what values of x is the height of the arch 100 ft? Round your answers to four decimal places.

(d) Approximate, to the nearest degree, the acute angle that the tangent line to the centerline makes with the ground at the ends of the arch.

74. Suppose that a hollow tube rotates with a constant angular velocity of ω rad/s about a horizontal axis at one end of the tube, as shown in the accompanying figure. Assume that an object is free to slide without friction in the tube while the tube is rotating. Let r be the distance from the object to the pivot point at time $t \geq 0$, and assume that the object is at rest and $r = 0$ when $t = 0$. It can be shown that if the tube is horizontal at time $t = 0$ and rotating as shown in the figure, then

$$r = \frac{g}{2\omega^2}[\sinh(\omega t) - \sin(\omega t)]$$

during the period that the object is in the tube. Assume that t is in seconds and r is in meters, and use $g = 9.8$ m/s^2 and $\omega = 2$ rad/s.

(a) Graph r versus t for $0 \leq t \leq 1$.

(b) Assuming that the tube has a length of 1 m, approximately how long does it take for the object to reach the end of the tube?

(c) Use the result of part (b) to approximate dr/dt at the instant that the object reaches the end of the tube.

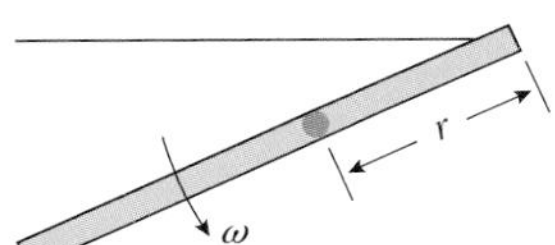

Figure Ex-74

75. The accompanying figure shows a person pulling a boat by holding a rope of length a attached to the bow and walking along the edge of a dock. If we assume that the rope is always tangent to the curve traced by the bow of the boat, then this curve, which is called a ***tractrix***, has the property that the segment of the tangent line between the curve and the y-axis has a constant length a. It can be proved that the equation of this tractrix is

$$y = a \operatorname{sech}^{-1}\frac{x}{a} - \sqrt{a^2 - x^2}$$

(a) Show that to move the bow of the boat to a point (x, y), the person must walk a distance

$$D = a \operatorname{sech}^{-1}\frac{x}{a}$$

from the origin.

(b) If the rope has a length of 15 m, how far must the person walk from the origin to bring the boat 10 m from the dock? Round your answer to two decimal places.

(c) Find the distance traveled by the bow along the tractrix as it moves from its initial position to the point where it is 5 m from the dock.

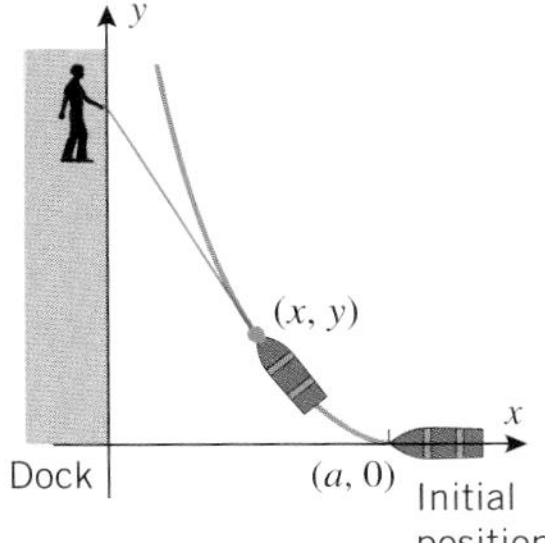

Figure Ex-75

✔ QUICK CHECK ANSWERS 7.8

1. $\dfrac{e^x + e^{-x}}{2}$; $\dfrac{e^x - e^{-x}}{2}$; $\dfrac{e^x - e^{-x}}{e^x + e^{-x}}$

2.

	$\cosh x$	$\sinh x$	$\tanh x$	$\coth x$	$\operatorname{sech} x$	$\operatorname{csch} x$
DOMAIN	$(-\infty, +\infty)$	$(-\infty, +\infty)$	$(-\infty, +\infty)$	$(-\infty, 0) \cup (0, +\infty)$	$(-\infty, +\infty)$	$(-\infty, 0) \cup (0, +\infty)$
RANGE	$[1, +\infty)$	$(-\infty, +\infty)$	$(-1, 1)$	$(-\infty, -1) \cup (1, +\infty)$	$(0, 1]$	$(-\infty, 0) \cup (0, +\infty)$

3. unit hyperbola; $x^2 - y^2 = 1$ 4. $\sinh x$; $\cosh x$; $\operatorname{sech}^2 x$ 5. $\sinh x + C$; $\cosh x + C$; $\ln(\cosh x) + C$

6. $\dfrac{1}{\sqrt{x^2 - 1}}$; $\dfrac{1}{\sqrt{1 + x^2}}$; $\dfrac{1}{1 - x^2}$

CHAPTER REVIEW EXERCISES

Graphing Utility

1. In each part, find $f^{-1}(x)$ if the inverse exists.
   (a) $f(x) = (e^x)^2 + 1$
   (b) $f(x) = \sin\left(\dfrac{1-2x}{x}\right), \quad \dfrac{2}{4+\pi} \le x \le \dfrac{2}{4-\pi}$
   (c) $f(x) = \dfrac{1}{1 + 3\tan^{-1} x}$

2. Let $f(x) = (ax+b)/(cx+d)$. What conditions on a, b, c, and d guarantee that f^{-1} exists? Find $f^{-1}(x)$.

3. In each part, find the exact numerical value of the given expression.
   (a) $\cos[\cos^{-1}(4/5) + \sin^{-1}(5/13)]$
   (b) $\sin[\sin^{-1}(4/5) + \cos^{-1}(5/13)]$

4. (a) State the restrictions on the domains of $\sin x$, $\cos x$, $\tan x$, and $\sec x$ that are imposed to make those functions one-to-one in the definitions of $\sin^{-1} x$, $\cos^{-1} x$, $\tan^{-1} x$, and $\sec^{-1} x$.
   (b) Sketch the graphs of the restricted trigonometric functions in part (a) and their inverses.

5. Suppose that the graph of $y = \log x$ is drawn with equal scales of 1 inch per unit in both the x- and y-directions. If a bug wants to walk along the graph until it reaches a height of 5 ft above the x-axis, how many miles to the right of the origin will it have to travel?

6. In each part, sketch the graph, and check your work with a graphing utility.
   (a) $f(x) = 3\sin^{-1}(x/2)$
   (b) $f(x) = \cos^{-1} x - \pi/2$
   (c) $f(x) = 2\tan^{-1}(-3x)$
   (d) $f(x) = \cos^{-1} x + \sin^{-1} x$

7. Express the following function as a rational function of x:
$$3\ln\left(e^{2x}(e^x)^3\right) + 2\exp(\ln 1)$$

8. Suppose that $y = Ce^{kt}$, where C and k are constants, and let $Y = \ln y$. Show that the graph of Y versus t is a line, and state its slope and Y-intercept.

9. (a) Sketch the curves $y = \pm e^{-x/2}$ and $y = e^{-x/2}\sin 2x$ for $-\pi/2 \le x \le 3\pi/2$ in the same coordinate system, and check your work using a graphing utility.
   (b) Find all x-intercepts of the curve $y = e^{-x/2}\sin 2x$ in the stated interval, and find the x-coordinates of all points where this curve intersects the curves $y = \pm e^{-x/2}$.

10. Suppose that a package of medical supplies is dropped from a helicopter straight down by parachute into a remote area. The velocity v (in feet per second) of the package t seconds after it is released is given by $v = 24.61(1 - e^{-1.3t})$.
   (a) Graph v versus t.
   (b) Show that the graph has a horizontal asymptote $v = c$.
   (c) The constant c is called the ***terminal velocity***. Explain what the terminal velocity means in practical terms.
   (d) Can the package actually reach its terminal velocity? Explain.
   (e) How long does it take for the package to reach 98% of its terminal velocity?

11. A breeding group of 20 bighorn sheep is released in a protected area in Colorado. It is expected that with careful management the number of sheep, N, after t years will be given by the formula
$$N = \frac{220}{1 + 10(0.83^t)}$$
and that the sheep population will be able to maintain itself without further supervision once the population reaches a size of 80.
   (a) Graph N versus t.
   (b) How many years must the state of Colorado maintain a program to care for the sheep?
   (c) How many bighorn sheep can the environment in the protected area support? [*Hint:* Examine the graph of N versus t for large values of t.]

12. An oven is preheated and then remains at a constant temperature. A potato is placed in the oven to bake. Suppose that the temperature T (in °F) of the potato t minutes later is given by $T = 400 - 325(0.97^t)$. The potato will be considered done when its temperature is anywhere between 260°F and 280°F.
   (a) During what interval of time would the potato be considered done?
   (b) How long does it take for the difference between the potato and oven temperatures to be cut in half?

13. (a) Show that the graphs of $y = \ln x$ and $y = x^{0.2}$ intersect.
   (b) Approximate the solution(s) of the equation $\ln x = x^{0.2}$ to three decimal places.

14. (a) Show that for $x > 0$ and $k \ne 0$ the equations
$$x^k = e^x \quad \text{and} \quad \frac{\ln x}{x} = \frac{1}{k}$$
have the same solutions.
   (b) Use the graph of $y = (\ln x)/x$ to determine the values of k for which the equation $x^k = e^x$ has two distinct positive solutions.
   (c) Estimate the positive solution(s) of $x^8 = e^x$.

15–18 Find the limits.

15. $\lim\limits_{t\to\pi/2^+} e^{\tan t}$

16. $\lim\limits_{\theta\to 0^+} \ln(\sin 2\theta) - \ln(\tan\theta)$

17. $\lim\limits_{x\to+\infty} \left(1 + \dfrac{3}{x}\right)^{-x}$

18. $\lim\limits_{x\to+\infty} \left(1 + \dfrac{a}{x}\right)^{bx}, \quad a, b > 0$

19–20 Find dy/dx by first using algebraic properties of the natural logarithm function.

19. $y = \ln\left(\frac{(x+1)(x+2)^2}{(x+3)^3(x+4)^4}\right)$

20. $y = \ln\left(\frac{\sqrt{x}\sqrt[3]{x+1}}{\sin x \sec x}\right)$

21–38 Find dy/dx.

21. $y = \ln 2x$

22. $y = (\ln x)^2$

23. $y = \sqrt[3]{\ln x + 1}$

24. $y = \ln(\sqrt[3]{x+1})$

25. $y = \log(\ln x)$

26. $y = \frac{1 + \log x}{1 - \log x}$

27. $y = \ln(x^{3/2}\sqrt{1+x^4})$

28. $y = \ln\left(\frac{\sqrt{x}\cos x}{1+x^2}\right)$

29. $y = e^{\ln(x^2+1)}$

30. $y = \ln\left(\frac{1+e^x+e^{2x}}{1-e^{3x}}\right)$

31. $y = 2xe^{\sqrt{x}}$

32. $y = \frac{a}{1+be^{-x}}$

33. $y = \frac{1}{\pi}\tan^{-1} 2x$

34. $y = 2^{\sin^{-1} x}$

35. $y = x^{(e^x)}$

36. $y = (1+x)^{1/x}$

37. $y = \sec^{-1}(2x+1)$

38. $y = \sqrt{\cos^{-1} x^2}$

39–40 Find dy/dx using logarithmic differentiation.

39. $y = \frac{x^3}{\sqrt{x^2+1}}$

40. $y = \sqrt[3]{\frac{x^2-1}{x^2+1}}$

41. (a) Make a conjecture about the shape of the graph of $y = \frac{1}{2}x - \ln x$, and draw a rough sketch.
(b) Check your conjecture by graphing the equation over the interval $0 < x < 5$ with a graphing utility.
(c) Show that the slopes of the tangent lines to the curve at $x = 1$ and $x = e$ have opposite signs.
(d) What does part (c) imply about the existence of a horizontal tangent line to the curve? Explain.
(e) Find the exact x-coordinates of all horizontal tangent lines to the curve.

42. The loudness β of a sound in decibels (dB) is given by $\beta = 10\log(I/I_0)$, where I is the intensity of the sound in watts per square meter (W/m^2) and I_0 is a constant that is approximately the intensity of a sound at the threshold of human hearing. Find the rate of change of β with respect to I at the point where
(a) $I/I_0 = 10$ (b) $I/I_0 = 100$ (c) $I/I_0 = 1000$.

43. A particle is moving along the curve $y = x\ln x$. Find all values of x at which the rate of change of y with respect to time is three times that of x. [Assume that dx/dt is never zero.]

44. Find the equation of the tangent line to the graph of $y = \ln(5 - x^2)$ at $x = 2$.

45. Find the value of b so that the line $y = x$ is tangent to the graph of $y = \log_b x$. Confirm your result by graphing both $y = x$ and $y = \log_b x$ in the same coordinate system.

46. In each part, find the value of k for which the graphs of $y = f(x)$ and $y = \ln x$ share a common tangent line at their point of intersection. Confirm your result by graphing $y = f(x)$ and $y = \ln x$ in the same coordinate system.
(a) $f(x) = \sqrt{x} + k$ (b) $f(x) = k\sqrt{x}$

47. If f and g are inverse functions and f is differentiable on its domain, must g be differentiable on its domain? Give a reasonable informal argument to support your answer.

48. In each part, find $(f^{-1})'(x)$ using Formula (1) of Section 7.3, and check your answer by differentiating f^{-1} directly.
(a) $f(x) = 3/(x+1)$ (b) $f(x) = \sqrt{e^x}$

49. Find a point on the graph of $y = e^{3x}$ at which the tangent line passes through the origin.

50. Show that the rate of change of $y = 5000e^{1.07x}$ is proportional to y.

51. Show that the function $y = e^{ax}\sin bx$ satisfies

$$y'' - 2ay' + (a^2 + b^2)y = 0$$

for any real constants a and b.

52. Show that the function $y = \tan^{-1} x$ satisfies

$$y'' = -2\sin y \cos^3 y$$

53. Suppose that the population of deer on an island is modeled by the equation

$$P(t) = \frac{95}{5 - 4e^{-t/4}}$$

where $P(t)$ is the number of deer t weeks after an initial observation at time $t = 0$.
(a) Use a graphing utility to graph the function $P(t)$.
(b) In words, explain what happens to the population over time. Check your conclusion by finding $\lim_{t\to+\infty} P(t)$.
(c) In words, what happens to the *rate* of population growth over time? Check your conclusion by graphing $P'(t)$.

54. In each part, find each limit by interpreting the expression as an appropriate derivative.
(a) $\lim_{h\to 0} \frac{(1+h)^\pi - 1}{h}$ (b) $\lim_{x\to e} \frac{1 - \ln x}{(x-e)\ln x}$

55. Suppose that $\lim f(x) = \pm\infty$ and $\lim g(x) = \pm\infty$. In each of the four possible cases, state whether $\lim[f(x) - g(x)]$ is an indeterminate form, and give a reasonable informal argument to support your answer.

56. (a) Under what conditions will a limit of the form

$$\lim_{x\to a} [f(x)/g(x)]$$

be an indeterminate form?

(b) If $\lim_{x\to a} g(x) = 0$, must $\lim_{x\to a}[f(x)/g(x)]$ be an indeterminate form? Give some examples to support your answer.

57–60 Evaluate the given limit.

57. $\lim_{x\to+\infty}(e^x - x^2)$ **58.** $\lim_{x\to 1}\sqrt{\dfrac{\ln x}{x^4-1}}$

59. $\lim_{x\to 0}\dfrac{x^2e^x}{\sin^2 3x}$ **60.** $\lim_{x\to 0}\dfrac{a^x-1}{x}$, $a > 0$

61–62 Find: (a) the intervals on which f is increasing, (b) the intervals on which f is decreasing, (c) the open intervals on which f is concave up, (d) the open intervals on which f is concave down, and (e) the x-coordinates of all inflection points.

61. $f(x) = 1/e^{x^2}$ **62.** $f(x) = \tan^{-1} x^2$

63–64 Use any method to find the relative extrema of the function f.

63. $f(x) = \ln(1 + x^2)$ **64.** $f(x) = x^2e^x$

65–66 Find the absolute minimum m and the absolute maximum M of f on the given interval (if they exist), and state where the absolute extrema occur.

65. $f(x) = e^x/x^2$; $(0, +\infty)$ **66.** $f(x) = x^x$; $(0, +\infty)$

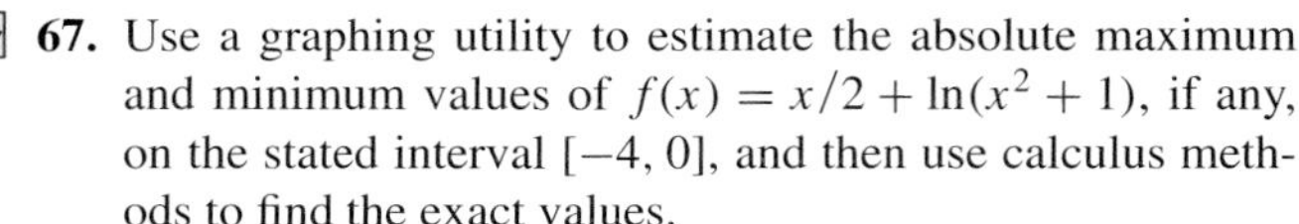

67. Use a graphing utility to estimate the absolute maximum and minimum values of $f(x) = x/2 + \ln(x^2 + 1)$, if any, on the stated interval $[-4, 0]$, and then use calculus methods to find the exact values.

68. Prove that $x \le \sin^{-1} x$ for all x in $[0, 1]$.

69–72 Evaluate the integrals.

69. $\displaystyle\int [x^{-2/3} - 5e^x]\,dx$ **70.** $\displaystyle\int \left[\frac{3}{4x} - \sec^2 x\right] dx$

71. $\displaystyle\int \left[\frac{1}{1+x^2} + \frac{2}{\sqrt{1-x^2}}\right] dx$

72. $\displaystyle\int \left[\frac{12}{x\sqrt{x^2-1}} + \frac{1-x^4}{1+x^2}\right] dx$

73–74 Use a calculating utility to find the left endpoint, right endpoint, and midpoint approximations to the area under the curve $y = f(x)$ over the stated interval using $n = 10$ subintervals.

73. $y = \ln x$; $[1, 2]$ **74.** $y = e^x$; $[0, 1]$

75. Give a convincing geometric argument to show that

$$\int_1^e \ln x\,dx + \int_0^1 e^x\,dx = e$$

76. Find the limit

$$\lim_{n\to+\infty}\frac{e^{1/n} + e^{2/n} + e^{3/n} + \cdots + e^{n/n}}{n}$$

by interpreting it as a limit of Riemann sums in which the interval $[0, 1]$ is divided into n subintervals of equal length.

77. Interpret the expression as a definite integral over $[0, 1]$, and then evaluate the limit by evaluating the integral.

$$\lim_{\max \Delta x_k\to 0}\sum_{k=1}^{n} e^{x_k^*}\,\Delta x_k$$

78. (a) Divide the interval $[1, 2]$ into 5 subintervals of equal length, and use appropriate Riemann sums to show that

$$0.2\left[\tfrac{1}{1.2} + \tfrac{1}{1.4} + \tfrac{1}{1.6} + \tfrac{1}{1.8} + \tfrac{1}{2.0}\right] < \ln 2$$
$$< 0.2\left[\tfrac{1}{1.0} + \tfrac{1}{1.2} + \tfrac{1}{1.4} + \tfrac{1}{1.6} + \tfrac{1}{1.8}\right]$$

(b) Show that if the interval $[1, 2]$ is divided into n subintervals of equal length, then

$$\sum_{k=1}^{n}\frac{1}{n+k} < \ln 2 < \sum_{k=0}^{n-1}\frac{1}{n+k}$$

(c) Show that the difference between the two sums in part (b) is $1/(2n)$, and use this result to show that the sums in part (a) approximate $\ln 2$ with an error of at most 0.1.

(d) How large must n be to ensure that the sums in part (b) approximate $\ln 2$ to three decimal places?

79–80 Find the area under the curve $y = f(x)$ over the stated interval.

79. $f(x) = e^x$; $[1, 3]$ **80.** $f(x) = \dfrac{1}{x}$; $[1, e^3]$

81. Solve the initial-value problems.

(a) $\dfrac{dy}{dx} = \cos x - 5e^x$, $y(0) = 0$

(b) $\dfrac{dy}{dx} = xe^{x^2}$, $y(0) = 0$

82–84 Evaluate the integrals.

82. $\displaystyle\int_e^{e^2}\frac{dx}{x\ln x}$ **83.** $\displaystyle\int_0^1\frac{dx}{\sqrt{e^x}}$

84. $\displaystyle\int_0^{2/\sqrt{3}}\frac{1}{4+9x^2}\,dx$

85. Find the volume of the solid whose base is the region bounded between the curves $y = \sqrt{x}$ and $y = 1/\sqrt{x}$ for $1 \le x \le 4$ and whose cross sections perpendicular to the x-axis are squares.

86. Find the average value of $f(x) = e^x + e^{-x}$ over the interval $\left[\ln \frac{1}{2}, \ln 2\right]$.

87. In each part, prove the identity.

(a) $\cosh 3x = 4\cosh^3 x - 3\cosh x$

(b) $\cosh \frac{1}{2}x = \sqrt{\frac{1}{2}(\cosh x + 1)}$

(c) $\sinh \frac{1}{2}x = \pm\sqrt{\frac{1}{2}(\cosh x - 1)}$

88. Show that for any constant a, the function $y = \sinh(ax)$ satisfies the equation $y'' = a^2 y$.

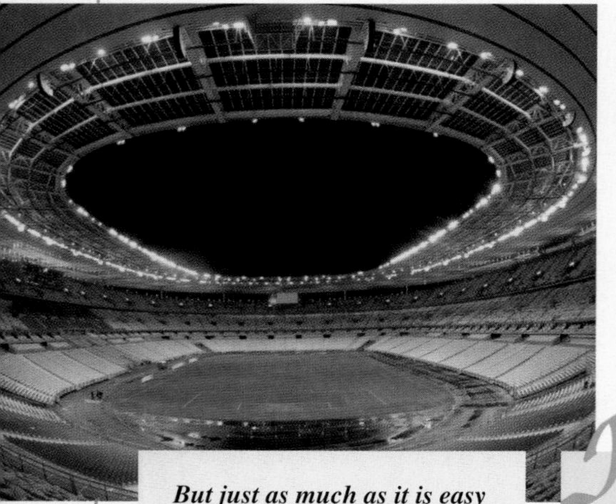

PRINCIPLES OF INTEGRAL EVALUATION

But just as much as it is easy to find the differential of a given quantity, so it is difficult to find the integral of a given differential. Moreover, sometimes we cannot say with certainty whether the integral of a given quantity can be found or not.

—Johann Bernoulli
Mathematician

In earlier chapters we obtained many basic integration formulas as an immediate consequence of the corresponding differentiation formulas. For example, knowing that the derivative of sin x is cos x enabled us to deduce that the integral of cos x is sin x. Subsequently, we expanded our integration repertoire by introducing the method of u-substitution. That method enabled us to integrate many functions by transforming the integrand of an unfamiliar integral into a familiar form. However, u-substitution alone is not adequate to handle the wide variety of integrals that arise in applications, so additional integration techniques are still needed. In this chapter we will discuss some of those techniques, and we will provide a more systematic procedure for attacking unfamiliar integrals. We will talk more about numerical approximations of definite integrals, and we will explore the idea of integrating over infinite intervals.

Photo: *The floating roof on the Stade de France sports complex is an ellipse. Finding the length of an ellipse involves numerical integration techniques introduced in this chapter.*

8.1 AN OVERVIEW OF INTEGRATION METHODS

In this section we will give a brief overview of methods for evaluating integrals, and we will review the integration formulas that were discussed in earlier sections.

METHODS FOR APPROACHING INTEGRATION PROBLEMS

There are three basic approaches for evaluating unfamiliar integrals:

- **Technology**—CAS programs such as *Mathematica*, *Maple*, and *Derive* are capable of evaluating extremely complicated integrals, and for both the computer and handheld calculator such programs are increasingly available.
- **Tables**—Prior to the development of CAS programs, scientists relied heavily on tables to evaluate difficult integrals arising in applications. Such tables were compiled over many years, incorporating the skills and experience of many people. One such table appears in the endpapers of this text, but more comprehensive tables appear in various reference books such as the *CRC Standard Mathematical Tables and Formulae*, CRC Press, Inc., 2002.

- **Transformation Methods**—Transformation methods are methods for converting unfamiliar integrals into familiar integrals. These include u-substitution, algebraic manipulation of the integrand, and other methods that we will discuss in this chapter.

None of the three methods is perfect; for example, CAS programs often encounter integrals that they cannot evaluate and they sometimes produce answers that are unnecessarily complicated, tables are not exhaustive and hence may not include a particular integral of interest, and transformation methods rely on human ingenuity that may prove to be inadequate in difficult problems.

In this chapter we will focus on transformation methods and tables, so it will *not be necessary* to have a CAS such as *Mathematica*, *Maple*, or *Derive*. However, if you have a CAS, then you can use it to confirm the results in the examples, and there are exercises that are designed to be solved with a CAS. If you have a CAS, keep in mind that many of the algorithms that it uses are based on the methods we will discuss here, so an understanding of these methods will help you to use your technology in a more informed way.

A REVIEW OF FAMILIAR INTEGRATION FORMULAS

The following is a list of basic integrals that we have encountered thus far:

CONSTANTS, POWERS, EXPONENTIALS

1. $\int du = u + C$

2. $\int a\,du = a\int du = au + C$

3. $\int u^r\,du = \dfrac{u^{r+1}}{r+1} + C,\ r \neq -1$

4. $\int \dfrac{du}{u} = \ln|u| + C$

5. $\int e^u\,du = e^u + C$

6. $\int b^u\,du = \dfrac{b^u}{\ln b} + C,\ b > 0, b \neq 1$

TRIGONOMETRIC FUNCTIONS

7. $\int \sin u\,du = -\cos u + C$

8. $\int \cos u\,du = \sin u + C$

9. $\int \sec^2 u\,du = \tan u + C$

10. $\int \csc^2 u\,du = -\cot u + C$

11. $\int \sec u \tan u\,du = \sec u + C$

12. $\int \csc u \cot u\,du = -\csc u + C$

13. $\int \tan u\,du = -\ln|\cos u| + C$

14. $\int \cot u\,du = \ln|\sin u| + C$

HYPERBOLIC FUNCTIONS

15. $\int \sinh u\,du = \cosh u + C$

16. $\int \cosh u\,du = \sinh u + C$

17. $\int \operatorname{sech}^2 u\,du = \tanh u + C$

18. $\int \operatorname{csch}^2 u\,du = -\coth u + C$

19. $\int \operatorname{sech} u \tanh u\,du = -\operatorname{sech} u + C$

20. $\int \operatorname{csch} u \coth u\,du = -\operatorname{csch} u + C$

ALGEBRAIC FUNCTIONS ($a > 0$)

21. $\int \dfrac{du}{\sqrt{a^2 - u^2}} = \sin^{-1}\dfrac{u}{a} + C \quad (|u| < a)$

22. $\int \dfrac{du}{a^2 + u^2} = \dfrac{1}{a}\tan^{-1}\dfrac{u}{a} + C$

23. $\displaystyle\int \frac{du}{u\sqrt{u^2-a^2}} = \frac{1}{a}\sec^{-1}\left|\frac{u}{a}\right| + C \qquad (0 < a < |u|)$

24. $\displaystyle\int \frac{du}{\sqrt{a^2+u^2}} = \ln(u+\sqrt{u^2+a^2}) + C$

25. $\displaystyle\int \frac{du}{\sqrt{u^2-a^2}} = \ln\left|u+\sqrt{u^2-a^2}\right| + C \qquad (0 < a < |u|)$

26. $\displaystyle\int \frac{du}{a^2-u^2} = \frac{1}{2a}\ln\left|\frac{a+u}{a-u}\right| + C$

27. $\displaystyle\int \frac{du}{u\sqrt{a^2-u^2}} = -\frac{1}{a}\ln\left|\frac{a+\sqrt{a^2-u^2}}{u}\right| + C \qquad (0 < |u| < a)$

28. $\displaystyle\int \frac{du}{u\sqrt{a^2+u^2}} = -\frac{1}{a}\ln\left|\frac{a+\sqrt{a^2+u^2}}{u}\right| + C$

Formula 25 is a generalization of a result in Theorem 7.8.6. Readers who did not cover Section 7.8 can ignore Formulas 24–28 for now, since we will develop other methods for obtaining them in this chapter.

✔QUICK CHECK EXERCISES 8.1 *(See page 513 for answers.)*

1. Use algebraic manipulation and (if necessary) u-substitution to integrate the function.

(a) $\displaystyle\int \frac{x+1}{x}\,dx =$ ______

(b) $\displaystyle\int \frac{x+2}{x+1}\,dx =$ ______

(c) $\displaystyle\int \frac{2x+1}{x^2+1}\,dx =$ ______

(d) $\displaystyle\int xe^{3\ln x}\,dx =$ ______

2. Use trigonometric identities and (if necessary) u-substitution to integrate the function.

(a) $\displaystyle\int \frac{1}{\csc x}\,dx =$ ______

(b) $\displaystyle\int \frac{1}{\cos^2 x}\,dx =$ ______

(c) $\displaystyle\int (\cot^2 x + 1)\,dx =$ ______

(d) $\displaystyle\int \frac{1}{\sec x + \tan x}\,dx =$ ______

3. Integrate the function.

(a) $\displaystyle\int \sqrt{x-1}\,dx =$ ______

(b) $\displaystyle\int e^{2x+1}\,dx =$ ______

(c) $\displaystyle\int (\sin^3 x\cos x + \sin x\cos^3 x)\,dx =$ ______

(d) $\displaystyle\int \frac{1}{(e^x+e^{-x})^2}\,dx =$ ______

EXERCISE SET 8.1

1–30 Evaluate the integrals by making appropriate u-substitutions and applying the formulas reviewed in this section.

1. $\displaystyle\int (4-2x)^3\,dx$

2. $\displaystyle\int 3\sqrt{4+2x}\,dx$

3. $\displaystyle\int x\sec^2(x^2)\,dx$

4. $\displaystyle\int 4x\tan(x^2)\,dx$

5. $\displaystyle\int \frac{\sin 3x}{2+\cos 3x}\,dx$

6. $\displaystyle\int \frac{1}{9+4x^2}\,dx$

7. $\displaystyle\int e^x\sinh(e^x)\,dx$

8. $\displaystyle\int \frac{\sec(\ln x)\tan(\ln x)}{x}\,dx$

9. $\displaystyle\int e^{\tan x}\sec^2 x\,dx$

10. $\displaystyle\int \frac{x}{\sqrt{1-x^4}}\,dx$

11. $\displaystyle\int \cos^5 5x\sin 5x\,dx$

12. $\displaystyle\int \frac{\cos x}{\sin x\sqrt{\sin^2 x+1}}\,dx$

13. $\displaystyle\int \frac{e^x}{\sqrt{4+e^{2x}}}\,dx$

14. $\displaystyle\int \frac{e^{\tan^{-1}x}}{1+x^2}\,dx$

15. $\int \frac{e^{\sqrt{x-1}}}{\sqrt{x-1}}\,dx$

16. $\int (x+1)\cot(x^2+2x)\,dx$

17. $\int \frac{\cosh\sqrt{x}}{\sqrt{x}}\,dx$

18. $\int \frac{dx}{x(\ln x)^2}$

19. $\int \frac{dx}{\sqrt{x}\,3^{\sqrt{x}}}$

20. $\int \sec(\sin\theta)\tan(\sin\theta)\cos\theta\,d\theta$

21. $\int \frac{\operatorname{csch}^2(2/x)}{x^2}\,dx$

22. $\int \frac{dx}{\sqrt{x^2-4}}$

23. $\int \frac{e^{-x}}{4-e^{-2x}}\,dx$

24. $\int \frac{\cos(\ln x)}{x}\,dx$

25. $\int \frac{e^x}{\sqrt{1-e^{2x}}}\,dx$

26. $\int \frac{\sinh(x^{-1/2})}{x^{3/2}}\,dx$

27. $\int \frac{x}{\csc(x^2)}\,dx$

28. $\int \frac{e^x}{\sqrt{4-e^{2x}}}\,dx$

29. $\int x4^{-x^2}\,dx$

30. $\int 2^{\pi x}\,dx$

FOCUS ON CONCEPTS

31. (a) Evaluate the integral $\int \sin x\cos x\,dx$ using the substitution $u=\sin x$.
(b) Evaluate the integral $\int \sin x\cos x\,dx$ using the identity $\sin 2x = 2\sin x\cos x$.
(c) Explain why your answers to parts (a) and (b) are consistent.

32. (a) Derive the identity

$$\frac{\operatorname{sech}^2 x}{1+\tanh^2 x} = \operatorname{sech} 2x$$

(b) Use the result in part (a) to evaluate $\int \operatorname{sech} x\,dx$.
(c) Derive the identity

$$\operatorname{sech} x = \frac{2e^x}{e^{2x}+1}$$

(d) Use the result in part (c) to evaluate $\int \operatorname{sech} x\,dx$.
(e) Explain why your answers to parts (b) and (d) are consistent.

33. (a) Derive the identity

$$\frac{\sec^2 x}{\tan x} = \frac{1}{\sin x\cos x}$$

(b) Use the identity $\sin 2x = 2\sin x\cos x$ along with the result in part (a) to evaluate $\int \csc x\,dx$.
(c) Use the identity $\cos x = \sin[(\pi/2)-x]$ along with your answer to part (a) to evaluate $\int \sec x\,dx$.

✔ QUICK CHECK ANSWERS 8.1

1. (a) $x+\ln|x|+C$ (b) $x+\ln|x+1|+C$ (c) $\ln(x^2+1)+\tan^{-1}x+C$ (d) $\frac{x^5}{5}+C$ 2. (a) $-\cos x + C$ (b) $\tan x + C$ (c) $-\cot x + C$ (d) $\ln(1+\sin x)+C$ 3. (a) $\frac{2}{3}(x-1)^{3/2}+C$ (b) $\frac{1}{2}e^{2x+1}+C$ (c) $\frac{1}{2}\sin^2 x + C$ (d) $\frac{1}{4}\tanh x + C$

8.2 INTEGRATION BY PARTS

In this section we will discuss an integration technique that is essentially an antiderivative formulation of the formula for differentiating a product of two functions.

THE PRODUCT RULE AND INTEGRATION BY PARTS

Our primary goal in this section is to develop a general method for attacking integrals of the form

$$\int f(x)g(x)\,dx$$

As a first step, let $G(x)$ be *any* antiderivative of $g(x)$. In this case $G'(x)=g(x)$, so the product rule for differentiating $f(x)G(x)$ can be expressed as

$$\frac{d}{dx}[f(x)G(x)] = f(x)g(x) + f'(x)G(x) \tag{1}$$

This implies that $f(x)G(x)$ is an antiderivative of the function on the right side of (1), so we can express (1) in integral form as

$$\int [f(x)g(x) + f'(x)G(x)]\,dx = f(x)G(x)$$

or, equivalently, as

$$\int f(x)g(x)\,dx = f(x)G(x) - \int f'(x)G(x)\,dx \tag{2}$$

This formula allows us to integrate $f(x)g(x)$ by integrating $f'(x)G(x)$ instead, and in many cases the net effect is to replace a difficult integration with an easier one. The application of this formula is called ***integration by parts***.

In practice, we usually rewrite (2) by letting

$$u = f(x), \quad du = f'(x)\,dx$$
$$v = G(x), \quad dv = G'(x)\,dx = g(x)\,dx$$

This yields the following alternative form for (2):

$$\int u\,dv = uv - \int v\,du \tag{3}$$

Note that in Example 1 we omitted the constant of integration in calculating v from dv. Had we included a constant of integration, it would have eventually dropped out. This is always the case in integration by parts [Exercise 62(b)], so it is common to omit the constant at this stage of the computation. However, there are certain cases in which making a clever choice of a constant of integration to include with v can simplify the computation of $\int v\,du$ (Exercises 63–65).

Perform the integration in Example 1 using Formula (2).

► **Example 1** Use integration by parts to evaluate $\int x \cos x\,dx$.

Solution. We will apply Formula (3). The first step is to make a choice for u and dv to put the given integral in the form $\int u\,dv$. We will let

$$u = x \quad \text{and} \quad dv = \cos x\,dx$$

(Other possibilities will be considered later.) The second step is to compute du from u and v from dv. This yields

$$du = dx \quad \text{and} \quad v = \int dv = \int \cos x\,dx = \sin x$$

The third step is to apply Formula (3). This yields

$$\int \underbrace{x}_{u}\,\underbrace{\cos x\,dx}_{dv} = \underbrace{x}_{u}\,\underbrace{\sin x}_{v} - \int \underbrace{\sin x}_{v}\,\underbrace{dx}_{du}$$
$$= x\sin x - (-\cos x) + C = x\sin x + \cos x + C \;◄$$

GUIDELINES FOR INTEGRATION BY PARTS

The main goal in integration by parts is to choose u and dv to obtain a new integral that is easier to evaluate than the original. In general, there are no hard and fast rules for doing this; it is mainly a matter of experience that comes from lots of practice. A strategy that often works is to choose u and dv so that u becomes "simpler" when differentiated, while leaving a dv that can be readily integrated to obtain v. Thus, for the integral $\int x\cos x\,dx$ in Example 1, both goals were achieved by letting $u = x$ and $dv = \cos x\,dx$. In contrast, $u = \cos x$ would not have been a good first choice in that example, since $du/dx = -\sin x$ is no simpler than u. Indeed, had we chosen

$$u = \cos x \qquad dv = x\,dx$$
$$du = -\sin x\,dx \qquad v = \int x\,dx = \frac{x^2}{2}$$

then we would have obtained

$$\int x\cos x\,dx = \frac{x^2}{2}\cos x - \int \frac{x^2}{2}(-\sin x)\,dx = \frac{x^2}{2}\cos x + \frac{1}{2}\int x^2 \sin x\,dx$$

For this choice of u and dv, the new integral is actually more complicated than the original.

The LIATE method is discussed in the article "A Technique for Integration by Parts," *American Mathematical Monthly*, Vol. 90, 1983, pp. 210–211, by Herbert Kasube.

There is another useful strategy for choosing u and dv that can be applied when the integrand is a product of two functions from *different* categories in the list

Logarithmic, Inverse trigonometric, Algebraic, Trigonometric, Exponential

In this case you will often be successful if you take u to be the function whose category occurs earlier in the list and take dv to the rest of the integrand. The acronym LIATE will help you to remember the order. The method does not work all the time, but it works often enough to be useful.

Note, for example, that the integrand in Example 1 consists of the product of the *algebraic* function x and the *trigonometric* function $\cos x$. Thus, the LIATE method suggests that we should let $u = x$ and $dv = \cos x\, dx$, which proved to be a successful choice.

► **Example 2** Evaluate $\int xe^x\, dx$.

Solution. In this case the integrand is the product of the algebraic function x with the exponential function e^x. According to LIATE we should let

$$u = x \quad \text{and} \quad dv = e^x\, dx$$

so that

$$du = dx \quad \text{and} \quad v = \int e^x\, dx = e^x$$

Thus, from (3)

$$\int xe^x\, dx = \int u\, dv = uv - \int v\, du = xe^x - \int e^x\, dx = xe^x - e^x + C \blacktriangleleft$$

► **Example 3** Evaluate $\int \ln x\, dx$.

Solution. One choice is to let $u = 1$ and $dv = \ln x\, dx$. But with this choice finding v is equivalent to evaluating $\int \ln x\, dx$ and we have gained nothing. Therefore, the only reasonable choice is to let

$$u = \ln x \qquad dv = dx$$
$$du = \frac{1}{x}\, dx \qquad v = \int dx = x$$

With this choice it follows from (3) that

$$\int \ln x\, dx = \int u\, dv = uv - \int v\, du = x \ln x - \int dx = x \ln x - x + C \blacktriangleleft$$

REPEATED INTEGRATION BY PARTS

It is sometimes necessary to use integration by parts more than once in the same problem.

► **Example 4** Evaluate $\int x^2 e^{-x}\, dx$.

Solution. Let

$$u = x^2, \quad dv = e^{-x}\, dx, \quad du = 2x\, dx, \quad v = \int e^{-x}\, dx = -e^{-x}$$

so that from (3)

$$\int x^2e^{-x}\,dx = \int u\,dv = uv - \int v\,du$$
$$= x^2(-e^{-x}) - \int -e^{-x}(2x)\,dx$$
$$= -x^2e^{-x} + 2\int xe^{-x}\,dx \tag{4}$$

The last integral is similar to the original except that we have replaced x^2 by x. Another integration by parts applied to $\int xe^{-x}\,dx$ will complete the problem. We let

$$u = x, \quad dv = e^{-x}\,dx, \quad du = dx, \quad v = \int e^{-x}\,dx = -e^{-x}$$

so that

$$\int xe^{-x}\,dx = x(-e^{-x}) - \int -e^{-x}\,dx = -xe^{-x} + \int e^{-x}\,dx = -xe^{-x} - e^{-x} + C$$

Finally, substituting this into the last line of (4) yields

$$\int x^2e^{-x}\,dx = -x^2e^{-x} + 2\int xe^{-x}\,dx = -x^2e^{-x} + 2(-xe^{-x} - e^{-x}) + C$$
$$= -(x^2 + 2x + 2)e^{-x} + C \blacktriangleleft$$

The LIATE method suggests that integrals of the form

$$\int e^{ax}\sin bx\,dx \quad \text{and} \quad \int e^{ax}\cos bx\,dx$$

can be evaluated by letting $u = \sin bx$ or $u = \cos bx$ and $dv = e^{ax}\,dx$. However, this will require a technique that deserves special attention.

▶ **Example 5** Evaluate $\int e^x \cos x\,dx$.

Solution. Let

$$u = \cos x, \quad dv = e^x\,dx, \quad du = -\sin x\,dx, \quad v = \int e^x\,dx = e^x$$

Thus,

$$\int e^x\cos x\,dx = \int u\,dv = uv - \int v\,du = e^x\cos x + \int e^x\sin x\,dx \tag{5}$$

Since the integral $\int e^x \sin x\,dx$ is similar in form to the original integral $\int e^x\cos x\,dx$, it seems that nothing has been accomplished. However, let us integrate this new integral by parts. We let

$$u = \sin x, \quad dv = e^x\,dx, \quad du = \cos x\,dx, \quad v = \int e^x\,dx = e^x$$

Thus,

$$\int e^x\sin x\,dx = \int u\,dv = uv - \int v\,du = e^x\sin x - \int e^x\cos x\,dx$$

Together with Equation (5) this yields

$$\int e^x\cos x\,dx = e^x\cos x + e^x\sin x - \int e^x\cos x\,dx \tag{6}$$

which is an equation we can solve for the unknown integral. We obtain

$$2\int e^x \cos x\,dx = e^x \cos x + e^x \sin x$$

and hence

$$\int e^x \cos x\,dx = \tfrac{1}{2}e^x \cos x + \tfrac{1}{2}e^x \sin x + C \blacktriangleleft$$

A TABULAR METHOD FOR REPEATED INTEGRATION BY PARTS

Integrals of the form

$$\int p(x)f(x)\,dx$$

where $p(x)$ is a polynomial, can sometimes be evaluated using repeated integration by parts in which u is taken to be $p(x)$ or one of its derivatives at each stage. Since du is computed by differentiating u, the repeated differentiation of $p(x)$ will eventually produce 0, at which point you may be left with a simplified integration problem. A convenient method for organizing the computations into two columns is called ***tabular integration by parts***.

More information on tabular integration by parts can be found in the articles "Tabular Integration by Parts," *College Mathematics Journal*, Vol. 21, 1990, pp. 307–311, by David Horowitz and "More on Tabular Integration by Parts," *College Mathematics Journal*, Vol. 22, 1991, pp. 407–410, by Leonard Gillman.

Tabular Integration by Parts

Step 1. Differentiate $p(x)$ repeatedly until you obtain 0, and list the results in the first column.

Step 2. Integrate $f(x)$ repeatedly and list the results in the second column.

Step 3. Draw an arrow from each entry in the first column to the entry that is one row down in the second column.

Step 4. Label the arrows with alternating $+$ and $-$ signs, starting with a $+$.

Step 5. For each arrow, form the product of the expressions at its tip and tail and then multiply that product by $+1$ or -1 in accordance with the sign on the arrow. Add the results to obtain the value of the integral.

This process is illustrated in Figure 8.2.1 for the integral $\int (x^2 - x)\cos x\,dx$.

REPEATED DIFFERENTIATION		REPEATED INTEGRATION
$x^2 - x$	$+$	$\cos x$
$2x - 1$	$-$	$\sin x$
2	$+$	$-\cos x$
0		$-\sin x$

$$\int (x^2 - x)\cos x\,dx = (x^2 - x)\sin x + (2x - 1)\cos x - 2\sin x + C$$
$$= (x^2 - x - 2)\sin x + (2x - 1)\cos x + C$$

Figure 8.2.1

▶ **Example 6** In Example 9 of Section 5.3 we evaluated $\int x^2\sqrt{x-1}\,dx$ using u-substitution. Evaluate this integral using tabular integration by parts.

Solution.

REPEATED DIFFERENTIATION		REPEATED INTEGRATION
x^2	+	$(x-1)^{1/2}$
$2x$	−	$\frac{2}{3}(x-1)^{3/2}$
2	+	$\frac{4}{15}(x-1)^{5/2}$
0		$\frac{8}{105}(x-1)^{7/2}$

Thus, it follows that

$$\int x^2\sqrt{x-1}\,dx = \tfrac{2}{3}x^2(x-1)^{3/2} - \tfrac{8}{15}x(x-1)^{5/2} + \tfrac{16}{105}(x-1)^{7/2} + C \blacktriangleleft$$

The result obtained in Example 6 looks quite different from that obtained in Example 9 of Section 5.3. Show that the two answers are equivalent.

INTEGRATION BY PARTS FOR DEFINITE INTEGRALS

For definite integrals the formula corresponding to (3) is

$$\int_a^b u\,dv = uv\Big]_a^b - \int_a^b v\,du \tag{7}$$

It is important to keep in mind that the variables u and v in this formula are functions of x and that the limits of integration in (7) are limits on the variable x. Sometimes it is helpful to emphasize this by writing (7) as

$$\int_{x=a}^b u\,dv = uv\Big]_{x=a}^b - \int_{x=a}^b v\,du \tag{8}$$

The next example illustrates how integration by parts can be used to integrate the inverse trigonometric functions.

▶ **Example 7** Evaluate $\displaystyle\int_0^1 \tan^{-1} x\,dx$.

Solution. Let

$$u = \tan^{-1} x,\quad dv = dx,\quad du = \frac{1}{1+x^2}\,dx,\quad v = x$$

Thus,

$$\int_0^1 \tan^{-1} x\,dx = \int_0^1 u\,dv = uv\Big]_0^1 - \int_0^1 v\,du$$

$$= x\tan^{-1} x\Big]_0^1 - \int_0^1 \frac{x}{1+x^2}\,dx$$

The limits of integration refer to x; that is, $x = 0$ and $x = 1$.

But

$$\int_0^1 \frac{x}{1+x^2}\,dx = \frac{1}{2}\int_0^1 \frac{2x}{1+x^2}\,dx = \frac{1}{2}\ln(1+x^2)\Big]_0^1 = \frac{1}{2}\ln 2$$

so

$$\int_0^1 \tan^{-1} x\,dx = x\tan^{-1} x\Big]_0^1 - \frac{1}{2}\ln 2 = \left(\frac{\pi}{4} - 0\right) - \frac{1}{2}\ln 2 = \frac{\pi}{4} - \ln\sqrt{2} \blacktriangleleft$$

REDUCTION FORMULAS

Integration by parts can be used to derive ***reduction formulas*** for integrals. These are formulas that express an integral involving a power of a function in terms of an integral that involves a *lower* power of that function. For example, if n is a positive integer and $n \geq 2$, then integration by parts can be used to obtain the reduction formulas

$$\int \sin^n x\,dx = -\frac{1}{n}\sin^{n-1} x\cos x + \frac{n-1}{n}\int \sin^{n-2} x\,dx \tag{9}$$

$$\int \cos^n x\,dx = \frac{1}{n}\cos^{n-1} x\sin x + \frac{n-1}{n}\int \cos^{n-2} x\,dx \tag{10}$$

To illustrate how such formulas can be obtained, let us derive (10). We begin by writing $\cos^n x$ as $\cos^{n-1} x \cdot \cos x$ and letting

$$\begin{aligned} u &= \cos^{n-1} x & dv &= \cos x\,dx \\ du &= (n-1)\cos^{n-2} x(-\sin x)\,dx & v &= \sin x \\ &= -(n-1)\cos^{n-2} x\sin x\,dx \end{aligned}$$

so that

$$\begin{aligned} \int \cos^n x\,dx = \int \cos^{n-1} x\cos x\,dx &= \int u\,dv = uv - \int v\,du \\ &= \cos^{n-1} x\sin x + (n-1)\int \sin^2 x\cos^{n-2} x\,dx \\ &= \cos^{n-1} x\sin x + (n-1)\int (1-\cos^2 x)\cos^{n-2} x\,dx \\ &= \cos^{n-1} x\sin x + (n-1)\int \cos^{n-2} x\,dx - (n-1)\int \cos^n x\,dx \end{aligned}$$

Moving the last term on the right to the left side yields

$$n\int \cos^n x\,dx = \cos^{n-1} x\sin x + (n-1)\int \cos^{n-2} x\,dx$$

from which (10) follows. The derivation of reduction formula (9) is similar (Exercise 57).

Reduction formulas (9) and (10) reduce the exponent of sine (or cosine) by 2. Thus, if the formulas are applied repeatedly, the exponent can eventually be reduced to 0 if n is even or 1 if n is odd, at which point the integration can be completed. We will discuss this method in more detail in the next section, but for now, here is an example that illustrates how reduction formulas work.

▸ **Example 8** Evaluate $\int \cos^4 x\,dx$.

Solution. From (10) with $n = 4$

$$\begin{aligned} \int \cos^4 x\,dx &= \tfrac{1}{4}\cos^3 x\sin x + \tfrac{3}{4}\textstyle\int \cos^2 x\,dx && \text{Now apply (10) with } n = 2. \\ &= \tfrac{1}{4}\cos^3 x\sin x + \tfrac{3}{4}\left(\tfrac{1}{2}\cos x\sin x + \tfrac{1}{2}\textstyle\int dx\right) \\ &= \tfrac{1}{4}\cos^3 x\sin x + \tfrac{3}{8}\cos x\sin x + \tfrac{3}{8}x + C \end{aligned}$$ ◂

✔QUICK CHECK EXERCISES 8.2
(See page 522 for answers.)

1. (a) If $G'(x) = g(x)$, then
$$\int f(x)g(x)\,dx = f(x)G(x) - \underline{\qquad}$$
(b) If $u = f(x)$ and $v = G(x)$, then the formula in part (a) can be written in the form $\int u\,dv = \underline{\qquad}$.

2. Find an appropriate choice of u and dv for integration by parts of each integral. Do not evaluate the integral.
(a) $\int x \ln x\,dx$; $u = \underline{\qquad}$, $dv = \underline{\qquad}$
(b) $\int (x-2)\sin x\,dx$; $u = \underline{\qquad}$, $dv = \underline{\qquad}$
(c) $\int \sin^{-1} x\,dx$; $u = \underline{\qquad}$, $dv = \underline{\qquad}$
(d) $\int \frac{x}{\sqrt{x-1}}\,dx$; $u = \underline{\qquad}$, $dv = \underline{\qquad}$

3. Use integration by parts to evaluate the integral.
(a) $\int xe^{2x}\,dx$ (b) $\int \ln(x-1)\,dx$
(c) $\int_0^{\pi/6} x \sin 3x\,dx$

4. Use a reduction formula to evaluate $\int \sin^3 x\,dx$.

EXERCISE SET 8.2

1–40 Evaluate the integral.

1. $\int xe^{-2x}\,dx$
2. $\int xe^{3x}\,dx$
3. $\int x^2e^x\,dx$
4. $\int x^2e^{-2x}\,dx$
5. $\int x\sin 3x\,dx$
6. $\int x\cos 2x\,dx$
7. $\int x^2\cos x\,dx$
8. $\int x^2\sin x\,dx$
9. $\int x\ln x\,dx$
10. $\int \sqrt{x}\ln x\,dx$
11. $\int (\ln x)^2\,dx$
12. $\int \frac{\ln x}{\sqrt{x}}\,dx$
13. $\int \ln(3x-2)\,dx$
14. $\int \ln(x^2+4)\,dx$
15. $\int \sin^{-1} x\,dx$
16. $\int \cos^{-1}(2x)\,dx$
17. $\int \tan^{-1}(3x)\,dx$
18. $\int x\tan^{-1} x\,dx$
19. $\int e^x\sin x\,dx$
20. $\int e^{3x}\cos 2x\,dx$
21. $\int e^{ax}\sin bx\,dx$
22. $\int e^{-3\theta}\sin 5\theta\,d\theta$
23. $\int \sin(\ln x)\,dx$
24. $\int \cos(\ln x)\,dx$
25. $\int x\sec^2 x\,dx$
26. $\int x\tan^2 x\,dx$
27. $\int x^3e^{x^2}\,dx$
28. $\int \frac{xe^x}{(x+1)^2}\,dx$
29. $\int_0^2 xe^{2x}\,dx$
30. $\int_0^1 xe^{-5x}\,dx$
31. $\int_1^e x^2\ln x\,dx$
32. $\int_{\sqrt{e}}^e \frac{\ln x}{x^2}\,dx$
33. $\int_{-1}^1 \ln(x+2)\,dx$
34. $\int_0^{\sqrt{3}/2} \sin^{-1} x\,dx$
35. $\int_2^4 \sec^{-1}\sqrt{\theta}\,d\theta$
36. $\int_1^2 x\sec^{-1} x\,dx$
37. $\int_0^\pi x\sin 2x\,dx$
38. $\int_0^\pi (x + x\cos x)\,dx$
39. $\int_1^3 \sqrt{x}\tan^{-1}\sqrt{x}\,dx$
40. $\int_0^2 \ln(x^2+1)\,dx$

41. In each part, evaluate the integral by making a u-substitution and then integrating by parts.
(a) $\int e^{\sqrt{x}}\,dx$ (b) $\int \cos\sqrt{x}\,dx$

42. Prove that tabular integration by parts gives the correct answer for
$$\int p(x)q(x)\,dx$$
where $p(x)$ is any *quadratic polynomial* and $q(x)$ is any function that can be repeatedly integrated.

43–46 Evaluate the integral using tabular integration by parts.

43. $\int (3x^2 - x + 2)e^{-x}\,dx$
44. $\int (x^2+x+1)\sin x\,dx$
45. $\int 4x^4\sin 2x\,dx$
46. $\int x^3\sqrt{2x+1}\,dx$

47. Evaluate the integral $\int \sin x\cos x\,dx$ using
(a) integration by parts (b) the substitution $u = \sin x$.

48. Evaluate the integral
$$\int_0^1 \frac{x^3}{\sqrt{x^2+1}}\,dx$$
using
(a) integration by parts
(b) the substitution $u=\sqrt{x^2+1}$.

49. (a) Find the area of the region enclosed by $y=\ln x$, the line $x=e$, and the x-axis.
(b) Find the volume of the solid generated when the region in part (a) is revolved about the x-axis.

50. Find the area of the region between $y=x\sin x$ and $y=x$ for $0\le x\le \pi/2$.

51. Find the volume of the solid generated when the region between $y=\sin x$ and $y=0$ for $0\le x\le\pi$ is revolved about the y-axis.

52. Find the volume of the solid generated when the region enclosed between $y=\cos x$ and $y=0$ for $0\le x\le\pi/2$ is revolved about the y-axis.

53. A particle moving along the x-axis has velocity function $v(t)=t^3\sin t$. How far does the particle travel from time $t=0$ to $t=\pi$?

54. The study of sawtooth waves in electrical engineering leads to integrals of the form
$$\int_{-\pi/\omega}^{\pi/\omega} t\sin(k\omega t)\,dt$$
where k is an integer and ω is a nonzero constant. Evaluate the integral.

55. Use reduction formula (9) to evaluate
(a) $\displaystyle\int \sin^4 x\,dx$ (b) $\displaystyle\int_0^{\pi/2} \sin^5 x\,dx$.

56. Use reduction formula (10) to evaluate
(a) $\displaystyle\int \cos^5 x\,dx$ (b) $\displaystyle\int_0^{\pi/2} \cos^6 x\,dx$.

57. Derive reduction formula (9).

58. In each part, use integration by parts or other methods to derive the reduction formula.
(a) $\displaystyle\int \sec^n x\,dx = \frac{\sec^{n-2}x\tan x}{n-1}+\frac{n-2}{n-1}\int \sec^{n-2}x\,dx$
(b) $\displaystyle\int \tan^n x\,dx = \frac{\tan^{n-1}x}{n-1}-\int \tan^{n-2}x\,dx$
(c) $\displaystyle\int x^n e^x\,dx = x^n e^x - n\int x^{n-1}e^x\,dx$

59–60 Use the reduction formulas in Exercise 58 to evaluate the integrals.

59. (a) $\displaystyle\int \tan^4 x\,dx$ (b) $\displaystyle\int \sec^4 x\,dx$ (c) $\displaystyle\int x^3 e^x\,dx$

60. (a) $\displaystyle\int x^2 e^{3x}\,dx$ (b) $\displaystyle\int_0^1 xe^{-\sqrt{x}}\,dx$
[*Hint:* First make a substitution.]

61. Let f be a function whose second derivative is continuous on $[-1,1]$. Show that
$$\int_{-1}^1 xf''(x)\,dx = f'(1)+f'(-1)-f(1)+f(-1)$$

FOCUS ON CONCEPTS

62. (a) In the integral $\int x\cos x\,dx$, let
$$u=x,\quad dv=\cos x\,dx,$$
$$du=dx,\quad v=\sin x + C_1$$
Show that the constant C_1 cancels out, thus giving the same solution obtained by omitting C_1.
(b) Show that in general
$$uv-\int v\,du = u(v+C_1)-\int (v+C_1)\,du$$
thereby justifying the omission of the constant of integration when calculating v in integration by parts.

63. Evaluate $\int \ln(x+1)\,dx$ using integration by parts. Simplify the computation of $\int v\,du$ by introducing a constant of integration $C_1=1$ when going from dv to v.

64. Evaluate $\int \ln(3x-2)\,dx$ using integration by parts. Simplify the computation of $\int v\,du$ by introducing a constant of integration $C_1=-\frac{2}{3}$ when going from dv to v. Compare your solution with your answer to Exercise 13.

65. Evaluate $\int x\tan^{-1}x\,dx$ using integration by parts. Simplify the computation of $\int v\,du$ by introducing a constant of integration $C_1=\frac{1}{2}$ when going from dv to v.

66. What equation results if integration by parts is applied to the integral
$$\int \frac{1}{x\ln x}\,dx$$
with the choices
$$u=\frac{1}{\ln x}\quad\text{and}\quad dv=\frac{1}{x}\,dx?$$
In what sense is this equation true? In what sense is it false?

67. Recall from Theorem 7.3.2 and the discussion preceding it that if $f'(x)>0$, then the function f is increasing and has an inverse function. Parts (a), (b), and (c) of this problem show that if this condition is satisfied and if f' is continuous, then a definite integral of f^{-1} can be expressed in terms of a definite integral of f.
(a) Use integration by parts to show that
$$\int_a^b f(x)\,dx = bf(b)-af(a)-\int_a^b xf'(x)\,dx$$
(b) Use the result in part (a) to show that if $y=f(x)$, then
$$\int_a^b f(x)\,dx = bf(b)-af(a)-\int_{f(a)}^{f(b)} f^{-1}(y)\,dy$$

(c) Show that if we let $\alpha = f(a)$ and $\beta = f(b)$, then the result in part (b) can be written as

$$\int_{\alpha}^{\beta} f^{-1}(x)\,dx = \beta f^{-1}(\beta) - \alpha f^{-1}(\alpha) - \int_{f^{-1}(\alpha)}^{f^{-1}(\beta)} f(x)\,dx$$

68. In each part, use the result in Exercise 67 to obtain the equation, and then confirm that the equation is correct by performing the integrations.

(a) $\displaystyle\int_0^{1/2} \sin^{-1} x\,dx = \tfrac{1}{2}\sin^{-1}\left(\tfrac{1}{2}\right) - \int_0^{\pi/6} \sin x\,dx$

(b) $\displaystyle\int_e^{e^2} \ln x\,dx = (2e^2 - e) - \int_1^2 e^x\,dx$

✔QUICK CHECK ANSWERS 8.2

1. (a) $\int f'(x)G(x)\,dx$ (b) $uv - \int v\,du$ **2.** (a) $\ln x$; $x\,dx$ (b) $x-2$; $\sin x\,dx$ (c) $\sin^{-1} x$; dx (d) x; $\dfrac{1}{\sqrt{x-1}}\,dx$

3. (a) $\left(\dfrac{x}{2} - \dfrac{1}{4}\right)e^{2x} + C$ (b) $(x-1)\ln(x-1) - x + C$ (c) $\frac{1}{9}$ **4.** $-\frac{1}{3}\sin^2 x\cos x - \frac{2}{3}\cos x + C$

8.3 TRIGONOMETRIC INTEGRALS

In the last section we derived reduction formulas for integrating positive integer powers of sine, cosine, tangent, and secant. In this section we will show how to work with those reduction formulas, and we will discuss methods for integrating other kinds of integrals that involve trigonometric functions.

INTEGRATING POWERS OF SINE AND COSINE

We begin by recalling two reduction formulas from the preceding section.

$$\int \sin^n x\,dx = -\frac{1}{n}\sin^{n-1} x\cos x + \frac{n-1}{n}\int \sin^{n-2} x\,dx \tag{1}$$

$$\int \cos^n x\,dx = \frac{1}{n}\cos^{n-1} x\sin x + \frac{n-1}{n}\int \cos^{n-2} x\,dx \tag{2}$$

In the case where $n = 2$, these formulas yield

$$\int \sin^2 x\,dx = -\tfrac{1}{2}\sin x\cos x + \tfrac{1}{2}\int dx = \tfrac{1}{2}x - \tfrac{1}{2}\sin x\cos x + C \tag{3}$$

$$\int \cos^2 x\,dx = \tfrac{1}{2}\cos x\sin x + \tfrac{1}{2}\int dx = \tfrac{1}{2}x + \tfrac{1}{2}\sin x\cos x + C \tag{4}$$

Alternative forms of these integration formulas can be derived from the trigonometric identities

$$\sin^2 x = \tfrac{1}{2}(1 - \cos 2x) \quad \text{and} \quad \cos^2 x = \tfrac{1}{2}(1 + \cos 2x) \tag{5–6}$$

which follow from the double-angle formulas

$$\cos 2x = 1 - 2\sin^2 x \quad \text{and} \quad \cos 2x = 2\cos^2 x - 1$$

These identities yield

$$\int \sin^2 x\,dx = \tfrac{1}{2}\int (1 - \cos 2x)\,dx = \tfrac{1}{2}x - \tfrac{1}{4}\sin 2x + C \tag{7}$$

$$\int \cos^2 x\,dx = \tfrac{1}{2}\int (1 + \cos 2x)\,dx = \tfrac{1}{2}x + \tfrac{1}{4}\sin 2x + C \tag{8}$$

Observe that the antiderivatives in Formulas (3) and (4) involve both sines and cosines, whereas those in (7) and (8) involve sines alone. However, the apparent discrepancy is easy to resolve by using the identity

$$\sin 2x = 2\sin x\cos x$$

to rewrite (7) and (8) in forms (3) and (4), or conversely.

In the case where $n = 3$, the reduction formulas for integrating $\sin^3 x$ and $\cos^3 x$ yield

$$\int \sin^3 x\,dx = -\tfrac{1}{3}\sin^2 x\cos x + \tfrac{2}{3}\int \sin x\,dx = -\tfrac{1}{3}\sin^2 x\cos x - \tfrac{2}{3}\cos x + C \tag{9}$$

$$\int \cos^3 x\,dx = \tfrac{1}{3}\cos^2 x\sin x + \tfrac{2}{3}\int \cos x\,dx = \tfrac{1}{3}\cos^2 x\sin x + \tfrac{2}{3}\sin x + C \tag{10}$$

If desired, Formula (9) can be expressed in terms of cosines alone by using the identity $\sin^2 x = 1 - \cos^2 x$, and Formula (10) can be expressed in terms of sines alone by using the identity $\cos^2 x = 1 - \sin^2 x$. We leave it for you to do this and confirm that

TECHNOLOGY MASTERY

The *Maple* CAS produces forms (11) and (12) when asked to integrate $\sin^3 x$ and $\cos^3 x$, but *Mathematica* produces

$$\int \sin^3 x\,dx = -\tfrac{3}{4}\cos x + \tfrac{1}{12}\cos 3x + C$$

$$\int \cos^3 x\,dx = \tfrac{3}{4}\sin x + \tfrac{1}{12}\sin 3x + C$$

Use trigonometric identities to reconcile the results of the two programs.

$$\int \sin^3 x\,dx = \tfrac{1}{3}\cos^3 x - \cos x + C \tag{11}$$

$$\int \cos^3 x\,dx = \sin x - \tfrac{1}{3}\sin^3 x + C \tag{12}$$

We leave it as an exercise to obtain the following formulas by first applying the reduction formulas, and then using appropriate trigonometric identities.

$$\int \sin^4 x\,dx = \tfrac{3}{8}x - \tfrac{1}{4}\sin 2x + \tfrac{1}{32}\sin 4x + C \tag{13}$$

$$\int \cos^4 x\,dx = \tfrac{3}{8}x + \tfrac{1}{4}\sin 2x + \tfrac{1}{32}\sin 4x + C \tag{14}$$

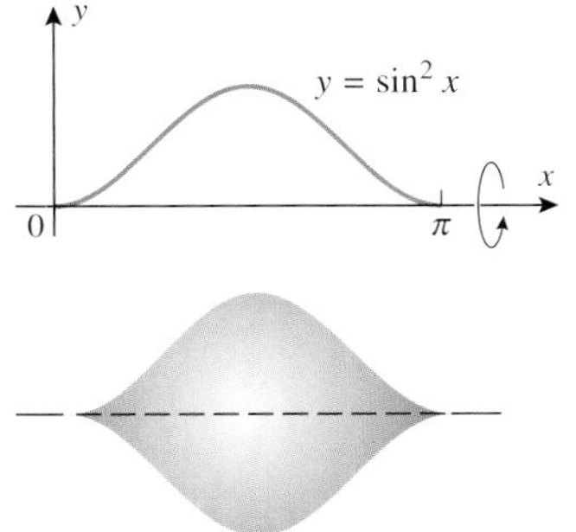

Figure 8.3.1

▶ **Example 1** Find the volume V of the solid that is obtained when the region under the curve $y = \sin^2 x$ over the interval $[0, \pi]$ is revolved about the x-axis (Figure 8.3.1).

Solution. Using the method of disks, Formula (5) of Section 6.2, and Formula (13) above yields

$$V = \int_0^{\pi} \pi\sin^4 x\,dx = \pi\left[\tfrac{3}{8}x - \tfrac{1}{4}\sin 2x + \tfrac{1}{32}\sin 4x\right]_0^{\pi} = \tfrac{3}{8}\pi^2 \blacktriangleleft$$

INTEGRATING PRODUCTS OF SINES AND COSINES

If m and n are positive integers, then the integral

$$\int \sin^m x\cos^n x\,dx$$

can be evaluated by one of the three procedures stated in Table 8.3.1, depending on whether m and n are odd or even.

▶ **Example 2** Evaluate

$$\text{(a)}\ \int \sin^4 x\cos^5 x\,dx \qquad \text{(b)}\ \int \sin^4 x\cos^4 x\,dx$$

Table 8.3.1

$\int \sin^m x \cos^n x\, dx$	PROCEDURE	RELEVANT IDENTITIES
n odd	• Split off a factor of $\cos x$. • Apply the relevant identity. • Make the substitution $u = \sin x$.	$\cos^2 x = 1 - \sin^2 x$
m odd	• Split off a factor of $\sin x$. • Apply the relevant identity. • Make the substitution $u = \cos x$.	$\sin^2 x = 1 - \cos^2 x$
$\begin{cases} m \text{ even} \\ n \text{ even} \end{cases}$	• Use the relevant identities to reduce the powers on $\sin x$ and $\cos x$.	$\begin{cases} \sin^2 x = \frac{1}{2}(1 - \cos 2x) \\ \cos^2 x = \frac{1}{2}(1 + \cos 2x) \end{cases}$

Solution (a). Since $n = 5$ is odd, we will follow the first procedure in Table 8.3.1:

$$\begin{aligned}
\int \sin^4 x \cos^5 x\, dx &= \int \sin^4 x \cos^4 x \cos x\, dx \\
&= \int \sin^4 x(1 - \sin^2 x)^2 \cos x\, dx \\
&= \int u^4(1 - u^2)^2\, du \\
&= \int (u^4 - 2u^6 + u^8)\, du \\
&= \tfrac{1}{5}u^5 - \tfrac{2}{7}u^7 + \tfrac{1}{9}u^9 + C \\
&= \tfrac{1}{5}\sin^5 x - \tfrac{2}{7}\sin^7 x + \tfrac{1}{9}\sin^9 x + C
\end{aligned}$$

Solution (b). Since $m = n = 4$, both exponents are even, so we will follow the third procedure in Table 8.3.1:

$$\begin{aligned}
\int \sin^4 x \cos^4 x\, dx &= \int (\sin^2 x)^2(\cos^2 x)^2\, dx \\
&= \int \left(\tfrac{1}{2}[1 - \cos 2x]\right)^2 \left(\tfrac{1}{2}[1 + \cos 2x]\right)^2 dx \\
&= \tfrac{1}{16}\int (1 - \cos^2 2x)^2\, dx \\
&= \tfrac{1}{16}\int \sin^4 2x\, dx
\end{aligned}$$

Note that this can be obtained more directly from the original integral using the identity $\sin x \cos x = \frac{1}{2}\sin 2x$.

$$= \tfrac{1}{32}\int \sin^4 u\, du$$

$u = 2x$
$du = 2\,dx$ or $dx = \frac{1}{2}\,du$

$$= \tfrac{1}{32}\left(\tfrac{3}{8}u - \tfrac{1}{4}\sin 2u + \tfrac{1}{32}\sin 4u\right) + C$$

Formula (13)

$$= \tfrac{3}{128}x - \tfrac{1}{128}\sin 4x + \tfrac{1}{1024}\sin 8x + C \blacktriangleleft$$

Integrals of the form

$$\int \sin mx \cos nx\, dx, \quad \int \sin mx \sin nx\, dx, \quad \int \cos mx \cos nx\, dx \tag{15}$$

can be found by using the trigonometric identities

$$\sin\alpha\cos\beta = \tfrac{1}{2}[\sin(\alpha-\beta)+\sin(\alpha+\beta)] \tag{16}$$

$$\sin\alpha\sin\beta = \tfrac{1}{2}[\cos(\alpha-\beta)-\cos(\alpha+\beta)] \tag{17}$$

$$\cos\alpha\cos\beta = \tfrac{1}{2}[\cos(\alpha-\beta)+\cos(\alpha+\beta)] \tag{18}$$

to express the integrand as a sum or difference of sines and cosines.

▶ **Example 3** Evaluate $\int \sin 7x \cos 3x\, dx$.

Solution. Using (16) yields

$$\int \sin 7x \cos 3x\, dx = \tfrac{1}{2}\int (\sin 4x + \sin 10x)\, dx = -\tfrac{1}{8}\cos 4x - \tfrac{1}{20}\cos 10x + C \quad ◀$$

INTEGRATING POWERS OF TANGENT AND SECANT

The procedures for integrating powers of tangent and secant closely parallel those for sine and cosine. The idea is to use the following reduction formulas (which were derived in Exercise 58 of Section 8.2) to reduce the exponent in the integrand until the resulting integral can be evaluated:

$$\int \tan^n x\, dx = \frac{\tan^{n-1} x}{n-1} - \int \tan^{n-2} x\, dx \tag{19}$$

$$\int \sec^n x\, dx = \frac{\sec^{n-2} x \tan x}{n-1} + \frac{n-2}{n-1}\int \sec^{n-2} x\, dx \tag{20}$$

In the case where n is odd, the exponent can be reduced to 1, leaving us with the problem of integrating $\tan x$ or $\sec x$. These integrals are given by

$$\int \tan x\, dx = \ln|\sec x| + C \tag{21}$$

$$\int \sec x\, dx = \ln|\sec x + \tan x| + C \tag{22}$$

Formula (21) can be obtained by writing

$$\begin{aligned}\int \tan x\, dx &= \int \frac{\sin x}{\cos x}\, dx \\ &= -\ln|\cos x| + C && \begin{bmatrix} u = \cos x \\ du = -\sin x\, dx \end{bmatrix} \\ &= \ln|\sec x| + C && \left[\ln|\cos x| = -\ln\frac{1}{|\cos x|}\right]\end{aligned}$$

To obtain Formula (22) we write

$$\begin{aligned}\int \sec x\, dx &= \int \sec x \left(\frac{\sec x + \tan x}{\sec x + \tan x}\right) dx = \int \frac{\sec^2 x + \sec x \tan x}{\sec x + \tan x}\, dx \\ &= \ln|\sec x + \tan x| + C && \begin{bmatrix} u = \sec x + \tan x \\ du = (\sec^2 x + \sec x \tan x)\, dx \end{bmatrix}\end{aligned}$$

The following basic integrals occur frequently and are worth noting:

$$\int \tan^2 x\,dx = \tan x - x + C \tag{23}$$

$$\int \sec^2 x\,dx = \tan x + C \tag{24}$$

Formula (24) is already known to us, since the derivative of $\tan x$ is $\sec^2 x$. Formula (23) can be obtained by applying reduction formula (19) with $n = 2$ (verify) or, alternatively, by using the identity

$$1 + \tan^2 x = \sec^2 x$$

to write

$$\int \tan^2 x\,dx = \int (\sec^2 x - 1)\,dx = \tan x - x + C$$

The formulas

$$\int \tan^3 x\,dx = \tfrac{1}{2}\tan^2 x - \ln|\sec x| + C \tag{25}$$

$$\int \sec^3 x\,dx = \tfrac{1}{2}\sec x\tan x + \tfrac{1}{2}\ln|\sec x + \tan x| + C \tag{26}$$

can be deduced from (21), (22), and reduction formulas (19) and (20) as follows:

$$\int \tan^3 x\,dx = \tfrac{1}{2}\tan^2 x - \int \tan x\,dx = \tfrac{1}{2}\tan^2 x - \ln|\sec x| + C$$

$$\int \sec^3 x\,dx = \tfrac{1}{2}\sec x\tan x + \tfrac{1}{2}\int \sec x\,dx = \tfrac{1}{2}\sec x\tan x + \tfrac{1}{2}\ln|\sec x + \tan x| + C$$

INTEGRATING PRODUCTS OF TANGENTS AND SECANTS

If m and n are positive integers, then the integral

$$\int \tan^m x \sec^n x\,dx$$

can be evaluated by one of the three procedures stated in Table 8.3.2, depending on whether m and n are odd or even.

Table 8.3.2

$\int \tan^m x \sec^n x\,dx$	PROCEDURE	RELEVANT IDENTITIES
n even	• Split off a factor of $\sec^2 x$. • Apply the relevant identity. • Make the substitution $u = \tan x$.	$\sec^2 x = \tan^2 x + 1$
m odd	• Split off a factor of $\sec x \tan x$. • Apply the relevant identity. • Make the substitution $u = \sec x$.	$\tan^2 x = \sec^2 x - 1$
$\begin{cases} m \text{ even} \\ n \text{ odd} \end{cases}$	• Use the relevant identities to reduce the integrand to powers of $\sec x$ alone. • Then use the reduction formula for powers of $\sec x$.	$\tan^2 x = \sec^2 x - 1$

▶ **Example 4** Evaluate

$$\text{(a)} \int \tan^2 x \sec^4 x\,dx \qquad \text{(b)} \int \tan^3 x \sec^3 x\,dx \qquad \text{(c)} \int \tan^2 x \sec x\,dx$$

Solution (a). Since $n = 4$ is even, we will follow the first procedure in Table 8.3.2:

$$\begin{aligned}\int \tan^2 x \sec^4 x\,dx &= \int \tan^2 x \sec^2 x \sec^2 x\,dx \\ &= \int \tan^2 x(\tan^2 x + 1)\sec^2 x\,dx \\ &= \int u^2(u^2+1)\,du \\ &= \tfrac{1}{5}u^5 + \tfrac{1}{3}u^3 + C = \tfrac{1}{5}\tan^5 x + \tfrac{1}{3}\tan^3 x + C\end{aligned}$$

Solution (b). Since $m = 3$ is odd, we will follow the second procedure in Table 8.3.2:

$$\begin{aligned}\int \tan^3 x \sec^3 x\,dx &= \int \tan^2 x \sec^2 x(\sec x \tan x)\,dx \\ &= \int (\sec^2 x - 1)\sec^2 x(\sec x \tan x)\,dx \\ &= \int (u^2-1)u^2\,du \\ &= \tfrac{1}{5}u^5 - \tfrac{1}{3}u^3 + C = \tfrac{1}{5}\sec^5 x - \tfrac{1}{3}\sec^3 x + C\end{aligned}$$

Solution (c). Since $m = 2$ is even and $n = 1$ is odd, we will follow the third procedure in Table 8.3.2:

$$\begin{aligned}\int \tan^2 x \sec x\,dx &= \int (\sec^2 x - 1)\sec x\,dx \\ &= \int \sec^3 x\,dx - \int \sec x\,dx \qquad \text{See (26) and (22).} \\ &= \tfrac{1}{2}\sec x \tan x + \tfrac{1}{2}\ln|\sec x + \tan x| - \ln|\sec x + \tan x| + C \\ &= \tfrac{1}{2}\sec x \tan x - \tfrac{1}{2}\ln|\sec x + \tan x| + C\end{aligned}$$ ◀

AN ALTERNATIVE METHOD FOR INTEGRATING POWERS OF SINE, COSINE, TANGENT, AND SECANT

The methods in Tables 8.3.1 and 8.3.2 can sometimes be applied if $m = 0$ or $n = 0$ to integrate positive integer powers of sine, cosine, tangent, and secant without reduction formulas. For example, instead of using the reduction formula to integrate $\sin^3 x$, we can apply the second procedure in Table 8.3.1:

$$\begin{aligned}\int \sin^3 x\,dx &= \int (\sin^2 x)\sin x\,dx \\ &= \int (1-\cos^2 x)\sin x\,dx \qquad \begin{matrix} u = \cos x \\ du = -\sin x\,dx \end{matrix} \\ &= -\int (1-u^2)\,du \\ &= \tfrac{1}{3}u^3 - u + C = \tfrac{1}{3}\cos^3 x - \cos x + C\end{aligned}$$

which agrees with (11).

With the aid of the identity

$$1 + \cot^2 x = \csc^2 x$$

the techniques in Table 8.3.2 can be adapted to evaluate integrals of the form

$$\int \cot^m x \csc^n x\,dx$$

It is also possible to derive reduction formulas for powers of cot and csc that are analogous to Formulas (19) and (20).

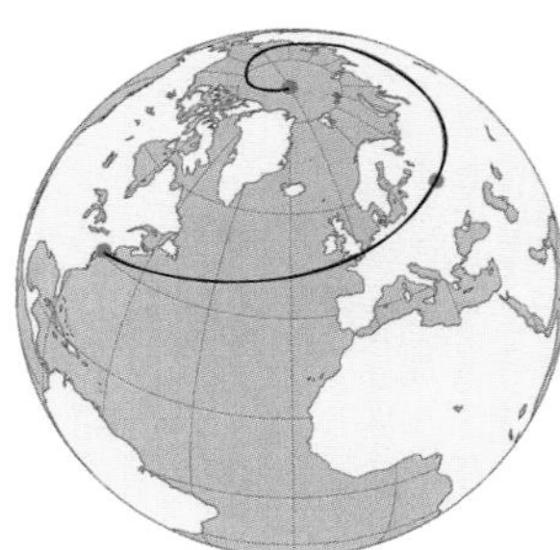

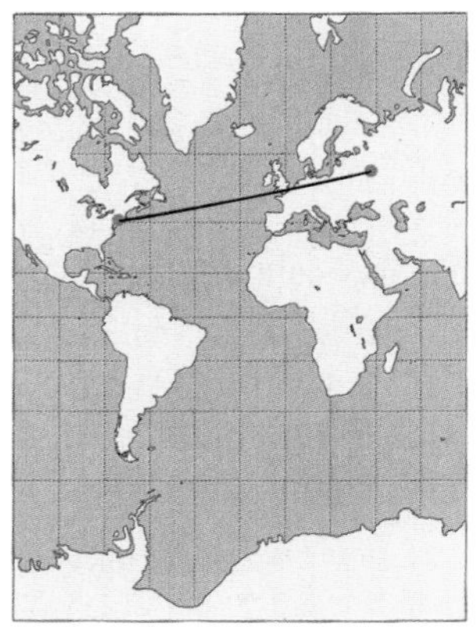

Figure 8.3.2 A flight with constant compass heading is shown from New York City to Moscow as it appears on a globe and on a Mercator projection.

MERCATOR'S MAP OF THE WORLD

The integral of $\sec x$ plays an important role in the design of navigational maps for charting nautical and aeronautical courses. Sailors and pilots usually chart their courses along paths with constant compass headings; for example, the course might be 30° northeast or 135° southeast. Except for courses that are parallel to the equator or run due north or south, a course with constant compass heading spirals around the Earth toward one of the poles (as in the top part of Figure 8.3.2). In 1569 the Flemish mathematician and geographer Gerhard Kramer (1512–1594) (better known by the Latin name Mercator) devised a world map, called the *Mercator projection*, in which spirals of constant compass headings appear as straight lines. This was extremely important because it enabled sailors to determine compass headings between two points by connecting them with a straight line on a map (as in the bottom part of Figure 8.3.2).

If the Earth is assumed to be a sphere of radius 4000 mi, then the lines of latitude at 1° increments are equally spaced about 70 mi apart (why?). However, in the Mercator projection, the lines of latitude become wider apart toward the poles, so that two widely spaced latitude lines near the poles may be actually the same distance apart on the Earth as two closely spaced latitude lines near the equator. It can be proved that on a Mercator map in which the equatorial line has length L, the vertical distance D_β on the map between the equator (latitude 0°) and the line of latitude $\beta°$ is

$$D_\beta = \frac{L}{2\pi}\int_0^{\beta\pi/180} \sec x\, dx \tag{27}$$

✔QUICK CHECK EXERCISES 8.3 (See page 530 for answers.)

1. Complete each trigonometric identity with an expression involving $\cos 2x$.
   (a) $\sin^2 x =$ ______
   (b) $\cos^2 x =$ ______
   (c) $\cos^2 x - \sin^2 x =$ ______
2. Evaluate the integral.
   (a) $\int \sec^2 x\, dx =$ ______
   (b) $\int \tan^2 x\, dx =$ ______
   (c) $\int \sec x\, dx =$ ______
   (d) $\int \tan x\, dx =$ ______
3. Use the indicated substitution to rewrite the integral in terms of u. Do not evaluate the integral.
   (a) $\int \sin^2 x \cos x\, dx;\ u = \sin x$
   (b) $\int \sin^3 x \cos^2 x\, dx;\ u = \cos x$
   (c) $\int \tan^3 x \sec^2 x\, dx;\ u = \tan x$
   (d) $\int \tan^3 x \sec x\, dx;\ u = \sec x$
4. Evaluate the integral.
   (a) $\int \sin^2 3x \cos 3x\, dx =$ ______
   (b) $\int \tan^2 x \sec^2 x\, dx =$ ______
   (c) $\int_0^{\pi/6} \sec 2x\, dx =$ ______
   (d) $\int_0^{\pi/8} \tan 2x\, dx =$ ______

EXERCISE SET 8.3

1–52 Evaluate the integral.

1. $\int \cos^3 x \sin x\, dx$
2. $\int \sin^5 3x \cos 3x\, dx$
3. $\int \sin^2 5\theta\, d\theta$
4. $\int \cos^2 3x\, dx$
5. $\int \sin^3 a\theta\, d\theta$
6. $\int \cos^3 at\, dt$
7. $\int \sin ax \cos ax\, dx$
8. $\int \sin^3 x \cos^3 x\, dx$
9. $\int \sin^2 t \cos^3 t\, dt$
10. $\int \sin^3 x \cos^2 x\, dx$
11. $\int \sin^2 x \cos^2 x\, dx$
12. $\int \sin^2 x \cos^4 x\, dx$
13. $\int \sin 2x \cos 3x\, dx$
14. $\int \sin 3\theta \cos 2\theta\, d\theta$
15. $\int \sin x \cos(x/2)\, dx$
16. $\int \cos^{1/3} x \sin x\, dx$
17. $\int_0^{\pi/2} \cos^3 x\, dx$
18. $\int_0^{\pi/2} \sin^2 \frac{x}{2} \cos^2 \frac{x}{2}\, dx$
19. $\int_0^{\pi/3} \sin^4 3x \cos^3 3x\, dx$
20. $\int_{-\pi}^{\pi} \cos^2 5\theta\, d\theta$
21. $\int_0^{\pi/6} \sin 4x \cos 2x\, dx$
22. $\int_0^{2\pi} \sin^2 kx\, dx$
23. $\int \sec^2(2x-1)\, dx$
24. $\int \tan 5x\, dx$
25. $\int e^{-x} \tan(e^{-x})\, dx$
26. $\int \cot 3x\, dx$
27. $\int \sec 4x\, dx$
28. $\int \frac{\sec(\sqrt{x})}{\sqrt{x}}\, dx$
29. $\int \tan^2 x \sec^2 x\, dx$
30. $\int \tan^5 x \sec^4 x\, dx$
31. $\int \tan 4x \sec^4 4x\, dx$
32. $\int \tan^4 \theta \sec^4 \theta\, d\theta$
33. $\int \sec^5 x \tan^3 x\, dx$
34. $\int \tan^5 \theta \sec \theta\, d\theta$
35. $\int \tan^4 x \sec x\, dx$
36. $\int \tan^2 x \sec^3 x\, dx$
37. $\int \tan t \sec^3 t\, dt$
38. $\int \tan x \sec^5 x\, dx$
39. $\int \sec^4 x\, dx$
40. $\int \sec^5 x\, dx$
41. $\int \tan^3 4x\, dx$
42. $\int \tan^4 x\, dx$
43. $\int \sqrt{\tan x} \sec^4 x\, dx$
44. $\int \tan x \sec^{3/2} x\, dx$
45. $\int_0^{\pi/8} \tan^2 2x\, dx$
46. $\int_0^{\pi/6} \sec^3 2\theta \tan 2\theta\, d\theta$
47. $\int_0^{\pi/2} \tan^5 \frac{x}{2}\, dx$
48. $\int_0^{1/4} \sec \pi x \tan \pi x\, dx$
49. $\int \cot^3 x \csc^3 x\, dx$
50. $\int \cot^2 3t \sec 3t\, dt$
51. $\int \cot^3 x\, dx$
52. $\int \csc^4 x\, dx$

53. Let m, n be distinct nonnegative integers. Use Formulas (16)–(18) to prove:
(a) $\int_0^{2\pi} \sin mx \cos nx\, dx = 0$
(b) $\int_0^{2\pi} \cos mx \cos nx\, dx = 0$
(c) $\int_0^{2\pi} \sin mx \sin nx\, dx = 0.$

54. Evaluate the integrals in Exercise 53 when m and n denote the *same* nonnegative integer.

55. Find the arc length of the curve $y = \ln(\cos x)$ over the interval $[0, \pi/4]$.

56. Find the volume of the solid generated when the region enclosed by $y = \tan x$, $y = 1$, and $x = 0$ is revolved about the x-axis.

57. Find the volume of the solid that results when the region enclosed by $y = \cos x$, $y = \sin x$, $x = 0$, and $x = \pi/4$ is revolved about the x-axis.

58. The region bounded below by the x-axis and above by the portion of $y = \sin x$ from $x = 0$ to $x = \pi$ is revolved about the x-axis. Find the volume of the resulting solid.

59. Use Formula (27) to show that if the length of the equatorial line on a Mercator projection is L, then the vertical distance D between the latitude lines at $\alpha°$ and $\beta°$ on the same side of the equator (where $\alpha < \beta$) is

$$D = \frac{L}{2\pi} \ln \left| \frac{\sec \beta° + \tan \beta°}{\sec \alpha° + \tan \alpha°} \right|$$

60. Suppose that the equator has a length of 100 cm on a Mercator projection. In each part, use the result in Exercise 59 to answer the question.
(a) What is the vertical distance on the map between the equator and the line at 25° north latitude?
(b) What is the vertical distance on the map between New Orleans, Louisiana, at 30° north latitude and Winnipeg, Canada, at 50° north latitude?

FOCUS ON CONCEPTS

61. (a) Show that

$$\int \csc x\, dx = -\ln|\csc x + \cot x| + C$$

(b) Show that the result in part (a) can also be written as

$$\int \csc x\, dx = \ln|\csc x - \cot x| + C$$

and

$$\int \csc x\,dx = \ln\left|\tan \tfrac{1}{2}x\right| + C$$

62. Rewrite $\sin x + \cos x$ in the form

$$A\sin(x+\phi)$$

and use your result together with Exercise 61 to evaluate

$$\int \frac{dx}{\sin x + \cos x}$$

63. Use the method of Exercise 62 to evaluate

$$\int \frac{dx}{a\sin x + b\cos x} \quad (a,\, b \text{ not both zero})$$

64. (a) Use Formula (9) in Section 8.2 to show that

$$\int_0^{\pi/2} \sin^n x\,dx = \frac{n-1}{n}\int_0^{\pi/2} \sin^{n-2} x\,dx \quad (n \geq 2)$$

(b) Use this result to derive the ***Wallis sine formulas***:

$$\int_0^{\pi/2} \sin^n x\,dx = \frac{\pi}{2}\cdot\frac{1\cdot 3\cdot 5\cdots(n-1)}{2\cdot 4\cdot 6\cdots n} \quad \begin{pmatrix} n \text{ even} \\ \text{and} \geq 2 \end{pmatrix}$$

$$\int_0^{\pi/2} \sin^n x\,dx = \frac{2\cdot 4\cdot 6\cdots(n-1)}{3\cdot 5\cdot 7\cdots n} \quad \begin{pmatrix} n \text{ odd} \\ \text{and} \geq 3 \end{pmatrix}$$

65. Use the Wallis formulas in Exercise 64 to evaluate

(a) $\int_0^{\pi/2} \sin^3 x\,dx$ (b) $\int_0^{\pi/2} \sin^4 x\,dx$

(c) $\int_0^{\pi/2} \sin^5 x\,dx$ (d) $\int_0^{\pi/2} \sin^6 x\,dx$.

66. Use Formula (10) in Section 8.2 and the method of Exercise 64 to derive the ***Wallis cosine formulas***:

$$\int_0^{\pi/2} \cos^n x\,dx = \frac{\pi}{2}\cdot\frac{1\cdot 3\cdot 5\cdots(n-1)}{2\cdot 4\cdot 6\cdots n} \quad \begin{pmatrix} n \text{ even} \\ \text{and} \geq 2 \end{pmatrix}$$

$$\int_0^{\pi/2} \cos^n x\,dx = \frac{2\cdot 4\cdot 6\cdots(n-1)}{3\cdot 5\cdot 7\cdots n} \quad \begin{pmatrix} n \text{ odd} \\ \text{and} \geq 3 \end{pmatrix}$$

✔ QUICK CHECK ANSWERS 8.3

1. (a) $\dfrac{1-\cos 2x}{2}$ (b) $\dfrac{1+\cos 2x}{2}$ (c) $\cos 2x$ **2.** (a) $\tan x + C$ (b) $\tan x - x + C$ (c) $\ln|\sec x + \tan x| + C$ (d) $\ln|\sec x| + C$ **3.** (a) $\int u^2\,du$ (b) $\int (u^2-1)u^2\,du$ (c) $\int u^3\,du$ (d) $\int (u^2-1)\,du$ **4.** (a) $\frac{1}{9}\sin^3 3x + C$ (b) $\frac{1}{3}\tan^3 x + C$ (c) $\frac{1}{2}\ln(2+\sqrt{3})$ (d) $\frac{1}{4}\ln 2$

8.4 TRIGONOMETRIC SUBSTITUTIONS

In this section we will discuss a method for evaluating integrals containing radicals by making substitutions involving trigonometric functions. We will also show how integrals containing quadratic polynomials can sometimes be evaluated by completing the square.

THE METHOD OF TRIGONOMETRIC SUBSTITUTION

To start, we will be concerned with integrals that contain expressions of the form

$$\sqrt{a^2-x^2}, \quad \sqrt{x^2+a^2}, \quad \sqrt{x^2-a^2}$$

in which a is a positive constant. The basic idea for evaluating such integrals is to make a substitution for x that will eliminate the radical. For example, to eliminate the radical in the expression $\sqrt{a^2-x^2}$, we can make the substitution

$$x = a\sin\theta, \quad -\pi/2 \leq \theta \leq \pi/2 \tag{1}$$

which yields

$$\begin{aligned}\sqrt{a^2-x^2} &= \sqrt{a^2-a^2\sin^2\theta} = \sqrt{a^2(1-\sin^2\theta)} \\ &= a\sqrt{\cos^2\theta} = a|\cos\theta| = a\cos\theta \qquad \boxed{\cos\theta \geq 0 \text{ since } -\pi/2 \leq \theta \leq \pi/2}\end{aligned}$$

The restriction on θ in (1) serves two purposes—it enables us to replace $|\cos\theta|$ by $\cos\theta$ to simplify the calculations, and it also ensures that the substitutions can be rewritten as $\theta = \sin^{-1}(x/a)$, if needed.

▶ Example 1 Evaluate $\displaystyle\int \frac{dx}{x^2\sqrt{4-x^2}}$.

Solution. To eliminate the radical we make the substitution

$$x = 2\sin\theta, \quad dx = 2\cos\theta\, d\theta$$

This yields

$$\begin{aligned}\int \frac{dx}{x^2\sqrt{4-x^2}} &= \int \frac{2\cos\theta\, d\theta}{(2\sin\theta)^2\sqrt{4-4\sin^2\theta}} \\ &= \int \frac{2\cos\theta\, d\theta}{(2\sin\theta)^2(2\cos\theta)} = \frac{1}{4}\int \frac{d\theta}{\sin^2\theta} \\ &= \frac{1}{4}\int \csc^2\theta\, d\theta = -\frac{1}{4}\cot\theta + C \qquad (2)\end{aligned}$$

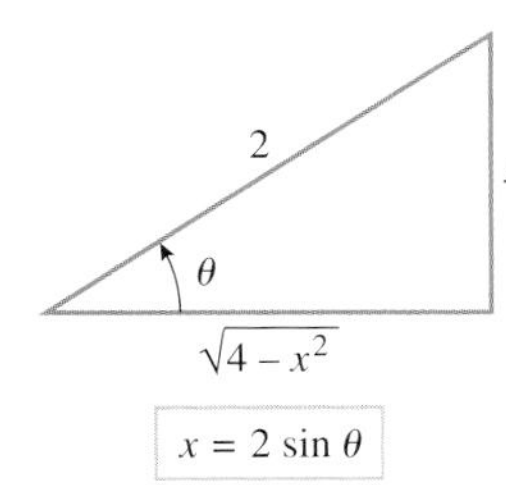

Figure 8.4.1

At this point we have completed the integration; however, because the original integral was expressed in terms of x, it is desirable to express $\cot\theta$ in terms of x as well. This can be done using trigonometric identities, but the expression can also be obtained by writing the substitution $x = 2\sin\theta$ as $\sin\theta = x/2$ and representing it geometrically as in Figure 8.4.1. From that figure we obtain

$$\cot\theta = \frac{\sqrt{4-x^2}}{x}$$

Substituting this in (2) yields

$$\int \frac{dx}{x^2\sqrt{4-x^2}} = -\frac{1}{4}\frac{\sqrt{4-x^2}}{x} + C \quad ◀$$

▶ Example 2 Evaluate $\displaystyle\int_1^{\sqrt{2}} \frac{dx}{x^2\sqrt{4-x^2}}$.

Solution. There are two possible approaches: we can make the substitution in the indefinite integral (as in Example 1) and then evaluate the definite integral using the x-limits of integration, or we can make the substitution in the definite integral and convert the x-limits to the corresponding θ-limits.

Method 1.

Using the result from Example 1 with the x-limits of integration yields

$$\int_1^{\sqrt{2}} \frac{dx}{x^2\sqrt{4-x^2}} = -\frac{1}{4}\left[\frac{\sqrt{4-x^2}}{x}\right]_1^{\sqrt{2}} = -\frac{1}{4}[1-\sqrt{3}\,] = \frac{\sqrt{3}-1}{4}$$

Method 2.

The substitution $x = 2\sin\theta$ can be expressed as $x/2 = \sin\theta$ or $\theta = \sin^{-1}(x/2)$, so the θ-limits that correspond to $x = 1$ and $x = \sqrt{2}$ are

$$x = 1: \quad \theta = \sin^{-1}(1/2) = \pi/6$$

$$x = \sqrt{2}: \quad \theta = \sin^{-1}(\sqrt{2}/2) = \pi/4$$

Thus, from (2) in Example 1 we obtain

$$\int_1^{\sqrt{2}} \frac{dx}{x^2\sqrt{4-x^2}} = -\frac{1}{4}\big[\cot\theta\big]_{\pi/6}^{\pi/4} = -\frac{1}{4}[1-\sqrt{3}\,] = \frac{\sqrt{3}-1}{4} \quad ◀$$

▶ Example 3 Find the area of the ellipse

$$\frac{x^2}{a^2} + \frac{y^2}{b^2} = 1$$

Solution. Because the ellipse is symmetric about both axes, its area A is four times the area in the first quadrant (Figure 8.4.2). If we solve the equation of the ellipse for y in terms of x, we obtain

$$y = \pm\frac{b}{a}\sqrt{a^2 - x^2}$$

where the positive square root gives the equation of the upper half. Thus, the area A is given by

$$A = 4\int_0^a \frac{b}{a}\sqrt{a^2 - x^2}\,dx = \frac{4b}{a}\int_0^a \sqrt{a^2 - x^2}\,dx$$

$\frac{x^2}{a^2} + \frac{y^2}{b^2} = 1$

Figure 8.4.2

To evaluate this integral, we will make the substitution $x = a\sin\theta$ $(dx = a\cos\theta\,d\theta)$ and convert the x-limits of integration to θ-limits. Since the substitution can be expressed as $\theta = \sin^{-1}(x/a)$, the θ-limits of integration are

$$x = 0\text{:} \quad \theta = \sin^{-1}(0) = 0$$

$$x = a\text{:} \quad \theta = \sin^{-1}(1) = \pi/2$$

Thus, we obtain

$$\begin{aligned} A &= \frac{4b}{a}\int_0^a \sqrt{a^2 - x^2}\,dx = \frac{4b}{a}\int_0^{\pi/2} a\cos\theta \cdot a\cos\theta\,d\theta \\ &= 4ab\int_0^{\pi/2} \cos^2\theta\,d\theta = 4ab\int_0^{\pi/2} \frac{1}{2}(1 + \cos 2\theta)\,d\theta \\ &= 2ab\left[\theta + \frac{1}{2}\sin 2\theta\right]_0^{\pi/2} = 2ab\left[\frac{\pi}{2} - 0\right] = \pi ab \end{aligned}$$ ◀

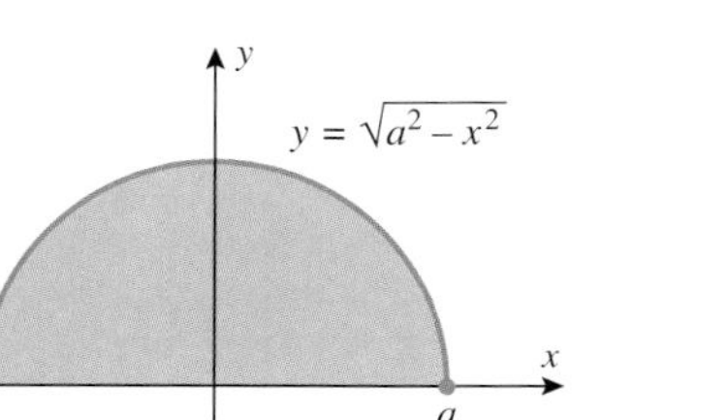

Figure 8.4.3

In the special case where $a = b$, the ellipse becomes a circle of radius a, and the area formula becomes $A = \pi a^2$, as expected. It is worth noting that

$$\int_{-a}^{a} \sqrt{a^2 - x^2}\,dx = \tfrac{1}{2}\pi a^2 \tag{3}$$

since this integral represents the area of the upper semicircle (Figure 8.4.3).

TECHNOLOGY MASTERY

If you have a calculating utility with a numerical integration capability, use it and Formula (3) to approximate π to three decimal places.

Thus far, we have focused on using the substitution $x = a\sin\theta$ to evaluate integrals involving radicals of the form $\sqrt{a^2 - x^2}$. Table 8.4.1 summarizes this method and describes some other substitutions of this type.

Table 8.4.1

EXPRESSION IN THE INTEGRAND	SUBSTITUTION	RESTRICTION ON θ	SIMPLIFICATION
$\sqrt{a^2 - x^2}$	$x = a\sin\theta$	$-\pi/2 \le \theta \le \pi/2$	$a^2 - x^2 = a^2 - a^2\sin^2\theta = a^2\cos^2\theta$
$\sqrt{a^2 + x^2}$	$x = a\tan\theta$	$-\pi/2 < \theta < \pi/2$	$a^2 + x^2 = a^2 + a^2\tan^2\theta = a^2\sec^2\theta$
$\sqrt{x^2 - a^2}$	$x = a\sec\theta$	$\begin{cases} 0 \le \theta < \pi/2 & (\text{if } x \ge a) \\ \pi/2 < \theta \le \pi & (\text{if } x \le -a) \end{cases}$	$x^2 - a^2 = a^2\sec^2\theta - a^2 = a^2\tan^2\theta$

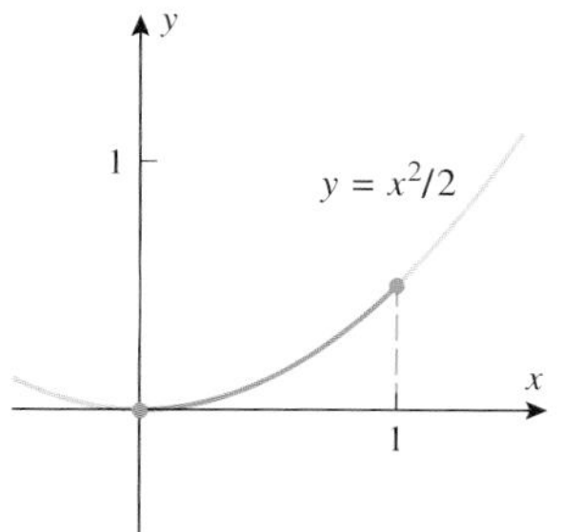

Figure 8.4.4

Example 4 Find the arc length of the curve $y = x^2/2$ from $x = 0$ to $x = 1$ (Figure 8.4.4).

Solution. From Formula (4) of Section 6.4 the arc length L of the curve is

$$L = \int_0^1 \sqrt{1 + \left(\frac{dy}{dx}\right)^2}\, dx = \int_0^1 \sqrt{1 + x^2}\, dx$$

The integrand involves a radical of the form $\sqrt{a^2 + x^2}$ with $a = 1$, so from Table 8.4.1 we make the substitution

$$x = \tan\theta, \quad -\pi/2 < \theta < \pi/2$$
$$\frac{dx}{d\theta} = \sec^2\theta \quad \text{or} \quad dx = \sec^2\theta\, d\theta$$

Since this substitution can be expressed as $\theta = \tan^{-1} x$, the θ-limits of integration that correspond to the x-limits, $x = 0$ and $x = 1$, are

$$x = 0{:} \quad \theta = \tan^{-1} 0 = 0$$
$$x = 1{:} \quad \theta = \tan^{-1} 1 = \pi/4$$

Thus,

$$\begin{aligned}
L = \int_0^1 \sqrt{1 + x^2}\, dx &= \int_0^{\pi/4} \sqrt{1 + \tan^2\theta}\, \sec^2\theta\, d\theta \\
&= \int_0^{\pi/4} \sqrt{\sec^2\theta}\, \sec^2\theta\, d\theta \\
&= \int_0^{\pi/4} |\sec\theta| \sec^2\theta\, d\theta \\
&= \int_0^{\pi/4} \sec^3\theta\, d\theta \qquad \boxed{\sec\theta > 0 \text{ since } -\pi/2 < \theta < \pi/2} \\
&= \left[\tfrac{1}{2}\sec\theta\tan\theta + \tfrac{1}{2}\ln|\sec\theta + \tan\theta|\right]_0^{\pi/4} \qquad \boxed{\text{Formula (26) of Section 8.3}} \\
&= \tfrac{1}{2}[\sqrt{2} + \ln(\sqrt{2} + 1)] \approx 1.148 \blacktriangleleft
\end{aligned}$$

Example 5 Evaluate $\displaystyle\int \frac{\sqrt{x^2 - 25}}{x}\, dx$, assuming that $x \geq 5$.

Solution. The integrand involves a radical of the form $\sqrt{x^2 - a^2}$ with $a = 5$, so from Table 8.4.1 we make the substitution

$$x = 5\sec\theta, \quad 0 \leq \theta < \pi/2$$
$$\frac{dx}{d\theta} = 5\sec\theta\tan\theta \quad \text{or} \quad dx = 5\sec\theta\tan\theta\, d\theta$$

Thus,

$$\begin{aligned}
\int \frac{\sqrt{x^2 - 25}}{x}\, dx &= \int \frac{\sqrt{25\sec^2\theta - 25}}{5\sec\theta}(5\sec\theta\tan\theta)\, d\theta \\
&= \int \frac{5|\tan\theta|}{5\sec\theta}(5\sec\theta\tan\theta)\, d\theta \\
&= 5\int \tan^2\theta\, d\theta \qquad \boxed{\tan\theta \geq 0 \text{ since } 0 \leq \theta < \pi/2} \\
&= 5\int (\sec^2\theta - 1)\, d\theta = 5\tan\theta - 5\theta + C
\end{aligned}$$

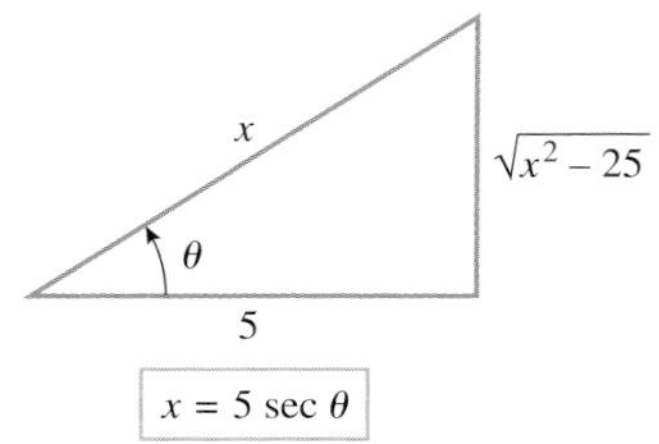

Figure 8.4.5

To express the solution in terms of x, we will represent the substitution $x = 5\sec\theta$ geometrically by the triangle in Figure 8.4.5, from which we obtain

$$\tan\theta = \frac{\sqrt{x^2 - 25}}{5}$$

From this and the fact that the substitution can be expressed as $\theta = \sec^{-1}(x/5)$, we obtain

$$\int \frac{\sqrt{x^2 - 25}}{x}\,dx = \sqrt{x^2 - 25} - 5\sec^{-1}\left(\frac{x}{5}\right) + C \blacktriangleleft$$

INTEGRALS INVOLVING $ax^2 + bx + c$

Integrals that involve a quadratic expression $ax^2 + bx + c$, where $a \neq 0$ and $b \neq 0$, can often be evaluated by first completing the square, then making an appropriate substitution. The following examples illustrate this idea.

▶ Example 6 Evaluate $\displaystyle\int \frac{x}{x^2 - 4x + 8}\,dx$.

Solution. Completing the square yields

$$x^2 - 4x + 8 = (x^2 - 4x + 4) + 8 - 4 = (x - 2)^2 + 4$$

Thus, the substitution

$$u = x - 2, \quad du = dx$$

yields

$$\begin{aligned}
\int \frac{x}{x^2 - 4x + 8}\,dx &= \int \frac{x}{(x-2)^2 + 4}\,dx = \int \frac{u + 2}{u^2 + 4}\,du \\
&= \int \frac{u}{u^2 + 4}\,du + 2\int \frac{du}{u^2 + 4} \\
&= \frac{1}{2}\int \frac{2u}{u^2 + 4}\,du + 2\int \frac{du}{u^2 + 4} \\
&= \frac{1}{2}\ln(u^2 + 4) + 2\left(\frac{1}{2}\right)\tan^{-1}\frac{u}{2} + C \\
&= \frac{1}{2}\ln[(x-2)^2 + 4] + \tan^{-1}\left(\frac{x-2}{2}\right) + C \blacktriangleleft
\end{aligned}$$

▶ Example 7 Evaluate $\displaystyle\int \frac{dx}{\sqrt{5 - 4x - 2x^2}}$.

Solution. Completing the square yields

$$\begin{aligned}
5 - 4x - 2x^2 &= 5 - 2(x^2 + 2x) = 5 - 2(x^2 + 2x + 1) + 2 \\
&= 5 - 2(x+1)^2 + 2 = 7 - 2(x+1)^2
\end{aligned}$$

Thus,

$$\int \frac{dx}{\sqrt{5-4x-2x^2}} = \int \frac{dx}{\sqrt{7-2(x+1)^2}}$$

$$= \int \frac{du}{\sqrt{7-2u^2}} \qquad \boxed{\begin{array}{l} u = x+1 \\ du = dx \end{array}}$$

$$= \frac{1}{\sqrt{2}} \int \frac{du}{\sqrt{(7/2)-u^2}}$$

$$= \frac{1}{\sqrt{2}} \sin^{-1}\left(\frac{u}{\sqrt{7/2}}\right) + C \qquad \boxed{\begin{array}{l} \text{Formula 21, Section 8.1} \\ \text{with } a = \sqrt{7/2} \end{array}}$$

$$= \frac{1}{\sqrt{2}} \sin^{-1}[\sqrt{2/7}(x+1)] + C \blacktriangleleft$$

✔QUICK CHECK EXERCISES 8.4 *(See page 536 for answers.)*

1. For each expression, give a trigonometric substitution that will eliminate the radical.
(a) $\sqrt{a^2 - x^2}$ ________ (b) $\sqrt{a^2 + x^2}$ ________
(c) $\sqrt{x^2 - a^2}$ ________

2. If $x = 2\sec\theta$ and $0 < \theta < \pi/2$, then
(a) $\sin\theta =$ ________ (b) $\cos\theta =$ ________
(c) $\tan\theta =$ ________.

3. In each part, state the trigonometric substitution that you would try first to evaluate the integral. Do not evaluate the integral.
(a) $\int \sqrt{9 + x^2}\, dx$ ________
(b) $\int \sqrt{9 - x^2}\, dx$ ________
(c) $\int \sqrt{1 - 9x^2}\, dx$ ________
(d) $\int \sqrt{x^2 - 9}\, dx$ ________
(e) $\int \sqrt{9 + 3x^2}\, dx$ ________
(f) $\int \sqrt{1 + (9x)^2}\, dx$ ________

4. In each part, determine the substitution u.
(a) $\int \frac{1}{x^2 - 2x + 10}\, dx = \int \frac{1}{u^2 + 3^2}\, du$;
$u =$ ________
(b) $\int \sqrt{x^2 - 6x + 8}\, dx = \int \sqrt{u^2 - 1}\, du$;
$u =$ ________
(c) $\int \sqrt{12 - 4x - x^2}\, dx = \int \sqrt{4^2 - u^2}\, du$;
$u =$ ________

EXERCISE SET 8.4  CAS

1–26 Evaluate the integral.

1. $\int \sqrt{4 - x^2}\, dx$

2. $\int \sqrt{1 - 4x^2}\, dx$

3. $\int \frac{x^2}{\sqrt{16 - x^2}}\, dx$

4. $\int \frac{dx}{x^2\sqrt{9 - x^2}}$

5. $\int \frac{dx}{(4 + x^2)^2}$

6. $\int \frac{x^2}{\sqrt{5 + x^2}}\, dx$

7. $\int \frac{\sqrt{x^2 - 9}}{x}\, dx$

8. $\int \frac{dx}{x^2\sqrt{x^2 - 16}}$

9. $\int \frac{3x^3}{\sqrt{1 - x^2}}\, dx$

10. $\int x^3\sqrt{5 - x^2}\, dx$

11. $\int \frac{dx}{x^2\sqrt{9x^2 - 4}}$

12. $\int \frac{\sqrt{1 + t^2}}{t}\, dt$

13. $\int \frac{dx}{(1 - x^2)^{3/2}}$

14. $\int \frac{dx}{x^2\sqrt{x^2 + 25}}$

15. $\int \frac{dx}{\sqrt{x^2 - 9}}$

16. $\int \frac{dx}{1 + 2x^2 + x^4}$

17. $\int \frac{dx}{(4x^2 - 9)^{3/2}}$

18. $\int \frac{3x^3}{\sqrt{x^2 - 25}}\, dx$

19. $\int e^x\sqrt{1 - e^{2x}}\, dx$

20. $\int \frac{\cos\theta}{\sqrt{2 - \sin^2\theta}}\, d\theta$

21. $\int_0^1 5x^3\sqrt{1 - x^2}\, dx$

22. $\int_0^{1/2} \frac{dx}{(1 - x^2)^2}$

23. $\int_{\sqrt{2}}^{2} \frac{dx}{x^2\sqrt{x^2-1}}$

24. $\int_{\sqrt{2}}^{2} \frac{\sqrt{2x^2-4}}{x}\,dx$

25. $\int_{1}^{3} \frac{dx}{x^4\sqrt{x^2+3}}$

26. $\int_{0}^{3} \frac{x^3}{(3+x^2)^{5/2}}\,dx$

FOCUS ON CONCEPTS

27. The integral
$$\int \frac{x}{x^2+4}\,dx$$
can be evaluated either by a trigonometric substitution or by the substitution $u = x^2 + 4$. Do it both ways and show that the results are equivalent.

28. The integral
$$\int \frac{x^2}{x^2+4}\,dx$$
can be evaluated either by a trigonometric substitution or by algebraically rewriting the numerator of the integrand as $(x^2+4) - 4$. Do it both ways and show that the results are equivalent.

29. Find the arc length of the curve $y = \ln x$ from $x = 1$ to $x = 2$.

30. Find the arc length of the curve $y = x^2$ from $x = 0$ to $x = 1$.

31. Find the area of the surface generated when the curve in Exercise 30 is revolved about the x-axis.

32. Find the volume of the solid generated when the region enclosed by $x = y(1-y^2)^{1/4}$, $y = 0$, $y = 1$, and $x = 0$ is revolved about the y-axis.

33–44 Evaluate the integral.

33. $\int \frac{dx}{x^2-4x+5}$

34. $\int \frac{dx}{\sqrt{2x-x^2}}$

35. $\int \frac{dx}{\sqrt{3+2x-x^2}}$

36. $\int \frac{dx}{16x^2+16x+5}$

37. $\int \frac{dx}{\sqrt{x^2-6x+10}}$

38. $\int \frac{x}{x^2+2x+2}\,dx$

39. $\int \sqrt{3-2x-x^2}\,dx$

40. $\int \frac{e^x}{\sqrt{1+e^x+e^{2x}}}\,dx$

41. $\int \frac{dx}{2x^2+4x+7}$

42. $\int \frac{2x+3}{4x^2+4x+5}\,dx$

43. $\int_{1}^{2} \frac{dx}{\sqrt{4x-x^2}}$

44. $\int_{0}^{4} \sqrt{x(4-x)}\,dx$

45–46 There is a good chance that your CAS will not be able to evaluate these integrals as stated. If this is so, make a substitution that converts the integral into one that your CAS can evaluate.

C **45.** $\int \cos x \sin x \sqrt{1-\sin^4 x}\,dx$

C **46.** $\int (x\cos x + \sin x)\sqrt{1+x^2\sin^2 x}\,dx$

FOCUS ON CONCEPTS

47. (a) Use the ***hyperbolic substitution*** $x = 3\sinh u$, the identity $\cosh^2 u - \sinh^2 u = 1$, and Theorem 7.8.4 to evaluate
$$\int \frac{dx}{\sqrt{x^2+9}}$$
(b) Evaluate the integral in part (a) using a trigonometric substitution and show that the result agrees with that obtained in part (a).

48. Use the hyperbolic substitutionn $x = \cosh u$, the identity $\sinh^2 u = \frac{1}{2}(\cosh 2u - 1)$, and the results referenced in Exercise 47 to evaluate
$$\int \sqrt{x^2-1}\,dx, \quad x \geq 1$$

✔ QUICK CHECK ANSWERS 8.4

1. (a) $x = a\sin\theta$ (b) $x = a\tan\theta$ (c) $x = a\sec\theta$ 2. (a) $\frac{\sqrt{x^2-4}}{x}$ (b) $\frac{2}{x}$ (c) $\frac{\sqrt{x^2-4}}{2}$ 3. (a) $x = 3\tan\theta$ (b) $x = 3\sin\theta$ (c) $x = \frac{1}{3}\sin\theta$ (d) $x = 3\sec\theta$ (e) $x = \sqrt{3}\tan\theta$ (f) $x = \frac{1}{9}\tan\theta$ 4. (a) $x - 1$ (b) $x - 3$ (c) $x + 2$

8.5 INTEGRATING RATIONAL FUNCTIONS BY PARTIAL FRACTIONS

Recall that a rational function is a ratio of two polynomials. In this section we will give a general method for integrating rational functions that is based on the idea of decomposing a rational function into a sum of simple rational functions that can be integrated by the methods studied in earlier sections.

PARTIAL FRACTIONS

In algebra, one learns to combine two or more fractions into a single fraction by finding a common denominator. For example,

$$\frac{2}{x-4}+\frac{3}{x+1}=\frac{2(x+1)+3(x-4)}{(x-4)(x+1)}=\frac{5x-10}{x^2-3x-4} \tag{1}$$

However, for purposes of integration, the left side of (1) is preferable to the right side since each of the terms is easy to integrate:

$$\int \frac{5x-10}{x^2-3x-4}\,dx=\int \frac{2}{x-4}\,dx+\int \frac{3}{x+1}\,dx=2\ln|x-4|+3\ln|x+1|+C$$

Thus, it is desirable to have some method that will enable us to obtain the left side of (1), starting with the right side. To illustrate how this can be done, we begin by noting that on the left side the numerators are constants and the denominators are the factors of the denominator on the right side. Thus, to find the left side of (1), starting from the right side, we could factor the denominator of the right side and look for constants A and B such that

$$\frac{5x-10}{(x-4)(x+1)}=\frac{A}{x-4}+\frac{B}{x+1} \tag{2}$$

One way to find the constants A and B is to multiply (2) through by $(x-4)(x+1)$ to clear fractions. This yields

$$5x-10=A(x+1)+B(x-4) \tag{3}$$

This relationship holds for all x, so it holds in particular if $x=4$ or $x=-1$. Substituting $x=4$ in (3) makes the second term on the right drop out and yields the equation $10=5A$ or $A=2$; and substituting $x=-1$ in (3) makes the first term on the right drop out and yields the equation $-15=-5B$ or $B=3$. Substituting these values in (2) we obtain

$$\frac{5x-10}{(x-4)(x+1)}=\frac{2}{x-4}+\frac{3}{x+1} \tag{4}$$

which agrees with (1).

A second method for finding the constants A and B is to multiply out the right side of (3) and collect like powers of x to obtain

$$5x-10=(A+B)x+(A-4B)$$

Since the polynomials on the two sides are identical, their corresponding coefficients must be the same. Equating the corresponding coefficients on the two sides yields the following system of equations in the unknowns A and B:

$$\begin{aligned} A+\ B &= \ \ 5 \\ A-4B &= -10 \end{aligned}$$

Solving this system yields $A=2$ and $B=3$ as before (verify).

The terms on the right side of (4) are called ***partial fractions*** of the expression on the left side because they each constitute *part* of that expression. To find those partial fractions we first had to make a guess about their form, and then we had to find the unknown constants. Our next objective is to extend this idea to general rational functions. For this purpose, suppose that $P(x)/Q(x)$ is a ***proper rational function***, by which we mean that the degree of the numerator is less than the degree of the denominator. There is a theorem in advanced algebra which states that every proper rational function can be expressed as a sum

$$\frac{P(x)}{Q(x)} = F_1(x) + F_2(x) + \cdots + F_n(x)$$

where $F_1(x), F_2(x), \ldots, F_n(x)$ are rational functions of the form

$$\frac{A}{(ax+b)^k} \quad \text{or} \quad \frac{Ax+B}{(ax^2+bx+c)^k}$$

in which the denominators are factors of $Q(x)$. The sum is called the ***partial fraction decomposition*** of $P(x)/Q(x)$, and the terms are called ***partial fractions***. As in our opening example, there are two parts to finding a partial fraction decomposition: determining the exact form of the decomposition and finding the unknown constants.

FINDING THE FORM OF A PARTIAL FRACTION DECOMPOSITION

The first step in finding the form of the partial fraction decomposition of a proper rational function $P(x)/Q(x)$ is to factor $Q(x)$ completely into linear and irreducible quadratic factors, and then collect all repeated factors so that $Q(x)$ is expressed as a product of *distinct* factors of the form

$$(ax+b)^m \quad \text{and} \quad (ax^2+bx+c)^m$$

From these factors we can determine the form of the partial fraction decomposition using two rules that we will now discuss.

LINEAR FACTORS

If all of the factors of $Q(x)$ are linear, then the partial fraction decomposition of $P(x)/Q(x)$ can be determined by using the following rule:

LINEAR FACTOR RULE. For each factor of the form $(ax+b)^m$, the partial fraction decomposition contains the following sum of m partial fractions:

$$\frac{A_1}{ax+b} + \frac{A_2}{(ax+b)^2} + \cdots + \frac{A_m}{(ax+b)^m}$$

where $A_1, A_2, \ldots, A_m$ are constants to be determined. In the case where $m = 1$, only the first term in the sum appears.

▶ **Example 1** Evaluate $\displaystyle\int \frac{dx}{x^2+x-2}$.

Solution. The integrand is a proper rational function that can be written as

$$\frac{1}{x^2+x-2} = \frac{1}{(x-1)(x+2)}$$

The factors $x-1$ and $x+2$ are both linear and appear to the first power, so each contributes one term to the partial fraction decomposition by the linear factor rule. Thus, the decomposition has the form

$$\frac{1}{(x-1)(x+2)} = \frac{A}{x-1} + \frac{B}{x+2} \tag{5}$$

where A and B are constants to be determined. Multiplying this expression through by $(x-1)(x+2)$ yields

$$1 = A(x+2) + B(x-1) \tag{6}$$

As discussed earlier, there are two methods for finding A and B: we can substitute values of x that are chosen to make terms on the right drop out, or we can multiply out on the right and equate corresponding coefficients on the two sides to obtain a system of equations that can be solved for A and B. We will use the first approach.

Setting $x = 1$ makes the second term in (6) drop out and yields $1 = 3A$ or $A = \frac{1}{3}$; and setting $x = -2$ makes the first term in (6) drop out and yields $1 = -3B$ or $B = -\frac{1}{3}$. Substituting these values in (5) yields the partial fraction decomposition

$$\frac{1}{(x-1)(x+2)} = \frac{\frac{1}{3}}{x-1} + \frac{-\frac{1}{3}}{x+2}$$

The integration can now be completed as follows:

$$\begin{aligned}\int \frac{dx}{(x-1)(x+2)} &= \frac{1}{3}\int \frac{dx}{x-1} - \frac{1}{3}\int \frac{dx}{x+2} \\ &= \frac{1}{3}\ln|x-1| - \frac{1}{3}\ln|x+2| + C = \frac{1}{3}\ln\left|\frac{x-1}{x+2}\right| + C \blacktriangleleft\end{aligned}$$

If the factors of $Q(x)$ are linear and none are repeated, as in the last example, then the recommended method for finding the constants in the partial fraction decomposition is to substitute appropriate values of x to make terms drop out. However, if some of the linear factors are repeated, then it will not be possible to find all of the constants in this way. In this case the recommended procedure is to find as many constants as possible by substitution and then find the rest by equating coefficients. This is illustrated in the next example.

▸ **Example 2** Evaluate $\displaystyle\int \frac{2x+4}{x^3-2x^2}\,dx$.

Solution. The integrand can be rewritten as

$$\frac{2x+4}{x^3-2x^2} = \frac{2x+4}{x^2(x-2)}$$

Although x^2 is a quadratic factor, it is *not* irreducible since $x^2 = xx$. Thus, by the linear factor rule, x^2 introduces two terms (since $m = 2$) of the form

$$\frac{A}{x} + \frac{B}{x^2}$$

and the factor $x - 2$ introduces one term (since $m = 1$) of the form

$$\frac{C}{x-2}$$

so the partial fraction decomposition is

$$\frac{2x+4}{x^2(x-2)} = \frac{A}{x} + \frac{B}{x^2} + \frac{C}{x-2} \tag{7}$$

Multiplying by $x^2(x-2)$ yields

$$2x + 4 = Ax(x-2) + B(x-2) + Cx^2 \tag{8}$$

which, after multiplying out and collecting like powers of x, becomes

$$2x + 4 = (A + C)x^2 + (-2A + B)x - 2B \tag{9}$$

Setting $x = 0$ in (8) makes the first and third terms drop out and yields $B = -2$, and setting $x = 2$ in (8) makes the first and second terms drop out and yields $C = 2$ (verify). However, there is no substitution in (8) that produces A directly, so we look to Equation (9) to find this value. This can be done by equating the coefficients of x^2 on the two sides to obtain

$$A + C = 0 \quad \text{or} \quad A = -C = -2$$

Substituting the values $A = -2$, $B = -2$, and $C = 2$ in (7) yields the partial fraction decomposition

$$\frac{2x+4}{x^2(x-2)} = \frac{-2}{x} + \frac{-2}{x^2} + \frac{2}{x-2}$$

Thus,

$$\int \frac{2x+4}{x^2(x-2)}\,dx = -2\int \frac{dx}{x} - 2\int \frac{dx}{x^2} + 2\int \frac{dx}{x-2}$$

$$= -2\ln|x| + \frac{2}{x} + 2\ln|x-2| + C = 2\ln\left|\frac{x-2}{x}\right| + \frac{2}{x} + C \blacktriangleleft$$

QUADRATIC FACTORS

If some of the factors of $Q(x)$ are irreducible quadratics, then the contribution of those factors to the partial fraction decomposition of $P(x)/Q(x)$ can be determined from the following rule:

> QUADRATIC FACTOR RULE. For each factor of the form $(ax^2 + bx + c)^m$, the partial fraction decomposition contains the following sum of m partial fractions:
>
> $$\frac{A_1x + B_1}{ax^2 + bx + c} + \frac{A_2x + B_2}{(ax^2 + bx + c)^2} + \cdots + \frac{A_mx + B_m}{(ax^2 + bx + c)^m}$$
>
> where $A_1, A_2, \ldots, A_m, B_1, B_2, \ldots, B_m$ are constants to be determined. In the case where $m = 1$, only the first term in the sum appears.

▶ **Example 3** Evaluate $\displaystyle\int \frac{x^2 + x - 2}{3x^3 - x^2 + 3x - 1}\,dx$.

Solution. The denominator in the integrand can be factored by grouping:

$$3x^3 - x^2 + 3x - 1 = x^2(3x-1) + (3x-1) = (3x-1)(x^2+1)$$

By the linear factor rule, the factor $3x - 1$ introduces one term, namely,

$$\frac{A}{3x-1}$$

and by the quadratic factor rule, the factor $x^2 + 1$ introduces one term, namely,

$$\frac{Bx + C}{x^2 + 1}$$

Thus, the partial fraction decomposition is

$$\frac{x^2 + x - 2}{(3x-1)(x^2+1)} = \frac{A}{3x-1} + \frac{Bx + C}{x^2+1} \tag{10}$$

Multiplying by $(3x-1)(x^2+1)$ yields

$$x^2 + x - 2 = A(x^2+1) + (Bx + C)(3x - 1) \tag{11}$$

We could find A by substituting $x = \frac{1}{3}$ to make the last term drop out, and then find the rest of the constants by equating corresponding coefficients. However, in this case it is just as easy to find *all* of the constants by equating coefficients and solving the resulting system. For this purpose we multiply out the right side of (11) and collect like terms:

$$x^2 + x - 2 = (A + 3B)x^2 + (-B + 3C)x + (A - C)$$

Equating corresponding coefficients gives

$$\begin{aligned} A + 3B \quad\quad &= 1 \\ -\ B + 3C &= 1 \\ A \quad\quad - C &= -2 \end{aligned}$$

To solve this system, subtract the third equation from the first to eliminate A. Then use the resulting equation together with the second equation to solve for B and C. Finally, determine A from the first or third equation. This yields (verify)

$$A = -\frac{7}{5}, \quad B = \frac{4}{5}, \quad C = \frac{3}{5}$$

Thus, (10) becomes

$$\frac{x^2 + x - 2}{(3x-1)(x^2+1)} = \frac{-\frac{7}{5}}{3x-1} + \frac{\frac{4}{5}x + \frac{3}{5}}{x^2+1}$$

and

TECHNOLOGY MASTERY

Computer algebra systems have built-in capabilities for finding partial fraction decompositions. If you have a CAS, use it to find the decompositions in Examples 1, 2, and 3.

$$\begin{aligned} \int \frac{x^2 + x - 2}{(3x-1)(x^2+1)}\,dx &= -\frac{7}{5}\int \frac{dx}{3x-1} + \frac{4}{5}\int \frac{x}{x^2+1}\,dx + \frac{3}{5}\int \frac{dx}{x^2+1} \\ &= -\frac{7}{15}\ln|3x-1| + \frac{2}{5}\ln(x^2+1) + \frac{3}{5}\tan^{-1}x + C \end{aligned}$$ ◀

▶ **Example 4** Evaluate $\displaystyle\int \frac{3x^4 + 4x^3 + 16x^2 + 20x + 9}{(x+2)(x^2+3)^2}\,dx.$

Solution. Observe that the integrand is a proper rational function since the numerator has degree 4 and the denominator has degree 5. Thus, the method of partial fractions is applicable. By the linear factor rule, the factor $x + 2$ introduces the single term

$$\frac{A}{x+2}$$

and by the quadratic factor rule, the factor $(x^2 + 3)^2$ introduces two terms (since $m = 2$):

$$\frac{Bx + C}{x^2+3} + \frac{Dx + E}{(x^2+3)^2}$$

Thus, the partial fraction decomposition of the integrand is

$$\frac{3x^4 + 4x^3 + 16x^2 + 20x + 9}{(x+2)(x^2+3)^2} = \frac{A}{x+2} + \frac{Bx + C}{x^2+3} + \frac{Dx + E}{(x^2+3)^2} \tag{12}$$

Multiplying by $(x+2)(x^2+3)^2$ yields

$$\begin{aligned} &3x^4 + 4x^3 + 16x^2 + 20x + 9 \\ &\quad = A(x^2+3)^2 + (Bx + C)(x^2+3)(x+2) + (Dx + E)(x+2) \end{aligned} \tag{13}$$

which, after multiplying out and collecting like powers of x, becomes

$$\begin{aligned} &3x^4 + 4x^3 + 16x^2 + 20x + 9 \\ &\quad = (A + B)x^4 + (2B + C)x^3 + (6A + 3B + 2C + D)x^2 \\ &\qquad + (6B + 3C + 2D + E)x + (9A + 6C + 2E) \end{aligned} \tag{14}$$

Equating corresponding coefficients in (14) yields the following system of five linear equations in five unknowns:

$$\begin{aligned} A + B &= 3 \\ 2B + C &= 4 \\ 6A + 3B + 2C + D &= 16 \\ 6B + 3C + 2D + E &= 20 \\ 9A + 6C + 2E &= 9 \end{aligned} \tag{15}$$

Efficient methods for solving systems of linear equations such as this are studied in a branch of mathematics called ***linear algebra***; those methods are outside the scope of this text. However, as a practical matter most linear systems of any size are solved by computer, and most computer algebra systems have commands that in many cases can solve linear systems exactly. In this particular case we can simplify the work by first substituting $x = -2$ in (13), which yields $A = 1$. Substituting this known value of A in (15) yields the simpler system

$$\begin{aligned} B &= 2 \\ 2B + C &= 4 \\ 3B + 2C + D &= 10 \\ 6B + 3C + 2D + E &= 20 \\ 6C + 2E &= 0 \end{aligned} \tag{16}$$

This system can be solved by starting at the top and working down, first substituting $B = 2$ in the second equation to get $C = 0$, then substituting the known values of B and C in the third equation to get $D = 4$, and so forth. This yields

$$A = 1, \quad B = 2, \quad C = 0, \quad D = 4, \quad E = 0$$

Thus, (12) becomes

$$\frac{3x^4 + 4x^3 + 16x^2 + 20x + 9}{(x + 2)(x^2 + 3)^2} = \frac{1}{x + 2} + \frac{2x}{x^2 + 3} + \frac{4x}{(x^2 + 3)^2}$$

and so

$$\begin{aligned} &\int \frac{3x^4 + 4x^3 + 16x^2 + 20x + 9}{(x + 2)(x^2 + 3)^2}\,dx \\ &\quad = \int \frac{dx}{x + 2} + \int \frac{2x}{x^2 + 3}\,dx + 4\int \frac{x}{(x^2 + 3)^2}\,dx \\ &\quad = \ln|x + 2| + \ln(x^2 + 3) - \frac{2}{x^2 + 3} + C \quad \blacktriangleleft \end{aligned}$$

INTEGRATING IMPROPER RATIONAL FUNCTIONS

Although the method of partial fractions only applies to proper rational functions, an improper rational function can be integrated by performing a long division and expressing the function as the quotient plus the remainder over the divisor. The remainder over the divisor will be a proper rational function, which can then be decomposed into partial fractions. This idea is illustrated in the following example.

▶ **Example 5** Evaluate $\displaystyle\int \frac{3x^4 + 3x^3 - 5x^2 + x - 1}{x^2 + x - 2}\,dx$.

Solution. The integrand is an improper rational function since the numerator has degree 4 and the denominator has degree 2. Thus, we first perform the long division

$$\begin{array}{r|l} & 3x^2 \qquad\quad + 1 \\ x^2 + x - 2 & 3x^4 + 3x^3 - 5x^2 + x - 1 \\ & 3x^4 + 3x^3 - 6x^2 \\ & \qquad\qquad\quad x^2 + x - 1 \\ & \qquad\qquad\quad x^2 + x - 2 \\ & \qquad\qquad\qquad\qquad 1 \end{array}$$

It follows that the integrand can be expressed as

$$\frac{3x^4 + 3x^3 - 5x^2 + x - 1}{x^2 + x - 2} = (3x^2 + 1) + \frac{1}{x^2 + x - 2}$$

and hence

$$\int \frac{3x^4+3x^3-5x^2+x-1}{x^2+x-2}\,dx = \int (3x^2+1)\,dx + \int \frac{dx}{x^2+x-2}$$

The second integral on the right now involves a proper rational function and can thus be evaluated by a partial fraction decomposition. Using the result of Example 1 we obtain

$$\int \frac{3x^4+3x^3-5x^2+x-1}{x^2+x-2}\,dx = x^3+x+\frac{1}{3}\ln\left|\frac{x-1}{x+2}\right|+C \blacktriangleleft$$

CONCLUDING REMARKS

There are some cases in which the method of partial fractions is inappropriate. For example, it would be inefficient to use partial fractions to perform the integration

$$\int \frac{3x^2+2}{x^3+2x-8}\,dx = \ln|x^3+2x-8|+C$$

since the substitution $u = x^3+2x-8$ is more direct. Similarly, the integration

$$\int \frac{2x-1}{x^2+1}\,dx = \int \frac{2x}{x^2+1}\,dx - \int \frac{dx}{x^2+1} = \ln(x^2+1) - \tan^{-1}x + C$$

requires only a little algebra since the integrand is already in partial fraction form.

✔QUICK CHECK EXERCISES 8.5

(See page 544 for answers.)

1. A partial fraction is a rational function of the form ________ or of the form ________.
2. (a) What is a proper rational function?
   (b) What condition must the degree of the numerator and the degree of the denominator of a rational function satisfy for the method of partial fractions to be applicable directly?
   (c) If the condition in part (b) is not satisfied, what must you do if you want to use partial fractions?
3. Suppose that the function $f(x) = P(x)/Q(x)$ is a proper rational function.
   (a) For each factor of $Q(x)$ of the form $(ax+b)^m$, the partial fraction decomposition of f contains the following sum of m partial fractions: ________
   (b) For each factor of $Q(x)$ of the form $(ax^2+bx+c)^m$, where ax^2+bx+c is an irreducible quadratic, the partial fraction decomposition of f contains the following sum of m partial fractions: ________
4. Find the partial fraction decomposition.
   (a) $\dfrac{3}{(x+1)(1-2x)}$ (b) $\dfrac{2x^2-3x}{(x^2+1)(3x+2)}$
5. Evaluate the integral.
   (a) $\displaystyle\int \frac{3}{(x+1)(1-2x)}\,dx$ (b) $\displaystyle\int \frac{2x^2-3x}{(x^2+1)(3x+2)}\,dx$

EXERCISE SET 8.5

C CAS

1–8 Write out the form of the partial fraction decomposition. (Do not find the numerical values of the coefficients.)

1. $\dfrac{3x-1}{(x-3)(x+4)}$
2. $\dfrac{5}{x(x^2-4)}$
3. $\dfrac{2x-3}{x^3-x^2}$
4. $\dfrac{x^2}{(x+2)^3}$
5. $\dfrac{1-x^2}{x^3(x^2+2)}$
6. $\dfrac{3x}{(x-1)(x^2+6)}$
7. $\dfrac{4x^3-x}{(x^2+5)^2}$
8. $\dfrac{1-3x^4}{(x-2)(x^2+1)^2}$

9–32 Evaluate the integral.

9. $\displaystyle\int \frac{dx}{x^2-3x-4}$
10. $\displaystyle\int \frac{dx}{x^2-6x-7}$
11. $\displaystyle\int \frac{11x+17}{2x^2+7x-4}\,dx$
12. $\displaystyle\int \frac{5x-5}{3x^2-8x-3}\,dx$
13. $\displaystyle\int \frac{2x^2-9x-9}{x^3-9x}\,dx$
14. $\displaystyle\int \frac{dx}{x(x^2-1)}$

15. $\int \frac{x^2-8}{x+3}\,dx$ 16. $\int \frac{x^2+1}{x-1}\,dx$

17. $\int \frac{3x^2-10}{x^2-4x+4}\,dx$ 18. $\int \frac{x^2}{x^2-3x+2}\,dx$

19. $\int \frac{x^5+x^2+2}{x^3-x}\,dx$ 20. $\int \frac{x^5-4x^3+1}{x^3-4x}\,dx$

21. $\int \frac{2x^2+3}{x(x-1)^2}\,dx$ 22. $\int \frac{3x^2-x+1}{x^3-x^2}\,dx$

23. $\int \frac{2x^2-10x+4}{(x+1)(x-3)^2}\,dx$ 24. $\int \frac{2x^2-2x-1}{x^3-x^2}\,dx$

25. $\int \frac{x^2}{(x+1)^3}\,dx$ 26. $\int \frac{2x^2+3x+3}{(x+1)^3}\,dx$

27. $\int \frac{2x^2-1}{(4x-1)(x^2+1)}\,dx$ 28. $\int \frac{dx}{x^3+2x}$

29. $\int \frac{x^3+3x^2+x+9}{(x^2+1)(x^2+3)}\,dx$ 30. $\int \frac{x^3+x^2+x+2}{(x^2+1)(x^2+2)}\,dx$

31. $\int \frac{x^3-2x^2+2x-2}{x^2+1}\,dx$

32. $\int \frac{x^4+6x^3+10x^2+x}{x^2+6x+10}\,dx$

33–34 Evaluate the integral by making a substitution that converts the integrand to a rational function.

33. $\int \frac{\cos\theta}{\sin^2\theta+4\sin\theta-5}\,d\theta$ 34. $\int \frac{e^t}{e^{2t}-4}\,dt$

35. Find the volume of the solid generated when the region enclosed by $y=x^2/(9-x^2)$, $y=0$, $x=0$, and $x=2$ is revolved about the x-axis.

36. Find the area of the region under the curve $y=1/(1+e^x)$, over the interval $[-\ln 5, \ln 5]$. [*Hint:* Make a substitution that converts the integrand to a rational function.]

37–38 Use a CAS to evaluate the integral in two ways: (i) integrate directly; (ii) use the CAS to find the partial fraction decomposition and integrate the decomposition. Integrate by hand to check the results.

C 37. $\int \frac{x^2+1}{(x^2+2x+3)^2}\,dx$

C 38. $\int \frac{x^5+x^4+4x^3+4x^2+4x+4}{(x^2+2)^3}\,dx$

39–40 Integrate by hand and check your answers using a CAS.

C 39. $\int \frac{dx}{x^4-3x^3-7x^2+27x-18}$

C 40. $\int \frac{dx}{16x^3-4x^2+4x-1}$

FOCUS ON CONCEPTS

41. Show that

$$\int_0^1 \frac{x}{x^4+1}\,dx = \frac{\pi}{8}$$

42. Use partial fractions to derive the integration formula

$$\int \frac{1}{a^2-x^2}\,dx = \frac{1}{2a}\ln\left|\frac{a+x}{a-x}\right| + C$$

43. Suppose that ax^2+bx+c is a quadratic polynomial and that the integration

$$\int \frac{1}{ax^2+bx+c}\,dx$$

produces a function with no inverse tangent terms. What does this tell you about the roots of the polynomial?

44. Suppose that ax^2+bx+c is a quadratic polynomial and that the integration

$$\int \frac{1}{ax^2+bx+c}\,dx$$

produces a function with neither logarithmic nor inverse tangent terms. What does this tell you about the roots of the polynomial?

45. Does there exist a quadratic polynomial ax^2+bx+c such that the integration

$$\int \frac{x}{ax^2+bx+c}\,dx$$

produces a function with no logarithmic terms? If so, give an example; if not, explain why no such polynomial can exist.

✔ QUICK CHECK ANSWERS 8.5

1. $\frac{A}{(ax+b)^k}$; $\frac{Ax+B}{(ax^2+bx+c)^k}$ 2. (a) A proper rational function is a rational function in which the degree of the numerator is less than the degree of the denominator. (b) The degree of the numerator must be less than the degree of the denominator. (c) Divide the denominator into the numerator, which results in the sum of a polynomial and a proper rational function.
3. (a) $\frac{A_1}{ax+b}+\frac{A_2}{(ax+b)^2}+\cdots+\frac{A_m}{(ax+b)^m}$ (b) $\frac{A_1x+B_1}{ax^2+bx+c}+\frac{A_2x+B_2}{(ax^2+bx+c)^2}+\cdots+\frac{A_mx+B_m}{(ax^2+bx+c)^m}$
4. (a) $\frac{1}{x+1}-\frac{2}{2x-1}$ (b) $\frac{2}{3x+2}-\frac{1}{x^2+1}$ 5. (a) $\int \frac{3}{(x+1)(1-2x)}\,dx = \ln\left|\frac{x+1}{1-2x}\right| + C$
(b) $\int \frac{2x^2-3x}{(x^2+1)(3x+2)}\,dx = \frac{2}{3}\ln|3x+2| - \tan^{-1}x + C$

8.6 USING COMPUTER ALGEBRA SYSTEMS AND TABLES OF INTEGRALS

In this section we will discuss how to integrate using tables, and we will address some of the issues that relate to using computer algebra systems for integration. Readers who are not using computer algebra systems can skip that material.

INTEGRAL TABLES

Tables of integrals are useful for eliminating tedious hand computation. The endpapers of this text contain a relatively brief table of integrals that we will refer to as the ***Endpaper Integral Table***; more comprehensive tables are published in standard reference books such as the *CRC Standard Mathematical Tables and Formulae*, CRC Press, Inc., 2002.

All integral tables have their own scheme for classifying integrals according to the form of the integrand. For example, the Endpaper Integral Table classifies the integrals into 15 categories; *Basic Functions*, *Reciprocals of Basic Functions*, *Powers of Trigonometric Functions*, *Products of Trigonometric Functions*, and so forth. The first step in working with tables is to read through the classifications so that you understand the classification scheme and know where to look in the table for integrals of different types.

PERFECT MATCHES

If you are lucky, the integral you are attempting to evaluate will match up perfectly with one of the forms in the table. However, when looking for matches you may have to make an adjustment for the variable of integration. For example, the integral

$$\int x^2 \sin x \, dx$$

is a perfect match with Formula (46) in the Endpaper Integral Table, except for the letter used for the variable of integration. Thus, to apply Formula (46) to the given integral we need to change the variable of integration in the formula from u to x. With that minor modification we obtain

$$\int x^2 \sin x \, dx = 2x \sin x + (2 - x^2) \cos x + C$$

Here are some more examples of perfect matches.

▶ Example 1 Use the Endpaper Integral Table to evaluate

(a) $\int \sin 7x \cos 2x \, dx$ (b) $\int x^2\sqrt{7+3x}\, dx$

(c) $\int \frac{\sqrt{2-x^2}}{x}\, dx$ (d) $\int (x^3 + 7x + 1) \sin \pi x \, dx$

Solution (a). The integrand can be classified as a product of trigonometric functions. Thus, from Formula (40) with $m = 7$ and $n = 2$ we obtain

$$\int \sin 7x \cos 2x \, dx = -\frac{\cos 9x}{18} - \frac{\cos 5x}{10} + C$$

Solution (b). The integrand can be classified as a power of x multiplying $\sqrt{a+bx}$. Thus, from Formula (103) with $a = 7$ and $b = 3$ we obtain

$$\int x^2\sqrt{7+3x}\, dx = \frac{2}{2835}(135x^2 - 252x + 392)(7+3x)^{3/2} + C$$

Solution (c). The integrand can be classified as a power of x dividing $\sqrt{a^2 - x^2}$. Thus, from Formula (79) with $a = \sqrt{2}$ we obtain

$$\int \frac{\sqrt{2-x^2}}{x}\,dx = \sqrt{2-x^2} - \sqrt{2}\ln\left|\frac{\sqrt{2}+\sqrt{2-x^2}}{x}\right| + C$$

Solution (d). The integrand can be classified as a polynomial multiplying a trigonometric function. Thus, we apply Formula (58) with $p(x) = x^3 + 7x + 1$ and $a = \pi$. The successive nonzero derivatives of $p(x)$ are

$$p'(x) = 3x^2 + 7, \quad p''(x) = 6x, \quad p'''(x) = 6$$

and hence

$$\int (x^3 + 7x + 1)\sin \pi x\,dx$$
$$= -\frac{x^3+7x+1}{\pi}\cos \pi x + \frac{3x^2+7}{\pi^2}\sin \pi x + \frac{6x}{\pi^3}\cos \pi x - \frac{6}{\pi^4}\sin \pi x + C \blacktriangleleft$$

MATCHES REQUIRING SUBSTITUTIONS

Sometimes an integral that does not match any table entry can be made to match by making an appropriate substitution. Here are some examples.

▶ **Example 2** Use the Endpaper Integral Table to evaluate $\int \sqrt{x - 4x^2}\,dx$.

Solution. The integrand does not match any of the forms in the table precisely. It comes closest to matching Formula (112), but it misses because of the factor of 4 multiplying x^2 inside the radical. However, if we make the substitution

$$u = 2x, \quad du = 2\,dx$$

then the $4x^2$ will become a u^2, and the transformed integral will be

$$\int \sqrt{x - 4x^2}\,dx = \frac{1}{2}\int \sqrt{\tfrac{1}{2}u - u^2}\,du$$

which matches Formula (112) with $a = \frac{1}{4}$. Thus, we obtain

$$\int \sqrt{x-4x^2}\,dx = \frac{1}{2}\left[\frac{u-\frac{1}{4}}{2}\sqrt{\tfrac{1}{2}u - u^2} + \frac{1}{32}\sin^{-1}\left(\frac{u-\frac{1}{4}}{\frac{1}{4}}\right)\right] + C$$
$$= \frac{1}{2}\left[\frac{2x-\frac{1}{4}}{2}\sqrt{x-4x^2} + \frac{1}{32}\sin^{-1}\left(\frac{2x-\frac{1}{4}}{\frac{1}{4}}\right)\right] + C$$
$$= \frac{8x-1}{16}\sqrt{x-4x^2} + \frac{1}{64}\sin^{-1}(8x-1) + C \blacktriangleleft$$

▶ **Example 3** Use the Endpaper Integral Table to evaluate

$$\text{(a)} \int e^{\pi x}\sin^{-1}(e^{\pi x})\,dx \qquad \text{(b)} \int x\sqrt{x^2 - 4x + 5}\,dx$$

Solution (a). The integrand does not even come close to matching any of the forms in the table. However, a little thought suggests the substitution

$$u = e^{\pi x}, \quad du = \pi e^{\pi x}\,dx$$

from which we obtain

$$\int e^{\pi x} \sin^{-1}(e^{\pi x})\, dx = \frac{1}{\pi} \int \sin^{-1} u\, du$$

The integrand is now a basic function, and Formula (7) yields

$$\begin{aligned} \int e^{\pi x} \sin^{-1}(e^{\pi x})\, dx &= \frac{1}{\pi}[u \sin^{-1} u + \sqrt{1-u^2}] + C \\ &= \frac{1}{\pi}[e^{\pi x} \sin^{-1}(e^{\pi x}) + \sqrt{1-e^{2\pi x}}] + C \end{aligned}$$

Solution (b). Again, the integrand does not closely match any of the forms in the table. However, a little thought suggests that it may be possible to bring the integrand closer to the form $x\sqrt{x^2+a^2}$ by completing the square to eliminate the term involving x inside the radical. Doing this yields

$$\int x\sqrt{x^2-4x+5}\, dx = \int x\sqrt{(x^2-4x+4)+1}\, dx = \int x\sqrt{(x-2)^2+1}\, dx \qquad (1)$$

At this point we are closer to the form $x\sqrt{x^2+a^2}$, but we are not quite there because of the $(x-2)^2$ rather than x^2 inside the radical. However, we can resolve that problem with the substitution

$$u = x - 2, \quad du = dx$$

With this substitution we have $x = u + 2$, so (1) can be expressed in terms of u as

$$\int x\sqrt{x^2-4x+5}\, dx = \int (u+2)\sqrt{u^2+1}\, du = \int u\sqrt{u^2+1}\, du + 2\int \sqrt{u^2+1}\, du$$

The first integral on the right is now a perfect match with Formula (84) with $a = 1$, and the second is a perfect match with Formula (72) with $a = 1$. Thus, applying these formulas we obtain

$$\int x\sqrt{x^2-4x+5}\, dx = \left[\tfrac{1}{3}(u^2+1)^{3/2}\right] + 2\left[\tfrac{1}{2}u\sqrt{u^2+1} + \tfrac{1}{2}\ln(u+\sqrt{u^2+1})\right] + C$$

If we now replace u by $x - 2$ (in which case $u^2 + 1 = x^2 - 4x + 5$), we obtain

$$\begin{aligned} \int x\sqrt{x^2-4x+5}\, dx = \tfrac{1}{3}(x^2-4x+5)^{3/2} &+ (x-2)\sqrt{x^2-4x+5} \\ &+ \ln(x-2+\sqrt{x^2-4x+5}) + C \end{aligned}$$

Although correct, this form of the answer has an unnecessary mixture of radicals and fractional exponents. If desired, we can "clean up" the answer by writing

$$(x^2-4x+5)^{3/2} = (x^2-4x+5)\sqrt{x^2-4x+5}$$

from which it follows that (verify)

$$\begin{aligned} \int x\sqrt{x^2-4x+5}\, dx = \tfrac{1}{3}(x^2-x-1)&\sqrt{x^2-4x+5} \\ &+ \ln(x-2+\sqrt{x^2-4x+5}) + C \quad \blacktriangleleft \end{aligned}$$

MATCHES REQUIRING REDUCTION FORMULAS

In cases where the entry in an integral table is a reduction formula, that formula will have to be applied first to reduce the given integral to a form in which it can be evaluated.

▶ **Example 4** Use the Endpaper Integral Table to evaluate $\int \frac{x^3}{\sqrt{1+x}}\,dx$.

Solution. The integrand can be classified as a power of x multiplying the reciprocal of $\sqrt{a+bx}$. Thus, from Formula (107) with $a = 1$, $b = 1$, and $n = 3$, followed by Formula (106), we obtain

$$\begin{aligned}\int \frac{x^3}{\sqrt{1+x}}\,dx &= \frac{2x^3\sqrt{1+x}}{7} - \frac{6}{7}\int \frac{x^2}{\sqrt{1+x}}\,dx \\ &= \frac{2x^3\sqrt{1+x}}{7} - \frac{6}{7}\left[\frac{2}{15}(3x^2 - 4x + 8)\sqrt{1+x}\right] + C \\ &= \left(\frac{2x^3}{7} - \frac{12x^2}{35} + \frac{16x}{35} - \frac{32}{35}\right)\sqrt{1+x} + C \quad \blacktriangleleft\end{aligned}$$

MATCHES REQUIRING SPECIAL SUBSTITUTIONS

The Endpaper Integral Table has numerous entries involving an exponent of $3/2$ or involving square roots (exponent $1/2$), but it has no entries with other fractional exponents. However, integrals involving fractional powers of x can often be simplified by making the substitution $u = x^{1/n}$ in which n is the least common multiple of the denominators of the exponents. The resulting integral will then involve integer powers of u. Here are some examples.

▶ **Example 5** Evaluate

$$\text{(a)}\ \int \frac{\sqrt{x}}{1+\sqrt[3]{x}}\,dx \qquad \text{(b)}\ \int \frac{dx}{2+2\sqrt{x}} \qquad \text{(c)}\ \int \sqrt{1+e^x}\,dx$$

Solution (a). The integrand contains $x^{1/2}$ and $x^{1/3}$, so we make the substitution $u = x^{1/6}$, from which we obtain

$$x = u^6, \quad dx = 6u^5\,du$$

Thus,

$$\int \frac{\sqrt{x}}{1+\sqrt[3]{x}}\,dx = \int \frac{(u^6)^{1/2}}{1+(u^6)^{1/3}}(6u^5)\,du = 6\int \frac{u^8}{1+u^2}\,du$$

By long division

$$\frac{u^8}{1+u^2} = u^6 - u^4 + u^2 - 1 + \frac{1}{1+u^2}$$

from which it follows that

$$\begin{aligned}\int \frac{\sqrt{x}}{1+\sqrt[3]{x}}\,dx &= 6\int \left(u^6 - u^4 + u^2 - 1 + \frac{1}{1+u^2}\right)du \\ &= \tfrac{6}{7}u^7 - \tfrac{6}{5}u^5 + 2u^3 - 6u + 6\tan^{-1}u + C \\ &= \tfrac{6}{7}x^{7/6} - \tfrac{6}{5}x^{5/6} + 2x^{1/2} - 6x^{1/6} + 6\tan^{-1}(x^{1/6}) + C\end{aligned}$$

Solution (b). The integrand contains $x^{1/2}$ but does not match any of the forms in the Endpaper Integral Table. Thus, we make the substitution $u = x^{1/2}$, from which we obtain

$$x = u^2, \quad dx = 2u\,du$$

Making this substitution yields

$$\begin{aligned}\int \frac{dx}{2+2\sqrt{x}} &= \int \frac{2u}{2+2u}\,du \\ &= \int \left(1 - \frac{1}{1+u}\right)du \qquad \boxed{\text{Long division}} \\ &= u - \ln|1+u| + C \\ &= \sqrt{x} - \ln(1+\sqrt{x}) + C \qquad \boxed{\text{Absolute value not needed}}\end{aligned}$$

Solution (c). Again, the integral does not match any of the forms in the Endpaper Integral Table. However, the integrand contains $(1+e^x)^{1/2}$, which is analogous to the situation in part (b), except that here it is $1+e^x$ rather than x that is raised to the $1/2$ power. This suggests the substitution $u=(1+e^x)^{1/2}$, from which we obtain (verify)

$$x=\ln(u^2-1), \quad dx=\frac{2u}{u^2-1}\,du$$

Thus,

$$\begin{aligned}
\int \sqrt{1+e^x}\,dx &= \int u\left(\frac{2u}{u^2-1}\right)du \\
&= \int \frac{2u^2}{u^2-1}\,du \\
&= \int \left(2+\frac{2}{u^2-1}\right)du && \text{Long division} \\
&= 2u+\int\left(\frac{1}{u-1}-\frac{1}{u+1}\right)du && \text{Partial fractions} \\
&= 2u+\ln|u-1|-\ln|u+1|+C \\
&= 2u+\ln\left|\frac{u-1}{u+1}\right|+C \\
&= 2\sqrt{1+e^x}+\ln\left[\frac{\sqrt{1+e^x}-1}{\sqrt{1+e^x}+1}\right]+C && \text{Absolute value not needed}
\end{aligned}$$

◀

Functions that consist of finitely many sums, differences, quotients, and products of $\sin x$ and $\cos x$ are called ***rational functions of sin x and cos x***. Some examples are

$$\frac{\sin x+3\cos^2 x}{\cos x+4\sin x}, \quad \frac{\sin x}{1+\cos x-\cos^2 x}, \quad \frac{3\sin^5 x}{1+4\sin x}$$

The Endpaper Integral Table gives a few formulas for integrating rational functions of $\sin x$ and $\cos x$ under the heading *Reciprocals of Basic Functions*. For example, it follows from Formula (18) that

$$\int \frac{1}{1+\sin x}\,dx=\tan x-\sec x+C \tag{2}$$

However, since the integrand is a rational function of $\sin x$, it may be desirable in a particular application to express the value of the integral in terms of $\sin x$ and $\cos x$ and rewrite (2) as

$$\int \frac{1}{1+\sin x}\,dx=\frac{\sin x-1}{\cos x}+C$$

Many rational functions of $\sin x$ and $\cos x$ can be evaluated by an ingenious method that was discovered by the mathematician Karl Weierstrass (see p. 117 for biography). The idea is to make the substitution

$$u=\tan(x/2), \quad -\pi/2<x/2<\pi/2$$

from which it follows that

$$x=2\tan^{-1}u, \quad dx=\frac{2}{1+u^2}\,du$$

To implement this substitution we need to express $\sin x$ and $\cos x$ in terms of u. For this purpose we will use the identities

$$\sin x=2\sin(x/2)\cos(x/2) \tag{3}$$

$$\cos x=\cos^2(x/2)-\sin^2(x/2) \tag{4}$$

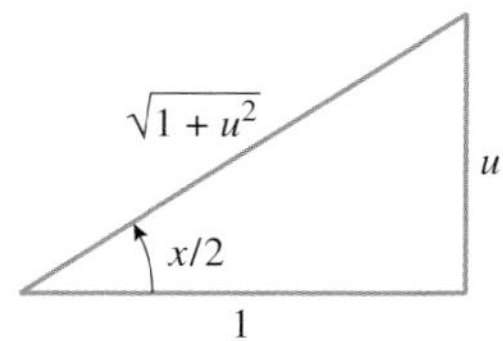

Figure 8.6.1

and the following relationships suggested by Figure 8.6.1:

$$\sin(x/2) = \frac{u}{\sqrt{1+u^2}} \quad \text{and} \quad \cos(x/2) = \frac{1}{\sqrt{1+u^2}}$$

Substituting these expressions in (3) and (4) yields

$$\sin x = 2\left(\frac{u}{\sqrt{1+u^2}}\right)\left(\frac{1}{\sqrt{1+u^2}}\right) = \frac{2u}{1+u^2}$$

$$\cos x = \left(\frac{1}{\sqrt{1+u^2}}\right)^2 - \left(\frac{u}{\sqrt{1+u^2}}\right)^2 = \frac{1-u^2}{1+u^2}$$

In summary, we have shown that the substitution $u = \tan(x/2)$ can be implemented in a rational function of $\sin x$ and $\cos x$ by letting

$$\sin x = \frac{2u}{1+u^2}, \quad \cos x = \frac{1-u^2}{1+u^2}, \quad dx = \frac{2}{1+u^2}\,du \tag{5}$$

► **Example 6** Evaluate $\displaystyle\int \frac{dx}{1 - \sin x + \cos x}$.

Solution. The integrand is a rational function of $\sin x$ and $\cos x$ that does not match any of the formulas in the Endpaper Integral Table, so we make the substitution $u = \tan(x/2)$. Thus, from (5) we obtain

$$\begin{aligned}\int \frac{dx}{1 - \sin x + \cos x} &= \int \frac{\dfrac{2\,du}{1+u^2}}{1 - \left(\dfrac{2u}{1+u^2}\right) + \left(\dfrac{1-u^2}{1+u^2}\right)} \\ &= \int \frac{2\,du}{(1+u^2) - 2u + (1-u^2)} \\ &= \int \frac{du}{1-u} = -\ln|1-u| + C = -\ln|1 - \tan(x/2)| + C \quad ◄\end{aligned}$$

> The substitution $u = \tan(x/2)$ will convert any rational function of $\sin x$ and $\cos x$ to an ordinary rational function of u. However, the method can lead to cumbersome partial fraction decompositions, so it may be worthwhile to consider other methods when hand computations are being used.

INTEGRATING WITH COMPUTER ALGEBRA SYSTEMS

Integration tables are rapidly giving way to computerized integration using computer algebra systems. However, as with many powerful tools, a knowledgeable operator is an important component of the system.

Sometimes computer algebra systems do not produce the most general form of the indefinite integral. For example, the integral formula

$$\int \frac{dx}{x-1} = \ln|x-1| + C$$

which can be obtained by inspection or by using the substitution $u = x - 1$, is valid for $x > 1$ or for $x < 1$. However, not all computer algebra systems produce this form of the answer. Some typical answers produced by various implementations of *Mathematica*, *Maple*, and *Derive* are

$$\ln(-1+x), \quad \ln(x-1), \quad \ln(|x-1|)$$

Observe that none of the systems include the constant of integration—the answer produced is a particular antiderivative and not the most general antiderivative (indefinite integral).

Observe also that only one of these answers includes the absolute value signs; the antiderivatives produced by the other systems are valid only for $x > 1$. All systems, however, are able to calculate the definite integral

$$\int_0^{1/2} \frac{dx}{x-1} = -\ln 2$$

correctly. Now let us examine how these systems handle the integral

$$\int x\sqrt{x^2-4x+5}\,dx = \tfrac{1}{3}(x^2-x-1)\sqrt{x^2-4x+5} + \ln(x-2+\sqrt{x^2-4x+5}) \quad (6)$$

which we obtained in Example 3(b) (with the constant of integration included). Some CAS implementations produce this result in slightly different algebraic forms, but a version of *Maple* produces the result

$$\int x\sqrt{x^2-4x+5}\,dx = \tfrac{1}{3}(x^2-4x+5)^{3/2} + \tfrac{1}{2}(2x-4)\sqrt{x^2-4x+5} + \sinh^{-1}(x-2)$$

This can be rewritten as (6) by expressing the fractional exponent in radical form and expressing $\sinh^{-1}(x-2)$ in logarithmic form using Theorem 7.8.4 (verify). A version of *Mathematica* produces the result

$$\int x\sqrt{x^2-4x+5}\,dx = \tfrac{1}{3}(x^2-x-1)\sqrt{x^2-4x+5} - \sinh^{-1}(2-x)$$

which can be rewritten in form (6) by using Theorem 7.8.4 together with the identity $\sinh^{-1}(-x) = -\sinh^{-1} x$ (verify).

Computer algebra systems can sometimes produce inconvenient or unnatural answers to integration problems. For example, various computer algebra systems produced the following results when asked to integrate $(x+1)^7$:

$$\frac{(x+1)^8}{8}, \quad \frac{1}{8}x^8 + x^7 + \frac{7}{2}x^6 + 7x^5 + \frac{35}{4}x^4 + 7x^3 + \frac{7}{2}x^2 + x \quad (7)$$

> Expanding the expression
>
> $$\frac{(x+1)^8}{8}$$
>
> produces a constant term of $\frac{1}{8}$, whereas the second expression in (7) has no constant term. What is the explanation?

The first form is in keeping with the hand computation

$$\int (x+1)^7\,dx = \frac{(x+1)^8}{8} + C$$

that uses the substitution $u = x+1$, whereas the second form is based on expanding $(x+1)^7$ and integrating term by term.

In Example 2(a) of Section 8.3 we showed that

$$\int \sin^4 x \cos^5 x\,dx = \tfrac{1}{5}\sin^5 x - \tfrac{2}{7}\sin^7 x + \tfrac{1}{9}\sin^9 x + C$$

However, a version of *Mathematica* integrates this as

$$\tfrac{3}{128}\sin x - \tfrac{1}{192}\sin 3x - \tfrac{1}{320}\sin 5x + \tfrac{1}{1792}\sin 7x + \tfrac{1}{2304}\sin 9x$$

whereas other computer algebra systems essentially integrate it as

$$-\tfrac{1}{9}\sin^3 x\cos^6 x - \tfrac{1}{21}\sin x\cos^6 x + \tfrac{1}{105}\cos^4 x\sin x + \tfrac{4}{315}\cos^2 x\sin x + \tfrac{8}{315}\sin x$$

Although these three results look quite different, they can be obtained from one another using appropriate trigonometric identities.

COMPUTER ALGEBRA SYSTEMS HAVE LIMITATIONS

A computer algebra system combines a set of integration rules (such as substitution) with a library of functions that it can use to construct antiderivatives. Such libraries contain elementary functions, such as polynomials, rational functions, trigonometric functions, as well as various nonelementary functions that arise in engineering, physics, and other applied fields. Just as our Endpaper Integral Table has only 121 indefinite integrals, these libraries are not exhaustive of all possible integrands. If the system cannot manipulate the integrand to a form matching one in its library, the program will give some indication that it cannot evaluate the integral. For example, when asked to evaluate the integral

$$\int (1 + \ln x)\sqrt{1 + (x \ln x)^2}\, dx \tag{8}$$

all of the systems mentioned above respond by displaying some form of the unevaluated integral as an answer, indicating that they could not perform the integration.

TECHNOLOGY MASTERY

Sometimes integrals that cannot be evaluated by a CAS in their given form can be evaluated by first rewriting them in a different form or by making a substitution. If you have a CAS, make a u-substitution in (8) that will enable you to evaluate the integral with your CAS. Then evaluate the integral.

Sometimes computer algebra systems respond by expressing an integral in terms of another integral. For example, if you try to integrate e^{x^2} using *Mathematica*, *Maple*, or *Derive*, you will obtain an expression involving erf (which stands for ***error function***). The function erf(x) is defined as

$$\operatorname{erf}(x) = \frac{2}{\sqrt{\pi}} \int_0^x e^{-t^2}\, dt$$

so all three programs essentially rewrite the given integral in terms of a closely related integral. From one point of view this is what we did in integrating $1/x$, since the natural logarithm function is (formally) defined as

$$\ln x = \int_1^x \frac{1}{t}\, dt$$

(see Section 7.6).

► **Example 7** A particle moves along an x-axis in such a way that its velocity $v(t)$ at time t is

$$v(t) = 30\cos^7 t \sin^4 t \quad (t \geq 0)$$

Graph the position versus time curve for the particle, given that the particle is at $x = 1$ when $t = 0$.

Solution. Since $dx/dt = v(t)$ and $x = 1$ when $t = 0$, the position function $x(t)$ is given by

$$x(t) = 1 + \int_0^t v(s)\, ds$$

Some computer algebra systems will allow this expression to be entered directly into a command for plotting functions, but it is often more efficient to perform the integration first. The authors' integration utility yields

$$\begin{aligned} x &= \int 30\cos^7 t \sin^4 t\, dt \\ &= -\tfrac{30}{11}\sin^{11} t + 10\sin^9 t - \tfrac{90}{7}\sin^7 t + 6\sin^5 t + C \end{aligned}$$

where we have added the required constant of integration. Using the initial condition $x(0) = 1$, we substitute the values $x = 1$ and $t = 0$ into this equation to find that $C = 1$, so

$$x(t) = -\tfrac{30}{11}\sin^{11} t + 10\sin^9 t - \tfrac{90}{7}\sin^7 t + 6\sin^5 t + 1 \quad (t \geq 0)$$

The graph of x versus t is shown in Figure 8.6.2. ◄

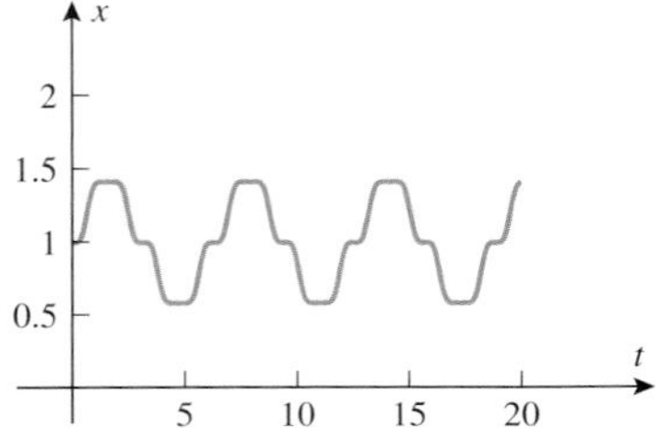

Figure 8.6.2

✔ QUICK CHECK EXERCISES 8.6

(See page 555 for answers.)

1. Find an integral formula in the Endpaper Integral Table that can be used to evaluate the integral. Do not evaluate the integral.

(a) $\int \frac{2x}{3x+4}\,dx$ ________

(b) $\int \frac{1}{x\sqrt{5x-4}}\,dx$ ________

(c) $\int x\sqrt{3x+2}\,dx$ ________

(d) $\int x^2 \ln x\,dx$ ________

2. In each part, make the indicated u-substitution, and then find an integral formula in the Endpaper Integral Table that can be used to evaluate the integral. Do not evaluate the integral.

(a) $\int \frac{x}{1+e^{x^2}}\,dx;\; u = x^2$ ________

(b) $\int e^{\sqrt{x}}\,dx;\; u = \sqrt{x}$ ________

(c) $\int \frac{e^x}{1+\sin(e^x)}\,dx;\; u = e^x$ ________

(d) $\int \frac{1}{(1-4x^2)^{3/2}}\,dx;\; u = 2x$ ________

3. In each part, use the Endpaper Integral Table to evaluate the integral. (If necessary, first make an appropriate substitution or complete the square.)

(a) $\int \frac{1}{4-x^2}\,dx =$ ________

(b) $\int \cos 2x \cos x\,dx =$ ________

(c) $\int \frac{e^{3x}}{\sqrt{1-e^{2x}}}\,dx =$ ________

(d) $\int \frac{x}{x^2-4x+8}\,dx =$ ________

EXERCISE SET 8.6

C CAS

1–24 (a) Use the Endpaper Integral Table to evaluate the given integral. (b) If you have a CAS, use it to evaluate the integral, and then confirm that the result is equivalent to the one that you found in part (a).

1. $\int \frac{4x}{3x-1}\,dx$

2. $\int \frac{x}{(4-5x)^2}\,dx$

3. $\int \frac{1}{x(2x+5)}\,dx$

4. $\int \frac{1}{x^2(1-5x)}\,dx$

5. $\int x\sqrt{2x+3}\,dx$

6. $\int \frac{x}{\sqrt{2-x}}\,dx$

7. $\int \frac{1}{x\sqrt{4-3x}}\,dx$

8. $\int \frac{1}{x\sqrt{3x-4}}\,dx$

9. $\int \frac{1}{16-x^2}\,dx$

10. $\int \frac{1}{x^2-9}\,dx$

11. $\int \sqrt{x^2-3}\,dx$

12. $\int \frac{\sqrt{x^2-5}}{x^2}\,dx$

13. $\int \frac{x^2}{\sqrt{x^2+4}}\,dx$

14. $\int \frac{1}{x^2\sqrt{x^2-2}}\,dx$

15. $\int \sqrt{9-x^2}\,dx$

16. $\int \frac{\sqrt{4-x^2}}{x^2}\,dx$

17. $\int \frac{\sqrt{4-x^2}}{x}\,dx$

18. $\int \frac{1}{x\sqrt{6x-x^2}}\,dx$

19. $\int \sin 3x \sin 4x\,dx$

20. $\int \sin 2x \cos 5x\,dx$

21. $\int x^3 \ln x\,dx$

22. $\int \frac{\ln x}{\sqrt{x^3}}\,dx$

23. $\int e^{-2x} \sin 3x\,dx$

24. $\int e^x \cos 2x\,dx$

25–36 (a) Make the indicated u-substitution, and then use the Endpaper Integral Table to evaluate the integral. (b) If you have a CAS, use it to evaluate the integral, and then confirm that the result is equivalent to the one that you found in part (a).

25. $\int \frac{e^{4x}}{(4-3e^{2x})^2}\,dx,\; u = e^{2x}$

26. $\int \frac{\sin 2x}{(\cos 2x)(3-\cos 2x)}\,dx,\; u = \cos 2x$

27. $\int \frac{1}{\sqrt{x}(9x+4)}\,dx,\; u = 3\sqrt{x}$

28. $\int \frac{\cos 4x}{9+\sin^2 4x}\,dx,\; u = \sin 4x$

29. $\int \frac{1}{\sqrt{4x^2-9}}\,dx,\ u = 2x$

30. $\int x\sqrt{2x^4+3}\,dx,\ u = \sqrt{2}x^2$

31. $\int \frac{4x^5}{\sqrt{2-4x^4}}\,dx,\ u = 2x^2$

32. $\int \frac{1}{x^2\sqrt{3-4x^2}}\,dx,\ u = 2x$

33. $\int \frac{\sin^2(\ln x)}{x}\,dx,\ u = \ln x$

34. $\int e^{-2x}\cos^2(e^{-2x})\,dx,\ u = e^{-2x}$

35. $\int xe^{-2x}\,dx,\ u = -2x$

36. $\int \ln(3x+1)\,dx,\ u = 3x+1$

37–48 (a) Make an appropriate u-substitution, and then use the Endpaper Integral Table to evaluate the integral. (b) If you have a CAS, use it to evaluate the integral (no substitution), and then confirm that the result is equivalent to that in part (a).

37. $\int \frac{\cos 3x}{(\sin 3x)(\sin 3x+1)^2}\,dx$

38. $\int \frac{\ln x}{x\sqrt{4\ln x-1}}\,dx$

39. $\int \frac{x}{16x^4-1}\,dx$

40. $\int \frac{e^x}{3-4e^{2x}}\,dx$

41. $\int e^x\sqrt{3-4e^{2x}}\,dx$

42. $\int \frac{\sqrt{4-9x^2}}{x^2}\,dx$

43. $\int \sqrt{5x-9x^2}\,dx$

44. $\int \frac{1}{x\sqrt{x-5x^2}}\,dx$

45. $\int 4x\sin 2x\,dx$

46. $\int \cos\sqrt{x}\,dx$

47. $\int e^{-\sqrt{x}}\,dx$

48. $\int x\ln(2+x^2)\,dx$

49–52 (a) Complete the square, make an appropriate u-substitution, and then use the Endpaper Integral Table to evaluate the integral. (b) If you have a CAS, use it to evaluate the integral (no substitution or square completion), and then confirm that the result is equivalent to that in part (a).

49. $\int \frac{1}{x^2+6x-7}\,dx$

50. $\int \sqrt{3-2x-x^2}\,dx$

51. $\int \frac{x}{\sqrt{5+4x-x^2}}\,dx$

52. $\int \frac{x}{x^2+6x+13}\,dx$

53–64 (a) Make an appropriate u-substitution of the form $u = x^{1/n}$ or $u = (x+a)^{1/n}$, and then evaluate the integral. (b) If you have a CAS, use it to evaluate the integral, and then confirm that the result is equivalent to the one that you found in part (a).

53. $\int x\sqrt{x-2}\,dx$

54. $\int \frac{x}{\sqrt{x+1}}\,dx$

55. $\int x^5\sqrt{x^3+1}\,dx$

56. $\int \frac{1}{x\sqrt{x^3-1}}\,dx$

57. $\int \frac{dx}{x-\sqrt[3]{x}}$

58. $\int \frac{dx}{\sqrt{x}+\sqrt[3]{x}}$

59. $\int \frac{dx}{x(1-x^{1/4})}$

60. $\int \frac{\sqrt{x}}{x+1}\,dx$

61. $\int \frac{dx}{x^{1/2}-x^{1/3}}$

62. $\int \frac{1+\sqrt{x}}{1-\sqrt{x}}\,dx$

63. $\int \frac{x^3}{\sqrt{1+x^2}}\,dx$

64. $\int \frac{x}{(x+3)^{1/5}}\,dx$

65–70 (a) Make u-substitution (5) to convert the integrand to a rational function of u, and then evaluate the integral. (b) If you have a CAS, use it to evaluate the integral (no substitution), and then confirm that the result is equivalent to that in part (a).

65. $\int \frac{dx}{1+\sin x+\cos x}$

66. $\int \frac{dx}{2+\sin x}$

67. $\int \frac{d\theta}{1-\cos\theta}$

68. $\int \frac{dx}{4\sin x-3\cos x}$

69. $\int \frac{dx}{\sin x+\tan x}$

70. $\int \frac{\sin x}{\sin x+\tan x}\,dx$

71–72 Use any method to solve for x.

71. $\int_2^x \frac{1}{t(4-t)}\,dt = 0.5,\ 2 < x < 4$

72. $\int_1^x \frac{1}{t\sqrt{2t-1}}\,dt = 1,\ x > \frac{1}{2}$

73–76 Use any method to find the area of the region enclosed by the curves.

73. $y = \sqrt{25-x^2},\ y = 0,\ x = 0,\ x = 4$

74. $y = \sqrt{9x^2-4},\ y = 0,\ x = 2$

75. $y = \frac{1}{25-16x^2},\ y = 0,\ x = 0,\ x = 1$

76. $y = \sqrt{x}\ln x,\ y = 0,\ x = 4$

77–80 Use any method to find the volume of the solid generated when the region enclosed by the curves is revolved about the y-axis.

77. $y = \cos x,\ y = 0,\ x = 0,\ x = \pi/2$

78. $y = \sqrt{x-4},\ y = 0,\ x = 8$

79. $y = e^{-x},\ y = 0,\ x = 0,\ x = 3$

80. $y = \ln x,\ y = 0,\ x = 5$

81–82 Use any method to find the arc length of the curve.

81. $y = 2x^2,\ 0 \le x \le 2$ **82.** $y = 3\ln x,\ 1 \le x \le 3$

83–84 Use any method to find the area of the surface generated by revolving the curve about the x-axis.

83. $y = \sin x,\ 0 \le x \le \pi$ **84.** $y = 1/x,\ 1 \le x \le 4$

85–86 Information is given about the motion of a particle moving along a coordinate line.
(a) Use a CAS to find the position function of the particle for $t \ge 0$.
(b) Graph the position versus time curve.

C **85.** $v(t) = 20\cos^6 t\sin^3 t,\ s(0) = 2$

C **86.** $a(t) = e^{-t}\sin 2t\sin 4t,\ v(0) = 0,\ s(0) = 10$

FOCUS ON CONCEPTS

87. (a) Use the substitution $u = \tan(x/2)$ to show that

$$\int \sec x\,dx = \ln\left|\frac{1+\tan(x/2)}{1-\tan(x/2)}\right| + C$$

and confirm that this is consistent with Formula (22) of Section 8.3.

(b) Use the result in part (a) to show that

$$\int \sec x\,dx = \ln\left|\tan\left(\frac{\pi}{4}+\frac{x}{2}\right)\right| + C$$

88. Use the substitution $u = \tan(x/2)$ to show that

$$\int \csc x\,dx = \frac{1}{2}\ln\left[\frac{1-\cos x}{1+\cos x}\right] + C$$

and confirm that this is consistent with the result in Exercise 61(a) of Section 8.3.

89. Find a substitution that can be used to integrate rational functions of $\sinh x$ and $\cosh x$ and use your substitution to evaluate

$$\int \frac{dx}{2\cosh x + \sinh x}$$

without expressing the integrand in terms of e^x and e^{-x}.

90–93 Some integrals that can be evaluated by hand cannot be evaluated by all computer algebra systems. Evaluate the integral by hand, and determine if it can be evaluated on your CAS.

C **90.** $\displaystyle\int \frac{x^3}{\sqrt{1-x^8}}\,dx$

C **91.** $\displaystyle\int (\cos^{32} x\sin^{30} x - \cos^{30} x\sin^{32} x)\,dx$

C **92.** $\displaystyle\int \sqrt{x - \sqrt{x^2-4}}\,dx$ [*Hint:* $\frac{1}{2}(\sqrt{x+2} - \sqrt{x-2})^2 = ?$]

C **93.** $\displaystyle\int \frac{1}{x^{10}+x}\,dx$
[*Hint:* Rewrite the denominator as $x^{10}(1+x^{-9})$.]

C **94.** Let

$$f(x) = \frac{-2x^5 + 26x^4 + 15x^3 + 6x^2 + 20x + 43}{x^6 - x^5 - 18x^4 - 2x^3 - 39x^2 - x - 20}$$

(a) Use a CAS to factor the denominator, and then write down the form of the partial fraction decomposition. You need not find the values of the constants.
(b) Check your answer in part (a) by using the CAS to find the partial fraction decomposition of f.
(c) Integrate f by hand, and then check your answer by integrating with the CAS.

✔ QUICK CHECK ANSWERS 8.6

1. (a) Formula (60) (b) Formula (108) (c) Formula (102) (d) Formula (50) **2.** (a) Formula (25) (b) Formula (51) (c) Formula (18) (d) Formula (97) **3.** (a) $\frac{1}{4}\ln\left|\frac{x+2}{x-2}\right| + C$ (b) $\frac{1}{6}\sin 3x + \frac{1}{2}\sin x + C$ (c) $-\frac{e^x}{2}\sqrt{1-e^{2x}} + \frac{1}{2}\sin^{-1} e^x + C$ (d) $\frac{1}{2}\ln\left(x^2 - 4x + 8\right) + \tan^{-1}\frac{x-2}{2} + C$

8.7 NUMERICAL INTEGRATION; SIMPSON'S RULE

If it is necessary to evaluate a definite integral of a function for which an antiderivative cannot be found, then one must settle for some kind of numerical approximation of the integral. In Section 5.4 we considered three such approximations in the context of areas—left endpoint approximation, right endpoint approximation, and midpoint approximation. In this section we will extend those methods to general definite integrals, and we will develop some new methods that often provide more accuracy with less computation. We will also discuss the errors that arise in integral approximations.

A REVIEW OF RIEMANN SUM APPROXIMATIONS

Recall from Section 5.5 that the definite integral of a continuous function f over an interval $[a, b]$ may be computed as

$$\int_a^b f(x)\,dx = \lim_{\max \Delta x_k \to 0} \sum_{k=1}^{n} f(x_k^*)\Delta x_k$$

where the sum that appears on the right side is called a Riemann sum. In this formula, Δx_k is the width of the kth subinterval of a partition $a = x_0 < x_1 < x_2 < \cdots < x_n = b$ of $[a, b]$ into n subintervals, and x_k^* denotes an arbitrary point in the kth subinterval. If we take all subintervals of the same width, so that $\Delta x_k = (b-a)/n$, then as n increases the Riemann sum will eventually be a good approximation to the definite integral. We denote this by writing

$$\int_a^b f(x)\,dx \approx \left(\frac{b-a}{n}\right)[f(x_1^*) + f(x_2^*) + \cdots + f(x_n^*)] \tag{1}$$

If we denote the values of f at the endpoints of the subintervals by

$$y_0 = f(a), \quad y_1 = f(x_1), \quad y_2 = f(x_2), \ldots, y_{n-1} = f(x_{n-1}), \quad y_n = f(x_n)$$

and the values of f at the midpoints of the subintervals by

$$y_{m_1}, y_{m_2}, \ldots, y_{m_n}$$

then it follows from (1) that the left endpoint, right endpoint, and midpoint approximations discussed in Section 5.4 can be expressed as shown in Table 8.7.1. Although we originally obtained these results for nonnegative functions in the context of approximating areas, they are applicable to any function that is continuous on $[a, b]$.

Table 8.7.1

LEFT ENDPOINT APPROXIMATION	RIGHT ENDPOINT APPROXIMATION	MIDPOINT APPROXIMATION
$\int_a^b f(x)\,dx \approx \left(\frac{b-a}{n}\right)[y_0 + y_1 + \cdots + y_{n-1}]$	$\int_a^b f(x)\,dx \approx \left(\frac{b-a}{n}\right)[y_1 + y_2 + \cdots + y_n]$	$\int_a^b f(x)\,dx \approx \left(\frac{b-a}{n}\right)[y_{m_1} + y_{m_2} + \cdots + y_{m_n}]$
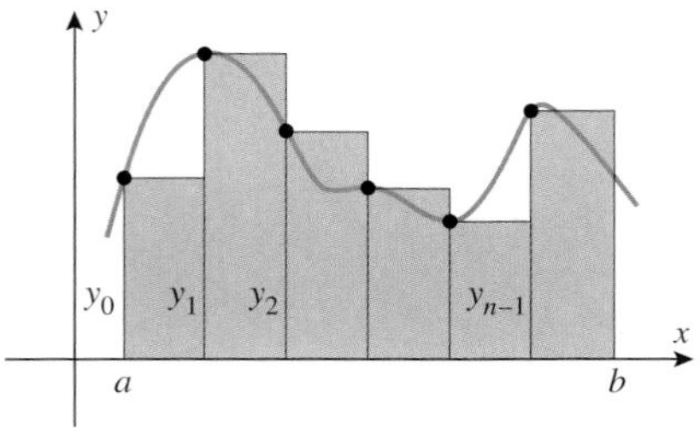 y; x; a; b; y_0; y_1; y_2; y_{n-1}	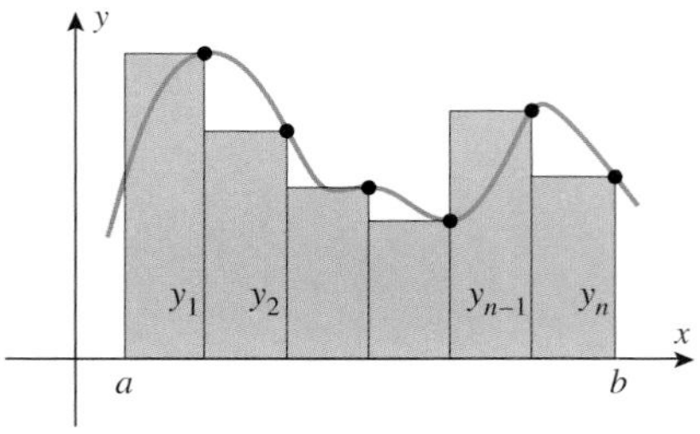 y; x; a; b; y_1; y_2; y_{n-1}; y_n	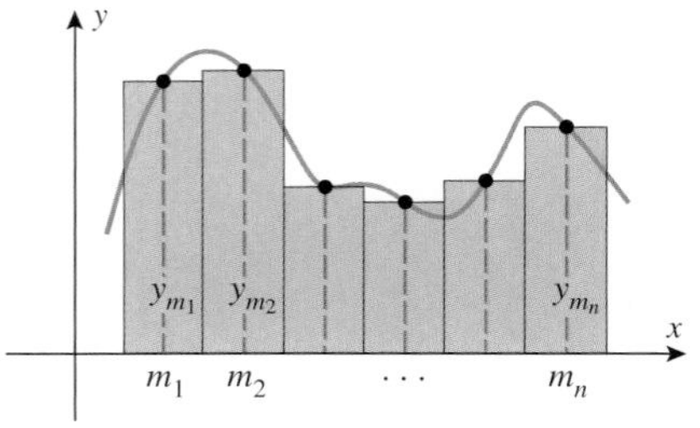 y; x; m_1; m_2; ...; m_n; y_{m_1}; y_{m_2}; y_{m_n}

TRAPEZOIDAL APPROXIMATION

It will be convenient in this section to denote the left endpoint, right endpoint, and midpoint approximations with n subintervals by L_n, R_n, and M_n, respectively. Of the three approximations, the midpoint approximation is most widely used in applications. If we take the average of L_n and R_n, then we obtain another important approximation denoted by

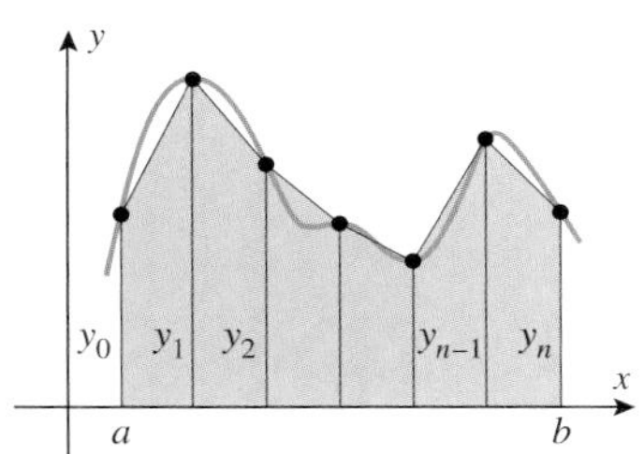

Trapezoidal approximation

Figure 8.7.1

$$T_n = \tfrac{1}{2}(L_n + R_n)$$

called the ***trapezoidal approximation***:

Trapezoidal Approximation

$$\int_a^b f(x)\,dx \approx T_n = \left(\frac{b-a}{2n}\right)[y_0 + 2y_1 + \cdots + 2y_{n-1} + y_n] \tag{2}$$

The name "trapezoidal approximation" results from the fact that in the case where f is nonnegative on the interval of integration, the approximation T_n is the sum of the trapezoidal areas shown in Figure 8.7.1 (see Exercise 47).

▶ **Example 1** In Table 8.7.2 we have approximated

$$\ln 2 = \int_1^2 \frac{1}{x}\,dx$$

using the midpoint approximation and the trapezoidal approximation.* In each case we used $n = 10$ subdivisions of the interval $[1, 2]$, so that

$$\underbrace{\frac{b-a}{n} = \frac{2-1}{10} = 0.1}_{\text{Midpoint}} \quad \text{and} \quad \underbrace{\frac{b-a}{2n} = \frac{2-1}{20} = 0.05}_{\text{Trapezoidal}}$$ ◀

Table 8.7.2

MIDPOINT APPROXIMATION

i	MIDPOINT m_i	$y_{m_i} = f(m_i) = 1/m_i$
1	1.05	0.952380952
2	1.15	0.869565217
3	1.25	0.800000000
4	1.35	0.740740741
5	1.45	0.689655172
6	1.55	0.645161290
7	1.65	0.606060606
8	1.75	0.571428571
9	1.85	0.540540541
10	1.95	0.512820513
		6.928353603

$$\int_1^2 \frac{1}{x}\,dx \approx (0.1)(6.928353603) \approx 0.692835360$$

TRAPEZOIDAL APPROXIMATION

i	ENDPOINT x_i	$y_i = f(x_i) = 1/x_i$	MULTIPLIER w_i	$w_i y_i$
0	1.0	1.000000000	1	1.000000000
1	1.1	0.909090909	2	1.818181818
2	1.2	0.833333333	2	1.666666667
3	1.3	0.769230769	2	1.538461538
4	1.4	0.714285714	2	1.428571429
5	1.5	0.666666667	2	1.333333333
6	1.6	0.625000000	2	1.250000000
7	1.7	0.588235294	2	1.176470588
8	1.8	0.555555556	2	1.111111111
9	1.9	0.526315789	2	1.052631579
10	2.0	0.500000000	1	0.500000000
				13.875428063

$$\int_1^2 \frac{1}{x}\,dx \approx (0.05)(13.875428063) \approx 0.693771403$$

*Throughout this section we will show numerical values to nine places to the right of the decimal point. If your calculating utility does not show this many places, then you will need to make the appropriate adjustments. What is important here is that you understand the principles being discussed.

By rewriting (3) and (4) in the form

$$\int_a^b f(x)\,dx = \text{approximation} + \text{error}$$

we see that positive values of E_M and E_T correspond to underestimates and negative values to overestimates.

COMPARISON OF THE MIDPOINT AND TRAPEZOIDAL APPROXIMATIONS

We define the ***errors*** in the midpoint and trapezoidal approximations to be

$$E_M = \int_a^b f(x)\,dx - M_n \quad \text{and} \quad E_T = \int_a^b f(x)\,dx - T_n \tag{3–4}$$

respectively, and we define $|E_M|$ and $|E_T|$ to be the ***absolute errors*** in these approximations. The absolute errors are nonnegative and do not distinguish between underestimates and overestimates.

► **Example 2** The value of ln 2 to nine decimal places is

$$\ln 2 = \int_1^2 \frac{1}{x}\,dx \approx 0.693147181 \tag{5}$$

so we see from Tables 8.7.2 and 8.7.3 that the absolute errors in approximating ln 2 by M_{10} and T_{10} are

$$|E_M| = |\ln 2 - M_{10}| \approx 0.000311821$$
$$|E_T| = |\ln 2 - T_{10}| \approx 0.000624222$$

Thus, the midpoint approximation is more accurate than the trapezoidal approximation in this case. ◄

Table 8.7.3

ln 2 (NINE DECIMAL PLACES)	APPROXIMATION	ERROR
0.693147181	$M_{10} \approx 0.692835360$	$E_M = \ln 2 - M_{10} \approx 0.000311821$
0.693147181	$T_{10} \approx 0.693771403$	$E_T = \ln 2 - T_{10} \approx -0.000624222$

It is not accidental in Example 2 that the midpoint approximation of ln 2 was more accurate than the trapezoidal approximation. To see why this is so, we first need to look at the midpoint approximation from another point of view. To simplify our explanation, we will assume that f is nonnegative on $[a, b]$, though the conclusions we reach will be true without this assumption.

If f is a differentiable function, then the midpoint approximation is sometimes called the ***tangent line approximation*** because for each subinterval of $[a, b]$ the area of the rectangle used in the midpoint approximation is equal to the area of the trapezoid whose upper boundary is the tangent line to $y = f(x)$ at the midpoint of the subinterval (Figure 8.7.2). The equality of these areas follows from the fact that the shaded areas in the figure are congruent. We will now show how this point of view about midpoint approximations can be used to establish useful criteria for determining which of M_n or T_n produces the better approximation of a given integral.

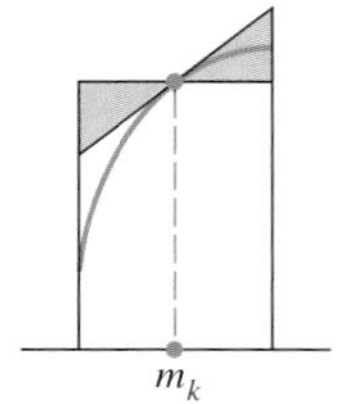

The shaded triangles have equal areas.

Figure 8.7.2

In Figure 8.7.3*a* we have isolated a subinterval of $[a, b]$ on which the graph of a function f is concave down, and we have shaded the areas that represent the errors in the midpoint and trapezoidal approximations over the subinterval. In Figure 8.7.3*b* we show a succession of four illustrations which make it evident that the error from the midpoint approximation is less than that from the trapezoidal approximation. If the graph of f were concave up, analogous figures would lead to the same conclusion. (This argument, due to Frank Buck, appeared in *The College Mathematics Journal*, Vol. 16, No. 1, 1985.)

Figure 8.7.3*a* also suggests that on a subinterval where the graph is concave down, the midpoint approximation is larger than the value of the integral and the trapezoidal

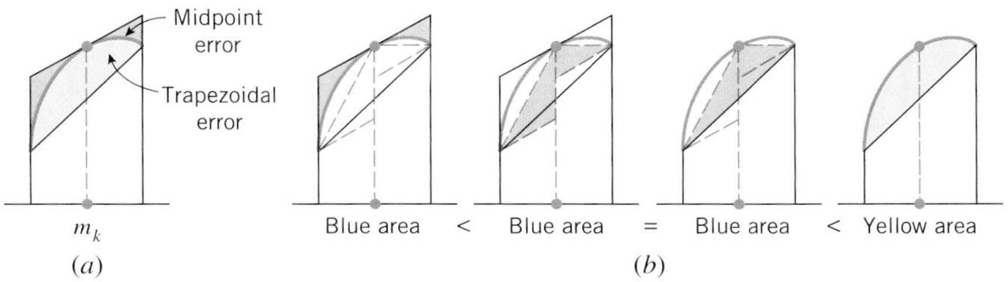

Figure 8.7.3

approximation is smaller. On an interval where the graph is concave up it is the other way around. In summary, we have the following result, which we state without formal proof:

8.7.1 THEOREM. *Let f be continuous on $[a, b]$, and let $|E_M|$ and $|E_T|$ be the absolute errors that result from the midpoint and trapezoidal approximations of $\int_a^b f(x)\,dx$ using n subintervals.*

(a) *If the graph of f is either concave up or concave down on (a, b), then $|E_M| < |E_T|$, that is, the absolute error from the midpoint approximation is less than that from the trapezoidal approximation.*

(b) *If the graph of f is concave down on (a, b), then*

$$T_n < \int_a^b f(x)\,dx < M_n$$

(c) *If the graph of f is concave up on (a, b), then*

$$M_n < \int_a^b f(x)\,dx < T_n$$

▶ **Example 3** Since the graph of $f(x) = 1/x$ is continuous on the interval $[1, 2]$ and concave up on the interval $(1, 2)$, it follows from part (a) of Theorem 8.7.1 that M_n will always provide a better approximation than T_n for

$$\int_1^2 \frac{1}{x}\,dx = \ln 2$$

Moreover, if follows from part (c) of Theorem 8.7.1 that $M_n < \ln 2 < T_n$ for every positive integer n. Note that this is consistent with our computations in Example 2. ◀

WARNING

Do not conclude that the midpoint approximation is always better than the trapezoidal approximation; the trapezoidal approximation may be better if the function changes concavity on the interval of integration.

▶ **Example 4** The midpoint and trapezoidal approximations can be used to approximate sin 1 by using the integral

$$\sin 1 = \int_0^1 \cos x\,dx$$

Since $f(x) = \cos x$ is continuous on $[0, 1]$ and concave down on $(0, 1)$, it follows from parts (a) and (b) of Theorem 8.7.1 that the absolute error in M_n will be less than that in T_n, and that $T_n < \sin 1 < M_n$ for every positive integer n. This is consistent with the results in Table 8.7.4 for $n = 5$ (intermediate computations are omitted). ◀

Table 8.7.4

sin 1 (NINE DECIMAL PLACES)	APPROXIMATION	ERROR
0.841470985	$M_5 \approx 0.842875074$	$E_M = \sin 1 - M_5 \approx -0.001404089$
0.841470985	$T_5 \approx 0.838664210$	$E_T = \sin 1 - T_5 \approx 0.002806775$

▶ **Example 5** Table 8.7.5 shows approximations for $\sin 3 = \int_0^3 \cos x\, dx$ using the midpoint and trapezoidal approximations with $n = 10$ subdivisions of the interval $[0, 3]$. Note that $|E_M| < |E_T|$ and $T_{10} < \sin 3 < M_{10}$, although these results are not guaranteed by Theorem 8.7.1 since $f(x) = \cos x$ changes concavity on the interval $[0, 3]$. ◀

Table 8.7.5

sin 3 (NINE DECIMAL PLACES)	APPROXIMATION	ERROR
0.141120008	$M_{10} \approx 0.141650601$	$E_M = \sin 3 - M_{10} \approx -0.000530592$
0.141120008	$T_{10} \approx 0.140060017$	$E_T = \sin 3 - T_{10} \approx 0.001059991$

SIMPSON'S RULE

Recall that the average of the left and right endpoint approximations yields the better trapezoidal approximation. We now see how a weighted average of the midpoint and trapezoidal approximations can yield an even better approximation.

The numerical evidence in Tables 8.7.3, 8.7.4, and 8.7.5 reveals that $E_T \approx -2E_M$ in these instances. This suggests that

$$\begin{aligned} 3\int_a^b f(x)\, dx &= 2\int_a^b f(x)\, dx + \int_a^b f(x)\, dx \\ &= 2(M_n + E_M) + (T_n + E_T) \\ &= (2M_n + T_n) + (2E_M + E_T) \\ &\approx 2M_n + T_n \end{aligned}$$

This gives

$$\int_a^b f(x)\, dx \approx \tfrac{1}{3}(2M_n + T_n)$$

We will use S_{2n} to denote the right side of this approximation. That is,

$$S_{2n} = \tfrac{1}{3}(2M_n + T_n)$$

Table 8.7.6 displays the approximations S_{2n} corresponding to the data in Tables 8.7.3 to 8.7.5.

Table 8.7.6

FUNCTION VALUE (NINE DECIMAL PLACES)	APPROXIMATION	ERROR
$\ln 2 \approx 0.693147181$	$\int_1^2 (1/x)\, dx \approx S_{20} = \frac{1}{3}(2M_{10} + T_{10}) \approx 0.693147375$	-0.000000194
$\sin 1 \approx 0.841470985$	$\int_0^1 \cos x\, dx \approx S_{10} = \frac{1}{3}(2M_5 + T_5) \approx 0.841471453$	-0.000000468
$\sin 3 \approx 0.141120008$	$\int_0^3 \cos x\, dx \approx S_{20} = \frac{1}{3}(2M_{10} + T_{10}) \approx 0.141120406$	-0.000000398

Using the midpoint approximation formula in Table 8.7.1 and Formula (2) for the trapezoidal approximation, we can derive a similar formula for S_{2n}. For convenience, we partition the interval $[a, b]$ into $2n$ subintervals, each of length $(b-a)/(2n)$. Label the endpoints of these subintervals by $a = x_0, x_1, x_2, \ldots, x_{2n} = b$. Then $x_0, x_2, x_4, \ldots, x_{2n}$ define a partition of $[a, b]$ into n equal subintervals, and the midpoints of these subintervals are $x_1, x_3, x_5, \ldots, x_{2n-1}$, respectively. Using $y_i = f(x_i)$, we have

$$\begin{aligned} M_n &= \left(\frac{b-a}{n}\right)[y_1 + y_3 + \cdots + y_{2n-1}] \\ &= \left(\frac{b-a}{2n}\right)[2y_1 + 2y_3 + \cdots + 2y_{2n-1}] \\ T_n &= \left(\frac{b-a}{2n}\right)[y_0 + 2y_2 + 2y_4 + \cdots + 2y_{2n-2} + y_{2n}] \end{aligned}$$

Thus, $S_{2n} = \frac{1}{3}(2M_n + T_n)$ can be expressed as

$$S_{2n} = \frac{1}{3}\left(\frac{b-a}{2n}\right)[y_0 + 4y_1 + 2y_2 + 4y_3 + 2y_4 + \cdots + 2y_{2n-2} + 4y_{2n-1} + y_{2n}] \tag{6}$$

The approximation

$$\int_a^b f(x)\,dx \approx S_{2n} \tag{7}$$

as given in (6) is known as ***Simpson's rule***. We denote the error in this approximation by

$$E_S = \int_a^b f(x)\,dx - S_{2n} \tag{8}$$

As before, the absolute error in approximation (7) is given by $|E_S|$.

▶ **Example 6** In Table 8.7.7 we have used Simpson's rule with $2n = 10$ subintervals to obtain the approximation

$$\ln 2 = \int_1^2 \frac{1}{x}\,dx \approx S_{10} = 0.693150231$$

For this approximation,

$$\frac{1}{3}\left(\frac{b-a}{2n}\right) = \frac{1}{3}\left(\frac{2-1}{10}\right) = \frac{1}{30}$$

Although S_{10} is a weighted average of M_5 and T_5, it makes sense to compare S_{10} to M_{10} and T_{10}, since the sums for these three approximations involve the same number of terms.

Thomas Simpson (1710–1761) English mathematician. Simpson was the son of a weaver. He was trained to follow in his father's footsteps and had little formal education in his early life. His interest in science and mathematics was aroused in 1724, when he witnessed an eclipse of the Sun and received two books from a peddler, one on astrology and the other on arithmetic. Simpson quickly absorbed their contents and soon became a successful local fortune teller. His improved financial situation enabled him to give up weaving and marry his landlady. Then in 1733 some mysterious "unfortunate incident" forced him to move. He settled in Derby, where he taught in an evening school and worked at weaving during the day. In 1736 he moved to London and published his first mathematical work in a periodical called the *Ladies' Diary* (of which he later became the editor). In 1737 he published a successful calculus textbook that enabled him to give up weaving completely and concentrate on textbook writing and teaching. His fortunes improved further in 1740 when one Robert Heath accused him of plagiarism. The publicity was marvelous, and Simpson proceeded to dash off a succession of best-selling textbooks: *Algebra* (ten editions plus translations), *Geometry* (twelve editions plus translations), *Trigonometry* (five editions plus translations), and numerous others. It is interesting to note that Simpson did not discover the rule that bears his name. It was a well-known result by Simpson's time.

Using the values for M_{10} and T_{10} from Example 2 and the value for S_{10} in Table 8.7.7, we have

$$|E_M| = |\ln 2 - M_{10}| \approx |0.693147181 - 0.692835360| = 0.000311821$$
$$|E_T| = |\ln 2 - T_{10}| \approx |0.693147181 - 0.693771403| = 0.000624222$$
$$|E_S| = |\ln 2 - S_{10}| \approx |0.693147181 - 0.693150231| = 0.000003050$$

Comparing these absolute errors, it is clear that S_{10} is a much more accurate approximation of ln 2 than either M_{10} or T_{10}. ◀

Table 8.7.7

i	ENDPOINT x_i	$y_i = f(x_i) = 1/x_i$	MULTIPLIER w_i	$w_i y_i$
0	1.0	1.000000000	1	1.000000000
1	1.1	0.909090909	4	3.636363636
2	1.2	0.833333333	2	1.666666667
3	1.3	0.769230769	4	3.076923077
4	1.4	0.714285714	2	1.428571429
5	1.5	0.666666667	4	2.666666667
6	1.6	0.625000000	2	1.250000000
7	1.7	0.588235294	4	2.352941176
8	1.8	0.555555556	2	1.111111111
9	1.9	0.526315789	4	2.105263158
10	2.0	0.500000000	1	0.500000000
				20.794506921

$$\int_1^2 \frac{1}{x}\,dx \approx \left(\tfrac{1}{30}\right)(20.794506921) \approx 0.693150231$$

GEOMETRIC INTERPRETATION OF SIMPSON'S RULE

The midpoint (or tangent line) approximation and the trapezoidal approximation for a definite integral are based on approximating a segment of the curve $y = f(x)$ by line segments. Intuition suggests that we might improve on these approximations using parabolic arcs rather than line segments, thereby accounting for concavity of the curve $y = f(x)$ more closely.

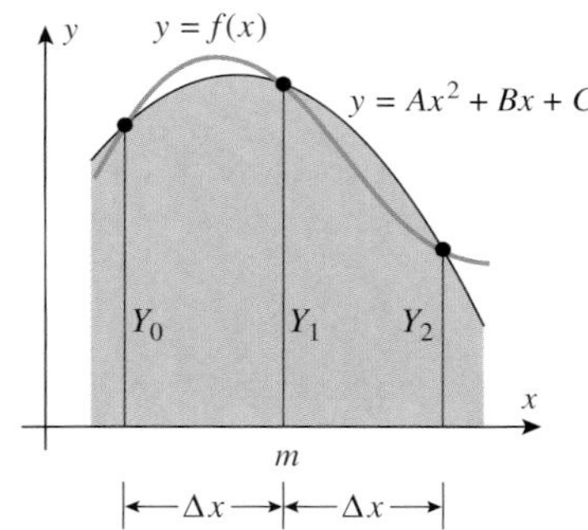

Figure 8.7.4

At the heart of this idea is a formula, sometimes called the ***one-third rule***. The one-third rule expresses a definite integral of a quadratic function $g(x) = Ax^2 + Bx + C$ in terms of the values Y_0, Y_1, and Y_2 of g at the left endpoint, midpoint, and right endpoint, respectively, of the interval of integration $[m - \Delta x, m + \Delta x]$ (see Figure 8.7.4).

$$\int_{m-\Delta x}^{m+\Delta x} (Ax^2 + Bx + C)\,dx = \frac{\Delta x}{3}[Y_0 + 4Y_1 + Y_2] \tag{9}$$

Verification of the one-third rule is left as an exercise. By applying the one-third rule to subintervals $[x_{2k-2}, x_{2k}]$, $k = 1, \ldots, n$, one arrives at Formula (6) for Simpson's rule (Exercises 49–50). Thus, Simpson's rule corresponds to the integral of a piecewise-quadratic approximation to $f(x)$.

ERROR BOUNDS

With all the methods studied in this section, there are two sources of error: the *intrinsic* or *truncation error* due to the approximation formula, and the *roundoff error* introduced in the calculations. In general, increasing n reduces the truncation error but increases the roundoff error, since more computations are required for larger n. In practical applications, it is important to know how large n must be taken to ensure that a specified degree of accuracy is obtained. The analysis of roundoff error is complicated and will not be considered here. However, the following theorems, which are proved in books on numerical analysis, provide upper bounds on the truncation errors in the midpoint, trapezoidal, and Simpson's rule approximations.

8.7.2 THEOREM (*Midpoint and Trapezoidal Error Bounds*). *If f'' is continuous on $[a, b]$ and if K_2 is the maximum value of $|f''(x)|$ on $[a, b]$, then*

$$(a)\ |E_M| = \left|\int_a^b f(x)\,dx - M_n\right| \le \frac{(b-a)^3 K_2}{24n^2} \tag{10}$$

$$(b)\ |E_T| = \left|\int_a^b f(x)\,dx - T_n\right| \le \frac{(b-a)^3 K_2}{12n^2} \tag{11}$$

8.7.3 THEOREM (*Simpson Error Bound*). *If $f^{(4)}$ is continuous on $[a, b]$ and if K_4 is the maximum value of $|f^{(4)}(x)|$ on $[a, b]$, then*

$$|E_S| = \left|\int_a^b f(x)\,dx - S_{2n}\right| \le \frac{(b-a)^5 K_4}{180(2n)^4} \tag{12}$$

▶ Example 7 Find an upper bound on the absolute error that results from approximating

$$\ln 2 = \int_1^2 \frac{1}{x}\,dx$$

using (a) the midpoint approximation M_{10} with $n = 10$ subintervals, (b) the trapezoidal approximation T_{10} with $n = 10$ subintervals, and (c) Simpson's rule S_{10} with $2n = 10$ subintervals.

Solution. We will apply Formulas (10), (11), and (12) with

$$f(x) = \frac{1}{x}, \quad a = 1, \quad \text{and} \quad b = 2$$

For (10) and (11) we use $n = 10$; for (12) we use $2n = 10$, or $n = 5$. We have

$$f'(x) = -\frac{1}{x^2}, \quad f''(x) = \frac{2}{x^3}, \quad f'''(x) = -\frac{6}{x^4}, \quad f^{(4)}(x) = \frac{24}{x^5}$$

Thus,

$$|f''(x)| = \left|\frac{2}{x^3}\right| = \frac{2}{x^3}, \quad |f^{(4)}(x)| = \left|\frac{24}{x^5}\right| = \frac{24}{x^5} \tag{13–14}$$

Note that the upper bounds calculated in Example 7 are consistent with the values $|E_M|$, $|E_T|$, and $|E_S|$ calculated in Example 6 but are considerably greater than those values. It is quite common that the upper bounds on the absolute errors given in Theorems 8.7.2 and 8.7.3 substantially exceed the actual absolute errors. However, that does not diminish the utility of these bounds.

where we have dropped the absolute values because $f''(x)$ and $f^{(4)}(x)$ have positive values for $1 \le x \le 2$. Since (13) and (14) are continuous and decreasing on $[1, 2]$, both functions have their maximum values at $x = 1$; for (13) this maximum value is 2 and for (14) the maximum value is 24. Thus we can take $K_2 = 2$ in (10) and (11), and $K_4 = 24$ in (12). This yields

$$|E_M| \le \frac{(b-a)^3 K_2}{24n^2} = \frac{1^3 \cdot 2}{24 \cdot 10^2} \approx 0.000833333$$

$$|E_T| \le \frac{(b-a)^3 K_2}{12n^2} = \frac{1^3 \cdot 2}{12 \cdot 10^2} \approx 0.001666667$$

$$|E_S| \le \frac{(b-a)^5 K_4}{180(2n)^4} = \frac{1^5 \cdot 24}{180 \cdot 10^4} \approx 0.000013333 \ ◀$$

▶ **Example 8** How many subintervals should be used in approximating

$$\ln 2 = \int_1^2 \frac{1}{x}\,dx$$

by Simpson's rule for five decimal-place accuracy?

Solution. To obtain five decimal-place accuracy, we must choose the number of subintervals so that

$$|E_S| \le 0.000005 = 5 \times 10^{-6}$$

From (12), this can be achieved by taking $2n$ in Simpson's rule to satisfy

$$\frac{(b-a)^5 K_4}{180(2n)^4} \le 5 \times 10^{-6}$$

Taking $a = 1$, $b = 2$, and $K_4 = 24$ (found in Example 7) in this inequality yields

$$\frac{1^5 \cdot 24}{180 \cdot (2n)^4} \le 5 \times 10^{-6}$$

which, on taking reciprocals, can be rewritten as

$$(2n)^4 \ge \frac{2 \times 10^6}{75} \quad \text{or} \quad n^4 \ge \frac{10^4}{6}$$

Thus,

$$n \ge \frac{10}{\sqrt[4]{6}} \approx 6.389$$

Since n must be an integer, the smallest value of n that satisfies this requirement is $n = 7$, or $2n = 14$. Thus, the approximation S_{14} using 14 subintervals will produce five decimal-place accuracy. ◀

In cases where it is difficult to find the values of K_2 and K_4 in Formulas (10), (11), and (12), these constants may be replaced by any larger constants. For example, suppose that a constant K can be easily found with the certainty that $|f''(x)| < K$ on the interval. Then $K_2 \le K$ and

$$|E_T| \le \frac{(b-a)^3 K_2}{12n^2} \le \frac{(b-a)^3 K}{12n^2} \tag{15}$$

so the right side of (15) is also an upper bound on the value of $|E_T|$. Using K, however, will likely increase the computed value of n needed for a given error tolerance. Many applications involve the resolution of competing practical issues, here illustrated through the trade-off between the convenience of finding a crude bound for $|f''(x)|$ versus the efficiency of using the smallest possible n for a desired accuracy.

▶ **Example 9** How many subintervals should be used in approximating

$$\int_0^1 \cos(x^2)\,dx$$

by the midpoint approximation for three decimal-place accuracy?

Solution. To obtain three decimal-place accuracy, we must choose n so that

$$|E_M| \le 0.0005 = 5 \times 10^{-4} \tag{16}$$

From (10) with $f(x) = \cos(x^2)$, $a = 0$, and $b = 1$, an upper bound on $|E_M|$ is given by

$$|E_M| \le \frac{K_2}{24n^2} \tag{17}$$

where $|K_2|$ is the maximum value of $|f''(x)|$ on the interval $[0, 1]$. But,

$$f'(x) = -2x\sin(x^2)$$
$$f''(x) = -4x^2\cos(x^2) - 2\sin(x^2) = -[4x^2\cos(x^2) + 2\sin(x^2)]$$

so that

$$|f''(x)| = |4x^2\cos(x^2) + 2\sin(x^2)| \tag{18}$$

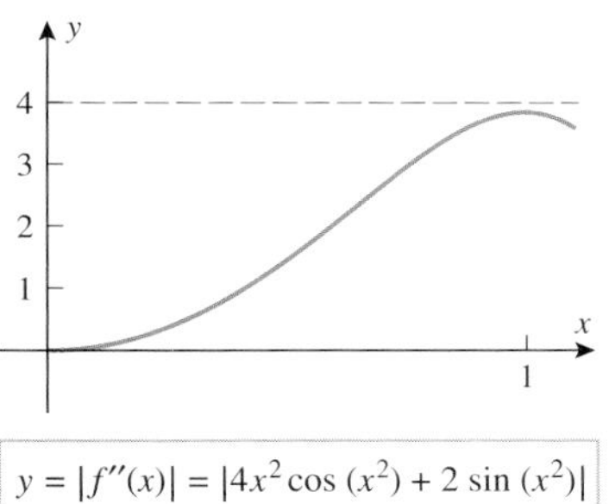

$y = |f''(x)| = |4x^2\cos(x^2) + 2\sin(x^2)|$

Figure 8.7.5

It would be tedious to look for the maximum value of this function on the interval $[0, 1]$. For x in $[0, 1]$, it is easy to see that each of the expressions x^2, $\cos(x^2)$, and $\sin(x^2)$ is bounded in absolute value by 1, so $|4x^2\cos(x^2) + 2\sin(x^2)| \le 4 + 2 = 6$ on $[0, 1]$. We can improve on this by using a graphing utility to sketch $|f''(x)|$, as shown in Figure 8.7.5. It is evident from the graph that

$$|f''(x)| < 4 \quad \text{for} \quad 0 \le x \le 1$$

Thus, it follows from (17) that

$$|E_M| \le \frac{K_2}{24n^2} < \frac{4}{24n^2} = \frac{1}{6n^2}$$

and hence we can satisfy (16) by choosing n so that

$$\frac{1}{6n^2} < 5 \times 10^{-4}$$

which, on taking reciprocals, can be written as

$$n^2 > \frac{10^4}{30} \quad \text{or} \quad n > \frac{10^2}{\sqrt{30}} \approx 18.257$$

The smallest integer value of n satisfying this inequality is $n = 19$. Thus, the midpoint approximation M_{19} using 19 subintervals will produce three decimal-place accuracy. ◀

A COMPARISON OF THE THREE METHODS

Of the three methods studied in this section, Simpson's rule generally produces more accurate results than the midpoint or trapezoidal approximations for the same amount of work. To make this plausible, let us express (10), (11), and (12) in terms of the subinterval width

$$\Delta x = \frac{b-a}{n} \quad \text{for } M_n \text{ and } T_n$$

and

$$\Delta x = \frac{b-a}{2n} \quad \text{for } S_{2n}$$

We obtain

$$|E_M| \le \frac{1}{24}K_2(b-a)(\Delta x)^2 \tag{19}$$

$$|E_T| \le \frac{1}{12}K_2(b-a)(\Delta x)^2 \tag{20}$$

$$|E_S| \le \frac{1}{180}K_4(b-a)(\Delta x)^4 \tag{21}$$

(verify). Thus, for Simpson's rule the upper bound on the absolute error is proportional to $(\Delta x)^4$, whereas the upper bound on the absolute error for the midpoint and trapezoidal approximations is proportional to $(\Delta x)^2$. Thus, reducing the interval width by a factor of 10, for example, reduces the error bound by a factor of 100 for the midpoint and trapezoidal approximations but reduces the error bound by a factor of 10,000 for Simpson's rule. This suggests that, as n increases, the accuracy of Simpson's rule improves much more rapidly than that of the other approximations.

As a final note, observe that if $f(x)$ is a polynomial of degree 3 or less, then we have $f^{(4)}(x) = 0$ for all x, so $K_4 = 0$ in (12) and consequently $|E_S| = 0$. Thus, Simpson's rule gives exact results for polynomials of degree 3 or less. Similarly, the midpoint and trapezoidal approximations give exact results for polynomials of degree 1 or less. (You should also be able to see that this is so geometrically.)

✔QUICK CHECK EXERCISES 8.7 *(See page 569 for answers.)*

1. Let T_n be the trapezoidal approximation for the definite integral of $f(x)$ over an interval $[a, b]$ using n subintervals.
(a) Expressed in terms of L_n and R_n (the left and right endpoint approximations), $T_n =$ ______.
(b) Expressed in terms of the function values $y_0, y_1, \ldots, y_n$ at the endpoints of the subintervals, $T_n =$ ______.

2. Let I denote the definite integral of f over an interval $[a, b]$ with T_n and M_n the respective trapezoidal and midpoint approximations of I for a given n. Assume that the graph of f is concave up on the interval $[a, b]$ and order the quantities T_n, M_n, and I from smallest to largest: ______ < ______ < ______.

3. Let S_{2n} be the Simpson's rule approximation for the definite integral of $f(x)$ over an interval $[a, b]$ using $2n$ subintervals.
(a) Expressed in terms of M_n and T_n (the midpoint and trapezoidal approximations), $S_{2n} =$ ______.
(b) Using the function values $y_0, y_1, \ldots, y_{2n}$ at the endpoints of the subintervals, $S_{2n} =$ ______.

4. Assume that $f^{(4)}$ is continuous on $[0, 1]$ and that $f^{(k)}(x)$ satisfies $|f^{(k)}(x)| \le 1$ on $[0, 1]$, $k = 1, 2, 3, 4$. Find an upper bound on the absolute error that results from approximating the integral of f over $[0, 1]$ using (a) the midpoint approximation M_{10}; (b) the trapezoidal approximation T_{10}; and (c) Simpson's rule S_{10}.

5. Approximate $\int_0^1 \sin x^2\, dx$ using the indicated method.
(a) $M_4 =$ ______ (b) $T_4 =$ ______
(c) $S_4 =$ ______ (d) $S_8 =$ ______

EXERCISE SET 8.7 [C] CAS

1–6 Use $n = 10$ subintervals to approximate the integral by (a) the midpoint approximation, (b) the trapezoidal approximation, and use $2n = 10$ subintervals to approximate the integral by (c) Simpson's rule. In each case, find the exact value of the integral and approximate the absolute error. Express your answers to at least four decimal places.

1. $\int_2^5 \sqrt{x-1}\, dx$ **2.** $\int_1^4 \frac{1}{\sqrt{x}}\, dx$ **3.** $\int_0^\pi \sin x\, dx$

4. $\int_0^1 \cos x\, dx$ **5.** $\int_1^4 e^{-x}\, dx$ **6.** $\int_0^2 \frac{1}{2x+1}\, dx$

7–12 Use inequalities (10), (11), and (12) to find upper bounds on the errors in parts (a), (b), or (c) of the indicated exercise.

7. Exercise 1 **8.** Exercise 2 **9.** Exercise 3
10. Exercise 4 **11.** Exercise 5 **12.** Exercise 6

13–18 Use inequalities (10), (11), and (12) to find a number n of subintervals for (a) the midpoint approximation and (b) the trapezoidal approximation to ensure that the absolute error will be less than the given value. Also, (c) find a number $2n$ of subintervals to ensure that the absolute error for the Simpson's rule approximation will be less than the given value.

13. Exercise 1; 5×10^{-4} **14.** Exercise 2; 5×10^{-4}
15. Exercise 3; 10^{-3} **16.** Exercise 4; 10^{-3}
17. Exercise 5; 10^{-6} **18.** Exercise 6; 10^{-6}

19–20 Find a function $g(x)$ of the form

$$g(x) = Ax^2 + Bx + C$$

whose graph contains the points $(m - \Delta x, f(m - \Delta x))$, $(m, f(m))$, and $(m + \Delta x, f(m + \Delta x))$, for the given function $f(x)$ and the given values of m and Δx. Then verify Formula (9):

$$\int_{m-\Delta x}^{m+\Delta x} g(x)\, dx = \frac{\Delta x}{3}[Y_0 + 4Y_1 + Y_2]$$

where $Y_0 = f(m - \Delta x)$, $Y_1 = f(m)$, and $Y_2 = f(m + \Delta x)$.

19. $f(x) = \frac{1}{x}$; $m = 3$, $\Delta x = 1$

20. $f(x) = \sin^2(\pi x)$; $m = \frac{1}{6}$, $\Delta x = \frac{1}{6}$

21–26 Approximate the integral using Simpson's rule with $2n = 10$ subintervals, and compare your answer to that produced by a calculating utility with a numerical integration capability. Express your answers to at least four decimal places.

21. $\int_0^1 e^{-x^2}\, dx$ **22.** $\int_0^2 \frac{x}{\sqrt{1+x^2}}\, dx$

23. $\int_1^2 \sqrt{x^3 - 1}\, dx$ **24.** $\int_{-\pi/2}^{\pi/2} \frac{1}{2 - \cos x}\, dx$

25. $\int_0^2 \sin(x^2)\,dx$ **26.** $\int_1^3 \sqrt{\ln x}\,dx$

27–28 The exact value of the given integral is π (verify). Use $n = 10$ subintervals to approximate the integral by (a) the midpoint approximation, (b) the trapezoidal approximation, and use $2n = 10$ subintervals to approximate the integral by (c) Simpson's rule. Estimate the absolute error, and express your answers to at least four decimal places.

27. $\int_0^1 \frac{4}{1+x^2}\,dx$ **28.** $\int_0^2 \sqrt{4-x^2}\,dx$

29. In Example 8 we showed that taking $2n = 14$ subdivisions ensures that the approximation of

$$\ln 2 = \int_1^2 \frac{1}{x}\,dx$$

by Simpson's rule is accurate to five decimal places. Confirm this by comparing the approximation of ln 2 produced by Simpson's rule with $2n = 14$ to the value produced directly by your calculating utility.

30. In each part, determine whether a trapezoidal approximation would be an underestimate or an overestimate for the definite integral.

(a) $\int_0^1 \cos(x^2)\,dx$ (b) $\int_{3/2}^2 \cos(x^2)\,dx$

31–32 Find a value of n to ensure that the absolute error in approximating the integral by the midpoint approximation will be less than 10^{-4}. Estimate the absolute error, and express your answers to at least four decimal places.

31. $\int_0^2 x \sin x\,dx$ **32.** $\int_0^1 e^{\cos x}\,dx$

33–34 Show that the inequalities (10) and (11) are of no value in finding an upper bound on the absolute error that results from approximating the integral using either the midpoint approximation or the trapezoidal approximation.

33. $\int_0^1 x\sqrt{x}\,dx$ **34.** $\int_0^1 \sin\sqrt{x}\,dx$

35–36 Use Simpson's rule with $2n = 10$ subintervals to approximate the length of the curve. Express your answer to at least four decimal places.

35. $y = \cos x,\ -\pi/2 \le x \le \pi/2$

36. $y = 1/x,\ 1 \le x \le 3$

FOCUS ON CONCEPTS

37. A graph of the speed v versus time t curve for a test run of a Mitsubishi Galant ES is shown in the accompanying figure. Estimate the speeds at times $t = 0$, 2.5, 5, 7.5, 10, 12.5, 15 s from the graph, convert to ft/s using 1 mi/h $= \frac{22}{15}$ ft/s, and use these speeds and Simpson's rule to approximate the number of feet traveled during the first 15 s. Round your answer to the nearest foot. [*Hint:* Distance traveled $= \int_0^{15} v(t)\,dt$.]

Source: Data from *Car and Driver*, November 2003.

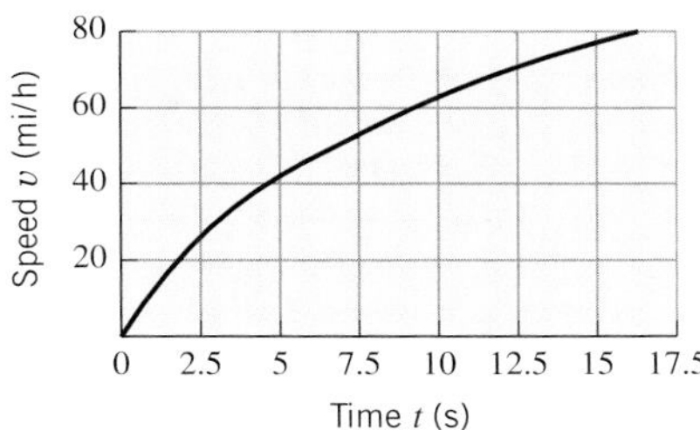

Figure Ex-37

38. A graph of the acceleration a versus time t for an object moving on a straight line is shown in the accompanying figure. Estimate the accelerations at $t = 0, 1, 2, \ldots, 8$ seconds (s) from the graph and use Simpson's rule to approximate the change in velocity from $t = 0$ to $t = 8$ s. Round your answer to the nearest tenth cm/s. [*Hint:* Change in velocity $= \int_0^8 a(t)\,dt$.]

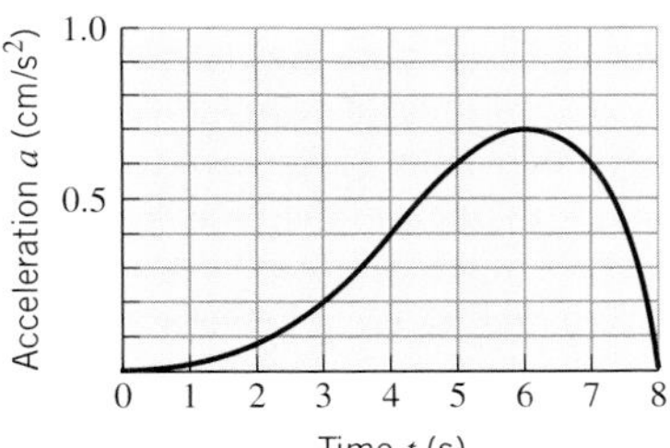

Figure Ex-38

39–42 Numerical integration methods can be used in problems where only measured or experimentally determined values of the integrand are available. Use Simpson's rule to estimate the value of the relevant integral in these exercises.

39. The accompanying table gives the speeds, in miles per second, at various times for a test rocket that was fired upward from the surface of the Earth. Use these values to approximate the number of miles traveled during the first 180 s. Round your answer to the nearest tenth of a mile. [*Hint:* Distance traveled $= \int_0^{180} v(t)\,dt$.]

TIME t (s)	SPEED v (mi/s)
0	0.00
30	0.03
60	0.08
90	0.16
120	0.27
150	0.42
180	0.65

Table Ex-39

40. The accompanying table gives the speeds of a bullet at various distances from the muzzle of a rifle. Use these values to approximate the number of seconds for the bullet to travel 1800 ft. Express your answer to the nearest hundredth of a second. [*Hint:* If v is the speed of the bullet and x is the distance traveled, then $v = dx/dt$ so that $dt/dx = 1/v$ and $t = \int_0^{1800} (1/v)\,dx$.]

DISTANCE x (ft)	SPEED v (ft/s)
0	3100
300	2908
600	2725
900	2549
1200	2379
1500	2216
1800	2059

Table Ex-40

41. Measurements of a pottery shard recovered from an archaeological dig reveal that the shard came from a pot with a flat bottom and circular cross sections (see the accompanying figure). The figure shows interior radius measurements of the shard made every 4 cm from the bottom of the pot to the top. Use those values to approximate the interior volume of the pot to the nearest tenth of a liter (1 L = 1000 cm^3). [*Hint:* Use 6.2.3 (volume by cross sections) to set up an appropriate integral for the volume.]

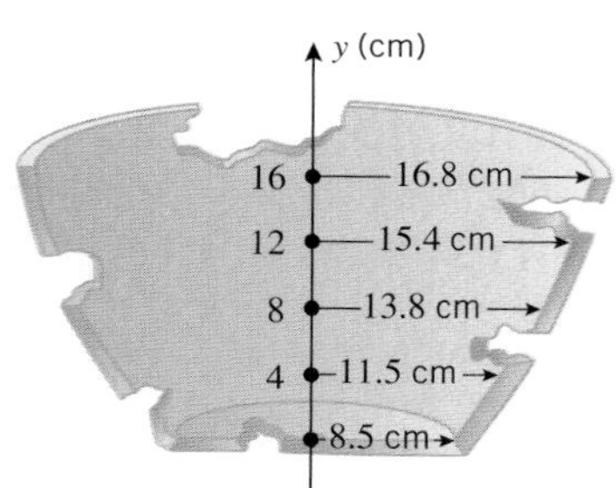

Figure Ex-41

42. Engineers want to construct a straight and level road 600 ft long and 75 ft wide by making a vertical cut through an intervening hill (see the accompanying figure). Heights of the hill above the centerline of the proposed road, as obtained at various points from a contour map of the region, are shown in the accompanying figure. To estimate the construction costs, the engineers need to know the volume of earth that must be removed. Approximate this volume, rounded to the nearest cubic foot. [*Hint:* First, set up an integral for the cross-sectional area of the cut along the centerline of the road, then assume that the height of the hill does not vary between the centerline and edges of the road.]

HORIZONTAL DISTANCE x (ft)	HEIGHT h (ft)
0	0
100	7
200	16
300	24
400	25
500	16
600	0

600 ft
x
Centerline
75 ft

Figure Ex-42

C **43.** Let $f(x) = \cos(x^2)$.

(a) Use a CAS to approximate the maximum value of $|f''(x)|$ on the interval $[0, 1]$.

(b) How large must n be in the midpoint approximation of $\int_0^1 f(x)\,dx$ to ensure that the absolute error is less than 5×10^{-4}? Compare your result with that obtained in Example 9.

(c) Estimate the integral using the midpoint approximation with the value of n obtained in part (b).

C **44.** Let $f(x) = \sqrt{1 + x^3}$.

(a) Use a CAS to approximate the maximum value of $|f''(x)|$ on the interval $[0, 1]$.

(b) How large must n be in the trapezoidal approximation of $\int_0^1 f(x)\,dx$ to ensure that the absolute error is less than 10^{-3}?

(c) Estimate the integral using the trapezoidal approximation with the value of n obtained in part (b).

C **45.** Let $f(x) = \cos(x^2)$.

(a) Use a CAS to approximate the maximum value of $|f^{(4)}(x)|$ on the interval $[0, 1]$.

(b) How large must the value of $2n$ be in the approximation of $\int_0^1 f(x)\,dx$ by Simpson's rule to ensure that the absolute error is less than 10^{-4}?

(c) Estimate the integral using Simpson's rule with the value of n obtained in part (b).

C **46.** Let $f(x) = \sqrt{1 + x^3}$.

(a) Use a CAS to approximate the maximum value of $|f^{(4)}(x)|$ on the interval $[0, 1]$.

(b) How large must the value of $2n$ be in the approximation of $\int_0^1 f(x)\,dx$ by Simpson's rule to ensure that the absolute error is less than 10^{-5}?

(c) Estimate the integral using Simpson's rule with the value of n obtained in part (b).

FOCUS ON CONCEPTS

47. (a) Verify that the average of the left and right endpoint approximations as given in Table 8.7.1 gives Formula (2) for the trapezoidal approximation.

(b) Suppose that f is a continuous nonnegative function on the interval $[a, b]$ and partition $[a, b]$ with equally spaced points, $a = x_0 < x_1 < \cdots < x_n = b$. Find the area of the trapezoid under the line segment joining points $(x_k, f(x_k))$ and $(x_{k+1}, f(x_{k+1}))$ and above the interval $[x_k, x_{k+1}]$. Show that the right side of Formula (2) is the sum of these trapezoidal areas (Figure 8.7.1).

48. Let f be a function that is positive, continuous, decreasing, and concave down on the interval $[a, b]$. Assuming that $[a, b]$ is subdivided into n equal subintervals, arrange the following approximations of $\int_a^b f(x)\,dx$ in order of increasing value: left endpoint, right endpoint, midpoint, and trapezoidal.

49. Suppose that $\Delta x > 0$ and $g(x) = Ax^2 + Bx + C$. Let m be a number and set $Y_0 = g(m - \Delta x)$, $Y_1 = g(m)$, and $Y_2 = g(m + \Delta x)$. Verify Formula (9):

$$\int_{m-\Delta x}^{m+\Delta x} g(x)\,dx = \frac{\Delta x}{3}[Y_0 + 4Y_1 + Y_2]$$

50. Suppose that f is a continuous nonnegative function on the interval $[a, b]$ and that $[a, b]$ is partitioned using equally spaced points, $a = x_0 < x_1 < \cdots < x_{2n} = b$. Set $y_k = f(x_k)$, $k = 0, 1, \ldots, 2n$. Let $g_i(x)$ be the function of the form $g_i(x) = Ax^2 + Bx + C$ that passes through the points (x_{2i}, y_{2i}), (x_{2i+1}, y_{2i+1}), and (x_{2i+2}, y_{2i+2}), $i = 0, 1, \ldots, n-1$. Verify that Formula (6) computes the area under a piecewise quadratic function by showing that

$$\sum_{i=0}^{n-1}\left(\int_{x_{2i}}^{x_{2i+2}} g_i(x)\,dx\right) = \frac{1}{3}\left(\frac{b-a}{2n}\right)[y_0 + 4y_1 + 2y_2 + 4y_3 + 2y_4 + \cdots + 2y_{2n-2} + 4y_{2n-1} + y_{2n}]$$

✔QUICK CHECK ANSWERS 8.7

1. (a) $\frac{1}{2}(L_n + R_n)$ (b) $\left(\frac{b-a}{2n}\right)[y_0 + 2y_1 + \cdots + 2y_{n-1} + y_n]$ **2.** $M_n < I < T_n$ **3.** (a) $\frac{1}{3}(2M_n + T_n)$ (b) $\frac{1}{3}\left(\frac{b-a}{2n}\right)[y_0 + 4y_1 + 2y_2 + 4y_3 + 2y_4 + \cdots + 2y_{2n-2} + 4y_{2n-1} + y_{2n}]$ **4.** (a) $\frac{1}{2400}$ (b) $\frac{1}{1200}$ (c) $\frac{1}{1{,}800{,}000}$
5. (a) 0.307385118 (b) 0.315975361 (c) 0.309943906 (d) 0.310248532

8.8 IMPROPER INTEGRALS

Up to now we have focused on definite integrals with continuous integrands and finite intervals of integration. In this section we will extend the concept of a definite integral to include infinite intervals of integration and integrands that become infinite within the interval of integration.

IMPROPER INTEGRALS

It is assumed in the definition of the definite integral

$$\int_a^b f(x)\,dx$$

that $[a, b]$ is a finite interval and that the limit that defines the integral exists; that is, the function f is integrable. We observed in Theorems 5.5.2 and 5.5.8 that continuous functions are integrable, as are bounded functions with finitely many points of discontinuity. We also observed in Theorem 5.5.8 that functions that are not bounded on the interval of integration are not integrable. Thus, for example, a function with a vertical asymptote within the interval of integration would not be integrable.

Our main objective in this section is to extend the concept of a definite integral to allow for infinite intervals of integration and integrands with vertical asymptotes within the interval of integration. We will call the vertical asymptotes ***infinite discontinuities***, and we will call integrals with infinite intervals of integration or infinite discontinuities within the interval of integration ***improper integrals***. Here are some examples:

- Improper integrals with infinite intervals of integration:

$$\int_1^{+\infty} \frac{dx}{x^2}, \quad \int_{-\infty}^{0} e^x\, dx, \quad \int_{-\infty}^{+\infty} \frac{dx}{1+x^2}$$

- Improper integrals with infinite discontinuities in the interval of integration:

$$\int_{-3}^{3} \frac{dx}{x^2}, \quad \int_1^2 \frac{dx}{x-1}, \quad \int_0^{\pi} \tan x\, dx$$

- Improper integrals with infinite discontinuities and infinite intervals of integration:

$$\int_0^{+\infty} \frac{dx}{\sqrt{x}}, \quad \int_{-\infty}^{+\infty} \frac{dx}{x^2-9}, \quad \int_1^{+\infty} \sec x\, dx$$

INTEGRALS OVER INFINITE INTERVALS

To motivate a reasonable definition for improper integrals of the form

$$\int_a^{+\infty} f(x)\, dx$$

let us begin with the case where f is continuous and nonnegative on $[a, +\infty)$, so we can think of the integral as the area under the curve $y = f(x)$ over the interval $[a, +\infty)$ (Figure 8.8.1). At first, you might be inclined to argue that this area is infinite because the region has infinite extent. However, such an argument would be based on vague intuition rather than precise mathematical logic, since the concept of area has only been defined over intervals of *finite extent*. Thus, before we can make any reasonable statements about the area of the region in Figure 8.8.1, we need to begin by defining what we mean by the area of this region. For that purpose, it will help to focus on a specific example.

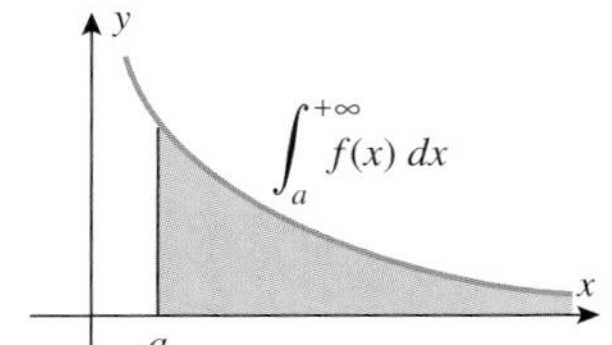

Figure 8.8.1

Suppose we are interested in the area A of the region that lies below the curve $y = 1/x^2$ and above the interval $[1, +\infty)$ on the x-axis. Instead of trying to find the entire area at once, let us begin by calculating the portion of the area that lies above a finite interval $[1, b]$ where $b > 1$ is arbitrary. That area is

$$\int_1^b \frac{dx}{x^2} = -\frac{1}{x}\bigg]_1^b = 1 - \frac{1}{b}$$

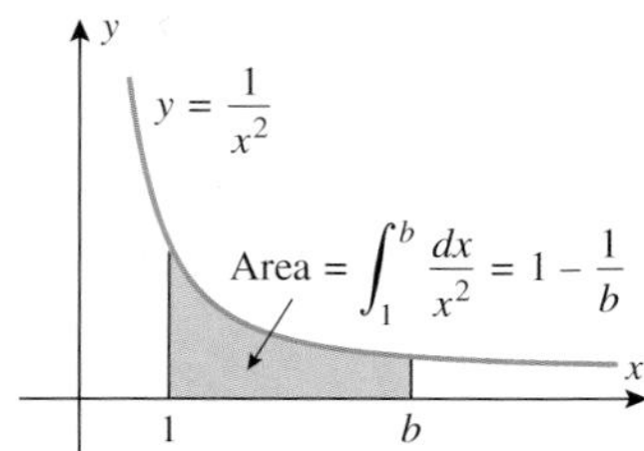

Figure 8.8.2

(Figure 8.8.2). If we now allow b to increase so that $b \to +\infty$, then the portion of the area over the interval $[1, b]$ will begin to fill out the area over the entire interval $[1, +\infty)$ (Figure 8.8.3), and hence we can reasonably define the area A under $y = 1/x^2$ over the interval $[1, +\infty)$ to be

$$A = \int_1^{+\infty} \frac{dx}{x^2} = \lim_{b\to+\infty} \int_1^b \frac{dx}{x^2} = \lim_{b\to+\infty} \left(1 - \frac{1}{b}\right) = 1 \tag{1}$$

Thus, the area has a finite value of 1 and is not infinite as we first conjectured.

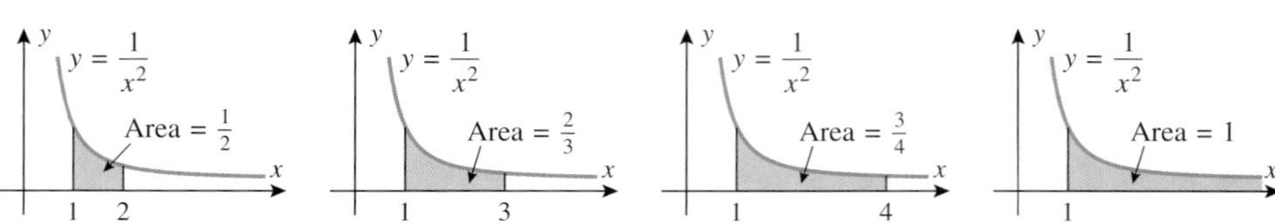

Figure 8.8.3

With the preceding discussion as our guide, we make the following definition (which is applicable to functions with both positive and negative values).

If f is nonnegative over the interval $[a, +\infty)$, then the improper integral in Definition 8.8.1 is interpreted to be the area under the graph of f over the interval $[a, +\infty)$. If the integral converges, then the area is finite and equal to the value of the integral, and if the integral diverges, then the area is regarded to be infinite.

8.8.1 DEFINITION. The ***improper integral of f over the interval*** $[a, +\infty)$ is defined to be

$$\int_a^{+\infty} f(x)\,dx = \lim_{b\to+\infty} \int_a^b f(x)\,dx$$

In the case where the limit exists, the improper integral is said to ***converge***, and the limit is defined to be the value of the integral. In the case where the limit does not exist, the improper integral is said to ***diverge***, and it is not assigned a value.

▶ **Example 1** Evaluate

$$\text{(a)} \int_1^{+\infty} \frac{dx}{x^3} \qquad \text{(b)} \int_1^{+\infty} \frac{dx}{x}$$

Solution (a). Following the definition, we replace the infinite upper limit by a finite upper limit b, and then take the limit of the resulting integral. This yields

$$\int_1^{+\infty} \frac{dx}{x^3} = \lim_{b\to+\infty} \int_1^b \frac{dx}{x^3} = \lim_{b\to+\infty} \left[-\frac{1}{2x^2}\right]_1^b = \lim_{b\to+\infty} \left(\frac{1}{2} - \frac{1}{2b^2}\right) = \frac{1}{2}$$

Solution (b).

$$\int_1^{+\infty} \frac{dx}{x} = \lim_{b\to+\infty} \int_1^b \frac{dx}{x} = \lim_{b\to+\infty} \left[\ln x\right]_1^b = \lim_{b\to+\infty} \ln b = +\infty$$

In this case the integral diverges and hence has no value. ◀

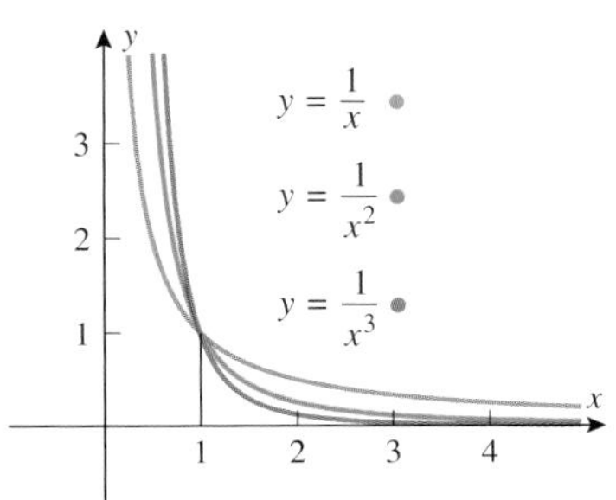

Figure 8.8.4

Because the functions $1/x^3$, $1/x^2$, and $1/x$ are nonnegative over the interval $[1, +\infty)$, it follows from (1) and the last example that over this interval the area under $y = 1/x^3$ is $\frac{1}{2}$, the area under $y = 1/x^2$ is 1, and the area under $y = 1/x$ is infinite. However, on the surface the graphs of the three functions seem very much alike (Figure 8.8.4), and there is nothing to suggest why one of the areas should be infinite and the other two finite. One explanation is that $1/x^3$ and $1/x^2$ approach zero more rapidly than $1/x$ as $x \to +\infty$, so that the area over the interval $[1, b]$ accumulates less rapidly under the curves $y = 1/x^3$ and $y = 1/x^2$ than under $y = 1/x$ as $b \to +\infty$, and the difference is just enough that the first two areas are finite and the third is infinite.

▶ **Example 2** For what values of p does the integral $\int_1^{+\infty} \frac{dx}{x^p}$ converge?

Solution. We know from the preceding example that the integral diverges if $p = 1$, so let us assume that $p \neq 1$. In this case we have

$$\int_1^{+\infty} \frac{dx}{x^p} = \lim_{b\to+\infty} \int_1^b x^{-p}\,dx = \lim_{b\to+\infty} \left.\frac{x^{1-p}}{1-p}\right]_1^b = \lim_{b\to+\infty} \left[\frac{b^{1-p}}{1-p} - \frac{1}{1-p}\right]$$

If $p > 1$, then the exponent $1 - p$ is negative and $b^{1-p} \to 0$ as $b \to +\infty$; and if $p < 1$, then the exponent $1 - p$ is positive and $b^{1-p} \to +\infty$ as $b \to +\infty$. Thus, the integral converges if $p > 1$ and diverges otherwise. In the convergent case the value of the integral is

$$\int_1^{+\infty} \frac{dx}{x^p} = \left[0 - \frac{1}{1-p}\right] = \frac{1}{p-1} \qquad (p > 1)$$ ◀

The following theorem summarizes this result.

8.8.2 THEOREM.
$$\int_1^{+\infty} \frac{dx}{x^p} = \begin{cases} \dfrac{1}{p-1} & \text{if } p > 1 \\ \text{diverges} & \text{if } p \le 1 \end{cases}$$

► **Example 3** Evaluate $\displaystyle\int_0^{+\infty} (1-x)e^{-x}\,dx$.

Solution. Integrating by parts with $u = 1 - x$ and $dv = e^{-x}\,dx$ yields

$$\int (1-x)e^{-x}\,dx = -e^{-x}(1-x) - \int e^{-x}\,dx = -e^{-x} + xe^{-x} + e^{-x} + C = xe^{-x} + C$$

Thus,
$$\int_0^{+\infty} (1-x)e^{-x}\,dx = \lim_{b\to+\infty} \left[xe^{-x}\right]_0^b = \lim_{b\to+\infty} \frac{b}{e^b}$$

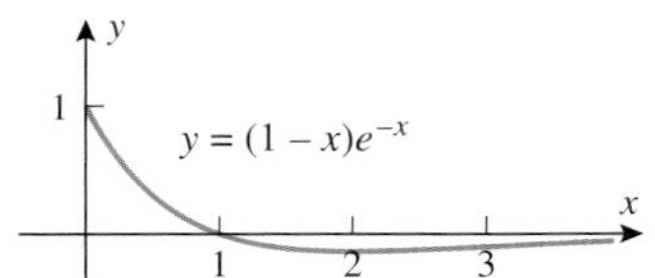

The net signed area between the graph and the interval $[0, +\infty)$ is zero.

Figure 8.8.5

The limit is an indeterminate form of type ∞/∞, so we will apply L'Hôpital's rule by differentiating the numerator and denominator with respect to b. This yields

$$\int_0^{+\infty} (1-x)e^{-x}\,dx = \lim_{b\to+\infty} \frac{1}{e^b} = 0$$

We can interpret this to mean that the net signed area between the graph of $y = (1-x)e^{-x}$ and the interval $[0, +\infty)$ is 0 (Figure 8.8.5). ◄

If f is nonnegative over the interval $(-\infty, +\infty)$, then the improper integral
$$\int_{-\infty}^{+\infty} f(x)\,dx$$
is interpreted to be the area under the graph of f over the interval $(-\infty, +\infty)$. The area is finite and equal to the value of the integral if the integral converges and is infinite if it diverges.

8.8.3 DEFINITION. The ***improper integral of f over the interval*** $(-\infty, b]$ is defined to be

$$\int_{-\infty}^{b} f(x)\,dx = \lim_{a\to-\infty} \int_a^b f(x)\,dx \tag{2}$$

The integral is said to ***converge*** if the limit exists and ***diverge*** if it does not. The ***improper integral of f over the interval*** $(-\infty, +\infty)$ is defined as

$$\int_{-\infty}^{+\infty} f(x)\,dx = \int_{-\infty}^{c} f(x)\,dx + \int_c^{+\infty} f(x)\,dx \tag{3}$$

where c is any real number. The improper integral is said to ***converge*** if both terms converge and ***diverge*** if either term diverges.

Although we usually choose $c = 0$ in (3), the choice does not matter because it can be proved that neither the convergence nor the value of the integral is affected by the choice of c.

► **Example 4** Evaluate $\displaystyle\int_{-\infty}^{+\infty} \frac{dx}{1+x^2}$.

Solution. We will evaluate the integral by choosing $c = 0$ in (3). With this value for c we obtain

$$\int_0^{+\infty} \frac{dx}{1+x^2} = \lim_{b\to+\infty} \int_0^b \frac{dx}{1+x^2} = \lim_{b\to+\infty} \left[\tan^{-1} x\right]_0^b = \lim_{b\to+\infty} (\tan^{-1} b) = \frac{\pi}{2}$$

$$\int_{-\infty}^{0} \frac{dx}{1+x^2} = \lim_{a\to-\infty} \int_a^0 \frac{dx}{1+x^2} = \lim_{a\to-\infty} \left[\tan^{-1} x\right]_a^0 = \lim_{a\to-\infty} (-\tan^{-1} a) = \frac{\pi}{2}$$

Thus, the integral converges and its value is

$$\int_{-\infty}^{+\infty} \frac{dx}{1+x^2} = \int_{-\infty}^{0} \frac{dx}{1+x^2} + \int_0^{+\infty} \frac{dx}{1+x^2} = \frac{\pi}{2} + \frac{\pi}{2} = \pi$$

Since the integrand is nonnegative on the interval $(-\infty, +\infty)$, the integral represents the area of the region shown in Figure 8.8.6. ◄

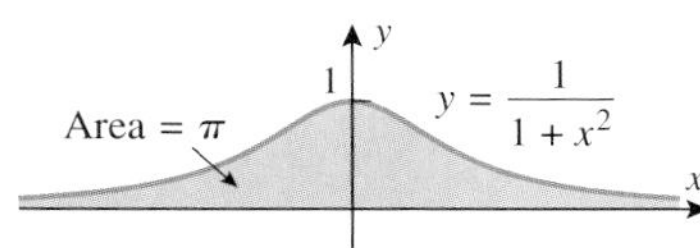

Figure 8.8.6

INTEGRALS WHOSE INTEGRANDS HAVE INFINITE DISCONTINUITIES

Next we will consider improper integrals whose integrands have infinite discontinuities. We will start with the case where the interval of integration is a finite interval $[a, b]$ and the infinite discontinuity occurs at the right-hand endpoint.

To motivate an appropriate definition for such an integral let us consider the case where f is nonnegative on $[a, b]$, so we can interpret the improper integral $\int_a^b f(x)\,dx$ as the area of the region in Figure 8.8.7*a*. The problem of finding the area of this region is complicated by the fact that it extends indefinitely in the positive y-direction. However, instead of trying to find the entire area at once, we can proceed indirectly by calculating the portion of the area over the interval $[a, k]$, where $a \le k < b$, and then letting k approach b to fill out the area of the entire region (Figure 8.8.7*b*). Motivated by this idea, we make the following definition.

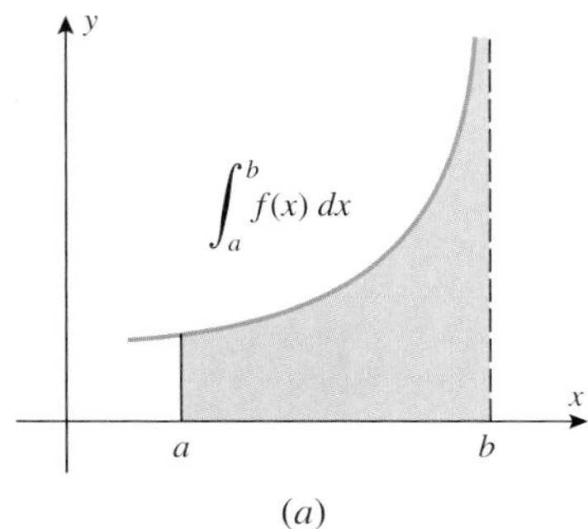

(*a*)

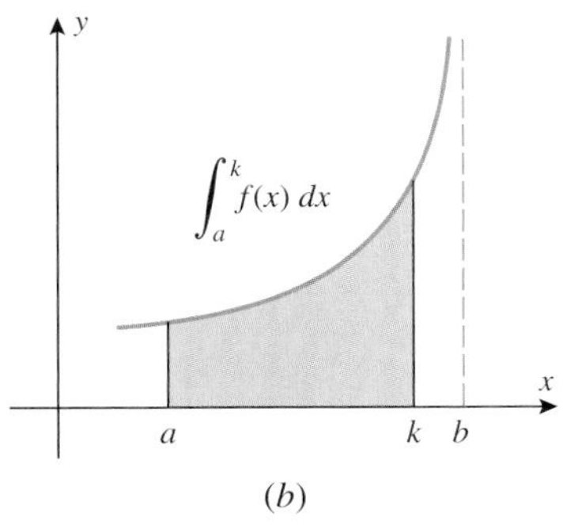

(*b*)

Figure 8.8.7

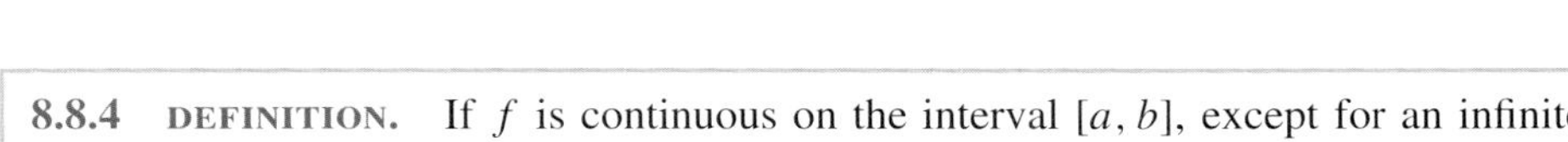

8.8.4 DEFINITION. If f is continuous on the interval $[a, b]$, except for an infinite discontinuity at b, then the ***improper integral of f over the interval* [*a*, *b*]** is defined as

$$\int_a^b f(x)\,dx = \lim_{k\to b^-} \int_a^k f(x)\,dx \tag{4}$$

In the case where the limit exists, the improper integral is said to ***converge***, and the limit is defined to be the value of the integral. In the case where the limit does not exist, the improper integral is said to ***diverge***, and it is not assigned a value.

► **Example 5** Evaluate $\displaystyle\int_0^1 \frac{dx}{\sqrt{1-x}}$.

Solution. The integral is improper because the integrand approaches $+\infty$ as x approaches the upper limit 1 from the left. From (4),

$$\int_0^1 \frac{dx}{\sqrt{1-x}} = \lim_{k\to 1^-} \int_0^k \frac{dx}{\sqrt{1-x}} = \lim_{k\to 1^-} \Big[-2\sqrt{1-x}\Big]_0^k$$
$$= \lim_{k\to 1^-} [-2\sqrt{1-k}+2] = 2 \text{ ◄}$$

Improper integrals with an infinite discontinuity at the left-hand endpoint or inside the interval of integration are defined as follows.

8.8.5 DEFINITION. If f is continuous on the interval $[a, b]$, except for an infinite discontinuity at a, then the ***improper integral of f over the interval* [*a*, *b*]** is defined as

$$\int_a^b f(x)\,dx = \lim_{k\to a^+} \int_k^b f(x)\,dx \tag{5}$$

The integral is said to ***converge*** if the limit exists and ***diverge*** if it does not. If f is continuous on the interval $[a, b]$, except for an infinite discontinuity at a point c in (a, b),

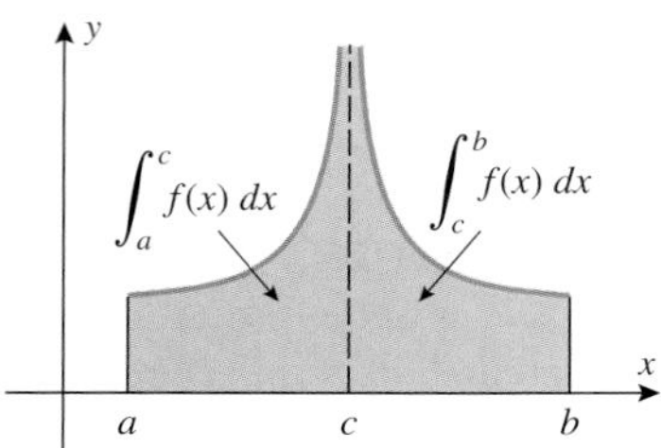

Figure 8.8.8

then the ***improper integral of f over the interval*** $[a, b]$ is defined as

$$\int_a^b f(x)\,dx = \int_a^c f(x)\,dx + \int_c^b f(x)\,dx \tag{6}$$

The improper integral is said to ***converge*** if both terms converge and ***diverge*** if either term diverges (Figure 8.8.8).

▶ **Example 6** Evaluate

$$\text{(a)} \int_1^2 \frac{dx}{1-x} \qquad \text{(b)} \int_1^4 \frac{dx}{(x-2)^{2/3}} \qquad \text{(c)} \int_0^{+\infty} \frac{dx}{\sqrt{x}(x+1)}$$

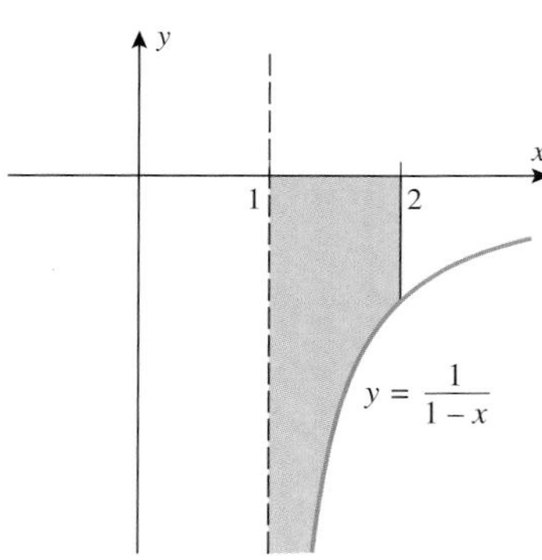

Figure 8.8.9

Solution (a). The integral is improper because the integrand approaches $-\infty$ as x approaches the lower limit 1 from the right (Figure 8.8.9). From Definition 8.8.5 we obtain

$$\begin{aligned}\int_1^2 \frac{dx}{1-x} &= \lim_{k\to 1^+} \int_k^2 \frac{dx}{1-x} = \lim_{k\to 1^+} \big[-\ln|1-x|\big]_k^2 \\ &= \lim_{k\to 1^+} \big[-\ln|-1| + \ln|1-k|\big] = \lim_{k\to 1^+} \ln|1-k| = -\infty\end{aligned}$$

so the integral diverges.

Solution (b). The integral is improper because the integrand approaches $+\infty$ at $x = 2$, which is inside the interval of integration. From Definition 8.8.5 we obtain

$$\int_1^4 \frac{dx}{(x-2)^{2/3}} = \int_1^2 \frac{dx}{(x-2)^{2/3}} + \int_2^4 \frac{dx}{(x-2)^{2/3}} \tag{7}$$

But

$$\int_1^2 \frac{dx}{(x-2)^{2/3}} = \lim_{k\to 2^-} \int_1^k \frac{dx}{(x-2)^{2/3}} = \lim_{k\to 2^-} [3(k-2)^{1/3} - 3(1-2)^{1/3}] = 3$$

$$\int_2^4 \frac{dx}{(x-2)^{2/3}} = \lim_{k\to 2^+} \int_k^4 \frac{dx}{(x-2)^{2/3}} = \lim_{k\to 2^+} [3(4-2)^{1/3} - 3(k-2)^{1/3}] = 3\sqrt[3]{2}$$

Thus, from (7)

$$\int_1^4 \frac{dx}{(x-2)^{2/3}} = 3 + 3\sqrt[3]{2}$$

Solution (c). This integral is improper for two reasons—the interval of integration is infinite, and there is an infinite discontinuity at $x = 0$. To evaluate this integral we will split the interval of integration at a convenient point, say $x = 1$, and write

$$\int_0^{+\infty} \frac{dx}{\sqrt{x}(x+1)} = \int_0^1 \frac{dx}{\sqrt{x}(x+1)} + \int_1^{+\infty} \frac{dx}{\sqrt{x}(x+1)}$$

The integrand in these two improper integrals does not match any of the forms in the Endpaper Integral Table, but the radical suggests the substitution $x = u^2$, $dx = 2u\,du$, from which we obtain

$$\begin{aligned}\int \frac{dx}{\sqrt{x}(x+1)} &= \int \frac{2u\,du}{u(u^2+1)} = 2\int \frac{du}{u^2+1} \\ &= 2\tan^{-1} u + C = 2\tan^{-1}\sqrt{x} + C\end{aligned}$$

Thus,

$$\begin{aligned}\int_0^{+\infty} \frac{dx}{\sqrt{x}(x+1)} &= 2\lim_{k\to 0^+} \big[\tan^{-1}\sqrt{x}\big]_k^1 + 2\lim_{k\to +\infty} \big[\tan^{-1}\sqrt{x}\big]_1^k \\ &= 2\Big[\frac{\pi}{4} - 0\Big] + 2\Big[\frac{\pi}{2} - \frac{\pi}{4}\Big] = \pi\end{aligned}$$ ◀

WARNING It is sometimes tempting to apply the Fundamental Theorem of Calculus directly to an improper integral without taking the appropriate limits. To illustrate what can go wrong with this procedure, suppose we ignore the fact that the integral

$$\int_0^2 \frac{dx}{(x-1)^2} \tag{8}$$

is improper and incorrectly evaluate this integral as

$$-\frac{1}{x-1}\bigg]_0^2 = -1 - (1) = -2$$

This result is clearly incorrect because the integrand is never negative and hence the integral cannot be negative! To evaluate (8) correctly we should first write

$$\int_0^2 \frac{dx}{(x-1)^2} = \int_0^1 \frac{dx}{(x-1)^2} + \int_1^2 \frac{dx}{(x-1)^2}$$

and then treat each term as an improper integral. For the first term,

$$\int_0^1 \frac{dx}{(x-1)^2} = \lim_{k\to 1^-} \int_0^k \frac{dx}{(x-1)^2} = \lim_{k\to 1^-}\left[-\frac{1}{k-1} - 1\right] = +\infty$$

so (8) diverges.

ARC LENGTH AND SURFACE AREA USING IMPROPER INTEGRALS

In Definitions 6.4.2 and 6.5.2 for arc length and surface area we required the function f to be smooth (continuous first derivative) to ensure the integrability in the resulting formula. However, smoothness is overly restrictive since some of the most basic formulas in geometry involve functions that are not smooth but lead to convergent improper integrals. Accordingly, let us agree to extend the definitions of arc length and surface area to allow functions that are not smooth, but for which the resulting integral in the formula converges.

▶ **Example 7** Derive the formula for the circumference of a circle of radius r.

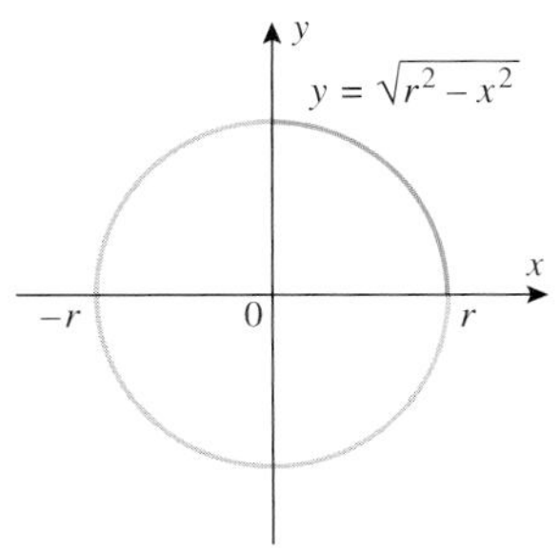

Figure 8.8.10

Solution. For convenience, let us assume that the circle is centered at the origin, in which case its equation is $x^2 + y^2 = r^2$. We will find the arc length of the portion of the circle that lies in the first quadrant and then multiply by 4 to obtain the total circumference (Figure 8.8.10).

Since the equation of the upper semicircle is $y = \sqrt{r^2 - x^2}$, it follows from Formula (4) of Section 7.4 that the circumference C is

$$\begin{aligned} C &= 4\int_0^r \sqrt{1 + (dy/dx)^2}\, dx = 4\int_0^r \sqrt{1 + \left(-\frac{x}{\sqrt{r^2 - x^2}}\right)^2}\, dx \\ &= 4r\int_0^r \frac{dx}{\sqrt{r^2 - x^2}} \end{aligned}$$

This integral is improper because of the infinite discontinuity at $x = r$, and hence we evaluate it by writing

$$\begin{aligned} C &= 4r \lim_{k\to r^-} \int_0^k \frac{dx}{\sqrt{r^2 - x^2}} \\ &= 4r \lim_{k\to r^-} \left[\sin^{-1}\left(\frac{x}{r}\right)\right]_0^k \\ &= 4r \lim_{k\to r^-} \left[\sin^{-1}\left(\frac{k}{r}\right) - \sin^{-1} 0\right] \\ &= 4r[\sin^{-1} 1 - \sin^{-1} 0] = 4r\left(\frac{\pi}{2} - 0\right) = 2\pi r \end{aligned}$$ ◀

Formula (77) in the Endpaper Integral Table

✔QUICK CHECK EXERCISES 8.8 *(See page 580 for answers.)*

1. In each part, determine whether the integral is improper, and if so, explain why. Do not evaluate the integrals.

(a) $\displaystyle\int_{\pi/4}^{3\pi/4} \cot x\,dx$ (b) $\displaystyle\int_{\pi/4}^{\pi} \cot x\,dx$

(c) $\displaystyle\int_{0}^{+\infty} \frac{1}{x^2+1}\,dx$ (d) $\displaystyle\int_{1}^{+\infty} \frac{1}{x^2-1}\,dx$

(e) $\displaystyle\int_{0}^{\pi/4} \tan x\,dx$

2. Express each improper integral in Quick Check Exercise 1 in terms of one or more appropriate limits. Do not evaluate the limits.

3. The improper integral

$$\int_1^{+\infty} x^{-p}\,dx$$

converges to ________ provided ________.

4. Evaluate the integrals that converge.

(a) $\displaystyle\int_0^{+\infty} e^{-x}\,dx$ (b) $\displaystyle\int_0^{+\infty} e^{x}\,dx$

(c) $\displaystyle\int_0^{1} \frac{1}{x^3}\,dx$ (d) $\displaystyle\int_0^{1} \frac{1}{\sqrt[3]{x^2}}\,dx$

EXERCISE SET 8.8  Graphing Utility  CAS

1. In each part, determine whether the integral is improper, and if so, explain why.

(a) $\displaystyle\int_1^5 \frac{dx}{x-3}$ (b) $\displaystyle\int_1^5 \frac{dx}{x+3}$ (c) $\displaystyle\int_0^1 \ln x\,dx$

(d) $\displaystyle\int_1^{+\infty} e^{-x}\,dx$ (e) $\displaystyle\int_{-\infty}^{+\infty} \frac{dx}{\sqrt[3]{x-1}}$ (f) $\displaystyle\int_0^{\pi/4} \tan x\,dx$

2. In each part, determine all values of p for which the integral is improper.

(a) $\displaystyle\int_0^1 \frac{dx}{x^p}$ (b) $\displaystyle\int_1^2 \frac{dx}{x-p}$ (c) $\displaystyle\int_0^1 e^{-px}\,dx$

3–30 Evaluate the integrals that converge.

3. $\displaystyle\int_0^{+\infty} e^{-2x}\,dx$

4. $\displaystyle\int_{-1}^{+\infty} \frac{x}{1+x^2}\,dx$

5. $\displaystyle\int_3^{+\infty} \frac{2}{x^2-1}\,dx$

6. $\displaystyle\int_0^{+\infty} xe^{-x^2}\,dx$

7. $\displaystyle\int_e^{+\infty} \frac{1}{x\ln^3 x}\,dx$

8. $\displaystyle\int_2^{+\infty} \frac{1}{x\sqrt{\ln x}}\,dx$

9. $\displaystyle\int_{-\infty}^{0} \frac{dx}{(2x-1)^3}$

10. $\displaystyle\int_{-\infty}^{3} \frac{dx}{x^2+9}$

11. $\displaystyle\int_{-\infty}^{0} e^{3x}\,dx$

12. $\displaystyle\int_{-\infty}^{0} \frac{e^x\,dx}{3-2e^x}$

13. $\displaystyle\int_{-\infty}^{+\infty} x\,dx$

14. $\displaystyle\int_{-\infty}^{+\infty} \frac{x}{\sqrt{x^2+2}}\,dx$

15. $\displaystyle\int_{-\infty}^{+\infty} \frac{x}{(x^2+3)^2}\,dx$

16. $\displaystyle\int_{-\infty}^{+\infty} \frac{e^{-t}}{1+e^{-2t}}\,dt$

17. $\displaystyle\int_0^4 \frac{dx}{(x-4)^2}$

18. $\displaystyle\int_0^8 \frac{dx}{\sqrt[3]{x}}$

19. $\displaystyle\int_0^{\pi/2} \tan x\,dx$

20. $\displaystyle\int_0^4 \frac{dx}{\sqrt{4-x}}$

21. $\displaystyle\int_0^1 \frac{dx}{\sqrt{1-x^2}}$

22. $\displaystyle\int_{-3}^1 \frac{x\,dx}{\sqrt{9-x^2}}$

23. $\displaystyle\int_{\pi/3}^{\pi/2} \frac{\sin x}{\sqrt{1-2\cos x}}\,dx$

24. $\displaystyle\int_0^{\pi/4} \frac{\sec^2 x}{1-\tan x}\,dx$

25. $\displaystyle\int_0^3 \frac{dx}{x-2}$

26. $\displaystyle\int_{-2}^2 \frac{dx}{x^2}$

27. $\displaystyle\int_{-1}^8 x^{-1/3}\,dx$

28. $\displaystyle\int_0^1 \frac{dx}{(x-1)^{2/3}}$

29. $\displaystyle\int_0^{+\infty} \frac{1}{x^2}\,dx$

30. $\displaystyle\int_1^{+\infty} \frac{dx}{x\sqrt{x^2-1}}$

31–34 Make the u-substitution and evaluate the resulting definite integral.

31. $\displaystyle\int_0^{+\infty} \frac{e^{-\sqrt{x}}}{\sqrt{x}}\,dx$; $u=\sqrt{x}$ [*Note:* $u\to+\infty$ as $x\to+\infty$.]

32. $\displaystyle\int_{12}^{+\infty} \frac{dx}{\sqrt{x}(x+4)}$; $u=\sqrt{x}$

33. $\displaystyle\int_0^{+\infty} \frac{e^{-x}}{\sqrt{1-e^{-x}}}\,dx$; $u=1-e^{-x}$
[*Note:* $u\to 1$ as $x\to+\infty$.]

34. $\displaystyle\int_0^{+\infty} \frac{e^{-x}}{\sqrt{1-e^{-2x}}}\,dx$; $u=e^{-x}$

35–36 Express the improper integral as a limit, and then evaluate that limit with a CAS. Confirm the answer by evaluating the integral directly with the CAS.

C 35. $\displaystyle\int_0^{+\infty} e^{-x}\cos x\,dx$

C 36. $\displaystyle\int_0^{+\infty} xe^{-3x}\,dx$

C **37.** In each part, try to evaluate the integral exactly with a CAS. If your result is not a simple numerical answer, then use the CAS to find a numerical approximation of the integral.

(a) $\displaystyle\int_{-\infty}^{+\infty} \frac{1}{x^8+x+1}\,dx$ (b) $\displaystyle\int_{0}^{+\infty} \frac{1}{\sqrt{1+x^3}}\,dx$

(c) $\displaystyle\int_{1}^{+\infty} \frac{\ln x}{e^x}\,dx$ (d) $\displaystyle\int_{1}^{+\infty} \frac{\sin x}{x^2}\,dx$

C **38.** In each part, confirm the result with a CAS.

(a) $\displaystyle\int_{0}^{+\infty} \frac{\sin x}{\sqrt{x}}\,dx = \sqrt{\frac{\pi}{2}}$ (b) $\displaystyle\int_{-\infty}^{+\infty} e^{-x^2}\,dx = \sqrt{\pi}$

(c) $\displaystyle\int_{0}^{1} \frac{\ln x}{1+x}\,dx = -\frac{\pi^2}{12}$

39. Find the length of the curve $y = (4 - x^{2/3})^{3/2}$ over the interval $[0, 8]$.

40. Find the length of the curve $y = \sqrt{4 - x^2}$ over the interval $[0, 2]$.

41–42 Use L'Hôpital's rule to help evaluate the improper integral.

41. $\displaystyle\int_{0}^{1} \ln x\,dx$ **42.** $\displaystyle\int_{1}^{+\infty} \frac{\ln x}{x^2}\,dx$

43. Find the area of the region between the x-axis and the curve $y = e^{-3x}$ for $x \geq 0$.

44. Find the area of the region between the x-axis and the curve $y = 8/(x^2 - 4)$ for $x \geq 4$.

45. Suppose that the region between the x-axis and the curve $y = e^{-x}$ for $x \geq 0$ is revolved about the x-axis.
(a) Find the volume of the solid that is generated.
(b) Find the surface area of the solid.

FOCUS ON CONCEPTS

46. Suppose that f and g are continuous functions and that

$$0 \leq f(x) \leq g(x)$$

if $x \geq a$. Give a reasonable informal argument using areas to explain why the following results are true.
(a) If $\int_a^{+\infty} f(x)\,dx$ diverges, then $\int_a^{+\infty} g(x)\,dx$ diverges.
(b) If $\int_a^{+\infty} g(x)\,dx$ converges, then $\int_a^{+\infty} f(x)\,dx$ converges and $\int_a^{+\infty} f(x)\,dx \leq \int_a^{+\infty} g(x)\,dx$.
[*Note:* The results in this exercise are sometimes called ***comparison tests*** for improper integrals.]

47–50 Use the results in Exercise 46.

47. (a) Confirm graphically and algebraically that

$$e^{-x^2} \leq e^{-x} \quad (x \geq 1)$$

(b) Evaluate the integral

$$\int_{1}^{+\infty} e^{-x}\,dx$$

(c) What does the result obtained in part (b) tell you about the integral

$$\int_{1}^{+\infty} e^{-x^2}\,dx?$$

48. (a) Confirm graphically and algebraically that

$$\frac{1}{2x+1} \leq \frac{e^x}{2x+1} \quad (x \geq 0)$$

(b) Evaluate the integral

$$\int_{0}^{+\infty} \frac{dx}{2x+1}$$

(c) What does the result obtained in part (b) tell you about the integral

$$\int_{0}^{+\infty} \frac{e^x}{2x+1}\,dx?$$

49. Let R be the region to the right of $x = 1$ that is bounded by the x-axis and the curve $y = 1/x$. When this region is revolved about the x-axis it generates a solid whose surface is known as ***Gabriel's Horn*** (for reasons that should be clear from the accompanying figure). Show that the solid has a finite volume but its surface has an infinite area. [*Note:* It has been suggested that if one could saturate the interior of the solid with paint and allow it to seep through to the surface, then one could paint an infinite surface with a finite amount of paint! What do you think?]

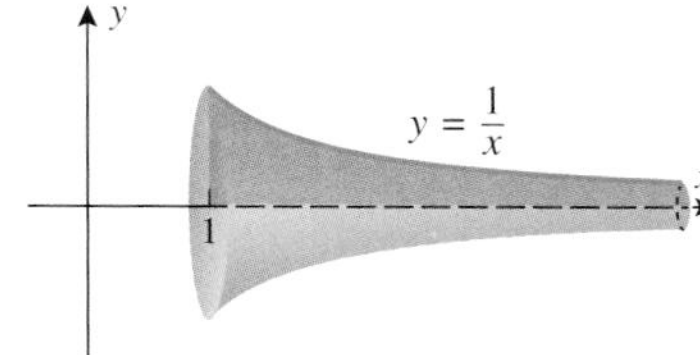

Figure Ex-49

50. In each part, use Exercise 46 to determine whether the integral converges or diverges. If it converges, then use part (b) of that exercise to find an upper bound on the value of the integral.

(a) $\displaystyle\int_{2}^{+\infty} \frac{\sqrt{x^3+1}}{x}\,dx$ (b) $\displaystyle\int_{2}^{+\infty} \frac{x}{x^5+1}\,dx$

(c) $\displaystyle\int_{0}^{+\infty} \frac{xe^x}{2x+1}\,dx$

FOCUS ON CONCEPTS

51. Sketch the region whose area is

$$\int_{0}^{+\infty} \frac{dx}{1+x^2}$$

and use your sketch to show that

$$\int_{0}^{+\infty} \frac{dx}{1+x^2} = \int_{0}^{1} \sqrt{\frac{1-y}{y}}\,dy$$

52. (a) Give a reasonable informal argument, based on areas, that explains why the integrals

$$\int_{0}^{+\infty} \sin x\,dx \quad \text{and} \quad \int_{0}^{+\infty} \cos x\,dx$$

diverge.

(b) Show that $\int_0^{+\infty} \frac{\cos\sqrt{x}}{\sqrt{x}}\,dx$ diverges.

53. In electromagnetic theory, the magnetic potential at a point on the axis of a circular coil is given by

$$u = \frac{2\pi NIr}{k}\int_a^{+\infty} \frac{dx}{(r^2+x^2)^{3/2}}$$

where N, I, r, k, and a are constants. Find u.

C **54.** The *average speed*, $\bar{v}$, of the molecules of an ideal gas is given by

$$\bar{v} = \frac{4}{\sqrt{\pi}}\left(\frac{M}{2RT}\right)^{3/2}\int_0^{+\infty} v^3 e^{-Mv^2/(2RT)}\,dv$$

and the *root-mean-square speed*, v_{rms}, by

$$v_{\text{rms}}^2 = \frac{4}{\sqrt{\pi}}\left(\frac{M}{2RT}\right)^{3/2}\int_0^{+\infty} v^4 e^{-Mv^2/(2RT)}\,dv$$

where v is the molecular speed, T is the gas temperature, M is the molecular weight of the gas, and R is the gas constant.

(a) Use a CAS to show that

$$\int_0^{+\infty} x^3 e^{-a^2x^2}\,dx = \frac{1}{2a^4}, \quad a > 0$$

and use this result to show that $\bar{v} = \sqrt{8RT/(\pi M)}$.

(b) Use a CAS to show that

$$\int_0^{+\infty} x^4 e^{-a^2x^2}\,dx = \frac{3\sqrt{\pi}}{8a^5}, \quad a > 0$$

and use this result to show that $v_{\text{rms}} = \sqrt{3RT/M}$.

55. In Exercise 23 of Section 6.7, we determined the work required to lift a 6000-lb satellite to an orbital position that is 1000 mi above the Earth's surface. The ideas discussed in that exercise will be needed here.

(a) Find a definite integral that represents the work required to lift a 6000-lb satellite to a position b miles above the Earth's surface.

(b) Find a definite integral that represents the work required to lift a 6000-lb satellite an "infinite distance" above the Earth's surface. Evaluate the integral. [*Note:* The result obtained here is sometimes called the work required to "escape" the Earth's gravity.]

56–57 A ***transform*** is a formula that converts or "transforms" one function into another. Transforms are used in applications to convert a difficult problem into an easier problem whose solution can then be used to solve the original difficult problem. The ***Laplace transform*** of a function $f(t)$, which plays an important role in the study of differential equations, is denoted by $\mathscr{L}\{f(t)\}$ and is defined by

$$\mathscr{L}\{f(t)\} = \int_0^{+\infty} e^{-st} f(t)\,dt$$

In this formula s is treated as a constant in the integration process; thus, the Laplace transform has the effect of transforming $f(t)$ into a function of s. Use this formula in these exercises.

56. Show that

(a) $\mathscr{L}\{1\} = \frac{1}{s}, \ s > 0$ (b) $\mathscr{L}\{e^{2t}\} = \frac{1}{s-2}, \ s > 2$

(c) $\mathscr{L}\{\sin t\} = \frac{1}{s^2+1}, \ s > 0$

(d) $\mathscr{L}\{\cos t\} = \frac{s}{s^2+1}, \ s > 0.$

57. In each part, find the Laplace transform.

(a) $f(t) = t, \ s > 0$ (b) $f(t) = t^2, \ s > 0$

(c) $f(t) = \begin{cases} 0, & t < 3 \\ 1, & t \ge 3 \end{cases}, \ s > 0$

C **58.** Later in the text, we will show that

$$\int_0^{+\infty} e^{-x^2}\,dx = \tfrac{1}{2}\sqrt{\pi}$$

Confirm that this is reasonable by using a CAS or a calculator with a numerical integration capability.

59. Use the result in Exercise 58 to show that

(a) $\int_{-\infty}^{+\infty} e^{-ax^2}\,dx = \sqrt{\frac{\pi}{a}}, \ a > 0$

(b) $\frac{1}{\sqrt{2\pi}\sigma}\int_{-\infty}^{+\infty} e^{-x^2/2\sigma^2}\,dx = 1, \ \sigma > 0.$

60–61 A convergent improper integral over an infinite interval can be approximated by first replacing the infinite limit(s) of integration by finite limit(s), then using a numerical integration technique, such as Simpson's rule, to approximate the integral with finite limit(s). This technique is illustrated in these exercises.

60. Suppose that the integral in Exercise 58 is approximated by first writing it as

$$\int_0^{+\infty} e^{-x^2}\,dx = \int_0^K e^{-x^2}\,dx + \int_K^{+\infty} e^{-x^2}\,dx$$

then dropping the second term, and then applying Simpson's rule to the integral

$$\int_0^K e^{-x^2}\,dx$$

The resulting approximation has two sources of error: the error from Simpson's rule and the error

$$E = \int_K^{+\infty} e^{-x^2}\,dx$$

that results from discarding the second term. We call E the ***truncation error***.

(a) Approximate the integral in Exercise 58 by applying Simpson's rule with $2n = 10$ subdivisions to the integral

$$\int_0^3 e^{-x^2}\,dx$$

Round your answer to four decimal places and compare it to $\frac{1}{2}\sqrt{\pi}$ rounded to four decimal places.

(b) Use the result that you obtained in Exercise 46 and the fact that $e^{-x^2} \le \frac{1}{3}xe^{-x^2}$ for $x \ge 3$ to show that the truncation error for the approximation in part (a) satisfies $0 < E < 2.1 \times 10^{-5}$.

61. (a) It can be shown that
$$\int_0^{+\infty} \frac{1}{x^6+1}\,dx = \frac{\pi}{3}$$
Approximate this integral by applying Simpson's rule with $2n = 20$ subdivisions to the integral
$$\int_0^4 \frac{1}{x^6+1}\,dx$$
Round your answer to three decimal places and compare it to $\pi/3$ rounded to three decimal places.

(b) Use the result that you obtained in Exercise 46 and the fact that $1/(x^6+1) < 1/x^6$ for $x \geq 4$ to show that the truncation error for the approximation in part (a) satisfies $0 < E < 2 \times 10^{-4}$.

62. For what values of p does $\int_0^{+\infty} e^{px}\,dx$ converge?

63. Show that $\int_0^1 dx/x^p$ converges if $p < 1$ and diverges if $p \geq 1$.

C **64.** It is sometimes possible to convert an improper integral into a "proper" integral having the same value by making an appropriate substitution. Evaluate the following integral by making the indicated substitution, and investigate what happens if you evaluate the integral directly using a CAS.
$$\int_0^1 \sqrt{\frac{1+x}{1-x}}\,dx; \; u = \sqrt{1-x}$$

65–66 Transform the given improper integral into a proper integral by making the stated u-substitution; then approximate the proper integral by Simpson's rule with $2n = 10$ subdivisions. Round your answer to three decimal places.

65. $\int_0^1 \frac{\cos x}{\sqrt{x}}\,dx; \; u = \sqrt{x}$

66. $\int_0^1 \frac{\sin x}{\sqrt{1-x}}\,dx; \; u = \sqrt{1-x}$

67. The ***Gamma function***, $\Gamma(x)$, is defined as
$$\Gamma(x) = \int_0^{+\infty} t^{x-1}e^{-t}\,dt$$
It can be shown that this improper integral converges if and only if $x > 0$.

(a) Find $\Gamma(1)$.

(b) Prove: $\Gamma(x+1) = x\Gamma(x)$ for all $x > 0$. [*Hint:* Use integration by parts.]

(c) Use the results in parts (a) and (b) to find $\Gamma(2)$, $\Gamma(3)$, and $\Gamma(4)$; and then make a conjecture about $\Gamma(n)$ for positive integer values of n.

(d) Show that $\Gamma\left(\frac{1}{2}\right) = \sqrt{\pi}$. [*Hint:* See Exercise 58.]

(e) Use the results obtained in parts (b) and (d) to show that $\Gamma\left(\frac{3}{2}\right) = \frac{1}{2}\sqrt{\pi}$ and $\Gamma\left(\frac{5}{2}\right) = \frac{3}{4}\sqrt{\pi}$.

68. Refer to the Gamma function defined in Exercise 67 to show that

(a) $\int_0^1 (\ln x)^n\,dx = (-1)^n\Gamma(n+1), \quad n > 0.$
[*Hint:* Let $t = -\ln x$.]

(b) $\int_0^{+\infty} e^{-x^n}\,dx = \Gamma\left(\frac{n+1}{n}\right), \quad n > 0.$
[*Hint:* Let $t = x^n$. Use the result in Exercise 67(b).]

C **69.** A ***simple pendulum*** consists of a mass that swings in a vertical plane at the end of a massless rod of length L, as shown in the accompanying figure. Suppose that a simple pendulum is displaced through an angle θ_0 and released from rest. It can be shown that in the absence of friction, the time T required for the pendulum to make one complete back-and-forth swing, called the ***period***, is given by
$$T = \sqrt{\frac{8L}{g}}\int_0^{\theta_0} \frac{1}{\sqrt{\cos\theta - \cos\theta_0}}\,d\theta \qquad (1)$$
where $\theta = \theta(t)$ is the angle the pendulum makes with the vertical at time t. The improper integral in (1) is difficult to evaluate numerically. By a substitution outlined below it can be shown that the period can be expressed as
$$T = 4\sqrt{\frac{L}{g}}\int_0^{\pi/2} \frac{1}{\sqrt{1-k^2\sin^2\phi}}\,d\phi \qquad (2)$$
where $k = \sin(\theta_0/2)$. The integral in (2) is called a ***complete elliptic integral of the first kind*** and is more easily evaluated by numerical methods.

(a) Obtain (2) from (1) by substituting
$$\cos\theta = 1 - 2\sin^2(\theta/2)$$
$$\cos\theta_0 = 1 - 2\sin^2(\theta_0/2)$$
$$k = \sin(\theta_0/2)$$
and then making the change of variable
$$\sin\phi = \frac{\sin(\theta/2)}{\sin(\theta_0/2)} = \frac{\sin(\theta/2)}{k}$$

(b) Use (2) and the numerical integration capability of your CAS to estimate the period of a simple pendulum for which $L = 1.5$ ft, $\theta_0 = 20°$, and $g = 32$ ft/s^2.

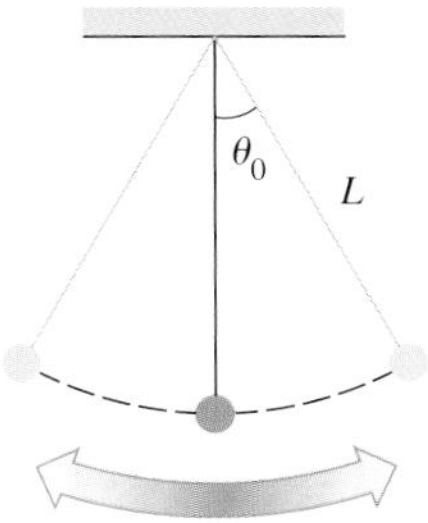

Figure Ex-69

✔QUICK CHECK ANSWERS 8.8

1. (a) proper (b) improper, since $\cot x$ has an infinite discontinuity at $x = \pi$ (c) improper, since there is an infinite interval of integration (d) improper, since there is an infinite interval of integration and the integrand has an infinite discontinuity at $x = 1$ (e) proper 2. (b) $\lim_{b\to\pi^-}\int_{\pi/4}^{b}\cot x\,dx$ (c) $\lim_{b\to+\infty}\int_0^b \frac{1}{x^2+1}\,dx$ (d) $\lim_{a\to 1^+}\int_a^2 \frac{1}{x^2+1}\,dx + \lim_{b\to+\infty}\int_2^b \frac{1}{x^2+1}\,dx$
3. $\frac{1}{p-1}$; $p>1$ 4. (a) 1 (b) diverges (c) diverges (d) 3

CHAPTER REVIEW EXERCISES

1–6 Evaluate the integral with the aid of an appropriate u-substitution.

1. $\int \sqrt{4+9x}\,dx$
2. $\int \frac{1}{\sec \pi x}\,dx$
3. $\int \sqrt{\cos x}\sin x\,dx$
4. $\int \frac{dx}{x\ln x}$
5. $\int x\tan^2(x^2)\sec^2(x^2)\,dx$
6. $\int_0^9 \frac{\sqrt{x}}{x+9}\,dx$
7. (a) Evaluate the integral
$$\int \frac{1}{\sqrt{2x-x^2}}\,dx$$
three ways: using the substitution $u=\sqrt{x}$, using the substitution $u=\sqrt{2-x}$, and completing the square.
(b) Show that the answers in part (a) are equivalent.
8. Evaluate the integral $\int_0^1 \frac{x^3}{\sqrt{x^2+1}}\,dx$
(a) using integration by parts
(b) using the substitution $u=\sqrt{x^2+1}$.

9–12 Use integration by parts to evaluate the integral.

9. $\int xe^{-x}\,dx$
10. $\int x\sin 2x\,dx$
11. $\int \ln(2x+3)\,dx$
12. $\int_0^{1/2}\tan^{-1}(2x)\,dx$
13. Evaluate $\int 8x^4\cos 2x\,dx$ using tabular integration by parts.
14. A particle moving along the x-axis has velocity function $v(t)=t^2e^{-t}$. How far does the particle travel from time $t=0$ to $t=5$?

15–20 Evaluate the integral.

15. $\int \sin^2 5\theta\,d\theta$
16. $\int \sin^3 2x\cos^2 2x\,dx$
17. $\int \sin x\cos 2x\,dx$
18. $\int_0^{\pi/6}\sin 2x\cos 4x\,dx$
19. $\int \sin^4 2x\,dx$
20. $\int x\cos^5(x^2)\,dx$

21–26 Evaluate the integral by making an appropriate trigonometric substitution.

21. $\int \frac{x^2}{\sqrt{9-x^2}}\,dx$
22. $\int \frac{dx}{x^2\sqrt{16-x^2}}$
23. $\int \frac{dx}{\sqrt{x^2-1}}$
24. $\int \frac{x^2}{\sqrt{x^2-25}}\,dx$
25. $\int \frac{x^2}{\sqrt{9+x^2}}\,dx$
26. $\int \frac{\sqrt{1+4x^2}}{x}\,dx$

27–32 Evaluate the integral using the method of partial fractions.

27. $\int \frac{dx}{x^2+3x-4}$
28. $\int \frac{dx}{x^2+8x+7}$
29. $\int \frac{x^2+2}{x+2}\,dx$
30. $\int \frac{x^2+x-16}{(x-1)(x-3)^2}\,dx$
31. $\int \frac{x^2}{(x+2)^3}\,dx$
32. $\int \frac{dx}{x^3+x}$
33. Consider the integral $\int \frac{1}{x^3-x}\,dx$.
(a) Evaluate the integral using the substitution $x=\sec\theta$. For what values of x is your result valid?
(b) Evaluate the integral using the substitution $x=\sin\theta$. For what values of x is your result valid?
(c) Evaluate the integral using the method of partial fractions. For what values of x is your result valid?
34. Find the area of the region that is enclosed by the curves $y=(x-3)/(x^3+x^2)$, $y=0$, $x=1$, and $x=2$.

35–40 Use the Endpaper Integral Table to evaluate the integral.

35. $\int \sin 7x\cos 9x\,dx$
36. $\int (x^3-x^2)e^{-x}\,dx$
37. $\int x\sqrt{x-x^2}\,dx$
38. $\int \frac{dx}{x\sqrt{4x+3}}$
39. $\int \tan^2 2x\,dx$
40. $\int \frac{3x-1}{2+x^2}\,dx$

41–42 Use $n = 10$ subintervals to approximate the integral by (a) the midpoint approximation, (b) the trapezoidal approximation, and use $2n = 10$ subintervals to approximate the integral by (c) Simpson's rule. In each case, find the exact value of the integral and approximate the absolute error. Express your answers to at least four decimal places.

41. $\int_0^3 \sqrt{x+1}\,dx$ **42.** $\int_{-1}^{1} \frac{1}{2x+3}\,dx$

43–44 Use inequalities (10), (11), and (12) of Section 8.7 to find upper bounds on the errors in parts (a), (b), or (c) of the indicated exercise.

43. Exercise 41 **44.** Exercise 42

45–46 Use inequalities (10), (11), and (12) of Section 8.7 to find a number n of subintervals for (a) the midpoint approximation and (b) the trapezoidal approximation to ensure that the absolute error will be less than the given value. Also, (c) find a number $2n$ of subintervals to ensure that the absolute error for Simpson's rule approximation will be less than the given value.

45. Exercise 41; 5×10^{-4} **46.** Exercise 42; 10^{-6}

47–50 Evaluate the integral if it converges.

47. $\int_0^{+\infty} e^{-x}\,dx$ **48.** $\int_{-\infty}^{2} \frac{dx}{x^2+4}$

49. $\int_0^9 \frac{dx}{\sqrt{9-x}}$ **50.** $\int_0^1 \frac{1}{2x-1}\,dx$

51. Find the area that is enclosed between the x-axis and the curve $y = (\ln x - 1)/x^2$ for $x \ge e$.

52. Find the volume of the solid that is generated when the region between the x-axis and the curve $y = e^{-x}$ for $x \ge 0$ is revolved about the y-axis.

53. Find a positive value of a that satisfies the equation

$$\int_0^{+\infty} \frac{1}{x^2+a^2}\,dx = 1$$

54. Consider the following methods for evaluating integrals: u-substitution, integration by parts, partial fractions, reduction formulas, and trigonometric substitutions. In each part, state the approach that you would try first to evaluate the integral. If none of them seems appropriate, then say so. You need not evaluate the integral.

(a) $\int x \sin x\,dx$ (b) $\int \cos x \sin x\,dx$

(c) $\int \tan^7 x\,dx$ (d) $\int \tan^7 x \sec^2 x\,dx$

(e) $\int \frac{3x^2}{x^3+1}\,dx$ (f) $\int \frac{3x^2}{(x+1)^3}\,dx$

(g) $\int \tan^{-1} x\,dx$ (h) $\int \sqrt{4-x^2}\,dx$

(i) $\int x\sqrt{4-x^2}\,dx$

55–74 Evaluate the integral.

55. $\int \frac{dx}{(3+x^2)^{3/2}}$ **56.** $\int x \cos 3x\,dx$

57. $\int_0^{\pi/4} \tan^7 \theta\,d\theta$ **58.** $\int \frac{\cos\theta}{\sin^2\theta - 6\sin\theta + 12}\,d\theta$

59. $\int \sin^2 2x \cos^3 2x\,dx$ **60.** $\int_0^4 \frac{1}{(x-3)^2}\,dx$

61. $\int e^{2x}\cos 3x\,dx$ **62.** $\int_{-1/\sqrt{2}}^{1/\sqrt{2}} (1-2x^2)^{3/2}\,dx$

63. $\int \frac{dx}{(x-1)(x+2)(x-3)}$ **64.** $\int_0^{1/3} \frac{dx}{(4-9x^2)^2}$

65. $\int_4^8 \frac{\sqrt{x-4}}{x}\,dx$ **66.** $\int_0^{\ln 2} \sqrt{e^x - 1}\,dx$

67. $\int \frac{1}{\sqrt{e^x+1}}\,dx$ **68.** $\int \frac{dx}{x(x^2+x+1)}$

69. $\int_0^{1/2} \sin^{-1} x\,dx$ **70.** $\int \tan^5 4x \sec^4 4x\,dx$

71. $\int \frac{x+3}{\sqrt{x^2+2x+2}}\,dx$ **72.** $\int \frac{\sec^2\theta}{\tan^3\theta - \tan^2\theta}\,d\theta$

73. $\int_a^{+\infty} \frac{x}{(x^2+1)^2}\,dx$

74. $\int_0^{+\infty} \frac{dx}{a^2+b^2x^2}, \quad a, b > 0$

Expanding the Calculus Horizon

As the chief engineer for a major railroad, you need to analyze the costs of trenches and tunnels needed to lay a track bed for a new rail line between two booming cities. To learn more about the mathematics needed to do this, and to build upon the mathematics learned in this chapter, go to

www.wiley.com/college/anton

MATHEMATICAL MODELING WITH DIFFERENTIAL EQUATIONS

Among all the mathematical disciplines the theory of differential equations is the most important...
—Marius Sophus Lie
Mathematician

Many of the principles in science and engineering concern relationships between changing quantities. Since rates of change are represented mathematically by derivatives, it should not be surprising that such principles are often expressed in terms of differential equations. We introduced the concept of a differential equation in Section 6.2, but in this chapter we will go into more detail. We will discuss some important mathematical models that involve differential equations, and we will discuss some methods for solving and approximating solutions of some of the basic types of differential equations. However, we will only be able to touch the surface of this topic, leaving many important topics in differential equations to courses that are devoted completely to the subject.

Photo: *Carbon dating of charred bison bones found in New Mexico near the "Folsom points" in 1950 confirmed that human hunters lived in the area between 9000 B.C. and 8000 B.C. We will study carbon dating in this chapter.*

9.1 FIRST-ORDER DIFFERENTIAL EQUATIONS AND APPLICATIONS

In this section we will introduce some basic terminology and concepts concerning differential equations. We will also discuss methods for solving certain basic types of differential equations, and we will give some applications of our work.

TERMINOLOGY

Recall from Section 6.2 that a ***differential equation*** is an equation involving one or more derivatives of an unknown function. In this section we will denote the unknown function by $y = y(x)$ unless the differential equation arises from an applied problem involving time, in which case we will denote it by $y = y(t)$. The ***order*** of a differential equation is the order of the highest derivative that it contains. Some examples are given in Table 9.1.1. The last two equations in that table are expressed in "prime" notation, which does not specify the independent variable explicitly. However, you will usually be able to tell from the equation itself or from the context in which it arises whether to interpret y' as dy/dx or dy/dt.

Table 9.1.1

DIFFERENTIAL EQUATION	ORDER
$\dfrac{dy}{dx} = 3y$	1
$\dfrac{d^2y}{dx^2} - 6\dfrac{dy}{dx} + 8y = 0$	2
$\dfrac{d^3y}{dt^3} - t\dfrac{dy}{dt} + (t^2 - 1)y = e^t$	3
$y' - y = e^{2x}$	1
$y'' + y' = \cos t$	2

SOLUTIONS OF DIFFERENTIAL EQUATIONS

A function $y = y(x)$ is a ***solution*** of a differential equation on an open interval I if the equation is satisfied identically on I when y and its derivatives are substituted into the equation. For example, $y = e^{2x}$ is a solution of the differential equation

$$\frac{dy}{dx} - y = e^{2x} \tag{1}$$

on the interval $I = (-\infty, +\infty)$, since substituting y and its derivative into the left side of this equation yields

$$\frac{dy}{dx} - y = \frac{d}{dx}[e^{2x}] - e^{2x} = 2e^{2x} - e^{2x} = e^{2x}$$

for all real values of x. However, this is not the only solution on I; for example, the function

$$y = Ce^{x} + e^{2x} \tag{2}$$

is also a solution for every real value of the constant C, since

$$\frac{dy}{dx} - y = \frac{d}{dx}[Ce^{x} + e^{2x}] - (Ce^{x} + e^{2x}) = (Ce^{x} + 2e^{2x}) - (Ce^{x} + e^{2x}) = e^{2x}$$

The first-order equation (1) has a single arbitrary constant in its general solution (2). Usually, the general solution of an nth-order differential equation will contain n arbitrary constants. This is plausible, since n integrations are needed to recover a function from its nth derivative.

After developing some techniques for solving equations such as (1), we will be able to show that *all* solutions of (1) on $(-\infty, +\infty)$ can be obtained by substituting values for the constant C in (2). On a given interval I, a solution of a differential equation from which all solutions on I can be derived by substituting values for arbitrary constants is called a ***general solution*** of the equation on I. Thus (2) is a general solution of (1) on the interval $I = (-\infty, +\infty)$.

The graph of a solution of a differential equation is called an ***integral curve*** for the equation, so the general solution of a differential equation produces a family of integral curves corresponding to the different possible choices for the arbitrary constants. For example, Figure 9.1.1 shows some integral curves for (1), which were obtained by assigning values to the arbitrary constant in (2).

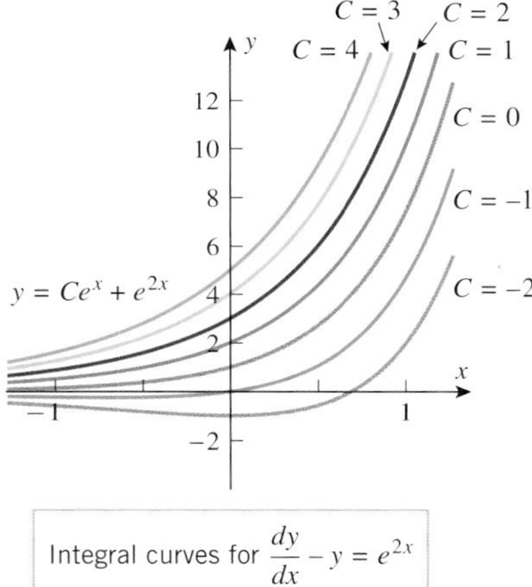

Figure 9.1.1

INITIAL-VALUE PROBLEMS

When an applied problem leads to a differential equation, there are usually conditions in the problem that determine specific values for the arbitrary constants. As a rule of thumb, it requires n conditions to determine values for all n arbitrary constants in the general solution of an nth-order differential equation (one condition for each constant). For a first-order equation, the single arbitrary constant can be determined by specifying the value of the unknown function $y(x)$ at an arbitrary x-value x_0, say $y(x_0) = y_0$. This is called an ***initial condition***, and the problem of solving a first-order equation subject to an initial condition is called a ***first-order initial-value problem***. Geometrically, the initial condition $y(x_0) = y_0$ has the effect of isolating the integral curve that passes through the point (x_0, y_0) from the complete family of integral curves.

▶ **Example 1** The solution of the initial-value problem

$$\frac{dy}{dx} - y = e^{2x}, \quad y(0) = 3$$

can be obtained by substituting the initial condition $x = 0$, $y = 3$ in the general solution (2) to find C. We obtain

$$3 = Ce^0 + e^0 = C + 1$$

Thus, $C = 2$, and the solution of the initial-value problem, which is obtained by substituting this value of C in (2), is

$$y = 2e^x + e^{2x}$$

Geometrically, this solution is realized as the integral curve in Figure 9.1.1 that passes through the point (0, 3). ◀

FIRST-ORDER LINEAR EQUATIONS

The simplest first-order equations are those that can be written in the form

$$\frac{dy}{dx} = q(x) \tag{3}$$

Such equations can often be solved by integration. For example, if

$$\frac{dy}{dx} = x^3 \tag{4}$$

then

$$y = \int x^3\,dx = \frac{x^4}{4} + C$$

is the general solution of (4) on the interval $I = (-\infty, +\infty)$. More generally, a first-order differential equation is called ***linear*** if it is expressible in the form

$$\frac{dy}{dx} + p(x)y = q(x) \tag{5}$$

Equation (3) is the special case of (5) that results when the function $p(x)$ is identically 0. Some other examples of first-order linear differential equations are

$$\frac{dy}{dx} + x^2y = e^x, \qquad \frac{dy}{dx} + (\sin x)y + x^3 = 0, \qquad \frac{dy}{dx} + 5y = 2$$

$p(x) = x^2,\ q(x) = e^x$ | $p(x) = \sin x,\ q(x) = -x^3$ | $p(x) = 5,\ q(x) = 2$

We will assume that the functions $p(x)$ and $q(x)$ in (5) are continuous on a common interval I, and we will look for a general solution that is valid on I. One method for doing this is based on the observation that if we define $\mu = \mu(x)$ by

$$\mu = e^{\int p(x)\,dx} \tag{6}$$

then

$$\frac{d\mu}{dx} = e^{\int p(x)\,dx} \cdot \frac{d}{dx}\int p(x)\,dx = \mu p(x)$$

Thus,

$$\frac{d}{dx}(\mu y) = \mu\frac{dy}{dx} + \frac{d\mu}{dx}y = \mu\frac{dy}{dx} + \mu p(x)y \tag{7}$$

If (5) is multiplied through by μ, it becomes

$$\mu\frac{dy}{dx} + \mu p(x)y = \mu q(x)$$

Combining this with (7) we have

$$\frac{d}{dx}(\mu y) = \mu q(x) \tag{8}$$

This equation can be solved for y by integrating both sides with respect to x and then dividing through by μ to obtain

$$y = \frac{1}{\mu}\int \mu q(x)\,dx \tag{9}$$

which is a general solution of (5) on I. The function μ in (6) is called an ***integrating factor*** for (5), and this method for finding a general solution of (5) is called the ***method of integrating factors***. Although one could simply memorize Formula (9), we recommend solving first-order linear equations by actually carrying out the steps used to derive this formula:

The Method of Integrating Factors

Step 1. Calculate the integrating factor

$$\mu = e^{\int p(x)\,dx}$$

Since any μ will suffice, we can take the constant of integration to be zero in this step.

Step 2. Multiply both sides of (5) by μ and express the result as

$$\frac{d}{dx}(\mu y) = \mu q(x)$$

Step 3. Integrate both sides of the equation obtained in Step 2 and then solve for y. Be sure to include a constant of integration in this step.

▶ **Example 2** Solve the differential equation

$$\frac{dy}{dx} - y = e^{2x}$$

Solution. Comparing the given equation to (5), we see that we have a first-order linear equation with $p(x) = -1$ and $q(x) = e^{2x}$. These coefficients are continuous on the interval $I = (-\infty, +\infty)$, so the method of integrating factors will produce a general solution on this interval. The first step is to compute the integrating factor. This yields

$$\mu = e^{\int p(x)\,dx} = e^{\int (-1)\,dx} = e^{-x}$$

Next we multiply both sides of the given equation by μ to obtain

$$e^{-x}\frac{dy}{dx} - e^{-x}y = e^{-x}e^{2x}$$

which we can rewrite as

$$\frac{d}{dx}[e^{-x}y] = e^{x}$$

Integrating both sides of this equation with respect to x we obtain

Confirm that the solution obtained in Example 2 agrees with that obtained by substituting the integrating factor into Formula (9).

$$e^{-x}y = e^{x} + C$$

Finally, solving for y yields the general solution

$$y = e^{2x} + Ce^{x}$$ ◀

A differential equation of the form

$$P(x)\frac{dy}{dx} + Q(x)y = R(x)$$

can be solved by dividing through by $P(x)$ to put the equation in form (5) and then applying the method of integrating factors. However, the resulting solution will only be valid on intervals where $p(x) = Q(x)/P(x)$ and $q(x) = R(x)/P(x)$ are both continuous.

▶ **Example 3** Solve the initial-value problem

$$x\frac{dy}{dx} - y = x, \quad y(1) = 2$$

Solution. This differential equation can be written in the form of (5) by dividing through by x. This yields

$$\frac{dy}{dx} - \frac{1}{x}y = 1 \tag{10}$$

where $q(x) = 1$ is continuous on $(-\infty, +\infty)$ and $p(x) = -1/x$ is continuous on $(-\infty, 0)$ and $(0, +\infty)$. Since we need $p(x)$ and $q(x)$ to be continuous on a common interval, and since our initial condition requires a solution for $x = 1$, we will find a general solution of (10) on the interval $(0, +\infty)$. On this interval we have $|x| = x$, so that

$$\int p(x)\,dx = -\int \frac{1}{x}\,dx = -\ln|x| = -\ln x$$

Taking the constant of integration to be 0

Thus, an integrating factor that will produce a general solution on the interval $(0, +\infty)$ is

$$\mu = e^{\int p(x)\,dx} = e^{-\ln x} = \frac{1}{x}$$

Multiplying both sides of Equation (10) by this integrating factor yields

$$\frac{1}{x}\frac{dy}{dx} - \frac{1}{x^2}y = \frac{1}{x}$$

or

$$\frac{d}{dx}\left[\frac{1}{x}y\right] = \frac{1}{x}$$

Therefore, on the interval $(0, +\infty)$,

$$\frac{1}{x}y = \int \frac{1}{x}\,dx = \ln x + C$$

from which it follows that

$$y = x\ln x + Cx \tag{11}$$

The initial condition $y(1) = 2$ requires that $y = 2$ if $x = 1$. Substituting these values into (11) and solving for C yields $C = 2$ (verify), so the solution of the initial-value problem is

$$y = x\ln x + 2x \quad ◀$$

It is not accidental that the initial-value problem in Example 3 has a unique solution. In general, if x_0 is any point in an open interval I on which the coefficients of (5) are continuous, then for any real number y_0, there will always exist a unique solution $y = y(x)$ of (5) on I for which $y(x_0) = y_0$ [Exercise 58(b)].

FIRST-ORDER SEPARABLE EQUATIONS

Although there is no general method for solving nonlinear first-order differential equations, we will now consider a method of solution that can often be applied to first-order equations that are expressible in the form

$$h(y)\frac{dy}{dx} = g(x) \tag{12}$$

Such first-order equations are said to be ***separable***. This name arises from the fact that this equation can be rewritten in the differential form

$$h(y)\,dy = g(x)\,dx \tag{13}$$

in which the expressions involving x appear on one side and those involving y appear on the other. The process of rewriting (12) in form (13) is called ***separating variables***.

To motivate a method for solving separable equations, assume that $h(y)$ and $g(x)$ are continuous functions of their respective variables, and let $H(y)$ and $G(x)$ denote antiderivatives of $h(y)$ and $g(x)$, respectively. Consider the equation that results if we integrate both sides of (13), the left side with respect to y and the right side with respect to x. We then have

$$\int h(y)\,dy = \int g(x)\,dx \tag{14}$$

or, equivalently,

$$H(y) = G(x) + C \tag{15}$$

where C denotes a constant. We claim that a differentiable function $y = y(x)$ is a solution to (12) if and only if y satisfies Equation (15) for some choice of the constant C.

Suppose that $y = y(x)$ is a solution to (12). It then follows from the chain rule that

$$\frac{d}{dx}[H(y)] = \frac{dH}{dy}\frac{dy}{dx} = h(y)\frac{dy}{dx} = g(x) = \frac{dG}{dx} \tag{16}$$

Since the functions $H(y)$ and $G(x)$ have the same derivative with respect to x, they must differ by a constant (Theorem 4.7.3). It then follows that y satisfies (15) for an appropriate choice of C. Conversely, if $y = y(x)$ is defined implicitly by Equation (15), then implicit differentiation shows that (16) is satisfied, and thus $y(x)$ is a solution to (12) (Exercise 59). Because of this, it is common practice to refer to Equation (15) as the "solution" to (12).

In summary, we have the following procedure for solving (12), called ***separation of variables***:

Separation of Variables

Step 1. Separate the variables in (12) by rewriting the equation in the differential form

$$h(y)\,dy = g(x)\,dx$$

Step 2. Integrate both sides of the equation in Step 1 (the left side with respect to y and the right side with respect to x):

$$\int h(y)\,dy = \int g(x)\,dx$$

Step 3. If $H(y)$ is any antiderivative of $h(y)$ and $G(x)$ is any antiderivative of $g(x)$, then the equation

$$H(y) = G(x) + C$$

will generally define a family of solutions implicitly. In some cases it may be possible to solve this equation explicitly for y.

▶ **Example 4** Solve the differential equation

$$\frac{dy}{dx} = -4xy^2$$

and then solve the initial-value problem

$$\frac{dy}{dx} = -4xy^2, \quad y(0) = 1$$

Solution. For $y \neq 0$ we can write this equation in form (12) as

$$\frac{1}{y^2}\frac{dy}{dx} = -4x$$

Separating variables and integrating yields

$$\frac{1}{y^2}\,dy = -4x\,dx$$

$$\int \frac{1}{y^2}\,dy = \int -4x\,dx$$

or

$$-\frac{1}{y} = -2x^2 + C$$

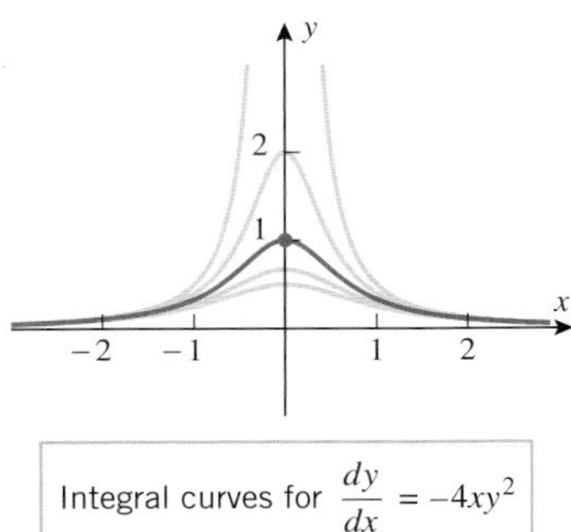

Figure 9.1.2

Solving for y as a function of x, we obtain

$$y = \frac{1}{2x^2 - C}$$

The initial condition $y(0) = 1$ requires that $y = 1$ when $x = 0$. Substituting these values into our solution yields $C = -1$ (verify). Thus, a solution to the initial-value problem is

$$y = \frac{1}{2x^2 + 1}$$

Some integral curves and our solution of the initial-value problem are graphed in Figure 9.1.2. ◄

One aspect of our solution to Example 4 deserves special comment. Had the initial condition been $y(0) = 0$ instead of $y(0) = 1$, the method we used would have failed to yield a solution to the resulting initial-value problem (Exercise 39). This is due to the fact that we assumed $y \neq 0$ in order to rewrite the equation $dy/dx = -4xy^2$ in the form

$$\frac{1}{y^2}\frac{dy}{dx} = -4x$$

It is important to be aware of such assumptions when manipulating a differential equation algebraically.

► **Example 5** Solve the initial-value problem

$$(4y - \cos y)\frac{dy}{dx} - 3x^2 = 0, \quad y(0) = 0$$

Solution. We can write this equation in form (12) as

$$(4y - \cos y)\frac{dy}{dx} = 3x^2$$

Separating variables and integrating yields

$$(4y - \cos y)\,dy = 3x^2\,dx$$

$$\int (4y - \cos y)\,dy = \int 3x^2\,dx$$

or

$$2y^2 - \sin y = x^3 + C \tag{17}$$

Equation (17) defines solutions of the differential equation implicitly; it cannot be solved explicitly for y as a function of x.

For the initial-value problem, the initial condition $y(0) = 0$ requires that $y = 0$ if $x = 0$. Substituting these values into (17) to determine the constant of integration yields $C = 0$ (verify). Thus, the solution of the initial-value problem is

$$2y^2 - \sin y = x^3 \quad ◄$$

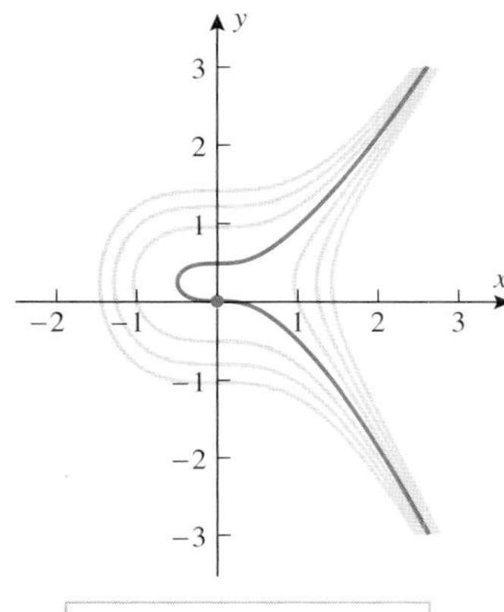

Integral curves for $(4y - \cos y)\dfrac{dy}{dx} - 3x^2 = 0$

Figure 9.1.3

Some integral curves and the solution of the initial-value problem in Example 5 are graphed in Figure 9.1.3.

We conclude this section with some applications of first-order differential equations.

TECHNOLOGY MASTERY

Some computer algebra systems can graph implicit equations. Figure 9.1.3 shows the graphs of (17) for $C = 0$, ± 1, ± 2, and ± 3. If you have a CAS that can graph implicit equations, try to duplicate this figure.

APPLICATIONS IN GEOMETRY

► Example 6 Find a curve in the xy-plane that passes through (0, 3) and whose tangent line at a point (x, y) has slope $2x/y^2$.

Solution. Since the slope of the tangent line is dy/dx, we have

$$\frac{dy}{dx} = \frac{2x}{y^2} \tag{18}$$

and, since the curve passes through (0, 3), we have the initial condition

$$y(0) = 3 \tag{19}$$

Equation (18) is separable and can be written as

$$y^2\,dy = 2x\,dx$$

so

$$\int y^2\,dy = \int 2x\,dx \quad \text{or} \quad \tfrac{1}{3}y^3 = x^2 + C$$

It follows from the initial condition (19) that $y = 3$ if $x = 0$. Substituting these values into the last equation yields $C = 9$ (verify), so the equation of the desired curve is

$$\tfrac{1}{3}y^3 = x^2 + 9 \quad \text{or} \quad y = (3x^2 + 27)^{1/3} \blacktriangleleft$$

MIXING PROBLEMS

In a typical mixing problem, a tank is filled to a specified level with a solution that contains a known amount of some soluble substance (say salt). The thoroughly stirred solution is allowed to drain from the tank at a known rate, and at the same time a solution with a known concentration of the soluble substance is added to the tank at a known rate that may or may not differ from the draining rate. As time progresses, the amount of the soluble substance in the tank will generally change, and the usual mixing problem seeks to determine the amount of the substance in the tank at a specified time. This type of problem serves as a model for many kinds of problems: discharge and filtration of pollutants in a river, injection and absorption of medication in the bloodstream, and migrations of species into and out of an ecological system, for example.

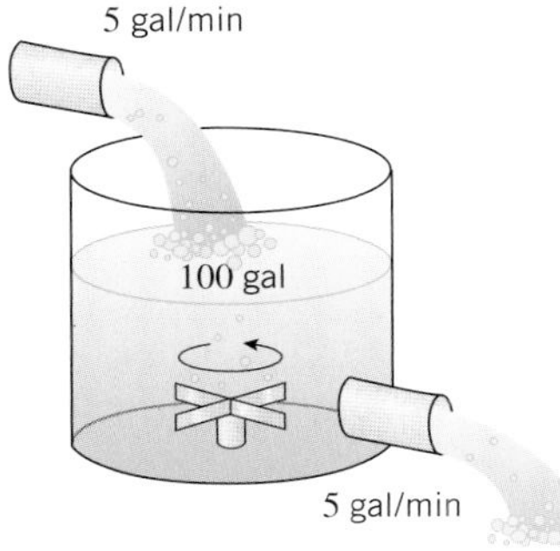

Figure 9.1.4

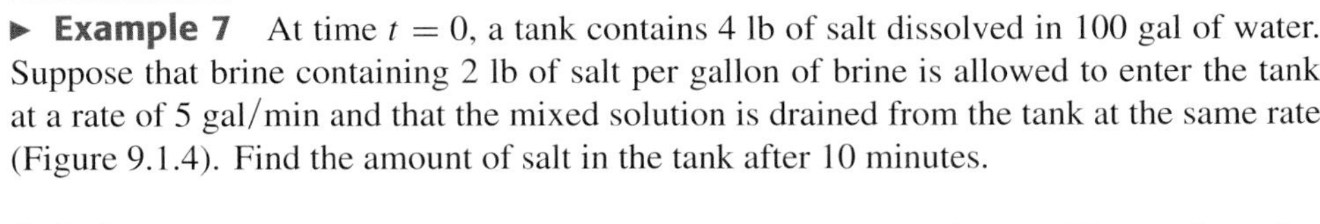

► Example 7 At time $t = 0$, a tank contains 4 lb of salt dissolved in 100 gal of water. Suppose that brine containing 2 lb of salt per gallon of brine is allowed to enter the tank at a rate of 5 gal/min and that the mixed solution is drained from the tank at the same rate (Figure 9.1.4). Find the amount of salt in the tank after 10 minutes.

Solution. Let $y(t)$ be the amount of salt (in pounds) after t minutes. We are given that $y(0) = 4$, and we want to find $y(10)$. We will begin by finding a differential equation that is satisfied by $y(t)$. To do this, observe that dy/dt, which is the rate at which the amount of salt in the tank changes with time, can be expressed as

$$\frac{dy}{dt} = \text{rate in} - \text{rate out} \tag{20}$$

where *rate in* is the rate at which salt enters the tank and *rate out* is the rate at which salt leaves the tank. But the rate at which salt enters the tank is

$$\text{rate in} = (2 \text{ lb/gal}) \cdot (5 \text{ gal/min}) = 10 \text{ lb/min}$$

Since brine enters and drains from the tank at the same rate, the volume of brine in the tank stays constant at 100 gal. Thus, after t minutes have elapsed, the tank contains $y(t)$ lb of salt per 100 gal of brine, and hence the rate at which salt leaves the tank at that instant is

$$\text{rate out} = \left(\frac{y(t)}{100} \text{ lb/gal}\right) \cdot (5 \text{ gal/min}) = \frac{y(t)}{20} \text{ lb/min}$$

Therefore, (20) can be written as

$$\frac{dy}{dt} = 10 - \frac{y}{20} \qquad \text{or} \qquad \frac{dy}{dt} + \frac{y}{20} = 10$$

which is a first-order linear differential equation satisfied by $y(t)$. Since we are given that $y(0) = 4$, the function $y(t)$ can be obtained by solving the initial-value problem

$$\frac{dy}{dt} + \frac{y}{20} = 10, \quad y(0) = 4$$

The integrating factor for the differential equation is

$$\mu = e^{\int (1/20)\, dt} = e^{t/20}$$

If we multiply the differential equation through by μ, then we obtain

$$\frac{d}{dt}(e^{t/20} y) = 10e^{t/20}$$

$$e^{t/20} y = \int 10e^{t/20}\, dt = 200e^{t/20} + C$$

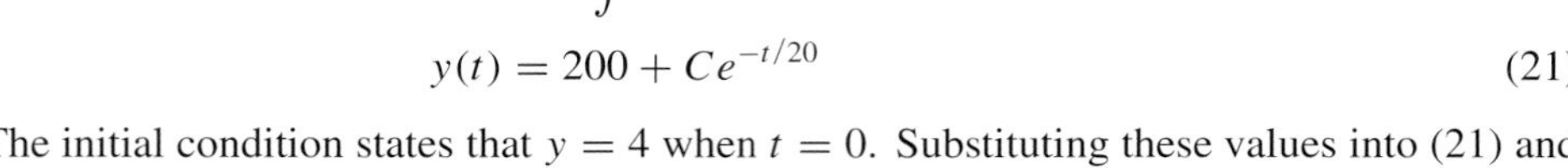

$$y(t) = 200 + Ce^{-t/20} \tag{21}$$

The initial condition states that $y = 4$ when $t = 0$. Substituting these values into (21) and solving for C yields $C = -196$ (verify), so

$$y(t) = 200 - 196e^{-t/20} \tag{22}$$

The graph of (22) is shown in Figure 9.1.5. At time $t = 10$ the amount of salt in the tank is

$$y(10) = 200 - 196e^{-0.5} \approx 81.1 \text{ lb} \blacktriangleleft$$

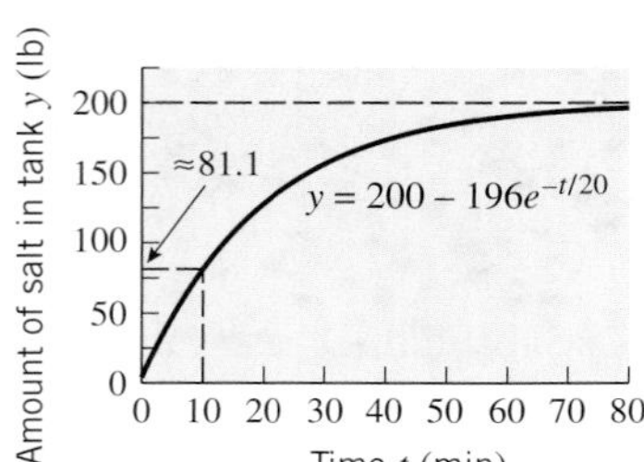

Figure 9.1.5

The graph shown in Figure 9.1.5 suggests that $y(t) \to 200$ as $t \to +\infty$. This means that over an extended period of time the amount of salt in the tank tends toward 200 lb. Give an informal physical argument to explain why this result is to be expected.

A MODEL OF FREE-FALL MOTION RETARDED BY AIR RESISTANCE

In Section 6.7 we considered the free-fall model of an object moving along a vertical axis near the surface of the Earth. It was assumed in that model that there is no air resistance and that the only force acting on the object is the Earth's gravity. Our goal here is to find a model that takes air resistance into account. For this purpose we make the following assumptions:

- The object moves along a vertical s-axis whose origin is at the surface of the Earth and whose positive direction is up (Figure 6.7.6).
- At time $t = 0$ the height of the object is s_0 and the velocity is v_0.
- The only forces on the object are the force $F_G = -mg$ of the Earth's gravity acting down and the force F_R of air resistance acting opposite to the direction of motion. The force F_R is called the ***drag force***.

We will also need the following result from physics.

9.1.1 NEWTON'S SECOND LAW OF MOTION. If an object with mass m is subjected to a force F, then the object undergoes an acceleration a that satisfies the equation

$$F = ma \tag{23}$$

In the case of free-fall motion retarded by air resistance, the net force acting on the object is

$$F_G + F_R = -mg + F_R$$

and the acceleration is d^2s/dt^2, so Newton's second law implies that

$$-mg + F_R = m\frac{d^2s}{dt^2} \tag{24}$$

Experimentation has shown that the force F_R of air resistance depends on the shape of the object and its speed—the greater the speed, the greater the drag force. There are many possible models for air resistance, but one of the most basic assumes that the drag force F_R is proportional to the velocity of the object, that is,

$$F_R = -cv$$

where c is a positive constant that depends on the object's shape and properties of the air.* (The minus sign ensures that the drag force is opposite to the direction of motion.) Substituting this in (24) and writing d^2s/dt^2 as dv/dt, we obtain

$$-mg - cv = m\frac{dv}{dt}$$

Dividing by m and rearranging we obtain

$$\frac{dv}{dt} + \frac{c}{m}v = -g$$

which is a first-order linear differential equation in the unknown function $v = v(t)$ with $p(t) = c/m$ and $q(t) = -g$ [see (5)]. For a specific object, the coefficient c can be determined experimentally, so we will assume that m, g, and c are known constants. Thus, the velocity function $v = v(t)$ can be obtained by solving the initial-value problem

$$\frac{dv}{dt} + \frac{c}{m}v = -g, \quad v(0) = v_0 \tag{25}$$

Once the velocity function is found, the position function $s = s(t)$ can be obtained by solving the initial-value problem

$$\frac{ds}{dt} = v(t), \quad s(0) = s_0 \tag{26}$$

In Exercise 47 we will ask you to solve (25) and show that

$$v(t) = e^{-ct/m}\left(v_0 + \frac{mg}{c}\right) - \frac{mg}{c} \tag{27}$$

Note that

$$\lim_{t\to+\infty} v(t) = -\frac{mg}{c} \tag{28}$$

(verify). Thus, the speed $|v(t)|$ does not increase indefinitely, as in free fall; rather, because of the air resistance, it approaches a finite limiting speed v_τ given by

$$v_\tau = \left|-\frac{mg}{c}\right| = \frac{mg}{c} \tag{29}$$

*Other common models assume that $F_R = -cv^2$ or, more generally, $F_R = -cv^p$ for some value of p.

This is called the ***terminal speed*** of the object, and (28) is called its ***terminal velocity***.

> Intuition suggests that near the limiting velocity, the velocity $v(t)$ changes very slowly; that is, $dv/dt \approx 0$. Thus, it should not be surprising that the limiting velocity can be obtained informally from (25) by setting $dv/dt = 0$ in the differential equation and solving for v. This yields
>
> $$v = -\frac{mg}{c}$$
>
> which agrees with (28).

✔QUICK CHECK EXERCISES 9.1 *(See page 595 for answers.)*

1. Match each differential equation with its family of solutions.

(a) $x\dfrac{dy}{dx} = y$ ______ (i) $y = x^2 + C$

(b) $y'' = 4y$ ______ (ii) $y = C_1 \sin 2x + C_2 \cos 2x$

(c) $\dfrac{dy}{dx} = 2x$ ______ (iii) $y = C_1e^{2x} + C_2e^{-2x}$

(d) $\dfrac{d^2y}{dx^2} = -4y$ ______ (iv) $y = Cx$

2. Solve the first-order linear differential equation

$$\frac{dy}{dx} + p(x)y = q(x)$$

by completing the following steps:

Step 1. Calculate the integrating factor $\mu =$ ______.

Step 2. Multiply both sides of the equation by the integrating factor and express the result as

$$\frac{d}{dx}[\_\_\_\_\_\_] = \_\_\_\_\_\_$$

Step 3. Integrate both sides of the equation obtained in Step 2 and solve for $y =$ ______.

3. If $y = C_1e^{2x} + C_2xe^{2x}$ is the general solution of a differential equation, then the order of the equation is ______, and a solution to the differential equation that satisfies the initial conditions $y(0) = 1$, $y'(0) = 4$ is given by $y =$ ______.

4. Solve the first-order separable equation

$$h(y)\frac{dy}{dx} = g(x)$$

by completing the following steps:

Step 1. Separate the variables by writing the equation in the differential form ______.

Step 2. Integrate both sides of the equation in Step 1: ______.

Step 3. If $H(y)$ is any antiderivative of $h(y)$, $G(x)$ is any antiderivative of $g(x)$, and C is an unspecified constant, then, as suggested by Step 2, the equation ______ will generally define a family of solutions to $h(y)\,dy/dx = g(x)$ implicitly.

5. (a) The graph of a differentiable function $y = y(x)$ passes through the point $(0, 1)$ and at every point $P(x, y)$ on the graph the tangent line is perpendicular to the line through P and the origin. Find an initial-value problem whose solution is $y(x)$.

(b) Explain why the differential equation in part (a) is separable. Solve the initial-value problem using either separation of variables or a geometric argument.

6. At time $t = 0$, a tank contains 30 oz of salt dissolved in 60 gal of water. Then brine containing 5 oz of salt per gallon of brine is allowed to enter the tank at a rate of 3 gal/min and the mixed solution is drained from the tank at the same rate. Give an initial-value problem satisfied by the amount of salt $y(t)$ in the tank at time t. Do not solve the problem.

EXERCISE SET 9.1 Graphing Utility [C] CAS

1. Confirm that $y = 3e^{x^3}$ is a solution of the initial-value problem $y' = 3x^2y$, $y(0) = 3$.

2. Confirm that $y = \frac{1}{4}x^4 + 2\cos x + 1$ is a solution of the initial-value problem $y' = x^3 - 2\sin x$, $y(0) = 3$.

3–4 State the order of the differential equation, and confirm that the functions in the given family are solutions.

3. (a) $(1+x)\dfrac{dy}{dx} = y;\ y = c(1+x)$

(b) $y'' + y = 0;\ y = c_1 \sin t + c_2 \cos t$

4. (a) $2\dfrac{dy}{dx} + y = x - 1;\ y = ce^{-x/2} + x - 3$

(b) $y'' - y = 0;\ y = c_1 e^t + c_2 e^{-t}$

5–6 Use implicit differentiation to confirm that the equation defines implicit solutions of the differential equation.

5. $\ln y = xy^2 + C;\ \dfrac{dy}{dx} = \dfrac{y^3}{1 - 2xy^2}$

6. $x^2 + xy^2 = C;\ 2x + y^2 + 2xy\dfrac{dy}{dx} = 0$

7–8 The first-order linear equations in these exercises can be rewritten as first-order separable equations. Solve the equations using both the method of integrating factors and the method of separation of variables, and determine whether the solutions produced are the same.

7. (a) $\dfrac{dy}{dx} + 3y = 0$ (b) $\dfrac{dy}{dt} - 2y = 0$

8. (a) $\dfrac{dy}{dx} - 4xy = 0$ (b) $\dfrac{dy}{dt} + y = 0$

9–14 Solve the differential equation by the method of integrating factors.

9. $\dfrac{dy}{dx} + 4y = e^{-3x}$

10. $\dfrac{dy}{dx} + 2xy = x$

11. $y' + y = \cos(e^x)$

12. $2\dfrac{dy}{dx} + 4y = 1$

13. $(x^2+1)\dfrac{dy}{dx} + xy = 0$

14. $\dfrac{dy}{dx} + y + \dfrac{1}{1 - e^x} = 0$

15–24 Solve the differential equation by separation of variables. Where reasonable, express the family of solutions as explicit functions of x.

15. $\dfrac{dy}{dx} = \dfrac{y}{x}$

16. $\dfrac{dy}{dx} = 2(1+y^2)x$

17. $\dfrac{\sqrt{1+x^2}}{1+y}\dfrac{dy}{dx} = -x$

18. $(1+x^4)\dfrac{dy}{dx} = \dfrac{x^3}{y}$

19. $(2+2y^2)y' = e^x y$

20. $y' = -xy$

21. $e^{-y}\sin x - y'\cos^2 x = 0$

22. $y' - (1+x)(1+y^2) = 0$

23. $\dfrac{dy}{dx} - \dfrac{y^2 - y}{\sin x} = 0$

24. $y - \dfrac{dy}{dx}\sec x = 0$

25. In each part, find the solution of the differential equation

$$x\frac{dy}{dx} + y = x$$

that satisfies the initial condition.

(a) $y(1) = 2$ (b) $y(-1) = 2$

26. In each part, find the solution of the differential equation

$$\frac{dy}{dx} = xy$$

that satisfies the initial condition.

(a) $y(0) = 1$ (b) $y(0) = \frac{1}{2}$

27–32 Solve the initial-value problem by any method.

27. $\dfrac{dy}{dx} - 2xy = 2x,\quad y(0) = 3$

28. $\dfrac{dy}{dt} + y = 2,\quad y(0) = 1$

29. $y' = \dfrac{3x^2}{2y + \cos y},\quad y(0) = \pi$

30. $y' - xe^y = 2e^y,\quad y(0) = 0$

31. $\dfrac{dy}{dt} = \dfrac{2t+1}{2y-2},\quad y(0) = -1$

32. $y'\cosh x + y\sinh x = \cosh^2 x,\quad y(0) = \frac{1}{4}$

33. (a) Sketch some typical integral curves of the differential equation $y' = y/2x$.

(b) Find an equation for the integral curve that passes through the point $(2, 1)$.

34. (a) Sketch some typical integral curves of the differential equation $y' = -x/y$.

(b) Find an equation for the integral curve that passes through the point $(3, 4)$.

35–36 Solve the differential equation and then use a graphing utility to generate five integral curves for the equation.

35. $(x^2+4)\dfrac{dy}{dx} + xy = 0$

36. $y' + 2y - 3e^t = 0$

37–38 Solve the differential equation. Then, if you have a CAS with implicit plotting capability, use the CAS to generate five integral curves for the equation.

C **37.** $y' = \dfrac{x^2}{1 - y^2}$

C **38.** $y' = \dfrac{y}{1 + y^2}$

39. Suppose that the initial condition in Example 4 had been $y(0) = 0$. Show that none of the solutions generated in Example 4 satisfy this initial condition, and then solve the initial-value problem

$$\frac{dy}{dx} = -4xy^2,\quad y(0) = 0$$

Why does the method of Example 4 fail to produce this particular solution?

40. Find all ordered pairs (x_0, y_0) such that if the initial condition in Example 4 is replaced by $y(x_0) = y_0$, the solution of the resulting initial-value problem is defined for all real numbers.

41. Find an equation of a curve with x-intercept 2 whose tangent line at any point (x, y) has slope xe^{-y}.

42. Use a graphing utility to generate a curve that passes through the point $(1, 1)$ and whose tangent line at (x, y) is perpendicular to the line through (x, y) with slope $-2y/(3x^2)$.

43. At time $t = 0$, a tank contains 25 oz of salt dissolved in 50 gal of water. Then brine containing 4 oz of salt per gallon of brine is allowed to enter the tank at a rate of 2 gal/min and the mixed solution is drained from the tank at the same rate.
(a) How much salt is in the tank at an arbitrary time t?
(b) How much salt is in the tank after 25 min?

44. A tank initially contains 200 gal of pure water. Then at time $t = 0$ brine containing 5 lb of salt per gallon of brine is allowed to enter the tank at a rate of 20 gal/min and the mixed solution is drained from the tank at the same rate.
(a) How much salt is in the tank at an arbitrary time t?
(b) How much salt is in the tank after 30 min?

45. A tank with a 1000-gal capacity initially contains 500 gal of water that is polluted with 50 lb of particulate matter. At time $t = 0$, pure water is added at a rate of 20 gal/min and the mixed solution is drained off at a rate of 10 gal/min. How much particulate matter is in the tank when it reaches the point of overflowing?

46. The water in a polluted lake initially contains 1 lb of mercury salts per 100,000 gal of water. The lake is circular with diameter 30 m and uniform depth 3 m. Polluted water is pumped from the lake at a rate of 1000 gal/h and is replaced with fresh water at the same rate. Construct a table that shows the amount of mercury in the lake (in lb) at the end of each hour over a 12-hour period. Discuss any assumptions you made. [Use 264 gal/m^3.]

47. (a) Use the method of integrating factors to derive solution (27) to the initial-value problem (25). [*Note:* Keep in mind that c, m, and g are constants.]
(b) Show that (27) can be expressed in terms of the terminal speed (29) as

$$v(t) = e^{-gt/v_\tau}(v_0 + v_\tau) - v_\tau$$

(c) Show that if $s(0) = s_0$, then the position function of the object can be expressed as

$$s(t) = s_0 - v_\tau t + \frac{v_\tau}{g}(v_0 + v_\tau)(1 - e^{-gt/v_\tau})$$

48. Suppose a fully equipped sky diver weighing 240 lb has a terminal speed of 120 ft/s with a closed parachute and 24 ft/s with an open parachute. Suppose further that this sky diver is dropped from an airplane at an altitude of 10,000 ft, falls for 25 s with a closed parachute, and then falls the rest of the way with an open parachute.
(a) Assuming that the sky diver's initial vertical velocity is zero, use Exercise 47 to find the sky diver's vertical velocity and height at the time the parachute opens. [Take $g = 32$ ft/s^2.]
(b) Use a calculating utility to find a numerical solution for the total time that the sky diver is in the air.

49. The accompanying figure is a schematic diagram of a basic RL series electrical circuit that contains a power source with a time-dependent voltage of $V(t)$ volts (V), a resistor with a constant resistance of R ohms (Ω), and an inductor with a constant inductance of L henrys (H). If you don't know anything about electrical circuits, don't worry; all you need to know is that electrical theory states that a current of $I(t)$ amperes (A) flows through the circuit where $I(t)$ satisfies the differential equation

$$L\frac{dI}{dt} + RI = V(t)$$

(a) Find $I(t)$ if $R = 10\,\Omega$, $L = 5$ H, V is a constant 20 V, and $I(0) = 0$ A.
(b) What happens to the current over a long period of time?

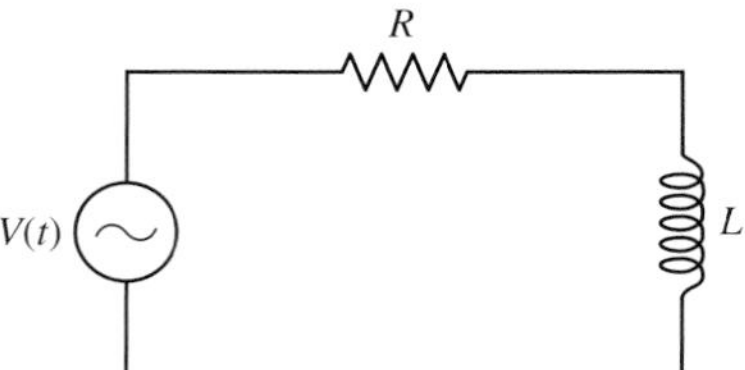

Figure Ex-49

50. Find $I(t)$ for the electrical circuit in Exercise 49 if $R = 6\,\Omega$, $L = 3$ H, $V(t) = 3\sin t$ V, and $I(0) = 15$ A.

51. A rocket, fired upward from rest at time $t = 0$, has an initial mass of m_0 (including its fuel). Assuming that the fuel is consumed at a constant rate k, the mass m of the rocket, while fuel is being burned, will be given by $m = m_0 - kt$. It can be shown that if air resistance is neglected and the fuel gases are expelled at a constant speed c relative to the rocket, then the velocity v of the rocket will satisfy the equation

$$m\frac{dv}{dt} = ck - mg$$

where g is the acceleration due to gravity.
(a) Find $v(t)$ keeping in mind that the mass m is a function of t.
(b) Suppose that the fuel accounts for 80% of the initial mass of the rocket and that all of the fuel is consumed in 100 s. Find the velocity of the rocket in meters per second at the instant the fuel is exhausted. [Take $g = 9.8$ m/s^2 and $c = 2500$ m/s.]

52. A bullet of mass m, fired straight up with an initial velocity of v_0, is slowed by the force of gravity and a drag force of air resistance kv^2, where g is the constant acceleration due to gravity and k is a positive constant. As the bullet moves upward, its velocity v satisfies the equation

$$m\frac{dv}{dt} = -(kv^2 + mg)$$

(a) Show that if $x = x(t)$ is the height of the bullet above the barrel opening at time t, then

$$mv\frac{dv}{dx} = -(kv^2 + mg)$$

(b) Express x in terms of v given that $x = 0$ when $v = v_0$.

(c) Assuming that

$$v_0 = 988 \text{ m/s}, \quad g = 9.8 \text{ m/s}^2$$
$$m = 3.56 \times 10^{-3} \text{ kg}, \quad k = 7.3 \times 10^{-6} \text{ kg/m}$$

use the result in part (b) to find out how high the bullet rises. [*Hint:* Find the velocity of the bullet at its highest point.]

53–54 Suppose that a tank containing a liquid is vented to the air at the top and has an outlet at the bottom through which the liquid can drain. It follows from ***Torricelli's law*** in physics that if the outlet is opened at time $t = 0$, then at each instant the depth of the liquid $h(t)$ and the area $A(h)$ of the liquid's surface are related by

$$A(h)\frac{dh}{dt} = -k\sqrt{h}$$

where k is a positive constant that depends on such factors as the viscosity of the liquid and the cross-sectional area of the outlet. Use this result in these exercises, assuming that h is in feet, $A(h)$ is in square feet, and t is in seconds.

53. Suppose that the cylindrical tank in the accompanying figure is filled to a depth of 4 feet at time $t = 0$ and that the constant in Torricelli's law is $k = 0.025$.
(a) Find $h(t)$.
(b) How many minutes will it take for the tank to drain completely?

54. Follow the directions of Exercise 53 for the cylindrical tank in the accompanying figure, assuming that the tank is filled to a depth of 4 feet at time $t = 0$ and that the constant in Torricelli's law is $k = 0.025$.

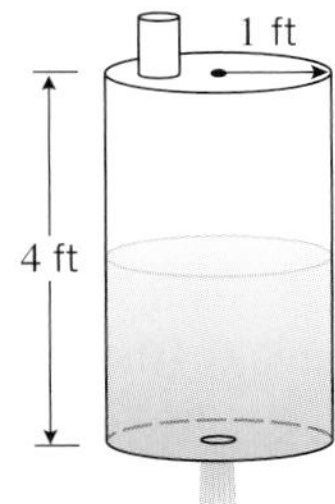

Figure Ex-53

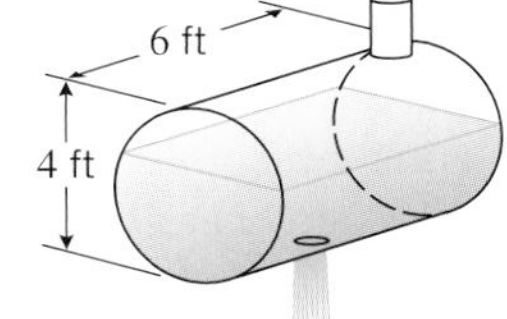

Figure Ex-54

55. Suppose that a particle moving along the x-axis encounters a resisting force that results in an acceleration of $a = dv/dt = -\frac{1}{32}v^2$. Given that $x = 0$ cm and $v = 128$ cm/s at time $t = 0$, find the velocity v and position x as a function of t for $t \geq 0$.

56. Suppose that a particle moving along the x-axis encounters a resisting force that results in an acceleration of $a = dv/dt = -0.02\sqrt{v}$. Given that $x = 0$ cm and $v = 9$ cm/s at time $t = 0$, find the velocity v and position x as a function of t for $t \geq 0$.

FOCUS ON CONCEPTS

57. Find an initial-value problem whose solution is

$$y = \cos x + \int_0^x e^{-t^2}\,dt$$

58. (a) Prove that any function $y = y(x)$ defined by Equation (9) will be a solution to (5) on the interval I.
(b) Consider the initial-value problem

$$\frac{dy}{dx} + p(x)y = q(x), \quad y(x_0) = y_0$$

where the functions $p(x)$ and $q(x)$ are both continuous on some open interval I. Using the general solution for a first-order linear equation, prove that this initial-value problem has a unique solution on I.

59. Use implicit differentiation to prove that any differentiable function defined implicitly by Equation (15) will be a solution to (12).

60. (a) Prove that solutions need not be unique for nonlinear initial-value problems by finding two solutions to

$$y\frac{dy}{dx} = x, \quad y(0) = 0$$

(b) Prove that solutions need not exist for nonlinear initial-value problems by showing that there is no solution for

$$y\frac{dy}{dx} = -x, \quad y(0) = 0$$

✔ QUICK CHECK ANSWERS 9.1

1. (a) (iv) (b) (iii) (c) (i) (d) (ii) **2.** Step 1: $e^{\int p(x)\,dx}$; Step 2: μy, $\mu q(x)$; Step 3: $\frac{1}{\mu}\int \mu q(x)\,dx$ **3.** 2; $e^{2x} + 2xe^{2x}$
4. Step 1: $h(y)\,dy = g(x)\,dx$; Step 2: $\int h(y)\,dy = \int g(x)\,dx$; Step 3: $H(y) = G(x) + C$ **5.** (a) $\frac{dy}{dx} = -\frac{x}{y}$, $y(0) = 1$
(b) The equation may be written in the form $y\frac{dy}{dx} = -x$ and it has the solution $y = \sqrt{1 - x^2}$, $-1 < x < 1$.
6. $\frac{dy}{dt} + \frac{y}{20} = 15$, $y(0) = 30$

9.2 SLOPE FIELDS; EULER'S METHOD

In this section we will reexamine the concept of a slope field and we will discuss a method for approximating solutions of first-order equations numerically. Numerical approximations are important in cases where the differential equation cannot be solved exactly.

FUNCTIONS OF TWO VARIABLES

We will be concerned here with first-order equations that are expressed with the derivative by itself on one side of the equation. For example,

$$y' = x^3 \quad \text{and} \quad y' = \sin(xy)$$

The first of these equations involves only x on the right side, so it has the form $y' = f(x)$. However, the second equation involves both x and y on the right side, so it has the form $y' = f(x, y)$, where the symbol $f(x, y)$ stands for a function of the two variables x and y. Later in the text we will study functions of two variables in more depth, but for now it will suffice to think of $f(x, y)$ as a formula that produces a unique output when values of x and y are given as inputs. For example, if

$$f(x, y) = x^2 + 3y$$

and if the inputs are $x = 2$ and $y = -4$, then the output is

$$f(2, -4) = 2^2 + 3(-4) = 4 - 12 = -8$$

In applied problems involving time, it is usual to use t as the independent variable, in which case one would be concerned with equations of the form $y' = f(t, y)$, where $y' = dy/dt$.

SLOPE FIELDS

In Section 6.2 we introduced the concept of a slope field in the context of differential equations of the form $y' = f(x)$; the same principles apply to differential equations of the form

$$y' = f(x, y)$$

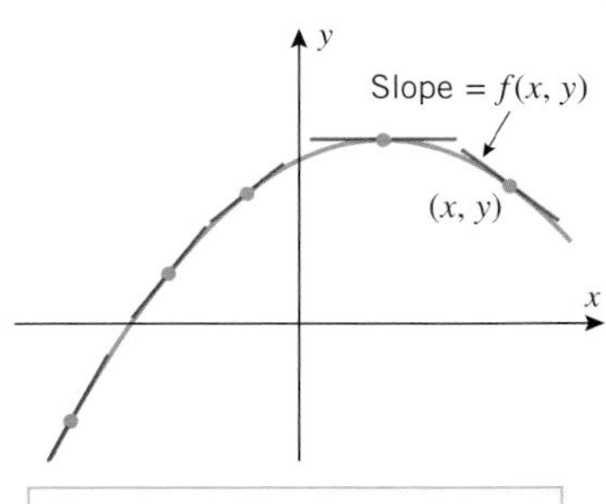

At each point (x, y) on an integral curve of $y' = f(x, y)$, the tangent line has slope $f(x, y)$.

Figure 9.2.1

To see why this is so, let us review the basic idea. If we interpret y' as the slope of a tangent line, then the differential equation states that at each point (x, y) on an integral curve, the slope of the tangent line is equal to the value of f at that point (Figure 9.2.1). For example, suppose that $f(x, y) = y - x$, in which case we have the differential equation

$$y' = y - x \tag{1}$$

A geometric description of the set of integral curves can be obtained by choosing a rectangular grid of points in the xy-plane, calculating the slopes of the tangent lines to the integral curves at the gridpoints, and drawing small segments of the tangent lines through those points. The resulting picture is called a ***slope field*** or a ***direction field*** for the differential equation because it shows the "slope" or "direction" of the integral curves at the gridpoints. The more gridpoints that are used, the better the description of the integral curves. For example, Figure 9.2.2 shows two slope fields for (1)—the first was obtained by hand calculation using the 49 gridpoints shown in the accompanying table, and the second, which gives a clearer picture of the integral curves, was obtained using 625 gridpoints and a CAS.

It so happens that Equation (1) can be solved exactly, since it can be written as

$$y' - y = -x$$

which, by comparison with Equation (5) in Section 9.1, is a first-order linear equation with $p(x) = -1$ and $q(x) = -x$. We leave it for you to use the method of integrating factors to show that the general solution of this equation is

$$y = x + 1 + Ce^x \tag{2}$$

Figure 9.2.3 shows some of the integral curves superimposed on the slope field. Note that it was not necessary to have the general solution to construct the slope field. Indeed, slope

VALUES OF $f(x, y) = y - x$

	$y=-3$	$y=-2$	$y=-1$	$y=0$	$y=1$	$y=2$	$y=3$
$x=-3$	0	1	2	3	4	5	6
$x=-2$	−1	0	1	2	3	4	5
$x=-1$	−2	−1	0	1	2	3	4
$x=0$	−3	−2	−1	0	1	2	3
$x=1$	−4	−3	−2	−1	0	1	2
$x=2$	−5	−4	−3	−2	−1	0	1
$x=3$	−6	−5	−4	−3	−2	−1	0

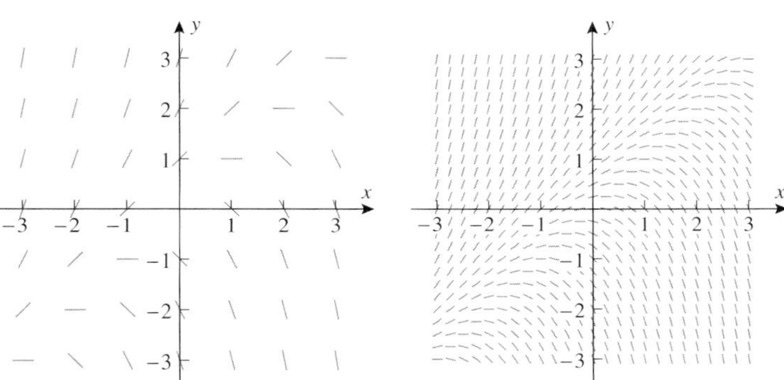

Figure 9.2.2

Confirm that the first slope field in Figure 9.2.2 is consistent with the accompanying table in that figure.

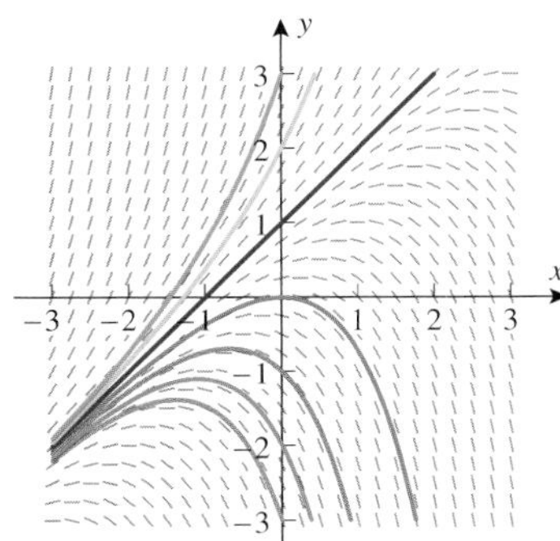

Figure 9.2.3

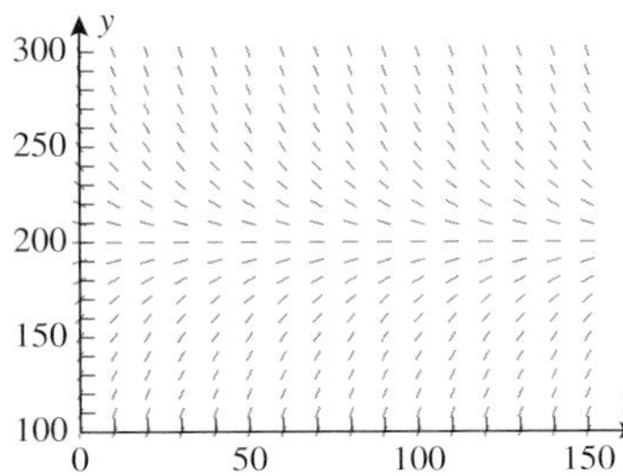

Figure 9.2.4

fields are important precisely because they can be constructed in cases where the differential equation cannot be solved exactly.

▶ **Example 1** In Example 7 of Section 9.1 we considered a mixing problem in which the amount of salt $y(t)$ in a tank at time t was shown to satisfy the differential equation

$$\frac{dy}{dt} + \frac{y}{20} = 10$$

which can be rewritten as

$$y' = 10 - \frac{y}{20} \tag{3}$$

We subsequently found the general solution of this equation to be

$$y(t) = 200 + Ce^{-t/20} \tag{4}$$

and then we found the value of the arbitrary constant C from the initial condition in the problem [the known amount of salt $y(0)$ at time $t = 0$]. However, it follows from (4) that

$$\lim_{t\to+\infty} y(t) = 200$$

for all values of C, so regardless of the amount of salt that is present in the tank initially, the amount of salt in the tank will eventually begin to stabilize at 200 lb. This can also be seen geometrically from the slope field for (3) shown in Figure 9.2.4. This slope field suggests that if the amount of salt present in the tank is greater than 200 lb initially, then the amount of salt will decrease steadily over time toward a limiting value of 200 lb; and if it is less than 200 lb initially, then it will increase steadily toward a limiting value of 200 lb. The slope field also suggests that if the amount present initially is exactly 200 lb, then the amount of salt in the tank will stay constant at 200 lb. This can also be seen from (4), since $C = 0$ in this case (verify). ◀

EULER'S METHOD

Our next objective is to develop a method for approximating the solution of an initial-value problem of the form

$$y' = f(x, y), \quad y(x_0) = y_0$$

We will not attempt to approximate $y(x)$ for all values of x; rather, we will choose some small increment Δx and focus on approximating the values of $y(x)$ at a succession of x-values spaced Δx units apart, starting from x_0. We will denote these x-values by

$$x_1 = x_0 + \Delta x, \quad x_2 = x_1 + \Delta x, \quad x_3 = x_2 + \Delta x, \quad x_4 = x_3 + \Delta x, \ldots$$

and we will denote the approximations of $y(x)$ at these points by

$$y_1 \approx y(x_1), \quad y_2 \approx y(x_2), \quad y_3 \approx y(x_3), \quad y_4 \approx y(x_4), \ldots$$

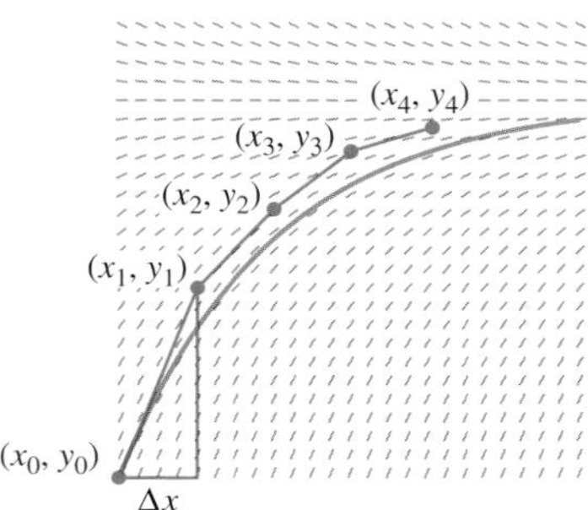

Figure 9.2.5

The technique that we will describe for obtaining these approximations is called ***Euler's Method***. Although there are better approximation methods available, many of them use Euler's Method as a starting point, so the underlying concepts are important to understand.

The basic idea behind Euler's Method is to start at the known initial point (x_0, y_0) and draw a line segment in the direction determined by the slope field until we reach the point (x_1, y_1) with x-coordinate $x_1 = x_0 + \Delta x$ (Figure 9.2.5). If Δx is small, then it is reasonable to expect that this line segment will not deviate much from the integral curve $y = y(x)$, and thus y_1 should closely approximate $y(x_1)$. To obtain the subsequent approximations, we repeat the process using the slope field as a guide at each step. Starting at the endpoint (x_1, y_1), we draw a line segment determined by the slope field until we reach the point (x_2, y_2) with x-coordinate $x_2 = x_1 + \Delta x$, and from that point we draw a line segment determined by the slope field to the point (x_3, y_3) with x-coordinate $x_3 = x_2 + \Delta x$, and so forth. As indicated in Figure 9.2.5, this procedure produces a polygonal path that tends to follow the integral curve closely, so it is reasonable to expect that the y-values $y_2, y_3, y_4, \ldots$ will closely approximate $y(x_2), y(x_3), y(x_4), \ldots$.

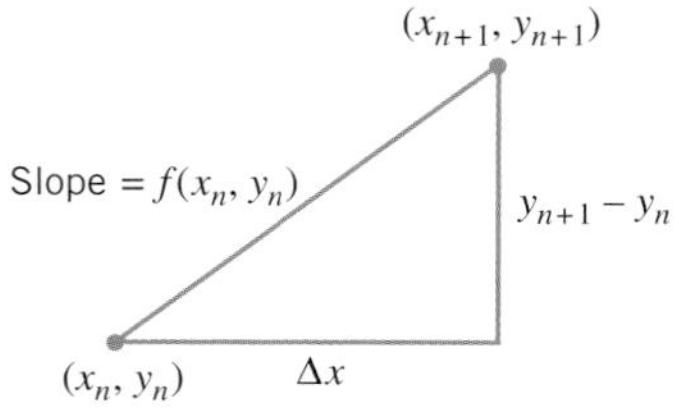

Figure 9.2.6

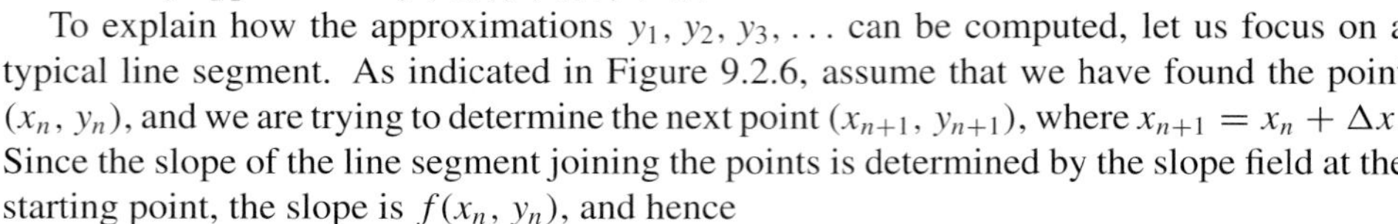

To explain how the approximations $y_1, y_2, y_3, \ldots$ can be computed, let us focus on a typical line segment. As indicated in Figure 9.2.6, assume that we have found the point (x_n, y_n), and we are trying to determine the next point (x_{n+1}, y_{n+1}), where $x_{n+1} = x_n + \Delta x$. Since the slope of the line segment joining the points is determined by the slope field at the starting point, the slope is $f(x_n, y_n)$, and hence

$$\frac{y_{n+1} - y_n}{x_{n+1} - x_n} = \frac{y_{n+1} - y_n}{\Delta x} = f(x_n, y_n)$$

which we can rewrite as

$$y_{n+1} = y_n + f(x_n, y_n)\Delta x$$

This formula, which is the heart of Euler's Method, tells us how to use each approximation to compute the next approximation.

Euler's Method

To approximate the solution of the initial-value problem

$$y' = f(x, y), \quad y(x_0) = y_0$$

proceed as follows:

Step 1. Choose a nonzero number Δx to serve as an ***increment*** or ***step size*** along the x-axis, and let

$$x_1 = x_0 + \Delta x, \quad x_2 = x_1 + \Delta x, \quad x_3 = x_2 + \Delta x, \ldots$$

Step 2. Compute successively

$$\begin{aligned} y_1 &= y_0 + f(x_0, y_0)\Delta x \\ y_2 &= y_1 + f(x_1, y_1)\Delta x \\ y_3 &= y_2 + f(x_2, y_2)\Delta x \\ &\vdots \\ y_{n+1} &= y_n + f(x_n, y_n)\Delta x \end{aligned}$$

The numbers $y_1, y_2, y_3, \ldots$ in these equations are the approximations of $y(x_1)$, $y(x_2), y(x_3), \ldots$.

▶ **Example 2** Use Euler's Method with a step size of 0.1 to make a table of approximate values of the solution of the initial-value problem

$$y' = y - x, \quad y(0) = 2 \tag{5}$$

over the interval $0 \le x \le 1$.

Solution. In this problem we have $f(x, y) = y - x$, $x_0 = 0$, and $y_0 = 2$. Moreover, since the step size is 0.1, the x-values at which the approximate values will be obtained are

$$x_1 = 0.1, \quad x_2 = 0.2, \quad x_3 = 0.3, \ldots, \quad x_9 = 0.9, \quad x_{10} = 1$$

The first three approximations are

$$y_1 = y_0 + f(x_0, y_0)\Delta x = 2 + (2 - 0)(0.1) = 2.2$$
$$y_2 = y_1 + f(x_1, y_1)\Delta x = 2.2 + (2.2 - 0.1)(0.1) = 2.41$$
$$y_3 = y_2 + f(x_2, y_2)\Delta x = 2.41 + (2.41 - 0.2)(0.1) = 2.631$$

Here is a way of organizing all 10 approximations rounded to five decimal places:

EULER'S METHOD FOR $y' = y - x$, $y(0) = 2$ WITH $\Delta x = 0.1$

n	x_n	y_n	$f(x_n, y_n)\Delta x$	$y_{n+1} = y_n + f(x_n, y_n)\Delta x$
0	0	2.00000	0.20000	2.20000
1	0.1	2.20000	0.21000	2.41000
2	0.2	2.41000	0.22100	2.63100
3	0.3	2.63100	0.23310	2.86410
4	0.4	2.86410	0.24641	3.11051
5	0.5	3.11051	0.26105	3.37156
6	0.6	3.37156	0.27716	3.64872
7	0.7	3.64872	0.29487	3.94359
8	0.8	3.94359	0.31436	4.25795
9	0.9	4.25795	0.33579	4.59374
10	1.0	4.59374	—	—

Observe that each entry in the last column becomes the next entry in the third column. ◀

ACCURACY OF EULER'S METHOD

It follows from (5) and the initial condition $y(0) = 2$ that the exact solution of the initial-value problem in Example 2 is

$$y = x + 1 + e^x$$

Thus, in this case we can compare the approximate values of $y(x)$ produced by Euler's Method with decimal approximations of the exact values (Table 9.2.1). In Table 9.2.1 the ***absolute error*** is calculated as

$$|\text{exact value} - \text{approximation}|$$

and the ***percentage error*** as

$$\frac{|\text{exact value} - \text{approximation}|}{|\text{exact value}|} \times 100\%$$

As a rule of thumb, the absolute error in an approximation produced by Euler's Method is proportional to the step size. Thus, reducing the step size by half reduces the absolute and percentage errors by roughly half. However, reducing the step size increases the amount of computation, thereby increasing the potential for more roundoff error. Such matters are discussed in differential equations or numerical analysis courses.

Table 9.2.1

x	EXACT SOLUTION	EULER APPROXIMATION	ABSOLUTE ERROR	PERCENTAGE ERROR
0	2.00000	2.00000	0.00000	0.00
0.1	2.20517	2.20000	0.00517	0.23
0.2	2.42140	2.41000	0.01140	0.47
0.3	2.64986	2.63100	0.01886	0.71
0.4	2.89182	2.86410	0.02772	0.96
0.5	3.14872	3.11051	0.03821	1.21
0.6	3.42212	3.37156	0.05056	1.48
0.7	3.71375	3.64872	0.06503	1.75
0.8	4.02554	3.94359	0.08195	2.04
0.9	4.35960	4.25795	0.10165	2.33
1.0	4.71828	4.59374	0.12454	2.64

✔ QUICK CHECK EXERCISES 9.2

(See page 602 for answers.)

1. Match each differential equation with its slope field.
   (a) $y' = y$ ______ (b) $y' = 2xy$ ______
   (c) $y' = e^{-y}$ ______ (d) $y' = 2xy^2$ ______

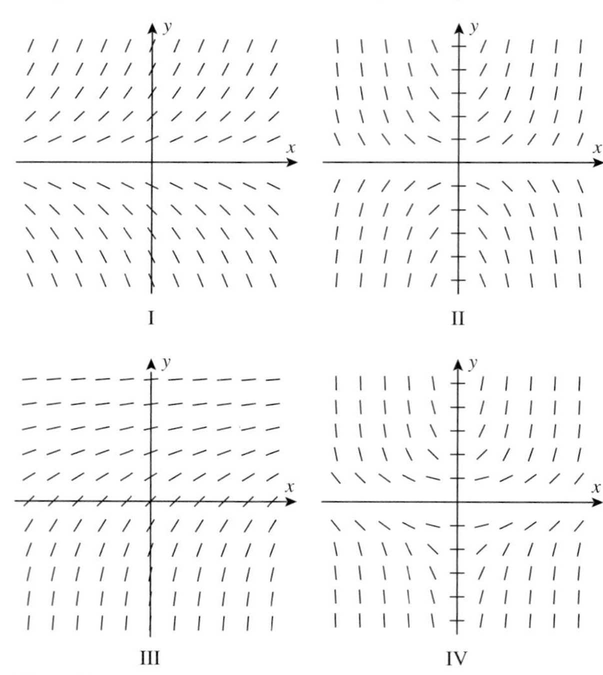

Figure Ex-1

2. The slope field for $y' = y/x$ at the 16 gridpoints (x, y), where $x = -2, -1, 1, 2$ and $y = -2, -1, 1, 2$ is shown in the accompanying figure. Use this slope field and geometric reasoning to find the integral curve that passes through the point $(1, 2)$.

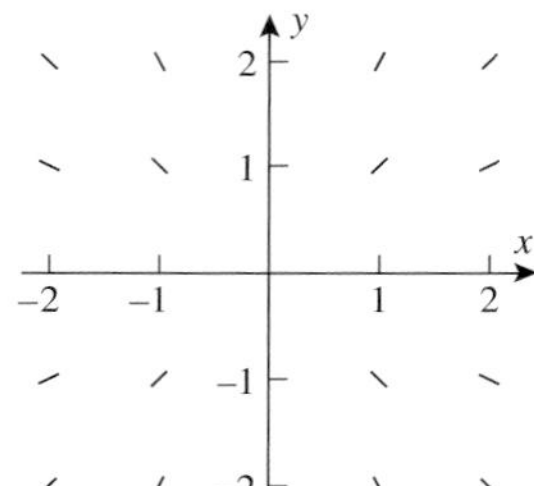

Figure Ex-2

3. When using Euler's Method on the initial-value problem $y' = f(x, y)$, $y(x_0) = y_0$, we obtain y_{n+1} from y_n, x_n, and Δx by means of the formula $y_{n+1} =$ ______.

4. Consider the initial-value problem $y' = y$, $y(0) = 1$.
   (a) Use Euler's Method with five steps to approximate $y(1)$.
   (b) What is the exact value of $y(1)$?

EXERCISE SET 9.2 Graphing Utility C CAS

1. Sketch the slope field for $y' = xy/4$ at the 25 gridpoints (x, y), where $x = -2, -1, \ldots, 2$ and $y = -2, -1, \ldots, 2$.

2. Sketch the slope field for $y' + y = 2$ at the 25 gridpoints (x, y), where $x = 0, 1, \ldots, 4$ and $y = 0, 1, \ldots, 4$.

3. A slope field for the differential equation $y' = 1 - y$ is shown in the accompanying figure. In each part, sketch the graph of the solution that satisfies the initial condition.
   (a) $y(0) = -1$ (b) $y(0) = 1$ (c) $y(0) = 2$

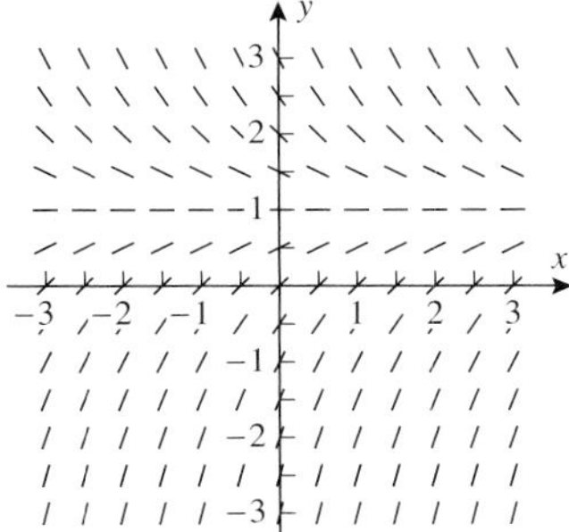
Figure Ex-3

4. Solve the initial-value problems in Exercise 3, and use a graphing utility to confirm that the integral curves for these solutions are consistent with the sketches you obtained from the slope field.

5. A slope field for the differential equation $y' = 2y - x$ is shown in the accompanying figure. In each part, sketch the graph of the solution that satisfies the initial condition.
(a) $y(1) = 1$ (b) $y(0) = -1$ (c) $y(-1) = 0$

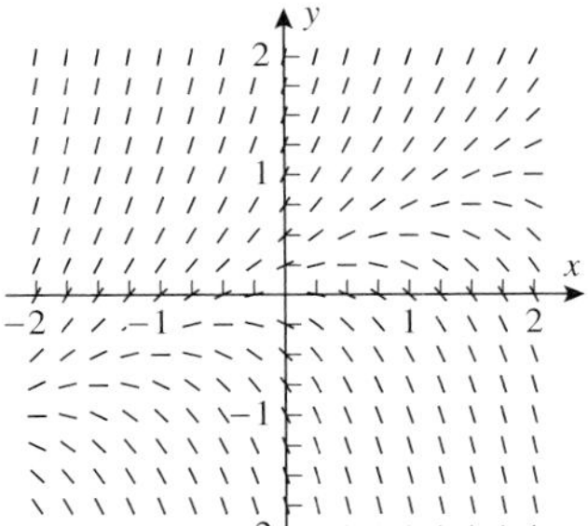
Figure Ex-5

6. Solve the initial-value problems in Exercise 5, and use a graphing utility to confirm that the integral curves for these solutions are consistent with the sketches you obtained from the slope field.

FOCUS ON CONCEPTS

7. Use the slope field in Exercise 3 to make a conjecture about the behavior of the solutions of $y' = 1 - y$ as $x \to +\infty$, and confirm your conjecture by examining the general solution of the equation.

8. Use the slope field in Exercise 5 to make a conjecture about the effect of y_0 on the behavior of the solution of the initial-value problem $y' = 2y - x$, $y(0) = y_0$ as $x \to +\infty$, and check your conjecture by examining the solution of the initial-value problem.

9. In each part, match the differential equation with the slope field, and explain your reasoning.
(a) $y' = 1/x$ (b) $y' = 1/y$ (c) $y' = e^{-x^2}$
(d) $y' = y^2 - 1$ (e) $y' = \dfrac{x+y}{x-y}$
(f) $y' = (\sin x)(\sin y)$

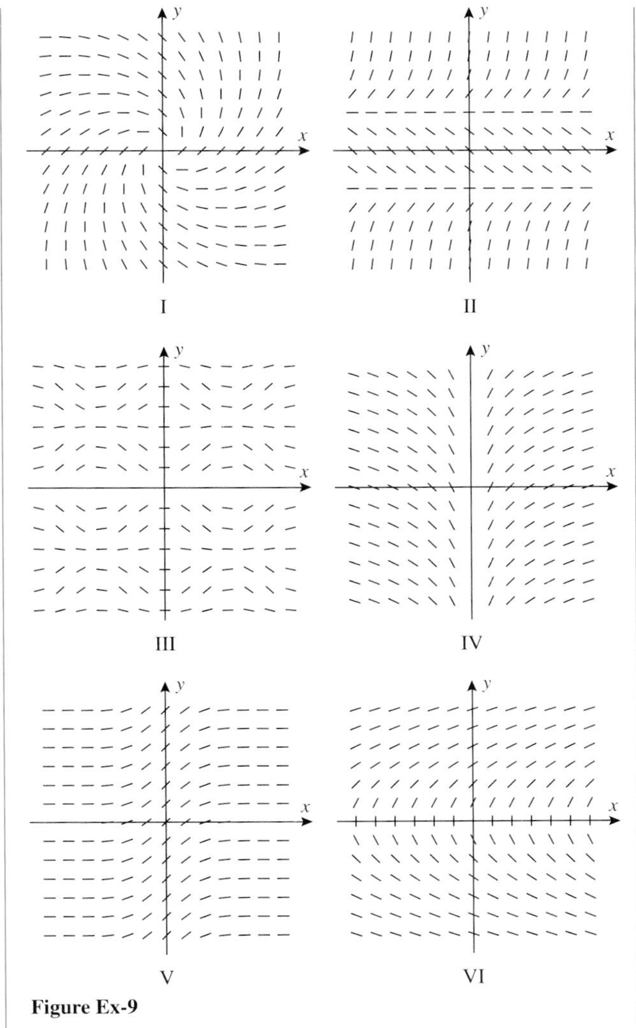

Figure Ex-9

10. If you have a CAS or a graphing utility that can generate slope fields, read the documentation on how to do it and check your answers in Exercise 9 by generating the slope fields for the differential equations.

11. (a) Use Euler's Method with a step size of $\Delta x = 0.2$ to approximate the solution of the initial-value problem

$$y' = x + y, \quad y(0) = 1$$

over the interval $0 \le x \le 1$.
(b) Solve the initial-value problem exactly, and calculate the error and the percentage error in each of the approximations in part (a).
(c) Sketch the exact solution and the approximate solution together.

12. It was stated at the end of this section that reducing the step size in Euler's Method by half reduces the error in each approximation by about half. Confirm that the error in $y(1)$ is reduced by about half if a step size of $\Delta x = 0.1$ is used in Exercise 11.

13–16 Use Euler's Method with the given step size Δx or Δt to approximate the solution of the initial-value problem over the stated interval. Present your answer as a table and as a graph.

13. $dy/dx = \sqrt[3]{y},\ y(0) = 1,\ 0 \le x \le 4,\ \Delta x = 0.5$

14. $dy/dx = x - y^2,\ y(0) = 1,\ 0 \le x \le 2,\ \Delta x = 0.25$

15. $dy/dt = \cos y,\ y(0) = 1,\ 0 \le t \le 2,\ \Delta t = 0.5$

16. $dy/dt = e^{-y},\ y(0) = 0,\ 0 \le t \le 1,\ \Delta t = 0.1$

17. Consider the initial-value problem

$$y' = \sin \pi t, \quad y(0) = 0$$

Use Euler's Method with five steps to approximate $y(1)$.

FOCUS ON CONCEPTS

18. (a) Show that the solution of the initial-value problem $y' = e^{-x^2}$, $y(0) = 0$ is

$$y(x) = \int_0^x e^{-t^2}\, dt$$

(b) Use Euler's Method with $\Delta x = 0.05$ to approximate the value of

$$y(1) = \int_0^1 e^{-t^2}\, dt$$

and compare the answer to that produced by a calculating utility with a numerical integration capability.

19. The accompanying figure shows a slope field for the differential equation $y' = -x/y$.

(a) Use the slope field to estimate $y(\frac{1}{2})$ for the solution that satisfies the given initial condition $y(0) = 1$.

(b) Compare your estimate to the exact value of $y(\frac{1}{2})$.

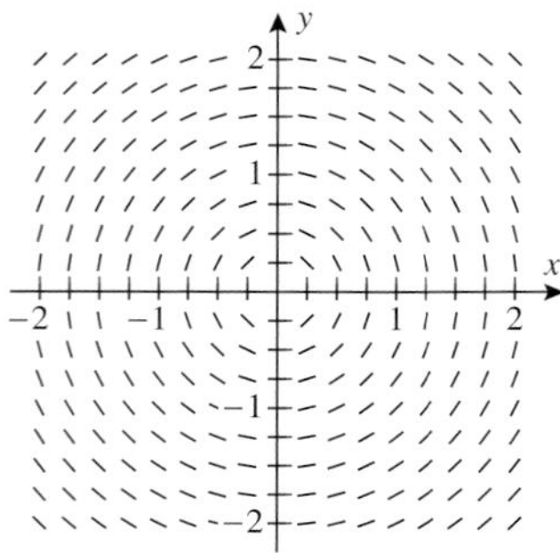

Figure Ex-19

20. Consider the initial-value problem

$$\frac{dy}{dx} = \frac{\sqrt{y}}{2}, \quad y(0) = 1$$

(a) Use Euler's Method with step sizes of $\Delta x = 0.2$, 0.1, and 0.05 to obtain three approximations of $y(1)$.

(b) Plot the three approximations versus Δx, and make a conjecture about the exact value of $y(1)$. Explain your reasoning.

(c) Check your conjecture by finding $y(1)$ exactly.

21. A slope field of the form $y' = f(y)$ is said to be ***autonomous***.

(a) Explain why the tangent segments along any horizontal line will be parallel for an autonomous slope field.

(b) The word *autonomous* means "independent." In what sense is an autonomous slope field independent?

(c) Suppose that $G(y)$ is an antiderivative of $1/[f(y)]$ and that C is a constant. Explain why any differentiable function defined implicitly by $G(y) - x = C$ will be a solution to the equation $y' = f(y)$.

22. (a) Solve the equation $y' = \sqrt{y}$ and show that every nonconstant solution has a graph that is everywhere concave up.

(b) Explain how the conclusion in part (a) may be obtained directly from the equation $y' = \sqrt{y}$ without solving.

23. (a) Use implicit differentiation to find a slope field whose integral curve through $(1, 1)$ is implicitly defined by the equation $xy^3 - x^2y = 0$.

(b) Prove that if $y(x)$ is any integral curve of the slope field in part (a), then $x[y(x)]^3 - x^2y(x)$ will be a constant function.

(c) Find an equation that implicitly defines the integral curve through $(-1, -1)$ of the slope field in part (a).

24. (a) Use implicit differentiation to find a slope field whose integral curve through $(0, 0)$ is implicitly defined by the equation $xe^y + ye^x = 0$.

(b) Prove that if $y(x)$ is any integral curve of the slope field in part (a), then $xe^{y(x)} + y(x)e^x$ will be a constant function.

(c) Find an equation that implicitly defines the integral curve through $(1, 1)$ of the slope field in part (a).

25. Explain the connection between Euler's Method and the local linear approximation discussed in Section 3.8.

26. Consider the initial-value problem $y' = y,\ y(0) = 1$, and let y_n denote the approximation of $y(1)$ using Euler's Method with n steps.

(a) What would you conjecture is the exact value of $\lim_{n\to+\infty} y_n$? Explain your reasoning.

(b) Find an explicit formula for y_n and use it to verify your conjecture in part (a).

✔ QUICK CHECK ANSWERS 9.2

1. (a) IV (b) III (c) I (d) II 2. $y = 2x, x > 0$ 3. $y_n + f(x_n, y_n)\Delta x$ 4. (a) 2.48832 (b) e

9.3 MODELING WITH FIRST-ORDER DIFFERENTIAL EQUATIONS

Since many of the fundamental laws of the physical and social sciences involve rates of change, it should not be surprising that such laws are modeled by differential equations. In this section we will discuss the general idea of modeling with differential equations, and we will investigate some important models that can be applied to population growth, carbon dating, medicine, and ecology.

POPULATION GROWTH

One of the simplest models of population growth is based on the observation that when populations (people, plants, bacteria, and fruit flies, for example) are not constrained by environmental limitations, they tend to grow at a rate that is proportional to the size of the population—the larger the population, the more rapidly it grows.

To translate this principle into a mathematical model, suppose that $y = y(t)$ denotes the population at time t. At each point in time, the rate of increase of the population with respect to time is dy/dt, so the assumption that the rate of growth is proportional to the population is described by the differential equation

$$\frac{dy}{dt} = ky \tag{1}$$

where k is a positive constant of proportionality that can usually be determined experimentally. Thus, if the population is known at some point in time, say $y = y_0$ at time $t = 0$, then a general formula for the population $y(t)$ can be obtained by solving the initial-value problem

$$\frac{dy}{dt} = ky, \quad y(0) = y_0$$

PHARMACOLOGY

When a drug (say, penicillin or aspirin) is administered to an individual, it enters the bloodstream and then is absorbed by the body over time. Medical research has shown that the amount of a drug that is present in the bloodstream tends to decrease at a rate that is proportional to the amount of the drug present—the more of the drug that is present in the bloodstream, the more rapidly it is absorbed by the body.

To translate this principle into a mathematical model, suppose that $y = y(t)$ is the amount of the drug present in the bloodstream at time t. At each point in time, the rate of change in y with respect to t is dy/dt, so the assumption that the rate of decrease is proportional to the amount y in the bloodstream translates into the differential equation

$$\frac{dy}{dt} = -ky \tag{2}$$

where k is a positive constant of proportionality that depends on the drug and can be determined experimentally. The negative sign is required because y decreases with time. Thus, if the initial dosage of the drug is known, say $y = y_0$ at time $t = 0$, then a general formula for $y(t)$ can be obtained by solving the initial-value problem

$$\frac{dy}{dt} = -ky, \quad y(0) = y_0$$

SPREAD OF DISEASE

Suppose that a disease begins to spread in a population of L individuals. Logic suggests that at each point in time the rate at which the disease spreads will depend on how many individuals are already affected and how many are not—as more individuals are affected, the opportunity to spread the disease tends to increase, but at the same time there are fewer individuals who are not affected, so the opportunity to spread the disease tends to decrease. Thus, there are two conflicting influences on the rate at which the disease spreads.

To translate this into a mathematical model, suppose that $y = y(t)$ is the number of individuals who have the disease at time t, so of necessity the number of individuals who do not have the disease at time t is $L - y$. As the value of y increases, the value of $L - y$ decreases, so the conflicting influences of the two factors on the rate of spread dy/dt are taken into account by the differential equation

$$\frac{dy}{dt} = ky(L - y)$$

where k is a positive constant of proportionality that depends on the nature of the disease and the behavior patterns of the individuals and can be determined experimentally. Thus, if the number of affected individuals is known at some point in time, say $y = y_0$ at time $t = 0$, then a general formula for $y(t)$ can be obtained by solving the initial-value problem

$$\frac{dy}{dt} = ky(L - y), \quad y(0) = y_0 \tag{3}$$

EXPONENTIAL GROWTH AND DECAY MODELS

Equations (1) and (2) are examples of a general class of models called *exponential models*. In general, exponential models arise in situations where a quantity increases or decreases at a rate that is proportional to the amount of the quantity present. More precisely, we make the following definition.

9.3.1 DEFINITION. A quantity $y = y(t)$ is said to have an ***exponential growth model*** if it increases at a rate that is proportional to the amount of the quantity present, and it is said to have an ***exponential decay model*** if it decreases at a rate that is proportional to the amount of the quantity present. Thus, for an exponential growth model, the quantity $y(t)$ satisfies an equation of the form

$$\frac{dy}{dt} = ky \quad (k > 0) \tag{4}$$

and for an exponential decay model, the quantity $y(t)$ satisfies an equation of the form

$$\frac{dy}{dt} = -ky \quad (k > 0) \tag{5}$$

The constant k is called the ***growth constant*** or the ***decay constant***, as appropriate.

Equations (4) and (5) are first-order linear equations, since they can be rewritten as

$$\frac{dy}{dt} - ky = 0 \quad \text{and} \quad \frac{dy}{dt} + ky = 0$$

both of which have the form of Equation (5) in Section 9.1 (but with t rather than x as the independent variable); in the first equation we have $p(t) = -k$ and $q(t) = 0$, and in the second we have $p(t) = k$ and $q(t) = 0$.

To illustrate how these equations can be solved, suppose that a quantity $y = y(t)$ has an exponential growth model and we know the amount of the quantity at some point in time, say $y = y_0$ when $t = 0$. Thus, a general formula for $y(t)$ can be obtained by solving the initial-value problem

$$\frac{dy}{dt} - ky = 0, \quad y(0) = y_0$$

Multiplying the differential equation through by the integrating factor

$$\mu = e^{\int (-k)\,dt} = e^{-kt}$$

yields

$$\frac{d}{dt}(e^{-kt}y) = 0$$

and then integrating with respect to t yields

$$e^{-kt}y = C \quad \text{or} \quad y = Ce^{kt}$$

The initial condition implies that $y = y_0$ when $t = 0$, from which it follows that $C = y_0$ (verify). Thus, the solution of the initial-value problem is

$$y = y_0 e^{kt} \tag{6}$$

We leave it for you to show that if $y = y(t)$ has an exponential decay model, and if $y(0) = y_0$, then

$$y = y_0 e^{-kt} \tag{7}$$

INTERPRETING THE GROWTH AND DECAY CONSTANTS

The significance of the constant k in Formulas (6) and (7) can be understood by reexamining the differential equations that gave rise to these formulas. For example, in the case of the exponential growth model, Equation (4) can be rewritten as

$$k = \frac{dy/dt}{y}$$

It is standard practice in applications to call the relative growth rate the *growth rate*, even though it is misleading (the growth rate is dy/dt). However, the practice is so common that we will follow it here.

which states that the growth rate as a fraction of the entire population remains constant over time, and this constant is k. For this reason, k is called the ***relative growth rate*** of the population. It is usual to express the relative growth rate as a percentage. Thus, a relative growth rate of 3% per unit of time in an exponential growth model means that $k = 0.03$. Similarly, the constant k in an exponential decay model is called the ***relative decay rate***.

▶ **Example 1** According to United Nations data, the world population in 1998 was approximately 5.9 billion and growing at a rate of about 1.33% per year. Assuming an exponential growth model, estimate the world population at the beginning of the year 2023.

Solution. We assume that the population at the beginning of 1998 was 5.9 billion and let

t = time elapsed from the beginning of 1998 (in years)

y = world population (in billions)

Since the beginning of 1998 corresponds to $t = 0$, it follows from the given data that

$$y_0 = y(0) = 5.9 \text{ (billion)}$$

Since the growth rate is 1.33% ($k = 0.0133$), it follows from (6) that the world population at time t will be

$$y(t) = y_0 e^{kt} = 5.9e^{0.0133t} \tag{8}$$

In Example 1 the growth rate was given, so there was no need to calculate it. If the growth rate or decay rate is unknown, then it can be calculated using the initial condition and the value of y at another point in time (Exercise 24).

Since the beginning of the year 2023 corresponds to an elapsed time of $t = 25$ years ($2023 - 1998 = 25$), it follows from (8) that the world population by the year 2023 will be

$$y(25) = 5.9e^{0.0133(25)} \approx 8.2$$

which is a population of approximately 8.2 billion. ◀

DOUBLING TIME AND HALF-LIFE

If a quantity y has an exponential growth model, then the time required for the original size to double is called the ***doubling time***, and if y has an exponential decay model, then the time required for the original size to reduce by half is called the ***half-life***. As it turns out, doubling time and half-life depend only on the growth or decay rate and not on the amount present initially. To see why this is so, suppose that $y = y(t)$ has an exponential growth model

$$y = y_0 e^{kt} \tag{9}$$

and let T denote the amount of time required for y to double in size. Thus, at time $t = T$

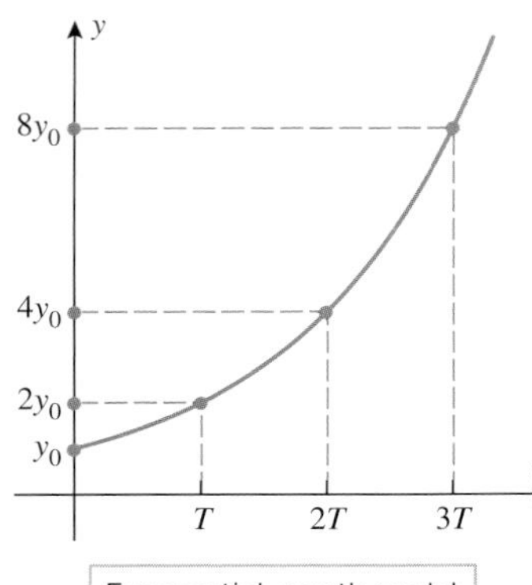

Exponential growth model with doubling time T

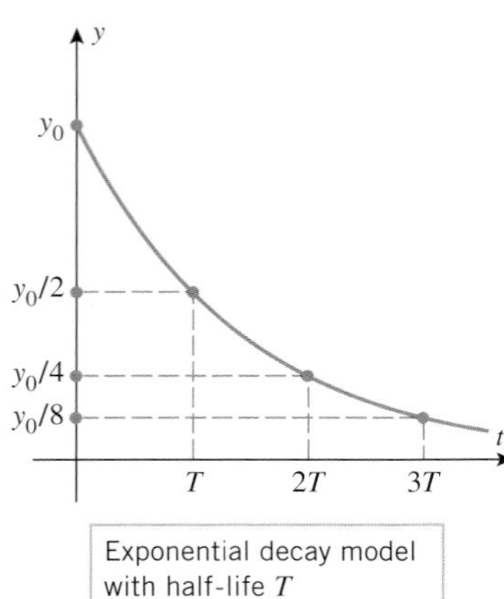

Exponential decay model with half-life T

Figure 9.3.1

the value of y will be $2y_0$, and hence from (9)

$$2y_0 = y_0e^{kT} \quad \text{or} \quad e^{kT} = 2$$

Taking the natural logarithm of both sides yields $kT = \ln 2$, which implies that the doubling time is

$$T = \frac{1}{k}\ln 2 \tag{10}$$

We leave it as an exercise to show that Formula (10) also gives the half-life of an exponential decay model. Observe that this formula does not involve the initial amount y_0, so that in an exponential growth or decay model, the quantity y doubles (or reduces by half) every T units (Figure 9.3.1).

▶ **Example 2** It follows from (10) that with a continued growth rate of 1.33% per year, the doubling time for the world population will be

$$T = \frac{1}{0.0133}\ln 2 \approx 52.116$$

or approximately 52 years. Thus, with a continued 1.33% annual growth rate the population of 5.9 billion in 1998 will double to 11.8 billion by the year 2050 and will double again to 23.6 billion by 2102. ◀

RADIOACTIVE DECAY

It is a fact of physics that radioactive elements disintegrate spontaneously in a process called ***radioactive decay***. Experimentation has shown that the rate of disintegration is proportional to the amount of the element present, which implies that the amount $y = y(t)$ of a radioactive element present as a function of time has an exponential decay model.

Every radioactive element has a specific half-life; for example, the half-life of radioactive carbon-14 is about 5730 years. Thus, from (10), the decay constant for this element is

$$k = \frac{1}{T}\ln 2 = \frac{\ln 2}{5730} \approx 0.000121$$

and this implies that if there are y_0 units of carbon-14 present at time $t = 0$, then the number of units present after t years will be approximately

$$y(t) = y_0e^{-0.000121t} \tag{11}$$

▶ **Example 3** If 100 grams of radioactive carbon-14 are stored in a cave for 1000 years, how many grams will be left at that time?

Solution. From (11) with $y_0 = 100$ and $t = 1000$, we obtain

$$y(1000) = 100e^{-0.000121(1000)} = 100e^{-0.121} \approx 88.6$$

Thus, about 88.6 grams will be left. ◀

CARBON DATING

When the nitrogen in the Earth's upper atmosphere is bombarded by cosmic radiation, the radioactive element carbon-14 is produced. This carbon-14 combines with oxygen to form carbon dioxide, which is ingested by plants, which in turn are eaten by animals. In this way all living plants and animals absorb quantities of radioactive carbon-14. In 1947 the American nuclear scientist W. F. Libby* proposed the theory that the percentage of

*W. F. Libby, "Radiocarbon Dating," *American Scientist*, Vol. 44, 1956, pp. 98–112.

carbon-14 in the atmosphere and in living tissues of plants is the same. When a plant or animal dies, the carbon-14 in the tissue begins to decay. Thus, the age of an artifact that contains plant or animal material can be estimated by determining what percentage of its original carbon-14 content remains. Various procedures, called ***carbon dating*** or ***carbon-14 dating***, have been developed for measuring this percentage.

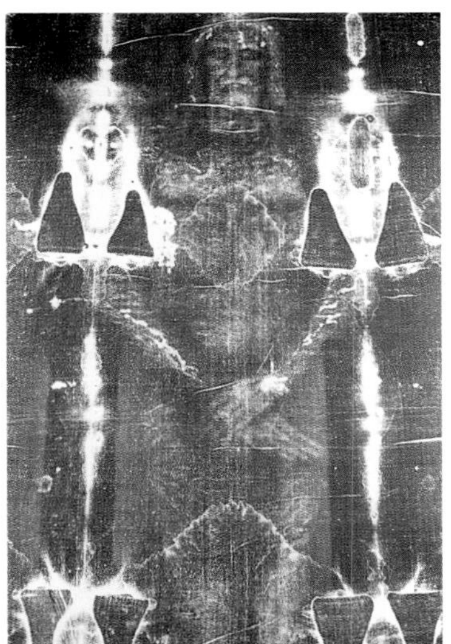

The Shroud of Turin

▶ **Example 4** In 1988 the Vatican authorized the British Museum to date a cloth relic known as the Shroud of Turin, possibly the burial shroud of Jesus of Nazareth. This cloth, which first surfaced in 1356, contains the negative image of a human body that was widely believed to be that of Jesus. The report of the British Museum showed that the fibers in the cloth contained between 92% and 93% of their original carbon-14. Use this information to estimate the age of the shroud.

Solution. From (11), the fraction of the original carbon-14 that remains after t years is

$$\frac{y(t)}{y_0} = e^{-0.000121t}$$

Taking the natural logarithm of both sides and solving for t, we obtain

$$t = -\frac{1}{0.000121} \ln\left(\frac{y(t)}{y_0}\right)$$

Thus, taking $y(t)/y_0$ to be 0.93 and 0.92, we obtain

$$t = -\frac{1}{0.000121} \ln(0.93) \approx 600$$

$$t = -\frac{1}{0.000121} \ln(0.92) \approx 689$$

This means that when the test was done in 1988, the shroud was between 600 and 689 years old, thereby placing its origin between 1299 A.D. and 1388 A.D. Thus, if one accepts the validity of carbon-14 dating, the Shroud of Turin cannot be the burial shroud of Jesus of Nazareth. ◀

✔ QUICK CHECK EXERCISES 9.3 *(See page 611 for answers.)*

1. Suppose that a quantity $y = y(t)$ has an exponential growth model with growth constant $k > 0$.
   (a) $y(t)$ satisfies a first-order differential equation of the form $dy/dt =$ ________.
   (b) In terms of k, the doubling time of the quantity is ________.
   (c) If $y_0 = y(0)$ is the initial amount of the quantity, then an explicit formula for $y(t)$ is given by $y(t) =$ ________.
2. Suppose that a quantity $y = y(t)$ has an exponential decay model with decay constant $k > 0$.
   (a) $y(t)$ satisfies a first-order differential equation of the form $dy/dt =$ ________.
   (b) In terms of k, the half-life of the quantity is ________.
   (c) If $y_0 = y(0)$ is the initial amount of the quantity, then an explicit formula for $y(t)$ is given by $y(t) =$ ________.
3. Suppose that the half-life of a radioactive element is 1 minute. If 32 g of the element are available in a container at 1:00 P.M., then the amount remaining at 1:05 P.M. will be ________.
4. A colony of fruit flies is growing exponentially at a rate of 2% per day. If the initial size of the colony is 100 fruit flies, then after t days the size of the colony will be $y(t) =$ ________.
5. Suppose that a square is growing in such a way that the rate of change of its area is equal in magnitude to its perimeter. If $A = A(t)$ is the area of the square at time t, then $A(t)$ satisfies the first-order equation $dA/dt =$ ________.

EXERCISE SET 9.3 Graphing Utility

FOCUS ON CONCEPTS

1. (a) Suppose that a quantity $y = y(t)$ increases at a rate that is proportional to the square of the amount present, and suppose that at time $t = 0$, the amount present is y_0. Find an initial-value problem whose solution is $y(t)$.
(b) Suppose that a quantity $y = y(t)$ decreases at a rate that is proportional to the square of the amount present, and suppose that at a time $t = 0$, the amount present is y_0. Find an initial-value problem whose solution is $y(t)$.

2. (a) Suppose that a quantity $y = y(t)$ changes in such a way that $dy/dt = k\sqrt{y}$, where $k > 0$. Describe how y changes in words.
(b) Suppose that a quantity $y = y(t)$ changes in such a way that $dy/dt = -ky^3$, where $k > 0$. Describe how y changes in words.

3. (a) Suppose that a particle moves along an s-axis in such a way that its velocity $v(t)$ is always half of $s(t)$. Find a differential equation whose solution is $s(t)$.
(b) Suppose that an object moves along an s-axis in such a way that its acceleration $a(t)$ is always twice the velocity. Find a differential equation whose solution is $s(t)$.

4. Suppose that a body moves along an s-axis through a resistive medium in such a way that the velocity $v = v(t)$ decreases at a rate that is twice the square of the velocity.
(a) Find a differential equation whose solution is the velocity $v(t)$.
(b) Find a differential equation whose solution is the position $s(t)$.

5. Suppose that an initial population of 10,000 bacteria grows exponentially at a rate of 2% per hour and that $y = y(t)$ is the number of bacteria present t hours later.
(a) Find an initial-value problem whose solution is $y(t)$.
(b) Find a formula for $y(t)$.
(c) How long does it take for the initial population of bacteria to double?
(d) How long does it take for the population of bacteria to reach 45,000?

6. A cell of the bacterium *E. coli* divides into two cells every 20 minutes when placed in a nutrient culture. Let $y = y(t)$ be the number of cells that are present t minutes after a single cell is placed in the culture. Assume that the growth of the bacteria is approximated by a continuous exponential growth model.
(a) Find an initial-value problem whose solution is $y(t)$.
(b) Find a formula for $y(t)$.
(c) How many cells are present after 2 hours?
(d) How long does it take for the number of cells to reach 1,000,000?

7. Radon-222 is a radioactive gas with a half-life of 3.83 days. This gas is a health hazard because it tends to get trapped in the basements of houses, and many health officials suggest that homeowners seal their basements to prevent entry of the gas. Assume that 5.0×10^7 radon atoms are trapped in a basement at the time it is sealed and that $y(t)$ is the number of atoms present t days later.
(a) Find an initial-value problem whose solution is $y(t)$.
(b) Find a formula for $y(t)$.
(c) How many atoms will be present after 30 days?
(d) How long will it take for 90% of the original quantity of gas to decay?

8. Polonium-210 is a radioactive element with a half-life of 140 days. Assume that 10 milligrams of the element are placed in a lead container and that $y(t)$ is the number of milligrams present t days later.
(a) Find an initial-value problem whose solution is $y(t)$.
(b) Find a formula for $y(t)$.
(c) How many milligrams will be present after 10 weeks?
(d) How long will it take for 70% of the original sample to decay?

9. Suppose that 100 fruit flies are placed in a breeding container that can support at most 10,000 flies. Assuming that the population grows exponentially at a rate of 2% per day, how long will it take for the container to reach capacity?

10. Suppose that the town of Grayrock had a population of 10,000 in 1998 and a population of 12,000 in 2003. Assuming an exponential growth model, in what year will the population reach 20,000?

11. A scientist wants to determine the half-life of a certain radioactive substance. She determines that in exactly 5 days a 10.0-milligram sample of the substance decays to 3.5 milligrams. Based on these data, what is the half-life?

12. Suppose that 30% of a certain radioactive substance decays in 5 years.
(a) What is the half-life of the substance in years?
(b) Suppose that a certain quantity of this substance is stored in a cave. What percentage of it will remain after t years?

13. In each part, find an exponential growth model $y = y_0e^{kt}$ that satisfies the stated conditions.
(a) $y_0 = 3$; doubling time $T = 6$
(b) $y(0) = 4$; growth rate 2%
(c) $y(1) = 1$; $y(10) = 200$
(d) $y(1) = 2$; doubling time $T = 6$

14. In each part, find an exponential decay model $y = y_0e^{-kt}$ that satisfies the stated conditions.
(a) $y_0 = 10$; half-life $T = 5$
(b) $y(0) = 10$; decay rate 1.5%
(c) $y(1) = 100$; $y(10) = 1$
(d) $y(1) = 10$; half-life $T = 5$

FOCUS ON CONCEPTS

15. (a) Make a conjecture about the effect on the graphs of $y = y_0e^{kt}$ and $y = y_0e^{-kt}$ of varying k and keeping y_0 fixed. Confirm your conjecture with a graphing utility.

(b) Make a conjecture about the effect on the graphs of $y = y_0e^{kt}$ and $y = y_0e^{-kt}$ of varying y_0 and keeping k fixed. Confirm your conjecture with a graphing utility.

16. (a) What effect does increasing y_0 and keeping k fixed have on the doubling time or half-life of an exponential model? Justify your answer.

(b) What effect does increasing k and keeping y_0 fixed have on the doubling time and half-life of an exponential model? Justify your answer.

17. (a) There is a trick, called the ***Rule of 70***, that can be used to get a quick estimate of the doubling time or half-life of an exponential model. According to this rule, the doubling time or half-life is roughly 70 divided by the percentage growth or decay rate. For example, we showed in Example 2 that with a continued growth rate of 1.33% per year the world population would double every 52 years. This result agrees with the Rule of 70, since $70/1.33 \approx 52.6$. Explain why this rule works.

(b) Use the Rule of 70 to estimate the doubling time of a population that grows exponentially at a rate of 1% per year.

(c) Use the Rule of 70 to estimate the half-life of a population that decreases exponentially at a rate of 3.5% per hour.

(d) Use the Rule of 70 to estimate the growth rate that would be required for a population growing exponentially to double every 10 years.

18. Find a formula for the tripling time of an exponential growth model.

19. In 1950, a research team digging near Folsom, New Mexico, found charred bison bones along with some leaf-shaped projectile points (called the "Folsom points") that had been made by a Paleo-Indian hunting culture. It was clear from the evidence that the bison had been cooked and eaten by the makers of the points, so that carbon-14 dating of the bones made it possible for the researchers to determine when the hunters roamed North America. Tests showed that the bones contained between 27% and 30% of their original carbon-14. Use this information to show that the hunters lived roughly between 9000 B.C. and 8000 B.C.

20. (a) Use a graphing utility to make a graph of p_{rem} versus t, where p_{rem} is the percentage of carbon-14 that remains in an artifact after t years.

(b) Use the graph to estimate the percentage of carbon-14 that would have to have been present in the 1988 test of the Shroud of Turin for it to have been the burial shroud of Jesus. [See Example 4.]

21. (a) It is currently accepted that the half-life of carbon-14 might vary ± 40 years from its nominal value of 5730 years. Does this variation make it possible that the Shroud of Turin dates to the time of Jesus of Nazareth? [See Example 4.]

(b) Review the subsection of Section 3.8 entitled Error Propagation in Applications, and then estimate the percentage error that results in the computed age of an artifact from an $r\%$ error in the half-life of carbon-14.

22. It has been observed experimentally that at a constant temperature the rate of change of the atmospheric pressure p with respect to the altitude h above sea level is proportional to the pressure.

(a) Assuming that the pressure at sea level is p_0, find an initial-value problem whose solution is $p(h)$. [*Note:* The differential equation in this case will involve a constant of proportionality.]

(b) Find a formula for $p(h)$ in atmospheres (atm) if the pressure at sea level is 1 atm and the pressure at 5000 ft above sea level is 0.83 atm.

23. (a) Show that if $b > 1$, then the equation $y = y_0b^t$ can be expressed as $y = y_0e^{kt}$ for some positive constant k. [*Note:* This shows that if $b > 1$, and if y grows in accordance with the equation $y = y_0b^t$, then y has an exponential growth model.]

(b) Show that if $0 < b < 1$, then the equation $y = y_0b^t$ can be expressed as $y = y_0e^{-kt}$ for some positive constant k. [*Note:* This shows that if $0 < b < 1$, and if y decays in accordance with the equation $y = y_0b^t$, then y has an exponential decay model.]

(c) Express $y = 4(2^t)$ in the form $y = y_0e^{kt}$.

(d) Express $y = 4(0.5^t)$ in the form $y = y_0e^{-kt}$.

24. Suppose that a quantity y has an exponential growth model $y = y_0e^{kt}$ or an exponential decay model $y = y_0e^{-kt}$, and it is known that $y = y_1$ if $t = t_1$. In each case find a formula for k in terms of y_0, y_1, and t_1, assuming that $t_1 \neq 0$.

25. (a) Show that if a quantity $y = y(t)$ has an exponential model, and if $y(t_1) = y_1$ and $y(t_2) = y_2$, then the doubling time or the half-life T is

$$T = \left| \frac{(t_2 - t_1)\ln 2}{\ln(y_2/y_1)} \right|$$

(b) In a certain 1-hour period the number of bacteria in a colony increases by 25%. Assuming an exponential growth model, what is the doubling time for the colony?

26. Suppose that P dollars is invested at an annual interest rate of $r \times 100\%$. If the accumulated interest is credited to the account at the end of the year, then the interest is said to be *compounded annually*; if it is credited at the end of each 6-month period, then it is said to be *compounded semiannually*; and if it is credited at the end of each 3-month period, then it is said to be *compounded quarterly*. The more frequently the interest is compounded, the better it is for the investor since more of the interest is itself earning interest.

(a) Show that if interest is compounded n times a year at equally spaced intervals, then the value A of the investment after t years is

$$A = P\left(1 + \frac{r}{n}\right)^{nt}$$

(b) One can imagine interest to be compounded each day, each hour, each minute, and so forth. Carried to the limit one can conceive of interest compounded at each instant of time; this is called ***continuous compounding***. Thus, from part (a), the value A of P dollars after t years when invested at an annual rate of $r \times 100\%$, compounded continuously, is

$$A = \lim_{n\to+\infty} P\left(1 + \frac{r}{n}\right)^{nt}$$

Use the fact that $\lim_{x\to 0}(1+x)^{1/x} = e$ to prove that $A = Pe^{rt}$.

(c) Use the result in part (b) to show that money invested at continuous compound interest increases at a rate proportional to the amount present.

27. (a) If \$1000 is invested at 8% per year compounded continuously (Exercise 26), what will the investment be worth after 5 years?

(b) If it is desired that an investment at 8% per year compounded continuously should have a value of \$10,000 after 10 years, how much should be invested now?

(c) How long does it take for an investment at 8% per year compounded continuously to double in value?

28. What is the effective annual interest rate for an interest rate of r% per year compounded continuously?

29–32 ***Newton's Law of Cooling*** states that the rate at which the temperature of a cooling object decreases and the rate at which a warming object increases are proportional to the difference between the temperature of the object and the temperature of the surrounding medium. Use this result in these exercises.

29. A cup of water with a temperature of 95°C is placed in a room with a constant temperature 21°C.

(a) Assuming that Newton's Law of Cooling applies, set up and solve an initial-value problem whose solution is the temperature of the water t minutes after it is placed in the room. [*Note:* The differential equation will involve a constant of proportionality.]

(b) How many minutes will it take for the water to reach a temperature of 51°C if it cools to 85°C in 1 minute?

30. A glass of lemonade with a temperature of 40°F is placed in a room with a constant temperature of 70°F, and 1 hour later its temperature is 52°F. Show that t hours after the lemonade is placed in the room its temperature is approximated by $T = 70 - 30e^{-0.5t}$.

31. The great detective Sherlock Holmes and his assistant Dr. Watson are discussing the murder of actor Cornelius McHam. McHam was shot in the head, and his understudy, Barry Moore, was found standing over the body with the murder weapon in hand. Let's listen in.

Watson: Open-and-shut case Holmes—Moore is the murderer.

Holmes: Not so fast Watson—you are forgetting Newton's Law of Cooling!

Watson: Huh?

Holmes: Elementary my dear Watson—Moore was found standing over McHam at 10:06 P.M., at which time the coroner recorded a body temperature of 77.9°F and noted that the room thermostat was set to 72°F. At 11:06 P.M. the coroner took another reading and recorded a body temperature of 75.6°F. Since McHam's normal temperature is 98.6°F, and since Moore was on stage between 6:00 P.M. and 8:00 P.M., Moore is obviously innocent.

Watson: Huh?

Holmes: Sometimes you are so dull Watson. Ask any calculus student to figure it out for you.

Watson: Hrrumph. . . .

How did Holmes know that Moore was innocent?

32. Suppose that at time $t = 0$ an object with temperature T_0 is placed in a room with constant temperature T_a. If $T_0 < T_a$, then the temperature of the object will increase, and if $T_0 > T_a$, then the temperature will decrease. Assuming that Newton's Law of Cooling applies, show that in both cases the temperature $T(t)$ at time t is given by

$$T(t) = T_a + (T_0 - T_a)e^{-kt}$$

where k is a positive constant.

FOCUS ON CONCEPTS

33–41 In population models, it is often important that the size of the population approach a positive constant L, called the ***carrying capacity*** of the system. One model with this property is provided by the ***logistic differential equation***

$$\frac{dy}{dt} = k\left(1 - \frac{y}{L}\right)y, \quad k > 0$$

Solutions to the logistic equation have applications in modeling population growth, the spread of disease, and ecology. These exercises develop some of the properties and applications of the logistic equation.

33. (a) Show that the constant functions $y = 0$ and $y = L$ are solutions to the logistic equation.

(b) Explain why the logistic model should behave much like an exponential growth model if y is very small relative to L.

(c) Explain why y increases when $y < L$ and y decreases when $y > L$.

(d) For what value of y is the rate of change of y with respect to t the greatest?

34. Assume that $y = y(t)$ satisfies the logistic equation with $y_0 = y(0)$ the initial value of y.

(a) Use separation of variables to derive the solution
$$y = \frac{y_0 L}{y_0 + (L - y_0)e^{-kt}}$$

(b) Use part (a) to show that $\lim_{t \to +\infty} y(t) = L$.

35. The graph of a solution to the logistic equation is known as a ***logistic curve***, and if $y_0 > 0$, it has one of the four general shapes, depending on the relationship between y_0 and L. In each part, assume that $k = 1$ and use a graphing utility to plot a logistic curve satisfying the given condition.

(a) $y_0 > L$ (b) $y_0 = L$

(c) $L/2 \le y_0 < L$ (d) $0 < y_0 < L/2$

36–37 The graph of a logistic model
$$y = \frac{y_0 L}{y_0 + (L - y_0)e^{-kt}}$$
is shown. Estimate y_0, L, and k.

36.

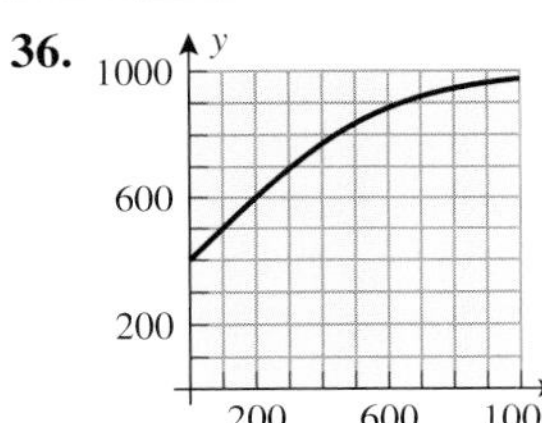

37.

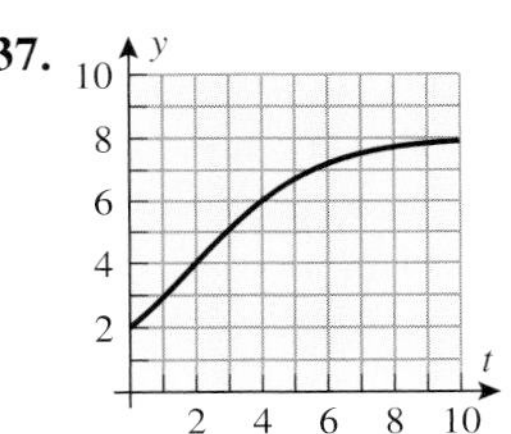

38. Plot a solution to the initial-value problem
$$\frac{dy}{dt} = 0.98\left(1 - \frac{y}{5}\right)y, \quad y_0 = 1$$

39. Suppose that the growth of a population $y = y(t)$ is given by the logistic equation
$$y = \frac{60}{5 + 7e^{-t}}$$

(a) What is the population at time $t = 0$?

(b) What is the carrying capacity L?

(c) What is the constant k?

(d) When does the population reach half of the carrying capacity?

(e) Find an initial-value problem whose solution is $y(t)$.

40. Suppose that the growth of a population $y = y(t)$ is given by the logistic equation
$$y = \frac{1000}{1 + 999e^{-0.9t}}$$

(a) What is the population at time $t = 0$?

(b) What is the carrying capacity L?

(c) What is the constant k?

(d) When does the population reach 75% of the carrying capacity?

(e) Find an initial-value problem whose solution is $y(t)$.

41. Suppose that a college residence hall houses 1000 students. Following the semester break, 20 students in the hall return with the flu, and 5 days later 35 students have the flu.

(a) Use a logistic model to set up an initial-value problem whose solution is the number of students who will have had the flu t days after the return from the break. [*Note:* The differential equation in this case will involve a constant of proportionality.]

(b) Solve the initial-value problem, and use the given data to find the constant of proportionality.

(c) Make a table that illustrates how the flu spreads day to day over a 2-week period.

(d) Use a graphing utility to generate a graph that illustrates how the flu spreads over a 2-week period.

✔ QUICK CHECK ANSWERS 9.3

1. (a) ky (b) $\frac{\ln 2}{k}$ (c) y_0e^{kt} **2.** (a) $-ky$ (b) $\frac{\ln 2}{k}$ (c) y_0e^{-kt} **3.** 1 g **4.** $100e^{0.02t}$ **5.** $4\sqrt{A}$

9.4 SECOND-ORDER LINEAR HOMOGENEOUS DIFFERENTIAL EQUATIONS; THE VIBRATING SPRING

In this section we will show how to solve an important collection of second-order differential equations. As an application, we will study the motion of a vibrating spring.

SECOND-ORDER LINEAR HOMOGENEOUS DIFFERENTIAL EQUATIONS WITH CONSTANT COEFFICIENTS

A ***second-order linear differential equation*** is one of the form

$$\frac{d^2y}{dx^2} + p(x)\frac{dy}{dx} + q(x)y = r(x) \tag{1}$$

or, in alternative notation,

$$y'' + p(x)y' + q(x)y = r(x)$$

If $r(x)$ is identically 0, then (1) reduces to

$$\frac{d^2y}{dx^2} + p(x)\frac{dy}{dx} + q(x)y = 0$$

which is called the second-order linear ***homogeneous*** differential equation.

In order to discuss the solutions to a second-order linear homogeneous differential equation, it will be useful to introduce some terminology. Two functions f and g are said to be ***linearly dependent*** if one is a *constant* multiple of the other. If neither is a constant multiple of the other, then they are called ***linearly independent***. Thus,

$$f(x) = \sin x \quad \text{and} \quad g(x) = 3\sin x$$

are linearly dependent, but

$$f(x) = x \quad \text{and} \quad g(x) = x^2$$

are linearly independent. The following theorem is central to the study of second-order linear homogeneous differential equations.

9.4.1 THEOREM. *Consider the homogeneous equation*

$$\frac{d^2y}{dx^2} + p(x)\frac{dy}{dx} + q(x)y = 0 \tag{2}$$

where the functions $p(x)$ and $q(x)$ are continuous on some common open interval I. Then there exist linearly independent solutions $y_1(x)$ and $y_2(x)$ to (2) on I. Furthermore, given any such pair of linearly independent solutions $y_1(x)$ and $y_2(x)$, a general solution of (2) on I is given by

$$y(x) = c_1y_1(x) + c_2y_2(x) \tag{3}$$

That is, every solution of (2) on I can be obtained from (3) by choosing appropriate values of the constants c_1 and c_2; conversely, (3) is a solution of (2) for all choices of c_1 and c_2.

A complete proof of Theorem 9.4.1 is best left for a course in differential equations. Portions of the argument can be found in Chapter 3 of *Elementary Differential Equations*, 8th ed., John Wiley & Sons, New York, 2004, by William E. Boyce and Richard C. DiPrima.

We will restrict our attention to second-order linear homogeneous equations of the form

$$\frac{d^2y}{dx^2} + p\frac{dy}{dx} + qy = 0 \tag{4}$$

where p and q are *constants*. Since the constant functions $p(x) = p$ and $q(x) = q$ are continuous on $I = (-\infty, +\infty)$, it follows from Theorem 9.4.1 that to determine a general solution to (4) we need only find two linearly independent solutions $y_1(x)$ and $y_2(x)$ on I. The general solution will then be given by $y(x) = c_1y_1(x) + c_2y_2(x)$, where c_1 and c_2 are arbitrary constants.

We will start by looking for solutions to (4) of the form $y = e^{mx}$. This is motivated by the fact that the first and second derivatives of this function are multiples of y, suggesting that a solution of (4) might result by choosing m appropriately. To find such an m, we substitute

$$y = e^{mx}, \quad \frac{dy}{dx} = me^{mx}, \quad \frac{d^2y}{dx^2} = m^2e^{mx} \tag{5}$$

into (4) to obtain

$$(m^2 + pm + q)e^{mx} = 0 \tag{6}$$

which is satisfied if and only if

$$m^2 + pm + q = 0 \tag{7}$$

since $e^{mx} \neq 0$ for every x.

Equation (7), which is called the ***auxiliary equation*** for (4), can be obtained from (4) by replacing d^2y/dx^2 by m^2, dy/dx by m $(= m^1)$, and y by 1 $(= m^0)$. The solutions, m_1 and m_2, of the auxiliary equation can be obtained by factoring or by the quadratic formula. These solutions are

Recall that the complex solutions of a polynomial equation, and in particular of a quadratic equation, occur as conjugate pairs $a + bi$ and $a - bi$, provided the coefficients in the equation are real.

$$m_1 = \frac{-p + \sqrt{p^2 - 4q}}{2}, \quad m_2 = \frac{-p - \sqrt{p^2 - 4q}}{2} \tag{8}$$

Depending on whether $p^2 - 4q$ is positive, zero, or negative, these roots will be distinct and real, equal and real, or complex conjugates. We will consider each of these cases separately.

DISTINCT REAL ROOTS

If m_1 and m_2 are distinct real roots, then (4) has the two solutions

$$y_1 = e^{m_1x}, \quad y_2 = e^{m_2x}$$

Neither of the functions e^{m_1x} and e^{m_2x} is a constant multiple of the other (Exercise 29), so the general solution of (4) in this case is

$$y(x) = c_1e^{m_1x} + c_2e^{m_2x} \tag{9}$$

▶ **Example 1** Find the general solution of $y'' - y' - 6y = 0$.

Solution. The auxiliary equation is

$$m^2 - m - 6 = 0 \quad \text{or equivalently} \quad (m + 2)(m - 3) = 0$$

so its roots are $m = -2$, $m = 3$. Thus, from (9) the general solution of the differential equation is

$$y = c_1e^{-2x} + c_2e^{3x}$$

where c_1 and c_2 are arbitrary constants. ◀

EQUAL REAL ROOTS

If m_1 and m_2 are equal real roots, say $m_1 = m_2\ (= m)$, then the auxiliary equation yields only one solution of (4):

$$y_1(x) = e^{mx}$$

We will now show that

$$y_2(x) = xe^{mx} \tag{10}$$

is a second linearly independent solution. To see that this is so, note that $p^2 - 4q = 0$ in (8) since the roots are equal. Thus,

$$m = m_1 = m_2 = -p/2$$

and (10) becomes

$$y_2(x) = xe^{(-p/2)x}$$

Differentiating yields

$$y_2'(x) = \left(1 - \frac{p}{2}x\right) e^{(-p/2)x} \quad \text{and} \quad y_2''(x) = \left(\frac{p^2}{4}x - p\right) e^{-(p/2)x}$$

so

$$\begin{aligned} y_2''(x) + py_2'(x) + qy_2(x) &= \left[\left(\frac{p^2}{4}x - p\right) + p\left(1 - \frac{p}{2}x\right) + qx\right] e^{(-p/2)x} \\ &= \left[-\frac{p^2}{4} + q\right] xe^{(-p/2)x} \end{aligned} \tag{11}$$

But $p^2 - 4q = 0$ implies that $(-p^2/4) + q = 0$, so (11) becomes

$$y_2''(x) + py_2'(x) + qy_2(x) = 0$$

which tells us that $y_2(x)$ is a solution of (4). It can be shown that

$$y_1(x) = e^{mx} \quad \text{and} \quad y_2(x) = xe^{mx}$$

are linearly independent (Exercise 29), so the general solution of (4) in this case is

$$y = c_1 e^{mx} + c_2 x e^{mx} \tag{12}$$

► **Example 2** Find the general solution of $y'' - 8y' + 16y = 0$.

Solution. The auxiliary equation is

$$m^2 - 8m + 16 = 0 \quad \text{or equivalently} \quad (m-4)^2 = 0$$

so $m = 4$ is the only root. Thus, from (12) the general solution of the differential equation is

$$y = c_1 e^{4x} + c_2 x e^{4x} \quad ◄$$

COMPLEX ROOTS

If the auxiliary equation has complex roots $m_1 = a + bi$ and $m_2 = a - bi$, then both $y_1(x) = e^{ax} \cos bx$ and $y_2(x) = e^{ax} \sin bx$ are linearly independent solutions of (4) and

$$y = e^{ax}(c_1 \cos bx + c_2 \sin bx) \tag{13}$$

is the general solution. The proof is discussed in the exercises (Exercise 30).

► **Example 3** Find the general solution of $y'' + y' + y = 0$.

Solution. The auxiliary equation $m^2 + m + 1 = 0$ has roots

$$m_1 = \frac{-1 + \sqrt{1-4}}{2} = -\frac{1}{2} + \frac{\sqrt{3}}{2}i$$

$$m_2 = \frac{-1 - \sqrt{1-4}}{2} = -\frac{1}{2} - \frac{\sqrt{3}}{2}i$$

Thus, from (13) with $a = -1/2$ and $b = \sqrt{3}/2$, the general solution of the differential equation is

$$y = e^{-x/2}\left(c_1 \cos \frac{\sqrt{3}}{2}x + c_2 \sin \frac{\sqrt{3}}{2}x\right) \blacktriangleleft$$

INITIAL-VALUE PROBLEMS

When a physical problem leads to a second-order differential equation, there are usually two conditions in the problem that determine specific values for the two arbitrary constants in the general solution of the equation. Conditions that specify the value of the solution $y(x)$ and its derivative $y'(x)$ at $x = x_0$ are called ***initial conditions***. A second-order differential equation with initial conditions is called a ***second-order initial-value problem***.

▶ **Example 4** Solve the initial-value problem

$$y'' - y = 0, \quad y(0) = 1, \quad y'(0) = 0$$

Solution. We must first solve the differential equation. The auxiliary equation

$$m^2 - 1 = 0$$

has distinct real roots $m_1 = 1$, $m_2 = -1$, so from (9) the general solution is

$$y(x) = c_1e^x + c_2e^{-x} \tag{14}$$

and the derivative of this solution is

$$y'(x) = c_1e^x - c_2e^{-x} \tag{15}$$

Substituting $x = 0$ in (14) and (15) and using the initial conditions $y(0) = 1$ and $y'(0) = 0$ yields the system of equations

$$\begin{aligned} c_1 + c_2 &= 1 \\ c_1 - c_2 &= 0 \end{aligned}$$

Solving this system yields $c_1 = \frac{1}{2}$, $c_2 = \frac{1}{2}$, so from (14) the solution of the initial-value problem is

$$y(x) = \tfrac{1}{2}e^x + \tfrac{1}{2}e^{-x} = \cosh x \blacktriangleleft$$

The following summary is included as a ready reference for the solution of second-order homogeneous linear differential equations with constant coefficients.

Summary

EQUATION: $y'' + py' + qy = 0$

AUXILIARY EQUATION: $m^2 + pm + q = 0$

CASE	GENERAL SOLUTION
Distinct real roots m_1, m_2 of the auxiliary equation	$y = c_1e^{m_1x} + c_2e^{m_2x}$
Equal real roots $m_1 = m_2$ $(= m)$ of the auxiliary equation	$y = c_1e^{mx} + c_2xe^{mx}$
Complex roots $m_1 = a + bi$, $m_2 = a - bi$ of the auxiliary equation	$y = e^{ax}(c_1 \cos bx + c_2 \sin bx)$

VIBRATIONS OF SPRINGS

We conclude this section with an engineering model that leads to a second-order differential equation of type (4).

As shown in Figure 9.4.1, consider a block of mass M that is suspended from a vertical spring and allowed to settle into an ***equilibrium position***. Assume that the block is then set into vertical vibratory motion by pulling or pushing on it and releasing it at time $t = 0$. We will be interested in finding a mathematical model that describes the vibratory motion of the block over time.

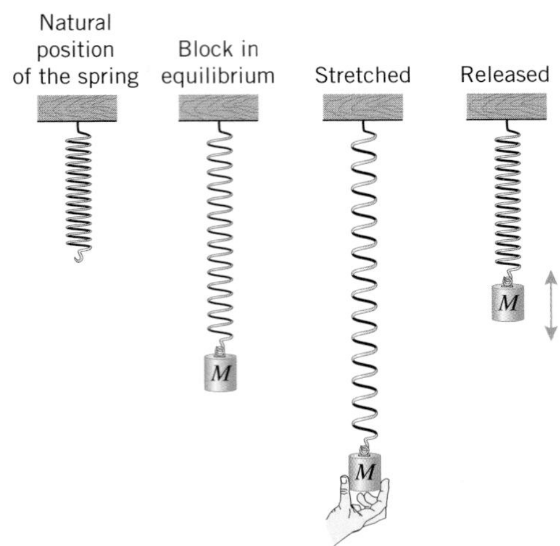

Figure 9.4.1

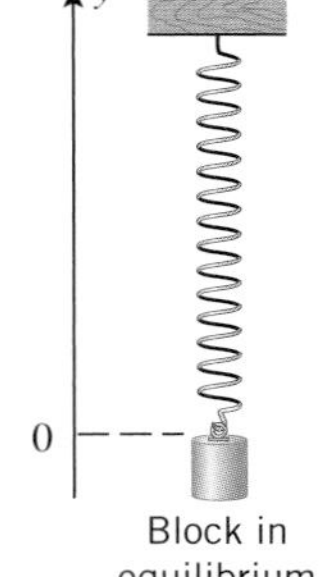

Figure 9.4.2

To translate this problem into mathematical form, we introduce a vertical y-axis whose positive direction is up and whose origin is at the connection of the spring to the block when the block is in equilibrium (Figure 9.4.2). Our goal is to find the coordinate $y = y(t)$ of the top of the block as a function of time. For this purpose we will need Newton's Second Law of Motion, which we will write as

$$F = Ma$$

rather than $F = ma$, as in Formula (23) of Section 9.1. This is to avoid a conflict with the letter "m" in the auxiliary equation. We will also need the following two results from physics.

9.4.2 HOOKE'S LAW. If a spring is stretched (or compressed) L units beyond its natural position, then it pulls (or pushes) with a force of magnitude

$$F = kL$$

where k is a positive constant, called the ***spring constant***. This constant, which is measured in units of force per unit length, depends on such factors as the thickness of the spring and its composition. The force exerted by the spring is called the ***restoring force***.

9.4.3 WEIGHT. The gravitational force exerted by the Earth on an object is called the object's ***weight*** (or, more precisely, its ***Earth weight***). It follows from Newton's Second Law of Motion that an object with mass M has a weight w of magnitude Mg, where g is the acceleration due to gravity. However, if the positive direction is up, as we are assuming here, then the force of the Earth's gravity is in the negative direction, so

$$w = -Mg$$

The weight of an object is measured in units of force.

The motion of the block in Figure 9.4.1 will depend on how far the spring is stretched or compressed initially and the forces that act on the block while it moves. In our model we will assume that there are only two such forces: the block's weight w and the restoring force F_s of the spring. In particular, we will ignore such forces as air resistance, internal frictional forces in the spring, forces due to movement of the spring support, and so forth. With these assumptions, the model is called the ***simple harmonic model*** and the motion of the block is called ***simple harmonic motion***.

Our goal is to produce a differential equation whose solution gives the position function $y(t)$ of the block as a function of time. We will do this by determining the net force $F(t)$ acting on the block at a general time t and then applying Newton's Second Law of Motion. Since the only forces acting on the block are its weight $w = -Mg$ and the restoring force F_s of the spring, and since the acceleration of the block at time t is $y''(t)$, it follows from Newton's Second Law of Motion that

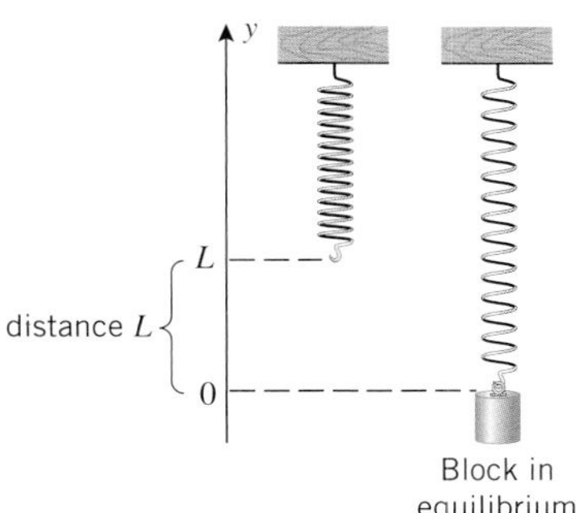

Figure 9.4.3

$$F_s(t) - Mg = My''(t) \tag{16}$$

To express $F_s(t)$ in terms of $y(t)$, we will begin by examining the forces on the block when it is in its equilibrium position. In this position the downward force of the weight is perfectly balanced by the upward restoring force of the spring, so that the sum of these two forces must be zero. Thus, if we assume that the spring constant is k and that the spring is stretched a distance of L units beyond its natural length when the block is in equilibrium (Figure 9.4.3), then

$$kL - Mg = 0 \tag{17}$$

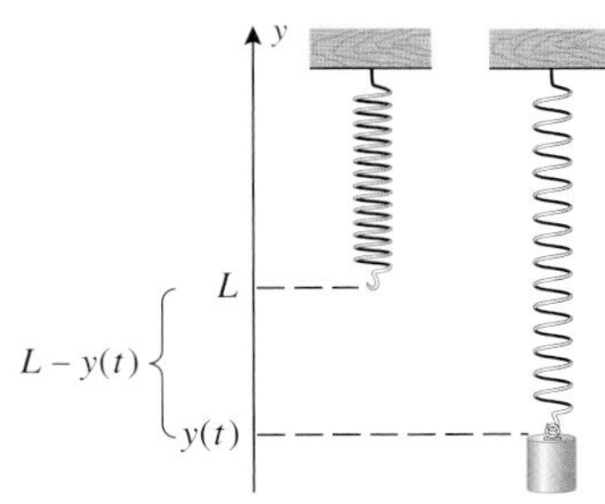

Figure 9.4.4

Now let us examine the restoring force acting on the block when the connection point has coordinate $y(t)$. At this point the end of the spring is displaced $L - y(t)$ units from its natural position (Figure 9.4.4), so Hooke's law implies that the restoring force is

$$F_s(t) = k(L - y(t)) = kL - ky(t)$$

which from (17) can be rewritten as

$$F_s(t) = Mg - ky(t)$$

Substituting this in (16) and canceling the Mg terms yields

$$-ky(t) = My''(t)$$

which we can rewrite as the homogeneous equation

$$y''(t) + \left(\frac{k}{M}\right) y(t) = 0 \tag{18}$$

The auxiliary equation for (18) is

$$m^2 + \frac{k}{M} = 0$$

which has imaginary roots $m_1 = \sqrt{k/M}\,i$, $m_2 = -\sqrt{k/M}\,i$ (since k and M are positive). It follows that the general solution of (18) is

$$y(t) = c_1 \cos\left(\sqrt{\frac{k}{M}}\,t\right) + c_2 \sin\left(\sqrt{\frac{k}{M}}\,t\right) \tag{19}$$

Confirm that the functions in family (19) are solutions of (18).

To determine the constants c_1 and c_2 in (19) we will take as our initial conditions the position and velocity at time $t = 0$. Specifically, we will ask you to show in Exercise 40 that if the position of the block at time $t = 0$ is y_0, and if the initial velocity of the block is zero (i.e., it is *released* from rest), then

$$y(t) = y_0 \cos\left(\sqrt{\frac{k}{M}}\, t\right) \tag{20}$$

This formula describes a periodic vibration with an amplitude of $|y_0|$, a period T given by

$$T = \frac{2\pi}{\sqrt{k/M}} = 2\pi\sqrt{M/k} \tag{21}$$

and a frequency f given by

$$f = \frac{1}{T} = \frac{\sqrt{k/M}}{2\pi} \tag{22}$$

(Figure 9.4.5).

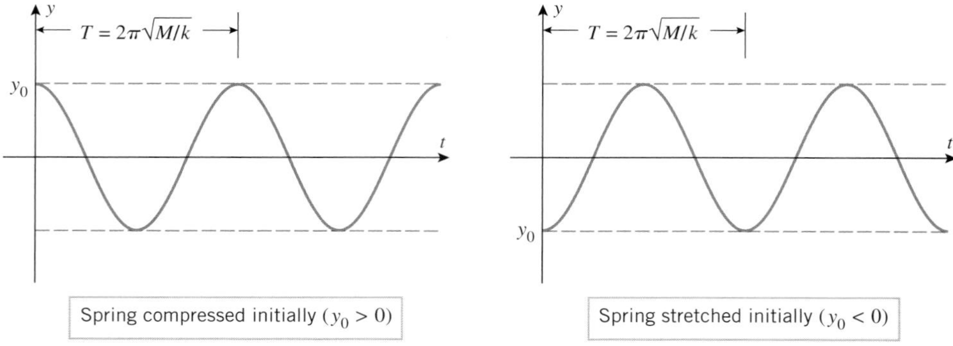

Figure 9.4.5

▶ **Example 5** Suppose that the block in Figure 9.4.2 stretches the spring 0.2 m in equilibrium. Suppose also that the block is pulled 0.5 m below its equilibrium position and released at time $t = 0$.

(a) Find the position function $y(t)$ of the block.

(b) Find the amplitude, period, and frequency of the vibration.

Solution (a). The appropriate formula is (20). Although we are not given the mass M of the block or the spring constant k, it does not matter because we can use the equilibrium condition (17) to find the ratio k/M without having values for k and M. Specifically, we are given that in equilibrium the block stretches the spring $L = 0.2$ m, and we know that $g = 9.8$ m/s^2. Thus, (17) implies that

$$\frac{k}{M} = \frac{g}{L} = \frac{9.8}{0.2} = 49 \text{ s}^{-2} \tag{23}$$

Substituting this in (20) yields

$$y(t) = y_0 \cos 7t$$

where y_0 is the coordinate of the block at time $t = 0$. However, we are given that the block is initially 0.5 m *below* the equilibrium position, so $y_0 = -0.5$ and hence the position function of the block is $y(t) = -0.5\cos 7t$.

Solution (b). The amplitude of the vibration is

$$\text{amplitude} = |y_0| = |-0.5| = 0.5 \text{ m}$$

and from (21), (22), and (23) the period and frequency are

$$\text{period} = T = 2\pi\sqrt{\frac{M}{k}} = 2\pi\sqrt{\frac{1}{49}} = \frac{2\pi}{7} \text{ s}, \quad \text{frequency} = f = \frac{1}{T} = \frac{7}{2\pi} \text{ Hz} \blacktriangleleft$$

✔QUICK CHECK EXERCISES 9.4 (See page 622 for answers.)

1. A second-order linear differential equation is one of the form ________. This equation is homogeneous provided ________.

2. (a) Two functions are said to be linearly independent provided ________.
(b) In each part, determine whether the functions f and g are linearly independent.
(i) $f(x) = \sin x$, $g(x) = \sin 2x$
(ii) $f(x) = \sin x \cos x$, $g(x) = \sin 2x$
(iii) $f(x) = \ln x$, $g(x) = \ln \sqrt{x}$
(iv) $f(x) = e^x$, $g(x) = e^{2x}$
(v) $f(x) = \sin^2 x$, $g(x) = 1 - \cos 2x$

3. If $y_1(x)$ and $y_2(x)$ are linearly independent solutions to a second-order linear homogeneous differential equation (with continuous coefficient functions), then the general solution to this differential equation is $y =$ ________.

4. In each part, use the information about the roots of the auxiliary equation to find the general solution to the corresponding second-order linear homogeneous differential equation with constant coefficients.
(a) The auxiliary equation has two distinct real roots m_1 and m_2.
(b) The auxiliary equation has a single real root m of multiplicity 2.
(c) The auxiliary equation has two distinct complex roots $m_1 = a + bi$ and $m_2 = a - bi$.

5. The differential equation for the simple harmonic motion of a mass M attached to a spring with spring constant k is ________.

EXERCISE SET 9.4  

1. Verify that the following are solutions of the differential equation $y'' + y' - 2y = 0$ by substituting these functions into the equation.
(a) e^{-2x} and e^x
(b) $c_1e^{-2x} + c_2e^x$ (c_1, c_2 constants)

2. Verify that the following are solutions of the differential equation $y'' + 4y' + 4y = 0$ by substituting these functions into the equation.
(a) e^{-2x} and xe^{-2x}
(b) $c_1e^{-2x} + c_2xe^{-2x}$ (c_1, c_2 constants)

3–16 Find the general solution of the differential equation.

3. $y'' + 3y' - 4y = 0$
4. $y'' + 5y' + 6y = 0$
5. $y'' - 2y' + y = 0$
6. $y'' - 6y' + 9y = 0$
7. $y'' + y = 0$
8. $y'' + 5y = 0$
9. $\dfrac{d^2y}{dx^2} - \dfrac{dy}{dx} = 0$
10. $\dfrac{d^2y}{dx^2} + 3\dfrac{dy}{dx} = 0$
11. $\dfrac{d^2y}{dt^2} - 4\dfrac{dy}{dt} + 4y = 0$
12. $\dfrac{d^2y}{dt^2} - 10\dfrac{dy}{dt} + 25y = 0$
13. $\dfrac{d^2y}{dx^2} + 4\dfrac{dy}{dx} + 13y = 0$
14. $\dfrac{d^2y}{dx^2} - 6\dfrac{dy}{dx} + 25y = 0$
15. $8y'' - 2y' - y = 0$
16. $9y'' - 6y' + y = 0$

17–22 Solve the initial-value problem.

17. $y'' + 2y' - 3y = 0$, $y(0) = 1$, $y'(0) = 9$
18. $y'' - 6y' - 7y = 0$, $y(0) = 5$, $y'(0) = 3$
19. $y'' + 6y' + 9y = 0$, $y(0) = 2$, $y'(0) = -5$
20. $y'' + 4y' + y = 0$, $y(0) = 5$, $y'(0) = 4$
21. $y'' + 4y' + 5y = 0$, $y(0) = -3$, $y'(0) = 0$
22. $y'' - 6y' + 13y = 0$, $y(0) = -2$, $y'(0) = 0$

23. In each part, find a second-order linear homogeneous differential equation with constant coefficients that has the given functions as solutions.
(a) $y_1 = e^{5x}$, $y_2 = e^{-2x}$
(b) $y_1 = e^{4x}$, $y_2 = xe^{4x}$
(c) $y_1 = e^{-x} \cos 4x$, $y_2 = e^{-x} \sin 4x$

24. Show that if e^x and e^{-x} are solutions of a second-order linear homogeneous differential equation, then so are $\cosh x$ and $\sinh x$.

25. Find all values of k for which the differential equation $y'' + ky' + ky = 0$ has a general solution of the given form.
(a) $y = c_1e^{ax} + c_2e^{bx}$
(b) $y = c_1e^{ax} + c_2xe^{ax}$
(c) $y = c_1e^{ax} \cos bx + c_2e^{ax} \sin bx$

26. The equation
$$x^2\frac{d^2y}{dx^2}+px\frac{dy}{dx}+qy=0 \quad (x>0)$$
where p and q are constants, is called *Euler's equidimensional equation*. Show that the substitution $x=e^z$ transforms this equation into the equation
$$\frac{d^2y}{dz^2}+(p-1)\frac{dy}{dz}+qy=0$$

27. Use the result in Exercise 26 to find the general solution of

(a) $x^2\dfrac{d^2y}{dx^2}+3x\dfrac{dy}{dx}+2y=0 \quad (x>0)$

(b) $x^2\dfrac{d^2y}{dx^2}-x\dfrac{dy}{dx}-2y=0 \quad (x>0)$.

FOCUS ON CONCEPTS

28. Let $y(x)$ be a solution of $y''+py'+qy=0$. Prove: If p and q are positive constants, then $\lim_{x\to+\infty} y(x)=0$.

29. Prove that the following functions are linearly independent.

(a) $y_1=e^{m_1x},\ y_2=e^{m_2x} \quad (m_1\neq m_2)$

(b) $y_1=e^{mx},\ y_2=xe^{mx}$

30. Prove: If the auxiliary equation of
$$y''+py'+qy=0$$
has complex roots $a+bi$ and $a-bi$, then the general solution of this differential equation is
$$y(x)=e^{ax}(c_1\cos bx+c_2\sin bx)$$
[*Hint:* Using substitution, verify that $y_1=e^{ax}\cos bx$ and $y_2=e^{ax}\sin bx$ are solutions of the differential equation. Then prove that y_1 and y_2 are linearly independent.]

31. Suppose that the auxiliary equation of the equation $y''+py'+qy=0$ has distinct real roots μ and m.

(a) Show that the function
$$g_\mu(x)=\frac{e^{\mu x}-e^{mx}}{\mu-m}$$
is a solution of the differential equation.

(b) Use L'Hôpital's rule to show that
$$\lim_{\mu\to m} g_\mu(x)=xe^{mx}$$
[*Note:* Can you see how the result in part (b) makes it plausible that the function $y(x)=xe^{mx}$ is a solution of $y''+py'+qy=0$ when m is a repeated root of the auxiliary equation?]

32. Consider the problem of solving the differential equation
$$y''+\lambda y=0$$
subject to the conditions $y(0)=0,\ y(\pi)=0$.

(a) Show that if $\lambda\le 0$, then $y=0$ is the only solution.

(b) Show that if $\lambda>0$, then the solution is
$$y=c\sin\sqrt{\lambda}x$$
where c is an arbitrary constant, if
$$\lambda=1,2^2,3^2,4^2,\ldots$$
and the only solution is $y=0$ otherwise.

33–38 These exercises involve vibrations of the block pictured in Figure 9.4.1. Assume that the y-axis is as shown in Figure 9.4.2 and that the simple harmonic model applies.

33. Suppose that the block has a mass of 2 kg, the spring constant is $k=0.5$ N/m, and the block is pushed 0.4 m above its equilibrium position and released at time $t=0$.

(a) Find the position function $y(t)$ of the block.

(b) Find the period and frequency of the vibration.

(c) Sketch the graph of $y(t)$.

(d) At what time does the block first pass through the equilibrium position?

(e) At what time does the block first reach its maximum distance below the equilibrium position?

34. Suppose that the block has a weight of 64 lb, the spring constant is $k=0.25$ lb/ft, and the block is pushed 1 ft above its equilibrium position and released at time $t=0$.

(a) Find the position function $y(t)$ of the block.

(b) Find the period and frequency of the vibration.

(c) Sketch the graph of $y(t)$.

(d) At what time does the block first pass through the equilibrium position?

(e) At what time does the block first reach its maximum distance below the equilibrium position?

35. Suppose that the block stretches the spring 0.05 m in equilibrium, and the block is pulled 0.12 m below the equilibrium position and released at time $t=0$.

(a) Find the position function $y(t)$ of the block.

(b) Find the period and frequency of the vibration.

(c) Sketch the graph of $y(t)$.

(d) At what time does the block first pass through the equilibrium position?

(e) At what time does the block first reach its maximum distance above the equilibrium position?

36. Suppose that the block stretches the spring 2 ft in equilibrium, and is pulled 2 ft below the equilibrium position and released at time $t=0$.

(a) Find the position function $y(t)$ of the block.

(b) Find the period and frequency of the vibration.

(c) Sketch the graph of $y(t)$.

(d) At what time does the block first pass through the equilibrium position?

(e) At what time does the block first reach its maximum distance above the equilibrium position?

37. (a) For what values of y would you expect the block in Exercise 36 to have its maximum speed? Confirm your answer to this question mathematically.

(b) For what values of y would you expect the block to have its minimum speed? Confirm your answer to this question mathematically.

38. Suppose that the block weighs w pounds and vibrates with a period of 3 s when it is pulled below the equilibrium position

and released. Suppose also that if the process is repeated with an additional 4 lb of weight, then the period is 5 s.
(a) Find the spring constant. (b) Find w.

39. As shown in the accompanying figure, suppose that a toy cart of mass M is attached to a wall by a spring with spring constant k, and let a horizontal x-axis be introduced with its origin at the connection point of the spring and cart when the cart is in equilibrium. Suppose that the cart is pulled or pushed horizontally to a point x_0 and then released at time $t = 0$. Find an initial-value problem whose solution is the position function of the cart, and state any assumptions you have made.

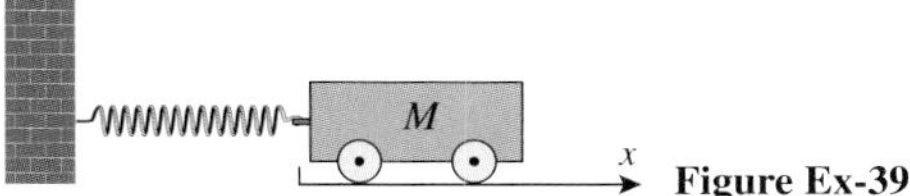

Figure Ex-39

40. Use the initial position $y(0) = y_0$ and the initial velocity $v(0) = 0$ to find the constants c_1 and c_2 in (19).

41. A block attached to a vertical spring is displaced from its equilibrium position and released, thereby causing it to vibrate with amplitude $|y_0|$ and period T.
(a) Show that the velocity of the block has maximum magnitude $2\pi|y_0|/T$ and that the maximum occurs when the block is at its equilibrium position.
(b) Show that the acceleration of the block has maximum magnitude $4\pi^2|y_0|/T^2$ and that the maximum occurs when the block is at a top or bottom point of its motion.

42. Assume that the motion of a block of mass M is governed by the simple harmonic model (18). Define the *potential energy* of the block at time t to be $\frac{1}{2}k[y(t)]^2$, and define the *kinetic energy* of the block at time t to be $\frac{1}{2}M[y'(t)]^2$. Prove that the sum of the potential energy of the block and the kinetic energy of the block is constant.

43–47 The accompanying figure shows a mass–spring system in which an object of mass M is suspended by a spring and linked to a piston that moves in a *dashpot* containing a viscous fluid. If there are no external forces acting on the system, then the object is said to have ***free motion*** and the motion of the object is completely determined by the displacement and velocity of the object at time $t = 0$, the stiffness of the spring as measured by the spring constant k, and the viscosity of the fluid in the dashpot as measured by a *damping constant* c. Mathematically, the displacement $y = y(t)$ of the object from its equilibrium position is the solution of an initial-value problem of the form

$$y'' + Ay' + By = 0, \quad y(0) = y_0, \quad y'(0) = v_0$$

where the coefficient A is determined by M and c and the coefficient B is determined by M and k. In our derivation of Equation (21) we considered only motion in which the coefficient A is zero and in which the object is released from rest, that is, $v_0 = 0$. In these exercises you are asked to consider initial-value problems for which both the coefficient A and the initial velocity v_0 are nonzero.

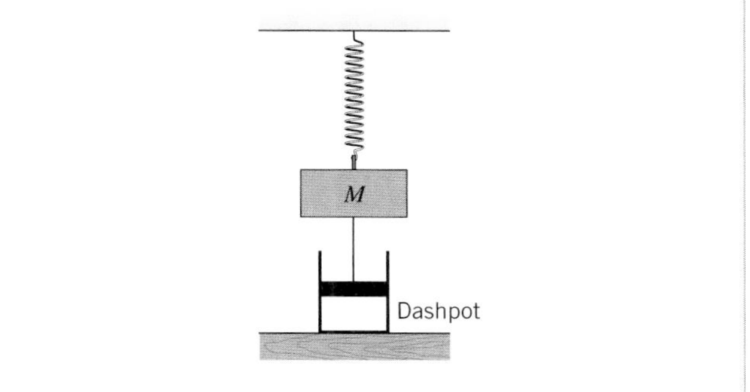

43. (a) Solve the initial-value problem $y'' + 2.4y' + 1.44y = 0$, $y(0) = 1$, $y'(0) = 2$ and graph $y = y(t)$ on the interval $[0, 5]$.
(b) Find the maximum distance above the equilibrium position attained by the object.
(c) The graph of $y(t)$ suggests that the object does not pass through the equilibrium position. Show that this is so.

C 44. (a) Solve the initial-value problem $y'' + 5y' + 2y = 0$, $y(0) = 1/2$, $y'(0) = -4$ and graph $y = y(t)$ on the interval $[0, 5]$.
(b) Find the maximum distance below the equilibrium position attained by the object.
(c) The graph of $y(t)$ suggests that the object passes through the equilibrium position exactly once. With what speed does the object pass through the equilibrium position?

C 45. (a) Solve the initial-value problem $y'' + y' + 5y = 0$, $y(0) = 1$, $y'(0) = -3.5$ and graph $y = y(t)$ on the interval $[0, 8]$.
(b) Find the maximum distance below the equilibrium position attained by the object.
(c) Find the velocity of the object when it passes through the equilibrium position the first time.
(d) Find, by inspection, the acceleration of the object when it passes through the equilibrium position the first time. [*Hint:* Examine the differential equation and use the result in part (c).]

C 46. (a) Solve the initial-value problem $y'' + y' + 3y = 0$, $y(0) = -2$, $y'(0) = v_0$.
(b) Find the largest positive value of v_0 for which the object will rise no higher than 1 unit above the equilibrium position. [*Hint:* Use a trial-and-error strategy. Estimate v_0 to the nearest hundredth.]
(c) Graph the solution of the initial-value problem on the interval $[0, 8]$ using the value of v_0 obtained in part (b).

47. (a) Solve the initial-value problem $y'' + 3.5y' + 3y = 0$, $y(0) = 1$, $y'(0) = v_0$.
(b) Use the result in part (a) to find the solutions for $v_0 = 2$, $v_0 = -1$, and $v_0 = -4$ and graph all three solutions on the interval $[0, 4]$ in the same coordinate system.
(c) Discuss the effect of the initial velocity on the motion of the object.

C **48.** If the block in Figure 9.4.1 is displaced y_0 units from its equilibrium position and given an initial velocity of v_0, rather than being released with an initial velocity of 0, then its position function $y(t)$ given in Equation (19) must satisfy the initial conditions $y(0) = y_0$ and $y'(0) = v_0$.

(a) Show that

$$y(t) = y_0 \cos\left(\sqrt{\frac{k}{M}}t\right) + v_0\sqrt{\frac{M}{k}} \sin\left(\sqrt{\frac{k}{M}}t\right)$$

(b) Suppose that a block with a mass of 1 kg stretches the spring 0.5 m in equilibrium. Use a graphing utility to graph the position function of the block if it is set in motion by pulling it down 1 m and imparting it an initial upward velocity of 0.25 m/s.

(c) What is the maximum displacement of the block from the equilibrium position?

FOCUS ON CONCEPTS

49. Consider the first-order linear homogeneous equation

$$\frac{dy}{dx} + p(x)y = 0$$

where $p(x)$ is a continuous function on some open interval I. By analogy to the results of Theorem 9.4.1, we might expect the general solution of this equation to be of the form

$$y = cy_1(x)$$

where $y_1(x)$ is a solution of the equation on the interval I and c is an arbitrary constant. Prove this to be the case.

50. Prove Theorem 9.4.1 in the special case where $q(x)$ is identically zero.

✔ QUICK CHECK ANSWERS 9.4

1. $\dfrac{d^2y}{dx^2} + p(x)\dfrac{dy}{dx} + q(x)y = r(x);\ r(x) = 0$ **2.** (a) neither is a constant multiple of the other (b) Only the pairs in (i) and (iv) are linearly independent. **3.** $c_1y_1(x) + c_2y_2(x)$ **4.** (a) $y(x) = c_1e^{m_1x} + c_2e^{m_2x}$ (b) $y(x) = c_1e^{mx} + c_2xe^{mx}$ (c) $y(x) = e^{ax}(c_1 \cos bx + c_2 \sin bx)$ **5.** $y''(t) + \left(\dfrac{k}{M}\right)y(t) = 0$

CHAPTER REVIEW EXERCISES

C CAS

1. We have seen that the general solution of a first-order linear equation involves a single arbitrary constant and that the general solution of a second-order linear differential equation involves two arbitrary constants. Give an informal explanation of why one might expect the number of arbitrary constants to equal the order of the equation.

2. (a) List the steps in the method of integrating factors for solving first-order linear differential equations.
(b) What would you do if you had to solve an initial-value problem involving a first-order linear differential equation whose integrating factor could not be obtained due to the complexity of the integration?

3. Classify the following first-order differential equations as separable, linear, both, or neither.

(a) $\dfrac{dy}{dx} - 3y = \sin x$ (b) $\dfrac{dy}{dx} + xy = x$

(c) $y\dfrac{dy}{dx} - x = 1$ (d) $\dfrac{dy}{dx} + xy^2 = \sin(xy)$

4. Which of the given differential equations are separable?

(a) $\dfrac{dy}{dx} = f(x)g(y)$ (b) $\dfrac{dy}{dx} = \dfrac{f(x)}{g(y)}$

(c) $\dfrac{dy}{dx} = f(x) + g(y)$ (d) $\dfrac{dy}{dx} = \sqrt{f(x)g(y)}$

5. Determine whether the methods of integrating factors and separation of variables produce the same solution of the differential equation

$$\frac{dy}{dx} - 4xy = x$$

6–10 Solve the differential equation either by the method of integrating factors or by separation of variables.

6. $\dfrac{dy}{dx} + 3y = e^{-2x}$ **7.** $\dfrac{dy}{dx} = (1 + y^2)x^2$

8. $\dfrac{dy}{dx} + y - \dfrac{1}{1 + e^x} = 0$ **9.** $(1 + y^2)y' = e^x y$

10. $3 \tan y - \dfrac{dy}{dx} \sec x = 0$

11–16 Solve the initial-value problem.

11. $y' - xy = x,\ y(0) = 3$ **12.** $y' = 1 + y^2,\ y(0) = 1$

13. $y' \cosh x + y \sinh x = \cosh^2 x,\ y(0) = 2$

14. $xy' + 2y = 4x^2,\ y(1) = 2$

15. $y' = \dfrac{y^5}{x(1 + y^4)},\ y(1) = 1$

16. $y' = 4y^2 \sec^2 2x,\ y(\pi/8) = 1$

C 17. (a) Solve the initial-value problem

$$y' - y = x \sin 3x, \quad y(0) = 1$$

by the method of integrating factors, using a CAS to perform any difficult integrations.

(b) Use the CAS to solve the initial-value problem directly, and confirm that the answer is consistent with that obtained in part (a).

(c) Graph the solution.

18. (a) Sketch the integral curve of $2yy' = 1$ that passes through the point $(0, 1)$ and the integral curve that passes through the point $(0, -1)$.

(b) Sketch the integral curve of $y' = -2xy^2$ that passes through the point $(0, 1)$.

19. A tank contains 1000 gal of fresh water. At time $t = 0$ min, brine containing 5 oz of salt per gallon of brine is poured into the tank at a rate of 10 gal/min, and the mixed solution is drained from the tank at the same rate. After 15 min that process is stopped and fresh water is poured into the tank at the rate of 5 gal/min, and the mixed solution is drained from the tank at the same rate. Find the amount of salt in the tank at time $t = 30$ min.

20. Suppose that a room containing 1200 ft^3 of air is free of carbon monoxide. At time $t = 0$ cigarette smoke containing 4% carbon monoxide is introduced at the rate of 0.1 ft^3/min, and the well-circulated mixture is vented from the room at the same rate.

(a) Find a formula for the percentage of carbon monoxide in the room at time t.

(b) Extended exposure to air containing 0.012% carbon monoxide is considered dangerous. How long will it take to reach this level?

Source: This is based on a problem from William E. Boyce and Richard C. DiPrima, *Elementary Differential Equations*, 8th ed., John Wiley & Sons, New York, 2004.

21. Sketch the slope field for $y' = xy/8$ at the 25 gridpoints (x, y), where $x = 0, 1, \ldots, 4$ and $y = 0, 1, \ldots, 4$.

22. Solve the differential equation $y' = xy/8$, and find a family of integral curves for the slope field in Exercise 21.

23–24 Use Euler's Method with the given step size Δx to approximate the solution of the initial-value problem over the stated interval. Present your answer as a table and as a graph.

23. $dy/dx = \sqrt{y},\ y(0) = 1,\ 0 \le x \le 4,\ \Delta x = 0.5$

24. $dy/dx = \sin y,\ y(0) = 1,\ 0 \le x \le 2,\ \Delta x = 0.5$

25. Consider the initial-value problem

$$y' = \cos 2\pi t, \quad y(0) = 1$$

Use Euler's Method with five steps to approximate $y(1)$.

26. (a) Use Euler's Method with a step size of $\Delta t = 0.1$ to approximate the solution of the initial-value problem

$$y' = 1 + 5t - y, \quad y(1) = 5$$

over the interval $[1, 2]$.

(b) Find the percentage error in the values computed.

27. In each part, find an exponential growth model $y = y_0 e^{kt}$ that satisfies the stated conditions.

(a) $y_0 = 2$; doubling time $T = 5$

(b) $y(0) = 5$; growth rate 1.5%

(c) $y(1) = 1$; $y(10) = 100$

(d) $y(1) = 1$; doubling time $T = 5$

28. Suppose that an initial population of 5000 bacteria grows exponentially at a rate of 1% per hour and that $y = y(t)$ is the number of bacteria present after t hours.

(a) Find an initial-value problem whose solution is $y(t)$.

(b) Find a formula for $y(t)$.

(c) What is the doubling time for the population?

(d) How long does it take for the population of bacteria to reach 30,000?

29. Cloth found in an Egyptian pyramid contains 78.5% of its original carbon-14. Estimate the age of the cloth.

30. The length and width of a rectangle are growing at the same constant rate. Prove that this implies that the rate of change of the area of the rectangle is proportional to the perimeter of the rectangle.

31. Find the general solution of each differential equation.

(a) $y'' - 3y' + 2y = 0$ (b) $4y'' - 4y' + y = 0$

(c) $y'' + y' + 2y = 0$

32. Solve the initial-value problem.

(a) $y'' + 2y' - 3y = 0,\ y(0) = 1,\ y'(0) = 5$

(b) $y'' - 6y' + 9y = 0,\ y(0) = 2,\ y'(0) = 1$

(c) $y'' - 4y' + 13y = 0,\ y(0) = 1,\ y'(0) = 5$

33–34 These exercises involve vibrations of the block pictured in Figure 9.4.1. Assume that the y-axis is as shown in Figure 9.4.2 and that the simple harmonic model applies.

33. Suppose that the block has a mass of 1 kg, the spring constant is $k = 0.25$ N/m, and the block is pushed 0.3 m above its equilibrium position and released at time $t = 0$.

(a) Find the position function $y(t)$ of the block.

(b) Find the period and frequency of the vibration.

(c) Sketch the graph of $y(t)$.

(d) At what time does the block first pass through the equilibrium position?

(e) At what time does the block first reach its maximum distance below the equilibrium position?

34. Suppose that the block stretches the spring 0.5 ft in equilibrium and is pulled 1.5 ft below the equilibrium position and released at time $t = 0$.

(a) Find the position function $y(t)$ of the block.

(b) Find the period and frequency of the vibration.

(c) Sketch the graph of $y(t)$.

(d) At what time does the block first pass through the equilibrium position?

(e) At what time does the block first reach its maximum distance above the equilibrium position?

c h a p t e r t e n

INFINITE SERIES

Great fleas have little fleas
upon their backs to bite 'em,
And little fleas have lesser
fleas, and so ad infinitum.
And the great fleas
themselves, in turn, have
greater fleas to go on;
While these again have
greater still, and greater still,
and so on.

—Augustus De Morgan
Mathematician

In this chapter we will be concerned with infinite series, which are sums that involve infinitely many terms. Infinite series play a fundamental role in both mathematics and science—they are used, for example, to approximate trigonometric functions and logarithms, to solve differential equations, to evaluate difficult integrals, to create new functions, and to construct mathematical models of physical laws. Since it is impossible to add up infinitely many numbers directly, one goal will be to define exactly what we mean by the sum of an infinite series. However, unlike finite sums, it turns out that not all infinite series actually have a sum, so we will need to develop tools for determining which infinite series have sums and which do not. Once the basic ideas have been developed we will begin to apply our work; we will show how infinite series are used to evaluate such quantities as ln 2, e, sin 3°, and π, how they are used to create functions, and finally, how they are used to model physical laws.

Photo: ***Perspective creates the illusion that the sequence of telephone poles continues indefinitely but converges toward a single point infinitely far away.***

10.1 SEQUENCES

In everyday language, the term "sequence" means a succession of things in a definite order—chronological order, size order, or logical order, for example. In mathematics, the term "sequence" is commonly used to denote a succession of numbers whose order is determined by a rule or a function. In this section, we will develop some of the basic ideas concerning sequences of numbers.

DEFINITION OF A SEQUENCE

Stated informally, an ***infinite sequence***, or more simply a ***sequence***, is an unending succession of numbers, called ***terms***. It is understood that the terms have a definite order; that is, there is a first term a_1, a second term a_2, a third term a_3, a fourth term a_4, and so forth. Such a sequence would typically be written as

$$a_1, a_2, a_3, a_4, \ldots$$

where the dots are used to indicate that the sequence continues indefinitely. Some specific examples are

$$1, 2, 3, 4, \ldots, \qquad 1, \tfrac{1}{2}, \tfrac{1}{3}, \tfrac{1}{4}, \ldots,$$
$$2, 4, 6, 8, \ldots, \qquad 1, -1, 1, -1, \ldots$$

Each of these sequences has a definite pattern that makes it easy to generate additional terms if we assume that those terms follow the same pattern as the displayed terms. However, such patterns can be deceiving, so it is better to have a rule or formula for generating the terms. One way of doing this is to look for a function that relates each term in the sequence to its term number. For example, in the sequence

$$2, 4, 6, 8, \ldots$$

each term is twice the term number; that is, the nth term in the sequence is given by the formula $2n$. We denote this by writing the sequence as

$$2, 4, 6, 8, \ldots, 2n, \ldots$$

We call the function $f(n) = 2n$ the *general term* of this sequence. Now, if we want to know a specific term in the sequence, we need only substitute its term number in the formula for the general term. For example, the 37th term in the sequence is $2 \cdot 37 = 74$.

► **Example 1** In each part, find the general term of the sequence.

(a) $\frac{1}{2}, \frac{2}{3}, \frac{3}{4}, \frac{4}{5}, \ldots$ (b) $\frac{1}{2}, \frac{1}{4}, \frac{1}{8}, \frac{1}{16}, \ldots$

(c) $\frac{1}{2}, -\frac{2}{3}, \frac{3}{4}, -\frac{4}{5}, \ldots$ (d) $1, 3, 5, 7, \ldots$

Table 10.1.1

TERM NUMBER	1	2	3	4	⋯	n	⋯
TERM	$\frac{1}{2}$	$\frac{2}{3}$	$\frac{3}{4}$	$\frac{4}{5}$	⋯	$\frac{n}{n+1}$	⋯

Solution (a). In Table 10.1.1, the four known terms have been placed below their term numbers, from which we see that the numerator is the same as the term number and the denominator is one greater than the term number. This suggests that the nth term has numerator n and denominator $n + 1$, as indicated in the table. Thus, the sequence can be expressed as

$$\frac{1}{2}, \frac{2}{3}, \frac{3}{4}, \frac{4}{5}, \ldots, \frac{n}{n+1}, \ldots$$

Table 10.1.2

TERM NUMBER	1	2	3	4	⋯	n	⋯
TERM	$\frac{1}{2}$	$\frac{1}{2^2}$	$\frac{1}{2^3}$	$\frac{1}{2^4}$	⋯	$\frac{1}{2^n}$	⋯

Solution (b). In Table 10.1.2, the denominators of the four known terms have been expressed as powers of 2 and the first four terms have been placed below their term numbers, from which we see that the exponent in the denominator is the same as the term number. This suggests that the denominator of the nth term is 2^n, as indicated in the table. Thus, the sequence can be expressed as

$$\frac{1}{2}, \frac{1}{4}, \frac{1}{8}, \frac{1}{16}, \ldots, \frac{1}{2^n}, \ldots$$

Solution (c). This sequence is identical to that in part (a), except for the alternating signs. Thus, the nth term in the sequence can be obtained by multiplying the nth term in part (a) by $(-1)^{n+1}$. This factor produces the correct alternating signs, since its successive values, starting with $n = 1$, are $1, -1, 1, -1, \ldots$. Thus, the sequence can be written as

$$\frac{1}{2}, -\frac{2}{3}, \frac{3}{4}, -\frac{4}{5}, \ldots, (-1)^{n+1}\frac{n}{n+1}, \ldots$$

Table 10.1.3

TERM NUMBER	1	2	3	4	⋯	n	⋯
TERM	1	3	5	7	⋯	$2n-1$	⋯

Solution (d). In Table 10.1.3, the four known terms have been placed below their term numbers, from which we see that each term is one less than twice its term number. This suggests that the nth term in the sequence is $2n - 1$, as indicated in the table. Thus, the sequence can be expressed as

$$1, 3, 5, 7, \ldots, 2n - 1, \ldots \blacktriangleleft$$

When the general term of a sequence

$$a_1, a_2, a_3, \ldots, a_n, \ldots \quad (1)$$

is known, there is no need to write out the initial terms, and it is common to write only the general term enclosed in braces. Thus, (1) might be written as

$$\{a_n\}_{n=1}^{+\infty} \quad \text{or as} \quad \{a_n\}_{n=1}^{\infty}$$

For example, here are the four sequences in Example 1 expressed in brace notation.

Consider the sequence whose general term is

$$f(n) = \tfrac{1}{3}(3 - 5n + 6n^2 - n^3)$$

Calculate the first three terms, and make a conjecture about the fourth term. Check your conjecture by calculating the fourth term. What message does this convey?

SEQUENCE	BRACE NOTATION
$\frac{1}{2}, \frac{2}{3}, \frac{3}{4}, \frac{4}{5}, \ldots, \frac{n}{n+1}, \ldots$	$\left\{\frac{n}{n+1}\right\}_{n=1}^{+\infty}$
$\frac{1}{2}, \frac{1}{4}, \frac{1}{8}, \frac{1}{16}, \ldots, \frac{1}{2^n}, \ldots$	$\left\{\frac{1}{2^n}\right\}_{n=1}^{+\infty}$
$\frac{1}{2}, -\frac{2}{3}, \frac{3}{4}, -\frac{4}{5}, \ldots, (-1)^{n+1}\frac{n}{n+1}, \ldots$	$\left\{(-1)^{n+1}\frac{n}{n+1}\right\}_{n=1}^{+\infty}$
$1, 3, 5, 7, \ldots, 2n-1, \ldots$	$\{2n-1\}_{n=1}^{+\infty}$

The letter n in (1) is called the ***index*** for the sequence. It is not essential to use n for the index; any letter not reserved for another purpose can be used. For example, we might view the general term of the sequence $a_1, a_2, a_3, \ldots$ to be the kth term, in which case we would denote this sequence as $\{a_k\}_{k=1}^{+\infty}$. Moreover, it is not essential to start the index at 1; sometimes it is more convenient to start it at 0 (or some other integer). For example, consider the sequence

$$1, \frac{1}{2}, \frac{1}{2^2}, \frac{1}{2^3}, \ldots$$

One way to write this sequence is

$$\left\{\frac{1}{2^{n-1}}\right\}_{n=1}^{+\infty}$$

However, the general term will be simpler if we think of the initial term in the sequence as the zeroth term, in which case we can write the sequence as

$$\left\{\frac{1}{2^n}\right\}_{n=0}^{+\infty}$$

We began this section by describing a sequence as an unending succession of numbers. Although this conveys the general idea, it is not a satisfactory mathematical definition because it relies on the term "succession," which is itself an undefined term. To motivate a precise definition, consider the sequence

$$2, 4, 6, 8, \ldots, 2n, \ldots$$

If we denote the general term by $f(n) = 2n$, then we can write this sequence as

$$f(1), f(2), f(3), \ldots, f(n), \ldots$$

which is a "list" of values of the function

$$f(n) = 2n, \quad n = 1, 2, 3, \ldots$$

whose domain is the set of positive integers. This suggests the following definition.

10.1.1 DEFINITION. A ***sequence*** is a function whose domain is a set of integers. Specifically, we will regard the expression $\{a_n\}_{n=1}^{+\infty}$ to be an alternative notation for the function $f(n) = a_n$, $n = 1, 2, 3, \ldots$.

GRAPHS OF SEQUENCES

When the starting value for the index of a sequence is not relevant to the discussion, it is common to use a notation such as $\{a_n\}$ in which there is no reference to the starting value of n. We can distinguish between different sequences by using different letters for their general terms; thus, $\{a_n\}$, $\{b_n\}$, and $\{c_n\}$ denote three different sequences.

Since sequences are functions, it makes sense to talk about the graph of a sequence. For example, the graph of the sequence $\{1/n\}_{n=1}^{+\infty}$ is the graph of the equation

$$y = \frac{1}{n}, \quad n = 1, 2, 3, \ldots$$

Because the right side of this equation is defined only for positive integer values of n, the graph consists of a succession of isolated points (Figure 10.1.1*a*). This is in distinction to the graph of

$$y = \frac{1}{x}, \quad x \geq 1$$

which is a continuous curve (Figure 10.1.1*b*).

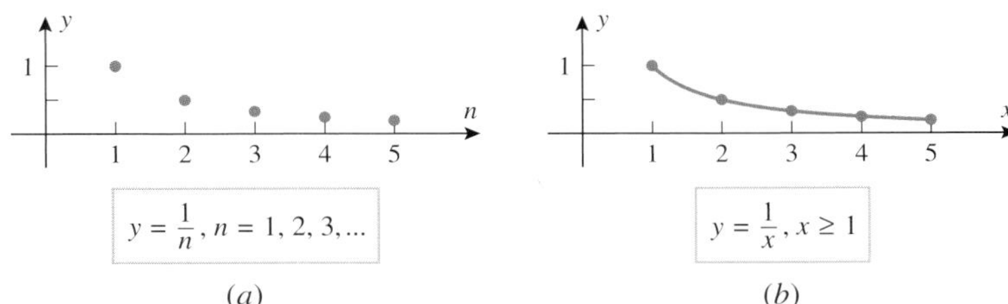

Figure 10.1.1 (*a*) (*b*)

LIMIT OF A SEQUENCE

Since sequences are functions, we can inquire about their limits. However, because a sequence $\{a_n\}$ is only defined for integer values of n, the only limit that makes sense is the limit of a_n as $n \to +\infty$. In Figure 10.1.2 we have shown the graphs of four sequences, each of which behaves differently as $n \to +\infty$:

- The terms in the sequence $\{n + 1\}$ increase without bound.
- The terms in the sequence $\{(-1)^{n+1}\}$ oscillate between -1 and 1.
- The terms in the sequence $\{n/(n + 1)\}$ increase toward a "limiting value" of 1.
- The terms in the sequence $\left\{1 + \left(-\frac{1}{2}\right)^n\right\}$ also tend toward a "limiting value" of 1, but do so in an oscillatory fashion.

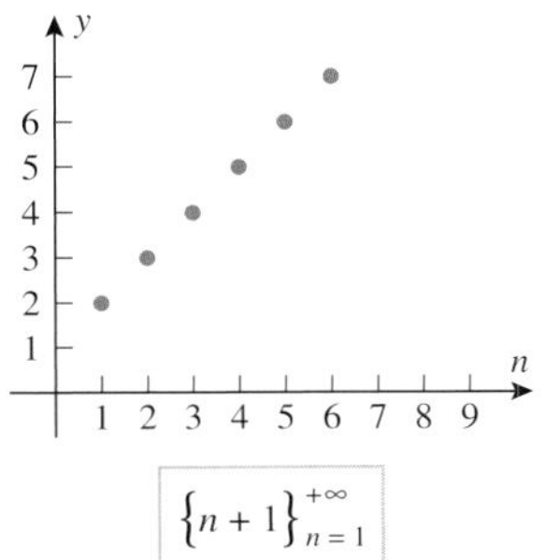

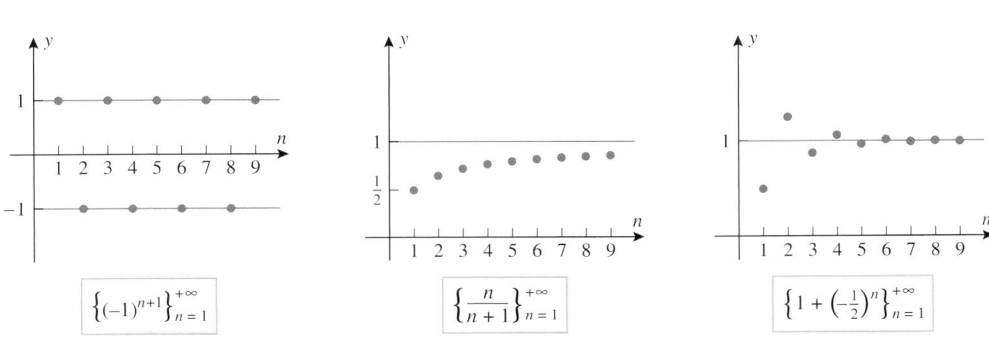

Figure 10.1.2

Informally speaking, the limit of a sequence $\{a_n\}$ is intended to describe how a_n behaves as $n \to +\infty$. To be more specific, we will say that *a sequence* $\{a_n\}$ *approaches a limit* L *if the terms in the sequence eventually become arbitrarily close to* L. Geometrically, this means that for any positive number ϵ there is a point in the sequence after which all terms lie between the lines $y = L - \epsilon$ and $y = L + \epsilon$ (Figure 10.1.3).

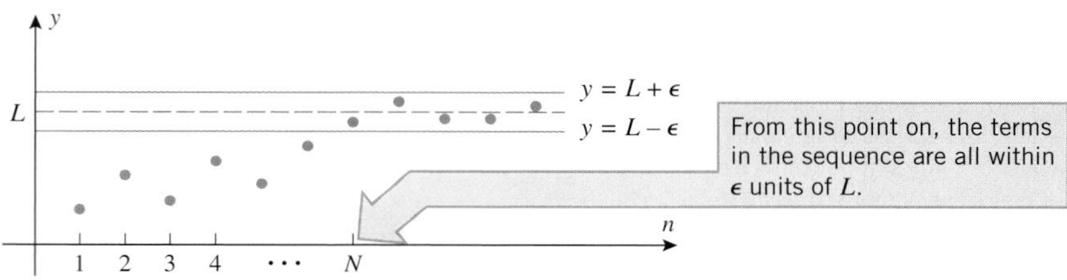

Figure 10.1.3

The following definition makes these ideas precise.

How would you define

$$\lim_{n\to+\infty} a_n = +\infty$$

and

$$\lim_{n\to+\infty} a_n = -\infty?$$

10.1.2 DEFINITION. A sequence $\{a_n\}$ is said to ***converge*** to the ***limit*** L if given any $\epsilon > 0$, there is a positive integer N such that $|a_n - L| < \epsilon$ for $n \geq N$. In this case we write

$$\lim_{n\to+\infty} a_n = L$$

A sequence that does not converge to some finite limit is said to ***diverge***.

▶ **Example 2** The first two sequences in Figure 10.1.2 diverge, and the second two converge to 1; that is,

$$\lim_{n\to+\infty} \frac{n}{n+1} = 1 \quad \text{and} \quad \lim_{n\to+\infty} \left[1 + \left(-\tfrac{1}{2}\right)^n\right] = 1 \; ◀$$

The following theorem, which we state without proof, shows that the familiar properties of limits apply to sequences. This theorem ensures that the algebraic techniques used to find limits of the form $\lim_{x\to+\infty}$ can also be used for limits of the form $\lim_{n\to+\infty}$.

10.1.3 THEOREM. *Suppose that the sequences $\{a_n\}$ and $\{b_n\}$ converge to limits L_1 and L_2, respectively, and c is a constant. Then:*

(a) $\lim_{n\to+\infty} c = c$

(b) $\lim_{n\to+\infty} ca_n = c \lim_{n\to+\infty} a_n = cL_1$

(c) $\lim_{n\to+\infty} (a_n + b_n) = \lim_{n\to+\infty} a_n + \lim_{n\to+\infty} b_n = L_1 + L_2$

(d) $\lim_{n\to+\infty} (a_n - b_n) = \lim_{n\to+\infty} a_n - \lim_{n\to+\infty} b_n = L_1 - L_2$

(e) $\lim_{n\to+\infty} (a_n b_n) = \lim_{n\to+\infty} a_n \cdot \lim_{n\to+\infty} b_n = L_1 L_2$

(f) $\lim_{n\to+\infty} \left(\frac{a_n}{b_n}\right) = \dfrac{\lim_{n\to+\infty} a_n}{\lim_{n\to+\infty} b_n} = \dfrac{L_1}{L_2}$ (if $L_2 \neq 0$)

Additional limit properties follow from those in Theorem 10.1.3. For example, use part (e) to show that if $a_n \to L$ and m is a positive integer, then

$$\lim_{n\to+\infty} (a_n)^m = L^m$$

▸ **Example 3** In each part, determine whether the sequence converges or diverges. If it converges, find the limit.

(a) $\left\{\dfrac{n}{2n+1}\right\}_{n=1}^{+\infty}$ (b) $\left\{(-1)^{n+1}\dfrac{n}{2n+1}\right\}_{n=1}^{+\infty}$

(c) $\left\{(-1)^{n+1}\dfrac{1}{n}\right\}_{n=1}^{+\infty}$ (d) $\{8-2n\}_{n=1}^{+\infty}$

Solution (a). Dividing numerator and denominator by n yields

$$\lim_{n\to+\infty}\frac{n}{2n+1}=\lim_{n\to+\infty}\frac{1}{2+1/n}=\frac{\lim\limits_{n\to+\infty}1}{\lim\limits_{n\to+\infty}(2+1/n)}=\frac{\lim\limits_{n\to+\infty}1}{\lim\limits_{n\to+\infty}2+\lim\limits_{n\to+\infty}1/n}$$

$$=\frac{1}{2+0}=\frac{1}{2}$$

Thus, the sequence converges to $\frac{1}{2}$.

Solution (b). This sequence is the same as that in part (a), except for the factor of $(-1)^{n+1}$, which oscillates between $+1$ and -1. Thus, the terms in this sequence oscillate between positive and negative values, with the odd-numbered terms being identical to those in part (a) and the even-numbered terms being the negatives of those in part (a). Since the sequence in part (a) has a limit of $\frac{1}{2}$, it follows that the odd-numbered terms in this sequence approach $\frac{1}{2}$, and the even-numbered terms approach $-\frac{1}{2}$. Therefore, this sequence has no limit—it diverges.

Solution (c). Since $\lim_{n\to+\infty}1/n=0$, the product $(-1)^{n+1}(1/n)$ oscillates between positive and negative values, with the odd-numbered terms approaching 0 through positive values and the even-numbered terms approaching 0 through negative values. Thus,

$$\lim_{n\to+\infty}(-1)^{n+1}\frac{1}{n}=0$$

so the sequence converges to 0.

Solution (d). $\lim\limits_{n\to+\infty}(8-2n)=-\infty$, so the sequence $\{8-2n\}_{n=1}^{+\infty}$ diverges. ◂

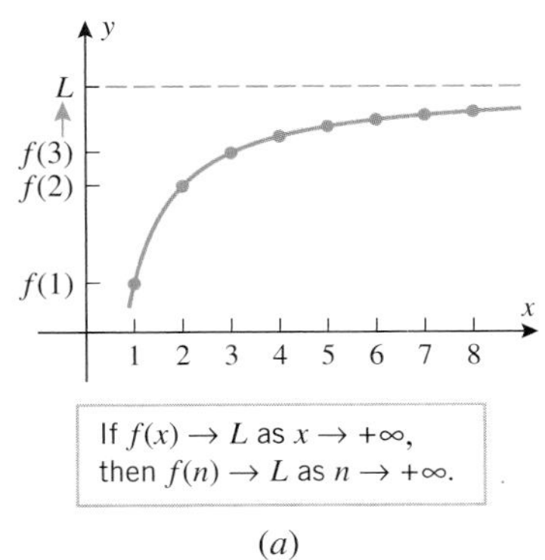

(*a*)

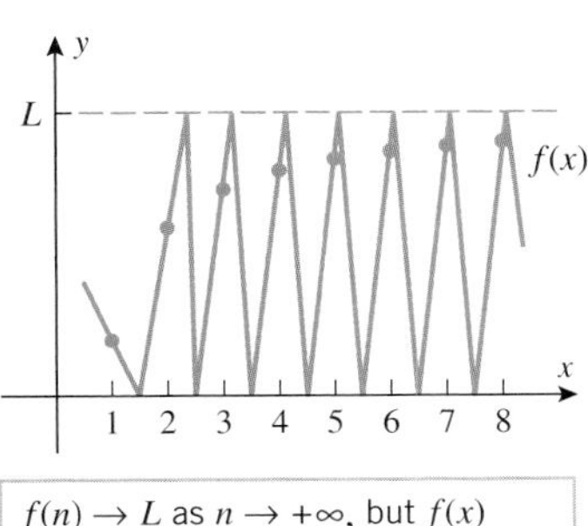

(*b*)

Figure 10.1.4

If the general term of a sequence is $f(n)$, where $f(x)$ is a function defined on the entire interval $[1,+\infty)$, then the values of $f(n)$ can be viewed as "sample values" of $f(x)$ taken at the positive integers. Thus,

$$\text{if } f(x)\to L \text{ as } x\to+\infty, \quad \text{then} \quad f(n)\to L \text{ as } n\to+\infty$$

(Figure 10.1.4*a*). However, the converse is not true; that is, one cannot infer that $f(x)\to L$ as $x\to+\infty$ from the fact that $f(n)\to L$ as $n\to+\infty$ (Figure 10.1.4*b*).

▸ **Example 4** In each part, determine whether the sequence converges, and if so, find its limit.

(a) $1, \dfrac{1}{2}, \dfrac{1}{2^2}, \dfrac{1}{2^3}, \ldots, \dfrac{1}{2^n}, \ldots$ (b) $1, 2, 2^2, 2^3, \ldots, 2^n, \ldots$

Solution. Replacing n by x in the first sequence produces the power function $(1/2)^x$, and replacing n by x in the second sequence produces the power function 2^x. Now recall that if

$0 < b < 1$, then $b^x \to 0$ as $x \to +\infty$, and if $b > 1$, then $b^x \to +\infty$ as $x \to +\infty$ (Figure 7.1.1). Thus,

$$\lim_{n\to+\infty} \frac{1}{2^n} = 0 \quad \text{and} \quad \lim_{n\to+\infty} 2^n = +\infty \blacktriangleleft$$

► **Example 5** Find the limit of the sequence $\left\{\frac{n}{e^n}\right\}_{n=1}^{+\infty}$.

Solution. The expression n/e^n is an indeterminate form of type ∞/∞ as $n \to +\infty$, so L'Hôpital's rule is indicated. However, we cannot apply this rule directly to n/e^n because the functions n and e^n have been defined here only at the positive integers, and hence are not differentiable functions. To circumvent this problem we extend the domains of these functions to all real numbers, here implied by replacing n by x, and apply L'Hôpital's rule to the limit of the quotient x/e^x. This yields

$$\lim_{x\to+\infty} \frac{x}{e^x} = \lim_{x\to+\infty} \frac{1}{e^x} = 0$$

from which we can conclude that

$$\lim_{n\to+\infty} \frac{n}{e^n} = 0 \blacktriangleleft$$

► **Example 6** Show that $\lim_{n\to+\infty} \sqrt[n]{n} = 1$.

Solution.

$$\lim_{n\to+\infty} \sqrt[n]{n} = \lim_{n\to+\infty} n^{1/n} = \lim_{n\to+\infty} e^{(1/n)\ln n} = e^0 = 1 \qquad \boxed{\text{By L'Hôpital's rule applied to } (1/x)\ln x} \blacktriangleleft$$

Sometimes the even-numbered and odd-numbered terms of a sequence behave sufficiently differently that it is desirable to investigate their convergence separately. The following theorem, whose proof is omitted, is helpful for that purpose.

> **10.1.4 THEOREM.** *A sequence converges to a limit L if and only if the sequences of even-numbered terms and odd-numbered terms both converge to L.*

► **Example 7** The sequence

$$\frac{1}{2}, \frac{1}{3}, \frac{1}{2^2}, \frac{1}{3^2}, \frac{1}{2^3}, \frac{1}{3^3}, \ldots$$

converges to 0, since the even-numbered terms and the odd-numbered terms both converge to 0, and the sequence

$$1, \tfrac{1}{2}, 1, \tfrac{1}{3}, 1, \tfrac{1}{4}, \ldots$$

diverges, since the odd-numbered terms converge to 1 and the even-numbered terms converge to 0. ◄

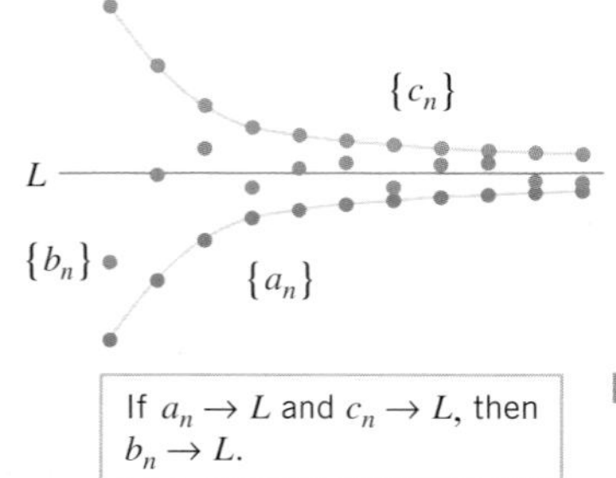

If $a_n \to L$ and $c_n \to L$, then $b_n \to L$.

Figure 10.1.5

THE SQUEEZING THEOREM FOR SEQUENCES

The following theorem, illustrated in Figure 10.1.5, is an adaptation of the Squeezing Theorem (2.6.2) to sequences. This theorem will be useful for finding limits of sequences that cannot be obtained directly. The proof is omitted.

> **10.1.5 THEOREM (*The Squeezing Theorem for Sequences*).** *Let $\{a_n\}$, $\{b_n\}$, and $\{c_n\}$ be sequences such that*
>
> $$a_n \le b_n \le c_n \quad (\text{for all values of } n \text{ beyond some index } N)$$
>
> *If the sequences $\{a_n\}$ and $\{c_n\}$ have a common limit L as $n \to +\infty$, then $\{b_n\}$ also has the limit L as $n \to +\infty$.*

Recall that if n is a positive integer, then $n!$ (read "n factorial") is the product of the first n positive integers. In addition, it is convenient to define $0! = 1$.

▶ Example 8 Use numerical evidence to make a conjecture about the limit of the sequence

$$\left\{\frac{n!}{n^n}\right\}_{n=1}^{+\infty}$$

and then confirm that your conjecture is correct.

Table 10.1.4

n	$\frac{n!}{n^n}$
1	1.0000000000
2	0.5000000000
3	0.2222222222
4	0.0937500000
5	0.0384000000
6	0.0154320988
7	0.0061198990
8	0.0024032593
9	0.0009366567
10	0.0003628800
11	0.0001399059
12	0.0000537232

Solution. Table 10.1.4, which was obtained with a calculating utility, suggests that the limit of the sequence may be 0. To confirm this we need to examine the limit of

$$a_n = \frac{n!}{n^n}$$

as $n \to +\infty$. Although this is an indeterminate form of type ∞/∞, L'Hôpital's rule is not helpful because we have no definition of $x!$ for values of x that are not integers. However, let us write out some of the initial terms and the general term in the sequence:

$$a_1 = 1, \quad a_2 = \frac{1 \cdot 2}{2 \cdot 2}, \quad a_3 = \frac{1 \cdot 2 \cdot 3}{3 \cdot 3 \cdot 3}, \ldots, \quad a_n = \frac{1 \cdot 2 \cdot 3 \cdots n}{n \cdot n \cdot n \cdots n}, \ldots$$

We can rewrite the general term as

$$a_n = \frac{1}{n}\left(\frac{2 \cdot 3 \cdots n}{n \cdot n \cdots n}\right)$$

from which it is evident that

$$0 \le a_n \le \frac{1}{n}$$

However, the two outside expressions have a limit of 0 as $n \to +\infty$; thus, the Squeezing Theorem for Sequences implies that $a_n \to 0$ as $n \to +\infty$, which confirms our conjecture. ◀

The following theorem is often useful for finding the limit of a sequence with both positive and negative terms—it states that if the sequence $\{|a_n|\}$ that is obtained by taking the absolute value of each term in the sequence $\{a_n\}$ converges to 0, then $\{a_n\}$ also converges to 0.

> **10.1.6 THEOREM.** *If* $\lim_{n\to+\infty} |a_n| = 0$, *then* $\lim_{n\to+\infty} a_n = 0$.

PROOF. Depending on the sign of a_n, either $a_n = |a_n|$ or $a_n = -|a_n|$. Thus, in all cases we have

$$-|a_n| \le a_n \le |a_n|$$

However, the limit of the two outside terms is 0, and hence the limit of a_n is 0 by the Squeezing Theorem for Sequences. ■

▶ **Example 9** Consider the sequence

$$1, -\frac{1}{2}, \frac{1}{2^2}, -\frac{1}{2^3}, \ldots, (-1)^n \frac{1}{2^n}, \ldots$$

If we take the absolute value of each term, we obtain the sequence

$$1, \frac{1}{2}, \frac{1}{2^2}, \frac{1}{2^3}, \ldots, \frac{1}{2^n}, \ldots$$

which, as shown in Example 4, converges to 0. Thus, from Theorem 10.1.6 we have

$$\lim_{n\to+\infty} \left[(-1)^n \frac{1}{2^n}\right] = 0 \blacktriangleleft$$

SEQUENCES DEFINED RECURSIVELY

Some sequences do not arise from a formula for the general term, but rather from a formula or set of formulas that specify how to generate each term in the sequence from terms that precede it; such sequences are said to be defined ***recursively***, and the defining formulas are called ***recursion formulas***. A good example is the mechanic's rule for approximating square roots. In Exercise 19 of Section 4.6 you were asked to show that

$$x_1 = 1, \quad x_{n+1} = \frac{1}{2}\left(x_n + \frac{a}{x_n}\right) \tag{2}$$

describes the sequence produced by Newton's Method to approximate $\sqrt{a}$ as a zero of the function $f(x) = x^2 - a$. Table 10.1.5 shows the first five terms in an application of the mechanic's rule to approximate $\sqrt{2}$.

Table 10.1.5

n	$x_1 = 1, \quad x_{n+1} = \frac{1}{2}\left(x_n + \frac{2}{x_n}\right)$	DECIMAL APPROXIMATION
	$x_1 = 1$ (Starting value)	1.00000000000
1	$x_2 = \frac{1}{2}\left[1 + \frac{2}{1}\right] = \frac{3}{2}$	1.50000000000
2	$x_3 = \frac{1}{2}\left[\frac{3}{2} + \frac{2}{3/2}\right] = \frac{17}{12}$	1.41666666667
3	$x_4 = \frac{1}{2}\left[\frac{17}{12} + \frac{2}{17/12}\right] = \frac{577}{408}$	1.41421568627
4	$x_5 = \frac{1}{2}\left[\frac{577}{408} + \frac{2}{577/408}\right] = \frac{665{,}857}{470{,}832}$	1.41421356237
5	$x_6 = \frac{1}{2}\left[\frac{665{,}857}{470{,}832} + \frac{2}{665{,}857/470{,}832}\right] = \frac{886{,}731{,}088{,}897}{627{,}013{,}566{,}048}$	1.41421356237

It would take us too far afield to investigate the convergence of sequences defined recursively, but we will conclude this section with a useful technique that can sometimes be used to compute limits of such sequences.

▶ **Example 10** Assuming that the sequence in Table 10.1.5 converges, show that the limit is $\sqrt{2}$.

Solution. Assume that $x_n \to L$, where L is to be determined. Since $n+1 \to +\infty$ as $n \to +\infty$, it is also true that $x_{n+1} \to L$ as $n \to +\infty$. Thus, if we take the limit of the expression

$$x_{n+1} = \frac{1}{2}\left(x_n + \frac{2}{x_n}\right)$$

as $n \to +\infty$, we obtain

$$L = \frac{1}{2}\left(L + \frac{2}{L}\right)$$

which can be rewritten as $L^2 = 2$. The negative solution of this equation is extraneous because $x_n > 0$ for all n, so $L = \sqrt{2}$. ◀

✔QUICK CHECK EXERCISES 10.1 *(See page 635 for answers.)*

1. Consider the sequence 4, 6, 8, 10, 12, ...
   (a) If $\{a_n\}_{n=1}^{+\infty}$ denotes this sequence, then $a_1 =$ ______, $a_4 =$ ______, and $a_7 =$ ______. The general term is $a_n =$ ______.
   (b) If $\{b_n\}_{n=0}^{+\infty}$ denotes this sequence, then $b_0 =$ ______, $b_4 =$ ______, and $b_8 =$ ______. The general term is $b_n =$ ______.

2. What does it mean to say that a sequence $\{a_n\}$ *converges*?

3. Consider the sequences $\{a_n\}$ and $\{b_n\}$, where
$$a_n = \frac{n(2n+1)}{n^2} \quad \text{and} \quad b_n = \frac{(-1)^n}{5}$$
Determine which of the following sequences converge and which diverge. If a sequence converges, indicate its limit.
(a) $\{a_n\}$ (b) $\{b_n\}$ (c) $\{3a_n - 1\}$ (d) $\{b_n^2\}$
(e) $\{a_n + b_n\}$ (f) $\{1/a_n\}$ (g) $\{a_n/b_n\}$

4. Let f be the function $f(x) = \cos\left(\frac{\pi}{2}x\right)$ and define sequences $\{a_n\}$ and $\{b_n\}$ by $a_n = f(2n)$ and $b_n = f(2n+1)$.
   (a) Does $\lim_{x \to +\infty} f(x)$ exist?
   (b) $a_1 =$ ______, $a_2 =$ ______, $a_3 =$ ______, $a_4 =$ ______.
   (c) Does $\{a_n\}$ converge?
   (d) $b_1 =$ ______, $b_2 =$ ______, $b_3 =$ ______, $b_4 =$ ______.
   (e) Does $\{b_n\}$ converge?
   (f) Does the sequence $\{f(n)\}$ converge?

5. Suppose that $\{a_n\}$, $\{b_n\}$, and $\{c_n\}$ are sequences such that $a_n \le b_n \le c_n$ for all $n \ge 10$, and that $\{a_n\}$ and $\{c_n\}$ both converge to 12. Then the ______ Theorem for Sequences implies that $\{b_n\}$ converges to ______.

EXERCISE SET 10.1  

1. In each part, find a formula for the general term of the sequence, starting with $n = 1$.
   (a) $1, \frac{1}{3}, \frac{1}{9}, \frac{1}{27}, \ldots$ (b) $1, -\frac{1}{3}, \frac{1}{9}, -\frac{1}{27}, \ldots$
   (c) $\frac{1}{2}, \frac{3}{4}, \frac{5}{6}, \frac{7}{8}, \ldots$ (d) $\frac{1}{\sqrt{\pi}}, \frac{4}{\sqrt[3]{\pi}}, \frac{9}{\sqrt[4]{\pi}}, \frac{16}{\sqrt[5]{\pi}}, \ldots$

2. In each part, find two formulas for the general term of the sequence, one starting with $n = 1$ and the other with $n = 0$.
   (a) $1, -r, r^2, -r^3, \ldots$ (b) $r, -r^2, r^3, -r^4, \ldots$

3. (a) Write out the first four terms of the sequence $\{1 + (-1)^n\}$, starting with $n = 0$.
   (b) Write out the first four terms of the sequence $\{\cos n\pi\}$, starting with $n = 0$.
   (c) Use the results in parts (a) and (b) to express the general term of the sequence 4, 0, 4, 0, ... in two different ways, starting with $n = 0$.

4. In each part, find a formula for the general term using factorials and starting with $n = 1$.
   (a) $1 \cdot 2, 1 \cdot 2 \cdot 3 \cdot 4, 1 \cdot 2 \cdot 3 \cdot 4 \cdot 5 \cdot 6, 1 \cdot 2 \cdot 3 \cdot 4 \cdot 5 \cdot 6 \cdot 7 \cdot 8, \ldots$
   (b) $1, 1 \cdot 2 \cdot 3, 1 \cdot 2 \cdot 3 \cdot 4 \cdot 5, 1 \cdot 2 \cdot 3 \cdot 4 \cdot 5 \cdot 6 \cdot 7, \ldots$

5–22 Write out the first five terms of the sequence, determine whether the sequence converges, and if so find its limit.

5. $\left\{\frac{n}{n+2}\right\}_{n=1}^{+\infty}$
6. $\left\{\frac{n^2}{2n+1}\right\}_{n=1}^{+\infty}$
7. $\{2\}_{n=1}^{+\infty}$
8. $\left\{\ln\left(\frac{1}{n}\right)\right\}_{n=1}^{+\infty}$
9. $\left\{\frac{\ln n}{n}\right\}_{n=1}^{+\infty}$
10. $\left\{n \sin \frac{\pi}{n}\right\}_{n=1}^{+\infty}$
11. $\{1 + (-1)^n\}_{n=1}^{+\infty}$
12. $\left\{\frac{(-1)^{n+1}}{n^2}\right\}_{n=1}^{+\infty}$

13. $\left\{(-1)^n \dfrac{2n^3}{n^3+1}\right\}_{n=1}^{+\infty}$ 14. $\left\{\dfrac{n}{2^n}\right\}_{n=1}^{+\infty}$

15. $\left\{\dfrac{(n+1)(n+2)}{2n^2}\right\}_{n=1}^{+\infty}$ 16. $\left\{\dfrac{\pi^n}{4^n}\right\}_{n=1}^{+\infty}$

17. $\left\{\cos \dfrac{3}{n}\right\}_{n=1}^{+\infty}$ 18. $\left\{\cos \dfrac{\pi n}{2}\right\}_{n=1}^{+\infty}$

19. $\{n^2 e^{-n}\}_{n=1}^{+\infty}$ 20. $\{\sqrt{n^2+3n} - n\}_{n=1}^{+\infty}$

21. $\left\{\left(\dfrac{n+3}{n+1}\right)^n\right\}_{n=1}^{+\infty}$ 22. $\left\{\left(1-\dfrac{2}{n}\right)^n\right\}_{n=1}^{+\infty}$

23–30 Find the general term of the sequence, starting with $n = 1$, determine whether the sequence converges, and if so find its limit.

23. $\dfrac{1}{2}, \dfrac{3}{4}, \dfrac{5}{6}, \dfrac{7}{8}, \ldots$ 24. $0, \dfrac{1}{2^2}, \dfrac{2}{3^2}, \dfrac{3}{4^2}, \ldots$

25. $\dfrac{1}{3}, -\dfrac{1}{9}, \dfrac{1}{27}, -\dfrac{1}{81}, \ldots$ 26. $-1, 2, -3, 4, -5, \ldots$

27. $\left(1-\dfrac{1}{2}\right), \left(\dfrac{1}{3}-\dfrac{1}{2}\right), \left(\dfrac{1}{3}-\dfrac{1}{4}\right), \left(\dfrac{1}{5}-\dfrac{1}{4}\right), \ldots$

28. $3, \dfrac{3}{2}, \dfrac{3}{2^2}, \dfrac{3}{2^3}, \ldots$

29. $(\sqrt{2}-\sqrt{3}), (\sqrt{3}-\sqrt{4}), (\sqrt{4}-\sqrt{5}), \ldots$

30. $\dfrac{1}{3^5}, -\dfrac{1}{3^6}, \dfrac{1}{3^7}, -\dfrac{1}{3^8}, \ldots$

FOCUS ON CONCEPTS

31. Give two examples of sequences, all of whose terms are between -10 and 10, that do not converge. Use graphs of your sequences to explain their properties.

32. (a) Suppose that f satisfies $\lim_{x\to 0^+} f(x) = +\infty$. Is it possible that the sequence $\{f(1/n)\}$ converges? Explain.
(b) Find a function f such that $\lim_{x\to 0^+} f(x)$ does not exist but the sequence $\{f(1/n)\}$ converges.

33. (a) Starting with $n = 1$, write out the first six terms of the sequence $\{a_n\}$, where

$$a_n = \begin{cases} 1, & \text{if } n \text{ is odd} \\ n, & \text{if } n \text{ is even} \end{cases}$$

(b) Starting with $n = 1$, and considering the even and odd terms separately, find a formula for the general term of the sequence

$$1, \frac{1}{2^2}, 3, \frac{1}{2^4}, 5, \frac{1}{2^6}, \ldots$$

(c) Starting with $n = 1$, and considering the even and odd terms separately, find a formula for the general term of the sequence

$$1, \frac{1}{3}, \frac{1}{3}, \frac{1}{5}, \frac{1}{5}, \frac{1}{7}, \frac{1}{7}, \frac{1}{9}, \frac{1}{9}, \ldots$$

(d) Determine whether the sequences in parts (a), (b), and (c) converge. For those that do, find the limit.

34. For what positive values of b does the sequence $b, 0, b^2, 0, b^3, 0, b^4, \ldots$ converge? Justify your answer.

C 35. (a) Use numerical evidence to make a conjecture about the limit of the sequence $\{\sqrt[n]{n^3}\}_{n=2}^{+\infty}$.
(b) Use a CAS to confirm your conjecture.

C 36. (a) Use numerical evidence to make a conjecture about the limit of the sequence $\{\sqrt[n]{3^n+n^3}\}_{n=2}^{+\infty}$.
(b) Use a CAS to confirm your conjecture.

37. Assuming that the sequence given in Formula (2) of this section converges, use the method of Example 10 to show that the limit of this sequence is $\sqrt{a}$.

38. Consider the sequence

$$a_1 = \sqrt{6}$$
$$a_2 = \sqrt{6+\sqrt{6}}$$
$$a_3 = \sqrt{6+\sqrt{6+\sqrt{6}}}$$
$$a_4 = \sqrt{6+\sqrt{6+\sqrt{6+\sqrt{6}}}}$$
$$\vdots$$

(a) Find a recursion formula for a_{n+1}.
(b) Assuming that the sequence converges, use the method of Example 10 to find the limit.

39. Consider the sequence $\{a_n\}_{n=1}^{+\infty}$, where

$$a_n = \frac{1}{n^2} + \frac{2}{n^2} + \cdots + \frac{n}{n^2}$$

(a) Find a_1, a_2, a_3, and a_4.
(b) Use numerical evidence to make a conjecture about the limit of the sequence.
(c) Confirm your conjecture by expressing a_n in closed form and calculating the limit.

40. Follow the directions in Exercise 39 with

$$a_n = \frac{1^2}{n^3} + \frac{2^2}{n^3} + \cdots + \frac{n^2}{n^3}$$

41–42 Use numerical evidence to make a conjecture about the limit of the sequence, and then use the Squeezing Theorem for Sequences (Theorem 10.1.5) to confirm that your conjecture is correct.

41. $\displaystyle\lim_{n\to+\infty} \frac{\sin^2 n}{n}$ 42. $\displaystyle\lim_{n\to+\infty} \left(\frac{1+n}{2n}\right)^n$

43. (a) A bored student enters the number 0.5 in a calculator display and then repeatedly computes the square of the number in the display. Taking $a_0 = 0.5$, find a formula for the general term of the sequence $\{a_n\}$ of numbers that appear in the display.

(b) Try this with a calculator and make a conjecture about the limit of a_n.
(c) Confirm your conjecture by finding the limit of a_n.
(d) For what values of a_0 will this procedure produce a convergent sequence?

44. Let
$$f(x) = \begin{cases} 2x, & 0 \le x < 0.5 \\ 2x - 1, & 0.5 \le x < 1 \end{cases}$$
Does the sequence $f(0.2)$, $f(f(0.2))$, $f(f(f(0.2)))$, ... converge? Justify your reasoning.

45. (a) Use a graphing utility to generate the graph of the equation $y = (2^x + 3^x)^{1/x}$, and then use the graph to make a conjecture about the limit of the sequence
$$\{(2^n + 3^n)^{1/n}\}_{n=1}^{+\infty}$$
(b) Confirm your conjecture by calculating the limit.

46. Consider the sequence $\{a_n\}_{n=1}^{+\infty}$ whose nth term is
$$a_n = \frac{1}{n}\sum_{k=1}^{n}\frac{1}{1+(k/n)}$$
Show that $\lim_{n\to+\infty} a_n = \ln 2$ by interpreting a_n as the Riemann sum of a definite integral.

47. Let a_n be the average value of $f(x) = 1/x$ over the interval $[1, n]$. Determine whether the sequence $\{a_n\}$ converges, and if so find its limit.

48. The sequence whose terms are 1, 1, 2, 3, 5, 8, 13, 21, ... is called the ***Fibonacci sequence*** in honor of the Italian mathematician Leonardo ("Fibonacci") da Pisa (c. 1170–1250). This sequence has the property that after starting with two 1's, each term is the sum of the preceding two.
(a) Denoting the sequence by $\{a_n\}$ and starting with $a_1 = 1$ and $a_2 = 1$, show that
$$\frac{a_{n+2}}{a_{n+1}} = 1 + \frac{a_n}{a_{n+1}} \quad \text{if } n \ge 1$$
(b) Give a reasonable informal argument to show that if the sequence $\{a_{n+1}/a_n\}$ converges to some limit L, then the sequence $\{a_{n+2}/a_{n+1}\}$ must also converge to L.
(c) Assuming that the sequence $\{a_{n+1}/a_n\}$ converges, show that its limit is $(1 + \sqrt{5})/2$.

49. If we accept the fact that the sequence $\{1/n\}_{n=1}^{+\infty}$ converges to the limit $L = 0$, then according to Definition 10.1.2, for every $\epsilon > 0$, there exists a positive integer N such that $|a_n - L| = |(1/n) - 0| < \epsilon$ when $n \ge N$. In each part, find the smallest possible value of N for the given value of ϵ.
(a) $\epsilon = 0.5$ (b) $\epsilon = 0.1$ (c) $\epsilon = 0.001$

50. If we accept the fact that the sequence
$$\left\{\frac{n}{n+1}\right\}_{n=1}^{+\infty}$$
converges to the limit $L = 1$, then according to Definition 10.1.2, for every $\epsilon > 0$ there exists an integer N such that
$$|a_n - L| = \left|\frac{n}{n+1} - 1\right| < \epsilon$$
when $n \ge N$. In each part, find the smallest value of N for the given value of ϵ.
(a) $\epsilon = 0.25$ (b) $\epsilon = 0.1$ (c) $\epsilon = 0.001$

51. Use Definition 10.1.2 to prove that
(a) the sequence $\{1/n\}_{n=1}^{+\infty}$ converges to 0
(b) the sequence $\left\{\frac{n}{n+1}\right\}_{n=1}^{+\infty}$ converges to 1.

52. Find $\lim_{n\to+\infty} r^n$, where r is a real number. [*Hint:* Consider the cases $|r| < 1$, $|r| > 1$, $r = 1$, and $r = -1$ separately.]

✔ QUICK CHECK ANSWERS 10.1

1. (a) 4; 10; 16; $2n + 2$ (b) 4; 12; 20; $2n + 4$ **2.** $\lim_{n\to+\infty} a_n$ exists **3.** (a) converges to 2 (b) diverges (c) converges to 5 (d) converges to $\frac{1}{25}$ (e) diverges (f) converges to $\frac{1}{2}$ (g) diverges **4.** (a) no (b) -1; 1; -1; 1 (c) no (d) 0; 0; 0; 0 (e) yes (to 0) (f) no **5.** Squeezing; 12

10.2 MONOTONE SEQUENCES

There are many situations in which it is important to know whether a sequence converges, but the value of the limit is not relevant to the problem at hand. In this section we will study several techniques that can be used to determine whether a sequence converges.

TERMINOLOGY

We begin with some terminology.

10.2.1 DEFINITION. A sequence $\{a_n\}_{n=1}^{+\infty}$ is called

strictly increasing if $a_1 < a_2 < a_3 < \cdots < a_n < \cdots$

increasing if $a_1 \le a_2 \le a_3 \le \cdots \le a_n \le \cdots$

strictly decreasing if $a_1 > a_2 > a_3 > \cdots > a_n > \cdots$

decreasing if $a_1 \ge a_2 \ge a_3 \ge \cdots \ge a_n \ge \cdots$

A sequence that is either increasing or decreasing is said to be ***monotone***, and a sequence that is either strictly increasing or strictly decreasing is said to be ***strictly monotone***.

Note that an increasing sequence need not be strictly increasing, and a decreasing sequence need not be strictly decreasing.

Some examples are given in Table 10.2.1 and their corresponding graphs are shown in Figure 10.2.1.

Table 10.2.1

SEQUENCE	DESCRIPTION
$\frac{1}{2}, \frac{2}{3}, \frac{3}{4}, \ldots, \frac{n}{n+1}, \ldots$	Strictly increasing
$1, \frac{1}{2}, \frac{1}{3}, \ldots, \frac{1}{n}, \ldots$	Strictly decreasing
$1, 1, 2, 2, 3, 3, \ldots$	Increasing; not strictly increasing
$1, 1, \frac{1}{2}, \frac{1}{2}, \frac{1}{3}, \frac{1}{3}, \ldots$	Decreasing; not strictly decreasing
$1, -\frac{1}{2}, \frac{1}{3}, -\frac{1}{4}, \ldots, (-1)^{n+1}\frac{1}{n}, \ldots$	Neither increasing nor decreasing

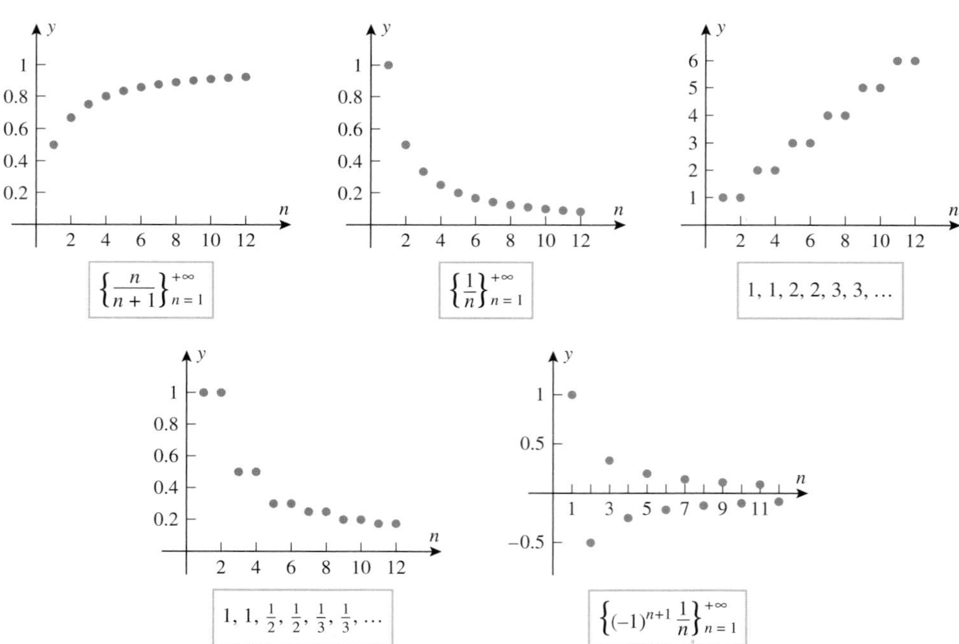

Figure 10.2.1

Can a sequence be both increasing and decreasing? Explain.

The first and second sequences in Table 10.2.1 are strictly monotone; the third and fourth sequences are monotone but not strictly monotone; and the fifth sequence is neither strictly monotone nor monotone.

TESTING FOR MONOTONICITY

Frequently, one can *guess* whether a sequence is monotone or strictly monotone by writing out some of the initial terms. However, to be certain that the guess is correct, one must give a precise mathematical argument. Table 10.2.2 provides two ways of doing this, one based on differences of successive terms and the other on ratios of successive terms. It is assumed in the latter case that the terms are positive. One must show that the specified conditions hold for *all* pairs of successive terms.

Table 10.2.2

DIFFERENCE BETWEEN SUCCESSIVE TERMS	RATIO OF SUCCESSIVE TERMS	CONCLUSION
$a_{n+1} - a_n > 0$	$a_{n+1}/a_n > 1$	Strictly increasing
$a_{n+1} - a_n < 0$	$a_{n+1}/a_n < 1$	Strictly decreasing
$a_{n+1} - a_n \geq 0$	$a_{n+1}/a_n \geq 1$	Increasing
$a_{n+1} - a_n \leq 0$	$a_{n+1}/a_n \leq 1$	Decreasing

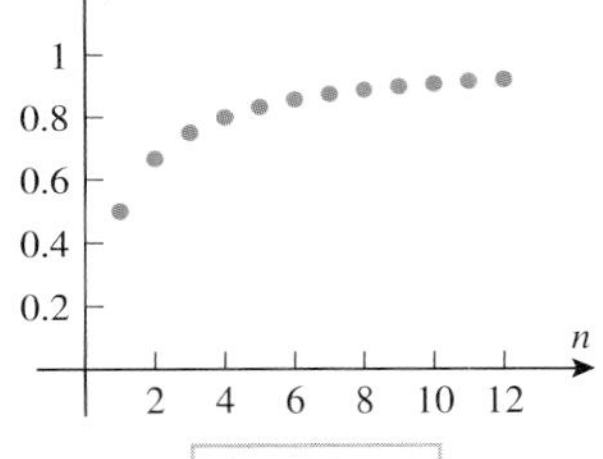

$\left\{\frac{n}{n+1}\right\}_{n=1}^{+\infty}$

Figure 10.2.2

▶ **Example 1** Use differences of successive terms to show that

$$\frac{1}{2}, \frac{2}{3}, \frac{3}{4}, \ldots, \frac{n}{n+1}, \ldots$$

(Figure 10.2.2) is a strictly increasing sequence.

Solution. The pattern of the initial terms suggests that the sequence is strictly increasing. To prove that this is so, let

$$a_n = \frac{n}{n+1}$$

We can obtain a_{n+1} by replacing n by $n+1$ in this formula. This yields

$$a_{n+1} = \frac{n+1}{(n+1)+1} = \frac{n+1}{n+2}$$

Thus, for $n \geq 1$

$$a_{n+1} - a_n = \frac{n+1}{n+2} - \frac{n}{n+1} = \frac{n^2+2n+1-n^2-2n}{(n+1)(n+2)} = \frac{1}{(n+1)(n+2)} > 0$$

which proves that the sequence is strictly increasing. ◀

▶ **Example 2** Use ratios of successive terms to show that the sequence in Example 1 is strictly increasing.

Solution. As shown in the solution of Example 1,

$$a_n = \frac{n}{n+1} \quad \text{and} \quad a_{n+1} = \frac{n+1}{n+2}$$

Thus,

$$\frac{a_{n+1}}{a_n} = \frac{(n+1)/(n+2)}{n/(n+1)} = \frac{n+1}{n+2} \cdot \frac{n+1}{n} = \frac{n^2+2n+1}{n^2+2n} \tag{1}$$

Since the numerator in (1) exceeds the denominator, it follows that $a_{n+1}/a_n > 1$ for $n \geq 1$. This proves that the sequence is strictly increasing. ◀

The following example illustrates still a third technique for determining whether a sequence is strictly monotone.

▶ **Example 3** In Examples 1 and 2 we proved that the sequence

$$\frac{1}{2}, \frac{2}{3}, \frac{3}{4}, \ldots, \frac{n}{n+1}, \ldots$$

is strictly increasing by considering the difference and ratio of successive terms. Alternatively, we can proceed as follows. Let

$$f(x) = \frac{x}{x+1}$$

so that the nth term in the given sequence is $a_n = f(n)$. The function f is increasing for $x \geq 1$ since

$$f'(x) = \frac{(x+1)(1) - x(1)}{(x+1)^2} = \frac{1}{(x+1)^2} > 0$$

Thus,

$$a_n = f(n) < f(n+1) = a_{n+1}$$

which proves that the given sequence is strictly increasing. ◀

Table 10.2.3

DERIVATIVE OF f FOR $x \geq 1$	CONCLUSION FOR THE SEQUENCE WITH $a_n = f(n)$
$f'(x) > 0$	Strictly increasing
$f'(x) < 0$	Strictly decreasing
$f'(x) \geq 0$	Increasing
$f'(x) \leq 0$	Decreasing

In general, if $f(n) = a_n$ is the nth term of a sequence, and if f is differentiable for $x \geq 1$, then the results in Table 10.2.3 can be used to investigate the monotonicity of the sequence.

PROPERTIES THAT HOLD EVENTUALLY

Sometimes a sequence will behave erratically at first and then settle down into a definite pattern. For example, the sequence

$$9, -8, -17, 12, 1, 2, 3, 4, \ldots \tag{2}$$

is strictly increasing from the fifth term on, but the sequence as a whole cannot be classified as strictly increasing because of the erratic behavior of the first four terms. To describe such sequences, we introduce the following terminology.

> **10.2.2 DEFINITION.** If discarding finitely many terms from the beginning of a sequence produces a sequence with a certain property, then the original sequence is said to have that property ***eventually***.

For example, although we cannot say that sequence (2) is strictly increasing, we can say that it is eventually strictly increasing.

▶ **Example 4** Show that the sequence $\left\{\dfrac{10^n}{n!}\right\}_{n=1}^{+\infty}$ is eventually strictly decreasing.

Solution. We have

$$a_n = \frac{10^n}{n!} \quad \text{and} \quad a_{n+1} = \frac{10^{n+1}}{(n+1)!}$$

so

$$\frac{a_{n+1}}{a_n} = \frac{10^{n+1}/(n+1)!}{10^n/n!} = \frac{10^{n+1}n!}{10^n(n+1)!} = 10\frac{n!}{(n+1)n!} = \frac{10}{n+1} \tag{3}$$

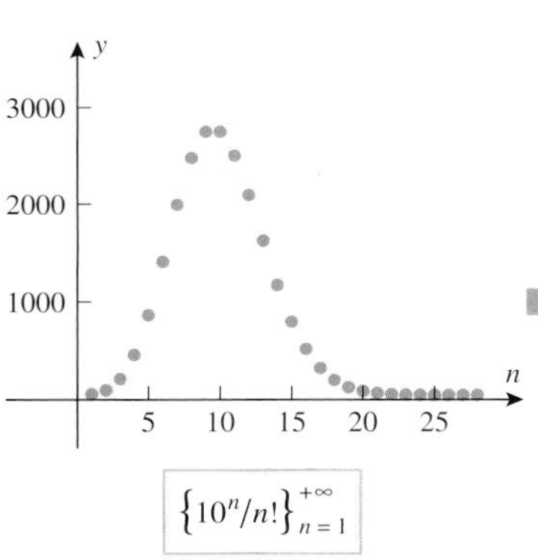

Figure 10.2.3

From (3), $a_{n+1}/a_n < 1$ for all $n \geq 10$, so the sequence is eventually strictly decreasing, as confirmed by the graph in Figure 10.2.3. ◄

AN INTUITIVE VIEW OF CONVERGENCE

Informally stated, the convergence or divergence of a sequence does not depend on the behavior of its *initial terms*, but rather on how the terms behave *eventually*. For example, the sequence

$$3, -9, -13, 17, 1, \frac{1}{2}, \frac{1}{3}, \frac{1}{4}, \ldots$$

eventually behaves like the sequence

$$1, \frac{1}{2}, \frac{1}{3}, \ldots, \frac{1}{n}, \ldots$$

and hence has a limit of 0.

CONVERGENCE OF MONOTONE SEQUENCES

The following two theorems, whose proofs are discussed at the end of this section, show that a monotone sequence either converges or becomes infinite—divergence by oscillation cannot occur.

10.2.3 THEOREM. *If a sequence $\{a_n\}$ is eventually increasing, then there are two possibilities:*

(*a*) *There is a constant M, called an* ***upper bound*** *for the sequence, such that $a_n \leq M$ for all n, in which case the sequence converges to a limit L satisfying $L \leq M$.*

(*b*) *No upper bound exists, in which case $\lim_{n\to+\infty} a_n = +\infty$.*

10.2.4 THEOREM. *If a sequence $\{a_n\}$ is eventually decreasing, then there are two possibilities:*

(*a*) *There is a constant M, called a* ***lower bound*** *for the sequence, such that $a_n \geq M$ for all n, in which case the sequence converges to a limit L satisfying $L \geq M$.*

(*b*) *No lower bound exists, in which case $\lim_{n\to+\infty} a_n = -\infty$.*

Theorems 10.2.3 and 10.2.4 are examples of ***existence theorems***; they tell us whether a limit exists, but they do not provide a method for finding it.

► **Example 5** Show that the sequence $\left\{\dfrac{10^n}{n!}\right\}_{n=1}^{+\infty}$ converges and find its limit.

Solution. We showed in Example 4 that the sequence is eventually strictly decreasing. Since all terms in the sequence are positive, it is bounded below by $M = 0$, and hence Theorem 10.2.4 guarantees that it converges to a nonnegative limit L. However, the limit is not evident directly from the formula $10^n/n!$ for the nth term, so we will need some ingenuity to obtain it.

Recall from Formula (3) of Example 4 that successive terms in the given sequence are related by the recursion formula

$$a_{n+1} = \frac{10}{n+1}a_n \tag{4}$$

where $a_n = 10^n/n!$. We will take the limit as $n \to +\infty$ of both sides of (4) and use the fact that

$$\lim_{n\to+\infty} a_{n+1} = \lim_{n\to+\infty} a_n = L$$

We obtain

$$L = \lim_{n\to+\infty} a_{n+1} = \lim_{n\to+\infty} \left(\frac{10}{n+1} a_n\right) = \lim_{n\to+\infty} \frac{10}{n+1} \lim_{n\to+\infty} a_n = 0 \cdot L = 0$$

so that

$$L = \lim_{n\to+\infty} \frac{10^n}{n!} = 0 \blacktriangleleft$$

In the exercises we will show that the technique illustrated in the last example can be adapted to obtain

$$\lim_{n\to+\infty} \frac{x^n}{n!} = 0 \qquad (5)$$

for any real value of x (Exercise 27). This result will be useful in our later work.

THE COMPLETENESS AXIOM

In this text we have accepted the familiar properties of real numbers without proof, and indeed, we have not even attempted to define the term *real number*. Although this is sufficient for many purposes, it was recognized by the late nineteenth century that the study of limits and functions in calculus requires a precise axiomatic formulation of the real numbers analogous to the axiomatic development of Euclidean geometry. Although we will not attempt to pursue this development, we will need to discuss one of the axioms about real numbers in order to prove Theorems 10.2.3 and 10.2.4. But first we will introduce some terminology.

If S is a nonempty set of real numbers, then we call u an ***upper bound*** for S if u is greater than or equal to every number in S, and we call l a ***lower bound*** for S if l is smaller than or equal to every number in S. For example, if S is the set of numbers in the interval $(1, 3)$, then $u = 4$, 10, and 100 are upper bounds for S and $l = -10$, 0, and $\frac{1}{2}$ are lower bounds for S. Observe also that $u = 3$ is the smallest of all upper bounds and $l = 1$ is the largest of all lower bounds. The existence of a smallest upper bound and a greatest lower bound for S is not accidental; it is a consequence of the following axiom.

10.2.5 AXIOM (*The Completeness Axiom*). *If a nonempty set S of real numbers has an upper bound, then it has a smallest upper bound (called the* ***least upper bound****), and if a nonempty set S of real numbers has a lower bound, then it has a largest lower bound (called the* ***greatest lower bound****).*

PROOF OF THEOREM 10.2.3.

(a) We will prove the result for increasing sequences, and leave it for the reader to adapt the argument to sequences that are eventually increasing. Assume there exists a number M such that $a_n \le M$ for $n = 1, 2, \ldots$. Then M is an upper bound for the set of terms in the sequence. By the Completeness Axiom there is a least upper bound for the terms; call it L. Now let ϵ be any positive number. Since L is the least upper bound for the terms, $L - \epsilon$ is not an upper bound for the terms, which means that there is at least one term a_N such that

$$a_N > L - \epsilon$$

Moreover, since $\{a_n\}$ is an increasing sequence, we must have

$$a_n \geq a_N > L - \epsilon \tag{6}$$

when $n \geq N$. But a_n cannot exceed L since L is an upper bound for the terms. This observation together with (6) tells us that $L \geq a_n > L - \epsilon$ for $n \geq N$, so all terms from the Nth on are within ϵ units of L. This is exactly the requirement to have

$$\lim_{n \to +\infty} a_n = L$$

Finally, $L \leq M$ since M is an upper bound for the terms and L is the least upper bound. This proves part (a).

(b) If there is no number M such that $a_n \leq M$ for $n = 1, 2, \ldots$, then no matter how large we choose M, there is a term a_N such that

$$a_N > M$$

and, since the sequence is increasing,

$$a_n \geq a_N > M$$

when $n \geq N$. Thus, the terms in the sequence become arbitrarily large as n increases. That is,

$$\lim_{n \to +\infty} a_n = +\infty$$ ■

The proof of Theorem 10.2.4 will be omitted since it is similar to that of 10.2.3.

✔QUICK CHECK EXERCISES 10.2 *(See page 642 for answers.)*

1. Classify each sequence as (I) increasing, (D) decreasing, or (N) neither increasing nor decreasing.
   ______ $\{2n\}$ ______ $\{2^{-n}\}$
   ______ $\left\{\dfrac{5-n}{n^2}\right\}$ ______ $\left\{\dfrac{-1}{n^2}\right\}$
   ______ $\left\{\dfrac{(-1)^n}{n^2}\right\}$

2. Classify each sequence as (M) monotonic, (S) strictly monotonic, or (N) not monotonic.
   ______ $\{n + (-1)^n\}$ ______ $\{2n + (-1)^n\}$
   ______ $\{3n + (-1)^n\}$

3. Since
$$\frac{n/[2(n+1)]}{(n-1)/(2n)} = \frac{n^2}{n^2-1} > \text{______}$$
the sequence $\{(n-1)/(2n)\}$ is strictly ______.

4. Since
$$\frac{d}{dx}[(x-8)^2] > 0 \text{ for } x > \text{______}$$
the sequence $\{(n-8)^2\}$ is ______ strictly ______.

EXERCISE SET 10.2

1–6 Use $a_{n+1} - a_n$ to show that the given sequence $\{a_n\}$ is strictly increasing or strictly decreasing.

1. $\left\{\dfrac{1}{n}\right\}_{n=1}^{+\infty}$
2. $\left\{1 - \dfrac{1}{n}\right\}_{n=1}^{+\infty}$
3. $\left\{\dfrac{n}{2n+1}\right\}_{n=1}^{+\infty}$
4. $\left\{\dfrac{n}{4n-1}\right\}_{n=1}^{+\infty}$
5. $\{n - 2^n\}_{n=1}^{+\infty}$
6. $\{n - n^2\}_{n=1}^{+\infty}$

7–12 Use a_{n+1}/a_n to show that the given sequence $\{a_n\}$ is strictly increasing or strictly decreasing.

7. $\left\{\dfrac{n}{2n+1}\right\}_{n=1}^{+\infty}$
8. $\left\{\dfrac{2^n}{1+2^n}\right\}_{n=1}^{+\infty}$
9. $\{ne^{-n}\}_{n=1}^{+\infty}$
10. $\left\{\dfrac{10^n}{(2n)!}\right\}_{n=1}^{+\infty}$
11. $\left\{\dfrac{n^n}{n!}\right\}_{n=1}^{+\infty}$
12. $\left\{\dfrac{5^n}{2^{(n^2)}}\right\}_{n=1}^{+\infty}$

13–18 Use differentiation to show that the given sequence is strictly increasing or strictly decreasing.

13. $\left\{\dfrac{n}{2n+1}\right\}_{n=1}^{+\infty}$ **14.** $\left\{3-\dfrac{1}{n}\right\}_{n=1}^{+\infty}$

15. $\left\{\dfrac{1}{n+\ln n}\right\}_{n=1}^{+\infty}$ **16.** $\{ne^{-2n}\}_{n=1}^{+\infty}$

17. $\left\{\dfrac{\ln(n+2)}{n+2}\right\}_{n=1}^{+\infty}$ **18.** $\{\tan^{-1} n\}_{n=1}^{+\infty}$

19–24 Show that the given sequence is eventually strictly increasing or eventually strictly decreasing.

19. $\{2n^2-7n\}_{n=1}^{+\infty}$ **20.** $\{n^3-4n^2\}_{n=1}^{+\infty}$

21. $\left\{\dfrac{n}{n^2+10}\right\}_{n=1}^{+\infty}$ **22.** $\left\{n+\dfrac{17}{n}\right\}_{n=1}^{+\infty}$

23. $\left\{\dfrac{n!}{3^n}\right\}_{n=1}^{+\infty}$ **24.** $\{n^5e^{-n}\}_{n=1}^{+\infty}$

FOCUS ON CONCEPTS

25. (a) Suppose that $\{a_n\}$ is a monotone sequence such that $1 \le a_n \le 2$ for all n. Must the sequence converge? If so, what can you say about the limit?
(b) Suppose that $\{a_n\}$ is a monotone sequence such that $a_n \le 2$ for all n. Must the sequence converge? If so, what can you say about the limit?

26. Give an example of a monotone sequence that is not eventually strictly monotone. What must be true of such a sequence?

27. The goal in this exercise is to prove Formula (5) in this section. The case where $x=0$ is obvious, so we will focus on the case where $x \ne 0$.
(a) Let $a_n = |x|^n/n!$. Show that

$$a_{n+1} = \frac{|x|}{n+1}a_n$$

(b) Show that the sequence $\{a_n\}$ is eventually strictly decreasing.
(c) Show that the sequence $\{a_n\}$ converges.
(d) Use the results in parts (a) and (c) to show that $a_n \to 0$ as $n \to +\infty$.
(e) Obtain Formula (5) from the result in part (d).

28. Let $\{a_n\}$ be the sequence defined recursively by $a_1 = \sqrt{2}$ and $a_{n+1} = \sqrt{2+a_n}$ for $n \ge 1$.
(a) List the first three terms of the sequence.
(b) Show that $a_n < 2$ for $n \ge 1$.
(c) Show that $a_{n+1}^2 - a_n^2 = (2-a_n)(1+a_n)$ for $n \ge 1$.
(d) Use the results in parts (b) and (c) to show that $\{a_n\}$ is a strictly increasing sequence. [*Hint:* If x and y are positive real numbers such that $x^2 - y^2 > 0$, then it follows by factoring that $x - y > 0$.]
(e) Show that $\{a_n\}$ converges and find its limit L.

29. Let $\{a_n\}$ be the sequence defined recursively by $a_1 = 1$ and $a_{n+1} = \frac{1}{2}[a_n + (3/a_n)]$ for $n \ge 1$.
(a) Show that $a_n \ge \sqrt{3}$ for $n \ge 2$. [*Hint:* What is the minimum value of $\frac{1}{2}[x + (3/x)]$ for $x > 0$?]
(b) Show that $\{a_n\}$ is eventually decreasing. [*Hint:* Examine $a_{n+1} - a_n$ or a_{n+1}/a_n and use the result in part (a).]
(c) Show that $\{a_n\}$ converges and find its limit L.

30. (a) Compare appropriate areas in the accompanying figure to deduce the following inequalities for $n \ge 2$:

$$\int_1^n \ln x\,dx < \ln n! < \int_1^{n+1} \ln x\,dx$$

(b) Use the result in part (a) to show that

$$\frac{n^n}{e^{n-1}} < n! < \frac{(n+1)^{n+1}}{e^n}, \quad n > 1$$

(c) Use the Squeezing Theorem for Sequences (Theorem 10.1.5) and the result in part (b) to show that

$$\lim_{n\to+\infty} \frac{\sqrt[n]{n!}}{n} = \frac{1}{e}$$

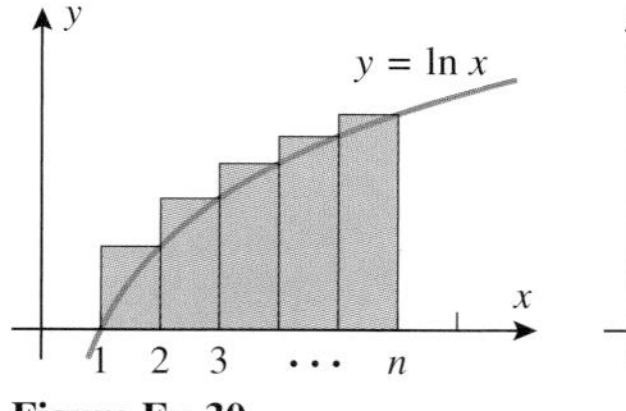

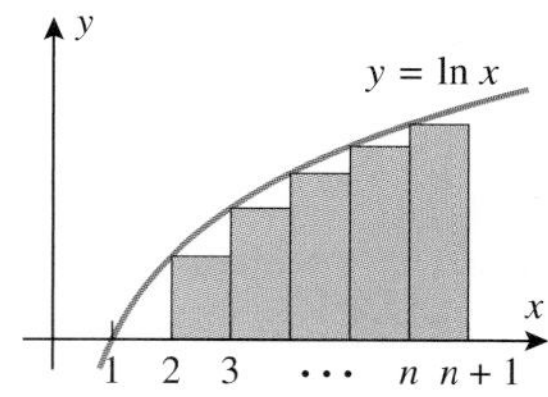

Figure Ex-30

31. Use the left inequality in Exercise 30(b) to show that

$$\lim_{n\to+\infty} \sqrt[n]{n!} = +\infty$$

✔ QUICK CHECK ANSWERS 10.2

1. I; D; N; I; N **2.** N; M; S **3.** 1; increasing **4.** 8; eventually; increasing

10.3 INFINITE SERIES

The purpose of this section is to discuss sums that contain infinitely many terms. The most familiar examples of such sums occur in the decimal representations of real numbers. For example, when we write $\frac{1}{3}$ in the decimal form $\frac{1}{3} = 0.3333\ldots$, we mean

$$\frac{1}{3} = 0.3 + 0.03 + 0.003 + 0.0003 + \cdots$$

which suggests that the decimal representation of $\frac{1}{3}$ can be viewed as a sum of infinitely many real numbers.

SUMS OF INFINITE SERIES

Our first objective is to define what is meant by the "sum" of infinitely many real numbers. We begin with some terminology.

10.3.1 DEFINITION. An ***infinite series*** is an expression that can be written in the form

$$\sum_{k=1}^{\infty} u_k = u_1 + u_2 + u_3 + \cdots + u_k + \cdots$$

The numbers $u_1, u_2, u_3, \ldots$ are called the ***terms*** of the series.

Since it is impossible to add infinitely many numbers together directly, sums of infinite series are defined and computed by an indirect limiting process. To motivate the basic idea, consider the decimal

$$0.3333\ldots \tag{1}$$

This can be viewed as the infinite series

$$0.3 + 0.03 + 0.003 + 0.0003 + \cdots$$

or, equivalently,

$$\frac{3}{10} + \frac{3}{10^2} + \frac{3}{10^3} + \frac{3}{10^4} + \cdots \tag{2}$$

Since (1) is the decimal expansion of $\frac{1}{3}$, any reasonable definition for the sum of an infinite series should yield $\frac{1}{3}$ for the sum of (2). To obtain such a definition, consider the following sequence of (finite) sums:

$$s_1 = \frac{3}{10} = 0.3$$

$$s_2 = \frac{3}{10} + \frac{3}{10^2} = 0.33$$

$$s_3 = \frac{3}{10} + \frac{3}{10^2} + \frac{3}{10^3} = 0.333$$

$$s_4 = \frac{3}{10} + \frac{3}{10^2} + \frac{3}{10^3} + \frac{3}{10^4} = 0.3333$$

$$\vdots$$

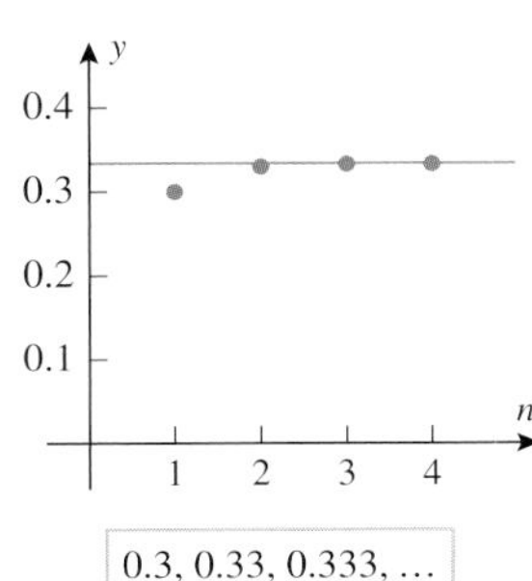

Figure 10.3.1

The sequence of numbers $s_1, s_2, s_3, s_4, \ldots$ (Figure 10.3.1) can be viewed as a succession of approximations to the "sum" of the infinite series, which we want to be $\frac{1}{3}$. As we progress through the sequence, more and more terms of the infinite series are used, and the approximations get better and better, suggesting that the desired sum of $\frac{1}{3}$ might be the *limit* of this sequence of approximations. To see that this is so, we must calculate the limit of the general term in the sequence of approximations, namely,

$$s_n = \frac{3}{10} + \frac{3}{10^2} + \cdots + \frac{3}{10^n} \tag{3}$$

The problem of calculating

$$\lim_{n\to+\infty} s_n = \lim_{n\to+\infty}\left(\frac{3}{10} + \frac{3}{10^2} + \cdots + \frac{3}{10^n}\right)$$

is complicated by the fact that both the last term and the number of terms in the sum change with n. It is best to rewrite such limits in a closed form in which the number of terms does not vary, if possible. (See the discussion of closed form and open form following Example 3 in Section 5.4.) To do this, we multiply both sides of (3) by $\frac{1}{10}$ to obtain

$$\frac{1}{10}s_n = \frac{3}{10^2} + \frac{3}{10^3} + \cdots + \frac{3}{10^n} + \frac{3}{10^{n+1}} \tag{4}$$

and then subtract (4) from (3) to obtain

$$\begin{aligned} s_n - \frac{1}{10}s_n &= \frac{3}{10} - \frac{3}{10^{n+1}} \\ \frac{9}{10}s_n &= \frac{3}{10}\left(1 - \frac{1}{10^n}\right) \\ s_n &= \frac{1}{3}\left(1 - \frac{1}{10^n}\right) \end{aligned}$$

Since $1/10^n \to 0$ as $n \to +\infty$, it follows that

$$\lim_{n\to+\infty} s_n = \lim_{n\to+\infty} \frac{1}{3}\left(1 - \frac{1}{10^n}\right) = \frac{1}{3}$$

which we denote by writing

$$\frac{1}{3} = \frac{3}{10} + \frac{3}{10^2} + \frac{3}{10^3} + \cdots + \frac{3}{10^n} + \cdots$$

Motivated by the preceding example, we are now ready to define the general concept of the "sum" of an infinite series

$$u_1 + u_2 + u_3 + \cdots + u_k + \cdots$$

We begin with some terminology: Let s_n denote the sum of the initial terms of the series, up to and including the term with index n. Thus,

$$\begin{aligned} s_1 &= u_1 \\ s_2 &= u_1 + u_2 \\ s_3 &= u_1 + u_2 + u_3 \\ &\vdots \\ s_n &= u_1 + u_2 + u_3 + \cdots + u_n = \sum_{k=1}^{n} u_k \end{aligned}$$

The number s_n is called the ***nth partial sum*** of the series and the sequence $\{s_n\}_{n=1}^{+\infty}$ is called the ***sequence of partial sums***.

As n increases, the partial sum $s_n = u_1 + u_2 + \cdots + u_n$ includes more and more terms of the series. Thus, if s_n tends toward a limit as $n \to +\infty$, it is reasonable to view this limit as the sum of *all* the terms in the series. This suggests the following definition.

10.3.2 DEFINITION. Let $\{s_n\}$ be the sequence of partial sums of the series

$$u_1 + u_2 + u_3 + \cdots + u_k + \cdots$$

If the sequence $\{s_n\}$ converges to a limit S, then the series is said to ***converge*** to S, and S is called the ***sum*** of the series. We denote this by writing

$$S = \sum_{k=1}^{\infty} u_k$$

If the sequence of partial sums diverges, then the series is said to ***diverge***. A divergent series has no sum.

WARNING

In everyday language the words "sequence" and "series" are often used interchangeably. However, in mathematics there is a difference between the two terms—a sequence is a *succession* whereas a series is a *sum*. It is essential that you keep this distinction in mind.

▶ Example 1 Determine whether the series

$$1 - 1 + 1 - 1 + 1 - 1 + \cdots$$

converges or diverges. If it converges, find the sum.

Solution. It is tempting to conclude that the sum of the series is zero by arguing that the positive and negative terms cancel one another. However, this is *not correct*; the problem is that algebraic operations that hold for finite sums do not carry over to infinite series in all cases. Later, we will discuss conditions under which familiar algebraic operations can be applied to infinite series, but for this example we turn directly to Definition 10.3.2. The partial sums are

$$s_1 = 1$$
$$s_2 = 1 - 1 = 0$$
$$s_3 = 1 - 1 + 1 = 1$$
$$s_4 = 1 - 1 + 1 - 1 = 0$$

and so forth. Thus, the sequence of partial sums is

$$1, 0, 1, 0, 1, 0, \ldots$$

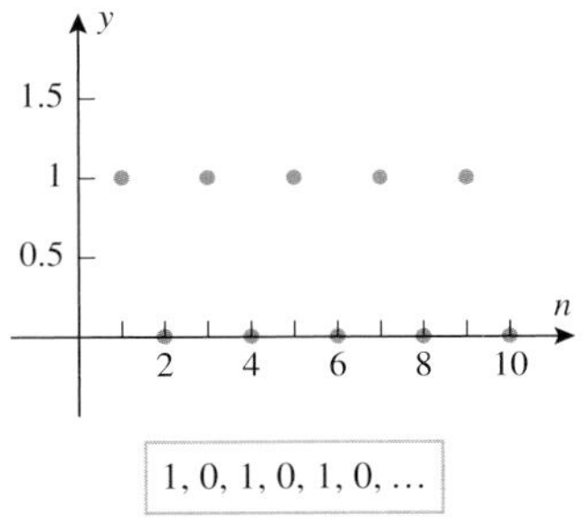

(Figure 10.3.2). Since this is a divergent sequence, the given series diverges and consequently has no sum. ◀

y
1.5
1
0.5
n
2 4 6 8 10
1, 0, 1, 0, 1, 0, ...

Figure 10.3.2

GEOMETRIC SERIES

In many important series, each term is obtained by multiplying the preceding term by some fixed constant. Thus, if the initial term of the series is a and each term is obtained by multiplying the preceding term by r, then the series has the form

$$\sum_{k=0}^{\infty} ar^k = a + ar + ar^2 + ar^3 + \cdots + ar^k + \cdots \quad (a \neq 0) \tag{5}$$

Such series are called ***geometric series***, and the number r is called the ***ratio*** for the series. Here are some examples:

$$1 + 2 + 4 + 8 + \cdots + 2^k + \cdots \qquad a = 1, r = 2$$

$$\frac{3}{10} + \frac{3}{10^2} + \frac{3}{10^3} + \cdots + \frac{3}{10^k} + \cdots \qquad a = \tfrac{3}{10}, r = \tfrac{1}{10}$$

$$\frac{1}{2} - \frac{1}{4} + \frac{1}{8} - \frac{1}{16} + \cdots + (-1)^{k+1}\frac{1}{2^k} + \cdots \qquad a = \tfrac{1}{2}, r = -\tfrac{1}{2}$$

$$1 + 1 + 1 + \cdots + 1 + \cdots \qquad a = 1, r = 1$$

$$1 - 1 + 1 - 1 + \cdots + (-1)^{k+1} + \cdots \qquad a = 1, r = -1$$

$$1 + x + x^2 + x^3 + \cdots + x^k + \cdots \qquad a = 1, r = x$$

The following theorem is the fundamental result on convergence of geometric series.

Sometimes it is desirable to start the index of summation of an infinite series at $k = 0$ rather than $k = 1$, in which case we would call u_0 the *zeroth term* and $s_0 = u_0$ the *zeroth partial sum*. One can prove that changing the starting value for the index of summation of an infinite series has no effect on the convergence, the divergence, or the sum. In the case of (5), the general term of the series would have been more complicated had we started the index at $k = 1$. What would it have been?

10.3.3 THEOREM. *A geometric series*

$$\sum_{k=0}^{\infty} ar^k = a + ar + ar^2 + \cdots + ar^k + \cdots \quad (a \neq 0)$$

converges if $|r| < 1$ and diverges if $|r| \geq 1$. If the series converges, then the sum is

$$\sum_{k=0}^{\infty} ar^k = \frac{a}{1-r}$$

PROOF. Let us treat the case $|r| = 1$ first. If $r = 1$, then the series is

$$a + a + a + a + \cdots$$

so the nth partial sum is $s_n = (n+1)a$ and $\lim_{n\to+\infty} s_n = \lim_{n\to+\infty}(n+1)a = \pm\infty$ (the sign depending on whether a is positive or negative). This proves divergence. If $r = -1$, the series is

$$a - a + a - a + \cdots$$

so the sequence of partial sums is

$$a, 0, a, 0, a, 0, \ldots$$

which diverges.

Now let us consider the case where $|r| \neq 1$. The nth partial sum of the series is

$$s_n = a + ar + ar^2 + \cdots + ar^n \tag{6}$$

Multiplying both sides of (6) by r yields

$$rs_n = ar + ar^2 + \cdots + ar^n + ar^{n+1} \tag{7}$$

and subtracting (7) from (6) gives

$$s_n - rs_n = a - ar^{n+1}$$

or

$$(1-r)s_n = a - ar^{n+1} \tag{8}$$

Since $r \neq 1$ in the case we are considering, this can be rewritten as

$$s_n = \frac{a - ar^{n+1}}{1-r} = \frac{a}{1-r}(1 - r^{n+1}) \tag{9}$$

If $|r| < 1$, then $\lim_{n\to+\infty} r^{n+1} = 0$ (can you see why?), so $\{s_n\}$ converges. From (9)

$$\lim_{n\to+\infty} s_n = \frac{a}{1-r}$$

If $|r| > 1$, then either $r > 1$ or $r < -1$. In the case $r > 1$, $\lim_{n\to+\infty} r^{n+1} = +\infty$, and in the case $r < -1$, r^{n+1} oscillates between positive and negative values that grow in magnitude, so $\{s_n\}$ diverges in both cases. ■

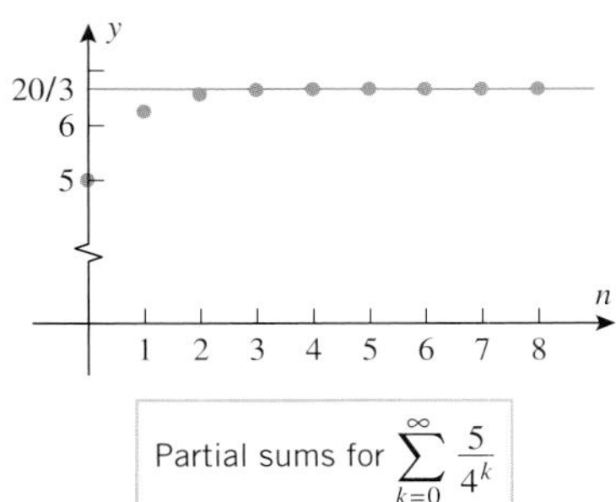

Partial sums for $\sum_{k=0}^{\infty} \frac{5}{4^k}$

Figure 10.3.3

▶ **Example 2** The series

$$\sum_{k=0}^{\infty} \frac{5}{4^k} = 5 + \frac{5}{4} + \frac{5}{4^2} + \cdots + \frac{5}{4^k} + \cdots$$

is a geometric series with $a = 5$ and $r = \frac{1}{4}$. Since $|r| = \frac{1}{4} < 1$, the series converges and the sum is

$$\frac{a}{1-r} = \frac{5}{1-\frac{1}{4}} = \frac{20}{3}$$

(Figure 10.3.3). ◀

TECHNOLOGY MASTERY

Computer algebra systems have commands for finding sums of convergent series. If you have a CAS, use it to compute the sums in Examples 2 and 3. Also, see what happens when you try to compute the sum in Example 4(a).

▶ **Example 3** Find the rational number represented by the repeating decimal

$$0.784784784\ldots$$

Solution. We can write

$$0.784784784\ldots = 0.784 + 0.000784 + 0.000000784 + \cdots$$

so the given decimal is the sum of a geometric series with $a = 0.784$ and $r = 0.001$. Thus,

$$0.784784784\ldots = \frac{a}{1-r} = \frac{0.784}{1-0.001} = \frac{0.784}{0.999} = \frac{784}{999} \quad ◀$$

▶ **Example 4** In each part, determine whether the series converges, and if so find its sum.

$$\text{(a)} \sum_{k=1}^{\infty} 3^{2k}5^{1-k} \qquad \text{(b)} \sum_{k=0}^{\infty} x^k$$

Solution (a). This is a geometric series in a concealed form, since we can rewrite it as

$$\sum_{k=1}^{\infty} 3^{2k}5^{1-k} = \sum_{k=1}^{\infty} \frac{9^k}{5^{k-1}} = \sum_{k=1}^{\infty} 9\left(\frac{9}{5}\right)^{k-1}$$

Since $r = \frac{9}{5} > 1$, the series diverges.

Solution (b). The expanded form of the series is

$$\sum_{k=0}^{\infty} x^k = 1 + x + x^2 + \cdots + x^k + \cdots$$

The series is a geometric series with $a = 1$ and $r = x$, so it converges if $|x| < 1$ and diverges otherwise. When the series converges its sum is

$$\sum_{k=0}^{\infty} x^k = \frac{1}{1-x} \quad ◀$$

TELESCOPING SUMS

▶ Example 5 Determine whether the series

$$\sum_{k=1}^{\infty} \frac{1}{k(k+1)} = \frac{1}{1\cdot 2} + \frac{1}{2\cdot 3} + \frac{1}{3\cdot 4} + \frac{1}{4\cdot 5} + \cdots$$

converges or diverges. If it converges, find the sum.

Solution. The nth partial sum of the series is

$$s_n = \sum_{k=1}^{n} \frac{1}{k(k+1)} = \frac{1}{1\cdot 2} + \frac{1}{2\cdot 3} + \frac{1}{3\cdot 4} + \cdots + \frac{1}{n(n+1)}$$

To calculate $\lim_{n\to+\infty} s_n$ we will rewrite s_n in closed form. This can be accomplished by using the method of partial fractions to obtain (verify)

$$\frac{1}{k(k+1)} = \frac{1}{k} - \frac{1}{k+1}$$

from which we obtain the sum

$$\begin{aligned} s_n &= \sum_{k=1}^{n} \left(\frac{1}{k} - \frac{1}{k+1}\right) \\ &= \left(1 - \frac{1}{2}\right) + \left(\frac{1}{2} - \frac{1}{3}\right) + \left(\frac{1}{3} - \frac{1}{4}\right) + \cdots + \left(\frac{1}{n} - \frac{1}{n+1}\right) \\ &= 1 + \left(-\frac{1}{2} + \frac{1}{2}\right) + \left(-\frac{1}{3} + \frac{1}{3}\right) + \cdots + \left(-\frac{1}{n} + \frac{1}{n}\right) - \frac{1}{n+1} \\ &= 1 - \frac{1}{n+1} \end{aligned} \tag{10}$$

The sum in (10) is an example of a ***telescoping sum***. The name is derived from the fact that in simplifying the sum, one term in each parenthetical expression cancels one term in the next parenthetical expression, until the entire sum collapses (like a folding telescope) into just two terms.

so

$$\sum_{k=1}^{\infty} \frac{1}{k(k+1)} = \lim_{n\to+\infty} s_n = \lim_{n\to+\infty} \left(1 - \frac{1}{n+1}\right) = 1 \blacktriangleleft$$

HARMONIC SERIES

One of the most important of all diverging series is the ***harmonic series***,

$$\sum_{k=1}^{\infty} \frac{1}{k} = 1 + \frac{1}{2} + \frac{1}{3} + \frac{1}{4} + \frac{1}{5} + \cdots$$

which arises in connection with the overtones produced by a vibrating musical string. It is not immediately evident that this series diverges. However, the divergence will become apparent when we examine the partial sums in detail. Because the terms in the series are all positive, the partial sums

$$s_1 = 1,\quad s_2 = 1 + \frac{1}{2},\quad s_3 = 1 + \frac{1}{2} + \frac{1}{3},\quad s_4 = 1 + \frac{1}{2} + \frac{1}{3} + \frac{1}{4}, \ldots$$

form a strictly increasing sequence

$$s_1 < s_2 < s_3 < \cdots < s_n < \cdots$$

(Figure 10.3.4*a*). Thus, by Theorem 10.2.3 we can prove divergence by demonstrating that there is no constant M that is greater than or equal to *every* partial sum. To this end, we will consider some selected partial sums, namely, $s_2, s_4, s_8, s_{16}, s_{32}, \ldots$. Note that the

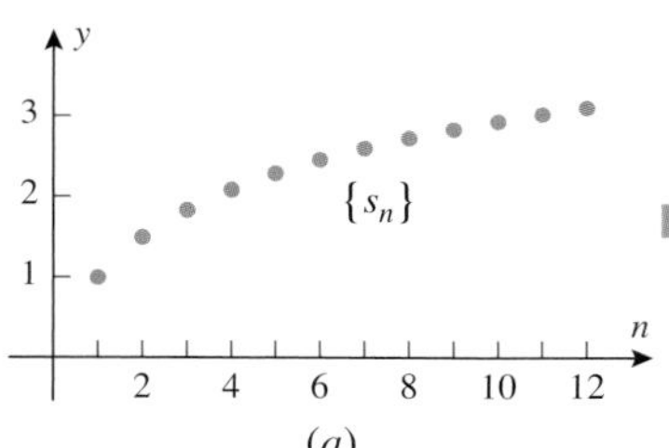

(*a*)

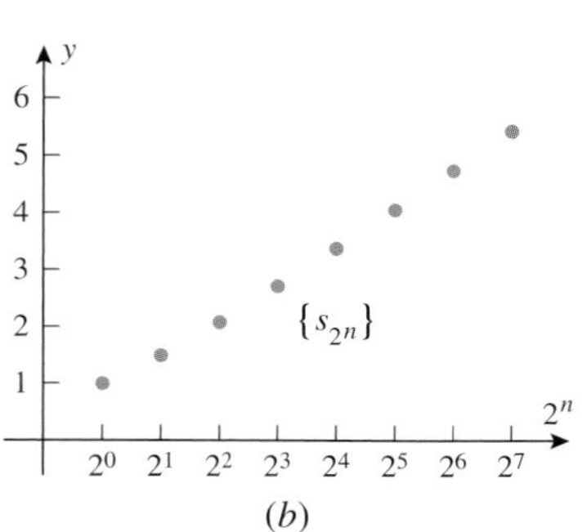

(*b*)

Partial sums for the harmonic series

Figure 10.3.4

subscripts are successive powers of 2, so that these are the partial sums of the form s_{2^n} (Figure 10.3.4*b*). These partial sums satisfy the inequalities

$$s_2 = 1 + \tfrac{1}{2} > \tfrac{1}{2} + \tfrac{1}{2} = \tfrac{2}{2}$$

$$s_4 = s_2 + \tfrac{1}{3} + \tfrac{1}{4} > s_2 + \left(\tfrac{1}{4} + \tfrac{1}{4}\right) = s_2 + \tfrac{1}{2} > \tfrac{3}{2}$$

$$s_8 = s_4 + \tfrac{1}{5} + \tfrac{1}{6} + \tfrac{1}{7} + \tfrac{1}{8} > s_4 + \left(\tfrac{1}{8} + \tfrac{1}{8} + \tfrac{1}{8} + \tfrac{1}{8}\right) = s_4 + \tfrac{1}{2} > \tfrac{4}{2}$$

$$s_{16} = s_8 + \tfrac{1}{9} + \tfrac{1}{10} + \tfrac{1}{11} + \tfrac{1}{12} + \tfrac{1}{13} + \tfrac{1}{14} + \tfrac{1}{15} + \tfrac{1}{16}$$

$$> s_8 + \left(\tfrac{1}{16} + \tfrac{1}{16} + \tfrac{1}{16} + \tfrac{1}{16} + \tfrac{1}{16} + \tfrac{1}{16} + \tfrac{1}{16} + \tfrac{1}{16}\right) = s_8 + \tfrac{1}{2} > \tfrac{5}{2}$$

$$\vdots$$

$$s_{2^n} > \frac{n+1}{2}$$

If M is any constant, we can find a positive integer n such that $(n+1)/2 > M$. But for this n

$$s_{2^n} > \frac{n+1}{2} > M$$

so that no constant M is greater than or equal to *every* partial sum of the harmonic series. This proves divergence.

This divergence proof, which predates the discovery of calculus, is due to a French bishop and teacher, Nicole Oresme (1323–1382). This series eventually attracted the interest of Johann and Jakob Bernoulli (p. 92) and led them to begin thinking about the general concept of convergence, which was a new idea at that time.

XVI. *Summa ſeriei infinitæ harmonicè progreſſionalium*, $\frac{1}{1} + \frac{1}{2} + \frac{1}{3} + \frac{1}{4} + \frac{1}{5}$ *&c. eſt infinita.*

Id primus deprehendit Frater: inventa namque per præced. ſumma ſeriei $\frac{1}{2} + \frac{1}{6} + \frac{1}{12} + \frac{1}{20} + \frac{1}{30}$, &c. viſurus porrò, quid emergeret ex iſta ſerie, $\frac{1}{2} + \frac{2}{6} + \frac{3}{12} + \frac{4}{20} + \frac{5}{30}$, &c. ſi reſolveretur methodo Prop. XIV. collegit propoſitionis veritatem ex abſurditate manifeſta, quæ ſequeretur, ſi ſumma ſeriei harmonicæ finita ſtatueretur. Animadvertit enim,

Seriem A, $\frac{1}{2} + \frac{1}{3} + \frac{1}{4} + \frac{1}{5} + \frac{1}{6} + \frac{1}{7}$, &c. ∞ (fractionibus ſingulis in alias, quarum numeratores ſunt 1, 2, 3, 4, &c. transmutatis) ſeriei B, $\frac{1}{2} + \frac{2}{6} + \frac{3}{12} + \frac{4}{20} + \frac{5}{30} + \frac{6}{42}$, &c. ∞ C + D + E + F, &c.

C. $\frac{1}{2} + \frac{1}{6} + \frac{1}{12} + \frac{1}{20} + \frac{1}{30} + \frac{1}{42}$, &c. ∞ per præc. $\frac{1}{1}$
D. . $+ \frac{1}{6} + \frac{1}{12} + \frac{1}{20} + \frac{1}{30} + \frac{1}{42}$, &c. ∞ C $- \frac{1}{2}$ ∞ $\frac{1}{2}$
E. . . $+ \frac{1}{12} + \frac{1}{20} + \frac{1}{30} + \frac{1}{42}$, &c. ∞ D $- \frac{1}{6}$ ∞ $\frac{1}{3}$
F. . . . . . $+ \frac{1}{20} + \frac{1}{30} + \frac{1}{42}$, &c. ∞ E $- \frac{1}{12}$ ∞ $\frac{1}{4}$
&c. ∞ &c.
∞ G; unde ſequitur, ſeriem G ∞ A, totum parti, ſi ſumma finita eſſet.

Ego

This is a proof of the divergence of the harmonic series, as it appeared in an appendix of Jakob Bernoulli's posthumous publication, *Ars Conjectandi*, which appeared in 1713.

✔ QUICK CHECK EXERCISES 10.3

(See page 652 for answers.)

1. In mathematics, the terms "sequence" and "series" have different meanings: a ________ is a succession, whereas a ________ is a sum.

2. Consider the series

$$\sum_{k=1}^{\infty} \frac{1}{2^k}$$

If $\{s_n\}$ is the sequence of partial sums for this series, then $s_1 =$ ________, $s_2 =$ ________, $s_3 =$ ________, $s_4 =$ ________, and $s_n =$ ________.

3. What does it mean to say that a series $\sum u_k$ *converges*?

4. A geometric series is a series of the form

$$\sum_{k=0}^{\infty} \text{\_\_\_\_\_\_\_\_}$$

This series converges to ________ if ________. This series diverges if ________.

5. The harmonic series has the form

$$\sum_{k=1}^{\infty} \text{\_\_\_\_\_\_\_\_}$$

Does the harmonic series converge or diverge?

EXERCISE SET 10.3 [C] CAS

1–2 In each part, find exact values for the first four partial sums, find a closed form for the nth partial sum, and determine whether the series converges by calculating the limit of the nth partial sum. If the series converges, then state its sum.

1. (a) $2 + \dfrac{2}{5} + \dfrac{2}{5^2} + \cdots + \dfrac{2}{5^{k-1}} + \cdots$

(b) $\dfrac{1}{4} + \dfrac{2}{4} + \dfrac{2^2}{4} + \cdots + \dfrac{2^{k-1}}{4} + \cdots$

(c) $\dfrac{1}{2\cdot 3} + \dfrac{1}{3\cdot 4} + \dfrac{1}{4\cdot 5} + \cdots + \dfrac{1}{(k+1)(k+2)} + \cdots$

2. (a) $\displaystyle\sum_{k=1}^{\infty}\left(\frac{1}{4}\right)^k$ (b) $\displaystyle\sum_{k=1}^{\infty} 4^{k-1}$ (c) $\displaystyle\sum_{k=1}^{\infty}\left(\frac{1}{k+3} - \frac{1}{k+4}\right)$

3–14 Determine whether the series converges, and if so find its sum.

3. $\sum_{k=1}^{\infty}\left(-\frac{3}{4}\right)^{k-1}$

4. $\sum_{k=1}^{\infty}\left(\frac{2}{3}\right)^{k+2}$

5. $\sum_{k=1}^{\infty}(-1)^{k-1}\frac{7}{6^{k-1}}$

6. $\sum_{k=1}^{\infty}\left(-\frac{3}{2}\right)^{k+1}$

7. $\sum_{k=1}^{\infty}\frac{1}{(k+2)(k+3)}$

8. $\sum_{k=1}^{\infty}\left(\frac{1}{2^k}-\frac{1}{2^{k+1}}\right)$

9. $\sum_{k=1}^{\infty}\frac{1}{9k^2+3k-2}$

10. $\sum_{k=2}^{\infty}\frac{1}{k^2-1}$

11. $\sum_{k=3}^{\infty}\frac{1}{k-2}$

12. $\sum_{k=5}^{\infty}\left(\frac{e}{\pi}\right)^{k-1}$

13. $\sum_{k=1}^{\infty}\frac{4^{k+2}}{7^{k-1}}$

14. $\sum_{k=1}^{\infty}5^{3k}7^{1-k}$

15. Match a series from one of Exercises 3, 5, 7, or 9 with the graph of its sequence of partial sums.

(a)

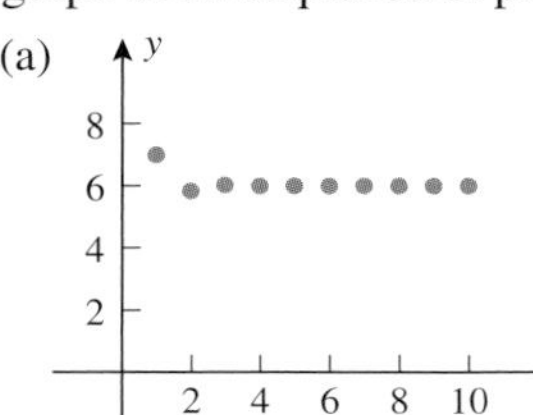

(b)

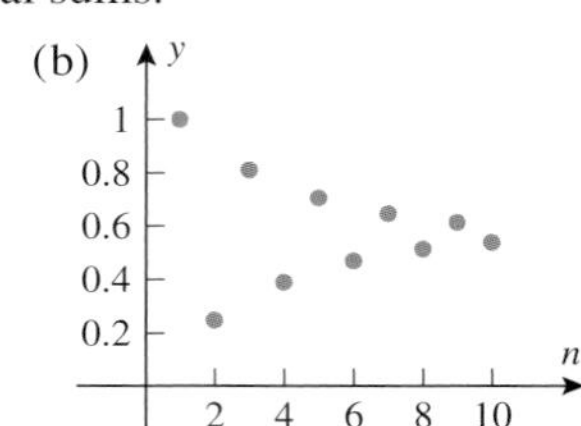

(c)

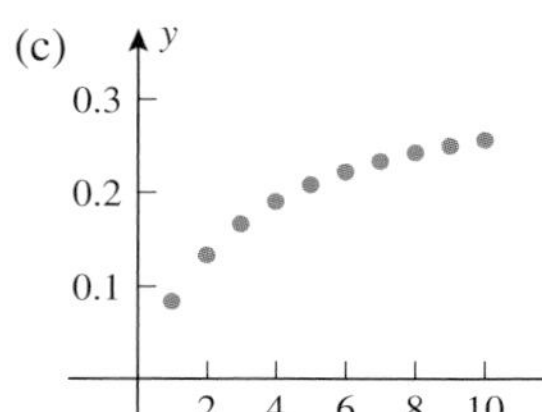

(d)

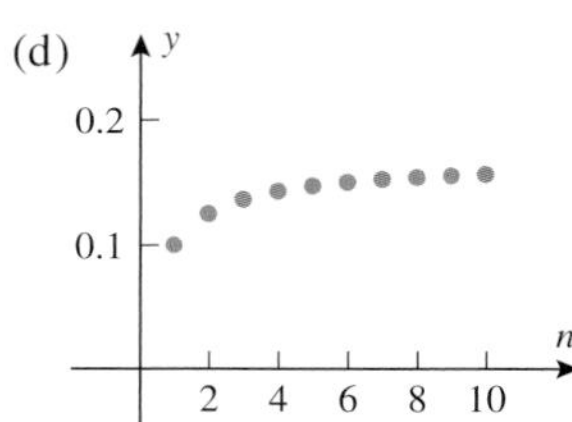

16. Match a series from one of Exercises 4, 6, 8, or 10 with the graph of its sequence of partial sums.

(a)

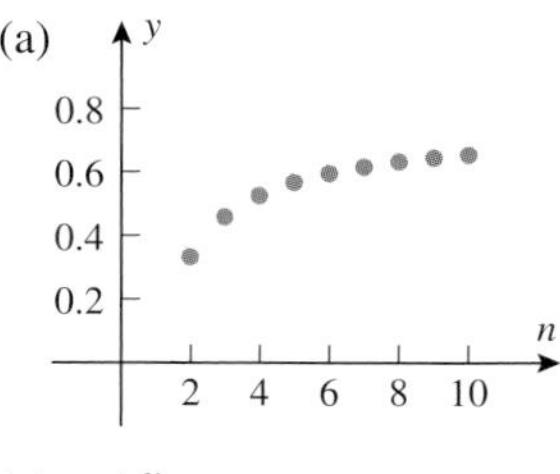

(b)

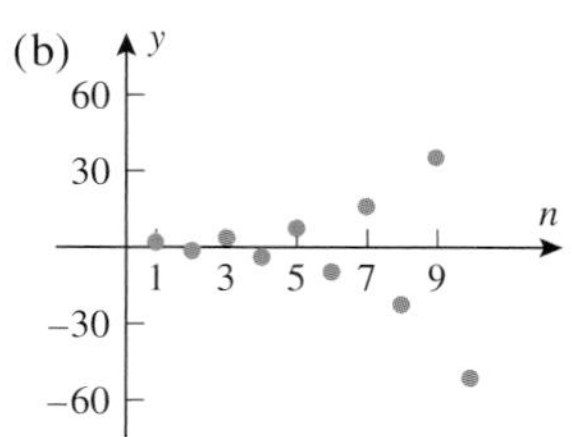

(c)

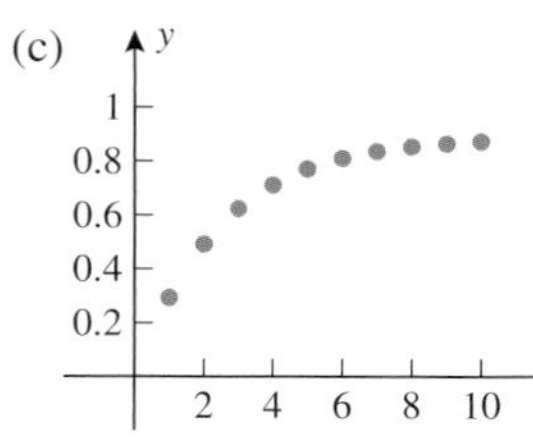

(d)

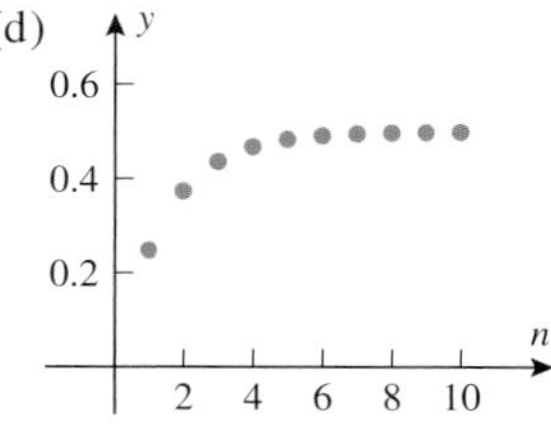

17–20 Express the repeating decimal as a fraction.

17. 0.4444...

18. 0.9999...

19. 5.373737...

20. 0.451141414...

21. Recall that a *terminating decimal* is a decimal whose digits are all 0 from some point on ($0.5 = 0.50000\ldots$, for example). Show that a decimal of the form $0.a_1a_2\ldots a_n9999\ldots$, where $a_n \neq 9$, can be expressed as a terminating decimal.

FOCUS ON CONCEPTS

22. The great Swiss mathematician Leonhard Euler (biography on p. 3) sometimes reached incorrect conclusions in his pioneering work on infinite series. For example, Euler deduced that

$$\tfrac{1}{2} = 1 - 1 + 1 - 1 + \cdots$$

and

$$-1 = 1 + 2 + 4 + 8 + \cdots$$

by substituting $x = -1$ and $x = 2$ in the formula

$$\frac{1}{1-x} = 1 + x + x^2 + x^3 + \cdots$$

What was the problem with his reasoning?

23. A ball is dropped from a height of 10 m. Each time it strikes the ground it bounces vertically to a height that is $\frac{3}{4}$ of the preceding height. Find the total distance the ball will travel if it is assumed to bounce infinitely often.

24. The accompanying figure shows an "infinite staircase" constructed from cubes. Find the total volume of the staircase, given that the largest cube has a side of length 1 and each successive cube has a side whose length is half that of the preceding cube.

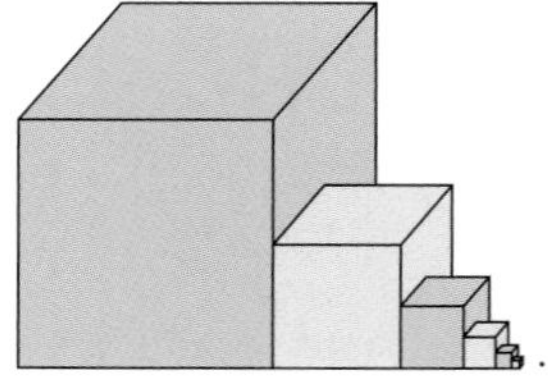

Figure Ex-24

25. In each part, find a closed form for the nth partial sum of the series, and determine whether the series converges. If so, find its sum.

(a) $\ln\frac{1}{2} + \ln\frac{2}{3} + \ln\frac{3}{4} + \cdots + \ln\frac{k}{k+1} + \cdots$

(b) $\ln\left(1-\frac{1}{4}\right) + \ln\left(1-\frac{1}{9}\right) + \ln\left(1-\frac{1}{16}\right) + \cdots + \ln\left(1-\frac{1}{(k+1)^2}\right) + \cdots$

26. Use geometric series to show that

(a) $\sum_{k=0}^{\infty}(-1)^k x^k = \frac{1}{1+x}$ if $-1 < x < 1$

(b) $\sum_{k=0}^{\infty}(x-3)^k = \frac{1}{4-x}$ if $2 < x < 4$

(c) $\sum_{k=0}^{\infty}(-1)^k x^{2k} = \frac{1}{1+x^2}$ if $-1 < x < 1$.

27. In each part, find all values of x for which the series converges, and find the sum of the series for those values of x.
(a) $x - x^3 + x^5 - x^7 + x^9 - \cdots$
(b) $\frac{1}{x^2} + \frac{2}{x^3} + \frac{4}{x^4} + \frac{8}{x^5} + \frac{16}{x^6} + \cdots$
(c) $e^{-x} + e^{-2x} + e^{-3x} + e^{-4x} + e^{-5x} + \cdots$

28. Show that for all real values of x

$$\sin x - \frac{1}{2}\sin^2 x + \frac{1}{4}\sin^3 x - \frac{1}{8}\sin^4 x + \cdots = \frac{2\sin x}{2+\sin x}$$

29. Let a_1 be any real number, and let $\{a_n\}$ be the sequence defined recursively by

$$a_{n+1} = \tfrac{1}{2}(a_n + 1)$$

Make a conjecture about the limit of the sequence, and confirm your conjecture by expressing a_n in terms of a_1 and taking the limit.

30. Show: $\sum_{k=1}^{\infty} \frac{\sqrt{k+1}-\sqrt{k}}{\sqrt{k^2+k}} = 1$.

31. Show: $\sum_{k=1}^{\infty} \left(\frac{1}{k} - \frac{1}{k+2}\right) = \frac{3}{2}$.

32. Show: $\frac{1}{1\cdot 3} + \frac{1}{2\cdot 4} + \frac{1}{3\cdot 5} + \cdots = \frac{3}{4}$.

33. Show: $\frac{1}{1\cdot 3} + \frac{1}{3\cdot 5} + \frac{1}{5\cdot 7} + \cdots = \frac{1}{2}$.

34. As shown in the accompanying figure, suppose that lines L_1 and L_2 form an angle θ, $0 < \theta < \pi/2$, at their point of intersection P. A point P_0 is chosen that is on L_1 and a units from P. Starting from P_0 a zig-zag path is constructed by successively going back and forth between L_1 and L_2 along a perpendicular from one line to the other. Find the following sums in terms of θ and a.
(a) $P_0P_1 + P_1P_2 + P_2P_3 + \cdots$
(b) $P_0P_1 + P_2P_3 + P_4P_5 + \cdots$
(c) $P_1P_2 + P_3P_4 + P_5P_6 + \cdots$

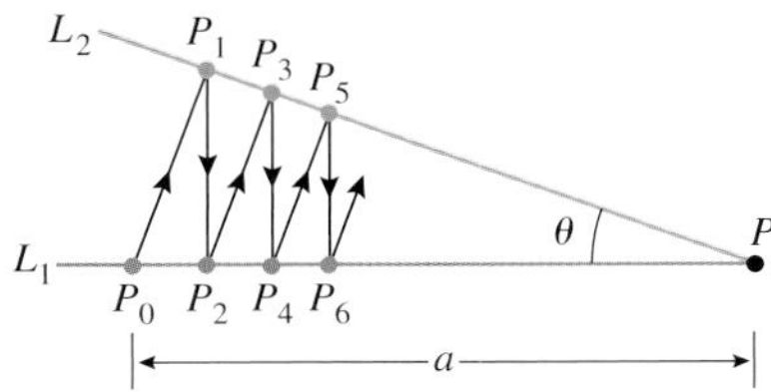

Figure Ex-34

35. As shown in the accompanying figure, suppose that an angle θ is bisected using a straightedge and compass to produce ray R_1, then the angle between R_1 and the initial side is bisected to produce ray R_2. Thereafter, rays $R_3, R_4, R_5, \ldots$ are constructed in succession by bisecting the angle between the preceding two rays. Show that the sequence of angles that these rays make with the initial side has a limit of $\theta/3$.

Source: This problem is based on *Trisection of an Angle in an Infinite Number of Steps* by Eric Kincannon, which appeared in *The College Mathematics Journal*, Vol. 21, No. 5, November 1990.

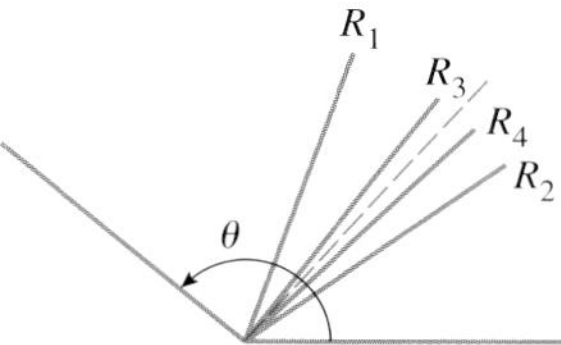

Figure Ex-35

36. In his *Treatise on the Configurations of Qualities and Motions* (written in the 1350s), the French Bishop of Lisieux, Nicole Oresme, used a geometric method to find the sum of the series

$$\sum_{k=1}^{\infty} \frac{k}{2^k} = \frac{1}{2} + \frac{2}{4} + \frac{3}{8} + \frac{4}{16} + \cdots$$

In part (*a*) of the accompanying figure, each term in the series is represented by the area of a rectangle, and in part (*b*) the configuration in part (*a*) has been divided into rectangles with areas $A_1, A_2, A_3, \ldots$. Find the sum $A_1 + A_2 + A_3 + \cdots$.

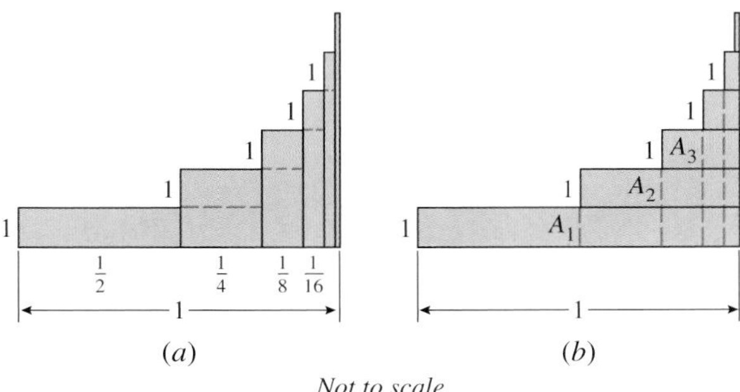

Figure Ex-36

C **37.** (a) See if your CAS can find the sum of the series

$$\sum_{k=1}^{\infty} \frac{6^k}{(3^{k+1}-2^{k+1})(3^k-2^k)}$$

(b) Find A and B such that

$$\frac{6^k}{(3^{k+1}-2^{k+1})(3^k-2^k)} = \frac{2^k A}{3^k-2^k} + \frac{2^k B}{3^{k+1}-2^{k+1}}$$

(c) Use the result in part (b) to find a closed form for the nth partial sum, and then find the sum of the series.

Source: This exercise is adapted from a problem that appeared in the Forty-Fifth Annual William Lowell Putnam Competition.

C **38.** In each part, use a CAS to find the sum of the series if it converges, and then confirm the result by hand calculation.

(a) $\sum_{k=1}^{\infty}(-1)^{k+1}2^k3^{2-k}$ (b) $\sum_{k=1}^{\infty}\frac{3^{3k}}{5^{k-1}}$ (c) $\sum_{k=1}^{\infty}\frac{1}{4k^2-1}$

✔QUICK CHECK ANSWERS 10.3

1. sequence; series 2. $\frac{1}{2}; \frac{3}{4}; \frac{7}{8}; \frac{15}{16}; 1 - \frac{1}{2^n}$ 3. The sequence of partial sums converges.
4. ar^k $(a \neq 0)$; $\frac{a}{1-r}$; $|r| < 1$; $|r| \geq 1$ 5. $\frac{1}{k}$; diverge

10.4 CONVERGENCE TESTS

In the last section we showed how to find the sum of a series by finding a closed form for the nth partial sum and taking its limit. However, it is relatively rare that one can find a closed form for the nth partial sum of a series, so alternative methods are needed for finding the sum of a series. One possibility is to prove that the series converges, and then to approximate the sum by a partial sum with sufficiently many terms to achieve the desired degree of accuracy. In this section we will develop various tests that can be used to determine whether a given series converges or diverges.

THE DIVERGENCE TEST

In stating general results about convergence or divergence of series, it is convenient to use the notation $\sum u_k$ as a generic template for a series, thus avoiding the issue of whether the sum begins with $k = 0$ or $k = 1$ or some other value. Indeed, we will see shortly that the starting index value is irrelevant to the issue of convergence. The kth term in an infinite series $\sum u_k$ is called the ***general term*** of the series. The following theorem establishes a relationship between the limit of the general term and the convergence properties of a series.

10.4.1 THEOREM (*The Divergence Test*).

(*a*) *If* $\lim_{k\to+\infty} u_k \neq 0$, *then the series* $\sum u_k$ *diverges.*

(*b*) *If* $\lim_{k\to+\infty} u_k = 0$, *then the series* $\sum u_k$ *may either converge or diverge.*

PROOF (*a*). To prove this result, it suffices to show that if the series converges, then $\lim_{k\to+\infty} u_k = 0$ (why?). We will prove this alternative form of (*a*).

Let us assume that the series converges. The general term u_k can be written as

$$u_k = s_k - s_{k-1} \tag{1}$$

where s_k is the sum of the terms through u_k and s_{k-1} is the sum of the terms through u_{k-1}. If S denotes the sum of the series, then $\lim_{k\to+\infty} s_k = S$, and since $(k-1) \to +\infty$ as $k \to +\infty$, we also have $\lim_{k\to+\infty} s_{k-1} = S$. Thus, from (1)

$$\lim_{k\to+\infty} u_k = \lim_{k\to+\infty} (s_k - s_{k-1}) = S - S = 0$$

PROOF (*b*). To prove this result, it suffices to produce both a convergent series and a divergent series for which $\lim_{k\to+\infty} u_k = 0$. The following series both have this property:

$$\frac{1}{2} + \frac{1}{2^2} + \cdots + \frac{1}{2^k} + \cdots \quad \text{and} \quad 1 + \frac{1}{2} + \frac{1}{3} + \cdots + \frac{1}{k} + \cdots$$

The first is a convergent geometric series and the second is the divergent harmonic series. ■

WARNING

The converse of Theorem 10.4.2 is false. Showing that

$$\lim_{k\to+\infty} u_k = 0$$

does not prove that $\sum u_k$ converges, since this property may hold for divergent as well as convergent series. This is illustrated in the proof of part (*b*) of Theorem 10.4.1.

The alternative form of part (a) given in the preceding proof is sufficiently important that we state it separately for future reference.

10.4.2 THEOREM. *If the series $\sum u_k$ converges, then $\lim_{k\to+\infty} u_k = 0$.*

▶ **Example 1** The series

$$\sum_{k=1}^{\infty} \frac{k}{k+1} = \frac{1}{2} + \frac{2}{3} + \frac{3}{4} + \cdots + \frac{k}{k+1} + \cdots$$

diverges since

$$\lim_{k\to+\infty} \frac{k}{k+1} = \lim_{k\to+\infty} \frac{1}{1+1/k} = 1 \neq 0 \ ◀$$

ALGEBRAIC PROPERTIES OF INFINITE SERIES

For brevity, the proof of the following result is omitted.

10.4.3 THEOREM.

(a) *If $\sum u_k$ and $\sum v_k$ are convergent series, then $\sum(u_k + v_k)$ and $\sum(u_k - v_k)$ are convergent series and the sums of these series are related by*

$$\sum_{k=1}^{\infty}(u_k + v_k) = \sum_{k=1}^{\infty} u_k + \sum_{k=1}^{\infty} v_k$$

$$\sum_{k=1}^{\infty}(u_k - v_k) = \sum_{k=1}^{\infty} u_k - \sum_{k=1}^{\infty} v_k$$

(b) *If c is a nonzero constant, then the series $\sum u_k$ and $\sum cu_k$ both converge or both diverge. In the case of convergence, the sums are related by*

$$\sum_{k=1}^{\infty} cu_k = c\sum_{k=1}^{\infty} u_k$$

(c) *Convergence or divergence is unaffected by deleting a finite number of terms from a series; in particular, for any positive integer K, the series*

$$\sum_{k=1}^{\infty} u_k = u_1 + u_2 + u_3 + \cdots$$

$$\sum_{k=K}^{\infty} u_k = u_K + u_{K+1} + u_{K+2} + \cdots$$

both converge or both diverge.

WARNING

Do not read too much into part (c) of Theorem 10.4.3. Although convergence is not affected when finitely many terms are deleted from the beginning of a convergent series, the *sum* of the series is changed by the removal of those terms.

▶ Example 2 Find the sum of the series

$$\sum_{k=1}^{\infty}\left(\frac{3}{4^k}-\frac{2}{5^{k-1}}\right)$$

Solution. The series

$$\sum_{k=1}^{\infty}\frac{3}{4^k}=\frac{3}{4}+\frac{3}{4^2}+\frac{3}{4^3}+\cdots$$

is a convergent geometric series $\left(a=\frac{3}{4}, r=\frac{1}{4}\right)$, and the series

$$\sum_{k=1}^{\infty}\frac{2}{5^{k-1}}=2+\frac{2}{5}+\frac{2}{5^2}+\frac{2}{5^3}+\cdots$$

is also a convergent geometric series $\left(a=2, r=\frac{1}{5}\right)$. Thus, from Theorems 10.4.3(a) and 10.3.3 the given series converges and

$$\sum_{k=1}^{\infty}\left(\frac{3}{4^k}-\frac{2}{5^{k-1}}\right)=\sum_{k=1}^{\infty}\frac{3}{4^k}-\sum_{k=1}^{\infty}\frac{2}{5^{k-1}}$$

$$=\frac{\frac{3}{4}}{1-\frac{1}{4}}-\frac{2}{1-\frac{1}{5}}=-\frac{3}{2} \blacktriangleleft$$

▶ Example 3 Determine whether the following series converge or diverge.

(a) $\displaystyle\sum_{k=1}^{\infty}\frac{5}{k}=5+\frac{5}{2}+\frac{5}{3}+\cdots+\frac{5}{k}+\cdots$ (b) $\displaystyle\sum_{k=10}^{\infty}\frac{1}{k}=\frac{1}{10}+\frac{1}{11}+\frac{1}{12}+\cdots$

Solution. The first series is a constant times the divergent harmonic series, and hence diverges by part (b) of Theorem 10.4.3. The second series results by deleting the first nine terms from the divergent harmonic series, and hence diverges by part (c) of Theorem 10.4.3. ◀

THE INTEGRAL TEST

The expressions

$$\sum_{k=1}^{\infty}\frac{1}{k^2} \quad \text{and} \quad \int_1^{+\infty}\frac{1}{x^2}\,dx$$

are related in that the integrand in the improper integral results when the index k in the general term of the series is replaced by x and the limits of summation in the series are replaced by the corresponding limits of integration. The following theorem shows that there is a relationship between the convergence of the series and the integral.

> **10.4.4 THEOREM (*The Integral Test*).** *Let $\sum u_k$ be a series with positive terms. If f is a function that is decreasing and continuous on an interval $[a,+\infty)$ and such that $u_k=f(k)$ for all $k \geq a$, then*
>
> $$\sum_{k=1}^{\infty}u_k \quad \textit{and} \quad \int_a^{+\infty}f(x)\,dx$$
>
> *both converge or both diverge.*

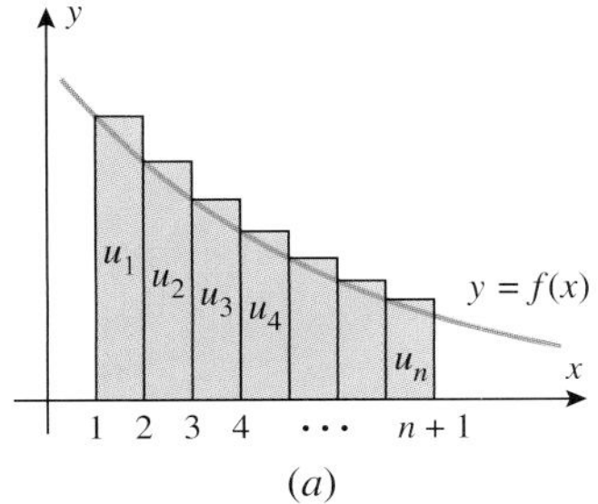

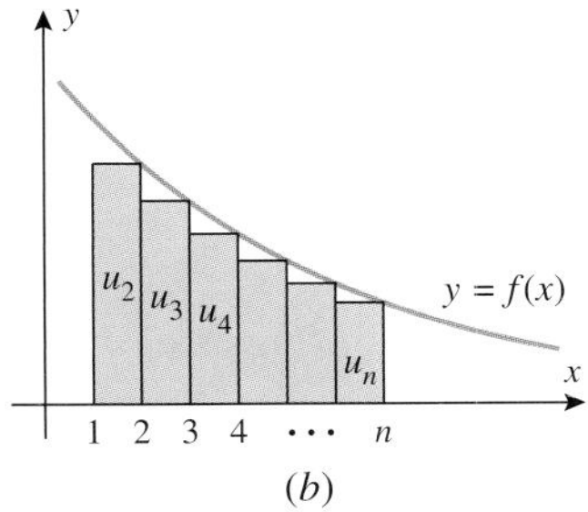

Figure 10.4.1

The proof of the integral test is deferred to the end of this section. However, the gist of the proof is captured in Figure 10.4.1: if the integral diverges, then so does the series (Figure 10.4.1*a*), and if the integral converges, then so does the series (Figure 10.4.1*b*).

▶ **Example 4** Use the integral test to determine whether the following series converge or diverge.

$$\text{(a)}\ \sum_{k=1}^{\infty}\frac{1}{k} \qquad \text{(b)}\ \sum_{k=1}^{\infty}\frac{1}{k^2}$$

Solution (a). We already know that this is the divergent harmonic series, so the integral test will simply provide another way of establishing the divergence. If we replace k by x in the general term $1/k$, we obtain the function $f(x) = 1/x$, which is decreasing and continuous for $x \geq 1$ (as required to apply the integral test with $a = 1$). Since

$$\int_1^{+\infty}\frac{1}{x}\,dx = \lim_{b\to+\infty}\int_1^b \frac{1}{x}\,dx = \lim_{b\to+\infty}[\ln b - \ln 1] = +\infty$$

the integral diverges and consequently so does the series.

Solution (b). If we replace k by x in the general term $1/k^2$, we obtain the function $f(x) = 1/x^2$, which is decreasing and continuous for $x \geq 1$. Since

$$\int_1^{+\infty}\frac{1}{x^2}\,dx = \lim_{b\to+\infty}\int_1^b \frac{dx}{x^2} = \lim_{b\to+\infty}\left[-\frac{1}{x}\right]_1^b = \lim_{b\to+\infty}\left[1 - \frac{1}{b}\right] = 1$$

the integral converges and consequently the series converges by the integral test with $a = 1$. ◀

WARNING

In part (b) of Example 4, do not erroneously conclude that the sum of the series is 1 because the value of the corresponding integral is 1. You can see that this is not so since the sum of the first two terms alone exceeds 1. Later, we will see that the sum of the series is actually $\pi^2/6$.

p-SERIES

The series in Example 4 are special cases of a class of series called ***p-series*** or ***hyperharmonic series***. A p-series is an infinite series of the form

$$\sum_{k=1}^{\infty}\frac{1}{k^p} = 1 + \frac{1}{2^p} + \frac{1}{3^p} + \cdots + \frac{1}{k^p} + \cdots$$

where $p > 0$. Examples of p-series are

$$\sum_{k=1}^{\infty}\frac{1}{k} = 1 + \frac{1}{2} + \frac{1}{3} + \cdots + \frac{1}{k} + \cdots \qquad p = 1$$

$$\sum_{k=1}^{\infty}\frac{1}{k^2} = 1 + \frac{1}{2^2} + \frac{1}{3^2} + \cdots + \frac{1}{k^2} + \cdots \qquad p = 2$$

$$\sum_{k=1}^{\infty}\frac{1}{\sqrt{k}} = 1 + \frac{1}{\sqrt{2}} + \frac{1}{\sqrt{3}} + \cdots + \frac{1}{\sqrt{k}} + \cdots \qquad p = \tfrac{1}{2}$$

The following theorem tells when a p-series converges.

10.4.5 THEOREM (*Convergence of p-Series*).

$$\sum_{k=1}^{\infty}\frac{1}{k^p} = 1 + \frac{1}{2^p} + \frac{1}{3^p} + \cdots + \frac{1}{k^p} + \cdots$$

converges if $p > 1$ and diverges if $0 < p \leq 1$.

PROOF. To establish this result when $p \neq 1$, we will use the integral test.

$$\int_1^{+\infty} \frac{1}{x^p}\,dx = \lim_{b\to+\infty} \int_1^b x^{-p}\,dx = \lim_{b\to+\infty} \frac{x^{1-p}}{1-p}\bigg]_1^b = \lim_{b\to+\infty}\left[\frac{b^{1-p}}{1-p} - \frac{1}{1-p}\right]$$

If $p > 1$, then $1 - p < 0$, so $b^{1-p} \to 0$ as $b \to +\infty$. Thus, the integral converges [its value is $-1/(1-p)$] and consequently the series also converges. For $0 < p < 1$, it follows that $1 - p > 0$ and $b^{1-p} \to +\infty$ as $b \to +\infty$, so the integral and the series diverge. The case $p = 1$ is the harmonic series, which was previously shown to diverge. ■

► **Example 5**

$$1 + \frac{1}{\sqrt[3]{2}} + \frac{1}{\sqrt[3]{3}} + \cdots + \frac{1}{\sqrt[3]{k}} + \cdots$$

diverges since it is a p-series with $p = \frac{1}{3} < 1$. ◄

PROOF OF THE INTEGRAL TEST

Before we can prove the integral test, we need a basic result about convergence of series with *nonnegative* terms. If $u_1 + u_2 + u_3 + \cdots + u_k + \cdots$ is such a series, then its sequence of partial sums is increasing, that is,

$$s_1 \le s_2 \le s_3 \le \cdots \le s_n \le \cdots$$

Thus, from Theorem 10.2.3 the sequence of partial sums converges to a limit S if and only if it has some upper bound M, in which case $S \le M$. If no upper bound exists, then the sequence of partial sums diverges. Since convergence of the sequence of partial sums corresponds to convergence of the series, we have the following theorem.

10.4.6 THEOREM. *If $\sum u_k$ is a series with nonnegative terms, and if there is a constant M such that*

$$s_n = u_1 + u_2 + \cdots + u_n \le M$$

for every n, then the series converges and the sum S satisfies $S \le M$. If no such M exists, then the series diverges.

In words, this theorem implies that *a series with nonnegative terms converges if and only if its sequence of partial sums is bounded above.*

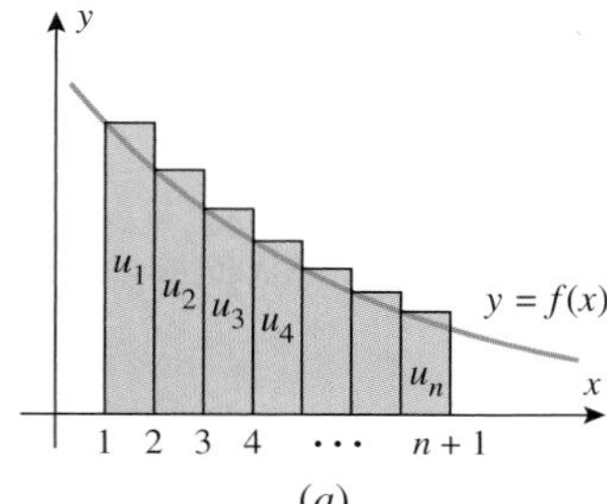

(a)

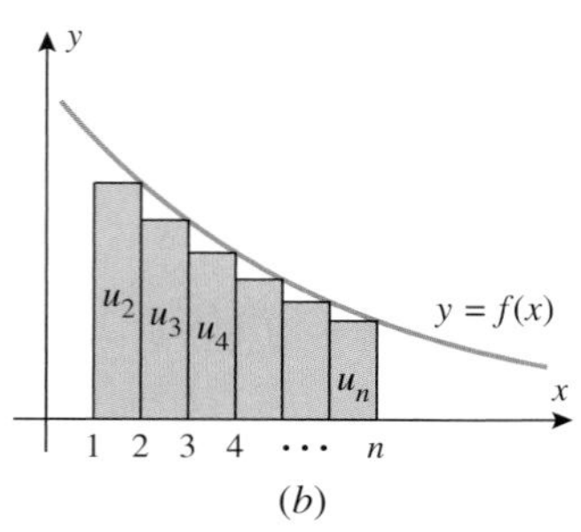

(b)

Figure 10.4.2

PROOF OF THEOREM 10.4.4. We need only show that the series converges when the integral converges and that the series diverges when the integral diverges. For simplicity, we will limit the proof to the case where $a = 1$. Assume that $f(x)$ satisfies the hypotheses of the theorem for $x \ge 1$. Since

$$f(1) = u_1,\ f(2) = u_2,\ \ldots,\ f(n) = u_n,\ \ldots$$

the values of $u_1, u_2, \ldots, u_n, \ldots$ can be interpreted as the areas of the rectangles shown in Figure 10.4.2.

The following inequalities result by comparing the areas under the curve $y = f(x)$ to the areas of the rectangles in Figure 10.4.2 for $n > 1$:

$$\int_1^{n+1} f(x)\,dx < u_1 + u_2 + \cdots + u_n = s_n \qquad \text{Figure 10.4.2}a$$

$$s_n - u_1 = u_2 + u_3 + \cdots + u_n < \int_1^n f(x)\,dx \qquad \text{Figure 10.4.2}b$$

These inequalities can be combined as

$$\int_1^{n+1} f(x)\,dx < s_n < u_1 + \int_1^n f(x)\,dx \tag{2}$$

If the integral $\int_1^{\infty} f(x)\,dx$ converges to a finite value L, then from the right-hand inequality in (2)

$$s_n < u_1 + \int_1^n f(x)\,dx < u_1 + \int_1^{+\infty} f(x)\,dx = u_1 + L$$

Thus, each partial sum is less than the finite constant $u_1 + L$, and the series converges by Theorem 10.4.6. On the other hand, if the integral $\int_1^{+\infty} f(x)\,dx$ diverges, then

$$\lim_{n\to+\infty} \int_1^{n+1} f(x)\,dx = +\infty$$

so that from the left-hand inequality in (2), $\lim_{n\to+\infty} s_n = +\infty$. This implies that the series also diverges. ■

✔QUICK CHECK EXERCISES 10.4

(See page 659 for answers.)

1. The divergence test says that if ______ $\neq 0$, then the series $\sum u_k$ diverges.

2. Given that

$$\sum_{k=1}^{\infty} \frac{1}{k(k+1)} = 1 \quad \text{and} \quad \sum_{k=1}^{\infty} \frac{1}{6^k} = \frac{1}{5}$$

it follows that

$$\sum_{k=1}^{\infty} \frac{7}{6^{k-1}} = \_\_\_\_\_\_$$

and

$$\sum_{k=1}^{\infty} \left(\frac{1}{2k(k+1)} - \frac{1}{6^k} \right) = \_\_\_\_\_\_$$

3. Since $\int_1^{+\infty} (1/\sqrt{x})\,dx = +\infty$, the ______ test applied to the series $\sum_{k=1}^{\infty}$ ______ shows that this series ______.

4. A p-series is a series of the form

$$\sum_{k=1}^{\infty} \_\_\_\_\_\_$$

This series converges if ______. This series diverges if ______.

EXERCISE SET 10.4

Graphing Utility   C CAS

1. Use Theorem 10.4.3 to find the sum of each series.

(a) $\left(\frac{1}{2} + \frac{1}{4}\right) + \left(\frac{1}{2^2} + \frac{1}{4^2}\right) + \cdots + \left(\frac{1}{2^k} + \frac{1}{4^k}\right) + \cdots$

(b) $\sum_{k=1}^{\infty} \left(\frac{1}{5^k} - \frac{1}{k(k+1)} \right)$

2. Use Theorem 10.4.3 to find the sum of each series.

(a) $\sum_{k=2}^{\infty} \left[\frac{1}{k^2-1} - \frac{7}{10^{k-1}} \right]$ (b) $\sum_{k=1}^{\infty} \left[7^{-k} 3^{k+1} - \frac{2^{k+1}}{5^k} \right]$

3–4 For each given p-series, identify p and determine whether the series converges.

3. (a) $\sum_{k=1}^{\infty} \frac{1}{k^3}$ (b) $\sum_{k=1}^{\infty} \frac{1}{\sqrt{k}}$ (c) $\sum_{k=1}^{\infty} k^{-1}$ (d) $\sum_{k=1}^{\infty} k^{-2/3}$

4. (a) $\sum_{k=1}^{\infty} k^{-4/3}$ (b) $\sum_{k=1}^{\infty} \frac{1}{\sqrt[4]{k}}$ (c) $\sum_{k=1}^{\infty} \frac{1}{\sqrt[3]{k^5}}$ (d) $\sum_{k=1}^{\infty} \frac{1}{k^{\pi}}$

5–6 Apply the divergence test and state what it tells you about the series.

5. (a) $\sum_{k=1}^{\infty} \frac{k^2+k+3}{2k^2+1}$ (b) $\sum_{k=1}^{\infty} \left(1 + \frac{1}{k}\right)^k$

(c) $\sum_{k=1}^{\infty} \cos k\pi$ (d) $\sum_{k=1}^{\infty} \frac{1}{k!}$

6. (a) $\sum_{k=1}^{\infty} \frac{k}{e^k}$ (b) $\sum_{k=1}^{\infty} \ln k$

(c) $\sum_{k=1}^{\infty} \frac{1}{\sqrt{k}}$ (d) $\sum_{k=1}^{\infty} \frac{\sqrt{k}}{\sqrt{k}+3}$

7–8 Confirm that the integral test is applicable and use it to determine whether the series converges.

7. (a) $\sum_{k=1}^{\infty} \frac{1}{5k+2}$ (b) $\sum_{k=1}^{\infty} \frac{1}{1+9k^2}$

8. (a) $\sum_{k=1}^{\infty} \frac{k}{1+k^2}$ (b) $\sum_{k=1}^{\infty} \frac{1}{(4+2k)^{3/2}}$

9–24 Determine whether the series converges.

9. $\sum_{k=1}^{\infty} \frac{1}{k+6}$ **10.** $\sum_{k=1}^{\infty} \frac{3}{5k}$ **11.** $\sum_{k=1}^{\infty} \frac{1}{\sqrt{k+5}}$

12. $\sum_{k=1}^{\infty} \frac{1}{\sqrt[k]{e}}$ **13.** $\sum_{k=1}^{\infty} \frac{1}{\sqrt[3]{2k-1}}$ **14.** $\sum_{k=3}^{\infty} \frac{\ln k}{k}$

15. $\sum_{k=1}^{\infty} \frac{k}{\ln(k+1)}$ **16.** $\sum_{k=1}^{\infty} ke^{-k^2}$ **17.** $\sum_{k=1}^{\infty} \left(1+\frac{1}{k}\right)^{-k}$

18. $\sum_{k=1}^{\infty} \frac{k^2+1}{k^2+3}$ **19.** $\sum_{k=1}^{\infty} \frac{\tan^{-1} k}{1+k^2}$ **20.** $\sum_{k=1}^{\infty} \frac{1}{\sqrt{k^2+1}}$

21. $\sum_{k=1}^{\infty} k^2 \sin^2\left(\frac{1}{k}\right)$ **22.** $\sum_{k=1}^{\infty} k^2 e^{-k^3}$

23. $\sum_{k=5}^{\infty} 7k^{-1.01}$ **24.** $\sum_{k=1}^{\infty} \operatorname{sech}^2 k$

25–26 Use the integral test to investigate the relationship between the value of p and the convergence of the series.

25. $\sum_{k=2}^{\infty} \frac{1}{k(\ln k)^p}$ **26.** $\sum_{k=3}^{\infty} \frac{1}{k(\ln k)[\ln(\ln k)]^p}$

FOCUS ON CONCEPTS

27. Suppose that the series $\sum u_k$ converges and the series $\sum v_k$ diverges. Show that the series $\sum(u_k+v_k)$ and $\sum(u_k-v_k)$ both diverge. [*Hint:* Assume that $\sum(u_k+v_k)$ converges and use Theorem 10.4.3 to obtain a contradiction.]

28. Find examples to show that if the series $\sum u_k$ and $\sum v_k$ both diverge, then the series $\sum(u_k+v_k)$ and $\sum(u_k-v_k)$ may either converge or diverge.

29–30 Use the results of Exercises 27 and 28, if needed, to determine whether each series converges or diverges.

29. (a) $\sum_{k=1}^{\infty} \left[\left(\frac{2}{3}\right)^{k-1} + \frac{1}{k}\right]$ (b) $\sum_{k=1}^{\infty} \left[\frac{1}{3k+2} - \frac{1}{k^{3/2}}\right]$

30. (a) $\sum_{k=2}^{\infty} \left[\frac{1}{k(\ln k)^2} - \frac{1}{k^2}\right]$ (b) $\sum_{k=2}^{\infty} \left[ke^{-k^2} + \frac{1}{k \ln k}\right]$

C **31.** Use a CAS to confirm that

$$\sum_{k=1}^{\infty} \frac{1}{k^2} = \frac{\pi^2}{6} \quad \text{and} \quad \sum_{k=1}^{\infty} \frac{1}{k^4} = \frac{\pi^4}{90}$$

and then use these results in each part to find the sum of the series.

(a) $\sum_{k=1}^{\infty} \frac{3k^2-1}{k^4}$ (b) $\sum_{k=3}^{\infty} \frac{1}{k^2}$ (c) $\sum_{k=2}^{\infty} \frac{1}{(k-1)^4}$

32–37 Exercise 32 will show how a partial sum can be used to obtain upper and lower bounds on the sum of a series when the hypotheses of the integral test are satisfied. This result will be needed in Exercises 33–37.

32. (a) Let $\sum_{k=1}^{\infty} u_k$ be a convergent series with positive terms, and let f be a function that is decreasing and continuous on $[n, +\infty)$ and such that $u_k = f(k)$ for $k \geq n$. Use an area argument and the accompanying figure to show that

$$\int_{n+1}^{+\infty} f(x)\,dx < \sum_{k=n+1}^{\infty} u_k < \int_{n}^{+\infty} f(x)\,dx$$

(b) Show that if S is the sum of the series $\sum_{k=1}^{\infty} u_k$ and s_n is the nth partial sum, then

$$s_n + \int_{n+1}^{+\infty} f(x)\,dx < S < s_n + \int_{n}^{+\infty} f(x)\,dx$$

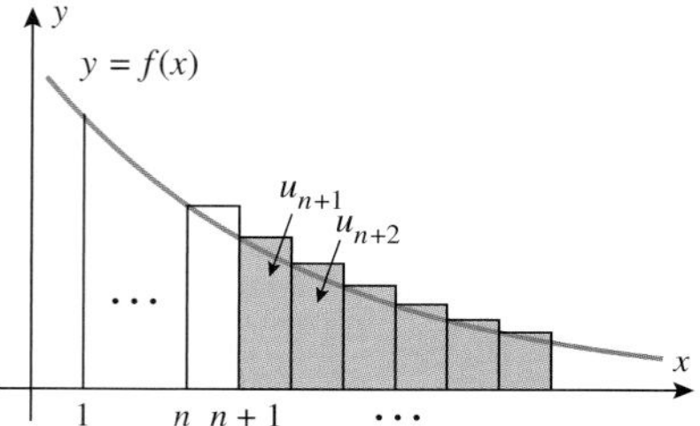

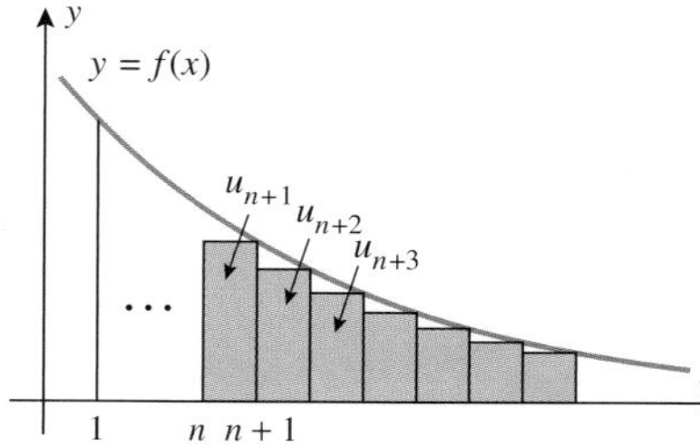

Figure Ex-32

33. (a) It was stated in Exercise 31 that

$$\sum_{k=1}^{\infty} \frac{1}{k^2} = \frac{\pi^2}{6}$$

Show that if s_n is the nth partial sum of this series, then

$$s_n + \frac{1}{n+1} < \frac{\pi^2}{6} < s_n + \frac{1}{n}$$

(b) Calculate s_3 exactly, and then use the result in part (a) to show that

$$\frac{29}{18} < \frac{\pi^2}{6} < \frac{61}{36}$$

(c) Use a calculating utility to confirm that the inequalities in part (b) are correct.

(d) Find upper and lower bounds on the error that results if the sum of the series is approximated by the 10th partial sum.

34. In each part, find upper and lower bounds on the error that results if the sum of the series is approximated by the 10th partial sum.

(a) $\sum_{k=1}^{\infty} \frac{1}{(2k+1)^2}$ (b) $\sum_{k=1}^{\infty} \frac{1}{k^2+1}$ (c) $\sum_{k=1}^{\infty} \frac{k}{e^k}$

35. Our objective in this problem is to approximate the sum of the series $\sum_{k=1}^{\infty} 1/k^3$ to two decimal-place accuracy.

(a) Show that if S is the sum of the series and s_n is the nth partial sum, then

$$s_n + \frac{1}{2(n+1)^2} < S < s_n + \frac{1}{2n^2}$$

(b) For two decimal-place accuracy, the error must be less than 0.005 (see Table 2.5.1 on p. 132). We can achieve this by finding an interval of length 0.01 (or less) that contains S and approximating S by the midpoint of that interval. Find the smallest value of n such that the interval containing S in part (a) has a length of 0.01 or less.

(c) Approximate S to two decimal-place accuracy.

36. (a) Use the method of Exercise 35 to approximate the sum of the series $\sum_{k=1}^{\infty} 1/k^4$ to two decimal-place accuracy.

(b) It was stated in Exercise 31 that the sum of this series is $\pi^4/90$. Use a calculating utility to confirm that your answer in part (a) is accurate to two decimal places.

37. We showed in Section 10.3 that the harmonic series $\sum_{k=1}^{\infty} 1/k$ diverges. Our objective in this problem is to demonstrate that although the partial sums of this series approach $+\infty$, they increase extremely slowly.

(a) Use inequality (2) to show that for $n \geq 2$

$$\ln(n+1) < s_n < 1 + \ln n$$

(b) Use the inequalities in part (a) to find upper and lower bounds on the sum of the first million terms in the series.

(c) Show that the sum of the first billion terms in the series is less than 22.

(d) Find a value of n so that the sum of the first n terms is greater than 100.

38. Investigate the relationship between the value of a and the convergence of the series $\sum_{k=1}^{\infty} k^{-\ln a}$.

39. Use a graphing utility to confirm that the integral test applies to the series $\sum_{k=1}^{\infty} k^2 e^{-k}$, and then determine whether the series converges.

C **40.** (a) Show that the hypotheses of the integral test are satisfied by the series $\sum_{k=1}^{\infty} 1/(k^3+1)$.

(b) Use a CAS and the integral test to confirm that the series converges.

(c) Construct a table of partial sums for $n = 10, 20, 30, \ldots, 100$, showing at least six decimal places.

(d) Based on your table, make a conjecture about the sum of the series to three decimal-place accuracy.

(e) Use part (b) of Exercise 32 to check your conjecture.

✔QUICK CHECK ANSWERS 10.4

1. $\lim_{k \to +\infty} u_k$ **2.** $\frac{42}{5}; \frac{3}{10}$ **3.** integral; $\frac{1}{\sqrt{k}}$; diverges **4.** $\frac{1}{k^p}$; $p > 1$; $0 < p \leq 1$

10.5 THE COMPARISON, RATIO, AND ROOT TESTS

In this section we will develop some more basic convergence tests for series with nonnegative terms. Later, we will use some of these tests to study the convergence of Taylor series.

THE COMPARISON TEST

We will begin with a test that is useful in its own right and is also the building block for other important convergence tests. The underlying idea of this test is to use the known convergence or divergence of a series to deduce the convergence or divergence of another series.

10.5.1 THEOREM (*The Comparison Test*). *Let $\sum_{k=1}^{\infty} a_k$ and $\sum_{k=1}^{\infty} b_k$ be series with nonnegative terms and suppose that*

$$a_1 \leq b_1, \ a_2 \leq b_2, \ a_3 \leq b_3, \ \ldots, \ a_k \leq b_k, \ \ldots$$

(a) *If the "bigger series" Σb_k converges, then the "smaller series" Σa_k also converges.*

(b) *If the "smaller series" Σa_k diverges, then the "bigger series" Σb_k also diverges.*

It is not essential in Theorem 10.5.1 that the condition $a_k \leq b_k$ hold for all k, as stated; the conclusions of the theorem remain true if this condition is eventually true.

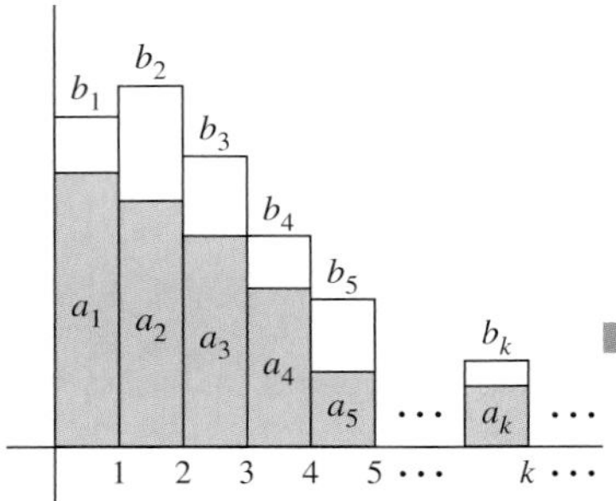

For each rectangle, b_k is the entire area and a_k is the area of the blue portion.

Figure 10.5.1

We have left the proof of this theorem for the exercises; however, it is easy to visualize why the theorem is true by interpreting the terms in the series as areas of rectangles (Figure 10.5.1). The comparison test states that if the total area $\sum b_k$ is finite, then the total area $\sum a_k$ must also be finite; and if the total area $\sum a_k$ is infinite, then the total area $\sum b_k$ must also be infinite.

USING THE COMPARISON TEST

There are two steps required for using the comparison test to determine whether a series $\sum u_k$ with positive terms converges:

- Guess at whether the series $\sum u_k$ converges or diverges.
- Find a series that proves the guess to be correct. That is, if the guess is divergence, we must find a divergent series whose terms are "smaller" than the corresponding terms of $\sum u_k$, and if the guess is convergence, we must find a convergent series whose terms are "bigger" than the corresponding terms of $\sum u_k$.

In most cases, the series $\sum u_k$ being considered will have its general term u_k expressed as a fraction. To help with the guessing process in the first step, we have formulated two principles that are based on the form of the denominator for u_k. These principles sometimes *suggest* whether a series is likely to converge or diverge. We have called these "informal principles" because they are not intended as formal theorems. In fact, we will not guarantee that they *always* work. However, they work often enough to be useful.

10.5.2 INFORMAL PRINCIPLE. *Constant terms in the denominator of u_k can usually be deleted without affecting the convergence or divergence of the series.*

10.5.3 INFORMAL PRINCIPLE. *If a polynomial in k appears as a factor in the numerator or denominator of u_k, all but the leading term in the polynomial can usually be discarded without affecting the convergence or divergence of the series.*

▶ Example 1 Use the comparison test to determine whether the following series converge or diverge.

$$\text{(a)} \sum_{k=1}^{\infty} \frac{1}{\sqrt{k}-\frac{1}{2}} \qquad \text{(b)} \sum_{k=1}^{\infty} \frac{1}{2k^2+k}$$

Solution (a). According to Principle 10.5.2, we should be able to drop the constant in the denominator without affecting the convergence or divergence. Thus, the given series is likely to behave like

$$\sum_{k=1}^{\infty} \frac{1}{\sqrt{k}} \tag{1}$$

which is a divergent p-series $\left(p=\frac{1}{2}\right)$. Thus, we will guess that the given series diverges and try to prove this by finding a divergent series that is "smaller" than the given series. However, series (1) does the trick since

$$\frac{1}{\sqrt{k}-\frac{1}{2}} > \frac{1}{\sqrt{k}} \quad \text{for } k=1,2,\ldots$$

Thus, we have proved that the given series diverges.

Solution (b). According to Principle 10.5.3, we should be able to discard all but the leading term in the polynomial without affecting the convergence or divergence. Thus, the given series is likely to behave like

$$\sum_{k=1}^{\infty} \frac{1}{2k^2} = \frac{1}{2} \sum_{k=1}^{\infty} \frac{1}{k^2} \tag{2}$$

which converges since it is a constant times a convergent p-series ($p = 2$). Thus, we will guess that the given series converges and try to prove this by finding a convergent series that is "bigger" than the given series. However, series (2) does the trick since

$$\frac{1}{2k^2 + k} < \frac{1}{2k^2} \quad \text{for } k = 1, 2, \ldots$$

Thus, we have proved that the given series converges. ◄

THE LIMIT COMPARISON TEST

In the last example, Principles 10.5.2 and 10.5.3 provided the guess about convergence or divergence as well as the series needed to apply the comparison test. Unfortunately, it is not always so straightforward to find the series required for comparison, so we will now consider an alternative to the comparison test that is usually easier to apply. The proof is given in Appendix C.

10.5.4 THEOREM (*The Limit Comparison Test*). *Let $\sum a_k$ and $\sum b_k$ be series with positive terms and suppose that*

$$\rho = \lim_{k \to +\infty} \frac{a_k}{b_k}$$

If ρ is finite and $\rho > 0$, then the series both converge or both diverge.

The cases where $\rho = 0$ or $\rho = +\infty$ are discussed in the exercises (Exercise 54).

► **Example 2** Use the limit comparison test to determine whether the following series converge or diverge.

$$\text{(a)} \sum_{k=1}^{\infty} \frac{1}{\sqrt{k} + 1} \qquad \text{(b)} \sum_{k=1}^{\infty} \frac{1}{2k^2 + k} \qquad \text{(c)} \sum_{k=1}^{\infty} \frac{3k^3 - 2k^2 + 4}{k^7 - k^3 + 2}$$

Solution (a). As in Example 1, Principle 10.5.2 suggests that the series is likely to behave like the divergent p-series (1). To prove that the given series diverges, we will apply the limit comparison test with

$$a_k = \frac{1}{\sqrt{k} + 1} \quad \text{and} \quad b_k = \frac{1}{\sqrt{k}}$$

We obtain

$$\rho = \lim_{k \to +\infty} \frac{a_k}{b_k} = \lim_{k \to +\infty} \frac{\sqrt{k}}{\sqrt{k} + 1} = \lim_{k \to +\infty} \frac{1}{1 + \dfrac{1}{\sqrt{k}}} = 1$$

Since ρ is finite and positive, it follows from Theorem 10.5.4 that the given series diverges.

Solution (b). As in Example 1, Principle 10.5.3 suggests that the series is likely to behave like the convergent series (2). To prove that the given series converges, we will apply the limit comparison test with

$$a_k = \frac{1}{2k^2 + k} \quad \text{and} \quad b_k = \frac{1}{2k^2}$$

We obtain

$$\rho = \lim_{k\to+\infty} \frac{a_k}{b_k} = \lim_{k\to+\infty} \frac{2k^2}{2k^2+k} = \lim_{k\to+\infty} \frac{2}{2+\dfrac{1}{k}} = 1$$

Since ρ is finite and positive, it follows from Theorem 10.5.4 that the given series converges, which agrees with the conclusion reached in Example 1 using the comparison test.

Solution (c). From Principle 10.5.3, the series is likely to behave like

$$\sum_{k=1}^{\infty} \frac{3k^3}{k^7} = \sum_{k=1}^{\infty} \frac{3}{k^4} \tag{3}$$

which converges since it is a constant times a convergent p-series. Thus, the given series is likely to converge. To prove this, we will apply the limit comparison test to series (3) and the given series. We obtain

$$\rho = \lim_{k\to+\infty} \frac{\dfrac{3k^3 - 2k^2 + 4}{k^7 - k^3 + 2}}{\dfrac{3}{k^4}} = \lim_{k\to+\infty} \frac{3k^7 - 2k^6 + 4k^4}{3k^7 - 3k^3 + 6} = 1$$

Since ρ is finite and nonzero, it follows from Theorem 10.5.4 that the given series converges, since (3) converges. ◀

THE RATIO TEST

The comparison test and the limit comparison test hinge on first making a guess about convergence and then finding an appropriate series for comparison, both of which can be difficult tasks in cases where Principles 10.5.2 and 10.5.3 cannot be applied. In such cases the next test can often be used, since it works exclusively with the terms of the given series—it requires neither an initial guess about convergence nor the discovery of a series for comparison. Its proof is given in Appendix C.

10.5.5 THEOREM (*The Ratio Test*). *Let $\sum u_k$ be a series with positive terms and suppose that*

$$\rho = \lim_{k\to+\infty} \frac{u_{k+1}}{u_k}$$

(*a*) *If $\rho < 1$, the series converges.*

(*b*) *If $\rho > 1$ or $\rho = +\infty$, the series diverges.*

(*c*) *If $\rho = 1$, the series may converge or diverge, so that another test must be tried.*

▶ **Example 3** Use the ratio test to determine whether the following series converge or diverge.

(a) $\displaystyle\sum_{k=1}^{\infty} \frac{1}{k!}$ (b) $\displaystyle\sum_{k=1}^{\infty} \frac{k}{2^k}$ (c) $\displaystyle\sum_{k=1}^{\infty} \frac{k^k}{k!}$ (d) $\displaystyle\sum_{k=3}^{\infty} \frac{(2k)!}{4^k}$ (e) $\displaystyle\sum_{k=1}^{\infty} \frac{1}{2k-1}$

Solution (a). The series converges, since

$$\rho = \lim_{k\to+\infty} \frac{u_{k+1}}{u_k} = \lim_{k\to+\infty} \frac{1/(k+1)!}{1/k!} = \lim_{k\to+\infty} \frac{k!}{(k+1)!} = \lim_{k\to+\infty} \frac{1}{k+1} = 0 < 1$$

Solution (b). The series converges, since

$$\rho = \lim_{k\to+\infty} \frac{u_{k+1}}{u_k} = \lim_{k\to+\infty} \frac{k+1}{2^{k+1}} \cdot \frac{2^k}{k} = \frac{1}{2} \lim_{k\to+\infty} \frac{k+1}{k} = \frac{1}{2} < 1$$

Solution (c). The series diverges, since

$$\rho = \lim_{k\to+\infty} \frac{u_{k+1}}{u_k} = \lim_{k\to+\infty} \frac{(k+1)^{k+1}}{(k+1)!} \cdot \frac{k!}{k^k} = \lim_{k\to+\infty} \frac{(k+1)^k}{k^k} = \lim_{k\to+\infty} \left(1 + \frac{1}{k}\right)^k = e > 1$$

See Formula (3) of Section 7.1

Solution (d). The series diverges, since

$$\rho = \lim_{k\to+\infty} \frac{u_{k+1}}{u_k} = \lim_{k\to+\infty} \frac{[2(k+1)]!}{4^{k+1}} \cdot \frac{4^k}{(2k)!} = \lim_{k\to+\infty} \left(\frac{(2k+2)!}{(2k)!} \cdot \frac{1}{4}\right)$$

$$= \frac{1}{4} \lim_{k\to+\infty} (2k+2)(2k+1) = +\infty$$

Solution (e). The ratio test is of no help since

$$\rho = \lim_{k\to+\infty} \frac{u_{k+1}}{u_k} = \lim_{k\to+\infty} \frac{1}{2(k+1)-1} \cdot \frac{2k-1}{1} = \lim_{k\to+\infty} \frac{2k-1}{2k+1} = 1$$

However, the integral test proves that the series diverges since

$$\int_1^{+\infty} \frac{dx}{2x-1} = \lim_{b\to+\infty} \int_1^b \frac{dx}{2x-1} = \lim_{b\to+\infty} \frac{1}{2} \ln(2x-1)\bigg]_1^b = +\infty$$

Both the comparison test and the limit comparison test would also have worked here (verify). ◀

THE ROOT TEST

In cases where it is difficult or inconvenient to find the limit required for the ratio test, the next test is sometimes useful. Since its proof is similar to the proof of the ratio test, we will omit it.

10.5.6 THEOREM (*The Root Test*). *Let $\sum u_k$ be a series with positive terms and suppose that*

$$\rho = \lim_{k\to+\infty} \sqrt[k]{u_k} = \lim_{k\to+\infty} (u_k)^{1/k}$$

(a) *If $\rho < 1$, the series converges.*

(b) *If $\rho > 1$ or $\rho = +\infty$, the series diverges.*

(c) *If $\rho = 1$, the series may converge or diverge, so that another test must be tried.*

▶ **Example 4** Use the root test to determine whether the following series converge or diverge.

$$\text{(a)}\ \sum_{k=2}^{\infty}\left(\frac{4k-5}{2k+1}\right)^k \qquad \text{(b)}\ \sum_{k=1}^{\infty}\frac{1}{(\ln(k+1))^k}$$

Solution (a). The series diverges, since

$$\rho = \lim_{k\to+\infty}(u_k)^{1/k} = \lim_{k\to+\infty}\frac{4k-5}{2k+1} = 2 > 1$$

Solution (b). The series converges, since

$$\rho = \lim_{k\to+\infty}(u_k)^{1/k} = \lim_{k\to+\infty}\frac{1}{\ln(k+1)} = 0 < 1 \blacktriangleleft$$

✔QUICK CHECK EXERCISES 10.5 (See page 665 for answers.)

1–4 Select between *converges* or *diverges* to fill the first blank.

1. The series

$$\sum_{k=1}^{\infty}\frac{2k^2+1}{2k^{8/3}-1}$$

_______ by comparison with the p-series $\sum_{k=1}^{\infty}$ _______.

2. Since

$$\lim_{k\to+\infty}\frac{(k+1)^3/3^{k+1}}{k^3/3^k} = \lim_{k\to+\infty}\frac{\left(1+\frac{1}{k}\right)^3}{3} = \frac{1}{3}$$

the series $\sum_{k=1}^{\infty}k^3/3^k$ _______ by the _______ test.

3. Since

$$\lim_{k\to+\infty}\frac{(k+1)!/3^{k+1}}{k!/3^k} = \lim_{k\to+\infty}\frac{k+1}{3} = +\infty$$

the series $\sum_{k=1}^{\infty}k!/3^k$ _______ by the _______ test.

4. Since

$$\lim_{k\to+\infty}\left(\frac{1}{k^{k/2}}\right)^{1/k} = \lim_{k\to+\infty}\frac{1}{k^{1/2}} = 0$$

the series $\sum_{k=1}^{\infty}1/k^{k/2}$ _______ by the _______ test.

EXERCISE SET 10.5 [C] CAS

1–2 Make a guess about the convergence or divergence of the series, and confirm your guess using the comparison test.

1. (a) $\displaystyle\sum_{k=1}^{\infty}\frac{1}{5k^2-k}$ (b) $\displaystyle\sum_{k=1}^{\infty}\frac{3}{k-\frac{1}{4}}$

2. (a) $\displaystyle\sum_{k=2}^{\infty}\frac{k+1}{k^2-k}$ (b) $\displaystyle\sum_{k=1}^{\infty}\frac{2}{k^4+k}$

3. In each part, use the comparison test to show that the series converges.

(a) $\displaystyle\sum_{k=1}^{\infty}\frac{1}{3^k+5}$ (b) $\displaystyle\sum_{k=1}^{\infty}\frac{5\sin^2 k}{k!}$

4. In each part, use the comparison test to show that the series diverges.

(a) $\displaystyle\sum_{k=1}^{\infty}\frac{\ln k}{k}$ (b) $\displaystyle\sum_{k=1}^{\infty}\frac{k}{k^{3/2}-\frac{1}{2}}$

5–10 Use the limit comparison test to determine whether the series converges.

5. $\displaystyle\sum_{k=1}^{\infty}\frac{4k^2-2k+6}{8k^7+k-8}$ **6.** $\displaystyle\sum_{k=1}^{\infty}\frac{1}{9k+6}$

7. $\displaystyle\sum_{k=1}^{\infty}\frac{5}{3^k+1}$ **8.** $\displaystyle\sum_{k=1}^{\infty}\frac{k(k+3)}{(k+1)(k+2)(k+5)}$

9. $\displaystyle\sum_{k=1}^{\infty}\frac{1}{\sqrt[3]{8k^2-3k}}$ **10.** $\displaystyle\sum_{k=1}^{\infty}\frac{1}{(2k+3)^{17}}$

11–16 Use the ratio test to determine whether the series converges. If the test is inconclusive, then say so.

11. $\displaystyle\sum_{k=1}^{\infty}\frac{3^k}{k!}$ **12.** $\displaystyle\sum_{k=1}^{\infty}\frac{4^k}{k^2}$ **13.** $\displaystyle\sum_{k=1}^{\infty}\frac{1}{5k}$

14. $\displaystyle\sum_{k=1}^{\infty}k\left(\frac{1}{2}\right)^k$ **15.** $\displaystyle\sum_{k=1}^{\infty}\frac{k!}{k^3}$ **16.** $\displaystyle\sum_{k=1}^{\infty}\frac{k}{k^2+1}$

17–20 Use the root test to determine whether the series converges. If the test is inconclusive, then say so.

17. $\sum_{k=1}^{\infty}\left(\frac{3k+2}{2k-1}\right)^k$

18. $\sum_{k=1}^{\infty}\left(\frac{k}{100}\right)^k$

19. $\sum_{k=1}^{\infty}\frac{k}{5^k}$

20. $\sum_{k=1}^{\infty}(1-e^{-k})^k$

21–44 Use any method to determine whether the series converges.

21. $\sum_{k=0}^{\infty}\frac{7^k}{k!}$

22. $\sum_{k=1}^{\infty}\frac{1}{2k+1}$

23. $\sum_{k=1}^{\infty}\frac{k^2}{5^k}$

24. $\sum_{k=1}^{\infty}\frac{k!10^k}{3^k}$

25. $\sum_{k=1}^{\infty}k^{50}e^{-k}$

26. $\sum_{k=1}^{\infty}\frac{k^2}{k^3+1}$

27. $\sum_{k=1}^{\infty}\frac{\sqrt{k}}{k^3+1}$

28. $\sum_{k=1}^{\infty}\frac{4}{2+3^k k}$

29. $\sum_{k=1}^{\infty}\frac{1}{\sqrt{k(k+1)}}$

30. $\sum_{k=1}^{\infty}\frac{2+(-1)^k}{5^k}$

31. $\sum_{k=1}^{\infty}\frac{2+\sqrt{k}}{(k+1)^3-1}$

32. $\sum_{k=1}^{\infty}\frac{4+|\cos k|}{k^3}$

33. $\sum_{k=1}^{\infty}\frac{1}{1+\sqrt{k}}$

34. $\sum_{k=1}^{\infty}\frac{k!}{k^k}$

35. $\sum_{k=1}^{\infty}\frac{\ln k}{e^k}$

36. $\sum_{k=1}^{\infty}\frac{k!}{e^{k^2}}$

37. $\sum_{k=0}^{\infty}\frac{(k+4)!}{4!k!4^k}$

38. $\sum_{k=1}^{\infty}\left(\frac{k}{k+1}\right)^{k^2}$

39. $\sum_{k=1}^{\infty}\frac{1}{4+2^{-k}}$

40. $\sum_{k=1}^{\infty}\frac{\sqrt{k}\ln k}{k^3+1}$

41. $\sum_{k=1}^{\infty}\frac{\tan^{-1}k}{k^2}$

42. $\sum_{k=1}^{\infty}\frac{5^k+k}{k!+3}$

43. $\sum_{k=0}^{\infty}\frac{(k!)^2}{(2k)!}$

44. $\sum_{k=1}^{\infty}\frac{(k!)^2 2^k}{(2k+2)!}$

45–46 Find the general term of the series and use the ratio test to show that the series converges.

45. $1+\frac{1\cdot 2}{1\cdot 3}+\frac{1\cdot 2\cdot 3}{1\cdot 3\cdot 5}+\frac{1\cdot 2\cdot 3\cdot 4}{1\cdot 3\cdot 5\cdot 7}+\cdots$

46. $1+\frac{1\cdot 3}{3!}+\frac{1\cdot 3\cdot 5}{5!}+\frac{1\cdot 3\cdot 5\cdot 7}{7!}+\cdots$

47–48 Use a CAS to investigate the convergence of the series.

C **47.** $\sum_{k=1}^{\infty}\frac{\ln k}{3^k}$

C **48.** $\sum_{k=1}^{\infty}\frac{[\pi(k+1)]^k}{k^{k+1}}$

FOCUS ON CONCEPTS

49. (a) Make a conjecture about the convergence of the series $\sum_{k=1}^{\infty}\sin(\pi/k)$ by considering the local linear approximation of $\sin x$ at $x=0$.
(b) Try to confirm your conjecture using the limit comparison test.

50. (a) We will see later that the polynomial $1-x^2/2$ is the "local quadratic" approximation for $\cos x$ at $x=0$. Make a conjecture about the convergence of the series

$$\sum_{k=1}^{\infty}\left[1-\cos\left(\frac{1}{k}\right)\right]$$

by considering this approximation.
(b) Try to confirm your conjecture using the limit comparison test.

51. Show that $\ln x<\sqrt{x}$ if $x>0$, and use this result to investigate the convergence of

(a) $\sum_{k=1}^{\infty}\frac{\ln k}{k^2}$ (b) $\sum_{k=2}^{\infty}\frac{1}{(\ln k)^2}$

52. For which positive values of α does the series $\sum_{k=1}^{\infty}(\alpha^k/k^\alpha)$ converge?

53. Use Theorem 10.4.6 to prove the comparison test (Theorem 10.5.1).

54. Let $\sum a_k$ and $\sum b_k$ be series with positive terms. Prove:
(a) If $\lim_{k\to+\infty}(a_k/b_k)=0$ and $\sum b_k$ converges, then $\sum a_k$ converges.
(b) If $\lim_{k\to+\infty}(a_k/b_k)=+\infty$ and $\sum b_k$ diverges, then $\sum a_k$ diverges.

✔ QUICK CHECK ANSWERS 10.5

1. diverges; $1/k^{2/3}$ 2. converges; ratio 3. diverges; ratio 4. converges; root

10.6 ALTERNATING SERIES; CONDITIONAL CONVERGENCE

Up to now we have focused exclusively on series with nonnegative terms. In this section we will discuss series that contain both positive and negative terms.

ALTERNATING SERIES

Series whose terms alternate between positive and negative, called ***alternating series***, are of special importance. Some examples are

$$\sum_{k=1}^{\infty}(-1)^{k+1}\frac{1}{k} = 1 - \frac{1}{2} + \frac{1}{3} - \frac{1}{4} + \frac{1}{5} - \cdots$$

$$\sum_{k=1}^{\infty}(-1)^{k}\frac{1}{k} = -1 + \frac{1}{2} - \frac{1}{3} + \frac{1}{4} - \frac{1}{5} + \cdots$$

In general, an alternating series has one of the following two forms:

$$\sum_{k=1}^{\infty}(-1)^{k+1}a_k = a_1 - a_2 + a_3 - a_4 + \cdots \tag{1}$$

$$\sum_{k=1}^{\infty}(-1)^{k}a_k = -a_1 + a_2 - a_3 + a_4 - \cdots \tag{2}$$

where the a_k's are assumed to be positive in both cases.

The following theorem is the key result on convergence of alternating series.

10.6.1 THEOREM (*Alternating Series Test*). *An alternating series of either form* (1) *or form* (2) *converges if the following two conditions are satisfied:*

(*a*) $a_1 \geq a_2 \geq a_3 \geq \cdots \geq a_k \geq \cdots$

(*b*) $\lim_{k\to+\infty} a_k = 0$

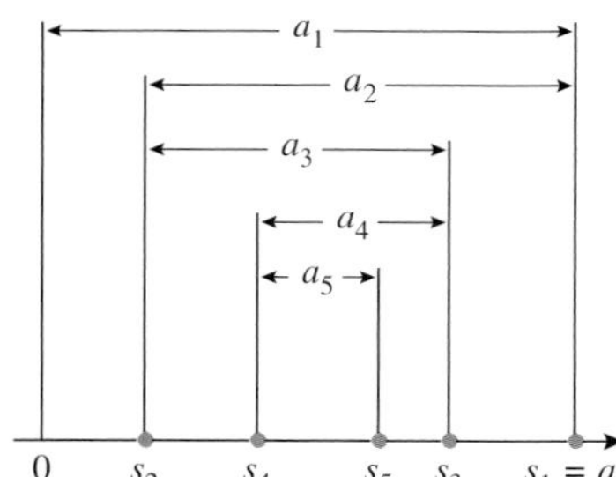

Figure 10.6.1

It is not essential for condition (*a*) in Theorem 10.6.1 to hold for all terms; an alternating series will converge if condition (*b*) is true and condition (*a*) holds eventually.

PROOF. We will consider only alternating series of form (1). The idea of the proof is to show that if conditions (*a*) and (*b*) hold, then the sequences of even-numbered and odd-numbered partial sums converge to a common limit S. It will then follow from Theorem 10.1.4 that the entire sequence of partial sums converges to S.

Figure 10.6.1 shows how successive partial sums satisfying conditions (*a*) and (*b*) appear when plotted on a horizontal axis. The even-numbered partial sums

$$s_2, s_4, s_6, s_8, \ldots, s_{2n}, \ldots$$

form an increasing sequence bounded above by a_1, and the odd-numbered partial sums

$$s_1, s_3, s_5, \ldots, s_{2n-1}, \ldots$$

form a decreasing sequence bounded below by 0. Thus, by Theorems 10.2.3 and 10.2.4, the even-numbered partial sums converge to some limit S_E and the odd-numbered partial sums converge to some limit S_O. To complete the proof we must show that $S_E = S_O$. But the $(2n)$-th term in the series is $-a_{2n}$, so that $s_{2n} - s_{2n-1} = -a_{2n}$, which can be written as

$$s_{2n-1} = s_{2n} + a_{2n}$$

However, $2n \to +\infty$ and $2n - 1 \to +\infty$ as $n \to +\infty$, so that

$$S_O = \lim_{n\to+\infty} s_{2n-1} = \lim_{n\to+\infty}(s_{2n} + a_{2n}) = S_E + 0 = S_E$$

which completes the proof. ■

▶ **Example 1** Use the alternating series test to show that the following series converge.

$$\text{(a)} \sum_{k=1}^{\infty}(-1)^{k+1}\frac{1}{k} \qquad \text{(b)} \sum_{k=1}^{\infty}(-1)^{k+1}\frac{k+3}{k(k+1)}$$

The series in part (a) of Example 1 is called the ***alternating harmonic series***. Note that this series converges, whereas the harmonic series diverges.

Solution (a). The two conditions in the alternating series test are satisfied since

$$a_k = \frac{1}{k} > \frac{1}{k+1} = a_{k+1} \quad \text{and} \quad \lim_{k\to+\infty} a_k = \lim_{k\to+\infty} \frac{1}{k} = 0$$

Solution (b). The two conditions in the alternating series test are satisfied since

$$\frac{a_{k+1}}{a_k} = \frac{k+4}{(k+1)(k+2)} \cdot \frac{k(k+1)}{k+3} = \frac{k^2+4k}{k^2+5k+6} = \frac{k^2+4k}{(k^2+4k)+(k+6)} < 1$$

so

$$a_k > a_{k+1}$$

and

$$\lim_{k\to+\infty} a_k = \lim_{k\to+\infty} \frac{k+3}{k(k+1)} = \lim_{k\to+\infty} \frac{\dfrac{1}{k}+\dfrac{3}{k^2}}{1+\dfrac{1}{k}} = 0 \ ◀$$

APPROXIMATING SUMS OF ALTERNATING SERIES

The following theorem is concerned with the error that results when the sum of an alternating series is approximated by a partial sum.

> **10.6.2 THEOREM.** *If an alternating series satisfies the hypotheses of the alternating series test, and if S is the sum of the series, then:*
>
> (*a*) *S lies between any two successive partial sums; that is, either*
>
> $$s_n \le S \le s_{n+1} \quad or \quad s_{n+1} \le S \le s_n \tag{3}$$
>
> *depending on which partial sum is larger.*
>
> (*b*) *If S is approximated by s_n, then the absolute error $|S - s_n|$ satisfies*
>
> $$|S - s_n| \le a_{n+1} \tag{4}$$
>
> *Moreover, the sign of the error $S - s_n$ is the same as that of the coefficient of a_{n+1}.*

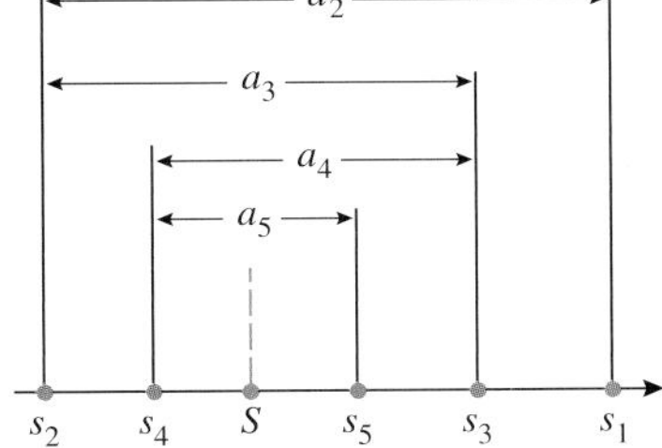

Figure 10.6.2

PROOF. We will prove the theorem for series of form (1). Referring to Figure 10.6.2 and keeping in mind our observation in the proof of Theorem 10.6.1 that the odd-numbered partial sums form a decreasing sequence converging to S and the even-numbered partial sums form an increasing sequence converging to S, we see that successive partial sums oscillate from one side of S to the other in smaller and smaller steps with the odd-numbered partial sums being larger than S and the even-numbered partial sums being smaller than S. Thus, depending on whether n is even or odd, we have

$$s_n \le S \le s_{n+1} \quad \text{or} \quad s_{n+1} \le S \le s_n$$

which proves (3). Moreover, in either case we have

$$|S - s_n| \le |s_{n+1} - s_n| \tag{5}$$

But $s_{n+1} - s_n = \pm a_{n+1}$ (the sign depending on whether n is even or odd). Thus, it follows from (5) that $|S - s_n| \le a_{n+1}$, which proves (4). Finally, since the odd-numbered partial sums are larger than S and the even-numbered partial sums are smaller than S, it follows that $S - s_n$ has the same sign as the coefficient of a_{n+1} (verify). ■

In words, inequality (4) states that for a series satisfying the hypotheses of the alternating series test, the magnitude of the error that results from approximating S by s_n is at most that of the first term that is *not* included in the partial sum. Also, note that if $a_1 > a_2 > \cdots > a_k > \cdots$, then inequality (4) can be strengthened to $|s - s_n| < a_{n+1}$.

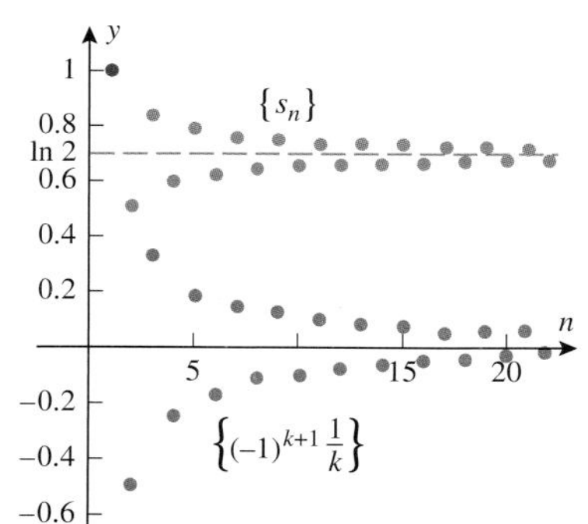

Graph of the sequences of terms and nth partial sums for the alternating harmonic series

Figure 10.6.3

► **Example 2** Later in this chapter we will show that the sum of the alternating harmonic series is

$$\ln 2 = 1 - \frac{1}{2} + \frac{1}{3} - \frac{1}{4} + \cdots + (-1)^{k+1}\frac{1}{k} + \cdots$$

This is illustrated in Figure 10.6.3.

(a) Accepting this to be so, find an upper bound on the magnitude of the error that results if ln 2 is approximated by the sum of the first eight terms in the series.

(b) Find a partial sum that approximates ln 2 to one decimal-place accuracy (the nearest tenth).

Solution (a). It follows from the strengthened form of (4) that

$$|\ln 2 - s_8| < a_9 = \frac{1}{9} < 0.12 \tag{6}$$

As a check, let us compute s_8 exactly. We obtain

$$s_8 = 1 - \frac{1}{2} + \frac{1}{3} - \frac{1}{4} + \frac{1}{5} - \frac{1}{6} + \frac{1}{7} - \frac{1}{8} = \frac{533}{840}$$

Thus, with the help of a calculator

$$|\ln 2 - s_8| = \left|\ln 2 - \frac{533}{840}\right| \approx 0.059$$

This shows that the error is well under the estimate provided by upper bound (6).

Solution (b). For one decimal-place accuracy, we must choose a value of n for which $|\ln 2 - s_n| \le 0.05$. However, it follows from the strengthened form of (4) that

$$|\ln 2 - s_n| < a_{n+1}$$

so it suffices to choose n so that $a_{n+1} \le 0.05$.

One way to find n is to use a calculating utility to obtain numerical values for $a_1, a_2, a_3, \ldots$ until you encounter the first value that is less than or equal to 0.05. If you do this, you will find that it is $a_{20} = 0.05$; this tells us that partial sum s_{19} will provide the desired accuracy. Another way to find n is to solve the inequality

$$\frac{1}{n+1} \le 0.05$$

algebraically. We can do this by taking reciprocals, reversing the sense of the inequality, and then simplifying to obtain $n \ge 19$. Thus, s_{19} will provide the required accuracy, which is consistent with the previous result.

With the help of a calculating utility, the value of s_{19} is approximately $s_{19} \approx 0.7$ and the value of ln 2 obtained directly is approximately $\ln 2 \approx 0.69$, which agrees with s_{19} when rounded to one decimal place. ◄

As Example 2 illustrates, the alternating harmonic series does not provide an efficient way to approximate ln 2, since too many terms and hence too much computation is required to achieve reasonable accuracy. Later, we will develop better ways to approximate logarithms.

ABSOLUTE CONVERGENCE

The series

$$1 - \frac{1}{2} - \frac{1}{2^2} + \frac{1}{2^3} + \frac{1}{2^4} - \frac{1}{2^5} - \frac{1}{2^6} + \cdots$$

does not fit in any of the categories studied so far—it has mixed signs but is not alternating. We will now develop some convergence tests that can be applied to such series.

10.6.3 DEFINITION. A series

$$\sum_{k=1}^{\infty} u_k = u_1 + u_2 + \cdots + u_k + \cdots$$

is said to ***converge absolutely*** if the series of absolute values

$$\sum_{k=1}^{\infty} |u_k| = |u_1| + |u_2| + \cdots + |u_k| + \cdots$$

converges and is said to ***diverge absolutely*** if the series of absolute values diverges.

▶ **Example 3** Determine whether the following series converge absolutely.

(a) $1 - \frac{1}{2} - \frac{1}{2^2} + \frac{1}{2^3} + \frac{1}{2^4} - \frac{1}{2^5} - \cdots$ (b) $1 - \frac{1}{2} + \frac{1}{3} - \frac{1}{4} + \frac{1}{5} - \cdots$

Solution (a). The series of absolute values is the convergent geometric series

$$1 + \frac{1}{2} + \frac{1}{2^2} + \frac{1}{2^3} + \frac{1}{2^4} + \frac{1}{2^5} + \cdots$$

so the given series converges absolutely.

Solution (b). The series of absolute values is the divergent harmonic series

$$1 + \frac{1}{2} + \frac{1}{3} + \frac{1}{4} + \frac{1}{5} + \cdots$$

so the given series diverges absolutely. ◀

It is important to distinguish between the notions of convergence and absolute convergence. For example, the series in part (b) of Example 3 converges, since it is the alternating harmonic series, yet we demonstrated that it does not converge absolutely. However, the following theorem shows that *if a series converges absolutely, then it converges.*

Theorem 10.6.4 provides a way of inferring convergence of a series with positive and negative terms from a related series with nonnegative terms (the series of absolute values). This is important because most of the convergence tests that we have developed apply only to series with nonnegative terms.

10.6.4 THEOREM. *If the series*

$$\sum_{k=1}^{\infty} |u_k| = |u_1| + |u_2| + \cdots + |u_k| + \cdots$$

converges, then so does the series

$$\sum_{k=1}^{\infty} u_k = u_1 + u_2 + \cdots + u_k + \cdots$$

PROOF. We will write the series $\sum u_k$ as

$$\sum_{k=1}^{\infty} u_k = \sum_{k=1}^{\infty} [(u_k + |u_k|) - |u_k|] \tag{7}$$

We are assuming that $\sum |u_k|$ converges, so that if we can show that $\sum (u_k + |u_k|)$ converges, then it will follow from (7) and Theorem 10.4.3(a) that $\sum u_k$ converges. However, the value of $u_k + |u_k|$ is either 0 or $2|u_k|$, depending on the sign of u_k. Thus, in all cases it is true that

$$0 \leq u_k + |u_k| \leq 2|u_k|$$

But $\sum 2|u_k|$ converges, since it is a constant times the convergent series $\sum |u_k|$; hence $\sum (u_k + |u_k|)$ converges by the comparison test. ■

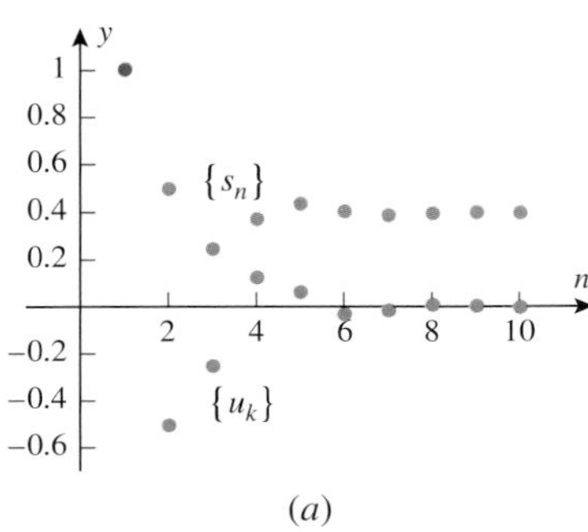

(a)

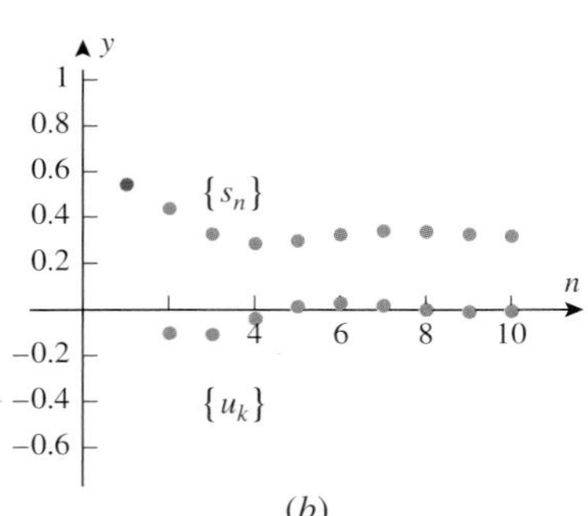

(b)

Graphs of the sequences of terms and nth partial sums for the series in Example 4

Figure 10.6.4

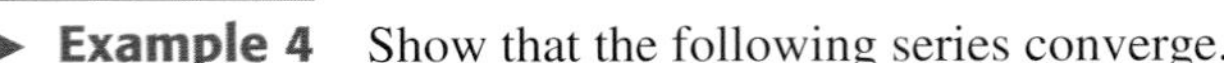

▶ **Example 4** Show that the following series converge.

$$\text{(a) } 1 - \frac{1}{2} - \frac{1}{2^2} + \frac{1}{2^3} + \frac{1}{2^4} - \frac{1}{2^5} - \frac{1}{2^6} + \cdots \qquad \text{(b) } \sum_{k=1}^{\infty} \frac{\cos k}{k^2}$$

Solution (a). Observe that this is not an alternating series because the signs alternate in pairs after the first term. Thus, we have no convergence test that can be applied directly. However, we showed in Example 3(a) that the series converges absolutely, so Theorem 10.6.4 implies that it converges (Figure 10.6.4a).

Solution (b). With the help of a calculating utility, you will be able to verify that the signs of the terms in this series vary irregularly. Thus, we will test for absolute convergence. The series of absolute values is

$$\sum_{k=1}^{\infty} \left| \frac{\cos k}{k^2} \right|$$

However,

$$\left| \frac{\cos k}{k^2} \right| \leq \frac{1}{k^2}$$

But $\sum 1/k^2$ is a convergent p-series ($p = 2$), so the series of absolute values converges by the comparison test. Thus, the given series converges absolutely and hence converges (Figure 10.6.4b). ◀

■ CONDITIONAL CONVERGENCE

Although Theorem 10.6.4 is a useful tool for series that converge absolutely, it provides no information about the convergence or divergence of a series that diverges absolutely. For example, consider the two series

$$1 - \frac{1}{2} + \frac{1}{3} - \frac{1}{4} + \cdots + (-1)^{k+1}\frac{1}{k} + \cdots \tag{8}$$

$$-1 - \frac{1}{2} - \frac{1}{3} - \frac{1}{4} - \cdots - \frac{1}{k} - \cdots \tag{9}$$

Both of these series diverge absolutely, since in each case the series of absolute values is the divergent harmonic series

$$1 + \frac{1}{2} + \frac{1}{3} + \cdots + \frac{1}{k} + \cdots$$

However, series (8) converges, since it is the alternating harmonic series, and series (9) diverges, since it is a constant times the divergent harmonic series. As a matter of terminology, a series that converges but diverges absolutely is said to ***converge conditionally*** (or to be ***conditionally convergent***). Thus, (8) is a conditionally convergent series.

THE RATIO TEST FOR ABSOLUTE CONVERGENCE

Although one cannot generally infer convergence or divergence of a series from absolute divergence, the following variation of the ratio test provides a way of deducing divergence from absolute divergence in certain situations. We omit the proof.

10.6.5 THEOREM (*Ratio Test for Absolute Convergence*). *Let $\sum u_k$ be a series with nonzero terms and suppose that*

$$\rho = \lim_{k\to+\infty} \frac{|u_{k+1}|}{|u_k|}$$

(*a*) *If $\rho < 1$, then the series $\sum u_k$ converges absolutely and therefore converges.*

(*b*) *If $\rho > 1$ or if $\rho = +\infty$, then the series $\sum u_k$ diverges.*

(*c*) *If $\rho = 1$, no conclusion about convergence or absolute convergence can be drawn from this test.*

► **Example 5** Use the ratio test for absolute convergence to determine whether the series converges.

$$\text{(a)}\ \sum_{k=1}^{\infty} (-1)^k \frac{2^k}{k!} \qquad \text{(b)}\ \sum_{k=1}^{\infty} (-1)^k \frac{(2k-1)!}{3^k}$$

Solution (a). Taking the absolute value of the general term u_k we obtain

$$|u_k| = \left|(-1)^k \frac{2^k}{k!}\right| = \frac{2^k}{k!}$$

Thus,

$$\rho = \lim_{k\to+\infty} \frac{|u_{k+1}|}{|u_k|} = \lim_{k\to+\infty} \frac{2^{k+1}}{(k+1)!} \cdot \frac{k!}{2^k} = \lim_{k\to+\infty} \frac{2}{k+1} = 0 < 1$$

which implies that the series converges absolutely and therefore converges.

Solution (b). Taking the absolute value of the general term u_k we obtain

$$|u_k| = \left|(-1)^k \frac{(2k-1)!}{3^k}\right| = \frac{(2k-1)!}{3^k}$$

Thus,

$$\rho = \lim_{k\to+\infty} \frac{|u_{k+1}|}{|u_k|} = \lim_{k\to+\infty} \frac{[2(k+1)-1]!}{3^{k+1}} \cdot \frac{3^k}{(2k-1)!}$$

$$= \lim_{k\to+\infty} \frac{1}{3} \cdot \frac{(2k+1)!}{(2k-1)!} = \frac{1}{3} \lim_{k\to+\infty} (2k)(2k+1) = +\infty$$

which implies that the series diverges. ◄

SUMMARY OF CONVERGENCE TESTS

We conclude this section with a summary of convergence tests that can be used for reference. The skill of selecting a good test is developed through lots of practice. In some instances a test may be inconclusive, so another test must be tried.

Summary of Convergence Tests

NAME	STATEMENT	COMMENTS
Divergence Test (10.4.1)	If $\lim_{k\to+\infty} u_k \neq 0$, then $\sum u_k$ diverges.	If $\lim_{k\to+\infty} u_k = 0$, then $\sum u_k$ may or may not converge.
Integral Test (10.4.4)	Let $\sum u_k$ be a series with positive terms. If f is a function that is decreasing and continuous on an interval $[a, +\infty)$ and such that $u_k = f(k)$ for all $k \geq a$, then $\sum_{k=1}^{\infty} u_k \quad \text{and} \quad \int_a^{+\infty} f(x)\,dx$ both converge or both diverge.	This test only applies to series that have positive terms. Try this test when $f(x)$ is easy to integrate.
Comparison Test (10.5.1)	Let $\sum_{k=1}^{\infty} a_k$ and $\sum_{k=1}^{\infty} b_k$ be series with nonnegative terms such that $a_1 \leq b_1,\ a_2 \leq b_2, \ldots, a_k \leq b_k, \ldots$ If $\sum b_k$ converges, then $\sum a_k$ converges, and if $\sum a_k$ diverges, then $\sum b_k$ diverges.	This test only applies to series with nonnegative terms. Try this test as a last resort; other tests are often easier to apply.
Limit Comparison Test (10.5.4)	Let $\sum a_k$ and $\sum b_k$ be series with positive terms and let $\rho = \lim_{k\to+\infty} \frac{a_k}{b_k}$ If $0 < \rho < +\infty$, then both series converge or both diverge.	This is easier to apply than the comparison test, but still requires some skill in choosing the series $\sum b_k$ for comparison.
Ratio Test (10.5.5)	Let $\sum u_k$ be a series with positive terms and suppose that $\rho = \lim_{k\to+\infty} \frac{u_{k+1}}{u_k}$ (a) Series converges if $\rho < 1$. (b) Series diverges if $\rho > 1$ or $\rho = +\infty$. (c) The test is inconclusive if $\rho = 1$.	Try this test when u_k involves factorials or kth powers.
Root Test (10.5.6)	Let $\sum u_k$ be a series with positive terms and suppose that $\rho = \lim_{k\to+\infty} \sqrt[k]{u_k}$ (a) The series converges if $\rho < 1$. (b) The series diverges if $\rho > 1$ or $\rho = +\infty$. (c) The test is inconclusive if $\rho = 1$.	Try this test when u_k involves kth powers.
Alternating Series Test (10.6.1)	If $a_k > 0$ for $k = 1, 2, 3, \ldots$, then the series $a_1 - a_2 + a_3 - a_4 + \cdots$ $-a_1 + a_2 - a_3 + a_4 - \cdots$ converge if the following conditions hold: (a) $a_1 \geq a_2 \geq a_3 \geq \cdots$ (b) $\lim_{k\to+\infty} a_k = 0$	This test applies only to alternating series.
Ratio Test for Absolute Convergence (10.6.5)	Let $\sum u_k$ be a series with nonzero terms and suppose that $\rho = \lim_{k\to+\infty} \frac{\lvert u_{k+1}\rvert}{\lvert u_k\rvert}$ (a) The series converges absolutely if $\rho < 1$. (b) The series diverges if $\rho > 1$ or $\rho = +\infty$. (c) The test is inconclusive if $\rho = 1$.	The series need not have positive terms and need not be alternating to use this test.

✔ QUICK CHECK EXERCISES 10.6 (See page 675 for answers.)

1. What characterizes an *alternating* series?

2. (a) The series
$$\sum_{k=1}^{\infty} \frac{(-1)^{k+1}}{k^2}$$
converges by the alternating series test since ________ and ________.

(b) If
$$S = \sum_{k=1}^{\infty} \frac{(-1)^{k+1}}{k^2} \quad \text{and} \quad s_9 = \sum_{k=1}^{9} \frac{(-1)^{k+1}}{k^2}$$
then $|S - s_9| <$ ________.

3. Classify each sequence as conditionally convergent, absolutely convergent, or divergent.

(a) $\sum_{k=1}^{\infty} (-1)^{k+1} \frac{1}{k}$: ________

(b) $\sum_{k=1}^{\infty} (-1)^{k} \frac{3k-1}{9k+15}$: ________

(c) $\sum_{k=1}^{\infty} (-1)^{k} \frac{1}{k(k+2)}$: ________

(d) $\sum_{k=1}^{\infty} (-1)^{k+1} \frac{1}{\sqrt[4]{k^3}}$: ________

4. Given that
$$\lim_{k\to+\infty} \frac{(k+1)^4/4^{k+1}}{k^4/4^k} = \lim_{k\to+\infty} \frac{\left(1+\frac{1}{k}\right)^4}{4} = \frac{1}{4}$$
is the series $\sum_{k=1}^{\infty} (-1)^k k^4/4^k$ conditionally convergent, absolutely convergent, or divergent?

EXERCISE SET 10.6

1–2 Show that the series converges by confirming that it satisfies the hypotheses of the alternating series test (Theorem 10.6.1).

1. $\sum_{k=1}^{\infty} \frac{(-1)^{k+1}}{2k+1}$ **2.** $\sum_{k=1}^{\infty} (-1)^{k+1} \frac{k}{3^k}$

3–6 Determine whether the alternating series converges; justify your answer.

3. $\sum_{k=1}^{\infty} (-1)^{k+1} \frac{k+1}{3k+1}$ **4.** $\sum_{k=1}^{\infty} (-1)^{k+1} \frac{k+1}{\sqrt{k}+1}$

5. $\sum_{k=1}^{\infty} (-1)^{k+1} e^{-k}$ **6.** $\sum_{k=3}^{\infty} (-1)^{k} \frac{\ln k}{k}$

7–12 Use the ratio test for absolute convergence (Theorem 10.6.5) to determine whether the series converges or diverges. If the test is inconclusive, say so.

7. $\sum_{k=1}^{\infty} \left(-\frac{3}{5}\right)^k$ **8.** $\sum_{k=1}^{\infty} (-1)^{k+1} \frac{2^k}{k!}$

9. $\sum_{k=1}^{\infty} (-1)^{k+1} \frac{3^k}{k^2}$ **10.** $\sum_{k=1}^{\infty} (-1)^{k} \frac{k}{5^k}$

11. $\sum_{k=1}^{\infty} (-1)^{k} \frac{k^3}{e^k}$ **12.** $\sum_{k=1}^{\infty} (-1)^{k+1} \frac{k^k}{k!}$

13–30 Classify each series as absolutely convergent, conditionally convergent, or divergent.

13. $\sum_{k=1}^{\infty} \frac{(-1)^{k+1}}{3k}$ **14.** $\sum_{k=1}^{\infty} \frac{(-1)^{k+1}}{k^{4/3}}$ **15.** $\sum_{k=1}^{\infty} \frac{(-4)^k}{k^2}$

16. $\sum_{k=1}^{\infty} \frac{(-1)^{k+1}}{k!}$ **17.** $\sum_{k=1}^{\infty} \frac{\cos k\pi}{k}$ **18.** $\sum_{k=3}^{\infty} \frac{(-1)^k \ln k}{k}$

19. $\sum_{k=1}^{\infty} (-1)^{k+1} \frac{k+2}{k(k+3)}$ **20.** $\sum_{k=1}^{\infty} \frac{(-1)^{k+1} k^2}{k^3+1}$

21. $\sum_{k=1}^{\infty} \sin \frac{k\pi}{2}$ **22.** $\sum_{k=1}^{\infty} \frac{\sin k}{k^3}$

23. $\sum_{k=2}^{\infty} \frac{(-1)^k}{k \ln k}$ **24.** $\sum_{k=1}^{\infty} \frac{(-1)^k}{\sqrt{k(k+1)}}$

25. $\sum_{k=2}^{\infty} \left(-\frac{1}{\ln k}\right)^k$ **26.** $\sum_{k=1}^{\infty} \frac{(-1)^{k+1}}{\sqrt{k+1}+\sqrt{k}}$

27. $\sum_{k=2}^{\infty} \frac{(-1)^k (k^2+1)}{k^3+2}$ **28.** $\sum_{k=1}^{\infty} \frac{k \cos k\pi}{k^2+1}$

29. $\sum_{k=1}^{\infty} \frac{(-1)^{k+1} k!}{(2k-1)!}$ **30.** $\sum_{k=1}^{\infty} (-1)^{k+1} \frac{3^{2k-1}}{k^2+1}$

31–34 Each series satisfies the hypotheses of the alternating series test. For the stated value of n, find an upper bound on the absolute error that results if the sum of the series is approximated by the nth partial sum.

31. $\sum_{k=1}^{\infty} \frac{(-1)^{k+1}}{k}$; $n = 7$ **32.** $\sum_{k=1}^{\infty} \frac{(-1)^{k+1}}{k!}$; $n = 5$

33. $\sum_{k=1}^{\infty} \frac{(-1)^{k+1}}{\sqrt{k}}$; $n = 99$

34. $\sum_{k=1}^{\infty} \frac{(-1)^{k+1}}{(k+1)\ln(k+1)}$; $n = 3$

35–38 Each series satisfies the hypotheses of the alternating series test. Find a value of n for which the nth partial sum is ensured to approximate the sum of the series to the stated accuracy.

35. $\sum_{k=1}^{\infty} \frac{(-1)^{k+1}}{k}$; $|\text{error}| < 0.0001$

36. $\sum_{k=1}^{\infty} \frac{(-1)^{k+1}}{k!}$; $|\text{error}| < 0.00001$

37. $\sum_{k=1}^{\infty} \frac{(-1)^{k+1}}{\sqrt{k}}$; two decimal places

38. $\sum_{k=1}^{\infty} \frac{(-1)^{k+1}}{(k+1)\ln(k+1)}$; one decimal place

39–40 Find an upper bound on the absolute error that results if s_{10} is used to approximate the sum of the given *geometric* series. Compute s_{10} rounded to four decimal places and compare this value with the exact sum of the series.

39. $\frac{3}{4} - \frac{3}{8} + \frac{3}{16} - \frac{3}{32} + \cdots$ **40.** $1 - \frac{2}{3} + \frac{4}{9} - \frac{8}{27} + \cdots$

41–44 Each series satisfies the hypotheses of the alternating series test. Approximate the sum of the series to two decimal-place accuracy.

41. $1 - \frac{1}{3!} + \frac{1}{5!} - \frac{1}{7!} + \cdots$ **42.** $1 - \frac{1}{2!} + \frac{1}{4!} - \frac{1}{6!} + \cdots$

43. $\frac{1}{1 \cdot 2} - \frac{1}{2 \cdot 2^2} + \frac{1}{3 \cdot 2^3} - \frac{1}{4 \cdot 2^4} + \cdots$

44. $\frac{1}{1^5 + 4 \cdot 1} - \frac{1}{3^5 + 4 \cdot 3} + \frac{1}{5^5 + 4 \cdot 5} - \frac{1}{7^5 + 4 \cdot 7} + \cdots$

FOCUS ON CONCEPTS

C **45.** The purpose of this exercise is to show that the error bound in part (*b*) of Theorem 10.6.2 can be overly conservative in certain cases.
(a) Use a CAS to confirm that
$$\frac{\pi}{4} = 1 - \frac{1}{3} + \frac{1}{5} - \frac{1}{7} + \cdots$$
(b) Use the CAS to show that $|(\pi/4) - s_{25}| < 10^{-2}$.
(c) According to the error bound in part (*b*) of Theorem 10.6.2, what value of n is required to ensure that $|(\pi/4) - s_n| < 10^{-2}$?

46. Prove: If a series $\sum a_k$ converges absolutely, then the series $\sum a_k^2$ converges.

47. (a) Find examples to show that if $\sum a_k$ converges, then $\sum a_k^2$ may diverge or converge.
(b) Find examples to show that if $\sum a_k^2$ converges, then $\sum a_k$ may diverge or converge.

48. Show that the alternating p-series
$$1 - \frac{1}{2^p} + \frac{1}{3^p} - \frac{1}{4^p} + \cdots + (-1)^{k+1}\frac{1}{k^p} + \cdots$$
converges absolutely if $p > 1$, converges conditionally if $0 < p \le 1$, and diverges if $p \le 0$.

49–51 It can be proved that any series that is constructed from an absolutely convergent series by rearranging the terms is absolutely convergent and has the same sum as the original series. Use this fact together with parts (*a*) and (*b*) of Theorem 10.4.3 in these exercises.

49. It was stated in Exercise 31 of Section 10.4 that
$$\frac{\pi^2}{6} = 1 + \frac{1}{2^2} + \frac{1}{3^2} + \frac{1}{4^2} + \cdots$$
Use this to show that
$$\frac{\pi^2}{8} = 1 + \frac{1}{3^2} + \frac{1}{5^2} + \frac{1}{7^2} + \cdots$$

50. Use the series for $\pi^2/6$ given in the preceding exercise to show that
$$\frac{\pi^2}{12} = 1 - \frac{1}{2^2} + \frac{1}{3^2} - \frac{1}{4^2} + \cdots$$

51. It was stated in Exercise 31 of Section 10.4 that
$$\frac{\pi^4}{90} = 1 + \frac{1}{2^4} + \frac{1}{3^4} + \frac{1}{4^4} + \cdots$$
Use this to show that
$$\frac{\pi^4}{96} = 1 + \frac{1}{3^4} + \frac{1}{5^4} + \frac{1}{7^4} + \cdots$$

FOCUS ON CONCEPTS

52. It can be proved that the terms of any conditionally convergent series can be rearranged to give either a divergent series or a conditionally convergent series whose sum is any given number S. For example, we stated in Example 2 that
$$\ln 2 = 1 - \frac{1}{2} + \frac{1}{3} - \frac{1}{4} + \frac{1}{5} - \frac{1}{6} + \cdots$$
Show that we can rearrange this series so that its sum is $\frac{1}{2}\ln 2$ by rewriting it as
$$\left(1 - \frac{1}{2} - \frac{1}{4}\right) + \left(\frac{1}{3} - \frac{1}{6} - \frac{1}{8}\right) + \left(\frac{1}{5} - \frac{1}{10} - \frac{1}{12}\right) + \cdots$$
[*Hint:* Add the first two terms in each grouping.]

53. Consider the series
$$1 - \frac{1}{2} + \frac{2}{3} - \frac{1}{3} + \frac{2}{4} - \frac{1}{4} + \frac{2}{5} - \frac{1}{5} + \cdots$$
(a) Show that this series diverges.
(b) Explain why the alternating series test does not apply to this series.

54. (a) Use a graphing utility to graph

$$f(x) = \frac{4x-1}{4x^2-2x}, \quad x \geq 1$$

(b) Based on your graph, do you think that the series

$$\sum_{k=1}^{\infty} (-1)^{k+1} \frac{4k-1}{4k^2-2k}$$

converges? Explain your reasoning.

55. As illustrated in the accompanying figure, a bug, starting at point A on a 180-cm wire, walks the length of the wire, stops and walks in the opposite direction for half the length of the wire, stops again and walks in the opposite direction for one-third the length of the wire, stops again and walks in the opposite direction for one-fourth the length of the wire, and so forth until it stops for the 1000th time.

(a) Give upper and lower bounds on the distance between the bug and point A when it finally stops. [*Hint:* As stated in Example 2, assume that the sum of the alternating harmonic series is ln 2.]

(b) Give upper and lower bounds on the total distance that the bug has traveled when it finally stops. [*Hint:* Use inequality (2) of Section 10.4.]

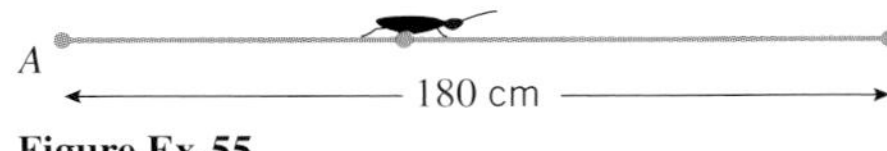

Figure Ex-55

✔ QUICK CHECK ANSWERS 10.6

1. Terms alternate between positive and negative. **2.** (a) $1 \geq \frac{1}{4} \geq \frac{1}{9} \geq \cdots \geq \frac{1}{k^2} \geq \frac{1}{(k+1)^2} \geq \cdots$; $\lim_{k \to +\infty} \frac{1}{k^2} = 0$ (b) $\frac{1}{100}$ **3.** (a) conditionally convergent (b) divergent (c) absolutely convergent (d) conditionally convergent **4.** absolutely convergent

10.7 MACLAURIN AND TAYLOR POLYNOMIALS

In a local linear approximation the tangent line to the graph of a function is used to obtain a linear approximation of the function near the point of tangency. In this section we will consider how one might improve on the accuracy of local linear approximations by using higher-order polynomials as approximations functions. We will also investigate the error associated with such approximations.

LOCAL QUADRATIC APPROXIMATIONS

Recall from Formula (1) in Section 3.9 that the local linear approximation of a function f at x_0 is

$$f(x) \approx f(x_0) + f'(x_0)(x - x_0) \tag{1}$$

In this formula, the approximating function

$$p(x) = f(x_0) + f'(x_0)(x - x_0)$$

is a first-degree polynomial satisfying $p(x_0) = f(x_0)$ and $p'(x_0) = f'(x_0)$ (verify). Thus, the local linear approximation of f at x_0 has the property that its value and the value of its first derivative match those of f at x_0.

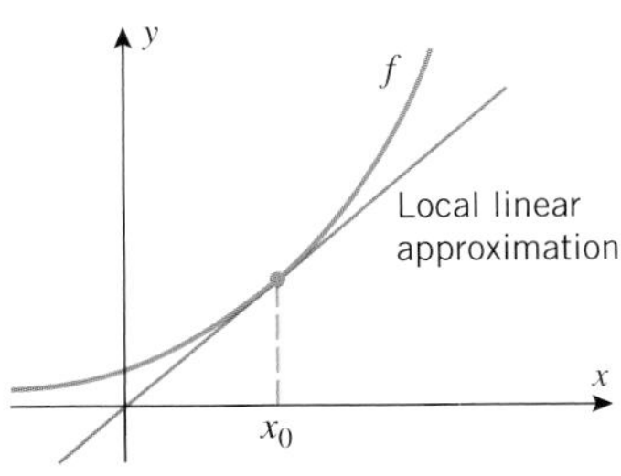

Figure 10.7.1

If the graph of a function f has a pronounced "bend" at x_0, then we can expect that the accuracy of the local linear approximation of f at x_0 will decrease rapidly as we progress away from x_0 (Figure 10.7.1). One way to deal with this problem is to approximate the function f at x_0 by a polynomial p of degree 2 with the property that the value of p and the values of its first two derivatives match those of f at x_0. This ensures that the graphs of f and p not only have the same tangent line at x_0, but they also bend in the same direction at x_0 (both concave up or concave down). As a result, we can expect that the graph of p will remain close to the graph of f over a larger interval around x_0 than the graph of the local linear approximation. The polynomial p is called the ***local quadratic approximation of f at $x = x_0$***.

Colin Maclaurin (1698–1746) Scottish mathematician. Maclaurin's father, a minister, died when the boy was only six months old, and his mother when he was nine years old. He was then raised by an uncle who was also a minister. Maclaurin entered Glasgow University as a divinity student but switched to mathematics after one year. He received his Master's degree at age 17 and, in spite of his youth, began teaching at Marischal College in Aberdeen, Scotland. He met Isaac Newton during a visit to London in 1719 and from that time on became Newton's disciple. During that era, some of Newton's analytic methods were bitterly attacked by major mathematicians and much of Maclaurin's important mathematical work resulted from his efforts to defend Newton's ideas geometrically. Maclaurin's work, *A Treatise of Fluxions* (1742), was the first systematic formulation of Newton's methods. The treatise was so carefully done that it was a standard of mathematical rigor in calculus until the work of Cauchy in 1821. Maclaurin was also an outstanding experimentalist; he devised numerous ingenious mechanical devices, made important astronomical observations, performed actuarial computations for insurance societies, and helped to improve maps of the islands around Scotland.

To illustrate this idea, let us try to find a formula for the local quadratic approximation of a function f at $x = 0$. This approximation has the form

$$f(x) \approx c_0 + c_1x + c_2x^2 \tag{2}$$

where c_0, c_1, and c_2 must be chosen so that the values of

$$p(x) = c_0 + c_1x + c_2x^2$$

and its first two derivatives match those of f at 0. Thus, we want

$$p(0) = f(0), \quad p'(0) = f'(0), \quad p''(0) = f''(0) \tag{3}$$

But the values of $p(0)$, $p'(0)$, and $p''(0)$ are as follows:

$$\begin{aligned} p(x) &= c_0 + c_1x + c_2x^2 & \qquad p(0) &= c_0 \\ p'(x) &= c_1 + 2c_2x & p'(0) &= c_1 \\ p''(x) &= 2c_2 & p''(0) &= 2c_2 \end{aligned}$$

Thus, it follows from (3) that

$$c_0 = f(0), \quad c_1 = f'(0), \quad c_2 = \frac{f''(0)}{2}$$

and substituting these in (2) yields the following formula for the local quadratic approximation of f at $x = 0$:

$$f(x) \approx f(0) + f'(0)x + \frac{f''(0)}{2}x^2 \tag{4}$$

► **Example 1** Find the local linear and quadratic approximations of e^x at $x = 0$, and graph e^x and the two approximations together.

Solution. If we let $f(x) = e^x$, then $f'(x) = f''(x) = e^x$; and hence

$$f(0) = f'(0) = f''(0) = e^0 = 1$$

Thus, from (4) the local quadratic approximation of e^x at $x = 0$ is

$$e^x \approx 1 + x + \frac{x^2}{2}$$

and the local linear approximation (which is the linear part of the local quadratic approximation) is

$$e^x \approx 1 + x$$

The graphs of e^x and the two approximations are shown in Figure 10.7.2. As expected, the local quadratic approximation is more accurate than the local linear approximation near $x = 0$. ◄

Figure 10.7.2

MACLAURIN POLYNOMIALS

It is natural to ask whether one can improve on the accuracy of a local quadratic approximation by using a polynomial of degree 3. Specifically, one might look for a polynomial of degree 3 with the property that its value and the values of its first three derivatives match those of f at a point; and if this provides an improvement in accuracy, why not go on to polynomials of even higher degree? Thus, we are led to consider the following general problem.

10.7.1 PROBLEM. Given a function f that can be differentiated n times at $x = x_0$, find a polynomial p of degree n with the property that the value of p and the values of its first n derivatives match those of f at x_0.

We will begin by solving this problem in the case where $x_0 = 0$. Thus, we want a polynomial

$$p(x) = c_0 + c_1x + c_2x^2 + c_3x^3 + \cdots + c_nx^n \tag{5}$$

such that

$$f(0) = p(0), \quad f'(0) = p'(0), \quad f''(0) = p''(0), \ldots, \quad f^{(n)}(0) = p^{(n)}(0) \tag{6}$$

But

$$\begin{aligned}
p(x) &= c_0 + c_1x + c_2x^2 + c_3x^3 + \cdots + c_nx^n \\
p'(x) &= c_1 + 2c_2x + 3c_3x^2 + \cdots + nc_nx^{n-1} \\
p''(x) &= 2c_2 + 3\cdot 2c_3x + \cdots + n(n-1)c_nx^{n-2} \\
p'''(x) &= 3\cdot 2c_3 + \cdots + n(n-1)(n-2)c_nx^{n-3} \\
&\vdots \\
p^{(n)}(x) &= n(n-1)(n-2)\cdots(1)c_n
\end{aligned}$$

Thus, to satisfy (6) we must have

$$\begin{aligned}
f(0) &= p(0) &&= c_0 \\
f'(0) &= p'(0) &&= c_1 \\
f''(0) &= p''(0) &&= 2c_2 = 2!c_2 \\
f'''(0) &= p'''(0) &&= 3\cdot 2c_3 = 3!c_3 \\
&\vdots \\
f^{(n)}(0) &= p^{(n)}(0) &&= n(n-1)(n-2)\cdots(1)c_n = n!c_n
\end{aligned}$$

which yields the following values for the coefficients of $p(x)$:

$$c_0 = f(0), \quad c_1 = f'(0), \quad c_2 = \frac{f''(0)}{2!}, \quad c_3 = \frac{f'''(0)}{3!}, \ldots, \quad c_n = \frac{f^{(n)}(0)}{n!}$$

The polynomial that results by using these coefficients in (5) is called the *nth Maclaurin polynomial for f*.

Augustin Louis Cauchy (1789–1857) French mathematician. Cauchy's early education was acquired from his father, a barrister and master of the classics. Cauchy entered L'Ecole Polytechnique in 1805 to study engineering, but because of poor health, was advised to concentrate on mathematics. His major mathematical work began in 1811 with a series of brilliant solutions to some difficult outstanding problems. In 1814 he wrote a treatise on integrals that was to become the basis for modern complex variable theory; in 1816 there followed a classic paper on wave propagation in liquids that won a prize from the French Academy; and in 1822 he wrote a paper that formed the basis of modern elasticity theory. Cauchy's mathematical contributions for the next 35 years were brilliant and staggering in quantity, over 700 papers filling 26 modern volumes. Cauchy's work initiated the era of modern analysis. He brought to mathematics standards of precision and rigor undreamed of by Leibniz and Newton.

Cauchy's life was inextricably tied to the political upheavals of the time. A strong partisan of the Bourbons, he left his wife and children in 1830 to follow the Bourbon king Charles X into exile. For his loyalty he was made a baron by the ex-king. Cauchy eventually returned to France, but refused to accept a university position until the government waived its requirement that he take a loyalty oath.

It is difficult to get a clear picture of the man. Devoutly Catholic, he sponsored charitable work for unwed mothers, criminals, and relief for Ireland. Yet other aspects of his life cast him in an unfavorable light. The Norwegian mathematician Abel described him as, "mad, infinitely Catholic, and bigoted." Some writers praise his teaching, yet others say he rambled incoherently and, according to a report of the day, he once devoted an entire lecture to extracting the square root of seventeen to ten decimal places by a method well known to his students. In any event, Cauchy is undeniably one of the greatest minds in the history of science.

Verify that $f(x) \approx p_1(x)$ is the local linear approximation of f at $x = 0$, and $f(x) \approx p_2(x)$ is the local quadratic approximation at $x = 0$. Thus, the polynomials in these approximations are special cases of the Maclaurin polynomials for f.

10.7.2 DEFINITION. If f can be differentiated n times at 0, then we define the ***nth Maclaurin polynomial for f*** to be

$$p_n(x) = f(0) + f'(0)x + \frac{f''(0)}{2!}x^2 + \frac{f'''(0)}{3!}x^3 + \cdots + \frac{f^{(n)}(0)}{n!}x^n \tag{7}$$

This polynomial has the property that its value and the values of its first n derivatives match the values of f and its first n derivatives at $x = 0$.

▶ **Example 2** Find the Maclaurin polynomials p_0, p_1, p_2, p_3, and p_n for e^x.

Solution. Let $f(x) = e^x$. Thus,

$$f'(x) = f''(x) = f'''(x) = \cdots = f^{(n)}(x) = e^x$$

and

$$f(0) = f'(0) = f''(0) = f'''(0) = \cdots = f^{(n)}(0) = e^0 = 1$$

Therefore,

$$\begin{aligned}
p_0(x) &= f(0) = 1 \\
p_1(x) &= f(0) + f'(0)x = 1 + x \\
p_2(x) &= f(0) + f'(0)x + \frac{f''(0)}{2!}x^2 = 1 + x + \frac{x^2}{2!} = 1 + x + \frac{1}{2}x^2 \\
p_3(x) &= f(0) + f'(0)x + \frac{f''(0)}{2!}x^2 + \frac{f'''(0)}{3!}x^3 \\
&= 1 + x + \frac{x^2}{2!} + \frac{x^3}{3!} = 1 + x + \frac{1}{2}x^2 + \frac{1}{6}x^3 \\
p_n(x) &= f(0) + f'(0)x + \frac{f''(0)}{2!}x^2 + \cdots + \frac{f^{(n)}(0)}{n!}x^n \\
&= 1 + x + \frac{x^2}{2!} + \cdots + \frac{x^n}{n!}
\end{aligned}$$ ◀

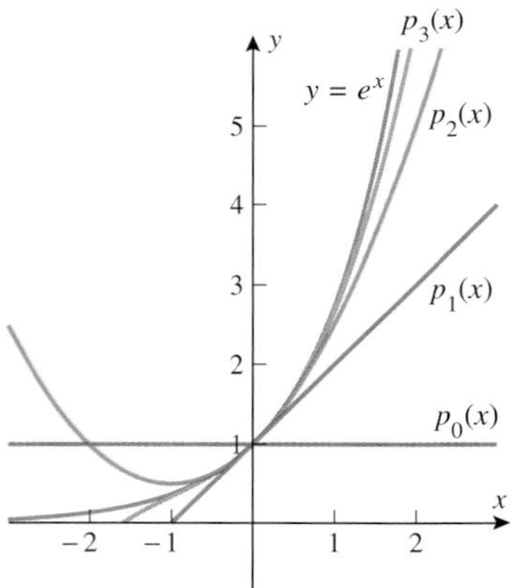

Figure 10.7.3

Figure 10.7.3 shows the graph of e^x (in blue) and the graph of the first four Maclaurin polynomials. Note that the graphs of $p_1(x)$, $p_2(x)$, and $p_3(x)$ are virtually indistinguishable from the graph of e^x near $x = 0$, so these polynomials are good approximations of e^x for x near 0. However, the farther x is from 0, the poorer these approximations become. This is typical of the Maclaurin polynomials for a function $f(x)$; they provide good approximations of $f(x)$ near 0, but the accuracy diminishes as x progresses away from 0. It is usually the case that the higher the degree of the polynomial, the larger the interval on which it provides a specified accuracy. Accuracy issues will be investigated later.

TAYLOR POLYNOMIALS

Up to now we have focused on approximating a function f in the vicinity of $x = 0$. Now we will consider the more general case of approximating f in the vicinity of an arbitrary domain value x_0. The basic idea is the same as before; we want to find an nth-degree polynomial p with the property that its value and the values of its first n derivatives match those of f at x_0. However, rather than expressing $p(x)$ in powers of x, it will simplify the computations if we express it in powers of $x - x_0$; that is,

$$p(x) = c_0 + c_1(x - x_0) + c_2(x - x_0)^2 + \cdots + c_n(x - x_0)^n \tag{8}$$

Verify that $f(x) \approx p_1(x)$ is the local linear approximation of f at $x = x_0$, and $f(x) \approx p_2(x)$ is the local quadratic approximation at $x = x_0$. Thus, the polynomials in these approximations are special cases of the Taylor polynomials for f at $x = x_0$.

We will leave it as an exercise for you to imitate the computations used in the case where $x_0 = 0$ to show that

$$c_0 = f(x_0), \quad c_1 = f'(x_0), \quad c_2 = \frac{f''(x_0)}{2!}, \quad c_3 = \frac{f'''(x_0)}{3!}, \ldots, \quad c_n = \frac{f^{(n)}(x_0)}{n!}$$

Substituting these values in (8) we obtain a polynomial called the *nth Taylor polynomial about $x = x_0$ for f*.

10.7.3 DEFINITION. If f can be differentiated n times at x_0, then we define the ***nth Taylor polynomial for f about $x = x_0$*** to be

$$p_n(x) = f(x_0) + f'(x_0)(x - x_0) + \frac{f''(x_0)}{2!}(x - x_0)^2 + \frac{f'''(x_0)}{3!}(x - x_0)^3 + \cdots + \frac{f^{(n)}(x_0)}{n!}(x - x_0)^n \tag{9}$$

The Maclaurin polynomials are the special cases of the Taylor polynomials in which $x_0 = 0$. Thus, theorems about Taylor polynomials also apply to Maclaurin polynomials.

▶ Example 3 Find the first four Taylor polynomials for $\ln x$ about $x = 2$.

Solution. Let $f(x) = \ln x$. Thus,

$$\begin{aligned} f(x) &= \ln x & f(2) &= \ln 2 \\ f'(x) &= 1/x & f'(2) &= 1/2 \\ f''(x) &= -1/x^2 & f''(2) &= -1/4 \\ f'''(x) &= 2/x^3 & f'''(2) &= 1/4 \end{aligned}$$

Substituting in (9) with $x_0 = 2$ yields

$$\begin{aligned} p_0(x) &= f(2) = \ln 2 \\ p_1(x) &= f(2) + f'(2)(x-2) = \ln 2 + \tfrac{1}{2}(x-2) \\ p_2(x) &= f(2) + f'(2)(x-2) + \frac{f''(2)}{2!}(x-2)^2 = \ln 2 + \tfrac{1}{2}(x-2) - \tfrac{1}{8}(x-2)^2 \\ p_3(x) &= f(2) + f'(2)(x-2) + \frac{f''(2)}{2!}(x-2)^2 + \frac{f'''(2)}{3!}(x-2)^3 \\ &= \ln 2 + \tfrac{1}{2}(x-2) - \tfrac{1}{8}(x-2)^2 + \tfrac{1}{24}(x-2)^3 \end{aligned}$$

Brook Taylor (1685–1731) English mathematician. Taylor was born of well-to-do parents. Musicians and artists were entertained frequently in the Taylor home, which undoubtedly had a lasting influence on him. In later years, Taylor published a definitive work on the mathematical theory of perspective and obtained major mathematical results about the vibrations of strings. There also exists an unpublished work, *On Musick*, that was intended to be part of a joint paper with Isaac Newton. Taylor's life was scarred with unhappiness, illness, and tragedy. Because his first wife was not rich enough to suit his father, the two men argued bitterly and parted ways. Subsequently, his wife died in childbirth. Then, after he remarried, his second wife also died in childbirth, though his daughter survived. Taylor's most productive period was from 1714 to 1719, during which time he wrote on a wide range of subjects—magnetism, capillary action, thermometers, perspective, and calculus. In his final years, Taylor devoted his writing efforts to religion and philosophy. According to Taylor, the results that bear his name were motivated by coffeehouse conversations about works of Newton on planetary motion and works of Halley ("Halley's comet") on roots of polynomials. Unfortunately, Taylor's writing style was so terse and hard to understand that he never received credit for many of his innovations.

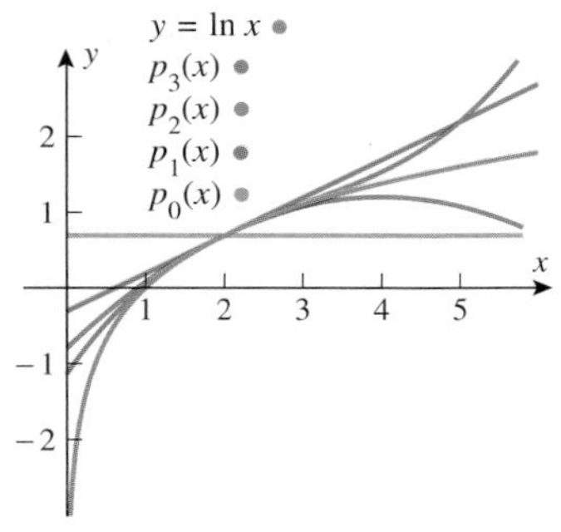

Figure 10.7.4

The graph of $\ln x$ (in blue) and its first four Taylor polynomials about $x = 2$ are shown in Figure 10.7.4. As expected, these polynomials produce their best approximations of $\ln x$ near 2. ◀

SIGMA NOTATION FOR TAYLOR AND MACLAURIN POLYNOMIALS

Frequently, we will want to express Formula (9) in sigma notation. To do this, we use the notation $f^{(k)}(x_0)$ to denote the kth derivative of f at $x = x_0$, and we make the convention that $f^{(0)}(x_0)$ denotes $f(x_0)$. This enables us to write

$$\sum_{k=0}^{n} \frac{f^{(k)}(x_0)}{k!}(x - x_0)^k = f(x_0) + f'(x_0)(x - x_0) + \frac{f''(x_0)}{2!}(x - x_0)^2 + \cdots + \frac{f^{(n)}(x_0)}{n!}(x - x_0)^n \tag{10}$$

In particular, we can write the nth Maclaurin polynomial for $f(x)$ as

$$\sum_{k=0}^{n} \frac{f^{(k)}(0)}{k!}x^k = f(0) + f'(0)x + \frac{f''(0)}{2!}x^2 + \cdots + \frac{f^{(n)}(0)}{n!}x^n \tag{11}$$

▶ **Example 4** Find the nth Maclaurin polynomials for

(a) $\sin x$ (b) $\cos x$ (c) $\dfrac{1}{1 - x}$

Solution (a). In the Maclaurin polynomials for $\sin x$, only the odd powers of x appear explicitly. To see this, let $f(x) = \sin x$; thus,

$$\begin{array}{llll} f(x) &= \sin x & f(0) &= 0 \\ f'(x) &= \cos x & f'(0) &= 1 \\ f''(x) &= -\sin x & f''(0) &= 0 \\ f'''(x) &= -\cos x & f'''(0) &= -1 \end{array}$$

Since $f^{(4)}(x) = \sin x = f(x)$, the pattern $0, 1, 0, -1$ will repeat as we evaluate successive derivatives at 0. Therefore, the successive Maclaurin polynomials for $\sin x$ are

$$\begin{aligned} p_0(x) &= 0 \\ p_1(x) &= 0 + x \\ p_2(x) &= 0 + x + 0 \\ p_3(x) &= 0 + x + 0 - \frac{x^3}{3!} \\ p_4(x) &= 0 + x + 0 - \frac{x^3}{3!} + 0 \\ p_5(x) &= 0 + x + 0 - \frac{x^3}{3!} + 0 + \frac{x^5}{5!} \\ p_6(x) &= 0 + x + 0 - \frac{x^3}{3!} + 0 + \frac{x^5}{5!} + 0 \\ p_7(x) &= 0 + x + 0 - \frac{x^3}{3!} + 0 + \frac{x^5}{5!} + 0 - \frac{x^7}{7!} \end{aligned}$$

Because of the zero terms, each even-order Maclaurin polynomial [after $p_0(x)$] is the same as the preceding odd-order Maclaurin polynomial. That is,

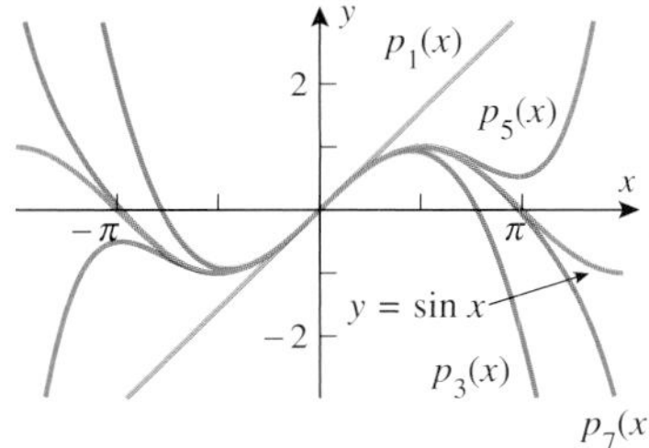

Figure 10.7.5

$$p_{2k+1}(x) = p_{2k+2}(x) = x - \frac{x^3}{3!} + \frac{x^5}{5!} - \frac{x^7}{7!} + \cdots + (-1)^k \frac{x^{2k+1}}{(2k+1)!} \quad (k = 0, 1, 2, \ldots)$$

The graphs of $\sin x$, $p_1(x)$, $p_3(x)$, $p_5(x)$, and $p_7(x)$ are shown in Figure 10.7.5.

Solution (b). In the Maclaurin polynomials for $\cos x$, only the even powers of x appear explicitly; the computations are similar to those in part (a). The reader should be able to show that

$$p_0(x) = p_1(x) = 1$$

$$p_2(x) = p_3(x) = 1 - \frac{x^2}{2!}$$

$$p_4(x) = p_5(x) = 1 - \frac{x^2}{2!} + \frac{x^4}{4!}$$

$$p_6(x) = p_7(x) = 1 - \frac{x^2}{2!} + \frac{x^4}{4!} - \frac{x^6}{6!}$$

In general, the Maclaurin polynomials for $\cos x$ are given by

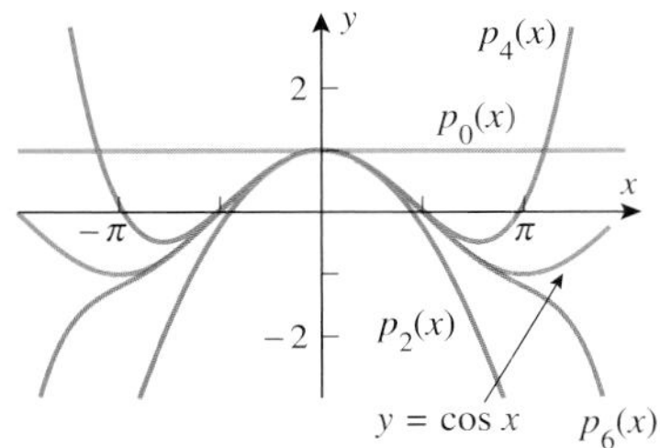

Figure 10.7.6

$$p_{2k}(x) = p_{2k+1}(x) = 1 - \frac{x^2}{2!} + \frac{x^4}{4!} - \frac{x^6}{6!} + \cdots + (-1)^k \frac{x^{2k}}{(2k)!} \quad (k = 0, 1, 2, \ldots)$$

The graphs of $\cos x$, $p_0(x)$, $p_2(x)$, $p_4(x)$, and $p_6(x)$ are shown in Figure 10.7.6.

Solution (c). Let $f(x) = 1/(1-x)$. The values of f and its first k derivatives at $x = 0$ are as follows:

$$\begin{aligned}
f(x) &= \frac{1}{1-x} & f(0) &= 1 = 0! \\
f'(x) &= \frac{1}{(1-x)^2} & f'(0) &= 1 = 1! \\
f''(x) &= \frac{2}{(1-x)^3} & f''(0) &= 2 = 2! \\
f'''(x) &= \frac{3 \cdot 2}{(1-x)^4} & f'''(0) &= 3! \\
f^{(4)}(x) &= \frac{4 \cdot 3 \cdot 2}{(1-x)^5} & f^{(4)}(0) &= 4! \\
&\vdots & &\vdots \\
f^{(k)}(x) &= \frac{k!}{(1-x)^{k+1}} & f^{(k)}(0) &= k!
\end{aligned}$$

Thus, substituting $f^{(k)}(0) = k!$ into Formula (11) yields the nth Maclaurin polynomial for $1/(1-x)$:

$$p_n(x) = \sum_{k=0}^{n} x^k = 1 + x + x^2 + \cdots + x^n \quad (n = 0, 1, 2, \ldots) \blacktriangleleft$$

TECHNOLOGY MASTERY

Computer algebra systems have commands for generating Taylor polynomials of any specified degree. If you have a CAS, use it to find some of the Maclaurin and Taylor polynomials in Examples 3 and 4.

▶ **Example 5** Find the nth Taylor polynomial for $1/x$ about $x = 1$.

Solution. Let $f(x) = 1/x$. The computations are similar to those in part (c) of Example 4. We leave it for you to show that

$$f(1) = 1, \quad f'(1) = -1, \quad f''(1) = 2!, \quad f'''(1) = -3!,$$
$$f^{(4)}(1) = 4!, \ldots, \quad f^{(k)}(1) = (-1)^k k!$$

Thus, substituting $f^{(k)}(1) = (-1)^k k!$ into Formula (10) with $x_0 = 1$ yields the nth Taylor polynomial for $1/x$:

$$\sum_{k=0}^{n} (-1)^k (x-1)^k = 1 - (x-1) + (x-1)^2 - (x-1)^3 + \cdots + (-1)^n (x-1)^n \quad ◀$$

THE nTH REMAINDER

It will be convenient to have a notation for the error in the approximation $f(x) \approx p_n(x)$. Accordingly, we will let $R_n(x)$ denote the difference between $f(x)$ and its nth Taylor polynomial; that is,

$$R_n(x) = f(x) - p_n(x) = f(x) - \sum_{k=0}^{n} \frac{f^{(k)}(x_0)}{k!}(x - x_0)^k \tag{12}$$

This can also be written as

$$f(x) = p_n(x) + R_n(x) = \sum_{k=0}^{n} \frac{f^{(k)}(x_0)}{k!}(x - x_0)^k + R_n(x) \tag{13}$$

The function $R_n(x)$ is called the ***nth remainder*** for the Taylor series of f, and Formula (13) is called ***Taylor's formula with remainder***.

Finding a bound for $R_n(x)$ gives an indication of the accuracy of the approximation $p_n(x) \approx f(x)$. The following theorem, which is proved in Appendix C, provides such a bound.

The bound for $|R_n(x)|$ in (14) is called the ***Lagrange error bound***.

10.7.4 THEOREM (*The Remainder Estimation Theorem*). *If the function f can be differentiated $n+1$ times on an interval I containing the number x_0, and if M is an upper bound for $|f^{(n+1)}(x)|$ on I, that is, $|f^{(n+1)}(x)| \le M$ for all x in I, then*

$$|R_n(x)| \le \frac{M}{(n+1)!}|x - x_0|^{n+1} \tag{14}$$

for all x in I.

▶ **Example 6** Use an nth Maclaurin polynomial for e^x to approximate e to five decimal-place accuracy.

Solution. We note first that the exponential function e^x has derivatives of all orders for every real number x. From Example 2, the nth Maclaurin polynomial for e^x is

$$\sum_{k=0}^{n} \frac{x^k}{k!} = 1 + x + \frac{x^2}{2!} + \cdots + \frac{x^n}{n!}$$

from which we have

$$e = e^1 \approx \sum_{k=0}^{n} \frac{1^k}{k!} = 1 + 1 + \frac{1}{2!} + \cdots + \frac{1}{n!}$$

Thus, our problem is to determine how many terms to include in a Maclaurin polynomial for e^x to achieve five decimal-place accuracy; that is, we want to choose n so that the absolute value of the nth remainder at $x = 1$ satisfies

$$|R_n(1)| \leq 0.000005$$

To determine n we use the Remainder Estimation Theorem with $f(x) = e^x$, $x = 1$, $x_0 = 0$, and I being the interval $[0, 1]$. In this case it follows from Formula (14) that

$$|R_n(1)| \leq \frac{M}{(n+1)!} \tag{15}$$

where M is an upper bound on the value of $f^{(n+1)}(x) = e^x$ for x in the interval $[0, 1]$. However, e^x is an increasing function, so its maximum value on the interval $[0, 1]$ occurs at $x = 1$; that is, $e^x \leq e$ on this interval. Thus, we can take $M = e$ in (15) to obtain

$$|R_n(1)| \leq \frac{e}{(n+1)!} \tag{16}$$

Unfortunately, this inequality is not very useful because it involves e, which is the very quantity we are trying to approximate. However, if we accept that $e < 3$, then we can replace (16) with the following less precise, but more easily applied, inequality:

$$|R_n(1)| \leq \frac{3}{(n+1)!}$$

Thus, we can achieve five decimal-place accuracy by choosing n so that

$$\frac{3}{(n+1)!} \leq 0.000005 \quad \text{or} \quad (n+1)! \geq 600{,}000$$

Since $9! = 362{,}880$ and $10! = 3{,}628{,}800$, the smallest value of n that meets this criterion is $n = 9$. Thus, to five decimal-place accuracy

$$e \approx 1 + 1 + \frac{1}{2!} + \frac{1}{3!} + \frac{1}{4!} + \frac{1}{5!} + \frac{1}{6!} + \frac{1}{7!} + \frac{1}{8!} + \frac{1}{9!} \approx 2.71828$$

As a check, a calculator's 12-digit representation of e is $e \approx 2.71828182846$, which agrees with the preceding approximation when rounded to five decimal places. ◀

✔QUICK CHECK EXERCISES 10.7 *(See page 685 for answers.)*

1. If f can be differentiated three times at 0, then the third Maclaurin polynomial for f is $p_3(x) =$ ______.
2. The third Maclaurin polynomial for $f(x) = e^{2x}$ is

$$p_3(x) = \underline{\qquad} + \underline{\qquad}\, x + \underline{\qquad}\, x^2 + \underline{\qquad}\, x^3$$

3. If $f(2) = 3$, $f'(2) = -4$, and $f''(2) = 10$, then the second Taylor polynomial for f about $x = 2$ is $p_2(x) =$ ______.
4. The third Taylor polynomial for $f(x) = x^5$ about $x = -1$ is

$$p_3(x) = \underline{\qquad} + \underline{\qquad}\,(x+1) + \underline{\qquad}\,(x+1)^2 + \underline{\qquad}\,(x+1)^3$$

5. (a) If a function f has nth Taylor polynomial $p_n(x)$ about $x = x_0$, then the nth remainder $R_n(x)$ is defined by $R_n(x) =$ ______.
   (b) Suppose that a function f can be differentiated five times on an interval I containing $x_0 = 2$ and that $|f^{(5)}(x)| \leq 20$ for all x in I. Then $|R_4(x)| \leq$ ______ for all x in I.

EXERCISE SET 10.7 Graphing Utility C CAS

1. In each part, find the local quadratic approximation of f at $x = x_0$, and use that approximation to find the local linear approximation of f at x_0. Use a graphing utility to graph f and the two approximations on the same screen.
(a) $f(x) = e^{-x}$; $x_0 = 0$ (b) $f(x) = \cos x$; $x_0 = 0$
(c) $f(x) = \sin x$; $x_0 = \pi/2$ (d) $f(x) = \sqrt{x}$; $x_0 = 1$

2. In each part, use a CAS to find the local quadratic approximation of f at $x = x_0$, and use that approximation to find the local linear approximation of f at $x = x_0$.
(a) $f(x) = e^{\sin x}$; $x_0 = 0$ (b) $f(x) = \sqrt{x}$; $x_0 = 9$
(c) $f(x) = \sec^{-1} x$; $x_0 = 2$
(d) $f(x) = \sin^{-1} x$; $x_0 = 0$

3. (a) Find the local quadratic approximation of $\sqrt{x}$ at $x_0 = 1$.
(b) Use the result obtained in part (a) to approximate $\sqrt{1.1}$, and compare your approximation to that produced directly by your calculating utility. [See Example 1 of Section 3.9.]

4. (a) Find the local quadratic approximation of $\cos x$ at $x_0 = 0$.
(b) Use the result obtained in part (a) to approximate $\cos 2°$, and compare the approximation to that produced directly by your calculating utility.

5. Use an appropriate local quadratic approximation to approximate $\tan 61°$, and compare the result to that produced directly by your calculating utility.

6. Use an appropriate local quadratic approximation to approximate $\sqrt{36.03}$, and compare the result to that produced directly by your calculating utility.

7–16 Find the Maclaurin polynomials of orders $n = 0, 1, 2, 3$, and 4, and then find the nth Maclaurin polynomials for the function in sigma notation.

7. e^{-x} 8. e^{ax} 9. $\cos \pi x$
10. $\sin \pi x$ 11. $\ln(1 + x)$ 12. $\dfrac{1}{1 + x}$
13. $\cosh x$ 14. $\sinh x$ 15. $x \sin x$
16. xe^x

17–24 Find the Taylor polynomials of orders $n = 0, 1, 2, 3$, and 4 about $x = x_0$, and then find the nth Taylor polynomial for the function in sigma notation.

17. e^x; $x_0 = 1$ 18. e^{-x}; $x_0 = \ln 2$
19. $\dfrac{1}{x}$; $x_0 = -1$ 20. $\dfrac{1}{x + 2}$; $x_0 = 3$
21. $\sin \pi x$; $x_0 = \dfrac{1}{2}$ 22. $\cos x$; $x_0 = \dfrac{\pi}{2}$
23. $\ln x$; $x_0 = 1$ 24. $\ln x$; $x_0 = e$

25. (a) Find the third Maclaurin polynomial for
$$f(x) = 1 + 2x - x^2 + x^3$$
(b) Find the third Taylor polynomial about $x = 1$ for
$$f(x) = 1 + 2(x - 1) - (x - 1)^2 + (x - 1)^3$$

26. (a) Find the nth Maclaurin polynomial for
$$f(x) = c_0 + c_1 x + c_2 x^2 + \cdots + c_n x^n$$
(b) Find the nth Taylor polynomial about $x = 1$ for
$$f(x) = c_0 + c_1(x - 1) + c_2(x - 1)^2 + \cdots + c_n(x - 1)^n$$

27–30 Find the first four distinct Taylor polynomials about $x = x_0$, and use a graphing utility to graph the given function and the Taylor polynomials on the same screen.

27. $f(x) = e^{-2x}$; $x_0 = 0$ 28. $f(x) = \sin x$; $x_0 = \pi/2$
29. $f(x) = \cos x$; $x_0 = \pi$ 30. $\ln(x + 1)$; $x_0 = 0$

31. Use the method of Example 6 to approximate $\sqrt{e}$ to four decimal-place accuracy, and check your work by comparing your answer to that produced directly by your calculating utility. [*Suggestion:* Write $\sqrt{e}$ as $e^{0.5}$.]

32. Use the method of Example 6 to approximate $1/e$ to three decimal-place accuracy, and check your work by comparing your answer to that produced directly by your calculating utility.

33. Show that the nth Taylor polynomial for $\sinh x$ about $x = \ln 4$ is
$$\sum_{k=0}^{n} \frac{16 - (-1)^k}{8k!}(x - \ln 4)^k$$

34. (a) The accompanying figure shows a sector of radius r and central angle 2α. Assuming that the angle α is small, use the local quadratic approximation of $\cos \alpha$ at $\alpha = 0$ to show that $x \approx r\alpha^2/2$.
(b) Assuming that the Earth is a sphere of radius 4000 mi, use the result in part (a) to approximate the maximum amount by which a 100-mi arc along the equator will diverge from its chord.

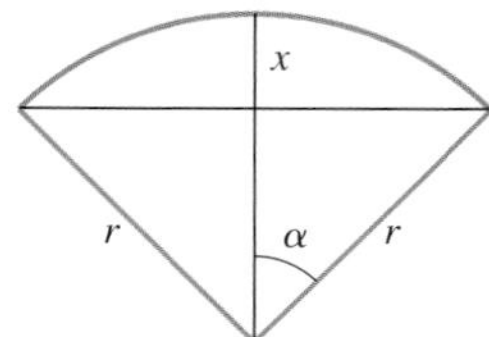

Figure Ex-34

FOCUS ON CONCEPTS

35. Which of the functions graphed in the following figure is most likely to have $p(x) = 1 - x + 2x^2$ as its second-order Maclaurin polynomial? Explain your reasoning.

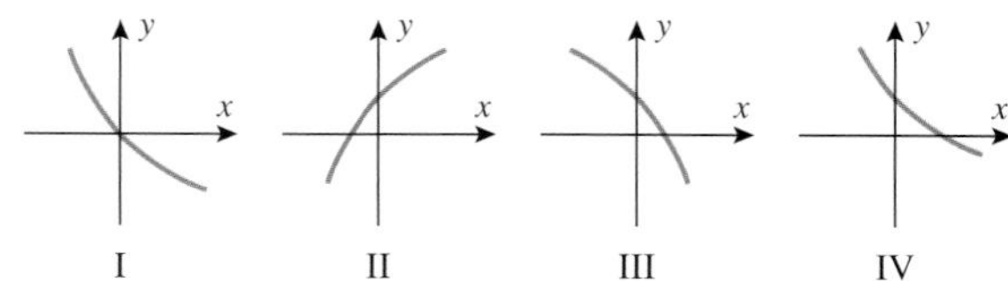

36. Suppose that the values of a function f and its first three derivatives at $x = 1$ are

$$f(1) = 2, \quad f'(1) = -3, \quad f''(1) = 0, \quad f'''(1) = 6$$

Find as many Taylor polynomials for f as you can about $x = 1$.

37. Let $p_1(x)$ and $p_2(x)$ be the local linear and local quadratic approximations of $f(x) = e^{\sin x}$ at $x = 0$.

(a) Use a graphing utility to generate the graphs of $f(x)$, $p_1(x)$, and $p_2(x)$ on the same screen for $-1 \le x \le 1$.

(b) Construct a table of values of $f(x)$, $p_1(x)$, and $p_2(x)$ for $x = -1.00, -0.75, -0.50, -0.25, 0, 0.25, 0.50, 0.75, 1.00$. Round the values to three decimal places.

(c) Generate the graph of $|f(x) - p_1(x)|$, and use the graph to determine an interval on which $p_1(x)$ approximates $f(x)$ with an error of at most ± 0.01. [*Suggestion:* Review the discussion relating to Figure 3.8.4.]

(d) Generate the graph of $|f(x) - p_2(x)|$, and use the graph to determine an interval on which $p_2(x)$ approximates $f(x)$ with an error of at most ± 0.01.

38. (a) Find an interval $[0, b]$ over which e^x can be approximated by $1 + x + (x^2/2!)$ to three decimal-place accuracy throughout the interval.

(b) Check your answer in part (a) by graphing

$$\left| e^x - \left(1 + x + \frac{x^2}{2!}\right)\right|$$

over the interval you obtained.

39–42 Use the Remainder Estimation Theorem to find an interval containing $x = 0$ over which $f(x)$ can be approximated by $p(x)$ to three decimal-place accuracy throughout the interval. Check your answer by graphing $|f(x) - p(x)|$ over the interval you obtained.

39. $f(x) = \sin x;\ p(x) = x - \dfrac{x^3}{3!}$

40. $f(x) = \cos x;\ p(x) = 1 - \dfrac{x^2}{2!} + \dfrac{x^4}{4!}$

41. $f(x) = \dfrac{1}{1 + x^2};\ p(x) = 1 - x^2 + x^4$

42. $f(x) = \ln(1 + x);\ p(x) = x - \dfrac{x^2}{2} + \dfrac{x^3}{3}$

✔ QUICK CHECK ANSWERS 10.7

1. $f(0) + f'(0)x + \dfrac{f''(0)}{2!}x^2 + \dfrac{f'''(0)}{3!}x^3$ 2. 1; 2; 2; $\frac{4}{3}$ 3. $3 - 4(x - 2) + 5(x - 2)^2$ 4. $-1; 5; -10; 10$
5. (a) $f(x) - p_n(x)$ (b) $\frac{1}{6}|x - 2|^5$

10.8 MACLAURIN AND TAYLOR SERIES; POWER SERIES

Recall from the last section that the nth Taylor polynomial $p_n(x)$ at $x = x_0$ for a function f was defined so its value and the values of its first n derivatives match those of f at x_0. This being the case, it is reasonable to expect that for values of x near x_0 the values of $p_n(x)$ will become better and better approximations of $f(x)$ as n increases, and may possibly converge to $f(x)$ as $n \to +\infty$. We will explore this idea in this section.

MACLAURIN AND TAYLOR SERIES

In Section 10.7 we defined the nth Maclaurin polynomial for a function f as

$$\sum_{k=0}^{n} \frac{f^{(k)}(0)}{k!}x^k = f(0) + f'(0)x + \frac{f''(0)}{2!}x^2 + \cdots + \frac{f^{(n)}(0)}{n!}x^n$$

and the nth Taylor polynomial for f about $x = x_0$ as

$$\sum_{k=0}^{n} \frac{f^{(k)}(x_0)}{k!}(x - x_0)^k = f(x_0) + f'(x_0)(x - x_0) + \frac{f''(x_0)}{2!}(x - x_0)^2 + \cdots + \frac{f^{(n)}(x_0)}{n!}(x - x_0)^n$$

It is not a big step to extend the notions of Maclaurin and Taylor polynomials to series by not stopping the summation index at n. Thus, we have the following definition.

10.8.1 DEFINITION. If f has derivatives of all orders at x_0, then we call the series

$$\sum_{k=0}^{\infty} \frac{f^{(k)}(x_0)}{k!}(x-x_0)^k = f(x_0) + f'(x_0)(x-x_0) + \frac{f''(x_0)}{2!}(x-x_0)^2 + \cdots + \frac{f^{(k)}(x_0)}{k!}(x-x_0)^k + \cdots \quad (1)$$

the ***Taylor series for f about x = x₀***. In the special case where $x_0 = 0$, this series becomes

$$\sum_{k=0}^{\infty} \frac{f^{(k)}(0)}{k!}x^k = f(0) + f'(0)x + \frac{f''(0)}{2!}x^2 + \cdots + \frac{f^{(k)}(0)}{k!}x^k + \cdots \quad (2)$$

in which case we call it the ***Maclaurin series for f***.

Note that the nth Maclaurin and Taylor polynomials are the nth partial sums for the corresponding Maclaurin and Taylor series.

► Example 1 Find the Maclaurin series for

(a) e^x (b) $\sin x$ (c) $\cos x$ (d) $\dfrac{1}{1-x}$

Solution (a). In Example 2 of Section 10.7 we found that the nth Maclaurin polynomial for e^x is

$$p_n(x) = \sum_{k=0}^{n} \frac{x^k}{k!} = 1 + x + \frac{x^2}{2!} + \cdots + \frac{x^n}{n!}$$

Thus, the Maclaurin series for e^x is

$$\sum_{k=0}^{\infty} \frac{x^k}{k!} = 1 + x + \frac{x^2}{2!} + \cdots + \frac{x^k}{k!} + \cdots$$

Solution (b). In Example 4(a) of Section 10.7 we found that the Maclaurin polynomials for $\sin x$ are given by

$$p_{2k+1}(x) = p_{2k+2}(x) = x - \frac{x^3}{3!} + \frac{x^5}{5!} - \frac{x^7}{7!} + \cdots + (-1)^k \frac{x^{2k+1}}{(2k+1)!} \quad (k = 0, 1, 2, \ldots)$$

Thus, the Maclaurin series for $\sin x$ is

$$\sum_{k=0}^{\infty} (-1)^k \frac{x^{2k+1}}{(2k+1)!} = x - \frac{x^3}{3!} + \frac{x^5}{5!} - \frac{x^7}{7!} + \cdots + (-1)^k \frac{x^{2k+1}}{(2k+1)!} + \cdots$$

Solution (c). In Example 4(b) of Section 10.7 we found that the Maclaurin polynomials for $\cos x$ are given by

$$p_{2k}(x) = p_{2k+1}(x) = 1 - \frac{x^2}{2!} + \frac{x^4}{4!} - \frac{x^6}{6!} + \cdots + (-1)^k \frac{x^{2k}}{(2k)!} \quad (k = 0, 1, 2, \ldots)$$

Thus, the Maclaurin series for $\cos x$ is

$$\sum_{k=0}^{\infty}(-1)^k\frac{x^{2k}}{(2k)!} = 1 - \frac{x^2}{2!} + \frac{x^4}{4!} - \frac{x^6}{6!} + \cdots + (-1)^k\frac{x^{2k}}{(2k)!} + \cdots$$

Solution (d). In Example 4(c) of Section 10.7 we found that the nth Maclaurin polynomial for $1/(1-x)$ is

$$p_n(x) = \sum_{k=0}^{n} x^k = 1 + x + x^2 + \cdots + x^n \quad (n = 0, 1, 2, \ldots)$$

Thus, the Maclaurin series for $1/(1-x)$ is

$$\sum_{k=0}^{\infty} x^k = 1 + x + x^2 + \cdots + x^k + \cdots \blacktriangleleft$$

▶ **Example 2** Find the Taylor series for $1/x$ about $x = 1$.

Solution. In Example 5 of Section 10.7 we found that the nth Taylor polynomial for $1/x$ about $x = 1$ is

$$\sum_{k=0}^{n}(-1)^k(x-1)^k = 1 - (x-1) + (x-1)^2 - (x-1)^3 + \cdots + (-1)^n(x-1)^n$$

Thus, the Taylor series for $1/x$ about $x = 1$ is

$$\sum_{k=0}^{\infty}(-1)^k(x-1)^k = 1 - (x-1) + (x-1)^2 - (x-1)^3 + \cdots + (-1)^k(x-1)^k + \cdots \blacktriangleleft$$

POWER SERIES IN x

Maclaurin and Taylor series differ from the series that we have considered in Sections 10.3 to 10.6 in that their terms are not merely constants, but instead involve a variable. These are examples of *power series*, which we now define.

If $c_0, c_1, c_2, \ldots$ are constants and x is a variable, then a series of the form

$$\sum_{k=0}^{\infty} c_k x^k = c_0 + c_1 x + c_2 x^2 + \cdots + c_k x^k + \cdots \tag{3}$$

is called a ***power series in x***. Some examples are

$$\sum_{k=0}^{\infty} x^k = 1 + x + x^2 + x^3 + \cdots$$

$$\sum_{k=0}^{\infty} \frac{x^k}{k!} = 1 + x + \frac{x^2}{2!} + \frac{x^3}{3!} + \cdots$$

$$\sum_{k=0}^{\infty}(-1)^k\frac{x^{2k}}{(2k)!} = 1 - \frac{x^2}{2!} + \frac{x^4}{4!} - \frac{x^6}{6!} + \cdots$$

From Example 1, these are the Maclaurin series for the functions $1/(1-x)$, e^x, and $\cos x$, respectively. Indeed, every Maclaurin series

$$\sum_{k=0}^{\infty} \frac{f^{(k)}(0)}{k!}x^k = f(0) + f'(0)x + \frac{f''(0)}{2!}x^2 + \cdots + \frac{f^{(k)}(0)}{k!}x^k + \cdots$$

is a power series in x.

RADIUS AND INTERVAL OF CONVERGENCE

If a numerical value is substituted for x in a power series $\sum c_k x^k$, then the resulting series of numbers may either converge or diverge. This leads to the problem of determining the set of x-values for which a given power series converges; this is called its ***convergence set***.

Observe that every power series in x converges at $x = 0$, since substituting this value in (3) produces the series

$$c_0 + 0 + 0 + 0 + \cdots + 0 + \cdots$$

whose sum is c_0. In some cases $x = 0$ may be the only number in the convergence set; in other cases the convergence set is some finite or infinite interval containing $x = 0$. This is the content of the following theorem, whose proof will be omitted.

> **10.8.2 THEOREM.** *For any power series in x, exactly one of the following is true:*
>
> (*a*) *The series converges only for $x = 0$.*
>
> (*b*) *The series converges absolutely (and hence converges) for all real values of x.*
>
> (*c*) *The series converges absolutely (and hence converges) for all x in some finite open interval $(-R, R)$ and diverges if $x < -R$ or $x > R$. At either of the values $x = R$ or $x = -R$, the series may converge absolutely, converge conditionally, or diverge, depending on the particular series.*

This theorem states that the convergence set for a power series in x is always an interval centered at $x = 0$ (possibly just the value $x = 0$ itself or possibly infinite). For this reason, the convergence set of a power series in x is called the ***interval of convergence***. In the case where the convergence set is the single value $x = 0$ we say that the series has ***radius of convergence* 0**, in the case where the convergence set is $(-\infty, +\infty)$ we say that the series has ***radius of convergence* $+\infty$**, and in the case where the convergence set extends between $-R$ and R we say that the series has ***radius of convergence R*** (Figure 10.8.1).

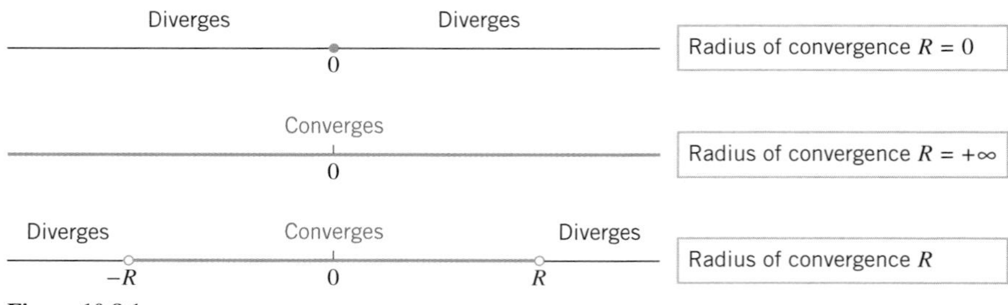

Figure 10.8.1

FINDING THE INTERVAL OF CONVERGENCE

The usual procedure for finding the interval of convergence of a power series is to apply the ratio test for absolute convergence (Theorem 10.6.5). The following example illustrates how this works.

► **Example 3** Find the interval of convergence and radius of convergence of the following power series.

$$\text{(a)}\ \sum_{k=0}^{\infty} x^k \qquad \text{(b)}\ \sum_{k=0}^{\infty} \frac{x^k}{k!} \qquad \text{(c)}\ \sum_{k=0}^{\infty} k!x^k \qquad \text{(d)}\ \sum_{k=0}^{\infty} \frac{(-1)^k x^k}{3^k(k+1)}$$

Solution (a). We apply the ratio test for absolute convergence. We have

$$\rho = \lim_{k\to+\infty}\left|\frac{u_{k+1}}{u_k}\right| = \lim_{k\to+\infty}\left|\frac{x^{k+1}}{x^k}\right| = \lim_{k\to+\infty}|x| = |x|$$

so the series converges absolutely if $\rho = |x| < 1$ and diverges if $\rho = |x| > 1$. The test is inconclusive if $|x| = 1$ (i.e., if $x = 1$ or $x = -1$), which means that we will have to investigate convergence at these values separately. At these values the series becomes

$$\sum_{k=0}^{\infty} 1^k = 1 + 1 + 1 + 1 + \cdots \qquad \boxed{x = 1}$$

$$\sum_{k=0}^{\infty} (-1)^k = 1 - 1 + 1 - 1 + \cdots \qquad \boxed{x = -1}$$

both of which diverge; thus, the interval of convergence for the given power series is $(-1, 1)$, and the radius of convergence is $R = 1$.

Solution (b). Applying the ratio test for absolute convergence, we obtain

$$\rho = \lim_{k\to+\infty}\left|\frac{u_{k+1}}{u_k}\right| = \lim_{k\to+\infty}\left|\frac{x^{k+1}}{(k+1)!}\cdot\frac{k!}{x^k}\right| = \lim_{k\to+\infty}\left|\frac{x}{k+1}\right| = 0$$

Since $\rho < 1$ for all x, the series converges absolutely for all x. Thus, the interval of convergence is $(-\infty, +\infty)$ and the radius of convergence is $R = +\infty$.

Solution (c). If $x \neq 0$, then the ratio test for absolute convergence yields

$$\rho = \lim_{k\to+\infty}\left|\frac{u_{k+1}}{u_k}\right| = \lim_{k\to+\infty}\left|\frac{(k+1)!x^{k+1}}{k!x^k}\right| = \lim_{k\to+\infty}|(k+1)x| = +\infty$$

Therefore, the series diverges for all nonzero values of x. Thus, the interval of convergence is the single value $x = 0$ and the radius of convergence is $R = 0$.

Solution (d). Since $|(-1)^k| = |(-1)^{k+1}| = 1$, we obtain

$$\begin{aligned}\rho = \lim_{k\to+\infty}\left|\frac{u_{k+1}}{u_k}\right| &= \lim_{k\to+\infty}\left|\frac{x^{k+1}}{3^{k+1}(k+2)}\cdot\frac{3^k(k+1)}{x^k}\right| \\ &= \lim_{k\to+\infty}\left[\frac{|x|}{3}\cdot\left(\frac{k+1}{k+2}\right)\right] \\ &= \frac{|x|}{3}\lim_{k\to+\infty}\left(\frac{1+(1/k)}{1+(2/k)}\right) = \frac{|x|}{3}\end{aligned}$$

The ratio test for absolute convergence implies that the series converges absolutely if $|x| < 3$ and diverges if $|x| > 3$. The ratio test fails to provide any information when $|x| = 3$, so the cases $x = -3$ and $x = 3$ need separate analyses. Substituting $x = -3$ in the given series yields

$$\sum_{k=0}^{\infty}\frac{(-1)^k(-3)^k}{3^k(k+1)} = \sum_{k=0}^{\infty}\frac{(-1)^k(-1)^k 3^k}{3^k(k+1)} = \sum_{k=0}^{\infty}\frac{1}{k+1}$$

which is the divergent harmonic series $1 + \frac{1}{2} + \frac{1}{3} + \frac{1}{4} + \cdots$. Substituting $x = 3$ in the given series yields

$$\sum_{k=0}^{\infty}\frac{(-1)^k 3^k}{3^k(k+1)} = \sum_{k=0}^{\infty}\frac{(-1)^k}{k+1} = 1 - \frac{1}{2} + \frac{1}{3} - \frac{1}{4} + \cdots$$

which is the conditionally convergent alternating harmonic series. Thus, the interval of convergence for the given series is $(-3, 3]$ and the radius of convergence is $R = 3$. ◄

POWER SERIES IN $x - x_0$

If x_0 is a constant, and if x is replaced by $x - x_0$ in (3), then the resulting series has the form

$$\sum_{k=0}^{\infty} c_k(x-x_0)^k = c_0 + c_1(x-x_0) + c_2(x-x_0)^2 + \cdots + c_k(x-x_0)^k + \cdots$$

This is called a ***power series in $x - x_0$***. Some examples are

$$\sum_{k=0}^{\infty} \frac{(x-1)^k}{k+1} = 1 + \frac{(x-1)}{2} + \frac{(x-1)^2}{3} + \frac{(x-1)^3}{4} + \cdots \qquad \boxed{x_0 = 1}$$

$$\sum_{k=0}^{\infty} \frac{(-1)^k(x+3)^k}{k!} = 1 - (x+3) + \frac{(x+3)^2}{2!} - \frac{(x+3)^3}{3!} + \cdots \qquad \boxed{x_0 = -3}$$

The first of these is a power series in $x - 1$ and the second is a power series in $x + 3$. Note that a power series in x is a power series in $x - x_0$ in which $x_0 = 0$. More generally, the Taylor series

$$\sum_{k=0}^{\infty} \frac{f^{(k)}(x_0)}{k!}(x-x_0)^k$$

is a power series in $x - x_0$.

The main result on convergence of a power series in $x - x_0$ can be obtained by substituting $x - x_0$ for x in Theorem 10.8.2. This leads to the following theorem.

10.8.3 THEOREM. *For a power series $\sum c_k(x-x_0)^k$, exactly one of the following statements is true:*

(*a*) *The series converges only for $x = x_0$.*

(*b*) *The series converges absolutely (and hence converges) for all real values of x.*

(*c*) *The series converges absolutely (and hence converges) for all x in some finite open interval $(x_0 - R, x_0 + R)$ and diverges if $x < x_0 - R$ or $x > x_0 + R$. At either of the values $x = x_0 - R$ or $x = x_0 + R$, the series may converge absolutely, converge conditionally, or diverge, depending on the particular series.*

It follows from this theorem that the set of values for which a power series in $x - x_0$ converges is always an interval centered at $x = x_0$; we call this the ***interval of convergence*** (Figure 10.8.2). In part (*a*) of Theorem 10.8.3 the interval of convergence reduces to the single value $x = x_0$, in which case we say that the series has ***radius of convergence $R = 0$***; in part (*b*) the interval of convergence is infinite (the entire real line), in which case we say that the series has ***radius of convergence $R = +\infty$***; and in part (*c*) the interval extends between $x_0 - R$ and $x_0 + R$, in which case we say that the series has ***radius of convergence R***.

▶ **Example 4** Find the interval of convergence and radius of convergence of the series

$$\sum_{k=1}^{\infty} \frac{(x-5)^k}{k^2}$$

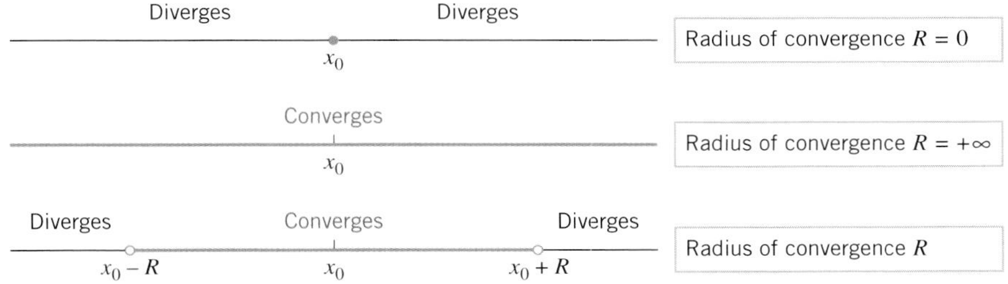

Figure 10.8.2

Solution. We apply the ratio test for absolute convergence.

$$\rho = \lim_{k\to+\infty}\left|\frac{u_{k+1}}{u_k}\right| = \lim_{k\to+\infty}\left|\frac{(x-5)^{k+1}}{(k+1)^2}\cdot\frac{k^2}{(x-5)^k}\right|$$

$$= \lim_{k\to+\infty}\left[|x-5|\left(\frac{k}{k+1}\right)^2\right]$$

$$= |x-5|\lim_{k\to+\infty}\left(\frac{1}{1+(1/k)}\right)^2 = |x-5|$$

Thus, the series converges absolutely if $|x-5| < 1$, or $-1 < x-5 < 1$, or $4 < x < 6$. The series diverges if $x < 4$ or $x > 6$.

To determine the convergence behavior at the endpoints $x = 4$ and $x = 6$, we substitute these values in the given series. If $x = 6$, the series becomes

$$\sum_{k=1}^{\infty}\frac{1^k}{k^2} = \sum_{k=1}^{\infty}\frac{1}{k^2} = 1 + \frac{1}{2^2} + \frac{1}{3^2} + \frac{1}{4^2} + \cdots$$

which is a convergent p-series ($p = 2$). If $x = 4$, the series becomes

$$\sum_{k=1}^{\infty}\frac{(-1)^k}{k^2} = -1 + \frac{1}{2^2} - \frac{1}{3^2} + \frac{1}{4^2} - \cdots$$

Since this series converges absolutely, the interval of convergence for the given series is $[4, 6]$. The radius of convergence is $R = 1$ (Figure 10.8.3). ◂

It will always be a waste of time to test for convergence at the endpoints of the interval of convergence using the ratio test, since ρ will always be 1 at those points if

$$\rho = \lim_{n\to+\infty}|a_{n+1}/a_n|$$

exists. Explain why this must be so.

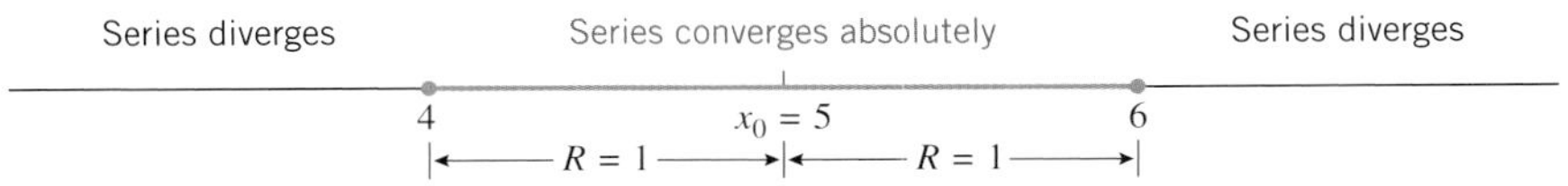

Series diverges | Series converges absolutely | Series diverges
4 | $x_0 = 5$ | 6
$R = 1$ | $R = 1$

Figure 10.8.3

FUNCTIONS DEFINED BY POWER SERIES

If a function f is expressed as a power series on some interval, then we say that the power series ***represents*** f on that interval. For example, we saw in Example 4(b) of Section 10.3 that

$$\frac{1}{1-x} = \sum_{k=0}^{\infty}x^k$$

if $|x| < 1$, so this power series represents the function $1/(1-x)$ on the interval $-1 < x < 1$.

Sometimes new functions actually originate as power series, and the properties of the functions are developed by working with their power series representations. For example, the functions

$$J_0(x) = \sum_{k=0}^{\infty} \frac{(-1)^k x^{2k}}{2^{2k}(k!)^2} = 1 - \frac{x^2}{2^2(1!)^2} + \frac{x^4}{2^4(2!)^2} - \frac{x^6}{2^6(3!)^2} + \cdots \tag{4}$$

and

$$J_1(x) = \sum_{k=0}^{\infty} \frac{(-1)^k x^{2k+1}}{2^{2k+1}(k!)(k+1)!} = \frac{x}{2} - \frac{x^3}{2^3(1!)(2!)} + \frac{x^5}{2^5(2!)(3!)} - \cdots \tag{5}$$

which are called ***Bessel functions*** in honor of the German mathematician and astronomer Friedrich Wilhelm Bessel (1784–1846), arise naturally in the study of planetary motion and in various problems that involve heat flow.

To find the domains of these functions, we must determine where their defining power series converge. For example, in the case of $J_0(x)$ we have

$$\rho = \lim_{k\to+\infty}\left|\frac{u_{k+1}}{u_k}\right| = \lim_{k\to+\infty}\left|\frac{x^{2(k+1)}}{2^{2(k+1)}[(k+1)!]^2}\cdot\frac{2^{2k}(k!)^2}{x^{2k}}\right|$$

$$= \lim_{k\to+\infty}\left|\frac{x^2}{4(k+1)^2}\right| = 0 < 1$$

so the series converges for all x; that is, the domain of $J_0(x)$ is $(-\infty, +\infty)$. We leave it as an exercise to show that the power series for $J_1(x)$ also converges for all x. Computer-generated graphs of $J_0(x)$ and $J_1(x)$ are shown in Figure 10.8.4.

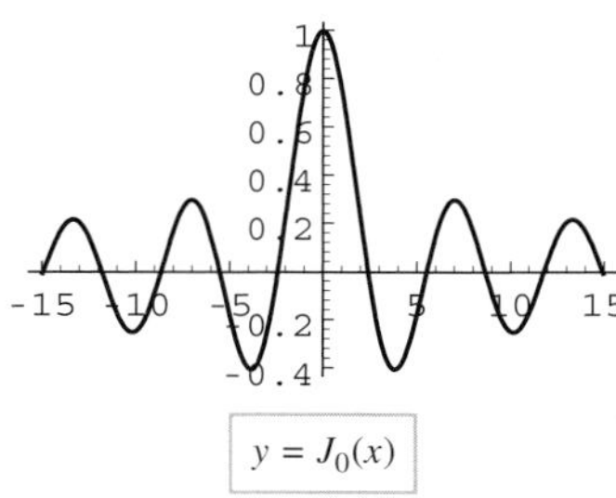

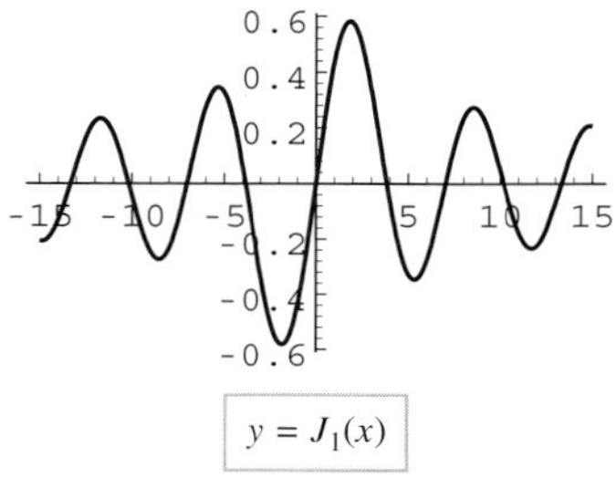

Figure 10.8.4

TECHNOLOGY MASTERY

Many computer algebra systems have the Bessel functions as part of their libraries. If you have a CAS with Bessel functions, use it to generate the graphs in Figure 10.8.4.

✔QUICK CHECK EXERCISES 10.8 *(See page 694 for answers.)*

1. If f has derivatives of all orders at x_0, then the Taylor series for f about $x = x_0$ is defined to be

$$\sum_{k=0}^{\infty} \_\_\_\_\_\_$$

2. Since

$$\lim_{k\to+\infty}\left|\frac{2^{k+1}x^{k+1}}{2^k x^k}\right| = 2|x|$$

the radius of convergence for the infinite series $\sum_{k=0}^{\infty} 2^k x^k$ is ______.

3. Since

$$\lim_{k\to+\infty}\left|\frac{(3^{k+1}x^{k+1})/(k+1)!}{(3^k x^k)/k!}\right| = \lim_{k\to+\infty}\left|\frac{3x}{k+1}\right| = 0$$

the interval of convergence for the series $\sum_{k=0}^{\infty}(3^k/k!)x^k$ is ______.

4. (a) Since

$$\lim_{k\to+\infty}\left|\frac{(x-4)^{k+1}/\sqrt{k+1}}{(x-4)^k/\sqrt{k}}\right| = \lim_{k\to+\infty}\left|\sqrt{\frac{k}{k+1}}(x-4)\right|$$
$$= |x-4|$$

the radius of convergence for the infinite series $\sum_{k=1}^{\infty}(1/\sqrt{k})(x-4)^k$ is ______.

(b) When $x = 3$,

$$\sum_{k=1}^{\infty}\frac{1}{\sqrt{k}}(x-4)^k = \sum_{k=1}^{\infty}\frac{1}{\sqrt{k}}(-1)^k$$

Does this series converge or diverge?

(c) When $x = 5$,

$$\sum_{k=1}^{\infty}\frac{1}{\sqrt{k}}(x-4)^k = \sum_{k=1}^{\infty}\frac{1}{\sqrt{k}}$$

Does this series converge or diverge?

(d) The interval of convergence for the infinite series $\sum_{k=1}^{\infty}(1/\sqrt{k})(x-4)^k$ is ______.

EXERCISE SET 10.8 Graphing Utility C CAS

1–10 Use sigma notation to write the Maclaurin series for the function.

1. e^{-x} 2. e^{ax} 3. $\cos \pi x$ 4. $\sin \pi x$

5. $\ln(1+x)$ 6. $\dfrac{1}{1+x}$ 7. $\cosh x$

8. $\sinh x$ 9. $x \sin x$ 10. xe^{x}

11–18 Use sigma notation to write the Taylor series about $x = x_0$ for the function.

11. e^{x}; $x_0 = 1$ 12. e^{-x}; $x_0 = \ln 2$

13. $\dfrac{1}{x}$; $x_0 = -1$ 14. $\dfrac{1}{x+2}$; $x_0 = 3$

15. $\sin \pi x$; $x_0 = \dfrac{1}{2}$ 16. $\cos x$; $x_0 = \dfrac{\pi}{2}$

17. $\ln x$; $x_0 = 1$ 18. $\ln x$; $x_0 = e$

19–22 Find the interval of convergence of the power series, and find a familiar function that is represented by the power series on that interval.

19. $1 - x + x^2 - x^3 + \cdots + (-1)^k x^k + \cdots$

20. $1 + x^2 + x^4 + \cdots + x^{2k} + \cdots$

21. $1 + (x-2) + (x-2)^2 + \cdots + (x-2)^k + \cdots$

22. $1 - (x+3) + (x+3)^2 - (x+3)^3 + \cdots + (-1)^k (x+3)^k| \ldots$

23. Suppose that the function f is represented by the power series

$$f(x) = 1 - \frac{x}{2} + \frac{x^2}{4} - \frac{x^3}{8} + \cdots + (-1)^k \frac{x^k}{2^k} + \cdots$$

(a) Find the domain of f. (b) Find $f(0)$ and $f(1)$.

24. Suppose that the function f is represented by the power series

$$f(x) = 1 - \frac{x-5}{3} + \frac{(x-5)^2}{3^2} - \frac{(x-5)^3}{3^3} + \cdots$$

(a) Find the domain of f. (b) Find $f(3)$ and $f(6)$.

25–48 Find the radius of convergence and the interval of convergence.

25. $\displaystyle\sum_{k=0}^{\infty} \frac{x^k}{k+1}$ 26. $\displaystyle\sum_{k=0}^{\infty} 3^k x^k$ 27. $\displaystyle\sum_{k=0}^{\infty} \frac{(-1)^k x^k}{k!}$

28. $\displaystyle\sum_{k=0}^{\infty} \frac{k!}{2^k} x^k$ 29. $\displaystyle\sum_{k=1}^{\infty} \frac{5^k}{k^2} x^k$ 30. $\displaystyle\sum_{k=2}^{\infty} \frac{x^k}{\ln k}$

31. $\displaystyle\sum_{k=1}^{\infty} \frac{x^k}{k(k+1)}$ 32. $\displaystyle\sum_{k=0}^{\infty} \frac{(-2)^k x^{k+1}}{k+1}$

33. $\displaystyle\sum_{k=1}^{\infty} (-1)^{k-1} \frac{x^k}{\sqrt{k}}$ 34. $\displaystyle\sum_{k=0}^{\infty} \frac{(-1)^k x^{2k}}{(2k)!}$

35. $\displaystyle\sum_{k=0}^{\infty} (-1)^k \frac{x^{2k+1}}{(2k+1)!}$ 36. $\displaystyle\sum_{k=1}^{\infty} (-1)^k \frac{x^{3k}}{k^{3/2}}$

37. $\displaystyle\sum_{k=0}^{\infty} \frac{3^k}{k!} x^k$ 38. $\displaystyle\sum_{k=2}^{\infty} (-1)^{k+1} \frac{x^k}{k(\ln k)^2}$

39. $\displaystyle\sum_{k=0}^{\infty} \frac{x^k}{1+k^2}$ 40. $\displaystyle\sum_{k=0}^{\infty} \frac{(x-3)^k}{2^k}$

41. $\displaystyle\sum_{k=1}^{\infty} (-1)^{k+1} \frac{(x+1)^k}{k}$ 42. $\displaystyle\sum_{k=0}^{\infty} (-1)^k \frac{(x-4)^k}{(k+1)^2}$

43. $\displaystyle\sum_{k=0}^{\infty} \left(\frac{3}{4}\right)^k (x+5)^k$ 44. $\displaystyle\sum_{k=1}^{\infty} \frac{(2k+1)!}{k^3} (x-2)^k$

45. $\displaystyle\sum_{k=1}^{\infty} (-1)^k \frac{(x+1)^{2k+1}}{k^2+4}$ 46. $\displaystyle\sum_{k=1}^{\infty} \frac{(\ln k)(x-3)^k}{k}$

47. $\displaystyle\sum_{k=0}^{\infty} \frac{\pi^k (x-1)^{2k}}{(2k+1)!}$ 48. $\displaystyle\sum_{k=0}^{\infty} \frac{(2x-3)^k}{4^{2k}}$

49. Use the root test to find the interval of convergence of

$$\sum_{k=2}^{\infty} \frac{x^k}{(\ln k)^k}$$

50. Find the domain of the function

$$f(x) = \sum_{k=1}^{\infty} \frac{1 \cdot 3 \cdot 5 \cdots (2k-1)}{(2k-2)!} x^k$$

51. Show that the series

$$1 - \frac{x}{2!} + \frac{x^2}{4!} - \frac{x^3}{6!} + \cdots$$

is the Maclaurin series for the function

$$f(x) = \begin{cases} \cos \sqrt{x}, & x \geq 0 \\ \cosh \sqrt{-x}, & x < 0 \end{cases}$$

[*Hint:* Use the Maclaurin series for $\cos x$ and $\cosh x$ to obtain series for $\cos \sqrt{x}$, where $x \geq 0$, and $\cosh \sqrt{-x}$, where $x \leq 0$.]

FOCUS ON CONCEPTS

52. If a function f is represented by a power series on an interval, then the graphs of the partial sums can be used as approximations to the graph of f.
   (a) Use a graphing utility to generate the graph of $1/(1-x)$ together with the graphs of the first four partial sums of its Maclaurin series over the interval $(-1, 1)$.
   (b) In general terms, where are the graphs of the partial sums the most accurate?

53. Prove:
   (a) If f is an even function, then all odd powers of x in its Maclaurin series have coefficient 0.
   (b) If f is an odd function, then all even powers of x in its Maclaurin series have coefficient 0.

54. Suppose that the power series $\sum c_k(x-x_0)^k$ has radius of convergence R and p is a nonzero constant. What can you say about the radius of convergence of the power series $\sum pc_k(x-x_0)^k$? Explain your reasoning. [*Hint:* See Theorem 10.4.3.]

55. Suppose that the power series $\sum c_k(x-x_0)^k$ has a finite radius of convergence R, and the power series $\sum d_k(x-x_0)^k$ has a radius of convergence of $+\infty$. What can you say about the radius of convergence of $\sum(c_k+d_k)(x-x_0)^k$? Explain your reasoning.

56. Suppose that the power series $\sum c_k(x-x_0)^k$ has a finite radius of convergence R_1 and the power series $\sum d_k(x-x_0)^k$ has a finite radius of convergence R_2. What can you say about the radius of convergence of $\sum(c_k+d_k)(x-x_0)^k$? Explain your reasoning. [*Hint:* The case $R_1 = R_2$ requires special attention.]

57. Show that if p is a positive integer, then the power series

$$\sum_{k=0}^{\infty} \frac{(pk)!}{(k!)^p} x^k$$

has a radius of convergence of $1/p^p$.

58. Show that if p and q are positive integers, then the power series

$$\sum_{k=0}^{\infty} \frac{(k+p)!}{k!(k+q)!} x^k$$

has a radius of convergence of $+\infty$.

59. Show that the power series representation of the Bessel function $J_1(x)$ converges for all x [Formula (5)].

C **60.** If the constant p in the general p-series is replaced by a variable x for $x > 1$, then the resulting function is called the ***Riemann zeta function*** and is denoted by

$$\zeta(x) = \sum_{k=1}^{\infty} \frac{1}{k^x}$$

(a) Let s_n be the nth partial sum of the series for $\zeta(3.7)$. Find n such that s_n approximates $\zeta(3.7)$ to two decimal-place accuracy, and calculate s_n using this value of n. [*Hint:* Use the right inequality in Exercise 32(b) of Section 10.4 with $f(x) = 1/x^{3.7}$.]

(b) Determine whether your CAS can evaluate the Riemann zeta function directly. If so, compare the value produced by the CAS to the value of s_n obtained in part (a).

61. Prove: If $\lim_{k\to+\infty} |c_k|^{1/k} = L$, where $L \neq 0$, then $1/L$ is the radius of convergence of the power series $\sum_{k=0}^{\infty} c_k x^k$.

62. Prove: If the power series $\sum_{k=0}^{\infty} c_k x^k$ has radius of convergence R, then the series $\sum_{k=0}^{\infty} c_k x^{2k}$ has radius of convergence $\sqrt{R}$.

63. Prove: If the interval of convergence of the series $\sum_{k=0}^{\infty} c_k(x-x_0)^k$ is $(x_0-R, x_0+R]$, then the series converges conditionally at x_0+R.

✔QUICK CHECK ANSWERS 10.8

1. $\dfrac{f^{(k)}(x_0)}{k!}(x-x_0)^k$ **2.** $\dfrac{1}{2}$ **3.** $(-\infty,+\infty)$ **4.** (a) 1 (b) converges (c) diverges (d) $[3, 5)$

10.9 CONVERGENCE OF TAYLOR SERIES

In this section we will investigate when a Taylor series for a function converges to that function on some interval, and we will consider how Taylor series can be used to approximate values of trigonometric, exponential, and logarithmic functions.

THE CONVERGENCE PROBLEM FOR TAYLOR SERIES

Recall that the nth Taylor polynomial for a function f about $x = x_0$ has the property that its value and the values of its first n derivatives match those of f at x_0. As n increases, more and more derivatives match up, so it is reasonable to hope that for values of x near x_0 the values of the Taylor polynomials might converge to the value of $f(x)$; that is,

$$f(x) = \lim_{n\to+\infty} \sum_{k=0}^{n} \frac{f^{(k)}(x_0)}{k!}(x-x_0)^k \tag{1}$$

However, the nth Taylor polynomial for f is the nth partial sum of the Taylor series for f, so (1) is equivalent to stating that the Taylor series for f converges at x, and its sum is $f(x)$. Thus, we are led to consider the following problem.

It is important to understand that Problem 10.9.1 is concerned with more than just convergence of the Taylor series for f; it is concerned with whether the series converges to the function f itself. Indeed, it is possible for a Taylor series of a function f to converge to values different from $f(x)$ for certain values of x (Exercise 14).

10.9.1 PROBLEM. Given a function f that has derivatives of all orders at $x = x_0$, determine whether there is an open interval containing x_0 such that $f(x)$ is the sum of its Taylor series about $x = x_0$ at each point in the interval; that is,

$$f(x) = \sum_{k=0}^{\infty} \frac{f^{(k)}(x_0)}{k!}(x - x_0)^k \tag{2}$$

for all values of x in the interval.

One way to show that (1) holds is to show that

$$\lim_{n \to +\infty} \left[f(x) - \sum_{k=0}^{n} \frac{f^{(k)}(x_0)}{k!}(x - x_0)^k \right] = 0$$

However, the difference appearing on the left side of this equation is the nth remainder for the Taylor series [Formula (12) of Section 10.7]. Thus, we have the following result.

10.9.2 THEOREM. *The equality*

$$f(x) = \sum_{k=0}^{\infty} \frac{f^{(k)}(x_0)}{k!}(x - x_0)^k$$

holds at a point x if and only if $\lim_{n \to +\infty} R_n(x) = 0$.

ESTIMATING THE nTH REMAINDER

It is relatively rare that one can prove directly that $R_n(x) \to 0$ as $n \to +\infty$. Usually, this is proved indirectly by finding appropriate bounds on $|R_n(x)|$ and applying the Squeezing Theorem for Sequences. The Remainder Estimation Theorem (Theorem 10.7.4) provides a useful bound for this purpose. Recall that this theorem asserts that if M is an upper bound for $|f^{(n+1)}(x)|$ on an interval I containing x_0, then

$$|R_n(x)| \le \frac{M}{(n+1)!}|x - x_0|^{n+1} \tag{3}$$

for all x in I.

The following example illustrates how the Remainder Estimation Theorem is applied.

▶ Example 1 Show that the Maclaurin series for $\cos x$ converges to $\cos x$ for all x; that is,

$$\cos x = \sum_{k=0}^{\infty} (-1)^k \frac{x^{2k}}{(2k)!} = 1 - \frac{x^2}{2!} + \frac{x^4}{4!} - \frac{x^6}{6!} + \cdots \qquad (-\infty < x < +\infty)$$

Solution. From Theorem 10.9.2 we must show that $R_n(x) \to 0$ for all x as $n \to +\infty$. For this purpose let $f(x) = \cos x$, so that for all x we have

$$f^{(n+1)}(x) = \pm \cos x \quad \text{or} \quad f^{(n+1)}(x) = \pm \sin x$$

In all cases we have $|f^{(n+1)}(x)| \le 1$, so we can apply (3) with $M = 1$ and $x_0 = 0$ to conclude that

$$0 \le |R_n(x)| \le \frac{|x|^{n+1}}{(n+1)!} \tag{4}$$

The method of Example 1 can be easily modified to prove that the Taylor series for $\sin x$ and $\cos x$ about any point $x = x_0$ converge to $\sin x$ and $\cos x$ respectively, for all x (Exercises 21 and 22). For reference, some of the most important Maclaurin series are listed in Table 10.9.1 at the end of this section.

However, it follows from Formula (5) of Section 10.2 with $n+1$ in place of n and $|x|$ in place of x that

$$\lim_{n\to+\infty} \frac{|x|^{n+1}}{(n+1)!} = 0 \tag{5}$$

Using this result and the Squeezing Theorem for Sequences (Theorem 10.1.5), it follows from (4) that $|R_n(x)| \to 0$ and hence that $R_n(x) \to 0$ as $n \to +\infty$ (Theorem 10.1.6). Since this is true for all x, we have proved that the Maclaurin series for $\cos x$ converges to $\cos x$ for all x. This is illustrated in Figure 10.9.1, where we can see how successive partial sums approximate the cosine curve more and more closely. ◄

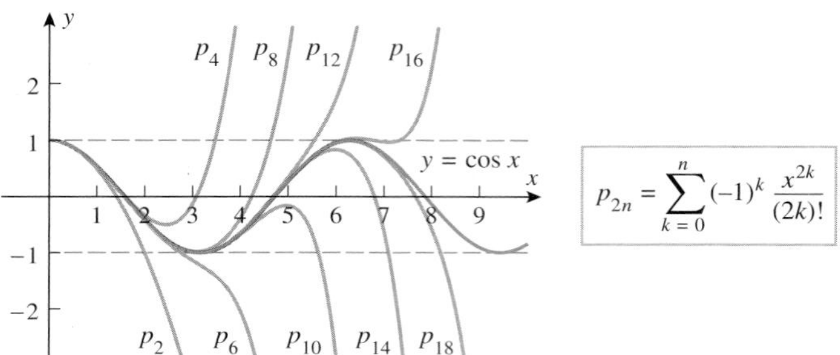

Figure 10.9.1

APPROXIMATING TRIGONOMETRIC FUNCTIONS

In general, to approximate the value of a function f at a point x using a Taylor series, there are two basic questions that must be answered:

- About what point x_0 should the Taylor series be expanded?
- How many terms in the series should be used to achieve the desired accuracy?

In response to the first question, x_0 needs to be a point at which the derivatives of f can be evaluated easily, since these values are needed for the coefficients in the Taylor series. Furthermore, if the function f is being evaluated at x, then x_0 should be chosen as close as possible to x, since Taylor series tend to converge more rapidly near x_0. For example, to approximate $\sin 3^\circ$ $(= \pi/60$ radians), it would be reasonable to take $x_0 = 0$, since $\pi/60$ is close to 0 and the derivatives of $\sin x$ are easy to evaluate at 0. On the other hand, to approximate $\sin 85^\circ$ $(= 17\pi/36$ radians), it would be more natural to take $x_0 = \pi/2$, since $17\pi/36$ is close to $\pi/2$ and the derivatives of $\sin x$ are easy to evaluate at $\pi/2$.

In response to the second question posed above, the number of terms required to achieve a specific accuracy needs to be determined on a problem-by-problem basis. The next example gives two methods for doing this.

► **Example 2** Use the Maclaurin series for $\sin x$ to approximate $\sin 3^\circ$ to five decimal-place accuracy.

Solution. In the Maclaurin series

$$\sin x = \sum_{k=0}^{\infty} (-1)^k \frac{x^{2k+1}}{(2k+1)!} = x - \frac{x^3}{3!} + \frac{x^5}{5!} - \frac{x^7}{7!} + \cdots \tag{6}$$

the angle x is assumed to be in radians (because the differentiation formulas for the trigonometric functions were derived with this assumption). Since $3^\circ = \pi/60$ radians, it follows from (6) that

$$\sin 3^\circ = \sin\frac{\pi}{60} = \left(\frac{\pi}{60}\right) - \frac{(\pi/60)^3}{3!} + \frac{(\pi/60)^5}{5!} - \frac{(\pi/60)^7}{7!} + \cdots \tag{7}$$

We must now determine how many terms in the series are required to achieve five decimal-place accuracy. We will consider two possible approaches, one using the Remainder Estimation Theorem (Theorem 10.7.4) and the other using the fact that (7) satisfies the hypotheses of the alternating series test (Theorem 10.6.1).

Method 1. (*The Remainder Estimation Theorem*)

Since we want to achieve five decimal-place accuracy, our goal is to choose n so that the absolute value of the nth remainder at $x = \pi/60$ does not exceed $0.000005 = 5 \times 10^{-6}$; that is,

$$\left|R_n\left(\frac{\pi}{60}\right)\right| \le 0.000005 \tag{8}$$

However, if we let $f(x) = \sin x$, then $f^{(n+1)}(x)$ is either $\pm\sin x$ or $\pm\cos x$, and in either case $|f^{(n+1)}(x)| \le 1$ for all x. Thus, it follows from the Remainder Estimation Theorem with $M = 1$, $x_0 = 0$, and $x = \pi/60$ that

$$\left|R_n\left(\frac{\pi}{60}\right)\right| \le \frac{(\pi/60)^{n+1}}{(n+1)!}$$

Thus, we can satisfy (8) by choosing n so that

$$\frac{(\pi/60)^{n+1}}{(n+1)!} \le 0.000005$$

With the help of a calculating utility you can verify that the smallest value of n that meets this criterion is $n = 3$. Thus, to achieve five decimal-place accuracy we need only keep terms up to the third power in (7). This yields

$$\sin 3^\circ \approx \left(\frac{\pi}{60}\right) - \frac{(\pi/60)^3}{3!} \approx 0.05234 \tag{9}$$

(verify). As a check, a calculator gives $\sin 3^\circ \approx 0.05233595624$, which agrees with (9) when rounded to five decimal places.

Method 2. (*The Alternating Series Test*)

We leave it for you to check that (7) satisfies the hypotheses of the alternating series test (Theorem 10.6.1).

Let s_n denote the sum of the terms in (7) up to and including the nth power of $\pi/60$. Since the exponents in the series are odd integers, the integer n must be odd, and the exponent of the first term *not* included in the sum s_n must be $n+2$. Thus, it follows from part (*b*) of Theorem 10.6.2 that

$$|\sin 3^\circ - s_n| < \frac{(\pi/60)^{n+2}}{(n+2)!}$$

This means that for five decimal-place accuracy we must look for the first positive odd integer n such that

$$\frac{(\pi/60)^{n+2}}{(n+2)!} \le 0.000005$$

With the help of a calculating utility you can verify that the smallest value of n that meets this criterion is $n = 3$. This agrees with the result obtained above using the Remainder Estimation Theorem and hence leads to approximation (9) as before. ◀

ROUNDOFF AND TRUNCATION ERROR

There are two types of errors that occur when computing with series. The first, called ***truncation error***, is the error that results when a series is approximated by a partial sum; and the second, called ***roundoff error***, is the error that arises from approximations in numerical computations. For example, in our derivation of (9) we took $n = 3$ to keep the truncation error below 0.000005. However, to evaluate the partial sum we had to approximate π, thereby introducing roundoff error. Had we not exercised some care in choosing this approximation, the roundoff error could easily have degraded the final result.

Methods for estimating and controlling roundoff error are studied in a branch of mathematics called ***numerical analysis***. However, as a rule of thumb, to achieve n decimal-place accuracy in a final result, all intermediate calculations must be accurate to at least $n + 1$ decimal places. Thus, in (9) at least six decimal-place accuracy in π is required to achieve the five decimal-place accuracy in the final numerical result. As a practical matter, a good working procedure is to perform all intermediate computations with the maximum number of digits that your calculating utility can handle and then round at the end.

APPROXIMATING EXPONENTIAL FUNCTIONS

► **Example 3** Show that the Maclaurin series for e^x converges to e^x for all x; that is,

$$e^x = \sum_{k=0}^{\infty} \frac{x^k}{k!} = 1 + x + \frac{x^2}{2!} + \frac{x^3}{3!} + \cdots + \frac{x^k}{k!} + \cdots \qquad (-\infty < x < +\infty)$$

Solution. Let $f(x) = e^x$, so that

$$f^{(n+1)}(x) = e^x$$

We want to show that $R_n(x) \to 0$ as $n \to +\infty$ for all x in the interval $-\infty < x < +\infty$. However, it will be helpful here to consider the cases $x \leq 0$ and $x > 0$ separately. If $x \leq 0$, then we will take the interval I in the Remainder Estimation Theorem (Theorem 10.7.4) to be $[x, 0]$, and if $x > 0$, then we will take it to be $[0, x]$. Since $f^{(n+1)}(x) = e^x$ is an increasing function, it follows that if c is in the interval $[x, 0]$, then

$$|f^{(n+1)}(c)| \leq |f^{(n+1)}(0)| = e^0 = 1$$

and if c is in the interval $[0, x]$, then

$$|f^{(n+1)}(c)| \leq |f^{(n+1)}(x)| = e^x$$

Thus, we can apply Theorem 10.7.4 with $M = 1$ in the case where $x \leq 0$ and with $M = e^x$ in the case where $x > 0$. This yields

$$0 \leq |R_n(x)| \leq \frac{|x|^{n+1}}{(n+1)!} \qquad \text{if } x \leq 0$$

$$0 \leq |R_n(x)| \leq e^x \frac{|x|^{n+1}}{(n+1)!} \qquad \text{if } x > 0$$

Thus, in both cases it follows from (5) and the Squeezing Theorem for Sequences that $|R_n(x)| \to 0$ as $n \to +\infty$, which in turn implies that $R_n(x) \to 0$ as $n \to +\infty$. Since this is true for all x, we have proved that the Maclaurin series for e^x converges to e^x for all x. ◄

Since the Maclaurin series for e^x converges to e^x for all x, we can use partial sums of the Maclaurin series to approximate powers of e to arbitrary precision. Recall that in Example 6 of Section 10.7 we were able to use the Remainder Estimation Theorem to determine that evaluating the ninth Maclaurin polynomial for e^x at $x = 1$ yields an approximation for e with five decimal-place accuracy:

$$e \approx 1 + 1 + \frac{1}{2!} + \frac{1}{3!} + \frac{1}{4!} + \frac{1}{5!} + \frac{1}{6!} + \frac{1}{7!} + \frac{1}{8!} + \frac{1}{9!} \approx 2.71828$$

APPROXIMATING LOGARITHMS

The Maclaurin series

$$\ln(1+x) = x - \frac{x^2}{2} + \frac{x^3}{3} - \frac{x^4}{4} + \cdots \qquad (-1 < x \le 1) \tag{10}$$

is the starting point for the approximation of natural logarithms. Unfortunately, the usefulness of this series is limited because of its slow convergence and the restriction $-1 < x \le 1$. However, if we replace x by $-x$ in this series, we obtain

$$\ln(1-x) = -x - \frac{x^2}{2} - \frac{x^3}{3} - \frac{x^4}{4} - \cdots \qquad (-1 \le x < 1) \tag{11}$$

and on subtracting (11) from (10) we obtain

$$\ln\left(\frac{1+x}{1-x}\right) = 2\left(x + \frac{x^3}{3} + \frac{x^5}{5} + \frac{x^7}{7} + \cdots\right) \qquad (-1 < x < 1) \tag{12}$$

In Example 2 of Section 10.6, we stated without proof that

$$\ln 2 = 1 - \frac{1}{2} + \frac{1}{3} - \frac{1}{4} + \frac{1}{5} - \cdots$$

This result can be obtained by letting $x = 1$ in (10), but as indicated in the text discussion, this series converges too slowly to be of practical use.

Series (12), first obtained by James Gregory in 1668, can be used to compute the natural logarithm of any positive number y by letting

$$y = \frac{1+x}{1-x}$$

or, equivalently,

$$x = \frac{y-1}{y+1} \tag{13}$$

and noting that $-1 < x < 1$. For example, to compute $\ln 2$ we let $y = 2$ in (13), which yields $x = \frac{1}{3}$. Substituting this value in (12) gives

$$\ln 2 = 2\left[\frac{1}{3} + \frac{\left(\frac{1}{3}\right)^3}{3} + \frac{\left(\frac{1}{3}\right)^5}{5} + \frac{\left(\frac{1}{3}\right)^7}{7} + \cdots\right] \tag{14}$$

In Exercise 19 we will ask you to show that five decimal-place accuracy can be achieved using the partial sum with terms up to and including the 13th power of $\frac{1}{3}$. Thus, to five decimal-place accuracy

$$\ln 2 \approx 2\left[\frac{1}{3} + \frac{\left(\frac{1}{3}\right)^3}{3} + \frac{\left(\frac{1}{3}\right)^5}{5} + \frac{\left(\frac{1}{3}\right)^7}{7} + \cdots + \frac{\left(\frac{1}{3}\right)^{13}}{13}\right] \approx 0.69315$$

(verify). As a check, a calculator gives $\ln 2 \approx 0.69314718056$, which agrees with the preceding approximation when rounded to five decimal places.

James Gregory (1638–1675) Scottish mathematician and astronomer. Gregory, the son of a minister, was famous in his time as the inventor of the Gregorian reflecting telescope, so named in his honor. Although he is not generally ranked with the great mathematicians, much of his work relating to calculus was studied by Leibniz and Newton and undoubtedly influenced some of their discoveries. There is a manuscript, discovered posthumously, which shows that Gregory had anticipated Taylor series well before Taylor.

APPROXIMATING π

In the next section we will show that

$$\tan^{-1} x = x - \frac{x^3}{3} + \frac{x^5}{5} - \frac{x^7}{7} + \cdots \qquad (-1 \le x \le 1) \tag{15}$$

Letting $x = 1$, we obtain

$$\frac{\pi}{4} = \tan^{-1} 1 = 1 - \frac{1}{3} + \frac{1}{5} - \frac{1}{7} + \cdots$$

or

$$\pi = 4\left[1 - \frac{1}{3} + \frac{1}{5} - \frac{1}{7} + \cdots\right]$$

This famous series, obtained by Leibniz in 1674, converges too slowly to be of computational value. A more practical procedure for approximating π uses the identity

$$\frac{\pi}{4} = \tan^{-1}\frac{1}{2} + \tan^{-1}\frac{1}{3} \tag{16}$$

which was derived in Exercise 89 of Section 7.7. By using this identity and series (15) to approximate $\tan^{-1}\frac{1}{2}$ and $\tan^{-1}\frac{1}{3}$, the value of π can be approximated efficiently to any degree of accuracy.

BINOMIAL SERIES

If m is a real number, then the Maclaurin series for $(1+x)^m$ is called the ***binomial series***; it is given by

$$1 + mx + \frac{m(m-1)}{2!}x^2 + \frac{m(m-1)(m-2)}{3!}x^3 + \cdots + \frac{m(m-1)\cdots(m-k+1)}{k!}x^k + \cdots$$

In the case where m is a nonnegative integer, the function $f(x) = (1+x)^m$ is a polynomial of degree m, so

$$f^{(m+1)}(0) = f^{(m+2)}(0) = f^{(m+3)}(0) = \cdots = 0$$

and the binomial series reduces to the familiar binomial expansion

$$(1+x)^m = 1 + mx + \frac{m(m-1)}{2!}x^2 + \frac{m(m-1)(m-2)}{3!}x^3 + \cdots + x^m$$

which is valid for $-\infty < x < +\infty$.

Let $f(x) = (1+x)^m$. Verify that

$$f(0) = 1$$
$$f'(0) = m$$
$$f''(0) = m(m-1)$$
$$f'''(0) = m(m-1)(m-2)$$
$$\vdots$$
$$f^{(k)}(0) = m(m-1)\cdots(m-k+1)$$

It can be proved that if m is not a nonnegative integer, then the binomial series converges to $(1+x)^m$ if $|x| < 1$. Thus, for such values of x

$$(1+x)^m = 1 + mx + \frac{m(m-1)}{2!}x^2 + \cdots + \frac{m(m-1)\cdots(m-k+1)}{k!}x^k + \cdots \tag{17}$$

or in sigma notation,

$$(1+x)^m = 1 + \sum_{k=1}^{\infty} \frac{m(m-1)\cdots(m-k+1)}{k!}x^k \quad \text{if } |x| < 1 \tag{18}$$

▶ **Example 4** Find binomial series for

$$\text{(a) } \frac{1}{(1+x)^2} \qquad \text{(b) } \frac{1}{\sqrt{1+x}}$$

Solution (a). Since the general term of the binomial series is complicated, you may find it helpful to write out some of the beginning terms of the series, as in Formula (17), to see developing patterns. Substituting $m = -2$ in this formula yields

$$\begin{aligned}
\frac{1}{(1+x)^2} = (1+x)^{-2} &= 1 + (-2)x + \frac{(-2)(-3)}{2!}x^2 \\
&\quad + \frac{(-2)(-3)(-4)}{3!}x^3 + \frac{(-2)(-3)(-4)(-5)}{4!}x^4 + \cdots \\
&= 1 - 2x + \frac{3!}{2!}x^2 - \frac{4!}{3!}x^3 + \frac{5!}{4!}x^4 - \cdots \\
&= 1 - 2x + 3x^2 - 4x^3 + 5x^4 - \cdots \\
&= \sum_{k=0}^{\infty} (-1)^k (k+1) x^k
\end{aligned}$$

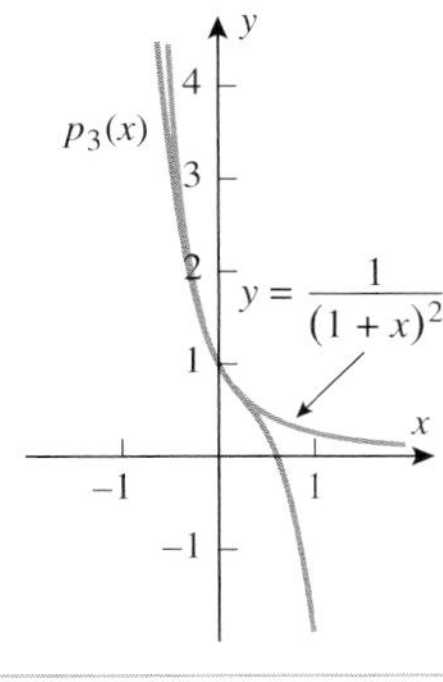

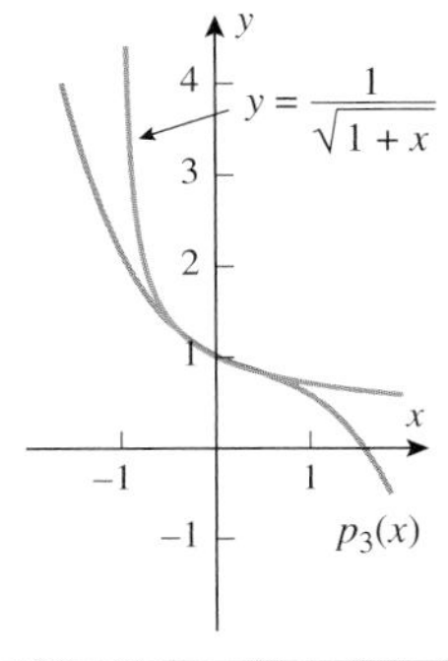

Figure 10.9.2

Solution (b). Substituting $m = -\frac{1}{2}$ in (17) yields

$$\begin{aligned}\frac{1}{\sqrt{1+x}} &= 1 - \frac{1}{2}x + \frac{\left(-\frac{1}{2}\right)\left(-\frac{1}{2}-1\right)}{2!}x^2 + \frac{\left(-\frac{1}{2}\right)\left(-\frac{1}{2}-1\right)\left(-\frac{1}{2}-2\right)}{3!}x^3 + \cdots \\ &= 1 - \frac{1}{2}x + \frac{1\cdot 3}{2^2\cdot 2!}x^2 - \frac{1\cdot 3\cdot 5}{2^3\cdot 3!}x^3 + \cdots \\ &= 1 + \sum_{k=1}^{\infty}(-1)^k\frac{1\cdot 3\cdot 5\cdots(2k-1)}{2^k k!}x^k \quad \blacktriangleleft\end{aligned}$$

Figure 10.9.2 shows the graphs of the functions in Example 4 compared to their third-degree Maclaurin polynomials.

SOME IMPORTANT MACLAURIN SERIES

For reference, Table 10.9.1 lists the Maclaurin series for some of the most important functions, together with a specification of the intervals over which the Maclaurin series converge to those functions. Some of these results are derived in the exercises and others will be derived in the next section using some special techniques that we will develop.

Table 10.9.1

MACLAURIN SERIES	INTERVAL OF CONVERGENCE
$\frac{1}{1-x} = \sum_{k=0}^{\infty} x^k = 1 + x + x^2 + x^3 + \cdots$	$-1 < x < 1$
$\frac{1}{1+x^2} = \sum_{k=0}^{\infty} (-1)^k x^{2k} = 1 - x^2 + x^4 - x^6 + \cdots$	$-1 < x < 1$
$e^x = \sum_{k=0}^{\infty} \frac{x^k}{k!} = 1 + x + \frac{x^2}{2!} + \frac{x^3}{3!} + \frac{x^4}{4!} + \cdots$	$-\infty < x < +\infty$
$\sin x = \sum_{k=0}^{\infty} (-1)^k \frac{x^{2k+1}}{(2k+1)!} = x - \frac{x^3}{3!} + \frac{x^5}{5!} - \frac{x^7}{7!} + \cdots$	$-\infty < x < +\infty$
$\cos x = \sum_{k=0}^{\infty} (-1)^k \frac{x^{2k}}{(2k)!} = 1 - \frac{x^2}{2!} + \frac{x^4}{4!} - \frac{x^6}{6!} + \cdots$	$-\infty < x < +\infty$
$\ln(1+x) = \sum_{k=1}^{\infty} (-1)^{k+1} \frac{x^k}{k} = x - \frac{x^2}{2} + \frac{x^3}{3} - \frac{x^4}{4} + \cdots$	$-1 < x \le 1$
$\tan^{-1} x = \sum_{k=0}^{\infty} (-1)^k \frac{x^{2k+1}}{2k+1} = x - \frac{x^3}{3} + \frac{x^5}{5} - \frac{x^7}{7} + \cdots$	$-1 \le x \le 1$
$\sinh x = \sum_{k=0}^{\infty} \frac{x^{2k+1}}{(2k+1)!} = x + \frac{x^3}{3!} + \frac{x^5}{5!} + \frac{x^7}{7!} + \cdots$	$-\infty < x < +\infty$
$\cosh x = \sum_{k=0}^{\infty} \frac{x^{2k}}{(2k)!} = 1 + \frac{x^2}{2!} + \frac{x^4}{4!} + \frac{x^6}{6!} + \cdots$	$-\infty < x < +\infty$
$(1+x)^m = 1 + \sum_{k=1}^{\infty} \frac{m(m-1)\cdots(m-k+1)}{k!} x^k$	$-1 < x < 1^*$ $(m \ne 0, 1, 2, \ldots)$

*The behavior at the endpoints depends on m: For $m > 0$ the series converges absolutely at both endpoints; for $m \le -1$ the series diverges at both endpoints; and for $-1 < m < 0$ the series converges conditionally at $x = 1$ and diverges at $x = -1$.

✔ QUICK CHECK EXERCISES 10.9 (See page 704 for answers.)

1. $\cos x = \sum_{k=0}^{\infty}$ ______

2. $e^x = \sum_{k=0}^{\infty}$ ______

3. $\ln(1+x) = \sum_{k=1}^{\infty}$ ______ for x in the interval ______.

4. If m is a real number but not a nonnegative integer, the *binomial series*
$$1 + \sum_{k=1}^{\infty} \text{\_\_\_\_\_\_}$$
converges to $(1+x)^m$ if $|x| <$ ______.

EXERCISE SET 10.9 Graphing Utility [C] CAS

1. Use both of the methods given in Example 2 to approximate $\sin 4°$ to five decimal-place accuracy, and check your work by comparing your answer to that produced directly by your calculating utility.

2. Use both of the methods given in Example 2 to approximate $\cos 3°$ to three decimal-place accuracy, and check your work by comparing your answer to that produced directly by your calculating utility.

3. Use the Maclaurin series for $\cos x$ to approximate $\cos 0.1$ to five decimal-place accuracy, and check your work by comparing your answer to that produced directly by your calculating utility.

4. Use the Maclaurin series for $\tan^{-1} x$ to approximate $\tan^{-1} 0.1$ to three decimal-place accuracy, and check your work by comparing your answer to that produced directly by your calculating utility.

5. Use an appropriate Taylor series to approximate $\sin 85°$ to four decimal-place accuracy, and check your work by comparing your answer to that produced directly by your calculating utility.

6. Use a Taylor series to approximate $\cos(-175°)$ to four decimal-place accuracy, and check your work by comparing your answer to that produced directly by your calculating utility.

7. Use the Maclaurin series for $\sinh x$ to approximate $\sinh 0.5$ to three decimal-place accuracy. Check your work by computing $\sinh 0.5$ with a calculating utility.

8. Use the Maclaurin series for $\cosh x$ to approximate $\cosh 0.1$ to three decimal-place accuracy. Check your work by computing $\cosh 0.1$ with a calculating utility.

9. Use the Remainder Estimation Theorem and the method of Example 1 to prove that the Taylor series for $\sin x$ about $x = \pi/4$ converges to $\sin x$ for all x.

10. Use the Remainder Estimation Theorem and the method of Example 3 to prove that the Taylor series for e^x about $x = 1$ converges to e^x for all x.

11. (a) Use Formula (12) in the text to find a series that converges to $\ln 1.25$.
(b) Approximate $\ln 1.25$ using the first two terms of the series. Round your answer to three decimal places, and compare the result to that produced directly by your calculating utility.

12. (a) Use Formula (12) to find a series that converges to $\ln 3$.
(b) Approximate $\ln 3$ using the first two terms of the series. Round your answer to three decimal places, and compare the result to that produced directly by your calculating utility.

FOCUS ON CONCEPTS

13. (a) Use the Maclaurin series for $\tan^{-1} x$ to approximate $\tan^{-1} \frac{1}{2}$ and $\tan^{-1} \frac{1}{3}$ to three decimal-place accuracy.
(b) Use the results in part (a) and Formula (16) to approximate π.
(c) Would you be willing to guarantee that your answer in part (b) is accurate to three decimal places? Explain your reasoning.
(d) Compare your answer in part (b) to that produced by your calculating utility.

14. The purpose of this exercise is to show that the Taylor series of a function f may possibly converge to a value different from $f(x)$ for certain values of x. Let
$$f(x) = \begin{cases} e^{-1/x^2}, & x \neq 0 \\ 0, & x = 0 \end{cases}$$
(a) Use the definition of a derivative to show that $f'(0) = 0$.
(b) With some difficulty it can be shown that if $n \geq 2$ then $f^{(n)}(0) = 0$. Accepting this fact, show that the Maclaurin series of f converges for all x, but converges to $f(x)$ only at $x = 0$.

15. (a) Find an upper bound on the error that can result if $\cos x$ is approximated by $1 - (x^2/2!) + (x^4/4!)$ over the interval $[-0.2, 0.2]$.

(b) Check your answer in part (a) by graphing

$$\left|\cos x - \left(1 - \frac{x^2}{2!} + \frac{x^4}{4!}\right)\right|$$

over the interval.

16. (a) Find an upper bound on the error that can result if $\ln(1+x)$ is approximated by x over the interval $[-0.01, 0.01]$.

(b) Check your answer in part (a) by graphing

$$|\ln(1+x) - x|$$

over the interval.

17. Use Formula (17) for the binomial series to obtain the Maclaurin series for

(a) $\dfrac{1}{1+x}$ (b) $\sqrt[3]{1+x}$ (c) $\dfrac{1}{(1+x)^3}$.

18. If m is any real number, and k is a nonnegative integer, then we define the ***binomial coefficient***

$$\binom{m}{k} \text{ by the formulas } \binom{m}{0} = 1 \text{ and}$$

$$\binom{m}{k} = \frac{m(m-1)(m-2)\cdots(m-k+1)}{k!}$$

for $k \geq 1$. Express Formula (17) in the text in terms of binomial coefficients.

19. In this exercise we will use the Remainder Estimation Theorem to determine the number of terms that are required in Formula (14) to approximate ln 2 to five decimal-place accuracy. For this purpose let

$$f(x) = \ln\frac{1+x}{1-x} = \ln(1+x) - \ln(1-x) \quad (-1 < x < 1)$$

(a) Show that

$$f^{(n+1)}(x) = n!\left[\frac{(-1)^n}{(1+x)^{n+1}} + \frac{1}{(1-x)^{n+1}}\right]$$

(b) Use the triangle inequality [Theorem 1.1.4(*d*)] to show that

$$|f^{(n+1)}(x)| \leq n!\left[\frac{1}{(1+x)^{n+1}} + \frac{1}{(1-x)^{n+1}}\right]$$

(c) Since we want to achieve five decimal-place accuracy, our goal is to choose n so that the absolute value of the nth remainder at $x = \frac{1}{3}$ does not exceed the value $0.000005 = 0.5 \times 10^{-5}$; that is, $\left|R_n\left(\frac{1}{3}\right)\right| \leq 0.000005$. Use the Remainder Estimation Theorem to show that this condition will be satisfied if n is chosen so that

$$\frac{M}{(n+1)!}\left(\frac{1}{3}\right)^{n+1} \leq 0.000005$$

where $|f^{(n+1)}(x)| \leq M$ on the interval $\left[0, \frac{1}{3}\right]$.

(d) Use the result in part (b) to show that M can be taken as

$$M = n!\left[1 + \frac{1}{\left(\frac{2}{3}\right)^{n+1}}\right]$$

(e) Use the results in parts (c) and (d) to show that five decimal-place accuracy will be achieved if n satisfies

$$\frac{1}{n+1}\left[\left(\frac{1}{3}\right)^{n+1} + \left(\frac{1}{2}\right)^{n+1}\right] \leq 0.000005$$

and then show that the smallest value of n that satisfies this condition is $n = 13$.

20. Use Formula (12) and the method of Exercise 19 to approximate $\ln\left(\frac{5}{3}\right)$ to five decimal-place accuracy. Then check your work by comparing your answer to that produced directly by your calculating utility.

21. Prove: The Taylor series for $\cos x$ about any value $x = x_0$ converges to $\cos x$ for all x.

22. Prove: The Taylor series for $\sin x$ about any value $x = x_0$ converges to $\sin x$ for all x.

23. Research has shown that the proportion p of the population with IQs (intelligence quotients) between α and β is approximately

$$p = \frac{1}{16\sqrt{2\pi}}\int_\alpha^\beta e^{-\frac{1}{2}\left(\frac{x-100}{16}\right)^2}\,dx$$

Use the first three terms of an appropriate Maclaurin series to estimate the proportion of the population that has IQs between 100 and 110.

24. In Section 6.7 we defined the kinetic energy K of a particle with mass m and velocity v to be $K = \frac{1}{2}mv^2$ [see Formula (6) of that section]. In this formula the mass m is assumed to be constant, and K is called the ***Newtonian kinetic energy***. However, in Albert Einstein's relativity theory the mass m increases with the velocity and the kinetic energy K is given by the formula

$$K = m_0c^2\left[\frac{1}{\sqrt{1-(v/c)^2}} - 1\right]$$

in which m_0 is the mass of the particle when its velocity is zero, and c is the speed of light. This is called the ***relativistic kinetic energy***. Use an appropriate binomial series to show that if the velocity is small compared to the speed of light (i.e., $v/c \approx 0$), then the Newtonian and relativistic kinetic energies are in close agreement.

C **25.** (a) In 1706 the British astronomer and mathematician John Machin discovered the following formula for $\pi/4$, called ***Machin's formula***:

$$\frac{\pi}{4} = 4\tan^{-1}\frac{1}{5} - \tan^{-1}\frac{1}{239}$$

Use a CAS to approximate $\pi/4$ using Machin's formula to 25 decimal places.

(b) In 1914 the brilliant Indian mathematician Srinivasa Ramanujan (1887–1920) showed that

$$\frac{1}{\pi} = \frac{\sqrt{8}}{9801}\sum_{k=0}^{\infty}\frac{(4k)!(1103 + 26{,}390k)}{(k!)^4 396^{4k}}$$

Use a CAS to compute the first four partial sums in ***Ramanujan's formula***.

✔ QUICK CHECK ANSWERS 10.9

1. $(-1)^k \dfrac{x^{2k}}{(2k)!}$ 2. $\dfrac{x^k}{k!}$ 3. $(-1)^{k+1}\dfrac{x^k}{k}$; $(-1, 1]$ 4. $\dfrac{m(m-1)\cdots(m-k+1)}{k!}x^k$; 1

10.10 DIFFERENTIATING AND INTEGRATING POWER SERIES; MODELING WITH TAYLOR SERIES

In this section we will discuss methods for finding power series for derivatives and integrals of functions, and we will discuss some practical methods for finding Taylor series that can be used in situations where it is difficult or impossible to find the series directly.

DIFFERENTIATING POWER SERIES

We begin by considering the following problem.

> **10.10.1 PROBLEM.** Suppose that a function f is represented by a power series on an open interval. How can we use the power series to find the derivative of f on that interval?

The solution to this problem can be motivated by considering the Maclaurin series for $\sin x$:

$$\sin x = x - \frac{x^3}{3!} + \frac{x^5}{5!} - \frac{x^7}{7!} + \cdots \qquad (-\infty < x < +\infty)$$

Of course, we already know that the derivative of $\sin x$ is $\cos x$; however, we are concerned here with using the Maclaurin series to deduce this. The solution is easy—all we need to do is differentiate the Maclaurin series term by term and observe that the resulting series is the Maclaurin series for $\cos x$:

$$\frac{d}{dx}\left[x - \frac{x^3}{3!} + \frac{x^5}{5!} - \frac{x^7}{7!} + \cdots\right] = 1 - 3\frac{x^2}{3!} + 5\frac{x^4}{5!} - 7\frac{x^6}{7!} + \cdots$$

$$= 1 - \frac{x^2}{2!} + \frac{x^4}{4!} - \frac{x^6}{6!} + \cdots = \cos x$$

Here is another example.

$$\frac{d}{dx}[e^x] = \frac{d}{dx}\left[1 + x + \frac{x^2}{2!} + \frac{x^3}{3!} + \frac{x^4}{4!} + \cdots\right]$$

$$= 1 + 2\frac{x}{2!} + 3\frac{x^2}{3!} + 4\frac{x^3}{4!} + \cdots = 1 + x + \frac{x^2}{2!} + \frac{x^3}{3!} + \cdots = e^x$$

The preceding computations suggest that if a function f is represented by a power series on an open interval, then a power series representation of f' on that interval can be obtained by differentiating the power series for f term by term. This is stated more precisely in the following theorem, which we give without proof.

> **10.10.2 THEOREM (*Differentiation of Power Series*).** *Suppose that a function f is represented by a power series in $x - x_0$ that has a nonzero radius of convergence R; that is,*
>
> $$f(x) = \sum_{k=0}^{\infty} c_k(x - x_0)^k \qquad (x_0 - R < x < x_0 + R)$$
>
> *Then:*

(*a*) *The function f is differentiable on the interval $(x_0 - R, x_0 + R)$.*

(*b*) *If the power series representation for f is differentiated term by term, then the resulting series has radius of convergence R and converges to f' on the interval $(x_0 - R, x_0 + R)$; that is,*

$$f'(x) = \sum_{k=0}^{\infty} \frac{d}{dx}[c_k(x - x_0)^k] \qquad (x_0 - R < x < x_0 + R)$$

This theorem has an important implication about the differentiability of functions that are represented by power series. According to the theorem, the power series for f' has the same radius of convergence as the power series for f, and this means that the theorem can be applied to f' as well as f. However, if we do this, then we conclude that f' is differentiable on the interval $(x_0 - R, x_0 + R)$, and the power series for f'' has the same radius of convergence as the power series for f and f'. We can now repeat this process ad infinitum, applying the theorem successively to $f'', f''', \dots, f^{(n)}, \dots$ to conclude that f has derivatives of all orders on the interval $(x_0 - R, x_0 + R)$. Thus, we have established the following result.

10.10.3 THEOREM. *If a function f can be represented by a power series in $x - x_0$ with a nonzero radius of convergence R, then f has derivatives of all orders on the interval $(x_0 - R, x_0 + R)$.*

In short, it is only the most "well-behaved" functions that can be represented by power series; that is, if a function f does not possess derivatives of all orders on an interval $(x_0 - R, x_0 + R)$, then it cannot be represented by a power series in $x - x_0$ on that interval.

▶ **Example 1** In Section 10.8, we showed that the Bessel function $J_0(x)$, represented by the power series

$$J_0(x) = \sum_{k=0}^{\infty} \frac{(-1)^k x^{2k}}{2^{2k}(k!)^2} \tag{1}$$

has radius of convergence $+\infty$ [see Formula (4) of that section and the related discussion]. Thus, $J_0(x)$ has derivatives of all orders on the interval $(-\infty, +\infty)$, and these can be obtained by differentiating the series term by term. For example, if we write (1) as

$$J_0(x) = 1 + \sum_{k=1}^{\infty} \frac{(-1)^k x^{2k}}{2^{2k}(k!)^2}$$

and differentiate term by term, we obtain

$$J_0'(x) = \sum_{k=1}^{\infty} \frac{(-1)^k (2k) x^{2k-1}}{2^{2k}(k!)^2} = \sum_{k=1}^{\infty} \frac{(-1)^k x^{2k-1}}{2^{2k-1} k!(k-1)!} \quad ◀$$

See Exercise 44 for a relationship between $J_0'(x)$ and $J_1(x)$.

The computations in this example use some techniques that are worth noting. First, when a power series is expressed in sigma notation, the formula for the general term of the series will often not be of a form that can be used for differentiating the constant term. Thus, if the series has a nonzero constant term, as here, it is usually a good idea to split it off from the summation before differentiating. Second, observe how we simplified the final formula by canceling the factor k from one of the factorials in the denominator. This is a standard simplification technique.

INTEGRATING POWER SERIES

Since the derivative of a function that is represented by a power series can be obtained by differentiating the series term by term, it should not be surprising that an antiderivative of a function represented by a power series can be obtained by integrating the series term by term. For example, we know that $\sin x$ is an antiderivative of $\cos x$. Here is how this result can be obtained by integrating the Maclaurin series for $\cos x$ term by term:

$$\begin{aligned}\int \cos x\,dx &= \int\left[1-\frac{x^2}{2!}+\frac{x^4}{4!}-\frac{x^6}{6!}+\cdots\right]dx\\ &= \left[x-\frac{x^3}{3(2!)}+\frac{x^5}{5(4!)}-\frac{x^7}{7(6!)}+\cdots\right]+C\\ &= \left[x-\frac{x^3}{3!}+\frac{x^5}{5!}-\frac{x^7}{7!}+\cdots\right]+C=\sin x+C\end{aligned}$$

The same idea applies to definite integrals. For example, by direct integration we have

$$\int_0^1 \frac{dx}{1+x^2}=\tan^{-1}x\Big]_0^1=\tan^{-1}1-\tan 0=\frac{\pi}{4}-0=\frac{\pi}{4}$$

and we will show later in this section that

$$\frac{\pi}{4}=1-\frac{1}{3}+\frac{1}{5}-\frac{1}{7}+\cdots \tag{2}$$

Thus,

$$\int_0^1 \frac{dx}{1+x^2}=1-\frac{1}{3}+\frac{1}{5}-\frac{1}{7}+\cdots$$

Here is how this result can be obtained by integrating the Maclaurin series for $1/(1+x^2)$ term by term (see Table 10.9.1):

$$\begin{aligned}\int_0^1 \frac{dx}{1+x^2} &= \int_0^1 [1-x^2+x^4-x^6+\cdots]\,dx\\ &= x-\frac{x^3}{3}+\frac{x^5}{5}-\frac{x^7}{7}+\cdots\Big]_0^1=1-\frac{1}{3}+\frac{1}{5}-\frac{1}{7}+\cdots\end{aligned}$$

The preceding computations are justified by the following theorem, which we give without proof.

10.10.4 THEOREM (*Integration of Power Series*). *Suppose that a function f is represented by a power series in $x-x_0$ that has a nonzero radius of convergence R; that is,*

$$f(x)=\sum_{k=0}^{\infty} c_k(x-x_0)^k \qquad (x_0-R<x<x_0+R)$$

(a) *If the power series representation of f is integrated term by term, then the resulting series has radius of convergence R and converges to an antiderivative for $f(x)$ on the interval (x_0-R, x_0+R); that is,*

$$\int f(x)\,dx=\sum_{k=0}^{\infty}\left[\frac{c_k}{k+1}(x-x_0)^{k+1}\right]+C \qquad (x_0-R<x<x_0+R)$$

(b) *If α and β are points in the interval (x_0-R, x_0+R), and if the power series representation of f is integrated term by term from α to β, then the resulting series converges absolutely on the interval (x_0-R, x_0+R) and*

$$\int_\alpha^\beta f(x)\,dx=\sum_{k=0}^{\infty}\left[\int_\alpha^\beta c_k(x-x_0)^k\,dx\right]$$

POWER SERIES REPRESENTATIONS MUST BE TAYLOR SERIES

For many functions it is difficult or impossible to find the derivatives that are required to obtain a Taylor series. For example, to find the Maclaurin series for $1/(1+x^2)$ directly would require some tedious derivative computations (try it). A more practical approach is to substitute $-x^2$ for x in the geometric series

$$\frac{1}{1-x} = 1 + x + x^2 + x^3 + x^4 + \cdots \qquad (-1 < x < 1)$$

to obtain

$$\frac{1}{1+x^2} = 1 - x^2 + x^4 - x^6 + x^8 - \cdots$$

However, there are two questions of concern with this procedure:

- Where does the power series that we obtained for $1/(1+x^2)$ actually converge to $1/(1+x^2)$?
- How do we know that the power series we have obtained is actually the Maclaurin series for $1/(1+x^2)$?

The first question is easy to resolve. Since the geometric series converges to $1/(1-x)$ if $|x| < 1$, the second series will converge to $1/(1+x^2)$ if $|-x^2| < 1$ or $|x^2| < 1$. However, this is true if and only if $|x| < 1$, so the power series we obtained for the function $1/(1+x^2)$ converges to this function if $-1 < x < 1$.

The second question is more difficult to answer and leads us to the following general problem.

10.10.5 PROBLEM. Suppose that a function f is represented by a power series in $x - x_0$ that has a nonzero radius of convergence. What relationship exists between the given power series and the Taylor series for f about $x = x_0$?

The answer is that they are the same; and here is the theorem that proves it.

Theorem 10.10.6 tells us that no matter how we arrive at a power series representation of a function f, be it by substitution, by differentiation, by integration, or by some algebraic process, that series will be the Taylor series for f about $x = x_0$, provided the series converges to f on some open interval containing x_0.

10.10.6 THEOREM. *If a function f is represented by a power series in $x - x_0$ on some open interval containing x_0, then that power series is the Taylor series for f about $x = x_0$.*

PROOF. Suppose that

$$f(x) = c_0 + c_1(x - x_0) + c_2(x - x_0)^2 + \cdots + c_k(x - x_0)^k + \cdots$$

for all x in some open interval containing x_0. To prove that this is the Taylor series for f about $x = x_0$, we must show that

$$c_k = \frac{f^{(k)}(x_0)}{k!} \quad \text{for} \quad k = 0, 1, 2, 3, \ldots$$

However, the assumption that the series converges to $f(x)$ on an open interval containing x_0 ensures that it has a nonzero radius of convergence R; hence we can differentiate term by term in accordance with Theorem 10.10.2. Thus,

$$\begin{aligned}
f(x) &= c_0 + c_1(x - x_0) + c_2(x - x_0)^2 + c_3(x - x_0)^3 + c_4(x - x_0)^4 + \cdots \\
f'(x) &= c_1 + 2c_2(x - x_0) + 3c_3(x - x_0)^2 + 4c_4(x - x_0)^3 + \cdots \\
f''(x) &= 2!c_2 + (3 \cdot 2)c_3(x - x_0) + (4 \cdot 3)c_4(x - x_0)^2 + \cdots \\
f'''(x) &= 3!c_3 + (4 \cdot 3 \cdot 2)c_4(x - x_0) + \cdots \\
&\vdots
\end{aligned}$$

On substituting $x = x_0$, all the powers of $x - x_0$ drop out, leaving

$$f(x_0) = c_0, \quad f'(x_0) = c_1, \quad f''(x_0) = 2!c_2, \quad f'''(x_0) = 3!c_3, \ldots$$

from which we obtain

$$c_0 = f(x_0), \quad c_1 = f'(x_0), \quad c_2 = \frac{f''(x_0)}{2!}, \quad c_3 = \frac{f'''(x_0)}{3!}, \ldots$$

which shows that the coefficients $c_0, c_1, c_2, c_3, \ldots$ are precisely the coefficients in the Taylor series about x_0 for $f(x)$. ■

SOME PRACTICAL WAYS TO FIND TAYLOR SERIES

▶ **Example 2** Find the Maclaurin series for $\tan^{-1} x$.

Solution. It would be tedious to find the Maclaurin series directly. A better approach is to start with the formula

$$\int \frac{1}{1 + x^2}\,dx = \tan^{-1} x + C$$

and integrate the Maclaurin series

$$\frac{1}{1 + x^2} = 1 - x^2 + x^4 - x^6 + x^8 - \cdots \qquad (-1 < x < 1)$$

term by term. This yields

$$\tan^{-1} x + C = \int \frac{1}{1 + x^2}\,dx = \int [1 - x^2 + x^4 - x^6 + x^8 - \cdots]\,dx$$

or

$$\tan^{-1} x = \left[x - \frac{x^3}{3} + \frac{x^5}{5} - \frac{x^7}{7} + \frac{x^9}{9} - \cdots\right] - C$$

The constant of integration can be evaluated by substituting $x = 0$ and using the condition $\tan^{-1} 0 = 0$. This gives $C = 0$, so that

$$\tan^{-1} x = x - \frac{x^3}{3} + \frac{x^5}{5} - \frac{x^7}{7} + \frac{x^9}{9} - \cdots \qquad (-1 < x < 1) \tag{3}$$

◀

Observe that neither Theorem 10.10.2 nor Theorem 10.10.3 addresses what happens at the endpoints of the interval of convergence. However, it can be proved that if the Taylor series for f about $x = x_0$ converges to $f(x)$ for all x in the interval $(x_0 - R, x_0 + R)$, and if the Taylor series converges at the right endpoint $x_0 + R$, then the value that it converges to at that point is the limit of $f(x)$ as $x \to x_0 + R$ from the left; and if the Taylor series converges at the left endpoint $x_0 - R$, then the value that it converges to at that point is the limit of $f(x)$ as $x \to x_0 - R$ from the right.

For example, the Maclaurin series for $\tan^{-1} x$ given in (3) converges at both $x = -1$ and $x = 1$, since the hypotheses of the alternating series test (Theorem 10.6.1) are satisfied at those points. Thus, the continuity of $\tan^{-1} x$ on the interval $[-1, 1]$ implies that at $x = 1$ the Maclaurin series converges to

$$\lim_{x \to 1^-} \tan^{-1} x = \tan^{-1} 1 = \frac{\pi}{4}$$

and at $x = -1$ it converges to

$$\lim_{x \to -1^+} \tan^{-1} x = \tan^{-1}(-1) = -\frac{\pi}{4}$$

This shows that the Maclaurin series for $\tan^{-1} x$ actually converges to $\tan^{-1} x$ on the closed interval $-1 \le x \le 1$. Moreover, the convergence at $x = 1$ establishes Formula (2).

Taylor series provide an alternative to Simpson's rule and other numerical methods for approximating definite integrals.

▶ **Example 3** Approximate the integral

$$\int_0^1 e^{-x^2}\,dx$$

to three decimal-place accuracy by expanding the integrand in a Maclaurin series and integrating term by term.

Solution. The simplest way to obtain the Maclaurin series for e^{-x^2} is to replace x by $-x^2$ in the Maclaurin series

$$e^x = 1 + x + \frac{x^2}{2!} + \frac{x^3}{3!} + \frac{x^4}{4!} + \cdots$$

to obtain

$$e^{-x^2} = 1 - x^2 + \frac{x^4}{2!} - \frac{x^6}{3!} + \frac{x^8}{4!} - \cdots$$

Therefore,

$$\begin{aligned}
\int_0^1 e^{-x^2}\,dx &= \int_0^1 \left[1 - x^2 + \frac{x^4}{2!} - \frac{x^6}{3!} + \frac{x^8}{4!} - \cdots\right] dx \\
&= \left[x - \frac{x^3}{3} + \frac{x^5}{5(2!)} - \frac{x^7}{7(3!)} + \frac{x^9}{9(4!)} - \cdots\right]_0^1 \\
&= 1 - \frac{1}{3} + \frac{1}{5 \cdot 2!} - \frac{1}{7 \cdot 3!} + \frac{1}{9 \cdot 4!} - \cdots \\
&= \sum_{k=0}^{\infty} \frac{(-1)^k}{(2k+1)k!}
\end{aligned}$$

Since this series clearly satisfies the hypotheses of the alternating series test (Theorem 10.6.1), it follows from Theorem 10.6.2 that if we approximate the integral by s_n (the nth partial sum of the series), then

$$\left|\int_0^1 e^{-x^2}\,dx - s_n\right| < \frac{1}{[2(n+1)+1](n+1)!} = \frac{1}{(2n+3)(n+1)!}$$

Thus, for three decimal-place accuracy we must choose n such that

$$\frac{1}{(2n+3)(n+1)!} \le 0.0005 = 5 \times 10^{-4}$$

With the help of a calculating utility you can show that the smallest value of n that satisfies this condition is $n = 5$. Thus, the value of the integral to three decimal-place accuracy is

$$\int_0^1 e^{-x^2}\,dx \approx 1 - \frac{1}{3} + \frac{1}{5\cdot 2!} - \frac{1}{7\cdot 3!} + \frac{1}{9\cdot 4!} - \frac{1}{11\cdot 5!} \approx 0.747$$

As a check, a calculator with a built-in numerical integration capability produced the approximation 0.746824, which agrees with our result when rounded to three decimal places. ◀

What advantages does the method of Example 3 have over Simpson's rule? What are its disadvantages?

FINDING MACLAURIN SERIES BY MULTIPLICATION AND DIVISION

The following examples illustrate some algebraic techniques that are sometimes useful for finding Taylor series.

$$\begin{array}{r} 1 - x^2 + \frac{x^4}{2} - \cdots \\ \times \quad x - \frac{x^3}{3} + \frac{x^5}{5} - \cdots \\ \hline x - x^3 + \frac{x^5}{2} - \cdots \\ -\frac{x^3}{3} + \frac{x^5}{3} - \frac{x^7}{6} + \cdots \\ \frac{x^5}{5} - \frac{x^7}{5} + \cdots \\ \hline x - \frac{4}{3}x^3 + \frac{31}{30}x^5 - \cdots \end{array}$$

▶ Example 4 Find the first three nonzero terms in the Maclaurin series for the function $f(x) = e^{-x^2}\tan^{-1}x$.

Solution. Using the series for e^{-x^2} and $\tan^{-1}x$ obtained in Examples 2 and 3 gives

$$e^{-x^2}\tan^{-1}x = \left(1 - x^2 + \frac{x^4}{2} - \cdots\right)\left(x - \frac{x^3}{3} + \frac{x^5}{5} - \cdots\right)$$

Multiplying, as shown in the margin, we obtain

$$e^{-x^2}\tan^{-1}x = x - \frac{4}{3}x^3 + \frac{31}{30}x^5 - \cdots$$

More terms in the series can be obtained by including more terms in the factors. Moreover, one can prove that a series obtained by this method converges at each point in the intersection of the intervals of convergence of the factors (and possibly on a larger interval). Thus, we can be certain that the series we have obtained converges for all x in the interval $-1 \le x \le 1$ (why?). ◀

$$\begin{array}{r|l} & x + \frac{x^3}{3} + \frac{2x^5}{15} + \cdots \\ \hline 1 - \frac{x^2}{2} + \frac{x^4}{24} - \cdots & x - \frac{x^3}{6} + \frac{x^5}{120} - \cdots \\ & x - \frac{x^3}{2} + \frac{x^5}{24} - \cdots \\ \hline & \frac{x^3}{3} - \frac{x^5}{30} + \cdots \\ & \frac{x^3}{3} - \frac{x^5}{6} + \cdots \\ \hline & \frac{2x^5}{15} + \cdots \end{array}$$

▶ Example 5 Find the first three nonzero terms in the Maclaurin series for $\tan x$.

Solution. Using the first three terms in the Maclaurin series for $\sin x$ and $\cos x$, we can express $\tan x$ as

$$\tan x = \frac{\sin x}{\cos x} = \frac{x - \frac{x^3}{3!} + \frac{x^5}{5!} - \cdots}{1 - \frac{x^2}{2!} + \frac{x^4}{4!} - \cdots}$$

Dividing, as shown in the margin, we obtain

$$\tan x = x + \frac{x^3}{3} + \frac{2x^5}{15} + \cdots$$ ◀

TECHNOLOGY MASTERY

If you have a CAS, use its capability for multiplying and dividing polynomials to perform the computations in Examples 4 and 5.

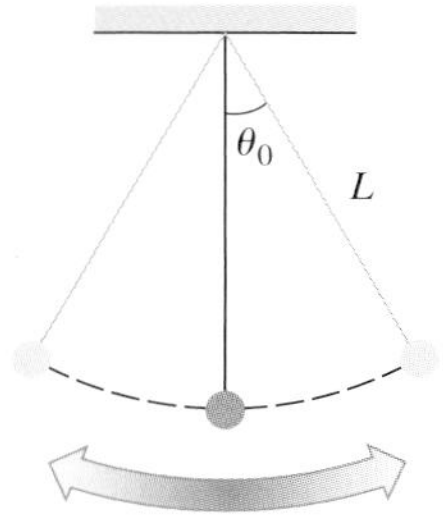

Figure 10.10.1

MODELING PHYSICAL LAWS WITH TAYLOR SERIES

Taylor series provide an important way of modeling physical laws. To illustrate the idea we will consider the problem of modeling the period of a simple pendulum (Figure 10.10.1). As explained in Exercise 69 of Section 8.8, the period T of such a pendulum is given by

$$T = 4\sqrt{\frac{L}{g}}\int_0^{\pi/2} \frac{1}{\sqrt{1-k^2\sin^2\phi}}\,d\phi \tag{4}$$

where

L = length of the supporting rod
g = acceleration due to gravity
$k = \sin(\theta_0/2)$, where θ_0 is the initial angle of displacement from the vertical

The integral, which is called a ***complete elliptic integral of the first kind***, cannot be expressed in terms of elementary functions and is often approximated by numerical methods. Unfortunately, numerical values are so specific that they often give little insight into general physical principles. However, if we expand the integrand of (4) in a series and integrate term by term, then we can generate an infinite series that can be used to construct various mathematical models for the period T that give a deeper understanding of the behavior of the pendulum.

To obtain a series for the integrand, we will substitute $-k^2\sin^2\phi$ for x in the binomial series for $1/\sqrt{1+x}$ that we derived in Example 4 of Section 10.9. If we do this, then we can rewrite (4) as

$$T = 4\sqrt{\frac{L}{g}}\int_0^{\pi/2}\left[1+\frac{1}{2}k^2\sin^2\phi+\frac{1\cdot 3}{2^2 2!}k^4\sin^4\phi+\frac{1\cdot 3\cdot 5}{2^3 3!}k^6\sin^6\phi+\cdots\right]d\phi \tag{5}$$

If we integrate term by term, then we can produce a series that converges to the period T. However, one of the most important cases of pendulum motion occurs when the initial displacement is small, in which case all subsequent displacements are small, and we can assume that $k = \sin(\theta_0/2) \approx 0$. In this case we expect the convergence of the series for T to be rapid, and we can approximate the sum of the series by dropping all but the constant term in (5). This yields

$$T = 2\pi\sqrt{\frac{L}{g}} \tag{6}$$

which is called the ***first-order model*** of T or the model for ***small vibrations***. This model can be improved on by using more terms in the series. For example, if we use the first two terms in the series, we obtain the ***second-order model***

$$T = 2\pi\sqrt{\frac{L}{g}}\left(1+\frac{k^2}{4}\right) \tag{7}$$

(verify).

✔QUICK CHECK EXERCISES 10.10 *(See page 714 for answers.)*

1. The Maclaurin series for e^{-x^2} obtained by substituting $-x^2$ for x in the series
$$e^x = \sum_{k=0}^{\infty}\frac{x^k}{k!}$$
is $e^{-x^2} = \sum_{k=0}^{\infty}$ ______.

2. $$\frac{d}{dx}\left[\sum_{k=1}^{\infty}(-1)^{k+1}\frac{x^k}{k}\right] = \text{\_\_\_\_} + \text{\_\_\_\_}\,x + \text{\_\_\_\_}\,x^2 + \text{\_\_\_\_}\,x^3 + \cdots = \sum_{k=0}^{\infty}\text{\_\_\_\_}$$

3. $\left(\sum_{k=0}^{\infty}\frac{x^k}{k!}\right)\left(\sum_{k=0}^{\infty}\frac{x^k}{k+1}\right)$

$$= \left(1 + x + \frac{x^2}{2!} + \cdots\right)\left(1 + \frac{x}{2} + \frac{x^2}{3} + \cdots\right)$$

$$= \underline{\qquad} + \underline{\qquad}\, x + \underline{\qquad}\, x^2 + \cdots$$

4. Suppose that $f(1) = 4$ and $f'(x) = \sum_{k=0}^{\infty}\frac{(-1)^k}{(k+1)!}(x-1)^k$

(a) $f''(1) = \underline{\qquad}$

(b) $f(x) = \underline{\qquad} + \underline{\qquad}\,(x-1)$

$+ \underline{\qquad}\,(x-1)^2 + \underline{\qquad}\,(x-1)^3 + \cdots$

$= \underline{\qquad} + \sum_{k=1}^{\infty}\underline{\qquad}$

EXERCISE SET 10.10 [C] CAS

1. In each part, obtain the Maclaurin series for the function by making an appropriate substitution in the Maclaurin series for $1/(1-x)$. Include the general term in your answer, and state the radius of convergence of the series.
(a) $\frac{1}{1+x}$ (b) $\frac{1}{1-x^2}$ (c) $\frac{1}{1-2x}$ (d) $\frac{1}{2-x}$

2. In each part, obtain the Maclaurin series for the function by making an appropriate substitution in the Maclaurin series for $\ln(1+x)$. Include the general term in your answer, and state the radius of convergence of the series.
(a) $\ln(1-x)$ (b) $\ln(1+x^2)$
(c) $\ln(1+2x)$ (d) $\ln(2+x)$

3. In each part, obtain the first four nonzero terms of the Maclaurin series for the function by making an appropriate substitution in one of the binomial series obtained in Example 4 of Section 10.9.
(a) $(2+x)^{-1/2}$ (b) $(1-x^2)^{-2}$

4. (a) Use the Maclaurin series for $1/(1-x)$ to find the Maclaurin series for $1/(a-x)$, where $a \neq 0$, and state the radius of convergence of the series.
(b) Use the binomial series for $1/(1+x)^2$ obtained in Example 4 of Section 10.9 to find the first four nonzero terms in the Maclaurin series for $1/(a+x)^2$, where $a \neq 0$, and state the radius of convergence of the series.

5–8 Find the first four nonzero terms of the Maclaurin series for the function by making an appropriate substitution in a known Maclaurin series and performing any algebraic operations that are required. State the radius of convergence of the series.

5. (a) $\sin 2x$ (b) e^{-2x} (c) e^{x^2} (d) $x^2\cos \pi x$

6. (a) $\cos 2x$ (b) $x^2 e^x$ (c) xe^{-x} (d) $\sin(x^2)$

7. (a) $\frac{x^2}{1+3x}$ (b) $x\sinh 2x$ (c) $x(1-x^2)^{3/2}$

8. (a) $\frac{x}{x-1}$ (b) $3\cosh(x^2)$ (c) $\frac{x}{(1+2x)^3}$

9–10 Find the first four nonzero terms of the Maclaurin series for the function by using an appropriate trigonometric identity or property of logarithms and then substituting in a known Maclaurin series.

9. (a) $\sin^2 x$ (b) $\ln[(1+x^3)^{12}]$

10. (a) $\cos^2 x$ (b) $\ln\left(\frac{1-x}{1+x}\right)$

11. (a) Use a known Maclaurin series to find the Taylor series of $1/x$ about $x = 1$ by expressing this function as
$$\frac{1}{x} = \frac{1}{1-(1-x)}$$
(b) Find the interval of convergence of the Taylor series.

12. Use the method of Exercise 11 to find the Taylor series of $1/x$ about $x = x_0$, and state the interval of convergence of the Taylor series.

13–14 Find the first four nonzero terms of the Maclaurin series for the function by multiplying the Maclaurin series of the factors.

13. (a) $e^x \sin x$ (b) $\sqrt{1+x}\,\ln(1+x)$

14. (a) $e^{-x^2}\cos x$ (b) $(1+x^2)^{4/3}(1+x)^{1/3}$

15–16 Find the first four nonzero terms of the Maclaurin series for the function by dividing appropriate Maclaurin series.

15. (a) $\sec x \left(= \frac{1}{\cos x}\right)$ (b) $\frac{\sin x}{e^x}$

16. (a) $\frac{\tan^{-1} x}{1+x}$ (b) $\frac{\ln(1+x)}{1-x}$

17. Use the Maclaurin series for e^x and e^{-x} to derive the Maclaurin series for $\sinh x$ and $\cosh x$. Include the general terms in your answers and state the radius of convergence of each series.

18. Use the Maclaurin series for $\sinh x$ and $\cosh x$ to obtain the first four nonzero terms in the Maclaurin series for $\tanh x$.

19–20 Find the first five nonzero terms of the Maclaurin series for the function by using partial fractions and a known Maclaurin series.

19. $\dfrac{4x-2}{x^2-1}$ **20.** $\dfrac{x^3+x^2+2x-2}{x^2-1}$

21–22 Confirm the derivative formula by differentiating the appropriate Maclaurin series term by term.

21. (a) $\dfrac{d}{dx}[\cos x] = -\sin x$ (b) $\dfrac{d}{dx}[\ln(1+x)] = \dfrac{1}{1+x}$

22. (a) $\dfrac{d}{dx}[\sinh x] = \cosh x$ (b) $\dfrac{d}{dx}[\tan^{-1} x] = \dfrac{1}{1+x^2}$

23–24 Confirm the integration formula by integrating the appropriate Maclaurin series term by term.

23. (a) $\int e^x\,dx = e^x + C$

(b) $\int \sinh x\,dx = \cosh x + C$

24. (a) $\int \sin x\,dx = -\cos x + C$

(b) $\int \dfrac{1}{1+x}\,dx = \ln(1+x) + C$

25. (a) Use the Maclaurin series for $1/(1-x)$ to find the Maclaurin series for

$$f(x) = \frac{x}{1-x^2}$$

(b) Use the Maclaurin series obtained in part (a) to find $f^{(5)}(0)$ and $f^{(6)}(0)$.

(c) What can you say about the value of $f^{(n)}(0)$?

26. Let $f(x) = x^2\cos 2x$. Use the method of Exercise 25 to find $f^{(99)}(0)$.

27–28 The limit of an indeterminate form as $x \to x_0$ can sometimes be found by expanding the functions involved in Taylor series about $x = x_0$ and taking the limit of the series term by term. Use this method to find the limits in these exercises.

27. (a) $\lim\limits_{x\to 0} \dfrac{\sin x}{x}$ (b) $\lim\limits_{x\to 0} \dfrac{\tan^{-1} x - x}{x^3}$

28. (a) $\lim\limits_{x\to 0} \dfrac{1-\cos x}{\sin x}$ (b) $\lim\limits_{x\to 0} \dfrac{\ln\sqrt{1+x} - \sin 2x}{x}$

29–32 Use Maclaurin series to approximate the integral to three decimal-place accuracy.

29. $\int_0^1 \sin(x^2)\,dx$ **30.** $\int_0^{1/2} \tan^{-1}(2x^2)\,dx$

31. $\int_0^{0.2} \sqrt[3]{1+x^4}\,dx$ **32.** $\int_0^{1/2} \dfrac{dx}{\sqrt[4]{x^2+1}}$

FOCUS ON CONCEPTS

33. (a) Find the Maclaurin series for e^{x^4}. What is the radius of convergence?

(b) Explain two different ways to use the Maclaurin series for e^x to find a series for $x^3e^{x^4}$. Confirm that both methods produce the same series.

34. (a) Differentiate the Maclaurin series for $1/(1-x)$, and use the result to show that

$$\sum_{k=1}^{\infty} kx^k = \frac{x}{(1-x)^2} \quad \text{for } -1 < x < 1$$

(b) Integrate the Maclaurin series for $1/(1-x)$, and use the result to show that

$$\sum_{k=1}^{\infty} \frac{x^k}{k} = -\ln(1-x) \quad \text{for } -1 < x < 1$$

(c) Use the result in part (b) to show that

$$\sum_{k=1}^{\infty} (-1)^{k+1}\frac{x^k}{k} = \ln(1+x) \quad \text{for } -1 < x < 1$$

(d) Show that the series in part (c) converges if $x = 1$.

(e) Use the remark following Example 2 to show that

$$\sum_{k=1}^{\infty} (-1)^{k+1}\frac{x^k}{k} = \ln(1+x) \quad \text{for } -1 < x \le 1$$

35. Use the results in Exercise 34 to find the sum of the series.

(a) $\displaystyle\sum_{k=1}^{\infty} \frac{k}{3^k} = \frac{1}{3} + \frac{2}{3^2} + \frac{3}{3^3} + \frac{4}{3^4} + \cdots$

(b) $\displaystyle\sum_{k=1}^{\infty} \frac{1}{k(4^k)} = \frac{1}{4} + \frac{1}{2(4^2)} + \frac{1}{3(4^3)} + \frac{1}{4(4^4)} + \cdots$

36. Use the results in Exercise 34 to find the sum of each series.

(a) $\displaystyle\sum_{k=1}^{\infty} (-1)^{k+1}\frac{1}{k} = 1 - \frac{1}{2} + \frac{1}{3} - \frac{1}{4} + \cdots$

(b) $\displaystyle\sum_{k=1}^{\infty} \frac{(e-1)^k}{ke^k} = \frac{e-1}{e} + \frac{(e-1)^2}{2(e^2)} - \frac{(e-1)^3}{3(e^3)} + \cdots$

37. (a) Use the relationship

$$\int \frac{1}{\sqrt{1+x^2}}\,dx = \sinh^{-1} x + C$$

to find the first four nonzero terms in the Maclaurin series for $\sinh^{-1} x$.

(b) Express the series in sigma notation.

(c) What is the radius of convergence?

38. (a) Use the relationship

$$\int \frac{1}{\sqrt{1-x^2}}\,dx = \sin^{-1} x + C$$

to find the first four nonzero terms in the Maclaurin series for $\sin^{-1} x$.

(b) Express the series in sigma notation.

(c) What is the radius of convergence?

39. We showed by Formula (11) of Section 9.3 that if there are y_0 units of radioactive carbon-14 present at time $t = 0$, then the number of units present t years later is

$$y(t) = y_0 e^{-0.000121t}$$

(a) Express $y(t)$ as a Maclaurin series.
(b) Use the first two terms in the series to show that the number of units present after 1 year is approximately $(0.999879)y_0$.
(c) Compare this to the value produced by the formula for $y(t)$.

40. In Section 9.1 we studied the motion of a falling object that has mass m and is retarded by air resistance. We showed that if the initial velocity is v_0 and the drag force F_R is proportional to the velocity, that is, $F_R = -cv$, then the velocity of the object at time t is

$$v(t) = e^{-ct/m}\left(v_0 + \frac{mg}{c}\right) - \frac{mg}{c}$$

where g is the acceleration due to gravity [see Formula (27) of Section 9.1].

(a) Use a Maclaurin series to show that if $ct/m \approx 0$, then the velocity can be approximated as

$$v(t) \approx v_0 - \left(\frac{cv_0}{m} + g\right)t$$

(b) Improve on the approximation in part (a).

C **41.** Suppose that a simple pendulum with a length of $L = 1$ meter is given an initial displacement of $\theta_0 = 5°$ from the vertical.
(a) Approximate the period of the pendulum using Formula (6) for the first-order model. [Take $g = 9.8\ \text{m/s}^2$.]
(b) Approximate the period of the pendulum using Formula (7) for the second-order model.
(c) Use the numerical integration capability of a CAS to approximate the period of the pendulum from Formula (4), and compare it to the values obtained in parts (a) and (b).

42. Use the first three nonzero terms in Formula (5) and the Wallis sine formula in the Endpaper Integral Table (Formula 122) to obtain a model for the period of a simple pendulum.

43. Recall that the gravitational force exerted by the Earth on an object is called the object's *weight* (or more precisely, its *Earth weight*). If an object of mass m is on the surface of the Earth (mean sea level), then the magnitude of its weight is mg, where g is the acceleration due to gravity at the Earth's surface. A more general formula for the magnitude of the gravitational force that the Earth exerts on an object of mass m is

$$F = \frac{mgR^2}{(R+h)^2}$$

where R is the radius of the Earth and h is the height of the object above the Earth's surface.
(a) Use the binomial series for $1/(1+x)^2$ obtained in Example 4 of Section 10.9 to express F as a Maclaurin series in powers of h/R.
(b) Show that if $h = 0$, then $F = mg$.
(c) Show that if $h/R \approx 0$, then $F \approx mg - (2mgh/R)$. [*Note:* The quantity $2mgh/R$ can be thought of as a "correction term" for the weight that takes the object's height above the Earth's surface into account.]
(d) If we assume that the Earth is a sphere of radius $R = 4000$ mi at mean sea level, by approximately what percentage does a person's weight change in going from mean sea level to the top of Mt. Everest (29,028 ft)?

44. (a) Show that the Bessel function $J_0(x)$ given by Formula (4) of Section 10.8 satisfies the differential equation $xy'' + y' + xy = 0$. (This is called the ***Bessel equation of order zero***.)
(b) Show that the Bessel function $J_1(x)$ given by Formula (5) of Section 10.8 satisfies the differential equation $x^2y'' + xy' + (x^2 - 1)y = 0$. (This is called the ***Bessel equation of order one***.)
(c) Show that $J_0'(x) = -J_1(x)$.

45. Prove: If the power series $\sum_{k=0}^{\infty} a_k x^k$ and $\sum_{k=0}^{\infty} b_k x^k$ have the same sum on an interval $(-r, r)$, then $a_k = b_k$ for all values of k.

✔QUICK CHECK ANSWERS 10.10

1. $(-1)^k \dfrac{x^{2k}}{k!}$ 2. $1; -1; 1; -1; (-1)^k x^k$ 3. $1; \dfrac{3}{2}; \dfrac{4}{3}$ 4. (a) $-\dfrac{1}{2}$ (b) $4; 1; -\dfrac{1}{4}; \dfrac{1}{18}; 4; (-1)^{k+1}\dfrac{(x-1)^k}{k\cdot(k!)}$

CHAPTER REVIEW EXERCISES

1. What is the difference between an infinite sequence and an infinite series?

2. What is meant by the sum of an infinite series?

3. (a) What is a geometric series? Give some examples of convergent and divergent geometric series.
(b) What is a p-series? Give some examples of convergent and divergent p-series.

4. State conditions under which an alternating series is guaranteed to converge.

5. (a) What does it mean to say that an infinite series converges absolutely?
(b) What relationship exists between convergence and absolute convergence of an infinite series?

6. State the Remainder Estimation Theorem, and describe some of its uses.

7. If a power series in $x - x_0$ has radius of convergence R, what can you say about the set of x-values at which the series converges?

8. (a) Write down the formula for the Maclaurin series for f in sigma notation.
(b) Write down the formula for the Taylor series for f about $x = x_0$ in sigma notation.

9. Are the following statements true or false? If true, state a theorem to justify your conclusion; if false, then give a counterexample.
(a) If $\sum u_k$ converges, then $u_k \to 0$ as $k \to +\infty$.
(b) If $u_k \to 0$ as $k \to +\infty$, then $\sum u_k$ converges.
(c) If $f(n) = a_n$ for $n = 1, 2, 3, \ldots$, and if $a_n \to L$ as $n \to +\infty$, then $f(x) \to L$ as $x \to +\infty$.
(d) If $f(n) = a_n$ for $n = 1, 2, 3, \ldots$, and if $f(x) \to L$ as $x \to +\infty$, then $a_n \to L$ as $n \to +\infty$.
(e) If $0 < a_n < 1$, then $\{a_n\}$ converges.
(f) If $0 < u_k < 1$, then $\sum u_k$ converges.
(g) If $\sum u_k$ and $\sum v_k$ converge, then $\sum(u_k + v_k)$ diverges.
(h) If $\sum u_k$ and $\sum v_k$ diverge, then $\sum(u_k - v_k)$ converges.
(i) If $0 \le u_k \le v_k$ and $\sum v_k$ converges, then $\sum u_k$ converges.
(j) If $0 \le u_k \le v_k$ and $\sum u_k$ diverges, then $\sum v_k$ diverges.
(k) If an infinite series converges, then it converges absolutely.
(l) If an infinite series diverges absolutely, then it diverges.

10. State whether each of the following is true or false. Justify your answers.
(a) The function $f(x) = x^{1/3}$ has a Maclaurin series.
(b) $1 + \frac{1}{2} - \frac{1}{2} + \frac{1}{3} - \frac{1}{3} + \frac{1}{4} - \frac{1}{4} + \cdots = 1$
(c) $1 + \frac{1}{2} - \frac{1}{2} + \frac{1}{2} - \frac{1}{2} + \frac{1}{2} - \frac{1}{2} + \cdots = 1$

11. Find the general term of the sequence, starting with $n = 1$, determine whether the sequence converges, and if so find its limit.
(a) $\dfrac{3}{2^2 - 1^2}, \dfrac{4}{3^2 - 2^2}, \dfrac{5}{4^2 - 3^2}, \ldots$
(b) $\dfrac{1}{3}, -\dfrac{2}{5}, \dfrac{3}{7}, -\dfrac{4}{9}, \ldots$

12. Suppose that the sequence $\{a_k\}$ is defined recursively by

$$a_0 = c, \quad a_{k+1} = \sqrt{a_k}$$

Assuming that the sequence converges, find its limit if
(a) $c = \frac{1}{2}$ (b) $c = \frac{3}{2}$.

13. Show that the sequence is eventually strictly monotone.
(a) $\left\{(n - 10)^4\right\}_{n=0}^{+\infty}$ (b) $\left\{\dfrac{100^n}{(2n)!(n!)}\right\}_{n=1}^{+\infty}$

14. (a) Give an example of a bounded sequence that diverges.
(b) Give an example of a monotonic sequence that diverges.

15–20 Use any method to determine whether the series converge.

15. (a) $\displaystyle\sum_{k=1}^{\infty} \frac{1}{5^k}$ (b) $\displaystyle\sum_{k=1}^{\infty} \frac{1}{5^k + 1}$

16. (a) $\displaystyle\sum_{k=1}^{\infty} (-1)^k \frac{k+4}{k^2 + k}$ (b) $\displaystyle\sum_{k=1}^{\infty} (-1)^{k+1} \left(\frac{k+2}{3k-1}\right)^k$

17. (a) $\displaystyle\sum_{k=1}^{\infty} \frac{1}{k^3 + 2k + 1}$ (b) $\displaystyle\sum_{k=1}^{\infty} \frac{1}{(3+k)^{2/5}}$

18. (a) $\displaystyle\sum_{k=1}^{\infty} \frac{\ln k}{k\sqrt{k}}$ (b) $\displaystyle\sum_{k=1}^{\infty} \frac{k^{4/3}}{8k^2 + 5k + 1}$

19. (a) $\displaystyle\sum_{k=1}^{\infty} \frac{9}{\sqrt{k} + 1}$ (b) $\displaystyle\sum_{k=1}^{\infty} \frac{\cos(1/k)}{k^2}$

20. (a) $\displaystyle\sum_{k=1}^{\infty} \frac{k^{-1/2}}{2 + \sin^2 k}$ (b) $\displaystyle\sum_{k=1}^{\infty} \frac{(-1)^{k+1}}{k^2 + 1}$

21. Find a formula for the exact error that results when the sum of the geometric series $\sum_{k=0}^{\infty}(1/5)^k$ is approximated by the sum of the first 100 terms in the series.

22. Suppose that $\displaystyle\sum_{k=1}^{n} u_k = 2 - \frac{1}{n}$. Find

(a) u_{100} (b) $\displaystyle\lim_{k \to +\infty} u_k$ (c) $\displaystyle\sum_{k=1}^{\infty} u_k$.

23. In each part, determine whether the series converges; if so, find its sum.
(a) $\displaystyle\sum_{k=1}^{\infty} \left(\frac{3}{2^k} - \frac{2}{3^k}\right)$ (b) $\displaystyle\sum_{k=1}^{\infty} [\ln(k+1) - \ln k]$
(c) $\displaystyle\sum_{k=1}^{\infty} \frac{1}{k(k+2)}$ (d) $\displaystyle\sum_{k=1}^{\infty} [\tan^{-1}(k+1) - \tan^{-1} k]$

24. It can be proved that

$$\lim_{n \to +\infty} \sqrt[n]{n!} = +\infty \quad \text{and} \quad \lim_{n \to +\infty} \frac{\sqrt[n]{n!}}{n} = \frac{1}{e}$$

In each part, use these limits and the root test to determine whether the series converges.
(a) $\displaystyle\sum_{k=0}^{\infty} \frac{2^k}{k!}$ (b) $\displaystyle\sum_{k=0}^{\infty} \frac{k^k}{k!}$

25. Let a, b, and p be positive constants. For which values of p does the series $\displaystyle\sum_{k=1}^{\infty} \frac{1}{(a+bk)^p}$ converge?

26. Find the interval of convergence of

$$\sum_{k=0}^{\infty} \frac{(x - x_0)^k}{b^k} \quad (b > 0)$$

27. (a) Show that $k^k \ge k!$.
(b) Use the comparison test to show that $\displaystyle\sum_{k=1}^{\infty} k^{-k}$ converges.
(c) Use the root test to show that the series converges.

28. Does the series $1 - \frac{2}{3} + \frac{3}{5} - \frac{4}{7} + \frac{5}{9} + \cdots$ converge? Justify your answer.

29. (a) Find the first five Maclaurin polynomials of the function $p(x) = 1 - 7x + 5x^2 + 4x^3$.
(b) Make a general statement about the Maclaurin polynomials of a polynomial of degree n.

30. Show that the approximation

$$\sin x \approx x - \frac{x^3}{3!} + \frac{x^5}{5!}$$

is accurate to four decimal places if $0 \le x \le \pi/4$.

31. Use a Maclaurin series and properties of alternating series to show that $|\ln(1+x) - x| \le x^2/2$ if $0 < x < 1$.

32. Use Maclaurin series to approximate the integral

$$\int_0^1 \frac{1-\cos x}{x}\,dx$$

to three decimal-place accuracy.

33. In parts (a)–(d), find the sum of the series by associating it with some Maclaurin series.

(a) $2 + \dfrac{4}{2!} + \dfrac{8}{3!} + \dfrac{16}{4!} + \cdots$

(b) $\pi - \dfrac{\pi^3}{3!} + \dfrac{\pi^5}{5!} - \dfrac{\pi^7}{7!} + \cdots$

(c) $1 - \dfrac{e^2}{2!} + \dfrac{e^4}{4!} - \dfrac{e^6}{6!} + \cdots$

(d) $1 - \ln 3 + \dfrac{(\ln 3)^2}{2!} - \dfrac{(\ln 3)^3}{3!} + \cdots$

34. In each part, write out the first four terms of the series, and then find the radius of convergence.

(a) $\displaystyle\sum_{k=1}^{\infty} \frac{1 \cdot 2 \cdot 3 \cdots k}{1 \cdot 4 \cdot 7 \cdots (3k-2)} x^k$

(b) $\displaystyle\sum_{k=1}^{\infty} (-1)^k \frac{1 \cdot 2 \cdot 3 \cdots k}{1 \cdot 3 \cdot 5 \cdots (2k-1)} x^{2k+1}$

35. Use an appropriate Taylor series for $\sqrt[3]{x}$ to approximate $\sqrt[3]{28}$ to three decimal-place accuracy, and check your answer by comparing it to that produced directly by your calculating utility.

36. Differentiate the Maclaurin series for xe^x and use the result to show that

$$\sum_{k=0}^{\infty} \frac{k+1}{k!} = 2e$$

37. Use the supplied Maclaurin series for $\sin x$ and $\cos x$ to find the first four nonzero terms of the Maclaurin series for the given functions.

$$\sin x = \sum_{k=0}^{\infty} (-1)^k \frac{x^{2k+1}}{(2k+1)!}$$

$$\cos x = \sum_{k=0}^{\infty} (-1)^k \frac{x^{2k}}{(2k)!}$$

(a) $\sin x \cos x$ (b) $\frac{1}{2}\sin 2x$

c h a p t e r   e l e v e n

ANALYTIC GEOMETRY IN CALCULUS

...without the treatises of the Greek geometers on the conic sections there could have been no Kepler, without Kepler no Newton, and without Newton no science in the modern sense of the term...

—Henry John Stephen Smith
Mathematician

In this chapter we will study aspects of analytic geometry that are important in applications of calculus. We will begin by introducing polar coordinate systems, which are used, for example, in tracking the motion of planets and satellites, in identifying the locations of objects from information on radar screens, and in the design of antennas. We will then discuss relationships between curves in polar coordinates and parametric curves in rectangular coordinates, and we will discuss methods for finding areas in polar coordinates and tangent lines to curves given in polar coordinates or parametrically in rectangular coordinates. We will then review the basic properties of parabolas, ellipses, and hyperbolas and discuss these curves in the context of polar coordinates. Finally, we will give some basic applications of these ideas to astronomy.

Photo: *Mathematical curves, such as the spirals in the sunflower blossom, can be described mathematically using analytic geometry.*

11.1 POLAR COORDINATES

Up to now we have specified the location of a point in the plane by means of coordinates relative to two perpendicular coordinate axes. However, sometimes a moving point has a special affinity for some fixed point, such as a planet moving in an orbit under the central attraction of the Sun. In such cases, the path of the particle is best described by its angular direction and its distance from the fixed point. In this section we will discuss a new kind of coordinate system that is based on this idea.

POLAR COORDINATE SYSTEMS

A ***polar coordinate system*** in a plane consists of a fixed point O, called the ***pole*** (or ***origin***), and a ray emanating from the pole, called the ***polar axis***. In such a coordinate system we can associate with each point P in the plane a pair of ***polar coordinates*** (r, θ), where r is the distance from P to the pole and θ is an angle from the polar axis to the ray OP (Figure 11.1.1). The number r is called the ***radial coordinate*** of P and the number θ the ***angular coordinate*** (or ***polar angle***) of P. In Figure 11.1.2, the points $(6, 45°)$, $(5, 120°)$, $(3, 225°)$, and $(4, 330°)$ are plotted in polar coordinate systems. If P is the pole, then $r = 0$, but there is no clearly defined polar angle. We will agree that an arbitrary angle can be used in this case; that is, $(0, \theta)$ are polar coordinates of the pole for all choices of θ.

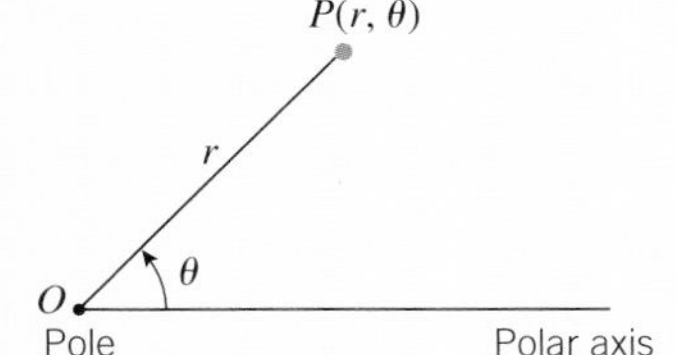

Figure 11.1.1

The polar coordinates of a point are not unique. For example, the polar coordinates

$$(1, 315°), \quad (1, -45°), \quad \text{and} \quad (1, 675°)$$

Figure 11.1.2

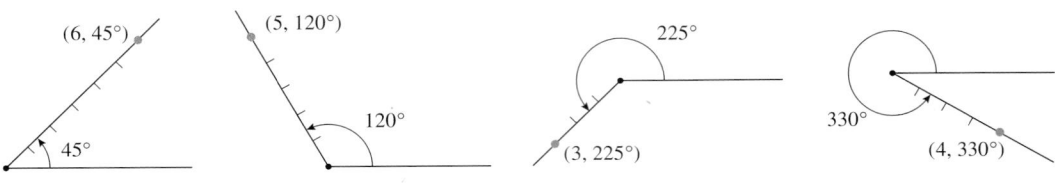

all represent the same point (Figure 11.1.3). In general, if a point P has polar coordinates (r, θ), then

$$(r, \theta + n \cdot 360^\circ) \quad \text{and} \quad (r, \theta - n \cdot 360^\circ)$$

are also polar coordinates of P for any nonnegative integer n. Thus, every point has infinitely many pairs of polar coordinates.

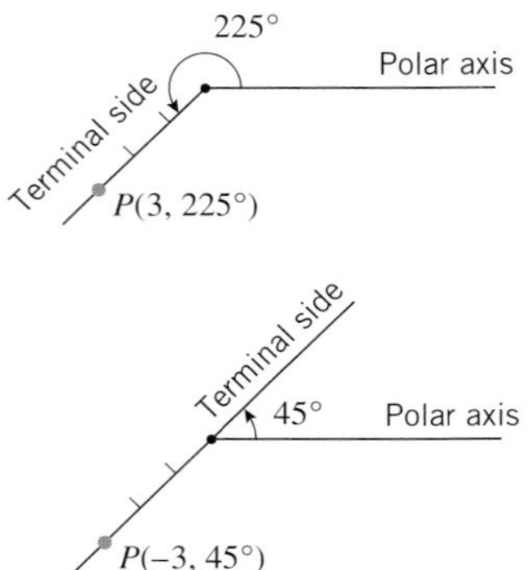

Figure 11.1.4

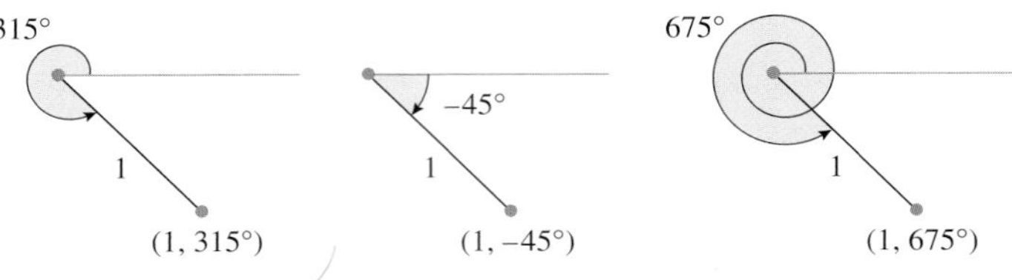

Figure 11.1.3

As defined above, the radial coordinate r of a point P is nonnegative, since it represents the distance from P to the pole. However, it will be convenient to allow for negative values of r as well. To motivate an appropriate definition, consider the point P with polar coordinates $(3, 225^\circ)$. As shown in Figure 11.1.4, we can reach this point by rotating the polar axis through an angle of 225° and then moving 3 units from the pole along the terminal side of the angle, or we can reach the point P by rotating the polar axis through an angle of 45° and then moving 3 units from the pole along the extension of the terminal side. This suggests that the point $(3, 225^\circ)$ might also be denoted by $(-3, 45^\circ)$, with the minus sign serving to indicate that the point is on the *extension* of the angle's terminal side rather than on the terminal side itself.

In general, the terminal side of the angle $\theta + 180^\circ$ is the extension of the terminal side of θ, so we define negative radial coordinates by agreeing that

$$(-r, \theta) \quad \text{and} \quad (r, \theta + 180^\circ)$$

are polar coordinates of the same point.

In problems involving derivative or integral formulas, angles must be in radians because those formulas were derived under that assumption. We will use radian measure for polar angles, except in applications where degree measure is more convenient and radian measure is not required.

RELATIONSHIP BETWEEN POLAR AND RECTANGULAR COORDINATES

Frequently, it will be useful to superimpose a rectangular xy-coordinate system on top of a polar coordinate system, making the positive x-axis coincide with the polar axis. If this is done, then every point P will have both rectangular coordinates (x, y) and polar coordinates (r, θ). As suggested by Figure 11.1.5, these coordinates are related by the equations

$$x = r\cos\theta, \quad y = r\sin\theta \tag{1}$$

These equations are well suited for finding x and y when r and θ are known. However, to find r and θ when x and y are known, it is preferable to use the identities $\sin^2\theta + \cos^2\theta = 1$ and $\tan\theta = \sin\theta / \cos\theta$ to rewrite (1) as

$$r^2 = x^2 + y^2, \quad \tan\theta = \frac{y}{x} \tag{2}$$

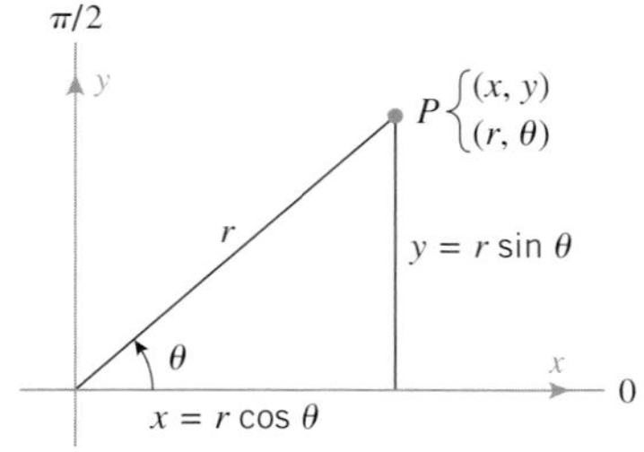

Figure 11.1.5

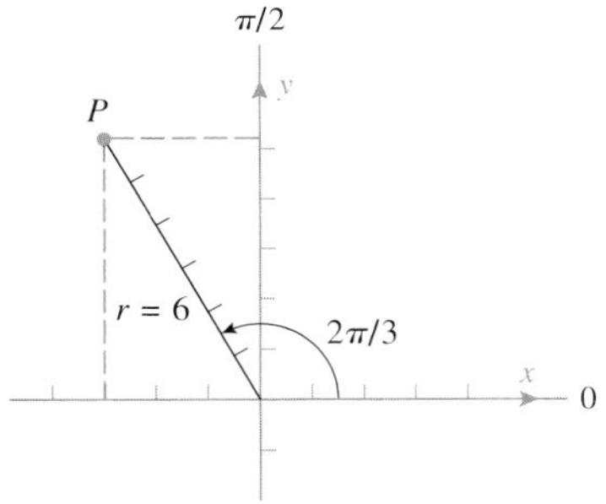

Figure 11.1.6

▸ **Example 1** Find the rectangular coordinates of the point P whose polar coordinates are $(6, 2\pi/3)$.

Solution. Substituting the polar coordinates $r = 6$ and $\theta = 2\pi/3$ in (1) yields

$$x = 6\cos\frac{2\pi}{3} = 6\left(-\frac{1}{2}\right) = -3$$

$$y = 6\sin\frac{2\pi}{3} = 6\left(\frac{\sqrt{3}}{2}\right) = 3\sqrt{3}$$

Thus, the rectangular coordinates of P are $(-3, 3\sqrt{3})$ (Figure 11.1.6). ◂

▸ **Example 2** Find polar coordinates of the point P whose rectangular coordinates are $(-2, 2\sqrt{3})$.

Solution. We will find the polar coordinates (r, θ) of P that satisfy the conditions $r > 0$ and $0 \le \theta < 2\pi$. From the first equation in (2),

$$r^2 = x^2 + y^2 = (-2)^2 + (2\sqrt{3})^2 = 4 + 12 = 16$$

so $r = 4$. From the second equation in (2),

$$\tan\theta = \frac{y}{x} = \frac{2\sqrt{3}}{-2} = -\sqrt{3}$$

From this and the fact that $(-2, 2\sqrt{3})$ lies in the second quadrant, it follows that the angle satisfying the requirement $0 \le \theta < 2\pi$ is $\theta = 2\pi/3$. Thus, $(4, 2\pi/3)$ are polar coordinates of P. All other polar coordinates of P are expressible in the form

$$\left(4, \frac{2\pi}{3} + 2n\pi\right) \quad \text{or} \quad \left(-4, \frac{5\pi}{3} + 2n\pi\right)$$

where n is an integer. ◂

GRAPHS IN POLAR COORDINATES

We will now consider the problem of graphing equations in r and θ, where θ is assumed to be measured in radians. Some examples of such equations are

$$r = 1, \quad \theta = \pi/4, \quad r = \theta, \quad r = \sin\theta, \quad r = \cos 2\theta$$

In a rectangular coordinate system the graph of an equation in x and y consists of all points whose coordinates (x, y) satisfy the equation. However, in a polar coordinate system, points have infinitely many different pairs of polar coordinates, so that a given point may have some polar coordinates that satisfy an equation and others that do not. Given an equation in r and θ, we define its ***graph in polar coordinates*** to consist of all points with *at least one* pair of coordinates (r, θ) that satisfy the equation.

▸ **Example 3** Sketch the graphs of

$$\text{(a) } r = 1 \qquad \text{(b) } \theta = \frac{\pi}{4}$$

in polar coordinates.

Solution (a). For all values of θ, the point $(1, \theta)$ is 1 unit away from the pole. Since θ is arbitrary, the graph is the circle of radius 1 centered at the pole (Figure 11.1.7*a*).

Solution (b). For all values of r, the point $(r, \pi/4)$ lies on a line that makes an angle of $\pi/4$ with the polar axis (Figure 11.1.7*b*). Positive values of r correspond to points on the line in the first quadrant and negative values of r to points on the line in the third quadrant. Thus, in absence of any restriction on r, the graph is the entire line. Observe, however, that had we imposed the restriction $r \geq 0$, the graph would have been just the ray in the first quadrant. ◀

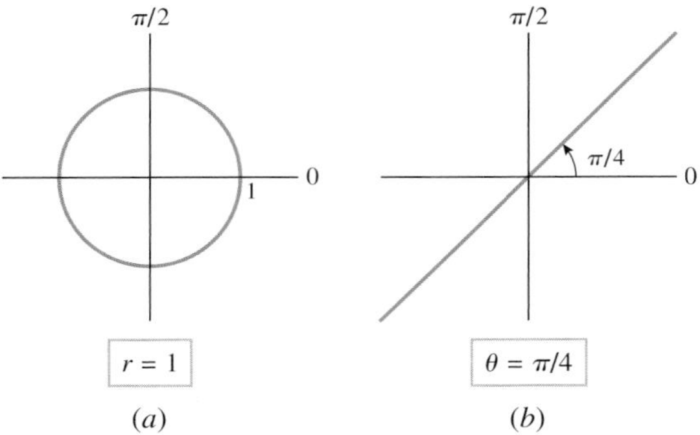

Figure 11.1.7

Equations $r = f(\theta)$ that express r as a function of θ are especially important. One way to graph such an equation is to choose some typical values of θ, calculate the corresponding values of r, and then plot the resulting pairs (r, θ) in a polar coordinate system. The next two examples illustrate this process.

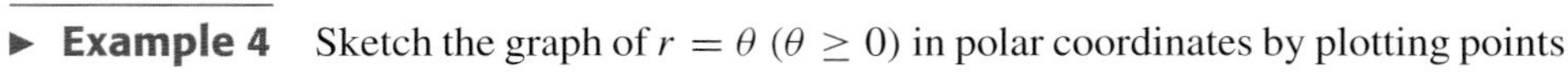

▶ **Example 4** Sketch the graph of $r = \theta$ $(\theta \geq 0)$ in polar coordinates by plotting points.

Solution. Observe that as θ increases, so does r; thus, the graph is a curve that spirals out from the pole as θ increases. A reasonably accurate sketch of the spiral can be obtained by plotting the points that correspond to values of θ that are integer multiples of $\pi/2$, keeping in mind that the value of r is always equal to the value of θ (Figure 11.1.8). ◀

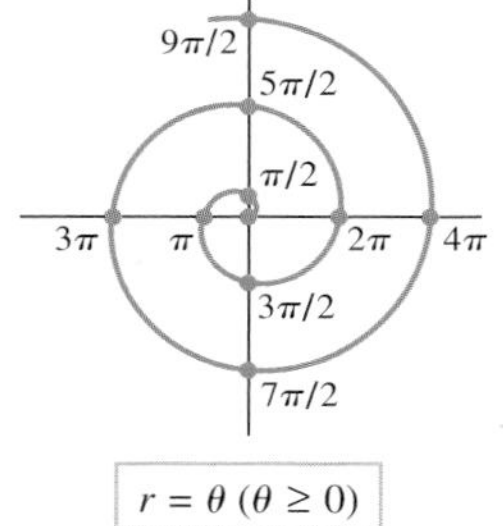

Figure 11.1.8

Graph the spiral $r = \theta$ $(\theta \leq 0)$. Compare your graph to that in Figure 11.1.8.

▶ **Example 5** Sketch the graph of the equation $r = \sin\theta$ in polar coordinates by plotting points.

Solution. Table 11.1.1 shows the coordinates of points on the graph at increments of $\pi/6$ $(= 30°)$.

These points are plotted in Figure 11.1.9. Note, however, that there are 13 points listed in the table but only 6 distinct plotted points. This is because the pairs from $\theta = \pi$ on yield duplicates of the preceding points. For example, $(-1/2, 7\pi/6)$ and $(1/2, \pi/6)$ represent the same point. ◀

Observe that the points in Figure 11.1.9 appear to lie on a circle. We can confirm that this is so by expressing the polar equation $r = \sin\theta$ in terms of x and y. To do this, we multiply the equation through by r to obtain

$$r^2 = r\sin\theta$$

Table 11.1.1

θ (RADIANS)	0	$\frac{\pi}{6}$	$\frac{\pi}{3}$	$\frac{\pi}{2}$	$\frac{2\pi}{3}$	$\frac{5\pi}{6}$	π	$\frac{7\pi}{6}$	$\frac{4\pi}{3}$	$\frac{3\pi}{2}$	$\frac{5\pi}{3}$	$\frac{11\pi}{6}$	2π
$r = \sin\theta$	0	$\frac{1}{2}$	$\frac{\sqrt{3}}{2}$	1	$\frac{\sqrt{3}}{2}$	$\frac{1}{2}$	0	$-\frac{1}{2}$	$-\frac{\sqrt{3}}{2}$	-1	$-\frac{\sqrt{3}}{2}$	$-\frac{1}{2}$	0
(r, θ)	$(0, 0)$	$\left(\frac{1}{2}, \frac{\pi}{6}\right)$	$\left(\frac{\sqrt{3}}{2}, \frac{\pi}{3}\right)$	$\left(1, \frac{\pi}{2}\right)$	$\left(\frac{\sqrt{3}}{2}, \frac{2\pi}{3}\right)$	$\left(\frac{1}{2}, \frac{5\pi}{6}\right)$	$(0, \pi)$	$\left(-\frac{1}{2}, \frac{7\pi}{6}\right)$	$\left(-\frac{\sqrt{3}}{2}, \frac{4\pi}{3}\right)$	$\left(-1, \frac{3\pi}{2}\right)$	$\left(-\frac{\sqrt{3}}{2}, \frac{5\pi}{3}\right)$	$\left(-\frac{1}{2}, \frac{11\pi}{6}\right)$	$(0, 2\pi)$

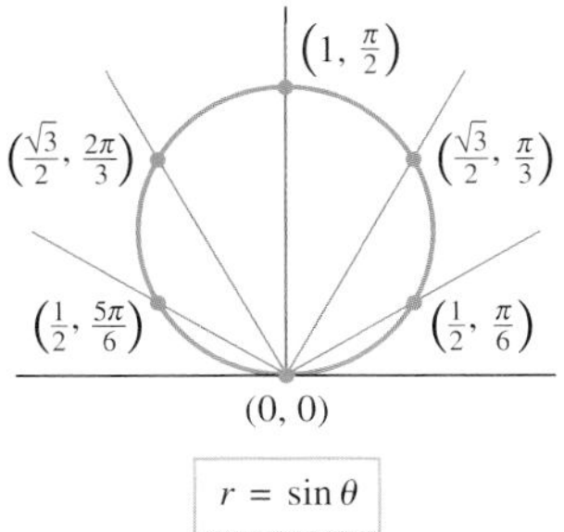

Figure 11.1.9

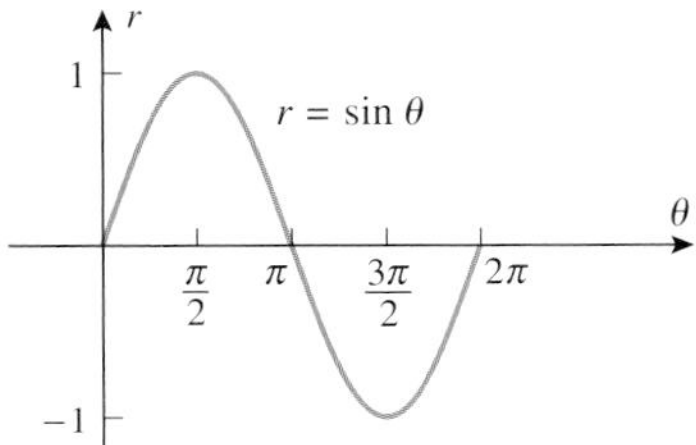

Figure 11.1.10

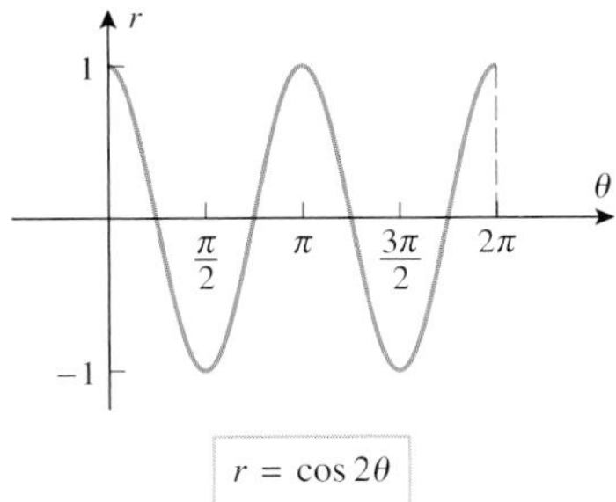

Figure 11.1.11

which now allows us to apply Formulas (1) and (2) to rewrite the equation as

$$x^2 + y^2 = y$$

Rewriting this equation as $x^2 + y^2 - y = 0$ and then completing the square yields

$$x^2 + \left(y - \tfrac{1}{2}\right)^2 = \tfrac{1}{4}$$

which is a circle of radius $\frac{1}{2}$ centered at the point $\left(0, \frac{1}{2}\right)$ in the xy-plane.

Just because an equation $r = f(\theta)$ involves the variables r and θ does not mean that it has to be graphed in a polar coordinate system. When useful, this equation can also be graphed in a rectangular coordinate system. For example, Figure 11.1.10 shows the graph of $r = \sin\theta$ in a rectangular θr-coordinate system. This graph can actually help to visualize how the polar graph in Figure 11.1.9 is generated:

- At $\theta = 0$ we have $r = 0$, which corresponds to the pole $(0, 0)$ on the polar graph.
- As θ varies from 0 to $\pi/2$, the value of r increases from 0 to 1, so the point (r, θ) moves along the circle from the pole to the high point at $(1, \pi/2)$.
- As θ varies from $\pi/2$ to π, the value of r decreases from 1 back to 0, so the point (r, θ) moves along the circle from the high point back to the pole.
- As θ varies from π to $3\pi/2$, the values of r are negative, varying from 0 to -1. Thus, the point (r, θ) moves along the circle from the pole to the high point at $(1, \pi/2)$, which is the same as the point $(-1, 3\pi/2)$. This duplicates the motion that occurred for $0 \le \theta \le \pi/2$.
- As θ varies from $3\pi/2$ to 2π, the value of r varies from -1 to 0. Thus, the point (r, θ) moves along the circle from the high point back to the pole, duplicating the motion that occurred for $\pi/2 \le \theta \le \pi$.

▶ **Example 6** Sketch the graph of $r = \cos 2\theta$ in polar coordinates.

Solution. Instead of plotting points, we will use the graph of $r = \cos 2\theta$ in rectangular coordinates (Figure 11.1.11) to visualize how the polar graph of this equation is generated. The analysis and the resulting polar graph are shown in Figure 11.1.12. This curve is called a *four-petal rose*. ◀

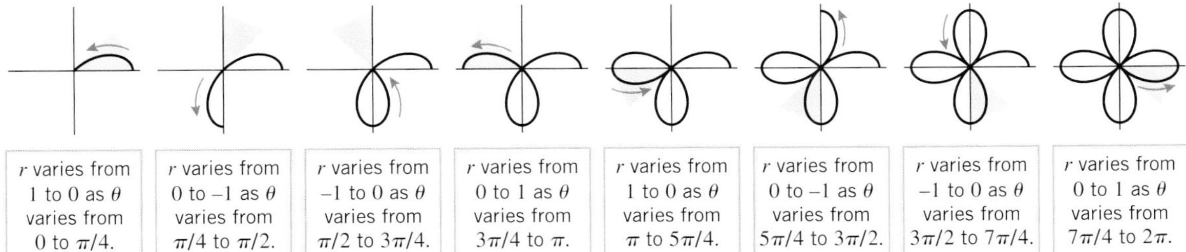

r varies from 1 to 0 as θ varies from 0 to $\pi/4$.
r varies from 0 to -1 as θ varies from $\pi/4$ to $\pi/2$.
r varies from -1 to 0 as θ varies from $\pi/2$ to $3\pi/4$.
r varies from 0 to 1 as θ varies from $3\pi/4$ to π.
r varies from 1 to 0 as θ varies from π to $5\pi/4$.
r varies from 0 to -1 as θ varies from $5\pi/4$ to $3\pi/2$.
r varies from -1 to 0 as θ varies from $3\pi/2$ to $7\pi/4$.
r varies from 0 to 1 as θ varies from $7\pi/4$ to 2π.

Figure 11.1.12

SYMMETRY TESTS

Observe that the polar graph of $r = \cos 2\theta$ in Figure 11.1.12 is symmetric about the x-axis and the y-axis. This symmetry could have been predicted from the following theorem, which is suggested by Figure 11.1.13 (we omit the proof).

The converse of each part of Theorem 11.1.1 is false. See Exercise 82.

11.1.1 THEOREM (*Symmetry Tests*).

(*a*) *A curve in polar coordinates is symmetric about the x-axis if replacing θ by $-\theta$ in its equation produces an equivalent equation* (Figure 11.1.13*a*).

(*b*) *A curve in polar coordinates is symmetric about the y-axis if replacing θ by $\pi - \theta$ in its equation produces an equivalent equation* (Figure 11.1.13*b*).

(*c*) *A curve in polar coordinates is symmetric about the origin if replacing θ by $\theta + \pi$, or replacing r by $-r$ in its equation produces an equivalent equation* (Figure 11.1.13*c*).

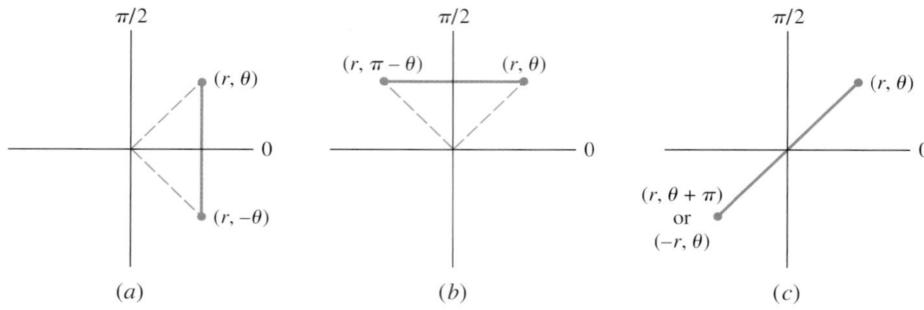

Figure 11.1.13

► **Example 7** Use Theorem 11.1.1 to confirm that the graph of $r = \cos 2\theta$ in Figure 11.1.12 is symmetric about the x-axis and y-axis.

A graph that is symmetric about both the x-axis and the y-axis is also symmetric about the origin. Use Theorem 11.1.1(c) to verify that the curve in Example 7 is symmetric about the origin.

Solution. To test for symmetry about the x-axis, we replace θ by $-\theta$. This yields

$$r = \cos(-2\theta) = \cos 2\theta$$

Thus, replacing θ by $-\theta$ does not alter the equation.

To test for symmetry about the y-axis, we replace θ by $\pi - \theta$. This yields

$$r = \cos 2(\pi - \theta) = \cos(2\pi - 2\theta) = \cos(-2\theta) = \cos 2\theta$$

Thus, replacing θ by $\pi - \theta$ does not alter the equation. ◄

► **Example 8** Sketch the graph of $r = a(1 - \cos\theta)$ in polar coordinates, assuming a to be a positive constant.

Solution. Observe first that replacing θ by $-\theta$ does not alter the equation, so we know in advance that the graph is symmetric about the polar axis. Thus, if we graph the upper half of the curve, then we can obtain the lower half by reflection about the polar axis.

As in our previous examples, we will first graph the equation in rectangular coordinates. This graph, which is shown in Figure 11.1.14*a*, can be obtained by rewriting the given equation as $r = a - a\cos\theta$, from which we see that the graph in rectangular coordinates can be obtained by first reflecting the graph of $r = a\cos\theta$ about the x-axis to obtain the graph of $r = -a\cos\theta$, and then translating that graph up a units to obtain the graph of $r = a - a\cos\theta$. Now we can see that:

- As θ varies from 0 to $\pi/3$, r increases from 0 to $a/2$.
- As θ varies from $\pi/3$ to $\pi/2$, r increases from $a/2$ to a.
- As θ varies from $\pi/2$ to $2\pi/3$, r increases from a to $3a/2$.
- As θ varies from $2\pi/3$ to π, r increases from $3a/2$ to $2a$.

This produces the polar curve shown in Figure 11.1.14*b*. The rest of the curve can be obtained by continuing the preceding analysis from π to 2π or, as noted above, by reflecting the portion already graphed about the x-axis (Figure 11.1.14*c*). This heart-shaped curve is called a ***cardioid*** (from the Greek word *kardia* meaning "heart"). ◄

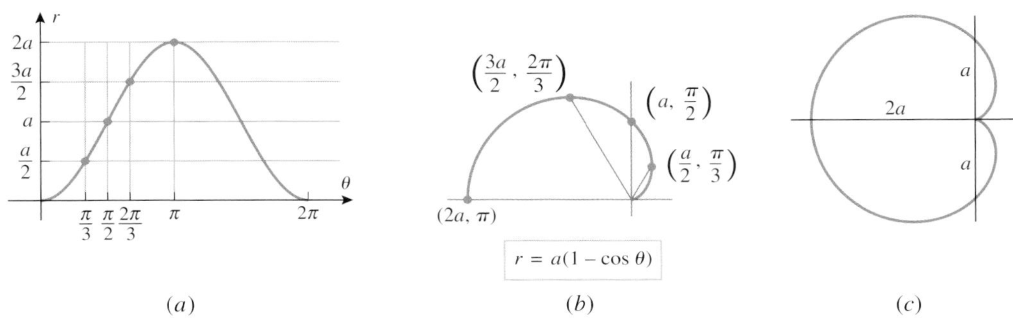

Figure 11.1.14

► **Example 9** Sketch the graph of $r^2 = 4\cos 2\theta$ in polar coordinates.

Solution. This equation does not express r as a function of θ, since solving for r in terms of θ yields two functions:

$$r = 2\sqrt{\cos 2\theta} \quad \text{and} \quad r = -2\sqrt{\cos 2\theta}$$

Thus, to graph the equation $r^2 = 4\cos 2\theta$ we will have to graph the two functions separately and then combine those graphs.

We will start with the graph of $r = 2\sqrt{\cos 2\theta}$. Observe first that this equation is not changed if we replace θ by $-\theta$ or if we replace θ by $\pi - \theta$. Thus, the graph is symmetric about the x-axis and the y-axis. This means that the entire graph can be obtained by graphing the portion in the first quadrant, reflecting that portion about the y-axis to obtain the portion in the second quadrant, and then reflecting those two portions about the x-axis to obtain the portions in the third and fourth quadrants.

To begin the analysis, we will graph the equation $r = 2\sqrt{\cos 2\theta}$ in rectangular coordinates (see Figure 11.1.15*a*). Note that there are gaps in that graph over the intervals $\pi/4 < \theta < 3\pi/4$ and $5\pi/4 < \theta < 7\pi/4$ because $\cos 2\theta$ is negative for those values of θ. From this graph we can see that:

- As θ varies from 0 to $\pi/4$, r decreases from 2 to 0.
- As θ varies from $\pi/4$ to $\pi/2$, no points are generated on the polar graph.

This produces the portion of the graph shown in Figure 11.1.15*b*. As noted above, we can complete the graph by a reflection about the y-axis followed by a reflection about the x-axis (11.1.15*c*). The resulting propeller-shaped graph is called a ***lemniscate*** (from the

Greek word *lemniscos* for a looped ribbon resembling the number 8). We leave it for you to verify that the equation $r = 2\sqrt{\cos 2\theta}$ has the same graph as $r = -2\sqrt{\cos 2\theta}$, but traced in a diagonally opposite manner. Thus, the graph of the equation $r^2 = 4\cos 2\theta$ consists of two identical superimposed lemniscates. ◀

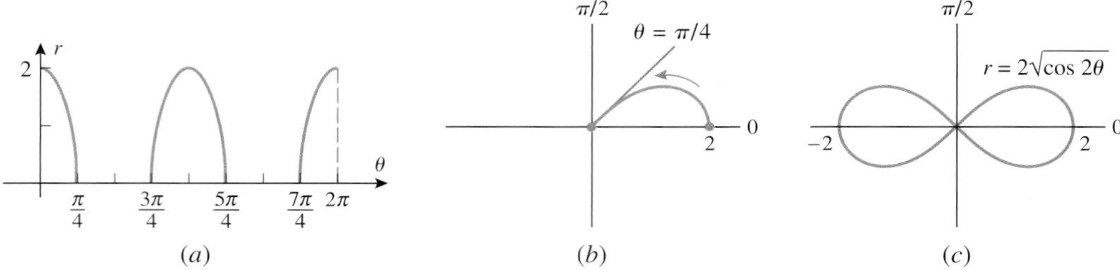

Figure 11.1.15

FAMILIES OF LINES AND RAYS THROUGH THE POLE

If θ_0 is a fixed angle, then for all values of r the point (r, θ_0) lies on the line that makes an angle of $\theta = \theta_0$ with the polar axis; and, conversely, every point on this line has a pair of polar coordinates of the form (r, θ_0). Thus, the equation $\theta = \theta_0$ represents the line that passes through the pole and makes an angle of θ_0 with the polar axis (Figure 11.1.16*a*). If r is restricted to be nonnegative, then the graph of the equation $\theta = \theta_0$ is the ray that emanates from the pole and makes an angle of θ_0 with the polar axis (Figure 11.1.16*b*). Thus, as θ_0 varies, the equation $\theta = \theta_0$ produces either a family of lines through the pole or a family of rays through the pole, depending on the restrictions on r.

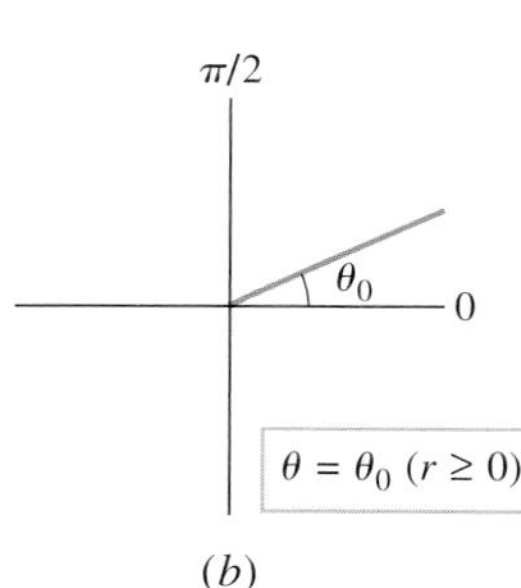

Figure 11.1.16

FAMILIES OF CIRCLES

We will consider three families of circles in which a is assumed to be a positive constant:

$$r = a \qquad r = 2a\cos\theta \qquad r = 2a\sin\theta \tag{3–5}$$

The equation $r = a$ represents a circle of radius a centered at the pole (Figure 11.1.17*a*). Thus, as a varies, this equation produces a family of circles centered at the pole. For families (4) and (5), recall from plane geometry that a triangle that is inscribed in a circle with a diameter of the circle for a side must be a right triangle. Thus, as indicated in Figures 11.1.17*b* and 11.1.17*c*, the equation $r = 2a\cos\theta$ represents a circle of radius a, centered on the x-axis and tangent to the y-axis at the origin; similarly, the equation $r = 2a\sin\theta$ represents a circle of radius a, centered on the y-axis and tangent to the x-axis at the origin. Thus, as a varies, Equations (4) and (5) produce the families illustrated in Figures 11.1.17*d* and 11.1.17*e*.

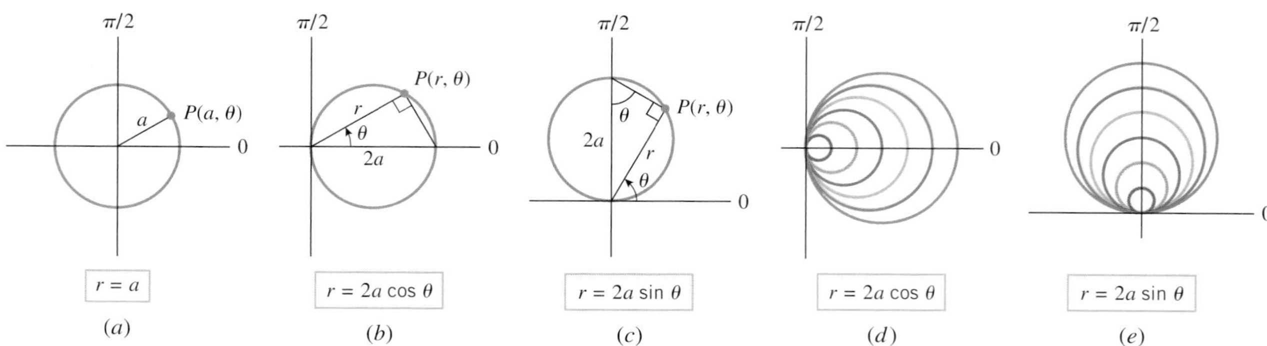

Figure 11.1.17

FAMILIES OF ROSE CURVES

In polar coordinates, equations of the form

$$r = a \sin n\theta \qquad r = a \cos n\theta \tag{6–7}$$

in which $a > 0$ and n is a positive integer represent families of flower-shaped curves called ***roses*** (Figure 11.1.18). The rose consists of n equally spaced petals of radius a if n is odd and $2n$ equally spaced petals of radius a if n is even. It can be shown that a rose with an even number of petals is traced out exactly once as θ varies over the interval $0 \le \theta < 2\pi$ and a rose with an odd number of petals is traced out exactly once as θ varies over the interval $0 \le \theta < \pi$ (Exercise 81). A four-petal rose of radius 1 was graphed in Example 6.

Observe that replacing θ by $-\theta$ does not change the equation $r = 2a\cos\theta$ and that replacing θ by $\pi - \theta$ does not change the equation $r = 2a\sin\theta$. This explains why the circles in Figure 11.1.17*d* are symmetric about the x-axis and those in Figure 11.1.17*e* are symmetric about the y-axis.

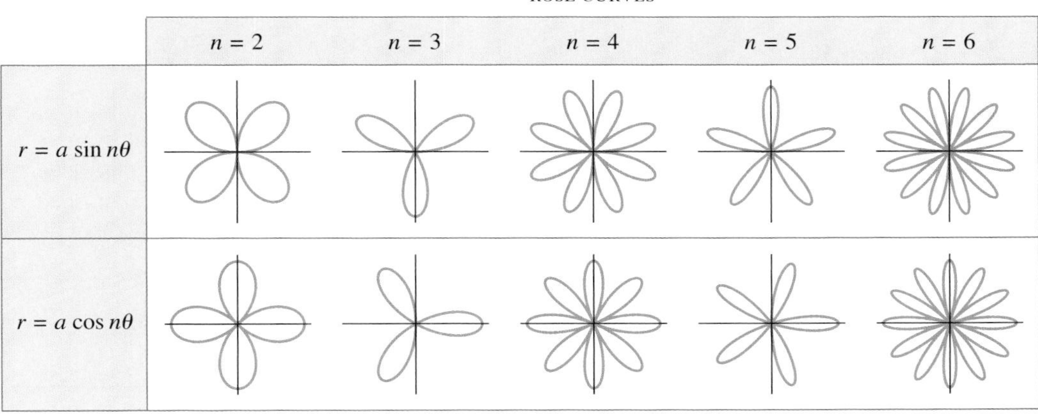

Figure 11.1.18

FAMILIES OF CARDIOIDS AND LIMAÇONS

Equations with any of the four forms

$$r = a \pm b\sin\theta \qquad r = a \pm b\cos\theta \tag{8–9}$$

in which $a > 0$ and $b > 0$ represent polar curves called ***limaçons*** (from the Latin word "limax" for a snail-like creature that is commonly called a slug). There are four possible shapes for a limaçon that are determined by the ratio a/b (Figure 11.1.19). If $a = b$ (the case $a/b = 1$), then the limaçon is called a ***cardioid*** because of its heart-shaped appearance, as noted in Example 8.

What do the graphs of one-petal roses look like?

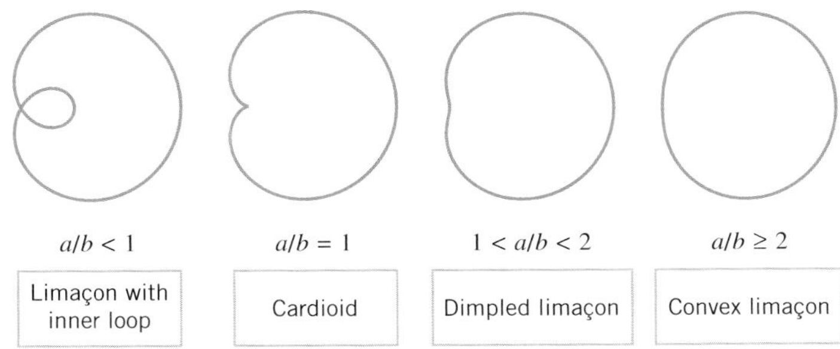

Figure 11.1.19

▶ **Example 10** Figure 11.1.20 shows the family of limaçons $r = a + \cos\theta$ with the constant a varying from 0.25 to 2.50 in steps of 0.25. In keeping with Figure 11.1.19, the limaçons evolve from the loop type to the convex type. As a increases from the starting value of 0.25, the loops get smaller and smaller until the cardioid is reached at $a = 1$. As a increases further, the limaçons evolve through the dimpled type into the convex type. ◀

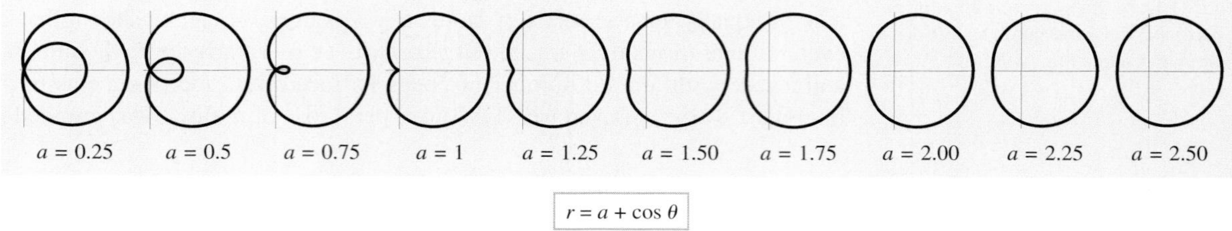

Figure 11.1.20

FAMILIES OF SPIRALS

A ***spiral*** is a curve that coils around a central point. Spirals generally have "left-hand" and "right-hand" versions that coil in opposite directions, depending on the restrictions on the polar angle and the signs of constants that appear in their equations. Some of the more common types of spirals are shown in Figure 11.1.21 for nonnegative values of θ, a, and b.

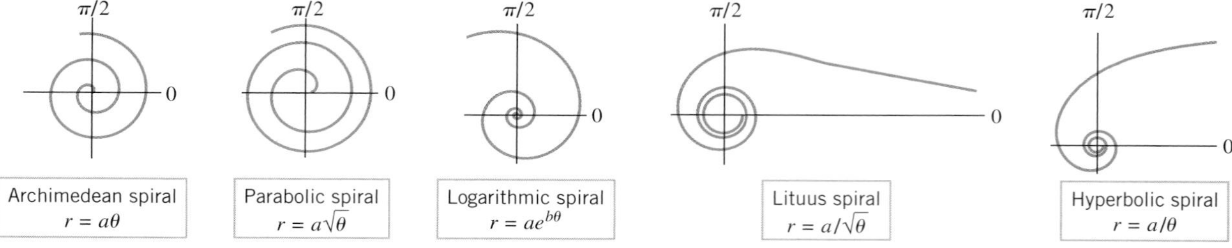

Figure 11.1.21

SPIRALS IN NATURE

Spirals of many kinds occur in nature. For example, the shell of the chambered nautilus (*below*) forms a logarithmic spiral, and a coiled sailor's rope forms an Archimedean spiral. Spirals also occur in flowers, the tusks of certain animals, and in the shapes of galaxies.

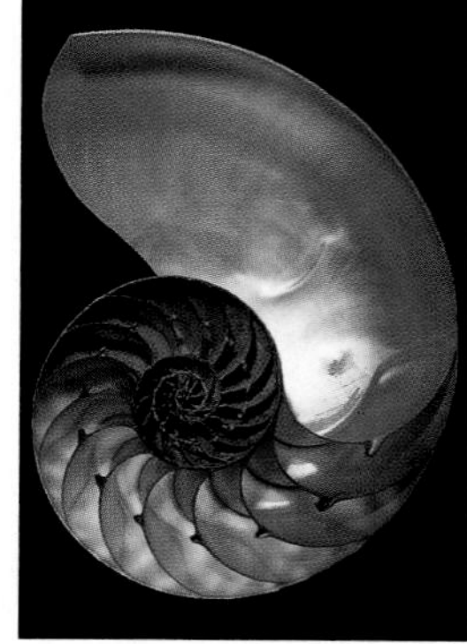

The shell of the chambered nautilus reveals a logarithmic spiral. The animal lives in the outermost chamber.

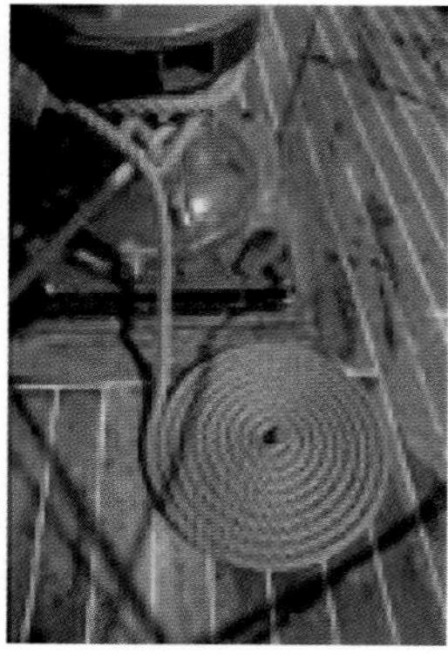

A sailor's coiled rope forms an Archimedean spiral.

A spiral galaxy

GENERATING POLAR CURVES WITH GRAPHING UTILITIES

For polar curves that are too complicated for hand computation, graphing utilities can be used. Although many graphing utilities are capable of graphing polar curves directly, some are not. However, if a graphing utility is capable of graphing parametric equations, then it can be used to graph a polar curve $r = f(\theta)$ by converting this equation to parametric form. This can be done by substituting $f(\theta)$ for r in (1). This yields

$$x = f(\theta)\cos\theta, \quad y = f(\theta)\sin\theta \tag{10}$$

which is a pair of parametric equations for the polar curve in terms of the parameter θ.

► **Example 11** Express the polar equation

$$r = 2 + \cos\frac{5\theta}{2}$$

parametrically, and generate the polar graph from the parametric equations using a graphing utility.

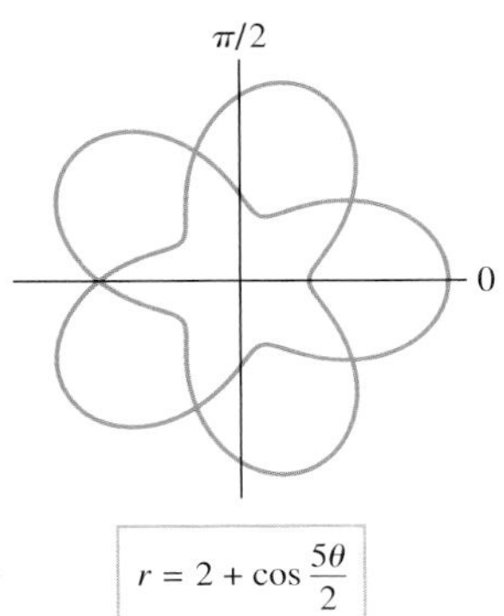

$r = 2 + \cos\dfrac{5\theta}{2}$

Figure 11.1.22

Solution. Substituting the given expression for r in $x = r\cos\theta$ and $y = r\sin\theta$ yields the parametric equations

$$x = \left[2 + \cos\frac{5\theta}{2}\right]\cos\theta, \quad y = \left[2 + \cos\frac{5\theta}{2}\right]\sin\theta$$

Next, we need to find an interval over which to vary θ to produce the entire graph. To find such an interval, we will look for the smallest number of complete revolutions that must occur until the value of r begins to repeat. Algebraically, this amounts to finding the smallest positive integer n such that

$$2 + \cos\left(\frac{5(\theta + 2n\pi)}{2}\right) = 2 + \cos\frac{5\theta}{2}$$

or

$$\cos\left(\frac{5\theta}{2} + 5n\pi\right) = \cos\frac{5\theta}{2}$$

For this equality to hold, the quantity $5n\pi$ must be an even multiple of π; the smallest n for which this occurs is $n = 2$. Thus, the entire graph will be traced in two revolutions, which means it can be generated from the parametric equations

$$x = \left[2 + \cos\frac{5\theta}{2}\right]\cos\theta, \quad y = \left[2 + \cos\frac{5\theta}{2}\right]\sin\theta \qquad (0 \le \theta \le 4\pi)$$

This yields the graph in Figure 11.1.22. ◄

TECHNOLOGY MASTERY

Use a graphing utility to duplicate the curve in Figure 11.1.22. If your graphing utility requires that t be used as the parameter, then you will have to replace θ by t in (10) to generate the graph.

✔QUICK CHECK EXERCISES 11.1 (See page 731 for answers.)

1. (a) Rectangular coordinates of a point (x, y) may be recovered from its polar coordinates (r, θ) by means of the equations $x =$ ______ and $y =$ ______.

(b) Polar coordinates (r, θ) may be recovered from rectangular coordinates (x, y) by means of the equations $r^2 =$ ______ and $\tan\theta =$ ______.

2. Find the rectangular coordinates of the points whose polar coordinates are given.
(a) $(4, \pi/3)$ (b) $(2, -\pi/6)$
(c) $(6, -2\pi/3)$ (d) $(4, 5\pi/4)$

3. In each part, find polar coordinates satisfying the stated conditions for the point whose rectangular coordinates are $(1, \sqrt{3})$.
(a) $r \ge 0$ and $0 \le \theta < 2\pi$
(b) $r \le 0$ and $0 \le \theta < 2\pi$

4. In each part, state the name that describes the polar curve most precisely: a rose, a line, a circle, a limaçon, a cardioid, a spiral, a lemniscate, or none of these.
(a) $r = 1 - \theta$ (b) $r = 1 + 2\sin\theta$
(c) $r = \sin 2\theta$ (d) $r = \cos^2\theta$
(e) $r = \csc\theta$ (f) $r = 2 + 2\cos\theta$
(g) $r = -2\sin\theta$

EXERCISE SET 11.1 Graphing Utility

1–2 Plot the points in polar coordinates.

1. (a) $(3, \pi/4)$ (b) $(5, 2\pi/3)$ (c) $(1, \pi/2)$
(d) $(4, 7\pi/6)$ (e) $(-6, -\pi)$ (f) $(-1, 9\pi/4)$

2. (a) $(2, -\pi/3)$ (b) $(3/2, -7\pi/4)$ (c) $(-3, 3\pi/2)$
(d) $(-5, -\pi/6)$ (e) $(2, 4\pi/3)$ (f) $(0, \pi)$

3–4 Find the rectangular coordinates of the points whose polar coordinates are given.

3. (a) $(6, \pi/6)$ (b) $(7, 2\pi/3)$ (c) $(-6, -5\pi/6)$
(d) $(0, -\pi)$ (e) $(7, 17\pi/6)$ (f) $(-5, 0)$

4. (a) $(-2, \pi/4)$ (b) $(6, -\pi/4)$ (c) $(4, 9\pi/4)$
(d) $(3, 0)$ (e) $(-4, -3\pi/2)$ (f) $(0, 3\pi)$

5. In each part, a point is given in rectangular coordinates. Find two pairs of polar coordinates for the point, one pair satisfying $r \geq 0$ and $0 \leq \theta < 2\pi$, and the second pair satisfying $r \geq 0$ and $-2\pi < \theta \leq 0$.
(a) $(-5, 0)$ (b) $(2\sqrt{3}, -2)$ (c) $(0, -2)$
(d) $(-8, -8)$ (e) $(-3, 3\sqrt{3})$ (f) $(1, 1)$

6. In each part, find polar coordinates satisfying the stated conditions for the point whose rectangular coordinates are $(-\sqrt{3}, 1)$.
(a) $r \geq 0$ and $0 \leq \theta < 2\pi$
(b) $r \leq 0$ and $0 \leq \theta < 2\pi$
(c) $r \geq 0$ and $-2\pi < \theta \leq 0$
(d) $r \leq 0$ and $-\pi < \theta \leq \pi$

7–8 Use a calculating utility, where needed, to approximate the polar coordinates of the points whose rectangular coordinates are given.

7. (a) $(3, 4)$ (b) $(6, -8)$ (c) $(-1, \tan^{-1} 1)$

8. (a) $(-3, 4)$ (b) $(-3, 1.7)$ (c) $\left(2, \sin^{-1} \frac{1}{2}\right)$

9–10 Identify the curve by transforming the given polar equation to rectangular coordinates.

9. (a) $r = 2$ (b) $r \sin\theta = 4$
(c) $r = 3\cos\theta$ (d) $r = \dfrac{6}{3\cos\theta + 2\sin\theta}$

10. (a) $r = 5\sec\theta$ (b) $r = 2\sin\theta$
(c) $r = 4\cos\theta + 4\sin\theta$ (d) $r = \sec\theta\tan\theta$

11–12 Express the given equations in polar coordinates.

11. (a) $x = 3$ (b) $x^2 + y^2 = 7$
(c) $x^2 + y^2 + 6y = 0$ (d) $9xy = 4$

12. (a) $y = -3$ (b) $x^2 + y^2 = 5$
(c) $x^2 + y^2 + 4x = 0$ (d) $x^2(x^2 + y^2) = y^2$

FOCUS ON CONCEPTS

13–16 A graph is given in a rectangular θr-coordinate system. Sketch the corresponding graph in polar coordinates.

13.
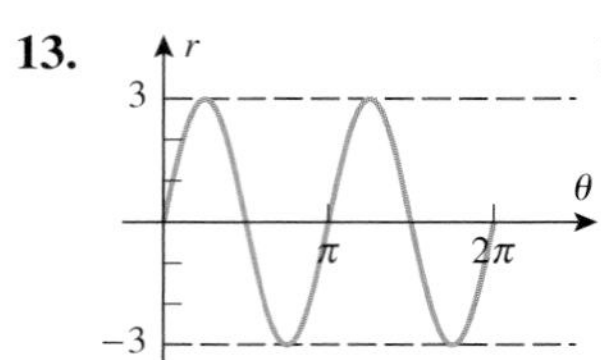

14.
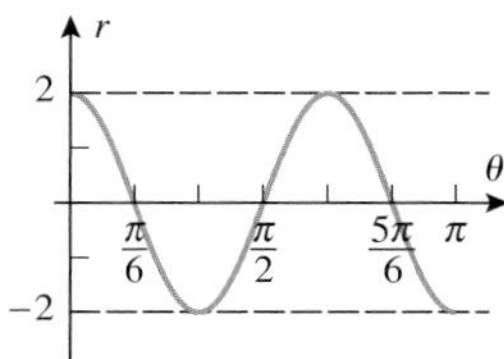

15.
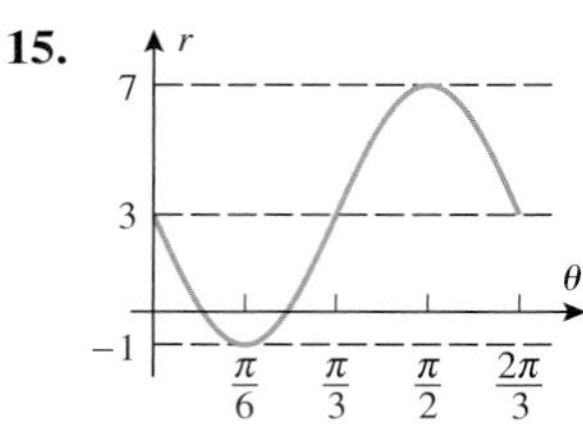

16.
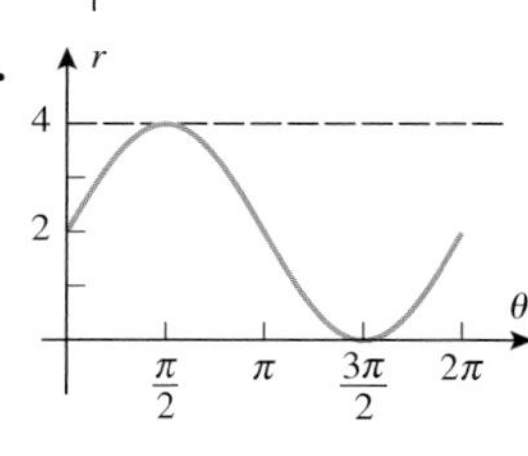

17–20 Find an equation for the given polar graph.

17. (a) (b)
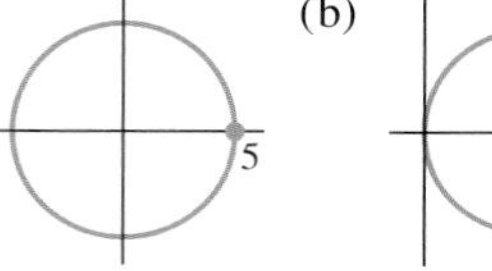

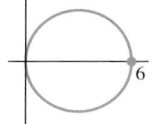

Circle Circle

(c)
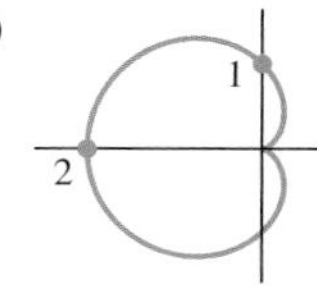

Cardioid

18. (a)
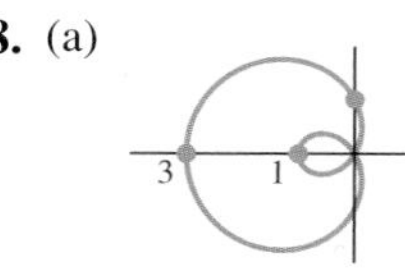

Limaçon

(b)
Circle

(c)
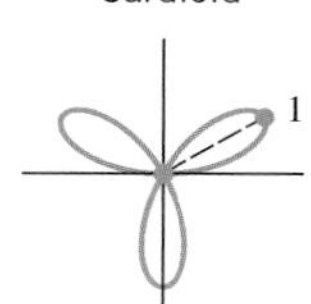

Three-petal rose

19. (a)
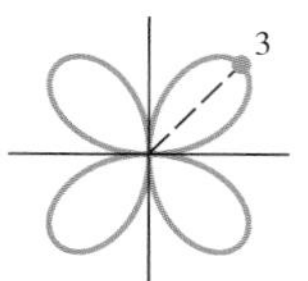

Four-petal rose

(b)
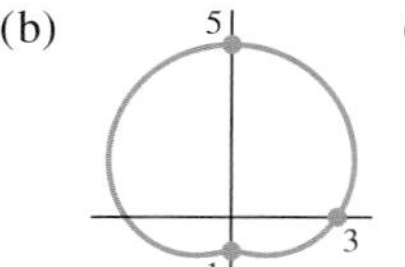

Limaçon

(c)
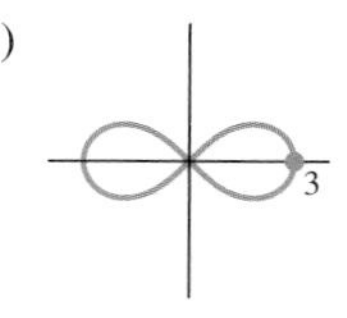

Lemniscate

20. (a)
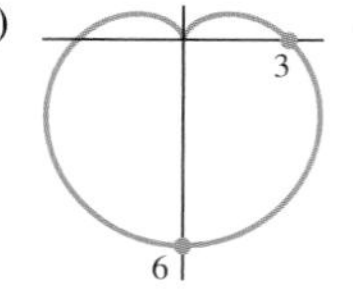

Cardioid

(b)
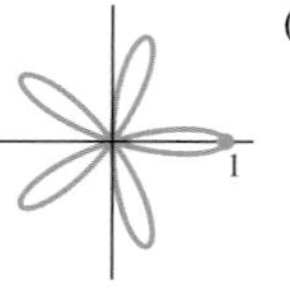

Five-petal rose

(c)
Circle

21–50 Sketch the curve in polar coordinates.

21. $\theta = \dfrac{\pi}{3}$ **22.** $\theta = -\dfrac{3\pi}{4}$ **23.** $r = 3$

24. $r = 4\cos\theta$ **25.** $r = 6\sin\theta$ **26.** $r = 1 + \sin\theta$

27. $2r = \cos\theta$ **28.** $r - 2 = 2\cos\theta$

29. $r = 3(1 + \sin\theta)$

30. $r = 5 - 5\sin\theta$

31. $r = 4 - 4\cos\theta$

32. $r = 1 + 2\sin\theta$

33. $r = -1 - \cos\theta$

34. $r = 4 + 3\cos\theta$

35. $r = 2 + \cos\theta$

36. $r = 3 - \sin\theta$

37. $r = 3 + 4\cos\theta$

38. $r - 5 = 3\sin\theta$

39. $r = 5 - 2\cos\theta$

40. $r = -3 - 4\sin\theta$

41. $r^2 = \cos 2\theta$

42. $r^2 = 9\sin 2\theta$

43. $r^2 = 16\sin 2\theta$

44. $r = 4\theta \quad (\theta \geq 0)$

45. $r = 4\theta \quad (\theta \leq 0)$

46. $r = 4\theta$

47. $r = -2\cos 2\theta$

48. $r = 3\sin 2\theta$

49. $r = 9\sin 4\theta$

50. $r = 2\cos 3\theta$

51–55 Use a graphing utility to generate the polar graph. Be sure to choose the parameter interval so that a complete graph is generated.

51. $r = \cos\dfrac{\theta}{2}$

52. $r = \sin\dfrac{\theta}{2}$

53. $r = 1 - 2\sin\dfrac{\theta}{4}$

54. $r = 0.5 + \cos\dfrac{\theta}{3}$

55. $r = \cos\dfrac{\theta}{5}$

56. The accompanying figure shows the graph of the "butterfly curve"

$$r = e^{\cos\theta} - 2\cos 4\theta + \sin^3\frac{\theta}{4}$$

Generate the complete butterfly with a graphing utility, and state the parameter interval you used.

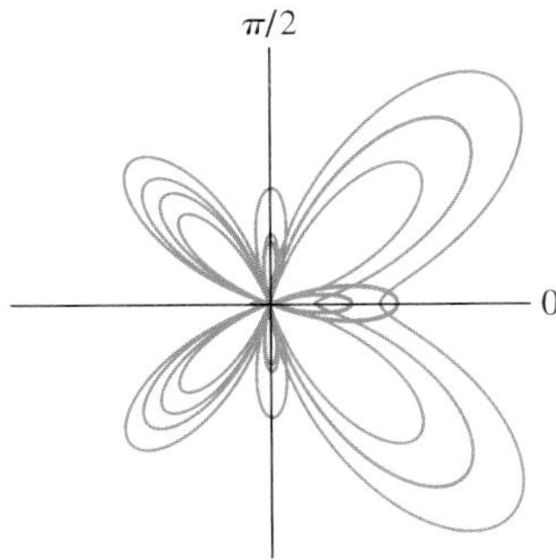

Figure Ex-56

57. The accompanying figure shows the Archimedean spiral $r = \theta/2$ produced with a graphing calculator.
(a) What interval of values for θ do you think was used to generate the graph?
(b) Duplicate the graph with your own graphing utility.

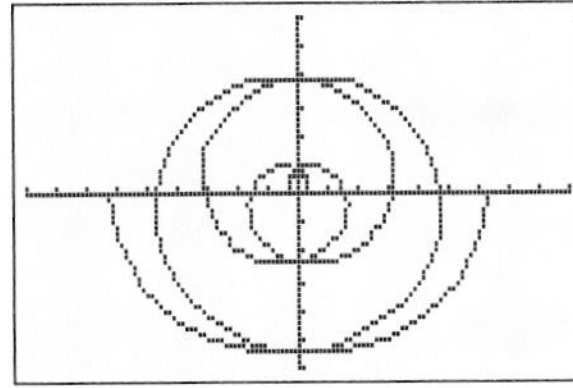

$[-9, 9] \times [-6, 6]$
$x\text{Scl} = 1, y\text{Scl} = 1$

Figure Ex-57

58. Find equations for the two families of circles in the accompanying figure.

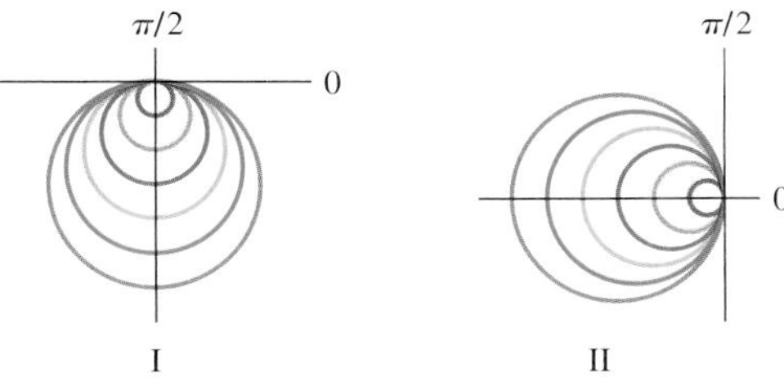

Figure Ex-58

59. (a) Show that if a varies, then the polar equation

$$r = a\sec\theta \quad (-\pi/2 < \theta < \pi/2)$$

describes a family of lines perpendicular to the polar axis.

(b) Show that if b varies, then the polar equation

$$r = b\csc\theta \quad (0 < \theta < \pi)$$

describes a family of lines parallel to the polar axis.

FOCUS ON CONCEPTS

60. The accompanying figure shows graphs of the Archimedean spiral $r = \theta$ and the parabolic spiral $r = \sqrt{\theta}$. Which is which? Explain your reasoning.

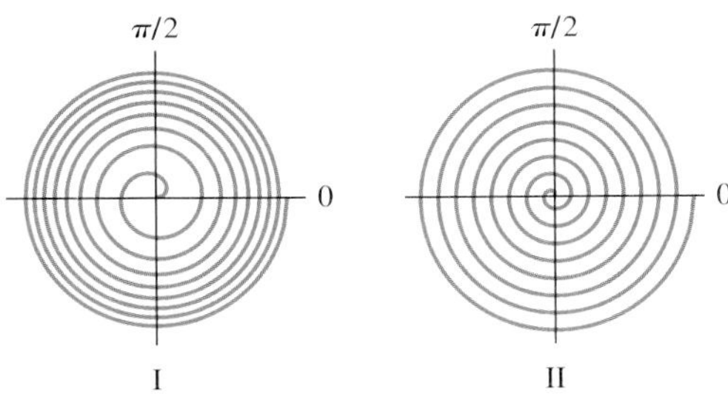

Figure Ex-60

61. The accompanying figure shows the polar graph of the equation $r = f(\theta)$, $(0 \leq \theta \leq \pi/2)$. Sketch the graph of
(a) $r = f(-\theta)$
(b) $r = f\left(\theta - \dfrac{\pi}{2}\right)$
(c) $r = f\left(\theta + \dfrac{\pi}{2}\right)$
(d) $r = -f(\theta)$
(e) $r = f(\theta) + 1$.

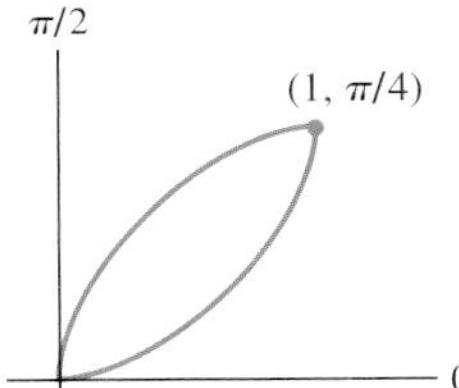

Figure Ex-61

62. Repeat Exercise 61 using the accompanying figure.

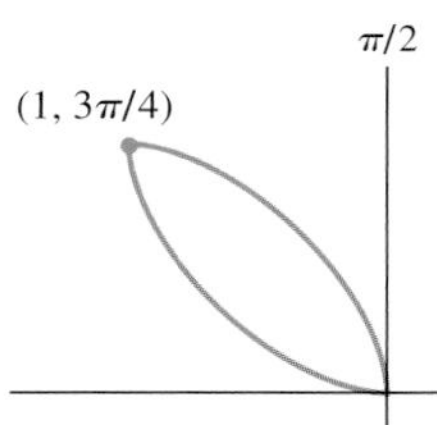

Figure Ex-62

63. Use a graphing utility to investigate how the family of polar curves $r = 1 + a\cos n\theta$ is affected by changing the values of a and n, where a is a positive real number and n is a positive integer. Write a brief paragraph to explain your conclusions.

64. Show that if the polar graph of $r = f(\theta)$ is rotated counterclockwise around the origin through an angle α, then $r = f(\theta - \alpha)$ is an equation for the rotated curve. [*Hint:* If (r_0, θ_0) is any point on the original graph, then $(r_0, \theta_0 + \alpha)$ is a point on the rotated graph.]

65. Use the result in Exercise 64 to find an equation for the cardioid $r = 1 + \cos\theta$ after it has been rotated through the given angle, and check your answer with a graphing utility.
(a) $\dfrac{\pi}{4}$ (b) $\dfrac{\pi}{2}$ (c) π (d) $\dfrac{5\pi}{4}$

66. Use the result in Exercise 64 to find an equation for the lemniscate that results when the lemniscate in Example 9 is rotated counterclockwise through an angle of $\pi/2$.

67. Sketch the polar graph of the equation $(r - 1)(\theta - 1) = 0$.

68. (a) Show that if A and B are not both zero, then the graph of the polar equation

$$r = A\sin\theta + B\cos\theta$$

is a circle. Find its radius.
(b) Derive Formulas (4) and (5) from the formula given in part (a).

69. Find the highest point on the cardioid $r = 1 + \cos\theta$.

70. Find the leftmost point on the upper half of the cardioid $r = 1 + \cos\theta$.

71. Define the width of a petal of a rose curve to be the dimension shown in the accompanying figure. Shown that the width w of a petal of the four-petal rose $r = \cos 2\theta$ is $w = 2\sqrt{6}/9$. [*Hint:* Express y in terms of θ, and investigate the maximum value of y.]

Figure Ex-71

72. Modify the argument in Exercise 71 to find the (vertical) width of the lemniscate $r^2 = \cos 2\theta$.

73. (a) Show that in a polar coordinate system the distance d between the points (r_1, θ_1) and (r_2, θ_2) is

$$d = \sqrt{r_1^2 + r_2^2 - 2r_1 r_2\cos(\theta_1 - \theta_2)}$$

(b) Show that if $0 \le \theta_1 < \theta_2 \le \pi$ and if r_1 and r_2 are positive, then the area A of the triangle with vertices $(0, 0)$, (r_1, θ_1), and (r_2, θ_2) is

$$A = \tfrac{1}{2}r_1 r_2\sin(\theta_2 - \theta_1)$$

(c) Find the distance between the points whose polar coordinates are $(3, \pi/6)$ and $(2, \pi/3)$.
(d) Find the area of the triangle whose vertices in polar coordinates are $(0, 0)$, $(1, 5\pi/6)$, and $(2, \pi/3)$.

74. Use the formula obtained in part (a) of Exercise 73 to find the distance between successive tips of the four-petal rose $r = \cos 2\theta$, and check your answer using geometry.

75. Use the formula obtained in part (a) of Exercise 73 to find the distance between successive tips of the three-petal rose $r = \sin 3\theta$, and check your answer using trigonometry.

76. In the late seventeenth century the Italian astronomer Giovanni Domenico Cassini (1625–1712) introduced the family of curves

$$(x^2 + y^2 + a^2)^2 - b^4 - 4a^2x^2 = 0 \quad (a > 0, b > 0)$$

in his studies of the relative motions of the Earth and the Sun. These curves, which are called ***Cassini ovals***, have one of the three basic shapes shown in the accompanying figure.
(a) Show that if $a = b$, then the polar equation of the Cassini oval is $r^2 = 2a^2\cos 2\theta$, which is a lemniscate.
(b) Use the formula in Exercise 73(a) to show that the lemniscate in part (a) is the curve traced by a point that moves in such a way that the product of its distances from the polar points $(a, 0)$ and (a, π) is a^2.

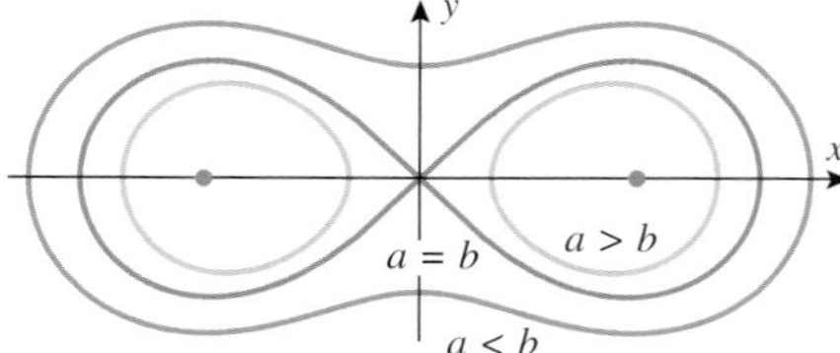

Figure Ex-76

77–80 Vertical and horizontal asymptotes of polar curves can sometimes be detected by investigating the behavior of $x = r\cos\theta$ and $y = r\sin\theta$ as θ varies. This idea is used in these exercises.

77. Show that the ***hyperbolic spiral*** $r = 1/\theta$ $(\theta > 0)$ has a horizontal asymptote at $y = 1$ by showing that $y \to 1$ and $x \to +\infty$ as $\theta \to 0^+$. Confirm this result by generating the spiral with a graphing utility.

78. Show that the spiral $r = 1/\theta^2$ does not have any horizontal asymptotes.

79. (a) Show that the ***kappa curve*** $r = 4\tan\theta$ $(0 \le \theta \le 2\pi)$ has a vertical asymptote at $x = 4$ by showing that $x \to 4$ and $y \to +\infty$ as $\theta \to \pi/2^-$ and that $x \to 4$ and $y \to -\infty$ as $\theta \to \pi/2^+$.
(b) Use the method in part (a) to show that the kappa curve also has a vertical asymptote at $x = -4$.
(c) Confirm the results in parts (a) and (b) by generating the kappa curve with a graphing utility.

80. Use a graphing utility to make a conjecture about the existence of asymptotes for the ***cissoid*** $r = 2\sin\theta\tan\theta$, and then confirm your conjecture by calculating appropriate limits.

81. Prove that a rose with an even number of petals is traced out exactly once as θ varies over the interval $0 \le \theta < 2\pi$ and a rose with an odd number of petals is traced out exactly once as θ varies over the interval $0 \le \theta < \pi$.

82. (a) Use a graphing utility to confirm that the graph of $r = 2 - \sin(\theta/2)$ $(0 \le \theta \le 4\pi)$ is symmetric about the x-axis.
(b) Show that replacing θ by $-\theta$ in the polar equation $r = 2 - \sin(\theta/2)$ does not produce an equivalent equation. Why does this not contradict the symmetry demonstrated in part (a)?

✔ QUICK CHECK ANSWERS 11.1

1. (a) $r\cos\theta$; $r\sin\theta$ (b) $x^2 + y^2$; y/x **2.** (a) $(2, 2\sqrt{3})$ (b) $(\sqrt{3}, -1)$ (c) $(-3, -3\sqrt{3})$ (d) $(-2\sqrt{2}, -2\sqrt{2})$
3. (a) $(2, \pi/3)$ (b) $(-2, 4\pi/3)$ **4.** (a) spiral (b) limaçon (c) rose (d) none of these (e) line (f) cardioid (g) circle

11.2 TANGENT LINES AND ARC LENGTH FOR PARAMETRIC AND POLAR CURVES

In this section we will derive the formulas required to find slopes, tangent lines, and arc lengths of parametric and polar curves.

TANGENT LINES TO PARAMETRIC CURVES

We will be concerned in this section with curves that are given by parametric equations

$$x = f(t), \quad y = g(t)$$

in which $f(t)$ and $g(t)$ have continuous first derivatives with respect to t. It can be proved that if $dx/dt \neq 0$, then y is a differentiable function of x, in which case the chain rule implies that

$$\frac{dy}{dx} = \frac{dy/dt}{dx/dt} \tag{1}$$

This formula makes it possible to find dy/dx directly from the parametric equations without eliminating the parameter.

► **Example 1** Find the slope of the tangent line to the unit circle

$$x = \cos t, \quad y = \sin t \quad (0 \le t \le 2\pi)$$

at the point where $t = \pi/6$ (Figure 11.2.1).

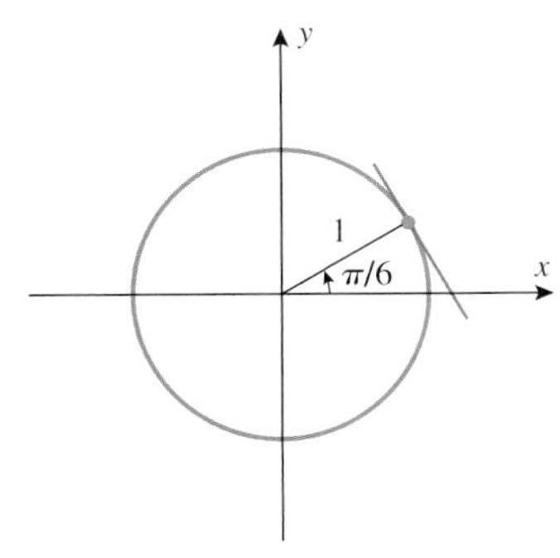

Figure 11.2.1

Solution. From (1), the slope at a general point on the circle is

$$\frac{dy}{dx} = \frac{dy/dt}{dx/dt} = \frac{\cos t}{-\sin t} = -\cot t \tag{2}$$

Thus, the slope at $t = \pi/6$ is

$$\left.\frac{dy}{dx}\right|_{t=\pi/6} = -\cot\frac{\pi}{6} = -\sqrt{3} \quad ◄$$

Note that Formula (2) makes sense geometrically because the radius from the origin to the point $P(\cos t, \sin t)$ has slope $m = \tan t$. Thus the tangent line at P, being perpendicular to the radius, has slope

$$-\frac{1}{m} = -\frac{1}{\tan t} = -\cot t$$

(Figure 11.2.2).

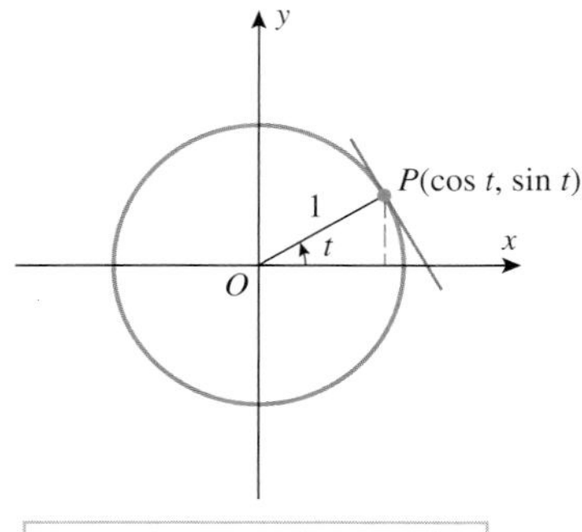

Radius OP has slope $m = \tan t$.

Figure 11.2.2

It follows from Formula (1) that the tangent line to a parametric curve will be horizontal at those points where $dy/dt = 0$ and $dx/dt \neq 0$, since $dy/dx = 0$ at such points. Two different situations occur when $dx/dt = 0$. At points where $dx/dt = 0$ and $dy/dt \neq 0$, the right side of (1) has a nonzero numerator and a zero denominator; we will agree that the curve has ***infinite slope*** and a ***vertical tangent line*** at such points. At points where dx/dt and dy/dt are both zero, the right side of (1) becomes an indeterminate form; we call such points ***singular points***. No general statement can be made about the behavior of parametric curves at singular points; they must be analyzed case by case.

▶ **Example 2** In a disastrous first flight, an experimental paper airplane follows the trajectory

$$x = t - 3\sin t, \quad y = 4 - 3\cos t \qquad (t \geq 0)$$

but crashes into a wall at time $t = 10$ (Figure 11.2.3).

(a) At what times was the airplane flying horizontally?

(b) At what times was it flying vertically?

Solution (a). The airplane was flying horizontally at those times when $dy/dt = 0$ and $dx/dt \neq 0$. From the given trajectory we have

$$\frac{dy}{dt} = 3\sin t \quad \text{and} \quad \frac{dx}{dt} = 1 - 3\cos t \tag{3}$$

Setting $dy/dt = 0$ yields the equation $3\sin t = 0$, or, more simply, $\sin t = 0$. This equation has four solutions in the time interval $0 \leq t \leq 10$:

$$t = 0, \quad t = \pi, \quad t = 2\pi, \quad t = 3\pi$$

Since $dx/dt = 1 - 3\cos t \neq 0$ for these values of t (verify), the airplane was flying horizontally at times

$$t = 0, \quad t = \pi \approx 3.14, \quad t = 2\pi \approx 6.28, \quad \text{and} \quad t = 3\pi \approx 9.42$$

which is consistent with Figure 11.2.3.

Solution (b). The airplane was flying vertically at those times when $dx/dt = 0$ and $dy/dt \neq 0$. Setting $dx/dt = 0$ in (3) yields the equation

$$1 - 3\cos t = 0 \quad \text{or} \quad \cos t = \tfrac{1}{3}$$

This equation has three solutions in the time interval $0 \leq t \leq 10$ (Figure 11.2.4):

$$t = \cos^{-1}\tfrac{1}{3}, \quad t = 2\pi - \cos^{-1}\tfrac{1}{3}, \quad t = 2\pi + \cos^{-1}\tfrac{1}{3}$$

Since $dy/dt = 3\sin t$ is not zero at these points (why?), it follows that the airplane was flying vertically at times

$$t = \cos^{-1}\tfrac{1}{3} \approx 1.23, \quad t \approx 2\pi - 1.23 \approx 5.05, \quad t \approx 2\pi + 1.23 \approx 7.51$$

which again is consistent with Figure 11.2.3. ◀

▶ **Example 3** The curve represented by the parametric equations

$$x = t^2, \quad y = t^3 \qquad (-\infty < t < +\infty)$$

is called a ***semicubical parabola***. The parameter t can be eliminated by cubing x and squaring y, from which it follows that $y^2 = x^3$. The graph of this equation, shown in Figure 11.2.5, consists of two branches: an upper branch obtained by graphing $y = x^{3/2}$ and a lower branch obtained by graphing $y = -x^{3/2}$. The two branches meet at the origin,

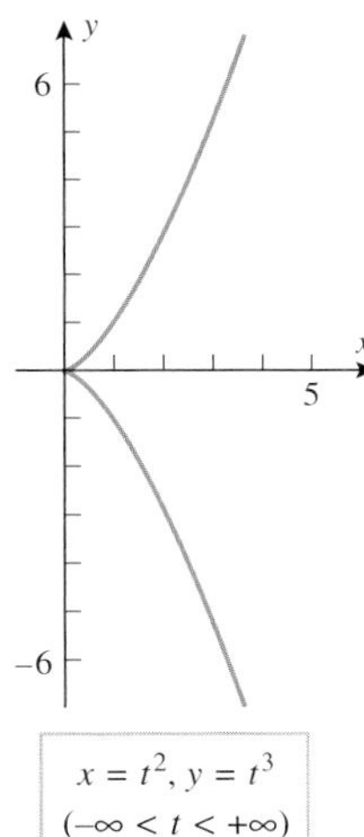

Figure 11.2.5

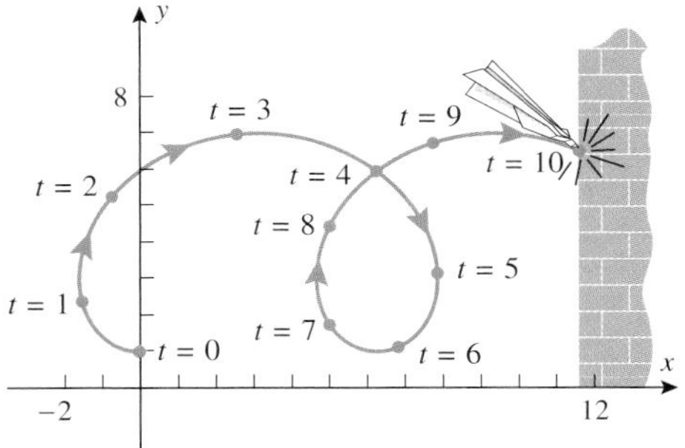

Figure 11.2.3

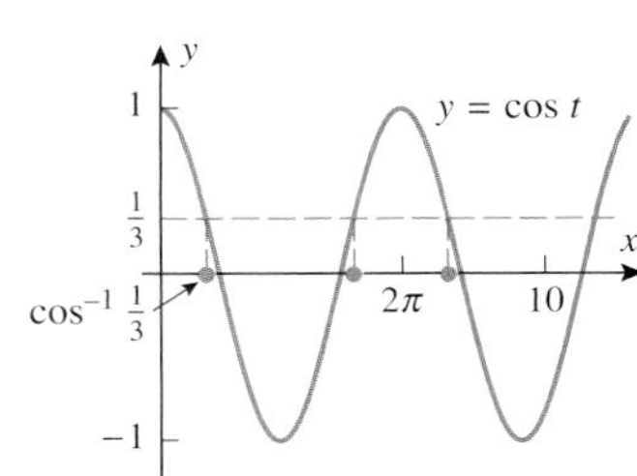

Figure 11.2.4

which corresponds to $t = 0$ in the parametric equations. This is a singular point because the derivatives $dx/dt = 2t$ and $dy/dt = 3t^2$ are both zero there. ◀

▶ **Example 4** Without eliminating the parameter, find dy/dx and d^2y/dx^2 at $(1, 1)$ and $(1, -1)$ on the semicubical parabola given by the parametric equations in Example 3.

Solution. From (1) we have

$$\frac{dy}{dx} = \frac{dy/dt}{dx/dt} = \frac{3t^2}{2t} = \frac{3}{2}t \qquad (t \neq 0) \tag{4}$$

and from (1) applied to $y' = dy/dx$ we have

$$\frac{d^2y}{dx^2} = \frac{dy'}{dx} = \frac{dy'/dt}{dx/dt} = \frac{3/2}{2t} = \frac{3}{4t} \tag{5}$$

Since the point $(1, 1)$ on the curve corresponds to $t = 1$ in the parametric equations, it follows from (4) and (5) that

$$\left.\frac{dy}{dx}\right|_{t=1} = \frac{3}{2} \quad \text{and} \quad \left.\frac{d^2y}{dx^2}\right|_{t=1} = \frac{3}{4}$$

Similarly, the point $(1, -1)$ corresponds to $t = -1$ in the parametric equations, so applying (4) and (5) again yields

$$\left.\frac{dy}{dx}\right|_{t=-1} = -\frac{3}{2} \quad \text{and} \quad \left.\frac{d^2y}{dx^2}\right|_{t=-1} = -\frac{3}{4}$$

Note that the values we obtained for the first and second derivatives are consistent with the graph in Figure 11.2.5, since at $(1, 1)$ on the upper branch the tangent line has positive slope and the curve is concave up, and at $(1, -1)$ on the lower branch the tangent line has negative slope and the curve is concave down.

Finally, observe that we were able to apply Formulas (4) and (5) for both $t = 1$ and $t = -1$, even though the points $(1, 1)$ and $(1, -1)$ lie on different branches. In contrast, had we chosen to perform the same computations by eliminating the parameter, we would have had to obtain separate derivative formulas for $y = x^{3/2}$ and $y = -x^{3/2}$. ◀

TANGENT LINES TO POLAR CURVES

Our next objective is to find a method for obtaining slopes of tangent lines to polar curves of the form $r = f(\theta)$ in which r is a differentiable function of θ. We showed in the last section that a curve of this form can be expressed parametrically in terms of the parameter θ by substituting $f(\theta)$ for r in the equations $x = r\cos\theta$ and $y = r\sin\theta$. This yields

$$x = f(\theta)\cos\theta, \quad y = f(\theta)\sin\theta$$

from which we obtain

$$\begin{aligned} \frac{dx}{d\theta} &= -f(\theta)\sin\theta + f'(\theta)\cos\theta = -r\sin\theta + \frac{dr}{d\theta}\cos\theta \\ \frac{dy}{d\theta} &= f(\theta)\cos\theta + f'(\theta)\sin\theta = r\cos\theta + \frac{dr}{d\theta}\sin\theta \end{aligned} \tag{6}$$

Thus, if $dx/d\theta$ and $dy/d\theta$ are continuous and if $dx/d\theta \neq 0$, then y is a differentiable function of x, and Formula (1) with θ in place of t yields

$$\frac{dy}{dx} = \frac{dy/d\theta}{dx/d\theta} = \frac{r\cos\theta + \sin\theta\dfrac{dr}{d\theta}}{-r\sin\theta + \cos\theta\dfrac{dr}{d\theta}} \tag{7}$$

▶ Example 5 Find the slope of the tangent line to the circle $r = 4\cos\theta$ at the point where $\theta = \pi/4$.

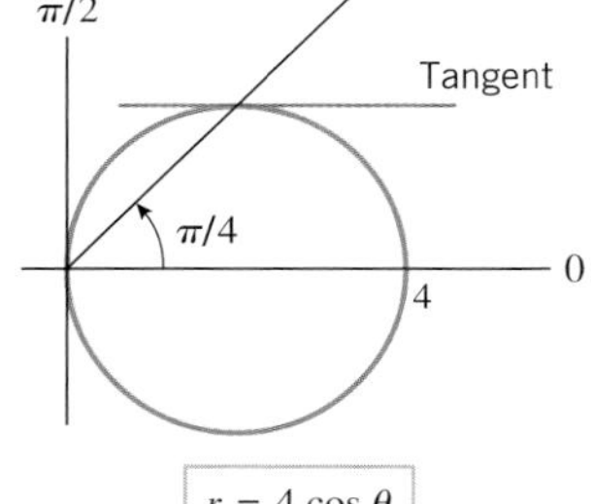

Figure 11.2.6

Solution. From (7) with $r = 4\cos\theta$ we obtain (verify)

$$\frac{dy}{dx} = \frac{4\cos^2\theta - 4\sin^2\theta}{-8\sin\theta\cos\theta} = \frac{4\cos 2\theta}{-4\sin 2\theta} = -\cot 2\theta$$

Thus, at the point where $\theta = \pi/4$ the slope of the tangent line is

$$m = \left.\frac{dy}{dx}\right|_{\theta=\pi/4} = -\cot\frac{\pi}{2} = 0$$

which implies that the circle has a horizontal tangent line at the point where $\theta = \pi/4$ (Figure 11.2.6). ◀

▶ Example 6 Find the points on the cardioid $r = 1 - \cos\theta$ at which there is a horizontal tangent line, a vertical tangent line, or a singular point.

Solution. A horizontal tangent line will occur where $dy/d\theta = 0$ and $dx/d\theta \neq 0$, a vertical tangent line where $dy/d\theta \neq 0$ and $dx/d\theta = 0$, and a singular point where $dy/d\theta = 0$ and $dx/d\theta = 0$. We could find these derivatives from the formulas in (6). However, an alternative approach is to go back to basic principles and express the cardioid parametrically by substituting $r = 1 - \cos\theta$ in the conversion formulas $x = r\cos\theta$ and $y = r\sin\theta$. This yields

$$x = (1 - \cos\theta)\cos\theta, \quad y = (1 - \cos\theta)\sin\theta \qquad (0 \le \theta \le 2\pi)$$

Differentiating these equations with respect to θ and then simplifying yields (verify)

$$\frac{dx}{d\theta} = \sin\theta(2\cos\theta - 1), \quad \frac{dy}{d\theta} = (1 - \cos\theta)(1 + 2\cos\theta)$$

Thus, $dx/d\theta = 0$ if $\sin\theta = 0$ or $\cos\theta = \frac{1}{2}$, and $dy/d\theta = 0$ if $\cos\theta = 1$ or $\cos\theta = -\frac{1}{2}$. We leave it for you to solve these equations and show that the solutions of $dx/d\theta = 0$ on the interval $0 \le \theta \le 2\pi$ are

$$\frac{dx}{d\theta} = 0: \quad \theta = 0, \ \frac{\pi}{3}, \ \pi, \ \frac{5\pi}{3}, \ 2\pi$$

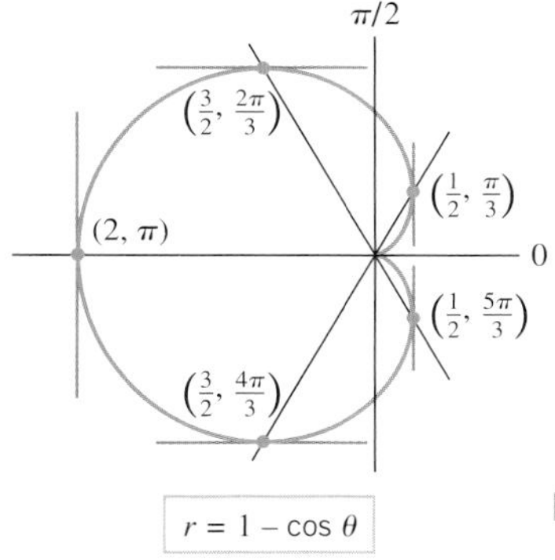

Figure 11.2.7

and the solutions of $dy/d\theta = 0$ on the interval $0 \le \theta \le 2\pi$ are

$$\frac{dy}{d\theta} = 0: \quad \theta = 0, \ \frac{2\pi}{3}, \ \frac{4\pi}{3}, \ 2\pi$$

Thus, horizontal tangent lines occur at $\theta = 2\pi/3$ and $\theta = 4\pi/3$; vertical tangent lines occur at $\theta = \pi/3$, π, and $5\pi/3$; and singular points occur at $\theta = 0$ and $\theta = 2\pi$ (Figure 11.2.7). Note, however, that $r = 0$ at both singular points, so there is really only one singular point on the cardioid—the pole. ◄

TANGENT LINES TO POLAR CURVES AT THE ORIGIN

Formula (7) reveals some useful information about the behavior of a polar curve $r = f(\theta)$ that passes through the origin. If we assume that $r = 0$ and $dr/d\theta \ne 0$ when $\theta = \theta_0$, then it follows from Formula (7) that the slope of the tangent line to the curve at $\theta = \theta_0$ is

$$\frac{dy}{dx} = \frac{0 + \sin\theta_0 \dfrac{dr}{d\theta}}{0 + \cos\theta_0 \dfrac{dr}{d\theta}} = \frac{\sin\theta_0}{\cos\theta_0} = \tan\theta_0$$

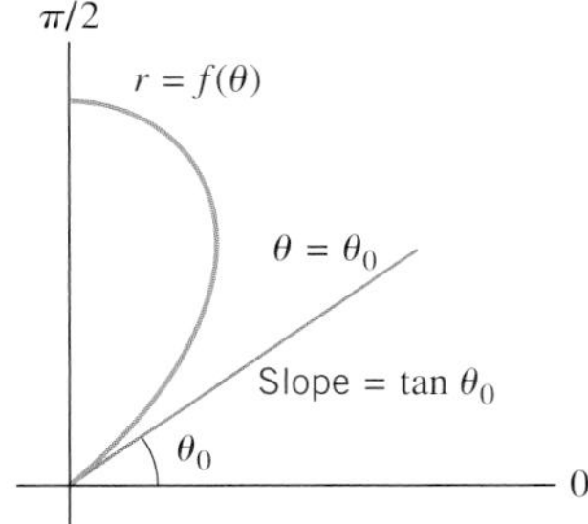

Figure 11.2.8

(Figure 11.2.8). However, $\tan\theta_0$ is also the slope of the line $\theta = \theta_0$, so we can conclude that this line is tangent to the curve at the origin. Thus, we have established the following result.

11.2.1 THEOREM. *If the polar curve $r = f(\theta)$ passes through the origin at $\theta = \theta_0$, and if $dr/d\theta \ne 0$ at $\theta = \theta_0$, then the line $\theta = \theta_0$ is tangent to the curve at the origin.*

This theorem tells us that equations of the tangent lines at the origin to the curve $r = f(\theta)$ can be obtained by solving the equation $f(\theta) = 0$. It is important to keep in mind, however, that $r = f(\theta)$ may be zero for more than one value of θ, so there may be more than one tangent line at the origin. This is illustrated in the next example.

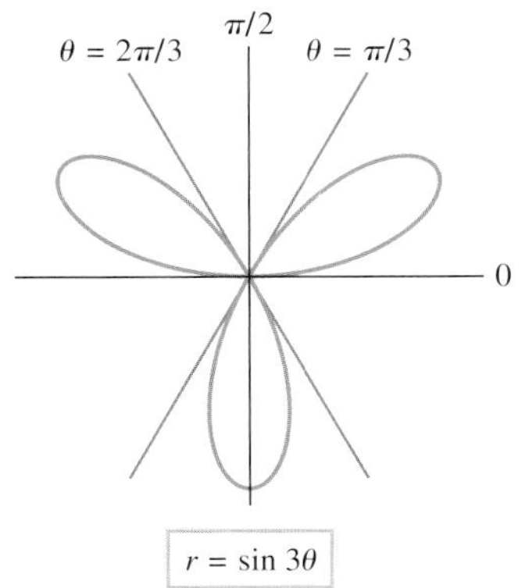

Figure 11.2.9

► **Example 7** The three-petal rose $r = \sin 3\theta$ in Figure 11.2.9 has three tangent lines at the origin, which can be found by solving the equation

$$\sin 3\theta = 0$$

It was shown in Exercise 81 of Section 11.1 that the complete rose is traced once as θ varies over the interval $0 \le \theta < \pi$, so we need only look for solutions in this interval. We leave it for you to confirm that these solutions are

$$\theta = 0, \quad \theta = \frac{\pi}{3}, \quad \text{and} \quad \theta = \frac{2\pi}{3}$$

Since $dr/d\theta = 3\cos 3\theta \ne 0$ for these values of θ, these three lines are tangent to the rose at the origin, which is consistent with the figure. ◄

ARC LENGTH OF A POLAR CURVE

A formula for the arc length of a polar curve $r = f(\theta)$ can be derived by expressing the curve in parametric form and applying Formula (6) of Section 6.4 for the arc length of a parametric curve. We leave it as an exercise to show the following.

11.2.2 ARC LENGTH FORMULA FOR POLAR CURVES. If no segment of the polar curve $r = f(\theta)$ is traced more than once as θ increases from α to β, and if $dr/d\theta$ is continuous for $\alpha \le \theta \le \beta$, then the arc length L from $\theta = \alpha$ to $\theta = \beta$ is

$$L = \int_{\alpha}^{\beta} \sqrt{[f(\theta)]^2 + [f'(\theta)]^2}\, d\theta = \int_{\alpha}^{\beta} \sqrt{r^2 + \left(\frac{dr}{d\theta}\right)^2}\, d\theta \tag{8}$$

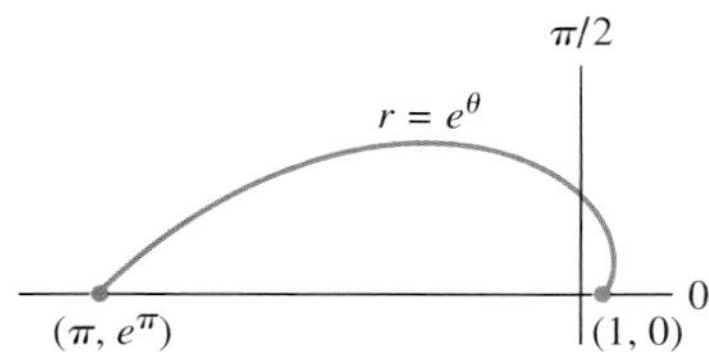

Figure 11.2.10

▶ **Example 8** Find the arc length of the spiral $r = e^{\theta}$ in Figure 11.2.10 between $\theta = 0$ and $\theta = \pi$.

Solution.

$$L = \int_{\alpha}^{\beta} \sqrt{r^2 + \left(\frac{dr}{d\theta}\right)^2}\, d\theta = \int_0^{\pi} \sqrt{(e^{\theta})^2 + (e^{\theta})^2}\, d\theta$$

$$= \int_0^{\pi} \sqrt{2}\, e^{\theta}\, d\theta = \sqrt{2}\, e^{\theta}\Big]_0^{\pi} = \sqrt{2}(e^{\pi} - 1) \approx 31.3 \blacktriangleleft$$

▶ **Example 9** Find the total arc length of the cardioid $r = 1 + \cos\theta$.

Solution. The cardioid is traced out once as θ varies from $\theta = 0$ to $\theta = 2\pi$. Thus,

$$L = \int_{\alpha}^{\beta} \sqrt{r^2 + \left(\frac{dr}{d\theta}\right)^2}\, d\theta = \int_0^{2\pi} \sqrt{(1 + \cos\theta)^2 + (-\sin\theta)^2}\, d\theta$$

$$= \sqrt{2} \int_0^{2\pi} \sqrt{1 + \cos\theta}\, d\theta$$

$$= 2 \int_0^{2\pi} \sqrt{\cos^2 \tfrac{1}{2}\theta}\, d\theta \qquad \text{Identity (45) of Appendix A}$$

$$= 2 \int_0^{2\pi} \left|\cos \tfrac{1}{2}\theta\right| d\theta$$

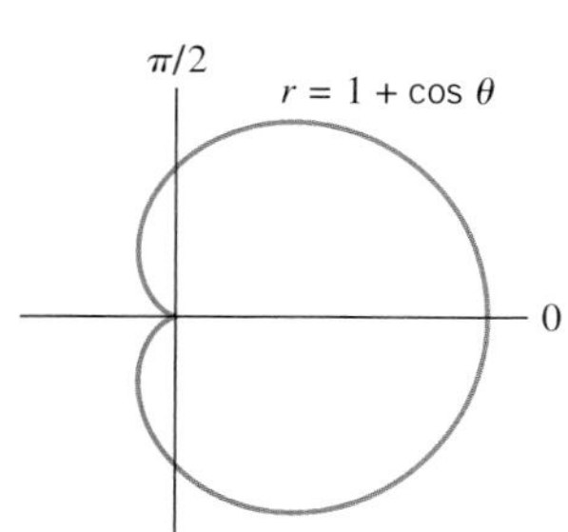

Figure 11.2.11

Since $\cos \frac{1}{2}\theta$ changes sign at π, we must split the last integral into the sum of two integrals: the integral from 0 to π plus the integral from π to 2π. However, the integral from π to 2π is equal to the integral from 0 to π, since the cardioid is symmetric about the polar axis (Figure 11.2.11). Thus,

$$L = 2 \int_0^{2\pi} \left|\cos \tfrac{1}{2}\theta\right| d\theta = 4 \int_0^{\pi} \cos \tfrac{1}{2}\theta\, d\theta = 8 \sin \tfrac{1}{2}\theta\Big]_0^{\pi} = 8 \blacktriangleleft$$

✔ QUICK CHECK EXERCISES 11.2 *(See page 740 for answers.)*

1. (a) To find dy/dx directly from the parametric equations

$$x = f(t), \quad y = g(t)$$

we can use the formula $dy/dx =$ ________.

(b) Find dy/dx and d^2y/dx^2 directly from the parametric equations $x = \cos^2 t$, $y = \sin^2 t$ $(0 < t < \pi/2)$.

2. (a) To obtain dy/dx directly from the polar equation $r = f(\theta)$, we can use the formula

$$\frac{dy}{dx} = \frac{dy/d\theta}{dx/d\theta} = \text{\_\_\_\_\_\_\_\_}$$

(b) Use the formula in part (a) to find dy/dx directly from the polar equation $r = \csc\theta$.

3. (a) What conditions on $f(\theta_0)$ and $f'(\theta_0)$ guarantee that the line $\theta = \theta_0$ is tangent to the polar curve $r = f(\theta)$ at the origin?
(b) What are the values of θ_0 in $[0, 2\pi]$ at which the lines $\theta = \theta_0$ are tangent at the origin to the four-petal rose $r = \cos 2\theta$?

4. (a) To find the arc length L of the polar curve $r = f(\theta)$ $(\alpha \le \theta \le \beta)$, we can use the formula $L =$ ______.
(b) The polar curve $r = \sec\theta$ $(0 \le \theta \le \pi/4)$ has arc length $L =$ ______.

EXERCISE SET 11.2 Graphing Utility C CAS

FOCUS ON CONCEPTS

1. (a) Find the slope of the tangent line to the parametric curve $x = t/2$, $y = t^2 + 1$ at $t = -1$ and at $t = 1$ without eliminating the parameter.
(b) Check your answers in part (a) by eliminating the parameter and differentiating an appropriate function of x.

2. (a) Find the slope of the tangent line to the parametric curve $x = 3\cos t$, $y = 4\sin t$ at $t = \pi/4$ and at $t = 7\pi/4$ without eliminating the parameter.
(b) Check your answers in part (a) by eliminating the parameter and differentiating an appropriate function of x.

3. For the parametric curve in Exercise 1, make a conjecture about the sign of d^2y/dx^2 at $t = -1$ and at $t = 1$, and confirm your conjecture without eliminating the parameter.

4. For the parametric curve in Exercise 2, make a conjecture about the sign of d^2y/dx^2 at $t = \pi/4$ and at $t = 7\pi/4$, and confirm your conjecture without eliminating the parameter.

5–10 Find dy/dx and d^2y/dx^2 at the given point without eliminating the parameter.

5. $x = \sqrt{t}$, $y = 2t + 4$; $t = 1$
6. $x = \frac{1}{2}t^2 + 1$, $y = \frac{1}{3}t^3 - t$; $t = 2$
7. $x = \sec t$, $y = \tan t$; $t = \pi/3$
8. $x = \sinh t$, $y = \cosh t$; $t = 0$
9. $x = \theta + \cos\theta$, $y = 1 + \sin\theta$; $\theta = \pi/6$
10. $x = \cos\phi$, $y = 3\sin\phi$; $\phi = 5\pi/6$

11. (a) Find the equation of the tangent line to the curve
$$x = e^t, \quad y = e^{-t}$$
at $t = 1$ without eliminating the parameter.
(b) Find the equation of the tangent line in part (a) by eliminating the parameter.

12. (a) Find the equation of the tangent line to the curve
$$x = 2t + 4, \quad y = 8t^2 - 2t + 4$$
at $t = 1$ without eliminating the parameter.
(b) Find the equation of the tangent line in part (a) by eliminating the parameter.

13–14 Find all values of t at which the parametric curve has (a) a horizontal tangent line and (b) a vertical tangent line.

13. $x = 2\sin t$, $y = 4\cos t$ $\quad (0 \le t \le 2\pi)$
14. $x = 2t^3 - 15t^2 + 24t + 7$, $y = t^2 + t + 1$

FOCUS ON CONCEPTS

15. The Lissajous curve
$$x = \sin t, \quad y = \sin 2t \qquad (0 \le t \le 2\pi)$$
crosses itself at the origin. (See the accompanying figure.) Find equations for the two tangent lines at the origin.

16. The ***prolate cycloid***
$$x = 2 - \pi\cos t, \quad y = 2t - \pi\sin t \qquad (-\pi \le t \le \pi)$$
crosses itself at a point on the x-axis. (See the accompanying figure.) Find equations for the two tangent lines at that point.

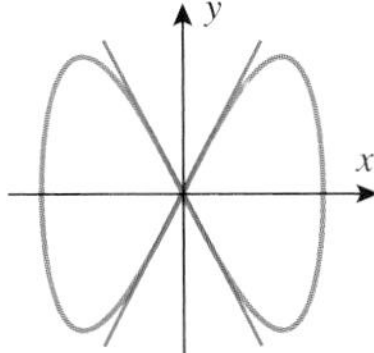

Figure Ex-15

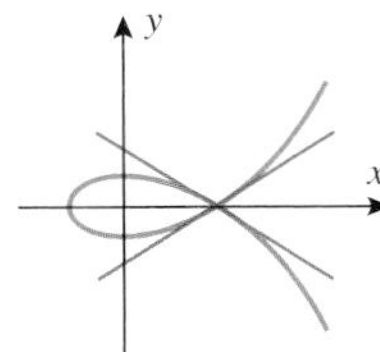

Figure Ex-16

17. Show that the curve $x = t^2$, $y = t^3 - 4t$ intersects itself at the point $(4, 0)$, and find equations for the two tangent lines to the curve at the point of intersection.

18. Show that the curve with parametric equations
$$x = t^2 - 3t + 5, \quad y = t^3 + t^2 - 10t + 9$$
intersects itself at the point $(3, 1)$, and find equations for the two tangent lines to the curve at the point of intersection.

19. (a) Use a graphing utility to generate the graph of the parametric curve
$$x = \cos^3 t, \quad y = \sin^3 t \qquad (0 \le t \le 2\pi)$$
and make a conjecture about the values of t at which singular points occur.

(b) Confirm your conjecture in part (a) by calculating appropriate derivatives.

20. (a) At what values of θ would you expect the cycloid in Figure 1.7.14 to have singular points?
(b) Confirm your answer in part (a) by calculating appropriate derivatives.

21–26 Find the slope of the tangent line to the polar curve for the given value of θ.

21. $r = 2\sin\theta;\ \theta = \pi/6$ **22.** $r = 1 + \cos\theta;\ \theta = \pi/2$

23. $r = 1/\theta;\ \theta = 2$ **24.** $r = a\sec 2\theta;\ \theta = \pi/6$

25. $r = \sin 3\theta;\ \theta = \pi/4$ **26.** $r = 4 - 3\sin\theta;\ \theta = \pi$

27–28 Calculate the slopes of the tangent lines indicated in the accompanying figures.

27. $r = 2 + 2\sin\theta$ **28.** $r = 1 - 2\sin\theta$

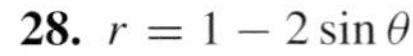

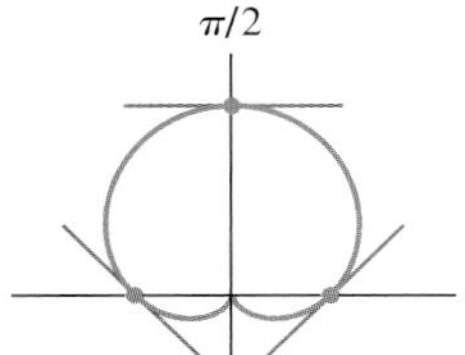

Figure Ex-27

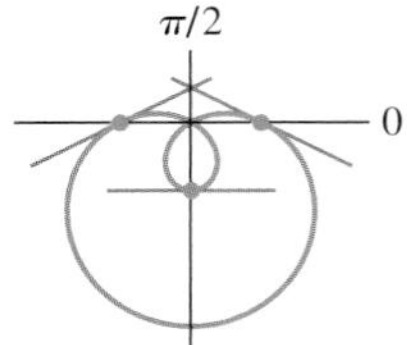

Figure Ex-28

29–30 Find polar coordinates of all points at which the polar curve has a horizontal or a vertical tangent line.

29. $r = a(1 + \cos\theta)$ **30.** $r = a\sin\theta$

31–32 Use a graphing utility to make a conjecture about the number of points on the polar curve at which there is a horizontal tangent line, and confirm your conjecture by finding appropriate derivatives.

 **31.** $r = \sin\theta\cos^2\theta$ **32.** $r = 1 - 2\sin\theta$

33–38 Sketch the polar curve and find polar equations of the tangent lines to the curve at the pole.

33. $r = 2\cos 3\theta$ **34.** $r = 4\sin\theta$ **35.** $r = 4\sqrt{\cos 2\theta}$

36. $r = \sin 2\theta$ **37.** $r = 1 - 2\cos\theta$ **38.** $r = 2\theta$

39–44 Use Formula (8) to calculate the arc length of the polar curve.

39. The entire circle $r = a$

40. The entire circle $r = 2a\cos\theta$

41. The entire cardioid $r = a(1 - \cos\theta)$

42. $r = \sin^2(\theta/2)$ from $\theta = 0$ to $\theta = \pi$

43. $r = e^{3\theta}$ from $\theta = 0$ to $\theta = 2$

44. $r = \sin^3(\theta/3)$ from $\theta = 0$ to $\theta = \pi/2$

45. (a) What is the slope of the tangent line at time t to the trajectory of the paper airplane in Example 2?
(b) What was the airplane's approximate angle of inclination when it crashed into the wall?

46. Suppose that a bee follows the trajectory

$$x = t - 2\cos t, \quad y = 2 - 2\sin t \quad (0 \le t \le 10)$$

(a) At what times was the bee flying horizontally?
(b) At what times was the bee flying vertically?

47. (a) Show that the arc length of one petal of the rose $r = \cos n\theta$ is given by

$$2\int_0^{\pi/(2n)} \sqrt{1 + (n^2 - 1)\sin^2 n\theta}\, d\theta$$

(b) Use the numerical integration capability of a calculating utility to approximate the arc length of one petal of the four-petal rose $r = \cos 2\theta$.
(c) Use the numerical integration capability of a calculating utility to approximate the arc length of one petal of the n-petal rose $r = \cos n\theta$ for $n = 2, 3, 4, \ldots, 20$; then make a conjecture about the limit of these arc lengths as $n \to +\infty$.

48. (a) Sketch the spiral $r = e^{-\theta/8}$ $(0 \le \theta < +\infty)$.
(b) Find an improper integral for the total arc length of the spiral.
(c) Show that the integral converges and find the total arc length of the spiral.

49–54 If $f'(t)$ and $g'(t)$ are continuous functions, and if no segment of the curve

$$x = f(t), \quad y = g(t) \quad (a \le t \le b)$$

is traced more than once, then it can be shown that the area of the surface generated by revolving this curve about the x-axis is

$$S = \int_a^b 2\pi y \sqrt{\left(\frac{dx}{dt}\right)^2 + \left(\frac{dy}{dt}\right)^2}\, dt$$

and the area of the surface generated by revolving the curve about the y-axis is

$$S = \int_a^b 2\pi x \sqrt{\left(\frac{dx}{dt}\right)^2 + \left(\frac{dy}{dt}\right)^2}\, dt$$

[The derivations are similar to those used to obtain Formulas (4) and (5) in Section 6.5.] Use the formulas above in these exercises.

49. Find the area of the surface generated by revolving $x = t^2$, $y = 3t$ $(0 \le t \le 2)$ about the x-axis.

50. Find the area of the surface generated by revolving the curve $x = e^t\cos t$, $y = e^t\sin t$ $(0 \le t \le \pi/2)$ about the x-axis.

51. Find the area of the surface generated by revolving the curve $x = \cos^2 t$, $y = \sin^2 t$ $(0 \le t \le \pi/2)$ about the y-axis.

52. Find the area of the surface generated by revolving $x = 6t$, $y = 4t^2$ $(0 \le t \le 1)$ about the y-axis.

53. By revolving the semicircle

$$x = r\cos t, \quad y = r\sin t \quad (0 \le t \le \pi)$$

about the x-axis, show that the surface area of a sphere of radius r is $4\pi r^2$.

54. The equations

$$x = a\phi - a\sin\phi, \quad y = a - a\cos\phi \quad (0 \le \phi \le 2\pi)$$

represent one arch of a cycloid. Show that the surface area generated by revolving this curve about the x-axis is given by $S = 64\pi a^2/3$.

FOCUS ON CONCEPTS

55. As illustrated in the accompanying figure, suppose that a rod with one end fixed at the pole of a polar coordinate system rotates counterclockwise at the constant rate of 1 rad/s. At time $t = 0$ a bug on the rod is 10 mm from the pole and is moving outward along the rod at the constant speed of 2 mm/s.

(a) Find an equation of the form $r = f(\theta)$ for the path of motion of the bug, assuming that $\theta = 0$ when $t = 0$.

(b) Find the distance the bug travels along the path in part (a) during the first 5 s. Round your answer to the nearest tenth of a millimeter.

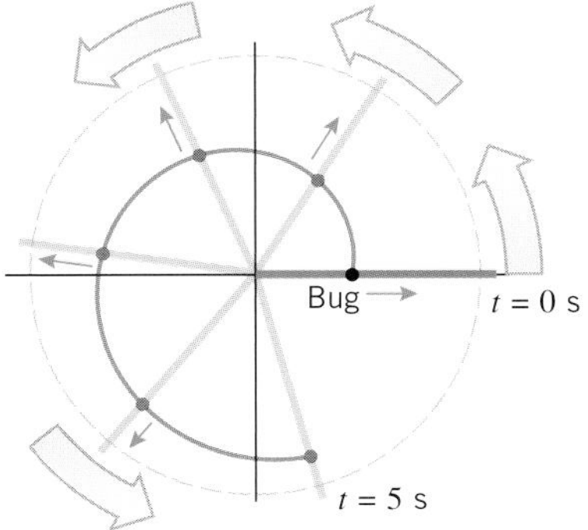

Figure Ex-55

56. Use Formula (6) of Section 6.4 to derive Formula (8).

57. The amusement park rides illustrated in the accompanying figure consist of two connected rotating arms of length 1—an inner arm that rotates counterclockwise at 1 radian per second and an outer arm that can be programmed to rotate either clockwise at 2 radians per second (the Scrambler ride) or counterclockwise at 2 radians per second (the Calypso ride). The center of the rider cage is at the end of the outer arm.

(a) Show that in the Scrambler ride the center of the cage has parametric equations

$$x = \cos t + \cos 2t, \quad y = \sin t - \sin 2t$$

(b) Find parametric equations for the center of the cage in the Calypso ride, and use a graphing utility to confirm that the center traces the curve shown in the accompanying figure.

(c) Do you think that a rider travels the same distance in one revolution of the Scrambler ride as in one revolution of the Calypso ride? Justify your conclusion.

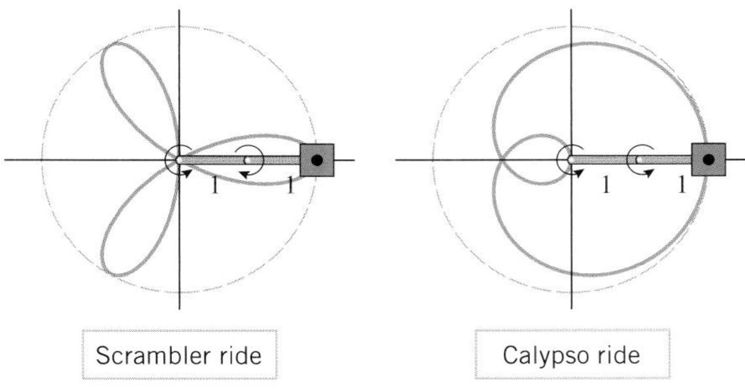

Figure Ex-57

58. Use a graphing utility to explore the effect of changing the rotation rates and the arm lengths in Exercise 57.

59. (a) If a thread is unwound from a fixed circle while being held taut (i.e., tangent to the circle), then the end of the thread traces a curve called an ***involute of a circle***. Show that if the circle is centered at the origin, has radius a, and the end of the thread is initially at the point $(a, 0)$, then the involute can be expressed parametrically as

$$x = a(\cos\theta + \theta\sin\theta), \quad y = a(\sin\theta - \theta\cos\theta)$$

where θ is the angle shown in part (a) of the accompanying figure.

(b) Assuming that the dog in part (b) of the accompanying figure unwinds its leash while keeping it taut, for what values of θ in the interval $0 \le \theta \le 2\pi$ will the dog be walking North? South? East? West?

(c) Use a graphing utility to generate the curve traced by the dog, and show that it is consistent with your answer in part (b).

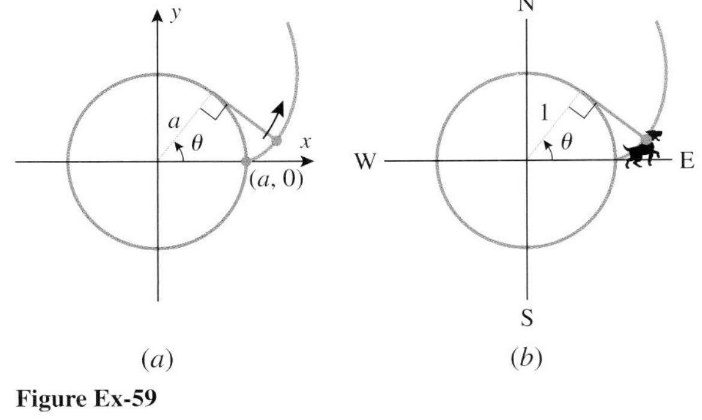

Figure Ex-59

60. Recall from Section 7.6 that the Fresnel sine and cosine functions are defined as

$$S(x) = \int_0^x \sin\left(\frac{\pi t^2}{2}\right) dt \quad \text{and} \quad C(x) = \int_0^x \cos\left(\frac{\pi t^2}{2}\right) dt$$

The following parametric curve, which is used to study amplitudes of light waves in optics, is called a ***clothoid*** or ***Cornu spiral*** in honor of the French scientist Marie Alfred Cornu (1841–1902):

$$\begin{aligned} x = C(t) &= \int_0^t \cos\left(\frac{\pi u^2}{2}\right) du \\ y = S(t) &= \int_0^t \sin\left(\frac{\pi u^2}{2}\right) du \end{aligned} \qquad (-\infty < t < +\infty)$$

(a) Use a CAS to graph the Cornu spiral.

(b) Describe the behavior of the spiral as $t \to +\infty$ and as $t \to -\infty$.

(c) Find the arc length of the spiral for $-1 \le t \le 1$.

61. As illustrated in the accompanying figure, let $P(r, \theta)$ be a point on the polar curve $r = f(\theta)$, let ψ be the smallest counterclockwise angle from the extended radius OP to the tangent line at P, and let ϕ be the angle of inclination of the tangent line. Derive the formula

$$\tan\psi = \frac{r}{dr/d\theta}$$

by substituting $\tan\phi$ for dy/dx in Formula (7) and applying the trigonometric identity

$$\tan(\phi - \theta) = \frac{\tan\phi - \tan\theta}{1 + \tan\phi\tan\theta}$$

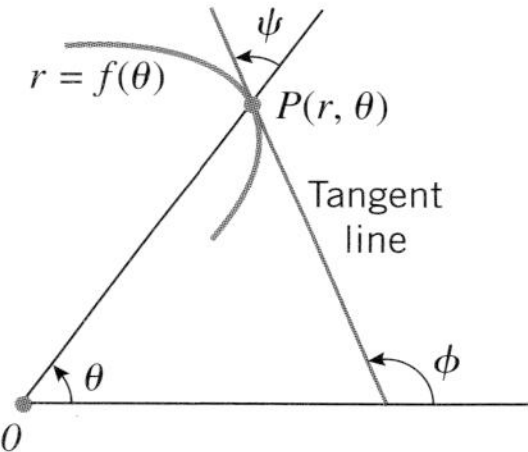

Figure Ex-61

62–63 Use the formula for ψ obtained in Exercise 61.

62. (a) Use the trigonometric identity

$$\tan\frac{\theta}{2} = \frac{1 - \cos\theta}{\sin\theta}$$

to show that if (r, θ) is a point on the cardioid

$$r = 1 - \cos\theta \qquad (0 \le \theta < 2\pi)$$

then $\psi = \theta/2$.

(b) Sketch the cardioid and show the angle ψ at the points where the cardioid crosses the y-axis.

(c) Find the angle ψ at the points where the cardioid crosses the y-axis.

63. Show that for a logarithmic spiral $r = ae^{b\theta}$, the angle from the radial line to the tangent line is constant along the spiral (see the accompanying figure). [*Note:* For this reason, logarithmic spirals are sometimes called ***equiangular spirals***.]

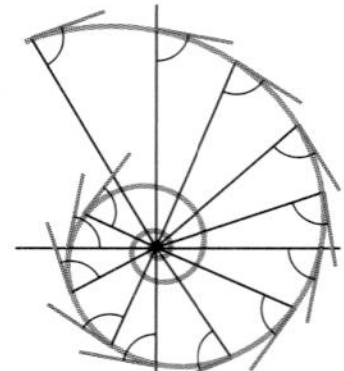

Figure Ex-63

✔QUICK CHECK ANSWERS 11.2

1. (a) $\dfrac{dy/dt}{dx/dt} = \dfrac{g'(t)}{f'(t)}$ (b) $\dfrac{dy}{dx} = -1,\ \dfrac{d^2y}{dx^2} = 0$ 2. (a) $\dfrac{r\cos\theta + \sin\theta\dfrac{dr}{d\theta}}{-r\sin\theta + \cos\theta\dfrac{dr}{d\theta}}$ (b) $\dfrac{dy}{dx} = 0$

3. (a) $f(\theta_0) = 0,\ f'(\theta_0) \neq 0$ (b) $\theta_0 = \dfrac{\pi}{4}, \dfrac{3\pi}{4}, \dfrac{5\pi}{4}, \dfrac{7\pi}{4}$ 4. (a) $\displaystyle\int_\alpha^\beta \sqrt{r^2 + \left(\frac{dr}{d\theta}\right)^2}\, d\theta$ (b) 1

11.3 AREA IN POLAR COORDINATES

In this section we will show how to find areas of regions that are bounded by polar curves.

AREA IN POLAR COORDINATES

We begin our investigation of area in polar coordinates with a simple case.

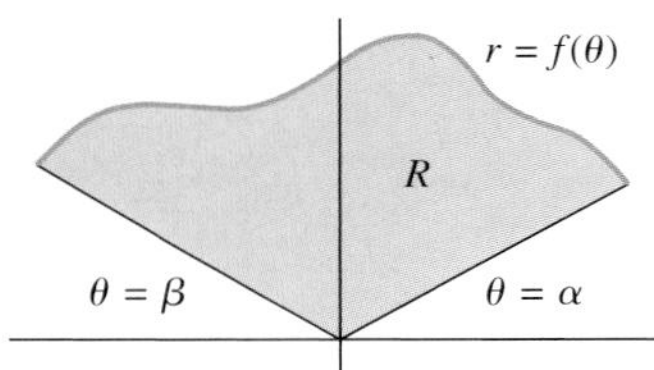

Figure 11.3.1

11.3.1 AREA PROBLEM IN POLAR COORDINATES. Suppose that α and β are angles that satisfy the condition

$$\alpha < \beta \le \alpha + 2\pi$$

and suppose that $f(\theta)$ is continuous and nonnegative for $\alpha \le \theta \le \beta$. Find the area of the region R enclosed by the polar curve $r = f(\theta)$ and the rays $\theta = \alpha$ and $\theta = \beta$ (Figure 11.3.1).

In rectangular coordinates we obtained areas under curves by dividing the region into an increasing number of vertical strips, approximating the strips by rectangles, and taking a limit. In polar coordinates rectangles are clumsy to work with, and it is better to divide the region into ***wedges*** by using rays

$$\theta = \theta_1,\ \theta = \theta_2, \ldots,\ \theta = \theta_{n-1}$$

such that

$$\alpha < \theta_1 < \theta_2 < \cdots < \theta_{n-1} < \beta$$

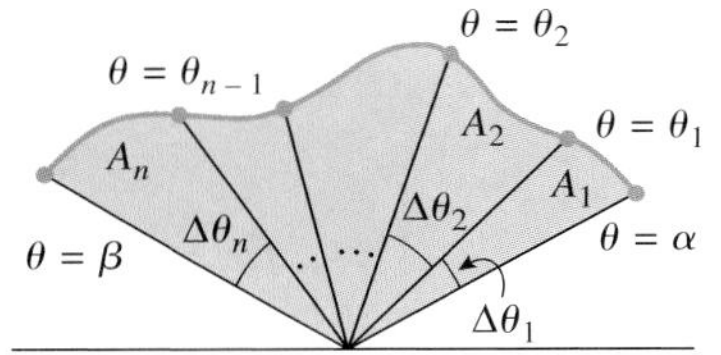

Figure 11.3.2

(Figure 11.3.2). As shown in that figure, the rays divide the region R into n wedges with areas $A_1, A_2, \ldots, A_n$ and central angles $\Delta\theta_1, \Delta\theta_2, \ldots, \Delta\theta_n$. The area of the entire region can be written as

$$A = A_1 + A_2 + \cdots + A_n = \sum_{k=1}^{n} A_k \tag{1}$$

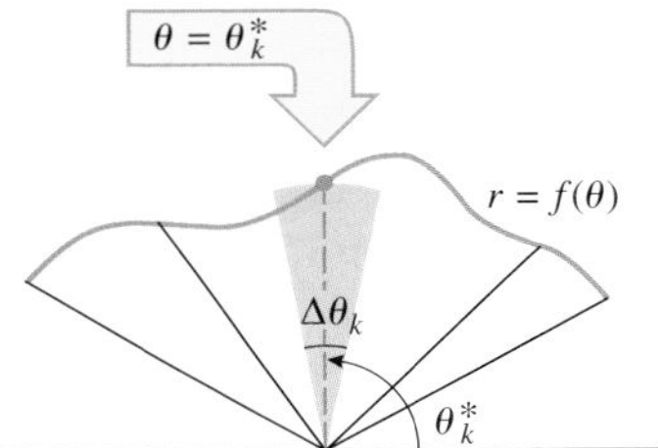

Figure 11.3.3

If $\Delta\theta_k$ is small, then we can approximate the area A_k of the kth wedge by the area of a sector with central angle $\Delta\theta_k$ and radius $f(\theta_k^*)$, where $\theta = \theta_k^*$ is any ray that lies in the kth wedge (Figure 11.3.3). Thus, from (1) and Formula (5) of Appendix A for the area of a sector, we obtain

$$A = \sum_{k=1}^{n} A_k \approx \sum_{k=1}^{n} \tfrac{1}{2}[f(\theta_k^*)]^2 \Delta\theta_k \tag{2}$$

If we now increase n in such a way that $\max \Delta\theta_k \to 0$, then the sectors will become better and better approximations of the wedges and it is reasonable to expect that (2) will approach the exact value of the area A (Figure 11.3.4); that is,

$$A = \lim_{\max \Delta\theta_k \to 0} \sum_{k=1}^{n} \tfrac{1}{2}[f(\theta_k^*)]^2 \Delta\theta_k = \int_{\alpha}^{\beta} \tfrac{1}{2}[f(\theta)]^2\, d\theta$$

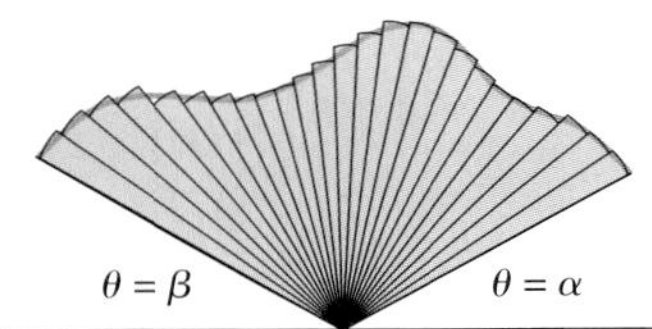

Figure 11.3.4

Note that the discussion above can easily be adapted to the case where $f(\theta)$ is nonpositive for $\alpha \le \theta \le \beta$. We summarize this result below.

11.3.2 AREA IN POLAR COORDINATES. If α and β are angles that satisfy the condition

$$\alpha < \beta \le \alpha + 2\pi$$

and if $f(\theta)$ is continuous and either nonnegative or nonpositive for $\alpha \le \theta \le \beta$, then the area A of the region R enclosed by the polar curve $r = f(\theta)$ $(\alpha \le \theta \le \beta)$ and the lines $\theta = \alpha$ and $\theta = \beta$ is

$$A = \int_{\alpha}^{\beta} \tfrac{1}{2}[f(\theta)]^2\, d\theta = \int_{\alpha}^{\beta} \tfrac{1}{2} r^2\, d\theta \tag{3}$$

The hardest part of applying (3) is determining the limits of integration. This can be done as follows:

Area in Polar Coordinates: Limits of Integration

Step 1. Sketch the region R whose area is to be determined.

Step 2. Draw an arbitrary "radial line" from the pole to the boundary curve $r = f(\theta)$.

Step 3. Ask, "Over what interval of values must θ vary in order for the radial line to sweep out the region R?"

Step 4. Your answer in Step 3 will determine the lower and upper limits of integration.

▶ **Example 1** Find the area of the region in the first quadrant that is within the cardioid $r = 1 - \cos\theta$.

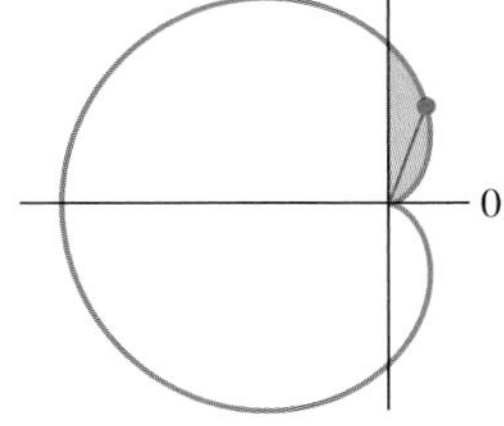

The shaded region is swept out by the radial line as θ varies from 0 to $\pi/2$.

Figure 11.3.5

Solution. The region and a typical radial line are shown in Figure 11.3.5. For the radial line to sweep out the region, θ must vary from 0 to $\pi/2$. Thus, from (3) with $\alpha = 0$ and $\beta = \pi/2$, we obtain

$$A = \int_0^{\pi/2} \frac{1}{2}r^2\,d\theta = \frac{1}{2}\int_0^{\pi/2} (1-\cos\theta)^2\,d\theta = \frac{1}{2}\int_0^{\pi/2} (1 - 2\cos\theta + \cos^2\theta)\,d\theta$$

With the help of the identity $\cos^2\theta = \frac{1}{2}(1+\cos 2\theta)$, this can be rewritten as

$$A = \frac{1}{2}\int_0^{\pi/2}\left(\frac{3}{2} - 2\cos\theta + \frac{1}{2}\cos 2\theta\right)d\theta = \frac{1}{2}\left[\frac{3}{2}\theta - 2\sin\theta + \frac{1}{4}\sin 2\theta\right]_0^{\pi/2} = \frac{3}{8}\pi - 1 \blacktriangleleft$$

▶ **Example 2** Find the entire area within the cardioid of Example 1.

Solution. For the radial line to sweep out the entire cardioid, θ must vary from 0 to 2π. Thus, from (3) with $\alpha = 0$ and $\beta = 2\pi$,

$$A = \int_0^{2\pi} \frac{1}{2}r^2\,d\theta = \frac{1}{2}\int_0^{2\pi} (1-\cos\theta)^2\,d\theta$$

If we proceed as in Example 1, this reduces to

$$A = \frac{1}{2}\int_0^{2\pi}\left(\frac{3}{2} - 2\cos\theta + \frac{1}{2}\cos 2\theta\right)d\theta = \frac{3\pi}{2}$$

Alternative Solution. Since the cardioid is symmetric about the x-axis, we can calculate the portion of the area above the x-axis and double the result. In the portion of the cardioid above the x-axis, θ ranges from 0 to π, so that

$$A = 2\int_0^{\pi} \frac{1}{2}r^2\,d\theta = \int_0^{\pi} (1-\cos\theta)^2\,d\theta = \frac{3\pi}{2} \blacktriangleleft$$

USING SYMMETRY

Although Formula (3) is applicable if $r = f(\theta)$ is negative, area computations can sometimes be simplified by using symmetry to restrict the limits of integration to intervals where $r \geq 0$. This is illustrated in the next example.

▶ Example 3 Find the area of the region enclosed by the rose curve $r = \cos 2\theta$.

Solution. Referring to Figure 11.1.12 and using symmetry, the area in the first quadrant that is swept out for $0 \le \theta \le \pi/4$ is one-eighth of the total area inside the rose. Thus, from Formula (3)

$$\begin{aligned} A &= 8\int_0^{\pi/4} \frac{1}{2}r^2\,d\theta = 4\int_0^{\pi/4} \cos^2 2\theta\,d\theta \\ &= 4\int_0^{\pi/4} \frac{1}{2}(1+\cos 4\theta)\,d\theta = 2\int_0^{\pi/4}(1+\cos 4\theta)\,d\theta \\ &= 2\theta + \frac{1}{2}\sin 4\theta\Big]_0^{\pi/4} = \frac{\pi}{2} \end{aligned}$$ ◀

Sometimes the most natural way to satisfy the restriction $\alpha < \beta \le \alpha + 2\pi$ required by Formula (3) is to use a negative value for α. For example, suppose that we are interested in finding the area of the shaded region in Figure 11.3.6*a*. The first step would be to determine the intersections of the cardioid $r = 4 + 4\cos\theta$ and the circle $r = 6$, since this information is needed for the limits of integration. To find the points of intersection, we can equate the two expressions for r. This yields

$$4 + 4\cos\theta = 6 \quad \text{or} \quad \cos\theta = \frac{1}{2}$$

which is satisfied by the positive angles

$$\theta = \frac{\pi}{3} \quad \text{and} \quad \theta = \frac{5\pi}{3}$$

However, there is a problem here because the radial lines to the circle and cardioid do not sweep through the shaded region shown in Figure 11.3.6*b* as θ varies over the interval $\pi/3 \le \theta \le 5\pi/3$. There are two ways to circumvent this problem—one is to take advantage of the symmetry by integrating over the interval $0 \le \theta \le \pi/3$ and doubling the result, and the second is to use a negative lower limit of integration and integrate over the interval $-\pi/3 \le \theta \le \pi/3$ (Figure 11.3.6*c*). The two methods are illustrated in the next example.

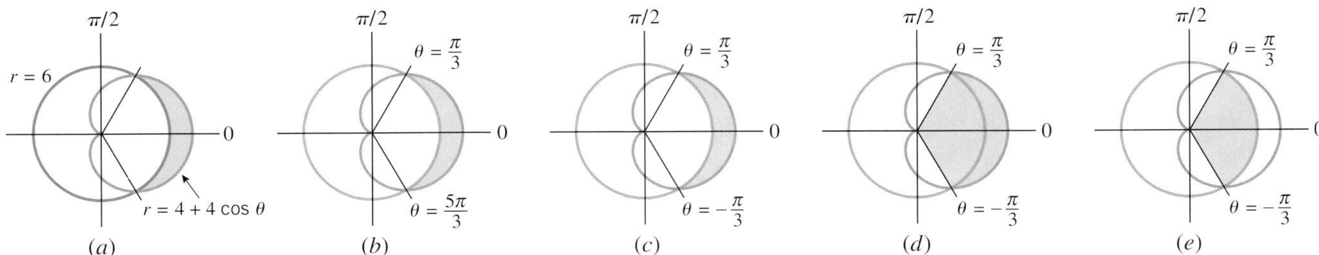

Figure 11.3.6

▶ Example 4 Find the area of the region that is inside of the cardioid $r = 4 + 4\cos\theta$ and outside of the circle $r = 6$.

Solution Using a Negative Angle. The area of the region can be obtained by subtracting the areas in Figures 11.3.6*d* and 11.3.6*e*:

$$\begin{aligned} A &= \int_{-\pi/3}^{\pi/3} \frac{1}{2}(4+4\cos\theta)^2\,d\theta - \int_{-\pi/3}^{\pi/3}\frac{1}{2}(6)^2\,d\theta \\ &= \int_{-\pi/3}^{\pi/3} \frac{1}{2}[(4+4\cos\theta)^2 - 36]\,d\theta = \int_{-\pi/3}^{\pi/3}(16\cos\theta + 8\cos^2\theta - 10)\,d\theta \\ &= \big[16\sin\theta + (4\theta + 2\sin 2\theta) - 10\theta\big]_{-\pi/3}^{\pi/3} = 18\sqrt{3} - 4\pi \end{aligned}$$

Area inside cardioid minus area inside circle.

Solution Using Symmetry. Using symmetry, we can calculate the area above the polar axis and double it. This yields (verify)

$$A = 2\int_0^{\pi/3} \frac{1}{2}[(4+4\cos\theta)^2 - 36]\,d\theta = 2(9\sqrt{3} - 2\pi) = 18\sqrt{3} - 4\pi$$

which agrees with the preceding result. ◄

INTERSECTIONS OF POLAR GRAPHS

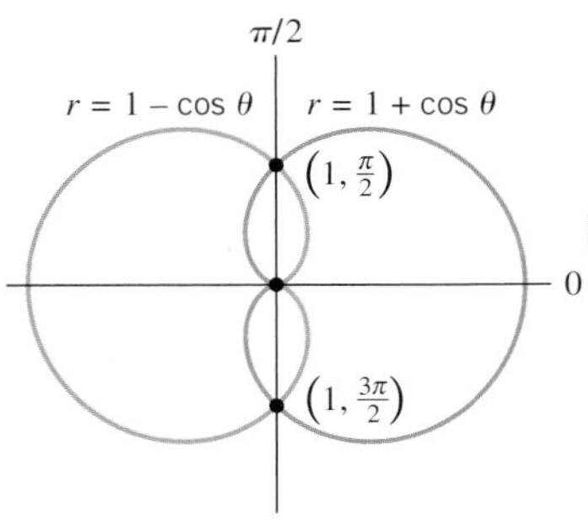

Figure 11.3.7

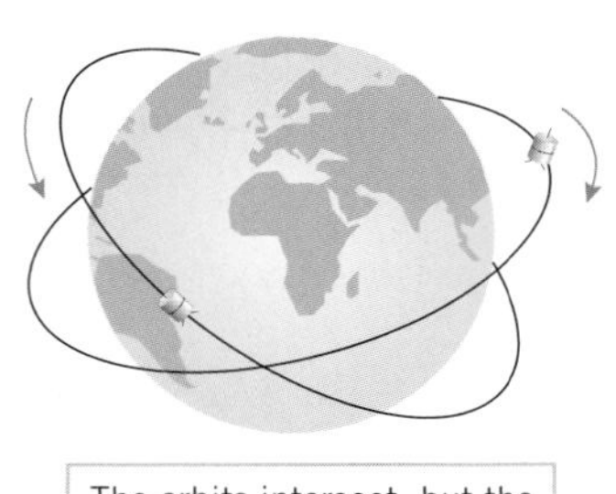

The orbits intersect, but the satellites do not collide.

Figure 11.3.8

In the last example we found the intersections of the cardioid and circle by equating their expressions for r and solving for θ. However, because a point can be represented in different ways in polar coordinates, this procedure will not always produce all of the intersections. For example, the cardioids

$$r = 1 - \cos\theta \quad \text{and} \quad r = 1 + \cos\theta \tag{4}$$

intersect at three points: the pole, the point $(1, \pi/2)$, and the point $(1, 3\pi/2)$ (Figure 11.3.7). Equating the right-hand sides of the equations in (4) yields $1 - \cos\theta = 1 + \cos\theta$ or $\cos\theta = 0$, so

$$\theta = \frac{\pi}{2} + k\pi, \quad k = 0, \pm 1, \pm 2, \ldots$$

Substituting any of these values in (4) yields $r = 1$, so that we have found only two distinct points of intersection, $(1, \pi/2)$ and $(1, 3\pi/2)$; the pole has been missed. This problem occurs because the two cardioids pass through the pole at different values of θ—the cardioid $r = 1 - \cos\theta$ passes through the pole at $\theta = 0$, and the cardioid $r = 1 + \cos\theta$ passes through the pole at $\theta = \pi$.

The situation with the cardioids is analogous to two satellites circling the Earth in intersecting orbits (Figure 11.3.8). The satellites will not collide unless they reach the same point at the same time. In general, when looking for intersections of polar curves, it is a good idea to graph the curves to determine how many intersections there should be.

✔QUICK CHECK EXERCISES 11.3 *(See page 746 for answers.)*

1. The area A enclosed by a nonnegative polar curve $r = f(\theta)$ $(\alpha \le \theta \le \beta)$ and the lines $\theta = \alpha$ and $\theta = \beta$ is given by the definite integral $A =$ ______.
2. Find the area of the circle $r = a$ by integration.
3. Write down, but do not evaluate, an integral for the area of each shaded region.

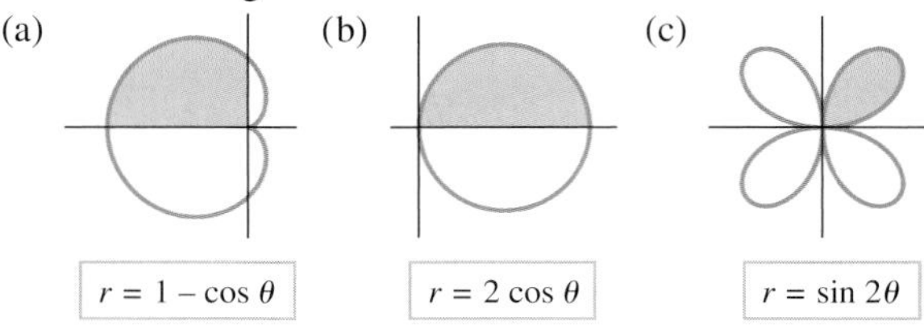

(d) $r = \theta$ (e) $r = 1 - \sin\theta$ (f) $r = \cos 2\theta$

4. Find the area of the shaded region in Quick Check Exercise 3(d).

EXERCISE SET 11.3

1. In each part, find the area of the circle by integration.
   (a) $r = 2a\sin\theta$ (b) $r = 2a\cos\theta$
2. (a) Show that $r = 2\sin\theta + 2\cos\theta$ is a circle.
   (b) Find the area of the circle using a geometric formula and then by integration.

3–8 Find the area of the region described.

3. The region that is enclosed by the cardioid $r = 2 + 2\sin\theta$.

4. The region in the first quadrant within the cardioid $r = 1 + \cos\theta$.

5. The region enclosed by the rose $r = 4\cos 3\theta$.

6. The region enclosed by the rose $r = 2\sin 2\theta$.

7. The region enclosed by the inner loop of the limaçon $r = 1 + 2\cos\theta$. [*Hint:* $r \le 0$ over the interval of integration.]

8. The region swept out by a radial line from the pole to the curve $r = 2/\theta$ as θ varies over the interval $1 \le \theta \le 3$.

9–12 Find the area of the shaded region.

9.

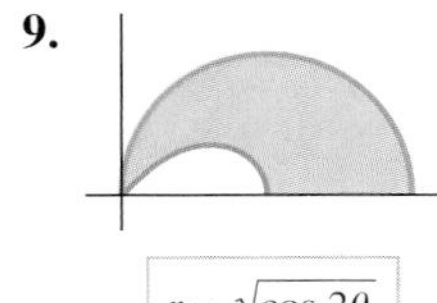

$r = \sqrt{\cos 2\theta}$
$r = 2\cos\theta$

10.

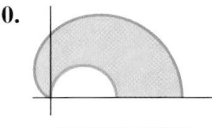

$r = 1 + \cos\theta$
$r = \cos\theta$

11.

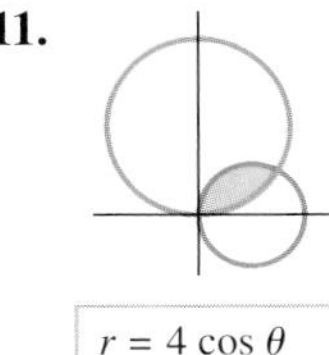

$r = 4\cos\theta$
$r = 4\sqrt{3}\sin\theta$

12.

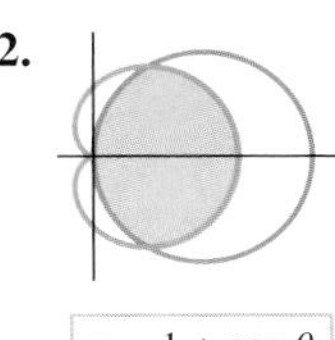

$r = 1 + \cos\theta$
$r = 3\cos\theta$

13–20 Find the area of the region described.

13. The region inside the circle $r = 3\sin\theta$ and outside the cardioid $r = 1 + \sin\theta$.

14. The region outside the cardioid $r = 2 - 2\cos\theta$ and inside the circle $r = 4$.

15. The region inside the cardioid $r = 2 + 2\cos\theta$ and outside the circle $r = 3$.

16. The region that is common to the circles $r = 2\cos\theta$ and $r = 2\sin\theta$.

17. The region between the loops of the limaçon $r = \frac{1}{2} + \cos\theta$.

18. The region inside the cardioid $r = 2 + 2\cos\theta$ and to the right of the line $r\cos\theta = \frac{3}{2}$.

19. The region inside the circle $r = 2$ and to the right of the line $r = \sqrt{2}\sec\theta$.

20. The region inside the rose $r = 2a\cos 2\theta$ and outside the circle $r = a\sqrt{2}$.

FOCUS ON CONCEPTS

21. (a) Find the error: The area that is inside the lemniscate $r^2 = a^2\cos 2\theta$ is

$$A = \int_0^{2\pi} \tfrac{1}{2}r^2\,d\theta = \int_0^{2\pi} \tfrac{1}{2}a^2\cos 2\theta\,d\theta$$
$$= \tfrac{1}{4}a^2\sin 2\theta\Big]_0^{2\pi} = 0$$

(b) Find the correct area.

(c) Find the area inside the lemniscate $r^2 = 4\cos 2\theta$ and outside the circle $r = \sqrt{2}$.

22. Find the area inside the curve $r^2 = \sin 2\theta$.

23. A radial line is drawn from the origin to the spiral $r = a\theta$ ($a > 0$ and $\theta \ge 0$). Find the area swept out during the second revolution of the radial line that was not swept out during the first revolution.

24. (a) In the discussion associated with Exercises 49–54 of Section 11.2, formulas were given for the area of the surface of revolution that is generated by revolving a parametric curve about the x-axis or y-axis. Use those formulas to derive the following formulas for the areas of the surfaces of revolution that are generated by revolving the portion of the polar curve $r = f(\theta)$ from $\theta = \alpha$ to $\theta = \beta$ about the polar axis and about the line $\theta = \pi/2$:

$$S = \int_\alpha^\beta 2\pi r\sin\theta\sqrt{r^2 + \left(\frac{dr}{d\theta}\right)^2}\,d\theta \qquad \text{About } \theta = 0$$

$$S = \int_\alpha^\beta 2\pi r\cos\theta\sqrt{r^2 + \left(\frac{dr}{d\theta}\right)^2}\,d\theta \qquad \text{About } \theta = \pi/2$$

(b) State conditions under which these formulas hold.

25–28 Sketch the surface, and use the formulas in Exercise 24 to find the surface area.

25. The surface generated by revolving the circle $r = \cos\theta$ about the line $\theta = \pi/2$.

26. The surface generated by revolving the spiral $r = e^\theta$ $(0 \le \theta \le \pi/2)$ about the line $\theta = \pi/2$.

27. The "apple" generated by revolving the upper half of the cardioid $r = 1 - \cos\theta$ $(0 \le \theta \le \pi)$ about the polar axis.

28. The sphere of radius a generated by revolving the semicircle $r = a$ in the upper half-plane about the polar axis.

C **29.** (a) Show that the Folium of Descartes $x^3 - 3xy + y^3 = 0$ can be expressed in polar coordinates as

$$r = \frac{3\sin\theta\cos\theta}{\cos^3\theta + \sin^3\theta}$$

(b) Use a CAS to show that the area inside of the loop is $\frac{3}{2}$ (Figure 4.1.3a).

C **30.** (a) What is the area that is enclosed by one petal of the rose $r = a\cos n\theta$ if n is an even integer?
(b) What is the area that is enclosed by one petal of the rose $r = a\cos n\theta$ if n is an odd integer?
(c) Use a CAS to show that the total area enclosed by the rose $r = a\cos n\theta$ is $\pi a^2/2$ if the number of petals is even. [*Hint:* See Exercise 81 of Section 11.1.]
(d) Use a CAS to show that the total area enclosed by the rose $r = a\cos n\theta$ is $\pi a^2/4$ if the number of petals is odd.

31. One of the most famous problems in Greek antiquity was "squaring the circle," that is, using a straightedge and compass to construct a square whose area is equal to that of a given circle. It was proved in the nineteenth century that no such construction is possible. However, show that the shaded areas in the accompanying figure are equal, thereby "squaring the crescent."

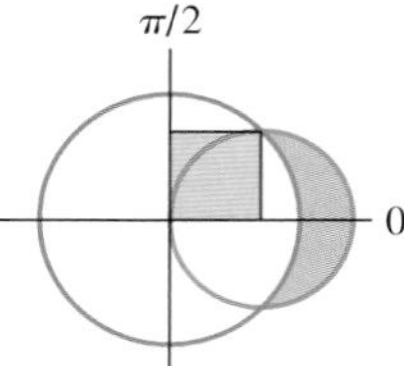

Figure Ex-31

32. Use a graphing utility to generate the polar graph of the equation $r = \cos 3\theta + 2$, and find the area that it encloses.

33. Use a graphing utility to generate the graph of the ***bifolium*** $r = 2\cos\theta\sin^2\theta$, and find the area of the upper loop.

✔QUICK CHECK ANSWERS 11.3

1. $\int_\alpha^\beta \frac{1}{2}[f(\theta)]^2\,d\theta = \int_\alpha^\beta \frac{1}{2}r^2\,d\theta$ 2. $\int_0^{2\pi} \frac{1}{2}a^2\,d\theta = \pi a^2$ 3. (a) $\int_{\pi/2}^{\pi} \frac{1}{2}(1-\cos\theta)^2\,d\theta$ (b) $\int_0^{\pi/2} 2\cos^2\theta\,d\theta$ (c) $\int_0^{\pi/2} \frac{1}{2}\sin^2 2\theta\,d\theta$ (d) $\int_0^{2\pi} \frac{1}{2}\theta^2\,d\theta$ (e) $\int_{-\pi/2}^{\pi/2} \frac{1}{2}(1-\sin\theta)^2\,d\theta$ (f) $\int_0^{\pi/4} \cos^2 2\theta\,d\theta$ 4. $\frac{4\pi^3}{3}$

11.4 CONIC SECTIONS IN CALCULUS

In this section we will discuss some of the basic geometric properties of parabolas, ellipses, and hyperbolas. These curves play an important role in calculus and also arise naturally in a broad range of applications in such fields as planetary motion, design of telescopes and antennas, geodetic positioning, and medicine, to name a few.

Some students may already be familiar with the material in this section, in which case it can be treated as a review. Instructors who want to spend some additional time on precalculus review may want to allocate more than one lecture on this material.

■ CONIC SECTIONS

Circles, ellipses, parabolas, and hyperbolas are called ***conic sections*** or ***conics*** because they can be obtained as intersections of a plane with a double-napped circular cone (Figure 11.4.1). If the plane passes through the vertex of the double-napped cone, then the intersection is a point, a pair of intersecting lines, or a single line. These are called ***degenerate conic sections***.

■ DEFINITIONS OF THE CONIC SECTIONS

Although we could derive properties of parabolas, ellipses, and hyperbolas by defining them as intersections with a double-napped cone, it will be better suited to calculus if we begin with equivalent definitions that are based on their geometric properties.

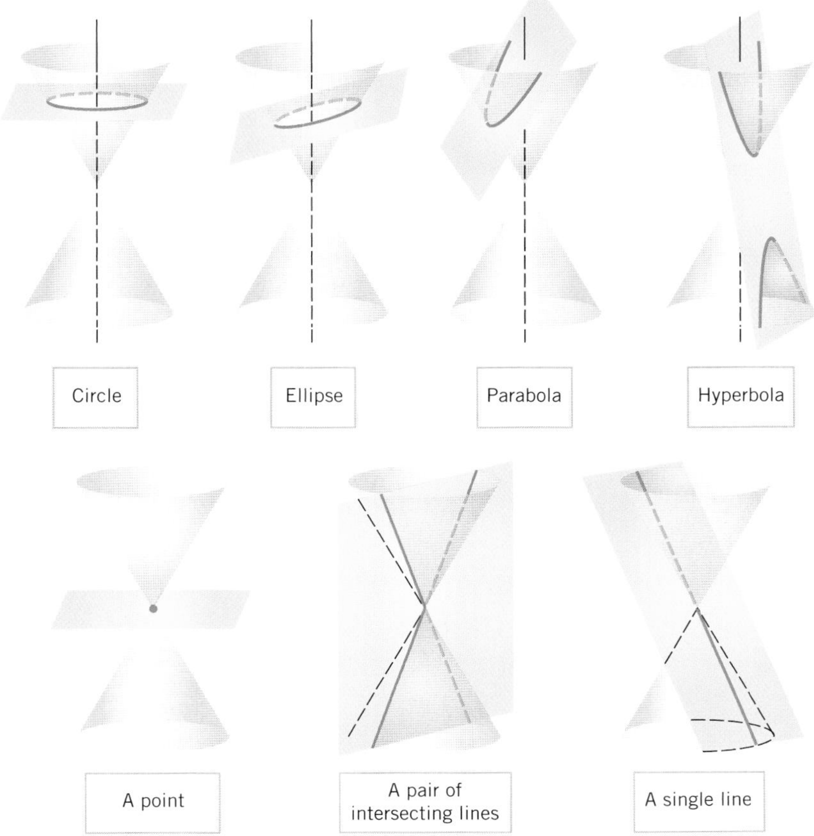

Figure 11.4.1

> **11.4.1 DEFINITION.** A ***parabola*** is the set of all points in the plane that are equidistant from a fixed line and a fixed point not on the line.

The line is called the ***directrix*** of the parabola, and the point is called the ***focus*** (Figure 11.4.2). A parabola is symmetric about the line that passes through the focus at right angles to the directrix. This line, called the ***axis*** or the ***axis of symmetry*** of the parabola, intersects the parabola at a point called the ***vertex***.

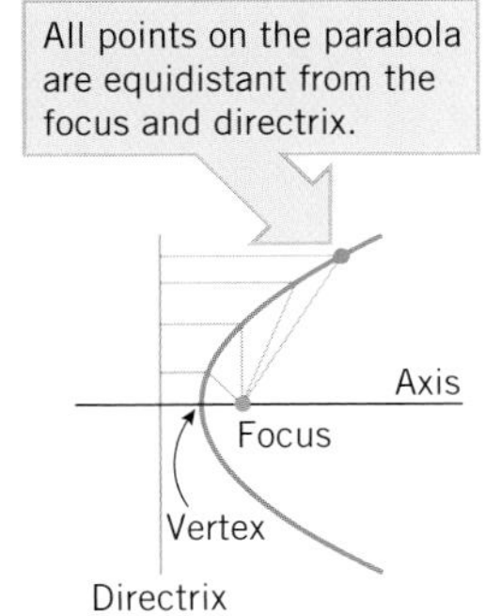

Figure 11.4.2

> **11.4.2 DEFINITION.** An ***ellipse*** is the set of all points in the plane, the sum of whose distances from two fixed points is a given positive constant that is greater than the distance between the fixed points.

The two fixed points are called the ***foci*** (plural of "focus") of the ellipse, and the midpoint of the line segment joining the foci is called the ***center*** (Figure 11.4.3a). To help visualize Definition 11.4.2, imagine that two ends of a string are tacked to the foci and a pencil traces a curve as it is held tight against the string (Figure 11.4.3b). The resulting curve will be an ellipse since the sum of the distances to the foci is a constant, namely, the total length of the string. Note that if the foci coincide, the ellipse reduces to a circle. For ellipses other than circles, the line segment through the foci and across the ellipse is called the ***major axis*** (Figure 11.4.3*c*), and the line segment across the ellipse, through the center, and perpendicular to the major axis is called the ***minor axis***. The endpoints of the major axis are called ***vertices***.

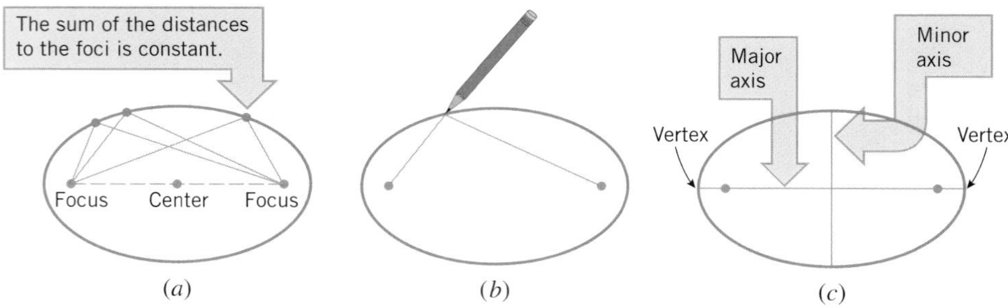

Figure 11.4.3

11.4.3 DEFINITION. A ***hyperbola*** is the set of all points in the plane, the difference of whose distances from two fixed distinct points is a given positive constant that is less than the distance between the fixed points.

The two fixed points are called the ***foci*** of the hyperbola, and the term "difference" that is used in the definition is understood to mean the distance to the farther focus minus the distance to the closer focus. As a result, the points on the hyperbola form two ***branches***, each "wrapping around" the closer focus (Figure 11.4.4*a*). The midpoint of the line segment joining the foci is called the ***center*** of the hyperbola, the line through the foci is called the ***focal axis***, and the line through the center that is perpendicular to the focal axis is called the ***conjugate axis***. The hyperbola intersects the focal axis at two points called the ***vertices***.

Associated with every hyperbola is a pair of lines, called the ***asymptotes*** of the hyperbola. These lines intersect at the center of the hyperbola and have the property that as a point P moves along the hyperbola away from the center, the vertical distance between P and one of the asymptotes approaches zero (Figure 11.4.4*b*).

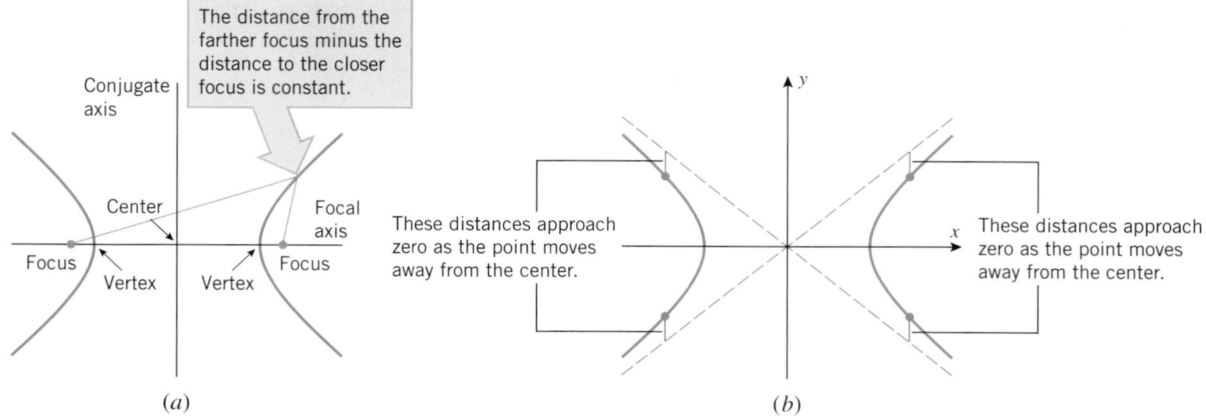

Figure 11.4.4

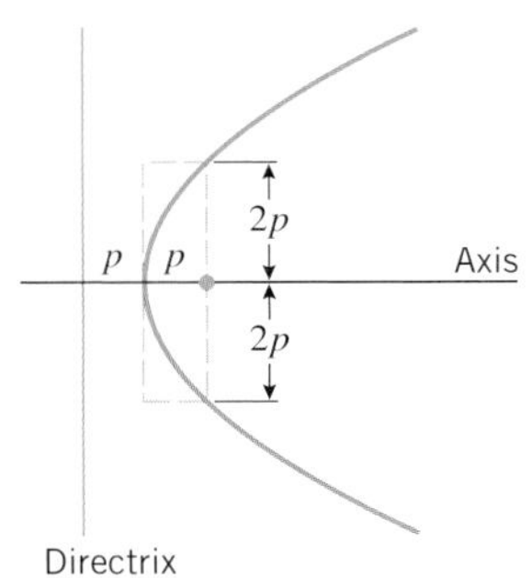

Figure 11.4.5

EQUATIONS OF PARABOLAS IN STANDARD POSITION

It is traditional in the study of parabolas to denote the distance between the focus and the vertex by p. The vertex is equidistant from the focus and the directrix, so the distance between the vertex and the directrix is also p; consequently, the distance between the focus and the directrix is $2p$ (Figure 11.4.5). As illustrated in that figure, the parabola passes through two of the corners of a box that extends from the vertex to the focus along the axis of symmetry and extends $2p$ units above and $2p$ units below the axis of symmetry.

The equation of a parabola is simplest if the vertex is the origin and the axis of symmetry is along the x-axis or y-axis. The four possible such orientations are shown in Figure 11.4.6. These are called the ***standard positions*** of a parabola, and the resulting equations are called the ***standard equations*** of a parabola.

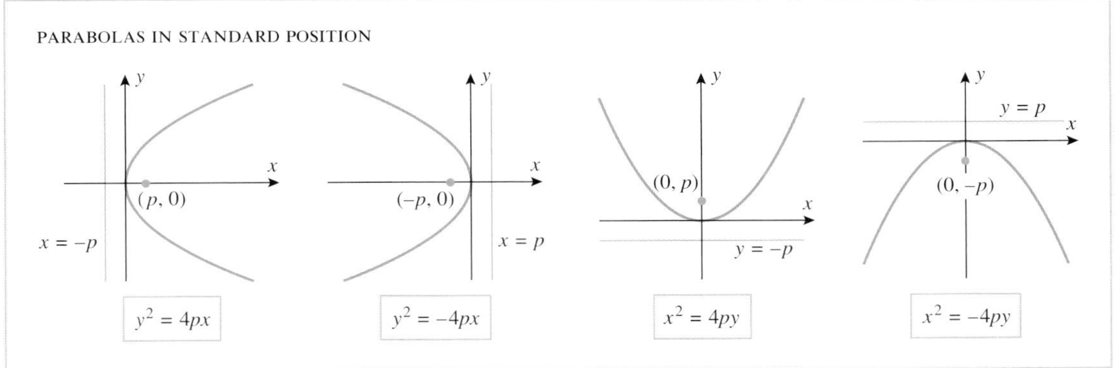

Figure 11.4.6

To illustrate how the equations in Figure 11.4.6 are obtained, we will derive the equation for the parabola with focus $(p, 0)$ and directrix $x = -p$. Let $P(x, y)$ be any point on the parabola. Since P is equidistant from the focus and directrix, the distances PF and PD in Figure 11.4.7 are equal; that is,

$$PF = PD \tag{1}$$

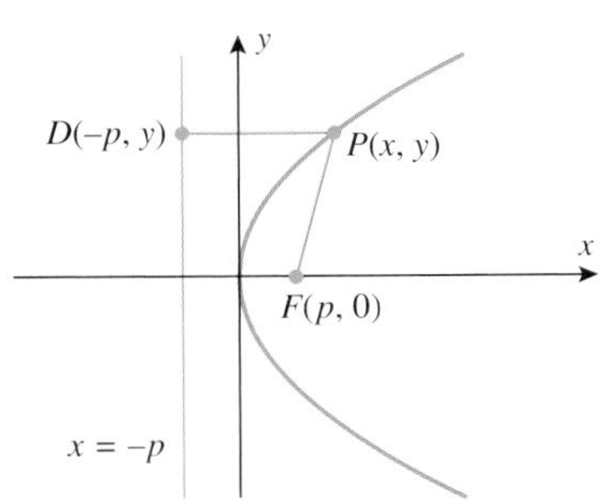

Figure 11.4.7

where $D(-p, y)$ is the foot of the perpendicular from P to the directrix. From the distance formula, the distances PF and PD are

$$PF = \sqrt{(x-p)^2 + y^2} \quad \text{and} \quad PD = \sqrt{(x+p)^2} \tag{2}$$

Substituting in (1) and squaring yields

$$(x-p)^2 + y^2 = (x+p)^2 \tag{3}$$

and after simplifying

$$y^2 = 4px \tag{4}$$

The derivations of the other equations in Figure 11.4.6 are similar.

A TECHNIQUE FOR SKETCHING PARABOLAS

Parabolas can be sketched from their *standard equations* using four basic steps:

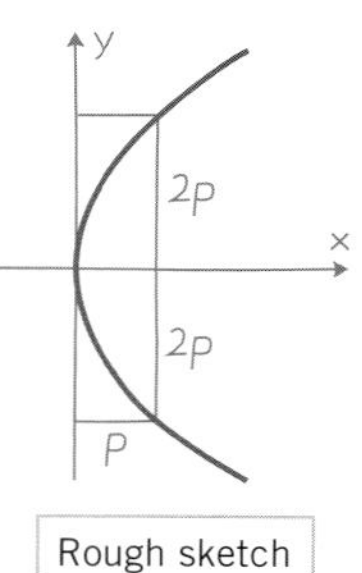

Figure 11.4.8

Sketching a Parabola from Its Standard Equation

Step 1. Determine whether the axis of symmetry is along the x-axis or the y-axis. Referring to Figure 11.4.6, the axis of symmetry is along the x-axis if the equation has a y^2-term, and it is along the y-axis if it has an x^2-term.

Step 2. Determine which way the parabola opens. If the axis of symmetry is along the x-axis, then the parabola opens to the right if the coefficient of x is positive, and it opens to the left if the coefficient is negative. If the axis of symmetry is along the y-axis, then the parabola opens up if the coefficient of y is positive, and it opens down if the coefficient is negative.

Step 3. Determine the value of p and draw a box extending p units from the origin along the axis of symmetry in the direction in which the parabola opens and extending $2p$ units on each side of the axis of symmetry.

Step 4. Using the box as a guide, sketch the parabola so that its vertex is at the origin and it passes through the corners of the box (Figure 11.4.8).

▶ **Example 1** Sketch the graphs of the parabolas

$$\text{(a) } x^2 = 12y \qquad \text{(b) } y^2 + 8x = 0$$

and show the focus and directrix of each.

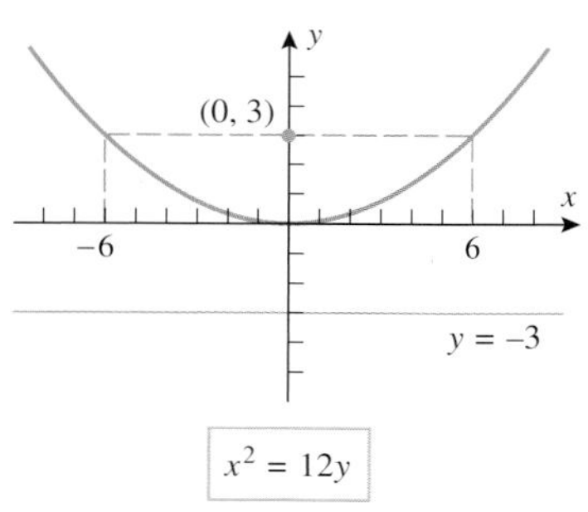

Figure 11.4.9

Solution (a). This equation involves x^2, so the axis of symmetry is along the y-axis, and the coefficient of y is positive, so the parabola opens upward. From the coefficient of y, we obtain $4p = 12$ or $p = 3$. Drawing a box extending $p = 3$ units up from the origin and $2p = 6$ units to the left and $2p = 6$ units to the right of the y-axis, then using corners of the box as a guide, yields the graph in Figure 11.4.9.

The focus is $p = 3$ units from the vertex along the axis of symmetry in the direction in which the parabola opens, so its coordinates are $(0, 3)$. The directrix is perpendicular to the axis of symmetry at a distance of $p = 3$ units from the vertex on the opposite side from the focus, so its equation is $y = -3$.

Solution (b). We first rewrite the equation in the standard form

$$y^2 = -8x$$

This equation involves y^2, so the axis of symmetry is along the x-axis, and the coefficient of x is negative, so the parabola opens to the left. From the coefficient of x we obtain $4p = 8$, so $p = 2$. Drawing a box extending $p = 2$ units left from the origin and $2p = 4$ units above and $2p = 4$ units below the x-axis, then using corners of the box as a guide, yields the graph in Figure 11.4.10. ◀

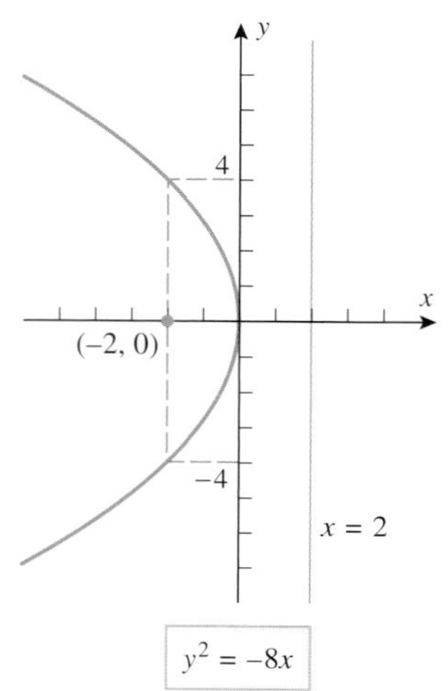

Figure 11.4.10

▶ **Example 2** Find an equation of the parabola that is symmetric about the y-axis, has its vertex at the origin, and passes through the point $(5, 2)$.

Solution. Since the parabola is symmetric about the y-axis and has its vertex at the origin, the equation is of the form

$$x^2 = 4py \quad \text{or} \quad x^2 = -4py$$

where the sign depends on whether the parabola opens up or down. But the parabola must open up since it passes through the point $(5, 2)$, which lies in the first quadrant. Thus, the equation is of the form

$$x^2 = 4py \tag{5}$$

Since the parabola passes through (5, 2), we must have $5^2 = 4p \cdot 2$ or $4p = \frac{25}{2}$. Therefore, (5) becomes

$$x^2 = \tfrac{25}{2}y \blacktriangleleft$$

EQUATIONS OF ELLIPSES IN STANDARD POSITION

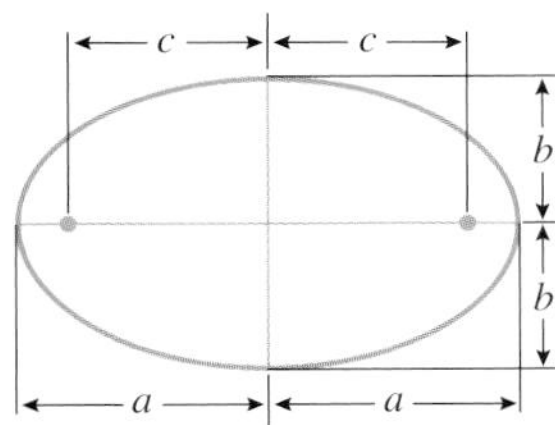

Figure 11.4.11

It is traditional in the study of ellipses to denote the length of the major axis by $2a$, the length of the minor axis by $2b$, and the distance between the foci by $2c$ (Figure 11.4.11). The number a is called the ***semimajor axis*** and the number b the ***semiminor axis*** (standard but odd terminology, since a and b are numbers, not geometric axes).

There is a basic relationship between the numbers a, b, and c that can be obtained by examining the sum of the distances to the foci from a point P at the end of the major axis and from a point Q at the end of the minor axis (Figure 11.4.12). From Definition 11.4.2, these sums must be equal, so we obtain

$$2\sqrt{b^2 + c^2} = (a - c) + (a + c)$$

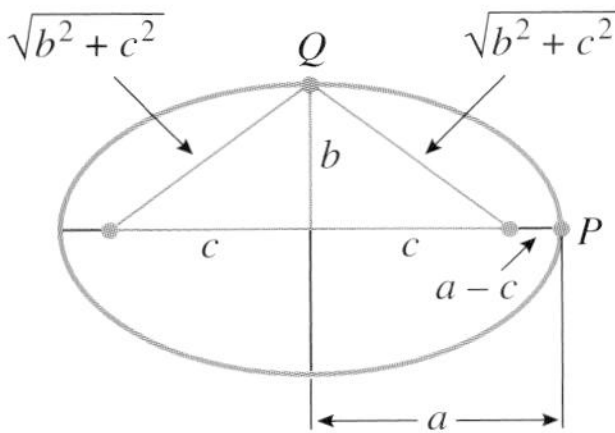

Figure 11.4.12

from which it follows that

$$a = \sqrt{b^2 + c^2} \tag{6}$$

or, equivalently,

$$c = \sqrt{a^2 - b^2} \tag{7}$$

From (6), the distance from a focus to an end of the minor axis is a (Figure 11.4.13), which implies that for *all* points on the ellipse the sum of the distances to the foci is $2a$.

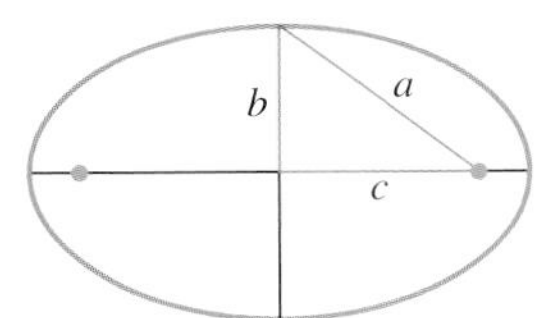

Figure 11.4.13

It also follows from (6) that $a \geq b$ with the equality holding only when $c = 0$. Geometrically, this means that the major axis of an ellipse is at least as large as the minor axis and that the two axes have equal length only when the foci coincide, in which case the ellipse is a circle.

The equation of an ellipse is simplest if the center of the ellipse is at the origin and the foci are on the x-axis or y-axis. The two possible such orientations are shown in Figure 11.4.14. These are called the ***standard positions*** of an ellipse, and the resulting equations are called the ***standard equations*** of an ellipse.

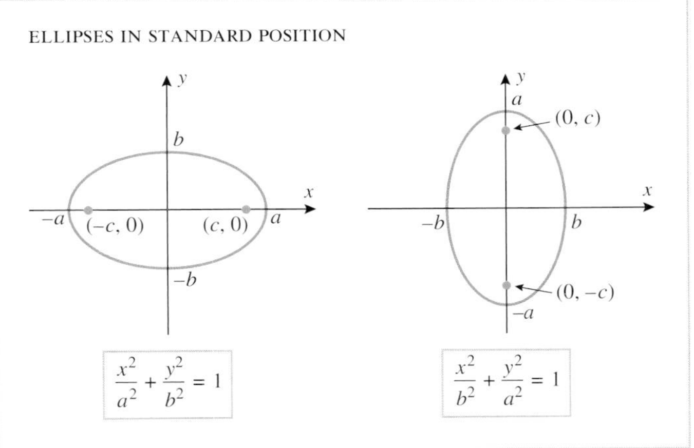

Figure 11.4.14

To illustrate how the equations in Figure 11.4.14 are obtained, we will derive the equation for the ellipse with foci on the x-axis. Let $P(x, y)$ be any point on that ellipse. Since the sum of the distances from P to the foci is $2a$, it follows (Figure 11.4.15) that

$$PF' + PF = 2a$$

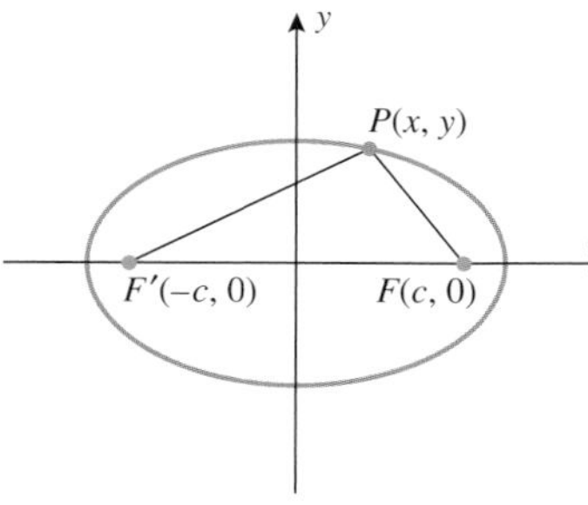

Figure 11.4.15

so

$$\sqrt{(x+c)^2+y^2}+\sqrt{(x-c)^2+y^2}=2a$$

Transposing the second radical to the right side of the equation and squaring yields

$$(x+c)^2+y^2=4a^2-4a\sqrt{(x-c)^2+y^2}+(x-c)^2+y^2$$

and, on simplifying,

$$\sqrt{(x-c)^2+y^2}=a-\frac{c}{a}x \tag{8}$$

Squaring again and simplifying yields

$$\frac{x^2}{a^2}+\frac{y^2}{a^2-c^2}=1$$

which, by virtue of (6), can be written as

$$\frac{x^2}{a^2}+\frac{y^2}{b^2}=1 \tag{9}$$

Conversely, it can be shown that any point whose coordinates satisfy (9) has $2a$ as the sum of its distances from the foci, so that such a point is on the ellipse.

A TECHNIQUE FOR SKETCHING ELLIPSES

Ellipses can be sketched from their *standard equations* using three basic steps:

Sketching an Ellipse from Its Standard Equation

Step 1. Determine whether the major axis is on the x-axis or the y-axis. This can be ascertained from the sizes of the denominators in the equation. Referring to Figure 11.4.14, and keeping in mind that $a^2 > b^2$ (since $a > b$), the major axis is along the x-axis if x^2 has the larger denominator, and it is along the y-axis if y^2 has the larger denominator. If the denominators are equal, the ellipse is a circle.

Step 2. Determine the values of a and b and draw a box extending a units on each side of the center along the major axis and b units on each side of the center along the minor axis.

Step 3. Using the box as a guide, sketch the ellipse so that its center is at the origin and it touches the sides of the box where the sides intersect the coordinate axes (Figure 11.4.16).

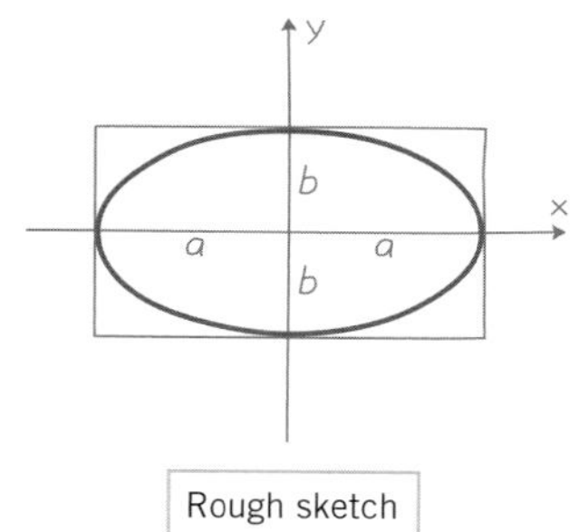

Figure 11.4.16

▶ **Example 3** Sketch the graphs of the ellipses

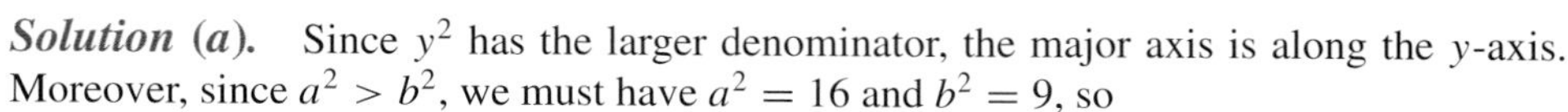

$$\text{(a)}\ \frac{x^2}{9}+\frac{y^2}{16}=1 \qquad \text{(b)}\ x^2+2y^2=4$$

showing the foci of each.

Solution (a). Since y^2 has the larger denominator, the major axis is along the y-axis. Moreover, since $a^2 > b^2$, we must have $a^2 = 16$ and $b^2 = 9$, so

$$a=4 \quad \text{and} \quad b=3$$

Drawing a box extending 4 units on each side of the origin along the y-axis and 3 units on each side of the origin along the x-axis as a guide yields the graph in Figure 11.4.17.

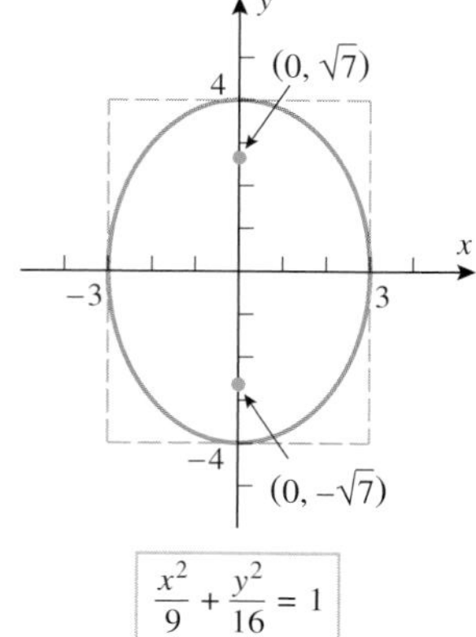

Figure 11.4.17

The foci lie c units on each side of the center along the major axis, where c is given by (7). From the values of a^2 and b^2 above, we obtain

$$c = \sqrt{a^2 - b^2} = \sqrt{16 - 9} = \sqrt{7} \approx 2.6$$

Thus, the coordinates of the foci are $(0, \sqrt{7})$ and $(0, -\sqrt{7})$, since they lie on the y-axis.

Solution (b). We first rewrite the equation in the standard form

$$\frac{x^2}{4} + \frac{y^2}{2} = 1$$

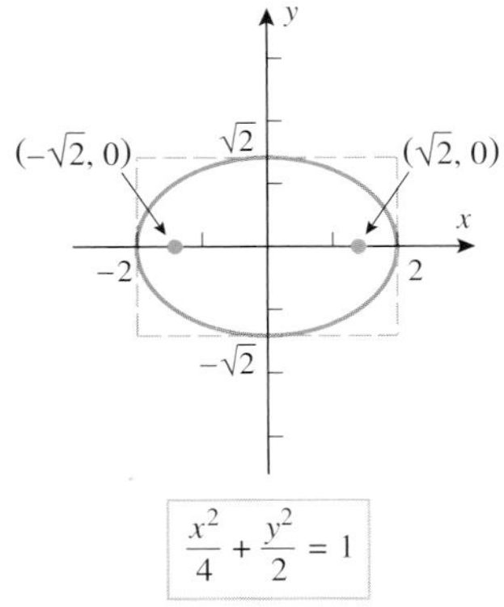

Since x^2 has the larger denominator, the major axis lies along the x-axis, and we have $a^2 = 4$ and $b^2 = 2$. Drawing a box extending $a = 2$ units on each side of the origin along the x-axis and extending $b = \sqrt{2} \approx 1.4$ units on each side of the origin along the y-axis as a guide yields the graph in Figure 11.4.18.

From (7), we obtain

$$c = \sqrt{a^2 - b^2} = \sqrt{2} \approx 1.4$$

Thus, the coordinates of the foci are $(\sqrt{2}, 0)$ and $(-\sqrt{2}, 0)$, since they lie on the x-axis. ◂

$(-\sqrt{2}, 0)$ $\sqrt{2}$ $(\sqrt{2}, 0)$ -2 2 $-\sqrt{2}$ x y

$$\frac{x^2}{4} + \frac{y^2}{2} = 1$$

Figure 11.4.18

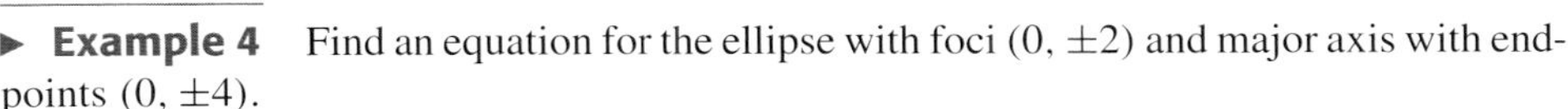

▸ **Example 4** Find an equation for the ellipse with foci $(0, \pm 2)$ and major axis with endpoints $(0, \pm 4)$.

Solution. From Figure 11.4.14, the equation has the form

$$\frac{x^2}{b^2} + \frac{y^2}{a^2} = 1$$

and from the given information, $a = 4$ and $c = 2$. It follows from (6) that

$$b^2 = a^2 - c^2 = 16 - 4 = 12$$

so the equation of the ellipse is

$$\frac{x^2}{12} + \frac{y^2}{16} = 1 \quad ◂$$

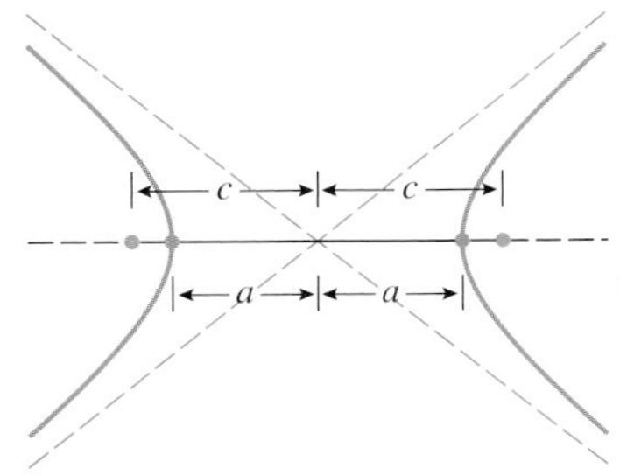

Figure 11.4.19

EQUATIONS OF HYPERBOLAS IN STANDARD POSITION

It is traditional in the study of hyperbolas to denote the distance between the vertices by $2a$, the distance between the foci by $2c$ (Figure 11.4.19), and to define the quantity b as

$$b = \sqrt{c^2 - a^2} \tag{10}$$

This relationship, which can also be expressed as

$$c = \sqrt{a^2 + b^2} \tag{11}$$

is pictured geometrically in Figure 11.4.20. As illustrated in that figure, and as we will show later in this section, the asymptotes pass through the corners of a box extending b units on each side of the center along the conjugate axis and a units on each side of the center along the focal axis. The number a is called the ***semifocal axis*** of the hyperbola and the number b the ***semiconjugate axis***. (As with the semimajor and semiminor axes of an ellipse, these are numbers, not geometric axes.)

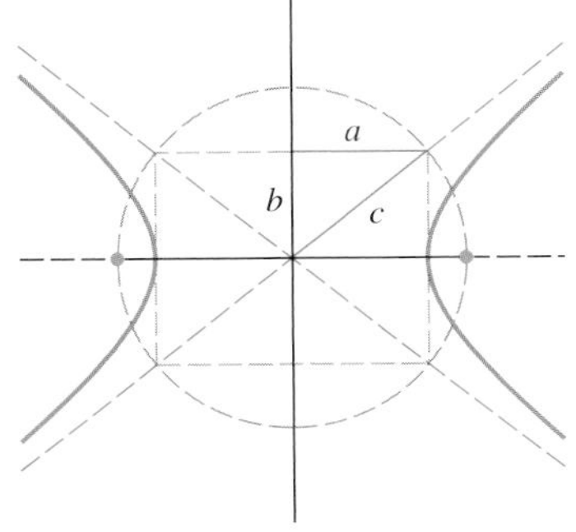

Figure 11.4.20

If V is one vertex of a hyperbola, then, as illustrated in Figure 11.4.21, the distance from V to the farther focus minus the distance from V to the closer focus is

$$[(c - a) + 2a] - (c - a) = 2a$$

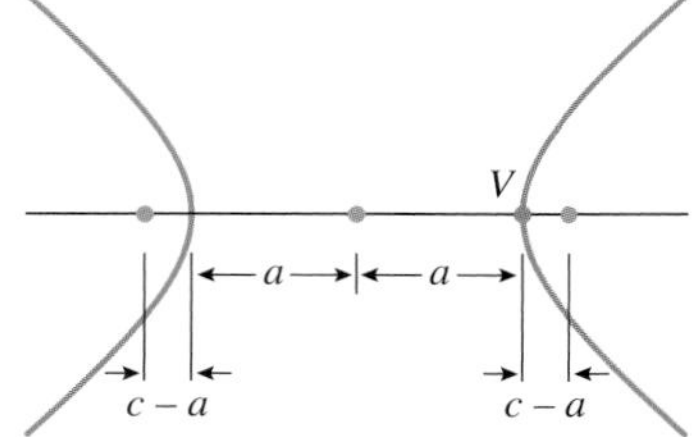

Figure 11.4.21

Thus, for *all* points on a hyperbola, the distance to the farther focus minus the distance to the closer focus is $2a$.

The equation of a hyperbola has an especially convenient form if the center of the hyperbola is at the origin and the foci are on the x-axis or y-axis. The two possible such orientations are shown in Figure 11.4.22. These are called the ***standard positions*** of a hyperbola, and the resulting equations are called the ***standard equations*** of a hyperbola.

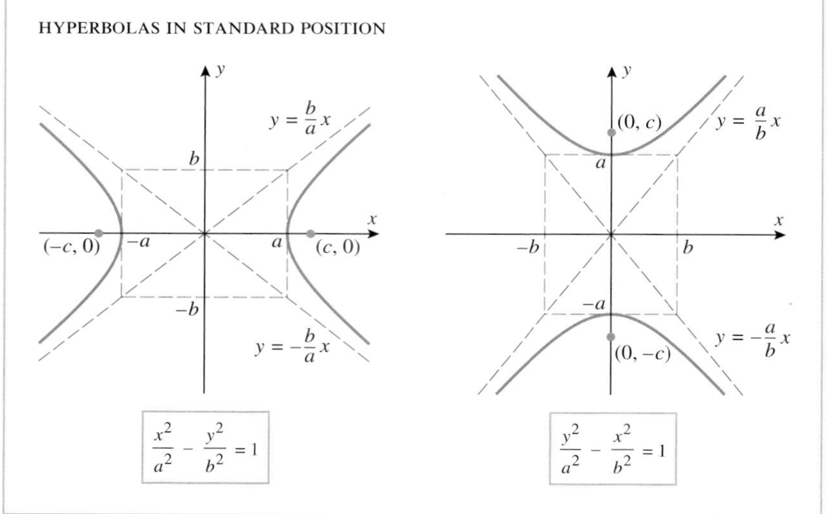

Figure 11.4.22

The derivations of these equations are similar to those already given for parabolas and ellipses, so we will leave them as exercises. However, to illustrate how the equations of the asymptotes are derived, we will derive those equations for the hyperbola

$$\frac{x^2}{a^2} - \frac{y^2}{b^2} = 1$$

We can rewrite this equation as

$$y^2 = \frac{b^2}{a^2}(x^2 - a^2)$$

which is equivalent to the pair of equations

$$y = \frac{b}{a}\sqrt{x^2 - a^2} \quad \text{and} \quad y = -\frac{b}{a}\sqrt{x^2 - a^2}$$

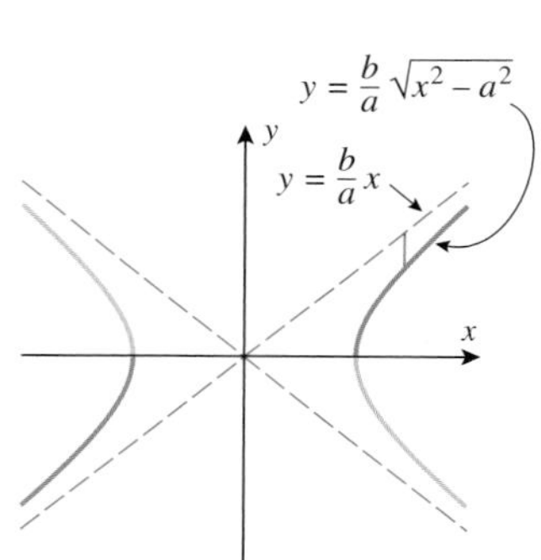

Figure 11.4.23

Thus, in the first quadrant, the vertical distance between the line $y = (b/a)x$ and the hyperbola can be written (Figure 11.4.23) as

$$\frac{b}{a}x - \frac{b}{a}\sqrt{x^2 - a^2}$$

But this distance tends to zero as $x \to +\infty$ since

$$\begin{aligned}
\lim_{x \to +\infty} \left(\frac{b}{a}x - \frac{b}{a}\sqrt{x^2 - a^2}\right) &= \lim_{x \to +\infty} \frac{b}{a}(x - \sqrt{x^2 - a^2}\,) \\
&= \lim_{x \to +\infty} \frac{b}{a}\frac{(x - \sqrt{x^2 - a^2}\,)(x + \sqrt{x^2 - a^2}\,)}{x + \sqrt{x^2 - a^2}} \\
&= \lim_{x \to +\infty} \frac{ab}{x + \sqrt{x^2 - a^2}} = 0
\end{aligned}$$

The analysis in the remaining quadrants is similar.

A QUICK WAY TO FIND ASYMPTOTES

There is a trick that can be used to avoid memorizing the equations of the asymptotes of a hyperbola. They can be obtained, when needed, by substituting 0 for the 1 on the right side of the hyperbola equation, and then solving for y in terms of x. For example, for the hyperbola

$$\frac{x^2}{a^2} - \frac{y^2}{b^2} = 1$$

we would write

$$\frac{x^2}{a^2} - \frac{y^2}{b^2} = 0 \quad \text{or} \quad y^2 = \frac{b^2}{a^2}x^2 \quad \text{or} \quad y = \pm\frac{b}{a}x$$

which are the equations for the asymptotes.

A TECHNIQUE FOR SKETCHING HYPERBOLAS

Hyperbolas can be sketched from their *standard equations* using four basic steps:

Sketching a Hyperbola from Its Standard Equation

Step 1. Determine whether the focal axis is on the x-axis or the y-axis. This can be ascertained from the location of the minus sign in the equation. Referring to Figure 11.4.22, the focal axis is along the x-axis when the minus sign precedes the y^2-term, and it is along the y-axis when the minus sign precedes the x^2-term.

Step 2. Determine the values of a and b and draw a box extending a units on either side of the center along the focal axis and b units on either side of the center along the conjugate axis. (The squares of a and b can be read directly from the equation.)

Step 3. Draw the asymptotes along the diagonals of the box.

Step 4. Using the box and the asymptotes as a guide, sketch the graph of the hyperbola (Figure 11.4.24).

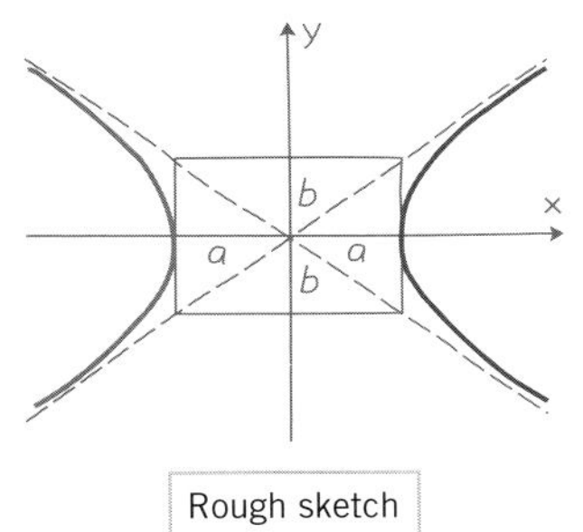

Figure 11.4.24

▶ **Example 5** Sketch the graphs of the hyperbolas

$$\text{(a)}\ \frac{x^2}{4} - \frac{y^2}{9} = 1 \qquad \text{(b)}\ y^2 - x^2 = 1$$

showing their vertices, foci, and asymptotes.

Solution (a). The minus sign precedes the y^2-term, so the focal axis is along the x-axis. From the denominators in the equation we obtain

$$a^2 = 4 \quad \text{and} \quad b^2 = 9$$

Since a and b are positive, we must have $a = 2$ and $b = 3$. Recalling that the vertices lie a units on each side of the center on the focal axis, it follows that their coordinates in this case are $(2, 0)$ and $(-2, 0)$. Drawing a box extending $a = 2$ units along the x-axis on each side of the origin and $b = 3$ units on each side of the origin along the y-axis, then drawing the asymptotes along the diagonals of the box as a guide, yields the graph in Figure 11.4.25.

To obtain equations for the asymptotes, we substitute 0 for 1 in the given equation; this yields

$$\frac{x^2}{4} - \frac{y^2}{9} = 0 \quad \text{or} \quad y = \pm\frac{3}{2}x$$

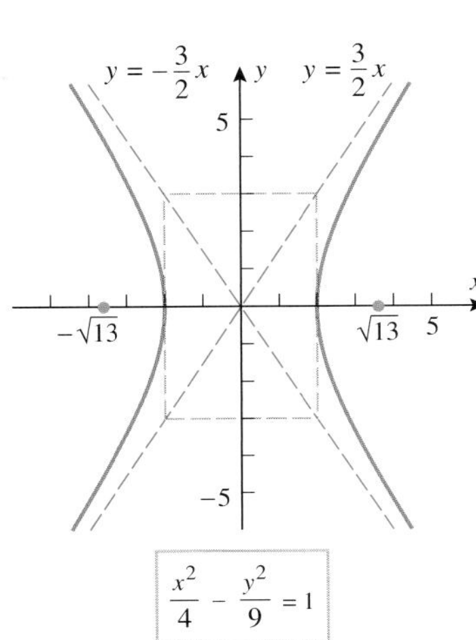

Figure 11.4.25

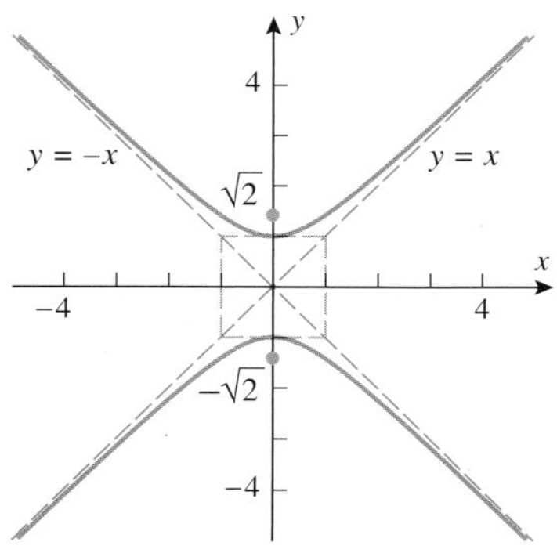

Figure 11.4.26

A hyperbola in which $a = b$, as in part (b) of Example 5, is called an ***equilateral hyperbola***. Such hyperbolas always have perpendicular asymptotes.

The foci lie c units on each side of the center along the focal axis, where c is given by (11). From the values of a^2 and b^2 above we obtain

$$c = \sqrt{a^2 + b^2} = \sqrt{4+9} = \sqrt{13} \approx 3.6$$

Since the foci lie on the x-axis in this case, their coordinates are $(\sqrt{13}, 0)$ and $(-\sqrt{13}, 0)$.

Solution (b). The minus sign precedes the x^2-term, so the focal axis is along the y-axis. From the denominators in the equation we obtain $a^2 = 1$ and $b^2 = 1$, from which it follows that

$$a = 1 \quad \text{and} \quad b = 1$$

Thus, the vertices are at $(0, -1)$ and $(0, 1)$. Drawing a box extending $a = 1$ unit on either side of the origin along the y-axis and $b = 1$ unit on either side of the origin along the x-axis, then drawing the asymptotes, yields the graph in Figure 11.4.26. Since the box is actually a square, the asymptotes are perpendicular and have equations $y = \pm x$. This can also be seen by substituting 0 for 1 in the given equation, which yields $y^2 - x^2 = 0$ or $y = \pm x$. Also,

$$c = \sqrt{a^2 + b^2} = \sqrt{1+1} = \sqrt{2}$$

so the foci, which lie on the y-axis, are $(0, -\sqrt{2})$ and $(0, \sqrt{2})$. ◀

▶ **Example 6** Find the equation of the hyperbola with vertices $(0, \pm 8)$ and asymptotes $y = \pm\frac{4}{3}x$.

Solution. Since the vertices are on the y-axis, the equation of the hyperbola has the form $(y^2/a^2) - (x^2/b^2) = 1$ and the asymptotes are

$$y = \pm\frac{a}{b}x$$

From the locations of the vertices we have $a = 8$, so the given equations of the asymptotes yield

$$y = \pm\frac{a}{b}x = \pm\frac{8}{b}x = \pm\frac{4}{3}x$$

from which it follows that $b = 6$. Thus, the hyperbola has the equation

$$\frac{y^2}{64} - \frac{x^2}{36} = 1 \quad ◀$$

TRANSLATED CONICS

Equations of conics that are translated from their standard positions can be obtained by replacing x by $x - h$ and y by $y - k$ in their standard equations. For a parabola, this translates the vertex from the origin to the point (h, k); and for ellipses and hyperbolas, this translates the center from the origin to the point (h, k).

Parabolas with vertex (h, k) and axis parallel to x-axis

$$(y-k)^2 = 4p(x-h) \quad \text{[Opens right]} \qquad (12)$$

$$(y-k)^2 = -4p(x-h) \quad \text{[Opens left]} \qquad (13)$$

Parabolas with vertex (h, k) and axis parallel to y-axis

$$(x-h)^2 = 4p(y-k) \quad \text{[Opens up]} \qquad (14)$$

$$(x-h)^2 = -4p(y-k) \quad \text{[Opens down]} \qquad (15)$$

Ellipse with center (h, k) and major axis parallel to x-axis

$$\frac{(x-h)^2}{a^2} + \frac{(y-k)^2}{b^2} = 1 \quad [b \le a] \qquad (16)$$

Ellipse with center (h, k) and major axis parallel to y-axis

$$\frac{(x-h)^2}{b^2}+\frac{(y-k)^2}{a^2}=1 \quad [b \le a] \tag{17}$$

Hyperbola with center (h, k) and focal axis parallel to x-axis

$$\frac{(x-h)^2}{a^2}-\frac{(y-k)^2}{b^2}=1 \tag{18}$$

Hyperbola with center (h, k) and focal axis parallel to y-axis

$$\frac{(y-k)^2}{a^2}-\frac{(x-h)^2}{b^2}=1 \tag{19}$$

▶ **Example 7** Find an equation for the parabola that has its vertex at $(1, 2)$ and its focus at $(4, 2)$.

Solution. Since the focus and vertex are on a horizontal line, and since the focus is to the right of the vertex, the parabola opens to the right and its equation has the form

$$(y-k)^2 = 4p(x-h)$$

Since the vertex and focus are 3 units apart, we have $p = 3$, and since the vertex is at $(h, k) = (1, 2)$, we obtain

$$(y-2)^2 = 12(x-1)$$ ◀

Sometimes the equations of translated conics occur in expanded form, in which case we are faced with the problem of identifying the graph of a quadratic equation in x and y:

$$Ax^2 + Cy^2 + Dx + Ey + F = 0 \tag{20}$$

The basic procedure for determining the nature of such a graph is to complete the squares of the quadratic terms and then try to match up the resulting equation with one of the forms of a translated conic.

▶ **Example 8** Describe the graph of the equation

$$y^2 - 8x - 6y - 23 = 0$$

Solution. The equation involves quadratic terms in y but none in x, so we first take all of the y-terms to one side:

$$y^2 - 6y = 8x + 23$$

Next, we complete the square on the y-terms by adding 9 to both sides:

$$(y-3)^2 = 8x + 32$$

Finally, we factor out the coefficient of the x-term to obtain

$$(y-3)^2 = 8(x+4)$$

This equation is of form (12) with $h = -4$, $k = 3$, and $p = 2$, so the graph is a parabola with vertex $(-4, 3)$ opening to the right. Since $p = 2$, the focus is 2 units to the right of the vertex, which places it at the point $(-2, 3)$; and the directrix is 2 units to the left of the vertex, which means that its equation is $x = -6$. The parabola is shown in Figure 11.4.27. ◀

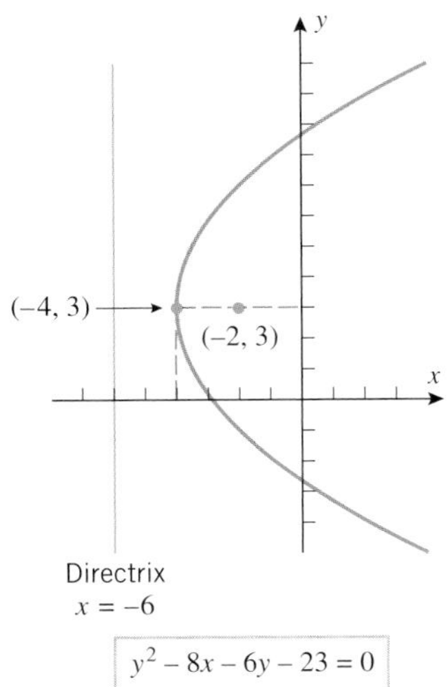

Figure 11.4.27

► Example 9 Describe the graph of the equation

$$16x^2 + 9y^2 - 64x - 54y + 1 = 0$$

Solution. This equation involves quadratic terms in both x and y, so we will group the x-terms and the y-terms on one side and put the constant on the other:

$$(16x^2 - 64x) + (9y^2 - 54y) = -1$$

Next, factor out the coefficients of x^2 and y^2 and complete the squares:

$$16(x^2 - 4x + 4) + 9(y^2 - 6y + 9) = -1 + 64 + 81$$

or

$$16(x - 2)^2 + 9(y - 3)^2 = 144$$

Finally, divide through by 144 to introduce a 1 on the right side:

$$\frac{(x-2)^2}{9} + \frac{(y-3)^2}{16} = 1$$

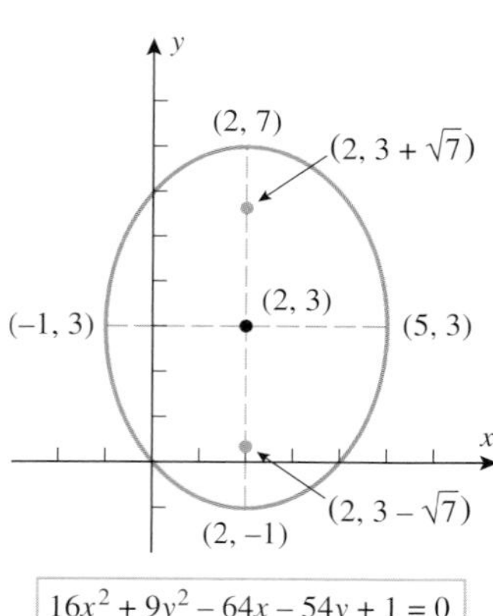

Figure 11.4.28

This is an equation of form (17), with $h = 2$, $k = 3$, $a^2 = 16$, and $b^2 = 9$. Thus, the graph of the equation is an ellipse with center $(2, 3)$ and major axis parallel to the y-axis. Since $a = 4$, the major axis extends 4 units above and 4 units below the center, so its endpoints are $(2, 7)$ and $(2, -1)$ (Figure 11.4.28). Since $b = 3$, the minor axis extends 3 units to the left and 3 units to the right of the center, so its endpoints are $(-1, 3)$ and $(5, 3)$. Since

$$c = \sqrt{a^2 - b^2} = \sqrt{16 - 9} = \sqrt{7}$$

the foci lie $\sqrt{7}$ units above and below the center, placing them at the points $(2, 3 + \sqrt{7})$ and $(2, 3 - \sqrt{7})$. ◄

► Example 10 Describe the graph of the equation

$$x^2 - y^2 - 4x + 8y - 21 = 0$$

Solution. This equation involves quadratic terms in both x and y, so we will group the x-terms and the y-terms on one side and put the constant on the other:

$$(x^2 - 4x) - (y^2 - 8y) = 21$$

We leave it for you to verify by completing the squares that this equation can be written as

$$\frac{(x-2)^2}{9} - \frac{(y-4)^2}{9} = 1 \tag{21}$$

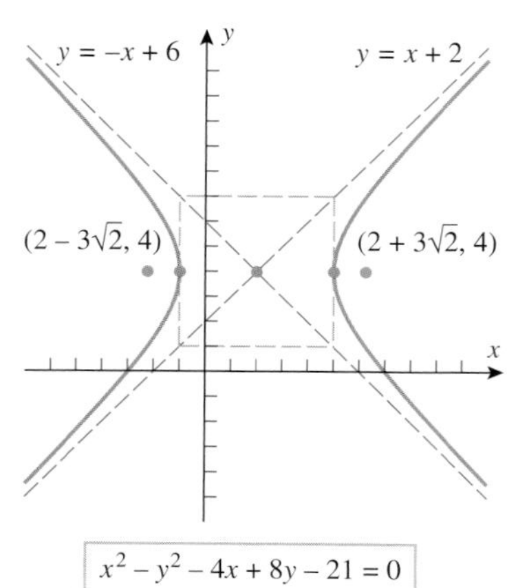

Figure 11.4.29

This is an equation of form (18) with $h = 2$, $k = 4$, $a^2 = 9$, and $b^2 = 9$. Thus, the equation represents a hyperbola with center $(2, 4)$ and focal axis parallel to the x-axis. Since $a = 3$, the vertices are located 3 units to the left and 3 units to the right of the center, or at the points $(-1, 4)$ and $(5, 4)$. From (11), $c = \sqrt{a^2 + b^2} = \sqrt{9 + 9} = 3\sqrt{2}$, so the foci are located $3\sqrt{2}$ units to the left and right of the center, or at the points $(2 - 3\sqrt{2}, 4)$ and $(2 + 3\sqrt{2}, 4)$.

The equations of the asymptotes may be found using the trick of substituting 0 for 1 in (21) to obtain

$$\frac{(x-2)^2}{9} - \frac{(y-4)^2}{9} = 0$$

This can be written as $y - 4 = \pm(x - 2)$, which yields the asymptotes

$$y = x + 2 \quad \text{and} \quad y = -x + 6$$

With the aid of a box extending $a = 3$ units left and right of the center and $b = 3$ units above and below the center, we obtain the sketch in Figure 11.4.29. ◄

REFLECTION PROPERTIES OF THE CONIC SECTIONS

Parabolas, ellipses, and hyperbolas have certain reflection properties that make them extremely valuable in various applications. In the exercises we will ask you to prove the following results.

> **11.4.4 THEOREM (*Reflection Property of Parabolas*).** *The tangent line at a point P on a parabola makes equal angles with the line through P parallel to the axis of symmetry and the line through P and the focus* (Figure 11.4.30*a*).

> **11.4.5 THEOREM (*Reflection Property of Ellipses*).** *A line tangent to an ellipse at a point P makes equal angles with the lines joining P to the foci* (Figure 11.4.30*b*).

> **11.4.6 THEOREM (*Reflection Property of Hyperbolas*).** *A line tangent to a hyperbola at a point P makes equal angles with the lines joining P to the foci* (Figure 11.4.30*c*).

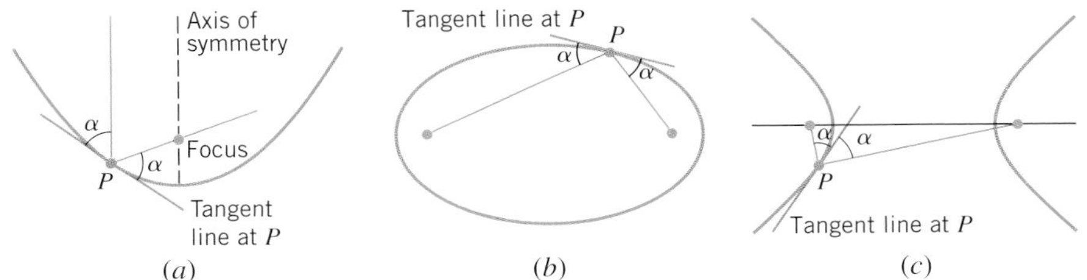

Figure 11.4.30

APPLICATIONS OF THE CONIC SECTIONS

Fermat's principle in optics implies that light reflects off of a surface at an angle equal to its angle of incidence. (See Exercise 65 in Section 4.5.) In particular, if a reflecting surface is generated by revolving a parabola about its axis of symmetry, it follows from Theorem 11.4.4 that all light rays entering parallel to the axis will be reflected to the focus (Figure 11.4.31*a*); conversely, if a light source is located at the focus, then the reflected rays will all be parallel to the axis (Figure 11.4.31*b*). This principle is used in certain telescopes to reflect the approximately parallel rays of light from the stars and planets off of a parabolic mirror to an eyepiece at the focus; and the parabolic reflectors in flashlights and automobile headlights utilize this principle to form a parallel beam of light rays from a bulb placed at the focus. The same optical principles apply to radar signals and sound waves, which explains the parabolic shape of many antennas.

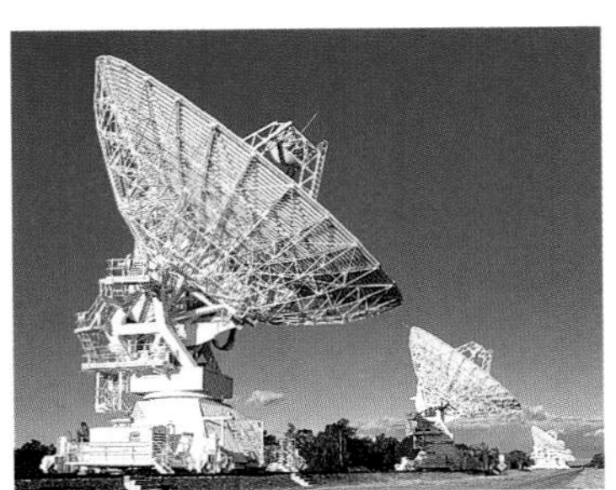

Incoming signals are reflected by the parabolic antenna to the receiver at the focus.

Visitors to various rooms in the United States Capitol Building and in St. Paul's Cathedral in Rome are often astonished by the "whispering gallery" effect in which two people at opposite ends of the room can hear one another's whispers very clearly. Such rooms have ceilings with elliptical cross sections and common foci. Thus, when the two people stand at the foci, their whispers are reflected directly to one another off of the elliptical ceiling.

Hyperbolic navigation systems, which were developed in World War II as navigational aids to ships, are based on the definition of a hyperbola. With these systems the ship receives synchronized radio signals from two widely spaced transmitters with known positions. The ship's electronic receiver measures the difference in reception times between the signals and then uses that difference to compute the difference $2a$ between its distances from the

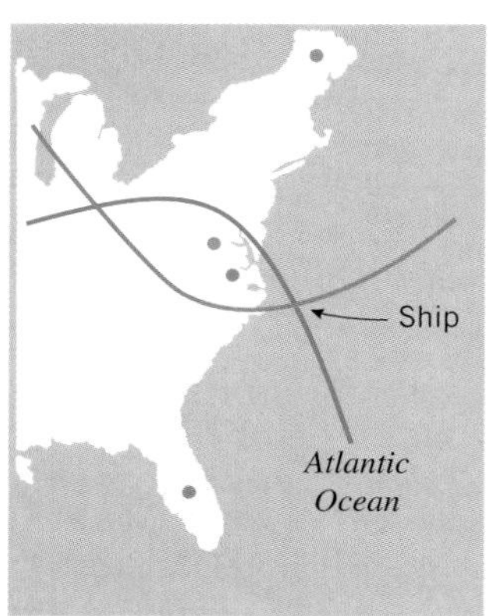

Figure 11.4.32

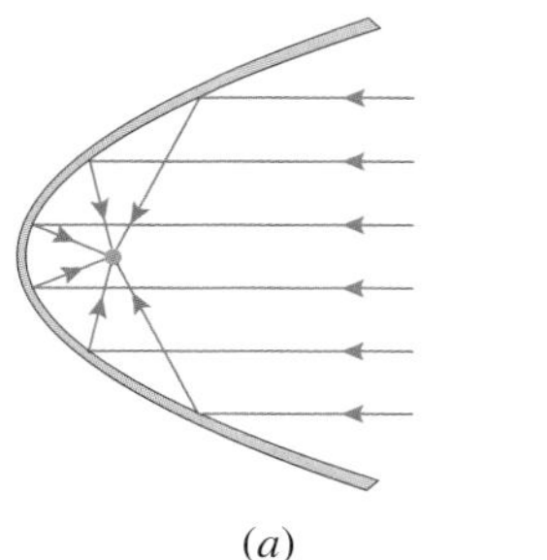
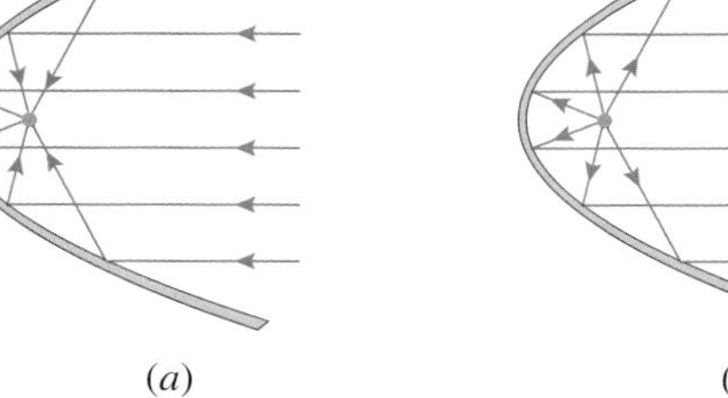

Figure 11.4.31 (a) (b)

two transmitters. This information places the ship somewhere on the hyperbola whose foci are at the transmitters and whose points have $2a$ as the difference in their distances from the foci. By repeating the process with a second set of transmitters, the position of the ship can be approximated as the intersection of two hyperbolas (Figure 11.4.32). (The modern "global positioning system" is based on the same principle.)

✔QUICK CHECK EXERCISES 11.4 *(See page 765 for answers.)*

1. Identify the conic.
   (a) The set of points in the plane, the sum of whose distances to two fixed points is a positive constant greater than the distance between the fixed points is ________.
   (b) The set of points in the plane, the difference of whose distances to two fixed points is a positive constant less than the distance between the fixed points is ________.
   (c) The set of points in the plane that are equidistant from a fixed line and a fixed point not on the line is ________.
2. (a) The equation of the parabola with focus $(p, 0)$ and directrix $x = -p$ is ________.
   (b) The equation of the parabola with focus $(0, p)$ and directrix $y = -p$ is ________.
3. (a) Suppose that an ellipse has semimajor axis a and semiminor axis b. Then for all points on the ellipse, the sum of the distances to the foci is equal to ________.
   (b) The two standard equations of an ellipse with semimajor axis a and semiminor axis b are ________ and ________.
   (c) Suppose that an ellipse has semimajor axis a, semiminor axis b, and foci $(\pm c, 0)$. Then c may be obtained from a and b by the equation $c =$ ________.
4. (a) Suppose that a hyperbola has semifocal axis a and semiconjugate axis b. Then for all points on the hyperbola, the difference of the distance to the farther focus minus the distance to the closer focus is equal to ________.
   (b) The two standard equations of a hyperbola with semifocal axis a and semiconjugate axis b are ________ and ________.
   (c) Suppose that a hyperbola in standard position has semifocal axis a, semiconjugate axis b, and foci $(\pm c, 0)$. Then c may be obtained from a and b by the equation $c =$ ________. The equations of the asymptotes of this hyperbola are $y = \pm$ ________.

EXERCISE SET 11.4 Graphing Utility C CAS

FOCUS ON CONCEPTS

1. In parts (a)–(f), find the equation of the conic.

(a)
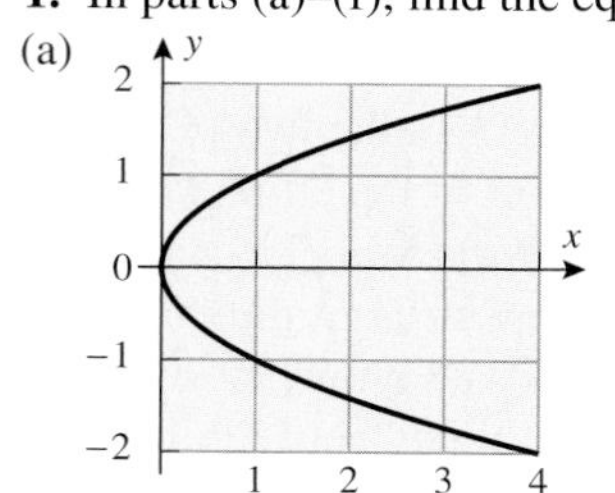

(b)
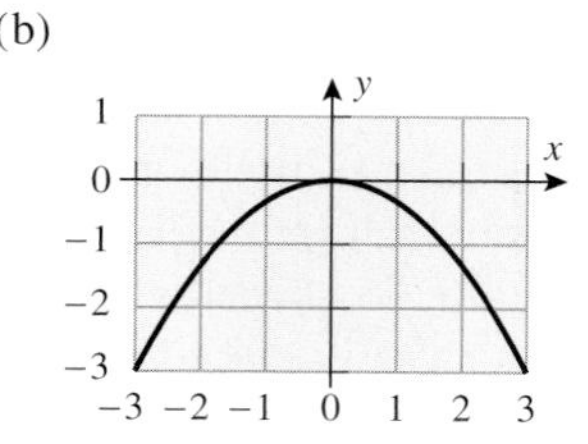

(c)
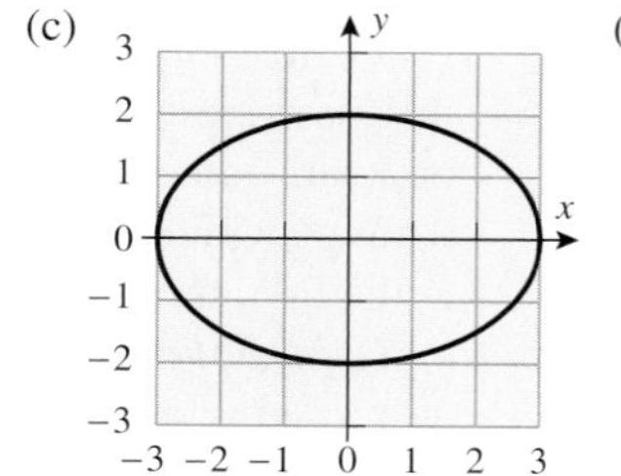

(d)
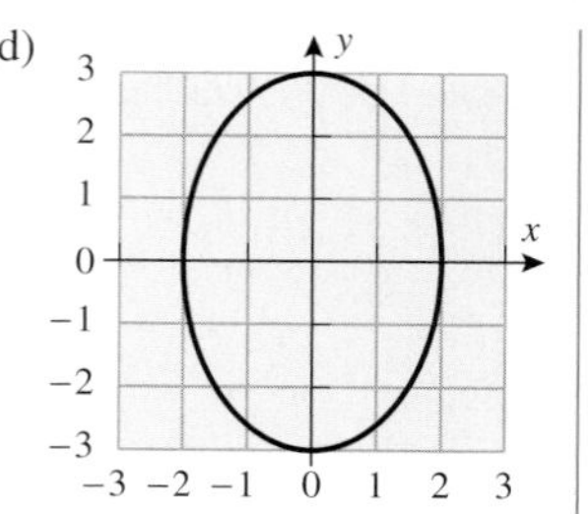

(e) 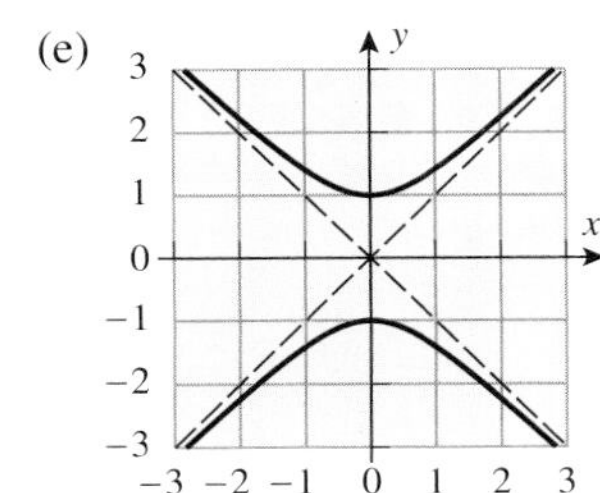

(f) 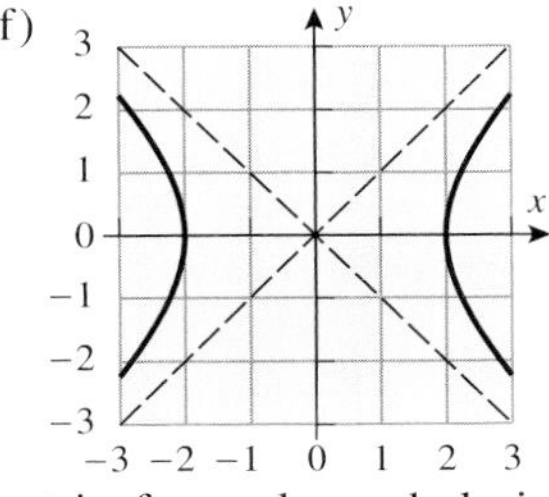

2. (a) Find the focus and directrix for each parabola in Exercise 1.
(b) Find the foci of the ellipses in Exercise 1.
(c) Find the foci and the equations of the asymptotes of the hyperbolas in Exercise 1.

3–8 Sketch the parabola, and label the focus, vertex, and directrix.

3. (a) $y^2 = 4x$ (b) $x^2 = -8y$

4. (a) $y^2 = -10x$ (b) $x^2 = 4y$

5. (a) $(y-3)^2 = 6(x-2)$ (b) $(x+2)^2 = -(y+2)$

6. (a) $(y-1)^2 = -12(x+4)$ (b) $(x-1)^2 = 2\left(y-\frac{1}{2}\right)$

7. (a) $x^2 - 4x + 2y = 1$ (b) $x = y^2 - 4y + 2$

8. (a) $y^2 - 6y - 2x + 1 = 0$ (b) $y = 4x^2 + 8x + 5$

9–14 Sketch the ellipse, and label the foci, the vertices, and the ends of the minor axis.

9. (a) $\dfrac{x^2}{16} + \dfrac{y^2}{9} = 1$ (b) $9x^2 + y^2 = 9$

10. (a) $\dfrac{x^2}{25} + \dfrac{y^2}{4} = 1$ (b) $4x^2 + y^2 = 36$

11. (a) $16(x-1)^2 + 9(y-3)^2 = 144$
(b) $9(x+2)^2 + 4(y+1)^2 = 36$

12. (a) $(x+3)^2 + 4(y-5)^2 = 16$
(b) $\frac{1}{4}x^2 + \frac{1}{9}(y+2)^2 - 1 = 0$

13. (a) $x^2 + 9y^2 + 2x - 18y + 1 = 0$
(b) $4x^2 + y^2 + 8x - 10y = -13$

14. (a) $9x^2 + 4y^2 - 18x + 24y + 9 = 0$
(b) $5x^2 + 9y^2 + 20x - 54y = -56$

15–20 Sketch the hyperbola, and label the vertices, foci, and asymptotes.

15. (a) $\dfrac{x^2}{16} - \dfrac{y^2}{9} = 1$ (b) $9y^2 - x^2 = 36$

16. (a) $\dfrac{y^2}{9} - \dfrac{x^2}{25} = 1$ (b) $16x^2 - 25y^2 = 400$

17. (a) $\dfrac{(x-1)^2}{9} - \dfrac{(y+2)^2}{4} = 1$
(b) $4(y-3)^2 - 9(x-2)^2 = 36$

18. (a) $\dfrac{(y+4)^2}{3} - \dfrac{(x-2)^2}{5} = 1$
(b) $16(x+1)^2 - 8(y-3)^2 = 16$

19. (a) $x^2 - 4y^2 + 2x + 8y - 7 = 0$
(b) $16x^2 - y^2 - 32x - 6y = 57$

20. (a) $4x^2 - 9y^2 - 16x - 54y - 29 = 0$
(b) $4y^2 - x^2 - 40y - 4x = -60$

21–26 Find an equation for the parabola that satisfies the given conditions.

21. (a) Vertex $(0, 0)$; focus $(3, 0)$.
(b) Vertex $(0, 0)$; directrix $x = 7$.

22. (a) Vertex $(0, 0)$; focus $(0, -3)$.
(b) Vertex $(0, 0)$; directrix $y = \frac{1}{4}$.

23. (a) Focus $(0, -3)$; directrix $y = 3$.
(b) Vertex $(1, 1)$; directrix $y = -2$.

24. (a) Focus $(6, 0)$; directrix $x = -6$.
(b) Focus $(-1, 4)$; directrix $x = 5$.

25. Axis $y = 0$; passes through $(3, 2)$ and $(2, -\sqrt{2})$.

26. Vertex $(5, -3)$; axis parallel to the y-axis; passes through $(9, 5)$.

27–32 Find an equation for the ellipse that satisfies the given conditions.

27. (a) Ends of major axis $(\pm 3, 0)$; ends of minor axis $(0, \pm 2)$.
(b) Length of major axis 26; foci $(\pm 5, 0)$.

28. (a) Ends of major axis $(0, \pm\sqrt{7})$; ends of minor axis $(\pm 1, 0)$.
(b) Length of minor axis 8; foci $(0, \pm 3)$.

29. (a) Foci $(\pm 1, 0)$; $b = \sqrt{2}$.
(b) $c = 2\sqrt{3}$; $a = 4$; center at the origin; foci on a coordinate axis (two answers).

30. (a) Foci $(\pm 3, 0)$; $a = 4$.
(b) $b = 3$; $c = 4$; center at the origin; foci on a coordinate axis (two answers).

31. (a) Ends of major axis $(0, \pm 6)$; passes through $(-3, 2)$.
(b) Foci $(-1, 1)$ and $(-1, 3)$; minor axis of length 4.

32. (a) Center at $(0, 0)$; major and minor axes along the coordinate axes; passes through $(3, 2)$ and $(1, 6)$.
(b) Foci $(2, 1)$ and $(2, -3)$; major axis of length 6.

33–38 Find an equation for a hyperbola that satisfies the given conditions. (In some cases there may be more than one hyperbola.)

33. (a) Vertices $(\pm 2, 0)$; foci $(\pm 3, 0)$.
(b) Vertices $(\pm 1, 0)$; asymptotes $y = \pm 2x$.

34. (a) Vertices $(0, \pm 4)$; foci $(0, \pm 5)$.
(b) Vertices $(0, \pm 2)$; asymptotes $y = \pm\frac{2}{3}x$.

35. (a) Asymptotes $y = \pm\frac{3}{2}x$; $b = 4$.
(b) Foci $(0, \pm 5)$; asymptotes $y = \pm 2x$.

36. (a) Asymptotes $y = \pm\frac{3}{4}x$; $c = 5$.
(b) Foci $(\pm 3, 0)$; asymptotes $y = \pm 2x$.

37. (a) Vertices $(0, 6)$ and $(6, 6)$; foci 10 units apart.
(b) Asymptotes $y = x - 2$ and $y = -x + 4$; passes through the origin.

38. (a) Foci $(1, 8)$ and $(1, -12)$; vertices 4 units apart.
(b) Vertices $(-3, -1)$ and $(5, -1)$; $b = 4$.

39. (a) As illustrated in the accompanying figure, a parabolic arch spans a road 40 ft wide. How high is the arch if a center section of the road 20 ft wide has a minimum clearance of 12 ft?
(b) How high would the center be if the arch were the upper half of an ellipse?

40. (a) Find an equation for the parabolic arch with base b and height h, shown in the accompanying figure.
(b) Find the area under the arch.

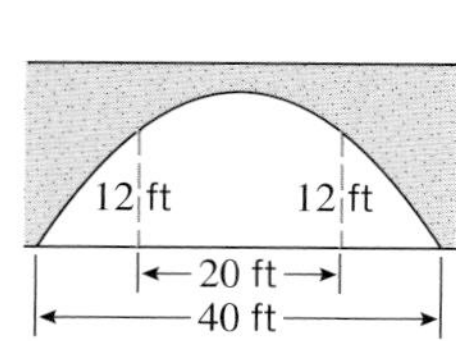

Figure Ex-39

$(\frac{1}{2}b, h)$ $(b, 0)$

Figure Ex-40

41. Show that the vertex is the closest point on a parabola to the focus. [*Suggestion:* Introduce a convenient coordinate system and use Definition 11.4.1.]

42. As illustrated in the accompanying figure, suppose that a comet moves in a parabolic orbit with the Sun at its focus and that the line from the Sun to the comet makes an angle of 60° with the axis of the parabola when the comet is 40 million miles from the center of the Sun. Use the result in Exercise 41 to determine how close the comet will come to the center of the Sun.

43. For the parabolic reflector in the accompanying figure, how far from the vertex should the light source be placed to produce a beam of parallel rays?

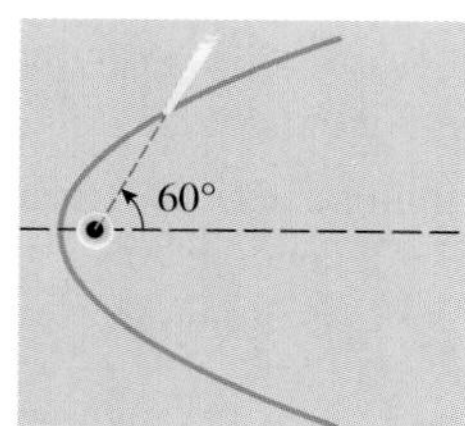

Figure Ex-42

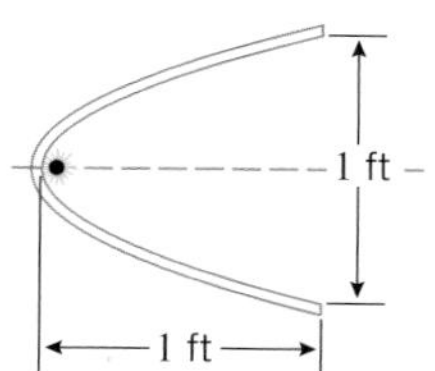

Figure Ex-43

44. In each part, find the shaded area in the figure.

(a)
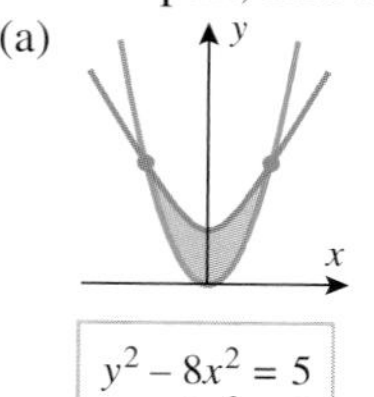

$y^2 - 8x^2 = 5$
$y - 2x^2 = 0$

(b)
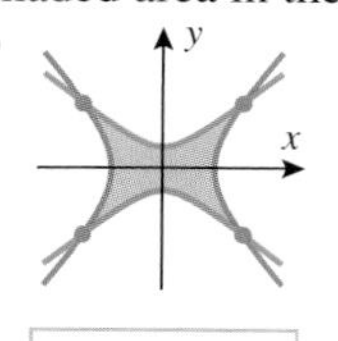

$3x^2 - 7y^2 = 5$
$9y^2 - 2x^2 = 1$

(c)
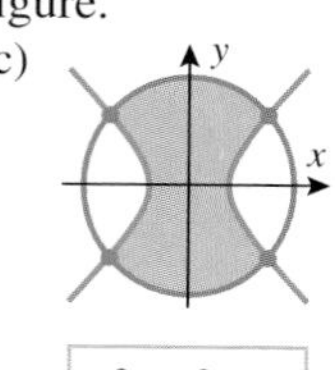

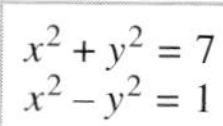

45. (a) The accompanying figure shows an ellipse with semimajor axis a and semiminor axis b. Express the coordinates of the points P, Q, and R in terms of t.
(b) How does the geometric interpretation of the parameter t differ between a circle

$$x = a\cos t, \quad y = a\sin t$$

and an ellipse

$$x = a\cos t, \quad y = b\sin t?$$

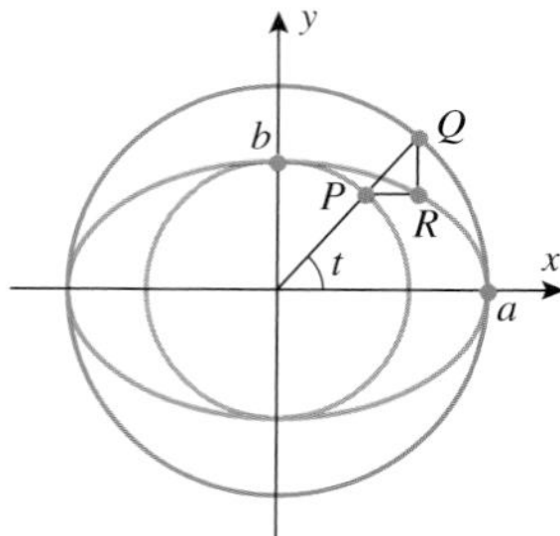

Figure Ex-45

46. (a) Show that the right and left branches of the hyperbola

$$\frac{x^2}{a^2} - \frac{y^2}{b^2} = 1$$

can be represented parametrically as

$$x = a\cosh t, \quad y = b\sinh t \quad (-\infty < t < +\infty)$$
$$x = -a\cosh t, \quad y = b\sinh t \quad (-\infty < t < +\infty)$$

(b) Use a graphing utility to generate both branches of the hyperbola $x^2 - y^2 = 1$ on the same screen.

47. (a) Show that the right and left branches of the hyperbola

$$\frac{x^2}{a^2} - \frac{y^2}{b^2} = 1$$

can be represented parametrically as

$$x = a\sec t, \quad y = b\tan t \quad (-\pi/2 < t < \pi/2)$$
$$x = -a\sec t, \quad y = b\tan t \quad (-\pi/2 < t < \pi/2)$$

(b) Use a graphing utility to generate both branches of the hyperbola $x^2 - y^2 = 1$ on the same screen.

48. Find an equation of the parabola traced by a point that moves so that its distance from $(2, 4)$ is the same as its distance to the x-axis.

49. Find an equation of the ellipse traced by a point that moves so that the sum of its distances to $(4, 1)$ and $(4, 5)$ is 12.

50. Find the equation of the hyperbola traced by a point that moves so that the difference between its distances to $(0, 0)$ and $(1, 1)$ is 1.

51. Suppose that the base of a solid is elliptical with a major axis of length 9 and a minor axis of length 4. Find the volume of the solid if the cross sections perpendicular to the major axis are squares (see the accompanying figure).

52. Suppose that the base of a solid is elliptical with a major axis of length 9 and a minor axis of length 4. Find the volume of the solid if the cross sections perpendicular to the minor axis are equilateral triangles (see the accompanying figure).

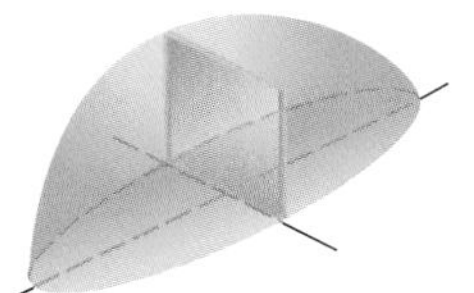
Figure Ex-51

Figure Ex-52

53. Show that an ellipse with semimajor axis a and semiminor axis b has area $A = \pi ab$.

54. (a) Show that the ellipsoid that results when an ellipse with semimajor axis a and semiminor axis b is revolved about the major axis has volume $V = \frac{4}{3}\pi ab^2$.
(b) Show that the ellipsoid that results when an ellipse with semimajor axis a and semiminor axis b is revolved about the minor axis has volume $V = \frac{4}{3}\pi a^2 b$.

55. Show that the ellipsoid that results when an ellipse with semimajor axis a and semiminor axis b is revolved about the major axis has surface area

$$S = 2\pi ab\left(\frac{b}{a} + \frac{a}{c}\sin^{-1}\frac{c}{a}\right)$$

where $c = \sqrt{a^2 - b^2}$.

56. Show that the ellipsoid that results when an ellipse with semimajor axis a and semiminor axis b is revolved about the minor axis has surface area

$$S = 2\pi ab\left(\frac{a}{b} + \frac{b}{c}\ln\frac{a+c}{b}\right)$$

where $c = \sqrt{a^2 - b^2}$.

FOCUS ON CONCEPTS

57. Suppose that you want to draw an ellipse that has given values for the lengths of the major and minor axes by using the method shown in Figure 11.4.3*b*. Assuming that the axes are drawn, explain how a compass can be used to locate the positions for the tacks.

58. The accompanying figure shows Kepler's method for constructing a parabola. A piece of string the length of the left edge of the drafting triangle is tacked to the vertex Q of the triangle and the other end to a fixed point F. A pencil holds the string taut against the base of the triangle as the edge opposite Q slides along a horizontal line L below F. Show that the pencil traces an arc of a parabola with focus F and directrix L.

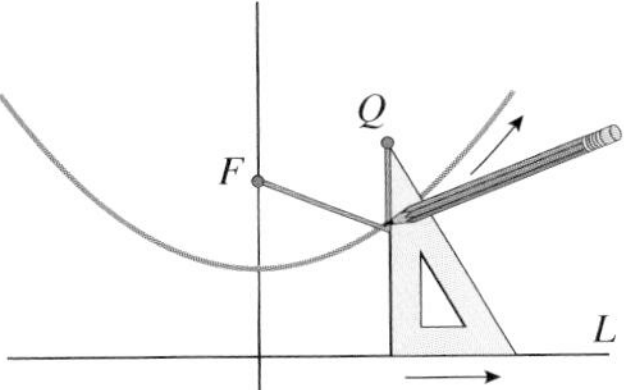

Figure Ex-58

59. The accompanying figure shows a method for constructing a hyperbola. A corner of a ruler is pinned to a fixed point F_1 and the ruler is free to rotate about that point. A piece of string whose length is less than that of the ruler is tacked to a point F_2 and to the free corner Q of the ruler on the same edge as F_1. A pencil holds the string taut against the top edge of the ruler as the ruler rotates about the point F_1. Show that the pencil traces an arc of a hyperbola with foci F_1 and F_2.

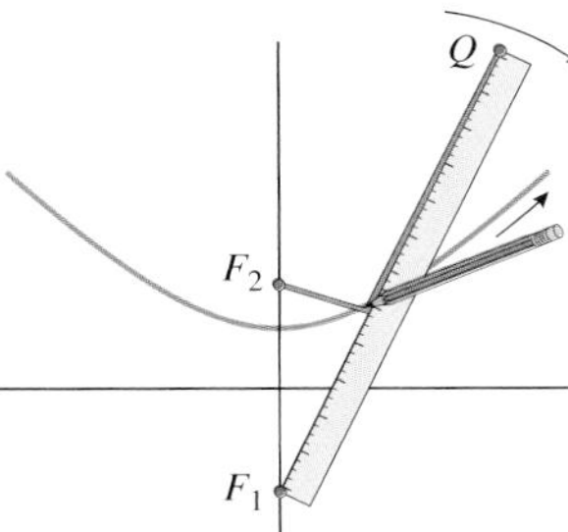

Figure Ex-59

60. Show that if a plane is not parallel to the axis of a right circular cylinder, then the intersection of the plane and cylinder is an ellipse (possibly a circle). [*Hint:* Let θ be the angle shown in the accompanying figure, introduce coordinate axes as shown, and express x' and y' in terms of x and y.]

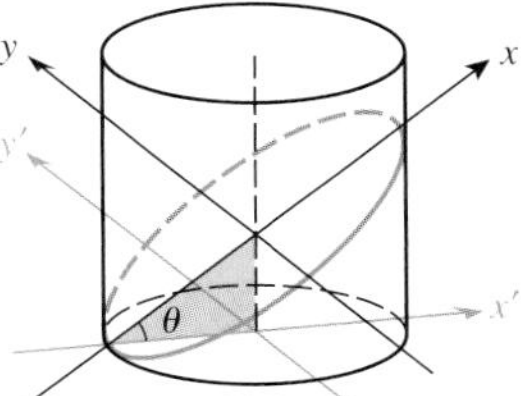

Figure Ex-60

61. As illustrated in the accompanying figure, a carpenter needs to cut an elliptical hole in a sloped roof through which a circular vent pipe of diameter D is to be inserted vertically. The carpenter wants to draw the outline of the hole on the roof using a pencil, two tacks, and a piece of string (as in Figure 11.4.3*b*). The center point of the ellipse is known, and common sense suggests that its major axis must be perpendicular to the drip line of the roof. The carpenter needs to determine the length L of the string and the distance T between a tack and the center point. The architect's plans show that the pitch of the roof is p (pitch = rise over run; see the accompanying figure). Find T and L in terms of D and p.

Source: This exercise is based on an article by William H. Enos, which appeared in the *Mathematics Teacher*, Feb. 1991, p. 148.

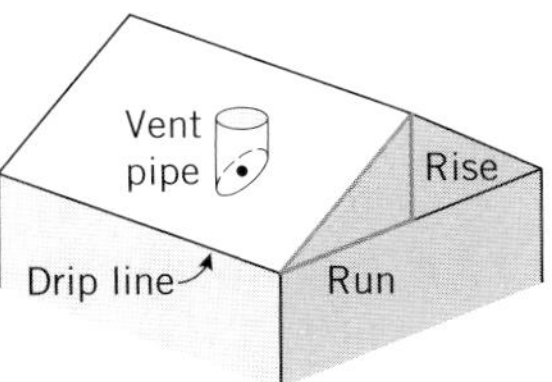

Figure Ex-61

62. Prove: The line tangent to the parabola $x^2 = 4py$ at the point (x_0, y_0) is $x_0 x = 2p(y + y_0)$.

63. Prove: The line tangent to the ellipse

$$\frac{x^2}{a^2} + \frac{y^2}{b^2} = 1$$

at the point (x_0, y_0) has the equation

$$\frac{xx_0}{a^2} + \frac{yy_0}{b^2} = 1$$

64. Prove: The line tangent to the hyperbola

$$\frac{x^2}{a^2} - \frac{y^2}{b^2} = 1$$

at the point (x_0, y_0) has the equation

$$\frac{xx_0}{a^2} - \frac{yy_0}{b^2} = 1$$

65. Use the results in Exercises 63 and 64 to show that if an ellipse and a hyperbola have the same foci, then at each point of intersection their tangent lines are perpendicular.

66. Find two values of k such that the line $x + 2y = k$ is tangent to the ellipse $x^2 + 4y^2 = 8$. Find the points of tangency.

67. Find the coordinates of all points on the hyperbola

$$4x^2 - y^2 = 4$$

where the two lines that pass through the point and the foci are perpendicular.

68. A line tangent to the hyperbola $4x^2 - y^2 = 36$ intersects the y-axis at the point $(0, 4)$. Find the point(s) of tangency.

FOCUS ON CONCEPTS

69. As illustrated in the accompanying figure, suppose that two observers are stationed at the points $F_1(c, 0)$ and $F_2(-c, 0)$ in an xy-coordinate system. Suppose also that the sound of an explosion in the xy-plane is heard by the F_1 observer t seconds before it is heard by the F_2 observer. Assuming that the speed of sound is a constant v, show that the explosion occurred somewhere on the hyperbola

$$\frac{x^2}{v^2t^2/4} - \frac{y^2}{c^2 - (v^2t^2/4)} = 1$$

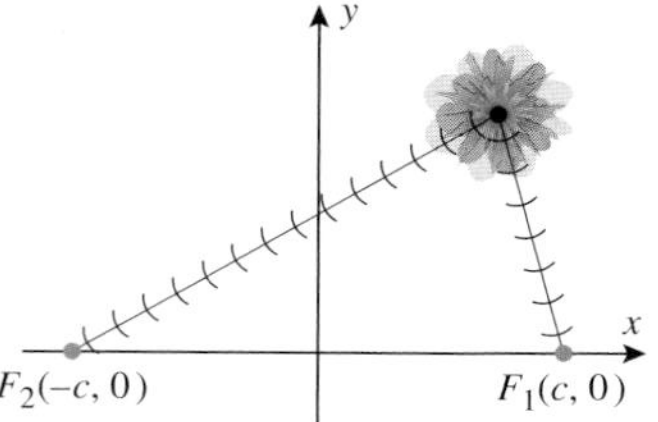

Figure Ex-69

70. As illustrated in the accompanying figure, suppose that two transmitting stations are positioned 100 km apart at points $F_1(50, 0)$ and $F_2(-50, 0)$ on a straight shoreline in an xy-coordinate system. Suppose also that a ship is traveling parallel to the shoreline but 200 km at sea. Find the coordinates of the ship if the stations transmit a pulse simultaneously, but the pulse from station F_1 is received by the ship 100 microseconds sooner than the pulse from station F_2. [*Hint:* Use the formula obtained in Exercise 69, assuming that the pulses travel at the speed of light (299,792,458 m/s).]

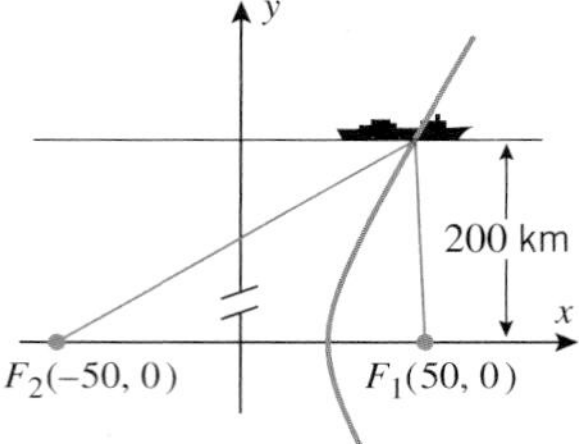

Figure Ex-70

C **71.** As illustrated in the accompanying figure, the tank of an oil truck is 18 ft long and has elliptical cross sections that are 6 ft wide and 4 ft high.

(a) Show that the volume V of oil in the tank (in cubic feet) when it is filled to a depth of h feet is

$$V = 27\left[4\sin^{-1}\frac{h-2}{2} + (h-2)\sqrt{4h-h^2} + 2\pi\right]$$

(b) Use the numerical root-finding capability of a CAS to determine how many inches from the bottom of a dipstick the calibration marks should be placed to indicate when the tank is $\frac{1}{4}$, $\frac{1}{2}$, and $\frac{3}{4}$ full.

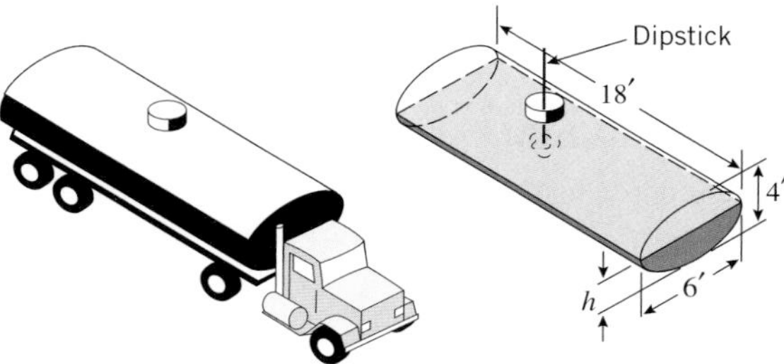

Figure Ex-71

72. Consider the second-degree equation

$$Ax^2 + Cy^2 + Dx + Ey + F = 0$$

where A and C are not both 0. Show by completing the square:

(a) If $AC > 0$, then the equation represents an ellipse, a circle, a point, or has no graph.

(b) If $AC < 0$, then the equation represents a hyperbola or a pair of intersecting lines.

(c) If $AC = 0$, then the equation represents a parabola, a pair of parallel lines, or has no graph.

73. In each part, use the result in Exercise 72 to make a statement about the graph of the equation, and then check your conclusion by completing the square and identifying the graph.

(a) $x^2 - 5y^2 - 2x - 10y - 9 = 0$

(b) $x^2 - 3y^2 - 6y - 3 = 0$

(c) $4x^2 + 8y^2 + 16x + 16y + 20 = 0$
(d) $3x^2 + y^2 + 12x + 2y + 13 = 0$
(e) $x^2 + 8x + 2y + 14 = 0$
(f) $5x^2 + 40x + 2y + 94 = 0$

74. Derive the equation $x^2 = 4py$ in Figure 11.4.6.

75. Derive the equation $(x^2/b^2) + (y^2/a^2) = 1$ given in Figure 11.4.14.

76. Derive the equation $(x^2/a^2) - (y^2/b^2) = 1$ given in Figure 11.4.22.

77. Prove Theorem 11.4.4. [*Hint:* Choose coordinate axes so that the parabola has the equation $x^2 = 4py$. Show that the tangent line at $P(x_0, y_0)$ intersects the y-axis at $Q(0, -y_0)$ and that the triangle whose three vertices are at P, Q, and the focus is isosceles.]

78. Given two intersecting lines, let L_2 be the line with the larger angle of inclination ϕ_2, and let L_1 be the line with the smaller angle of inclination ϕ_1. We define the ***angle θ between L_1 and L_2*** by $\theta = \phi_2 - \phi_1$. (See the accompanying figure.)

(a) Prove: If L_1 and L_2 are not perpendicular, then

$$\tan\theta = \frac{m_2 - m_1}{1 + m_1 m_2}$$

where L_1 and L_2 have slopes m_1 and m_2.

(b) Prove Theorem 11.4.5. [*Hint:* Introduce coordinates so that the equation $x^2/a^2 + y^2/b^2 = 1$ describes the ellipse, and use part (a).]

(c) Prove Theorem 11.4.6. [*Hint:* Introduce coordinates so that the equation $x^2/a^2 - y^2/b^2 = 1$ describes the hyperbola, and use part (a).]

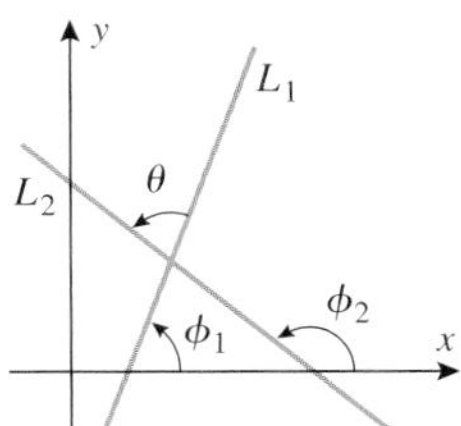

Figure Ex-78

✔ QUICK CHECK ANSWERS 11.4

1. (a) an ellipse (b) a hyperbola (c) a parabola **2.** (a) $y^2 = 4px$ (b) $x^2 = 4py$
3. (a) $2a$ (b) $\dfrac{x^2}{a^2} + \dfrac{y^2}{b^2} = 1$; $\dfrac{x^2}{b^2} + \dfrac{y^2}{a^2} = 1$ (c) $\sqrt{a^2 - b^2}$ **4.** (a) $2a$ (b) $\dfrac{x^2}{a^2} - \dfrac{y^2}{b^2} = 1$; $\dfrac{y^2}{a^2} - \dfrac{x^2}{b^2} = 1$ (c) $\sqrt{a^2 + b^2}$; $\dfrac{b}{a}x$

11.5 ROTATION OF AXES; SECOND-DEGREE EQUATIONS

In the preceding section we obtained equations of conic sections with axes parallel to the coordinate axes. In this section we will study the equations of conics that are "tilted" relative to the coordinate axes. This will lead us to investigate rotations of coordinate axes.

QUADRATIC EQUATIONS IN x AND y

We saw in Examples 8 to 10 of the preceding section that equations of the form

$$Ax^2 + Cy^2 + Dx + Ey + F = 0 \tag{1}$$

can represent conic sections. Equation (1) is a special case of the more general equation

$$Ax^2 + Bxy + Cy^2 + Dx + Ey + F = 0 \tag{2}$$

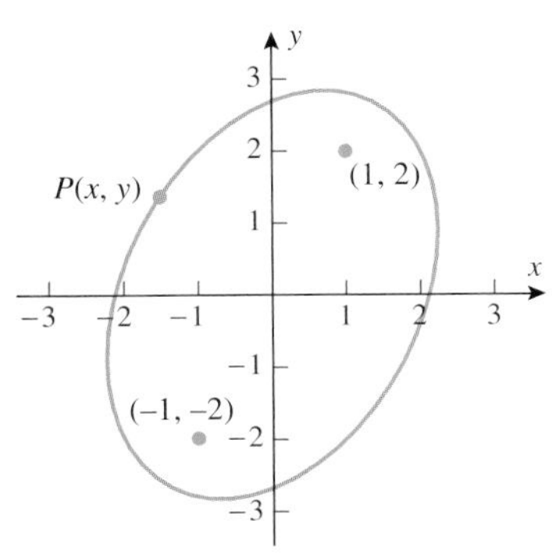

Figure 11.5.1

which, if A, B, and C are not all zero, is called a ***quadratic equation*** in x and y. It is usually the case that the graph of any second-degree equation is a conic section. If $B = 0$, then (2) reduces to (1) and the conic section has its axis or axes parallel to the coordinate axes. However, if $B \neq 0$, then (2) contains a "cross-product" term Bxy, and the graph of the conic section represented by the equation has its axis or axes "tilted" relative to the coordinate axes. As an illustration, consider the ellipse with foci $F_1(1, 2)$ and $F_2(-1, -2)$ and such that the sum of the distances from each point $P(x, y)$ on the ellipse to the foci is 6 units. Expressing this condition as an equation, we obtain (Figure 11.5.1)

$$\sqrt{(x-1)^2 + (y-2)^2} + \sqrt{(x+1)^2 + (y+2)^2} = 6$$

Squaring both sides, then isolating the remaining radical, then squaring again ultimately yields

$$8x^2 - 4xy + 5y^2 = 36$$

as the equation of the ellipse. This is of form (2) with $A = 8$, $B = -4$, $C = 5$, $D = 0$, $E = 0$, and $F = -36$.

ROTATION OF AXES

To study conics that are tilted relative to the coordinate axes it is frequently helpful to rotate the coordinate axes, so that the rotated coordinate axes are parallel to the axes of the conic. Before we can discuss the details, we need to develop some ideas about rotation of coordinate axes.

In Figure 11.5.2*a* the axes of an xy-coordinate system have been rotated about the origin through an angle θ to produce a new $x'y'$-coordinate system. As shown in the figure, each point P in the plane has coordinates (x', y') as well as coordinates (x, y). To see how the two are related, let r be the distance from the common origin to the point P, and let α be the angle shown in Figure 11.5.2*b*. It follows that

$$x = r\cos(\theta + \alpha), \quad y = r\sin(\theta + \alpha) \tag{3}$$

and

$$x' = r\cos\alpha, \quad y' = r\sin\alpha \tag{4}$$

Using familiar trigonometric identities, the relationships in (3) can be written as

$$x = r\cos\theta\cos\alpha - r\sin\theta\sin\alpha$$
$$y = r\sin\theta\cos\alpha + r\cos\theta\sin\alpha$$

and on substituting (4) in these equations we obtain the following relationships called the ***rotation equations***:

$$\begin{aligned} x &= x'\cos\theta - y'\sin\theta \\ y &= x'\sin\theta + y'\cos\theta \end{aligned} \tag{5}$$

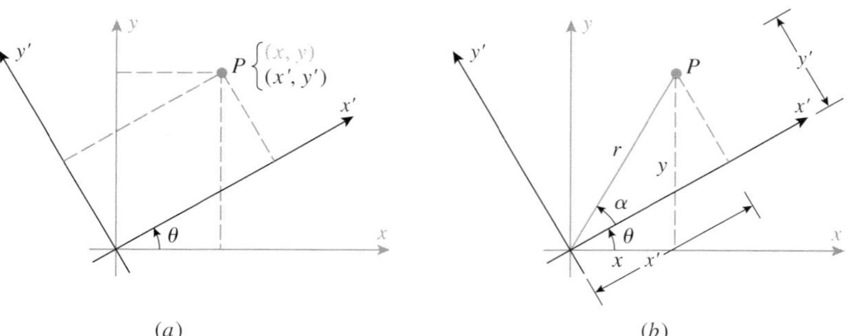

Figure 11.5.2

▶ **Example 1** Suppose that the axes of an xy-coordinate system are rotated through an angle of $\theta = 45°$ to obtain an $x'y'$-coordinate system. Find the equation of the curve

$$x^2 - xy + y^2 - 6 = 0$$

in $x'y'$-coordinates.

Solution. Substituting $\sin\theta = \sin 45° = 1/\sqrt{2}$ and $\cos\theta = \cos 45° = 1/\sqrt{2}$ in (5) yields the rotation equations

$$x = \frac{x'}{\sqrt{2}} - \frac{y'}{\sqrt{2}} \quad \text{and} \quad y = \frac{x'}{\sqrt{2}} + \frac{y'}{\sqrt{2}}$$

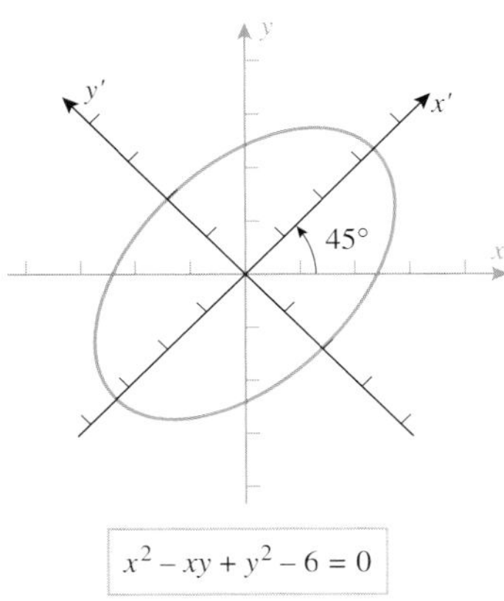

Figure 11.5.3

Substituting these into the given equation yields

$$\left(\frac{x'}{\sqrt{2}} - \frac{y'}{\sqrt{2}}\right)^2 - \left(\frac{x'}{\sqrt{2}} - \frac{y'}{\sqrt{2}}\right)\left(\frac{x'}{\sqrt{2}} + \frac{y'}{\sqrt{2}}\right) + \left(\frac{x'}{\sqrt{2}} + \frac{y'}{\sqrt{2}}\right)^2 - 6 = 0$$

or

$$\frac{x'^2 - 2x'y' + y'^2 - x'^2 + y'^2 + x'^2 + 2x'y' + y'^2}{2} = 6$$

or

$$\frac{x'^2}{12} + \frac{y'^2}{4} = 1$$

which is the equation of an ellipse (Figure 11.5.3). ◀

If the rotation equations (5) are solved for x' and y' in terms of x and y, one obtains (Exercise 16):

$$\begin{aligned} x' &= x\cos\theta + y\sin\theta \\ y' &= -x\sin\theta + y\cos\theta \end{aligned} \tag{6}$$

▶ **Example 2** Find the new coordinates of the point $(2, 4)$ if the coordinate axes are rotated through an angle of $\theta = 30^\circ$.

Solution. Using the rotation equations in (6) with $x = 2$, $y = 4$, $\cos\theta = \cos 30^\circ = \sqrt{3}/2$, and $\sin\theta = \sin 30^\circ = 1/2$, we obtain

$$\begin{aligned} x' &= 2(\sqrt{3}/2) + 4(1/2) = \sqrt{3} + 2 \\ y' &= -2(1/2) + 4(\sqrt{3}/2) = -1 + 2\sqrt{3} \end{aligned}$$

Thus, the new coordinates are $(\sqrt{3} + 2, -1 + 2\sqrt{3})$. ◀

ELIMINATING THE CROSS-PRODUCT TERM

In Example 1 we were able to identify the curve $x^2 - xy + y^2 - 6 = 0$ as an ellipse because the rotation of axes eliminated the xy-term, thereby reducing the equation to a familiar form. This occurred because the new $x'y'$-axes were aligned with the axes of the ellipse. The following theorem tells how to determine an appropriate rotation of axes to eliminate the cross-product term of a second-degree equation in x and y.

It is always possible to satisfy (8) with an angle θ in the interval

$$0 < \theta < \pi/2$$

We will always choose θ in this way.

11.5.1 THEOREM. *If the equation*

$$Ax^2 + Bxy + Cy^2 + Dx + Ey + F = 0 \tag{7}$$

is such that $B \neq 0$, and if an $x'y'$-coordinate system is obtained by rotating the xy-axes through an angle θ satisfying

$$\cot 2\theta = \frac{A - C}{B} \tag{8}$$

then, in $x'y'$-coordinates, Equation (7) will have the form

$$A'x'^2 + C'y'^2 + D'x' + E'y' + F' = 0$$

PROOF. Substituting (5) into (7) and simplifying yields

$$A'x'^2 + B'x'y' + C'y'^2 + D'x' + E'y' + F' = 0$$

where

$$\begin{aligned} A' &= A\cos^2\theta + B\cos\theta\sin\theta + C\sin^2\theta \\ B' &= B(\cos^2\theta - \sin^2\theta) + 2(C - A)\sin\theta\cos\theta \\ C' &= A\sin^2\theta - B\sin\theta\cos\theta + C\cos^2\theta \\ D' &= D\cos\theta + E\sin\theta \\ E' &= -D\sin\theta + E\cos\theta \\ F' &= F \end{aligned} \tag{9}$$

(Verify.) To complete the proof we must show that $B' = 0$ if

$$\cot 2\theta = \frac{A - C}{B}$$

or, equivalently,

$$\frac{\cos 2\theta}{\sin 2\theta} = \frac{A - C}{B} \tag{10}$$

However, by using the trigonometric double-angle formulas, we can rewrite B' in the form

$$B' = B\cos 2\theta - (A - C)\sin 2\theta$$

Thus, $B' = 0$ if θ satisfies (10). ■

▶ **Example 3** Identify and sketch the curve $xy = 1$.

Solution. As a first step, we will rotate the coordinate axes to eliminate the cross-product term. Comparing the given equation to (7), we have

$$A = 0, \quad B = 1, \quad C = 0$$

Thus, the desired angle of rotation must satisfy

$$\cot 2\theta = \frac{A - C}{B} = \frac{0 - 0}{1} = 0$$

This condition can be met by taking $2\theta = \pi/2$ or $\theta = \pi/4 = 45°$. Making the substitutions $\cos\theta = \cos 45° = 1/\sqrt{2}$ and $\sin\theta = \sin 45° = 1/\sqrt{2}$ in (5) yields

$$x = \frac{x'}{\sqrt{2}} - \frac{y'}{\sqrt{2}} \quad \text{and} \quad y = \frac{x'}{\sqrt{2}} + \frac{y'}{\sqrt{2}}$$

Substituting these in the equation $xy = 1$ yields

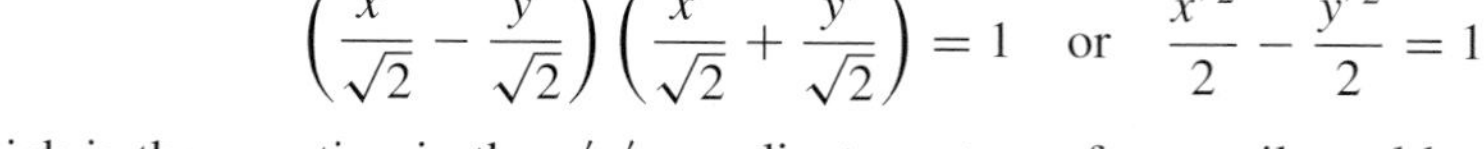

$$\left(\frac{x'}{\sqrt{2}} - \frac{y'}{\sqrt{2}}\right)\left(\frac{x'}{\sqrt{2}} + \frac{y'}{\sqrt{2}}\right) = 1 \quad \text{or} \quad \frac{x'^2}{2} - \frac{y'^2}{2} = 1$$

which is the equation in the $x'y'$-coordinate system of an equilateral hyperbola with vertices at $(\sqrt{2}, 0)$ and $(-\sqrt{2}, 0)$ in that coordinate system (Figure 11.5.4). ◀

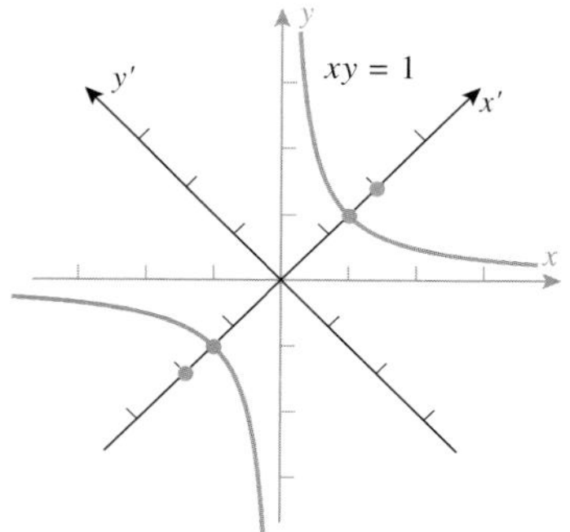

Figure 11.5.4

In problems where it is inconvenient to solve

$$\cot 2\theta = \frac{A - C}{B}$$

for θ, the values of $\sin\theta$ and $\cos\theta$ needed for the rotation equations can be obtained by first calculating $\cos 2\theta$ and then computing $\sin\theta$ and $\cos\theta$ from the identities

$$\sin\theta = \sqrt{\frac{1 - \cos 2\theta}{2}} \quad \text{and} \quad \cos\theta = \sqrt{\frac{1 + \cos 2\theta}{2}}$$

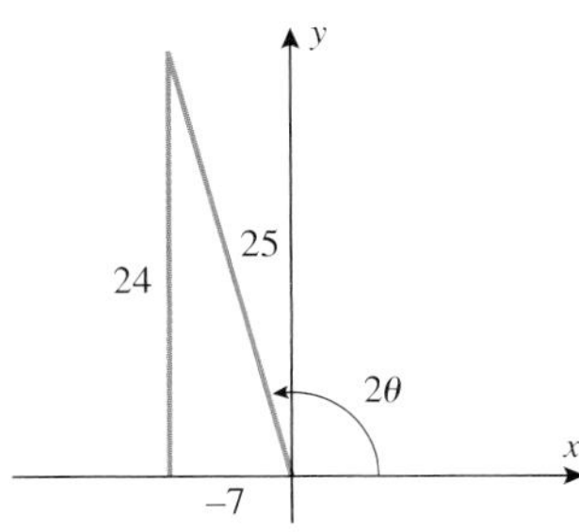

Figure 11.5.5

▶ **Example 4** Identify and sketch the curve

$$153x^2 - 192xy + 97y^2 - 30x - 40y - 200 = 0$$

Solution. We have $A = 153$, $B = -192$, and $C = 97$, so

$$\cot 2\theta = \frac{A - C}{B} = -\frac{56}{192} = -\frac{7}{24}$$

Since θ is to be chosen in the range $0 < \theta < \pi/2$, this relationship is represented by the triangle in Figure 11.5.5. From that triangle we obtain $\cos 2\theta = -\frac{7}{25}$, which implies that

$$\cos\theta = \sqrt{\frac{1 + \cos 2\theta}{2}} = \sqrt{\frac{1 - \frac{7}{25}}{2}} = \frac{3}{5}$$

$$\sin\theta = \sqrt{\frac{1 - \cos 2\theta}{2}} = \sqrt{\frac{1 + \frac{7}{25}}{2}} = \frac{4}{5}$$

Substituting these values in (5) yields the rotation equations

$$x = \tfrac{3}{5}x' - \tfrac{4}{5}y' \quad \text{and} \quad y = \tfrac{4}{5}x' + \tfrac{3}{5}y'$$

and substituting these in turn in the given equation yields

$$\tfrac{153}{25}(3x' - 4y')^2 - \tfrac{192}{25}(3x' - 4y')(4x' + 3y') + \tfrac{97}{25}(4x' + 3y')^2$$
$$- \tfrac{30}{5}(3x' - 4y') - \tfrac{40}{5}(4x' + 3y') - 200 = 0$$

which simplifies to

$$25x'^2 + 225y'^2 - 50x' - 200 = 0$$

or

$$x'^2 + 9y'^2 - 2x' - 8 = 0$$

Completing the square yields

$$\frac{(x' - 1)^2}{9} + y'^2 = 1$$

which is the equation in the $x'y'$-coordinate system of an ellipse with center $(1, 0)$ in that coordinate system and semiaxes $a = 3$ and $b = 1$ (Figure 11.5.6). ◀

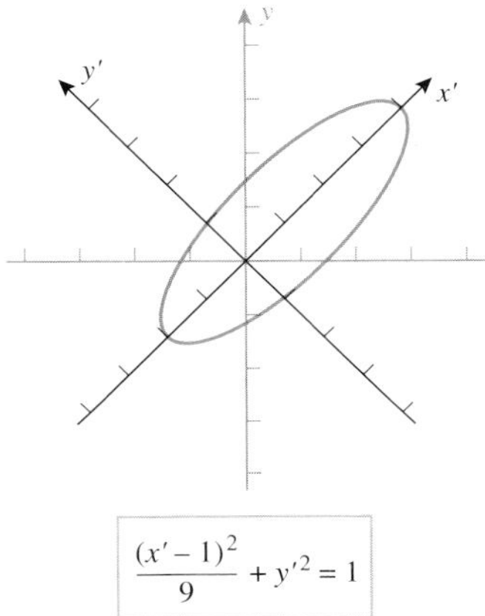

Figure 11.5.6

There is a method for deducing the kind of curve represented by a second-degree equation directly from the equation itself without rotating coordinate axes. For a discussion of this topic, see the section on the ***discriminant*** that appears in Web Appendix H.

✔QUICK CHECK EXERCISES 11.5 *(See page 771 for answers.)*

1. Suppose that an xy-coordinate system is rotated θ radians to produce a new $x'y'$-coordinate system.
   (a) x and y may be obtained from x', y', and θ using the rotation equations $x =$ ______ and $y =$ ______.
   (b) x' and y' may be obtained from x, y, and θ using the equations $x' =$ ______ and $y' =$ ______.
2. If the equation
$$Ax^2 + Bxy + Cy^2 + Dx + Ey + F = 0$$
is such that $B \neq 0$, then the xy-term in this equation can be eliminated by a rotation of axes through an angle θ satisfying $\cot 2\theta =$ ______.
3. In each part, determine a rotation angle θ that will eliminate the xy-term.
   (a) $2x^2 + xy + 2y^2 + x - y = 0$
   (b) $x^2 + 2\sqrt{3}xy + 3y^2 - 2x + y = 1$
   (c) $3x^2 + \sqrt{3}xy + 2y^2 + y = 0$
4. Express $2x^2 + xy + 2y^2 = 1$ in the $x'y'$-coordinate system obtained by rotating the xy-coordinate system through the angle $\theta = \pi/4$.

EXERCISE SET 11.5

1. Let an $x'y'$-coordinate system be obtained by rotating an xy-coordinate system through an angle of $\theta = 60°$.
   (a) Find the $x'y'$-coordinates of the point whose xy-coordinates are $(-2, 6)$.
   (b) Find an equation of the curve $\sqrt{3}xy + y^2 = 6$ in $x'y'$-coordinates.
   (c) Sketch the curve in part (b), showing both xy-axes and $x'y'$-axes.
2. Let an $x'y'$-coordinate system be obtained by rotating an xy-coordinate system through an angle of $\theta = 30°$.
   (a) Find the $x'y'$-coordinates of the point whose xy-coordinates are $(1, -\sqrt{3})$.
   (b) Find an equation of the curve $2x^2 + 2\sqrt{3}xy = 3$ in $x'y'$-coordinates.
   (c) Sketch the curve in part (b), showing both xy-axes and $x'y'$-axes.

3–12 Rotate the coordinate axes to remove the xy-term. Then identify the type of conic and sketch its graph.

3. $xy = -9$
4. $x^2 - xy + y^2 - 2 = 0$
5. $x^2 + 4xy - 2y^2 - 6 = 0$
6. $31x^2 + 10\sqrt{3}xy + 21y^2 - 144 = 0$
7. $x^2 + 2\sqrt{3}xy + 3y^2 + 2\sqrt{3}x - 2y = 0$
8. $34x^2 - 24xy + 41y^2 - 25 = 0$
9. $9x^2 - 24xy + 16y^2 - 80x - 60y + 100 = 0$
10. $5x^2 - 6xy + 5y^2 - 8\sqrt{2}x + 8\sqrt{2}y = 8$
11. $52x^2 - 72xy + 73y^2 + 40x + 30y - 75 = 0$
12. $6x^2 + 24xy - y^2 - 12x + 26y + 11 = 0$
13. Let an $x'y'$-coordinate system be obtained by rotating an xy-coordinate system through an angle of $45°$. Use (6) to find an equation of the curve $3x'^2 + y'^2 = 6$ in xy-coordinates.
14. Let an $x'y'$-coordinate system be obtained by rotating an xy-coordinate system through an angle of $30°$. Use (5) to find an equation in $x'y'$-coordinates of the curve $y = x^2$.

FOCUS ON CONCEPTS

15. Let an $x'y'$-coordinate system be obtained by rotating an xy-coordinate system through an angle θ. Prove: For every value of θ, the equation $x^2 + y^2 = r^2$ becomes the equation $x'^2 + y'^2 = r^2$. Give a geometric explanation.
16. Derive (6) by solving the rotation equations in (5) for x' and y' in terms of x and y.
17. Let an $x'y'$-coordinate system be obtained by rotating an xy-coordinate system through an angle θ. Explain how to find the xy-coordinates of a point whose $x'y'$-coordinates are known.
18. Let an $x'y'$-coordinate system be obtained by rotating an xy-coordinate system through an angle θ. Explain how to find the xy-equation of a line whose $x'y'$-equation is known.

19–22 Show that the graph of the given equation is a parabola. Find its vertex, focus, and directrix.

19. $x^2 + 2xy + y^2 + 4\sqrt{2}x - 4\sqrt{2}y = 0$
20. $x^2 - 2\sqrt{3}xy + 3y^2 - 8\sqrt{3}x - 8y = 0$
21. $9x^2 - 24xy + 16y^2 - 80x - 60y + 100 = 0$
22. $x^2 + 2\sqrt{3}xy + 3y^2 + 16\sqrt{3}x - 16y - 96 = 0$

23–26 Show that the graph of the given equation is an ellipse. Find its foci, vertices, and the ends of its minor axis.

23. $288x^2 - 168xy + 337y^2 - 3600 = 0$
24. $25x^2 - 14xy + 25y^2 - 288 = 0$
25. $31x^2 + 10\sqrt{3}xy + 21y^2 - 32x + 32\sqrt{3}y - 80 = 0$
26. $43x^2 - 14\sqrt{3}xy + 57y^2 - 36\sqrt{3}x - 36y - 540 = 0$

27–30 Show that the graph of the given equation is a hyperbola. Find its foci, vertices, and asymptotes.

27. $x^2 - 10\sqrt{3}xy + 11y^2 + 64 = 0$
28. $17x^2 - 312xy + 108y^2 - 900 = 0$
29. $32y^2 - 52xy - 7x^2 + 72\sqrt{5}x - 144\sqrt{5}y + 900 = 0$
30. $2\sqrt{2}y^2 + 5\sqrt{2}xy + 2\sqrt{2}x^2 + 18x + 18y + 36\sqrt{2} = 0$
31. Show that the graph of the equation
$$\sqrt{x} + \sqrt{y} = 1$$
is a portion of a parabola. [*Hint:* First rationalize the equation and then perform a rotation of axes.]

FOCUS ON CONCEPTS

32. Derive the expression for B' in (9).
33. Use (9) to prove that $B^2 - 4AC = B'^2 - 4A'C'$ for all values of θ.
34. Use (9) to prove that $A + C = A' + C'$ for all values of θ.
35. Prove: If $A = C$ in (7), then the cross-product term can be eliminated by rotating through $45°$.
36. Prove: If $B \neq 0$, then the graph of $x^2 + Bxy + F = 0$ is a hyperbola if $F \neq 0$ and two intersecting lines if $F = 0$.

✔ QUICK CHECK ANSWERS 11.5

1. (a) $x' \cos\theta - y' \sin\theta;\ x' \sin\theta + y' \cos\theta$ (b) $x \cos\theta + y \sin\theta;\ -x \sin\theta + y \cos\theta$ 2. $\dfrac{A-C}{B}$ 3. (a) $\dfrac{\pi}{4}$ (b) $\dfrac{\pi}{3}$ (c) $\dfrac{\pi}{6}$
4. $5x'^2 + 3y'^2 = 2$

11.6 CONIC SECTIONS IN POLAR COORDINATES

It will be shown later in the text that if an object moves in a gravitational field that is directed toward a fixed point (such as the center of the Sun), then the path of that object must be a conic section with the fixed point at a focus. For example, planets in our solar system move along elliptical paths with the Sun at a focus, and the comets move along parabolic, elliptical, or hyperbolic paths with the Sun at a focus, depending on the conditions under which they were born. For applications of this type it is usually desirable to express the equations of the conic sections in polar coordinates with the pole at a focus. In this section we will show how to do this.

THE FOCUS–DIRECTRIX CHARACTERIZATION OF CONICS

To obtain polar equations for the conic sections we will need the following theorem.

It is an unfortunate historical accident that the letter e is used for the base of the natural logarithm as well as for the eccentricity of conic sections. However, as a practical matter the appropriate interpretation will usually be clear from the context in which the letter is used.

11.6.1 THEOREM (*Focus–Directrix Property of Conics*). *Suppose that a point P moves in the plane determined by a fixed point (called the **focus**) and a fixed line (called the **directrix**), where the focus does not lie on the directrix. If the point moves in such a way that its distance to the focus divided by its distance to the directrix is some constant e (called the **eccentricity**), then the curve traced by the point is a conic section. Moreover, the conic is a parabola if $e = 1$, an ellipse if $0 < e < 1$, and a hyperbola if $e > 1$.*

We will not give a formal proof of this theorem; rather, we will use the specific cases in Figure 11.6.1 to illustrate the basic ideas. For the parabola, we will take the directrix to be $x = -p$, as usual; and for the ellipse and the hyperbola we will take the directrix to be $x = a^2/c$. We want to show in all three cases that if P is a point on the graph, F is the focus, and D is the directrix, then the ratio PF/PD is some constant e, where $e = 1$ for the parabola, $0 < e < 1$ for the ellipse, and $e > 1$ for the hyperbola. We will give the arguments for the parabola and ellipse and leave the argument for the hyperbola as an exercise.

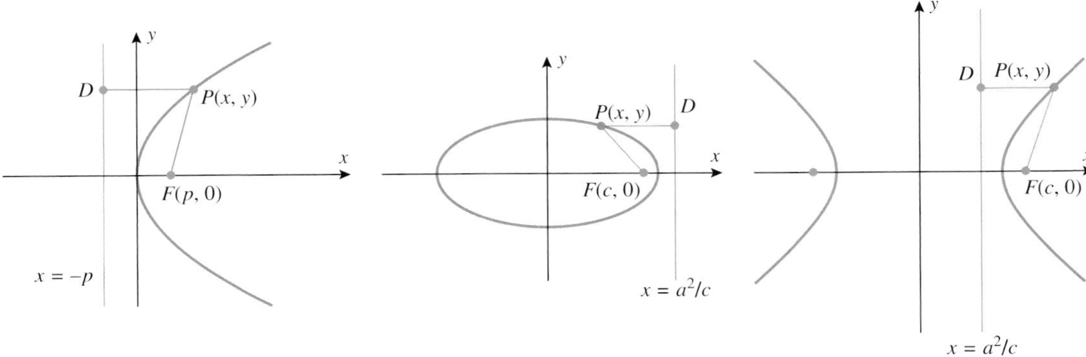

Figure 11.6.1

For the parabola, the distance PF to the focus is equal to the distance PD to the directrix, so that $PF/PD = 1$, which is what we wanted to show. For the ellipse, we rewrite Equation (8) of Section 11.4 as

$$\sqrt{(x-c)^2+y^2} = a - \frac{c}{a}x = \frac{c}{a}\left(\frac{a^2}{c} - x\right)$$

But the expression on the left side is the distance PF, and the expression in the parentheses on the right side is the distance PD, so we have shown that

$$PF = \frac{c}{a}PD$$

Thus, PF/PD is constant, and the eccentricity is

$$e = \frac{c}{a} \tag{1}$$

If we rule out the degenerate case where $a = 0$ or $c = 0$, then it follows from Formula (7) of Section 11.4 that $0 < c < a$, so $0 < e < 1$, which is what we wanted to show.

We will leave it as an exercise to show that the eccentricity of the hyperbola in Figure 11.6.1 is also given by Formula (1), but in this case it follows from Formula (11) of Section 11.4 that $c > a$, so $e > 1$.

ECCENTRICITY OF AN ELLIPSE AS A MEASURE OF FLATNESS

The eccentricity of an ellipse can be viewed as a measure of its flatness—as e approaches 0 the ellipses become more and more circular, and as e approaches 1 they become more and more flat (Figure 11.6.2). Table 11.6.1 shows the orbital eccentricities of various celestial objects. Note that most of the planets actually have fairly circular orbits.

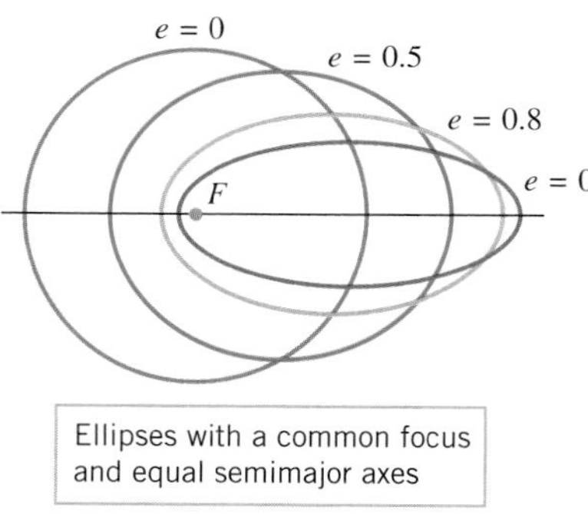

Figure 11.6.2

Table 11.6.1

CELESTIAL BODY	ECCENTRICITY
Mercury	0.206
Venus	0.007
Earth	0.017
Mars	0.093
Jupiter	0.048
Saturn	0.056
Uranus	0.046
Neptune	0.010
Pluto	0.249
Halley's comet	0.970

POLAR EQUATIONS OF CONICS

Our next objective is to derive polar equations for the conic sections from their focus–directrix characterizations. We will assume that the focus is at the pole and the directrix is either parallel or perpendicular to the polar axis. If the directrix is parallel to the polar axis, then it can be above or below the pole; and if the directrix is perpendicular to the polar axis, then it can be to the left or right of the pole. Thus, there are four cases to consider. We will derive the formulas for the case in which the directrix is perpendicular to the polar axis and to the right of the pole.

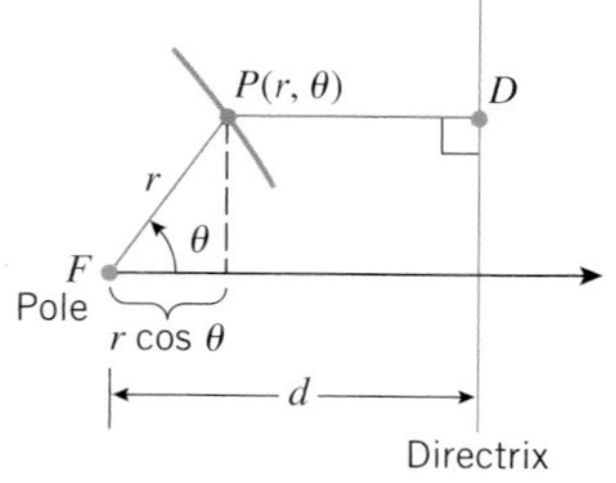

Figure 11.6.3

As illustrated in Figure 11.6.3, let us assume that the directrix is perpendicular to the polar axis and d units to the right of the pole, where the constant d is known. If P is a point on the conic and if the eccentricity of the conic is e, then it follows from Theorem 11.6.1 that $PF/PD = e$ or, equivalently, that

$$PF = ePD \tag{2}$$

However, it is evident from Figure 11.6.3 that $PF = r$ and $PD = d - r\cos\theta$. Thus, (2) can be written as

$$r = e(d - r\cos\theta)$$

which can be solved for r and expressed as

$$r = \frac{ed}{1 + e\cos\theta}$$

(verify). Observe that this single polar equation can represent a parabola, an ellipse, or a hyperbola, depending on the value of e. In contrast, the rectangular equations for these conics all have different forms. The derivations in the other three cases are similar.

11.6.2 THEOREM. *If a conic section with eccentricity e is positioned in a polar coordinate system so that its focus is at the pole and the corresponding directrix is d units from the pole and is either parallel or perpendicular to the polar axis, then the equation of the conic has one of four possible forms, depending on its orientation:*

$$r = \frac{ed}{1 + e\cos\theta} \qquad r = \frac{ed}{1 - e\cos\theta} \tag{3–4}$$

Directrix right of pole — Directrix left of pole

$$r = \frac{ed}{1 + e\sin\theta} \qquad r = \frac{ed}{1 - e\sin\theta} \tag{5–6}$$

Directrix above pole — Directrix below pole

SKETCHING CONICS IN POLAR COORDINATES

Precise graphs of conic sections in polar coordinates can be generated with graphing utilities. However, it is often useful to be able to make quick sketches of these graphs that show their orientations and give some sense of their dimensions. The orientation of a conic relative to the polar axis can be deduced by matching its equation with one of the four forms in Theorem 11.6.2. The key dimensions of a parabola are determined by the constant p (Figure 11.4.5) and those of ellipses and hyperbolas by the constants a, b, and c (Figures 11.4.11 and 11.4.20). Thus, we need to show how these constants can be obtained from the polar equations.

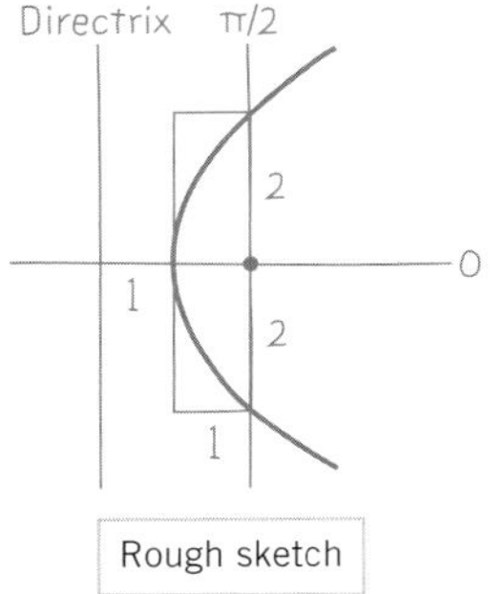

Figure 11.6.4

▶ **Example 1** Sketch the graph of $r = \dfrac{2}{1 - \cos\theta}$ in polar coordinates.

Solution. The equation is an exact match to (4) with $d = 2$ and $e = 1$. Thus, the graph is a parabola with the focus at the pole and the directrix 2 units to the left of the pole. This tells us that the parabola opens to the right along the polar axis and $p = 1$. Thus, the parabola looks roughly like that sketched in Figure 11.6.4. ◀

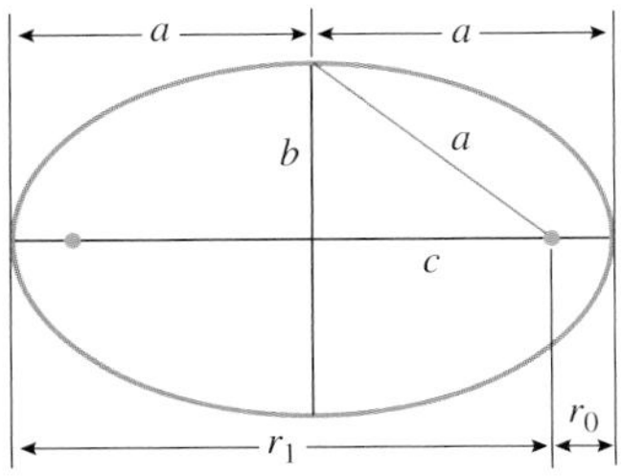

Figure 11.6.5

All of the important geometric information about an ellipse can be obtained from the values of a, b, and c in Figure 11.6.5. One way to find these values from the polar equation of an ellipse is based on finding the distances from the focus to the vertices. As shown in the figure, let r_0 be the distance from the focus to the closest vertex and r_1 the distance to the farthest vertex. Thus,

$$r_0 = a - c \quad \text{and} \quad r_1 = a + c \tag{7}$$

from which it follows that

$$a = \tfrac{1}{2}(r_1 + r_0) \qquad c = \tfrac{1}{2}(r_1 - r_0) \tag{8–9}$$

In words, Formula (8) states that a is the ***arithmetic average*** (also called the ***arithmetic mean***) of r_0 and r_1, and Formula (10) states that b is the ***geometric mean*** of r_0 and r_1.

Moreover, it also follows from (7) that

$$r_0 r_1 = a^2 - c^2 = b^2$$

Thus,

$$b = \sqrt{r_0 r_1} \tag{10}$$

▶ **Example 2** Sketch the graph of $r = \dfrac{6}{2 + \cos\theta}$ in polar coordinates.

Solution. This equation does not match any of the forms in Theorem 11.6.2 because they all require a constant term of 1 in the denominator. However, we can put the equation into one of these forms by dividing the numerator and denominator by 2 to obtain

$$r = \frac{3}{1 + \frac{1}{2}\cos\theta}$$

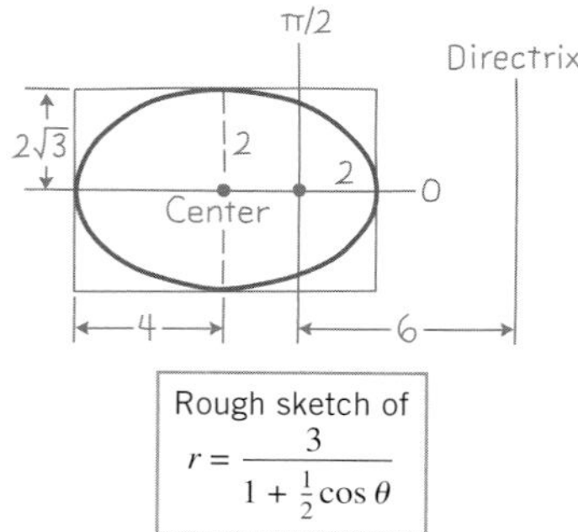

Figure 11.6.6

This is an exact match to (3) with $d = 6$ and $e = \frac{1}{2}$, so the graph is an ellipse with the directrix 6 units to the right of the pole. The distance r_0 from the focus to the closest vertex can be obtained by setting $\theta = 0$ in this equation, and the distance r_1 to the farthest vertex can be obtained by setting $\theta = \pi$. This yields

$$r_0 = \frac{3}{1 + \frac{1}{2}\cos 0} = \frac{3}{\frac{3}{2}} = 2, \quad r_1 = \frac{3}{1 + \frac{1}{2}\cos\pi} = \frac{3}{\frac{1}{2}} = 6$$

Thus, from Formulas (8), (10), and (9), respectively, we obtain

$$a = \tfrac{1}{2}(r_1 + r_0) = 4, \quad b = \sqrt{r_0 r_1} = 2\sqrt{3}, \quad c = \tfrac{1}{2}(r_1 - r_0) = 2$$

Thus, the ellipse looks roughly like that sketched in Figure 11.6.6. ◀

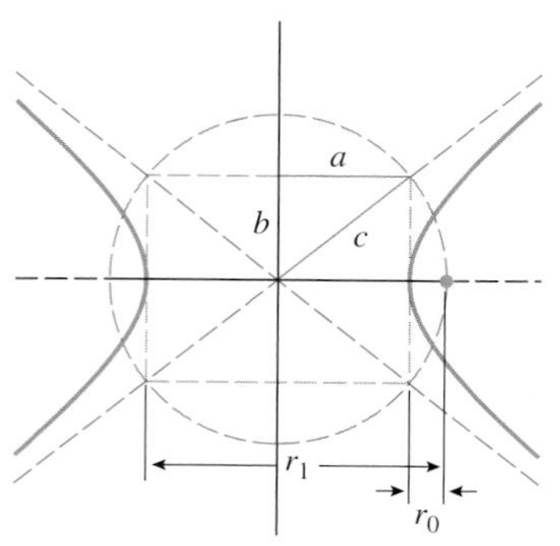

Figure 11.6.7

All of the important information about a hyperbola can be obtained from the values of a, b, and c in Figure 11.6.7. As with the ellipse, one way to find these values from the polar equation of a hyperbola is based on finding the distances from the focus to the vertices. As shown in the figure, let r_0 be the distance from the focus to the closest vertex and r_1 the distance to the farthest vertex. Thus,

$$r_0 = c - a \quad \text{and} \quad r_1 = c + a \tag{11}$$

from which it follows that

$$a = \tfrac{1}{2}(r_1 - r_0) \qquad c = \tfrac{1}{2}(r_1 + r_0) \tag{12–13}$$

In words, Formula (13) states that c is the ***arithmetic mean*** of r_0 and r_1, and Formula (14) states that b is the ***geometric mean*** of r_0 and r_1.

Moreover, it also follows from (11) that

$$r_0 r_1 = c^2 - a^2 = b^2$$

from which it follows that

$$b = \sqrt{r_0 r_1} \tag{14}$$

▶ **Example 3** Sketch the graph of $r = \dfrac{2}{1 + 2\sin\theta}$ in polar coordinates.

Solution. This equation is an exact match to (5) with $d = 1$ and $e = 2$. Thus, the graph is a hyperbola with its directrix 1 unit above the pole. However, it is not so straightforward to compute the values of r_0 and r_1, since hyperbolas in polar coordinates are generated in a strange way as θ varies from 0 to 2π. This can be seen from Figure 11.6.8a, which is the graph of the given equation in rectangular coordinates. It follows from this graph that the corresponding polar graph is generated in pieces (see Figure 11.6.8b):

- As θ varies over the interval $0 \le \theta < 7\pi/6$, the value of r is positive and varies from 2 down to $2/3$ and then to $+\infty$, which generates part of the lower branch.
- As θ varies over the interval $7\pi/6 < \theta \le 3\pi/2$, the value of r is negative and varies from $-\infty$ to -2, which generates the right part of the upper branch.
- As θ varies over the interval $3\pi/2 \le \theta < 11\pi/6$, the value of r is negative and varies from -2 to $-\infty$, which generates the left part of the upper branch.
- As θ varies over the interval $11\pi/6 < \theta \le 2\pi$, the value of r is positive and varies from $+\infty$ to 2, which fills in the missing piece of the lower right branch.

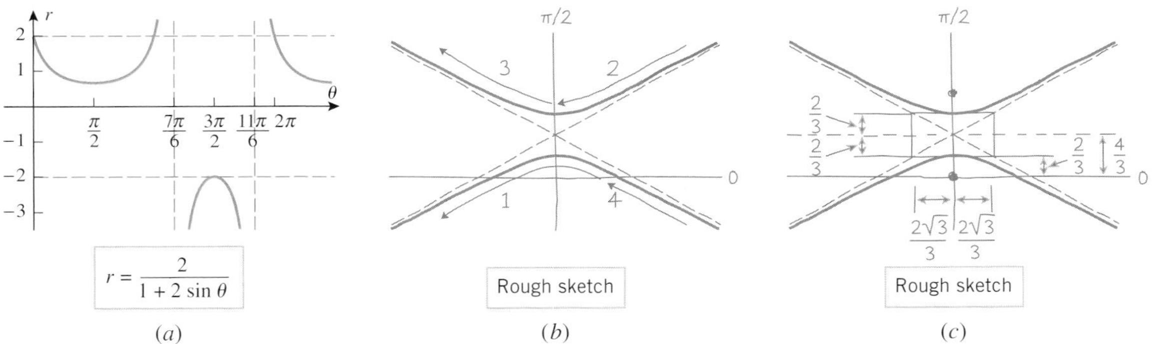

Figure 11.6.8

It is now clear that we can obtain r_0 by setting $\theta = \pi/2$ and r_1 by setting $\theta = 3\pi/2$. Keeping in mind that r_0 and r_1 are positive, this yields

$$r_0 = \frac{2}{1 + 2\sin(\pi/2)} = \frac{2}{3}, \quad r_1 = \left|\frac{2}{1 + 2\sin(3\pi/2)}\right| = \left|\frac{2}{-1}\right| = 2$$

Thus, from Formulas (12), (14), and (13), respectively, we obtain

$$a = \frac{1}{2}(r_1 - r_0) = \frac{2}{3}, \quad b = \sqrt{r_0 r_1} = \frac{2\sqrt{3}}{3}, \quad c = \frac{1}{2}(r_1 + r_0) = \frac{4}{3}$$

Thus, the hyperbola looks roughly like that sketched in Figure 11.6.8c. ◀

APPLICATIONS IN ASTRONOMY

In 1609 Johannes Kepler published a book known as *Astronomia Nova* (or sometimes *Commentaries on the Motions of Mars*) in which he succeeded in distilling thousands of years of observational astronomy into three beautiful laws of planetary motion (Figure 11.6.9).

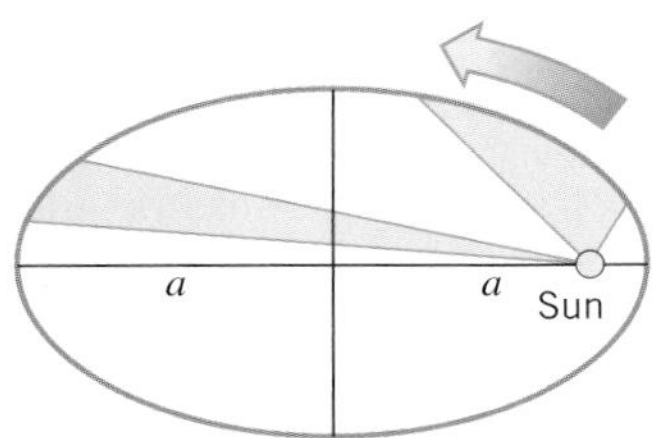

Equal areas are swept out in equal times, and the square of the period T is proportional to a^3.

Figure 11.6.9

> **11.6.3 KEPLER'S LAWS.**
>
> - First law (***Law of Orbits***). Each planet moves in an elliptical orbit with the Sun at a focus.
> - Second law (***Law of Areas***). The radial line from the center of the Sun to the center of a planet sweeps out equal areas in equal times.
> - Third law (***Law of Periods***). The square of a planet's period (the time it takes the planet to complete one orbit about the Sun) is proportional to the cube of the semimajor axis of its orbit.

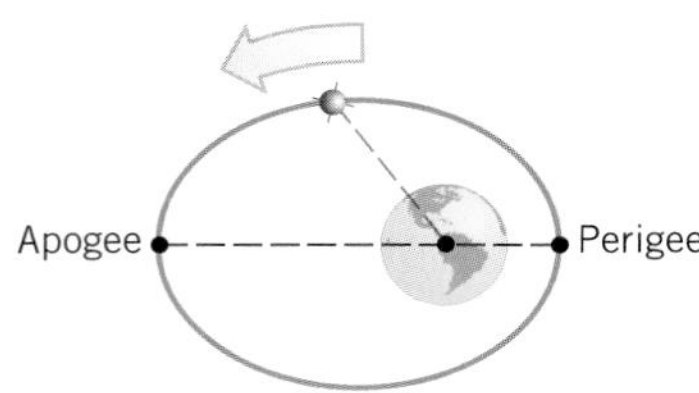

Figure 11.6.10

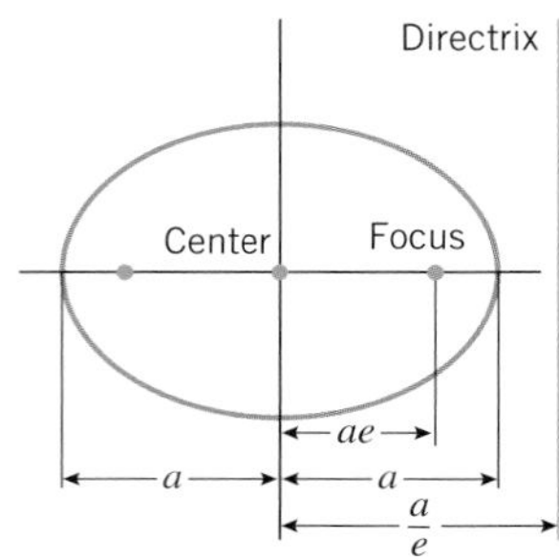

Figure 11.6.11

Kepler's laws, although stated for planetary motion around the Sun, apply to all orbiting celestial bodies that are subjected to a *single* central gravitational force—artificial satellites subjected only to the central force of Earth's gravity and moons subjected only to the central gravitational force of a planet, for example. Later in the text we will derive Kepler's laws from basic principles, but for now we will show how they can be used in basic astronomical computations.

In an elliptical orbit, the closest point to the focus is called the ***perigee*** and the farthest point the ***apogee*** (Figure 11.6.10). The distances from the focus to the perigee and apogee are called the ***perigee distance*** and ***apogee distance***, respectively. For orbits around the Sun, it is more common to use the terms ***perihelion*** and ***aphelion***, rather than perigee and apogee, and to measure time in Earth years and distances in astronomical units (AU), where 1 AU is the semimajor axis a of the Earth's orbit (approximately 150×10^6 km or 92.9×10^6 mi). With this choice of units, the constant of proportionality in Kepler's third law is 1, since $a = 1$ AU produces a period of $T = 1$ Earth year. In this case Kepler's third law can be expressed as

$$T = a^{3/2} \tag{15}$$

Shapes of elliptical orbits are often specified by giving the eccentricity e and the semimajor axis a, so it is useful to express the polar equations of an ellipse in terms of these constants. Figure 11.6.11, which can be obtained from the ellipse in Figure 11.6.1 and the

Johannes Kepler (1571–1630) German astronomer and physicist. Kepler, whose work provided our contemporary view of planetary motion, led a fascinating but ill-starred life. His alcoholic father made him work in a family-owned tavern as a child, later withdrawing him from elementary school and hiring him out as a field laborer, where the boy contracted smallpox, permanently crippling his hands and impairing his eyesight. In later years, Kepler's first wife and several children died, his mother was accused of witchcraft, and being a Protestant he was often subjected to persecution by Catholic authorities. He was often impoverished, eking out a living as an astrologer and prognosticator. Looking back on his unhappy childhood, Kepler described his father as "criminally inclined" and "quarrelsome" and his mother as "garrulous" and "bad-tempered." However, it was his mother who left an indelible mark on the six-year-old Kepler by showing him the comet of 1577; and in later life he personally prepared her defense against the witchcraft charges. Kepler became acquainted with the work of Copernicus as a student at the University of Tübingen, where he received his master's degree in 1591. He continued on as a theological student, but at the urging of the university officials he abandoned his clerical studies and accepted a position as a mathematician and teacher in Graz, Austria. However, he was expelled from the city when it came under Catholic control, and in 1600 he finally moved on to Prague, where he became an assistant at the observatory of the famous Danish astronomer Tycho Brahe. Brahe was a brilliant and meticulous astronomical observer who amassed the most accurate astronomical data known at that time; and when Brahe died in 1601 Kepler inherited the treasure-trove of data. After eight years of intense labor, Kepler deciphered the underlying principles buried in the data and in 1609 published his monumental work, *Astronomia Nova*, in which he stated his first two laws of planetary motion. Commenting on his discovery of elliptical orbits, Kepler wrote, "I was almost driven to madness in considering and calculating this matter. I could not find out why the planet would rather go on an elliptical orbit (rather than a circle). Oh ridiculous me!" It ultimately remained for Isaac Newton to discover the laws of gravitation that explained the reason for elliptical orbits.

relationship $c = ea$, implies that the distance d between the focus and the directrix is

$$d = \frac{a}{e} - c = \frac{a}{e} - ea = \frac{a(1-e^2)}{e} \tag{16}$$

from which it follows that $ed = a(1-e^2)$. Thus, depending on the orientation of the ellipse, the formulas in Theorem 11.6.2 can be expressed in terms of a and e as

$$r = \frac{a(1-e^2)}{1 \pm e\cos\theta} \qquad r = \frac{a(1-e^2)}{1 \pm e\sin\theta} \tag{17–18}$$

+: Directrix right of pole   +: Directrix above pole
−: Directrix left of pole   −: Directrix below pole

Moreover, it is evident from Figure 11.6.11 that the distances from the focus to the closest and farthest vertices can be expressed in terms of a and e as

$$r_0 = a - ea = a(1-e) \quad \text{and} \quad r_1 = a + ea = a(1+e) \tag{19–20}$$

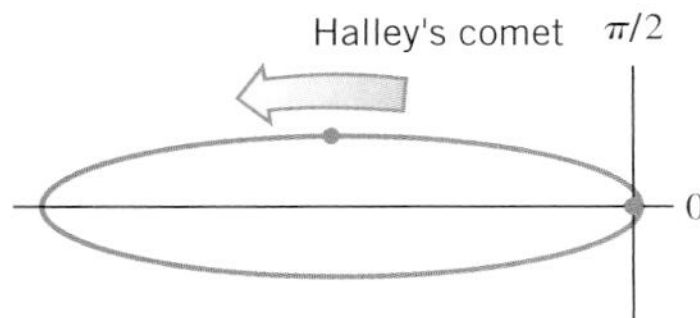

Figure 11.6.12

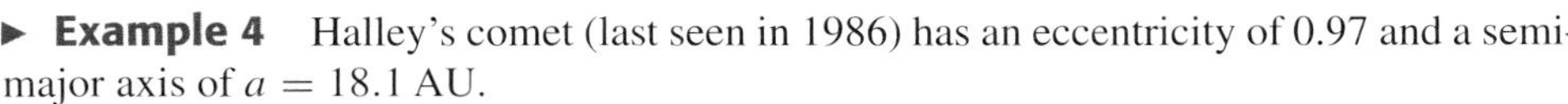

▶ **Example 4** Halley's comet (last seen in 1986) has an eccentricity of 0.97 and a semimajor axis of $a = 18.1$ AU.

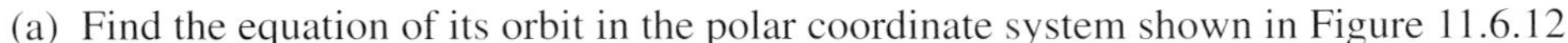

(a) Find the equation of its orbit in the polar coordinate system shown in Figure 11.6.12.
(b) Find the period of its orbit.
(c) Find its perihelion and aphelion distances.

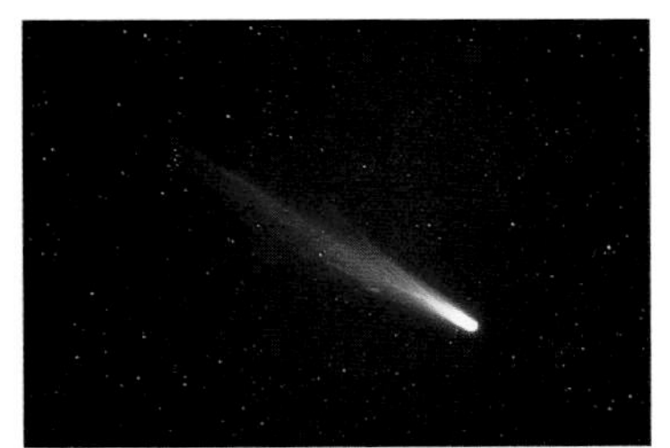
Halley's comet photographed April 21, 1910 in Peru

Solution (a). From (17), the polar equation of the orbit has the form

$$r = \frac{a(1-e^2)}{1 + e\cos\theta}$$

But $a(1-e^2) = 18.1[1 - (0.97)^2] \approx 1.07$. Thus, the equation of the orbit is

$$r = \frac{1.07}{1 + 0.97\cos\theta}$$

Solution (b). From (15), with $a = 18.1$, the period of the orbit is

$$T = (18.1)^{3/2} \approx 77 \text{ years}$$

Solution (c). Since the perihelion and aphelion distances are the distances to the closest and farthest vertices, respectively, it follows from (19) and (20) that

$$r_0 = a - ea = a(1-e) = 18.1(1 - 0.97) \approx 0.543 \text{ AU}$$
$$r_1 = a + ea = a(1+e) = 18.1(1 + 0.97) \approx 35.7 \text{ AU}$$

or since 1 AU $\approx 150 \times 10^6$ km, the perihelion and aphelion distances in kilometers are

$$r_0 = 18.1(1 - 0.97)(150 \times 10^6) \approx 81{,}500{,}000 \text{ km}$$
$$r_1 = 18.1(1 + 0.97)(150 \times 10^6) \approx 5{,}350{,}000{,}000 \text{ km} \;◀$$

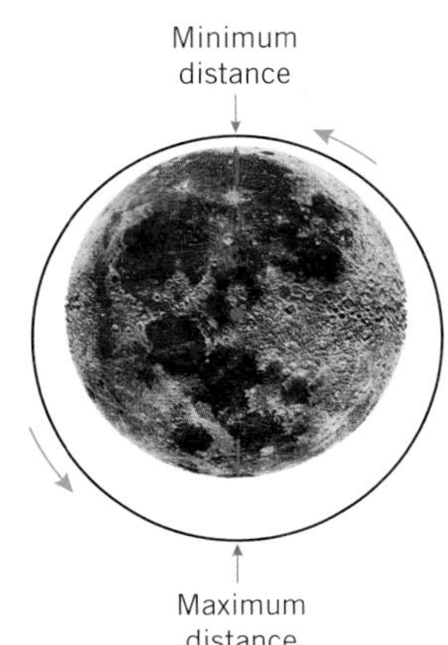

Figure 11.6.13

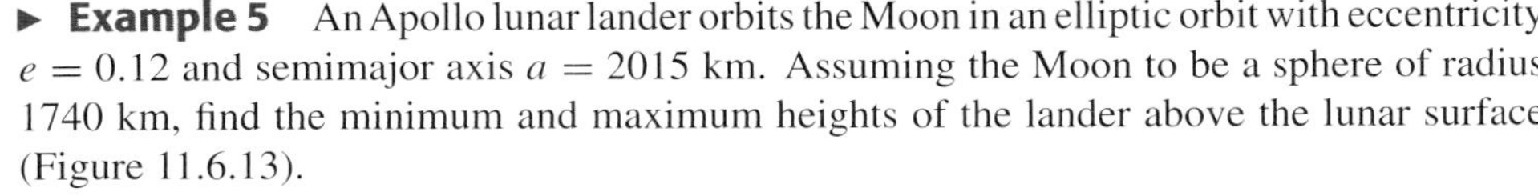

▶ **Example 5** An Apollo lunar lander orbits the Moon in an elliptic orbit with eccentricity $e = 0.12$ and semimajor axis $a = 2015$ km. Assuming the Moon to be a sphere of radius 1740 km, find the minimum and maximum heights of the lander above the lunar surface (Figure 11.6.13).

Solution. If we let r_0 and r_1 denote the minimum and maximum distances from the center of the Moon, then the minimum and maximum distances from the surface of the Moon will be

$$d_{\min} = r_0 - 1740$$
$$d_{\max} = r_1 - 1740$$

or from Formulas (19) and (20)

$$d_{\min} = r_0 - 1740 = a(1-e) - 1740 = 2015(0.88) - 1740 = 33.2 \text{ km}$$
$$d_{\max} = r_1 - 1740 = a(1+e) - 1740 = 2015(1.12) - 1740 = 516.8 \text{ km}$$ ◄

✔QUICK CHECK EXERCISES 11.6 *(See page 780 for answers.)*

1. In each part, name the conic section described.
(a) The set of points whose distance to the point (2, 3) is half the distance to the line $x + y = 1$ is ________.
(b) The set of points whose distance to the point (2, 3) is equal to the distance to the line $x + y = 1$ is ________.
(c) The set of points whose distance to the point (2, 3) is twice the distance to the line $x + y = 1$ is ________.

2. In each part: (i) Identify the polar graph as a parabola, an ellipse, or a hyperbola; (ii) state whether the directrix is above, below, to the left, or to the right of the pole; and (iii) find the distance from the pole to the directrix.
(a) $r = \dfrac{1}{4+\cos\theta}$ (b) $r = \dfrac{1}{1-4\cos\theta}$
(c) $r = \dfrac{1}{4+4\sin\theta}$ (d) $r = \dfrac{4}{1-\sin\theta}$

3. If the distance from a vertex of an ellipse to the nearest focus is r_0, and if the distance from that vertex to the farthest focus is r_1, then the semimajor axis is $a =$ ________ and the semiminor axis is $b =$ ________.

4. If the distance from a vertex of a hyperbola to the nearest focus is r_0, and if the distance from that vertex to the farthest focus is r_1, then the semifocal axis is $a =$ ________ and the semiconjugate axis is $b =$ ________.

EXERCISE SET 11.6 Graphing Utility

1–2 Find the eccentricity and the distance from the pole to the directrix, and sketch the graph in polar coordinates.

1. (a) $r = \dfrac{3}{2-2\cos\theta}$ (b) $r = \dfrac{3}{2+\sin\theta}$
(c) $r = \dfrac{4}{2+3\cos\theta}$ (d) $r = \dfrac{5}{3+3\sin\theta}$

2. (a) $r = \dfrac{3}{4-2\cos\theta}$ (b) $r = \dfrac{3}{3-6\sin\theta}$
(c) $r = \dfrac{1}{2+2\cos\theta}$ (d) $r = \dfrac{1}{2+8\cos\theta}$

3–4 Use Formulas (3)–(6) to identify the type of conic and its orientation. Check your answer by generating the graph with a graphing utility.

3. (a) $r = \dfrac{8}{1-\sin\theta}$ (b) $r = \dfrac{16}{4+3\sin\theta}$
(c) $r = \dfrac{4}{2-3\sin\theta}$ (d) $r = \dfrac{12}{4+\cos\theta}$

4. (a) $r = \dfrac{15}{1+\cos\theta}$ (b) $r = \dfrac{2}{3+3\cos\theta}$
(c) $r = \dfrac{64}{7-12\sin\theta}$ (d) $r = \dfrac{12}{3-2\cos\theta}$

5–8 Find a polar equation for the conic that has its focus at the pole and satisfies the stated conditions. Points are in polar coordinates and directrices in rectangular coordinates for simplicity. (In some cases there may be more than one conic that satisfies the conditions.)

5. (a) Ellipse; $e = \frac{3}{4}$; directrix $x = 2$.
(b) Parabola; directrix $x = 1$.
(c) Hyperbola; $e = \frac{4}{3}$; directrix $y = 3$.

6. (a) Ellipse; $e = \frac{2}{3}$; directrix $y = -1$.
(b) Parabola; directrix $y = 1$.
(c) Hyperbola; $e = \frac{4}{3}$; directrix $x = -1$.

7. (a) Ellipse; vertices $(6, 0)$ and $(4, \pi)$.
(b) Parabola; vertex $(1, 3\pi/2)$.
(c) Hyperbola; vertices $(3, \pi/2)$ and $(-7, 3\pi/2)$.

8. (a) Ellipse; ends of major axis $(2, \pi/2)$ and $(6, 3\pi/2)$.
(b) Parabola; vertex $(2, \pi)$.
(c) Hyperbola; $e = \sqrt{2}$; vertex $(2, 0)$.

9–10 Find the distances from the pole to the vertices, and then apply Formulas (8)–(10) to find the equation of the ellipse in rectangular coordinates.

9. (a) $r = \dfrac{6}{2 + \sin\theta}$ (b) $r = \dfrac{1}{2 - \cos\theta}$

10. (a) $r = \dfrac{6}{5 + 2\cos\theta}$ (b) $r = \dfrac{8}{4 - 3\sin\theta}$

11–12 Find the distances from the pole to the vertices, and then apply Formulas (12)–(14) to find the equation of the hyperbola in rectangular coordinates.

11. (a) $r = \dfrac{3}{1 + 2\sin\theta}$ (b) $r = \dfrac{5}{2 - 3\cos\theta}$

12. (a) $r = \dfrac{4}{1 - 2\sin\theta}$ (b) $r = \dfrac{15}{2 + 8\cos\theta}$

13–14 Find a polar equation for the ellipse that has its focus at the pole and satisfies the stated conditions.

13. (a) Directrix to the right of the pole; $a = 8$; $e = \frac{1}{2}$.
(b) Directrix below the pole; $a = 4$; $e = \frac{3}{5}$.
(c) Directrix to the left of the pole; $b = 4$; $e = \frac{3}{5}$.
(d) Directrix above the pole; $c = 5$; $e = \frac{1}{5}$.

14. (a) Directrix above the pole; $a = 6$; $e = \frac{1}{2}$.
(b) Directrix to the left of the pole; $a = 5$; $e = \frac{1}{5}$.
(c) Directrix below the pole; $b = 4$; $e = \frac{4}{5}$.
(d) Directrix to the right of the pole; $c = 9$; $e = \frac{3}{4}$.

FOCUS ON CONCEPTS

15. Prove that a hyperbola is an equilateral hyperbola if and only if $e = \sqrt{2}$.

16. How is the shape of a hyperbola affected as its eccentricity approaches 1? As it approaches $+\infty$? Draw some pictures to illustrate your conclusions.

17. What happens to the distance between the directrix and the center of an ellipse if the foci remain fixed and the eccentricity approaches 0?

18. (a) Show that the coordinates of the point P on the hyperbola in Figure 11.6.1 satisfy the equation

$$\sqrt{(x - c)^2 + y^2} = \frac{c}{a}x - a$$

(b) Use the result obtained in part (a) to show that $PF/PD = c/a$.

19. (a) Show that the eccentricity of an ellipse can be expressed in terms of r_0 and r_1 as

$$e = \frac{r_1 - r_0}{r_1 + r_0}$$

(b) Show that

$$\frac{r_1}{r_0} = \frac{1 + e}{1 - e}$$

20. (a) Show that the eccentricity of a hyperbola can be expressed in terms of r_0 and r_1 as

$$e = \frac{r_1 + r_0}{r_1 - r_0}$$

(b) Show that

$$\frac{r_1}{r_0} = \frac{e + 1}{e - 1}$$

21. Find the polar equation of an equilateral hyperbola with a focus at the pole and vertex (5, 0).

22. (a) Sketch the curves

$$r = \frac{1}{1 + \cos\theta} \quad \text{and} \quad r = \frac{1}{1 - \cos\theta}$$

(b) Find polar coordinates of the intersections of the curves in part (a).
(c) Show that the curves are *orthogonal*, that is, their tangent lines are perpendicular at the points of intersection.

23–28 Use the following values, where needed:

radius of the Earth = 4000 mi = 6440 km
1 year (Earth year) = 365 days (Earth days)
1 AU = 92.9×10^6 mi = 150×10^6 km

23. The planet Pluto has eccentricity $e = 0.249$ and semimajor axis $a = 39.5$ AU.
(a) Find the period T in years.
(b) Find the perihelion and aphelion distances.
(c) Choose a polar coordinate system with the center of the Sun at the pole, and find a polar equation of Pluto's orbit in that coordinate system.
(d) Make a sketch of the orbit with reasonably accurate proportions.

24. (a) Let a be the semimajor axis of a planet's orbit around the Sun, and let T be its period. Show that if T is measured in days and a is measured in kilometers, then $T = (365 \times 10^{-9})(a/150)^{3/2}$.
(b) Use the result in part (a) to find the period of the planet Mercury in days, given that its semimajor axis is $a = 57.95 \times 10^6$ km.
(c) Choose a polar coordinate system with the Sun at the pole, and find an equation for the orbit of Mercury in that coordinate system given that the eccentricity of the orbit is $e = 0.206$.
(d) Use a graphing utility to generate the orbit of Mercury from the equation obtained in part (c).

25. The Hale–Bopp comet, discovered independently on July 23, 1995 by Alan Hale and Thomas Bopp, has an orbital eccentricity of $e = 0.9951$ and a period of 2380 years.
(a) Find its semimajor axis in astronomical units (AU).
(b) Find its perihelion and aphelion distances.
(c) Choose a polar coordinate system with the center of the Sun at the pole, and find an equation for the Hale–Bopp orbit in that coordinate system.

(d) Make a sketch of the Hale–Bopp orbit with reasonably accurate proportions.

26. Mars has a perihelion distance of 204,520,000 km and an aphelion distance of 246,280,000 km.
(a) Use these data to calculate the eccentricity, and compare your answer to the value given in Table 11.6.1.
(b) Find the period of Mars.
(c) Choose a polar coordinate system with the center of the Sun at the pole, and find an equation for the orbit of Mars in that coordinate system.
(d) Use a graphing utility to generate the orbit of Mars from the equation obtained in part (c).

27. *Vanguard 1* was launched in March 1958 into an orbit around the Earth with eccentricity $e = 0.21$ and semimajor axis 8864.5 km. Find the minimum and maximum heights of *Vanguard 1* above the surface of the Earth.

28. The planet Jupiter is believed to have a rocky core of radius 10,000 km surrounded by two layers of hydrogen—a 40,000-km-thick layer of compressed metallic-like hydrogen and a 20,000-km-thick layer of ordinary molecular hydrogen. The visible features, such as the Great Red Spot, are at the outer surface of the molecular hydrogen layer. On November 6, 1997 the spacecraft *Galileo* was placed in a Jovian orbit to study the moon Europa. The orbit had eccentricity 0.814580 and semimajor axis 3,514,918.9 km. Find *Galileo*'s minimum and maximum heights above the molecular hydrogen layer (see the accompanying figure).

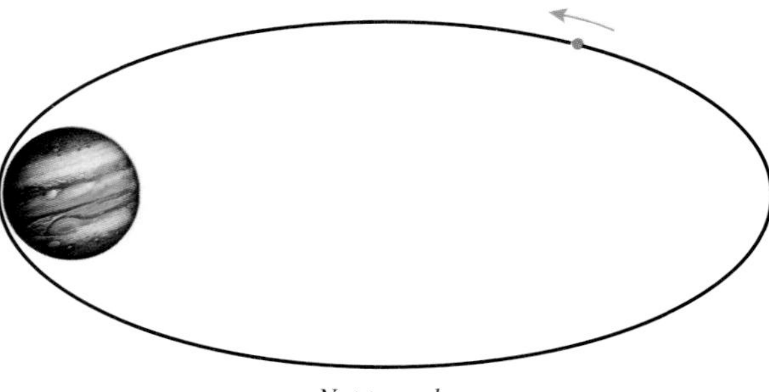

Figure Ex-28

✔ QUICK CHECK ANSWERS 11.6

1. (a) an ellipse (b) a parabola (c) a hyperbola **2.** (a) (i) ellipse (ii) to the right of the pole (iii) distance = 1 (b) (i) hyperbola (ii) to the left of the pole (iii) distance = $\frac{1}{4}$ (c) (i) parabola (ii) above the pole (iii) distance = $\frac{1}{4}$ (d) (i) parabola (ii) below the pole (iii) distance = 4 **3.** $\frac{1}{2}(r_1 + r_0)$; $\sqrt{r_0 r_1}$ **4.** $\frac{1}{2}(r_1 - r_0)$; $\sqrt{r_0 r_1}$

CHAPTER REVIEW EXERCISES  

Graphing Utility C CAS

1. In each part, find the rectangular coordinates of the point whose polar coordinates are given.
(a) $(-8, \pi/4)$ (b) $(7, -\pi/4)$ (c) $(8, 9\pi/4)$
(d) $(5, 0)$ (e) $(-2, -3\pi/2)$ (f) $(0, \pi)$

2. Express the point whose xy-coordinates are $(-1, 1)$ in polar coordinates with
(a) $r > 0,\ 0 \le \theta < 2\pi$ (b) $r < 0,\ 0 \le \theta < 2\pi$
(c) $r > 0,\ -\pi < \theta \le \pi$ (d) $r < 0,\ -\pi < \theta \le \pi$.

3. In each part, use a calculating utility to approximate the polar coordinates of the point whose rectangular coordinates are given.
(a) $(4, 3)$ (b) $(2, -5)$ (c) $(1, \tan^{-1} 1)$

4. In each part, state the name that describes the polar curve most precisely: a rose, a line, a circle, a limaçon, a cardioid, a spiral, a lemniscate, or none of these.
(a) $r = 3\cos\theta$ (b) $r = \cos 3\theta$
(c) $r = \dfrac{3}{\cos\theta}$ (d) $r = 3 - \cos\theta$
(e) $r = 1 - 3\cos\theta$ (f) $r^2 = 3\cos\theta$
(g) $r = (3\cos\theta)^2$ (h) $r = 1 + 3\theta$

5. In each part, identify the curve by converting the polar equation to rectangular coordinates. Assume that $a > 0$.
(a) $r = a\sec^2\dfrac{\theta}{2}$ (b) $r^2\cos 2\theta = a^2$
(c) $r = 4\csc\left(\theta - \dfrac{\pi}{4}\right)$ (d) $r = 4\cos\theta + 8\sin\theta$

6. In each part, express the given equation in polar coordinates.
(a) $x = 7$ (b) $x^2 + y^2 = 9$
(c) $x^2 + y^2 - 6y = 0$ (d) $4xy = 9$

7–12 Sketch the curve in polar coordinates.

7. $\theta = \dfrac{\pi}{6}$ **8.** $r = 6\cos\theta$

9. $r = 3(1 - \sin\theta)$ **10.** $r = 2 + \sin\theta$

11. $r = 3 - \cos\theta$ **12.** $r^2 = \sin 2\theta$

13. (a) Find the minimum and maximum x-coordinates of points on the cardioid $r = 1 - \cos\theta$.
(b) Find the minimum and maximum y-coordinates of points on the cardioid in part (a).

14. (a) Show that the maximum value of the y-coordinate of points on the curve $r = 1/\sqrt{\theta}$ for θ in the interval $(0, \pi]$ occurs when $\tan\theta = 2\theta$.
(b) Use a calculating utility to solve the equation in part (a) to at least four decimal-place accuracy.
(c) Use the result of part (b) to approximate the maximum value of y for $0 < \theta \leq \pi$.

15. (a) Find the slope of the tangent line to the parametric curve $x = t^2 + 1$, $y = t/2$ at $t = -1$ and $t = 1$ without eliminating the parameter.
(b) Check your answers in part (a) by eliminating the parameter and differentiating a function of x.

16. Find dy/dx and d^2y/dx^2 at $t = 2$ for the parametric curve $x = \frac{1}{2}t^2$, $y = \frac{1}{3}t^3$.

17. Find all values of t at which a tangent line to the parametric curve $x = 2\cos t$, $y = 4\sin t$ is
(a) horizontal (b) vertical.

18. Determine the slope of the tangent line to the polar curve $r = 1 + \sin\theta$ at $\theta = \pi/4$.

19. A parametric curve of the form

$$x = a\cot t + b\cos t, \quad y = a + b\sin t \quad (0 < t < 2\pi)$$

is called a ***conchoid of Nicomedes*** (see the accompanying figure for the case $0 < a < b$).
(a) Describe how the conchoid

$$x = \cot t + 4\cos t, \quad y = 1 + 4\sin t$$

is generated as t varies over the interval $0 < t < 2\pi$.
(b) Find the horizontal asymptote of the conchoid given in part (a).
(c) For what values of t does the conchoid in part (a) have a horizontal tangent line? A vertical tangent line?
(d) Find a polar equation $r = f(\theta)$ for the conchoid in part (a), and then find polar equations for the tangent lines to the conchoid at the pole.

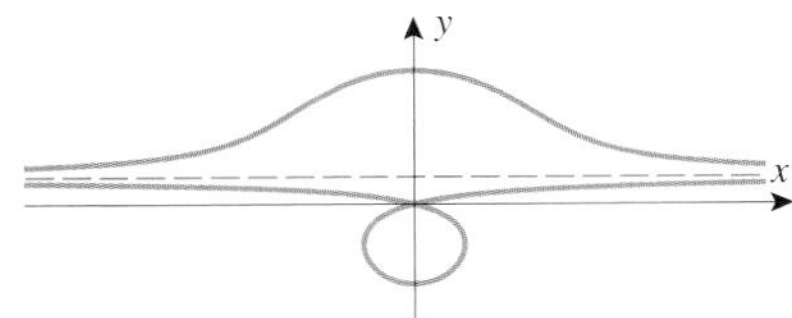

Figure Ex-19

20. (a) Find the arc length of the polar curve $r = 1/\theta$ for $\pi/4 \leq \theta \leq \pi/2$.
(b) What can you say about the arc length of the portion of the curve that lies inside the circle $r = 1$?

21. Find the area of the region that is enclosed by the cardioid $r = 2 + 2\cos\theta$.

22. Find the area of the region in the first quadrant within the cardioid $r = 1 + \sin\theta$.

23. Find the area of the region that is common to the circles $r = 1$, $r = 2\cos\theta$, and $r = 2\sin\theta$.

24. Find the area of the region that is inside the cardioid $r = a(1 + \sin\theta)$ and outside the circle $r = a\sin\theta$.

25–28 Sketch the parabola, and label the focus, vertex, and directrix.

25. $y^2 = 6x$
26. $x^2 = -9y$
27. $(y + 1)^2 = -7(x - 4)$
28. $\left(x - \frac{1}{2}\right)^2 = 2(y - 1)$

29–32 Sketch the ellipse, and label the foci, the vertices, and the ends of the minor axis.

29. $\dfrac{x^2}{4} + \dfrac{y^2}{25} = 1$
30. $4x^2 + 9y^2 = 36$
31. $9(x - 1)^2 + 16(y - 3)^2 = 144$
32. $3(x + 2)^2 + 4(y + 1)^2 = 12$

33–36 Sketch the hyperbola, and label the vertices, foci, and asymptotes.

33. $\dfrac{x^2}{16} - \dfrac{y^2}{4} = 1$
34. $9y^2 - 4x^2 = 36$
35. $\dfrac{(x - 2)^2}{9} - \dfrac{(y - 4)^2}{4} = 1$
36. $(y + 3)^2 - 9(x + 2)^2 = 36$

37. In each part, sketch the graph of the conic section with reasonably accurate proportions.
(a) $x^2 - 4x + 8y + 36 = 0$
(b) $3x^2 + 4y^2 - 30x - 8y + 67 = 0$
(c) $4x^2 - 5y^2 - 8x - 30y - 21 = 0$

C 38. If you have a CAS with implicit plotting capability, use it to check your work in Exercise 37.

39–41 Find an equation for the conic described.

39. A parabola with vertex $(0, 0)$ and focus $(0, -4)$.

40. An ellipse with the ends of the major axis $(0, \pm\sqrt{5})$ and the ends of the minor axis $(\pm 1, 0)$.

41. A hyperbola with vertices $(0, \pm 3)$ and asymptotes $y = \pm x$.

42. It can be shown in the accompanying figure (next page) that hanging cables form parabolic arcs rather than catenaries if they are subjected to uniformly distributed downward forces along their length. For example, if the weight of the roadway in a suspension bridge is assumed to be uniformly distributed along the supporting cables, then the cables can be modeled by parabolas.
(a) Assuming a parabolic model, find an equation for the cable in the accompanying figure, taking the y-axis to be vertical and the origin at the low point of the cable.
(b) Find the length of the cable between the supports.

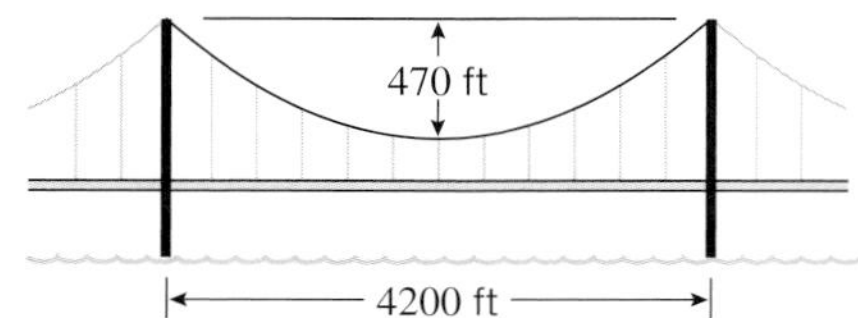

Figure Ex-42

43. It will be shown later in this text that if a projectile is launched with speed v_0 at an angle α with the horizontal and at a height y_0 above ground level, then the resulting trajectory relative to the coordinate system in the accompanying figure will have parametric equations

$$x = (v_0 \cos\alpha)t, \quad y = y_0 + (v_0 \sin\alpha)t - \tfrac{1}{2}gt^2$$

where g is the acceleration due to gravity.
(a) Show that the trajectory is a parabola.
(b) Find the coordinates of the vertex.

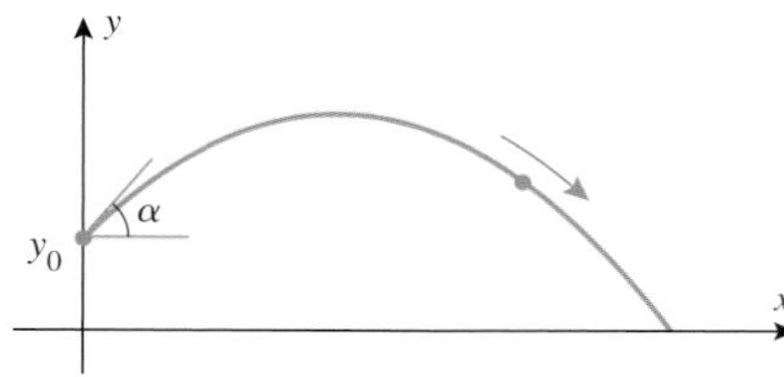

Figure Ex-43

44. Mickey Mantle is recognized as baseball's unofficial king of long home runs. On April 17, 1953 Mantle blasted a pitch by Chuck Stobbs of the hapless Washington Senators out of Griffith Stadium, just clearing the 50-ft wall at the 391-ft marker in left center. Assuming that the ball left the bat at a height of 3 ft above the ground and at an angle of 45°, use the parametric equations in Exercise 43 with $g = 32$ ft/s^2 to find
(a) the speed of the ball as it left the bat
(b) the maximum height of the ball
(c) the distance along the ground from home plate to where the ball struck the ground.

45. Let R be the region that is above the x-axis and enclosed between the curve $b^2x^2 - a^2y^2 = a^2b^2$ and the line $x = \sqrt{a^2 + b^2}$.
(a) Sketch the solid generated by revolving R about the x-axis, and find its volume.
(b) Sketch the solid generated by revolving R about the y-axis, and find its volume.

46. A nuclear cooling tower is to have a height of h feet and the shape of the solid that is generated by revolving the region R enclosed by the right branch of the hyperbola $1521x^2 - 225y^2 = 342{,}225$ and the lines $x = 0$, $y = -h/2$, and $y = h/2$ about the y-axis.
(a) Find the volume of the tower.
(b) Find the lateral surface area of the tower.

47–49 Rotate the coordinate axes to remove the xy-term, and then name the conic.

47. $x^2 + y^2 - 3xy - 3 = 0$

48. $7x^2 + 2\sqrt{3}xy + 5y^2 - 4 = 0$

49. $4\sqrt{5}x^2 + 4\sqrt{5}xy + \sqrt{5}y^2 + 5x - 10y = 0$

50. Rotate the coordinate axes to show that the graph of

$$17x^2 - 312xy + 108y^2 + 1080x - 1440y + 4500 = 0$$

is a hyperbola. Then find its vertices, foci, and asymptotes.

51. In each part: (i) Identify the polar graph as a parabola, an ellipse, or a hyperbola; (ii) state whether the directrix is above, below, to the left, or to the right of the pole; and (iii) find the distance from the pole to the directrix.
(a) $r = \dfrac{1}{3 + \cos\theta}$ (b) $r = \dfrac{1}{1 - 3\cos\theta}$
(c) $r = \dfrac{1}{3(1 + \sin\theta)}$ (d) $r = \dfrac{3}{1 - \sin\theta}$

52–53 Find an equation in xy-coordinates for the conic section that satisfies the given conditions.

52. (a) Ellipse with eccentricity $e = \frac{2}{7}$ and ends of the minor axis at the points $(0, \pm 3)$.
(b) Parabola with vertex at the origin, focus on the y-axis, and directrix passing through the point $(7, 4)$.
(c) Hyperbola that has the same foci as the ellipse $3x^2 + 16y^2 = 48$ and asymptotes $y = \pm 2x/3$.

53. (a) Ellipse with center $(-3, 2)$, vertex $(2, 2)$, and eccentricity $e = \frac{4}{5}$.
(b) Parabola with focus $(-2, -2)$ and vertex $(-2, 0)$.
(c) Hyperbola with vertex $(-1, 7)$ and asymptotes $y - 5 = \pm 8(x + 1)$.

54. Use the parametric equations $x = a\cos t$, $y = b\sin t$ to show that the circumference C of an ellipse with semimajor axis a and eccentricity e is

$$C = 4a\int_0^{\pi/2} \sqrt{1 - e^2 \sin^2 u}\, du$$

55. Use Simpson's rule or the numerical integration capability of a graphing utility to approximate the circumference of the ellipse $4x^2 + 9y^2 = 36$ from the integral obtained in Exercise 54.

56. (a) Calculate the eccentricity of the Earth's orbit, given that the ratio of the distance between the center of the Earth and the center of the Sun at perihelion to the distance between the centers at aphelion is $\frac{59}{61}$.
(b) Find the distance between the center of the Earth and the center of the Sun at perihelion, given that the average value of the perihelion and aphelion distances between the centers is 93 million miles.
(c) Use the result in Exercise 54 and Simpson's rule or the numerical integration capability of a graphing utility to approximate the distance that the Earth travels in 1 year (one revolution around the Sun).

EXPANDING THE CALCULUS HORIZON

Comet Collision

The Earth lives in a cosmic shooting gallery of comets and asteroids. Although the probability that the Earth will be hit by a comet or asteroid in any given year is small, the consequences of such a collision are so catastrophic that the international community is now beginning to track **near Earth objects** *(NEOs). Your job, as part of the international NEO tracking team, is to compute the orbits of incoming comets and asteroids, determine how close they will come to colliding with the Earth, and issue a notification if there is danger of a collision or near miss.*

At the time when the Earth is at its *aphelion* (its farthest point from the Sun), your NEO tracking team receives a notification from the NASA/Caltech Jet Propulsion Laboratory that a previously unknown comet (designation Rogue 2000) is traveling in the plane of Earth's orbit and hurtling in the direction of the Earth. You immediately transmit a request to NASA for the orbital parameters and the current positions of the Earth and Rogue 2000 and receive the following report:

ORBITAL PARAMETERS

EARTH	ROGUE 2000
Eccentricity: $e_1 = 0.017$	Eccentricity: $e_2 = 0.98$
Semimajor axis: $a_1 = 1$ AU $= 1.496 \times 10^8$ km	Semimajor axis: $a_2 = 5$ AU $= 7.48 \times 10^8$ km
Period: $T_1 = 1$ year	Period: $T_2 = 5\sqrt{5}$ years

INITIAL POSITION INFORMATION

The major axes of Earth and Rogue 2000 lie on the same line.
The aphelions of Earth and Rogue 2000 are on the same side of the Sun.
Initial polar angle of Earth: $\theta = 0$ radians.
Initial polar angle of Rogue 2000: $\theta = 0.45$ radian.

The Calculation Strategy

Since the immediate concern is a possible collision at intersection A in Figure 1, your team works out the following plan:

Step 1. Find the polar equations for Earth and Rogue 2000.
Step 2. Find the polar coordinates of intersection A.
Step 3. Determine how long it will take the Earth to reach intersection A.
Step 4. Determine where Rogue 2000 will be when the Earth reaches intersection A.
Step 5. Determine how far Rogue 2000 will be from the Earth when the Earth is at intersection A.

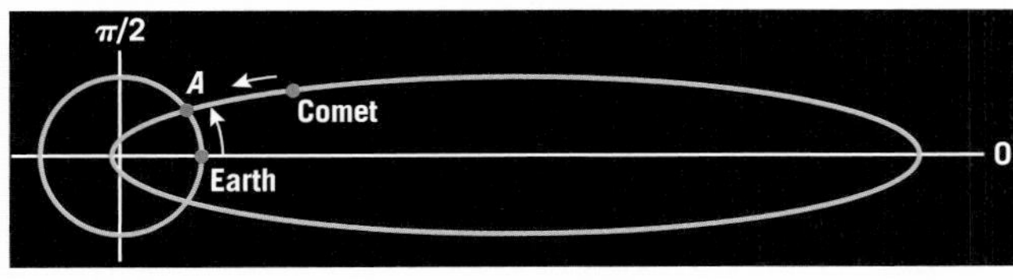

Initial configuration of Earth and Rogue 2000

Figure 1

Polar Equations of the Orbits

Exercise 1 Write polar equations of the form

$$r = \frac{a(1-e^2)}{1-e\cos\theta}$$

for the orbits of Earth and Rogue 2000 using AU units for r.

Exercise 2 Use a graphing utility to generate the two orbits on the same screen.

Intersection of the Orbits

The second step in your team's calculation plan is to find the polar coordinates of intersection A in Figure 1.

Exercise 3 For simplicity, let $k_1 = a_1(1-e_1^2)$ and $k_2 = a_2(1-e_2^2)$, and use the polar equations obtained in Exercise 1 to show that the angle θ at intersection A satisfies the equation

$$\cos\theta = \frac{k_1 - k_2}{k_1e_2 - k_2e_1}$$

Exercise 4 Use the result in Exercise 3 and the inverse cosine capability of a calculating utility to show that the angle θ at intersection A in Figure 1 is $\theta = 0.607$ radian.

Exercise 5 Use the result in Exercise 4 and either polar equation obtained in Exercise 1 to show that if r is in AU units, then the polar coordinates of intersection A are $(r, \theta) = (1.014, 0.607)$.

Time Required for Earth to Reach Intersection *A*

According to Kepler's second law (see 11.6.3), the radial line from the center of the Sun to the center of an object orbiting around it sweeps out equal areas in equal times. Thus, if t is the time that it takes for the radial line to sweep out an "elliptic sector" from some initial angle θ_{I} to some final angle θ_{F} (Figure 2), and if T is the period of the object (the time for one complete revolution), then

$$\frac{t}{T} = \frac{\text{area of the "elliptic sector"}}{\text{area of the entire ellipse}} \tag{1}$$

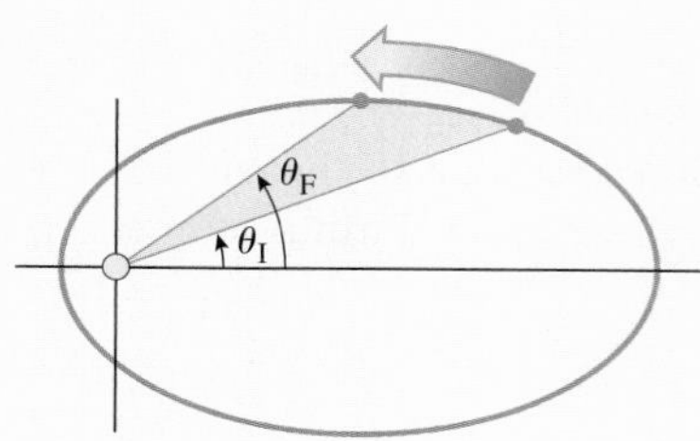

Figure 2

Exercise 6 Use Formula (1) to show that

$$t = \frac{T\displaystyle\int_{\theta_{\mathrm{I}}}^{\theta_{\mathrm{F}}} r^2\,d\theta}{2\pi a^2\sqrt{1-e^2}} \tag{2}$$

Exercise 7 Use a calculating utility with a numerical integration capability, Formula (2), and the polar equation for the orbit of the Earth obtained in Exercise 1 to find the time t (in years) required for the Earth to move from its initial position to intersection A.

Position of Rogue 2000 When the Earth Is at Intersection A

The fourth step in your team's calculation strategy is to determine the position of Rogue 2000 when the Earth reaches intersection A.

Exercise 8 During the time that it takes for the Earth to move from its initial position to intersection A, the polar angle of Rogue 2000 will change from its initial value $\theta_I = 0.45$ radian to some final value θ_F that remains to be determined. Apply Formula (2) using the orbital data for Rogue 2000 and the time t obtained in Exercise 7 to show that θ_F satisfies the equation

$$\int_{0.45}^{\theta_F} \left[\frac{a_2(1-e_2^2)}{1-e_2\cos\theta}\right]^2 d\theta = \frac{2t\pi a_2^2\sqrt{1-e_2^2}}{5\sqrt{5}} \tag{3}$$

Your team is now faced with the problem of solving Equation (3) for the unknown upper limit θ_F. Some members of the team plan to use a CAS to perform the integration, some plan to use integration tables, and others plan to use hand calculation by making the substitution $u = \tan(\theta/2)$ and applying the formulas in (5) of Section 8.6.

Exercise 9

(a) Evaluate the integral in (3) using a CAS or by hand calculation.

(b) Use the root-finding capability of a calculating utility to find the polar angle of Rogue 2000 when the Earth is at intersection A.

Calculating the Critical Distance

It is the policy of your NEO tracking team to issue a notification to various governmental agencies for any asteroid or comet that will be within 4 million kilometers of the Earth at an orbital intersection. (This distance is roughly 10 times that between the Earth and the Moon.) Accordingly, the final step in your team's plan is to calculate the distance between the Earth and Rogue 2000 when the Earth is at intersection A, and then determine whether a notification should be issued.

Exercise 10 Use the polar equation of Rogue 2000 obtained in Exercise 1 and the result in Exercise 9(b) to find polar coordinates of Rogue 2000 with r in AU units when the Earth is at intersection A.

Exercise 11 Use the distance formula in Exercise 67(a) of Section 11.1 to calculate the distance between the Earth and Rogue 2000 in AU units when the Earth is at intersection A, and then use the conversion factor 1 AU $= 1.496 \times 10^8$ km to determine whether a government notification should be issued.

Note: One of the closest near misses in recent history occurred on October 30, 1937 when the asteroid Hermes passed within 900,000 km of the Earth. More recently, on June 14, 1968 the asteroid Icarus passed within 23,000,000 km of the Earth.

Module by Mary Ann Connors, USMA, West Point, and Howard Anton, Drexel University

c h a p t e r   t w e l v e

THREE-DIMENSIONAL SPACE; VECTORS

What if angry vectors veer
Round your sleeping head,
and form. There's never
need to fear Violence of the
poor world's abstract storm.
—Robert Penn Warren
Poet

In this chapter we will discuss rectangular coordinate systems in three dimensions, and we will study the analytic geometry of lines, planes, and other basic surfaces. The second theme of this chapter is the study of vectors. These are the mathematical objects that physicists and engineers use to study forces, displacements, and velocities of objects moving on curved paths. More generally, vectors are used to represent all physical entities that involve both a magnitude and a direction for their complete description. We will introduce various algebraic operations on vectors, and we will apply these operations to problems involving force, work, and rotational tendencies in two and three dimensions. Finally, we will discuss cylindrical and spherical coordinate systems, which are appropriate in problems that involve various kinds of symmetries and also have specific applications in navigation and celestial mechanics.

Photo: *To fully describe the motion of a boat one must specify its speed and direction of motion at each instant. Speed and direction together describe a "vector" quantity. We will study vectors in this chapter.*

12.1 RECTANGULAR COORDINATES IN 3-SPACE; SPHERES; CYLINDRICAL SURFACES

In this section we will discuss coordinate systems in three-dimensional space and some basic facts about surfaces in three dimensions.

RECTANGULAR COORDINATE SYSTEMS

In the remainder of this text we will call three-dimensional space ***3-space***, two-dimensional space (a plane) ***2-space***, and one-dimensional space (a line) ***1-space***. Just as points in 2-space can be placed in one-to-one correspondence with pairs of real numbers using two perpendicular coordinate lines, so points in 3-space can be placed in one-to-one correspondence with triples of real numbers by using three mutually perpendicular coordinate lines, called the ***x-axis***, the ***y-axis***, and the ***z-axis***, positioned so that their origins coincide (Figure 12.1.1). The three coordinate axes form a three-dimensional ***rectangular coordinate system*** (or ***Cartesian coordinate system***). The point of intersection of the coordinate axes is called the ***origin*** of the coordinate system.

Figure 12.1.1

Rectangular coordinate systems in 3-space fall into two categories: ***left-handed*** and ***right-handed***. A right-handed system has the property that when the fingers of the right hand are cupped so that they curve from the positive x-axis toward the positive y-axis, the thumb points (roughly) in the direction of the positive z-axis (Figure 12.1.2). Similarly for

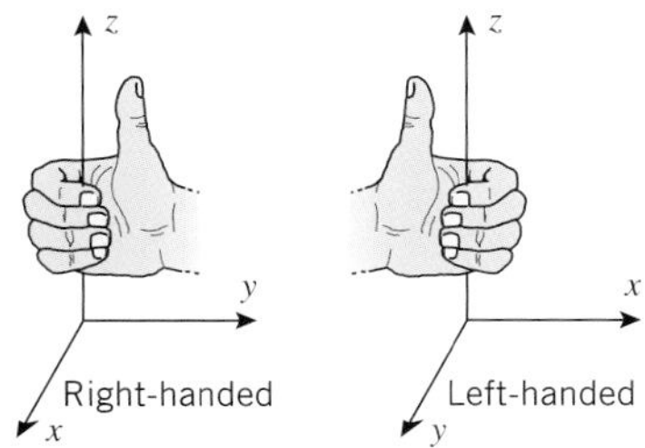

Figure 12.1.2

a left-handed coordinate system (Figure 12.1.2). We will use only right-handed coordinate systems in this text.

The coordinate axes, taken in pairs, determine three ***coordinate planes***: the ***xy-plane***, the ***xz-plane***, and the ***yz-plane*** (Figure 12.1.3). To each point P in 3-space we can assign a triple of real numbers by passing three planes through P parallel to the coordinate planes and letting a, b, and c be the coordinates of the intersections of those planes with the x-axis, y-axis, and z-axis, respectively (Figure 12.1.4). We call a, b, and c the ***x-coordinate***, ***y-coordinate***, and ***z-coordinate*** of P, respectively, and we denote the point P by (a, b, c) or by $P(a, b, c)$. Figure 12.1.5 shows the points $(4, 5, 6)$ and $(-3, 2, -4)$.

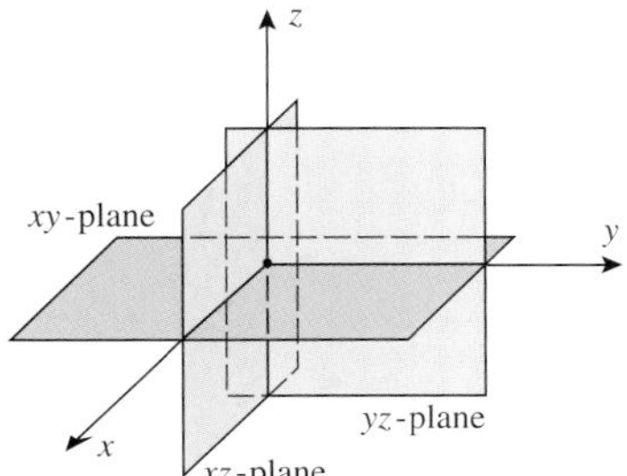

Figure 12.1.3

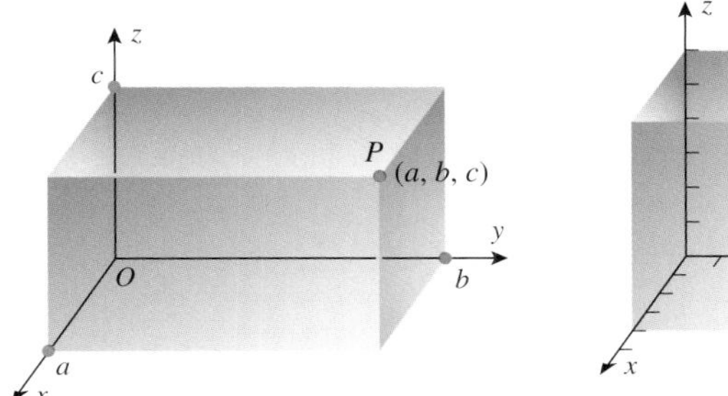

Figure 12.1.4

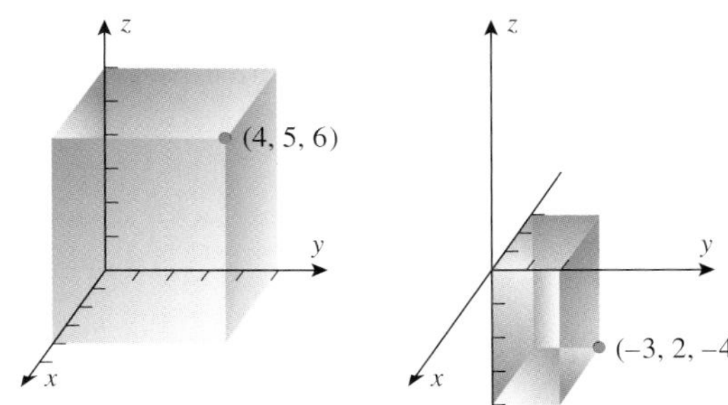

Figure 12.1.5

Just as the coordinate axes in a two-dimensional coordinate system divide 2-space into four quadrants, so the coordinate planes of a three-dimensional coordinate system divide 3-space into eight parts, called ***octants***. The set of points with three positive coordinates forms the ***first octant***; the remaining octants have no standard numbering.

You should be able to visualize the following facts about three-dimensional rectangular coordinate systems:

REGION	DESCRIPTION
xy-plane	Consists of all points of the form $(x, y, 0)$
xz-plane	Consists of all points of the form $(x, 0, z)$
yz-plane	Consists of all points of the form $(0, y, z)$
x-axis	Consists of all points of the form $(x, 0, 0)$
y-axis	Consists of all points of the form $(0, y, 0)$
z-axis	Consists of all points of the form $(0, 0, z)$

DISTANCE IN 3-SPACE

To derive a formula for the distance between two points in 3-space, we start by considering a box whose sides have lengths a, b, and c (Figure 12.1.6). The length d of a diagonal of the box can be obtained by applying the Theorem of Pythagoras twice: first to show that a diagonal of the base has length $\sqrt{a^2 + b^2}$, then again to show that a diagonal of the box has length

$$d = \sqrt{\left(\sqrt{a^2 + b^2}\right)^2 + c^2} = \sqrt{a^2 + b^2 + c^2} \tag{1}$$

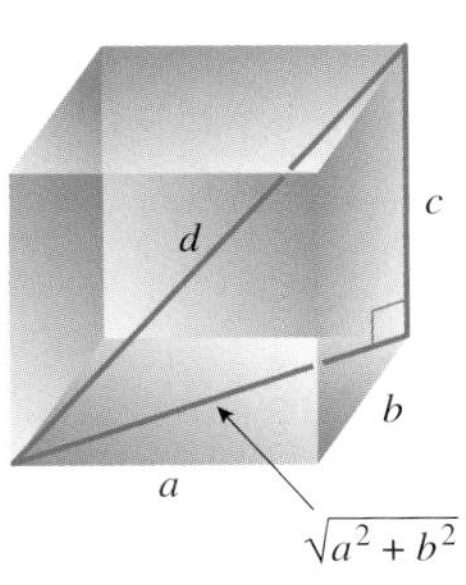

Figure 12.1.6

We can now obtain a formula for the distance d between two points $P_1(x_1, y_1, z_1)$ and $P_2(x_2, y_2, z_2)$ in 3-space by finding the length of the diagonal of a box that has these points as diagonal corners (Figure 12.1.7). The sides of such a box have lengths

$$|x_2 - x_1|, \quad |y_2 - y_1|, \quad \text{and} \quad |z_2 - z_1|$$

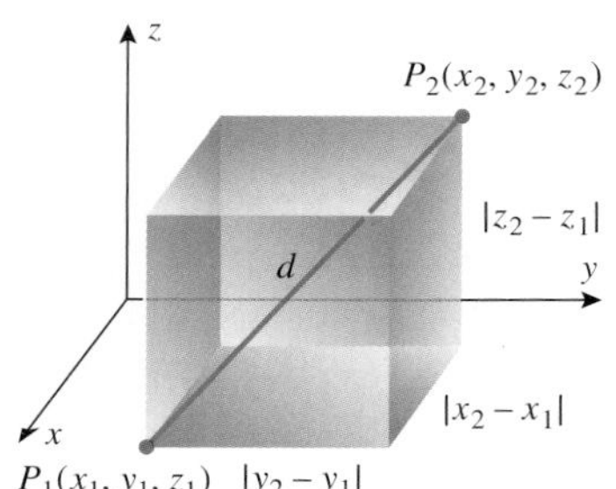

Figure 12.1.7

and hence from (1) the distance d between the points P_1 and P_2 is

$$d = \sqrt{(x_2 - x_1)^2 + (y_2 - y_1)^2 + (z_2 - z_1)^2} \tag{2}$$

(where we have omitted the unnecessary absolute value signs).

> Recall that in 2-space the distance d between points $P_1(x_1, y_1)$ and $P_2(x_2, y_2)$ is
>
> $$d = \sqrt{(x_2 - x_1)^2 + (y_2 - y_1)^2}$$
>
> Thus, the distance formula in 3-space has the same form as the formula in 2-space, but it has a third term to account for the additional dimension. We will see that this is a common occurrence in extending formulas from 2-space to 3-space.

► **Example 1** Find the distance d between the points $(2, 3, -1)$ and $(4, -1, 3)$.

Solution. From Formula (2)

$$d = \sqrt{(4-2)^2 + (-1-3)^2 + (3+1)^2} = \sqrt{36} = 6 \blacktriangleleft$$

SPHERES

Recall that in an xy-coordinate system, the set of points (x, y) whose coordinates satisfy an equation in x and y is called the *graph* of the equation. Analogously, in an xyz-coordinate system, the set of points (x, y, z) whose coordinates satisfy an equation in x, y, and z is called the ***graph*** of the equation. For example, consider the equation

$$x^2 + y^2 + z^2 = 25 \tag{3}$$

This equation can be rewritten as

$$\sqrt{x^2 + y^2 + z^2} = 5$$

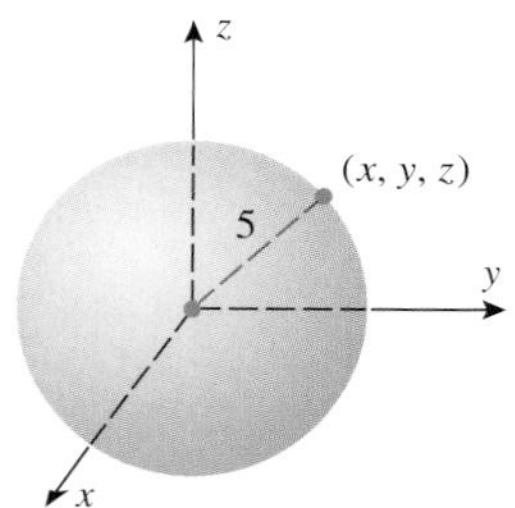

Figure 12.1.8

so the graph of (3) consists of all points that are at a distance of 5 units from the origin. Thus, the graph is a sphere of radius 5 centered at the origin (Figure 12.1.8).

In general, the sphere with center (x_0, y_0, z_0) and radius r consists of those points (x, y, z) whose coordinates satisfy

$$\sqrt{(x - x_0)^2 + (y - y_0)^2 + (z - z_0)^2} = r$$

or, equivalently,

$$(x - x_0)^2 + (y - y_0)^2 + (z - z_0)^2 = r^2 \tag{4}$$

> Recall that in 2-space the standard equation of the circle with center (x_0, y_0) and radius r is
>
> $$(x - x_0)^2 + (y - y_0)^2 = r^2$$
>
> Comparing this to (4) we see that the standard equation for the sphere in 3-space has the same form as the standard equation for the circle in 2-space, but with an additional term to account for the third coordinate.

This is called the ***standard equation of the sphere*** with center (x_0, y_0, z_0) and radius r. Some examples are given in the following table:

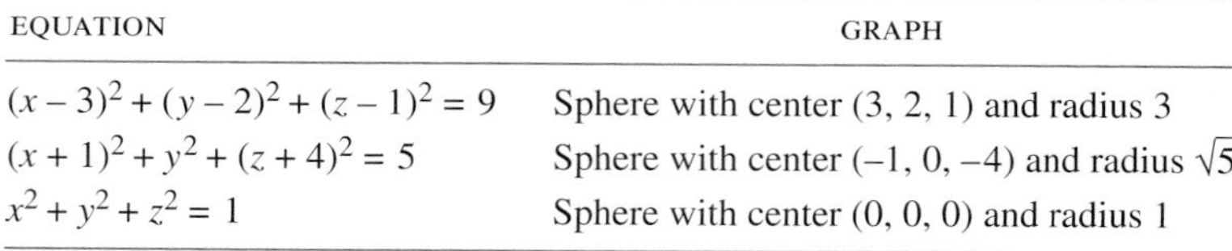

EQUATION	GRAPH
$(x-3)^2 + (y-2)^2 + (z-1)^2 = 9$	Sphere with center (3, 2, 1) and radius 3
$(x+1)^2 + y^2 + (z+4)^2 = 5$	Sphere with center (–1, 0, –4) and radius $\sqrt{5}$
$x^2 + y^2 + z^2 = 1$	Sphere with center (0, 0, 0) and radius 1

If the terms in (4) are expanded and like terms are then collected, then the resulting equation has the form

$$x^2 + y^2 + z^2 + Gx + Hy + Iz + J = 0 \tag{5}$$

The following example shows how the center and radius of a sphere that is expressed in this form can be obtained by completing the squares.

▶ Example 2 Find the center and radius of the sphere

$$x^2 + y^2 + z^2 - 2x - 4y + 8z + 17 = 0$$

Solution. We can put the equation in the form of (4) by completing the squares:

$$(x^2 - 2x) + (y^2 - 4y) + (z^2 + 8z) = -17$$
$$(x^2 - 2x + 1) + (y^2 - 4y + 4) + (z^2 + 8z + 16) = -17 + 21$$
$$(x - 1)^2 + (y - 2)^2 + (z + 4)^2 = 4$$

which is the equation of the sphere with center $(1, 2, -4)$ and radius 2. ◀

In general, completing the squares in (5) produces an equation of the form

$$(x - x_0)^2 + (y - y_0)^2 + (z - z_0)^2 = k$$

If $k > 0$, then the graph of this equation is a sphere with center (x_0, y_0, z_0) and radius $\sqrt{k}$. If $k = 0$, then the sphere has radius zero, so the graph is the single point (x_0, y_0, z_0). If $k < 0$, the equation is not satisfied by any values of x, y, and z (why?), so it has no graph.

12.1.1 THEOREM. *An equation of the form*

$$x^2 + y^2 + z^2 + Gx + Hy + Iz + J = 0$$

represents a sphere, a point, or has no graph.

CYLINDRICAL SURFACES

Although it is natural to graph equations in two variables in 2-space and equations in three variables in 3-space, it is also possible to graph equations in two variables in 3-space. For example, the graph of the equation $y = x^2$ in an xy-coordinate system is a parabola; however, there is nothing to prevent us from writing this equation as $y = x^2 + 0z$ and inquiring about its graph in an xyz-coordinate system. To obtain this graph we need only observe that the equation $y = x^2$ does not impose any restrictions on z. Thus, if we find values of x and y that satisfy this equation, then the coordinates of the point (x, y, z) will also satisfy the equation for *arbitrary* values of z. Geometrically, the point (x, y, z) lies on the vertical line through the point $(x, y, 0)$ in the xy-plane, which means that we can obtain the graph of $y = x^2$ in an xyz-coordinate system by first graphing the equation in the xy-plane and then translating that graph parallel to the z-axis to generate the entire graph (Figure 12.1.9).

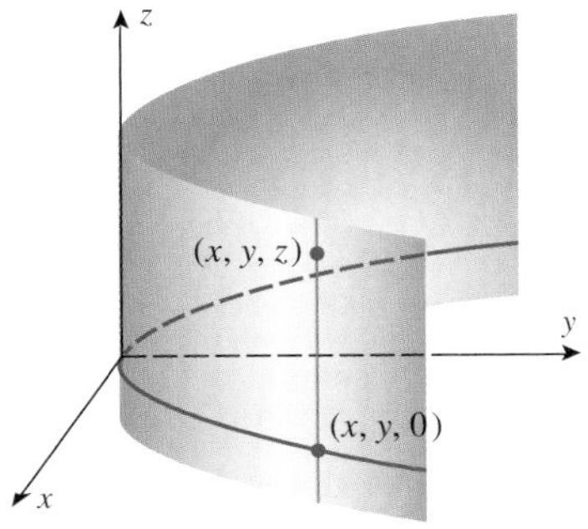

z
(x, y, z)
y
(x, y, 0)
x

Figure 12.1.9

The process of generating a surface by translating a plane curve parallel to some line is called ***extrusion***, and surfaces that are generated by extrusion are called ***cylindrical surfaces***. A familiar example is the surface of a right circular cylinder, which can be generated by translating a circle parallel to the axis of the cylinder. The following theorem provides basic information about graphing equations in two variables in 3-space:

12.1.2 THEOREM. *An equation that contains only two of the variables x, y, and z represents a cylindrical surface in an xyz-coordinate system. The surface can be obtained by graphing the equation in the coordinate plane of the two variables that appear in the equation and then translating that graph parallel to the axis of the missing variable.*

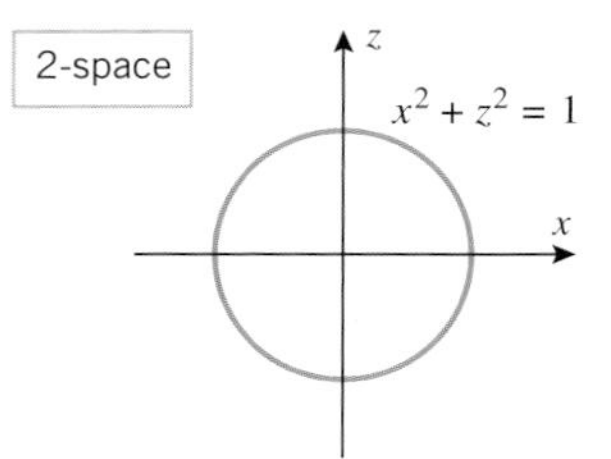

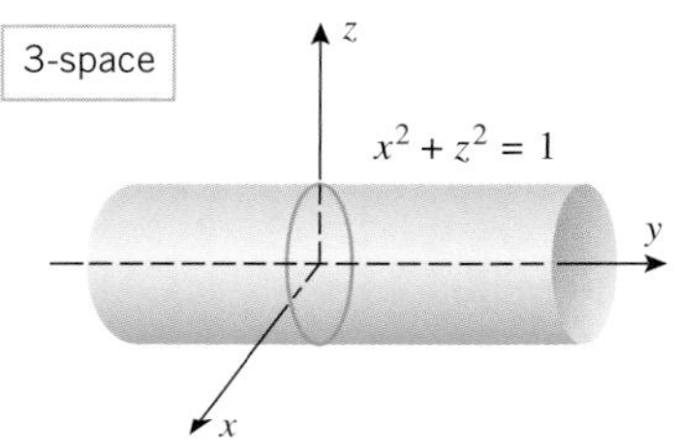

Figure 12.1.10

In an xy-coordinate system, the graph of the equation $x = 1$ is a line parallel to the y-axis. What is the graph of this equation in an xyz-coordinate system?

▶ **Example 3** Sketch the graph of $x^2 + z^2 = 1$ in 3-space.

Solution. Since y does not appear in this equation, the graph is a cylindrical surface generated by extrusion parallel to the y-axis. In the xz-plane the graph of the equation $x^2 + z^2 = 1$ is a circle (Figure 12.1.10). Thus, in 3-space the graph is a right circular cylinder along the y-axis. ◀

▶ **Example 4** Sketch the graph of $z = \sin y$ in 3-space.

Solution. (See Figure 12.1.11.) ◀

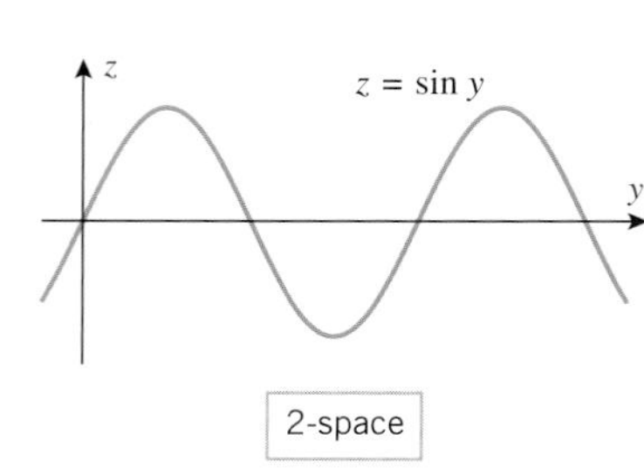

z

z = sin y

y

x

3-space

Figure 12.1.11

✔ QUICK CHECK EXERCISES 12.1 *(See page 792 for answers.)*

1. The distance between the points $(1, -2, 0)$ and $(4, 0, 5)$ is ________.

2. The graph of $(x-3)^2 + (y-2)^2 + (z+1)^2 = 16$ is a ________ of radius ________ centered at ________.

3. The shortest distance from the point $(4, 0, 5)$ to the sphere $(x-1)^2 + (y+2)^2 + z^2 = 36$ is ________.

4. Let S be the graph of $x^2 + z^2 + 6z = 16$ in 3-space.
   (a) The intersection of S with the xz-plane is a circle with center ________ and radius ________.
   (b) The intersection of S with the xy-plane is two lines, $x =$ ________ and $x =$ ________.
   (c) The intersection of S with the yz-plane is two lines, $z =$ ________ and $z =$ ________.

EXERCISE SET 12.1 Graphing Utility

1. In each part, find the coordinates of the eight corners of the box.

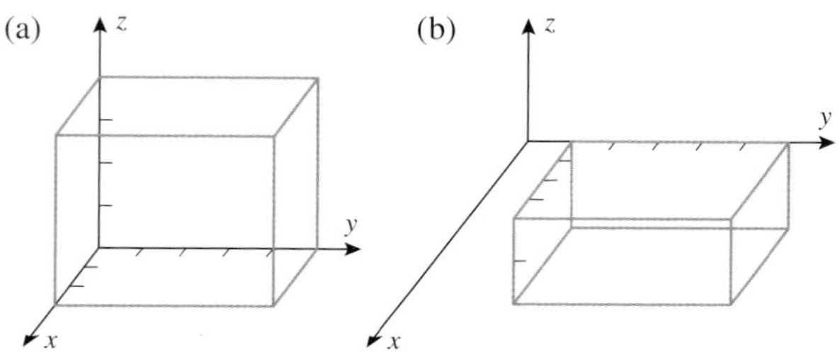

2. A cube of side 4 has its geometric center at the origin and its faces parallel to the coordinate planes. Sketch the cube and give the coordinates of the corners.

FOCUS ON CONCEPTS

3. Suppose that a box has its faces parallel to the coordinate planes and the points $(4, 2, -2)$ and $(-6, 1, 1)$ are endpoints of a diagonal. Sketch the box and give the coordinates of the remaining six corners.

4. Suppose that a box has its faces parallel to the coordinate planes and the points (x_1, y_1, z_1) and (x_2, y_2, z_2) are endpoints of a diagonal.
(a) Find the coordinates of the remaining six corners.
(b) Show that the midpoint of the line segment joining (x_1, y_1, z_1) and (x_2, y_2, z_2) is
$$\left(\tfrac{1}{2}(x_1 + x_2), \tfrac{1}{2}(y_1 + y_2), \tfrac{1}{2}(z_1 + z_2)\right)$$
[*Suggestion:* Apply Theorem G.2 in Web Appendix G to three appropriate edges of the box.]

5. Interpret the graph of $x = 1$ in the contexts of
(a) a number line (b) 2-space (c) 3-space.

6. Consider the points $P(3, 1, 0)$ and $Q(1, 4, 4)$.
(a) Sketch the triangle with vertices P, Q, and $(1, 4, 0)$. Without computing distances, explain why this triangle is a right triangle, and then apply the Theorem of Pythagoras twice to find the distance from P to Q.
(b) Repeat part (a) using the points P, Q, and $(3, 4, 0)$.
(c) Repeat part (a) using the points P, Q, and $(1, 1, 4)$.

7. Find the center and radius of the sphere that has $(1, -2, 4)$ and $(3, 4, -12)$ as endpoints of a diameter. [See Exercise 4.]

8. Show that $(4, 5, 2)$, $(1, 7, 3)$, and $(2, 4, 5)$ are vertices of an equilateral triangle.

9. (a) Show that $(2, 1, 6)$, $(4, 7, 9)$, and $(8, 5, -6)$ are the vertices of a right triangle.
(b) Which vertex is at the 90° angle?
(c) Find the area of the triangle.

10. Find the distance from the point $(-5, 2, -3)$ to the
(a) xy-plane (b) xz-plane (c) yz-plane
(d) x-axis (e) y-axis (f) z-axis.

11. In each part, find the standard equation of the sphere that satisfies the stated conditions.
(a) Center $(1, 0, -1)$; diameter $= 8$.
(b) Center $(-1, 3, 2)$ and passing through the origin.
(c) A diameter has endpoints $(-1, 2, 1)$ and $(0, 2, 3)$.

12. Find equations of two spheres that are centered at the origin and are tangent to the sphere of radius 1 centered at $(3, -2, 4)$.

13. In each part, find an equation of the sphere with center $(2, -1, -3)$ and satisfying the given condition.
(a) Tangent to the xy-plane
(b) Tangent to the xz-plane
(c) Tangent to the yz-plane

14. (a) Find an equation of the sphere that is inscribed in the cube that is centered at the point $(-2, 1, 3)$ and has sides of length 1 that are parallel to the coordinate planes.
(b) Find an equation of the sphere that is circumscribed about the cube in part (a).

15. A sphere has center in the first octant and is tangent to each of the three coordinate planes. Show that the center of the sphere is at a point of the form (r, r, r), where r is the radius of the sphere.

16. A sphere has center in the first octant and is tangent to each of the three coordinate planes. The distance from the origin to the sphere is $3 - \sqrt{3}$ units. Find an equation for the sphere.

17–22 Describe the surface whose equation is given.

17. $x^2 + y^2 + z^2 + 10x + 4y + 2z - 19 = 0$
18. $x^2 + y^2 + z^2 - y = 0$
19. $2x^2 + 2y^2 + 2z^2 - 2x - 3y + 5z - 2 = 0$
20. $x^2 + y^2 + z^2 + 2x - 2y + 2z + 3 = 0$
21. $x^2 + y^2 + z^2 - 3x + 4y - 8z + 25 = 0$
22. $x^2 + y^2 + z^2 - 2x - 6y - 8z + 1 = 0$

23. In each part, sketch the portion of the surface that lies in the first octant.
(a) $y = x$ (b) $y = z$ (c) $x = z$

24. In each part, sketch the graph of the equation in 3-space.
(a) $x = 1$ (b) $y = 1$ (c) $z = 1$

25. In each part, sketch the graph of the equation in 3-space.
(a) $x^2 + y^2 = 25$ (b) $y^2 + z^2 = 25$ (c) $x^2 + z^2 = 25$

26. In each part, sketch the graph of the equation in 3-space.
(a) $x = y^2$ (b) $z = x^2$ (c) $y = z^2$

27. In each part, write an equation for the surface.
(a) The plane that contains the x-axis and the point $(0, 1, 2)$.
(b) The plane that contains the y-axis and the point $(1, 0, 2)$.
(c) The right circular cylinder that has radius 1 and is centered on the line parallel to the z-axis that passes through the point $(1, 1, 0)$.
(d) The right circular cylinder that has radius 1 and is centered on the line parallel to the y-axis that passes through the point $(1, 0, 1)$.

28. Find equations for the following right circular cylinders. Each cylinder has radius a and is "tangent" to two coordinate planes.

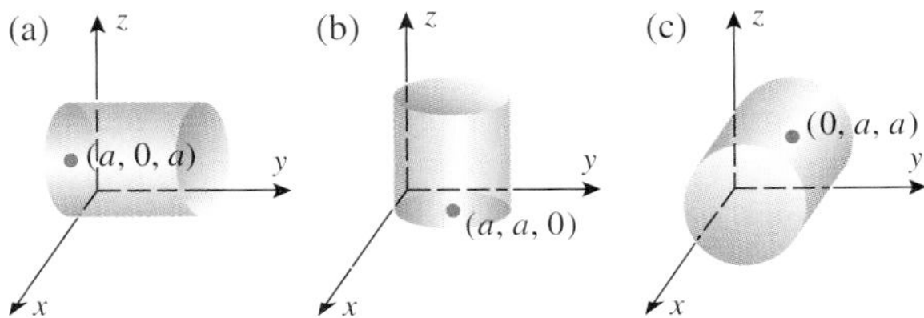

29–38 Sketch the surface in 3-space.

29. $y = \sin x$
30. $y = e^x$
31. $z = 1 - y^2$
32. $z = \cos x$
33. $2x + z = 3$
34. $2x + 3y = 6$
35. $4x^2 + 9z^2 = 36$
36. $z = \sqrt{3 - x}$
37. $y^2 - 4z^2 = 4$
38. $yz = 1$

39. Use a graphing utility to generate the curve $y = x^3/(1 + x^2)$ in the xy-plane, and then use the graph to help sketch the surface $z = y^3/(1 + y^2)$ in 3-space.

40. Use a graphing utility to generate the curve $y = x/(1 + x^4)$ in the xy-plane, and then use the graph to help sketch the surface $z = y/(1 + y^4)$ in 3-space.

41. If a bug walks on the sphere

$$x^2 + y^2 + z^2 + 2x - 2y - 4z - 3 = 0$$

how close and how far can it get from the origin?

42. Describe the set of all points in 3-space whose coordinates satisfy the inequality $x^2 + y^2 + z^2 - 2x + 8z \leq 8$.

43. Describe the set of all points in 3-space whose coordinates satisfy the inequality $y^2 + z^2 + 6y - 4z > 3$.

44. The distance between a point $P(x, y, z)$ and the point $A(1, -2, 0)$ is twice the distance between P and the point $B(0, 1, 1)$. Show that the set of all such points is a sphere, and find the center and radius of the sphere.

45. As shown in the accompanying figure, a bowling ball of radius R is placed inside a box just large enough to hold it, and it is secured for shipping by packing a Styrofoam sphere into each corner of the box. Find the radius of the largest Styrofoam sphere that can be used. [*Hint:* Take the origin of a Cartesian coordinate system at a corner of the box with the coordinate axes along the edges.]

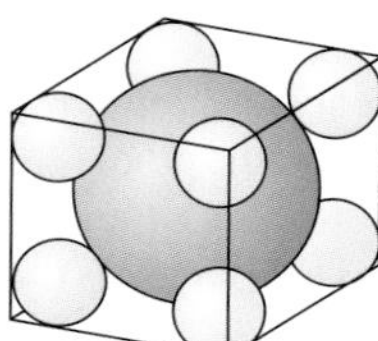

Figure Ex-45

46. Consider the equation

$$x^2 + y^2 + z^2 + Gx + Hy + Iz + J = 0$$

and let $K = G^2 + H^2 + I^2 - 4J$.

(a) Prove that the equation represents a sphere if $K > 0$, a point if $K = 0$, and has no graph if $K < 0$.

(b) In the case where $K > 0$, find the center and radius of the sphere.

47. (a) The accompanying figure shows a surface of revolution that is generated by revolving the curve $y = f(x)$ in the xy-plane about the x-axis. Show that the equation of this surface is $y^2 + z^2 = [f(x)]^2$. [*Hint:* Each point on the curve traces a circle as it revolves about the x-axis.]

(b) Find an equation of the surface of revolution that is generated by revolving the curve $y = e^x$ in the xy-plane about the x-axis.

(c) Show that the ellipsoid $3x^2 + 4y^2 + 4z^2 = 16$ is a surface of revolution about the x-axis by finding a curve $y = f(x)$ in the xy-plane that generates it.

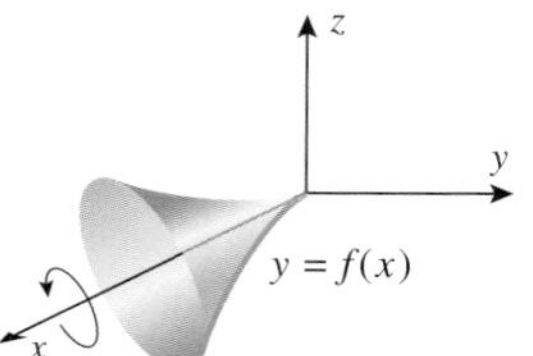

Figure Ex-47

48. In each part, use the idea in Exercise 47(a) to derive a formula for the stated surface of revolution.

(a) The surface generated by revolving the curve $x = f(y)$ in the xy-plane about the y-axis.

(b) The surface generated by revolving the curve $y = f(z)$ in the yz-plane about the z-axis.

(c) The surface generated by revolving the curve $z = f(x)$ in the xz-plane about the x-axis.

49. Show that for all values of θ and ϕ, the point

$$(a \sin\phi \cos\theta, a \sin\phi \sin\theta, a \cos\phi)$$

lies on the sphere $x^2 + y^2 + z^2 = a^2$.

✔ QUICK CHECK ANSWERS 12.1

1. $\sqrt{38}$ **2.** sphere; 4; $(3, 2, -1)$ **3.** $\sqrt{38} - 6$ **4.** (a) $(0, 0, -3)$; 5 (b) 4; -4 (c) 2; -8

12.2 VECTORS

Many physical quantities such as area, length, mass, and temperature are completely described once the magnitude of the quantity is given. Such quantities are called "scalars." Other physical quantities, called "vectors," are not completely determined until both a magnitude and a direction are specified. For example, winds are usually described by giving their speed and direction, say 20 mi/h northeast. The wind speed and wind direction together form a vector quantity called the wind velocity. Other examples of vectors are force and displacement. In this section we will develop the basic mathematical properties of vectors.

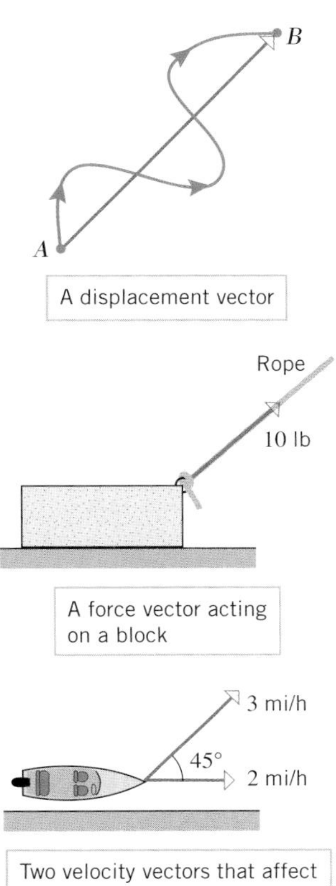

Figure 12.2.1

VECTORS IN PHYSICS AND ENGINEERING

A particle that moves along a line can move in only two directions, so its direction of motion can be described by taking one direction to be positive and the other negative. Thus, the *displacement* or *change in position* of the point can be described by a signed real number. For example, a displacement of 3 (= +3) describes a position change of 3 units in the positive direction, and a displacement of −3 describes a position change of 3 units in the negative direction. However, for a particle that moves in two dimensions or three dimensions, a plus or minus sign is no longer sufficient to specify the direction of motion—other methods are required. One method is to use an arrow, called a ***vector***, that points in the direction of motion and whose length represents the distance from the starting point to the ending point; this is called the ***displacement vector*** for the motion. For example, the first part of Figure 12.2.1 shows the displacement vector of a particle that moves from point A to point B along a circuitous path. Note that the length of the arrow describes the distance between the starting and ending points and not the actual distance traveled by the particle.

Arrows are not limited to describing displacements—they can be used to describe any physical quantity that involves both a magnitude and a direction. Two important examples are forces and velocities. For example, the arrow in the second part of Figure 12.2.1 represents a force vector of 10 lb acting in a specific direction on a block, and the arrows in the third part of that figure show the velocity vector of a boat whose motor propels it parallel to the shore at 2 mi/h and the velocity vector of a 3 mi/h wind acting at an angle of 45° with the shoreline. Intuition suggests that the two velocity vectors will combine to produce some net velocity for the boat at an angle to the shoreline. Thus, our first objective in this section is to define mathematical operations on vectors that can be used to determine the combined effect of vectors.

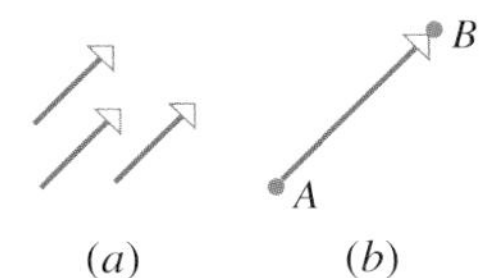

Figure 12.2.2

VECTORS VIEWED GEOMETRICALLY

Vectors can be represented geometrically by arrows in 2-space or 3-space; the direction of the arrow specifies the direction of the vector and the length of the arrow describes its magnitude. The tail of the arrow is called the ***initial point*** of the vector, and the tip of the arrow the ***terminal point***. We will denote vectors with lowercase boldface type such as $\mathbf{a}$, $\mathbf{k}$, $\mathbf{v}$, $\mathbf{w}$, and $\mathbf{x}$. When discussing vectors, we will refer to real numbers as ***scalars***. Scalars will be denoted by lowercase italic type such as a, k, v, w, and x. Two vectors, $\mathbf{v}$ and $\mathbf{w}$, are considered to be ***equal*** (also called ***equivalent***) if they have the same length and same direction, in which case we write $\mathbf{v} = \mathbf{w}$. Geometrically, two vectors are equal if they are translations of one another; thus, the three vectors in Figure 12.2.2*a* are equal, even though they are in different positions.

Because vectors are not affected by translation, the initial point of a vector $\mathbf{v}$ can be moved to any convenient point A by making an appropriate translation. If the initial point of $\mathbf{v}$ is A and the terminal point is B, then we write $\mathbf{v} = \overrightarrow{AB}$ when we want to emphasize the initial and terminal points (Figure 12.2.2*b*). If the initial and terminal points of a vector coincide, then the vector has length zero; we call this the ***zero vector*** and denote it by $\mathbf{0}$. The zero vector does not have a specific direction, so we will agree that it can be assigned any convenient direction in a specific problem.

There are various algebraic operations that are performed on vectors, all of whose definitions originated in physics. We begin with vector addition.

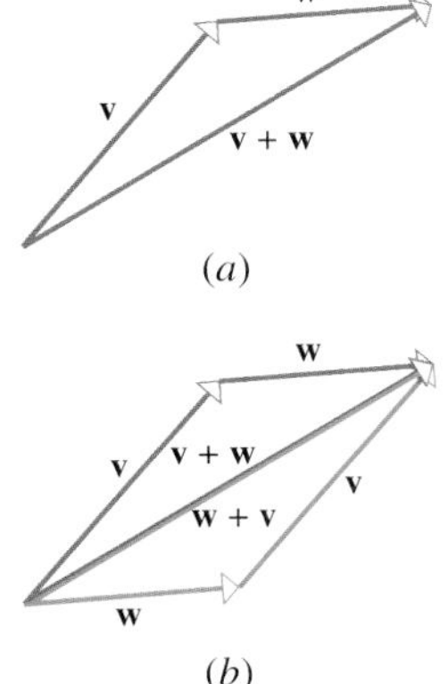

Figure 12.2.3

> **12.2.1 DEFINITION.** If $\mathbf{v}$ and $\mathbf{w}$ are vectors, then the ***sum*** $\mathbf{v} + \mathbf{w}$ is the vector from the initial point of $\mathbf{v}$ to the terminal point of $\mathbf{w}$ when the vectors are positioned so the initial point of $\mathbf{w}$ is at the terminal point of $\mathbf{v}$ (Figure 12.2.3*a*).

In Figure 12.2.3*b* we have constructed two sums, $\mathbf{v}+\mathbf{w}$ (purple arrows) and $\mathbf{w}+\mathbf{v}$ (green arrows). It is evident that

$$\mathbf{v}+\mathbf{w}=\mathbf{w}+\mathbf{v}$$

and that the sum coincides with the diagonal of the parallelogram determined by $\mathbf{v}$ and $\mathbf{w}$ when these vectors are positioned so they have the same initial point.

Since the initial and terminal points of $\mathbf{0}$ coincide, it follows that

$$\mathbf{0}+\mathbf{v}=\mathbf{v}+\mathbf{0}=\mathbf{v}$$

12.2.2 DEFINITION. If $\mathbf{v}$ is a nonzero vector and k is a nonzero real number (a scalar), then the ***scalar multiple*** $k\mathbf{v}$ is defined to be the vector whose length is $|k|$ times the length of $\mathbf{v}$ and whose direction is the same as that of $\mathbf{v}$ if $k>0$ and opposite to that of $\mathbf{v}$ if $k<0$. We define $k\mathbf{v}=\mathbf{0}$ if $k=0$ or $\mathbf{v}=\mathbf{0}$.

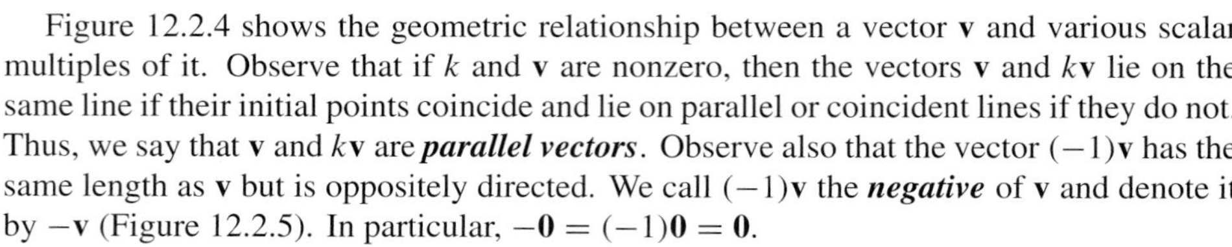

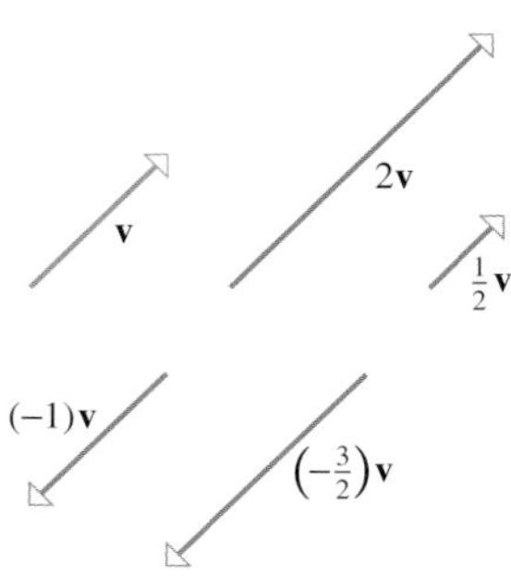

Figure 12.2.4

Figure 12.2.4 shows the geometric relationship between a vector $\mathbf{v}$ and various scalar multiples of it. Observe that if k and $\mathbf{v}$ are nonzero, then the vectors $\mathbf{v}$ and $k\mathbf{v}$ lie on the same line if their initial points coincide and lie on parallel or coincident lines if they do not. Thus, we say that $\mathbf{v}$ and $k\mathbf{v}$ are ***parallel vectors***. Observe also that the vector $(-1)\mathbf{v}$ has the same length as $\mathbf{v}$ but is oppositely directed. We call $(-1)\mathbf{v}$ the ***negative*** of $\mathbf{v}$ and denote it by $-\mathbf{v}$ (Figure 12.2.5). In particular, $-\mathbf{0}=(-1)\mathbf{0}=\mathbf{0}$.

Vector subtraction is defined in terms of addition and scalar multiplication by

$$\mathbf{v}-\mathbf{w}=\mathbf{v}+(-\mathbf{w})$$

The difference $\mathbf{v}-\mathbf{w}$ can be obtained geometrically by first constructing the vector $-\mathbf{w}$ and then adding $\mathbf{v}$ and $-\mathbf{w}$, say by the parallelogram method (Figure 12.2.6*a*). However, if $\mathbf{v}$ and $\mathbf{w}$ are positioned so their initial points coincide, then $\mathbf{v}-\mathbf{w}$ can be formed more directly, as shown in Figure 12.2.6*b*, by drawing the vector from the terminal point of $\mathbf{w}$ (the second term) to the terminal point of $\mathbf{v}$ (the first term). In the special case where $\mathbf{v}=\mathbf{w}$ the terminal points of the vectors coincide, so their difference is $\mathbf{0}$; that is,

$$\mathbf{v}+(-\mathbf{v})=\mathbf{v}-\mathbf{v}=\mathbf{0}$$

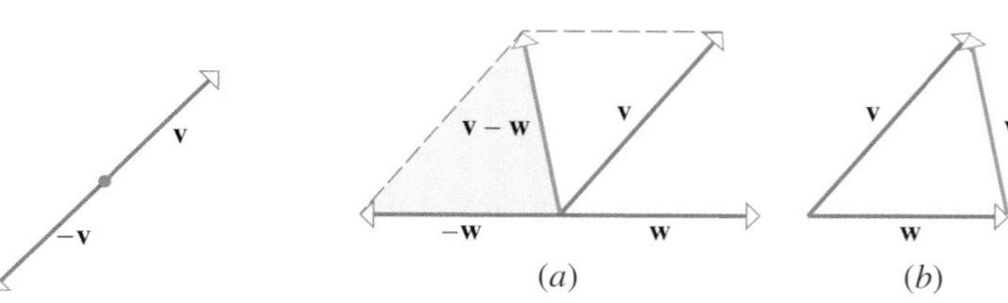

Figure 12.2.5

Figure 12.2.6

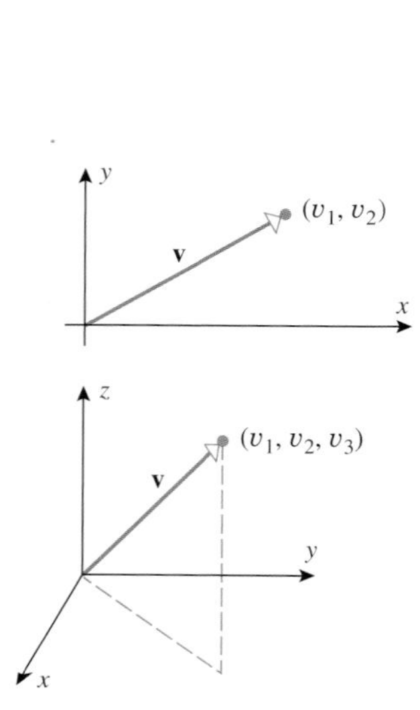

Figure 12.2.7

VECTORS IN COORDINATE SYSTEMS

Problems involving vectors are often best solved by introducing a rectangular coordinate system. If a vector $\mathbf{v}$ is positioned with its initial point at the origin of a rectangular coordinate system, then its terminal point will have coordinates of the form (v_1, v_2) or (v_1, v_2, v_3), depending on whether the vector is in 2-space or 3-space (Figure 12.2.7). We call these coordinates the ***components*** of $\mathbf{v}$, and we write $\mathbf{v}$ in *component form* as

$$\underbrace{\mathbf{v}=\langle v_1, v_2\rangle}_{\text{2-space}} \quad \text{or} \quad \underbrace{\mathbf{v}=\langle v_1, v_2, v_3\rangle}_{\text{3-space}}$$

In particular, the zero vectors in 2-space and 3-space are

$$\mathbf{0} = \langle 0, 0 \rangle \quad \text{and} \quad \mathbf{0} = \langle 0, 0, 0 \rangle$$

respectively.

Components provide a simple way of identifying equivalent vectors. For example, consider the vectors $\mathbf{v} = \langle v_1, v_2 \rangle$ and $\mathbf{w} = \langle w_1, w_2 \rangle$ in 2-space. If $\mathbf{v} = \mathbf{w}$, then the vectors have the same length and same direction, and this means that their terminal points coincide when their initial points are placed at the origin. It follows that $v_1 = w_1$ and $v_2 = w_2$, so we have shown that equivalent vectors have the same components. Conversely, if $v_1 = w_1$ and $v_2 = w_2$, then the terminal points of the vectors coincide when their initial points are placed at the origin. It follows that the vectors have the same length and same direction, so we have shown that vectors with the same components are equivalent. A similar argument holds for vectors in 3-space, so we have the following result.

12.2.3 THEOREM. *Two vectors are equivalent if and only if their corresponding components are equal.*

For example,

$$\langle a, b, c \rangle = \langle 1, -4, 2 \rangle$$

if and only if $a = 1$, $b = -4$, and $c = 2$.

ARITHMETIC OPERATIONS ON VECTORS

The next theorem shows how to perform arithmetic operations on vectors using components.

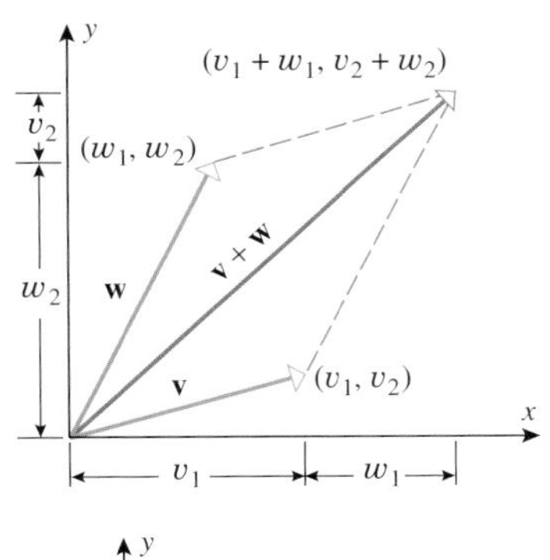

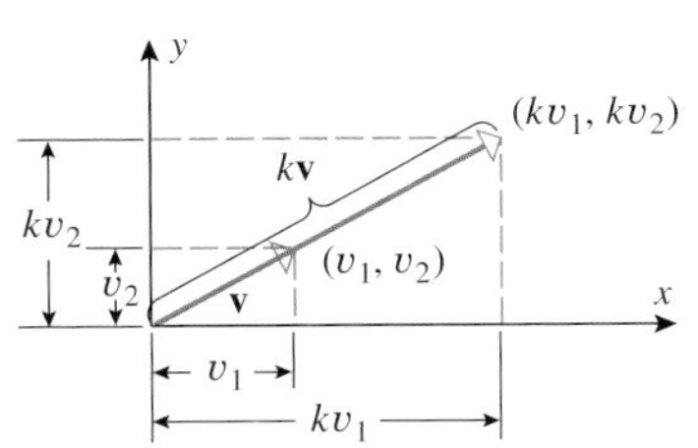

Figure 12.2.8

12.2.4 THEOREM. *If $\mathbf{v} = \langle v_1, v_2 \rangle$ and $\mathbf{w} = \langle w_1, w_2 \rangle$ are vectors in 2-space and k is any scalar, then*

$$\mathbf{v} + \mathbf{w} = \langle v_1 + w_1, v_2 + w_2 \rangle \tag{1}$$
$$\mathbf{v} - \mathbf{w} = \langle v_1 - w_1, v_2 - w_2 \rangle \tag{2}$$
$$k\mathbf{v} = \langle kv_1, kv_2 \rangle \tag{3}$$

Similarly, if $\mathbf{v} = \langle v_1, v_2, v_3 \rangle$ and $\mathbf{w} = \langle w_1, w_2, w_3 \rangle$ are vectors in 3-space and k is any scalar, then

$$\mathbf{v} + \mathbf{w} = \langle v_1 + w_1, v_2 + w_2, v_3 + w_3 \rangle \tag{4}$$
$$\mathbf{v} - \mathbf{w} = \langle v_1 - w_1, v_2 - w_2, v_3 - w_3 \rangle \tag{5}$$
$$k\mathbf{v} = \langle kv_1, kv_2, kv_3 \rangle \tag{6}$$

We will not prove this theorem. However, results (1) and (3) should be evident from Figure 12.2.8. Similar figures in 3-space can be used to motivate (4) and (6). Formulas (2) and (5) can be obtained by writing $\mathbf{v} + \mathbf{w} = \mathbf{v} + (-1)\mathbf{w}$.

▶ **Example 1** If $\mathbf{v} = \langle -2, 0, 1 \rangle$ and $\mathbf{w} = \langle 3, 5, -4 \rangle$, then

$$\begin{aligned} \mathbf{v} + \mathbf{w} &= \langle -2, 0, 1 \rangle + \langle 3, 5, -4 \rangle = \langle 1, 5, -3 \rangle \\ 3\mathbf{v} &= \langle -6, 0, 3 \rangle \\ -\mathbf{w} &= \langle -3, -5, 4 \rangle \\ \mathbf{w} - 2\mathbf{v} &= \langle 3, 5, -4 \rangle - \langle -4, 0, 2 \rangle = \langle 7, 5, -6 \rangle \end{aligned}$$ ◀

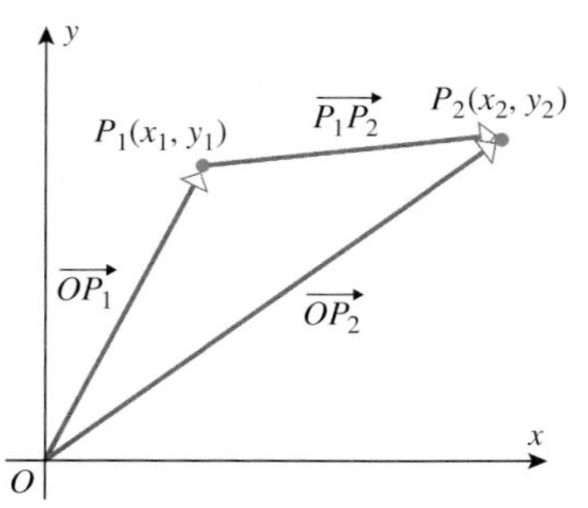

Figure 12.2.9

VECTORS WITH INITIAL POINT NOT AT THE ORIGIN

Recall that we defined the components of a vector to be the coordinates of its terminal point when its initial point is at the origin. We will now consider the problem of finding the components of a vector whose initial point is not at the origin. To be specific, suppose that $P_1(x_1, y_1)$ and $P_2(x_2, y_2)$ are points in 2-space and we are interested in finding the components of the vector $\overrightarrow{P_1P_2}$. As illustrated in Figure 12.2.9, we can write this vector as

$$\overrightarrow{P_1P_2} = \overrightarrow{OP_2} - \overrightarrow{OP_1} = \langle x_2, y_2\rangle - \langle x_1, y_1\rangle = \langle x_2 - x_1, y_2 - y_1\rangle$$

Thus, we have shown that the components of the vector $\overrightarrow{P_1P_2}$ can be obtained by subtracting the coordinates of its initial point from the coordinates of its terminal point. Similar computations hold in 3-space, so we have established the following result.

12.2.5 THEOREM. *If $\overrightarrow{P_1P_2}$ is a vector in 2-space with initial point $P_1(x_1, y_1)$ and terminal point $P_2(x_2, y_2)$, then*

$$\overrightarrow{P_1P_2} = \langle x_2 - x_1, y_2 - y_1\rangle \tag{7}$$

Similarly, if $\overrightarrow{P_1P_2}$ is a vector in 3-space with initial point $P_1(x_1, y_1, z_1)$ and terminal point $P_2(x_2, y_2, z_2)$, then

$$\overrightarrow{P_1P_2} = \langle x_2 - x_1, y_2 - y_1, z_2 - z_1\rangle \tag{8}$$

► **Example 2** In 2-space the vector from $P_1(1, 3)$ to $P_2(4, -2)$ is

$$\overrightarrow{P_1P_2} = \langle 4 - 1, -2 - 3\rangle = \langle 3, -5\rangle$$

and in 3-space the vector from $A(0, -2, 5)$ to $B(3, 4, -1)$ is

$$\overrightarrow{AB} = \langle 3 - 0, 4 - (-2), -1 - 5\rangle = \langle 3, 6, -6\rangle$$ ◄

RULES OF VECTOR ARITHMETIC

The following theorem shows that many of the familiar rules of ordinary arithmetic also hold for vector arithmetic.

12.2.6 THEOREM. *For any vectors* **u**, **v**, *and* **w** *and any scalars k and l, the following relationships hold:*

(*a*) $\mathbf{u} + \mathbf{v} = \mathbf{v} + \mathbf{u}$	(*e*) $k(l\mathbf{u}) = (kl)\mathbf{u}$
(*b*) $(\mathbf{u} + \mathbf{v}) + \mathbf{w} = \mathbf{u} + (\mathbf{v} + \mathbf{w})$	(*f*) $k(\mathbf{u} + \mathbf{v}) = k\mathbf{u} + k\mathbf{v}$
(*c*) $\mathbf{u} + \mathbf{0} = \mathbf{0} + \mathbf{u} = \mathbf{u}$	(*g*) $(k + l)\mathbf{u} = k\mathbf{u} + l\mathbf{u}$
(*d*) $\mathbf{u} + (-\mathbf{u}) = \mathbf{0}$	(*h*) $1\mathbf{u} = \mathbf{u}$

It follows from part (*b*) of Theorem 12.2.6 that the expression

$$\mathbf{u} + \mathbf{v} + \mathbf{w}$$

is unambiguous since the same vector results no matter how the terms are grouped.

The results in this theorem can be proved either algebraically by using components or geometrically by treating the vectors as arrows. We will prove part (*b*) both ways and leave some of the remaining proofs as exercises.

PROOF (*b*) (ALGEBRAIC IN 2-SPACE). Let $\mathbf{u} = \langle u_1, u_2 \rangle$, $\mathbf{v} = \langle v_1, v_2 \rangle$, and $\mathbf{w} = \langle w_1, w_2 \rangle$. Then

$$\begin{aligned}(\mathbf{u}+\mathbf{v})+\mathbf{w} &= (\langle u_1, u_2\rangle + \langle v_1, v_2\rangle) + \langle w_1, w_2\rangle \\ &= \langle u_1+v_1, u_2+v_2\rangle + \langle w_1, w_2\rangle \\ &= \langle (u_1+v_1)+w_1, (u_2+v_2)+w_2\rangle \\ &= \langle u_1+(v_1+w_1), u_2+(v_2+w_2)\rangle \\ &= \langle u_1, u_2\rangle + \langle v_1+w_1, v_2+w_2\rangle \\ &= \mathbf{u}+(\mathbf{v}+\mathbf{w})\end{aligned}$$

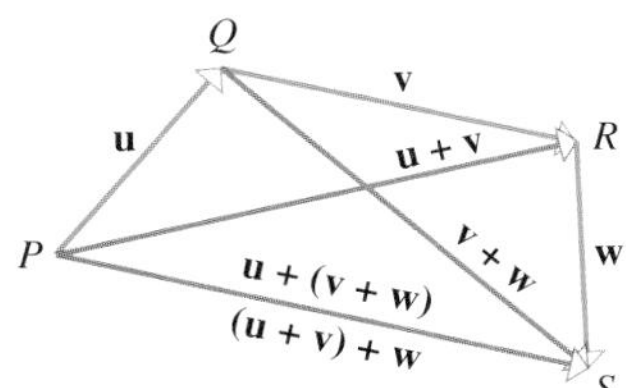

Figure 12.2.10

PROOF (*b*) (GEOMETRIC). Let **u**, **v**, and **w** be represented by $\overrightarrow{PQ}$, $\overrightarrow{QR}$, and $\overrightarrow{RS}$ as shown in Figure 12.2.10. Then

$$\mathbf{v}+\mathbf{w} = \overrightarrow{QS} \quad \text{and} \quad \mathbf{u}+(\mathbf{v}+\mathbf{w}) = \overrightarrow{PS}$$
$$\mathbf{u}+\mathbf{v} = \overrightarrow{PR} \quad \text{and} \quad (\mathbf{u}+\mathbf{v})+\mathbf{w} = \overrightarrow{PS}$$

Therefore,

$$(\mathbf{u}+\mathbf{v})+\mathbf{w} = \mathbf{u}+(\mathbf{v}+\mathbf{w}) \qquad \blacksquare$$

Observe that in Figure 12.2.10 the vectors **u**, **v**, and **w** are positioned "tip to tail" and that

$$\mathbf{u}+\mathbf{v}+\mathbf{w}$$

is the vector from the initial point of **u** (the first term in the sum) to the terminal point of **w** (the last term in the sum). This "tip to tail" method of vector addition also works for four or more vectors (Figure 12.2.11).

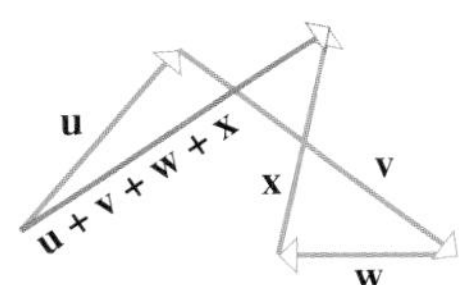

Figure 12.2.11

NORM OF A VECTOR

The distance between the initial and terminal points of a vector **v** is called the ***length***, the ***norm***, or the ***magnitude*** of **v** and is denoted by $\|\mathbf{v}\|$. This distance does not change if the vector is translated, so for purposes of calculating the norm we can assume that the vector is positioned with its initial point at the origin (Figure 12.2.12). This makes it evident that the norm of a vector $\mathbf{v} = \langle v_1, v_2\rangle$ in 2-space is given by

$$\|\mathbf{v}\| = \sqrt{v_1^2 + v_2^2} \tag{9}$$

and the norm of a vector $\mathbf{v} = \langle v_1, v_2, v_3\rangle$ in 3-space is given by

$$\|\mathbf{v}\| = \sqrt{v_1^2 + v_2^2 + v_3^2} \tag{10}$$

▶ **Example 3** Find the norms of $\mathbf{v} = \langle -2, 3\rangle$, $10\mathbf{v} = \langle -20, 30\rangle$, and $\mathbf{w} = \langle 2, 3, 6\rangle$.

Solution. From (9) and (10)

$$\begin{aligned}\|\mathbf{v}\| &= \sqrt{(-2)^2+3^2} = \sqrt{13} \\ \|10\mathbf{v}\| &= \sqrt{(-20)^2+30^2} = \sqrt{1300} = 10\sqrt{13} \\ \|\mathbf{w}\| &= \sqrt{2^2+3^2+6^2} = \sqrt{49} = 7 \quad \blacktriangleleft\end{aligned}$$

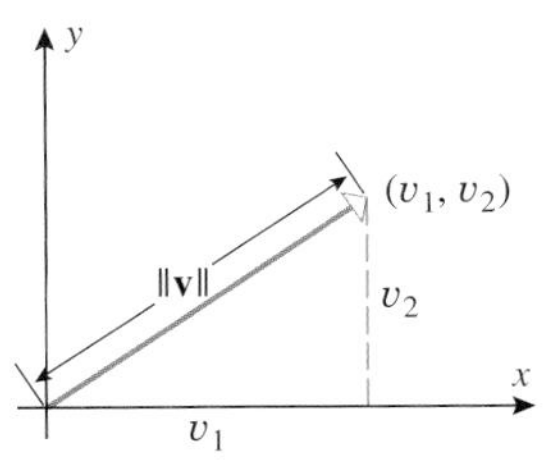

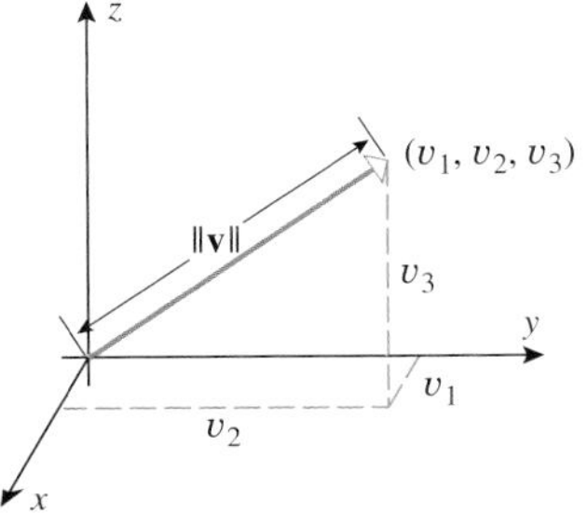

Figure 12.2.12

Note that $\|10\mathbf{v}\| = 10\|\mathbf{v}\|$ in Example 3. This is consistent with Definition 12.2.2, which stipulated that for any vector **v** and scalar k, the length of $k\mathbf{v}$ must be $|k|$ times the length of **v**; that is,

$$\|k\mathbf{v}\| = |k|\|\mathbf{v}\| \tag{11}$$

Thus, for example,

$$\begin{aligned}\|3\mathbf{v}\| &= |3|\|\mathbf{v}\| = 3\|\mathbf{v}\| \\ \|-2\mathbf{v}\| &= |-2|\|\mathbf{v}\| = 2\|\mathbf{v}\| \\ \|-1\mathbf{v}\| &= |-1|\|\mathbf{v}\| = \|\mathbf{v}\|\end{aligned}$$

This applies to vectors in 2-space and 3-space.

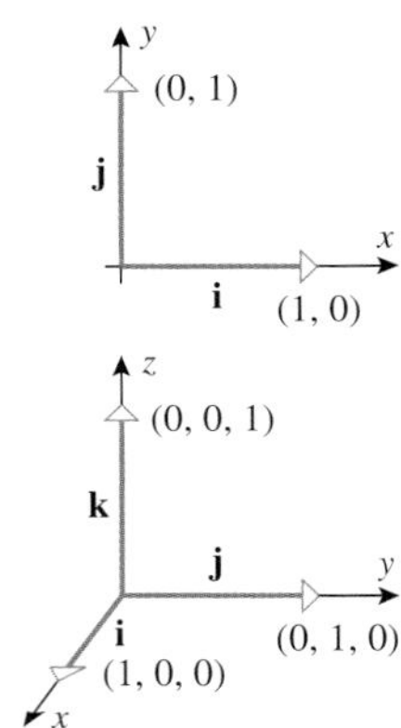

Figure 12.2.13

UNIT VECTORS

A vector of length 1 is called a ***unit vector***. In an xy-coordinate system the unit vectors along the x- and y-axes are denoted by $\mathbf{i}$ and $\mathbf{j}$, respectively; and in an xyz-coordinate system the unit vectors along the x-, y-, and z-axes are denoted by $\mathbf{i}$, $\mathbf{j}$, and $\mathbf{k}$, respectively (Figure 12.2.13). Thus,

$$\mathbf{i} = \langle 1, 0 \rangle, \quad \mathbf{j} = \langle 0, 1 \rangle \qquad \text{In 2-space}$$

$$\mathbf{i} = \langle 1, 0, 0 \rangle, \quad \mathbf{j} = \langle 0, 1, 0 \rangle, \quad \mathbf{k} = \langle 0, 0, 1 \rangle \qquad \text{In 3-space}$$

Every vector in 2-space is expressible uniquely in terms of $\mathbf{i}$ and $\mathbf{j}$, and every vector in 3-space is expressible uniquely in terms of $\mathbf{i}$, $\mathbf{j}$, and $\mathbf{k}$ as follows:

$$\mathbf{v} = \langle v_1, v_2 \rangle = \langle v_1, 0 \rangle + \langle 0, v_2 \rangle = v_1\langle 1, 0 \rangle + v_2\langle 0, 1 \rangle = v_1\mathbf{i} + v_2\mathbf{j}$$

$$\mathbf{v} = \langle v_1, v_2, v_3 \rangle = v_1\langle 1, 0, 0 \rangle + v_2\langle 0, 1, 0 \rangle + v_3\langle 0, 0, 1 \rangle = v_1\mathbf{i} + v_2\mathbf{j} + v_3\mathbf{k}$$

► Example 4

The two notations for vectors illustrated in Example 4 are completely interchangeable, the choice being a matter of convenience or personal preference.

2-SPACE	3-SPACE
$\langle 2, 3 \rangle = 2\mathbf{i} + 3\mathbf{j}$	$\langle 2, -3, 4 \rangle = 2\mathbf{i} - 3\mathbf{j} + 4\mathbf{k}$
$\langle -4, 0 \rangle = -4\mathbf{i} + 0\mathbf{j} = -4\mathbf{i}$	$\langle 0, 3, 0 \rangle = 3\mathbf{j}$
$\langle 0, 0 \rangle = 0\mathbf{i} + 0\mathbf{j} = \mathbf{0}$	$\langle 0, 0, 0 \rangle = 0\mathbf{i} + 0\mathbf{j} + 0\mathbf{k} = \mathbf{0}$
$(3\mathbf{i} + 2\mathbf{j}) + (4\mathbf{i} + \mathbf{j}) = 7\mathbf{i} + 3\mathbf{j}$	$(3\mathbf{i} + 2\mathbf{j} - \mathbf{k}) - (4\mathbf{i} - \mathbf{j} + 2\mathbf{k}) = -\mathbf{i} + 3\mathbf{j} - 3\mathbf{k}$
$5(6\mathbf{i} - 2\mathbf{j}) = 30\mathbf{i} - 10\mathbf{j}$	$2(\mathbf{i} + \mathbf{j} - \mathbf{k}) + 4(\mathbf{i} - \mathbf{j}) = 6\mathbf{i} - 2\mathbf{j} - 2\mathbf{k}$
$\lVert 2\mathbf{i} - 3\mathbf{j} \rVert = \sqrt{2^2 + (-3)^2} = \sqrt{13}$	$\lVert \mathbf{i} + 2\mathbf{j} - 3\mathbf{k} \rVert = \sqrt{1^2 + 2^2 + (-3)^2} = \sqrt{14}$
$\lVert v_1\mathbf{i} + v_2\mathbf{j} \rVert = \sqrt{v_1^2 + v_2^2}$	$\lVert \langle v_1, v_2, v_3 \rangle \rVert = \sqrt{v_1^2 + v_2^2 + v_3^2}$ ◄

NORMALIZING A VECTOR

A common problem in applications is to find a unit vector $\mathbf{u}$ that has the same direction as some given nonzero vector $\mathbf{v}$. This can be done by multiplying $\mathbf{v}$ by the reciprocal of its length; that is,

$$\mathbf{u} = \frac{1}{\lVert \mathbf{v} \rVert}\mathbf{v} = \frac{\mathbf{v}}{\lVert \mathbf{v} \rVert}$$

is a unit vector with the same direction as $\mathbf{v}$—the direction is the same because $k = 1/\lVert \mathbf{v} \rVert$ is a positive scalar, and the length is 1 because

$$\lVert \mathbf{u} \rVert = \lVert k\mathbf{v} \rVert = |k|\lVert \mathbf{v} \rVert = k\lVert \mathbf{v} \rVert = \frac{1}{\lVert \mathbf{v} \rVert}\lVert \mathbf{v} \rVert = 1$$

The process of multiplying a vector $\mathbf{v}$ by the reciprocal of its length to obtain a unit vector with the same direction is called ***normalizing*** $\mathbf{v}$.

TECHNOLOGY MASTERY

Many calculating utilities can perform vector operations, and some have built-in norm and normalization operations. If your calculator has these capabilities, use it to check the computations in Examples 1, 3, and 5.

► Example 5

Find the unit vector that has the same direction as $\mathbf{v} = 2\mathbf{i} + 2\mathbf{j} - \mathbf{k}$.

Solution. The vector $\mathbf{v}$ has length

$$\lVert \mathbf{v} \rVert = \sqrt{2^2 + 2^2 + (-1)^2} = 3$$

so the unit vector $\mathbf{u}$ in the same direction as $\mathbf{v}$ is

$$\mathbf{u} = \tfrac{1}{3}\mathbf{v} = \tfrac{2}{3}\mathbf{i} + \tfrac{2}{3}\mathbf{j} - \tfrac{1}{3}\mathbf{k} \ ◄$$

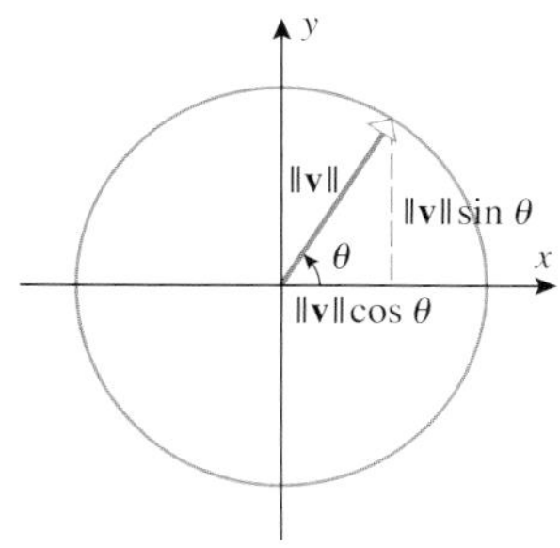

Figure 12.2.14

VECTORS DETERMINED BY LENGTH AND ANGLE

If $\mathbf{v}$ is a nonzero vector with its initial point at the origin of an xy-coordinate system, and if θ is the angle from the positive x-axis to the radial line through $\mathbf{v}$, then the x-component of $\mathbf{v}$ can be written as $\|\mathbf{v}\|\cos\theta$ and the y-component as $\|\mathbf{v}\|\sin\theta$ (Figure 12.2.14); and hence $\mathbf{v}$ can be expressed in trigonometric form as

$$\mathbf{v} = \|\mathbf{v}\|\langle\cos\theta, \sin\theta\rangle \quad \text{or} \quad \mathbf{v} = \|\mathbf{v}\|\cos\theta\,\mathbf{i} + \|\mathbf{v}\|\sin\theta\,\mathbf{j} \tag{12}$$

In the special case of a unit vector $\mathbf{u}$ this simplifies to

$$\mathbf{u} = \langle\cos\theta, \sin\theta\rangle \quad \text{or} \quad \mathbf{u} = \cos\theta\,\mathbf{i} + \sin\theta\,\mathbf{j} \tag{13}$$

► **Example 6**

(a) Find the vector of length 2 that makes an angle of $\pi/4$ with the positive x-axis.

(b) Find the angle that the vector $\mathbf{v} = -\sqrt{3}\,\mathbf{i} + \mathbf{j}$ makes with the positive x-axis.

Solution (a). From (12)

$$\mathbf{v} = 2\cos\frac{\pi}{4}\mathbf{i} + 2\sin\frac{\pi}{4}\mathbf{j} = \sqrt{2}\,\mathbf{i} + \sqrt{2}\,\mathbf{j}$$

Solution (b). We will normalize $\mathbf{v}$, then use (13) to find $\sin\theta$ and $\cos\theta$, and then use these values to find θ. Normalizing $\mathbf{v}$ yields

$$\frac{\mathbf{v}}{\|\mathbf{v}\|} = \frac{-\sqrt{3}\,\mathbf{i} + \mathbf{j}}{\sqrt{(-\sqrt{3}\,)^2 + 1^2}} = -\frac{\sqrt{3}}{2}\mathbf{i} + \frac{1}{2}\mathbf{j}$$

Thus, $\cos\theta = -\sqrt{3}/2$ and $\sin\theta = \frac{1}{2}$, from which we conclude that $\theta = 5\pi/6$. ◄

VECTORS DETERMINED BY LENGTH AND A VECTOR IN THE SAME DIRECTION

It is a common problem in many applications that a direction in 2-space or 3-space is determined by some known unit vector $\mathbf{u}$, and it is of interest to find the components of a vector $\mathbf{v}$ that has the same direction as $\mathbf{u}$ and some specified length $\|\mathbf{v}\|$. This can be done by expressing $\mathbf{v}$ as

$$\mathbf{v} = \|\mathbf{v}\|\mathbf{u}$$

v is equal to its length times a unit vector in the same direction.

and then reading off the components of $\|\mathbf{v}\|\mathbf{u}$.

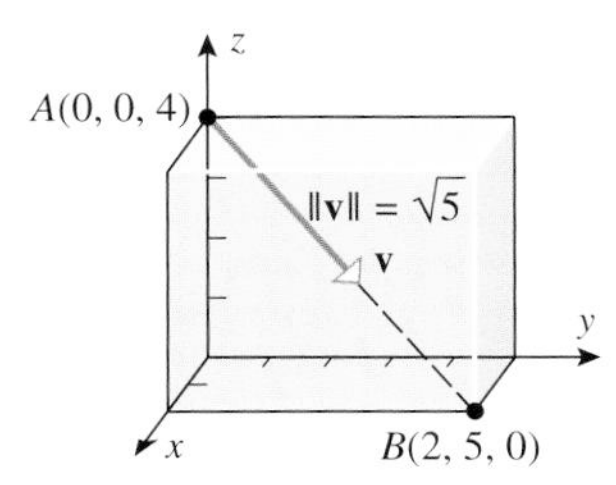

Figure 12.2.15

► **Example 7** Figure 12.2.15 shows a vector $\mathbf{v}$ of length $\sqrt{5}$ that extends along the line through A and B. Find the components of $\mathbf{v}$.

Solution. First we will find the components of the vector $\overrightarrow{AB}$, then we will normalize this vector to obtain a unit vector in the direction of $\mathbf{v}$, and then we will multiply this unit vector by $\|\mathbf{v}\|$ to obtain the vector $\mathbf{v}$. The computations are as follows:

$$\overrightarrow{AB} = \langle 2, 5, 0\rangle - \langle 0, 0, 4\rangle = \langle 2, 5, -4\rangle$$

$$\|\overrightarrow{AB}\| = \sqrt{2^2 + 5^2 + (-4)^2} = \sqrt{45} = 3\sqrt{5}$$

$$\frac{\overrightarrow{AB}}{\|\overrightarrow{AB}\|} = \left\langle \frac{2}{3\sqrt{5}}, \frac{5}{3\sqrt{5}}, -\frac{4}{3\sqrt{5}} \right\rangle$$

$$\mathbf{v} = \|\mathbf{v}\|\left(\frac{\overrightarrow{AB}}{\|\overrightarrow{AB}\|}\right) = \sqrt{5}\left\langle \frac{2}{3\sqrt{5}}, \frac{5}{3\sqrt{5}}, -\frac{4}{3\sqrt{5}} \right\rangle = \left\langle \frac{2}{3}, \frac{5}{3}, -\frac{4}{3} \right\rangle$$ ◄

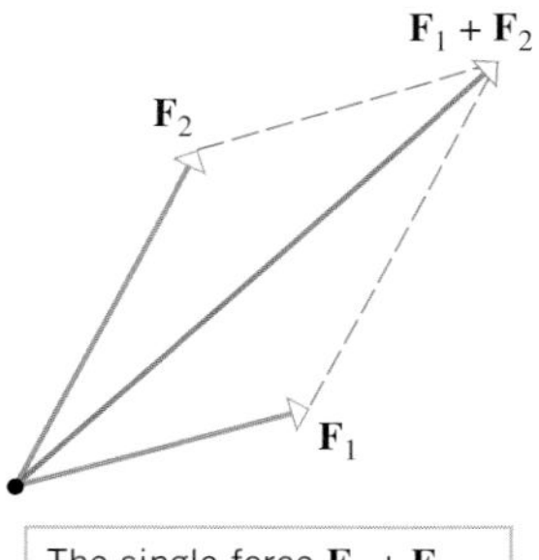

Figure 12.2.16

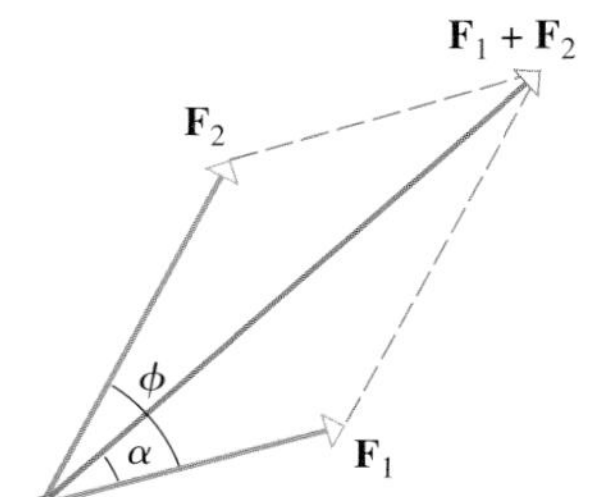

Figure 12.2.17

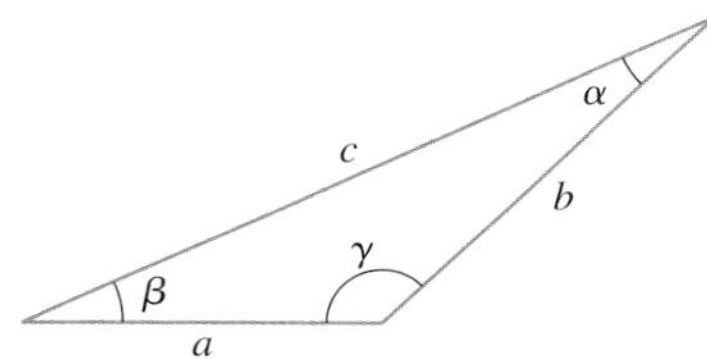

Figure 12.2.18

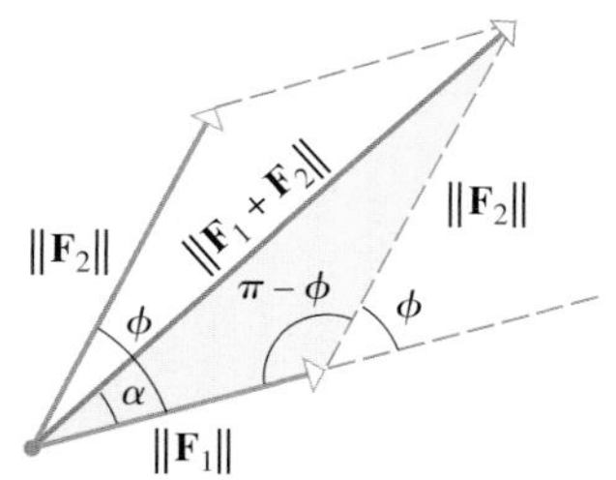

Figure 12.2.19

RESULTANT OF TWO CONCURRENT FORCES

The effect that a force has on an object depends on the magnitude and direction of the force and the point at which it is applied. Thus, forces are regarded to be vector quantities and, indeed, the algebraic operations on vectors that we have defined in this section have their origin in the study of forces. For example, it is a fact of physics that if two forces $\mathbf{F}_1$ and $\mathbf{F}_2$ are applied at the same point on an object, then the two forces have the same effect on the object as the single force $\mathbf{F}_1 + \mathbf{F}_2$ applied at the point (Figure 12.2.16). Physicists and engineers call $\mathbf{F}_1 + \mathbf{F}_2$ the ***resultant*** of $\mathbf{F}_1$ and $\mathbf{F}_2$, and they say that the forces $\mathbf{F}_1$ and $\mathbf{F}_2$ are ***concurrent*** to indicate that they are applied at the same point.

In many applications, the magnitudes of two concurrent forces and the angle between them are known, and the problem is to find the magnitude and direction of the resultant. For example, referring to Figure 12.2.17, suppose that we know the magnitudes of the forces $\mathbf{F}_1$ and $\mathbf{F}_2$ and the angle ϕ between them, and we are interested in finding the magnitude of the resultant $\mathbf{F}_1 + \mathbf{F}_2$ and the angle α that the resultant makes with the force $\mathbf{F}_1$. This can be done by trigonometric methods based on the laws of sines and cosines. For this purpose, recall that the law of sines applied to the triangle in Figure 12.2.18 states that

$$\frac{a}{\sin\alpha} = \frac{b}{\sin\beta} = \frac{c}{\sin\gamma}$$

and the law of cosines implies that

$$c^2 = a^2 + b^2 - 2ab\cos\gamma$$

Referring to Figure 12.2.19, and using the fact that $\cos(\pi - \phi) = -\cos\phi$, it follows from the law of cosines that

$$\|\mathbf{F}_1 + \mathbf{F}_2\|^2 = \|\mathbf{F}_1\|^2 + \|\mathbf{F}_2\|^2 + 2\|\mathbf{F}_1\|\|\mathbf{F}_2\|\cos\phi \tag{14}$$

Moreover, it follows from the law of sines that

$$\frac{\|\mathbf{F}_2\|}{\sin\alpha} = \frac{\|\mathbf{F}_1 + \mathbf{F}_2\|}{\sin(\pi - \phi)}$$

which, with the help of the identity $\sin(\pi - \phi) = \sin\phi$, can be expressed as

$$\sin\alpha = \frac{\|\mathbf{F}_2\|}{\|\mathbf{F}_1 + \mathbf{F}_2\|}\sin\phi \tag{15}$$

▸ **Example 8** Suppose that two forces are applied to an eye bracket, as shown in Figure 12.2.20. Find the magnitude of the resultant and the angle θ that it makes with the positive x-axis.

Solution. We are given that $\|\mathbf{F}_1\| = 200$ N and $\|\mathbf{F}_2\| = 300$ N and that the angle between the vectors $\mathbf{F}_1$ and $\mathbf{F}_2$ is $\phi = 40°$. Thus, it follows from (14) that the magnitude of the resultant is

$$\begin{aligned}\|\mathbf{F}_1 + \mathbf{F}_2\| &= \sqrt{\|\mathbf{F}_1\|^2 + \|\mathbf{F}_2\|^2 + 2\|\mathbf{F}_1\|\|\mathbf{F}_2\|\cos\phi} \\ &= \sqrt{(200)^2 + (300)^2 + 2(200)(300)\cos 40°} \\ &\approx 471 \text{ N}\end{aligned}$$

Moreover, it follows from (15) that the angle α between $\mathbf{F}_1$ and the resultant is

$$\alpha = \sin^{-1}\left(\frac{\|\mathbf{F}_2\|}{\|\mathbf{F}_1 + \mathbf{F}_2\|}\sin\phi\right) \approx \sin^{-1}\left(\frac{300}{471}\sin 40°\right) \approx 24.2°$$

Thus, the angle θ that the resultant makes with the positive x-axis is

$$\theta = \alpha + 30^\circ \approx 24.2^\circ + 30^\circ = 54.2^\circ$$

(Figure 12.2.21). ◀

The resultant of three or more concurrent forces can be found by working in pairs. For example, the resultant of three forces can be found by finding the resultant of any two of the forces and then finding the resultant of that resultant with the third force.

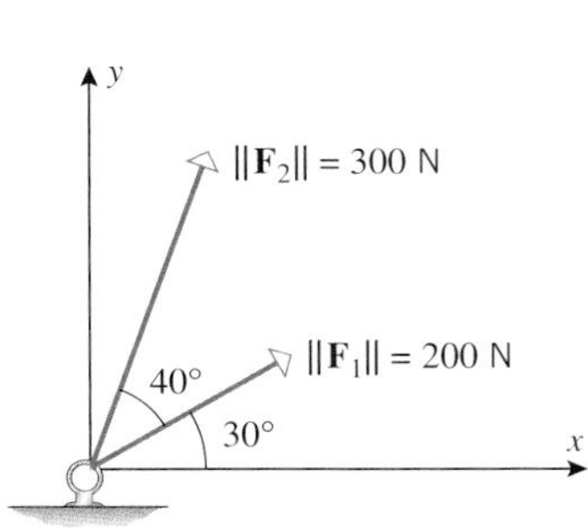

Figure 12.2.20

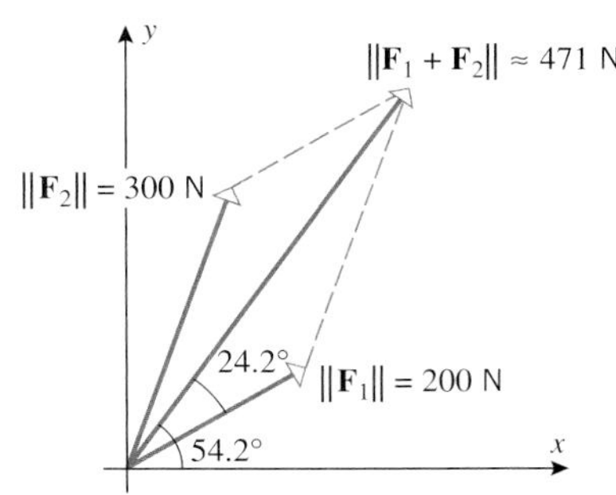

Figure 12.2.21

✔QUICK CHECK EXERCISES 12.2 *(See page 804 for answers.)*

1. If $\mathbf{v} = \langle 3, -1, 7\rangle$ and $\mathbf{w} = \langle 4, 10, -5\rangle$, then
(a) $\|\mathbf{v}\| =$ ________ (b) $\mathbf{v} + \mathbf{w} =$ ________
(c) $\mathbf{v} - \mathbf{w} =$ ________ (d) $2\mathbf{v} =$ ________.

2. The unit vector in the direction of $\mathbf{v} = \langle 3, -1, 7\rangle$ is ________.

3. The unit vector in 2-space that makes an angle of $\pi/3$ with the positive x-axis is ________.

4. Consider points $A(3, 4, 0)$ and $B(0, 0, 5)$.
(a) $\overrightarrow{AB} =$ ________
(b) If $\mathbf{v}$ is a vector in the same direction as $\overrightarrow{AB}$ and the length of $\mathbf{v}$ is $\sqrt{2}$, then $\mathbf{v} =$ ________.

EXERCISE SET 12.2

1–4 Sketch the vectors with their initial points at the origin.

1. (a) $\langle 2, 5\rangle$ (b) $\langle -5, -4\rangle$ (c) $\langle 2, 0\rangle$
(d) $-5\mathbf{i} + 3\mathbf{j}$ (e) $3\mathbf{i} - 2\mathbf{j}$ (f) $-6\mathbf{j}$

2. (a) $\langle -3, 7\rangle$ (b) $\langle 6, -2\rangle$ (c) $\langle 0, -8\rangle$
(d) $4\mathbf{i} + 2\mathbf{j}$ (e) $-2\mathbf{i} - \mathbf{j}$ (f) $4\mathbf{i}$

3. (a) $\langle 1, -2, 2\rangle$ (b) $\langle 2, 2, -1\rangle$
(c) $-\mathbf{i} + 2\mathbf{j} + 3\mathbf{k}$ (d) $2\mathbf{i} + 3\mathbf{j} - \mathbf{k}$

4. (a) $\langle -1, 3, 2\rangle$ (b) $\langle 3, 4, 2\rangle$
(c) $2\mathbf{j} - \mathbf{k}$ (d) $\mathbf{i} - \mathbf{j} + 2\mathbf{k}$

5–6 Find the components of the vector, and sketch an equivalent vector with its initial point at the origin.

5. (a)

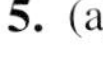

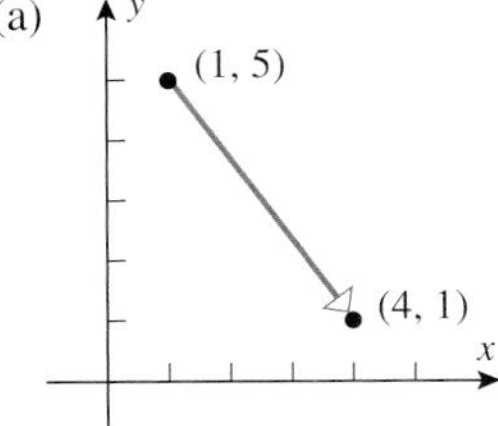

(b)

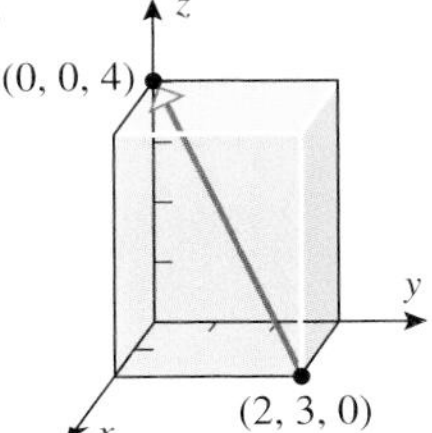

6. (a)

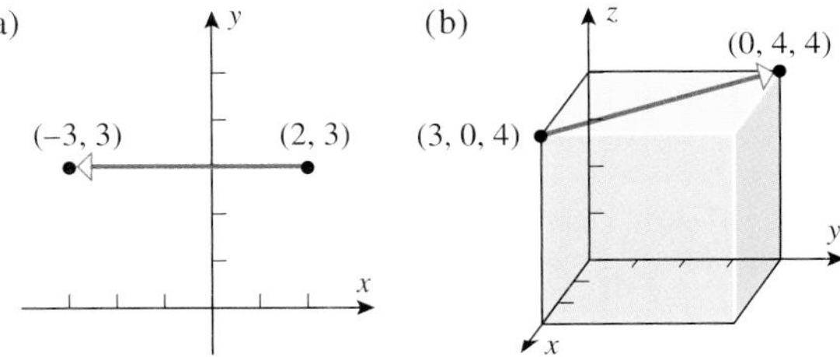

7–8 Find the components of the vector $\overrightarrow{P_1P_2}$.

7. (a) $P_1(3, 5)$, $P_2(2, 8)$ (b) $P_1(7, -2)$, $P_2(0, 0)$
(c) $P_1(5, -2, 1)$, $P_2(2, 4, 2)$

8. (a) $P_1(-6, -2)$, $P_2(-4, -1)$
(b) $P_1(0, 0, 0)$, $P_2(-1, 6, 1)$
(c) $P_1(4, 1, -3)$, $P_2(9, 1, -3)$

9. (a) Find the terminal point of $\mathbf{v} = 3\mathbf{i} - 2\mathbf{j}$ if the initial point is $(1, -2)$.
(b) Find the initial point of $\mathbf{v} = \langle -3, 1, 2\rangle$ if the terminal point is $(5, 0, -1)$.

10. (a) Find the terminal point of $\mathbf{v} = \langle 7, 6\rangle$ if the initial point is $(2, -1)$.

(b) Find the terminal point of $\mathbf{v} = \mathbf{i} + 2\mathbf{j} - 3\mathbf{k}$ if the initial point is $(-2, 1, 4)$.

11–12 Perform the stated operations on the vectors $\mathbf{u}$, $\mathbf{v}$, and $\mathbf{w}$.

11. $\mathbf{u} = 3\mathbf{i} - \mathbf{k}$, $\mathbf{v} = \mathbf{i} - \mathbf{j} + 2\mathbf{k}$, $\mathbf{w} = 3\mathbf{j}$
(a) $\mathbf{w} - \mathbf{v}$ (b) $6\mathbf{u} + 4\mathbf{w}$
(c) $-\mathbf{v} - 2\mathbf{w}$ (d) $4(3\mathbf{u} + \mathbf{v})$
(e) $-8(\mathbf{v} + \mathbf{w}) + 2\mathbf{u}$ (f) $3\mathbf{w} - (\mathbf{v} - \mathbf{w})$

12. $\mathbf{u} = \langle 2, -1, 3\rangle$, $\mathbf{v} = \langle 4, 0, -2\rangle$, $\mathbf{w} = \langle 1, 1, 3\rangle$
(a) $\mathbf{u} - \mathbf{w}$ (b) $7\mathbf{v} + 3\mathbf{w}$ (c) $-\mathbf{w} + \mathbf{v}$
(d) $3(\mathbf{u} - 7\mathbf{v})$ (e) $-3\mathbf{v} - 8\mathbf{w}$ (f) $2\mathbf{v} - (\mathbf{u} + \mathbf{w})$

13–14 Find the norm of $\mathbf{v}$.

13. (a) $\mathbf{v} = \langle 1, -1\rangle$ (b) $\mathbf{v} = -\mathbf{i} + 7\mathbf{j}$
(c) $\mathbf{v} = \langle -1, 2, 4\rangle$ (d) $\mathbf{v} = -3\mathbf{i} + 2\mathbf{j} + \mathbf{k}$

14. (a) $\mathbf{v} = \langle 3, 4\rangle$ (b) $\mathbf{v} = \sqrt{2}\mathbf{i} - \sqrt{7}\mathbf{j}$
(c) $\mathbf{v} = \langle 0, -3, 0\rangle$ (d) $\mathbf{v} = \mathbf{i} + \mathbf{j} + \mathbf{k}$

15. Let $\mathbf{u} = \mathbf{i} - 3\mathbf{j} + 2\mathbf{k}$, $\mathbf{v} = \mathbf{i} + \mathbf{j}$, and $\mathbf{w} = 2\mathbf{i} + 2\mathbf{j} - 4\mathbf{k}$. Find
(a) $\|\mathbf{u} + \mathbf{v}\|$ (b) $\|\mathbf{u}\| + \|\mathbf{v}\|$
(c) $\|-2\mathbf{u}\| + 2\|\mathbf{v}\|$ (d) $\|3\mathbf{u} - 5\mathbf{v} + \mathbf{w}\|$
(e) $\dfrac{1}{\|\mathbf{w}\|}\mathbf{w}$ (f) $\left\|\dfrac{1}{\|\mathbf{w}\|}\mathbf{w}\right\|$.

16. Is it possible to have $\|\mathbf{u}\| + \|\mathbf{v}\| = \|\mathbf{u} + \mathbf{v}\|$ if $\mathbf{u}$ and $\mathbf{v}$ are nonzero vectors? Justify your conclusion geometrically.

17–18 Find the unit vectors that satisfy the stated conditions.

17. (a) Same direction as $-\mathbf{i} + 4\mathbf{j}$.
(b) Oppositely directed to $6\mathbf{i} - 4\mathbf{j} + 2\mathbf{k}$.
(c) Same direction as the vector from the point $A(-1, 0, 2)$ to the point $B(3, 1, 1)$.

18. (a) Oppositely directed to $3\mathbf{i} - 4\mathbf{j}$.
(b) Same direction as $2\mathbf{i} - \mathbf{j} - 2\mathbf{k}$.
(c) Same direction as the vector from the point $A(-3, 2)$ to the point $B(1, -1)$.

19–20 Find the vectors that satisfy the stated conditions.

19. (a) Oppositely directed to $\mathbf{v} = \langle 3, -4\rangle$ and half the length of $\mathbf{v}$.
(b) Length $\sqrt{17}$ and same direction as $\mathbf{v} = \langle 7, 0, -6\rangle$.

20. (a) Same direction as $\mathbf{v} = -2\mathbf{i} + 3\mathbf{j}$ and three times the length of $\mathbf{v}$.
(b) Length 2 and oppositely directed to $\mathbf{v} = -3\mathbf{i} + 4\mathbf{j} + \mathbf{k}$.

21. In each part, find the component form of the vector $\mathbf{v}$ in 2-space that has the stated length and makes the stated angle θ with the positive x-axis.
(a) $\|\mathbf{v}\| = 3$; $\theta = \pi/4$ (b) $\|\mathbf{v}\| = 2$; $\theta = 90°$
(c) $\|\mathbf{v}\| = 5$; $\theta = 120°$ (d) $\|\mathbf{v}\| = 1$; $\theta = \pi$

22. Find the component forms of $\mathbf{v} + \mathbf{w}$ and $\mathbf{v} - \mathbf{w}$ in 2-space, given that $\|\mathbf{v}\| = 1$, $\|\mathbf{w}\| = 1$, $\mathbf{v}$ makes an angle of $\pi/6$ with the positive x-axis, and $\mathbf{w}$ makes an angle of $3\pi/4$ with the positive x-axis.

23–24 Find the component form of $\mathbf{v} + \mathbf{w}$, given that $\mathbf{v}$ and $\mathbf{w}$ are unit vectors.

23.

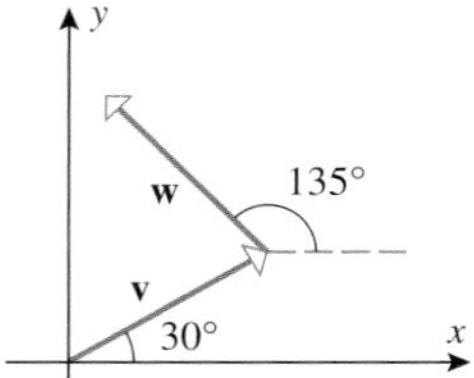

24.

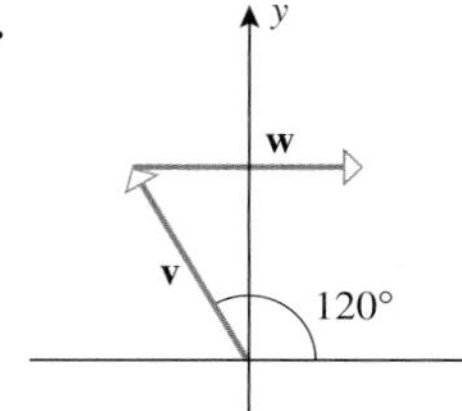

25. In each part, sketch the vector $\mathbf{u} + \mathbf{v} + \mathbf{w}$ and express it in component form.

(a)

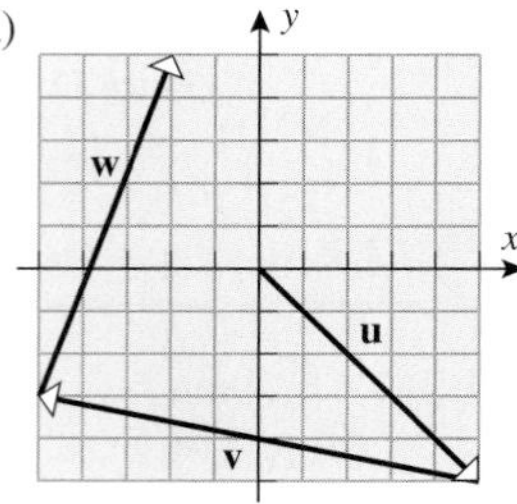

(b)

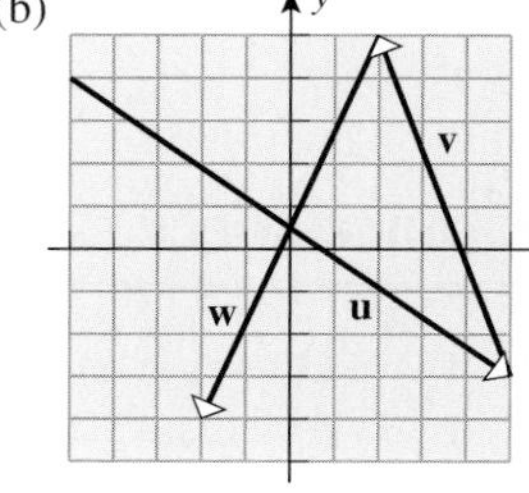

26. In each part of Exercise 25, sketch the vector $\mathbf{u} - \mathbf{v} + \mathbf{w}$ and express it in component form.

27. Let $\mathbf{u} = \langle 1, 3\rangle$, $\mathbf{v} = \langle 2, 1\rangle$, $\mathbf{w} = \langle 4, -1\rangle$. Find the vector $\mathbf{x}$ that satisfies $2\mathbf{u} - \mathbf{v} + \mathbf{x} = 7\mathbf{x} + \mathbf{w}$.

28. Let $\mathbf{u} = \langle -1, 1\rangle$, $\mathbf{v} = \langle 0, 1\rangle$, and $\mathbf{w} = \langle 3, 4\rangle$. Find the vector $\mathbf{x}$ that satisfies $\mathbf{u} - 2\mathbf{x} = \mathbf{x} - \mathbf{w} + 3\mathbf{v}$.

29. Find $\mathbf{u}$ and $\mathbf{v}$ if $\mathbf{u} + 2\mathbf{v} = 3\mathbf{i} - \mathbf{k}$ and $3\mathbf{u} - \mathbf{v} = \mathbf{i} + \mathbf{j} + \mathbf{k}$.

30. Find $\mathbf{u}$ and $\mathbf{v}$ if $\mathbf{u} + \mathbf{v} = \langle 2, -3\rangle$ and $3\mathbf{u} + 2\mathbf{v} = \langle -1, 2\rangle$.

31. Use vectors to find the lengths of the diagonals of the parallelogram that has $\mathbf{i} + \mathbf{j}$ and $\mathbf{i} - 2\mathbf{j}$ as adjacent sides.

32. Use vectors to find the fourth vertex of a parallelogram, three of whose vertices are $(0, 0)$, $(1, 3)$, and $(2, 4)$. [*Note:* There is more than one answer.]

33. (a) Given that $\|\mathbf{v}\| = 3$, find all values of k such that $\|k\mathbf{v}\| = 5$.
(b) Given that $k = -2$ and $\|k\mathbf{v}\| = 6$, find $\|\mathbf{v}\|$.

34. What do you know about k and $\mathbf{v}$ if $\|k\mathbf{v}\| = 0$?

35. In each part, find two unit vectors in 2-space that satisfy the stated condition.
(a) Parallel to the line $y = 3x + 2$
(b) Parallel to the line $x + y = 4$
(c) Perpendicular to the line $y = -5x + 1$

36. In each part, find two unit vectors in 3-space that satisfy the stated condition.
(a) Perpendicular to the xy-plane
(b) Perpendicular to the xz-plane
(c) Perpendicular to the yz-plane

FOCUS ON CONCEPTS

37. Let $\mathbf{r} = \langle x, y \rangle$ be an arbitrary vector. In each part, describe the set of all points (x, y) in 2-space that satisfy the stated condition.
(a) $\|\mathbf{r}\| = 1$ (b) $\|\mathbf{r}\| \le 1$ (c) $\|\mathbf{r}\| > 1$

38. Let $\mathbf{r} = \langle x, y \rangle$ and $\mathbf{r}_0 = \langle x_0, y_0 \rangle$. In each part, describe the set of all points (x, y) in 2-space that satisfy the stated condition.
(a) $\|\mathbf{r}-\mathbf{r}_0\| = 1$ (b) $\|\mathbf{r}-\mathbf{r}_0\| \le 1$ (c) $\|\mathbf{r}-\mathbf{r}_0\| > 1$

39. Let $\mathbf{r} = \langle x, y, z \rangle$ be an arbitrary vector. In each part, describe the set of all points (x, y, z) in 3-space that satisfy the stated condition.
(a) $\|\mathbf{r}\| = 1$ (b) $\|\mathbf{r}\| \le 1$ (c) $\|\mathbf{r}\| > 1$

40. Let $\mathbf{r}_1 = \langle x_1, y_1 \rangle$, $\mathbf{r}_2 = \langle x_2, y_2 \rangle$, and $\mathbf{r} = \langle x, y \rangle$. Assuming that $k > \|\mathbf{r}_2 - \mathbf{r}_1\|$, describe the set of all points (x, y) for which $\|\mathbf{r} - \mathbf{r}_1\| + \|\mathbf{r} - \mathbf{r}_2\| = k$.

41–46 Find the magnitude of the resultant force and the angle that it makes with the positive x-axis.

41.

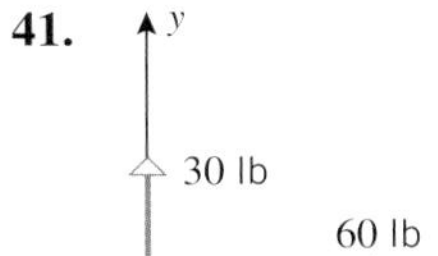

42.

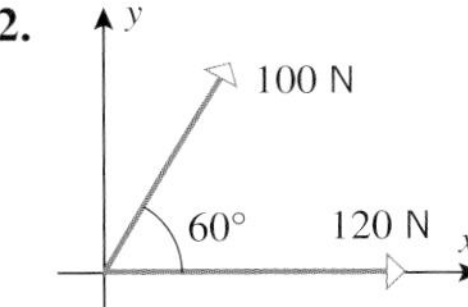

43.

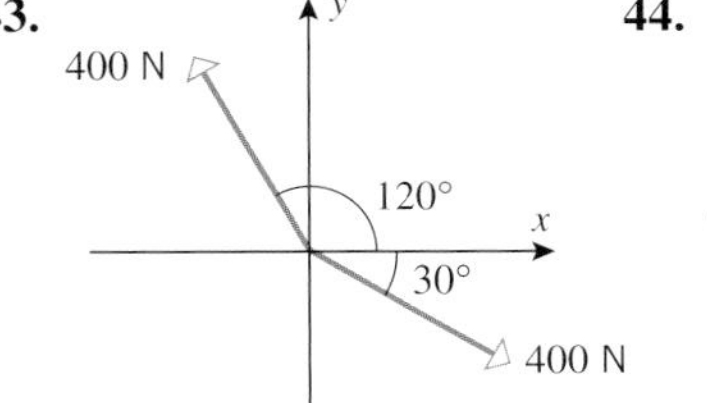

44.

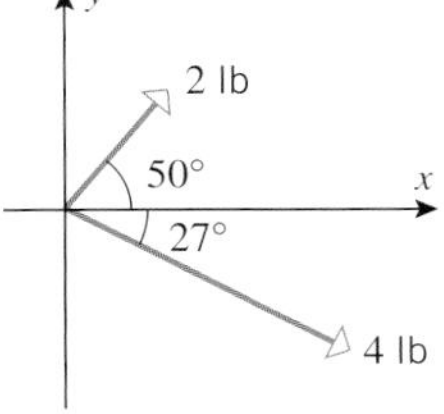

45.

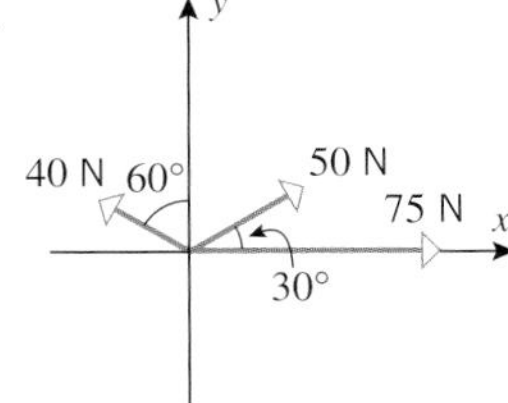

46.

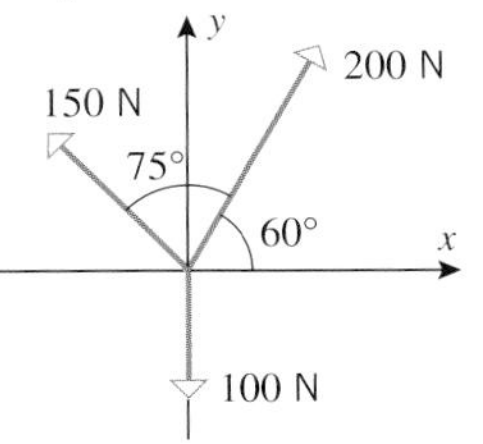

47–48 A particle is said to be in ***static equilibrium*** if the resultant of all forces applied to it is zero. In these exercises, find the force $\mathbf{F}$ that must be applied to the point to produce static equilibrium. Describe $\mathbf{F}$ by specifying its magnitude and the angle that it makes with the positive x-axis.

47.

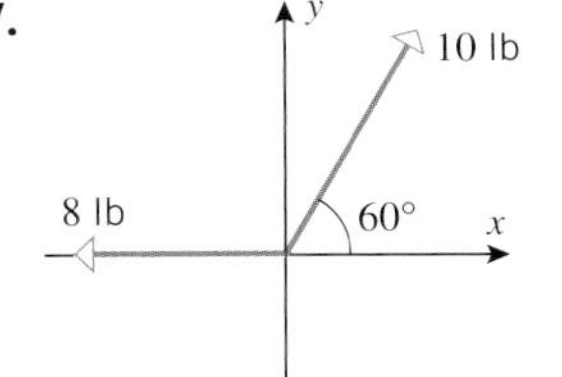

48.

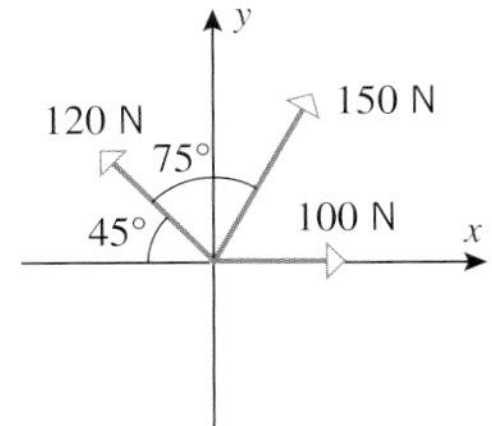

49. The accompanying figure shows a 250-lb traffic light supported by two flexible cables. The magnitudes of the forces that the cables apply to the eye ring are called the cable ***tensions***. Find the tensions in the cables if the traffic light is in static equilibrium (defined above Exercise 47).

50. Find the tensions in the cables shown in the accompanying figure if the block is in static equilibrium (see Exercise 49).

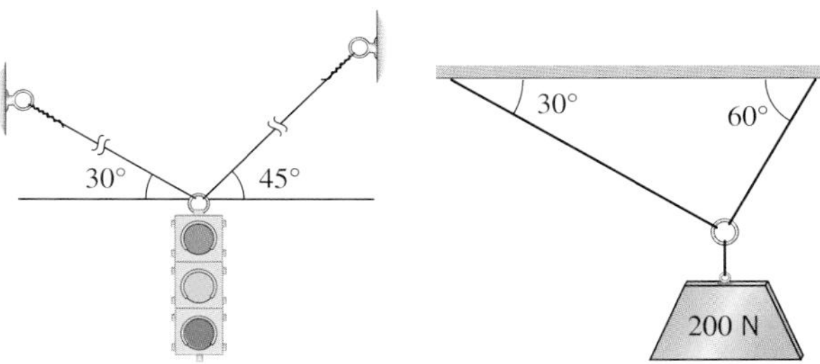

Figure Ex-49 **Figure Ex-50**

51. A vector $\mathbf{w}$ is said to be a ***linear combination*** of the vectors $\mathbf{v}_1$ and $\mathbf{v}_2$ if $\mathbf{w}$ can be expressed as $\mathbf{w} = c_1\mathbf{v}_1 + c_2\mathbf{v}_2$, where c_1 and c_2 are scalars.
(a) Find scalars c_1 and c_2 to express the vector $4\mathbf{j}$ as a linear combination of the vectors $\mathbf{v}_1 = 2\mathbf{i} - \mathbf{j}$ and $\mathbf{v}_2 = 4\mathbf{i} + 2\mathbf{j}$.
(b) Show that the vector $\langle 3, 5 \rangle$ cannot be expressed as a linear combination of the vectors $\mathbf{v}_1 = \langle 1, -3 \rangle$ and $\mathbf{v}_2 = \langle -2, 6 \rangle$.

52. A vector $\mathbf{w}$ is a ***linear combination*** of the vectors $\mathbf{v}_1$, $\mathbf{v}_2$, and $\mathbf{v}_3$ if $\mathbf{w}$ can be expressed as $\mathbf{w} = c_1\mathbf{v}_1 + c_2\mathbf{v}_2 + c_3\mathbf{v}_3$, where c_1, c_2, and c_3 are scalars.
(a) Find scalars c_1, c_2, and c_3 to express $\langle -1, 1, 5 \rangle$ as a linear combination of $\mathbf{v}_1 = \langle 1, 0, 1 \rangle$, $\mathbf{v}_2 = \langle 3, 2, 0 \rangle$, and $\mathbf{v}_3 = \langle 0, 1, 1 \rangle$.
(b) Show that the vector $2\mathbf{i} + \mathbf{j} - \mathbf{k}$ cannot be expressed as a linear combination of $\mathbf{v}_1 = \mathbf{i} - \mathbf{j}$, $\mathbf{v}_2 = 3\mathbf{i} + \mathbf{k}$, and $\mathbf{v}_3 = 4\mathbf{i} - \mathbf{j} + \mathbf{k}$.

53. Use a theorem from plane geometry to show that if $\mathbf{u}$ and $\mathbf{v}$ are vectors in 2-space or 3-space, then

$$\|\mathbf{u} + \mathbf{v}\| \le \|\mathbf{u}\| + \|\mathbf{v}\|$$

which is called the ***triangle inequality for vectors***. Give some examples to illustrate this inequality.

54. Prove parts (a), (c), and (e) of Theorem 12.2.6 algebraically in 2-space.

55. Prove parts (d), (g), and (h) of Theorem 12.2.6 algebraically in 2-space.

56. Prove part (f) of Theorem 12.2.6 geometrically.

FOCUS ON CONCEPTS

57. Use vectors to prove that the line segment joining the midpoints of two sides of a triangle is parallel to the third side and half as long.

58. Use vectors to prove that the midpoints of the sides of a quadrilateral are the vertices of a parallelogram.

✔QUICK CHECK ANSWERS 12.2

1. (a) $\sqrt{59}$ (b) $\langle 7, 9, 2\rangle$ (c) $\langle -1, -11, 12\rangle$ (d) $\langle 6, -2, 14\rangle$ 2. $\dfrac{1}{\sqrt{59}}\mathbf{v} = \left\langle \dfrac{3}{\sqrt{59}}, -\dfrac{1}{\sqrt{59}}, \dfrac{7}{\sqrt{59}}\right\rangle$ 3. $\left\langle \dfrac{1}{2}, \dfrac{\sqrt{3}}{2}\right\rangle = \dfrac{1}{2}\mathbf{i} + \dfrac{\sqrt{3}}{2}\mathbf{j}$
4. (a) $\langle -3, -4, 5\rangle$ (b) $\frac{1}{5}\overrightarrow{AB} = \langle -\frac{3}{5}, -\frac{4}{5}, 1\rangle$

12.3 DOT PRODUCT; PROJECTIONS

In the last section we defined three operations on vectors—addition, subtraction, and scalar multiplication. In scalar multiplication a vector is multiplied by a scalar and the result is a vector. In this section we will define a new kind of multiplication in which two vectors are multiplied to produce a scalar. This multiplication operation has many uses, some of which we will also discuss in this section.

■ DEFINITION OF THE DOT PRODUCT

In words, the dot product of two vectors is formed by multiplying their corresponding components and adding the resulting products. Note that the dot product of two vectors is a scalar.

12.3.1 DEFINITION. If $\mathbf{u} = \langle u_1, u_2\rangle$ and $\mathbf{v} = \langle v_1, v_2\rangle$ are vectors in 2-space, then the ***dot product*** of $\mathbf{u}$ and $\mathbf{v}$ is written as $\mathbf{u}\cdot\mathbf{v}$ and is defined as

$$\mathbf{u}\cdot\mathbf{v} = u_1v_1 + u_2v_2$$

Similarly, if $\mathbf{u} = \langle u_1, u_2, u_3\rangle$ and $\mathbf{v} = \langle v_1, v_2, v_3\rangle$ are vectors in 3-space, then their dot product is defined as

$$\mathbf{u}\cdot\mathbf{v} = u_1v_1 + u_2v_2 + u_3v_3$$

TECHNOLOGY MASTERY

Many calculating utilities have a built-in dot product operation. If your calculating utility has this capability, use it to check the computations in Example 1.

▶ Example 1

$$\langle 3, 5\rangle \cdot \langle -1, 2\rangle = 3(-1) + 5(2) = 7$$
$$\langle 2, 3\rangle \cdot \langle -3, 2\rangle = 2(-3) + 3(2) = 0$$
$$\langle 1, -3, 4\rangle \cdot \langle 1, 5, 2\rangle = 1(1) + (-3)(5) + 4(2) = -6$$

Here are the same computations expressed another way:

$$(3\mathbf{i} + 5\mathbf{j}) \cdot (-\mathbf{i} + 2\mathbf{j}) = 3(-1) + 5(2) = 7$$
$$(2\mathbf{i} + 3\mathbf{j}) \cdot (-3\mathbf{i} + 2\mathbf{j}) = 2(-3) + 3(2) = 0$$
$$(\mathbf{i} - 3\mathbf{j} + 4\mathbf{k}) \cdot (\mathbf{i} + 5\mathbf{j} + 2\mathbf{k}) = 1(1) + (-3)(5) + 4(2) = -6$$ ◀

■ ALGEBRAIC PROPERTIES OF THE DOT PRODUCT

The following theorem provides some of the basic algebraic properties of the dot product.

Note the difference between the two zeros that appear in part (e) of Theorem 12.3.2—the zero on the left side is the *zero vector* (boldface), whereas the zero on the right side is the *zero scalar* (lightface).

12.3.2 THEOREM. *If* $\mathbf{u}$, $\mathbf{v}$, *and* $\mathbf{w}$ *are vectors in 2- or 3-space and k is a scalar, then*

(a) $\mathbf{u}\cdot\mathbf{v} = \mathbf{v}\cdot\mathbf{u}$

(b) $\mathbf{u}\cdot(\mathbf{v} + \mathbf{w}) = \mathbf{u}\cdot\mathbf{v} + \mathbf{u}\cdot\mathbf{w}$

(c) $k(\mathbf{u}\cdot\mathbf{v}) = (k\mathbf{u})\cdot\mathbf{v} = \mathbf{u}\cdot(k\mathbf{v})$

(d) $\mathbf{v}\cdot\mathbf{v} = \|\mathbf{v}\|^2$

(e) $\mathbf{0}\cdot\mathbf{v} = 0$

We will prove parts (c) and (d) for vectors in 3-space and leave some of the others as exercises.

PROOF (c). Let $\mathbf{u} = \langle u_1, u_2, u_3 \rangle$ and $\mathbf{v} = \langle v_1, v_2, v_3 \rangle$. Then

$$k(\mathbf{u} \cdot \mathbf{v}) = k(u_1v_1 + u_2v_2 + u_3v_3) = (ku_1)v_1 + (ku_2)v_2 + (ku_3)v_3 = (k\mathbf{u}) \cdot \mathbf{v}$$

Similarly, $k(\mathbf{u} \cdot \mathbf{v}) = \mathbf{u} \cdot (k\mathbf{v})$.

PROOF (d). $\mathbf{v} \cdot \mathbf{v} = v_1v_1 + v_2v_2 + v_3v_3 = v_1^2 + v_2^2 + v_3^2 = \|\mathbf{v}\|^2$. ■

The following alternative form of the formula in part (d) of Theorem 12.3.2 provides a useful way of expressing the norm of a vector in terms of a dot product:

$$\|\mathbf{v}\| = \sqrt{\mathbf{v} \cdot \mathbf{v}} \tag{1}$$

ANGLE BETWEEN VECTORS

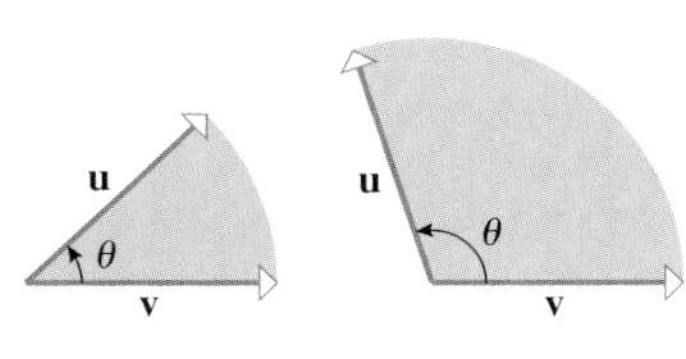

Suppose that $\mathbf{u}$ and $\mathbf{v}$ are nonzero vectors in 2-space or 3-space that are positioned so their initial points coincide. We define the ***angle between* u *and* v** to be the angle θ determined by the vectors that satisfies the condition $0 \le \theta \le \pi$ (Figure 12.3.1). In 2-space, θ is the smallest counterclockwise angle through which one of the vectors can be rotated until it aligns with the other.

The next theorem provides a way of calculating the angle between two vectors from their components.

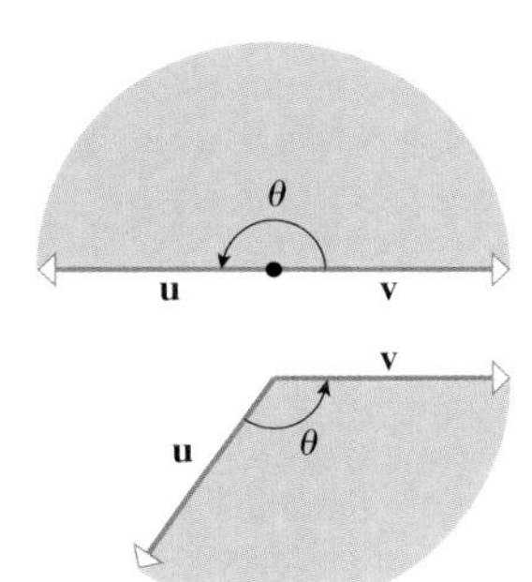

θ is the angle between **u** and **v**.

Figure 12.3.1

12.3.3 THEOREM. *If* $\mathbf{u}$ *and* $\mathbf{v}$ *are nonzero vectors in 2-space or 3-space, and if* θ *is the angle between them, then*

$$\cos\theta = \frac{\mathbf{u} \cdot \mathbf{v}}{\|\mathbf{u}\|\,\|\mathbf{v}\|} \tag{2}$$

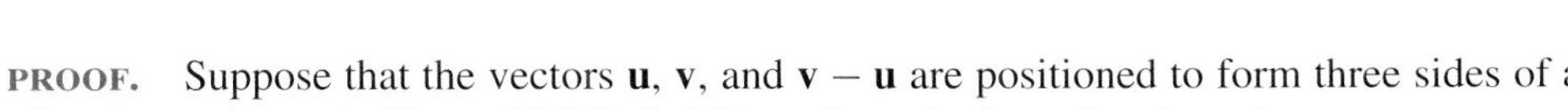

PROOF. Suppose that the vectors $\mathbf{u}$, $\mathbf{v}$, and $\mathbf{v} - \mathbf{u}$ are positioned to form three sides of a triangle, as shown in Figure 12.3.2. It follows from the law of cosines that

$$\|\mathbf{v} - \mathbf{u}\|^2 = \|\mathbf{u}\|^2 + \|\mathbf{v}\|^2 - 2\|\mathbf{u}\|\,\|\mathbf{v}\|\cos\theta \tag{3}$$

Using the properties of the dot product in Theorem 12.3.2, we can rewrite the left side of this equation as

$$\begin{aligned}
\|\mathbf{v} - \mathbf{u}\|^2 &= (\mathbf{v} - \mathbf{u}) \cdot (\mathbf{v} - \mathbf{u}) \\
&= (\mathbf{v} - \mathbf{u}) \cdot \mathbf{v} - (\mathbf{v} - \mathbf{u}) \cdot \mathbf{u} \\
&= \mathbf{v} \cdot \mathbf{v} - \mathbf{u} \cdot \mathbf{v} - \mathbf{v} \cdot \mathbf{u} + \mathbf{u} \cdot \mathbf{u} \\
&= \|\mathbf{v}\|^2 - 2\mathbf{u} \cdot \mathbf{v} + \|\mathbf{u}\|^2
\end{aligned}$$

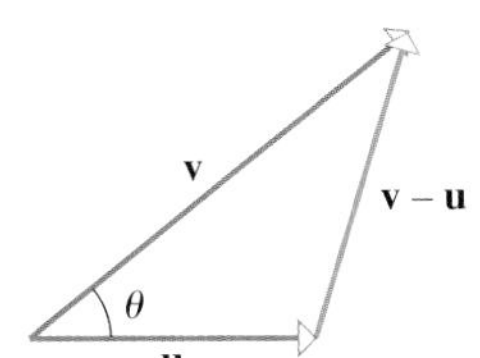

Figure 12.3.2

Substituting this back into (3) yields

$$\|\mathbf{v}\|^2 - 2\mathbf{u} \cdot \mathbf{v} + \|\mathbf{u}\|^2 = \|\mathbf{u}\|^2 + \|\mathbf{v}\|^2 - 2\|\mathbf{u}\|\,\|\mathbf{v}\|\cos\theta$$

which we can simplify and rewrite as

$$\mathbf{u} \cdot \mathbf{v} = \|\mathbf{u}\|\,\|\mathbf{v}\|\cos\theta$$

Finally, dividing both sides of this equation by $\|\mathbf{u}\|\,\|\mathbf{v}\|$ yields (2). ■

▶ **Example 2** Find the angle between the vector $\mathbf{u} = \mathbf{i} - 2\mathbf{j} + 2\mathbf{k}$ and

(a) $\mathbf{v} = -3\mathbf{i} + 6\mathbf{j} + 2\mathbf{k}$ (b) $\mathbf{w} = 2\mathbf{i} + 7\mathbf{j} + 6\mathbf{k}$ (c) $\mathbf{z} = -3\mathbf{i} + 6\mathbf{j} - 6\mathbf{k}$

Solution (a).

$$\cos\theta = \frac{\mathbf{u} \cdot \mathbf{v}}{\|\mathbf{u}\|\,\|\mathbf{v}\|} = \frac{-11}{(3)(7)} = -\frac{11}{21}$$

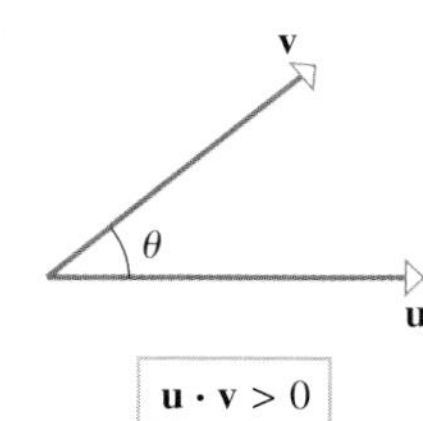

Thus,

$$\theta = \cos^{-1}\left(-\tfrac{11}{21}\right) \approx 2.12 \text{ radians} \approx 121.6^\circ$$

Solution (b).

$$\cos\theta = \frac{\mathbf{u}\cdot\mathbf{w}}{\|\mathbf{u}\|\,\|\mathbf{w}\|} = \frac{0}{\|\mathbf{u}\|\,\|\mathbf{w}\|} = 0$$

Thus, $\theta = \pi/2$, which means that the vectors are perpendicular.

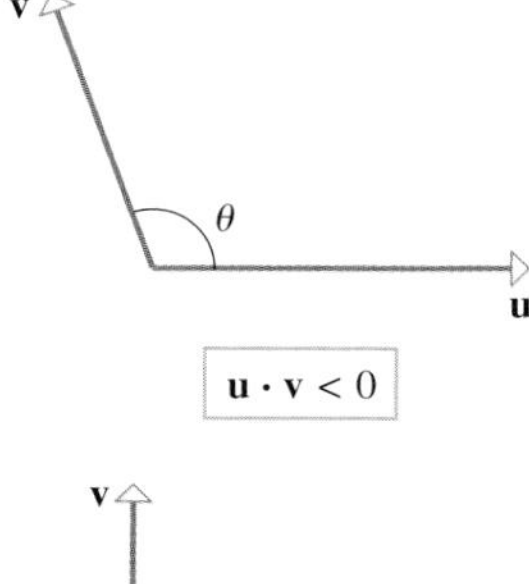

Solution (c).

$$\cos\theta = \frac{\mathbf{u}\cdot\mathbf{z}}{\|\mathbf{u}\|\,\|\mathbf{z}\|} = \frac{-27}{(3)(9)} = -1$$

Thus, $\theta = \pi$, which means that the vectors are oppositely directed. In retrospect, we could have seen this without computing θ, since $\mathbf{z} = -3\mathbf{u}$. ◄

INTERPRETING THE SIGN OF THE DOT PRODUCT

It will often be convenient to express Formula (2) as

$$\mathbf{u}\cdot\mathbf{v} = \|\mathbf{u}\|\,\|\mathbf{v}\|\cos\theta \tag{4}$$

which expresses the dot product of **u** and **v** in terms of the lengths of these vectors and the angle between them. Since **u** and **v** are assumed to be nonzero vectors, this version of the formula makes it clear that the sign of $\mathbf{u}\cdot\mathbf{v}$ is the same as the sign of $\cos\theta$. Thus, we can tell from the dot product whether the angle between two vectors is acute or obtuse or whether the vectors are perpendicular (Figure 12.3.3).

Figure 12.3.3

> The terms "perpendicular," "orthogonal," and "normal" are all commonly used to describe geometric objects that meet at right angles. For consistency, we will say that two vectors are *orthogonal*, a vector is *normal* to a plane, and two planes are *perpendicular*. Moreover, although the zero vector does not make a well-defined angle with other vectors, we will consider **0** to be orthogonal to *all* vectors. This convention allows us to say that **u** and **v** are orthogonal vectors if and only if $\mathbf{u}\cdot\mathbf{v} = 0$, and makes Formula (4) valid if **u** or **v** (or both) is zero.

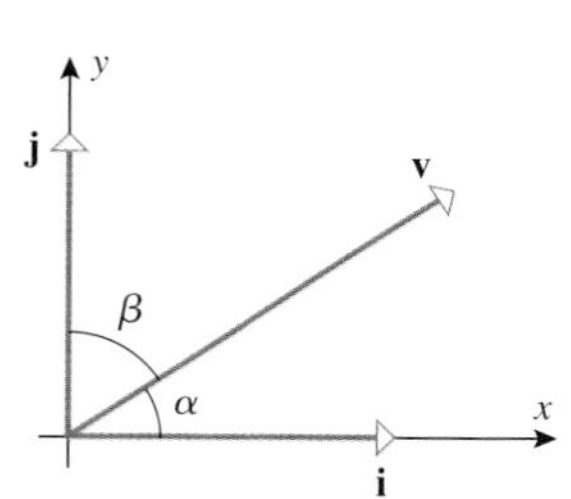

Figure 12.3.4

DIRECTION ANGLES

In an xy-coordinate system, the direction of a nonzero vector **v** is completely determined by the angles α and β between **v** and the unit vectors **i** and **j** (Figure 12.3.4), and in an xyz-coordinate system the direction is completely determined by the angles α, β, and γ between **v** and the unit vectors **i**, **j**, and **k** (Figure 12.3.5). In both 2-space and 3-space the angles between a nonzero vector **v** and the vectors **i**, **j**, and **k** are called the ***direction angles*** of **v**, and the cosines of those angles are called the ***direction cosines*** of **v**. Formulas for the direction cosines of a vector can be obtained from Formula (2). For example, if $\mathbf{v} = v_1\mathbf{i} + v_2\mathbf{j} + v_3\mathbf{k}$, then

$$\cos\alpha = \frac{\mathbf{v}\cdot\mathbf{i}}{\|\mathbf{v}\|\,\|\mathbf{i}\|} = \frac{v_1}{\|\mathbf{v}\|}, \quad \cos\beta = \frac{\mathbf{v}\cdot\mathbf{j}}{\|\mathbf{v}\|\,\|\mathbf{j}\|} = \frac{v_2}{\|\mathbf{v}\|}, \quad \cos\gamma = \frac{\mathbf{v}\cdot\mathbf{k}}{\|\mathbf{v}\|\,\|\mathbf{k}\|} = \frac{v_3}{\|\mathbf{v}\|}$$

Thus, we have the following theorem.

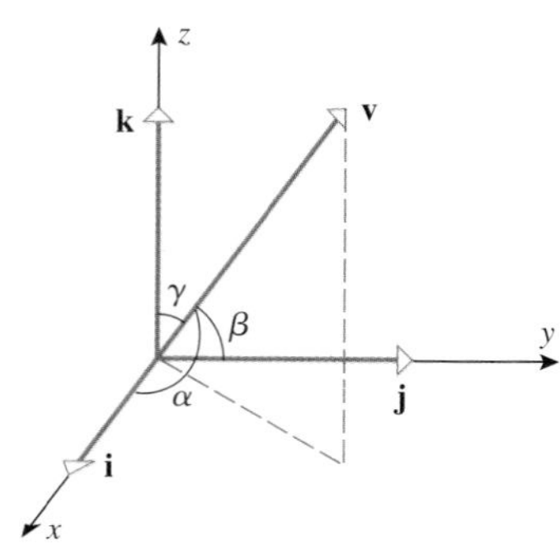

Figure 12.3.5

> **12.3.4 THEOREM.** *The direction cosines of a nonzero vector* $\mathbf{v} = v_1\mathbf{i} + v_2\mathbf{j} + v_3\mathbf{k}$ *are*
>
> $$\cos\alpha = \frac{v_1}{\|\mathbf{v}\|}, \quad \cos\beta = \frac{v_2}{\|\mathbf{v}\|}, \quad \cos\gamma = \frac{v_3}{\|\mathbf{v}\|}$$

The direction cosines of a vector $\mathbf{v} = v_1\mathbf{i} + v_2\mathbf{j} + v_3\mathbf{k}$ can be computed by normalizing $\mathbf{v}$ and reading off the components of $\mathbf{v}/\|\mathbf{v}\|$, since

$$\frac{\mathbf{v}}{\|\mathbf{v}\|} = \frac{v_1}{\|\mathbf{v}\|}\mathbf{i} + \frac{v_2}{\|\mathbf{v}\|}\mathbf{j} + \frac{v_3}{\|\mathbf{v}\|}\mathbf{k} = (\cos\alpha)\mathbf{i} + (\cos\beta)\mathbf{j} + (\cos\gamma)\mathbf{k}$$

We leave it as an exercise for you to show that the direction cosines of a vector satisfy the equation

$$\cos^2\alpha + \cos^2\beta + \cos^2\gamma = 1 \tag{5}$$

► **Example 3** Find the direction cosines of the vector $\mathbf{v} = 2\mathbf{i} - 4\mathbf{j} + 4\mathbf{k}$, and approximate the direction angles to the nearest degree.

Solution. First we will normalize the vector $\mathbf{v}$ and then read off the components. We have $\|\mathbf{v}\| = \sqrt{4 + 16 + 16} = 6$, so that $\mathbf{v}/\|\mathbf{v}\| = \frac{1}{3}\mathbf{i} - \frac{2}{3}\mathbf{j} + \frac{2}{3}\mathbf{k}$. Thus,

$$\cos\alpha = \tfrac{1}{3}, \quad \cos\beta = -\tfrac{2}{3}, \quad \cos\gamma = \tfrac{2}{3}$$

With the help of a calculating utility we obtain

$$\alpha = \cos^{-1}\left(\tfrac{1}{3}\right) \approx 71^\circ, \quad \beta = \cos^{-1}\left(-\tfrac{2}{3}\right) \approx 132^\circ, \quad \gamma = \cos^{-1}\left(\tfrac{2}{3}\right) \approx 48^\circ$$ ◄

► **Example 4** Find the angle between a diagonal of a cube and one of its edges.

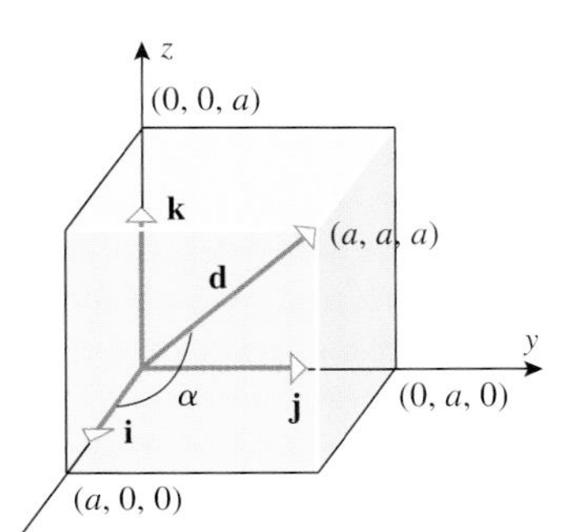

Figure 12.3.6

Solution. Assume that the cube has side a, and introduce a coordinate system as shown in Figure 12.3.6. In this coordinate system the vector

$$\mathbf{d} = a\mathbf{i} + a\mathbf{j} + a\mathbf{k}$$

is a diagonal of the cube and the unit vectors $\mathbf{i}$, $\mathbf{j}$, and $\mathbf{k}$ run along the edges. By symmetry, the diagonal makes the same angle with each edge, so it is sufficient to find the angle between $\mathbf{d}$ and $\mathbf{i}$ (the direction angle α). Thus,

$$\cos\alpha = \frac{\mathbf{d}\cdot\mathbf{i}}{\|\mathbf{d}\|\|\mathbf{i}\|} = \frac{a}{\|\mathbf{d}\|} = \frac{a}{\sqrt{3a^2}} = \frac{1}{\sqrt{3}}$$

and hence

$$\alpha = \cos^{-1}\left(\frac{1}{\sqrt{3}}\right) \approx 0.955 \text{ radian} \approx 54.7^\circ$$ ◄

DECOMPOSING VECTORS INTO ORTHOGONAL COMPONENTS

In many applications it is desirable to "decompose" a vector into a sum of two orthogonal vectors with convenient specified directions. For example, Figure 12.3.7 shows a block on an inclined plane. The downward force $\mathbf{F}$ that gravity exerts on the block can be decomposed into the sum

$$\mathbf{F} = \mathbf{F}_1 + \mathbf{F}_2$$

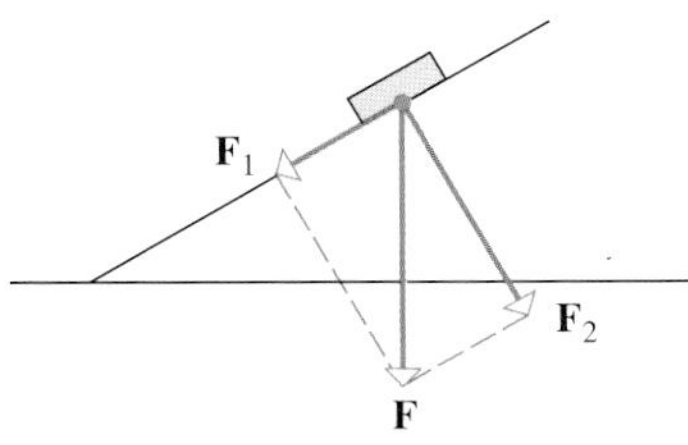

The force of gravity pulls the block against the ramp and down the ramp.

Figure 12.3.7

where the force $\mathbf{F}_1$ is parallel to the ramp and the force $\mathbf{F}_2$ is perpendicular to the ramp. The forces $\mathbf{F}_1$ and $\mathbf{F}_2$ are useful because $\mathbf{F}_1$ is the force that pulls the block *along* the ramp, and $\mathbf{F}_2$ is the force that the block exerts *against* the ramp.

Thus, our next objective is to develop a computational procedure for decomposing a vector into a sum of orthogonal vectors. For this purpose, suppose that $\mathbf{e}_1$ and $\mathbf{e}_2$ are two orthogonal *unit* vectors in 2-space, and suppose that we want to express a given vector $\mathbf{v}$ as a sum

$$\mathbf{v} = \mathbf{w}_1 + \mathbf{w}_2$$

so that $\mathbf{w}_1$ is a scalar multiple of $\mathbf{e}_1$ and $\mathbf{w}_2$ is a scalar multiple of $\mathbf{e}_2$ (Figure 12.3.8*a*). That is, we want to find scalars k_1 and k_2 such that

$$\mathbf{v} = k_1\mathbf{e}_1 + k_2\mathbf{e}_2 \tag{6}$$

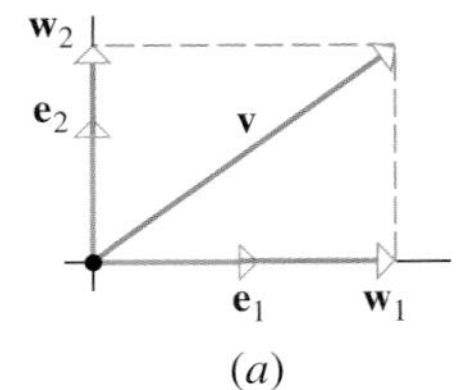

(*a*)

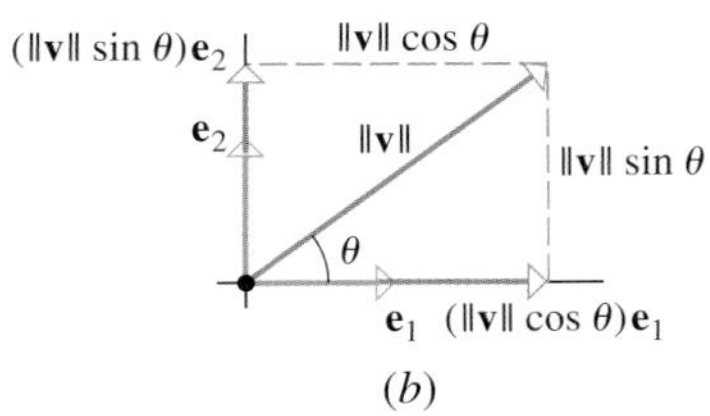

(*b*)

Figure 12.3.8

We can find k_1 by taking the dot product of $\mathbf{v}$ with $\mathbf{e}_1$. This yields

$$\begin{aligned}\mathbf{v} \cdot \mathbf{e}_1 &= (k_1\mathbf{e}_1 + k_2\mathbf{e}_2) \cdot \mathbf{e}_1 \\ &= k_1(\mathbf{e}_1 \cdot \mathbf{e}_1) + k_2(\mathbf{e}_2 \cdot \mathbf{e}_1) \\ &= k_1\|\mathbf{e}_1\|^2 + 0 = k_1\end{aligned}$$

Similarly,

$$\mathbf{v} \cdot \mathbf{e}_2 = (k_1\mathbf{e}_1 + k_2\mathbf{e}_2) \cdot \mathbf{e}_2 = k_1(\mathbf{e}_1 \cdot \mathbf{e}_2) + k_2(\mathbf{e}_2 \cdot \mathbf{e}_2) = 0 + k_2\|\mathbf{e}_2\|^2 = k_2$$

Substituting these expressions for k_1 and k_2 in (6) yields

$$\mathbf{v} = (\mathbf{v} \cdot \mathbf{e}_1)\mathbf{e}_1 + (\mathbf{v} \cdot \mathbf{e}_2)\mathbf{e}_2 \tag{7}$$

In this formula we call $(\mathbf{v} \cdot \mathbf{e}_1)\mathbf{e}_1$ and $(\mathbf{v} \cdot \mathbf{e}_2)\mathbf{e}_2$ the ***vector components*** of $\mathbf{v}$ along $\mathbf{e}_1$ and $\mathbf{e}_2$, respectively; and we call $\mathbf{v} \cdot \mathbf{e}_1$ and $\mathbf{v} \cdot \mathbf{e}_2$ the ***scalar components*** of $\mathbf{v}$ along $\mathbf{e}_1$ and $\mathbf{e}_2$, respectively. If θ denotes the angle between $\mathbf{v}$ and $\mathbf{e}_1$, then the scalar components of $\mathbf{v}$ can be written in trigonometric form as

$$\mathbf{v} \cdot \mathbf{e}_1 = \|\mathbf{v}\|\cos\theta \quad \text{and} \quad \mathbf{v} \cdot \mathbf{e}_2 = \|\mathbf{v}\|\sin\theta \tag{8}$$

(Figure 12.3.8*b*). Moreover, the vector components of $\mathbf{v}$ can be expressed as

$$(\mathbf{v} \cdot \mathbf{e}_1)\mathbf{e}_1 = (\|\mathbf{v}\|\cos\theta)\mathbf{e}_1 \quad \text{and} \quad (\mathbf{v} \cdot \mathbf{e}_2)\mathbf{e}_2 = (\|\mathbf{v}\|\sin\theta)\mathbf{e}_2 \tag{9}$$

and the decomposition (6) can be expressed as

$$\mathbf{v} = (\|\mathbf{v}\|\cos\theta)\mathbf{e}_1 + (\|\mathbf{v}\|\sin\theta)\mathbf{e}_2 \tag{10}$$

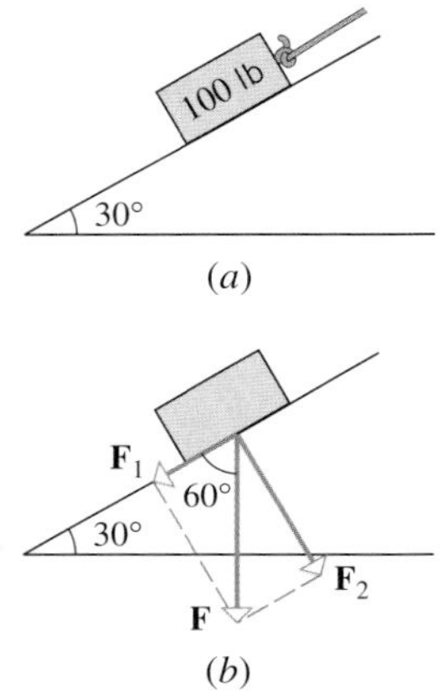

(*a*)

(*b*)

Figure 12.3.9

► **Example 5** A rope is attached to a 100-lb block on a ramp that is inclined at an angle of 30° with the ground (Figure 12.3.9*a*). How much force does the block exert against the ramp, and how much force must be applied to the rope in a direction parallel to the ramp to prevent the block from sliding down the ramp? (Assume that the ramp is smooth, that is, exerts no frictional forces.)

Solution. Let $\mathbf{F}$ denote the downward force of gravity on the block (so $\|\mathbf{F}\| = 100$ lb), and let $\mathbf{F}_1$ and $\mathbf{F}_2$ be the vector components of $\mathbf{F}$ parallel and perpendicular to the ramp (as shown in Figure 12.3.9*b*). The lengths of $\mathbf{F}_1$ and $\mathbf{F}_2$ are

$$\|\mathbf{F}_1\| = \|\mathbf{F}\|\cos 60^\circ = 100\left(\frac{1}{2}\right) = 50 \text{ lb}$$

$$\|\mathbf{F}_2\| = \|\mathbf{F}\|\sin 60^\circ = 100\left(\frac{\sqrt{3}}{2}\right) \approx 86.6 \text{ lb}$$

Thus, the block exerts a force of approximately 86.6 lb against the ramp, and it requires a force of 50 lb to prevent the block from sliding down the ramp. ◄

ORTHOGONAL PROJECTIONS

The vector components of $\mathbf{v}$ along $\mathbf{e}_1$ and $\mathbf{e}_2$ in (7) are also called the *orthogonal projections* of $\mathbf{v}$ on $\mathbf{e}_1$ and $\mathbf{e}_2$ and are commonly denoted by

$$\text{proj}_{\mathbf{e}_1}\mathbf{v} = (\mathbf{v} \cdot \mathbf{e}_1)\mathbf{e}_1 \quad \text{and} \quad \text{proj}_{\mathbf{e}_2}\mathbf{v} = (\mathbf{v} \cdot \mathbf{e}_2)\mathbf{e}_2$$

In general, if $\mathbf{e}$ is a unit vector, then we define the ***orthogonal projection of*** $\mathbf{v}$ ***on*** $\mathbf{e}$ to be

$$\text{proj}_{\mathbf{e}}\mathbf{v} = (\mathbf{v} \cdot \mathbf{e})\mathbf{e} \tag{11}$$

The orthogonal projection of $\mathbf{v}$ on an arbitrary nonzero vector $\mathbf{b}$ can be obtained by normalizing $\mathbf{b}$ and then applying Formula (11); that is,

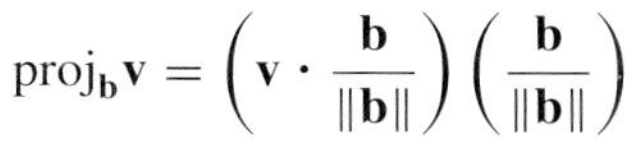

$$\text{proj}_{\mathbf{b}}\mathbf{v} = \left(\mathbf{v} \cdot \frac{\mathbf{b}}{\|\mathbf{b}\|}\right)\left(\frac{\mathbf{b}}{\|\mathbf{b}\|}\right)$$

which can be rewritten as

$$\text{proj}_{\mathbf{b}}\mathbf{v} = \frac{\mathbf{v} \cdot \mathbf{b}}{\|\mathbf{b}\|^2}\mathbf{b} \tag{12}$$

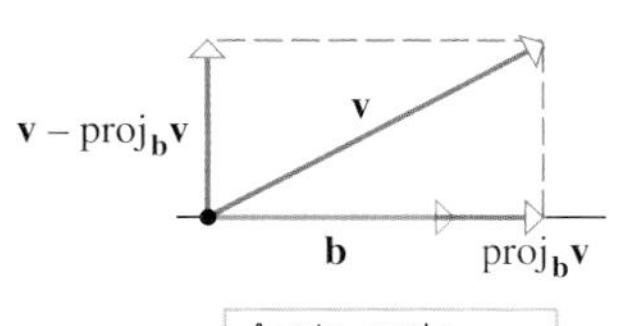

Acute angle between **v** and **b**

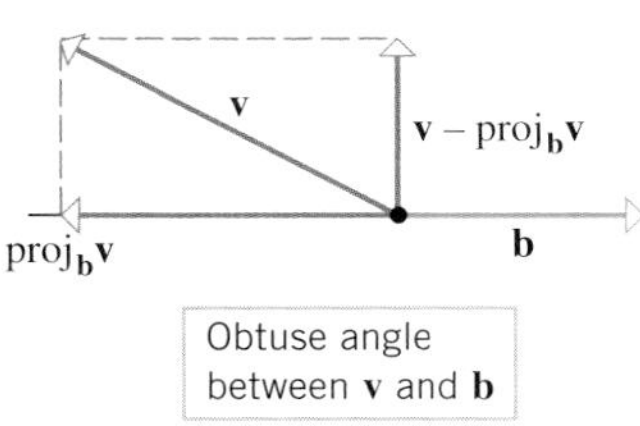

Obtuse angle between **v** and **b**

Figure 12.3.10

Geometrically, if $\mathbf{b}$ and $\mathbf{v}$ have a common initial point, then $\text{proj}_{\mathbf{b}}\mathbf{v}$ is the vector that is determined when a perpendicular is dropped from the terminal point of $\mathbf{v}$ to the line through $\mathbf{b}$ (illustrated in Figure 12.3.10 in two cases). Moreover, it is evident from Figure 12.3.10 that if we subtract $\text{proj}_{\mathbf{b}}\mathbf{v}$ from $\mathbf{v}$, then the resulting vector

$$\mathbf{v} - \text{proj}_{\mathbf{b}}\mathbf{v}$$

will be orthogonal to $\mathbf{b}$; we call this the ***vector component of*** $\mathbf{v}$ ***orthogonal to*** $\mathbf{b}$.

Example 6 Find the orthogonal projection of $\mathbf{v} = \mathbf{i} + \mathbf{j} + \mathbf{k}$ on $\mathbf{b} = 2\mathbf{i} + 2\mathbf{j}$, and then find the vector component of $\mathbf{v}$ orthogonal to $\mathbf{b}$.

Solution. We have

$$\mathbf{v} \cdot \mathbf{b} = (\mathbf{i} + \mathbf{j} + \mathbf{k}) \cdot (2\mathbf{i} + 2\mathbf{j}) = 2 + 2 + 0 = 4$$
$$\|\mathbf{b}\|^2 = 2^2 + 2^2 = 8$$

Thus, the orthogonal projection of $\mathbf{v}$ on $\mathbf{b}$ is

$$\text{proj}_{\mathbf{b}}\mathbf{v} = \frac{\mathbf{v} \cdot \mathbf{b}}{\|\mathbf{b}\|^2}\mathbf{b} = \frac{4}{8}(2\mathbf{i} + 2\mathbf{j}) = \mathbf{i} + \mathbf{j}$$

and the vector component of $\mathbf{v}$ orthogonal to $\mathbf{b}$ is

$$\mathbf{v} - \text{proj}_{\mathbf{b}}\mathbf{v} = (\mathbf{i} + \mathbf{j} + \mathbf{k}) - (\mathbf{i} + \mathbf{j}) = \mathbf{k}$$

These results are consistent with Figure 12.3.11. ◀

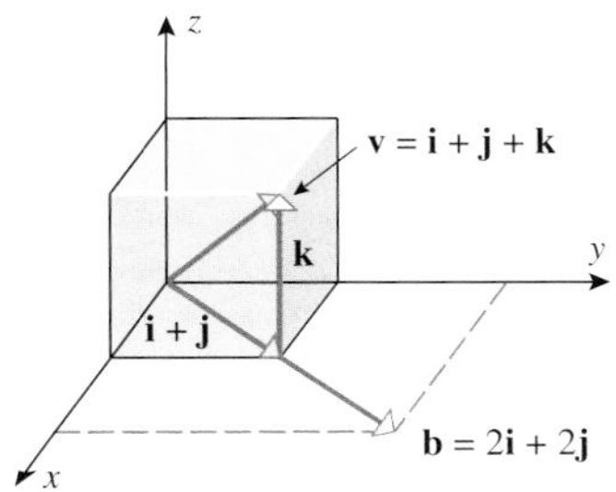

Figure 12.3.11

WORK

In Section 6.7 we discussed the work done by a constant force acting on an object that moves along a line. We defined the work W done on the object by a constant force of magnitude F acting in the direction of motion over a distance d to be

$$W = Fd = \text{force} \times \text{distance} \tag{13}$$

If we let $\mathbf{F}$ denote a force vector of magnitude $\|\mathbf{F}\| = F$ *acting in the direction of motion*, then we can write (13) as

$$W = \|\mathbf{F}\|d$$

Furthermore, if we assume that the object moves along a line from point P to point Q, then $d = \|\overrightarrow{PQ}\|$, so that the work can be expressed entirely in vector form as

$$W = \|\mathbf{F}\|\|\overrightarrow{PQ}\|$$

(Figure 12.3.12a). The vector $\overrightarrow{PQ}$ is called the ***displacement vector*** for the object. In the case where a constant force $\mathbf{F}$ is not in the direction of motion, but rather makes an angle θ with the displacement vector, then we *define* the work W done by $\mathbf{F}$ to be

$$W = (\|\mathbf{F}\|\cos\theta)\|\overrightarrow{PQ}\| = \mathbf{F} \cdot \overrightarrow{PQ} \tag{14}$$

(Figure 12.3.12b).

Note that in Formula (14) the quantity $\|\mathbf{F}\|\cos\theta$ is the scalar component of force along the displacement vector. Thus, in the case where $\cos\theta > 0$, a force of magnitude $\|\mathbf{F}\|$ acting at an angle θ does the same work as a force of magnitude $\|\mathbf{F}\|\cos\theta$ acting in the direction of motion.

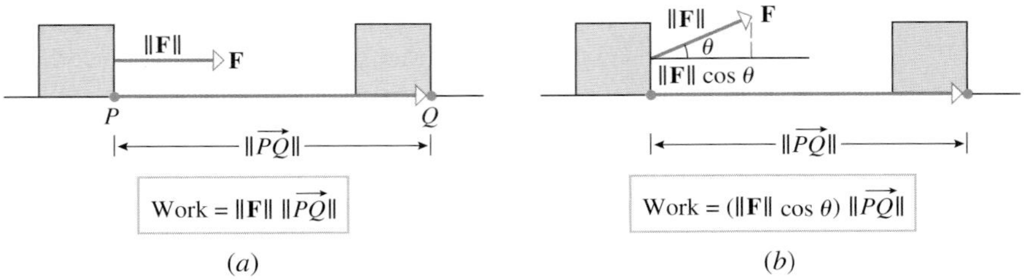

Work = $\|\mathbf{F}\| \, \|\overrightarrow{PQ}\|$

(a)

Work = $(\|\mathbf{F}\| \cos\theta) \, \|\overrightarrow{PQ}\|$

(b)

Figure 12.3.12

▶ **Example 7** A wagon is pulled horizontally by exerting a constant force of 10 lb on the handle at an angle of 60° with the horizontal. How much work is done in moving the wagon 50 ft?

Solution. Introduce an xy-coordinate system so that the wagon moves from $P(0, 0)$ to $Q(50, 0)$ along the x-axis (Figure 12.3.13). In this coordinate system

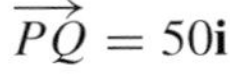

$$\overrightarrow{PQ} = 50\mathbf{i}$$

and

$$\mathbf{F} = (10\cos 60^\circ)\mathbf{i} + (10\sin 60^\circ)\mathbf{j} = 5\mathbf{i} + 5\sqrt{3}\mathbf{j}$$

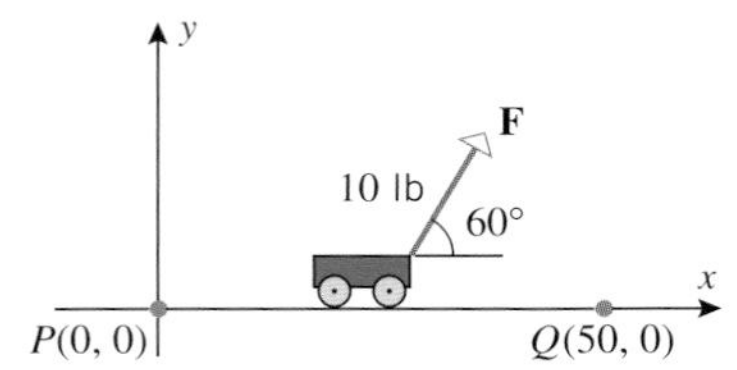

Figure 12.3.13

so the work done is

$$W = \mathbf{F} \cdot \overrightarrow{PQ} = (5\mathbf{i} + 5\sqrt{3}\mathbf{j}) \cdot (50\mathbf{i}) = 250 \text{ (foot-pounds)}$$ ◀

✔ QUICK CHECK EXERCISES 12.3 *(See page 813 for answers.)*

1. $\langle 3, 1, -2\rangle \cdot \langle 6, 0, 5\rangle =$ ________
2. Suppose that **u**, **v**, and **w** are vectors in 3-space such that $\|\mathbf{u}\| = 5$, $\mathbf{u} \cdot \mathbf{v} = 7$, and $\mathbf{u} \cdot \mathbf{w} = -3$.
   (a) $\mathbf{u} \cdot \mathbf{u} =$ ________ (b) $\mathbf{v} \cdot \mathbf{u} =$ ________
   (c) $\mathbf{u} \cdot (\mathbf{v} - \mathbf{w}) =$ ________ (d) $\mathbf{u} \cdot (2\mathbf{w}) =$ ________
3. For the vectors **u** and **v** in the preceding exercise, if the angle between **u** and **v** is $\pi/3$, then $\|\mathbf{v}\| =$ ________.
4. The direction cosines of $\langle 2, -1, 3\rangle$ are $\cos\alpha =$ ________, $\cos\beta =$ ________, and $\cos\gamma =$ ________.
5. The orthogonal projection of $\mathbf{v} = 10\mathbf{i}$ on $\mathbf{b} = -3\mathbf{i} + \mathbf{j}$ is ________.

EXERCISE SET 12.3 Graphing Utility C CAS

1. In each part, find the dot product of the vectors and the cosine of the angle between them.
   (a) $\mathbf{u} = \mathbf{i} + 2\mathbf{j}$, $\mathbf{v} = 6\mathbf{i} - 8\mathbf{j}$
   (b) $\mathbf{u} = \langle -7, -3\rangle$, $\mathbf{v} = \langle 0, 1\rangle$
   (c) $\mathbf{u} = \mathbf{i} - 3\mathbf{j} + 7\mathbf{k}$, $\mathbf{v} = 8\mathbf{i} - 2\mathbf{j} - 2\mathbf{k}$
   (d) $\mathbf{u} = \langle -3, 1, 2\rangle$, $\mathbf{v} = \langle 4, 2, -5\rangle$
2. In each part use the given information to find $\mathbf{u} \cdot \mathbf{v}$.
   (a) $\|\mathbf{u}\| = 1$, $\|\mathbf{v}\| = 2$, the angle between **u** and **v** is $\pi/6$.
   (b) $\|\mathbf{u}\| = 2$, $\|\mathbf{v}\| = 3$, the angle between **u** and **v** is 135°.
3. In each part, determine whether **u** and **v** make an acute angle, an obtuse angle, or are orthogonal.
   (a) $\mathbf{u} = 7\mathbf{i} + 3\mathbf{j} + 5\mathbf{k}$, $\mathbf{v} = -8\mathbf{i} + 4\mathbf{j} + 2\mathbf{k}$
   (b) $\mathbf{u} = 6\mathbf{i} + \mathbf{j} + 3\mathbf{k}$, $\mathbf{v} = 4\mathbf{i} - 6\mathbf{k}$
   (c) $\mathbf{u} = \langle 1, 1, 1\rangle$, $\mathbf{v} = \langle -1, 0, 0\rangle$
   (d) $\mathbf{u} = \langle 4, 1, 6\rangle$, $\mathbf{v} = \langle -3, 0, 2\rangle$

FOCUS ON CONCEPTS

4. Does the triangle in 3-space with vertices $(-1, 2, 3)$, $(2, -2, 0)$, and $(3, 1, -4)$ have an obtuse angle? Justify your answer.
5. The accompanying figure shows eight vectors that are equally spaced around a circle of radius 1. Find the dot product of $\mathbf{v}_0$ with each of the other seven vectors.
6. The accompanying figure shows six vectors that are equally spaced around a circle of radius 5. Find the dot product of $\mathbf{v}_0$ with each of the other five vectors.

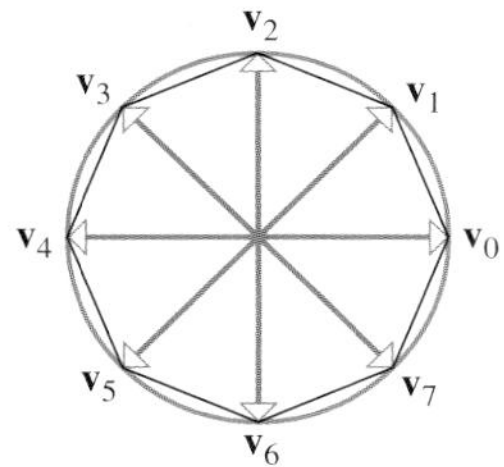

Figure Ex-5

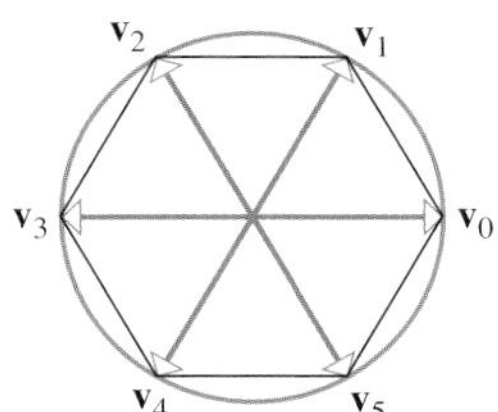

Figure Ex-6

7. (a) Use vectors to show that $A(2, -1, 1)$, $B(3, 2, -1)$, and $C(7, 0, -2)$ are vertices of a right triangle. At which vertex is the right angle?
(b) Use vectors to find the interior angles of the triangle with vertices $(-1, 0)$, $(2, -1)$, and $(1, 4)$. Express your answers to the nearest degree.

8. (a) Show that if $\mathbf{v} = a\mathbf{i} + b\mathbf{j}$ is a vector in 2-space, then the vectors

$$\mathbf{v}_1 = -b\mathbf{i} + a\mathbf{j} \quad \text{and} \quad \mathbf{v}_2 = b\mathbf{i} - a\mathbf{j}$$

are both orthogonal to $\mathbf{v}$.
(b) Use the result in part (a) to find two unit vectors that are orthogonal to the vector $\mathbf{v} = 3\mathbf{i} - 2\mathbf{j}$. Sketch the vectors $\mathbf{v}$, $\mathbf{v}_1$, and $\mathbf{v}_2$.

9. Explain why each of the following expressions makes no sense.
(a) $\mathbf{u} \cdot (\mathbf{v} \cdot \mathbf{w})$ (b) $(\mathbf{u} \cdot \mathbf{v}) + \mathbf{w}$
(c) $\|\mathbf{u} \cdot \mathbf{v}\|$ (d) $k \cdot (\mathbf{u} + \mathbf{v})$

10. True or false? If $\mathbf{a} \cdot \mathbf{b} = \mathbf{a} \cdot \mathbf{c}$ and if $\mathbf{a} \neq \mathbf{0}$, then $\mathbf{b} = \mathbf{c}$. Justify your conclusion.

11. Verify parts (*b*) and (*c*) of Theorem 12.3.2 for the vectors $\mathbf{u} = 6\mathbf{i} - \mathbf{j} + 2\mathbf{k}$, $\mathbf{v} = 2\mathbf{i} + 7\mathbf{j} + 4\mathbf{k}$, $\mathbf{w} = \mathbf{i} + \mathbf{j} - 3\mathbf{k}$ and $k = -5$.

12. Let $\mathbf{u} = \langle 1, 2 \rangle$, $\mathbf{v} = \langle 4, -2 \rangle$, and $\mathbf{w} = \langle 6, 0 \rangle$. Find
(a) $\mathbf{u} \cdot (7\mathbf{v} + \mathbf{w})$ (b) $\|(\mathbf{u} \cdot \mathbf{w})\mathbf{w}\|$
(c) $\|\mathbf{u}\|(\mathbf{v} \cdot \mathbf{w})$ (d) $(\|\mathbf{u}\|\mathbf{v}) \cdot \mathbf{w}$.

13. Find r so that the vector from the point $A(1, -1, 3)$ to the point $B(3, 0, 5)$ is orthogonal to the vector from A to the point $P(r, r, r)$.

14. Find two unit vectors in 2-space that make an angle of 45° with $4\mathbf{i} + 3\mathbf{j}$.

15–16 Find the direction cosines of $\mathbf{v}$ and confirm that they satisfy Equation (5). Then use the direction cosines to approximate the direction angles to the nearest degree.

15. (a) $\mathbf{v} = \mathbf{i} + \mathbf{j} - \mathbf{k}$ (b) $\mathbf{v} = 2\mathbf{i} - 2\mathbf{j} + \mathbf{k}$

16. (a) $\mathbf{v} = 3\mathbf{i} - 2\mathbf{j} - 6\mathbf{k}$ (b) $\mathbf{v} = 3\mathbf{i} - 4\mathbf{k}$

FOCUS ON CONCEPTS

17. Show that the direction cosines of a vector satisfy

$$\cos^2\alpha + \cos^2\beta + \cos^2\gamma = 1$$

18. Let θ and λ be the angles shown in the accompanying figure. Show that the direction cosines of $\mathbf{v}$ can be expressed as

$$\cos\alpha = \cos\lambda\cos\theta$$
$$\cos\beta = \cos\lambda\sin\theta$$
$$\cos\gamma = \sin\lambda$$

[*Hint:* Express $\mathbf{v}$ in component form and normalize.]

19. The accompanying figure shows a cube.
(a) Find the angle between the vectors $\mathbf{d}$ and $\mathbf{u}$ to the nearest degree.
(b) Make a conjecture about the angle between the vectors $\mathbf{d}$ and $\mathbf{v}$, and confirm your conjecture by computing the angle.

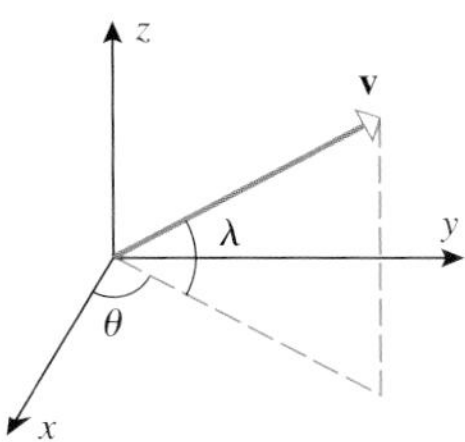

Figure Ex-18

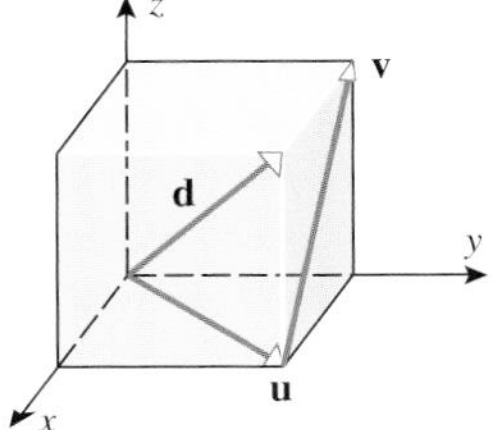

Figure Ex-19

20. Show that two nonzero vectors $\mathbf{v}_1$ and $\mathbf{v}_2$ are orthogonal if and only if their direction cosines satisfy

$$\cos\alpha_1\cos\alpha_2 + \cos\beta_1\cos\beta_2 + \cos\gamma_1\cos\gamma_2 = 0$$

21. Use the result in Exercise 18 to find the direction angles of the vector shown in the accompanying figure to the nearest degree.

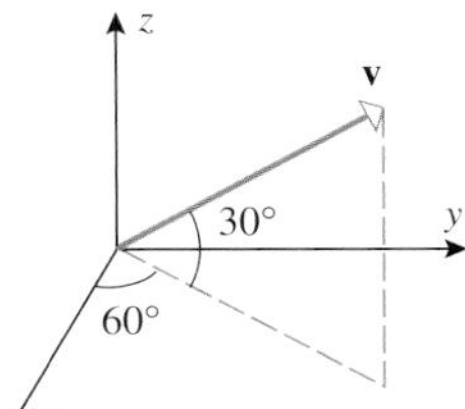

Figure Ex-21

22. Find, to the nearest degree, the acute angle formed by two diagonals of a cube.

23. Find, to the nearest degree, the angles that a diagonal of a box with dimensions 10 cm by 15 cm by 25 cm makes with the edges of the box.

24. In each part, find the vector component of $\mathbf{v}$ along $\mathbf{b}$ and the vector component of $\mathbf{v}$ orthogonal to $\mathbf{b}$. Then sketch the vectors $\mathbf{v}$, $\text{proj}_{\mathbf{b}}\mathbf{v}$, and $\mathbf{v} - \text{proj}_{\mathbf{b}}\mathbf{v}$.
(a) $\mathbf{v} = 2\mathbf{i} - \mathbf{j}$, $\mathbf{b} = 3\mathbf{i} + 4\mathbf{j}$
(b) $\mathbf{v} = \langle 4, 5 \rangle$, $\mathbf{b} = \langle 1, -2 \rangle$
(c) $\mathbf{v} = -3\mathbf{i} - 2\mathbf{j}$, $\mathbf{b} = 2\mathbf{i} + \mathbf{j}$

25. In each part, find the vector component of $\mathbf{v}$ along $\mathbf{b}$ and the vector component of $\mathbf{v}$ orthogonal to $\mathbf{b}$.
(a) $\mathbf{v} = 2\mathbf{i} - \mathbf{j} + 3\mathbf{k}$, $\mathbf{b} = \mathbf{i} + 2\mathbf{j} + 2\mathbf{k}$
(b) $\mathbf{v} = \langle 4, -1, 7 \rangle$, $\mathbf{b} = \langle 2, 3, -6 \rangle$

26–27 Express the vector $\mathbf{v}$ as the sum of a vector parallel to $\mathbf{b}$ and a vector orthogonal to $\mathbf{b}$.

26. (a) $\mathbf{v} = 2\mathbf{i} - 4\mathbf{j},\ \mathbf{b} = \mathbf{i} + \mathbf{j}$
(b) $\mathbf{v} = 3\mathbf{i} + \mathbf{j} - 2\mathbf{k},\ \mathbf{b} = 2\mathbf{i} - \mathbf{k}$
(c) $\mathbf{v} = 4\mathbf{i} - 2\mathbf{j} + 6\mathbf{k},\ \mathbf{b} = -2\mathbf{i} + \mathbf{j} - 3\mathbf{k}$

27. (a) $\mathbf{v} = \langle -3, 5\rangle,\ \mathbf{b} = \langle 1, 1\rangle$
(b) $\mathbf{v} = \langle -2, 1, 6\rangle,\ \mathbf{b} = \langle 0, -2, 1\rangle$
(c) $\mathbf{v} = \langle 1, 4, 1\rangle,\ \mathbf{b} = \langle 3, -2, 5\rangle$

28. If L is a line in 2-space or 3-space that passes through the points A and B, then the distance from a point P to the line L is equal to the length of the component of the vector $\overrightarrow{AP}$ that is orthogonal to the vector $\overrightarrow{AB}$ (see the accompanying figure). Use this result to find the distance from the point $P(1, 0)$ to the line through $A(2, -3)$ and $B(5, 1)$.

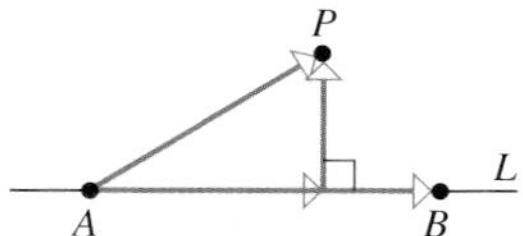

Figure Ex-28

29. Use the method of Exercise 28 to find the distance from the point $P(-3, 1, 2)$ to the line through $A(1, 1, 0)$ and $B(-2, 3, -4)$.

30. As shown in the accompanying figure, a child with mass 34 kg is seated on a smooth (frictionless) playground slide that is inclined at an angle of 27° with the horizontal. How much force does the child exert on the slide, and how much force must be applied in the direction of $\mathbf{P}$ to prevent the child from sliding down the slide? Take the acceleration due to gravity to be 9.8 m/s^2.

31. For the child in Exercise 30, how much force must be applied in the direction of $\mathbf{Q}$ (shown in the accompanying figure) to prevent the child from sliding down the slide?

Figure Ex-30

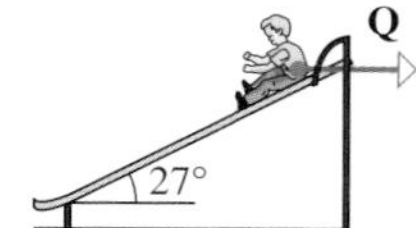

Figure Ex-31

32. A block weighing 300 lb is suspended by cables A and B, as shown in the accompanying figure. Determine the forces that the block exerts along the cables.

33. A block weighing 100 N is suspended by cables A and B, as shown in the accompanying figure.
(a) Use a graphing utility to graph the forces that the block exerts along cables A and B as functions of the "sag" d.
(b) Does increasing the sag increase or decrease the forces on the cables?
(c) How much sag is required if the cables cannot tolerate forces in excess of 150 N?

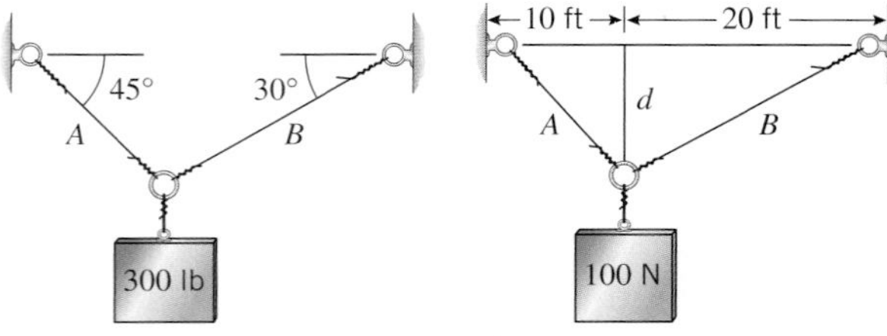

Figure Ex-32 **Figure Ex-33**

34. Find the work done by a force $\mathbf{F} = -3\mathbf{j}$ (pounds) applied to a point that moves on a line from $(1, 3)$ to $(4, 7)$. Assume that distance is measured in feet.

35. A force of $\mathbf{F} = 4\mathbf{i} - 6\mathbf{j} + \mathbf{k}$ newtons is applied to a point that moves a distance of 15 meters in the direction of the vector $\mathbf{i} + \mathbf{j} + \mathbf{k}$. How much work is done?

36. A boat travels 100 meters due north while the wind exerts a force of 500 newtons toward the northeast. How much work does the wind do?

37. A box is dragged along the floor by a rope that applies a force of 50 lb at an angle of 60° with the floor. How much work is done in moving the box 15 ft?

38. As shown in the accompanying figure, a force of 250 N is applied to a boat at an angle of 38° with the positive x-axis. What force $\mathbf{F}$ should be applied to the boat to produce a resultant force of 1000 N acting in the positive x-direction? State your answer by giving the magnitude of the force and its angle with the positive x-axis to the nearest degree.

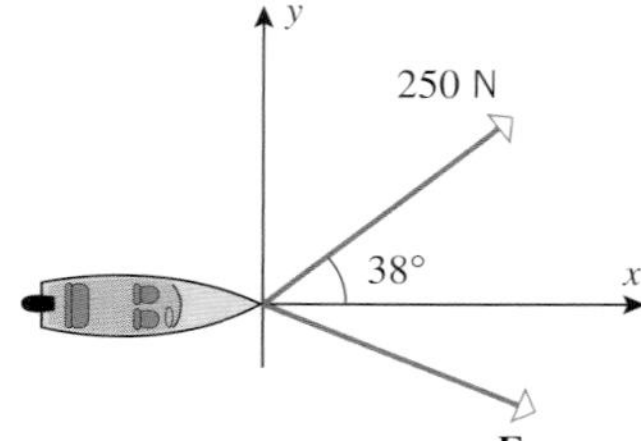

Figure Ex-38

FOCUS ON CONCEPTS

39. Let $\mathbf{u}$ and $\mathbf{v}$ be adjacent sides of a parallelogram. Use vectors to prove that the diagonals of the parallelogram are perpendicular if the sides are equal in length.

40. Let $\mathbf{u}$ and $\mathbf{v}$ be adjacent sides of a parallelogram. Use vectors to prove that the parallelogram is a rectangle if the diagonals are equal in length.

41. Prove that

$$\|\mathbf{u} + \mathbf{v}\|^2 + \|\mathbf{u} - \mathbf{v}\|^2 = 2\|\mathbf{u}\|^2 + 2\|\mathbf{v}\|^2$$

and interpret the result geometrically by translating it into a theorem about parallelograms.

42. Prove: $\mathbf{u} \cdot \mathbf{v} = \frac{1}{4}\|\mathbf{u} + \mathbf{v}\|^2 - \frac{1}{4}\|\mathbf{u} - \mathbf{v}\|^2$.

43. Show that if $\mathbf{v}_1$, $\mathbf{v}_2$, and $\mathbf{v}_3$ are mutually orthogonal nonzero vectors in 3-space, and if a vector $\mathbf{v}$ in 3-space is expressed as

$$\mathbf{v} = c_1\mathbf{v}_1 + c_2\mathbf{v}_2 + c_3\mathbf{v}_3$$

then the scalars c_1, c_2, and c_3 are given by the formulas

$$c_i = (\mathbf{v} \cdot \mathbf{v}_i)/\|\mathbf{v}_i\|^2, \quad i = 1, 2, 3$$

44. Show that the three vectors

$$\mathbf{v}_1 = 3\mathbf{i} - \mathbf{j} + 2\mathbf{k}, \quad \mathbf{v}_2 = \mathbf{i} + \mathbf{j} - \mathbf{k}, \quad \mathbf{v}_3 = \mathbf{i} - 5\mathbf{j} - 4\mathbf{k}$$

are mutually orthogonal, and then use the result of Exercise 43 to find scalars c_1, c_2, and c_3 so that

$$c_1\mathbf{v}_1 + c_2\mathbf{v}_2 + c_3\mathbf{v}_3 = \mathbf{i} - \mathbf{j} + \mathbf{k}$$

C **45.** For each x in $(-\infty, +\infty)$, let $\mathbf{u}(x)$ be the vector from the origin to the point $P(x, y)$ on the curve $y = x^2 + 1$, and $\mathbf{v}(x)$ the vector from the origin to the point $Q(x, y)$ on the line $y = -x - 1$.

(a) Use a CAS to find, to the nearest degree, the minimum angle between $\mathbf{u}(x)$ and $\mathbf{v}(x)$ for x in $(-\infty, +\infty)$.

(b) Determine whether there are any real values of x for which $\mathbf{u}(x)$ and $\mathbf{v}(x)$ are orthogonal.

C **46.** Let $\mathbf{u}$ be a unit vector in the xy-plane of an xyz-coordinate system, and let $\mathbf{v}$ be a unit vector in the yz-plane. Let θ_1 be the angle between $\mathbf{u}$ and $\mathbf{i}$, let θ_2 be the angle between $\mathbf{v}$ and $\mathbf{k}$, and let θ be the angle between $\mathbf{u}$ and $\mathbf{v}$.

(a) Show that $\cos\theta = \pm\sin\theta_1 \sin\theta_2$.

(b) Find θ if θ is acute and $\theta_1 = \theta_2 = 45°$.

(c) Use a CAS to find, to the nearest degree, the maximum and minimum values of θ if θ is acute and $\theta_2 = 2\theta_1$.

47. Prove parts (*b*) and (*e*) of Theorem 12.3.2 for vectors in 3-space.

✔ QUICK CHECK ANSWERS 12.3

1. 8 2. (a) 25 (b) 7 (c) 10 (d) −6 3. $\frac{14}{5}$ 4. $\frac{2}{\sqrt{14}}; -\frac{1}{\sqrt{14}}; \frac{3}{\sqrt{14}}$ 5. $9\mathbf{i} - 3\mathbf{j}$

12.4 CROSS PRODUCT

In many applications of vectors in mathematics, physics, and engineering, there is a need to find a vector that is orthogonal to two given vectors. In this section we will discuss a new type of vector multiplication that can be used for this purpose.

DETERMINANTS

Some of the concepts that we will develop in this section require basic ideas about ***determinants***, which are functions that assign numerical values to square arrays of numbers. For example, if a_1, a_2, b_1, and b_2 are real numbers, then we define a **2 × 2 *determinant*** by

$$\begin{vmatrix} a_1 & a_2 \\ b_1 & b_2 \end{vmatrix} = a_1b_2 - a_2b_1 \tag{1}$$

The purpose of the arrows is to help you remember the formula—the determinant is the product of the entries on the rightward arrow minus the product of the entries on the leftward arrow. For example,

$$\begin{vmatrix} 3 & -2 \\ 4 & 5 \end{vmatrix} = (3)(5) - (-2)(4) = 15 + 8 = 23$$

A **3 × 3 *determinant*** is defined in terms of 2 × 2 determinants by

$$\begin{vmatrix} a_1 & a_2 & a_3 \\ b_1 & b_2 & b_3 \\ c_1 & c_2 & c_3 \end{vmatrix} = a_1\begin{vmatrix} b_2 & b_3 \\ c_2 & c_3 \end{vmatrix} - a_2\begin{vmatrix} b_1 & b_3 \\ c_1 & c_3 \end{vmatrix} + a_3\begin{vmatrix} b_1 & b_2 \\ c_1 & c_2 \end{vmatrix} \tag{2}$$

The right side of this formula is easily remembered by noting that a_1, a_2, and a_3 are the entries in the first "row" of the left side, and the 2 × 2 determinants on the right side arise by deleting the first row and an appropriate column from the left side. The pattern is as follows:

$$\begin{vmatrix} a_1 & a_2 & a_3 \\ b_1 & b_2 & b_3 \\ c_1 & c_2 & c_3 \end{vmatrix} = a_1 \begin{vmatrix} a_1 & a_2 & a_3 \\ b_1 & b_2 & b_3 \\ c_1 & c_2 & c_3 \end{vmatrix} - a_2 \begin{vmatrix} a_1 & a_2 & a_3 \\ b_1 & b_2 & b_3 \\ c_1 & c_2 & c_3 \end{vmatrix} + a_3 \begin{vmatrix} a_1 & a_2 & a_3 \\ b_1 & b_2 & b_3 \\ c_1 & c_2 & c_3 \end{vmatrix}$$

For example,

$$\begin{vmatrix} 3 & -2 & -5 \\ 1 & 4 & -4 \\ 0 & 3 & 2 \end{vmatrix} = 3 \begin{vmatrix} 4 & -4 \\ 3 & 2 \end{vmatrix} - (-2) \begin{vmatrix} 1 & -4 \\ 0 & 2 \end{vmatrix} + (-5) \begin{vmatrix} 1 & 4 \\ 0 & 3 \end{vmatrix}$$

$$= 3(20) + 2(2) - 5(3) = 49$$

There are also definitions of 4×4 determinants, 5×5 determinants, and higher, but we will not need them in this text. Properties of determinants are studied in a branch of mathematics called ***linear algebra***, but we will only need the two properties stated in the following theorem.

12.4.1 THEOREM.

(*a*) *If two rows in the array of a determinant are the same, then the value of the determinant is* 0.

(*b*) *Interchanging two rows in the array of a determinant multiplies its value by* -1.

We will give the proofs of parts (*a*) and (*b*) for 2×2 determinants and leave the proofs for 3×3 determinants as exercises.

PROOF (*a*).

$$\begin{vmatrix} a_1 & a_2 \\ a_1 & a_2 \end{vmatrix} = a_1 a_2 - a_2 a_1 = 0$$

PROOF (*b*).

$$\begin{vmatrix} b_1 & b_2 \\ a_1 & a_2 \end{vmatrix} = b_1 a_2 - b_2 a_1 = -(a_1 b_2 - a_2 b_1) = - \begin{vmatrix} a_1 & a_2 \\ b_1 & b_2 \end{vmatrix} \quad \blacksquare$$

CROSS PRODUCT

We now turn to the main concept in this section.

12.4.2 DEFINITION. If $\mathbf{u} = \langle u_1, u_2, u_3 \rangle$ and $\mathbf{v} = \langle v_1, v_2, v_3 \rangle$ are vectors in 3-space, then the ***cross product*** $\mathbf{u} \times \mathbf{v}$ is the vector defined by

$$\mathbf{u} \times \mathbf{v} = \begin{vmatrix} u_2 & u_3 \\ v_2 & v_3 \end{vmatrix} \mathbf{i} - \begin{vmatrix} u_1 & u_3 \\ v_1 & v_3 \end{vmatrix} \mathbf{j} + \begin{vmatrix} u_1 & u_2 \\ v_1 & v_2 \end{vmatrix} \mathbf{k} \tag{3}$$

or, equivalently,

$$\mathbf{u} \times \mathbf{v} = (u_2 v_3 - u_3 v_2)\mathbf{i} - (u_1 v_3 - u_3 v_1)\mathbf{j} + (u_1 v_2 - u_2 v_1)\mathbf{k} \tag{4}$$

Observe that the right side of Formula (3) has the same form as the right side of Formula (2), the difference being notation and the order of the factors in the three terms. Thus, we can rewrite (3) as

$$\mathbf{u} \times \mathbf{v} = \begin{vmatrix} \mathbf{i} & \mathbf{j} & \mathbf{k} \\ u_1 & u_2 & u_3 \\ v_1 & v_2 & v_3 \end{vmatrix} \tag{5}$$

However, this is just a mnemonic device and not a true determinant since the entries in a determinant are numbers, not vectors.

▶ Example 1 Let $\mathbf{u} = \langle 1, 2, -2 \rangle$ and $\mathbf{v} = \langle 3, 0, 1 \rangle$. Find

(a) $\mathbf{u} \times \mathbf{v}$ (b) $\mathbf{v} \times \mathbf{u}$

Solution (a).

$$\mathbf{u} \times \mathbf{v} = \begin{vmatrix} \mathbf{i} & \mathbf{j} & \mathbf{k} \\ 1 & 2 & -2 \\ 3 & 0 & 1 \end{vmatrix}$$

$$= \begin{vmatrix} 2 & -2 \\ 0 & 1 \end{vmatrix}\mathbf{i} - \begin{vmatrix} 1 & -2 \\ 3 & 1 \end{vmatrix}\mathbf{j} + \begin{vmatrix} 1 & 2 \\ 3 & 0 \end{vmatrix}\mathbf{k} = 2\mathbf{i} - 7\mathbf{j} - 6\mathbf{k}$$

Solution (b). We could use the method of part (a), but it is really not necessary to perform any computations. We need only observe that reversing $\mathbf{u}$ and $\mathbf{v}$ interchanges the second and third rows in (5), which in turn interchanges the rows in the arrays for the 2×2 determinants in (3). But interchanging the rows in the array of a 2×2 determinant reverses its sign, so the net effect of reversing the factors in a cross product is to reverse the signs of the components. Thus, by inspection

$$\mathbf{v} \times \mathbf{u} = -(\mathbf{u} \times \mathbf{v}) = -2\mathbf{i} + 7\mathbf{j} + 6\mathbf{k} \blacktriangleleft$$

▶ Example 2 Show that $\mathbf{u} \times \mathbf{u} = \mathbf{0}$ for any vector $\mathbf{u}$ in 3-space.

Solution. We could let $\mathbf{u} = u_1\mathbf{i} + u_2\mathbf{j} + u_3\mathbf{k}$ and apply the method in part (a) of Example 1 to show that

$$\mathbf{u} \times \mathbf{u} = \begin{vmatrix} \mathbf{i} & \mathbf{j} & \mathbf{k} \\ u_1 & u_2 & u_3 \\ u_1 & u_2 & u_3 \end{vmatrix} = 0$$

However, the actual computations are unnecessary. We need only observe that if the two factors in a cross product are the same, then each 2×2 determinant in (3) is zero because its array has identical rows. Thus, $\mathbf{u} \times \mathbf{u} = \mathbf{0}$ by inspection. ◀

ALGEBRAIC PROPERTIES OF THE CROSS PRODUCT

Our next goal is to establish some of the basic algebraic properties of the cross product. As you read the discussion, keep in mind the essential differences between the cross product and the dot product:

- The cross product is defined only for vectors in 3-space, whereas the dot product is defined for vectors in 2-space and 3-space.
- The cross product of two vectors is a vector, whereas the dot product of two vectors is a scalar.

The main algebraic properties of the cross product are listed in the next theorem.

Unlike for ordinary multiplication for dot products, the order of the factors matters for cross products. Specifically, part (*a*) of Theorem 12.4.3 shows that reversing the order of the factors in a cross product reverses the direction of the resulting vector.

12.4.3 THEOREM. *If* $\mathbf{u}$, $\mathbf{v}$, *and* $\mathbf{w}$ *are any vectors in 3-space and* k *is any scalar, then*

(*a*) $\mathbf{u} \times \mathbf{v} = -(\mathbf{v} \times \mathbf{u})$

(*b*) $\mathbf{u} \times (\mathbf{v} + \mathbf{w}) = (\mathbf{u} \times \mathbf{v}) + (\mathbf{u} \times \mathbf{w})$

(*c*) $(\mathbf{u} + \mathbf{v}) \times \mathbf{w} = (\mathbf{u} \times \mathbf{w}) + (\mathbf{v} \times \mathbf{w})$

(*d*) $k(\mathbf{u} \times \mathbf{v}) = (k\mathbf{u}) \times \mathbf{v} = \mathbf{u} \times (k\mathbf{v})$

(*e*) $\mathbf{u} \times \mathbf{0} = \mathbf{0} \times \mathbf{u} = \mathbf{0}$

(*f*) $\mathbf{u} \times \mathbf{u} = \mathbf{0}$

Parts (*a*) and (*f*) were addressed in Examples 1 and 2. The other proofs are left as exercises.

The following cross products occur so frequently that it is helpful to be familiar with them:

$$\begin{array}{lll} \mathbf{i} \times \mathbf{j} = \mathbf{k} & \mathbf{j} \times \mathbf{k} = \mathbf{i} & \mathbf{k} \times \mathbf{i} = \mathbf{j} \\ \mathbf{j} \times \mathbf{i} = -\mathbf{k} & \mathbf{k} \times \mathbf{j} = -\mathbf{i} & \mathbf{i} \times \mathbf{k} = -\mathbf{j} \end{array} \tag{6}$$

These results are easy to obtain; for example,

$$\mathbf{i} \times \mathbf{j} = \begin{vmatrix} \mathbf{i} & \mathbf{j} & \mathbf{k} \\ 1 & 0 & 0 \\ 0 & 1 & 0 \end{vmatrix} = \begin{vmatrix} 0 & 0 \\ 1 & 0 \end{vmatrix}\mathbf{i} - \begin{vmatrix} 1 & 0 \\ 0 & 0 \end{vmatrix}\mathbf{j} + \begin{vmatrix} 1 & 0 \\ 0 & 1 \end{vmatrix}\mathbf{k} = \mathbf{k}$$

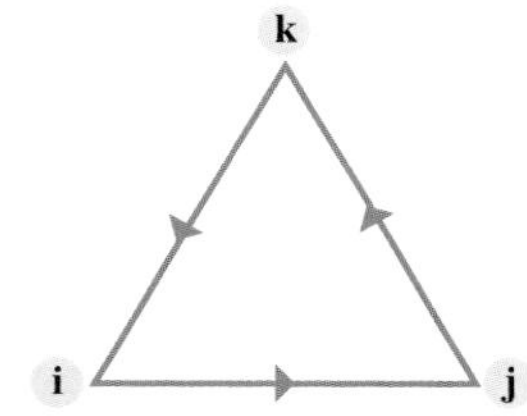

Figure 12.4.1

However, rather than computing these cross products each time you need them, you can use the diagram in Figure 12.4.1. In this diagram, the cross product of two consecutive vectors in the counterclockwise direction is the next vector around, and the cross product of two consecutive vectors in the clockwise direction is the negative of the next vector around.

WARNING We can write a product of three real numbers as uvw since the associative law $u(vw) = (uv)w$ ensures that the same value for the product results no matter how the factors are grouped. However, the associative law *does not* hold for cross products. For example,

$$\mathbf{i} \times (\mathbf{j} \times \mathbf{j}) = \mathbf{i} \times \mathbf{0} = \mathbf{0} \quad \text{and} \quad (\mathbf{i} \times \mathbf{j}) \times \mathbf{j} = \mathbf{k} \times \mathbf{j} = -\mathbf{i}$$

so that $\mathbf{i} \times (\mathbf{j} \times \mathbf{j}) \neq (\mathbf{i} \times \mathbf{j}) \times \mathbf{j}$. Thus, we cannot write a cross product with three vectors as $\mathbf{u} \times \mathbf{v} \times \mathbf{w}$, since this expression is ambiguous without parentheses.

GEOMETRIC PROPERTIES OF THE CROSS PRODUCT

The following theorem shows that the cross product of two vectors is orthogonal to both factors.

12.4.4 THEOREM. *If* $\mathbf{u}$ *and* $\mathbf{v}$ *are vectors in 3-space, then:*

(*a*) $\mathbf{u} \cdot (\mathbf{u} \times \mathbf{v}) = 0$ ($\mathbf{u} \times \mathbf{v}$ *is orthogonal to* $\mathbf{u}$)

(*b*) $\mathbf{v} \cdot (\mathbf{u} \times \mathbf{v}) = 0$ ($\mathbf{u} \times \mathbf{v}$ *is orthogonal to* $\mathbf{v}$)

We will prove part (a). The proof of part (b) is similar.

PROOF (a). Let $\mathbf{u} = \langle u_1, u_2, u_3\rangle$ and $\mathbf{v} = \langle v_1, v_2, v_3\rangle$. Then from (4)

$$\mathbf{u}\times\mathbf{v} = \langle u_2v_3 - u_3v_2,\, u_3v_1 - u_1v_3,\, u_1v_2 - u_2v_1\rangle \tag{7}$$

so that

$$\mathbf{u}\cdot(\mathbf{u}\times\mathbf{v}) = u_1(u_2v_3 - u_3v_2) + u_2(u_3v_1 - u_1v_3) + u_3(u_1v_2 - u_2v_1) = 0 \quad ■$$

► **Example 3** In Example 1 we showed that the cross product $\mathbf{u}\times\mathbf{v}$ of $\mathbf{u} = \langle 1, 2, -2\rangle$ and $\mathbf{v} = \langle 3, 0, 1\rangle$ is

$$\mathbf{u}\times\mathbf{v} = 2\mathbf{i} - 7\mathbf{j} - 6\mathbf{k} = \langle 2, -7, -6\rangle$$

Theorem 12.4.4 guarantees that this vector is orthogonal to both $\mathbf{u}$ and $\mathbf{v}$; this is confirmed by the computations

$$\mathbf{u}\cdot(\mathbf{u}\times\mathbf{v}) = \langle 1, 2, -2\rangle\cdot\langle 2, -7, -6\rangle = (1)(2) + (2)(-7) + (-2)(-6) = 0$$
$$\mathbf{v}\cdot(\mathbf{u}\times\mathbf{v}) = \langle 3, 0, 1\rangle\cdot\langle 2, -7, -6\rangle = (3)(2) + (0)(-7) + (1)(-6) = 0 \quad ◄$$

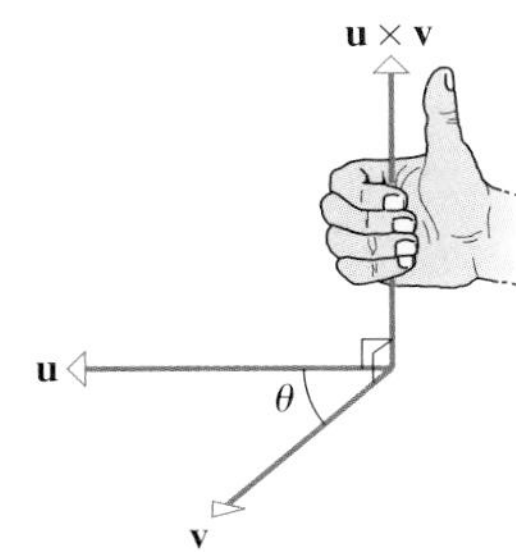

Figure 12.4.2

It can be proved that if $\mathbf{u}$ and $\mathbf{v}$ are nonzero and nonparallel vectors, then the direction of $\mathbf{u}\times\mathbf{v}$ relative to $\mathbf{u}$ and $\mathbf{v}$ is determined by a right-hand rule;* that is, if the fingers of the right hand are cupped so they curl from $\mathbf{u}$ toward $\mathbf{v}$ in the direction of rotation that takes $\mathbf{u}$ into $\mathbf{v}$ in less than 180°, then the thumb will point (roughly) in the direction of $\mathbf{u}\times\mathbf{v}$ (Figure 12.4.2). For example, we stated in (6) that

$$\mathbf{i}\times\mathbf{j} = \mathbf{k}, \quad \mathbf{j}\times\mathbf{k} = \mathbf{i}, \quad \mathbf{k}\times\mathbf{i} = \mathbf{j}$$

all of which are consistent with the right-hand rule (verify).

The next theorem lists some more important geometric properties of the cross product.

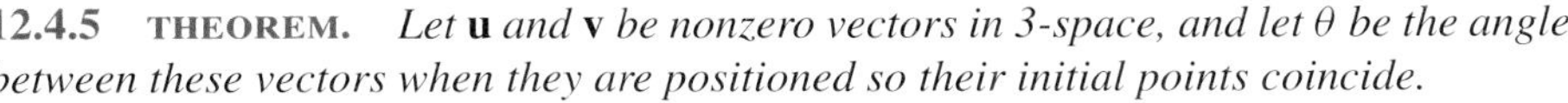

12.4.5 THEOREM. *Let* $\mathbf{u}$ *and* $\mathbf{v}$ *be nonzero vectors in 3-space, and let* θ *be the angle between these vectors when they are positioned so their initial points coincide.*

(a) $\|\mathbf{u}\times\mathbf{v}\| = \|\mathbf{u}\|\|\mathbf{v}\|\sin\theta$

(b) *The area A of the parallelogram that has* $\mathbf{u}$ *and* $\mathbf{v}$ *as adjacent sides is*

$$A = \|\mathbf{u}\times\mathbf{v}\| \tag{8}$$

(c) $\mathbf{u}\times\mathbf{v} = \mathbf{0}$ *if and only if* $\mathbf{u}$ *and* $\mathbf{v}$ *are parallel vectors, that is, if and only if they are scalar multiples of one another.*

PROOF (a).

$$\begin{aligned}
\|\mathbf{u}\|\|\mathbf{v}\|\sin\theta &= \|\mathbf{u}\|\|\mathbf{v}\|\sqrt{1-\cos^2\theta} \\
&= \|\mathbf{u}\|\|\mathbf{v}\|\sqrt{1-\frac{(\mathbf{u}\cdot\mathbf{v})^2}{\|\mathbf{u}\|^2\|\mathbf{v}\|^2}} \qquad \boxed{\text{Theorem 12.3.3}} \\
&= \sqrt{\|\mathbf{u}\|^2\|\mathbf{v}\|^2 - (\mathbf{u}\cdot\mathbf{v})^2} \\
&= \sqrt{(u_1^2+u_2^2+u_3^2)(v_1^2+v_2^2+v_3^2) - (u_1v_1+u_2v_2+u_3v_3)^2} \\
&= \sqrt{(u_2v_3-u_3v_2)^2 + (u_1v_3-u_3v_1)^2 + (u_1v_2-u_2v_1)^2} \\
&= \|\mathbf{u}\times\mathbf{v}\| \qquad \boxed{\text{See Formula (4).}}
\end{aligned}$$

---

*Recall that we agreed to consider only right-handed coordinate systems in this text. Had we used left-handed systems instead, a "left-hand rule" would apply here.

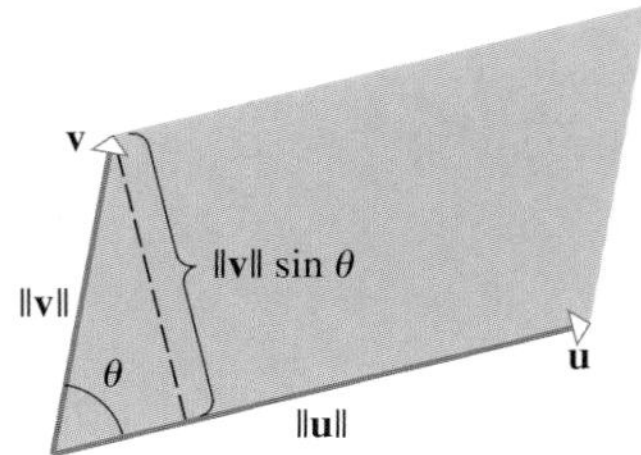

Figure 12.4.3

PROOF (*b*). Referring to Figure 12.4.3, the parallelogram that has **u** and **v** as adjacent sides can be viewed as having base $\|\mathbf{u}\|$ and altitude $\|\mathbf{v}\|\sin\theta$. Thus, its area A is

$$A = (\text{base})(\text{altitude}) = \|\mathbf{u}\|\,\|\mathbf{v}\|\sin\theta = \|\mathbf{u}\times\mathbf{v}\|$$

PROOF (*c*). Since **u** and **v** are assumed to be nonzero vectors, it follows from part (*a*) that $\mathbf{u}\times\mathbf{v} = \mathbf{0}$ if and only if $\sin\theta = 0$; this is true if and only if $\theta = 0$ or $\theta = \pi$ (since $0 \le \theta \le \pi$). Geometrically, this means that $\mathbf{u}\times\mathbf{v} = \mathbf{0}$ if and only if **u** and **v** are parallel vectors. ■

▶ **Example 4** Find the area of the triangle that is determined by the points $P_1(2, 2, 0)$, $P_2(-1, 0, 2)$, and $P_3(0, 4, 3)$.

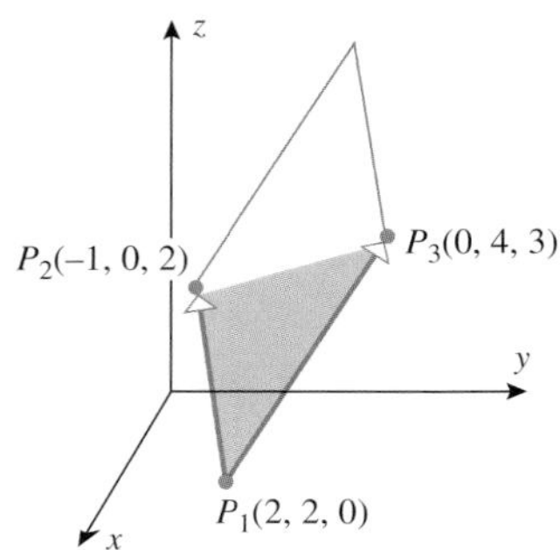

Figure 12.4.4

Solution. The area A of the triangle is half the area of the parallelogram determined by the vectors $\overrightarrow{P_1P_2}$ and $\overrightarrow{P_1P_3}$ (Figure 12.4.4). But $\overrightarrow{P_1P_2} = \langle -3, -2, 2\rangle$ and $\overrightarrow{P_1P_3} = \langle -2, 2, 3\rangle$, so

$$\overrightarrow{P_1P_2} \times \overrightarrow{P_1P_3} = \langle -10, 5, -10\rangle$$

(verify), and consequently

$$A = \tfrac{1}{2}\|\overrightarrow{P_1P_2} \times \overrightarrow{P_1P_3}\| = \tfrac{15}{2} \quad ◀$$

SCALAR TRIPLE PRODUCTS

If $\mathbf{u} = \langle u_1, u_2, u_3\rangle$, $\mathbf{v} = \langle v_1, v_2, v_3\rangle$, and $\mathbf{w} = \langle w_1, w_2, w_3\rangle$ are vectors in 3-space, then the number

$$\mathbf{u}\cdot(\mathbf{v}\times\mathbf{w})$$

is called the ***scalar triple product*** of **u**, **v**, and **w**. It is not necessary to compute the dot product and cross product to evaluate a scalar triple product—the value can be obtained directly from the formula

$$\mathbf{u}\cdot(\mathbf{v}\times\mathbf{w}) = \begin{vmatrix} u_1 & u_2 & u_3 \\ v_1 & v_2 & v_3 \\ w_1 & w_2 & w_3 \end{vmatrix} \tag{9}$$

the validity of which can be seen by writing

$$\begin{aligned}
\mathbf{u}\cdot(\mathbf{v}\times\mathbf{w}) &= \mathbf{u}\cdot\left(\begin{vmatrix} v_2 & v_3 \\ w_2 & w_3 \end{vmatrix}\mathbf{i} - \begin{vmatrix} v_1 & v_3 \\ w_1 & w_3 \end{vmatrix}\mathbf{j} + \begin{vmatrix} v_1 & v_2 \\ w_1 & w_2 \end{vmatrix}\mathbf{k}\right) \\
&= u_1\begin{vmatrix} v_2 & v_3 \\ w_2 & w_3 \end{vmatrix} - u_2\begin{vmatrix} v_1 & v_3 \\ w_1 & w_3 \end{vmatrix} + u_3\begin{vmatrix} v_1 & v_2 \\ w_1 & w_2 \end{vmatrix} \\
&= \begin{vmatrix} u_1 & u_2 & u_3 \\ v_1 & v_2 & v_3 \\ w_1 & w_2 & w_3 \end{vmatrix}
\end{aligned}$$

▶ **Example 5** Calculate the scalar triple product $\mathbf{u}\cdot(\mathbf{v}\times\mathbf{w})$ of the vectors

$$\mathbf{u} = 3\mathbf{i} - 2\mathbf{j} - 5\mathbf{k}, \quad \mathbf{v} = \mathbf{i} + 4\mathbf{j} - 4\mathbf{k}, \quad \mathbf{w} = 3\mathbf{j} + 2\mathbf{k}$$

Solution.

$$\mathbf{u}\cdot(\mathbf{v}\times\mathbf{w}) = \begin{vmatrix} 3 & -2 & -5 \\ 1 & 4 & -4 \\ 0 & 3 & 2 \end{vmatrix} = 49 \quad ◀$$

TECHNOLOGY MASTERY

Many calculating utilities have built-in cross product and determinant operations. If your calculating utility has these capabilities, use it to check the computations in Examples 1 and 5.

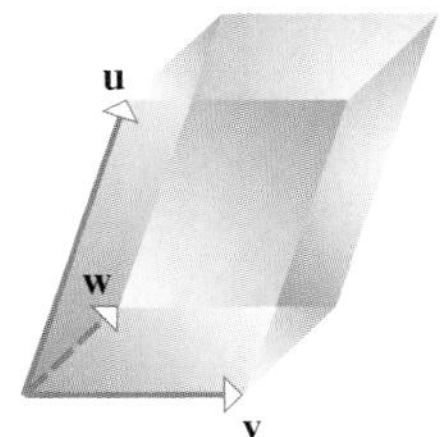

Figure 12.4.5

GEOMETRIC PROPERTIES OF THE SCALAR TRIPLE PRODUCT

If **u**, **v**, and **w** are nonzero vectors in 3-space that are positioned so their initial points coincide, then these vectors form the adjacent sides of a parallelepiped (Figure 12.4.5). The following theorem establishes a relationship between the volume of this parallelepiped and the scalar triple product of the sides.

12.4.6 THEOREM. *Let* **u**, **v**, *and* **w** *be nonzero vectors in 3-space.*

(*a*) *The volume V of the parallelepiped that has* **u**, **v**, *and* **w** *as adjacent edges is*

$$V = |\mathbf{u} \cdot (\mathbf{v} \times \mathbf{w})| \tag{10}$$

(*b*) $\mathbf{u} \cdot (\mathbf{v} \times \mathbf{w}) = 0$ *if and only if* **u**, **v**, *and* **w** *lie in the same plane.*

It follows from Formula (10) that

$$\mathbf{u} \cdot (\mathbf{v} \times \mathbf{w}) = \pm V$$

The + occurs when **u** makes an acute angle with $\mathbf{v} \times \mathbf{w}$ and the − occurs when it makes an obtuse angle.

PROOF (*a*). Referring to Figure 12.4.6, let us regard the base of the parallelepiped with **u**, **v**, and **w** as adjacent sides to be the parallelogram determined by **v** and **w**. Thus, the area of the base is $\|\mathbf{v} \times \mathbf{w}\|$, and the altitude h of the parallelepiped (shown in the figure) is the length of the orthogonal projection of **u** on the vector $\mathbf{v} \times \mathbf{w}$. Therefore, from Formula (12) of Section 12.3 we have

$$h = \|\text{proj}_{\mathbf{v}\times\mathbf{w}}\mathbf{u}\| = \frac{|\mathbf{u} \cdot (\mathbf{v} \times \mathbf{w})|}{\|\mathbf{v} \times \mathbf{w}\|^2}\|\mathbf{v} \times \mathbf{w}\| = \frac{|\mathbf{u} \cdot (\mathbf{v} \times \mathbf{w})|}{\|\mathbf{v} \times \mathbf{w}\|}$$

It now follows that the volume of the parallelepiped is

$$V = (\text{area of base})(\text{height}) = \|\mathbf{v} \times \mathbf{w}\|h = |\mathbf{u} \cdot (\mathbf{v} \times \mathbf{w})|$$

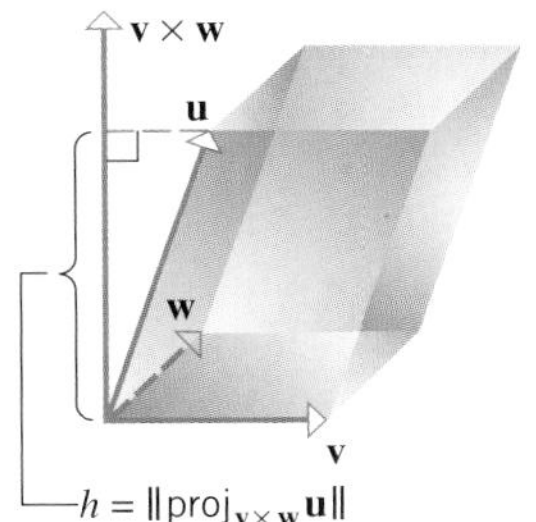

Figure 12.4.6

PROOF (*b*). The vectors **u**, **v**, and **w** lie in the same plane if and only if the parallelepiped with these vectors as adjacent sides has volume zero (why?). Thus, from part (*a*) the vectors lie in the same plane if and only if $\mathbf{u} \cdot (\mathbf{v} \times \mathbf{w}) = 0$. ■

ALGEBRAIC PROPERTIES OF THE SCALAR TRIPLE PRODUCT

We observed earlier in this section that the expression $\mathbf{u} \times \mathbf{v} \times \mathbf{w}$ must be avoided because it is ambiguous without parentheses. However, the expression $\mathbf{u} \cdot \mathbf{v} \times \mathbf{w}$ is not ambiguous—it has to mean $\mathbf{u} \cdot (\mathbf{v} \times \mathbf{w})$ and not $(\mathbf{u} \cdot \mathbf{v}) \times \mathbf{w}$ because we cannot form the cross product of a scalar and a vector. Similarly, the expression $\mathbf{u} \times \mathbf{v} \cdot \mathbf{w}$ must mean $(\mathbf{u} \times \mathbf{v}) \cdot \mathbf{w}$ and not $\mathbf{u} \times (\mathbf{v} \cdot \mathbf{w})$. Thus, when you see an expression of the form $\mathbf{u} \cdot \mathbf{v} \times \mathbf{w}$ or $\mathbf{u} \times \mathbf{v} \cdot \mathbf{w}$, the cross product is formed first and the dot product second.

Since interchanging two rows of a determinant multiplies its value by −1, making two row interchanges in a determinant has no effect on its value. This being the case, it follows that

$$\mathbf{u} \cdot (\mathbf{v} \times \mathbf{w}) = \mathbf{w} \cdot (\mathbf{u} \times \mathbf{v}) = \mathbf{v} \cdot (\mathbf{w} \times \mathbf{u}) \tag{11}$$

since the 3×3 determinants that are used to compute these scalar triple products can be obtained from one another by two row interchanges (verify).

A good way to remember Formula (11) is to observe that the second expression in the formula can be obtained from the first by leaving the dot, cross, and parentheses fixed, moving the first two vectors to the right, and bringing the third vector to the first position. The same procedure produces the third expression from the second and the first expression from the third (verify).

Another useful formula can be obtained by rewriting the first equality in (11) as

$$\mathbf{u} \cdot (\mathbf{v} \times \mathbf{w}) = (\mathbf{u} \times \mathbf{v}) \cdot \mathbf{w}$$

and then omitting the superfluous parentheses to obtain

$$\mathbf{u} \cdot \mathbf{v} \times \mathbf{w} = \mathbf{u} \times \mathbf{v} \cdot \mathbf{w} \tag{12}$$

In words, this formula states that the dot and cross in a scalar triple product can be interchanged (provided the factors are grouped appropriately).

DOT AND CROSS PRODUCTS ARE COORDINATE INDEPENDENT

In Definitions 12.3.1 and 12.4.2 we defined the dot product and the cross product of two vectors in terms of the components of those vectors in a coordinate system. Thus, it is theoretically possible that changing the coordinate system might change $\mathbf{u}\cdot\mathbf{v}$ or $\mathbf{u}\times\mathbf{v}$, since the components of a vector depend on the coordinate system that is chosen. However, the relationships

$$\mathbf{u}\cdot\mathbf{v} = \|\mathbf{u}\|\,\|\mathbf{v}\|\cos\theta \tag{13}$$

$$\|\mathbf{u}\times\mathbf{v}\| = \|\mathbf{u}\|\,\|\mathbf{v}\|\sin\theta \tag{14}$$

that were obtained in Theorems 12.3.3 and 12.4.5 show that this is not the case. Formula (13) shows that the value of $\mathbf{u}\cdot\mathbf{v}$ depends only on the lengths of the vectors and the angle between them—not on the coordinate system. Similarly, Formula (14), in combination with the right-hand rule and Theorem 12.4.4, shows that $\mathbf{u}\times\mathbf{v}$ does not depend on the coordinate system (as long as it is right-handed). These facts are important in applications because they allow us to choose any convenient coordinate system for solving a problem with full confidence that the choice will not affect computations that involve dot products or cross products.

MOMENTS AND ROTATIONAL MOTION IN 3-SPACE

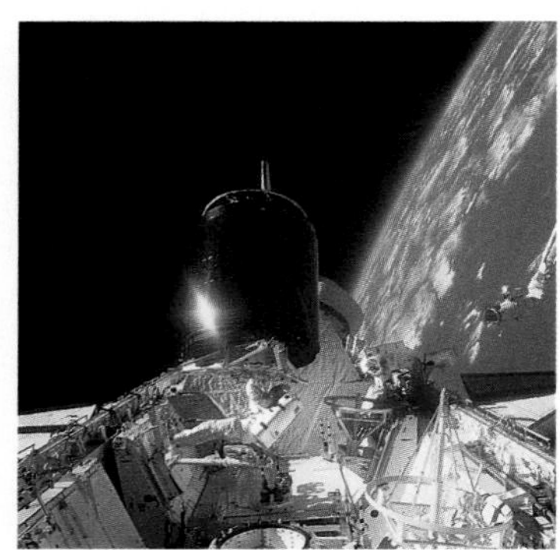
Astronauts use tools that are designed to limit forces that would impart unintended rotational motion to a satellite.

Cross products play an important role in describing rotational motion in 3-space. For example, suppose that an astronaut on a satellite repair mission in space applies a force $\mathbf{F}$ at a point Q on the surface of a spherical satellite. If the force is directed along a line that passes through the center P of the satellite, then Newton's Second Law of Motion implies that the force will accelerate the satellite in the direction of $\mathbf{F}$. However, if the astronaut applies the same force at an angle θ with the vector $\overrightarrow{PQ}$, then $\mathbf{F}$ will tend to cause a rotation, as well as an acceleration in the direction of $\mathbf{F}$. To see why this is so, let us resolve $\mathbf{F}$ into a sum of orthogonal components $\mathbf{F} = \mathbf{F}_1 + \mathbf{F}_2$, where $\mathbf{F}_1$ is the orthogonal projection of $\mathbf{F}$ on the vector $\overrightarrow{PQ}$ and $\mathbf{F}_2$ is the component of $\mathbf{F}$ orthogonal to $\overrightarrow{PQ}$ (Figure 12.4.7). Since the force $\mathbf{F}_1$ acts along the line through the center of the satellite, it contributes to the linear acceleration of the satellite but does not cause any rotation. However, the force $\mathbf{F}_2$ is tangent to the circle around the satellite in the plane of $\mathbf{F}$ and $\overrightarrow{PQ}$, so it causes the satellite to rotate about an axis that is perpendicular to that plane.

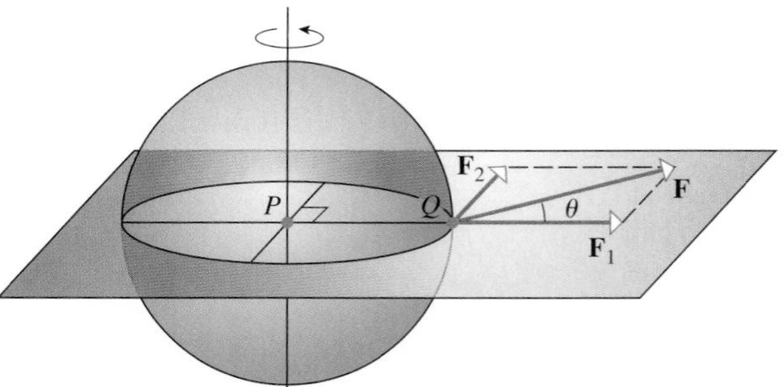

Figure 12.4.7

You know from your own experience that the "tendency" for rotation about an axis depends both on the amount of force and how far from the axis it is applied. For example, it is easier to close a door by pushing on its outer edge than applying the same force close to the hinges. In fact, the tendency of rotation of the satellite can be measured by

$$\|\overrightarrow{PQ}\|\,\|\mathbf{F}_2\| \quad \text{distance from the center} \times \text{magnitude of the force} \tag{15}$$

However, $\|\mathbf{F}_2\| = \|\mathbf{F}\| \sin\theta$, so we can rewrite (15) as

$$\|\overrightarrow{PQ}\|\|\mathbf{F}\|\sin\theta = \|\overrightarrow{PQ} \times \mathbf{F}\|$$

This is called the ***scalar moment*** or ***torque*** of $\mathbf{F}$ about the point P. Scalar moments have units of force times distance—pound-feet or newton-meters, for example. The vector $\overrightarrow{PQ} \times \mathbf{F}$ is called the ***vector moment*** or ***torque vector*** of $\mathbf{F}$ about P.

Recalling that the direction of $\overrightarrow{PQ} \times \mathbf{F}$ is determined by the right-hand rule, it follows that the direction of rotation about P that results by applying the force $\mathbf{F}$ at the point Q is counterclockwise looking down the axis of $\overrightarrow{PQ} \times \mathbf{F}$ (Figure 12.4.7). Thus, the vector moment $\overrightarrow{PQ} \times \mathbf{F}$ captures the essential information about the rotational effect of the force—the magnitude of the cross product provides the scalar moment of the force, and the cross product vector itself provides the axis and direction of rotation.

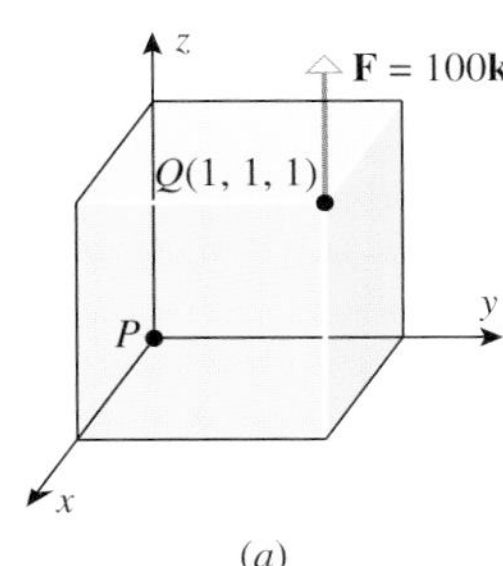

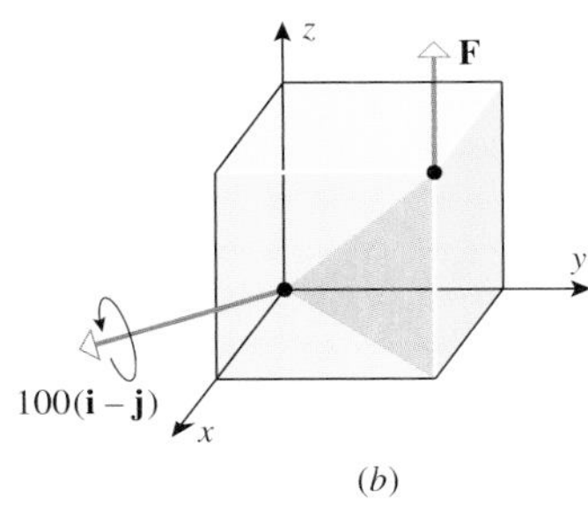

Figure 12.4.8

▶ **Example 6** Figure 12.4.8a shows a force $\mathbf{F}$ of 100 N applied in the positive z-direction at the point $Q(1, 1, 1)$ of a cube whose sides have a length of 1 m. Assuming that the cube is free to rotate about the point $P(0, 0, 0)$ (the origin), find the scalar moment of the force about P, and describe the direction of rotation.

Solution. The force vector is $\mathbf{F} = 100\mathbf{k}$, and the vector from P to Q is $\overrightarrow{PQ} = \mathbf{i} + \mathbf{j} + \mathbf{k}$, so the vector moment of $\mathbf{F}$ about P is

$$\overrightarrow{PQ} \times \mathbf{F} = \begin{vmatrix} \mathbf{i} & \mathbf{j} & \mathbf{k} \\ 1 & 1 & 1 \\ 0 & 0 & 100 \end{vmatrix} = 100\mathbf{i} - 100\mathbf{j}$$

Thus, the scalar moment of $\mathbf{F}$ about P is $\|100\mathbf{i} - 100\mathbf{j}\| = 100\sqrt{2} \approx 141$ N·m, and the direction of rotation is counterclockwise looking along the vector $100\mathbf{i} - 100\mathbf{j} = 100(\mathbf{i} - \mathbf{j})$ toward its initial point (Figure 12.4.8b). ◀

✔ QUICK CHECK EXERCISES 12.4 (See page 823 for answers.)

1. (a) $\begin{vmatrix} 3 & 2 \\ 4 & 5 \end{vmatrix} =$ ______ (b) $\begin{vmatrix} 3 & 2 & 1 \\ 3 & 2 & 1 \\ 5 & 5 & 5 \end{vmatrix} =$ ______

2. $\langle 1, 2, 0\rangle \times \langle 3, 0, 4\rangle =$ ______

3. Suppose that $\mathbf{u}$, $\mathbf{v}$, and $\mathbf{w}$ are vectors in 3-space such that $\mathbf{u} \times \mathbf{v} = \langle 2, 7, 3\rangle$ and $\mathbf{u} \times \mathbf{w} = \langle -5, 4, 0\rangle$.
 (a) $\mathbf{u} \times \mathbf{u} =$ ______ (b) $\mathbf{v} \times \mathbf{u} =$ ______
 (c) $\mathbf{u} \times (\mathbf{v} + \mathbf{w}) =$ ______
 (d) $\mathbf{u} \times (2\mathbf{w}) =$ ______

4. Let $\mathbf{u} = \mathbf{i} - 5\mathbf{k}$, $\mathbf{v} = 2\mathbf{i} - 4\mathbf{j} + \mathbf{k}$, and $\mathbf{w} = 3\mathbf{i} - 2\mathbf{j} + 5\mathbf{k}$.
 (a) $\mathbf{u} \cdot (\mathbf{v} \times \mathbf{w}) =$ ______
 (b) The volume of the parallelepiped that has $\mathbf{u}$, $\mathbf{v}$, and $\mathbf{w}$ as adjacent edges is $V =$ ______.

EXERCISE SET 12.4 [C] CAS

1. (a) Use a determinant to find the cross product

$$\mathbf{i} \times (\mathbf{i} + \mathbf{j} + \mathbf{k})$$

 (b) Check your answer in part (a) by rewriting the cross product as

$$\mathbf{i} \times (\mathbf{i} + \mathbf{j} + \mathbf{k}) = (\mathbf{i} \times \mathbf{i}) + (\mathbf{i} \times \mathbf{j}) + (\mathbf{i} \times \mathbf{k})$$

 and evaluating each term.

2. In each part, use the two methods in Exercise 1 to find
 (a) $\mathbf{j} \times (\mathbf{i} + \mathbf{j} + \mathbf{k})$ (b) $\mathbf{k} \times (\mathbf{i} + \mathbf{j} + \mathbf{k})$.

3–6 Find $\mathbf{u} \times \mathbf{v}$ and check that it is orthogonal to both $\mathbf{u}$ and $\mathbf{v}$.

3. $\mathbf{u} = \langle 1, 2, -3\rangle$, $\mathbf{v} = \langle -4, 1, 2\rangle$

4. $\mathbf{u} = 3\mathbf{i} + 2\mathbf{j} - \mathbf{k}$, $\mathbf{v} = -\mathbf{i} - 3\mathbf{j} + \mathbf{k}$

5. $\mathbf{u} = \langle 0, 1, -2\rangle$, $\mathbf{v} = \langle 3, 0, -4\rangle$

6. $\mathbf{u} = 4\mathbf{i} + \mathbf{k}$, $\mathbf{v} = 2\mathbf{i} - \mathbf{j}$

7. Let $\mathbf{u} = \langle 2, -1, 3\rangle$, $\mathbf{v} = \langle 0, 1, 7\rangle$, and $\mathbf{w} = \langle 1, 4, 5\rangle$. Find
(a) $\mathbf{u} \times (\mathbf{v} \times \mathbf{w})$
(b) $(\mathbf{u} \times \mathbf{v}) \times \mathbf{w}$
(c) $(\mathbf{u} \times \mathbf{v}) \times (\mathbf{v} \times \mathbf{w})$
(d) $(\mathbf{v} \times \mathbf{w}) \times (\mathbf{u} \times \mathbf{v})$.

C **8.** Use a CAS or a calculating utility that can compute determinants or cross products to solve Exercise 7.

9. Find the direction cosines of $\mathbf{u} \times \mathbf{v}$ for the vectors $\mathbf{u}$ and $\mathbf{v}$ in the accompanying figure.

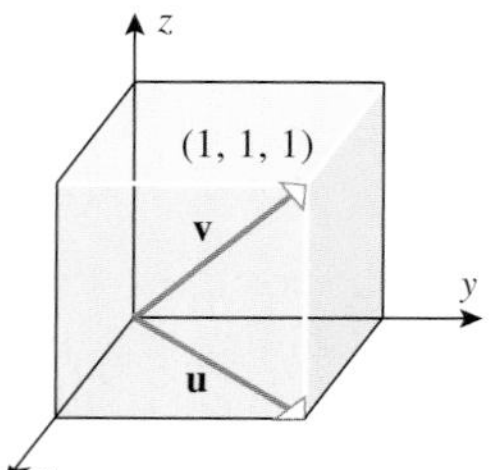

Figure Ex-9

10. Find two unit vectors that are orthogonal to both

$$\mathbf{u} = -7\mathbf{i} + 3\mathbf{j} + \mathbf{k}, \quad \mathbf{v} = 2\mathbf{i} + 4\mathbf{k}$$

11. Find two unit vectors that are normal to the plane determined by the points $A(0, -2, 1)$, $B(1, -1, -2)$, and $C(-1, 1, 0)$.

12. Find two unit vectors that are parallel to the yz-plane and are orthogonal to the vector $3\mathbf{i} - \mathbf{j} + 2\mathbf{k}$.

13–14 Find the area of the parallelogram that has $\mathbf{u}$ and $\mathbf{v}$ as adjacent sides.

13. $\mathbf{u} = \mathbf{i} - \mathbf{j} + 2\mathbf{k}$, $\mathbf{v} = 3\mathbf{j} + \mathbf{k}$

14. $\mathbf{u} = 2\mathbf{i} + 3\mathbf{j}$, $\mathbf{v} = -\mathbf{i} + 2\mathbf{j} - 2\mathbf{k}$

15–16 Find the area of the triangle with vertices P, Q, and R.

15. $P(1, 5, -2)$, $Q(0, 0, 0)$, $R(3, 5, 1)$

16. $P(2, 0, -3)$, $Q(1, 4, 5)$, $R(7, 2, 9)$

17–20 Find $\mathbf{u} \cdot (\mathbf{v} \times \mathbf{w})$.

17. $\mathbf{u} = 2\mathbf{i} - 3\mathbf{j} + \mathbf{k}$, $\mathbf{v} = 4\mathbf{i} + \mathbf{j} - 3\mathbf{k}$, $\mathbf{w} = \mathbf{j} + 5\mathbf{k}$

18. $\mathbf{u} = \langle 1, -2, 2\rangle$, $\mathbf{v} = \langle 0, 3, 2\rangle$, $\mathbf{w} = \langle -4, 1, -3\rangle$

19. $\mathbf{u} = \langle 2, 1, 0\rangle$, $\mathbf{v} = \langle 1, -3, 1\rangle$, $\mathbf{w} = \langle 4, 0, 1\rangle$

20. $\mathbf{u} = \mathbf{i}$, $\mathbf{v} = \mathbf{i} + \mathbf{j}$, $\mathbf{w} = \mathbf{i} + \mathbf{j} + \mathbf{k}$

21–22 Use a scalar triple product to find the volume of the parallelepiped that has $\mathbf{u}$, $\mathbf{v}$, and $\mathbf{w}$ as adjacent edges.

21. $\mathbf{u} = \langle 2, -6, 2\rangle$, $\mathbf{v} = \langle 0, 4, -2\rangle$, $\mathbf{w} = \langle 2, 2, -4\rangle$

22. $\mathbf{u} = 3\mathbf{i} + \mathbf{j} + 2\mathbf{k}$, $\mathbf{v} = 4\mathbf{i} + 5\mathbf{j} + \mathbf{k}$, $\mathbf{w} = \mathbf{i} + 2\mathbf{j} + 4\mathbf{k}$

23. In each part, use a scalar triple product to determine whether the vectors lie in the same plane.
(a) $\mathbf{u} = \langle 1, -2, 1\rangle$, $\mathbf{v} = \langle 3, 0, -2\rangle$, $\mathbf{w} = \langle 5, -4, 0\rangle$
(b) $\mathbf{u} = 5\mathbf{i} - 2\mathbf{j} + \mathbf{k}$, $\mathbf{v} = 4\mathbf{i} - \mathbf{j} + \mathbf{k}$, $\mathbf{w} = \mathbf{i} - \mathbf{j}$
(c) $\mathbf{u} = \langle 4, -8, 1\rangle$, $\mathbf{v} = \langle 2, 1, -2\rangle$, $\mathbf{w} = \langle 3, -4, 12\rangle$

24. Suppose that $\mathbf{u} \cdot (\mathbf{v} \times \mathbf{w}) = 3$. Find
(a) $\mathbf{u} \cdot (\mathbf{w} \times \mathbf{v})$
(b) $(\mathbf{v} \times \mathbf{w}) \cdot \mathbf{u}$
(c) $\mathbf{w} \cdot (\mathbf{u} \times \mathbf{v})$
(d) $\mathbf{v} \cdot (\mathbf{u} \times \mathbf{w})$
(e) $(\mathbf{u} \times \mathbf{w}) \cdot \mathbf{v}$
(f) $\mathbf{v} \cdot (\mathbf{w} \times \mathbf{w})$.

25. Consider the parallelepiped with adjacent edges

$$\mathbf{u} = 3\mathbf{i} + 2\mathbf{j} + \mathbf{k}$$
$$\mathbf{v} = \mathbf{i} + \mathbf{j} + 2\mathbf{k}$$
$$\mathbf{w} = \mathbf{i} + 3\mathbf{j} + 3\mathbf{k}$$

(a) Find the volume.
(b) Find the area of the face determined by $\mathbf{u}$ and $\mathbf{w}$.
(c) Find the angle between $\mathbf{u}$ and the plane containing the face determined by $\mathbf{v}$ and $\mathbf{w}$.

26. Show that in 3-space the distance d from a point P to the line L through points A and B can be expressed as

$$d = \frac{\|\overrightarrow{AP} \times \overrightarrow{AB}\|}{\|\overrightarrow{AB}\|}$$

27. Use the result in Exercise 26 to find the distance between the point P and the line through the points A and B.
(a) $P(-3, 1, 2)$, $A(1, 1, 0)$, $B(-2, 3, -4)$
(b) $P(4, 3)$, $A(2, 1)$, $B(0, 2)$

28. It is a theorem of solid geometry that the volume of a tetrahedron is $\frac{1}{3}$(area of base) $\cdot$ (height). Use this result to prove that the volume of a tetrahedron with adjacent edges given by the vectors $\mathbf{u}$, $\mathbf{v}$, and $\mathbf{w}$ is $\frac{1}{6}|\mathbf{u} \cdot (\mathbf{v} \times \mathbf{w})|$.

29. Use the result of Exercise 28 to find the volume of the tetrahedron with vertices

$$P(-1, 2, 0), \quad Q(2, 1, -3), \quad R(1, 0, 1), \quad S(3, -2, 3)$$

30. Let θ be the angle between the vectors $\mathbf{u} = 2\mathbf{i} + 3\mathbf{j} - 6\mathbf{k}$ and $\mathbf{v} = 2\mathbf{i} + 3\mathbf{j} + 6\mathbf{k}$.
(a) Use the dot product to find $\cos\theta$.
(b) Use the cross product to find $\sin\theta$.
(c) Confirm that $\sin^2\theta + \cos^2\theta = 1$.

FOCUS ON CONCEPTS

31. Let A, B, C and D be four distinct points in 3-space. If $\overrightarrow{AB} \times \overrightarrow{CD} \neq \mathbf{0}$ and $\overrightarrow{AC} \cdot (\overrightarrow{AB} \times \overrightarrow{CD}) = 0$, explain why the line through A and B must intersect the line through C and D.

32. Let A, B, and C be three distinct noncollinear points in 3-space. Describe the set of all points P that satisfy the vector equation $\overrightarrow{AP} \cdot (\overrightarrow{AB} \times \overrightarrow{AC}) = 0$.

33. What can you say about the angle between nonzero vectors $\mathbf{u}$ and $\mathbf{v}$ if $\mathbf{u} \cdot \mathbf{v} = \|\mathbf{u} \times \mathbf{v}\|$?

34. Show that if $\mathbf{u}$ and $\mathbf{v}$ are vectors in 3-space, then

$$\|\mathbf{u} \times \mathbf{v}\|^2 = \|\mathbf{u}\|^2\|\mathbf{v}\|^2 - (\mathbf{u} \cdot \mathbf{v})^2$$

[*Note:* This result is sometimes called ***Lagrange's identity***.]

35. The accompanying figure shows a force **F** of 10 lb applied in the positive y-direction to the point $Q(1, 1, 1)$ of a cube whose sides have a length of 1 ft. In each part, find the scalar moment of **F** about the point P, and describe the direction of rotation, if any, if the cube is free to rotate about P.
(a) P is the point $(0, 0, 0)$. (b) P is the point $(1, 0, 0)$.
(c) P is the point $(1, 0, 1)$.

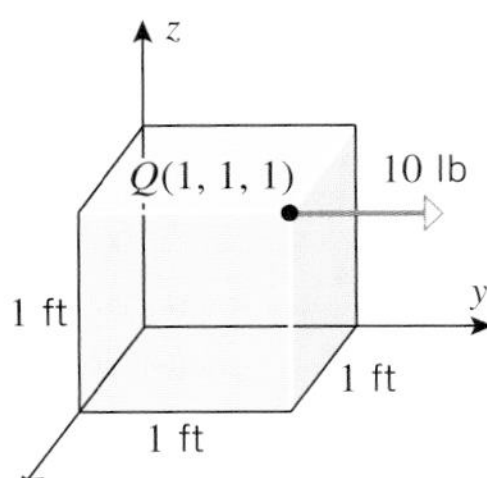

Figure Ex-35

36. The accompanying figure shows a force **F** of 1000 N applied to the corner of a box.
(a) Find the scalar moment of **F** about the point P.
(b) Find the direction angles of the vector moment of **F** about the point P to the nearest degree.

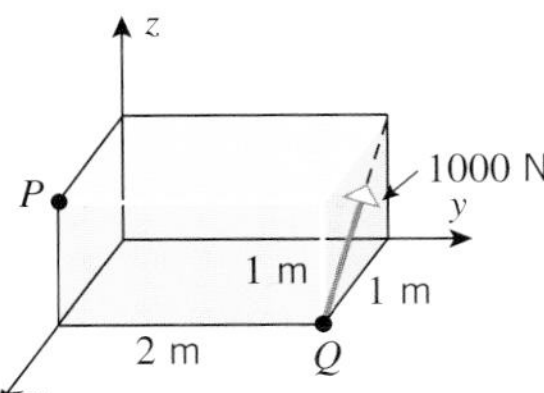

Figure Ex-36

37. As shown in the accompanying figure, a force of 200 N is applied at an angle of 18° to a point near the end of a monkey wrench. Find the scalar moment of the force about the center of the bolt. [Treat this as a problem in two dimensions.]

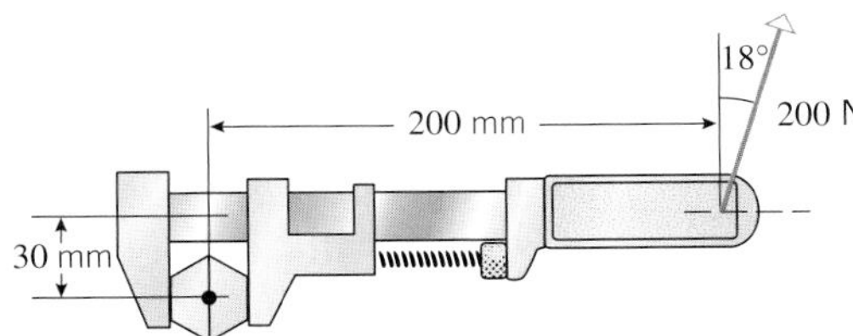

Figure Ex-37

38. Prove parts (*b*) and (*c*) of Theorem 12.4.3.

39. Prove parts (*d*) and (*e*) of Theorem 12.4.3.

40. Prove part (*b*) of Theorem 12.4.1 for 3×3 determinants. [Just give the proof for the first two rows.] Then use (*b*) to prove (*a*).

FOCUS ON CONCEPTS

41. Expressions of the form

$$\mathbf{u} \times (\mathbf{v} \times \mathbf{w}) \quad \text{and} \quad (\mathbf{u} \times \mathbf{v}) \times \mathbf{w}$$

are called ***vector triple products***. It can be proved with some effort that

$$\mathbf{u} \times (\mathbf{v} \times \mathbf{w}) = (\mathbf{u} \cdot \mathbf{w})\mathbf{v} - (\mathbf{u} \cdot \mathbf{v})\mathbf{w}$$
$$(\mathbf{u} \times \mathbf{v}) \times \mathbf{w} = (\mathbf{w} \cdot \mathbf{u})\mathbf{v} - (\mathbf{w} \cdot \mathbf{v})\mathbf{u}$$

These expressions can be summarized with the following mnemonic rule:

vector triple product = (outer · remote)adjacent − (outer · adjacent)remote

See if you can figure out what the expressions "outer," "remote," and "adjacent" mean in this rule, and then use the rule to find the two vector triple products of the vectors

$$\mathbf{u} = \mathbf{i} + 3\mathbf{j} - \mathbf{k}, \quad \mathbf{v} = \mathbf{i} + \mathbf{j} + 2\mathbf{k}, \quad \mathbf{w} = 3\mathbf{i} - \mathbf{j} + 2\mathbf{k}$$

42. (a) Use the result in Exercise 41 to show that $\mathbf{u} \times (\mathbf{v} \times \mathbf{w})$ lies in the same plane as **v** and **w**, and $(\mathbf{u} \times \mathbf{v}) \times \mathbf{w}$ lies in the same plane as **u** and **v**.
(b) Use a geometrical argument to justify the results in part (a).

43. In each part, use the result in Exercise 41 to prove the vector identity.
(a) $(\mathbf{a} \times \mathbf{b}) \times (\mathbf{c} \times \mathbf{d}) = (\mathbf{a} \times \mathbf{b} \cdot \mathbf{d})\mathbf{c} - (\mathbf{a} \times \mathbf{b} \cdot \mathbf{c})\mathbf{d}$
(b) $(\mathbf{a} \times \mathbf{b}) \times \mathbf{c} + (\mathbf{b} \times \mathbf{c}) \times \mathbf{a} + (\mathbf{c} \times \mathbf{a}) \times \mathbf{b} = 0$

44. Prove: If **a**, **b**, **c**, and **d** lie in the same plane when positioned with a common initial point, then

$$(\mathbf{a} \times \mathbf{b}) \times (\mathbf{c} \times \mathbf{d}) = \mathbf{0}$$

C **45.** Use a CAS to approximate the minimum area of a triangle if two of its vertices are $(2, -1, 0)$ and $(3, 2, 2)$ and its third vertex is on the curve $y = \ln x$ in the xy-plane.

46. If a force **F** is applied to an object at a point Q, then the line through Q parallel to **F** is called the ***line of action*** of the force. We defined the vector moment of **F** about a point P to be $\overrightarrow{PQ} \times \mathbf{F}$. Show that if Q' is any point on the line of action of **F**, then $\overrightarrow{PQ} \times \mathbf{F} = \overrightarrow{PQ'} \times \mathbf{F}$; that is, it is not essential to use the point of application to compute the vector moment—any point on the line of action will do. [*Hint:* Write $\overrightarrow{PQ'} = \overrightarrow{PQ} + \overrightarrow{QQ'}$ and use properties of the cross product.]

✔QUICK CHECK ANSWERS 12.4

1. (a) 7 (b) 0 **2.** $8\mathbf{i} - 4\mathbf{j} - 6\mathbf{k}$ **3.** (a) $\langle 0, 0, 0\rangle$ (b) $\langle -2, -7, -3\rangle$ (c) $\langle -3, 11, 3\rangle$ (d) $\langle -10, 8, 0\rangle$ **4.** (a) -58 (b) 58

12.5 PARAMETRIC EQUATIONS OF LINES

In this section we will discuss parametric equations of lines in 2-space and 3-space. In 3-space, parametric equations of lines are especially important because they generally provide the most convenient form for representing lines algebraically.

LINES DETERMINED BY A POINT AND A VECTOR

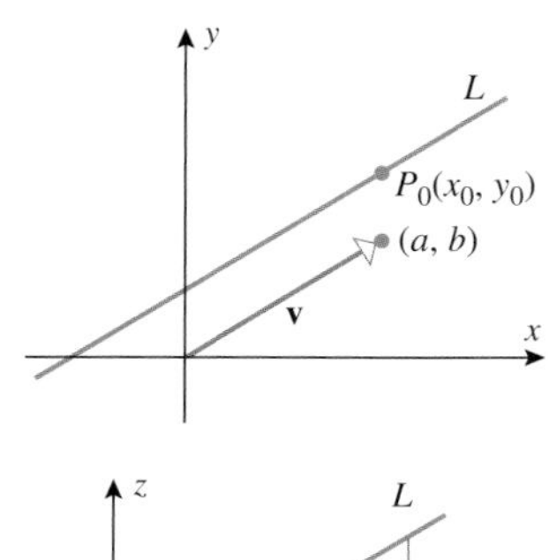

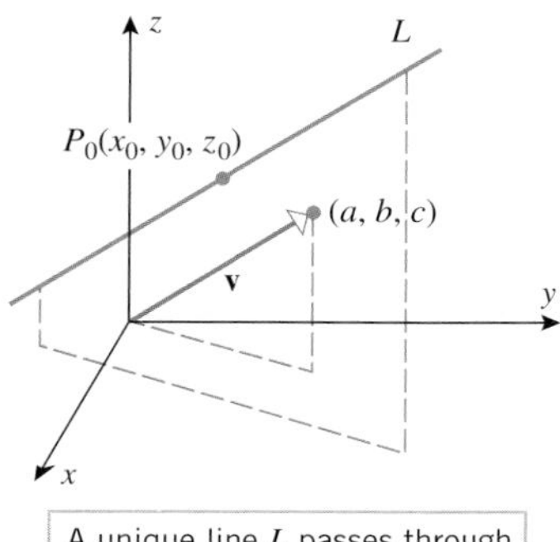

A unique line L passes through P_0 and is parallel to $\mathbf{v}$.

Figure 12.5.1

A line in 2-space or 3-space can be determined uniquely by specifying a point on the line and a nonzero vector parallel to the line (Figure 12.5.1). For example, consider a line L in 3-space that passes through the point $P_0(x_0, y_0, z_0)$ and is parallel to the nonzero vector $\mathbf{v} = \langle a, b, c \rangle$. Then L consists precisely of those points $P(x, y, z)$ for which the vector $\overrightarrow{P_0P}$ is parallel to $\mathbf{v}$ (Figure 12.5.2). In other words, the point $P(x, y, z)$ is on L if and only if $\overrightarrow{P_0P}$ is a scalar multiple of $\mathbf{v}$, say

$$\overrightarrow{P_0P} = t\mathbf{v}$$

This equation can be written as

$$\langle x - x_0, y - y_0, z - z_0 \rangle = \langle ta, tb, tc \rangle$$

which implies that

$$x - x_0 = ta, \quad y - y_0 = tb, \quad z - z_0 = tc$$

Thus, L can be described by the parametric equations

$$x = x_0 + at, \quad y = y_0 + bt, \quad z = z_0 + ct$$

A similar description applies to lines in 2-space. We summarize these descriptions in the following theorem.

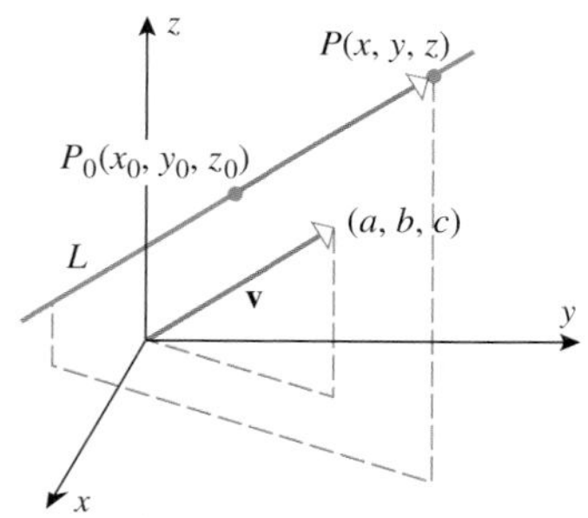

Figure 12.5.2

12.5.1 THEOREM.

(a) *The line in 2-space that passes through the point $P_0(x_0, y_0)$ and is parallel to the nonzero vector $\mathbf{v} = \langle a, b \rangle = a\mathbf{i} + b\mathbf{j}$ has parametric equations*

$$x = x_0 + at, \quad y = y_0 + bt \tag{1}$$

(b) *The line in 3-space that passes through the point $P_0(x_0, y_0, z_0)$ and is parallel to the nonzero vector $\mathbf{v} = \langle a, b, c \rangle = a\mathbf{i} + b\mathbf{j} + c\mathbf{k}$ has parametric equations*

$$x = x_0 + at, \quad y = y_0 + bt, \quad z = z_0 + ct \tag{2}$$

Although it is not stated explicitly, it is understood in Equations (1) and (2) that $-\infty < t < +\infty$, which reflects the fact that lines extend indefinitely.

► **Example 1** Find parametric equations of the line

(a) passing through (4, 2) and parallel to $\mathbf{v} = \langle -1, 5 \rangle$;

(b) passing through $(1, 2, -3)$ and parallel to $\mathbf{v} = 4\mathbf{i} + 5\mathbf{j} - 7\mathbf{k}$;

(c) passing through the origin in 3-space and parallel to $\mathbf{v} = \langle 1, 1, 1 \rangle$.

Solution (a). From (1) with $x_0 = 4$, $y_0 = 2$, $a = -1$, and $b = 5$ we obtain

$$x = 4 - t, \quad y = 2 + 5t$$

Solution (b). From (2) we obtain

$$x = 1 + 4t, \quad y = 2 + 5t, \quad z = -3 - 7t$$

Solution (c). From (2) with $x_0 = 0$, $y_0 = 0$, $z_0 = 0$, $a = 1$, $b = 1$, and $c = 1$ we obtain

$$x = t, \quad y = t, \quad z = t \blacktriangleleft$$

► Example 2

(a) Find parametric equations of the line L passing through the points $P_1(2, 4, -1)$ and $P_2(5, 0, 7)$.

(b) Where does the line intersect the xy-plane?

Solution (a). The vector $\overrightarrow{P_1P_2} = \langle 3, -4, 8\rangle$ is parallel to L and the point $P_1(2, 4, -1)$ lies on L, so it follows from (2) that L has parametric equations

$$x = 2 + 3t, \quad y = 4 - 4t, \quad z = -1 + 8t \tag{3}$$

Had we used P_2 as the point on L rather than P_1, we would have obtained the equations

$$x = 5 + 3t, \quad y = -4t, \quad z = 7 + 8t$$

Although these equations look different from those obtained using P_1, the two sets of equations are actually equivalent in that both generate L as t varies from $-\infty$ to $+\infty$. To see this, note that if t_1 gives a point

$$(x, y, z) = (2 + 3t_1, 4 - 4t_1, -1 + 8t_1)$$

on L using the first set of equations, then $t_2 = t_1 - 1$ gives the *same* point

$$\begin{aligned}(x, y, z) &= (5 + 3t_2, -4t_2, 7 + 8t_2)\\ &= (5 + 3(t_1 - 1), -4(t_1 - 1), 7 + 8(t_1 - 1))\\ &= (2 + 3t_1, 4 - 4t_1, -1 + 8t_1)\end{aligned}$$

on L using the second set of equations. Conversely, if t_2 gives a point on L using the second set of equations, then $t_1 = t_2 + 1$ gives the same point using the first set.

Solution (b). It follows from (3) in part (a) that the line intersects the xy-plane at the point where $z = -1 + 8t = 0$, that is, when $t = \frac{1}{8}$. Substituting this value of t in (3) yields the point of intersection $(x, y, z) = \left(\frac{19}{8}, \frac{7}{2}, 0\right)$. ◄

► Example 3

Let L_1 and L_2 be the lines

$$\begin{aligned}L_1\colon x &= 1 + 4t, \quad y = 5 - 4t, \quad z = -1 + 5t\\ L_2\colon x &= 2 + 8t, \quad y = 4 - 3t, \quad z = 5 + t\end{aligned}$$

(a) Are the lines parallel?

(b) Do the lines intersect?

Solution (a). The line L_1 is parallel to the vector $4\mathbf{i} - 4\mathbf{j} + 5\mathbf{k}$, and the line L_2 is parallel to the vector $8\mathbf{i} - 3\mathbf{j} + \mathbf{k}$. These vectors are not parallel since neither is a scalar multiple of the other. Thus, the lines are not parallel.

Solution (b). For L_1 and L_2 to intersect at some point (x_0, y_0, z_0) these coordinates would have to satisfy the equations of both lines. In other words, there would have to exist values t_1 and t_2 for the parameters such that

$$x_0 = 1 + 4t_1, \quad y_0 = 5 - 4t_1, \quad z_0 = -1 + 5t_1$$

and

$$x_0 = 2 + 8t_2, \quad y_0 = 4 - 3t_2, \quad z_0 = 5 + t_2$$

This leads to three conditions on t_1 and t_2,

$$\begin{aligned} 1 + 4t_1 &= 2 + 8t_2 \\ 5 - 4t_1 &= 4 - 3t_2 \\ -1 + 5t_1 &= 5 + t_2 \end{aligned} \tag{4}$$

Thus, the lines intersect if there are values of t_1 and t_2 that satisfy all three equations, and the lines do not intersect if there are no such values. You should be familiar with methods for solving systems of two linear equations in two unknowns; however, this is a system of three linear equations in two unknowns. To determine whether this system has a solution we will solve the first two equations for t_1 and t_2 and then check whether these values satisfy the third equation.

We will solve the first two equations by the method of elimination. We can eliminate the unknown t_1 by adding the equations. This yields the equation

$$6 = 6 + 5t_2$$

from which we obtain $t_2 = 0$. We can now find t_1 by substituting this value of t_2 in either the first or second equation. This yields $t_1 = \frac{1}{4}$. However, the values $t_1 = \frac{1}{4}$ and $t_2 = 0$ do not satisfy the third equation in (4), so the lines do not intersect. ◄

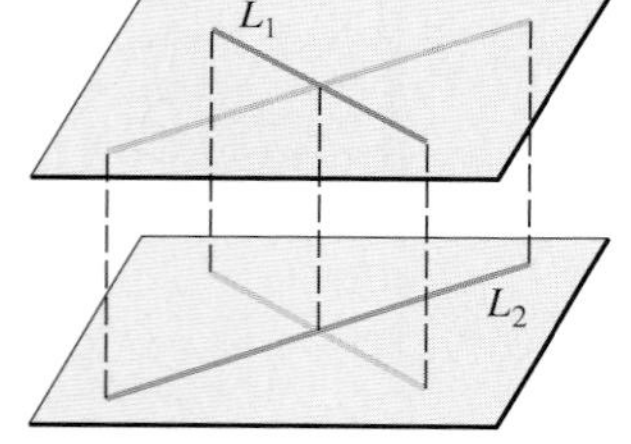

Parallel planes containing skew lines L_1 and L_2 can be determined by translating each line until it intersects the other.

Figure 12.5.3

Two lines in 3-space that are not parallel and do not intersect (such as those in Example 3) are called ***skew*** lines. As illustrated in Figure 12.5.3, any two skew lines lie in parallel planes.

LINE SEGMENTS

Sometimes one is not interested in an entire line, but rather some *segment* of a line. Parametric equations of a line segment can be obtained by finding parametric equations for the entire line, then restricting the parameter appropriately so that only the desired segment is generated.

► **Example 4** Find parametric equations describing the line segment joining the points $P_1(2, 4, -1)$ and $P_2(5, 0, 7)$.

Solution. From Example 2, the line through the points P_1 and P_2 has parametric equations $x = 2 + 3t$, $y = 4 - 4t$, $z = -1 + 8t$. With these equations, the point P_1 corresponds to $t = 0$ and P_2 to $t = 1$. Thus, the line segment that joins P_1 and P_2 is given by

$$x = 2 + 3t, \quad y = 4 - 4t, \quad z = -1 + 8t \qquad (0 \le t \le 1)$$ ◄

VECTOR EQUATIONS OF LINES

We will now show how vector notation can be used to express the parametric equations of a line more compactly. Because two vectors are equal if and only if their components are equal, (1) and (2) can be written in vector form as

$$\langle x, y\rangle = \langle x_0 + at, y_0 + bt\rangle$$
$$\langle x, y, z\rangle = \langle x_0 + at, y_0 + bt, z_0 + ct\rangle$$

or, equivalently, as

$$\langle x, y\rangle = \langle x_0, y_0\rangle + t\langle a, b\rangle \tag{5}$$
$$\langle x, y, z\rangle = \langle x_0, y_0, z_0\rangle + t\langle a, b, c\rangle \tag{6}$$

For the equation in 2-space we define the vectors $\mathbf{r}$, $\mathbf{r}_0$, and $\mathbf{v}$ as

$$\mathbf{r} = \langle x, y\rangle, \quad \mathbf{r}_0 = \langle x_0, y_0\rangle, \quad \mathbf{v} = \langle a, b\rangle \tag{7}$$

and for the equation in 3-space we define them as

$$\mathbf{r} = \langle x, y, z\rangle, \quad \mathbf{r}_0 = \langle x_0, y_0, z_0\rangle, \quad \mathbf{v} = \langle a, b, c\rangle \tag{8}$$

Substituting (7) and (8) in (5) and (6), respectively, yields the equation

$$\mathbf{r} = \mathbf{r}_0 + t\mathbf{v} \tag{9}$$

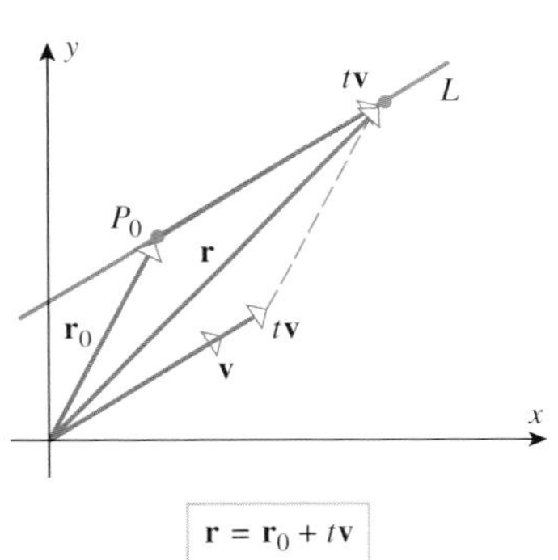

Figure 12.5.4

in both cases. We call this the ***vector equation of a line*** in 2-space or 3-space. In this equation, $\mathbf{v}$ is a nonzero vector parallel to the line, and $\mathbf{r}_0$ is a vector whose components are the coordinates of a point on the line.

We can interpret Equation (9) geometrically by positioning the vectors $\mathbf{r}_0$ and $\mathbf{v}$ with their initial points at the origin and the vector $t\mathbf{v}$ with its initial point at P_0 (Figure 12.5.4). The vector $t\mathbf{v}$ is a scalar multiple of $\mathbf{v}$ and hence is parallel to $\mathbf{v}$ and L. Moreover, since the initial point of $t\mathbf{v}$ is at the point P_0 on L, this vector actually runs along L; hence, the vector $\mathbf{r} = \mathbf{r}_0 + t\mathbf{v}$ can be interpreted as the vector from the origin to a point on L. As the parameter t varies from 0 to $+\infty$, the terminal point of $\mathbf{r}$ traces out the portion of L that extends from P_0 in the direction of $\mathbf{v}$, and as t varies from 0 to $-\infty$, the terminal point of $\mathbf{r}$ traces out the portion of L that extends from P_0 in the direction that is opposite to $\mathbf{v}$. Thus, the entire line is traced as t varies over the interval $(-\infty, +\infty)$, and it is traced in the direction of $\mathbf{v}$ as t increases.

▶ **Example 5** The equation

$$\langle x, y, z\rangle = \langle -1, 0, 2\rangle + t\langle 1, 5, -4\rangle$$

is of form (9) with

$$\mathbf{r}_0 = \langle -1, 0, 2\rangle \quad \text{and} \quad \mathbf{v} = \langle 1, 5, -4\rangle$$

Thus, the equation represents the line in 3-space that passes through the point $(-1, 0, 2)$ and is parallel to the vector $\langle 1, 5, -4\rangle$. ◀

▶ **Example 6** Find an equation of the line in 3-space that passes through the points $P_1(2, 4, -1)$ and $P_2(5, 0, 7)$.

Solution. The vector

$$\overrightarrow{P_1P_2} = \langle 3, -4, 8\rangle$$

is parallel to the line, so it can be used as $\mathbf{v}$ in (9). For $\mathbf{r}_0$ we can use either the vector from the origin to P_1 or the vector from the origin to P_2. Using the former yields

$$\mathbf{r}_0 = \langle 2, 4, -1\rangle$$

Thus, a vector equation of the line through P_1 and P_2 is

$$\langle x, y, z\rangle = \langle 2, 4, -1\rangle + t\langle 3, -4, 8\rangle$$

If needed, we can express the line parametrically by equating corresponding components on the two sides of this vector equation, in which case we obtain the parametric equations in Example 2 (verify). ◄

✔QUICK CHECK EXERCISES 12.5 (See page 830 for answers.)

1. Let L be the line through (2, 5) and parallel to $\mathbf{v} = \langle 3, -1\rangle$.
   (a) Parametric equations of L are
   $x =$ ______ $y =$ ______
   (b) A vector equation of L is $\langle x, y\rangle =$ ______.
2. Parametric equations for the line through (5, 3, 7) and parallel to the line $x = 3 - t$, $y = 2$, $z = 8 + 4t$ are
   $x =$ ______, $y =$ ______, $z =$ ______
3. Parametric equations for the line segment joining the points (3, 0, 11) and (2, 6, 7) are
   $x =$ ______, $y =$ ______, $z =$ ______ (______)
4. The line through the points $(-3, 8, -4)$ and $(1, 0, 8)$ intersects the yz-plane at ______.

EXERCISE SET 12.5 Graphing Utility C CAS

1. (a) Find parametric equations for the lines through the corner of the unit square shown in part (a) of the accompanying figure.
   (b) Find parametric equations for the lines through the corner of the unit cube shown in part (b) of the accompanying figure.

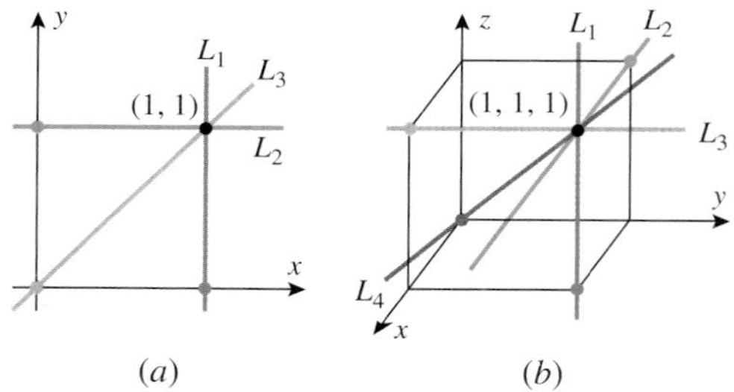

Figure Ex-1

2. (a) Find parametric equations for the line segments in the unit square in part (a) of the accompanying figure.
   (b) Find parametric equations for the line segments in the unit cube shown in part (b) of the accompanying figure.

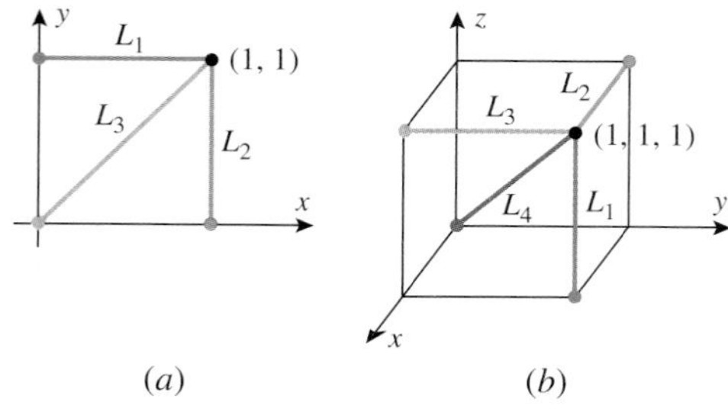

Figure Ex-2

3–4 Find parametric equations for the line through P_1 and P_2 and also for the line segment joining those points.

3. (a) $P_1(3, -2)$, $P_2(5, 1)$ (b) $P_1(5, -2, 1)$, $P_2(2, 4, 2)$
4. (a) $P_1(0, 1)$, $P_2(-3, -4)$
   (b) $P_1(-1, 3, 5)$, $P_2(-1, 3, 2)$

5–6 Find parametric equations for the line whose vector equation is given.

5. (a) $\langle x, y\rangle = \langle 2, -3\rangle + t\langle 1, -4\rangle$
   (b) $x\mathbf{i} + y\mathbf{j} + z\mathbf{k} = \mathbf{k} + t(\mathbf{i} - \mathbf{j} + \mathbf{k})$
6. (a) $x\mathbf{i} + y\mathbf{j} = (3\mathbf{i} - 4\mathbf{j}) + t(2\mathbf{i} + \mathbf{j})$
   (b) $\langle x, y, z\rangle = \langle -1, 0, 2\rangle + t\langle -1, 3, 0\rangle$

7–8 Find a point P on the line and a vector $\mathbf{v}$ parallel to the line by inspection.

7. (a) $x\mathbf{i} + y\mathbf{j} = (2\mathbf{i} - \mathbf{j}) + t(4\mathbf{i} - \mathbf{j})$
   (b) $\langle x, y, z\rangle = \langle -1, 2, 4\rangle + t\langle 5, 7, -8\rangle$
8. (a) $\langle x, y\rangle = \langle -1, 5\rangle + t\langle 2, 3\rangle$
   (b) $x\mathbf{i} + y\mathbf{j} + z\mathbf{k} = (\mathbf{i} + \mathbf{j} - 2\mathbf{k}) + t\mathbf{j}$

9–10 Express the given parametric equations of a line using bracket notation and also using $\mathbf{i}, \mathbf{j}, \mathbf{k}$ notation.

9. (a) $x = -3 + t$, $y = 4 + 5t$
   (b) $x = 2 - t$, $y = -3 + 5t$, $z = t$
10. (a) $x = t$, $y = -2 + t$
    (b) $x = 1 + t$, $y = -7 + 3t$, $z = 4 - 5t$

11–18 Find parametric equations of the line that satisfies the stated conditions.

11. The line through $(-5, 2)$ that is parallel to $2\mathbf{i} - 3\mathbf{j}$.
12. The line through $(0, 3)$ that is parallel to the line $x = -5 + t$, $y = 1 - 2t$.
13. The line that is tangent to the circle $x^2 + y^2 = 25$ at the point $(3, -4)$.
14. The line that is tangent to the parabola $y = x^2$ at the point $(-2, 4)$.
15. The line through $(-1, 2, 4)$ that is parallel to $3\mathbf{i} - 4\mathbf{j} + \mathbf{k}$.
16. The line through $(2, -1, 5)$ that is parallel to $\langle -1, 2, 7\rangle$.
17. The line through $(-2, 0, 5)$ that is parallel to the line given by $x = 1 + 2t$, $y = 4 - t$, $z = 6 + 2t$.
18. The line through the origin that is parallel to the line given by $x = t$, $y = -1 + t$, $z = 2$.
19. Where does the line $x = 1 + 3t$, $y = 2 - t$ intersect
    (a) the x-axis (b) the y-axis
    (c) the parabola $y = x^2$?
20. Where does the line $\langle x, y\rangle = \langle 4t, 3t\rangle$ intersect the circle $x^2 + y^2 = 25$?

21–22 Find the intersections of the lines with the xy-plane, the xz-plane, and the yz-plane.

21. $x = -2,\ y = 4 + 2t,\ z = -3 + t$
22. $x = -1 + 2t,\ y = 3 + t,\ z = 4 - t$
23. Where does the line $x = 1 + t$, $y = 3 - t$, $z = 2t$ intersect the cylinder $x^2 + y^2 = 16$?
24. Where does the line $x = 2 - t$, $y = 3t$, $z = -1 + 2t$ intersect the plane $2y + 3z = 6$?

25–26 Show that the lines L_1 and L_2 intersect, and find their point of intersection.

25. $L_1\colon x = 2 + t,\ y = 2 + 3t,\ z = 3 + t$
    $L_2\colon x = 2 + t,\ y = 3 + 4t,\ z = 4 + 2t$
26. $L_1\colon x + 1 = 4t,\ y - 3 = t,\ z - 1 = 0$
    $L_2\colon x + 13 = 12t,\ y - 1 = 6t,\ z - 2 = 3t$

27–28 Show that the lines L_1 and L_2 are skew.

27. $L_1\colon x = 1 + 7t,\ y = 3 + t,\ z = 5 - 3t$
    $L_2\colon x = 4 - t,\ y = 6,\ z = 7 + 2t$
28. $L_1\colon x = 2 + 8t,\ y = 6 - 8t,\ z = 10t$
    $L_2\colon x = 3 + 8t,\ y = 5 - 3t,\ z = 6 + t$

29–30 Determine whether the lines L_1 and L_2 are parallel.

29. $L_1\colon x = 3 - 2t,\ y = 4 + t,\ z = 6 - t$
    $L_2\colon x = 5 - 4t,\ y = -2 + 2t,\ z = 7 - 2t$
30. $L_1\colon x = 5 + 3t,\ y = 4 - 2t,\ z = -2 + 3t$
    $L_2\colon x = -1 + 9t,\ y = 5 - 6t,\ z = 3 + 8t$

31–32 Determine whether the points P_1, P_2, and P_3 lie on the same line.

31. $P_1(6, 9, 7),\ P_2(9, 2, 0),\ P_3(0, -5, -3)$
32. $P_1(1, 0, 1),\ P_2(3, -4, -3),\ P_3(4, -6, -5)$

33–34 Show that the lines L_1 and L_2 are the same.

33. $L_1\colon x = 3 - t,\ y = 1 + 2t$
    $L_2\colon x = -1 + 3t,\ y = 9 - 6t$
34. $L_1\colon x = 1 + 3t,\ y = -2 + t,\ z = 2t$
    $L_2\colon x = 4 - 6t,\ y = -1 - 2t,\ z = 2 - 4t$

FOCUS ON CONCEPTS

35. Sketch the vectors $\mathbf{r}_0 = \langle -1, 2\rangle$ and $\mathbf{v} = \langle 1, 1\rangle$, and then sketch the six vectors $\mathbf{r}_0 \pm \mathbf{v}$, $\mathbf{r}_0 \pm 2\mathbf{v}$, $\mathbf{r}_0 \pm 3\mathbf{v}$. Draw the line $L : x = -1 + t,\ y = 2 + t$, and describe the relationship between L and the vectors you sketched. What is the vector equation of L?
36. Sketch the vectors $\mathbf{r}_0 = \langle 0, 2, 1\rangle$ and $\mathbf{v} = \langle 1, 0, 1\rangle$, and then sketch the vectors $\mathbf{r}_0 + \mathbf{v}$, $\mathbf{r}_0 + 2\mathbf{v}$, and $\mathbf{r}_0 + 3\mathbf{v}$. Draw the line $L : x = t,\ y = 2,\ z = 1 + t$, and describe the relationship between L and the vectors you sketched. What is the vector equation of L?
37. Sketch the vectors $\mathbf{r}_0 = \langle -2, 0\rangle$ and $\mathbf{r}_1 = \langle 1, 3\rangle$, and then sketch the vectors
    $$\tfrac{1}{3}\mathbf{r}_0 + \tfrac{2}{3}\mathbf{r}_1,\quad \tfrac{1}{2}\mathbf{r}_0 + \tfrac{1}{2}\mathbf{r}_1,\quad \tfrac{2}{3}\mathbf{r}_0 + \tfrac{1}{3}\mathbf{r}_1$$
    Draw the line segment $(1 - t)\mathbf{r}_0 + t\mathbf{r}_1$ $(0 \le t \le 1)$. If n is a positive integer, what is the position of the point on this line segment corresponding to $t = 1/n$, relative to the points $(-2, 0)$ and $(1, 3)$?
38. Sketch the vectors $\mathbf{r}_0 = \langle 2, 0, 4\rangle$ and $\mathbf{r}_1 = \langle 0, 4, 0\rangle$, and then sketch the vectors
    $$\tfrac{1}{4}\mathbf{r}_0 + \tfrac{3}{4}\mathbf{r}_1,\quad \tfrac{1}{2}\mathbf{r}_0 + \tfrac{1}{2}\mathbf{r}_1,\quad \tfrac{3}{4}\mathbf{r}_0 + \tfrac{1}{4}\mathbf{r}_1$$
    Draw the line segment $(1 - t)\mathbf{r}_0 + t\mathbf{r}_1$ $(0 \le t \le 1)$. If n is a positive integer, what is the position of the point on this line segment corresponding to $t = 1/n$, relative to the points $(2, 0, 4)$ and $(0, 4, 0)$?

39–40 Describe the line segment represented by the vector equation.

39. $\langle x, y\rangle = \langle 1, 0\rangle + t\langle -2, 3\rangle \quad (0 \le t \le 2)$
40. $\langle x, y, z\rangle = \langle -2, 1, 4\rangle + t\langle 3, 0, -1\rangle \quad (0 \le t \le 3)$
41. Find the point on the line segment joining $P_1(3, 6)$ and $P_2(8, -4)$ that is $\frac{2}{5}$ of the way from P_1 to P_2.
42. Find the point on the line segment joining $P_1(1, 4, -3)$ and $P_2(1, 5, -1)$ that is $\frac{2}{3}$ of the way from P_1 to P_2.

43–44 Use the method in Exercise 28 of Section 12.3 to find the distance from the point P to the line L, and then check your answer using the method in Exercise 26 of Section 12.4.

43. $P(-2, 1, 1)$
$L: x = 3 - t,\ y = t,\ z = 1 + 2t$

44. $P(1, 4, -3)$
$L: x = 2 + t,\ y = -1 - t,\ z = 3t$

45–46 Show that the lines L_1 and L_2 are parallel, and find the distance between them.

45. $L_1: x = 2 - t,\ y = 2t,\ z = 1 + t$
$L_2: x = 1 + 2t,\ y = 3 - 4t,\ z = 5 - 2t$

46. $L_1: x = 2t,\ y = 3 + 4t,\ z = 2 - 6t$
$L_2: x = 1 + 3t,\ y = 6t,\ z = -9t$

47. (a) Find parametric equations for the line through the points (x_0, y_0, z_0) and (x_1, y_1, z_1).
(b) Find parametric equations for the line through the point (x_1, y_1, z_1) and parallel to the line

$$x = x_0 + at, \quad y = y_0 + bt, \quad z = z_0 + ct$$

48. Let L be the line that passes through the point (x_0, y_0, z_0) and is parallel to the vector $\mathbf{v} = \langle a, b, c\rangle$, where a, b, and c are nonzero. Show that a point (x, y, z) lies on the line L if and only if

$$\frac{x - x_0}{a} = \frac{y - y_0}{b} = \frac{z - z_0}{c}$$

These equations, which are called the ***symmetric equations*** of L, provide a nonparametric representation of L.

49. (a) Describe the line whose symmetric equations are

$$\frac{x - 1}{2} = \frac{y + 3}{4} = z - 5$$

[See Exercise 48.]
(b) Find parametric equations for the line in part (a).

50. Consider the lines L_1 and L_2 whose symmetric equations are

$$L_1: \frac{x - 1}{2} = \frac{y + \frac{3}{2}}{1} = \frac{z + 1}{2}$$

$$L_2: \frac{x - 4}{-1} = \frac{y - 3}{-2} = \frac{z + 4}{2}$$

[See Exercise 48.]
(a) Are L_1 and L_2 parallel? Perpendicular?
(b) Find parametric equations for L_1 and L_2.
(c) Do L_1 and L_2 intersect? If so, where?

51. Let L_1 and L_2 be the lines whose parametric equations are

$$L_1: x = 1 + 2t, \quad y = 2 - t, \quad z = 4 - 2t$$
$$L_2: x = 9 + t, \quad y = 5 + 3t, \quad z = -4 - t$$

(a) Show that L_1 and L_2 intersect at the point $(7, -1, -2)$.
(b) Find, to the nearest degree, the acute angle between L_1 and L_2 at their intersection.
(c) Find parametric equations for the line that is perpendicular to L_1 and L_2 and passes through their point of intersection.

52. Let L_1 and L_2 be the lines whose parametric equations are

$$L_1: x = 4t, \quad y = 1 - 2t, \quad z = 2 + 2t$$
$$L_2: x = 1 + t, \quad y = 1 - t, \quad z = -1 + 4t$$

(a) Show that L_1 and L_2 intersect at the point $(2, 0, 3)$.
(b) Find, to the nearest degree, the acute angle between L_1 and L_2 at their intersection.
(c) Find parametric equations for the line that is perpendicular to L_1 and L_2 and passes through their point of intersection.

53–54 Find parametric equations of the line that contains the point P and intersects the line L at a right angle, and find the distance between P and L.

53. $P(0, 2, 1)$
$L: x = 2t,\ y = 1 - t,\ z = 2 + t$

54. $P(3, 1, -2)$
$L: x = -2 + 2t,\ y = 4 + 2t,\ z = 2 + t$

55. Two bugs are walking along lines in 3-space. At time t bug 1 is at the point (x, y, z) on the line

$$x = 4 - t, \quad y = 1 + 2t, \quad z = 2 + t$$

and at the same time t bug 2 is at the point (x, y, z) on the line

$$x = t, \quad y = 1 + t, \quad z = 1 + 2t$$

Assume that distance is in centimeters and that time is in minutes.
(a) Find the distance between the bugs at time $t = 0$.
(b) Use a graphing utility to graph the distance between the bugs as a function of time from $t = 0$ to $t = 5$.
(c) What does the graph tell you about the distance between the bugs?
(d) How close do the bugs get?

C **56.** Suppose that the temperature T at a point (x, y, z) on the line $x = t$, $y = 1 + t$, $z = 3 - 2t$ is $T = 25x^2yz$. Use a CAS or a calculating utility with a root-finding capability to approximate the maximum temperature on that portion of the line that extends from the xz-plane to the xy-plane.

✔ QUICK CHECK ANSWERS 12.5

1. (a) $2 + 3t;\ 5 - t$ (b) $\langle 2, 5\rangle + t\langle 3, -1\rangle$ **2.** $5 - t;\ 3;\ 7 + 4t$ **3.** $3 - t;\ 6t;\ 11 - 4t;\ 0 \le t \le 1$ **4.** $(0, 2, 5)$

12.6 PLANES IN 3-SPACE

In this section we will use vectors to derive equations of planes in 3-space, and then we will use these equations to solve various geometric problems.

PLANES PARALLEL TO THE COORDINATE PLANES

The graph of the equation $x = a$ in an xyz-coordinate system consists of all points of the form (a, y, z), where y and z are arbitrary. One such point is $(a, 0, 0)$, and all others are in the plane that passes through this point and is parallel to the yz-plane (Figure 12.6.1). Similarly, the graph of $y = b$ is the plane through $(0, b, 0)$ that is parallel to the xz-plane, and the graph of $z = c$ is the plane through $(0, 0, c)$ that is parallel to the xy-plane.

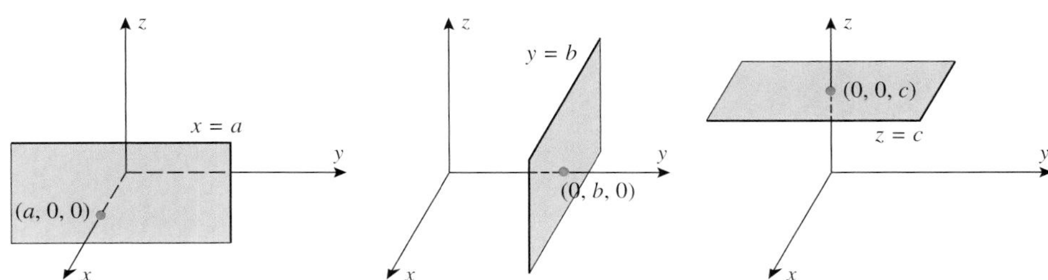

Figure 12.6.1

PLANES DETERMINED BY A POINT AND A NORMAL VECTOR

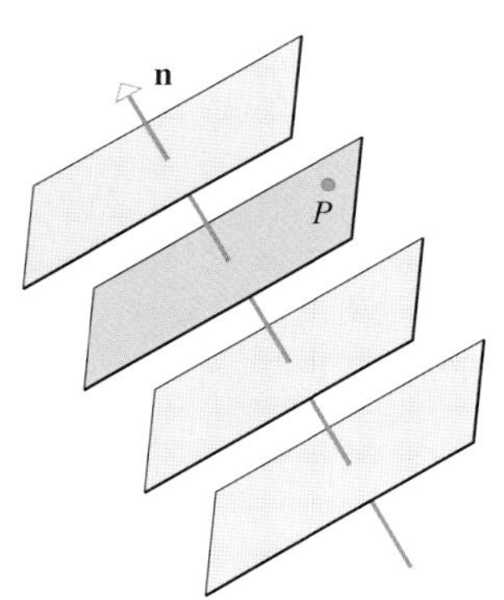

The colored plane is determined uniquely by the point P and the vector $\mathbf{n}$ perpendicular to the plane.

Figure 12.6.2

A plane in 3-space can be determined uniquely by specifying a point in the plane and a vector perpendicular to the plane (Figure 12.6.2). A vector perpendicular to a plane is called a ***normal*** to the plane.

Suppose that we want to find an equation of the plane passing through $P_0(x_0, y_0, z_0)$ and perpendicular to the vector $\mathbf{n} = \langle a, b, c\rangle$. Define the vectors $\mathbf{r}_0$ and $\mathbf{r}$ as

$$\mathbf{r}_0 = \langle x_0, y_0, z_0\rangle \quad \text{and} \quad \mathbf{r} = \langle x, y, z\rangle$$

It should be evident from Figure 12.6.3 that the plane consists precisely of those points $P(x, y, z)$ for which the vector $\mathbf{r} - \mathbf{r}_0$ is orthogonal to $\mathbf{n}$; or, expressed as an equation,

$$\mathbf{n} \cdot (\mathbf{r} - \mathbf{r}_0) = 0 \qquad (1)$$

If preferred, we can express this vector equation in terms of components as

$$\langle a, b, c\rangle \cdot \langle x - x_0, y - y_0, z - z_0\rangle = 0 \qquad (2)$$

from which we obtain

$$a(x - x_0) + b(y - y_0) + c(z - z_0) = 0 \qquad (3)$$

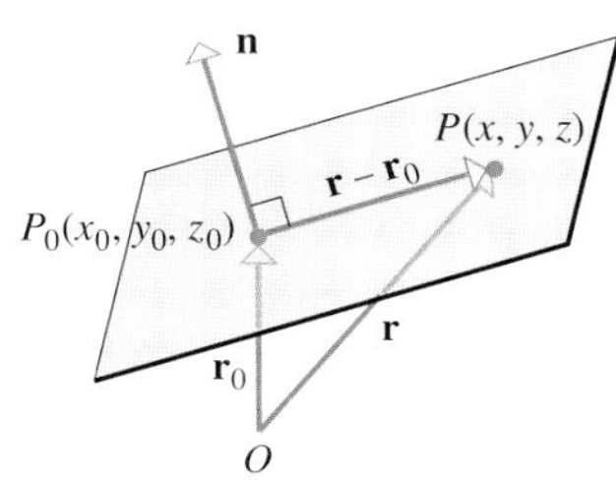

Figure 12.6.3

This is called the ***point-normal form*** of the equation of a plane. Formulas (1) and (2) are vector versions of this formula.

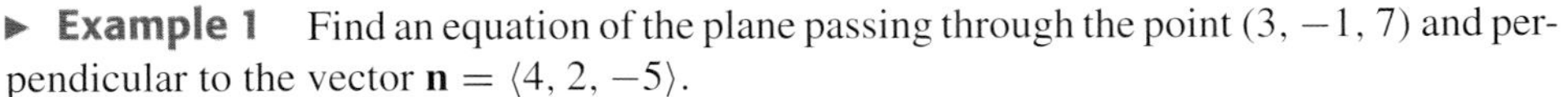

▶ **Example 1** Find an equation of the plane passing through the point $(3, -1, 7)$ and perpendicular to the vector $\mathbf{n} = \langle 4, 2, -5\rangle$.

Solution. From (3), a point-normal form of the equation is

$$4(x - 3) + 2(y + 1) - 5(z - 7) = 0 \qquad (4)$$

What does Equation (1) represent if

$\mathbf{n} = \langle a, b\rangle,\ \mathbf{r}_0 = \langle x_0, y_0\rangle,\ \mathbf{r} = \langle x, y\rangle$

are vectors in an xy-plane in 2-space? Draw a picture.

If preferred, this equation can be written in vector form as

$$\langle 4, 2, -5\rangle \cdot \langle x-3, y+1, z-7\rangle = 0 \blacktriangleleft$$

Observe that if we multiply out the terms in (3) and simplify, we obtain an equation of the form

$$ax + by + cz + d = 0 \tag{5}$$

For example, Equation (4) in Example 1 can be rewritten as

$$4x + 2y - 5z + 25 = 0$$

The following theorem shows that every equation of form (5) represents a plane in 3-space.

12.6.1 THEOREM. *If a, b, c, and d are constants, and a, b, and c are not all zero, then the graph of the equation*

$$ax + by + cz + d = 0 \tag{6}$$

is a plane that has the vector $\mathbf{n} = \langle a, b, c\rangle$ as a normal.

PROOF. Since a, b, and c are not all zero, there is at least one point (x_0, y_0, z_0) whose coordinates satisfy Equation (6). For example, if $a \neq 0$, then such a point is $(-d/a, 0, 0)$, and similarly if $b \neq 0$ or $c \neq 0$ (verify). Thus, let (x_0, y_0, z_0) be any point whose coordinates satisfy (6); that is,

$$ax_0 + by_0 + cz_0 + d = 0$$

Subtracting this equation from (6) yields

$$a(x - x_0) + b(y - y_0) + c(z - z_0) = 0$$

which is the point-normal form of a plane with normal $\mathbf{n} = \langle a, b, c\rangle$. ■

Equation (6) is called the ***general form*** of the equation of a plane.

▶ Example 2 Determine whether the planes

$$3x - 4y + 5z = 0 \quad \text{and} \quad -6x + 8y - 10z - 4 = 0$$

are parallel.

Solution. It is clear geometrically that two planes are parallel if and only if their normals are parallel vectors. A normal to the first plane is

$$\mathbf{n}_1 = \langle 3, -4, 5\rangle$$

and a normal to the second plane is

$$\mathbf{n}_2 = \langle -6, 8, -10\rangle$$

Since $\mathbf{n}_2$ is a scalar multiple of $\mathbf{n}_1$, the normals are parallel, and hence so are the planes. ◀

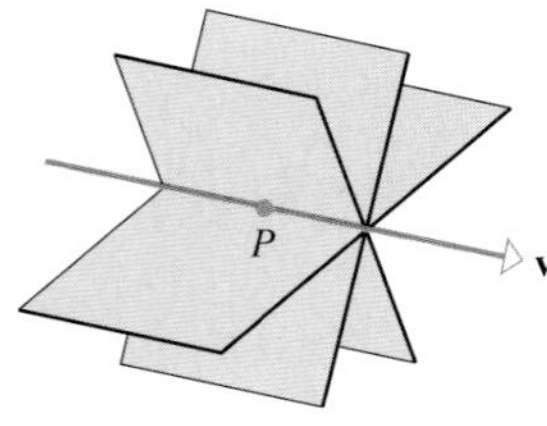

There are infinitely many planes containing P and parallel to $\mathbf{v}$.

Figure 12.6.4

We have seen that a unique plane is determined by a point in the plane and a nonzero vector normal to the plane. In contrast, a unique plane is not determined by a point in the plane and a nonzero vector *parallel* to the plane (Figure 12.6.4). However, a unique plane is determined by a point in the plane and two nonparallel vectors that are parallel to the plane (Figure 12.6.5). A unique plane is also determined by three noncollinear points that lie in the plane (Figure 12.6.6).

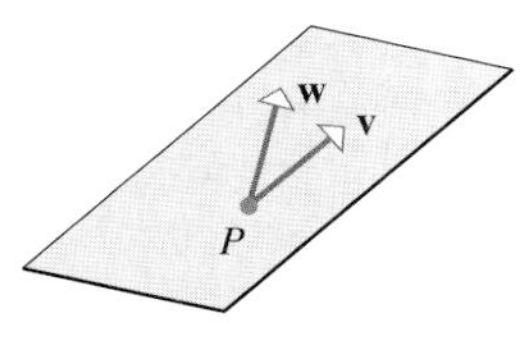

There is a unique plane through P that is parallel to both **v** and **w**.

Figure 12.6.5

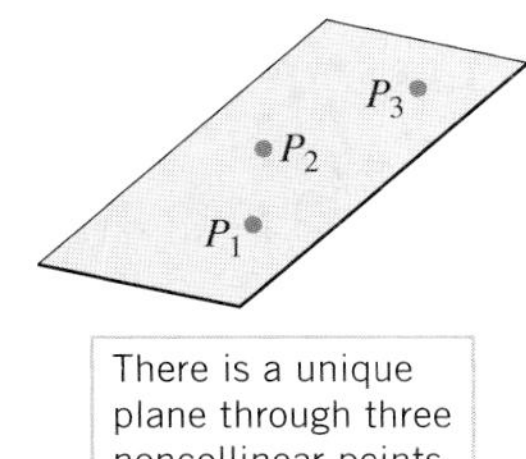

There is a unique plane through three noncollinear points.

Figure 12.6.6

▶ **Example 3** Find an equation of the plane through the points $P_1(1, 2, -1)$, $P_2(2, 3, 1)$, and $P_3(3, -1, 2)$.

Solution. Since the points P_1, P_2, and P_3 lie in the plane, the vectors $\overrightarrow{P_1P_2} = \langle 1, 1, 2\rangle$ and $\overrightarrow{P_1P_3} = \langle 2, -3, 3\rangle$ are parallel to the plane. Therefore,

$$\overrightarrow{P_1P_2} \times \overrightarrow{P_1P_3} = \begin{vmatrix} \mathbf{i} & \mathbf{j} & \mathbf{k} \\ 1 & 1 & 2 \\ 2 & -3 & 3 \end{vmatrix} = 9\mathbf{i} + \mathbf{j} - 5\mathbf{k}$$

is normal to the plane, since it is orthogonal to both $\overrightarrow{P_1P_2}$ and $\overrightarrow{P_1P_3}$. By using this normal and the point $P_1(1, 2, -1)$ in the plane, we obtain the point-normal form

$$9(x-1) + (y-2) - 5(z+1) = 0$$

which can be rewritten as

$$9x + y - 5z - 16 = 0 \quad ◀$$

▶ **Example 4** Determine whether the line

$$x = 3 + 8t, \quad y = 4 + 5t, \quad z = -3 - t$$

is parallel to the plane $x - 3y + 5z = 12$.

Solution. The vector $\mathbf{v} = \langle 8, 5, -1\rangle$ is parallel to the line and the vector $\mathbf{n} = \langle 1, -3, 5\rangle$ is normal to the plane. For the line and plane to be parallel, the vectors **v** and **n** must be orthogonal. But this is not so, since the dot product

$$\mathbf{v} \cdot \mathbf{n} = (8)(1) + (5)(-3) + (-1)(5) = -12$$

is nonzero. Thus, the line and plane are not parallel. ◀

▶ **Example 5** Find the intersection of the line and plane in Example 4.

Solution. If we let (x_0, y_0, z_0) be the point of intersection, then the coordinates of this point satisfy both the equation of the plane and the parametric equations of the line. Thus,

$$x_0 - 3y_0 + 5z_0 = 12 \tag{7}$$

and for some value of t, say $t = t_0$,

$$x_0 = 3 + 8t_0, \quad y_0 = 4 + 5t_0, \quad z_0 = -3 - t_0 \tag{8}$$

Substituting (8) in (7) yields

$$(3 + 8t_0) - 3(4 + 5t_0) + 5(-3 - t_0) = 12$$

Solving for t_0 yields $t_0 = -3$ and on substituting this value in (8), we obtain

$$(x_0, y_0, z_0) = (-21, -11, 0) \quad ◀$$

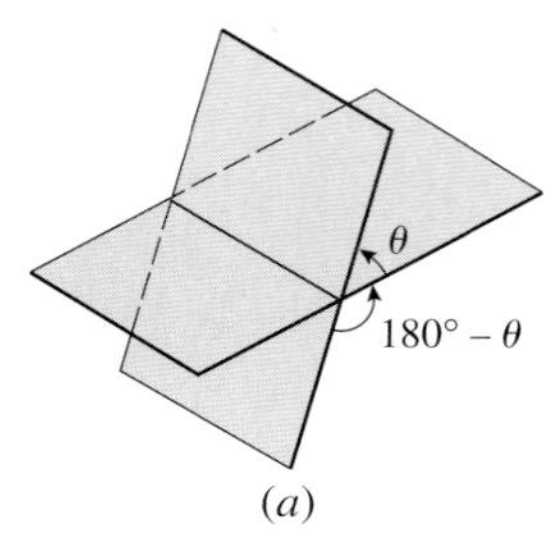

(a)

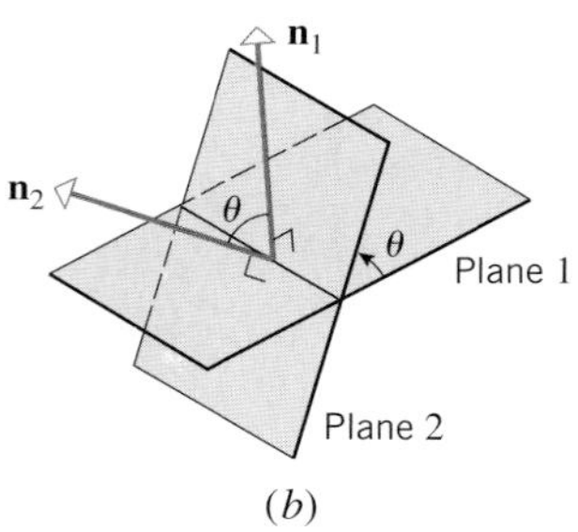

(b)

Figure 12.6.7

INTERSECTING PLANES

Two distinct intersecting planes determine two positive angles of intersection—an (acute) angle θ that satisfies the condition $0 \le \theta \le \pi/2$ and the supplement of that angle (Figure 12.6.7a). If $\mathbf{n}_1$ and $\mathbf{n}_2$ are normals to the planes, then depending on the directions of $\mathbf{n}_1$ and $\mathbf{n}_2$, the angle θ is either the angle between $\mathbf{n}_1$ and $\mathbf{n}_2$ or the angle between $\mathbf{n}_1$ and $-\mathbf{n}_2$ (Figure 12.6.7b). In both cases, Theorem 12.3.3 yields the following formula for the acute angle θ between the planes:

$$\cos\theta = \frac{|\mathbf{n}_1 \cdot \mathbf{n}_2|}{\|\mathbf{n}_1\|\,\|\mathbf{n}_2\|} \tag{9}$$

► **Example 6** Find the acute angle of intersection between the two planes

$$2x - 4y + 4z = 6 \quad \text{and} \quad 6x + 2y - 3z = 4$$

Solution. The given equations yield the normals $\mathbf{n}_1 = \langle 2, -4, 4\rangle$ and $\mathbf{n}_2 = \langle 6, 2, -3\rangle$. Thus, Formula (9) yields

$$\cos\theta = \frac{|\mathbf{n}_1 \cdot \mathbf{n}_2|}{\|\mathbf{n}_1\|\,\|\mathbf{n}_2\|} = \frac{|-8|}{\sqrt{36}\sqrt{49}} = \frac{4}{21}$$

from which we obtain

$$\theta = \cos^{-1}\left(\frac{4}{21}\right) \approx 79^\circ \blacktriangleleft$$

► **Example 7** Find an equation for the line L of intersection of the planes in Example 6.

Solution. First compute $\mathbf{v} = \mathbf{n}_1 \times \mathbf{n}_2 = \langle 2, -4, 4\rangle \times \langle 6, 2, -3\rangle = \langle 4, 30, 28\rangle$. Since $\mathbf{v}$ is orthogonal to $\mathbf{n}_1$, it is parallel to the first plane, and since $\mathbf{v}$ is orthogonal to $\mathbf{n}_2$, it is parallel to the second plane. That is, $\mathbf{v}$ is parallel to L, the intersection of the two planes. To find a point on L we observe that L must intersect the xy-plane, $z = 0$, since $\mathbf{v} \cdot \langle 0, 0, 1\rangle = 28 \ne 0$. Substituting $z = 0$ in the equations of both planes yields

$$\begin{aligned} 2x - 4y &= 6 \\ 6x + 2y &= 4 \end{aligned}$$

with solution $x = 1$, $y = -1$. Thus, $P(1, -1, 0)$ is a point on L. A vector equation for L is

$$\langle x, y, z\rangle = \langle 1, -1, 0\rangle + t\langle 4, 30, 28\rangle \blacktriangleleft$$

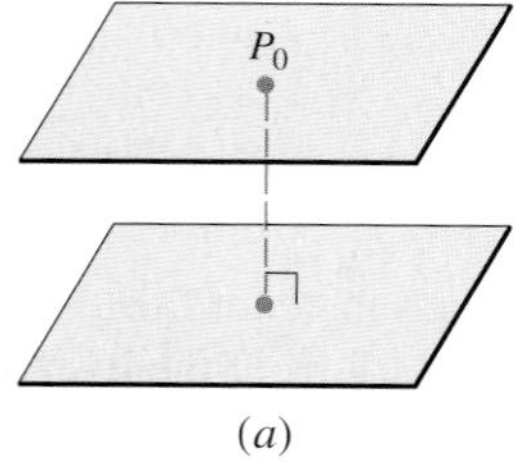

(a)

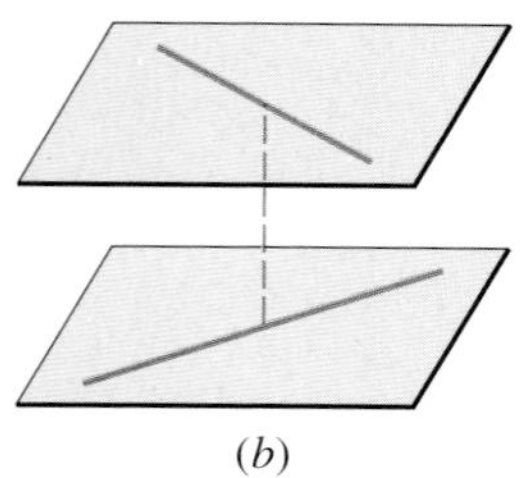
(b)

Figure 12.6.8

DISTANCE PROBLEMS INVOLVING PLANES

Next we will consider three basic "distance problems" in 3-space:

- Find the distance between a point and a plane.
- Find the distance between two parallel planes.
- Find the distance between two skew lines.

The three problems are related. If we can find the distance between a point and a plane, then we can find the distance between parallel planes by computing the distance between one of the planes and an arbitrary point P_0 in the other plane (Figure 12.6.8a). Moreover, we can find the distance between two skew lines by computing the distance between parallel planes containing them (Figure 12.6.8b).

12.6.2 THEOREM. *The distance D between a point $P_0(x_0, y_0, z_0)$ and the plane $ax + by + cz + d = 0$ is*

$$D = \frac{|ax_0 + by_0 + cz_0 + d|}{\sqrt{a^2 + b^2 + c^2}} \tag{10}$$

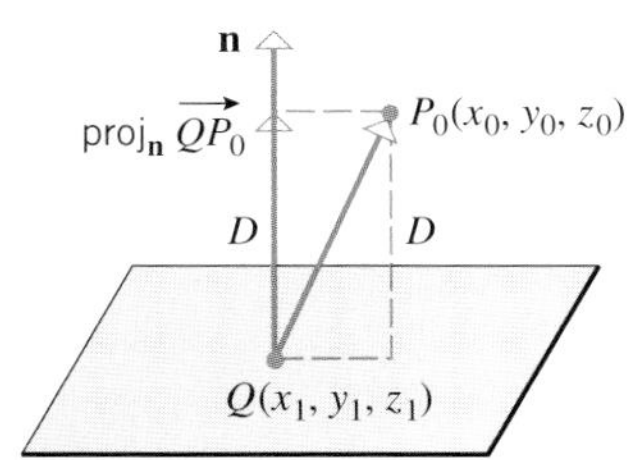

Figure 12.6.9

PROOF. Let $Q(x_1, y_1, z_1)$ be any point in the plane, and position the normal $\mathbf{n} = \langle a, b, c\rangle$ so that its initial point is at Q. As illustrated in Figure 12.6.9, the distance D is equal to the length of the orthogonal projection of $\overrightarrow{QP_0}$ on $\mathbf{n}$. Thus, from (12) of Section 12.3,

$$D = \|\text{proj}_{\mathbf{n}} \overrightarrow{QP_0}\| = \left\| \frac{\overrightarrow{QP_0} \cdot \mathbf{n}}{\|\mathbf{n}\|^2}\mathbf{n} \right\| = \frac{|\overrightarrow{QP_0} \cdot \mathbf{n}|}{\|\mathbf{n}\|^2}\|\mathbf{n}\| = \frac{|\overrightarrow{QP_0} \cdot \mathbf{n}|}{\|\mathbf{n}\|}$$

But

$$\overrightarrow{QP_0} = \langle x_0 - x_1, y_0 - y_1, z_0 - z_1\rangle$$
$$\overrightarrow{QP_0} \cdot \mathbf{n} = a(x_0 - x_1) + b(y_0 - y_1) + c(z_0 - z_1)$$
$$\|\mathbf{n}\| = \sqrt{a^2 + b^2 + c^2}$$

Thus,

$$D = \frac{|a(x_0 - x_1) + b(y_0 - y_1) + c(z_0 - z_1)|}{\sqrt{a^2 + b^2 + c^2}} \tag{11}$$

Since the point $Q(x_1, y_1, z_1)$ lies in the plane, its coordinates satisfy the equation of the plane; that is,

$$ax_1 + by_1 + cz_1 + d = 0$$

or

$$d = -ax_1 - by_1 - cz_1$$

Combining this expression with (11) yields (10). ■

▶ **Example 8** Find the distance D between the point $(1, -4, -3)$ and the plane

$$2x - 3y + 6z = -1$$

There is an analog of Formula (10) in 2-space that can be used to compute the distance between a point and a line. (See Exercise 50).

Solution. Formula (10) requires the plane be rewritten in the form $ax + by + cz + d = 0$. Thus, we rewrite the equation of the given plane as

$$2x - 3y + 6z + 1 = 0$$

from which we obtain $a = 2$, $b = -3$, $c = 6$, and $d = 1$. Substituting these values and the coordinates of the given point in (10), we obtain

$$D = \frac{|(2)(1) + (-3)(-4) + 6(-3) + 1|}{\sqrt{2^2 + (-3)^2 + 6^2}} = \frac{|-3|}{7} = \frac{3}{7}$$ ◀

▶ **Example 9** The planes

$$x + 2y - 2z = 3 \quad \text{and} \quad 2x + 4y - 4z = 7$$

are parallel since their normals, $\langle 1, 2, -2\rangle$ and $\langle 2, 4, -4\rangle$, are parallel vectors. Find the distance between these planes.

Solution. To find the distance D between the planes, we can select an arbitrary point in one of the planes and compute its distance to the other plane. By setting $y = z = 0$ in the equation $x + 2y - 2z = 3$, we obtain the point $P_0(3, 0, 0)$ in this plane. From (10), the distance from P_0 to the plane $2x + 4y - 4z = 7$ is

$$D = \frac{|(2)(3) + 4(0) + (-4)(0) - 7|}{\sqrt{2^2 + 4^2 + (-4)^2}} = \frac{1}{6} \blacktriangleleft$$

▶ **Example 10** It was shown in Example 3 of Section 12.5 that the lines

$$L_1: x = 1 + 4t, \quad y = 5 - 4t, \quad z = -1 + 5t$$
$$L_2: x = 2 + 8t, \quad y = 4 - 3t, \quad z = 5 + t$$

are skew. Find the distance between them.

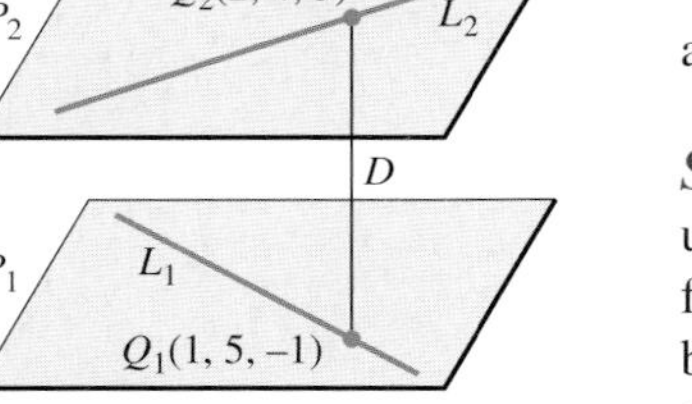

Figure 12.6.10

Solution. Let P_1 and P_2 denote parallel planes containing L_1 and L_2, respectively (Figure 12.6.10). To find the distance D between L_1 and L_2, we will calculate the distance from a point in P_1 to the plane P_2. Since L_1 lies in plane P_1, we can find a point in P_1 by finding a point on the line L_1; we can do this by substituting any convenient value of t in the parametric equations of L_1. The simplest choice is $t = 0$, which yields the point $Q_1(1, 5, -1)$.

The next step is to find an equation for the plane P_2. For this purpose, observe that the vector $\mathbf{u}_1 = \langle 4, -4, 5\rangle$ is parallel to line L_1, and therefore also parallel to planes P_1 and P_2. Similarly, $\mathbf{u}_2 = \langle 8, -3, 1\rangle$ is parallel to L_2 and hence parallel to P_1 and P_2. Therefore, the cross product

$$\mathbf{n} = \mathbf{u}_1 \times \mathbf{u}_2 = \begin{vmatrix} \mathbf{i} & \mathbf{j} & \mathbf{k} \\ 4 & -4 & 5 \\ 8 & -3 & 1 \end{vmatrix} = 11\mathbf{i} + 36\mathbf{j} + 20\mathbf{k}$$

is normal to both P_1 and P_2. Using this normal and the point $Q_2(2, 4, 5)$ found by setting $t = 0$ in the equations of L_2, we obtain an equation for P_2:

$$11(x - 2) + 36(y - 4) + 20(z - 5) = 0$$

or

$$11x + 36y + 20z - 266 = 0$$

The distance between $Q_1(1, 5, -1)$ and this plane is

$$D = \frac{|(11)(1) + (36)(5) + (20)(-1) - 266|}{\sqrt{11^2 + 36^2 + 20^2}} = \frac{95}{\sqrt{1817}}$$

which is also the distance between L_1 and L_2. ◀

✔ QUICK CHECK EXERCISES 12.6 (See page 839 for answers.)

1. The point-normal form of the equation of the plane through $(0, 3, 5)$ and perpendicular to $\langle -4, 1, 7\rangle$ is ______.
2. A normal vector for the plane $4x - 2y + 7z - 11 = 0$ is ______.
3. A normal vector for the plane through the points $(2, 5, 1)$, $(3, 7, 0)$, and $(2, 5, 2)$ is ______.
4. The acute angle of intersection of the planes $x + y - 2z = 5$ and $3y - 4z = 6$ is ______.
5. The distance between the point $(9, 8, 3)$ and the plane $x + y - 2z = 5$ is ______.

EXERCISE SET 12.6

1. Find equations of the planes P_1, P_2, and P_3 that are parallel to the coordinate planes and pass through the corner $(3, 4, 5)$ of the box shown in the accompanying figure.

2. Find equations of the planes P_1, P_2, and P_3 that are parallel to the coordinate planes and pass through the corner (x_0, y_0, z_0) of the box shown in the accompanying figure.

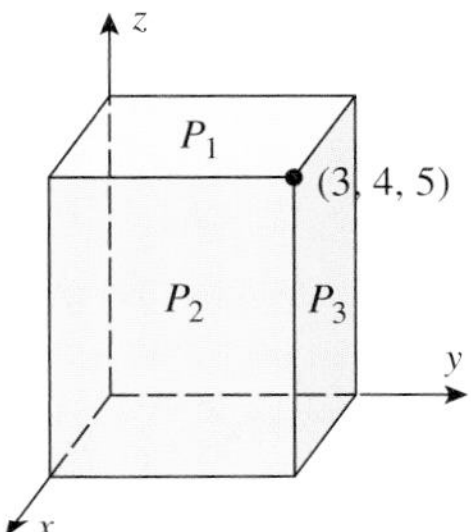

Figure Ex-1

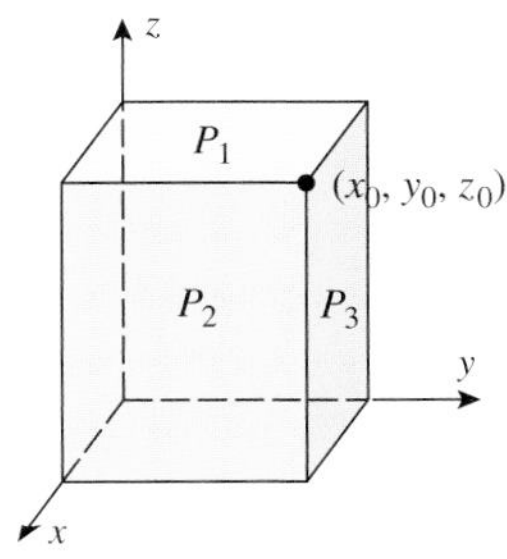

Figure Ex-2

3–6 Find an equation of the plane that passes through the point P and has the vector $\mathbf{n}$ as a normal.

3. $P(2, 6, 1)$; $\mathbf{n} = \langle 1, 4, 2\rangle$

4. $P(-1, -1, 2)$; $\mathbf{n} = \langle -1, 7, 6\rangle$

5. $P(1, 0, 0)$; $\mathbf{n} = \langle 0, 0, 1\rangle$

6. $P(0, 0, 0)$; $\mathbf{n} = \langle 2, -3, -4\rangle$

7–10 Find an equation of the plane indicated in the figure.

7.
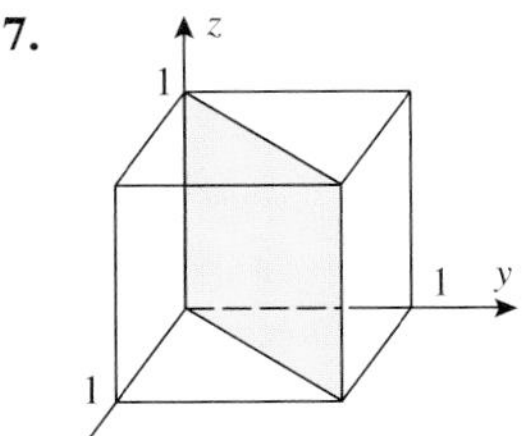

8.
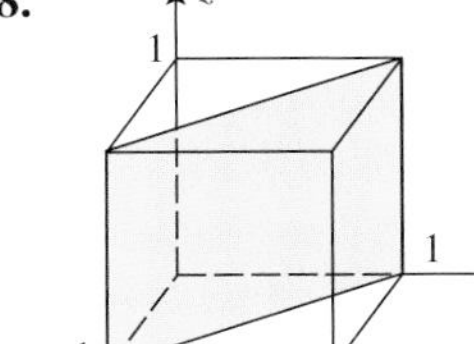

9.
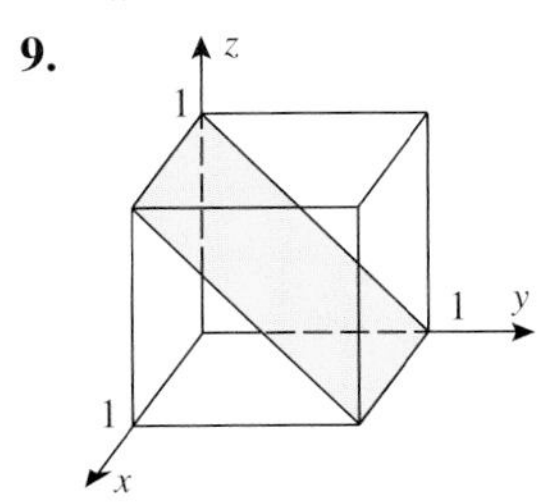

10.
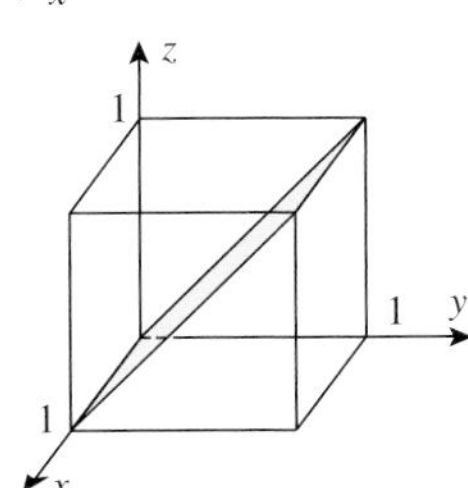

11–12 Find an equation of the plane that passes through the given points.

11. $(-2, 1, 1)$, $(0, 2, 3)$, and $(1, 0, -1)$

12. $(3, 2, 1)$, $(2, 1, -1)$, and $(-1, 3, 2)$

13–14 Determine whether the planes are parallel, perpendicular, or neither.

13. (a) $2x - 8y - 6z - 2 = 0$
$-x + 4y + 3z - 5 = 0$
(b) $3x - 2y + z = 1$
$4x + 5y - 2z = 4$
(c) $x - y + 3z - 2 = 0$
$2x + z = 1$

14. (a) $3x - 2y + z = 4$
$6x - 4y + 3z = 7$
(b) $y = 4x - 2z + 3$
$x = \frac{1}{4}y + \frac{1}{2}z$
(c) $x + 4y + 7z = 3$
$5x - 3y + z = 0$

15–16 Determine whether the line and plane are parallel, perpendicular, or neither.

15. (a) $x = 4 + 2t,\ y = -t,\ z = -1 - 4t;$
$3x + 2y + z - 7 = 0$
(b) $x = t,\ y = 2t,\ z = 3t;$
$x - y + 2z = 5$
(c) $x = -1 + 2t,\ y = 4 + t,\ z = 1 - t;$
$4x + 2y - 2z = 7$

16. (a) $x = 3 - t,\ y = 2 + t,\ z = 1 - 3t;$
$2x + 2y - 5 = 0$
(b) $x = 1 - 2t,\ y = t,\ z = -t;$
$6x - 3y + 3z = 1$
(c) $x = t,\ y = 1 - t,\ z = 2 + t;$
$x + y + z = 1$

17–18 Determine whether the line and plane intersect; if so, find the coordinates of the intersection.

17. (a) $x = t,\ y = t,\ z = t;$
$3x - 2y + z - 5 = 0$
(b) $x = 2 - t,\ y = 3 + t,\ z = t;$
$2x + y + z = 1$

18. (a) $x = 3t,\ y = 5t,\ z = -t;$
$2x - y + z + 1 = 0$
(b) $x = 1 + t,\ y = -1 + 3t,\ z = 2 + 4t;$
$x - y + 4z = 7$

19–20 Find the acute angle of intersection of the planes to the nearest degree.

19. $x = 0$ and $2x - y + z - 4 = 0$

20. $x + 2y - 2z = 5$ and $6x - 3y + 2z = 8$

21–30 Find an equation of the plane that satisfies the stated conditions.

21. The plane through the origin that is parallel to the plane $4x - 2y + 7z + 12 = 0$.

22. The plane that contains the line $x = -2 + 3t$, $y = 4 + 2t$, $z = 3 - t$ and is perpendicular to the plane $x - 2y + z = 5$.

23. The plane through the point $(-1, 4, 2)$ that contains the line of intersection of the planes $4x - y + z - 2 = 0$ and $2x + y - 2z - 3 = 0$.

24. The plane through $(-1, 4, -3)$ that is perpendicular to the line $x - 2 = t$, $y + 3 = 2t$, $z = -t$.

25. The plane through $(1, 2, -1)$ that is perpendicular to the line of intersection of the planes $2x + y + z = 2$ and $x + 2y + z = 3$.

26. The plane through the points $P_1(-2, 1, 4)$, $P_2(1, 0, 3)$ that is perpendicular to the plane $4x - y + 3z = 2$.

27. The plane through $(-1, 2, -5)$ that is perpendicular to the planes $2x - y + z = 1$ and $x + y - 2z = 3$.

28. The plane that contains the point $(2, 0, 3)$ and the line $x = -1 + t$, $y = t$, $z = -4 + 2t$.

29. The plane whose points are equidistant from $(2, -1, 1)$ and $(3, 1, 5)$.

30. The plane that contains the line $x = 3t$, $y = 1 + t$, $z = 2t$ and is parallel to the intersection of the planes $y + z = -1$ and $2x - y + z = 0$.

31. Find parametric equations of the line through the point $(5, 0, -2)$ that is parallel to the planes $x - 4y + 2z = 0$ and $2x + 3y - z + 1 = 0$.

32. Let L be the line $x = 3t + 1$, $y = -5t$, $z = t$.
(a) Show that L lies in the plane $2x + y - z = 2$.
(b) Show that L is parallel to the plane $x + y + 2z = 0$. Is the line above, below, or on this plane?

33. Show that the lines

$$\begin{aligned} x &= -2 + t, & y &= 3 + 2t, & z &= 4 - t \\ x &= 3 - t, & y &= 4 - 2t, & z &= t \end{aligned}$$

are parallel and find an equation of the plane they determine.

34. Show that the lines

$$\begin{aligned} L_1&: x + 1 = 4t, & y - 3 &= t, & z - 1 &= 0 \\ L_2&: x + 13 = 12t, & y - 1 &= 6t, & z - 2 &= 3t \end{aligned}$$

intersect and find an equation of the plane they determine.

FOCUS ON CONCEPTS

35. Do the points $(1, 0, -1)$, $(0, 2, 3)$, $(-2, 1, 1)$, and $(4, 2, 3)$ lie in the same plane? Justify your answer two different ways.

36. Show that if a, b, and c are nonzero, then the plane whose intercepts with the coordinate axes are $x = a$, $y = b$, and $z = c$ is given by the equation

$$\frac{x}{a} + \frac{y}{b} + \frac{z}{c} = 1$$

37. If L is a line in 3-space, must L lie in some vertical plane? Explain.

38. If L is a line in 3-space, must L lie in some horizontal plane? Explain.

39–40 Find parametric equations of the line of intersection of the planes.

39. $-2x + 3y + 7z + 2 = 0$
$x + 2y - 3z + 5 = 0$

40. $3x - 5y + 2z = 0$
$z = 0$

41–42 Find the distance between the point and the plane.

41. $(1, -2, 3)$; $2x - 2y + z = 4$

42. $(0, 1, 5)$; $3x + 6y - 2z - 5 = 0$

43–44 Find the distance between the given parallel planes.

43. $-2x + y + z = 0$
$6x - 3y - 3z - 5 = 0$

44. $x + y + z = 1$
$x + y + z = -1$

45–46 Find the distance between the given skew lines.

45. $x = 1 + 7t$, $y = 3 + t$, $z = 5 - 3t$
$x = 4 - t$, $y = 6$, $z = 7 + 2t$

46. $x = 3 - t$, $y = 4 + 4t$, $z = 1 + 2t$
$x = t$, $y = 3$, $z = 2t$

47. Find an equation of the sphere with center $(2, 1, -3)$ that is tangent to the plane $x - 3y + 2z = 4$.

48. Locate the point of intersection of the plane $2x + y - z = 0$ and the line through $(3, 1, 0)$ that is perpendicular to the plane.

49. Show that the line $x = -1 + t$, $y = 3 + 2t$, $z = -t$ and the plane $2x - 2y - 2z + 3 = 0$ are parallel, and find the distance between them.

FOCUS ON CONCEPTS

50. Formulas (1), (2), (3), (5), and (10), which apply to planes in 3-space, have analogs for lines in 2-space.
(a) Draw an analog of Figure 12.6.3 in 2-space to illustrate that the equation of the line that passes through the point $P(x_0, y_0)$ and is perpendicular to the vector $\mathbf{n} = \langle a, b\rangle$ can be expressed as

$$\mathbf{n} \cdot (\mathbf{r} - \mathbf{r}_0) = 0$$

where $\mathbf{r} = \langle x, y\rangle$ and $\mathbf{r}_0 = \langle x_0, y_0\rangle$.
(b) Show that the vector equation in part (a) can be expressed as

$$a(x - x_0) + b(y - y_0) = 0$$

This is called the ***point-normal form of a line***.
(c) Using the proof of Theorem 12.6.1 as a guide, show that if a and b are not both zero, then the graph of the equation

$$ax + by + c = 0$$

is a line that has $\mathbf{n} = \langle a, b\rangle$ as a normal.

(d) Using the proof of Theorem 12.6.2 as a guide, show that the distance D between a point $P(x_0, y_0)$ and the line $ax + by + c = 0$ is

$$D = \frac{|ax_0 + by_0 + c|}{\sqrt{a^2 + b^2}}$$

(e) Use the formula in part (d) to find the distance between the point $P(-3, 5)$ and the line $y = -2x + 1$.

51. (a) Show that the distance D between parallel planes

$$ax + by + cz + d_1 = 0$$
$$ax + by + cz + d_2 = 0$$

is

$$D = \frac{|d_1 - d_2|}{\sqrt{a^2 + b^2 + c^2}}$$

(b) Use the formula in part (a) to solve Exercise 43.

✔QUICK CHECK ANSWERS 12.6

1. $-4x + (y - 3) + 7(z - 5) = 0$ **2.** $\langle 4, -2, 7\rangle$ **3.** $\langle 2, -1, 0\rangle$ **4.** $\cos^{-1}\dfrac{11}{5\sqrt{6}} \approx 26°$ **5.** $\sqrt{6}$

12.7 QUADRIC SURFACES

In this section we will study an important class of surfaces that are the three-dimensional analogs of the conic sections.

TRACES OF SURFACES

Although the general shape of a curve in 2-space can be obtained by plotting points, this method is not usually helpful for surfaces in 3-space because too many points are required. It is more common to build up the shape of a surface with a network of ***mesh lines***, which are curves obtained by cutting the surface with well-chosen planes. For example, Figure 12.7.1, which was generated by a CAS, shows the graph of $z = x^3 - 3xy^2$ rendered with a combination of mesh lines and colorization to produce the surface detail. This surface is called a "monkey saddle" because a monkey sitting astride the surface has a place for its two legs and tail.

A monkey saddle

Figure 12.7.1

The mesh line that results when a surface is cut by a plane is called the ***trace*** of the surface in the plane (Figure 12.7.2). Usually, surfaces are built up from traces in planes that are parallel to the coordinate planes, so we will begin by showing how the equations of such traces can be obtained. For this purpose, we will consider the surface

$$z = x^2 + y^2 \tag{1}$$

shown in Figure 12.7.3a.

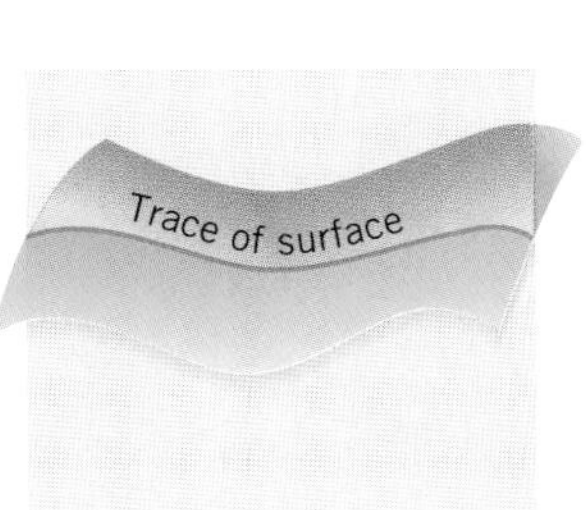

Figure 12.7.2

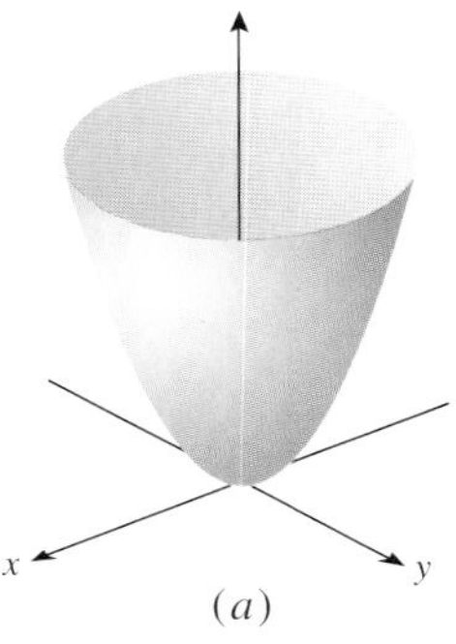

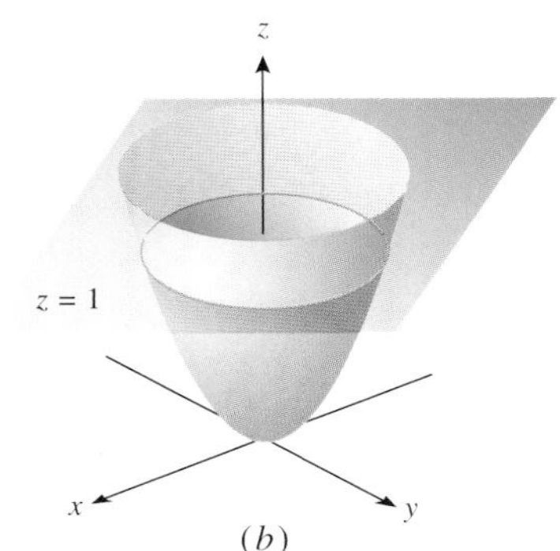

Figure 12.7.3

The parenthetical part of Equation (2) is a reminder that the z-coordinate of each point in the trace is $z = 1$. This needs to be stated explicitly because the variable z does not appear in the equation $x^2 + y^2 = 1$.

The basic procedure for finding the equation of a trace is to substitute the equation of the plane into the equation of the surface. For example, to find the trace of the surface $z = x^2 + y^2$ in the plane $z = 1$, we substitute $z = 1$ in (1), which yields

$$x^2 + y^2 = 1 \qquad (z = 1) \tag{2}$$

This is a circle of radius 1 centered at the point (0, 0, 1) (Figure 12.7.3*b*).

Figure 12.7.4*a* suggests that the traces of (1) in planes that are parallel to and above the xy-plane form a family of circles that are centered on the z-axis and whose radii increase with z. To confirm this, let us consider the trace in a general plane $z = k$ that is parallel to the xy-plane. The equation of the trace is

$$x^2 + y^2 = k \qquad (z = k)$$

If $k \geq 0$, then the trace is a circle of radius $\sqrt{k}$ centered at the point $(0, 0, k)$. In particular, if $k = 0$, then the radius is zero, so the trace in the xy-plane is the single point (0, 0, 0). Thus, for nonnegative values of k the traces parallel to the xy-plane form a family of circles, centered on the z-axis, whose radii start at zero and increase with k. This confirms our conjecture. If $k < 0$, then the equation $x^2 + y^2 = k$ has no graph, which means that there is no trace.

Now let us examine the traces of (1) in planes parallel to the yz-plane. Such planes have equations of the form $x = k$, so we substitute this in (1) to obtain

$$z = k^2 + y^2 \qquad (x = k)$$

which we can rewrite as

$$z - k^2 = y^2 \qquad (x = k) \tag{3}$$

For simplicity, let us start with the case where $k = 0$ (the trace in the yz-plane), in which case the trace has the equation

$$z = y^2 \qquad (x = 0)$$

You should be able to recognize that this is a parabola that has its vertex at the origin, opens in the positive z-direction, and is symmetric about the z-axis (Figure 12.7.4*b* shows a two-dimensional view). You should also be able to recognize that the $-k^2$ term in (3) has the effect of translating the parabola $z = y^2$ in the positive z-direction, so the new vertex falls at $(k, 0, k^2)$. Thus, the traces parallel to the yz-plane form a family of parabolas whose vertices move upward as k^2 increases. This is consistent with Figure 12.7.4*c*. Similarly, the traces in planes parallel to the xz-plane have equations of the form

$$z - k^2 = x^2 \qquad (y = k)$$

which again is a family of parabolas whose vertices move upward as k^2 increases (Figure 12.7.4*d*).

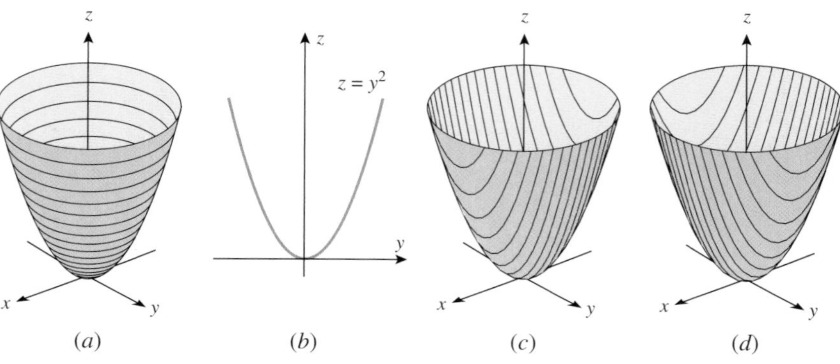

Figure 12.7.4

THE QUADRIC SURFACES

In the discussion of Formula (2) in Section 11.5 we noted that a second-degree equation

$$Ax^2 + Bxy + Cy^2 + Dx + Ey + F = 0$$

represents a conic section (possibly degenerate). The analog of this equation in an xyz-coordinate system is

$$Ax^2 + By^2 + Cz^2 + Dxy + Exz + Fyz + Gx + Hy + Iz + J = 0 \qquad (4)$$

which is called a ***second-degree equation in x, y, and z***. The graphs of such equations are called ***quadric surfaces*** or sometimes ***quadrics***.

Six common types of quadric surfaces are shown in Table 12.7.1—*ellipsoids, hyperboloids of one sheet, hyperboloids of two sheets, elliptic cones, elliptic paraboloids,* and *hyperbolic paraboloids*. (The constants a, b, and c that appear in the equations in the table are assumed to be positive.) Observe that none of the quadric surfaces in the table have cross-product terms in their equations. This is because of their orientations relative to the coordinate axes. Later in this section we will discuss other possible orientations that produce equations of the quadric surfaces with no cross-product terms. In the special case where the elliptic cross sections of an elliptic cone or an elliptic paraboloid are circles, the terms *circular cone* and *circular paraboloid* are used.

TECHNIQUES FOR GRAPHING QUADRIC SURFACES

Accurate graphs of quadric surfaces are best left for graphing utilities. However, the techniques that we will now discuss can be used to generate rough sketches of these surfaces that are useful for various purposes.

A rough sketch of an ellipsoid

$$\frac{x^2}{a^2} + \frac{y^2}{b^2} + \frac{z^2}{c^2} = 1 \qquad (a > 0, b > 0, c > 0) \qquad (5)$$

can be obtained by first plotting the intersections with the coordinate axes, then sketching the elliptical traces in the coordinate planes, and then sketching the surface itself using the traces as a guide. Example 1 illustrates this technique.

▶ Example 1 Sketch the ellipsoid

$$\frac{x^2}{4} + \frac{y^2}{16} + \frac{z^2}{9} = 1 \qquad (6)$$

Solution. The x-intercepts can be obtained by setting $y = 0$ and $z = 0$ in (6). This yields $x = \pm 2$. Similarly, the y-intercepts are $y = \pm 4$, and the z-intercepts are $z = \pm 3$. From these intercepts we obtain the elliptical traces and the ellipsoid sketched in Figure 12.7.5. ◀

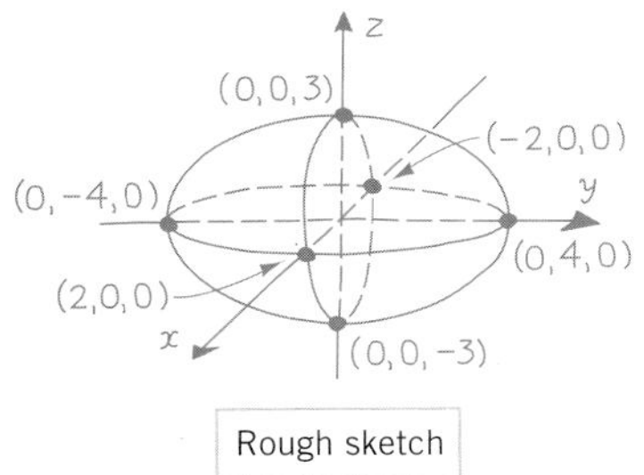

Figure 12.7.5

A rough sketch of a hyperboloid of one sheet

$$\frac{x^2}{a^2} + \frac{y^2}{b^2} - \frac{z^2}{c^2} = 1 \qquad (a > 0, b > 0, c > 0) \qquad (7)$$

can be obtained by first sketching the elliptical trace in the xy-plane, then the elliptical traces in the planes $z = \pm c$, and then the hyperbolic curves that join the endpoints of the axes of these ellipses. The next example illustrates this technique.

Table 12.7.1

SURFACE	EQUATION	SURFACE	EQUATION
ELLIPSOID	$\frac{x^2}{a^2}+\frac{y^2}{b^2}+\frac{z^2}{c^2}=1$ The traces in the coordinate planes are ellipses, as are the traces in those planes that are parallel to the coordinate planes and intersect the surface in more than one point.	ELLIPTIC CONE	$z^2=\frac{x^2}{a^2}+\frac{y^2}{b^2}$ The trace in the xy-plane is a point (the origin), and the traces in planes parallel to the xy-plane are ellipses. The traces in the yz- and xz-planes are pairs of lines intersecting at the origin. The traces in planes parallel to these are hyperbolas.
HYPERBOLOID OF ONE SHEET	$\frac{x^2}{a^2}+\frac{y^2}{b^2}-\frac{z^2}{c^2}=1$ The trace in the xy-plane is an ellipse, as are the traces in planes parallel to the xy-plane. The traces in the yz-plane and xz-plane are hyperbolas, as are the traces in those planes that are parallel to these and do not pass through the x- or y-intercepts. At these intercepts the traces are pairs of intersecting lines.	ELLIPTIC PARABOLOID	$z=\frac{x^2}{a^2}+\frac{y^2}{b^2}$ The trace in the xy-plane is a point (the origin), and the traces in planes parallel to and above the xy-plane are ellipses. The traces in the yz- and xz-planes are parabolas, as are the traces in planes parallel to these.
HYPERBOLOID OF TWO SHEETS	$\frac{z^2}{c^2}-\frac{x^2}{a^2}-\frac{y^2}{b^2}=1$ There is no trace in the xy-plane. In planes parallel to the xy-plane that intersect the surface in more than one point the traces are ellipses. In the yz- and xz-planes, the traces are hyperbolas, as are the traces in those planes that are parallel to these.	HYPERBOLIC PARABOLOID	$z=\frac{y^2}{b^2}-\frac{x^2}{a^2}$ The trace in the xy-plane is a pair of lines intersecting at the origin. The traces in planes parallel to the xy-plane are hyperbolas. The hyperbolas above the xy-plane open in the y-direction, and those below in the x-direction. The traces in the yz- and xz-planes are parabolas, as are the traces in planes parallel to these.

► **Example 2** Sketch the graph of the hyperboloid of one sheet

$$x^2+y^2-\frac{z^2}{4}=1 \tag{8}$$

Solution. The trace in the xy-plane, obtained by setting $z=0$ in (8), is

$$x^2+y^2=1 \qquad (z=0)$$

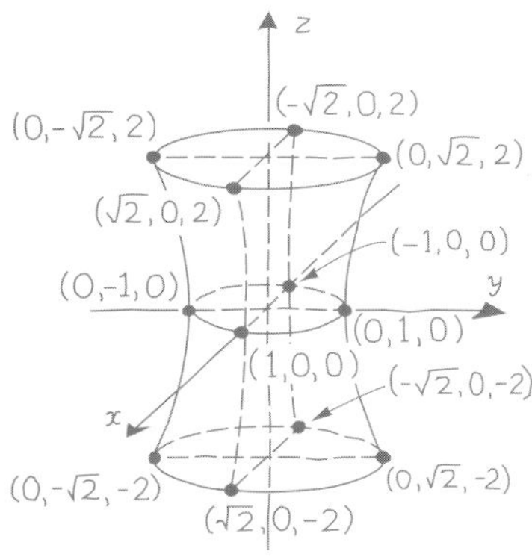

Figure 12.7.6

which is a circle of radius 1 centered on the z-axis. The traces in the planes $z = 2$ and $z = -2$, obtained by setting $z = \pm 2$ in (8), are given by

$$x^2 + y^2 = 2 \qquad (z = \pm 2)$$

which are circles of radius $\sqrt{2}$ centered on the z-axis. Joining these circles by the hyperbolic traces in the vertical coordinate planes yields the graph in Figure 12.7.6. ◀

A rough sketch of the hyperboloid of two sheets

$$\frac{z^2}{c^2} - \frac{x^2}{a^2} - \frac{y^2}{b^2} = 1 \qquad (a > 0, b > 0, c > 0) \tag{9}$$

can be obtained by first plotting the intersections with the z-axis, then sketching the elliptical traces in the planes $z = \pm 2c$, and then sketching the hyperbolic traces that connect the z-axis intersections and the endpoints of the axes of the ellipses. (It is not essential to use the planes $z = \pm 2c$, but these are good choices since they simplify the calculations slightly and have the right spacing for a good sketch.) The next example illustrates this technique.

▶ **Example 3** Sketch the graph of the hyperboloid of two sheets

$$z^2 - x^2 - \frac{y^2}{4} = 1 \tag{10}$$

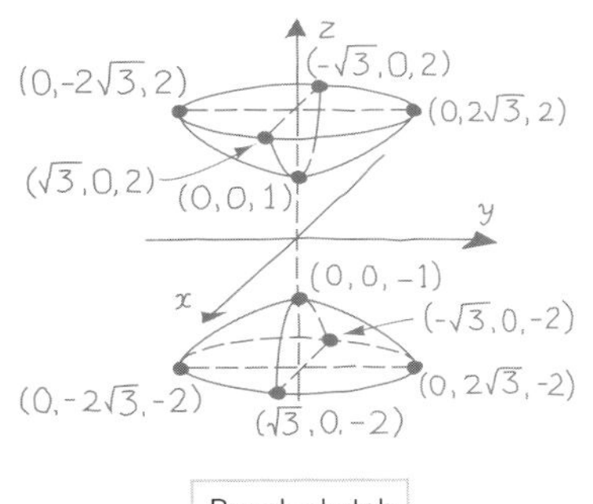

Figure 12.7.7

Solution. The z-intercepts, obtained by setting $x = 0$ and $y = 0$ in (10), are $z = \pm 1$. The traces in the planes $z = 2$ and $z = -2$, obtained by setting $z = \pm 2$ in (10), are given by

$$\frac{x^2}{3} + \frac{y^2}{12} = 1 \qquad (z = \pm 2)$$

Sketching these ellipses and the hyperbolic traces in the vertical coordinate planes yields Figure 12.7.7. ◀

A rough sketch of the elliptic cone

$$z^2 = \frac{x^2}{a^2} + \frac{y^2}{b^2} \qquad (a > 0, b > 0) \tag{11}$$

can be obtained by first sketching the elliptical traces in the planes $z = \pm 1$ and then sketching the linear traces that connect the endpoints of the axes of the ellipses. The next example illustrates this technique.

▶ **Example 4** Sketch the graph of the elliptic cone

$$z^2 = x^2 + \frac{y^2}{4} \tag{12}$$

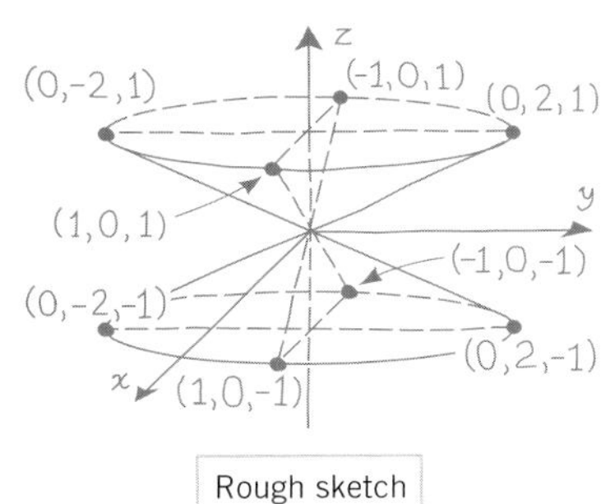

Figure 12.7.8

Solution. The traces of (12) in the planes $z = \pm 1$ are given by

$$x^2 + \frac{y^2}{4} = 1 \qquad (z = \pm 1)$$

Sketching these ellipses and the linear traces in the vertical coordinate planes yields the graph in Figure 12.7.8. ◀

In the special cases of (11) and (13) where $a = b$, the traces parallel to the xy-plane are circles. In these cases, we call (11) a ***circular cone*** and (13) a ***circular paraboloid.***

A rough sketch of the elliptic paraboloid

$$z = \frac{x^2}{a^2} + \frac{y^2}{b^2} \qquad (a > 0, b > 0) \tag{13}$$

can be obtained by first sketching the elliptical trace in the plane $z = 1$ and then sketching the parabolic traces in the vertical coordinate planes to connect the origin to the ends of the axes of the ellipse. The next example illustrates this technique.

▶ Example 5 Sketch the graph of the elliptic paraboloid

$$z = \frac{x^2}{4} + \frac{y^2}{9} \tag{14}$$

Solution. The trace of (14) in the plane $z = 1$ is

$$\frac{x^2}{4} + \frac{y^2}{9} = 1 \qquad (z = 1)$$

Sketching this ellipse and the parabolic traces in the vertical coordinate planes yields the graph in Figure 12.7.9. ◀

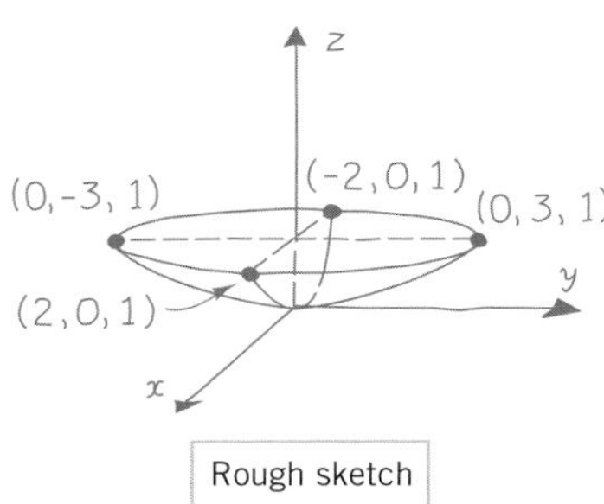

Figure 12.7.9

A rough sketch of the hyperbolic paraboloid

$$z = \frac{y^2}{b^2} - \frac{x^2}{a^2} \qquad (a > 0, b > 0) \tag{15}$$

can be obtained by first sketching the two parabolic traces that pass through the origin (one in the plane $x = 0$ and the other in the plane $y = 0$). After the parabolic traces are drawn, sketch the hyperbolic traces in the planes $z = \pm 1$ and then fill in any missing edges. The next example illustrates this technique.

▶ Example 6 Sketch the graph of the hyperbolic paraboloid

$$z = \frac{y^2}{4} - \frac{x^2}{9} \tag{16}$$

Solution. Setting $x = 0$ in (16) yields

$$z = \frac{y^2}{4} \qquad (x = 0)$$

which is a parabola in the yz-plane with vertex at the origin and opening in the positive z-direction (since $z \geq 0$), and setting $y = 0$ yields

$$z = -\frac{x^2}{9} \qquad (y = 0)$$

which is a parabola in the xz-plane with vertex at the origin and opening in the negative z-direction.

The trace in the plane $z = 1$ is

$$\frac{y^2}{4} - \frac{x^2}{9} = 1 \qquad (z = 1)$$

which is a hyperbola that opens along a line parallel to the y-axis (verify), and the trace in the plane $z = -1$ is

$$\frac{x^2}{9} - \frac{y^2}{4} = 1 \qquad (z = -1)$$

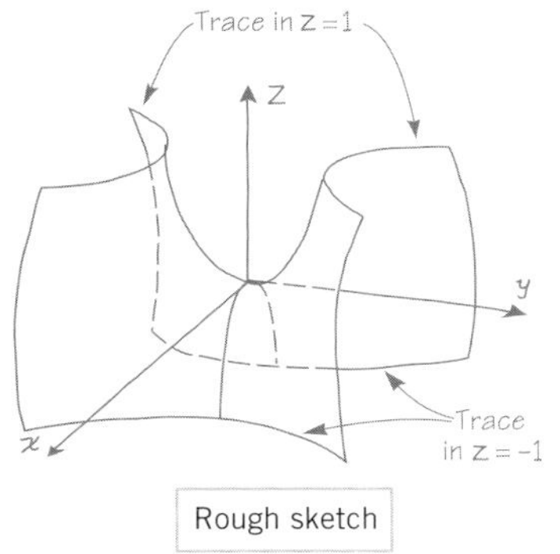

Figure 12.7.10

which is a hyperbola that opens along a line parallel to the x-axis. Combining all of the above information leads to the sketch in Figure 12.7.10. ◀

> The hyperbolic paraboloid in Figure 12.7.10 has an interesting behavior at the origin—the trace in the xz-plane has a relative maximum at $(0, 0, 0)$, and the trace in the yz-plane has a relative minimum at $(0, 0, 0)$. Thus, a bug walking on the surface may view the origin as a highest point if traveling along one path, or may view the origin as a lowest point if traveling along a different path. A point with this property is commonly called a ***saddle point*** or a ***minimax point***.

Figure 12.7.11 shows two computer-generated views of the hyperbolic paraboloid in Example 6. The first view, which is much like our rough sketch in Figure 12.7.10, has cuts at the top and bottom that are hyperbolic traces parallel to the xy-plane. In the second view the top horizontal cut has been omitted; this helps to emphasize the parabolic traces parallel to the xz-plane.

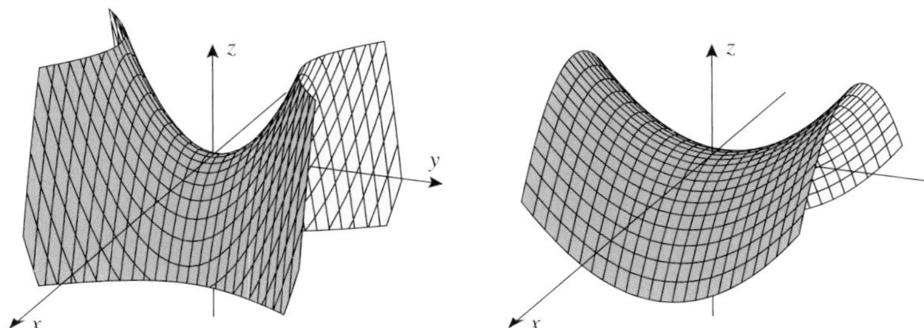

Figure 12.7.11

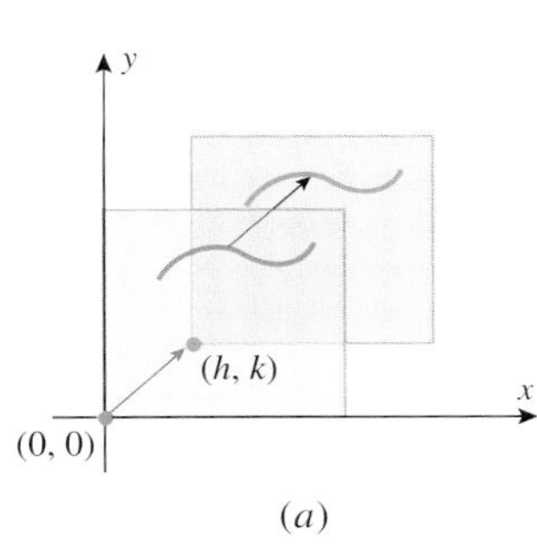

Figure 12.7.12

TRANSLATIONS OF QUADRIC SURFACES

In Section 11.4 we saw that a conic in an xy-coordinate system can be translated by substituting $x - h$ for x and $y - k$ for y in its equation. To understand why this works, think of the xy-axes as fixed and think of the plane as a transparent sheet of plastic on which all graphs are drawn. When the coordinates of points are modified by substituting $(x - h, y - k)$ for (x, y), the geometric effect is to translate the sheet of plastic (and hence all curves) so that the point on the plastic that was initially at $(0, 0)$ is moved to the point (h, k) (see Figure 12.7.12a).

For the analog in three dimensions, think of the xyz-axes as fixed and think of 3-space as a transparent block of plastic in which all surfaces are embedded. When the coordinates of points are modified by substituting $(x - h, y - k, z - l)$ for (x, y, z), the geometric effect is to translate the block of plastic (and hence all surfaces) so that the point in the plastic block that was initially at $(0, 0, 0)$ is moved to the point (h, k, l) (see Figure 12.7.12b).

▶ **Example 7** Describe the surface $z = (x - 1)^2 + (y + 2)^2 + 3$.

Solution. The equation can be rewritten as

$$z - 3 = (x - 1)^2 + (y + 2)^2$$

This surface is the paraboloid that results by translating the paraboloid

$$z = x^2 + y^2$$

in Figure 12.7.3 so that the new "vertex" is at the point $(1, -2, 3)$. A rough sketch of this paraboloid is shown in Figure 12.7.13. ◀

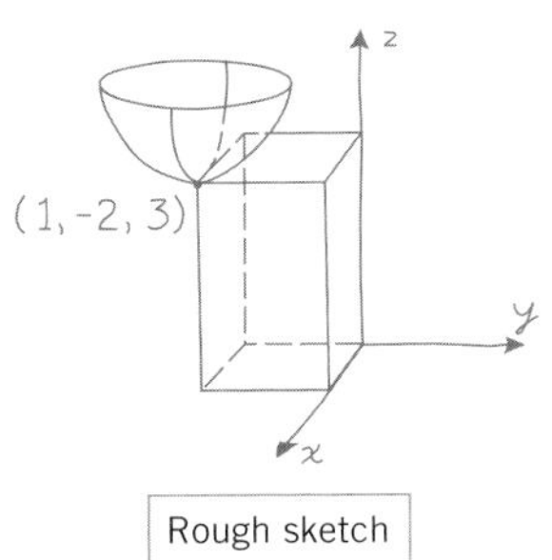

Figure 12.7.13

▶ Example 8 Describe the surface

$$4x^2 + 4y^2 + z^2 + 8y - 4z = -4$$

Solution. Completing the squares yields

$$4x^2 + 4(y+1)^2 + (z-2)^2 = -4 + 4 + 4$$

or

$$x^2 + (y+1)^2 + \frac{(z-2)^2}{4} = 1$$

Thus, the surface is the ellipsoid that results when the ellipsoid

$$x^2 + y^2 + \frac{z^2}{4} = 1$$

is translated so that the new "center" is at the point $(0, -1, 2)$. A rough sketch of this ellipsoid is shown in Figure 12.7.14. ◀

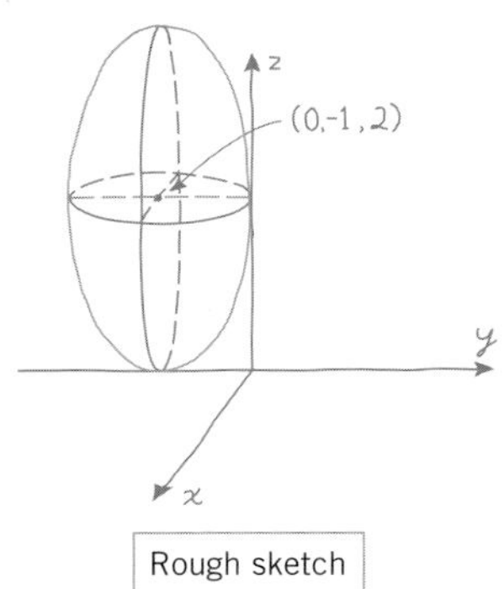

Figure 12.7.14

In Figure 12.7.14, the cross section in the yz-plane is shown tangent to both the y- and z-axes. Confirm that this is correct.

REFLECTIONS OF SURFACES IN 3-SPACE

Recall that in an xy-coordinate system a point (x, y) is reflected about the x-axis if y is replaced by $-y$, and it is reflected about the y-axis if x is replaced by $-x$. In an xyz-coordinate system, a point (x, y, z) is reflected about the xy-plane if z is replaced by $-z$, it is reflected about the yz-plane if x is replaced by $-x$, and it is reflected about the xz-plane if y is replaced by $-y$ (Figure 12.7.15). It follows that *replacing a variable by its negative in the equation of a surface causes that surface to be reflected about a coordinate plane.*

Recall also that in an xy-coordinate system a point (x, y) is reflected about the line $y = x$ if x and y are interchanged. However, in an xyz-coordinate system, interchanging x and y reflects the point (x, y, z) about the plane $y = x$ (Figure 12.7.16). Similarly, interchanging x and z reflects the point about the plane $x = z$, and interchanging y and z reflects it about the plane $y = z$. Thus, it follows that *interchanging two variables in the equation of a surface reflects that surface about a plane that makes a* 45° *angle with two of the coordinate planes.*

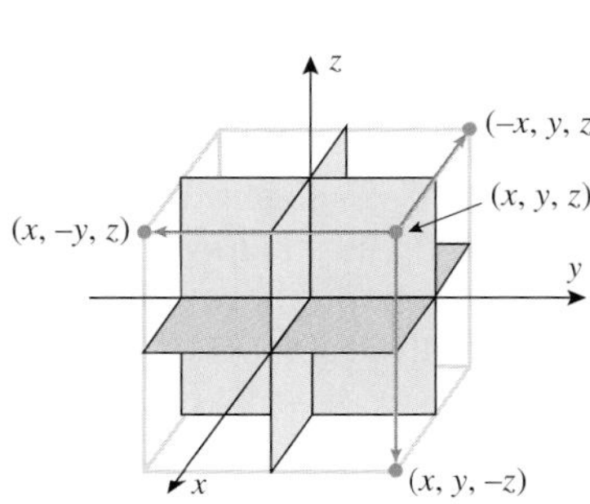

Figure 12.7.15

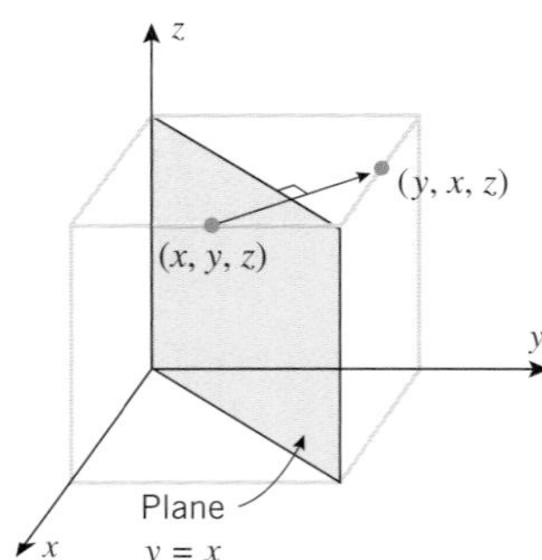

Figure 12.7.16

▶ Example 9 Describe the surfaces

$$\text{(a) } y^2 = x^2 + z^2 \qquad \text{(b) } z = -(x^2 + y^2)$$

Solution (a). The graph of the equation $y^2 = x^2 + z^2$ results from interchanging y and z in the equation $z^2 = x^2 + y^2$. Thus, the graph of the equation $y^2 = x^2 + z^2$ can be obtained by reflecting the graph of $z^2 = x^2 + y^2$ about the plane $y = z$. Since the graph of

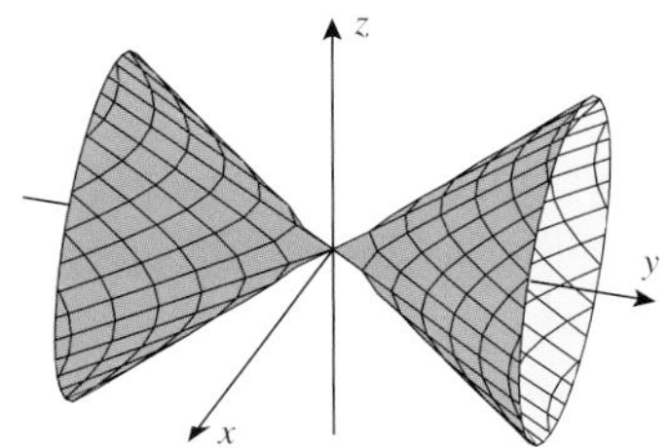

Figure 12.7.17

$z^2 = x^2 + y^2$ is a circular cone opening along the z-axis (see Table 12.7.1), it follows that the graph of $y^2 = x^2 + z^2$ is a circular cone opening along the y-axis (Figure 12.7.17).

Solution (b). The graph of the equation $z = -(x^2 + y^2)$ can be written as $-z = x^2 + y^2$, which can be obtained by replacing z with $-z$ in the equation $z = x^2 + y^2$. Since the graph of $z = x^2 + y^2$ is a circular paraboloid opening in the positive z-direction (see Table 12.7.1), it follows that the graph of $z = -(x^2 + y^2)$ is a circular paraboloid opening in the negative z-direction (Figure 12.7.18). ◄

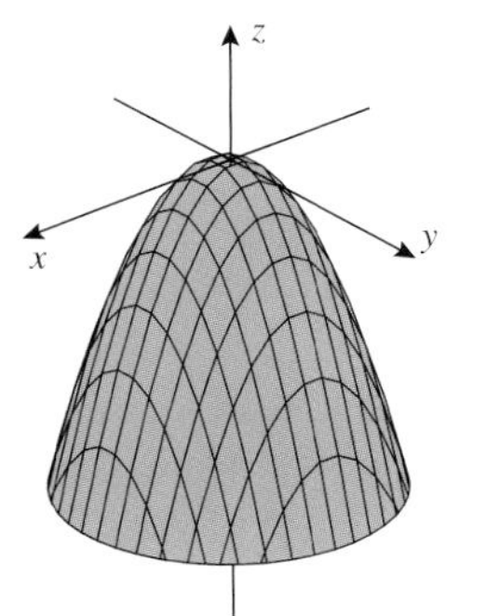

Figure 12.7.18

A TECHNIQUE FOR IDENTIFYING QUADRIC SURFACES

The equations of the quadric surfaces in Table 12.7.1 have certain characteristics that make it possible to identify quadric surfaces that are derived from these equations by reflections. These identifying characteristics, which are shown in Table 12.7.2, are based on writing the equation of the quadric surface so that all of the variable terms are on the left side of the equation and there is a 1 or a 0 on the right side. When there is a 1 on the right side the surface is an ellipsoid, hyperboloid of one sheet, or a hyperboloid of two sheets, and when there is a 0 on the right side it is an elliptic cone, an elliptic paraboloid, or a hyperbolic paraboloid. Within the group with a 1 on the right side, ellipsoids have no minus signs, hyperboloids of one sheet have one minus sign, and hyperboloids of two sheets have two minus signs. Within the group with a 0 on the right side, elliptic cones have no linear terms, elliptic paraboloids have one linear term and two quadratic terms with the same sign, and hyperbolic paraboloids have one linear term and two quadratic terms with opposite signs. These characteristics do not change when the surface is reflected about a coordinate plane or planes of the form $x = y$, $x = z$, or $y = z$, thereby making it possible to identify the reflected quadric surface from the form of its equation.

Table 12.7.2

EQUATION	$\frac{x^2}{a^2}+\frac{y^2}{b^2}+\frac{z^2}{c^2}=1$	$\frac{x^2}{a^2}+\frac{y^2}{b^2}-\frac{z^2}{c^2}=1$	$\frac{z^2}{c^2}-\frac{x^2}{a^2}-\frac{y^2}{b^2}=1$	$z^2-\frac{x^2}{a^2}-\frac{y^2}{b^2}=0$	$z-\frac{x^2}{a^2}-\frac{y^2}{b^2}=0$	$z-\frac{y^2}{b^2}+\frac{x^2}{a^2}=0$
CHARACTERISTIC	No minus signs	One minus sign	Two minus signs	No linear terms	One linear term; two quadratic terms with the same sign	One linear term; two quadratic terms with opposite signs
CLASSIFICATION	Ellipsoid	Hyperboloid of one sheet	Hyperboloid of two sheets	Elliptic cone	Elliptic paraboloid	Hyperbolic paraboloid

► **Example 10** Identify the surfaces

(a) $3x^2 - 4y^2 + 12z^2 + 12 = 0$ (b) $4x^2 - 4y + z^2 = 0$

Solution (a). The equation can be rewritten as

$$\frac{y^2}{3} - \frac{x^2}{4} - z^2 = 1$$

This equation has a 1 on the right side and two negative terms on the left side, so its graph is a hyperboloid of two sheets.

Solution (b). The equation has one linear term and two quadratic terms with the same sign, so its graph is an elliptic paraboloid. ◄

✔QUICK CHECK EXERCISES 12.7

(See page 850 for answers.)

1. For the surface $4x^2 + y^2 + z^2 = 9$, classify the indicated trace as an ellipse, hyperbola, or parabola.
(a) $x = 0$ (b) $y = 0$ (c) $z = 1$

2. For the surface $4x^2 + z^2 - y^2 = 9$, classify the indicated trace as an ellipse, hyperbola, or parabola.
(a) $x = 0$ (b) $y = 0$ (c) $z = 1$

3. For the surface $4x^2 + y^2 - z = 0$, classify the indicated trace as an ellipse, hyperbola, or parabola.
(a) $x = 0$ (b) $y = 0$ (c) $z = 1$

4. Classify each surface as an ellipsoid, hyperboloid of one sheet, hyperboloid of two sheets, elliptic cone, elliptic paraboloid, or hyperbolic paraboloid.
(a) $\frac{x^2}{36} + \frac{y^2}{25} - z = 0$ (b) $\frac{x^2}{36} + \frac{y^2}{25} + z^2 = 1$
(c) $\frac{x^2}{36} - \frac{y^2}{25} + z = 0$ (d) $\frac{x^2}{36} + \frac{y^2}{25} - z^2 = 1$
(e) $\frac{x^2}{36} + \frac{y^2}{25} - z^2 = 0$ (f) $z^2 - \frac{x^2}{36} - \frac{y^2}{25} = 1$

EXERCISE SET 12.7

1–2 Identify the quadric surface as an ellipsoid, hyperboloid of one sheet, hyperboloid of two sheets, elliptic cone, elliptic paraboloid, or hyperbolic paraboloid by matching the equation with one of the forms given in Table 12.7.1. State the values of a, b, and c in each case.

1. (a) $z = \frac{x^2}{4} + \frac{y^2}{9}$ (b) $z = \frac{y^2}{25} - x^2$
(c) $x^2 + y^2 - z^2 = 16$ (d) $x^2 + y^2 - z^2 = 0$
(e) $4z = x^2 + 4y^2$ (f) $z^2 - x^2 - y^2 = 1$

2. (a) $6x^2 + 3y^2 + 4z^2 = 12$ (b) $y^2 - x^2 - z = 0$
(c) $9x^2 + y^2 - 9z^2 = 9$ (d) $4x^2 + y^2 - 4z^2 = -4$
(e) $2z - x^2 - 4y^2 = 0$ (f) $12z^2 - 3x^2 = 4y^2$

3. Find an equation for and sketch the surface that results when the circular paraboloid $z = x^2 + y^2$ is reflected about the plane
(a) $z = 0$ (b) $x = 0$ (c) $y = 0$
(d) $y = x$ (e) $x = z$ (f) $y = z$.

4. Find an equation for and sketch the surface that results when the hyperboloid of one sheet $x^2 + y^2 - z^2 = 1$ is reflected about the plane
(a) $z = 0$ (b) $x = 0$ (c) $y = 0$
(d) $y = x$ (e) $x = z$ (f) $y = z$.

FOCUS ON CONCEPTS

5. The given equations represent quadric surfaces whose orientations are different from those in Table 12.7.1. In each part, identify the quadric surface, and give a verbal description of its orientation (e.g., an elliptic cone opening along the z-axis or a hyperbolic paraboloid straddling the y-axis).
(a) $\frac{z^2}{c^2} - \frac{y^2}{b^2} + \frac{x^2}{a^2} = 1$ (b) $\frac{x^2}{a^2} - \frac{y^2}{b^2} - \frac{z^2}{c^2} = 1$
(c) $x = \frac{y^2}{b^2} + \frac{z^2}{c^2}$ (d) $x^2 = \frac{y^2}{b^2} + \frac{z^2}{c^2}$
(e) $y = \frac{z^2}{c^2} - \frac{x^2}{a^2}$ (f) $y = -\left(\frac{x^2}{a^2} + \frac{z^2}{c^2}\right)$

6. For each of the surfaces in Exercise 5, find the equation of the surface that results if the given surface is reflected about the xz-plane and that surface is then reflected about the plane $z = 0$.

7–8 Find equations of the traces in the coordinate planes and sketch the traces in an xyz-coordinate system. [*Suggestion:* If you have trouble sketching a trace directly in three dimensions, start with a sketch in two dimensions by placing the coordinate plane in the plane of the paper, then transfer the sketch to three dimensions.]

7. (a) $\frac{x^2}{9} + \frac{y^2}{25} + \frac{z^2}{4} = 1$ (b) $z = x^2 + 4y^2$
(c) $\frac{x^2}{9} + \frac{y^2}{16} - \frac{z^2}{4} = 1$

8. (a) $y^2 + 9z^2 = x$ (b) $4x^2 - y^2 + 4z^2 = 4$
(c) $z^2 = x^2 + \frac{y^2}{4}$

9–10 In these exercises, traces of the surfaces in the planes are conic sections. In each part, find an equation of the trace, and state whether it is an ellipse, a parabola, or a hyperbola.

9. (a) $4x^2 + y^2 + z^2 = 4;\ y = 1$
(b) $4x^2 + y^2 + z^2 = 4;\ x = \frac{1}{2}$
(c) $9x^2 - y^2 - z^2 = 16;\ x = 2$
(d) $9x^2 - y^2 - z^2 = 16;\ z = 2$
(e) $z = 9x^2 + 4y^2;\ y = 2$
(f) $z = 9x^2 + 4y^2;\ z = 4$

10. (a) $9x^2 - y^2 + 4z^2 = 9;\ x = 2$
(b) $9x^2 - y^2 + 4z^2 = 9;\ y = 4$
(c) $x^2 + 4y^2 - 9z^2 = 0;\ y = 1$
(d) $x^2 + 4y^2 - 9z^2 = 0;\ z = 1$
(e) $z = x^2 - 4y^2;\ x = 1$
(f) $z = x^2 - 4y^2;\ z = 4$

11–22 Identify and sketch the quadric surface.

11. $x^2 + \frac{y^2}{4} + \frac{z^2}{9} = 1$

12. $x^2 + 4y^2 + 9z^2 = 36$

13. $\frac{x^2}{4} + \frac{y^2}{9} - \frac{z^2}{16} = 1$

14. $x^2 + y^2 - z^2 = 9$

15. $4z^2 = x^2 + 4y^2$

16. $9x^2 + 4y^2 - 36z^2 = 0$

17. $9z^2 - 4y^2 - 9x^2 = 36$

18. $y^2 - \frac{x^2}{4} - \frac{z^2}{9} = 1$

19. $z = y^2 - x^2$

20. $16z = y^2 - x^2$

21. $4z = x^2 + 2y^2$

22. $z - 3x^2 - 3y^2 = 0$

23–28 The given equation represents a quadric surface whose orientation is different from that in Table 12.7.1. Identify and sketch the surface.

23. $x^2 - 3y^2 - 3z^2 = 0$

24. $x - y^2 - 4z^2 = 0$

25. $2y^2 - x^2 + 2z^2 = 8$

26. $x^2 - 3y^2 - 3z^2 = 9$

27. $z = \frac{x^2}{4} - \frac{y^2}{9}$

28. $4x^2 - y^2 + 4z^2 = 16$

29–32 Sketch the surface.

29. $z = \sqrt{x^2 + y^2}$

30. $z = \sqrt{1 - x^2 - y^2}$

31. $z = \sqrt{x^2 + y^2 - 1}$

32. $z = \sqrt{1 + x^2 + y^2}$

33–36 Identify the surface and make a rough sketch that shows its position and orientation.

33. $z = (x + 2)^2 + (y - 3)^2 - 9$

34. $4x^2 - y^2 + 16(z - 2)^2 = 100$

35. $9x^2 + y^2 + 4z^2 - 18x + 2y + 16z = 10$

36. $z^2 = 4x^2 + y^2 + 8x - 2y + 4z$

37–38 Use the ellipsoid $4x^2 + 9y^2 + 18z^2 = 72$ in these exercises.

37. (a) Find an equation of the elliptical trace in the plane $z = \sqrt{2}$.
(b) Find the lengths of the major and minor axes of the ellipse in part (a).
(c) Find the coordinates of the foci of the ellipse in part (a).
(d) Describe the orientation of the focal axis of the ellipse in part (a) relative to the coordinate axes.

38. (a) Find an equation of the elliptical trace in the plane $x = 3$.
(b) Find the lengths of the major and minor axes of the ellipse in part (a).
(c) Find the coordinates of the foci of the ellipse in part (a).
(d) Describe the orientation of the focal axis of the ellipse in part (a) relative to the coordinate axes.

39–42 These exercises refer to the hyperbolic paraboloid $z = y^2 - x^2$.

39. (a) Find an equation of the hyperbolic trace in the plane $z = 4$.
(b) Find the vertices of the hyperbola in part (a).
(c) Find the foci of the hyperbola in part (a).
(d) Describe the orientation of the focal axis of the hyperbola in part (a) relative to the coordinate axes.

40. (a) Find an equation of the hyperbolic trace in the plane $z = -4$.
(b) Find the vertices of the hyperbola in part (a).
(c) Find the foci of the hyperbola in part (a).
(d) Describe the orientation of the focal axis of the hyperbola in part (a) relative to the coordinate axes.

41. (a) Find an equation of the parabolic trace in the plane $x = 2$.
(b) Find the vertices of the parabola in part (a).
(c) Find the focus of the parabola in part (a).
(d) Describe the orientation of the focal axis of the parabola in part (a) relative to the coordinate axes.

42. (a) Find an equation of the parabolic trace in the plane $y = 2$.
(b) Find the vertex of the parabola in part (a).
(c) Find the focus of the parabola in part (a).
(d) Describe the orientation of the focal axis of the parabola in part (a) relative to the coordinate axes.

43–44 Sketch the region enclosed between the surfaces and describe their curve of intersection.

43. The paraboloids $z = x^2 + y^2$ and $z = 4 - x^2 - y^2$

44. The hyperbolic paraboloid $x^2 = y^2 + z$ and the ellipsoid $x^2 = 4 - 2y^2 - 2z$

45–46 Find an equation for the surface generated by revolving the curve about the y-axis.

45. $y = 4x^2$ $(z = 0)$

46. $y = 2x$ $(z = 0)$

47. Find an equation of the surface consisting of all points $P(x, y, z)$ that are equidistant from the point $(0, 0, 1)$ and the plane $z = -1$. Identify the surface.

48. Find an equation of the surface consisting of all points $P(x, y, z)$ that are twice as far from the plane $z = -1$ as from the point $(0, 0, 1)$. Identify the surface.

49. If a sphere

$$\frac{x^2}{a^2} + \frac{y^2}{a^2} + \frac{z^2}{a^2} = 1$$

of radius a is compressed in the z-direction, then the resulting surface, called an ***oblate spheroid***, has an equation of the form

$$\frac{x^2}{a^2} + \frac{y^2}{a^2} + \frac{z^2}{c^2} = 1$$

where $c < a$. Show that the oblate spheroid has a circular trace of radius a in the xy-plane and an elliptical trace in the

xz-plane with major axis of length $2a$ along the x-axis and minor axis of length $2c$ along the z-axis.

50. The Earth's rotation causes a flattening at the poles, so its shape is often modeled as an oblate spheroid rather than a sphere (see Exercise 49 for terminology). One of the models used by global positioning satellites is the ***World Geodetic System of 1984*** (WGS-84), which treats the Earth as an oblate spheroid whose equatorial radius is 6378.1370 km and whose polar radius (the distance from the Earth's center to the poles) is 6356.5231 km. Use the WGS-84 model to find an equation for the surface of the Earth relative to the coordinate system shown in the accompanying figure.

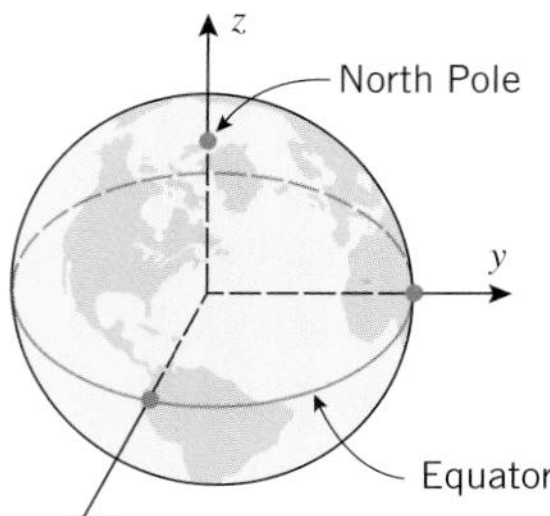

Figure Ex-50

51. Use the method of slicing to show that the volume of the ellipsoid

$$\frac{x^2}{a^2} + \frac{y^2}{b^2} + \frac{z^2}{c^2} = 1$$

is $\frac{4}{3}\pi abc$.

QUICK CHECK ANSWERS 12.7

1. (a) ellipse (b) ellipse (c) ellipse **2.** (a) hyperbola (b) ellipse (c) hyperbola **3.** (a) parabola (b) parabola (c) ellipse **4.** (a) elliptic paraboloid (b) ellipsoid (c) hyperbolic paraboloid (d) hyperboloid of one sheet (e) elliptic cone (f) hyperboloid of two sheets

12.8 CYLINDRICAL AND SPHERICAL COORDINATES

In this section we will discuss two new types of coordinate systems in 3-space that are often more useful than rectangular coordinate systems for studying surfaces with symmetries. These new coordinate systems also have important applications in navigation, astronomy, and the study of rotational motion about an axis.

CYLINDRICAL AND SPHERICAL COORDINATE SYSTEMS

Three coordinates are required to establish the location of a point in 3-space. We have already done this using rectangular coordinates. However, Figure 12.8.1 shows two other possibilities: part (a) of the figure shows the ***rectangular coordinates*** (x, y, z) of a point P, part (b) shows the ***cylindrical coordinates*** (r, θ, z) of P, and part (c) shows the ***spherical coordinates*** (ρ, θ, ϕ) of P. In a rectangular coordinate system the coordinates can be any real numbers, but in cylindrical and spherical coordinate systems there are restrictions on the allowable values of the coordinates (as indicated in Figure 12.8.1).

CONSTANT SURFACES

In rectangular coordinates the surfaces represented by equations of the form

$$x = x_0, \quad y = y_0, \quad \text{and} \quad z = z_0$$

where x_0, y_0, and z_0 are constants, are planes parallel to the yz-plane, xz-plane, and xy-plane, respectively (Figure 12.8.2). In cylindrical coordinates the surfaces represented by equations of the form

$$r = r_0, \quad \theta = \theta_0, \quad \text{and} \quad z = z_0$$

where r_0, θ_0, and z_0 are constants, are shown in Figure 12.8.3:

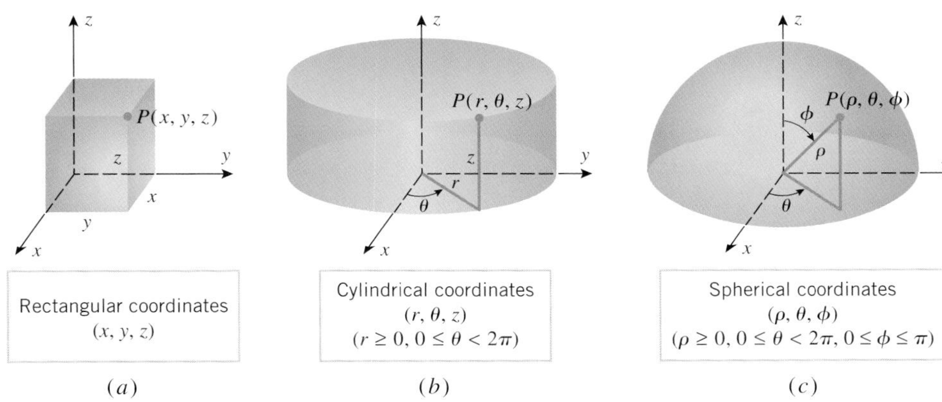

Figure 12.8.1

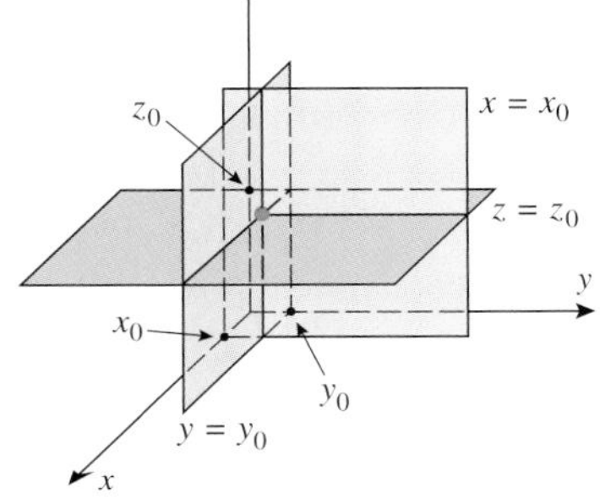

Figure 12.8.2

- The surface $r = r_0$ is a right circular cylinder of radius r_0 centered on the z-axis. At each point (r, θ, z) on this cylinder, r has the value r_0, but θ and z are unrestricted except for our general restriction that $0 \le \theta < 2\pi$.
- The surface $\theta = \theta_0$ is a half-plane attached along the z-axis and making an angle θ_0 with the positive x-axis. At each point (r, θ, z) on this surface, θ has the value θ_0, but r and z are unrestricted except for our general restriction that $r \ge 0$.
- The surface $z = z_0$ is a horizontal plane. At each point (r, θ, z) on this plane, z has the value z_0, but r and θ are unrestricted except for the general restrictions.

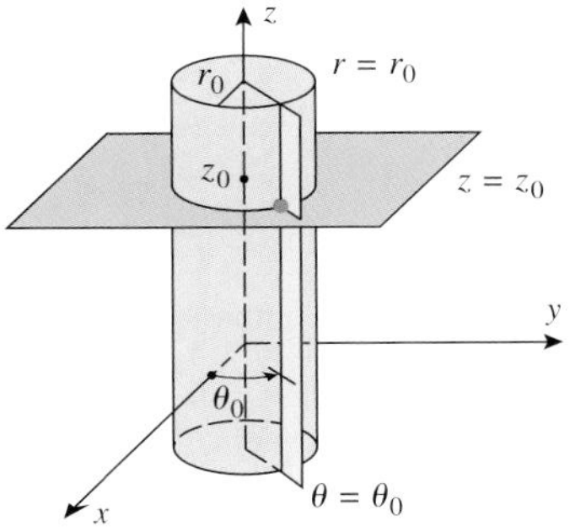

Figure 12.8.3

In spherical coordinates the surfaces represented by equations of the form

$$\rho = \rho_0, \quad \theta = \theta_0, \quad \text{and} \quad \phi = \phi_0$$

where ρ_0, θ_0, and ϕ_0 are constants, are shown in Figure 12.8.4:

- The surface $\rho = \rho_0$ consists of all points whose distance ρ from the origin is ρ_0. Assuming ρ_0 to be nonnegative, this is a sphere of radius ρ_0 centered at the origin.
- As in cylindrical coordinates, the surface $\theta = \theta_0$ is a half-plane attached along the z-axis, making an angle of θ_0 with the positive x-axis.
- The surface $\phi = \phi_0$ consists of all points from which a line segment to the origin makes an angle of ϕ_0 with the positive z-axis. Depending on whether $0 < \phi_0 < \pi/2$ or $\pi/2 < \phi_0 < \pi$, this will be the nappe of a cone opening up or opening down. (If $\phi_0 = \pi/2$, then the cone is flat, and the surface is the xy-plane.)

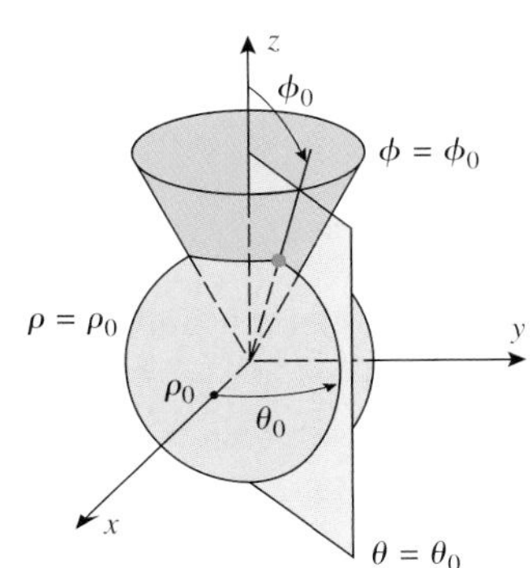

Figure 12.8.4

CONVERTING COORDINATES

Just as we needed to convert between rectangular and polar coordinates in 2-space, so we will need to be able to convert between rectangular, cylindrical, and spherical coordinates in 3-space. Table 12.8.1 provides formulas for making these conversions.

The diagrams in Figure 12.8.5 will help you to understand how the formulas in Table 12.8.1 are derived. For example, part (a) of the figure shows that in converting between rectangular coordinates (x, y, z) and cylindrical coordinates (r, θ, z), we can interpret (r, θ) as polar coordinates of (x, y). Thus, the polar-to-rectangular and rectangular-to-polar conversion formulas (1) and (2) of Section 11.1 provide the conversion formulas between rectangular and cylindrical coordinates in the table.

Table 12.8.1

CONVERSION		FORMULAS	RESTRICTIONS
Cylindrical to rectangular Rectangular to cylindrical	$(r, \theta, z) \to (x, y, z)$ $(x, y, z) \to (r, \theta, z)$	$x = r\cos\theta,\quad y = r\sin\theta,\quad z = z$ $r = \sqrt{x^2+y^2},\quad \tan\theta = y/x,\quad z = z$	$r \geq 0, \rho \geq 0$ $0 \leq \theta < 2\pi$ $0 \leq \phi \leq \pi$
Spherical to cylindrical Cylindrical to spherical	$(\rho, \theta, \phi) \to (r, \theta, z)$ $(r, \theta, z) \to (\rho, \theta, \phi)$	$r = \rho\sin\phi,\quad \theta = \theta,\quad z = \rho\cos\phi$ $\rho = \sqrt{r^2+z^2},\quad \theta = \theta,\quad \tan\phi = r/z$	
Spherical to rectangular Rectangular to spherical	$(\rho, \theta, \phi) \to (x, y, z)$ $(x, y, z) \to (\rho, \theta, \phi)$	$x = \rho\sin\phi\cos\theta,\quad y = \rho\sin\phi\sin\theta,\quad z = \rho\cos\phi$ $\rho = \sqrt{x^2+y^2+z^2},\quad \tan\theta = y/x,\quad \cos\phi = z/\sqrt{x^2+y^2+z^2}$	

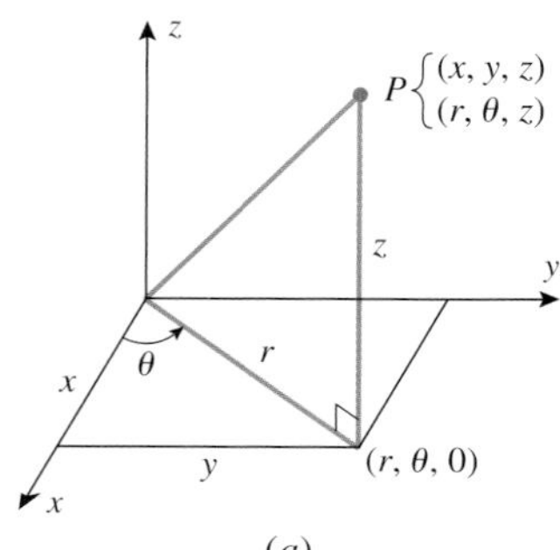

(*a*)

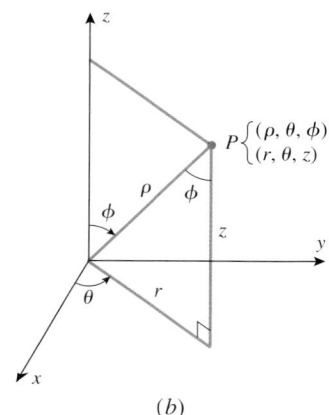

(*b*)

Figure 12.8.5

Part (*b*) of Figure 12.8.5 suggests that the spherical coordinates (ρ, θ, ϕ) of a point P can be converted to cylindrical coordinates (r, θ, z) by the conversion formulas

$$r = \rho\sin\phi,\quad \theta = \theta,\quad z = \rho\cos\phi \tag{1}$$

Moreover, since the cylindrical coordinates (r, θ, z) of P can be converted to rectangular coordinates (x, y, z) by the conversion formulas

$$x = r\cos\theta,\quad y = r\sin\theta,\quad z = z \tag{2}$$

we can obtain direct conversion formulas from spherical coordinates to rectangular coordinates by substituting (1) in (2). This yields

$$x = \rho\sin\phi\cos\theta,\quad y = \rho\sin\phi\sin\theta,\quad z = \rho\cos\phi \tag{3}$$

The other conversion formulas in Table 12.8.1 are left as exercises.

▶ Example 1

(a) Find the rectangular coordinates of the point with cylindrical coordinates

$$(r, \theta, z) = (4, \pi/3, -3)$$

(b) Find the rectangular coordinates of the point with spherical coordinates

$$(\rho, \theta, \phi) = (4, \pi/3, \pi/4)$$

Solution (a). Applying the cylindrical-to-rectangular conversion formulas in Table 12.8.1 yields

$$x = r\cos\theta = 4\cos\frac{\pi}{3} = 2,\quad y = r\sin\theta = 4\sin\frac{\pi}{3} = 2\sqrt{3},\quad z = -3$$

Thus, the rectangular coordinates of the point are $(x, y, z) = (2, 2\sqrt{3}, -3)$ (Figure 12.8.6).

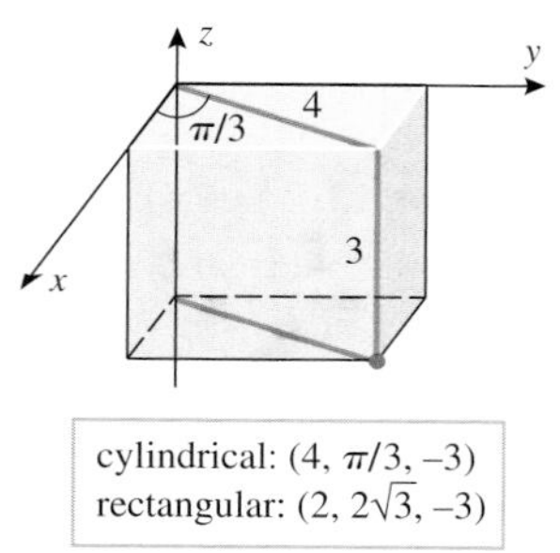

Figure 12.8.6

Solution (b). Applying the spherical-to-rectangular conversion formulas in Table 12.8.1 yields

$$x = \rho\sin\phi\cos\theta = 4\sin\frac{\pi}{4}\cos\frac{\pi}{3} = \sqrt{2}$$

$$y = \rho\sin\phi\sin\theta = 4\sin\frac{\pi}{4}\sin\frac{\pi}{3} = \sqrt{6}$$

$$z = \rho\cos\phi = 4\cos\frac{\pi}{4} = 2\sqrt{2}$$

The rectangular coordinates of the point are $(x, y, z) = (\sqrt{2}, \sqrt{6}, 2\sqrt{2})$ (Figure 12.8.7). ◀

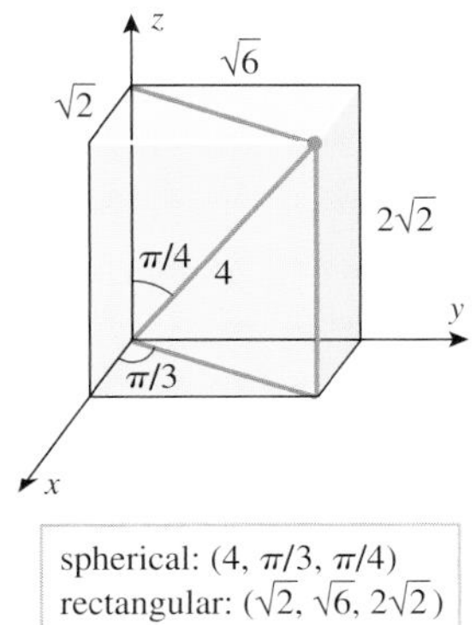

Figure 12.8.7

How should θ be chosen if $x = 0$? How should θ be chosen if $y = 0$?

Since the interval $0 \le \theta < 2\pi$ covers two periods of the tangent function, the conversion formula $\tan\theta = y/x$ does not completely determine θ. When converting from rectangular to cylindrical or spherical coordinates, it is evident from parts (*b*) and (*c*) of Figure 12.8.1 that we should select θ so that

$$0 < \theta < \pi \quad \text{if } y > 0 \quad \text{and} \quad \pi < \theta < 2\pi \quad \text{if } y < 0$$

This is illustrated in the following example.

▶ **Example 2** Find the spherical coordinates of the point that has rectangular coordinates

$$(x, y, z) = (4, -4, 4\sqrt{6})$$

Solution. From the rectangular-to-spherical conversion formulas in Table 12.8.1 we obtain

$$\rho = \sqrt{x^2 + y^2 + z^2} = \sqrt{16 + 16 + 96} = \sqrt{128} = 8\sqrt{2}$$

$$\tan\theta = \frac{y}{x} = -1$$

$$\cos\phi = \frac{z}{\sqrt{x^2 + y^2 + z^2}} = \frac{4\sqrt{6}}{8\sqrt{2}} = \frac{\sqrt{3}}{2}$$

From the restriction $0 \le \theta < 2\pi$ and the computed value of $\tan\theta$, the possibilities for θ are $\theta = 3\pi/4$ and $\theta = 7\pi/4$. However, the given point has a negative y-coordinate, so we must have $\theta = 7\pi/4$. Moreover, from the restriction $0 \le \phi \le \pi$ and the computed value of $\cos\phi$, the only possibility for ϕ is $\phi = \pi/6$. Thus, the spherical coordinates of the point are $(\rho, \theta, \phi) = (8\sqrt{2}, 7\pi/4, \pi/6)$ (Figure 12.8.8). ◀

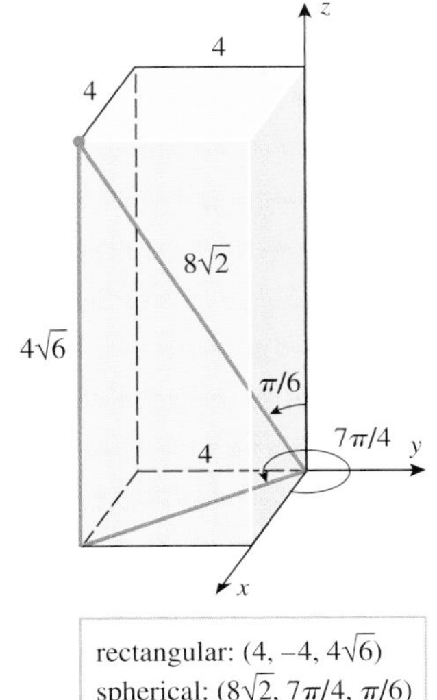

Figure 12.8.8

EQUATIONS OF SURFACES IN CYLINDRICAL AND SPHERICAL COORDINATES

Surfaces of revolution about the z-axis of a rectangular coordinate system usually have simpler equations in cylindrical coordinates than in rectangular coordinates, and the equations of surfaces with symmetry about the origin are usually simpler in spherical coordinates than in rectangular coordinates. For example, consider the upper nappe of the circular cone whose equation in rectangular coordinates is

$$z = \sqrt{x^2 + y^2}$$

(Table 12.8.2). The corresponding equation in cylindrical coordinates can be obtained from the cylindrical-to-rectangular conversion formulas in Table 12.8.1. This yields

$$z = \sqrt{(r\cos\theta)^2 + (r\sin\theta)^2} = \sqrt{r^2} = |r| = r$$

so the equation of the cone in cylindrical coordinates is $z = r$. Going a step further, the equation of the cone in spherical coordinates can be obtained from the spherical-to-cylindrical conversion formulas from Table 12.8.1. This yields

$$\rho\cos\phi = \rho\sin\phi$$

which, if $\rho \ne 0$, can be rewritten as

$$\tan\phi = 1 \quad \text{or} \quad \phi = \frac{\pi}{4}$$

Geometrically, this tells us that the radial line from the origin to any point on the cone makes an angle of $\pi/4$ with the z-axis.

Table 12.8.2

	CONE	CYLINDER	SPHERE	PARABOLOID	HYPERBOLOID
RECTANGULAR	$z = \sqrt{x^2 + y^2}$	$x^2 + y^2 = 1$	$x^2 + y^2 + z^2 = 1$	$z = x^2 + y^2$	$x^2 + y^2 - z^2 = 1$
CYLINDRICAL	$z = r$	$r = 1$	$z^2 = 1 - r^2$	$z = r^2$	$z^2 = r^2 - 1$
SPHERICAL	$\phi = \pi/4$	$\rho = \csc\phi$	$\rho = 1$	$\rho = \cos\phi \csc^2\phi$	$\rho^2 = -\sec 2\phi$

▶ **Example 3** Find equations of the paraboloid $z = x^2 + y^2$ in cylindrical and spherical coordinates.

Verify the equations given in Table 12.8.2 for the cylinder and hyperboloid in cylindrical and spherical coordinates.

Solution. The rectangular-to-cylindrical conversion formulas in Table 12.8.1 yield

$$z = r^2 \tag{4}$$

which is the equation in cylindrical coordinates. Now applying the spherical-to-cylindrical conversion formulas to (4) yields

$$\rho \cos\phi = \rho^2 \sin^2\phi$$

which we can rewrite as

$$\rho = \cos\phi \csc^2\phi$$

Alternatively, we could have obtained this equation directly from the equation in rectangular coordinates by applying the spherical-to-rectangular conversion formulas (verify). ◀

SPHERICAL COORDINATES IN NAVIGATION

Spherical coordinates are related to longitude and latitude coordinates used in navigation. To see why this is so, let us construct a right-hand rectangular coordinate system with its origin at the center of the Earth, its positive z-axis passing through the North Pole, and its positive x-axis passing through the prime meridian (Figure 12.8.9). If we assume the Earth to be a sphere of radius $\rho = 4000$ miles, then each point on the Earth has spherical coordinates of the form $(4000, \theta, \phi)$, where ϕ and θ determine the latitude and longitude of the point. It is common to specify longitudes in degrees east or west of the prime meridian and latitudes in degrees north or south of the equator. However, the next example shows that it is a simple matter to determine ϕ and θ from such data.

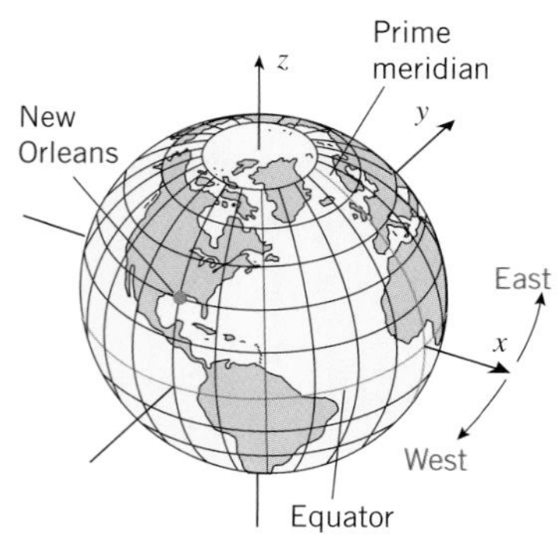

Figure 12.8.9

▶ **Example 4** The city of New Orleans is located at 90° west longitude and 30° north latitude. Find its spherical and rectangular coordinates relative to the coordinate axes of Figure 12.8.9. (Assume that distance is in miles.)

Solution. A longitude of 90° west corresponds to $\theta = 360° - 90° = 270°$ or $\theta = 3\pi/2$ radians; and a latitude of 30° north corresponds to $\phi = 90° - 30° = 60°$ or $\phi = \pi/3$ radians. Thus, the spherical coordinates (ρ, θ, ϕ) of New Orleans are $(4000, 3\pi/2, \pi/3)$.

To find the rectangular coordinates we apply the spherical-to-rectangular conversion formulas in Table 12.8.1. This yields

$$x = 4000 \sin\frac{\pi}{3}\cos\frac{3\pi}{2} = 4000\frac{\sqrt{3}}{2}(0) = 0 \text{ mi}$$

$$y = 4000 \sin\frac{\pi}{3}\sin\frac{3\pi}{2} = 4000\frac{\sqrt{3}}{2}(-1) = -2000\sqrt{3} \text{ mi}$$

$$z = 4000 \cos\frac{\pi}{3} = 4000\left(\frac{1}{2}\right) = 2000 \text{ mi} \blacktriangleleft$$

✔QUICK CHECK EXERCISES 12.8

(See page 856 for answers.)

1. The conversion formulas from cylindrical coordinates (r, θ, z) to rectangular coordinates (x, y, z) are

$x =$ ______, $y =$ ______, $z =$ ______

2. The conversion formulas from spherical coordinates (ρ, θ, ϕ) to rectangular coordinates (x, y, z) are

$x =$ ______, $y =$ ______, $z =$ ______

3. The conversion formulas from spherical coordinates (ρ, θ, ϕ) to cylindrical coordinates (r, θ, z) are

$r =$ ______, $\theta =$ ______, $z =$ ______

4. Let P be the point in 3-space with rectangular coordinates $(\sqrt{2}, -\sqrt{2}, 2\sqrt{3})$.
(a) Cylindrical coordinates for P are $(r, \theta, z) =$ ______.
(b) Spherical coordinates for P are $(\rho, \theta, \phi) =$ ______.

5. Give an equation of a sphere of radius 5, centered at the origin, in
(a) rectangular coordinates
(b) cylindrical coordinates
(c) spherical coordinates.

EXERCISE SET 12.8

Graphing Utility   C CAS

1–2 Convert from rectangular to cylindrical coordinates.

1. (a) $(4\sqrt{3}, 4, -4)$ (b) $(-5, 5, 6)$
(c) $(0, 2, 0)$ (d) $(4, -4\sqrt{3}, 6)$

2. (a) $(\sqrt{2}, -\sqrt{2}, 1)$ (b) $(0, 1, 1)$
(c) $(-4, 4, -7)$ (d) $(2, -2, -2)$

3–4 Convert from cylindrical to rectangular coordinates.

3. (a) $(4, \pi/6, 3)$ (b) $(8, 3\pi/4, -2)$
(c) $(5, 0, 4)$ (d) $(7, \pi, -9)$

4. (a) $(6, 5\pi/3, 7)$ (b) $(1, \pi/2, 0)$
(c) $(3, \pi/2, 5)$ (d) $(4, \pi/2, -1)$

5–6 Convert from rectangular to spherical coordinates.

5. (a) $(1, \sqrt{3}, -2)$ (b) $(1, -1, \sqrt{2})$
(c) $(0, 3\sqrt{3}, 3)$ (d) $(-5\sqrt{3}, 5, 0)$

6. (a) $(4, 4, 4\sqrt{6})$ (b) $(1, -\sqrt{3}, -2)$
(c) $(2, 0, 0)$ (d) $(\sqrt{3}, 1, 2\sqrt{3})$

7–8 Convert from spherical to rectangular coordinates.

7. (a) $(5, \pi/6, \pi/4)$ (b) $(7, 0, \pi/2)$
(c) $(1, \pi, 0)$ (d) $(2, 3\pi/2, \pi/2)$

8. (a) $(1, 2\pi/3, 3\pi/4)$ (b) $(3, 7\pi/4, 5\pi/6)$
(c) $(8, \pi/6, \pi/4)$ (d) $(4, \pi/2, \pi/3)$

9–10 Convert from cylindrical to spherical coordinates.

9. (a) $(\sqrt{3}, \pi/6, 3)$ (b) $(1, \pi/4, -1)$
(c) $(2, 3\pi/4, 0)$ (d) $(6, 1, -2\sqrt{3})$

10. (a) $(4, 5\pi/6, 4)$ (b) $(2, 0, -2)$
(c) $(4, \pi/2, 3)$ (d) $(6, \pi, 2)$

11–12 Convert from spherical to cylindrical coordinates.

11. (a) $(5, \pi/4, 2\pi/3)$ (b) $(1, 7\pi/6, \pi)$
(c) $(3, 0, 0)$ (d) $(4, \pi/6, \pi/2)$

12. (a) $(5, \pi/2, 0)$ (b) $(6, 0, 3\pi/4)$
(c) $(\sqrt{2}, 3\pi/4, \pi)$ (d) $(5, 2\pi/3, 5\pi/6)$

C **13.** Use a CAS or a programmable calculating utility to set up the conversion formulas in Table 12.8.1, and then use the CAS or calculating utility to solve the problems in Exercises 1, 3, 5, 7, 9, and 11.

C **14.** Use a CAS or a programmable calculating utility to set up the conversion formulas in Table 12.8.1, and then use the CAS or calculating utility to solve the problems in Exercises 2, 4, 6, 8, 10, and 12.

15–22 An equation is given in cylindrical coordinates. Express the equation in rectangular coordinates and sketch the graph.

15. $r = 3$ **16.** $\theta = \pi/4$ **17.** $z = r^2$

18. $z = r\cos\theta$ **19.** $r = 4\sin\theta$ **20.** $r = 2\sec\theta$

21. $r^2 + z^2 = 1$ **22.** $r^2\cos 2\theta = z$

23–30 An equation is given in spherical coordinates. Express the equation in rectangular coordinates and sketch the graph.

23. $\rho = 3$ **24.** $\theta = \pi/3$ **25.** $\phi = \pi/4$

26. $\rho = 2\sec\phi$ **27.** $\rho = 4\cos\phi$ **28.** $\rho\sin\phi = 1$

29. $\rho\sin\phi = 2\cos\theta$ **30.** $\rho - 2\sin\phi\cos\theta = 0$

31–42 An equation of a surface is given in rectangular coordinates. Find an equation of the surface in (a) cylindrical coordinates and (b) spherical coordinates.

31. $z = 3$ **32.** $y = 2$

33. $z = 3x^2 + 3y^2$ **34.** $z = \sqrt{3x^2 + 3y^2}$

35. $x^2 + y^2 = 4$ **36.** $x^2 + y^2 - 6y = 0$

37. $x^2 + y^2 + z^2 = 9$ **38.** $z^2 = x^2 - y^2$

39. $2x + 3y + 4z = 1$ **40.** $x^2 + y^2 - z^2 = 1$

41. $x^2 = 16 - z^2$ **42.** $x^2 + y^2 + z^2 = 2z$

FOCUS ON CONCEPTS

43–46 Describe the region in 3-space that satisfies the given inequalities.

43. $r^2 \le z \le 4$ **44.** $0 \le r \le 2\sin\theta,\quad 0 \le z \le 3$

45. $1 \le \rho \le 3$ **46.** $0 \le \phi \le \pi/6,\quad 0 \le \rho \le 2$

47. St. Petersburg (Leningrad), Russia, is located at 30° east longitude and 60° north latitude. Find its spherical and rectangular coordinates relative to the coordinate axes of Figure 12.8.9. Take miles as the unit of distance and assume the Earth to be a sphere of radius 4000 miles.

48. (a) Show that the curve of intersection of the surfaces $z = \sin\theta$ and $r = a$ (cylindrical coordinates) is an ellipse.
(b) Sketch the surface $z = \sin\theta$ for $0 \le \theta \le \pi/2$.

49. The accompanying figure shows a right circular cylinder of radius 10 cm spinning at 3 revolutions per minute about the z-axis. At time $t = 0$ s, a bug at the point $(0, 10, 0)$ begins walking straight up the face of the cylinder at the rate of 0.5 cm/min.
(a) Find the cylindrical coordinates of the bug after 2 min.
(b) Find the rectangular coordinates of the bug after 2 min.
(c) Find the spherical coordinates of the bug after 2 min.

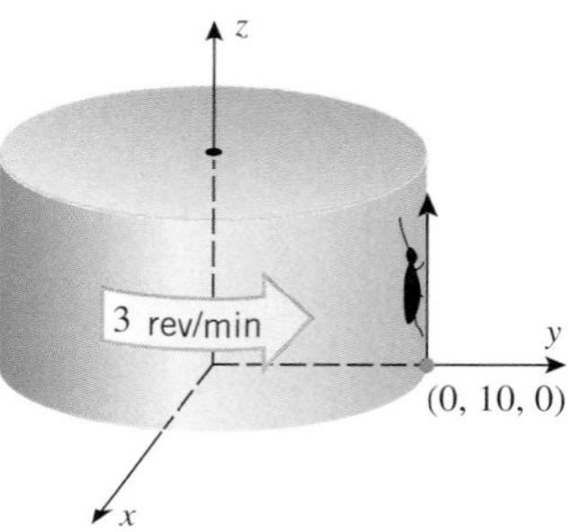

Figure Ex-49

50. Referring to Exercise 49, use a graphing utility to graph the bug's distance from the origin as a function of time.

51. A ship at sea is at point A that is 60° west longitude and 40° north latitude. The ship travels to point B that is 40° west longitude and 20° north latitude. Assuming that the Earth is a sphere with radius 6370 kilometers, find the shortest distance the ship can travel in going from A to B, given that the shortest distance between two points on a sphere is along the arc of the great circle joining the points. [*Suggestion:* Introduce an xyz-coordinate system as in Figure 12.8.9, and consider the angle between the vectors from the center of the Earth to the points A and B. If you are not familiar with the term "great circle," consult a dictionary.]

✔ QUICK CHECK ANSWERS 12.8

1. $r\cos\theta$; $r\sin\theta$; z 2. $\rho\sin\phi\cos\theta$; $\rho\sin\phi\sin\theta$; $\rho\cos\phi$ 3. $\rho\sin\phi$; θ; $\rho\cos\theta$
4. (a) $(2, 7\pi/4, 2\sqrt{3})$ (b) $(4, 7\pi/4, \pi/6)$ 5. (a) $x^2 + y^2 + z^2 = 25$ (b) $r^2 + z^2 = 25$ (c) $\rho = 5$

CHAPTER REVIEW EXERCISES

1. (a) What is the difference between a vector and a scalar? Give a physical example of each.
(b) How can you determine whether or not two vectors are orthogonal?
(c) How can you determine whether or not two vectors are parallel?
(d) How can you determine whether or not three vectors with a common initial point lie in the same plane in 3-space?

2. (a) Sketch vectors $\mathbf{u}$ and $\mathbf{v}$ for which $\mathbf{u}+\mathbf{v}$ and $\mathbf{u}-\mathbf{v}$ are orthogonal.
(b) How can you use vectors to determine whether four points in 3-space lie in the same plane?
(c) If forces $\mathbf{F}_1 = \mathbf{i}$ and $\mathbf{F}_2 = \mathbf{j}$ are applied at a point in 2-space, what force would you apply at that point to cancel the combined effect of $\mathbf{F}_1$ and $\mathbf{F}_2$?
(d) Write an equation of the sphere with center $(1, -2, 2)$ that passes through the origin.

3. (a) Draw a picture that shows the direction angles α, β, and γ of a vector.
(b) What are the components of a unit vector in 2-space that makes an angle of 120° with the vector $\mathbf{i}$ (two answers)?
(c) How can you use vectors to determine whether a triangle with known vertices P_1, P_2, and P_3 has an obtuse angle?
(d) True or false: The cross product of orthogonal unit vectors is a unit vector. Explain your reasoning.

4. (a) Make a table that shows all possible cross products of the vectors $\mathbf{i}$, $\mathbf{j}$, and $\mathbf{k}$.
(b) Give a geometric interpretation of $\|\mathbf{u}\times\mathbf{v}\|$.
(c) Give a geometric interpretation of $|\mathbf{u}\cdot(\mathbf{v}\times\mathbf{w})|$.
(d) Write an equation of the plane that passes through the origin and is perpendicular to the line $x = t$, $y = 2t$, $z = -t$.

5. In each part, find an equation of the sphere with center $(-3, 5, -4)$ and satisfying the given condition.
(a) Tangent to the xy-plane
(b) Tangent to the xz-plane
(c) Tangent to the yz-plane

6. Find the largest and smallest distances between the point $P(1, 1, 1)$ and the sphere

$$x^2 + y^2 + z^2 - 2y + 6z - 6 = 0$$

7. Given the points $P(3, 4)$, $Q(1, 1)$, and $R(5, 2)$, use vector methods to find the coordinates of the fourth vertex of the parallelogram whose adjacent sides are $\overrightarrow{PQ}$ and $\overrightarrow{QR}$.

8. Let $\mathbf{u} = \langle 3, 5, -1\rangle$ and $\mathbf{v} = \langle 2, -2, 3\rangle$. Find
(a) $2\mathbf{u} + 5\mathbf{v}$
(b) $\dfrac{1}{\|\mathbf{v}\|}\mathbf{v}$
(c) $\|\mathbf{u}\|$
(d) $\|\mathbf{u}-\mathbf{v}\|$.

9. Let $\mathbf{a} = c\mathbf{i} + \mathbf{j}$ and $\mathbf{b} = 4\mathbf{i} + 3\mathbf{j}$. Find c so that
(a) $\mathbf{a}$ and $\mathbf{b}$ are orthogonal
(b) the angle between $\mathbf{a}$ and $\mathbf{b}$ is $\pi/4$
(c) the angle between $\mathbf{a}$ and $\mathbf{b}$ is $\pi/6$
(d) $\mathbf{a}$ and $\mathbf{b}$ are parallel.

10. Let $\mathbf{r}_0 = \langle x_0, y_0, z_0\rangle$ and $\mathbf{r} = \langle x, y, z\rangle$. Describe the set of all points (x, y, z) for which
(a) $\mathbf{r}\cdot\mathbf{r}_0 = 0$
(b) $(\mathbf{r}-\mathbf{r}_0)\cdot\mathbf{r}_0 = 0$.

11. Show that if $\mathbf{u}$ and $\mathbf{v}$ are unit vectors and θ is the angle between them, then $\|\mathbf{u}-\mathbf{v}\| = 2\sin\frac{1}{2}\theta$.

12. Find the vector with length 5 and direction angles $\alpha = 60°$, $\beta = 120°$, $\gamma = 135°$.

13. Assuming that force is in pounds and distance is in feet, find the work done by a constant force $\mathbf{F} = 3\mathbf{i} - 4\mathbf{j} + \mathbf{k}$ acting on a particle that moves on a straight line from $P(5, 7, 0)$ to $Q(6, 6, 6)$.

14. Assuming that force is in newtons and distance is in meters, find the work done by the resultant of the constant forces $\mathbf{F}_1 = \mathbf{i} - 3\mathbf{j} + \mathbf{k}$ and $\mathbf{F}_2 = \mathbf{i} + 2\mathbf{j} + 2\mathbf{k}$ acting on a particle that moves on a straight line from $P(-1, -2, 3)$ to $Q(0, 2, 0)$.

15. (a) Find the area of the triangle with vertices $A(1, 0, 1)$, $B(0, 2, 3)$, and $C(2, 1, 0)$.
(b) Use the result in part (a) to find the length of the altitude from vertex C to side AB.

16. True or false? Explain your reasoning.
(a) If $\mathbf{u}\cdot\mathbf{v} = 0$, then $\mathbf{u} = \mathbf{0}$ or $\mathbf{v} = \mathbf{0}$.
(b) If $\mathbf{u}\times\mathbf{v} = \mathbf{0}$, then $\mathbf{u} = \mathbf{0}$ or $\mathbf{v} = \mathbf{0}$.
(c) If $\mathbf{u}\cdot\mathbf{v} = 0$ and $\mathbf{u}\times\mathbf{v} = \mathbf{0}$, then $\mathbf{u} = \mathbf{0}$ or $\mathbf{v} = \mathbf{0}$.

17. Consider the points

$$A(1, -1, 2),\quad B(2, -3, 0),\quad C(-1, -2, 0),\quad D(2, 1, -1)$$

(a) Find the volume of the parallelepiped that has the vectors $\overrightarrow{AB}$, $\overrightarrow{AC}$, $\overrightarrow{AD}$ as adjacent edges.
(b) Find the distance from D to the plane containing A, B, and C.

18. Suppose that a force $\mathbf{F}$ with a magnitude of 9 lb is applied to the lever–shaft assembly shown in the accompanying figure.
(a) Express the force $\mathbf{F}$ in component form.
(b) Find the vector moment of $\mathbf{F}$ about the origin.

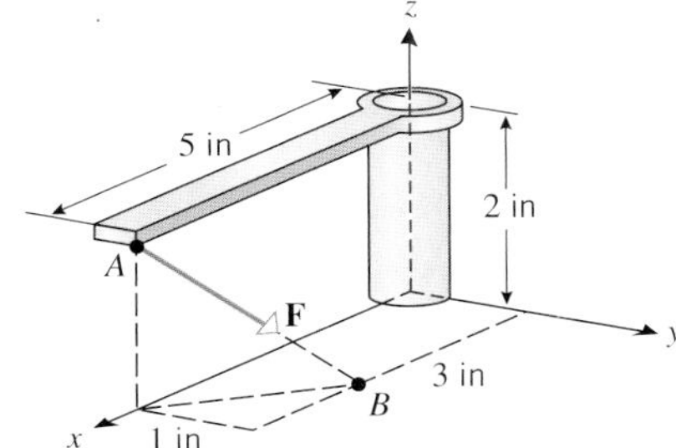

Figure Ex-18

19. Let P be the point $(4, 1, 2)$. Find parametric equations for the line through P and parallel to the vector $\langle 1, -1, 0\rangle$.

20. (a) Find parametric equations for the intersection of the planes $2x + y - z = 3$ and $x + 2y + z = 3$.
(b) Find the acute angle between the two planes.

21. Find an equation of the plane that is parallel to the plane $x + 5y - z + 8 = 0$ and contains the point $(1, 1, 4)$.

22. Find an equation of the plane through the point $(4, 3, 0)$ and parallel to the vectors $\mathbf{i} + \mathbf{k}$ and $2\mathbf{j} - \mathbf{k}$.

23. What condition must the constants satisfy for the planes

$$a_1x + b_1y + c_1z = d_1 \quad\text{and}\quad a_2x + b_2y + c_2z = d_2$$

to be perpendicular?

24. (a) List six common types of quadric surfaces, and describe their traces in planes parallel to the coordinate planes.
(b) Give the coordinates of the points that result when the point (x, y, z) is reflected about the plane $y = x$, the plane $y = z$, and the plane $x = z$.
(c) Describe the intersection of the surfaces $r = 5$ and $z = 1$ in cylindrical coordinates.
(d) Describe the intersection of the surfaces $\phi = \pi/4$ and $\theta = 0$ in spherical coordinates.

25. In each part, identify the surface by completing the squares.
(a) $x^2 + 4y^2 - z^2 - 6x + 8y + 4z = 0$
(b) $x^2 + y^2 + z^2 + 6x - 4y + 12z = 0$
(c) $x^2 + y^2 - z^2 - 2x + 4y + 5 = 0$

26. In each part, express the equation in cylindrical and spherical coordinates.
(a) $x^2 + y^2 = z$ (b) $x^2 - y^2 - z^2 = 0$

27. In each part, express the equation in rectangular coordinates.
(a) $z = r^2 \cos 2\theta$ (b) $\rho^2 \sin\phi \cos\phi \cos\theta = 1$

28–29 Sketch the solid in 3-space that is described in cylindrical coordinates by the stated inequalities.

28. (a) $1 \le r \le 2$ (b) $2 \le z \le 3$ (c) $\pi/6 \le \theta \le \pi/3$
(d) $1 \le r \le 2,\ 2 \le z \le 3,$ and $\pi/6 \le \theta \le \pi/3$

29. (a) $r^2 + z^2 \le 4$ (b) $r \le 1$
(c) $r^2 + z^2 \le 4$ and $r > 1$

30–31 Sketch the solid in 3-space that is described in spherical coordinates by the stated inequalities.

30. (a) $0 \le \rho \le 2$ (b) $0 \le \phi \le \pi/6$
(c) $0 \le \rho \le 2$ and $0 \le \phi \le \pi/6$

31. (a) $0 \le \rho \le 5,\ 0 \le \phi \le \pi/2,$ and $0 \le \theta \le \pi/2$
(b) $0 \le \phi \le \pi/3$ and $0 \le \rho \le 2\sec\phi$
(c) $0 \le \rho \le 2$ and $\pi/6 \le \phi \le \pi/3$

32. Sketch the surface whose equation in spherical coordinates is $\rho = a(1 - \cos\phi)$. [*Hint:* The surface is shaped like a familiar fruit.]

chapter thirteen

VECTOR-VALUED FUNCTIONS

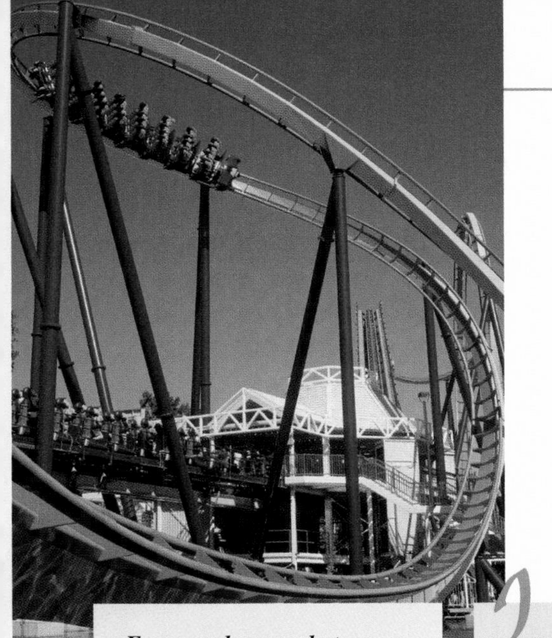

Everyone knows what a curve is, until he has studied enough mathematics to become confused through the countless number of possible exceptions.

—Felix Klein
Mathematician

In this chapter we will consider functions whose values are vectors. Such functions provide a unified way of studying parametric curves in 2-space and 3-space and are a basic tool for analyzing the motion of particles along curved paths. We will begin by developing the calculus of vector-valued functions—we will show how to differentiate and integrate such functions, and we will develop some of the basic properties of these operations. We will then apply these calculus tools to define three fundamental vectors that can be used to describe such basic characteristics of curves as curvature and twisting tendencies. Once this is done, we will develop the concepts of velocity and acceleration for such motion, and we will apply these concepts to explain various physical phenomena. Finally, we will use the calculus of vector-valued functions to develop basic principles of gravitational attraction and to derive Kepler's laws of planetary motion.

Photo: *A roller coaster moves with variable velocity and direction. We will study this kind of motion in this chapter.*

13.1 INTRODUCTION TO VECTOR-VALUED FUNCTIONS

In Section 12.5 we discussed parametric equations of lines in 3-space. In this section we will discuss more general parametric curves in 3-space, and we will show how vector notation can be used to express parametric equations in 2-space and 3-space in a more compact form. This will lead us to consider a new kind of function—namely, functions that associate vectors with real numbers. Such functions have many important applications in physics and engineering.

PARAMETRIC CURVES IN 3-SPACE

Recall from Section 1.7 that if f and g are well-behaved functions, then the pair of parametric equations

$$x = f(t), \quad y = g(t) \tag{1}$$

generates a curve in 2-space that is traced in a specific direction as the parameter t increases. We defined this direction to be the *orientation* of the curve or the *direction of increasing parameter*, and we called the curve together with its orientation the *graph* of the parametric equations or the *parametric curve* represented by the equations. Analogously, if f, g, and h are three well-behaved functions, then the parametric equations

$$x = f(t), \quad y = g(t), \quad z = h(t) \tag{2}$$

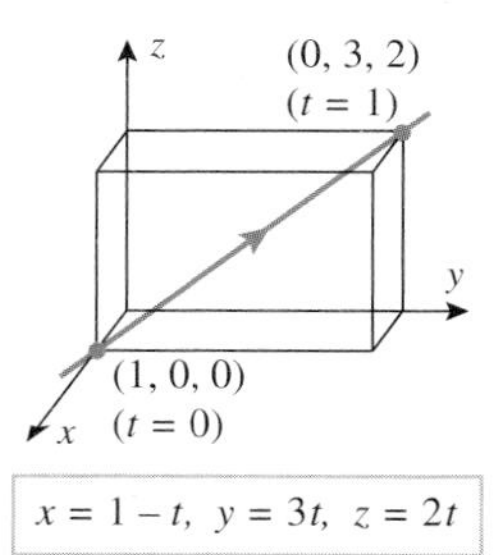

Figure 13.1.1

generate a curve in 3-space that is traced in a specific direction as t increases. As in 2-space, this direction is called the ***orientation*** or ***direction of increasing parameter***, and the curve together with its orientation is called the ***graph*** of the parametric equations or the ***parametric curve*** represented by the equations. If no restrictions are stated explicitly or are implied by the equations, then it will be understood that t varies over the interval $(-\infty, +\infty)$.

▶ Example 1 The parametric equations

$$x = 1 - t, \quad y = 3t, \quad z = 2t$$

represent a line in 3-space that passes through the point $(1, 0, 0)$ and is parallel to the vector $\langle -1, 3, 2\rangle$. Since x decreases as t increases, the line has the orientation shown in Figure 13.1.1. ◀

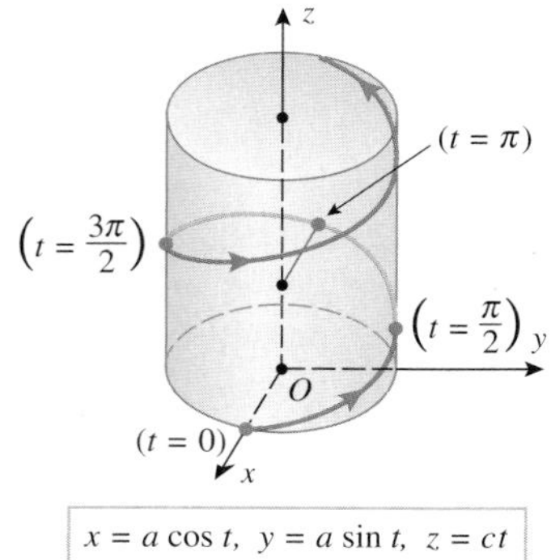

Figure 13.1.2

▶ Example 2 Describe the parametric curve represented by the equations

$$x = a\cos t, \quad y = a\sin t, \quad z = ct$$

where a and c are positive constants.

Solution. As the parameter t increases, the value of $z = ct$ also increases, so the point (x, y, z) moves upward. However, as t increases, the point (x, y, z) also moves in a path directly over the circle

$$x = a\cos t, \quad y = a\sin t$$

in the xy-plane. The combination of these upward and circular motions produces a corkscrew-shaped curve that wraps around a right circular cylinder of radius a centered on the z-axis (Figure 13.1.2). This curve is called a ***circular helix***. ◀

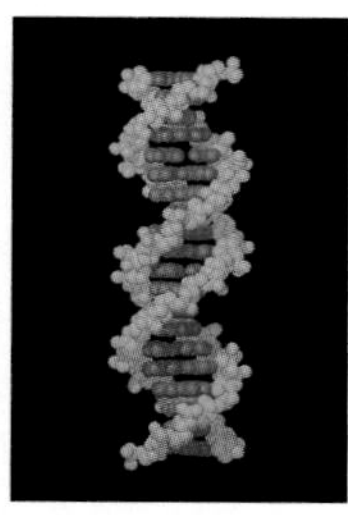

The circular helix described in Example 2 occurs in nature. Above is a computer representation of the twin helix DNA molecule (deoxyribonucleic acid). This structure contains all the inherited instructions necessary for the development of a living organism.

PARAMETRIC CURVES GENERATED WITH TECHNOLOGY

Except in the simplest cases, parametric curves in 3-space can be difficult to visualize and draw without the help of a graphing utility. For example, Figure 13.1.3a shows the graph of the parametric curve called a *torus knot* that was produced by a CAS. However, even this computer rendering is difficult to visualize because it is unclear whether the points of overlap are intersections or whether one portion of the curve is in front of the other. To resolve this visualization problem, some graphing utilities provide the capability of enclosing the curve within a thin tube, as in Figure 13.1.3b. Such graphs are called ***tube plots***.

TECHNOLOGY MASTERY

If you have a CAS, use it to generate the line in Example 1 and the helix

$$x = 4\cos t,\quad y = 4\sin t,\quad z = t \qquad (0 \le t \le 3\pi)$$

in Figure 13.1.4.

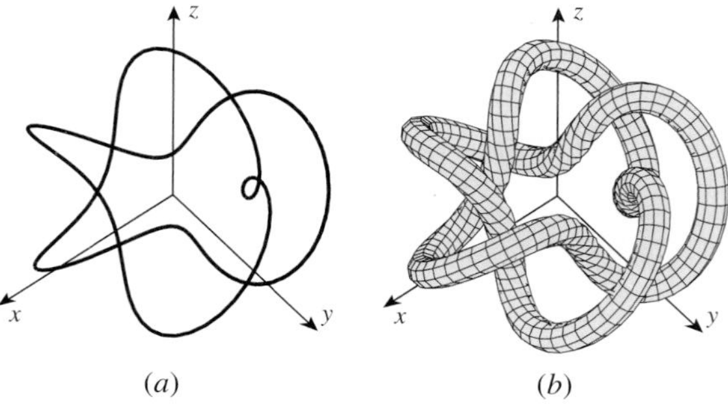

Figure 13.1.3

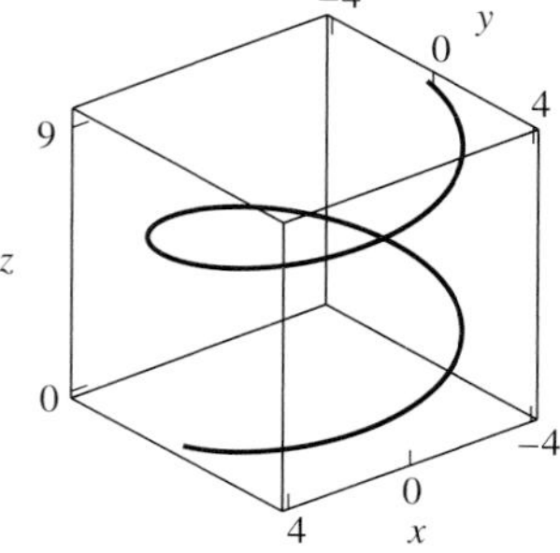

Figure 13.1.4

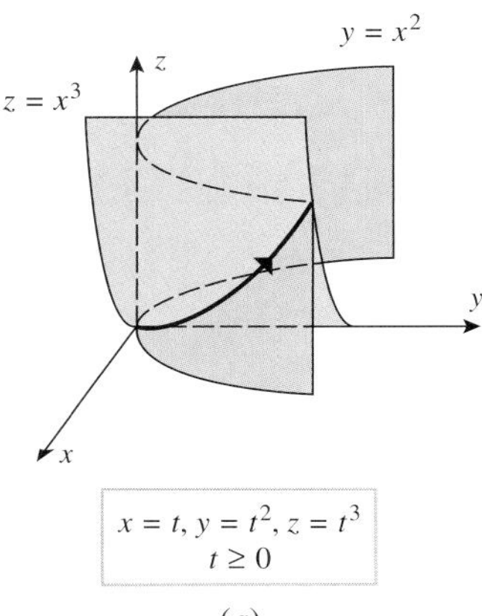

(a)

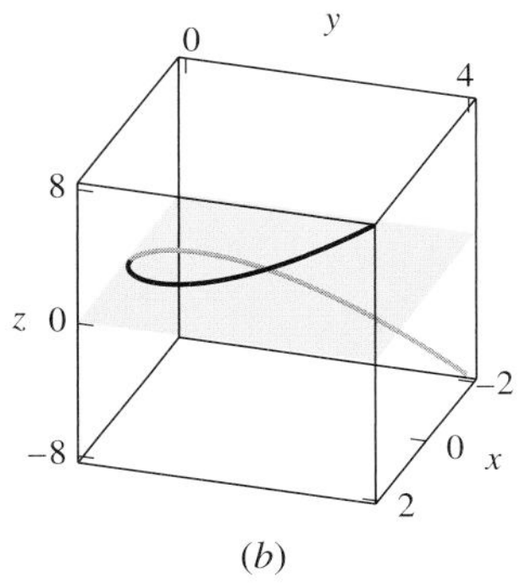

(b)

Figure 13.1.5

Find the vector-valued function in 2-space whose component functions are $x(t) = t$ and $y(t) = t^2$.

PARAMETRIC EQUATIONS FOR INTERSECTIONS OF SURFACES

Curves in 3-space often arise as intersections of surfaces. For example, Figure 13.1.5*a* shows a portion of the intersection of the cylinders $z = x^3$ and $y = x^2$. One method for finding parametric equations for the curve of intersection is to choose one of the variables as the parameter and use the two equations to express the remaining two variables in terms of that parameter. In particular, if we choose $x = t$ as the parameter and substitute this into the equations $z = x^3$ and $y = x^2$, we obtain the parametric equations

$$x = t, \quad y = t^2, \quad z = t^3 \tag{3}$$

This curve is called a ***twisted cubic***. The portion of the twisted cubic shown in Figure 13.1.5*a* corresponds to $t \geq 0$; a computer-generated graph of the twisted cubic for positive and negative values of t is shown in Figure 13.1.5*b*. Some other examples and techniques for finding intersections of surfaces are discussed in the exercises.

VECTOR-VALUED FUNCTIONS

The twisted cubic defined by the equations in (3) is the set of points of the form (t, t^2, t^3) for real values of t. If we view each of these points as a terminal point for a vector **r** whose initial point is at the origin,

$$\mathbf{r} = \langle x, y, z\rangle = \langle t, t^2, t^3\rangle = t\mathbf{i} + t^2\mathbf{j} + t^3\mathbf{k}$$

then we obtain **r** as a function of the parameter t, that is, $\mathbf{r} = \mathbf{r}(t)$. Since this function produces a *vector*, we say that $\mathbf{r} = \mathbf{r}(t)$ defines **r** as a ***vector-valued function of a real variable***, or more simply, a ***vector-valued function***. The vectors that we will consider in this text are either in 2-space or 3-space, so we will say that a vector-valued function is in 2-space or in 3-space according to the kind of vectors that it produces.

If $\mathbf{r}(t)$ is a vector-valued function in 2-space, then for each allowable value of t the vector $\mathbf{r} = \mathbf{r}(t)$ can be represented in terms of components as

$$\mathbf{r} = \mathbf{r}(t) = \langle x(t), y(t)\rangle = x(t)\mathbf{i} + y(t)\mathbf{j}$$

The functions $x(t)$ and $y(t)$ are called the ***component functions*** or the ***components*** of $\mathbf{r}(t)$. Similarly, the component functions of a vector-valued function

$$\mathbf{r}(t) = \langle x(t), y(t), z(t)\rangle = x(t)\mathbf{i} + y(t)\mathbf{j} + z(t)\mathbf{k}$$

in 3-space are $x(t)$, $y(t)$, and $z(t)$.

▶ Example 3 The component functions of

$$\mathbf{r}(t) = \langle t, t^2, t^3\rangle = t\mathbf{i} + t^2\mathbf{j} + t^3\mathbf{k}$$

are

$$x(t) = t, \quad y(t) = t^2, \quad z(t) = t^3 \quad ◀$$

The ***domain*** of a vector-valued function $\mathbf{r}(t)$ is the set of allowable values for t. If $\mathbf{r}(t)$ is defined in terms of component functions and the domain is not specified explicitly, then it will be understood that the domain is the intersection of the natural domains of the component functions; this is called the ***natural domain*** of $\mathbf{r}(t)$.

▶ Example 4 Find the natural domain of

$$\mathbf{r}(t) = \langle \ln|t-1|, e^t, \sqrt{t}\rangle = (\ln|t-1|)\mathbf{i} + e^t\mathbf{j} + \sqrt{t}\mathbf{k}$$

Solution. The natural domains of the component functions

$$x(t) = \ln|t-1|, \quad y(t) = e^t, \quad z(t) = \sqrt{t}$$

are

$$(-\infty, 1) \cup (1, +\infty), \quad (-\infty, +\infty), \quad [0, +\infty)$$

respectively. The intersection of these sets is

$$[0, 1) \cup (1, +\infty)$$

(verify), so the natural domain of $\mathbf{r}(t)$ consists of all values of t such that

$$0 \le t < 1 \quad \text{or} \quad t > 1 \blacktriangleleft$$

GRAPHS OF VECTOR-VALUED FUNCTIONS

If $\mathbf{r}(t)$ is a vector-valued function in 2-space or 3-space, then we define the ***graph*** of $\mathbf{r}(t)$ to be the parametric curve described by the component functions for $\mathbf{r}(t)$. For example, if

$$\mathbf{r}(t) = \langle 1 - t, 3t, 2t \rangle = (1 - t)\mathbf{i} + 3t\mathbf{j} + 2t\mathbf{k} \tag{4}$$

then the graph of $\mathbf{r} = \mathbf{r}(t)$ is the graph of the parametric equations

$$x = 1 - t, \quad y = 3t, \quad z = 2t$$

Thus, the graph of (4) is the line in Figure 13.1.1.

Strictly speaking, we should write $(\cos t)\mathbf{i}$ and $(\sin t)\mathbf{j}$ rather than $\cos t\mathbf{i}$ and $\sin t\mathbf{j}$ for clarity. However, it is a common practice to omit the parentheses in such cases, since no misinterpretation is possible. Why?

▶ **Example 5** Describe the graph of the vector-valued function

$$\mathbf{r}(t) = \langle \cos t, \sin t, t \rangle = \cos t\mathbf{i} + \sin t\mathbf{j} + t\mathbf{k}$$

Solution. The corresponding parametric equations are

$$x = \cos t, \quad y = \sin t, \quad z = t$$

Thus, as we saw in Example 2, the graph is a circular helix wrapped around a cylinder of radius 1. ◀

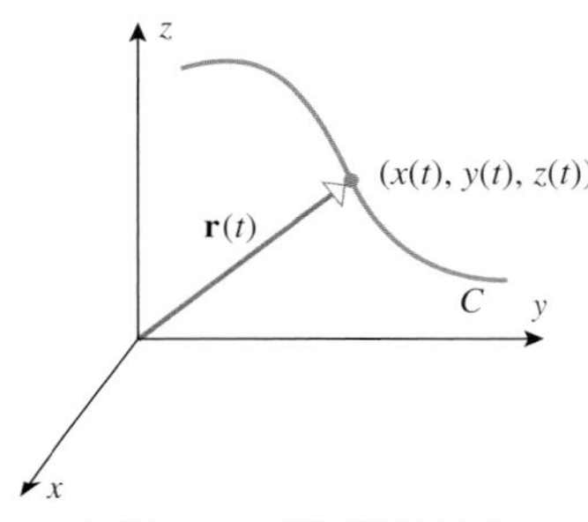

Figure 13.1.6

Up to now we have considered parametric curves to be paths traced by moving points. However, if a parametric curve is viewed as the graph of a vector-valued function, then we can also imagine the graph to be traced by the tip of a moving vector. For example, if the curve C in 3-space is the graph of

$$\mathbf{r}(t) = x(t)\mathbf{i} + y(t)\mathbf{j} + z(t)\mathbf{k}$$

and if we position $\mathbf{r}(t)$ so its initial point is at the origin, then its terminal point will fall on the curve C (as shown in Figure 13.1.6). Thus, when $\mathbf{r}(t)$ is positioned with its initial point at the origin, its terminal point will trace out the curve C as the parameter t varies, in which case we call $\mathbf{r}(t)$ the ***radius vector*** or the ***position vector*** for C. For simplicity, we will sometimes let the dependence on t be understood and write $\mathbf{r}$ rather than $\mathbf{r}(t)$ for a radius vector.

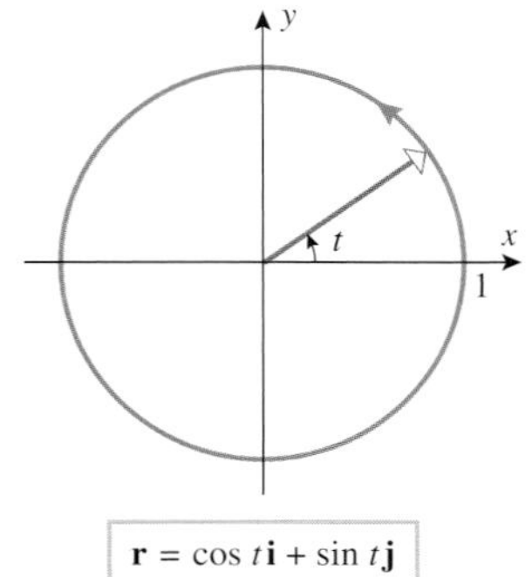

Figure 13.1.7

▶ **Example 6** Sketch the graph and a radius vector of

(a) $\mathbf{r}(t) = \cos t\mathbf{i} + \sin t\mathbf{j}, \quad 0 \le t \le 2\pi$

(b) $\mathbf{r}(t) = \cos t\mathbf{i} + \sin t\mathbf{j} + 2\mathbf{k}, \quad 0 \le t \le 2\pi$

Solution (a). The corresponding parametric equations are

$$x = \cos t, \quad y = \sin t \qquad (0 \le t \le 2\pi)$$

so the graph is a circle of radius 1, centered at the origin, and oriented counterclockwise. The graph and a radius vector are shown in Figure 13.1.7.

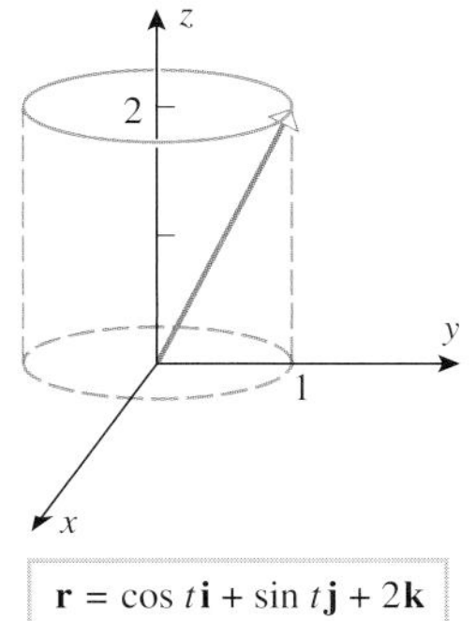

Figure 13.1.8

Solution (b). The corresponding parametric equations are

$$x = \cos t, \quad y = \sin t, \quad z = 2 \qquad (0 \le t \le 2\pi)$$

From the third equation, the tip of the radius vector traces a curve in the plane $z = 2$, and from the first two equations, the curve is a circle of radius 1 centered at the point $(0, 0, 2)$ and traced counterclockwise looking down the z-axis. The graph and a radius vector are shown in Figure 13.1.8. ◀

VECTOR FORM OF A LINE SEGMENT

Recall from Formula (9) of Section 12.5 that if $\mathbf{r}_0$ is a vector in 2-space or 3-space with its initial point at the origin, then the line that passes through the terminal point of $\mathbf{r}_0$ and is parallel to the vector $\mathbf{v}$ can be expressed in vector form as

$$\mathbf{r} = \mathbf{r}_0 + t\mathbf{v}$$

In particular, if $\mathbf{r}_0$ and $\mathbf{r}_1$ are vectors in 2-space or 3-space with their initial points at the origin, then the line that passes through the terminal points of these vectors can be expressed in vector form as

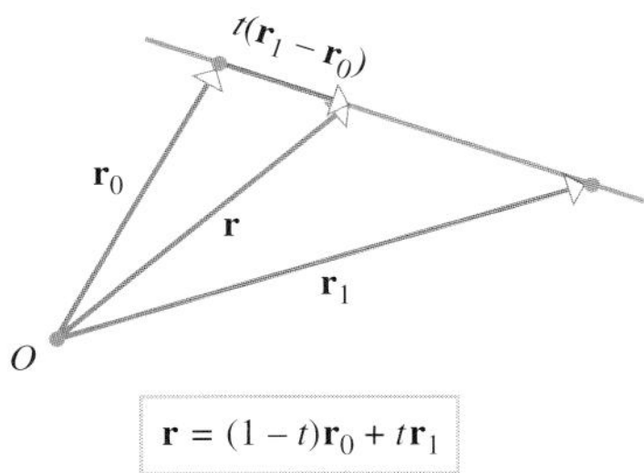

Figure 13.1.9

$$\mathbf{r} = \mathbf{r}_0 + t(\mathbf{r}_1 - \mathbf{r}_0) \qquad \text{or} \qquad \mathbf{r} = (1 - t)\mathbf{r}_0 + t\mathbf{r}_1 \tag{5–6}$$

as indicated in Figure 13.1.9.

It is common to call either (5) or (6) the ***two-point vector form of a line*** and to say, for simplicity, that the line passes through the *points* $\mathbf{r}_0$ and $\mathbf{r}_1$ (as opposed to saying that it passes through the *terminal points* of $\mathbf{r}_0$ and $\mathbf{r}_1$).

It is understood in (5) and (6) that t varies from $-\infty$ to $+\infty$. However, if we restrict t to vary over the interval $0 \le t \le 1$, then $\mathbf{r}$ will vary from $\mathbf{r}_0$ to $\mathbf{r}_1$. Thus, the equation

$$\mathbf{r} = (1 - t)\mathbf{r}_0 + t\mathbf{r}_1 \qquad (0 \le t \le 1) \tag{7}$$

represents the line segment in 2-space or 3-space that is traced from $\mathbf{r}_0$ to $\mathbf{r}_1$.

✔QUICK CHECK EXERCISES 13.1

(See page 865 for answers.)

1. (a) Express the parametric equations

$$x = \frac{1}{t}, \quad y = \sqrt{t}, \quad z = \sin^{-1} t$$

as a single vector equation of the form

$$\mathbf{r} = x(t)\mathbf{i} + y(t)\mathbf{j} + z(t)\mathbf{k}$$

(b) The vector equation in part (a) defines $\mathbf{r} = \mathbf{r}(t)$ as a vector-valued function. The domain of $\mathbf{r}(t)$ is ________ and $\mathbf{r}\left(\frac{1}{2}\right) =$ ________.

2. Describe the graph of $\mathbf{r}(t) = \langle 1 + 2t, -1 + 3t \rangle$.
3. Describe the graph of $\mathbf{r}(t) = \sin^2 t\mathbf{i} + \cos^2 t\mathbf{j}$.
4. Find a vector equation for the curve of intersection of the surfaces $y = x^2$ and $z = y$ in terms of the parameter $x = t$.

EXERCISE SET 13.1

Graphing Utility

1–4 Find the domain of $\mathbf{r}(t)$ and the value of $\mathbf{r}(t_0)$.

1. $\mathbf{r}(t) = \cos t\mathbf{i} - 3t\mathbf{j}; \ t_0 = \pi$
2. $\mathbf{r}(t) = \langle \sqrt{3t + 1}, t^2 \rangle; \ t_0 = 1$
3. $\mathbf{r}(t) = \cos \pi t\mathbf{i} - \ln t\mathbf{j} + \sqrt{t - 2}\mathbf{k}; \ t_0 = 3$
4. $\mathbf{r}(t) = \langle 2e^{-t}, \sin^{-1} t, \ln(1 - t) \rangle; \ t_0 = 0$

5–8 Express the parametric equations as a single vector equation of the form

$$\mathbf{r} = x(t)\mathbf{i} + y(t)\mathbf{j} \quad \text{or} \quad \mathbf{r} = x(t)\mathbf{i} + y(t)\mathbf{j} + z(t)\mathbf{k}$$

5. $x = 3\cos t, \ y = t + \sin t$
6. $x = t^2 + 1, \ y = e^{-2t}$
7. $x = 2t, \ y = 2\sin 3t, \ z = 5\cos 3t$

8. $x = t\sin t,\ y = \ln t,\ z = \cos^2 t$

9–12 Find the parametric equations that correspond to the given vector equation.

9. $\mathbf{r} = 3t^2\mathbf{i} - 2\mathbf{j}$

10. $\mathbf{r} = \sin^2 t\mathbf{i} + (1 - \cos 2t)\mathbf{j}$

11. $\mathbf{r} = (2t - 1)\mathbf{i} - 3\sqrt{t}\mathbf{j} + \sin 3t\mathbf{k}$

12. $\mathbf{r} = te^{-t}\mathbf{i} - 5t^2\mathbf{k}$

13–18 Describe the graph of the equation.

13. $\mathbf{r} = (3 - 2t)\mathbf{i} + 5t\mathbf{j}$

14. $\mathbf{r} = 2\sin 3t\mathbf{i} - 2\cos 3t\mathbf{j}$

15. $\mathbf{r} = 2t\mathbf{i} - 3\mathbf{j} + (1 + 3t)\mathbf{k}$

16. $\mathbf{r} = 3\mathbf{i} + 2\cos t\mathbf{j} + 2\sin t\mathbf{k}$

17. $\mathbf{r} = 2\cos t\mathbf{i} - 3\sin t\mathbf{j} + \mathbf{k}$

18. $\mathbf{r} = -3\mathbf{i} + (1 - t^2)\mathbf{j} + t\mathbf{k}$

19. (a) Find the slope of the line in 2-space that is represented by the vector equation $\mathbf{r} = (1 - 2t)\mathbf{i} - (2 - 3t)\mathbf{j}$.
(b) Find the coordinates of the point where the line
$$\mathbf{r} = (2 + t)\mathbf{i} + (1 - 2t)\mathbf{j} + 3t\mathbf{k}$$
intersects the xz-plane.

20. (a) Find the y-intercept of the line in 2-space that is represented by the vector equation $\mathbf{r} = (3 + 2t)\mathbf{i} + 5t\mathbf{j}$.
(b) Find the coordinates of the point where the line
$$\mathbf{r} = t\mathbf{i} + (1 + 2t)\mathbf{j} - 3t\mathbf{k}$$
intersects the plane $3x - y - z = 2$.

21–22 Sketch the line segment represented by the vector equation.

21. (a) $\mathbf{r} = (1 - t)\mathbf{i} + t\mathbf{j};\ 0 \le t \le 1$
(b) $\mathbf{r} = (1 - t)(\mathbf{i} + \mathbf{j}) + t(\mathbf{i} - \mathbf{j});\ 0 \le t \le 1$

22. (a) $\mathbf{r} = (1 - t)(\mathbf{i} + \mathbf{j}) + t\mathbf{k};\ 0 \le t \le 1$
(b) $\mathbf{r} = (1 - t)(\mathbf{i} + \mathbf{j} + \mathbf{k}) + t(\mathbf{i} + \mathbf{j});\ 0 \le t \le 1$

23–24 Write a vector equation for the line segment from P to Q.

23.
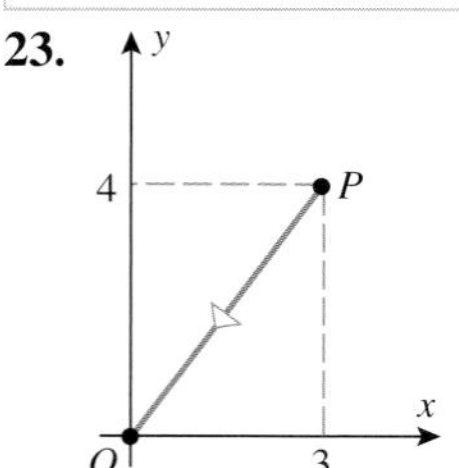

24.
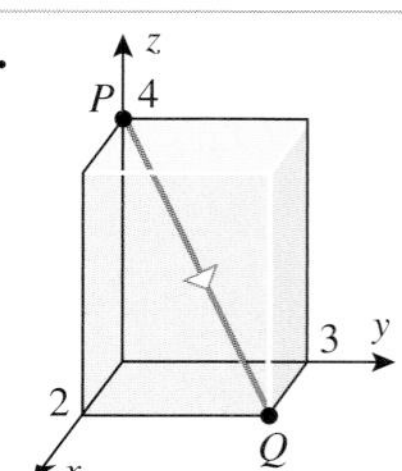

25–34 Sketch the graph of $\mathbf{r}(t)$ and show the direction of increasing t.

25. $\mathbf{r}(t) = 2\mathbf{i} + t\mathbf{j}$

26. $\mathbf{r}(t) = \langle 3t - 4, 6t + 2\rangle$

27. $\mathbf{r}(t) = (1 + \cos t)\mathbf{i} + (3 - \sin t)\mathbf{j};\ 0 \le t \le 2\pi$

28. $\mathbf{r}(t) = \langle 2\cos t, 5\sin t\rangle;\ 0 \le t \le 2\pi$

29. $\mathbf{r}(t) = \cosh t\mathbf{i} + \sinh t\mathbf{j}$

30. $\mathbf{r}(t) = \sqrt{t}\mathbf{i} + (2t + 4)\mathbf{j}$

31. $\mathbf{r}(t) = 2\cos t\mathbf{i} + 2\sin t\mathbf{j} + t\mathbf{k}$

32. $\mathbf{r}(t) = 9\cos t\mathbf{i} + 4\sin t\mathbf{j} + t\mathbf{k}$

33. $\mathbf{r}(t) = t\mathbf{i} + t^2\mathbf{j} + 2\mathbf{k}$

34. $\mathbf{r}(t) = t\mathbf{i} + t\mathbf{j} + \sin t\mathbf{k};\ 0 \le t \le 2\pi$

35–36 Sketch the curve of intersection of the surfaces, and find parametric equations for the intersection in terms of parameter $x = t$. Check your work with a graphing utility by generating the parametric curve over the interval $-1 \le t \le 1$.

35. $z = x^2 + y^2,\ x - y = 0$

36. $y + x = 0,\ z = \sqrt{2 - x^2 - y^2}$

37–38 Sketch the curve of intersection of the surfaces, and find a vector equation for the curve in terms of the parameter $x = t$.

37. $9x^2 + y^2 + 9z^2 = 81,\ y = x^2\quad (z > 0)$

38. $y = x,\ x + y + z = 1$

39. Show that the graph of
$$\mathbf{r} = t\sin t\mathbf{i} + t\cos t\mathbf{j} + t^2\mathbf{k}$$
lies on the paraboloid $z = x^2 + y^2$.

40. Show that the graph of
$$\mathbf{r} = t\mathbf{i} + \frac{1+t}{t}\mathbf{j} + \frac{1-t^2}{t}\mathbf{k},\quad t > 0$$
lies in the plane $x - y + z + 1 = 0$.

FOCUS ON CONCEPTS

41. Show that the graph of
$$\mathbf{r} = \sin t\mathbf{i} + 2\cos t\mathbf{j} + \sqrt{3}\sin t\mathbf{k}$$
is a circle, and find its center and radius. [*Hint:* Show that the curve lies on both a sphere and a plane.]

42. Show that the graph of
$$\mathbf{r} = 3\cos t\mathbf{i} + 3\sin t\mathbf{j} + 3\sin t\mathbf{k}$$
is an ellipse, and find the lengths of the major and minor axes. [*Hint:* Show that the graph lies on both a circular cylinder and a plane and use the result in Exercise 60 of Section 11.4.]

43. For the helix $\mathbf{r} = a\cos t\mathbf{i} + a\sin t\mathbf{j} + ct\mathbf{k}$, find the value of c ($c > 0$) so that the helix will make one complete turn in a distance of 3 units measured along the z-axis.

44. How many revolutions will the circular helix
$$\mathbf{r} = a\cos t\mathbf{i} + a\sin t\mathbf{j} + 0.2t\mathbf{k}$$
make in a distance of 10 units measured along the z-axis?

45. Show that the curve $\mathbf{r} = t\cos t\mathbf{i} + t\sin t\mathbf{j} + t\mathbf{k}$, $t \ge 0$, lies on the cone $z = \sqrt{x^2 + y^2}$. Describe the curve.

46. Describe the curve $\mathbf{r} = a\cos t\mathbf{i} + b\sin t\mathbf{j} + ct\mathbf{k}$, where a, b, and c are positive constants such that $a \ne b$.

47. In each part, match the vector equation with one of the accompanying graphs, and explain your reasoning.
(a) $\mathbf{r} = t\mathbf{i} - t\mathbf{j} + \sqrt{2 - t^2}\,\mathbf{k}$
(b) $\mathbf{r} = \sin \pi t\mathbf{i} - t\mathbf{j} + t\mathbf{k}$
(c) $\mathbf{r} = \sin t\mathbf{i} + \cos t\,\mathbf{j} + \sin 2t\mathbf{k}$
(d) $\mathbf{r} = \frac{1}{2}t\mathbf{i} + \cos 3t\,\mathbf{j} + \sin 3t\mathbf{k}$

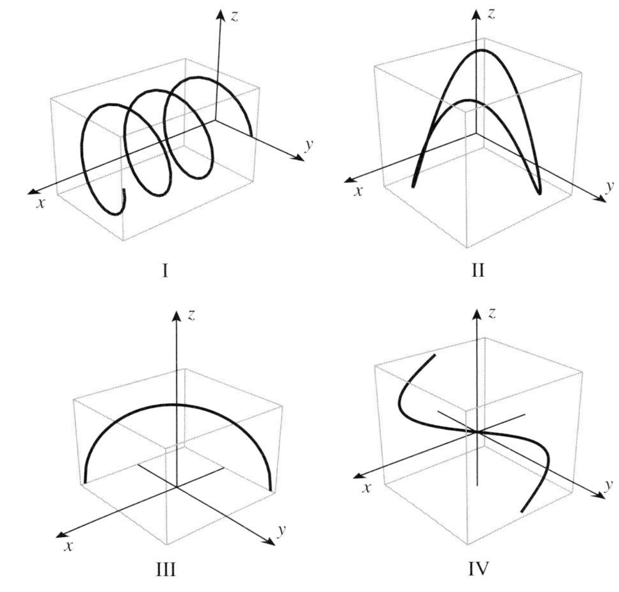

48. Check your conclusions in Exercise 47 by generating the curves with a graphing utility. [*Note:* Your graphing utility may look at the curve from a different viewpoint. Read the documentation for your graphing utility to determine how to control the viewpoint, and see if you can generate a reasonable facsimile of the graphs shown in the figure by adjusting the viewpoint and choosing the interval of t-values appropriately.]

49. (a) Find parametric equations for the curve of intersection of the circular cylinder $x^2 + y^2 = 9$ and the parabolic cylinder $z = x^2$ in terms of a parameter t for which $x = 3\cos t$.
(b) Use a graphing utility to generate the curve of intersection in part (a).

50. Use a graphing utility to generate the intersection of the cone $z = \sqrt{x^2 + y^2}$ and the plane $z = y + 2$. Identify the curve and explain your reasoning.

51. (a) Sketch the graph of

$$\mathbf{r}(t) = \left\langle 2t, \frac{2}{1+t^2} \right\rangle$$

(b) Prove that the curve in part (a) is also the graph of the function

$$y = \frac{8}{4 + x^2}$$

[The graphs of $y = a^3/(a^2 + x^2)$, where a denotes a constant, were first studied by the French mathematician Pierre de Fermat, and later by the Italian mathematicians Guido Grandi and Maria Agnesi. Any such curve is now known as a "witch of Agnesi." There are a number of theories for the origin of this name. Some suggest there was a mistranslation by either Grandi or Agnesi of some less colorful Latin name into Italian. Others lay the blame on a translation into English of Agnesi's 1748 treatise, *Analytical Institutions*.]

✔ QUICK CHECK ANSWERS 13.1

1. (a) $\mathbf{r} = \dfrac{1}{t}\mathbf{i} + \sqrt{t}\mathbf{j} + \sin^{-1} t\mathbf{k}$ (b) $0 < t \le 1$; $2\mathbf{i} + \dfrac{\sqrt{2}}{2}\mathbf{j} + \dfrac{\pi}{6}\mathbf{k}$ **2.** The graph is a line through $(1, -1)$ with direction vector $2\mathbf{i} + 3\mathbf{j}$. **3.** The graph is the line segment in the xy-plane from $(0, 1)$ to $(1, 0)$. **4.** $\mathbf{r} = \langle t, t^2, t^2 \rangle$

13.2 CALCULUS OF VECTOR-VALUED FUNCTIONS

In this section we will define limits, derivatives, and integrals of vector-valued functions and discuss their properties.

LIMITS AND CONTINUITY

Our first goal in this section is to develop a notion of what it means for a vector-valued function $\mathbf{r}(t)$ in 2-space or 3-space to approach a limiting vector $\mathbf{L}$ as t approaches a number a. That is, we want to define

$$\lim_{t \to a} \mathbf{r}(t) = \mathbf{L} \tag{1}$$

One way to motivate a reasonable definition of (1) is to position $\mathbf{r}(t)$ and $\mathbf{L}$ with their initial points at the origin and interpret this limit to mean that the terminal point of $\mathbf{r}(t)$ approaches the terminal point of $\mathbf{L}$ as t approaches a or, equivalently, that the vector $\mathbf{r}(t)$ approaches the vector $\mathbf{L}$ in both length and direction at t approaches a (Figure 13.2.1). Algebraically, this is equivalent to stating that

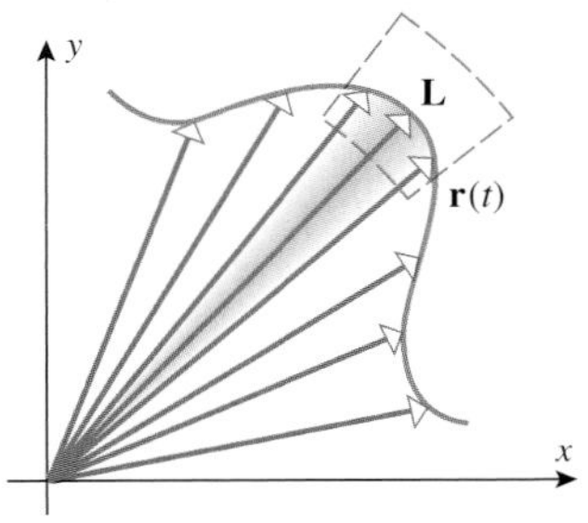

$$\lim_{t\to a} \|\mathbf{r}(t) - \mathbf{L}\| = 0 \tag{2}$$

(Figure 13.2.2). Thus, we make the following definition.

$\mathbf{r}(t)$ approaches $\mathbf{L}$ in length and direction if $\lim_{t\to a} \mathbf{r}(t) = \mathbf{L}$.

Figure 13.2.1

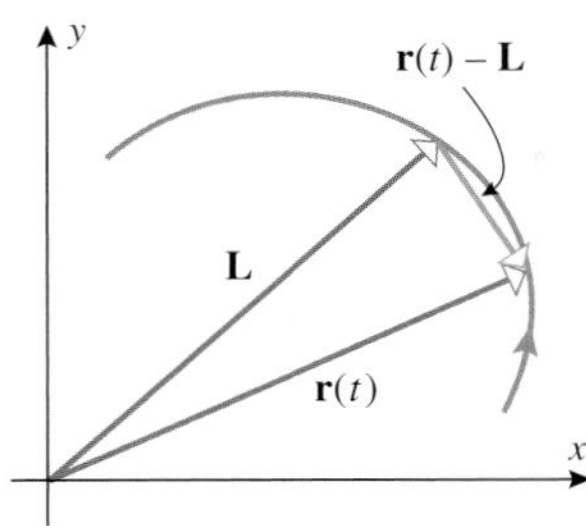

$\|\mathbf{r}(t) - \mathbf{L}\|$ is the distance between terminal points for vectors $\mathbf{r}(t)$ and $\mathbf{L}$ when positioned with the same initial points.

Figure 13.2.2

> **13.2.1 DEFINITION.** Let $\mathbf{r}(t)$ be a vector-valued function that is defined for all t in some open interval containing the number a, except that $\mathbf{r}(t)$ need not be defined at a. We will write
>
> $$\lim_{t\to a} \mathbf{r}(t) = \mathbf{L}$$
>
> if and only if
>
> $$\lim_{t\to a} \|\mathbf{r}(t) - \mathbf{L}\| = 0$$

Note that $\|\mathbf{r}(t) - \mathbf{L}\|$ is a real number for each value of t, so even though this expression involves a vector-valued function, the limit

$$\lim_{t\to a} \|\mathbf{r}(t) - \mathbf{L}\|$$

is an ordinary limit of a real-valued function.

It is clear intuitively that $\mathbf{r}(t)$ will approach a limiting vector $\mathbf{L}$ as t approaches a if and only if the component functions of $\mathbf{r}(t)$ approach the corresponding components of $\mathbf{L}$. This suggests the following theorem, whose formal proof is omitted.

> **13.2.2 THEOREM.**
>
> (a) *If* $\mathbf{r}(t) = \langle x(t), y(t)\rangle = x(t)\mathbf{i} + y(t)\mathbf{j}$, *then*
>
> $$\lim_{t\to a} \mathbf{r}(t) = \left\langle \lim_{t\to a} x(t), \lim_{t\to a} y(t)\right\rangle = \lim_{t\to a} x(t)\mathbf{i} + \lim_{t\to a} y(t)\mathbf{j}$$
>
> *provided the limits of the component functions exist. Conversely, the limits of the component functions exist provided* $\mathbf{r}(t)$ *approaches a limiting vector as* t *approaches* a.
>
> (b) *If* $\mathbf{r}(t) = \langle x(t), y(t), z(t)\rangle = x(t)\mathbf{i} + y(t)\mathbf{j} + z(t)\mathbf{k}$, *then*
>
> $$\lim_{t\to a} \mathbf{r}(t) = \left\langle \lim_{t\to a} x(t), \lim_{t\to a} y(t), \lim_{t\to a} z(t)\right\rangle = \lim_{t\to a} x(t)\mathbf{i} + \lim_{t\to a} y(t)\mathbf{j} + \lim_{t\to a} z(t)\mathbf{k}$$
>
> *provided the limits of the component functions exist. Conversely, the limits of the component functions exist provided* $\mathbf{r}(t)$ *approaches a limiting vector as* t *approaches* a.

How would you define the one-sided limits

$$\lim_{t\to a^+} \mathbf{r}(t) \quad\text{and}\quad \lim_{t\to a^-} \mathbf{r}(t)?$$

▶ Example 1 Let $\mathbf{r}(t) = t^2\mathbf{i} + e^t\mathbf{j} - (2\cos \pi t)\mathbf{k}$. Then

$$\lim_{t\to 0} \mathbf{r}(t) = \left(\lim_{t\to 0} t^2\right)\mathbf{i} + \left(\lim_{t\to 0} e^t\right)\mathbf{j} - \left(\lim_{t\to 0} 2\cos \pi t\right)\mathbf{k} = \mathbf{j} - 2\mathbf{k}$$

Alternatively, using the angle bracket notation for vectors,

$$\lim_{t\to 0} \mathbf{r}(t) = \lim_{t\to 0} \langle t^2, e^t, -2\cos \pi t\rangle = \left\langle \lim_{t\to 0} t^2, \lim_{t\to 0} e^t, \lim_{t\to 0}(-2\cos \pi t)\right\rangle = \langle 0, 1, -2\rangle \quad ◀$$

Limits of vector-valued functions have many of the same properties as limits of real-valued functions. For example, assuming that the limits exist, the limit of a sum is the sum of the limits, the limit of a difference is the difference of the limits, and a constant scalar factor can be moved through a limit symbol.

Motivated by the definition of continuity for real-valued functions, we define a vector-valued function $\mathbf{r}(t)$ to be ***continuous*** at $t = a$ if

$$\lim_{t\to a} \mathbf{r}(t) = \mathbf{r}(a) \tag{3}$$

That is, $\mathbf{r}(a)$ is defined, the limit of $\mathbf{r}(t)$ as $t \to a$ exists, and the two are equal. As in the case for real-valued functions, we say that $\mathbf{r}(t)$ is ***continuous on an interval*** I if it is continuous at each point of I [with the understanding that at an endpoint in I the two-sided limit in (3) is replaced by the appropriate one-sided limit]. It follows from Theorem 13.2.2 that a vector-valued function is continuous at $t = a$ if and only if its component functions are continuous at $t = a$.

DERIVATIVES

The derivative of a vector-valued function is defined by a limit similar to that for the derivative of a real-valued function.

13.2.3 DEFINITION. If $\mathbf{r}(t)$ is a vector-valued function, we define the ***derivative of* r *with respect to t*** to be the vector-valued function $\mathbf{r}'$ given by

$$\mathbf{r}'(t) = \lim_{h\to 0} \frac{\mathbf{r}(t+h) - \mathbf{r}(t)}{h} \tag{4}$$

The domain of $\mathbf{r}'$ consists of all values of t in the domain of $\mathbf{r}(t)$ for which the limit exists.

The function $\mathbf{r}(t)$ is ***differentiable*** at t if the limit in (4) exists. All of the standard notations for derivatives continue to apply. For example, the derivative of $\mathbf{r}(t)$ can be expressed as

$$\frac{d}{dt}[\mathbf{r}(t)], \quad \frac{d\mathbf{r}}{dt}, \quad \mathbf{r}'(t), \quad \text{or} \quad \mathbf{r}'$$

It is important to keep in mind that $\mathbf{r}'(t)$ is a vector, not a number, and hence has a magnitude and a direction for each value of t [except if $\mathbf{r}'(t) = \mathbf{0}$, in which case $\mathbf{r}'(t)$ has magnitude zero but no specific direction]. In the next section we will consider the significance of the magnitude of $\mathbf{r}'(t)$, but for now our goal is to obtain a geometric interpretation of the direction of $\mathbf{r}'(t)$. For this purpose, consider parts (a) and (b) of Figure 13.2.3. These illustrations show the graph C of $\mathbf{r}(t)$ (with its orientation) and the vectors $\mathbf{r}(t)$, $\mathbf{r}(t+h)$, and $\mathbf{r}(t+h) - \mathbf{r}(t)$ for positive h and for negative h. In both cases, the vector $\mathbf{r}(t+h) - \mathbf{r}(t)$ runs along the secant line joining the terminal points of $\mathbf{r}(t+h)$ and $\mathbf{r}(t)$, but with opposite directions in the two cases. In the case where h is positive the vector $\mathbf{r}(t+h) - \mathbf{r}(t)$ points in the direction of increasing parameter, and in the case where h is negative it points in the opposite direction. However, in the case where h is negative the direction gets reversed when we multiply by $1/h$, so in both cases the vector

$$\frac{1}{h}[\mathbf{r}(t+h) - \mathbf{r}(t)] = \frac{\mathbf{r}(t+h) - \mathbf{r}(t)}{h}$$

points in the direction of increasing parameter and runs along the secant line. As $h \to 0$, the secant line approaches the tangent line at the terminal point of $\mathbf{r}(t)$, so we can conclude that the limit

$$\mathbf{r}'(t) = \lim_{h\to 0} \frac{\mathbf{r}(t+h) - \mathbf{r}(t)}{h}$$

(if it exists and is nonzero) is a vector that is tangent to the curve C at the tip of $\mathbf{r}(t)$ and points in the direction of increasing parameter (Figure 13.2.3).

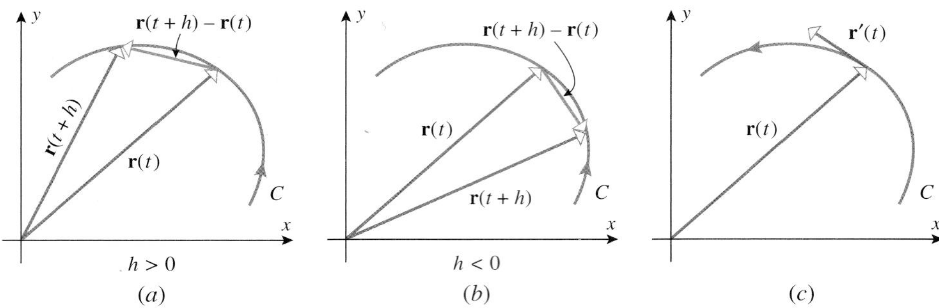

Figure 13.2.3

We can summarize all of this as follows.

> **13.2.4 GEOMETRIC INTERPRETATION OF THE DERIVATIVE.** Suppose that C is the graph of a vector-valued function $\mathbf{r}(t)$ in 2-space or 3-space and that $\mathbf{r}'(t)$ exists and is nonzero for a given value of t. If the vector $\mathbf{r}'(t)$ is positioned with its initial point at the terminal point of the radius vector $\mathbf{r}(t)$, then $\mathbf{r}'(t)$ is tangent to C and points in the direction of increasing parameter.

Since limits of vector-valued functions can be computed componentwise, it seems reasonable that we should be able to compute derivatives in terms of component functions as well. This is the result of the next theorem.

> **13.2.5 THEOREM.** *If $\mathbf{r}(t)$ is a vector-valued function, then $\mathbf{r}$ is differentiable at t if and only if each of its component functions is differentiable at t, in which case the component functions of $\mathbf{r}'(t)$ are the derivatives of the corresponding component functions of $\mathbf{r}(t)$.*

PROOF. For simplicity, we give the proof in 2-space; the proof in 3-space is identical, except for the additional component. Assume that $\mathbf{r}(t) = x(t)\mathbf{i} + y(t)\mathbf{j}$. Then

$$\begin{aligned}
\mathbf{r}'(t) &= \lim_{h\to 0} \frac{\mathbf{r}(t+h) - \mathbf{r}(t)}{h} \\
&= \lim_{h\to 0} \frac{[x(t+h)\mathbf{i} + y(t+h)\mathbf{j}] - [x(t)\mathbf{i} + y(t)\mathbf{j}]}{h} \\
&= \left(\lim_{h\to 0} \frac{x(t+h) - x(t)}{h}\right)\mathbf{i} + \left(\lim_{h\to 0} \frac{y(t+h) - y(t)}{h}\right)\mathbf{j} \\
&= x'(t)\mathbf{i} + y'(t)\mathbf{j}
\end{aligned}$$

■

▶ Example 2 Let $\mathbf{r}(t) = t^2\mathbf{i} + e^t\mathbf{j} - (2\cos \pi t)\mathbf{k}$. Then

$$\begin{aligned}
\mathbf{r}'(t) &= \frac{d}{dt}(t^2)\mathbf{i} + \frac{d}{dt}(e^t)\mathbf{j} - \frac{d}{dt}(2\cos \pi t)\mathbf{k} \\
&= 2t\mathbf{i} + e^t\mathbf{j} + (2\pi \sin \pi t)\mathbf{k} \quad ◀
\end{aligned}$$

DERIVATIVE RULES

Many of the rules for differentiating real-valued functions have analogs in the context of differentiating vector-valued functions. We state some of these in the following theorem.

13.2.6 THEOREM (*Rules of Differentiation*). *Let $\mathbf{r}(t)$, $\mathbf{r}_1(t)$, and $\mathbf{r}_2(t)$ be vector-valued functions that are all in 2-space or all in 3-space, and let $f(t)$ be a real-valued function, k a scalar, and $\mathbf{c}$ a constant vector (that is, a vector whose value does not depend on t). Then the following rules of differentiation hold:*

(a) $\dfrac{d}{dt}[\mathbf{c}] = \mathbf{0}$

(b) $\dfrac{d}{dt}[k\mathbf{r}(t)] = k\dfrac{d}{dt}[\mathbf{r}(t)]$

(c) $\dfrac{d}{dt}[\mathbf{r}_1(t) + \mathbf{r}_2(t)] = \dfrac{d}{dt}[\mathbf{r}_1(t)] + \dfrac{d}{dt}[\mathbf{r}_2(t)]$

(d) $\dfrac{d}{dt}[\mathbf{r}_1(t) - \mathbf{r}_2(t)] = \dfrac{d}{dt}[\mathbf{r}_1(t)] - \dfrac{d}{dt}[\mathbf{r}_2(t)]$

(e) $\dfrac{d}{dt}[f(t)\mathbf{r}(t)] = f(t)\dfrac{d}{dt}[\mathbf{r}(t)] + \dfrac{d}{dt}[f(t)]\mathbf{r}(t)$

The proofs of most of these rules are immediate consequences of Definition 13.2.3, although the last rule can be seen more easily by application of the product rule for real-valued functions to the component functions. The proof of Theorem 13.2.6 is left as an exercise.

TANGENT LINES TO GRAPHS OF VECTOR-VALUED FUNCTIONS

Motivated by the discussion of the geometric interpretation of the derivative of a vector-valued function, we make the following definition.

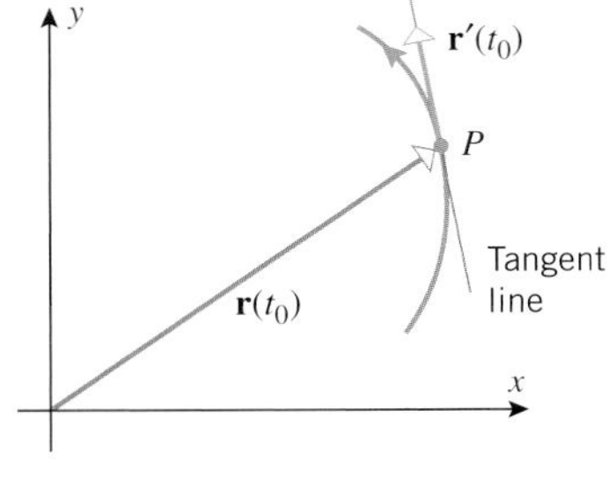

Figure 13.2.4

13.2.7 DEFINITION. Let P be a point on the graph of a vector-valued function $\mathbf{r}(t)$, and let $\mathbf{r}(t_0)$ be the radius vector from the origin to P (Figure 13.2.4). If $\mathbf{r}'(t_0)$ exists and $\mathbf{r}'(t_0) \neq \mathbf{0}$, then we call $\mathbf{r}'(t_0)$ a ***tangent vector*** to the graph of $\mathbf{r}(t)$ at $\mathbf{r}(t_0)$, and we call the line through P that is parallel to the tangent vector the ***tangent line*** to the graph of $\mathbf{r}(t)$ at $\mathbf{r}(t_0)$.

Let $\mathbf{r}_0 = \mathbf{r}(t_0)$ and $\mathbf{v}_0 = \mathbf{r}'(t_0)$. It follows from Formula (9) of Section 12.5 that the tangent line to the graph of $\mathbf{r}(t)$ at $\mathbf{r}_0$ is given by the vector equation

$$\mathbf{r} = \mathbf{r}_0 + t\mathbf{v}_0 \tag{5}$$

▶ **Example 3** Find parametric equations of the tangent line to the circular helix

$$x = \cos t, \quad y = \sin t, \quad z = t$$

where $t = t_0$, and use that result to find parametric equations for the tangent line at the point where $t = \pi$.

Solution. The vector equation of the helix is

$$\mathbf{r}(t) = \cos t\mathbf{i} + \sin t\mathbf{j} + t\mathbf{k}$$

so we have

$$\mathbf{r}_0 = \mathbf{r}(t_0) = \cos t_0\mathbf{i} + \sin t_0\mathbf{j} + t_0\mathbf{k}$$
$$\mathbf{v}_0 = \mathbf{r}'(t_0) = (-\sin t_0)\mathbf{i} + \cos t_0\mathbf{j} + \mathbf{k}$$

It follows from (5) that the vector equation of the tangent line at $t = t_0$ is

$$\begin{aligned}\mathbf{r} &= \cos t_0\mathbf{i} + \sin t_0\mathbf{j} + t_0\mathbf{k} + t[(-\sin t_0)\mathbf{i} + \cos t_0\mathbf{j} + \mathbf{k}] \\ &= (\cos t_0 - t\sin t_0)\mathbf{i} + (\sin t_0 + t\cos t_0)\mathbf{j} + (t_0 + t)\mathbf{k}\end{aligned}$$

Thus, the parametric equations of the tangent line at $t = t_0$ are

$$x = \cos t_0 - t\sin t_0, \quad y = \sin t_0 + t\cos t_0, \quad z = t_0 + t$$

In particular, the tangent line at the point where $t = \pi$ has parametric equations

$$x = -1, \quad y = -t, \quad z = \pi + t$$

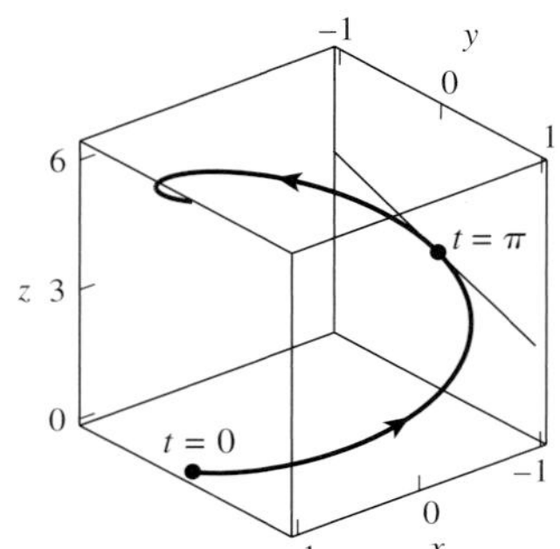

Figure 13.2.5

The graph of the helix and this tangent line are shown in Figure 13.2.5. ◄

▶ **Example 4** Let

$$\mathbf{r}_1(t) = (\tan^{-1} t)\mathbf{i} + (\sin t)\mathbf{j} + t^2\mathbf{k}$$

and

$$\mathbf{r}_2(t) = (t^2 - t)\mathbf{i} + (2t - 2)\mathbf{j} + (\ln t)\mathbf{k}$$

The graphs of $\mathbf{r}_1(t)$ and $\mathbf{r}_2(t)$ intersect at the origin. Find the degree measure of the acute angle between the tangent lines to the graphs of $\mathbf{r}_1(t)$ and $\mathbf{r}_2(t)$ at the origin.

Solution. The graph of $\mathbf{r}_1(t)$ passes through the origin at $t = 0$, where its tangent vector is

$$\mathbf{r}_1'(0) = \left\langle \frac{1}{1+t^2}, \cos t, 2t \right\rangle\bigg|_{t=0} = \langle 1, 1, 0\rangle$$

The graph of $\mathbf{r}_2(t)$ passes through the origin at $t = 1$ (verify), where its tangent vector is

$$\mathbf{r}_2'(1) = \left\langle 2t - 1, 2, \frac{1}{t} \right\rangle\bigg|_{t=1} = \langle 1, 2, 1\rangle$$

By Theorem 12.3.3, the angle θ between these two tangent vectors satisfies

$$\cos\theta = \frac{\langle 1, 1, 0\rangle \cdot \langle 1, 2, 1\rangle}{\|\langle 1, 1, 0\rangle\|\,\|\langle 1, 2, 1\rangle\|} = \frac{3}{\sqrt{12}} = \frac{\sqrt{3}}{2}$$

It follows that $\theta = \pi/6$ radians, or $30°$. ◄

DERIVATIVES OF DOT AND CROSS PRODUCTS

The following rules, which are derived in the exercises, provide a method for differentiating dot products in 2-space and 3-space and cross products in 3-space.

Note that in (6) the order of the factors in each term on the right does not matter, but in (7) it does.

$$\frac{d}{dt}[\mathbf{r}_1(t) \cdot \mathbf{r}_2(t)] = \mathbf{r}_1(t) \cdot \frac{d\mathbf{r}_2}{dt} + \frac{d\mathbf{r}_1}{dt} \cdot \mathbf{r}_2(t) \tag{6}$$

$$\frac{d}{dt}[\mathbf{r}_1(t) \times \mathbf{r}_2(t)] = \mathbf{r}_1(t) \times \frac{d\mathbf{r}_2}{dt} + \frac{d\mathbf{r}_1}{dt} \times \mathbf{r}_2(t) \tag{7}$$

In plane geometry one learns that a tangent line to a circle is perpendicular to the radius at the point of tangency. Consequently, if a point moves along a circle in 2-space that is centered at the origin, then one would expect the radius vector and the tangent vector at any point on the circle to be orthogonal. This is the motivation for the following useful theorem, which is applicable in both 2-space and 3-space.

13.2.8 THEOREM. *If $\mathbf{r}(t)$ is a vector-valued function in 2-space or 3-space and $\|\mathbf{r}(t)\|$ is constant for all t, then*

$$\mathbf{r}(t) \cdot \mathbf{r}'(t) = 0 \tag{8}$$

that is, $\mathbf{r}(t)$ and $\mathbf{r}'(t)$ are orthogonal vectors for all t.

PROOF. It follows from (6) with $\mathbf{r}_1(t) = \mathbf{r}_2(t) = \mathbf{r}(t)$ that

$$\frac{d}{dt}[\mathbf{r}(t) \cdot \mathbf{r}(t)] = \mathbf{r}(t) \cdot \frac{d\mathbf{r}}{dt} + \frac{d\mathbf{r}}{dt} \cdot \mathbf{r}(t)$$

or, equivalently,

$$\frac{d}{dt}[\|\mathbf{r}(t)\|^2] = 2\mathbf{r}(t) \cdot \frac{d\mathbf{r}}{dt} \tag{9}$$

But $\|\mathbf{r}(t)\|^2$ is constant, so its derivative is zero. Thus

$$2\mathbf{r}(t) \cdot \frac{d\mathbf{r}}{dt} = 0$$

from which (8) follows. ■

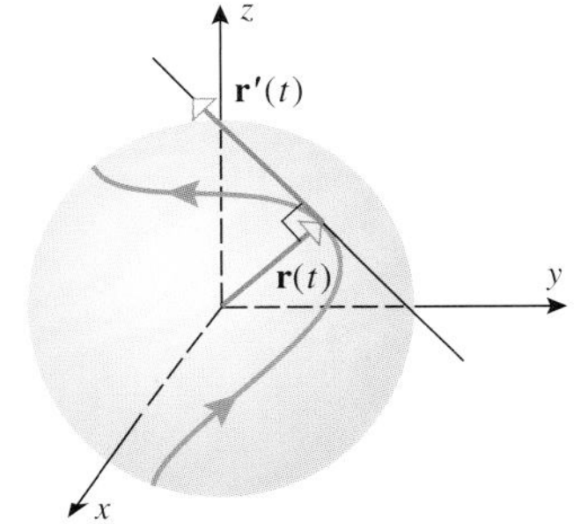

Figure 13.2.6

▶ **Example 5** Just as a tangent line to a circle in 2-space is perpendicular to the radius at the point of tangency, so a tangent vector to a curve on the surface of a sphere in 3-space that is centered at the origin is orthogonal to the radius vector at the point of tangency (Figure 13.2.6). To see that this is so, suppose that the graph of $\mathbf{r}(t)$ lies on the surface of a sphere of positive radius k centered at the origin. For each value of t we have $\|\mathbf{r}(t)\| = k$, so by Theorem 13.2.8

$$\mathbf{r}(t) \cdot \mathbf{r}'(t) = 0$$

and hence the radius vector $\mathbf{r}(t)$ and the tangent vector $\mathbf{r}'(t)$ are orthogonal. ◀

DEFINITE INTEGRALS OF VECTOR-VALUED FUNCTIONS

If $\mathbf{r}(t)$ is a vector-valued function that is continuous on the interval $a \le t \le b$, then we define the ***definite integral*** of $\mathbf{r}(t)$ over this interval as a limit of Riemann sums, just as in Definition 5.5.1, except here the integrand is a vector-valued function. Specifically, we define

$$\int_a^b \mathbf{r}(t)\, dt = \lim_{\max \Delta t_k \to 0} \sum_{k=1}^{n} \mathbf{r}(t_k^*)\Delta t_k \tag{10}$$

It follows from (10) that the definite integral of $\mathbf{r}(t)$ over the interval $a \le t \le b$ can be expressed as a vector whose components are the definite integrals of the component functions of $\mathbf{r}(t)$. For example, if $\mathbf{r}(t) = x(t)\mathbf{i} + y(t)\mathbf{j}$, then

$$\begin{aligned}\int_a^b \mathbf{r}(t)\,dt &= \lim_{\max \Delta t_k \to 0} \sum_{k=1}^{n} \mathbf{r}(t_k^*)\Delta t_k \\ &= \lim_{\max \Delta t_k \to 0} \left[\left(\sum_{k=1}^{n} x(t_k^*)\Delta t_k\right)\mathbf{i} + \left(\sum_{k=1}^{n} y(t_k^*)\Delta t_k\right)\mathbf{j}\right] \\ &= \left(\lim_{\max \Delta t_k \to 0} \sum_{k=1}^{n} x(t_k^*)\Delta t_k\right)\mathbf{i} + \left(\lim_{\max \Delta t_k \to 0} \sum_{k=1}^{n} y(t_k^*)\Delta t_k\right)\mathbf{j} \\ &= \left(\int_a^b x(t)\,dt\right)\mathbf{i} + \left(\int_a^b y(t)\,dt\right)\mathbf{j}\end{aligned}$$

In general, we have

$$\int_a^b \mathbf{r}(t)\,dt = \left(\int_a^b x(t)\,dt\right)\mathbf{i} + \left(\int_a^b y(t)\,dt\right)\mathbf{j} \qquad \text{2-space} \qquad (11)$$

$$\int_a^b \mathbf{r}(t)\,dt = \left(\int_a^b x(t)\,dt\right)\mathbf{i} + \left(\int_a^b y(t)\,dt\right)\mathbf{j} + \left(\int_a^b z(t)\,dt\right)\mathbf{k} \qquad \text{3-space} \qquad (12)$$

Rewrite Formulas (11) and (12) in bracket notation with

$$\mathbf{r}(t) = \langle x(t), y(t)\rangle$$

and

$$\mathbf{r}(t) = \langle x(t), y(t), z(t)\rangle$$

respectively.

▶ **Example 6** Let $\mathbf{r}(t) = t^2\mathbf{i} + e^t\mathbf{j} - (2\cos \pi t)\mathbf{k}$. Then

$$\begin{aligned}\int_0^1 \mathbf{r}(t)\,dt &= \left(\int_0^1 t^2\,dt\right)\mathbf{i} + \left(\int_0^1 e^t\,dt\right)\mathbf{j} - \left(\int_0^1 2\cos \pi t\,dt\right)\mathbf{k} \\ &= \frac{t^3}{3}\bigg]_0^1 \mathbf{i} + e^t\bigg]_0^1 \mathbf{j} - \frac{2}{\pi}\sin \pi t\bigg]_0^1 \mathbf{k} = \frac{1}{3}\mathbf{i} + (e-1)\mathbf{j} \;◀\end{aligned}$$

RULES OF INTEGRATION

As with differentiation, many of the rules for integrating real-valued functions have analogs for vector-valued functions.

13.2.9 THEOREM (*Rules of Integration*). *Let $\mathbf{r}(t)$, $\mathbf{r}_1(t)$, and $\mathbf{r}_2(t)$ be vector-valued functions in 2-space or 3-space that are continuous on the interval $a \le t \le b$, and let k be a scalar. Then the following rules of integration hold:*

(a) $\displaystyle\int_a^b k\mathbf{r}(t)\,dt = k\int_a^b \mathbf{r}(t)\,dt$

(b) $\displaystyle\int_a^b [\mathbf{r}_1(t) + \mathbf{r}_2(t)]\,dt = \int_a^b \mathbf{r}_1(t)\,dt + \int_a^b \mathbf{r}_2(t)\,dt$

(c) $\displaystyle\int_a^b [\mathbf{r}_1(t) - \mathbf{r}_2(t)]\,dt = \int_a^b \mathbf{r}_1(t)\,dt - \int_a^b \mathbf{r}_2(t)\,dt$

We omit the proof.

ANTIDERIVATIVES OF VECTOR-VALUED FUNCTIONS

An ***antiderivative*** for a vector-valued function $\mathbf{r}(t)$ is a vector-valued function $\mathbf{R}(t)$ such that

$$\mathbf{R}'(t) = \mathbf{r}(t) \tag{13}$$

As in Chapter 5, we express Equation (13) using integral notation as

$$\int \mathbf{r}(t)\,dt = \mathbf{R}(t) + \mathbf{C} \tag{14}$$

where $\mathbf{C}$ represents an arbitrary constant *vector*.

Since differentiation of vector-valued functions can be performed componentwise, it follows that antidifferentiation can be done this way as well. This is illustrated in the next example.

▸ **Example 7**

$$\begin{aligned}\int (2t\mathbf{i} + 3t^2\mathbf{j})\,dt &= \left(\int 2t\,dt\right)\mathbf{i} + \left(\int 3t^2\,dt\right)\mathbf{j}\\ &= (t^2 + C_1)\mathbf{i} + (t^3 + C_2)\mathbf{j}\\ &= (t^2\mathbf{i} + t^3\mathbf{j}) + (C_1\mathbf{i} + C_2\mathbf{j}) = (t^2\mathbf{i} + t^3\mathbf{j}) + \mathbf{C}\end{aligned}$$

where $\mathbf{C} = C_1\mathbf{i} + C_2\mathbf{j}$ is an arbitrary vector constant of integration. ◂

Most of the familiar integration properties have vector counterparts. For example, vector differentiation and integration are inverse operations in the sense that

$$\frac{d}{dt}\left[\int \mathbf{r}(t)\,dt\right] = \mathbf{r}(t) \qquad \text{and} \qquad \int \mathbf{r}'(t)\,dt = \mathbf{r}(t) + \mathbf{C} \tag{15–16}$$

Moreover, if $\mathbf{R}(t)$ is an antiderivative of $\mathbf{r}(t)$ on an interval containing $t = a$ and $t = b$, then we have the following vector form of the Fundamental Theorem of Calculus:

$$\int_a^b \mathbf{r}(t)\,dt = \mathbf{R}(t)\Big]_a^b = \mathbf{R}(b) - \mathbf{R}(a) \tag{17}$$

▸ **Example 8** Evaluate the definite integral $\displaystyle\int_0^2 (2t\mathbf{i} + 3t^2\mathbf{j})\,dt$.

Solution. Integrating the components yields

$$\int_0^2 (2t\mathbf{i} + 3t^2\mathbf{j})\,dt = t^2\Big]_0^2 \mathbf{i} + t^3\Big]_0^2 \mathbf{j} = 4\mathbf{i} + 8\mathbf{j}$$

Alternative Solution. The function $\mathbf{R}(t) = t^2\mathbf{i} + t^3\mathbf{j}$ is an antiderivative of the integrand since $\mathbf{R}'(t) = 2t\mathbf{i} + 3t^2\mathbf{j}$. Thus, it follows from (17) that

$$\int_0^2 (2t\mathbf{i} + 3t^2\mathbf{j})\,dt = \mathbf{R}(t)\Big]_0^2 = t^2\mathbf{i} + t^3\mathbf{j}\Big]_0^2 = (4\mathbf{i} + 8\mathbf{j}) - (0\mathbf{i} + 0\mathbf{j}) = 4\mathbf{i} + 8\mathbf{j} \ ◂$$

▶ **Example 9** Find $\mathbf{r}(t)$ given that $\mathbf{r}'(t) = \langle 3, 2t \rangle$ and $\mathbf{r}(1) = \langle 2, 5 \rangle$.

Solution. Integrating $\mathbf{r}'(t)$ to obtain $\mathbf{r}(t)$ yields

$$\mathbf{r}(t) = \int \mathbf{r}'(t)\,dt = \int \langle 3, 2t \rangle\,dt = \langle 3t, t^2 \rangle + \mathbf{C}$$

where $\mathbf{C}$ is a vector constant of integration. To find the value of $\mathbf{C}$ we substitute $t = 1$ and use the given value of $\mathbf{r}(1)$ to obtain

$$\mathbf{r}(1) = \langle 3, 1 \rangle + \mathbf{C} = \langle 2, 5 \rangle$$

so that $\mathbf{C} = \langle -1, 4 \rangle$. Thus,

$$\mathbf{r}(t) = \langle 3t, t^2 \rangle + \langle -1, 4 \rangle = \langle 3t - 1, t^2 + 4 \rangle \quad ◀$$

✔ QUICK CHECK EXERCISES 13.2 *(See page 876 for answers.)*

1. (a) $\lim_{t \to 3} (t^2\mathbf{i} + 2t\mathbf{j}) =$ ________
   (b) $\lim_{t \to \pi/4} \langle \cos t, \sin t \rangle =$ ________

2. Find $\mathbf{r}'(t)$.
   (a) $\mathbf{r}(t) = (4 + 5t)\mathbf{i} + (t - t^2)\mathbf{j}$
   (b) $\mathbf{r}(t) = \left\langle \dfrac{1}{t}, \tan t, e^{2t} \right\rangle$

3. Suppose that $\mathbf{r}_1(0) = \langle 3, 2, 1 \rangle$, $\mathbf{r}_2(0) = \langle 1, 2, 3 \rangle$, $\mathbf{r}_1'(0) = \langle 0, 0, 0 \rangle$, and $\mathbf{r}_2'(0) = \langle -6, -4, -2 \rangle$. Use this information to evaluate the derivative of each function at $t = 0$.
   (a) $\mathbf{r}(t) = 2\mathbf{r}_1(t) - \mathbf{r}_2(t)$
   (b) $\mathbf{r}(t) = \cos t\,\mathbf{r}_1(t) + e^{2t}\mathbf{r}_2(t)$
   (c) $\mathbf{r}(t) = \mathbf{r}_1(t) \times \mathbf{r}_2(t)$
   (d) $f(t) = \mathbf{r}_1(t) \cdot \mathbf{r}_2(t)$

4. (a) $\int_0^1 \langle 2t, t^2, \sin \pi t \rangle\,dt =$ ________
   (b) $\int (t\mathbf{i} - 3t^2\mathbf{j} + e^t\mathbf{k})\,dt =$ ________

EXERCISE SET 13.2 ☐ Graphing Utility

1–4 Find the limit.

1. $\lim_{t \to +\infty} \left\langle \dfrac{t^2 + 1}{3t^2 + 2}, \dfrac{1}{t} \right\rangle$

2. $\lim_{t \to 0^+} \left(\sqrt{t}\,\mathbf{i} + \dfrac{\sin t}{t}\mathbf{j} \right)$

3. $\lim_{t \to 2} (t\mathbf{i} - 3\mathbf{j} + t^2\mathbf{k})$

4. $\lim_{t \to 1} \left\langle \dfrac{3}{t^2}, \dfrac{\ln t}{t^2 - 1}, \sin 2t \right\rangle$

5–6 Determine whether $\mathbf{r}(t)$ is continuous at $t = 0$. Explain your reasoning.

5. (a) $\mathbf{r}(t) = 3\sin t\,\mathbf{i} - 2t\mathbf{j}$ (b) $\mathbf{r}(t) = t^2\mathbf{i} + \dfrac{1}{t}\mathbf{j} + t\mathbf{k}$

6. (a) $\mathbf{r}(t) = e^t\mathbf{i} + \mathbf{j} + \csc t\,\mathbf{k}$
   (b) $\mathbf{r}(t) = 5\mathbf{i} - \sqrt{3t + 1}\,\mathbf{j} + e^{2t}\mathbf{k}$

7. Sketch the circle $\mathbf{r}(t) = \cos t\,\mathbf{i} + \sin t\,\mathbf{j}$, and in each part draw the vector with its correct length.
   (a) $\mathbf{r}'(\pi/4)$ (b) $\mathbf{r}''(\pi)$ (c) $\mathbf{r}(2\pi) - \mathbf{r}(3\pi/2)$

8. Sketch the circle $\mathbf{r}(t) = \cos t\,\mathbf{i} - \sin t\,\mathbf{j}$, and in each part draw the vector with its correct length.
   (a) $\mathbf{r}'(\pi/4)$ (b) $\mathbf{r}''(\pi)$ (c) $\mathbf{r}(2\pi) - \mathbf{r}(3\pi/2)$

9–10 Find $\mathbf{r}'(t)$.

9. $\mathbf{r}(t) = 4\mathbf{i} - \cos t\,\mathbf{j}$

10. $\mathbf{r}(t) = (\tan^{-1} t)\mathbf{i} + t\cos t\,\mathbf{j} - \sqrt{t}\,\mathbf{k}$

11–14 Find the vector $\mathbf{r}'(t_0)$; then sketch the graph of $\mathbf{r}(t)$ in 2-space and draw the tangent vector $\mathbf{r}'(t_0)$.

11. $\mathbf{r}(t) = \langle t, t^2 \rangle$; $t_0 = 2$

12. $\mathbf{r}(t) = t^3\mathbf{i} + t^2\mathbf{j}$; $t_0 = 1$

13. $\mathbf{r}(t) = \sec t\,\mathbf{i} + \tan t\,\mathbf{j}$; $t_0 = 0$

14. $\mathbf{r}(t) = 2\sin t\,\mathbf{i} + 3\cos t\,\mathbf{j}$; $t_0 = \pi/6$

15–16 Find the vector $\mathbf{r}'(t_0)$; then sketch the graph of $\mathbf{r}(t)$ in 3-space and draw the tangent vector $\mathbf{r}'(t_0)$.

15. $\mathbf{r}(t) = 2\sin t\,\mathbf{i} + \mathbf{j} + 2\cos t\,\mathbf{k}$; $t_0 = \pi/2$

16. $\mathbf{r}(t) = \cos t\,\mathbf{i} + \sin t\,\mathbf{j} + t\mathbf{k}$; $t_0 = \pi/4$

17–18 Use a graphing utility to generate the graph of $\mathbf{r}(t)$ and the graph of the tangent line at t_0 on the same screen.

☐ 17. $\mathbf{r}(t) = \sin \pi t\,\mathbf{i} + t^2\mathbf{j}$; $t_0 = \frac{1}{2}$

18. $\mathbf{r}(t) = 3\sin t\mathbf{i} + 4\cos t\mathbf{j};\ t_0 = \pi/4$

19–22 Find parametric equations of the line tangent to the graph of $\mathbf{r}(t)$ at the point where $t = t_0$.

19. $\mathbf{r}(t) = t^2\mathbf{i} + (2 - \ln t)\mathbf{j};\ t_0 = 1$
20. $\mathbf{r}(t) = e^{2t}\mathbf{i} - 2\cos 3t\mathbf{j};\ t_0 = 0$
21. $\mathbf{r}(t) = 2\cos \pi t\mathbf{i} + 2\sin \pi t\mathbf{j} + 3t\mathbf{k};\ t_0 = \frac{1}{3}$
22. $\mathbf{r}(t) = \ln t\mathbf{i} + e^{-t}\mathbf{j} + t^3\mathbf{k};\ t_0 = 2$

23–26 Find a vector equation of the line tangent to the graph of $\mathbf{r}(t)$ at the point P_0 on the curve.

23. $\mathbf{r}(t) = (2t - 1)\mathbf{i} + \sqrt{3t + 4}\mathbf{j};\ P_0(-1, 2)$
24. $\mathbf{r}(t) = 4\cos t\mathbf{i} - 3t\mathbf{j};\ P_0(2, -\pi)$
25. $\mathbf{r}(t) = t^2\mathbf{i} - \dfrac{1}{t+1}\mathbf{j} + (4 - t^2)\mathbf{k};\ P_0(4, 1, 0)$
26. $\mathbf{r}(t) = \sin t\mathbf{i} + \cosh t\mathbf{j} + (\tan^{-1} t)\mathbf{k};\ P_0(0, 1, 0)$
27. Let $\mathbf{r}(t) = \cos t\mathbf{i} + \sin t\mathbf{j} + \mathbf{k}$. Find
    (a) $\lim_{t\to 0} (\mathbf{r}(t) - \mathbf{r}'(t))$ (b) $\lim_{t\to 0} (\mathbf{r}(t) \times \mathbf{r}'(t))$
    (c) $\lim_{t\to 0} (\mathbf{r}(t) \cdot \mathbf{r}'(t))$.
28. Let $\mathbf{r}(t) = t\mathbf{i} + t^2\mathbf{j} + t^3\mathbf{k}$. Find
$$\lim_{t\to 1} \mathbf{r}(t) \cdot (\mathbf{r}'(t) \times \mathbf{r}''(t))$$

29–30 Calculate
$$\frac{d}{dt}[\mathbf{r}_1(t) \cdot \mathbf{r}_2(t)] \quad\text{and}\quad \frac{d}{dt}[\mathbf{r}_1(t) \times \mathbf{r}_2(t)]$$
first by differentiating the product directly and then by applying Formulas (6) and (7).

29. $\mathbf{r}_1(t) = 2t\mathbf{i} + 3t^2\mathbf{j} + t^3\mathbf{k},\ \mathbf{r}_2(t) = t^4\mathbf{k}$
30. $\mathbf{r}_1(t) = \cos t\mathbf{i} + \sin t\mathbf{j} + t\mathbf{k},\ \mathbf{r}_2(t) = \mathbf{i} + t\mathbf{k}$

31–36 Evaluate the indefinite integral.

31. $\int (3\mathbf{i} + 4t\mathbf{j})\,dt$
32. $\int (\sin t\mathbf{i} - \cos t\mathbf{j})\,dt$
33. $\int (t\sin t\mathbf{i} + \mathbf{j})\,dt$
34. $\int \langle te^t, \ln t\rangle\,dt$
35. $\int \left(t^2\mathbf{i} - 2t\mathbf{j} + \frac{1}{t}\mathbf{k}\right)dt$
36. $\int \langle e^{-t}, e^t, 3t^2\rangle\,dt$

37–42 Evaluate the definite integral.

37. $\int_0^{\pi/2} \langle \cos 2t, \sin 2t\rangle\,dt$
38. $\int_0^1 (t^2\mathbf{i} + t^3\mathbf{j})\,dt$
39. $\int_0^2 \|t\mathbf{i} + t^2\mathbf{j}\|\,dt$
40. $\int_{-3}^3 \langle (3-t)^{3/2}, (3+t)^{3/2}, 1\rangle\,dt$
41. $\int_1^9 (t^{1/2}\mathbf{i} + t^{-1/2}\mathbf{j})\,dt$
42. $\int_0^1 (e^{2t}\mathbf{i} + e^{-t}\mathbf{j} + t\mathbf{k})\,dt$

43–46 Solve the vector initial-value problem for $\mathbf{y}(t)$ by integrating and using the initial conditions to find the constants of integration.

43. $\mathbf{y}'(t) = 2t\mathbf{i} + 3t^2\mathbf{j},\ \mathbf{y}(0) = \mathbf{i} - \mathbf{j}$
44. $\mathbf{y}'(t) = \cos t\mathbf{i} + \sin t\mathbf{j},\ \mathbf{y}(0) = \mathbf{i} - \mathbf{j}$
45. $\mathbf{y}''(t) = \mathbf{i} + e^t\mathbf{j},\ \mathbf{y}(0) = 2\mathbf{i},\ \mathbf{y}'(0) = \mathbf{j}$
46. $\mathbf{y}''(t) = 12t^2\mathbf{i} - 2t\mathbf{j},\ \mathbf{y}(0) = 2\mathbf{i} - 4\mathbf{j},\ \mathbf{y}'(0) = \mathbf{0}$

47–48 Let $\theta(t)$ be the angle between $\mathbf{r}(t)$ and $\mathbf{r}'(t)$. Use a graphing calculator to generate the graph of θ versus t, and make rough estimates of the t-values at which t-intercepts or relative extrema occur. What do these values tell you about the vectors $\mathbf{r}(t)$ and $\mathbf{r}'(t)$?

47. $\mathbf{r}(t) = 4\cos t\mathbf{i} + 3\sin t\mathbf{j};\ 0 \le t \le 2\pi$
48. $\mathbf{r}(t) = t^2\mathbf{i} + t^3\mathbf{j};\ 0 \le t \le 1$
49. (a) Find the points where the curve
$$\mathbf{r} = t\mathbf{i} + t^2\mathbf{j} - 3t\mathbf{k}$$
intersects the plane $2x - y + z = -2$.
    (b) For the curve and plane in part (a), find, to the nearest degree, the acute angle that the tangent line to the curve makes with a line normal to the plane at each point of intersection.
50. Find where the tangent line to the curve
$$\mathbf{r} = e^{-2t}\mathbf{i} + \cos t\mathbf{j} + 3\sin t\mathbf{k}$$
at the point $(1, 1, 0)$ intersects the yz-plane.

51–52 Show that the graphs of $\mathbf{r}_1(t)$ and $\mathbf{r}_2(t)$ intersect at the point P. Find, to the nearest degree, the acute angle between the tangent lines to the graphs of $\mathbf{r}_1(t)$ and $\mathbf{r}_2(t)$ at the point P.

51. $\mathbf{r}_1(t) = t^2\mathbf{i} + t\mathbf{j} + 3t^3\mathbf{k}$
    $\mathbf{r}_2(t) = (t-1)\mathbf{i} + \frac{1}{4}t^2\mathbf{j} + (5-t)\mathbf{k};\ P(1, 1, 3)$
52. $\mathbf{r}_1(t) = 2e^{-t}\mathbf{i} + \cos t\mathbf{j} + (t^2 + 3)\mathbf{k}$
    $\mathbf{r}_2(t) = (1-t)\mathbf{i} + t^2\mathbf{j} + (t^3 + 4)\mathbf{k};\ P(2, 1, 3)$

FOCUS ON CONCEPTS

53. Use Formula (7) to derive the differentiation formula
$$\frac{d}{dt}[\mathbf{r}(t) \times \mathbf{r}'(t)] = \mathbf{r}(t) \times \mathbf{r}''(t)$$
54. Let $\mathbf{u} = \mathbf{u}(t)$, $\mathbf{v} = \mathbf{v}(t)$, and $\mathbf{w} = \mathbf{w}(t)$ be differentiable vector-valued functions. Use Formulas (6) and (7) to show that
$$\frac{d}{dt}[\mathbf{u} \cdot (\mathbf{v} \times \mathbf{w})] = \frac{d\mathbf{u}}{dt} \cdot [\mathbf{v} \times \mathbf{w}] + \mathbf{u} \cdot \left[\frac{d\mathbf{v}}{dt} \times \mathbf{w}\right] + \mathbf{u} \cdot \left[\mathbf{v} \times \frac{d\mathbf{w}}{dt}\right]$$

55. Let u_1, u_2, u_3, v_1, v_2, v_3, w_1, w_2, and w_3 be differentiable functions of t. Use Exercise 54 to show that

$$\frac{d}{dt}\begin{vmatrix} u_1 & u_2 & u_3 \\ v_1 & v_2 & v_3 \\ w_1 & w_2 & w_3 \end{vmatrix} = \begin{vmatrix} u_1' & u_2' & u_3' \\ v_1 & v_2 & v_3 \\ w_1 & w_2 & w_3 \end{vmatrix} + \begin{vmatrix} u_1 & u_2 & u_3 \\ v_1' & v_2' & v_3' \\ w_1 & w_2 & w_3 \end{vmatrix} + \begin{vmatrix} u_1 & u_2 & u_3 \\ v_1 & v_2 & v_3 \\ w_1' & w_2' & w_3' \end{vmatrix}$$

56. Prove Theorem 13.2.6 for 2-space.

57. Derive Formulas (6) and (7) for 3-space.

58. Prove Theorem 13.2.9 for 2-space.

✔QUICK CHECK ANSWERS 13.2

1. (a) $9\mathbf{i}+6\mathbf{j}$ (b) $\left\langle \frac{\sqrt{2}}{2}, \frac{\sqrt{2}}{2} \right\rangle$ **2.** (a) $\mathbf{r}'(t) = 5\mathbf{i} + (1-2t)\mathbf{j}$ (b) $\mathbf{r}'(t) = \left\langle -\frac{1}{t^2}, \sec^2 t, 2e^{2t} \right\rangle$ **3.** (a) $\langle 6, 4, 2\rangle$ (b) $\langle -4, 0, 4\rangle$ (c) $\mathbf{0}$ (d) -28 **4.** (a) $\left\langle 1, \frac{1}{3}, \frac{2}{\pi} \right\rangle$ (b) $\frac{t^2}{2}\mathbf{i} - t^3\mathbf{j} + e^t\mathbf{k} + \mathbf{C}$

13.3 CHANGE OF PARAMETER; ARC LENGTH

We observed in earlier sections that a curve in 2-space or 3-space can be represented parametrically in more than one way. For example, in Section 1.7 we gave two parametric representations of a circle—one in which the circle was traced clockwise and the other in which it was traced counterclockwise. Sometimes it will be desirable to change the parameter for a parametric curve to a different parameter that is better suited for the problem at hand. In this section we will investigate issues associated with changes of parameter, and we will show that arc length plays a special role in parametric representations of curves.

SMOOTH PARAMETRIZATIONS

Graphs of vector-valued functions range from continuous and smooth to discontinuous and wildly erratic. In this text we will not be concerned with graphs of the latter type, so we will need to impose restrictions to eliminate the unwanted behavior. We will say that $\mathbf{r}(t)$ is ***smoothly parametrized*** or that $\mathbf{r}(t)$ is a ***smooth function*** of t if $\mathbf{r}'(t)$ is continuous and $\mathbf{r}'(t) \neq \mathbf{0}$ for any allowable value of t. Algebraically, smoothness implies that the components of $\mathbf{r}(t)$ have continuous derivatives that are not all zero for the same value of t, and geometrically, it implies that the tangent vector $\mathbf{r}'(t)$ varies continuously along the curve. For this reason a smoothly parametrized function is said to have a ***continuously turning tangent vector***.

► **Example 1** Determine whether the following vector-valued functions have continuously turning tangent vectors.

(a) $\mathbf{r}(t) = a\cos t\mathbf{i} + a\sin t\mathbf{j} + ct\mathbf{k} \quad (a > 0, c > 0)$

(b) $\mathbf{r}(t) = t^2\mathbf{i} + t^3\mathbf{j}$

Solution (a). We have

$$\mathbf{r}'(t) = -a\sin t\mathbf{i} + a\cos t\mathbf{j} + c\mathbf{k}$$

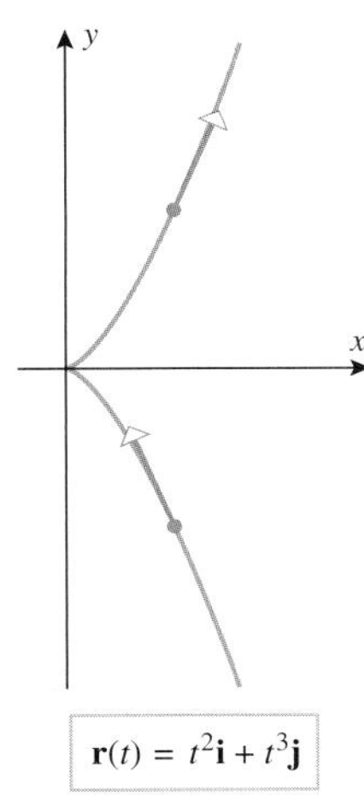

Figure 13.3.1

The components are continuous functions, and there is no value of t for which all three of them are zero (verify), so $\mathbf{r}(t)$ has a continuously turning tangent vector. The graph of $\mathbf{r}(t)$ is the circular helix in Figure 13.1.2.

Solution (b). We have

$$\mathbf{r}'(t) = 2t\mathbf{i} + 3t^2\mathbf{j}$$

Although the components are continuous functions, they are both equal to zero if $t = 0$, so $\mathbf{r}(t)$ does not have a continuously turning tangent vector. The graph of $\mathbf{r}(t)$, which is shown in Figure 13.3.1, is a semicubical parabola traced in the upward direction (see Example 3 of Section 11.2). Observe that for values of t slightly less than zero the angle between $\mathbf{r}'(t)$ and $\mathbf{i}$ is near π, and for values of t slightly larger than zero the angle is near 0; hence there is a sudden reversal in the direction of the tangent vector as t increases through $t = 0$. ◀

ARC LENGTH FROM THE VECTOR VIEWPOINT

Recall from Theorem 6.4.3 that the arc length L of a parametric curve

$$x = x(t), \quad y = y(t) \qquad (a \le t \le b) \tag{1}$$

is given by the formula

$$L = \int_a^b \sqrt{\left(\frac{dx}{dt}\right)^2 + \left(\frac{dy}{dt}\right)^2}\, dt \tag{2}$$

Analogously, the arc length L of a parametric curve

$$x = x(t), \quad y = y(t), \quad z = z(t) \qquad (a \le t \le b) \tag{3}$$

in 3-space is given by the formula

$$L = \int_a^b \sqrt{\left(\frac{dx}{dt}\right)^2 + \left(\frac{dy}{dt}\right)^2 + \left(\frac{dz}{dt}\right)^2}\, dt \tag{4}$$

Formulas (2) and (4) have vector forms that we can obtain by letting

$$\underbrace{\mathbf{r}(t) = x(t)\mathbf{i} + y(t)\mathbf{j}}_{\text{2-space}} \quad \text{or} \quad \underbrace{\mathbf{r}(t) = x(t)\mathbf{i} + y(t)\mathbf{j} + z(t)\mathbf{k}}_{\text{3-space}}$$

It follows that

$$\underbrace{\frac{d\mathbf{r}}{dt} = \frac{dx}{dt}\mathbf{i} + \frac{dy}{dt}\mathbf{j}}_{\text{2-space}} \quad \text{or} \quad \underbrace{\frac{d\mathbf{r}}{dt} = \frac{dx}{dt}\mathbf{i} + \frac{dy}{dt}\mathbf{j} + \frac{dz}{dt}\mathbf{k}}_{\text{3-space}}$$

and hence

$$\underbrace{\left\|\frac{d\mathbf{r}}{dt}\right\| = \sqrt{\left(\frac{dx}{dt}\right)^2 + \left(\frac{dy}{dt}\right)^2}}_{\text{2-space}} \quad \text{or} \quad \underbrace{\left\|\frac{d\mathbf{r}}{dt}\right\| = \sqrt{\left(\frac{dx}{dt}\right)^2 + \left(\frac{dy}{dt}\right)^2 + \left(\frac{dz}{dt}\right)^2}}_{\text{3-space}}$$

Substituting these expressions in (2) and (4) leads us to the following theorem.

> **13.3.1 THEOREM.** *If C is the graph in 2-space or 3-space of a smooth vector-valued function $\mathbf{r}(t)$, then its arc length L from $t = a$ to $t = b$ is*
>
> $$L = \int_a^b \left\| \frac{d\mathbf{r}}{dt} \right\| dt \tag{5}$$

▶ **Example 2** Find the arc length of that portion of the circular helix

$$x = \cos t, \quad y = \sin t, \quad z = t$$

from $t = 0$ to $t = \pi$.

Solution. Set $\mathbf{r}(t) = (\cos t)\mathbf{i} + (\sin t)\mathbf{j} + t\mathbf{k} = \langle \cos t, \sin t, t \rangle$. Then

$$\mathbf{r}'(t) = \langle -\sin t, \cos t, 1 \rangle \quad \text{and} \quad \|\mathbf{r}'(t)\| = \sqrt{(-\sin t)^2 + (\cos t)^2 + 1} = \sqrt{2}$$

From Theorem 13.3.1 the arc length of the helix is

$$L = \int_0^\pi \left\| \frac{d\mathbf{r}}{dt} \right\| dt = \int_0^\pi \sqrt{2}\, dt = \sqrt{2}\pi \;◀$$

ARC LENGTH AS A PARAMETER

For many purposes the best parameter to use for representing a curve in 2-space or 3-space parametrically is the length of arc measured along the curve from some fixed reference point. This can be done as follows:

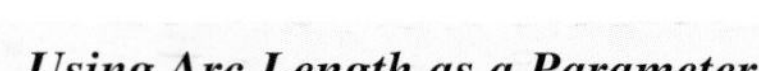

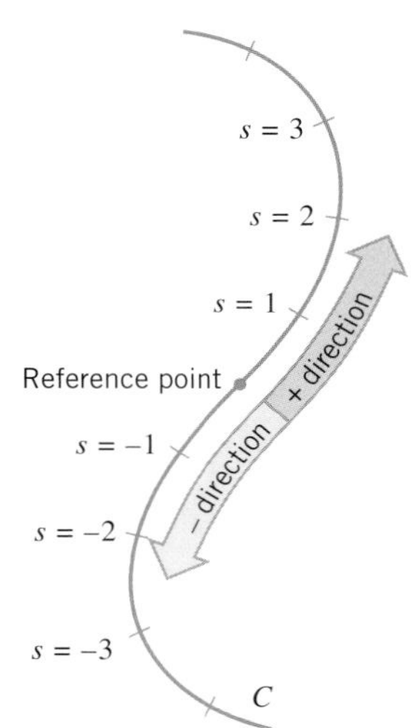

Figure 13.3.2

> ***Using Arc Length as a Parameter***
>
> **Step 1.** Select an arbitrary point on the curve C to serve as a ***reference point***.
>
> **Step 2.** Starting from the reference point, choose one direction along the curve to be the ***positive direction*** and the other to be the ***negative direction***.
>
> **Step 3.** If P is a point on the curve, let s be the "signed" arc length along C from the reference point to P, where s is positive if P is in the positive direction from the reference point and s is negative if P is in the negative direction. Figure 13.3.2 illustrates this idea.

By this procedure, a unique point P on the curve is determined when a value for s is given. For example, $s = 2$ determines the point that is 2 units along the curve in the positive direction from the reference point, and $s = -\frac{3}{2}$ determines the point that is $\frac{3}{2}$ units along the curve in the negative direction from the reference point.

Let us now treat s as a variable. As the value of s changes, the corresponding point P moves along C and the coordinates of P become functions of s. Thus, in 2-space the coordinates of P are $(x(s), y(s))$, and in 3-space they are $(x(s), y(s), z(s))$. Therefore, in 2-space or 3-space the curve C is given by the parametric equations

$$x = x(s), \quad y = y(s) \quad \text{or} \quad x = x(s), \quad y = y(s), \quad z = z(s)$$

A parametric representation of a curve with arc length as the parameter is called an ***arc length parametrization*** of the curve. Note that a given curve will generally have infinitely many different arc length parametrizations, since the reference point and orientation can be chosen arbitrarily.

▶ **Example 3** Find the arc length parametrization of the circle $x^2 + y^2 = a^2$ with counterclockwise orientation and $(a, 0)$ as the reference point.

Solution. The circle with counterclockwise orientation can be represented by the parametric equations

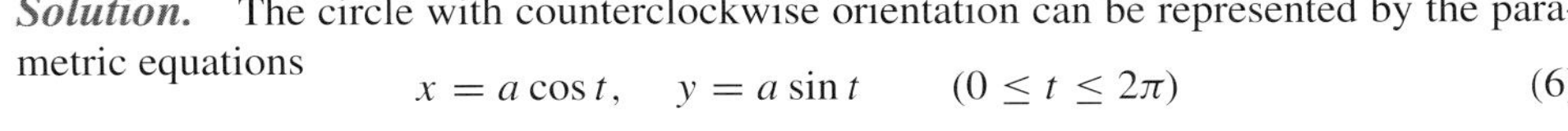

$$x = a\cos t, \quad y = a\sin t \quad (0 \le t \le 2\pi) \tag{6}$$

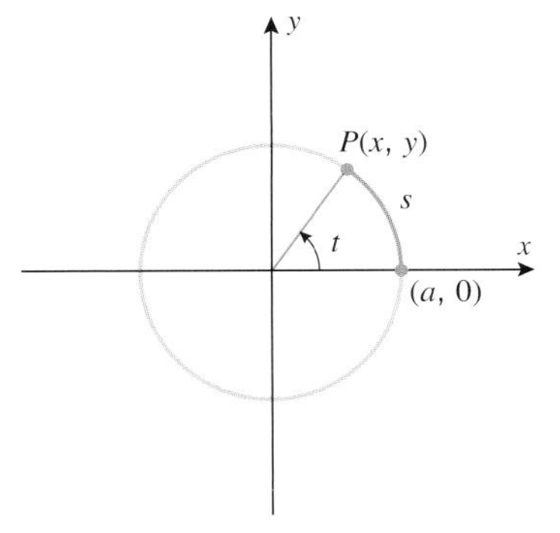

Figure 13.3.3

in which t can be interpreted as the angle in radian measure from the positive x-axis to the radius from the origin to the point $P(x, y)$ (Figure 13.3.3). If we take the positive direction for measuring the arc length to be counterclockwise, and we take $(a, 0)$ to be the reference point, then s and t are related by

$$s = at \quad \text{or} \quad t = s/a$$

Making this change of variable in (6) and noting that s increases from 0 to $2\pi a$ as t increases from 0 to 2π yields the following arc length parametrization of the circle:

$$x = a\cos(s/a), \quad y = a\sin(s/a) \quad (0 \le s \le 2\pi a)$$ ◀

CHANGE OF PARAMETER

In many situations the solution of a problem can be simplified by choosing the parameter in a vector-valued function or a parametric curve in the right way. The two most common parameters for curves in 2-space or 3-space are time and arc length. However, there are other useful possibilities as well. For example, in analyzing the motion of a particle in 2-space, it is often desirable to parametrize its trajectory in terms of the angle ϕ between the tangent vector and the positive x-axis (Figure 13.3.4). Thus, our next objective is to develop methods for changing the parameter in a vector-valued function or parametric curve. This will allow us to move freely between different possible parametrizations.

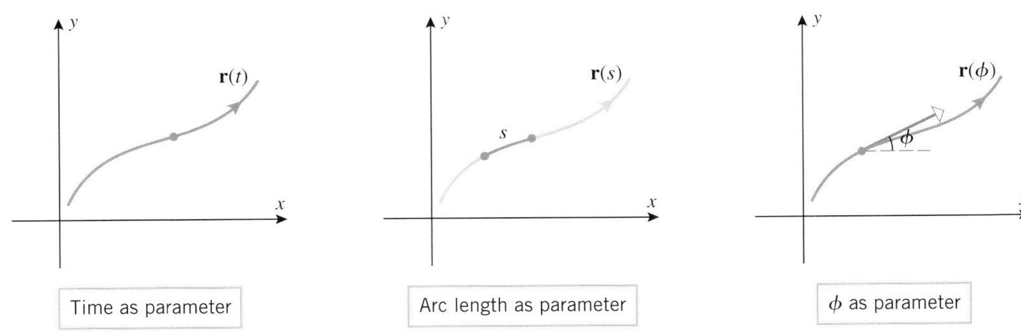

Figure 13.3.4

A ***change of parameter*** in a vector-valued function $\mathbf{r}(t)$ is a substitution $t = g(\tau)$ that produces a new vector-valued function $\mathbf{r}(g(\tau))$ having the same graph as $\mathbf{r}(t)$, but possibly traced differently as the parameter τ increases.

▶ **Example 4** Find a change of parameter $t = g(\tau)$ for the circle

$$\mathbf{r}(t) = \cos t\,\mathbf{i} + \sin t\,\mathbf{j} \quad (0 \le t \le 2\pi)$$

such that

(a) the circle is traced counterclockwise as τ increases over the interval [0, 1];

(b) the circle is traced clockwise as τ increases over the interval [0, 1].

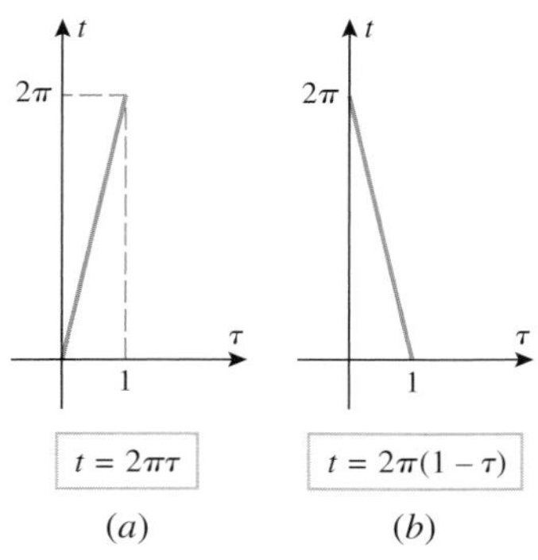

Figure 13.3.5

Solution (a). The given circle is traced counterclockwise as t increases. Thus, if we choose g to be an increasing function, then it will follow from the relationship $t = g(\tau)$ that t increases when τ increases, thereby ensuring that the circle will be traced counterclockwise as τ increases. We also want to choose g so that t increases from 0 to 2π as τ increases from 0 to 1. A simple choice of g that satisfies all of the required criteria is the linear function graphed in Figure 13.3.5*a*. The equation of this line is

$$t = g(\tau) = 2\pi\tau \tag{7}$$

which is the desired change of parameter. The resulting representation of the circle in terms of the parameter τ is

$$\mathbf{r}(g(\tau)) = \cos 2\pi\tau\,\mathbf{i} + \sin 2\pi\tau\,\mathbf{j} \qquad (0 \le \tau \le 1)$$

Solution (b). To ensure that the circle is traced clockwise, we will choose g to be a decreasing function such that t decreases from 2π to 0 as τ increases from 0 to 1. A simple choice of g that achieves this is the linear function

$$t = g(\tau) = 2\pi(1 - \tau) \tag{8}$$

graphed in Figure 13.3.5*b*. The resulting representation of the circle in terms of the parameter τ is

$$\mathbf{r}(g(\tau)) = \cos(2\pi(1 - \tau))\mathbf{i} + \sin(2\pi(1 - \tau))\mathbf{j} \qquad (0 \le \tau \le 1)$$

which simplifies to (verify)

$$\mathbf{r}(g(\tau)) = \cos 2\pi\tau\,\mathbf{i} - \sin 2\pi\tau\,\mathbf{j} \qquad (0 \le \tau \le 1)$$ ◄

When making a change of parameter $t = g(\tau)$ in a vector-valued function $\mathbf{r}(t)$, it will be important to ensure that the new vector-valued function $\mathbf{r}(g(\tau))$ is smooth if $\mathbf{r}(t)$ is smooth. To establish conditions under which this happens, we will need the following version of the chain rule for vector-valued functions. The proof is left as an exercise.

Strictly speaking, since $d\mathbf{r}/dt$ is a vector and $dt/d\tau$ is a scalar, Formula (9) should be written in the form

$$\frac{d\mathbf{r}}{d\tau} = \frac{dt}{d\tau}\frac{d\mathbf{r}}{dt}$$

However, reversing the order of the factors makes the formula easier to remember, and we will continue to do so.

13.3.2 THEOREM (*Chain Rule*). *Let $\mathbf{r}(t)$ be a vector-valued function in 2-space or 3-space that is differentiable with respect to t. If $t = g(\tau)$ is a change of parameter in which g is differentiable with respect to τ, then $\mathbf{r}(g(\tau))$ is differentiable with respect to τ and*

$$\frac{d\mathbf{r}}{d\tau} = \frac{d\mathbf{r}}{dt}\frac{dt}{d\tau} \tag{9}$$

A change of parameter $t = g(\tau)$ in which $\mathbf{r}(g(\tau))$ is smooth if $\mathbf{r}(t)$ is smooth is called a ***smooth change of parameter***. It follows from (9) that $t = g(\tau)$ will be a smooth change of parameter if $dt/d\tau$ is continuous and $dt/d\tau \ne 0$ for all values of τ, since these conditions imply that $d\mathbf{r}/d\tau$ is continuous and nonzero if $d\mathbf{r}/dt$ is continuous and nonzero. Smooth changes of parameter fall into two categories—those for which $dt/d\tau > 0$ for all τ (called ***positive changes of parameter***) and those for which $dt/d\tau < 0$ for all τ (called ***negative changes of parameter***). A positive change of parameter preserves the orientation of a parametric curve, and a negative change of parameter reverses it.

► **Example 5** In Example 4 the change of parameter in Formula (7) is positive since $dt/d\tau = 2\pi > 0$, and the change of parameter given by Formula (8) is negative since $dt/d\tau = -2\pi < 0$. The positive change of parameter preserved the orientation of the circle, and the negative change of parameter reversed it. ◄

FINDING ARC LENGTH PARAMETRIZATIONS

Next we will consider the problem of finding an arc length parametrization of a vector-valued function that is expressed initially in terms of some other parameter t. The following theorem will provide a general method for doing this.

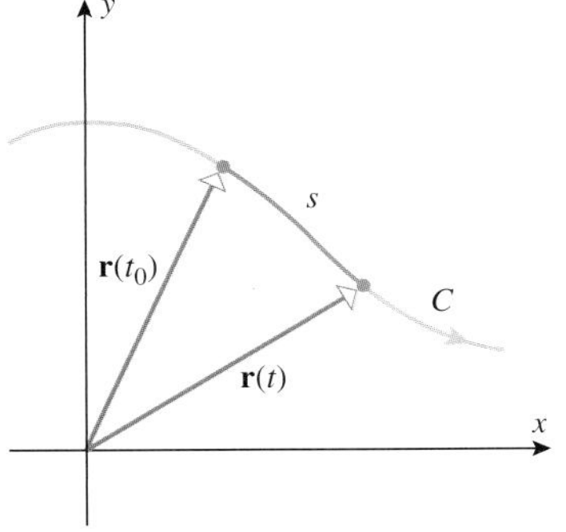

Figure 13.3.6

13.3.3 THEOREM. *Let C be the graph of a smooth vector-valued function $\mathbf{r}(t)$ in 2-space or 3-space, and let $\mathbf{r}(t_0)$ be any point on C. Then the following formula defines a positive change of parameter from t to s, where s is an arc length parameter having $\mathbf{r}(t_0)$ as its reference point* (Figure 13.3.6):

$$s = \int_{t_0}^{t} \left\| \frac{d\mathbf{r}}{du} \right\| du \tag{10}$$

PROOF. From (5) with u as the variable of integration instead of t, the integral represents the arc length of that portion of C between $\mathbf{r}(t_0)$ and $\mathbf{r}(t)$ if $t > t_0$ and the negative of that arc length if $t < t_0$. Thus, s is the arc length parameter with $\mathbf{r}(t_0)$ as its reference point and its positive direction in the direction of increasing t. ■

When needed, Formula (10) can be expressed in component form as

$$s = \int_{t_0}^{t} \sqrt{\left(\frac{dx}{du}\right)^2 + \left(\frac{dy}{du}\right)^2}\, du \qquad \text{2-space} \tag{11}$$

$$s = \int_{t_0}^{t} \sqrt{\left(\frac{dx}{du}\right)^2 + \left(\frac{dy}{du}\right)^2 + \left(\frac{dz}{du}\right)^2}\, du \qquad \text{3-space} \tag{12}$$

► **Example 6** Find the arc length parametrization of the circular helix

$$\mathbf{r} = \cos t\mathbf{i} + \sin t\mathbf{j} + t\mathbf{k} \tag{13}$$

that has reference point $\mathbf{r}(0) = (1, 0, 0)$ and the same orientation as the given helix.

Solution. Replacing t by u in $\mathbf{r}$ for integration purposes and taking $t_0 = 0$ in Formula (10), we obtain

$$\mathbf{r} = \cos u\mathbf{i} + \sin u\mathbf{j} + u\mathbf{k}$$

$$\frac{d\mathbf{r}}{du} = (-\sin u)\mathbf{i} + \cos u\mathbf{j} + \mathbf{k}$$

$$\left\| \frac{d\mathbf{r}}{du} \right\| = \sqrt{(-\sin u)^2 + \cos^2 u + 1} = \sqrt{2}$$

$$s = \int_0^t \left\| \frac{d\mathbf{r}}{du} \right\| du = \int_0^t \sqrt{2}\, du = \sqrt{2}u \Big]_0^t = \sqrt{2}t$$

Thus, $t = s/\sqrt{2}$, so (13) can be reparametrized in terms of s as

$$\mathbf{r} = \cos\left(\frac{s}{\sqrt{2}}\right)\mathbf{i} + \sin\left(\frac{s}{\sqrt{2}}\right)\mathbf{j} + \frac{s}{\sqrt{2}}\mathbf{k}$$

We are guaranteed that this reparametrization preserves the orientation of the helix since Formula (10) produces a positive change of parameter. ◀

▶ **Example 7** A bug, starting at the reference point $(1, 0, 0)$ of the helix in Example 6, walks up the helix for a distance of 10 units. What are the bug's final coordinates?

Solution. From Example 6, the arc length parametrization of the helix relative to the reference point $(1, 0, 0)$ is

$$\mathbf{r} = \cos\left(\frac{s}{\sqrt{2}}\right)\mathbf{i} + \sin\left(\frac{s}{\sqrt{2}}\right)\mathbf{j} + \frac{s}{\sqrt{2}}\mathbf{k}$$

or, expressed parametrically,

$$x = \cos\left(\frac{s}{\sqrt{2}}\right), \quad y = \sin\left(\frac{s}{\sqrt{2}}\right), \quad z = \frac{s}{\sqrt{2}}$$

Thus, at $s = 10$ the coordinates are

$$\left(\cos\left(\frac{10}{\sqrt{2}}\right), \sin\left(\frac{10}{\sqrt{2}}\right), \frac{10}{\sqrt{2}}\right) \approx (0.705, 0.709, 7.07) \quad ◀$$

▶ **Example 8** Recall from Formula (9) of Section 12.5 that the equation

$$\mathbf{r} = \mathbf{r}_0 + t\mathbf{v} \tag{14}$$

is the vector form of the line that passes through the terminal point of $\mathbf{r}_0$ and is parallel to the vector $\mathbf{v}$. Find the arc length parametrization of the line that has reference point $\mathbf{r}_0$ and the same orientation as the given line.

Solution. Replacing t by u in (14) for integration purposes and taking $t_0 = 0$ in Formula (10), we obtain

$$\mathbf{r} = \mathbf{r}_0 + u\mathbf{v} \quad \text{and} \quad \frac{d\mathbf{r}}{du} = \mathbf{v}$$

Since $\mathbf{r}_0$ is constant

It follows from this that

$$s = \int_0^t \left\|\frac{d\mathbf{r}}{du}\right\| du = \int_0^t \|\mathbf{v}\|\, du = \|\mathbf{v}\|u\Big]_0^t = t\|\mathbf{v}\|$$

This implies that $t = s/\|\mathbf{v}\|$, so (14) can be reparametrized in terms of s as

$$\mathbf{r} = \mathbf{r}_0 + s\left(\frac{\mathbf{v}}{\|\mathbf{v}\|}\right) \quad ◀ \tag{15}$$

In words, Formula (15) tells us that the line represented by Equation (14) can be reparametrized in terms of arc length with $\mathbf{r}_0$ as the reference point by normalizing $\mathbf{v}$ and then replacing t by s.

▶ **Example 9** Find the arc length parametrization of the line

$$x = 2t + 1, \quad y = 3t - 2$$

that has the same orientation as the given line and uses $(1, -2)$ as the reference point.

Solution. The line passes through the point $(1, -2)$ and is parallel to $\mathbf{v} = 2\mathbf{i} + 3\mathbf{j}$. To find the arc length parametrization of the line, we need only rewrite the given equations using $\mathbf{v}/\|\mathbf{v}\|$ rather than $\mathbf{v}$ to determine the direction and replace t by s. Since

$$\frac{\mathbf{v}}{\|\mathbf{v}\|} = \frac{2\mathbf{i} + 3\mathbf{j}}{\sqrt{13}} = \frac{2}{\sqrt{13}}\mathbf{i} + \frac{3}{\sqrt{13}}\mathbf{j}$$

it follows that the parametric equations for the line in terms of s are

$$x = \frac{2}{\sqrt{13}}s + 1, \quad y = \frac{3}{\sqrt{13}}s - 2 \blacktriangleleft$$

PROPERTIES OF ARC LENGTH PARAMETRIZATIONS

Because arc length parameters for a curve C are intimately related to the geometric characteristics of C, arc length parametrizations have properties that are not enjoyed by other parametrizations. For example, the following theorem shows that if a smooth curve is represented parametrically using an arc length parameter, then the tangent vectors all have length 1.

13.3.4 THEOREM.

(*a*) *If C is the graph of a smooth vector-valued function $\mathbf{r}(t)$ in 2-space or 3-space, where t is a general parameter, and if s is the arc length parameter for C defined by Formula* (10), *then for every value of t the tangent vector has length*

$$\left\|\frac{d\mathbf{r}}{dt}\right\| = \frac{ds}{dt} \tag{16}$$

(*b*) *If C is the graph of a smooth vector-valued function $\mathbf{r}(s)$ in 2-space or 3-space, where s is an arc length parameter, then for every value of s the tangent vector to C has length*

$$\left\|\frac{d\mathbf{r}}{ds}\right\| = 1 \tag{17}$$

(*c*) *If C is the graph of a smooth vector-valued function $\mathbf{r}(t)$ in 2-space or 3-space, and if $\|d\mathbf{r}/dt\| = 1$ for every value of t, then for any value of t_0 in the domain of $\mathbf{r}$, the parameter $s = t - t_0$ is an arc length parameter that has its reference point at the point on C where $t = t_0$.*

PROOF (*a*). This result follows by applying the Fundamental Theorem of Calculus (Theorem 5.6.3) to Formula (10).

PROOF (*b*). Let $t = s$ in part (*a*).

PROOF (*c*). It follows from Theorem 13.3.3 that the formula

$$s = \int_{t_0}^{t} \left\|\frac{d\mathbf{r}}{du}\right\| du$$

defines an arc length parameter for C with reference point $\mathbf{r}(0)$. However, $\|d\mathbf{r}/du\| = 1$ by hypothesis, so we can rewrite the formula for s as

$$s = \int_{t_0}^{t} du = u\Big]_{t_0}^{t} = t - t_0 \qquad \blacksquare$$

The component forms of Formulas (16) and (17) will be of sufficient interest in later sections that we provide them here for reference:

$$\frac{ds}{dt} = \left\|\frac{d\mathbf{r}}{dt}\right\| = \sqrt{\left(\frac{dx}{dt}\right)^2 + \left(\frac{dy}{dt}\right)^2} \qquad \text{2-space} \tag{18}$$

Note that Formulas (18) and (19) do not involve t_0, and hence do not depend on where the reference point for s is chosen. This is to be expected since changing the reference point shifts s by a constant (the arc length between the two reference points), and this constant drops out on differentiating.

$$\frac{ds}{dt} = \left\|\frac{d\mathbf{r}}{dt}\right\| = \sqrt{\left(\frac{dx}{dt}\right)^2 + \left(\frac{dy}{dt}\right)^2 + \left(\frac{dz}{dt}\right)^2} \quad \text{3-space} \qquad (19)$$

$$\left\|\frac{d\mathbf{r}}{ds}\right\| = \sqrt{\left(\frac{dx}{ds}\right)^2 + \left(\frac{dy}{ds}\right)^2} = 1 \quad \text{2-space} \qquad (20)$$

$$\left\|\frac{d\mathbf{r}}{ds}\right\| = \sqrt{\left(\frac{dx}{ds}\right)^2 + \left(\frac{dy}{ds}\right)^2 + \left(\frac{dz}{ds}\right)^2} = 1 \quad \text{3-space} \qquad (21)$$

✔QUICK CHECK EXERCISES 13.3 (See page 886 for answers.)

1. If $\mathbf{r}(t)$ is a smooth vector-valued function, then the integral
$$\int_a^b \left\|\frac{d\mathbf{r}}{dt}\right\| dt$$
may be interpreted geometrically as the ______.

2. If $\mathbf{r}(s)$ is a smooth vector-valued function parametrized by arc length s, then
$$\left\|\frac{d\mathbf{r}}{ds}\right\| = \underline{\qquad}$$
and the arc length of the graph of $\mathbf{r}$ over the interval $a \le s \le b$ is ______.

3. If $\mathbf{r}(t)$ is a smooth vector-valued function, then the arc length parameter s having $\mathbf{r}(t_0)$ as the reference point may be defined by the integral
$$s = \int_{t_0}^{t} \underline{\qquad}\, du$$

4. Suppose that $\mathbf{r}(t)$ is a smooth vector-valued function of t with $\mathbf{r}'(1) = \langle\sqrt{3}, -\sqrt{3}, -1\rangle$, and let $\mathbf{r}_1(t)$ be defined by the equation $\mathbf{r}_1(t) = \mathbf{r}(2\cos t)$. Then $\mathbf{r}_1'(\pi/3) =$ ______.

EXERCISE SET 13.3

1–4 Determine whether $\mathbf{r}(t)$ is a smooth function of the parameter t.

1. $\mathbf{r}(t) = t^3\mathbf{i} + (3t^2 - 2t)\mathbf{j} + t^2\mathbf{k}$
2. $\mathbf{r}(t) = \cos t^2\mathbf{i} + \sin t^2\mathbf{j} + e^{-t}\mathbf{k}$
3. $\mathbf{r}(t) = te^{-t}\mathbf{i} + (t^2 - 2t)\mathbf{j} + \cos \pi t\mathbf{k}$
4. $\mathbf{r}(t) = \sin \pi t\mathbf{i} + (2t - \ln t)\mathbf{j} + (t^2 - t)\mathbf{k}$

5–8 Find the arc length of the parametric curve.

5. $x = \cos^3 t,\ y = \sin^3 t,\ z = 2;\ \ 0 \le t \le \pi/2$
6. $x = 3\cos t,\ y = 3\sin t,\ z = 4t;\ \ 0 \le t \le \pi$
7. $x = e^t,\ y = e^{-t},\ z = \sqrt{2}t;\ \ 0 \le t \le 1$
8. $x = \frac{1}{2}t,\ y = \frac{1}{3}(1-t)^{3/2},\ z = \frac{1}{3}(1+t)^{3/2};\ \ -1 \le t \le 1$

9–12 Find the arc length of the graph of $\mathbf{r}(t)$.

9. $\mathbf{r}(t) = t^3\mathbf{i} + t\mathbf{j} + \frac{1}{2}\sqrt{6}t^2\mathbf{k};\ \ 1 \le t \le 3$
10. $\mathbf{r}(t) = (4+3t)\mathbf{i} + (2-2t)\mathbf{j} + (5+t)\mathbf{k};\ \ 3 \le t \le 4$
11. $\mathbf{r}(t) = 3\cos t\mathbf{i} + 3\sin t\mathbf{j} + t\mathbf{k};\ \ 0 \le t \le 2\pi$
12. $\mathbf{r}(t) = t^2\mathbf{i} + (\cos t + t\sin t)\mathbf{j} + (\sin t - t\cos t)\mathbf{k};$ $0 \le t \le \pi$

13–16 Calculate $d\mathbf{r}/d\tau$ by the chain rule, and then check your result by expressing $\mathbf{r}$ in terms of τ and differentiating.

13. $\mathbf{r} = t\mathbf{i} + t^2\mathbf{j};\ \ t = 4\tau + 1$
14. $\mathbf{r} = \langle 3\cos t, 3\sin t\rangle;\ \ t = \pi\tau$
15. $\mathbf{r} = e^t\mathbf{i} + 4e^{-t}\mathbf{j};\ \ t = \tau^2$
16. $\mathbf{r} = \mathbf{i} + 3t^{3/2}\mathbf{j} + t\mathbf{k};\ \ t = 1/\tau$

FOCUS ON CONCEPTS

17. The accompanying figure shows the graph of the *four-cusped hypocycloid*
$$\mathbf{r}(t) = \cos^3 t\mathbf{i} + \sin^3 t\mathbf{j} \qquad (0 \le t \le 2\pi)$$
(a) Give an informal explanation of why $\mathbf{r}(t)$ is not smooth.
(b) Confirm that $\mathbf{r}(t)$ is not smooth by examining $\mathbf{r}'(t)$.

18. The accompanying figure shows the graph of the vector-valued function
$$\mathbf{r}(t) = \sin t\mathbf{i} + \sin^2 t\mathbf{j} \qquad (0 \le t \le 2\pi)$$
Show that this parametric curve is not smooth, even though it has no corners. Give an informal explanation of what causes the lack of smoothness.

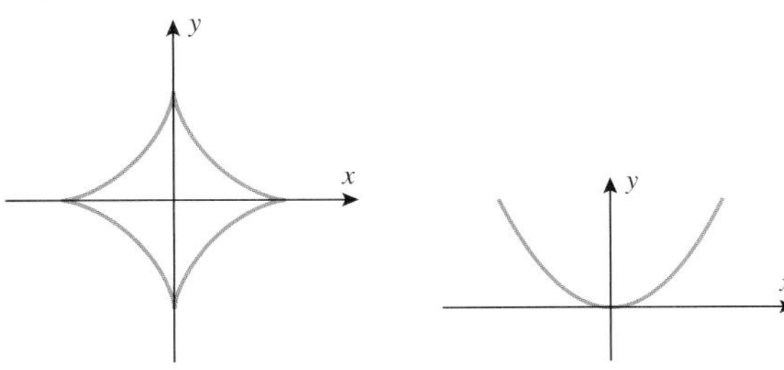

Figure Ex-17 Figure Ex-18

19. (a) Find the arc length parametrization of the line
$$x = t, \quad y = t$$
that has the same orientation as the given line and has reference point $(0, 0)$.
(b) Find the arc length parametrization of the line
$$x = t, \quad y = t, \quad z = t$$
that has the same orientation as the given line and has reference point $(0, 0, 0)$.

20. Find arc length parametrizations of the lines in Exercise 19 that have the stated reference points but are oriented opposite to the given lines.

21. (a) Find the arc length parametrization of the line
$$x = 1 + t, \quad y = 3 - 2t, \quad z = 4 + 2t$$
that has the same direction as the given line and has reference point $(1, 3, 4)$.
(b) Use the parametric equations obtained in part (a) to find the point on the line that is 25 units from the reference point in the direction of increasing parameter.

22. (a) Find the arc length parametrization of the line
$$x = -5 + 3t, \quad y = 2t, \quad z = 5 + t$$
that has the same direction as the given line and has reference point $(-5, 0, 5)$.
(b) Use the parametric equations obtained in part (a) to find the point on the line that is 10 units from the reference point in the direction of increasing parameter.

23–28 Find an arc length parametrization of the curve that has the same orientation as the given curve and has $t = 0$ as the reference point.

23. $\mathbf{r}(t) = (3 + \cos t)\mathbf{i} + (2 + \sin t)\mathbf{j};\ 0 \le t \le 2\pi$

24. $\mathbf{r}(t) = \cos^3 t\mathbf{i} + \sin^3 t\mathbf{j};\ 0 \le t \le \pi/2$

25. $\mathbf{r}(t) = \frac{1}{3}t^3\mathbf{i} + \frac{1}{2}t^2\mathbf{j};\ t \ge 0$

26. $\mathbf{r}(t) = (1 + t)^2\mathbf{i} + (1 + t)^3\mathbf{j};\ 0 \le t \le 1$

27. $\mathbf{r}(t) = e^t \cos t\mathbf{i} + e^t \sin t\mathbf{j};\ 0 \le t \le \pi/2$

28. $\mathbf{r}(t) = \sin e^t\mathbf{i} + \cos e^t\mathbf{j} + \sqrt{3}e^t\mathbf{k};\ t \ge 0$

29. Show that the arc length of the circular helix $x = a\cos t$, $y = a\sin t$, $z = ct$ for $0 \le t \le t_0$ is $t_0\sqrt{a^2 + c^2}$.

30. Use the result in Exercise 29 to show the circular helix
$$\mathbf{r} = a\cos t\mathbf{i} + a\sin t\mathbf{j} + ct\mathbf{k}$$
can be expressed as
$$\mathbf{r} = \left(a\cos\frac{s}{w}\right)\mathbf{i} + \left(a\sin\frac{s}{w}\right)\mathbf{j} + \frac{cs}{w}\mathbf{k}$$
where $w = \sqrt{a^2 + c^2}$ and s is an arc length parameter with reference point at $(a, 0, 0)$.

31. Find an arc length parametrization of the cycloid
$$x = at - a\sin t, \quad y = a - a\cos t \quad (0 \le t \le 2\pi)$$
with $(0, 0)$ as the reference point.

32. Show that in cylindrical coordinates a curve given by the parametric equations $r = r(t)$, $\theta = \theta(t)$, $z = z(t)$ for $a \le t \le b$ has arc length
$$L = \int_a^b \sqrt{\left(\frac{dr}{dt}\right)^2 + r^2\left(\frac{d\theta}{dt}\right)^2 + \left(\frac{dz}{dt}\right)^2}\, dt$$
[*Hint:* Use the relationships $x = r\cos\theta$, $y = r\sin\theta$.]

33. In each part, use the formula in Exercise 32 to find the arc length of the curve.
(a) $r = e^{2t}, \theta = t, z = e^{2t};\ 0 \le t \le \ln 2$
(b) $r = t^2, \theta = \ln t, z = \frac{1}{3}t^3;\ 1 \le t \le 2$

34. Show that in spherical coordinates a curve given by the parametric equations $\rho = \rho(t)$, $\theta = \theta(t)$, $\phi = \phi(t)$ for $a \le t \le b$ has arc length
$$L = \int_a^b \sqrt{\left(\frac{d\rho}{dt}\right)^2 + \rho^2\sin^2\phi\left(\frac{d\theta}{dt}\right)^2 + \rho^2\left(\frac{d\phi}{dt}\right)^2}\, dt$$
[*Hint:* $x = \rho\sin\phi\cos\theta$, $y = \rho\sin\phi\sin\theta$, $z = \rho\cos\phi$.]

35. In each part, use the formula in Exercise 34 to find the arc length of the curve.
(a) $\rho = e^{-t}, \theta = 2t, \phi = \pi/4;\ 0 \le t \le 2$
(b) $\rho = 2t, \theta = \ln t, \phi = \pi/6;\ 1 \le t \le 5$

FOCUS ON CONCEPTS

36. (a) Show that $\mathbf{r}(t) = t\mathbf{i} + t^2\mathbf{j}\ (-1 \le t \le 1)$ is a smooth vector-valued function, but the change of parameter $t = \tau^3$ produces a vector-valued function that is not smooth, yet has the same graph as $\mathbf{r}(t)$.
(b) Examine how the two vector-valued functions are traced and see if you can explain what causes the problem.

37. Find a change of parameter $t = g(\tau)$ for the semicircle
$$\mathbf{r}(t) = \cos t\mathbf{i} + \sin t\mathbf{j} \quad (0 \le t \le \pi)$$
such that
(a) the semicircle is traced counterclockwise as τ varies over the interval $[0, 1]$
(b) the semicircle is traced clockwise as τ varies over the interval $[0, 1]$.

38. What change of parameter $t = g(\tau)$ would you make if you wanted to trace the graph of $\mathbf{r}(t)$ $(0 \le t \le 1)$ in the opposite direction with τ varying from 0 to 1?

39. As illustrated in the accompanying figure, copper cable with a diameter of $\frac{1}{2}$ inch is to be wrapped in a circular helix around a cylinder that has a 12-inch diameter. What length of cable (measured along its centerline) will make one complete turn around the cylinder in a distance of 20 inches (between centerlines) measured parallel to the axis of the cylinder?

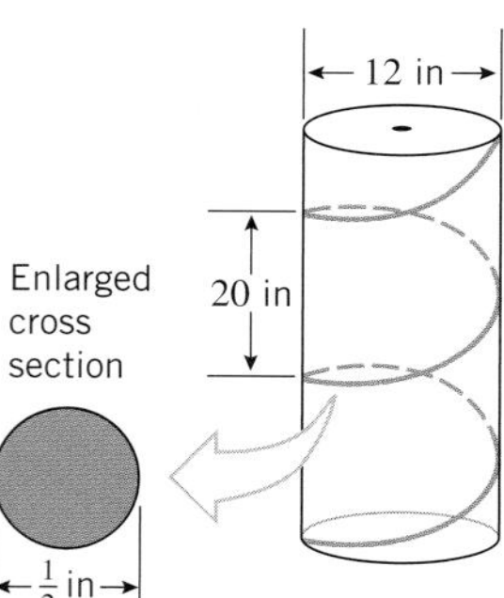

Figure Ex-39

40. Let $x = \cos t,\ y = \sin t,\ z = t^{3/2}$. Find

(a) $\|\mathbf{r}'(t)\|$ (b) $\dfrac{ds}{dt}$ (c) $\displaystyle\int_0^2 \|\mathbf{r}'(t)\|\,dt$.

41. Let $\mathbf{r}(t) = \ln t\,\mathbf{i} + 2t\mathbf{j} + t^2\mathbf{k}$. Find

(a) $\|\mathbf{r}'(t)\|$ (b) $\dfrac{ds}{dt}$ (c) $\displaystyle\int_1^3 \|\mathbf{r}'(t)\|\,dt$.

42. Prove: If $\mathbf{r}(t)$ is a smoothly parametrized function, then the angles between $\mathbf{r}'(t)$ and the vectors $\mathbf{i}$, $\mathbf{j}$, and $\mathbf{k}$ are continuous functions of t.

43. Prove the vector form of the chain rule for 2-space (Theorem 13.3.2) by expressing $\mathbf{r}(t)$ in terms of components.

✔QUICK CHECK ANSWERS 13.3

1. arc length of the graph of $\mathbf{r}(t)$ from $t = a$ to $t = b$ 2. 1; $b - a$ 3. $\left\|\dfrac{d\mathbf{r}}{du}\right\|$ 4. $\langle -3, 3, \sqrt{3}\rangle$

13.4 UNIT TANGENT, NORMAL, AND BINORMAL VECTORS

In this section we will discuss some of the fundamental geometric properties of vector-valued functions. Our work here will have important applications to the study of motion along a curved path in 2-space or 3-space and to the study of the geometric properties of curves and surfaces.

UNIT TANGENT VECTORS

Recall that if C is the graph of a *smooth* vector-valued function $\mathbf{r}(t)$ in 2-space or 3-space, then the vector $\mathbf{r}'(t)$ is nonzero, tangent to C, and points in the direction of increasing parameter. Thus, by normalizing $\mathbf{r}'(t)$ we obtain a unit vector

$$\mathbf{T}(t) = \frac{\mathbf{r}'(t)}{\|\mathbf{r}'(t)\|} \tag{1}$$

that is tangent to C and points in the direction of increasing parameter. We call $\mathbf{T}(t)$ the ***unit tangent vector*** to C at t.

Unless stated otherwise, we will assume that $\mathbf{T}(t)$ is positioned with its initial point at the terminal point of $\mathbf{r}(t)$, as in Figure 13.4.1. This will ensure that $\mathbf{T}(t)$ is actually tangent to the graph of $\mathbf{r}(t)$ and not simply parallel to the tangent line.

► **Example 1** Find the unit tangent vector to the graph of $\mathbf{r}(t) = t^2\mathbf{i} + t^3\mathbf{j}$ at the point where $t = 2$.

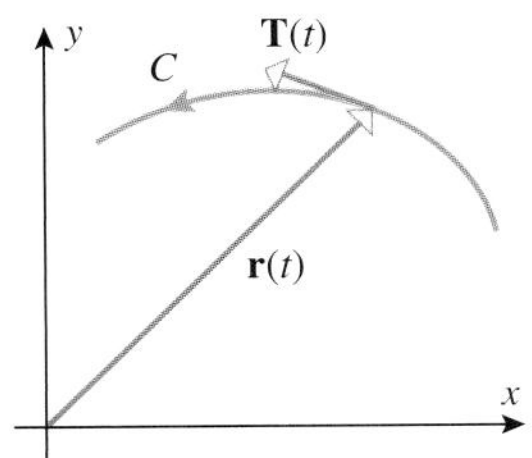

Figure 13.4.1

Solution. Since

$$\mathbf{r}'(t) = 2t\mathbf{i} + 3t^2\mathbf{j}$$

we obtain

$$\mathbf{T}(2) = \frac{\mathbf{r}'(2)}{\|\mathbf{r}'(2)\|} = \frac{4\mathbf{i} + 12\mathbf{j}}{\sqrt{160}} = \frac{4\mathbf{i} + 12\mathbf{j}}{4\sqrt{10}} = \frac{1}{\sqrt{10}}\mathbf{i} + \frac{3}{\sqrt{10}}\mathbf{j}$$

The graph of $\mathbf{r}(t)$ and the vector $\mathbf{T}(2)$ are shown in Figure 13.4.2. ◀

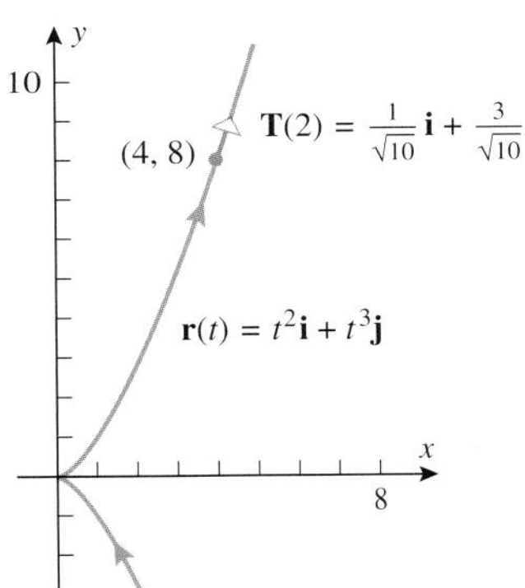

Figure 13.4.2

UNIT NORMAL VECTORS

Recall from Theorem 13.2.8 that if a vector-valued function $\mathbf{r}(t)$ has constant norm, then $\mathbf{r}(t)$ and $\mathbf{r}'(t)$ are orthogonal vectors. In particular, $\mathbf{T}(t)$ has constant norm 1, so $\mathbf{T}(t)$ and $\mathbf{T}'(t)$ are orthogonal vectors. This implies that $\mathbf{T}'(t)$ is perpendicular to the tangent line to C at t, so we say that $\mathbf{T}'(t)$ is ***normal*** to C at t. It follows that if $\mathbf{T}'(t) \neq \mathbf{0}$, and if we normalize $\mathbf{T}'(t)$, then we obtain a unit vector

$$\mathbf{N}(t) = \frac{\mathbf{T}'(t)}{\|\mathbf{T}'(t)\|} \tag{2}$$

that is normal to C and points in the same direction as $\mathbf{T}'(t)$. We call $\mathbf{N}(t)$ the ***principal unit normal vector*** to C at t, or more simply, the ***unit normal vector***. Observe that the unit normal vector is only defined at points where $\mathbf{T}'(t) \neq \mathbf{0}$. Unless stated otherwise, we will assume that this condition is satisfied. In particular, this *excludes* straight lines.

In 2-space there are two unit vectors that are orthogonal to $\mathbf{T}(t)$, and in 3-space there are infinitely many such vectors (Figure 13.4.3). In both cases the principal unit normal is that particular normal that points in the direction of $\mathbf{T}'(t)$. After the next example we will show that for a nonlinear parametric curve in 2-space the principal unit normal is the one that points "inward" toward the concave side of the curve.

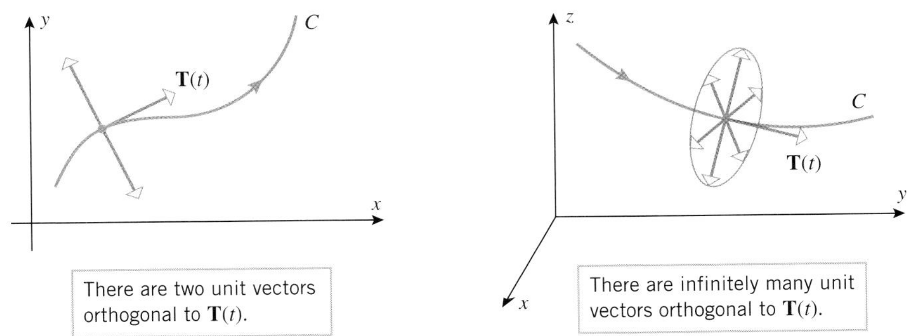

Figure 13.4.3

▶ **Example 2** Find $\mathbf{T}(t)$ and $\mathbf{N}(t)$ for the circular helix

$$x = a\cos t, \quad y = a\sin t, \quad z = ct$$

where $a > 0$.

Solution. The radius vector for the helix is

$$\mathbf{r}(t) = a\cos t\,\mathbf{i} + a\sin t\,\mathbf{j} + ct\mathbf{k}$$

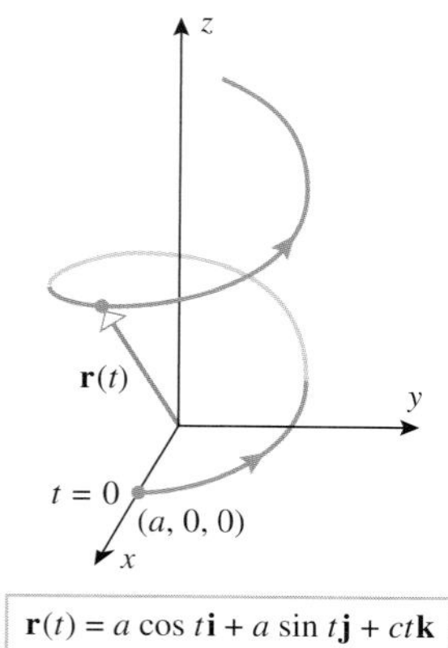

Figure 13.4.4

(Figure 13.4.4). Thus,

$$\mathbf{r}'(t) = (-a\sin t)\mathbf{i} + a\cos t\,\mathbf{j} + c\mathbf{k}$$

$$\|\mathbf{r}'(t)\| = \sqrt{(-a\sin t)^2 + (a\cos t)^2 + c^2} = \sqrt{a^2+c^2}$$

$$\mathbf{T}(t) = \frac{\mathbf{r}'(t)}{\|\mathbf{r}'(t)\|} = -\frac{a\sin t}{\sqrt{a^2+c^2}}\mathbf{i} + \frac{a\cos t}{\sqrt{a^2+c^2}}\mathbf{j} + \frac{c}{\sqrt{a^2+c^2}}\mathbf{k}$$

$$\mathbf{T}'(t) = -\frac{a\cos t}{\sqrt{a^2+c^2}}\mathbf{i} - \frac{a\sin t}{\sqrt{a^2+c^2}}\mathbf{j}$$

$$\|\mathbf{T}'(t)\| = \sqrt{\left(-\frac{a\cos t}{\sqrt{a^2+c^2}}\right)^2 + \left(-\frac{a\sin t}{\sqrt{a^2+c^2}}\right)^2} = \sqrt{\frac{a^2}{a^2+c^2}} = \frac{a}{\sqrt{a^2+c^2}}$$

$$\mathbf{N}(t) = \frac{\mathbf{T}'(t)}{\|\mathbf{T}'(t)\|} = (-\cos t)\mathbf{i} - (\sin t)\mathbf{j} = -(\cos t\mathbf{i} + \sin t\mathbf{j})$$

Note that the **k** component of the principal unit normal $\mathbf{N}(t)$ is zero for every value of t, so this vector always lies in a horizontal plane, as illustrated in Figure 13.4.5. We leave it as an exercise to show that this vector actually always points toward the z-axis. ◀

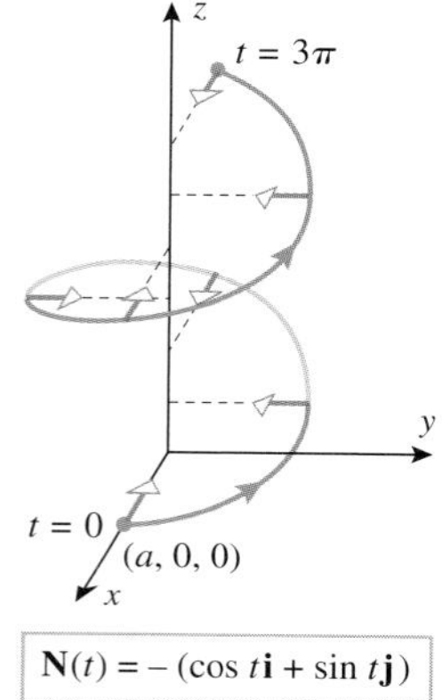

Figure 13.4.5

INWARD UNIT NORMAL VECTORS IN 2-SPACE

Our next objective is to show that for a nonlinear parametric curve C in 2-space the unit normal vector always points toward the concave side of C. For this purpose, let $\phi(t)$ be the angle from the positive x-axis to $\mathbf{T}(t)$, and let $\mathbf{n}(t)$ be the unit vector that results when $\mathbf{T}(t)$ is rotated counterclockwise through an angle of $\pi/2$ (Figure 13.4.6). Since $\mathbf{T}(t)$ and $\mathbf{n}(t)$ are unit vectors, it follows from Formula (12) of Section 12.2 that these vectors can be expressed as

$$\mathbf{T}(t) = \cos\phi(t)\mathbf{i} + \sin\phi(t)\mathbf{j} \tag{3}$$

and

$$\mathbf{n}(t) = \cos[\phi(t) + \pi/2]\mathbf{i} + \sin[\phi(t) + \pi/2]\mathbf{j} = -\sin\phi(t)\mathbf{i} + \cos\phi(t)\mathbf{j} \tag{4}$$

Observe that on intervals where $\phi(t)$ is increasing the vector $\mathbf{n}(t)$ points *toward* the concave side of C, and on intervals where $\phi(t)$ is decreasing it points *away* from the concave side (Figure 13.4.7).

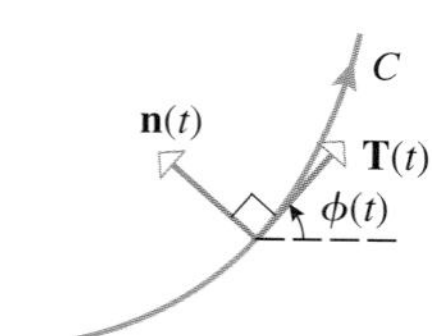

Figure 13.4.6

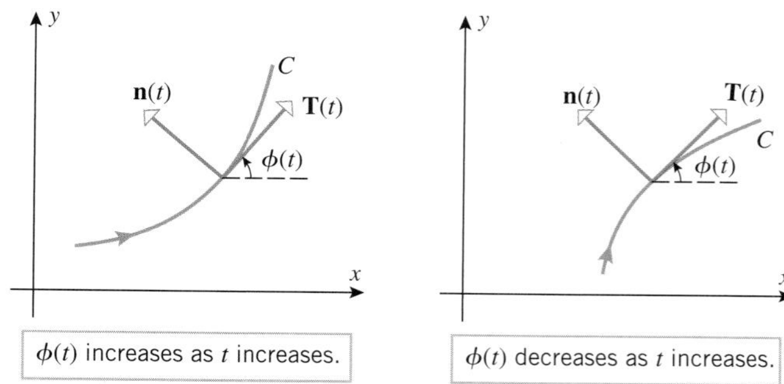

Figure 13.4.7

Now let us differentiate $\mathbf{T}(t)$ by using Formula (3) and applying the chain rule. This yields

$$\frac{d\mathbf{T}}{dt} = \frac{d\mathbf{T}}{d\phi}\frac{d\phi}{dt} = [(-\sin\phi)\mathbf{i} + (\cos\phi)\mathbf{j}]\frac{d\phi}{dt}$$

and thus from (4)

$$\frac{d\mathbf{T}}{dt} = \mathbf{n}(t)\frac{d\phi}{dt} \tag{5}$$

But $d\phi/dt > 0$ on intervals where $\phi(t)$ is increasing and $d\phi/dt < 0$ on intervals where $\phi(t)$ is decreasing. Thus, it follows from (5) that $d\mathbf{T}/dt$ has the same direction as $\mathbf{n}(t)$ on intervals where $\phi(t)$ is increasing and the opposite direction on intervals where $\phi(t)$ is decreasing. Therefore, $\mathbf{T}'(t) = d\mathbf{T}/dt$ points "inward" toward the concave side of the curve in all cases, and hence so does $\mathbf{N}(t)$. For this reason, $\mathbf{N}(t)$ is also called the ***inward unit normal*** when applied to curves in 2-space.

COMPUTING T AND N FOR CURVES PARAMETRIZED BY ARC LENGTH

In the case where $\mathbf{r}(s)$ is parametrized by arc length, the procedures for computing the unit tangent vector $\mathbf{T}(s)$ and the unit normal vector $\mathbf{N}(s)$ are simpler than in the general case. For example, we showed in Theorem 13.3.4 that if s is an arc length parameter, then $\|\mathbf{r}'(s)\| = 1$. Thus, Formula (1) for the unit tangent vector simplifies to

$$\mathbf{T}(s) = \mathbf{r}'(s) \tag{6}$$

and consequently Formula (2) for the unit normal vector simplifies to

$$\mathbf{N}(s) = \frac{\mathbf{r}''(s)}{\|\mathbf{r}''(s)\|} \tag{7}$$

WARNING

Formulas (6) and (7) are only applicable when the curve is parametrized by an arc length parameter s. For other parametrizations Formulas (1) and (2) can be used.

▶ **Example 3** The circle of radius a with counterclockwise orientation and centered at the origin can be represented by the vector-valued function

$$\mathbf{r} = a\cos t\mathbf{i} + a\sin t\mathbf{j} \qquad (0 \le t \le 2\pi) \tag{8}$$

In this representation we can interpret t as the angle in radian measure from the positive x-axis to the radius vector (Figure 13.4.8). This angle subtends an arc of length $s = at$ on the circle, so we can reparametrize the circle in terms of s by substituting s/a for t in (8). This yields

$$\mathbf{r}(s) = a\cos(s/a)\mathbf{i} + a\sin(s/a)\mathbf{j} \qquad (0 \le s \le 2\pi a)$$

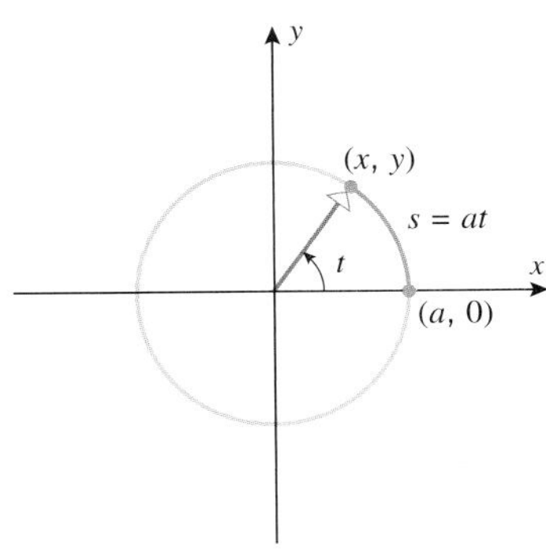

Figure 13.4.8

To find $\mathbf{T}(s)$ and $\mathbf{N}(s)$ from Formulas (6) and (7), we must compute $\mathbf{r}'(s)$, $\mathbf{r}''(s)$, and $\|\mathbf{r}''(s)\|$. Doing so, we obtain

$$\mathbf{r}'(s) = -\sin(s/a)\mathbf{i} + \cos(s/a)\mathbf{j}$$
$$\mathbf{r}''(s) = -(1/a)\cos(s/a)\mathbf{i} - (1/a)\sin(s/a)\mathbf{j}$$
$$\|\mathbf{r}''(s)\| = \sqrt{(-1/a)^2\cos^2(s/a) + (-1/a)^2\sin^2(s/a)} = 1/a$$

Thus,

$$\mathbf{T}(s) = \mathbf{r}'(s) = -\sin(s/a)\mathbf{i} + \cos(s/a)\mathbf{j}$$
$$\mathbf{N}(s) = \mathbf{r}''(s)/\|\mathbf{r}''(s)\| = -\cos(s/a)\mathbf{i} - \sin(s/a)\mathbf{j}$$

so $\mathbf{N}(s)$ points toward the center of the circle for all s (Figure 13.4.9). This makes sense geometrically and is also consistent with our earlier observation that in 2-space the unit normal vector is the inward normal. ◀

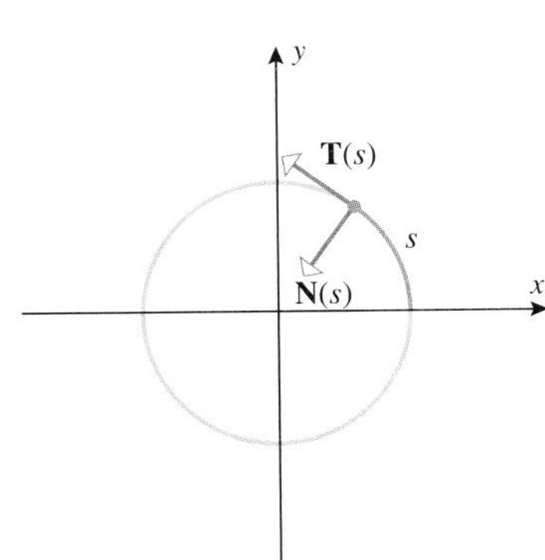

Figure 13.4.9

BINORMAL VECTORS IN 3-SPACE

If C is the graph of a vector-valued function $\mathbf{r}(t)$ in 3-space, then we define the ***binormal vector*** to C at t to be

$$\mathbf{B}(t) = \mathbf{T}(t) \times \mathbf{N}(t) \tag{9}$$

It follows from properties of the cross product that $\mathbf{B}(t)$ is orthogonal to both $\mathbf{T}(t)$ and $\mathbf{N}(t)$ and is oriented relative to $\mathbf{T}(t)$ and $\mathbf{N}(t)$ by the right-hand rule. Moreover, $\mathbf{T}(t) \times \mathbf{N}(t)$ is a unit vector since

$$\|\mathbf{T}(t) \times \mathbf{N}(t)\| = \|\mathbf{T}(t)\|\|\mathbf{N}(t)\|\sin(\pi/2) = 1$$

Thus, $\{\mathbf{T}(t), \mathbf{N}(t), \mathbf{B}(t)\}$ is a set of three mutually orthogonal unit vectors.

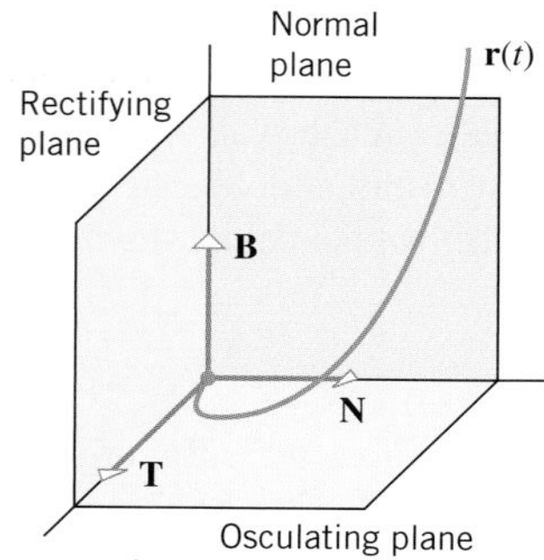

Figure 13.4.10

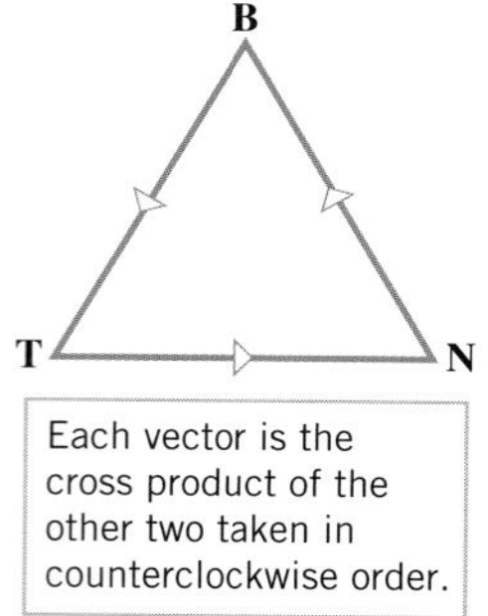

B
T
N
Each vector is the cross product of the other two taken in counterclockwise order.

Figure 13.4.11

Just as the vectors **i**, **j**, and **k** determine a right-handed coordinate system in 3-space, so do the vectors $\mathbf{T}(t)$, $\mathbf{N}(t)$, and $\mathbf{B}(t)$. At each point on a smooth parametric curve C in 3-space, these vectors determine three mutually perpendicular planes that pass through the point—the **TB**-plane (called the ***rectifying plane***), the **TN**-plane (called the ***osculating plane***), and the **NB**-plane (called the ***normal plane***) (Figure 13.4.10). Moreover, one can show that a coordinate system determined by $\mathbf{T}(t)$, $\mathbf{N}(t)$, and $\mathbf{B}(t)$ is right-handed in the sense that each of these vectors is related to the other two by the right-hand rule (Figure 13.4.11):

$$\mathbf{B}(t) = \mathbf{T}(t) \times \mathbf{N}(t), \quad \mathbf{N}(t) = \mathbf{B}(t) \times \mathbf{T}(t), \quad \mathbf{T}(t) = \mathbf{N}(t) \times \mathbf{B}(t) \tag{10}$$

The coordinate system determined by $\mathbf{T}(t)$, $\mathbf{N}(t)$, and $\mathbf{B}(t)$ is called the **TNB-*frame*** or sometimes the ***Frenet frame*** in honor of the French mathematician Jean Frédéric Frenet (1816–1900) who pioneered its application to the study of space curves. Typically, the xyz-coordinate system determined by the unit vectors **i**, **j**, and **k** remains fixed, whereas the **TNB**-frame changes as its origin moves along the curve C (Figure 13.4.12).

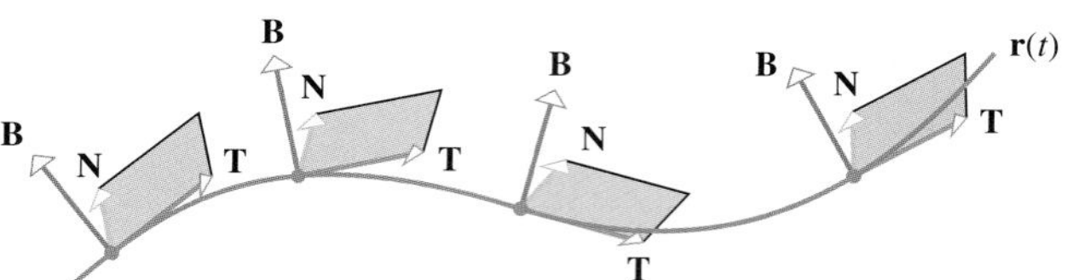

Figure 13.4.12

Formula (9) expresses $\mathbf{B}(t)$ in terms of $\mathbf{T}(t)$ and $\mathbf{N}(t)$. Alternatively, the binormal $\mathbf{B}(t)$ can be expressed directly in terms of $\mathbf{r}(t)$ as

$$\mathbf{B}(t) = \frac{\mathbf{r}'(t) \times \mathbf{r}''(t)}{\|\mathbf{r}'(t) \times \mathbf{r}''(t)\|} \tag{11}$$

and in the case where the parameter is arc length it can be expressed in terms of $\mathbf{r}(s)$ as

$$\mathbf{B}(s) = \frac{\mathbf{r}'(s) \times \mathbf{r}''(s)}{\|\mathbf{r}''(s)\|} \tag{12}$$

We omit the proof.

✔QUICK CHECK EXERCISES 13.4

(See page 892 for answers.)

1. If C is the graph of a smooth vector-valued function $\mathbf{r}(t)$, then the unit tangent, unit normal, and binormal to C at t are defined, respectively, by

   $\mathbf{T}(t) =$ ______, $\mathbf{N}(t) =$ ______, $\mathbf{B}(t) =$ ______

2. If C is the graph of a smooth vector-valued function $\mathbf{r}(s)$ parametrized by arc length, then the definitions of the unit tangent and unit normal to C at s simplify, respectively, to

   $\mathbf{T}(s) =$ ______ and $\mathbf{N}(s) =$ ______

3. If C is the graph of a smooth vector-valued function $\mathbf{r}(t)$, then the unit binormal vector to C at t may be computed directly in terms of $\mathbf{r}'(t)$ and $\mathbf{r}''(t)$ by the formula $\mathbf{B}(t) =$ ______. When $t = s$ is the arc length parameter, this formula simplifies to $\mathbf{B}(s) =$ ______.

4. Suppose that C is the graph of a smooth vector-valued function $\mathbf{r}(s)$ parametrized by arc length with $\mathbf{r}'(0) = \langle \frac{2}{3}, \frac{1}{3}, \frac{2}{3} \rangle$ and $\mathbf{r}''(0) = \langle -3, 12, -3 \rangle$. Then

   $\mathbf{T}(0) =$ ______, $\mathbf{N}(0) =$ ______, $\mathbf{B}(0) =$ ______

EXERCISE SET 13.4

FOCUS ON CONCEPTS

1. In each part, sketch the unit tangent and normal vectors at the points P, Q, and R, taking into account the orientation of the curve C.

(a) 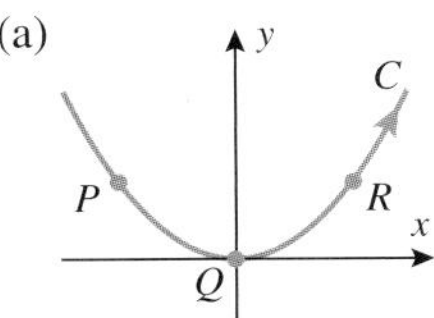

(b) 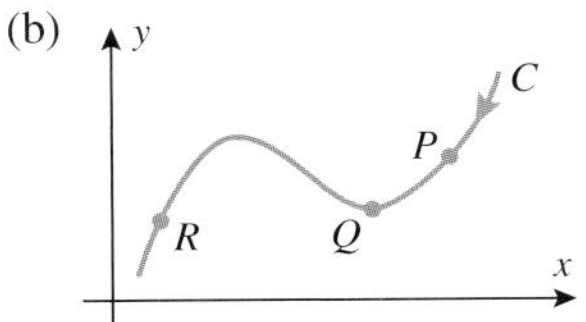

2. Make a rough sketch that shows the ellipse

$$\mathbf{r}(t) = 3\cos t\,\mathbf{i} + 2\sin t\,\mathbf{j}$$

for $0 \le t \le 2\pi$ and the unit tangent and normal vectors at the points $t = 0$, $t = \pi/4$, $t = \pi/2$, and $t = \pi$.

3. In the marginal note associated with Example 8 of Section 13.3, we observed that a line $\mathbf{r} = \mathbf{r}_0 + t\mathbf{v}$ can be parametrized in terms of an arc length parameter s with reference point $\mathbf{r}_0$ by normalizing $\mathbf{v}$. Use this result to show that the tangent line to the graph of $\mathbf{r}(t)$ at the point t_0 can be expressed as

$$\mathbf{r} = \mathbf{r}(t_0) + s\mathbf{T}(t_0)$$

where s is an arc length parameter with reference point $\mathbf{r}(t_0)$.

4. Use the result in Exercise 3 to show that the tangent line to the parabola

$$x = t, \quad y = t^2$$

at the point $(1, 1)$ can be expressed parametrically as

$$x = 1 + \frac{s}{\sqrt{5}}, \quad y = 1 + \frac{2s}{\sqrt{5}}$$

5–12 Find $\mathbf{T}(t)$ and $\mathbf{N}(t)$ at the given point.

5. $\mathbf{r}(t) = (t^2 - 1)\mathbf{i} + t\mathbf{j};\ t = 1$
6. $\mathbf{r}(t) = \frac{1}{2}t^2\mathbf{i} + \frac{1}{3}t^3\mathbf{j};\ t = 1$
7. $\mathbf{r}(t) = 5\cos t\,\mathbf{i} + 5\sin t\,\mathbf{j};\ t = \pi/3$
8. $\mathbf{r}(t) = \ln t\,\mathbf{i} + t\mathbf{j};\ t = e$
9. $\mathbf{r}(t) = 4\cos t\,\mathbf{i} + 4\sin t\,\mathbf{j} + t\mathbf{k};\ t = \pi/2$
10. $\mathbf{r}(t) = t\mathbf{i} + \frac{1}{2}t^2\mathbf{j} + \frac{1}{3}t^3\mathbf{k};\ t = 0$
11. $x = e^t\cos t,\ y = e^t\sin t,\ z = e^t;\ t = 0$
12. $x = \cosh t,\ y = \sinh t,\ z = t;\ t = \ln 2$

13–14 Use the result in Exercise 3 to find parametric equations for the tangent line to the graph of $\mathbf{r}(t)$ at t_0 in terms of an arc length parameter s.

13. $\mathbf{r}(t) = \sin t\,\mathbf{i} + \cos t\,\mathbf{j} + \frac{1}{2}t^2\mathbf{k};\ t_0 = 0$
14. $\mathbf{r}(t) = t\mathbf{i} + t\mathbf{j} + \sqrt{9 - t^2}\,\mathbf{k};\ t_0 = 1$

15–18 Use the formula $\mathbf{B}(t) = \mathbf{T}(t) \times \mathbf{N}(t)$ to find $\mathbf{B}(t)$, and then check your answer by using Formula (11) to find $\mathbf{B}(t)$ directly from $\mathbf{r}(t)$.

15. $\mathbf{r}(t) = 3\sin t\,\mathbf{i} + 3\cos t\,\mathbf{j} + 4t\mathbf{k}$
16. $\mathbf{r}(t) = e^t\sin t\,\mathbf{i} + e^t\cos t\,\mathbf{j} + 3\mathbf{k}$
17. $\mathbf{r}(t) = (\sin t - t\cos t)\mathbf{i} + (\cos t + t\sin t)\mathbf{j} + \mathbf{k}$
18. $\mathbf{r}(t) = a\cos t\,\mathbf{i} + a\sin t\,\mathbf{j} + ct\mathbf{k} \quad (a \ne 0, c \ne 0)$

19–20 Find $\mathbf{T}(t)$, $\mathbf{N}(t)$, and $\mathbf{B}(t)$ for the given value of t. Then find equations for the osculating, normal, and rectifying planes at the point that corresponds to that value of t.

19. $\mathbf{r}(t) = \cos t\,\mathbf{i} + \sin t\,\mathbf{j} + \mathbf{k};\ t = \pi/4$
20. $\mathbf{r}(t) = e^t\mathbf{i} + e^t\cos t\,\mathbf{j} + e^t\sin t\,\mathbf{k};\ t = 0$

21. (a) Use the formula $\mathbf{N}(t) = \mathbf{B}(t) \times \mathbf{T}(t)$ and Formulas (1) and (11) to show that $\mathbf{N}(t)$ can be expressed in terms of $\mathbf{r}(t)$ as

$$\mathbf{N}(t) = \frac{\mathbf{r}'(t) \times \mathbf{r}''(t)}{\|\mathbf{r}'(t) \times \mathbf{r}''(t)\|} \times \frac{\mathbf{r}'(t)}{\|\mathbf{r}'(t)\|}$$

(b) Use properties of cross products to show that the formula in part (a) can be expressed as

$$\mathbf{N}(t) = \frac{(\mathbf{r}'(t) \times \mathbf{r}''(t)) \times \mathbf{r}'(t)}{\|(\mathbf{r}'(t) \times \mathbf{r}''(t)) \times \mathbf{r}'(t)\|}$$

(c) Use the result in part (b) and Exercise 41 of Section 12.4 to show that $\mathbf{N}(t)$ can be expressed directly in terms of $\mathbf{r}(t)$ as

$$\mathbf{N}(t) = \frac{\mathbf{u}(t)}{\|\mathbf{u}(t)\|}$$

where

$$\mathbf{u}(t) = \|\mathbf{r}'(t)\|^2\mathbf{r}''(t) - (\mathbf{r}'(t) \cdot \mathbf{r}''(t))\mathbf{r}'(t)$$

22. Use the result in part (b) of Exercise 21 to find the unit normal vector requested in
(a) Exercise 5 (b) Exercise 9.

23–24 Use the result in part (c) of Exercise 21 to find $\mathbf{N}(t)$.

23. $\mathbf{r}(t) = \sin t\,\mathbf{i} + \cos t\,\mathbf{j} + t\mathbf{k}$
24. $\mathbf{r}(t) = t\mathbf{i} + t^2\mathbf{j} + t^3\mathbf{k}$

✔QUICK CHECK ANSWERS 13.4

1. $\dfrac{\mathbf{r}'(t)}{\|\mathbf{r}'(t)\|}$; $\dfrac{\mathbf{T}'(t)}{\|\mathbf{T}'(t)\|}$; $\mathbf{T}(t)\times\mathbf{N}(t)$ 2. $\mathbf{r}'(s)$; $\dfrac{\mathbf{r}''(s)}{\|\mathbf{r}''(s)\|}$ 3. $\dfrac{\mathbf{r}'(t)\times\mathbf{r}''(t)}{\|\mathbf{r}'(t)\times\mathbf{r}''(t)\|}$; $\dfrac{\mathbf{r}'(s)\times\mathbf{r}''(s)}{\|\mathbf{r}''(s)\|}$
4. $\left\langle \frac{2}{3}, \frac{1}{3}, \frac{2}{3}\right\rangle$; $\left\langle -\frac{1}{3\sqrt{2}}, \frac{4}{3\sqrt{2}}, -\frac{1}{3\sqrt{2}}\right\rangle$; $\left\langle -\frac{1}{\sqrt{2}}, 0, \frac{1}{\sqrt{2}}\right\rangle$

13.5 CURVATURE

In this section we will consider the problem of obtaining a numerical measure of how sharply a curve in 2-space or 3-space bends. Our results will have applications in geometry and in the study of motion along a curved path.

DEFINITION OF CURVATURE

Suppose that C is the graph of a smooth vector-valued function in 2-space or 3-space that is parametrized in terms of arc length. Figure 13.5.1 suggests that for a curve in 2-space the "sharpness" of the bend in C is closely related to $d\mathbf{T}/ds$, which is the rate of change of the unit tangent vector $\mathbf{T}$ with respect to s. (Keep in mind that $\mathbf{T}$ has constant length, so only its direction changes.) If C is a straight line (no bend), then the direction of $\mathbf{T}$ remains constant (Figure 13.5.1a); if C bends slightly, then $\mathbf{T}$ undergoes a gradual change of direction (Figure 13.5.1b); and if C bends sharply, then $\mathbf{T}$ undergoes a rapid change of direction (Figure 13.5.1c).

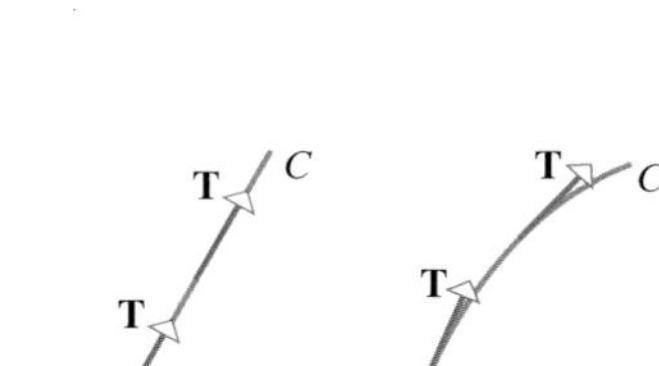

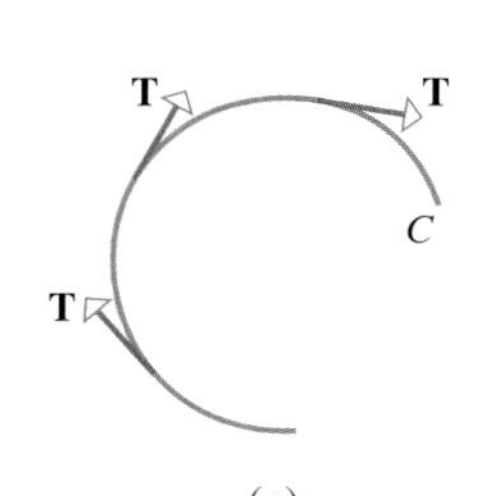

Figure 13.5.1

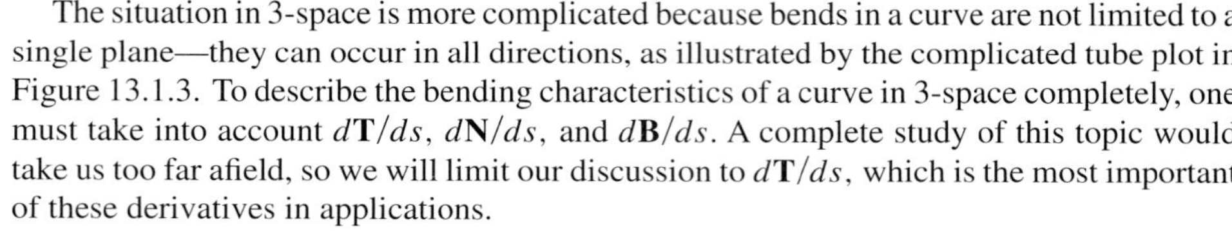

The situation in 3-space is more complicated because bends in a curve are not limited to a single plane—they can occur in all directions, as illustrated by the complicated tube plot in Figure 13.1.3. To describe the bending characteristics of a curve in 3-space completely, one must take into account $d\mathbf{T}/ds$, $d\mathbf{N}/ds$, and $d\mathbf{B}/ds$. A complete study of this topic would take us too far afield, so we will limit our discussion to $d\mathbf{T}/ds$, which is the most important of these derivatives in applications.

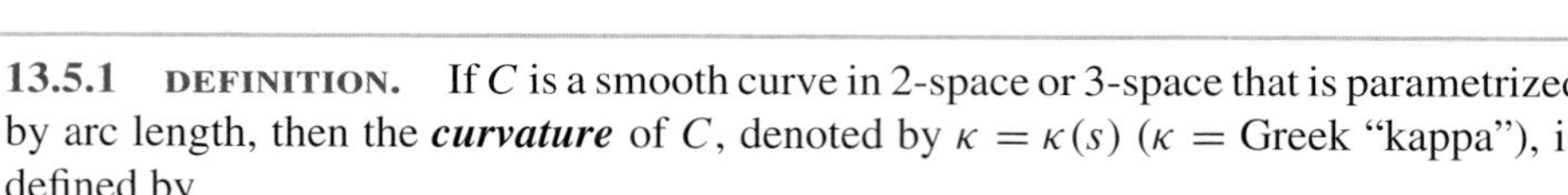

13.5.1 DEFINITION. If C is a smooth curve in 2-space or 3-space that is parametrized by arc length, then the ***curvature*** of C, denoted by $\kappa = \kappa(s)$ ($\kappa =$ Greek "kappa"), is defined by

$$\kappa(s) = \left\| \frac{d\mathbf{T}}{ds} \right\| = \|\mathbf{r}''(s)\| \tag{1}$$

Observe that $\kappa(s)$ is a real-valued function of s, since it is the *length* of $d\mathbf{T}/ds$ that measures the curvature. In general, the curvature will vary from point to point along a curve; however, the following example shows that the curvature is constant for circles in 2-space, as you might expect.

► **Example 1** In Example 3 of Section 13.4 we showed that the circle of radius a, centered at the origin, can be parametrized in terms of arc length as

$$\mathbf{r}(s) = a\cos\left(\frac{s}{a}\right)\mathbf{i} + a\sin\left(\frac{s}{a}\right)\mathbf{j} \qquad (0 \le s \le 2\pi a)$$

Thus,

$$\mathbf{r}''(s) = -\frac{1}{a}\cos\left(\frac{s}{a}\right)\mathbf{i} - \frac{1}{a}\sin\left(\frac{s}{a}\right)\mathbf{j}$$

and hence from (1)

$$\kappa(s) = \|\mathbf{r}''(s)\| = \sqrt{\left[-\frac{1}{a}\cos\left(\frac{s}{a}\right)\right]^2 + \left[-\frac{1}{a}\sin\left(\frac{s}{a}\right)\right]^2} = \frac{1}{a}$$

so the circle has constant curvature $1/a$. ◀

The next example shows that lines have zero curvature, which is consistent with the fact that they do not bend.

▶ **Example 2** Recall from Formula (15) of Section 13.3 that a line in 2-space or 3-space can be parametrized in terms of arc length as

$$\mathbf{r} = \mathbf{r}_0 + s\mathbf{u}$$

where the terminal point of $\mathbf{r}_0$ is a point on the line and $\mathbf{u}$ is a unit vector parallel to the line. Since $\mathbf{u}$ and $\mathbf{r}_0$ are constant, their derivatives with respect to s are zero, and hence

$$\mathbf{r}'(s) = \frac{d\mathbf{r}}{ds} = \frac{d}{ds}[\mathbf{r}_0 + s\mathbf{u}] = \mathbf{0} + \mathbf{u} = \mathbf{u}$$

$$\mathbf{r}''(s) = \frac{d\mathbf{r}'}{ds} = \frac{d}{ds}[\mathbf{u}] = \mathbf{0}$$

Thus,

$$\kappa(s) = \|\mathbf{r}''(s)\| = 0 \quad ◀$$

FORMULAS FOR CURVATURE

Formula (1) is only applicable if the curve is parametrized in terms of arc length. The following theorem provides two formulas for curvature in terms of a general parameter t.

13.5.2 THEOREM. *If $\mathbf{r}(t)$ is a smooth vector-valued function in 2-space or 3-space, then for each value of t at which $\mathbf{T}'(t)$ and $\mathbf{r}''(t)$ exist, the curvature κ can be expressed as*

(a) $$\kappa(t) = \frac{\|\mathbf{T}'(t)\|}{\|\mathbf{r}'(t)\|} \tag{2}$$

(b) $$\kappa(t) = \frac{\|\mathbf{r}'(t) \times \mathbf{r}''(t)\|}{\|\mathbf{r}'(t)\|^3} \tag{3}$$

PROOF (*a*). It follows from Formula (1) and Formulas (16) and (17) of Section 13.3 that

$$\kappa(t) = \left\|\frac{d\mathbf{T}}{ds}\right\| = \left\|\frac{d\mathbf{T}/dt}{ds/dt}\right\| = \left\|\frac{d\mathbf{T}/dt}{\|d\mathbf{r}/dt\|}\right\| = \frac{\|\mathbf{T}'(t)\|}{\|\mathbf{r}'(t)\|}$$

PROOF (*b*). It follows from Formula (1) of Section 13.4 that

$$\mathbf{r}'(t) = \|\mathbf{r}'(t)\|\mathbf{T}(t) \tag{4}$$

$$\mathbf{r}''(t) = \|\mathbf{r}'(t)\|'\mathbf{T}(t) + \|\mathbf{r}'(t)\|\mathbf{T}'(t) \tag{5}$$

But from Formula (2) of Section 13.4 and part (*a*) of this theorem we have

$$\mathbf{T}'(t) = \|\mathbf{T}'(t)\|\mathbf{N}(t) \quad \text{and} \quad \|\mathbf{T}'(t)\| = \kappa(t)\|\mathbf{r}'(t)\|$$

so

$$\mathbf{T}'(t) = \kappa(t)\|\mathbf{r}'(t)\|\mathbf{N}(t)$$

Substituting this into (5) yields

$$\mathbf{r}''(t) = \|\mathbf{r}'(t)\|'\mathbf{T}(t) + \kappa(t)\|\mathbf{r}'(t)\|^2\mathbf{N}(t) \tag{6}$$

Thus, from (4) and (6)

$$\mathbf{r}'(t) \times \mathbf{r}''(t) = \|\mathbf{r}'(t)\|\|\mathbf{r}'(t)\|'(\mathbf{T}(t) \times \mathbf{T}(t)) + \kappa(t)\|\mathbf{r}'(t)\|^3(\mathbf{T}(t) \times \mathbf{N}(t))$$

But the cross product of a vector with itself is zero, so this equation simplifies to

$$\mathbf{r}'(t) \times \mathbf{r}''(t) = \kappa(t)\|\mathbf{r}'(t)\|^3(\mathbf{T}(t) \times \mathbf{N}(t)) = \kappa(t)\|\mathbf{r}'(t)\|^3\mathbf{B}(t)$$

It follows from this equation and the fact that $\mathbf{B}(t)$ is a unit vector that

$$\|\mathbf{r}'(t) \times \mathbf{r}''(t)\| = \kappa(t)\|\mathbf{r}'(t)\|^3$$

Formula (3) now follows. ■

Formula (2) is useful if $\mathbf{T}(t)$ is known or is easy to obtain; however, Formula (3) will usually be easier to apply, since it involves only $\mathbf{r}(t)$ and its derivatives. We also note that cross products were defined only for vectors in 3-space, so to use Formula (3) in 2-space we must first write the 2-space function $\mathbf{r}(t) = x(t)\mathbf{i} + y(t)\mathbf{j}$ as the 3-space function $\mathbf{r}(t) = x(t)\mathbf{i} + y(t)\mathbf{j} + 0\mathbf{k}$ with a zero $\mathbf{k}$ component.

► **Example 3** Find $\kappa(t)$ for the circular helix

$$x = a\cos t, \quad y = a\sin t, \quad z = ct$$

where $a > 0$.

Solution. The radius vector for the helix is

$$\mathbf{r}(t) = a\cos t\mathbf{i} + a\sin t\mathbf{j} + ct\mathbf{k}$$

Thus,

$$\mathbf{r}'(t) = (-a\sin t)\mathbf{i} + a\cos t\mathbf{j} + c\mathbf{k}$$
$$\mathbf{r}''(t) = (-a\cos t)\mathbf{i} + (-a\sin t)\mathbf{j}$$
$$\mathbf{r}'(t) \times \mathbf{r}''(t) = \begin{vmatrix} \mathbf{i} & \mathbf{j} & \mathbf{k} \\ -a\sin t & a\cos t & c \\ -a\cos t & -a\sin t & 0 \end{vmatrix} = (ac\sin t)\mathbf{i} - (ac\cos t)\mathbf{j} + a^2\mathbf{k}$$

Therefore,

$$\|\mathbf{r}'(t)\| = \sqrt{(-a\sin t)^2 + (a\cos t)^2 + c^2} = \sqrt{a^2 + c^2}$$

and

$$\|\mathbf{r}'(t) \times \mathbf{r}''(t)\| = \sqrt{(ac\sin t)^2 + (-ac\cos t)^2 + a^4}$$
$$= \sqrt{a^2c^2 + a^4} = a\sqrt{a^2 + c^2}$$

so

$$\kappa(t) = \frac{\|\mathbf{r}'(t) \times \mathbf{r}''(t)\|}{\|\mathbf{r}'(t)\|^3} = \frac{a\sqrt{a^2 + c^2}}{(\sqrt{a^2 + c^2})^3} = \frac{a}{a^2 + c^2}$$

Note that κ does not depend on t, which tells us that the helix has constant curvature. ◄

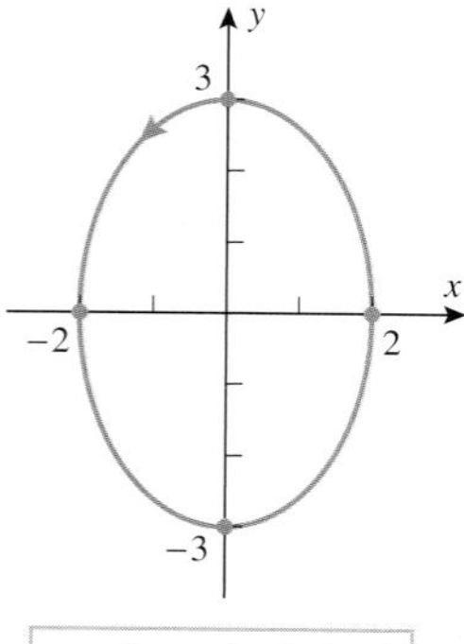

Figure 13.5.2

► **Example 4** The graph of the vector equation

$$\mathbf{r} = 2\cos t\mathbf{i} + 3\sin t\mathbf{j} \quad (0 \le t \le 2\pi)$$

is the ellipse in Figure 13.5.2. Find the curvature of the ellipse at the endpoints of the major and minor axes, and use a graphing utility to generate the graph of $\kappa(t)$.

Solution. To apply Formula (3), we must treat the ellipse as a curve in the xy-plane of an xyz-coordinate system by adding a zero $\mathbf{k}$ component and writing its equation as

$$\mathbf{r} = 2\cos t\mathbf{i} + 3\sin t\mathbf{j} + 0\mathbf{k}$$

It is not essential to write the zero $\mathbf{k}$ component explicitly as long as you assume it to be there when you calculate a cross product. Thus,

$$\mathbf{r}'(t) = (-2\sin t)\mathbf{i} + 3\cos t\mathbf{j}$$
$$\mathbf{r}''(t) = (-2\cos t)\mathbf{i} + (-3\sin t)\mathbf{j}$$
$$\mathbf{r}'(t) \times \mathbf{r}''(t) = \begin{vmatrix} \mathbf{i} & \mathbf{j} & \mathbf{k} \\ -2\sin t & 3\cos t & 0 \\ -2\cos t & -3\sin t & 0 \end{vmatrix} = [(6\sin^2 t) + (6\cos^2 t)]\mathbf{k} = 6\mathbf{k}$$

Therefore,

$$\|\mathbf{r}'(t)\| = \sqrt{(-2\sin t)^2 + (3\cos t)^2} = \sqrt{4\sin^2 t + 9\cos^2 t}$$
$$\|\mathbf{r}'(t) \times \mathbf{r}''(t)\| = 6$$

so

$$\kappa(t) = \frac{\|\mathbf{r}'(t) \times \mathbf{r}''(t)\|}{\|\mathbf{r}'(t)\|^3} = \frac{6}{[4\sin^2 t + 9\cos^2 t]^{3/2}} \tag{7}$$

The endpoints of the minor axis are $(2, 0)$ and $(-2, 0)$, which correspond to $t = 0$ and $t = \pi$, respectively. Substituting these values in (7) yields the same curvature at both points, namely,

$$\kappa = \kappa(0) = \kappa(\pi) = \frac{6}{9^{3/2}} = \frac{6}{27} = \frac{2}{9}$$

The endpoints of the major axis are $(0, 3)$ and $(0, -3)$, which correspond to $t = \pi/2$ and $t = 3\pi/2$, respectively; from (7) the curvature at these points is

$$\kappa = \kappa\left(\frac{\pi}{2}\right) = \kappa\left(\frac{3\pi}{2}\right) = \frac{6}{4^{3/2}} = \frac{3}{4}$$

Observe that the curvature is greater at the ends of the major axis than at the ends of the minor axis, as you might expect. Figure 13.5.3 shows the graph of κ versus t. This graph illustrates clearly that the curvature is minimum at $t = 0$ (the right end of the minor axis), increases to a maximum at $t = \pi/2$ (the top of the major axis), decreases to a minimum again at $t = \pi$ (the left end of the minor axis), and continues cyclically in this manner. Figure 13.5.4 provides another way of picturing the curvature. ◀

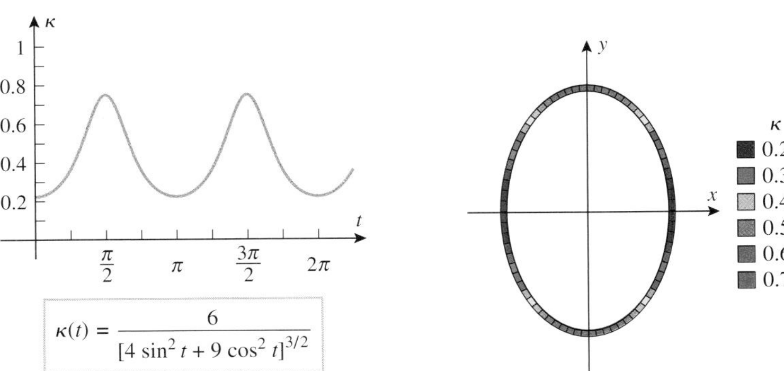

Figure 13.5.3

Figure 13.5.4

RADIUS OF CURVATURE

In the last example we found the curvature at the ends of the minor axis to be $\frac{2}{9}$ and the curvature at the ends of the major axis to be $\frac{3}{4}$. To obtain a better understanding of the meaning of these numbers, recall from Example 1 that a circle of radius a has a constant

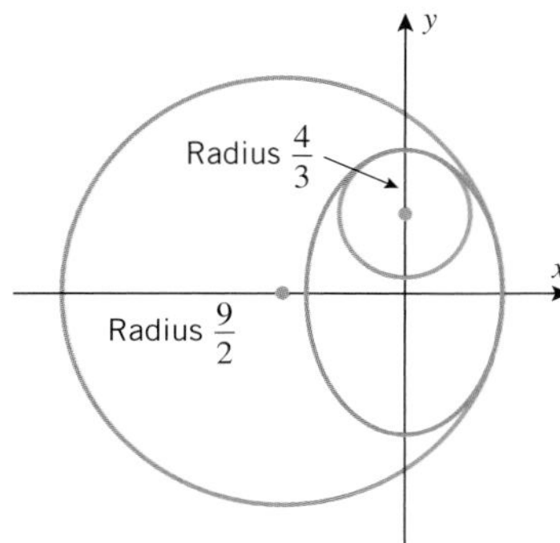

Figure 13.5.5

curvature of $1/a$; thus, the curvature of the ellipse at the ends of the minor axis is the same as that of a circle of radius $\frac{9}{2}$, and the curvature at the ends of the major axis is the same as that of a circle of radius $\frac{4}{3}$ (Figure 13.5.5).

In general, if a curve C in 2-space has nonzero curvature κ at a point P, then the circle of radius $\rho = 1/\kappa$ sharing a common tangent with C at P, and centered on the concave side of the curve at P, is called the ***circle of curvature*** or ***osculating circle*** at P (Figure 13.5.6). The osculating circle and the curve C not only touch at P but they have equal curvatures at that point. In this sense, the osculating circle is the circle that best approximates the curve C near P. The radius ρ of the osculating circle at P is called the ***radius of curvature*** at P, and the center of the circle is called the ***center of curvature*** at P (Figure 13.5.6).

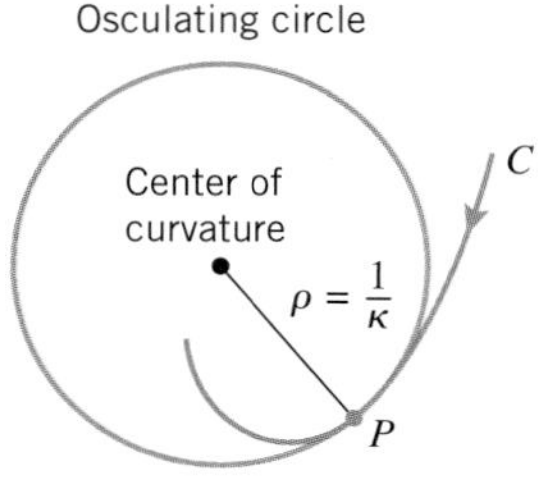

Figure 13.5.6

AN INTERPRETATION OF CURVATURE IN 2-SPACE

A useful geometric interpretation of curvature in 2-space can be obtained by considering the angle ϕ measured counterclockwise from the direction of the positive x-axis to the unit tangent vector $\mathbf{T}$ (Figure 13.5.7). By Formula (12) of Section 12.2, we can express $\mathbf{T}$ in terms of ϕ as

$$\mathbf{T}(\phi) = \cos\phi\mathbf{i} + \sin\phi\mathbf{j}$$

Thus,

$$\frac{d\mathbf{T}}{d\phi} = (-\sin\phi)\mathbf{i} + \cos\phi\mathbf{j}$$

$$\frac{d\mathbf{T}}{ds} = \frac{d\mathbf{T}}{d\phi}\frac{d\phi}{ds}$$

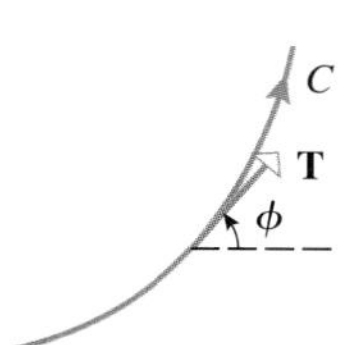

Figure 13.5.7

from which we obtain

$$\kappa(s) = \left\|\frac{d\mathbf{T}}{ds}\right\| = \left|\frac{d\phi}{ds}\right|\left\|\frac{d\mathbf{T}}{d\phi}\right\| = \left|\frac{d\phi}{ds}\right|\sqrt{(-\sin\phi)^2 + \cos^2\phi} = \left|\frac{d\phi}{ds}\right|$$

In summary, we have shown that

$$\kappa(s) = \left|\frac{d\phi}{ds}\right| \tag{8}$$

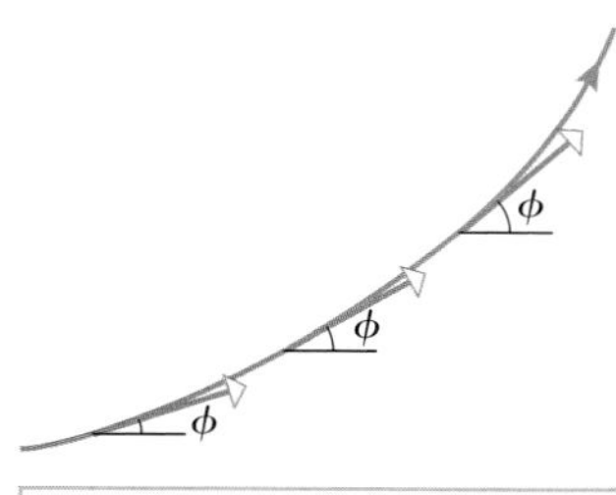

In 2-space, $\kappa(s)$ is the magnitude of the rate of change of ϕ with respect to s.

Figure 13.5.8

which tells us that curvature in 2-space can be interpreted as the magnitude of the rate of change of ϕ with respect to s—the greater the curvature, the more rapidly ϕ changes with s (Figure 13.5.8). In the case of a straight line, the angle ϕ is constant (Figure 13.5.9) and consequently $\kappa(s) = |d\phi/ds| = 0$, which is consistent with the fact that a straight line has zero curvature at every point.

FORMULA SUMMARY

We conclude this section with a summary of formulas for $\mathbf{T}$, $\mathbf{N}$, and $\mathbf{B}$. These formulas have either been derived in the text or are easily derivable from formulas we have already established.

$$\mathbf{T}(s) = \mathbf{r}'(s) \tag{9}$$

$$\mathbf{N}(s) = \frac{1}{\kappa(s)}\frac{d\mathbf{T}}{ds} = \frac{\mathbf{r}''(s)}{\|\mathbf{r}''(s)\|} = \frac{\mathbf{r}''(s)}{\kappa(s)} \tag{10}$$

$$\mathbf{B}(s) = \frac{\mathbf{r}'(s)\times\mathbf{r}''(s)}{\|\mathbf{r}''(s)\|} = \frac{\mathbf{r}'(s)\times\mathbf{r}''(s)}{\kappa(s)} \tag{11}$$

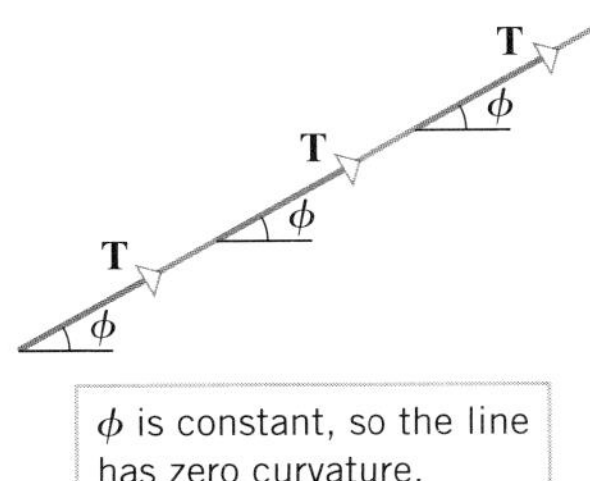

Figure 13.5.9

$$\mathbf{T}(t) = \frac{\mathbf{r}'(t)}{\|\mathbf{r}'(t)\|} \tag{12}$$

$$\mathbf{B}(t) = \frac{\mathbf{r}'(t) \times \mathbf{r}''(t)}{\|\mathbf{r}'(t) \times \mathbf{r}''(t)\|} \tag{13}$$

$$\mathbf{N}(t) = \mathbf{B}(t) \times \mathbf{T}(t) \tag{14}$$

✔QUICK CHECK EXERCISES 13.5 *(See page 900 for answers.)*

1. If C is a smooth curve parametrized by arc length, then the curvature is defined by $\kappa(s) =$ ________.
2. Let $\mathbf{r}(t)$ be a smooth vector-valued function with curvature $\kappa(t)$.
   (a) The curvature may be expressed in terms of $\mathbf{T}'(t)$ and $\mathbf{r}'(t)$ as $\kappa(t) =$ ________.
   (b) The curvature may be expressed directly in terms of $\mathbf{r}'(t)$ and $\mathbf{r}''(t)$ as $\kappa(t) =$ ________.
3. Suppose that C is the graph of a smooth vector-valued function $\mathbf{r}(s) = \langle x(s), y(s)\rangle$ parametrized by arc length and that the unit tangent $\mathbf{T}(s) = \langle \cos\phi(s), \sin\phi(s)\rangle$. Then the curvature may be expressed in terms of $\phi(s)$ as $\kappa(s) =$ ________.
4. Suppose that C is a smooth curve and that $x^2 + y^2 = 4$ is the osculating circle to C at $P(1, \sqrt{3})$. Then the curvature of C at P is ________.

EXERCISE SET 13.5  Graphing Utility  CAS

FOCUS ON CONCEPTS

1–2 Use the osculating circle shown in the figure to estimate the curvature at the indicated point.

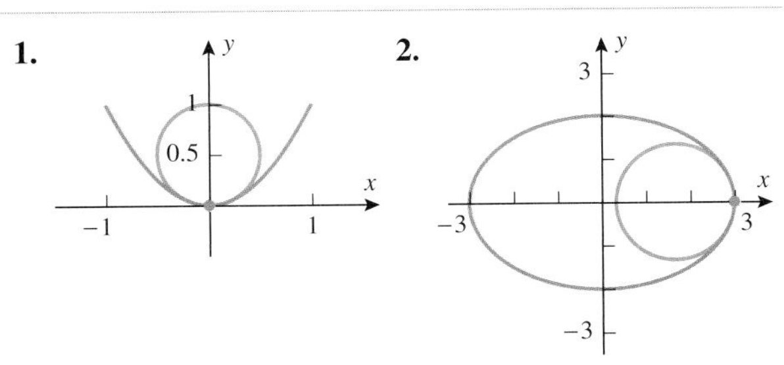

3–4 For a plane curve $y = f(x)$ the curvature at $(x, f(x))$ is a function $\kappa(x)$. In these exercises the graphs of $f(x)$ and $\kappa(x)$ are shown. Determine which is which and explain your reasoning.

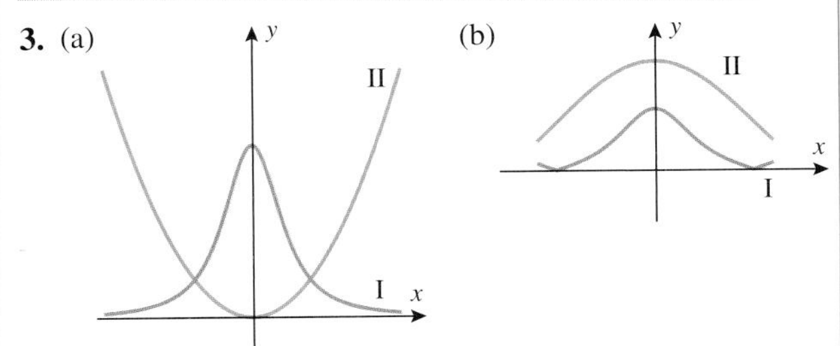

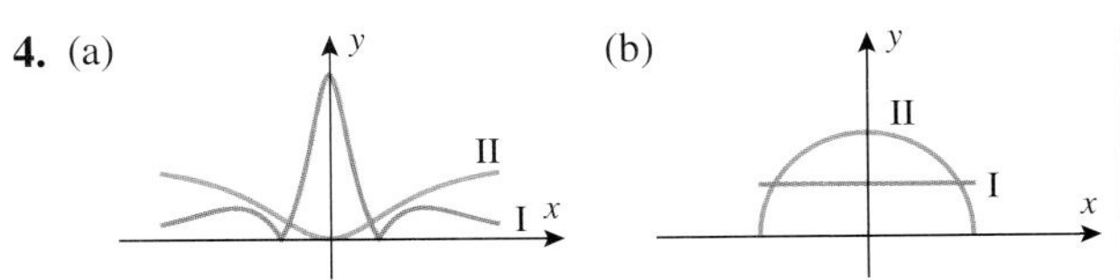

5–12 Use Formula (3) to find $\kappa(t)$.

5. $\mathbf{r}(t) = t^2\mathbf{i} + t^3\mathbf{j}$
6. $\mathbf{r}(t) = 4\cos t\,\mathbf{i} + \sin t\,\mathbf{j}$
7. $\mathbf{r}(t) = e^{3t}\mathbf{i} + e^{-t}\mathbf{j}$
8. $x = 1 - t^3,\ \ y = t - t^2$
9. $\mathbf{r}(t) = 4\cos t\,\mathbf{i} + 4\sin t\,\mathbf{j} + t\mathbf{k}$
10. $\mathbf{r}(t) = t\mathbf{i} + \frac{1}{2}t^2\mathbf{j} + \frac{1}{3}t^3\mathbf{k}$
11. $x = \cosh t,\ \ y = \sinh t,\ \ z = t$
12. $\mathbf{r}(t) = \mathbf{i} + t\mathbf{j} + t^2\mathbf{k}$

13–16 Find the curvature and the radius of curvature at the stated point.

13. $\mathbf{r}(t) = 3\cos t\,\mathbf{i} + 4\sin t\,\mathbf{j} + t\mathbf{k};\ \ t = \pi/2$
14. $\mathbf{r}(t) = e^t\mathbf{i} + e^{-t}\mathbf{j} + t\mathbf{k};\ \ t = 0$
15. $x = e^t\cos t,\ y = e^t\sin t,\ z = e^t;\ \ t = 0$
16. $x = \sin t,\ y = \cos t,\ z = \frac{1}{2}t^2;\ \ t = 0$

17–18 Confirm that s is an arc length parameter by showing that $\|d\mathbf{r}/ds\| = 1$, and then apply Formula (1) to find $\kappa(s)$.

17. $\mathbf{r} = \sin\left(1+\frac{s}{2}\right)\mathbf{i} + \cos\left(1+\frac{s}{2}\right)\mathbf{j} + \sqrt{3}\left(1+\frac{s}{2}\right)\mathbf{k}$

18. $\mathbf{r} = \left(1-\frac{2}{3}s\right)^{3/2}\mathbf{i} + \left(\frac{2}{3}s\right)^{3/2}\mathbf{j} \quad \left(0 \le s \le \frac{3}{2}\right)$

19. (a) Use Formula (3) to show that in 2-space the curvature of a smooth parametric curve

$$x = x(t), \quad y = y(t)$$

is

$$\kappa(t) = \frac{|x'y'' - y'x''|}{(x'^2 + y'^2)^{3/2}}$$

where primes denote differentiation with respect to t.

(b) Use the result in part (a) to show that in 2-space the curvature of the plane curve given by $y = f(x)$ is

$$\kappa(x) = \frac{|d^2y/dx^2|}{[1 + (dy/dx)^2]^{3/2}}$$

[*Hint:* Express $y = f(x)$ parametrically with $x = t$ as the parameter.]

20. Use part (b) of Exercise 19 to show that the curvature of $y = f(x)$ can be expressed in terms of the angle of inclination of the tangent line as

$$\kappa(\phi) = \left|\frac{d^2y}{dx^2}\cos^3\phi\right|$$

[*Hint:* $\tan\phi = dy/dx$.]

21–26 Use the result in Exercise 19(b) to find the curvature at the stated point.

21. $y = \sin x;\ x = \pi/2$

22. $y = x^3/3;\ x = 0$

23. $y = 1/x;\ x = 1$

24. $y = e^{-x};\ x = 1$

25. $y = \tan x;\ x = \pi/4$

26. $y^2 - 4x^2 = 9;\ (2, 5)$

27–32 Use the result in Exercise 19(a) to find the curvature at the stated point.

27. $x = t^2,\ y = t^3;\ t = \frac{1}{2}$

28. $x = 4\cos t,\ y = \sin t;\ t = \pi/2$

29. $x = e^{3t},\ y = e^{-t};\ t = 0$

30. $x = 1 - t^3,\ y = t - t^2;\ t = 1$

31. $x = t,\ y = 1/t;\ t = 1$

32. $x = 2\sin 2t,\ y = 3\sin t;\ t = \pi/2$

33. In each part, use the formulas in Exercise 19 to help find the radius of curvature at the stated points. Then sketch the graph together with the osculating circles at those points.
(a) $y = \cos x$ at $x = 0$ and $x = \pi$
(b) $x = 2\cos t,\ y = \sin t\ \ (0 \le t \le 2\pi)$ at $t = 0$ and $t = \pi/2$

34. Use the formula in Exercise 19(a) to find $\kappa(t)$ for the curve $x = e^{-t}\cos t,\ y = e^{-t}\sin t$. Then sketch the graph of $\kappa(t)$.

35–36 Generate the graph of $y = f(x)$ using a graphing utility, and then make a conjecture about the shape of the graph of $y = \kappa(x)$. Check your conjecture by generating the graph of $y = \kappa(x)$.

35. $f(x) = xe^{-x}$ for $0 \le x \le 5$

36. $f(x) = x^3 - x$ for $-1 \le x \le 1$

C **37.** (a) If you have a CAS, read the documentation on calculating higher-order derivatives. Then use the CAS and part (b) of Exercise 19 to find $\kappa(x)$ for $f(x) = x^4 - 2x^2$.
(b) Use the CAS to generate the graphs of $f(x) = x^4 - 2x^2$ and $\kappa(x)$ on the same screen for $-2 \le x \le 2$.
(c) Find the radius of curvature at each relative extremum.
(d) Make a reasonably accurate hand-drawn sketch that shows the graph of $f(x) = x^4 - 2x^2$ and the osculating circles in their correct proportions at the relative extrema.

C **38.** (a) Use a CAS to graph the parametric curve $x = t\cos t$, $y = t\sin t$ for $t \ge 0$.
(b) Make a conjecture about the behavior of the curvature $\kappa(t)$ as $t \to +\infty$.
(c) Use the CAS and part (a) of Exercise 19 to find $\kappa(t)$.
(d) Check your conjecture by finding the limit of $\kappa(t)$ as $t \to +\infty$.

39. Use the formula in Exercise 19(a) to show that for a curve in polar coordinates described by $r = f(\theta)$ the curvature is

$$\kappa(\theta) = \frac{\left|r^2 + 2\left(\frac{dr}{d\theta}\right)^2 - r\frac{d^2r}{d\theta^2}\right|}{\left[r^2 + \left(\frac{dr}{d\theta}\right)^2\right]^{3/2}}$$

[*Hint:* Let θ be the parameter and use the relationships $x = r\cos\theta,\ y = r\sin\theta$.]

40. Use the result in Exercise 39 to show that a circle has constant curvature.

41–44 Use the formula in Exercise 39 to find the curvature at the indicated point.

41. $r = 1 + \cos\theta;\ \theta = \pi/2$

42. $r = e^{2\theta};\ \theta = 1$

43. $r = \sin 3\theta;\ \theta = 0$

44. $r = \theta;\ \theta = 1$

45. The accompanying figure is the graph of the radius of curvature versus θ in rectangular coordinates for the cardioid $r = 1 + \cos\theta$. In words, explain what the graph tells you about the cardioid.

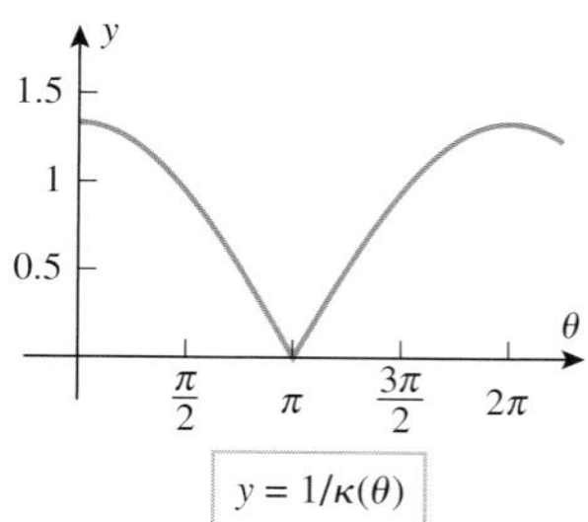

Figure Ex-45

46. Use the formula in Exercise 39 and a graphing utility to generate the graph in Exercise 45.

47. Find the radius of curvature of the parabola $y^2 = 4px$ at $(0, 0)$.

48. At what point(s) does $y = e^x$ have maximum curvature?

49. At what point(s) does $4x^2 + 9y^2 = 36$ have minimum radius of curvature?

50. Find the value of $x, x > 0$, where $y = x^3$ has maximum curvature.

51. Find the maximum and minimum values of the radius of curvature for the curve $x = \cos t,\ y = \sin t,\ z = \cos t$.

52. Find the minimum value of the radius of curvature for the curve $x = e^t,\ y = e^{-t},\ z = \sqrt{2}t$.

53. Use the formula in Exercise 39 to show that the curvature of the polar curve $r = e^{a\theta}$ is inversely proportional to r.

54. Use the formula in Exercise 39 and a CAS to show that the curvature of the lemniscate $r = \sqrt{a \cos 2\theta}$ is directly proportional to r.

55. (a) Use the result in Exercise 20 to show that for the parabola $y = x^2$ the curvature $\kappa(\phi)$ at points where the tangent line has an angle of inclination of ϕ is

$$\kappa(\phi) = |2\cos^3 \phi|$$

(b) Use the result in part (a) to find the radius of curvature of the parabola at the point on the parabola where the tangent line has slope 1.

(c) Make a sketch with reasonably accurate proportions that shows the osculating circle at the point on the parabola where the tangent line has slope 1.

56. The ***evolute*** of a smooth parametric curve C in 2-space is the curve formed from the centers of curvature of C. The accompanying figure shows the ellipse $x = 3\cos t$, $y = 2\sin t$ $(0 \le t \le 2\pi)$ and its evolute graphed together.

(a) Which points on the evolute correspond to $t = 0$ and $t = \pi/2$?

(b) In what direction is the evolute traced as t increases from 0 to 2π?

(c) What does the evolute of a circle look like? Explain your reasoning.

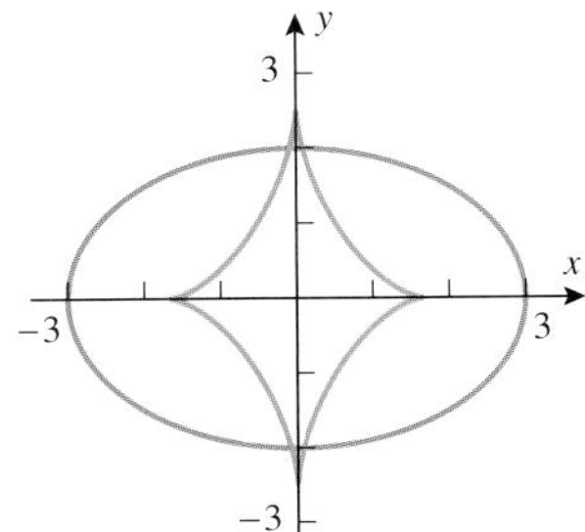

Figure Ex-56

FOCUS ON CONCEPTS

57–62 These exercises are concerned with the problem of creating a single smooth curve by piecing together two separate smooth curves. If two smooth curves C_1 and C_2 are joined at a point P to form a curve C, then we will say that C_1 and C_2 make a ***smooth transition*** at P if the curvature of C is continuous at P.

57. Show that the transition at $x = 0$ from the horizontal line $y = 0$ for $x \le 0$ to the parabola $y = x^2$ for $x > 0$ is not smooth, whereas the transition to $y = x^3$ for $x > 0$ is smooth.

58. (a) Sketch the graph of the curve defined piecewise by $y = x^2$ for $x < 0$, $y = x^4$ for $x \ge 0$.

(b) Show that for the curve in part (a) the transition at $x = 0$ is not smooth.

59. The accompanying figure shows the arc of a circle of radius r with center at $(0, r)$. Find the value of a so that there is a smooth transition from the circle to the parabola $y = ax^2$ at the point where $x = 0$.

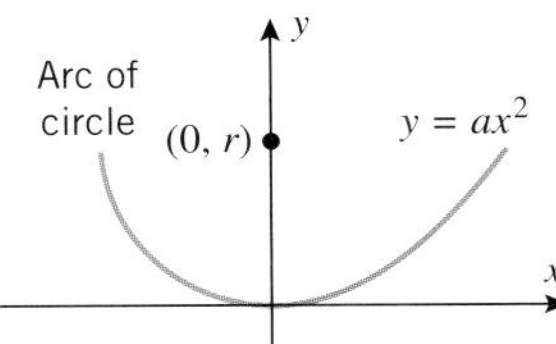

Figure Ex-59

60. Find a, b, and c so that there is a smooth transition at $x = 0$ from the curve $y = e^x$ for $x \le 0$ to the parabola $y = ax^2 + bx + c$ for $x > 0$. [*Hint:* The curvature is continuous at those points where y'' is continuous.]

61. Assume that f is a function for which $f'''(x)$ is defined for all $x \le 0$. Explain why it is always possible to find numbers a, b, and c such that there is a smooth transition at $x = 0$ from the curve $y = f(x)$, $x \le 0$, to the parabola $y = ax^2 + bx + c$.

62. In Exercise 60 of Section 11.2 we defined the Cornu spiral parametrically as

$$x = \int_0^t \cos\left(\frac{\pi u^2}{2}\right) du, \quad y = \int_0^t \sin\left(\frac{\pi u^2}{2}\right) du$$

This curve, which is graphed in the accompanying figure (next page), is used in highway design to create a gradual transition from a straight road (zero curvature) to an exit ramp with positive curvature.

(a) Express the Cornu spiral as a vector-valued function $\mathbf{r}(t)$, and then use Theorem 13.3.4 to show that $s = t$ is the arc length parameter with reference point $(0, 0)$.

(b) Replace t by s and use Formula (1) to show that $\kappa(s) = \pi|s|$. [*Note:* If $s \ge 0$, then the curvature $\kappa(s) = \pi s$ increases from 0 at a constant rate with respect to s. This makes the spiral ideal for joining a curved road to a straight road.]

(c) What happens to the curvature of the Cornu spiral as $s \to +\infty$? In words, explain why this is consistent with the graph.

Figure Ex-62

63–66 Assume that s is an arc length parameter for a smooth vector-valued function $\mathbf{r}(s)$ in 3-space and that $d\mathbf{T}/ds$ and $d\mathbf{N}/ds$ exist at each point on the curve. (This implies that $d\mathbf{B}/ds$ exists as well, since $\mathbf{B} = \mathbf{T} \times \mathbf{N}$.)

63. Show that
$$\frac{d\mathbf{T}}{ds} = \kappa(s)\mathbf{N}(s)$$
and use this result to obtain the formulas in (10).

64. (a) Show that $d\mathbf{B}/ds$ is perpendicular to $\mathbf{B}(s)$.

(b) Show that $d\mathbf{B}/ds$ is perpendicular to $\mathbf{T}(s)$. [*Hint:* Use the fact that $\mathbf{B}(s)$ is perpendicular to both $\mathbf{T}(s)$ and $\mathbf{N}(s)$, and differentiate $\mathbf{B} \cdot \mathbf{T}$ with respect to s.]

(c) Use the results in parts (a) and (b) to show that $d\mathbf{B}/ds$ is a scalar multiple of $\mathbf{N}(s)$. The *negative* of this scalar is called the ***torsion*** of $\mathbf{r}(s)$ and is denoted by $\tau(s)$. Thus,
$$\frac{d\mathbf{B}}{ds} = -\tau(s)\mathbf{N}(s)$$

(d) Show that $\tau(s) = 0$ for all s if the graph of $\mathbf{r}(s)$ lies in a plane. [*Note:* For reasons that we cannot discuss here, the torsion is related to the "twisting" properties of the curve, and $\tau(s)$ is regarded as a numerical measure of the tendency for the curve to twist out of the osculating plane.]

65. Let κ be the curvature of C and τ the torsion (defined in Exercise 64). By differentiating $\mathbf{N} = \mathbf{B} \times \mathbf{T}$ with respect to s, show that $d\mathbf{N}/ds = -\kappa\mathbf{T} + \tau\mathbf{B}$.

66. The following derivatives, known as the ***Frenet–Serret formulas***, are fundamental in the theory of curves in 3-space:

$d\mathbf{T}/ds = \kappa\mathbf{N}$ [Exercise 63]

$d\mathbf{N}/ds = -\kappa\mathbf{T} + \tau\mathbf{B}$ [Exercise 65]

$d\mathbf{B}/ds = -\tau\mathbf{N}$ [Exercise 64(c)]

Use the first two Frenet–Serret formulas and the fact that $\mathbf{r}'(s) = \mathbf{T}$ if $\mathbf{r} = \mathbf{r}(s)$ to show that
$$\tau = \frac{[\mathbf{r}'(s) \times \mathbf{r}''(s)] \cdot \mathbf{r}'''(s)}{\|\mathbf{r}''(s)\|^2} \quad \text{and} \quad \mathbf{B} = \frac{\mathbf{r}'(s) \times \mathbf{r}''(s)}{\|\mathbf{r}''(s)\|}$$

67. Use the results in Exercise 66 and the results in Exercise 30 of Section 13.3 to show that for the circular helix
$$\mathbf{r} = a\cos t\mathbf{i} + a\sin t\mathbf{j} + ct\mathbf{k}$$
with $a > 0$ the torsion and the binormal vector are
$$\tau = \frac{c}{w^2}$$
and
$$\mathbf{B} = \left(\frac{c}{w}\sin\frac{s}{w}\right)\mathbf{i} - \left(\frac{c}{w}\cos\frac{s}{w}\right)\mathbf{j} + \left(\frac{a}{w}\right)\mathbf{k}$$
where $w = \sqrt{a^2 + c^2}$ and s has reference point $(a, 0, 0)$.

68. (a) Use the chain rule and the first two Frenet–Serret formulas in Exercise 66 to show that
$$\mathbf{T}' = \kappa s'\mathbf{N} \quad \text{and} \quad \mathbf{N}' = -\kappa s'\mathbf{T} + \tau s'\mathbf{B}$$
where primes denote differentiation with respect to t.

(b) Show that Formulas (4) and (6) can be written in the form
$$\mathbf{r}'(t) = s'\mathbf{T} \quad \text{and} \quad \mathbf{r}''(t) = s''\mathbf{T} + \kappa(s')^2\mathbf{N}$$

(c) Use the results in parts (a) and (b) to show that
$$\mathbf{r}'''(t) = [s''' - \kappa^2(s')^3]\mathbf{T} + [3\kappa s's'' + \kappa'(s')^2]\mathbf{N} + \kappa\tau(s')^3\mathbf{B}$$

(d) Use the results in parts (b) and (c) to show that
$$\tau(t) = \frac{[\mathbf{r}'(t) \times \mathbf{r}''(t)] \cdot \mathbf{r}'''(t)}{\|\mathbf{r}'(t) \times \mathbf{r}''(t)\|^2}$$

69–72 Use the formula in Exercise 68(d) to find the torsion $\tau = \tau(t)$.

69. The twisted cubic $\mathbf{r}(t) = 2t\mathbf{i} + t^2\mathbf{j} + \frac{1}{3}t^3\mathbf{k}$

70. The circular helix $\mathbf{r}(t) = a\cos t\mathbf{i} + a\sin t\mathbf{j} + ct\mathbf{k}$

71. $\mathbf{r}(t) = e^t\mathbf{i} + e^{-t}\mathbf{j} + \sqrt{2}t\mathbf{k}$

72. $\mathbf{r}(t) = (t - \sin t)\mathbf{i} + (1 - \cos t)\mathbf{j} + t\mathbf{k}$

✔ QUICK CHECK ANSWERS 13.5

1. $\left\|\frac{d\mathbf{T}}{ds}\right\| = \|\mathbf{r}''(s)\|$ 2. (a) $\frac{\|\mathbf{T}'(t)\|}{\|\mathbf{r}'(t)\|}$ (b) $\frac{\|\mathbf{r}'(t) \times \mathbf{r}''(t)\|}{\|\mathbf{r}'(t)\|^3}$ 3. $\left|\frac{d\phi}{ds}\right|$ 4. $\frac{1}{2}$

13.6 MOTION ALONG A CURVE

In earlier sections we considered the motion of a particle along a line. In that situation there are only two directions in which the particle can move—the positive direction or the negative direction. Motion in 2-space or 3-space is more complicated because there are infinitely many directions in which a particle can move. In this section we will show how vectors can be used to analyze motion along curves in 2-space or 3-space.

VELOCITY, ACCELERATION, AND SPEED

Let us assume that the motion of a particle in 2-space or 3-space is described by a smooth vector-valued function $\mathbf{r}(t)$ in which the parameter t denotes time; we will call this the ***position function*** or ***trajectory*** of the particle. As the particle moves along its trajectory, its direction of motion and its speed can vary from instant to instant. Thus, before we can undertake any analysis of such motion, we must have clear answers to the following questions:

- What is the direction of motion of the particle at an instant of time?
- What is the speed of the particle at an instant of time?

We will define the direction of motion at time t to be the direction of the unit tangent vector $\mathbf{T}(t)$, and we will define the speed to be ds/dt—the instantaneous rate of change of the arc length traveled by the particle from an arbitrary reference point. Taking this a step further, we will combine the speed and the direction of motion to form the vector

$$\mathbf{v}(t) = \frac{ds}{dt}\mathbf{T}(t) \tag{1}$$

which we call the ***velocity*** of the particle at time t. Thus, at each instant of time the velocity vector $\mathbf{v}(t)$ points in the direction of motion and has a magnitude that is equal to the speed of the particle (Figure 13.6.1).

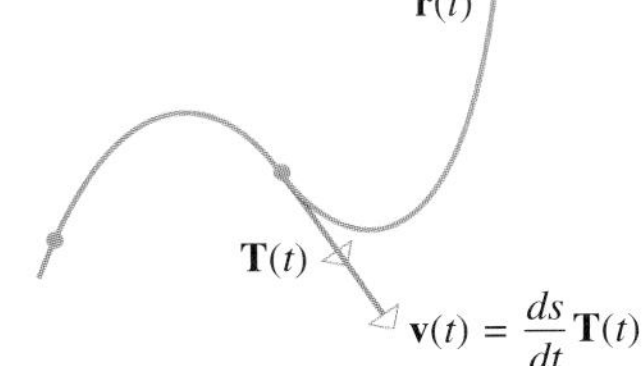

The length of the velocity vector is the speed of the particle, and the direction of the velocity vector is the direction of motion.

Figure 13.6.1

Recall that for motion along a coordinate line the velocity function is the derivative of the position function. The same is true for motion along a curve, since

$$\frac{d\mathbf{r}}{dt} = \frac{d\mathbf{r}}{ds}\frac{ds}{dt} = \frac{ds}{dt}\mathbf{T}(t) = \mathbf{v}(t)$$

For motion along a coordinate line, the acceleration function was defined to be the derivative of the velocity function. The definition is the same for motion along a curve.

13.6.1 DEFINITION. If $\mathbf{r}(t)$ is the position function of a particle moving along a curve in 2-space or 3-space, then the ***instantaneous velocity***, ***instantaneous acceleration***, and ***instantaneous speed*** of the particle at time t are defined by

$$\text{velocity} = \mathbf{v}(t) = \frac{d\mathbf{r}}{dt} \tag{2}$$

$$\text{acceleration} = \mathbf{a}(t) = \frac{d\mathbf{v}}{dt} = \frac{d^2\mathbf{r}}{dt^2} \tag{3}$$

$$\text{speed} = \|\mathbf{v}(t)\| = \frac{ds}{dt} \tag{4}$$

As shown in Table 13.6.1, the position, velocity, acceleration, and speed can also be expressed in component form.

Table 13.6.1

	2-SPACE	3-SPACE
POSITION	$\mathbf{r}(t) = x(t)\mathbf{i} + y(t)\mathbf{j}$	$\mathbf{r}(t) = x(t)\mathbf{i} + y(t)\mathbf{j} + z(t)\mathbf{k}$
VELOCITY	$\mathbf{v}(t) = \dfrac{dx}{dt}\mathbf{i} + \dfrac{dy}{dt}\mathbf{j}$	$\mathbf{v}(t) = \dfrac{dx}{dt}\mathbf{i} + \dfrac{dy}{dt}\mathbf{j} + \dfrac{dz}{dt}\mathbf{k}$
ACCELERATION	$\mathbf{a}(t) = \dfrac{d^2x}{dt^2}\mathbf{i} + \dfrac{d^2y}{dt^2}\mathbf{j}$	$\mathbf{a}(t) = \dfrac{d^2x}{dt^2}\mathbf{i} + \dfrac{d^2y}{dt^2}\mathbf{j} + \dfrac{d^2z}{dt^2}\mathbf{k}$
SPEED	$\|\mathbf{v}(t)\| = \sqrt{\left(\dfrac{dx}{dt}\right)^2 + \left(\dfrac{dy}{dt}\right)^2}$	$\|\mathbf{v}(t)\| = \sqrt{\left(\dfrac{dx}{dt}\right)^2 + \left(\dfrac{dy}{dt}\right)^2 + \left(\dfrac{dz}{dt}\right)^2}$

► **Example 1** A particle moves along a circular path in such a way that its x- and y-coordinates at time t are

$$x = 2\cos t, \quad y = 2\sin t$$

(a) Find the instantaneous velocity and speed of the particle at time t.

(b) Sketch the path of the particle, and show the position and velocity vectors at time $t = \pi/4$ with the velocity vector drawn so that its initial point is at the tip of the position vector.

(c) Show that at each instant the acceleration vector is perpendicular to the velocity vector.

Solution (a). At time t, the position vector is

$$\mathbf{r}(t) = 2\cos t\mathbf{i} + 2\sin t\mathbf{j}$$

so the instantaneous velocity and speed are

$$\mathbf{v}(t) = \frac{d\mathbf{r}}{dt} = -2\sin t\mathbf{i} + 2\cos t\mathbf{j}$$

$$\|\mathbf{v}(t)\| = \sqrt{(-2\sin t)^2 + (2\cos t)^2} = 2$$

Solution (b). The graph of the parametric equations is a circle of radius 2 centered at the origin. At time $t = \pi/4$ the position and velocity vectors of the particle are

$$\mathbf{r}(\pi/4) = 2\cos(\pi/4)\mathbf{i} + 2\sin(\pi/4)\mathbf{j} = \sqrt{2}\mathbf{i} + \sqrt{2}\mathbf{j}$$

$$\mathbf{v}(\pi/4) = -2\sin(\pi/4)\mathbf{i} + 2\cos(\pi/4)\mathbf{j} = -\sqrt{2}\mathbf{i} + \sqrt{2}\mathbf{j}$$

These vectors and the circle are shown in Figure 13.6.2.

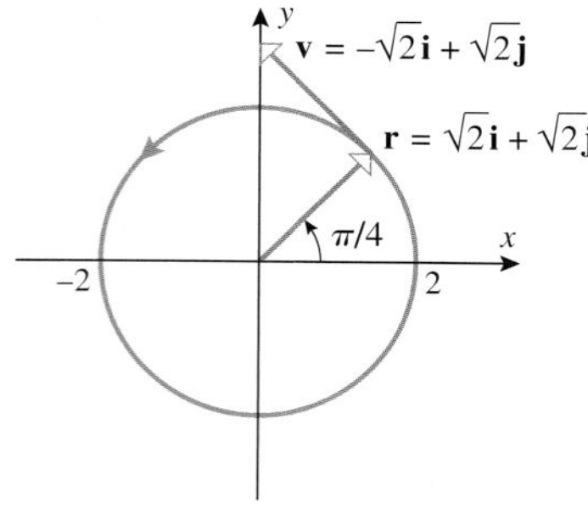

Figure 13.6.2

Solution (c). At time t, the acceleration vector is

$$\mathbf{a}(t) = \frac{d\mathbf{v}}{dt} = -2\cos t\mathbf{i} - 2\sin t\mathbf{j}$$

One way of showing that $\mathbf{v}(t)$ and $\mathbf{a}(t)$ are perpendicular is to show that their dot product is zero (try it). However, it is easier to observe that $\mathbf{a}(t)$ is the negative of $\mathbf{r}(t)$, which implies that $\mathbf{v}(t)$ and $\mathbf{a}(t)$ are perpendicular, since at each point on a circle the radius and tangent line are perpendicular. ◄

Since $\mathbf{v}(t)$ can be obtained by differentiating $\mathbf{r}(t)$, and since $\mathbf{a}(t)$ can be obtained by differentiating $\mathbf{v}(t)$, it follows that $\mathbf{r}(t)$ can be obtained by integrating $\mathbf{v}(t)$, and $\mathbf{v}(t)$ can be obtained by integrating $\mathbf{a}(t)$. However, such integrations do not produce unique functions because constants of integration occur. Typically, initial conditions are required to determine these constants.

Example 2 A particle moves through 3-space in such a way that its velocity is

$$\mathbf{v}(t) = \mathbf{i} + t\mathbf{j} + t^2\mathbf{k}$$

Find the coordinates of the particle at time $t = 1$ given that the particle is at the point $(-1, 2, 4)$ at time $t = 0$.

Solution. Integrating the velocity function to obtain the position function yields

$$\mathbf{r}(t) = \int \mathbf{v}(t)\,dt = \int (\mathbf{i} + t\mathbf{j} + t^2\mathbf{k})\,dt = t\mathbf{i} + \frac{t^2}{2}\mathbf{j} + \frac{t^3}{3}\mathbf{k} + \mathbf{C} \tag{5}$$

where $\mathbf{C}$ is a vector constant of integration. Since the coordinates of the particle at time $t = 0$ are $(-1, 2, 4)$, the position vector at time $t = 0$ is

$$\mathbf{r}(0) = -\mathbf{i} + 2\mathbf{j} + 4\mathbf{k} \tag{6}$$

It follows on substituting $t = 0$ in (5) and equating the result with (6) that

$$\mathbf{C} = -\mathbf{i} + 2\mathbf{j} + 4\mathbf{k}$$

Substituting this value of $\mathbf{C}$ in (5) and simplifying yields

$$\mathbf{r}(t) = (t - 1)\mathbf{i} + \left(\frac{t^2}{2} + 2\right)\mathbf{j} + \left(\frac{t^3}{3} + 4\right)\mathbf{k}$$

Thus, at time $t = 1$ the position vector of the particle is

$$\mathbf{r}(1) = 0\mathbf{i} + \frac{5}{2}\mathbf{j} + \frac{13}{3}\mathbf{k}$$

so its coordinates at that instant are $\left(0, \frac{5}{2}, \frac{13}{3}\right)$. ◀

DISPLACEMENT AND DISTANCE TRAVELED

If a particle travels along a curve C in 2-space or 3-space, the ***displacement*** of the particle over the time interval $t_1 \le t \le t_2$ is commonly denoted by $\Delta\mathbf{r}$ and is defined as

$$\Delta\mathbf{r} = \mathbf{r}(t_2) - \mathbf{r}(t_1) \tag{7}$$

(Figure 13.6.3). The displacement vector, which describes the change in position of the particle during the time interval, can be obtained by integrating the velocity function from t_1 to t_2:

$$\Delta\mathbf{r} = \int_{t_1}^{t_2} \mathbf{v}(t)\,dt = \int_{t_1}^{t_2} \frac{d\mathbf{r}}{dt}\,dt = \mathbf{r}(t)\Big]_{t_1}^{t_2} = \mathbf{r}(t_2) - \mathbf{r}(t_1) \quad \text{Displacement} \tag{8}$$

It follows from Theorem 13.3.1 that we can find the distance s traveled by a particle over a time interval $t_1 \le t \le t_2$ by integrating the speed over that interval, since

$$s = \int_{t_1}^{t_2} \left\|\frac{d\mathbf{r}}{dt}\right\| dt = \int_{t_1}^{t_2} \|\mathbf{v}(t)\|\,dt \quad \text{Distance traveled} \tag{9}$$

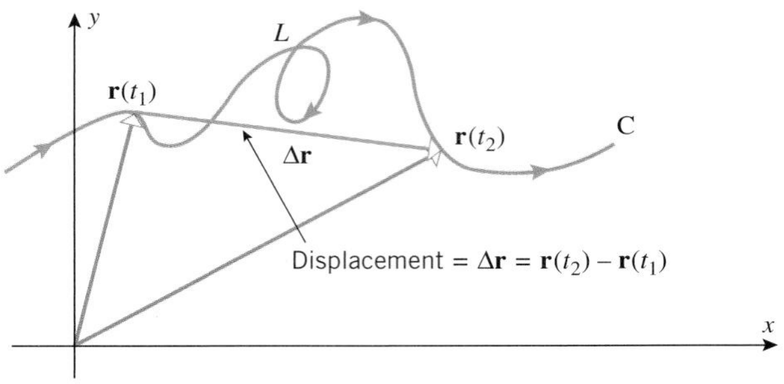

Figure 13.6.3

▶ **Example 3** Suppose that a particle moves along a circular helix in 3-space so that its position vector at time t is

$$\mathbf{r}(t) = (4\cos \pi t)\mathbf{i} + (4\sin \pi t)\mathbf{j} + t\mathbf{k}$$

Find the distance traveled and the displacement of the particle during the time interval $1 \le t \le 5$.

Solution. We have

$$\mathbf{v}(t) = \frac{d\mathbf{r}}{dt} = (-4\pi \sin \pi t)\mathbf{i} + (4\pi \cos \pi t)\mathbf{j} + \mathbf{k}$$

$$\|\mathbf{v}(t)\| = \sqrt{(-4\pi \sin \pi t)^2 + (4\pi \cos \pi t)^2 + 1} = \sqrt{16\pi^2 + 1}$$

Thus, it follows from (9) that the distance traveled by the particle from time $t = 1$ to $t = 5$ is

$$s = \int_1^5 \sqrt{16\pi^2 + 1}\, dt = 4\sqrt{16\pi^2 + 1}$$

Moreover, it follows from (8) that the displacement over the time interval is

$$\begin{aligned}\Delta\mathbf{r} &= \mathbf{r}(5) - \mathbf{r}(1)\\ &= (4\cos 5\pi\mathbf{i} + 4\sin 5\pi\mathbf{j} + 5\mathbf{k}) - (4\cos \pi\mathbf{i} + 4\sin \pi\mathbf{j} + \mathbf{k})\\ &= (-4\mathbf{i} + 5\mathbf{k}) - (-4\mathbf{i} + \mathbf{k}) = 4\mathbf{k}\end{aligned}$$

which tells us that the change in the position of the particle over the time interval was 4 units straight up. ◀

NORMAL AND TANGENTIAL COMPONENTS OF ACCELERATION

You know from your experience as an automobile passenger that if a car speeds up rapidly, then your body is thrown back against the backrest of the seat. You also know that if the car rounds a turn in the road, then your body is thrown toward the outside of the curve—the greater the curvature in the road, the greater this effect. The explanation of these effects can be understood by resolving the velocity and acceleration components of the motion into vector components that are parallel to the unit tangent and unit normal vectors. The following theorem explains how to do this.

13.6.2 THEOREM. *If a particle moves along a smooth curve C in 2-space or 3-space, then at each point on the curve velocity and acceleration vectors can be written as*

$$\mathbf{v} = \frac{ds}{dt}\mathbf{T} \qquad \mathbf{a} = \frac{d^2s}{dt^2}\mathbf{T} + \kappa\left(\frac{ds}{dt}\right)^2\mathbf{N} \tag{10–11}$$

where s is an arc length parameter for the curve, and **T**, **N**, *and* κ *denote the unit tangent vector, unit normal vector, and curvature at the point* (Figure 13.6.4).

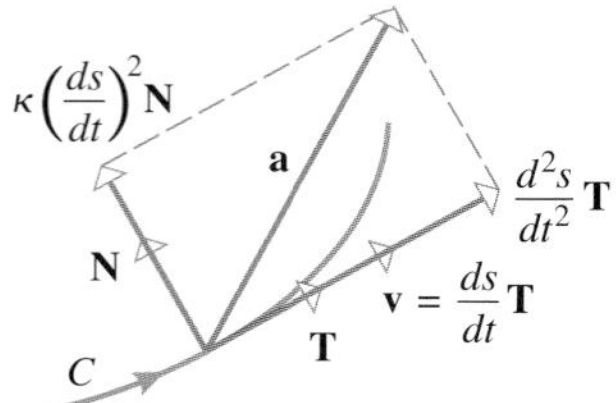

Figure 13.6.4

PROOF. Formula (10) is just a restatement of (1). To obtain (11), we differentiate both sides of (10) with respect to t; this yields

$$\begin{aligned}\mathbf{a} = \frac{d}{dt}\left(\frac{ds}{dt}\mathbf{T}\right) &= \frac{d^2s}{dt^2}\mathbf{T} + \frac{ds}{dt}\frac{d\mathbf{T}}{dt} \\ &= \frac{d^2s}{dt^2}\mathbf{T} + \frac{ds}{dt}\frac{d\mathbf{T}}{ds}\frac{ds}{dt} \\ &= \frac{d^2s}{dt^2}\mathbf{T} + \left(\frac{ds}{dt}\right)^2\frac{d\mathbf{T}}{ds} \\ &= \frac{d^2s}{dt^2}\mathbf{T} + \left(\frac{ds}{dt}\right)^2\kappa\mathbf{N}\end{aligned}$$

Formula (10) of Section 13.5

from which (11) follows. ■

The coefficients of $\mathbf{T}$ and $\mathbf{N}$ in (11) are commonly denoted by

$$a_T = \frac{d^2s}{dt^2} \qquad a_N = \kappa\left(\frac{ds}{dt}\right)^2 \tag{12–13}$$

Formula (14) applies to motion in both 2-space and 3-space. What is interesting is that the 3-space formula does not involve the binormal vector $\mathbf{B}$, so the acceleration vector always lies in the plane of $\mathbf{T}$ and $\mathbf{N}$ (the osculating plane), even for highly twisting paths of motion (Figure 13.6.5).

in which case Formula (11) is expressed as

$$\mathbf{a} = a_T\mathbf{T} + a_N\mathbf{N} \tag{14}$$

In this formula the scalars a_T and a_N are called the ***tangential scalar component of acceleration*** and the ***normal scalar component of acceleration***, and the vectors $a_T\mathbf{T}$ and $a_N\mathbf{N}$ are called the ***tangential vector component of acceleration*** and the ***normal vector component of acceleration***.

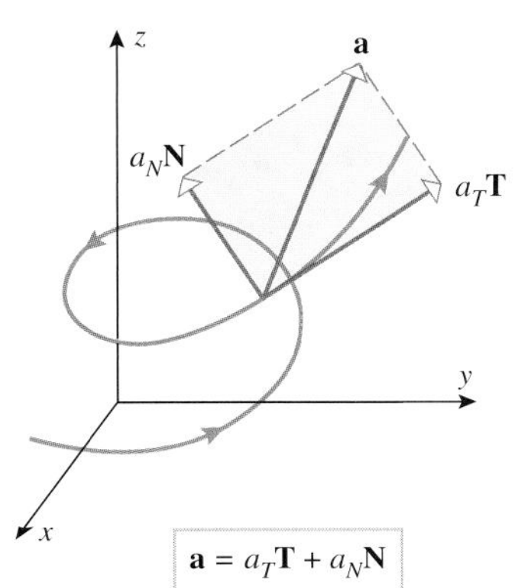

Figure 13.6.5

The scalar components of acceleration explain the effect that you experience when a car speeds up rapidly or rounds a turn. The rapid increase in speed produces a large value for d^2s/dt^2, which results in a large tangential scalar component of acceleration; and by Newton's second law this corresponds to a large tangential force on the car in the direction of motion. To understand the effect of rounding a turn, observe that the normal scalar component of acceleration has the curvature κ and the square of the speed ds/dt as factors. Thus, sharp turns or turns taken at high speed both correspond to large normal forces on the car.

Although Formulas (12) and (13) provide useful insight into the behavior of particles moving along curved paths, they are not always the best formulas for computations. The following theorem provides some more useful formulas that relate a_T, a_N, and κ to the velocity $\mathbf{v}$ and acceleration $\mathbf{a}$.

Theorem 13.6.3 applies to motion in 2-space and 3-space, but for motion in 2-space you will have to add a zero $\mathbf{k}$ component to $\mathbf{v}$ to calculate the cross product.

13.6.3 THEOREM. *If a particle moves along a smooth curve C in 2-space or 3-space, then at each point on the curve the velocity $\mathbf{v}$ and the acceleration $\mathbf{a}$ are related to a_T, a_N, and κ by the formulas*

$$a_T = \frac{\mathbf{v}\cdot\mathbf{a}}{\|\mathbf{v}\|} \qquad a_N = \frac{\|\mathbf{v}\times\mathbf{a}\|}{\|\mathbf{v}\|} \qquad \kappa = \frac{\|\mathbf{v}\times\mathbf{a}\|}{\|\mathbf{v}\|^3} \tag{15–17}$$

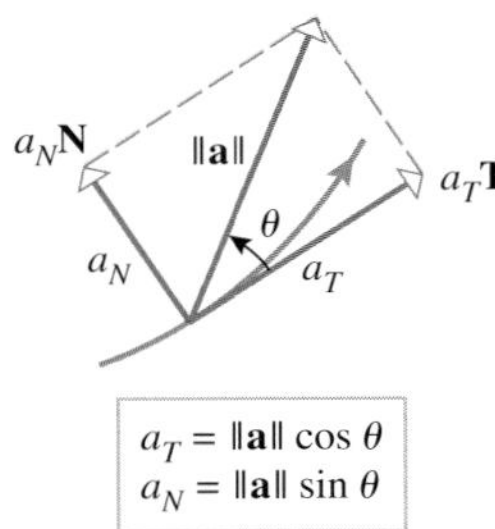

Figure 13.6.6

PROOF. As illustrated in Figure 13.6.6, let θ be the angle between the vector $\mathbf{a}$ and the vector $a_T\mathbf{T}$. Thus,

$$a_T = \|\mathbf{a}\|\cos\theta \quad \text{and} \quad a_N = \|\mathbf{a}\|\sin\theta$$

from which we obtain

$$a_T = \|\mathbf{a}\|\cos\theta = \frac{\|\mathbf{v}\|\|\mathbf{a}\|\cos\theta}{\|\mathbf{v}\|} = \frac{\mathbf{v}\cdot\mathbf{a}}{\|\mathbf{v}\|}$$

$$a_N = \|\mathbf{a}\|\sin\theta = \frac{\|\mathbf{v}\|\|\mathbf{a}\|\sin\theta}{\|\mathbf{v}\|} = \frac{\|\mathbf{v}\times\mathbf{a}\|}{\|\mathbf{v}\|}$$

$$\kappa = \frac{a_N}{(ds/dt)^2} = \frac{a_N}{\|\mathbf{v}\|^2} = \frac{1}{\|\mathbf{v}\|^2}\frac{\|\mathbf{v}\times\mathbf{a}\|}{\|\mathbf{v}\|} = \frac{\|\mathbf{v}\times\mathbf{a}\|}{\|\mathbf{v}\|^3} \qquad \blacksquare$$

Recall that for nonlinear smooth curves in 2-space the unit normal vector $\mathbf{N}$ is the inward normal (points toward the concave side of the curve). Explain why the same is true for $a_N\mathbf{N}$.

▸ Example 4 Suppose that a particle moves through 3-space so that its position vector at time t is

$$\mathbf{r}(t) = t\mathbf{i} + t^2\mathbf{j} + t^3\mathbf{k}$$

(The path is the twisted cubic shown in Figure 13.1.5.)

(a) Find the scalar tangential and normal components of acceleration at time t.

(b) Find the scalar tangential and normal components of acceleration at time $t = 1$.

(c) Find the vector tangential and normal components of acceleration at time $t = 1$.

(d) Find the curvature of the path at the point where the particle is located at time $t = 1$.

Solution (a). We have

$$\mathbf{v}(t) = \mathbf{r}'(t) = \mathbf{i} + 2t\mathbf{j} + 3t^2\mathbf{k}$$
$$\mathbf{a}(t) = \mathbf{v}'(t) = 2\mathbf{j} + 6t\mathbf{k}$$
$$\|\mathbf{v}(t)\| = \sqrt{1 + 4t^2 + 9t^4}$$
$$\mathbf{v}(t)\cdot\mathbf{a}(t) = 4t + 18t^3$$
$$\mathbf{v}(t)\times\mathbf{a}(t) = \begin{vmatrix} \mathbf{i} & \mathbf{j} & \mathbf{k} \\ 1 & 2t & 3t^2 \\ 0 & 2 & 6t \end{vmatrix} = 6t^2\mathbf{i} - 6t\mathbf{j} + 2\mathbf{k}$$

Thus, from (15) and (16)

$$a_T = \frac{\mathbf{v}\cdot\mathbf{a}}{\|\mathbf{v}\|} = \frac{4t + 18t^3}{\sqrt{1 + 4t^2 + 9t^4}}$$

$$a_N = \frac{\|\mathbf{v}\times\mathbf{a}\|}{\|\mathbf{v}\|} = \frac{\sqrt{36t^4 + 36t^2 + 4}}{\sqrt{1 + 4t^2 + 9t^4}} = 2\sqrt{\frac{9t^4 + 9t^2 + 1}{9t^4 + 4t^2 + 1}}$$

Solution (b). At time $t = 1$, the components a_T and a_N in part (a) are

$$a_T = \frac{22}{\sqrt{14}} \approx 5.88 \quad \text{and} \quad a_N = 2\sqrt{\frac{19}{14}} \approx 2.33$$

Solution (c). Since $\mathbf{T}$ and $\mathbf{v}$ have the same direction, $\mathbf{T}$ can be obtained by normalizing $\mathbf{v}$, that is,

$$\mathbf{T}(t) = \frac{\mathbf{v}(t)}{\|\mathbf{v}(t)\|}$$

At time $t = 1$ we have

$$\mathbf{T}(1) = \frac{\mathbf{v}(1)}{\|\mathbf{v}(1)\|} = \frac{\mathbf{i} + 2\mathbf{j} + 3\mathbf{k}}{\|\mathbf{i} + 2\mathbf{j} + 3\mathbf{k}\|} = \frac{1}{\sqrt{14}}(\mathbf{i} + 2\mathbf{j} + 3\mathbf{k})$$

From this and part (b) we obtain the vector tangential component of acceleration:

$$a_T(1)\mathbf{T}(1) = \frac{22}{\sqrt{14}}\mathbf{T}(1) = \frac{11}{7}(\mathbf{i}+2\mathbf{j}+3\mathbf{k}) = \frac{11}{7}\mathbf{i}+\frac{22}{7}\mathbf{j}+\frac{33}{7}\mathbf{k}$$

To find the normal vector component of acceleration, we rewrite $\mathbf{a} = a_T\mathbf{T} + a_N\mathbf{N}$ as

$$a_N\mathbf{N} = \mathbf{a} - a_T\mathbf{T}$$

Thus, at time $t = 1$ the normal vector component of acceleration is

$$\begin{aligned} a_N(1)\mathbf{N}(1) &= \mathbf{a}(1) - a_T(1)\mathbf{T}(1) \\ &= (2\mathbf{j}+6\mathbf{k}) - \left(\frac{11}{7}\mathbf{i}+\frac{22}{7}\mathbf{j}+\frac{33}{7}\mathbf{k}\right) \\ &= -\frac{11}{7}\mathbf{i}-\frac{8}{7}\mathbf{j}+\frac{9}{7}\mathbf{k} \end{aligned}$$

Solution (d). We will apply Formula (17) with $t = 1$. From part (a)

$$\|\mathbf{v}(1)\| = \sqrt{14} \quad \text{and} \quad \mathbf{v}(1)\times\mathbf{a}(1) = 6\mathbf{i}-6\mathbf{j}+2\mathbf{k}$$

Thus, at time $t = 1$

$$\kappa = \frac{\|\mathbf{v}\times\mathbf{a}\|}{\|\mathbf{v}\|^3} = \frac{\sqrt{76}}{(\sqrt{14})^3} = \frac{1}{14}\sqrt{\frac{38}{7}} \approx 0.17 \blacktriangleleft$$

In the case where $\|\mathbf{a}\|$ and a_T are known, there is a useful alternative to Formula (16) for a_N that does not require the calculation of a cross product. It follows algebraically from Formula (14) or geometrically from Figure 13.6.6 and the Theorem of Pythagoras that

Use Formula (18) to confirm the value of a_N found in Example 4.

$$a_N = \sqrt{\|\mathbf{a}\|^2 - a_T^2} \tag{18}$$

A MODEL OF PROJECTILE MOTION

Earlier in this text we examined various problems concerned with objects moving *vertically* in the Earth's gravitational field (see the subsection of Section 5.7 entitled Free-Fall Model and the subsection of Section 9.1 entitled A Model of Free-Fall Motion Retarded by Air Resistance). Now we will consider the motion of a projectile launched along a *curved* path in the Earth's gravitational field. For this purpose we will need the following *vector version* of Newton's Second Law of Motion (9.1.1)

$$\mathbf{F} = m\mathbf{a} \tag{19}$$

and we will need to make three modeling assumptions:

- The mass m of the object is constant.
- The only force acting on the object after it is launched is the force of the Earth's gravity. (Thus, air resistance and the gravitational effect of other planets and celestial objects are ignored.)
- The object remains sufficiently close to the Earth that we can assume the force of gravity to be constant.

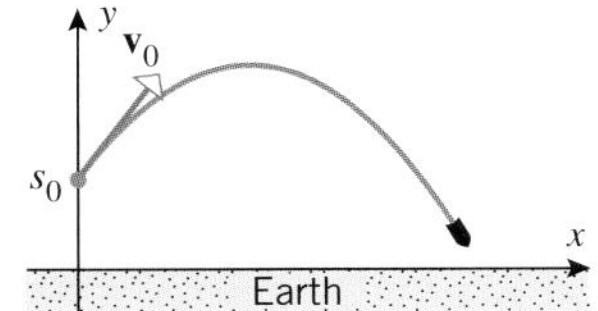

Figure 13.6.7

Let us assume that at time $t = 0$ an object of mass m is launched from a height of s_0 above the Earth with an initial velocity vector of $\mathbf{v}_0$. Furthermore, let us introduce an xy-coordinate system as shown in Figure 13.6.7. In this coordinate system the positive y-direction is up, the origin is at the surface of the Earth, and the initial location of the object is $(0, s_0)$. Our objective is to use basic principles of physics to derive the velocity function $\mathbf{v}(t)$ and the position function $\mathbf{r}(t)$ from the acceleration function $\mathbf{a}(t)$ of the object. Our starting point is the physical observation that the downward force $\mathbf{F}$ of the Earth's gravity on an object of mass m is

$$\mathbf{F} = -mg\mathbf{j}$$

where g is the acceleration due to gravity (see 9.4.3). It follows from this fact and Newton's second law (19) that

$$m\mathbf{a} = -mg\mathbf{j}$$

or on canceling m from both sides

$$\mathbf{a} = -g\mathbf{j} \tag{20}$$

Observe that this acceleration function does not involve t and hence is constant. We can now obtain the velocity function $\mathbf{v}(t)$ by integrating this acceleration function and using the initial condition $\mathbf{v}(0) = \mathbf{v}_0$ to find the constant of integration. Integrating (20) with respect to t and keeping in mind that $-g\mathbf{j}$ is constant yields

$$\mathbf{v}(t) = \int -g\mathbf{j}\,dt = -gt\mathbf{j} + \mathbf{c}_1$$

where $\mathbf{c}_1$ is a vector constant of integration. Substituting $t = 0$ in this equation and using the initial condition $\mathbf{v}(0) = \mathbf{v}_0$ yields $\mathbf{v}_0 = \mathbf{c}_1$. Thus, the velocity function of the object is

$$\mathbf{v}(t) = -gt\mathbf{j} + \mathbf{v}_0 \tag{21}$$

To obtain the position function $\mathbf{r}(t)$ of the object, we will integrate the velocity function and use the known initial position of the object to find the constant of integration. For this purpose observe that the object has coordinates $(0, s_0)$ at time $t = 0$, so the position vector at that time is

$$\mathbf{r}(0) = 0\mathbf{i} + s_0\mathbf{j} = s_0\mathbf{j} \tag{22}$$

This is the initial condition that we will need to find the constant of integration. Integrating (21) with respect to t yields

$$\mathbf{r}(t) = \int (-gt\mathbf{j} + \mathbf{v}_0)\,dt = -\tfrac{1}{2}gt^2\mathbf{j} + t\mathbf{v}_0 + \mathbf{c}_2 \tag{23}$$

Observe that the mass m does not appear in Formulas (21) and (24) and hence has no influence on the velocity or the trajectory of the object. This explains the famous observation of Galileo that two objects of different mass that are released from the same height reach the ground at the same time if air resistance is neglected.

where $\mathbf{c}_2$ is another vector constant of integration. Substituting $t = 0$ in (23) and using initial condition (22) yields

$$s_0\mathbf{j} = \mathbf{c}_2$$

so that (23) can be written as

$$\mathbf{r}(t) = \left(-\tfrac{1}{2}gt^2 + s_0\right)\mathbf{j} + t\mathbf{v}_0 \tag{24}$$

This formula expresses the position function of the object in terms of its known initial position and velocity.

PARAMETRIC EQUATIONS OF PROJECTILE MOTION

Formulas (21) and (24) can be used to obtain parametric equations for the position and velocity in terms of the initial speed of the object and the angle that the initial velocity vector makes with the positive x-axis. For this purpose, let $v_0 = \|\mathbf{v}_0\|$ be the initial speed, let α be the angle that the initial velocity vector $\mathbf{v}_0$ makes with the positive x-axis, let v_x and v_y be the horizontal and vertical scalar components of $\mathbf{v}(t)$ at time t, and let x and y be the horizontal and vertical components of $\mathbf{r}(t)$ at time t. As illustrated in Figure 13.6.8, the initial velocity vector can be expressed as

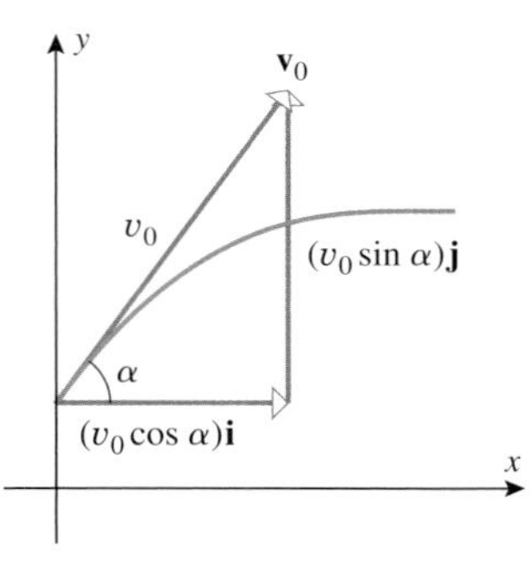

Figure 13.6.8

$$\mathbf{v}_0 = (v_0 \cos\alpha)\mathbf{i} + (v_0 \sin\alpha)\mathbf{j} \tag{25}$$

Substituting this expression in (24) and combining like components yields (verify)

$$\mathbf{r}(t) = (v_0 \cos\alpha)t\mathbf{i} + \left(s_0 + (v_0 \sin\alpha)t - \tfrac{1}{2}gt^2\right)\mathbf{j} \tag{26}$$

which is equivalent to the parametric equations

$$x = (v_0 \cos\alpha)t, \quad y = s_0 + (v_0 \sin\alpha)t - \tfrac{1}{2}gt^2 \tag{27}$$

Similarly, substituting (25) in (21) and combining like components yields

$$\mathbf{v}(t) = (v_0 \cos\alpha)\mathbf{i} + (v_0 \sin\alpha - gt)\mathbf{j}$$

which is equivalent to the parametric equations

$$v_x = v_0 \cos\alpha, \quad v_y = v_0 \sin\alpha - gt \tag{28}$$

The parameter t can be eliminated in (27) by solving the first equation for t and substituting in the second equation. We leave it for you to show that this yields

$$y = s_0 + (\tan\alpha)x - \left(\frac{g}{2v_0^2 \cos^2\alpha}\right)x^2 \tag{29}$$

which is the equation of a parabola, since the right side is a quadratic polynomial in x. Thus, we have shown that the trajectory of the projectile is a parabolic arc.

▶ **Example 5** A shell, fired from a cannon, has a muzzle speed (the speed as it leaves the barrel) of 800 ft/s. The barrel makes an angle of 45° with the horizontal and, for simplicity, the barrel opening is assumed to be at ground level.

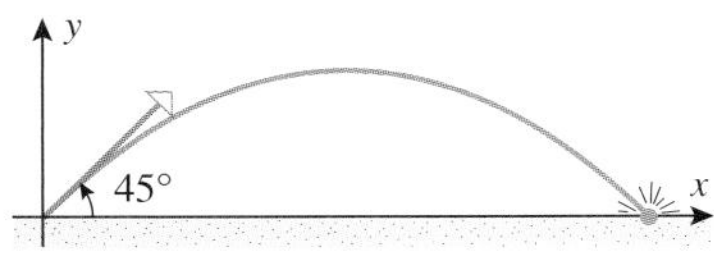

Figure 13.6.9

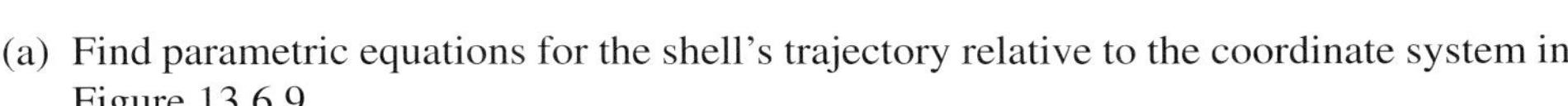

(a) Find parametric equations for the shell's trajectory relative to the coordinate system in Figure 13.6.9.
(b) How high does the shell rise?
(c) How far does the shell travel horizontally?
(d) What is the speed of the shell at its point of impact with the ground?

Solution (a). From (27) with $v_0 = 800$ ft/s, $\alpha = 45°$, $s_0 = 0$ ft (since the shell starts at ground level), and $g = 32$ ft/s^2, we obtain the parametric equations

$$x = (800 \cos 45°)t, \quad y = (800 \sin 45°)t - 16t^2 \quad (t \geq 0)$$

which simplify to

$$x = 400\sqrt{2}t, \quad y = 400\sqrt{2}t - 16t^2 \quad (t \geq 0) \tag{30}$$

Solution (b). The maximum height of the shell is the maximum value of y in (30), which occurs when $dy/dt = 0$, that is, when

$$400\sqrt{2} - 32t = 0 \quad \text{or} \quad t = \frac{25\sqrt{2}}{2}$$

Substituting this value of t in (30) yields

$$y = 5000 \text{ ft}$$

as the maximum height of the shell.

Solution (c). The shell will hit the ground when $y = 0$. From (30), this occurs when

$$400\sqrt{2}t - 16t^2 = 0 \quad \text{or} \quad t(400\sqrt{2} - 16t) = 0$$

The solution $t = 0$ corresponds to the initial position of the shell and the solution $t = 25\sqrt{2}$ to the time of impact. Substituting the latter value in the equation for x in (30) yields

$$x = 20{,}000 \text{ ft}$$

as the horizontal distance traveled by the shell.

Solution (d). From (30), the position function of the shell is

$$\mathbf{r}(t) = 400\sqrt{2}t\mathbf{i} + (400\sqrt{2}t - 16t^2)\mathbf{j}$$

so that the velocity function is

$$\mathbf{v}(t) = \mathbf{r}'(t) = 400\sqrt{2}\mathbf{i} + (400\sqrt{2} - 32t)\mathbf{j}$$

From part (c), impact occurs when $t = 25\sqrt{2}$, so the velocity vector at this point is

$$\mathbf{v}(25\sqrt{2}) = 400\sqrt{2}\mathbf{i} + [400\sqrt{2} - 32(25\sqrt{2})]\mathbf{j} = 400\sqrt{2}\mathbf{i} - 400\sqrt{2}\mathbf{j}$$

Thus, the speed at impact is

$$\|\mathbf{v}(25\sqrt{2})\| = \sqrt{(400\sqrt{2})^2 + (-400\sqrt{2})^2} = 800 \text{ ft/s} \blacktriangleleft$$

✔ QUICK CHECK EXERCISES 13.6 (See page 914 for answers.)

1. If $\mathbf{r}(t)$ is the position function of a particle, then the velocity, acceleration, and speed of the particle at time t are given, respectively, by

$$\mathbf{v}(t) = \_\_\_\_\_\_,\quad \mathbf{a}(t) = \_\_\_\_\_\_,\quad \frac{ds}{dt} = \_\_\_\_\_\_$$

2. If $\mathbf{r}(t)$ is the position function of a particle, then the displacement of the particle over the time interval $t_1 \le t \le t_2$ is ______, and the distance s traveled by the particle during this time interval is given by the integral ______.

3. The tangential scalar component of acceleration is given by the formula ______, and the normal scalar component of acceleration is given by the formula ______.

4. The projectile motion model

$$\mathbf{r}(t) = \left(-\tfrac{1}{2}gt^2 + s_0\right)\mathbf{j} + t\mathbf{v}_0$$

describes the motion of an object with constant acceleration $\mathbf{a} =$ ______ and velocity function $\mathbf{v}(t) =$ ______. The initial position of the object is ______ and its initial velocity is ______.

EXERCISE SET 13.6  Graphing Utility C CAS

1–4 In these exercises $\mathbf{r}(t)$ is the position vector of a particle moving in the plane. Find the velocity, acceleration, and speed at an arbitrary time t. Then sketch the path of the particle together with the velocity and acceleration vectors at the indicated time t.

1. $\mathbf{r}(t) = 3\cos t\mathbf{i} + 3\sin t\mathbf{j};\ t = \pi/3$

2. $\mathbf{r}(t) = t\mathbf{i} + t^2\mathbf{j};\ t = 2$

3. $\mathbf{r}(t) = e^t\mathbf{i} + e^{-t}\mathbf{j};\ t = 0$

4. $\mathbf{r}(t) = (2 + 4t)\mathbf{i} + (1 - t)\mathbf{j};\ t = 1$

5–8 Find the velocity, speed, and acceleration at the given time t of a particle moving along the given curve.

5. $\mathbf{r}(t) = t\mathbf{i} + \frac{1}{2}t^2\mathbf{j} + \frac{1}{3}t^3\mathbf{k};\ t = 1$

6. $x = 1 + 3t,\ y = 2 - 4t,\ z = 7 + t;\ t = 2$

7. $x = 2\cos t,\ y = 2\sin t,\ z = t;\ t = \pi/4$

8. $\mathbf{r}(t) = e^t\sin t\mathbf{i} + e^t\cos t\mathbf{j} + t\mathbf{k};\ t = \pi/2$

FOCUS ON CONCEPTS

9. As illustrated in the accompanying figure, suppose that the equations of motion of a particle moving along an elliptic path are $x = a\cos\omega t,\ y = b\sin\omega t$.
   (a) Show that the acceleration is directed toward the origin.
   (b) Show that the magnitude of the acceleration is proportional to the distance from the particle to the origin.

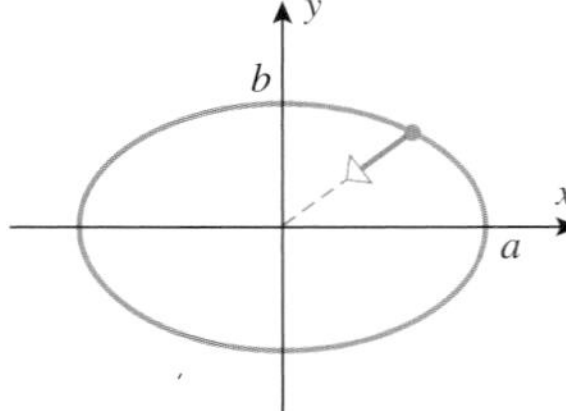

Figure Ex-9

10. Suppose that a particle vibrates in such a way that its position function is $\mathbf{r}(t) = 16\sin\pi t\mathbf{i} + 4\cos 2\pi t\,\mathbf{j}$, where distance is in millimeters and t is in seconds.

(a) Find the velocity and acceleration at time $t = 1$ s.
(b) Show that the particle moves along a parabolic curve.
(c) Show that the particle moves back and forth along the curve.

11. What can you say about the trajectory of a particle that moves in 2-space or 3-space with zero acceleration? Justify your answer.

12. Recall from Theorem 13.2.8 that if $\mathbf{r}(t)$ is a vector-valued function in 2-space or 3-space, and if $\|\mathbf{r}(t)\|$ is constant for all t, then $\mathbf{r}(t) \cdot \mathbf{r}'(t) = 0$.
(a) Translate this theorem into a statement about the motion of a particle in 2-space or 3-space.
(b) Replace $\mathbf{r}(t)$ by $\mathbf{r}'(t)$ in the theorem, and translate the result into a statement about the motion of a particle in 2-space or 3-space.

13. Suppose that the position vector of a particle moving in the plane is $\mathbf{r} = 12\sqrt{t}\,\mathbf{i} + t^{3/2}\mathbf{j}$, $t > 0$. Find the minimum speed of the particle and its location when it has this speed.

14. Suppose that the motion of a particle is described by the position vector $\mathbf{r} = (t - t^2)\mathbf{i} - t^2\mathbf{j}$. Find the minimum speed of the particle and its location when it has this speed.

15. Suppose that the position function of a particle moving in 2-space is $\mathbf{r} = \sin 3t\mathbf{i} - 2\cos 3t\mathbf{j}$.
(a) Use a graphing utility to graph the speed of the particle versus time from $t = 0$ to $t = 2\pi/3$.
(b) What are the maximum and minimum speeds of the particle?
(c) Use the graph to estimate the time at which the maximum speed first occurs.
(d) Find the exact time at which the maximum speed first occurs.

16. Suppose that the position function of a particle moving in 3-space is $\mathbf{r} = 3\cos 2t\mathbf{i} + \sin 2t\mathbf{j} + 4t\mathbf{k}$.
(a) Use a graphing utility to graph the speed of the particle versus time from $t = 0$ to $t = \pi$.
(b) Use the graph to estimate the maximum and minimum speeds of the particle.
(c) Use the graph to estimate the time at which the maximum speed first occurs.
(d) Find the exact values of the maximum and minimum speeds and the exact time at which the maximum speed first occurs.

17–20 Use the given information to find the position and velocity vectors of the particle.

17. $\mathbf{a}(t) = -\cos t\mathbf{i} - \sin t\mathbf{j}$; $\mathbf{v}(0) = \mathbf{i}$; $\mathbf{r}(0) = \mathbf{j}$

18. $\mathbf{a}(t) = \mathbf{i} + e^{-t}\mathbf{j}$; $\mathbf{v}(0) = 2\mathbf{i} + \mathbf{j}$; $\mathbf{r}(0) = \mathbf{i} - \mathbf{j}$

19. $\mathbf{a}(t) = \sin t\mathbf{i} + \cos t\mathbf{j} + e^t\mathbf{k}$; $\mathbf{v}(0) = \mathbf{k}$; $\mathbf{r}(0) = -\mathbf{i} + \mathbf{k}$

20. $\mathbf{a}(t) = (t + 1)^{-2}\mathbf{j} - e^{-2t}\mathbf{k}$; $\mathbf{v}(0) = 3\mathbf{i} - \mathbf{j}$; $\mathbf{r}(0) = 2\mathbf{k}$

21. Find, to the nearest degree, the angle between $\mathbf{v}$ and $\mathbf{a}$ for $\mathbf{r} = t^3\mathbf{i} + t^2\mathbf{j}$ when $t = 1$.

22. Show that the angle between $\mathbf{v}$ and $\mathbf{a}$ is constant for the position vector $\mathbf{r} = e^t \cos t\mathbf{i} + e^t \sin t\mathbf{j}$. Find the angle.

23. (a) Suppose that at time $t = t_0$ an electron has a position vector of $\mathbf{r} = 3.5\mathbf{i} - 1.7\mathbf{j} + \mathbf{k}$, and at a later time $t = t_1$ it has a position vector of $\mathbf{r} = 4.2\mathbf{i} + \mathbf{j} - 2.4\mathbf{k}$. What is the displacement of the electron during the time interval from t_0 to t_1?
(b) Suppose that during a certain time interval a proton has a displacement of $\Delta\mathbf{r} = 0.7\mathbf{i} + 2.9\mathbf{j} - 1.2\mathbf{k}$ and its final position vector is known to be $\mathbf{r} = 3.6\mathbf{k}$. What was the initial position vector of the proton?

24. Suppose that the position function of a particle moving along a circle in the xy-plane is $\mathbf{r} = 5\cos 2\pi t\mathbf{i} + 5\sin 2\pi t\mathbf{j}$.
(a) Sketch some typical displacement vectors over the time interval from $t = 0$ to $t = 1$.
(b) What is the distance traveled by the particle during the time interval?

25–28 Find the displacement and the distance traveled over the indicated time interval.

25. $\mathbf{r} = t^2\mathbf{i} + \frac{1}{3}t^3\mathbf{j}$; $1 \le t \le 3$

26. $\mathbf{r} = (1 - 3\sin t)\mathbf{i} + 3\cos t\mathbf{j}$; $0 \le t \le 3\pi/2$

27. $\mathbf{r} = e^t\mathbf{i} + e^{-t}\mathbf{j} + \sqrt{2}t\mathbf{k}$; $0 \le t \le \ln 3$

28. $\mathbf{r} = \cos 2t\mathbf{i} + (1 - \cos 2t)\mathbf{j} + \left(3 + \frac{1}{2}\cos 2t\right)\mathbf{k}$; $0 \le t \le \pi$

29–30 The position vectors of two particles are given. Show that the particles move along the same path but the speed of the first is constant and the speed of the second is not.

29. $\mathbf{r}_1 = 2\cos 3t\mathbf{i} + 2\sin 3t\mathbf{j}$
$\mathbf{r}_2 = 2\cos(t^2)\mathbf{i} + 2\sin(t^2)\mathbf{j} \quad (t \ge 0)$

30. $\mathbf{r}_1 = (3 + 2t)\mathbf{i} + t\mathbf{j} + (1 - t)\mathbf{k}$
$\mathbf{r}_2 = (5 - 2t^3)\mathbf{i} + (1 - t^3)\mathbf{j} + t^3\mathbf{k}$

31–38 The position function of a particle is given. Use Theorem 13.6.3 to find
(a) the scalar tangential and normal components of acceleration at the stated time t;
(b) the vector tangential and normal components of acceleration at the stated time t;
(c) the curvature of the path at the point where the particle is located at the stated time t.

31. $\mathbf{r} = e^{-t}\mathbf{i} + e^t\mathbf{j}$; $t = 0$

32. $\mathbf{r} = \cos(t^2)\mathbf{i} + \sin(t^2)\mathbf{j}$; $t = \sqrt{\pi}/2$

33. $\mathbf{r} = (t^3 - 2t)\mathbf{i} + (t^2 - 4)\mathbf{j}$; $t = 1$

34. $\mathbf{r} = e^t \cos t\mathbf{i} + e^t \sin t\mathbf{j}$; $t = \pi/4$

35. $\mathbf{r} = (1/t)\mathbf{i} + t^2\mathbf{j} + t^3\mathbf{k}$; $t = 1$

36. $\mathbf{r} = e^t\mathbf{i} + e^{-2t}\mathbf{j} + t\mathbf{k}$; $t = 0$

37. $\mathbf{r} = 3\sin t\mathbf{i} + 2\cos t\mathbf{j} - \sin 2t\mathbf{k}$; $t = \pi/2$

38. $\mathbf{r} = 2\mathbf{i} + t^3\mathbf{j} - 16\ln t\mathbf{k}$; $t = 1$

39–42 In these exercises **v** and **a** are given at a certain instant of time. Find a_T, a_N, **T**, and **N** at this instant.

39. $\mathbf{v} = -4\mathbf{j},\ \mathbf{a} = 2\mathbf{i} + 3\mathbf{j}$ **40.** $\mathbf{v} = \mathbf{i} + 2\mathbf{j},\ \mathbf{a} = 3\mathbf{i}$

41. $\mathbf{v} = 2\mathbf{i} + 2\mathbf{j} + \mathbf{k},\ \mathbf{a} = \mathbf{i} + 2\mathbf{k}$

42. $\mathbf{v} = 3\mathbf{i} - 4\mathbf{k},\ \mathbf{a} = \mathbf{i} - \mathbf{j} + 2\mathbf{k}$

43–46 The speed $\|\mathbf{v}\|$ of a particle at an arbitrary time t is given. Find the scalar tangential component of acceleration at the indicated time.

43. $\|\mathbf{v}\| = \sqrt{3t^2 + 4};\ t = 2$ **44.** $\|\mathbf{v}\| = \sqrt{t^2 + e^{-3t}};\ t = 0$

45. $\|\mathbf{v}\| = \sqrt{(4t-1)^2 + \cos^2 \pi t};\ t = \frac{1}{4}$

46. $\|\mathbf{v}\| = \sqrt{t^4 + 5t^2 + 3};\ t = 1$

47. The nuclear accelerator at the Enrico Fermi Laboratory is circular with a radius of 1 km. Find the scalar normal component of acceleration of a proton moving around the accelerator with a constant speed of 2.9×10^5 km/s.

48. Suppose that a particle moves with nonzero acceleration along the curve $y = f(x)$. Use part (b) of Exercise 19 in Section 13.5 to show that the acceleration vector is tangent to the curve at each point where $f''(x) = 0$.

49–50 Use the given information and Exercise 19 of Section 13.5 to find the normal scalar component of acceleration as a function of x.

49. A particle moves along the parabola $y = x^2$ with a constant speed of 3 units per second.

50. A particle moves along the curve $x = \ln y$ with a constant speed of 2 units per second.

51–52 Use the given information to find the normal scalar component of acceleration at time $t = 1$.

51. $\mathbf{a}(1) = \mathbf{i} + 2\mathbf{j} - 2\mathbf{k};\ a_T(1) = 3$

52. $\|\mathbf{a}(1)\| = 9;\ a_T(1)\mathbf{T}(1) = 2\mathbf{i} - 2\mathbf{j} + \mathbf{k}$

53. An automobile travels at a constant speed around a curve whose radius of curvature is 1000 m. What is the maximum allowable speed if the maximum acceptable value for the normal scalar component of acceleration is 1.5 m/s²?

54. If an automobile of mass m rounds a curve, then its inward vector component of acceleration $a_N\mathbf{N}$ is caused by the frictional force **F** of the road. Thus, it follows from the vector form of Newton's second law [Equation (19)] that the frictional force and the normal scalar component of acceleration are related by the equation $\mathbf{F} = ma_N\mathbf{N}$. Thus,

$$\|\mathbf{F}\| = m\kappa\left(\frac{ds}{dt}\right)^2$$

Use this result to find the magnitude of the frictional force in newtons exerted by the road on a 500-kg go-cart driven at a speed of 10 km/h around a circular track of radius 15 m. [*Note:* 1 N = 1 kg·m/s²]

55. A shell is fired from ground level with a muzzle speed of 320 ft/s and elevation angle of 60°. Find
(a) parametric equations for the shell's trajectory
(b) the maximum height reached by the shell
(c) the horizontal distance traveled by the shell
(d) the speed of the shell at impact.

56. Solve Exercise 55 assuming that the muzzle speed is 980 m/s and the elevation angle is 45°.

57. A rock is thrown downward from the top of a building, 168 ft high, at an angle of 60° with the horizontal. How far from the base of the building will the rock land if its initial speed is 80 ft/s?

58. Solve Exercise 57 assuming that the rock is thrown horizontally at a speed of 80 ft/s.

59. A shell is to be fired from ground level at an elevation angle of 30°. What should the muzzle speed be in order for the maximum height of the shell to be 2500 ft?

60. A shell, fired from ground level at an elevation angle of 45°, hits the ground 24,500 m away. Calculate the muzzle speed of the shell.

61. Find two elevation angles that will enable a shell, fired from ground level with a muzzle speed of 800 ft/s, to hit a ground-level target 10,000 ft away.

62. A ball rolls off a table 4 ft high while moving at a constant speed of 5 ft/s.
(a) How long does it take for the ball to hit the floor after it leaves the table?
(b) At what speed does the ball hit the floor?
(c) If a ball were dropped from rest at table height just as the rolling ball leaves the table, which ball would hit the ground first? Justify your answer.

63. As illustrated in the accompanying figure, a fire hose sprays water with an initial velocity of 40 ft/s at an angle of 60° with the horizontal.
(a) Confirm that the water will clear corner point A.
(b) Confirm that the water will hit the roof.
(c) How far from corner point A will the water hit the roof?

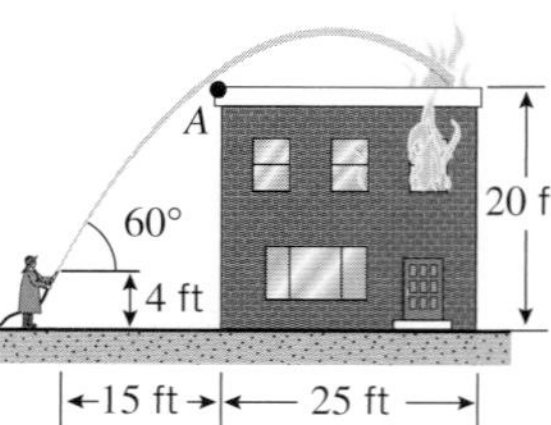

Figure Ex-63

64. What is the minimum initial velocity that will allow the water in Exercise 63 to hit the roof?

65. As shown in the accompanying figure, water is sprayed from a hose with an initial velocity of 35 m/s at an angle of 45° with the horizontal.
(a) What is the radius of curvature of the stream at the point where it leaves the hose?

(b) What is the maximum height of the stream above the nozzle of the hose?

66. As illustrated in the accompanying figure, a train is traveling on a curved track. At a point where the train is traveling at a speed of 132 ft/s and the radius of curvature of the track is 3000 ft, the engineer hits the brakes to make the train slow down at a constant rate of 7.5 ft/s^2.

(a) Find the magnitude of the acceleration vector at the instant the engineer hits the brakes.

(b) Approximate the angle between the acceleration vector and the unit tangent vector **T** at the instant the engineer hits the brakes.

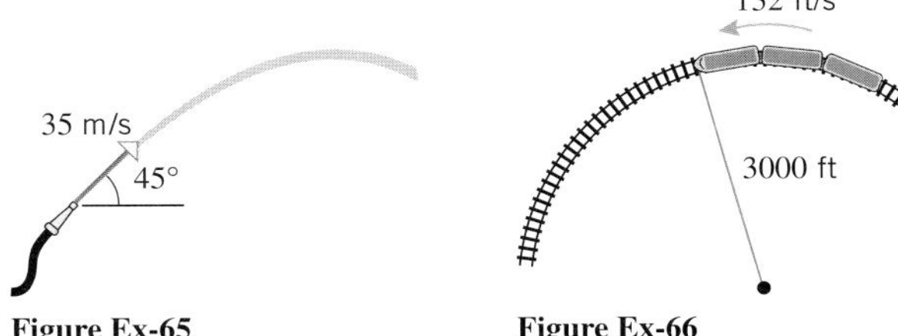

Figure Ex-65 **Figure Ex-66**

67. A shell is fired from ground level at an elevation angle of α and a muzzle speed of v_0.

(a) Show that the maximum height reached by the shell is

$$\text{maximum height} = \frac{(v_0 \sin \alpha)^2}{2g}$$

(b) The ***horizontal range*** R of the shell is the horizontal distance traveled when the shell returns to ground level. Show that $R = (v_0^2 \sin 2\alpha)/g$. For what elevation angle will the range be maximum? What is the maximum range?

68. A shell is fired from ground level with an elevation angle α and a muzzle speed of v_0. Find two angles that can be used to hit a target at ground level that is a distance of three-fourths the maximum range of the shell. Express your answer to the nearest tenth of a degree. [*Hint:* See Exercise 67(b).]

69. At time $t = 0$ a baseball that is 5 ft above the ground is hit with a bat. The ball leaves the bat with a speed of 80 ft/s at an angle of 30° above the horizontal.

(a) How long will it take for the baseball to hit the ground? Express your answer to the nearest hundredth of a second.

(b) Use the result in part (a) to find the horizontal distance traveled by the ball. Express your answer to the nearest tenth of a foot.

70. Repeat Exercise 69, assuming that the ball leaves the bat with a speed of 70 ft/s at an angle of 60° above the horizontal.

C **71.** At time $t = 0$ a skier leaves the end of a ski jump with a speed of v_0 ft/s at an angle α with the horizontal (see the accompanying figure). The skier lands 259 ft down the incline 2.9 s later.

(a) Approximate v_0 to the nearest ft/s and α to the nearest degree.

(b) Use a CAS or a calculating utility with a numerical integration capability to approximate the distance traveled by the skier.

(Use $g = 32$ ft/s^2 as the acceleration due to gravity.)

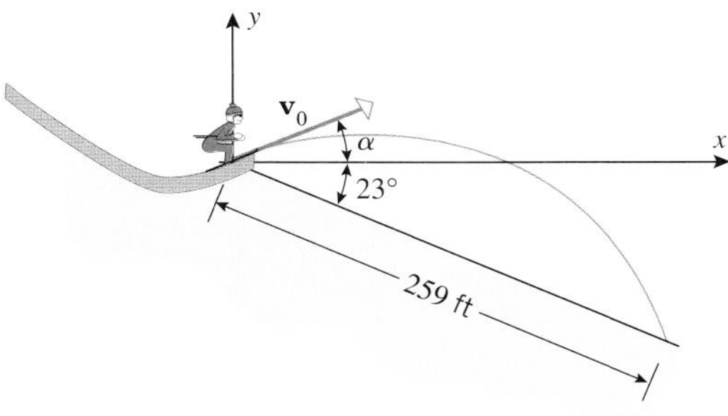

Figure Ex-71

FOCUS ON CONCEPTS

72. At time $t = 0$ a projectile is fired from a height h above level ground at an elevation angle of α with a speed v. Let R be the horizontal distance to the point where the projectile hits the ground.

(a) Show that α and R must satisfy the equation

$$g(\sec^2 \alpha)R^2 - 2v^2(\tan \alpha)R - 2v^2 h = 0$$

(b) If g, h, and v are constant, then the equation in part (a) defines R implicitly as a function of α. Let R_0 be the maximum value of R and α_0 the value of α when $R = R_0$. Use implicit differentiation to find $dR/d\alpha$ and show that

$$\tan \alpha_0 = \frac{v^2}{gR_0}$$

[*Hint:* Assume that $dR/d\alpha = 0$ when R attains a maximum.]

(c) Use the results in parts (a) and (b) to show that

$$R_0 = \frac{v}{g}\sqrt{v^2 + 2gh}$$

and

$$\alpha_0 = \tan^{-1} \frac{v}{\sqrt{v^2 + 2gh}}$$

73. Suppose that the position function of a point moving in the xy-plane is

$$\mathbf{r} = x(t)\mathbf{i} + y(t)\mathbf{j}$$

This equation can be expressed in polar coordinates by making the substitution

$$x(t) = r(t)\cos\theta(t), \quad y(t) = r(t)\sin\theta(t)$$

This yields

$$\mathbf{r} = r(t)\cos\theta(t)\mathbf{i} + r(t)\sin\theta(t)\mathbf{j}$$

which can be expressed as

$$\mathbf{r} = r(t)\mathbf{e}_r(t)$$

where $\mathbf{e}_r(t) = \cos\theta(t)\mathbf{i} + \sin\theta(t)\mathbf{j}$.

(a) Show that $\mathbf{e}_r(t)$ is a unit vector that has the same direction as the radius vector $\mathbf{r}$ if $r(t) > 0$ and that $\mathbf{e}_\theta(t) = -\sin\theta(t)\mathbf{i} + \cos\theta(t)\mathbf{j}$ is the unit vector that results when $\mathbf{e}_r(t)$ is rotated counterclockwise through an angle of $\pi/2$. The vector $\mathbf{e}_r(t)$ is called the ***radial unit vector*** and the vector $\mathbf{e}_\theta(t)$ is called the ***transverse unit vector*** (see the accompanying figure).

(b) Show that the velocity function $\mathbf{v} = \mathbf{v}(t)$ can be expressed in terms of radial and transverse components as

$$\mathbf{v} = \frac{dr}{dt}\mathbf{e}_r + r\frac{d\theta}{dt}\mathbf{e}_\theta$$

(c) Show that the acceleration function $\mathbf{a} = \mathbf{a}(t)$ can be expressed in terms of radial and transverse components as

$$\mathbf{a} = \left[\frac{d^2r}{dt^2} - r\left(\frac{d\theta}{dt}\right)^2\right]\mathbf{e}_r + \left[r\frac{d^2\theta}{dt^2} + 2\frac{dr}{dt}\frac{d\theta}{dt}\right]\mathbf{e}_\theta$$

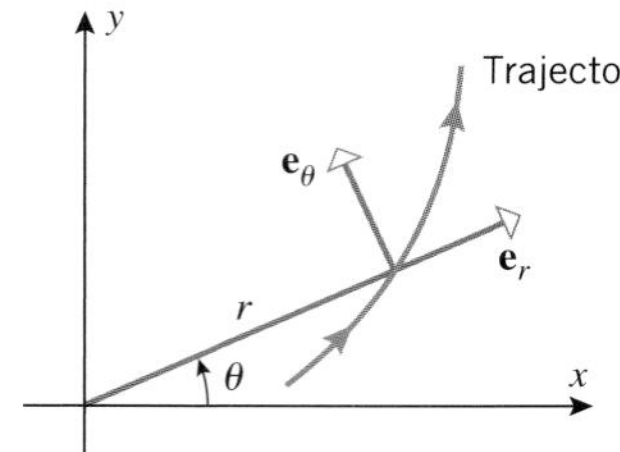

Figure Ex-73

✔QUICK CHECK ANSWERS 13.6

1. $\dfrac{d\mathbf{r}}{dt}$; $\dfrac{d\mathbf{v}}{dt} = \dfrac{d^2\mathbf{r}}{dt^2}$; $\|\mathbf{v}(t)\|$ 2. $\mathbf{r}(t_2) - \mathbf{r}(t_1)$; $\displaystyle\int_{t_1}^{t_2} \|\mathbf{v}(t)\|\,dt$ 3. $\dfrac{d^2s}{dt^2}$; $\kappa(ds/dt)^2$ 4. $-g\mathbf{j}$; $-gt\mathbf{j} + \mathbf{v}_0$; $s_0\mathbf{j}$; $\mathbf{v}_0$

13.7 KEPLER'S LAWS OF PLANETARY MOTION

One of the great advances in the history of astronomy occurred in the early 1600s when Johannes Kepler deduced from empirical data that all planets in our solar system move in elliptical orbits with the Sun at a focus. Subsequently, Isaac Newton showed mathematically that such planetary motion is the consequence of an inverse-square law of gravitational attraction. In this section we will use the concepts developed in the preceding sections of this chapter to derive three basic laws of planetary motion, known as **Kepler's laws**.*

KEPLER'S LAWS

In Section 11.6 we stated the following laws of planetary motion that were published by Johannes Kepler in 1609 in his book known as *Astronomia Nova*.

13.7.1 KEPLER'S LAWS.

- First law (***Law of Orbits***). Each planet moves in an elliptical orbit with the Sun at a focus.
- Second law (***Law of Areas***). Equal areas are swept out in equal times by the line from the Sun to a planet.
- Third law (***Law of Periods***). The square of a planet's period (the time it takes the planet to complete one orbit about the Sun) is proportional to the cube of the semimajor axis of its orbit.

---

* See biography on p. 780.

CENTRAL FORCES

If a particle moves under the influence of a *single* force that is always directed toward a fixed point O, then the particle is said to be moving in a ***central force field***. The force is called a ***central force***, and the point O is called the ***center of force***. For example, in the simplest model of planetary motion, it is assumed that the only force acting on a planet is the force of the Sun's gravity, directed toward the center of the Sun. This model, which produces Kepler's laws, ignores the forces that other celestial objects exert on the planet as well as the minor effect that the planet's gravity has on the Sun. Central force models are also used to study the motion of comets, asteroids, planetary moons, and artificial satellites. They also have important applications in electromagnetics. Our objective in this section is to develop some basic principles about central force fields and then use those results to derive Kepler's laws.

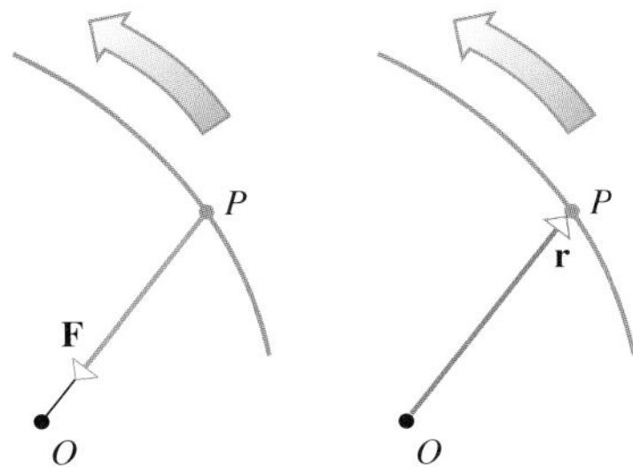

Figure 13.7.1

Suppose that a particle P of mass m moves in a central force field due to a force $\mathbf{F}$ that is directed toward a fixed point O, and let $\mathbf{r} = \mathbf{r}(t)$ be the position vector from O to P (Figure 13.7.1). Let $\mathbf{v} = \mathbf{v}(t)$ and $\mathbf{a} = \mathbf{a}(t)$ be the velocity and acceleration functions of the particle, and assume that $\mathbf{F}$ and $\mathbf{a}$ are related by Newton's second law ($\mathbf{F} = m\mathbf{a}$).

Our first objective is to show that the particle P moves in a plane containing the point O. For this purpose observe that $\mathbf{a}$ has the same direction as $\mathbf{F}$ by Newton's second law, and this implies that $\mathbf{a}$ and $\mathbf{r}$ are oppositely directed vectors. Thus, it follows from part (c) of Theorem 12.4.5 that

$$\mathbf{r} \times \mathbf{a} = \mathbf{0}$$

Since the velocity and acceleration of the particle are given by $\mathbf{v} = d\mathbf{r}/dt$ and $\mathbf{a} = d\mathbf{v}/dt$, respectively, we have

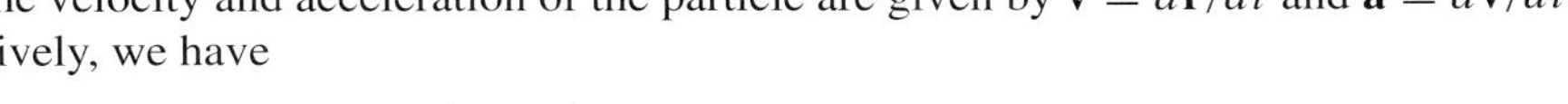

$$\frac{d}{dt}(\mathbf{r} \times \mathbf{v}) = \mathbf{r} \times \frac{d\mathbf{v}}{dt} + \frac{d\mathbf{r}}{dt} \times \mathbf{v} = (\mathbf{r} \times \mathbf{a}) + (\mathbf{v} \times \mathbf{v}) = \mathbf{0} + \mathbf{0} = \mathbf{0} \qquad (1)$$

Integrating the left and right sides of this equation with respect to t yields

$$\mathbf{r} \times \mathbf{v} = \mathbf{b} \qquad (2)$$

where $\mathbf{b}$ is a constant (independent of t). However, $\mathbf{b}$ is orthogonal to both $\mathbf{r}$ and $\mathbf{v}$, so we can conclude that $\mathbf{r} = \mathbf{r}(t)$ and $\mathbf{v} = \mathbf{v}(t)$ lie in a fixed plane containing the point O.

> Astronomers call the plane containing the orbit of a planet the ***ecliptic*** of the planet.

NEWTON'S LAW OF UNIVERSAL GRAVITATION

Our next objective is to derive the position function of a particle moving under a central force in a polar coordinate system. For this purpose we will need the following result, known as ***Newton's Law of Universal Gravitation***.

> **13.7.2 NEWTON'S LAW OF UNIVERSAL GRAVITATION.** Every particle of matter in the Universe attracts every other particle of matter in the Universe with a force that is proportional to the product of their masses and inversely proportional to the square of the distance between them. Specifically, if a particle of mass M and a particle of mass m are at a distance r from one another, then they attract each other with equal and opposite forces, $\mathbf{F}$ and $-\mathbf{F}$, of magnitude
>
> $$\|\mathbf{F}\| = \frac{GMm}{r^2} \qquad (3)$$
>
> where G is a constant called the ***universal gravitational constant***.

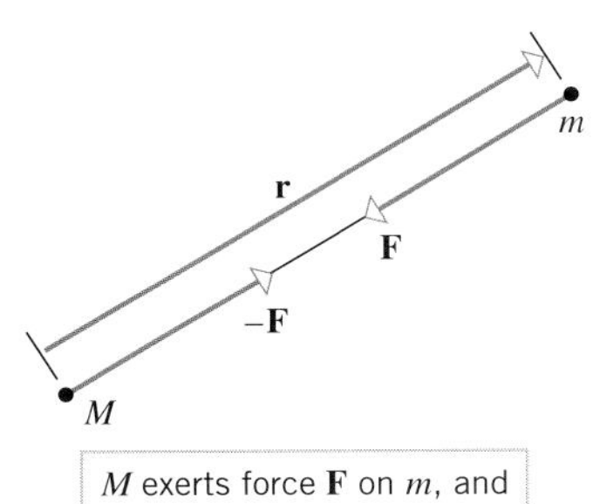

M exerts force $\mathbf{F}$ on m, and m exerts force $-\mathbf{F}$ on M.

Figure 13.7.2

To obtain a formula for the vector force $\mathbf{F}$ that mass M exerts on mass m, we will let $\mathbf{r}$ be the radius vector from mass M to mass m (Figure 13.7.2). Thus, the distance r between the masses is $\|\mathbf{r}\|$, and the force $\mathbf{F}$ can be expressed in terms of $\mathbf{r}$ as

$$\mathbf{F} = \|\mathbf{F}\|\left(-\frac{\mathbf{r}}{\|\mathbf{r}\|}\right) = \|\mathbf{F}\|\left(-\frac{\mathbf{r}}{r}\right)$$

which from (3) can be expressed as

$$\mathbf{F} = -\frac{GMm}{r^3}\mathbf{r} \tag{4}$$

We start by finding a formula for the acceleration function. To do this we use Formula (4) and Newton's second law to obtain

$$m\mathbf{a} = -\frac{GMm}{r^3}\mathbf{r}$$

Observe in Formula (5) that the acceleration $\mathbf{a}$ does not involve m. Thus, the mass of a planet has no effect on its acceleration.

from which we obtain

$$\mathbf{a} = -\frac{GM}{r^3}\mathbf{r} \tag{5}$$

To obtain a formula for the position function of the mass m, we will need to introduce a coordinate system and make some assumptions about the initial conditions. Let us assume:

- The distance r from m to M is minimum at time $t = 0$.
- The mass m has nonzero position and velocity vectors $\mathbf{r}_0$ and $\mathbf{v}_0$ at time $t = 0$.
- A polar coordinate system is introduced with its pole at mass M and oriented so $\theta = 0$ at time $t = 0$.
- The vector $\mathbf{v}_0$ is perpendicular to the polar axis at time $t = 0$.

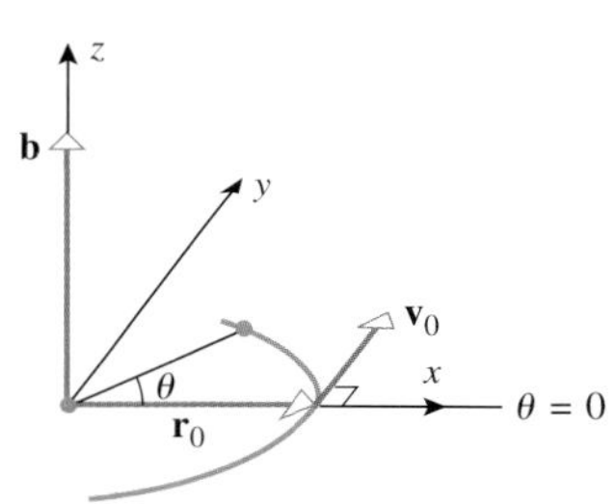

Figure 13.7.3

Moreover, to ensure that the polar angle θ increases with t, let us agree to observe this polar coordinate system looking toward the pole from the terminal point of the vector $\mathbf{b} = \mathbf{r}_0 \times \mathbf{v}_0$. We will also find it useful to superimpose an xyz-coordinate system on the polar coordinate system with the positive z-axis in the direction of $\mathbf{b}$ (Figure 13.7.3).

For computational purposes, it will be helpful to denote $\|\mathbf{r}_0\|$ by r_0 and $\|\mathbf{v}_0\|$ by v_0, in which case we can express the vectors $\mathbf{r}_0$ and $\mathbf{v}_0$ in xyz-coordinates as

$$\mathbf{r}_0 = r_0\mathbf{i} \quad \text{and} \quad \mathbf{v}_0 = v_0\mathbf{j}$$

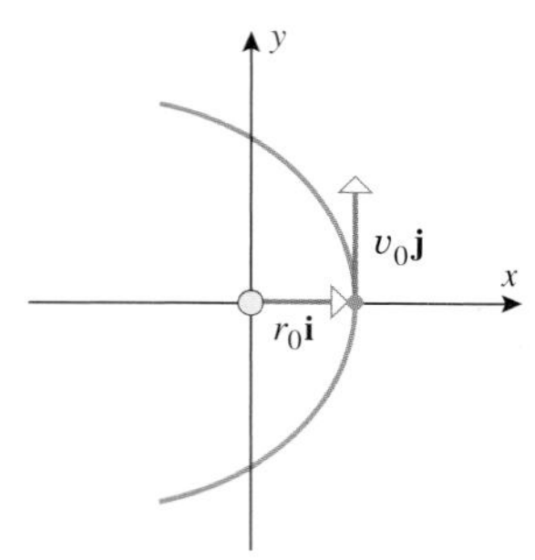

Figure 13.7.4

and the vector $\mathbf{b}$ as

$$\mathbf{b} = \mathbf{r}_0 \times \mathbf{v}_0 = r_0\mathbf{i} \times v_0\mathbf{j} = r_0v_0\mathbf{k} \tag{6}$$

(Figure 13.7.4). It will also be useful to introduce the unit vector

$$\mathbf{u} = \cos\theta\mathbf{i} + \sin\theta\mathbf{j} \tag{7}$$

which will allow us to express the polar form of the position vector $\mathbf{r}$ as

$$\mathbf{r} = r\cos\theta\mathbf{i} + r\sin\theta\mathbf{j} = r(\cos\theta\mathbf{i} + \sin\theta\mathbf{j}) = r\mathbf{u} \tag{8}$$

and to express the acceleration vector $\mathbf{a}$ in terms of $\mathbf{u}$ by rewriting (5) as

$$\mathbf{a} = -\frac{GM}{r^2}\mathbf{u} \tag{9}$$

We are now ready to derive the position function of the mass m in polar coordinates. For this purpose, recall from (2) that the vector $\mathbf{b} = \mathbf{r} \times \mathbf{v}$ is constant, so it follows from (6) that the relationship

$$\mathbf{b} = \mathbf{r} \times \mathbf{v} = r_0v_0\mathbf{k} \tag{10}$$

holds for *all* values of t. Now let us examine $\mathbf{b}$ from another point of view. It follows from (8) that

$$\mathbf{v} = \frac{d\mathbf{r}}{dt} = \frac{d}{dt}(r\mathbf{u}) = r\frac{d\mathbf{u}}{dt} + \frac{dr}{dt}\mathbf{u}$$

and hence

$$\mathbf{b} = \mathbf{r} \times \mathbf{v} = (r\mathbf{u}) \times \left(r\frac{d\mathbf{u}}{dt} + \frac{dr}{dt}\mathbf{u}\right) = r^2\mathbf{u} \times \frac{d\mathbf{u}}{dt} + r\frac{dr}{dt}\mathbf{u} \times \mathbf{u} = r^2\mathbf{u} \times \frac{d\mathbf{u}}{dt} \tag{11}$$

But (7) implies that

$$\frac{d\mathbf{u}}{dt} = \frac{d\mathbf{u}}{d\theta}\frac{d\theta}{dt} = (-\sin\theta\,\mathbf{i} + \cos\theta\,\mathbf{j})\frac{d\theta}{dt}$$

so

$$\mathbf{u} \times \frac{d\mathbf{u}}{dt} = \frac{d\theta}{dt}\mathbf{k} \tag{12}$$

Substituting (12) in (11) yields

$$\mathbf{b} = r^2\frac{d\theta}{dt}\mathbf{k} \tag{13}$$

Thus, it follows from (7), (9), and (13) that

$$\begin{aligned} \mathbf{a} \times \mathbf{b} &= -\frac{GM}{r^2}(\cos\theta\,\mathbf{i} + \sin\theta\,\mathbf{j}) \times \left(r^2\frac{d\theta}{dt}\mathbf{k}\right) \\ &= GM(-\sin\theta\,\mathbf{i} + \cos\theta\,\mathbf{j})\frac{d\theta}{dt} = GM\frac{d\mathbf{u}}{dt} \end{aligned} \tag{14}$$

From this formula and the fact that $d\mathbf{b}/dt = \mathbf{0}$ (since $\mathbf{b}$ is constant), we obtain

$$\frac{d}{dt}(\mathbf{v} \times \mathbf{b}) = \mathbf{v} \times \frac{d\mathbf{b}}{dt} + \frac{d\mathbf{v}}{dt} \times \mathbf{b} = \mathbf{a} \times \mathbf{b} = GM\frac{d\mathbf{u}}{dt}$$

Integrating both sides of this equation with respect to t yields

$$\mathbf{v} \times \mathbf{b} = GM\mathbf{u} + \mathbf{C} \tag{15}$$

where $\mathbf{C}$ is a vector constant of integration. This constant can be obtained by evaluating both sides of the equation at $t = 0$. We leave it as an exercise to show that

$$\mathbf{C} = (r_0v_0^2 - GM)\mathbf{i} \tag{16}$$

from which it follows that

$$\mathbf{v} \times \mathbf{b} = GM\mathbf{u} + (r_0v_0^2 - GM)\mathbf{i} \tag{17}$$

We can now obtain the position function by computing the scalar triple product $\mathbf{r} \cdot (\mathbf{v} \times \mathbf{b})$ in two ways. First we use (10) and property (11) of Section 12.4 to obtain

$$\mathbf{r} \cdot (\mathbf{v} \times \mathbf{b}) = (\mathbf{r} \times \mathbf{v}) \cdot \mathbf{b} = \mathbf{b} \cdot \mathbf{b} = r_0^2v_0^2 \tag{18}$$

and next we use (17) to obtain

$$\begin{aligned} \mathbf{r} \cdot (\mathbf{v} \times \mathbf{b}) &= \mathbf{r} \cdot (GM\mathbf{u}) + \mathbf{r} \cdot (r_0v_0^2 - GM)\mathbf{i} \\ &= \mathbf{r} \cdot \left(GM\frac{\mathbf{r}}{r}\right) + r\mathbf{u} \cdot (r_0v_0^2 - GM)\mathbf{i} \\ &= GMr + r(r_0v_0^2 - GM)\cos\theta \end{aligned}$$

If we now equate this to (18), we obtain

$$r_0^2v_0^2 = GMr + r(r_0v_0^2 - GM)\cos\theta$$

which when solved for r gives

$$r = \frac{r_0^2v_0^2}{GM + (r_0v_0^2 - GM)\cos\theta} = \frac{\dfrac{r_0^2v_0^2}{GM}}{1 + \left(\dfrac{r_0v_0^2}{GM} - 1\right)\cos\theta} \tag{19}$$

or more simply

$$r = \frac{k}{1 + e\cos\theta} \tag{20}$$

where

$$k = \frac{r_0^2 v_0^2}{GM} \quad \text{and} \quad e = \frac{r_0 v_0^2}{GM} - 1 \tag{21–22}$$

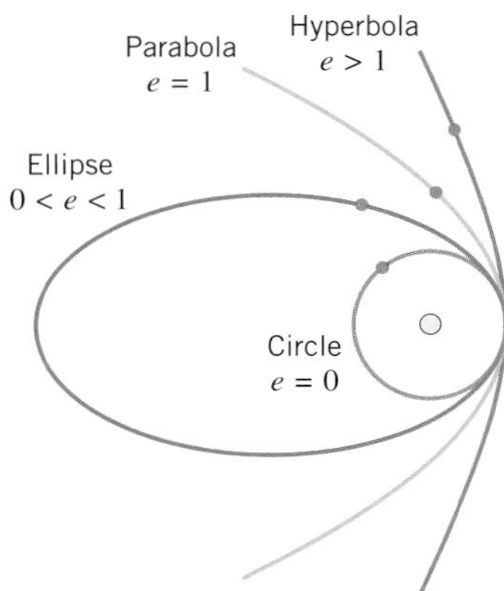

Figure 13.7.5

We will leave it as an exercise to show that $e \geq 0$. Accepting this to be so, it follows by comparing (20) to Formula (3) of Section 11.6 that the trajectory is a conic section with eccentricity e, the focus at the pole, and $d = k/e$. Thus, depending on whether $e < 1$, $e = 1$, or $e > 1$, the trajectory will be, respectively, an ellipse, a parabola, or a hyperbola (Figure 13.7.5).

Note from Formula (22) that e depends on r_0 and v_0, so the exact form of the trajectory is determined by the mass M and the initial conditions. If the initial conditions are such that $e < 1$, then the mass m becomes trapped in an elliptical orbit; otherwise the mass m "escapes" and never returns to its initial position. Accordingly, the initial velocity that produces an eccentricity of $e = 1$ is called the ***escape speed*** and is denoted by v_{esc}. Thus, it follows from (22) that

$$v_{\text{esc}} = \sqrt{\frac{2GM}{r_0}} \tag{23}$$

(verify).

KEPLER'S FIRST AND SECOND LAWS

It follows from our general discussion of central force fields that the planets have elliptical orbits with the Sun at the focus, which is Kepler's first law. To derive Kepler's second law, we begin by equating (10) and (13) to obtain

$$r^2 \frac{d\theta}{dt} = r_0 v_0 \tag{24}$$

To prove that the radial line from the center of the Sun to the center of a planet sweeps out equal areas in equal times, let $r = f(\theta)$ denote the polar equation of the planet, and let A denote the area swept out by the radial line as it varies from any fixed angle θ_0 to an angle θ. It follows from the area formula in 11.3.2 that A can be expressed as

$$A = \int_{\theta_0}^{\theta} \frac{1}{2}[f(\phi)]^2 \, d\phi$$

where the dummy variable ϕ is introduced for the integration to reserve θ for the upper limit. It now follows from Part 2 of the Fundamental Theorem of Calculus and the chain rule that

$$\frac{dA}{dt} = \frac{dA}{d\theta}\frac{d\theta}{dt} = \frac{1}{2}[f(\theta)]^2 \frac{d\theta}{dt} = \frac{1}{2} r^2 \frac{d\theta}{dt}$$

Thus, it follows from (24) that

$$\frac{dA}{dt} = \frac{1}{2} r_0 v_0 \tag{25}$$

which shows that A changes at a constant rate. This implies that equal areas are swept out in equal times.

KEPLER'S THIRD LAW

To derive Kepler's third law, we let a and b be the semimajor and semiminor axes of the elliptical orbit, and we recall that the area of this ellipse is πab. It follows by integrating (25) that in t units of time the radial line will sweep out an area of $A = \frac{1}{2} r_0 v_0 t$. Thus, if T denotes the time required for the planet to make one revolution around the Sun (the period), then the radial line will sweep out the area of the entire ellipse during that time and hence

$$\pi ab = \frac{1}{2} r_0 v_0 T$$

from which we obtain

$$T^2 = \frac{4\pi^2 a^2 b^2}{r_0^2 v_0^2} \tag{26}$$

However, it follows from Formula (1) of Section 11.6 and the relationship $c^2 = a^2 - b^2$ for an ellipse that

$$e = \frac{c}{a} = \frac{\sqrt{a^2 - b^2}}{a}$$

Thus, $b^2 = a^2(1 - e^2)$ and hence (26) can be written as

$$T^2 = \frac{4\pi^2 a^4 (1 - e^2)}{r_0^2 v_0^2} \tag{27}$$

But comparing Equation (20) to Equation (17) of Section 11.6 shows that

$$k = a(1 - e^2)$$

Finally, substituting this expression and (21) in (27) yields

$$T^2 = \frac{4\pi^2 a^3}{r_0^2 v_0^2} k = \frac{4\pi^2 a^3}{r_0^2 v_0^2} \frac{r_0^2 v_0^2}{GM} = \frac{4\pi^2}{GM} a^3 \tag{28}$$

Thus, we have proved that T^2 is proportional to a^3, which is Kepler's third law. When convenient, Formula (28) can also be expressed as

$$T = \frac{2\pi}{\sqrt{GM}} a^{3/2} \tag{29}$$

ARTIFICIAL SATELLITES

Kepler's second and third laws and Formula (23) also apply to satellites that orbit a celestial body; we need only interpret M to be the mass of the body exerting the force and m to be the mass of the satellite. Values of GM that are required in many of the formulas in this section have been determined experimentally for various attracting bodies (Table 13.7.1).

Table 13.7.1

ATTRACTING BODY	INTERNATIONAL SYSTEM	BRITISH ENGINEERING SYSTEM
Earth	$GM = 3.99 \times 10^{14}$ m^3/s^2 $GM = 3.99 \times 10^5$ km^3/s^2	$GM = 1.41 \times 10^{16}$ ft^3/s^2 $GM = 1.24 \times 10^{12}$ mi^3/h^2
Sun	$GM = 1.33 \times 10^{20}$ m^3/s^2 $GM = 1.33 \times 10^{11}$ km^3/s^2	$GM = 4.69 \times 10^{21}$ ft^3/s^2 $GM = 4.13 \times 10^{17}$ mi^3/h^2
Moon	$GM = 4.90 \times 10^{12}$ m^3/s^2 $GM = 4.90 \times 10^3$ km^3/s^2	$GM = 1.73 \times 10^{14}$ ft^3/s^2 $GM = 1.53 \times 10^{10}$ mi^3/h^2

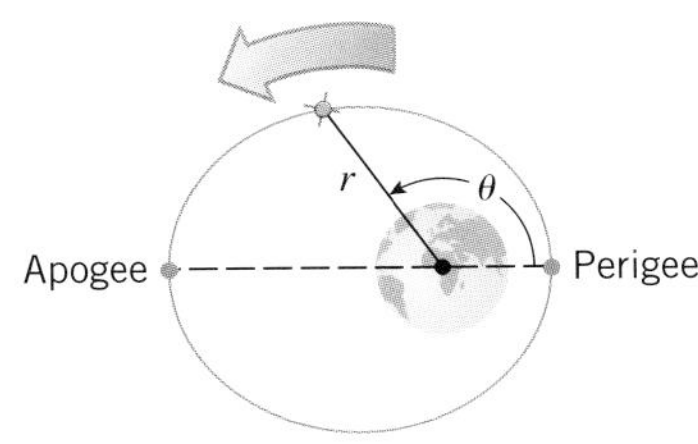

Figure 13.7.6

Recall that for orbits of planets around the Sun, the point at which the distance between the center of the planet and the center of the Sun is maximum is called the *aphelion* and the point at which it is minimum the *perihelion*. For satellites around the Earth the point at which the maximum distance occurs is called the ***apogee*** and the point at which the minimum distance occurs is called the ***perigee*** (Figure 13.7.6). The actual distances between the centers at apogee and perigee are called the ***apogee distance*** and the ***perigee distance***.

▶ **Example 1** A geosynchronous orbit for a satellite is a circular orbit about the equator of the Earth in which the satellite stays fixed over a point on the equator. Use the fact that the Earth makes one revolution about its axis every 24 hours to find the altitude in miles of a communications satellite in geosynchronous orbit. Assume the Earth to be a sphere of radius 4000 mi.

Solution. To remain fixed over a point on the equator, the satellite must have a period of $T = 24$ h. It follows from (28) or (29) and the Earth value of $GM = 1.24 \times 10^{12}$ mi^3/h^2 from Table 13.7.1 that

$$a = \sqrt[3]{\frac{GMT^2}{4\pi^2}} = \sqrt[3]{\frac{(1.24 \times 10^{12})(24)^2}{4\pi^2}} \approx 26{,}250 \text{ mi}$$

and hence the altitude h of the satellite is

$$h \approx 26{,}250 - 4000 = 22{,}250 \text{ mi} \blacktriangleleft$$

✔QUICK CHECK EXERCISES 13.7 *(See page 921 for answers.)*

1. Let G denote the universal gravitational constant and let M and m denote masses a distance r apart.
   (a) According to Newton's Law of Universal Gravitation, M and m attract each other with a force of magnitude ________.
   (b) If $\mathbf{r}$ is the radius vector from M to m, then the force of attraction that mass M exerts on mass m is ________.
2. Suppose that a mass m is in an orbit about a mass M and that r_0 is the minimum distance from m to M. If G is the universal gravitational constant, then the "escape" speed of m is ________.
3. For a planet in an elliptical orbit about the Sun, the square of the planet's period is proportional to what power of the semimajor axis of its orbit?
4. Suppose that a mass m is in an orbit about a mass M and that r_0 is the minimum distance from m to M. If v_0 is the speed of mass m when it is a distance r_0 from M, and if G denotes the universal gravitational constant, then the eccentricity of the orbit is ________.

EXERCISE SET 13.7

In exercises that require numerical values, use Table 13.7.1 and the following values, where needed:

radius of Earth = 4000 mi = 6440 km
radius of Moon = 1080 mi = 1740 km
1 year (Earth year) = 365 days

FOCUS ON CONCEPTS

1. (a) Obtain the value of $\mathbf{C}$ given in Formula (16) by setting $t = 0$ in (15).
   (b) Use Formulas (7), (17), and (22) to show that
   $$\mathbf{v} \times \mathbf{b} = GM[(e + \cos\theta)\mathbf{i} + \sin\theta\,\mathbf{j}]$$
   (c) Show that $\|\mathbf{v} \times \mathbf{b}\| = \|\mathbf{v}\|\|\mathbf{b}\|$.
   (d) Use the results in parts (b) and (c) to show that the speed of a particle in an elliptical orbit is
   $$v = \frac{v_0}{1+e}\sqrt{e^2 + 2e\cos\theta + 1}$$
   (e) Suppose that a particle is in an elliptical orbit. Use part (d) to conclude that the distance from the particle to the center of force is a minimum if and only if the speed of the particle is a maximum. Similarly, argue that the distance from the particle to the center of force is a maximum if and only if the speed of the particle is a minimum.
2. Use the result in Exercise 1(d) to show that when a particle in an elliptical orbit with eccentricity e reaches an end of the minor axis, its speed is
   $$v = v_0\sqrt{\frac{1-e}{1+e}}$$
3. Use the result in Exercise 1(d) to show that for a particle in an elliptical orbit with eccentricity e, the maximum and minimum speeds are related by
   $$v_{\max} = v_{\min}\frac{1+e}{1-e}$$
4. Use Formula (22) and the result in Exercise 1(d) to show that the speed v of a particle in a circular orbit of radius r_0 is constant and is given by
   $$v = \sqrt{\frac{GM}{r_0}}$$

5. Suppose that a particle is in an elliptical orbit in a central force field in which the center of force is at a focus, and let $\mathbf{r} = \mathbf{r}(t)$ and $\mathbf{v} = \mathbf{v}(t)$ be the position and velocity functions of the particle respectively. Let $r_{\min}$ and $r_{\max}$ denote the minimum and maximum distances from the particle to the center of force, and let $v_{\min}$ and $v_{\max}$ denote the minimum and maximum speeds of the particle.
   (a) Review the discussion of ellipses in polar coordinates in Section 11.6, and show that if the ellipse has eccentricity e and semimajor axis a, then $r_{\min} = a(1 - e)$ and $r_{\max} = a(1 + e)$.
   (b) Explain why $r_{\min}$ and $r_{\max}$ occur at points at which $\mathbf{r}$ and $\mathbf{v}$ are orthogonal. [*Hint:* First argue that the extreme values of $\|\mathbf{r}\|$ occur at critical points of the function $\|\mathbf{r}\|^2 = \mathbf{r} \cdot \mathbf{r}$.]
   (c) Explain why $v_{\min}$ and $v_{\max}$ occur at points at which $\mathbf{r}$ and $\mathbf{v}$ are orthogonal. [*Hint:* First argue that the extreme values of $\|\mathbf{v}\|$ occur at critical points of the function $\|\mathbf{v}\|^2 = \mathbf{v} \cdot \mathbf{v}$. Then use Equation (5).]
   (d) Use Equation (2) and parts (b) and (c) to conclude that $r_{\max} v_{\min} = r_{\min} v_{\max}$.

6. Use the results in parts (a) and (d) of Exercise 5 to give a derivation of the equation in Exercise 3.

7. Use the result in Exercise 4 to find the speed in km/s of a satellite in a circular orbit that is 200 km above the surface of the Earth.

8. Use the result in Exercise 4 to find the speed in mi/h of a communications satellite that is in geosynchronous orbit around the Earth. [See Example 1.]

9. Find the escape speed in km/s for a space probe in a circular orbit that is 300 km above the surface of the Earth.

10. The universal gravitational constant is approximately
$$G = 6.67 \times 10^{-11}\ \text{m}^3/\text{kg}\cdot\text{s}^2$$
and the semimajor axis of the Earth's orbit is approximately
$$a = 149.6 \times 10^6\ \text{km}$$
Estimate the mass of the Sun in kg.

11. (a) The eccentricity of the Moon's orbit around the Earth is 0.055, and its semimajor axis is $a = 238{,}900$ mi. Find the maximum and minimum distances between the surface of the Earth and the surface of the Moon.
   (b) Find the period of the Moon's orbit in days.

12. (a) *Vanguard 1* was launched in March 1958 with perigee and apogee altitudes above the Earth of 649 km and 4340 km, respectively. Find the length of the semimajor axis of its orbit.
   (b) Use the result in part (a) of Exercise 19 in Section 11.6 to find the eccentricity of its orbit.
   (c) Find the period of *Vanguard 1* in minutes.

13. (a) Suppose that a space probe is in a circular orbit at an altitude of 180 mi above the surface of the Earth. Use the result in Exercise 4 to find its speed.
   (b) During a very short period of time, a thruster rocket on the space probe is fired to increase the speed of the probe by 600 mi/h in its direction of motion. Find the eccentricity of the resulting elliptical orbit, and use the result in part (a) of Exercise 5 to find the apogee altitude.

14. Show that the quantity e defined by Formula (22) is nonnegative. [*Hint:* The polar axis was chosen so that r is minimum when $\theta = 0$.]

✔ QUICK CHECK ANSWERS 13.7

1. (a) $\dfrac{GMm}{r^2}$ (b) $-\dfrac{GMm}{r^3}\mathbf{r}$ 2. $\sqrt{\dfrac{2GM}{r_0}}$ 3. 3 4. $e = \dfrac{r_0 v_0^2}{GM} - 1$

CHAPTER REVIEW EXERCISES

1. In words, what is meant by the graph of a vector-valued function?

2–5 Describe the graph of the equation.

2. $\mathbf{r} = (2 - 3t)\mathbf{i} - 4t\mathbf{j}$
3. $\mathbf{r} = 3\sin 2t\mathbf{i} + 3\cos 2t\mathbf{j}$
4. $\mathbf{r} = 3\cos t\mathbf{i} + 2\sin t\mathbf{j} - \mathbf{k}$
5. $\mathbf{r} = -2\mathbf{i} + t\mathbf{j} + (t^2 - 1)\mathbf{k}$

6. Describe the graph of the vector-valued function.
   (a) $\mathbf{r} = \mathbf{r}_0 + t(\mathbf{r}_1 - \mathbf{r}_0)$
   (b) $\mathbf{r} = \mathbf{r}_0 + t(\mathbf{r}_1 - \mathbf{r}_0) \quad (0 \le t \le 1)$
   (c) $\mathbf{r} = \mathbf{r}_0 + t\mathbf{r}'(t_0)$

7. Show that the graph of $\mathbf{r}(t) = t\sin \pi t\mathbf{i} + t\mathbf{j} + t\cos \pi t\mathbf{k}$ lies on the surface of a cone, and sketch the cone.

8. Find parametric equations for the intersection of the surfaces
$$y = x^2 \quad \text{and} \quad 2x^2 + y^2 + 6z^2 = 24$$
and sketch the intersection.

9. In words, give a geometric description of the statement $\lim_{t \to a} \mathbf{r}(t) = \mathbf{L}$.

10. Evaluate $\lim_{t\to 0}\left(e^{-t}\mathbf{i} + \frac{1-\cos t}{t}\mathbf{j} + t^2\mathbf{k}\right)$.

11. Find parametric equations of the line tangent to the graph of

$$\mathbf{r}(t) = (t + \cos 2t)\mathbf{i} - (t^2 + t)\mathbf{j} + \sin t\mathbf{k}$$

at the point where $t = 0$.

12. Suppose that $\mathbf{r}_1(t)$ and $\mathbf{r}_2(t)$ are smooth vector-valued functions such that $\mathbf{r}_1(0) = \langle -1, 1, 2\rangle$, $\mathbf{r}_2(0) = \langle 1, 2, 1\rangle$, $\mathbf{r}_1'(0) = \langle 1, 0, 1\rangle$, and $\mathbf{r}_2'(0) = \langle 4, 0, 2\rangle$. Use this information to evaluate the derivative at $t = 0$ of each function.

(a) $\mathbf{r}(t) = 3\mathbf{r}_1(t) + 2\mathbf{r}_2(t)$ (b) $\mathbf{r}(t) = [\ln(t+1)]\mathbf{r}_1(t)$
(c) $\mathbf{r}(t) = \mathbf{r}_1(t) \times \mathbf{r}_2(t)$ (d) $f(t) = \mathbf{r}_1(t) \cdot \mathbf{r}_2(t)$

13. Evaluate $\int (\cos t\mathbf{i} + \sin t\mathbf{j})\,dt$.

14. Evaluate $\int_0^{\pi/3} \langle \cos 3t, -\sin 3t\rangle\,dt$.

15. Solve the vector initial-value problem

$$\mathbf{y}'(t) = t^2\mathbf{i} + 2t\mathbf{j}, \quad \mathbf{y}(0) = \mathbf{i} + \mathbf{j}$$

16. Solve the vector initial-value problem

$$\frac{d\mathbf{r}}{dt} = \mathbf{r}, \quad \mathbf{r}(0) = \mathbf{r}_0$$

for the unknown vector-valued function $\mathbf{r}(t)$.

17. Find the arc length of the graph of

$$\mathbf{r}(t) = e^{\sqrt{2}t}\mathbf{i} + e^{-\sqrt{2}t}\mathbf{j} + 2t\mathbf{k} \quad (0 \le t \le \sqrt{2}\ln 2)$$

18. Suppose that $\mathbf{r}(t)$ is a smooth vector-valued function of t with $\mathbf{r}'(0) = 3\mathbf{i} - \mathbf{j} + \mathbf{k}$ and that $\mathbf{r}_1(t) = \mathbf{r}(2 - e^{t\ln 2})$. Find $\mathbf{r}_1'(1)$.

19. Find the arc length parametrization of the line through $P(-1, 4, 3)$ and $Q(0, 2, 5)$ that has reference point P and orients the line in the direction from P to Q.

20. Find an arc length parametrization of the curve

$$\mathbf{r}(t) = \langle e^t\cos t, -e^t\sin t\rangle \quad (0 \le t \le \pi/2)$$

which has the same orientation and has $\mathbf{r}(0)$ as the reference point.

21. Suppose that $\mathbf{r}(t)$ is a smooth vector-valued function. State the definitions of $\mathbf{T}(t)$, $\mathbf{N}(t)$, and $\mathbf{B}(t)$.

22. Find $\mathbf{T}(0)$, $\mathbf{N}(0)$, and $\mathbf{B}(0)$ for the curve

$$\mathbf{r}(t) = \left\langle 2\cos t, 2\cos t + \frac{3}{\sqrt{5}}\sin t, \cos t - \frac{6}{\sqrt{5}}\sin t\right\rangle$$

23. State the definition of "curvature" and explain what it means geometrically.

24. Suppose that $\mathbf{r}(t)$ is a smooth curve with $\mathbf{r}'(0) = \mathbf{i}$ and $\mathbf{r}''(0) = \mathbf{i} + 2\mathbf{j}$. Find the curvature at $t = 0$.

25–28 Find the curvature of the curve at the stated point.

25. $\mathbf{r}(t) = 2\cos t\mathbf{i} + 3\sin t\mathbf{j} - t\mathbf{k}$; $t = \pi/2$

26. $\mathbf{r}(t) = \langle 2t, e^{2t}, e^{-2t}\rangle$; $t = 0$

27. $y = \cos x$; $x = \pi/2$ **28.** $y = \ln x$; $x = 1$

29. Suppose that $\mathbf{r}(t)$ is the position function of a particle moving in 2-space or 3-space. In each part, explain what the given quantity represents physically.

(a) $\left\|\frac{d\mathbf{r}}{dt}\right\|$ (b) $\int_{t_0}^{t_1}\left\|\frac{d\mathbf{r}}{dt}\right\|dt$ (c) $\|\mathbf{r}(t)\|$

30. (a) What does Theorem 13.2.8 tell you about the velocity vector of a particle that moves over a sphere?
(b) What does Theorem 13.2.8 tell you about the acceleration vector of a particle that moves with constant speed?
(c) Show that the particle with position function

$$\mathbf{r}(t) = \sqrt{1 - \tfrac{1}{4}\cos^2 t}\,\cos t\mathbf{i} + \sqrt{1 - \tfrac{1}{4}\cos^2 t}\,\sin t\mathbf{j} + \tfrac{1}{2}\cos t\mathbf{k}$$

moves over a sphere.

31. As illustrated in the accompanying figure, suppose that a particle moves counterclockwise around a circle of radius R centered at the origin at a constant rate of ω radians per second. This is called ***uniform circular motion***. If we assume that the particle is at the point $(R, 0)$ at time $t = 0$, then its position function will be

$$\mathbf{r}(t) = R\cos\omega t\mathbf{i} + R\sin\omega t\mathbf{j}$$

(a) Show that the velocity vector $\mathbf{v}(t)$ is always tangent to the circle and that the particle has constant speed v given by

$$v = R\omega$$

(b) Show that the acceleration vector $\mathbf{a}(t)$ is always directed toward the center of the circle and has constant magnitude a given by

$$a = R\omega^2$$

(c) Show that the time T required for the particle to make one complete revolution is

$$T = \frac{2\pi}{\omega} = \frac{2\pi R}{v}$$

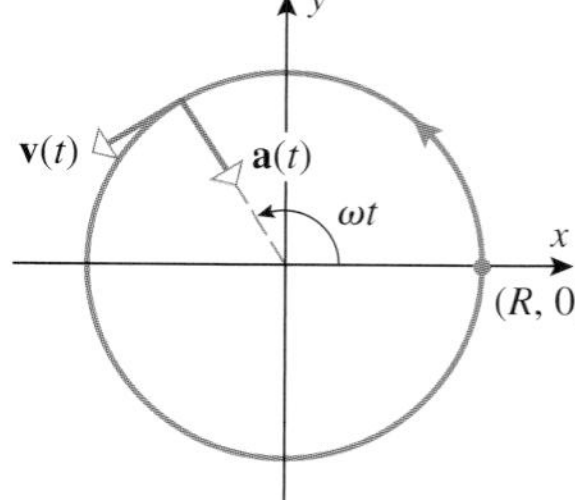

Figure Ex-31

32. If a particle of mass m has uniform circular motion (see Exercise 31), then the acceleration vector $\mathbf{a}(t)$ is called the ***centripetal acceleration***. According to Newton's second law, this acceleration must be produced by some force $\mathbf{F}(t)$, called the ***centripetal force***, that is related to $\mathbf{a}(t)$ by the equation $\mathbf{F}(t) = m\mathbf{a}(t)$. If this force is not present, then the particle cannot undergo uniform circular motion.

(a) Show that the direction of the centripetal force varies with time but that it has constant magnitude F given by

$$F = \frac{mv^2}{R}$$

(b) An astronaut with a mass of $m = 60$ kg orbits the Earth at an altitude of $h = 3200$ km with a constant speed of $v = 6.43$ km/s. Find her centripetal acceleration assuming that the radius of the Earth is 6440 km.

(c) What centripetal gravitational force in newtons does the Earth exert on the astronaut?

33. At time $t = 0$ a particle at the origin of an xyz-coordinate system has a velocity vector of $\mathbf{v}_0 = \mathbf{i} + 2\mathbf{j} - \mathbf{k}$. The acceleration function of the particle is $\mathbf{a}(t) = 2t^2\mathbf{i} + \mathbf{j} + \cos 2t\mathbf{k}$.
(a) Find the position function of the particle.
(b) Find the speed of the particle at time $t = 1$.

34. As illustrated in the accompanying figure, the polar coordinates of a rocket are tracked by radar from a point that is b units from the launching pad. Show that the speed v of the rocket can be expressed in terms b, θ, and $d\theta/dt$ as

$$v = b\sec^2\theta\frac{d\theta}{dt}$$

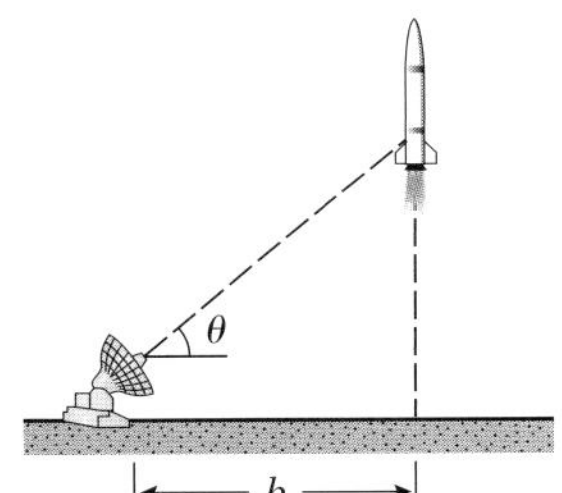

Figure Ex-34

35. A player throws a ball with an initial speed of 60 ft/s at an unknown angle α with the horizontal from a point that is 4 ft above the floor of a gymnasium. Given that the ceiling of the gymnasium is 25 ft high, determine the maximum height h at which the ball can hit a wall that is 60 ft away (see the accompanying figure).

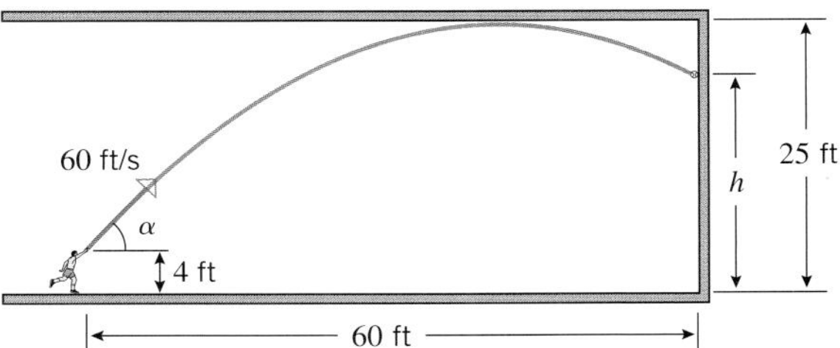

Figure Ex-35

36. Let $\mathbf{v} = \mathbf{v}(t)$ and $\mathbf{a} = \mathbf{a}(t)$ be the velocity and acceleration vectors for a particle moving in 2-space or 3-space. Show that the rate of change of its speed can be expressed as

$$\frac{d}{dt}(\|\mathbf{v}\|) = \frac{1}{\|\mathbf{v}\|}(\mathbf{v}\cdot\mathbf{a})$$

37. Use Formula (23) in Section 13.7 to find the escape speed (in km/s) for a space probe in a circular orbit 600 km above the surface of the Earth.

38. The universal gravitational constant is approximately 6.67×10^{-11} m^3/kg·s^2 and the semimajor axis of the Moon's orbit about the Earth is $a = 384{,}629$ km. Estimate the mass of the Earth in kilograms.

c h a p t e r f o u r t e e n

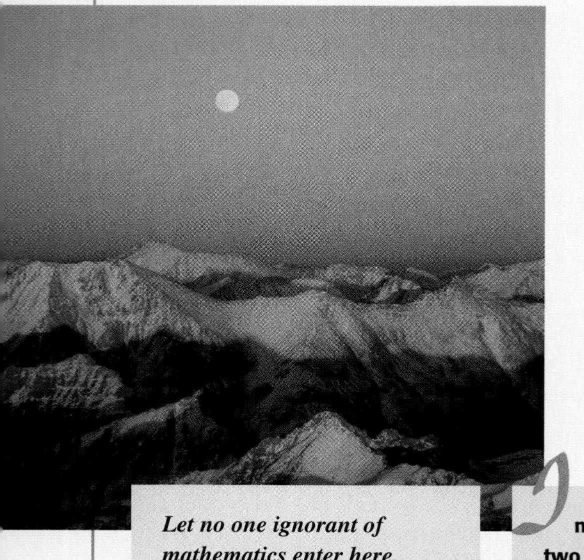

PARTIAL DERIVATIVES

Let no one ignorant of mathematics enter here.
—Plato
Ancient Greek Philosopher and Scholar

In this chapter we will extend many of the basic concepts of calculus to functions of two or more variables, commonly called functions of several variables. We will begin by discussing limits and continuity for functions of two and three variables, then we will define derivatives of such functions, and then we will use these derivatives to study tangent planes, rates of change, slopes of surfaces, and maximization and minimization problems. Although many of the basic ideas that we developed for functions of one variable will carry over in a natural way, functions of several variables are intrinsically more complicated than functions of one variable, so we will need to develop new tools and new ideas to deal with such functions.

Photo: ***Three-dimensional surfaces are analogous to mountain ranges. In this chapter we will use derivatives to analyze steepness and other features of such surfaces.***

14.1 FUNCTIONS OF TWO OR MORE VARIABLES

In previous sections we studied real-valued functions of a real variable and vector-valued functions of a real variable. In this section we will consider real-valued functions of two or more real variables.

NOTATION AND TERMINOLOGY

There are many familiar formulas in which a given variable depends on two or more other variables. For example, the area A of a triangle depends on the base length b and height h by the formula $A = \frac{1}{2}bh$; the volume V of a rectangular box depends on the length l, the width w, and the height h by the formula $V = lwh$; and the arithmetic average $\bar{x}$ of n real numbers, $x_1, x_2, \ldots, x_n$, depends on those numbers by the formula

$$\bar{x} = \frac{1}{n}(x_1 + x_2 + \cdots + x_n)$$

Thus, we say that

A is a function of the two variables b and h;
V is a function of the three variables l, w, and h;
$\bar{x}$ is a function of the n variables $x_1, x_2, \ldots, x_n$.

The terminology and notation for functions of two or more variables is similar to that for functions of one variable. For example, the expression

$$z = f(x, y)$$

means that z is a function of x and y in the sense that a unique value of the dependent variable z is determined by specifying values for the independent variables x and y. Similarly,

$$w = f(x, y, z)$$

expresses w as a function of x, y, and z, and

$$u = f(x_1, x_2, \ldots, x_n)$$

expresses u as a function of $x_1, x_2, \ldots, x_n$.

We will find it useful to think of functions of two or three independent variables in geometric terms. For example, if $z = f(x, y)$, then we can view (x, y) as a point in the xy-plane and think of f as a rule that associates a unique numerical value z with the point (x, y); similarly, we can think of $w = f(x, y, z)$ as a rule that associates a unique numerical value w with a point (x, y, z) in an xyz-coordinate system (Figure 14.1.1).

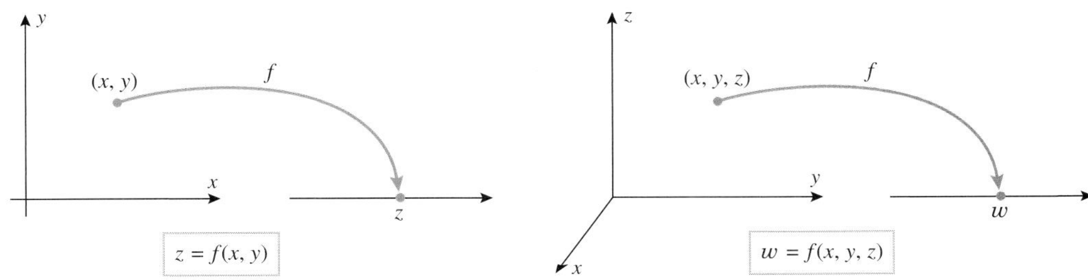

Figure 14.1.1

As with functions of one variable, the independent variables of a function of two or more variables may be restricted to lie in some set D, which we call the ***domain*** of f. Sometimes the domain will be determined by physical restrictions on the variables. If the function is defined by a formula and if there are no physical restrictions or other restrictions stated explicitly, then it is understood that the domain consists of all points for which the formula yields a real value for the dependent variable. We call this the ***natural domain*** of the function. The following definitions summarize this discussion.

14.1.1 DEFINITION. A ***function f of two variables***, x and y, is a rule that assigns a unique real number $f(x, y)$ to each point (x, y) in some set D in the xy-plane.

By extension, one can define the notion of "n-dimensional space" in which a "point" is a sequence of n real numbers $(x_1, x_2, \ldots, x_n)$. Then a function of n real variables is a rule that assigns a unique real number $f(x_1, x_2, \ldots, x_n)$ to each point in some set in this space.

14.1.2 DEFINITION. A ***function f of three variables***, x, y, and z, is a rule that assigns a unique real number $f(x, y, z)$ to each point (x, y, z) in some set D in three-dimensional space.

▶ Example 1 Let

$$f(x, y) = 3x^2\sqrt{y} - 1$$

Find $f(1, 4)$, $f(0, 9)$, $f(t^2, t)$, $f(ab, 9b)$, and the natural domain of f.

Solution. By substitution

$$f(1, 4) = 3(1)^2\sqrt{4} - 1 = 5$$
$$f(0, 9) = 3(0)^2\sqrt{9} - 1 = -1$$
$$f(t^2, t) = 3(t^2)^2\sqrt{t} - 1 = 3t^4\sqrt{t} - 1$$
$$f(ab, 9b) = 3(ab)^2\sqrt{9b} - 1 = 9a^2b^2\sqrt{b} - 1$$

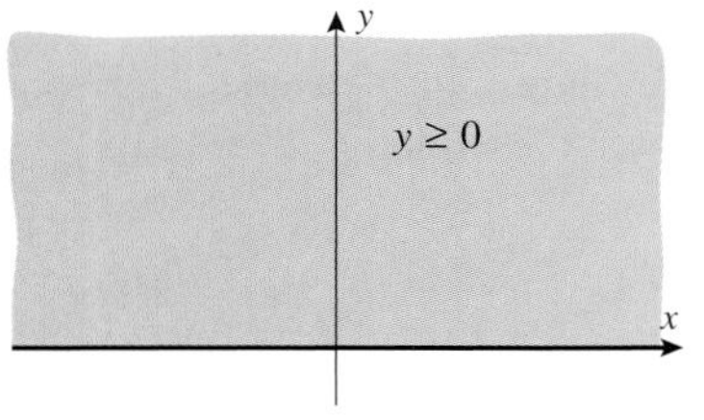

The solid boundary line is included in the domain.

Figure 14.1.2

Because of the radical $\sqrt{y}$ in the formula for f, we must have $y \geq 0$ to avoid non-real values for $f(x, y)$. Thus, the natural domain of f consists of all points in the xy-plane that are on or above the x-axis. (See Figure 14.1.2.) ◀

▶ Example 2 Sketch the natural domain of the function $f(x, y) = \ln(x^2 - y)$.

Solution. $\ln(x^2 - y)$ is defined only when $0 < x^2 - y$ or $y < x^2$. We first sketch the parabola $y = x^2$ as a "dashed" curve. The region $y < x^2$ then consists of all points below this curve (Figure 14.1.3). ◀

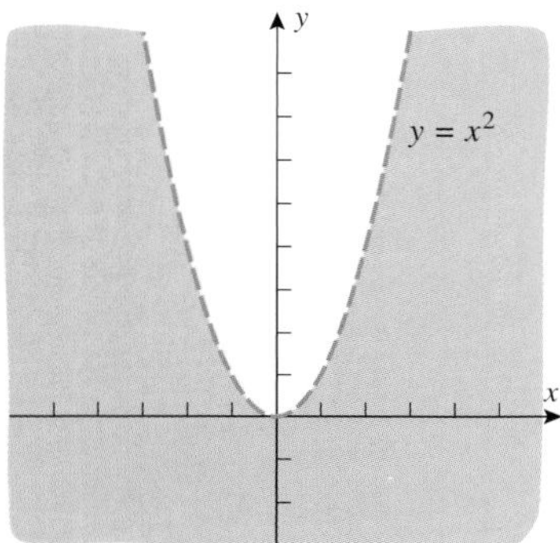

The dashed boundary does not belong to the domain.

Figure 14.1.3

▶ Example 3 Let

$$f(x, y, z) = \sqrt{1 - x^2 - y^2 - z^2}$$

Find $f\left(0, \frac{1}{2}, -\frac{1}{2}\right)$ and the natural domain of f.

Solution. By substitution,

$$f\left(0, \tfrac{1}{2}, -\tfrac{1}{2}\right) = \sqrt{1 - (0)^2 - \left(\tfrac{1}{2}\right)^2 - \left(-\tfrac{1}{2}\right)^2} = \sqrt{\tfrac{1}{2}}$$

Because of the square root sign, we must have $0 \leq 1 - x^2 - y^2 - z^2$ in order to have a real value for $f(x, y, z)$. Rewriting this inequality in the form

$$x^2 + y^2 + z^2 \leq 1$$

we see that the natural domain of f consists of all points on or within the sphere

$$x^2 + y^2 + z^2 = 1 \quad ◀$$

FUNCTIONS DESCRIBED BY TABLES

Sometimes it is either desirable or necessary to represent a function of two variables in table form, rather than as an explicit formula. For example, the U.S. National Weather Service uses the formula

$$W = 35.74 + 0.6215T + (0.4275T - 35.75)v^{0.16} \tag{1}$$

to model the wind chill index W (in °F) as a function of the temperature T (in °F) and the wind speed v (in mi/h) for wind speeds greater than 3 mi/h. This formula is sufficiently complex that it is difficult to get an intuitive feel for the relationship between the variables. One can get a clearer sense of the relationship by selecting sample values of T and v and constructing a table, such as Table 14.1.1, in which we have rounded the values of W to the

> The wind chill index is that temperature (in °F) which would produce the same sensation on exposed skin at a wind speed of 3 mi/h as the temperature and wind speed combination in current weather conditions.

Table 14.1.1

WIND SPEED v (mi/h) \ TEMPERATURE T (°F)	20	25	30	35
5	13	19	25	31
15	6	13	19	25
25	3	9	16	23
35	0	7	14	21
45	−2	5	12	19

nearest integer. For example, if the temperature is 30°F and the wind speed is 5 mi/h, it feels as if the temperature is 25°F. If the wind speed increases to 15 mi/h, the temperature then feels as if it has dropped to 19°F. Note that in this case, an increase in wind speed of 10 mi/h causes a 6°F decrease in the wind chill index. To estimate wind chill values not displayed in the table, we can use ***linear interpolation***. For example, suppose that the temperature is 30°F and the wind speed is 7 mi/h. A reasonable estimate for the drop in the wind chill index from its value when the wind speed is 5 mi/h would be $\frac{2}{10} \cdot 6°\text{F} = 1.2°\text{F}$. (Why?) The resulting estimate in wind chill would then be $25° - 1.2° = 23.8°$F.

In some cases, tables for functions of two variables arise directly from experimental data, in which case one must either work directly with the table or else use some technique to construct a formula that models the data in the table. We will not discuss such modeling techniques in this text.

GRAPHS OF FUNCTIONS OF TWO VARIABLES

Recall that for a function f of one variable, the graph of $f(x)$ in the xy-plane was defined to be the graph of the equation $y = f(x)$. Similarly, if f is a function of two variables, we define the ***graph*** of $f(x, y)$ in xyz-space to be the graph of the equation $z = f(x, y)$. In general, such a graph will be a surface in 3-space.

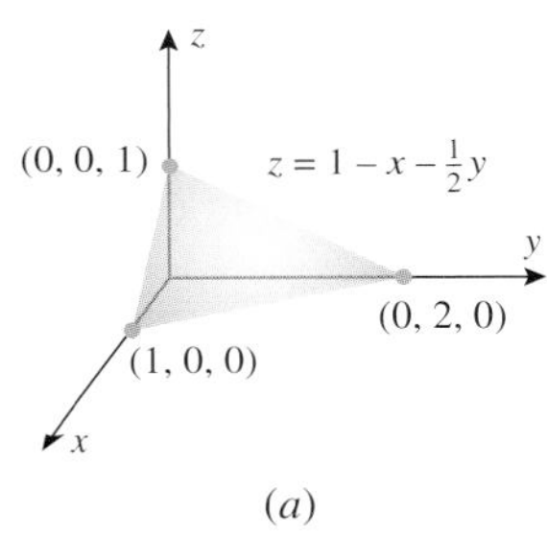

(a)

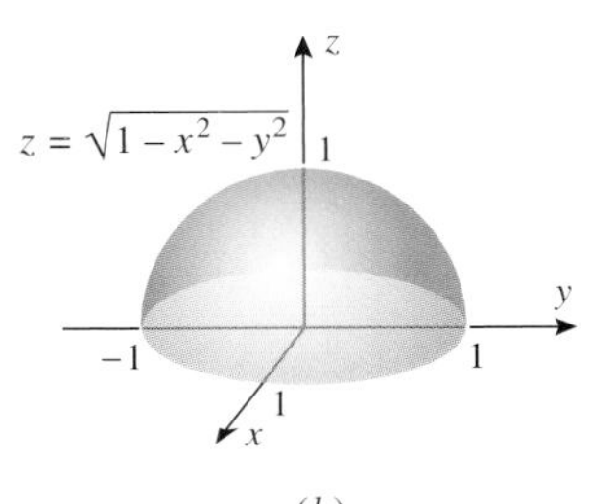

(b)

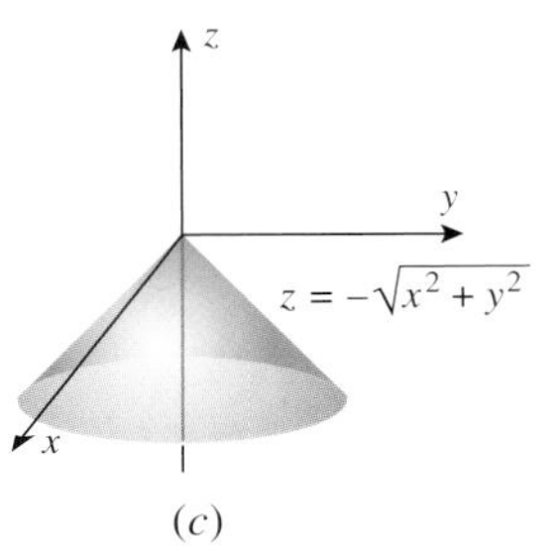

(c)

Figure 14.1.4

▶ **Example 4** In each part, describe the graph of the function in an xyz-coordinate system.

(a) $f(x, y) = 1 - x - \frac{1}{2}y$ (b) $f(x, y) = \sqrt{1 - x^2 - y^2}$

(c) $f(x, y) = -\sqrt{x^2 + y^2}$

Solution (a). By definition, the graph of the given function is the graph of the equation

$$z = 1 - x - \tfrac{1}{2}y$$

which is a plane. A triangular portion of the plane can be sketched by plotting the intersections with the coordinate axes and joining them with line segments (Figure 14.1.4*a*).

Solution (b). By definition, the graph of the given function is the graph of the equation

$$z = \sqrt{1 - x^2 - y^2} \tag{2}$$

After squaring both sides, this can be rewritten as

$$x^2 + y^2 + z^2 = 1$$

which represents a sphere of radius 1, centered at the origin. Since (2) imposes the added condition that $z \geq 0$, the graph is just the upper hemisphere (Figure 14.1.4*b*).

Solution (c). The graph of the given function is the graph of the equation

$$z = -\sqrt{x^2 + y^2} \tag{3}$$

After squaring, we obtain

$$z^2 = x^2 + y^2$$

which is the equation of a circular cone (see Table 12.7.1). Since (3) imposes the condition that $z \leq 0$, the graph is just the lower nappe of the cone (Figure 14.1.4*c*). ◀

LEVEL CURVES

We are all familiar with the topographic (or contour) maps in which a three-dimensional landscape, such as a mountain range, is represented by two-dimensional contour lines or curves of constant elevation. Consider, for example, the model hill and its contour map shown in Figure 14.1.5. The contour map is constructed by passing planes of constant elevation through the hill, projecting the resulting contours onto a flat surface, and labeling the contours with their elevations. In Figure 14.1.5, note how the two gullies appear as indentations in the contour lines and how the curves are close together on the contour map where the hill has a steep slope and become more widely spaced where the slope is gradual.

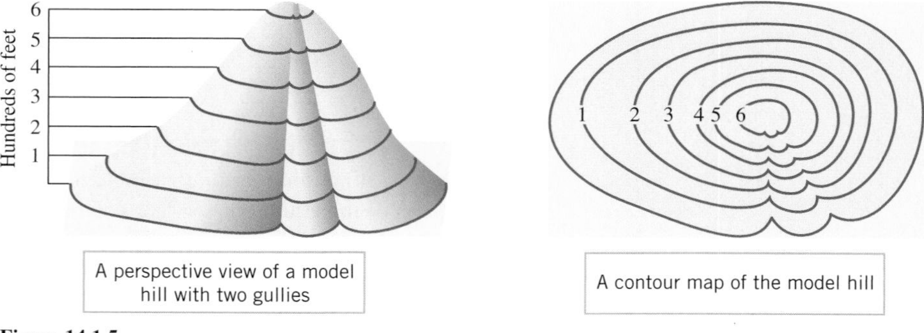

A perspective view of a model hill with two gullies

A contour map of the model hill

Figure 14.1.5

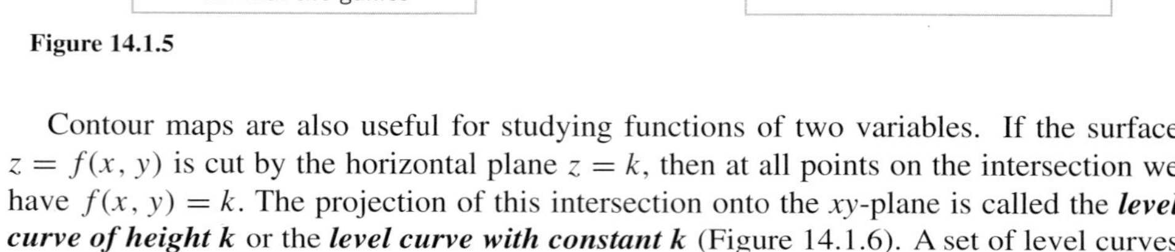

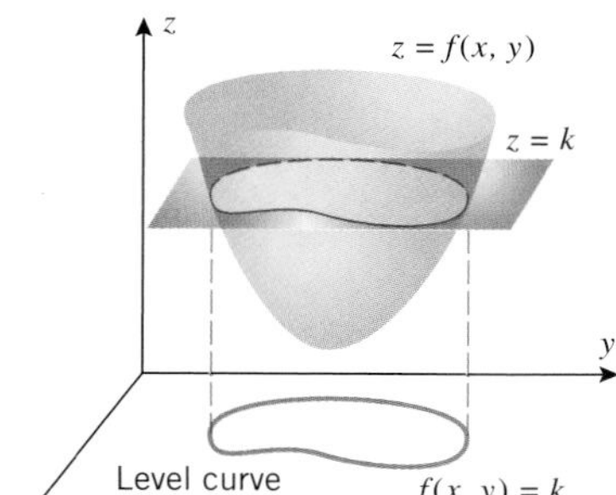

Figure 14.1.6

Contour maps are also useful for studying functions of two variables. If the surface $z = f(x, y)$ is cut by the horizontal plane $z = k$, then at all points on the intersection we have $f(x, y) = k$. The projection of this intersection onto the xy-plane is called the ***level curve of height k*** or the ***level curve with constant k*** (Figure 14.1.6). A set of level curves for $z = f(x, y)$ is called a ***contour plot*** or ***contour map*** of f.

► **Example 5** The graph of the function $f(x, y) = y^2 - x^2$ in xyz-space is the hyperbolic paraboloid (saddle surface) shown in Figure 14.1.7a. The level curves have equations of the form $y^2 - x^2 = k$. For $k > 0$ these curves are hyperbolas opening along lines parallel to the y-axis; for $k < 0$ they are hyperbolas opening along lines parallel to the x-axis; and for $k = 0$ the level curve consists of the intersecting lines $y + x = 0$ and $y - x = 0$ (Figure 14.1.7b). ◄

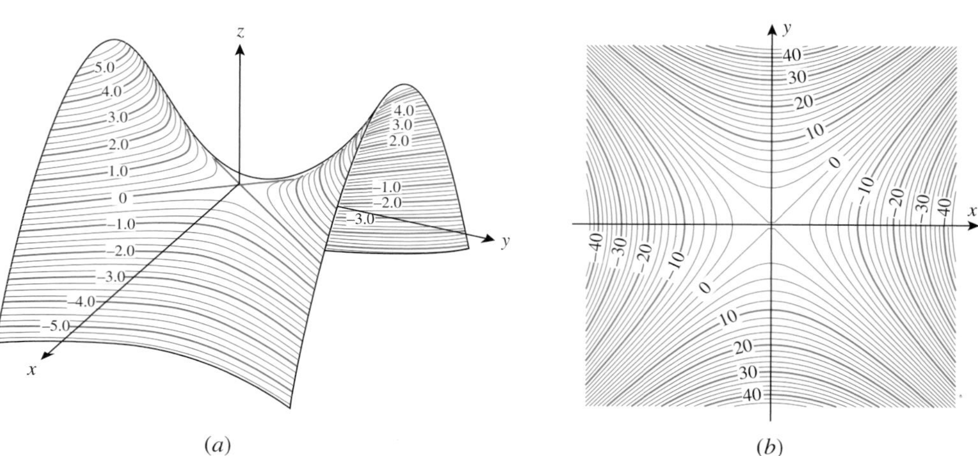

Figure 14.1.7

▶ **Example 6** Sketch the contour plot of $f(x, y) = 4x^2 + y^2$ using level curves of height $k = 0, 1, 2, 3, 4, 5$.

Solution. The graph of the surface $z = 4x^2 + y^2$ is the paraboloid shown in the left part of Figure 14.1.8, so we can reasonably expect the contour plot to be a family of ellipses centered at the origin. The level curve of height k has the equation $4x^2 + y^2 = k$. If $k = 0$, then the graph is the single point $(0, 0)$. For $k > 0$ we can rewrite the equation as

$$\frac{x^2}{k/4} + \frac{y^2}{k} = 1$$

which represents a family of ellipses with x-intercepts $\pm\sqrt{k}/2$ and y-intercepts $\pm\sqrt{k}$. The contour plot for the specified values of k is shown in the right part of Figure 14.1.8. ◀

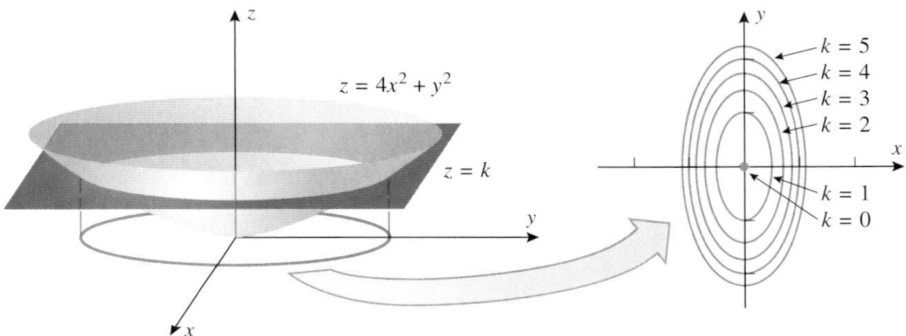

Figure 14.1.8

▶ **Example 7** Sketch the contour plot of $f(x, y) = 2 - x - y$ using level curves of height $k = -6, -4, -2, 0, 2, 4, 6$.

Solution. The graph of the surface $z = 2 - x - y$ is the plane shown in the left part of Figure 14.1.9, so we can reasonably expect the contour plot to be a family of parallel lines. The level curve of height k has the equation $2 - x - y = k$, which we can rewrite as

$$y = -x + (2 - k)$$

This represents a family of parallel lines of slope -1. The contour plot for the specified values of k is shown in the right part of Figure 14.1.9. ◀

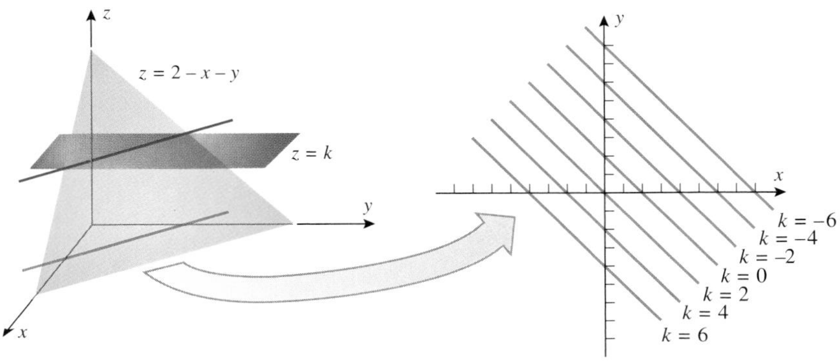

Figure 14.1.9

CONTOUR PLOTS USING TECHNOLOGY

Except in the simplest cases, contour plots can be difficult to produce without the help of a graphing utility. Figure 14.1.10 illustrates how graphing technology can be used to display level curves. The table shows two graphical representations of the level curves of the function $f(x, y) = |\sin x \sin y|$ produced with a CAS over the domain $0 \le x \le 2\pi$, $0 \le y \le 2\pi$.

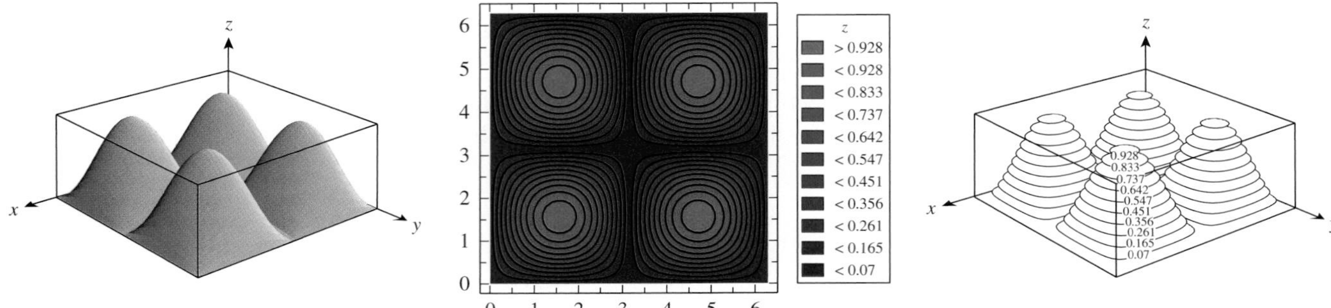

Figure 14.1.10

LEVEL SURFACES

The term "level surface" is standard but confusing, since a level surface need *not* be level in the sense of being horizontal—it is simply a surface on which all values of f are the same.

Observe that the graph of $y = f(x)$ is a curve in 2-space, and the graph of $z = f(x, y)$ is a surface in 3-space, so the number of dimensions required for these graphs is one greater than the number of independent variables. Accordingly, there is no "direct" way to graph a function of three variables since four dimensions are required. However, if k is a constant, then the graph of the equation $f(x, y, z) = k$ will generally be a surface in 3-space (e.g., the graph of $x^2 + y^2 + z^2 = 1$ is a sphere), which we call the ***level surface with constant k***. Some geometric insight into the behavior of the function f can sometimes be obtained by graphing these level surfaces for various values of k.

Example 8 Describe the level surfaces of

$$\text{(a)}\ f(x, y, z) = x^2 + y^2 + z^2 \qquad \text{(b)}\ f(x, y, z) = z^2 - x^2 - y^2$$

Solution (a). The level surfaces have equations of the form

$$x^2 + y^2 + z^2 = k$$

For $k > 0$ the graph of this equation is a sphere of radius $\sqrt{k}$, centered at the origin; for $k = 0$ the graph is the single point $(0, 0, 0)$; and for $k < 0$ there is no level surface (Figure 14.1.11).

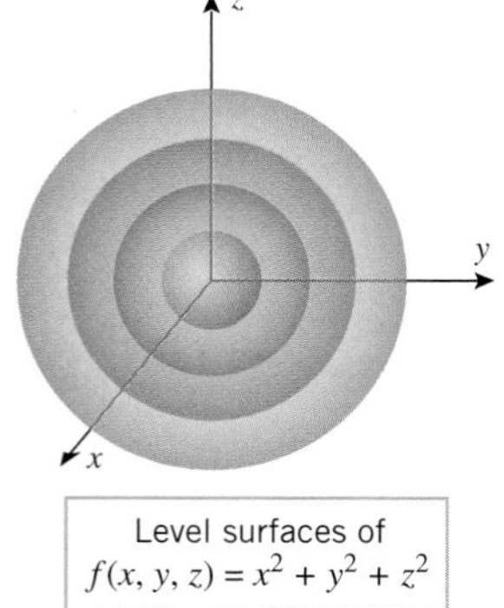

Level surfaces of $f(x, y, z) = x^2 + y^2 + z^2$

Figure 14.1.11

Solution (b). The level surfaces have equations of the form

$$z^2 - x^2 - y^2 = k$$

As discussed in Section 12.7, this equation represents a cone if $k = 0$, a hyperboloid of two sheets if $k > 0$, and a hyperboloid of one sheet if $k < 0$ (Figure 14.1.12). ◀

GRAPHING FUNCTIONS OF TWO VARIABLES USING TECHNOLOGY

Generating surfaces with a graphing utility is more complicated than generating plane curves because there are more factors that must be taken into account. We can only touch on the ideas here, so if you want to use a graphing utility, its documentation will be your main source of information.

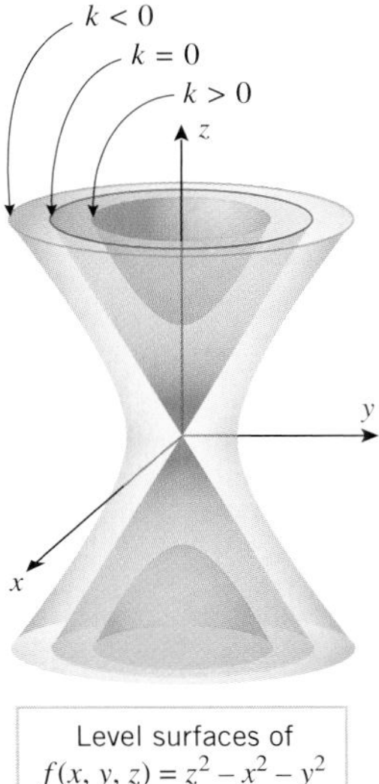

Figure 14.1.12

TECHNOLOGY MASTERY

If you have a graphing utility that can generate surfaces in 3-space, read the documentation and try to duplicate some of the surfaces in Figures 14.1.13 and 14.1.14 and Table 14.1.2.

Graphing utilities can only show a portion of xyz-space in a viewing screen, so the first step in graphing a surface is to determine which portion of xyz-space you want to display. This region is called the ***viewing box*** or ***viewing window***. For example, Figure 14.1.13 shows the effect of graphing the paraboloid $z = x^2 + y^2$ in three different viewing windows. However, within a fixed viewing box, the appearance of the surface is also affected by the ***viewpoint***, that is, the direction from which the surface is viewed, and the distance from the viewer to the surface. For example, Figure 14.1.14 shows the graph of the paraboloid $z = x^2 + y^2$ from three different viewpoints using the first viewing box in Figure 14.1.13.

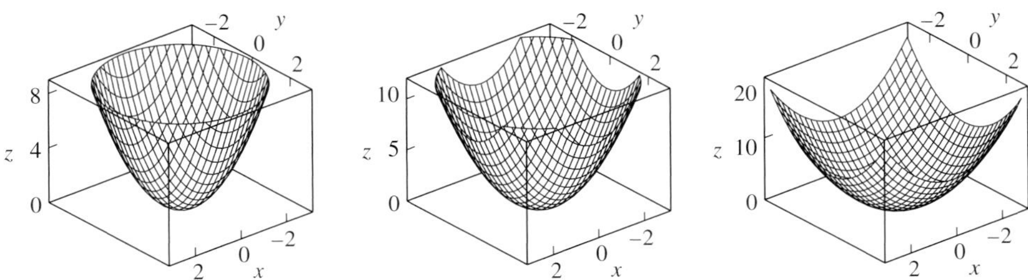

Figure 14.1.13 Varying the viewing box.

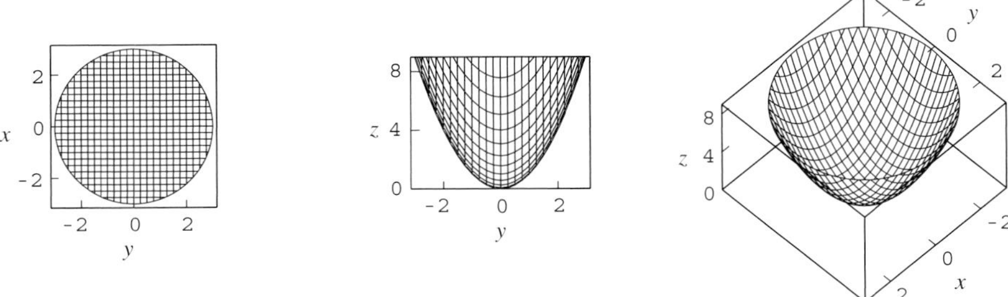

Figure 14.1.14 Viewing the viewpoint.

Table 14.1.2 shows six surfaces in 3-space along with their associated contour plots. Note that the mesh lines on the surface are traces in vertical planes, whereas the level curves correspond to traces in horizontal planes. In these contour plots the color gradation is from dark to light as z increases.

Table 14.1.2

SURFACE	CONTOUR PLOT	SURFACE	CONTOUR PLOT
$z = \cos y$ 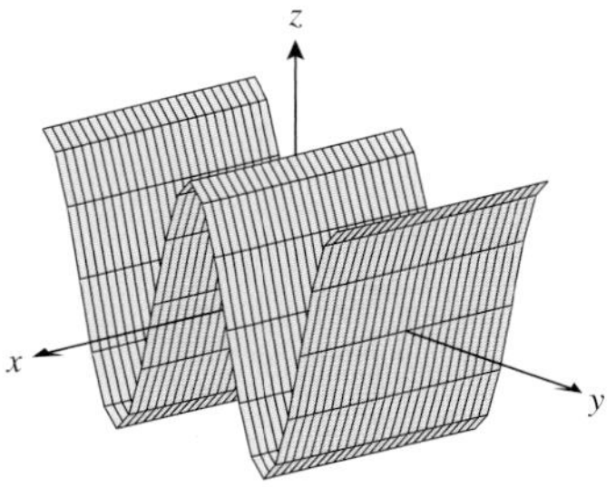	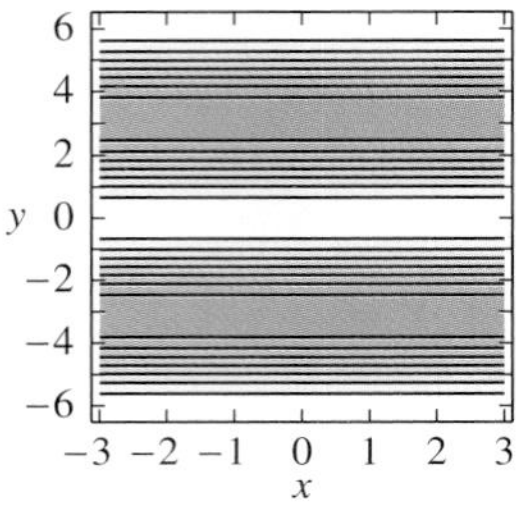	$z = e^x \sin y$ 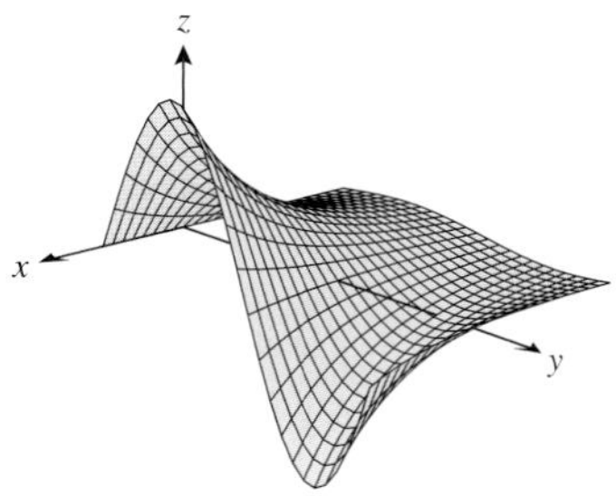	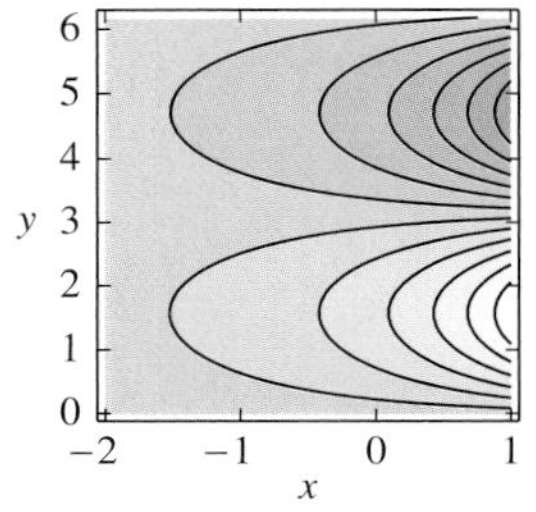
$z = \sin(\sqrt{x^2 + y^2})$ 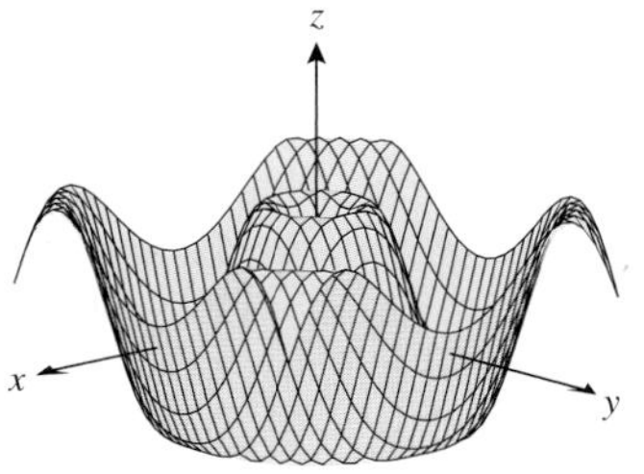	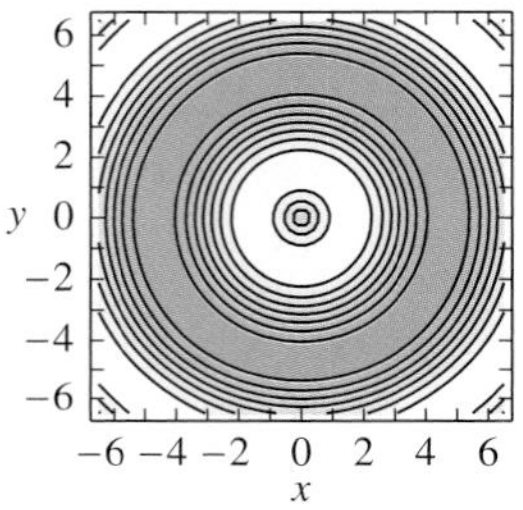	$z = xye^{-\frac{1}{2}(x^2 + y^2)}$ 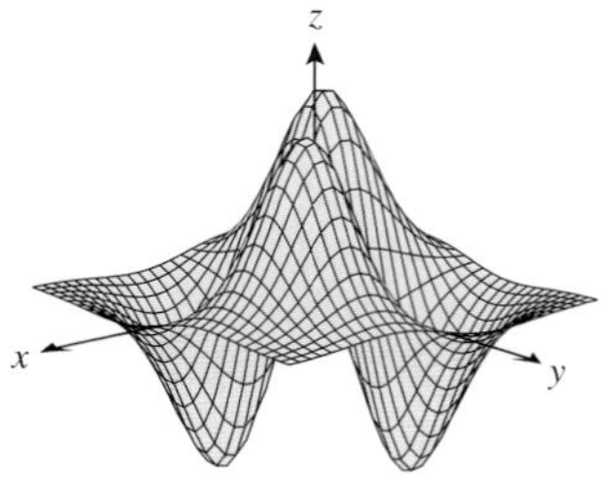	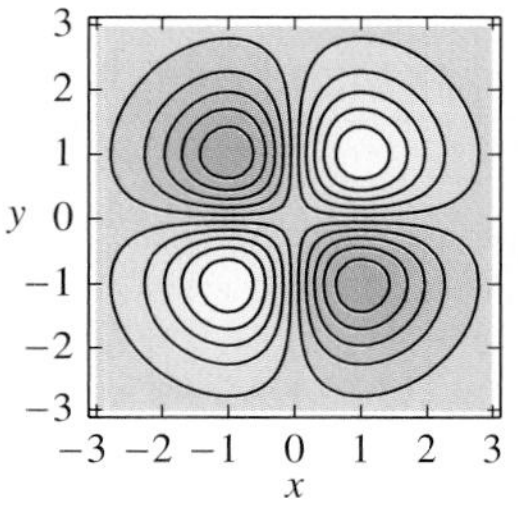
$z = \cos(xy)$ 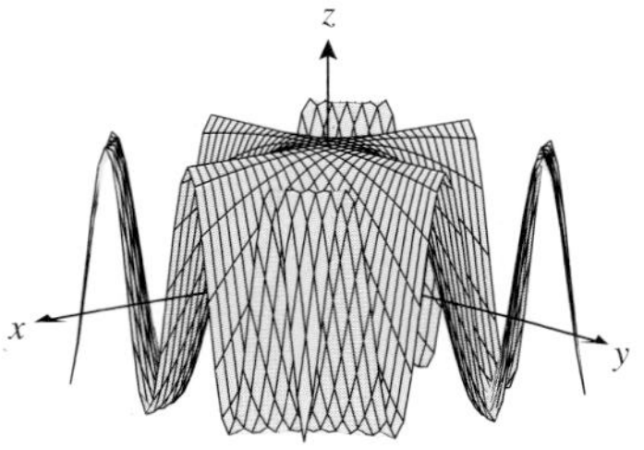	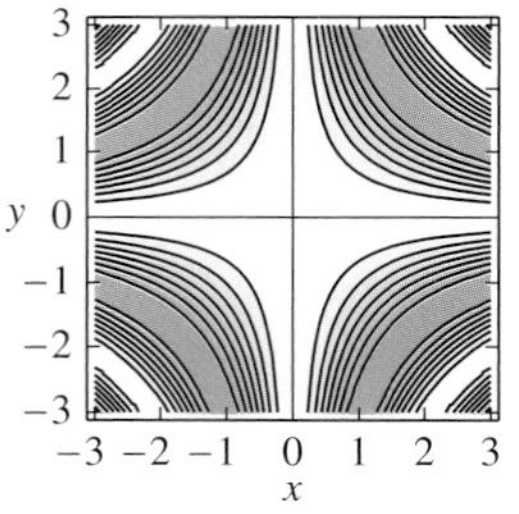	$z = \lvert xy \rvert$ 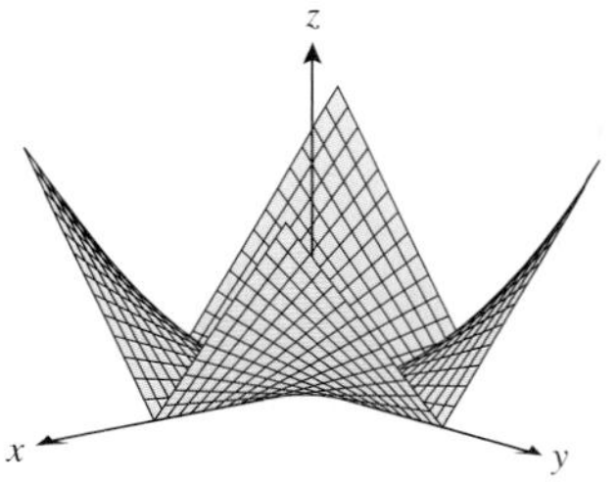	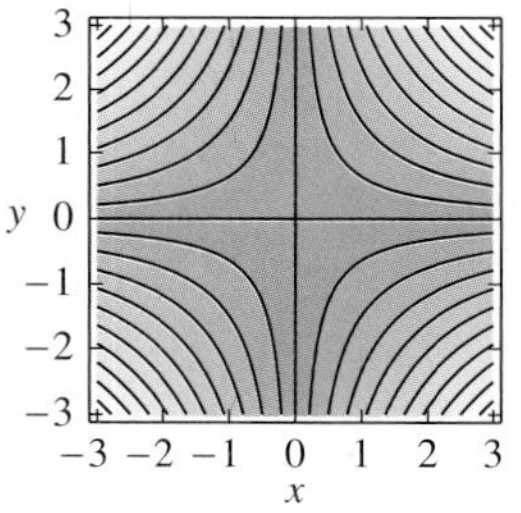

✔ QUICK CHECK EXERCISES 14.1

(See page 936 for answers.)

1. The domain of $f(x, y) = \ln xy$ is ________ and the domain of $g(x, y) = \ln x + \ln y$ is ________.

2. Let $f(x, y) = \dfrac{x - y}{x + y + 1}$.
   (a) $f(2, 1) =$ ________ (b) $f(1, 2) =$ ________
   (c) $f(a, a) =$ ________ (d) $f(y + 1, y) =$ ________

3. Let $f(x, y) = e^{x+y}$.
   (a) for what values of k does the level curve $f(x, y) = k$ contain at least one point?
   (b) Describe the level curves $f(x, y) = k$ for the values of k obtained in part (a).

4. Let $f(x, y, z) = \dfrac{1}{x^2 + y^2 + z^2 + 1}$.
   (a) Determine all values of k such that the level surface $f(x, y, z) = k$ contains at least one point.
   (b) Describe the level surfaces $f(x, y, z) = k$ for the values of k obtained in part (a).

EXERCISE SET 14.1 Graphing Utility C CAS

1–8 These exercises are concerned with functions of two variables.

1. Let $f(x, y) = x^2y + 1$. Find
(a) $f(2, 1)$ (b) $f(1, 2)$ (c) $f(0, 0)$
(d) $f(1, -3)$ (e) $f(3a, a)$ (f) $f(ab, a - b)$.

2. Let $f(x, y) = x + \sqrt[3]{xy}$. Find
(a) $f(t, t^2)$ (b) $f(x, x^2)$ (c) $f(2y^2, 4y)$.

3. Let $f(x, y) = xy + 3$. Find
(a) $f(x + y, x - y)$ (b) $f(xy, 3x^2y^3)$.

4. Let $g(x) = x \sin x$. Find
(a) $g(x/y)$ (b) $g(xy)$ (c) $g(x - y)$.

5. Find $F(g(x), h(y))$ if $F(x, y) = xe^{xy}$, $g(x) = x^3$, and $h(y) = 3y + 1$.

6. Find $g(u(x, y), v(x, y))$ if $g(x, y) = y \sin(x^2y)$, $u(x, y) = x^2y^3$, and $v(x, y) = \pi xy$.

7. Let $f(x, y) = x + 3x^2y^2$, $x(t) = t^2$, and $y(t) = t^3$. Find
(a) $f(x(t), y(t))$ (b) $f(x(0), y(0))$
(c) $f(x(2), y(2))$.

8. Let $g(x, y) = ye^{-3x}$, $x(t) = \ln(t^2 + 1)$, and $y(t) = \sqrt{t}$. Find $g(x(t), y(t))$.

9. Refer to Table 14.1.1 to estimate the wind chill index when
(a) the temperature is 25°F and the wind speed is 7 mi/h
(b) the temperature is 28°F and the wind speed is 5 mi/h.

10. Refer to Table 14.1.1 to estimate the wind chill index when
(a) the temperature is 35°F and the wind speed is 14 mi/h
(b) the temperature is 32°F and the wind speed is 15 mi/h.

11. One method for determining relative humidity is to wet the bulb of a thermometer, whirl it through the air, and then compare the thermometer reading with the actual air temperature. If the relative humidity is less than 100%, the reading on the thermometer will be less than the temperature of the air. This difference in temperature is known as the *wet-bulb depression*. The accompanying table gives the relative humidity as a function of the air temperature and the wet-bulb depression. Use the table to complete parts (a)–(c).
(a) What is the relative humidity if the air temperature is 20°C and the wet-bulb thermometer reads 16°C?
(b) Estimate the relative humidity if the air temperature is 25°C and the wet-bulb depression is 3.5°C.
(c) Estimate the relative humidity if the air temperature is 22°C and the wet-bulb depression is 5°C.

WET-BULB DEPRESSION (°C) \ AIR TEMPERATURE (°C)	15	20	25	30
3	71	74	77	79
4	62	66	70	73
5	53	59	63	67

Table Ex-11

12. Use the table in Exercise 11 to complete parts (a)–(c).
(a) What is the wet-bulb depression if the air temperature is 30°C and the relative humidity is 73%?
(b) Estimate the relative humidity if the air temperature is 15°C and the wet-bulb depression is 4.25°C.
(c) Estimate the relative humidity if the air temperature is 26°C and the wet-bulb depression is 3°C.

13–16 These exercises involve functions of three variables.

13. Let $f(x, y, z) = xy^2z^3 + 3$. Find
(a) $f(2, 1, 2)$ (b) $f(-3, 2, 1)$
(c) $f(0, 0, 0)$ (d) $f(a, a, a)$
(e) $f(t, t^2, -t)$ (f) $f(a + b, a - b, b)$.

14. Let $f(x, y, z) = zxy + x$. Find
(a) $f(x + y, x - y, x^2)$ (b) $f(xy, y/x, xz)$.

15. Find $F(f(x), g(y), h(z))$ if $F(x, y, z) = ye^{xyz}$, $f(x) = x^2$, $g(y) = y + 1$, and $h(z) = z^2$.

16. Find $g(u(x, y, z), v(x, y, z), w(x, y, z))$ if $g(x, y, z) = z \sin xy$, $u(x, y, z) = x^2z^3$, $v(x, y, z) = \pi xyz$, and $w(x, y, z) = xy/z$.

17–18 These exercises are concerned with functions of four or more variables.

17. (a) Let $f(x, y, z, t) = x^2y^3\sqrt{z + t}$. Find $f(\sqrt{5}, 2, \pi, 3\pi)$.
(b) Let $f(x_1, x_2, \ldots, x_n) = \sum_{k=1}^{n} kx_k$. Find $f(1, 1, \ldots, 1)$.

18. (a) Let $f(u, v, \lambda, \phi) = e^{u+v} \cos \lambda \tan \phi$. Find $f(-2, 2, 0, \pi/4)$.
(b) Let $f(x_1, x_2, \ldots, x_n) = x_1^2 + x_2^2 + \cdots + x_n^2$. Find $f(1, 2, \ldots, n)$.

19–22 Sketch the domain of f. Use solid lines for portions of the boundary included in the domain and dashed lines for portions not included.

19. $f(x, y) = \ln(1 - x^2 - y^2)$ **20.** $f(x, y) = \sqrt{x^2 + y^2 - 4}$

21. $f(x, y) = \dfrac{1}{x - y^2}$ **22.** $f(x, y) = \ln xy$

23–24 Describe the domain of f in words.

23. (a) $f(x, y) = xe^{-\sqrt{y+2}}$
(b) $f(x, y, z) = \sqrt{25 - x^2 - y^2 - z^2}$
(c) $f(x, y, z) = e^{xyz}$

24. (a) $f(x, y) = \dfrac{\sqrt{4 - x^2}}{y^2 + 3}$ (b) $f(x, y) = \ln(y - 2x)$
(c) $f(x, y, z) = \dfrac{xyz}{x + y + z}$

25–34 Sketch the graph of f.

25. $f(x, y) = 3$
26. $f(x, y) = \sqrt{9 - x^2 - y^2}$
27. $f(x, y) = \sqrt{x^2 + y^2}$
28. $f(x, y) = x^2 + y^2$
29. $f(x, y) = x^2 - y^2$
30. $f(x, y) = 4 - x^2 - y^2$
31. $f(x, y) = \sqrt{x^2 + y^2 + 1}$
32. $f(x, y) = \sqrt{x^2 + y^2 - 1}$
33. $f(x, y) = y + 1$
34. $f(x, y) = x^2$

FOCUS ON CONCEPTS

35. In each part, match the contour plot with one of the functions

$$f(x, y) = \sqrt{x^2 + y^2}, \quad f(x, y) = x^2 + y^2, \\ f(x, y) = 1 - x^2 - y^2$$

by inspection, and explain your reasoning. Larger values of z are indicated by lighter colors in the contour plot, and the concentric contours correspond to equally spaced values of z.

(a)
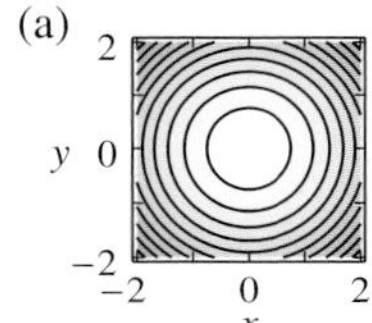

(b)
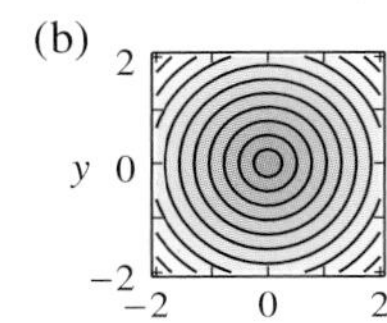

(c)
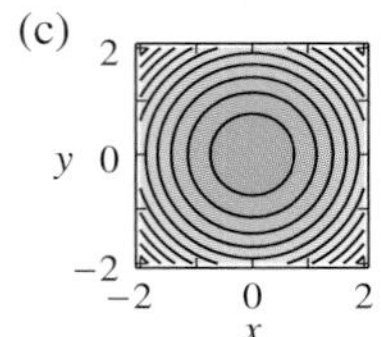

36. In each part, match the contour plot with one of the surfaces in the accompanying figure by inspection, and explain your reasoning. The larger the value of z, the lighter the color in the contour plot.

(a)
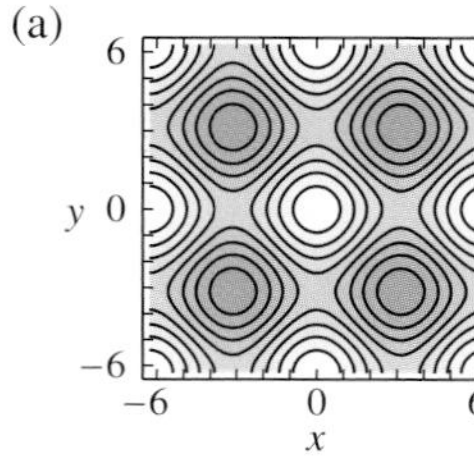

(b)
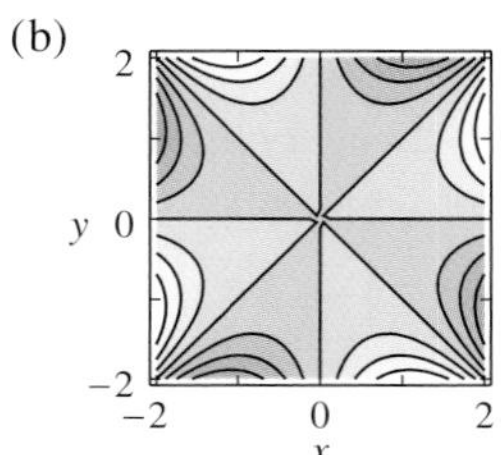

(c)
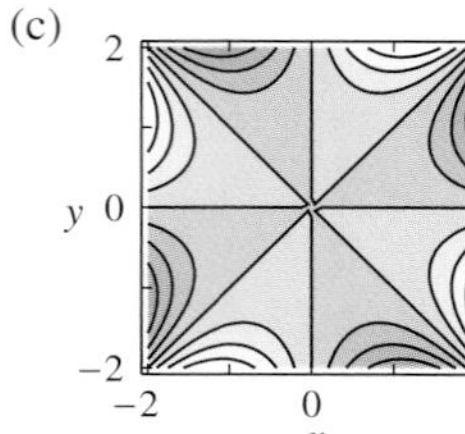

(d)
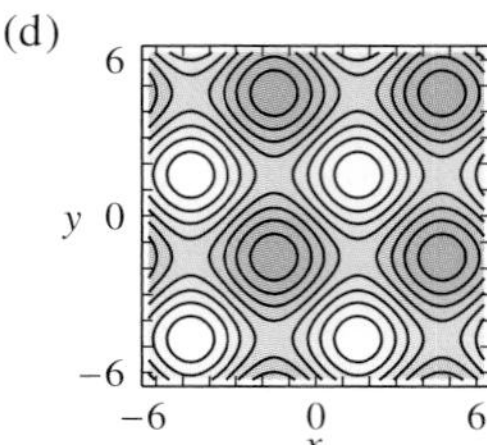

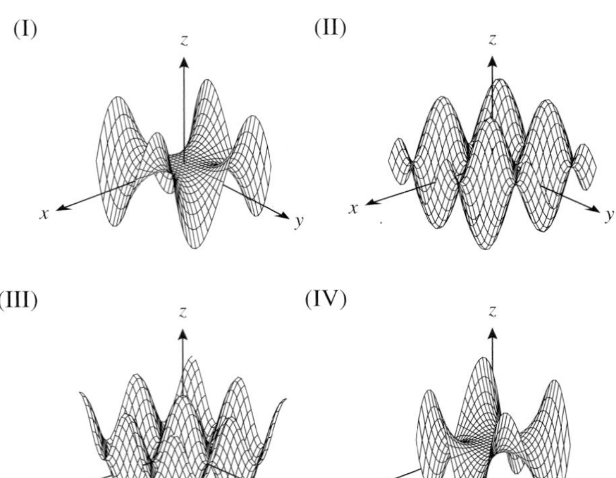

Figure Ex-36

37. In each part, the questions refer to the contour map in the accompanying figure.
(a) Is A or B the higher point? Explain your reasoning.
(b) Is the slope steeper at point A or at point B? Explain your reasoning.
(c) Starting at A and moving so that y remains constant and x increases, will the elevation begin to increase or decrease?
(d) Starting at B and moving so that y remains constant and x increases, will the elevation begin to increase or decrease?
(e) Starting at A and moving so that x remains constant and y decreases, will the elevation begin to increase or decrease?
(f) Starting at B and moving so that x remains constant and y decreases, will the elevation begin to increase or decrease?

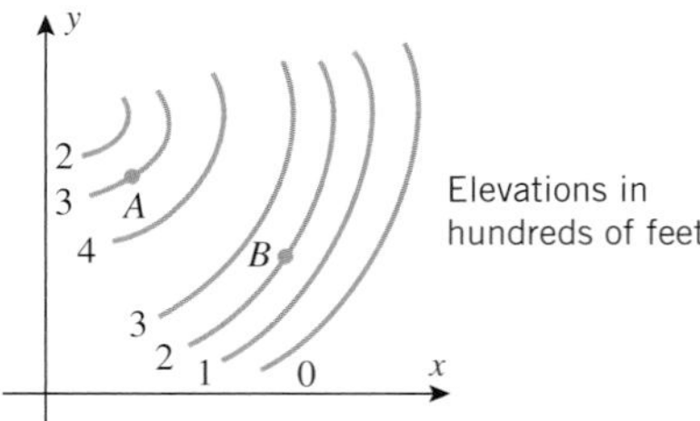

Figure Ex-37

38. A curve connecting points of equal atmospheric pressure on a weather map is called an ***isobar***. On a typical weather map the isobars refer to pressure at mean sea level and are given in units of ***millibars*** (mb). Mathematically, isobars are level curves for the pressure function $p(x, y)$ defined at the geographic points (x, y) represented on the map. Tightly packed isobars correspond to steep slopes on the graph of the pressure function, and these are usually associated with strong winds—the steeper the slope, the greater the speed of the wind.

(a) Referring to the accompanying weather map, is the wind speed greater in Medicine Hat, Alberta or in Chicago? Explain your reasoning.

(b) Estimate the average rate of change in atmospheric pressure (in mb/mi) from Medicine Hat to Chicago, given that the distance between the two cities is approximately 1400 mi.

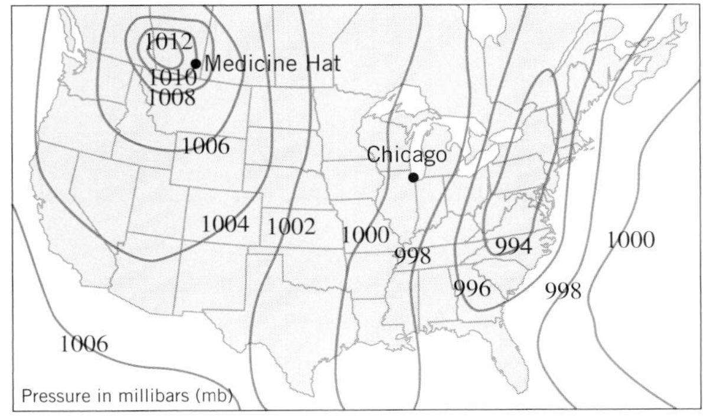

Figure Ex-38

39–44 Sketch the level curve $z = k$ for the specified values of k.

39. $z = x^2 + y^2$; $k = 0, 1, 2, 3, 4$

40. $z = y/x$; $k = -2, -1, 0, 1, 2$

41. $z = x^2 + y$; $k = -2, -1, 0, 1, 2$

42. $z = x^2 + 9y^2$; $k = 0, 1, 2, 3, 4$

43. $z = x^2 - y^2$; $k = -2, -1, 0, 1, 2$

44. $z = y \csc x$; $k = -2, -1, 0, 1, 2$

45–48 Sketch the level surface $f(x, y, z) = k$.

45. $f(x, y, z) = 4x^2 + y^2 + 4z^2$; $k = 16$

46. $f(x, y, z) = x^2 + y^2 - z^2$; $k = 0$

47. $f(x, y, z) = z - x^2 - y^2 + 4$; $k = 7$

48. $f(x, y, z) = 4x - 2y + z$; $k = 1$

49–52 Describe the level surfaces in words.

49. $f(x, y, z) = (x - 2)^2 + y^2 + z^2$

50. $f(x, y, z) = 3x - y + 2z$ **51.** $f(x, y, z) = x^2 + z^2$

52. $f(x, y, z) = z - x^2 - y^2$

53. Let $f(x, y) = x^2 - 2x^3 + 3xy$. Find an equation of the level curve that passes through the point
(a) $(-1, 1)$ (b) $(0, 0)$ (c) $(2, -1)$.

54. Let $f(x, y) = ye^x$. Find an equation of the level curve that passes through the point
(a) $(\ln 2, 1)$ (b) $(0, 3)$ (c) $(1, -2)$.

55. Let $f(x, y, z) = x^2 + y^2 - z$. Find an equation of the level surface that passes through the point
(a) $(1, -2, 0)$ (b) $(1, 0, 3)$ (c) $(0, 0, 0)$.

56. Let $f(x, y, z) = xyz + 3$. Find an equation of the level surface that passes through the point
(a) $(1, 0, 2)$ (b) $(-2, 4, 1)$ (c) $(0, 0, 0)$.

57. If $T(x, y)$ is the temperature at a point (x, y) on a thin metal plate in the xy-plane, then the level curves of T are called ***isothermal curves***. All points on such a curve are at the same temperature. Suppose that a plate occupies the first quadrant and $T(x, y) = xy$.

(a) Sketch the isothermal curves on which $T = 1$, $T = 2$, and $T = 3$.

(b) An ant, initially at $(1, 4)$, wants to walk on the plate so that the temperature along its path remains constant. What path should the ant take and what is the temperature along that path?

58. If $V(x, y)$ is the voltage or potential at a point (x, y) in the xy-plane, then the level curves of V are called ***equipotential curves***. Along such a curve, the voltage remains constant. Given that

$$V(x, y) = \frac{8}{\sqrt{16 + x^2 + y^2}}$$

sketch the equipotential curves at which $V = 2.0$, $V = 1.0$, and $V = 0.5$.

59. Let $f(x, y) = x^2 + y^3$.

(a) Use a graphing utility to generate the level curve that passes through the point $(2, -1)$.

(b) Generate the level curve of height 1.

60. Let $f(x, y) = 2\sqrt{xy}$.

(a) Use a graphing utility to generate the level curve that passes through the point $(2, 2)$.

(b) Generate the level curve of height 8.

C **61.** Let $f(x, y) = xe^{-(x^2+y^2)}$.

(a) Use a CAS to generate the graph of f for $-2 \le x \le 2$ and $-2 \le y \le 2$.

(b) Generate a contour plot for the surface, and confirm visually that it is consistent with the surface obtained in part (a).

(c) Read the appropriate documentation and explore the effect of generating the graph of f from various viewpoints.

C **62.** Let $f(x, y) = \frac{1}{10}e^x \sin y$.

(a) Use a CAS to generate the graph of f for $0 \le x \le 4$ and $0 \le y \le 2\pi$.

(b) Generate a contour plot for the surface, and confirm visually that it is consistent with the surface obtained in part (a).

(c) Read the appropriate documentation and explore the effect of generating the graph of f from various viewpoints.

63. In each part, describe in words how the graph of g is related to the graph of f.
(a) $g(x, y) = f(x - 1, y)$ (b) $g(x, y) = 1 + f(x, y)$
(c) $g(x, y) = -f(x, y + 1)$

64. (a) Sketch the graph of $f(x, y) = e^{-(x^2+y^2)}$.
(b) Describe in words how the graph of the function $g(x, y) = e^{-a(x^2+y^2)}$ is related to the graph of f for positive values of a.

✔ QUICK CHECK ANSWERS 14.1

1. points (x, y) in the first or third quadrants; points (x, y) in the first quadrant **2.** (a) $\frac{1}{4}$ (b) $-\frac{1}{4}$ (c) 0 (d) $1/(2y + 2)$ **3.** (a) $k > 0$ (b) the lines $x + y = \ln k$ **4.** (a) $0 < k \leq 1$ (b) spheres of radius $\sqrt{(1 - k)/k}$ for $0 < k < 1$, the single point $(0, 0, 0)$ for $k = 1$

14.2 LIMITS AND CONTINUITY

In this section we will introduce the notions of limit and continuity for functions of two or more variables. We will not go into great detail—our objective is to develop the basic concepts accurately and to obtain results needed in later sections. A more extensive study of these topics is usually given in advanced calculus.

LIMITS ALONG CURVES

For a function of one variable there are two one-sided limits at a point x_0, namely,

$$\lim_{x \to x_0^+} f(x) \quad \text{and} \quad \lim_{x \to x_0^-} f(x)$$

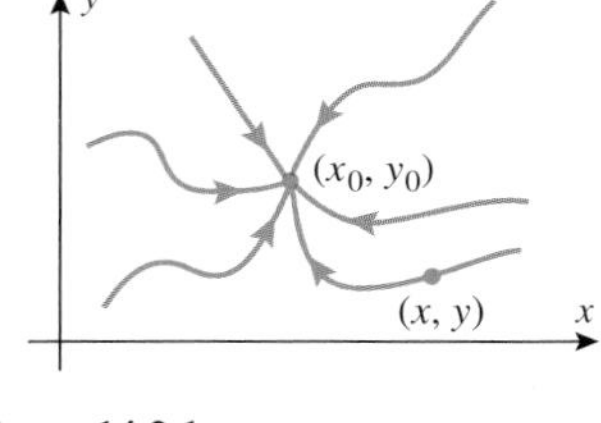

Figure 14.2.1

reflecting the fact that there are only two directions from which x can approach x_0, the right or the left. For functions of two or three variables the situation is more complicated because there are infinitely many different curves along which one point can approach another (Figure 14.2.1). Our first objective in this section is to define the limit of $f(x, y)$ as (x, y) approaches a point (x_0, y_0) along a curve C (and similarly for functions of three variables).

If C is a smooth parametric curve in 2-space or 3-space that is represented by the equations

$$x = x(t), \quad y = y(t) \qquad \text{or} \qquad x = x(t), \quad y = y(t), \quad z = z(t)$$

and if $x_0 = x(t_0)$, $y_0 = y(t_0)$, and $z_0 = z(t_0)$, then the limits

$$\lim_{\substack{(x,y) \to (x_0,y_0) \\ (\text{along } C)}} f(x, y) \quad \text{and} \quad \lim_{\substack{(x,y,z) \to (x_0,y_0,z_0) \\ (\text{along } C)}} f(x, y, z)$$

are defined by

$$\lim_{\substack{(x,y) \to (x_0,y_0) \\ (\text{along } C)}} f(x, y) = \lim_{t \to t_0} f(x(t), y(t)) \tag{1}$$

$$\lim_{\substack{(x,y,z) \to (x_0,y_0,z_0) \\ (\text{along } C)}} f(x, y, z) = \lim_{t \to t_0} f(x(t), y(t), z(t)) \tag{2}$$

In words, Formulas (1) and (2) state that a limit of a function f along a parametric curve can be obtained by substituting the parametric equations for the curve into the formula for the function and then computing the limit of the resulting function of one variable at the appropriate point.

In these formulas the limit of the function of t must be treated as a one-sided limit if (x_0, y_0) or (x_0, y_0, z_0) is an endpoint of C.

A geometric interpretation of the limit along a curve for a function of two variables is shown in Figure 14.2.2: As the point $(x(t), y(t))$ moves along the curve C in the xy-plane toward (x_0, y_0), the point $(x(t), y(t), f(x(t), y(t)))$ moves directly above it along the graph of $z = f(x, y)$ with $f(x(t), y(t))$ approaching the limiting value L. In the figure we followed a common practice of omitting the zero z-coordinate for points in the xy-plane.

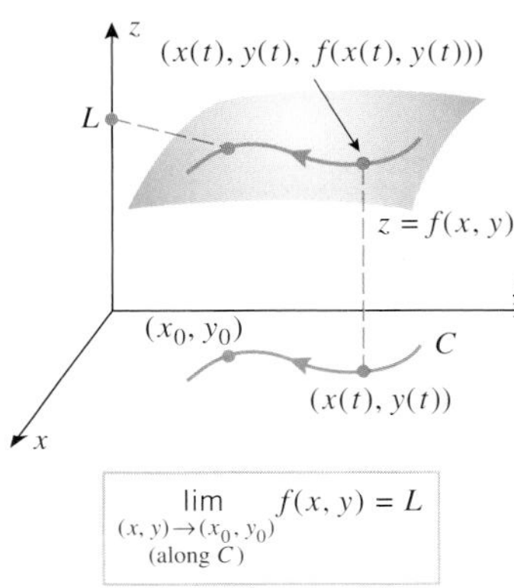

Figure 14.2.2

▶ Example 1 Figure 14.2.3*a* shows a computer-generated graph of the function

$$f(x, y) = -\frac{xy}{x^2 + y^2}$$

The graph reveals that the surface has a ridge above the line $y = -x$, which is to be expected since $f(x, y)$ has a constant value of $\frac{1}{2}$ for $y = -x$, except at $(0, 0)$ where f is undefined (verify). Moreover, the graph suggests that the limit of $f(x, y)$ as $(x, y) \to (0, 0)$ along a line through the origin varies with the direction of the line. Find this limit along

(a) the x-axis (b) the y-axis (c) the line $y = x$
(d) the line $y = -x$ (e) the parabola $y = x^2$

Solution (a). The x-axis has parametric equations $x = t$, $y = 0$, with $(0, 0)$ corresponding to $t = 0$, so

$$\lim_{\substack{(x,y)\to(0,0)\\ (\text{along } y=0)}} f(x, y) = \lim_{t\to 0} f(t, 0) = \lim_{t\to 0}\left(-\frac{0}{t^2}\right) = \lim_{t\to 0} 0 = 0$$

which is consistent with Figure 14.2.3*b*.

Solution (b). The y-axis has parametric equations $x = 0$, $y = t$, with $(0, 0)$ corresponding to $t = 0$, so

$$\lim_{\substack{(x,y)\to(0,0)\\ (\text{along } x=0)}} f(x, y) = \lim_{t\to 0} f(0, t) = \lim_{t\to 0}\left(-\frac{0}{t^2}\right) = \lim_{t\to 0} 0 = 0$$

which is consistent with Figure 14.2.3*b*.

Solution (c). The line $y = x$ has parametric equations $x = t$, $y = t$, with $(0, 0)$ corresponding to $t = 0$, so

$$\lim_{\substack{(x,y)\to(0,0)\\ (\text{along } y=x)}} f(x, y) = \lim_{t\to 0} f(t, t) = \lim_{t\to 0}\left(-\frac{t^2}{2t^2}\right) = \lim_{t\to 0}\left(-\frac{1}{2}\right) = -\frac{1}{2}$$

which is consistent with Figure 14.2.3*b*.

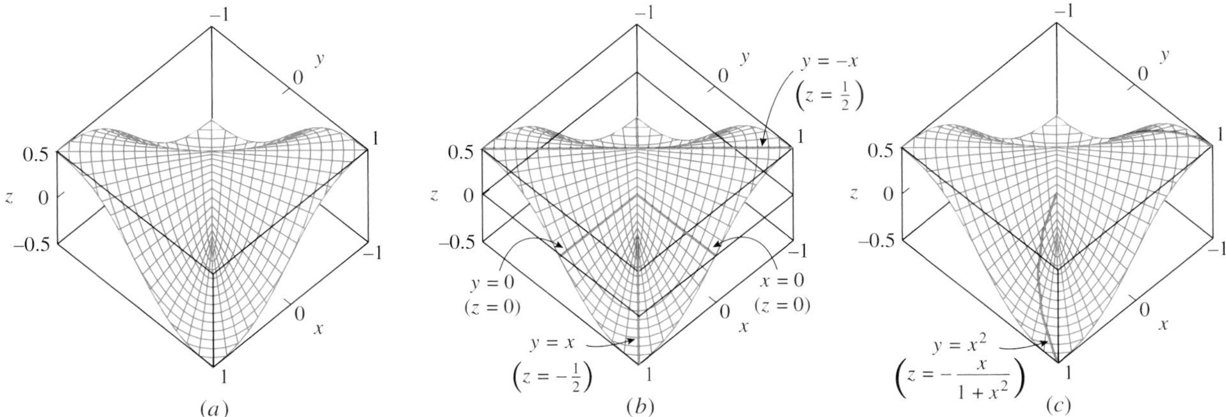

Figure 14.2.3

Solution (d). The line $y = -x$ has parametric equations $x = t$, $y = -t$, with $(0, 0)$ corresponding to $t = 0$, so

$$\lim_{\substack{(x,y)\to(0,0)\\ (\text{along } y=-x)}} f(x, y) = \lim_{t\to 0} f(t, -t) = \lim_{t\to 0} \frac{t^2}{2t^2} = \lim_{t\to 0} \frac{1}{2} = \frac{1}{2}$$

which is consistent with Figure 14.2.3*b*.

Solution (e). The parabola $y = x^2$ has parametric equations $x = t$, $y = t^2$, with $(0, 0)$ corresponding to $t = 0$, so

$$\lim_{\substack{(x,y)\to(0,0)\\ (\text{along } y=x^2)}} f(x, y) = \lim_{t\to 0} f(t, t^2) = \lim_{t\to 0} \left(-\frac{t^3}{t^2 + t^4}\right) = \lim_{t\to 0} \left(-\frac{t}{1 + t^2}\right) = 0$$

This is consistent with Figure 14.2.3*c*, which shows the parametric curve

$$x = t, \quad y = t^2, \quad z = -\frac{t}{1 + t^2}$$

superimposed on the surface. ◀

OPEN AND CLOSED SETS

Although limits along specific curves are useful for many purposes, they do not always tell the complete story about the limiting behavior of a function at a point; what is required is a limit concept that accounts for the behavior of the function in an *entire vicinity* of a point, not just along smooth curves passing through the point. For this purpose, we start by introducing some terminology.

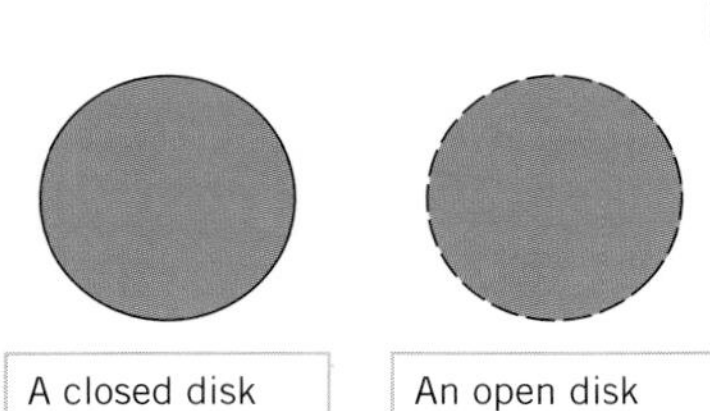

A closed disk includes all of the points on its bounding circle.
An open disk contains none of the points on its bounding circle.

Figure 14.2.4

Let C be a circle in 2-space that is centered at (x_0, y_0) and has positive radius δ. The set of points that are enclosed by the circle, but do not lie on the circle, is called the ***open disk*** of radius δ centered at (x_0, y_0), and the set of points that lie on the circle together with those enclosed by the circle is called the ***closed disk*** of radius δ centered at (x_0, y_0) (Figure 14.2.4). Analogously, if S is a sphere in 3-space that is centered at (x_0, y_0, z_0) and has positive radius δ, then the set of points that are enclosed by the sphere, but do not lie on the sphere, is called the ***open ball*** of radius δ centered at (x_0, y_0, z_0), and the set of points that lie on the sphere together with those enclosed by the sphere is called the ***closed ball*** of radius δ centered at (x_0, y_0, z_0). Disks and balls are the two-dimensional and three-dimensional analogs of intervals on a line.

The notions of "open" and "closed" can be extended to more general sets in 2-space and 3-space. If D is a set of points in 2-space, then a point (x_0, y_0) is called an ***interior point*** of D if there is *some* open disk centered at (x_0, y_0) that contains only points of D, and (x_0, y_0) is called a ***boundary point*** of D if *every* open disk centered at (x_0, y_0) contains both points in D and points not in D. The same terminology applies to sets in 3-space, but in that case the definitions use balls rather than disks (Figure 14.2.5).

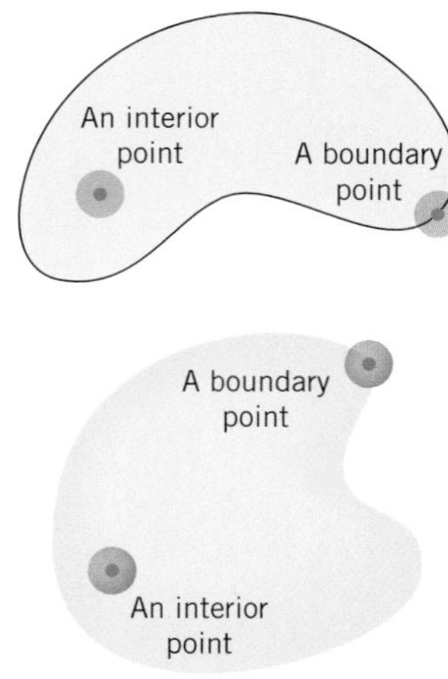

Figure 14.2.5

For a set D in either 2-space or 3-space, the set of all interior points is called the ***interior*** of D and the set of all boundary points is called the ***boundary*** of D. Moreover, just as for disks, we say that D is ***closed*** if it contains all of its boundary points and ***open*** if it contains *none* of its boundary points. The set of all points in 2-space and the set of all points in 3-space have no boundary points (why?), so by agreement they are regarded to be both open and closed.

GENERAL LIMITS OF FUNCTIONS OF TWO VARIABLES

The statement

$$\lim_{(x,y)\to(x_0,y_0)} f(x, y) = L$$

is intended to convey the idea that the value of $f(x, y)$ can be made as close as we like to the number L by restricting the point (x, y) to be sufficiently close to (but different from) the point (x_0, y_0). This idea has a formal expression in the following definition and is illustrated in Figure 14.2.6.

14.2.1 DEFINITION. Let f be a function of two variables, and assume that f is defined at all points of some open disk centered at (x_0, y_0), except possibly at (x_0, y_0). We will write

$$\lim_{(x,y)\to(x_0,y_0)} f(x, y) = L \tag{3}$$

if given any number $\epsilon > 0$, we can find a number $\delta > 0$ such that $f(x, y)$ satisfies

$$|f(x, y) - L| < \epsilon$$

whenever the distance between (x, y) and (x_0, y_0) satisfies

$$0 < \sqrt{(x - x_0)^2 + (y - y_0)^2} < \delta$$

When convenient, (3) can also be written as

$$\lim_{\substack{x\to x_0 \\ y\to y_0}} f(x, y) = L$$

or as

$$f(x, y) \to L \quad \text{as} \quad (x, y) \to (x_0, y_0)$$

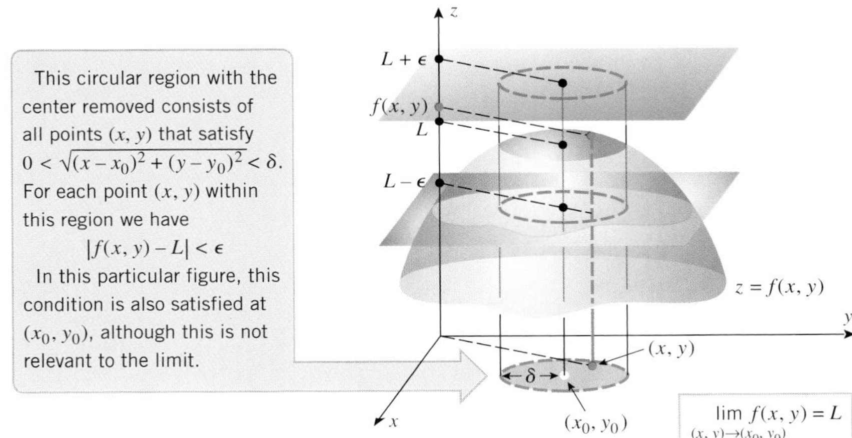
This circular region with the center removed consists of all points (x, y) that satisfy $0 < \sqrt{(x - x_0)^2 + (y - y_0)^2} < \delta$. For each point (x, y) within this region we have $|f(x, y) - L| < \epsilon$
In this particular figure, this condition is also satisfied at (x_0, y_0), although this is not relevant to the limit.
z
$L + \epsilon$
$f(x, y)$
L
$L - \epsilon$
$z = f(x, y)$
y
(x, y)
δ
x
(x_0, y_0)
$\lim_{(x,y)\to(x_0,y_0)} f(x, y) = L$

Figure 14.2.6

Another illustration of Definition 14.2.1 is shown in the "arrow diagram" of Figure 14.2.7. As in Figure 14.2.6, this figure is intended to convey the idea that the values of $f(x, y)$ can be forced within ϵ units of L on the z-axis by restricting (x, y) to lie within δ units of (x_0, y_0) in the xy-plane. We used a white dot at (x_0, y_0) to suggest that the epsilon condition need not hold at this point.

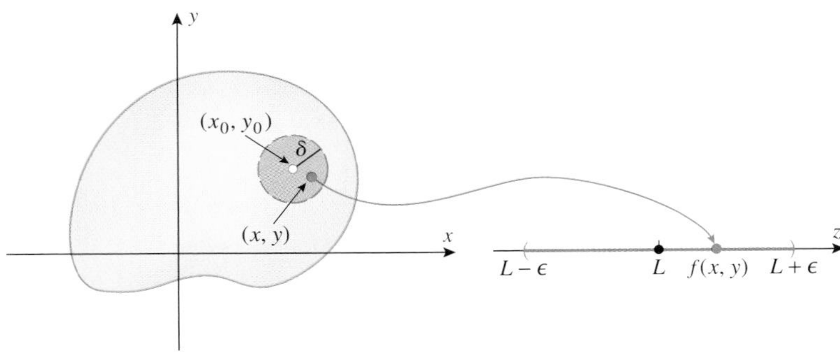
y
(x_0, y_0)
δ
(x, y)
x
z
$L - \epsilon$
L
$f(x, y)$
$L + \epsilon$

Figure 14.2.7

We note without proof that the standard properties of limits hold for limits along curves and for general limits of functions of two variables, so that computations involving such limits can be performed in the usual way.

▶ **Example 2**

$$\lim_{(x,y)\to(1,4)} [5x^3y^2 - 9] = \lim_{(x,y)\to(1,4)} [5x^3y^2] - \lim_{(x,y)\to(1,4)} 9$$

$$= 5\left[\lim_{(x,y)\to(1,4)} x\right]^3\left[\lim_{(x,y)\to(1,4)} y\right]^2 - 9$$

$$= 5(1)^3(4)^2 - 9 = 71 \blacktriangleleft$$

RELATIONSHIPS BETWEEN GENERAL LIMITS AND LIMITS ALONG SMOOTH CURVES

Stated informally, if $f(x, y)$ has limit L as (x, y) approaches (x_0, y_0), then the value of $f(x, y)$ gets closer and closer to L as the distance between (x, y) and (x_0, y_0) approaches zero. Since this statement imposes no restrictions on the direction in which (x, y) approaches (x_0, y_0), it is plausible that the function $f(x, y)$ will also have the limit L as (x, y) approaches (x_0, y_0) along *any* smooth curve C. This is the implication of the following theorem, which we state without proof.

WARNING

In general, one cannot show that

$$\lim_{(x,y)\to(x_0,y_0)} f(x, y) = L$$

by showing that this limit holds along a specific curve, or even some specific family of curves. The problem is there may be some other curve along which the limit does not exist or has a value different from L (see Exercise 26, for example).

14.2.2 THEOREM.

(*a*) *If $f(x, y) \to L$ as $(x, y) \to (x_0, y_0)$, then $f(x, y) \to L$ as $(x, y) \to (x_0, y_0)$ along any smooth curve.*

(*b*) *If the limit of $f(x, y)$ fails to exist as $(x, y) \to (x_0, y_0)$ along some smooth curve, or if $f(x, y)$ has different limits as $(x, y) \to (x_0, y_0)$ along two different smooth curves, then the limit of $f(x, y)$ does not exist as $(x, y) \to (x_0, y_0)$.*

▶ **Example 3** The limit

$$\lim_{(x,y)\to(0,0)} \frac{xy}{x^2 + y^2}$$

does not exist because in Example 1 we found two different smooth curves along which this limit had different values. Specifically,

$$\lim_{\substack{(x,y)\to(0,0)\\(\text{along } x=0)}} \frac{xy}{x^2 + y^2} = 0 \quad \text{and} \quad \lim_{\substack{(x,y)\to(0,0)\\(\text{along } y=x)}} \frac{xy}{x^2 + y^2} = -\frac{1}{2} \blacktriangleleft$$

CONTINUITY

Stated informally, a function of one variable is continuous if its graph is an unbroken curve without jumps or holes. To extend this idea to functions of two variables, imagine that the graph of $z = f(x, y)$ is molded from a thin sheet of clay that has been hollowed or pinched into peaks and valleys. We will regard f as being continuous if the clay surface has no tears or holes. The functions graphed in Figure 14.2.8 fail to be continuous because of their behavior at $(0, 0)$.

The precise definition of continuity at a point for functions of two variables is similar to that for functions of one variable—we require the limit of the function and the value of the function to be the same at the point.

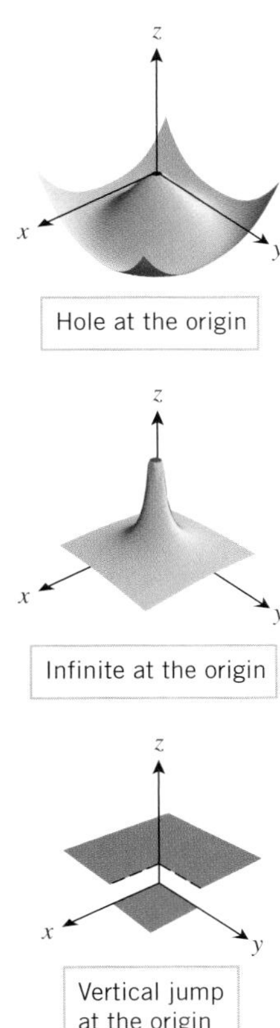

Figure 14.2.8

14.2.3 DEFINITION. A function $f(x, y)$ is said to be ***continuous at*** (x_0, y_0) if $f(x_0, y_0)$ is defined and if

$$\lim_{(x,y)\to(x_0,y_0)} f(x, y) = f(x_0, y_0)$$

In addition, if f is continuous at every point in an open set D, then we say that f is ***continuous on D***, and if f is continuous at every point in the xy-plane, then we say that f is ***continuous everywhere***.

The following theorem, which we state without proof, illustrates some of the ways in which continuous functions can be combined to produce new continuous functions.

14.2.4 THEOREM.

(*a*) *If $g(x)$ is continuous at x_0 and $h(y)$ is continuous at y_0, then $f(x, y) = g(x)h(y)$ is continuous at (x_0, y_0).*

(*b*) *If $h(x, y)$ is continuous at (x_0, y_0) and $g(u)$ is continuous at $u = h(x_0, y_0)$, then the composition $f(x, y) = g(h(x, y))$ is continuous at (x_0, y_0).*

(*c*) *If $f(x, y)$ is continuous at (x_0, y_0), and if $x(t)$ and $y(t)$ are continuous at t_0 with $x(t_0) = x_0$ and $y(t_0) = y_0$, then the composition $f(x(t), y(t))$ is continuous at t_0.*

▶ **Example 4** Use Theorem 14.2.4 to show that the functions $f(x, y) = 3x^2y^5$ and $f(x, y) = \sin(3x^2y^5)$ are continuous everywhere.

Solution. The polynomials $g(x) = 3x^2$ and $h(y) = y^5$ are continuous at every real number, and therefore by part (*a*) of Theorem 14.2.4, the function $f(x, y) = 3x^2y^5$ is continuous at every point (x, y) in the xy-plane. Since $3x^2y^5$ is continuous at every point in the xy-plane and $\sin u$ is continuous at every real number u, it follows from part (*b*) of Theorem 14.2.4 that the composition $f(x, y) = \sin(3x^2y^5)$ is continuous everywhere. ◀

Theorem 14.2.4 is one of a whole class of theorems about continuity of functions in two or more variables. The content of these theorems can be summarized informally with three basic principles:

Recognizing Continuous Functions

- A composition of continuous functions is continuous.
- A sum, difference, or product of continuous functions is continuous.
- A quotient of continuous functions is continuous, except where the denominator is zero.

By using these principles and Theorem 14.2.4, you should be able to confirm that the following functions are all continuous everywhere:

$$xe^{xy} + y^{2/3}, \quad \cosh(xy^3) - |xy|, \quad \frac{xy}{1 + x^2 + y^2}$$

▶ **Example 5** Evaluate $\displaystyle\lim_{(x,y)\to(-1,2)} \frac{xy}{x^2+y^2}$.

Solution. Since $f(x, y) = xy/(x^2 + y^2)$ is continuous at $(-1, 2)$ (why?), it follows from the definition of continuity for functions of two variables that

$$\lim_{(x,y)\to(-1,2)} \frac{xy}{x^2+y^2} = \frac{(-1)(2)}{(-1)^2+(2)^2} = -\frac{2}{5} \blacktriangleleft$$

▶ **Example 6** Since the function

$$f(x, y) = \frac{x^3y^2}{1-xy}$$

is a quotient of continuous functions, it is continuous except where $1 - xy = 0$. Thus, $f(x, y)$ is continuous everywhere except on the hyperbola $xy = 1$. ◀

LIMITS AT DISCONTINUITIES

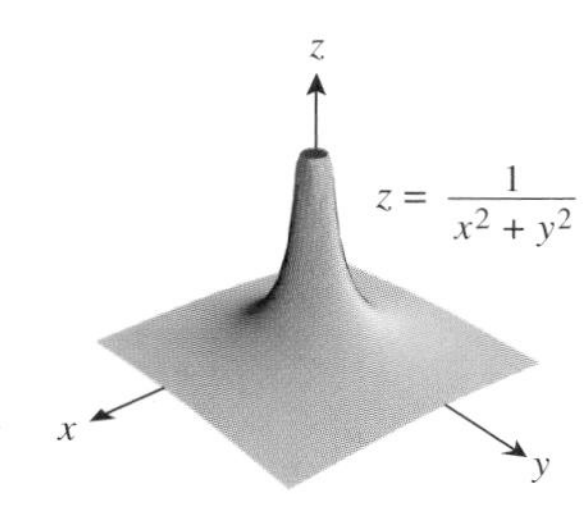

Figure 14.2.9

Sometimes it is easy to recognize when a limit does not exist. For example, it is evident that

$$\lim_{(x,y)\to(0,0)} \frac{1}{x^2+y^2} = +\infty$$

which implies that the values of the function approach $+\infty$ as $(x, y)\to(0, 0)$ along any smooth curve (Figure 14.2.9). However, it is not evident whether the limit

$$\lim_{(x,y)\to(0,0)} (x^2+y^2)\ln(x^2+y^2)$$

exists because it is an indeterminate form of type $0 \cdot \infty$. Although L'Hôpital's rule cannot be applied directly, the following example illustrates a method for finding this limit by converting to polar coordinates.

▶ **Example 7** Find $\displaystyle\lim_{(x,y)\to(0,0)} (x^2+y^2)\ln(x^2+y^2)$.

Solution. Let (r, θ) be polar coordinates of the point (x, y) with $r \ge 0$. Then we have

$$x = r\cos\theta, \quad y = r\sin\theta, \quad r^2 = x^2 + y^2$$

Moreover, since $r \ge 0$ we have $r = \sqrt{x^2+y^2}$, so that $r \to 0^+$ if and only if $(x, y) \to (0, 0)$. Thus, we can rewrite the given limit as

$$\begin{aligned}
\lim_{(x,y)\to(0,0)} (x^2+y^2)\ln(x^2+y^2) &= \lim_{r\to0^+} r^2 \ln r^2 \\
&= \lim_{r\to0^+} \frac{2\ln r}{1/r^2} && \text{This converts the limit to an indeterminate form of type } \infty/\infty. \\
&= \lim_{r\to0^+} \frac{2/r}{-2/r^3} && \text{L'Hôpital's rule} \\
&= \lim_{r\to0^+} (-r^2) = 0 \blacktriangleleft
\end{aligned}$$

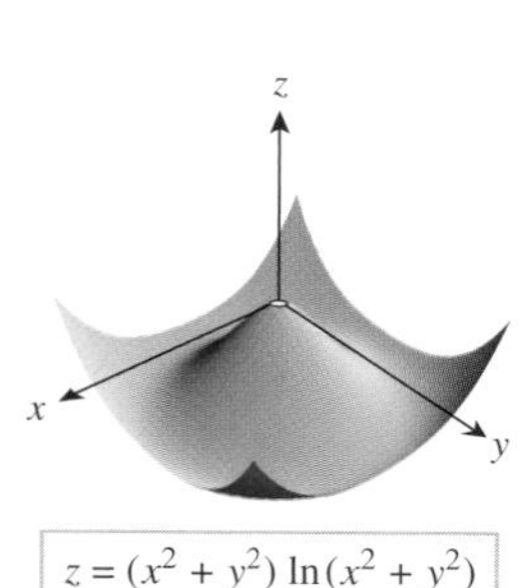

Figure 14.2.10

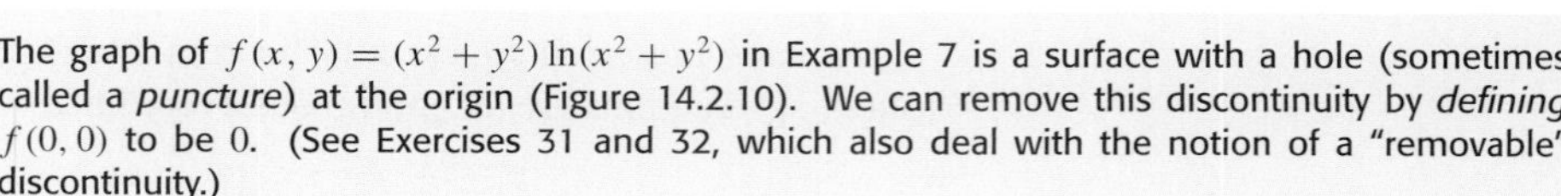

The graph of $f(x, y) = (x^2 + y^2)\ln(x^2 + y^2)$ in Example 7 is a surface with a hole (sometimes called a *puncture*) at the origin (Figure 14.2.10). We can remove this discontinuity by *defining* $f(0, 0)$ to be 0. (See Exercises 31 and 32, which also deal with the notion of a "removable" discontinuity.)

CONTINUITY AT BOUNDARY POINTS

Recall that in our study of continuity for functions of one variable, we first defined continuity at a point, then continuity on an open interval, and then, by using one-sided limits, we extended the notion of continuity to include the boundary points of the interval. Similarly, for functions of two variables one can extend the notion of continuity of $f(x, y)$ to the boundary of its domain by modifying Definition 14.2.1 appropriately so that (x, y) is restricted to approach (x_0, y_0) through points lying wholly in the domain of f. We will omit the details.

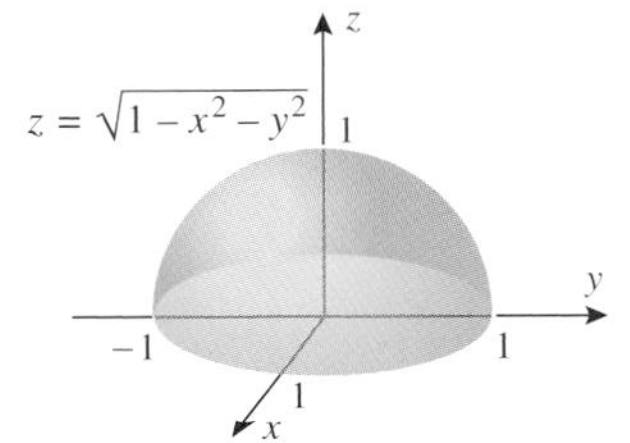

Figure 14.2.11

▶ **Example 8** The graph of the function $f(x, y) = \sqrt{1 - x^2 - y^2}$ is the upper hemisphere shown in Figure 14.2.11, and the natural domain of f is the closed unit disk

$$x^2 + y^2 \leq 1$$

The graph of f has no tears or holes, so it passes our "intuitive test" of continuity. In this case the continuity at a point (x_0, y_0) on the boundary reflects the fact that

$$\lim_{(x,y)\to(x_0,y_0)} \sqrt{1 - x^2 - y^2} = \sqrt{1 - x_0^2 - y_0^2} = 0$$

when (x, y) is restricted to points on the closed unit disk $x^2 + y^2 \leq 1$. It follows that f is continuous on its domain. ◀

EXTENSIONS TO THREE VARIABLES

All of the results in this section can be extended to functions of three or more variables. For example, the distance between the points (x, y, z) and (x_0, y_0, z_0) in 3-space is

$$\sqrt{(x - x_0)^2 + (y - y_0)^2 + (z - z_0)^2}$$

so the natural extension of Definition 14.2.1 to 3-space is as follows:

14.2.5 DEFINITION. Let f be a function of three variables, and assume that f is defined at all points within a ball centered at (x_0, y_0, z_0), except possibly at (x_0, y_0, z_0). We will write

$$\lim_{(x,y,z)\to(x_0,y_0,z_0)} f(x, y, z) = L \tag{4}$$

if given any number $\epsilon > 0$, we can find a number $\delta > 0$ such that $f(x, y, z)$ satisfies

$$|f(x, y, z) - L| < \epsilon$$

whenever the distance between (x, y, z) and (x_0, y_0, z_0) satisfies

$$0 < \sqrt{(x - x_0)^2 + (y - y_0)^2 + (z - z_0)^2} < \delta$$

As with functions of one and two variables, we define a function $f(x, y, z)$ of three variables to be continuous at a point (x_0, y_0, z_0) if the limit of the function and the value of the function are the same at this point; that is,

$$\lim_{(x,y,z)\to(x_0,y_0,z_0)} f(x, y, z) = f(x_0, y_0, z_0)$$

Although we will omit the details, the properties of limits and continuity that we discussed for functions of two variables, including the notion of continuity at boundary points, carry over to functions of three variables.

✔ QUICK CHECK EXERCISES 14.2

(See page 945 for answers.)

1. Let
$$f(x, y) = \frac{x^2 - y^2}{x^2 + y^2}$$
Determine the limit of $f(x, y)$ as (x, y) approaches $(0, 0)$ along the curve C.
(a) $C: x = 0$ (b) $C: y = 0$
(c) $C: y = x$ (d) $C: y = x^2$

2. (a) $\lim_{(x,y)\to(3,2)} x \cos \pi y =$ ______
(b) $\lim_{(x,y)\to(0,1)} e^{xy^2} =$ ______
(c) $\lim_{(x,y)\to(0,0)} (x^2 + y^2) \sin\left(\frac{1}{x^2 + y^2}\right) =$ ______

3. A function $f(x, y)$ is continuous at (x_0, y_0) provided $f(x_0, y_0)$ exists and provided $f(x, y)$ has limit ______ as (x, y) approaches ______.

4. Determine all values of the constant a such that the function $f(x, y) = \sqrt{x^2 - ay^2 + 1}$ is continuous everywhere.

EXERCISE SET 14.2

1–6 Use limit laws and continuity properties to evaluate the limit.

1. $\lim_{(x,y)\to(1,3)} (4xy^2 - x)$
2. $\lim_{(x,y)\to(1/2,\pi)} (xy^2 \sin xy)$
3. $\lim_{(x,y)\to(-1,2)} \frac{xy^3}{x + y}$
4. $\lim_{(x,y)\to(1,-3)} e^{2x-y^2}$
5. $\lim_{(x,y)\to(0,0)} \ln(1 + x^2y^3)$
6. $\lim_{(x,y)\to(4,-2)} x\sqrt[3]{y^3 + 2x}$

7–8 Show that the limit does not exist by considering the limits as $(x, y)\to(0, 0)$ along the coordinate axes.

7. (a) $\lim_{(x,y)\to(0,0)} \frac{3}{x^2 + 2y^2}$ (b) $\lim_{(x,y)\to(0,0)} \frac{x + y}{2x^2 + y^2}$
8. (a) $\lim_{(x,y)\to(0,0)} \frac{x - y}{x^2 + y^2}$ (b) $\lim_{(x,y)\to(0,0)} \frac{\cos xy}{x^2 + y^2}$

9–12 Evaluate the limit using the substitution $z = x^2 + y^2$ and observing that $z\to 0^+$ if and only if $(x, y)\to(0, 0)$.

9. $\lim_{(x,y)\to(0,0)} \frac{\sin(x^2 + y^2)}{x^2 + y^2}$
10. $\lim_{(x,y)\to(0,0)} \frac{1 - \cos(x^2 + y^2)}{x^2 + y^2}$
11. $\lim_{(x,y)\to(0,0)} e^{-1/(x^2+y^2)}$
12. $\lim_{(x,y)\to(0,0)} \frac{e^{-1/\sqrt{x^2+y^2}}}{\sqrt{x^2 + y^2}}$

13–20 Determine whether the limit exists. If so, find its value.

13. $\lim_{(x,y)\to(0,0)} \frac{x^4 - y^4}{x^2 + y^2}$
14. $\lim_{(x,y)\to(0,0)} \frac{x^4 - 16y^4}{x^2 + 4y^2}$
15. $\lim_{(x,y)\to(0,0)} \frac{xy}{3x^2 + 2y^2}$
16. $\lim_{(x,y)\to(0,0)} \frac{1 - x^2 - y^2}{x^2 + y^2}$
17. $\lim_{(x,y,z)\to(2,-1,2)} \frac{xz^2}{\sqrt{x^2 + y^2 + z^2}}$
18. $\lim_{(x,y,z)\to(2,0,-1)} \ln(2x + y - z)$
19. $\lim_{(x,y,z)\to(0,0,0)} \frac{\sin(x^2 + y^2 + z^2)}{\sqrt{x^2 + y^2 + z^2}}$
20. $\lim_{(x,y,z)\to(0,0,0)} \frac{\sin\sqrt{x^2 + y^2 + z^2}}{x^2 + y^2 + z^2}$

21–22 Evaluate the limit, if it exists, by converting to polar coordinates, as in Example 7.

21. $\lim_{(x,y)\to(0,0)} y \ln(x^2 + y^2)$
22. $\lim_{(x,y)\to(0,0)} \frac{x^2y^2}{\sqrt{x^2 + y^2}}$

23–24 Evaluate the limit, if it exists, by converting to spherical coordinates (ρ, θ, ϕ) and observe that $\rho\to 0^+$ if and only if $(x, y, z)\to(0, 0, 0)$, since $\rho = \sqrt{x^2 + y^2 + z^2}$.

23. $\lim_{(x,y,z)\to(0,0,0)} \frac{e^{\sqrt{x^2+y^2+z^2}}}{\sqrt{x^2 + y^2 + z^2}}$
24. $\lim_{(x,y,z)\to(0,0,0)} \tan^{-1}\left[\frac{1}{x^2 + y^2 + z^2}\right]$

FOCUS ON CONCEPTS

25. The accompanying figure shows a portion of the graph of
$$f(x, y) = \frac{x^2y}{x^4 + y^2}$$
(a) Based on the graph in the figure, does $f(x, y)$ have a limit as $(x, y)\to(0, 0)$? Explain your reasoning.
(b) Show that $f(x, y)\to 0$ as $(x, y)\to(0, 0)$ along any line $y = mx$. Does this imply that $f(x, y)\to 0$ as $(x, y)\to(0, 0)$? Explain.
(c) Show that $f(x, y)\to\frac{1}{2}$ as $(x, y)\to(0, 0)$ along the parabola $y = x^2$, and confirm visually that this is consistent with the graph of $f(x, y)$.
(d) Based on parts (b) and (c), does $f(x, y)$ have a limit as $(x, y)\to(0, 0)$? Is this consistent with your answer to part (a)?

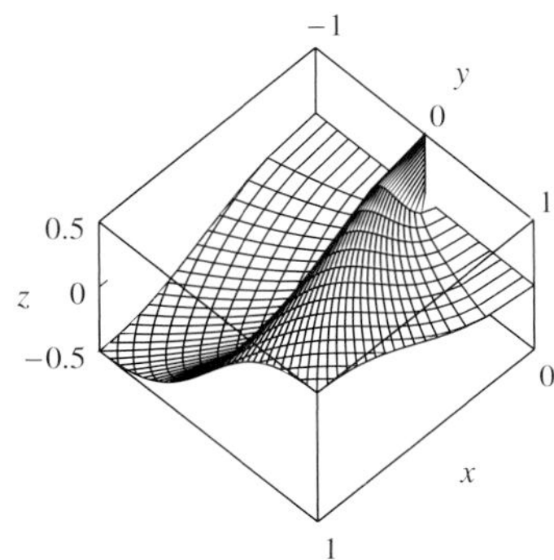

Figure Ex-25

26. (a) Show that the value of
$$\frac{x^3y}{2x^6+y^2}$$
approaches 0 as $(x, y) \to (0, 0)$ along any straight line $y = mx$, or along any parabola $y = kx^2$.

(b) Show that
$$\lim_{(x,y)\to(0,0)} \frac{x^3y}{2x^6+y^2}$$
does not exist by letting $(x, y) \to (0, 0)$ along the curve $y = x^3$.

27. (a) Show that the value of
$$\frac{xyz}{x^2+y^4+z^4}$$
approaches 0 as $(x, y, z) \to (0, 0, 0)$ along any line $x = at$, $y = bt$, $z = ct$.

(b) Show that the limit
$$\lim_{(x,y,z)\to(0,0,0)} \frac{xyz}{x^2+y^4+z^4}$$
does not exist by letting $(x, y, z) \to (0, 0, 0)$ along the curve $x = t^2$, $y = t$, $z = t$.

28. Find $\displaystyle\lim_{(x,y)\to(0,1)} \tan^{-1}\left[\frac{x^2+1}{x^2+(y-1)^2}\right]$.

29. Find $\displaystyle\lim_{(x,y)\to(0,1)} \tan^{-1}\left[\frac{x^2-1}{x^2+(y-1)^2}\right]$.

30. Let $f(x, y) = \begin{cases} \dfrac{\sin(x^2+y^2)}{x^2+y^2}, & (x, y) \neq (0, 0) \\ 1, & (x, y) = (0, 0). \end{cases}$

Show that f is continuous at $(0, 0)$.

31–32 A function $f(x, y)$ is said to have a ***removable discontinuity*** at (x_0, y_0) if $\lim_{(x,y)\to(x_0,y_0)} f(x, y)$ exists but f is not continuous at (x_0, y_0), either because f is not defined at (x_0, y_0) or because $f(x_0, y_0)$ differs from the value of the limit. Determine whether $f(x, y)$ has a removable discontinuity at $(0, 0)$.

31. $f(x, y) = \dfrac{x^2}{x^2+y^2}$

32. $f(x, y) = xy\ln(x^2+y^2)$

33–40 Sketch the largest region on which the function f is continuous.

33. $f(x, y) = y\ln(1+x)$

34. $f(x, y) = \sqrt{x-y}$

35. $f(x, y) = \dfrac{x^2y}{\sqrt{25-x^2-y^2}}$

36. $f(x, y) = \ln(2x-y+1)$

37. $f(x, y) = \cos\left(\dfrac{xy}{1+x^2+y^2}\right)$

38. $f(x, y) = e^{1-xy}$

39. $f(x, y) = \sin^{-1}(xy)$

40. $f(x, y) = \tan^{-1}(y-x)$

41–44 Describe the largest region on which the function f is continuous.

41. $f(x, y, z) = 3x^2e^{yz}\cos(xyz)$

42. $f(x, y, z) = \ln(4-x^2-y^2-z^2)$

43. $f(x, y, z) = \dfrac{y+1}{x^2+z^2-1}$

44. $f(x, y, z) = \sin\sqrt{x^2+y^2+3z^2}$

✔ QUICK CHECK ANSWERS 14.2

1. (a) −1 (b) 1 (c) 0 (d) 1 2. (a) 3 (b) 1 (c) 0 3. $f(x_0, y_0)$; (x_0, y_0) 4. $a \le 0$

14.3 PARTIAL DERIVATIVES

In this section we will develop the mathematical tools for studying rates of change that involve two or more independent variables.

PARTIAL DERIVATIVES OF FUNCTIONS OF TWO VARIABLES

If $z = f(x, y)$, then one can inquire how the value of z changes if y is held fixed and x is allowed to vary, or if x is held fixed and y is allowed to vary. For example, the ideal gas law in physics states that under appropriate conditions the pressure exerted by a gas is a function of the volume of the gas and its temperature. Thus, a physicist studying gases might be interested in the rate of change of the pressure if the volume is held fixed and the temperature is allowed to vary, or if the temperature is held fixed and the volume is allowed to vary. We now define a derivative that describes such rates of change.

Suppose that (x_0, y_0) is a point in the domain of a function $f(x, y)$. If we fix $y = y_0$, then $f(x, y_0)$ is a function of the variable x alone. The value of the derivative

$$\frac{d}{dx}[f(x, y_0)]$$

at x_0 then gives us a measure of the instantaneous rate of change of f with respect to x at the point (x_0, y_0). Similarly, the value of the derivative

$$\frac{d}{dy}[f(x_0, y)]$$

at y_0 gives us a measure of the instantaneous rate of change of f with respect to y at the point (x_0, y_0). These derivatives are so basic to the study of differential calculus of multivariable functions that they have their own name and notation.

14.3.1 DEFINITION. If $z = f(x, y)$ and (x_0, y_0) is a point in the domain of f, then the ***partial derivative of f with respect to x*** at (x_0, y_0) [also called the ***partial derivative of z with respect to x*** at (x_0, y_0)] is the derivative at x_0 of the function that results when $y = y_0$ is held fixed and x is allowed to vary. This partial derivative is denoted by $f_x(x_0, y_0)$ and is given by

$$f_x(x_0, y_0) = \frac{d}{dx}[f(x, y_0)]\bigg|_{x=x_0} = \lim_{\Delta x \to 0} \frac{f(x_0 + \Delta x, y_0) - f(x_0, y_0)}{\Delta x} \qquad (1)$$

Similarly, the ***partial derivative of f with respect to y*** at (x_0, y_0) [also called the ***partial derivative of z with respect to y*** at (x_0, y_0)] is the derivative at y_0 of the function that results when $x = x_0$ is held fixed and y is allowed to vary. This partial derivative is denoted by $f_y(x_0, y_0)$ and is given by

$$f_y(x_0, y_0) = \frac{d}{dy}[f(x_0, y)]\bigg|_{y=y_0} = \lim_{\Delta y \to 0} \frac{f(x_0, y_0 + \Delta y) - f(x_0, y_0)}{\Delta y} \qquad (2)$$

The limits in (1) and (2) show the relationship between partial derivatives and derivatives of functions of one variable. In practice, our usual method for computing partial derivatives is to hold one variable fixed and then differentiate the resulting function using the derivative rules for functions of one variable.

▶ **Example 1** Find $f_x(1, 3)$ and $f_y(1, 3)$ for the function $f(x, y) = 2x^3y^2 + 2y + 4x$.

Solution. Since

$$f_x(x, 3) = \frac{d}{dx}[f(x, 3)] = \frac{d}{dx}[18x^3 + 4x + 6] = 54x^2 + 4$$

we have $f_x(1, 3) = 54 + 4 = 58$. Also, since

$$f_y(1, y) = \frac{d}{dy}[f(1, y)] = \frac{d}{dy}[2y^2 + 2y + 4] = 4y + 2$$

we have $f_y(1, 3) = 4(3) + 2 = 14$. ◀

THE PARTIAL DERIVATIVE FUNCTIONS

Formulas (1) and (2) define the partial derivatives of a function at a specific point (x_0, y_0). However, often it will be desirable to omit the subscripts and think of the partial derivatives as functions of the variables x and y. These functions are

$$f_x(x, y) = \lim_{\Delta x \to 0} \frac{f(x + \Delta x, y) - f(x, y)}{\Delta x} \qquad f_y(x, y) = \lim_{\Delta y \to 0} \frac{f(x, y + \Delta y) - f(x, y)}{\Delta y}$$

The following example gives an alternative way of performing the computations in Example 1.

▶ **Example 2** Find $f_x(x, y)$ and $f_y(x, y)$ for $f(x, y) = 2x^3y^2 + 2y + 4x$, and use those partial derivatives to compute $f_x(1, 3)$ and $f_y(1, 3)$.

Solution. Keeping y fixed and differentiating with respect x yields

$$f_x(x, y) = \frac{d}{dx}[2x^3y^2 + 2y + 4x] = 6x^2y^2 + 4$$

and keeping x fixed and differentiating with respect to y yields

$$f_y(x, y) = \frac{d}{dy}[2x^3y^2 + 2y + 4x] = 4x^3y + 2$$

Thus,

$$f_x(1, 3) = 6(1^2)(3^2) + 4 = 58 \quad \text{and} \quad f_y(1, 3) = 4(1^3)3 + 2 = 14$$

which agree with the results in Example 1. ◀

TECHNOLOGY MASTERY

Computer algebra systems have specific commands for calculating partial derivatives. If you have a CAS, use it to find the partial derivatives $f_x(x, y)$ and $f_y(x, y)$ in Example 2.

PARTIAL DERIVATIVES VIEWED AS RATES OF CHANGE AND SLOPES

Recall that if $y = f(x)$, then the value of $f'(x_0)$ can be interpreted either as the rate of change of y with respect to x at x_0 or as the slope of the tangent line to the graph of f at x_0. Partial derivatives have analogous interpretations. To see that this is so, suppose that C_1 is the intersection of the surface $z = f(x, y)$ with the plane $y = y_0$ and that C_2 is its intersection with the plane $x = x_0$ (Figure 14.3.1). Thus, $f_x(x, y_0)$ can be interpreted as the rate of change of z with respect to x along the curve C_1, and $f_y(x_0, y)$ can be interpreted as the rate of change of z with respect to y along the curve C_2. In particular, $f_x(x_0, y_0)$ is the rate of change of z with respect to x along the curve C_1 at the point (x_0, y_0), and $f_y(x_0, y_0)$ is the rate of change of z with respect to y along the curve C_2 at the point (x_0, y_0).

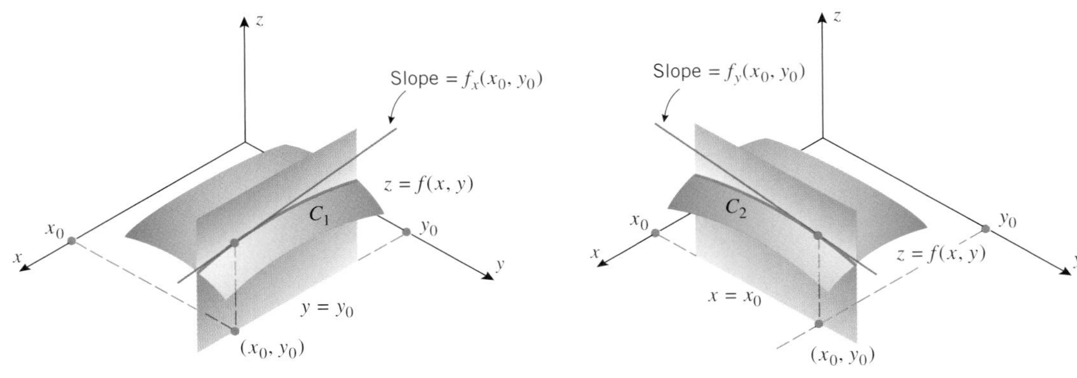

Figure 14.3.1

▶ **Example 3** Recall that the wind chill temperature index is given by the formula

$$W = 35.74 + 0.6215T + (0.4275T - 35.75)v^{0.16}$$

Compute the partial derivative of W with respect to v at the point $(T, v) = (25, 10)$ and interpret this partial derivative as a rate of change.

In an applied problem, the interpretations of $f_x(x_0, y_0)$ and $f_y(x_0, y_0)$ must be accompanied by the proper units. See Example 3.

Solution. Holding T fixed and differentiating with respect to v yields

$$\frac{\partial W}{\partial v}(T, v) = 0 + 0 + (0.4275T - 35.75)(0.16)v^{0.16-1} = (0.4275T - 35.75)(0.16)v^{-0.84}$$

Since W is in degrees Fahrenheit and v is in miles per hour, a rate of change of W with respect to v will have units °F/(mi/h) (which may also be written as °F·h/mi). Substituting $T = 25$ and $v = 10$ gives

$$\frac{\partial W}{\partial v}(25, 10) = (-4.01)10^{-0.84} \approx -0.58 \frac{°\text{F}}{\text{mi/h}}$$

as the instantaneous rate of change of W with respect to v at $(T, v) = (25, 10)$. We conclude that if the air temperature is a constant 25°F and the wind speed changes by a small amount from an initial speed of 10 mi/h, then the ratio of the change in the wind chill index to the change in wind speed should be about −0.58°F/(mi/h). ◀

Confirm the conclusion of Example 3 by calculating

$$\frac{W(25, 10 + \Delta v) - W(25, 10)}{\Delta v}$$

for values of Δv near 0.

Geometrically, $f_x(x_0, y_0)$ can be viewed as the slope of the tangent line to the curve C_1 at the point (x_0, y_0), and $f_y(x_0, y_0)$ can be viewed as the slope of the tangent line to the curve C_2 at the point (x_0, y_0) (Figure 14.3.1). We will call $f_x(x_0, y_0)$ the ***slope of the surface in the x-direction*** at (x_0, y_0) and $f_y(x_0, y_0)$ the ***slope of the surface in the y-direction*** at (x_0, y_0).

▶ **Example 4** Let $f(x, y) = x^2y + 5y^3$.

(a) Find the slope of the surface $z = f(x, y)$ in the x-direction at the point $(1, -2)$.

(b) Find the slope of the surface $z = f(x, y)$ in the y-direction at the point $(1, -2)$.

Solution (a). Differentiating f with respect to x with y held fixed yields

$$f_x(x, y) = 2xy$$

Thus, the slope in the x-direction is $f_x(1, -2) = -4$; that is, z is decreasing at the rate of 4 units per unit increase in x.

Solution (b). Differentiating f with respect to y with x held fixed yields

$$f_y(x, y) = x^2 + 15y^2$$

Thus, the slope in the y-direction is $f_y(1, -2) = 61$; that is, z is increasing at the rate of 61 units per unit increase in y. ◀

▶ **Example 5** Let

$$f(x, y) = \begin{cases} -\dfrac{xy}{x^2 + y^2}, & (x, y) \neq (0, 0) \\ 0, & (x, y) = (0, 0) \end{cases} \tag{3}$$

(a) Show that $f_x(x, y)$ and $f_y(x, y)$ exist at all points (x, y).

(b) Explain why f is not continuous at $(0, 0)$.

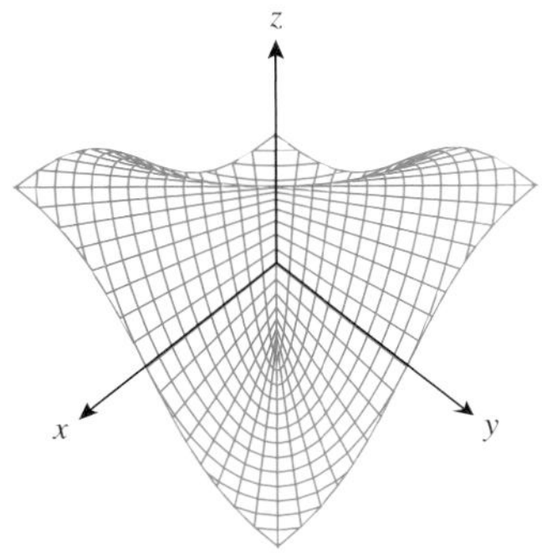

Figure 14.3.2

Solution (a). Figure 14.3.2 shows the graph of f. Note that f is similar to the function considered in Example 1 of Section 14.2, except that here we have assigned f a value of 0 at (0, 0). Except at this point, the partial derivatives of f are

$$f_x(x, y) = -\frac{(x^2+y^2)y - xy(2x)}{(x^2+y^2)^2} = \frac{x^2y - y^3}{(x^2+y^2)^2} \quad (4)$$

$$f_y(x, y) = -\frac{(x^2+y^2)x - xy(2y)}{(x^2+y^2)^2} = \frac{xy^2 - x^3}{(x^2+y^2)^2} \quad (5)$$

It is not evident from Formula (3) whether f has partial derivatives at (0, 0), and if so, what the values of those derivatives are. To answer that question we will have to use the definitions of the partial derivatives (Definition 14.3.1). Applying Formulas (1) and (2) to (3) we obtain

$$f_x(0, 0) = \lim_{\Delta x \to 0} \frac{f(\Delta x, 0) - f(0, 0)}{\Delta x} = \lim_{\Delta x \to 0} \frac{0 - 0}{\Delta x} = 0$$

$$f_y(0, 0) = \lim_{\Delta y \to 0} \frac{f(0, \Delta y) - f(0, 0)}{\Delta y} = \lim_{\Delta y \to 0} \frac{0 - 0}{\Delta y} = 0$$

This shows that f has partial derivatives at (0, 0) and the values of both partial derivatives are 0 at that point.

Solution (b). We saw in Example 3 of Section 14.2 that

$$\lim_{(x,y)\to(0,0)} -\frac{xy}{x^2+y^2}$$

does not exist. Thus, f is not continuous at (0, 0). ◀

Example 5 shows that, in contrast to the case of functions of a single variable, the existence of partial derivatives for a multivariable function does not guarantee the continuity of the function. We will return to this issue in the next section.

PARTIAL DERIVATIVE NOTATION

If $z = f(x, y)$, then the partial derivatives f_x and f_y are also denoted by the symbols

$$\frac{\partial f}{\partial x}, \quad \frac{\partial z}{\partial x} \quad \text{and} \quad \frac{\partial f}{\partial y}, \quad \frac{\partial z}{\partial y}$$

The symbol ∂ is called a partial derivative sign. It is derived from the Cyrillic alphabet.

Some typical notations for the partial derivatives of $z = f(x, y)$ at a point (x_0, y_0) are

$$\left.\frac{\partial f}{\partial x}\right|_{x=x_0,\, y=y_0}, \quad \left.\frac{\partial z}{\partial x}\right|_{(x_0, y_0)}, \quad \left.\frac{\partial f}{\partial x}\right|_{(x_0, y_0)}, \quad \frac{\partial f}{\partial x}(x_0, y_0), \quad \frac{\partial z}{\partial x}(x_0, y_0)$$

► **Example 6** Find $\partial z/\partial x$ and $\partial z/\partial y$ if $z = x^4 \sin(xy^3)$.

Solution.

$$\begin{aligned}\frac{\partial z}{\partial x} &= \frac{\partial}{\partial x}[x^4 \sin(xy^3)] = x^4 \frac{\partial}{\partial x}[\sin(xy^3)] + \sin(xy^3) \cdot \frac{\partial}{\partial x}(x^4) \\ &= x^4 \cos(xy^3) \cdot y^3 + \sin(xy^3) \cdot 4x^3 = x^4y^3 \cos(xy^3) + 4x^3 \sin(xy^3)\end{aligned}$$

$$\begin{aligned}\frac{\partial z}{\partial y} &= \frac{\partial}{\partial y}[x^4 \sin(xy^3)] = x^4 \frac{\partial}{\partial y}[\sin(xy^3)] + \sin(xy^3) \cdot \frac{\partial}{\partial y}(x^4) \\ &= x^4 \cos(xy^3) \cdot 3xy^2 + \sin(xy^3) \cdot 0 = 3x^5y^2 \cos(xy^3)\end{aligned}$$ ◀

For functions that are presented in tabular form, we can estimate partial derivatives by using adjacent entries within the table.

Table 14.3.1

TEMPERATURE T (°F)

WIND SPEED v (mi/h)	20	25	30	35
5	13	19	25	31
10	9	15	21	27
15	6	13	19	25
20	4	11	17	24

▶ **Example 7** Use the values of the wind chill index function $W(T, v)$ displayed in Table 14.3.1 to estimate the partial derivative of W with respect to v at $(T, v) = (25, 10)$. Compare this estimate with the value of the partial derivative obtained in Example 3.

Solution. Since

$$\frac{\partial W}{\partial v}(25, 10) = \lim_{\Delta v \to 0} \frac{W(25, 10 + \Delta v) - W(25, 10)}{\Delta v} = \lim_{\Delta v \to 0} \frac{W(25, 10 + \Delta v) - 15}{\Delta v}$$

we can approximate the partial derivative by

$$\frac{\partial W}{\partial v}(25, 10) \approx \frac{W(25, 10 + \Delta v) - 15}{\Delta v}$$

With $\Delta v = 5$ this approximation is

$$\frac{\partial W}{\partial v}(25, 10) \approx \frac{W(25, 10 + 5) - 15}{5} = \frac{W(25, 15) - 15}{5} = \frac{13 - 15}{5} = -\frac{2}{5}\ \frac{°\text{F}}{\text{mi/h}}$$

and with $\Delta v = -5$ this approximation is

$$\frac{\partial W}{\partial v}(25, 10) \approx \frac{W(25, 10 - 5) - 15}{-5} = \frac{W(25, 5) - 15}{-5} = \frac{19 - 15}{-5} = -\frac{4}{5}\ \frac{°\text{F}}{\text{mi/h}}$$

We will take the average, $-\frac{3}{5} = -0.6°\text{F/(mi/h)}$, of these two approximations as our estimate of $(\partial W/\partial v)(25, 10)$. This is close to the value

$$\frac{\partial W}{\partial v}(25, 10) = (-4.01)10^{-0.84} \approx -0.58\ \frac{°\text{F}}{\text{mi/h}}$$

found in Example 3. ◀

IMPLICIT PARTIAL DIFFERENTIATION

▶ **Example 8** Find the slope of the sphere $x^2 + y^2 + z^2 = 1$ in the y-direction at the points $\left(\frac{2}{3}, \frac{1}{3}, \frac{2}{3}\right)$ and $\left(\frac{2}{3}, \frac{1}{3}, -\frac{2}{3}\right)$ (Figure 14.3.3).

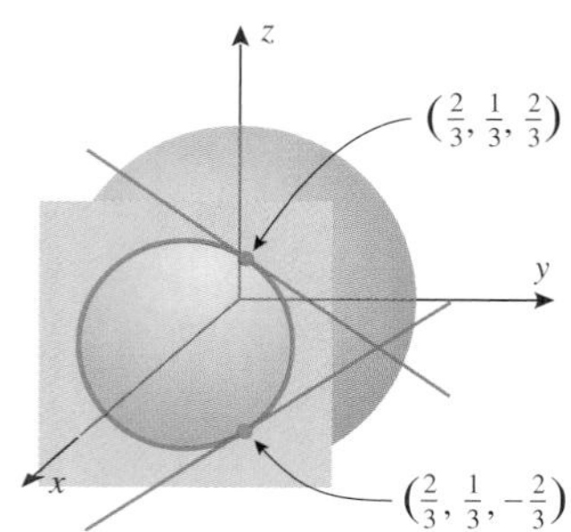

Figure 14.3.3

Solution. The point $\left(\frac{2}{3}, \frac{1}{3}, \frac{2}{3}\right)$ lies on the upper hemisphere $z = \sqrt{1 - x^2 - y^2}$, and the point $\left(\frac{2}{3}, \frac{1}{3}, -\frac{2}{3}\right)$ lies on the lower hemisphere $z = -\sqrt{1 - x^2 - y^2}$. We could find the slopes by differentiating each expression for z separately with respect to y and then evaluating the derivatives at $x = \frac{2}{3}$ and $y = \frac{1}{3}$. However, it is more efficient to differentiate the given equation

$$x^2 + y^2 + z^2 = 1$$

implicitly with respect to y, since this will give us both slopes with one differentiation. To perform the implicit differentiation, we view z as a function of x and y and differentiate both sides with respect to y, taking x to be fixed. The computations are as follows:

$$\frac{\partial}{\partial y}[x^2 + y^2 + z^2] = \frac{\partial}{\partial y}[1]$$

$$0 + 2y + 2z\frac{\partial z}{\partial y} = 0$$

$$\frac{\partial z}{\partial y} = -\frac{y}{z}$$

Substituting the y- and z-coordinates of the points $\left(\frac{2}{3}, \frac{1}{3}, \frac{2}{3}\right)$ and $\left(\frac{2}{3}, \frac{1}{3}, -\frac{2}{3}\right)$ in this expression, we find that the slope at the point $\left(\frac{2}{3}, \frac{1}{3}, \frac{2}{3}\right)$ is $-\frac{1}{2}$ and the slope at $\left(\frac{2}{3}, \frac{1}{3}, -\frac{2}{3}\right)$ is $\frac{1}{2}$. ◀

Check the results in Example 8 by differentiating the functions

$$z = \sqrt{1 - x^2 - y^2}$$

and

$$z = -\sqrt{1 - x^2 - y^2}$$

directly.

▶ **Example 9** Suppose that $D = \sqrt{x^2 + y^2}$ is the length of the diagonal of a rectangle whose sides have lengths x and y that are allowed to vary. Find a formula for the rate of change of D with respect to x if x varies with y held constant, and use this formula to find the rate of change of D with respect to x at the point where $x = 3$ and $y = 4$.

Solution. Differentiating both sides of the equation $D^2 = x^2 + y^2$ with respect to x yields

$$2D\frac{\partial D}{\partial x} = 2x \quad \text{and thus} \quad D\frac{\partial D}{\partial x} = x$$

Since $D = 5$ when $x = 3$ and $y = 4$, it follows that

$$5\left.\frac{\partial D}{\partial x}\right|_{x=3,\,y=4} = 3 \quad \text{or} \quad \left.\frac{\partial D}{\partial x}\right|_{x=3,\,y=4} = \frac{3}{5}$$

Thus, D is increasing at a rate of $\frac{3}{5}$ unit per unit increase in x at the point $(3, 4)$. ◀

PARTIAL DERIVATIVES OF FUNCTIONS WITH MORE THAN TWO VARIABLES

For a function $f(x, y, z)$ of three variables, there are three ***partial derivatives***:

$$f_x(x, y, z), \quad f_y(x, y, z), \quad f_z(x, y, z)$$

The partial derivative f_x is calculated by holding y and z constant and differentiating with respect to x. For f_y the variables x and z are held constant, and for f_z the variables x and y are held constant. If a dependent variable

$$w = f(x, y, z)$$

is used, then the three partial derivatives of f can be denoted by

$$\frac{\partial w}{\partial x}, \quad \frac{\partial w}{\partial y}, \quad \text{and} \quad \frac{\partial w}{\partial z}$$

▶ **Example 10** If $f(x, y, z) = x^3y^2z^4 + 2xy + z$, then

$$\begin{aligned}
f_x(x, y, z) &= 3x^2y^2z^4 + 2y \\
f_y(x, y, z) &= 2x^3yz^4 + 2x \\
f_z(x, y, z) &= 4x^3y^2z^3 + 1 \\
f_z(-1, 1, 2) &= 4(-1)^3(1)^2(2)^3 + 1 = -31
\end{aligned}$$ ◀

▶ **Example 11** If $f(\rho, \theta, \phi) = \rho^2 \cos\phi \sin\theta$, then

$$\begin{aligned}
f_\rho(\rho, \theta, \phi) &= 2\rho \cos\phi \sin\theta \\
f_\theta(\rho, \theta, \phi) &= \rho^2 \cos\phi \cos\theta \\
f_\phi(\rho, \theta, \phi) &= -\rho^2 \sin\phi \sin\theta
\end{aligned}$$ ◀

In general, if $f(v_1, v_2, \ldots, v_n)$ is a function of n variables, there are n partial derivatives of f, each of which is obtained by holding $n - 1$ of the variables fixed and differentiating the function f with respect to the remaining variable. If $w = f(v_1, v_2, \dots, v_n)$, then these partial derivatives are denoted by

$$\frac{\partial w}{\partial v_1}, \frac{\partial w}{\partial v_2}, \dots, \frac{\partial w}{\partial v_n}$$

where $\partial w/\partial v_i$ is obtained by holding all variables except v_i fixed and differentiating with respect to v_i.

▶ **Example 12** Find

$$\frac{\partial}{\partial x_i}\left[\sqrt{x_1^2 + x_2^2 + \cdots + x_n^2}\right]$$

for $i = 1, 2, \dots, n$.

Solution. For each $i = 1, 2, \dots, n$ we obtain

$$\begin{aligned}\frac{\partial}{\partial x_i}\left[\sqrt{x_1^2 + x_2^2 + \cdots + x_n^2}\right] &= \frac{1}{2\sqrt{x_1^2 + x_2^2 + \cdots + x_n^2}} \cdot \frac{\partial}{\partial x_i}[x_1^2 + x_2^2 + \cdots + x_n^2] \\ &= \frac{1}{2\sqrt{x_1^2 + x_2^2 + \cdots + x_n^2}}[2x_i] \\ &= \frac{x_i}{\sqrt{x_1^2 + x_2^2 + \cdots + x_n^2}}\end{aligned}$$ ◀

All terms except x_i^2 are constant.

HIGHER-ORDER PARTIAL DERIVATIVES

Suppose that f is a function of two variables x and y. Since the partial derivatives $\partial f/\partial x$ and $\partial f/\partial y$ are also functions of x and y, these functions may themselves have partial derivatives. This gives rise to four possible ***second-order*** partial derivatives of f, which are defined by

$$\frac{\partial^2 f}{\partial x^2} = \frac{\partial}{\partial x}\left(\frac{\partial f}{\partial x}\right) = f_{xx} \qquad \frac{\partial^2 f}{\partial y^2} = \frac{\partial}{\partial y}\left(\frac{\partial f}{\partial y}\right) = f_{yy}$$

Differentiate twice with respect to x.

Differentiate twice with respect to y.

$$\frac{\partial^2 f}{\partial y \partial x} = \frac{\partial}{\partial y}\left(\frac{\partial f}{\partial x}\right) = f_{xy} \qquad \frac{\partial^2 f}{\partial x \partial y} = \frac{\partial}{\partial x}\left(\frac{\partial f}{\partial y}\right) = f_{yx}$$

Differentiate first with respect to x and then with respect to y.

Differentiate first with respect to y and then with respect to x.

The last two cases are called the ***mixed second-order partial derivatives*** or the ***mixed second partials***. Also, the derivatives $\partial f/\partial x$ and $\partial f/\partial y$ are often called the ***first-order partial derivatives*** when it is necessary to distinguish them from higher-order partial derivatives. Similar conventions apply to the second-order partial derivatives of a function of three variables.

WARNING

Observe that the two notations for the mixed second partials have opposite conventions for the order of differentiation. In the "∂" notation the derivatives are taken right to left, and in the "subscript" notation they are taken left to right. The conventions are logical if you insert parentheses:

$$\frac{\partial^2 f}{\partial y \partial x} = \frac{\partial}{\partial y}\left(\frac{\partial f}{\partial x}\right)$$

Right to left. Differentiate inside the parentheses first.

$$f_{xy} = (f_x)_y$$

Left to right. Differentiate inside the parentheses first.

▶ **Example 13** Find the second-order partial derivatives of $f(x, y) = x^2y^3 + x^4y$.

Solution. We have

$$\frac{\partial f}{\partial x} = 2xy^3 + 4x^3y \quad \text{and} \quad \frac{\partial f}{\partial y} = 3x^2y^2 + x^4$$

so that

$$\frac{\partial^2 f}{\partial x^2} = \frac{\partial}{\partial x}\left(\frac{\partial f}{\partial x}\right) = \frac{\partial}{\partial x}(2xy^3 + 4x^3y) = 2y^3 + 12x^2y$$

$$\frac{\partial^2 f}{\partial y^2} = \frac{\partial}{\partial y}\left(\frac{\partial f}{\partial y}\right) = \frac{\partial}{\partial y}(3x^2y^2 + x^4) = 6x^2y$$

$$\frac{\partial^2 f}{\partial x \partial y} = \frac{\partial}{\partial x}\left(\frac{\partial f}{\partial y}\right) = \frac{\partial}{\partial x}(3x^2y^2 + x^4) = 6xy^2 + 4x^3$$

$$\frac{\partial^2 f}{\partial y \partial x} = \frac{\partial}{\partial y}\left(\frac{\partial f}{\partial x}\right) = \frac{\partial}{\partial y}(2xy^3 + 4x^3y) = 6xy^2 + 4x^3 \blacktriangleleft$$

Third-order, fourth-order, and higher-order partial derivatives can be obtained by successive differentiation. Some possibilities are

$$\frac{\partial^3 f}{\partial x^3} = \frac{\partial}{\partial x}\left(\frac{\partial^2 f}{\partial x^2}\right) = f_{xxx} \qquad \frac{\partial^4 f}{\partial y^4} = \frac{\partial}{\partial y}\left(\frac{\partial^3 f}{\partial y^3}\right) = f_{yyyy}$$

$$\frac{\partial^3 f}{\partial y^2 \partial x} = \frac{\partial}{\partial y}\left(\frac{\partial^2 f}{\partial y \partial x}\right) = f_{xyy} \qquad \frac{\partial^4 f}{\partial y^2 \partial x^2} = \frac{\partial}{\partial y}\left(\frac{\partial^3 f}{\partial y \partial x^2}\right) = f_{xxyy}$$

▶ **Example 14** Let $f(x, y) = y^2e^x + y$. Find f_{xyy}.

Solution.

$$f_{xyy} = \frac{\partial^3 f}{\partial y^2 \partial x} = \frac{\partial^2}{\partial y^2}\left(\frac{\partial f}{\partial x}\right) = \frac{\partial^2}{\partial y^2}(y^2e^x) = \frac{\partial}{\partial y}(2ye^x) = 2e^x \blacktriangleleft$$

EQUALITY OF MIXED PARTIALS

For a function $f(x, y)$ it might be expected that there would be four distinct second-order partial derivatives: f_{xx}, f_{xy}, f_{yx}, and f_{yy}. However, observe that the mixed second-order partial derivatives in Example 13 are equal. The following theorem (proved in advanced courses) explains why this is so.

If f is a function of three variables, then the analog of Theorem 14.3.2 holds for each pair of mixed second-order partials if we replace "open disk" by "open ball." How many second-order partials does $f(x, y, z)$ have?

14.3.2 THEOREM. *Let f be a function of two variables. If f_{xy} and f_{yx} are continuous on some open disk, then $f_{xy} = f_{yx}$ on that disk.*

It follows from this theorem that if $f_{xy}(x, y)$ and $f_{yx}(x, y)$ are continuous everywhere, then $f_{xy}(x, y) = f_{yx}(x, y)$ for all values of x and y. Since polynomials are continuous everywhere, this explains why the mixed second-order partials in Example 13 are equal.

THE WAVE EQUATION

Consider a string of length L that is stretched taut between $x = 0$ and $x = L$ on an x-axis, and suppose that the string is set into vibratory motion by "plucking" it at time $t = 0$ (Figure

14.3.4*a*). The displacement of a point on the string depends both on its coordinate x and the elapsed time t, and hence is described by a function $u(x, t)$ of two variables. For a fixed value t, the function $u(x, t)$ depends on x alone, and the graph of u versus x describes the shape of the string—think of it as a "snapshot" of the string at time t (Figure 14.3.4*b*). It follows that at a fixed time t, the partial derivative $\partial u/\partial x$ represents the slope of the string at x, and the sign of the second partial derivative $\partial^2 u/\partial x^2$ tells us whether the string is concave up or concave down at x (Figure 14.3.4*c*).

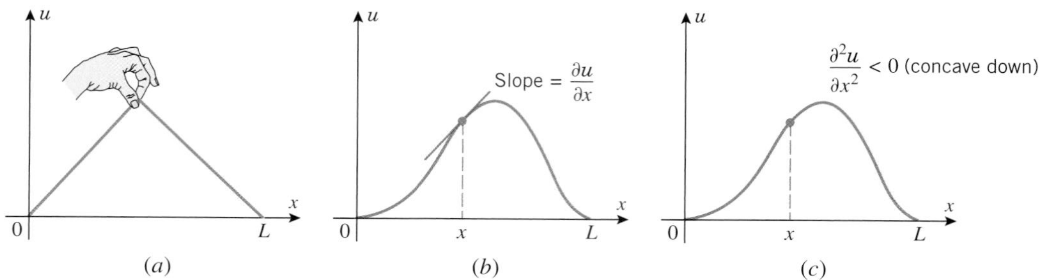

Figure 14.3.4

For a fixed value of x, the function $u(x, t)$ depends on t alone, and the graph of u versus t is the position versus time curve of the point on the string with coordinate x. Thus, for a fixed value of x, the partial derivative $\partial u/\partial t$ is the velocity of the point with coordinate x, and $\partial^2 u/\partial t^2$ is the acceleration of that point.

The vibration of a plucked string is governed by the wave equation.

It can be proved that under appropriate conditions the function $u(x, t)$ satisfies an equation of the form

$$\frac{\partial^2 u}{\partial t^2} = c^2 \frac{\partial^2 u}{\partial x^2} \tag{6}$$

where c is a positive constant that depends on the physical characteristics of the string. This equation, which is called the ***one-dimensional wave equation***, involves partial derivatives of the unknown function $u(x, t)$ and hence is classified as a ***partial differential equation***. Techniques for solving partial differential equations are studied in advanced courses and will not be discussed in this text.

▶ **Example 15** Show that the function $u(x, t) = \sin(x - ct)$ is a solution of Equation (6).

Solution. We have

$$\frac{\partial u}{\partial x} = \cos(x - ct), \qquad \frac{\partial^2 u}{\partial x^2} = -\sin(x - ct)$$

$$\frac{\partial u}{\partial t} = -c\cos(x - ct), \qquad \frac{\partial^2 u}{\partial t^2} = -c^2 \sin(x - ct)$$

Thus, $u(x, t)$ satisfies (6). ◀

✔ QUICK CHECK EXERCISES 14.3 *(See page 959 for answers.)*

1. Let $f(x, y) = x \sin xy$. Then $f_x(x, y) =$ ________ and $f_y(x, y) =$ ________.

2. The slope of the surface $z = xy^2$ in the x-direction at the point $(2, 3)$ is ________, and the slope of this surface in the y-direction at the point $(2, 3)$ is ________.

3. The volume V of a right circular cone of radius r and height h is given by $V = \frac{1}{3}\pi r^2 h$.
 (a) Find a formula for the instantaneous rate of change of V with respect to r if r changes and h remains constant.
 (b) Find a formula for the instantaneous rate of change of V with respect to h if h changes and r remains constant.

4. Find all second-order partial derivatives for the function $f(x, y) = x^2y^3$.

EXERCISE SET 14.3 Graphing Utility

1. Let $f(x, y) = 3x^3y^2$. Find
 (a) $f_x(x, y)$ (b) $f_y(x, y)$ (c) $f_x(1, y)$
 (d) $f_x(x, 1)$ (e) $f_y(1, y)$ (f) $f_y(x, 1)$
 (g) $f_x(1, 2)$ (h) $f_y(1, 2)$.

2. Let $z = e^{2x} \sin y$. Find
 (a) $\partial z/\partial x$ (b) $\partial z/\partial y$ (c) $\partial z/\partial x|_{(0,y)}$
 (d) $\partial z/\partial x|_{(x,0)}$ (e) $\partial z/\partial y|_{(0,y)}$ (f) $\partial z/\partial y|_{(x,0)}$
 (g) $\partial z/\partial x|_{(\ln 2, 0)}$ (h) $\partial z/\partial y|_{(\ln 2, 0)}$.

3. Let $f(x, y) = \sqrt{3x + 2y}$.
 (a) Find the slope of the surface $z = f(x, y)$ in the x-direction at the point $(4, 2)$.
 (b) Find the slope of the surface $z = f(x, y)$ in the y-direction at the point $(4, 2)$.

4. Let $f(x, y) = xe^{-y} + 5y$.
 (a) Find the slope of the surface $z = f(x, y)$ in the x-direction at the point $(3, 0)$.
 (b) Find the slope of the surface $z = f(x, y)$ in the y-direction at the point $(3, 0)$.

5. Let $z = \sin(y^2 - 4x)$.
 (a) Find the rate of change of z with respect to x at the point $(2, 1)$ with y held fixed.
 (b) Find the rate of change of z with respect to y at the point $(2, 1)$ with x held fixed.

6. Let $z = (x + y)^{-1}$.
 (a) Find the rate of change of z with respect to x at the point $(-2, 4)$ with y held fixed.
 (b) Find the rate of change of z with respect to y at the point $(-2, 4)$ with x held fixed.

FOCUS ON CONCEPTS

7. Use the information in the accompanying figure to find the values of the first-order partial derivatives of f at the point $(1, 2)$.

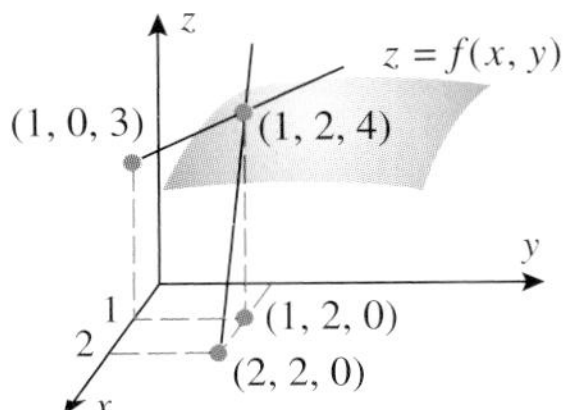

Figure Ex-7

8. The accompanying figure shows a contour plot for an unspecified function $f(x, y)$. Make a conjecture about the signs of the partial derivatives $f_x(x_0, y_0)$ and $f_y(x_0, y_0)$, and explain your reasoning.

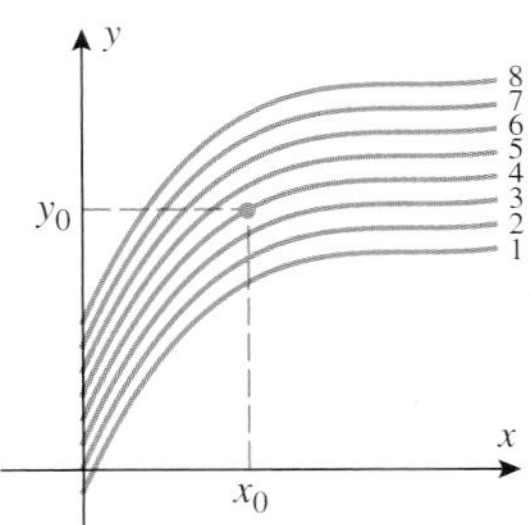

Figure Ex-8

9. Suppose that Nolan throws a baseball to Ryan and that the baseball leaves Nolan's hand at the same height at which it is caught by Ryan. It we ignore air resistance, the horizontal range r of the baseball is a function of the initial speed v of the ball when it leaves Nolan's hand and the angle θ above the horizontal at which it is thrown. Use the accompanying table and the method of Example 7 to estimate
 (a) the partial derivative of r with respect to v when $v = 80$ ft/s and $\theta = 40°$
 (b) the partial derivative of r with respect to θ when $v = 80$ ft/s and $\theta = 40°$.

ANGLE θ (degrees) \ SPEED v (ft/s)	75	80	85	90
35	165	188	212	238
40	173	197	222	249
45	176	200	226	253
50	173	197	222	249

Table Ex-9

10. Use the table in Exercise 9 and the method of Example 7 to estimate
 (a) the partial derivative of r with respect to v when $v = 85$ ft/s and $\theta = 45°$
 (b) the partial derivative of r with respect to θ when $v = 85$ ft/s and $\theta = 45°$.

11. The accompanying figure (next page) shows the graphs of an unspecified function $f(x, y)$ and its partial derivatives $f_x(x, y)$ and $f_y(x, y)$. Determine which is which, and explain your reasoning.

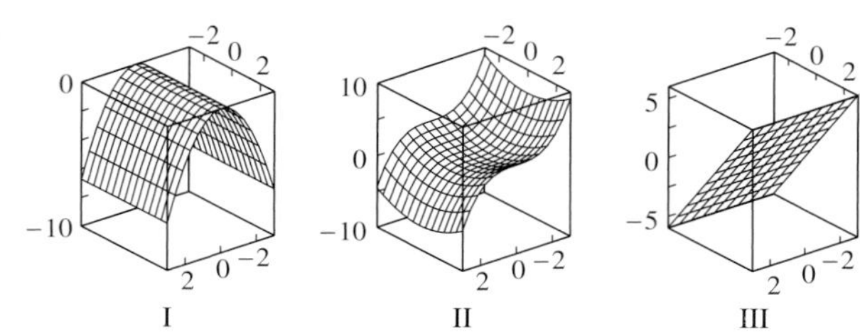

Figure Ex-11

12. What can you say about the signs of $\partial z/\partial x$, $\partial^2 z/\partial x^2$, $\partial z/\partial y$, and $\partial^2 z/\partial y^2$ at the point P in the accompanying figure? Explain your reasoning.

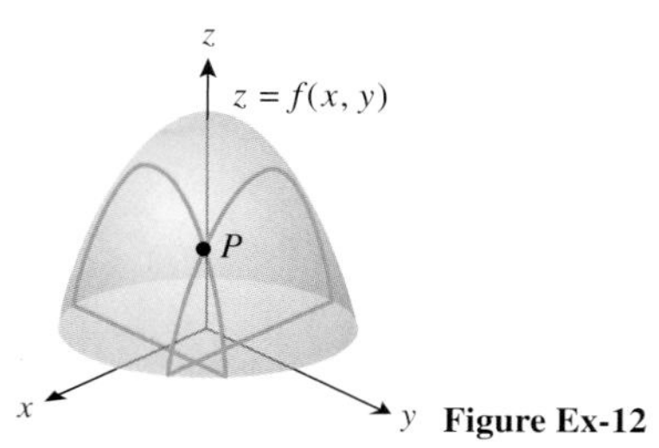

Figure Ex-12

13–18 Find $\partial z/\partial x$ and $\partial z/\partial y$.

13. $z = 4e^{x^2y^3}$

14. $z = \cos(x^5y^4)$

15. $z = x^3 \ln(1 + xy^{-3/5})$

16. $z = e^{xy} \sin 4y^2$

17. $z = \dfrac{xy}{x^2 + y^2}$

18. $z = \dfrac{x^2y^3}{\sqrt{x + y}}$

19–24 Find $f_x(x, y)$ and $f_y(x, y)$.

19. $f(x, y) = \sqrt{3x^5y - 7x^3y}$

20. $f(x, y) = \dfrac{x + y}{x - y}$

21. $f(x, y) = y^{-3/2} \tan^{-1}(x/y)$

22. $f(x, y) = x^3e^{-y} + y^3 \sec \sqrt{x}$

23. $f(x, y) = (y^2 \tan x)^{-4/3}$

24. $f(x, y) = \cosh(\sqrt{x}) \sinh^2(xy^2)$

25–28 Evaluate the indicated partial derivatives.

25. $f(x, y) = 9 - x^2 - 7y^3$; $f_x(3, 1)$, $f_y(3, 1)$

26. $f(x, y) = x^2ye^{xy}$; $\partial f/\partial x(1, 1)$, $\partial f/\partial y(1, 1)$

27. $z = \sqrt{x^2 + 4y^2}$; $\partial z/\partial x(1, 2)$, $\partial z/\partial y(1, 2)$

28. $w = x^2 \cos xy$; $\partial w/\partial x\left(\frac{1}{2}, \pi\right)$, $\partial w/\partial y\left(\frac{1}{2}, \pi\right)$

29. Let $f(x, y, z) = x^2y^4z^3 + xy + z^2 + 1$. Find
(a) $f_x(x, y, z)$ (b) $f_y(x, y, z)$ (c) $f_z(x, y, z)$
(d) $f_x(1, y, z)$ (e) $f_y(1, 2, z)$ (f) $f_z(1, 2, 3)$.

30. Let $w = x^2y \cos z$. Find
(a) $\partial w/\partial x(x, y, z)$ (b) $\partial w/\partial y(x, y, z)$
(c) $\partial w/\partial z(x, y, z)$ (d) $\partial w/\partial x(2, y, z)$
(e) $\partial w/\partial y(2, 1, z)$ (f) $\partial w/\partial z(2, 1, 0)$.

31–34 Find f_x, f_y, and f_z.

31. $f(x, y, z) = z \ln(x^2y \cos z)$

32. $f(x, y, z) = y^{-3/2} \sec\left(\dfrac{xz}{y}\right)$

33. $f(x, y, z) = \tan^{-1}\left(\dfrac{1}{xy^2z^3}\right)$

34. $f(x, y, z) = \cosh(\sqrt{z}) \sinh^2(x^2yz)$

35–38 Find $\partial w/\partial x$, $\partial w/\partial y$, and $\partial w/\partial z$.

35. $w = ye^z \sin xz$

36. $w = \dfrac{x^2 - y^2}{y^2 + z^2}$

37. $w = \sqrt{x^2 + y^2 + z^2}$

38. $w = y^3e^{2x+3z}$

39. Let $f(x, y, z) = y^2e^{xz}$. Find
(a) $\partial f/\partial x|_{(1,1,1)}$ (b) $\partial f/\partial y|_{(1,1,1)}$ (c) $\partial f/\partial z|_{(1,1,1)}$.

40. Let $w = \sqrt{x^2 + 4y^2 - z^2}$. Find
(a) $\partial w/\partial x|_{(2,1,-1)}$ (b) $\partial w/\partial y|_{(2,1,-1)}$
(c) $\partial w/\partial z|_{(2,1,-1)}$.

41. Let $f(x, y) = e^x \cos y$. Use a graphing utility to graph the functions $f_x(0, y)$ and $f_y(x, \pi/2)$.

42. Let $f(x, y) = e^x \sin y$. Use a graphing utility to graph the functions $f_x(0, y)$ and $f_y(x, 0)$.

43. A point moves along the intersection of the elliptic paraboloid $z = x^2 + 3y^2$ and the plane $y = 1$. At what rate is z changing with respect to x when the point is at $(2, 1, 7)$?

44. A point moves along the intersection of the elliptic paraboloid $z = x^2 + 3y^2$ and the plane $x = 2$. At what rate is z changing with respect to y when the point is at $(2, 1, 7)$?

45. A point moves along the intersection of the plane $y = 3$ and the surface $z = \sqrt{29 - x^2 - y^2}$. At what rate is z changing with respect to x when the point is at $(4, 3, 2)$?

46. Find the slope of the tangent line at $(-1, 1, 5)$ to the curve of intersection of the surface $z = x^2 + 4y^2$ and
(a) the plane $x = -1$ (b) the plane $y = 1$.

47. The volume V of a right circular cylinder is given by the formula $V = \pi r^2 h$, where r is the radius and h is the height.
(a) Find a formula for the instantaneous rate of change of V with respect to r if r changes and h remains constant.
(b) Find a formula for the instantaneous rate of change of V with respect to h if h changes and r remains constant.
(c) Suppose that h has aconstant value of 4 in, but r varies. Find the rate of change of V with respect to r at the point where $r = 6$ in.
(d) Suppose that r has a constant value of 8 in, but h varies. Find the instantaneous rate of change of V with respect to h at the point where $h = 10$ in.

48. The volume V of a right circular cone is given by

$$V = \frac{\pi}{24}d^2\sqrt{4s^2 - d^2}$$

where s is the slant height and d is the diameter of the base.

(a) Find a formula for the instantaneous rate of change of V with respect to s if d remains constant.
(b) Find a formula for the instantaneous rate of change of V with respect to d if s remains constant.
(c) Suppose that d has a constant value of 16 cm, but s varies. Find the rate of change of V with respect to s when $s = 10$ cm.
(d) Suppose that s has a constant value of 10 cm, but d varies. Find the rate of change of V with respect to d when $d = 16$ cm.

49. According to the ideal gas law, the pressure, temperature, and volume of a gas are related by $P = kT/V$, where k is a constant of proportionality. Suppose that V is measured in cubic inches (in^3), T is measured in kelvins (K), and that for a certain gas the constant of proportionality is $k = 10$ in·lb/K.
(a) Find the instantaneous rate of change of pressure with respect to temperature if the temperature is 80 K and the volume remains fixed at 50 in^3.
(b) Find the instantaneous rate of change of volume with respect to pressure if the volume is 50 in^3 and the temperature remains fixed at 80 K.

50. The temperature at a point (x, y) on a metal plate in the xy-plane is $T(x, y) = x^3 + 2y^2 + x$ degrees Celsius. Assume that distance is measured in centimeters and find the rate at which temperature changes with respect to distance if we start at the point $(1, 2)$ and move
(a) to the right and parallel to the x-axis
(b) upward and parallel to the y-axis.

51. The length, width, and height of a rectangular box are $l = 5$, $w = 2$, and $h = 3$, respectively.
(a) Find the instantaneous rate of change of the volume of the box with respect to the length if w and h are held constant.
(b) Find the instantaneous rate of change of the volume of the box with respect to the width if l and h are held constant.
(c) Find the instantaneous rate of change of the volume of the box with respect to the height if l and w are held constant.

52. The area A of a triangle is given by $A = \frac{1}{2}ab\sin\theta$, where a and b are the lengths of two sides and θ is the angle between these sides. Suppose that $a = 5$, $b = 10$, and $\theta = \pi/3$.
(a) Find the rate at which A changes with respect to a if b and θ are held constant.
(b) Find the rate at which A changes with respect to θ if a and b are held constant.
(c) Find the rate at which b changes with respect to a if A and θ are held constant.

53. The volume of a right circular cone of radius r and height h is $V = \frac{1}{3}\pi r^2 h$. Show that if the height remains constant while the radius changes, then the volume satisfies

$$\frac{\partial V}{\partial r} = \frac{2V}{r}$$

54. Find parametric equations for the tangent line at $(1, 3, 3)$ to the curve of intersection of the surface $z = x^2 y$ and
(a) the plane $x = 1$ (b) the plane $y = 3$.

55. (a) By differentiating implicitly, find the slope of the hyperboloid $x^2 + y^2 - z^2 = 1$ in the x-direction at the points $(3, 4, 2\sqrt{6})$ and $(3, 4, -2\sqrt{6})$.
(b) Check the results in part (a) by solving for z and differentiating the resulting functions directly.

56. (a) By differentiating implicitly, find the slope of the hyperboloid $x^2 + y^2 - z^2 = 1$ in the y-direction at the points $(3, 4, 2\sqrt{6})$ and $(3, 4, -2\sqrt{6})$.
(b) Check the results in part (a) by solving for z and differentiating the resulting functions directly.

57–60 Calculate $\partial z/\partial x$ and $\partial z/\partial y$ using implicit differentiation. Leave your answers in terms of x, y, and z.

57. $(x^2 + y^2 + z^2)^{3/2} = 1$ **58.** $\ln(2x^2 + y - z^3) = x$

59. $x^2 + z\sin xyz = 0$ **60.** $e^{xy}\sinh z - z^2 x + 1 = 0$

61–64 Find $\partial w/\partial x$, $\partial w/\partial y$, and $\partial w/\partial z$ using implicit differentiation. Leave your answers in terms of x, y, z, and w.

61. $(x^2 + y^2 + z^2 + w^2)^{3/2} = 4$

62. $\ln(2x^2 + y - z^3 + 3w) = z$

63. $w^2 + w\sin xyz = 1$

64. $e^{xy}\sinh w - z^2 w + 1 = 0$

65–66 Find f_x and f_y.

65. $f(x, y) = \int_y^x e^{t^2}\,dt$ **66.** $f(x, y) = \int_1^{xy} e^{t^2}\,dt$

67. Let $z = \sqrt{x}\cos y$. Find
(a) $\partial^2 z/\partial x^2$ (b) $\partial^2 z/\partial y^2$
(c) $\partial^2 z/\partial x\partial y$ (d) $\partial^2 z/\partial y\partial x$.

68. Let $f(x, y) = 4x^2 - 2y + 7x^4 y^5$. Find
(a) f_{xx} (b) f_{yy} (c) f_{xy} (d) f_{yx}.

69–76 Confirm that the mixed second-order partial derivatives of f are the same.

69. $f(x, y) = 4x^2 - 8xy^4 + 7y^5 - 3$

70. $f(x, y) = \sqrt{x^2 + y^2}$ **71.** $f(x, y) = e^x\cos y$

72. $f(x, y) = e^{x-y^2}$ **73.** $f(x, y) = \ln(4x - 5y)$

74. $f(x, y) = \ln(x^2 + y^2)$

75. $f(x, y) = (x - y)/(x + y)$

76. $f(x, y) = (x^2 - y^2)/(x^2 + y^2)$

77. Express the following derivatives in "∂" notation.
(a) f_{xxx} (b) f_{xyy} (c) f_{yyxx} (d) f_{xyyy}

78. Express the derivatives in "subscript" notation.
(a) $\dfrac{\partial^3 f}{\partial y^2\partial x}$ (b) $\dfrac{\partial^4 f}{\partial x^4}$ (c) $\dfrac{\partial^4 f}{\partial y^2\partial x^2}$ (d) $\dfrac{\partial^5 f}{\partial x^2\partial y^3}$

79. Given $f(x, y) = x^3y^5 - 2x^2y + x$, find
(a) f_{xxy} (b) f_{yxy} (c) f_{yyy}.

80. Given $z = (2x - y)^5$, find
(a) $\dfrac{\partial^3 z}{\partial y \partial x \partial y}$ (b) $\dfrac{\partial^3 z}{\partial x^2 \partial y}$ (c) $\dfrac{\partial^4 z}{\partial x^2 \partial y^2}$.

81. Given $f(x, y) = y^3 e^{-5x}$, find
(a) $f_{xyy}(0, 1)$ (b) $f_{xxx}(0, 1)$ (c) $f_{yyxx}(0, 1)$.

82. Given $w = e^y \cos x$, find
(a) $\left.\dfrac{\partial^3 w}{\partial y^2 \partial x}\right|_{(\pi/4,0)}$ (b) $\left.\dfrac{\partial^3 w}{\partial x^2 \partial y}\right|_{(\pi/4,0)}$

83. Let $f(x, y, z) = x^3y^5z^7 + xy^2 + y^3z$. Find
(a) f_{xy} (b) f_{yz} (c) f_{xz} (d) f_{zz}
(e) f_{zyy} (f) f_{xxy} (g) f_{zyx} (h) f_{xxyz}.

84. Let $w = (4x - 3y + 2z)^5$. Find
(a) $\dfrac{\partial^2 w}{\partial x \partial z}$ (b) $\dfrac{\partial^3 w}{\partial x \partial y \partial z}$ (c) $\dfrac{\partial^4 w}{\partial z^2 \partial y \partial x}$.

85. Show that the function satisfies ***Laplace's equation***

$$\frac{\partial^2 z}{\partial x^2} + \frac{\partial^2 z}{\partial y^2} = 0$$

(a) $z = x^2 - y^2 + 2xy$
(b) $z = e^x \sin y + e^y \cos x$
(c) $z = \ln(x^2 + y^2) + 2\tan^{-1}(y/x)$

86. Show that the function satisfies the ***heat equation***

$$\frac{\partial z}{\partial t} = c^2 \frac{\partial^2 z}{\partial x^2} \quad (c > 0, \text{ constant})$$

(a) $z = e^{-t}\sin(x/c)$ (b) $z = e^{-t}\cos(x/c)$

87. Show that the function $u(x, t) = \sin c\omega t \sin \omega x$ satisfies the wave equation [Equation (6)] for all real values of ω.

88. In each part, show that $u(x, y)$ and $v(x, y)$ satisfy the ***Cauchy–Riemann equations***

$$\frac{\partial u}{\partial x} = \frac{\partial v}{\partial y} \quad \text{and} \quad \frac{\partial u}{\partial y} = -\frac{\partial v}{\partial x}$$

(a) $u = x^2 - y^2$, $v = 2xy$
(b) $u = e^x \cos y$, $v = e^x \sin y$
(c) $u = \ln(x^2 + y^2)$, $v = 2\tan^{-1}(y/x)$

89. Show that if $u(x, y)$ and $v(x, y)$ each have equal mixed second partials, and if u and v satisfy the Cauchy–Riemann equations (Exercise 88), then u, v, and $u + v$ satisfy Laplace's equation (Exercise 85).

90. When two resistors having resistances R_1 ohms and R_2 ohms are connected in parallel, their combined resistance R in ohms is $R = R_1R_2/(R_1 + R_2)$. Show that

$$\frac{\partial^2 R}{\partial R_1^2}\frac{\partial^2 R}{\partial R_2^2} = \frac{4R^2}{(R_1 + R_2)^4}$$

91–94 Find the indicated partial derivatives.

91. $f(v, w, x, y) = 4v^2w^3x^4y^5$;
$\partial f/\partial v$, $\partial f/\partial w$, $\partial f/\partial x$, $\partial f/\partial y$

92. $w = r\cos st + e^u \sin ur$;
$\partial w/\partial r$, $\partial w/\partial s$, $\partial w/\partial t$, $\partial w/\partial u$

93. $f(v_1, v_2, v_3, v_4) = \dfrac{v_1^2 - v_2^2}{v_3^2 + v_4^2}$;
$\partial f/\partial v_1$, $\partial f/\partial v_2$, $\partial f/\partial v_3$, $\partial f/\partial v_4$

94. $V = xe^{2x-y} + we^{zw} + yw$;
$\partial V/\partial x$, $\partial V/\partial y$, $\partial V/\partial z$, $\partial V/\partial w$

95. Let $u(w, x, y, z) = xe^{yw}\sin^2 z$. Find
(a) $\dfrac{\partial u}{\partial x}(0, 0, 1, \pi)$ (b) $\dfrac{\partial u}{\partial y}(0, 0, 1, \pi)$
(c) $\dfrac{\partial u}{\partial w}(0, 0, 1, \pi)$ (d) $\dfrac{\partial u}{\partial z}(0, 0, 1, \pi)$
(e) $\dfrac{\partial^4 u}{\partial x \partial y \partial w \partial z}$ (f) $\dfrac{\partial^4 u}{\partial w \partial z \partial y^2}$.

96. Let $f(v, w, x, y) = 2v^{1/2}w^4x^{1/2}y^{2/3}$. Find $f_v(1, -2, 4, 8)$, $f_w(1, -2, 4, 8)$, $f_x(1, -2, 4, 8)$, and $f_y(1, -2, 4, 8)$.

97–98 Find $\partial w/\partial x_i$ for $i = 1, 2, \ldots, n$.

97. $w = \cos(x_1 + 2x_2 + \cdots + nx_n)$

98. $w = \left(\displaystyle\sum_{k=1}^{n} x_k\right)^{1/n}$

99–100 Describe the largest set on which Theorem 14.3.2 may be used to prove that f_{xy} and f_{yx} are equal on that set. Then confirm by direct computation that $f_{xy} = f_{yx}$ on the given set.

99. (a) $f(x, y) = 4x^3y + 3x^2y$ (b) $f(x, y) = x^3/y$

100. (a) $f(x, y) = \sqrt{x^2 + y^2 - 1}$
(b) $f(x, y) = \sin(x^2 + y^3)$

101. Let $f(x, y) = 2x^2 - 3xy + y^2$. Find $f_x(2, -1)$ and $f_y(2, -1)$ by evaluating the limits in Definition 14.3.1. Then check your work by calculating the derivative in the usual way.

102. Let $f(x, y) = (x^2 + y^2)^{2/3}$. Show that

$$f_x(x, y) = \begin{cases} \dfrac{4x}{3(x^2 + y^2)^{1/3}}, & (x, y) \neq (0, 0) \\ 0, & (x, y) = (0, 0) \end{cases}$$

Source: This problem, due to Don Cohen, appeared in *Mathematics and Computer Education*, Vol. 25, No. 2, 1991, p. 179.

103. Let $f(x, y) = (x^3 + y^3)^{1/3}$.
(a) Show that $f_y(0, 0) = 1$.
(b) At what points, if any, does $f_y(x, y)$ fail to exist?

✔ QUICK CHECK ANSWERS 14.3

1. $\sin xy + xy\cos xy$; $x^2\cos xy$ **2.** 9; 12 **3.** (a) $\frac{2}{3}\pi rh$ (b) $\frac{1}{3}\pi r^2 s$
4. $f_{xx}(x, y) = 2y^3$, $f_{yy}(x, y) = 6x^2y$, $f_{xy}(x, y) = f_{yx}(x, y) = 6xy^2$

14.4 DIFFERENTIABILITY, DIFFERENTIALS, AND LOCAL LINEARITY

In this section we will extend the notion of differentiability to functions of two or three variables. Our definition of differentiability will be based on the idea that a function is differentiable at a point provided it can be very closely approximated by a linear function near that point. In the process, we will expand the concept of a "differential" to functions of more than one variable and define the "local linear approximation" of a function.

DIFFERENTIABILITY

Recall that a function f of one variable is called differentiable at x_0 if it has a derivative at x_0, that is, if the limit

$$f'(x_0) = \lim_{\Delta x \to 0} \frac{f(x_0 + \Delta x) - f(x_0)}{\Delta x} \tag{1}$$

exists. As a consequence of (1) a differentiable function enjoys a number of other important properties:

- The graph of $y = f(x)$ has a nonvertical tangent line at the point $(x_0, f(x_0))$;
- f may be closely approximated by a linear function near x_0 (Section 3.9);
- f is continuous at x_0.

Our primary objective in this section is to extend the notion of differentiability to functions of two or three variables in such a way that the natural analogs of these properties hold. For example, if a function $f(x, y)$ of two variables is differentiable at a point (x_0, y_0), we want it to be the case that

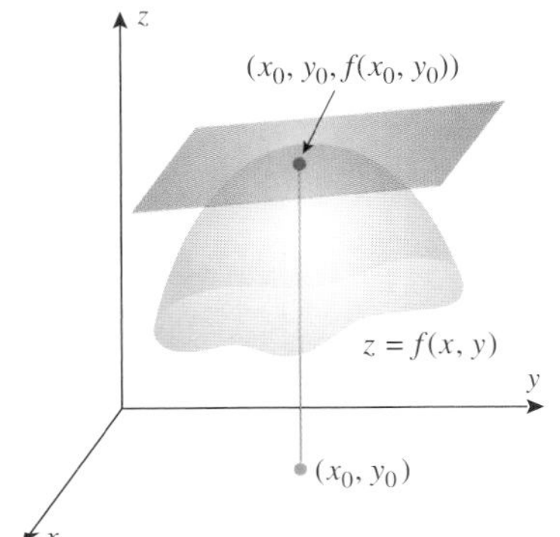

Figure 14.4.1

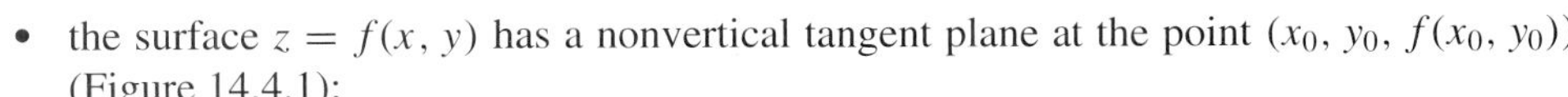

- the surface $z = f(x, y)$ has a nonvertical tangent plane at the point $(x_0, y_0, f(x_0, y_0))$ (Figure 14.4.1);
- the values of f at points near (x_0, y_0) can be very closely approximated by the values of a linear function;
- f is continuous at (x_0, y_0).

One could reasonably conjecture that a function f of two or three variables should be called differentiable at a point if all the first-order partial derivatives of the function exist at that point. Unfortunately, this condition is not strong enough to guarantee that the properties above hold. For instance, we saw in Example 5 of Section 14.3 that the mere existence of both first-order partial derivatives for a function is not sufficient to guarantee the continuity of the function. To determine what else we should include in our definition, it will be helpful to reexamine one of the consequences of differentiability for a *single-variable* function $f(x)$. Suppose that $f(x)$ is differentiable at $x = x_0$ and let

$$\Delta f = f(x_0 + \Delta x) - f(x_0)$$

denote the change in f that corresponds to the change Δx in x from x_0 to $x_0 + \Delta x$. We saw in Section 3.9 that

$$\Delta f \approx f'(x_0)\Delta x$$

provide Δx is close to 0. In fact, for Δx close to 0 the error $\Delta f - f'(x_0)\Delta x$ in this approximation will have magnitude much smaller than that of Δx because

$$\lim_{\Delta x \to 0} \frac{\Delta f - f'(x_0)\Delta x}{\Delta x} = \lim_{\Delta x \to 0} \left(\frac{f(x_0 + \Delta x) - f(x_0)}{\Delta x} - f'(x_0) \right) = f'(x_0) - f'(x_0) = 0$$

Since the magnitude of Δx is just the distance between the points x_0 and $x_0 + \Delta x$, we see that when the two points are close together, the magnitude of the error in the approximation will be much smaller than the distance between the two points (Figure 14.4.2). The extension of this idea to functions of two or three variables is the "extra ingredient" needed in our definition of differentiability for multivariable functions.

For a function $f(x, y)$, the symbol Δf, called the ***increment*** of f, denotes the change in the value of $f(x, y)$ that results when (x, y) varies from some initial position (x_0, y_0) to some new position $(x_0 + \Delta x, y_0 + \Delta y)$; thus

$$\Delta f = f(x_0 + \Delta x, y_0 + \Delta y) - f(x_0, y_0) \tag{2}$$

(see Figure 14.4.3). [If a dependent variable $z = f(x, y)$ is used, then we will sometimes write Δz rather than Δf.] Let us assume that both $f_x(x_0, y_0)$ and $f_y(x_0, y_0)$ exist and (by analogy with the one-variable case) make the approximation

$$\Delta f \approx f_x(x_0, y_0)\Delta x + f_y(x_0, y_0)\Delta y \tag{3}$$

Show that if $f(x, y)$ is a linear function, then (3) becomes an equality.

For Δx and Δy close to 0, we would like the error

$$\Delta f - f_x(x_0, y_0)\Delta x - f_y(x_0, y_0)\Delta y$$

in this approximation to be much smaller than the distance $\sqrt{(\Delta x)^2 + (\Delta y)^2}$ between (x_0, y_0) and $(x_0 + \Delta x, y_0 + \Delta y)$. We can guarantee this by requiring that

$$\lim_{(\Delta x, \Delta y) \to (0,0)} \frac{\Delta f - f_x(x_0, y_0)\Delta x - f_y(x_0, y_0)\Delta y}{\sqrt{(\Delta x)^2 + (\Delta y)^2}} = 0$$

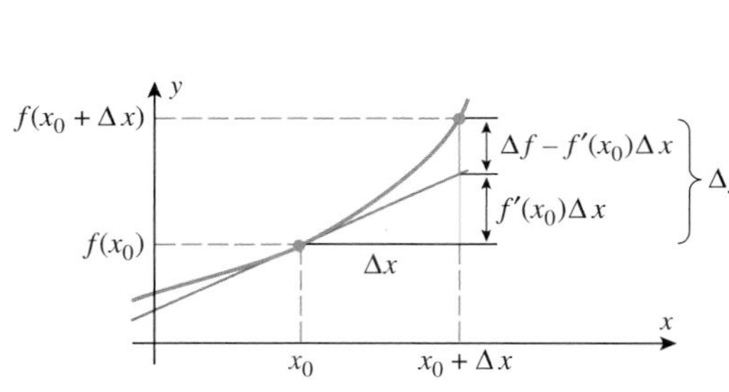

Figure 14.4.2

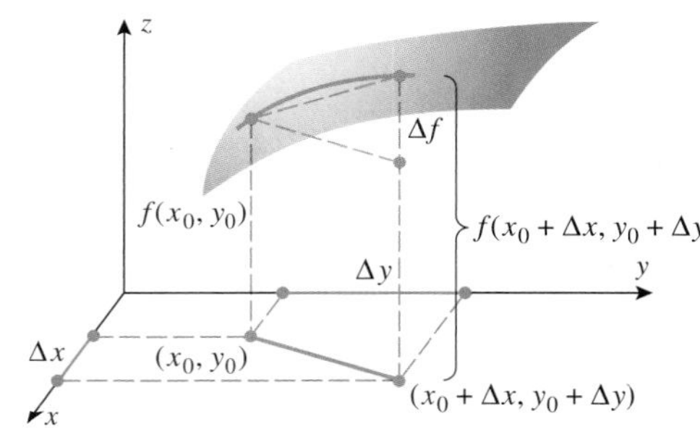

Figure 14.4.3

Based on these ideas, we can now give our definition of differentiability for functions of two variables.

14.4.1 DEFINITION. A function f of two variables is said to be ***differentiable*** at (x_0, y_0) provided $f_x(x_0, y_0)$ and $f_y(x_0, y_0)$ both exist and

$$\lim_{(\Delta x, \Delta y) \to (0,0)} \frac{\Delta f - f_x(x_0, y_0)\Delta x - f_y(x_0, y_0)\Delta y}{\sqrt{(\Delta x)^2 + (\Delta y)^2}} = 0 \tag{4}$$

As with the one-variable case, verification of differentiability using this definition involves the computation of a limit.

▶ **Example 1** Use Definition 14.4.1 to prove that $f(x, y) = x^2 + y^2$ is differentiable at $(0, 0)$.

Solution. The increment is

$$\Delta f = f(0 + \Delta x, 0 + \Delta y) - f(0, 0) = (\Delta x)^2 + (\Delta y)^2$$

Since $f_x(x, y) = 2x$ and $f_y(x, y) = 2y$, we have $f_x(0, 0) = f_y(0, 0) = 0$, and (4) becomes

$$\lim_{(\Delta x, \Delta y) \to (0,0)} \frac{(\Delta x)^2 + (\Delta y)^2}{\sqrt{(\Delta x)^2 + (\Delta y)^2}} = \lim_{(\Delta x, \Delta y) \to (0,0)} \sqrt{(\Delta x)^2 + (\Delta y)^2} = 0$$

Therefore, f is differentiable at $(0, 0)$. ◀

We now derive an important consequence of limit (4). Define a function

$$\epsilon = \epsilon(\Delta x, \Delta y) = \frac{\Delta f - f_x(x_0, y_0)\Delta x - f_y(x_0, y_0)\Delta y}{\sqrt{(\Delta x)^2 + (\Delta y)^2}} \quad \text{for } (\Delta x, \Delta y) \neq (0, 0)$$

and define $\epsilon(0, 0)$ to be 0. Equation (4) then implies that

$$\lim_{(\Delta x, \Delta y) \to (0,0)} \epsilon(\Delta x, \Delta y) = 0$$

Furthermore, it immediately follows from the definition of ϵ that

$$\Delta f = f_x(x_0, y_0)\Delta x + f_y(x_0, y_0)\Delta y + \epsilon\sqrt{(\Delta x)^2 + (\Delta y)^2} \tag{5}$$

In other words, if f is differentiable at (x_0, y_0), then Δf may be expressed as shown in (5), where $\epsilon \to 0$ as $(\Delta x, \Delta y) \to 0$ and where $\epsilon = 0$ if $(\Delta x, \Delta y) = (0, 0)$.

For functions of three variables we have an analogous definition of differentiability in terms of the increment

$$\Delta f = f(x_0 + \Delta x, y_0 + \Delta y, z_0 + \Delta z) - f(x_0, y_0, z_0)$$

14.4.2 DEFINITION. A function f of three variables is said to be ***differentiable*** at (x_0, y_0, z_0) provided $f_x(x_0, y_0, z_0)$, $f_y(x_0, y_0, z_0)$, and $f_z(x_0, y_0, z_0)$ exist and

$$\lim_{(\Delta x, \Delta y, \Delta z) \to (0,0,0)} \frac{\Delta f - f_x(x_0, y_0, z_0)\Delta x - f_y(x_0, y_0, z_0)\Delta y - f_z(x_0, y_0, z_0)\Delta z}{\sqrt{(\Delta x)^2 + (\Delta y)^2 + (\Delta z)^2}} = 0 \tag{6}$$

In a manner similar to the two-variable case, we can express the limit (6) in terms of a function $\epsilon(\Delta x, \Delta y, \Delta z)$ that vanishes at $(\Delta x, \Delta y, \Delta z) = (0, 0, 0)$ and is continuous there. The details are left as an exercise for the reader.

If a function f of two variables is differentiable at each point of a region R in the xy-plane, then we say that f is ***differentiable on R***; and if f is differentiable at every point in the xy-plane, then we say that f is ***differentiable everywhere***. For a function f of three variables we have corresponding conventions.

DIFFERENTIABILITY AND CONTINUITY

Recall that we want a function to be continuous at every point at which it is differentiable. The next result shows this to be the case.

14.4.3 THEOREM. *If a function is differentiable at a point, then it is continuous at that point.*

PROOF. We will give the proof for $f(x, y)$, a function of two variables, since that will reveal the essential ideas. Assume that f is differentiable at (x_0, y_0). To prove that f is continuous at (x_0, y_0) we must show that

$$\lim_{(x,y)\to(x_0,y_0)} f(x, y) = f(x_0, y_0)$$

which, on letting $x = x_0 + \Delta x$ and $y = y_0 + \Delta y$, is equivalent to

$$\lim_{(\Delta x,\Delta y)\to(0,0)} f(x_0 + \Delta x, y_0 + \Delta y) = f(x_0, y_0)$$

By Equation (2) this is equivalent to

$$\lim_{(\Delta x,\Delta y)\to(0,0)} \Delta f = 0$$

However, from Equation (5)

$$\begin{aligned}\lim_{(\Delta x,\Delta y)\to(0,0)} \Delta f &= \lim_{(\Delta x,\Delta y)\to(0,0)} [f_x(x_0, y_0)\Delta x + f_y(x_0, y_0)\Delta y \\ &\qquad + \epsilon(\Delta x, \Delta y)\sqrt{(\Delta x)^2 + (\Delta y)^2}\,] \\ &= 0 + 0 + 0 \cdot 0 = 0\end{aligned}$$

■

The converse of Theorem 14.4.3 is false. For example, explain why

$$f(x, y) = \sqrt{x^2 + y^2}$$

is continuous at $(0, 0)$ but not differentiable at $(0, 0)$.

It can be difficult to verify that a function is differentiable at a point directly from the definition. The next theorem, whose proof is usually studied in more advanced courses, provides simple conditions for a function to be differentiable at a point.

14.4.4 THEOREM. *If all first-order partial derivatives of f exist and are continuous at a point, then f is differentiable at that point.*

For example, consider the function

$$f(x, y, z) = x + yz$$

Since $f_x(x, y, z) = 1$, $f_y(x, y, z) = z$, and $f_z(x, y, z) = y$ are defined and continuous everywhere, we conclude from Theorem 14.4.4 that f is differentiable everywhere.

DIFFERENTIALS

As with the one-variable case, the approximations

$$\Delta f \approx f_x(x_0, y_0)\Delta x + f_y(x_0, y_0)\Delta y$$

for a function of two variables and the approximation

$$\Delta f \approx f_x(x_0, y_0, z_0)\Delta x + f_y(x_0, y_0, z_0)\Delta y + f_z(x_0, y_0, z_0)\Delta z$$

for a function of three variables have a convenient formulation in the language of differentials. If $z = f(x, y)$ is differentiable at a point (x_0, y_0), we let

$$dz = f_x(x_0, y_0)\,dx + f_y(x_0, y_0)\,dy \tag{7}$$

denote a new function with dependent variable dz and independent variables dx and dy. We refer to this function (also denoted df) as the ***total differential of z*** at (x_0, y_0) or as the

total differential of f at (x_0, y_0). Similarly, for a function $w = f(x, y, z)$ of three variables we have the ***total differential of w*** at (x_0, y_0, z_0),

$$dw = f_x(x_0, y_0, z_0)\,dx + f_y(x_0, y_0, z_0)\,dy + f_z(x_0, y_0, z_0)\,dz \tag{8}$$

which is also referred to as the ***total differential of f*** at (x_0, y_0, z_0). It is common practice to omit the subscripts and write Equations (7) and (8) as

$$dz = f_x(x, y)\,dx + f_y(x, y)\,dy \tag{9}$$

and

$$dw = f_x(x, y, z)\,dx + f_y(x, y, z)\,dy + f_z(x, y, z)\,dz \tag{10}$$

In the two-variable case, the approximation

$$\Delta f \approx f_x(x_0, y_0)\Delta x + f_y(x_0, y_0)\Delta y$$

can be written in the form

$$\Delta f \approx df \tag{11}$$

for $dx = \Delta x$ and $dy = \Delta y$. Equivalently, we can write approximation (11) as

$$\Delta z \approx dz \tag{12}$$

In other words, we can estimate the change Δz in z by the value of the differential dz where dx is the change in x and dy is the change in y. Furthermore, it follows from (4) that if Δx and Δy are close to 0, then the magnitude of the error in approximation (12) will be much smaller than the distance $\sqrt{(\Delta x)^2 + (\Delta y)^2}$ between (x_0, y_0) and $(x_0 + \Delta x, y_0 + \Delta y)$.

▶ **Example 2** Use (12) to approximate the change in $z = xy^2$ from its value at (0.5, 1.0) to its value at (0.503, 1.004). Compare the magnitude of the error in this approximation with the distance between the points (0.5, 1.0) and (0.503, 1.004).

Solution. For $z = xy^2$ we have $dz = y^2\,dx + 2xy\,dy$. Evaluating this differential at $(x, y) = (0.5, 1.0)$, $dx = \Delta x = 0.503 - 0.5 = 0.003$, and $dy = \Delta y = 1.004 - 1.0 = 0.004$ yields

$$dz = 1.0^2(0.003) + 2(0.5)(1.0)(0.004) = 0.007$$

Since $z = 0.5$ at $(x, y) = (0.5, 1.0)$ and $z = 0.507032048$ at $(x, y) = (0.503, 1.004)$, we have

$$\Delta z = 0.507032048 - 0.5 = 0.007032048$$

and the error in approximating Δz by dz has magnitude

$$|dz - \Delta z| = |0.007 - 0.007032048| = 0.000032048$$

Since the distance between (0.5, 1.0) and $(0.503, 1.004) = (0.5 + \Delta x, 1.0 + \Delta y)$ is

$$\sqrt{(\Delta x)^2 + (\Delta y)^2} = \sqrt{(0.003)^2 + (0.004)^2} = \sqrt{0.000025} = 0.005$$

we have

$$\frac{|dz - \Delta z|}{\sqrt{(\Delta x)^2 + (\Delta y)^2}} = \frac{0.000032048}{0.005} = 0.0064096 < \frac{1}{150}$$

Thus, the magnitude of the error in our approximation is less than $\frac{1}{150}$ of the distance between the two points. ◀

With the appropriate changes in notation, the preceding analysis can be extended to functions of three or more variables.

▶ **Example 3** The length, width, and height of a rectangular box are measured with an error of at most 5%. Use a total differential to estimate the maximum percentage error that results if these quantities are used to calculate the diagonal of the box.

Solution. The diagonal D of a box with length x, width y, and height z is given by

$$D = \sqrt{x^2 + y^2 + z^2}$$

Let x_0, y_0, z_0, and $D_0 = \sqrt{x_0^2 + y_0^2 + z_0^2}$ denote the actual values of the length, width, height, and diagonal of the box. The total differential dD of D at (x_0, y_0, z_0) is given by

$$dD = \frac{x_0}{\sqrt{x_0^2 + y_0^2 + z_0^2}}\,dx + \frac{y_0}{\sqrt{x_0^2 + y_0^2 + z_0^2}}\,dy + \frac{z_0}{\sqrt{x_0^2 + y_0^2 + z_0^2}}\,dz$$

If x, y, z, and $D = \sqrt{x^2 + y^2 + z^2}$ are the measured and computed values of the length, width, height, and diagonal, respectively, then $\Delta x = x - x_0$, $\Delta y = y - y_0$, $\Delta z = z - z_0$, and

$$\left|\frac{\Delta x}{x_0}\right| \le 0.05, \quad \left|\frac{\Delta y}{y_0}\right| \le 0.05, \quad \left|\frac{\Delta z}{z_0}\right| \le 0.05$$

We are seeking an estimate for the maximum size of $\Delta D / D_0$. With the aid of Equation (10) we have

$$\begin{aligned}\frac{\Delta D}{D_0} \approx \frac{dD}{D_0} &= \frac{1}{x_0^2 + y_0^2 + z_0^2}[x_0\Delta x + y_0\Delta y + z_0\Delta z] \\ &= \frac{1}{x_0^2 + y_0^2 + z_0^2}\left[x_0^2\frac{\Delta x}{x_0} + y_0^2\frac{\Delta y}{y_0} + z_0^2\frac{\Delta z}{z_0}\right]\end{aligned}$$

Since

$$\begin{aligned}\left|\frac{dD}{D_0}\right| &= \frac{1}{x_0^2 + y_0^2 + z_0^2}\left|x_0^2\frac{\Delta x}{x_0} + y_0^2\frac{\Delta y}{y_0} + z_0^2\frac{\Delta z}{z_0}\right| \\ &\le \frac{1}{x_0^2 + y_0^2 + z_0^2}\left(x_0^2\left|\frac{\Delta x}{x_0}\right| + y_0^2\left|\frac{\Delta y}{y_0}\right| + z_0^2\left|\frac{\Delta z}{z_0}\right|\right) \\ &\le \frac{1}{x_0^2 + y_0^2 + z_0^2}\left(x_0^2(0.05) + y_0^2(0.05) + z_0^2(0.05)\right) = 0.05\end{aligned}$$

we estimate the maximum percentage error in D to be 5%. ◀

LOCAL LINEAR APPROXIMATIONS

We now show that if a function f is differentiable at a point, then it can be very closely approximated by a linear function near that point. For example, suppose that $f(x, y)$ is differentiable at the point (x_0, y_0). Then approximation (3) can be written in the form

$$f(x_0 + \Delta x, y_0 + \Delta y) \approx f(x_0, y_0) + f_x(x_0, y_0)\Delta x + f_y(x_0, y_0)\Delta y$$

If we let $x = x_0 + \Delta x$ and $y = x_0 + \Delta y$, this approximation becomes

$$f(x, y) \approx f(x_0, y_0) + f_x(x_0, y_0)(x - x_0) + f_y(x_0, y_0)(y - y_0) \tag{13}$$

> Show that if $f(x, y)$ is a linear function, then (13) becomes an equality.

which yields a linear approximation of $f(x, y)$. Since the error in this approximation is equal to the error in the approximation (3), we conclude that for (x, y) close to (x_0, y_0), the error in (13) will be much smaller than the distance between these two points. When $f(x, y)$ is differentiable at (x_0, y_0) we get

> Explain why the error in approximation (13) is the same as the error in approximation (3).

$$L(x, y) = f(x_0, y_0) + f_x(x_0, y_0)(x - x_0) + f_y(x_0, y_0)(y - y_0) \tag{14}$$

and refer to $L(x, y)$ as the ***local linear approximation to f at*** (x_0, y_0).

▶ **Example 4** Let $L(x, y)$ denote the local linear approximation to $f(x, y) = \sqrt{x^2 + y^2}$ at the point (3, 4). Compare the error in approximating

$$f(3.04, 3.98) = \sqrt{(3.04)^2 + (3.98)^2}$$

by $L(3.04, 3.98)$ with the distance between the points (3, 4) and (3.04, 3.98).

Solution. We have

$$f_x(x, y) = \frac{x}{\sqrt{x^2 + y^2}} \quad \text{and} \quad f_y(x, y) = \frac{y}{\sqrt{x^2 + y^2}}$$

with $f_x(3, 4) = \frac{3}{5}$ and $f_y(3, 4) = \frac{4}{5}$. Therefore, the local linear approximation to f at (3, 4) is given by

$$L(x, y) = 5 + \tfrac{3}{5}(x - 3) + \tfrac{4}{5}(y - 4)$$

Consequently,

$$f(3.04, 3.98) \approx L(3.04, 3.98) = 5 + \tfrac{3}{5}(0.04) + \tfrac{4}{5}(-0.02) = 5.008$$

Since

$$f(3.04, 3.98) = \sqrt{(3.04)^2 + (3.98)^2} \approx 5.00819$$

the error in the approximation is about $5.00819 - 5.008 = 0.00019$. This is less than $\frac{1}{200}$ of the distance

$$\sqrt{(3.04 - 3)^2 + (3.98 - 4)^2} \approx 0.045$$

between the points (3, 4) and (3.04, 3.98). ◀

Similarly, for a function $f(x, y, z)$ that is differentiable at (x_0, y_0, z_0), the local linear approximation is

$$\begin{aligned} L(x, y, z) = f(x_0, y_0, z_0) &+ f_x(x_0, y_0, z_0)(x - x_0) \\ &+ f_y(x_0, y_0, z_0)(y - y_0) + f_z(x_0, y_0, z_0)(z - z_0) \end{aligned} \tag{15}$$

We have formulated our definitions in this section in such a way that continuity and local linearity are consequences of differentiability. In Section 14.7 we will show that if a function $f(x, y)$ is differentiable at a point (x_0, y_0), then the graph of $L(x, y)$ is a nonvertical tangent plane to the graph of f at the point $(x_0, y_0, f(x_0, y_0))$.

✔QUICK CHECK EXERCISES 14.4 *(See page 967 for answers.)*

1. Assume that $f(x, y)$ is differentiable at (x_0, y_0) and let Δf denote the change in f from its value at (x_0, y_0) to its value at $(x_0 + \Delta x, y_0 + \Delta y)$.
   (a) $\Delta f \approx$ ________
   (b) The limit that guarantees the error in the approximation in part (a) is very small when both Δx and Δy are close to 0 is ________.
2. Compute the differential of each function.
   (a) $z = xe^{y^2}$ (b) $w = x \sin(yz)$
3. If f is differentiable at (x_0, y_0), then the local linear approximation to f at (x_0, y_0) is $L(x) =$ ________.
4. Assume that $f(1, -2) = 4$ and $f(x, y)$ is differentiable at $(1, -2)$ with $f_x(1, -2) = 2$ and $f_y(1, -2) = -3$. Estimate the value of $f(0.9, -1.950)$.

EXERCISE SET 14.4

FOCUS ON CONCEPTS

1. Suppose that a function $f(x, y)$ is differentiable at the point (3, 4) with $f_x(3, 4) = 2$ and $f_y(3, 4) = -1$. If $f(3, 4) = 5$, estimate the value of $f(3.01, 3.98)$.
2. Suppose that a function $f(x, y)$ is differentiable at the point $(-1, 2)$ with $f_x(-1, 2) = 1$ and $f_y(-1, 2) = 3$. If $f(-1, 2) = 2$, estimate the value of $f(-0.99, 2.02)$.

3. Suppose that a function $f(x, y, z)$ is differentiable at the point $(1, 2, 3)$ with $f_x(1, 2, 3) = 1$, $f_y(1, 2, 3) = 2$, and $f_z(1, 2, 3) = 3$. If $f(1, 2, 3) = 4$, estimate the value of $f(1.01, 2.02, 3.03)$.

4. Suppose that a function $f(x, y, z)$ is differentiable at the point $(2, 1, -2)$, $f_x(2, 1, -2) = -1$, $f_y(2, 1, -2) = 1$, and $f_z(2, 1, -2) = -2$. If $f(2, 1, -2) = 0$, estimate the value of $f(1.98, 0.99, -1.97)$.

5. Use Definitions 14.4.1 and 14.4.2 to prove that a constant function of two or three variables is differentiable everywhere.

6. Use Definitions 14.4.1 and 14.4.2 to prove that a linear function of two or three variables is differentiable everywhere.

7. Use Definition 14.4.2 to prove that
$$f(x, y, z) = x^2 + y^2 + z^2$$
is differentiable at $(0, 0, 0)$.

8. Use Definition 14.4.2 to determine all values of r such that $f(x, y, z) = (x^2 + y^2 + z^2)^r$ is differentiable at $(0, 0, 0)$.

9–20 Compute the differential dz or dw of the specified function.

9. $z = 7x - 2y$
10. $z = e^{xy}$
11. $z = x^3y^2$
12. $z = 5x^2y^5 - 2x + 4y + 7$
13. $z = \tan^{-1} xy$
14. $z = \sec^2(x - 3y)$
15. $w = 8x - 3y + 4z$
16. $w = e^{xyz}$
17. $w = x^3y^2z$
18. $w = 4x^2y^3z^7 - 3xy + z + 5$
19. $w = \tan^{-1}(xyz)$
20. $w = \sqrt{x} + \sqrt{y} + \sqrt{z}$

21–26 Use a total differential to approximate the change in the values of f from P to Q. Compare your estimate with the actual change in f.

21. $f(x, y) = x^2 + 2xy - 4x$; $P(1, 2)$, $Q(1.01, 2.04)$
22. $f(x, y) = x^{1/3}y^{1/2}$; $P(8, 9)$, $Q(7.78, 9.03)$
23. $f(x, y) = \dfrac{x + y}{xy}$; $P(-1, -2)$, $Q(-1.02, -2.04)$
24. $f(x, y) = \ln\sqrt{1 + xy}$; $P(0, 2)$, $Q(-0.09, 1.98)$
25. $f(x, y, z) = 2xy^2z^3$; $P(1, -1, 2)$, $Q(0.99, -1.02, 2.02)$
26. $f(x, y, z) = \dfrac{xyz}{x + y + z}$; $P(-1, -2, 4)$, $Q(-1.04, -1.98, 3.97)$

27. In the accompanying figure a rectangle with initial length x_0 and initial width y_0 has been enlarged, resulting in a rectangle with length $x_0 + \Delta x$ and width $y_0 + \Delta y$. What portion of the figure represents the increase in the area of the rectangle? What portion of the figure represents an approximation of the increase in area by a total differential?

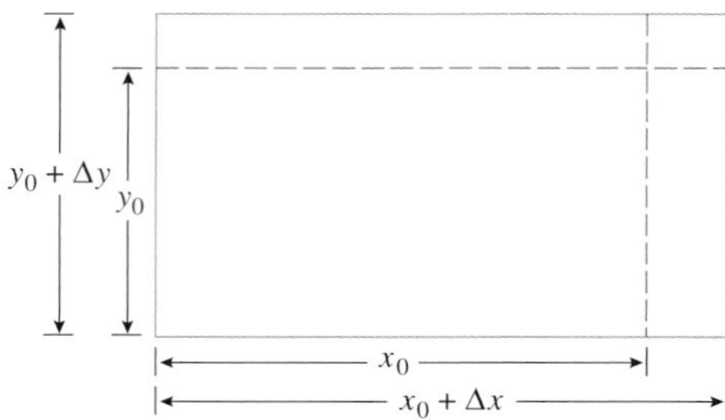

Figure Ex-27

28. The volume V of a right circular cone of radius r and height h is given by $V = \frac{1}{3}\pi r^2 h$. Suppose that the height decreases from 20 in to 19.95 in and the radius increases from 4 in to 4.05 in. Compare the change in volume of the cone with an approximation of this change using a total differential.

29–36 (a) Find the local linear approximation L to the specified function f at the designated point P. (b) Compare the error in approximating f by L at the specified point Q with the distance between P and Q.

29. $f(x, y) = \dfrac{1}{\sqrt{x^2 + y^2}}$; $P(4, 3)$, $Q(3.92, 3.01)$
30. $f(x, y) = x^{0.5}y^{0.3}$; $P(1, 1)$, $Q(1.05, 0.97)$
31. $f(x, y) = x \sin y$; $P(0, 0)$, $Q(0.003, 0.004)$
32. $f(x, y) = \ln xy$; $P(1, 2)$, $Q(1.01, 2.02)$
33. $f(x, y, z) = xyz$; $P(1, 2, 3)$, $Q(1.001, 2.002, 3.003)$
34. $f(x, y, z) = \dfrac{x + y}{y + z}$; $P(-1, 1, 1)$, $Q(-0.99, 0.99, 1.01)$
35. $f(x, y, z) = xe^{yz}$; $P(1, -1, -1)$, $Q(0.99, -1.01, -0.99)$
36. $f(x, y, z) = \ln(x + yz)$; $P(2, 1, -1)$, $Q(2.02, 0.97, -1.01)$

37. In each part, confirm that the stated formula is the local linear approximation at $(0, 0)$.
(a) $e^x \sin y \approx y$ (b) $\dfrac{2x + 1}{y + 1} \approx 1 + 2x - y$

38. Show that the local linear approximation of the function $f(x, y) = x^\alpha y^\beta$ at $(1, 1)$ is
$$x^\alpha y^\beta \approx 1 + \alpha(x - 1) + \beta(y - 1)$$

39. In each part, confirm that the stated formula is the local linear approximation at $(1, 1, 1)$.
(a) $xyz + 2 \approx x + y + z$ (b) $\dfrac{4x}{y + z} \approx 2x - y - z + 2$

40. Based on Exercise 38, what would you conjecture is the local linear approximation to $x^\alpha y^\beta z^\gamma$ at $(1, 1, 1)$? Verify your conjecture by finding this local linear approximation.

41. Suppose that a function $f(x, y)$ is differentiable at the point $(1, 1)$ with $f_x(1, 1) = 2$ and $f(1, 1) = 3$. Let $L(x, y)$ denote the local linear approximation of f at $(1, 1)$. If $L(1.1, 0.9) = 3.15$, find the value of $f_y(1, 1)$.

42. Suppose that a function $f(x, y)$ is differentiable at the point $(0, -1)$ with $f_y(0, -1) = -2$ and $f(0, -1) = 3$. Let

$L(x, y)$ denote the local linear approximation of f at $(0, -1)$. If $L(0.1, -1.1) = 3.3$, find the value of $f_x(0, -1)$.

43. Suppose that a function $f(x, y, z)$ is differentiable at the point $(3, 2, 1)$ and $L(x, y, z) = x - y + 2z - 2$ is the local linear approximation to f at $(3, 2, 1)$. Find $f(3, 2, 1)$, $f_x(3, 2, 1)$, $f_y(3, 2, 1)$, and $f_z(3, 2, 1)$.

44. Suppose that a function $f(x, y, z)$ is differentiable at the point $(0, -1, -2)$ and $L(x, y, z) = x + 2y + 3z + 4$ is the local linear approximation to f at $(0, -1, -2)$. Find $f(0, -1, -2)$, $f_x(0, -1, -2)$, $f_y(0, -1, -2)$, and $f_z(0, -1, -2)$.

45–48 A function f is given along with a local linear approximation L to f at a point P. Use the information given to determine point P.

45. $f(x, y) = x^2 + y^2$; $L(x, y) = 2y - 2x - 2$

46. $f(x, y) = x^2y$; $L(x, y) = 4y - 4x + 8$

47. $f(x, y, z) = xy + z^2$; $L(x, y, z) = y + 2z - 1$

48. $f(x, y, z) = xyz$; $L(x, y, z) = x - y - z - 2$

49. The length and width of a rectangle are measured with errors of at most 3% and 5%, respectively. Use differentials to approximate the maximum percentage error in the calculated area.

50. The radius and height of a right circular cone are measured with errors of at most 1% and 4%, respectively. Use differentials to approximate the maximum percentage error in the calculated volume.

51. The length and width of a rectangle are measured with errors of at most r%, where r is small. Use differentials to approximate the maximum percentage error in the calculated length of the diagonal.

52. The legs of a right triangle are measured to be 3 cm and 4 cm, with a maximum error of 0.05 cm in each measurement. Use differentials to approximate the maximum possible error in the calculated value of (a) the hypotenuse and (b) the area of the triangle.

53. The period T of a simple pendulum with small oscillations is calculated from the formula $T = 2\pi\sqrt{L/g}$, where L is the length of the pendulum and g is the acceleration due to gravity. Suppose that measured values of L and g have errors of at most 0.5% and 0.1%, respectively. Use differentials to approximate the maximum percentage error in the calculated value of T.

54. According to the ideal gas law, the pressure, temperature, and volume of a confined gas are related by $P = kT/V$, where k is a constant. Use differentials to approximate the percentage change in pressure if the temperature of a gas is increased 3% and the volume is increased 5%.

55. Suppose that certain measured quantities x and y have errors of at most r% and s%, respectively. For each of the following formulas in x and y, use differentials to approximate the maximum possible error in the calculated result.
(a) xy (b) x/y (c) x^2y^3 (d) $x^3\sqrt{y}$

56. The total resistance R of three resistances R_1, R_2, and R_3, connected in parallel, is given by

$$\frac{1}{R} = \frac{1}{R_1} + \frac{1}{R_2} + \frac{1}{R_3}$$

Suppose that R_1, R_2, and R_3 are measured to be 100 ohms, 200 ohms, and 500 ohms, respectively, with a maximum error of 10% in each. Use differentials to approximate the maximum percentage error in the calculated value of R.

57. The area of a triangle is to be computed from the formula $A = \frac{1}{2}ab\sin\theta$, where a and b are the lengths of two sides and θ is the included angle. Suppose that a, b, and θ are measured to be 40 ft, 50 ft, and 30°, respectively. Use differentials to approximate the maximum error in the calculated value of A if the maximum errors in a, b, and θ are $\frac{1}{2}$ ft, $\frac{1}{4}$ ft, and 2°, respectively.

58. The length, width, and height of a rectangular box are measured with errors of at most r% (where r is small). Use differentials to approximate the maximum percentage error in the computed value of the volume.

59. Use Theorem 14.4.4 to prove that $f(x, y) = x^2 \sin y$ is differentiable everywhere.

60. Use Theorem 14.4.4 to prove that $f(x, y, z) = xy \sin z$ is differentiable everywhere.

61. Suppose that $f(x, y)$ is differentiable at the point (x_0, y_0) and let $z_0 = f(x_0, y_0)$. Prove that $g(x, y, z) = z - f(x, y)$ is differentiable at (x_0, y_0, z_0).

62. Suppose that Δf satisfies an equation in the form of (5), where $\epsilon(\Delta x, \Delta y)$ is continuous at $(\Delta x, \Delta y) = (0, 0)$ with $\epsilon(0, 0) = 0$. Prove that f is differentiable at (x_0, y_0).

✔ QUICK CHECK ANSWERS 14.4

1. (a) $f_x(x_0, y_0)\Delta x + f_y(x_0, y_0)\Delta y$ (b) $\displaystyle\lim_{(\Delta x, \Delta y)\to(0,0)} \frac{\Delta f - f_x(x_0, y_0)\Delta x - f_y(x_0, y_0)\Delta y}{\sqrt{(\Delta x)^2 + (\Delta y)^2}} = 0$ 2. (a) $dz = e^{y^2}dx + 2xye^{y^2}dy$ (b) $dw = \sin(yz)\,dx + xz\cos(yz)\,dy + xy\cos(yz)\,dz$ 3. $f(x_0, y_0) + f_x(x_0, y_0)(x - x_0) + f_y(x_0, y_0)(y - y_0)$ 4. 3.65

14.5 THE CHAIN RULE

In this section we will derive versions of the chain rule for functions of two or three variables. These new versions will allow us to generate useful relationships among the derivatives and partial derivatives of various functions.

THE CHAIN RULE FOR DERIVATIVES

If y is a differentiable function of x and x is a differentiable function of t, then the chain rule for functions of one variable states that, under composition, y becomes a differentiable function of t with

$$\frac{dy}{dt} = \frac{dy}{dx}\frac{dx}{dt}$$

We will now derive a version of the chain rule for functions of two variables.

Assume that $z = f(x, y)$ is a function of x and y, and suppose that x and y are in turn functions of a single variable t, say

$$x = x(t), \quad y = y(t)$$

The composition $z = f(x(t), y(t))$ then expresses z as a function of the single variable t. Thus, we can ask for the derivative dz/dt and we can inquire about its relationship to the derivatives $\partial z/\partial x$, $\partial z/\partial y$, dx/dt, and dy/dt. Letting Δx, Δy, and Δz denote the changes in x, y, and z, respectively, that correspond to a change of Δt in t, we have

$$\frac{dz}{dt} = \lim_{\Delta t \to 0} \frac{\Delta z}{\Delta t}, \quad \frac{dx}{dt} = \lim_{\Delta t \to 0} \frac{\Delta x}{\Delta t}, \quad \text{and} \quad \frac{dy}{dt} = \lim_{\Delta t \to 0} \frac{\Delta y}{\Delta t}$$

It follows from (3) of Section 14.4 that

$$\Delta z \approx \frac{\partial z}{\partial x}\Delta x + \frac{\partial z}{\partial y}\Delta y \tag{1}$$

where the partial derivatives are evaluated at $(x(t), y(t))$. Dividing both sides of (1) by Δt yields

$$\frac{\Delta z}{\Delta t} \approx \frac{\partial z}{\partial x}\frac{\Delta x}{\Delta t} + \frac{\partial z}{\partial y}\frac{\Delta y}{\Delta t} \tag{2}$$

Taking the limit as $\Delta t \to 0$ of both sides of (2) suggests the following result (whose complete proof can be found in Appendix C).

14.5.1 THEOREM (*Two-Variable Chain Rule*). *If $x = x(t)$ and $y = y(t)$ are differentiable at t, and if $z = f(x, y)$ is differentiable at the point $(x, y) = (x(t), y(t))$, then $z = f(x(t), y(t))$ is differentiable at t and*

$$\frac{dz}{dt} = \frac{\partial z}{\partial x}\frac{dx}{dt} + \frac{\partial z}{\partial y}\frac{dy}{dt} \tag{3}$$

where the ordinary derivatives are evaluated at t and the partial derivatives are evaluated at (x, y).

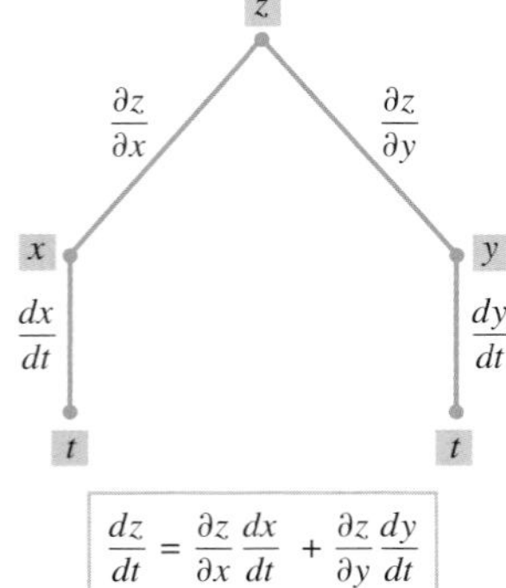

Figure 14.5.1

Formula (3) can be represented schematically by a "tree diagram" that is constructed as follows (Figure 14.5.1). Starting with z at the top of the tree and moving downward, join each variable by lines (or branches) to those variables on which it depends *directly*. Thus, z is joined to x and y and these in turn are joined to t. Next, label each branch with a derivative whose "numerator" contains the variable at the top end of that branch and whose "denominator" contains the variable at the bottom end of that branch. This completes the "tree." To find the formula for dz/dt, follow the two paths through the tree that start with z and end with t. Each such path corresponds to a term in Formula (3).

▶ **Example 1** Suppose that

$$z = x^2y, \quad x = t^2, \quad y = t^3$$

Use the chain rule to find dz/dt, and check the result by expressing z as a function of t and differentiating directly.

Solution. By the chain rule

$$\begin{aligned}\frac{dz}{dt} &= \frac{\partial z}{\partial x}\frac{dx}{dt} + \frac{\partial z}{\partial y}\frac{dy}{dt} = (2xy)(2t) + (x^2)(3t^2)\\ &= (2t^5)(2t) + (t^4)(3t^2) = 7t^6\end{aligned}$$

Alternatively, we can express z directly as a function of t,

$$z = x^2y = (t^2)^2(t^3) = t^7$$

and then differentiate to obtain $dz/dt = 7t^6$. However, this procedure may not always be convenient. ◀

▶ **Example 2** Suppose that

$$z = \sqrt{xy + y}, \quad x = \cos\theta, \quad y = \sin\theta$$

Use the chain rule to find $dz/d\theta$ when $\theta = \pi/2$.

Solution. From the chain rule with θ in place of t,

$$\frac{dz}{d\theta} = \frac{\partial z}{\partial x}\frac{dx}{d\theta} + \frac{\partial z}{\partial y}\frac{dy}{d\theta}$$

we obtain

$$\frac{dz}{d\theta} = \frac{1}{2}(xy + y)^{-1/2}(y)(-\sin\theta) + \frac{1}{2}(xy + y)^{-1/2}(x + 1)(\cos\theta)$$

When $\theta = \pi/2$, we have

$$x = \cos\frac{\pi}{2} = 0, \quad y = \sin\frac{\pi}{2} = 1$$

Substituting $x = 0$, $y = 1$, $\theta = \pi/2$ in the formula for $dz/d\theta$ yields

$$\left.\frac{dz}{d\theta}\right|_{\theta=\pi/2} = \frac{1}{2}(1)(1)(-1) + \frac{1}{2}(1)(1)(0) = -\frac{1}{2}$$ ◀

Confirm the result of Example 2 by expressing z directly as a function of θ.

There are many variations in derivative notations, each of which gives the chain rule a different look. If $z = f(x, y)$, where x and y are functions of t, then some possibilities are

$$\frac{dz}{dt} = f_x\frac{dx}{dt} + f_y\frac{dy}{dt}$$

$$\frac{df}{dt} = \frac{\partial f}{\partial x}\frac{dx}{dt} + \frac{\partial f}{\partial y}\frac{dy}{dt}$$

$$\frac{df}{dt} = f_x x'(t) + f_y y'(t)$$

Theorem 14.5.1 has a natural extension to functions $w = f(x, y, z)$ of three variables, which we state without proof.

14.5.2 THEOREM (*Three-Variable Chain Rule*). *If each of the functions $x = x(t)$, $y = y(t)$, and $z = z(t)$ is differentiable at t, and if $w = f(x, y, z)$ is differentiable at the point $(x, y, z) = (x(t), y(t), z(t))$, then $w = f(x(t), y(t), z(t))$ is differentiable at t and*

$$\frac{dw}{dt} = \frac{\partial w}{\partial x}\frac{dx}{dt} + \frac{\partial w}{\partial y}\frac{dy}{dt} + \frac{\partial w}{\partial z}\frac{dz}{dt} \tag{4}$$

where the ordinary derivatives are evaluated at t and the partial derivatives are evaluated at (x, y, z).

One of the principal uses of the chain rule for functions of a *single* variable was to compute formulas for the derivatives of compositions of functions. Theorems 14.5.1 and 14.5.2 are important not so much for the computation of formulas but because they allow us to express *relationships* among various derivatives. As illustrations, we revisit the topics of implicit differentiation and related rates problems.

IMPLICIT DIFFERENTIATION

Consider the special case where $z = f(x, y)$ is a function of x and y and y is a differentiable function of x. Equation (3) then becomes

$$\frac{dz}{dx} = \frac{\partial f}{\partial x}\frac{dx}{dx} + \frac{\partial f}{\partial y}\frac{dy}{dx} = \frac{\partial f}{\partial x} + \frac{\partial f}{\partial y}\frac{dy}{dx} \tag{5}$$

This result can be used to find derivatives of functions that are defined implicitly. For example, suppose that the equation

$$f(x, y) = c \tag{6}$$

defines y implicitly as a differentiable function of x and we are interested in finding dy/dx. Differentiating both sides of (6) with respect to x and applying (5) yields

$$\frac{\partial f}{\partial x} + \frac{\partial f}{\partial y}\frac{dy}{dx} = 0$$

Thus, if $\partial f/\partial y \neq 0$, we obtain

$$\frac{dy}{dx} = -\frac{\partial f/\partial x}{\partial f/\partial y}$$

In summary, we have the following result.

Show that the function $y = x$ is defined implicitly by the equation

$$x^2 - 2xy + y^2 = 0$$

but that Theorem 14.5.3 is not applicable for finding dy/dx.

14.5.3 THEOREM. *If the equation $f(x, y) = c$ defines y implicitly as a differentiable function of x, and if $\partial f/\partial y \neq 0$, then*

$$\frac{dy}{dx} = -\frac{\partial f/\partial x}{\partial f/\partial y} \tag{7}$$

► **Example 3** Given that

$$x^3 + y^2x - 3 = 0$$

find dy/dx using (7), and check the result using implicit differentiation.

Solution. By (7) with $f(x, y) = x^3 + y^2x - 3$,

$$\frac{dy}{dx} = -\frac{\partial f/\partial x}{\partial f/\partial y} = -\frac{3x^2 + y^2}{2yx}$$

Alternatively, differentiating the given equation implicitly yields

$$3x^2 + y^2 + x\left(2y\frac{dy}{dx}\right) - 0 = 0 \quad \text{or} \quad \frac{dy}{dx} = -\frac{3x^2 + y^2}{2yx}$$

which agrees with the result obtained by (7). ◀

RELATED RATES PROBLEMS

Theorems 14.5.1 and 14.5.2 provide us with additional perspective on related rates problems such as those in Section 3.8.

▶ Example 4 At what rate is the volume of a box changing if its length is 8 ft and increasing at 3 ft/s, its width is 6 ft and increasing at 2 ft/s, and its height is 4 ft and increasing at 1 ft/s?

Solution. Let x, y, and z denote the length, width, and height of the box, respectively, and let t denote time in seconds. We can interpret the given rates to mean that

$$\frac{dx}{dt} = 3, \quad \frac{dy}{dt} = 2, \quad \text{and} \quad \frac{dz}{dt} = 1 \tag{8}$$

at the instant when

$$x = 8, \quad y = 6, \quad \text{and} \quad z = 4 \tag{9}$$

We want to find dV/dt at that instant. For this purpose we use the volume formula $V = xyz$ to obtain

$$\frac{dV}{dt} = \frac{\partial V}{\partial x}\frac{dx}{dt} + \frac{\partial V}{\partial y}\frac{dy}{dt} + \frac{\partial V}{\partial z}\frac{dz}{dt} = yz\frac{dx}{dt} + xz\frac{dy}{dt} + xy\frac{dz}{dt}$$

Substituting (8) and (9) into this equation yields

$$\frac{dV}{dt} = (6)(4)(3) + (8)(4)(2) + (8)(6)(1) = 184$$

Thus, the volume is increasing at a rate of 184 ft^3/s at the given instant. ◀

Redo Example 4 using the methods of Section 3.8. What derivative rule replaces the chain rule in your solution?

THE CHAIN RULE FOR PARTIAL DERIVATIVES

In Theorem 14.5.1 the variables x and y are each functions of a single variable t. We now consider the case where x and y are each functions of two variables. Let

$$z = f(x, y) \tag{10}$$

and suppose that x and y are functions of u and v, say

$$x = x(u, v), \quad y = y(u, v)$$

On substituting these functions of u and v into (10), we obtain the relationship

$$z = f(x(u, v), y(u, v))$$

which expresses z as a function of the two variables u and v. Thus, we can ask for the partial derivatives $\partial z/\partial u$ and $\partial z/\partial v$; and we can inquire about the relationship between these derivatives and the derivatives $\partial z/\partial x$, $\partial z/\partial y$, $\partial x/\partial u$, $\partial x/\partial v$, $\partial y/\partial u$, and $\partial y/\partial v$.

14.5.4 THEOREM (*Two-Variable Chain Rule*). *If $x = x(u, v)$ and $y = y(u, v)$ have first-order partial derivatives at the point (u, v), and if $z = f(x, y)$ is differentiable at the point $(x(u, v), y(u, v))$, then $z = f(x(u, v), y(u, v))$ has first-order partial derivatives at (u, v) given by*

$$\frac{\partial z}{\partial u} = \frac{\partial z}{\partial x}\frac{\partial x}{\partial u} + \frac{\partial z}{\partial y}\frac{\partial y}{\partial u} \quad \text{and} \quad \frac{\partial z}{\partial v} = \frac{\partial z}{\partial x}\frac{\partial x}{\partial v} + \frac{\partial z}{\partial y}\frac{\partial y}{\partial v}$$

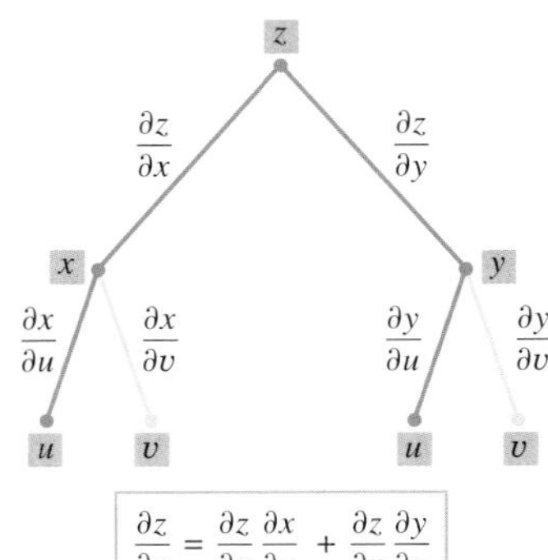

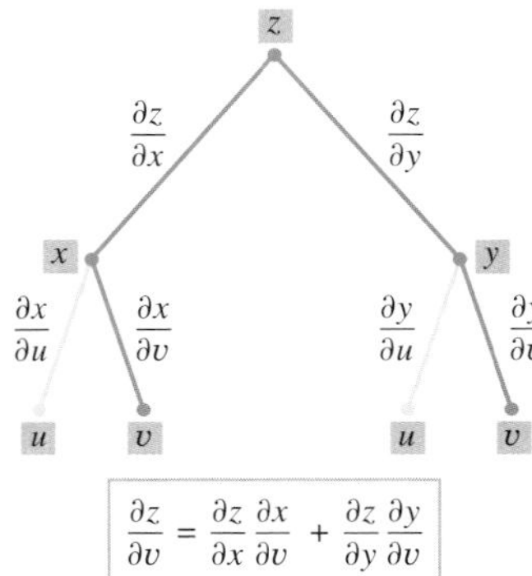

Figure 14.5.2

PROOF. If v is held fixed, then $x = x(u, v)$ and $y = y(u, v)$ become functions of u alone. Thus, we are back to the case of Theorem 14.5.1. If we apply that theorem with u in place of t, and if we use ∂ rather than d to indicate that the variable v is fixed, we obtain

$$\frac{\partial z}{\partial u} = \frac{\partial z}{\partial x}\frac{\partial x}{\partial u} + \frac{\partial z}{\partial y}\frac{\partial y}{\partial u}$$

The formula for $\partial z/\partial v$ is derived similarly. ■

Figure 14.5.2 shows tree diagrams for the formulas in Theorem 14.5.4. The formula for $\partial z/\partial u$ can be obtained by tracing all paths through the tree that start with z and end with u, and the formula for $\partial z/\partial v$ can be obtained by tracing all paths through the tree that start with z and end with v.

▸ **Example 5** Given that

$$z = e^{xy}, \quad x = 2u + v, \quad y = u/v$$

find $\partial z/\partial u$ and $\partial z/\partial v$ using the chain rule.

Solution.

$$\begin{aligned}
\frac{\partial z}{\partial u} &= \frac{\partial z}{\partial x}\frac{\partial x}{\partial u} + \frac{\partial z}{\partial y}\frac{\partial y}{\partial u} = (ye^{xy})(2) + (xe^{xy})\left(\frac{1}{v}\right) = \left[2y + \frac{x}{v}\right]e^{xy} \\
&= \left[\frac{2u}{v} + \frac{2u+v}{v}\right]e^{(2u+v)(u/v)} = \left[\frac{4u}{v} + 1\right]e^{(2u+v)(u/v)} \\
\frac{\partial z}{\partial v} &= \frac{\partial z}{\partial x}\frac{\partial x}{\partial v} + \frac{\partial z}{\partial y}\frac{\partial y}{\partial v} = (ye^{xy})(1) + (xe^{xy})\left(-\frac{u}{v^2}\right) \\
&= \left[y - x\left(\frac{u}{v^2}\right)\right]e^{xy} = \left[\frac{u}{v} - (2u+v)\left(\frac{u}{v^2}\right)\right]e^{(2u+v)(u/v)} \\
&= -\frac{2u^2}{v^2}e^{(2u+v)(u/v)}
\end{aligned}$$ ◂

Theorem 14.5.4 has a natural extension to functions $w = f(x, y, z)$ of three variables, which we state without proof.

14.5.5 THEOREM (*Three-Variable Chain Rule*). *If $x = x(u, v)$, $y = y(u, v)$, and $z = z(u, v)$ have first-order partial derivatives at the point (u, v), and if the function $w = f(x, y, z)$ is differentiable at the point $(x(u, v), y(u, v), z(u, v))$, then the function $w = f(x(u, v), y(u, v), z(u, v))$ has first-order partial derivatives at (u, v) given by*

$$\frac{\partial w}{\partial u} = \frac{\partial w}{\partial x}\frac{\partial x}{\partial u} + \frac{\partial w}{\partial y}\frac{\partial y}{\partial u} + \frac{\partial w}{\partial z}\frac{\partial z}{\partial u} \quad \text{and} \quad \frac{\partial w}{\partial v} = \frac{\partial w}{\partial x}\frac{\partial x}{\partial v} + \frac{\partial w}{\partial y}\frac{\partial y}{\partial v} + \frac{\partial w}{\partial z}\frac{\partial z}{\partial v}$$

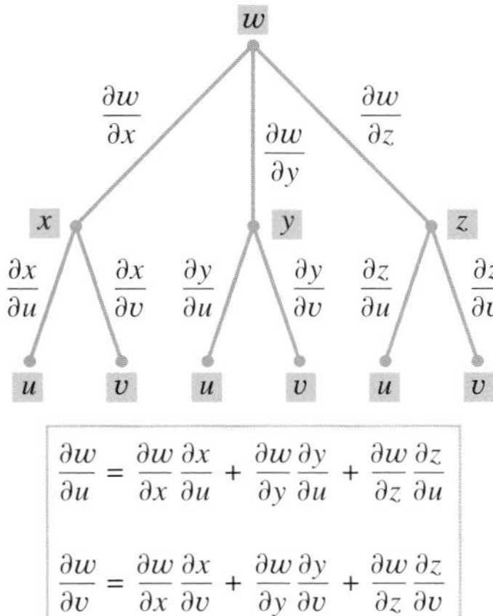

Figure 14.5.3

▶ **Example 6** Suppose that

$$w = e^{xyz}, \quad x = 3u + v, \quad y = 3u - v, \quad z = u^2 v$$

Use appropriate forms of the chain rule to find $\partial w/\partial u$ and $\partial w/\partial v$.

Solution. From the tree diagram and corresponding formulas in Figure 14.5.3 we obtain

$$\frac{\partial w}{\partial u} = yze^{xyz}(3) + xze^{xyz}(3) + xye^{xyz}(2uv) = e^{xyz}(3yz + 3xz + 2xyuv)$$

and

$$\frac{\partial w}{\partial v} = yze^{xyz}(1) + xze^{xyz}(-1) + xye^{xyz}(u^2) = e^{xyz}(yz - xz + xyu^2)$$

If desired, we can express $\partial w/\partial u$ and $\partial w/\partial v$ in terms of u and v alone by replacing x, y, and z by their expressions in terms of u and v. ◀

OTHER VERSIONS OF THE CHAIN RULE

Although we will not prove it, the chain rule extends to functions $w = f(v_1, v_2, \ldots, v_n)$ of n variables. For example, if each v_i is a function of t, $i = 1, 2, \ldots, n$, the relevant formula is

$$\frac{dw}{dt} = \frac{\partial w}{\partial v_1}\frac{dv_1}{dt} + \frac{\partial w}{\partial v_2}\frac{dv_2}{dt} + \cdots + \frac{\partial w}{\partial v_n}\frac{dv_n}{dt} \tag{11}$$

Note that (11) is a natural extension of Formula (3) in Theorem 14.5.1 and Formula (4) in Theorem 14.5.2.

There are infinitely many variations of the chain rule, depending on the number of variables and the choice of independent and dependent variables. A good working procedure is to use tree diagrams to derive new versions of the chain rule as needed. This approach will give correct results for the functions that we will usually encounter.

▶ **Example 7** Suppose that $w = x^2 + y^2 - z^2$ and

$$x = \rho \sin\phi \cos\theta, \quad y = \rho \sin\phi \sin\theta, \quad z = \rho\cos\phi$$

Use appropriate forms of the chain rule to find $\partial w/\partial \rho$ and $\partial w/\partial \theta$.

Solution. From the tree diagram and corresponding formulas in Figure 14.5.4 we obtain

$$\begin{aligned}
\frac{\partial w}{\partial \rho} &= 2x \sin\phi\cos\theta + 2y\sin\phi\sin\theta - 2z\cos\phi \\
&= 2\rho\sin^2\phi\cos^2\theta + 2\rho\sin^2\phi\sin^2\theta - 2\rho\cos^2\phi \\
&= 2\rho\sin^2\phi(\cos^2\theta + \sin^2\theta) - 2\rho\cos^2\phi \\
&= 2\rho(\sin^2\phi - \cos^2\phi) \\
&= -2\rho\cos 2\phi \\
\frac{\partial w}{\partial \theta} &= (2x)(-\rho\sin\phi\sin\theta) + (2y)\rho\sin\phi\cos\theta \\
&= -2\rho^2\sin^2\phi\sin\theta\cos\theta + 2\rho^2\sin^2\phi\sin\theta\cos\theta \\
&= 0
\end{aligned}$$

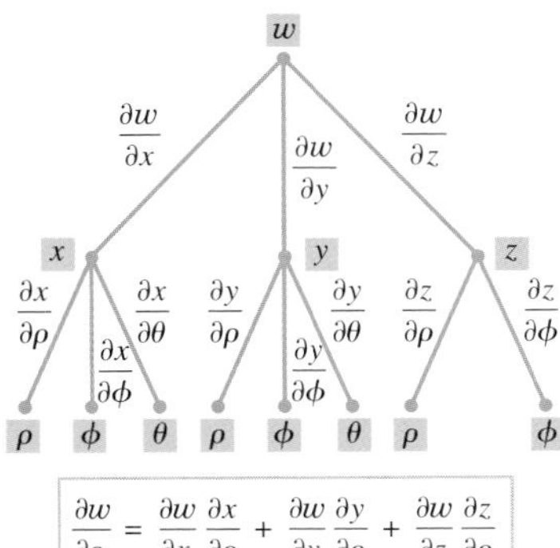

Figure 14.5.4

This result is explained by the fact that w does not vary with θ. You can see this directly by expressing the variables x, y, and z in terms of ρ, ϕ, and θ in the formula for w. (Verify that $w = -\rho^2\cos 2\phi$.) ◀

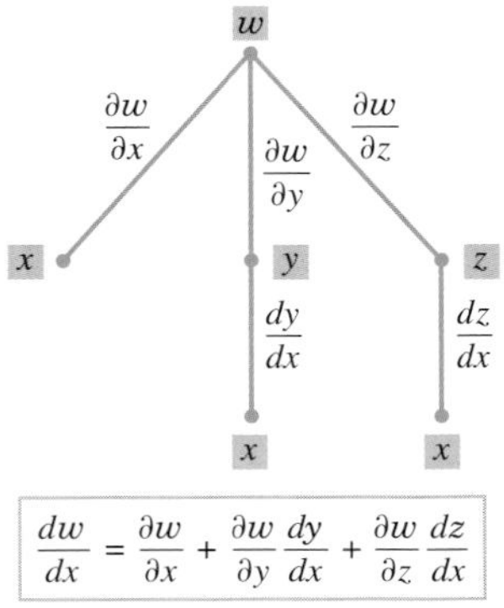

$$\frac{dw}{dx} = \frac{\partial w}{\partial x} + \frac{\partial w}{\partial y}\frac{dy}{dx} + \frac{\partial w}{\partial z}\frac{dz}{dx}$$

Figure 14.5.5

▶ Example 8 Suppose that

$$w = xy + yz, \quad y = \sin x, \quad z = e^x$$

Use an appropriate form of the chain rule to find dw/dx.

Solution. From the tree diagram and corresponding formulas in Figure 14.5.5 we obtain

$$\begin{aligned} \frac{dw}{dx} &= y + (x + z)\cos x + ye^x \\ &= \sin x + (x + e^x)\cos x + e^x \sin x \end{aligned}$$

This result can also be obtained by first expressing w explicitly in terms of x as

$$w = x \sin x + e^x \sin x$$

and then differentiating with respect to x; however, such direct substitution is not always possible. ◀

WARNING The symbol ∂z, unlike the differential dz, has no meaning of its own. For example, if we were to "cancel" partial symbols in the chain-rule formula

$$\frac{\partial z}{\partial u} = \frac{\partial z}{\partial x}\frac{\partial x}{\partial u} + \frac{\partial z}{\partial y}\frac{\partial y}{\partial u}$$

we would obtain

$$\frac{\partial z}{\partial u} = \frac{\partial z}{\partial u} + \frac{\partial z}{\partial u}$$

which is false in cases where $\partial z/\partial u \neq 0$.

In each of the expressions

$$z = \sin xy, \quad z = \frac{xy}{1 + xy}, \quad z = e^{xy}$$

the independent variables occur only in the combination xy, so the substitution $t = xy$ reduces the expression to a function of one variable:

$$z = \sin t, \quad z = \frac{t}{1 + t}, \quad z = e^t$$

Conversely, if we begin with a function of one variable $z = f(t)$ and substitute $t = xy$, we obtain a function $z = f(xy)$ in which the variables appear only in the combination xy. Functions whose variables occur in fixed combinations arise frequently in applications.

▶ Example 9 Show that when f is differentiable, a function of the form $z = f(xy)$ satisfies the equation

$$x\frac{\partial z}{\partial x} - y\frac{\partial z}{\partial y} = 0$$

Solution. Let $t = xy$, so that $z = f(t)$. From the tree diagram in Figure 14.5.6 we obtain the formulas

$$\frac{\partial z}{\partial x} = \frac{dz}{dt}\frac{\partial t}{\partial x} = y\frac{dz}{dt} \quad \text{and} \quad \frac{\partial z}{\partial y} = \frac{dz}{dt}\frac{\partial t}{\partial y} = x\frac{dz}{dt}$$

from which it follows that

$$x\frac{\partial z}{\partial x} - y\frac{\partial z}{\partial y} = xy\frac{dz}{dt} - yx\frac{dz}{dt} = 0 \ ◀$$

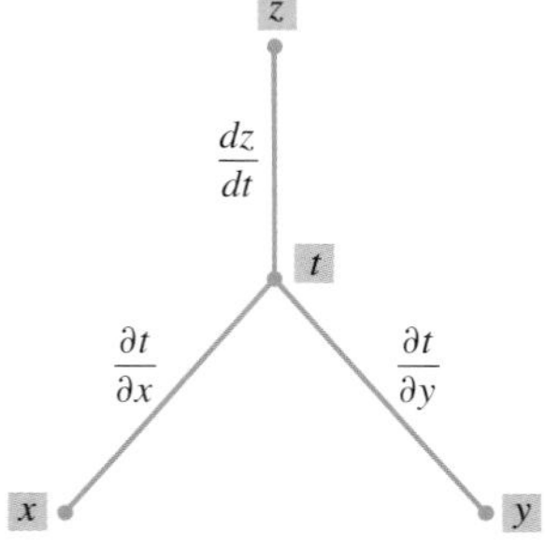

Figure 14.5.6

✔ QUICK CHECK EXERCISES 14.5

(See page 978 for answers.)

1. Suppose that $z = xy^2$ and x and y are differentiable functions of t with $x = 1$, $y = -1$, $dx/dt = -2$, and $dy/dt = 3$ when $t = -1$. Then $dz/dt =$ ________ when $t = -1$.
2. Suppose that C is the graph of the equation $f(x, y) = 1$ and that this equation defines y implicitly as a differentiable function of x. If the point $(2, 1)$ belongs to C with $f_x(2, 1) = 3$ and $f_y(2, 1) = -1$, then the tangent line to C at the point $(2, 1)$ has slope ________.
3. A rectangle is growing in such a way that when its length is 5 ft and its width is 2 ft, the length is increasing at a rate of 3 ft/s and its width is increasing at a rate of 4 ft/s. At this instant the area of the rectangle is growing at a rate of ________.
4. Suppose that $z = x/y$, where x and y are differentiable functions of u and v such that $x = 3$, $y = 1$, $\partial x/\partial u = 4$, $\partial x/\partial v = -2$, $\partial y/\partial u = 1$, and $\partial y/\partial v = -1$ when $u = 2$ and $v = 1$. When $u = 2$ and $v = 1$, $\partial z/\partial u =$ ________ and $\partial z/\partial v =$ ________.

EXERCISE SET 14.5

1–6 Use an appropriate form of the chain rule to find dz/dt.

1. $z = 3x^2y^3$; $x = t^4$, $y = t^2$
2. $z = \ln(2x^2 + y)$; $x = \sqrt{t}$, $y = t^{2/3}$
3. $z = 3\cos x - \sin xy$; $x = 1/t$, $y = 3t$
4. $z = \sqrt{1 + x - 2xy^4}$; $x = \ln t$, $y = t$
5. $z = e^{1-xy}$; $x = t^{1/3}$, $y = t^3$
6. $z = \cosh^2 xy$; $x = t/2$, $y = e^t$

7–10 Use an appropriate form of the chain rule to find dw/dt.

7. $w = 5x^2y^3z^4$; $x = t^2$, $y = t^3$, $z = t^5$
8. $w = \ln(3x^2 - 2y + 4z^3)$; $x = t^{1/2}$, $y = t^{2/3}$, $z = t^{-2}$
9. $w = 5\cos xy - \sin xz$; $x = 1/t$, $y = t$, $z = t^3$
10. $w = \sqrt{1 + x - 2yz^4x}$; $x = \ln t$, $y = t$, $z = 4t$

FOCUS ON CONCEPTS

11. Suppose that
$$w = x^3y^2z^4; \quad x = t^2, \quad y = t + 2, \quad z = 2t^4$$
Find the rate of change of w with respect to t at $t = 1$ by using the chain rule, and then check your work by expressing w as a function of t and differentiating.
12. Suppose that
$$w = x\sin yz^2; \quad x = \cos t, \quad y = t^2, \quad z = e^t$$
Find the rate of change of w with respect to t at $t = 0$ by using the chain rule, and then check your work by expressing w as a function of t and differentiating.
13. Suppose that $z = f(x, y)$ is differentiable at the point $(4, 8)$ with $f_x(4, 8) = 3$ and $f_y(4, 8) = -1$. If $x = t^2$ and $y = t^3$, find dz/dt when $t = 2$.
14. Suppose that $w = f(x, y, z)$ is differentiable at the point $(1, 0, 2)$ with $f_x(1, 0, 2) = 1$, $f_y(1, 0, 2) = 2$, and $f_z(1, 0, 2) = 3$. If $x = t$, $y = \sin(\pi t)$, and $z = t^2 + 1$, find dw/dt when $t = 1$.
15. Explain how the product rule for functions of a single variable may be viewed as a consequence of the chain rule applied to a particular function of two variables.
16. A student attempts to differentiate the function x^x using the power rule, mistakenly getting $x \cdot x^{x-1}$. A second student attempts to differentiate x^x by treating it as an exponential function, mistakenly getting $(\ln x)x^x$. Use the chain rule to explain why the correct derivative is the sum of these two incorrect results.

17–22 Use appropriate forms of the chain rule to find $\partial z/\partial u$ and $\partial z/\partial v$.

17. $z = 8x^2y - 2x + 3y$; $x = uv$, $y = u - v$
18. $z = x^2 - y\tan x$; $x = u/v$, $y = u^2v^2$
19. $z = x/y$; $x = 2\cos u$, $y = 3\sin v$
20. $z = 3x - 2y$; $x = u + v\ln u$, $y = u^2 - v\ln v$
21. $z = e^{x^2y}$; $x = \sqrt{uv}$, $y = 1/v$
22. $z = \cos x \sin y$; $x = u - v$, $y = u^2 + v^2$

23–30 Use appropriate forms of the chain rule to find the derivatives.

23. Let $T = x^2y - xy^3 + 2$; $x = r\cos\theta$, $y = r\sin\theta$. Find $\partial T/\partial r$ and $\partial T/\partial \theta$.
24. Let $R = e^{2s-t^2}$; $s = 3\phi$, $t = \phi^{1/2}$. Find $dR/d\phi$.
25. Let $t = u/v$; $u = x^2 - y^2$, $v = 4xy^3$. Find $\partial t/\partial x$ and $\partial t/\partial y$.
26. Let $w = rs/(r^2 + s^2)$; $r = uv$, $s = u - 2v$. Find $\partial w/\partial u$ and $\partial w/\partial v$.
27. Let $z = \ln(x^2 + 1)$, where $x = r\cos\theta$. Find $\partial z/\partial r$ and $\partial z/\partial \theta$.

28. Let $u = rs^2 \ln t$, $r = x^2$, $s = 4y + 1$, $t = xy^3$. Find $\partial u/\partial x$ and $\partial u/\partial y$.

29. Let $w = 4x^2 + 4y^2 + z^2$, $x = \rho \sin\phi \cos\theta$, $y = \rho \sin\phi \sin\theta$, $z = \rho\cos\phi$. Find $\partial w/\partial \rho$, $\partial w/\partial \phi$, and $\partial w/\partial \theta$.

30. Let $w = 3xy^2z^3$, $y = 3x^2 + 2$, $z = \sqrt{x-1}$. Find dw/dx.

31. Use a chain rule to find the value of $\left.\dfrac{dw}{ds}\right|_{s=1/4}$ if $w = r^2 - r\tan\theta$; $r = \sqrt{s}$, $\theta = \pi s$.

32. Use a chain rule to find the values of

$$\left.\frac{\partial f}{\partial u}\right|_{u=1, v=-2} \quad \text{and} \quad \left.\frac{\partial f}{\partial v}\right|_{u=1, v=-2}$$

if $f(x, y) = x^2y^2 - x + 2y$; $x = \sqrt{u}$, $y = uv^3$.

33. Use a chain rule to find the values of

$$\left.\frac{\partial z}{\partial r}\right|_{r=2, \theta=\pi/6} \quad \text{and} \quad \left.\frac{\partial z}{\partial \theta}\right|_{r=2, \theta=\pi/6}$$

if $z = xye^{x/y}$; $x = r\cos\theta$, $y = r\sin\theta$.

34. Use a chain rule to find $\left.\dfrac{dz}{dt}\right|_{t=3}$ if $z = x^2y$; $x = t^2$, $y = t + 7$.

35–38 Use Theorem 14.5.3 to find dy/dx and check your result using implicit differentiation.

35. $x^2y^3 + \cos y = 0$

36. $x^3 - 3xy^2 + y^3 = 5$

37. $e^{xy} + ye^y = 1$

38. $x - \sqrt{xy} + 3y = 4$

39. Assume that $F(x, y, z) = 0$ defines z implicitly as a function of x and y. Show that if $\partial F/\partial z \neq 0$, then

$$\frac{\partial z}{\partial x} = -\frac{\partial F/\partial x}{\partial F/\partial z}$$

40. Assume that $F(x, y, z) = 0$ defines z implicitly as a function of x and y. Show that if $\partial F/\partial z \neq 0$, then

$$\frac{\partial z}{\partial y} = -\frac{\partial F/\partial y}{\partial F/\partial z}$$

41–44 Find $\partial z/\partial x$ and $\partial z/\partial y$ by implicit differentiation, and confirm that the results obtained agree with those predicted by the formulas in Exercises 39 and 40.

41. $x^2 - 3yz^2 + xyz - 2 = 0$

42. $\ln(1 + z) + xy^2 + z = 1$

43. $ye^x - 5\sin 3z = 3z$

44. $e^{xy}\cos yz - e^{yz}\sin xz + 2 = 0$

45. Two straight roads intersect at right angles. Car A, moving on one of the roads, approaches the intersection at 25 mi/h and car B, moving on the other road, approaches the intersection at 30 mi/h. At what rate is the distance between the cars changing when A is 0.3 mile from the intersection and B is 0.4 mile from the intersection?

46. Use the ideal gas law $P = kT/V$ with V in cubic inches (in^3), T in kelvins (K), and $k = 10$ in·lb/K to find the rate at which the temperature of a gas is changing when the volume is 200 in^3 and increasing at the rate of 4 in^3/s, while the pressure is 5 lb/in^2 and decreasing at the rate of 1 $\text{lb/in}^2/\text{s}$.

47. Two sides of a triangle have lengths $a = 4$ cm and $b = 3$ cm but are increasing at the rate of 1 cm/s. If the area of the triangle remains constant, at what rate is the angle θ between a and b changing when $\theta = \pi/6$?

48. Two sides of a triangle have lengths $a = 5$ cm and $b = 10$ cm, and the included angle is $\theta = \pi/3$. If a is increasing at a rate of 2 cm/s, b is increasing at a rate of 1 cm/s, and θ remains constant, at what rate is the third side changing? Is it increasing or decreasing? [*Hint:* Use the law of cosines.]

49. Suppose that the portion of a tree that is usable for lumber is a right circular cylinder. If the usable height of a tree increases 2 ft per year and the usable diameter increases 3 in per year, how fast is the volume of usable lumber increasing when the usable height of the tree is 20 ft and the usable diameter is 30 in?

50. Suppose that a particle moving along a metal plate in the xy-plane has velocity $\mathbf{v} = \mathbf{i} - 4\mathbf{j}$ (cm/s) at the point (3, 2). Given that the temperature of the plate at points in the xy-plane is $T(x, y) = y^2 \ln x$, $x \geq 1$, in degrees Celsius, find dT/dt at the point (3, 2).

51. The length, width, and height of a rectangular box are increasing at rates of 1 in/s, 2 in/s, and 3 in/s, respectively.

(a) At what rate is the volume increasing when the length is 2 in, the width is 3 in, and the height is 6 in?

(b) At what rate is the length of the diagonal increasing at that instant?

52. Consider the box in Exercise 51. At what rate is the surface area of the box increasing at the given instant?

53–54 A function $f(x, y)$ is said to be ***homogeneous of degree n*** if $f(tx, ty) = t^n f(x, y)$ for $t > 0$. This terminology is needed in these exercises.

53. In each part, show that the function is homogeneous, and find its degree.

(a) $f(x, y) = 3x^2 + y^2$

(b) $f(x, y) = \sqrt{x^2 + y^2}$

(c) $f(x, y) = x^2y - 2y^3$

(d) $f(x, y) = \dfrac{5}{(x^2 + 2y^2)^2}$

54. (a) Show that if $f(x, y)$ is a homogeneous function of degree n, then

$$x\frac{\partial f}{\partial x} + y\frac{\partial f}{\partial y} = nf$$

[*Hint:* Let $u = tx$ and $v = ty$ in $f(tx, ty)$, and differentiate both sides of $f(u, v) = t^n f(x, y)$ with respect to t.]

(b) Confirm that the functions in Exercise 53 satisfy the equation in part (a).

55. (a) Suppose that $z = f(u)$ and $u = g(x, y)$. Draw a tree diagram, and use it to construct chain rules that express $\partial z/\partial x$ and $\partial z/\partial y$ in terms of dz/du, $\partial u/\partial x$, and $\partial u/\partial y$.

(b) Show that

$$\frac{\partial^2 z}{\partial x^2} = \frac{dz}{du}\frac{\partial^2 u}{\partial x^2} + \frac{d^2 z}{du^2}\left(\frac{\partial u}{\partial x}\right)^2$$

$$\frac{\partial^2 z}{\partial y^2} = \frac{dz}{du}\frac{\partial^2 u}{\partial y^2} + \frac{d^2 z}{du^2}\left(\frac{\partial u}{\partial y}\right)^2$$

$$\frac{\partial^2 z}{\partial y \partial x} = \frac{dz}{du}\frac{\partial^2 u}{\partial y \partial x} + \frac{d^2 z}{du^2}\frac{\partial u}{\partial x}\frac{\partial u}{\partial y}$$

56. (a) Let $z = f(x^2 - y^2)$. Use the result in Exercise 55(a) to show that

$$y\frac{\partial z}{\partial x} + x\frac{\partial z}{\partial y} = 0$$

(b) Let $z = f(xy)$. Use the result in Exercise 55(a) to show that

$$x\frac{\partial z}{\partial x} - y\frac{\partial z}{\partial y} = 0$$

(c) Confirm the result in part (a) in the case where $z = \sin(x^2 - y^2)$.

(d) Confirm the result in part (b) in the case where $z = e^{xy}$.

57. Let f be a differentiable function of one variable, and let $z = f(x + 2y)$. Show that

$$2\frac{\partial z}{\partial x} - \frac{\partial z}{\partial y} = 0$$

58. Let f be a differentiable function of one variable, and let $z = f(x^2 + y^2)$. Show that

$$y\frac{\partial z}{\partial x} - x\frac{\partial z}{\partial y} = 0$$

59. Let f be a differentiable function of one variable, and let $w = f(u)$, where $u = x + 2y + 3z$. Show that

$$\frac{\partial w}{\partial x} + \frac{\partial w}{\partial y} + \frac{\partial w}{\partial z} = 6\frac{dw}{du}$$

60. Let f be a differentiable function of one variable, and let $w = f(\rho)$, where $\rho = (x^2 + y^2 + z^2)^{1/2}$. Show that

$$\left(\frac{\partial w}{\partial x}\right)^2 + \left(\frac{\partial w}{\partial y}\right)^2 + \left(\frac{\partial w}{\partial z}\right)^2 = \left(\frac{dw}{d\rho}\right)^2$$

61. Let $z = f(x - y, y - x)$. Show that $\partial z/\partial x + \partial z/\partial y = 0$.

62. Let f be a differentiable function of three variables and suppose that $w = f(x - y, y - z, z - x)$. Show that

$$\frac{\partial w}{\partial x} + \frac{\partial w}{\partial y} + \frac{\partial w}{\partial z} = 0$$

63. In parts (a)–(e), suppose that the equation $z = f(x, y)$ is expressed in the polar form $z = g(r, \theta)$ by making the substitution $x = r\cos\theta$ and $y = r\sin\theta$.

(a) View r and θ as functions of x and y and use implicit differentiation to show that

$$\frac{\partial r}{\partial x} = \cos\theta \quad\text{and}\quad \frac{\partial \theta}{\partial x} = -\frac{\sin\theta}{r}$$

(b) View r and θ as functions of x and y and use implicit differentiation to show that

$$\frac{\partial r}{\partial y} = \sin\theta \quad\text{and}\quad \frac{\partial \theta}{\partial y} = \frac{\cos\theta}{r}$$

(c) Use the results in parts (a) and (b) to show that

$$\frac{\partial z}{\partial x} = \frac{\partial z}{\partial r}\cos\theta - \frac{1}{r}\frac{\partial z}{\partial \theta}\sin\theta$$

$$\frac{\partial z}{\partial y} = \frac{\partial z}{\partial r}\sin\theta + \frac{1}{r}\frac{\partial z}{\partial \theta}\cos\theta$$

(d) Use the result in part (c) to show that

$$\left(\frac{\partial z}{\partial x}\right)^2 + \left(\frac{\partial z}{\partial y}\right)^2 = \left(\frac{\partial z}{\partial r}\right)^2 + \frac{1}{r^2}\left(\frac{\partial z}{\partial \theta}\right)^2$$

(e) Use the result in part (c) to show that if $z = f(x, y)$ satisfies Laplace's equation

$$\frac{\partial^2 z}{\partial x^2} + \frac{\partial^2 z}{\partial y^2} = 0$$

then $z = g(r, \theta)$ satisfies the equation

$$\frac{\partial^2 z}{\partial r^2} + \frac{1}{r^2}\frac{\partial^2 z}{\partial \theta^2} + \frac{1}{r}\frac{\partial z}{\partial r} = 0$$

and conversely. The latter equation is called the ***polar form of Laplace's equation***.

64. Show that the function

$$z = \tan^{-1}\frac{2xy}{x^2 - y^2}$$

satisfies Laplace's equation; then make the substitution $x = r\cos\theta$, $y = r\sin\theta$, and show that the resulting function of r and θ satisfies the polar form of Laplace's equation given in part (e) of Exercise 63.

65. (a) Show that if $u(x, y)$ and $v(x, y)$ satisfy the Cauchy–Riemann equations (Exercise 88, Section 14.3), and if $x = r\cos\theta$ and $y = r\sin\theta$, then

$$\frac{\partial u}{\partial r} = \frac{1}{r}\frac{\partial v}{\partial \theta} \quad\text{and}\quad \frac{\partial v}{\partial r} = -\frac{1}{r}\frac{\partial u}{\partial \theta}$$

This is called the ***polar form of the Cauchy–Riemann equations***.

(b) Show that the functions

$$u = \ln(x^2 + y^2), \quad v = 2\tan^{-1}(y/x)$$

satisfy the Cauchy–Riemann equations; then make the substitution $x = r\cos\theta$, $y = r\sin\theta$, and show that the resulting functions of r and θ satisfy the polar form of the Cauchy–Riemann equations.

66. In parts (a)–(d), recall from Formula (6) of Section 14.3 that under appropriate conditions a plucked string satisfies the wave equation

$$\frac{\partial^2 u}{\partial t^2} = c^2\frac{\partial^2 u}{\partial x^2}$$

where c is a positive constant.

(a) Show that a function of the form $u(x, t) = f(x + ct)$ satisfies the wave equation.

(b) Show that a function of the form $u(x, t) = g(x - ct)$ satisfies the wave equation.

(c) Show that a function of the form

$$u(x, t) = f(x + ct) + g(x - ct)$$

satisfies the wave equation.

(d) It can be proved that every solution of the wave equation is expressible in the form stated in part (c). Confirm that $u(x, t) = \sin t \sin x$ satisfies the wave equation in which $c = 1$, and then use appropriate trigonometric identities to express this function in the form $f(x + t) + g(x - t)$.

67. Let f be a differentiable function of three variables, and let $w = f(x, y, z)$, $x = \rho \sin\phi \cos\theta$, $y = \rho \sin\phi \sin\theta$, and $z = \rho \cos\phi$. Express $\partial w/\partial\rho$, $\partial w/\partial\phi$, and $\partial w/\partial\theta$ in terms of $\partial w/\partial x$, $\partial w/\partial y$, and $\partial w/\partial z$.

68. Let $w = f(x, y, z)$ be differentiable, where $z = g(x, y)$. Taking x and y as the independent variables, express each of the following in terms of $\partial f/\partial x$, $\partial f/\partial y$, $\partial f/\partial z$, $\partial z/\partial x$, and $\partial z/\partial y$.
(a) $\partial w/\partial x$ (b) $\partial w/\partial y$

69. Let $w = \ln(e^r + e^s + e^t + e^u)$. Show that

$$w_{rstu} = -6e^{r+s+t+u-4w}$$

[*Hint:* Take advantage of the relationship $e^w = e^r + e^s + e^t + e^u$.]

70. Suppose that w is a differentiable function of x_1, x_2, and x_3, and

$$x_1 = a_1 y_1 + b_1 y_2$$
$$x_2 = a_2 y_1 + b_2 y_2$$
$$x_3 = a_3 y_1 + b_3 y_2$$

where the a's and b's are constants. Express $\partial w/\partial y_1$ and $\partial w/\partial y_2$ in terms of $\partial w/\partial x_1$, $\partial w/\partial x_2$, and $\partial w/\partial x_3$.

71. (a) Let w be a differentiable function of x_1, x_2, x_3, and x_4, and let each x_i be a differentiable function of t. Find a chain-rule formula for dw/dt.

(b) Let w be a differentiable function of x_1, x_2, x_3, and x_4, and let each x_i be a differentiable function of v_1, v_2, and v_3. Find chain-rule formulas for $\partial w/\partial v_1$, $\partial w/\partial v_2$, and $\partial w/\partial v_3$.

72. Let $w = (x_1^2 + x_2^2 + \cdots + x_n^2)^k$, where $n \geq 2$. For what values of k does

$$\frac{\partial^2 w}{\partial x_1^2} + \frac{\partial^2 w}{\partial x_2^2} + \cdots + \frac{\partial^2 w}{\partial x_n^2} = 0$$

hold?

73. We showed in Exercise 24 of Section 7.6 that

$$\frac{d}{dx}\int_{h(x)}^{g(x)} f(t)\,dt = f(g(x))g'(x) - f(h(x))h'(x)$$

Derive this same result by letting $u = g(x)$ and $v = h(x)$ and then differentiating the function

$$F(u, v) = \int_v^u f(t)\,dt$$

with respect to x.

74. Prove: If f, f_x, and f_y are continuous on a circular region containing $A(x_0, y_0)$ and $B(x_1, y_1)$, then there is a point (x^*, y^*) on the line segment joining A and B such that

$$f(x_1, y_1) - f(x_0, y_0)$$
$$= f_x(x^*, y^*)(x_1 - x_0) + f_y(x^*, y^*)(y_1 - y_0)$$

This result is the two-dimensional version of the Mean-Value Theorem. [*Hint:* Express the line segment joining A and B in parametric form and use the Mean-Value Theorem for functions of one variable.]

75. Prove: If $f_x(x, y) = 0$ and $f_y(x, y) = 0$ throughout a circular region, then $f(x, y)$ is constant on that region. [*Hint:* Use the result of Exercise 74.]

✔ QUICK CHECK ANSWERS 14.5

1. -8 **2.** 3 **3.** 26 ft^2/s **4.** 1; 1

14.6 DIRECTIONAL DERIVATIVES AND GRADIENTS

The partial derivatives $f_x(x, y)$ and $f_y(x, y)$ represent the rates of change of $f(x, y)$ in directions parallel to the x- and y-axes. In this section we will investigate rates of change of $f(x, y)$ in other directions.

DIRECTIONAL DERIVATIVES

In this section we extend the concept of a *partial* derivative to the more general notion of a *directional* derivative. We have seen that the partial derivatives of a function give the instantaneous rates of change of that function in directions parallel to the coordinate axes. Directional derivatives allow us to compute the rates of change of a function with respect to distance in *any* direction.

Suppose that we wish to compute the instantaneous rate of change of a function $f(x, y)$ with respect to distance from a point (x_0, y_0) in some direction. Since there are infinitely many different directions from (x_0, y_0) in which we could move, we need a convenient method for describing a specific direction starting at (x_0, y_0). One way to do this is to use a unit vector

$$\mathbf{u} = u_1\mathbf{i} + u_2\mathbf{j}$$

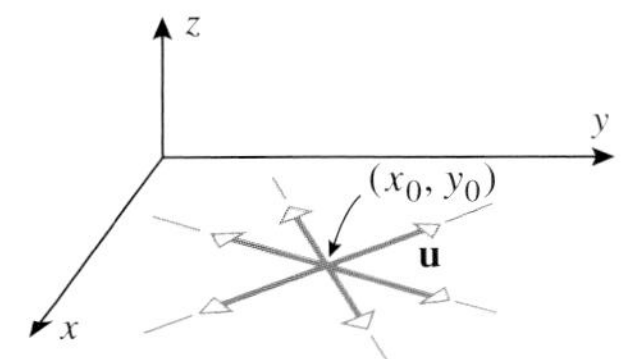

Figure 14.6.1

that has its initial point at (x_0, y_0) and points in the desired direction (Figure 14.6.1). This vector determines a line l in the xy-plane that can be expressed parametrically as

$$x = x_0 + su_1, \quad y = y_0 + su_2 \tag{1}$$

where s is the arc length parameter that has its reference point at (x_0, y_0) and has positive values in the direction of $\mathbf{u}$. For $s = 0$, the point (x, y) is at the reference point (x_0, y_0), and as s increases, the point (x, y) moves along l in the direction of $\mathbf{u}$. On the line l the variable $z = f(x_0 + su_1, y_0 + su_2)$ is a function of the parameter s. The value of the derivative dz/ds at $s = 0$ then gives an instantaneous rate of change of $f(x, y)$ with respect to distance from (x_0, y_0) in the direction of $\mathbf{u}$.

14.6.1 DEFINITION. If $f(x, y)$ is a function of x and y, and if $\mathbf{u} = u_1\mathbf{i} + u_2\mathbf{j}$ is a unit vector, then the ***directional derivative of f in the direction of* u** at (x_0, y_0) is denoted by $D_{\mathbf{u}}f(x_0, y_0)$ and is defined by

$$D_{\mathbf{u}}f(x_0, y_0) = \frac{d}{ds}[f(x_0 + su_1, y_0 + su_2)]_{s=0} \tag{2}$$

provided this derivative exists.

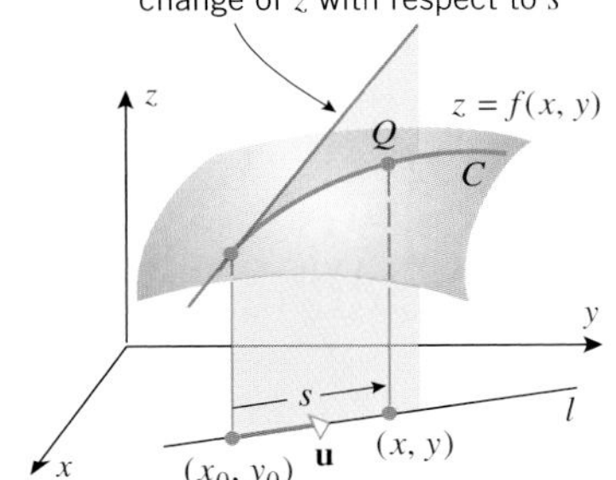

Figure 14.6.2

Geometrically, $D_{\mathbf{u}}f(x_0, y_0)$ can be interpreted as the ***slope of the surface z = f(x, y) in the direction of* u** at the point $(x_0, y_0, f(x_0, y_0))$ (Figure 14.6.2). Usually the value of $D_{\mathbf{u}}f(x_0, y_0)$ will depend on both the point (x_0, y_0) and the direction $\mathbf{u}$. Thus, at a fixed point the slope of the surface may vary with the direction (Figure 14.6.3). Analytically, the directional derivative represents the ***instantaneous rate of change of sfmar f(x, y) with respect to distance in the direction of* u** at the point (x_0, y_0).

▶ Example 1 Let $f(x, y) = xy$ and find $D_{\mathbf{u}}f(1, 2)$, where $\mathbf{u} = \frac{\sqrt{3}}{2}\mathbf{i} + \frac{1}{2}\mathbf{j}$.

Solution. It follows from Equation (2) that

$$D_{\mathbf{u}}f(1, 2) = \frac{d}{ds}\left[f\left(1 + \frac{\sqrt{3}s}{2}, 2 + \frac{s}{2}\right)\right]_{s=0}$$

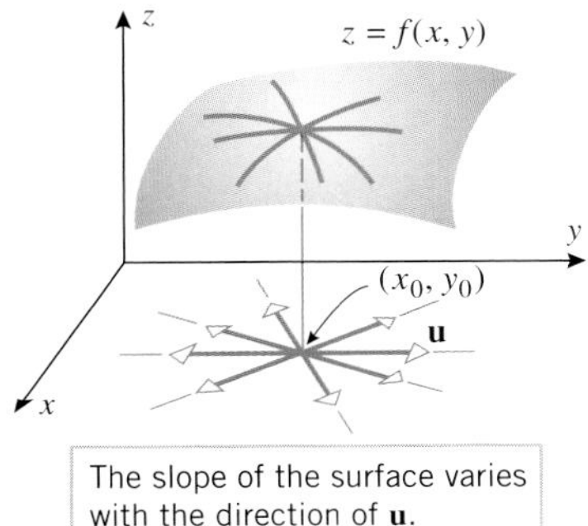

The slope of the surface varies with the direction of **u**.

Figure 14.6.3

Since

$$f\left(1 + \frac{\sqrt{3}s}{2}, 2 + \frac{s}{2}\right) = \left(1 + \frac{\sqrt{3}s}{2}\right)\left(2 + \frac{s}{2}\right) = \frac{\sqrt{3}}{4}s^2 + \left(\frac{1}{2} + \sqrt{3}\right)s + 2$$

we have

$$\frac{d}{ds}\left[f\left(1 + \frac{\sqrt{3}s}{2}, 2 + \frac{s}{2}\right)\right] = \frac{\sqrt{3}}{2}s + \frac{1}{2} + \sqrt{3}$$

and thus

$$D_{\mathbf{u}}f(1, 2) = \frac{d}{ds}\left[f\left(1 + \frac{\sqrt{3}s}{2}, 2 + \frac{s}{2}\right)\right]_{s=0} = \frac{1}{2} + \sqrt{3}$$

Since $\frac{1}{2} + \sqrt{3} \approx 2.23$, we conclude that if we move a small distance from the point $(1, 2)$ in the direction of $\mathbf{u}$, the function $f(x, y) = xy$ will increase by about 2.23 times the distance moved. ◀

The definition of a directional derivative for a function $f(x, y, z)$ of three variables is similar to Definition 14.6.1.

14.6.2 DEFINITION. If $\mathbf{u} = u_1\mathbf{i} + u_2\mathbf{j} + u_3\mathbf{k}$ is a unit vector, and if $f(x, y, z)$ is a function of x, y, and z, then the ***directional derivative of f in the direction of*** $\mathbf{u}$ at (x_0, y_0, z_0) is denoted by $D_{\mathbf{u}}f(x_0, y_0, z_0)$ and is defined by

$$D_{\mathbf{u}}f(x_0, y_0, z_0) = \frac{d}{ds}[f(x_0 + su_1, y_0 + su_2, z_0 + su_3)]_{s=0} \tag{3}$$

provided this derivative exists.

What are the difficulties in interpreting (3) as a "slope"?

Although Equation (3) does not have a convenient geometric interpretation, we can still interpret directional derivatives for functions of three variables in terms of instantaneous rates of change in a specified direction.

For a function that is differentiable at a point, directional derivatives exist in every direction from the point and can be computed directly in terms of the first-order partial derivatives of the function.

14.6.3 THEOREM.

(*a*) *If $f(x, y)$ is differentiable at (x_0, y_0), and if $\mathbf{u} = u_1\mathbf{i} + u_2\mathbf{j}$ is a unit vector, then the directional derivative $D_{\mathbf{u}}f(x_0, y_0)$ exists and is given by*

$$D_{\mathbf{u}}f(x_0, y_0) = f_x(x_0, y_0)u_1 + f_y(x_0, y_0)u_2 \tag{4}$$

(*b*) *If $f(x, y, z)$ is differentiable at (x_0, y_0, z_0), and if $\mathbf{u} = u_1\mathbf{i} + u_2\mathbf{j} + u_3\mathbf{k}$ is a unit vector, then the directional derivative $D_{\mathbf{u}}f(x_0, y_0, z_0)$ exists and is given by*

$$D_{\mathbf{u}}f(x_0, y_0, z_0) = f_x(x_0, y_0, z_0)u_1 + f_y(x_0, y_0, z_0)u_2 + f_z(x_0, y_0, z_0)u_3 \tag{5}$$

PROOF. We will give the proof of part (*a*); the proof of part (*b*) is similar and will be omitted. The function $z = f(x_0 + su_1, y_0 + su_2)$ is the composition of the function $z = f(x, y)$ with the functions

$$x = x(s) = x_0 + su_1 \quad \text{and} \quad y = y(s) = y_0 + su_2$$

As such, the chain rule in Theorem 14.5.1 immediately gives

$$\begin{aligned} D_{\mathbf{u}}f(x_0, y_0) &= \frac{d}{ds}[f(x_0 + su_1, y_0 + su_2)]_{s=0} \\ &= \frac{dz}{ds}(0) = f_x(x_0, y_0)u_1 + f_y(x_0, y_0)u_2 \end{aligned}$$ ■

We can use Theorem 14.6.3 to confirm the result of Example 1. For $f(x, y) = xy$ we have $f_x(1, 2) = 2$ and $f_y(1, 2) = 1$ (verify). With

$$\mathbf{u} = \frac{\sqrt{3}}{2}\mathbf{i} + \frac{1}{2}\mathbf{j}$$

Equation (4) becomes

$$D_{\mathbf{u}}f(1,2) = 2\left(\frac{\sqrt{3}}{2}\right) + \frac{1}{2} = \sqrt{3} + \frac{1}{2}$$

which agrees with our solution in Example 1.

Recall from Formula (13) of Section 12.2 that a unit vector $\mathbf{u}$ in the xy-plane can be expressed as

$$\mathbf{u} = \cos\phi\,\mathbf{i} + \sin\phi\,\mathbf{j} \tag{6}$$

where ϕ is the angle from the positive x-axis to $\mathbf{u}$. Thus, Formula (4) can also be expressed as

$$D_{\mathbf{u}}f(x_0, y_0) = f_x(x_0, y_0)\cos\phi + f_y(x_0, y_0)\sin\phi \tag{7}$$

► **Example 2** Find the directional derivative of $f(x, y) = e^{xy}$ at $(-2, 0)$ in the direction of the unit vector that makes an angle of $\pi/3$ with the positive x-axis.

Solution. The partial derivatives of f are

$$f_x(x,y) = ye^{xy}, \quad f_y(x,y) = xe^{xy}$$
$$f_x(-2,0) = 0, \quad f_y(-2,0) = -2$$

The unit vector $\mathbf{u}$ that makes an angle of $\pi/3$ with the positive x-axis is

$$\mathbf{u} = \cos(\pi/3)\mathbf{i} + \sin(\pi/3)\mathbf{j} = \frac{1}{2}\mathbf{i} + \frac{\sqrt{3}}{2}\mathbf{j}$$

Thus, from (7)

$$\begin{aligned} D_{\mathbf{u}}f(-2,0) &= f_x(-2,0)\cos(\pi/3) + f_y(-2,0)\sin(\pi/3) \\ &= 0(1/2) + (-2)(\sqrt{3}/2) = -\sqrt{3} \end{aligned}$$ ◄

It is important that the direction of a directional derivative be specified by a *unit vector* when applying either Equation (4) or Equation (5).

► **Example 3** Find the directional derivative of $f(x, y, z) = x^2y - yz^3 + z$ at the point $(1, -2, 0)$ in the direction of the vector $\mathbf{a} = 2\mathbf{i} + \mathbf{j} - 2\mathbf{k}$.

Solution. The partial derivatives of f are

$$f_x(x,y,z) = 2xy, \quad f_y(x,y,z) = x^2 - z^3, \quad f_z(x,y,z) = -3yz^2 + 1$$
$$f_x(1,-2,0) = -4, \quad f_y(1,-2,0) = 1, \quad f_z(1,-2,0) = 1$$

Since $\mathbf{a}$ is not a unit vector, we normalize it, getting

$$\mathbf{u} = \frac{\mathbf{a}}{\|\mathbf{a}\|} = \frac{1}{\sqrt{9}}(2\mathbf{i} + \mathbf{j} - 2\mathbf{k}) = \frac{2}{3}\mathbf{i} + \frac{1}{3}\mathbf{j} - \frac{2}{3}\mathbf{k}$$

Formula (5) then yields

$$D_{\mathbf{u}}f(1,-2,0) = (-4)\left(\frac{2}{3}\right) + \frac{1}{3} - \frac{2}{3} = -3$$ ◄

THE GRADIENT

Formula (4) can be expressed in the form of a dot product as

$$\begin{aligned} D_{\mathbf{u}}f(x_0, y_0) &= (f_x(x_0, y_0)\mathbf{i} + f_y(x_0, y_0)\mathbf{j}) \cdot (u_1\mathbf{i} + u_2\mathbf{j}) \\ &= (f_x(x_0, y_0)\mathbf{i} + f_y(x_0, y_0)\mathbf{j}) \cdot \mathbf{u} \end{aligned}$$

Similarly, Formula (5) can be expressed as

$$D_{\mathbf{u}}f(x_0, y_0, z_0) = (f_x(x_0, y_0, z_0)\mathbf{i} + f_y(x_0, y_0, z_0)\mathbf{j} + f_z(x_0, y_0, z_0)\mathbf{k}) \cdot \mathbf{u}$$

In both cases the directional derivative is obtained by dotting the direction vector **u** with a new vector constructed from the first-order partial derivatives of f.

Remember that ∇f is not a product of ∇ and f. Think of ∇ as an "operator" that acts on a function f to produce the gradient ∇f.

14.6.4 DEFINITION.

(a) If f is a function of x and y, then the ***gradient of f*** is defined by

$$\nabla f(x, y) = f_x(x, y)\mathbf{i} + f_y(x, y)\mathbf{j} \tag{8}$$

(b) If f is a function of x, y, and z, then the ***gradient of f*** is defined by

$$\nabla f(x, y, z) = f_x(x, y, z)\mathbf{i} + f_y(x, y, z)\mathbf{j} + f_z(x, y, z)\mathbf{k} \tag{9}$$

The symbol ∇ (read "del") is an inverted delta. (It is sometimes called a "nabla" because of its similarity in form to an ancient Hebrew ten-stringed harp of that name.)

Formulas (4) and (5) can now be written as

$$D_{\mathbf{u}}f(x_0, y_0) = \nabla f(x_0, y_0) \cdot \mathbf{u} \tag{10}$$

and

$$D_{\mathbf{u}}f(x_0, y_0, z_0) = \nabla f(x_0, y_0, z_0) \cdot \mathbf{u} \tag{11}$$

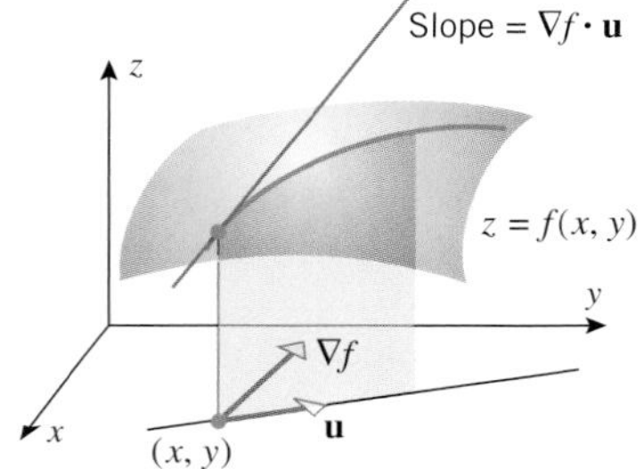

Figure 14.6.4

respectively. For example, using Formula (11) our solution to Example 3 would take the form

$$\begin{aligned} D_{\mathbf{u}}f(1, -2, 0) = \nabla f(1, -2, 0) \cdot \mathbf{u} &= (-4\mathbf{i} + \mathbf{j} + \mathbf{k}) \cdot \left(\tfrac{2}{3}\mathbf{i} + \tfrac{1}{3}\mathbf{j} - \tfrac{2}{3}\mathbf{k}\right) \\ &= (-4)\left(\tfrac{2}{3}\right) + \tfrac{1}{3} - \tfrac{2}{3} = -3 \end{aligned}$$

Formula (10) can be interpreted to mean that the slope of the surface $z = f(x, y)$ at the point (x_0, y_0) in the direction of **u** is the dot product of the gradient with **u** (Figure 14.6.4).

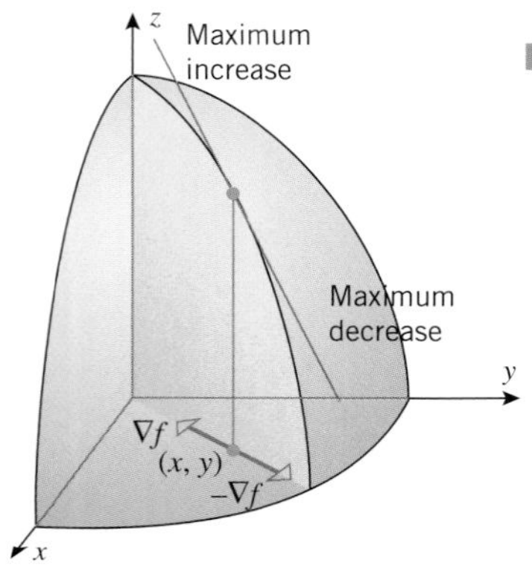

Figure 14.6.5

PROPERTIES OF THE GRADIENT

The gradient is not merely a notational device to simplify the formula for the directional derivative; we will see that the length and direction of the gradient ∇f provide important information about the function f and the surface $z = f(x, y)$. For example, suppose that $\nabla f(x, y) \neq \mathbf{0}$, and let us use Formula (4) of Section 12.3 to rewrite (10) as

$$D_{\mathbf{u}}f(x, y) = \nabla f(x, y) \cdot \mathbf{u} = \|\nabla f(x, y)\|\|\mathbf{u}\|\cos\theta = \|\nabla f(x, y)\|\cos\theta \tag{12}$$

where θ is the angle between $\nabla f(x, y)$ and **u**. This equation tells us that the maximum value of $D_{\mathbf{u}}f(x, y)$ is $\|\nabla f(x, y)\|$, and this maximum occurs when $\theta = 0$, that is, when **u** is in the direction of $\nabla f(x, y)$. Geometrically, this means that *the surface $z = f(x, y)$ has its maximum slope at a point (x, y) in the direction of the gradient, and the maximum slope is* $\|\nabla f(x, y)\|$ (Figure 14.6.5). Similarly, (12) tells us that the minimum value of $D_{\mathbf{u}}f(x, y)$ is $-\|\nabla f(x, y)\|$, and this minimum occurs when $\theta = \pi$, that is, when **u** is oppositely directed to $\nabla f(x, y)$. Geometrically, this means that *the surface $z = f(x, y)$ has its minimum slope at a point (x, y) in the direction that is opposite to the gradient, and the minimum slope is* $-\|\nabla f(x, y)\|$ (Figure 14.6.5).

Finally, in the case where $\nabla f(x, y) = \mathbf{0}$, it follows from (12) that $D_{\mathbf{u}} f(x, y) = 0$ in all directions at the point (x, y). This typically occurs where the surface $z = f(x, y)$ has a "relative maximum," a "relative minimum," or a saddle point.

A similar analysis applies to functions of three variables. As a consequence, we have the following result.

14.6.5 THEOREM. *Let f be a function of either two variables or three variables, and let P denote the point $P(x_0, y_0)$ or $P(x_0, y_0, z_0)$, respectively. Assume that f is differentiable at P.*

(*a*) *If $\nabla f = \mathbf{0}$ at P, then all directional derivatives of f at P are zero.*

(*b*) *If $\nabla f \neq \mathbf{0}$ at P, then among all possible directional derivatives of f at P, the derivative in the direction of ∇f at P has the largest value. The value of this largest directional derivative is $\|\nabla f\|$ at P.*

(*c*) *If $\nabla f \neq \mathbf{0}$ at P, then among all possible directional derivatives of f at P, the derivative in the direction opposite to that of ∇f at P has the smallest value. The value of this smallest directional derivative is $-\|\nabla f\|$ at P.*

▶ **Example 4** Let $f(x, y) = x^2 e^y$. Find the maximum value of a directional derivative at $(-2, 0)$, and find the unit vector in the direction in which the maximum value occurs.

Solution. Since

$$\nabla f(x, y) = f_x(x, y)\mathbf{i} + f_y(x, y)\mathbf{j} = 2xe^y\mathbf{i} + x^2e^y\mathbf{j}$$

the gradient of f at $(-2, 0)$ is

$$\nabla f(-2, 0) = -4\mathbf{i} + 4\mathbf{j}$$

By Theorem 14.6.5, the maximum value of the directional derivative is

$$\|\nabla f(-2, 0)\| = \sqrt{(-4)^2 + 4^2} = \sqrt{32} = 4\sqrt{2}$$

This maximum occurs in the direction of $\nabla f(-2, 0)$. The unit vector in this direction is

$$\mathbf{u} = \frac{\nabla f(-2, 0)}{\|\nabla f(-2, 0)\|} = \frac{1}{4\sqrt{2}}(-4\mathbf{i} + 4\mathbf{j}) = -\frac{1}{\sqrt{2}}\mathbf{i} + \frac{1}{\sqrt{2}}\mathbf{j} \quad ◀$$

What would be the minimum value of a directional derivative of

$$f(x, y) = x^2 e^y$$

at $(-2, 0)$?

GRADIENTS ARE NORMAL TO LEVEL CURVES

We have seen that the gradient points in the direction in which a function increases most rapidly. For a function $f(x, y)$ of two variables, we will now consider how this direction of maximum rate of increase can be determined from a contour map of the function. Suppose that (x_0, y_0) is a point on a level curve $f(x, y) = c$ of f, and assume that this curve can be smoothly parametrized as

$$x = x(s), \quad y = y(s) \tag{13}$$

where s is an arc length parameter. Recall from Formula (6) of Section 13.4 that the unit tangent vector to (13) is

$$\mathbf{T} = \mathbf{T}(s) = \left(\frac{dx}{ds}\right)\mathbf{i} + \left(\frac{dy}{ds}\right)\mathbf{j}$$

Since $\mathbf{T}$ gives a direction along which f is nearly constant, we would expect the instantaneous rate of change of f with respect to distance in the direction of $\mathbf{T}$ to be 0. That is, we would expect that

$$D_{\mathbf{T}} f(x, y) = \nabla f(x, y) \cdot \mathbf{T}(s) = 0$$

To show this to be the case, we differentiate both sides of the equation $f(x, y) = c$ with respect to s. Assuming that f is differentiable at (x, y), we can use the chain rule to obtain

$$\frac{\partial f}{\partial x}\frac{dx}{ds} + \frac{\partial f}{\partial y}\frac{dy}{ds} = 0$$

which we can rewrite as

$$\left(\frac{\partial f}{\partial x}\mathbf{i} + \frac{\partial f}{\partial y}\mathbf{j}\right) \cdot \left(\frac{dx}{ds}\mathbf{i} + \frac{dy}{ds}\mathbf{j}\right) = 0$$

or, alternatively, as

$$\nabla f(x, y) \cdot \mathbf{T} = 0$$

Therefore, if $\nabla f(x, y) \neq \mathbf{0}$, then $\nabla f(x, y)$ should be normal to the level curve $f(x, y) = c$ at any point (x, y) on the curve.

It is proved in advanced courses that if $f(x, y)$ has continuous first-order partial derivatives, and if $\nabla f(x_0, y_0) \neq \mathbf{0}$, then near (x_0, y_0) the graph of $f(x, y) = c$ is indeed a smooth curve through (x_0, y_0). Furthermore, we also know from Theorem 14.4.4 that f will be differentiable at (x_0, y_0). We therefore have the following result.

Verify Theorem 14.6.6 for

$$f(x, y) = x^2 + y^2$$

and $(x_0, y_0) = (3, 4)$.

14.6.6 THEOREM. *Assume that $f(x, y)$ has continuous first-order partial derivatives in an open disk centered at (x_0, y_0) and that $\nabla f(x_0, y_0) \neq \mathbf{0}$. Then $\nabla f(x_0, y_0)$ is normal to the level curve of f through (x_0, y_0).*

When we examine a contour map, we instinctively regard the distance between adjacent contours to be measured in a normal direction. If the contours correspond to equally spaced values of f, then the closer together the contours appear to be, the more rapidly the values of f will be changing in that normal direction. It follows from Theorems 14.6.5 and 14.6.6 that this rate of change of f is given by $\|\nabla f(x, y)\|$. Thus, the closer together the contours appear to be, the greater the length of the gradient of f.

► **Example 5** A contour plot of a function f is given in Figure 14.6.6*a*. Sketch the directions of the gradient of f at the points P, Q, and R. At which of these three points does the gradient vector have maximum length? Minimum length?

Solution. It follows from Theorems 14.6.5 and 14.6.6 that the directions of the gradient vectors will be as given in Figure 14.6.6*b*. Based on the density of the contour lines, we would guess that the gradient of f has maximum length at R and minimum length at P, with the length at Q somewhere in between. ◄

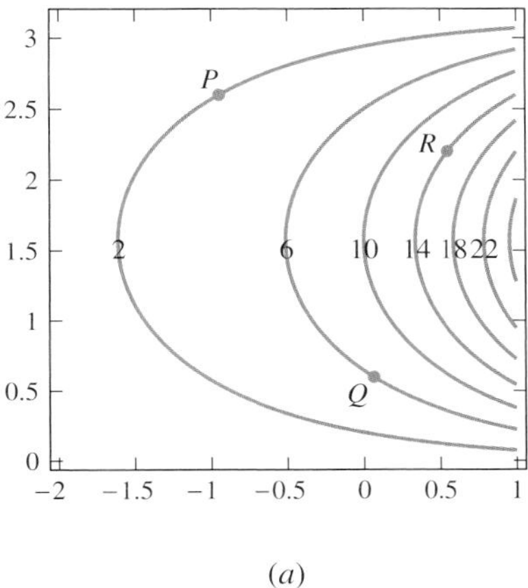

(a)

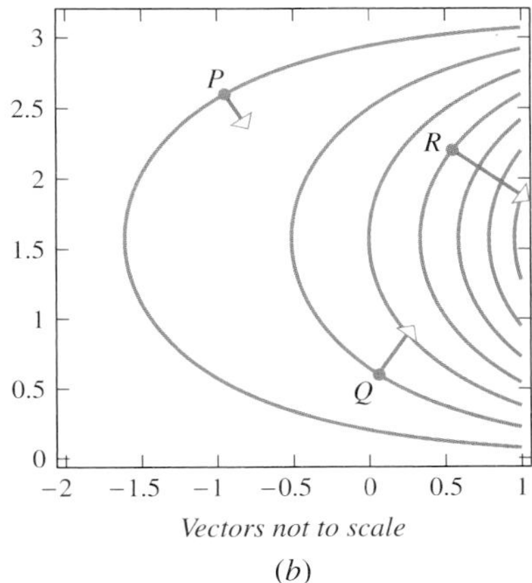

(b)

Figure 14.6.6

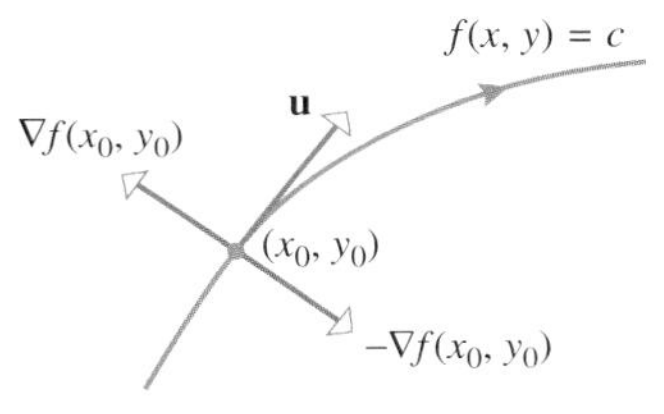

Figure 14.6.7

If (x_0, y_0) is a point on the level curve $f(x, y) = c$, then the slope of the surface $z = f(x, y)$ at that point in the direction of **u** is

$$D_{\mathbf{u}} f(x_0, y_0) = \nabla f(x_0, y_0) \cdot \mathbf{u}$$

If **u** is tangent to the level curve at (x_0, y_0), then $f(x, y)$ is neither increasing nor decreasing in that direction, so $D_{\mathbf{u}} f(x_0, y_0) = 0$. Thus, $\nabla f(x_0, y_0)$, $-\nabla f(x_0, y_0)$, and the tangent vector **u** mark the directions of maximum slope, minimum slope, and zero slope at a point (x_0, y_0) on a level curve (Figure 14.6.7). Good skiers use these facts intuitively to control their speed by zigzagging down ski slopes—they ski across the slope with their skis tangential to a level curve to stop their downhill motion, and they point their skis down the slope and normal to the level curve to obtain the most rapid descent.

AN APPLICATION OF GRADIENTS

There are numerous applications in which the motion of an object must be controlled so that it moves toward a heat source. For example, in medical applications the operation of certain diagnostic equipment is designed to locate heat sources generated by tumors or infections, and in military applications the trajectories of heat-seeking missiles are controlled to seek and destroy enemy aircraft. The following example illustrates how gradients are used to solve such problems.

► **Example 6** A heat-seeking particle is located at the point (2, 3) on a flat metal plate whose temperature at a point (x, y) is

$$T(x, y) = 10 - 8x^2 - 2y^2$$

Find an equation for the trajectory of the particle if it moves continuously in the direction of maximum temperature increase.

Solution. Assume that the trajectory is represented parametrically by the equations

$$x = x(t), \quad y = y(t)$$

where the particle is at the point (2, 3) at time $t = 0$. Because the particle moves in the direction of maximum temperature increase, its direction of motion at time t is in the direction of the gradient of $T(x, y)$, and hence its velocity vector $\mathbf{v}(t)$ at time t points in the direction of the gradient. Thus, there is a scalar k that depends on t such that

$$\mathbf{v}(t) = k\nabla T(x, y)$$

from which we obtain

$$\frac{dx}{dt}\mathbf{i} + \frac{dy}{dt}\mathbf{j} = k(-16x\mathbf{i} - 4y\mathbf{j})$$

Equating components yields

$$\frac{dx}{dt} = -16kx, \qquad \frac{dy}{dt} = -4ky$$

and dividing to eliminate k yields

$$\frac{dy}{dx} = \frac{-4ky}{-16kx} = \frac{y}{4x}$$

Thus, we can obtain the trajectory by solving the initial-value problem

$$\frac{dy}{dx} - \frac{y}{4x} = 0, \quad y(2) = 3$$

The differential equation is a separable first-order linear equation and hence can be solved by separating the variables or by the method of integrating factors discussed in Section 9.1. We leave it for you to show that the solution of the initial-value problem is

$$y = \frac{3}{\sqrt[4]{2}}x^{1/4}$$

The graph of the trajectory and a contour plot of the temperature function are shown in Figure 14.6.8. ◀

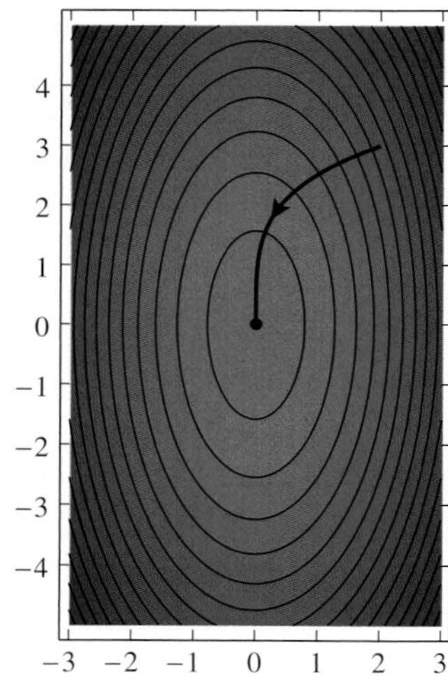

Figure 14.6.8

✔QUICK CHECK EXERCISES 14.6 (See page 989 for answers.)

1. The gradient of $f(x, y, z) = xy^2z^3$ at the point $(1, 1, 1)$ is ________.

2. Suppose that the differentiable function $f(x, y)$ has the property that

$$f\left(2 + \frac{s\sqrt{3}}{2}, 1 + \frac{s}{2}\right) = 3se^s$$

The directional derivative of f in the direction of

$$\mathbf{u} = \frac{\sqrt{3}}{2}\mathbf{i} + \frac{1}{2}\mathbf{j}$$

at $(2, 1)$ is ________.

3. If the gradient of $f(x, y)$ at the origin is $6\mathbf{i} + 8\mathbf{j}$, then the directional derivative of f in the direction of $\mathbf{a} = 3\mathbf{i} + 4\mathbf{j}$ at the origin is ________. The slope of the tangent line to the level curve of f through the origin at $(0, 0)$ is ________.

4. If the gradient of $f(x, y, z)$ at $(1, 2, 3)$ is $2\mathbf{i} - 2\mathbf{j} + \mathbf{k}$, then the maximum value for a directional derivative of f at $(1, 2, 3)$ is ________ and the minimum value for a directional derivative at this point is ________.

EXERCISE SET 14.6 Graphing Utility C CAS

1–8 Find $D_\mathbf{u}f$ at P.

1. $f(x, y) = (1 + xy)^{3/2}$; $P(3, 1)$; $\mathbf{u} = \frac{1}{\sqrt{2}}\mathbf{i} + \frac{1}{\sqrt{2}}\mathbf{j}$

2. $f(x, y) = e^{2xy}$; $P(4, 0)$; $\mathbf{u} = -\frac{3}{5}\mathbf{i} + \frac{4}{5}\mathbf{j}$

3. $f(x, y) = \ln(1 + x^2 + y)$; $P(0, 0)$; $\mathbf{u} = -\frac{1}{\sqrt{10}}\mathbf{i} - \frac{3}{\sqrt{10}}\mathbf{j}$

4. $f(x, y) = \dfrac{cx + dy}{x - y}$; $P(3, 4)$; $\mathbf{u} = \frac{4}{5}\mathbf{i} + \frac{3}{5}\mathbf{j}$

5. $f(x, y, z) = 4x^5y^2z^3$; $P(2, -1, 1)$; $\mathbf{u} = \frac{1}{3}\mathbf{i} + \frac{2}{3}\mathbf{j} - \frac{2}{3}\mathbf{k}$

6. $f(x, y, z) = ye^{xz} + z^2$; $P(0, 2, 3)$; $\mathbf{u} = \frac{2}{7}\mathbf{i} - \frac{3}{7}\mathbf{j} + \frac{6}{7}\mathbf{k}$

7. $f(x, y, z) = \ln(x^2 + 2y^2 + 3z^2)$; $P(-1, 2, 4)$; $\mathbf{u} = -\frac{3}{13}\mathbf{i} - \frac{4}{13}\mathbf{j} - \frac{12}{13}\mathbf{k}$

8. $f(x, y, z) = \sin xyz$; $P\left(\frac{1}{2}, \frac{1}{3}, \pi\right)$;
$\mathbf{u} = \frac{1}{\sqrt{3}}\mathbf{i} - \frac{1}{\sqrt{3}}\mathbf{j} + \frac{1}{\sqrt{3}}\mathbf{k}$

9–18 Find the directional derivative of f at P in the direction of $\mathbf{a}$.

9. $f(x, y) = 4x^3y^2$; $P(2, 1)$; $\mathbf{a} = 4\mathbf{i} - 3\mathbf{j}$

10. $f(x, y) = x^2 - 3xy + 4y^3$; $P(-2, 0)$; $\mathbf{a} = \mathbf{i} + 2\mathbf{j}$

11. $f(x, y) = y^2 \ln x$; $P(1, 4)$; $\mathbf{a} = -3\mathbf{i} + 3\mathbf{j}$

12. $f(x, y) = e^x \cos y$; $P(0, \pi/4)$; $\mathbf{a} = 5\mathbf{i} - 2\mathbf{j}$

13. $f(x, y) = \tan^{-1}(y/x)$; $P(-2, 2)$; $\mathbf{a} = -\mathbf{i} - \mathbf{j}$

14. $f(x, y) = xe^y - ye^x$; $P(0, 0)$; $\mathbf{a} = 5\mathbf{i} - 2\mathbf{j}$

15. $f(x, y, z) = x^3z - yx^2 + z^2$; $P(2, -1, 1)$;
$\mathbf{a} = 3\mathbf{i} - \mathbf{j} + 2\mathbf{k}$

16. $f(x, y, z) = y - \sqrt{x^2 + z^2}$; $P(-3, 1, 4)$;
$\mathbf{a} = 2\mathbf{i} - 2\mathbf{j} - \mathbf{k}$

17. $f(x, y, z) = \dfrac{z - x}{z + y}$; $P(1, 0, -3)$; $\mathbf{a} = -6\mathbf{i} + 3\mathbf{j} - 2\mathbf{k}$

18. $f(x, y, z) = e^{x+y+3z}$; $P(-2, 2, -1)$; $\mathbf{a} = 20\mathbf{i} - 4\mathbf{j} + 5\mathbf{k}$

19–22 Find the directional derivative of f at P in the direction of a vector making the counterclockwise angle θ with the positive x-axis.

19. $f(x, y) = \sqrt{xy}$; $P(1, 4)$; $\theta = \pi/3$

20. $f(x, y) = \dfrac{x - y}{x + y}$; $P(-1, -2)$; $\theta = \pi/2$

21. $f(x, y) = \tan(2x + y)$; $P(\pi/6, \pi/3)$; $\theta = 7\pi/4$

22. $f(x, y) = \sinh x \cosh y$; $P(0, 0)$; $\theta = \pi$

23. Find the directional derivative of

$$f(x, y) = \frac{x}{x + y}$$

at $P(1, 0)$ in the direction of $Q(-1, -1)$.

24. Find the directional derivative of $f(x, y) = e^{-x} \sec y$ at $P(0, \pi/4)$ in the direction of the origin.

25. Find the directional derivative of $f(x, y) = \sqrt{xy}e^y$ at $P(1, 1)$ in the direction of the negative y-axis.

26. Let

$$f(x, y) = \frac{y}{x + y}$$

Find a unit vector $\mathbf{u}$ for which $D_{\mathbf{u}}f(2, 3) = 0$.

27. Find the directional derivative of

$$f(x, y, z) = \frac{y}{x + z}$$

at $P(2, 1, -1)$ in the direction from P to $Q(-1, 2, 0)$.

28. Find the directional derivative of the function

$$f(x, y, z) = x^3y^2z^5 - 2xz + yz + 3x$$

at $P(-1, -2, 1)$ in the direction of the negative z-axis.

FOCUS ON CONCEPTS

29. Suppose that $D_{\mathbf{u}}f(1, 2) = -5$ and $D_{\mathbf{v}}f(1, 2) = 10$, where $\mathbf{u} = \frac{3}{5}\mathbf{i} - \frac{4}{5}\mathbf{j}$ and $\mathbf{v} = \frac{4}{5}\mathbf{i} + \frac{3}{5}\mathbf{j}$. Find
(a) $f_x(1, 2)$ (b) $f_y(1, 2)$
(c) the directional derivative of f at $(1, 2)$ in the direction of the origin.

30. Given that $f_x(-5, 1) = -3$ and $f_y(-5, 1) = 2$, find the directional derivative of f at $P(-5, 1)$ in the direction of the vector from P to $Q(-4, 3)$.

31. The accompanying figure shows some level curves of an unspecified function $f(x, y)$. Which of the three vectors shown in the figure is most likely to be ∇f? Explain.

32. The accompanying figure shows some level curves of an unspecified function $f(x, y)$. Of the gradients at P and Q, which probably has the greater length? Explain.

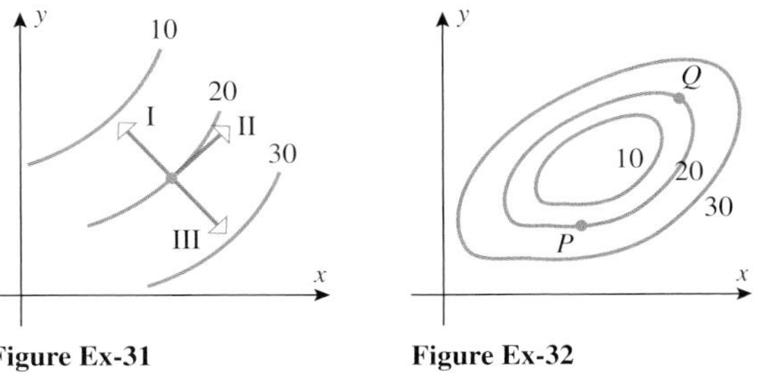

Figure Ex-31 Figure Ex-32

33–36 Find ∇z or ∇w.

33. $z = 4x - 8y$

34. $z = e^{-3y} \cos 4x$

35. $w = \ln \sqrt{x^2 + y^2 + z^2}$

36. $w = e^{-5x} \sec x^2yz$

37–40 Find the gradient of f at the indicated point.

37. $f(x, y) = (x^2 + xy)^3$; $(-1, -1)$

38. $f(x, y) = (x^2 + y^2)^{-1/2}$; $(3, 4)$

39. $f(x, y, z) = y \ln(x + y + z)$; $(-3, 4, 0)$

40. $f(x, y, z) = y^2z \tan^3 x$; $(\pi/4, -3, 1)$

41–44 Sketch the level curve of $f(x, y)$ that passes through P and draw the gradient vector at P.

41. $f(x, y) = 4x - 2y + 3$; $P(1, 2)$

42. $f(x, y) = y/x^2$; $P(-2, 2)$

43. $f(x, y) = x^2 + 4y^2$; $P(-2, 0)$

44. $f(x, y) = x^2 - y^2$; $P(2, -1)$

45. Find a unit vector $\mathbf{u}$ that is normal at $P(1, -2)$ to the level curve of $f(x, y) = 4x^2y$ through P.

46. Find a unit vector $\mathbf{u}$ that is normal at $P(2, -3)$ to the level curve of $f(x, y) = 3x^2y - xy$ through P.

47–54 Find a unit vector in the direction in which f increases most rapidly at P, and find the rate of change of f at P in that direction.

47. $f(x, y) = 4x^3y^2$; $P(-1, 1)$

48. $f(x, y) = 3x - \ln y$; $P(2, 4)$

49. $f(x, y) = \sqrt{x^2 + y^2}$; $P(4, -3)$

50. $f(x, y) = \dfrac{x}{x+y}$; $P(0, 2)$

51. $f(x, y, z) = x^3z^2 + y^3z + z - 1$; $P(1, 1, -1)$

52. $f(x, y, z) = \sqrt{x - 3y + 4z}$; $P(0, -3, 0)$

53. $f(x, y, z) = \dfrac{x}{z} + \dfrac{z}{y^2}$; $P(1, 2, -2)$

54. $f(x, y, z) = \tan^{-1}\left(\dfrac{x}{y+z}\right)$; $P(4, 2, 2)$

55–60 Find a unit vector in the direction in which f decreases most rapidly at P, and find the rate of change of f at P in that direction.

55. $f(x, y) = 20 - x^2 - y^2$; $P(-1, -3)$

56. $f(x, y) = e^{xy}$; $P(2, 3)$

57. $f(x, y) = \cos(3x - y)$; $P(\pi/6, \pi/4)$

58. $f(x, y) = \sqrt{\dfrac{x-y}{x+y}}$; $P(3, 1)$

59. $f(x, y, z) = \dfrac{x+z}{z-y}$; $P(5, 7, 6)$

60. $f(x, y, z) = 4e^{xy}\cos z$; $P(0, 1, \pi/4)$

FOCUS ON CONCEPTS

61. Given that $\nabla f(4, -5) = 2\mathbf{i} - \mathbf{j}$, find the directional derivative of the function f at the point $(4, -5)$ in the direction of $\mathbf{a} = 5\mathbf{i} + 2\mathbf{j}$.

62. Given that $\nabla f(x_0, y_0) = \mathbf{i} - 2\mathbf{j}$ and $D_{\mathbf{u}} f(x_0, y_0) = -2$, find $\mathbf{u}$ (two answers).

63. The accompanying figure shows some level curves of an unspecified function $f(x, y)$.
(a) Use the available information to approximate the length of the vector $\nabla f(1, 2)$, and sketch the approximation. Explain how you approximated the length and determined the direction of the vector.
(b) Sketch an approximation of the vector $-\nabla f(4, 4)$.

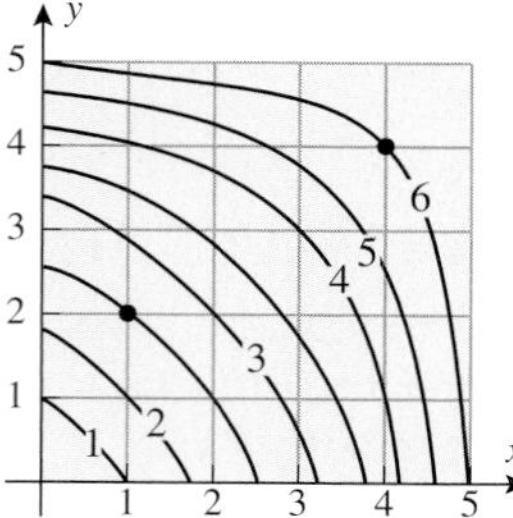

Figure Ex-63

64. The accompanying figure shows a topographic map of a hill and a point P at the bottom of the hill. Suppose that you want to climb from the point P toward the top of the hill in such a way that you are always ascending in the direction of steepest slope. Sketch the projection of your path on the contour map. This is called the ***path of steepest ascent***. Explain how you determined the path.

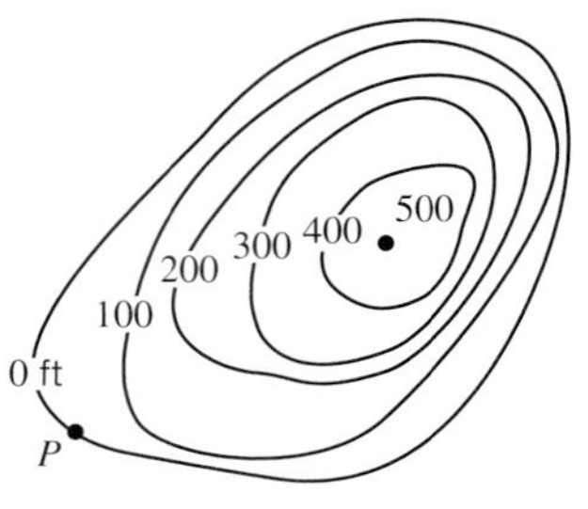

Figure Ex-64

65. Let $z = 3x^2 - y^2$. Find all points at which $\|\nabla z\| = 6$.

66. Given that $z = 3x + y^2$, find $\nabla\|\nabla z\|$ at the point $(5, 2)$.

67. A particle moves along a path C given by the equations $x = t$ and $y = -t^2$. If $z = x^2 + y^2$, find dz/ds along C at the instant when the particle is at the point $(2, -4)$.

68. The temperature (in degrees Celsius) at a point (x, y) on a metal plate in the xy-plane is
$$T(x, y) = \frac{xy}{1 + x^2 + y^2}$$
(a) Find the rate of change of temperature at $(1, 1)$ in the direction of $\mathbf{a} = 2\mathbf{i} - \mathbf{j}$.
(b) An ant at $(1, 1)$ wants to walk in the direction in which the temperature drops most rapidly. Find a unit vector in that direction.

69. If the electric potential at a point (x, y) in the xy-plane is $V(x, y)$, then the ***electric intensity vector*** at the point (x, y) is $\mathbf{E} = -\nabla V(x, y)$. Suppose that $V(x, y) = e^{-2x}\cos 2y$.
(a) Find the electric intensity vector at $(\pi/4, 0)$.
(b) Show that at each point in the plane, the electric potential decreases most rapidly in the direction of the vector $\mathbf{E}$.

70. On a certain mountain, the elevation z above a point (x, y) in an xy-plane at sea level is $z = 2000 - 0.02x^2 - 0.04y^2$, where x, y, and z are in meters. The positive x-axis points east, and the positive y-axis north. A climber is at the point $(-20, 5, 1991)$.
(a) If the climber uses a compass reading to walk due west, will she begin to ascend or descend?
(b) If the climber uses a compass reading to walk northeast, will she ascend or descend? At what rate?
(c) In what compass direction should the climber begin walking to travel a level path (two answers)?

71. Given that the directional derivative of $f(x, y, z)$ at the point $(3, -2, 1)$ in the direction of $\mathbf{a} = 2\mathbf{i} - \mathbf{j} - 2\mathbf{k}$ is -5 and that $\|\nabla f(3, -2, 1)\| = 5$, find $\nabla f(3, -2, 1)$.

72. The temperature (in degrees Celsius) at a point (x, y, z) in a metal solid is
$$T(x, y, z) = \frac{xyz}{1 + x^2 + y^2 + z^2}$$

(a) Find the rate of change of temperature with respect to distance at $(1, 1, 1)$ in the direction of the origin.
(b) Find the direction in which the temperature rises most rapidly at the point $(1, 1, 1)$. (Express your answer as a unit vector.)
(c) Find the rate at which the temperature rises moving from $(1, 1, 1)$ in the direction obtained in part (b).

73. Let $r = \sqrt{x^2 + y^2}$.
(a) Show that $\nabla r = \dfrac{\mathbf{r}}{r}$, where $\mathbf{r} = x\mathbf{i} + y\mathbf{j}$.
(b) Show that $\nabla f(r) = f'(r)\nabla r = \dfrac{f'(r)}{r}\mathbf{r}$.

74. Use the formula in part (b) of Exercise 73 to find
(a) $\nabla f(r)$ if $f(r) = re^{-3r}$
(b) $f(r)$ if $\nabla f(r) = 3r^2\mathbf{r}$ and $f(2) = 1$.

75. Let $\mathbf{u}_r$ be a unit vector whose counterclockwise angle from the positive x-axis is θ, and let $\mathbf{u}_\theta$ be a unit vector $90°$ counterclockwise from $\mathbf{u}_r$. Show that if $z = f(x, y)$, $x = r\cos\theta$, and $y = r\sin\theta$, then

$$\nabla z = \frac{\partial z}{\partial r}\mathbf{u}_r + \frac{1}{r}\frac{\partial z}{\partial \theta}\mathbf{u}_\theta$$

[*Hint:* Use part (c) of Exercise 63, Section 14.5.]

76. Prove: If f and g are differentiable, then
(a) $\nabla(f + g) = \nabla f + \nabla g$
(b) $\nabla(cf) = c\nabla f$ (c constant)
(c) $\nabla(fg) = f\nabla g + g\nabla f$
(d) $\nabla\left(\dfrac{f}{g}\right) = \dfrac{g\nabla f - f\nabla g}{g^2}$
(e) $\nabla(f^n) = nf^{n-1}\nabla f$.

77–78 A heat-seeking particle is located at the point P on a flat metal plate whose temperature at a point (x, y) is $T(x, y)$. Find parametric equations for the trajectory of the particle if it moves continuously in the direction of maximum temperature increase.

77. $T(x, y) = 5 - 4x^2 - y^2$; $P(1, 4)$

78. $T(x, y) = 100 - x^2 - 2y^2$; $P(5, 3)$

79. Use a graphing utility to generate the trajectory of the particle together with some representative level curves of the temperature function in Exercise 77.

80. Use a graphing utility to generate the trajectory of the particle together with some representative level curves of the temperature function in Exercise 78.

C **81.** (a) Use a CAS to graph $f(x, y) = (x^2 + 3y^2)e^{-(x^2+y^2)}$.
(b) At how many points do you think it is true that $D_\mathbf{u} f(x, y) = 0$ for all unit vectors $\mathbf{u}$?
(c) Use a CAS to find ∇f.
(d) Use a CAS to solve the equation $\nabla f(x, y) = 0$ for x and y.
(e) Use the result in part (d) together with Theorem 14.6.5 to check your conjecture in part (b).

82. Prove: If $x = x(t)$ and $y = y(t)$ are differentiable at t, and if $z = f(x, y)$ is differentiable at the point $(x(t), y(t))$, then

$$\frac{dz}{dt} = \nabla z \cdot \mathbf{r}'(t)$$

where $\mathbf{r}(t) = x(t)\mathbf{i} + y(t)\mathbf{j}$.

83. Prove: If f, f_x, and f_y are continuous on a circular region, and if $\nabla f(x, y) = \mathbf{0}$ throughout the region, then $f(x, y)$ is constant on the region. [*Hint:* See Exercise 75, Section 14.5.]

84. Prove: If the function f is differentiable at the point (x, y) and if $D_\mathbf{u} f(x, y) = 0$ in two nonparallel directions, then $D_\mathbf{u} f(x, y) = 0$ in all directions.

85. Given that the functions $u = u(x, y, z)$, $v = v(x, y, z)$, $w = w(x, y, z)$, and $f(u, v, w)$ are all differentiable, show that

$$\nabla f(u, v, w) = \frac{\partial f}{\partial u}\nabla u + \frac{\partial f}{\partial v}\nabla v + \frac{\partial f}{\partial w}\nabla w$$

✔ QUICK CHECK ANSWERS 14.6

1. $\langle 1, 2, 3\rangle$ **2.** 3 **3.** 10; $-\frac{3}{4}$ **4.** 3; -3

14.7 TANGENT PLANES AND NORMAL VECTORS

In this section we will discuss tangent planes to surfaces in three-dimensional space. We will be concerned with three main questions: What is a tangent plane? When do tangent planes exist? How do we find equations of tangent planes?

TANGENT PLANES

Recall from Section 14.4 that if a function $f(x, y)$ is differentiable at a point (x_0, y_0), then we want it to be the case that the surface $z = f(x, y)$ has a nonvertical tangent plane at the point $P_0(x_0, y_0, f(x_0, y_0))$. We also saw in Section 14.4 that the linear function

$$L(x, y) = f(x_0, y_0) + f_x(x_0, y_0)(x - x_0) + f_y(x_0, y_0)(y - y_0)$$

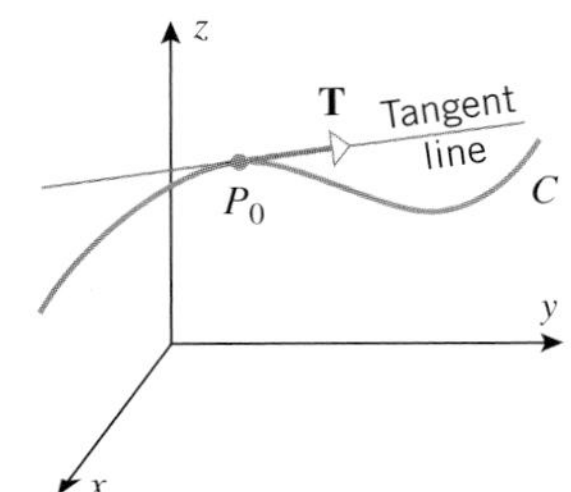

Figure 14.7.1

approximates $f(x, y)$ very closely near (x_0, y_0) and the graph of L is a nonvertical plane through the point P_0. This suggests that the graph of L is the tangent plane we seek. We can now provide some *geometric* justification for this conclusion.

We will base our concept of a tangent plane to a surface $S : z = f(x, y)$ on the more elementary notion of a tangent line to a curve C in 3-space (Figure 14.7.1). Intuitively, we would expect a tangent plane to S at a point P_0 to be composed of the tangent lines at P_0 of all curves on S that pass through P_0 (Figure 14.7.2). The following theorem shows that the graph of the local linear approximation is indeed tangent to the surface $z = f(x, y)$ in this geometric sense.

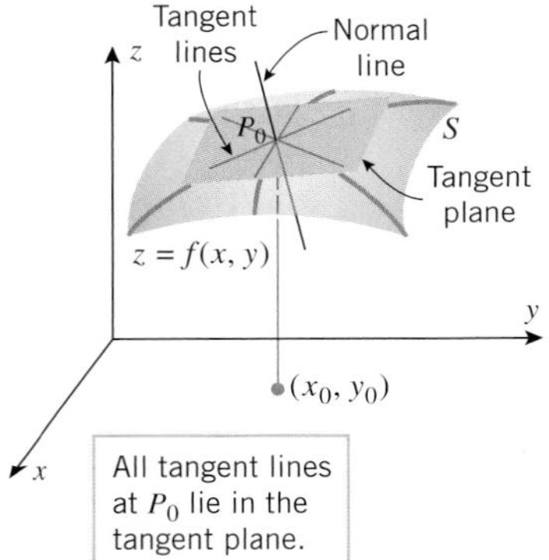

Figure 14.7.2

14.7.1 THEOREM. *Assume that the function $f(x, y)$ is differentiable at (x_0, y_0) and let $P_0(x_0, y_0, f(x_0, y_0))$ denote the corresponding point on the graph of f. Let T denote the graph of the local linear approximation*

$$L(x, y) = f(x_0, y_0) + f_x(x_0, y_0)(x - x_0) + f_y(x_0, y_0)(y - y_0) \tag{1}$$

to f at (x_0, y_0). Then a line is tangent at P_0 to a curve C on the surface $z = f(x, y)$ if and only if the line is contained in T.

PROOF. The graph T of (1) is the plane

$$z = f(x_0, y_0) + f_x(x_0, y_0)(x - x_0) + f_y(x_0, y_0)(y - y_0)$$

for which

$$\mathbf{n} = f_x(x_0, y_0)\mathbf{i} + f_y(x_0, y_0)\mathbf{j} - \mathbf{k}$$

is a normal vector (verify). Let C denote a curve on the surface $z = f(x, y)$ through P_0 and assume that C is parametrized by

$$x = x(t), \quad y = y(t), \quad z = z(t)$$

with

$$x_0 = x(t_0), \quad y_0 = y(t_0), \quad f(x_0, y_0) = z(t_0)$$

The tangent line l to C through P_0 is then parallel to the vector

$$\mathbf{r}' = x'(t_0)\mathbf{i} + y'(t_0)\mathbf{j} + z'(t_0)\mathbf{k}$$

where we assume that $\mathbf{r}' \neq \mathbf{0}$ (Definition 13.2.7). To prove that l is contained in T, it suffices to prove that $\mathbf{n} \cdot \mathbf{r}' = 0$. Since C lies on the graph of f, we have

$$z(t) = f(x(t), y(t))$$

Using the chain rule to compute the derivative of $z(t)$ at t_0 yields

$$z'(t_0) = f_x(x_0, y_0)x'(t_0) + f_y(x_0, y_0)y'(t_0)$$

or, equivalently, that

$$(f_x(x_0, y_0)\mathbf{i} + f_y(x_0, y_0)\mathbf{j} - \mathbf{k}) \cdot (x'(t_0)\mathbf{i} + y'(t_0)\mathbf{j} + z'(t_0)\mathbf{k}) = 0$$

But this is just the equation $\mathbf{n} \cdot \mathbf{r}' = 0$, which completes the proof that l is contained in T.

Conversely, let $\mathbf{a} = a_1\mathbf{i} + a_2\mathbf{j} + a_3\mathbf{k}$ denote the direction vector for a line l through P_0 contained in T. Then

$$0 = \mathbf{n} \cdot \mathbf{a} = a_1 f_x(x_0, y_0) + a_2 f_y(x_0, y_0) - a_3$$

which implies that

$$a_3 = a_1 f_x(x_0, y_0) + a_2 f_y(x_0, y_0)$$

Let C denote the curve with parametric equations

$$x = x(t) = x_0 + a_1 t, \quad y = y(t) = y_0 + a_2 t, \quad z = z(t) = f(x(t), y(t))$$

The curve C passes through P_0 when $t = 0$ and the tangent line to C at P_0 has direction vector

$$\mathbf{r}' = x'(0)\mathbf{i} + y'(0)\mathbf{j} + z'(0)\mathbf{k} = a_1\mathbf{i} + a_2\mathbf{j} + z'(0)\mathbf{k}$$

It follows from the chain rule that

$$z'(0) = a_1 f_x(x_0, y_0) + a_2 f_y(x_0, y_0) = a_3$$

and therefore

$$\mathbf{r}' = x'(0)\mathbf{i} + y'(0)\mathbf{j} + z'(0)\mathbf{k} = a_1\mathbf{i} + a_2\mathbf{j} + a_3\mathbf{k} = \mathbf{a}$$

Thus, the vector $\mathbf{a} = a_1\mathbf{i} + a_2\mathbf{j} + a_3\mathbf{k}$ is the direction vector $\mathbf{r}'$ for the line through P_0 tangent to C. Therefore, this line is l, which completes the proof that l is tangent at P_0 to a curve C on the surface $z = f(x, y)$. ■

Use Theorem 14.7.1 to show that if a curve is contained in a plane, then so is any tangent line to the curve.

Based on Theorem 14.7.1 we make the following definition.

14.7.2 DEFINITION. If $f(x, y)$ is differentiable at the point (x_0, y_0), then the ***tangent plane*** to the surface $z = f(x, y)$ at the point $P_0(x_0, y_0, f(x_0, y_0))$ [or (x_0, y_0)] is the plane

$$z = f(x_0, y_0) + f_x(x_0, y_0)(x - x_0) + f_y(x_0, y_0)(y - y_0) \tag{2}$$

The line through the point P_0 parallel to the vector $\mathbf{n}$ is perpendicular to the tangent plane (2). We will refer to this line as the ***normal line*** to the surface $z = f(x, y)$ at P_0. It follows that this normal line can be expressed parametrically as

$$x = x_0 + f_x(x_0, y_0)t, \quad y = y_0 + f_y(x_0, y_0)t, \quad z = f(x_0, y_0) - t \tag{3}$$

▶ **Example 1** Find an equation for the tangent plane and parametric equations for the normal line to the surface $z = x^2 y$ at the point $(2, 1, 4)$.

Solution. The partial derivatives of f are

$$f_x(x, y) = 2xy, \quad f_y(x, y) = x^2$$
$$f_x(2, 1) = 4, \quad f_y(2, 1) = 4$$

Therefore, the tangent plane has equation

$$z = 4 + 4(x - 2) + 4(y - 1) = 4x + 4y - 8$$

and the normal line has equations

$$x = 2 + 4t, \quad y = 1 + 4t, \quad z = 4 - t \blacktriangleleft$$

TANGENT PLANES AND TOTAL DIFFERENTIALS

Recall that for a function $z = f(x, y)$ of two variables, the approximation by differentials is

$$\Delta z = \Delta f = f(x, y) - f(x_0, y_0) \approx dz = f_x(x_0, y_0)(x - x_0) + f_y(x_0, y_0)(y - y_0)$$

The tangent plane provides a geometric interpretation of this approximation. We see in Figure 14.7.3 that Δz is the change in z *along the surface* $z = f(x, y)$ from the point $P_0(x_0, y_0, f(x_0, y_0))$ to the point $P(x, y, f(x, y))$, and dz is the change in z *along the tangent plane* from P_0 to $Q(x, y, L(x, y))$. The small vertical displacement at (x, y) between the surface and the plane represents the error in the local linear approximation to

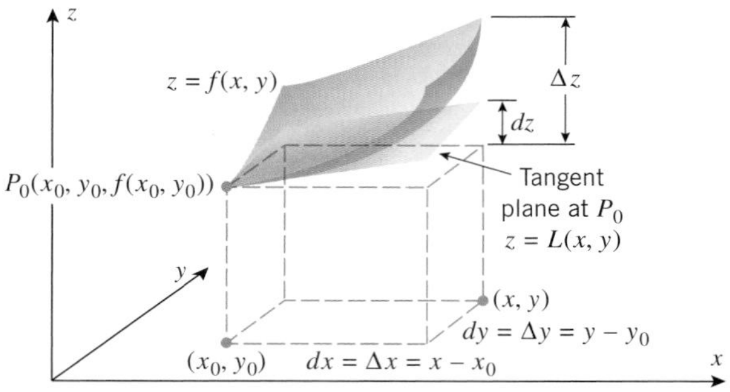

Note that the tangent plane in Figure 14.7.3 is analogous to the tangent line in Figure 14.7.2.

Figure 14.7.3

f at (x_0, y_0). We have seen that near (x_0, y_0) this error term has magnitude much smaller than the distance between (x, y) and (x_0, y_0).

TANGENT PLANES TO LEVEL SURFACES

We now consider the problem of finding tangent planes to surfaces that can be represented implicitly by equations of the form $F(x, y, z) = c$. We will assume that F has continuous first-order partial derivatives. This assumption poses no real restriction on the functions we will routinely encounter and has an important geometric consequence. In advanced courses it is proved that if F has continuous first-order partial derivatives, and if $\nabla F(x_0, y_0, z_0) \neq \mathbf{0}$, then near $P_0(x_0, y_0, z_0)$ the graph of $F(x, y, z) = c$ is actually the graph of an implicitly defined differentiable function of (at least) one of the following forms:

$$z = f(x, y), \quad y = g(x, z), \quad x = h(y, z) \tag{4}$$

This guarantees that near P_0 the graph of $F(x, y, z) = c$ is indeed a "surface" (rather than some possibly exotic-looking set of points in 3-space), and it follows from Theorem 14.7.1 that there is a tangent plane to the surface at the point P_0.

Fortunately, we do not need to solve the equation $F(x, y, z) = c$ for one of the functions in (4) in order to find the tangent plane at P_0. (In practice, this may be impossible.) We know from Theorem 14.7.1 that a line through P_0 will belong to this tangent plane if and only if it is a tangent line at P_0 of a curve C on the surface $F(x, y, z) = c$. Suppose that C is parametrized by

$$x = x(t), \quad y = y(t), \quad z = z(t)$$

with

$$x_0 = x(t_0), \quad y_0 = y(t_0), \quad z_0 = z(t_0)$$

The tangent line l to C through P_0 is then parallel to the vector

$$\mathbf{r}' = x'(t_0)\mathbf{i} + y'(t_0)\mathbf{j} + z'(t_0)\mathbf{k}$$

where we assume that $\mathbf{r}' \neq \mathbf{0}$ (Definition 13.2.7). Since C is on the surface $F(x, y, z) = c$, we have

$$c = F(x(t), y(t), z(t)) \tag{5}$$

Computing the derivative at t_0 of both sides of (5), we have by the chain rule that

$$0 = F_x(x_0, y_0, z_0)x'(t_0) + F_y(x_0, y_0, z_0)y'(t_0) + F_z(x_0, y_0, z_0)z'(t_0)$$

We can write this equation in vector form as

$$0 = (F_x(x_0, y_0, z_0)\mathbf{i} + F_y(x_0, y_0, z_0)\mathbf{j} + F_z(x_0, y_0, z_0)\mathbf{k}) \cdot (x'(t_0)\mathbf{i} + y'(t_0)\mathbf{j} + z'(t_0)\mathbf{k})$$

or

$$0 = \nabla F(x_0, y_0, z_0) \cdot \mathbf{r}'$$

It follows that if $\nabla F(x_0, y_0, z_0) \neq \mathbf{0}$, then $\nabla F(x_0, y_0, z_0)$ is normal to line l. We conclude that if $\nabla F(x_0, y_0, z_0) \neq \mathbf{0}$, then $\nabla F(x_0, y_0, z_0)$ is normal to any line through P_0 that is contained in the tangent plane to the surface $F(x, y, z) = c$ at P_0. It follows that $\nabla F(x_0, y_0, z_0)$ is a normal vector to this plane and hence is normal to the level surface (Figure 14.7.4).

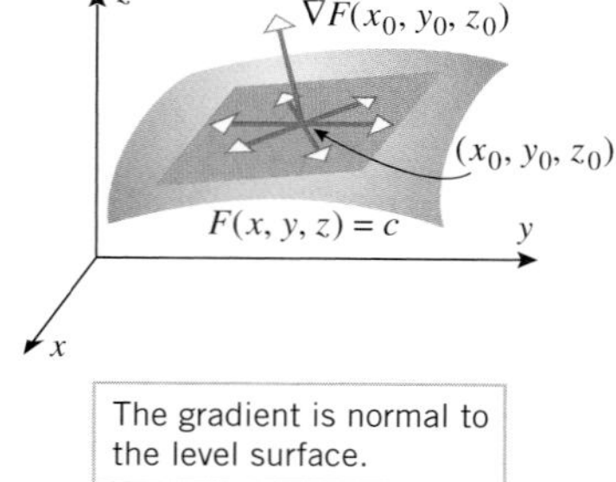

The gradient is normal to the level surface.

Figure 14.7.4

We can now express the equation of the tangent plane to the level surface at P_0 in point-normal form as

$$F_x(x_0, y_0, z_0)(x - x_0) + F_y(x_0, y_0, z_0)(y - y_0) + F_z(x_0, y_0, z_0)(z - z_0) = 0$$

[see Formula (3) of Section 12.6]. Based on this analysis we have the following theorem.

Theorem 14.7.3 can be viewed as an extension of Theorem 14.6.6 from curves to surfaces.

14.7.3 THEOREM. *Assume that $F(x, y, z)$ has continuous first-order partial derivatives and let $c = F(x_0, y_0, z_0)$. If $\nabla F(x_0, y_0, z_0) \neq \mathbf{0}$, then $\nabla F(x_0, y_0, z_0)$ is a* ***normal vector*** *to the surface $F(x, y, z) = c$ at the point $P_0(x_0, y_0, z_0)$, and the* ***tangent plane*** *to this surface at P_0 is the plane with equation*

$$F_x(x_0, y_0, z_0)(x - x_0) + F_y(x_0, y_0, z_0)(y - y_0) + F_z(x_0, y_0, z_0)(z - z_0) = 0 \quad (6)$$

▶ **Example 2** Find an equation of the tangent plane to the ellipsoid $x^2 + 4y^2 + z^2 = 18$ at the point $(1, 2, 1)$, and determine the acute angle that this plane makes with the xy-plane.

Solution. The ellipsoid is a level surface of the function $F(x, y, z) = x^2 + 4y^2 + z^2$, so we begin by finding the gradient of this function at the point $(1, 2, 1)$. The computations are

$$\nabla F(x, y, z) = \frac{\partial F}{\partial x}\mathbf{i} + \frac{\partial F}{\partial y}\mathbf{j} + \frac{\partial F}{\partial z}\mathbf{k} = 2x\mathbf{i} + 8y\mathbf{j} + 2z\mathbf{k}$$

$$\nabla F(1, 2, 1) = 2\mathbf{i} + 16\mathbf{j} + 2\mathbf{k}$$

Thus,

$$F_x(1, 2, 1) = 2, \quad F_y(1, 2, 1) = 16, \quad F_z(1, 2, 1) = 2$$

and hence from (6) the equation of the tangent plane is

$$2(x - 1) + 16(y - 2) + 2(z - 1) = 0 \quad \text{or} \quad x + 8y + z = 18$$

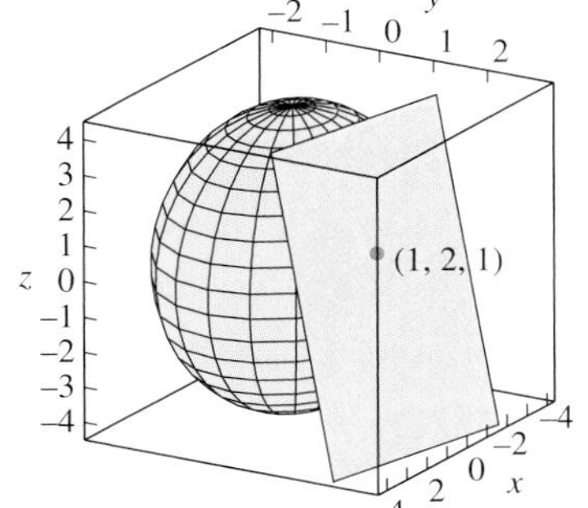

Figure 14.7.5

To find the acute angle θ between the tangent plane and the xy-plane, we will apply Formula (9) of Section 12.6 with $\mathbf{n}_1 = \nabla F(1, 2, 1) = 2\mathbf{i} + 16\mathbf{j} + 2\mathbf{k}$ and $\mathbf{n}_2 = \mathbf{k}$. This yields

$$\cos\theta = \frac{|\nabla F(1, 2, 1) \cdot \mathbf{k}|}{\|\nabla F(1, 2, 1)\|\|\mathbf{k}\|} = \frac{2}{(2\sqrt{66})(1)} = \frac{1}{\sqrt{66}}$$

Thus,

$$\theta = \cos^{-1}\left(\frac{1}{\sqrt{66}}\right) \approx 83^\circ$$

(Figure 14.7.5). ◀

USING GRADIENTS TO FIND TANGENT LINES TO INTERSECTIONS OF SURFACES

In general, the intersection of two surfaces $F(x, y, z) = 0$ and $G(x, y, z) = 0$ will be a curve in 3-space. If (x_0, y_0, z_0) is a point on this curve, then $\nabla F(x_0, y_0, z_0)$ will be normal to the surface $F(x, y, z) = 0$ at (x_0, y_0, z_0) and $\nabla G(x_0, y_0, z_0)$ will be normal to the surface $G(x, y, z) = 0$ at (x_0, y_0, z_0). Thus, if the curve of intersection can be smoothly parametrized, then its unit tangent vector $\mathbf{T}$ at (x_0, y_0, z_0) will be orthogonal to both $\nabla F(x_0, y_0, z_0)$ and $\nabla G(x_0, y_0, z_0)$ (Figure 14.7.6). Consequently, if

$$\nabla F(x_0, y_0, z_0) \times \nabla G(x_0, y_0, z_0) \neq \mathbf{0}$$

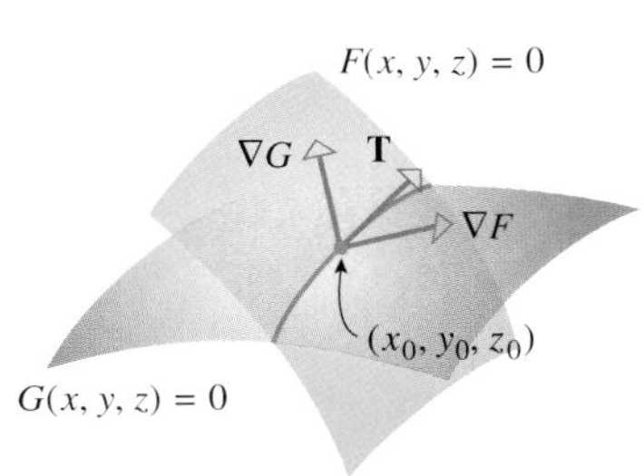

Figure 14.7.6

then this cross product will be parallel to $\mathbf{T}$ and hence will be tangent to the curve of intersection. This tangent vector can be used to determine the direction of the tangent line to the curve of intersection at the point (x_0, y_0, z_0).

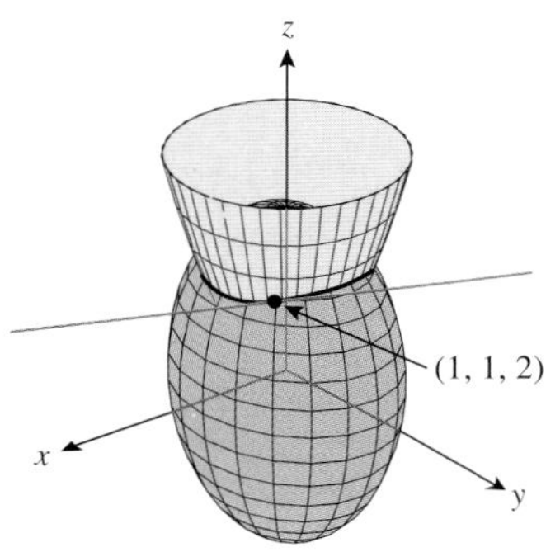

Figure 14.7.7

▶ **Example 3** Find parametric equations of the tangent line to the curve of intersection of the paraboloid $z = x^2 + y^2$ and the ellipsoid $3x^2 + 2y^2 + z^2 = 9$ at the point $(1, 1, 2)$ (Figure 14.7.7).

Solution. We begin by rewriting the equations of the surfaces as

$$x^2 + y^2 - z = 0 \quad \text{and} \quad 3x^2 + 2y^2 + z^2 - 9 = 0$$

and we take

$$F(x, y, z) = x^2 + y^2 - z \quad \text{and} \quad G(x, y, z) = 3x^2 + 2y^2 + z^2 - 9$$

We will need the gradients of these functions at the point $(1, 1, 2)$. The computations are

$$\begin{aligned} \nabla F(x, y, z) &= 2x\mathbf{i} + 2y\mathbf{j} - \mathbf{k}, \quad & \nabla G(x, y, z) &= 6x\mathbf{i} + 4y\mathbf{j} + 2z\mathbf{k} \\ \nabla F(1, 1, 2) &= 2\mathbf{i} + 2\mathbf{j} - \mathbf{k}, \quad & \nabla G(1, 1, 2) &= 6\mathbf{i} + 4\mathbf{j} + 4\mathbf{k} \end{aligned}$$

Thus, a tangent vector at $(1, 1, 2)$ to the curve of intersection is

$$\nabla F(1, 1, 2) \times \nabla G(1, 1, 2) = \begin{vmatrix} \mathbf{i} & \mathbf{j} & \mathbf{k} \\ 2 & 2 & -1 \\ 6 & 4 & 4 \end{vmatrix} = 12\mathbf{i} - 14\mathbf{j} - 4\mathbf{k}$$

Since any scalar multiple of this vector will do just as well, we can multiply by $\frac{1}{2}$ to reduce the size of the coefficients and use the vector of $6\mathbf{i} - 7\mathbf{j} - 2\mathbf{k}$ to determine the direction of the tangent line. This vector and the point $(1, 1, 2)$ yield the parametric equations

$$x = 1 + 6t, \quad y = 1 - 7t, \quad z = 2 - 2t \; ◀$$

✔ QUICK CHECK EXERCISES 14.7 *(See page 996 for answers.)*

1. Suppose that $f(x, y)$ is differentiable at the point $(3, 1)$ with $f(3, 1) = 4$, $f_x(3, 1) = 2$, and $f_y(3, 1) = -3$. An equation for the tangent plane to the graph of f at the point $(3, 1, 4)$ is ________, and parametric equations for the normal line to the graph of f through the point $(3, 1, 4)$ are

   $x =$ ________, $y =$ ________, $z =$ ________

2. An equation for the tangent plane to the graph of $z = x^2\sqrt{y}$ at the point $(2, 4, 8)$ is ________, and parametric equations for the normal line to the graph of $z = x^2\sqrt{y}$ through the point $(2, 4, 8)$ are

   $x =$ ________, $y =$ ________, $z =$ ________

3. Suppose that $f(1, 0, -1) = 2$, and $f(x, y, z)$ is differentiable at $(1, 0, -1)$ with $\nabla f(1, 0, -1) = \langle 2, 1, 1\rangle$. An equation for the tangent plane to the level surface $f(x, y, z) = 2$ at the point $(1, 0, -1)$ is ________, and parametric equations for the normal line to the level surface through the point $(1, 0, -1)$ are

   $x =$ ________, $y =$ ________, $z =$ ________

4. The sphere $x^2 + y^2 + z^2 = 9$ and the plane $x + y + z = 5$ intersect in a circle that passes through the point $(2, 1, 2)$. Parametric equations for the tangent line to this circle at $(2, 1, 2)$ are

   $x =$ ________, $y =$ ________, $z =$ ________

EXERCISE SET 14.7 [C] CAS

1–8 Find an equation for the tangent plane and parametric equations for the normal line to the surface at the point P.

1. $z = 4x^3y^2 + 2y$; $P(1, -2, 12)$
2. $z = \frac{1}{2}x^7y^{-2}$; $P(2, 4, 4)$
3. $z = xe^{-y}$; $P(1, 0, 1)$
4. $z = \ln\sqrt{x^2 + y^2}$; $P(-1, 0, 0)$
5. $z = e^{3y}\sin 3x$; $P(\pi/6, 0, 1)$
6. $z = x^{1/2} + y^{1/2}$; $P(4, 9, 5)$
7. $x^2 + y^2 + z^2 = 25$; $P(-3, 0, 4)$
8. $x^2y - 4z^2 = -7$; $P(-3, 1, -2)$

FOCUS ON CONCEPTS

9. Find all points on the surface at which the tangent plane is horizontal.
(a) $z = x^3y^2$
(b) $z = x^2 - xy + y^2 - 2x + 4y$

10. Find a point on the surface $z = 3x^2 - y^2$ at which the tangent plane is parallel to the plane $6x + 4y - z = 5$.

11. Find a point on the surface $z = 8 - 3x^2 - 2y^2$ at which the tangent plane is perpendicular to the line $x = 2 - 3t$, $y = 7 + 8t$, $z = 5 - t$.

12. Show that the surfaces

$$z = \sqrt{x^2 + y^2} \quad \text{and} \quad z = \tfrac{1}{10}(x^2 + y^2) + \tfrac{5}{2}$$

intersect at $(3, 4, 5)$ and have a common tangent plane at that point.

13. (a) Find all points of intersection of the line

$$x = -1 + t, \quad y = 2 + t, \quad z = 2t + 7$$

and the surface

$$z = x^2 + y^2$$

(b) At each point of intersection, find the cosine of the acute angle between the given line and the line normal to the surface.

14. Show that if f is differentiable and $z = xf(x/y)$, then all tangent planes to the graph of this equation pass through the origin.

15. Consider the ellipsoid $x^2 + y^2 + 4z^2 = 12$.
(a) Use the method of Example 2 to find an equation of the tangent plane to the ellipsoid at the point $(2, 2, 1)$.
(b) Find parametric equations of the line that is normal to the ellipsoid at the point $(2, 2, 1)$.
(c) Find the acute angle that the tangent plane at the point $(2, 2, 1)$ makes with the xy-plane.

16. Consider the surface $xz - yz^3 + yz^2 = 2$.
(a) Use the method of Example 2 to find an equation of the tangent plane to the surface at the point $(2, -1, 1)$.
(b) Find parametric equations of the line that is normal to the surface at the point $(2, -1, 1)$.
(c) Find the acute angle that the tangent plane at the point $(2, -1, 1)$ makes with the xy-plane.

17–18 Find two unit vectors that are normal to the given surface at the point P.

17. $\sqrt{\dfrac{z + x}{y - 1}} = z^2$; $P(3, 5, 1)$

18. $\sin xz - 4\cos yz = 4$; $P(\pi, \pi, 1)$

19. Show that every line that is normal to the sphere

$$x^2 + y^2 + z^2 = 1$$

passes through the origin.

20. Find all points on the ellipsoid $2x^2 + 3y^2 + 4z^2 = 9$ at which the plane tangent to the ellipsoid is parallel to the plane $x - 2y + 3z = 5$.

21. Find all points on the surface $x^2 + y^2 - z^2 = 1$ at which the normal line is parallel to the line through $P(1, -2, 1)$ and $Q(4, 0, -1)$.

22. Show that the ellipsoid $2x^2 + 3y^2 + z^2 = 9$ and the sphere

$$x^2 + y^2 + z^2 - 6x - 8y - 8z + 24 = 0$$

have a common tangent plane at the point $(1, 1, 2)$.

23. Find parametric equations for the tangent line to the curve of intersection of the paraboloid $z = x^2 + y^2$ and the ellipsoid $x^2 + 4y^2 + z^2 = 9$ at the point $(1, -1, 2)$.

24. Find parametric equations for the tangent line to the curve of intersection of the cone $z = \sqrt{x^2 + y^2}$ and the plane $x + 2y + 2z = 20$ at the point $(4, 3, 5)$.

25. Find parametric equations for the tangent line to the curve of intersection of the cylinders $x^2 + z^2 = 25$ and $y^2 + z^2 = 25$ at the point $(3, -3, 4)$.

C **26.** The accompanying figure shows the intersection of the surfaces $z = 8 - x^2 - y^2$ and $4x + 2y - z = 0$.
(a) Find parametric equations for the tangent line to the curve of intersection at the point $(0, 2, 4)$.
(b) Use a CAS to generate a reasonable facsimile of the figure. You need not generate the colors, but try to obtain a similar viewpoint.

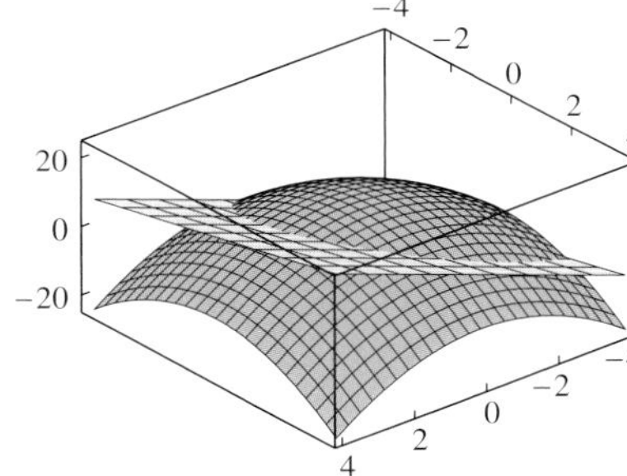

Figure Ex-26

27. Show that the equation of the plane that is tangent to the ellipsoid

$$\frac{x^2}{a^2} + \frac{y^2}{b^2} + \frac{z^2}{c^2} = 1$$

at (x_0, y_0, z_0) can be written in the form

$$\frac{x_0x}{a^2} + \frac{y_0y}{b^2} + \frac{z_0z}{c^2} = 1$$

28. Show that the equation of the plane that is tangent to the paraboloid

$$z = \frac{x^2}{a^2} + \frac{y^2}{b^2}$$

at (x_0, y_0, z_0) can be written in the form

$$z + z_0 = \frac{2x_0x}{a^2} + \frac{2y_0y}{b^2}$$

29. Prove: If the surfaces $z = f(x, y)$ and $z = g(x, y)$ intersect at $P(x_0, y_0, z_0)$, and if f and g are differentiable at (x_0, y_0), then the normal lines at P are perpendicular if and only if

$$f_x(x_0, y_0)g_x(x_0, y_0) + f_y(x_0, y_0)g_y(x_0, y_0) = -1$$

30. Use the result in Exercise 29 to show that the normal lines to the cones $z = \sqrt{x^2 + y^2}$ and $z = -\sqrt{x^2 + y^2}$ are perpendicular to the normal lines to the sphere $x^2 + y^2 + z^2 = a^2$ at every point of intersection (see Figure Ex-32).

31. Two surfaces $f(x, y, z) = 0$ and $g(x, y, z) = 0$ are said to be ***orthogonal*** at a point P of intersection if ∇f and ∇g are nonzero at P and the normal lines to the surfaces are perpendicular at P. Show that if $\nabla f(x_0, y_0, z_0) \neq \mathbf{0}$ and $\nabla g(x_0, y_0, z_0) \neq \mathbf{0}$, then the surfaces $f(x, y, z) = 0$ and $g(x, y, z) = 0$ are orthogonal at the point (x_0, y_0, z_0) if and only if

$$f_x g_x + f_y g_y + f_z g_z = 0$$

at this point. [*Note:* This is a more general version of the result in Exercise 29.]

32. Use the result of Exercise 31 to show that the sphere $x^2 + y^2 + z^2 = a^2$ and the cone $z^2 = x^2 + y^2$ are orthogonal at every point of intersection (see the accompanying figure).

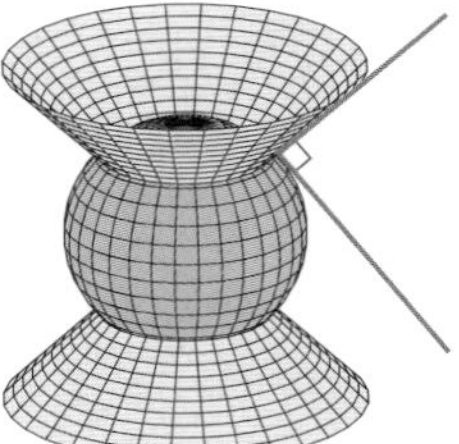

Figure Ex-32

33. Show that the volume of the solid bounded by the coordinate planes and a plane tangent to the portion of the surface $xyz = k$, $k > 0$, in the first octant does not depend on the point of tangency.

✔ QUICK CHECK ANSWERS 14.7

1. $z = 4 + 2(x - 3) - 3(y - 1)$; $x = 3 + 2t$; $y = 1 - 3t$; $z = 4 - t$
2. $z = 8 + 8(x - 2) + (y - 4)$; $x = 2 + 8t$; $y = 4 + t$; $z = 8 - t$
3. $2(x - 1) + y + (z + 1) = 0$; $x = 1 + 2t$; $y = t$; $z = -1 + t$ **4.** $x = 2 + t$; $y = 1$; $z = 2 - t$

14.8 MAXIMA AND MINIMA OF FUNCTIONS OF TWO VARIABLES

Earlier in this text we learned how to find maximum and minimum values of a function of one variable. In this section we will develop similar techniques for functions of two variables.

EXTREMA

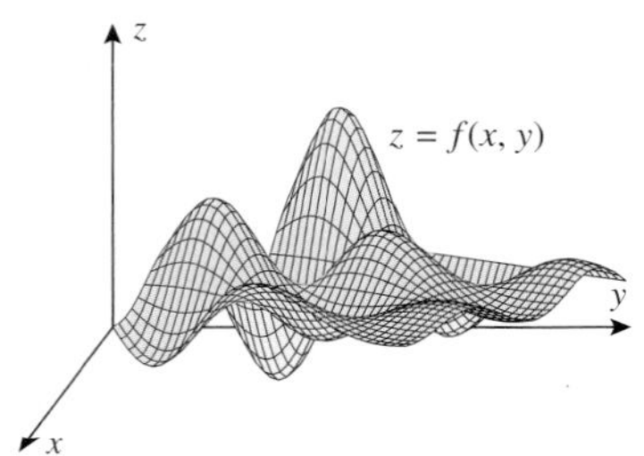

Figure 14.8.1

If we imagine the graph of a function f of two variables to be a mountain range (Figure 14.8.1), then the mountaintops, which are the high points in their immediate vicinity, are called *relative maxima* of f, and the valley bottoms, which are the low points in their immediate vicinity, are called *relative minima* of f.

Just as a geologist might be interested in finding the highest mountain and deepest valley in an entire mountain range, so a mathematician might be interested in finding the largest and smallest values of $f(x, y)$ over the *entire* domain of f. These are called the *absolute maximum* and *absolute minimum values* of f. The following definitions make these informal ideas precise.

14.8.1 DEFINITION. A function f of two variables is said to have a ***relative maximum*** at a point (x_0, y_0) if there is a disk centered at (x_0, y_0) such that $f(x_0, y_0) \geq f(x, y)$ for all points (x, y) that lie inside the disk, and f is said to have an ***absolute maximum*** at (x_0, y_0) if $f(x_0, y_0) \geq f(x, y)$ for all points (x, y) in the domain of f.

14.8.2 DEFINITION. A function f of two variables is said to have a ***relative minimum*** at a point (x_0, y_0) if there is a disk centered at (x_0, y_0) such that $f(x_0, y_0) \le f(x, y)$ for all points (x, y) that lie inside the disk, and f is said to have an ***absolute minimum*** at (x_0, y_0) if $f(x_0, y_0) \le f(x, y)$ for all points (x, y) in the domain of f.

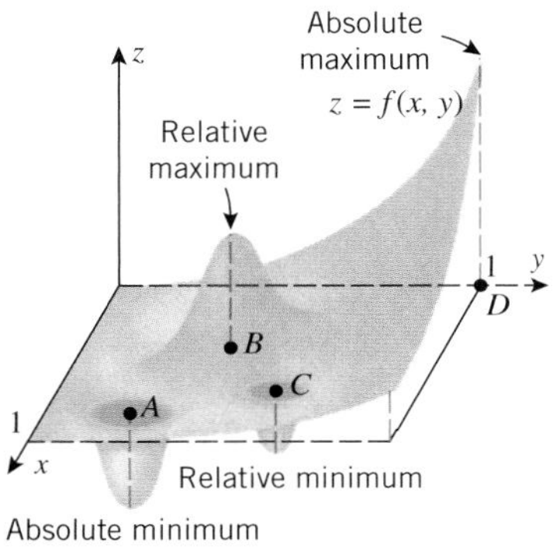

Absolute maximum

$z = f(x, y)$

Relative maximum

Relative minimum

Absolute minimum

Figure 14.8.2

If f has a relative maximum or a relative minimum at (x_0, y_0), then we say that f has a ***relative extremum*** at (x_0, y_0), and if f has an absolute maximum or absolute minimum at (x_0, y_0), then we say that f has an ***absolute extremum*** at (x_0, y_0).

Figure 14.8.2 shows the graph of a function f whose domain is the square region in the xy-plane whose points satisfy the inequalities $0 \le x \le 1, 0 \le y \le 1$. The function f has relative minima at the points A and C and a relative maximum at B. There is an absolute minimum at A and an absolute maximum at D.

For functions of two variables we will be concerned with two important questions:

- Are there any relative or absolute extrema?
- If so, where are they located?

BOUNDED SETS

Just as we distinguished between finite intervals and infinite intervals on the real line, so we will want to distinguish between regions of "finite extent" and regions of "infinite extent" in 2-space and 3-space. A set of points in 2-space is called ***bounded*** if the entire set can be contained within some rectangle, and is called ***unbounded*** if there is no rectangle that contains all the points of the set. Similarly, a set of points in 3-space is ***bounded*** if the entire set can be contained within some box, and is unbounded otherwise (Figure 14.8.3).

Explain why any subset of a bounded set is also bounded.

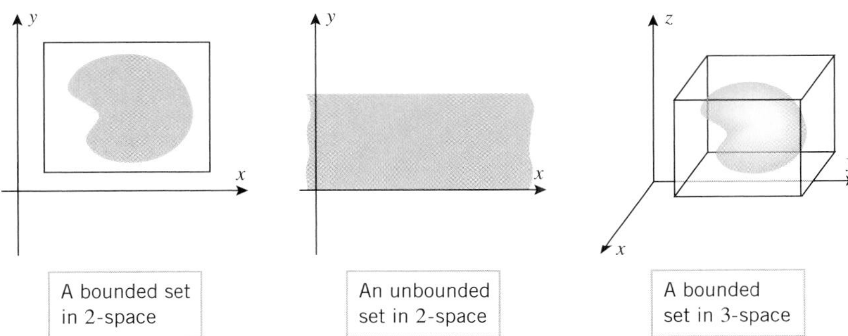

Figure 14.8.3

THE EXTREME-VALUE THEOREM

For functions of one variable that are continuous on a closed interval, the Extreme-Value Theorem (Theorem 4.4.2) answered the existence question for absolute extrema. The following theorem, which we state without proof, is the corresponding result for functions of two variables.

14.8.3 THEOREM (*Extreme-Value Theorem*). *If $f(x, y)$ is continuous on a closed and bounded set R, then f has both an absolute maximum and an absolute minimum on R.*

▶ Example 1 The square region R whose points satisfy the inequalities

$$0 \le x \le 1 \quad \text{and} \quad 0 \le y \le 1$$

is a closed and bounded set in the xy-plane. The function f whose graph is shown in Figure 14.8.2 is continuous on R; thus, it is guaranteed to have an absolute maximum and minimum on R by the last theorem. These occur at points D and A that are shown in the figure. ◀

> If any of the conditions in the Extreme-Value Theorem fail to hold, then there is no guarantee that an absolute maximum or absolute minimum exists on the region R. Thus, a discontinuous function on a closed and bounded set need not have any absolute extrema, and a continuous function on a set that is not closed and bounded also need not have any absolute extrema.

FINDING RELATIVE EXTREMA

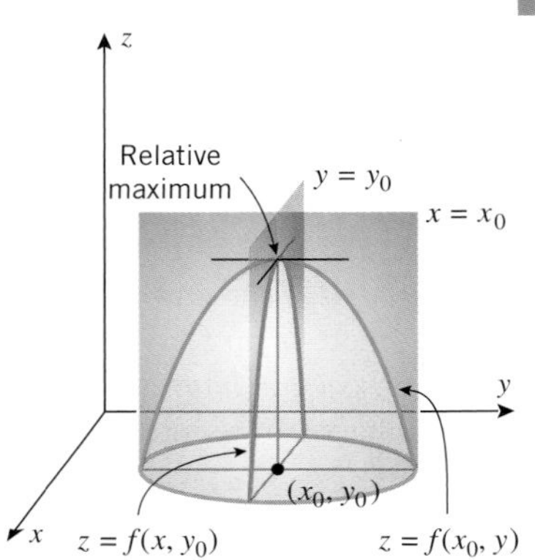

Figure 14.8.4

Recall that if a function g of one variable has a relative extremum at a point x_0 where g is differentiable, then $g'(x_0) = 0$. To obtain the analog of this result for functions of two variables, suppose that $f(x, y)$ has a relative maximum at a point (x_0, y_0) and that the partial derivatives of f exist at (x_0, y_0). It seems plausible geometrically that the traces of the surface $z = f(x, y)$ on the planes $x = x_0$ and $y = y_0$ have horizontal tangent lines at (x_0, y_0) (Figure 14.8.4), so

$$f_x(x_0, y_0) = 0 \quad \text{and} \quad f_y(x_0, y_0) = 0$$

The same conclusion holds if f has a relative minimum at (x_0, y_0), all of which suggests the following result, which we state without formal proof.

14.8.4 THEOREM. *If f has a relative extremum at a point (x_0, y_0), and if the first-order partial derivatives of f exist at this point, then*

$$f_x(x_0, y_0) = 0 \quad \textit{and} \quad f_y(x_0, y_0) = 0$$

Recall that the *critical points* of a function f of one variable are those values of x in the domain of f at which $f'(x) = 0$ or f is not differentiable. The following definition is the analog for functions of two variables.

> Explain why
>
> $$D_{\mathbf{u}} f(x_0, y_0) = 0$$
>
> for all $\mathbf{u}$ if (x_0, y_0) is a critical point of f and f is differentiable at (x_0, y_0).

14.8.5 DEFINITION. A point (x_0, y_0) in the domain of a function $f(x, y)$ is called a ***critical point*** of the function if $f_x(x_0, y_0) = 0$ and $f_y(x_0, y_0) = 0$ or if one or both partial derivatives do not exist at (x_0, y_0).

It follows from this definition and Theorem 14.8.4 that relative extrema occur at critical points, just as for a function of one variable. However, recall that for a function of one variable a relative extremum need not occur at *every* critical point. For example, the function might have an inflection point with a horizontal tangent line at the critical point (see Figure 4.2.6). Similarly, a function of two variables need not have a relative extremum at every critical point. For example, consider the function

$$f(x, y) = y^2 - x^2$$

This function, whose graph is the hyperbolic paraboloid shown in Figure 14.8.5, has a

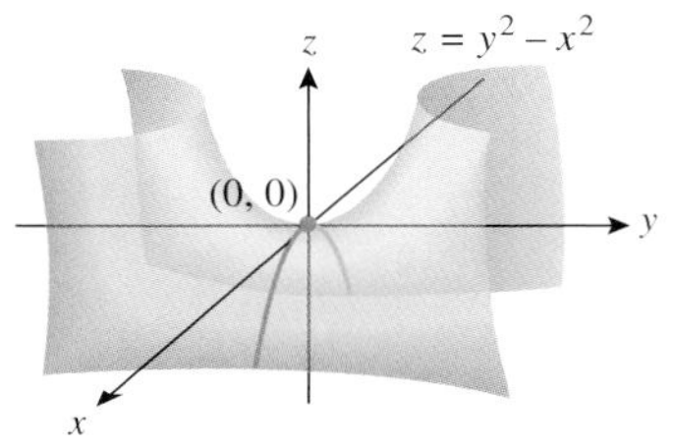

The function $f(x, y) = y^2 - x^2$ has neither a relative maximum nor a relative minimum at the critical point $(0, 0)$.

Figure 14.8.5

critical point at $(0, 0)$, since

$$f_x(x, y) = -2x \quad \text{and} \quad f_y(x, y) = 2y$$

from which it follows that

$$f_x(0, 0) = 0 \quad \text{and} \quad f_y(0, 0) = 0$$

However, the function f has neither a relative maximum nor a relative minimum at $(0, 0)$. For obvious reasons, the point $(0, 0)$ is called a *saddle point* of f. In general, we will say that a surface $z = f(x, y)$ has a ***saddle point*** at (x_0, y_0) if there are two distinct vertical planes through this point such that the trace of the surface in one of the planes has a relative maximum at (x_0, y_0) and the trace in the other has a relative minimum at (x_0, y_0).

▶ **Example 2** The three functions graphed in Figure 14.8.6 all have critical points at $(0, 0)$. For the paraboloids, the partial derivatives at the origin are zero. You can check this algebraically by evaluating the partial derivatives at $(0, 0)$, but you can see it geometrically by observing that the traces in the xz-plane and yz-plane have horizontal tangent lines at $(0, 0)$. For the cone neither partial derivative exists at the origin because the traces in the xz-plane and the yz-plane have corners there. The paraboloid in part (a) and the cone in part (c) have a relative minimum and absolute minimum at the origin, and the paraboloid in part (b) has a relative maximum and an absolute maximum at the origin. ◀

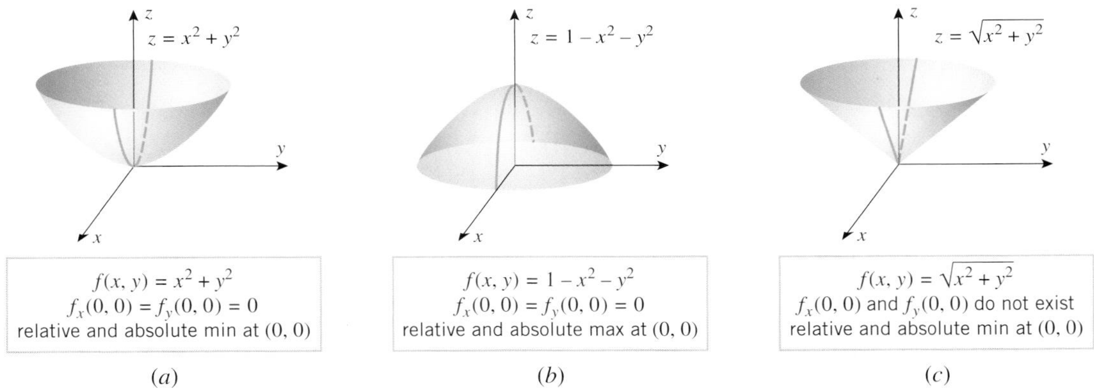

$f(x, y) = x^2 + y^2$
$f_x(0, 0) = f_y(0, 0) = 0$
relative and absolute min at $(0, 0)$

(a)

$f(x, y) = 1 - x^2 - y^2$
$f_x(0, 0) = f_y(0, 0) = 0$
relative and absolute max at $(0, 0)$

(b)

$f(x, y) = \sqrt{x^2 + y^2}$
$f_x(0, 0)$ and $f_y(0, 0)$ do not exist
relative and absolute min at $(0, 0)$

(c)

Figure 14.8.6

THE SECOND PARTIALS TEST

For functions of one variable the second derivative test (Theorem 4.2.4) was used to determine the behavior of a function at a critical point. The following theorem, which is usually proved in advanced calculus, is the analog of that theorem for functions of two variables.

With the notation of Theorem 14.8.6, show that if $D > 0$, then $f_{xx}(x_0, y_0)$ and $f_{yy}(x_0, y_0)$ have the same sign. Thus, we can replace $f_{xx}(x_0, y_0)$ by $f_{yy}(x_0, y_0)$ in parts (a) and (b) of the theorem.

14.8.6 THEOREM (*The Second Partials Test*). *Let f be a function of two variables with continuous second-order partial derivatives in some disk centered at a critical point (x_0, y_0), and let*

$$D = f_{xx}(x_0, y_0) f_{yy}(x_0, y_0) - f_{xy}^2(x_0, y_0)$$

(a) *If $D > 0$ and $f_{xx}(x_0, y_0) > 0$, then f has a relative minimum at (x_0, y_0).*

(b) *If $D > 0$ and $f_{xx}(x_0, y_0) < 0$, then f has a relative maximum at (x_0, y_0).*

(c) *If $D < 0$, then f has a saddle point at (x_0, y_0).*

(d) *If $D = 0$, then no conclusion can be drawn.*

▶ **Example 3** Locate all relative extrema and saddle points of

$$f(x, y) = 3x^2 - 2xy + y^2 - 8y$$

Solution. Since $f_x(x, y) = 6x - 2y$ and $f_y(x, y) = -2x + 2y - 8$, the critical points of f satisfy the equations

$$6x - 2y = 0$$
$$-2x + 2y - 8 = 0$$

Solving these for x and y yields $x = 2$, $y = 6$ (verify), so $(2, 6)$ is the only critical point. To apply Theorem 14.8.6 we need the second-order partial derivatives

$$f_{xx}(x, y) = 6, \quad f_{yy}(x, y) = 2, \quad f_{xy}(x, y) = -2$$

At the point $(2, 6)$ we have

$$D = f_{xx}(2, 6) f_{yy}(2, 6) - f_{xy}^2(2, 6) = (6)(2) - (-2)^2 = 8 > 0$$

and

$$f_{xx}(2, 6) = 6 > 0$$

so f has a relative minimum at $(2, 6)$ by part (a) of the second partials test. Figure 14.8.7 shows a graph of f in the vicinity of the relative minimum. ◀

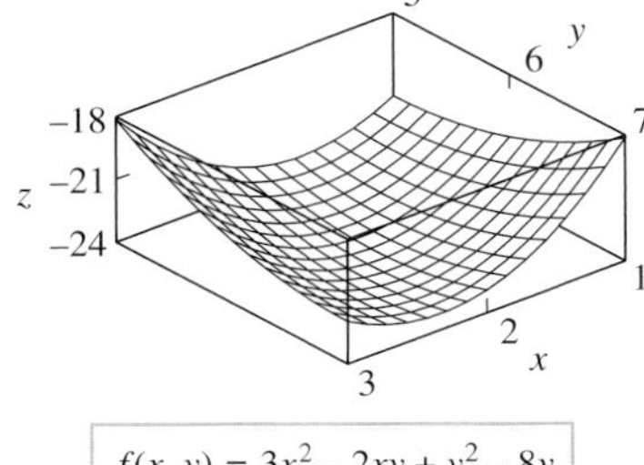

$f(x, y) = 3x^2 - 2xy + y^2 - 8y$

Figure 14.8.7

▶ **Example 4** Locate all relative extrema and saddle points of

$$f(x, y) = 4xy - x^4 - y^4$$

Solution. Since

$$f_x(x, y) = 4y - 4x^3$$
$$f_y(x, y) = 4x - 4y^3 \qquad (1)$$

the critical points of f have coordinates satisfying the equations

$$4y - 4x^3 = 0 \qquad y = x^3$$
$$\text{or} \qquad (2)$$
$$4x - 4y^3 = 0 \qquad x = y^3$$

Substituting the top equation in the bottom yields $x = (x^3)^3$ or equivalently $x^9 - x = 0$ or $x(x^8 - 1) = 0$, which has solutions $x = 0, x = 1, x = -1$. Substituting these values in the top equation of (2), we obtain the corresponding y-values $y = 0, y = 1, y = -1$. Thus, the critical points of f are $(0, 0)$, $(1, 1)$, and $(-1, -1)$.

From (1),

$$f_{xx}(x, y) = -12x^2, \quad f_{yy}(x, y) = -12y^2, \quad f_{xy}(x, y) = 4$$

which yields the following table:

CRITICAL POINT (x_0, y_0)	$f_{xx}(x_0, y_0)$	$f_{yy}(x_0, y_0)$	$f_{xy}(x_0, y_0)$	$D = f_{xx}f_{yy} - f_{xy}^2$
(0, 0)	0	0	4	−16
(1, 1)	−12	−12	4	128
(−1, −1)	−12	−12	4	128

At the points $(1, 1)$ and $(-1, -1)$, we have $D > 0$ and $f_{xx} < 0$, so relative maxima occur at these critical points. At $(0, 0)$ there is a saddle point since $D < 0$. The surface and a contour plot are shown in Figure 14.8.8. ◀

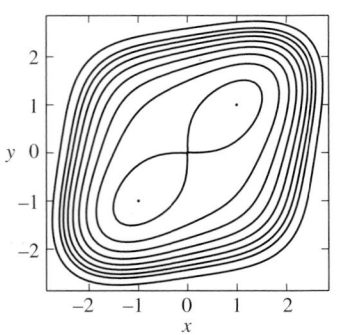

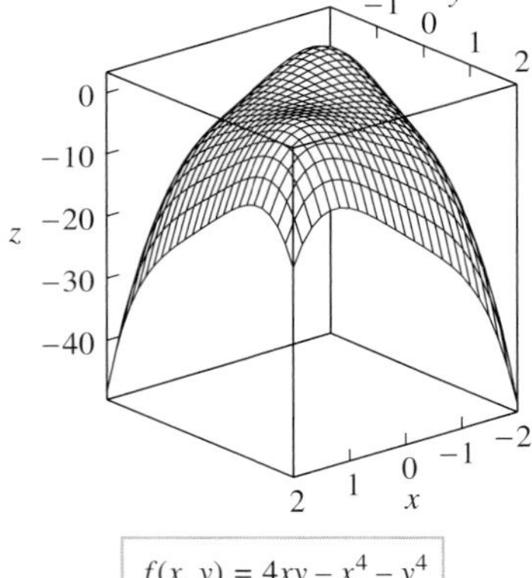

$f(x, y) = 4xy - x^4 - y^4$

Figure 14.8.8

The "figure eight" pattern at $(0, 0)$ in the contour plot for the surface in Figure 14.8.8 is typical for level curves that pass through a saddle point. If a bug starts at the point $(0, 0, 0)$ on the surface, in how many directions can it walk and remain in the xy-plane?

The following theorem, which is the analog for functions of two variables of Theorem 4.4.3, will lead to an important method for finding absolute extrema.

14.8.7 THEOREM. *If a function f of two variables has an absolute extremum (either an absolute maximum or an absolute minimum) at an interior point of its domain, then this extremum occurs at a critical point.*

PROOF. If f has an absolute maximum at the point (x_0, y_0) in the interior of the domain of f, then f has a relative maximum at (x_0, y_0). If both partial derivatives exist at (x_0, y_0), then

$$f_x(x_0, y_0) = 0 \quad \text{and} \quad f_y(x_0, y_0) = 0$$

by Theorem 14.8.4, so (x_0, y_0) is a critical point of f. If either partial derivative does not exist, then again (x_0, y_0) is a critical point, so (x_0, y_0) is a critical point in all cases. The proof for an absolute minimum is similar. ■

FINDING ABSOLUTE EXTREMA ON CLOSED AND BOUNDED SETS

If $f(x, y)$ is continuous on a closed and bounded set R, then the Extreme-Value Theorem (Theorem 14.8.3) guarantees the existence of an absolute maximum and an absolute minimum of f on R. These absolute extrema can occur either on the boundary of R or in the interior of R, but if an absolute extremum occurs in the interior, then it occurs at a critical point by Theorem 14.8.7. Thus, we are led to the following procedure for finding absolute extrema:

Compare this procedure with that in Section 4.4 for finding the extreme values of $f(x)$ on a closed interval.

How to Find the Absolute Extrema of a Continuous Function f of Two Variables on a Closed and Bounded Set R

Step 1. Find the critical points of f that lie in the interior of R.

Step 2. Find all boundary points at which the absolute extrema can occur.

Step 3. Evaluate $f(x, y)$ at the points obtained in the preceding steps. The largest of these values is the absolute maximum and the smallest the absolute minimum.

▶ **Example 5** Find the absolute maximum and minimum values of

$$f(x, y) = 3xy - 6x - 3y + 7 \tag{3}$$

on the closed triangular region R with vertices $(0, 0)$, $(3, 0)$, and $(0, 5)$.

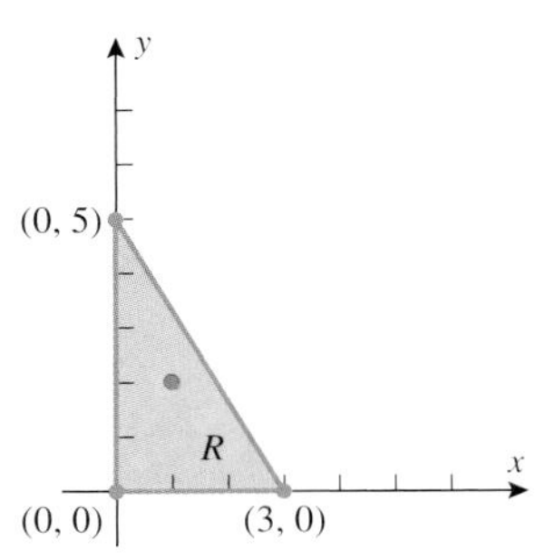

Figure 14.8.9

Solution. The region R is shown in Figure 14.8.9. We have

$$\frac{\partial f}{\partial x} = 3y - 6 \quad \text{and} \quad \frac{\partial f}{\partial y} = 3x - 3$$

so all critical points occur where

$$3y - 6 = 0 \quad \text{and} \quad 3x - 3 = 0$$

Solving these equations yields $x = 1$ and $y = 2$, so $(1, 2)$ is the only critical point. As shown in Figure 14.8.9, this critical point is in the interior of R.

Next we want to determine the locations of the points on the boundary of R at which the absolute extrema might occur. The boundary of R consists of three line segments, each of which we will treat separately:

The line segment between $(0, 0)$ *and* $(3, 0)$: On this line segment we have $y = 0$, so (3) simplifies to a function of the single variable x,

$$u(x) = f(x, 0) = -6x + 7, \quad 0 \le x \le 3$$

This function has no critical points because $u'(x) = -6$ is nonzero for all x. Thus the extreme values of $u(x)$ occur at the endpoints $x = 0$ and $x = 3$, which correspond to the points $(0, 0)$ and $(3, 0)$ of R.

The line segment between $(0, 0)$ *and* $(0, 5)$: On this line segment we have $x = 0$, so (3) simplifies to a function of the single variable y,

$$v(y) = f(0, y) = -3y + 7, \quad 0 \le y \le 5$$

This function has no critical points because $v'(y) = -3$ is nonzero for all y. Thus, the extreme values of $v(y)$ occur at the endpoints $y = 0$ and $y = 5$, which correspond to the points $(0, 0)$ and $(0, 5)$ of R.

The line segment between $(3, 0)$ *and* $(0, 5)$: In the xy-plane, an equation for this line segment is

$$y = -\tfrac{5}{3}x + 5, \quad 0 \le x \le 3 \tag{4}$$

so (3) simplifies to a function of the single variable x,

$$\begin{aligned} w(x) &= f\left(x, -\tfrac{5}{3}x + 5\right) = 3x\left(-\tfrac{5}{3}x + 5\right) - 6x - 3\left(-\tfrac{5}{3}x + 5\right) + 7 \\ &= -5x^2 + 14x - 8, \qquad 0 \le x \le 3 \end{aligned}$$

Since $w'(x) = -10x + 14$, the equation $w'(x) = 0$ yields $x = \frac{7}{5}$ as the only critical point of w. Thus, the extreme values of w occur either at the critical point $x = \frac{7}{5}$ or at the endpoints $x = 0$ and $x = 3$. The endpoints correspond to the points $(0, 5)$ and $(3, 0)$ of R, and from (4) the critical point corresponds to $\left(\frac{7}{5}, \frac{8}{3}\right)$.

Finally, Table 14.8.1 lists the values of $f(x, y)$ at the interior critical point and at the points on the boundary where an absolute extremum can occur. From the table we conclude that the absolute maximum value of f is $f(0, 0) = 7$ and the absolute minimum value is $f(3, 0) = -11$. ◄

Table 14.8.1

(x, y)	$(0, 0)$	$(3, 0)$	$(0, 5)$	$\left(\frac{7}{5}, \frac{8}{3}\right)$	$(1, 2)$
$f(x, y)$	7	−11	−8	$\frac{9}{5}$	1

► **Example 6** Determine the dimensions of a rectangular box, open at the top, having a volume of 32 ft^3, and requiring the least amount of material for its construction.

Solution. Let

x = length of the box (in feet)
y = width of the box (in feet)
z = height of the box (in feet)
S = surface area of the box (in square feet)

We may reasonably assume that the box with least surface area requires the least amount of material, so our objective is to minimize the surface area

$$S = xy + 2xz + 2yz \tag{5}$$

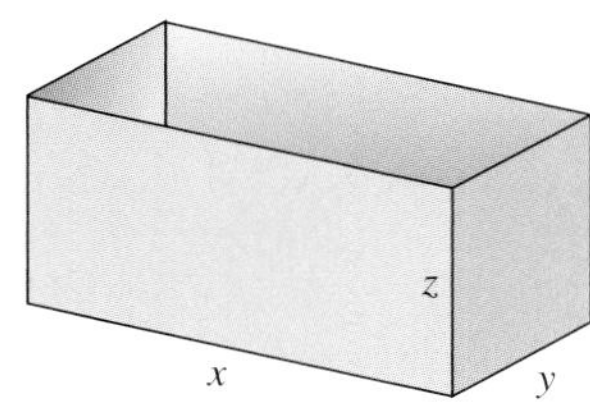

Figure 14.8.10

(Figure 14.8.10) subject to the volume requirement

$$xyz = 32 \tag{6}$$

From (6) we obtain $z = 32/xy$, so (5) can be rewritten as

$$S = xy + \frac{64}{y} + \frac{64}{x} \tag{7}$$

which expresses S as a function of two variables. The dimensions x and y in this formula must be positive, but otherwise have no limitation, so our problem reduces to finding the absolute minimum value of S over the first quadrant: $x > 0$, $y > 0$ (Figure 14.8.11). Because this region is neither closed nor bounded, we have no mathematical guarantee at this stage that an absolute minimum exists. However, note that S will have a large value at any point (x, y) in the first quadrant for which the product xy is large or for which either x or y is close to 0. We can use this observation to prove the existence of an absolute minimum value of S.

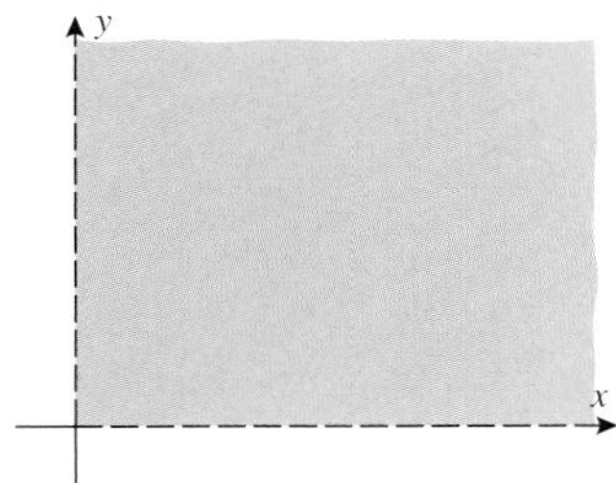

Figure 14.8.11

Let R denote the region in the first quadrant defined by the inequalities

$$\tfrac{1}{2} \le x, \quad \tfrac{1}{2} \le y, \quad \text{and} \quad xy \le 128$$

This region is both closed and bounded (verify) and the function S is continuous on R. It follows from Theorem 14.8.3 that S has an absolute minimum on R. Furthermore, note that $S > 128$ at any point (x, y) not in R and that the point $(8, 8)$ belongs to R with $S = 80 < 128$ at this point. We conclude that the minimum value of S on R is also the minimum value of S on the entire first quadrant.

Since S has an absolute minimum value in the first quadrant, it must occur at a critical point of S. Differentiating (7) we obtain

$$\frac{\partial S}{\partial x} = y - \frac{64}{x^2}, \quad \frac{\partial S}{\partial y} = x - \frac{64}{y^2} \tag{8}$$

so the coordinates of the critical points of S satisfy

$$y - \frac{64}{x^2} = 0, \quad x - \frac{64}{y^2} = 0$$

Solving the first equation for y yields

$$y = \frac{64}{x^2} \tag{9}$$

and substituting this expression in the second equation yields

$$x - \frac{64}{(64/x^2)^2} = 0$$

which can be rewritten as

$$x\left(1 - \frac{x^3}{64}\right) = 0$$

The solutions of this equation are $x = 0$ and $x = 4$. Since we require $x > 0$, the only solution of significance is $x = 4$. Substituting this value into (9) yields $y = 4$. Substituting $x = 4$ and $y = 4$ into (6) yields $z = 2$, so the box using least material has a height of 2 ft and a square base whose edges are 4 ft long. ◀

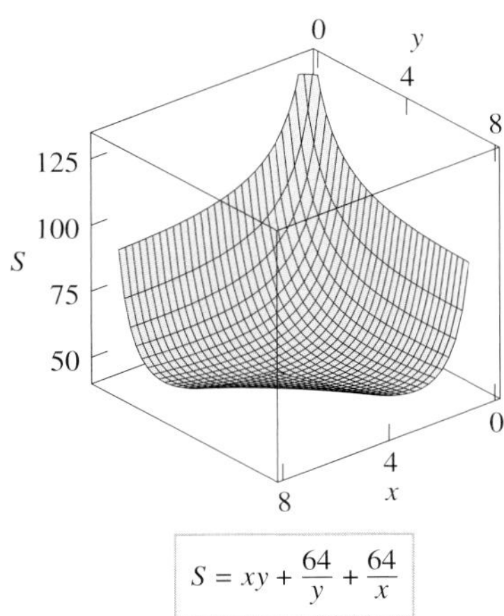

Figure 14.8.12

Fortunately, in our solution to Example 6 we were able to prove the existence of an absolute minimum of S in the first quadrant. The general problem of finding the absolute extrema of a function on an unbounded region, or on a region that is not closed, can be difficult and will not be considered in this text. However, in applied problems we can sometimes use physical considerations to deduce that an absolute extremum has been found. For example, the graph of Equation (7) in Figure 14.8.12 strongly suggests that the relative minimum at $x = 4$ and $y = 4$ is also an absolute minimum.

✔QUICK CHECK EXERCISES 14.8

(See page 1005 for answers.)

1. The critical points of the function $f(x, y) = x^3 + xy + y^2$ are ________.

2. Suppose that $f(x, y)$ has continuous second-order partial derivatives everywhere and that the origin is a critical point for f. State what information (if any) is provided by the second partials test if
   (a) $f_{xx}(0, 0) = 2,\ f_{xy}(0, 0) = 2,\ f_{yy}(0, 0) = 2$
   (b) $f_{xx}(0, 0) = -2,\ f_{xy}(0, 0) = 2,\ f_{yy}(0, 0) = 2$
   (c) $f_{xx}(0, 0) = 3,\ f_{xy}(0, 0) = 2,\ f_{yy}(0, 0) = 2$
   (d) $f_{xx}(0, 0) = -3,\ f_{xy}(0, 0) = 2,\ f_{yy}(0, 0) = -2$.

3. For the function $f(x, y) = x^3 - 3xy + y^3$, state what information (if any) is provided by the second partials test at the point
   (a) $(0, 0)$ (b) $(-1, -1)$ (c) $(1, 1)$.

4. A rectangular box has total surface area of 2 ft^2. Express the volume of the box as a function of the dimensions x and y of the base of the box.

EXERCISE SET 14.8

Graphing Utility   C CAS

1–2 Locate all absolute maxima and minima, if any, by inspection. Then check your answers using calculus.

1. (a) $f(x, y) = (x - 2)^2 + (y + 1)^2$
   (b) $f(x, y) = 1 - x^2 - y^2$
   (c) $f(x, y) = x + 2y - 5$

2. (a) $f(x, y) = 1 - (x + 1)^2 - (y - 5)^2$
   (b) $f(x, y) = e^{xy}$
   (c) $f(x, y) = x^2 - y^2$

3–4 Complete the squares and locate all absolute maxima and minima, if any, by inspection. Then check your answers using calculus.

3. $f(x, y) = 13 - 6x + x^2 + 4y + y^2$

4. $f(x, y) = 1 - 2x - x^2 + 4y - 2y^2$

FOCUS ON CONCEPTS

5–8 The contour plots show all significant features of the function. Make a conjecture about the number and the location of all relative extrema and saddle points, and then use calculus to check your conjecture.

5.
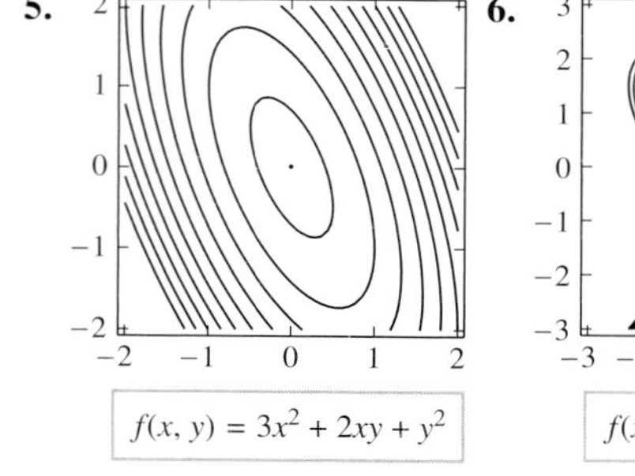

$f(x, y) = 3x^2 + 2xy + y^2$

6.
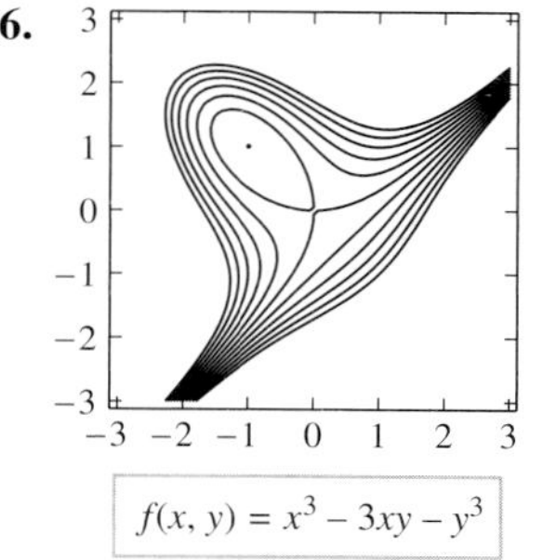

$f(x, y) = x^3 - 3xy - y^3$

7. 8.
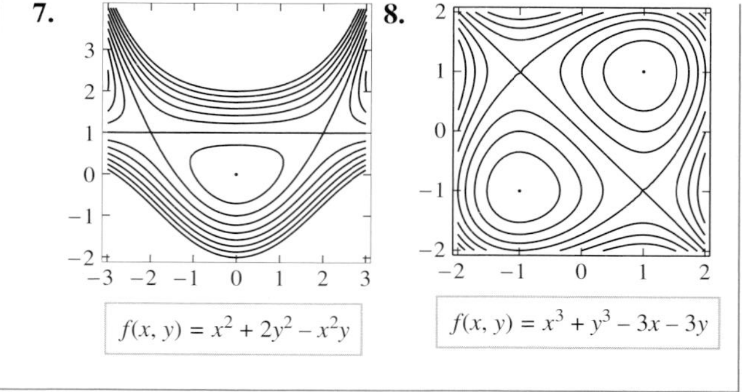

$f(x, y) = x^2 + 2y^2 - x^2y$ $f(x, y) = x^3 + y^3 - 3x - 3y$

9–20 Locate all relative maxima, relative minima, and saddle points, if any.

9. $f(x, y) = y^2 + xy + 3y + 2x + 3$

10. $f(x, y) = x^2 + xy - 2y - 2x + 1$

11. $f(x, y) = x^2 + xy + y^2 - 3x$

12. $f(x, y) = xy - x^3 - y^2$

13. $f(x, y) = x^2 + y^2 + \dfrac{2}{xy}$

14. $f(x, y) = xe^y$

15. $f(x, y) = x^2 + y - e^y$

16. $f(x, y) = xy + \dfrac{2}{x} + \dfrac{4}{y}$

17. $f(x, y) = e^x \sin y$

18. $f(x, y) = y \sin x$

19. $f(x, y) = e^{-(x^2+y^2+2x)}$

20. $f(x, y) = xy + \dfrac{a^3}{x} + \dfrac{b^3}{y} \quad (a \neq 0, b \neq 0)$

C 21. Use a CAS to generate a contour plot of
$$f(x, y) = 2x^2 - 4xy + y^4 + 2$$
for $-2 \le x \le 2$ and $-2 \le y \le 2$, and use the plot to approximate the locations of all relative extrema and saddle points in the region. Check your answer using calculus, and identify the relative extrema as relative maxima or minima.

C 22. Use a CAS to generate a contour plot of
$$f(x, y) = 2y^2x - yx^2 + 4xy$$
for $-5 \le x \le 5$ and $-5 \le y \le 5$, and use the plot to ap-

proximate the locations of all relative extrema and saddle points in the region. Check your answer using calculus, and identify the relative extrema as relative maxima or minima.

FOCUS ON CONCEPTS

23. (a) Show that the second partials test provides no information about the critical points of the function $f(x, y) = x^4 + y^4$.
(b) Classify all critical points of f as relative maxima, relative minima, or saddle points.

24. (a) Show that the second partials test provides no information about the critical points of the function $f(x, y) = x^4 - y^4$.
(b) Classify all critical points of f as relative maxima, relative minima, or saddle points.

25. Recall from Theorem 4.4.4 that if a continuous function of one variable has exactly one relative extremum on an interval, then that relative extremum is an absolute extremum on the interval. This exercise shows that this result does not extend to functions of two variables.
(a) Show that $f(x, y) = 3xe^y - x^3 - e^{3y}$ has only one critical point and that a relative maximum occurs there. (See the accompanying figure.)
(b) Show that f does not have an absolute maximum.

Source: This exercise is based on the article "The Only Critical Point in Town Test" by Ira Rosenholtz and Lowell Smylie, *Mathematics Magazine*, Vol. 58, No. 3, May 1985, pp. 149–150.

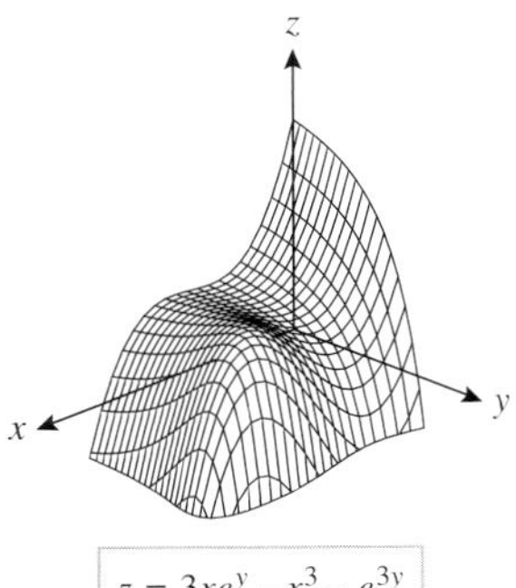

Figure Ex-25

26. If f is a continuous function of one variable with two relative maxima on an interval, then there must be a relative minimum between the relative maxima. (Convince yourself of this by drawing some pictures.) The purpose of this exercise is to show that this result does not extend to functions of two variables. Show that $f(x, y) = 4x^2e^y - 2x^4 - e^{4y}$ has two relative maxima but no other critical points (see Figure Ex-26).

Source: This exercise is based on the problem "Two Mountains Without a Valley" proposed and solved by Ira Rosenholtz, *Mathematics Magazine*, Vol. 60, No. 1, February 1987, p. 48.

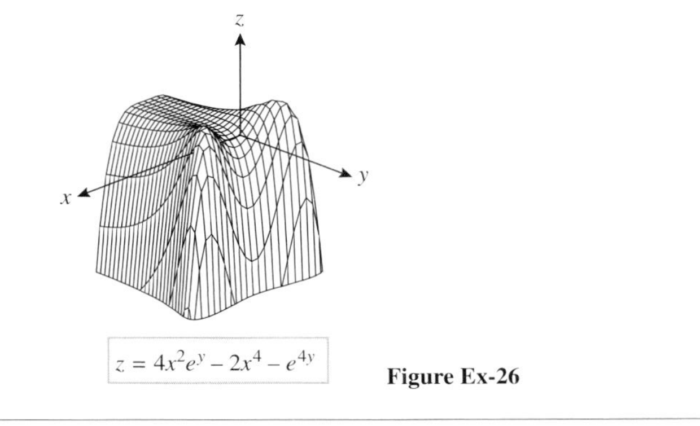

Figure Ex-26

27–32 Find the absolute extrema of the given function on the indicated closed and bounded set R.

27. $f(x, y) = xy - x - 3y$; R is the triangular region with vertices (0, 0), (0, 4), and (5, 0).

28. $f(x, y) = xy - 2x$; R is the triangular region with vertices (0, 0), (0, 4), and (4, 0).

29. $f(x, y) = x^2 - 3y^2 - 2x + 6y$; R is the region bounded by the square with vertices (0, 0), (0, 2), (2, 2), and (2, 0).

30. $f(x, y) = xe^y - x^2 - e^y$; R is the rectangular region with vertices (0, 0), (0, 1), (2, 1), and (2, 0).

31. $f(x, y) = x^2 + 2y^2 - x$; R is the disk $x^2 + y^2 \leq 4$.

32. $f(x, y) = xy^2$; R is the region that satisfies the inequalities $x \geq 0$, $y \geq 0$, and $x^2 + y^2 \leq 1$.

33. Find three positive numbers whose sum is 48 and such that their product is as large as possible.

34. Find three positive numbers whose sum is 27 and such that the sum of their squares is as small as possible.

35. Find all points on the portion of the plane $x + y + z = 5$ in the first octant at which $f(x, y, z) = xy^2z^2$ has a maximum value.

36. Find the points on the surface $x^2 - yz = 5$ that are closest to the origin.

37. Find the dimensions of the rectangular box of maximum volume that can be inscribed in a sphere of radius a.

38. Find the maximum volume of a rectangular box with three faces in the coordinate planes and a vertex in the first octant on the plane $x + y + z = 1$.

39. A closed rectangular box with a volume of 16 ft^3 is made from two kinds of materials. The top and bottom are made of material costing 10¢ per square foot and the sides from material costing 5¢ per square foot. Find the dimensions of the box so that the cost of materials is minimized.

40. A manufacturer makes two models of an item, standard and deluxe. It costs $40 to manufacture the standard model and $60 for the deluxe. A market research firm estimates that if the standard model is priced at x dollars and the deluxe at y dollars, then the manufacturer will sell $500(y - x)$ of the standard items and $45{,}000 + 500(x - 2y)$ of the deluxe each year. How should the items be priced to maximize the profit?

41. Consider the function
$$f(x, y) = 4x^2 - 3y^2 + 2xy$$
over the unit square $0 \le x \le 1, 0 \le y \le 1$.
(a) Find the maximum and minimum values of f on each edge of the square.
(b) Find the maximum and minimum values of f on each diagonal of the square.
(c) Find the maximum and minimum values of f on the entire square.

42. Show that among all parallelograms with perimeter l, a square with sides of length $l/4$ has maximum area. [*Hint:* The area of a parallelogram is given by the formula $A = ab \sin \alpha$, where a and b are the lengths of two adjacent sides and α is the angle between them.]

43. Determine the dimensions of a rectangular box, open at the top, having volume V, and requiring the least amount of material for its construction.

44. A length of sheet metal 27 inches wide is to be made into a water trough by bending up two sides as shown in the accompanying figure. Find x and ϕ so that the trapezoid-shaped cross section has a maximum area.

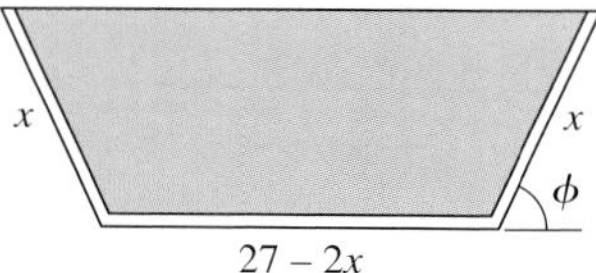

Figure Ex-44

45–46 A common problem in experimental work is to obtain a mathematical relationship $y = f(x)$ between two variables x and y by "fitting" a curve to points in the plane that correspond to experimentally determined values of x and y, say
$$(x_1, y_1), (x_2, y_2), \ldots, (x_n, y_n)$$
The curve $y = f(x)$ is called a ***mathematical model*** of the data. The general form of the function f is commonly determined by some underlying physical principle, but sometimes it is just determined by the pattern of the data. We are concerned with fitting a straight line $y = mx + b$ to data. Usually, the data will not lie on a line (possibly due to experimental error or variations in experimental conditions), so the problem is to find a line that fits the data "best" according to some criterion. One criterion for selecting the line of best fit is to choose m and b to minimize the function
$$g(m, b) = \sum_{i=1}^{n} (mx_i + b - y_i)^2$$
This is called the ***method of least squares***, and the resulting line is called the ***regression line*** or the ***least squares line of best fit***. Geometrically, $|mx_i + b - y_i|$ is the vertical distance between the data point (x_i, y_i) and the line $y = mx + b$.

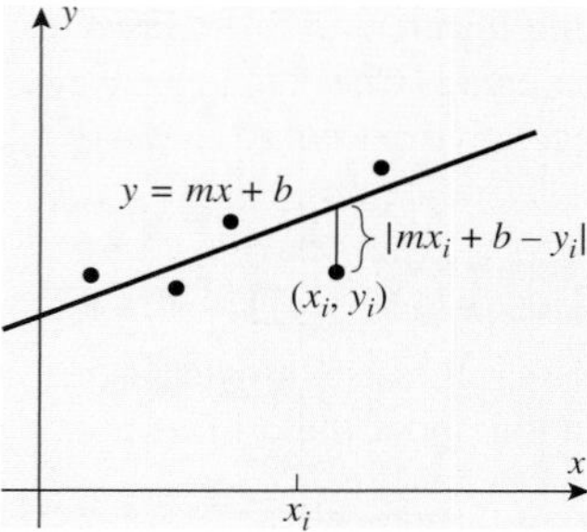

These vertical distances are called the ***residuals*** of the data points, so the effect of minimizing $g(m, b)$ is to minimize the sum of the squares of the residuals. In these exercises, we will derive a formula for the regression line. More on this topic can be found in Section 1.6.

45. The purpose of this exercise is to find the values of m and b that produce the regression line.
(a) To minimize $g(m, b)$, we start by finding values of m and b such that $\partial g/\partial m = 0$ and $\partial g/\partial b = 0$. Show that these equations are satisfied if m and b satisfy the conditions
$$\left(\sum_{i=1}^{n} x_i^2\right) m + \left(\sum_{i=1}^{n} x_i\right) b = \sum_{i=1}^{n} x_i y_i$$
$$\left(\sum_{i=1}^{n} x_i\right) m + nb = \sum_{i=1}^{n} y_i$$
(b) Let $\bar{x} = (x_1 + x_2 + \cdots + x_n)/n$ denote the arithmetic average of $x_1, x_2, \ldots, x_n$. Use the fact that
$$\sum_{i=1}^{n} (x_i - \bar{x})^2 \ge 0$$
to show that
$$n\left(\sum_{i=1}^{n} x_i^2\right) - \left(\sum_{i=1}^{n} x_i\right)^2 \ge 0$$
with equality if and only if all the x_i's are the same.
(c) Assuming that not all the x_i's are the same, prove that the equations in part (a) have the unique solution
$$m = \frac{n\sum_{i=1}^{n} x_i y_i - \sum_{i=1}^{n} x_i \sum_{i=1}^{n} y_i}{n\sum_{i=1}^{n} x_i^2 - \left(\sum_{i=1}^{n} x_i\right)^2}$$
$$b = \frac{1}{n}\left(\sum_{i=1}^{n} y_i - m\sum_{i=1}^{n} x_i\right)$$
[*Note:* We have shown that g has a critical point at these values of m and b. In the next exercise we will show that g has an absolute minimum at this critical point. Accepting this to be so, we have shown that the line $y = mx + b$ is the regression line for these values of m and b.]

46. Assume that not all the x_i's are the same, so that $g(m, b)$ has a unique critical point at the values of m and b obtained in Exercise 45(c). The purpose of this exercise is to show that g has an absolute minimum value at this point.

(a) Find the partial derivatives $g_{mm}(m, b)$, $g_{bb}(m, b)$, and $g_{mb}(m, b)$, and then apply the second partials test to show that g has a relative minimum at the critical point obtained in Exercise 45.

(b) Show that the graph of the equation $z = g(m, b)$ is a quadric surface. [*Hint:* See Formula (4) of Section 12.7.]

(c) It can be proved that the graph of $z = g(m, b)$ is an elliptic paraboloid. Accepting this to be so, show that this paraboloid opens in the positive z-direction, and explain how this shows that g has an absolute minimum at the critical point obtained in Exercise 45.

47–50 Use the formulas obtained in Exercise 45 to find and draw the regression line. If you have a calculating utility that can calculate regression lines, use it to check your work.

47. **48.**

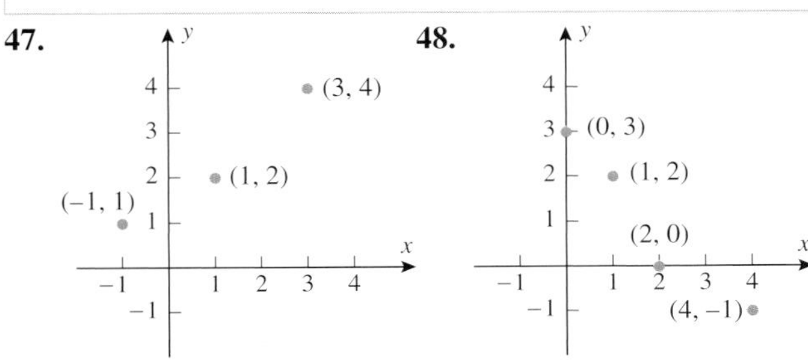

49.

x	1	2	3	4
y	1.5	1.6	2.1	3.0

50.

x	1	2	3	4	5
y	4.2	3.5	3.0	2.4	2.0

51. The following table shows the life expectancy by year of birth of females in the United States:

YEAR OF BIRTH	1930	1940	1950	1960	1970	1980	1990	2000
LIFE EXPECTANCY	61.6	65.2	71.1	73.1	74.7	77.5	78.8	79.7

(a) Take $t = 0$ to be the year 1930, and let y be the life expectancy for birth year t. Use the regression capability of a calculating utility to find the regression line of y as a function of t.

(b) Use a graphing utility to make a graph that shows the data points and the regression line.

(c) Use the regression line to make a conjecture about the life expectancy of females born in the year 2010.

52. A company manager wants to establish a relationship between the sales of a certain product and the price. The company research department provides the following data:

PRICE (x) IN DOLLARS	\$35.00	\$40.00	\$45.00	\$48.00	\$50.00
DAILY SALES VOLUME (y) IN UNITS	80	75	68	66	63

(a) Use a calculating utility to find the regression line of y as a function of x.

(b) Use a graphing utility to make a graph that shows the data points and the regression line.

(c) Use the regression line to make a conjecture about the number of units that would be sold at a price of \$60.00.

53. If a gas is cooled with its volume held constant, then it follows from the ***ideal gas law*** in physics that its pressure drops proportionally to the drop in temperature. The temperature that, in theory, corresponds to a pressure of zero is called ***absolute zero***. Suppose that an experiment produces the following data for pressure P versus temperature T with the volume held constant:

P (KILOPASCALS)	134	142	155	160	171	184
T (°CELSIUS)	0	20	40	60	80	100

(a) Use a calculating utility to find the regression line of P as a function of T.

(b) Use a graphing utility to make a graph that shows the data points and the regression line.

(c) Use the regression line to estimate the value of absolute zero in degrees Celsius.

54. Find

(a) a continuous function $f(x, y)$ that is defined on the entire xy-plane and has no absolute extrema on the xy-plane;

(b) a function $f(x, y)$ that is defined everywhere on the rectangle $0 \le x \le 1, 0 \le y \le 1$ and has no absolute extrema on the rectangle.

55. Show that if f has a relative maximum at (x_0, y_0), then $G(x) = f(x, y_0)$ has a relative maximum at $x = x_0$ and $H(y) = f(x_0, y)$ has a relative maximum at $y = y_0$.

✔ QUICK CHECK ANSWERS 14.8

1. $(0, 0)$ and $\left(\frac{1}{6}, -\frac{1}{12}\right)$ **2.** (a) no information (b) a saddle point at $(0, 0)$ (c) a relative minimum at $(0, 0)$ (d) a relative maximum at $(0, 0)$ **3.** (a) a saddle point at $(0, 0)$ (b) no information, since $(-1, -1)$ is not a critical point (c) a relative minimum at $(1, 1)$ **4.** $V = \dfrac{xy(1 - xy)}{x + y}$

14.9 LAGRANGE MULTIPLIERS

In this section we will study a powerful new method for maximizing or minimizing a function subject to constraints on the variables. This method will help us to solve certain optimization problems that are difficult or impossible to solve using the methods studied in the last section.

EXTREMUM PROBLEMS WITH CONSTRAINTS

In Example 6 of the last section, we solved the problem of minimizing

$$S = xy + 2xz + 2yz \tag{1}$$

subject to the constraint

$$xyz - 32 = 0 \tag{2}$$

This is a special case of the following general problem:

14.9.1 ***Three-Variable Extremum Problem with One Constraint***
Maximize or minimize the function $f(x, y, z)$ subject to the constraint $g(x, y, z) = 0$.

We will also be interested in the following two-variable version of this problem:

14.9.2 ***Two-Variable Extremum Problem with One Constraint***
Maximize or minimize the function $f(x, y)$ subject to the constraint $g(x, y) = 0$.

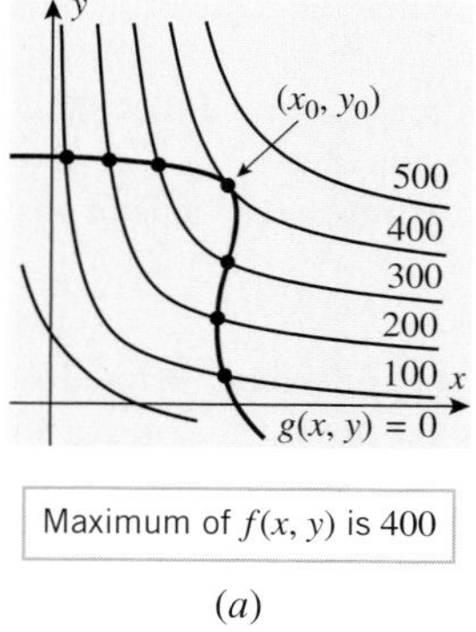

(*a*)

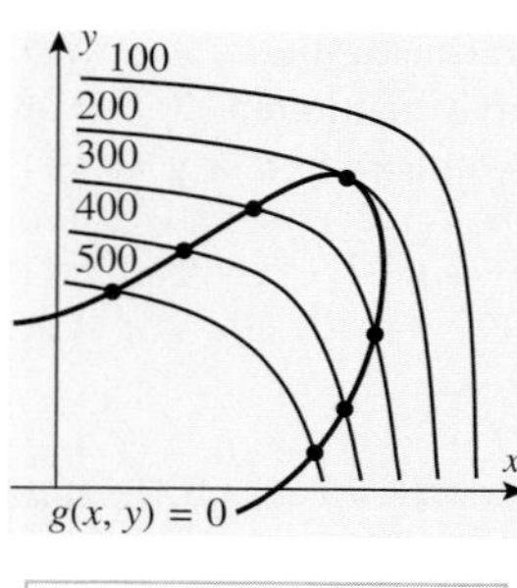

(*b*)

Figure 14.9.1

LAGRANGE MULTIPLIERS

One way to attack problems of these types is to solve the constraint equation for one of the variables in terms of the others and substitute the result into f. This produces a new function of one or two variables that incorporates the constraint and can be maximized or minimized by applying standard methods. For example, to solve the problem in Example 6 of the last section we substituted (2) into (1) to obtain

$$S = xy + \frac{64}{y} + \frac{64}{x}$$

which we then minimized by finding the critical points and applying the second partials test. However, this approach hinges on our ability to solve the constraint equation for one of the variables in terms of the others. If this cannot be done, then other methods must be used. One such method, called the *method of Lagrange multipliers*, will be discussed in this section.

To motivate the method of Lagrange multipliers, suppose that we are trying to maximize a function $f(x, y)$ subject to the constraint $g(x, y) = 0$. Geometrically, this means that we are looking for a point (x_0, y_0) on the graph of the constraint curve at which $f(x, y)$ is as large as possible. To help locate such a point, let us construct a contour plot of $f(x, y)$ in the same coordinate system as the graph of $g(x, y) = 0$. For example, Figure 14.9.1*a* shows some typical level curves of $f(x, y) = c$, which we have labeled $c = 100, 200, 300, 400,$ and 500 for purposes of illustration. In this figure, each point of intersection of $g(x, y) = 0$ with a level curve is a candidate for a solution, since these points lie on the constraint curve. Among the seven such intersections shown in the figure, the maximum value of $f(x, y)$ occurs at the intersection (x_0, y_0), where $f(x, y)$ has a value of 400. Note that at (x_0, y_0) the constraint curve and the level curve just touch and thus have a *common* tangent line at this point. Since $\nabla f(x_0, y_0)$ is normal to the level curve $f(x, y) = 400$ at (x_0, y_0), and since $\nabla g(x_0, y_0)$ is normal to the constraint curve $g(x, y) = 0$ at (x_0, y_0), we conclude that the vectors $\nabla f(x_0, y_0)$ and $\nabla g(x_0, y_0)$ must be parallel. That is,

$$\nabla f(x_0, y_0) = \lambda \nabla g(x_0, y_0) \tag{3}$$

for some scalar λ. The same condition holds at points on the constraint curve where $f(x, y)$ has a minimum. For example, if the level curves are as shown in Figure 14.9.1*b*, then the minimum value of $f(x, y)$ occurs where the constraint curve just touches a level curve. Thus, to find the maximum or minimum of $f(x, y)$ subject to the constraint $g(x, y) = 0$, we look for points at which (3) holds—this is the method of Lagrange multipliers.

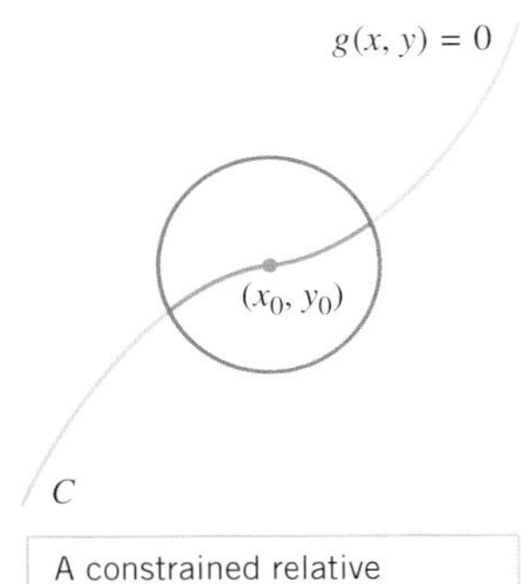

A constrained relative maximum occurs at (x_0, y_0) if $f(x_0, y_0) \geq f(x, y)$ on some segment of C that extends on both sides of (x_0, y_0).

Figure 14.9.2

Our next objective in this section is to make the preceding intuitive argument more precise. For this purpose it will help to begin with some terminology about the problem of maximizing or minimizing a function $f(x, y)$ subject to a constraint $g(x, y) = 0$. As with other kinds of maximization and minimization problems, we need to distinguish between relative and absolute extrema. We will say that f has a ***constrained absolute maximum (minimum)*** at (x_0, y_0) if $f(x_0, y_0)$ is the largest (smallest) value of f on the constraint curve, and we will say that f has a ***constrained relative maximum (minimum)*** at (x_0, y_0) if $f(x_0, y_0)$ is the largest (smallest) value of f on some segment of the constraint curve that extends on both sides of the point (x_0, y_0) (Figure 14.9.2).

Let us assume that a constrained relative maximum or minimum occurs at the point (x_0, y_0) and for simplicity, let us further assume that the equation $g(x, y) = 0$ can be smoothly parametrized as

$$x = x(s), \quad y = y(s)$$

where s is an arc length parameter with reference point (x_0, y_0) at $s = 0$. Thus, the quantity

$$z = f(x(s), y(s))$$

has a relative maximum or minimum at $s = 0$, and this implies that $dz/ds = 0$ at that point. From the chain rule, this equation can be expressed as

$$\frac{dz}{ds} = \frac{\partial f}{\partial x}\frac{dx}{ds} + \frac{\partial f}{\partial y}\frac{dy}{ds} = \left(\frac{\partial f}{\partial x}\mathbf{i} + \frac{\partial f}{\partial y}\mathbf{j}\right) \cdot \left(\frac{dx}{ds}\mathbf{i} + \frac{dy}{ds}\mathbf{j}\right) = 0$$

where the derivatives are all evaluated at $s = 0$. However, the first factor in the dot product is the gradient of f, and the second factor is the unit tangent vector to the constraint curve.

Joseph Louis Lagrange (1736–1813) French–Italian mathematician and astronomer. Lagrange, the son of a public official, was born in Turin, Italy. (Baptismal records list his name as Giuseppe Lodovico Lagrangia.) Although his father wanted him to be a lawyer, Lagrange was attracted to mathematics and astronomy after reading a memoir by the astronomer Halley. At age 16 he began to study mathematics on his own and by age 19 was appointed to a professorship at the Royal Artillery School in Turin. The following year Lagrange sent Euler solutions to some famous problems using new methods that eventually blossomed into a branch of mathematics called calculus of variations. These methods and Lagrange's applications of them to problems in celestial mechanics were so monumental that by age 25 he was regarded by many of his contemporaries as the greatest living mathematician.

In 1776, on the recommendations of Euler, he was chosen to succeed Euler as the director of the Berlin Academy. During his stay in Berlin, Lagrange distinguished himself not only in celestial mechanics, but also in algebraic equations and the theory of numbers. After twenty years in Berlin, he moved to Paris at the invitation of Louis XVI. He was given apartments in the Louvre and treated with great honor, even during the revolution.

Napoleon was a great admirer of Lagrange and showered him with honors—count, senator, and Legion of Honor. The years Lagrange spent in Paris were devoted primarily to didactic treatises summarizing his mathematical conceptions. One of Lagrange's most famous works is a memoir, *Mécanique Analytique*, in which he reduced the theory of mechanics to a few general formulas from which all other necessary equations could be derived.

It is an interesting historical fact that Lagrange's father speculated unsuccessfully in several financial ventures, so his family was forced to live quite modestly. Lagrange himself stated that if his family had money, he would not have made mathematics his vocation. In spite of his fame, Lagrange was always a shy and modest man. On his death, he was buried with honor in the Pantheon.

Since the point (x_0, y_0) corresponds to $s = 0$, it follows from this equation that

$$\nabla f(x_0, y_0) \cdot \mathbf{T}(0) = 0$$

which implies that the gradient is either $\mathbf{0}$ or is normal to the constraint curve at a constrained relative extremum. However, the constraint curve $g(x, y) = 0$ is a level curve for the function $g(x, y)$, so that if $\nabla g(x_0, y_0) \neq \mathbf{0}$, then $\nabla g(x_0, y_0)$ is normal to this curve at (x_0, y_0). It then follows that there is some scalar λ such that

$$\nabla f(x_0, y_0) = \lambda \nabla g(x_0, y_0) \tag{4}$$

This scalar is called a ***Lagrange multiplier***. Thus, the ***method of Lagrange multipliers*** for finding constrained relative extrema is to look for points on the constraint curve $g(x, y) = 0$ at which Equation (4) is satisfied for some scalar λ.

14.9.3 THEOREM *(Constrained-Extremum Principle for Two Variables and One Constraint).* *Let f and g be functions of two variables with continuous first partial derivatives on some open set containing the constraint curve $g(x, y) = 0$, and assume that $\nabla g \neq \mathbf{0}$ at any point on this curve. If f has a constrained relative extremum, then this extremum occurs at a point (x_0, y_0) on the constraint curve at which the gradient vectors $\nabla f(x_0, y_0)$ and $\nabla g(x_0, y_0)$ are parallel; that is, there is some number λ such that*

$$\nabla f(x_0, y_0) = \lambda \nabla g(x_0, y_0)$$

► **Example 1** At what point or points on the circle $x^2 + y^2 = 1$ does $f(x, y) = xy$ have an absolute maximum, and what is that maximum?

Solution. The circle $x^2 + y^2 = 1$ is a closed and bounded set and $f(x, y) = xy$ is a continuous function, so it follows from the Extreme-Value Theorem (Theorem 14.8.3) that f has an absolute maximum and an absolute minimum on the circle. To find these extrema, we will use Lagrange multipliers to find the constrained relative extrema, and then we will evaluate f at those relative extrema to find the absolute extrema.

We want to maximize $f(x, y) = xy$ subject to the constraint

$$g(x, y) = x^2 + y^2 - 1 = 0 \tag{5}$$

First we will look for constrained *relative* extrema. For this purpose we will need the gradients

$$\nabla f = y\mathbf{i} + x\mathbf{j} \quad \text{and} \quad \nabla g = 2x\mathbf{i} + 2y\mathbf{j}$$

From the formula for ∇g we see that $\nabla g = \mathbf{0}$ if and only if $x = 0$ and $y = 0$, so $\nabla g \neq \mathbf{0}$ at any point on the circle $x^2 + y^2 = 1$. Thus, at a constrained relative extremum we must have

$$\nabla f = \lambda \nabla g \quad \text{or} \quad y\mathbf{i} + x\mathbf{j} = \lambda(2x\mathbf{i} + 2y\mathbf{j})$$

which is equivalent to the pair of equations

$$y = 2x\lambda \quad \text{and} \quad x = 2y\lambda$$

It follows from these equations that if $x = 0$, then $y = 0$, and if $y = 0$, then $x = 0$. In either case we have $x^2 + y^2 = 0$, so the constraint equation $x^2 + y^2 = 1$ is not satisfied. Thus, we can assume that x and y are nonzero, and we can rewrite the equations as

$$\lambda = \frac{y}{2x} \quad \text{and} \quad \lambda = \frac{x}{2y}$$

from which we obtain

$$\frac{y}{2x} = \frac{x}{2y}$$

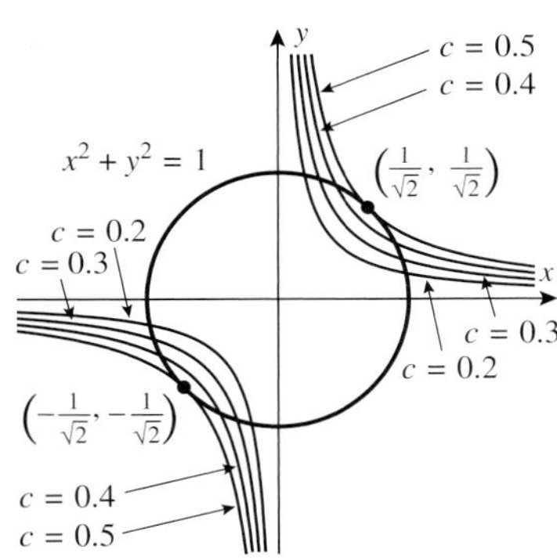

Figure 14.9.3

Give another solution to Example 1 using the parametrization

$$x = \cos\theta, \quad y = \sin\theta$$

and the identity

$$\sin 2\theta = 2\sin\theta\cos\theta$$

or

$$y^2 = x^2 \tag{6}$$

Substituting this in (5) yields

$$2x^2 - 1 = 0$$

from which we obtain $x = \pm 1/\sqrt{2}$. Each of these values, when substituted in Equation (6), produces y-values of $y = \pm 1/\sqrt{2}$. Thus, constrained relative extrema occur at the points $(1/\sqrt{2}, 1/\sqrt{2})$, $(1/\sqrt{2}, -1/\sqrt{2})$, $(-1/\sqrt{2}, 1/\sqrt{2})$, and $(-1/\sqrt{2}, -1/\sqrt{2})$. The values of xy at these points are as follows:

(x, y)	$(1/\sqrt{2}, 1/\sqrt{2})$	$(1/\sqrt{2}, -1/\sqrt{2})$	$(-1/\sqrt{2}, 1/\sqrt{2})$	$(-1/\sqrt{2}, -1/\sqrt{2})$
xy	1/2	−1/2	−1/2	1/2

Thus, the function $f(x, y) = xy$ has an absolute maximum of $\frac{1}{2}$ occurring at the two points $(1/\sqrt{2}, 1/\sqrt{2})$ and $(-1/\sqrt{2}, -1/\sqrt{2})$. Although it was not asked for, we can also see that f has an absolute minimum of $-\frac{1}{2}$ occurring at the points $(1/\sqrt{2}, -1/\sqrt{2})$ and $(-1/\sqrt{2}, 1/\sqrt{2})$. Figure 14.9.3 shows some level curves $xy = c$ and the constraint curve in the vicinity of the maxima. A similar figure for the minima can be obtained using negative values of c for the level curves $xy = c$. ◄

If c is a constant, then the functions $g(x, y)$ and $g(x, y) - c$ have the same gradient since the constant c drops out when we differentiate. Consequently, it is *not* essential to rewrite a constraint of the form $g(x, y) = c$ as $g(x, y) - c = 0$ in order to apply the constrained-extremum principle. Thus, in the last example, we could have kept the constraint in the form $x^2 + y^2 = 1$ and then taken $g(x, y) = x^2 + y^2$ rather than $g(x, y) = x^2 + y^2 - 1$.

► **Example 2** Use the method of Lagrange multipliers to find the dimensions of a rectangle with perimeter p and maximum area.

Solution. Let

x = length of the rectangle, y = width of the rectangle, A = area of the rectangle

We want to maximize $A = xy$ on the line segment

$$2x + 2y = p, \quad 0 \le x, y \tag{7}$$

that corresponds to the perimeter constraint. This segment is a closed and bounded set, and since $f(x, y) = xy$ is a continuous function, it follows from the Extreme-Value Theorem (Theorem 14.8.3) that f has an absolute maximum on this segment. This absolute maximum must also be a constrained relative maximum since f is 0 at the endpoints of the segment and positive elsewhere on the segment. If $g(x, y) = 2x + 2y$, then we have

$$\nabla f = y\mathbf{i} + x\mathbf{j} \quad \text{and} \quad \nabla g = 2\mathbf{i} + 2\mathbf{j}$$

Noting that $\nabla g \neq \mathbf{0}$, it follows from (4) that

$$y\mathbf{i} + x\mathbf{j} = \lambda(2\mathbf{i} + 2\mathbf{j})$$

at a constrained relative maximum. This is equivalent to the two equations

$$y = 2\lambda \quad \text{and} \quad x = 2\lambda$$

Eliminating λ from these equations we obtain $x = y$, which shows that the rectangle is actually a square. Using this condition and constraint (7), we obtain $x = p/4$, $y = p/4$. ◄

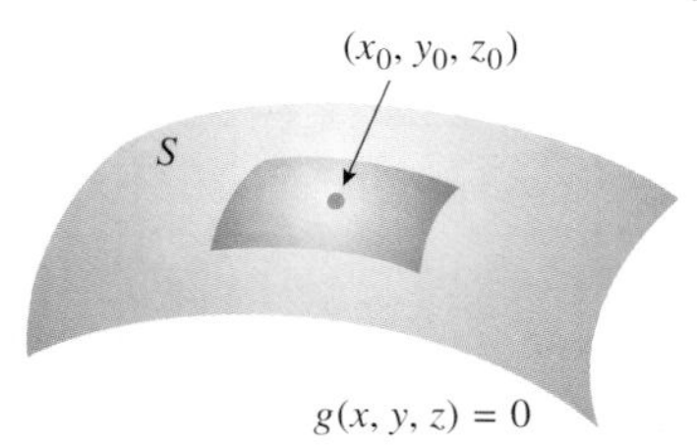

A constrained relative maximum occurs at (x_0, y_0, z_0) if $f(x_0, y_0, z_0) \geq f(x, y, z)$ at all points of S near (x_0, y_0, z_0).

Figure 14.9.4

THREE VARIABLES AND ONE CONSTRAINT

The method of Lagrange multipliers can also be used to maximize or minimize a function of three variables $f(x, y, z)$ subject to a constraint $g(x, y, z) = 0$. As a rule, the graph of $g(x, y, z) = 0$ will be some surface S in 3-space. Thus, from a geometric viewpoint, the problem is to maximize or minimize $f(x, y, z)$ as (x, y, z) varies over the surface S (Figure 14.9.4). As usual, we distinguish between relative and absolute extrema. We will say that f has a ***constrained absolute maximum (minimum)*** at (x_0, y_0, z_0) if $f(x_0, y_0, z_0)$ is the largest (smallest) value of $f(x, y, z)$ on S, and we will say that f has a ***constrained relative maximum (minimum)*** at (x_0, y_0, z_0) if $f(x_0, y_0, z_0)$ is the largest (smallest) value of $f(x, y, z)$ at all points of S "near" (x_0, y_0, z_0).

The following theorem, which we state without proof, is the three-variable analog of Theorem 14.9.3.

14.9.4 THEOREM *(Constrained-Extremum Principle for Three Variables and One Constraint).* *Let f and g be functions of three variables with continuous first partial derivatives on some open set containing the constraint surface $g(x, y, z) = 0$, and assume that $\nabla g \neq \mathbf{0}$ at any point on this surface. If f has a constrained relative extremum, then this extremum occurs at a point (x_0, y_0, z_0) on the constraint surface at which the gradient vectors $\nabla f(x_0, y_0, z_0)$ and $\nabla g(x_0, y_0, z_0)$ are parallel; that is, there is some number λ such that*

$$\nabla f(x_0, y_0, z_0) = \lambda \nabla g(x_0, y_0, z_0)$$

▶ **Example 3** Find the points on the sphere $x^2 + y^2 + z^2 = 36$ that are closest to and farthest from the point $(1, 2, 2)$.

Solution. To avoid radicals, we will find points on the sphere that minimize and maximize the *square* of the distance to $(1, 2, 2)$. Thus, we want to find the relative extrema of

$$f(x, y, z) = (x-1)^2 + (y-2)^2 + (z-2)^2$$

subject to the constraint

$$x^2 + y^2 + z^2 = 36 \tag{8}$$

If we let $g(x, y, z) = x^2 + y^2 + z^2$, then $\nabla g = 2x\mathbf{i} + 2y\mathbf{j} + 2z\mathbf{k}$. Thus, $\nabla g = \mathbf{0}$ if and only if $x = y = z = 0$. It follows that $\nabla g \neq \mathbf{0}$ at any point of the sphere (8), and hence the constrained relative extrema must occur at points where

$$\nabla f(x, y, z) = \lambda \nabla g(x, y, z)$$

That is,

$$2(x-1)\mathbf{i} + 2(y-2)\mathbf{j} + 2(z-2)\mathbf{k} = \lambda(2x\mathbf{i} + 2y\mathbf{j} + 2z\mathbf{k})$$

which leads to the equations

$$2(x-1) = 2x\lambda, \quad 2(y-2) = 2y\lambda, \quad 2(z-2) = 2z\lambda \tag{9}$$

We may assume that x, y, and z are nonzero since $x = 0$ does not satisfy the first equation, $y = 0$ does not satisfy the second, and $z = 0$ does not satisfy the third. Thus, we can rewrite (9) as

$$\frac{x-1}{x} = \lambda, \quad \frac{y-2}{y} = \lambda, \quad \frac{z-2}{z} = \lambda$$

The first two equations imply that

$$\frac{x-1}{x} = \frac{y-2}{y}$$

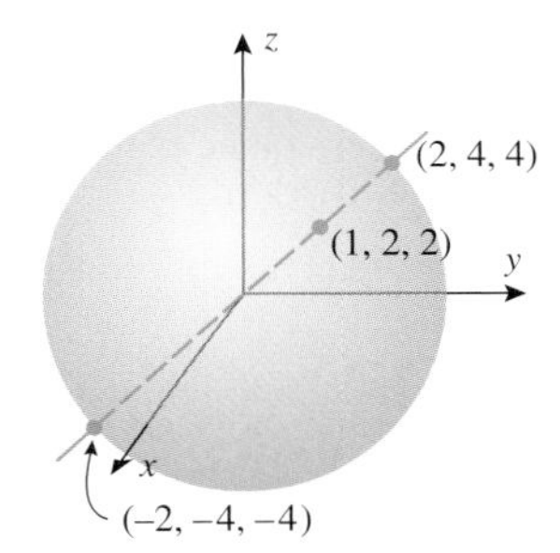

Figure 14.9.5

from which it follows that

$$y = 2x \tag{10}$$

Similarly, the first and third equations imply that

$$z = 2x \tag{11}$$

Substituting (10) and (11) in the constraint equation (8), we obtain

$$9x^2 = 36 \quad \text{or} \quad x = \pm 2$$

Substituting these values in (10) and (11) yields two points:

$$(2, 4, 4) \quad \text{and} \quad (-2, -4, -4)$$

Since $f(2, 4, 4) = 9$ and $f(-2, -4, -4) = 81$, it follows that $(2, 4, 4)$ is the point on the sphere closest to $(1, 2, 2)$, and $(-2, -4, -4)$ is the point that is farthest (Figure 14.9.5). ◀

Next we will use Lagrange multipliers to solve the problem of Example 6 in the last section.

▶ **Example 4** Use Lagrange multipliers to determine the dimensions of a rectangular box, open at the top, having a volume of 32 ft^3, and requiring the least amount of material for its construction.

Solution. With the notation introduced in Example 6 of the last section, the problem is to minimize the surface area

$$S = xy + 2xz + 2yz$$

subject to the volume constraint

$$xyz = 32 \tag{12}$$

If we let $f(x, y, z) = xy + 2xz + 2yz$ and $g(x, y, z) = xyz$, then

$$\nabla f = (y + 2z)\mathbf{i} + (x + 2z)\mathbf{j} + (2x + 2y)\mathbf{k} \quad \text{and} \quad \nabla g = yz\mathbf{i} + xz\mathbf{j} + xy\mathbf{k}$$

It follows that $\nabla g \neq \mathbf{0}$ at any point on the surface $xyz = 32$, since x, y, and z are all nonzero on this surface. Thus, at a constrained relative extremum we must have $\nabla f = \lambda \nabla g$, that is,

$$(y + 2z)\mathbf{i} + (x + 2z)\mathbf{j} + (2x + 2y)\mathbf{k} = \lambda(yz\mathbf{i} + xz\mathbf{j} + xy\mathbf{k})$$

This condition yields the three equations

$$y + 2z = \lambda yz, \quad x + 2z = \lambda xz, \quad 2x + 2y = \lambda xy$$

Because x, y, and z are nonzero, these equations can be rewritten as

$$\frac{1}{z} + \frac{2}{y} = \lambda, \quad \frac{1}{z} + \frac{2}{x} = \lambda, \quad \frac{2}{y} + \frac{2}{x} = \lambda$$

From the first two equations,

$$y = x \tag{13}$$

and from the first and third equations,

$$z = \tfrac{1}{2}x \tag{14}$$

Substituting (13) and (14) in the volume constraint (12) yields

$$\tfrac{1}{2}x^3 = 32$$

This equation, together with (13) and (14), yields

$$x = 4, \quad y = 4, \quad z = 2$$

which agrees with the result that was obtained in Example 6 of the last section. ◀

There are variations in the method of Lagrange multipliers that can be used to solve problems with two or more constraints. However, we will not discuss that topic here.

✔QUICK CHECK EXERCISES 14.9

(See page 1015 for answers.)

1. (a) Suppose that $f(x, y)$ and $g(x, y)$ are differentiable at the origin and have nonzero gradients there, and that $g(0, 0) = 0$. If the maximum value of $f(x, y)$ subject to the constraint $g(x, y) = 0$ occurs at the origin, how is the tangent line to the graph of $g(x, y) = 0$ related to the tangent line at the origin to the level curve of f through $(0, 0)$?
   (b) Suppose that $f(x, y, z)$ and $g(x, y, z)$ are differentiable at the origin and have nonzero gradients there, and that $g(0, 0, 0) = 0$. If the maximum value of $f(x, y, z)$ subject to the constraint $g(x, y, z) = 0$ occurs at the origin, how is the tangent plane to the graph of the constraint $g(x, y, z) = 0$ related to the tangent plane at the origin to the level surface of f through $(0, 0, 0)$?
2. The maximum value of $x + y$ subject to the constraint $x^2 + y^2 = 1$ is ______.
3. The maximum value of $x + y + z$ subject to the constraint $x^2 + y^2 + z^2 = 1$ is ______.
4. The maximum and minimum values of $2x + 3y$ subject to the constraint $x + y = 1$, where $0 \le x, 0 \le y$, are ______ and ______, respectively.

EXERCISE SET 14.9

Graphing Utility C CAS

FOCUS ON CONCEPTS

1. The accompanying figure shows graphs of the line $x + y = 4$ and the level curves of height $c = 2, 4, 6$, and 8 for the function $f(x, y) = xy$.
   (a) Use the figure to find the maximum value of the function $f(x, y) = xy$ subject to $x + y = 4$, and explain your reasoning.
   (b) How can you tell from the figure that your answer to part (a) is not the minimum value of f subject to the constraint?
   (c) Use Lagrange multipliers to check your work.
2. The accompanying figure shows the graphs of the line $3x + 4y = 25$ and the level curves of height $c = 9, 16, 25, 36$, and 49 for the function $f(x, y) = x^2 + y^2$.
   (a) Use the accompanying figure to find the minimum value of the function $f(x, y) = x^2 + y^2$ subject to $3x + 4y = 25$, and explain your reasoning.
   (b) How can you tell from the accompanying figure that your answer to part (a) is not the maximum value of f subject to the constraint?
   (c) Use Lagrange multipliers to check your work.

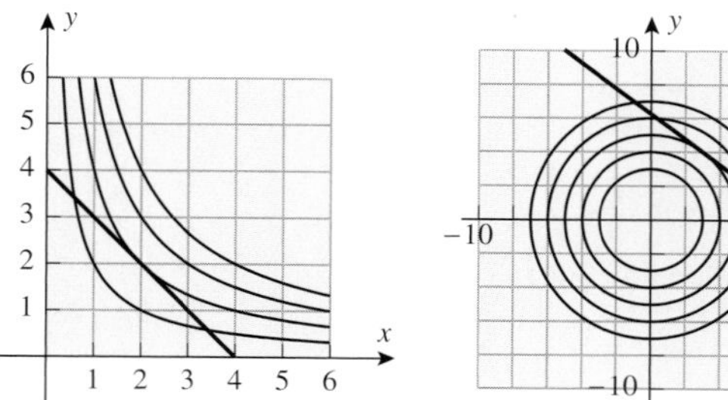

Figure Ex-1 Figure Ex-2

3. (a) On a graphing utility, graph the circle $x^2 + y^2 = 25$ and two distinct level curves of $f(x, y) = x^2 - y$ that just touch the circle in a single point.
   (b) Use the results you obtained in part (a) to approximate the maximum and minimum values of f subject to the constraint $x^2 + y^2 = 25$.
   (c) Check your approximations in part (b) using Lagrange multipliers.

C 4. (a) If you have a CAS with implicit plotting capability, use it to graph the circle $(x - 4)^2 + (y - 4)^2 = 4$ and two level curves of $f(x, y) = x^3 + y^3 - 3xy$ that just touch the circle.
   (b) Use the result you obtained in part (a) to approximate the minimum value of f subject to the constraint $(x - 4)^2 + (y - 4)^2 = 4$.
   (c) Confirm graphically that you have found a minimum and not a maximum.
   (d) Check your approximation using Lagrange multipliers and solving the required equations numerically.

5–12 Use Lagrange multipliers to find the maximum and minimum values of f subject to the given constraint. Also, find the points at which these extreme values occur.

5. $f(x, y) = xy;\ 4x^2 + 8y^2 = 16$
6. $f(x, y) = x^2 - y^2;\ x^2 + y^2 = 25$
7. $f(x, y) = 4x^3 + y^2;\ 2x^2 + y^2 = 1$
8. $f(x, y) = x - 3y - 1;\ x^2 + 3y^2 = 16$
9. $f(x, y, z) = 2x + y - 2z;\ x^2 + y^2 + z^2 = 4$
10. $f(x, y, z) = 3x + 6y + 2z;\ 2x^2 + 4y^2 + z^2 = 70$
11. $f(x, y, z) = xyz;\ x^2 + y^2 + z^2 = 1$
12. $f(x, y, z) = x^4 + y^4 + z^4;\ x^2 + y^2 + z^2 = 1$

13–20 Solve using Lagrange multipliers.

13. Find the point on the line $2x - 4y = 3$ that is closest to the origin.
14. Find the point on the line $y = 2x + 3$ that is closest to $(4, 2)$.
15. Find the point on the plane $x + 2y + z = 1$ that is closest to the origin.
16. Find the point on the plane $4x + 3y + z = 2$ that is closest to $(1, -1, 1)$.

17. Find the points on the circle $x^2 + y^2 = 45$ that are closest to and farthest from $(1, 2)$.

18. Find the points on the surface $xy - z^2 = 1$ that are closest to the origin.

19. Find a vector in 3-space whose length is 5 and whose components have the largest possible sum.

20. Suppose that the temperature at a point (x, y) on a metal plate is $T(x, y) = 4x^2 - 4xy + y^2$. An ant, walking on the plate, traverses a circle of radius 5 centered at the origin. What are the highest and lowest temperatures encountered by the ant?

21–28 Use Lagrange multipliers to solve the indicated problems from Section 14.8.

21. Exercise 34 **22.** Exercise 35
23. Exercise 36 **24.** Exercise 37
25. Exercise 39 **26.** Exercises 41(a) and (b)
27. Exercise 42 **28.** Exercise 43

C **29.** Let α, β, and γ be the angles of a triangle.
(a) Use Lagrange multipliers to find the maximum value of $f(\alpha, \beta, \gamma) = \cos\alpha \cos\beta \cos\gamma$, and determine the angles for which the maximum occurs.
(b) Express $f(\alpha, \beta, \gamma)$ as a function of α and β alone, and use a CAS to graph this function of two variables. Confirm that the result obtained in part (a) is consistent with the graph.

30. The accompanying figure shows the intersection of the elliptic paraboloid $z = x^2 + 4y^2$ and the right circular cylinder $x^2 + y^2 = 1$. Use Lagrange multipliers to find the highest and lowest points on the curve of intersection.

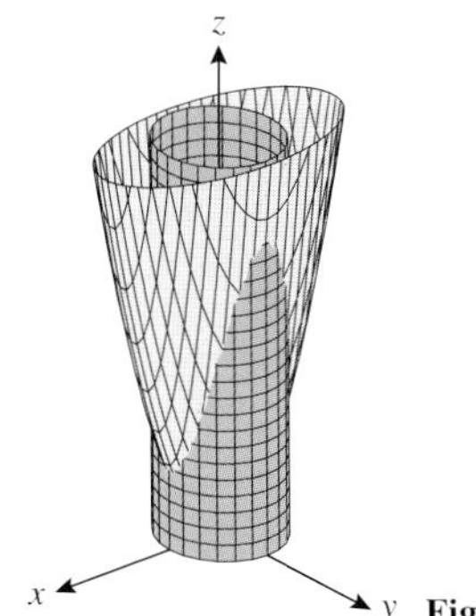

Figure Ex-30

✔QUICK CHECK ANSWERS 14.9

1. (a) They are the same line. (b) They are the same plane. 2. $\sqrt{2}$ 3. $\sqrt{3}$ 4. 3; 2

CHAPTER REVIEW EXERCISES

Graphing Utility

1. Let $f(x, y) = e^x \ln y$. Find
(a) $f(\ln y, e^x)$ (b) $f(r + s, rs)$.

2. Sketch the domain of f using solid lines for portions of the boundary included in the domain and dashed lines for portions not included.
(a) $f(x, y) = \ln(xy - 1)$ (b) $f(x, y) = (\sin^{-1} x)/e^y$

3. Show that the level curves of the cone $z = \sqrt{x^2 + y^2}$ and the paraboloid $z = x^2 + y^2$ are circles, and make a sketch that illustrates the difference between the contour plots of the two functions.

4. (a) In words, describe the level surfaces of the function $f(x, y, z) = a^2x^2 + a^2y^2 + z^2$, where $a > 0$.
(b) Find a function $f(x, y, z)$ whose level surfaces form a family of circular paraboloids that open in the positive z-direction.

5–6 (a) Find the limit of $f(x, y)$ as $(x, y) \to (0, 0)$ if it exists, and (b) determine whether f is continuous at $(0, 0)$.

5. $f(x, y) = \dfrac{x^4 - x + y - x^3y}{x - y}$

6. $f(x, y) = \begin{cases} \dfrac{x^4 - y^4}{x^2 + y^2} & \text{if } (x, y) \neq (0, 0) \\ 0 & \text{if } (x, y) = (0, 0) \end{cases}$

7. (a) A company manufactures two types of computer monitors: standard monitors and high resolution monitors. Suppose that $P(x, y)$ is the profit that results from producing and selling x standard monitors and y high-resolution monitors. What do the two partial derivatives $\partial P/\partial x$ and $\partial P/\partial y$ represent?
(b) Suppose that the temperature at time t at a point (x, y) on the surface of a lake is $T(x, y, t)$. What do the partial derivatives $\partial T/\partial x$, $\partial T/\partial y$, and $\partial T/\partial t$ represent?

8. Let $z = f(x, y)$.
(a) Express $\partial z/\partial x$ and $\partial z/\partial y$ as limits.
(b) In words, what do the derivatives $f_x(x_0, y_0)$ and $f_y(x_0, y_0)$ tell you about the surface $z = f(x, y)$?
(c) In words, what do the derivatives $\partial z/\partial x(x_0, y_0)$ and $\partial z/\partial y(x_0, y_0)$ tell you about the rates of change of z with respect to x and y?

9. The pressure in newtons per square meter (N/m^2) of a gas in a cylinder is given by $P = 10T/V$ with T in kelvins (K) and V in cubic meters (m^3).
(a) If T is increasing at a rate of 3 K/min with V held fixed at 2.5 m^3, find the rate at which the pressure is changing when $T = 50$ K.
(b) If T is held fixed at 50 K while V is decreasing at the rate of 3 m^3/min, find the rate at which the pressure is changing when $V = 2.5$ m^3.

10. Find the slope of the tangent line at the point $(1, -2, -3)$ on the curve of intersection of the surface $z = 5 - 4x^2 - y^2$ with
(a) the plane $x = 1$ (b) the plane $y = -2$.

11–14 Verify the assertion.

11. If $w = \tan(x^2 + y^2) + x\sqrt{y}$, then $w_{xy} = w_{yx}$.

12. If $w = \ln(3x - 3y) + \cos(x + y)$, then $\partial^2 w/\partial x^2 = \partial^2 w/\partial y^2$.

13. If $F(x, y, z) = 2z^3 - 3(x^2 + y^2)z$, then F satisfies the equation $F_{xx} + F_{yy} + F_{zz} = 0$.

14. If $f(x, y, z) = xyz + x^2 + \ln(y/z)$, then $f_{xyzx} = f_{zxxy}$.

15. What do Δf and df represent, and how are they related?

16. If $w = x^2y - 2xy + y^2x$, find the increment Δw and the differential dw if (x, y) varies from $(1, 0)$ to $(1.1, -0.1)$.

17. Use differentials to estimate the change in the volume $V = \frac{1}{3}x^2h$ of a pyramid with a square base when its height h is increased from 2 to 2.2 m and its base dimension x is decreased from 1 to 0.9 m. Compare this to ΔV.

18. Find the local linear approximation of $f(x, y) = \sin(xy)$ at $\left(\frac{1}{3}, \pi\right)$.

19. Suppose that z is a differentiable function of x and y with

$$\frac{\partial z}{\partial x}(1, 2) = 4 \quad \text{and} \quad \frac{\partial z}{\partial y}(1, 2) = 2$$

If $x = x(t)$ and $y = y(t)$ are differentiable functions of t with $x(0) = 1$, $y(0) = 2$, $x'(0) = -\frac{1}{2}$, and (under composition) $z'(0) = 2$, find $y'(0)$.

20. In each part, use Theorem 14.5.3 to find dy/dx.
(a) $3x^2 - 5xy + \tan xy = 0$
(b) $x \ln y + \sin(x - y) = \pi$

21. Given that $f(x, y) = 0$, use Theorem 14.5.3 to express d^2y/dx^2 in terms of partial derivatives of f.

22. Let $z = f(x, y)$, where $x = g(t)$ and $y = h(t)$.
(a) Show that

$$\frac{d}{dt}\left(\frac{\partial z}{\partial x}\right) = \frac{\partial^2 z}{\partial x^2}\frac{dx}{dt} + \frac{\partial^2 z}{\partial y \partial x}\frac{dy}{dt}$$

and

$$\frac{d}{dt}\left(\frac{\partial z}{\partial y}\right) = \frac{\partial^2 z}{\partial x \partial y}\frac{dx}{dt} + \frac{\partial^2 z}{\partial y^2}\frac{dy}{dt}$$

(b) Use the formulas in part (a) to help find a formula for d^2z/dt^2.

23. (a) How are the directional derivative and the gradient of a function related?
(b) Under what conditions is the directional derivative of a differentiable function 0?
(c) In what direction does the directional derivative of a differentiable function have its maximum value? Its minimum value?

24. In words, what does the derivative $D_{\mathbf{u}}f(x_0, y_0)$ tell you about the surface $z = f(x, y)$?

25. Find $D_{\mathbf{u}}f(-3, 5)$ for $f(x, y) = y\ln(x + y)$ if $\mathbf{u} = \frac{3}{5}\mathbf{i} + \frac{4}{5}\mathbf{j}$.

26. Suppose that $\nabla f(0, 0) = 2\mathbf{i} + \frac{3}{2}\mathbf{j}$.
(a) Find a unit vector $\mathbf{u}$ such that $D_{\mathbf{u}}f(0, 0)$ is a maximum. What is this maximum value?
(b) Find a unit vector $\mathbf{u}$ such that $D_{\mathbf{u}}f(0, 0)$ is a minimum. What is this minimum value?

27. At the point $(1, 2)$, the directional derivative $D_{\mathbf{u}}f$ is $2\sqrt{2}$ toward $P_1(2, 3)$ and -3 toward $P_2(1, 0)$. Find $D_{\mathbf{u}}f(1, 2)$ toward the origin.

28. Find equations for the tangent plane and normal line to the given surface at P_0.
(a) $z = x^2e^{2y}$; $P_0(1, \ln 2, 4)$
(b) $x^2y^3z^4 + xyz = 2$; $P_0(2, 1, -1)$

29. Find all points P_0 on the surface $z = 2 - xy$ at which the normal line passes through the origin.

30. Show that for all tangent planes to the surface

$$x^{2/3} + y^{2/3} + z^{2/3} = 1$$

the sum of the squares of the x-, y-, and z-intercepts is 1.

31. Find all points on the paraboloid $z = 9x^2 + 4y^2$ at which the normal line is parallel to the line through the points $P(4, -2, 5)$ and $Q(-2, -6, 4)$.

32. Suppose the equations of motion of a particle are $x = t - 1$, $y = 4e^{-t}$, $z = 2 - \sqrt{t}$, where $t > 0$. Find, to the nearest tenth of a degree, the acute angle between the velocity vector and the normal line to the surface $(x^2/4) + y^2 + z^2 = 1$ at the points where the particle collides with the surface. Use a calculating utility with a root-finding capability where needed.

33–36 Locate all relative minima, relative maxima, and saddle points.

33. $f(x, y) = x^2 + 3xy + 3y^2 - 6x + 3y$

34. $f(x, y) = x^2y - 6y^2 - 3x^2$

35. $f(x, y) = x^3 - 3xy + \frac{1}{2}y^2$

36. $f(x, y) = 4x^2 - 12xy + 9y^2$

37–39 Solve these exercises two ways:
(a) Use the constraint to eliminate a variable.
(b) Use Lagrange multipliers.

37. Find all relative extrema of x^2y^2 subject to the constraint $4x^2 + y^2 = 8$.

38. Find the dimensions of the rectangular box of maximum volume that can be inscribed in the ellipsoid

$$(x/a)^2 + (y/b)^2 + (z/c)^2 = 1$$

39. As illustrated in the accompanying figure, suppose that a current I branches into currents I_1, I_2, and I_3 through resistors R_1, R_2, and R_3 in such a way that the total power dissipated in the three resistors is a minimum. Find the ratios $I_1 : I_2 : I_3$ if the power dissipated in R_i is $I_i^2 R_i (i = 1, 2, 3)$ and $I_1 + I_2 + I_3 = I$.

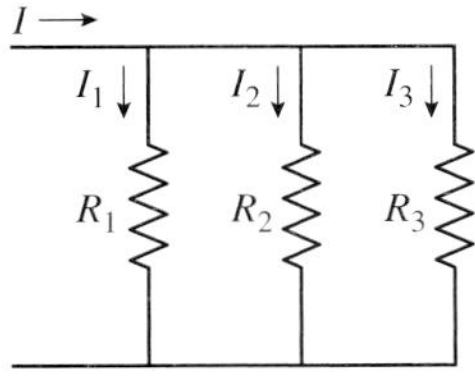

Figure Ex-39

40–42 In economics, a ***production model*** is a mathematical relationship between the output of a company or a country and the labor and capital equipment required to produce that output. Much of the pioneering work in the field of production models occurred in the 1920s when Paul Douglas of the University of Chicago and his collaborator Charles Cobb proposed that the output P can be expressed in terms of the labor L and the capital equipment K by an equation of the form

$$P = cL^{\alpha}K^{\beta}$$

where c is a constant of proportionality and α and β are constants such that $0 < \alpha < 1$ and $0 < \beta < 1$. This is called the ***Cobb–Douglas production model***. Typically, P, L, and K are all expressed in terms of their equivalent monetary values. These exercises explore properties of this model.

40. (a) Consider the Cobb–Douglas production model given by the formula $P = L^{0.75}K^{0.25}$. Sketch the level curves $P(L, K) = 1$, $P(L, K) = 2$, and $P(L, K) = 3$ in an LK-coordinate system (L horizontal and K vertical). Your sketch need not be accurate numerically, but it should show the general shape of the curves and their relative positions.

(b) Use a graphing utility to make a more extensive contour plot of the model.

41. (a) Find $\partial P/\partial L$ and $\partial P/\partial K$ for the Cobb–Douglas production model $P = cL^{\alpha}K^{\beta}$.

(b) The derivative $\partial P/\partial L$ is called the ***marginal productivity of labor***, and the derivative $\partial P/\partial K$ is called the ***marginal productivity of capital***. Explain what these quantities mean in practical terms.

(c) Show that if $\beta = 1 - \alpha$, then P satisfies the partial differential equation

$$K\frac{\partial P}{\partial K} + L\frac{\partial P}{\partial L} = P$$

42. Consider the Cobb–Douglas production model

$$P = 1000\, L^{0.6}K^{0.4}$$

(a) Find the maximum output value of P if labor costs \$50.00 per unit, capital costs \$100.00 per unit, and the total cost of labor and capital is set at \$200,000.

(b) How should the \$200,000 be allocated between labor and capital to achieve the maximum?

c h a p t e r f i f t e e n

MULTIPLE INTEGRALS

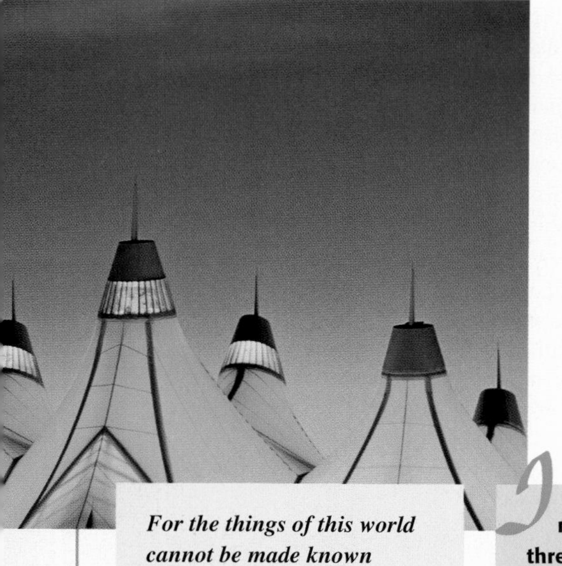

For the things of this world cannot be made known without a knowledge of mathematics.

—Roger Bacon
Mathematician and Scientist

In this chapter we will extend the concept of a definite integral to functions of two and three variables. Whereas functions of one variable are usually integrated over intervals, functions of two variables are usually integrated over regions in 2-space and functions of three variables over regions in 3-space. Calculating such integrals will require some new techniques that will be a central focus in this chapter. Once we have developed the basic methods for integrating functions of two and three variables, we will show how such integrals can be used to calculate surface areas and volumes of solids; and we will also show how they can be used to find masses and centers of gravity of flat plates and three-dimensional solids. In addition to our study of integration, we will generalize the concept of a parametic curve in 2-space to a parametric surface in 3-space. This will allow us to work with a wider variety of surfaces than previously possible and will provide a powerful tool for generating surfaces using computers and other graphing utilities.

Photo: ***Finding the area of complex surfaces such as those used in the design of the Denver International Airport require integration methods studied in this chapter.***

15.1 DOUBLE INTEGRALS

The notion of a definite integral can be extended to functions of two or more variables. In this section we will discuss the double integral, which is the extension to functions of two variables.

VOLUME

Recall that the definite integral of a function of one variable

$$\int_a^b f(x)\,dx = \lim_{\max \Delta x_k \to 0} \sum_{k=1}^{n} f(x_k^*)\Delta x_k = \lim_{n \to +\infty} \sum_{k=1}^{n} f(x_k^*)\Delta x_k \tag{1}$$

arose from the problem of finding areas under curves. [In the rightmost expression in (1), we use the "limit as $n \to +\infty$" to encapsulate the process by which we increase the number of subintervals of $[a, b]$ in such a way that the lengths of the subintervals approach zero.] Integrals of functions of two variables arise from the problem of finding volumes under surfaces.

> **15.1.1 THE VOLUME PROBLEM.** Given a function f of two variables that is continuous and nonnegative on a region R in the xy-plane, find the volume of the solid enclosed between the surface $z = f(x, y)$ and the region R (Figure 15.1.1).

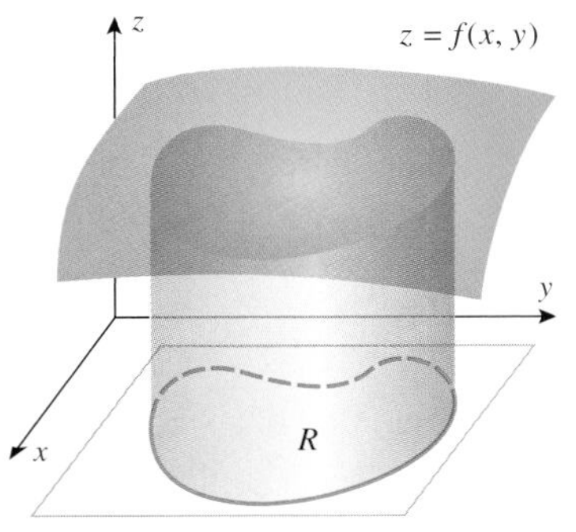

Figure 15.1.1

Later, we will place more restrictions on the region R, but for now we will just assume that the entire region can be enclosed within some suitably large rectangle with sides parallel to the coordinate axes. This ensures that R does not extend indefinitely in any direction.

The procedure for finding the volume V of the solid in Figure 15.1.1 will be similar to the limiting process used for finding areas, except that now the approximating elements will be rectangular parallelepipeds rather than rectangles. We proceed as follows:

- Using lines parallel to the coordinate axes, divide the rectangle enclosing the region R into subrectangles, and exclude from consideration all those subrectangles that contain any points outside of R. This leaves only rectangles that are subsets of R (Figure 15.1.2). Assume that there are n such rectangles, and denote the area of the kth such rectangle by ΔA_k.
- Choose any arbitrary point in each subrectangle, and denote the point in the kth subrectangle by (x_k^*, y_k^*). As shown in Figure 15.1.3, the product $f(x_k^*, y_k^*)\Delta A_k$ is the volume of a rectangular parallelepiped with base area ΔA_k and height $f(x_k^*, y_k^*)$, so the sum

$$\sum_{k=1}^{n} f(x_k^*, y_k^*)\Delta A_k$$

can be viewed as an approximation to the volume V of the entire solid.

- There are two sources of error in the approximation: first, the parallelepipeds have flat tops, whereas the surface $z = f(x, y)$ may be curved; second, the rectangles that form the bases of the parallelepipeds may not completely cover the region R. However, if we repeat the above process with more and more subdivisions in such a way that both the lengths and the widths of the subrectangles approach zero, then it is plausible that the errors of both types approach zero, and the exact volume of the solid will be

$$V = \lim_{n \to +\infty} \sum_{k=1}^{n} f(x_k^*, y_k^*)\Delta A_k$$

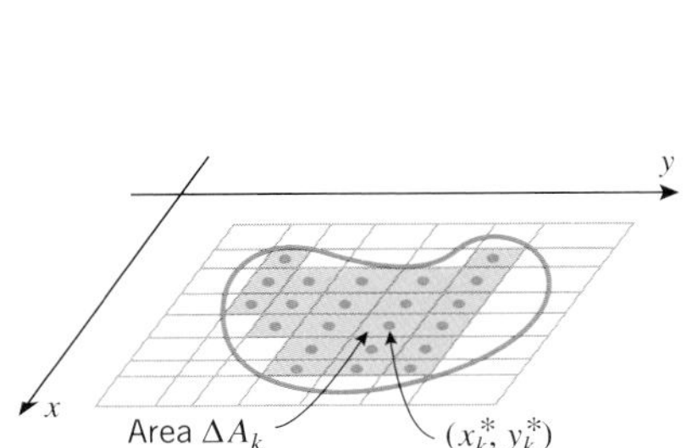

Figure 15.1.2

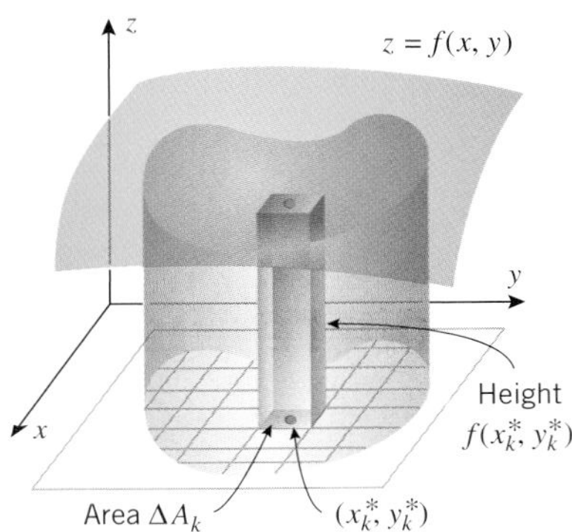

Figure 15.1.3

This suggests the following definition.

Definition 15.1.2 is satisfactory for our present purposes, but some issues would have to be resolved before it could be regarded as rigorous. For example, we would have to prove that the limit actually exists and that its value does not depend on how the points $(x_1^*, y_1^*), (x_2^*, y_2^*), \ldots, (x_n^*, y_n^*)$ are chosen. This can be verified if the region R is not too "complicated" and if f is continuous on R. The details are beyond the scope of this text.

15.1.2 DEFINITION (*Volume Under a Surface*). If f is a function of two variables that is continuous and nonnegative on a region R in the xy-plane, then the volume of the solid enclosed between the surface $z = f(x, y)$ and the region R is defined by

$$V = \lim_{n\to+\infty} \sum_{k=1}^{n} f(x_k^*, y_k^*)\Delta A_k \tag{2}$$

Here, $n \to +\infty$ indicates the process of increasing the number of subrectangles of the rectangle enclosing R in such a way that both the lengths and the widths of the subrectangles approach zero.

It is assumed in Definition 15.1.2 that f is nonnegative on the region R. If f is continuous on R and has both positive and negative values, then the limit

$$\lim_{n\to+\infty} \sum_{k=1}^{n} f(x_k^*, y_k^*)\Delta A_k \tag{3}$$

no longer represents the volume between R and the surface $z = f(x, y)$; rather, it represents a *difference* of volumes—the volume between R and the portion of the surface that is above the xy-plane minus the volume between R and the portion of the surface below the xy-plane. We call this the ***net signed volume*** between the region R and the surface $z = f(x, y)$.

DEFINITION OF A DOUBLE INTEGRAL

As in Definition 15.1.2, the notation $n \to +\infty$ in (3) encapsulates a process in which the enclosing rectangle for R is repeatedly subdivided in such a way that both the lengths and the widths of the subrectangles approach zero. Note that subdividing so that the subrectangle lengths approach zero forces the mesh of the partition of the length of the enclosing rectangle for R to approach zero. Similarly, subdividing so that the subrectangle widths approach zero forces the mesh of the partition of the width of the enclosing rectangle for R to approach zero. Thus, we have extended the notion conveyed by Formula (1) where the definite integral of a one-variable function is expressed as a limit of Riemann sums. By extension, the sums in (3) are also called ***Riemann sums***, and the limit of the Riemann sums is denoted by

$$\iint_R f(x, y)\,dA = \lim_{n\to+\infty} \sum_{k=1}^{n} f(x_k^*, y_k^*)\Delta A_k \tag{4}$$

which is called the ***double integral*** of $f(x, y)$ over R.

If f is continuous and nonnegative on the region R, then the volume formula in (2) can be expressed as

$$V = \iint_R f(x, y)\,dA \tag{5}$$

If f has both positive and negative values on R, then a positive value for the double integral of f over R means that there is more volume above R than below, a negative value for the double integral means that there is more volume below R than above, and a value of zero means that the volume above R is the same as the volume below R.

EVALUATING DOUBLE INTEGRALS

Except in the simplest cases, it is impractical to obtain the value of a double integral from the limit in (4). However, we will now show how to evaluate double integrals by calculating two successive single integrals. For the rest of this section we will limit our discussion to the case where R is a rectangle; in the next section we will consider double integrals over more complicated regions.

The partial derivatives of a function $f(x, y)$ are calculated by holding one of the variables fixed and differentiating with respect to the other variable. Let us consider the reverse of this process, ***partial integration***. The symbols

$$\int_a^b f(x, y)\,dx \quad \text{and} \quad \int_c^d f(x, y)\,dy$$

denote ***partial definite integrals***; the first integral, called the ***partial definite integral with respect to x***, is evaluated by holding y fixed and integrating with respect to x, and the second integral, called the ***partial definite integral with respect to y***, is evaluated by holding x fixed and integrating with respect to y. As the following example shows, the partial definite integral with respect to x is a function of y, and the partial definite integral with respect to y is a function of x.

▶ Example 1

$$\int_0^1 xy^2\,dx = y^2\int_0^1 x\,dx = \left.\frac{y^2x^2}{2}\right]_{x=0}^1 = \frac{y^2}{2}$$

$$\int_0^1 xy^2\,dy = x\int_0^1 y^2\,dy = \left.\frac{xy^3}{3}\right]_{y=0}^1 = \frac{x}{3} \quad ◀$$

A partial definite integral with respect to x is a function of y and hence can be integrated with respect to y; similarly, a partial definite integral with respect to y can be integrated with respect to x. This two-stage integration process is called ***iterated*** (or ***repeated***) ***integration***. We introduce the following notation:

$$\int_c^d\int_a^b f(x, y)\,dx\,dy = \int_c^d\left[\int_a^b f(x, y)\,dx\right]dy \tag{6}$$

$$\int_a^b\int_c^d f(x, y)\,dy\,dx = \int_a^b\left[\int_c^d f(x, y)\,dy\right]dx \tag{7}$$

These integrals are called ***iterated integrals***.

▶ Example 2 Evaluate

(a) $\displaystyle\int_1^3\int_2^4 (40 - 2xy)\,dy\,dx$ (b) $\displaystyle\int_2^4\int_1^3 (40 - 2xy)\,dx\,dy$

Solution (a).

$$\begin{aligned}\int_1^3\int_2^4 (40 - 2xy)\,dy\,dx &= \int_1^3\left[\int_2^4 (40 - 2xy)\,dy\right]dx \\ &= \int_1^3 (40y - xy^2)\Big]_{y=2}^4\,dx \\ &= \int_1^3 [(160 - 16x) - (80 - 4x)]\,dx \\ &= \int_1^3 (80 - 12x)\,dx \\ &= (80x - 6x^2)\Big]_1^3 = 112\end{aligned}$$

Solution (b).

$$\int_2^4 \int_1^3 (40 - 2xy)\,dx\,dy = \int_2^4 \left[\int_1^3 (40 - 2xy)\,dx\right] dy$$
$$= \int_2^4 (40x - x^2 y)\Big]_{x=1}^{3}\,dy$$
$$= \int_2^4 [(120 - 9y) - (40 - y)]\,dy$$
$$= \int_2^4 (80 - 8y)\,dy$$
$$= (80y - 4y^2)\Big]_2^4 = 112 \blacktriangleleft$$

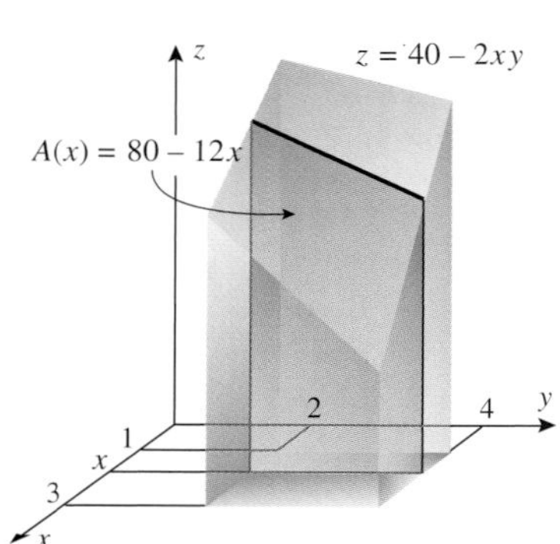

Figure 15.1.4

It is no accident that both parts of Example 2 produced the same answer. Consider the solid S bounded above by the surface $z = 40 - 2xy$ and below by the rectangle R defined by $1 \le x \le 3$ and $2 \le y \le 4$. By the method of slicing discussed in Section 6.2, the volume of S is given by

$$V = \int_1^3 A(x)\,dx$$

where $A(x)$ is the area of a vertical cross section of S taken perpendicular to the x-axis (Figure 15.1.4). For a fixed value of x, $1 \le x \le 3$, $z = 40 - 2xy$ is a function of y, so the integral

$$A(x) = \int_2^4 (40 - 2xy)\,dy$$

represents the area under the graph of this function of y. Thus,

$$V = \int_1^3 \left[\int_2^4 (40 - 2xy)\,dy\right] dx = \int_1^3 \int_2^4 (40 - 2xy)\,dy\,dx$$

is the volume of S. Similarly, by the method of slicing with cross sections of S taken perpendicular to the y-axis, the volume of S is given by

$$V = \int_2^4 A(y)\,dy = \int_2^4 \left[\int_1^3 (40 - 2xy)\,dx\right] dy = \int_2^4 \int_1^3 (40 - 2xy)\,dx\,dy$$

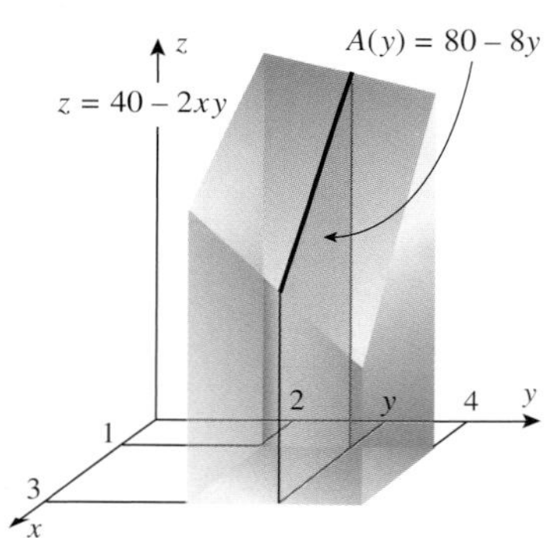

Figure 15.1.5

(Figure 15.1.5). Thus, the iterated integrals in parts (a) and (b) of Example 2 both measure the volume of S, which by Formula (5) is the double integral of $z = 40 - 2xy$ over R. That is,

$$\int_1^3 \int_2^4 (40 - 2xy)\,dy\,dx = \iint_R (40 - 2xy)\,dA = \int_2^4 \int_1^3 (40 - 2xy)\,dx\,dy$$

We will often denote the rectangle
$$\{(x, y) : a \le x \le b, c \le y \le d\}$$
as $[a, b] \times [c, d]$ for simplicity.

The geometric argument above applies to any continuous function $f(x, y)$ that is nonnegative on a rectangle $R = [a, b] \times [c, d]$, as is the case for $f(x, y) = 40 - 2xy$ on $[1, 3] \times [2, 4]$. The conclusion that the double integral of $f(x, y)$ over R has the same value as either of the two possible iterated integrals is true even when f is negative at some points in R. We state this result in the following theorem and omit a formal proof.

15.1.3 THEOREM. *Let R be the rectangle defined by the inequalities*

$$a \le x \le b, \quad c \le y \le d$$

If $f(x, y)$ is continuous on this rectangle, then

$$\iint_R f(x, y)\,dA = \int_c^d \int_a^b f(x, y)\,dx\,dy = \int_a^b \int_c^d f(x, y)\,dy\,dx$$

Theorem 15.1.3 allows us to evaluate a double integral over a rectangle by converting it to an iterated integral. This can be done in two ways, both of which produce the value of the double integral.

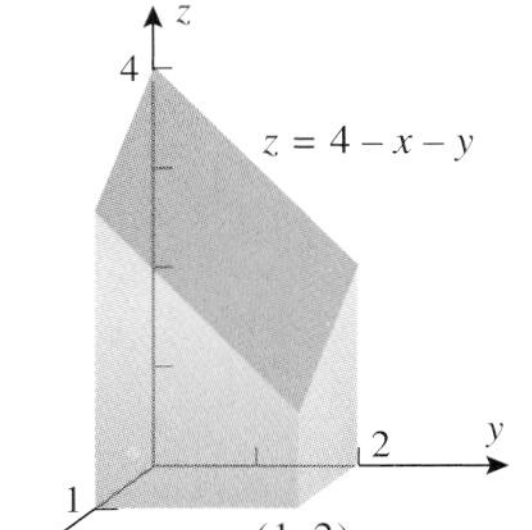

Figure 15.1.6

▶ **Example 3** Use a double integral to find the volume of the solid that is bounded above by the plane $z = 4 - x - y$ and below by the rectangle $R = [0, 1] \times [0, 2]$ (Figure 15.1.6).

Solution. The volume is the double integral of $z = 4 - x - y$ over R. Using Theorem 15.1.3, this can be obtained from either of the iterated integrals

$$\int_0^2 \int_0^1 (4 - x - y)\, dx\, dy \quad \text{or} \quad \int_0^1 \int_0^2 (4 - x - y)\, dy\, dx \tag{8}$$

Using the first of these, we obtain

$$\begin{aligned} V &= \iint_R (4 - x - y)\, dA = \int_0^2 \int_0^1 (4 - x - y)\, dx\, dy \\ &= \int_0^2 \left[4x - \frac{x^2}{2} - xy \right]_{x=0}^{1} dy = \int_0^2 \left(\frac{7}{2} - y \right) dy \\ &= \left[\frac{7}{2}y - \frac{y^2}{2} \right]_0^2 = 5 \end{aligned}$$

You can check this result by evaluating the second integral in (8). ◀

TECHNOLOGY MASTERY

If you have a CAS with a built-in capability for computing iterated double integrals, use it to check Example 3.

Theorem 15.1.3 guarantees that the double integral in Example 4 can also be evaluated by integrating first with respect to y and then with respect to x. Verify this.

▶ **Example 4** Evaluate the double integral

$$\iint_R y^2 x\, dA$$

over the rectangle $R = \{(x, y) : -3 \le x \le 2, 0 \le y \le 1\}$.

Solution. In view of Theorem 15.1.3, the value of the double integral can be obtained by evaluating one of two possible iterated double integrals. We choose to integrate first with respect to x and then with respect to y.

$$\begin{aligned} \iint_R y^2 x\, dA &= \int_0^1 \int_{-3}^2 y^2 x\, dx\, dy = \int_0^1 \left[\frac{1}{2} y^2 x^2 \right]_{x=-3}^{2} dy \\ &= \int_0^1 \left(-\frac{5}{2} y^2 \right) dy = -\frac{5}{6} y^3 \bigg]_0^1 = -\frac{5}{6} \end{aligned}$$ ◀

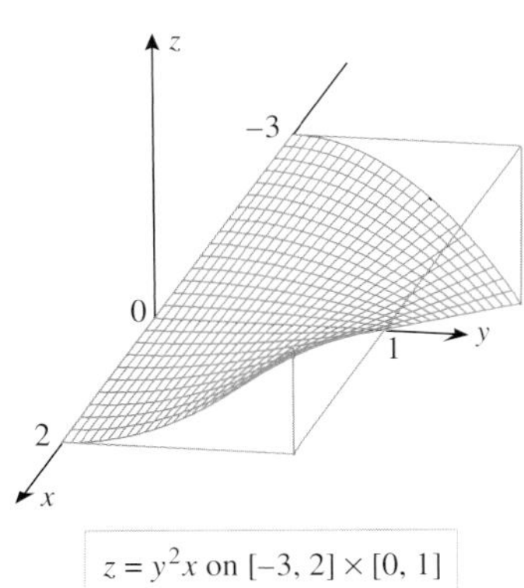

$z = y^2x$ on $[-3, 2] \times [0, 1]$

Figure 15.1.7

The integral in Example 4 can be interpreted as the net signed volume between the rectangle $[-3, 2] \times [0, 1]$ and the surface $z = y^2x$. That is, it is the volume below $z = y^2x$ and above $[0, 2] \times [0, 1]$ minus the volume above $z = y^2x$ and below $[-3, 0] \times [0, 1]$ (Figure 15.1.7).

PROPERTIES OF DOUBLE INTEGRALS

To distinguish between double integrals of functions of two variables and definite integrals of functions of one variable, we will refer to the latter as ***single integrals***. Because double integrals, like single integrals, are defined as limits, they inherit many of the properties of limits. The following results, which we state without proof, are analogs of those in Theorem 6.5.4.

$$\iint_R cf(x, y)\, dA = c \iint_R f(x, y)\, dA \quad (c \text{ a constant}) \tag{9}$$

$$\iint_R [f(x, y) + g(x, y)]\, dA = \iint_R f(x, y)\, dA + \iint_R g(x, y)\, dA \tag{10}$$

$$\iint_R [f(x, y) - g(x, y)]\, dA = \iint_R f(x, y)\, dA - \iint_R g(x, y)\, dA \tag{11}$$

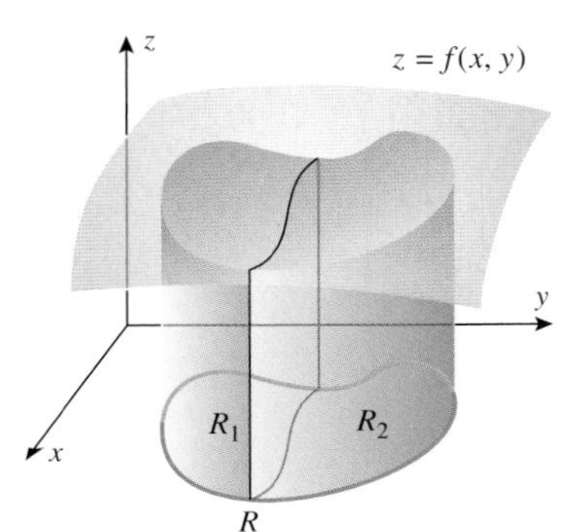

The volume of the entire solid is the sum of the volumes of the solids above R_1 and R_2.

Figure 15.1.8

It is evident intuitively that if $f(x, y)$ is nonnegative on a region R, then subdividing R into two regions R_1 and R_2 has the effect of subdividing the solid between R and $z = f(x, y)$ into two solids, the sum of whose volumes is the volume of the entire solid (Figure 15.1.8). This suggests the following result, which holds even if f has negative values:

$$\iint_R f(x, y)\, dA = \iint_{R_1} f(x, y)\, dA + \iint_{R_2} f(x, y)\, dA \tag{12}$$

The proof of this result will be omitted.

✔QUICK CHECK EXERCISES 15.1 *(See page 1026 for answers.)*

1. The double integral is defined as a limit of Riemann sums by
$$\iint_R f(x, y)\, dA = \_\_\_\_\_\_$$

2. The iterated integral
$$\int_1^5 \int_2^4 f(x, y)\, dx\, dy$$
integrates f over the rectangle defined by
$\_\_\_\_ \le x \le \_\_\_\_, \quad \_\_\_\_ \le y \le \_\_\_\_$

3. Supply the missing integrand and limits of integration.
$$\int_1^5 \int_2^4 (3x^2 - 2xy + y^2)\, dx\, dy = \int_\square^\square \_\_\_\_\_\_ \, dy$$

4. The volume of the solid enclosed by the surface $z = x/y$ and the rectangle $0 \le x \le 4$, $1 \le y \le e^2$ in the xy-plane is ______.

EXERCISE SET 15.1 [C] CAS

1–12 Evaluate the iterated integrals.

1. $\int_0^1 \int_0^2 (x + 3)\, dy\, dx$
2. $\int_1^3 \int_{-1}^1 (2x - 4y)\, dy\, dx$
3. $\int_2^4 \int_0^1 x^2 y\, dx\, dy$
4. $\int_{-2}^0 \int_{-1}^2 (x^2 + y^2)\, dx\, dy$
5. $\int_0^{\ln 3} \int_0^{\ln 2} e^{x+y}\, dy\, dx$
6. $\int_0^2 \int_0^1 y \sin x\, dy\, dx$
7. $\int_{-1}^0 \int_2^5 dx\, dy$
8. $\int_4^6 \int_{-3}^7 dy\, dx$
9. $\int_0^1 \int_0^1 \frac{x}{(xy + 1)^2}\, dy\, dx$
10. $\int_{\pi/2}^{\pi} \int_1^2 x \cos xy\, dy\, dx$
11. $\int_0^{\ln 2} \int_0^1 xy e^{y^2 x}\, dy\, dx$
12. $\int_3^4 \int_1^2 \frac{1}{(x + y)^2}\, dy\, dx$

13–16 Evaluate the double integral over the rectangular region R.

13. $\iint_R 4xy^3\, dA$; $R = \{(x, y) : -1 \le x \le 1, -2 \le y \le 2\}$

14. $\iint\limits_R \frac{xy}{\sqrt{x^2+y^2+1}}\, dA$;
$R = \{(x, y) : 0 \le x \le 1, 0 \le y \le 1\}$

15. $\iint\limits_R x\sqrt{1-x^2}\, dA$; $R = \{(x, y) : 0 \le x \le 1, 2 \le y \le 3\}$

16. $\iint\limits_R (x \sin y - y \sin x)\, dA$;
$R = \{(x, y) : 0 \le x \le \pi/2, 0 \le y \le \pi/3\}$

FOCUS ON CONCEPTS

17. (a) Let $f(x, y) = x^2 + y$, and as shown in the accompanying figure, let the rectangle $R = [0, 2] \times [0, 2]$ be subdivided into 16 subrectangles. Take (x_k^*, y_k^*) to be the center of the kth rectangle, and approximate the double integral of f over R by the resulting Riemann sum.
(b) Compare the result in part (a) to the exact value of the integral.

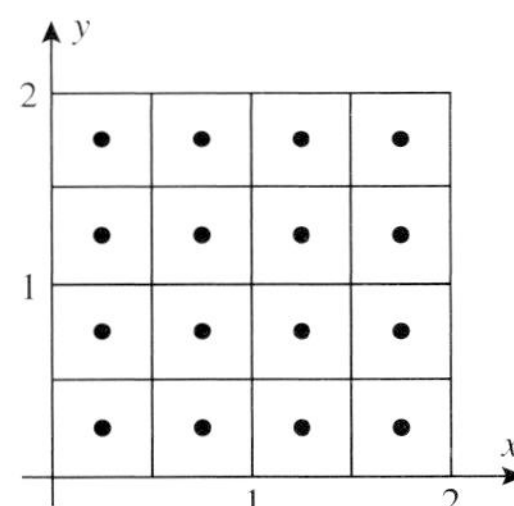

Figure Ex-17

18. (a) Let $f(x, y) = x - 2y$, and as shown in Exercise 17, let the rectangle $R = [0, 2] \times [0, 2]$ be subdivided into 16 subrectangles. Take (x_k^*, y_k^*) to be the center of the kth rectangle, and approximate the double integral of f over R by the resulting Riemann sum.
(b) Compare the result in part (a) to the exact value of the integral.

19–20 Each iterated integral represents the volume of a solid. Make a sketch of the solid. (You do *not* have to find the volume.)

19. (a) $\int_0^5 \int_1^2 4\, dx\, dy$

(b) $\int_0^3 \int_0^4 \sqrt{25 - x^2 - y^2}\, dy\, dx$

20. (a) $\int_0^1 \int_0^1 (2 - x - y)\, dy\, dx$

(b) $\int_{-2}^2 \int_{-2}^2 (x^2 + y^2)\, dx\, dy$

21–24 Use a double integral to find the volume.

21. The volume under the plane $z = 2x + y$ and over the rectangle $R = \{(x, y) : 3 \le x \le 5, 1 \le y \le 2\}$.

22. The volume under the surface $z = 3x^3 + 3x^2y$ and over the rectangle $R = \{(x, y) : 1 \le x \le 3, 0 \le y \le 2\}$.

23. The volume of the solid enclosed by the surface $z = x^2$ and the planes $x = 0$, $x = 2$, $y = 3$, $y = 0$, and $z = 0$.

24. The volume in the first octant bounded by the coordinate planes, the plane $y = 4$, and the plane $(x/3) + (z/5) = 1$.

25. Evaluate the integral by choosing a convenient order of integration:

$$\iint\limits_R x \cos(xy) \cos^2 \pi x\, dA;\ R = \left[0, \tfrac{1}{2}\right] \times [0, \pi]$$

26. (a) Sketch the solid in the first octant that is enclosed by the planes $x = 0$, $z = 0$, $x = 5$, $z - y = 0$, and $z = -2y + 6$.
(b) Find the volume of the solid by breaking it into two parts.

27–30 The ***average value*** or ***mean value*** of a continuous function $f(x, y)$ over a rectangle $R = [a, b] \times [c, d]$ is defined as

$$f_{\text{ave}} = \frac{1}{A(R)} \iint\limits_R f(x, y)\, dA$$

where $A(R) = (b - a)(d - c)$ is the area of the rectangle R (compare to Definition 6.6.1). Use this definition in these exercises.

27. Find the average value of $f(x, y) = y \sin xy$ over the rectangle $[0, 1] \times [0, \pi/2]$.

28. Find the average value of $f(x, y) = x(x^2 + y)^{1/2}$ over the interval $[0, 1] \times [0, 3]$.

29. Suppose that the temperature in degrees Celsius at a point (x, y) on a flat metal plate is $T(x, y) = 10 - 8x^2 - 2y^2$, where x and y are in meters. Find the average temperature of the rectangular portion of the plate for which $0 \le x \le 1$ and $0 \le y \le 2$.

30. Show that if $f(x, y)$ is constant on the rectangle $R = [a, b] \times [c, d]$, say $f(x, y) = k$, then $f_{\text{ave}} = k$ over R.

31–32 Most computer algebra systems have commands for approximating double integrals numerically. Read the relevant documentation and use a CAS to find a numerical approximation of the double integral in these exercises.

C **31.** $\int_0^2 \int_0^1 \sin\sqrt{x^3 + y^3}\, dx\, dy$

C **32.** $\int_{-1}^1 \int_{-1}^1 e^{-(x^2+y^2)}\, dx\, dy$

33. In this exercise, suppose that $f(x, y) = g(x)h(y)$ and $R = \{(x, y) : a \le x \le b, c \le y \le d\}$. Show that

$$\iint\limits_R f(x, y)\, dA = \left[\int_a^b g(x)\, dx\right]\left[\int_c^d h(y)\, dy\right]$$

34. Use the result in Exercise 33 to evaluate the integral

$$\int_0^{\ln 2} \int_{-1}^{1} \sqrt{e^y + 1} \tan x \, dx \, dy$$

by inspection. Explain your reasoning.

C **35.** Use a CAS to evaluate the iterated integrals

$$\int_0^1 \int_0^1 \frac{y-x}{(x+y)^3} \, dx \, dy \quad \text{and} \quad \int_0^1 \int_0^1 \frac{y-x}{(x+y)^3} \, dy \, dx$$

Does this violate Theorem 15.1.3? Explain.

C **36.** Use a CAS to show that the volume V under the surface $z = xy^3 \sin xy$ over the rectangle shown in the accompanying figure is $V = 3/\pi$.

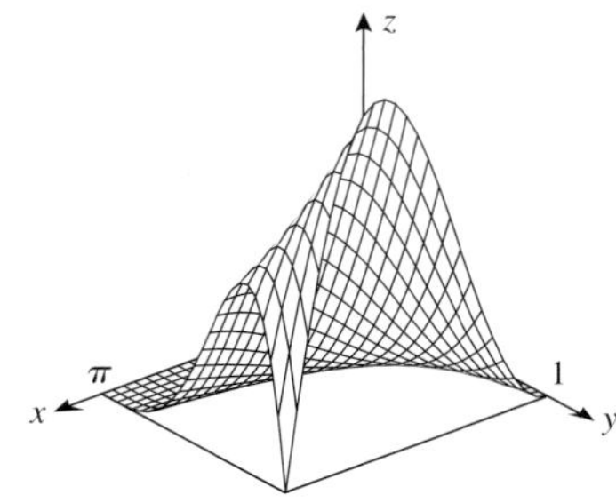

Figure Ex-36

✔QUICK CHECK ANSWERS 15.1

1. $\lim_{n \to +\infty} \sum_{k=1}^{n} f(x_k^*, y_k^*) \Delta A_k$ **2.** $2 \le x \le 4;\ 1 \le y \le 5$ **3.** $\int_1^5 (56 - 12y + 2y^2) \, dy$ **4.** 16

15.2 DOUBLE INTEGRALS OVER NONRECTANGULAR REGIONS

In this section we will show how to evaluate double integrals over regions other than rectangles.

ITERATED INTEGRALS WITH NONCONSTANT LIMITS OF INTEGRATION

Later in this section we will see that double integrals over nonrectangular regions can often be evaluated as iterated integrals of the following types:

$$\int_a^b \int_{g_1(x)}^{g_2(x)} f(x, y) \, dy \, dx = \int_a^b \left[\int_{g_1(x)}^{g_2(x)} f(x, y) \, dy \right] dx \tag{1}$$

$$\int_c^d \int_{h_1(y)}^{h_2(y)} f(x, y) \, dx \, dy = \int_c^d \left[\int_{h_1(y)}^{h_2(y)} f(x, y) \, dx \right] dy \tag{2}$$

We begin with an example that illustrates how to evaluate such integrals.

► **Example 1** Evaluate

$$\text{(a)} \int_0^1 \int_{-x}^{x^2} y^2 x \, dy \, dx \qquad \text{(b)} \int_0^{\pi/3} \int_0^{\cos y} x \sin y \, dx \, dy$$

Solution (a).

$$\int_0^1 \int_{-x}^{x^2} y^2 x \, dy \, dx = \int_0^1 \left[\int_{-x}^{x^2} y^2 x \, dy \right] dx = \int_0^1 \frac{y^3 x}{3} \bigg]_{y=-x}^{x^2} dx$$

$$= \int_0^1 \left[\frac{x^7}{3} + \frac{x^4}{3} \right] dx = \left(\frac{x^8}{24} + \frac{x^5}{15} \right) \bigg]_0^1 = \frac{13}{120}$$

Solution (b).

$$\int_0^{\pi/3}\int_0^{\cos y} x\sin y\,dx\,dy = \int_0^{\pi/3}\left[\int_0^{\cos y} x\sin y\,dx\right]dy = \int_0^{\pi/3}\frac{x^2}{2}\sin y\bigg]_{x=0}^{\cos y}dy$$

$$= \int_0^{\pi/3}\left[\frac{1}{2}\cos^2 y\sin y\right]dy = -\frac{1}{6}\cos^3 y\bigg]_0^{\pi/3} = \frac{7}{48} \blacktriangleleft$$

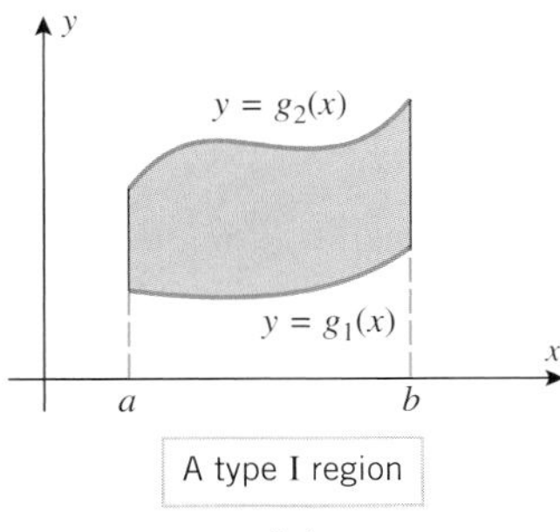

(a)

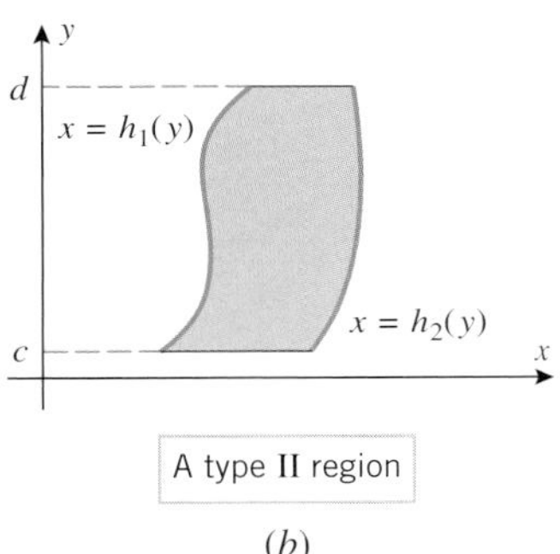

(b)

Figure 15.2.1

DOUBLE INTEGRALS OVER NONRECTANGULAR REGIONS

Plane regions can be extremely complex, and the theory of double integrals over very general regions is a topic for advanced courses in mathematics. We will limit our study of double integrals to two basic types of regions, which we will call *type I* and *type II*; they are defined as follows.

15.2.1 DEFINITION.

(a) A ***type I region*** is bounded on the left and right by vertical lines $x = a$ and $x = b$ and is bounded below and above by continuous curves $y = g_1(x)$ and $y = g_2(x)$, where $g_1(x) \le g_2(x)$ for $a \le x \le b$ (Figure 15.2.1*a*).

(b) A ***type II region*** is bounded below and above by horizontal lines $y = c$ and $y = d$ and is bounded on the left and right by continuous curves $x = h_1(y)$ and $x = h_2(y)$ satisfying $h_1(y) \le h_2(y)$ for $c \le y \le d$ (Figure 15.2.1*b*).

The following theorem will enable us to evaluate double integrals over type I and type II regions using iterated integrals.

15.2.2 THEOREM.

(*a*) *If R is a type I region on which $f(x, y)$ is continuous, then*

$$\iint_R f(x,y)\,dA = \int_a^b\int_{g_1(x)}^{g_2(x)} f(x,y)\,dy\,dx \tag{3}$$

(*b*) *If R is a type II region on which $f(x, y)$ is continuous, then*

$$\iint_R f(x,y)\,dA = \int_c^d\int_{h_1(y)}^{h_2(y)} f(x,y)\,dx\,dy \tag{4}$$

Using Theorem 15.2.2, the integral in Example 1(a) is the double integral of the function $f(x, y) = y^2x$ over the type I region bounded on the left and right by the vertical lines $x = 0$ and $x = 1$ and bounded below and above by the curves $y = -x$ and $y = x^2$ (Figure 15.2.2). Also, the integral in Example 1(b) is the double integral of $f(x, y) = x \sin y$ over the type II region bounded below and above by the horizontal lines $y = 0$ and $y = \pi/3$ and bounded on the left and right by the curves $x = 0$ and $x = \cos y$ (Figure 15.2.3).

We will not prove Theorem 15.2.2, but for the case where $f(x, y)$ is nonnegative on the region R, it can be made plausible by a geometric argument that is similar to that given for Theorem 15.1.3. Since $f(x, y)$ is nonnegative, the double integral can be interpreted as the volume of the solid S that is bounded above by the surface $z = f(x, y)$ and below

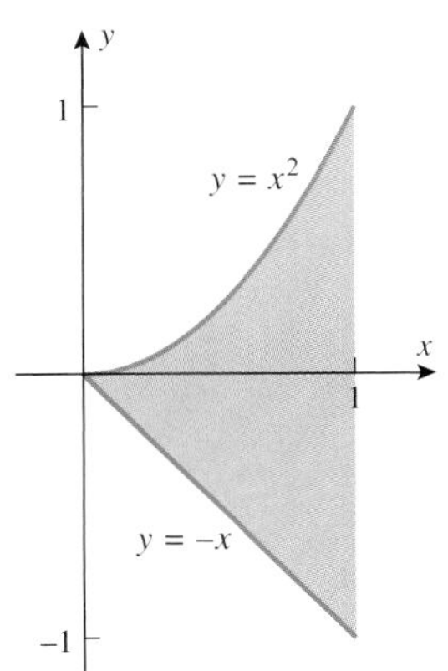

Figure 15.2.2

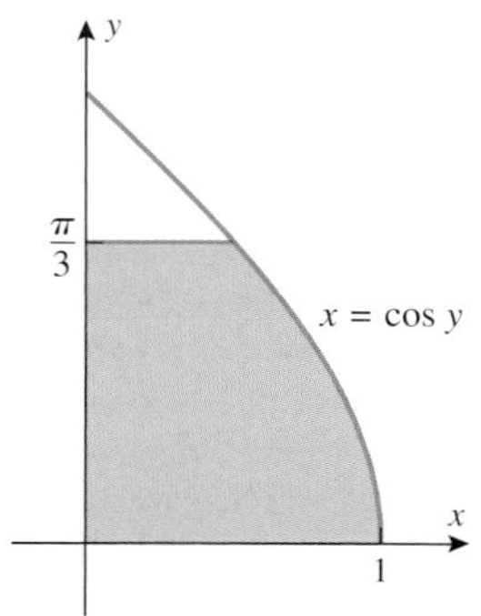

Figure 15.2.3

by the region R, so it suffices to show that the iterated integrals also represent this volume. Consider the iterated integral in (3), for example. For a fixed value of x, the function $f(x, y)$ is a function of y, and hence the integral

$$A(x) = \int_{g_1(x)}^{g_2(x)} f(x, y)\, dy$$

represents the area under the graph of this function of y between $y = g_1(x)$ and $y = g_2(x)$. This area, shown in yellow in Figure 15.2.4, is the cross-sectional area at x of the solid S, and hence by the method of slicing, the volume V of the solid S is

$$V = \int_a^b \int_{g_1(x)}^{g_2(x)} f(x, y)\, dy\, dx$$

which shows that in (3) the iterated integral is equal to the double integral. Similarly for (4).

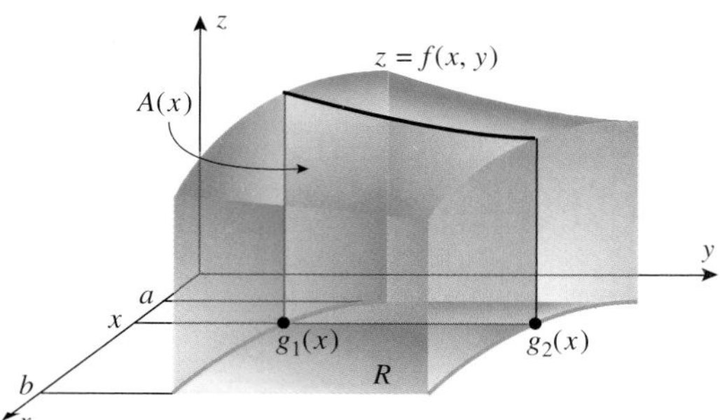

Figure 15.2.4

SETTING UP LIMITS OF INTEGRATION FOR EVALUATING DOUBLE INTEGRALS

To apply Theorem 15.2.2, it is helpful to start with a two-dimensional sketch of the region R. [It is not necessary to graph $f(x, y)$.] For a type I region, the limits of integration in Formula (3) can be obtained as follows:

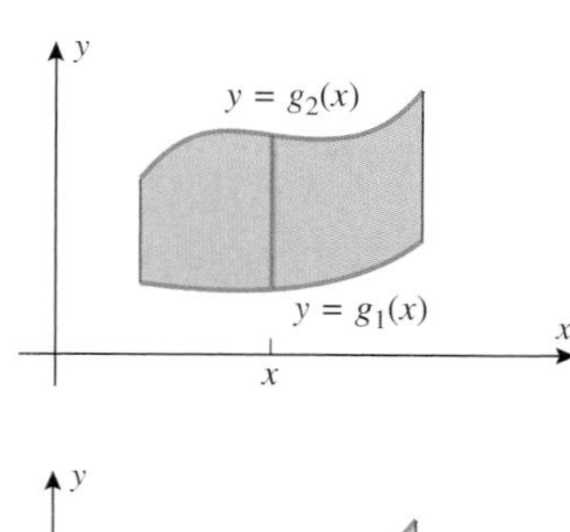

Figure 15.2.5

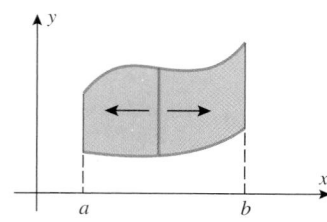

Determining Limits of Integration: Type I Region

Step 1. Since x is held fixed for the first integration, we draw a vertical line through the region R at an arbitrary fixed value x (Figure 15.2.5). This line crosses the boundary of R twice. The lower point of intersection is on the curve $y = g_1(x)$ and the higher point is on the curve $y = g_2(x)$. These two intersections determine the lower and upper y-limits of integration in Formula (3).

Step 2. Imagine moving the line drawn in Step 1 first to the left and then to the right (Figure 15.2.5). The leftmost position where the line intersects the region R is $x = a$, and the rightmost position where the line intersects the region R is $x = b$. This yields the limits for the x-integration in Formula (3).

▶ **Example 2** Evaluate

$$\iint_R xy\, dA$$

over the region R enclosed between $y = \frac{1}{2}x$, $y = \sqrt{x}$, $x = 2$, and $x = 4$.

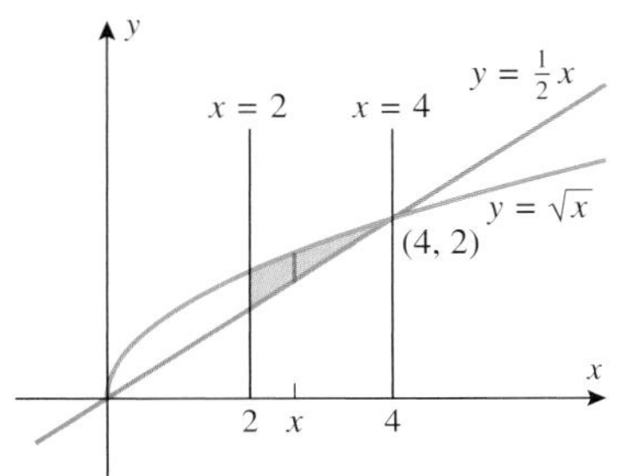

Figure 15.2.6

Solution. We view R as a type I region. The region R and a vertical line corresponding to a fixed x are shown in Figure 15.2.6. This line meets the region R at the lower boundary $y = \frac{1}{2}x$ and the upper boundary $y = \sqrt{x}$. These are the y-limits of integration. Moving this line first left and then right yields the x-limits of integration, $x = 2$ and $x = 4$. Thus,

$$\iint_R xy\,dA = \int_2^4 \int_{x/2}^{\sqrt{x}} xy\,dy\,dx = \int_2^4 \left[\frac{xy^2}{2}\right]_{y=x/2}^{\sqrt{x}} dx = \int_2^4 \left(\frac{x^2}{2} - \frac{x^3}{8}\right) dx$$

$$= \left[\frac{x^3}{6} - \frac{x^4}{32}\right]_2^4 = \left(\frac{64}{6} - \frac{256}{32}\right) - \left(\frac{8}{6} - \frac{16}{32}\right) = \frac{11}{6} \blacktriangleleft$$

If R is a type II region, then the limits of integration in Formula (4) can be obtained as follows:

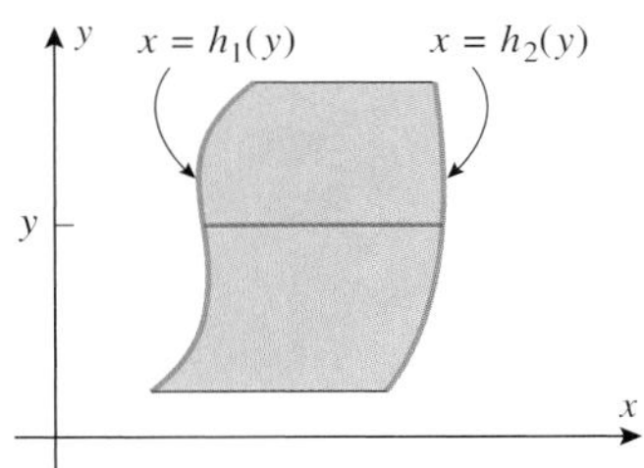

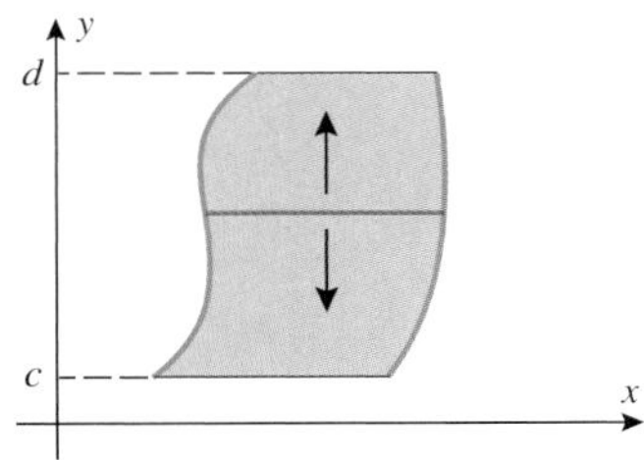

Figure 15.2.7

Determining Limits of Integration: Type II Region

Step 1. Since y is held fixed for the first integration, we draw a horizontal line through the region R at a fixed value y (Figure 15.2.7). This line crosses the boundary of R twice. The leftmost point of intersection is on the curve $x = h_1(y)$ and the rightmost point is on the curve $x = h_2(y)$. These intersections determine the x-limits of integration in (4).

Step 2. Imagine moving the line drawn in Step 1 first down and then up (Figure 15.2.7). The lowest position where the line intersects the region R is $y = c$, and the highest position where the line intersects the region R is $y = d$. This yields the y-limits of integration in (4).

▶ **Example 3** Evaluate

$$\iint_R (2x - y^2)\,dA$$

over the triangular region R enclosed between the lines $y = -x + 1$, $y = x + 1$, and $y = 3$.

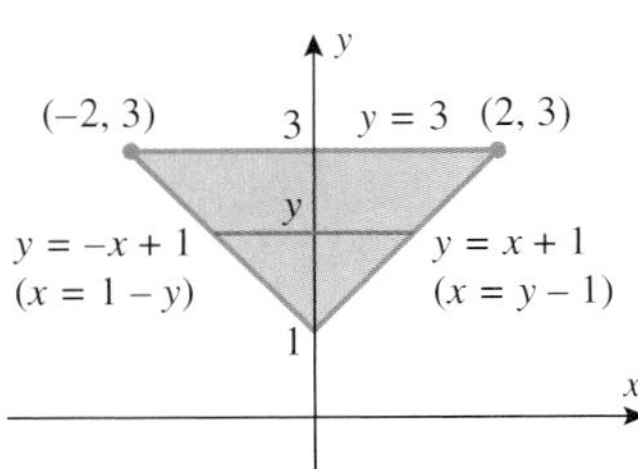

Figure 15.2.8

Solution. We view R as a type II region. The region R and a horizontal line corresponding to a fixed y are shown in Figure 15.2.8. This line meets the region R at its left-hand boundary $x = 1 - y$ and its right-hand boundary $x = y - 1$. These are the x-limits of integration. Moving this line first down and then up yields the y-limits, $y = 1$ and $y = 3$. Thus,

$$\iint_R (2x - y^2)\,dA = \int_1^3 \int_{1-y}^{y-1} (2x - y^2)\,dx\,dy = \int_1^3 \left[x^2 - y^2x\right]_{x=1-y}^{y-1} dy$$

$$= \int_1^3 [(1 - 2y + 2y^2 - y^3) - (1 - 2y + y^3)]\,dy$$

$$= \int_1^3 (2y^2 - 2y^3)\,dy = \left[\frac{2y^3}{3} - \frac{y^4}{2}\right]_1^3 = -\frac{68}{3} \blacktriangleleft$$

To integrate over a type II region, the left- and right-hand boundaries must be expressed in the form $x = h_1(y)$ and $x = h_2(y)$. This is why we rewrote the boundary equations

$$y = -x + 1 \quad \text{and} \quad y = x + 1$$

as

$$x = 1 - y \quad \text{and} \quad x = y - 1$$

in Example 3.

In Example 3 we could have treated R as a type I region, but with an added complication. Viewed as a type I region, the upper boundary of R is the line $y = 3$ (Figure 15.2.9) and the lower boundary consists of two parts, the line $y = -x + 1$ to the left of the y-axis and

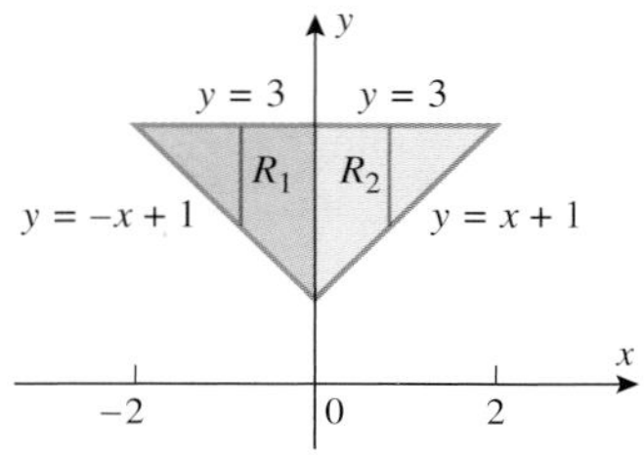

Figure 15.2.9

the line $y = x + 1$ to the right of the y-axis. To carry out the integration it is necessary to decompose the region R into two parts, R_1 and R_2, as shown in Figure 15.2.9, and write

$$\iint_R (2x - y^2)\,dA = \iint_{R_1} (2x - y^2)\,dA + \iint_{R_2} (2x - y^2)\,dA$$
$$= \int_{-2}^{0}\int_{-x+1}^{3} (2x - y^2)\,dy\,dx + \int_{0}^{2}\int_{x+1}^{3} (2x - y^2)\,dy\,dx$$

This will yield the same result that was obtained in Example 3. (Verify.)

▶ **Example 4** Use a double integral to find the volume of the tetrahedron bounded by the coordinate planes and the plane $z = 4 - 4x - 2y$.

Solution. The tetrahedron in question is bounded above by the plane

$$z = 4 - 4x - 2y \tag{5}$$

and below by the triangular region R shown in Figure 15.2.10. Thus, the volume is given by

$$V = \iint_R (4 - 4x - 2y)\,dA$$

The region R is bounded by the x-axis, the y-axis, and the line $y = 2 - 2x$ [set $z = 0$ in (5)], so that treating R as a type I region yields

$$V = \iint_R (4 - 4x - 2y)\,dA = \int_0^1\int_0^{2-2x} (4 - 4x - 2y)\,dy\,dx$$
$$= \int_0^1 \left[4y - 4xy - y^2\right]_{y=0}^{2-2x} dx = \int_0^1 (4 - 8x + 4x^2)\,dx = \frac{4}{3} \blacktriangleleft$$

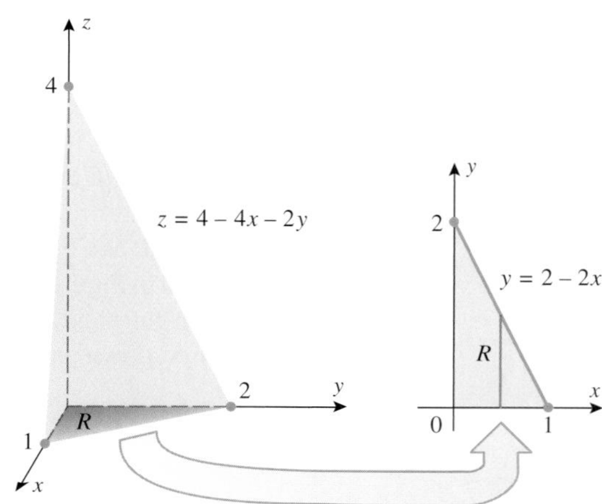

Figure 15.2.10

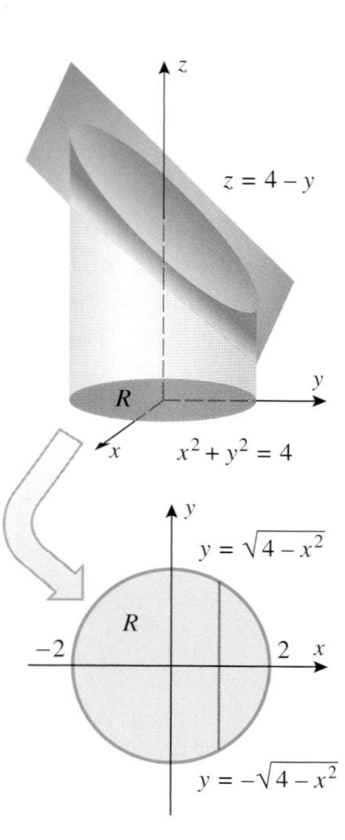

Figure 15.2.11

▶ **Example 5** Find the volume of the solid bounded by the cylinder $x^2 + y^2 = 4$ and the planes $y + z = 4$ and $z = 0$.

Solution. The solid shown in Figure 15.2.11 is bounded above by the plane $z = 4 - y$ and below by the region R within the circle $x^2 + y^2 = 4$. The volume is given by

$$V = \iint_R (4 - y)\,dA$$

Treating R as a type I region we obtain

$$V = \int_{-2}^{2} \int_{-\sqrt{4-x^2}}^{\sqrt{4-x^2}} (4-y)\,dy\,dx = \int_{-2}^{2} \left[4y - \frac{1}{2}y^2\right]_{y=-\sqrt{4-x^2}}^{\sqrt{4-x^2}} dx$$

$$= \int_{-2}^{2} 8\sqrt{4-x^2}\,dx = 8(2\pi) = 16\pi$$

See Formula (3) of Section 8.4. ◀

REVERSING THE ORDER OF INTEGRATION

Sometimes the evaluation of an iterated integral can be simplified by reversing the order of integration. The next example illustrates how this is done.

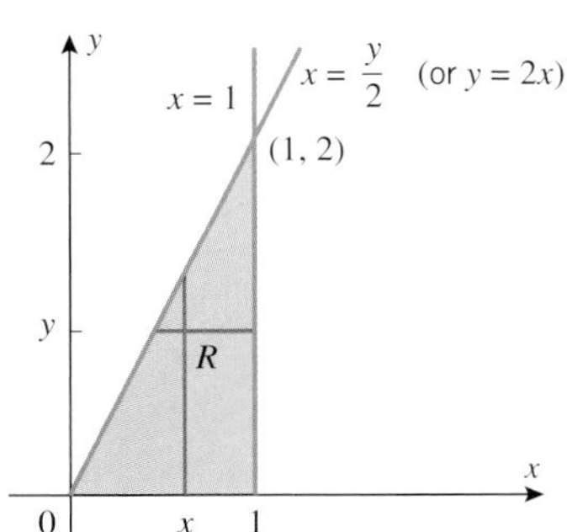

Figure 15.2.12

▶ **Example 6** Since there is no elementary antiderivative of e^{x^2}, the integral

$$\int_{0}^{2} \int_{y/2}^{1} e^{x^2}\,dx\,dy$$

cannot be evaluated by performing the x-integration first. Evaluate this integral by expressing it as an equivalent iterated integral with the order of integration reversed.

Solution. For the inside integration, y is fixed and x varies from the line $x = y/2$ to the line $x = 1$ (Figure 15.2.12). For the outside integration, y varies from 0 to 2, so the given iterated integral is equal to a double integral over the triangular region R in Figure 15.2.12.

To reverse the order of integration, we treat R as a type I region, which enables us to write the given integral as

$$\int_{0}^{2} \int_{y/2}^{1} e^{x^2}\,dx\,dy = \iint_R e^{x^2}\,dA = \int_{0}^{1} \int_{0}^{2x} e^{x^2}\,dy\,dx = \int_{0}^{1} \left[e^{x^2} y\right]_{y=0}^{2x} dx$$

$$= \int_{0}^{1} 2xe^{x^2}\,dx = e^{x^2}\Big]_{0}^{1} = e - 1 \quad ◀$$

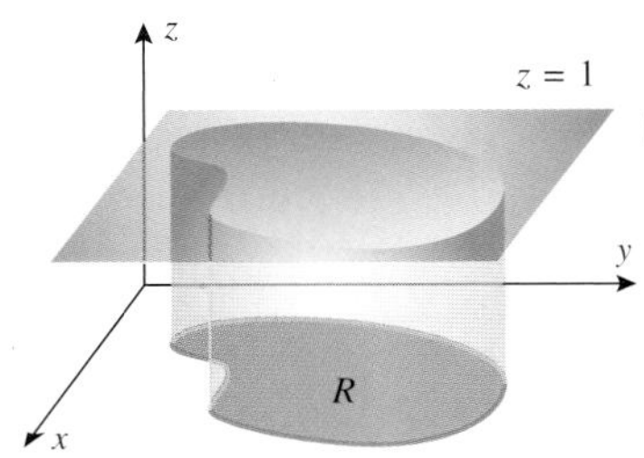

Cylinder with base R and height 1

Figure 15.2.13

AREA CALCULATED AS A DOUBLE INTEGRAL

Although double integrals arose in the context of calculating volumes, they can also be used to calculate areas. To see why this is so, recall that a *right cylinder* is a solid that is generated when a plane region is translated along a line that is perpendicular to the region. In Formula (2) of Section 6.2 we stated that the volume V of a right cylinder with cross-sectional area A and height h is

$$V = A \cdot h \tag{6}$$

Now suppose that we are interested in finding the area A of a region R in the xy-plane. If we translate the region R upward 1 unit, then the resulting solid will be a right cylinder that has cross-sectional area A, base R, and the plane $z = 1$ as its top (Figure 15.2.13). Thus, it follows from (6) that

$$\iint_R 1\,dA = (\text{area of } R) \cdot 1$$

which we can rewrite as

$$\text{area of } R = \iint_R 1\,dA = \iint_R dA \tag{7}$$

Formula (7) can be confusing because it equates an area and a volume; the formula is intended to equate only the *numerical values* of the area and volume and not the units, which must, of course, be different.

▶ **Example 7** Use a double integral to find the area of the region R enclosed between the parabola $y = \frac{1}{2}x^2$ and the line $y = 2x$.

Solution. The region R may be treated equally well as type I (Figure 15.2.14a) or type II (Figure 15.2.14b). Treating R as type I yields

$$\text{area of } R = \iint_R dA = \int_0^4 \int_{x^2/2}^{2x} dy\, dx = \int_0^4 \left[y\right]_{y=x^2/2}^{2x} dx$$

$$= \int_0^4 \left(2x - \frac{1}{2}x^2\right) dx = \left[x^2 - \frac{x^3}{6}\right]_0^4 = \frac{16}{3}$$

Treating R as type II yields

$$\text{area of } R = \iint_R dA = \int_0^8 \int_{y/2}^{\sqrt{2y}} dx\, dy = \int_0^8 \left[x\right]_{x=y/2}^{\sqrt{2y}} dy$$

$$= \int_0^8 \left(\sqrt{2y} - \frac{1}{2}y\right) dy = \left[\frac{2\sqrt{2}}{3}y^{3/2} - \frac{y^2}{4}\right]_0^8 = \frac{16}{3} \blacktriangleleft$$

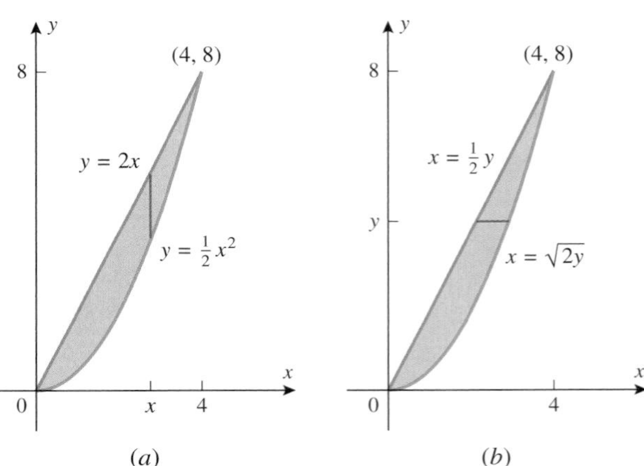

Figure 15.2.14

✔QUICK CHECK EXERCISES 15.2 *(See page 1035 for answers.)*

1. Supply the missing integrand and limits of integration.

(a) $\displaystyle\int_1^5 \int_2^{y/2} 6x^2y\, dx\, dy = \int_\square^\square \underline{\hspace{3em}}\, dy$

(b) $\displaystyle\int_1^5 \int_2^{x/2} 6x^2y\, dy\, dx = \int_\square^\square \underline{\hspace{3em}}\, dx$

2. Let R be the triangular region in the xy-plane with vertices $(0, 0)$, $(3, 0)$, and $(0, 4)$. Supply the missing portions of the integrals.

(a) Treating R as a type I region,

$$\iint_R f(x, y)\, dA = \int_\square^\square \int_\square^\square f(x, y) \underline{\hspace{3em}}$$

(b) Treating R as a type II region,

$$\iint_R f(x, y)\, dA = \int_\square^\square \int_\square^\square f(x, y) \underline{\hspace{3em}}$$

3. Let R be the triangular region in the xy-plane with vertices $(0, 0)$, $(3, 3)$, and $(0, 4)$. Expressed as an iterated double integral, the area of R is $A(R) = \underline{\hspace{3em}}$.

4. The line $y = 2 - x$ and the parabola $y = x^2$ intersect at the points $(-2, 4)$ and $(1, 1)$. If R is the region enclosed by $y = 2 - x$ and $y = x^2$, then

$$\iint_R (1 + 2y)\, dA = \underline{\hspace{3em}}$$

EXERCISE SET 15.2 Graphing Utility C CAS

1–10 Evaluate the iterated integral.

1. $\int_0^1 \int_{x^2}^x xy^2\,dy\,dx$

2. $\int_1^{3/2} \int_y^{3-y} y\,dx\,dy$

3. $\int_0^3 \int_0^{\sqrt{9-y^2}} y\,dx\,dy$

4. $\int_{1/4}^1 \int_{x^2}^x \sqrt{\frac{x}{y}}\,dy\,dx$

5. $\int_{\sqrt{\pi}}^{\sqrt{2\pi}} \int_0^{x^3} \sin\frac{y}{x}\,dy\,dx$

6. $\int_{-1}^1 \int_{-x^2}^{x^2} (x^2 - y)\,dy\,dx$

7. $\int_{\pi/2}^{\pi} \int_0^{x^2} \frac{1}{x}\cos\frac{y}{x}\,dy\,dx$

8. $\int_0^1 \int_0^x e^{x^2}\,dy\,dx$

9. $\int_0^1 \int_0^x y\sqrt{x^2 - y^2}\,dy\,dx$

10. $\int_1^2 \int_0^{y^2} e^{x/y^2}\,dx\,dy$

FOCUS ON CONCEPTS

11. Let R be the region shown in the accompanying figure. Fill in the missing limits of integration.

(a) $\iint_R f(x, y)\,dA = \int_\square^\square \int_\square^\square f(x, y)\,dy\,dx$

(b) $\iint_R f(x, y)\,dA = \int_\square^\square \int_\square^\square f(x, y)\,dx\,dy$

12. Let R be the region shown in the accompanying figure. Fill in the missing limits of integration.

(a) $\iint_R f(x, y)\,dA = \int_\square^\square \int_\square^\square f(x, y)\,dy\,dx$

(b) $\iint_R f(x, y)\,dA = \int_\square^\square \int_\square^\square f(x, y)\,dx\,dy$

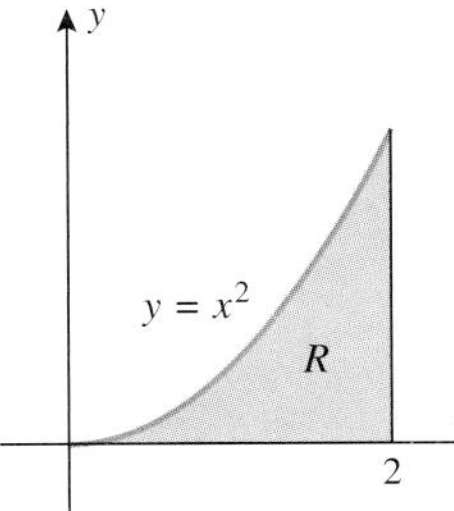

Figure Ex-11

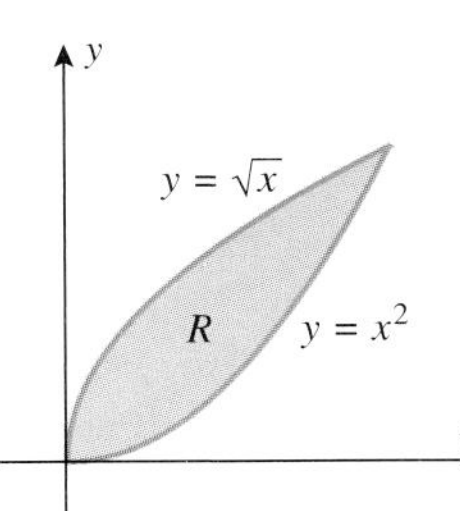

Figure Ex-12

13. Let R be the region shown in the accompanying figure. Fill in the missing limits of integration.

(a) $\iint_R f(x, y)\,dA = \int_1^2 \int_\square^\square f(x, y)\,dy\,dx + \int_2^4 \int_\square^\square f(x, y)\,dy\,dx + \int_4^5 \int_\square^\square f(x, y)\,dy\,dx$

(b) $\iint_R f(x, y)\,dA = \int_\square^\square \int_\square^\square f(x, y)\,dx\,dy$

14. Let R be the region shown in the accompanying figure. Fill in the missing limits of integration.

(a) $\iint_R f(x, y)\,dA = \int_\square^\square \int_\square^\square f(x, y)\,dy\,dx$

(b) $\iint_R f(x, y)\,dA = \int_\square^\square \int_\square^\square f(x, y)\,dx\,dy$

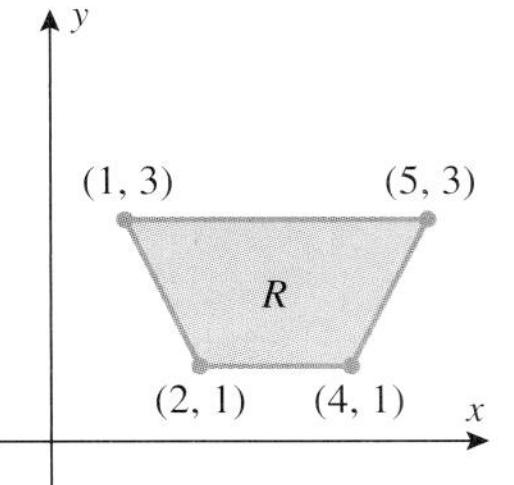

Figure Ex-13

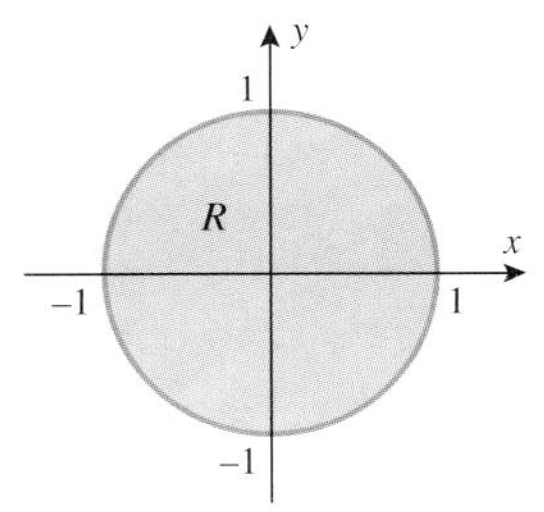

Figure Ex-14

15. Evaluate $\iint_R xy\,dA$, where R is the region in

(a) Exercise 11 (b) Exercise 13.

16. Evaluate $\iint_R (x + y)\,dA$, where R is the region in

(a) Exercise 12 (b) Exercise 14.

17–20 Evaluate the double integral in two ways using iterated integrals: (a) viewing R as a type I region, and (b) viewing R as a type II region.

17. $\iint_R x^2\,dA$; R is the region bounded by $y = 16/x$, $y = x$, and $x = 8$.

18. $\iint_R xy^2\,dA$; R is the region enclosed by $y = 1$, $y = 2$, $x = 0$, and $y = x$.

19. $\iint_R (3x - 2y)\,dA$; R is the region enclosed by the circle $x^2 + y^2 = 1$.

20. $\iint_R y\,dA$; R is the region in the first quadrant enclosed between the circle $x^2 + y^2 = 25$ and the line $x + y = 5$.

21–26 Evaluate the double integral.

21. $\iint_R x(1 + y^2)^{-1/2}\,dA$; R is the region in the first quadrant enclosed by $y = x^2$, $y = 4$, and $x = 0$.

22. $\iint_R x \cos y \, dA$; R is the triangular region bounded by the lines $y = x$, $y = 0$, and $x = \pi$.

23. $\iint_R xy \, dA$; R is the region enclosed by $y = \sqrt{x}$, $y = 6 - x$, and $y = 0$.

24. $\iint_R x \, dA$; R is the region enclosed by $y = \sin^{-1} x$, $x = 1/\sqrt{2}$, and $y = 0$.

25. $\iint_R (x - 1) \, dA$; R is the region in the first quadrant enclosed between $y = x$ and $y = x^3$.

26. $\iint_R x^2 \, dA$; R is the region in the first quadrant enclosed by $xy = 1$, $y = x$, and $y = 2x$.

27. (a) By hand or with the help of a graphing utility, make a sketch of the region R enclosed between the curves $y = x + 2$ and $y = e^x$.
(b) Estimate the intersections of the curves in part (a).
(c) Viewing R as a type I region, estimate $\iint_R x \, dA$.
(d) Viewing R as a type II region, estimate $\iint_R x \, dA$.

28. (a) By hand or with the help of a graphing utility, make a sketch of the region R enclosed between the curves $y = 4x^3 - x^4$ and $y = 3 - 4x + 4x^2$.
(b) Find the intersections of the curves in part (a).
(c) Find $\iint_R x \, dA$.

29–32 Use double integration to find the area of the plane region enclosed by the given curves.

29. $y = \sin x$ and $y = \cos x$, for $0 \le x \le \pi/4$.

30. $y^2 = -x$ and $3y - x = 4$.

31. $y^2 = 9 - x$ and $y^2 = 9 - 9x$.

32. $y = \cosh x$, $y = \sinh x$, $x = 0$, and $x = 1$.

33–34 Use double integration to find the volume of the solid.

33.

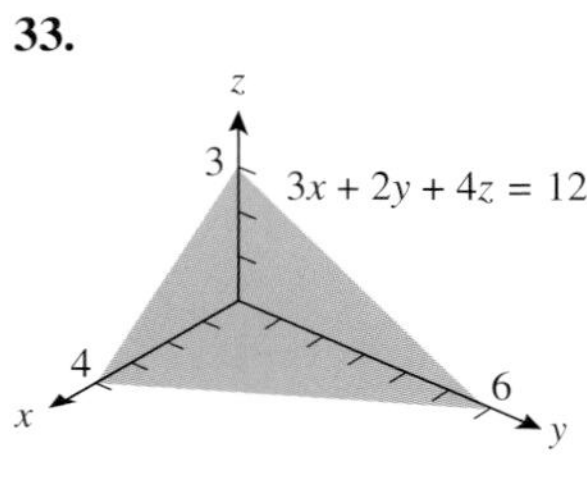

34.

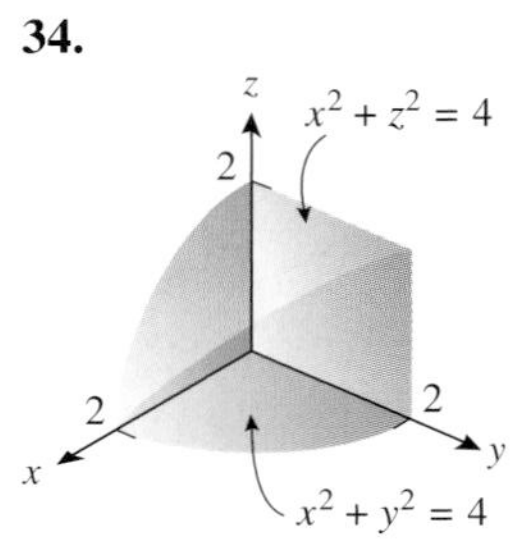

35–42 Use double integration to find the volume of each solid.

35. The solid bounded by the cylinder $x^2 + y^2 = 9$ and the planes $z = 0$ and $z = 3 - x$.

36. The solid in the first octant bounded above by the paraboloid $z = x^2 + 3y^2$, below by the plane $z = 0$, and laterally by $y = x^2$ and $y = x$.

37. The solid bounded above by the paraboloid $z = 9x^2 + y^2$, below by the plane $z = 0$, and laterally by the planes $x = 0$, $y = 0$, $x = 3$, and $y = 2$.

38. The solid enclosed by $y^2 = x$, $z = 0$, and $x + z = 1$.

39. The wedge cut from the cylinder $4x^2 + y^2 = 9$ by the planes $z = 0$ and $z = y + 3$.

40. The solid in the first octant bounded above by $z = 9 - x^2$, below by $z = 0$, and laterally by $y^2 = 3x$.

41. The solid that is common to the cylinders $x^2 + y^2 = 25$ and $x^2 + z^2 = 25$.

42. The solid bounded above by the paraboloid $z = x^2 + y^2$, below by the xy-plane, and laterally by the circular cylinder $x^2 + (y - 1)^2 = 1$.

43–44 Use a double integral and a CAS to find the volume of the solid.

C 43. The solid bounded above by the paraboloid $z = 1 - x^2 - y^2$ and below by the xy-plane.

C 44. The solid in the first octant that is bounded by the paraboloid $z = x^2 + y^2$, the cylinder $x^2 + y^2 = 4$ and the coordinate planes.

45–50 Express the integral as an equivalent integral with the order of integration reversed.

45. $\displaystyle\int_0^2 \int_0^{\sqrt{x}} f(x, y) \, dy \, dx$

46. $\displaystyle\int_0^4 \int_{2y}^{8} f(x, y) \, dx \, dy$

47. $\displaystyle\int_0^2 \int_1^{e^y} f(x, y) \, dx \, dy$

48. $\displaystyle\int_1^e \int_0^{\ln x} f(x, y) \, dy \, dx$

49. $\displaystyle\int_0^1 \int_{\sin^{-1} y}^{\pi/2} f(x, y) \, dx \, dy$

50. $\displaystyle\int_0^1 \int_{y^2}^{\sqrt{y}} f(x, y) \, dx \, dy$

51–54 Evaluate the integral by first reversing the order of integration.

51. $\displaystyle\int_0^1 \int_{4x}^{4} e^{-y^2} \, dy \, dx$

52. $\displaystyle\int_0^2 \int_{y/2}^{1} \cos(x^2) \, dx \, dy$

53. $\displaystyle\int_0^4 \int_{\sqrt{y}}^{2} e^{x^3} \, dx \, dy$

54. $\displaystyle\int_1^3 \int_0^{\ln x} x \, dy \, dx$

55. Evaluate $\iint_R \sin(y^3) \, dA$, where R is the region bounded by $y = \sqrt{x}$, $y = 2$, and $x = 0$. [*Hint:* Choose the order of integration carefully.]

56. Evaluate $\iint_R x\,dA$, where R is the region bounded by $x = \ln y$, $x = 0$, and $y = e$.

C **57.** Try to evaluate the integral with a CAS using the stated order of integration, and then by reversing the order of integration.

(a) $\int_0^4 \int_{\sqrt{x}}^2 \sin \pi y^3\,dy\,dx$

(b) $\int_0^1 \int_{\sin^{-1} y}^{\pi/2} \sec^2(\cos x)\,dx\,dy$

58. Use the appropriate Wallis formula (see Exercise Set 8.3) to find the volume of the solid enclosed between the circular paraboloid $z = x^2 + y^2$, the right circular cylinder $x^2 + y^2 = 4$, and the xy-plane (see the accompanying figure for cut view).

59. Evaluate $\iint_R xy^2\,dA$ over the region R shown in the accompanying figure.

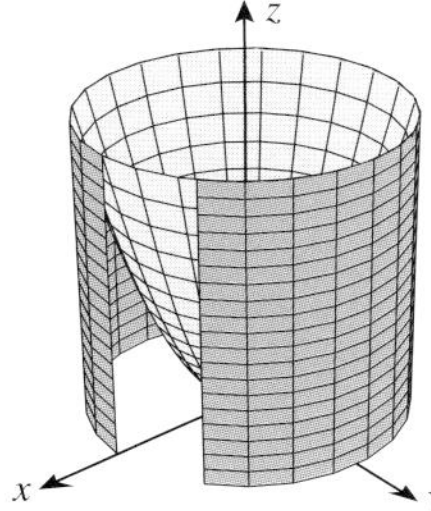

Figure Ex-58

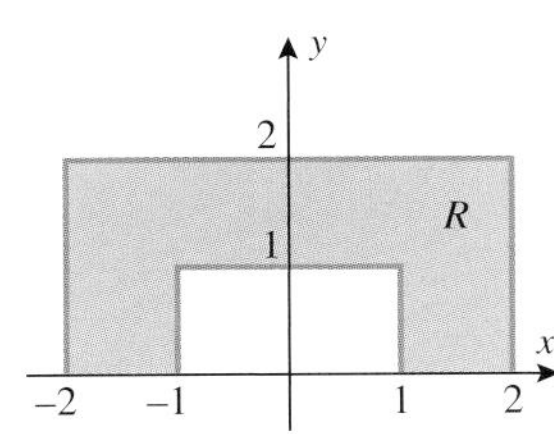

Figure Ex-59

60. Give a geometric argument to show that

$$\int_0^1 \int_0^{\sqrt{1-y^2}} \sqrt{1 - x^2 - y^2}\,dx\,dy = \frac{\pi}{6}$$

61–62 The ***average value*** or ***mean value*** of a continuous function $f(x, y)$ over a region R in the xy-plane is defined as

$$f_{\text{ave}} = \frac{1}{A(R)} \iint_R f(x, y)\,dA$$

where $A(R)$ is the area of the region R (compare to the definition preceding Exercise 27 in Section 15.1). Use this definition in these exercises.

61. Find the average value of $1/(1 + x^2)$ over the triangular region with vertices $(0, 0)$, $(1, 1)$, and $(0, 1)$.

62. Find the average value of $f(x, y) = x^2 - xy$ over the region enclosed by $y = x$ and $y = 3x - x^2$.

63. Suppose that the temperature in degrees Celsius at a point (x, y) on a flat metal plate is $T(x, y) = 5xy + x^2$, where x and y are in meters. Find the average temperature of the diamond-shaped portion of the plate for which $|2x + y| \leq 4$ and $|2x - y| \leq 4$.

64. A circular lens of radius 2 inches has thickness $1 - (r^2/4)$ inches at all points r inches from the center of the lens. Find the average thickness of the lens.

C **65.** Use a CAS to approximate the intersections of the curves $y = \sin x$ and $y = x/2$, and then approximate the volume of the solid in the first octant that is below the surface $z = \sqrt{1 + x + y}$ and above the region in the xy-plane that is enclosed by the curves.

✔QUICK CHECK ANSWERS 15.2

1. (a) $\int_1^5 \left(\frac{1}{4}y^4 - 16y\right) dy$ (b) $\int_1^5 \left(\frac{3}{4}x^4 - 12x^2\right) dx$ **2.** (a) $\int_0^3 \int_0^{-\frac{4}{3}x+4} f(x, y)\,dy\,dx$ (b) $\int_0^4 \int_0^{-\frac{3}{4}y+3} f(x, y)\,dx\,dy$

3. $\int_0^3 \int_x^{-\frac{1}{3}x+4} dy\,dx$ **4.** $\int_{-2}^1 \int_{x^2}^{2-x} (1 + 2y)\,dy\,dx = 18.9$

15.3 DOUBLE INTEGRALS IN POLAR COORDINATES

In this section we will study double integrals in which the integrand and the region of integration are expressed in polar coordinates. Such integrals are important for two reasons: first, they arise naturally in many applications, and second, many double integrals in rectangular coordinates can be evaluated more easily if they are converted to polar coordinates.

SIMPLE POLAR REGIONS

Some double integrals are easier to evaluate if the region of integration is expressed in polar coordinates. This is usually true if the region is bounded by a cardioid, a rose curve, a spiral, or, more generally, by any curve whose equation is simpler in polar coordinates than in rectangular coordinates. Moreover, double integrals whose integrands involve $x^2 + y^2$ also tend to be easier to evaluate in polar coordinates because this sum simplifies to r^2 when the conversion formulas $x = r\cos\theta$ and $y = r\sin\theta$ are applied.

An overview of polar coordinates can be found in Section 11.1.

Figure 15.3.1a shows a region R in a polar coordinate system that is enclosed between two rays, $\theta = \alpha$ and $\theta = \beta$, and two polar curves, $r = r_1(\theta)$ and $r = r_2(\theta)$. If, as shown in the figure, the functions $r_1(\theta)$ and $r_2(\theta)$ are continuous and their graphs do not cross, then the region R is called a *simple polar region*. If $r_1(\theta)$ is identically zero, then the boundary $r = r_1(\theta)$ reduces to a point (the origin), and the region has the general shape shown in Figure 15.3.1b. If, in addition, $\beta = \alpha + 2\pi$, then the rays coincide, and the region has the general shape shown in Figure 15.3.1c. The following definition expresses these geometric ideas algebraically.

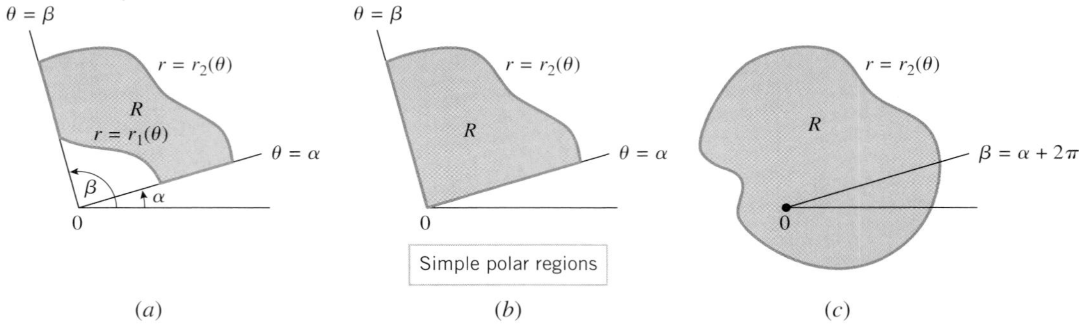

Figure 15.3.1

> **15.3.1 DEFINITION.** A ***simple polar region*** in a polar coordinate system is a region that is enclosed between two rays, $\theta = \alpha$ and $\theta = \beta$, and two continuous polar curves, $r = r_1(\theta)$ and $r = r_2(\theta)$, where the equations of the rays and the polar curves satisfy the following conditions:
>
> (i) $\alpha \le \beta$ (ii) $\beta - \alpha \le 2\pi$ (iii) $0 \le r_1(\theta) \le r_2(\theta)$

Conditions (i) and (ii) together imply that the ray $\theta = \beta$ can be obtained by rotating the ray $\theta = \alpha$ counterclockwise through an angle that is at most 2π radians. This is consistent with Figure 15.3.1. Condition (iii) implies that the boundary curves $r = r_1(\theta)$ and $r = r_2(\theta)$ can touch but cannot actually cross over one another (why?). Thus, in keeping with Figure 15.3.1, it is appropriate to describe $r = r_1(\theta)$ as the ***inner boundary*** of the region and $r = r_2(\theta)$ as the ***outer boundary***.

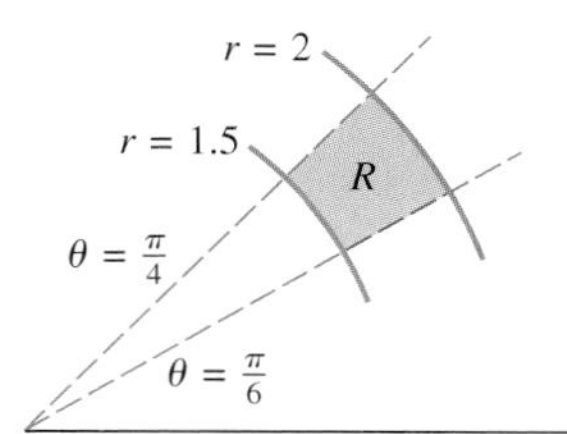

Figure 15.3.2

A ***polar rectangle*** is a simple polar region for which the bounding polar curves are circular arcs. For example, Figure 15.3.2 shows the polar rectangle R given by

$$1.5 \le r \le 2, \quad \frac{\pi}{6} \le \theta \le \frac{\pi}{4}$$

DOUBLE INTEGRALS IN POLAR COORDINATES

Next we will consider the polar version of Problem 15.1.1.

> **15.3.2 THE VOLUME PROBLEM IN POLAR COORDINATES.** Given a function $f(r, \theta)$ that is continuous and nonnegative on a simple polar region R, find the volume of the solid that is enclosed between the region R and the surface whose equation in cylindrical coordinates is $z = f(r, \theta)$ (Figure 15.3.3).

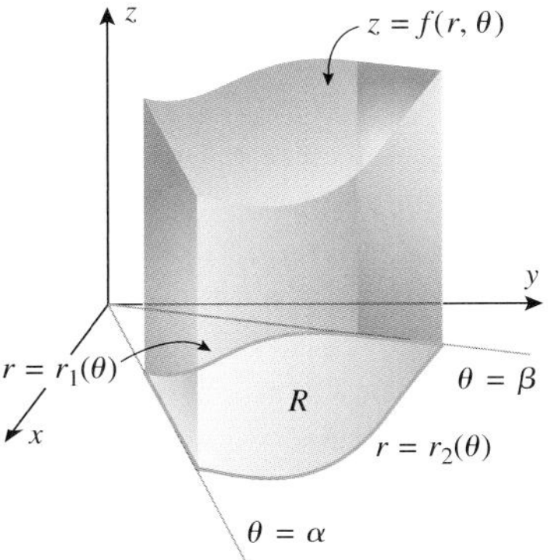

Figure 15.3.3

To motivate a formula for the volume V of the solid in Figure 15.3.3, we will use a limit process similar to that used to obtain Formula (2) of Section 15.1, except that here we will use circular arcs and rays to subdivide the region R into polar rectangles. As shown in Figure 15.3.4, we will exclude from consideration all polar rectangles that contain any points outside of R, leaving only polar rectangles that are subsets of R. Assume that there are n such polar rectangles, and denote the area of the kth polar rectangle by ΔA_k. Let (r_k^*, θ_k^*) be any point in this polar rectangle. As shown in Figure 15.3.5, the product $f(r_k^*, \theta_k^*)\Delta A_k$ is the volume of a solid with base area ΔA_k and height $f(r_k^*, \theta_k^*)$, so the sum

$$\sum_{k=1}^{n} f(r_k^*, \theta_k^*)\Delta A_k$$

can be viewed as an approximation to the volume V of the entire solid.

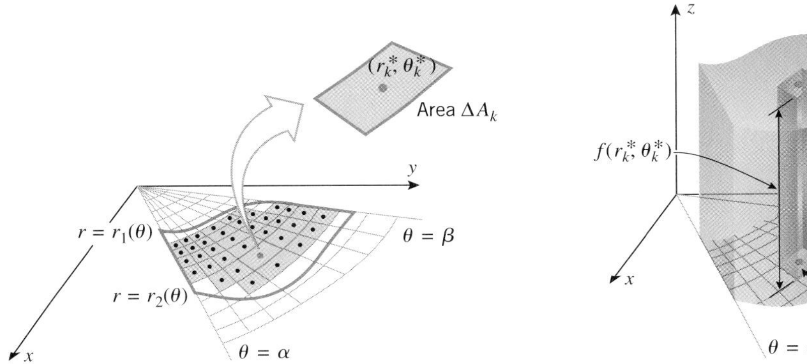

Figure 15.3.4

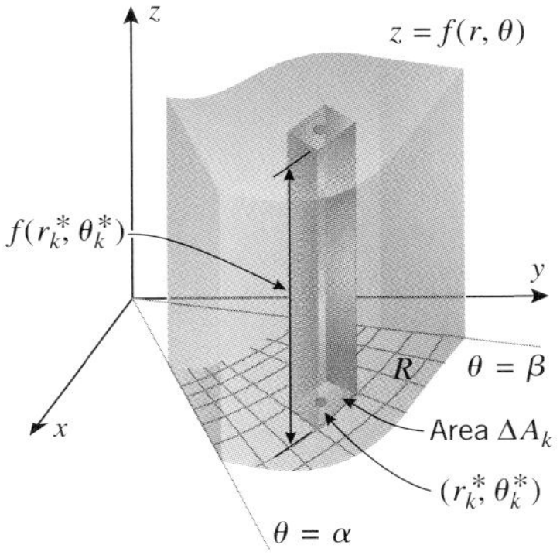

Figure 15.3.5

If we now increase the number of subdivisions in such a way that the dimensions of the polar rectangles approach zero, then it seems plausible that the errors in the approximations approach zero, and the exact volume of the solid is

$$V = \lim_{n \to +\infty} \sum_{k=1}^{n} f(r_k^*, \theta_k^*)\Delta A_k \tag{1}$$

If $f(r, \theta)$ is continuous on R and has both positive and negative values, then the limit

$$\lim_{n \to +\infty} \sum_{k=1}^{n} f(r_k^*, \theta_k^*)\Delta A_k \tag{2}$$

represents the net signed volume between the region R and the surface $z = f(r, \theta)$ (as with double integrals in rectangular coordinates). The sums in (2) are called ***polar Riemann sums***, and the limit of the polar Riemann sums is denoted by

$$\iint_R f(r, \theta)\, dA = \lim_{n \to +\infty} \sum_{k=1}^{n} f(r_k^*, \theta_k^*)\Delta A_k \tag{3}$$

Polar double integrals are also called ***double integrals in polar coordinates*** to distinguish them from double integrals over regions in the xy-plane; the latter are called ***double integrals in rectangular coordinates***. Double integrals in polar coordinates have the usual integral properties, such as those stated in Formulas (9), (10), and (11) of Section 15.1.

which is called the ***polar double integral*** of $f(r, \theta)$ over R. If $f(r, \theta)$ is continuous and nonnegative on R, then the volume formula (1) can be expressed as

$$V = \iint_R f(r, \theta)\, dA \tag{4}$$

EVALUATING POLAR DOUBLE INTEGRALS

In Sections 15.1 and 15.2 we evaluated double integrals in rectangular coordinates by expressing them as iterated integrals. Polar double integrals are evaluated the same way. To motivate the formula that expresses a double polar integral as an iterated integral, we will assume that $f(r, \theta)$ is nonnegative so that we can interpret (3) as a volume. However, the results that we will obtain will also be applicable if f has negative values. To begin, let us choose the arbitrary point (r_k^*, θ_k^*) in (3) to be at the "center" of the kth polar rectangle as shown in Figure 15.3.6. Suppose also that this polar rectangle has a central angle $\Delta\theta_k$ and a "radial thickness" Δr_k. Thus, the inner radius of this polar rectangle is $r_k^* - \frac{1}{2}\Delta r_k$ and the outer radius is $r_k^* + \frac{1}{2}\Delta r_k$. Treating the area ΔA_k of this polar rectangle as the difference in area of two sectors, we obtain

$$\Delta A_k = \tfrac{1}{2}\left(r_k^* + \tfrac{1}{2}\Delta r_k\right)^2 \Delta\theta_k - \tfrac{1}{2}\left(r_k^* - \tfrac{1}{2}\Delta r_k\right)^2 \Delta\theta_k$$

which simplifies to

$$\Delta A_k = r_k^* \Delta r_k \Delta\theta_k \tag{5}$$

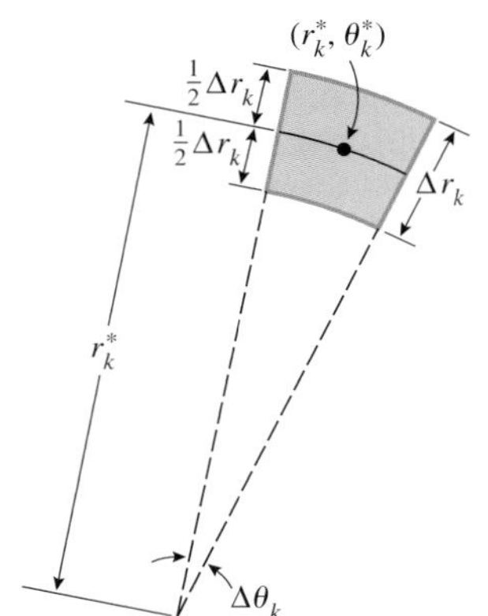

Figure 15.3.6

Thus, from (3) and (4)

$$V = \iint_R f(r, \theta)\, dA = \lim_{n \to +\infty} \sum_{k=1}^{n} f(r_k^*, \theta_k^*) r_k^* \Delta r_k \Delta\theta_k$$

which suggests that the volume V can be expressed as the iterated integral

$$V = \iint_R f(r, \theta)\, dA = \int_\alpha^\beta \int_{r_1(\theta)}^{r_2(\theta)} f(r, \theta) r\, dr\, d\theta \tag{6}$$

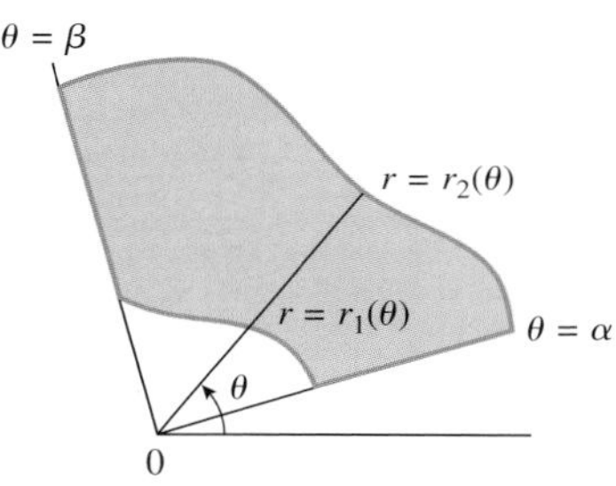

Figure 15.3.7

in which the limits of integration are chosen to cover the region R; that is, with θ fixed between α and β, the value of r varies from $r_1(\theta)$ to $r_2(\theta)$ (Figure 15.3.7).

Although we assumed $f(r, \theta)$ to be nonnegative in deriving Formula (6), it can be proved that the relationship between the polar double integral and the iterated integral in this formula also holds if f has negative values. Accepting this to be so, we obtain the following theorem, which we state without formal proof.

Note the extra factor of r that appears in the integrand when expressing a polar double integral as an iterated integral in polar coordinates.

15.3.3 THEOREM. *If R is a simple polar region whose boundaries are the rays $\theta = \alpha$ and $\theta = \beta$ and the curves $r = r_1(\theta)$ and $r = r_2(\theta)$ shown in Figure 15.3.7, and if $f(r, \theta)$ is continuous on R, then*

$$\iint_R f(r, \theta)\, dA = \int_\alpha^\beta \int_{r_1(\theta)}^{r_2(\theta)} f(r, \theta) r\, dr\, d\theta \tag{7}$$

To apply this theorem you will need to be able to find the rays and the curves that form the boundary of the region R, since these determine the limits of integration in the iterated integral. This can be done as follows:

Determining Limits of Integration for a Polar Double Integral: Simple Polar Region

Step 1. Since θ is held fixed for the first integration, draw a radial line from the origin through the region R at a fixed angle θ (Figure 15.3.8*a*). This line crosses the boundary of R at most twice. The innermost point of intersection is on the inner boundary curve $r = r_1(\theta)$ and the outermost point is on the outer boundary curve $r = r_2(\theta)$. These intersections determine the r-limits of integration in (7).

Step 2. Imagine rotating a ray along the polar x-axis one revolution counterclockwise about the origin. The smallest angle at which this ray intersects the region R is $\theta = \alpha$ and the largest angle is $\theta = \beta$ (Figure 15.3.8*b*). This determines the θ-limits of integration.

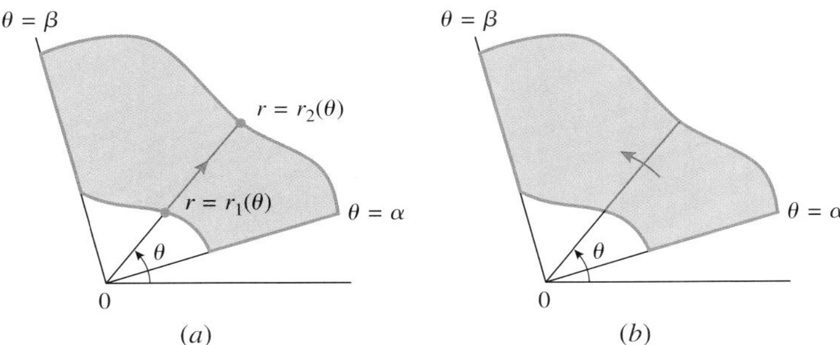

Figure 15.3.8

▶ Example 1 Evaluate

$$\iint_R \sin\theta \, dA$$

where R is the region in the first quadrant that is outside the circle $r = 2$ and inside the cardioid $r = 2(1 + \cos\theta)$.

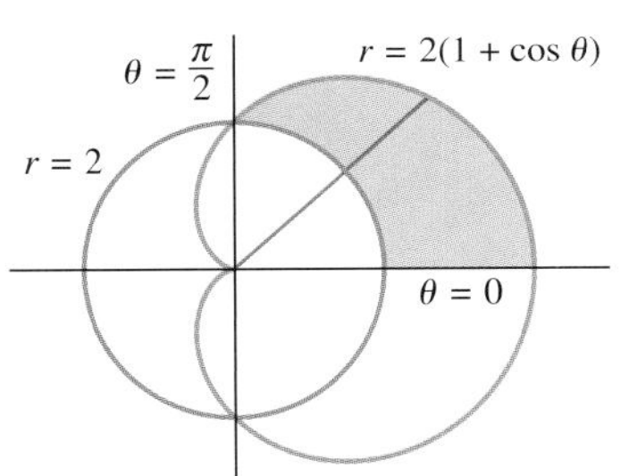

Figure 15.3.9

Solution. The region R is sketched in Figure 15.3.9. Following the two steps outlined above we obtain

$$\begin{aligned}
\iint_R \sin\theta \, dA &= \int_0^{\pi/2} \int_2^{2(1+\cos\theta)} (\sin\theta) r \, dr \, d\theta \\
&= \int_0^{\pi/2} \left[\frac{1}{2} r^2 \sin\theta\right]_{r=2}^{2(1+\cos\theta)} d\theta \\
&= 2 \int_0^{\pi/2} [(1+\cos\theta)^2 \sin\theta - \sin\theta] \, d\theta \\
&= 2\left[-\frac{1}{3}(1+\cos\theta)^3 + \cos\theta\right]_0^{\pi/2} \\
&= 2\left[-\frac{1}{3} - \left(-\frac{5}{3}\right)\right] = \frac{8}{3} \; ◀
\end{aligned}$$

▶ Example 2 The sphere of radius a centered at the origin is expressed in rectangular coordinates as $x^2 + y^2 + z^2 = a^2$, and hence its equation in cylindrical coordinates is $r^2 + z^2 = a^2$. Use this equation and a polar double integral to find the volume of the sphere.

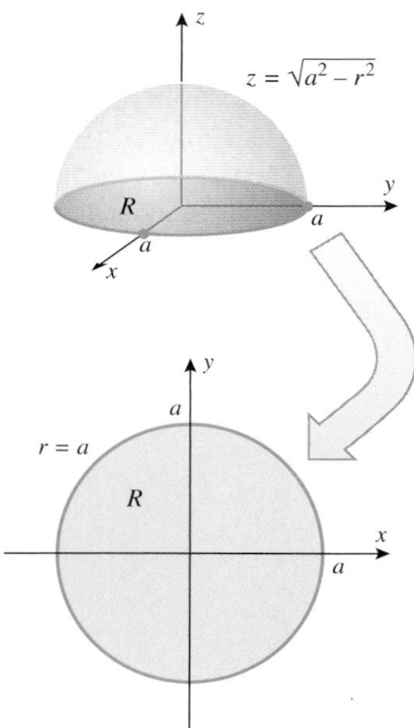

Figure 15.3.10

Solution. In cylindrical coordinates the upper hemisphere is given by the equation

$$z = \sqrt{a^2 - r^2}$$

so the volume enclosed by the entire sphere is

$$V = 2 \iint_R \sqrt{a^2 - r^2}\, dA$$

where R is the circular region shown in Figure 15.3.10. Thus,

$$V = 2 \iint_R \sqrt{a^2 - r^2}\, dA = \int_0^{2\pi} \int_0^a \sqrt{a^2 - r^2}(2r)\, dr\, d\theta$$

$$= \int_0^{2\pi} \left[-\frac{2}{3}(a^2 - r^2)^{3/2} \right]_{r=0}^{a} d\theta = \int_0^{2\pi} \frac{2}{3} a^3\, d\theta$$

$$= \left[\frac{2}{3} a^3 \theta \right]_0^{2\pi} = \frac{4}{3}\pi a^3 \blacktriangleleft$$

FINDING AREAS USING POLAR DOUBLE INTEGRALS

Recall from Formula (7) of Section 15.2 that the area of a region R in the xy-plane can be expressed as

$$\text{area of } R = \iint_R 1\, dA = \iint_R dA \tag{8}$$

The argument used to derive this result can also be used to show that the formula applies to polar double integrals over regions in polar coordinates.

▶ **Example 3** Use a polar double integral to find the area enclosed by the three-petaled rose $r = \sin 3\theta$.

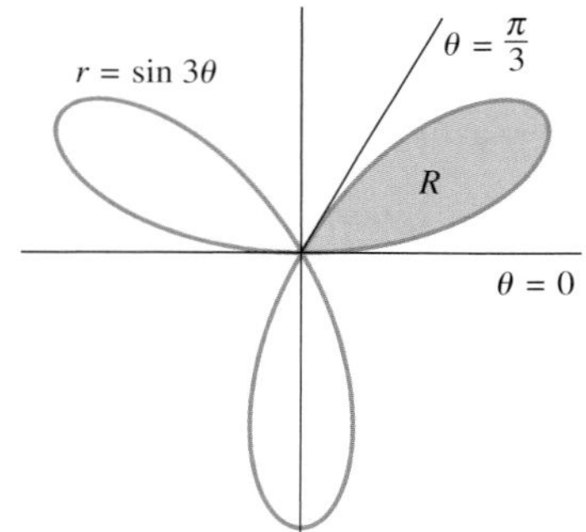

Figure 15.3.11

Solution. The rose is sketched in Figure 15.3.11. We will use Formula (8) to calculate the area of the petal R in the first quadrant and multiply by three.

$$A = 3 \iint_R dA = 3 \int_0^{\pi/3} \int_0^{\sin 3\theta} r\, dr\, d\theta$$

$$= \frac{3}{2} \int_0^{\pi/3} \sin^2 3\theta\, d\theta = \frac{3}{4} \int_0^{\pi/3} (1 - \cos 6\theta)\, d\theta$$

$$= \frac{3}{4} \left[\theta - \frac{\sin 6\theta}{6} \right]_0^{\pi/3} = \frac{1}{4}\pi \blacktriangleleft$$

CONVERTING DOUBLE INTEGRALS FROM RECTANGULAR TO POLAR COORDINATES

Sometimes a double integral that is difficult to evaluate in rectangular coordinates can be evaluated more easily in polar coordinates by making the substitution $x = r\cos\theta$, $y = r\sin\theta$ and expressing the region of integration in polar form; that is, we rewrite the double integral in rectangular coordinates as

$$\iint_R f(x, y)\, dA = \iint_R f(r\cos\theta, r\sin\theta)\, dA = \iint_{\substack{\text{appropriate}\\ \text{limits}}} f(r\cos\theta, r\sin\theta) r\, dr\, d\theta \tag{9}$$

▶ **Example 4** Use polar coordinates to evaluate $\displaystyle\int_{-1}^{1}\int_{0}^{\sqrt{1-x^2}} (x^2+y^2)^{3/2}\,dy\,dx$.

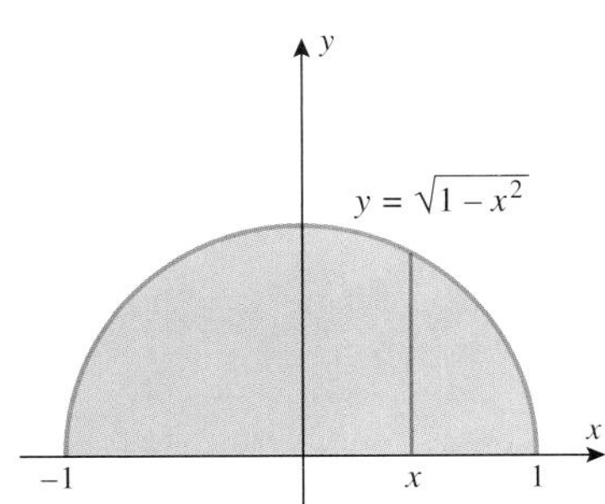

Figure 15.3.12

Solution. In this problem we are starting with an iterated integral in rectangular coordinates rather than a double integral, so before we can make the conversion to polar coordinates we will have to identify the region of integration. To do this, we observe that for fixed x the y-integration runs from $y = 0$ to $y = \sqrt{1-x^2}$, which tells us that the lower boundary of the region is the x-axis and the upper boundary is a semicircle of radius 1 centered at the origin. From the x-integration we see that x varies from -1 to 1, so we conclude that the region of integration is as shown in Figure 15.3.12. In polar coordinates, this is the region swept out as r varies between 0 and 1 and θ varies between 0 and π. Thus,

$$\int_{-1}^{1}\int_{0}^{\sqrt{1-x^2}} (x^2+y^2)^{3/2}\,dy\,dx = \iint_R (x^2+y^2)^{3/2}\,dA$$

$$= \int_0^{\pi}\int_0^1 (r^3)r\,dr\,d\theta = \int_0^{\pi} \frac{1}{5}\,d\theta = \frac{\pi}{5} \quad \blacktriangleleft$$

> The conversion to polar coordinates worked so nicely in Example 4 because the substitution $x = r\cos\theta$, $y = r\sin\theta$ collapsed the sum x^2+y^2 into the single term r^2, thereby simplifying the integrand. Whenever you see an expression involving x^2+y^2 in the integrand, you should consider the possibility of converting to polar coordinates.

✔ QUICK CHECK EXERCISES 15.3 *(See page 1043 for answers.)*

1. The polar region inside the circle $r = 2\sin\theta$ and outside the circle $r = 1$ is a simple polar region given by the inequalities

   ______ $\le r \le$ ______, ______ $\le \theta \le$ ______

2. Let R be the region in the first quadrant enclosed between the circles $x^2+y^2=9$ and $x^2+y^2=100$. Supply the missing limits of integration.

   $$\iint_R f(r,\theta)\,dA = \int_{\square}^{\square}\int_{\square}^{\square} f(r,\theta)r\,dr\,d\theta$$

3. Let V be the volume of the solid bounded above by the hemisphere $z = \sqrt{1-r^2}$ and bounded below by the disk enclosed within the circle $r = \sin\theta$. Expressed as a double integral in polar coordinates, $V =$ ______.

4. Express the iterated integral as a double integral in polar coordinates.

   $$\int_{1/\sqrt{2}}^{1}\int_{\sqrt{1-x^2}}^{x} \left(\frac{1}{x^2+y^2}\right) dy\,dx = \underline{\qquad}$$

EXERCISE SET 15.3 [C] CAS

1–6 Evaluate the iterated integral.

1. $\displaystyle\int_0^{\pi/2}\int_0^{\sin\theta} r\cos\theta\,dr\,d\theta$

2. $\displaystyle\int_0^{\pi}\int_0^{1+\cos\theta} r\,dr\,d\theta$

3. $\displaystyle\int_0^{\pi/2}\int_0^{a\sin\theta} r^2\,dr\,d\theta$

4. $\displaystyle\int_0^{\pi/6}\int_0^{\cos 3\theta} r\,dr\,d\theta$

5. $\displaystyle\int_0^{\pi}\int_0^{1-\sin\theta} r^2\cos\theta\,dr\,d\theta$

6. $\displaystyle\int_0^{\pi/2}\int_0^{\cos\theta} r^3\,dr\,d\theta$

7–10 Use a double integral in polar coordinates to find the area of the region described.

7. The region enclosed by the cardioid $r = 1-\cos\theta$.

8. The region enclosed by the rose $r = \sin 2\theta$.

9. The region in the first quadrant bounded by $r = 1$ and $r = \sin 2\theta$, with $\pi/4 \le \theta \le \pi/2$.

10. The region inside the circle $x^2+y^2=4$ and to the right of the line $x = 1$.

FOCUS ON CONCEPTS

11–12 Let R be the region described. Sketch the region R and fill in the missing limits of integration.

$$\iint_R f(r,\theta)\,dA = \int_{\square}^{\square}\int_{\square}^{\square} f(r,\theta)r\,dr\,d\theta$$

11. The region inside the circle $r = 4\sin\theta$ and outside the circle $r = 2$.

12. The region inside the circle $r = 1$ and outside the cardioid $r = 1 + \cos\theta$.

13–16 Express the volume of the solid described as a double integral in polar coordinates.

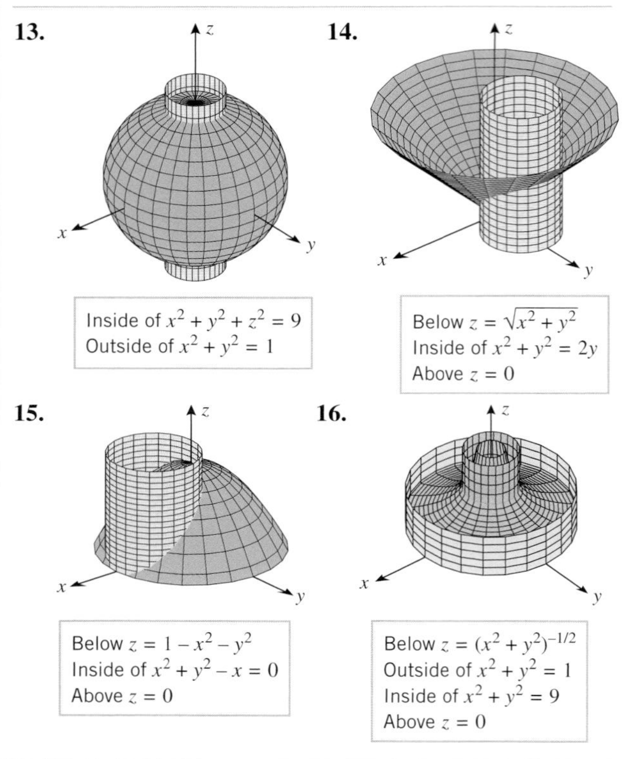

13.
14.
Inside of $x^2 + y^2 + z^2 = 9$
Outside of $x^2 + y^2 = 1$
Below $z = \sqrt{x^2 + y^2}$
Inside of $x^2 + y^2 = 2y$
Above $z = 0$
15.
16.
Below $z = 1 - x^2 - y^2$
Inside of $x^2 + y^2 - x = 0$
Above $z = 0$
Below $z = (x^2 + y^2)^{-1/2}$
Outside of $x^2 + y^2 = 1$
Inside of $x^2 + y^2 = 9$
Above $z = 0$

17. Find the volume of the solid described in Exercise 13.

18. Find the volume of the solid described in Exercise 14.

19. Find the volume of the solid described in Exercise 15.

20. Find the volume of the solid described in Exercise 16.

21. Find the volume of the solid in the first octant bounded above by the surface $z = r\sin\theta$, below by the xy-plane, and laterally by the plane $x = 0$ and the surface $r = 3\sin\theta$.

22. Find the volume of the solid inside the surface $r^2 + z^2 = 4$ and outside the surface $r = 2\cos\theta$.

23–26 Use polar coordinates to evaluate the double integral.

23. $\displaystyle\iint_R e^{-(x^2+y^2)}\,dA$, where R is the region enclosed by the circle $x^2 + y^2 = 1$.

24. $\displaystyle\iint_R \sqrt{9 - x^2 - y^2}\,dA$, where R is the region in the first quadrant within the circle $x^2 + y^2 = 9$.

25. $\displaystyle\iint_R \frac{1}{1 + x^2 + y^2}\,dA$, where R is the sector in the first quadrant bounded by $y = 0$, $y = x$, and $x^2 + y^2 = 4$.

26. $\displaystyle\iint_R 2y\,dA$, where R is the region in the first quadrant bounded above by the circle $(x - 1)^2 + y^2 = 1$ and below by the line $y = x$.

27–34 Evaluate the iterated integral by converting to polar coordinates.

27. $\displaystyle\int_0^1\int_0^{\sqrt{1-x^2}} (x^2 + y^2)\,dy\,dx$

28. $\displaystyle\int_{-2}^2\int_{-\sqrt{4-y^2}}^{\sqrt{4-y^2}} e^{-(x^2+y^2)}\,dx\,dy$

29. $\displaystyle\int_0^2\int_0^{\sqrt{2x-x^2}} \sqrt{x^2 + y^2}\,dy\,dx$

30. $\displaystyle\int_0^1\int_0^{\sqrt{1-y^2}} \cos(x^2 + y^2)\,dx\,dy$

31. $\displaystyle\int_0^a\int_0^{\sqrt{a^2-x^2}} \frac{dy\,dx}{(1 + x^2 + y^2)^{3/2}} \quad (a > 0)$

32. $\displaystyle\int_0^1\int_y^{\sqrt{y}} \sqrt{x^2 + y^2}\,dx\,dy$

33. $\displaystyle\int_0^{\sqrt{2}}\int_y^{\sqrt{4-y^2}} \frac{1}{\sqrt{1 + x^2 + y^2}}\,dx\,dy$

34. $\displaystyle\int_0^4\int_3^{\sqrt{25-x^2}} dy\,dx$

35. Use a double integral in polar coordinates to find the volume of a cylinder of radius a and height h.

36. (a) Use a double integral in polar coordinates to find the volume of the oblate spheroid

$$\frac{x^2}{a^2} + \frac{y^2}{a^2} + \frac{z^2}{c^2} = 1 \quad (0 < c < a)$$

(b) Use the result in part (a) and the World Geodetic System of 1984 (WGS-84) discussed in Exercise 50 of Section 12.7 to find the volume of the Earth in cubic meters.

37. Use polar coordinates to find the volume of the solid that is above the xy-plane, inside the cylinder $x^2 + y^2 - ay = 0$, and inside the ellipsoid

$$\frac{x^2}{a^2} + \frac{y^2}{a^2} + \frac{z^2}{c^2} = 1$$

38. Find the area of the region enclosed by the lemniscate $r^2 = 2a^2\cos 2\theta$.

39. Find the area in the first quadrant that is inside the circle $r = 4\sin\theta$ and outside the lemniscate $r^2 = 8\cos 2\theta$.

40. Show that the shaded area in the accompanying figure is $a^2\phi - \frac{1}{2}a^2 \sin 2\phi$.

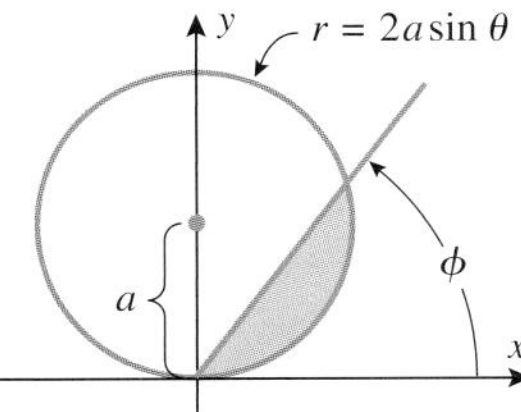

Figure Ex-40

41. The integral $\int_0^{+\infty} e^{-x^2}\,dx$, which arises in probability theory, can be evaluated using the following method. Let the value of the integral be I. Thus,

$$I = \int_0^{+\infty} e^{-x^2}\,dx = \int_0^{+\infty} e^{-y^2}\,dy$$

since the letter used for the variable of integration in a definite integral does not matter.

(a) Give a reasonable argument to show that

$$I^2 = \int_0^{+\infty}\int_0^{+\infty} e^{-(x^2+y^2)}\,dx\,dy$$

(b) Evaluate the iterated integral in part (a) by converting to polar coordinates.

(c) Use the result in part (b) to show that $I = \sqrt{\pi}/2$.

42. Show that

$$\int_0^{+\infty}\int_0^{+\infty} \frac{1}{(1+x^2+y^2)^2}\,dx\,dy = \frac{\pi}{4}$$

[*Hint:* See Exercise 41.]

C **43.** (a) Use the numerical integration capability of a CAS to approximate the value of the double integral

$$\int_{-1}^{1}\int_0^{\sqrt{1-x^2}} e^{-(x^2+y^2)^2}\,dy\,dx$$

(b) Compare the approximation obtained in part (a) to the approximation that results if the integral is first converted to polar coordinates.

44. Suppose that a geyser, centered at the origin of a polar coordinate system, sprays water in a circular pattern in such a way that the depth D of water that reaches a point at a distance of r feet from the origin in 1 hour is $D = ke^{-r}$. Find the total volume of water that the geyser sprays inside a circle of radius R centered at the origin.

45. Evaluate $\iint_R x^2\,dA$ over the region R shown in the accompanying figure.

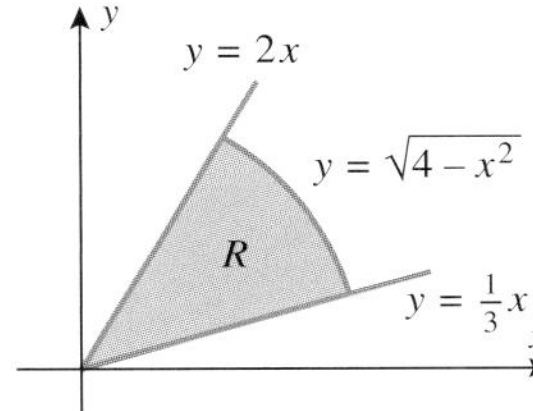

Figure Ex-45

✔ QUICK CHECK ANSWERS 15.3

1. $1 \le r \le 2\sin\theta,\ \pi/6 \le \theta \le 5\pi/6$ 2. $\int_0^{\pi/2}\int_3^{10} f(r,\theta)r\,dr\,d\theta$ 3. $\int_0^{\pi}\int_0^{\sin\theta} r\sqrt{1-r^2}\,dr\,d\theta$ 4. $\int_0^{\pi/4}\int_1^{\sec\theta} \frac{1}{r}\,dr\,d\theta$

15.4 PARAMETRIC SURFACES; SURFACE AREA

In previous sections we considered parametric curves in 2-space and 3-space. In this section we will discuss parametric surfaces in 3-space. As we will see, parametric representations of surfaces are not only important in computer graphics but also allow us to study more general kinds of surfaces than those encountered so far. In Section 6.5 we showed how to find the surface area of a surface of revolution. Our work on parametric surfaces will enable us to derive area formulas for more general kinds of surfaces.

PARAMETRIC REPRESENTATION OF SURFACES

We have seen that curves in 3-space can be represented by three equations involving one parameter, say

$$x = x(t), \quad y = y(t), \quad z = z(t)$$

Surfaces in 3-space can be represented parametrically by three equations involving two parameters, say

$$x = x(u, v), \quad y = y(u, v), \quad z = z(u, v) \tag{1}$$

To visualize why such equations represent a surface, think of (u, v) as a point that varies over some region in a uv-plane. If u is held constant, then v is the only varying parameter in (1), and hence these equations represent a curve in 3-space. We call this a ***constant u-curve*** (Figure 15.4.1). Similarly, if v is held constant, then u is the only varying parameter in (1), so again these equations represent a curve in 3-space. We call this a ***constant v-curve***. By varying the constants we generate a family of u-curves and a family of v-curves that together form a surface.

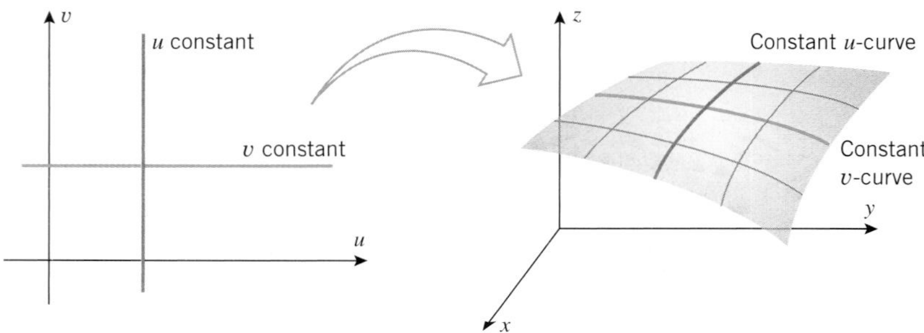

Figure 15.4.1

▶ **Example 1** Consider the paraboloid $z = 4 - x^2 - y^2$. One way to parametrize this surface is to take $x = u$ and $y = v$ as the parameters, in which case the surface is represented by the parametric equations

$$x = u, \quad y = v, \quad z = 4 - u^2 - v^2 \tag{2}$$

Figure 15.4.2a shows a computer-generated graph of this surface. The constant u-curves correspond to constant x-values and hence appear on the surface as traces parallel to the yz-plane. Similarly, the constant v-curves correspond to constant y-values and hence appear on the surface as traces parallel to the xz-plane. ◀

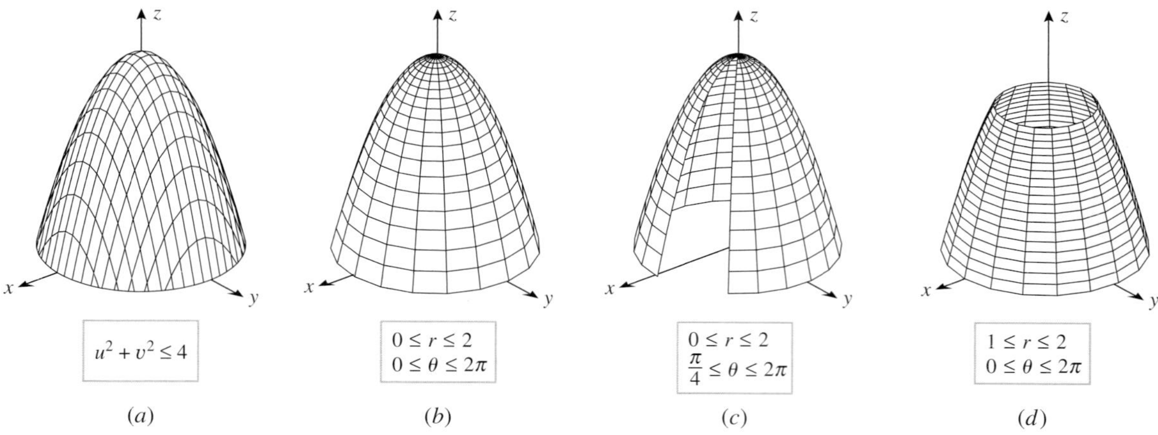

Figure 15.4.2

▶ **Example 2** The paraboloid $z = 4 - x^2 - y^2$ that was considered in Example 1 can also be parametrized by first expressing the equation in cylindrical coordinates. For this

TECHNOLOGY MASTERY

If you have a graphing utility that can generate parametric surfaces, consult the relevant documentation and then try to generate the surfaces in Figure 15.4.2.

purpose, we make the substitution $x = r\cos\theta$, $y = r\sin\theta$, which yields $z = 4 - r^2$. Thus, the paraboloid can be represented parametrically in terms of r and θ as

$$x = r\cos\theta, \quad y = r\sin\theta, \quad z = 4 - r^2 \tag{3}$$

A computer-generated graph of this surface for $0 \le r \le 2$ and $0 \le \theta \le 2\pi$ is shown in Figure 15.4.2*b*. The constant r-curves correspond to constant z-values and hence appear on the surface as traces parallel to the xy-plane. The constant θ-curves appear on the surface as traces from vertical planes through the origin at varying angles with the x-axis. Parts (*c*) and (*d*) of Figure 15.4.2 show the effect of restrictions on the parameters r and θ. ◀

▶ **Example 3** One way to generate the sphere $x^2 + y^2 + z^2 = 1$ with a graphing utility is to graph the upper and lower hemispheres

$$z = \sqrt{1 - x^2 - y^2} \quad \text{and} \quad z = -\sqrt{1 - x^2 - y^2}$$

on the same screen. However, this usually produces a fragmented sphere (Figure 15.4.3*a*) because roundoff error sporadically produces negative values inside the radical when $1 - x^2 - y^2$ is near zero. A better graph can be generated by first expressing the sphere in spherical coordinates as $\rho = 1$ and then using the spherical-to-rectangular conversion formulas in Table 12.8.1 to obtain the parametric equations

$$x = \sin\phi\cos\theta, \quad y = \sin\phi\sin\theta, \quad z = \cos\phi$$

with parameters θ and ϕ. Figure 15.4.3*b* shows the graph of this parametric surface for $0 \le \theta \le 2\pi$ and $0 \le \phi \le \pi$. In the language of cartographers, the constant ϕ-curves are the ***lines of latitude*** and the constant θ-curves are the ***lines of longitude***. ◀

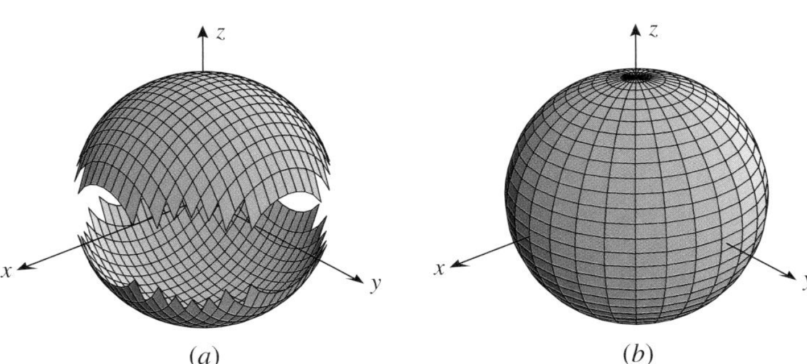

Figure 15.4.3

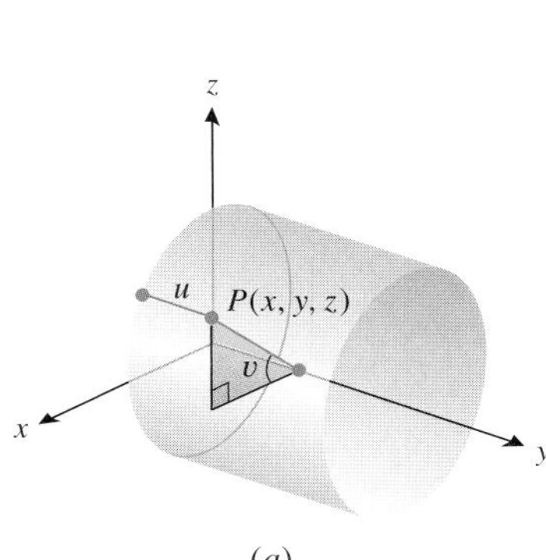

(*a*)

(*b*)

Figure 15.4.4

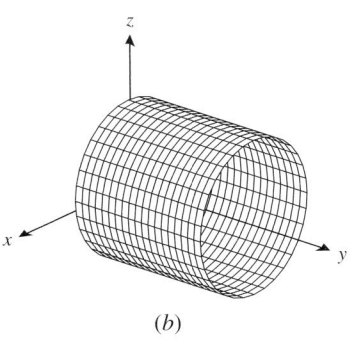

▶ **Example 4** Find parametric equations for the portion of the right circular cylinder

$$x^2 + z^2 = 9 \quad \text{for which} \quad 0 \le y \le 5$$

in terms of the parameters u and v shown in Figure 15.4.4*a*. The parameter u is the y-coordinate of a point $P(x, y, z)$ on the surface, and v is the angle shown in the figure.

Solution. The radius of the cylinder is 3, so it is evident from the figure that $y = u$, $x = 3\cos v$, and $z = 3\sin v$. Thus, the surface can be represented parametrically as

$$x = 3\cos v, \quad y = u, \quad z = 3\sin v$$

To obtain the portion of the surface from $y = 0$ to $y = 5$, we let the parameter u vary over the interval $0 \le u \le 5$, and to ensure that the entire lateral surface is covered, we let the parameter v vary over the interval $0 \le v \le 2\pi$. Figure 15.4.4*b* shows a computer-generated graph of the surface in which u and v vary over these intervals. Constant u-curves appear as circular traces parallel to the xz-plane, and constant v-curves appear as lines parallel to the y-axis. ◀

REPRESENTING SURFACES OF REVOLUTION PARAMETRICALLY

The basic idea of Example 4 can be adapted to obtain parametric equations for surfaces of revolution. For example, suppose that we want to find parametric equations for the surface generated by revolving the plane curve $y = f(x)$ about the x-axis. Figure 15.4.5 suggests that the surface can be represented parametrically as

$$x = u, \quad y = f(u)\cos v, \quad z = f(u)\sin v \tag{4}$$

where v is the angle shown. In the exercises we will discuss analogous formulas for surfaces of revolution about other axes.

▶ **Example 5** Find parametric equations for the surface generated by revolving the curve $y = 1/x$ about the x-axis.

Solution. From (4) this surface can be represented parametrically as

$$x = u, \quad y = \frac{1}{u}\cos v, \quad z = \frac{1}{u}\sin v$$

Figure 15.4.6 shows a computer-generated graph of the surface in which $0.7 \le u \le 5$ and $0 \le v \le 2\pi$. This surface is a portion of Gabriel's horn, which was discussed in Exercise 49 of Section 8.8. ◀

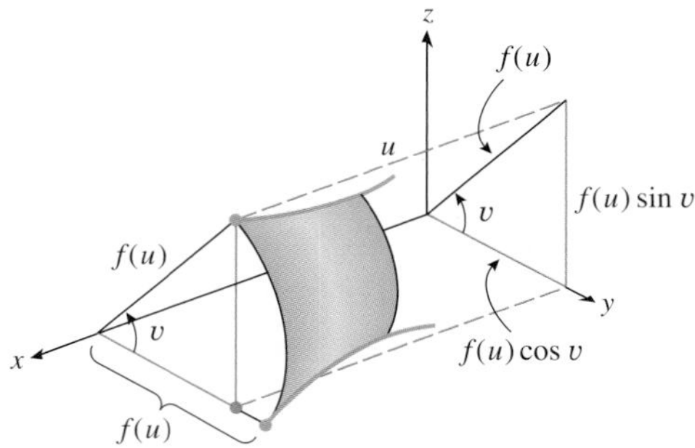

Figure 15.4.5

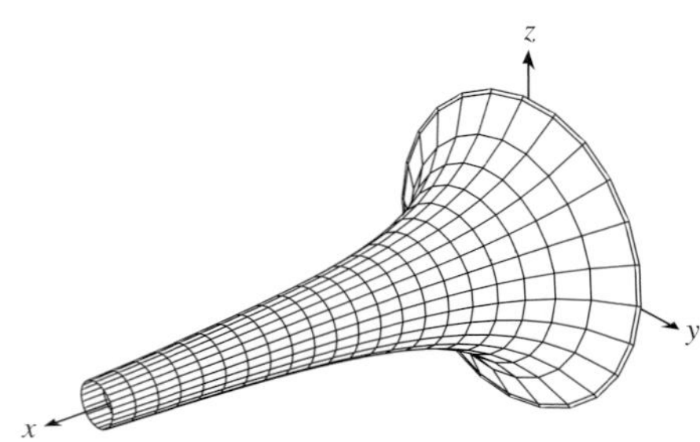

Figure 15.4.6

VECTOR-VALUED FUNCTIONS OF TWO VARIABLES

Recall that the parametric equations

$$x = x(t), \quad y = y(t), \quad z = z(t)$$

can be expressed in vector form as

$$\mathbf{r} = x(t)\mathbf{i} + y(t)\mathbf{j} + z(t)\mathbf{k}$$

where $\mathbf{r} = x\mathbf{i} + y\mathbf{j} + z\mathbf{k}$ is the radius vector and $\mathbf{r}(t) = x(t)\mathbf{i} + y(t)\mathbf{j} + z(t)\mathbf{k}$ is a vector-valued function of one variable. Similarly, the parametric equations

$$x = x(u, v), \quad y = y(u, v), \quad z = z(u, v)$$

can be expressed in vector form as

$$\mathbf{r} = x(u, v)\mathbf{i} + y(u, v)\mathbf{j} + z(u, v)\mathbf{k}$$

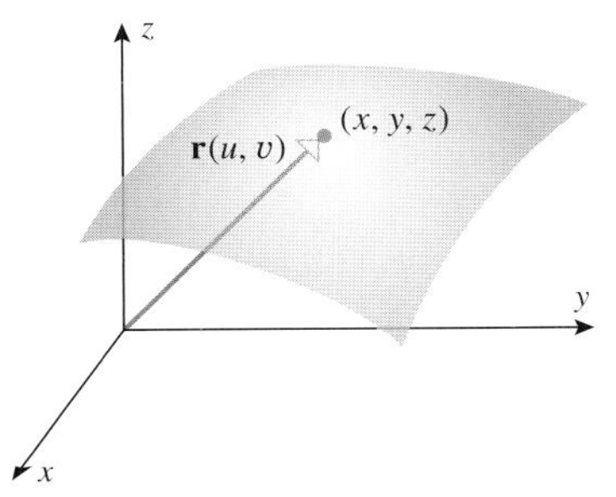

Figure 15.4.7

Here the function $\mathbf{r}(u, v) = x(u, v)\mathbf{i} + y(u, v)\mathbf{j} + z(u, v)\mathbf{k}$ is a ***vector-valued function of two variables***. We define the ***graph*** of $\mathbf{r}(u, v)$ to be the graph of the corresponding parametric equations. Geometrically, we can view $\mathbf{r}$ as a vector from the origin to a point (x, y, z) that moves over the surface $\mathbf{r} = \mathbf{r}(u, v)$ as u and v vary (Figure 15.4.7). As with vector-valued functions of one variable, we say that $\mathbf{r}(u, v)$ is ***continuous*** if each component is continuous.

▶ **Example 6** The paraboloid in Example 1 was expressed parametrically as

$$x = u, \quad y = v, \quad z = 4 - u^2 - v^2$$

These equations can be expressed in vector form as

$$\mathbf{r} = u\mathbf{i} + v\mathbf{j} + (4 - u^2 - v^2)\mathbf{k} \quad ◀$$

PARTIAL DERIVATIVES OF VECTOR-VALUED FUNCTIONS

Partial derivatives of vector-valued functions of two variables are obtained by taking partial derivatives of the components. For example, if

$$\mathbf{r}(u, v) = x(u, v)\mathbf{i} + y(u, v)\mathbf{j} + z(u, v)\mathbf{k}$$

then

$$\frac{\partial \mathbf{r}}{\partial u} = \frac{\partial x}{\partial u}\mathbf{i} + \frac{\partial y}{\partial u}\mathbf{j} + \frac{\partial z}{\partial u}\mathbf{k}$$

$$\frac{\partial \mathbf{r}}{\partial v} = \frac{\partial x}{\partial v}\mathbf{i} + \frac{\partial y}{\partial v}\mathbf{j} + \frac{\partial z}{\partial v}\mathbf{k}$$

These derivatives can also be written as $\mathbf{r}_u$ and $\mathbf{r}_v$ or $\mathbf{r}_u(u, v)$ and $\mathbf{r}_v(u, v)$ and can be expressed as the limits

$$\frac{\partial \mathbf{r}}{\partial u} = \lim_{\Delta u \to 0} \frac{\mathbf{r}(u + \Delta u, v) - \mathbf{r}(u, v)}{\Delta u} = \lim_{w \to u} \frac{\mathbf{r}(w, v) - \mathbf{r}(u, v)}{w - u} \tag{5}$$

$$\frac{\partial \mathbf{r}}{\partial v} = \lim_{\Delta v \to 0} \frac{\mathbf{r}(u, v + \Delta v) - \mathbf{r}(u, v)}{\Delta v} = \lim_{w \to v} \frac{\mathbf{r}(u, w) - \mathbf{r}(u, v)}{w - v} \tag{6}$$

▶ **Example 7** Find the partial derivatives of the vector-valued function $\mathbf{r}$ in Example 6.

Solution.

$$\frac{\partial \mathbf{r}}{\partial u} = \frac{\partial}{\partial u}[u\mathbf{i} + v\mathbf{j} + (4 - u^2 - v^2)\mathbf{k}] = \mathbf{i} - 2u\mathbf{k}$$

$$\frac{\partial \mathbf{r}}{\partial v} = \frac{\partial}{\partial v}[u\mathbf{i} + v\mathbf{j} + (4 - u^2 - v^2)\mathbf{k}] = \mathbf{j} - 2v\mathbf{k} \quad ◀$$

TANGENT PLANES TO PARAMETRIC SURFACES

Our next objective is to show how to find tangent planes to parametric surfaces. Let σ denote a parametric surface in 3-space, with P_0 a point on σ. We will say that a plane is ***tangent*** to σ at P_0 provided a line through P_0 lies in the plane if and only if it is a tangent line at P_0 to a curve on σ. We showed in Section 14.7 that if $z = f(x, y)$, then the graph of f has a tangent plane at a point if f is differentiable at that point. It is beyond the scope of this text to obtain precise conditions under which a parametric surface has a tangent plane at a point, so we will simply assume the existence of tangent planes at points of interest and focus on finding their equations.

Suppose that the parametric surface σ is the graph of the vector-valued function $\mathbf{r}(u, v)$ and that we are interested in the tangent plane at the point (x_0, y_0, z_0) on the surface that corresponds to the parameter values $u = u_0$ and $v = v_0$; that is,

$$\mathbf{r}(u_0, v_0) = x_0\mathbf{i} + y_0\mathbf{j} + z_0\mathbf{k}$$

If $v = v_0$ is kept fixed and u is allowed to vary, then $\mathbf{r}(u, v_0)$ is a vector-valued function of one variable whose graph is the constant v-curve through the point (u_0, v_0); similarly, if $u = u_0$ is kept fixed and v is allowed to vary, then $\mathbf{r}(u_0, v)$ is a vector-valued function of one variable whose graph is the constant u-curve through the point (u_0, v_0). Moreover, it follows from the geometric interpretation of the derivative developed in Section 13.2 that if $\partial\mathbf{r}/\partial u \neq \mathbf{0}$ at (u_0, v_0), then this vector is tangent to the constant v-curve through (u_0, v_0); and if $\partial\mathbf{r}/\partial v \neq \mathbf{0}$ at (u_0, v_0), then this vector is tangent to the constant u-curve through (u_0, v_0) (Figure 15.4.8). Thus, if $\partial\mathbf{r}/\partial u \times \partial\mathbf{r}/\partial v \neq \mathbf{0}$ at (u_0, v_0), then the vector

$$\frac{\partial\mathbf{r}}{\partial u} \times \frac{\partial\mathbf{r}}{\partial v} = \begin{vmatrix} \mathbf{i} & \mathbf{j} & \mathbf{k} \\ \dfrac{\partial x}{\partial u} & \dfrac{\partial y}{\partial u} & \dfrac{\partial z}{\partial u} \\ \dfrac{\partial x}{\partial v} & \dfrac{\partial y}{\partial v} & \dfrac{\partial z}{\partial v} \end{vmatrix} \tag{7}$$

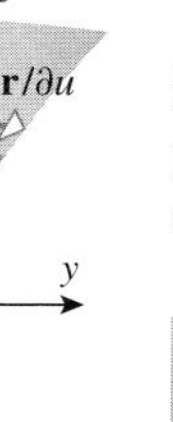

Figure 15.4.8

is orthogonal to both tangent vectors at the point (u_0, v_0) and hence is normal to the tangent plane and the surface at this point (Figure 15.4.8). Accordingly, we make the following definition.

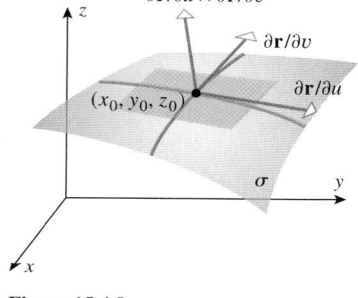

15.4.1 DEFINITION. If a parametric surface σ is the graph of $\mathbf{r} = \mathbf{r}(u, v)$, and if $\partial\mathbf{r}/\partial u \times \partial\mathbf{r}/\partial v \neq \mathbf{0}$ at a point on the surface, then the ***principal unit normal vector*** to the surface at that point is denoted by $\mathbf{n}$ or $\mathbf{n}(u, v)$ and is defined as

$$\mathbf{n} = \frac{\dfrac{\partial\mathbf{r}}{\partial u} \times \dfrac{\partial\mathbf{r}}{\partial v}}{\left\| \dfrac{\partial\mathbf{r}}{\partial u} \times \dfrac{\partial\mathbf{r}}{\partial v} \right\|} \tag{8}$$

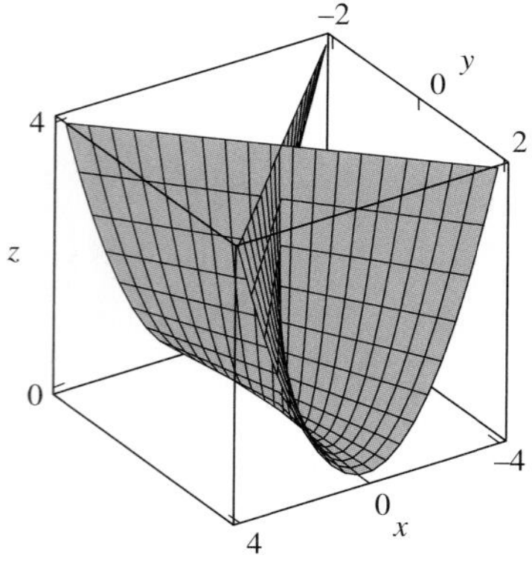

Figure 15.4.9

► **Example 8** Find an equation of the tangent plane to the parametric surface

$$x = uv, \quad y = u, \quad z = v^2$$

at the point where $u = 2$ and $v = -1$. This surface, called *Whitney's umbrella*, is an example of a self-intersecting parametric surface (Figure 15.4.9).

Solution. We start by writing the equations in the vector form

$$\mathbf{r} = uv\mathbf{i} + u\mathbf{j} + v^2\mathbf{k}$$

The partial derivatives of $\mathbf{r}$ are

$$\frac{\partial\mathbf{r}}{\partial u}(u, v) = v\mathbf{i} + \mathbf{j}$$

$$\frac{\partial\mathbf{r}}{\partial v}(u, v) = u\mathbf{i} + 2v\mathbf{k}$$

and at $u = 2$ and $v = -1$ these partial derivatives are

$$\frac{\partial\mathbf{r}}{\partial u}(2, -1) = -\mathbf{i} + \mathbf{j}$$

$$\frac{\partial\mathbf{r}}{\partial v}(2, -1) = 2\mathbf{i} - 2\mathbf{k}$$

Thus, from (7) and (8) a normal to the surface at this point is

$$\frac{\partial \mathbf{r}}{\partial u}(2,-1) \times \frac{\partial \mathbf{r}}{\partial v}(2,-1) = \begin{vmatrix} \mathbf{i} & \mathbf{j} & \mathbf{k} \\ -1 & 1 & 0 \\ 2 & 0 & -2 \end{vmatrix} = -2\mathbf{i} - 2\mathbf{j} - 2\mathbf{k}$$

Convince yourself that the result obtained in Example 8 is consistent with Figure 15.4.9.

Since any normal will suffice to find the tangent plane, it makes sense to multiply this vector by $-\frac{1}{2}$ and use the simpler normal $\mathbf{i}+\mathbf{j}+\mathbf{k}$. It follows from the given parametric equations that the point on the surface corresponding to $u = 2$ and $v = -1$ is $(-2, 2, 1)$, so the tangent plane at this point can be expressed in point-normal form as

$$(x+2) + (y-2) + (z-1) = 0 \quad \text{or} \quad x + y + z = 1 \blacktriangleleft$$

▶ **Example 9** The sphere $x^2 + y^2 + z^2 = a^2$ can be expressed in spherical coordinates as $\rho = a$, and the spherical-to-rectangular conversion formulas in Table 12.8.1 can then be used to express the sphere as the graph of the vector-valued function

$$\mathbf{r}(\phi, \theta) = a \sin\phi \cos\theta\mathbf{i} + a \sin\phi \sin\theta\mathbf{j} + a\cos\phi\mathbf{k}$$

where $0 \le \phi \le \pi$ and $0 \le \theta \le 2\pi$ (verify). Use this function to show that the radius vector is normal to the tangent plane at each point on the sphere.

Solution. We will show that at each point of the sphere the unit normal vector $\mathbf{n}$ is a scalar multiple of $\mathbf{r}$ (and hence is parallel to $\mathbf{r}$). We have

$$\frac{\partial \mathbf{r}}{\partial \phi} \times \frac{\partial \mathbf{r}}{\partial \theta} = \begin{vmatrix} \mathbf{i} & \mathbf{j} & \mathbf{k} \\ \dfrac{\partial x}{\partial \phi} & \dfrac{\partial y}{\partial \phi} & \dfrac{\partial z}{\partial \phi} \\ \dfrac{\partial x}{\partial \theta} & \dfrac{\partial y}{\partial \theta} & \dfrac{\partial z}{\partial \theta} \end{vmatrix} = \begin{vmatrix} \mathbf{i} & \mathbf{j} & \mathbf{k} \\ a\cos\phi\cos\theta & a\cos\phi\sin\theta & -a\sin\phi \\ -a\sin\phi\sin\theta & a\sin\phi\cos\theta & 0 \end{vmatrix}$$
$$= a^2\sin^2\phi\cos\theta\mathbf{i} + a^2\sin^2\phi\sin\theta\mathbf{j} + a^2\sin\phi\cos\phi\mathbf{k}$$

and hence

$$\begin{aligned} \left\| \frac{\partial \mathbf{r}}{\partial \phi} \times \frac{\partial \mathbf{r}}{\partial \theta} \right\| &= \sqrt{a^4\sin^4\phi\cos^2\theta + a^4\sin^4\phi\sin^2\theta + a^4\sin^2\phi\cos^2\phi} \\ &= \sqrt{a^4\sin^4\phi + a^4\sin^2\phi\cos^2\phi} \\ &= a^2\sqrt{\sin^2\phi} = a^2|\sin\phi| = a^2\sin\phi \end{aligned}$$

For $\phi \ne 0$ or π, it follows from (8) that

$$\mathbf{n} = \sin\phi\cos\theta\mathbf{i} + \sin\phi\sin\theta\mathbf{j} + \cos\phi\mathbf{k} = \frac{1}{a}\mathbf{r}$$

Furthermore, the tangent planes at $\phi \ne 0$ or π are horizontal, to which $\mathbf{r} = \pm a\mathbf{k}$ is clearly normal. ◀

SURFACE AREA OF PARAMETRIC SURFACES

In Section 6.5 we obtained formulas for the surface area of a surface of revolution [see Formulas (4) and (5) and Exercise 33 in that section]. We will now obtain a formula for the surface area S of a parametric surface σ and from that formula we will then derive a formula for the surface area of a surface of the form $z = f(x, y)$.

Let σ be a parametric surface whose vector equation is

$$\mathbf{r} = x(u, v)\mathbf{i} + y(u, v)\mathbf{j} + z(u, v)\mathbf{k}$$

We will say that σ is a ***smooth parametric surface*** on a region R of the uv-plane if $\partial\mathbf{r}/\partial u$ and $\partial\mathbf{r}/\partial v$ are continuous on R and $\partial\mathbf{r}/\partial u \times \partial\mathbf{r}/\partial v \neq \mathbf{0}$ on R. Geometrically, this means that σ has a principal unit normal vector (and hence a tangent plane) for all (u, v) in R and $\mathbf{n} = \mathbf{n}(u, v)$ is a continuous function on R. Thus, on a smooth parametric surface the unit normal vector $\mathbf{n}$ varies continuously and has no abrupt changes in direction. We will derive a surface area formula for parametric surfaces that have no self-intersections and are smooth on a region R, with the possible exception that $\partial\mathbf{r}/\partial u \times \partial\mathbf{r}/\partial v$ may equal $\mathbf{0}$ on the boundary of R.

We begin by subdividing R into rectangular regions by lines parallel to the u- and v-axes and discarding any nonrectangular portions that contain points of the boundary. Assume that there are n rectangles, and let R_k denote the kth rectangle. Let (u_k, v_k) be the lower left corner of R_k, and assume that R_k has area $\Delta A_k = \Delta u_k \Delta v_k$, where Δu_k and Δv_k are the dimensions of R_k (Figure 15.4.10a). The image of R_k will be some *curvilinear patch* σ_k on the surface σ that has a corner at $\mathbf{r}(u_k, v_k)$; denote the area of this patch by ΔS_k (Figure 15.4.10b).

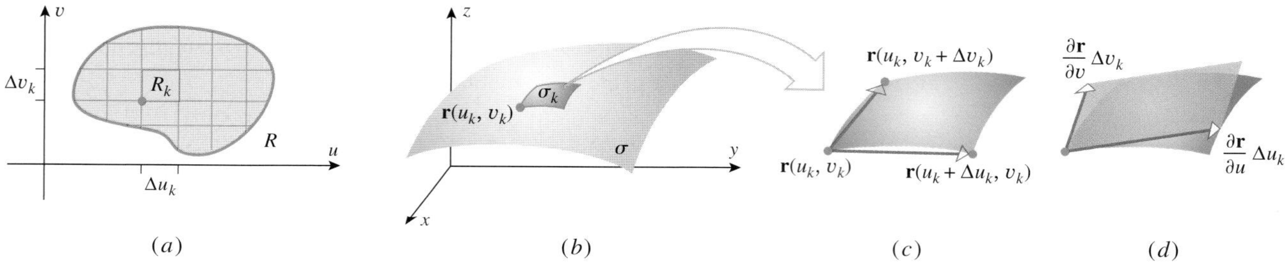

Figure 15.4.10

As suggested by Figure 15.4.10c, the two edges of the patch that meet at $\mathbf{r}(u_k, v_k)$ can be approximated by the "secant" vectors

$$\mathbf{r}(u_k + \Delta u_k, v_k) - \mathbf{r}(u_k, v_k)$$
$$\mathbf{r}(u_k, v_k + \Delta v_k) - \mathbf{r}(u_k, v_k)$$

and hence the area of σ_k can be approximated by the area of the parallelogram determined by these vectors. However, it follows from Formulas (5) and (6) that if Δu_k and Δv_k are small, then these secant vectors can in turn be approximated by the tangent vectors

$$\frac{\partial \mathbf{r}}{\partial u}\Delta u_k \quad \text{and} \quad \frac{\partial \mathbf{r}}{\partial v}\Delta v_k$$

where the partial derivatives are evaluated at (u_k, v_k). Thus, the area of the patch σ_k can be approximated by the area of the parallelogram determined by these vectors (Figure 15.4.10d); that is,

$$\Delta S_k \approx \left\| \frac{\partial \mathbf{r}}{\partial u}\Delta u_k \times \frac{\partial \mathbf{r}}{\partial v}\Delta v_k \right\| = \left\| \frac{\partial \mathbf{r}}{\partial u} \times \frac{\partial \mathbf{r}}{\partial v} \right\| \Delta u_k \Delta v_k = \left\| \frac{\partial \mathbf{r}}{\partial u} \times \frac{\partial \mathbf{r}}{\partial v} \right\| \Delta A_k \tag{9}$$

It follows that the surface area S of the entire surface σ can be approximated as

$$S \approx \sum_{k=1}^{n} \left\| \frac{\partial \mathbf{r}}{\partial u} \times \frac{\partial \mathbf{r}}{\partial v} \right\| \Delta A_k$$

Thus, if we assume that the errors in the approximations approach zero as n increases in such a way that the dimensions of the rectangles approach zero, then it is plausible that the exact value of S is

$$S = \lim_{n \to +\infty} \sum_{k=1}^{n} \left\| \frac{\partial \mathbf{r}}{\partial u} \times \frac{\partial \mathbf{r}}{\partial v} \right\| \Delta A_k$$

or, equivalently,

$$S = \iint_R \left\| \frac{\partial \mathbf{r}}{\partial u} \times \frac{\partial \mathbf{r}}{\partial v} \right\| dA \tag{10}$$

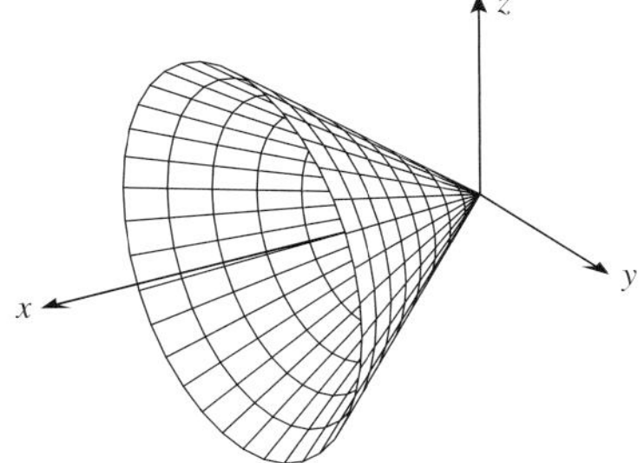

Figure 15.4.11

► **Example 10** It follows from (4) that the parametric equations

$$x = u, \quad y = u\cos v, \quad z = u\sin v$$

represent the cone that results when the line $y = x$ in the xy-plane is revolved about the x-axis. Use Formula (10) to find the surface area of that portion of the cone for which $0 \le u \le 2$ and $0 \le v \le 2\pi$ (Figure 15.4.11).

Solution. The surface can be expressed in vector form as

$$\mathbf{r} = u\mathbf{i} + u\cos v\mathbf{j} + u\sin v\mathbf{k} \quad (0 \le u \le 2, \;\; 0 \le v \le 2\pi)$$

Thus,

$$\frac{\partial \mathbf{r}}{\partial u} = \mathbf{i} + \cos v\mathbf{j} + \sin v\mathbf{k}$$

$$\frac{\partial \mathbf{r}}{\partial v} = -u\sin v\mathbf{j} + u\cos v\mathbf{k}$$

$$\frac{\partial \mathbf{r}}{\partial u} \times \frac{\partial \mathbf{r}}{\partial v} = \begin{vmatrix} \mathbf{i} & \mathbf{j} & \mathbf{k} \\ 1 & \cos v & \sin v \\ 0 & -u\sin v & u\cos v \end{vmatrix} = u\mathbf{i} - u\cos v\mathbf{j} - u\sin v\mathbf{k}$$

$$\left\| \frac{\partial \mathbf{r}}{\partial u} \times \frac{\partial \mathbf{r}}{\partial v} \right\| = \sqrt{u^2 + (-u\cos v)^2 + (-u\sin v)^2} = |u|\sqrt{2} = u\sqrt{2}$$

Thus, from (10)

$$S = \iint_R \left\| \frac{\partial \mathbf{r}}{\partial u} \times \frac{\partial \mathbf{r}}{\partial v} \right\| dA = \int_0^{2\pi}\int_0^2 \sqrt{2}u\,du\,dv = 2\sqrt{2}\int_0^{2\pi} dv = 4\pi\sqrt{2} \;◄$$

SURFACE AREA OF SURFACES OF THE FORM $z = f(x, y)$

In the case where σ is a surface of the form $z = f(x, y)$, we can take $x = u$ and $y = v$ as parameters and express the surface parametrically as

$$x = u, \quad y = v, \quad z = f(u, v)$$

or in vector form as

$$\mathbf{r} = u\mathbf{i} + v\mathbf{j} + f(u, v)\mathbf{k}$$

Thus,

$$\frac{\partial \mathbf{r}}{\partial u} = \mathbf{i} + \frac{\partial f}{\partial u}\mathbf{k} = \mathbf{i} + \frac{\partial z}{\partial x}\mathbf{k}$$

$$\frac{\partial \mathbf{r}}{\partial v} = \mathbf{j} + \frac{\partial f}{\partial v}\mathbf{k} = \mathbf{j} + \frac{\partial z}{\partial y}\mathbf{k}$$

$$\frac{\partial \mathbf{r}}{\partial u} \times \frac{\partial \mathbf{r}}{\partial v} = \begin{vmatrix} \mathbf{i} & \mathbf{j} & \mathbf{k} \\ 1 & 0 & \dfrac{\partial z}{\partial x} \\ 0 & 1 & \dfrac{\partial z}{\partial y} \end{vmatrix} = -\frac{\partial z}{\partial x}\mathbf{i} - \frac{\partial z}{\partial y}\mathbf{j} + \mathbf{k}$$

$$\left\| \frac{\partial \mathbf{r}}{\partial u} \times \frac{\partial \mathbf{r}}{\partial v} \right\| = \sqrt{\left(\frac{\partial z}{\partial x}\right)^2 + \left(\frac{\partial z}{\partial y}\right)^2 + 1}$$

In Formula (11) the region R lies in the xy-plane because the parameters are x and y. Geometrically, this region is the projection on the xy-plane of that portion of the surface $z = f(x, y)$ whose area is being determined by the formula (Figure 15.4.12).

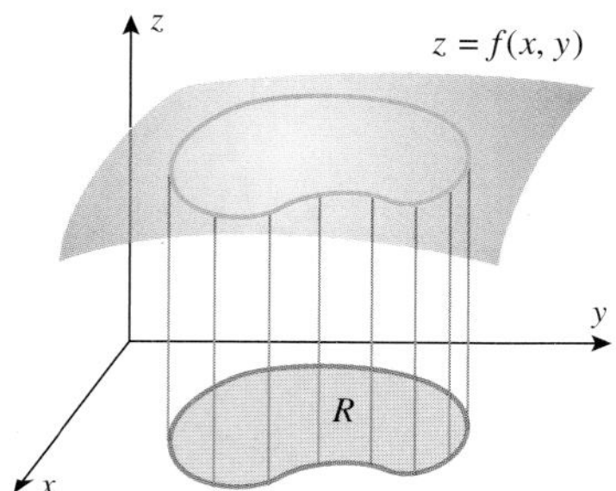

Figure 15.4.12

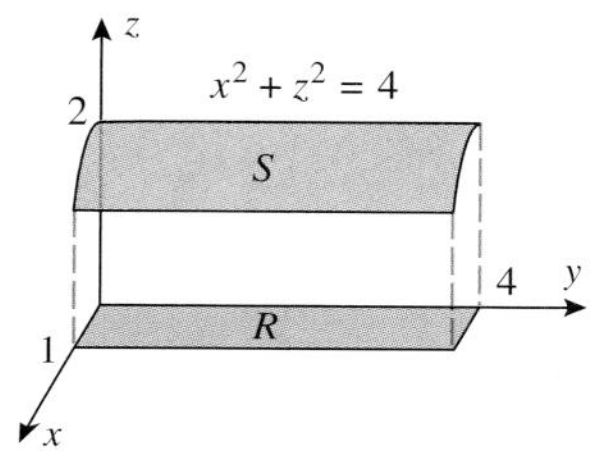

Figure 15.4.13

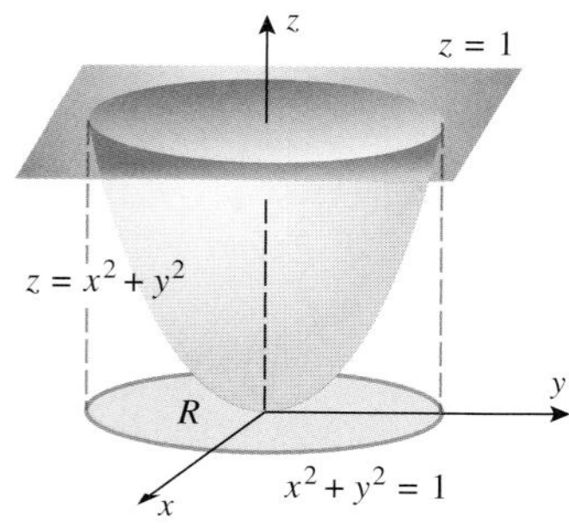

Figure 15.4.14

Thus, it follows from (10) that

$$S = \iint_R \sqrt{\left(\frac{\partial z}{\partial x}\right)^2 + \left(\frac{\partial z}{\partial y}\right)^2 + 1}\, dA \tag{11}$$

▶ **Example 11** Find the surface area of that portion of the surface $z = \sqrt{4 - x^2}$ that lies above the rectangle R in the xy-plane whose coordinates satisfy $0 \le x \le 1$ and $0 \le y \le 4$.

Solution. As shown in Figure 15.4.13, the surface is a portion of the cylinder $x^2 + z^2 = 4$. It follows from (11) that the surface area is

$$\begin{aligned} S &= \iint_R \sqrt{\left(\frac{\partial z}{\partial x}\right)^2 + \left(\frac{\partial z}{\partial y}\right)^2 + 1}\, dA \\ &= \iint_R \sqrt{\left(-\frac{x}{\sqrt{4 - x^2}}\right)^2 + 0 + 1}\, dA = \int_0^4 \int_0^1 \frac{2}{\sqrt{4 - x^2}}\, dx\, dy \\ &= 2\int_0^4 \left[\sin^{-1}\left(\frac{1}{2}x\right)\right]_{x=0}^1 dy = 2\int_0^4 \frac{\pi}{6}\, dy = \frac{4}{3}\pi \quad ◀ \end{aligned}$$

Formula 21 of Section 8.1

▶ **Example 12** Find the surface area of the portion of the paraboloid $z = x^2 + y^2$ below the plane $z = 1$.

Solution. The surface $z = x^2 + y^2$ is the circular paraboloid shown in Figure 15.4.14. The trace of the paraboloid in the plane $z = 1$ projects onto the circle $x^2 + y^2 = 1$ in the xy-plane, and the portion of the paraboloid that lies below the plane $z = 1$ projects onto the region R that is enclosed by this circle. Thus, it follows from (11) that the surface area is

$$S = \iint_R \sqrt{4x^2 + 4y^2 + 1}\, dA$$

The expression $4x^2 + 4y^2 + 1 = 4(x^2 + y^2) + 1$ in the integrand suggests that we evaluate the integral in polar coordinates. In accordance with Formula (9) of Section 15.3, we substitute $x = r\cos\theta$ and $y = r\sin\theta$ in the integrand, replace dA by $r\, dr\, d\theta$, and find the limits of integration by expressing the region R in polar coordinates. This yields

$$\begin{aligned} S &= \int_0^{2\pi} \int_0^1 \sqrt{4r^2 + 1}\, r\, dr\, d\theta = \int_0^{2\pi} \left[\frac{1}{12}(4r^2 + 1)^{3/2}\right]_{r=0}^1 d\theta \\ &= \int_0^{2\pi} \frac{1}{12}(5\sqrt{5} - 1)\, d\theta = \frac{1}{6}\pi(5\sqrt{5} - 1) \quad ◀ \end{aligned}$$

✔ QUICK CHECK EXERCISES 15.4 *(See page 1056 for answers.)*

1. Consider the surface represented parametrically by

$$\begin{aligned} x &= 1 - u \\ y &= (1 - u)\cos v \qquad (0 \le u \le 1,\ 0 \le v \le 2\pi) \\ z &= (1 - u)\sin v \end{aligned}$$

(a) Describe the constant u-curves.
(b) Describe the constant v-curves.

2. If

$$\mathbf{r}(u, v) = (1 - u)\mathbf{i} + [(1 - u)\cos v]\mathbf{j} + [(1 - u)\sin v]\mathbf{k}$$

then

$$\frac{\partial \mathbf{r}}{\partial u} = \_\_\_\_\_\_ \quad \text{and} \quad \frac{\partial \mathbf{r}}{\partial v} = \_\_\_\_\_\_$$

3. If

$$\mathbf{r}(u, v) = (1 - u)\mathbf{i} + [(1 - u)\cos v]\mathbf{j} + [(1 - u)\sin v]\mathbf{k}$$

the principal unit normal to the graph of **r** at the point where $u = 1/2$ and $v = \pi/6$ is given by ______.

4. Suppose σ is a parametric surface with vector equation

$$\mathbf{r}(u, v) = x(u, v)\mathbf{i} + y(u, v)\mathbf{j} + z(u, v)\mathbf{k}$$

If σ has no self-intersections and σ is smooth on a region R in the uv-plane, then the surface area of σ is given by

$$S = \iint_R \_\_\_\_\_\_ \, dA$$

5. The surface area of a surface of the form $z = f(x, y)$ over a region R in the xy-plane is given by

$$S = \iint_R \_\_\_\_\_\_ \, dA$$

EXERCISE SET 15.4  Graphing Utility C CAS

1–2 Sketch the parametric surface.

1. (a) $x = u,\ y = v,\ z = \sqrt{u^2 + v^2}$
   (b) $x = u,\ y = \sqrt{u^2 + v^2},\ z = v$
   (c) $x = \sqrt{u^2 + v^2},\ y = u,\ z = v$

2. (a) $x = u,\ y = v,\ z = u^2 + v^2$
   (b) $x = u,\ y = u^2 + v^2,\ z = v$
   (c) $x = u^2 + v^2,\ y = u,\ z = v$

3–4 Find a parametric representation of the surface in terms of the parameters $u = x$ and $v = y$.

3. (a) $2z - 3x + 4y = 5$ (b) $z = x^2$
4. (a) $z + zx^2 - y = 0$ (b) $y^2 - 3z = 5$
5. (a) Find parametric equations for the portion of the cylinder $x^2 + y^2 = 5$ that extends between the planes $z = 0$ and $z = 1$.
   (b) Find parametric equations for the portion of the cylinder $x^2 + z^2 = 4$ that extends between the planes $y = 1$ and $y = 3$.
6. (a) Find parametric equations for the portion of the plane $x + y = 1$ that extends between the planes $z = -1$ and $z = 1$.
   (b) Find parametric equations for the portion of the plane $y - 2z = 5$ that extends between the planes $x = 0$ and $x = 3$.
7. Find parametric equations for the surface generated by revolving the curve $y = \sin x$ about the x-axis.
8. Find parametric equations for the surface generated by revolving the curve $y - e^x = 0$ about the x-axis.

9–14 Find a parametric representation of the surface in terms of the parameters r and θ, where (r, θ, z) are the cylindrical coordinates of a point on the surface.

9. $z = \dfrac{1}{1 + x^2 + y^2}$
10. $z = e^{-(x^2+y^2)}$
11. $z = 2xy$
12. $z = x^2 - y^2$
13. The portion of the sphere $x^2 + y^2 + z^2 = 9$ on or above the plane $z = 2$.
14. The portion of the cone $z = \sqrt{x^2 + y^2}$ on or below the plane $z = 3$.
15. Find a parametric representation of the cone

$$z = \sqrt{3x^2 + 3y^2}$$

in terms of parameters ρ and θ, where (ρ, θ, ϕ) are spherical coordinates of a point on the surface.

16. Describe the cylinder $x^2 + y^2 = 9$ in terms of parameters θ and ϕ, where (ρ, θ, ϕ) are spherical coordinates of a point on the surface.

FOCUS ON CONCEPTS

17–22 Eliminate the parameters to obtain an equation in rectangular coordinates, and describe the surface.

17. $x = 2u + v,\ y = u - v,\ z = 3v$ for $-\infty < u < +\infty$ and $-\infty < v < +\infty$.
18. $x = u\cos v,\ y = u^2,\ z = u\sin v$ for $0 \le u \le 2$ and $0 \le v < 2\pi$.
19. $x = 3\sin u,\ y = 2\cos u,\ z = 2v$ for $0 \le u < 2\pi$ and $1 \le v \le 2$.
20. $x = \sqrt{u}\cos v,\ y = \sqrt{u}\sin v,\ z = u$ for $0 \le u \le 4$ and $0 \le v < 2\pi$.
21. $\mathbf{r}(u, v) = 3u\cos v\mathbf{i} + 4u\sin v\mathbf{j} + u\mathbf{k}$ for $0 \le u \le 1$ and $0 \le v < 2\pi$.
22. $r(u, v) = \sin u\cos v\mathbf{i} + 2\sin u\sin v\mathbf{j} + 3\cos u\mathbf{k}$ for $0 \le u \le \pi$ and $0 \le v < 2\pi$.
23. The accompanying figure shows the graphs of two parametric representations of the cone $z = \sqrt{x^2 + y^2}$ for $0 \le z \le 2$.

(a) Find parametric equations that produce reasonable facsimiles of these surfaces.
(b) Use a graphing utility to check your answer in part (a).

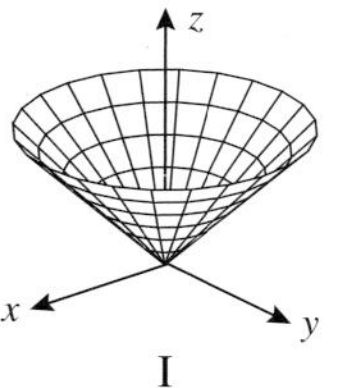
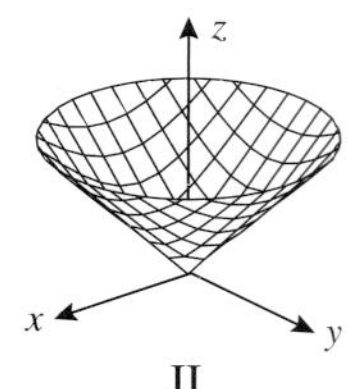

I II Figure Ex-23

24. The accompanying figure shows the graphs of two parametric representations of the paraboloid $z = x^2 + y^2$ for $0 \le z \le 2$.
(a) Find parametric equations that produce reasonable facsimiles of these surfaces.
(b) Use a graphing utility to check your answer in part (a).

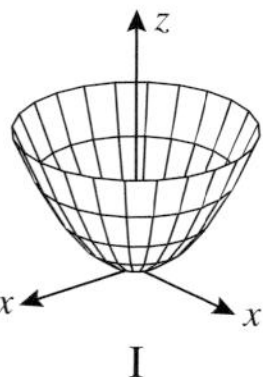
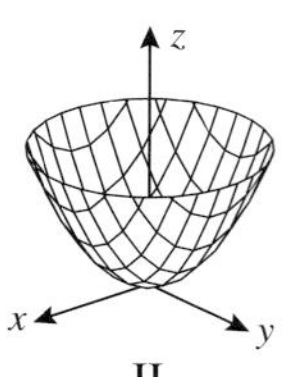

I II Figure Ex-24

25. In each part, the figure shows a portion of the parametric surface $x = 3\cos v$, $y = u$, $z = 3\sin v$. Find restrictions on u and v that produce the surface, and check your answer with a graphing utility.

(a) 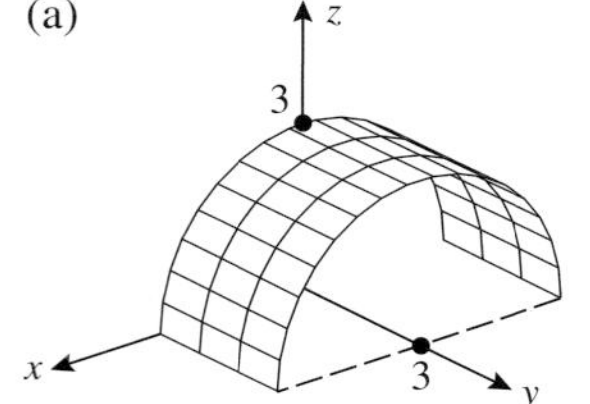
(b) 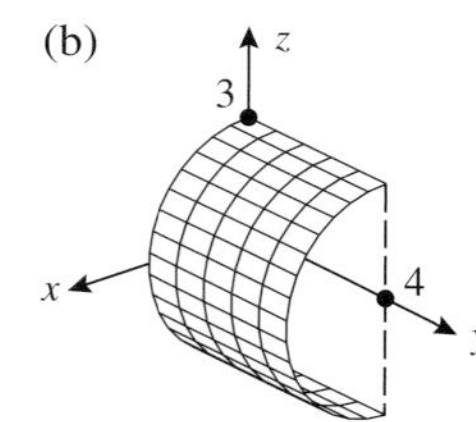

26. In each part, the figure shows a portion of the parametric surface $x = 3\cos v$, $y = 3\sin v$, $z = u$. Find restrictions on u and v that produce the surface, and check your answer with a graphing utility.

(a) 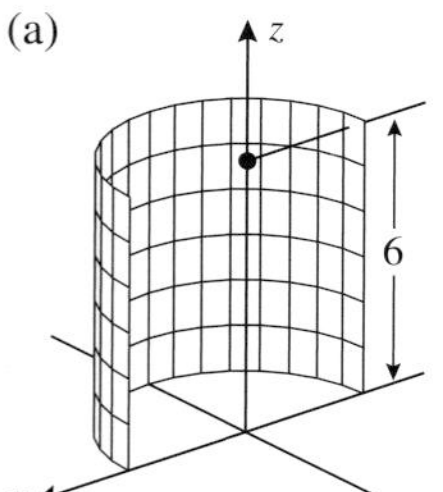
(b) 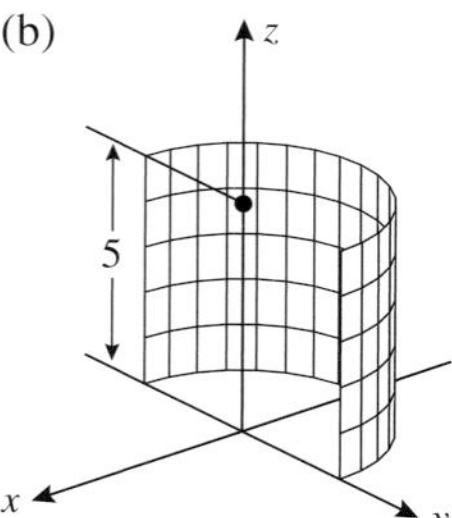

27. In each part, the figure shows a hemisphere that is a portion of the sphere $x = \sin\phi\cos\theta$, $y = \sin\phi\sin\theta$, $z = \cos\phi$. Find restrictions on ϕ and θ that produce the hemisphere, and check your answer with a graphing utility.

(a) 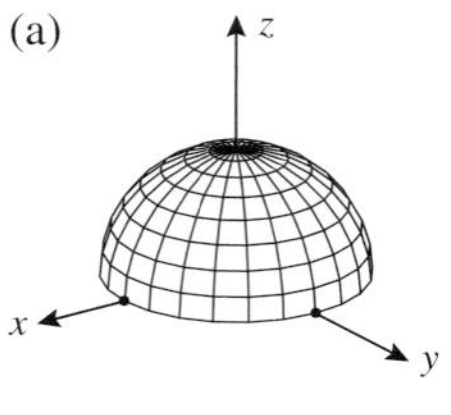
(b) 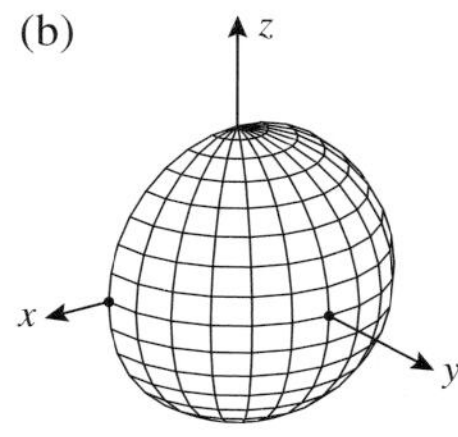

28. In each part, the figure shows a portion of the sphere $x = \sin\phi\cos\theta$, $y = \sin\phi\sin\theta$, $z = \cos\phi$. Find restrictions on ϕ and θ that produce the surface, and check your answer with a graphing utility.

(a) 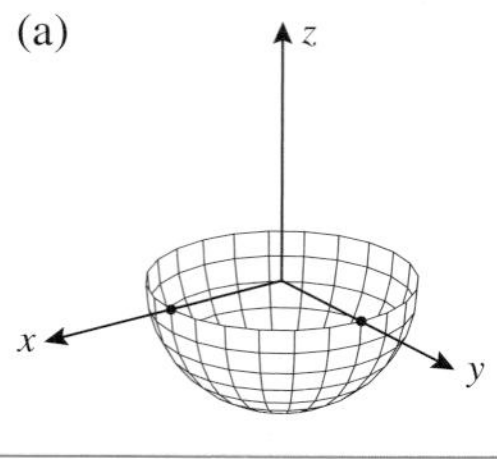
(b) 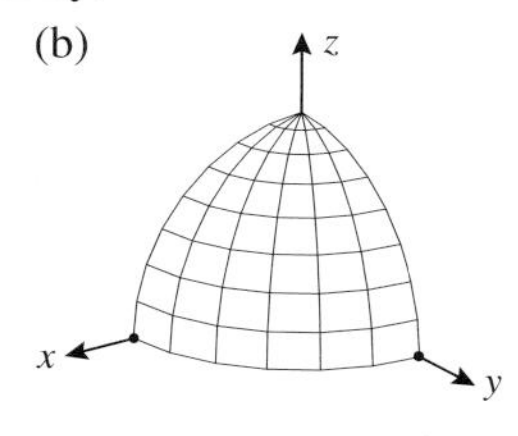

29–34 Find an equation of the tangent plane to the parametric surface at the stated point.

29. $x = u$, $y = v$, $z = u^2 + v^2$; $(1, 2, 5)$

30. $x = u^2$, $y = v^2$, $z = u + v$; $(1, 4, 3)$

31. $x = 3v\sin u$, $y = 2v\cos u$, $z = u^2$; $(0, 2, 0)$

32. $\mathbf{r} = uv\mathbf{i} + (u - v)\mathbf{j} + (u + v)\mathbf{k}$; $u = 1, v = 2$

33. $\mathbf{r} = u\cos v\mathbf{i} + u\sin v\mathbf{j} + v\mathbf{k}$; $u = 1/2, v = \pi/4$

34. $\mathbf{r} = uv\mathbf{i} + ue^v\mathbf{j} + ve^u\mathbf{k}$; $u = \ln 2, v = 0$

35–46 Find the area of the given surface.

35. The portion of the cylinder $y^2 + z^2 = 9$ that is above the rectangle $R = \{(x, y) : 0 \le x \le 2, -3 \le y \le 3\}$.

36. The portion of the plane $2x + 2y + z = 8$ in the first octant.

37. The portion of the cone $z^2 = 4x^2 + 4y^2$ that is above the region in the first quadrant bounded by the line $y = x$ and the parabola $y = x^2$.

38. The portion of the cone $z = \sqrt{x^2 + y^2}$ that lies inside the cylinder $x^2 + y^2 = 2x$.

39. The portion of the paraboloid $z = 1 - x^2 - y^2$ that is above the xy-plane.

40. The portion of the surface $z = 2x + y^2$ that is above the triangular region with vertices $(0, 0)$, $(0, 1)$, and $(1, 1)$.

41. The portion of the paraboloid

$$\mathbf{r}(u, v) = u\cos v\mathbf{i} + u\sin v\mathbf{j} + u^2\mathbf{k}$$

for which $1 \le u \le 2$, $0 \le v \le 2\pi$.

42. The portion of the cone

$$\mathbf{r}(u, v) = u\cos v\mathbf{i} + u\sin v\mathbf{j} + u\mathbf{k}$$

for which $0 \le u \le 2v$, $0 \le v \le \pi/2$.

43. The portion of the surface $z = xy$ that is above the sector in the first quadrant bounded by the lines $y = x/\sqrt{3}$, $y = 0$, and the circle $x^2 + y^2 = 9$.

44. The portion of the paraboloid $2z = x^2 + y^2$ that is inside the cylinder $x^2 + y^2 = 8$.

45. The portion of the sphere $x^2 + y^2 + z^2 = 16$ between the planes $z = 1$ and $z = 2$.

46. The portion of the sphere $x^2 + y^2 + z^2 = 8$ that is inside the cone $z = \sqrt{x^2 + y^2}$.

47. Use parametric equations to derive the formula for the surface area of a sphere of radius a.

48. Use parametric equations to derive the formula for the lateral surface area of a right circular cylinder of radius r and height h.

49. The portion of the surface

$$z = \frac{h}{a}\sqrt{x^2 + y^2} \quad (a, h > 0)$$

between the xy-plane and the plane $z = h$ is a right circular cone of height h and radius a. Use a double integral to show that the lateral surface area of this cone is $S = \pi a\sqrt{a^2 + h^2}$.

50. The accompanying figure shows the ***torus*** that is generated by revolving the circle

$$(x - a)^2 + z^2 = b^2 \quad (0 < b < a)$$

in the xz-plane about the z-axis.

(a) Show that this torus can be expressed parametrically as

$$x = (a + b\cos v)\cos u$$
$$y = (a + b\cos v)\sin u$$
$$z = b\sin v$$

where u and v are the parameters shown in the figure and $0 \le u \le 2\pi$, $0 \le v \le 2\pi$.

(b) Use a graphing utility to generate a torus.

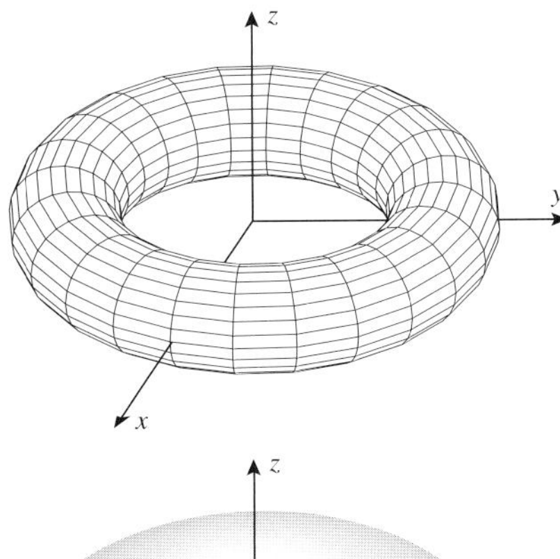

Figure Ex-50

51. Find the surface area of the torus in Exercise 50(a).

52. Use a CAS to graph the ***helicoid***

$$x = u\cos v, \quad y = u\sin v, \quad z = v$$

for $0 \le u \le 5$ and $0 \le v \le 4\pi$ (see the accompanying figure), and then use the numerical double integration operation of the CAS to approximate the surface area.

53. Use a CAS to graph the ***pseudosphere***

$$x = \cos u \sin v$$
$$y = \sin u \sin v$$
$$z = \cos v + \ln\left(\tan\frac{v}{2}\right)$$

for $0 \le u \le 2\pi$, $0 < v < \pi$ (see the accompanying figure), and then use the numerical double integration operation of the CAS to approximate the surface area between the planes $z = -1$ and $z = 1$.

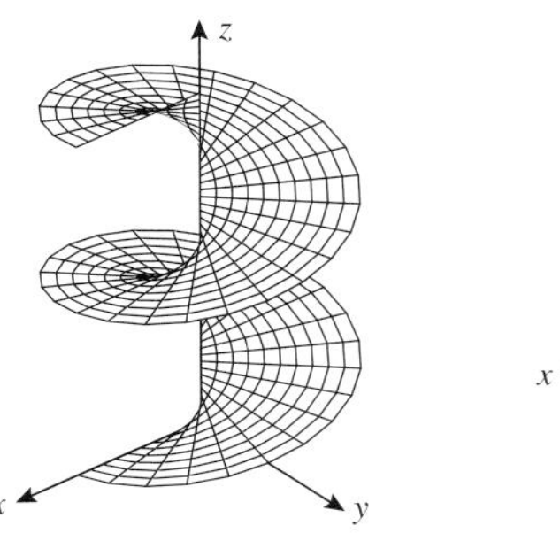

Figure Ex-52 Figure Ex-53

54. The accompanying figure shows the graph of an ***astroidal sphere***

$$x^{2/3} + y^{2/3} + z^{2/3} = a^{2/3}$$

(a) Show that this surface can be represented parametrically as

$$x = a(\sin u \cos v)^3$$
$$y = a(\sin u \sin v)^3 \quad (0 \le u \le \pi,\ 0 \le v \le 2\pi)$$
$$z = a(\cos u)^3$$

(b) Use a CAS to approximate the surface area in the case where $a = 1$.

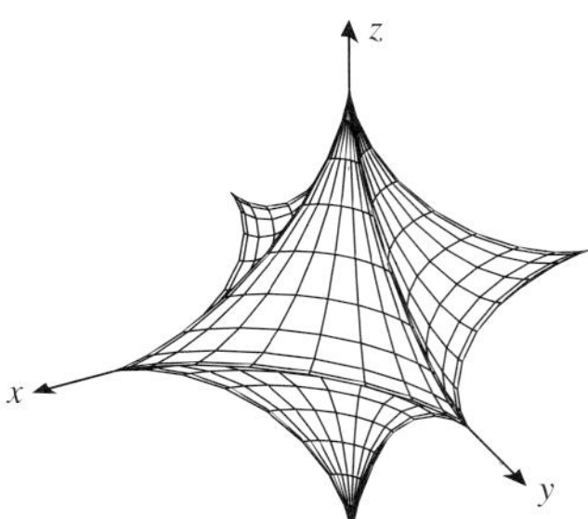

Figure Ex-54

55. (a) Describe the surface that is represented by the parametric equations

$$x = a\sin\phi\cos\theta$$
$$y = b\sin\phi\sin\theta \quad (0 \le \phi \le \pi,\ 0 \le \theta \le 2\pi)$$
$$z = c\cos\phi$$

where $a > 0$, $b > 0$, and $c > 0$.

(b) Use a CAS to approximate the area of the surface for $a = 2, b = 3, c = 4$.

56. (a) Find parametric equations for the surface of revolution that is generated by revolving the curve $z = f(x)$ in the xz-plane about the z-axis.
(b) Use the result obtained in part (a) to find parametric equations for the surface of revolution that is generated by revolving the curve $z = 1/x^2$ in the xz-plane about the z-axis.
(c) Use a graphing utility to check your work by graphing the parametric surface.

57–59 The parametric equations in these exercises represent a quadric surface for positive values of a, b, and c. Identify the type of surface by eliminating the parameters u and v. Check your conclusion by choosing specific values for the constants and generating the surface with a graphing utility.

57. $x = a\cos u\cos v,\ \ y = b\sin u\cos v,\ \ z = c\sin v$

58. $x = a\cos u\cosh v,\ \ y = b\sin u\cosh v,\ \ z = c\sinh v$

59. $x = a\sinh v,\ \ y = b\sinh u\cosh v,\ \ z = c\cosh u\cosh v$

✔ QUICK CHECK ANSWERS 15.4

1. (a) The constant u-curves are circles of radius $1 - u$ centered at $(1 - u, 0, 0)$ and parallel to the yz-plane. (b) The constant v-curves are line segments joining the points $(1, \cos v, \sin v)$ and $(0, 0, 0)$. 2. $\dfrac{\partial \mathbf{r}}{\partial u} = -\mathbf{i} - (\cos v)\mathbf{j} - (\sin v)\mathbf{k}$; $\dfrac{\partial \mathbf{r}}{\partial v} = -[(1-u)\sin v]\mathbf{j} + [(1-u)\cos v]\mathbf{k}$ 3. $\dfrac{1}{\sqrt{8}}(-2\mathbf{i} + \sqrt{3}\mathbf{j} + \mathbf{k})$ 4. $\left\| \dfrac{\partial \mathbf{r}}{\partial u} \times \dfrac{\partial \mathbf{r}}{\partial v} \right\|$ 5. $\sqrt{\left(\dfrac{\partial z}{\partial x}\right)^2 + \left(\dfrac{\partial z}{\partial y}\right)^2 + 1}$

15.5 TRIPLE INTEGRALS

In the preceding sections we defined and discussed properties of double integrals for functions of two variables. In this section we will define triple integrals for functions of three variables.

DEFINITION OF A TRIPLE INTEGRAL

A single integral of a function $f(x)$ is defined over a finite closed interval on the x-axis, and a double integral of a function $f(x, y)$ is defined over a finite closed region R in the xy-plane. Our first goal in this section is to define what is meant by a *triple integral* of $f(x, y, z)$ over a closed solid region G in an xyz-coordinate system. To ensure that G does not extend indefinitely in some direction, we will assume that it can be enclosed in a suitably large box whose sides are parallel to the coordinate planes (Figure 15.5.1). In this case we say that G is a ***finite solid***.

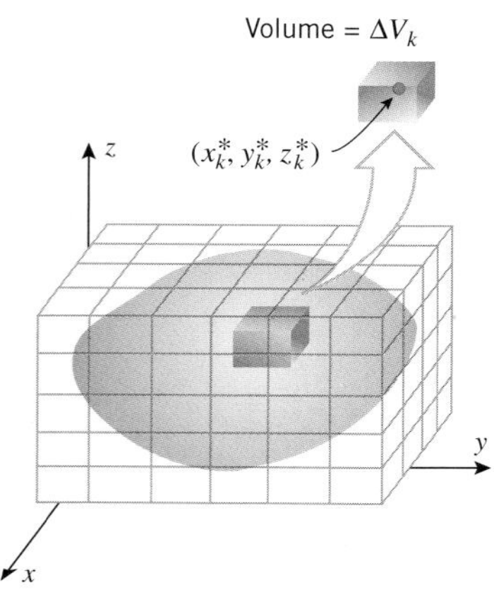

Figure 15.5.1

To define the triple integral of $f(x, y, z)$ over G, we first divide the box into n "subboxes" by planes parallel to the coordinate planes. We then discard those subboxes that contain any points outside of G and choose an arbitrary point in each of the remaining subboxes. As shown in Figure 15.5.1, we denote the volume of the kth remaining subbox by ΔV_k and the point selected in the kth subbox by (x_k^*, y_k^*, z_k^*). Next we form the product

$$f(x_k^*, y_k^*, z_k^*)\Delta V_k$$

for each subbox, then add the products for all of the subboxes to obtain the ***Riemann sum***

$$\sum_{k=1}^{n} f(x_k^*, y_k^*, z_k^*)\Delta V_k$$

Finally, we repeat this process with more and more subdivisions in such a way that the length, width, and height of each subbox approach zero, and n approaches $+\infty$. The limit

$$\iiint_G f(x, y, z)\, dV = \lim_{n\to+\infty} \sum_{k=1}^{n} f(x_k^*, y_k^*, z_k^*)\Delta V_k \tag{1}$$

is called the ***triple integral*** of $f(x, y, z)$ over the region G. Conditions under which the triple integral exists are studied in advanced calculus. However, for our purposes it suffices to say that existence is ensured when f is continuous on G and the region G is not too "complicated."

PROPERTIES OF TRIPLE INTEGRALS

Triple integrals enjoy many properties of single and double integrals:

$$\iiint_G cf(x, y, z)\, dV = c\iiint_G f(x, y, z)\, dV \quad (c \text{ a constant})$$

$$\iiint_G [f(x, y, z) + g(x, y, z)]\, dV = \iiint_G f(x, y, z)\, dV + \iiint_G g(x, y, z)\, dV$$

$$\iiint_G [f(x, y, z) - g(x, y, z)]\, dV = \iiint_G f(x, y, z)\, dV - \iiint_G g(x, y, z)\, dV$$

Moreover, if the region G is subdivided into two subregions G_1 and G_2 (Figure 15.5.2), then

$$\iiint_G f(x, y, z)\, dV = \iiint_{G_1} f(x, y, z)\, dV + \iiint_{G_2} f(x, y, z)\, dV$$

We omit the proofs.

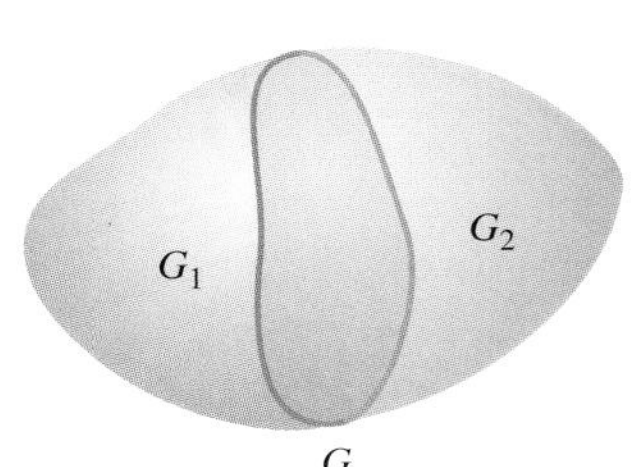

Figure 15.5.2

EVALUATING TRIPLE INTEGRALS OVER RECTANGULAR BOXES

Just as a double integral can be evaluated by two successive single integrations, so a triple integral can be evaluated by three successive integrations. The following theorem, which we state without proof, is the analog of Theorem 15.1.3.

> **15.5.1 THEOREM.** *Let G be the rectangular box defined by the inequalities*
>
> $$a \le x \le b, \quad c \le y \le d, \quad k \le z \le l$$
>
> *If f is continuous on the region G, then*
>
> $$\iiint_G f(x, y, z)\, dV = \int_a^b \int_c^d \int_k^l f(x, y, z)\, dz\, dy\, dx \tag{2}$$
>
> *Moreover, the iterated integral on the right can be replaced with any of the five other iterated integrals that result by altering the order of integration.*

There are two possible orders of integration for the iterated integrals in Theorem 15.1.3:

$$dx\,dy, \quad dy\,dx$$

Six orders of integration are possible for the iterated integral in Theorem 15.5.1:

$$dx\,dy\,dz, \quad dy\,dz\,dx, \quad dz\,dx\,dy$$
$$dx\,dz\,dy, \quad dz\,dy\,dx, \quad dy\,dx\,dz$$

► **Example 1** Evaluate the triple integral

$$\iiint_G 12xy^2z^3\, dV$$

over the rectangular box G defined by the inequalities $-1 \le x \le 2, 0 \le y \le 3, 0 \le z \le 2$.

Solution. Of the six possible iterated integrals we might use, we will choose the one in (2). Thus, we will first integrate with respect to z, holding x and y fixed, then with respect to y, holding x fixed, and finally with respect to x.

$$\begin{aligned}\iiint_G 12xy^2z^3\,dV &= \int_{-1}^{2}\int_{0}^{3}\int_{0}^{2} 12xy^2z^3\,dz\,dy\,dx \\ &= \int_{-1}^{2}\int_{0}^{3}\left[3xy^2z^4\right]_{z=0}^{2} dy\,dx = \int_{-1}^{2}\int_{0}^{3} 48xy^2\,dy\,dx \\ &= \int_{-1}^{2}\left[16xy^3\right]_{y=0}^{3} dx = \int_{-1}^{2} 432x\,dx \\ &= 216x^2\Big]_{-1}^{2} = 648 \blacktriangleleft\end{aligned}$$

EVALUATING TRIPLE INTEGRALS OVER MORE GENERAL REGIONS

Next we will consider how triple integrals can be evaluated over solids that are not rectangular boxes. For the moment we will limit our discussion to solids of the type shown in Figure 15.5.3. Specifically, we will assume that the solid G is bounded above by a surface $z = g_2(x, y)$ and below by a surface $z = g_1(x, y)$ and that the projection of the solid on the xy-plane is a type I or type II region R (see Definition 15.2.1). In addition, we will assume that $g_1(x, y)$ and $g_2(x, y)$ are continuous on R and that $g_1(x, y) \le g_2(x, y)$ on R. Geometrically, this means that the surfaces may touch but cannot cross. We call a solid of this type a ***simple xy-solid***.

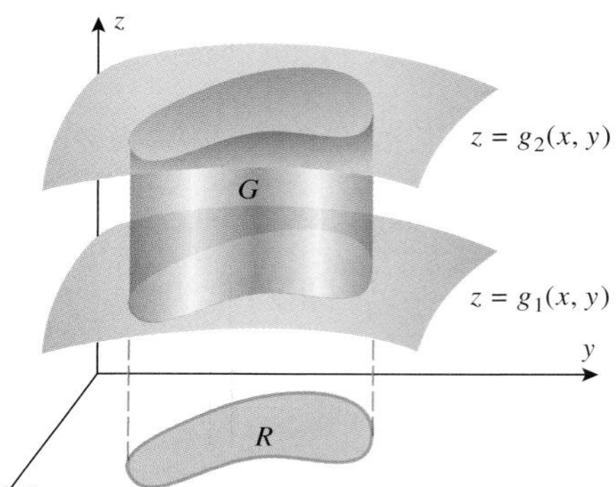

Figure 15.5.3

The following theorem, which we state without proof, will enable us to evaluate triple integrals over simple xy-solids.

> **15.5.2 THEOREM.** *Let G be a simple xy-solid with upper surface $z = g_2(x, y)$ and lower surface $z = g_1(x, y)$, and let R be the projection of G on the xy-plane. If $f(x, y, z)$ is continuous on G, then*
>
> $$\iiint_G f(x, y, z)\,dV = \iint_R \left[\int_{g_1(x,y)}^{g_2(x,y)} f(x, y, z)\,dz\right] dA \qquad (3)$$

In (3), the first integration is with respect to z, after which a function of x and y remains. This function of x and y is then integrated over the region R in the xy-plane. To apply (3), it is helpful to begin with a three-dimensional sketch of the solid G. The limits of integration can be obtained from the sketch as follows:

Determining Limits of Integration: Simple xy-Solid

Step 1. Find an equation $z = g_2(x, y)$ for the upper surface and an equation $z = g_1(x, y)$ for the lower surface of G. The functions $g_1(x, y)$ and $g_2(x, y)$ determine the lower and upper z-limits of integration.

Step 2. Make a two-dimensional sketch of the projection R of the solid on the xy-plane. From this sketch determine the limits of integration for the double integral over R in (3).

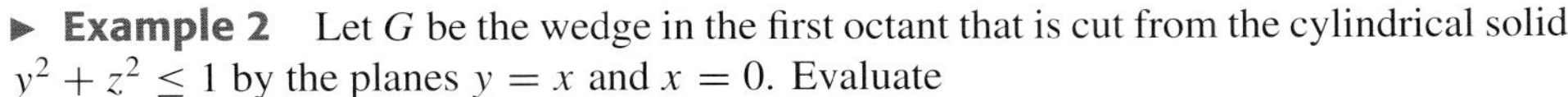

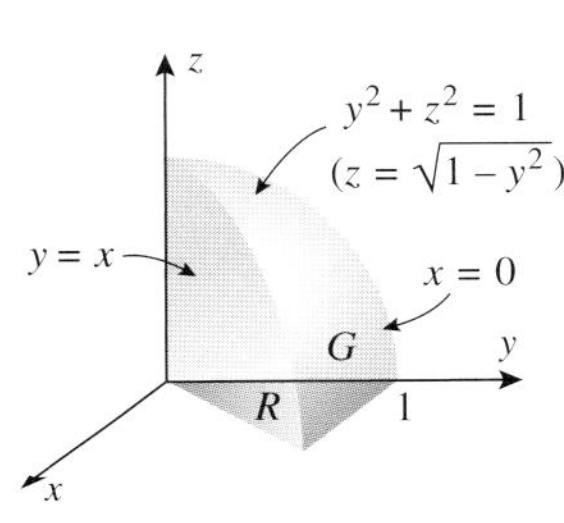

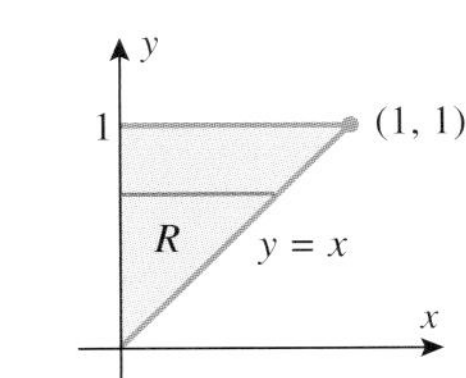

Figure 15.5.4

▶ **Example 2** Let G be the wedge in the first octant that is cut from the cylindrical solid $y^2 + z^2 \le 1$ by the planes $y = x$ and $x = 0$. Evaluate

$$\iiint_G z\, dV$$

Solution. The solid G and its projection R on the xy-plane are shown in Figure 15.5.4. The upper surface of the solid is formed by the cylinder and the lower surface by the xy-plane. Since the portion of the cylinder $y^2 + z^2 = 1$ that lies above the xy-plane has the equation $z = \sqrt{1 - y^2}$, and the xy-plane has the equation $z = 0$, it follows from (3) that

$$\iiint_G z\, dV = \iint_R \left[\int_0^{\sqrt{1-y^2}} z\, dz \right] dA \tag{4}$$

For the double integral over R, the x- and y-integrations can be performed in either order, since R is both a type I and type II region. We will integrate with respect to x first. With this choice, (4) yields

$$\begin{aligned}
\iiint_G z\, dV &= \int_0^1 \int_0^y \int_0^{\sqrt{1-y^2}} z\, dz\, dx\, dy = \int_0^1 \int_0^y \frac{1}{2} z^2 \bigg]_{z=0}^{\sqrt{1-y^2}} dx\, dy \\
&= \int_0^1 \int_0^y \frac{1}{2}(1 - y^2)\, dx\, dy = \frac{1}{2} \int_0^1 (1 - y^2)x \bigg]_{x=0}^{y} dy \\
&= \frac{1}{2} \int_0^1 (y - y^3)\, dy = \frac{1}{2} \left[\frac{1}{2} y^2 - \frac{1}{4} y^4 \right]_0^1 = \frac{1}{8}
\end{aligned}$$ ◀

TECHNOLOGY MASTERY

Most computer algebra systems have a built-in capability for computing iterated triple integrals. If you have a CAS, consult the relevant documentation and use the CAS to check Examples 1 and 2.

VOLUME CALCULATED AS A TRIPLE INTEGRAL

Triple integrals have many physical interpretations, some of which we will consider in the next section. However, in the special case where $f(x, y, z) = 1$, Formula (1) yields

$$\iiint_G dV = \lim_{n \to +\infty} \sum_{k=1}^{n} \Delta V_k$$

which Figure 15.5.1 suggests is the volume of G; that is,

$$\text{volume of } G = \iiint\limits_G dV \tag{5}$$

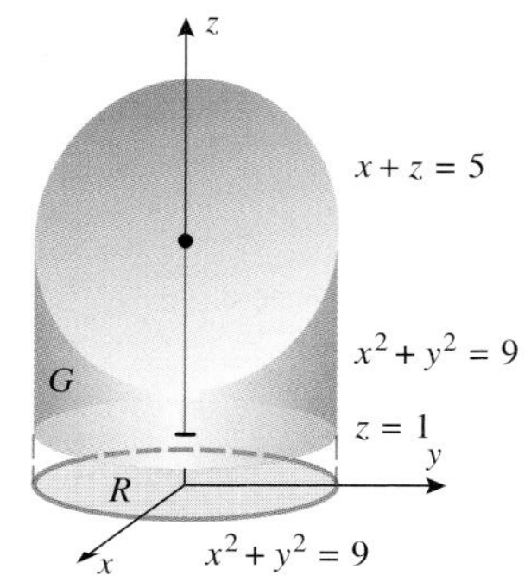

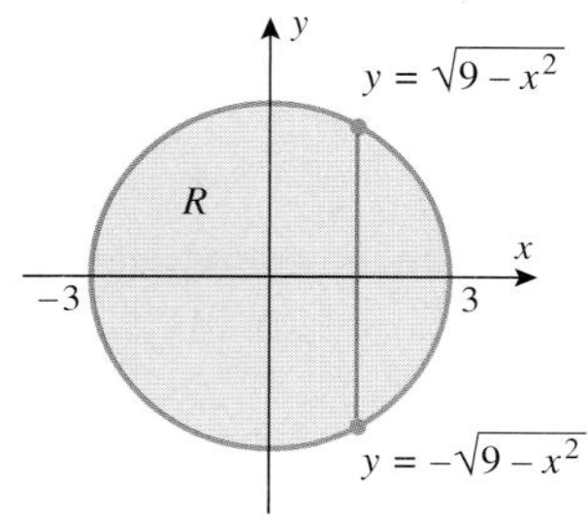

Figure 15.5.5

▶ **Example 3** Use a triple integral to find the volume of the solid within the cylinder $x^2 + y^2 = 9$ and between the planes $z = 1$ and $x + z = 5$.

Solution. The solid G and its projection R on the xy-plane are shown in Figure 15.5.5. The lower surface of the solid is the plane $z = 1$ and the upper surface is the plane $x + z = 5$ or, equivalently, $z = 5 - x$. Thus, from (3) and (5)

$$\text{volume of } G = \iiint\limits_G dV = \iint\limits_R \left[\int_1^{5-x} dz\right] dA \tag{6}$$

For the double integral over R, we will integrate with respect to y first. Thus, (6) yields

$$\begin{aligned}
\text{volume of } G &= \int_{-3}^{3}\int_{-\sqrt{9-x^2}}^{\sqrt{9-x^2}}\int_1^{5-x} dz\,dy\,dx = \int_{-3}^{3}\int_{-\sqrt{9-x^2}}^{\sqrt{9-x^2}} z\Big]_{z=1}^{5-x} dy\,dx \\
&= \int_{-3}^{3}\int_{-\sqrt{9-x^2}}^{\sqrt{9-x^2}} (4-x)\,dy\,dx = \int_{-3}^{3}(8-2x)\sqrt{9-x^2}\,dx \\
&= 8\int_{-3}^{3}\sqrt{9-x^2}\,dx - \int_{-3}^{3} 2x\sqrt{9-x^2}\,dx \\
&= 8\left(\frac{9}{2}\pi\right) - \int_{-3}^{3} 2x\sqrt{9-x^2}\,dx \\
&= 8\left(\frac{9}{2}\pi\right) - 0 = 36\pi
\end{aligned}$$ ◀

For the first integral, see Formula (3) of Section 8.4.

The second integral is 0 because the integrand is an odd function.

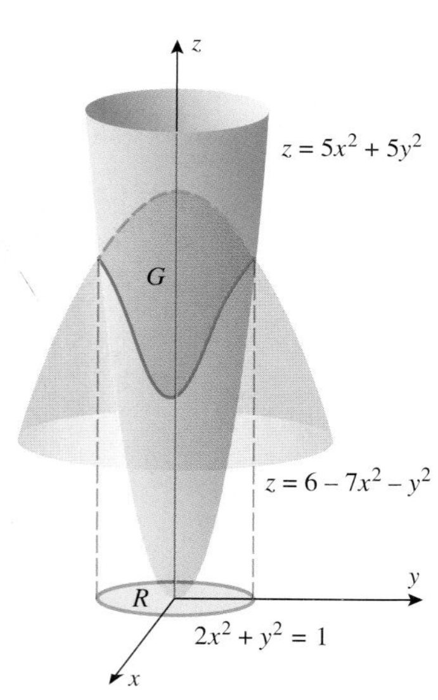

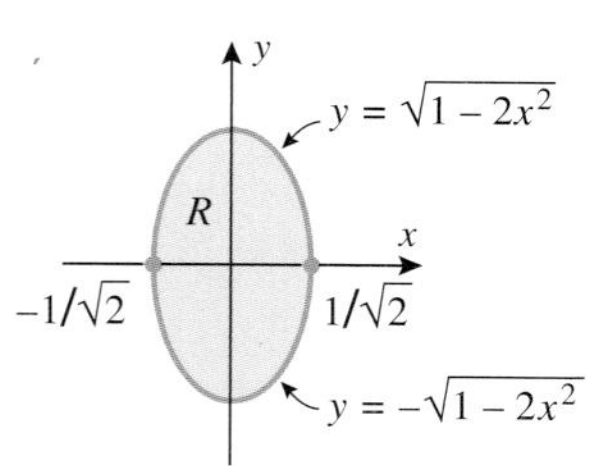

Figure 15.5.6

▶ **Example 4** Find the volume of the solid enclosed between the paraboloids

$$z = 5x^2 + 5y^2 \quad \text{and} \quad z = 6 - 7x^2 - y^2$$

Solution. The solid G and its projection R on the xy-plane are shown in Figure 15.5.6. The projection R is obtained by solving the given equations simultaneously to determine where the paraboloids intersect. We obtain

$$5x^2 + 5y^2 = 6 - 7x^2 - y^2$$

or

$$2x^2 + y^2 = 1 \tag{7}$$

which tells us that the paraboloids intersect in a curve on the elliptic cylinder given by (7).

The projection of this intersection on the xy-plane is an ellipse with this same equation. Therefore,

$$\text{volume of } G = \iiint\limits_G dV = \iint\limits_R \left[\int_{5x^2+5y^2}^{6-7x^2-y^2} dz\right] dA$$

$$= \int_{-1/\sqrt{2}}^{1/\sqrt{2}} \int_{-\sqrt{1-2x^2}}^{\sqrt{1-2x^2}} \int_{5x^2+5y^2}^{6-7x^2-y^2} dz\,dy\,dx$$

$$= \int_{-1/\sqrt{2}}^{1/\sqrt{2}} \int_{-\sqrt{1-2x^2}}^{\sqrt{1-2x^2}} (6-12x^2-6y^2)\,dy\,dx$$

$$= \int_{-1/\sqrt{2}}^{1/\sqrt{2}} \left[6(1-2x^2)y - 2y^3\right]_{y=-\sqrt{1-2x^2}}^{\sqrt{1-2x^2}} dx$$

$$= 8\int_{-1/\sqrt{2}}^{1/\sqrt{2}} (1-2x^2)^{3/2}\,dx = \frac{8}{\sqrt{2}}\int_{-\pi/2}^{\pi/2} \cos^4\theta\,d\theta = \frac{3\pi}{\sqrt{2}} \blacktriangleleft$$

Let $x = \frac{1}{\sqrt{2}}\sin\theta$.

Use the Wallis cosine formula in Exercise 66 of Section 8.3.

INTEGRATION IN OTHER ORDERS

In Formula (3) for integrating over a simple xy-solid, the z-integration was performed first. However, there are situations in which it is preferable to integrate in a different order. For example, Figure 15.5.7a shows a ***simple xz-solid***, and Figure 15.5.7b shows a ***simple yz-solid***. For a simple xz-solid it is usually best to integrate with respect to y first, and for a simple yz-solid it is usually best to integrate with respect to x first:

$$\underset{\text{simple } xz\text{-solid}}{\iiint\limits_G} f(x,y,z)\,dV = \iint\limits_R \left[\int_{g_1(x,z)}^{g_2(x,z)} f(x,y,z)\,dy\right] dA \qquad (8)$$

$$\underset{\text{simple } yz\text{-solid}}{\iiint\limits_G} f(x,y,z)\,dV = \iint\limits_R \left[\int_{g_1(y,z)}^{g_2(y,z)} f(x,y,z)\,dx\right] dA \qquad (9)$$

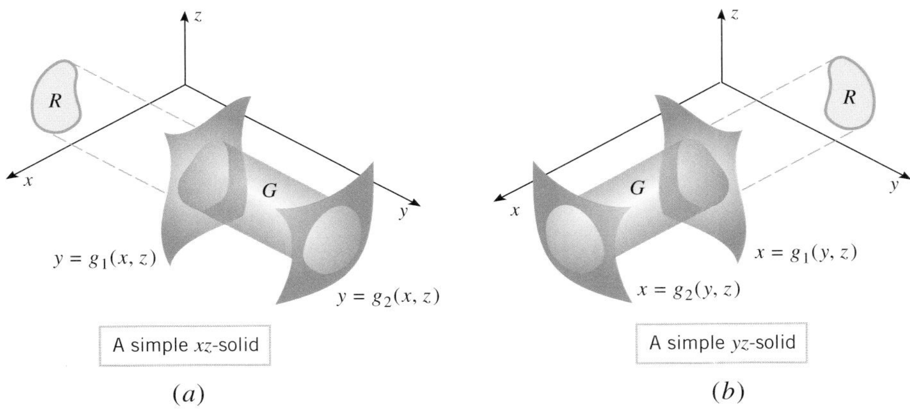

Figure 15.5.7

Sometimes a solid G can be viewed as a simple xy-solid, a simple xz-solid, and a simple yz-solid, in which case the order of integration can be chosen to simplify the computations.

▶ Example 5 In Example 2, we evaluated

$$\iiint_G z\,dV$$

over the wedge in Figure 15.5.4 by integrating first with respect to z. Evaluate this integral by integrating first with respect to x.

Solution. The solid is bounded in the back by the plane $x = 0$ and in the front by the plane $x = y$, so

$$\iiint_G z\,dV = \iint_R \left[\int_0^y z\,dx\right] dA$$

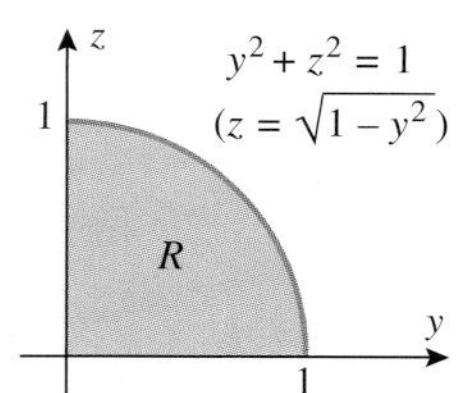

Figure 15.5.8

where R is the projection of G on the yz-plane (Figure 15.5.8). The integration over R can be performed first with respect to z and then y or vice versa. Performing the z-integration first yields

$$\iiint_G z\,dV = \int_0^1 \int_0^{\sqrt{1-y^2}} \int_0^y z\,dx\,dz\,dy = \int_0^1 \int_0^{\sqrt{1-y^2}} zx\Big]_{x=0}^{y} dz\,dy$$

$$= \int_0^1 \int_0^{\sqrt{1-y^2}} zy\,dz\,dy = \int_0^1 \frac{1}{2}z^2 y\Big]_{z=0}^{\sqrt{1-y^2}} dy = \int_0^1 \frac{1}{2}(1-y^2)y\,dy = \frac{1}{8}$$

which agrees with the result in Example 2. ◀

✔ QUICK CHECK EXERCISES 15.5 *(See page 1064 for answers.)*

1. The iterated integral
$$\int_1^5 \int_2^4 \int_3^6 f(x, y, z)\,dx\,dz\,dy$$
integrates f over the rectangular box defined by
$\_\_\_\_ \le x \le \_\_\_\_$, $\_\_\_\_ \le y \le \_\_\_\_$,
$\_\_\_\_ \le z \le \_\_\_\_$

2. Let G be the solid in the first octant bounded below by the surface $z = y + x^2$ and bounded above by the plane $z = 4$. Supply the missing limits of integration.

(a) $\iiint_G f(x, y, z)\,dA = \int_\square^\square \int_\square^\square \int_{y+x^2}^4 f(x, y, z)\,dz\,dx\,dy$

(b) $\iiint_G f(x, y, z)\,dA = \int_\square^\square \int_\square^\square \int_{y+x^2}^4 f(x, y, z)\,dz\,dy\,dx$

(c) $\iiint_G f(x, y, z)\,dA = \int_\square^\square \int_\square^\square \int_\square^\square f(x, y, z)\,dy\,dz\,dx$

3. The volume of the solid G in Quick Check Exercises 2 is ____.

EXERCISE SET 15.5 [C] CAS

1–8 Evaluate the iterated integral.

1. $\int_{-1}^1 \int_0^2 \int_0^1 (x^2 + y^2 + z^2)\,dx\,dy\,dz$

2. $\int_{1/3}^{1/2} \int_0^\pi \int_0^1 zx \sin xy\,dz\,dy\,dx$

3. $\int_0^2 \int_{-1}^{y^2} \int_{-1}^z yz\,dx\,dz\,dy$

4. $\int_0^{\pi/4} \int_0^1 \int_0^{x^2} x \cos y\,dz\,dx\,dy$

5. $\int_0^3 \int_0^{\sqrt{9-z^2}} \int_0^x xy\,dy\,dx\,dz$

6. $\int_1^3 \int_x^{x^2} \int_0^{\ln z} xe^y\,dy\,dz\,dx$

7. $\int_0^2 \int_0^{\sqrt{4-x^2}} \int_{-5+x^2+y^2}^{3-x^2-y^2} x\,dz\,dy\,dx$

8. $\int_1^2 \int_z^2 \int_0^{\sqrt{3}y} \frac{y}{x^2+y^2}\,dx\,dy\,dz$

9–12 Evaluate the triple integral.

9. $\iiint_G xy \sin yz\, dV$, where G is the rectangular box defined by the inequalities $0 \le x \le \pi$, $0 \le y \le 1$, $0 \le z \le \pi/6$.

10. $\iiint_G y\, dV$, where G is the solid enclosed by the plane $z = y$, the xy-plane, and the parabolic cylinder $y = 1 - x^2$.

11. $\iiint_G xyz\, dV$, where G is the solid in the first octant that is bounded by the parabolic cylinder $z = 2 - x^2$ and the planes $z = 0$, $y = x$, and $y = 0$.

12. $\iiint_G \cos(z/y)\, dV$, where G is the solid defined by the inequalities $\pi/6 \le y \le \pi/2$, $y \le x \le \pi/2$, $0 \le z \le xy$.

C **13.** Use the numerical triple integral operation of a CAS to approximate

$$\iiint_G \frac{\sqrt{x + z^2}}{y}\, dV$$

where G is the rectangular box defined by the inequalities $0 \le x \le 3$, $1 \le y \le 2$, $-2 \le z \le 1$.

C **14.** Use the numerical triple integral operation of a CAS to approximate

$$\iiint_G e^{-x^2-y^2-z^2}\, dV$$

where G is the spherical region $x^2 + y^2 + z^2 \le 1$.

15–18 Use a triple integral to find the volume of the solid.

15. The solid in the first octant bounded by the coordinate planes and the plane $3x + 6y + 4z = 12$.

16. The solid bounded by the surface $z = \sqrt{y}$ and the planes $x + y = 1$, $x = 0$, and $z = 0$.

17. The solid bounded by the surface $y = x^2$ and the planes $y + z = 4$ and $z = 0$.

18. The wedge in the first octant that is cut from the solid cylinder $y^2 + z^2 \le 1$ by the planes $y = x$ and $x = 0$.

FOCUS ON CONCEPTS

19. Let G be the solid enclosed by the surfaces in the accompanying figure. Fill in the missing limits of integration.

(a) $\iiint_G f(x, y, z)\, dV$

$$= \int_{\square}^{\square} \int_{\square}^{\square} \int_{\square}^{\square} f(x, y, z)\, dz\, dy\, dx$$

(b) $\iiint_G f(x, y, z)\, dV$

$$= \int_{\square}^{\square} \int_{\square}^{\square} \int_{\square}^{\square} f(x, y, z)\, dz\, dx\, dy$$

20. Let G be the solid enclosed by the surfaces in the accompanying figure. Fill in the missing limits of integration.

(a) $\iiint_G f(x, y, z)\, dV$

$$= \int_{\square}^{\square} \int_{\square}^{\square} \int_{\square}^{\square} f(x, y, z)\, dz\, dy\, dx$$

(b) $\iiint_G f(x, y, z)\, dV$

$$= \int_{\square}^{\square} \int_{\square}^{\square} \int_{\square}^{\square} f(x, y, z)\, dz\, dx\, dy$$

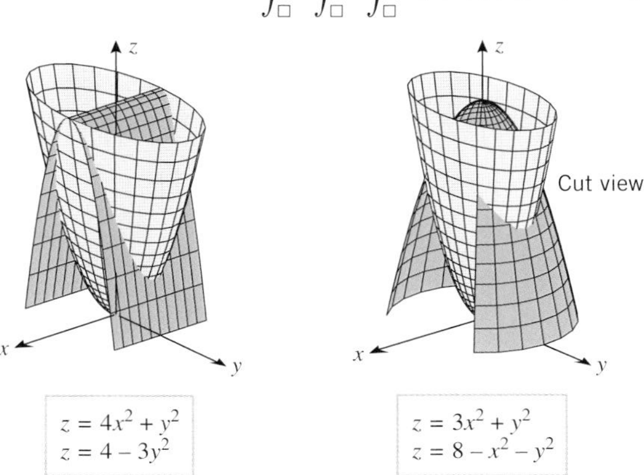

Figure Ex-19 **Figure Ex-20**

21–24 Set up (but do not evaluate) an iterated triple integral for the volume of the solid enclosed between the given surfaces.

21. The surfaces in Exercise 19.

22. The surfaces in Exercise 20.

23. The elliptic cylinder $x^2 + 9y^2 = 9$ and the planes $z = 0$ and $z = x + 3$.

24. The cylinders $x^2 + y^2 = 1$ and $x^2 + z^2 = 1$.

25–26 In each part, sketch the solid whose volume is given by the integral.

25. (a) $\displaystyle\int_{-1}^{1} \int_{-\sqrt{1-x^2}}^{\sqrt{1-x^2}} \int_{0}^{y+1} dz\, dy\, dx$

(b) $\displaystyle\int_{0}^{9} \int_{0}^{y/3} \int_{0}^{\sqrt{y^2-9x^2}} dz\, dx\, dy$

(c) $\displaystyle\int_{0}^{1} \int_{0}^{\sqrt{1-x^2}} \int_{0}^{2} dy\, dz\, dx$

26. (a) $\displaystyle\int_{0}^{3} \int_{x^2}^{9} \int_{0}^{2} dz\, dy\, dx$

(b) $\displaystyle\int_{0}^{2} \int_{0}^{2-y} \int_{0}^{2-x-y} dz\, dx\, dy$

(c) $\displaystyle\int_{-2}^{2} \int_{0}^{4-y^2} \int_{0}^{2} dx\, dz\, dy$

27–30 The ***average value*** or ***mean value*** of a continuous function $f(x, y, z)$ over a solid G is defined as

$$f_{\text{ave}} = \frac{1}{V(G)} \iiint_G f(x, y, z)\, dV$$

where $V(G)$ is the volume of the solid G (compare to the definition preceding Exercise 61 of Section 15.2). Use this definition in these exercises.

27. Find the average value of $f(x, y, z) = x + y + z$ over the tetrahedron shown in the accompanying figure.

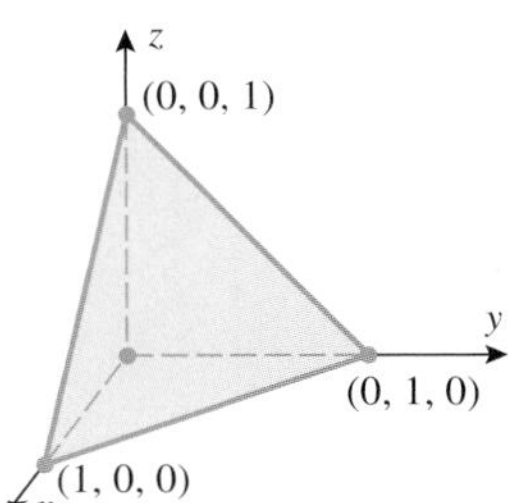

Figure Ex-27

28. Find the average value of $f(x, y, z) = xyz$ over the spherical region $x^2 + y^2 + z^2 \le 1$.

C **29.** Use the numerical triple integral operation of a CAS to approximate the average distance from the origin to a point in the solid of Example 4.

C **30.** Let $d(x, y, z)$ be the distance from the point (z, z, z) to the point $(x, y, 0)$. Use the numerical triple integral operation of a CAS to approximate the average value of d for $0 \le x \le 1$, $0 \le y \le 1$, and $0 \le z \le 1$. Write a short explanation as to why this value may be considered to be the average distance between a point on the diagonal from $(0, 0, 0)$ to $(1, 1, 1)$ and a point on the face in the xy-plane for the unit cube $0 \le x \le 1$, $0 \le y \le 1$, and $0 \le z \le 1$.

31. Let G be the tetrahedron in the first octant bounded by the coordinate planes and the plane

$$\frac{x}{a} + \frac{y}{b} + \frac{z}{c} = 1 \quad (a > 0, b > 0, c > 0)$$

(a) List six different iterated integrals that represent the volume of G.

(b) Evaluate any one of the six to show that the volume of G is $\frac{1}{6}abc$.

32. Use a triple integral to derive the formula for the volume of the ellipsoid

$$\frac{x^2}{a^2} + \frac{y^2}{b^2} + \frac{z^2}{c^2} = 1$$

FOCUS ON CONCEPTS

33–34 Express each integral as an equivalent integral in which the z-integration is performed first, the y-integration second, and the x-integration last.

33. (a) $\displaystyle\int_0^5 \int_0^2 \int_0^{\sqrt{4-y^2}} f(x, y, z)\, dx\, dy\, dz$

(b) $\displaystyle\int_0^9 \int_0^{3-\sqrt{x}} \int_0^z f(x, y, z)\, dy\, dz\, dx$

(c) $\displaystyle\int_0^4 \int_y^{8-y} \int_0^{\sqrt{4-y}} f(x, y, z)\, dx\, dz\, dy$

34. (a) $\displaystyle\int_0^3 \int_0^{\sqrt{9-z^2}} \int_0^{\sqrt{9-y^2-z^2}} f(x, y, z)\, dx\, dy\, dz$

(b) $\displaystyle\int_0^4 \int_0^2 \int_0^{x/2} f(x, y, z)\, dy\, dz\, dx$

(c) $\displaystyle\int_0^4 \int_0^{4-y} \int_0^{\sqrt{z}} f(x, y, z)\, dx\, dz\, dy$

C **35.** (a) Find the region G over which the triple integral

$$\iiint_G (1 - x^2 - y^2 - z^2)\, dV$$

has its maximum value.

(b) Use the numerical triple integral operation of a CAS to approximate the maximum value.

(c) Find the exact maximum value.

36. Let G be the rectangular box defined by the inequalities $a \le x \le b$, $c \le y \le d$, $k \le z \le l$. Show that

$$\iiint_G f(x)g(y)h(z)\, dV = \left[\int_a^b f(x)\, dx\right]\left[\int_c^d g(y)\, dy\right]\left[\int_k^l h(z)\, dz\right]$$

37. Use the result of Exercise 36 to evaluate

(a) $\displaystyle\iiint_G xy^2 \sin z\, dV$, where G is the set of points satisfying $-1 \le x \le 1$, $0 \le y \le 1$, $0 \le z \le \pi/2$;

(b) $\displaystyle\iiint_G e^{2x+y-z}\, dV$, where G is the set of points satisfying $0 \le x \le 1$, $0 \le y \le \ln 3$, $0 \le z \le \ln 2$.

✔ QUICK CHECK ANSWERS 15.5

1. $3 \le x \le 6$, $1 \le y \le 5$, $2 \le z \le 4$ **2.** (a) $\displaystyle\int_0^4 \int_0^{\sqrt{4-y}} \int_{y+x^2}^4 f(x, y, z)\, dz\, dx\, dy$ (b) $\displaystyle\int_0^2 \int_0^{4-x^2} \int_{y+x^2}^4 f(x, y, z)\, dz\, dy\, dx$ (c) $\displaystyle\int_0^2 \int_{x^2}^4 \int_0^{z-x^2} f(x, y, z)\, dy\, dz\, dx$ **3.** $\dfrac{128}{15}$

15.6 CENTROID, CENTER OF GRAVITY, THEOREM OF PAPPUS

Suppose that a rigid physical body is acted on by a gravitational field. Because the body is composed of many particles, each of which is affected by gravity, the action of a constant gravitational field on the body consists of a large number of forces distributed over the entire body. However, these individual forces can be replaced by a single force acting at a point called the ***center of gravity*** *of the body. In this section we will show how double and triple integrals can be used to locate centers of gravity.*

DENSITY OF A LAMINA

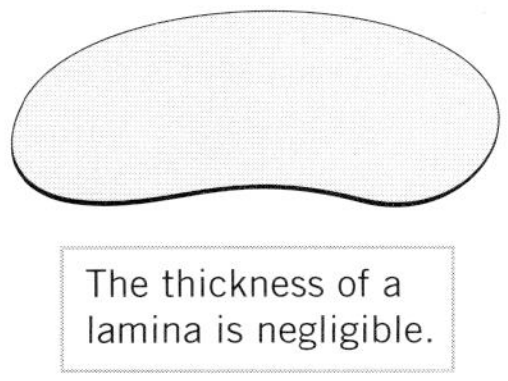

The thickness of a lamina is negligible.

Figure 15.6.1

Let us consider an idealized flat object that is thin enough to be viewed as a two-dimensional plane region (Figure 15.6.1). Such an object is called a ***lamina***. A lamina is called ***homogeneous*** if its composition is uniform throughout and ***inhomogeneous*** otherwise. The ***density*** of a *homogeneous* lamina is defined to be its mass per unit area. Thus, the density δ of a homogeneous lamina of mass M and area A is given by $\delta = M/A$.

For an inhomogeneous lamina the composition may vary from point to point, and hence an appropriate definition of "density" must reflect this. To motivate such a definition, suppose that the lamina is placed in an xy-plane. The density at a point (x, y) can be specified by a function $\delta(x, y)$, called the ***density function***, which can be interpreted as follows. Construct a small rectangle centered at (x, y) and let ΔM and ΔA be the mass and area of the portion of the lamina enclosed by this rectangle (Figure 15.6.2). If the ratio $\Delta M/\Delta A$ approaches a limiting value as the dimensions (and hence the area) of the rectangle approach zero, then this limit is considered to be the density of the lamina at (x, y). Symbolically,

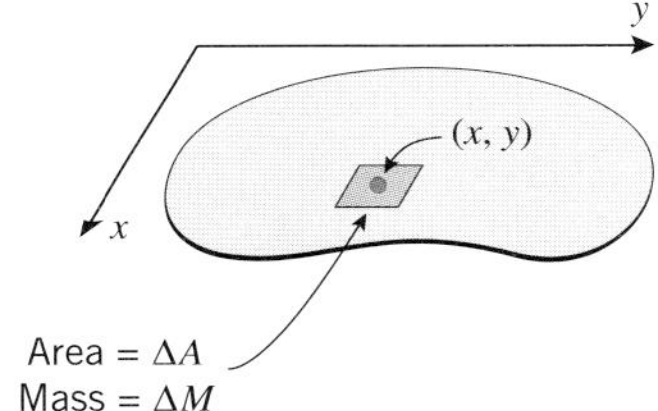

Figure 15.6.2

$$\delta(x, y) = \lim_{\Delta A \to 0} \frac{\Delta M}{\Delta A} \tag{1}$$

From this relationship we obtain the approximation

$$\Delta M \approx \delta(x, y)\Delta A \tag{2}$$

which relates the mass and area of a small rectangular portion of the lamina centered at (x, y). It is assumed that as the dimensions of the rectangle tend to zero, the error in this approximation also tends to zero.

MASS OF A LAMINA

The following result shows how to find the mass of a lamina from its density function.

15.6.1 MASS OF A LAMINA. If a lamina with a continuous density function $\delta(x, y)$ occupies a region R in the xy-plane, then its total mass M is given by

$$M = \iint_R \delta(x, y)\, dA \tag{3}$$

This formula can be motivated by a familiar limiting process that can be outlined as follows: Imagine the lamina to be subdivided into rectangular pieces using lines parallel to the coordinate axes and excluding from consideration any nonrectangular parts at the boundary (Figure 15.6.3). Assume that there are n such rectangular pieces, and suppose that the kth piece has area ΔA_k. If we let (x_k^*, y_k^*) denote the center of the kth piece, then from Formula (2), the mass ΔM_k of this piece can be approximated by

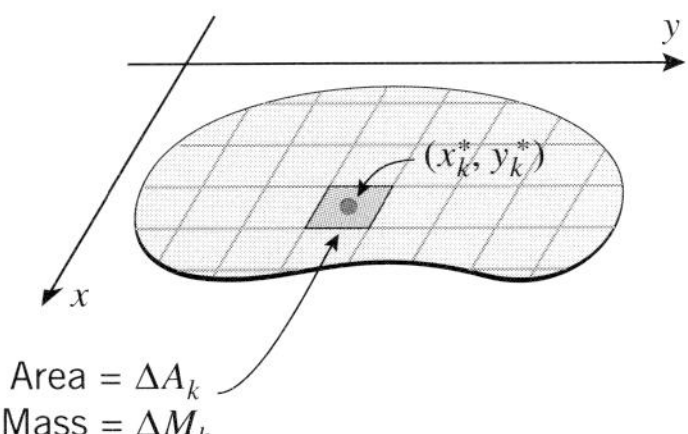

Figure 15.6.3

$$\Delta M_k \approx \delta(x_k^*, y_k^*)\Delta A_k \tag{4}$$

and hence the mass M of the entire lamina can be approximated by

$$M \approx \sum_{k=1}^{n} \delta(x_k^*, y_k^*)\Delta A_k$$

If we now increase n in such a way that the dimensions of the rectangles tend to zero, then it is plausible that the errors in our approximations will approach zero, so

$$M = \lim_{n\to+\infty} \sum_{k=1}^{n} \delta(x_k^*, y_k^*)\Delta A_k = \iint_R \delta(x, y)\, dA$$

▶ **Example 1** A triangular lamina with vertices (0, 0), (0, 1), and (1, 0) has density function $\delta(x, y) = xy$. Find its total mass.

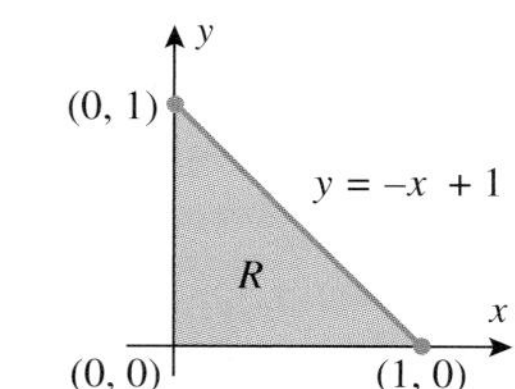

Figure 15.6.4

Solution. Referring to (3) and Figure 15.6.4, the mass M of the lamina is

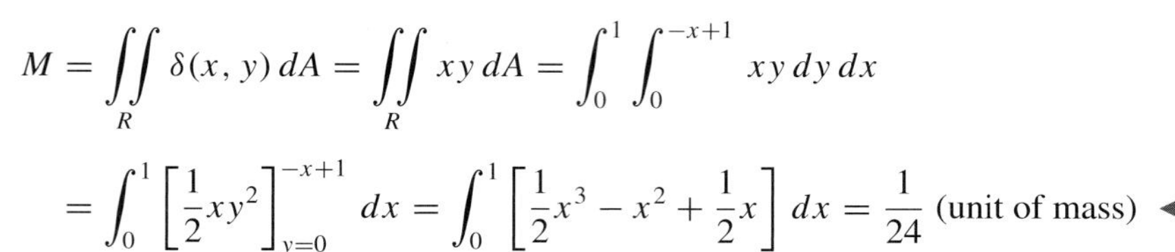

$$M = \iint_R \delta(x, y)\, dA = \iint_R xy\, dA = \int_0^1 \int_0^{-x+1} xy\, dy\, dx$$

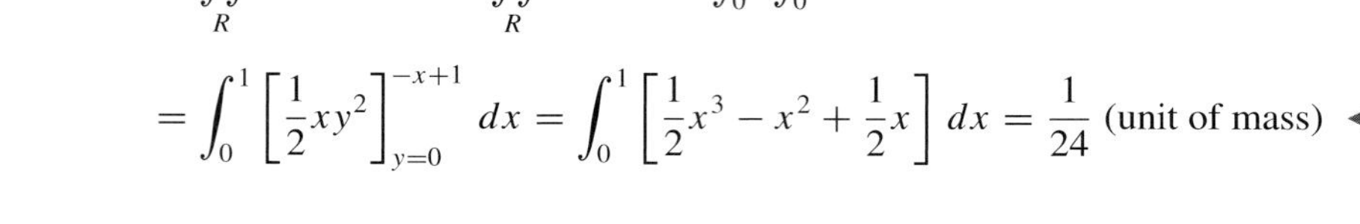

$$= \int_0^1 \left[\frac{1}{2}xy^2\right]_{y=0}^{-x+1} dx = \int_0^1 \left[\frac{1}{2}x^3 - x^2 + \frac{1}{2}x\right] dx = \frac{1}{24} \text{ (unit of mass)}$$ ◀

CENTER OF GRAVITY OF A LAMINA

Assume that the acceleration due to the force of gravity is constant and acts downward, and suppose that a lamina occupies a region R in a horizontal xy-plane. It can be shown that there exists a unique point $(\bar{x}, \bar{y})$ (which may or may not belong to R) such that the effect of gravity on the lamina is "equivalent" to that of a single force acting at the point $(\bar{x}, \bar{y})$. This point is called the ***center of gravity*** of the lamina, and if it is in R, then the lamina will balance horizontally on the point of a support placed at $(\bar{x}, \bar{y})$. For example, the center of gravity of a disk of uniform density is at the center of the disk, and the center of gravity of a rectangular region of uniform density is at the center of the rectangle. For an irregularly shaped lamina or for a lamina in which the density varies from point to point, locating the center of gravity requires calculus.

> **15.6.2 PROBLEM.** Suppose that a lamina with a continuous density function $\delta(x, y)$ occupies a region R in a horizontal xy-plane. Find the coordinates $(\bar{x}, \bar{y})$ of the center of gravity.

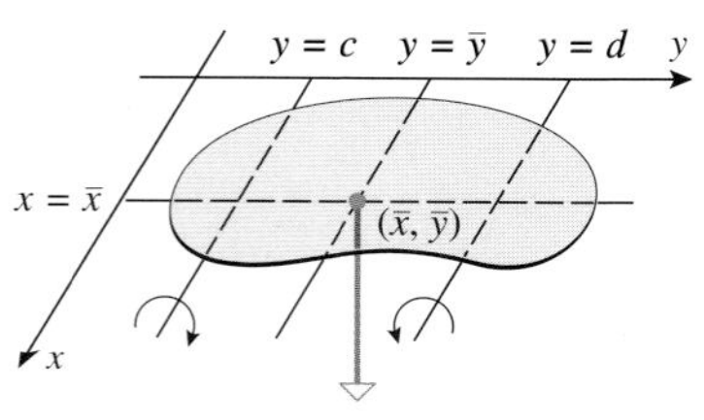

Force of gravity acting on the center of gravity of the lamina

Figure 15.6.5

To motivate the solution, consider what happens if we try to balance the lamina on a knife-edge parallel to the x-axis. Suppose the lamina in Figure 15.6.5 is placed on a knife-edge along a line $y = c$ that does not pass through the center of gravity. Because the lamina behaves as if its entire mass is concentrated at the center of gravity $(\bar{x}, \bar{y})$, the lamina will be rotationally unstable and the force of gravity will cause a rotation about $y = c$. Similarly, the lamina will undergo a rotation if placed on a knife-edge along $y = d$. However, if the knife-edge runs along the line $y = \bar{y}$ through the center of gravity, the lamina will be in perfect balance. Similarly, the lamina will be in perfect balance on a knife-edge along the line $x = \bar{x}$ through the center of gravity. This suggests that the center of gravity of a lamina can be determined as the intersection of two lines of balance, one parallel to the x-axis and the other parallel to the y-axis. In order to find these lines of balance, we will need some preliminary results about rotations.

Children on a seesaw learn by experience that a lighter child can balance a heavier one by sitting farther from the fulcrum or pivot point. This is because the tendency for an object to produce rotation is proportional not only to its mass but also to the distance between the object and the fulcrum. To make this more precise, consider an x-axis, which we view as a weightless beam. If a point-mass m is located on the axis at x, then the tendency for that mass to produce a rotation of the beam about a point a on the axis is measured by the following quantity, called the ***moment of m about x = a***:

$$\begin{bmatrix}\text{moment of } m\\ \text{about } a\end{bmatrix} = m(x-a)$$

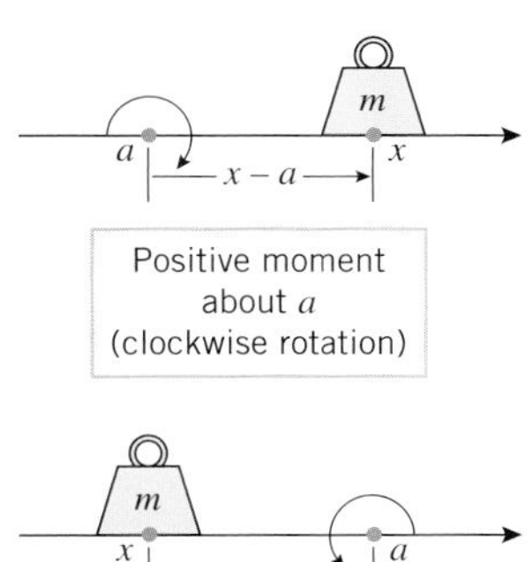

Figure 15.6.6

The number $x - a$ is called the ***lever arm***. Depending on whether the mass is to the right or left of a, the lever arm is either the distance between x and a or the negative of this distance (Figure 15.6.6). Positive lever arms result in positive moments and clockwise rotations, and negative lever arms result in negative moments and counterclockwise rotations.

Suppose that masses $m_1, m_2, \ldots, m_n$ are located at $x_1, x_2, \ldots, x_n$ on a coordinate axis and a fulcrum is positioned at the point a (Figure 15.6.7). Depending on whether the sum of the moments about a,

$$\sum_{k=1}^{n} m_k(x_k - a) = m_1(x_1 - a) + m_2(x_2 - a) + \cdots + m_n(x_n - a)$$

is positive, negative, or zero, a weightless beam along the axis will rotate clockwise about a, rotate counterclockwise about a, or balance perfectly. In the last case, the system of masses is said to be in ***equilibrium***.

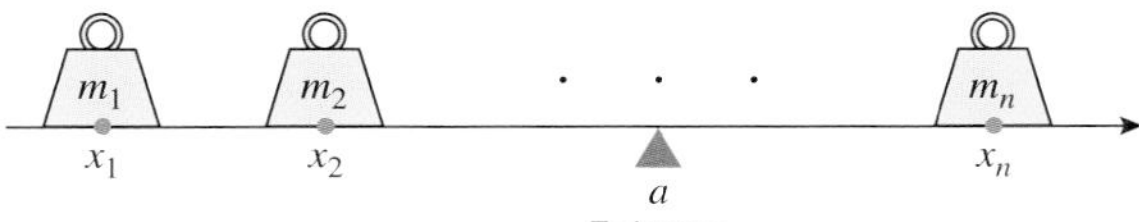

Figure 15.6.7

The preceding ideas can be extended to masses distributed in two-dimensional space. If we imagine the xy-plane to be a weightless sheet supporting a point-mass m located at a point (x, y), then the tendency for the mass to produce a rotation of the sheet about the line $x = a$ is $m(x - a)$, called the ***moment of m about x = a***, and the tendency for the mass to produce a rotation about the line $y = c$ is $m(y - c)$, called the ***moment of m about y = c*** (Figure 15.6.8). In summary,

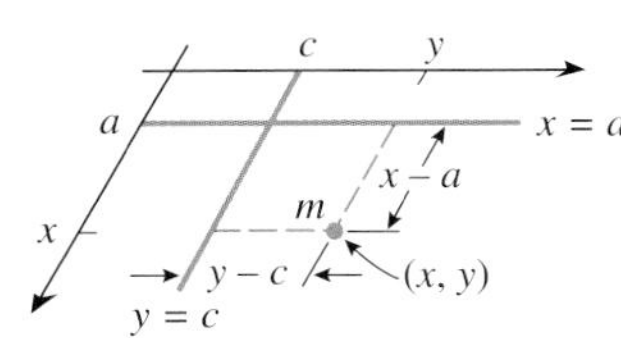

Figure 15.6.8

$$\begin{bmatrix}\text{moment of } m\\ \text{about the}\\ \text{line } x = a\end{bmatrix} = m(x-a) \quad\text{and}\quad \begin{bmatrix}\text{moment of } m\\ \text{about the}\\ \text{line } y = c\end{bmatrix} = m(y-c) \qquad (5\text{–}6)$$

If a number of masses are distributed throughout the xy-plane, then the plane (viewed as a weightless sheet) will balance on a knife-edge along the line $x = a$ if the sum of the moments about the line is zero. Similarly for the line $y = c$.

We are now ready to solve Problem 15.6.2. We imagine the lamina to be subdivided into rectangular pieces using lines parallel to the coordinate axes and excluding from consideration any nonrectangular pieces at the boundary (Figure 15.6.3). We assume that there are n such rectangular pieces and that the kth piece has area ΔA_k and mass ΔM_k. We will let (x_k^*, y_k^*) be the center of the kth piece, and we will assume that the entire mass of the kth piece is concentrated at its center. From (4), the mass of the kth piece can be approximated by

$$\Delta M_k \approx \delta(x_k^*, y_k^*)\Delta A_k$$

Since the lamina balances on the lines $x = \bar{x}$ and $y = \bar{y}$, the sum of the moments of the rectangular pieces about those lines should be close to zero; that is,

$$\sum_{k=1}^{n}(x_k^* - \bar{x})\Delta M_k = \sum_{k=1}^{n}(x_k^* - \bar{x})\delta(x_k^*, y_k^*)\Delta A_k \approx 0$$

$$\sum_{k=1}^{n}(y_k^* - \bar{y})\Delta M_k = \sum_{k=1}^{n}(y_k^* - \bar{y})\delta(x_k^*, y_k^*)\Delta A_k \approx 0$$

If we now increase n in such a way that the dimensions of the rectangles tend to zero, then it is plausible that the errors in our approximations will approach zero, so that

$$\lim_{n\to+\infty}\sum_{k=1}^{n}(x_k^* - \bar{x})\delta(x_k^*, y_k^*)\Delta A_k = 0$$

$$\lim_{n\to+\infty}\sum_{k=1}^{n}(y_k^* - \bar{y})\delta(x_k^*, y_k^*)\Delta A_k = 0$$

from which we obtain

$$\iint_R (x - \bar{x})\delta(x, y)\, dA = 0$$

$$\iint_R (y - \bar{y})\delta(x, y)\, dA = 0$$

Since $\bar{x}$ and $\bar{y}$ are constant, these equations can be rewritten as

$$\iint_R x\delta(x, y)\, dA = \bar{x}\iint_R \delta(x, y)\, dA$$

$$\iint_R y\delta(x, y)\, dA = \bar{y}\iint_R \delta(x, y)\, dA$$

from which we obtain the following formulas for the center of gravity of the lamina:

Center of Gravity* $(\bar{x}, \bar{y})$ *of a Lamina

$$\bar{x} = \frac{\displaystyle\iint_R x\delta(x, y)\, dA}{\displaystyle\iint_R \delta(x, y)\, dA}, \qquad \bar{y} = \frac{\displaystyle\iint_R y\delta(x, y)\, dA}{\displaystyle\iint_R \delta(x, y)\, dA} \tag{7–8}$$

Observe that in both formulas the denominator is the mass M of the lamina [see (3)]. The numerator in the formula for $\bar{x}$ is denoted by M_y and is called the ***first moment of the lamina about the y-axis***; the numerator of the formula for $\bar{y}$ is denoted by M_x and is called the ***first moment of the lamina about the x-axis***. Thus, Formulas (7) and (8) can be expressed as

$$\bar{x} = \frac{M_y}{M} = \frac{1}{\text{mass of } R}\iint_R x\delta(x, y)\, dA \tag{9}$$

$$\bar{y} = \frac{M_x}{M} = \frac{1}{\text{mass of } R}\iint_R y\delta(x, y)\, dA \tag{10}$$

▶ **Example 2** Find the center of gravity of the triangular lamina with vertices $(0, 0)$, $(0, 1)$, and $(1, 0)$ and density function $\delta(x, y) = xy$.

Solution. The lamina is shown in Figure 15.6.4. In Example 1 we found the mass of the lamina to be

$$M = \iint_R \delta(x, y)\, dA = \iint_R xy\, dA = \frac{1}{24}$$

The moment of the lamina about the y-axis is

$$\begin{aligned} M_y &= \iint_R x\delta(x, y)\, dA = \iint_R x^2 y\, dA = \int_0^1 \int_0^{-x+1} x^2 y\, dy\, dx \\ &= \int_0^1 \left[\frac{1}{2}x^2y^2\right]_{y=0}^{-x+1} dx = \int_0^1 \left(\frac{1}{2}x^4 - x^3 + \frac{1}{2}x^2\right) dx = \frac{1}{60} \end{aligned}$$

and the moment about the x-axis is

$$\begin{aligned} M_x &= \iint_R y\delta(x, y)\, dA = \iint_R xy^2\, dA = \int_0^1 \int_0^{-x+1} xy^2\, dy\, dx \\ &= \int_0^1 \left[\frac{1}{3}xy^3\right]_{y=0}^{-x+1} dx = \int_0^1 \left(-\frac{1}{3}x^4 + x^3 - x^2 + \frac{1}{3}x\right) dx = \frac{1}{60} \end{aligned}$$

From (9) and (10),

$$\bar{x} = \frac{M_y}{M} = \frac{1/60}{1/24} = \frac{2}{5}, \quad \bar{y} = \frac{M_x}{M} = \frac{1/60}{1/24} = \frac{2}{5}$$

so the center of gravity is $\left(\frac{2}{5}, \frac{2}{5}\right)$. ◀

CENTROIDS

In the special case of a *homogeneous* lamina, the center of gravity is called the ***centroid of the lamina*** or sometimes the ***centroid of the region R***. Because the density function δ is constant for a homogeneous lamina, the factor δ may be moved through the integral signs in (7) and (8) and canceled. Thus, the centroid $(\bar{x}, \bar{y})$ is a geometric property of the region R and is given by the following formulas:

Centroid of a Region R

$$\bar{x} = \frac{\iint_R x\, dA}{\iint_R dA} = \frac{1}{\text{area of } R} \iint_R x\, dA \tag{11}$$

$$\bar{y} = \frac{\iint_R y\, dA}{\iint_R dA} = \frac{1}{\text{area of } R} \iint_R y\, dA \tag{12}$$

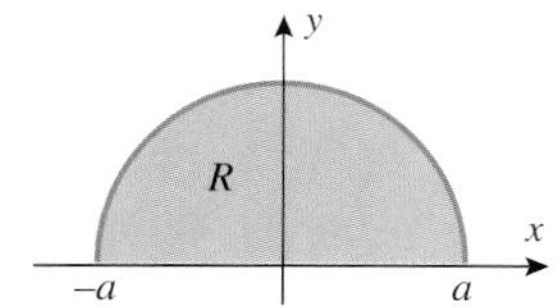

Figure 15.6.9

► **Example 3** Find the centroid of the semicircular region in Figure 15.6.9.

Solution. By symmetry, $\bar{x} = 0$ since the y-axis is obviously a line of balance. From (12),

$$\bar{y} = \frac{1}{\text{area of } R}\iint_R y\,dA = \frac{1}{\frac{1}{2}\pi a^2}\iint_R y\,dA$$

$$= \frac{1}{\frac{1}{2}\pi a^2}\int_0^{\pi}\int_0^{a}(r\sin\theta)r\,dr\,d\theta \qquad \text{Evaluating in polar coordinates}$$

$$= \frac{1}{\frac{1}{2}\pi a^2}\int_0^{\pi}\left[\frac{1}{3}r^3\sin\theta\right]_{r=0}^{a}d\theta$$

$$= \frac{1}{\frac{1}{2}\pi a^2}\left(\frac{1}{3}a^3\right)\int_0^{\pi}\sin\theta\,d\theta = \frac{1}{\frac{1}{2}\pi a^2}\left(\frac{2}{3}a^3\right) = \frac{4a}{3\pi}$$

so the centroid is $\left(0, \dfrac{4a}{3\pi}\right)$. ◄

CENTER OF GRAVITY AND CENTROID OF A SOLID

For a three-dimensional solid G, the formulas for moments, center of gravity, and centroid are similar to those for laminas. If G is *homogeneous*, then its ***density*** is defined to be its mass per unit volume. Thus, if G is a homogeneous solid of mass M and volume V, then its density δ is given by $\delta = M/V$. If G is inhomogeneous and is in an xyz-coordinate system, then its density at a general point (x, y, z) is specified by a ***density function*** $\delta(x, y, z)$ whose value at a point can be viewed as a limit:

$$\delta(x, y, z) = \lim_{\Delta V \to 0}\frac{\Delta M}{\Delta V}$$

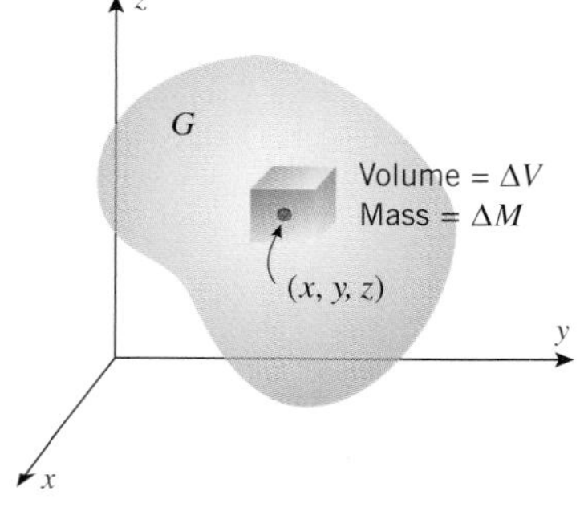

Figure 15.6.10

where ΔM and ΔV represent the mass and volume of a rectangular parallelepiped, centered at (x, y, z), whose dimensions tend to zero (Figure 15.6.10).

Using the discussion of laminas as a model, you should be able to show that the mass M of a solid with a continuous density function $\delta(x, y, z)$ is

$$M = \text{mass of } G = \iiint_G \delta(x, y, z)\,dV \tag{13}$$

The formulas for center of gravity and centroid are as follows:

Center of Gravity $(\bar{x}, \bar{y}, \bar{z})$ ***of a Solid G***

$$\bar{x} = \frac{1}{M}\iiint_G x\delta(x, y, z)\,dV \qquad \bar{y} = \frac{1}{M}\iiint_G y\delta(x, y, z)\,dV \qquad \bar{z} = \frac{1}{M}\iiint_G z\delta(x, y, z)\,dV$$

Centroid $(\bar{x}, \bar{y}, \bar{z})$ ***of a Solid G***

$$\bar{x} = \frac{1}{V}\iiint_G x\,dV \qquad \bar{y} = \frac{1}{V}\iiint_G y\,dV \qquad \bar{z} = \frac{1}{V}\iiint_G z\,dV \tag{14–15}$$

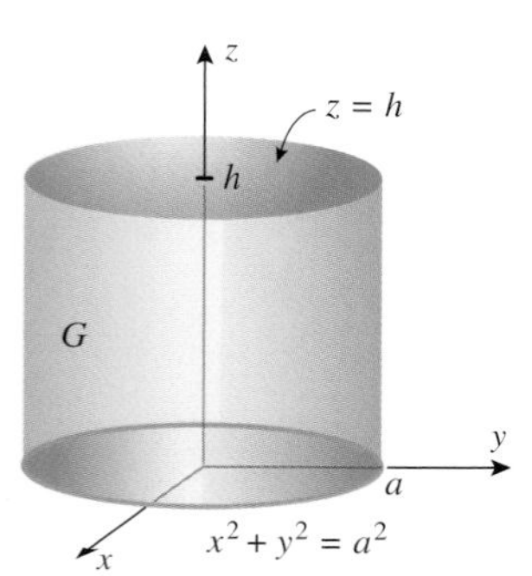

Figure 15.6.11

► **Example 4** Find the mass and the center of gravity of a cylindrical solid of height h and radius a (Figure 15.6.11), assuming that the density at each point is proportional to the distance between the point and the base of the solid.

Solution. Since the density is proportional to the distance z from the base, the density function has the form $\delta(x, y, z) = kz$, where k is some (unknown) positive constant of proportionality. From (13) the mass of the solid is

$$M = \iiint_G \delta(x, y, z)\, dV = \int_{-a}^{a} \int_{-\sqrt{a^2-x^2}}^{\sqrt{a^2-x^2}} \int_0^h kz\, dz\, dy\, dx$$

$$= k \int_{-a}^{a} \int_{-\sqrt{a^2-x^2}}^{\sqrt{a^2-x^2}} \frac{1}{2}h^2\, dy\, dx$$

$$= kh^2 \int_{-a}^{a} \sqrt{a^2 - x^2}\, dx$$

$$= \tfrac{1}{2}kh^2\pi a^2$$

Interpret the integral as the area of a semicircle.

Without additional information, the constant k cannot be determined. However, as we will now see, the value of k does not affect the center of gravity.

From (14),

$$\bar{z} = \frac{1}{M}\iiint_G z\delta(x, y, z)\, dV = \frac{1}{\frac{1}{2}kh^2\pi a^2}\iiint_G z\delta(x, y, z)\, dV$$

$$= \frac{1}{\frac{1}{2}kh^2\pi a^2} \int_{-a}^{a} \int_{-\sqrt{a^2-x^2}}^{\sqrt{a^2-x^2}} \int_0^h z(kz)\, dz\, dy\, dx$$

$$= \frac{k}{\frac{1}{2}kh^2\pi a^2} \int_{-a}^{a} \int_{-\sqrt{a^2-x^2}}^{\sqrt{a^2-x^2}} \frac{1}{3}h^3\, dy\, dx$$

$$= \frac{\frac{1}{3}kh^3}{\frac{1}{2}kh^2\pi a^2} \int_{-a}^{a} 2\sqrt{a^2 - x^2}\, dx$$

$$= \frac{\frac{1}{3}kh^3\pi a^2}{\frac{1}{2}kh^2\pi a^2} = \frac{2}{3}h$$

Similar calculations using (14) will yield $\bar{x} = \bar{y} = 0$. However, this is evident by inspection, since it follows from the symmetry of the solid and the form of its density function that the center of gravity is on the z-axis. Thus, the center of gravity is $\left(0, 0, \frac{2}{3}h\right)$. ◄

THEOREM OF PAPPUS

The following theorem, due to the Greek mathematician Pappus, gives an important relationship between the centroid of a plane region R and the volume of the solid generated when the region is revolved about a line.

Pappus of Alexandria (4th century A.D.) Greek mathematician. Pappus lived during the early Christian era when mathematical activity was in a period of decline. His main contributions to mathematics appeared in a series of eight books called *The Collection* (written about 340 A.D.). This work, which survives only partially, contained some original results but was devoted mostly to statements, refinements, and proofs of results by earlier mathematicians. Pappus' Theorem, stated without proof in Book VII of *The Collection*, was probably known and proved in earlier times. This result is sometimes called Guldin's Theorem in recognition of the Swiss mathematician, Paul Guldin (1577–1643), who rediscovered it independently.

15.6.3 THEOREM (*Theorem of Pappus*). *If R is a bounded plane region and L is a line that lies in the plane of R such that R is entirely on one side of L, then the volume of the solid formed by revolving R about L is given by*

$$\textit{volume} = (\textit{area of } R) \cdot \begin{pmatrix} \textit{distance traveled} \\ \textit{by the centroid} \end{pmatrix}$$

PROOF. Introduce an xy-coordinate system so that L is along the y-axis and the region R is in the first quadrant (Figure 15.6.12). Let R be partitioned into subregions in the usual way and let R_k be a typical rectangle interior to R. If (x_k^*, y_k^*) is the center of R_k, and if the area of R_k is $\Delta A_k = \Delta x_k \Delta y_k$, then from Formula (1) of Section 6.3 the volume generated by R_k as it revolves about L is

$$2\pi x_k^* \Delta x_k \Delta y_k = 2\pi x_k^* \Delta A_k$$

Therefore, the total volume of the solid is approximately

$$V \approx \sum_{k=1}^{n} 2\pi x_k^* \Delta A_k$$

from which it follows that the exact volume is

$$V = \iint_R 2\pi x \, dA = 2\pi \iint_R x \, dA$$

Thus, it follows from (11) that

$$V = 2\pi \cdot \bar{x} \cdot [\text{area of } R]$$

This completes the proof since $2\pi\bar{x}$ is the distance traveled by the centroid when R is revolved about the y-axis. ■

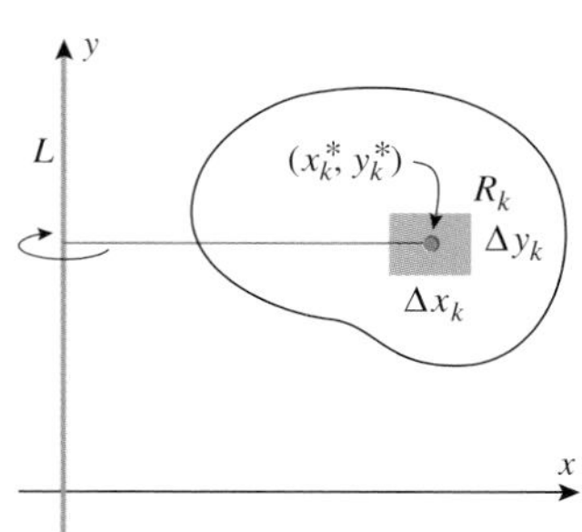

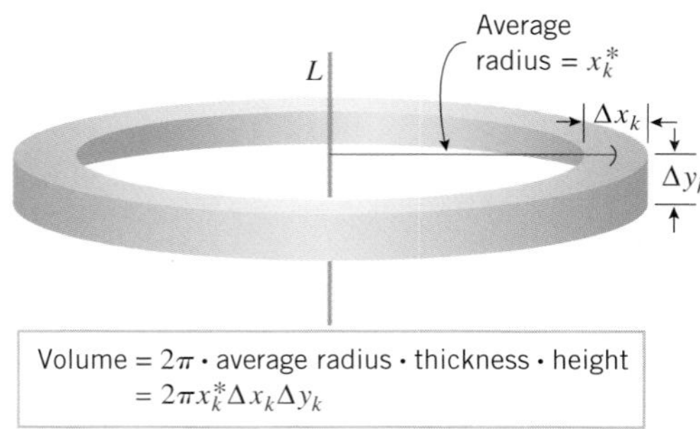

Figure 15.6.12

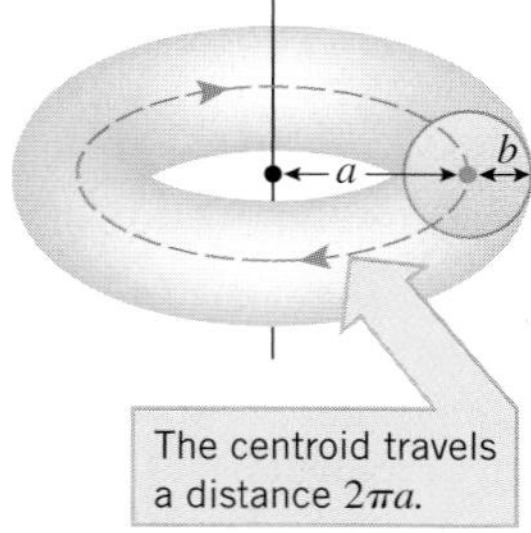

Figure 15.6.13

► **Example 5** Use Pappus' Theorem to find the volume V of the torus generated by revolving a circular region of radius b about a line at a distance a (greater than b) from the center of the circle (Figure 15.6.13).

Solution. By symmetry, the centroid of a circular region is its center. Thus, the distance traveled by the centroid is $2\pi a$. Since the area of a circle of radius b is πb^2, it follows from Pappus' Theorem that the volume of the torus is

$$V = (2\pi a)(\pi b^2) = 2\pi^2 ab^2 \quad ◄$$

✔ QUICK CHECK EXERCISES 15.6 (See page 1075 for answers.)

1. The total mass of a lamina with continuous density function $\delta(x, y)$ that occupies a region R in the xy-plane is given by $M =$ ________.

2. Consider a lamina with mass M and continuous density function $\delta(x, y)$ that occupies a region R in the xy-plane. The x-coordinate of the center of gravity of the lamina is M_y/M, where M_y is called the ________ and is given by the double integral ________.

3. Let R be the region between the graphs of $y = x^2$ and $y = 2 - x$ for $0 \le x \le 1$. The area of R is $\frac{7}{6}$ and the centroid of R is ________.

4. If the region R in Quick Check Exercise 3 is used to generate a solid G by rotating R about a horizontal line 6 units above its centroid, then the volume of G is ________.

EXERCISE SET 15.6 Graphing Utility C CAS

FOCUS ON CONCEPTS

1. Masses $m_1 = 5$, $m_2 = 10$, and $m_3 = 20$ are positioned on a weightless beam as shown in the accompanying figure.
   (a) Suppose that the fulcrum is positioned at $x = 5$. Without computing the sum of moments about 5, determine whether the sum is positive, zero, or negative. Explain.
   (b) Where should the fulcrum be placed so that the beam is in equilibrium?

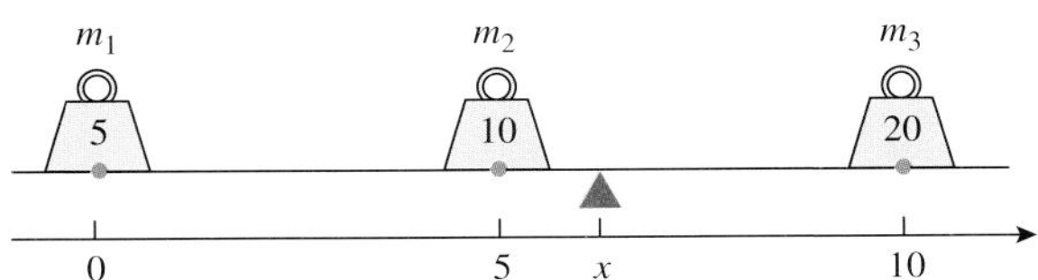

Figure Ex-1

2. Masses $m_1 = 10$, $m_2 = 3$, $m_3 = 4$, and m are positioned on a weightless beam, with the fulcrum positioned at point 4, as shown in the accompanying figure.
   (a) Suppose that $m = 14$. Without computing the sum of the moments about 4, determine whether the sum is positive, zero, or negative. Explain.
   (b) For what value of m is the beam in equilibrium?

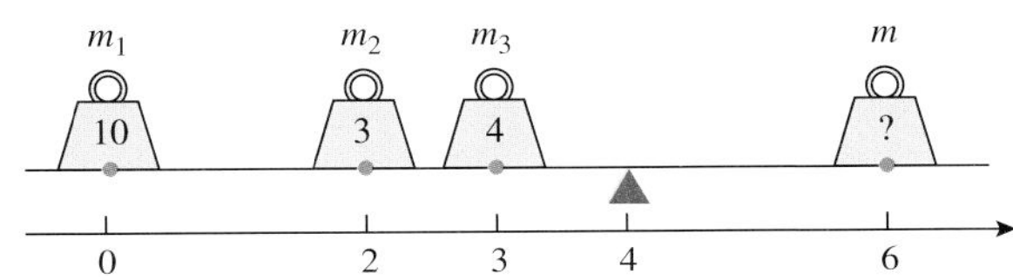

Figure Ex-2

3–4 Make a conjecture about the coordinates of the centroid of the region and confirm your conjecture by integrating.

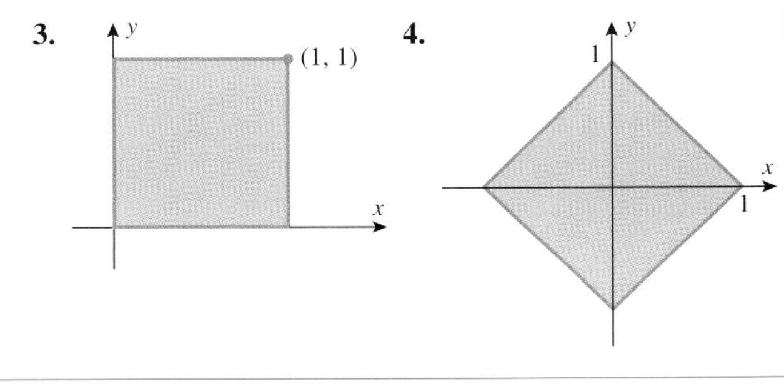

5–10 Find the centroid of the region.

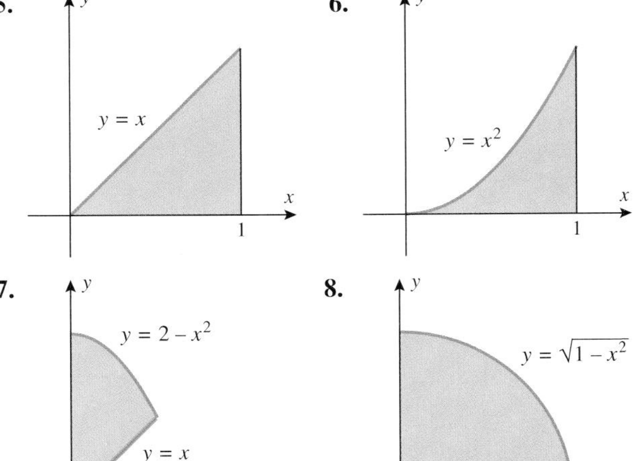

9. The region above the x-axis and between the circles $x^2 + y^2 = a^2$ and $x^2 + y^2 = b^2$ $(a < b)$.

10. The region enclosed between the y-axis and the right half of the circle $x^2 + y^2 = a^2$.

FOCUS ON CONCEPTS

11–12 Make a conjecture about the coordinates of the center of gravity and confirm your conjecture by integrating.

11. The lamina of Exercise 3 with density function $\delta(x, y) = |x + y - 1|$.

12. The lamina of Exercise 4 with density function $\delta(x, y) = 1 + x^2 + y^2$.

13–16 Find the mass and center of gravity of the lamina.

13. A lamina with density $\delta(x, y) = x + y$ is bounded by the x-axis, the line $x = 1$, and the curve $y = \sqrt{x}$.

14. A lamina with density $\delta(x, y) = y$ is bounded by $y = \sin x$, $y = 0$, $x = 0$, and $x = \pi$.

15. A lamina with density $\delta(x, y) = xy$ is in the first quadrant and is bounded by the circle $x^2 + y^2 = a^2$ and the coordinate axes.

16. A lamina with density $\delta(x, y) = x^2 + y^2$ is bounded by the x-axis and the upper half of the circle $x^2 + y^2 = 1$.

FOCUS ON CONCEPTS

17–18 Make a conjecture about the coordinates of the centroid and confirm your conjecture by integrating.

17. **18.**

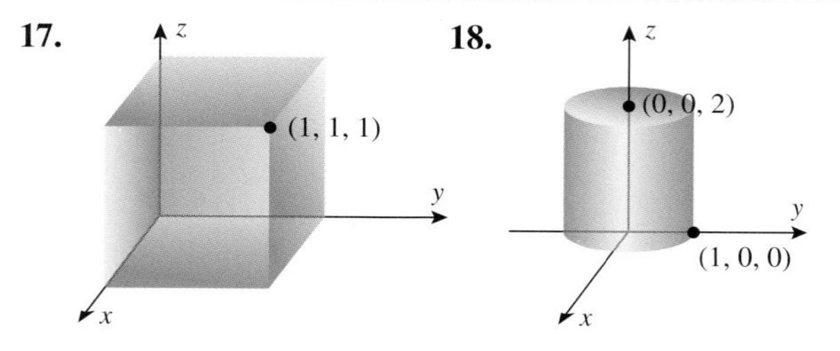

19–24 Find the centroid of the solid.

19. The tetrahedron in the first octant enclosed by the coordinate planes and the plane $x + y + z = 1$.

20. The solid bounded by the parabolic cylinder $z = 1 - y^2$ and the planes $x + z = 1$, $x = 0$, and $z = 0$.

21. The solid bounded by the surface $z = y^2$ and the planes $x = 0$, $x = 1$, and $z = 1$.

22. The solid in the first octant bounded by the surface $z = xy$ and the planes $z = 0$, $x = 2$, and $y = 2$.

23. The solid in the first octant that is bounded by the sphere $x^2 + y^2 + z^2 = a^2$ and the coordinate planes.

24. The solid enclosed by the xy-plane and the hemisphere $z = \sqrt{a^2 - x^2 - y^2}$.

25–28 Find the mass and center of gravity of the solid.

25. The cube that has density $\delta(x, y, z) = a - x$ and is defined by the inequalities $0 \le x \le a$, $0 \le y \le a$, and $0 \le z \le a$.

26. The cylindrical solid that has density $\delta(x, y, z) = h - z$ and is enclosed by $x^2 + y^2 = a^2$, $z = 0$, and $z = h$.

27. The solid that has density $\delta(x, y, z) = yz$ and is enclosed by $z = 1 - y^2$ (for $y \ge 0$), $z = 0$, $y = 0$, $x = -1$, and $x = 1$.

28. The solid that has density $\delta(x, y, z) = xz$ and is enclosed by $y = 9 - x^2$ (for $x \ge 0$), $x = 0$, $y = 0$, $z = 0$, and $z = 1$.

29. Find the center of gravity of the square lamina with vertices $(0, 0)$, $(1, 0)$, $(0, 1)$, and $(1, 1)$ if
(a) the density is proportional to the square of the distance from the origin;
(b) the density is proportional to the distance from the y-axis.

30. Find the center of gravity of the cube that is determined by the inequalities $0 \le x \le 1$, $0 \le y \le 1$, $0 \le z \le 1$ if
(a) the density is proportional to the square of the distance to the origin;
(b) the density is proportional to the sum of the distances to the faces that lie in the coordinate planes.

C **31.** Use the numerical triple integral capability of a CAS to approximate the location of the centroid of the solid that is bounded above by the surface $z = 1/(1 + x^2 + y^2)$, below by the xy-plane, and laterally by the plane $y = 0$ and the surface $y = \sin x$ for $0 \le x \le \pi$ (see the accompanying figure).

32. The accompanying figure shows the solid that is bounded above by the surface $z = 1/(x^2 + y^2 + 1)$, below by the xy-plane, and laterally by the surface $x^2 + y^2 = a^2$.
(a) By symmetry, the centroid of the solid lies on the z-axis. Make a conjecture about the behavior of the z-coordinate of the centroid as $a \to 0^+$ and as $a \to +\infty$.
(b) Find the z-coordinate of the centroid, and check your conjecture by calculating the appropriate limits.
(c) Use a graphing utility to plot the z-coordinate of the centroid versus a, and use the graph to estimate the value of a for which the centroid is $(0, 0, 0.25)$.

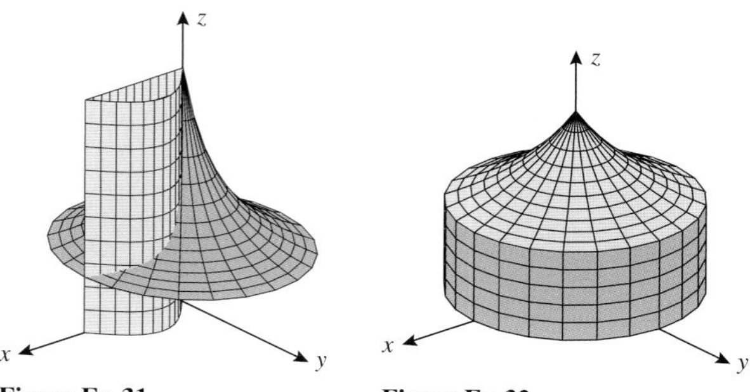

Figure Ex-31 **Figure Ex-32**

33. Show that in polar coordinates the formulas for the centroid $(\bar{x}, \bar{y})$ of a region R are

$$\bar{x} = \frac{1}{\text{area of } R} \iint_R r^2 \cos\theta \, dr \, d\theta$$

$$\bar{y} = \frac{1}{\text{area of } R} \iint_R r^2 \sin\theta \, dr \, d\theta$$

34. Use the result of Exercise 33 to find the centroid $(\bar{x}, \bar{y})$ of the region enclosed by the cardioid $r = a(1 + \sin\theta)$.

35. Use the result of Exercise 33 to find the centroid $(\bar{x}, \bar{y})$ of the petal of the rose $r = \sin 2\theta$ in the first quadrant.

36. Let R be the rectangle bounded by the lines $x = 0$, $x = 3$, $y = 0$, and $y = 2$. By inspection, find the centroid of R and use it to evaluate

$$\iint_R x \, dA \quad \text{and} \quad \iint_R y \, dA$$

37. Use the Theorem of Pappus and the fact that the volume of a sphere of radius a is $V = \frac{4}{3}\pi a^3$ to show that the centroid of the lamina that is bounded by the x-axis and the semicircle $y = \sqrt{a^2 - x^2}$ is $(0, 4a/(3\pi))$. (This problem was solved directly in Example 3.)

38. Use the Theorem of Pappus and the result of Exercise 37 to find the volume of the solid generated when the region bounded by the x-axis and the semicircle $y = \sqrt{a^2 - x^2}$ is revolved about

(a) the line $y = -a$ (b) the line $y = x - a$.

39. Use the Theorem of Pappus and the fact that the area of an ellipse with semiaxes a and b is πab to find the volume of the elliptical torus generated by revolving the ellipse

$$\frac{(x-k)^2}{a^2} + \frac{y^2}{b^2} = 1$$

about the y-axis. Assume that $k > a$.

40. Use the Theorem of Pappus to find the volume of the solid that is generated when the region enclosed by $y = x^2$ and $y = 8 - x^2$ is revolved about the x-axis.

41. Use the Theorem of Pappus to find the centroid of the triangular region with vertices $(0, 0)$, $(a, 0)$, and $(0, b)$, where $a > 0$ and $b > 0$. [*Hint:* Revolve the region about the x-axis to obtain $\bar{y}$ and about the y-axis to obtain $\bar{x}$.]

42. It can be proved that if a bounded plane region slides along a helix in such a way that the region is always orthogonal to the helix (i.e., orthogonal to the unit tangent vector to the helix), then the volume swept out by the region is equal to the area of the region times the distance traveled by its centroid. Use this result to find the volume of the "tube" in the accompanying figure that is swept out by sliding a circle of radius $\frac{1}{2}$ along the helix

$$x = \cos t, \quad y = \sin t, \quad z = \frac{t}{4} \qquad (0 \le t \le 4\pi)$$

in such a way that the circle is always centered on the helix and lies in the plane perpendicular to the helix.

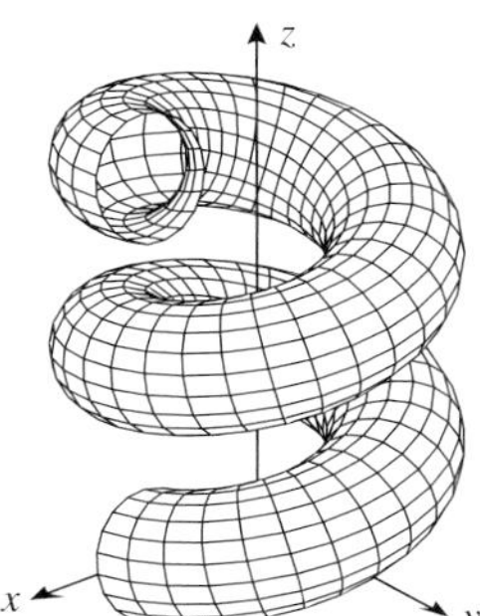

Figure Ex-42

43–44 The tendency of a lamina to resist a change in rotational motion about an axis is measured by its ***moment of inertia*** about that axis. If a lamina occupies a region R of the xy-plane, and if its density function $\delta(x, y)$ is continuous on R, then the moments of inertia about the x-axis, the y-axis, and the z-axis are denoted by I_x, I_y, and I_z, respectively, and are defined by

$$I_x = \iint_R y^2 \, \delta(x, y) \, dA, \quad I_y = \iint_R x^2 \, \delta(x, y) \, dA,$$

$$I_z = \iint_R (x^2 + y^2) \, \delta(x, y) \, dA$$

Use these definitions in Exercises 43 and 44.

43. Consider the rectangular lamina that occupies the region described by the inequalities $0 \le x \le a$ and $0 \le y \le b$. Assuming that the lamina has constant density δ, show that

$$I_x = \frac{\delta ab^3}{3}, \quad I_y = \frac{\delta a^3 b}{3}, \quad I_z = \frac{\delta ab(a^2 + b^2)}{3}$$

44. Consider the circular lamina that occupies the region described by the inequalities $0 \le x^2 + y^2 \le a^2$. Assuming that the lamina has constant density δ, show that

$$I_x = I_y = \frac{\delta \pi a^4}{4}, \quad I_z = \frac{\delta \pi a^4}{2}$$

✔ QUICK CHECK ANSWERS 15.6

1. $\iint_R \delta(x, y) \, dA$ **2.** first moment about the y-axis; $\iint_R x\delta(x, y) \, dA$ **3.** $\left(\frac{5}{14}, \frac{32}{35}\right)$ **4.** 14π

15.7 TRIPLE INTEGRALS IN CYLINDRICAL AND SPHERICAL COORDINATES

Earlier we saw that some double integrals are easier to evaluate in polar coordinates than in rectangular coordinates. Similarly, some triple integrals are easier to evaluate in cylindrical or spherical coordinates than in rectangular coordinates. In this section we will study triple integrals in these coordinate systems.

TRIPLE INTEGRALS IN CYLINDRICAL COORDINATES

Recall that in rectangular coordinates the triple integral of a continuous function f over a solid region G is defined as

$$\iiint_G f(x, y, z)\, dV = \lim_{n \to +\infty} \sum_{k=1}^{n} f(x_k^*, y_k^*, z_k^*)\Delta V_k$$

where ΔV_k denotes the volume of a rectangular parallelepiped interior to G and (x_k^*, y_k^*, z_k^*) is a point in this parallelepiped (see Figure 15.5.1). Triple integrals in cylindrical and spherical coordinates are defined similarly, except that the region G is divided not into rectangular parallelepipeds but into regions more appropriate to these coordinate systems.

In cylindrical coordinates, the simplest equations are of the form

$$r = \text{constant}, \quad \theta = \text{constant}, \quad z = \text{constant}$$

The first equation represents a right circular cylinder centered on the z-axis, the second a vertical half-plane hinged on the z-axis, and the third a horizontal plane. (See Figure 12.8.3.) These surfaces can be paired up to determine solids called ***cylindrical wedges*** or ***cylindrical elements of volume***. To be precise, a cylindrical wedge is a solid enclosed between six surfaces of the following form:

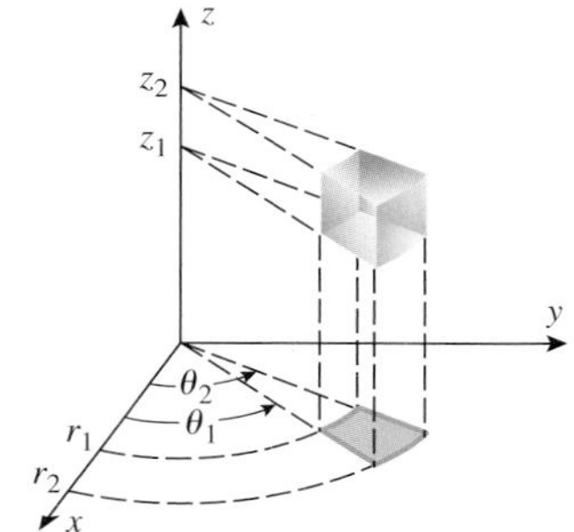

Figure 15.7.1

two cylinders	$r = r_1,$	$r = r_2$	$(r_1 < r_2)$
two half-planes	$\theta = \theta_1,$	$\theta = \theta_2$	$(\theta_1 < \theta_2)$
two planes	$z = z_1,$	$z = z_2$	$(z_1 < z_2)$

(Figure 15.7.1). The dimensions $\theta_2 - \theta_1$, $r_2 - r_1$, and $z_2 - z_1$ are called the ***central angle***, ***thickness***, and ***height*** of the wedge.

To define the triple integral over G of a function $f(r, \theta, z)$ in cylindrical coordinates we proceed as follows:

- Subdivide G into pieces by a three-dimensional grid consisting of concentric circular cylinders centered on the z-axis, half-planes hinged on the z-axis, and horizontal planes. Exclude from consideration all pieces that contain any points outside of G, thereby leaving only cylindrical wedges that are subsets of G.
- Assume that there are n such cylindrical wedges, and denote the volume of the kth cylindrical wedge by ΔV_k. As indicated in Figure 15.7.2, let $(r_k^*, \theta_k^*, z_k^*)$ be any point in the kth cylindrical wedge.
- Repeat this process with more and more subdivisions so that as n increases, the height, thickness, and central angle of the cylindrical wedges approach zero. Define

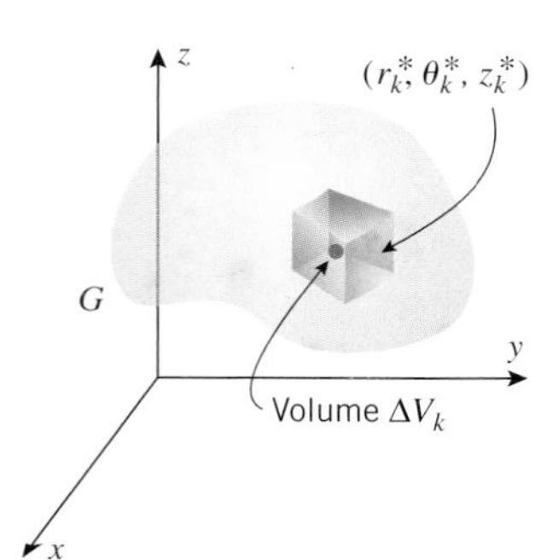

Figure 15.7.2

$$\iiint_G f(r, \theta, z)\, dV = \lim_{n \to +\infty} \sum_{k=1}^{n} f(r_k^*, \theta_k^*, z_k^*)\Delta V_k \tag{1}$$

For computational purposes, it will be helpful to express (1) as an iterated integral. Toward this end we note that the volume ΔV_k of the kth cylindrical wedge can be expressed as

$$\Delta V_k = [\text{area of base}] \cdot [\text{height}] \tag{2}$$

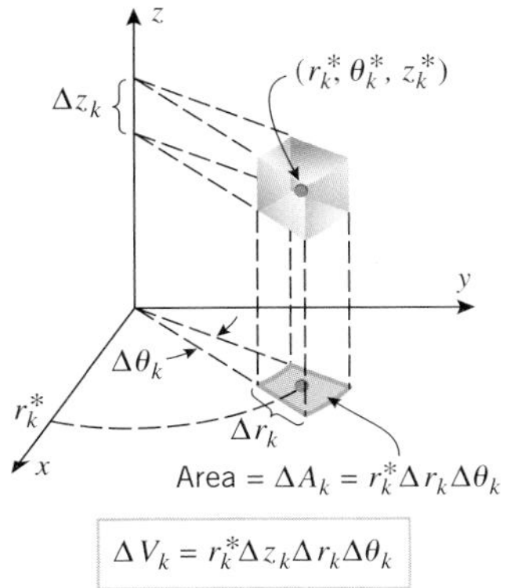

Figure 15.7.3

Note the extra factor of r that appears in the integrand on converting a triple integral to an iterated integral in cylindrical coordinates.

If we denote the thickness, central angle, and height of this wedge by Δr_k, $\Delta\theta_k$, and Δz_k, and if we choose the arbitrary point $(r_k^*, \theta_k^*, z_k^*)$ to lie above the "center" of the base (Figures 15.3.5 and 15.7.3), then it follows from (5) of Section 15.3 that the base has area $\Delta A_k = r_k^* \Delta r_k \Delta\theta_k$. Thus, (2) can be written as

$$\Delta V_k = r_k^* \Delta r_k \Delta\theta_k \Delta z_k = r_k^* \Delta z_k \Delta r_k \Delta\theta_k$$

Substituting this expression in (1) yields

$$\iiint_G f(r, \theta, z)\, dV = \lim_{n\to+\infty} \sum_{k=1}^{n} f(r_k^*, \theta_k^*, z_k^*) r_k^* \Delta z_k \Delta r_k \Delta\theta_k$$

which suggests that a triple integral in cylindrical coordinates can be evaluated as an iterated integral of the form

$$\iiint_G f(r, \theta, z)\, dV = \iiint_{\substack{\text{appropriate} \\ \text{limits}}} f(r, \theta, z) r\, dz\, dr\, d\theta \tag{3}$$

In this formula the integration with respect to z is done first, then with respect to r, and then with respect to θ, but any order of integration is allowable.

The following theorem, which we state without proof, makes the preceding ideas more precise.

15.7.1 THEOREM. *Let G be a solid region whose upper surface has the equation $z = g_2(r, \theta)$ and whose lower surface has the equation $z = g_1(r, \theta)$ in cylindrical coordinates. If the projection of the solid on the xy-plane is a simple polar region R, and if $f(r, \theta, z)$ is continuous on G, then*

$$\iiint_G f(r, \theta, z)\, dV = \iint_R \left[\int_{g_1(r,\theta)}^{g_2(r,\theta)} f(r, \theta, z)\, dz\right] dA \tag{4}$$

where the double integral over R is evaluated in polar coordinates. In particular, if the projection R is as shown in Figure 15.7.4, then (4) can be written as

$$\iiint_G f(r, \theta, z)\, dV = \int_{\theta_1}^{\theta_2} \int_{r_1(\theta)}^{r_2(\theta)} \int_{g_1(r,\theta)}^{g_2(r,\theta)} f(r, \theta, z) r\, dz\, dr\, d\theta \tag{5}$$

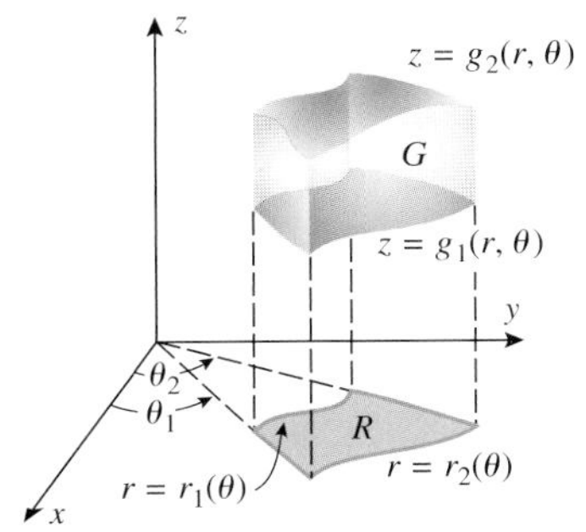

Figure 15.7.4

The type of solid to which Formula (5) applies is illustrated in Figure 15.7.4. To apply (4) and (5) it is best to begin with a three-dimensional sketch of the solid G, from which the limits of integration can be obtained as follows:

Determining Limits of Integration: Cylindrical Coordinates

Step 1. Identify the upper surface $z = g_2(r, \theta)$ and the lower surface $z = g_1(r, \theta)$ of the solid. The functions $g_1(r, \theta)$ and $g_2(r, \theta)$ determine the z-limits of integration. (If the upper and lower surfaces are given in rectangular coordinates, convert them to cylindrical coordinates.)

Step 2. Make a two-dimensional sketch of the projection R of the solid on the xy-plane. From this sketch the r- and θ-limits of integration may be obtained exactly as with double integrals in polar coordinates.

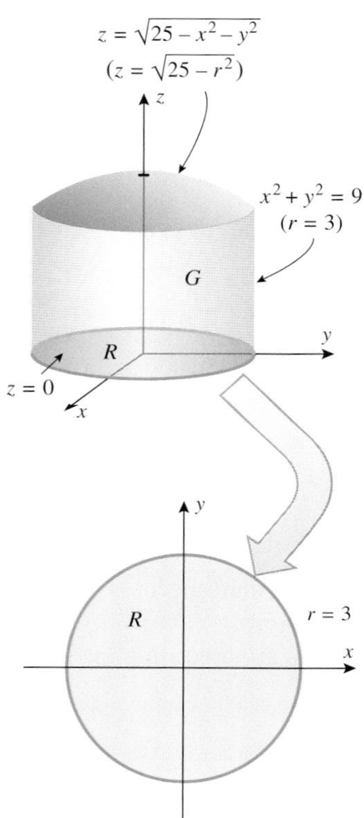

Figure 15.7.5

► **Example 1** Use triple integration in cylindrical coordinates to find the volume and the centroid of the solid G that is bounded above by the hemisphere $z = \sqrt{25 - x^2 - y^2}$, below by the xy-plane, and laterally by the cylinder $x^2 + y^2 = 9$.

Solution. The solid G and its projection R on the xy-plane are shown in Figure 15.7.5. In cylindrical coordinates, the upper surface of G is the hemisphere $z = \sqrt{25 - r^2}$ and the lower surface is the plane $z = 0$. Thus, from (4), the volume of G is

$$V = \iiint_G dV = \iint_R \left[\int_0^{\sqrt{25-r^2}} dz\right] dA$$

For the double integral over R, we use polar coordinates:

$$\begin{aligned} V &= \int_0^{2\pi}\int_0^3\int_0^{\sqrt{25-r^2}} r\,dz\,dr\,d\theta = \int_0^{2\pi}\int_0^3 \big[rz\big]_{z=0}^{\sqrt{25-r^2}}\,dr\,d\theta \\ &= \int_0^{2\pi}\int_0^3 r\sqrt{25-r^2}\,dr\,d\theta = \int_0^{2\pi}\left[-\frac{1}{3}(25-r^2)^{3/2}\right]_{r=0}^{3} d\theta \\ &= \int_0^{2\pi}\frac{61}{3}\,d\theta = \frac{122}{3}\pi \end{aligned}$$

$u = 25 - r^2$
$du = -2r\,dr$

From this result and (15) of Section 15.6,

$$\begin{aligned} \bar{z} &= \frac{1}{V}\iiint_G z\,dV = \frac{3}{122\pi}\iiint_G z\,dV = \frac{3}{122\pi}\iint_R\left[\int_0^{\sqrt{25-r^2}} z\,dz\right]dA \\ &= \frac{3}{122\pi}\int_0^{2\pi}\int_0^3\int_0^{\sqrt{25-r^2}} zr\,dz\,dr\,d\theta = \frac{3}{122\pi}\int_0^{2\pi}\int_0^3\left[\frac{1}{2}rz^2\right]_{z=0}^{\sqrt{25-r^2}} dr\,d\theta \\ &= \frac{3}{244\pi}\int_0^{2\pi}\int_0^3 (25r - r^3)\,dr\,d\theta = \frac{3}{244\pi}\int_0^{2\pi}\frac{369}{4}\,d\theta = \frac{1107}{488} \end{aligned}$$

By symmetry, the centroid $(\bar{x}, \bar{y}, \bar{z})$ of G lies on the z-axis, so $\bar{x} = \bar{y} = 0$. Thus, the centroid is at the point $(0, 0, 1107/488)$. ◄

CONVERTING TRIPLE INTEGRALS FROM RECTANGULAR TO CYLINDRICAL COORDINATES

Sometimes a triple integral that is difficult to integrate in rectangular coordinates can be evaluated more easily by making the substitution $x = r\cos\theta$, $y = r\sin\theta$, $z = z$ to convert it to an integral in cylindrical coordinates. Under such a substitution, a rectangular triple integral can be expressed as an iterated integral in cylindrical coordinates as

$$\iiint_G f(x, y, z)\,dV = \iiint_{\substack{\text{appropriate}\\ \text{limits}}} f(r\cos\theta, r\sin\theta, z)r\,dz\,dr\,d\theta \tag{6}$$

The order of integration on the right side of (6) can be changed, provided the limits of integration are adjusted accordingly.

► **Example 2** Use cylindrical coordinates to evaluate

$$\int_{-3}^{3}\int_{-\sqrt{9-x^2}}^{\sqrt{9-x^2}}\int_0^{9-x^2-y^2} x^2\,dz\,dy\,dx$$

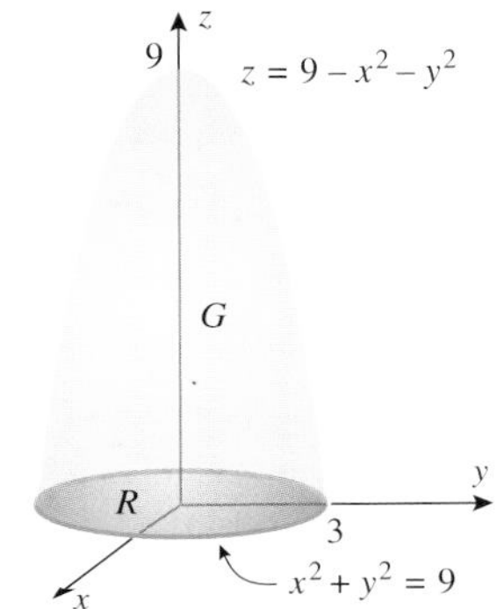

Figure 15.7.6

Solution. In problems of this type, it is helpful to sketch the region of integration G and its projection R on the xy-plane. From the z-limits of integration, the upper surface of G is the paraboloid $z = 9 - x^2 - y^2$ and the lower surface is the xy-plane $z = 0$. From the x- and y-limits of integration, the projection R is the region in the xy-plane enclosed by the circle $x^2 + y^2 = 9$ (Figure 15.7.6). Thus,

$$\int_{-3}^{3}\int_{-\sqrt{9-x^2}}^{\sqrt{9-x^2}}\int_{0}^{9-x^2-y^2} x^2\,dz\,dy\,dx = \iiint_G x^2\,dV$$

$$= \iint_R \left[\int_0^{9-r^2} r^2\cos^2\theta\,dz\right] dA = \int_0^{2\pi}\int_0^3\int_0^{9-r^2} (r^2\cos^2\theta)r\,dz\,dr\,d\theta$$

$$= \int_0^{2\pi}\int_0^3\int_0^{9-r^2} r^3\cos^2\theta\,dz\,dr\,d\theta = \int_0^{2\pi}\int_0^3 \left[zr^3\cos^2\theta\right]_{z=0}^{9-r^2} dr\,d\theta$$

$$= \int_0^{2\pi}\int_0^3 (9r^3 - r^5)\cos^2\theta\,dr\,d\theta = \int_0^{2\pi}\left[\left(\frac{9r^4}{4} - \frac{r^6}{6}\right)\cos^2\theta\right]_{r=0}^{3} d\theta$$

$$= \frac{243}{4}\int_0^{2\pi}\cos^2\theta\,d\theta = \frac{243}{4}\int_0^{2\pi}\frac{1}{2}(1+\cos 2\theta)\,d\theta = \frac{243\pi}{4} \blacktriangleleft$$

TRIPLE INTEGRALS IN SPHERICAL COORDINATES

In spherical coordinates, the simplest equations are of the form

$$\rho = \text{constant}, \quad \theta = \text{constant}, \quad \phi = \text{constant}$$

As indicated in Figure 12.8.4, the first equation represents a sphere centered at the origin and the second a half-plane hinged on the z-axis. The graph of the third equation is a right circular cone nappe with its vertex at the origin and its line of symmetry along the z-axis for $\phi \neq \pi/2$, and is the xy-plane if $\phi = \pi/2$. By a ***spherical wedge*** or ***spherical element of volume*** we mean a solid enclosed between six surfaces of the following form:

two spheres	$\rho = \rho_1$,	$\rho = \rho_2$	$(\rho_1 < \rho_2)$
two half-planes	$\theta = \theta_1$,	$\theta = \theta_2$	$(\theta_1 < \theta_2)$
nappes of two right circular cones	$\phi = \phi_1$,	$\phi = \phi_2$	$(\phi_1 < \phi_2)$

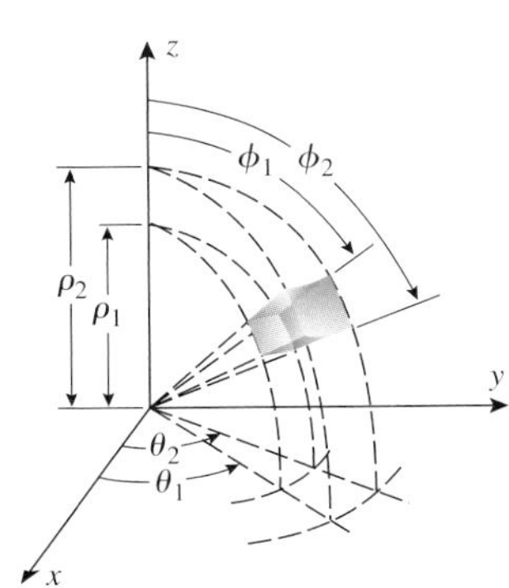

Figure 15.7.7

(Figure 15.7.7). We will refer to the numbers $\rho_2 - \rho_1$, $\theta_2 - \theta_1$, and $\phi_2 - \phi_1$ as the ***dimensions*** of a spherical wedge.

If G is a solid region in three-dimensional space, then the triple integral over G of a continuous function $f(\rho, \theta, \phi)$ in spherical coordinates is similar in definition to the triple integral in cylindrical coordinates, except that the solid G is partitioned into *spherical* wedges by a three-dimensional grid consisting of spheres centered at the origin, half-planes hinged on the z-axis, and nappes of right circular cones with vertices at the origin and lines of symmetry along the z-axis (Figure 15.7.8).

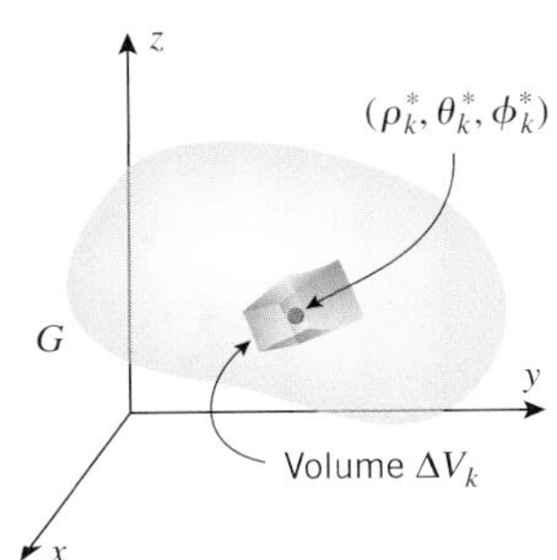

Figure 15.7.8

The defining equation of a triple integral in spherical coordinates is

$$\iiint_G f(\rho, \theta, \phi)\,dV = \lim_{n\to+\infty}\sum_{k=1}^{n} f(\rho_k^*, \theta_k^*, \phi_k^*)\Delta V_k \tag{7}$$

where ΔV_k is the volume of the kth spherical wedge that is interior to G, $(\rho_k^*, \theta_k^*, \phi_k^*)$ is an arbitrary point in this wedge, and n increases in such a way that the dimensions of each interior spherical wedge tend to zero.

For computational purposes, it will be desirable to express (7) as an iterated integral. In the exercises we will help you to show that if the point $(\rho_k^*, \theta_k^*, \phi_k^*)$ is suitably chosen, then the volume ΔV_k in (7) can be written as

$$\Delta V_k = \rho_k^{*2} \sin \phi_k^* \Delta\rho_k \Delta\phi_k \Delta\theta_k \tag{8}$$

where $\Delta\rho_k$, $\Delta\phi_k$, and $\Delta\theta_k$ are the dimensions of the wedge (Exercise 42). Substituting this in (7) we obtain

$$\iiint\limits_G f(\rho, \theta, \phi)\, dV = \lim_{n\to+\infty} \sum_{k=1}^{n} f(\rho_k^*, \theta_k^*, \phi_k^*)\rho_k^{*2} \sin \phi_k^* \Delta\rho_k \Delta\phi_k \Delta\theta_k$$

which suggests that a triple integral in spherical coordinates can be evaluated as an iterated integral of the form

$$\iiint\limits_G f(\rho, \theta, \phi)\, dV = \iiint\limits_{\substack{\text{appropriate}\\ \text{limits}}} f(\rho, \theta, \phi)\rho^2 \sin \phi \, d\rho\, d\phi\, d\theta \tag{9}$$

Note the extra factor of $\rho^2 \sin\phi$ that appears in the integrand on converting a triple integral to an iterated integral in spherical coordinates. This is analogous to the extra factor of r that appears in an iterated integral in cylindrical coordinates.

The analog of Theorem 15.7.1 for triple integrals in spherical coordinates is tedious to state, so instead we will give some examples that illustrate techniques for obtaining the limits of integration. In all of our examples we will use the same order of integration—first with respect to ρ, then ϕ, and then θ. Once you have mastered the basic ideas, there should be no trouble using other orders of integration.

Suppose that we want to integrate $f(\rho, \theta, \phi)$ over the spherical solid G enclosed by the sphere $\rho = \rho_0$. The basic idea is to choose the limits of integration so that every point of the solid is accounted for in the integration process. Figure 15.7.9 illustrates one way of doing this. Holding θ and ϕ fixed for the first integration, we let ρ vary from 0 to ρ_0. This covers a radial line from the origin to the surface of the sphere. Next, keeping θ fixed, we let ϕ vary from 0 to π so that the radial line sweeps out a fan-shaped region. Finally, we let θ vary from 0 to 2π so that the fan-shaped region makes a complete revolution, thereby sweeping out the entire sphere. Thus, the triple integral of $f(\rho, \theta, \phi)$ over the spherical solid G may be evaluated by writing

$$\iiint\limits_G f(\rho, \theta, \phi)\, dV = \int_0^{2\pi} \int_0^{\pi} \int_0^{\rho_0} f(\rho, \theta, \phi)\rho^2 \sin \phi \, d\rho\, d\phi\, d\theta$$

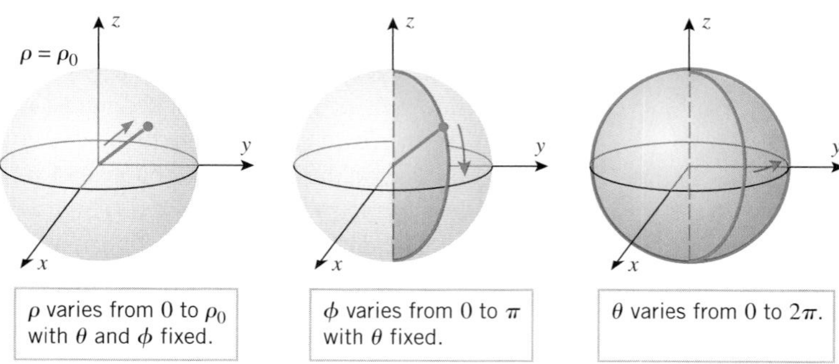

Figure 15.7.9

Table 15.7.1 suggests how the limits of integration in spherical coordinates can be obtained for some other common solids.

Table 15.7.1

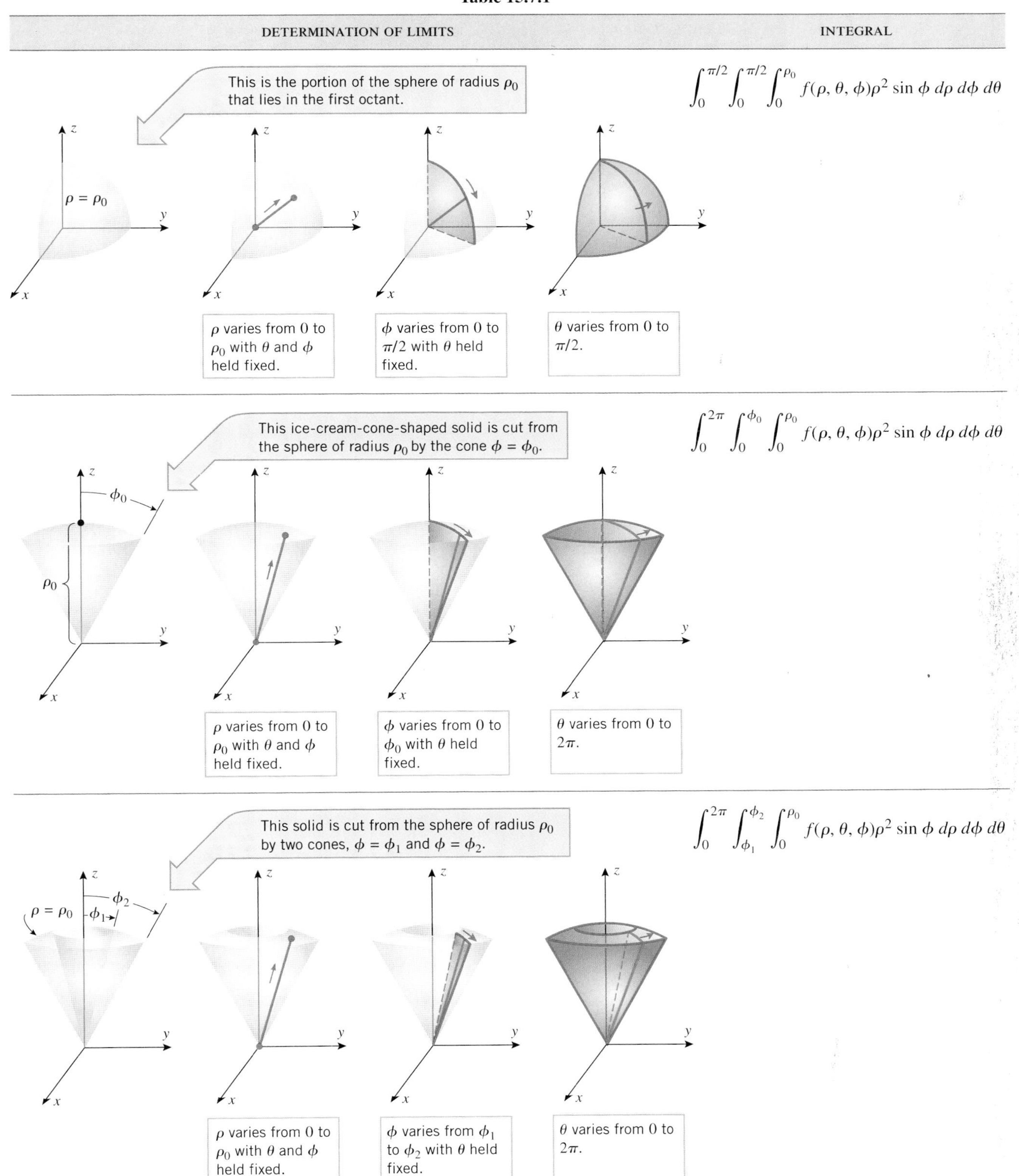

DETERMINATION OF LIMITS	INTEGRAL
This is the portion of the sphere of radius ρ_0 that lies in the first octant. $\rho = \rho_0$. ρ varies from 0 to ρ_0 with θ and ϕ held fixed. ϕ varies from 0 to $\pi/2$ with θ held fixed. θ varies from 0 to $\pi/2$.	$\int_0^{\pi/2}\int_0^{\pi/2}\int_0^{\rho_0} f(\rho, \theta, \phi)\rho^2 \sin\phi \, d\rho \, d\phi \, d\theta$
This ice-cream-cone-shaped solid is cut from the sphere of radius ρ_0 by the cone $\phi = \phi_0$. ρ varies from 0 to ρ_0 with θ and ϕ held fixed. ϕ varies from 0 to ϕ_0 with θ held fixed. θ varies from 0 to 2π.	$\int_0^{2\pi}\int_0^{\phi_0}\int_0^{\rho_0} f(\rho, \theta, \phi)\rho^2 \sin\phi \, d\rho \, d\phi \, d\theta$
This solid is cut from the sphere of radius ρ_0 by two cones, $\phi = \phi_1$ and $\phi = \phi_2$. $\rho = \rho_0$. ρ varies from 0 to ρ_0 with θ and ϕ held fixed. ϕ varies from ϕ_1 to ϕ_2 with θ held fixed. θ varies from 0 to 2π.	$\int_0^{2\pi}\int_{\phi_1}^{\phi_2}\int_0^{\rho_0} f(\rho, \theta, \phi)\rho^2 \sin\phi \, d\rho \, d\phi \, d\theta$

Table 15.7.1 (*continued*)

DETERMINATION OF LIMITS	INTEGRAL
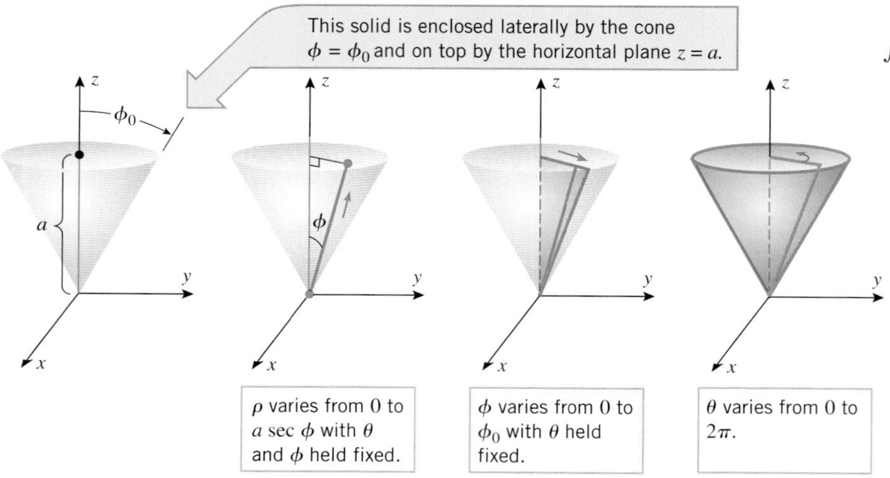 This solid is enclosed laterally by the cone $\phi = \phi_0$ and on top by the horizontal plane $z = a$. ρ varies from 0 to $a \sec \phi$ with θ and ϕ held fixed. ϕ varies from 0 to ϕ_0 with θ held fixed. θ varies from 0 to 2π.	$\int_0^{2\pi} \int_0^{\phi_0} \int_0^{a \sec \phi} f(\rho, \theta, \phi)\rho^2 \sin \phi \, d\rho \, d\phi \, d\theta$
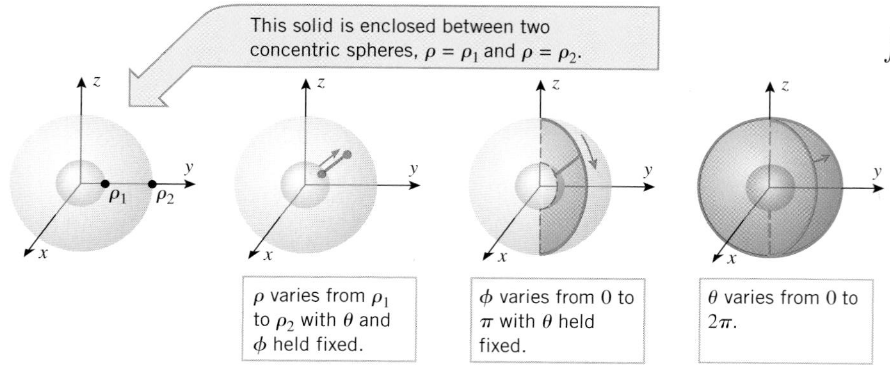 This solid is enclosed between two concentric spheres, $\rho = \rho_1$ and $\rho = \rho_2$. ρ varies from ρ_1 to ρ_2 with θ and ϕ held fixed. ϕ varies from 0 to π with θ held fixed. θ varies from 0 to 2π.	$\int_0^{2\pi} \int_0^{\pi} \int_{\rho_1}^{\rho_2} f(\rho, \theta, \phi)\rho^2 \sin \phi \, d\rho \, d\phi \, d\theta$

► **Example 3** Use spherical coordinates to find the volume and the centroid of the solid G bounded above by the sphere $x^2 + y^2 + z^2 = 16$ and below by the cone $z = \sqrt{x^2 + y^2}$.

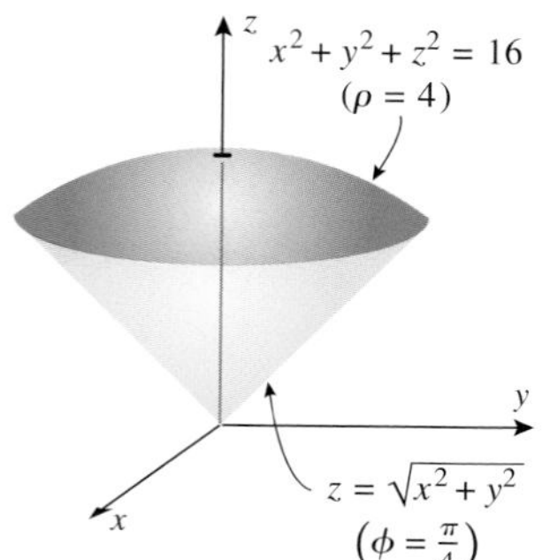

Figure 15.7.10

Solution. The solid G is sketched in Figure 15.7.10.

In spherical coordinates, the equation of the sphere $x^2 + y^2 + z^2 = 16$ is $\rho = 4$ and the equation of the cone $z = \sqrt{x^2 + y^2}$ is

$$\rho \cos \phi = \sqrt{\rho^2 \sin^2 \phi \cos^2 \theta + \rho^2 \sin^2 \phi \sin^2 \theta}$$

which simplifies to

$$\rho \cos \phi = \rho \sin \phi$$

or, on dividing both sides by $\rho \cos \phi$,

$$\tan \phi = 1$$

Thus $\phi = \pi/4$, and using the second entry in Table 15.7.1, the volume of G is

$$\begin{aligned}
V = \iiint_G dV &= \int_0^{2\pi}\int_0^{\pi/4}\int_0^4 \rho^2 \sin\phi \, d\rho \, d\phi \, d\theta \\
&= \int_0^{2\pi}\int_0^{\pi/4}\left[\frac{\rho^3}{3}\sin\phi\right]_{\rho=0}^4 d\phi \, d\theta \\
&= \int_0^{2\pi}\int_0^{\pi/4}\frac{64}{3}\sin\phi \, d\phi \, d\theta \\
&= \frac{64}{3}\int_0^{2\pi}\left[-\cos\phi\right]_{\phi=0}^{\pi/4} d\theta = \frac{64}{3}\int_0^{2\pi}\left(1 - \frac{\sqrt{2}}{2}\right) d\theta \\
&= \frac{64\pi}{3}(2-\sqrt{2})
\end{aligned}$$

By symmetry, the centroid $(\bar{x}, \bar{y}, \bar{z})$ is on the z-axis, so $\bar{x} = \bar{y} = 0$. From (15) of Section 15.6 and the volume calculated above,

$$\begin{aligned}
\bar{z} = \frac{1}{V}\iiint_G z \, dV &= \frac{1}{V}\int_0^{2\pi}\int_0^{\pi/4}\int_0^4 (\rho\cos\phi)\rho^2 \sin\phi \, d\rho \, d\phi \, d\theta \\
&= \frac{1}{V}\int_0^{2\pi}\int_0^{\pi/4}\left[\frac{\rho^4}{4}\cos\phi\sin\phi\right]_{\rho=0}^4 d\phi \, d\theta \\
&= \frac{64}{V}\int_0^{2\pi}\int_0^{\pi/4}\sin\phi\cos\phi \, d\phi \, d\theta = \frac{64}{V}\int_0^{2\pi}\left[\frac{1}{2}\sin^2\phi\right]_{\phi=0}^{\pi/4} d\theta \\
&= \frac{16}{V}\int_0^{2\pi} d\theta = \frac{32\pi}{V} = \frac{3}{2(2-\sqrt{2})}
\end{aligned}$$

With the help of a calculator, $\bar{z} \approx 2.56$ (to two decimal places), so the approximate location of the centroid in the xyz-coordinate system is $(0, 0, 2.56)$. ◄

CONVERTING TRIPLE INTEGRALS FROM RECTANGULAR TO SPHERICAL COORDINATES

Referring to Table 12.8.1, triple integrals can be converted from rectangular coordinates to spherical coordinates by making the substitution $x = \rho\sin\phi\cos\theta$, $y = \rho\sin\phi\sin\theta$, $z = \rho\cos\phi$. The two integrals are related by the equation

$$\iiint_G f(x, y, z) \, dV = \iiint_{\substack{\text{appropriate}\\ \text{limits}}} f(\rho\sin\phi\cos\theta, \rho\sin\phi\sin\theta, \rho\cos\phi)\rho^2\sin\phi \, d\rho \, d\phi \, d\theta \tag{10}$$

► **Example 4** Use spherical coordinates to evaluate

$$\int_{-2}^{2}\int_{-\sqrt{4-x^2}}^{\sqrt{4-x^2}}\int_0^{\sqrt{4-x^2-y^2}} z^2\sqrt{x^2+y^2+z^2} \, dz \, dy \, dx$$

Solution. In problems like this, it is helpful to begin (when possible) with a sketch of the region G of integration. From the z-limits of integration, the upper surface of G is the hemisphere $z = \sqrt{4 - x^2 - y^2}$ and the lower surface is the xy-plane $z = 0$. From the x- and y-limits of integration, the projection of the solid G on the xy-plane is the region

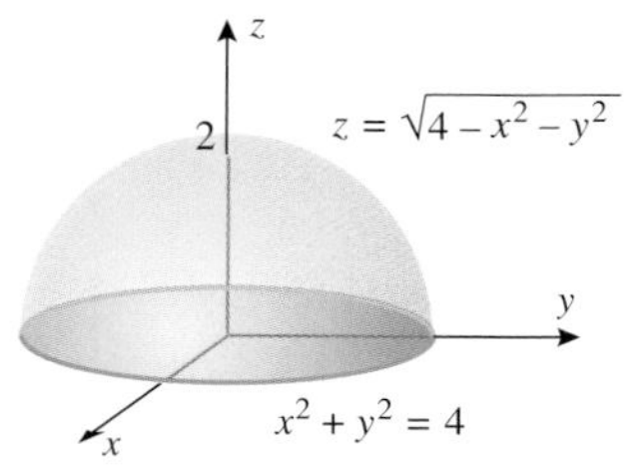

Figure 15.7.11

enclosed by the circle $x^2 + y^2 = 4$. From this information we obtain the sketch of G in Figure 15.7.11. Thus,

$$\begin{aligned}\int_{-2}^{2}\int_{-\sqrt{4-x^2}}^{\sqrt{4-x^2}}\int_{0}^{\sqrt{4-x^2-y^2}} & z^2\sqrt{x^2+y^2+z^2}\,dz\,dy\,dx \\ &= \iiint_G z^2\sqrt{x^2+y^2+z^2}\,dV \\ &= \int_0^{2\pi}\int_0^{\pi/2}\int_0^2 \rho^5\cos^2\phi\sin\phi\,d\rho\,d\phi\,d\theta \\ &= \int_0^{2\pi}\int_0^{\pi/2}\frac{32}{3}\cos^2\phi\sin\phi\,d\phi\,d\theta \\ &= \frac{32}{3}\int_0^{2\pi}\left[-\frac{1}{3}\cos^3\phi\right]_{\phi=0}^{\pi/2} d\theta = \frac{32}{9}\int_0^{2\pi} d\theta = \frac{64}{9}\pi \blacktriangleleft\end{aligned}$$

✔QUICK CHECK EXERCISES 15.7 *(See page 1086 for answers.)*

1. (a) The cylindrical wedge $1 \le r \le 3$, $\pi/6 \le \theta \le \pi/2$, $0 \le z \le 5$ has volume $V =$ ______.
   (b) The spherical wedge $1 \le \rho \le 3$, $\pi/6 \le \theta \le \pi/2$, $0 \le \phi \le \pi/3$ has volume $V =$ ______.
2. Let G be the solid region inside the sphere of radius 2 centered at the origin and above the plane $z = 1$. In each part, supply the missing integrand and limits of integration for the iterated integral in cylindrical coordinates.
   (a) The volume of G is
   $$\iiint_G dV = \int_\square^\square\int_\square^\square\int_\square^\square \underline{\qquad}\,dz\,dr\,d\theta$$
   (b) $$\iiint_G \frac{z}{x^2+y^2+z^2}\,dV = \int_\square^\square\int_\square^\square\int_\square^\square \underline{\qquad}\,dz\,dr\,d\theta$$
3. Let G be the solid region described in Quick Check Exercise 2. In each part, supply the missing integrand and limits of integration for the iterated integral in spherical coordinates.
   (a) The volume of G is
   $$\iiint_G dV = \int_\square^\square\int_\square^\square\int_\square^\square \underline{\qquad}\,d\rho\,d\phi\,d\theta$$
   (b) $$\iiint_G \frac{z}{x^2+y^2+z^2}\,dV = \int_\square^\square\int_\square^\square\int_\square^\square \underline{\qquad}\,d\rho\,d\phi\,d\theta$$

EXERCISE SET 15.7 [C] CAS

1–4 Evaluate the iterated integral.

1. $\displaystyle\int_0^{2\pi}\int_0^1\int_0^{\sqrt{1-r^2}} zr\,dz\,dr\,d\theta$
2. $\displaystyle\int_0^{\pi/2}\int_0^{\cos\theta}\int_0^{r^2} r\sin\theta\,dz\,dr\,d\theta$
3. $\displaystyle\int_0^{\pi/2}\int_0^{\pi/2}\int_0^1 \rho^3\sin\phi\cos\phi\,d\rho\,d\phi\,d\theta$
4. $\displaystyle\int_0^{2\pi}\int_0^{\pi/4}\int_0^{a\sec\phi} \rho^2\sin\phi\,d\rho\,d\phi\,d\theta \quad (a > 0)$

FOCUS ON CONCEPTS

5. Sketch the region G and identify the function f so that
   $$\iiint_G f(r, \theta, z)\,dV$$
   corresponds to the iterated integral in Exercise 1.

6. Sketch the region G and identify the function f so that

$$\iiint_G f(r, \theta, z)\, dV$$

corresponds to the iterated integral in Exercise 2.

7. Sketch the region G and identify the function f so that

$$\iiint_G f(\rho, \theta, \phi)\, dV$$

corresponds to the iterated integral in Exercise 3.

8. Sketch the region G and identify the function f so that

$$\iiint_G f(\rho, \theta, \phi)\, dV$$

corresponds to the iterated integral in Exercise 4.

9–12 Use cylindrical coordinates to find the volume of the solid.

9. The solid enclosed by the paraboloid $z = x^2 + y^2$ and the plane $z = 9$.

10. The solid that is bounded above and below by the sphere $x^2 + y^2 + z^2 = 9$ and inside the cylinder $x^2 + y^2 = 4$.

11. The solid that is inside the surface $r^2 + z^2 = 20$ but not above the surface $z = r^2$.

12. The solid enclosed between the cone $z = (hr)/a$ and the plane $z = h$.

13–16 Use spherical coordinates to find the volume of the solid.

13. The solid bounded above by the sphere $\rho = 4$ and below by the cone $\phi = \pi/3$.

14. The solid within the cone $\phi = \pi/4$ and between the spheres $\rho = 1$ and $\rho = 2$.

15. The solid enclosed by the sphere $x^2 + y^2 + z^2 = 4a^2$ and the planes $z = 0$ and $z = a$.

16. The solid within the sphere $x^2 + y^2 + z^2 = 9$, outside the cone $z = \sqrt{x^2 + y^2}$, and above the xy-plane.

17–20 Use cylindrical or spherical coordinates to evaluate the integral.

17. $\displaystyle\int_0^a \int_0^{\sqrt{a^2-x^2}} \int_0^{a^2-x^2-y^2} x^2\, dz\, dy\, dx \quad (a > 0)$

18. $\displaystyle\int_{-1}^1 \int_0^{\sqrt{1-x^2}} \int_0^{\sqrt{1-x^2-y^2}} e^{-(x^2+y^2+z^2)^{3/2}}\, dz\, dy\, dx$

19. $\displaystyle\int_0^2 \int_0^{\sqrt{4-y^2}} \int_{\sqrt{x^2+y^2}}^{\sqrt{8-x^2-y^2}} z^2\, dz\, dx\, dy$

20. $\displaystyle\int_{-3}^3 \int_{-\sqrt{9-y^2}}^{\sqrt{9-y^2}} \int_{-\sqrt{9-x^2-y^2}}^{\sqrt{9-x^2-y^2}} \sqrt{x^2 + y^2 + z^2}\, dz\, dx\, dy$

C 21. (a) Use a CAS to evaluate

$$\int_{-2}^2 \int_1^4 \int_{\pi/6}^{\pi/3} \frac{r \tan^3 \theta}{\sqrt{1+z^2}}\, d\theta\, dr\, dz$$

(b) Find a function $f(x, y, z)$ and sketch a region G in 3-space so that the triple integral in rectangular coordinates

$$\iiint_G f(x, y, z)\, dV$$

matches the iterated integral in cylindrical coordinates given in part (a).

C 22. Use a CAS to evaluate

$$\int_0^{\pi/2} \int_0^{\pi/4} \int_0^{\cos\theta} \rho^{17} \cos\phi \cos^{19}\theta\, d\rho\, d\phi\, d\theta$$

23. Find the volume enclosed by $x^2 + y^2 + z^2 = a^2$ using
(a) cylindrical coordinates
(b) spherical coordinates.

24. Let G be the solid in the first octant bounded by the sphere $x^2 + y^2 + z^2 = 4$ and the coordinate planes. Evaluate

$$\iiint_G xyz\, dV$$

(a) using rectangular coordinates
(b) using cylindrical coordinates
(c) using spherical coordinates.

25–26 Use cylindrical coordinates.

25. Find the mass of the solid with density $\delta(x, y, z) = 3 - z$ that is bounded by the cone $z = \sqrt{x^2 + y^2}$ and the plane $z = 3$.

26. Find the mass of a right circular cylinder of radius a and height h if the density is proportional to the distance from the base. (Let k be the constant of proportionality.)

27–28 Use spherical coordinates.

27. Find the mass of a spherical solid of radius a if the density is proportional to the distance from the center. (Let k be the constant of proportionality.)

28. Find the mass of the solid enclosed between the spheres $x^2 + y^2 + z^2 = 1$ and $x^2 + y^2 + z^2 = 4$ if the density is $\delta(x, y, z) = (x^2 + y^2 + z^2)^{-1/2}$.

29–30 Use cylindrical coordinates to find the centroid of the solid.

29. The solid that is bounded above by the sphere

$$x^2 + y^2 + z^2 = 2$$

and below by the paraboloid $z = x^2 + y^2$.

30. The solid that is bounded by the cone $z = \sqrt{x^2 + y^2}$ and the plane $z = 2$.

31–32 Use spherical coordinates to find the centroid of the solid.

31. The solid in the first octant bounded by the coordinate planes and the sphere $x^2 + y^2 + z^2 = a^2$.

32. The solid bounded above by the sphere $\rho = 4$ and below by the cone $\phi = \pi/3$.

33–34 Use the Wallis formulas in Exercises 64 and 66 of Section 8.3.

33. Find the centroid of the solid bounded above by the paraboloid $z = x^2 + y^2$, below by the plane $z = 0$, and laterally by the cylinder $(x-1)^2 + y^2 = 1$.

34. Find the mass of the solid in the first octant bounded above by the paraboloid $z = 4 - x^2 - y^2$, below by the plane $z = 0$, and laterally by the cylinder $x^2 + y^2 = 2x$ and the plane $y = 0$, assuming the density to be $\delta(x, y, z) = z$.

35–40 Solve the problem using either cylindrical or spherical coordinates (whichever seems appropriate).

35. Find the volume of the solid in the first octant bounded by the sphere $\rho = 2$, the coordinate planes, and the cones $\phi = \pi/6$ and $\phi = \pi/3$.

36. Find the mass of the solid that is enclosed by the sphere $x^2 + y^2 + z^2 = 1$ and lies within the cone $z = \sqrt{x^2 + y^2}$ if the density is $\delta(x, y, z) = \sqrt{x^2 + y^2 + z^2}$.

37. Find the center of gravity of the solid bounded by the paraboloid $z = 1 - x^2 - y^2$ and the xy-plane, assuming the density to be $\delta(x, y, z) = x^2 + y^2 + z^2$.

38. Find the center of gravity of the solid that is bounded by the cylinder $x^2 + y^2 = 1$, the cone $z = \sqrt{x^2 + y^2}$, and the xy-plane if the density is $\delta(x, y, z) = z$.

39. Find the center of gravity of the solid hemisphere bounded by $z = \sqrt{a^2 - x^2 - y^2}$ and $z = 0$ if the density is proportional to the distance from the origin.

40. Find the centroid of the solid that is enclosed by the hemispheres $y = \sqrt{9 - x^2 - z^2}$, $y = \sqrt{4 - x^2 - z^2}$, and the plane $y = 0$.

41. Suppose that the density at a point in a gaseous spherical star is modeled by the formula

$$\delta = \delta_0 e^{-(\rho/R)^3}$$

where δ_0 is a positive constant, R is the radius of the star, and ρ is the distance from the point to the star's center. Find the mass of the star.

42. In this exercise we will obtain a formula for the volume of the spherical wedge in Figure 15.7.7.

(a) Use a triple integral in cylindrical coordinates to show that the volume of the solid bounded above by a sphere $\rho = \rho_0$, below by a cone $\phi = \phi_0$, and on the sides by $\theta = \theta_1$ and $\theta = \theta_2$ $(\theta_1 < \theta_2)$ is

$$V = \tfrac{1}{3}\rho_0^3(1 - \cos\phi_0)(\theta_2 - \theta_1)$$

[*Hint:* In cylindrical coordinates, the sphere has the equation $r^2 + z^2 = \rho_0^2$ and the cone has the equation $z = r\cot\phi_0$. For simplicity, consider only the case $0 < \phi_0 < \pi/2$.]

(b) Subtract appropriate volumes and use the result in part (a) to deduce that the volume ΔV of the spherical wedge is

$$\Delta V = \frac{\rho_2^3 - \rho_1^3}{3}(\cos\phi_1 - \cos\phi_2)(\theta_2 - \theta_1)$$

(c) Apply the Mean-Value Theorem to the functions $\cos\phi$ and ρ^3 to deduce that the formula in part (b) can be written as

$$\Delta V = \rho^{*2}\sin\phi^*\,\Delta\rho\,\Delta\phi\,\Delta\theta$$

where ρ^* is between ρ_1 and ρ_2, ϕ^* is between ϕ_1 and ϕ_2, and $\Delta\rho = \rho_2 - \rho_1$, $\Delta\phi = \phi_2 - \phi_1$, $\Delta\theta = \theta_2 - \theta_1$.

43–46 The tendency of a solid to resist a change in rotational motion about an axis is measured by its ***moment of inertia*** about that axis. If the solid occupies a region G in an xyz-coordinate system, and if its density function $\delta(x, y, z)$ is continuous on G, then the moments of inertia about the x-axis, the y-axis, and the z-axis are denoted by I_x, I_y, and I_z, respectively, and are defined by

$$I_x = \iiint_G (y^2 + z^2)\,\delta(x, y, z)\,dV$$

$$I_y = \iiint_G (x^2 + z^2)\,\delta(x, y, z)\,dV$$

$$I_z = \iiint_G (x^2 + y^2)\,\delta(x, y, z)\,dV$$

(compare with the discussion preceding Exercise 43 in Section 15.6). In these exercises, find the indicated moments of inertia of the solid, assuming that it has constant density δ.

43. I_z for the solid cylinder $x^2 + y^2 \le a^2$, $0 \le z \le h$.

44. I_y for the solid cylinder $x^2 + y^2 \le a^2$, $0 \le z \le h$.

45. I_z for the hollow cylinder $a_1^2 \le x^2 + y^2 \le a_2^2$, $0 \le z \le h$.

46. I_z for the solid sphere $x^2 + y^2 + z^2 \le a^2$.

✔ QUICK CHECK ANSWERS 15.7

1. (a) $\dfrac{20}{3}\pi$ (b) $\dfrac{13}{9}\pi$ **2.** (a) $\displaystyle\int_0^{2\pi}\int_0^{\sqrt{3}}\int_1^{\sqrt{4-r^2}} r\,dz\,dr\,d\theta$ (b) $\displaystyle\int_0^{2\pi}\int_0^{\sqrt{3}}\int_1^{\sqrt{4-r^2}} \frac{rz}{r^2+z^2}\,dz\,dr\,d\theta$

3. (a) $\displaystyle\int_0^{2\pi}\int_0^{\pi/3}\int_{\sec\phi}^{2} \rho^2\sin\phi\,d\rho\,d\phi\,d\theta$ (b) $\displaystyle\int_0^{2\pi}\int_0^{\pi/3}\int_{\sec\phi}^{2} \rho\cos\phi\sin\phi\,d\rho\,d\phi\,d\theta$

15.8 CHANGE OF VARIABLES IN MULTIPLE INTEGRALS; JACOBIANS

In this section we will discuss a general method for evaluating double and triple integrals by substitution. Most of the results in this section are very difficult to prove, so our approach will be informal and motivational. Our goal is to provide a geometric understanding of the basic principles and an exposure to computational techniques.

CHANGE OF VARIABLE IN A SINGLE INTEGRAL

To motivate techniques for evaluating double and triple integrals by substitution, it will be helpful to consider the effect of a substitution $x = g(u)$ on a single integral over an interval $[a, b]$. If g is differentiable and either increasing or decreasing, then g is one-to-one and

$$\int_a^b f(x)\,dx = \int_{g^{-1}(a)}^{g^{-1}(b)} f(g(u))g'(u)\,du$$

In this relationship $f(x)$ and dx are expressed in terms of u, and the u-limits of integration result from solving the equations

$$a = g(u) \quad \text{and} \quad b = g(u)$$

In the case where g is decreasing we have $g^{-1}(b) < g^{-1}(a)$, which is contrary to our usual convention of writing definite integrals with the larger limit of integration at the top. We can remedy this by reversing the limits of integration and writing

$$\int_a^b f(x)\,dx = -\int_{g^{-1}(b)}^{g^{-1}(a)} f(g(u))g'(u)\,du = \int_{g^{-1}(b)}^{g^{-1}(a)} f(g(u))|g'(u)|\,du$$

where the absolute value results from the fact that $g'(u)$ is negative. Thus, regardless of whether g is increasing or decreasing we can write

$$\int_a^b f(x)\,dx = \int_\alpha^\beta f(g(u))|g'(u)|\,du \tag{1}$$

where α and β are the u-limits of integration and $\alpha < \beta$.

The expression $g'(u)$ that appears in (1) is called the ***Jacobian*** of the change of variable $x = g(u)$ in honor of C. G. J. Jacobi, who made the first serious study of change of variables in multiple integrals in the mid-1800s. Formula (1) reveals three effects of the change of variable $x = g(u)$:

- The new integrand becomes $f(g(u))$ times the absolute value of the Jacobian.
- dx becomes du.
- The x-interval of integration is transformed into a u-interval of integration.

Our goal in this section is to show that analogous results hold for changing variables in double and triple integrals.

TRANSFORMATIONS OF THE PLANE

In earlier sections we considered parametric equations of three kinds:

$$x = x(t), \quad y = y(t)$$ A curve in the plane

$$x = x(t), \quad y = y(t), \quad z = z(t)$$ A curve in 3-space

$$x = x(u, v), \quad y = y(u, v), \quad z = z(u, v)$$ A surface in 3-space

Now we will consider parametric equations of the form

$$x = x(u, v), \quad y = y(u, v) \tag{2}$$

Parametric equations of this type associate points in the xy-plane with points in the uv-plane. These equations can be written in vector form as

$$\mathbf{r} = \mathbf{r}(u, v) = x(u, v)\mathbf{i} + y(u, v)\mathbf{j}$$

where $\mathbf{r} = x\mathbf{i} + y\mathbf{j}$ is a position vector in the xy-plane and $\mathbf{r}(u, v)$ is a vector-valued function of the variables u and v.

It will also be useful in this section to think of the parametric equations in (2) in terms of inputs and outputs. If we think of the pair of numbers (u, v) as an input, then the two equations, in combination, produce a unique output (x, y), and hence define a function T that associates points in the xy-plane with points in the uv-plane. This function is described by the formula

$$T(u, v) = (x(u, v), y(u, v))$$

We call T a ***transformation*** from the uv-plane to the xy-plane and (x, y) the ***image*** of (u, v) under the transformation T. We also say that T ***maps*** (u, v) into (x, y). The set R of all images in the xy-plane of a set S in the uv-plane is called the ***image of S under T***. If distinct points in the uv-plane have distinct images in the xy-plane, then T is said to be ***one-to-one***. In this case the equations in (2) define u and v as functions of x and y, say

$$u = u(x, y), \quad v = v(x, y)$$

These equations, which can often be obtained by solving (2) for u and v in terms of x and y, define a transformation from the xy-plane to the uv-plane that maps the image of (u, v) under T back into (u, v). This transformation is denoted by T^{-1} and is called the ***inverse of T*** (Figure 15.8.1).

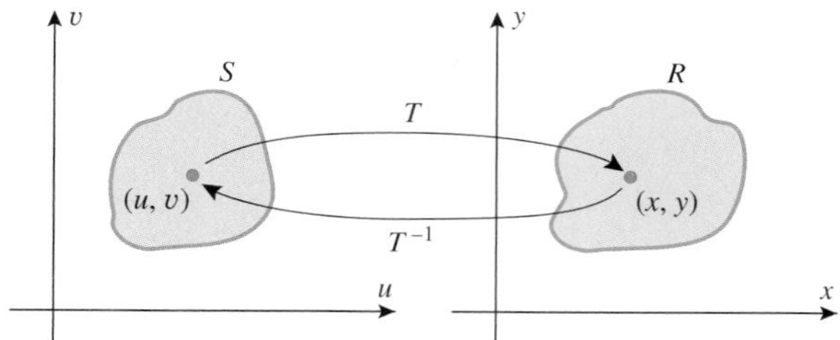

Figure 15.8.1

Carl Gustav Jacob Jacobi (1804–1851) German mathematician. Jacobi, the son of a banker, grew up in a background of wealth and culture and showed brilliance in mathematics early. He resisted studying mathematics by rote, preferring instead to learn general principles from the works of the masters, Euler and Lagrange. He entered the University of Berlin at age 16 as a student of mathematics and classical studies. However, he soon realized that he could not do both and turned fully to mathematics with a blazing intensity that he would maintain throughout his life. He received his Ph.D. in 1825 and was able to secure a position as a lecturer at the University of Berlin by giving up Judaism and becoming a Christian. However, his promotion opportunities remained limited and he moved on to the University of Königsberg. Jacobi was born to teach—he had a dynamic personality and delivered his lectures with a clarity and enthusiasm that frequently left his audience spellbound. In spite of extensive teaching commitments, he was able to publish volumes of revolutionary mathematical research that eventually made him the leading European mathematician after Gauss. His main body of research was in the area of elliptic functions, a branch of mathematics with important applications in astronomy and physics as well as in other fields of mathematics. Because of his family wealth, Jacobi was not dependent on his teaching salary in his early years. However, his comfortable world eventually collapsed. In 1840 his family went bankrupt and he was wiped out financially. In 1842 he had a nervous breakdown from overwork. In 1843 he became seriously ill with diabetes and moved to Berlin with the help of a government grant to defray his medical expenses. In 1848 he made an injudicious political speech that caused the government to withdraw the grant, eventually resulting in the loss of his home. His health continued to decline and in 1851 he finally succumbed to successive bouts of influenza and smallpox. In spite of all his problems, Jacobi was a tireless worker to the end. When a friend expressed concern about the effect of the hard work on his health, Jacobi replied, "Certainly, I have sometimes endangered my health by overwork, but what of it? Only cabbages have no nerves, no worries. And what do they get out of their perfect well-being?"

One way to visualize the geometric effect of a transformation T is to determine the images in the xy-plane of the vertical and horizontal lines in the uv-plane. Following the discussion on page 1048 in Section 15.4, sets of points in the xy-plane that are images of horizontal lines (v constant) are called ***constant v-curves***, and sets of points that are images of vertical lines (u constant) are called ***constant u-curves*** (Figure 15.8.2).

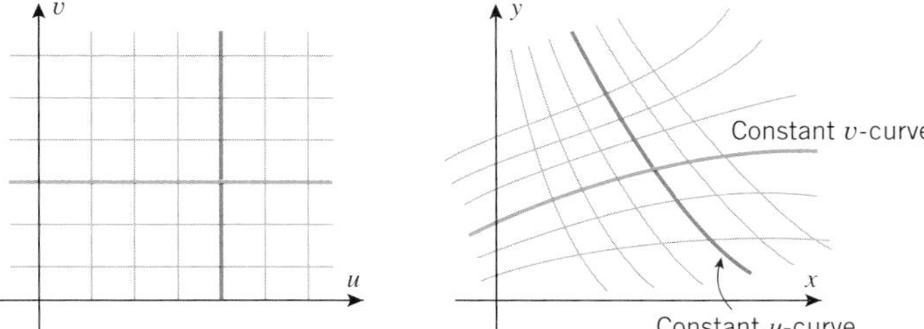

Figure 15.8.2

▶ **Example 1** Let T be the transformation from the uv-plane to the xy-plane defined by the equations

$$x = \tfrac{1}{4}(u + v), \quad y = \tfrac{1}{2}(u - v) \tag{3}$$

(a) Find $T(1, 3)$.

(b) Sketch the constant v-curves corresponding to $v = -2, -1, 0, 1, 2$.

(c) Sketch the constant u-curves corresponding to $u = -2, -1, 0, 1, 2$.

(d) Sketch the image under T of the square region in the uv-plane bounded by the lines $u = -2$, $u = 2$, $v = -2$, and $v = 2$.

Solution (a). Substituting $u = 1$ and $v = 3$ in (3) yields $T(1, 3) = (1, -1)$.

Solutions (b and c). In these parts it will be convenient to express the transformation equations with u and v as functions of x and y. We leave it for you to show that

$$u = 2x + y, \quad v = 2x - y$$

Thus, the constant v-curves corresponding to $v = -2, -1, 0, 1$, and 2 are

$$2x - y = -2, \quad 2x - y = -1, \quad 2x - y = 0, \quad 2x - y = 1, \quad 2x - y = 2$$

and the constant u-curves corresponding to $u = -2, -1, 0, 1$, and 2 are

$$2x + y = -2, \quad 2x + y = -1, \quad 2x + y = 0, \quad 2x + y = 1, \quad 2x + y = 2$$

In Figure 15.8.3 the constant v-curves are shown in green and the constant u-curves in red.

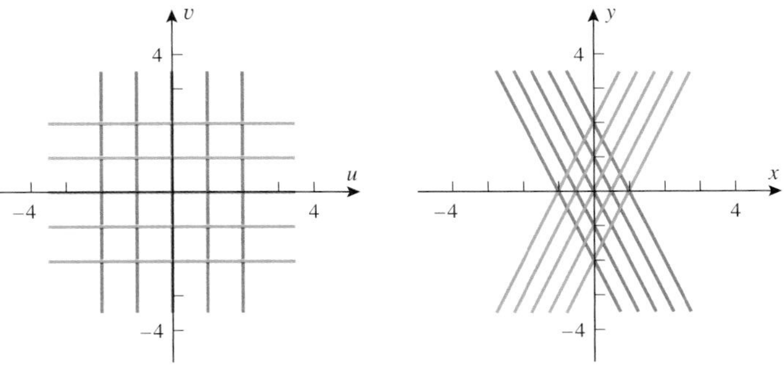

Figure 15.8.3

Solution (d). The image of a region can often be found by finding the image of its boundary. In this case the images of the boundary lines $u = -2$, $u = 2$, $v = -2$, and $v = 2$ enclose the diamond-shaped region in the xy-plane shown in Figure 15.8.4. ◄

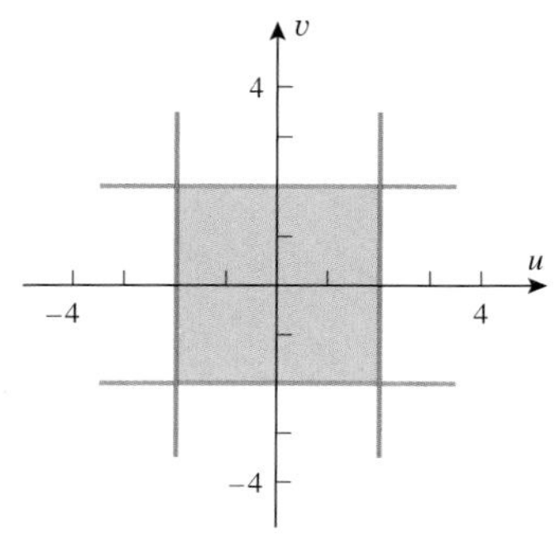

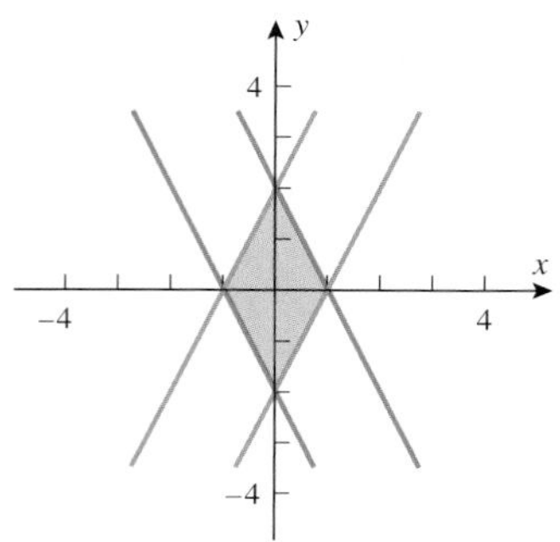

Figure 15.8.4

JACOBIANS IN TWO VARIABLES

To derive the change of variables formula for double integrals, we will need to understand the relationship between the area of a *small* rectangular region in the uv-plane and the area of its image in the xy-plane under a transformation T given by the equations

$$x = x(u, v), \quad y = y(u, v)$$

For this purpose, suppose that Δu and Δv are positive, and consider a rectangular region S in the uv-plane enclosed by the lines

$$u = u_0, \quad u = u_0 + \Delta u, \quad v = v_0, \quad v = v_0 + \Delta v$$

If the functions $x(u, v)$ and $y(u, v)$ are continuous, and if Δu and Δv are not too large, then the image of S in the xy-plane will be a region R that looks like a slightly distorted parallelogram (Figure 15.8.5). The sides of R are the constant u-curves and v-curves that correspond to the sides of S.

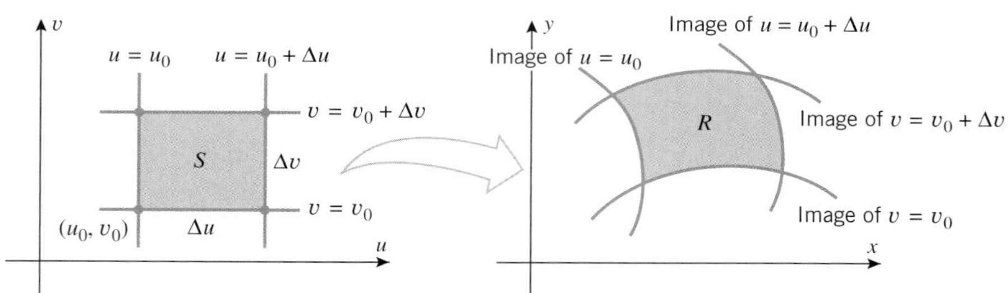

Figure 15.8.5

If we let

$$\mathbf{r} = \mathbf{r}(u, v) = x(u, v)\mathbf{i} + y(u, v)\mathbf{j}$$

be the position vector of the point in the xy-plane that corresponds to the point (u, v) in the uv-plane, then the constant v-curve corresponding to $v = v_0$ and the constant u-curve corresponding to $u = u_0$ can be represented in vector form as

$$\mathbf{r}(u, v_0) = x(u, v_0)\mathbf{i} + y(u, v_0)\mathbf{j} \quad \text{Constant } v\text{-curve}$$

$$\mathbf{r}(u_0, v) = x(u_0, v)\mathbf{i} + y(u_0, v)\mathbf{j} \quad \text{Constant } u\text{-curve}$$

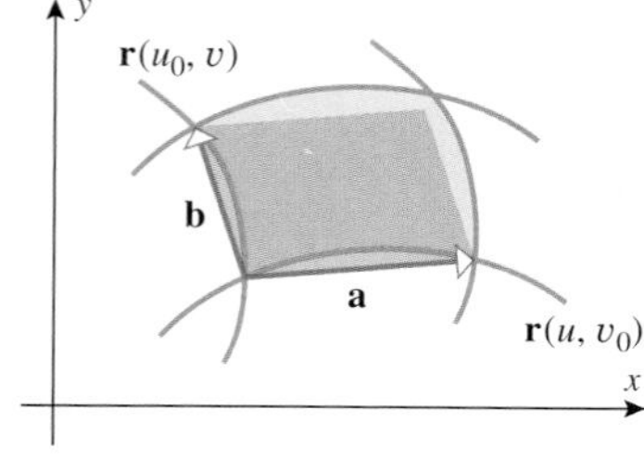

Figure 15.8.6

Since we are assuming Δu and Δv to be small, the region R can be approximated by a parallelogram determined by the "secant vectors"

$$\mathbf{a} = \mathbf{r}(u_0 + \Delta u, v_0) - \mathbf{r}(u_0, v_0) \tag{4}$$

$$\mathbf{b} = \mathbf{r}(u_0, v_0 + \Delta v) - \mathbf{r}(u_0, v_0) \tag{5}$$

shown in Figure 15.8.6. A more useful approximation of R can be obtained by using Formulas (5) and (6) of Section 15.4 to approximate these secant vectors by tangent vectors as follows:

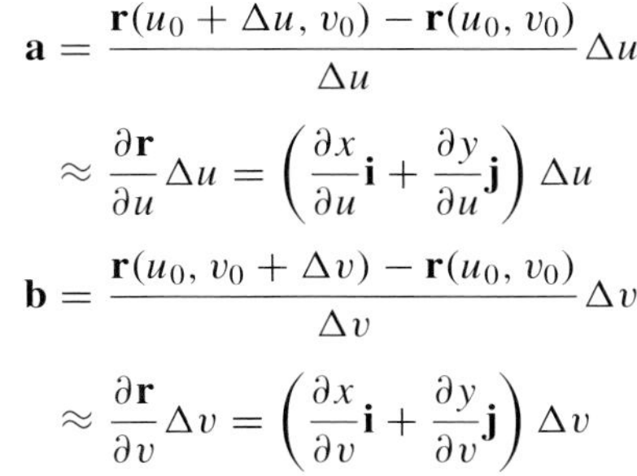

$$\mathbf{a} = \frac{\mathbf{r}(u_0 + \Delta u, v_0) - \mathbf{r}(u_0, v_0)}{\Delta u}\Delta u$$

$$\approx \frac{\partial \mathbf{r}}{\partial u}\Delta u = \left(\frac{\partial x}{\partial u}\mathbf{i} + \frac{\partial y}{\partial u}\mathbf{j}\right)\Delta u$$

$$\mathbf{b} = \frac{\mathbf{r}(u_0, v_0 + \Delta v) - \mathbf{r}(u_0, v_0)}{\Delta v}\Delta v$$

$$\approx \frac{\partial \mathbf{r}}{\partial v}\Delta v = \left(\frac{\partial x}{\partial v}\mathbf{i} + \frac{\partial y}{\partial v}\mathbf{j}\right)\Delta v$$

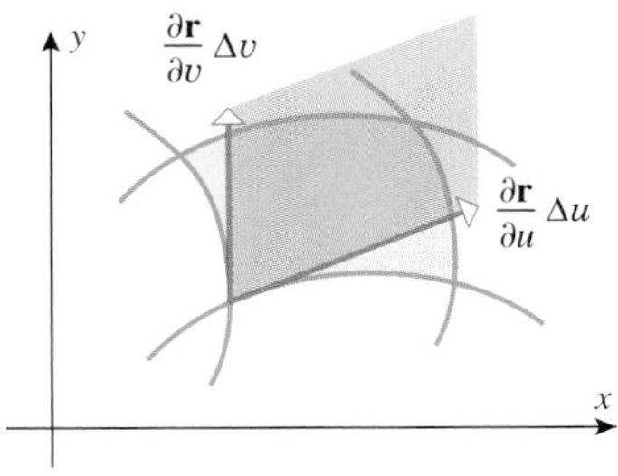

Figure 15.8.7

where the partial derivatives are evaluated at (u_0, v_0) (Figure 15.8.7). Hence, it follows that the area of the region R, which we will denote by ΔA, can be approximated by the area of the parallelogram determined by these vectors. Thus, from Formula (8) of Section 12.4 we have

$$\Delta A \approx \left\| \frac{\partial \mathbf{r}}{\partial u}\Delta u \times \frac{\partial \mathbf{r}}{\partial v}\Delta v \right\| = \left\| \frac{\partial \mathbf{r}}{\partial u} \times \frac{\partial \mathbf{r}}{\partial v} \right\| \Delta u\, \Delta v \tag{6}$$

where the derivatives are evaluated at (u_0, v_0). Computing the cross product, we obtain

$$\frac{\partial \mathbf{r}}{\partial u} \times \frac{\partial \mathbf{r}}{\partial v} = \begin{vmatrix} \mathbf{i} & \mathbf{j} & \mathbf{k} \\ \dfrac{\partial x}{\partial u} & \dfrac{\partial y}{\partial u} & 0 \\ \dfrac{\partial x}{\partial v} & \dfrac{\partial y}{\partial v} & 0 \end{vmatrix} = \begin{vmatrix} \dfrac{\partial x}{\partial u} & \dfrac{\partial y}{\partial u} \\ \dfrac{\partial x}{\partial v} & \dfrac{\partial y}{\partial v} \end{vmatrix}\mathbf{k} = \begin{vmatrix} \dfrac{\partial x}{\partial u} & \dfrac{\partial x}{\partial v} \\ \dfrac{\partial y}{\partial u} & \dfrac{\partial y}{\partial v} \end{vmatrix}\mathbf{k} \tag{7}$$

The determinant in (7) is sufficiently important that it has its own terminology and notation.

15.8.1 DEFINITION. If T is the transformation from the uv-plane to the xy-plane defined by the equations $x = x(u, v)$, $y = y(u, v)$, then the ***Jacobian of T*** is denoted by $J(u, v)$ or by $\partial(x, y)/\partial(u, v)$ and is defined by

$$J(u, v) = \frac{\partial(x, y)}{\partial(u, v)} = \begin{vmatrix} \dfrac{\partial x}{\partial u} & \dfrac{\partial x}{\partial v} \\ \dfrac{\partial y}{\partial u} & \dfrac{\partial y}{\partial v} \end{vmatrix} = \frac{\partial x}{\partial u}\frac{\partial y}{\partial v} - \frac{\partial y}{\partial u}\frac{\partial x}{\partial v}$$

Using the notation in this definition, it follows from (6) and (7) that

$$\Delta A \approx \left\| \frac{\partial(x, y)}{\partial(u, v)}\mathbf{k} \right\| \Delta u\, \Delta v$$

or, since $\mathbf{k}$ is a unit vector,

$$\Delta A \approx \left| \frac{\partial(x, y)}{\partial(u, v)} \right| \Delta u\, \Delta v \tag{8}$$

At the point (u_0, v_0) this important formula relates the areas of the regions R and S in Figure 15.8.5; it tells us that *for small values of Δu and Δv, the area of R is approximately the absolute value of the Jacobian times the area of S.* Moreover, it is proved in advanced calculus courses that the relative error in the approximation approaches zero as $\Delta u \to 0$ and $\Delta v \to 0$.

CHANGE OF VARIABLES IN DOUBLE INTEGRALS

Our next objective is to provide a geometric motivation for the following result.

A precise statement of conditions under which Formula (9) holds is beyond the scope of this course. Suffice it to say that the formula holds if T is a one-to-one transformation, $f(x, y)$ is continuous on R, the partial derivatives of $x(u, v)$ and $y(u, v)$ exist and are continuous on S, and the regions R and S are not complicated.

15.8.2 CHANGE OF VARIABLES FORMULA FOR DOUBLE INTEGRALS. If the transformation $x = x(u, v)$, $y = y(u, v)$ maps the region S in the uv-plane into the region R in the xy-plane, and if the Jacobian $\partial(x, y)/\partial(u, v)$ is nonzero and does not change sign on S, then with appropriate restrictions on the transformation and the regions it follows that

$$\iint_R f(x, y)\, dA_{xy} = \iint_S f(x(u, v), y(u, v)) \left| \frac{\partial(x, y)}{\partial(u, v)} \right| dA_{uv} \qquad (9)$$

where we have attached subscripts to the dA's to help identify the associated variables.

To motivate Formula (9), we proceed as follows:

- Subdivide the region S in the uv-plane into pieces by lines parallel to the coordinate axes, and exclude from consideration any pieces that contain points outside of S. This leaves only rectangular regions that are subsets of S. Assume that there are n such regions and denote the kth such region by S_k. Assume that S_k has dimensions Δu_k by Δv_k and, as shown in Figure 15.8.8a, let (u_k^*, v_k^*) be its "lower left corner."
- As shown in Figure 15.8.8b, the transformation T defined by the equations $x = x(u, v)$, $y = y(u, v)$ maps S_k into a curvilinear parallelogram R_k in the xy-plane and maps the point (u_k^*, v_k^*) into the point $(x_k^*, y_k^*) = (x(u_k^*, v_k^*), y(u_k^*, v_k^*))$ in R_k. Denote the area of R_k by ΔA_k.
- In rectangular coordinates the double integral of $f(x, y)$ over a region R is defined as a limit of Riemann sums in which R is subdivided into *rectangular* subregions. It is proved in advanced calculus courses that under appropriate conditions subdivisions into *curvilinear* parallelograms can be used instead. Accepting this to be so, we can approximate the double integral of $f(x, y)$ over R as

$$\begin{aligned}\iint_R f(x, y)\, dA_{xy} &\approx \sum_{k=1}^{n} f(x_k^*, y_k^*)\, \Delta A_k \\ &\approx \sum_{k=1}^{n} f(x(u_k^*, v_k^*), y(u_k^*, v_k^*)) \left| \frac{\partial(x, y)}{\partial(u, v)} \right| \Delta u_k\, \Delta v_k\end{aligned}$$

where the Jacobian is evaluated at (u_k^*, v_k^*). But the last expression is a Riemann sum for the integral

$$\iint_S f(x(u, v), y(u, v)) \left| \frac{\partial(x, y)}{\partial(u, v)} \right| dA_{uv}$$

so Formula (9) follows if we assume that the errors in the approximations approach zero as $n \to +\infty$.

► **Example 2** Evaluate

$$\iint_R \frac{x - y}{x + y}\, dA$$

where R is the region enclosed by $x - y = 0$, $x - y = 1$, $x + y = 1$, and $x + y = 3$ (Figure 15.8.9a).

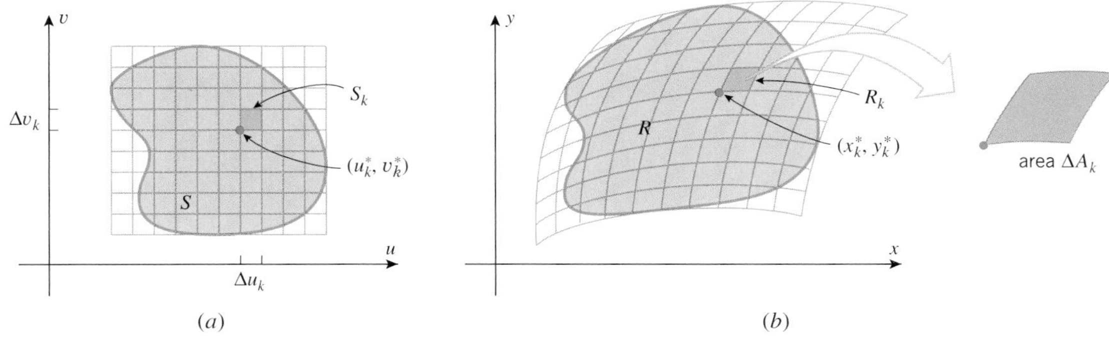

Figure 15.8.8

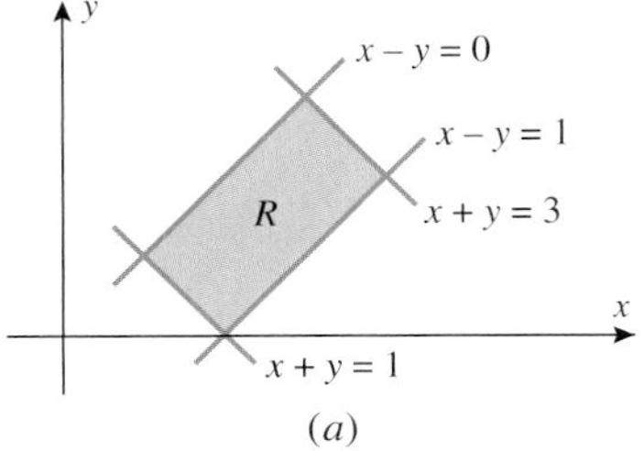

(a)

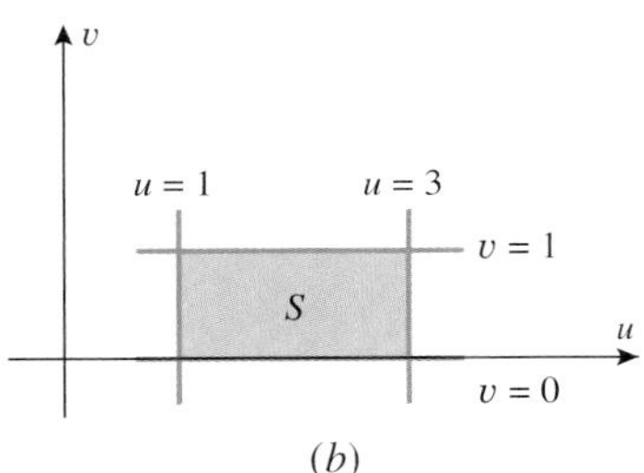

(b)

Figure 15.8.9

Solution. This integral would be tedious to evaluate directly because the region R is oriented in such a way that we would have to subdivide it and integrate over each part separately. However, the occurrence of the expressions $x - y$ and $x + y$ in the equations of the boundary suggests that the transformation

$$u = x + y, \quad v = x - y \tag{10}$$

would be helpful, since with this transformation the boundary lines

$$x + y = 1, \quad x + y = 3, \quad x - y = 0, \quad x - y = 1$$

are constant u-curves and constant v-curves corresponding to the lines

$$u = 1, \quad u = 3, \quad v = 0, \quad v = 1$$

in the uv-plane. These lines enclose the rectangular region S shown in Figure 15.8.9b. To find the Jacobian $\partial(x, y)/\partial(u, v)$ of this transformation, we first solve (10) for x and y in terms of u and v. This yields

$$x = \tfrac{1}{2}(u + v), \quad y = \tfrac{1}{2}(u - v)$$

from which we obtain

$$\frac{\partial(x, y)}{\partial(u, v)} = \begin{vmatrix} \dfrac{\partial x}{\partial u} & \dfrac{\partial x}{\partial v} \\[2mm] \dfrac{\partial y}{\partial u} & \dfrac{\partial y}{\partial v} \end{vmatrix} = \begin{vmatrix} \frac{1}{2} & \frac{1}{2} \\ \frac{1}{2} & -\frac{1}{2} \end{vmatrix} = -\tfrac{1}{4} - \tfrac{1}{4} = -\tfrac{1}{2}$$

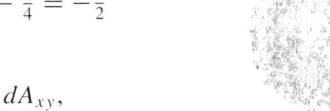

Thus, from Formula (9), but with the notation dA rather than dA_{xy},

$$\begin{aligned} \iint_R \frac{x - y}{x + y}\, dA &= \iint_S \frac{v}{u} \left| \frac{\partial(x, y)}{\partial(u, v)} \right| dA_{uv} \\ &= \iint_S \frac{v}{u} \left| -\frac{1}{2} \right| dA_{uv} = \frac{1}{2} \int_0^1 \int_1^3 \frac{v}{u}\, du\, dv \\ &= \frac{1}{2} \int_0^1 v \ln|u| \bigg]_{u=1}^{3} dv \\ &= \frac{1}{2} \ln 3 \int_0^1 v\, dv = \frac{1}{4} \ln 3 \quad \blacktriangleleft \end{aligned}$$

The underlying idea illustrated in Example 2 is to find a one-to-one transformation that maps a rectangle S in the uv-plane into the region R of integration, and then use that transformation as a substitution in the integral to produce an equivalent integral over S.

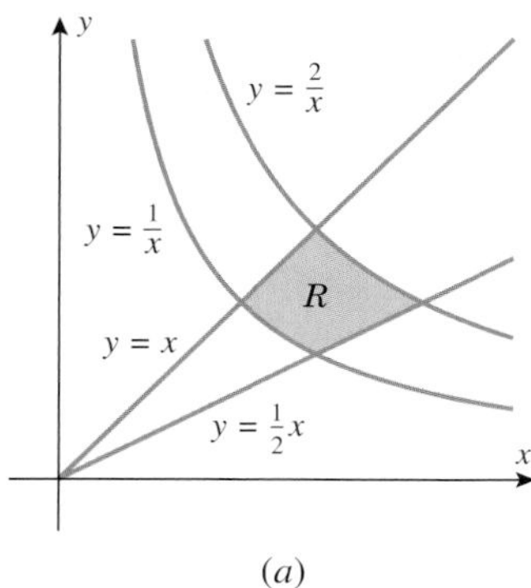

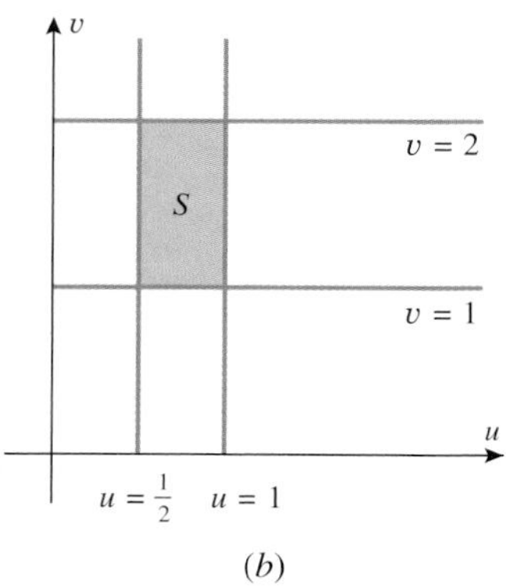

Figure 15.8.10

► **Example 3** Evaluate

$$\iint_R e^{xy}\, dA$$

where R is the region enclosed by the lines $y = \frac{1}{2}x$ and $y = x$ and the hyperbolas $y = 1/x$ and $y = 2/x$ (Figure 15.8.10a).

Solution. As in the last example, we look for a transformation in which the boundary curves in the xy-plane become constant v-curves and constant u-curves. For this purpose we rewrite the four boundary curves as

$$\frac{y}{x} = \frac{1}{2}, \quad \frac{y}{x} = 1, \quad xy = 1, \quad xy = 2$$

which suggests the transformation

$$u = \frac{y}{x}, \quad v = xy \tag{11}$$

With this transformation the boundary curves in the xy-plane are constant u-curves and constant v-curves corresponding to the lines

$$u = \tfrac{1}{2}, \quad u = 1, \quad v = 1, \quad v = 2$$

in the uv-plane. These lines enclose the region S shown in Figure 15.8.10b. To find the Jacobian $\partial(x, y)/\partial(u, v)$ of this transformation, we first solve (11) for x and y in terms of u and v. This yields

$$x = \sqrt{v/u}, \quad y = \sqrt{uv}$$

from which we obtain

$$\frac{\partial(x,y)}{\partial(u,v)} = \begin{vmatrix} \dfrac{\partial x}{\partial u} & \dfrac{\partial x}{\partial v} \\ \dfrac{\partial y}{\partial u} & \dfrac{\partial y}{\partial v} \end{vmatrix} = \begin{vmatrix} -\dfrac{1}{2u}\sqrt{\dfrac{v}{u}} & \dfrac{1}{2\sqrt{uv}} \\ \dfrac{1}{2}\sqrt{\dfrac{v}{u}} & \dfrac{1}{2}\sqrt{\dfrac{u}{v}} \end{vmatrix} = -\frac{1}{4u} - \frac{1}{4u} = -\frac{1}{2u}$$

Thus, from Formula (9), but with the notation dA rather than dA_{xy},

$$\begin{aligned} \iint_R e^{xy}\, dA &= \iint_S e^v \left| -\frac{1}{2u} \right| dA_{uv} = \frac{1}{2}\iint_S \frac{1}{u} e^v\, dA_{uv} \\ &= \frac{1}{2}\int_1^2 \int_{1/2}^1 \frac{1}{u} e^v\, du\, dv = \frac{1}{2}\int_1^2 e^v \ln|u| \bigg]_{u=1/2}^1 dv \\ &= \frac{1}{2}\ln 2 \int_1^2 e^v\, dv = \frac{1}{2}(e^2 - e)\ln 2 \; ◄ \end{aligned}$$

CHANGE OF VARIABLES IN TRIPLE INTEGRALS

Equations of the form

$$x = x(u, v, w), \quad y = y(u, v, w), \quad z = z(u, v, w) \tag{12}$$

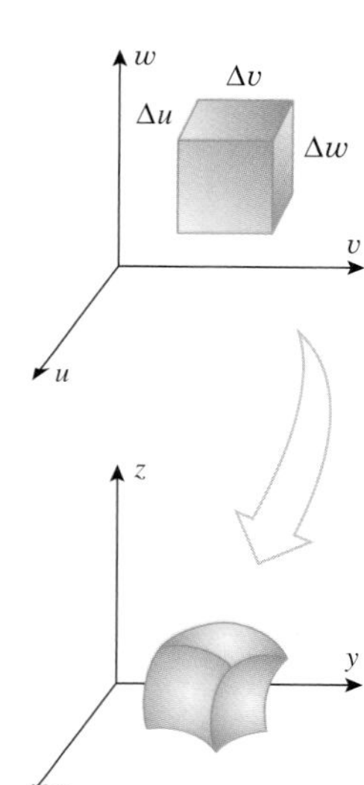

Figure 15.8.11

define a ***transformation*** T from uvw-space to xyz-space. Just as a transformation $x = x(u, v)$, $y = y(u, v)$ in two variables maps small rectangles in the uv-plane into curvilinear parallelograms in the xy-plane, so (12) maps small rectangular parallelepipeds in uvw-space into curvilinear parallelepipeds in xyz-space (Figure 15.8.11). The definition of the Jacobian of (12) is similar to Definition 15.8.1.

15.8.3 DEFINITION. If T is the transformation from uvw-space to xyz-space defined by the equations $x = x(u, v, w)$, $y = y(u, v, w)$, $z = z(u, v, w)$, then the ***Jacobian of T*** is denoted by $J(u, v, w)$ or $\partial(x, y, z)/\partial(u, v, w)$ and is defined by

$$J(u, v, w) = \frac{\partial(x, y, z)}{\partial(u, v, w)} = \begin{vmatrix} \dfrac{\partial x}{\partial u} & \dfrac{\partial x}{\partial v} & \dfrac{\partial x}{\partial w} \\ \dfrac{\partial y}{\partial u} & \dfrac{\partial y}{\partial v} & \dfrac{\partial y}{\partial w} \\ \dfrac{\partial z}{\partial u} & \dfrac{\partial z}{\partial v} & \dfrac{\partial z}{\partial w} \end{vmatrix}$$

For small values of Δu, Δv, and Δw, the volume ΔV of the curvilinear parallelepiped in Figure 15.8.11 is related to the volume $\Delta u\, \Delta v\, \Delta w$ of the rectangular parallelepiped by

$$\Delta V \approx \left| \frac{\partial(x, y, z)}{\partial(u, v, w)} \right| \Delta u\, \Delta v\, \Delta w \tag{13}$$

which is the analog of Formula (8). Using this relationship and an argument similar to the one that led to Formula (9), we can obtain the following result.

15.8.4 CHANGE OF VARIABLES FORMULA FOR TRIPLE INTEGRALS. If the transformation $x = x(u, v, w)$, $y = y(u, v, w)$, $z = z(u, v, w)$ maps the region S in uvw-space into the region R in xyz-space, and if the Jacobian $\partial(x, y, z)/\partial(u, v, w)$ is nonzero and does not change sign on S, then with appropriate restrictions on the transformation and the regions it follows that

$$\iiint_R f(x, y, z)\, dV_{xyz} = \iiint_S f(x(u, v, w), y(u, v, w), z(u, v, w)) \left| \frac{\partial(x, y, z)}{\partial(u, v, w)} \right| dV_{uvw} \tag{14}$$

► **Example 4** Find the volume of the region G enclosed by the ellipsoid

$$\frac{x^2}{a^2} + \frac{y^2}{b^2} + \frac{z^2}{c^2} = 1$$

Solution. The volume V is given by the triple integral

$$V = \iiint_G dV$$

To evaluate this integral, we make the change of variables

$$x = au, \quad y = bv, \quad z = cw \tag{15}$$

which maps the region S in uvw-space enclosed by a sphere of radius 1 into the region G in xyz-space. This can be seen from (15) by noting that

$$\frac{x^2}{a^2} + \frac{y^2}{b^2} + \frac{z^2}{c^2} = 1 \quad \text{becomes} \quad u^2 + v^2 + w^2 = 1$$

The Jacobian of (15) is

$$\frac{\partial(x, y, z)}{\partial(u, v, w)} = \begin{vmatrix} \frac{\partial x}{\partial u} & \frac{\partial x}{\partial v} & \frac{\partial x}{\partial w} \\ \frac{\partial y}{\partial u} & \frac{\partial y}{\partial v} & \frac{\partial y}{\partial w} \\ \frac{\partial z}{\partial u} & \frac{\partial z}{\partial v} & \frac{\partial z}{\partial w} \end{vmatrix} = \begin{vmatrix} a & 0 & 0 \\ 0 & b & 0 \\ 0 & 0 & c \end{vmatrix} = abc$$

Thus, from Formula (14), but with the notation dV rather than dV_{xyz},

$$V = \iiint_G dV = \iiint_S \left| \frac{\partial(x, y, z)}{\partial(u, v, w)} \right| dV_{uvw} = abc \iiint_S dV_{uvw}$$

The last integral is the volume enclosed by a sphere of radius 1, which we know to be $\frac{4}{3}\pi$. Thus, the volume enclosed by the ellipsoid is $V = \frac{4}{3}\pi abc$. ◀

Jacobians also arise in converting triple integrals in rectangular coordinates to iterated integrals in cylindrical and spherical coordinates. For example, we will ask you to show in Exercise 47 that the Jacobian of the transformation

$$x = r\cos\theta, \quad y = r\sin\theta, \quad z = z$$

is

$$\frac{\partial(x, y, z)}{\partial(r, \theta, z)} = r$$

and the Jacobian of the transformation

$$x = \rho\sin\phi\cos\theta, \quad y = \rho\sin\phi\sin\theta, \quad z = \rho\cos\phi$$

is

$$\frac{\partial(x, y, z)}{\partial(\rho, \phi, \theta)} = \rho^2\sin\phi$$

Thus, Formulas (6) and (10) of Section 15.7 can be expressed in terms of Jacobians as

$$\iiint_G f(x, y, z)\, dV = \iiint_{\substack{\text{appropriate}\\ \text{limits}}} f(r\cos\theta, r\sin\theta, z) \frac{\partial(x, y, z)}{\partial(r, \theta, z)}\, dz\, dr\, d\theta \tag{16}$$

$$\iiint_G f(x, y, z)\, dV = \iiint_{\substack{\text{appropriate}\\ \text{limits}}} f(\rho\sin\phi\cos\theta, \rho\sin\phi\sin\theta, \rho\cos\phi) \frac{\partial(x, y, z)}{\partial(\rho, \phi, \theta)}\, d\rho\, d\phi\, d\theta \tag{17}$$

The absolute-value signs are omitted from Formulas (16) and (17) because the Jacobians are nonnegative (see the restrictions in Table 12.8.1).

✔ QUICK CHECK EXERCISES 15.8

(See page 1099 for answers.)

1. Let T be the transformation from the uv-plane to the xy-plane defined by the equations

$$x = u - 2v, \quad y = 3u + v$$

(a) Sketch the image under T of the rectangle $1 \le u \le 3$, $0 \le v \le 2$.

(b) Solve for u and v in terms of x and y:

$u =$ ________, $v =$ ________

2. State the relationship between R and S in the change of variables formula

$$\iint_R f(x, y)\, dA_{xy} = \iint_S f(x(u, v), y(u, v)) \left| \frac{\partial(x, y)}{\partial(u, v)} \right| dA_{uv}$$

3. Let T be the transformation in Quick Check Exercise 1.

(a) The Jacobian $\partial(x, y)/\partial(u, v)$ of T is ________.

(b) Let R be the region in Quick Check Exercise 1(a). Fill in the missing integrand and limits of integration for the change of variables given by T.

$$\iint_R e^{x+2y}\,dA = \int_{\square}^{\square}\int_{\square}^{\square} \rule{2cm}{0.4pt}\; du\,dv$$

4. The Jacobian of the transformation

$$x = uv, \quad y = vw, \quad z = 2w$$

is

$$\frac{\partial(x, y, z)}{\partial(u, v, w)} = \rule{2cm}{0.4pt}$$

EXERCISE SET 15.8

1–4 Find the Jacobian $\partial(x, y)/\partial(u, v)$.

1. $x = u + 4v,\ y = 3u - 5v$
2. $x = u + 2v^2,\ y = 2u^2 - v$
3. $x = \sin u + \cos v,\ y = -\cos u + \sin v$
4. $x = \dfrac{2u}{u^2 + v^2},\ y = -\dfrac{2v}{u^2 + v^2}$

5–8 Solve for x and y in terms of u and v, and then find the Jacobian $\partial(x, y)/\partial(u, v)$.

5. $u = 2x - 5y,\ v = x + 2y$
6. $u = e^x,\ v = ye^{-x}$
7. $u = x^2 - y^2,\ v = x^2 + y^2 \quad (x > 0, y > 0)$
8. $u = xy,\ v = xy^3 \quad (x > 0, y > 0)$

9–12 Find the Jacobian $\partial(x, y, z)/\partial(u, v, w)$.

9. $x = 3u + v,\ y = u - 2w,\ z = v + w$
10. $x = u - uv,\ y = uv - uvw,\ z = uvw$
11. $u = xy,\ v = y,\ w = x + z$
12. $u = x + y + z,\ v = x + y - z,\ w = x - y + z$

FOCUS ON CONCEPTS

13–16 Sketch the image in the xy-plane of the set S under the given transformation.

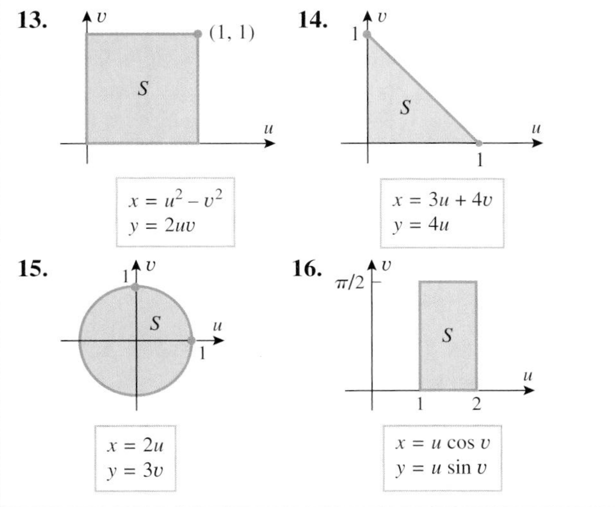

17. Use the transformation $u = x - 2y,\ v = 2x + y$ to find

$$\iint_R \frac{x - 2y}{2x + y}\,dA$$

where R is the rectangular region enclosed by the lines $x - 2y = 1,\ x - 2y = 4,\ 2x + y = 1,\ 2x + y = 3$.

18. Use the transformation $u = x + y,\ v = x - y$ to find

$$\iint_R (x - y)e^{x^2 - y^2}\,dA$$

over the rectangular region R enclosed by the lines $x + y = 0,\ x + y = 1,\ x - y = 1,\ x - y = 4$.

19. Use the transformation $u = \frac{1}{2}(x + y),\ v = \frac{1}{2}(x - y)$ to find

$$\iint_R \sin \tfrac{1}{2}(x + y) \cos \tfrac{1}{2}(x - y)\,dA$$

over the triangular region R with vertices $(0, 0)$, $(2, 0)$, $(1, 1)$.

20. Use the transformation $u = y/x,\ v = xy$ to find

$$\iint_R xy^3\,dA$$

over the region R in the first quadrant enclosed by $y = x$, $y = 3x$, $xy = 1$, $xy = 4$.

21–24 The transformation $x = au,\ y = bv\ (a > 0, b > 0)$ can be rewritten as $x/a = u,\ y/b = v$, and hence it maps the circular region

$$u^2 + v^2 \le 1$$

into the elliptical region

$$\frac{x^2}{a^2} + \frac{y^2}{b^2} \le 1$$

In these exercises, perform the integration by transforming the elliptical region of integration into a circular region of integration and then evaluating the transformed integral in polar coordinates.

21. $\displaystyle\iint_R \sqrt{16x^2 + 9y^2}\,dA$, where R is the region enclosed by the ellipse $(x^2/9) + (y^2/16) = 1$.

22. $\displaystyle\iint_R e^{-(x^2 + 4y^2)}\,dA$, where R is the region enclosed by the ellipse $(x^2/4) + y^2 = 1$.

23. $\displaystyle\iint_R \sin(4x^2+9y^2)\,dA$, where R is the region in the first quadrant enclosed by the ellipse $4x^2+9y^2=1$ and the coordinate axes.

24. Show that the area of the ellipse

$$\frac{x^2}{a^2}+\frac{y^2}{b^2}=1$$

is πab.

25–26 If a, b, and c are positive constants, then the transformation $x=au$, $y=bv$, $z=cw$ can be rewritten as $x/a=u$, $y/b=v$, $z/c=w$, and hence it maps the spherical region

$$u^2+v^2+w^2\le 1$$

into the ellipsoidal region

$$\frac{x^2}{a^2}+\frac{y^2}{b^2}+\frac{z^2}{c^2}\le 1$$

In these exercises, perform the integration by transforming the ellipsoidal region of integration into a spherical region of integration and then evaluating the transformed integral in spherical coordinates.

25. $\displaystyle\iiint_G x^2\,dV$, where G is the region enclosed by the ellipsoid $9x^2+4y^2+z^2=36$.

26. Find the moment of inertia about the x-axis of the solid ellipsoid bounded by

$$\frac{x^2}{a^2}+\frac{y^2}{b^2}+\frac{z^2}{c^2}=1$$

given that $\delta(x,y,z)=1$. [See the definition preceding Exercise 43 of Section 15.7.]

FOCUS ON CONCEPTS

27–30 Find a transformation

$$u=f(x,y),\quad v=g(x,y)$$

that when applied to the region R in the xy-plane has as its image the region S in the uv-plane.

27.

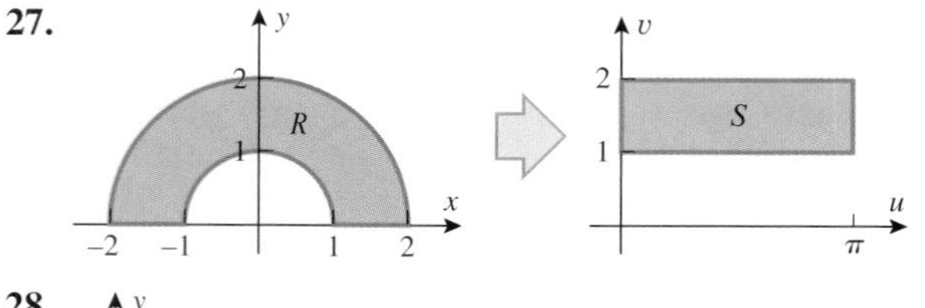

28.

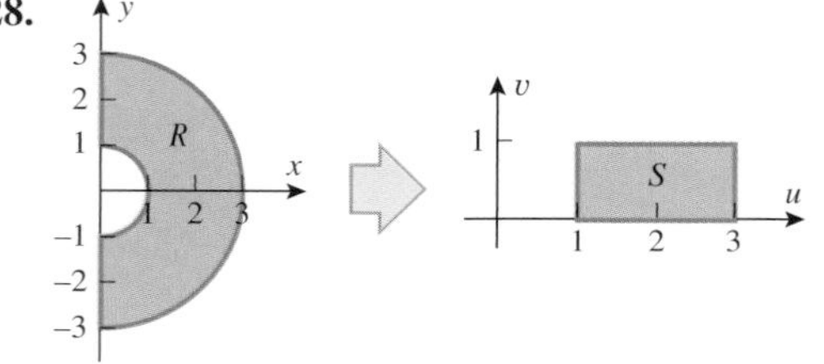

29.

30.

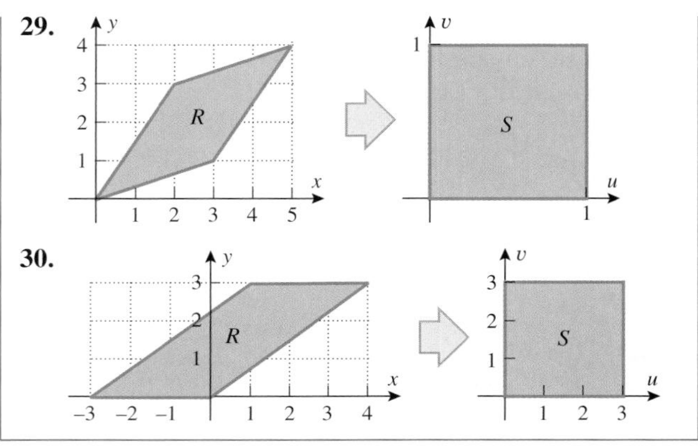

31–34 Evaluate the integral by making an appropriate change of variables.

31. $\displaystyle\iint_R \frac{y-4x}{y+4x}\,dA$, where R is the region enclosed by the lines $y=4x$, $y=4x+2$, $y=2-4x$, $y=5-4x$.

32. $\displaystyle\iint_R (x^2-y^2)\,dA$, where R is the rectangular region enclosed by the lines $y=-x$, $y=1-x$, $y=x$, $y=x+2$.

33. $\displaystyle\iint_R \frac{\sin(x-y)}{\cos(x+y)}\,dA$, where R is the triangular region enclosed by the lines $y=0$, $y=x$, $x+y=\pi/4$.

34. $\displaystyle\iint_R e^{(y-x)/(y+x)}\,dA$, where R is the region in the first quadrant enclosed by the trapezoid with vertices $(0,1)$, $(1,0)$, $(0,4)$, $(4,0)$.

35. Use an appropriate change of variables to find the area of the region in the first quadrant enclosed by the curves $y=x$, $y=2x$, $x=y^2$, $x=4y^2$.

36. Use an appropriate change of variables to find the volume of the solid bounded above by the plane $x+y+z=9$, below by the xy-plane, and laterally by the elliptic cylinder $4x^2+9y^2=36$. [*Hint:* Express the volume as a double integral in xy-coordinates, then use polar coordinates to evaluate the transformed integral.]

37. Use the transformation $u=x$, $v=z-y$, $w=xy$ to find

$$\iiint_G (z-y)^2xy\,dV$$

where G is the region enclosed by the surfaces $x=1$, $x=3$, $z=y$, $z=y+1$, $xy=2$, $xy=4$.

38. Use the transformation $u=xy$, $v=yz$, $w=xz$ to find the volume of the region in the first octant that is enclosed by the hyperbolic cylinders $xy=1$, $xy=2$, $yz=1$, $yz=3$, $xz=1$, $xz=4$.

39. An astroidal sphere has equation $x^{2/3} + y^{2/3} + x^{2/3} = a^{2/3}$ (see Exercise 54 in Section 15.4). Find the volume of the astroidal sphere using a triple integral and the transformation
$$x = \rho(\sin\phi\cos\theta)^3$$
$$y = \rho(\sin\phi\sin\theta)^3$$
$$z = \rho(\cos\phi)^3$$
for which $0 \le \rho \le a, 0 \le \phi \le \pi, 0 \le \theta \le 2\pi$.

40. (a) Verify that
$$\begin{vmatrix} a_1 & b_1 \\ c_1 & d_1 \end{vmatrix}\begin{vmatrix} a_2 & b_2 \\ c_2 & d_2 \end{vmatrix} = \begin{vmatrix} a_1a_2 + b_1c_2 & a_1b_2 + b_1d_2 \\ c_1a_2 + d_1c_2 & c_1b_2 + d_1d_2 \end{vmatrix}$$
(b) If $x = x(u, v)$, $y = y(u, v)$ is a one-to-one transformation, then $u = u(x, y)$, $v = v(x, y)$. Assuming the necessary differentiability, use the result in part (a) and the chain rule to show that
$$\frac{\partial(x, y)}{\partial(u, v)} \cdot \frac{\partial(u, v)}{\partial(x, y)} = 1$$

41. In each part, confirm that the formula obtained in part (b) of Exercise 40 holds for the given transformation.
(a) $x = u - uv,\ y = uv$
(b) $x = uv,\ y = v^2 \quad (v > 0)$
(c) $x = \frac{1}{2}(u^2 + v^2),\ y = \frac{1}{2}(u^2 - v^2) \quad (u > 0, v > 0)$

42–44 The formula obtained in part (b) of Exercise 40 is useful in integration problems where it is inconvenient or impossible to solve the transformation equations $u = f(x, y)$, $v = g(x, y)$ explicitly for x and y in terms of u and v. In these exercises, use the relationship
$$\frac{\partial(x, y)}{\partial(u, v)} = \frac{1}{\partial(u, v)/\partial(x, y)}$$
to avoid solving for x and y in terms of u and v.

42. Use the transformation $u = xy$, $v = xy^4$ to find
$$\iint_R \sin(xy)\, dA$$
where R is the region enclosed by the curves $xy = \pi$, $xy = 2\pi$, $xy^4 = 1$, $xy^4 = 2$.

43. Use the transformation $u = x^2 - y^2$, $v = x^2 + y^2$ to find
$$\iint_R xy\, dA$$
where R is the region in the first quadrant that is enclosed by the hyperbolas $x^2 - y^2 = 1$, $x^2 - y^2 = 4$ and the circles $x^2 + y^2 = 9$, $x^2 + y^2 = 16$.

44. Use the transformation $u = xy$, $v = x^2 - y^2$ to find
$$\iint_R (x^4 - y^4)e^{xy}\, dA$$
where R is the region in the first quadrant enclosed by the hyperbolas $xy = 1$, $xy = 3$, $x^2 - y^2 = 3$, $x^2 - y^2 = 4$.

45. The three-variable analog of the formula derived in part (b) of Exercise 40 is
$$\frac{\partial(x, y, z)}{\partial(u, v, w)} \cdot \frac{\partial(u, v, w)}{\partial(x, y, z)} = 1$$
Use this result to show that the volume V of the oblique parallelepiped that is bounded by the planes $x + y + 2z = \pm 3$, $x - 2y + z = \pm 2$, $4x + y + z = \pm 6$ is $V = 16$.

46. (a) Show that if R is the triangular region with vertices $(0, 0)$, $(1, 0)$, and $(0, 1)$, then
$$\iint_R f(x + y)\, dA = \int_0^1 uf(u)\, du$$
(b) Use the result in part (a) to evaluate the integral
$$\iint_R e^{x+y}\, dA$$

47. (a) Consider the transformation
$$x = r\cos\theta, \quad y = r\sin\theta, \quad z = z$$
from cylindrical to rectangular coordinates, where $r \ge 0$. Show that
$$\frac{\partial(x, y, z)}{\partial(r, \theta, z)} = r$$
(b) Consider the transformation
$$x = \rho\sin\phi\cos\theta, \quad y = \rho\sin\phi\sin\theta, \quad z = \rho\cos\phi$$
from spherical to rectangular coordinates, where $0 \le \phi \le \pi$. Show that
$$\frac{\partial(x, y, z)}{\partial(\rho, \phi, \theta)} = \rho^2\sin\phi$$

✔ QUICK CHECK ANSWERS 15.8

1. (a) The image is the region in the xy-plane enclosed by the parallelogram with vertices $(1, 3)$, $(-3, 5)$, $(-1, 11)$, and $(3, 9)$. (b) $u = \frac{1}{7}(x + 2y)$; $v = \frac{1}{7}(y - 3x)$ **2.** S is a region in the uv-plane and R is the image of S in the xy-plane under the transformation $x = x(u, v)$, $y = y(u, v)$ **3.** (a) 7 (b) $\int_0^2 \int_1^3 7e^{7u}\, du\, dv$ **4.** $2vw$

CHAPTER REVIEW EXERCISES

1. The double integral over a region R in the xy-plane is defined as

$$\iint_R f(x, y)\, dA = \lim_{n \to +\infty} \sum_{k=1}^{n} f(x_k^*, y_k^*)\, \Delta A_k$$

Describe the procedure on which this definition is based.

2. The triple integral over a solid G in an xyz-coordinate system is defined as

$$\iiint_G f(x, y, z)\, dV = \lim_{n \to +\infty} \sum_{k=1}^{n} f(x_k^*, y_k^*, z_k^*)\, \Delta V_k$$

Describe the procedure on which this definition is based.

3. (a) Express the area of a region R in the xy-plane as a double integral.
(b) Express the volume of a region G in an xyz-coordinate system as a triple integral.
(c) Express the area of the portion of the surface $z = f(x, y)$ that lies above the region R in the xy-plane as a double integral.

4. (a) Write down parametric equations for a sphere of radius a centered at the origin.
(b) Write down parametric equations for the right circular cylinder of radius a and height h that is centered on the z-axis, has its base in the xy-plane, and extends in the positive z-direction.

5. (a) In physical terms, what is meant by the center of gravity of a lamina?
(b) What is meant by the centroid of a lamina?
(c) Write down formulas for the coordinates of the center of gravity of a lamina in the xy-plane.
(d) Write down formulas for the coordinates of the centroid of a lamina in the xy-plane.

6. Suppose that you have a double integral over a region R in the xy-plane and you want to transform that integral into an equivalent double integral over a region S in the uv-plane. Describe the procedure you would use.

7. Let R be the region in the accompanying figure. Fill in the missing limits of integration in the iterated integral

$$\int_{\square}^{\square} \int_{\square}^{\square} f(x, y)\, dx\, dy$$

over R.

8. Let R be the region shown in the accompanying figure. Fill in the missing limits of integration in the sum of the iterated integrals

$$\int_{0}^{2} \int_{\square}^{\square} f(x, y)\, dy\, dx + \int_{2}^{3} \int_{\square}^{\square} f(x, y)\, dy\, dx$$

over R.

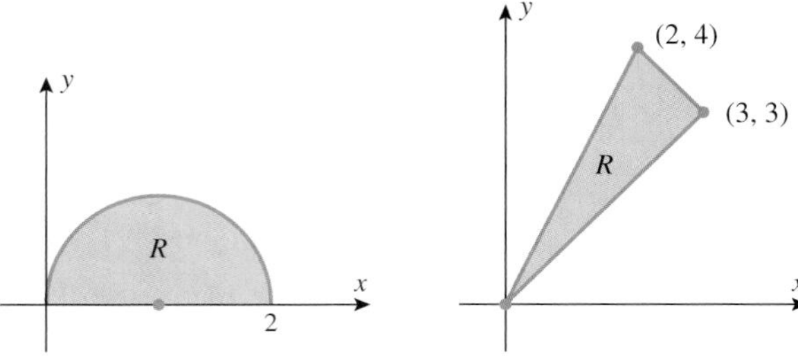

Figure Ex-7 **Figure Ex-8**

9. (a) Find constants a, b, c, and d such that the transformation $x = au + bv$, $y = cu + dv$ maps the region S in the accompanying figure into the region R.
(b) Find the area of the parallelogram R by integrating over the region S, and check your answer using a formula from geometry.

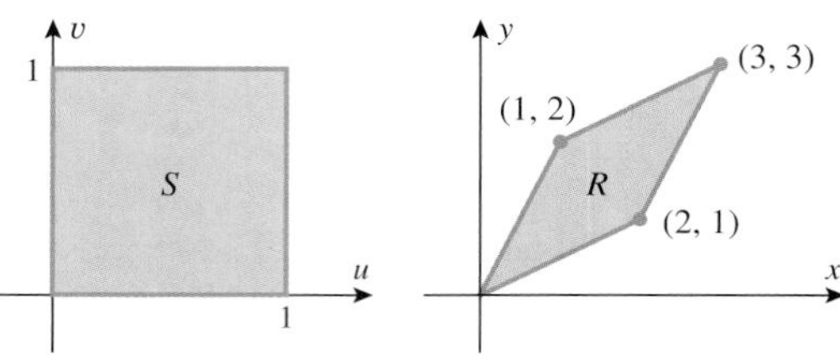

Figure Ex-9

10. Give a geometric argument to show that

$$0 < \int_0^{\pi} \int_0^{\pi} \sin \sqrt{xy}\, dy\, dx < \pi^2$$

11–12 Evaluate the iterated integral.

11. $\displaystyle\int_{1/2}^{1} \int_0^{2x} \cos(\pi x^2)\, dy\, dx$ **12.** $\displaystyle\int_0^{2} \int_{-y}^{2y} x e^{y^3}\, dx\, dy$

13–14 Express the iterated integral as an equivalent integral with the order of integration reversed.

13. $\displaystyle\int_0^{2} \int_0^{x/2} e^x e^y\, dy\, dx$ **14.** $\displaystyle\int_0^{\pi} \int_y^{\pi} \frac{\sin x}{x}\, dx\, dy$

15–16 Sketch the region whose area is represented by the iterated integral.

15. $\displaystyle\int_0^{\pi/2} \int_{\tan(x/2)}^{\sin x} dy\, dx$

16. $\displaystyle\int_{\pi/6}^{\pi/2} \int_a^{a(1+\cos\theta)} r\, dr\, d\theta \quad (a > 0)$

17–18 Evaluate the double integral.

17. $\displaystyle\iint_R x^2 \sin y^2\, dA$; R is the region that is bounded by $y = x^3$, $y = -x^3$, and $y = 8$.

18. $\iint_R (4 - x^2 - y^2)\, dA$; R is the sector in the first quadrant bounded by the circle $x^2 + y^2 = 4$ and the coordinate axes.

19. Convert to rectangular coordinates and evaluate:

$$\int_0^{\pi/2} \int_0^{2a \sin\theta} r \sin 2\theta \, dr\, d\theta$$

20. Convert to polar coordinates and evaluate:

$$\int_0^{\sqrt{2}} \int_x^{\sqrt{4-x^2}} 4xy\, dy\, dx$$

21–22 Find the area of the region using a double integral.

21. The region bounded by $y = 2x^3$, $2x + y = 4$, and the x-axis.

22. The region enclosed by the rose $r = \cos 3\theta$.

23. Convert to cylindrical coordinates and evaluate:

$$\int_{-2}^{2} \int_{-\sqrt{4-x^2}}^{\sqrt{4-x^2}} \int_{(x^2+y^2)^2}^{16} x^2\, dz\, dy\, dx$$

24. Convert to spherical coordinates and evaluate:

$$\int_0^1 \int_0^{\sqrt{1-x^2}} \int_0^{\sqrt{1-x^2-y^2}} \frac{1}{1 + x^2 + y^2 + z^2}\, dz\, dy\, dx$$

25. Let G be the region bounded above by the sphere $\rho = a$ and below by the cone $\phi = \pi/3$. Express

$$\iiint_G (x^2 + y^2)\, dV$$

as an iterated integral in
(a) spherical coordinates (b) cylindrical coordinates
(c) rectangular coordinates.

26. Let $G = \{(x, y, z) : x^2 + y^2 \le z \le 4x\}$. Express the volume of G as an iterated integral in
(a) rectangular coordinates (b) cylindrical coordinates.

27–28 Find the volume of the solid using a triple integral.

27. The solid bounded below by the cone $\phi = \pi/6$ and above by the plane $z = a$.

28. The solid enclosed between the surfaces $x = y^2 + z^2$ and $x = 1 - y^2$.

29. Find the surface area of the portion of the hyperbolic paraboloid

$$\mathbf{r}(u, v) = (u + v)\mathbf{i} + (u - v)\mathbf{j} + uv\mathbf{k}$$

for which $u^2 + v^2 \le 4$.

30. Find the surface area of the portion of the spiral ramp

$$\mathbf{r}(u, v) = u \cos v\mathbf{i} + u \sin v\mathbf{j} + v\mathbf{k}$$

for which $0 \le u \le 2$, $0 \le v \le 3u$.

31–32 Find the equation of the tangent plane to the surface at the specified point.

31. $\mathbf{r} = u\mathbf{i} + v\mathbf{j} + (u^2 + v^2)\mathbf{k}$; $u = 1, v = 2$

32. $x = u \cosh v$, $y = u \sinh v$, $z = u^2$; $(-3, 0, 9)$

33–34 Find the centroid of the region.

33. The region bounded by $y^2 = 4x$ and $y^2 = 8(x - 2)$.

34. The upper half of the ellipse $(x/a)^2 + (y/b)^2 = 1$.

35–36 Find the centroid of the solid.

35. The solid cone with vertex $(0, 0, h)$ and with base the disk $x^2 + y^2 \le a^2$ in the xy-plane.

36. The solid bounded by $y = x^2$, $z = 0$, and $y + z = 4$.

37. Find the average distance from a point inside a sphere of radius a to the center. [See the definition preceding Exercise 27 of Section 15.5.]

38. Use the transformation $u = x - 3y$, $v = 3x + y$ to find

$$\iint_R \frac{x - 3y}{(3x + y)^2}\, dA$$

where R is the rectangular region enclosed by the lines $x - 3y = 0$, $x - 3y = 4$, $3x + y = 1$, and $3x + y = 3$.

39. Let G be the solid in 3-space defined by the inequalities

$$1 - e^x \le y \le 3 - e^x, \quad 1 - y \le 2z \le 2 - y, \quad y \le e^x \le y + 4$$

(a) Using the coordinate transformation

$$u = e^x + y, \quad v = y + 2z, \quad w = e^x - y$$

calculate the Jacobian $\partial(x, y, z)/\partial(u, v, w)$. Express your answer in terms of u, v, and w.

(b) Using a triple integral and the change of variables given in part (a), find the volume of G.

c h a p t e r   s i x t e e n

TOPICS IN VECTOR CALCULUS

I'm very good at integral and differential calculus, I know the scientific names of beings animalculous; In short, in matters vegetable, animal, and mineral, I am the very model of a modern Major-General.

—W. S. Gilbert
Librettist of the operetta The Mikado

The main theme of this chapter is the concept of a "flow." The body of mathematics that we will study here is concerned with analyzing flows of various types—the flow of a fluid or the flow of electricity, for example. Indeed, the early writings of Isaac Newton on calculus are replete with such nouns as "fluxion" and "fluent," which are rooted in the Latin *fluens* (to flow). We will begin this chapter by introducing the concept of a vector field, which is the mathematical description of a flow. In subsequent sections, we will introduce two new kinds of integrals that are used in a variety of applications to analyze properties of vector fields and flows. Finally, we conclude with three major theorems, Green's Theorem, the Divergence Theorem, and Stokes' Theorem. These theorems provide a deep insight into the nature of flows and are the basis for many of the most important principles in physics and engineering.

Photo: *Results in this chapter provide tools for analyzing and understanding the behavior of hurricanes and other fluid flows.*

16.1 VECTOR FIELDS

In this section we will consider functions that associate vectors with points in 2-space or 3-space. We will see that such functions play an important role in the study of fluid flow, gravitational force fields, electromagnetic force fields, and a wide range of other applied problems.

VECTOR FIELDS

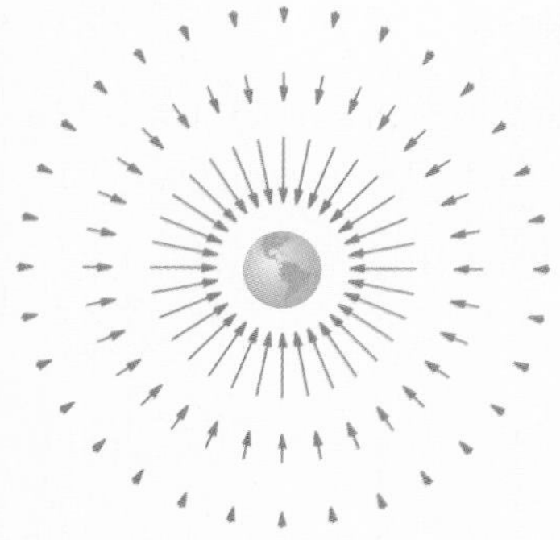

Figure 16.1.1

To motivate the mathematical ideas in this section, consider a *unit* point mass located at any point in the universe. According to Newton's Law of Universal Gravitation, the Earth exerts an attractive force on the mass that is directed toward the center of the Earth and has a magnitude that is inversely proportional to the square of the distance from the mass to the Earth's center (Figure 16.1.1). This association of force vectors with points in space is called the Earth's *gravitational field*. A similar idea arises in fluid flow. Imagine a stream in which the water flows horizontally at every level, and consider the layer of water at a specific depth. At each point of the layer, the water has a certain velocity, which we can represent by a vector at that point (Figure 16.1.2). This association of velocity vectors with points in the two-dimensional layer is called the *velocity field* at that layer. These ideas are captured in the following definition.

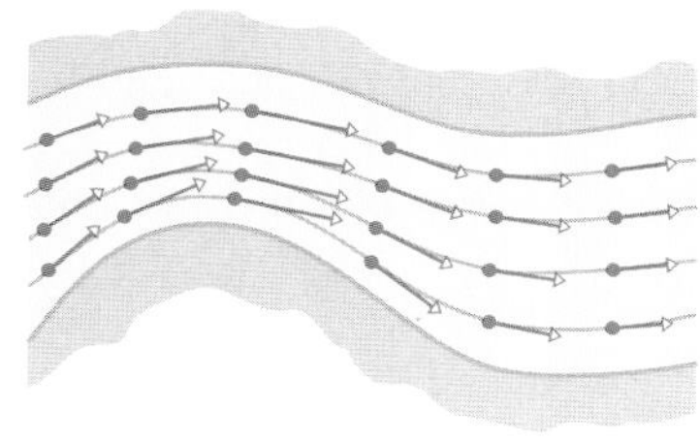

Figure 16.1.2

> **16.1.1 DEFINITION.** A ***vector field*** in a plane is a function that associates with each point P in the plane a unique vector $\mathbf{F}(P)$ parallel to the plane. Similarly, a vector field in 3-space is a function that associates with each point P in 3-space a unique vector $\mathbf{F}(P)$ in 3-space.

Observe that in this definition there is no reference to a coordinate system. However, for computational purposes it is usually desirable to introduce a coordinate system so that vectors can be assigned components. Specifically, if $\mathbf{F}(P)$ is a vector field in an xy-coordinate system, then the point P will have some coordinates (x, y) and the associated vector will have components that are functions of x and y. Thus, the vector field $\mathbf{F}(P)$ can be expressed as

$$\mathbf{F}(x, y) = f(x, y)\mathbf{i} + g(x, y)\mathbf{j}$$

Similarly, in 3-space with an xyz-coordinate system, a vector field $\mathbf{F}(P)$ can be expressed as

$$\mathbf{F}(x, y, z) = f(x, y, z)\mathbf{i} + g(x, y, z)\mathbf{j} + h(x, y, z)\mathbf{k}$$

GRAPHICAL REPRESENTATIONS OF VECTOR FIELDS

A vector field in 2-space can be pictured geometrically by drawing representative field vectors $\mathbf{F}(x, y)$ at some well-chosen points in the xy-plane. But, just as it is usually not possible to describe a plane curve completely by plotting finitely many points, so it is usually not possible to describe a vector field completely by drawing finitely many vectors. Nevertheless, such graphical representations can provide useful information about the general behavior of the field if the vectors are chosen appropriately. However, graphical representations of vector fields require a substantial amount of computation, so they are usually created using computers. Figure 16.1.3 shows four computer-generated vector fields. The vector field in part (a) might describe the velocity of the current in a stream at various depths. At the bottom of the stream the velocity is zero, but the speed of the current increases as the depth decreases. Points at the same depth have the same speed. The vector field in part (b) might describe the velocity at points on a rotating wheel. At the center of the wheel the velocity is zero, but the speed increases with the distance from the center. Points at the same distance from the center have the same speed. The vector field in part (c) might describe the repulsive force of an electrical charge—the closer to the charge, the greater the force of repulsion. Part (d) shows a vector field in 3-space. Such pictures tend to be cluttered and hence are of lesser value than graphical representations of vector fields in 2-space. Note also that the vectors in parts (b) and (c) are not to scale—their lengths have been compressed for clarity. We will follow this procedure throughout this chapter.

TECHNOLOGY MASTERY

If you have a graphing utility that can generate vector fields, read the relevant documentation and try to make reasonable duplicates of parts (a) and (b) of Figure 16.1.3.

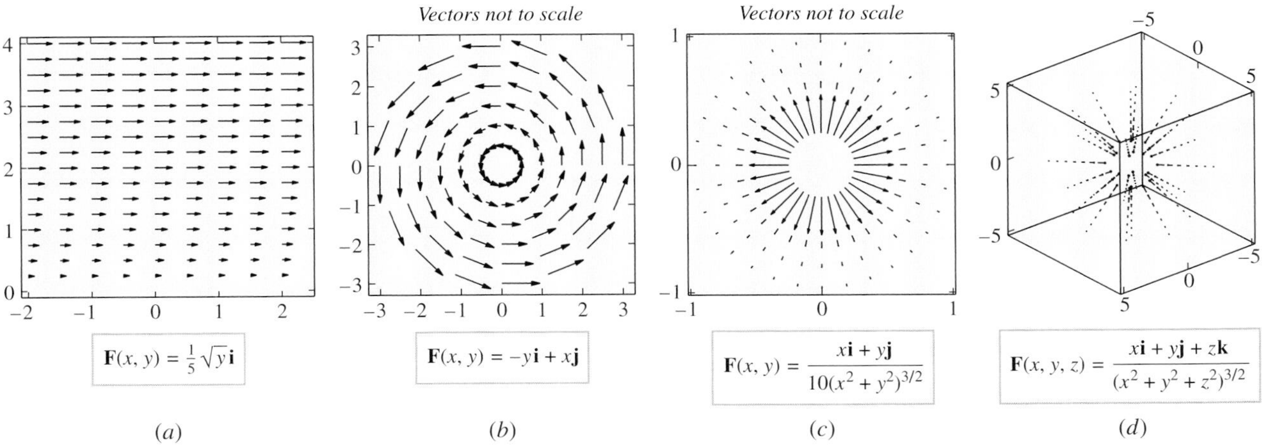

Figure 16.1.3

A COMPACT NOTATION FOR VECTOR FIELDS

Sometimes it is helpful to denote the vector fields $\mathbf{F}(x, y)$ and $\mathbf{F}(x, y, z)$ entirely in vector notation by identifying (x, y) with the radius vector $\mathbf{r} = x\mathbf{i} + y\mathbf{j}$ and (x, y, z) with the radius vector $\mathbf{r} = x\mathbf{i} + y\mathbf{j} + z\mathbf{k}$. With this notation a vector field in either 2-space or 3-space can be written as $\mathbf{F}(\mathbf{r})$. When no confusion is likely to arise, we will sometimes omit the $\mathbf{r}$ altogether and denote the vector field as $\mathbf{F}$.

INVERSE-SQUARE FIELDS

According to Newton's Law of Universal Gravitation, particles with masses m and M attract each other with a force $\mathbf{F}$ of magnitude

$$\|\mathbf{F}\| = \frac{GmM}{r^2} \tag{1}$$

where r is the distance between the particles and G is a constant. If we assume that the particle of mass M is located at the origin of an xyz-coordinate system and $\mathbf{r}$ is the radius vector to the particle of mass m, then $r = \|\mathbf{r}\|$, and the force $\mathbf{F}(\mathbf{r})$ exerted by the particle of mass M on the particle of mass m is in the direction of the unit vector $-\mathbf{r}/\|\mathbf{r}\|$. Thus, it follows from (1) that

$$\mathbf{F}(\mathbf{r}) = -\frac{GmM}{\|\mathbf{r}\|^2}\frac{\mathbf{r}}{\|\mathbf{r}\|} = -\frac{GmM}{\|\mathbf{r}\|^3}\mathbf{r} \tag{2}$$

If m and M are constant, and we let $c = -GmM$, then this formula can be expressed as

$$\mathbf{F}(\mathbf{r}) = \frac{c}{\|\mathbf{r}\|^3}\mathbf{r}$$

Vector fields of this form arise in electromagnetic as well as gravitational problems. Such fields are so important that they have their own terminology.

16.1.2 DEFINITION. If $\mathbf{r}$ is a radius vector in 2-space or 3-space, and if c is a constant, then a vector field of the form

$$\mathbf{F}(\mathbf{r}) = \frac{c}{\|\mathbf{r}\|^3}\mathbf{r} \tag{3}$$

is called an ***inverse-square field***.

Observe that if $c > 0$ in (3), then $\mathbf{F}(\mathbf{r})$ has the same direction as $\mathbf{r}$, so each vector in the field is directed away from the origin; and if $c < 0$, then $\mathbf{F}(\mathbf{r})$ is oppositely directed to $\mathbf{r}$, so each vector in the field is directed toward the origin. In either case the magnitude of $\mathbf{F}(\mathbf{r})$ is inversely proportional to the square of the distance from the terminal point of $\mathbf{r}$ to the origin, since

$$\|\mathbf{F}(\mathbf{r})\| = \frac{|c|}{\|\mathbf{r}\|^3}\|\mathbf{r}\| = \frac{|c|}{\|\mathbf{r}\|^2}$$

We leave it for you to verify that in 2-space Formula (3) can be written in component form as

$$\mathbf{F}(x, y) = \frac{c}{(x^2 + y^2)^{3/2}}(x\mathbf{i} + y\mathbf{j}) \tag{4}$$

and in 3-space as

$$\mathbf{F}(x, y, z) = \frac{c}{(x^2 + y^2 + z^2)^{3/2}}(x\mathbf{i} + y\mathbf{j} + z\mathbf{k}) \tag{5}$$

[see parts (c) and (d) of Figure 16.1.3].

▶ **Example 1** ***Coulomb's law*** states that *the electrostatic force exerted by one charged particle on another is directly proportional to the product of the charges and inversely proportional to the square of the distance between them.* This has the same form as Newton's Law of Universal Gravitation, so the electrostatic force field exerted by a charged particle is an inverse-square field. Specifically, if a particle of charge Q is at the origin of a coordinate system, and if $\mathbf{r}$ is the radius vector to a particle of charge q, then the force $\mathbf{F}(\mathbf{r})$ that the particle of charge Q exerts on the particle of charge q is of the form

$$\mathbf{F}(\mathbf{r}) = \frac{qQ}{4\pi\epsilon_0\|\mathbf{r}\|^3}\mathbf{r}$$

where ϵ_0 is a positive constant (called the ***permittivity constant***). This formula is of form (3) with $c = qQ/4\pi\epsilon_0$. ◀

GRADIENT FIELDS

An important class of vector fields arises from the process of finding gradients. Recall that if ϕ is a function of three variables, then the gradient of ϕ is defined as

$$\nabla\phi = \frac{\partial\phi}{\partial x}\mathbf{i} + \frac{\partial\phi}{\partial y}\mathbf{j} + \frac{\partial\phi}{\partial z}\mathbf{k}$$

This formula defines a vector field in 3-space called the ***gradient field of*** $\boldsymbol{\phi}$. Similarly, the gradient of a function of two variables defines a gradient field in 2-space. At each point in a gradient field where the gradient is nonzero, the vector points in the direction in which the rate of increase of ϕ is maximum.

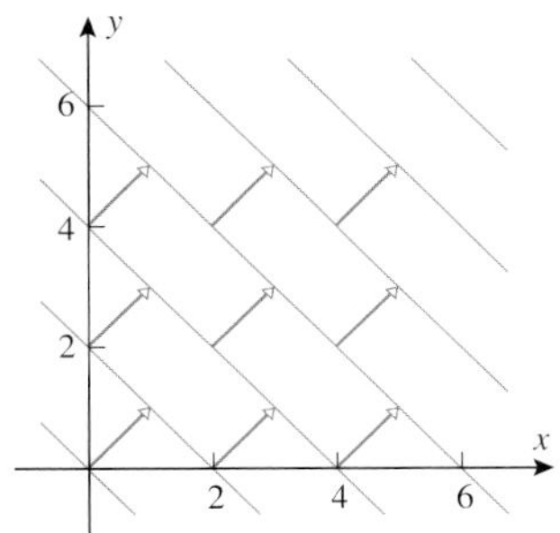

Figure 16.1.4

▶ **Example 2** Sketch the gradient field of $\phi(x, y) = x + y$.

Solution. The gradient of ϕ is

$$\nabla\phi = \frac{\partial\phi}{\partial x}\mathbf{i} + \frac{\partial\phi}{\partial y}\mathbf{j} = \mathbf{i} + \mathbf{j}$$

which is the same at each point. A portion of the vector field is sketched in Figure 16.1.4 together with some level curves of ϕ. Note that at each point, $\nabla\phi$ is normal to the level curve of ϕ through the point (Theorem 14.6.6). ◀

CONSERVATIVE FIELDS AND POTENTIAL FUNCTIONS

If $\mathbf{F}(\mathbf{r})$ is an arbitrary vector field in 2-space or 3-space, we can ask whether it is the gradient field of some function ϕ, and if so, how we can find ϕ. This is an important problem in various applications, and we will study it in more detail later. However, there is some terminology for such fields that we will introduce now.

> **16.1.3 DEFINITION.** A vector field $\mathbf{F}$ in 2-space or 3-space is said to be ***conservative*** in a region if it is the gradient field for some function ϕ in that region; that is, if
>
> $$\mathbf{F} = \nabla\phi$$
>
> The function ϕ is called a ***potential function*** for $\mathbf{F}$ in the region.

► **Example 3** Inverse-square fields are conservative in any region that does not contain the origin. For example, in the two-dimensional case the function

$$\phi(x, y) = -\frac{c}{(x^2 + y^2)^{1/2}} \tag{6}$$

is a potential function for (4) in any region not containing the origin, since

$$\begin{aligned}
\nabla\phi(x, y) &= \frac{\partial\phi}{\partial x}\mathbf{i} + \frac{\partial\phi}{\partial y}\mathbf{j} \\
&= \frac{cx}{(x^2 + y^2)^{3/2}}\mathbf{i} + \frac{cy}{(x^2 + y^2)^{3/2}}\mathbf{j} \\
&= \frac{c}{(x^2 + y^2)^{3/2}}(x\mathbf{i} + y\mathbf{j}) \\
&= \mathbf{F}(x, y)
\end{aligned}$$

In a later section we will discuss methods for finding potential functions for conservative vector fields. ◄

DIVERGENCE AND CURL

We will now define two important operations on vector fields in 3-space—the *divergence* and the *curl* of the field. These names originate in the study of fluid flow, in which case the divergence relates to the way in which fluid flows toward or away from a point and the curl relates to the rotational properties of the fluid at a point. We will investigate the physical interpretations of these operations in more detail later, but for now we will focus only on their computation.

16.1.4 DEFINITION. If $\mathbf{F}(x, y, z) = f(x, y, z)\mathbf{i} + g(x, y, z)\mathbf{j} + h(x, y, z)\mathbf{k}$, then we define the ***divergence of*** **F**, written div **F**, to be the function given by

$$\text{div}\,\mathbf{F} = \frac{\partial f}{\partial x} + \frac{\partial g}{\partial y} + \frac{\partial h}{\partial z} \tag{7}$$

16.1.5 DEFINITION. If $\mathbf{F}(x, y, z) = f(x, y, z)\mathbf{i} + g(x, y, z)\mathbf{j} + h(x, y, z)\mathbf{k}$, then we define the ***curl of*** **F**, written curl **F**, to be the vector field given by

$$\text{curl}\,\mathbf{F} = \left(\frac{\partial h}{\partial y} - \frac{\partial g}{\partial z}\right)\mathbf{i} + \left(\frac{\partial f}{\partial z} - \frac{\partial h}{\partial x}\right)\mathbf{j} + \left(\frac{\partial g}{\partial x} - \frac{\partial f}{\partial y}\right)\mathbf{k} \tag{8}$$

Observe that div **F** and curl **F** depend on the point at which they are computed, and hence are more properly written as div $\mathbf{F}(x, y, z)$ and curl $\mathbf{F}(x, y, z)$. However, even though these functions are expressed in terms of x, y, and z, it can be proved that their values at a fixed point depend only on the point and not on the coordinate system selected. This is important in applications, since it allows physicists and engineers to compute the curl and divergence in any convenient coordinate system.

Before proceeding to some examples, we note that div **F** has scalar values, whereas curl **F** has vector values (i.e., curl **F** is itself a vector field). Moreover, for computational purposes it is useful to note that the formula for the curl can be expressed in the determinant form

$$\operatorname{curl} \mathbf{F} = \begin{vmatrix} \mathbf{i} & \mathbf{j} & \mathbf{k} \\ \dfrac{\partial}{\partial x} & \dfrac{\partial}{\partial y} & \dfrac{\partial}{\partial z} \\ f & g & h \end{vmatrix} \tag{9}$$

You should verify that Formula (8) results if the determinant is computed by interpreting a "product" such as $(\partial/\partial x)(g)$ to mean $\partial g/\partial x$. Keep in mind, however, that (9) is just a mnemonic device and not a true determinant, since the entries in a determinant must be numbers, not vectors and partial derivative symbols.

▸ **Example 4** Find the divergence and the curl of the vector field

$$\mathbf{F}(x, y, z) = x^2 y\mathbf{i} + 2y^3 z\mathbf{j} + 3z\mathbf{k}$$

Solution. From (7)

$$\begin{aligned} \operatorname{div} \mathbf{F} &= \frac{\partial}{\partial x}(x^2 y) + \frac{\partial}{\partial y}(2y^3 z) + \frac{\partial}{\partial z}(3z) \\ &= 2xy + 6y^2 z + 3 \end{aligned}$$

and from (9)

TECHNOLOGY MASTERY

Most computer algebra systems can compute gradient fields, divergence, and curl. If you have a CAS with these capabilities, read the relevant documentation and use your CAS to check the computations in Examples 2 and 4.

$$\begin{aligned} \operatorname{curl} \mathbf{F} &= \begin{vmatrix} \mathbf{i} & \mathbf{j} & \mathbf{k} \\ \dfrac{\partial}{\partial x} & \dfrac{\partial}{\partial y} & \dfrac{\partial}{\partial z} \\ x^2 y & 2y^3 z & 3z \end{vmatrix} \\ &= \left[\frac{\partial}{\partial y}(3z) - \frac{\partial}{\partial z}(2y^3 z)\right]\mathbf{i} + \left[\frac{\partial}{\partial z}(x^2 y) - \frac{\partial}{\partial x}(3z)\right]\mathbf{j} + \left[\frac{\partial}{\partial x}(2y^3 z) - \frac{\partial}{\partial y}(x^2 y)\right]\mathbf{k} \\ &= -2y^3\mathbf{i} - x^2\mathbf{k} \end{aligned}$$ ◂

▸ **Example 5** Show that the divergence of the inverse-square field

$$\mathbf{F}(x, y, z) = \frac{c}{(x^2 + y^2 + z^2)^{3/2}}(x\mathbf{i} + y\mathbf{j} + z\mathbf{k})$$

is zero.

Solution. The computations can be simplified by letting $r = (x^2 + y^2 + z^2)^{1/2}$, in which case **F** can be expressed as

$$\mathbf{F}(x, y, z) = \frac{cx\mathbf{i} + cy\mathbf{j} + cz\mathbf{k}}{r^3} = \frac{cx}{r^3}\mathbf{i} + \frac{cy}{r^3}\mathbf{j} + \frac{cz}{r^3}\mathbf{k}$$

We leave it for you to show that

$$\frac{\partial r}{\partial x} = \frac{x}{r}, \quad \frac{\partial r}{\partial y} = \frac{y}{r}, \quad \frac{\partial r}{\partial z} = \frac{z}{r}$$

Thus

$$\operatorname{div}\mathbf{F} = c\left[\frac{\partial}{\partial x}\left(\frac{x}{r^3}\right) + \frac{\partial}{\partial y}\left(\frac{y}{r^3}\right) + \frac{\partial}{\partial z}\left(\frac{z}{r^3}\right)\right] \tag{10}$$

But

$$\begin{aligned}
\frac{\partial}{\partial x}\left(\frac{x}{r^3}\right) &= \frac{r^3 - x(3r^2)(x/r)}{(r^3)^2} = \frac{1}{r^3} - \frac{3x^2}{r^5} \\
\frac{\partial}{\partial y}\left(\frac{y}{r^3}\right) &= \frac{1}{r^3} - \frac{3y^2}{r^5} \\
\frac{\partial}{\partial z}\left(\frac{z}{r^3}\right) &= \frac{1}{r^3} - \frac{3z^2}{r^5}
\end{aligned}$$

Substituting these expressions in (10) yields

$$\operatorname{div}\mathbf{F} = c\left[\frac{3}{r^3} - \frac{3x^2 + 3y^2 + 3z^2}{r^5}\right] = c\left[\frac{3}{r^3} - \frac{3r^2}{r^5}\right] = 0 \blacktriangleleft$$

THE ∇ OPERATOR

Thus far, the symbol $\boldsymbol{\nabla}$ that appears in the gradient expression $\boldsymbol{\nabla}\phi$ has not been given a meaning of its own. However, it is often convenient to view $\boldsymbol{\nabla}$ as an operator

$$\boldsymbol{\nabla} = \frac{\partial}{\partial x}\mathbf{i} + \frac{\partial}{\partial y}\mathbf{j} + \frac{\partial}{\partial z}\mathbf{k} \tag{11}$$

which when applied to $\phi(x, y, z)$ produces the gradient

$$\boldsymbol{\nabla}\phi = \frac{\partial \phi}{\partial x}\mathbf{i} + \frac{\partial \phi}{\partial y}\mathbf{j} + \frac{\partial \phi}{\partial z}\mathbf{k}$$

We call (11) the ***del operator***. This is analogous to the derivative operator d/dx, which when applied to $f(x)$ produces the derivative $f'(x)$.

The del operator allows us to express the divergence of a vector field

$$\mathbf{F} = f(x, y, z)\mathbf{i} + g(x, y, z)\mathbf{j} + h(x, y, z)\mathbf{k}$$

in dot product notation as

$$\operatorname{div}\mathbf{F} = \boldsymbol{\nabla}\cdot\mathbf{F} = \frac{\partial f}{\partial x} + \frac{\partial g}{\partial y} + \frac{\partial h}{\partial z} \tag{12}$$

and the curl of this field in cross-product notation as

$$\operatorname{curl}\mathbf{F} = \boldsymbol{\nabla}\times\mathbf{F} = \begin{vmatrix} \mathbf{i} & \mathbf{j} & \mathbf{k} \\ \dfrac{\partial}{\partial x} & \dfrac{\partial}{\partial y} & \dfrac{\partial}{\partial z} \\ f & g & h \end{vmatrix} \tag{13}$$

THE LAPLACIAN ∇^2

The operator that results by taking the dot product of the del operator with itself is denoted by $\boldsymbol{\nabla}^2$ and is called the ***Laplacian operator***. This operator has the form

$$\boldsymbol{\nabla}^2 = \boldsymbol{\nabla}\cdot\boldsymbol{\nabla} = \frac{\partial^2}{\partial x^2} + \frac{\partial^2}{\partial y^2} + \frac{\partial^2}{\partial z^2} \tag{14}$$

When applied to $\phi(x, y, z)$ the Laplacian operator produces the function

$$\nabla^2\phi = \frac{\partial^2\phi}{\partial x^2} + \frac{\partial^2\phi}{\partial y^2} + \frac{\partial^2\phi}{\partial z^2}$$

Note that $\nabla^2\phi$ can also be expressed as div $(\nabla\phi)$. The equation $\nabla^2\phi = 0$ or, equivalently,

$$\frac{\partial^2\phi}{\partial x^2} + \frac{\partial^2\phi}{\partial y^2} + \frac{\partial^2\phi}{\partial z^2} = 0$$

is known as ***Laplace's equation***. This partial differential equation plays an important role in a wide variety of applications, resulting from the fact that it is satisfied by the potential function for the inverse-square field.

✔QUICK CHECK EXERCISES 16.1

(See page 1111 for answers.)

1. The function $\phi(x, y, z) = xy + yz + xz$ is a potential for the vector field $\mathbf{F} =$ ________.
2. The vector field $\mathbf{F}(x, y, z) =$ ________, defined for $(x, y, z) \neq (0, 0, 0)$, is always directed toward the origin and is of length equal to the distance from (x, y, z) to the origin.
3. An inverse-square field is one that can be written in the form $\mathbf{F}(\mathbf{r}) =$ ________.
4. The vector field

$$\mathbf{F}(x, y, z) = yz\mathbf{i} + xy^2\mathbf{j} + yz^2\mathbf{k}$$

has divergence ________ and curl ________.

Pierre-Simon de Laplace (1749–1827) French mathematician and physicist. Laplace is sometimes referred to as the French Isaac Newton because of his work in celestial mechanics. In a five-volume treatise entitled *Traité de Mécanique Céleste*, he solved extremely difficult problems involving gravitational interactions between the planets. In particular, he was able to show that our solar system is stable and not prone to catastrophic collapse as a result of these interactions. This was an issue of major concern at the time because Jupiter's orbit appeared to be shrinking and Saturn's expanding; Laplace showed that these were expected periodic anomalies. In addition to his work in celestial mechanics, he founded modern probability theory, showed with Lavoisier that respiration is a form of combustion, and developed methods that fostered many new branches of pure mathematics.

Laplace was born to moderately successful parents in Normandy, his father being a farmer and cider merchant. He matriculated in the theology program at the University of Caen at age 16 but left for Paris at age 18 with a letter of introduction to the influential mathematician d'Alembert, who eventually helped him undertake a career in mathematics. Laplace was a prolific writer, and after his election to the Academy of Sciences in 1773, the secretary wrote that the Academy had never received so many important research papers by so young a person in such a short time. Laplace had little interest in pure mathematics—he regarded mathematics merely as a tool for solving applied problems. In his impatience with mathematical detail, he frequently omitted complicated arguments with the statement, "It is easy to show that. . . ." He admitted, however, that as time passed he often had trouble reconstructing the omitted details himself!

At the height of his fame, Laplace served on many government committees and held the posts of Minister of the Interior and Chancellor of the Senate. He barely escaped imprisonment and execution during the period of the Revolution, probably because he was able to convince each opposing party that he sided with them. Napoleon described him as a great mathematician but a poor administrator who "sought subtleties everywhere, had only doubtful ideas, and . . . carried the spirit of the infinitely small into administration." In spite of his genius, Laplace was both egotistic and insecure, attempting to ensure his place in history by conveniently failing to credit mathematicians whose work he used—an unnecessary pettiness since his own work was so brilliant. However, on the positive side he was supportive of young mathematicians, often treating them as his own children. Laplace ranks as one of the most influential mathematicians in history.

EXERCISE SET 16.1 Graphing Utility [C] CAS

FOCUS ON CONCEPTS

1–2 Match the vector field $\mathbf{F}(x, y)$ with one of the plots, and explain your reasoning.

1. (a) $\mathbf{F}(x, y) = x\mathbf{i}$ (b) $\mathbf{F}(x, y) = \sin x\mathbf{i} + \mathbf{j}$

2. (a) $\mathbf{F}(x, y) = \mathbf{i} + \mathbf{j}$
(b) $\mathbf{F}(x, y) = \dfrac{x}{\sqrt{x^2 + y^2}}\mathbf{i} + \dfrac{y}{\sqrt{x^2 + y^2}}\mathbf{j}$

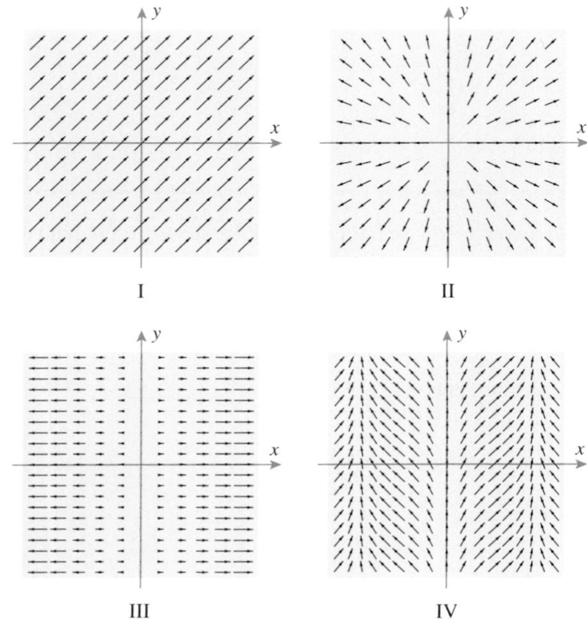

3–4 Determine whether the statement about the vector field $\mathbf{F}(x, y)$ is true or false. If false, explain why.

3. $\mathbf{F}(x, y) = x^2\mathbf{i} - y\mathbf{j}$.
(a) $\|\mathbf{F}(x, y)\| \to 0$ as $(x, y) \to (0, 0)$.
(b) If (x, y) is on the positive y-axis, then the vector points in the negative y-direction.
(c) If (x, y) is in the first quadrant, then the vector points down and to the right.

4. $\mathbf{F}(x, y) = \dfrac{x}{\sqrt{x^2 + y^2}}\mathbf{i} - \dfrac{y}{\sqrt{x^2 + y^2}}\mathbf{j}$.
(a) As (x, y) moves away from the origin, the lengths of the vectors decrease.
(b) If (x, y) is a point on the positive x-axis, then the vector points up.
(c) If (x, y) is a point on the positive y-axis, the vector points to the right.

5–8 Sketch the vector field by drawing some representative nonintersecting vectors. The vectors need not be drawn to scale, but they should be in reasonably correct proportion relative to each other.

5. $\mathbf{F}(x, y) = 2\mathbf{i} - \mathbf{j}$

6. $\mathbf{F}(x, y) = y\mathbf{j}, \quad y > 0$

7. $\mathbf{F}(x, y) = y\mathbf{i} - x\mathbf{j}$. [*Note:* Each vector in the field is perpendicular to the position vector $\mathbf{r} = x\mathbf{i} + y\mathbf{j}$.]

8. $\mathbf{F}(x, y) = \dfrac{x\mathbf{i} + y\mathbf{j}}{\sqrt{x^2 + y^2}}$. [*Note:* Each vector in the field is a unit vector in the same direction as the position vector $\mathbf{r} = x\mathbf{i} + y\mathbf{j}$.]

9–10 Use a graphing utility to generate a plot of the vector field.

9. $\mathbf{F}(x, y) = \mathbf{i} + \cos y\mathbf{j}$

10. $\mathbf{F}(x, y) = y\mathbf{i} - x\mathbf{j}$

11–12 Confirm that ϕ is a potential function for $\mathbf{F}(\mathbf{r})$ on some region, and state the region.

11. (a) $\phi(x, y) = \tan^{-1} xy$
$\mathbf{F}(x, y) = \dfrac{y}{1 + x^2y^2}\mathbf{i} + \dfrac{x}{1 + x^2y^2}\mathbf{j}$
(b) $\phi(x, y, z) = x^2 - 3y^2 + 4z^2$
$\mathbf{F}(x, y, z) = 2x\mathbf{i} - 6y\mathbf{j} + 8z\mathbf{k}$

12. (a) $\phi(x, y) = 2y^2 + 3x^2y - xy^3$
$\mathbf{F}(x, y) = (6xy - y^3)\mathbf{i} + (4y + 3x^2 - 3xy^2)\mathbf{j}$
(b) $\phi(x, y, z) = x \sin z + y \sin x + z \sin y$
$\mathbf{F}(x, y, z) = (\sin z + y \cos x)\mathbf{i} + (\sin x + z \cos y)\mathbf{j} + (\sin y + x \cos z)\mathbf{k}$

13–18 Find div **F** and curl **F**.

13. $\mathbf{F}(x, y, z) = x^2\mathbf{i} - 2\mathbf{j} + yz\mathbf{k}$

14. $\mathbf{F}(x, y, z) = xz^3\mathbf{i} + 2y^4x^2\mathbf{j} + 5z^2y\mathbf{k}$

15. $\mathbf{F}(x, y, z) = 7y^3z^2\mathbf{i} - 8x^2z^5\mathbf{j} - 3xy^4\mathbf{k}$

16. $\mathbf{F}(x, y, z) = e^{xy}\mathbf{i} - \cos y\mathbf{j} + \sin^2 z\mathbf{k}$

17. $\mathbf{F}(x, y, z) = \dfrac{1}{\sqrt{x^2 + y^2 + z^2}}(x\mathbf{i} + y\mathbf{j} + z\mathbf{k})$

18. $\mathbf{F}(x, y, z) = \ln x\mathbf{i} + e^{xyz}\mathbf{j} + \tan^{-1}(z/x)\mathbf{k}$

19–20 Find $\nabla \cdot (\mathbf{F} \times \mathbf{G})$.

19. $\mathbf{F}(x, y, z) = 2x\mathbf{i} + \mathbf{j} + 4y\mathbf{k}$
$\mathbf{G}(x, y, z) = x\mathbf{i} + y\mathbf{j} - z\mathbf{k}$

20. $\mathbf{F}(x, y, z) = yz\mathbf{i} + xz\mathbf{j} + xy\mathbf{k}$
$\mathbf{G}(x, y, z) = xy\mathbf{j} + xyz\mathbf{k}$

21–22 Find $\nabla \cdot (\nabla \times \mathbf{F})$.

21. $\mathbf{F}(x, y, z) = \sin x\mathbf{i} + \cos(x - y)\mathbf{j} + z\mathbf{k}$

22. $\mathbf{F}(x, y, z) = e^{xz}\mathbf{i} + 3xe^y\mathbf{j} - e^{yz}\mathbf{k}$

23–24 Find $\nabla \times (\nabla \times \mathbf{F})$.

23. $\mathbf{F}(x, y, z) = xy\mathbf{j} + xyz\mathbf{k}$

24. $\mathbf{F}(x, y, z) = y^2x\mathbf{i} - 3yz\mathbf{j} + xy\mathbf{k}$

C **25.** Use a CAS to check the calculations in Exercises 19, 21, and 23.

C **26.** Use a CAS to check the calculations in Exercises 20, 22, and 24.

27–34 Let k be a constant, $\mathbf{F} = \mathbf{F}(x, y, z)$, $\mathbf{G} = \mathbf{G}(x, y, z)$, and $\phi = \phi(x, y, z)$. Prove the following identities, assuming that all derivatives involved exist and are continuous.

27. $\operatorname{div}(k\mathbf{F}) = k \operatorname{div} \mathbf{F}$ **28.** $\operatorname{curl}(k\mathbf{F}) = k \operatorname{curl} \mathbf{F}$

29. $\operatorname{div}(\mathbf{F} + \mathbf{G}) = \operatorname{div} \mathbf{F} + \operatorname{div} \mathbf{G}$

30. $\operatorname{curl}(\mathbf{F} + \mathbf{G}) = \operatorname{curl} \mathbf{F} + \operatorname{curl} \mathbf{G}$

31. $\operatorname{div}(\phi\mathbf{F}) = \phi \operatorname{div} \mathbf{F} + \nabla\phi \cdot \mathbf{F}$

32. $\operatorname{curl}(\phi\mathbf{F}) = \phi \operatorname{curl} \mathbf{F} + \nabla\phi \times \mathbf{F}$

33. $\operatorname{div}(\operatorname{curl} \mathbf{F}) = 0$ **34.** $\operatorname{curl}(\nabla\phi) = \mathbf{0}$

35. Rewrite the identities in Exercises 27, 29, 31, and 33 in an equivalent form using the notation $\nabla \cdot$ for divergence and $\nabla \times$ for curl.

36. Rewrite the identities in Exercises 28, 30, 32, and 34 in an equivalent form using the notation $\nabla \cdot$ for divergence and $\nabla \times$ for curl.

37–38 Verify that the radius vector $\mathbf{r} = x\mathbf{i} + y\mathbf{j} + z\mathbf{k}$ has the stated property.

37. (a) $\operatorname{curl} \mathbf{r} = \mathbf{0}$ (b) $\nabla\|\mathbf{r}\| = \dfrac{\mathbf{r}}{\|\mathbf{r}\|}$

38. (a) $\operatorname{div} \mathbf{r} = 3$ (b) $\nabla \dfrac{1}{\|\mathbf{r}\|} = -\dfrac{\mathbf{r}}{\|\mathbf{r}\|^3}$

39–40 Let $\mathbf{r} = x\mathbf{i} + y\mathbf{j} + z\mathbf{k}$, let $r = \|\mathbf{r}\|$, let f be a differentiable function of one variable, and let $\mathbf{F}(\mathbf{r}) = f(r)\mathbf{r}$.

39. (a) Use the chain rule and Exercise 37(b) to show that

$$\nabla f(r) = \frac{f'(r)}{r}\mathbf{r}$$

(b) Use the result in part (a) and Exercises 31 and 38(a) to show that $\operatorname{div} \mathbf{F} = 3f(r) + rf'(r)$.

40. (a) Use part (a) of Exercise 39, Exercise 32, and Exercise 37(a) to show that $\operatorname{curl} \mathbf{F} = \mathbf{0}$.

(b) Use the result in part (a) of Exercise 39 and Exercises 31 and 38(a) to show that

$$\nabla^2 f(r) = 2\frac{f'(r)}{r} + f''(r)$$

41. Use the result in Exercise 39(b) to show that the divergence of the inverse-square field $\mathbf{F} = \mathbf{r}/\|\mathbf{r}\|^3$ is zero.

42. Use the result of Exercise 39(b) to show that if $\mathbf{F}$ is a vector field of the form $\mathbf{F} = f(\|\mathbf{r}\|)\mathbf{r}$ and if $\operatorname{div} \mathbf{F} = 0$, then $\mathbf{F}$ is an inverse-square field. [*Suggestion:* Let $r = \|\mathbf{r}\|$ and multiply $3f(r) + rf'(r) = 0$ through by r^2. Then write the result as a derivative of a product.

43. A curve C is called a ***flow line*** of a vector field $\mathbf{F}$ if $\mathbf{F}$ is a tangent vector to C at each point along C (see the accompanying figure).

(a) Let C be a flow line for $\mathbf{F}(x, y) = -y\mathbf{i} + x\mathbf{j}$, and let (x, y) be a point on C for which $y \neq 0$. Show that the flow lines satisfy the differential equation

$$\frac{dy}{dx} = -\frac{x}{y}$$

(b) Solve the differential equation in part (a) by separation of variables, and show that the flow lines are concentric circles centered at the origin.

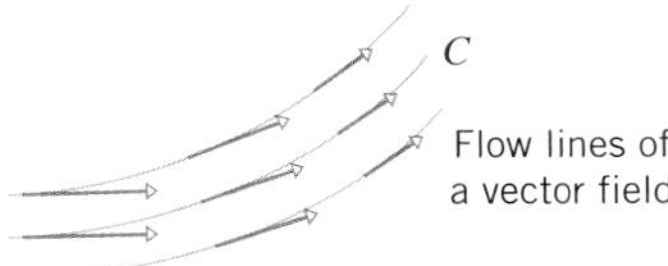

Figure Ex-43

44–46 Find a differential equation satisfied by the flow lines of $\mathbf{F}$ (see Exercise 43), and solve it to find equations for the flow lines of $\mathbf{F}$. Sketch some typical flow lines and tangent vectors.

44. $\mathbf{F}(x, y) = \mathbf{i} + x\mathbf{j}$ **45.** $\mathbf{F}(x, y) = x\mathbf{i} + \mathbf{j}, \quad x > 0$

46. $\mathbf{F}(x, y) = x\mathbf{i} - y\mathbf{j}, \quad x > 0$ and $y > 0$

✔ QUICK CHECK ANSWERS 16.1

1. $(y + z)\mathbf{i} + (x + z)\mathbf{j} + (x + y)\mathbf{k}$ 2. $-\mathbf{r} = -x\mathbf{i} - y\mathbf{j} - z\mathbf{k}$ 3. $\dfrac{c}{\|\mathbf{r}\|^3}\mathbf{r}$ 4. $2xy + 2yz$; $z^2\mathbf{i} + y\mathbf{j} + (y^2 - z)\mathbf{k}$

16.2 LINE INTEGRALS

In earlier chapters we considered three kinds of integrals in rectangular coordinates: single integrals over intervals, double integrals over two-dimensional regions, and triple integrals over three-dimensional regions. In this section we will discuss integrals along curves in two- or three-dimensional space.

LINE INTEGRALS

The first goal of this section is to define what it means to integrate a function along a curve. To motivate the definition we will consider the problem of finding the mass of a very thin wire whose linear density function (mass per unit length) is known. We assume that we can model the wire by a smooth curve C between two points P and Q in 3-space (Figure 16.2.1). Given any point (x, y, z) on C, we let $f(x, y, z)$ denote the corresponding value of the density function. To compute the mass of the wire, we proceed as follows:

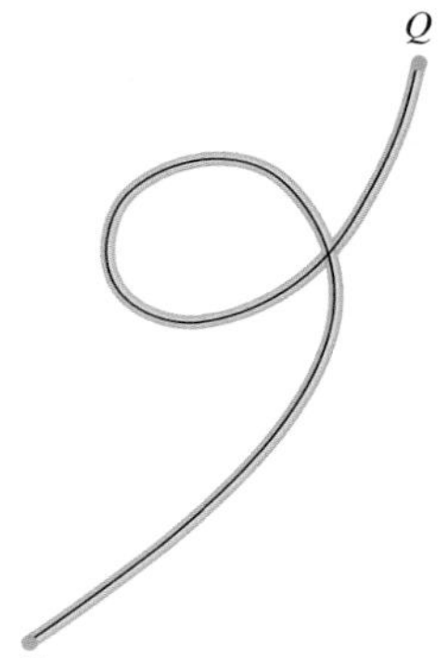

Figure 16.2.1

- Divide C into n very small sections using a succession of distinct partition points

$$P = P_0, P_1, P_2, \dots, P_{n-1}, P_n = Q$$

as illustrated on the left side of Figure 16.2.2. Let ΔM_k be the mass of the kth section, and let Δs_k be the length of the arc between P_{k-1} and P_k.

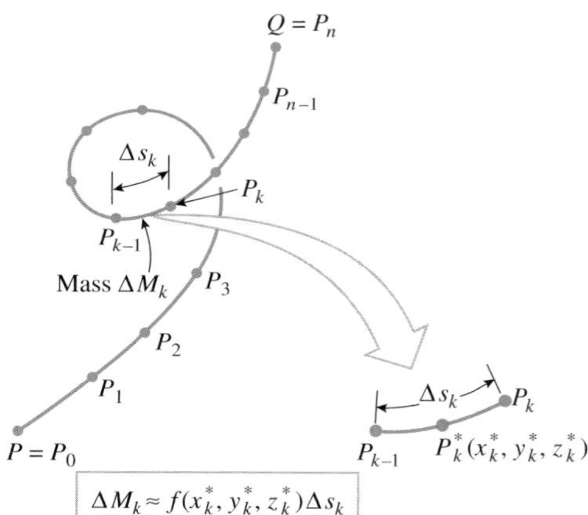

Figure 16.2.2

- Choose an arbitrary sampling point $P_k^*(x_k^*, y_k^*, z_k^*)$ on the kth arc, as illustrated on the right side of Figure 16.2.2. If Δs_k is very small, the value of f will not vary much along the kth section and we can approximate f along this section by the value $f(x_k^*, y_k^*, z_k^*)$. It follows that the mass of the kth section can be approximated by

$$\Delta M_k \approx f(x_k^*, y_k^*, z_k^*)\Delta s_k$$

- The mass M of the entire wire can then be approximated by

$$M = \sum_{k=1}^{n} \Delta M_k \approx \sum_{k=1}^{n} f(x_k^*, y_k^*, z_k^*)\Delta s_k \tag{1}$$

- We will use the expression max $\Delta s_k \to 0$ to indicate the process of increasing n in such a way that the lengths of all the sections approach 0. It is plausible that the error in (1) will approach 0 as max $\Delta s_k \to 0$ and the exact value of M will be given by

$$M = \lim_{\max \Delta s_k \to 0} \sum_{k=1}^{n} f(x_k^*, y_k^*, z_k^*)\Delta s_k \tag{2}$$

The limit in (2) is similar to the limit of Riemann sums used to define the definite integral of a function over an interval (Definition 5.5.1). With this similarity in mind, we make the following definition.

Although the term "curve integrals" is more descriptive, the integrals in Definition 16.2.1 are called "line integrals" for historical reasons.

16.2.1 DEFINITION. If C is a smooth curve in 2-space or 3-space, then the ***line integral of f with respect to s along C*** is

$$\int_C f(x, y)\, ds = \lim_{\max \Delta s_k \to 0} \sum_{k=1}^{n} f(x_k^*, y_k^*)\Delta s_k \qquad \text{2-space} \tag{3}$$

or

$$\int_C f(x, y, z)\, ds = \lim_{\max \Delta s_k \to 0} \sum_{k=1}^{n} f(x_k^*, y_k^*, z_k^*)\Delta s_k \qquad \text{3-space} \tag{4}$$

provided this limit exists and does not depend on the choice of partition or on the choice of sample points.

It is usually impractical to evaluate line integrals directly from Definition 16.2.1. However, the definition is important in the application and interpretation of line integrals. For example:

- If C is a curve in 3-space that models a thin wire, and if $f(x, y, z)$ is the linear density function of the wire, then it follows from (2) and Definition 16.2.1 that the mass M of the wire is given by

$$M = \int_C f(x, y, z)\, ds \tag{5}$$

  That is, to obtain the mass of a thin wire, we integrate the linear density function over the smooth curve that models the wire.

- If C is a smooth curve of arc length L, and f is identically 1, then it immediately follows from Definition 16.2.1 that

$$\int_C ds = \lim_{\max \Delta s_k \to 0} \sum_{k=1}^{n} \Delta s_k = \lim_{\max \Delta s_k \to 0} L = L \tag{6}$$

- If C is a curve in the xy-plane and $f(x, y)$ is a nonnegative continuous function defined on C, then $\int_C f(x, y)\, ds$ can be interpreted as the area A of the "sheet" that is swept out by a vertical line segment that extends upward from the point (x, y) to a height of $f(x, y)$ and moves along C from one endpoint to the other (Figure 16.2.3). To see why this is so, refer to Figure 16.2.4 and note the approximation

$$\Delta A_k \approx f(x_k^*, y_k^*)\Delta s_k$$

  It follows that

$$A = \sum_{k=1}^{n} \Delta A_k \approx \sum_{k=1}^{n} f(x_k^*, y_k^*)\Delta s_k$$

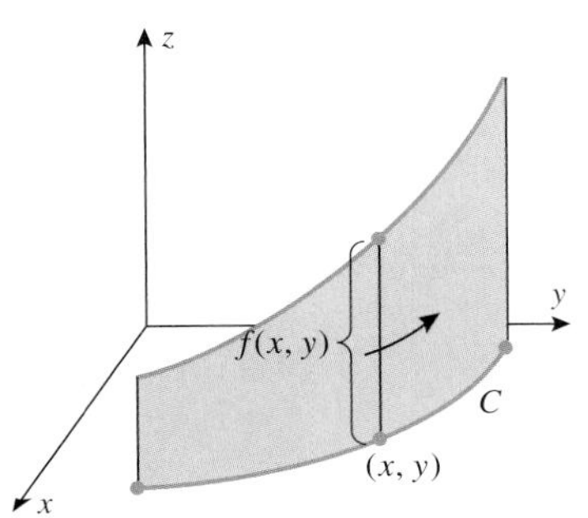

Figure 16.2.3

It is then plausible that

$$A = \lim_{\max \Delta s_k \to 0} \sum_{k=1}^{n} f(x_k^*, y_k^*)\Delta s_k = \int_C f(x, y)\, ds \tag{7}$$

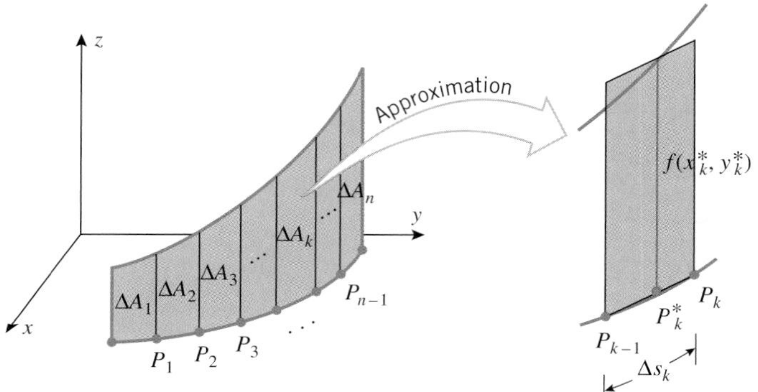

Figure 16.2.4

Since Definition 16.2.1 is closely modeled on Definition 5.5.1, it should come as no surprise that line integrals share many of the common properties of ordinary definite integrals. For example, we have

$$\int_C [f(x, y) + g(x, y)]\, ds = \int_C f(x, y)\, ds + \int_C g(x, y)\, ds$$

provided both line integrals on the right-hand side of this equation exist. Similarly, it can be shown that if f is continuous on C, then the line integral of f with respect to s along C exists.

EVALUATING LINE INTEGRALS

Except in simple cases, it will not be feasible to evaluate a line integral directly from (3) or (4). However, we will now show that it is possible to express a line integral as an ordinary definite integral, so that no special methods of evaluation are required. For example, suppose that C is a curve in the xy-plane that is smoothly parametrized by

$$\mathbf{r}(t) = x(t)\mathbf{i} + y(t)\mathbf{j} \qquad (a \le t \le b)$$

Moreover, suppose that each partition point P_k of C corresponds to a parameter value of t_k in $[a, b]$. The arc length of C between points P_{k-1} and P_k is then given by

$$\Delta s_k = \int_{t_{k-1}}^{t_k} \|\mathbf{r}'(t)\|\, dt \tag{8}$$

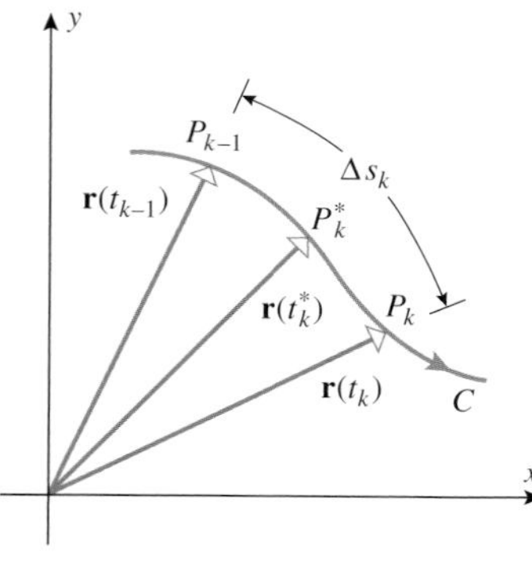

Figure 16.2.5

(Theorem 13.3.1). If we let $\Delta t_k = t_k - t_{k-1}$, then it follows from (8) and the Mean-Value Theorem for Integrals (Theorem 5.6.2) that there exists a point t_k^* in $[t_{k-1}, t_k]$ such that

$$\Delta s_k = \int_{t_{k-1}}^{t_k} \|\mathbf{r}'(t)\|\, dt = \|\mathbf{r}'(t_k^*)\|\Delta t_k$$

We let $P_k^*(x_k^*, y_k^*) = P_k^*(x(t_k^*), y(t_k^*))$ correspond to the parameter value t_k^* (Figure 16.2.5).

Since the parametrization of C is smooth, it can be shown that $\max \Delta s_k \to 0$ if and only if $\max \Delta t_k \to 0$ (Exercise 49). Furthermore, the composition $f(x(t), y(t))$ is a real-valued function defined on $[a, b]$ and we have

$$\begin{aligned}
\int_C f(x, y)\, ds &= \lim_{\max \Delta s_k \to 0} \sum_{k=1}^{n} f(x_k^*, y_k^*)\, \Delta s_k && \boxed{\text{Definition 16.2.1}} \\
&= \lim_{\max \Delta s_k \to 0} \sum_{k=1}^{n} f(x(t_k^*), y(t_k^*)) \|\mathbf{r}'(t_k^*)\| \Delta t_k && \boxed{\text{Substitution}} \\
&= \lim_{\max \Delta t_k \to 0} \sum_{k=1}^{n} f(x(t_k^*), y(t_k^*)) \|\mathbf{r}'(t_k^*)\| \Delta t_k \\
&= \int_a^b f(x(t), y(t)) \|\mathbf{r}'(t)\|\, dt && \boxed{\text{Definition 5.5.1}}
\end{aligned}$$

Therefore, if C is smoothly parametrized by

$$\mathbf{r}(t) = x(t)\mathbf{i} + y(t)\mathbf{j} \qquad (a \le t \le b)$$

then

$$\int_C f(x, y)\, ds = \int_a^b f(x(t), y(t)) \|\mathbf{r}'(t)\|\, dt \tag{9}$$

Similarly, if C is a curve in 3-space that is smoothly parametrized by

$$\mathbf{r}(t) = x(t)\mathbf{i} + y(t)\mathbf{j} + z(t)\mathbf{k} \qquad (a \le t \le b)$$

then

$$\int_C f(x, y, z)\, ds = \int_a^b f(x(t), y(t), z(t)) \|\mathbf{r}'(t)\|\, dt \tag{10}$$

Explain how Formulas (9) and (10) confirm Formula (6) for arc length.

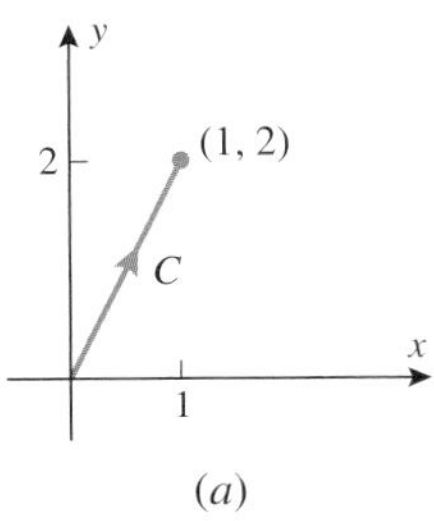

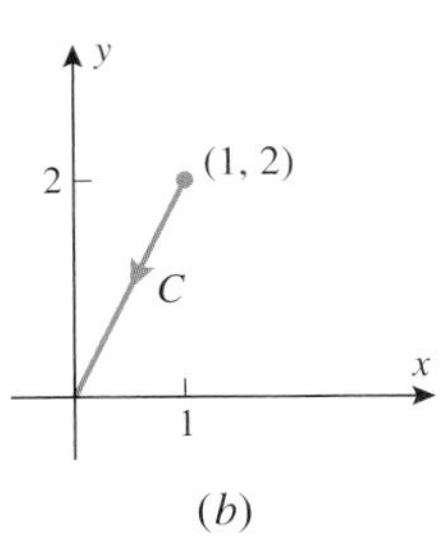

Figure 16.2.6

▶ **Example 1** Using the given parametrization, evaluate the line integral $\int_C (1 + xy^2)\, ds$.

(a) $C : \mathbf{r}(t) = t\mathbf{i} + 2t\mathbf{j} \quad (0 \le t \le 1)$ (see Figure 16.2.6*a*)

(b) $C : \mathbf{r}(t) = (1 - t)\mathbf{i} + (2 - 2t)\mathbf{j} \quad (0 \le t \le 1)$ (see Figure 16.2.6*b*)

Solution (a). Since $\mathbf{r}'(t) = \mathbf{i} + 2\mathbf{j}$, we have $\|\mathbf{r}'(t)\| = \sqrt{5}$ and it follows from Formula (9) that

$$\begin{aligned}
\int_C (1 + xy^2)\, ds &= \int_0^1 [1 + t(2t)^2]\sqrt{5}\, dt \\
&= \int_0^1 (1 + 4t^3)\sqrt{5}\, dt \\
&= \sqrt{5}\left[t + t^4\right]_0^1 = 2\sqrt{5}
\end{aligned}$$

Solution (b). Since $\mathbf{r}'(t) = -\mathbf{i} - 2\mathbf{j}$, we have $\|\mathbf{r}'(t)\| = \sqrt{5}$ and it follows from Formula (9) that

$$\begin{aligned}
\int_C (1 + xy^2)\, ds &= \int_0^1 [1 + (1 - t)(2 - 2t)^2]\sqrt{5}\, dt \\
&= \int_0^1 [1 + 4(1 - t)^3]\sqrt{5}\, dt \\
&= \sqrt{5}\left[t - (1 - t)^4\right]_0^1 = 2\sqrt{5}
\end{aligned}$$ ◀

Note that the integrals in parts (a) and (b) of Example 1 agree, even though the corresponding parametrizations of C have opposite orientations. This illustrates the important result that the value of a line integral of f with respect to s along C does not depend on an orientation of C. (This is because Δs_k is always positive; therefore, it does not matter in which *direction* along C we list the partition points of the curve in Definition 16.2.1.) Later in this section we will discuss line integrals that are defined only for oriented curves.

Formula (9) has an alternative expression for a curve C in the xy-plane that is given by parametric equations

$$x = x(t), \quad y = y(t) \qquad (a \le t \le b)$$

In this case, we write (9) in the expanded form

$$\int_C f(x, y)\, ds = \int_a^b f(x(t), y(t))\sqrt{\left(\frac{dx}{dt}\right)^2 + \left(\frac{dy}{dt}\right)^2}\, dt \tag{11}$$

Similarly, if C is a curve in 3-space that is parametrized by

$$x = x(t), \quad y = y(t), \quad z = z(t) \qquad (a \le t \le b)$$

then we write (10) in the form

$$\int_C f(x, y, z)\, ds = \int_a^b f(x(t), y(t), z(t))\sqrt{\left(\frac{dx}{dt}\right)^2 + \left(\frac{dy}{dt}\right)^2 + \left(\frac{dz}{dt}\right)^2}\, dt \tag{12}$$

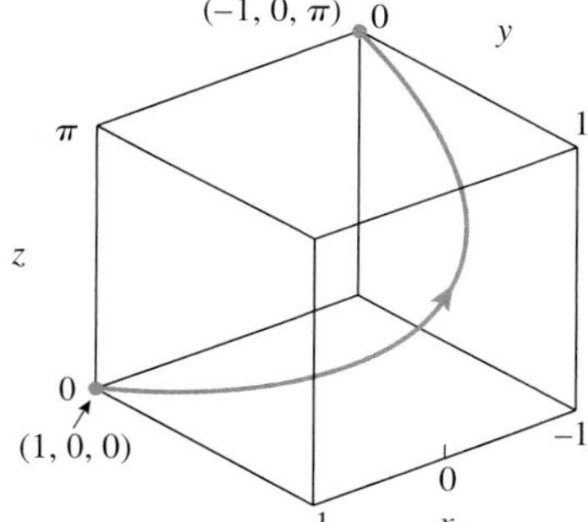

Figure 16.2.7

► **Example 2** Evaluate the line integral $\int_C (xy + z^3)\, ds$ from $(1, 0, 0)$ to $(-1, 0, \pi)$ along the helix C that is represented by the parametric equations

$$x = \cos t, \quad y = \sin t, \quad z = t \qquad (0 \le t \le \pi)$$

(Figure 16.2.7).

Solution. From (12)

$$\begin{aligned}
\int_C (xy + z^3)\, ds &= \int_0^\pi (\cos t \sin t + t^3)\sqrt{\left(\frac{dx}{dt}\right)^2 + \left(\frac{dy}{dt}\right)^2 + \left(\frac{dz}{dt}\right)^2}\, dt \\
&= \int_0^\pi (\cos t \sin t + t^3)\sqrt{(-\sin t)^2 + (\cos t)^2 + 1}\, dt \\
&= \sqrt{2}\int_0^\pi (\cos t \sin t + t^3)\, dt \\
&= \sqrt{2}\left[\frac{\sin^2 t}{2} + \frac{t^4}{4}\right]_0^\pi = \frac{\sqrt{2}\pi^4}{4} \blacktriangleleft
\end{aligned}$$

If $\delta(x, y)$ is the linear density function of a wire that is modeled by a smooth curve C in the xy-plane, then an argument similar to the derivation of Formula (5) shows that the mass of the wire is given by $\int_C \delta(x, y)\, ds$.

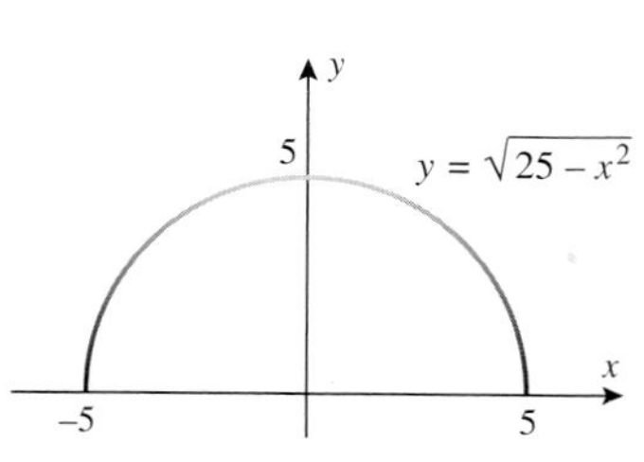

Figure 16.2.8

► **Example 3** Suppose that a semicircular wire has the equation $y = \sqrt{25 - x^2}$ and that its mass density is $\delta(x, y) = 15 - y$ (Figure 16.2.8). Physically, this means the wire has a maximum density of 15 units at the base ($y = 0$) and that the density of the wire decreases linearly with respect to y to a value of 10 units at the top ($y = 5$). Find the mass of the wire.

Solution. The mass M of the wire can be expressed as the line integral

$$M = \int_C \delta(x, y)\, ds = \int_C (15 - y)\, ds$$

along the semicircle C. To evaluate this integral we will express C parametrically as

$$x = 5\cos t, \quad y = 5\sin t \qquad (0 \le t \le \pi)$$

Thus, it follows from (11) that

$$\begin{aligned} M = \int_C (15 - y)\, ds &= \int_0^\pi (15 - 5\sin t)\sqrt{\left(\frac{dx}{dt}\right)^2 + \left(\frac{dy}{dt}\right)^2}\, dt \\ &= \int_0^\pi (15 - 5\sin t)\sqrt{(-5\sin t)^2 + (5\cos t)^2}\, dt \\ &= 5\int_0^\pi (15 - 5\sin t)\, dt \\ &= 5\big[15t + 5\cos t\big]_0^\pi \\ &= 75\pi - 50 \approx 185.6 \text{ units of mass} \end{aligned}$$ ◀

In the special case where t is an arc length parameter, say $t = s$, it follows from Formulas (20) and (21) in Section 13.3 that the radicals in (11) and (12) reduce to 1 and the equations simplify to

$$\int_C f(x, y)\, ds = \int_a^b f(x(s), y(s))\, ds \tag{13}$$

and

$$\int_C f(x, y, z)\, ds = \int_a^b f(x(s), y(s), z(s))\, ds \tag{14}$$

respectively.

▶ **Example 4** Find the area of the surface extending upward from the circle $x^2 + y^2 = 1$ in the xy-plane to the parabolic cylinder $z = 1 - x^2$ (Figure 16.2.9).

Solution. It follows from (7) that the area A of the surface can be expressed as the line integral

$$A = \int_C (1 - x^2)\, ds \tag{15}$$

where C is the circle $x^2 + y^2 = 1$. This circle can be parametrized in terms of arc length as

$$x = \cos s, \quad y = \sin s \qquad (0 \le s \le 2\pi)$$

Thus, it follows from (13) and (15) that

$$\begin{aligned} A &= \int_C (1 - x^2)\, ds = \int_0^{2\pi} (1 - \cos^2 s)\, ds \\ &= \int_0^{2\pi} \sin^2 s\, ds = \frac{1}{2}\int_0^{2\pi} (1 - \cos 2s)\, ds = \pi \end{aligned}$$ ◀

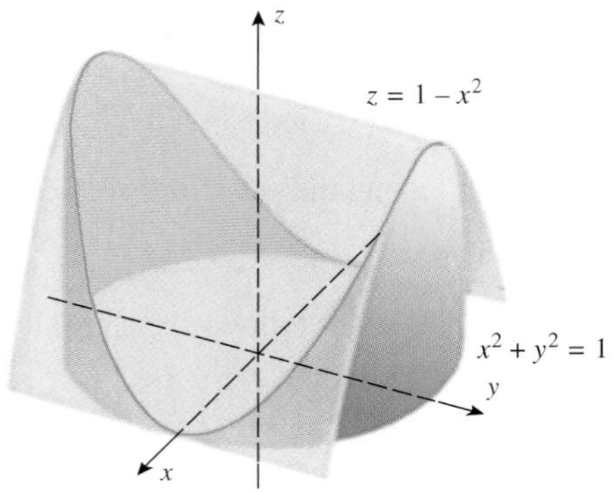

Figure 16.2.9

LINE INTEGRALS WITH RESPECT TO *x*, *y*, AND *z*

We now describe a second type of line integral in which we replace the "ds" in the integral by dx, dy, or dz. For example, suppose that f is a function defined on a smooth curve C in the xy-plane and that partition points of C are denoted by $P_k(x_k, y_k)$. Letting

$$\Delta x_k = x_k - x_{k-1} \quad \text{and} \quad \Delta y_k = y_k - y_{k-1}$$

we would like to define

$$\int_C f(x, y)\, dx = \lim_{\max \Delta s_k \to 0} \sum_{k=1}^{n} f(x_k^*, y_k^*)\Delta x_k \tag{16}$$

$$\int_C f(x, y)\, dy = \lim_{\max \Delta s_k \to 0} \sum_{k=1}^{n} f(x_k^*, y_k^*)\Delta y_k \tag{17}$$

However, unlike Δs_k, the values of Δx_k and Δy_k change sign if the order of the partition points along C is reversed. Therefore, in order to define the line integrals using Formulas (16) and (17), we must restrict ourselves to *oriented* curves C and to partitions of C in which the partition points are ordered in the direction of the curve. With this restriction, if the limit in (16) exists and does not depend on the choice of partition or sampling points, then we refer to (16) as the ***line integral of f with respect to x along C***. Similarly, (17) defines the ***line integral of f with respect to y along C***. If C is a smooth curve in 3-space, we can have ***line integrals of f with respect to x, y, and z along C***. For example,

> Explain why Formula (16) implies that $\int_C dx = x_1 - x_0$, where x_1 and x_0 are the respective x-coordinates of the final and initial points of C. What about $\int_C dy$?

$$\int_C f(x, y, z)\, dx = \lim_{\max \Delta s_k \to 0} \sum_{k=1}^{n} f(x_k^*, y_k^*, z_k^*)\Delta x_k$$

and so forth. As was the case with line integrals with respect to s, line integrals of f with respect to x, y, and z exist if f is continuous on C.

> Explain why Formula (16) implies that $\int_C f(x, y)\, dx = 0$ on any oriented segment parallel to the y-axis. What can you say about $\int_C f(x, y)\, dy$ on any oriented segment parallel to the x-axis?

The basic procedure for evaluating these line integrals is to find parametric equations for C, say

$$x = x(t), \quad y = y(t), \quad z = z(t) \qquad (a \le t \le b)$$

in which the orientation of C is in the direction of increasing t, and then express the integrand in terms of t. For example,

$$\int_C f(x, y, z)\, dz = \int_a^b f(x(t), y(t), z(t))z'(t)\, dt$$

[Such a formula is easy to remember—just substitute for x, y, and z using the parametric equations and recall that $dz = z'(t)\,dt$.]

▶ **Example 5** Evaluate $\int_C 3xy\,dx$, where C is the line segment joining $(0, 0)$ and $(1, 2)$ with the given orientation.

(a) Oriented from $(0, 0)$ to $(1, 2)$ as in Figure 16.2.6a.

(b) Oriented from $(1, 2)$ to $(0, 0)$ as in Figure 16.2.6b.

Solution (a). Using the parametrization

$$x = t, \quad y = 2t \qquad (0 \le t \le 1)$$

we have

$$\int_C 3xy\,dy = \int_0^1 3(t)(2t)(2)\,dt = \int_0^1 12t^2\,dt = 4t^3\Big]_0^1 = 4$$

Solution (b). Using the parametrization

$$x = 1 - t, \quad y = 2 - 2t \qquad (0 \le t \le 1)$$

we have

$$\int_C 3xy\,dy = \int_0^1 3(1-t)(2-2t)(-2)\,dt = \int_0^1 -12(1-t)^2\,dt = 4(1-t)^3\Big]_0^1 = -4$$ ◀

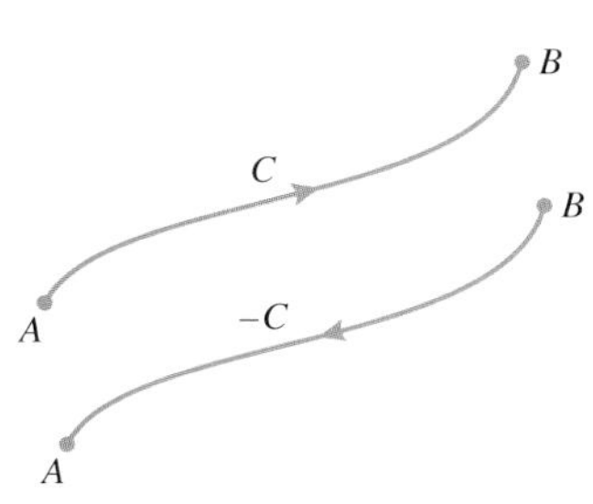

Figure 16.2.10

In Example 5, note that reversing the orientation of the curve changed the sign of the line integral. This is because reversing the orientation of a curve changes the sign of Δx_k in definition (16). Thus, unlike line integrals of functions with respect to s along C, reversing the orientation of C changes the sign of a line integral with respect to x, y, and z. If C is a smooth oriented curve, we will let $-C$ denote the oriented curve consisting of the same points as C but with the opposite orientation (Figure 16.2.10). We then have

$$\int_{-C} f(x, y)\,dx = -\int_C f(x, y)\,dx \quad \text{and} \quad \int_{-C} g(x, y)\,dy = -\int_C g(x, y)\,dy \tag{18–19}$$

while

$$\int_{-C} f(x, y)\,ds = \int_C f(x, y)\,ds \tag{20}$$

and similarly for line integrals in 3-space. Unless indicated otherwise, we will assume that parametric curves are oriented in the direction of increasing parameter.

Frequently, the line integrals with respect to x and y occur in combination, in which case we will dispense with one of the integral signs and write

$$\int_C f(x, y)\,dx + g(x, y)\,dy = \int_C f(x, y)\,dx + \int_C g(x, y)\,dy \tag{21}$$

We will use a similar convention for combinations of line integrals with respect to x, y, and z along curves in 3-space.

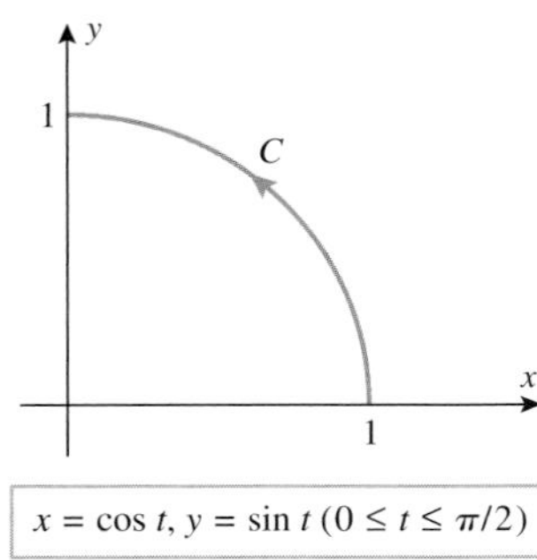

$x = \cos t,\ y = \sin t\ (0 \le t \le \pi/2)$

Figure 16.2.11

▶ **Example 6** Evaluate

$$\int_C 2xy\,dx + (x^2 + y^2)\,dy$$

along the circular arc C given by $x = \cos t$, $y = \sin t$ $(0 \le t \le \pi/2)$ (Figure 16.2.11).

Solution. We have

$$\begin{aligned}\int_C 2xy\,dx &= \int_0^{\pi/2} (2\cos t \sin t)\left[\frac{d}{dt}(\cos t)\right] dt \\ &= -2\int_0^{\pi/2} \sin^2 t \cos t\,dt = -\frac{2}{3}\sin^3 t\Big]_0^{\pi/2} = -\frac{2}{3}\end{aligned}$$

$$\begin{aligned}\int_C (x^2 + y^2)\,dy &= \int_0^{\pi/2} (\cos^2 t + \sin^2 t)\left[\frac{d}{dt}(\sin t)\right] dt \\ &= \int_0^{\pi/2} \cos t\,dt = \sin t\Big]_0^{\pi/2} = 1\end{aligned}$$

Thus, from (21)

$$\begin{aligned}\int_C 2xy\,dx + (x^2 + y^2)\,dy &= \int_C 2xy\,dx + \int_C (x^2 + y^2)\,dy \\ &= -\frac{2}{3} + 1 = \frac{1}{3}\end{aligned}$$ ◀

It can be shown that if f and g are continuous functions on C, then combinations of line integrals with respect to x and y can be expressed in terms of a limit and can be evaluated together in a single step. For example, we have

$$\int_C f(x, y)\,dx + g(x, y)\,dy = \lim_{\max \Delta s_k \to 0} \sum_{k=1}^{n} [f(x_k^*, y_k^*)\Delta x_k + g(x_k^*, y_k^*)\,\Delta y_k] \tag{22}$$

and

$$\int_C f(x, y)\,dx + g(x, y)\,dy = \int_a^b [f(x(t), y(t))x'(t) + g(x(t), y(t))y'(t)]\,dt \tag{23}$$

Similar results hold for line integrals in 3-space. The evaluation of a line integral can sometimes be simplified by using Formula (23).

▶ **Example 7** Evaluate

$$\int_C (3x^2 + y^2)\,dx + 2xy\,dy$$

along the circular arc C given by $x = \cos t$, $y = \sin t$ $(0 \le t \le \pi/2)$ (Figure 16.2.11).

Solution. From (23) we have

$$\begin{aligned}\int_C (3x^2 + y^2)\,dx + 2xy\,dy &= \int_0^{\pi/2} [(3\cos^2 t + \sin^2 t)(-\sin t) + 2(\cos t)(\sin t)(\cos t)]\,dt \\ &= \int_0^{\pi/2} (-3\cos^2 t \sin t - \sin^3 t + 2\cos^2 t \sin t)\,dt \\ &= \int_0^{\pi/2} (-\cos^2 t - \sin^2 t)(\sin t)\,dt = \int_0^{\pi/2} -\sin t\,dt \\ &= \cos t\Big]_0^{\pi/2} = -1\end{aligned}$$ ◀

Compare the computations in Example 7 with those involved in computing

$$\int_C (3x^2 + y^2)\,dx + \int_C 2xy\,dy$$

It follows from (18) and (19) that

$$\int_{-C} f(x, y)\,dx + g(x, y)\,dy = -\int_{C} f(x, y)\,dx + g(x, y)\,dy \tag{24}$$

so that reversing the orientation of C changes the sign of a line integral in which x and y occur in combination. Similarly,

$$\begin{aligned}\int_{-C} & f(x, y, z)\,dx + g(x, y, z)\,dy + h(x, y, z)\,dz \\ &= -\int_{C} f(x, y, z)\,dx + g(x, y, z)\,dy + h(x, y, z)\,dz\end{aligned} \tag{25}$$

INTEGRATING A VECTOR FIELD ALONG A CURVE

There is an alternative notation for line integrals with respect to x, y, and z that is particularly appropriate for dealing with problems involving vector fields. We will interpret $d\mathbf{r}$ as

$$d\mathbf{r} = dx\mathbf{i} + dy\mathbf{j} \quad \text{or} \quad d\mathbf{r} = dx\mathbf{i} + dy\mathbf{j} + dz\mathbf{k}$$

depending on whether C is in 2-space or 3-space. For an oriented curve C in 2-space and a vector field

$$\mathbf{F}(x, y) = f(x, y)\mathbf{i} + g(x, y)\mathbf{j}$$

we will write

$$\int_C \mathbf{F}\cdot d\mathbf{r} = \int_C (f(x, y)\mathbf{i} + g(x, y)\mathbf{j})\cdot(dx\mathbf{i} + dy\mathbf{j}) = \int_C f(x, y)\,dx + g(x, y)\,dy \tag{26}$$

Similarily, for a curve C in 3-space and vector field

$$\mathbf{F}(x, y, z) = f(x, y, z)\mathbf{i} + g(x, y, z)\mathbf{j} + h(x, y, z)\mathbf{k}$$

we will write

$$\begin{aligned}\int_C \mathbf{F}\cdot d\mathbf{r} &= \int_C (f(x, y, z)\mathbf{i} + g(x, y, z)\mathbf{j} + h(x, y, z)\mathbf{k})\cdot(dx\mathbf{i} + dy\mathbf{j} + dz\mathbf{k}) \\ &= \int_C f(x, y, z)\,dx + g(x, y, z)\,dy + h(x, y, z)\,dz\end{aligned} \tag{27}$$

With these conventions, we are led to the following definition.

16.2.2 DEFINITION. If $\mathbf{F}$ is a continuous vector field and C is a smooth oriented curve, then the ***line integral of* F *along C*** is

$$\int_C \mathbf{F}\cdot d\mathbf{r} \tag{28}$$

The notation in Definition 16.2.2 makes it easy to remember the formula for evaluating the line integral of $\mathbf{F}$ along C. For example, suppose that C is an oriented curve in the plane given in vector form by

$$\mathbf{r} = \mathbf{r}(t) = x(t)\mathbf{i} + y(t)\mathbf{j} \qquad (a \le t \le b)$$

If we write

$$\mathbf{F}(\mathbf{r}(t)) = f(x(t), y(t))\mathbf{i} + g(x(t), y(t))\mathbf{j}$$

then

$$\int_C \mathbf{F} \cdot d\mathbf{r} = \int_a^b \mathbf{F}(\mathbf{r}(t)) \cdot \mathbf{r}'(t)\, dt \tag{29}$$

Formula (29) is also valid for oriented curves in 3-space.

Vectors not to scale

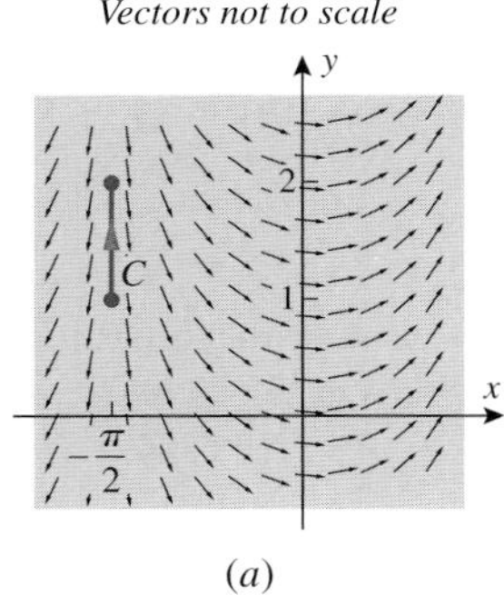

(a)

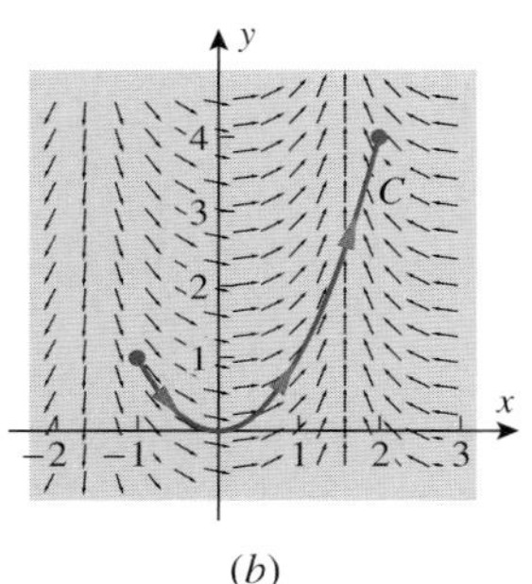

(b)

Figure 16.2.12

▶ **Example 8** Evaluate $\int_C \mathbf{F} \cdot d\mathbf{r}$ where $\mathbf{F}(x, y) = \cos x\mathbf{i} + \sin x\mathbf{j}$ and where C is the given oriented curve.

(a) $C : \mathbf{r}(t) = -\dfrac{\pi}{2}\mathbf{i} + t\mathbf{j} \quad (1 \le t \le 2)$ (see Figure 16.2.12*a*)

(b) $C : \mathbf{r}(t) = t\mathbf{i} + t^2\mathbf{j} \quad (-1 \le t \le 2)$ (see Figure 16.2.12*b*)

Solution (a). Using (27) we have

$$\int_C \mathbf{F} \cdot d\mathbf{r} = \int_1^2 \mathbf{F}(\mathbf{r}(t)) \cdot \mathbf{r}'(t)\, dt = \int_1^2 (-\mathbf{j}) \cdot \mathbf{j}\, dt = \int_1^2 (-1)\, dt = -1$$

Solution (b). Using (27) we have

$$\begin{aligned}\int_C \mathbf{F} \cdot d\mathbf{r} &= \int_{-1}^2 \mathbf{F}(\mathbf{r}(t)) \cdot \mathbf{r}'(t)\, dt = \int_{-1}^2 (\cos t\mathbf{i} + \sin t\mathbf{j}) \cdot (\mathbf{i} + 2t\mathbf{j})\, dt \\ &= \int_{-1}^2 (\cos t + 2t \sin t)\, dt = -2t \cos t + 3 \sin t\Big]_{-1}^2 \\ &= -2\cos 1 - 4\cos 2 + 3(\sin 1 + \sin 2) \approx 5.83629\end{aligned}$$ ◀

If we let t denote an arc length parameter, say $t = s$, with $\mathbf{T} = \mathbf{r}'(s)$ the unit tangent vector field along C, then

$$\int_C \mathbf{F} \cdot d\mathbf{r} = \int_a^b \mathbf{F}(\mathbf{r}(s)) \cdot \mathbf{r}'(s)\, ds = \int_a^b \mathbf{F}(\mathbf{r}(s)) \cdot \mathbf{T}\, ds = \int_C \mathbf{F} \cdot \mathbf{T}\, ds$$

which shows that

$$\int_C \mathbf{F} \cdot d\mathbf{r} = \int_C \mathbf{F} \cdot \mathbf{T}\, ds \tag{30}$$

In words, the integral of a vector field along a curve has the same value as the integral of the tangential component of the vector field along the curve.

We can use (30) to interpret $\int_C \mathbf{F} \cdot d\mathbf{r}$ geometrically. If θ is the angle between $\mathbf{F}$ and $\mathbf{T}$ at a point on C, then at this point

$$\begin{aligned}\mathbf{F} \cdot \mathbf{T} &= \|\mathbf{F}\|\|\mathbf{T}\| \cos\theta && \text{Formula (4) in Section 12.3} \\ &= \|\mathbf{F}\| \cos\theta && \text{Since } \|\mathbf{T}\| = 1\end{aligned}$$

Thus,

$$-\|\mathbf{F}\| \le \mathbf{F} \cdot \mathbf{T} \le \|\mathbf{F}\|$$

and if $\mathbf{F} \ne \mathbf{0}$, then the sign of $\mathbf{F} \cdot \mathbf{T}$ will depend on the angle between the direction of $\mathbf{F}$ and the direction of C (Figure 16.2.13). That is, $\mathbf{F} \cdot \mathbf{T}$ will be positive where $\mathbf{F}$ has the same general direction as C, it will be 0 if $\mathbf{F}$ is normal to C, and it will be negative where $\mathbf{F}$ and C have more or less opposite directions. The line integral of $\mathbf{F}$ along C can be interpreted as the accumulated effect of the magnitude of **F** along C, the extent to which **F** and C have the same direction, and the arc length of C.

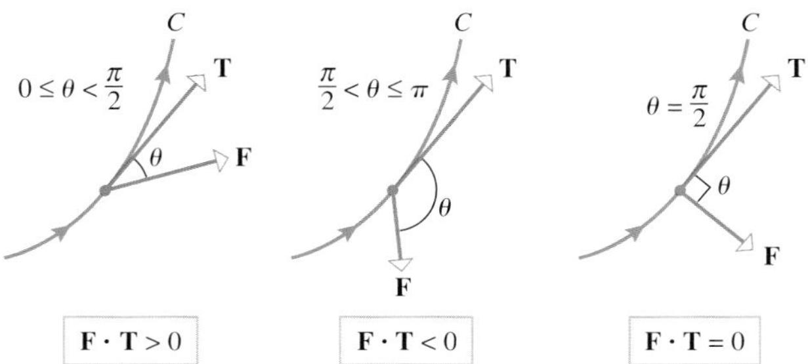

Figure 16.2.13

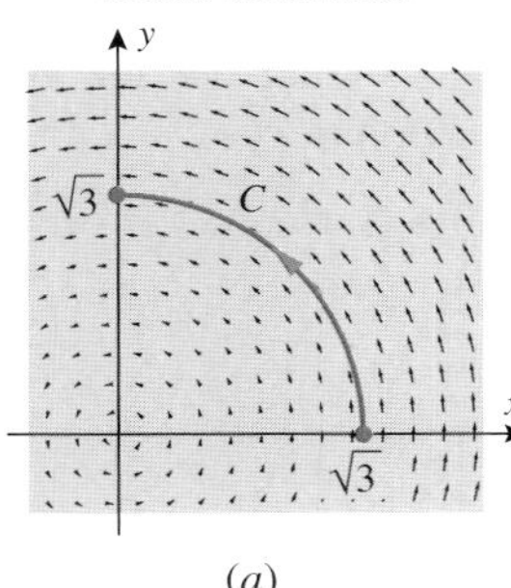

(a)

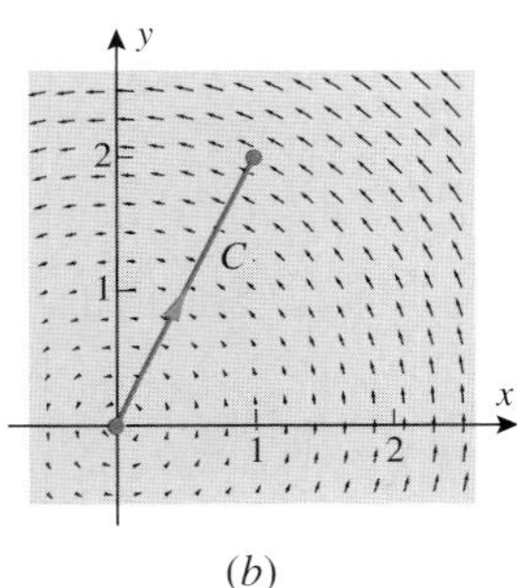

(b)

Figure 16.2.14

▶ **Example 9** Use (30) to evaluate $\int_C \mathbf{F}\cdot d\mathbf{r}$ where $\mathbf{F}(x, y) = -y\mathbf{i} + x\mathbf{j}$ and where C is the given oriented curve.

(a) $C : x^2 + y^2 = 3 \quad (0 \le x, y$; oriented as in Figure 16.2.14a)

(b) $C : \mathbf{r}(t) = t\mathbf{i} + 2t\mathbf{j} \quad (0 \le t \le 1$; see Figure 16.2.14$b$)

Solution (a). At every point on C the direction of **F** and the direction of C are the same. (Why?) In addition, at every point on C

$$\|\mathbf{F}\| = \sqrt{(-y)^2 + x^2} = \sqrt{x^2 + y^2} = \sqrt{3}$$

Therefore, $\mathbf{F}\cdot\mathbf{T} = \|\mathbf{F}\|\cos(0) = \|\mathbf{F}\| = \sqrt{3}$, and

$$\int_C \mathbf{F}\cdot d\mathbf{r} = \int_C \mathbf{F}\cdot\mathbf{T}\,ds = \int_C \sqrt{3}\,ds = \sqrt{3}\int_C ds = \frac{3\pi}{2}$$

Solution (b). The vector field **F** is normal to C at every point. (Why?) Therefore,

$$\int_C \mathbf{F}\cdot d\mathbf{r} = \int_C \mathbf{F}\cdot\mathbf{T}\,ds = \int_C 0\,ds = 0 \blacktriangleleft$$

Refer to Figure 16.2.12 and explain the sign of each line integral in Example 8 geometrically. Exercises 5 and 6 take this geometric analysis further.

In light of (20) and (30), you might expect that reversing the orientation of C in $\int_C \mathbf{F}\cdot d\mathbf{r}$ would have no effect on the value of the line integral. However, reversing the orientation of C reverses the orientation of **T** in the integrand and hence reverses the sign of the integral; that is,

$$\int_{-C} \mathbf{F}\cdot\mathbf{T}\,ds = -\int_C \mathbf{F}\cdot\mathbf{T}\,ds \tag{31}$$

$$\int_{-C} \mathbf{F}\cdot d\mathbf{r} = -\int_C \mathbf{F}\cdot d\mathbf{r} \tag{32}$$

WORK AS A LINE INTEGRAL

An important application of line integrals with respect to x, y, and z is to the problem of defining the work performed by a variable force moving a particle along a curved path. In Section 6.7 we defined the work W performed by a force of constant magnitude acting on an object in the direction of motion (Definition 6.7.1), and later in that section we extended the definition to allow for a force of variable magnitude acting in the direction of motion (Definition 6.7.3). In Section 12.3 we took the concept of work a step further by defining the work W performed by a constant force $\mathbf{F}$ moving a particle in a straight line from point P to point Q. We defined the work to be

$$W = \mathbf{F} \cdot \overrightarrow{PQ} \tag{33}$$

[Formula (14) in Section 12.3]. Our next goal is to define a more general concept of work —the work performed by a variable force acting on a particle that moves along a curved path in 2-space or 3-space.

In many applications variable forces arise from force fields (gravitational fields, electromagnetic fields, and so forth), so we will consider the problem of work in that context. To motivate an appropriate definition for work performed by a force field, we will use a limit process, and since the procedure is the same for 2-space and 3-space, we will discuss it in detail for 2-space only. The idea is as follows:

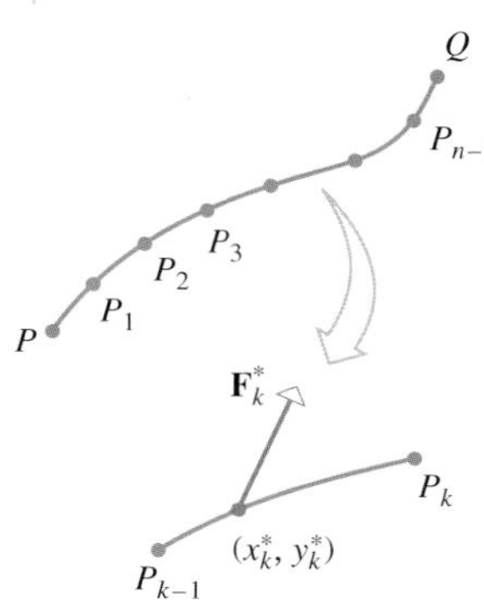

Figure 16.2.15

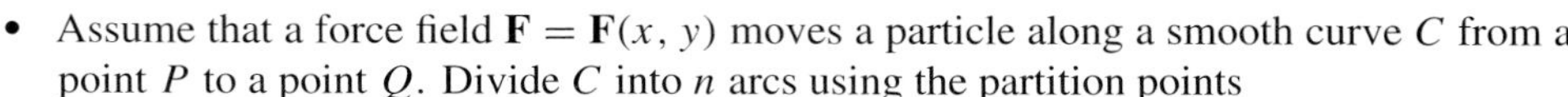

- Assume that a force field $\mathbf{F} = \mathbf{F}(x, y)$ moves a particle along a smooth curve C from a point P to a point Q. Divide C into n arcs using the partition points

$$P = P_0(x_0, y_0),\ P_1(x_1, y_1),\ P_2(x_2, y_2), \ldots,\ P_{n-1}(x_{n-1}, y_{n-1}),\ P_n(x_n, y_n) = Q$$

  directed along C from P to Q, and denote the length of the kth arc by Δs_k. Let (x_k^*, y_k^*) be any point on the kth arc, and let

$$\mathbf{F}_k^* = \mathbf{F}(x_k^*, y_k^*) = f(x_k^*, y_k^*)\mathbf{i} + g(x_k^*, y_k^*)\mathbf{j}$$

  be the force vector at this point (Figure 16.2.15).

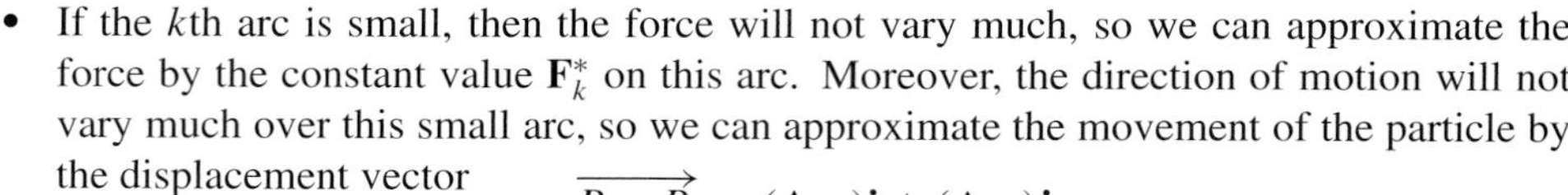

- If the kth arc is small, then the force will not vary much, so we can approximate the force by the constant value $\mathbf{F}_k^*$ on this arc. Moreover, the direction of motion will not vary much over this small arc, so we can approximate the movement of the particle by the displacement vector

$$\overrightarrow{P_{k-1}P_k} = (\Delta x_k)\mathbf{i} + (\Delta y_k)\mathbf{j}$$

  where $\Delta x_k = x_k - x_{k-1}$ and $\Delta y_k = y_k - y_{k-1}$.

- Since the work done by a constant force $\mathbf{F}_k^*$ moving a particle along a straight line from P_{k-1} to P_k is

$$\begin{aligned}\mathbf{F}_k^* \cdot \overrightarrow{P_{k-1}P_k} &= (f(x_k^*, y_k^*)\mathbf{i} + g(x_k^*, y_k^*)\mathbf{j}) \cdot ((\Delta x_k)\mathbf{i} + (\Delta y_k)\mathbf{j}) \\ &= f(x_k^*, y_k^*)\Delta x_k + g(x_k^*, y_k^*)\Delta y_k\end{aligned}$$

  [Formula (33)], the work ΔW_k performed by the force field along the kth arc of C can be approximated by

$$\Delta W_k \approx f(x_k^*, y_k^*)\Delta x_k + g(x_k^*, y_k^*)\Delta y_k$$

  The total work W performed by the force moving the particle over the entire curve C can then be approximated as

$$W = \sum_{k=1}^{n} \Delta W_k \approx \sum_{k=1}^{n} [f(x_k^*, y_k^*)\Delta x_k + g(x_k^*, y_k^*)\Delta y_k]$$

- As max $\Delta s_k \to 0$, it is plausible that the error in this approximation approaches 0 and the exact work performed by the force field is

$$\begin{aligned}W &= \lim_{\max \Delta s_k \to 0} \sum_{k=1}^{n} [f(x_k^*, y_k^*)\Delta x_k + g(x_k^*, y_k^*)\Delta y_k] \\ &= \int_C f(x, y)\,dx + g(x, y)\,dy \qquad \text{Formula (22)} \\ &= \int_C \mathbf{F} \cdot d\mathbf{r} \qquad \text{Formula (26)}\end{aligned}$$

  Thus, we are led to the following definition.

Note from Formula (30) that the work performed by a force field on a particle moving along a smooth curve C is obtained by integrating the scalar tangential component of force along C. This implies that the component of force orthogonal to the direction of motion of the particle has no effect on the work done.

16.2.3 DEFINITION. Suppose that under the influence of a continuous force field $\mathbf{F}$ a particle moves along a smooth curve C and that C is oriented in the direction of motion of the particle. Then the ***work performed by the force field*** on the particle is

$$\int_C \mathbf{F} \cdot d\mathbf{r} \tag{34}$$

For example suppose that force is measured in pounds and distance is measured in feet. It follows from part (a) of Example 9 that the work done by a force $\mathbf{F}(x, y) = -y\mathbf{i} + x\mathbf{j}$ acting on a particle moving along the circle $x^2 + y^2 = 3$ from $(\sqrt{3}, 0)$ to $(0, \sqrt{3})$ is $3\pi/2$ foot-pounds.

LINE INTEGRALS ALONG PIECEWISE SMOOTH CURVES

Thus far, we have only considered line integrals along smooth curves. However, the notion of a line integral can be extended to curves formed from finitely many smooth curves $C_1, C_2, \ldots, C_n$ joined end to end. Such a curve is called ***piecewise smooth*** (Figure 16.2.16). We define a line integral along a piecewise smooth curve C to be the sum of the integrals along the sections:

$$\int_C = \int_{C_1} + \int_{C_2} + \cdots + \int_{C_n}$$

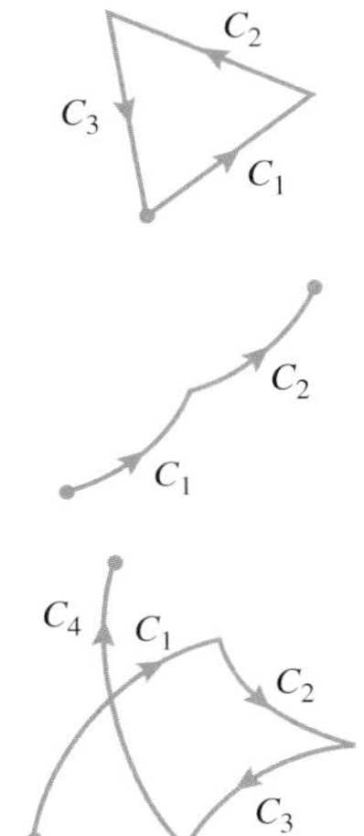

Figure 16.2.16

Example 10 Evaluate

$$\int_C x^2 y\, dx + x\, dy$$

in a counterclockwise direction around the triangular path shown in Figure 16.2.17.

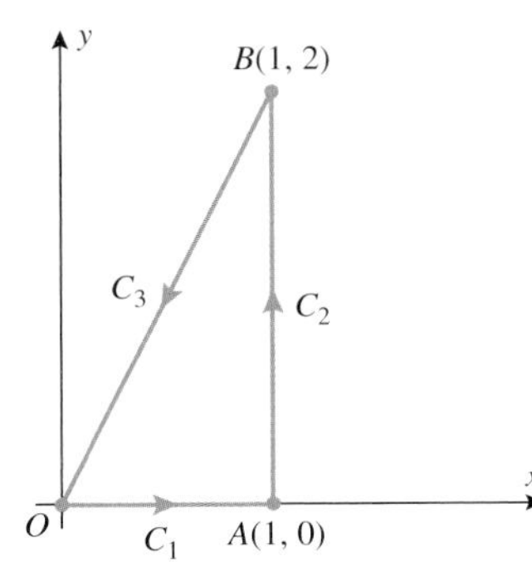

Figure 16.2.17

Solution. We will integrate over C_1, C_2, and C_3 separately and add the results. For each of the three integrals we must find parametric equations that trace the path of integration in the correct direction. For this purpose recall from Formula (7) of Section 13.1 that the graph of the vector-valued function

$$\mathbf{r}(t) = (1 - t)\mathbf{r}_0 + t\mathbf{r}_1 \qquad (0 \le t \le 1)$$

is the line segment joining $\mathbf{r}_0$ and $\mathbf{r}_1$, oriented in the direction from $\mathbf{r}_0$ to $\mathbf{r}_1$. Thus, the line segments C_1, C_2, and C_3 can be represented in vector notation as

$$\begin{aligned}
C_1&: \mathbf{r}(t) = (1-t)\langle 0, 0\rangle + t\langle 1, 0\rangle = \langle t, 0\rangle \\
C_2&: \mathbf{r}(t) = (1-t)\langle 1, 0\rangle + t\langle 1, 2\rangle = \langle 1, 2t\rangle \\
C_3&: \mathbf{r}(t) = (1-t)\langle 1, 2\rangle + t\langle 0, 0\rangle = \langle 1-t, 2-2t\rangle
\end{aligned}$$

where t varies from 0 to 1 in each case. From these equations we obtain

$$\begin{aligned}
\int_{C_1} x^2 y\, dx + x\, dy &= \int_{C_1} x^2 y\, dx = \int_0^1 (t^2)(0)\frac{d}{dt}[t]\, dt = 0 \\
\int_{C_2} x^2 y\, dx + x\, dy &= \int_{C_2} x\, dy = \int_0^1 (1)\frac{d}{dt}[2t]\, dt = 2 \\
\int_{C_3} x^2 y\, dx + x\, dy &= \int_0^1 (1-t)^2(2-2t)\frac{d}{dt}[1-t]\, dt + \int_0^1 (1-t)\frac{d}{dt}[2-2t]\, dt \\
&= 2\int_0^1 (t-1)^3\, dt + 2\int_0^1 (t-1)\, dt = -\tfrac{1}{2} - 1 = -\tfrac{3}{2}
\end{aligned}$$

Thus,

$$\int_C x^2 y\, dx + x\, dy = 0 + 2 + \left(-\tfrac{3}{2}\right) = \tfrac{1}{2} \blacktriangleleft$$

✔ QUICK CHECK EXERCISES 16.2 *(See page 1128 for answers.)*

1. The area of the surface extending upward from the line segment $y = x$ $(0 \le x \le 1)$ in the xy-plane to the plane $z = 2x + 1$ is ______.

2. Suppose that a wire has equation $y = 1 - x$ $(0 \le x \le 1)$ and that its mass density is $\delta(x, y) = 2 - x$. The mass of the wire is ______.

3. If C is the curve represented by the equations

$$x = \sin t, \quad y = \cos t, \quad z = t \qquad (0 \le t \le 2\pi)$$

then $\int_C y\,dx - x\,dy + dz =$ ______.

4. If C is the unit circle $x^2 + y^2 = 1$ oriented counterclockwise and $\mathbf{F}(x, y) = x\mathbf{i} + y\mathbf{j}$, then

$$\int_C \mathbf{F} \cdot d\mathbf{r} = \text{______}$$

EXERCISE SET 16.2  Graphing Utility C CAS

FOCUS ON CONCEPTS

1. Let C be the line segment from $(0, 0)$ to $(0, 1)$. In each part, evaluate the line integral along C by inspection, and explain your reasoning.

(a) $\int_C ds$ (b) $\int_C \sin xy\,dy$

2. Let C be the line segment from $(0, 2)$ to $(0, 4)$. In each part, evaluate the line integral along C by inspection, and explain your reasoning.

(a) $\int_C ds$ (b) $\int_C e^{xy}\,dx$

3–4 Evaluate $\int_C \mathbf{F} \cdot d\mathbf{r}$ by inspection for the force field $\mathbf{F}(x, y) = \mathbf{i} + \mathbf{j}$ and the curve C shown in the figure. Explain your reasoning. [For clarity, the vectors in the force field are shown at less than true scale.]

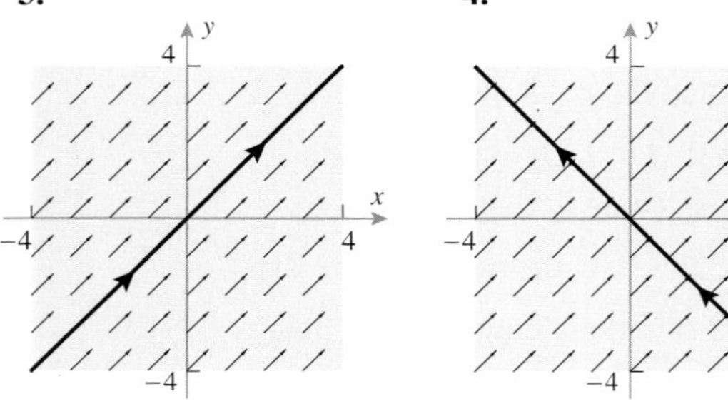

5. Use (30) to explain why the line integral in part (a) of Example 8 can be found by multiplying the length of the line segment C by -1.

6. (a) Use (30) to explain why the line integral in part (b) of Example 8 should be close to, but somewhat less than, the length of the parabolic curve C.

(b) Verify the conclusion in part (a) of this exercise by computing the length of C and comparing the length with the value of the line integral.

7. Let C be the curve represented by the equations

$$x = 2t, \quad y = 3t^2 \qquad (0 \le t \le 1)$$

In each part, evaluate the line integral along C.

(a) $\int_C (x - y)\,ds$ (b) $\int_C (x - y)\,dx$

(c) $\int_C (x - y)\,dy$

8. Let C be the curve represented by the equations

$$x = t, \quad y = 3t^2, \quad z = 6t^3 \qquad (0 \le t \le 1)$$

In each part, evaluate the line integral along C.

(a) $\int_C xyz^2\,ds$ (b) $\int_C xyz^2\,dx$

(c) $\int_C xyz^2\,dy$ (d) $\int_C xyz^2\,dz$

9. In each part, evaluate the integral

$$\int_C (3x + 2y)\,dx + (2x - y)\,dy$$

along the stated curve.

(a) The line segment from $(0, 0)$ to $(1, 1)$.
(b) The parabolic arc $y = x^2$ from $(0, 0)$ to $(1, 1)$.
(c) The curve $y = \sin(\pi x/2)$ from $(0, 0)$ to $(1, 1)$.
(d) The curve $x = y^3$ from $(0, 0)$ to $(1, 1)$.

10. In each part, evaluate the integral

$$\int y\,dx + z\,dy - x\,dz$$

along the stated curve.

(a) The line segment from $(0, 0, 0)$ to $(1, 1, 1)$.
(b) The twisted cubic $x = t$, $y = t^2$, $z = t^3$ from $(0, 0, 0)$ to $(1, 1, 1)$.
(c) The helix $x = \cos \pi t$, $y = \sin \pi t$, $z = t$ from $(1, 0, 0)$ to $(-1, 0, 1)$.

11–14 Evaluate the line integral with respect to s along the curve C.

11. $\displaystyle\int_C \frac{1}{1+x}\,ds$
$C: \mathbf{r}(t) = t\mathbf{i} + \frac{2}{3}t^{3/2}\mathbf{j} \quad (0 \le t \le 3)$

12. $\displaystyle\int_C \frac{x}{1+y^2}\,ds$
$C: x = 1+2t,\ y = t \quad (0 \le t \le 1)$

13. $\displaystyle\int_C 3x^2yz\,ds$
$C: x = t,\ y = t^2,\ z = \frac{2}{3}t^3 \quad (0 \le t \le 1)$

14. $\displaystyle\int_C \frac{e^{-z}}{x^2+y^2}\,ds$
$C: \mathbf{r}(t) = 2\cos t\,\mathbf{i} + 2\sin t\,\mathbf{j} + t\mathbf{k} \quad (0 \le t \le 2\pi)$

15–22 Evaluate the line integral along the curve C.

15. $\displaystyle\int_C (x+2y)\,dx + (x-y)\,dy$
$C: x = 2\cos t,\ y = 4\sin t \quad (0 \le t \le \pi/4)$

16. $\displaystyle\int_C (x^2-y^2)\,dx + x\,dy$
$C: x = t^{2/3},\ y = t \quad (-1 \le t \le 1)$

17. $\displaystyle\int_C -y\,dx + x\,dy$
$C: y^2 = 3x$ from $(3, 3)$ to $(0, 0)$

18. $\displaystyle\int_C (y-x)\,dx + x^2y\,dy$
$C: y^2 = x^3$ from $(1, -1)$ to $(1, 1)$

19. $\displaystyle\int_C (x^2+y^2)\,dx - x\,dy$
$C: x^2+y^2 = 1$, counterclockwise from $(1, 0)$ to $(0, 1)$

20. $\displaystyle\int_C (y-x)\,dx + xy\,dy$
C: the line segment from $(3, 4)$ to $(2, 1)$

21. $\displaystyle\int_C yz\,dx - xz\,dy + xy\,dz$
$C: x = e^t,\ y = e^{3t},\ z = e^{-t} \quad (0 \le t \le 1)$

22. $\displaystyle\int_C x^2\,dx + xy\,dy + z^2\,dz$
$C: x = \sin t,\ y = \cos t,\ z = t^2 \quad (0 \le t \le \pi/2)$

23–24 Use a CAS to evaluate the line integrals along the given curves.

C 23. (a) $\displaystyle\int_C (x^3+y^3)\,ds$
$C: \mathbf{r}(t) = e^t\mathbf{i} + e^{-t}\mathbf{j} \quad (0 \le t \le \ln 2)$

(b) $\displaystyle\int_C xe^z\,dx + (x-z)\,dy + (x^2+y^2+z^2)\,dz$
$C: x = \sin t,\ y = \cos t,\ z = t \quad (0 \le t \le \pi/2)$

C 24. (a) $\displaystyle\int_C x^7y^3\,ds$
$C: x = \cos^3 t,\ y = \sin^3 t \quad (0 \le t \le \pi/2)$

(b) $\displaystyle\int_C x^5z\,dx + 7y\,dy + y^2z\,dz$
$C: \mathbf{r}(t) = t\mathbf{i} + t^2\mathbf{j} + \ln t\,\mathbf{k} \quad (1 \le t \le e)$

25–26 Evaluate $\int_C y\,dx - x\,dy$ along the curve C shown in the figure.

25. (a) (b)
26. (a) (b)

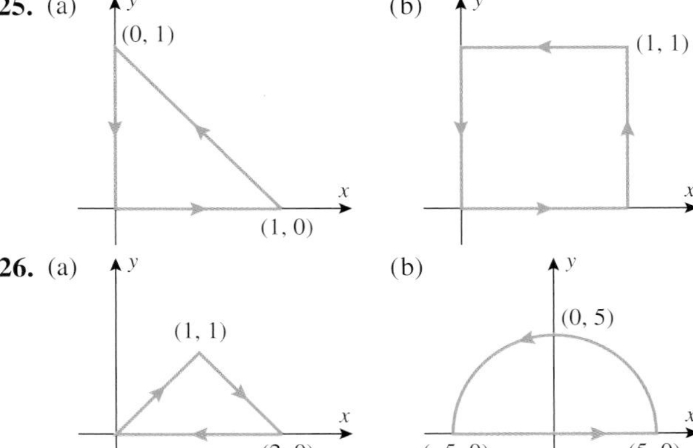

27–28 Evaluate $\int_C x^2z\,dx - yx^2\,dy + 3\,dz$ along the curve C shown in the figure.

27. 28.

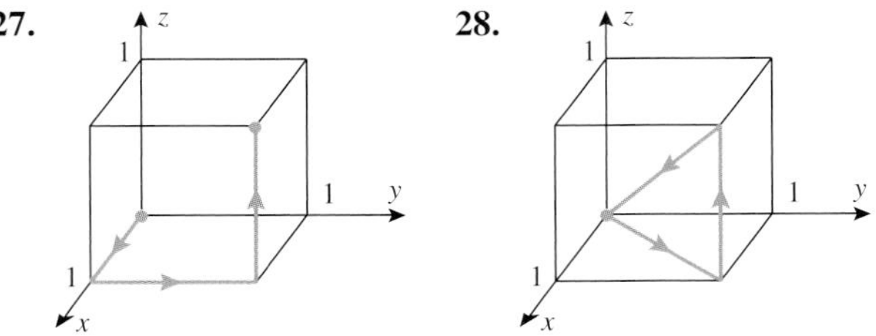

29–32 Evaluate $\int_C \mathbf{F}\cdot d\mathbf{r}$ along the curve C.

29. $\mathbf{F}(x, y) = x^2\mathbf{i} + xy\mathbf{j}$
$C: \mathbf{r}(t) = 2\cos t\,\mathbf{i} + 2\sin t\,\mathbf{j} \quad (0 \le t \le \pi)$

30. $\mathbf{F}(x, y) = x^2y\mathbf{i} + 4\mathbf{j}$
$C: \mathbf{r}(t) = e^t\mathbf{i} + e^{-t}\mathbf{j} \quad (0 \le t \le 1)$

31. $\mathbf{F}(x, y) = (x^2+y^2)^{-3/2}(x\mathbf{i} + y\mathbf{j})$
$C: \mathbf{r}(t) = e^t \sin t\,\mathbf{i} + e^t \cos t\,\mathbf{j} \quad (0 \le t \le 1)$

32. $\mathbf{F}(x, y, z) = z\mathbf{i} + x\mathbf{j} + y\mathbf{k}$
$C: \mathbf{r}(t) = \sin t\,\mathbf{i} + 3\sin t\,\mathbf{j} + \sin^2 t\,\mathbf{k} \quad (0 \le t \le \pi/2)$

33. Find the mass of a thin wire shaped in the form of the circular arc $y = \sqrt{9-x^2}$ $(0 \le x \le 3)$ if the density function is $\delta(x, y) = x\sqrt{y}$.

34. Find the mass of a thin wire shaped in the form of the curve $x = e^t \cos t$, $y = e^t \sin t$ $(0 \le t \le 1)$ if the density function δ is proportional to the distance from the origin.

35. Find the mass of a thin wire shaped in the form of the helix $x = 3\cos t$, $y = 3\sin t$, $z = 4t$ $(0 \le t \le \pi/2)$ if the density function is $\delta = kx/(1+y^2)$ $(k > 0)$.

36. Find the mass of a thin wire shaped in the form of the curve $x = 2t$, $y = \ln t$, $z = 4\sqrt{t}$ $(1 \le t \le 4)$ if the density function is proportional to the distance above the xy-plane.

37–40 Find the work done by the force field $\mathbf{F}$ on a particle that moves along the curve C.

37. $\mathbf{F}(x, y) = xy\mathbf{i} + x^2\mathbf{j}$
C: $x = y^2$ from $(0, 0)$ to $(1, 1)$

38. $\mathbf{F}(x, y) = (x^2 + xy)\mathbf{i} + (y - x^2y)\mathbf{j}$
C: $x = t$, $y = 1/t$ $\quad (1 \le t \le 3)$

39. $\mathbf{F}(x, y, z) = xy\mathbf{i} + yz\mathbf{j} + xz\mathbf{k}$
C: $\mathbf{r}(t) = t\mathbf{i} + t^2\mathbf{j} + t^3\mathbf{k}$ $\quad (0 \le t \le 1)$

40. $\mathbf{F}(x, y, z) = (x + y)\mathbf{i} + xy\mathbf{j} - z^2\mathbf{k}$
C: along line segments from $(0, 0, 0)$ to $(1, 3, 1)$ to $(2, -1, 4)$

41–42 Find the work done by the force field

$$\mathbf{F}(x, y) = \frac{1}{x^2 + y^2}\mathbf{i} + \frac{4}{x^2 + y^2}\mathbf{j}$$

on a particle that moves along the curve C shown in the figure.

41. **42.**

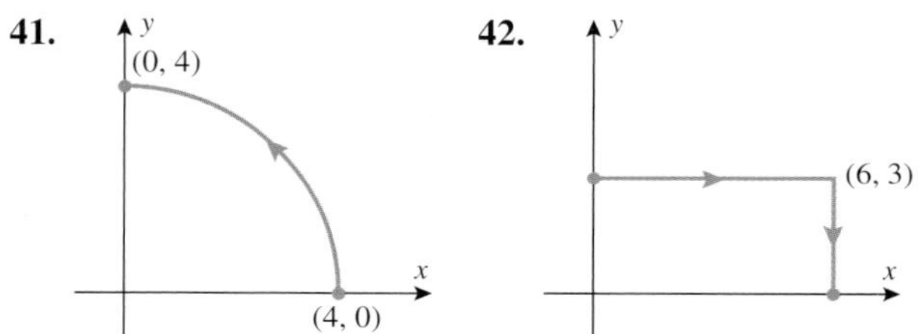

43–44 Use a line integral to find the area of the surface.

43. The surface that extends upward from the parabola $y = x^2$ $(0 \le x \le 2)$ in the xy-plane to the plane $z = 3x$.

44. The surface that extends upward from the semicircle $y = \sqrt{4 - x^2}$ in the xy-plane to the surface $z = x^2y$.

45. As illustrated in the accompanying figure, a sinusoidal cut is made in the top of a cylindrical tin can. Suppose that the base is modeled by the parametric equations $x = \cos t$, $y = \sin t$, $z = 0$ $(0 \le t \le 2\pi)$, and the height of the cut as a function of t is $z = 2 + 0.5 \sin 3t$.

(a) Use a geometric argument to find the lateral surface area of the cut can.
(b) Write down a line integral for the surface area.
(c) Use the line integral to calculate the surface area.

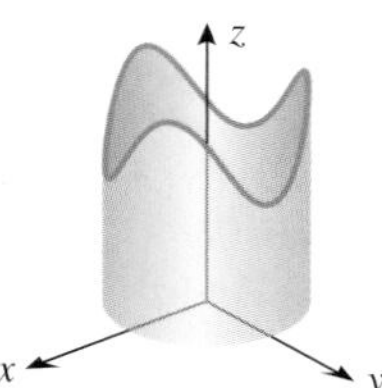

Figure Ex-45

46. Evaluate the integral $\displaystyle\int_{-C} \frac{x\,dy - y\,dx}{x^2 + y^2}$, where C is the circle $x^2 + y^2 = a^2$ traversed counterclockwise.

47. Suppose that a particle moves through the force field $\mathbf{F}(x, y) = xy\mathbf{i} + (x - y)\mathbf{j}$ from the point $(0, 0)$ to the point $(1, 0)$ along the curve $x = t$, $y = \lambda t(1 - t)$. For what value of λ will the work done by the force field be 1?

48. A farmer weighing 150 lb carries a sack of grain weighing 20 lb up a circular helical staircase around a silo of radius 25 ft. As the farmer climbs, grain leaks from the sack at a rate of 1 lb per 10 ft of ascent. How much work is performed by the farmer in climbing through a vertical distance of 60 ft in exactly four revolutions? [*Hint:* Find a vector field that represents the force exerted by the farmer in lifting his own weight plus the weight of the sack upward at each point along his path.]

49. Suppose that a curve C in the xy-plane is smoothly parametrized by

$$\mathbf{r}(t) = x(t)\mathbf{i} + y(t)\mathbf{j} \qquad (a \le t \le b)$$

In each part, refer to the notation used in the derivation of Formula (9).

(a) Let m and M denote the respective minimum and maximum values of $\|\mathbf{r}'(t)\|$ on $[a, b]$. Prove that

$$0 \le m(\max \Delta t_k) \le \max \Delta s_k \le M(\max \Delta t_k)$$

(b) Use part (a) to prove that $\max \Delta s_k \to 0$ if and only if $\max \Delta t_k \to 0$.

✔ QUICK CHECK ANSWERS 16.2

1. $2\sqrt{2}$ 2. $\dfrac{3\sqrt{2}}{2}$ 3. 4π 4. 0

16.3 INDEPENDENCE OF PATH; CONSERVATIVE VECTOR FIELDS

In this section we will show that for certain kinds of vector fields the line integral of **F** *along a curve depends only on the endpoints of the curve and not on the curve itself. Such vector fields are of special importance in physics and engineering.*

WORK INTEGRALS

We saw in the last section that if **F** is a force field in 2-space or 3-space, then the work performed by the field on a particle moving along a parametric curve C from an initial point P to a final point Q is given by the integral

$$\int_C \mathbf{F}\cdot d\mathbf{r} \quad \text{or equivalently} \quad \int_C \mathbf{F}\cdot\mathbf{T}\,ds$$

Accordingly, we call an integral of this type a ***work integral***. Recall that a work integral can also be expressed in scalar form as

$$\int_C \mathbf{F}\cdot d\mathbf{r} = \int_C f(x,y)\,dx + g(x,y)\,dy \qquad \text{2-space} \tag{1}$$

$$\int_C \mathbf{F}\cdot d\mathbf{r} = \int_C f(x,y,z)\,dx + g(x,y,z)\,dy + h(x,y,z)\,dz \qquad \text{3-space} \tag{2}$$

where f, g, and h are the component functions of **F**.

INDEPENDENCE OF PATH

The parametric curve C in a work integral is called the ***path of integration***. One of the important problems in applications is to determine how the path of integration affects the work performed by a force field on a particle that moves from a fixed point P to a fixed point Q. We will show shortly that if the force field **F** is conservative (i.e., is the gradient of some potential function ϕ), then the work that the field performs on a particle that moves from P to Q does not depend on the particular path C that the particle follows. This is illustrated in the following example.

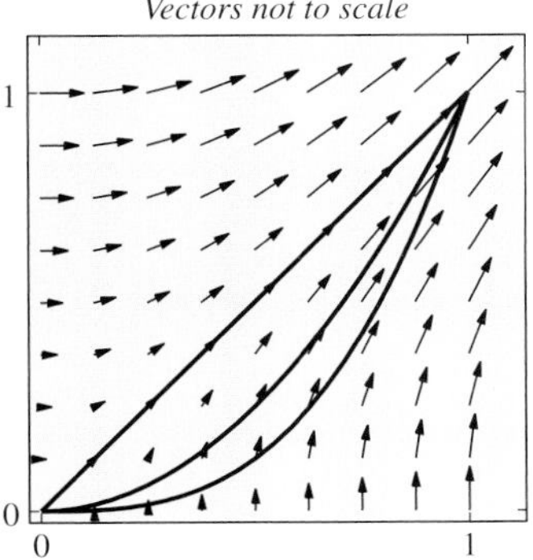

Figure 16.3.1

▶ **Example 1** The force field $\mathbf{F}(x,y) = y\mathbf{i} + x\mathbf{j}$ is conservative since it is the gradient of $\phi(x,y) = xy$ (verify). Thus, the preceding discussion suggests that the work performed by the field on a particle that moves from the point (0, 0) to the point (1, 1) should be the same along different paths. Confirm that the value of the work integral

$$\int_C \mathbf{F}\cdot d\mathbf{r}$$

is the same along the following paths (Figure 16.3.1):

(a) The line segment $y = x$ from (0, 0) to (1, 1).

(b) The parabola $y = x^2$ from (0, 0) to (1, 1).

(c) The cubic $y = x^3$ from (0, 0) to (1, 1).

Solution (a). With $x = t$ as the parameter, the path of integration is given by

$$x = t, \quad y = t \qquad (0 \le t \le 1)$$

Thus,

$$\int_C \mathbf{F}\cdot d\mathbf{r} = \int_C (y\mathbf{i} + x\mathbf{j})\cdot(dx\mathbf{i} + dy\mathbf{j}) = \int_C y\,dx + x\,dy$$

$$= \int_0^1 2t\,dt = 1$$

Solution (b). With $x = t$ as the parameter, the path of integration is given by

$$x = t, \quad y = t^2 \qquad (0 \le t \le 1)$$

Thus,

$$\int_C \mathbf{F} \cdot d\mathbf{r} = \int_C y\,dx + x\,dy = \int_0^1 3t^2\,dt = 1$$

Solution (c). With $x = t$ as the parameter, the path of integration is given by

$$x = t, \quad y = t^3 \qquad (0 \le t \le 1)$$

Thus,

$$\int_C \mathbf{F} \cdot d\mathbf{r} = \int_C y\,dx + x\,dy = \int_0^1 4t^3\,dt = 1 \blacktriangleleft$$

THE FUNDAMENTAL THEOREM OF LINE INTEGRALS

Recall from the Fundamental Theorem of Calculus (Theorem 5.6.1) that if F is an antiderivative of f, then

$$\int_a^b f(x)\,dx = F(b) - F(a)$$

The following result is the analog of that theorem for line integrals in 2-space.

16.3.1 THEOREM (*The Fundamental Theorem of Line Integrals*). *Suppose that*

$$\mathbf{F}(x, y) = f(x, y)\mathbf{i} + g(x, y)\mathbf{j}$$

is a conservative vector field in some open region D containing the points (x_0, y_0) and (x_1, y_1) and that $f(x, y)$ and $g(x, y)$ are continuous in this region. If

$$\mathbf{F}(x, y) = \nabla\phi(x, y)$$

and if C is any piecewise smooth parametric curve that starts at (x_0, y_0), ends at (x_1, y_1), and lies in the region D, then

$$\int_C \mathbf{F}(x, y) \cdot d\mathbf{r} = \phi(x_1, y_1) - \phi(x_0, y_0) \tag{3}$$

or, equivalently,

$$\int_C \nabla\phi \cdot d\mathbf{r} = \phi(x_1, y_1) - \phi(x_0, y_0) \tag{4}$$

The value of

$$\int_C \mathbf{F} \cdot d\mathbf{r} = \int_C \mathbf{F} \cdot \mathbf{T}\,ds$$

depends on the magnitude of $\mathbf{F}$ along C, the alignment of $\mathbf{F}$ with the direction of C at each point, and the length of C. If $\mathbf{F}$ is conservative, these various factors always "balance out" so that the value of $\int_C \mathbf{F} \cdot d\mathbf{r}$ depends only on the initial and final points of C.

PROOF. We will give the proof for a smooth curve C. The proof for a piecewise smooth curve, which is left as an exercise, can be obtained by applying the theorem to each individual smooth piece and adding the results. Suppose that C is given parametrically by $x = x(t)$, $y = y(t)$ $(a \le t \le b)$, so that the initial and final points of the curve are

$$(x_0, y_0) = (x(a), y(a)) \quad \text{and} \quad (x_1, y_1) = (x(b), y(b))$$

Since $\mathbf{F}(x, y) = \nabla\phi$, it follows that

$$\mathbf{F}(x, y) = \frac{\partial\phi}{\partial x}\mathbf{i} + \frac{\partial\phi}{\partial y}\mathbf{j}$$

so

$$\begin{aligned}\int_C \mathbf{F}(x, y) \cdot d\mathbf{r} &= \int_C \frac{\partial \phi}{\partial x}\, dx + \frac{\partial \phi}{\partial y}\, dy = \int_a^b \left[\frac{\partial \phi}{\partial x}\frac{dx}{dt} + \frac{\partial \phi}{\partial y}\frac{dy}{dt}\right] dt \\ &= \int_a^b \frac{d}{dt}[\phi(x(t), y(t))]\, dt = \phi(x(t), y(t))\bigg]_{t=a}^{b} \\ &= \phi(x(b), y(b)) - \phi(x(a), y(a)) \\ &= \phi(x_1, y_1) - \phi(x_0, y_0)\end{aligned}$$

■

Stated informally, this theorem shows that *the value of a line integral along a piecewise smooth path in a conservative vector field is* ***independent of the path***; that is, the value of the integral depends on the endpoints and not on the actual path C. Accordingly, for line integrals along paths in conservative vector fields, it is common to express (3) and (4) as

$$\int_{(x_0, y_0)}^{(x_1, y_1)} \mathbf{F} \cdot d\mathbf{r} = \int_{(x_0, y_0)}^{(x_1, y_1)} \nabla\phi \cdot d\mathbf{r} = \phi(x_1, y_1) - \phi(x_0, y_0) \tag{5}$$

▶ Example 2

(a) Confirm that the force field $\mathbf{F}(x, y) = y\mathbf{i} + x\mathbf{j}$ in Example 1 is conservative by showing that $\mathbf{F}(x, y)$ is the gradient of $\phi(x, y) = xy$.

(b) Use the Fundamental Theorem of Line Integrals to evaluate $\int_{(0,0)}^{(1,1)} \mathbf{F} \cdot d\mathbf{r}$.

If $\mathbf{F}$ is conservative, then you have a choice of methods for evaluating $\int_C \mathbf{F} \cdot d\mathbf{r}$. You can work directly with the curve C, you can replace C with another curve that has the same endpoints as C, or you can apply (3).

Solution (a).

$$\nabla\phi = \frac{\partial \phi}{\partial x}\mathbf{i} + \frac{\partial \phi}{\partial y}\mathbf{j} = y\mathbf{i} + x\mathbf{j}$$

Solution (b). From (5) we obtain

$$\int_{(0,0)}^{(1,1)} \mathbf{F} \cdot d\mathbf{r} = \phi(1, 1) - \phi(0, 0) = 1 - 0 = 1$$

which agrees with the results obtained in Example 1 by integrating from (0, 0) to (1, 1) along specific paths. ◀

LINE INTEGRALS ALONG CLOSED PATHS

Parametric curves that begin and end at the same point play an important role in the study of vector fields, so there is some special terminology associated with them. A parametric curve C that is represented by the vector-valued function $\mathbf{r}(t)$ for $a \le t \le b$ is said to be ***closed*** if the initial point $\mathbf{r}(a)$ and the terminal point $\mathbf{r}(b)$ coincide; that is, $\mathbf{r}(a) = \mathbf{r}(b)$ (Figure 16.3.2).

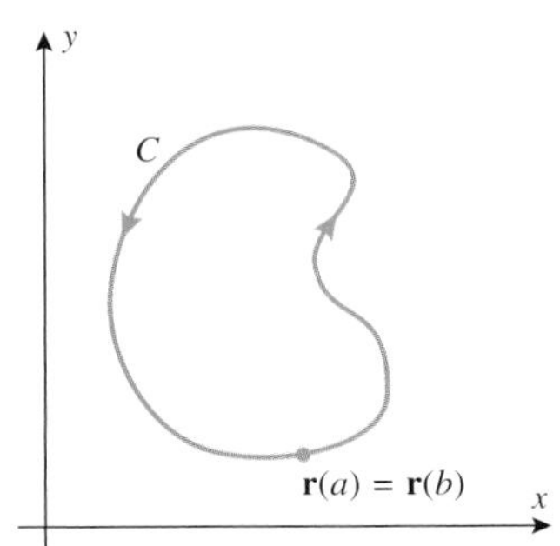

Figure 16.3.2

It follows from (5) that the line integral of a conservative vector field along a closed path C that begins and ends at (x_0, y_0) is zero. This is because the point (x_1, y_1) in (5) is the same as (x_0, y_0) and hence

$$\int_C \mathbf{F} \cdot d\mathbf{r} = \phi(x_1, y_1) - \phi(x_0, y_0) = 0$$

Our next objective is to show that the converse of this result is also true. That is, we want to show that under appropriate conditions a vector field whose line integral is zero along *all* closed paths must be conservative. For this to be true we will need to require that

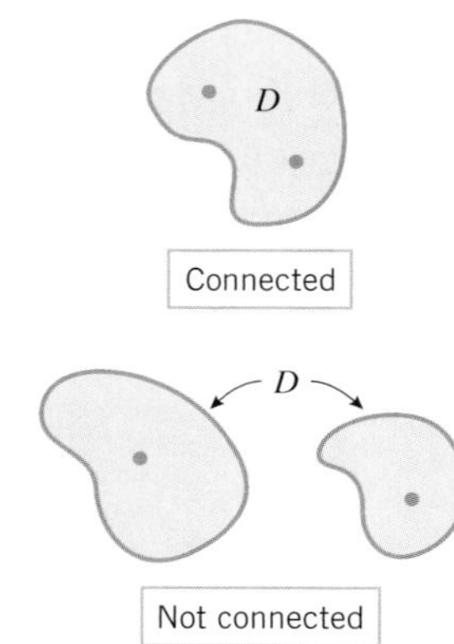

Figure 16.3.3

the domain D of the vector field be ***connected***, by which we mean that any two points in D can be joined by some piecewise smooth curve that lies entirely in D. Stated informally, D is connected if it does not consist of two or more separate pieces (Figure 16.3.3).

16.3.2 THEOREM. *If $f(x, y)$ and $g(x, y)$ are continuous on some open connected region D, then the following statements are equivalent (all true or all false):*

(*a*) $\mathbf{F}(x, y) = f(x, y)\mathbf{i} + g(x, y)\mathbf{j}$ *is a conservative vector field on the region D.*

(*b*) $\int_C \mathbf{F} \cdot d\mathbf{r} = 0$ *for every piecewise smooth closed curve C in D.*

(*c*) $\int_C \mathbf{F} \cdot d\mathbf{r}$ *is independent of the path from any point P in D to any point Q in D for every piecewise smooth curve C in D.*

This theorem can be established by proving three implications: $(a) \Rightarrow (b)$, $(b) \Rightarrow (c)$, and $(c) \Rightarrow (a)$. Since we showed above that $(a) \Rightarrow (b)$, we need only prove the last two implications. We will prove $(c) \Rightarrow (a)$ and leave the other implication as an exercise.

PROOF. $(c) \Rightarrow (a)$. We are assuming that $\int_C \mathbf{F} \cdot d\mathbf{r}$ is independent of the path for every piecewise smooth curve C in the region, and we want to show that there is a function $\phi = \phi(x, y)$ such that $\nabla\phi = \mathbf{F}(x, y)$ at each point of the region; that is,

$$\frac{\partial \phi}{\partial x} = f(x, y) \quad \text{and} \quad \frac{\partial \phi}{\partial y} = g(x, y) \tag{6}$$

Now choose a fixed point (a, b) in D, let (x, y) be any point in D, and define

$$\phi(x, y) = \int_{(a,b)}^{(x,y)} \mathbf{F} \cdot d\mathbf{r} \tag{7}$$

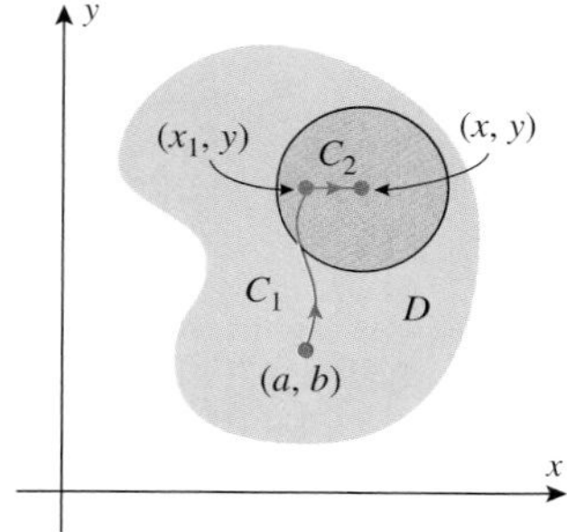

Figure 16.3.4

This is an unambiguous definition because we have assumed that the integral is independent of the path. We will show that $\nabla\phi = \mathbf{F}$. Since D is open, we can find a circular disk centered at (x, y) whose points lie entirely in D. As shown in Figure 16.3.4, choose any point (x_1, y) in this disk that lies on the same horizontal line as (x, y) such that $x_1 < x$. Because the integral in (7) is independent of path, we can evaluate it by first integrating from (a, b) to (x_1, y) along an arbitrary piecewise smooth curve C_1 in D, and then continuing along the horizontal line segment C_2 from (x_1, y) to (x, y). This yields

$$\phi(x, y) = \int_{C_1} \mathbf{F} \cdot d\mathbf{r} + \int_{C_2} \mathbf{F} \cdot d\mathbf{r} = \int_{(a,b)}^{(x_1,y)} \mathbf{F} \cdot d\mathbf{r} + \int_{C_2} \mathbf{F} \cdot d\mathbf{r}$$

Since the first term does not depend on x, its partial derivative with respect to x is zero and hence

$$\frac{\partial \phi}{\partial x} = \frac{\partial}{\partial x} \int_{C_2} \mathbf{F} \cdot d\mathbf{r} = \frac{\partial}{\partial x} \int_{C_2} f(x, y)\,dx + g(x, y)\,dy$$

However, the line integral with respect to y is zero along the horizontal line segment C_2, so this equation simplifies to

$$\frac{\partial \phi}{\partial x} = \frac{\partial}{\partial x} \int_{C_2} f(x, y)\,dx \tag{8}$$

To evaluate the integral in this expression, we treat y as a constant and express the line C_2 parametrically as

$$x = t, \quad y = y \qquad (x_1 \le t \le x)$$

At the risk of confusion, but to avoid complicating the notation, we have used x both as the dependent variable in the parametric equations and as the endpoint of the line segment. With the latter interpretation of x, it follows that (8) can be expressed as

$$\frac{\partial \phi}{\partial x} = \frac{\partial}{\partial x} \int_{x_1}^{x} f(t, y)\, dt$$

Now we apply Part 2 of the Fundamental Theorem of Calculus (Theorem 5.6.3), treating y as constant. This yields

$$\frac{\partial \phi}{\partial x} = f(x, y)$$

which proves the first part of (6). The proof that $\partial\phi/\partial y = g(x, y)$ can be obtained in a similar manner by joining (x, y) to a point (x, y_1) with a vertical line segment (Exercise 35). ■

A TEST FOR CONSERVATIVE VECTOR FIELDS

Although Theorem 16.3.2 is an important characterization of conservative vector fields, it is not an effective computational tool because it is usually not possible to evaluate the line integral over all possible piecewise smooth curves in D, as required in parts (b) and (c). To develop a method for determining whether a vector field is conservative, we will need to introduce some new concepts about parametric curves and connected sets. We will say that a parametric curve is ***simple*** if it does not intersect itself between its endpoints. A simple parametric curve may or may not be closed (Figure 16.3.5). In addition, we will say that a connected set D in 2-space is ***simply connected*** if no simple closed curve in D encloses points that are not in D. Stated informally, a connected set D is simply connected if it has no holes; a connected set with one or more holes is said to be ***multiply connected*** (Figure 16.3.6).

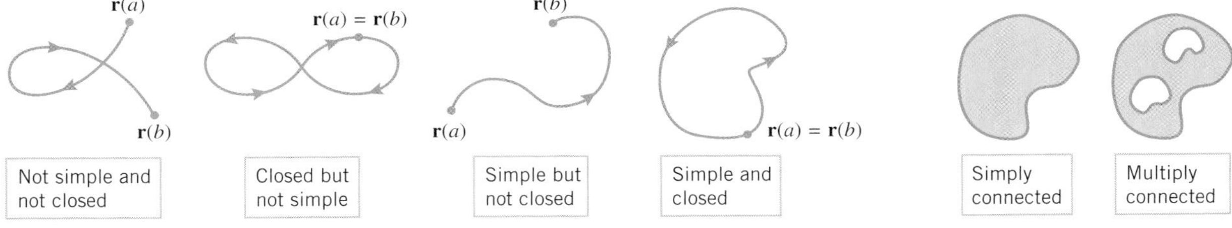

Figure 16.3.5

Figure 16.3.6

The following theorem is the primary tool for determining whether a vector field in 2-space is conservative.

16.3.3 THEOREM (*Conservative Field Test*). *If $f(x, y)$ and $g(x, y)$ are continuous and have continuous first partial derivatives on some open region D, and if the vector field $\mathbf{F}(x, y) = f(x, y)\mathbf{i} + g(x, y)\mathbf{j}$ is conservative on D, then*

$$\frac{\partial f}{\partial y} = \frac{\partial g}{\partial x} \qquad (9)$$

at each point in D. Conversely, if D is simply connected and (9) holds at each point in D, then $\mathbf{F}(x, y) = f(x, y)\mathbf{i} + g(x, y)\mathbf{j}$ is conservative.

WARNING

In (9), the **i**-component of **F** is differentiated with respect to y and the **j**-component with respect to x. It is easy to get this backwards by mistake.

A complete proof of this theorem requires results from advanced calculus and will be omitted. However, it is not hard to see why (9) must hold if **F** is conservative. For this purpose suppose that $\mathbf{F} = \nabla\phi$, in which case we can express the functions f and g as

$$\frac{\partial \phi}{\partial x} = f \quad \text{and} \quad \frac{\partial \phi}{\partial y} = g \tag{10}$$

Thus,

$$\frac{\partial f}{\partial y} = \frac{\partial}{\partial y}\left(\frac{\partial \phi}{\partial x}\right) = \frac{\partial^2 \phi}{\partial y \partial x} \quad \text{and} \quad \frac{\partial g}{\partial x} = \frac{\partial}{\partial x}\left(\frac{\partial \phi}{\partial y}\right) = \frac{\partial^2 \phi}{\partial x \partial y}$$

But the mixed partial derivatives in these equations are equal (Theorem 14.3.2), so (9) follows.

▶ **Example 3** Use Theorem 16.3.3 to determine whether the vector field

$$\mathbf{F}(x, y) = (y + x)\mathbf{i} + (y - x)\mathbf{j}$$

is conservative on some open set.

Solution. Let $f(x, y) = y + x$ and $g(x, y) = y - x$. Then

$$\frac{\partial f}{\partial y} = 1 \quad \text{and} \quad \frac{\partial g}{\partial x} = -1$$

Thus, there are no points in the xy-plane at which condition (9) holds, and hence $\mathbf{F}$ is not conservative on any open set. ◀

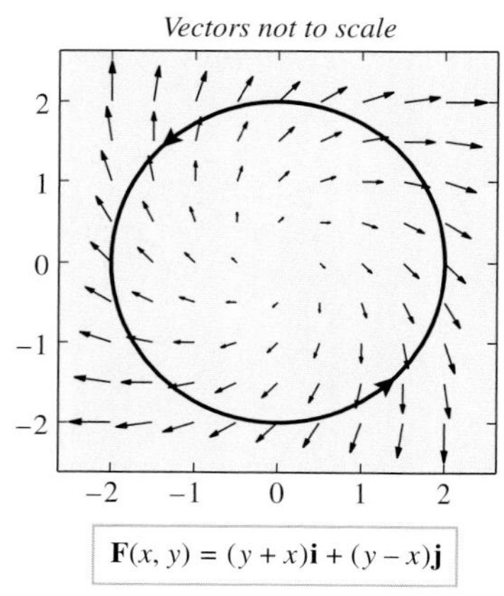

Figure 16.3.7

Since the vector field $\mathbf{F}$ in Example 3 is not conservative, it follows from Theorem 16.3.2 that there must exist piecewise smooth closed curves in every open connected set in the xy-plane on which

$$\int_C \mathbf{F} \cdot d\mathbf{r} = \int_C \mathbf{F} \cdot \mathbf{T}\, ds \neq 0$$

One such curve is the oriented circle shown in Figure 16.3.7. The figure suggests that $\mathbf{F} \cdot \mathbf{T} < 0$ at each point of C (why?), so $\int_C \mathbf{F} \cdot \mathbf{T}\, ds < 0$.

Once it is established that a vector field is conservative, a potential function for the field can be obtained by first integrating either of the equations in (10). This is illustrated in the following example.

▶ **Example 4** Let $\mathbf{F}(x, y) = 2xy^3\mathbf{i} + (1 + 3x^2y^2)\mathbf{j}$.

(a) Show that $\mathbf{F}$ is a conservative vector field on the entire xy-plane.

(b) Find ϕ by first integrating $\partial\phi/\partial x$.

(c) Find ϕ by first integrating $\partial\phi/\partial y$.

Solution (a). Since $f(x, y) = 2xy^3$ and $g(x, y) = 1 + 3x^2y^2$, we have

$$\frac{\partial f}{\partial y} = 6xy^2 = \frac{\partial g}{\partial x}$$

so (9) holds for all (x, y).

Solution (b). Since the field $\mathbf{F}$ is conservative, there is a potential function ϕ such that

$$\frac{\partial \phi}{\partial x} = 2xy^3 \quad \text{and} \quad \frac{\partial \phi}{\partial y} = 1 + 3x^2y^2 \tag{11}$$

Integrating the first of these equations with respect to x (and treating y as a constant) yields

$$\phi = \int 2xy^3\,dx = x^2y^3 + k(y) \tag{12}$$

where $k(y)$ represents the "constant" of integration. We are justified in treating the constant of integration as a function of y, since y is held constant in the integration process. To find $k(y)$ we differentiate (12) with respect to y and use the second equation in (11) to obtain

$$\frac{\partial \phi}{\partial y} = 3x^2y^2 + k'(y) = 1 + 3x^2y^2$$

from which it follows that $k'(y) = 1$. Thus,

$$k(y) = \int k'(y)\,dy = \int 1\,dy = y + K$$

where K is a (numerical) constant of integration. Substituting in (12) we obtain

$$\phi = x^2y^3 + y + K$$

The appearance of the arbitrary constant K tells us that ϕ is not unique. As a check on the computations, you may want to verify that $\nabla\phi = \mathbf{F}$.

You can also use (7) to find a potential function for a conservative vector field. For example, find a potential function for the vector field in Example 4 by evaluating (7) on the line segment

$$\mathbf{r}(t) = t(x\mathbf{i}) + t(y\mathbf{j}) \quad (0 \le t \le 1)$$

from $(0, 0)$ to (x, y).

Solution (c). Integrating the second equation in (11) with respect to y (and treating x as a constant) yields

$$\phi = \int (1 + 3x^2y^2)\,dy = y + x^2y^3 + k(x) \tag{13}$$

where $k(x)$ is the "constant" of integration. Differentiating (13) with respect to x and using the first equation in (11) yields

$$\frac{\partial \phi}{\partial x} = 2xy^3 + k'(x) = 2xy^3$$

from which it follows that $k'(x) = 0$ and consequently that $k(x) = K$, where K is a numerical constant of integration. Substituting this in (13) yields

$$\phi = y + x^2y^3 + K$$

which agrees with the solution in part (b). ◀

▶ **Example 5** Use the potential function obtained in Example 4 to evaluate the integral

$$\int_{(1,4)}^{(3,1)} 2xy^3\,dx + (1 + 3x^2y^2)\,dy$$

Solution. The integrand can be expressed as $\mathbf{F}\cdot d\mathbf{r}$, where $\mathbf{F}$ is the vector field in Example 4. Thus, using Formula (3) and the potential function $\phi = y + x^2y^3 + K$ for $\mathbf{F}$, we obtain

$$\begin{aligned}\int_{(1,4)}^{(3,1)} 2xy^3\,dx + (1 + 3x^2y^2)\,dy &= \int_{(1,4)}^{(3,1)} \mathbf{F}\cdot d\mathbf{r} = \phi(3, 1) - \phi(1, 4)\\ &= (10 + K) - (68 + K) = -58\end{aligned}$$ ◀

In the solution to Example 5, note that the constant K drops out. In future integration problems we will sometimes omit K from the computations.

▶ **Example 6** Let $\mathbf{F}(x, y) = e^y\mathbf{i} + xe^y\mathbf{j}$ denote a force field in the xy-plane.

(a) Verify that the force field $\mathbf{F}$ is conservative on the entire xy-plane.

(b) Find the work done by the field on a particle that moves from $(1, 0)$ to $(-1, 0)$ along the semicircular path C shown in Figure 16.3.8.

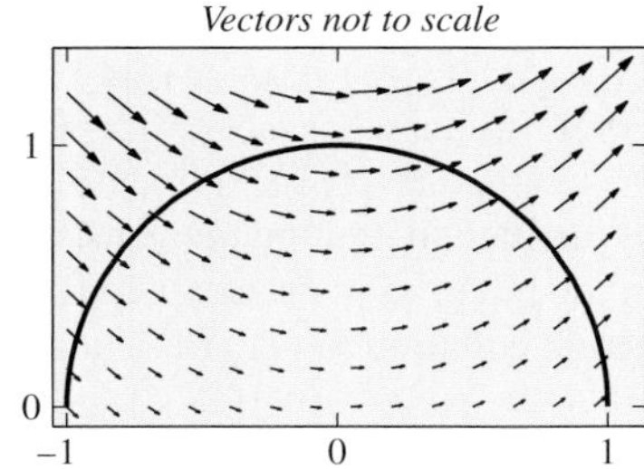

Figure 16.3.8

Solution (a). For the given field we have $f(x, y) = e^y$ and $g(x, y) = xe^y$. Thus,

$$\frac{\partial}{\partial y}(e^y) = e^y = \frac{\partial}{\partial x}(xe^y)$$

so (9) holds for all (x, y) and hence $\mathbf{F}$ is conservative on the entire xy-plane.

Solution (b). From Formula (34) of Section 16.2, the work done by the field is

$$W = \int_C \mathbf{F} \cdot d\mathbf{r} = \int_C e^y\,dx + xe^y\,dy \tag{14}$$

However, the calculations involved in integrating along C are tedious, so it is preferable to apply Theorem 16.3.1, taking advantage of the fact that the field is conservative and the integral is independent of path. Thus, we write (14) as

$$W = \int_{(1,0)}^{(-1,0)} e^y\,dx + xe^y\,dy = \phi(-1, 0) - \phi(1, 0) \tag{15}$$

As illustrated in Example 4, we can find ϕ by integrating either of the equations

$$\frac{\partial \phi}{\partial x} = e^y \quad \text{and} \quad \frac{\partial \phi}{\partial y} = xe^y \tag{16}$$

We will integrate the first. We obtain

$$\phi = \int e^y\,dx = xe^y + k(y) \tag{17}$$

Differentiating this equation with respect to y and using the second equation in (16) yields

$$\frac{\partial \phi}{\partial y} = xe^y + k'(y) = xe^y$$

from which it follows that $k'(y) = 0$ or $k(y) = K$. Thus, from (17)

$$\phi = xe^y + K$$

and hence from (15)

$$W = \phi(-1, 0) - \phi(1, 0) = (-1)e^0 - 1e^0 = -2 \blacktriangleleft$$

CONSERVATIVE VECTOR FIELDS IN 3-SPACE

All of the results in this section have analogs in 3-space: Theorems 16.3.1 and 16.3.2 can be extended to vector fields in 3-space simply by adding a third variable and modifying the hypotheses appropriately. For example, in 3-space, Formula (3) becomes

$$\int_C \mathbf{F}(x, y, z) \cdot d\mathbf{r} = \phi(x_1, y_1, z_1) - \phi(x_0, y_0, z_0) \tag{18}$$

Theorem 16.3.3 can also be extended to vector fields in 3-space. We leave it for the exercises to show that if $\mathbf{F}(x, y, z) = f(x, y, z)\mathbf{i} + g(x, y, z)\mathbf{j} + h(x, y, z)\mathbf{k}$ is a conservative field, then

$$\frac{\partial f}{\partial y} = \frac{\partial g}{\partial x}, \quad \frac{\partial f}{\partial z} = \frac{\partial h}{\partial x}, \quad \frac{\partial g}{\partial z} = \frac{\partial h}{\partial y} \tag{19}$$

that is, curl $\mathbf{F} = \mathbf{0}$. Conversely, a vector field satisfying these conditions on a suitably restricted region is conservative on that region if f, g, and h are continuous and have continuous first partial derivatives in the region. Some problems involving Formulas (18) and (19) are given in the review exercises at the end of this chapter.

CONSERVATION OF ENERGY

If $\mathbf{F}(x, y, z)$ is a conservative force field with a potential function $\phi(x, y, z)$, then we call $V(x, y, z) = -\phi(x, y, z)$ the ***potential energy*** of the field at the point (x, y, z). Thus, it follows from the 3-space version of Theorem 16.3.1 that the work W done by $\mathbf{F}$ on a particle that moves along any path C from a point (x_0, y_0, z_0) to a point (x_1, y_1, z_1) is related to the potential energy by the equation

$$W = \int_C \mathbf{F} \cdot d\mathbf{r} = \phi(x_1, y_1, z_1) - \phi(x_0, y_0, z_0) = -[V(x_1, y_1, z_1) - V(x_0, y_0, z_0)] \quad (20)$$

That is, the work done by the field is the negative of the change in potential energy. In particular, it follows from the 3-space analog of Theorem 16.3.2 that if a particle traverses a piecewise smooth closed path in a conservative vector field, then the work done by the field is zero, and there is no change in potential energy. To take this a step further, suppose that a particle of mass m moves along any piecewise smooth curve (not necessarily closed) in a conservative force field $\mathbf{F}$, starting at (x_0, y_0, z_0) with velocity v_i and ending at (x_1, y_1, z_1) with velocity v_f. If $\mathbf{F}$ is the only force acting on the particle, then an argument similar to the derivation of Equation (5) in Section 6.7 shows that the work done on the particle by $\mathbf{F}$ is equal to the change in kinetic energy $\frac{1}{2}mv_f^2 - \frac{1}{2}mv_i^2$ of the particle. If we let V_i denote the potential energy at the starting point and V_f the potential energy at the final point, then it follows from (20)

$$\tfrac{1}{2}mv_f^2 - \tfrac{1}{2}mv_i^2 = -[V_f - V_i]$$

which we can rewrite as

$$\tfrac{1}{2}mv_f^2 + V_f = \tfrac{1}{2}mv_i^2 + V_i$$

This equation states that the total energy of the particle (kinetic energy + potential energy) does not change as the particle moves along a path in a conservative vector field. This result, called the ***conservation of energy principle***, explains the origin of the term "conservative vector field."

✔ QUICK CHECK EXERCISES 16.3

(See page 1139 for answers.)

1. If C is a piecewise smooth curve from $(1, 2, 3)$ to $(4, 5, 6)$, then
$$\int_C dx + 2\,dy + 3\,dz = \underline{\qquad}$$

2. If C is the portion of the circle $x^2 + y^2 = 1$ where $0 \le x$, oriented counterclockwise, and $f(x, y) = ye^x$ then
$$\int_C \nabla f \cdot d\mathbf{r} = \underline{\qquad}$$

3. A potential function for the vector field
$$\mathbf{F}(x, y, z) = yz\mathbf{i} + (xz + z)\mathbf{j} + (xy + y + 1)\mathbf{k}$$
is $\phi(x, y, z) = \underline{\qquad}$.

4. If a, b, and c are nonzero real numbers such that the vector field $x^5y^a\mathbf{i} + x^by^c\mathbf{j}$ is a conservative vector field, then
$$a = \underline{\qquad}, \quad b = \underline{\qquad}, \quad c = \underline{\qquad}$$

EXERCISE SET 16.3 ⓒ CAS

1–6 Determine whether $\mathbf{F}$ is a conservative vector field. If so, find a potential function for it.

1. $\mathbf{F}(x, y) = x\mathbf{i} + y\mathbf{j}$
2. $\mathbf{F}(x, y) = 3y^2\mathbf{i} + 6xy\mathbf{j}$
3. $\mathbf{F}(x, y) = x^2y\mathbf{i} + 5xy^2\mathbf{j}$
4. $\mathbf{F}(x, y) = e^x \cos y\mathbf{i} - e^x \sin y\mathbf{j}$
5. $\mathbf{F}(x, y) = (\cos y + y\cos x)\mathbf{i} + (\sin x - x\sin y)\mathbf{j}$
6. $\mathbf{F}(x, y) = x\ln y\mathbf{i} + y\ln x\mathbf{j}$
7. (a) Show that the line integral $\int_C y^2\,dx + 2xy\,dy$ is independent of the path.
   (b) Evaluate the integral in part (a) along the line segment from $(-1, 2)$ to $(1, 3)$.
   (c) Evaluate the integral $\int_{(-1,2)}^{(1,3)} y^2\,dx + 2xy\,dy$ using Theorem 16.3.1, and confirm that the value is the same as that obtained in part (b).

8. (a) Show that the line integral $\int_C y \sin x\,dx - \cos x\,dy$ is independent of the path.
(b) Evaluate the integral in part (a) along the line segment from $(0, 1)$ to $(\pi, -1)$.
(c) Evaluate the integral $\int_{(0,1)}^{(\pi,-1)} y \sin x\,dx - \cos x\,dy$ using Theorem 16.3.1, and confirm that the value is the same as that obtained in part (b).

9–14 Show that the integral is independent of the path, and use Theorem 16.3.1 to find its value.

9. $\displaystyle\int_{(1,2)}^{(4,0)} 3y\,dx + 3x\,dy$

10. $\displaystyle\int_{(0,0)}^{(1,\pi/2)} e^x \sin y\,dx + e^x \cos y\,dy$

11. $\displaystyle\int_{(0,0)}^{(3,2)} 2xe^y\,dx + x^2e^y\,dy$

12. $\displaystyle\int_{(-1,2)}^{(0,1)} (3x - y + 1)\,dx - (x + 4y + 2)\,dy$

13. $\displaystyle\int_{(2,-2)}^{(-1,0)} 2xy^3\,dx + 3y^2x^2\,dy$

14. $\displaystyle\int_{(1,1)}^{(3,3)} \left(e^x \ln y - \frac{e^y}{x}\right)dx + \left(\frac{e^x}{y} - e^y \ln x\right)dy$, where x and y are positive.

15–18 Confirm that the force field $\mathbf{F}$ is conservative in some open connected region containing the points P and Q, and then find the work done by the force field on a particle moving along an arbitrary smooth curve in the region from P to Q.

15. $\mathbf{F}(x, y) = xy^2\mathbf{i} + x^2y\mathbf{j}$; $P(1, 1)$, $Q(0, 0)$

16. $\mathbf{F}(x, y) = 2xy^3\mathbf{i} + 3x^2y^2\mathbf{j}$; $P(-3, 0)$, $Q(4, 1)$

17. $\mathbf{F}(x, y) = ye^{xy}\mathbf{i} + xe^{xy}\mathbf{j}$; $P(-1, 1)$, $Q(2, 0)$

18. $\mathbf{F}(x, y) = e^{-y} \cos x\mathbf{i} - e^{-y} \sin x\mathbf{j}$; $P(\pi/2, 1)$, $Q(-\pi/2, 0)$

19–20 Find the exact value of $\int_C \mathbf{F} \cdot d\mathbf{r}$ using any method.

19. $\mathbf{F}(x, y) = (e^y + ye^x)\mathbf{i} + (xe^y + e^x)\mathbf{j}$
$C: \mathbf{r}(t) = \sin(\pi t/2)\mathbf{i} + \ln t\,\mathbf{j} \quad (1 \le t \le 2)$

20. $\mathbf{F}(x, y) = 2xy\mathbf{i} + (x^2 + \cos y)\mathbf{j}$
$C: \mathbf{r}(t) = t\mathbf{i} + t\cos(t/3)\mathbf{j} \quad (0 \le t \le \pi)$

C **21.** Use the numerical integration capability of a CAS or other calculating utility to approximate the value of the integral in Exercise 19 by direct integration. Confirm that the numerical approximation is consistent with the exact value.

C **22.** Use the numerical integration capability of a CAS or other calculating utility to approximate the value of the integral in Exercise 20 by direct integration. Confirm that the numerical approximation is consistent with the exact value.

FOCUS ON CONCEPTS

23–24 Is the vector field conservative? Explain.

23. **24.**

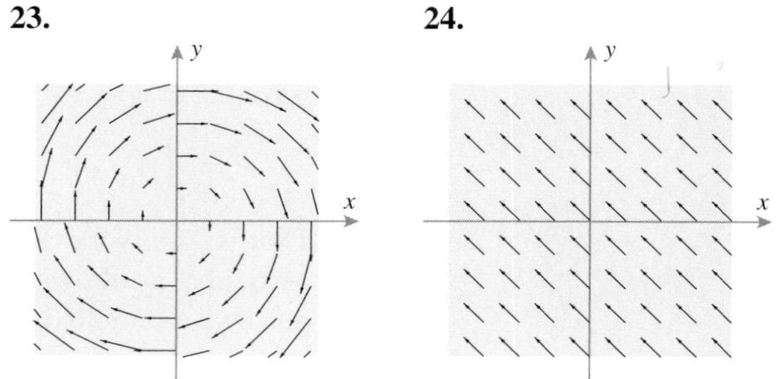

25. Suppose that C is a circle in the domain of a conservative vector field in the xy-plane whose component functions are continuous. Explain why there must be at least two points on C at which the vector field is normal to the circle.

26. Does the result in Exercise 25 remain true if the circle C is replaced by a square? Explain.

27. Prove: If

$$\mathbf{F}(x, y, z) = f(x, y, z)\mathbf{i} + g(x, y, z)\mathbf{j} + h(x, y, z)\mathbf{k}$$

is a conservative field and f, g, and h are continuous and have continuous first partial derivatives in a region, then

$$\frac{\partial f}{\partial y} = \frac{\partial g}{\partial x}, \quad \frac{\partial f}{\partial z} = \frac{\partial h}{\partial x}, \quad \frac{\partial g}{\partial z} = \frac{\partial h}{\partial y}$$

in the region.

28. Use the result in Exercise 27 to show that the integral

$$\int_C yz\,dx + xz\,dy + yx^2\,dz$$

is not independent of the path.

29. Find a nonzero function h for which

$$\mathbf{F}(x, y) = h(x)[x \sin y + y \cos y]\mathbf{i} + h(x)[x \cos y - y \sin y]\mathbf{j}$$

is conservative.

30. (a) In Example 3 of Section 16.1 we showed that

$$\phi(x, y) = -\frac{c}{(x^2 + y^2)^{1/2}}$$

is a potential function for the two-dimensional inverse-square field

$$\mathbf{F}(x, y) = \frac{c}{(x^2 + y^2)^{3/2}}(x\mathbf{i} + y\mathbf{j})$$

but we did not explain how the potential function $\phi(x, y)$ was obtained. Use Theorem 16.3.3 to show that the two-dimensional inverse-square field is conservative everywhere except at the origin, and then use the method of Example 4 to derive the formula for $\phi(x, y)$.

(b) Use an appropriate generalization of the method of Example 4 to derive the potential function

$$\phi(x, y, z) = -\frac{c}{(x^2 + y^2 + z^2)^{1/2}}$$

for the three-dimensional inverse-square field given by Formula (5) of Section 16.1.

31–32 Use the result in Exercise 30(b).

31. In each part, find the work done by the three-dimensional inverse-square field

$$\mathbf{F}(\mathbf{r}) = \frac{1}{\|\mathbf{r}\|^3}\mathbf{r}$$

on a particle that moves along the curve C.

(a) C is the line segment from $P(1, 1, 2)$ to $Q(3, 2, 1)$.

(b) C is the curve

$$\mathbf{r}(t) = (2t^2 + 1)\mathbf{i} + (t^3 + 1)\mathbf{j} + (2 - \sqrt{t})\mathbf{k}$$

where $0 \le t \le 1$.

(c) C is the circle in the xy-plane of radius 1 centered at $(2, 0, 0)$ traversed counterclockwise.

32. Let $\mathbf{F}(x, y) = \dfrac{y}{x^2 + y^2}\mathbf{i} - \dfrac{x}{x^2 + y^2}\mathbf{j}$.

(a) Show that

$$\int_{C_1} \mathbf{F} \cdot d\mathbf{r} \neq \int_{C_2} \mathbf{F} \cdot d\mathbf{r}$$

if C_1 and C_2 are the semicircular paths from $(1, 0)$ to $(-1, 0)$ given by

$$C_1\colon\ x = \cos t,\quad y = \sin t \qquad (0 \le t \le \pi)$$

$$C_2\colon\ x = \cos t,\quad y = -\sin t \qquad (0 \le t \le \pi)$$

(b) Show that the components of $\mathbf{F}$ satisfy Formula (9).

(c) Do the results in parts (a) and (b) violate Theorem 16.3.3? Explain.

33. Prove Theorem 16.3.1 if C is a piecewise smooth curve composed of smooth curves $C_1, C_2, \ldots, C_n$.

34. Prove that (*b*) implies (*c*) in Theorem 16.3.2. [*Hint:* Consider any two piecewise smooth oriented curves C_1 and C_2 in the region from a point P to a point Q, and integrate around the closed curve consisting of C_1 and $-C_2$.]

35. Complete the proof of Theorem 16.3.2 by showing that $\partial\phi/\partial y = g(x, y)$, where $\phi(x, y)$ is the function in (7).

✔ QUICK CHECK ANSWERS 16.3

1. 18 **2.** 2 **3.** $xyz + yz + z$ **4.** 6; 6; 5

16.4 GREEN'S THEOREM

In this section we will discuss a remarkable and beautiful theorem that expresses a double integral over a plane region in terms of a line integral around its boundary.

GREEN'S THEOREM

16.4.1 THEOREM (*Green's Theorem*). *Let R be a simply connected plane region whose boundary is a simple, closed, piecewise smooth curve C oriented counterclockwise. If $f(x, y)$ and $g(x, y)$ are continuous and have continuous first partial derivatives on some open set containing R, then*

$$\int_C f(x, y)\,dx + g(x, y)\,dy = \iint_R \left(\frac{\partial g}{\partial x} - \frac{\partial f}{\partial y}\right) dA \tag{1}$$

PROOF. For simplicity, we will prove the theorem for regions that are simultaneously type I and type II (see Definition 15.2.1). Such a region is shown in Figure 16.4.1. The crux of the proof is to show that

$$\int_C f(x, y)\,dx = -\iint_R \frac{\partial f}{\partial y}\,dA \quad \text{and} \quad \int_C g(x, y)\,dy = \iint_R \frac{\partial g}{\partial x}\,dA \tag{2–3}$$

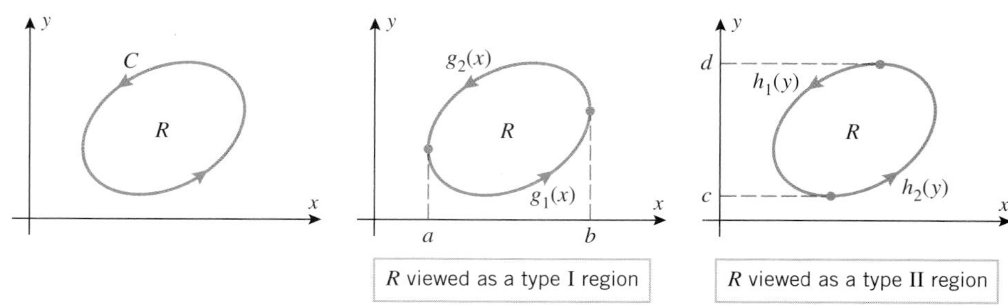

Figure 16.4.1

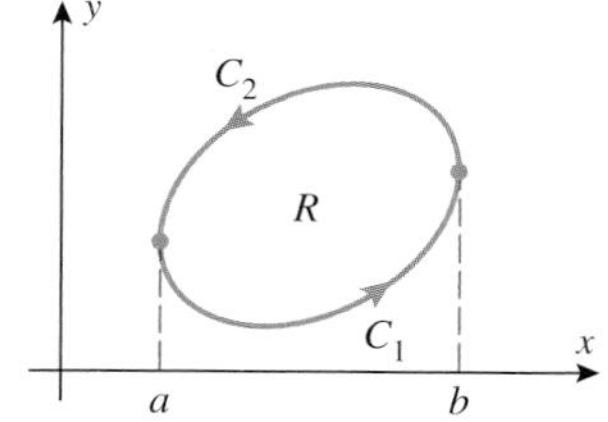

Figure 16.4.2

To prove (2), view R as a type I region and let C_1 and C_2 be the lower and upper boundary curves, oriented as in Figure 16.4.2. Then

$$\int_C f(x, y)\,dx = \int_{C_1} f(x, y)\,dx + \int_{C_2} f(x, y)\,dx$$

or, equivalently,

$$\int_C f(x, y)\,dx = \int_{C_1} f(x, y)\,dx - \int_{-C_2} f(x, y)\,dx \tag{4}$$

(This step will help simplify our calculations since C_1 and $-C_2$ are then both oriented left to right.) The curves C_1 and $-C_2$ can be expressed parametrically as

$$\begin{aligned} C_1\colon x &= t, \quad y = g_1(t) \quad (a \le t \le b) \\ -C_2\colon x &= t, \quad y = g_2(t) \quad (a \le t \le b) \end{aligned}$$

Thus, we can rewrite (4) as

$$\begin{aligned} \int_C f(x, y)\,dx &= \int_a^b f(t, g_1(t))x'(t)\,dt - \int_a^b f(t, g_2(t))x'(t)\,dt \\ &= \int_a^b f(t, g_1(t))\,dt - \int_a^b f(t, g_2(t))\,dt \\ &= -\int_a^b [f(t, g_2(t)) - f(t, g_1(t))]\,dt \\ &= -\int_a^b \Big[f(t, y)\Big]_{y=g_1(t)}^{y=g_2(t)} dt = -\int_a^b \left[\int_{g_1(t)}^{g_2(t)} \frac{\partial f}{\partial y}\,dy\right] dt \\ &= -\int_a^b \int_{g_1(x)}^{g_2(x)} \frac{\partial f}{\partial y}\,dy\,dx = -\iint_R \frac{\partial f}{\partial y}\,dA \end{aligned}$$

Since $x = t$

Supply the details for the proof of (3).

The proof of (3) is obtained similarly by treating R as a type II region. We omit the details. ■

▶ Example 1 Use Green's Theorem to evaluate

$$\int_C x^2 y\,dx + x\,dy$$

along the triangular path shown in Figure 16.4.3.

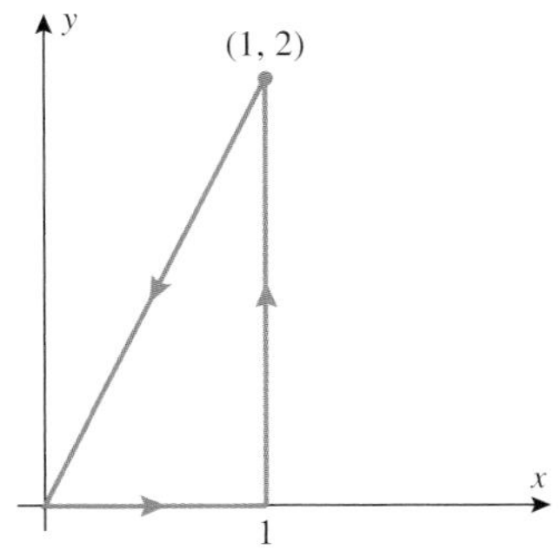

Figure 16.4.3

Solution. Since $f(x, y) = x^2 y$ and $g(x, y) = x$, it follows from (1) that

$$\int_C x^2 y\,dx + x\,dy = \iint_R \left[\frac{\partial}{\partial x}(x) - \frac{\partial}{\partial y}(x^2 y)\right] dA = \int_0^1 \int_0^{2x} (1 - x^2)\,dy\,dx$$

$$= \int_0^1 (2x - 2x^3)\,dx = \left[x^2 - \frac{x^4}{2}\right]_0^1 = \frac{1}{2}$$

This agrees with the result obtained in Example 10 of Section 16.2, where we evaluated the line integral directly. Note how much simpler this solution is. ◄

A NOTATION FOR LINE INTEGRALS AROUND SIMPLE CLOSED CURVES

It is common practice to denote a line integral around a simple closed curve by an integral sign with a superimposed circle. With this notation Formula (1) would be written as

$$\oint_C f(x, y)\,dx + g(x, y)\,dy = \iint_R \left(\frac{\partial g}{\partial x} - \frac{\partial f}{\partial y}\right) dA$$

Sometimes a direction arrow is added to the circle to indicate whether the integration is clockwise or counterclockwise. Thus, if we wanted to emphasize the counterclockwise direction of integration required by Theorem 16.4.1, we could express (1) as

$$\ointctrclockwise_C f(x, y)\,dx + g(x, y)\,dy = \iint_R \left(\frac{\partial g}{\partial x} - \frac{\partial f}{\partial y}\right) dA \tag{5}$$

FINDING WORK USING GREEN'S THEOREM

It follows from Formula (26) of Section 16.2 that the integral on the left side of (5) is the work performed by the force field $\mathbf{F}(x, y) = f(x, y)\mathbf{i} + g(x, y)\mathbf{j}$ on a particle moving counterclockwise around the simple closed curve C. In the case where this vector field is conservative, it follows from Theorem 16.3.2 that the integrand in the double integral on the right side of (5) is zero, so the work performed by the field is zero, as expected. For vector fields that are not conservative, it is often more efficient to calculate the work around simple closed curves by using Green's Theorem than by parametrizing the curve.

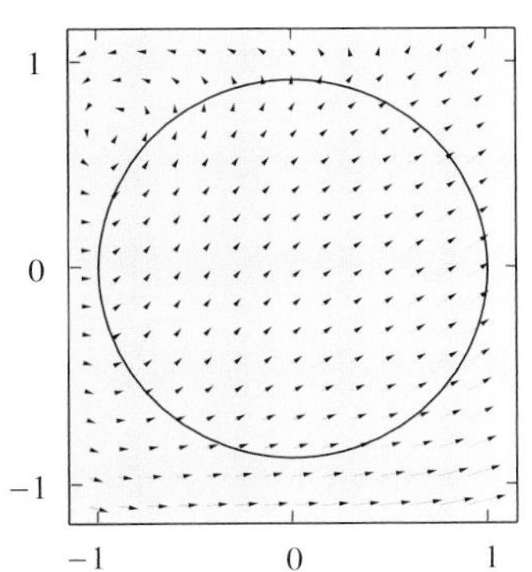

Figure 16.4.4

► **Example 2** Find the work done by the force field

$$\mathbf{F}(x, y) = (e^x - y^3)\mathbf{i} + (\cos y + x^3)\mathbf{j}$$

on a particle that travels once around the unit circle $x^2 + y^2 = 1$ in the counterclockwise direction (Figure 16.4.4).

George Green (1793–1841) English mathematician and physicist. Green left school at an early age to work in his father's bakery and consequently had little early formal education. When his father opened a mill, the boy used the top room as a study in which he taught himself physics and mathematics from library books. In 1828 Green published his most important work, *An Essay on the Application of Mathematical Analysis to the Theories of Electricity and Magnetism.* Although Green's Theorem appeared in that paper, the result went virtually unnoticed because of the small pressrun and local distribution. Following the death of his father in 1829, Green was urged by friends to seek a college education. In 1833, after four years of self-study to close the gaps in his elementary education, Green was admitted to Caius College, Cambridge. He graduated four years later, but with a disappointing performance on his final examinations—possibly because he was more interested in his own research. After a succession of works on light and sound, he was named to be Perse Fellow at Caius College. Two years later he died. In 1845, four years after his death, his paper of 1828 was published and the theories developed therein by this obscure, self-taught baker's son helped pave the way to the modern theories of electricity and magnetism.

Solution. The work W performed by the field is

$$W = \oint_C \mathbf{F} \cdot d\mathbf{r} = \oint_C (e^x - y^3)\,dx + (\cos y + x^3)\,dy$$

$$= \iint_R \left[\frac{\partial}{\partial x}(\cos y + x^3) - \frac{\partial}{\partial y}(e^x - y^3)\right] dA$$

Green's Theorem

$$= \iint_R (3x^2 + 3y^2)\,dA = 3\iint_R (x^2 + y^2)\,dA$$

$$= 3\int_0^{2\pi}\int_0^1 (r^2)r\,dr\,d\theta = \frac{3}{4}\int_0^{2\pi} d\theta = \frac{3\pi}{2} \blacktriangleleft$$

We converted to polar coordinates.

FINDING AREAS USING GREEN'S THEOREM

Green's Theorem leads to some useful new formulas for the area A of a region R that satisfies the conditions of the theorem. Two such formulas can be obtained as follows:

$$A = \iint_R dA = \oint_C x\,dy \quad \text{and} \quad A = \iint_R dA = \oint_C (-y)\,dx$$

Set $f(x, y) = 0$ and $g(x, y) = x$ in (1).

Set $f(x, y) = -y$ and $g(x, y) = 0$ in (1).

A third formula can be obtained by adding these two equations together. Thus, we have the following three formulas that express the area A of a region R in terms of line integrals around the boundary:

Although the third formula in (6) looks more complicated than the other two, it often leads to simpler integrations. Each has advantages in certain situations.

$$A = \oint_C x\,dy = -\oint_C y\,dx = \frac{1}{2}\oint_C -y\,dx + x\,dy \tag{6}$$

► **Example 3** Use a line integral to find the area enclosed by the ellipse

$$\frac{x^2}{a^2} + \frac{y^2}{b^2} = 1$$

Solution. The ellipse, with counterclockwise orientation, can be represented parametrically by

$$x = a\cos t, \quad y = b\sin t \qquad (0 \le t \le 2\pi)$$

If we denote this curve by C, then from the third formula in (6) the area A enclosed by the ellipse is

$$A = \frac{1}{2}\oint_C -y\,dx + x\,dy$$

$$= \frac{1}{2}\int_0^{2\pi} [(-b\sin t)(-a\sin t) + (a\cos t)(b\cos t)]\,dt$$

$$= \frac{1}{2}ab\int_0^{2\pi} (\sin^2 t + \cos^2 t)\,dt = \frac{1}{2}ab\int_0^{2\pi} dt = \pi ab \blacktriangleleft$$

GREEN'S THEOREM FOR MULTIPLY CONNECTED REGIONS

Recall that a plane region is said to be simply connected if it has no holes and is said to be multiply connected if it has one or more holes (see Figure 16.3.6). At the beginning of this section we stated Green's Theorem for a counterclockwise integration around the

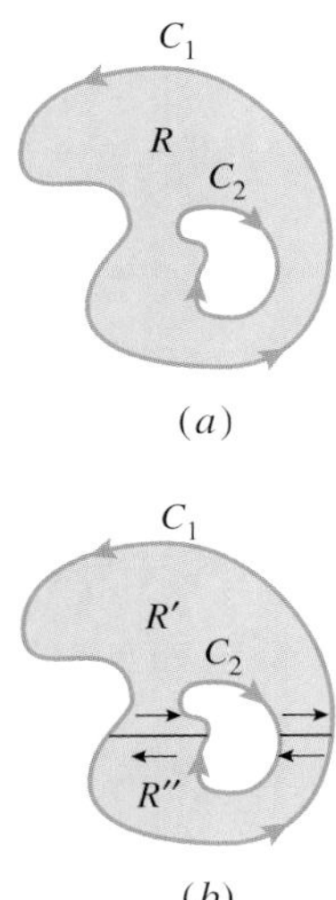

Figure 16.4.5

boundary of a simply connected region R (Theorem 16.4.1). Our next goal is to extend this theorem to multiply connected regions. To make this extension we will need to assume that *the region lies on the left when any portion of the boundary is traversed in the direction of its orientation*. This implies that the outer boundary curve of the region is oriented counterclockwise and the boundary curves that enclose holes have clockwise orientation (Figure 16.4.5*a*). If all portions of the boundary of a multiply connected region R are oriented in this way, then we say that the boundary of R has ***positive orientation***.

We will now derive a version of Green's Theorem that applies to multiply connected regions with positively oriented boundaries. For simplicity, we will consider a multiply connected region R with one hole, and we will assume that $f(x, y)$ and $g(x, y)$ have continuous first partial derivatives on some open set containing R. As shown in Figure 16.4.5*b*, let us divide R into two regions R' and R'' by introducing two "cuts" in R. The cuts are shown as line segments, but any piecewise smooth curves will suffice. If we assume that f and g satisfy the hypotheses of Green's Theorem on R (and hence on R' and R''), then we can apply this theorem to both R' and R'' to obtain

$$\iint_R \left(\frac{\partial g}{\partial x} - \frac{\partial f}{\partial y}\right) dA = \iint_{R'} \left(\frac{\partial g}{\partial x} - \frac{\partial f}{\partial y}\right) dA + \iint_{R''} \left(\frac{\partial g}{\partial x} - \frac{\partial f}{\partial y}\right) dA$$

$$= \oint_{\substack{\text{Boundary}\\ \text{of } R'}} f(x, y)\,dx + g(x, y)\,dy + \oint_{\substack{\text{Boundary}\\ \text{of } R''}} f(x, y)\,dx + g(x, y)\,dy$$

However, the two line integrals are taken in opposite directions along the cuts, and hence cancel there, leaving only the contributions along C_1 and C_2. Thus,

$$\iint_R \left(\frac{\partial g}{\partial x} - \frac{\partial f}{\partial y}\right) dA = \oint_{C_1} f(x, y)\,dx + g(x, y)\,dy + \oint_{C_2} f(x, y)\,dx + g(x, y)\,dy \tag{7}$$

which is an extension of Green's Theorem to a multiply connected region with one hole. Observe that the integral around the outer boundary is taken counterclockwise and the integral around the hole is taken clockwise. More generally, if R is a multiply connected region with n holes, then the analog of (7) involves a sum of $n + 1$ integrals, one taken counterclockwise around the outer boundary of R and the rest taken clockwise around the holes.

Sketch a proof of the version of Green's Theorem that applies to a multiply connected region with two holes.

▶ **Example 4** Evaluate the integral

$$\oint_C \frac{-y\,dx + x\,dy}{x^2 + y^2}$$

if C is a piecewise smooth simple closed curve oriented counterclockwise such that (a) C does not enclose the origin and (b) C encloses the origin.

Solution (a). Let

$$f(x, y) = -\frac{y}{x^2 + y^2}, \quad g(x, y) = \frac{x}{x^2 + y^2} \tag{8}$$

so that

$$\frac{\partial g}{\partial x} = \frac{y^2 - x^2}{(x^2 + y^2)^2} = \frac{\partial f}{\partial y}$$

if x and y are not both zero. Thus, if C does not enclose the origin, we have

$$\frac{\partial g}{\partial x} - \frac{\partial f}{\partial y} = 0 \tag{9}$$

on the simply connected region enclosed by C, and hence the given integral is zero by Green's Theorem.

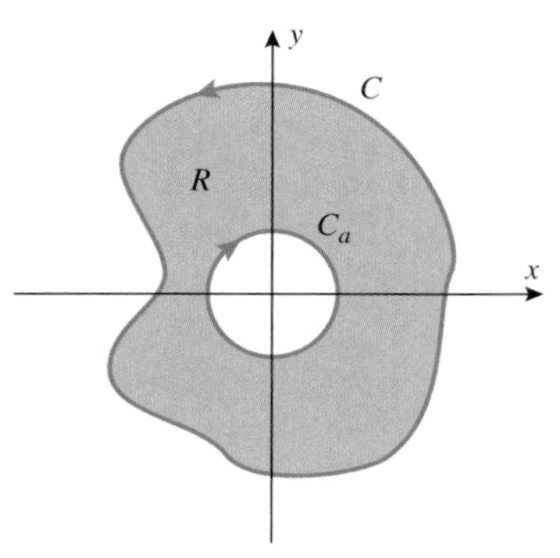

Figure 16.4.6

Solution (b). Unlike the situation in part (a), we cannot apply Green's Theorem directly because the functions $f(x, y)$ and $g(x, y)$ in (8) are discontinuous at the origin. Our problems are further compounded by the fact that we do not have a specific curve C that we can parametrize to evaluate the integral. Our strategy for circumventing these problems will be to replace C with a specific curve that produces the same value for the integral and then use that curve for the evaluation. To obtain such a curve, we will apply Green's Theorem for multiply connected regions to a region that does not contain the origin. For this purpose we construct a circle C_a with *clockwise* orientation, centered at the origin, and with sufficiently small radius a that it lies inside the region enclosed by C (Figure 16.4.6). This creates a multiply connected region R whose boundary curves C and C_a have the orientations required by Formula (7) and such that within R the functions $f(x, y)$ and $g(x, y)$ in (8) satisfy the hypotheses of Green's Theorem (the origin is outside of R). Thus, it follows from (7) and (9) that

$$\oint_C \frac{-y\,dx + x\,dy}{x^2 + y^2} + \oint_{C_a} \frac{-y\,dx + x\,dy}{x^2 + y^2} = \iint_R 0\,dA = 0$$

It follows from this equation that

$$\oint_C \frac{-y\,dx + x\,dy}{x^2 + y^2} = -\oint_{C_a} \frac{-y\,dx + x\,dy}{x^2 + y^2}$$

which we can rewrite as

$$\oint_C \frac{-y\,dx + x\,dy}{x^2 + y^2} = \oint_{-C_a} \frac{-y\,dx + x\,dy}{x^2 + y^2}$$

Reversing the orientation of C_a reverses the sign of the integral.

But C_a has clockwise orientation, so $-C_a$ has counterclockwise orientation. Thus, we have shown that the original integral can be evaluated by integrating counterclockwise around a circle of radius a that is centered at the origin and lies within the region enclosed by C. Such a circle can be expressed parametrically as $x = a\cos t$, $y = a\sin t$ $(0 \le t \le 2\pi)$; and hence

$$\oint_C \frac{-y\,dx + x\,dy}{x^2 + y^2} = \int_0^{2\pi} \frac{(-a\sin t)(-a\sin t)\,dt + (a\cos t)(a\cos t)\,dt}{(a\cos t)^2 + (a\sin t)^2}$$

$$= \int_0^{2\pi} 1\,dt = 2\pi$$

✔QUICK CHECK EXERCISES 16.4 *(See page 1147 for answers.)*

1. If C is the square with vertices $(\pm 1, \pm 1)$ oriented counterclockwise, then

$$\int_C -y\,dx + x\,dy = \underline{\qquad}$$

2. If C is the triangle with vertices $(0, 0)$, $(1, 0)$, and $(1, 1)$ oriented counterclockwise, then

$$\int_C 2xy\,dx + (x^2 + x)\,dy = \underline{\qquad}$$

3. Sometimes symmetry considerations can simplify an application of Green's Theorem. For example, if C is the unit circle centered at the origin and oriented counterclockwise, then

$$\int_C (y^3 - y - x)\,dx + (x^3 + x + y)\,dy = \underline{\qquad}$$

4. What region R and choice of functions $f(x, y)$ and $g(x, y)$ allow us to use Formula (1) of Theorem 16.4.1 to claim that

$$\int_0^1 \int_0^{\sqrt{1-x^2}} (2x + 2y)\,dy\,dx = \int_0^{\pi/2} (\sin^3 t + \cos^3 t)\,dt?$$

EXERCISE SET 16.4 [C] CAS

1–2 Evaluate the line integral using Green's Theorem and check the answer by evaluating it directly.

1. $\oint_C y^2\,dx + x^2\,dy$, where C is the square with vertices $(0, 0)$, $(1, 0)$, $(1, 1)$, and $(0, 1)$ oriented counterclockwise.

2. $\oint_C y\,dx + x\,dy$, where C is the unit circle oriented counterclockwise.

3–13 Use Green's Theorem to evaluate the integral. In each exercise, assume that the curve C is oriented counterclockwise.

3. $\oint_C 3xy\,dx + 2xy\,dy$, where C is the rectangle bounded by $x = -2$, $x = 4$, $y = 1$, and $y = 2$.

4. $\oint_C (x^2 - y^2)\,dx + x\,dy$, where C is the circle $x^2 + y^2 = 9$.

5. $\oint_C x\cos y\,dx - y\sin x\,dy$, where C is the square with vertices $(0, 0)$, $(\pi/2, 0)$, $(\pi/2, \pi/2)$, and $(0, \pi/2)$.

6. $\oint_C y\tan^2 x\,dx + \tan x\,dy$, where C is the circle $x^2 + (y+1)^2 = 1$.

7. $\oint_C (x^2 - y)\,dx + x\,dy$, where C is the circle $x^2 + y^2 = 4$.

8. $\oint_C (e^x + y^2)\,dx + (e^y + x^2)\,dy$, where C is the boundary of the region between $y = x^2$ and $y = x$.

9. $\oint_C \ln(1+y)\,dx - \dfrac{xy}{1+y}\,dy$, where C is the triangle with vertices $(0, 0)$, $(2, 0)$, and $(0, 4)$.

10. $\oint_C x^2y\,dx - y^2x\,dy$, where C is the boundary of the region in the first quadrant, enclosed between the coordinate axes and the circle $x^2 + y^2 = 16$.

11. $\oint_C \tan^{-1} y\,dx - \dfrac{y^2x}{1+y^2}\,dy$, where C is the square with vertices $(0, 0)$, $(1, 0)$, $(1, 1)$, and $(0, 1)$.

12. $\oint_C \cos x\sin y\,dx + \sin x\cos y\,dy$, where C is the triangle with vertices $(0, 0)$, $(3, 3)$, and $(0, 3)$.

13. $\oint_C x^2y\,dx + (y + xy^2)\,dy$, where C is the boundary of the region enclosed by $y = x^2$ and $x = y^2$.

14. Let C be the boundary of the region enclosed between $y = x^2$ and $y = 2x$. Assuming that C is oriented counterclockwise, evaluate the following integrals by Green's Theorem:

(a) $\oint_C (6xy - y^2)\,dx$ (b) $\oint_C (6xy - y^2)\,dy$

[C] **15.** Use a CAS to check Green's Theorem by evaluating both integrals in the equation

$$\oint_C e^y\,dx + ye^x\,dy = \iint_R \left[\frac{\partial}{\partial x}(ye^x) - \frac{\partial}{\partial y}(e^y)\right] dA$$

where

(a) C is the circle $x^2 + y^2 = 1$

(b) C is the boundary of the region enclosed by $y = x^2$ and $x = y^2$.

16. In Example 3, we used Green's Theorem to obtain the area of an ellipse. Obtain this area using the first and then the second formula in (6).

17. Use a line integral to find the area of the region enclosed by the astroid

$$x = a\cos^3\phi, \quad y = a\sin^3\phi \qquad (0 \le \phi \le 2\pi)$$

[See Exercise 25 of Section 6.4.]

18. Use a line integral to find the area of the triangle with vertices $(0, 0)$, $(a, 0)$, and $(0, b)$, where $a > 0$ and $b > 0$.

19. Use the formula

$$A = \frac{1}{2}\oint_C -y\,dx + x\,dy$$

to find the area of the region swept out by the line from the origin to the ellipse $x = a\cos t$, $y = b\sin t$ if t varies from $t = 0$ to $t = t_0$ $(0 \le t_0 \le 2\pi)$.

20. Use the formula

$$A = \frac{1}{2}\oint_C -y\,dx + x\,dy$$

to find the area of the region swept out by the line from the origin to the hyperbola $x = a\cosh t$, $y = b\sinh t$ if t varies from $t = 0$ to $t = t_0$ $(t_0 \ge 0)$.

FOCUS ON CONCEPTS

21. Suppose that $\mathbf{F}(x, y) = f(x, y)\mathbf{i} + g(x, y)\mathbf{j}$ is a vector field whose component functions f and g have continuous first partial derivatives. Let C denote a simple, closed, piecewise smooth curve oriented counterclockwise that bounds a region R contained in the domain of $\mathbf{F}$. We can think of $\mathbf{F}$ as a vector field in 3-space by writing it as

$$\mathbf{F}(x, y, z) = f(x, y)\mathbf{i} + g(x, y)\mathbf{j} + 0\mathbf{k}$$

With this convention, explain why

$$\int_C \mathbf{F}\cdot d\mathbf{r} = \iint_R \operatorname{curl}\mathbf{F}\cdot\mathbf{k}\,dA$$

22. Suppose that $\mathbf{F}(x, y) = f(x, y)\mathbf{i} + g(x, y)\mathbf{j}$ is a vector field on the xy-plane and that f and g have continuous first partial derivatives with $f_y = g_x$ everywhere. Use Green's Theorem to explain why

$$\int_{C_1} \mathbf{F}\cdot d\mathbf{r} = \int_{C_2} \mathbf{F}\cdot d\mathbf{r}$$

where C_1 and C_2 are the oriented curves in the accompanying figure. (Compare this result with Theorems 16.3.2 and 16.3.3.)

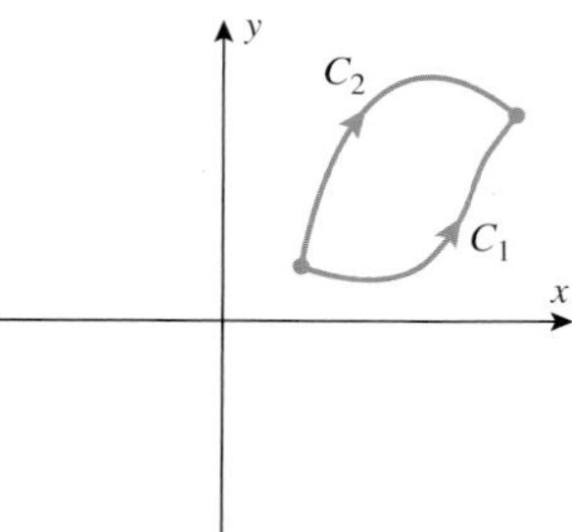

Figure Ex-22

23. Suppose that $f(x)$ and $g(x)$ are continuous functions with $g(x) \le f(x)$. Let R denote the region bounded by the graph of f, the graph of g, and the vertical lines $x = a$ and $x = b$. Let C denote the boundary of R oriented counterclockwise. What familiar formula results from applying Green's Theorem to $\int_C (-y)\,dx$?

24. In the accompanying figure, C is a smooth oriented curve from $P(x_0, y_0)$ to $Q(x_1, y_1)$ that is contained inside the rectangle with corners at the origin and Q and outside the rectangle with corners at the origin and P.

(a) What region in the figure has area $\int_C x\,dy$?

(b) What region in the figure has area $\int_C y\,dx$?

(c) Express $\int_C x\,dy + \int_C y\,dx$ in terms of the coordinates of P and Q.

(d) Interpret the result of part (c) in terms of the Fundamental Theorem of Line Integrals.

(e) Interpret the result in part (c) in terms of integration by parts.

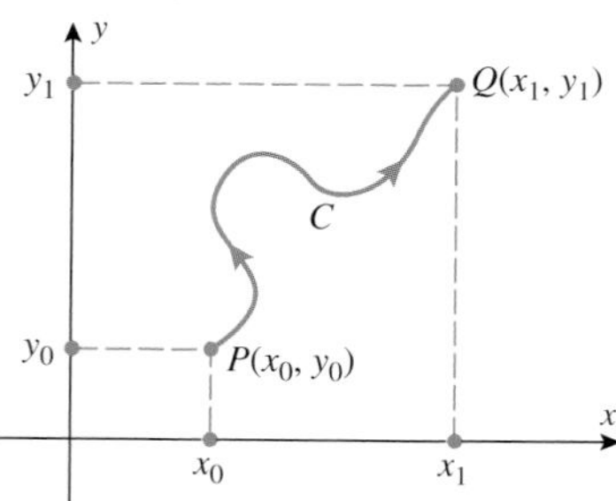

Figure Ex-24

25–26 Use Green's Theorem to find the work done by the force field **F** on a particle that moves along the stated path.

25. $\mathbf{F}(x, y) = xy\mathbf{i} + \left(\frac{1}{2}x^2 + xy\right)\mathbf{j}$; the particle starts at $(5, 0)$, traverses the upper semicircle $x^2 + y^2 = 25$, and returns to its starting point along the x-axis.

26. $\mathbf{F}(x, y) = \sqrt{y}\,\mathbf{i} + \sqrt{x}\,\mathbf{j}$; the particle moves counterclockwise one time around the closed curve given by the equations $y = 0$, $x = 2$, and $y = x^3/4$.

27. Evaluate $\oint_C y\,dx - x\,dy$, where C is the cardioid

$$r = a(1 + \cos\theta) \quad (0 \le \theta \le 2\pi)$$

28. Let R be a plane region with area A whose boundary is a piecewise smooth simple closed curve C. Use Green's Theorem to prove that the centroid $(\bar{x}, \bar{y})$ of R is given by

$$\bar{x} = \frac{1}{2A}\oint_C x^2\,dy, \quad \bar{y} = -\frac{1}{2A}\oint_C y^2\,dx$$

29–32 Use the result in Exercise 28 to find the centroid of the region.

29. **30.** **31.** **32.**

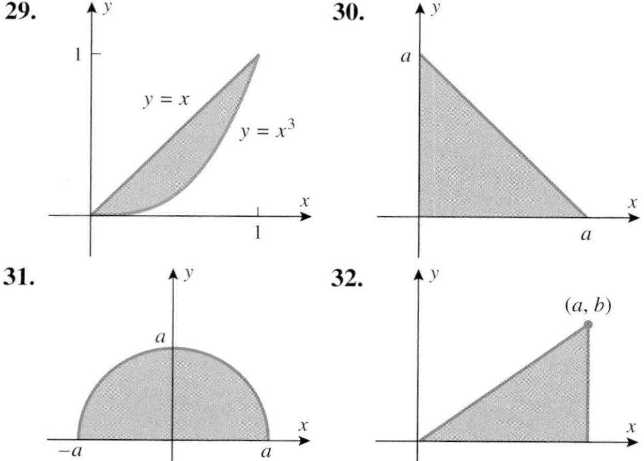

33. Find a simple closed curve C with counterclockwise orientation that maximizes the value of

$$\oint_C \tfrac{1}{3}y^3\,dx + \left(x - \tfrac{1}{3}x^3\right)dy$$

and explain your reasoning.

34. (a) Let C be the line segment from a point (a, b) to a point (c, d). Show that

$$\int_C -y\,dx + x\,dy = ad - bc$$

(b) Use the result in part (a) to show that the area A of a triangle with successive vertices (x_1, y_1), (x_2, y_2), and (x_3, y_3) going counterclockwise is

$$A = \tfrac{1}{2}[(x_1y_2 - x_2y_1) + (x_2y_3 - x_3y_2) + (x_3y_1 - x_1y_3)]$$

(c) Find a formula for the area of a polygon with successive vertices $(x_1, y_1), (x_2, y_2), \ldots, (x_n, y_n)$ going counterclockwise.

(d) Use the result in part (c) to find the area of a quadrilateral with vertices $(0, 0)$, $(3, 4)$, $(-2, 2)$, $(-1, 0)$.

35–36 Evaluate the integral $\int_C \mathbf{F} \cdot d\mathbf{r}$, where C is the boundary of the region R and C is oriented so that the region is on the left when the boundary is traversed in the direction of its orientation.

35. $\mathbf{F}(x, y) = (x^2 + y)\mathbf{i} + (4x - \cos y)\mathbf{j}$; C is the boundary of the region R that is inside the square with vertices $(0, 0)$, $(5, 0)$, $(5, 5)$, $(0, 5)$ but is outside the rectangle with vertices $(1, 1)$, $(3, 1)$, $(3, 2)$, $(1, 2)$.

36. $\mathbf{F}(x, y) = (e^{-x} + 3y)\mathbf{i} + x\mathbf{j}$; C is the boundary of the region R inside the circle $x^2 + y^2 = 16$ and outside the circle $x^2 - 2x + y^2 = 3$.

QUICK CHECK ANSWERS 16.4

1. 8 2. $\frac{1}{2}$ 3. 2π 4. R is the region $x^2 + y^2 \leq 1$ $(0 \leq x, 0 \leq y)$ and $f(x, y) = -y^2$, $g(x, y) = x^2$

16.5 SURFACE INTEGRALS

In previous sections we considered four kinds of integrals—integrals over intervals, double integrals over two-dimensional regions, triple integrals over three-dimensional solids, and line integrals along curves in two- or three-dimensional space. In this section we will discuss integrals over surfaces in three-dimensional space. Such integrals occur in problems involving fluid and heat flow, electricity, magnetism, mass, and center of gravity.

DEFINITION OF A SURFACE INTEGRAL

In this section we will define what it means to integrate a function $f(x, y, z)$ over a smooth parametric surface σ. To motivate the definition we will consider the problem of finding the mass of a curved lamina whose density function (mass per unit area) is known. Recall that in Section 15.6 we defined a *lamina* to be an idealized flat object that is thin enough to be viewed as a plane region. Analogously, a ***curved lamina*** is an idealized object that is thin enough to be viewed as a surface in 3-space. A curved lamina may look like a bent plate, as in Figure 16.5.1, or it may enclose a region in 3-space, like the shell of an egg. We will model the lamina by a smooth parametric surface σ. Given any point (x, y, z) on σ, we let $f(x, y, z)$ denote the corresponding value of the density function. To compute the mass of the lamina, we proceed as follows:

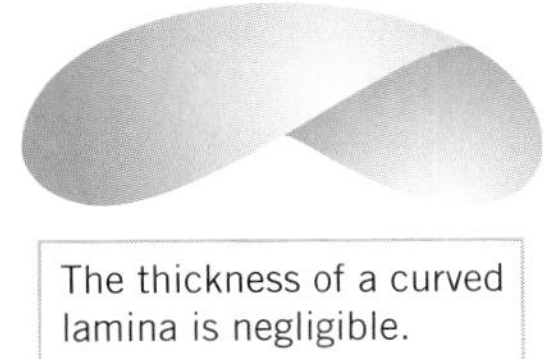

Figure 16.5.1

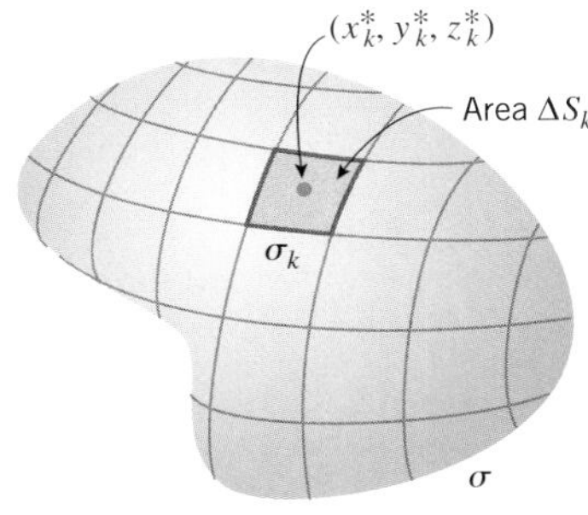

Figure 16.5.2

- As shown in Figure 16.5.2, we divide σ into n very small patches $\sigma_1, \sigma_2, \ldots, \sigma_n$ with areas $\Delta S_1, \Delta S_2, \ldots, \Delta S_n$, respectively. Let (x_k^*, y_k^*, z_k^*) be a sample point in the kth patch with ΔM_k the mass of the corresponding section.
- If the dimensions of σ_k are very small, the value of f will not vary much along the kth section and we can approximate f along this section by the value $f(x_k^*, y_k^*, z_k^*)$. It follows that the mass of the kth section can be approximated by

$$\Delta M_k \approx f(x_k^*, y_k^*, z_k^*)\Delta S_k$$

- The mass M of the entire lamina can then be approximated by

$$M = \sum_{k=1}^{n} \Delta M_k \approx \sum_{k=1}^{n} f(x_k^*, y_k^*, z_k^*)\Delta S_k \tag{1}$$

- We will use the expression $n \to \infty$ to indicate the process of increasing n in such a way that the maximum dimension of each patch approaches 0. It is plausible that the error in (1) will approach 0 as $n \to \infty$ and the exact value of M will be given by

$$M = \lim_{n \to \infty} \sum_{k=1}^{n} f(x_k^*, y_k^*, z_k^*)\Delta S_k \tag{2}$$

The limit in (2) is very similar to the limit used to find the mass of a thin wire [Formula (2) in Section 16.2]. By analogy to Definition 16.2.1, we make the following definition.

16.5.1 DEFINITION. If σ is a smooth parametric surface, then the ***surface integral*** of $f(x, y, z)$ over σ is

$$\iint_{\sigma} f(x, y, z)\, dS = \lim_{n\to\infty} \sum_{k=1}^{n} f(x_k^*, y_k^*, z_k^*)\Delta S_k \tag{3}$$

provided this limit exists and does not depend on the way the subdivisions of σ are made or how the sample points (x_k^*, y_k^*, z_k^*) are chosen.

It can be shown that the integral of f over σ exists if f is continuous on σ.

We see from (2) and Definition 16.5.1 that if σ models a lamina and if $f(x, y, z)$ is the density function of the lamina, then the mass M of the lamina is given by

$$M = \iint_{\sigma} f(x, y, z)\, dS \tag{4}$$

That is, to obtain the mass of a lamina, we integrate the density function over the smooth surface that models the lamina.

Note that if σ is a smooth surface of surface area S, and f is identically 1, then it immediately follows from Definition 16.5.1 that

$$\iint_{\sigma} dS = \lim_{n\to\infty} \sum_{k=1}^{n} \Delta S_k = \lim_{n\to\infty} S = S \tag{5}$$

EVALUATING SURFACE INTEGRALS

There are various procedures for evaluating surface integrals that depend on how the surface σ is represented. The following theorem provides a method for evaluating a surface integral when σ is represented parametrically.

16.5.2 THEOREM. *Let σ be a smooth parametric surface whose vector equation is*

$$\mathbf{r} = x(u, v)\mathbf{i} + y(u, v)\mathbf{j} + z(u, v)\mathbf{k}$$

where (u, v) varies over a region R in the uv-plane. If $f(x, y, z)$ is continuous on σ, then

$$\iint_{\sigma} f(x, y, z)\, dS = \iint_{R} f(x(u, v), y(u, v), z(u, v)) \left\| \frac{\partial \mathbf{r}}{\partial u} \times \frac{\partial \mathbf{r}}{\partial v} \right\| dA \tag{6}$$

Explain how to use Formula (6) to confirm Formula (5).

To motivate this result, suppose that the parameter domain R is subdivided as in Figure 15.4.10, and suppose that the point (x_k^*, y_k^*, z_k^*) in (3) corresponds to parameter values of u_k^* and v_k^*. If we use Formula (9) of Section 15.4 to approximate ΔS_k, and if we assume that the errors in the approximations approach zero as $n \to +\infty$, then it follows from (3) that

$$\iint_\sigma f(x, y, z)\, dS = \lim_{n \to +\infty} \sum_{k=1}^{n} f(x(u_k^*, v_k^*), y(u_k^*, v_k^*), z(u_k^*, v_k^*)) \left\| \frac{\partial \mathbf{r}}{\partial u} \times \frac{\partial \mathbf{r}}{\partial v} \right\| \Delta A_k$$

which suggests Formula (6).

Although Theorem 16.5.2 is stated for *smooth* parametric surfaces, Formula (6) remains valid even if $\partial \mathbf{r}/\partial u \times \partial \mathbf{r}/\partial v$ is allowed to equal $\mathbf{0}$ on the boundary of R.

▶ **Example 1** Evaluate the surface integral $\iint_\sigma x^2\, dS$ over the sphere $x^2 + y^2 + z^2 = 1$.

Solution. As in Example 9 of Section 15.4 (with $a = 1$), the sphere is the graph of the vector-valued function

$$\mathbf{r}(\phi, \theta) = \sin\phi \cos\theta\, \mathbf{i} + \sin\phi \sin\theta\, \mathbf{j} + \cos\phi\, \mathbf{k} \quad (0 \le \phi \le \pi,\ \ 0 \le \theta \le 2\pi) \tag{7}$$

and

$$\left\| \frac{\partial \mathbf{r}}{\partial \phi} \times \frac{\partial \mathbf{r}}{\partial \theta} \right\| = \sin\phi$$

Explain why the function $\mathbf{r}(\phi, \theta)$ given in (7) fails to be smooth on its domain.

From the $\mathbf{i}$-component of $\mathbf{r}$, the integrand in the surface integral can be expressed in terms of ϕ and θ as $x^2 = \sin^2\phi \cos^2\theta$. Thus, it follows from (6) with ϕ and θ in place of u and v and R as the rectangular region in the $\phi\theta$-plane determined by the inequalities in (7) that

$$\begin{aligned}
\iint_\sigma x^2\, dS &= \iint_R (\sin^2\phi \cos^2\theta) \left\| \frac{\partial \mathbf{r}}{\partial \phi} \times \frac{\partial \mathbf{r}}{\partial \theta} \right\| dA \\
&= \int_0^{2\pi} \int_0^{\pi} \sin^3\phi \cos^2\theta\, d\phi\, d\theta \\
&= \int_0^{2\pi} \left[\int_0^{\pi} \sin^3\phi\, d\phi \right] \cos^2\theta\, d\theta \\
&= \int_0^{2\pi} \left[\frac{1}{3}\cos^3\phi - \cos\phi \right]_0^{\pi} \cos^2\theta\, d\theta && \text{Formula (11), Section 8.3} \\
&= \frac{4}{3} \int_0^{2\pi} \cos^2\theta\, d\theta \\
&= \frac{4}{3} \left[\frac{1}{2}\theta + \frac{1}{4}\sin 2\theta \right]_0^{2\pi} = \frac{4\pi}{3} && \text{Formula (8), Section 8.3}
\end{aligned}$$
◀

SURFACE INTEGRALS OVER $z = g(x, y)$, $y = g(x, z)$, AND $x = g(y, z)$

In the case where σ is a surface of the form $z = g(x, y)$, we can take $x = u$ and $y = v$ as parameters and express the equation of the surface as

$$\mathbf{r} = u\mathbf{i} + v\mathbf{j} + g(u, v)\mathbf{k}$$

in which case we obtain

$$\left\| \frac{\partial \mathbf{r}}{\partial u} \times \frac{\partial \mathbf{r}}{\partial v} \right\| = \sqrt{\left(\frac{\partial z}{\partial x}\right)^2 + \left(\frac{\partial z}{\partial y}\right)^2 + 1}$$

[see the derivation of Formula (11) in Section 15.4]. Thus, it follows from (6) that

$$\iint_{\sigma} f(x, y, z)\, dS = \iint_{R} f(x, y, g(x, y))\sqrt{\left(\frac{\partial z}{\partial x}\right)^2 + \left(\frac{\partial z}{\partial y}\right)^2 + 1}\, dA$$

Note that in this formula the region R lies in the xy-plane because the parameters are x and y. Geometrically, this region is the projection of σ on the xy-plane. The following theorem summarizes this result and gives analogous formulas for surface integrals over surfaces of the form $y = g(x, z)$ and $x = g(y, z)$.

16.5.3 THEOREM.

(*a*) *Let σ be a surface with equation $z = g(x, y)$ and let R be its projection on the xy-plane. If g has continuous first partial derivatives on R and $f(x, y, z)$ is continuous on σ, then*

$$\iint_{\sigma} f(x, y, z)\, dS = \iint_{R} f(x, y, g(x, y))\sqrt{\left(\frac{\partial z}{\partial x}\right)^2 + \left(\frac{\partial z}{\partial y}\right)^2 + 1}\, dA \tag{8}$$

(*b*) *Let σ be a surface with equation $y = g(x, z)$ and let R be its projection on the xz-plane. If g has continuous first partial derivatives on R and $f(x, y, z)$ is continuous on σ, then*

$$\iint_{\sigma} f(x, y, z)\, dS = \iint_{R} f(x, g(x, z), z)\sqrt{\left(\frac{\partial y}{\partial x}\right)^2 + \left(\frac{\partial y}{\partial z}\right)^2 + 1}\, dA \tag{9}$$

(*c*) *Let σ be a surface with equation $x = g(y, z)$ and let R be its projection on the yz-plane. If g has continuous first partial derivatives on R and $f(x, y, z)$ is continuous on σ, then*

$$\iint_{\sigma} f(x, y, z)\, dS = \iint_{R} f(g(y, z), y, z)\sqrt{\left(\frac{\partial x}{\partial y}\right)^2 + \left(\frac{\partial x}{\partial z}\right)^2 + 1}\, dA \tag{10}$$

Formulas (9) and (10) can be recovered from Formula (8). Explain how.

▸ **Example 2** Evaluate the surface integral

$$\iint_{\sigma} xz\, dS$$

where σ is the part of the plane $x + y + z = 1$ that lies in the first octant.

Solution. The equation of the plane can be written as

$$z = 1 - x - y$$

Consequently, we can apply Formula (8) with $z = g(x, y) = 1 - x - y$ and $f(x, y, z) = xz$. We have

$$\frac{\partial z}{\partial x} = -1 \quad \text{and} \quad \frac{\partial z}{\partial y} = -1$$

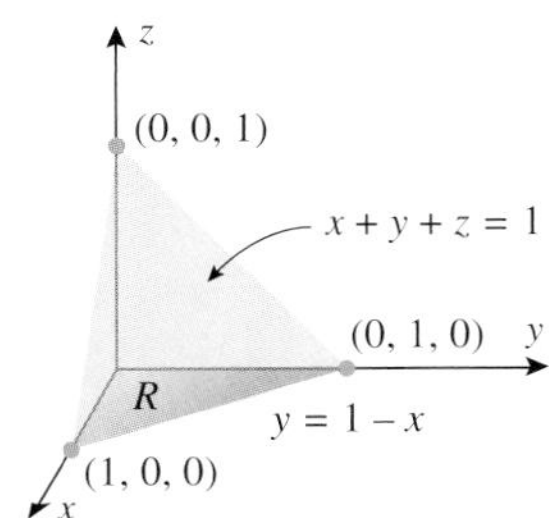

Figure 16.5.3

so (8) becomes

$$\iint_{\sigma} xz\,dS = \iint_{R} x(1-x-y)\sqrt{(-1)^2+(-1)^2+1}\,dA \tag{11}$$

where R is the projection of σ on the xy-plane (Figure 16.5.3). Rewriting the double integral in (11) as an iterated integral yields

$$\begin{aligned}\iint_{\sigma} xz\,dS &= \sqrt{3}\int_0^1\int_0^{1-x}(x-x^2-xy)\,dy\,dx\\ &= \sqrt{3}\int_0^1\left[xy - x^2y - \frac{xy^2}{2}\right]_{y=0}^{1-x}dx\\ &= \sqrt{3}\int_0^1\left(\frac{x}{2}-x^2+\frac{x^3}{2}\right)dx\\ &= \sqrt{3}\left[\frac{x^2}{4}-\frac{x^3}{3}+\frac{x^4}{8}\right]_0^1 = \frac{\sqrt{3}}{24} \blacktriangleleft\end{aligned}$$

▶ **Example 3** Evaluate the surface integral

$$\iint_{\sigma} y^2z^2\,dS$$

where σ is the part of the cone $z=\sqrt{x^2+y^2}$ that lies between the planes $z=1$ and $z=2$ (Figure 16.5.4).

Solution. We will apply Formula (8) with

$$z = g(x,y) = \sqrt{x^2+y^2} \quad \text{and} \quad f(x,y,z) = y^2z^2$$

Thus,

$$\frac{\partial z}{\partial x} = \frac{x}{\sqrt{x^2+y^2}} \quad \text{and} \quad \frac{\partial z}{\partial y} = \frac{y}{\sqrt{x^2+y^2}}$$

so

$$\sqrt{\left(\frac{\partial z}{\partial x}\right)^2+\left(\frac{\partial z}{\partial y}\right)^2+1} = \sqrt{2}$$

(verify), and (8) yields

$$\iint_{\sigma} y^2z^2\,dS = \iint_{R} y^2\left(\sqrt{x^2+y^2}\right)^2\sqrt{2}\,dA = \sqrt{2}\iint_{R} y^2(x^2+y^2)\,dA$$

where R is the annulus enclosed between $x^2+y^2=1$ and $x^2+y^2=4$ (Figure 16.5.4). Using polar coordinates to evaluate this double integral over the annulus R yields

$$\begin{aligned}\iint_{\sigma} y^2z^2\,dS &= \sqrt{2}\int_0^{2\pi}\int_1^2 (r\sin\theta)^2(r^2)r\,dr\,d\theta\\ &= \sqrt{2}\int_0^{2\pi}\int_1^2 r^5\sin^2\theta\,dr\,d\theta\\ &= \sqrt{2}\int_0^{2\pi}\left[\frac{r^6}{6}\sin^2\theta\right]_{r=1}^{2}d\theta = \frac{21}{\sqrt{2}}\int_0^{2\pi}\sin^2\theta\,d\theta\\ &= \frac{21}{\sqrt{2}}\left[\frac{1}{2}\theta - \frac{1}{4}\sin 2\theta\right]_0^{2\pi} = \frac{21\pi}{\sqrt{2}} \blacktriangleleft\end{aligned}$$

Formula (7), Section 8.3

Evaluate the integral in Example 3 with the help of Formula (6) and the parametrization

$$\mathbf{r} = \langle r\cos\theta, r\sin\theta, r\rangle$$
$$(1 \le r \le 2, 0 \le \theta \le 2\pi)$$

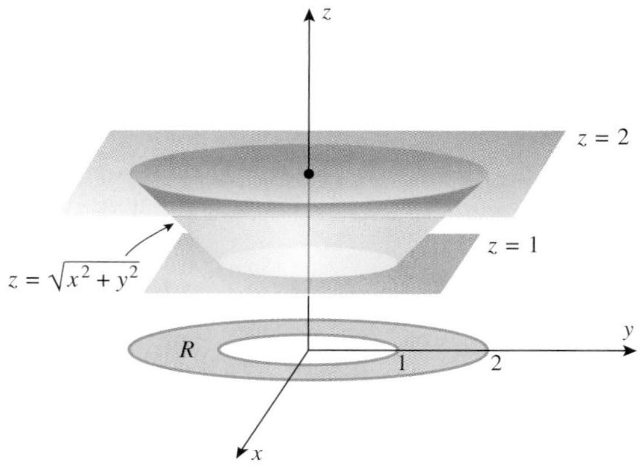

Figure 16.5.4

▶ **Example 4** Suppose that a curved lamina σ with constant density $\delta(x, y, z) = \delta_0$ is the portion of the paraboloid $z = x^2 + y^2$ below the plane $z = 1$ (Figure 16.5.5). Find the mass of the lamina.

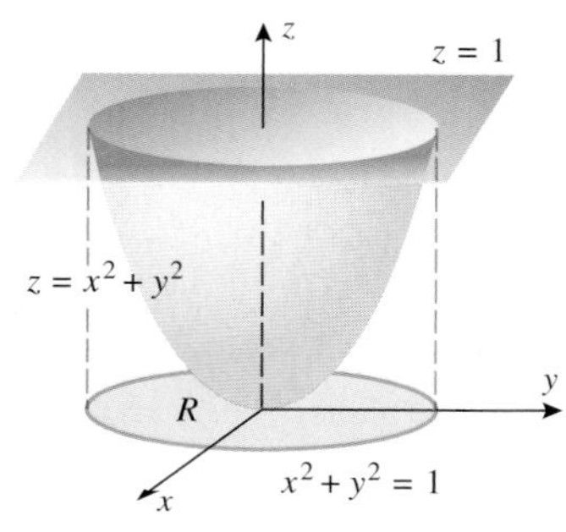

Figure 16.5.5

Solution. Since $z = g(x, y) = x^2 + y^2$, it follows that

$$\frac{\partial z}{\partial x} = 2x \quad \text{and} \quad \frac{\partial z}{\partial y} = 2y$$

Therefore,

$$M = \iint_{\sigma} \delta_0 \, dS = \iint_{R} \delta_0 \sqrt{(2x)^2 + (2y)^2 + 1} \, dA = \delta_0 \iint_{R} \sqrt{4x^2 + 4y^2 + 1} \, dA \tag{12}$$

where R is the circular region enclosed by $x^2 + y^2 = 1$. To evaluate (12) we use polar coordinates:

$$\begin{aligned} M &= \delta_0 \int_0^{2\pi} \int_0^1 \sqrt{4r^2 + 1} \, r \, dr \, d\theta = \frac{\delta_0}{12} \int_0^{2\pi} (4r^2 + 1)^{3/2} \Big]_{r=0}^1 d\theta \\ &= \frac{\delta_0}{12} \int_0^{2\pi} (5^{3/2} - 1) \, d\theta = \frac{\pi \delta_0}{6} (5\sqrt{5} - 1) \end{aligned}$$ ◀

✔ QUICK CHECK EXERCISES 16.5 *(See page 1155 for answers.)*

1. Consider the surface integral $\iint_{\sigma} f(x, y, z) \, dS$.
   (a) If σ is a parametric surface whose vector equation is
   $$\mathbf{r} = x(u, v)\mathbf{i} + y(u, v)\mathbf{j} + z(u, v)\mathbf{k}$$
   to evaluate the integral replace dS by ________.
   (b) If σ is the graph of a function $z = g(x, y)$ with continuous first partial derivatives, to evaluate the integral replace dS by ________.

2. If σ is the triangular region with vertices $(1, 0, 0)$, $(0, 1, 0)$, and $(0, 0, 1)$, then
$$\iint_{\sigma} (x + y + z) \, dS = \_\_\_\_\_\_\_\_$$

3. If σ is the sphere of radius 2 centered at the origin, then
$$\iint_{\sigma} (x^2 + y^2 + z^2) \, dS = \_\_\_\_\_\_\_\_$$

4. If $f(x, y, z)$ is the mass density function of a curved lamina σ, then the mass of σ is given by the integral ________.

EXERCISE SET 16.5 [C] CAS

1–8 Evaluate the surface integral

$$\iint_{\sigma} f(x, y, z)\, dS$$

1. $f(x, y, z) = z^2$; σ is the portion of the cone $z = \sqrt{x^2 + y^2}$ between the planes $z = 1$ and $z = 2$.
2. $f(x, y, z) = xy$; σ is the portion of the plane $x + y + z = 1$ lying in the first octant.
3. $f(x, y, z) = x^2 y$; σ is the portion of the cylinder $x^2 + z^2 = 1$ between the planes $y = 0$, $y = 1$, and above the xy-plane.
4. $f(x, y, z) = (x^2 + y^2)z$; σ is the portion of the sphere $x^2 + y^2 + z^2 = 4$ above the plane $z = 1$.
5. $f(x, y, z) = x - y - z$; σ is the portion of the plane $x + y = 1$ in the first octant between $z = 0$ and $z = 1$.
6. $f(x, y, z) = x + y$; σ is the portion of the plane $z = 6 - 2x - 3y$ in the first octant.
7. $f(x, y, z) = x + y + z$; σ is the surface of the cube defined by the inequalities $0 \le x \le 1, 0 \le y \le 1, 0 \le z \le 1$. [*Hint:* Integrate over each face separately.]
8. $f(x, y, z) = x^2 + y^2$; σ is the surface of the sphere $x^2 + y^2 + z^2 = a^2$.

9–10 Sometimes evaluating a surface integral results in an improper integral. When this happens, one can either attempt to determine the value of the integral using an appropriate limit or one can try another method. These exercises explore both approaches.

9. Consider the integral of $f(x, y, z) = z + 1$ over the upper hemisphere σ: $z = \sqrt{1 - x^2 - y^2}$ $(0 \le x^2 + y^2 \le 1)$.
   (a) Explain why evaluating this surface integral using (8) results in an improper integral.
   (b) Use (8) to evaluate the integral of f over the surface $\sigma_r : z = \sqrt{1 - x^2 - y^2}$ $(0 \le x^2 + y^2 \le r^2 < 1)$. Take the limit of this result as $r \to 1^-$ to determine the integral of f over σ.
   (c) Parametrize σ using spherical coordinates and evaluate the integral of f over σ using (6). Verify that your answer agrees with the result in part (b).
10. Consider the integral of $f(x, y, z) = \sqrt{x^2 + y^2 + z^2}$ over the cone $\sigma : z = \sqrt{x^2 + y^2}$ $(0 \le z \le 1)$.
    (a) Explain why evaluating this surface integral using (8) results in an improper integral.
    (b) Use (8) to evaluate the integral of f over the surface $\sigma_r : z = \sqrt{x^2 + y^2}$ $(0 < r^2 \le x^2 + y^2 \le 1)$. Take the limit of this result as $r \to 0^+$ to determine the integral of f over σ.
    (c) Parametrize σ using spherical coordinates and evaluate the integral of f over σ using (6). Verify that your answer agrees with the result in part (b).

FOCUS ON CONCEPTS

11–14 In some cases it is possible to use Definition 16.5.1 along with symmetry considerations to evaluate a surface integral without reference to a parametrization of the surface. In these exercises, σ denotes the unit sphere centered at the origin.

11. (a) Explain why it is possible to subdivide σ into patches and choose corresponding sample points (x_k^*, y_k^*, z_k^*) such that (i) the dimensions of each patch are as small as desired and (ii) for each sample point (x_k^*, y_k^*, z_k^*), there exists a sample point (x_j^*, y_j^*, z_j^*) with
    $$x_k = -x_j, \quad y_k = y_j, \quad z_k = z_j$$
    and with $\Delta S_k = \Delta S_j$.
    (b) Use Definition 16.5.1, the result in part (a), and the fact that surface integrals exist for continuous functions to prove that $\iint_\sigma x^n\, dS = 0$ for n an odd positive integer.
12. Use the argument in Exercise 11 to prove that if $f(x)$ is a continuous odd function of x and if $g(y, z)$ is a continuous function, then
    $$\iint_{\sigma} f(x)g(y, z)\, dS = 0$$
13. (a) Explain why
    $$\iint_{\sigma} x^2\, dS = \iint_{\sigma} y^2\, dS = \iint_{\sigma} z^2\, dS$$
    (b) Conclude from part (a) that
    $$\iint_{\sigma} x^2\, dS = \frac{1}{3}\left[\iint_{\sigma} x^2\, dS + \iint_{\sigma} y^2\, dS + \iint_{\sigma} y^2\, dS\right]$$
    (c) Use part (b) to evaluate
    $$\iint_{\sigma} x^2\, dS$$
    without performing an integration.
14. Use the results of Exercises 12 and 13 to evaluate
    $$\iint_{\sigma} (x - y)^2\, dS$$
    without performing an integration.

15–16 Set up, but do not evaluate, an iterated integral equal to the given surface integral by projecting σ on (a) the xy-plane, (b) the yz-plane, and (c) the xz-plane.

15. $\displaystyle\iint_{\sigma} xyz\, dS$, where σ is the portion of the plane $2x + 3y + 4z = 12$ in the first octant.

16. $\iint_\sigma xz\,dS$, where σ is the portion of the sphere $x^2+y^2+z^2=a^2$ in the first octant.

C 17. Use a CAS to confirm that the three integrals you obtained in Exercise 15 are equal, and find the exact value of the surface integral.

C 18. Try to confirm with a CAS that the three integrals you obtained in Exercise 16 are equal. If you did not succeed, what was the difficulty?

19–20 Set up, but do not evaluate, two different iterated integrals equal to the given integral.

19. $\iint_\sigma xyz\,dS$, where σ is the portion of the surface $y^2=x$ between the planes $z=0$, $z=4$, $y=1$, and $y=2$.

20. $\iint_\sigma x^2y\,dS$, where σ is the portion of the cylinder $y^2+z^2=a^2$ in the first octant between the planes $x=0$, $x=9$, $z=y$, and $z=2y$.

C 21. Use a CAS to confirm that the two integrals you obtained in Exercise 19 are equal, and find the exact value of the surface integral.

C 22. Use a CAS to find the value of the surface integral

$$\iint_\sigma x^2yz\,dS$$

where the surface σ is the portion of the elliptic paraboloid $z=5-3x^2-2y^2$ that lies above the xy-plane.

23–24 Find the mass of the lamina with constant density δ_0.

23. The lamina that is the portion of the circular cylinder $x^2+z^2=4$ that lies directly above the rectangle $R=\{(x,y): 0\le x\le 1, 0\le y\le 4\}$ in the xy-plane.

24. The lamina that is the portion of the paraboloid $2z=x^2+y^2$ inside the cylinder $x^2+y^2=8$.

25. Find the mass of the lamina that is the portion of the surface $y^2=4-z$ between the planes $x=0$, $x=3$, $y=0$, and $y=3$ if the density is $\delta(x,y,z)=y$.

26. Find the mass of the lamina that is the portion of the cone $z=\sqrt{x^2+y^2}$ between $z=1$ and $z=4$ if the density is $\delta(x,y,z)=x^2z$.

27. If a curved lamina has constant density δ_0, what relationship must exist between its mass and surface area? Explain your reasoning.

28. Show that if the density of the lamina $x^2+y^2+z^2=a^2$ at each point is equal to the distance between that point and the xy-plane, then the mass of the lamina is $2\pi a^3$.

29–30 The centroid of a surface σ is defined by

$$\bar{x}=\frac{\iint_\sigma x\,dS}{\text{area of }\sigma},\quad \bar{y}=\frac{\iint_\sigma y\,dS}{\text{area of }\sigma},\quad \bar{z}=\frac{\iint_\sigma z\,dS}{\text{area of }\sigma}$$

Find the centroid of the surface.

29. The portion of the paraboloid $z=\frac{1}{2}(x^2+y^2)$ below the plane $z=4$.

30. The portion of the sphere $x^2+y^2+z^2=4$ above the plane $z=1$.

31–34 Evaluate the integral $\iint_\sigma f(x,y,z)\,dS$ over the surface σ represented by the vector-valued function $\mathbf{r}(u,v)$.

31. $f(x,y,z)=xyz$; $\mathbf{r}(u,v)=u\cos v\mathbf{i}+u\sin v\mathbf{j}+3u\mathbf{k}$ $(1\le u\le 2,\ 0\le v\le \pi/2)$

32. $f(x,y,z)=\dfrac{x^2+z^2}{y}$; $\mathbf{r}(u,v)=2\cos v\mathbf{i}+u\mathbf{j}+2\sin v\mathbf{k}$ $(1\le u\le 3,\ 0\le v\le 2\pi)$

33. $f(x,y,z)=\dfrac{1}{\sqrt{1+4x^2+4y^2}}$; $\mathbf{r}(u,v)=u\cos v\mathbf{i}+u\sin v\mathbf{j}+u^2\mathbf{k}$ $(0\le u\le \sin v,\ 0\le v\le \pi)$

34. $f(x,y,z)=e^{-z}$; $\mathbf{r}(u,v)=2\sin u\cos v\mathbf{i}+2\sin u\sin v\mathbf{j}+2\cos u\mathbf{k}$ $(0\le u\le \pi/2, 0\le v\le 2\pi)$

C 35. Use a CAS to approximate the mass of the curved lamina $z=e^{-x^2-y^2}$ that lies above the region in the xy-plane enclosed by $x^2+y^2=9$ given that the density function is $\delta(x,y,z)=\sqrt{x^2+y^2}$.

C 36. The surface σ shown in the accompanying figure, called a ***Möbius strip***, is represented by the parametric equations

$$x=(5+u\cos(v/2))\cos v$$
$$y=(5+u\cos(v/2))\sin v$$
$$z=u\sin(v/2)$$

where $-1\le u\le 1$ and $0\le v\le 2\pi$.

(a) Use a CAS to generate a reasonable facsimile of this surface.

(b) Use a CAS to approximate the location of the centroid of σ (see the definition preceding Exercise 29).

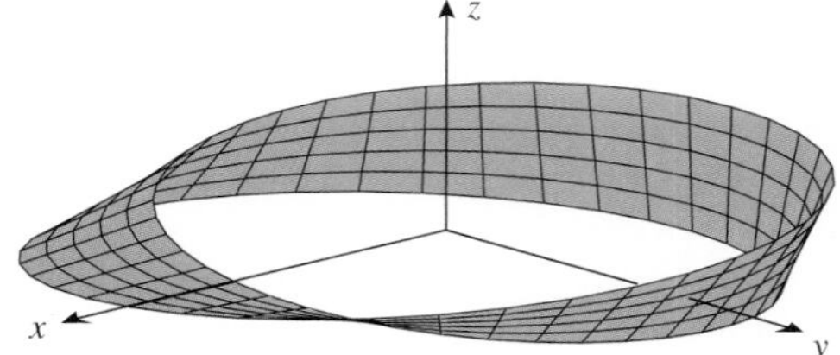

Figure Ex-36

✔ QUICK CHECK ANSWERS 16.5

1. (a) $\left\|\dfrac{\partial \mathbf{r}}{\partial u} \times \dfrac{\partial \mathbf{r}}{\partial v}\right\| dA$ (b) $\sqrt{\left(\dfrac{\partial z}{\partial x}\right)^2 + \left(\dfrac{\partial z}{\partial y}\right)^2 + 1}\, dA$ 2. $\dfrac{\sqrt{3}}{2}$ 3. 64π 4. $\displaystyle\iint_\sigma f(x, y, z)\, dS$

16.6 APPLICATIONS OF SURFACE INTEGRALS; FLUX

In this section we will discuss applications of surface integrals in vector fields associated with fluid flow and electrostatic forces. However, the ideas that we will develop will be general in nature and applicable to other kinds of vector fields as well.

FLOW FIELDS

We will be concerned in this section with vector fields in 3-space that involve some type of "flow"—the flow of a fluid or the flow of charged particles in an electrostatic field, for example. In the case of fluid flow, the vector field $\mathbf{F}(x, y, z)$ represents the velocity of a fluid particle at the point (x, y, z), and the fluid particles flow along "streamlines" that are tangential to the velocity vectors (Figure 16.6.1a). In the case of an electrostatic field, $\mathbf{F}(x, y, z)$ is the force that the field exerts on a small unit of positive charge at the point (x, y, z), and such charges have acceleration in the directions of "electric lines" that are tangential to the force vectors (Figures 16.6.1b and 16.6.1c).

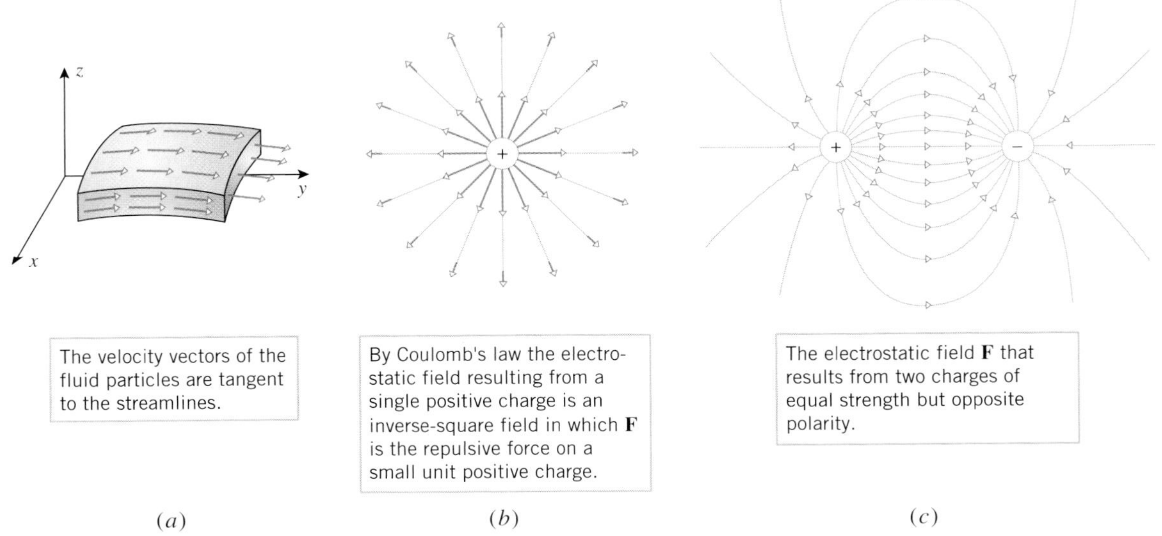

z
y
x
+
+
−
The velocity vectors of the fluid particles are tangent to the streamlines.
By Coulomb's law the electrostatic field resulting from a single positive charge is an inverse-square field in which **F** is the repulsive force on a small unit positive charge.
The electrostatic field **F** that results from two charges of equal strength but opposite polarity.
(a)
(b)
(c)

Figure 16.6.1

ORIENTED SURFACES

Our main goal in this section is to study flows of vector fields through permeable surfaces placed in the field. For this purpose we will need to consider some basic ideas about surfaces. Most surfaces that we encounter in applications have two sides—a sphere has an inside and an outside, and an infinite horizontal plane has a top side and a bottom side, for example. However, there exist mathematical surfaces with only one side. For example, Figure 16.6.2a

shows the construction of a surface called a ***Möbius strip*** [in honor of the German mathematician August Möbius (1790–1868)]. The Möbius strip has only one side in the sense that a bug can traverse the *entire* surface without crossing an edge (Figure 16.6.2*b*). In contrast, a sphere is two-sided in the sense that a bug walking on the sphere can traverse the inside surface or the outside surface but cannot traverse both without somehow passing through the sphere. A two-sided surface is said to be ***orientable***, and a one-sided surface is said to be ***nonorientable***. In the rest of this text we will only be concerned with orientable surfaces.

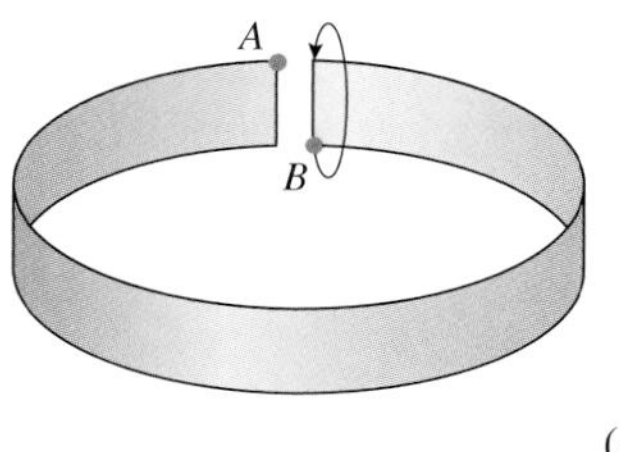

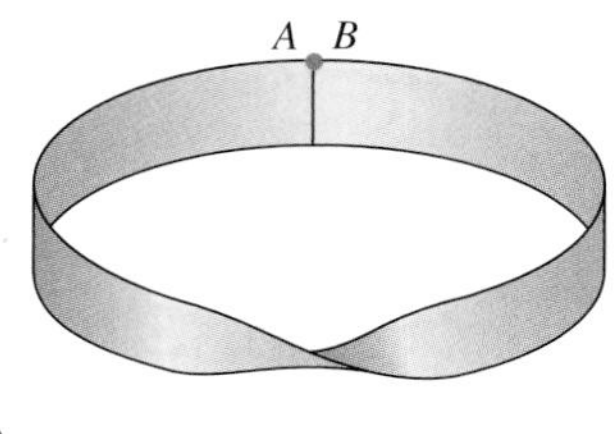

(*a*)

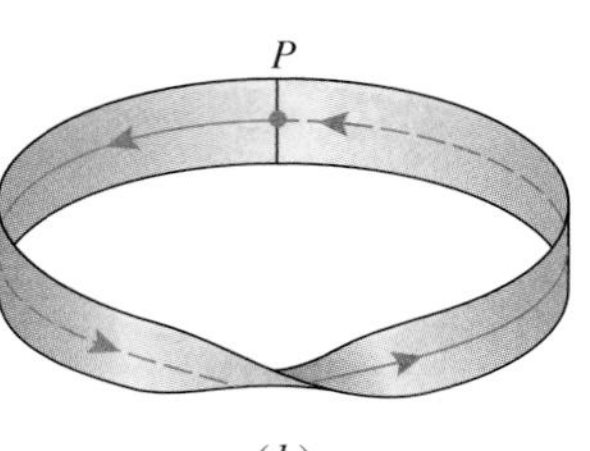

(*b*)

If an ant starts at P with its back facing you and makes one circuit around the strip, then its back will face away from you when it returns to P.

Figure 16.6.2

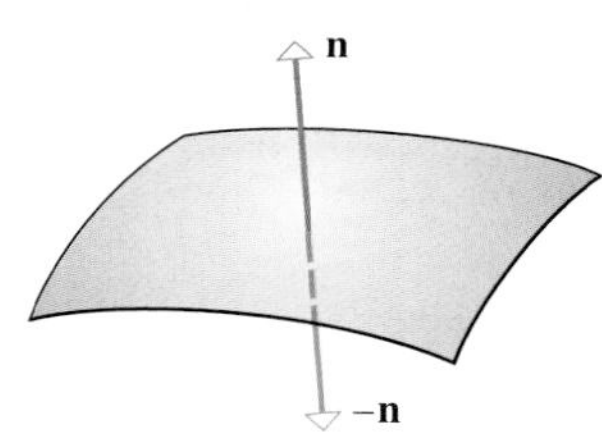

Figure 16.6.3

In applications, it is important to have some way of distinguishing between the two sides of an orientable surface. For this purpose let us suppose that σ is an orientable surface that has a unit normal vector $\mathbf{n}$ at each point. As illustrated in Figure 16.6.3, the vectors $\mathbf{n}$ and $-\mathbf{n}$ point to opposite sides of the surface and hence serve to distinguish between the two sides. It can be proved that if σ is a smooth orientable surface, then it is always possible to choose the direction of $\mathbf{n}$ at each point so that $\mathbf{n} = \mathbf{n}(x, y, z)$ varies continuously over the surface. These unit vectors are then said to form an ***orientation*** of the surface. It can also be proved that a smooth orientable surface has only two possible orientations. For example, the surface in Figure 16.6.4 is oriented up by the purple vectors and down by the green vectors. However, we cannot create a third orientation by mixing the two since this produces points on the surface at which there is an abrupt change in direction (across the black curve in the figure, for example).

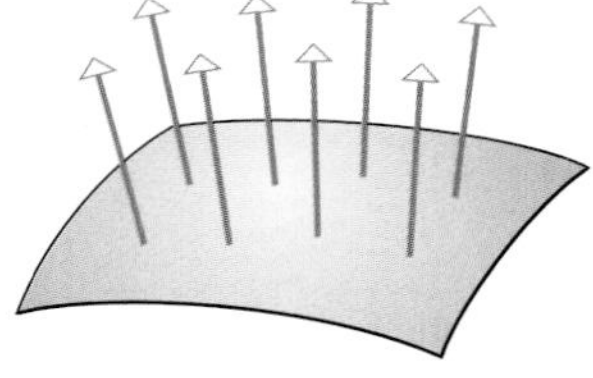

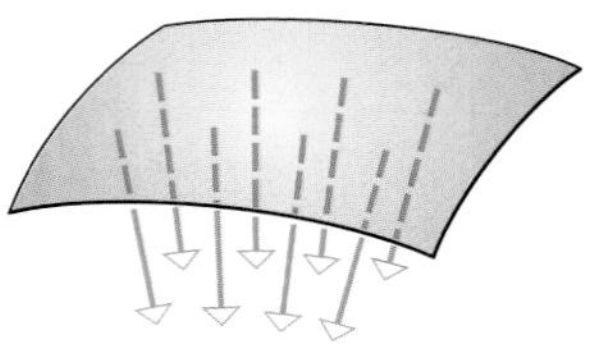

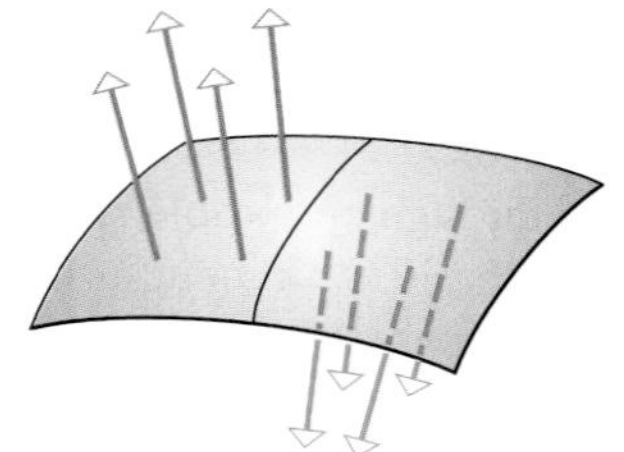

Figure 16.6.4

ORIENTATION OF A SMOOTH PARAMETRIC SURFACE

When a surface is expressed parametrically, the parametric equations create a natural orientation of the surface. To see why this is so, recall from Section 15.4 that if a smooth parametric surface σ is given by the vector equation

$$\mathbf{r} = x(u, v)\mathbf{i} + y(u, v)\mathbf{j} + z(u, v)\mathbf{k}$$

then the unit normal

$$\mathbf{n} = \mathbf{n}(u, v) = \frac{\dfrac{\partial \mathbf{r}}{\partial u} \times \dfrac{\partial \mathbf{r}}{\partial v}}{\left\| \dfrac{\partial \mathbf{r}}{\partial u} \times \dfrac{\partial \mathbf{r}}{\partial v} \right\|} \tag{1}$$

is a continuous vector-valued function of u and v. Thus, Formula (1) defines an orientation of the surface; we call this the ***positive orientation*** of the parametric surface and we say that $\mathbf{n}$ points in the ***positive direction*** from the surface. The orientation determined by $-\mathbf{n}$ is called the ***negative orientation*** of the surface and we say that $-\mathbf{n}$ points in the ***negative direction*** from the surface. For example, consider the cylinder that is represented parametrically by the vector equation

$$\mathbf{r}(u, v) = \cos u\mathbf{i} + v\mathbf{j} - \sin u\mathbf{k} \qquad (0 \le u \le 2\pi,\ 0 \le v \le 1)$$

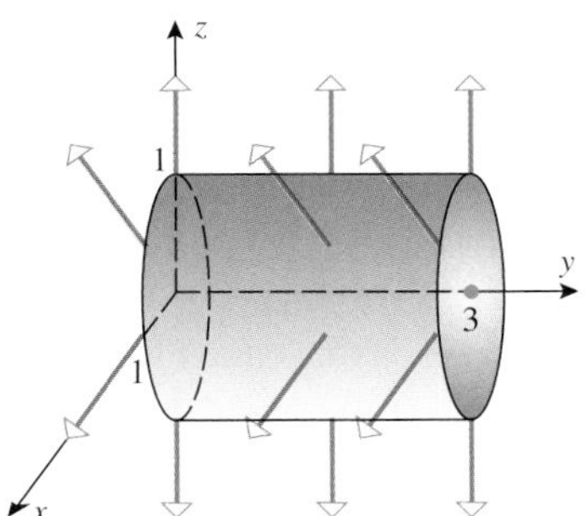

Figure 16.6.5

See if you can find a parametrization of the cylinder in which the positive direction is inward.

Then

$$\frac{\partial \mathbf{r}}{\partial u} \times \frac{\partial \mathbf{r}}{\partial v} = \cos u\mathbf{i} - \sin u\mathbf{k}$$

has unit length, so that Formula (1) becomes

$$\mathbf{n} = \frac{\partial \mathbf{r}}{\partial u} \times \frac{\partial \mathbf{r}}{\partial v} = \cos u\mathbf{i} - \sin u\mathbf{k}$$

Since **n** has the same **i**- and **k**-components as **r**, the positive orientation of the cylinder is *outward* and the negative orientation is *inward* (Figure 16.6.5).

FLUX

In physics, the term *fluid* is used to describe both liquids and gases. Liquids are usually regarded to be ***incompressible***, meaning that the liquid has a uniform density (mass per unit volume) that cannot be altered by compressive forces. Gases are regarded to be ***compressible***, meaning that the density may vary from point to point and can be altered by compressive forces. In this text we will be concerned primarily with incompressible fluids. Moreover, we will assume that the velocity of the fluid at a fixed point does not vary with time. Fluid flows with this property are said to be in a ***steady state***.

Our next goal in this section is to define a fundamental concept of physics known as *flux* (from the Latin word *fluxus*, meaning "flow"). This concept is applicable in any vector field, but we will motivate it in the context of steady-state flow of an incompressible fluid. We consider the following problem.

16.6.1 PROBLEM. Suppose that an oriented surface σ is immersed in an incompressible, steady-state fluid flow and that the surface is permeable so that the fluid can flow through it freely in either direction. Find the net volume of fluid Φ that passes through the surface per unit of time, where the net volume is interpreted to mean the volume that passes through the surface in the positive direction minus the volume that passes through the surface in the negative direction.

To solve this problem, suppose that the velocity of the fluid at a point (x, y, z) on the surface σ is given by

$$\mathbf{F}(x, y, z) = f(x, y, z)\mathbf{i} + g(x, y, z)\mathbf{j} + h(x, y, z)\mathbf{k}$$

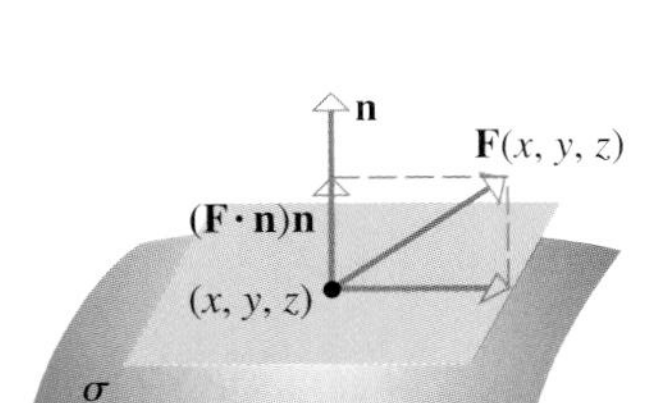

Figure 16.6.6

Let **n** be the unit normal toward the positive side of σ at the point (x, y, z). As illustrated in Figure 16.6.6, the velocity vector **F** can be resolved into two orthogonal components—a component $(\mathbf{F} \cdot \mathbf{n})\mathbf{n}$ that is perpendicular to the surface σ and a second component that is along the "face" of σ. The component of velocity along the face of the surface does not contribute to the flow through σ and hence can be ignored in our computations. Moreover, observe that the sign of $\mathbf{F} \cdot \mathbf{n}$ determines the direction of flow—a positive value means the flow is in the direction of **n** and a negative value means that it is opposite to **n**.

To solve Problem 16.6.1, we subdivide σ into n patches $\sigma_1, \sigma_2, \ldots, \sigma_n$ with areas

$$\Delta S_1, \Delta S_2, \ldots, \Delta S_n$$

If the patches are small and the flow is not too erratic, it is reasonable to assume that the velocity does not vary much on each patch. Thus, if (x_k^*, y_k^*, z_k^*) is any point in the kth patch, we can assume that $\mathbf{F}(x, y, z)$ is constant and equal to $\mathbf{F}(x_k^*, y_k^*, z_k^*)$ throughout the patch and that the component of velocity across the surface σ_k is

$$\mathbf{F}(x_k^*, y_k^*, z_k^*) \cdot \mathbf{n}(x_k^*, y_k^*, z_k^*) \tag{2}$$

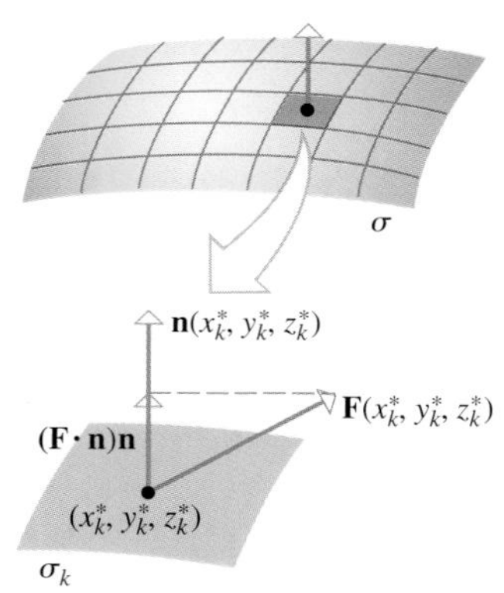

Figure 16.6.7

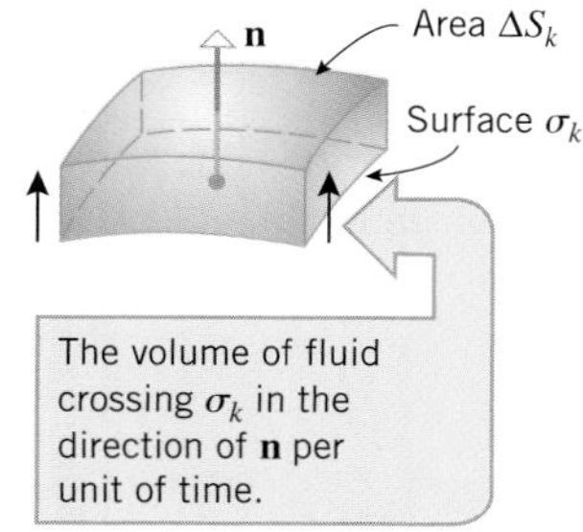

Figure 16.6.8

(Figure 16.6.7). Thus, we can interpret

$$\mathbf{F}(x_k^*, y_k^*, z_k^*) \cdot \mathbf{n}(x_k^*, y_k^*, z_k^*)\Delta S_k$$

as the approximate volume of fluid crossing the patch σ_k in the direction of **n** per unit of time (Figure 16.6.8). For example, if the component of velocity in the direction of **n** is $\mathbf{F}(x_k^*, y_k^*, z_k^*) \cdot \mathbf{n} = 25$ cm/s, and the area of the patch is $\Delta S_k = 2\ \text{cm}^2$, then the volume of fluid ΔV_k crossing the patch in the direction of **n** per unit of time is approximately

$$\Delta V_k \approx \mathbf{F}(x_k^*, y_k^*, z_k^*) \cdot \mathbf{n}(x_k^*, y_k^*, z_k^*)\Delta S_k = 25\ \text{cm/s} \cdot 2\ \text{cm}^2 = 50\ \text{cm}^3/\text{s}$$

In the case where the velocity component $\mathbf{F}(x_k^*, y_k^*, z_k^*) \cdot \mathbf{n}(x_k^*, y_k^*, z_k^*)$ is negative, the flow is in the direction opposite to **n**, so that $-\Delta V_k$ is the approximate volume of fluid crossing the patch σ_k in the direction opposite to **n** per unit time. Thus, the sum

$$\sum_{k=1}^{n} \mathbf{F}(x_k^*, y_k^*, z_k^*) \cdot \mathbf{n}(x_k^*, y_k^*, z_k^*)\Delta S_k$$

measures the approximate net volume of fluid that crosses the surface σ in the direction of its orientation **n** per unit of time.

If we now increase n in such a way that the maximum dimension of each patch approaches zero, then it is plausible that the errors in the approximations approach zero, and the limit

$$\Phi = \lim_{n\to+\infty} \sum_{k=1}^{n} \mathbf{F}(x_k^*, y_k^*, z_k^*) \cdot \mathbf{n}(x_k^*, y_k^*, z_k^*)\Delta S_k \tag{3}$$

represents the exact net volume of fluid that crosses the surface σ in the direction of its orientation **n** per unit of time. The quantity Φ defined by Equation (3) is called the ***flux of F across σ***. The flux can also be expressed as the surface integral

$$\Phi = \iint_{\sigma} \mathbf{F}(x, y, z) \cdot \mathbf{n}(x, y, z)\, dS \tag{4}$$

If the fluid has mass density δ, then $\Phi\delta$ (volume × density) represents the net mass of fluid that passes through σ per unit of time.

A positive flux means that in one unit of time a greater volume of fluid passes through σ in the positive direction than in the negative direction, a negative flux means that a greater volume passes through the surface in the negative direction than in the positive direction, and a zero flux means that the same volume passes through the surface in each direction. Integrals of form (4) arise in other contexts as well and are called ***flux integrals***.

EVALUATING FLUX INTEGRALS

An effective formula for evaluating flux integrals can be obtained by applying Theorem 16.5.2 and using Formula (1) for **n**. This yields

$$\begin{aligned}\iint_{\sigma} \mathbf{F}\cdot\mathbf{n}\, dS &= \iint_{R} \mathbf{F}\cdot\mathbf{n} \left\|\frac{\partial \mathbf{r}}{\partial u} \times \frac{\partial \mathbf{r}}{\partial v}\right\| dA \\ &= \iint_{R} \mathbf{F}\cdot \frac{\dfrac{\partial \mathbf{r}}{\partial u} \times \dfrac{\partial \mathbf{r}}{\partial v}}{\left\|\dfrac{\partial \mathbf{r}}{\partial u} \times \dfrac{\partial \mathbf{r}}{\partial v}\right\|} \left\|\frac{\partial \mathbf{r}}{\partial u} \times \frac{\partial \mathbf{r}}{\partial v}\right\| dA \\ &= \iint_{R} \mathbf{F}\cdot\left(\frac{\partial \mathbf{r}}{\partial u} \times \frac{\partial \mathbf{r}}{\partial v}\right) dA\end{aligned}$$

In summary, we have the following result.

16.6.2 THEOREM. *Let σ be a smooth parametric surface represented by the vector equation $\mathbf{r} = \mathbf{r}(u, v)$ in which (u, v) varies over a region R in the uv-plane. If the component functions of the vector field $\mathbf{F}$ are continuous on σ, and if $\mathbf{n}$ determines the positive orientation of σ, then*

$$\Phi = \iint_{\sigma} \mathbf{F} \cdot \mathbf{n}\, dS = \iint_{R} \mathbf{F} \cdot \left(\frac{\partial \mathbf{r}}{\partial u} \times \frac{\partial \mathbf{r}}{\partial v} \right) dA \qquad (5)$$

where it is understood that the integrand on the right side of the equation is expressed in terms of u and v.

Although Theorem 16.6.2 was derived for smooth parametric surfaces, Formula (5) is valid more generally. For example, as long as σ has a continuous normal vector field $\mathbf{n}$ and the component functions of $\mathbf{r}(u, v)$ have continuous first partial derivatives, Formula (5) can be applied whenever $\partial \mathbf{r}/\partial u \times \partial \mathbf{r}/\partial v$ is a positive multiple of $\mathbf{n}$ in the *interior* of R. (That is, $\partial \mathbf{r}/\partial u \times \partial \mathbf{r}/\partial v$ is allowed to equal $\mathbf{0}$ on the boundary of R.)

▶ **Example 1** Find the flux of the vector field $\mathbf{F}(x, y, z) = z\mathbf{k}$ across the outward-oriented sphere $x^2 + y^2 + z^2 = a^2$.

Solution. The sphere with outward positive orientation can be represented by the vector-valued function

$$\mathbf{r}(\phi, \theta) = a \sin\phi \cos\theta\mathbf{i} + a \sin\phi \sin\theta\mathbf{j} + a \cos\phi\mathbf{k} \qquad (0 \le \phi \le \pi,\ \ 0 \le \theta \le 2\pi)$$

From this formula we obtain (see Example 9 of Section 15.4 for the computations)

$$\frac{\partial \mathbf{r}}{\partial \phi} \times \frac{\partial \mathbf{r}}{\partial \theta} = a^2 \sin^2\phi \cos\theta\mathbf{i} + a^2 \sin^2\phi \sin\theta\mathbf{j} + a^2 \sin\phi \cos\phi\mathbf{k}$$

Moreover, for points on the sphere we have $\mathbf{F} = z\mathbf{k} = a\cos\phi\mathbf{k}$; hence,

$$\mathbf{F} \cdot \left(\frac{\partial \mathbf{r}}{\partial \phi} \times \frac{\partial \mathbf{r}}{\partial \theta} \right) = a^3 \sin\phi \cos^2\phi$$

Thus, it follows from (5) with the parameters u and v replaced by ϕ and θ that

$$\begin{aligned}
\Phi &= \iint_{\sigma} \mathbf{F} \cdot \mathbf{n}\, dS \\
&= \iint_{R} \mathbf{F} \cdot \left(\frac{\partial \mathbf{r}}{\partial \phi} \times \frac{\partial \mathbf{r}}{\partial \theta} \right) dA \\
&= \int_0^{2\pi} \int_0^{\pi} a^3 \sin\phi \cos^2\phi \, d\phi \, d\theta \\
&= a^3 \int_0^{2\pi} \left[-\frac{\cos^3\phi}{3} \right]_0^{\pi} d\theta \\
&= \frac{2a^3}{3} \int_0^{2\pi} d\theta = \frac{4\pi a^3}{3}
\end{aligned}$$ ◀

Solve Example 1 using symmetry: First argue that the vector fields $x\mathbf{i}$, $y\mathbf{j}$, and $z\mathbf{k}$ will have the same flux across the sphere. Then define

$$\mathbf{H} = x\mathbf{i} + y\mathbf{j} + z\mathbf{k}$$

and explain why

$$\mathbf{H} \cdot \mathbf{n} = a$$

Use this to compute Φ.

Reversing the orientation of the surface σ in (5) reverses the sign of $\mathbf{n}$, hence the sign of $\mathbf{F} \cdot \mathbf{n}$, and hence reverses the sign of Φ. This can also be seen physically by interpreting the flux integral as the volume of fluid per unit time that crosses σ in the positive direction minus the volume per unit time that crosses in the negative direction—reversing the orientation of σ changes the sign of the difference. Thus, in Example 1 an inward orientation of the sphere would produce a flux of $-4\pi a^3/3$.

ORIENTATION OF NONPARAMETRIC SURFACES

Nonparametric surfaces of the form $z = g(x, y)$, $y = g(z, x)$, and $x = g(y, z)$ can be expressed parametrically using the independent variables as parameters. More precisely, these surfaces can be represented by the vector equations

$$\mathbf{r} = u\mathbf{i} + v\mathbf{j} + g(u, v)\mathbf{k}, \quad \mathbf{r} = v\mathbf{i} + g(u, v)\mathbf{j} + u\mathbf{k}, \quad \mathbf{r} = g(u, v)\mathbf{i} + u\mathbf{j} + v\mathbf{k} \qquad (6\text{–}8)$$

$z = g(x, y)$ $\qquad$ $y = g(z, x)$ $\qquad$ $x = g(y, z)$

These representations impose positive and negative orientations on the surfaces in accordance with Formula (1). We leave it as an exercise to calculate $\mathbf{n}$ and $-\mathbf{n}$ in each case and to show that the positive and negative orientations are as shown in Table 16.6.1. (To assist with perspective, each graph is pictured as a portion of the surface of a small solid region.)

Table 16.6.1

$z = g(x, y)$	$y = g(z, x)$	$x = g(y, z)$
$\mathbf{n} = \dfrac{-\frac{\partial z}{\partial x}\mathbf{i} - \frac{\partial z}{\partial y}\mathbf{j} + \mathbf{k}}{\sqrt{\left(\frac{\partial z}{\partial x}\right)^2 + \left(\frac{\partial z}{\partial y}\right)^2 + 1}}$ Positive k-component; Positive orientation	$\mathbf{n} = \dfrac{-\frac{\partial y}{\partial x}\mathbf{i} + \mathbf{j} - \frac{\partial y}{\partial z}\mathbf{k}}{\sqrt{\left(\frac{\partial y}{\partial x}\right)^2 + \left(\frac{\partial y}{\partial z}\right)^2 + 1}}$ Positive j-component; Positive orientation	$\mathbf{n} = \dfrac{\mathbf{i} - \frac{\partial x}{\partial y}\mathbf{j} - \frac{\partial x}{\partial z}\mathbf{k}}{\sqrt{\left(\frac{\partial x}{\partial y}\right)^2 + \left(\frac{\partial x}{\partial z}\right)^2 + 1}}$ Positive i-component; Positive orientation
$-\mathbf{n} = \dfrac{\frac{\partial z}{\partial x}\mathbf{i} + \frac{\partial z}{\partial y}\mathbf{j} - \mathbf{k}}{\sqrt{\left(\frac{\partial z}{\partial x}\right)^2 + \left(\frac{\partial z}{\partial y}\right)^2 + 1}}$ Negative k-component; Negative orientation	$-\mathbf{n} = \dfrac{\frac{\partial y}{\partial x}\mathbf{i} - \mathbf{j} + \frac{\partial y}{\partial z}\mathbf{k}}{\sqrt{\left(\frac{\partial y}{\partial x}\right)^2 + \left(\frac{\partial y}{\partial z}\right)^2 + 1}}$ Negative j-component; Negative orientation	$-\mathbf{n} = \dfrac{-\mathbf{i} + \frac{\partial x}{\partial y}\mathbf{j} + \frac{\partial x}{\partial z}\mathbf{k}}{\sqrt{\left(\frac{\partial x}{\partial y}\right)^2 + \left(\frac{\partial x}{\partial z}\right)^2 + 1}}$ Negative i-component; Negative orientation

The dependent variable will increase as you move away from a surface

$$z = g(x, y), \quad y = g(x, z)$$

or

$$x = g(y, z)$$

in the direction of positive orientation.

The results in Table 16.6.1 can also be obtained using gradients. To see how this can be done, rewrite the equations of the surfaces as

$$z - g(x, y) = 0, \quad y - g(z, x) = 0, \quad x - g(y, z) = 0$$

Each of these equations has the form $G(x, y, z) = 0$ and hence can be viewed as a level surface of a function $G(x, y, z)$. Since the gradient of G is normal to the level surface, it follows that the unit normal $\mathbf{n}$ is either $\nabla G/\|\nabla G\|$ or $-\nabla G/\|\nabla G\|$. However, if $G(x, y, z) = z - g(x, y)$, then ∇G has a $\mathbf{k}$-component of 1; if $G(x, y, z) = y - g(z, x)$, then ∇G has a $\mathbf{j}$-component of 1; and if $G(x, y, z) = x - g(y, z)$, then ∇G has an $\mathbf{i}$-component of 1. Thus, it is evident from Table 16.6.1 that in all three cases we have

$$\mathbf{n} = \frac{\nabla G}{\|\nabla G\|} \tag{9}$$

Moreover, we leave it as an exercise to show that if the surfaces $z = g(x, y)$, $y = g(z, x)$, and $x = g(y, z)$ are expressed in vector forms (6), (7), and (8), then

$$\nabla G = \frac{\partial \mathbf{r}}{\partial u} \times \frac{\partial \mathbf{r}}{\partial v} \tag{10}$$

[compare (1) and (9)]. Thus, we are led to the following version of Theorem 16.6.2 for non-parametric surfaces.

16.6.3 THEOREM. *Let σ be a smooth surface of the form $z = g(x, y)$, $y = g(z, x)$, or $x = g(y, z)$, and suppose that the component functions of the vector field $\mathbf{F}$ are continuous on σ. Suppose also that the equation for σ is rewritten as $G(x, y, z) = 0$ by taking g to the left side of the equation, and let R be the projection of σ on the coordinate plane determined by the independent variables of g. If σ has positive orientation, then*

$$\Phi = \iint_{\sigma} \mathbf{F} \cdot \mathbf{n}\, dS = \iint_{R} \mathbf{F} \cdot \nabla G\, dA \tag{11}$$

Formula (11) can either be used directly for computations or to derive some more specific formulas for each of the three surface types. For example, if $z = g(x, y)$, then we have $G(x, y, z) = z - g(x, y)$, so

$$\nabla G = -\frac{\partial g}{\partial x}\mathbf{i} - \frac{\partial g}{\partial y}\mathbf{j} + \mathbf{k} = -\frac{\partial z}{\partial x}\mathbf{i} - \frac{\partial z}{\partial y}\mathbf{j} + \mathbf{k}$$

Substituting this expression for ∇G in (11) and taking R to be the projection of the surface $z = g(x, y)$ on the xy-plane yields

$$\iint_{\sigma} \mathbf{F} \cdot \mathbf{n}\, dS = \iint_{R} \mathbf{F} \cdot \left(-\frac{\partial z}{\partial x}\mathbf{i} - \frac{\partial z}{\partial y}\mathbf{j} + \mathbf{k}\right) dA \tag{12}$$

σ of the form $z = g(x, y)$ and oriented up

$$\iint_{\sigma} \mathbf{F} \cdot \mathbf{n}\, dS = \iint_{R} \mathbf{F} \cdot \left(\frac{\partial z}{\partial x}\mathbf{i} + \frac{\partial z}{\partial y}\mathbf{j} - \mathbf{k}\right) dA \tag{13}$$

σ of the form $z = g(x, y)$ and oriented down

The derivations of the corresponding formulas when $y = g(z, x)$ and $x = g(y, z)$ are left as exercises.

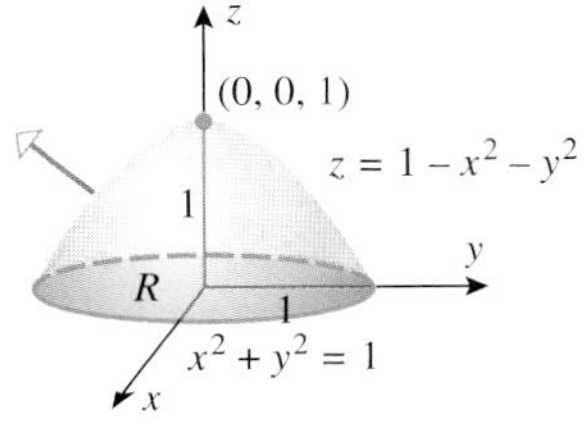

Figure 16.6.9

► **Example 2** Let σ be the portion of the surface $z = 1 - x^2 - y^2$ that lies above the xy-plane, and suppose that σ is oriented up, as shown in Figure 16.6.9. Find the flux of the vector field $\mathbf{F}(x, y, z) = x\mathbf{i} + y\mathbf{j} + z\mathbf{k}$ across σ.

Solution. From (12) the flux Φ is given by

$$\Phi = \iint_{\sigma} \mathbf{F}\cdot\mathbf{n}\,dS = \iint_{R} \mathbf{F}\cdot\left(-\frac{\partial z}{\partial x}\mathbf{i} - \frac{\partial z}{\partial y}\mathbf{j} + \mathbf{k}\right) dA$$

$$= \iint_{R} (x\mathbf{i} + y\mathbf{j} + z\mathbf{k})\cdot(2x\mathbf{i} + 2y\mathbf{j} + \mathbf{k})\,dA$$

$$= \iint_{R} (x^2 + y^2 + 1)\,dA \qquad \text{Since } z = 1 - x^2 - y^2 \text{ on the surface}$$

$$= \int_0^{2\pi}\int_0^1 (r^2 + 1)r\,dr\,d\theta \qquad \text{Using polar coordinates to evaluate the integral}$$

$$= \int_0^{2\pi} \left(\frac{3}{4}\right) d\theta = \frac{3\pi}{2}$$ ◀

✔QUICK CHECK EXERCISES 16.6 *(See page 1164 for answers.)*

In these exercises, let $\mathbf{F}(x, y, z)$ denote a vector field defined on a surface σ that is oriented by a unit normal vector field $\mathbf{n}(x, y, z)$, and let Φ denote the flux of $\mathbf{F}$ across σ.

1. (a) Φ is the value of the surface integral ________.
   (b) If σ is the unit sphere and $\mathbf{n}$ is the outward unit normal, then the flux of
   $$\mathbf{F}(x, y, z) = x\mathbf{i} + y\mathbf{j} + z\mathbf{k}$$
   across σ is $\Phi =$ ________.
2. (a) Assume that σ is parametrized by a vector-valued function $\mathbf{r}(u, v)$ whose domain is a region R in the uv-plane and that $\mathbf{n}$ is a positive multiple of
   $$\frac{\partial \mathbf{r}}{\partial u} \times \frac{\partial \mathbf{r}}{\partial v}$$
   Then the double integral over R whose value is Φ is ________.
   (b) Suppose that σ is the parametric surface
   $$\mathbf{r}(u, v) = u\mathbf{i} + v\mathbf{j} + (u + v)\mathbf{k} \qquad (0 \le u^2 + v^2 \le 1)$$
   and that $\mathbf{n}$ is a positive multiple of
   $$\frac{\partial \mathbf{r}}{\partial u} \times \frac{\partial \mathbf{r}}{\partial v}$$
   Then the flux of $\mathbf{F}(x, y, z) = x\mathbf{i} + y\mathbf{j} + z\mathbf{k}$ across σ is $\Phi =$ ________.
3. (a) Assume that σ is the graph of a function $z = g(x, y)$ over a region R in the xy-plane and that $\mathbf{n}$ has a positive $\mathbf{k}$-component for every point on σ. Then a double integral over R whose value is Φ is ________.
   (b) Suppose that σ is the triangular region with vertices $(1, 0, 0)$, $(0, 1, 0)$, and $(0, 0, 1)$ with upward orientation. Then the flux of
   $$\mathbf{F}(x, y, z) = x\mathbf{i} + y\mathbf{j} + z\mathbf{k}$$
   across σ is $\Phi =$ ________.
4. In the case of steady-state incompressible fluid flow, with $\mathbf{F}(x, y, z)$ the fluid velocity at (x, y, z) on σ, Φ can be interpreted as ________.

EXERCISE SET 16.6

FOCUS ON CONCEPTS

1. Suppose that the surface σ of the unit cube in the accompanying figure has an outward orientation. In each part, determine whether the flux of the vector field $\mathbf{F}(x, y, z) = z\mathbf{j}$ across the specified face is positive, negative, or zero.
   (a) The face $x = 1$ (b) The face $x = 0$
   (c) The face $y = 1$ (d) The face $y = 0$
   (e) The face $z = 1$ (f) The face $z = 0$

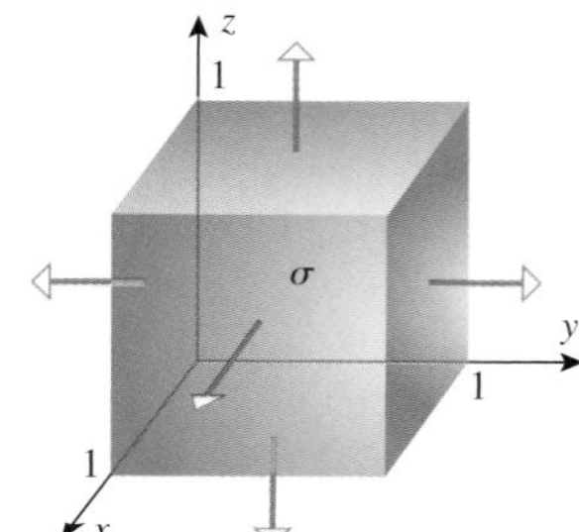

Figure Ex-1

2. Answer the questions posed in Exercise 1 for the vector field $\mathbf{F}(x, y, z) = x\mathbf{i} - z\mathbf{k}$.

3. Answer the questions posed in Exercise 1 for the vector field $\mathbf{F}(x, y, z) = x\mathbf{i} + y\mathbf{j} + z\mathbf{k}$.

4. Find the flux of the constant vector field $\mathbf{F}(x, y, z) = \mathbf{i}$ across the entire surface σ in Figure Ex-1. Explain your reasoning.

5. Let σ be the cylindrical surface that is represented by the vector-valued function $\mathbf{r}(u, v) = \cos v\mathbf{i} + \sin v\mathbf{j} + u\mathbf{k}$ with $0 \le u \le 1$ and $0 \le v \le 2\pi$.
   (a) Find the unit normal $\mathbf{n} = \mathbf{n}(u, v)$ that defines the positive orientation of σ.
   (b) Is the positive orientation inward or outward? Justify your answer.

6. Let σ be the conical surface that is represented by the parametric equations $x = r\cos\theta$, $y = r\sin\theta$, $z = r$ with $1 \le r \le 2$ and $0 \le \theta \le 2\pi$.
   (a) Find the unit normal $\mathbf{n} = \mathbf{n}(r, \theta)$ that defines the positive orientation of σ.
   (b) Is the positive orientation upward or downward? Justify your answer.

7–12 Find the flux of the vector field $\mathbf{F}$ across σ.

7. $\mathbf{F}(x, y, z) = x\mathbf{i} + y\mathbf{j} + 2z\mathbf{k}$; σ is the portion of the surface $z = 1 - x^2 - y^2$ above the xy-plane, oriented by upward normals.

8. $\mathbf{F}(x, y, z) = (x + y)\mathbf{i} + (y + z)\mathbf{j} + (z + x)\mathbf{k}$; σ is the portion of the plane $x + y + z = 1$ in the first octant, oriented by unit normals with positive components.

9. $\mathbf{F}(x, y, z) = x\mathbf{i} + y\mathbf{j} + 2z\mathbf{k}$; σ is the portion of the cone $z^2 = x^2 + y^2$ between the planes $z = 1$ and $z = 2$, oriented by upward unit normals.

10. $\mathbf{F}(x, y, z) = y\mathbf{j} + \mathbf{k}$; σ is the portion of the paraboloid $z = x^2 + y^2$ below the plane $z = 4$, oriented by downward unit normals.

11. $\mathbf{F}(x, y, z) = x\mathbf{k}$; the surface σ is the portion of the paraboloid $z = x^2 + y^2$ below the plane $z = y$, oriented by downward unit normals.

12. $\mathbf{F}(x, y, z) = x^2\mathbf{i} + yx\mathbf{j} + zx\mathbf{k}$; σ is the portion of the plane $6x + 3y + 2z = 6$ in the first octant, oriented by unit normals with positive components.

13–16 Find the flux of the vector field $\mathbf{F}$ across σ in the direction of positive orientation.

13. $\mathbf{F}(x, y, z) = x\mathbf{i} + y\mathbf{j} + \mathbf{k}$; σ is the portion of the paraboloid
$$\mathbf{r}(u, v) = u\cos v\mathbf{i} + u\sin v\mathbf{j} + (1 - u^2)\mathbf{k}$$
with $1 \le u \le 2$, $0 \le v \le 2\pi$.

14. $\mathbf{F}(x, y, z) = e^{-y}\mathbf{i} - y\mathbf{j} + x\sin z\mathbf{k}$; σ is the portion of the elliptic cylinder
$$\mathbf{r}(u, v) = 2\cos v\mathbf{i} + \sin v\mathbf{j} + u\mathbf{k}$$
with $0 \le u \le 5$, $0 \le v \le 2\pi$.

15. $\mathbf{F}(x, y, z) = \sqrt{x^2 + y^2}\,\mathbf{k}$; σ is the portion of the cone
$$\mathbf{r}(u, v) = u\cos v\mathbf{i} + u\sin v\mathbf{j} + 2u\mathbf{k}$$
with $0 \le u \le \sin v$, $0 \le v \le \pi$.

16. $\mathbf{F}(x, y, z) = x\mathbf{i} + y\mathbf{j} + z\mathbf{k}$; σ is the portion of the sphere
$$\mathbf{r}(u, v) = 2\sin u\cos v\mathbf{i} + 2\sin u\sin v\mathbf{j} + 2\cos u\mathbf{k}$$
with $0 \le u \le \pi/3$, $0 \le v \le 2\pi$.

17. Let σ be the surface of the cube bounded by the planes $x = \pm 1$, $y = \pm 1$, $z = \pm 1$, oriented by outward unit normals. In each part, find the flux of $\mathbf{F}$ across σ.
   (a) $\mathbf{F}(x, y, z) = x\mathbf{i}$
   (b) $\mathbf{F}(x, y, z) = x\mathbf{i} + y\mathbf{j} + z\mathbf{k}$
   (c) $\mathbf{F}(x, y, z) = x^2\mathbf{i} + y^2\mathbf{j} + z^2\mathbf{k}$

18. Let σ be the closed surface consisting of the portion of the paraboloid $z = x^2 + y^2$ for which $0 \le z \le 1$ and capped by the disk $x^2 + y^2 \le 1$ in the plane $z = 1$. Find the flux of the vector field $\mathbf{F}(x, y, z) = z\mathbf{j} - y\mathbf{k}$ in the outward direction across σ.

19–20 Find the flux of $\mathbf{F}$ across the surface σ by expressing σ parametrically.

19. $\mathbf{F}(x, y, z) = \mathbf{i} + \mathbf{j} + \mathbf{k}$; the surface σ is the portion of the cone $z = \sqrt{x^2 + y^2}$ between the planes $z = 1$ and $z = 2$, oriented by downward unit normals.

20. $\mathbf{F}(x, y, z) = x\mathbf{i} + y\mathbf{j} + z\mathbf{k}$; σ is the portion of the cylinder $x^2 + z^2 = 1$ between the planes $y = 1$ and $y = -2$, oriented by outward unit normals.

21. Let x, y, and z be measured in meters, and suppose that $\mathbf{F}(x, y, z) = 2x\mathbf{i} - 3y\mathbf{j} + z\mathbf{k}$ is the velocity vector (in m/s) of a fluid particle at the point (x, y, z) in a steady-state fluid flow.
   (a) Find the net volume of fluid that passes in the upward direction through the portion of the plane $x + y + z = 1$ in the first octant in 1 s.
   (b) Assuming that the fluid has a mass density of 806 kg/m^3, find the net mass of fluid that passes in the upward direction through the surface in part (a) in 1 s.

22. Let x, y, and z be measured in meters, and suppose that $\mathbf{F}(x, y, z) = -y\mathbf{i} + z\mathbf{j} + 3x\mathbf{k}$ is the velocity vector (in m/s) of a fluid particle at the point (x, y, z) in a steady-state incompressible fluid flow.
   (a) Find the net volume of fluid that passes in the upward direction through the hemisphere $z = \sqrt{9 - x^2 - y^2}$ in 1 s.
   (b) Assuming that the fluid has a mass density of 1060 kg/m^3, find the net mass of fluid that passes in the upward direction through the surface in part (a) in 1 s.

23. (a) Derive the analogs of Formulas (12) and (13) for surfaces of the form $x = g(y, z)$.
(b) Let σ be the portion of the paraboloid $x = y^2 + z^2$ for $x \leq 1$ and $z \geq 0$ oriented by unit normals with negative x-components. Use the result in part (a) to find the flux of
$$\mathbf{F}(x, y, z) = y\mathbf{i} - z\mathbf{j} + 8\mathbf{k}$$
across σ.

24. (a) Derive the analogs of Formulas (12) and (13) for surfaces of the form $y = g(z, x)$.
(b) Let σ be the portion of the paraboloid $y = z^2 + x^2$ for $y \leq 1$ and $z \geq 0$ oriented by unit normals with positive y-components. Use the result in part (a) to find the flux of
$$\mathbf{F}(x, y, z) = x\mathbf{i} + y\mathbf{j} + z\mathbf{k}$$
across σ.

25. Let $\mathbf{F} = \|\mathbf{r}\|^k\mathbf{r}$, where $\mathbf{r} = x\mathbf{i} + y\mathbf{j} + z\mathbf{k}$ and k is a constant. (Note that if $k = -3$, this is an inverse-square field.) Let σ be the sphere of radius a centered at the origin and oriented by the outward normal $\mathbf{n} = \mathbf{r}/\|\mathbf{r}\| = \mathbf{r}/a$.
(a) Find the flux of $\mathbf{F}$ across σ without performing any integrations. [*Hint:* The surface area of a sphere of radius a is $4\pi a^2$.]
(b) For what value of k is the flux independent of the radius of the sphere?

26. Let
$$\mathbf{F}(x, y, z) = a^2x\mathbf{i} + (y/a)\mathbf{j} + az^2\mathbf{k}$$
and let σ be the sphere of radius 1, centered at the origin and oriented outward. Approximate all values of a such that the flux of $\mathbf{F}$ across σ is 10.

✔QUICK CHECK ANSWERS 16.6

1. (a) $\iint_\sigma \mathbf{F} \cdot \mathbf{n}\, dS$ (b) 4π 2. (a) $\iint_R \mathbf{F} \cdot \left(\frac{\partial \mathbf{r}}{\partial u} \times \frac{\partial \mathbf{r}}{\partial v}\right) dA$ (b) 0 3. (a) $\iint_R \mathbf{F} \cdot \left(-\frac{\partial z}{\partial x}\mathbf{i} - \frac{\partial z}{\partial y}\mathbf{j} + \mathbf{k}\right) dA$ (b) $\frac{1}{2}$
4. the net volume of fluid crossing σ in the positive direction per unit time

16.7 THE DIVERGENCE THEOREM

In this section we will be concerned with flux across surfaces, such as spheres, that "enclose" a region of space. We will show that the flux across such surfaces can be expressed in terms of the divergence of the vector field, and we will use this result to give a physical interpretation of the concept of divergence.

ORIENTATION OF PIECEWISE SMOOTH CLOSED SURFACES

In the last section we studied flux across general surfaces. Here we will be concerned exclusively with surfaces that are boundaries of finite solids—the surface of a solid sphere, the surface of a solid box, or the surface of a solid cylinder, for example. Such surfaces are said to be ***closed***. A closed surface may or may not be smooth, but most of the surfaces that arise in applications are ***piecewise smooth***; that is, they consist of finitely many smooth surfaces joined together at the edges (a box, for example). We will limit our discussion to piecewise smooth surfaces that can be assigned an ***inward orientation*** (toward the interior of the solid) and an ***outward orientation*** (away from the interior). It is very difficult to make this concept mathematically precise, but the basic idea is that each piece of the surface is orientable, and oriented pieces fit together in such a way that the entire surface can be assigned an orientation (Figure 16.7.1).

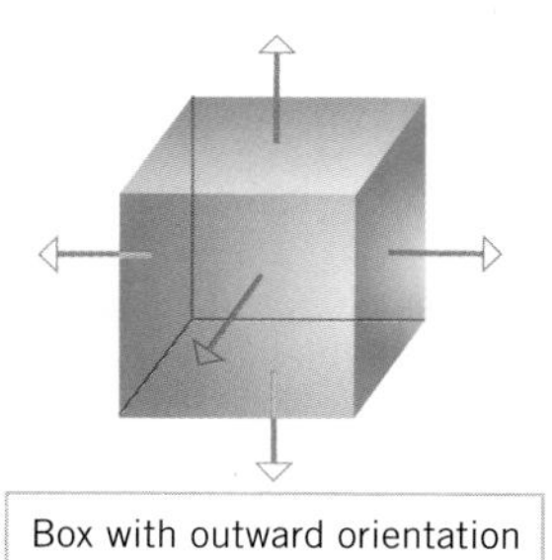

Box with outward orientation

Figure 16.7.1

THE DIVERGENCE THEOREM

In Section 16.1 we defined the divergence of a vector field
$$\mathbf{F}(x, y, z) = f(x, y, z)\mathbf{i} + g(x, y, z)\mathbf{j} + h(x, y, z)\mathbf{k}$$
as
$$\operatorname{div} \mathbf{F} = \frac{\partial f}{\partial x} + \frac{\partial g}{\partial y} + \frac{\partial h}{\partial z}$$
but we did not attempt to give a physical explanation of its meaning at that time. The following result, known as the ***Divergence Theorem*** or ***Gauss's Theorem***, will provide us with a physical interpretation of divergence in the context of fluid flow.

16.7.1 THEOREM (*The Divergence Theorem*). *Let G be a solid whose surface σ is oriented outward. If*

$$\mathbf{F}(x, y, z) = f(x, y, z)\mathbf{i} + g(x, y, z)\mathbf{j} + h(x, y, z)\mathbf{k}$$

where f, g, and h have continuous first partial derivatives on some open set containing G, and if $\mathbf{n}$ is the outward unit normal on σ, then

$$\iint_{\sigma} \mathbf{F} \cdot \mathbf{n}\, dS = \iiint_{G} \operatorname{div} \mathbf{F}\, dV \tag{1}$$

Carl Friedrich Gauss (1777–1855) German mathematician and scientist. Sometimes called the "prince of mathematicians," Gauss ranks with Newton and Archimedes as one of the three greatest mathematicians who ever lived. His father, a laborer, was an uncouth but honest man who would have liked Gauss to take up a trade such as gardening or bricklaying; but the boy's genius for mathematics was not to be denied. In the entire history of mathematics there may never have been a child so precocious as Gauss—by his own account he worked out the rudiments of arithmetic before he could talk. One day, before he was even three years old, his genius became apparent to his parents in a very dramatic way. His father was preparing the weekly payroll for the laborers under his charge while the boy watched quietly from a corner. At the end of the long and tedious calculation, Gauss informed his father that there was an error in the result and stated the answer, which he had worked out in his head. To the astonishment of his parents, a check of the computations showed Gauss to be correct!

For his elementary education Gauss was enrolled in a squalid school run by a man named Büttner whose main teaching technique was thrashing. Büttner was in the habit of assigning long addition problems which, unknown to his students, were arithmetic progressions that he could sum up using formulas. On the first day that Gauss entered the arithmetic class, the students were asked to sum the numbers from 1 to 100. But no sooner had Büttner stated the problem than Gauss turned over his slate and exclaimed in his peasant dialect, "Ligget se'." (Here it lies.) For nearly an hour Büttner glared at Gauss, who sat with folded hands while his classmates toiled away. When Büttner examined the slates at the end of the period, Gauss's slate contained a single number, 5050—the only correct solution in the class. To his credit, Büttner recognized the genius of Gauss and with the help of his assistant, John Bartels, had him brought to the attention of Karl Wilhelm Ferdinand, Duke of Brunswick. The shy and awkward boy, who was then fourteen, so captivated the Duke that he subsidized him through preparatory school, college, and the early part of his career.

From 1795 to 1798 Gauss studied mathematics at the University of Göttingen, receiving his degree in absentia from the University of Helmstadt. For his dissertation, he gave the first complete proof of the fundamental theorem of algebra, which states that every polynomial equation has as many solutions as its degree. At age 19 he solved a problem that baffled Euclid, inscribing a regular polygon of 17 sides in a circle using straightedge and compass; and in 1801, at age 24, he published his first masterpiece, *Disquisitiones Arithmeticae*, considered by many to be one of the most brilliant achievements in mathematics. In that book Gauss systematized the study of number theory (properties of the integers) and formulated the basic concepts that form the foundation of that subject.

In the same year that the *Disquisitiones* was published, Gauss again applied his phenomenal computational skills in a dramatic way. The astronomer Giuseppi Piazzi had observed the asteroid Ceres for $\frac{1}{40}$ of its orbit, but lost it in the Sun. Using only three observations and the "method of least squares" that he had developed in 1795, Gauss computed the orbit with such accuracy that astronomers had no trouble relocating it the following year. This achievement brought him instant recognition as the premier mathematician in Europe, and in 1807 he was made Professor of Astronomy and head of the astronomical observatory at Göttingen.

In the years that followed, Gauss revolutionized mathematics by bringing to it standards of precision and rigor undreamed of by his predecessors. He had a passion for perfection that drove him to polish and rework his papers rather than publish less finished work in greater numbers—his favorite saying was "Pauca, sed matura" (Few, but ripe). As a result, many of his important discoveries were squirreled away in diaries that remained unpublished until years after his death.

Among his myriad achievements, Gauss discovered the Gaussian or "bell-shaped" error curve fundamental in probability, gave the first geometric interpretation of complex numbers and established their fundamental role in mathematics, developed methods of characterizing surfaces intrinsically by means of the curves that they contain, developed the theory of conformal (angle-preserving) maps, and discovered non-Euclidean geometry 30 years before the ideas were published by others. In physics he made major contributions to the theory of lenses and capillary action, and with Wilhelm Weber he did fundamental work in electromagnetism. Gauss invented the heliotrope, bifilar magnetometer, and an electrotelegraph.

Gauss was deeply religious and aristocratic in demeanor. He mastered foreign languages with ease, read extensively, and enjoyed mineralogy and botany as hobbies. He disliked teaching and was usually cool and discouraging to other mathematicians, possibly because he had already anticipated their work. It has been said that if Gauss had published all of his discoveries, the current state of mathematics would be advanced by 50 years. He was without a doubt the greatest mathematician of the modern era.

The proof of this theorem for a general solid G is too difficult to present here. However, we can give a proof for the special case where G is simultaneously a simple xy-solid, a simple yz-solid, and a simple zx-solid (see Figure 15.5.3 and the related discussion for terminology).

PROOF. Formula (1) can be expressed as

$$\iint_{\sigma} [f(x, y, z)\mathbf{i} + g(x, y, z)\mathbf{j} + h(x, y, z)\mathbf{k}] \cdot \mathbf{n}\, dS = \iiint_{G} \left(\frac{\partial f}{\partial x} + \frac{\partial g}{\partial y} + \frac{\partial h}{\partial z}\right) dV$$

so it suffices to prove the three equalities

$$\iint_{\sigma} f(x, y, z)\mathbf{i} \cdot \mathbf{n}\, dS = \iiint_{G} \frac{\partial f}{\partial x}\, dV \tag{2a}$$

$$\iint_{\sigma} g(x, y, z)\mathbf{j} \cdot \mathbf{n}\, dS = \iiint_{G} \frac{\partial g}{\partial y}\, dV \tag{2b}$$

$$\iint_{\sigma} h(x, y, z)\mathbf{k} \cdot \mathbf{n}\, dS = \iiint_{G} \frac{\partial h}{\partial z}\, dV \tag{2c}$$

Since the proofs of all three equalities are similar, we will prove only the third.

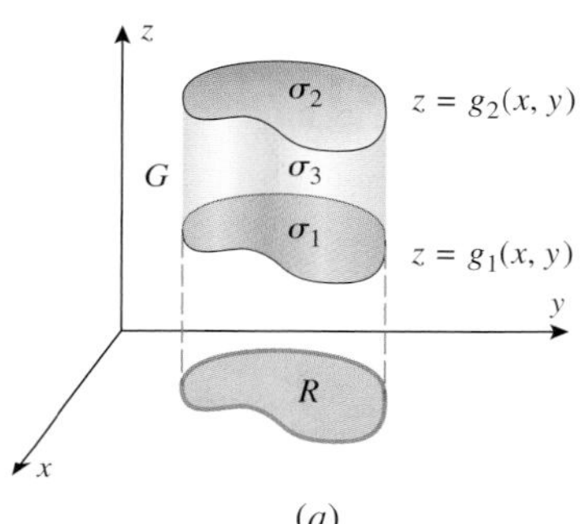

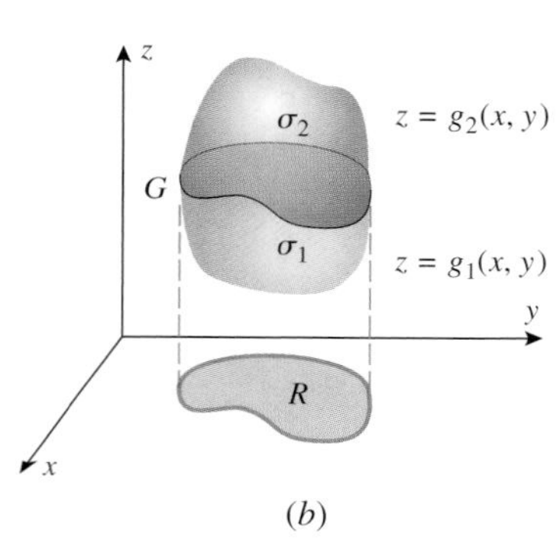

Figure 16.7.2

Suppose that G has upper surface $z = g_2(x, y)$, lower surface $z = g_1(x, y)$, and projection R on the xy-plane. Let σ_1 denote the lower surface, σ_2 the upper surface, and σ_3 the lateral surface (Figure 16.7.2a). If the upper surface and lower surface meet as in Figure 16.7.2b, then there is no lateral surface σ_3. Our proof will allow for both cases shown in those figures.

It follows from Theorem 15.5.2 that

$$\iiint_{G} \frac{\partial h}{\partial z}\, dV = \iint_{R} \left[\int_{g_1(x,y)}^{g_2(x,y)} \frac{\partial h}{\partial z}\, dz\right] dA = \iint_{R} \left[h(x, y, z)\right]_{z=g_1(x,y)}^{g_2(x,y)} dA$$

so

$$\iiint_{G} \frac{\partial h}{\partial z}\, dV = \iint_{R} [h(x, y, g_2(x, y)) - h(x, y, g_1(x, y))]\, dA \tag{3}$$

Next we will evaluate the surface integral in (2c) by integrating over each surface of G separately. If there is a lateral surface σ_3, then at each point of this surface $\mathbf{k} \cdot \mathbf{n} = 0$ since $\mathbf{n}$ is horizontal and $\mathbf{k}$ is vertical. Thus,

$$\iint_{\sigma_3} h(x, y, z)\mathbf{k} \cdot \mathbf{n}\, dS = 0$$

Therefore, regardless of whether G has a lateral surface, we can write

$$\iint_{\sigma} h(x, y, z)\mathbf{k} \cdot \mathbf{n}\, dS = \iint_{\sigma_1} h(x, y, z)\mathbf{k} \cdot \mathbf{n}\, dS + \iint_{\sigma_2} h(x, y, z)\mathbf{k} \cdot \mathbf{n}\, dS \tag{4}$$

On the upper surface σ_2, the outer normal is an upward normal, and on the lower surface σ_1, the outer normal is a downward normal. Thus, Formulas (12) and (13) of Section 16.6 imply that

$$\begin{aligned}\iint_{\sigma_2} h(x, y, z)\mathbf{k} \cdot \mathbf{n}\, dS &= \iint_{R} h(x, y, g_2(x, y))\mathbf{k} \cdot \left(-\frac{\partial z}{\partial x}\mathbf{i} - \frac{\partial z}{\partial y}\mathbf{j} + \mathbf{k}\right) dA \\ &= \iint_{R} h(x, y, g_2(x, y))\, dA\end{aligned} \tag{5}$$

and

$$\begin{aligned}\iint_{\sigma_1} h(x, y, z)\mathbf{k} \cdot \mathbf{n}\, dS &= \iint_R h(x, y, g_1(x, y))\mathbf{k} \cdot \left(\frac{\partial z}{\partial x}\mathbf{i} + \frac{\partial z}{\partial y}\mathbf{j} - \mathbf{k}\right) dA \\ &= -\iint_R h(x, y, g_1(x, y))\, dA\end{aligned} \tag{6}$$

Substituting (5) and (6) into (4) and combining the terms into a single integral yields

Explain how the derivation of (2c) should be modified to yield a proof of (2a) or (2b).

$$\iint_{\sigma} h(x, y, z)\mathbf{k} \cdot \mathbf{n}\, dS = \iint_R [h(x, y, g_2(x, y)) - h(x, y, g_1(x, y))]\, dA \tag{7}$$

Equation (2c) now follows from (3) and (7). ■

In words, the Divergence Theorem states:

The flux of a vector field across a closed surface with outward orientation is equal to the triple integral of the divergence over the region enclosed by the surface.

This is sometimes called the ***outward flux*** across the surface.

USING THE DIVERGENCE THEOREM TO FIND FLUX

Sometimes it is easier to find the flux across a closed surface by using the Divergence Theorem than by evaluating the flux integral directly. This is illustrated in the following example.

▶ **Example 1** Use the Divergence Theorem to find the outward flux of the vector field $\mathbf{F}(x, y, z) = z\mathbf{k}$ across the sphere $x^2 + y^2 + z^2 = a^2$.

Solution. Let σ denote the outward-oriented spherical surface and G the region that it encloses. The divergence of the vector field is

$$\operatorname{div} \mathbf{F} = \frac{\partial z}{\partial z} = 1$$

so from (1) the flux across σ is

$$\Phi = \iint_{\sigma} \mathbf{F} \cdot \mathbf{n}\, dS = \iiint_G dV = \text{volume of } G = \frac{4\pi a^3}{3}$$

Note how much simpler this calculation is than that in Example 1 of Section 16.6. ◀

The Divergence Theorem is usually the method of choice for finding the flux across closed piecewise smooth surfaces with multiple sections, since it eliminates the need for a separate integral evaluation over each section. This is illustrated in the next three examples.

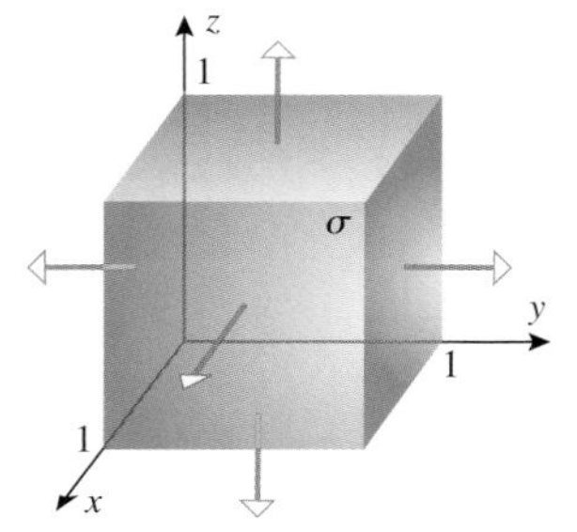

Figure 16.7.3

Let $\mathbf{F}(x, y, z)$ be the vector field in Example 2 and show that $\mathbf{F} \cdot \mathbf{n}$ is constant on each of the six faces of the cube in Figure 16.7.3. Use your computations to confirm the result in Example 2.

▶ **Example 2** Use the Divergence Theorem to find the outward flux of the vector field

$$\mathbf{F}(x, y, z) = 2x\mathbf{i} + 3y\mathbf{j} + z^2\mathbf{k}$$

across the unit cube in Figure 16.7.3.

Solution. Let σ denote the outward-oriented surface of the cube and G the region that it encloses. The divergence of the vector field is

$$\operatorname{div} \mathbf{F} = \frac{\partial}{\partial x}(2x) + \frac{\partial}{\partial y}(3y) + \frac{\partial}{\partial z}(z^2) = 5 + 2z$$

so from (1) the flux across σ is

$$\begin{aligned}\Phi = \iint_{\sigma} \mathbf{F} \cdot \mathbf{n}\, dS &= \iiint_{G} (5 + 2z)\, dV = \int_0^1 \int_0^1 \int_0^1 (5 + 2z)\, dz\, dy\, dx \\ &= \int_0^1 \int_0^1 \left[5z + z^2\right]_{z=0}^{1} dy\, dx = \int_0^1 \int_0^1 6\, dy\, dx = 6 \blacktriangleleft\end{aligned}$$

▶ **Example 3** Use the Divergence Theorem to find the outward flux of the vector field

$$\mathbf{F}(x, y, z) = x^3\mathbf{i} + y^3\mathbf{j} + z^2\mathbf{k}$$

across the surface of the region that is enclosed by the circular cylinder $x^2 + y^2 = 9$ and the planes $z = 0$ and $z = 2$ (Figure 16.7.4).

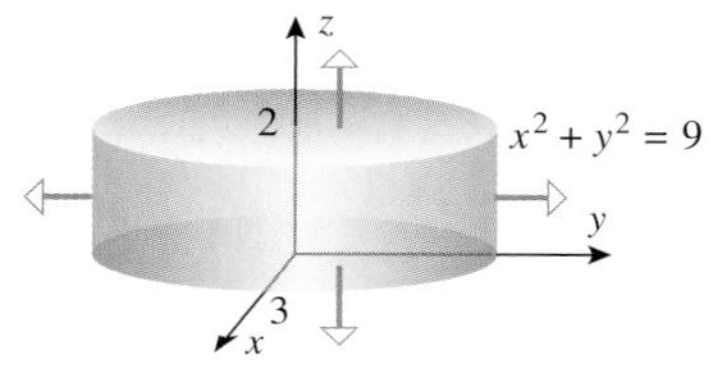

Figure 16.7.4

Solution. Let σ denote the outward-oriented surface and G the region that it encloses. The divergence of the vector field is

$$\operatorname{div} \mathbf{F} = \frac{\partial}{\partial x}(x^3) + \frac{\partial}{\partial y}(y^3) + \frac{\partial}{\partial z}(z^2) = 3x^2 + 3y^2 + 2z$$

so from (1) the flux across σ is

$$\begin{aligned}\Phi = \iint_{\sigma} \mathbf{F} \cdot \mathbf{n}\, dS &= \iiint_{G} (3x^2 + 3y^2 + 2z)\, dV \\ &= \int_0^{2\pi} \int_0^3 \int_0^2 (3r^2 + 2z) r\, dz\, dr\, d\theta \qquad \text{Using cylindrical coordinates} \\ &= \int_0^{2\pi} \int_0^3 \left[3r^3 z + z^2 r\right]_{z=0}^{2} dr\, d\theta \\ &= \int_0^{2\pi} \int_0^3 (6r^3 + 4r)\, dr\, d\theta \\ &= \int_0^{2\pi} \left[\frac{3r^4}{2} + 2r^2\right]_0^3 d\theta \\ &= \int_0^{2\pi} \frac{279}{2}\, d\theta = 279\pi \blacktriangleleft\end{aligned}$$

▶ **Example 4** Use the Divergence Theorem to find the outward flux of the vector field

$$\mathbf{F}(x, y, z) = x^3\mathbf{i} + y^3\mathbf{j} + z^3\mathbf{k}$$

across the surface of the region that is enclosed by the hemisphere $z = \sqrt{a^2 - x^2 - y^2}$ and the plane $z = 0$ (Figure 16.7.5).

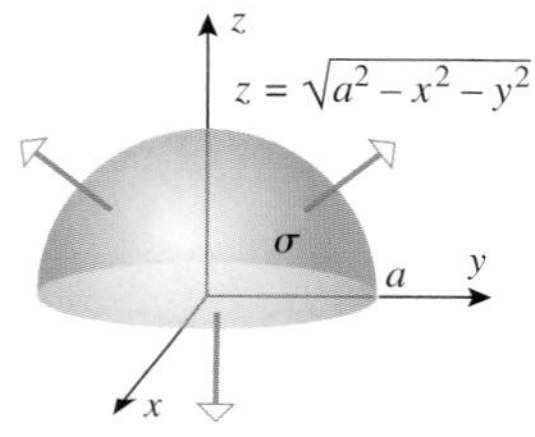

Figure 16.7.5

Solution. Let σ denote the outward-oriented surface and G the region that it encloses. The divergence of the vector field is

$$\operatorname{div} \mathbf{F} = \frac{\partial}{\partial x}(x^3) + \frac{\partial}{\partial y}(y^3) + \frac{\partial}{\partial z}(z^3) = 3x^2 + 3y^2 + 3z^2$$

so from (1) the flux across σ is

$$\begin{aligned}
\Phi = \iint_{\sigma} \mathbf{F} \cdot \mathbf{n}\, dS &= \iiint_G (3x^2 + 3y^2 + 3z^2)\, dV \\
&= \int_0^{2\pi} \int_0^{\pi/2} \int_0^a (3\rho^2)\rho^2 \sin\phi \, d\rho \, d\phi \, d\theta \\
&= 3 \int_0^{2\pi} \int_0^{\pi/2} \int_0^a \rho^4 \sin\phi \, d\rho \, d\phi \, d\theta \\
&= 3 \int_0^{2\pi} \int_0^{\pi/2} \left[\frac{\rho^5}{5} \sin\phi\right]_{\rho=0}^a d\phi \, d\theta \\
&= \frac{3a^5}{5} \int_0^{2\pi} \int_0^{\pi/2} \sin\phi \, d\phi \, d\theta \\
&= \frac{3a^5}{5} \int_0^{2\pi} \big[-\cos\phi\big]_0^{\pi/2} d\theta \\
&= \frac{3a^5}{5} \int_0^{2\pi} d\theta = \frac{6\pi a^5}{5} \blacktriangleleft
\end{aligned}$$

Using spherical coordinates

DIVERGENCE VIEWED AS FLUX DENSITY

The Divergence Theorem provides a way of interpreting the divergence of a vector field $\mathbf{F}$. Suppose that G is a *small* spherical region centered at the point P_0 and that its surface, denoted by $\sigma(G)$, is oriented outward. Denote the volume of the region by $\operatorname{vol}(G)$ and the flux of $\mathbf{F}$ across $\sigma(G)$ by $\Phi(G)$. If $\operatorname{div} \mathbf{F}$ is continuous on G, then across the small region G the value of $\operatorname{div} \mathbf{F}$ will not vary much from its value $\operatorname{div} \mathbf{F}(P_0)$ at the center, and we can reasonably approximate $\operatorname{div} \mathbf{F}$ by the constant $\operatorname{div} \mathbf{F}(P_0)$ on G. Thus, the Divergence Theorem implies that the flux $\Phi(G)$ of $\mathbf{F}$ across $\sigma(G)$ can be approximated by

$$\Phi(G) = \iint_{\sigma(G)} \mathbf{F} \cdot \mathbf{n}\, dS = \iiint_G \operatorname{div} \mathbf{F}\, dV \approx \operatorname{div} \mathbf{F}(P_0) \iiint_G dV = \operatorname{div} \mathbf{F}(P_0) \operatorname{vol}(G)$$

from which we obtain the approximation

$$\operatorname{div} \mathbf{F}(P_0) \approx \frac{\Phi(G)}{\operatorname{vol}(G)} \tag{8}$$

The expression on the right side of (8) is called the ***outward flux density of* F *across* G.** If we now let the radius of the sphere approach zero [so that $\operatorname{vol}(G)$ approaches zero], then it is plausible that the error in this approximation will approach zero, and the divergence of $\mathbf{F}$ at the point P_0 will be given exactly by

$$\operatorname{div} \mathbf{F}(P_0) = \lim_{\operatorname{vol}(G) \to 0} \frac{\Phi(G)}{\operatorname{vol}(G)}$$

which we can express as

$$\operatorname{div} \mathbf{F}(P_0) = \lim_{\operatorname{vol}(G) \to 0} \frac{1}{\operatorname{vol}(G)} \iint_{\sigma(G)} \mathbf{F} \cdot \mathbf{n}\, dS \tag{9}$$

Formula (9) is sometimes taken as the definition of divergence. This is a useful alternative to Definition 16.1.4 because it does not require a coordinate system.

This limit, which is called the ***outward flux density of* F *at* P_0**, tells us that *in a steady-state fluid flow*, div **F** *can be interpreted as the limiting flux per unit volume at a point*. Moreover, it follows from (8) that for a small spherical region G centered at a point P_0 in the flow, the outward flux across the surface of G can be approximated by

$$\Phi(G) \approx (\text{div}\,\mathbf{F}(P_0))(\text{vol}(G)) \tag{10}$$

SOURCES AND SINKS

If P_0 is a point in an incompressible fluid at which div $\mathbf{F}(P_0) > 0$, then it follows from (8) that $\Phi(G) > 0$ for a sufficiently small sphere G centered at P_0. Thus, there is a greater volume of fluid going out through the surface of G than coming in. But this can only happen if there is some point *inside* the sphere at which fluid is entering the flow (say by condensation, melting of a solid, or a chemical reaction); otherwise the net outward flow through the surface would result in a decrease in density within the sphere, contradicting the incompressibility assumption. Similarly, if div $\mathbf{F}(P_0) < 0$, there would have to be a point *inside* the sphere at which fluid is leaving the flow (say by evaporation); otherwise the net inward flow through the surface would result in an increase in density within the sphere. In an incompressible fluid, points at which div $\mathbf{F}(P_0) > 0$ are called ***sources*** and points at which div $\mathbf{F}(P_0) < 0$ are called ***sinks***. Fluid enters the flow at a source and drains out at a sink. In an incompressible fluid without sources or sinks we must have

$$\text{div}\,\mathbf{F}(P) = 0$$

at every point P. In hydrodynamics this is called the ***continuity equation for incompressible fluids*** and is sometimes taken as the defining characteristic of an incompressible fluid.

GAUSS'S LAW FOR INVERSE-SQUARE FIELDS

The Divergence Theorem applied to inverse-square fields (see Definition 16.1.2) produces a result called ***Gauss's Law for Inverse-Square Fields***. This result is the basis for many important principles in physics.

> **16.7.2 GAUSS'S LAW FOR INVERSE-SQUARE FIELDS.** If
>
> $$\mathbf{F}(\mathbf{r}) = \frac{c}{\|\mathbf{r}\|^3}\mathbf{r}$$
>
> is an inverse-square field in 3-space, and if σ is a closed orientable surface that surrounds the origin, then the outward flux of **F** across σ is
>
> $$\Phi = \iint_\sigma \mathbf{F}\cdot\mathbf{n}\,dS = 4\pi c \tag{11}$$

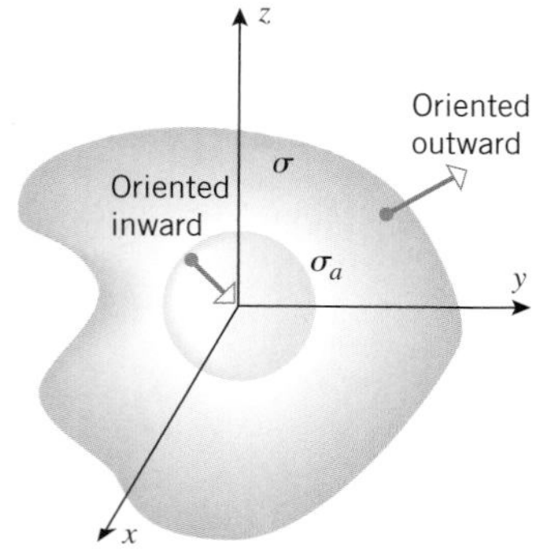

Figure 16.7.6

To derive this result, recall from Formula (5) of Section 16.1 that **F** can be expressed in component form as

$$\mathbf{F}(x, y, z) = \frac{c}{(x^2 + y^2 + z^2)^{3/2}}(x\mathbf{i} + y\mathbf{j} + z\mathbf{k}) \tag{12}$$

Since the components of **F** are not continuous at the origin, we cannot apply the Divergence Theorem across the solid enclosed by σ. However, we can circumvent this difficulty by constructing a sphere of radius a centered at the origin, where the radius is sufficiently small that the sphere lies entirely within the region enclosed by σ (Figure 16.7.6). We will denote the surface of this sphere by σ_a. The solid G enclosed between σ_a and σ is an example of a three-dimensional solid with an internal "cavity." Just as we were able to extend Green's Theorem to multiply connected regions in the plane (regions with holes), so it is possible to extend the Divergence Theorem to solids in 3-space with internal cavities, provided the surface integral in the theorem is taken over the *entire* boundary with the outside boundary of the solid oriented outward and the boundaries of the cavities oriented inward. Thus, if $\mathbf{F}$ is the inverse-square field in (12), and if σ_a is oriented inward, then the Divergence Theorem yields

$$\iiint_G \operatorname{div} \mathbf{F}\, dV = \iint_\sigma \mathbf{F} \cdot \mathbf{n}\, dS + \iint_{\sigma_a} \mathbf{F} \cdot \mathbf{n}\, dS \tag{13}$$

But we showed in Example 5 of Section 16.1 that div $\mathbf{F} = 0$, so (13) yields

$$\iint_\sigma \mathbf{F} \cdot \mathbf{n}\, dS = -\iint_{\sigma_a} \mathbf{F} \cdot \mathbf{n}\, dS \tag{14}$$

We can evaluate the surface integral over σ_a by expressing the integrand in terms of components; however, it is easier to leave it in vector form. At each point on the sphere the unit normal $\mathbf{n}$ points inward along a radius from the origin, and hence $\mathbf{n} = -\mathbf{r}/\|\mathbf{r}\|$. Thus, (14) yields

$$\begin{aligned}
\iint_\sigma \mathbf{F} \cdot \mathbf{n}\, dS &= -\iint_{\sigma_a} \frac{c}{\|\mathbf{r}\|^3}\mathbf{r} \cdot \left(-\frac{\mathbf{r}}{\|\mathbf{r}\|}\right) dS \\
&= \iint_{\sigma_a} \frac{c}{\|\mathbf{r}\|^4}(\mathbf{r} \cdot \mathbf{r})\, dS \\
&= \iint_{\sigma_a} \frac{c}{\|\mathbf{r}\|^2}\, dS \\
&= \frac{c}{a^2}\iint_{\sigma_a} dS && \boxed{\|\mathbf{r}\| = a \text{ on } \sigma_a} \\
&= \frac{c}{a^2}(4\pi a^2) && \boxed{\text{The integral is the surface area of the sphere.}} \\
&= 4\pi c
\end{aligned}$$

which establishes (11).

GAUSS'S LAW IN ELECTROSTATICS

It follows from Example 1 of Section 16.1 with $q = 1$ that a single charged particle of charge Q located at the origin creates an inverse-square field

$$\mathbf{F}(\mathbf{r}) = \frac{Q}{4\pi\epsilon_0\|\mathbf{r}\|^3}\mathbf{r}$$

in which $\mathbf{F}(\mathbf{r})$ is the electrical force exerted by Q on a unit positive charge ($q = 1$) located at the point with position vector $\mathbf{r}$. In this case Gauss's law (16.7.2) states that the outward flux Φ across any closed orientable surface σ that surrounds Q is

$$\Phi = \iint_\sigma \mathbf{F} \cdot \mathbf{n}\, dS = 4\pi\left(\frac{Q}{4\pi\epsilon_0}\right) = \frac{Q}{\epsilon_0}$$

This result, which is called ***Gauss's Law for Electric Fields***, can be extended to more than one charge. It is one of the fundamental laws in electricity and magnetism.

✔ QUICK CHECK EXERCISES 16.7 (See page 1173 for answers.)

1. Let G be a solid whose surface σ is oriented outward by the unit normal $\mathbf{n}$, and let $\mathbf{F}(x, y, z)$ denote a vector field whose component functions have continuous first partial derivatives on some open set containing G. The Divergence Theorem states that the surface integral ______ and the triple integral ______ have the same value.
2. The outward flux of $\mathbf{F}(x, y, z) = x\mathbf{i} + y\mathbf{j} + z\mathbf{k}$ across any unit cube is ______.
3. If $\mathbf{F}(x, y, z)$ is the velocity vector field for a steady-state incompressible fluid flow, then a point at which div $\mathbf{F}$ is positive is called a ______ and a point at which $\mathbf{F}$ is negative is called a ______. The continuity equation for an incompressible fluid states that ______.
4. If
$$\mathbf{F}(\mathbf{r}) = \frac{c}{\|\mathbf{r}\|^3}\mathbf{r}$$
is an inverse-square field, and if σ is a closed orientable surface that surrounds the origin, then Gauss's law states that the outward flux of $\mathbf{F}$ across σ is ______. On the other hand, if σ does not surround the origin, then it follows from the Divergence Theorem that the outward flux of $\mathbf{F}$ across σ is ______.

EXERCISE SET 16.7 [C] CAS

1–4 Verify Formula (1) in the Divergence Theorem by evaluating the surface integral and the triple integral.

1. $\mathbf{F}(x, y, z) = x\mathbf{i} + y\mathbf{j} + z\mathbf{k}$; σ is the surface of the cube bounded by the planes $x = 0$, $x = 1$, $y = 0$, $y = 1$, $z = 0$, $z = 1$.
2. $\mathbf{F}(x, y, z) = x\mathbf{i} + y\mathbf{j} + z\mathbf{k}$; σ is the spherical surface $x^2 + y^2 + z^2 = 1$.
3. $\mathbf{F}(x, y, z) = 2x\mathbf{i} - yz\mathbf{j} + z^2\mathbf{k}$; the surface σ is the paraboloid $z = x^2 + y^2$ capped by the disk $x^2 + y^2 \le 1$ in the plane $z = 1$.
4. $\mathbf{F}(x, y, z) = xy\mathbf{i} + yz\mathbf{j} + xz\mathbf{k}$; σ is the surface of the cube bounded by the planes $x = 0$, $x = 2$, $y = 0$, $y = 2$, $z = 0$, $z = 2$.

5–15 Use the Divergence Theorem to find the flux of $\mathbf{F}$ across the surface σ with outward orientation.

5. $\mathbf{F}(x, y, z) = (x^2 + y)\mathbf{i} + z^2\mathbf{j} + (e^y - z)\mathbf{k}$; σ is the surface of the rectangular solid bounded by the coordinate planes and the planes $x = 3$, $y = 1$, and $z = 2$.
6. $\mathbf{F}(x, y, z) = z^3\mathbf{i} - x^3\mathbf{j} + y^3\mathbf{k}$, where σ is the sphere $x^2 + y^2 + z^2 = a^2$.
7. $\mathbf{F}(x, y, z) = (x - z)\mathbf{i} + (y - x)\mathbf{j} + (z - y)\mathbf{k}$; σ is the surface of the cylindrical solid bounded by $x^2 + y^2 = a^2$, $z = 0$, and $z = 1$.
8. $\mathbf{F}(x, y, z) = x\mathbf{i} + y\mathbf{j} + z\mathbf{k}$; σ is the surface of the solid bounded by the paraboloid $z = 1 - x^2 - y^2$ and the xy-plane.
9. $\mathbf{F}(x, y, z) = x^3\mathbf{i} + y^3\mathbf{j} + z^3\mathbf{k}$; σ is the surface of the cylindrical solid bounded by $x^2 + y^2 = 4$, $z = 0$, and $z = 3$.
10. $\mathbf{F}(x, y, z) = (x^2 + y)\mathbf{i} + xy\mathbf{j} - (2xz + y)\mathbf{k}$; σ is the surface of the tetrahedron in the first octant bounded by $x + y + z = 1$ and the coordinate planes.
11. $\mathbf{F}(x, y, z) = (x^3 - e^y)\mathbf{i} + (y^3 + \sin z)\mathbf{j} + (z^3 - xy)\mathbf{k}$, where σ is the surface of the solid bounded above by $z = \sqrt{4 - x^2 - y^2}$ and below by the xy-plane. [*Hint:* Use spherical coordinates.]
12. $\mathbf{F}(x, y, z) = 2xz\mathbf{i} + yz\mathbf{j} + z^2\mathbf{k}$, where σ is the surface of the solid bounded above by $z = \sqrt{a^2 - x^2 - y^2}$ and below by the xy-plane.
13. $\mathbf{F}(x, y, z) = x^2\mathbf{i} + y^2\mathbf{j} + z^2\mathbf{k}$; σ is the surface of the conical solid bounded by $z = \sqrt{x^2 + y^2}$ and $z = 1$.
14. $\mathbf{F}(x, y, z) = x^2y\mathbf{i} - xy^2\mathbf{j} + (z + 2)\mathbf{k}$; σ is the surface of the solid bounded above by the plane $z = 2x$ and below by the paraboloid $z = x^2 + y^2$.
15. $\mathbf{F}(x, y, z) = x^3\mathbf{i} + x^2y\mathbf{j} + xy\mathbf{k}$; σ is the surface of the solid bounded by $z = 4 - x^2$, $y + z = 5$, $z = 0$, and $y = 0$.
16. Prove that if $\mathbf{r} = x\mathbf{i} + y\mathbf{j} + z\mathbf{k}$ and σ is the surface of a solid G oriented by outward unit normals, then
$$\text{vol}(G) = \frac{1}{3}\iint_{\sigma} \mathbf{r} \cdot \mathbf{n}\, dS$$
where vol(G) is the volume of G.
17. Use the result in Exercise 16 to find the outward flux of the vector field $\mathbf{F}(x, y, z) = x\mathbf{i} + y\mathbf{j} + z\mathbf{k}$ across the surface σ of the cylindrical solid bounded by $x^2 + 4x + y^2 = 5$, $z = -1$, and $z = 4$.

FOCUS ON CONCEPTS

18. Let $\mathbf{F}(x, y, z) = a\mathbf{i} + b\mathbf{j} + c\mathbf{k}$ be a constant vector field and let σ be the surface of a solid G. Use the Divergence Theorem to show that the flux of $\mathbf{F}$ across σ is zero. Give an informal physical explanation of this result.
19. Find a vector field $\mathbf{F}(x, y, z)$ that has
 (a) positive divergence everywhere
 (b) negative divergence everywhere.

20. In each part, the figure shows a horizontal layer of the vector field of a fluid flow in which the flow is parallel to the xy-plane at every point and is identical in each layer (i.e., is independent of z). For each flow, what can you say about the sign of the divergence at the origin? Explain your reasoning.

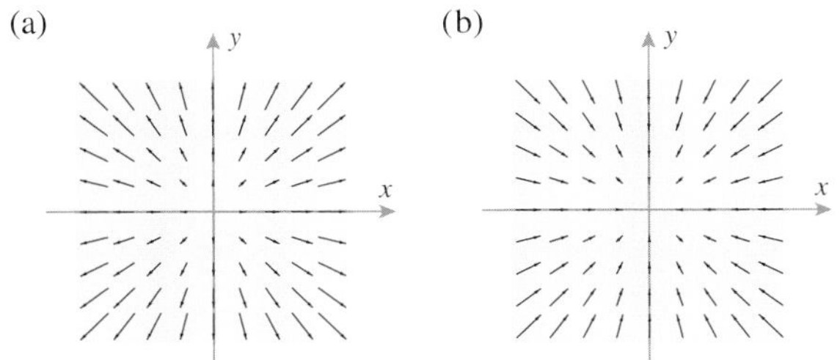

21. Let $\mathbf{F}(x, y, z)$ be a nonzero vector field in 3-space whose component functions have continuous first partial derivatives, and assume that div $\mathbf{F} = 0$ everywhere. If σ is any sphere in 3-space, explain why there are infinitely many points on σ at which $\mathbf{F}$ is tangent to the sphere.

22. Does the result in Exercise 21 remain true if the sphere σ is replaced by a cube? Explain.

23–27 Prove the identity, assuming that $\mathbf{F}$, σ, and G satisfy the hypotheses of the Divergence Theorem and that all necessary differentiability requirements for the functions $f(x, y, z)$ and $g(x, y, z)$ are met.

23. $\displaystyle\iint_\sigma \operatorname{curl} \mathbf{F} \cdot \mathbf{n}\, dS = 0$ [*Hint:* See Exercise 33, Section 16.1.]

24. $\displaystyle\iint_\sigma \nabla f \cdot \mathbf{n}\, dS = \iiint_G \nabla^2 f\, dV$

$\left(\nabla^2 f = \dfrac{\partial^2 f}{\partial x^2} + \dfrac{\partial^2 f}{\partial y^2} + \dfrac{\partial^2 f}{\partial z^2}\right)$

25. $\displaystyle\iint_\sigma (f\nabla g) \cdot \mathbf{n}\, dS = \iiint_G (f\nabla^2 g + \nabla f \cdot \nabla g)\, dV$

26. $\displaystyle\iint_\sigma (f\nabla g - g\nabla f) \cdot \mathbf{n}\, dS = \iiint_G (f\nabla^2 g - g\nabla^2 f)\, dV$

[*Hint:* Interchange f and g in 25.]

27. $\displaystyle\iint_\sigma (f\mathbf{n}) \cdot \mathbf{v}\, dS = \iiint_G \nabla f \cdot \mathbf{v}\, dV$ ($\mathbf{v}$ a fixed vector)

28. Use the Divergence Theorem to find all positive values of k such that

$$\mathbf{F}(\mathbf{r}) = \frac{\mathbf{r}}{\|\mathbf{r}\|^k}$$

satisfies the condition div $\mathbf{F} = 0$ when $\mathbf{r} \neq \mathbf{0}$. [*Hint:* Modify the proof of (11).]

29–32 Determine whether the vector field $\mathbf{F}(x, y, z)$ is free of sources and sinks. If it is not, locate them.

29. $\mathbf{F}(x, y, z) = (y + z)\mathbf{i} - xz^3\mathbf{j} + (x^2 \sin y)\mathbf{k}$

30. $\mathbf{F}(x, y, z) = xy\mathbf{i} - xy\mathbf{j} + y^2\mathbf{k}$

31. $\mathbf{F}(x, y, z) = x^3\mathbf{i} + y^3\mathbf{j} + z^3\mathbf{k}$

32. $\mathbf{F}(x, y, z) = (x^3 - x)\mathbf{i} + (y^3 - y)\mathbf{j} + (z^3 - z)\mathbf{k}$

C **33.** Let σ be the surface of the solid G that is enclosed by the paraboloid $z = 1 - x^2 - y^2$ and the plane $z = 0$. Use a CAS to verify Formula (1) in the Divergence Theorem for the vector field

$$\mathbf{F} = (x^2 y - z^2)\mathbf{i} + (y^3 - x)\mathbf{j} + (2x + 3z - 1)\mathbf{k}$$

by evaluating the surface integral and the triple integral.

✔ QUICK CHECK ANSWERS 16.7

1. $\displaystyle\iint_\sigma \mathbf{F} \cdot \mathbf{n}\, dS;\ \iiint_G \operatorname{div} \mathbf{F}\, dV$ 2. 3 3. source; sink; div $\mathbf{F} = 0$ 4. $4\pi c$; 0

16.8 STOKES' THEOREM

In this section we will discuss a generalization of Green's Theorem to three dimensions that has important applications in the study of vector fields, particularly in the analysis of rotational motion of fluids. This theorem will also provide us with a physical interpretation of the curl of a vector field.

■ RELATIVE ORIENTATION OF CURVES AND SURFACES

We will be concerned in this section with oriented surfaces in 3-space that are bounded by simple closed parametric curves (Figure 16.8.1a). If σ is an oriented surface bounded by a simple closed parametric curve C, then there are two possible relationships between the orientations of σ and C, which can be described as follows. Imagine a person walking along the curve C with his or her head in the direction of the orientation of σ. The person is said to be walking in the ***positive direction*** of C relative to the orientation of σ if the surface is on the person's left (Figure 16.8.1*b*), and the person is said to be walking in the ***negative direction*** of C relative to the orientation of σ if the surface is on the person's right (Figure 16.8.1*c*). The positive direction of C establishes a right-hand relationship between the orientations of σ and C in the sense that if the fingers of the right hand are curled from the direction of C towards σ, then the thumb points (roughly) in the direction of the orientation of σ.

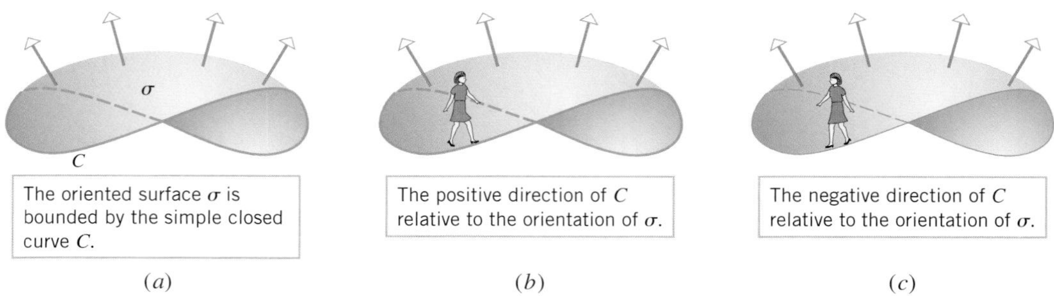

σ
C
The oriented surface σ is bounded by the simple closed curve C.
The positive direction of C relative to the orientation of σ.
The negative direction of C relative to the orientation of σ.
(a)
(b)
(c)

Figure 16.8.1

STOKES' THEOREM

In Section 16.1 we defined the curl of a vector field

$$\mathbf{F}(x, y, z) = f(x, y, z)\mathbf{i} + g(x, y, z)\mathbf{j} + h(x, y, z)\mathbf{k}$$

as

$$\text{curl } \mathbf{F} = \left(\frac{\partial h}{\partial y} - \frac{\partial g}{\partial z}\right)\mathbf{i} + \left(\frac{\partial f}{\partial z} - \frac{\partial h}{\partial x}\right)\mathbf{j} + \left(\frac{\partial g}{\partial x} - \frac{\partial f}{\partial y}\right)\mathbf{k} = \begin{vmatrix} \mathbf{i} & \mathbf{j} & \mathbf{k} \\ \dfrac{\partial}{\partial x} & \dfrac{\partial}{\partial y} & \dfrac{\partial}{\partial z} \\ f & g & h \end{vmatrix} \qquad (1)$$

but we did not attempt to give a physical explanation of its meaning at that time. The following result, known as ***Stokes' Theorem***, will provide us with a physical interpretation of the curl in the context of fluid flow.

16.8.1 THEOREM (*Stokes' Theorem*). *Let σ be a piecewise smooth oriented surface that is bounded by a simple, closed, piecewise smooth curve C with positive orientation. If the components of the vector field*

$$\mathbf{F}(x, y, z) = f(x, y, z)\mathbf{i} + g(x, y, z)\mathbf{j} + h(x, y, z)\mathbf{k}$$

are continuous and have continuous first partial derivatives on some open set containing σ, and if $\mathbf{T}$ is the unit tangent vector to C, then

$$\oint_C \mathbf{F} \cdot \mathbf{T}\, ds = \iint_\sigma (\text{curl } \mathbf{F}) \cdot \mathbf{n}\, dS \qquad (2)$$

The proof of this theorem is beyond the scope of this text, so we will focus on its applications.

Recall from Formulas (30) and (34) in Section 16.2 that if $\mathbf{F}$ is a force field, the integral on the left side of (2) represents the work performed by the force field on a particle that traverses the curve C. Thus, loosely phrased, Stokes' Theorem states:

The work performed by a force field on a particle that traverses a simple, closed, piecewise smooth curve C in the positive direction can be obtained by integrating the normal component of the curl over an oriented surface σ bounded by C.

USING STOKES' THEOREM TO CALCULATE WORK

For computational purposes it is usually preferable to use Formula (30) in Section 16.2 to rewrite the formula in Stokes' Theorem as

$$\oint_C \mathbf{F} \cdot d\mathbf{r} = \iint_\sigma (\text{curl } \mathbf{F}) \cdot \mathbf{n}\, dS \tag{3}$$

Stokes' Theorem is usually the method of choice for calculating work around piecewise smooth curves with multiple sections, since it eliminates the need for a separate integral evaluation over each section. This is illustrated in the following example.

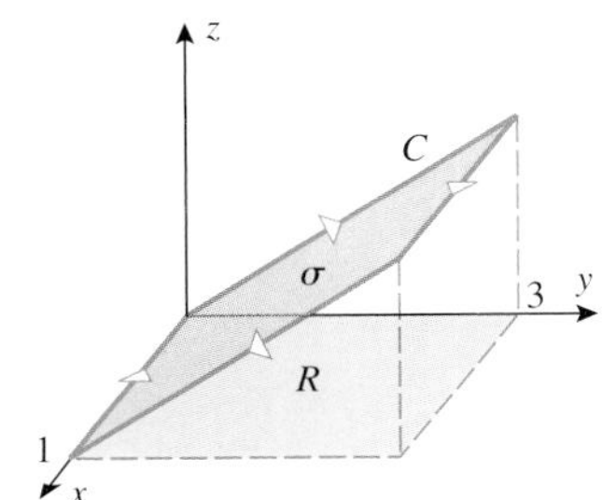

Figure 16.8.2

Example 1 Find the work performed by the force field

$$\mathbf{F}(x, y, z) = x^2\mathbf{i} + 4xy^3\mathbf{j} + y^2x\mathbf{k}$$

on a particle that traverses the rectangle C in the plane $z = y$ shown in Figure 16.8.2.

Solution. The work performed by the field is

$$W = \oint_C \mathbf{F} \cdot d\mathbf{r}$$

However, to evaluate this integral directly would require four separate integrations, one over each side of the rectangle. Instead, we will use Formula (3) to express the work as the surface integral

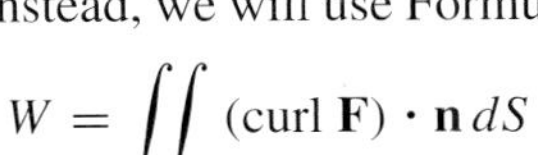

$$W = \iint_\sigma (\text{curl } \mathbf{F}) \cdot \mathbf{n}\, dS$$

George Gabriel Stokes (1819–1903) Irish mathematician and physicist. Born in Skreen, Ireland, Stokes came from a family deeply rooted in the Church of Ireland. His father was a rector, his mother the daughter of a rector, and three of his brothers took holy orders. He received his early education from his father and a local parish clerk. In 1837, he entered Pembroke College and after graduating with top honors accepted a fellowship at the college. In 1847 he was appointed Lucasian professor of mathematics at Cambridge, a position once held by Isaac Newton, but one that had lost its esteem through the years. By virtue of his accomplishments, Stokes ultimately restored the position to the eminence it once held. Unfortunately, the position paid very little and Stokes was forced to teach at the Government School of Mines during the 1850s to supplement his income.

Stokes was one of several outstanding nineteenth century scientists who helped turn the physical sciences in a more empirical direction. He systematically studied hydrodynamics, elasticity of solids, behavior of waves in elastic solids, and diffraction of light. For Stokes, mathematics was a tool for his physical studies. He wrote classic papers on the motion of viscous fluids that laid the foundation for modern hydrodynamics; he elaborated on the wave theory of light; and he wrote papers on gravitational variation that established him as a founder of the modern science of geodesy.

Stokes was honored in his later years with degrees, medals, and memberships in foreign societies. He was knighted in 1889. Throughout his life, Stokes gave generously of his time to learned societies and readily assisted those who sought his help in solving problems. He was deeply religious and vitally concerned with the relationship between science and religion.

in which the plane surface σ enclosed by C is assigned a *downward* orientation to make the orientation of C positive, as required by Stokes' Theorem.

Since the surface σ has equation $z = y$ and

$$\operatorname{curl} \mathbf{F} = \begin{vmatrix} \mathbf{i} & \mathbf{j} & \mathbf{k} \\ \dfrac{\partial}{\partial x} & \dfrac{\partial}{\partial y} & \dfrac{\partial}{\partial z} \\ x^2 & 4xy^3 & xy^2 \end{vmatrix} = 2xy\mathbf{i} - y^2\mathbf{j} + 4y^3\mathbf{k}$$

it follows from Formula (13) of Section 16.6 with curl $\mathbf{F}$ replacing $\mathbf{F}$ that

$$\begin{aligned} W = \iint_{\sigma} (\operatorname{curl} \mathbf{F}) \cdot \mathbf{n}\, dS &= \iint_{R} (\operatorname{curl} \mathbf{F}) \cdot \left(\frac{\partial z}{\partial x}\mathbf{i} + \frac{\partial z}{\partial y}\mathbf{j} - \mathbf{k} \right) dA \\ &= \iint_{R} (2xy\mathbf{i} - y^2\mathbf{j} + 4y^3\mathbf{k}) \cdot (0\mathbf{i} + \mathbf{j} - \mathbf{k})\, dA \\ &= \int_0^1 \int_0^3 (-y^2 - 4y^3)\, dy\, dx \\ &= -\int_0^1 \left[\frac{y^3}{3} + y^4 \right]_{y=0}^{3} dx \\ &= -\int_0^1 90\, dx = -90 \blacktriangleleft \end{aligned}$$

Explain how the result in Example 1 shows that the given force field is not conservative.

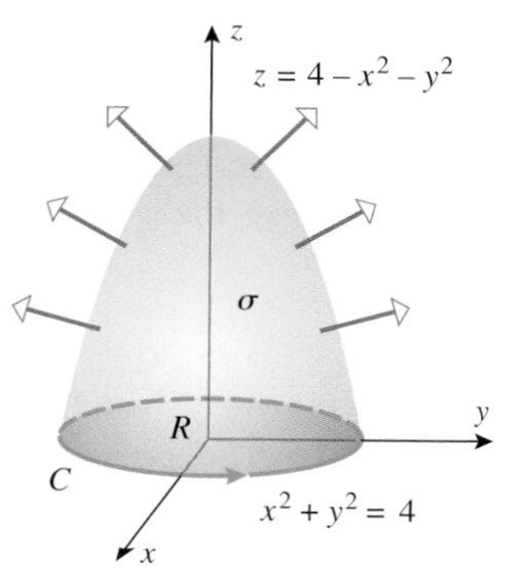

Figure 16.8.3

▶ Example 2 Verify Stokes' Theorem for the vector field $\mathbf{F}(x, y, z) = 2z\mathbf{i} + 3x\mathbf{j} + 5y\mathbf{k}$, taking σ to be the portion of the paraboloid $z = 4 - x^2 - y^2$ for which $z \geq 0$ with upward orientation, and C to be the positively oriented circle $x^2 + y^2 = 4$ that forms the boundary of σ in the xy-plane (Figure 16.8.3).

Solution. We will verify Formula (3). Since σ is oriented up, the positive orientation of C is counterclockwise looking down the positive z-axis. Thus, C can be represented parametrically (with positive orientation) by

$$x = 2\cos t, \quad y = 2\sin t, \quad z = 0 \qquad (0 \leq t \leq 2\pi) \tag{4}$$

Therefore,

$$\begin{aligned} \oint_C \mathbf{F} \cdot d\mathbf{r} &= \oint_C 2z\, dx + 3x\, dy + 5y\, dz \\ &= \int_0^{2\pi} [0 + (6\cos t)(2\cos t) + 0]\, dt \\ &= \int_0^{2\pi} 12\cos^2 t\, dt = 12\left[\frac{1}{2}t + \frac{1}{4}\sin 2t \right]_0^{2\pi} = 12\pi \end{aligned}$$

To evaluate the right side of (3), we start by finding curl $\mathbf{F}$. We obtain

$$\operatorname{curl} \mathbf{F} = \begin{vmatrix} \mathbf{i} & \mathbf{j} & \mathbf{k} \\ \dfrac{\partial}{\partial x} & \dfrac{\partial}{\partial y} & \dfrac{\partial}{\partial z} \\ 2z & 3x & 5y \end{vmatrix} = 5\mathbf{i} + 2\mathbf{j} + 3\mathbf{k}$$

Since σ is oriented up and is expressed in the form $z = g(x, y) = 4 - x^2 - y^2$, it follows from Formula (12) of Section 16.6 with curl $\mathbf{F}$ replacing $\mathbf{F}$ that

$$\begin{aligned}
\iint_{\sigma} (\text{curl } \mathbf{F}) \cdot \mathbf{n}\, dS &= \iint_{R} (\text{curl } \mathbf{F}) \cdot \left(-\frac{\partial z}{\partial x}\mathbf{i} - \frac{\partial z}{\partial y}\mathbf{j} + \mathbf{k}\right) dA \\
&= \iint_{R} (5\mathbf{i} + 2\mathbf{j} + 3\mathbf{k}) \cdot (2x\mathbf{i} + 2y\mathbf{j} + \mathbf{k})\, dA \\
&= \iint_{R} (10x + 4y + 3)\, dA \\
&= \int_0^{2\pi} \int_0^2 (10r\cos\theta + 4r\sin\theta + 3)r\, dr\, d\theta \\
&= \int_0^{2\pi} \left[\frac{10r^3}{3}\cos\theta + \frac{4r^3}{3}\sin\theta + \frac{3r^2}{2}\right]_{r=0}^{2} d\theta \\
&= \int_0^{2\pi} \left(\frac{80}{3}\cos\theta + \frac{32}{3}\sin\theta + 6\right) d\theta \\
&= \left[\frac{80}{3}\sin\theta - \frac{32}{3}\cos\theta + 6\theta\right]_0^{2\pi} = 12\pi
\end{aligned}$$

As guaranteed by Stokes' Theorem, the value of this surface integral is the same as the value of the line integral obtained above. Note, however, that the line integral was simpler to evaluate and hence would be the method of choice in this case. ◀

Observe that in Formula (3) the only relationships required between σ and C are that C be the boundary of σ and that C be positively oriented relative to the orientation of σ. Thus, if σ_1 and σ_2 are *different* oriented surfaces but have the *same* positively oriented boundary curve C, then it follows from (3) that

$$\iint_{\sigma_1} \text{curl } \mathbf{F} \cdot \mathbf{n}\, dS = \iint_{\sigma_2} \text{curl } \mathbf{F} \cdot \mathbf{n}\, dS$$

For example, the parabolic surface in Example 2 has the same positively oriented boundary C as the disk R in Figure 16.8.3 with upper orientation. Thus, the value of the surface integral in that example would not change if σ is replaced by R (or by any other oriented surface that has the positively oriented circle C as its boundary). This can be useful in computations because it is sometimes possible to circumvent a difficult integration by changing the surface of integration.

RELATIONSHIP BETWEEN GREEN'S THEOREM AND STOKES' THEOREM

It is sometimes convenient to regard a vector field

$$\mathbf{F}(x, y) = f(x, y)\mathbf{i} + g(x, y)\mathbf{j}$$

in 2-space as a vector field in 3-space by expressing it as

$$\mathbf{F}(x, y) = f(x, y)\mathbf{i} + g(x, y)\mathbf{j} + 0\mathbf{k} \tag{5}$$

If R is a region in the xy-plane enclosed by a simple, closed, piecewise smooth curve C, then we can treat R as a *flat* surface, and we can treat a surface integral over R as an ordinary double integral over R. Thus, if we orient R up and C counterclockwise looking down the positive z-axis, then Formula (3) applied to (5) yields

$$\oint_C \mathbf{F} \cdot d\mathbf{r} = \iint_R \operatorname{curl} \mathbf{F} \cdot \mathbf{k}\, dA \tag{6}$$

But

$$\operatorname{curl} \mathbf{F} = \begin{vmatrix} \mathbf{i} & \mathbf{j} & \mathbf{k} \\ \dfrac{\partial}{\partial x} & \dfrac{\partial}{\partial y} & \dfrac{\partial}{\partial z} \\ f & g & 0 \end{vmatrix} = -\frac{\partial g}{\partial z}\mathbf{i} + \frac{\partial f}{\partial z}\mathbf{j} + \left(\frac{\partial g}{\partial x} - \frac{\partial f}{\partial y}\right)\mathbf{k} = \left(\frac{\partial g}{\partial x} - \frac{\partial f}{\partial y}\right)\mathbf{k}$$

since $\partial g/\partial z = \partial f/\partial z = 0$. Substituting this expression in (6) and expressing the integrals in terms of components yields

$$\oint_C f\,dx + g\,dy = \iint_R \left(\frac{\partial g}{\partial x} - \frac{\partial f}{\partial y}\right) dA$$

which is Green's Theorem [Formula (1) of Section 16.4]. Thus, we have shown that Green's Theorem can be viewed as a special case of Stokes' Theorem.

CURL VIEWED AS CIRCULATION

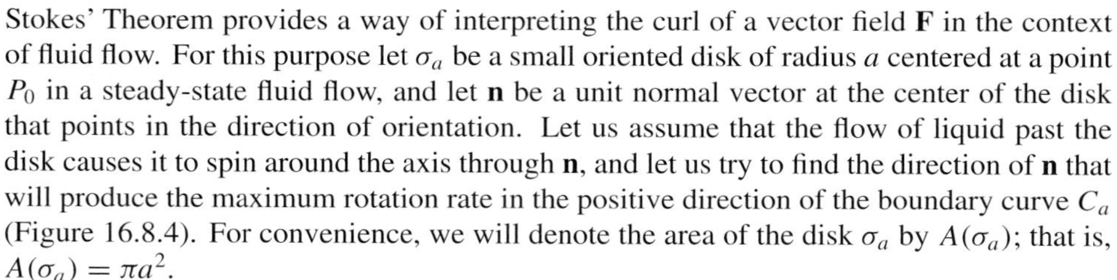

Stokes' Theorem provides a way of interpreting the curl of a vector field $\mathbf{F}$ in the context of fluid flow. For this purpose let σ_a be a small oriented disk of radius a centered at a point P_0 in a steady-state fluid flow, and let $\mathbf{n}$ be a unit normal vector at the center of the disk that points in the direction of orientation. Let us assume that the flow of liquid past the disk causes it to spin around the axis through $\mathbf{n}$, and let us try to find the direction of $\mathbf{n}$ that will produce the maximum rotation rate in the positive direction of the boundary curve C_a (Figure 16.8.4). For convenience, we will denote the area of the disk σ_a by $A(\sigma_a)$; that is, $A(\sigma_a) = \pi a^2$.

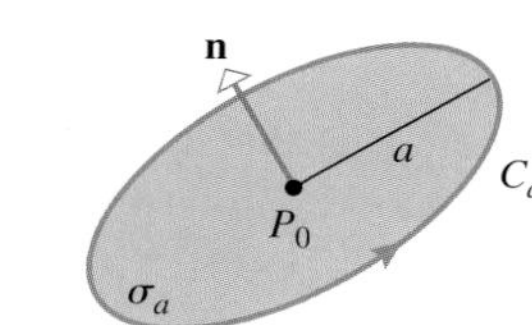

Figure 16.8.4

If the direction of $\mathbf{n}$ is fixed, then at each point of C_a the only component of $\mathbf{F}$ that contributes to the rotation of the disk about $\mathbf{n}$ is the component $\mathbf{F} \cdot \mathbf{T}$ tangent to C_a (Figure 16.8.5). Thus, for a fixed $\mathbf{n}$ the integral

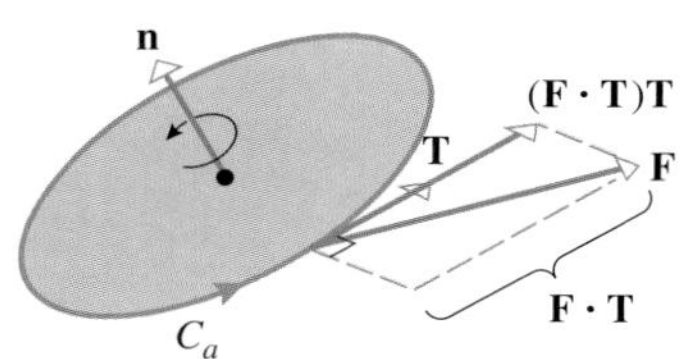

Figure 16.8.5

$$\oint_{C_a} \mathbf{F} \cdot \mathbf{T}\, ds \tag{7}$$

can be viewed as a measure of the tendency for the fluid to flow in the positive direction around C_a. Accordingly, (7) is called the ***circulation of* F *around* C_a**. For example, in the extreme case where the flow is normal to the circle at each point, the circulation around C_a is zero, since $\mathbf{F} \cdot \mathbf{T} = 0$ at each point. The more closely that $\mathbf{F}$ aligns with $\mathbf{T}$ along the circle, the larger the value of $\mathbf{F} \cdot \mathbf{T}$ and the larger the value of the circulation.

To see the relationship between circulation and curl, suppose that curl $\mathbf{F}$ is continuous on σ_a, so that when σ_a is small the value of curl $\mathbf{F}$ at any point of σ_a will not vary much from the value of curl $\mathbf{F}(P_0)$ at the center. Thus, for a small disk σ_a we can reasonably approximate curl $\mathbf{F}$ by the constant value curl $\mathbf{F}(P_0)$ on σ_a. Moreover, because the surface σ_a is flat, the unit normal vectors that orient σ_a are all equal. Thus, the vector quantity $\mathbf{n}$ in Formula (3) can be treated as a constant, and we can write

$$\oint_{C_a} \mathbf{F} \cdot \mathbf{T}\, ds = \iint_{\sigma_a} (\operatorname{curl} \mathbf{F}) \cdot \mathbf{n}\, dS \approx \operatorname{curl} \mathbf{F}(P_0) \cdot \mathbf{n} \iint_{\sigma_a} dS$$

where the line integral is taken in the positive direction of C_a. But the last double integral in this equation represents the surface area of σ_a, so

$$\oint_{C_a} \mathbf{F} \cdot \mathbf{T}\, ds \approx [\operatorname{curl} \mathbf{F}(P_0) \cdot \mathbf{n}] A(\sigma_a)$$

from which we obtain

$$\text{curl } \mathbf{F}(P_0) \cdot \mathbf{n} \approx \frac{1}{A(\sigma_a)} \oint_{C_a} \mathbf{F} \cdot \mathbf{T}\, ds \tag{8}$$

Formula (9) is sometimes taken as a definition of curl. This is a useful alternative to Definition 16.1.5 because it does not require a coordinate system.

The quantity on the right side of (8) is called the ***circulation density of* F *around* C_a**. If we now let the radius a of the disk approach zero (with $\mathbf{n}$ fixed), then it is plausible that the error in this approximation will approach zero and the exact value of curl $\mathbf{F}(P_0) \cdot \mathbf{n}$ will be given by

$$\text{curl } \mathbf{F}(P_0) \cdot \mathbf{n} = \lim_{a \to 0} \frac{1}{A(\sigma_a)} \oint_{C_a} \mathbf{F} \cdot \mathbf{T}\, ds \tag{9}$$

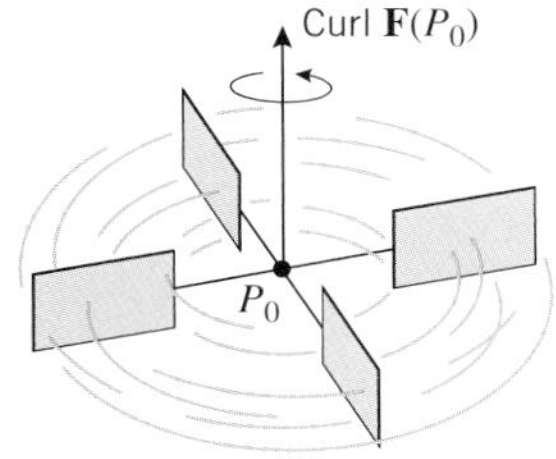

Figure 16.8.6

We call curl $\mathbf{F}(P_0) \cdot \mathbf{n}$ the ***circulation density of* F *at* P_0 *in the direction of* n**. This quantity has its maximum value when $\mathbf{n}$ is in the same direction as curl $\mathbf{F}(P_0)$; this tells us that *at each point in a steady-state fluid flow the maximum circulation density occurs in the direction of the curl*. Physically, this means that if a small paddle wheel is immersed in the fluid so that the pivot point is at P_0, then the paddles will turn most rapidly when the spindle is aligned with curl $\mathbf{F}(P_0)$ (Figure 16.8.6). If curl $\mathbf{F} = \mathbf{0}$ at each point of a region, then $\mathbf{F}$ is said to be ***irrotational*** in that region, since no circulation occurs about any point of the region.

✔QUICK CHECK EXERCISES 16.8 *(See page 1181 for answers.)*

1. Let σ be a piecewise smooth oriented surface that is bounded by a simple, closed, piecewise smooth curve C with positive orientation. If the component functions of the vector field $\mathbf{F}(x, y, z)$ have continuous first partial derivatives on some open set containing σ, and if $\mathbf{T}$ is the unit tangent vector to C, then Stokes' Theorem states that the line integral ________ and the surface integral ________ are equal.

2. We showed in Example 2 that the vector field
$$\mathbf{F}(x, y, z) = 2z\mathbf{i} + 3x\mathbf{j} + 5y\mathbf{k}$$
satisfies the equation curl $\mathbf{F} = 5\mathbf{i} + 2\mathbf{j} + 3\mathbf{k}$. It follows from Stokes' Theorem that if C is any circle of radius a in the xy-plane that is oriented counterclockwise when viewed from the positive z-axis, then
$$\int_C \mathbf{F} \cdot \mathbf{T}\, ds = \underline{\qquad\qquad}$$
where $\mathbf{T}$ denotes the unit tangent vector to C.

3. (a) If σ_1 and σ_2 are two oriented surfaces that have the same positively oriented boundary curve C, and if the vector field $\mathbf{F}(x, y, z)$ has continuous first partial derivatives on some open set containing σ_1 and σ_2, then it follows from Stokes' Theorem that the surface integrals ________ and ________ are equal.
(b) Let $\mathbf{F}(x, y, z) = 2z\mathbf{i} + 3x\mathbf{j} + 5y\mathbf{k}$, let a be a positive number, and let σ be the portion of the paraboloid $z = a^2 - x^2 - y^2$ for which $z \geq 0$ with upward orientation. Using part (a) and Quick Check Exercise 2, it follows that
$$\iint_\sigma (\text{curl } \mathbf{F}) \cdot \mathbf{n}\, dS = \underline{\qquad\qquad}$$

4. For steady-state flow, the maximum circulation density occurs in the direction of the ________ of the velocity vector field for the flow.

EXERCISE SET 16.8  CAS

1–4 Verify Formula (2) in Stokes' Theorem by evaluating the line integral and the double integral. Assume that the surface has an upward orientation.

1. $\mathbf{F}(x, y, z) = (x - y)\mathbf{i} + (y - z)\mathbf{j} + (z - x)\mathbf{k}$; σ is the portion of the plane $x + y + z = 1$ in the first octant.

2. $\mathbf{F}(x, y, z) = x^2\mathbf{i} + y^2\mathbf{j} + z^2\mathbf{k}$; σ is the portion of the cone $z = \sqrt{x^2 + y^2}$ below the plane $z = 1$.

3. $\mathbf{F}(x, y, z) = x\mathbf{i} + y\mathbf{j} + z\mathbf{k}$; σ is the upper hemisphere $z = \sqrt{a^2 - x^2 - y^2}$.

4. $\mathbf{F}(x, y, z) = (z - y)\mathbf{i} + (z + x)\mathbf{j} - (x + y)\mathbf{k}$; σ is the portion of the paraboloid $z = 9 - x^2 - y^2$ above the xy-plane.

5–12 Use Stokes' Theorem to evaluate $\oint_C \mathbf{F} \cdot d\mathbf{r}$.

5. $\mathbf{F}(x, y, z) = z^2\mathbf{i} + 2x\mathbf{j} - y^3\mathbf{k}$; C is the circle $x^2 + y^2 = 1$ in the xy-plane with counterclockwise orientation looking down the positive z-axis.

6. $\mathbf{F}(x, y, z) = xz\mathbf{i} + 3x^2y^2\mathbf{j} + yx\mathbf{k}$; C is the rectangle in the plane $z = y$ shown in Figure 16.8.2.

7. $\mathbf{F}(x, y, z) = 3z\mathbf{i} + 4x\mathbf{j} + 2y\mathbf{k}$; C is the boundary of the paraboloid shown in Figure 16.8.3.

8. $\mathbf{F}(x, y, z) = -3y^2\mathbf{i} + 4z\mathbf{j} + 6x\mathbf{k}$; C is the triangle in the plane $z = \frac{1}{2}y$ with vertices $(2, 0, 0)$, $(0, 2, 1)$, and $(0, 0, 0)$ with a counterclockwise orientation looking down the positive z-axis.

9. $\mathbf{F}(x, y, z) = xy\mathbf{i} + x^2\mathbf{j} + z^2\mathbf{k}$; C is the intersection of the paraboloid $z = x^2 + y^2$ and the plane $z = y$ with a counterclockwise orientation looking down the positive z-axis.

10. $\mathbf{F}(x, y, z) = xy\mathbf{i} + yz\mathbf{j} + zx\mathbf{k}$; C is the triangle in the plane $x + y + z = 1$ with vertices $(1, 0, 0)$, $(0, 1, 0)$, and $(0, 0, 1)$ with a counterclockwise orientation looking from the first octant toward the origin.

11. $\mathbf{F}(x, y, z) = (x - y)\mathbf{i} + (y - z)\mathbf{j} + (z - x)\mathbf{k}$; C is the circle $x^2 + y^2 = a^2$ in the xy-plane with counterclockwise orientation looking down the positive z-axis.

12. $\mathbf{F}(x, y, z) = (z + \sin x)\mathbf{i} + (x + y^2)\mathbf{j} + (y + e^z)\mathbf{k}$; C is the intersection of the sphere $x^2 + y^2 + z^2 = 1$ and the cone $z = \sqrt{x^2 + y^2}$ with counterclockwise orientation looking down the positive z-axis.

13. Consider the vector field given by the formula

$$\mathbf{F}(x, y, z) = (x - z)\mathbf{i} + (y - x)\mathbf{j} + (z - xy)\mathbf{k}$$

(a) Use Stokes' Theorem to find the circulation around the triangle with vertices $A(1, 0, 0)$, $B(0, 2, 0)$, and $C(0, 0, 1)$ oriented counterclockwise looking from the origin toward the first octant.
(b) Find the circulation density of $\mathbf{F}$ at the origin in the direction of $\mathbf{k}$.
(c) Find the unit vector $\mathbf{n}$ such that the circulation density of $\mathbf{F}$ at the origin is maximum in the direction of $\mathbf{n}$.

FOCUS ON CONCEPTS

14. (a) Let σ denote the surface of a solid G with $\mathbf{n}$ the outward unit normal vector field to σ. Assume that $\mathbf{F}$ is a vector field with continuous first-order partial derivatives on σ. Prove that

$$\iint_\sigma (\text{curl } \mathbf{F}) \cdot \mathbf{n}\, dS = 0$$

[*Hint:* Let C denote a simple closed curve on σ that separates the surface into two subsurfaces σ_1 and σ_2 that share C as their common boundary. Apply Stokes' Theorem to σ_1 and to σ_2 and add the results.]
(b) The vector field curl($\mathbf{F}$) is called the ***curl field*** of $\mathbf{F}$. In words, interpret the formula in part (a) as a statement about the flux of the curl field.

15–16 The figures in these exercises show a horizontal layer of the vector field of a fluid flow in which the flow is parallel to the xy-plane at every point and is identical in each layer (i.e., is independent of z). For each flow, state whether you believe that the curl is nonzero at the origin, and explain your reasoning. If you believe that it is nonzero, then state whether it points in the positive or negative z-direction.

15. (a) (b)

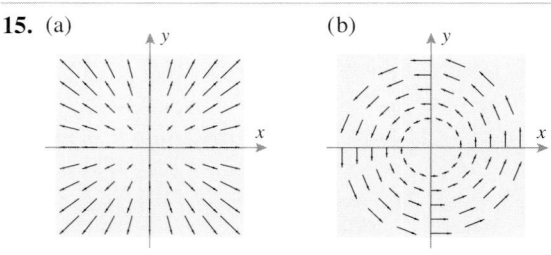

16. (a) (b)

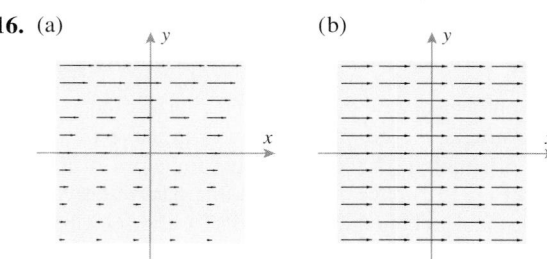

17. Let $\mathbf{F}(x, y, z)$ be a conservative vector field in 3-space whose component functions have continuous first partial derivatives. Explain how to use Formula (9) to prove that curl $\mathbf{F} = \mathbf{0}$.

18. In 1831 the physicist Michael Faraday discovered that an electric current can be produced by varying the magnetic flux through a conducting loop. His experiments showed that the electromotive force $\mathbf{E}$ is related to the magnetic induction $\mathbf{B}$ by the equation

$$\oint_C \mathbf{E} \cdot d\mathbf{r} = -\iint_\sigma \frac{\partial \mathbf{B}}{\partial t} \cdot \mathbf{n}\, dS$$

Use this result to make a conjecture about the relationship between curl $\mathbf{E}$ and $\mathbf{B}$, and explain your reasoning.

C 19. Let σ be the portion of the paraboloid $z = 1 - x^2 - y^2$ for which $z \geq 0$, and let C be the circle $x^2 + y^2 = 1$ that forms the boundary of σ in the xy-plane. Assuming that σ is oriented up, use a CAS to verify Formula (2) in Stokes' Theorem for the vector field

$$\mathbf{F} = (x^2y - z^2)\mathbf{i} + (y^3 - x)\mathbf{j} + (2x + 3z - 1)\mathbf{k}$$

by evaluating the line integral and the surface integral.

✔QUICK CHECK ANSWERS 16.8

1. $\int_C \mathbf{F}\cdot\mathbf{T}\,ds$; $\iint_\sigma (\text{curl } \mathbf{F})\cdot\mathbf{n}\,dS$ 2. $3\pi a^2$ 3. (a) $\iint_{\sigma_1} (\text{curl } \mathbf{F})\cdot\mathbf{n}\,dS$; $\iint_{\sigma_2} (\text{curl } \mathbf{F})\cdot\mathbf{n}\,dS$ (b) $3\pi a^2$ 4. curl

CHAPTER REVIEW EXERCISES

1. In words, what is a vector field? Give some physical examples of vector fields.

2. (a) Give a physical example of an inverse-square field $\mathbf{F}(\mathbf{r})$ in 3-space.
(b) Write a formula for a general inverse-square field $\mathbf{F}(\mathbf{r})$ in terms of the radius vector $\mathbf{r}$.
(c) Write a formula for a general inverse-square field $\mathbf{F}(x, y, z)$ in 3-space using rectangular coordinates.

3. Find an explicit coordinate expression for the vector field $\mathbf{F}(x, y)$ that at every point $(x, y) \neq (1, 2)$ is the unit vector directed from (x, y) to $(1, 2)$.

4. Find $\nabla\left(\dfrac{x+y}{x-y}\right)$.

5. Find $\text{curl}(z\mathbf{i} + x\mathbf{j} + y\mathbf{k})$.

6. Let

$$\mathbf{F}(x, y, z) = \frac{x}{x^2+y^2}\mathbf{i} + \frac{y}{x^2+y^2}\mathbf{j} + \frac{z}{x^2+y^2}\mathbf{k}$$

Sketch the level surface $\text{div } \mathbf{F} = 1$.

7. Assume that C is the parametric curve $x = x(t)$, $y = y(t)$, where t varies from a to b. In each part, express the line integral as a definite integral with variable of integration t.
(a) $\displaystyle\int_C f(x, y)\,dx + g(x, y)\,dy$ (b) $\displaystyle\int_C f(x, y)\,ds$

8. (a) Express the mass M of a thin wire in 3-space as a line integral.
(b) Express the length of a curve as a line integral.

9. Give a physical interpretation of $\int_C \mathbf{F}\cdot\mathbf{T}\,ds$.

10. State some alternative notations for $\int_C \mathbf{F}\cdot\mathbf{T}\,ds$.

11–13 Evaluate the line integral.

11. $\displaystyle\int_C (x - y)\,ds$; $C : x^2 + y^2 = 1$

12. $\displaystyle\int_C x\,dx + z\,dy - 2y^2\,dz$;
$C : x = \cos t,\ y = \sin t,\ z = t \quad (0 \le t \le 2\pi)$

13. $\displaystyle\int_C \mathbf{F}\cdot d\mathbf{r}$ where $\mathbf{F}(x, y) = (x/y)\mathbf{i} - (y/x)\mathbf{j}$;
$\mathbf{r}(t) = t\mathbf{i} + 2t\mathbf{j} \quad (1 \le t \le 2)$

14. Find the work done by the force field

$$\mathbf{F}(x, y) = y^2\mathbf{i} + xy\mathbf{j}$$

moving a particle from $(0, 0)$ to $(1, 1)$ along the parabola $y = x^2$.

15. State the Fundamental Theorem of Line Integrals, including all required hypotheses.

16. Evaluate $\int_C \nabla f\cdot d\mathbf{r}$ where $f(x, y, z) = xy^2z^3$ and

$$\mathbf{r}(t) = t\mathbf{i} + (t^2 + t)\mathbf{j} + \sin(3\pi t/2)\mathbf{k} \quad (0 \le t \le 1)$$

17. Let $\mathbf{F}(x, y) = y\mathbf{i} - 2x\mathbf{j}$.
(a) Find a nonzero function $h(x)$ such that $h(x)\mathbf{F}(x, y)$ is a conservative vector field.
(b) Find a nonzero function $g(y)$ such that $g(y)\mathbf{F}(x, y)$ is a conservative vector field.

18. Let $\mathbf{F}(x, y) = (ye^{xy} - 1)\mathbf{i} + xe^{xy}\mathbf{j}$.
(a) Show that $\mathbf{F}$ is a conservative vector field.
(b) Find a potential function for $\mathbf{F}$.
(c) Find the work performed by the force field on a particle that moves along the sawtooth curve represented by the parametric equations

$$x = t + \sin^{-1}(\sin t), \quad y = (2/\pi)\sin^{-1}(\sin t) \quad (0 \le t \le 8\pi)$$

(see the accompanying figure).

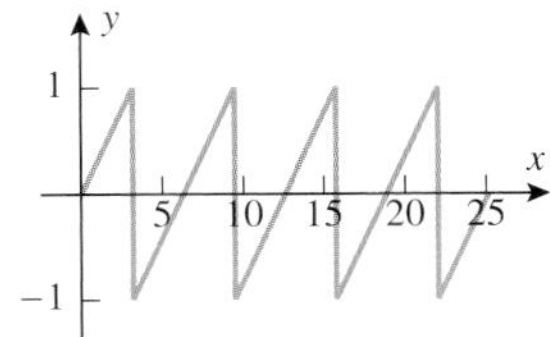

Figure Ex-18

19. State Green's Theorem, including all of the required hypotheses.

20. Express the area of a plane region as a line integral.

21. Let α and β denote angles that satisfy $0 < \beta - \alpha \le 2\pi$ and assume that $r = f(\theta)$ is a smooth polar curve with $f(\theta) > 0$ on the interval $[\alpha, \beta]$. Use the formula

$$A = \frac{1}{2}\int_C -y\,dx + x\,dy$$

to find the area of the region R enclosed by the curve $r = f(\theta)$ and the rays $\theta = \alpha$ and $\theta = \beta$.

22. (a) Use Green's Theorem to prove that

$$\int_C f(x)\,dx + g(y)\,dy = 0$$

if f and g are differentiable functions and C is a simple, closed, piecewise smooth curve.
(b) What does this tell you about the vector field $\mathbf{F}(x, y) = f(x)\mathbf{i} + g(y)\mathbf{j}$?

23. Assume that σ is the parametric surface

$$\mathbf{r} = x(u, v)\mathbf{i} + y(u, v)\mathbf{j} + z(u, v)\mathbf{k}$$

where (u, v) varies over a region R. Express the surface integral

$$\iint_{\sigma} f(x, y, z)\, dS$$

as a double integral with variables of integration u and v.

24. Evaluate $\iint_{\sigma} z\, dS$; $\sigma : x^2 + y^2 = 1 (0 \le z \le 1)$.

25. Do you think that the surface in the accompanying figure is orientable? Explain your reasoning.

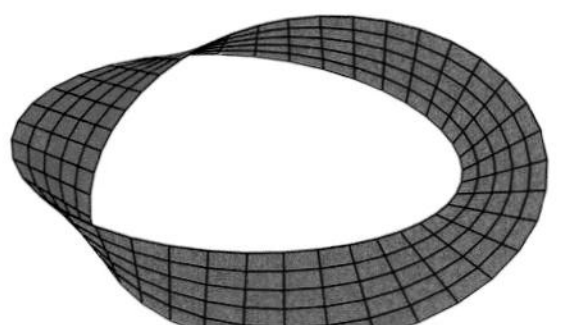

Figure Ex-25

26. Give a physical interpretation of $\iint_{\sigma} \mathbf{F} \cdot \mathbf{n}\, dS$.

27. Find the flux of $\mathbf{F}(x, y, z) = x\mathbf{i} + y\mathbf{j} + 2z\mathbf{k}$ through the portion of the paraboloid $z = 1 - x^2 - y^2$ that is on or above the xy-plane, with upward orientation.

28. Find the flux of $\mathbf{F}(x, y, z) = x\mathbf{i} + 2y\mathbf{j} + 3z\mathbf{k}$ through the unit sphere centered at the origin with outward orientation.

29. State the Divergence Theorem and Stokes' Theorem, including all required hypotheses.

30. Let G be a solid with the surface σ oriented by outward unit normals, suppose that ϕ has continuous first and second partial derivatives in some open set containing G, and let $D_{\mathbf{n}}\phi$ be the directional derivative of ϕ, where $\mathbf{n}$ is an outward unit normal to σ. Show that

$$\iint_{\sigma} D_{\mathbf{n}}\phi\, dS = \iiint_{G} \left[\frac{\partial^2\phi}{\partial x^2} + \frac{\partial^2\phi}{\partial y^2} + \frac{\partial^2\phi}{\partial z^2}\right] dV$$

31. Let σ be the sphere $x^2 + y^2 + z^2 = 1$, let $\mathbf{n}$ be an inward unit normal, and let $D_{\mathbf{n}}f$ be the directional derivative of $f(x, y, z) = x^2 + y^2 + z^2$. Use the result in Exercise 30 to evaluate the surface integral

$$\iint_{\sigma} D_{\mathbf{n}}f\, dS$$

32. Use Stokes' Theorem to evaluate $\iint_{\sigma} \operatorname{curl} \mathbf{F} \cdot \mathbf{n}\, dS$ where $\mathbf{F}(x, y, z) = (z - y)\mathbf{i} + (x + z)\mathbf{j} - (x + y)\mathbf{k}$ and σ is the portion of the paraboloid $z = 2 - x^2 - y^2$ on or above the plane $z = 1$, with upward orientation.

33. Let $\mathbf{F}(x, y, z) = f(x, y, z)\mathbf{i} + g(x, y, z)\mathbf{j} + h(x, y, z)\mathbf{k}$ and suppose that f, g, and h are continuous and have continuous first partial derivatives in a region. It was shown in Exercise 27 of Section 16.3 that if $\mathbf{F}$ is conservative in the region, then

$$\frac{\partial f}{\partial y} = \frac{\partial g}{\partial x}, \quad \frac{\partial f}{\partial z} = \frac{\partial h}{\partial x}, \quad \frac{\partial g}{\partial z} = \frac{\partial h}{\partial y}$$

there. Use this result to show that if $\mathbf{F}$ is conservative in an open spherical region, then $\operatorname{curl} \mathbf{F} = \mathbf{0}$ in that region.

34–35 With the aid of Exercise 33, determine whether $\mathbf{F}$ is conservative.

34. (a) $\mathbf{F}(x, y, z) = z^2\mathbf{i} + e^{-y}\mathbf{j} + 2xz\mathbf{k}$
(b) $\mathbf{F}(x, y, z) = xy\mathbf{i} + x^2\mathbf{j} + \sin z\mathbf{k}$

35. (a) $\mathbf{F}(x, y, z) = \sin x\mathbf{i} + z\mathbf{j} + y\mathbf{k}$
(b) $\mathbf{F}(x, y, z) = z\mathbf{i} + 2yz\mathbf{j} + y^2\mathbf{k}$

36. As discussed in Example 1 of Section 16.1, *Coulomb's law* states that the electrostatic force $\mathbf{F}(\mathbf{r})$ that a particle of charge Q exerts on a particle of charge q is given by the formula

$$\mathbf{F}(\mathbf{r}) = \frac{qQ}{4\pi\epsilon_0\|\mathbf{r}\|^3}\mathbf{r}$$

where $\mathbf{r}$ is the radius vector from Q to q and ϵ_0 is the permittivity constant.
(a) Express the vector field $\mathbf{F}(\mathbf{r})$ in coordinate form $\mathbf{F}(x, y, z)$ with Q at the origin.
(b) Find the work performed by the force field $\mathbf{F}$ on a charge q that moves along a straight line from $(3, 0, 0)$ to $(3, 1, 5)$.

EXPANDING THE CALCULUS HORIZON

Hurricane Modeling

Each year population centers throughout the world are ravaged by hurricanes, and it is the mission of the National Hurricane Center to minimize the damage and loss of life by issuing warnings and forecasts of hurricanes developing in the Caribbean, Atlantic, Gulf of Mexico, and Eastern Pacific regions. Your assignment as a trainee at the Center is to construct a simple mathematical model of a hurricane using basic principles of fluid flow and properties of vector fields.

Modeling Assumptions

You have been notified of a developing hurricane in the Bahamas (designated hurricane *Isaac*) and have been asked to construct a model of its velocity field. Because hurricanes are complicated three-dimensional fluid flows, you will have to make many simplifying assumptions about the structure of a hurricane and the properties of the fluid flow. Accordingly, you decide to model the moisture in Isaac as an ***ideal fluid***, meaning that it is ***incompressible*** and its ***viscosity*** can be ignored. An incompressible fluid is one in which the density of the fluid is the same at all points and cannot be altered by compressive forces. Experience has shown that water can be regarded as incompressible but water vapor cannot. However, incompressibility is a reasonable assumption for a basic hurricane model because a hurricane is not restricted to a closed container that would produce compressive forces.

All fluids have a certain amount of viscosity, which is a resistance to flow—oil and molasses have a high viscosity, whereas water has almost none at subsonic speeds. Thus, it is reasonable to ignore viscosity in a basic model. Next, you decide to assume that the flow is in a ***steady state***, meaning that the velocity of the fluid at any point does not vary with time. This is reasonable over very short time periods for hurricanes that move and change slowly. Finally, although hurricanes are three-dimensional flows, you decide to model a two-dimensional horizontal cross section, so you make the simplifying assumption that the fluid in the cross section flows horizontally.

The photograph of Isaac shown at the beginning of this module reveals a typical pattern of a Caribbean hurricane—a counterclockwise swirl of fluid around the ***eye*** through which the fluid exits the flow in the form of rain. The lower pressure in the eye causes an inward-rushing air mass, and circular winds around the eye contribute to the swirling effect.

Your first objective is to find an explicit formula for Isaac's velocity field $\mathbf{F}(x, y)$, so you begin by introducing a rectangular coordinate system with its origin at the eye and its y-axis pointing north. Moreover, based on the hurricane picture and your knowledge of meteorological theory, you decide to build up the velocity field for Isaac from the velocity fields of simpler flows—a counterclockwise "vortex flow" $\mathbf{F}_1(x, y)$ in which fluid flows counterclockwise in concentric circles around the eye and a "sink flow" $\mathbf{F}_2(x, y)$ in which the fluid flows in straight lines toward a sink at the eye. Once you find explicit formulas for $\mathbf{F}_1(x, y)$ and $\mathbf{F}_2(x, y)$, your plan is to use the ***superposition principle*** from fluid dynamics to express the velocity field for Isaac as $\mathbf{F}(x, y) = \mathbf{F}_1(x, y) + \mathbf{F}_2(x, y)$.

Modeling a Vortex Flow

A ***counterclockwise vortex flow*** of an ideal fluid around the origin has four defining characteristics (Figure 1a on the following page):

- The velocity vector at a point (x, y) is tangent to the circle that is centered at the origin and passes through the point (x, y).

- The direction of the velocity vector at a point (x, y) indicates a counterclockwise motion.
- The speed of the fluid is constant on circles centered at the origin.
- The speed of the fluid along a circle is inversely proportional to the radius of the circle (and hence the speed approaches $+\infty$ as the radius of the circle approaches 0).

In fluid dynamics, the ***strength*** k of a vortex flow is defined to be 2π times the speed of the fluid along the unit circle. If the strength of a vortex flow is known, then the speed of the fluid along any other circle can be found from the fact that speed is inversely proportional to the radius of the circle. Thus, your first objective is to find a formula for a vortex flow $\mathbf{F}_1(x, y)$ with a specified strength k.

Exercise 1 Show that

$$\mathbf{F}_1(x, y) = -\frac{k}{2\pi(x^2 + y^2)}(y\mathbf{i} - x\mathbf{j})$$

is a model for the velocity field of a counterclockwise vortex flow around the origin of strength k by confirming that

(a) $\mathbf{F}_1(x, y)$ has the four properties required of a counterclockwise vortex flow around the origin;

(b) k is 2π times the speed of the fluid along the unit circle.

Exercise 2 Use a graphing utility that can generate vector fields to generate a vortex flow of strength 2π.

Modeling a Sink Flow

A ***uniform sink flow*** of an ideal fluid toward the origin has four defining characteristics (Figure 1*b*):

- The velocity vector at every point (x, y) is directed toward the origin.
- The speed of the fluid is the same at all points on a circle centered at the origin.
- The speed of the fluid at a point is inversely proportional to its distance from the origin (from which it follows that the speed approaches $+\infty$ as the distance from the origin approaches 0).
- There is a sink at the origin at which fluid leaves the flow.

As with a vortex flow, the ***strength*** q of a uniform sink flow is defined to be 2π times the speed of the fluid at points on the unit circle. If the strength of a sink flow is known, then the speed of the fluid at any point in the flow can be found using the fact that the speed is inversely proportional to the distance from the origin. Thus, your next objective is to find a formula for a uniform sink flow $\mathbf{F}_2(x, y)$ with a specified strength q.

Exercise 3 Show that

$$\mathbf{F}_2(x, y) = -\frac{q}{2\pi(x^2 + y^2)}(x\mathbf{i} + y\mathbf{j})$$

is a model for the velocity field of a uniform sink flow toward the origin of strength q by confirming the following facts:

(a) $\mathbf{F}_2(x, y)$ has the four properties required of a uniform sink flow toward the origin. [A reasonable physical argument to confirm the existence of the sink will suffice.]

(b) q is 2π times the speed of the fluid at points on the unit circle.

Exercise 4 Use a graphing utility that can generate vector fields to generate a uniform sink flow of strength 2π.

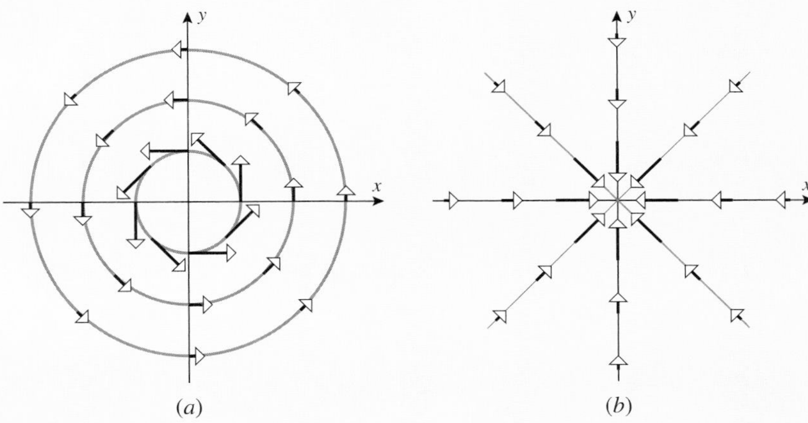

Figure 1

A Basic Hurricane Model

It now follows from Exercises 1 and 3 that the vector field $\mathbf{F}(x, y)$ for a hurricane model that combines a vortex flow around the origin of strength k and a uniform sink flow toward the origin of strength q is

$$\mathbf{F}(x, y) = -\frac{1}{2\pi(x^2 + y^2)}[(qx + ky)\mathbf{i} + (qy - kx)\mathbf{j}] \tag{1}$$

Exercise 5

(a) Figure 2 shows a vector field for a hurricane with vortex strength $k = 2\pi$ and sink strength $q = 2\pi$. Use a graphing utility that can generate vector fields to produce a reasonable facsimile of this figure.

(b) Make a conjecture about the effect of increasing k and keeping q fixed, and check your conjecture using a graphing utility.

(c) Make a conjecture about the effect of increasing q and keeping k fixed, and check your conjecture using a graphing utility.

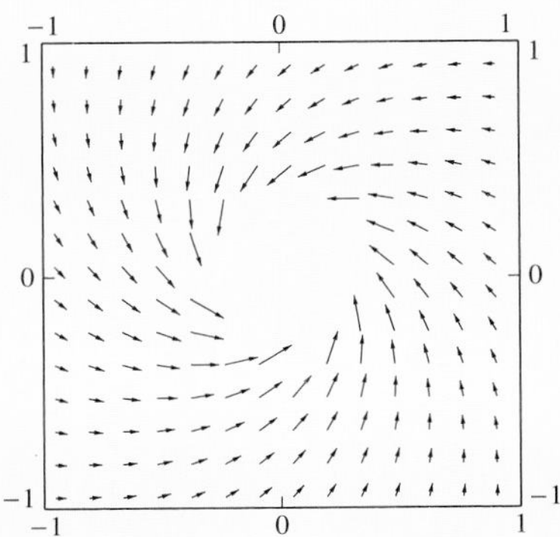

Figure 2

Modeling Hurricane Isaac

You are now ready to apply Formula (1) to obtain a model of the vector field $\mathbf{F}(x, y)$ of hurricane Isaac. You need some observational data to determine the constants k and q, so you call the Technical Support Branch of the Center for the latest information on hurricane Isaac. They report that 20 km from the eye the wind velocity has a component of 15 km/h toward the eye and a counterclockwise tangential component of 45 km/h.

Exercise 6

(a) Find the strengths k and q of the vortex and sink for hurricane Isaac.

(b) Find the vector field $\mathbf{F}(x, y)$ for hurricane Isaac.

(c) Estimate the size of hurricane Isaac by finding a radius beyond which the wind speed is less than 5 km/h.

Streamlines for the Basic Hurricane Model

The paths followed by the fluid particles in a fluid flow are called the ***streamlines*** of the flow. Thus, the vectors $\mathbf{F}(x, y)$ in the velocity field of a fluid flow are tangent to the streamlines. If the streamlines can be represented as the level curves of some function $\psi(x, y)$, then the function ψ is called a ***stream function*** for the flow. Since $\nabla\psi$ is normal to the level curves $\psi(x, y) = c$, it follows that $\nabla\psi$ is normal to the streamlines; and this in turn implies that

$$\nabla\psi \cdot \mathbf{F} = 0 \tag{2}$$

Your plan is to use this equation to find the stream function and then the streamlines of the basic hurricane model.

Since the vortex and sink flows that produce the basic hurricane model have a central symmetry, intuition suggests that polar coordinates may lead to simpler equations for the streamlines than rectangular coordinates. Thus, you decide to express the velocity vector $\mathbf{F}$ at a point (r, θ) in terms of the orthogonal unit vectors

$$\mathbf{u}_r = \cos\theta\,\mathbf{i} + \sin\theta\,\mathbf{j} \quad \text{and} \quad \mathbf{u}_\theta = -\sin\theta\,\mathbf{i} + \cos\theta\,\mathbf{j}$$

The vector $\mathbf{u}_r$, called the ***radial unit vector***, points away from the origin, and the vector $\mathbf{u}_\theta$, called the ***transverse unit vector***, is obtained by rotating $\mathbf{u}_r$ counterclockwise 90° (Figure 3).

Exercise 7 Show that the vector field for the basic hurricane model given in (1) can be expressed in terms of $\mathbf{u}_r$ and $\mathbf{u}_\theta$ as

$$\mathbf{F} = -\frac{1}{2\pi r}(q\mathbf{u}_r - k\mathbf{u}_\theta)$$

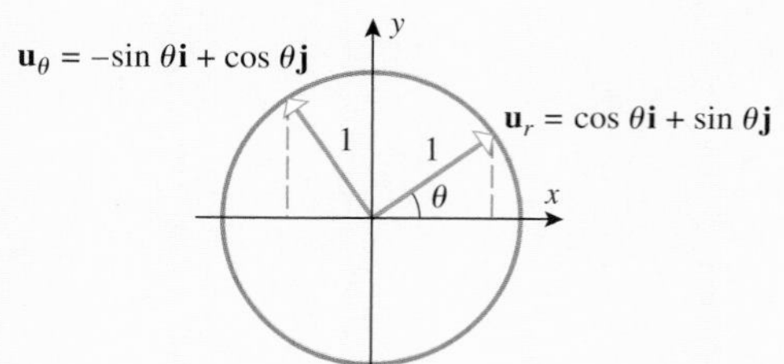

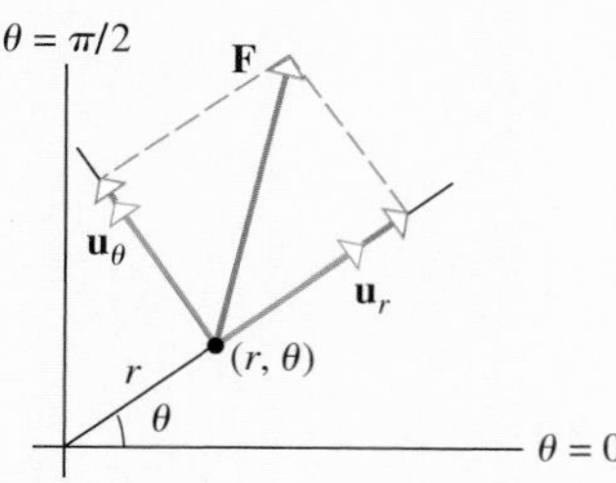

Figure 3

It follows from Exercise 75 of Section 14.6 that the gradient of the stream function can be expressed in terms of $\mathbf{u}_r$ and $\mathbf{u}_\theta$ as

$$\nabla\psi = \frac{\partial\psi}{\partial r}\mathbf{u}_r + \frac{1}{r}\frac{\partial\psi}{\partial\theta}\mathbf{u}_\theta$$

Exercise 8 Confirm that for the basic hurricane model the orthogonality condition in (2) is satisfied if

$$\frac{\partial \psi}{\partial r} = \frac{k}{r} \quad \text{and} \quad \frac{\partial \psi}{\partial \theta} = q$$

Exercise 9 By integrating the equations in Exercise 8, show that

$$\psi = k \ln r + q\theta$$

is a stream function for the basic hurricane model.

Exercise 10 Show that the streamlines for the basic hurricane model are logarithmic spirals of the form

$$r = Ke^{-q\theta/k} \quad (K > 0)$$

Exercise 11 Use a graphing utility to generate some typical streamlines for the basic hurricane model with vortex strength 2π and sink strength 2π.

Streamlines for Hurricane Isaac

Exercise 12 In Exercise 6 you found the strengths k and q of the vortex and sink for hurricane Isaac. Use that information to find a formula for the family of streamlines for Isaac; and then use a graphing utility to graph the streamline that passes through the point that is 20 km from the eye in the direction that is 45° NE from the eye.

Module by: *Josef S. Torok, Rochester Institute of Technology*
Howard Anton, Drexel University

a p p e n d i x a

TRIGONOMETRY REVIEW

TRIGONOMETRIC FUNCTIONS AND IDENTITIES

ANGLES

Angles in the plane can be generated by rotating a ray about its endpoint. The starting position of the ray is called the ***initial side*** of the angle, the final position is called the ***terminal side*** of the angle, and the point at which the initial and terminal sides meet is called the ***vertex*** of the angle. We allow for the possibility that the ray may make more than one complete revolution. Angles are considered to be ***positive*** if generated counterclockwise and ***negative*** if generated clockwise (Figure A.1).

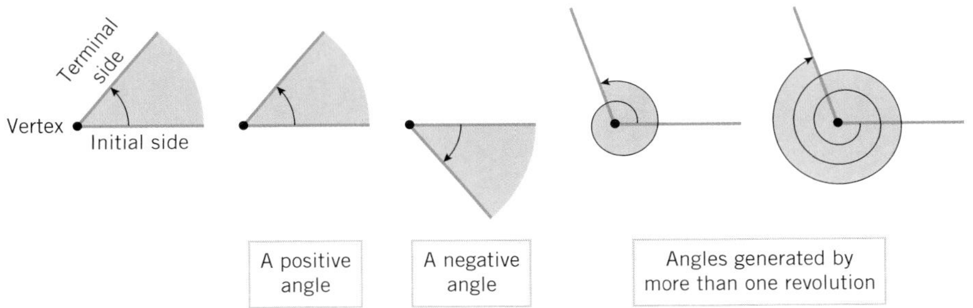

Figure A.1

There are two standard measurement systems for describing the size of an angle: ***degree measure*** and ***radian measure***. In degree measure, one degree (written $1°$) is the measure of an angle generated by $1/360$ of one revolution. Thus, there are $360°$ in an angle of one revolution, $180°$ in an angle of one-half revolution, $90°$ in an angle of one-quarter revolution (a *right angle*), and so forth. Degrees are divided into sixty equal parts, called ***minutes***, and minutes are divided into sixty equal parts, called ***seconds***. Thus, one minute (written $1'$) is $1/60$ of a degree, and one second (written $1''$) is $1/60$ of a minute. Smaller subdivisions of a degree are expressed as fractions of a second.

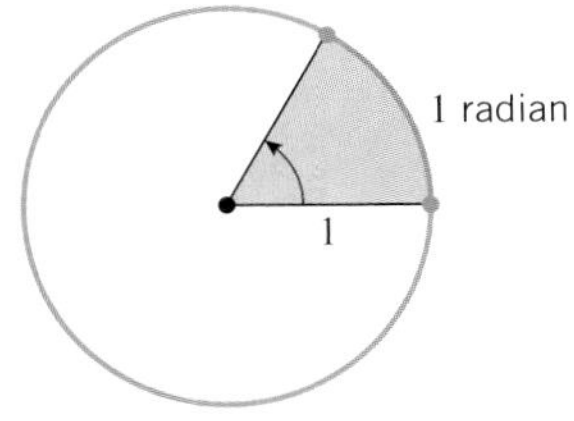

Figure A.2

In radian measure, angles are measured by the length of the arc that the angle subtends on a circle of radius 1 when the vertex is at the center. One unit of arc on a circle of radius 1 is called one ***radian*** (written 1 radian or 1 rad) (Figure A.2), and hence the entire circumference of a circle of radius 1 is 2π radians. It follows that an angle of $360°$ subtends an arc of 2π radians, an angle of $180°$ subtends an arc of π radians, an angle of $90°$ subtends an arc of $\pi/2$ radians, and so forth. Figure A.3 and Table 1 show the relationship between degree measure and radian measure for some important positive angles.

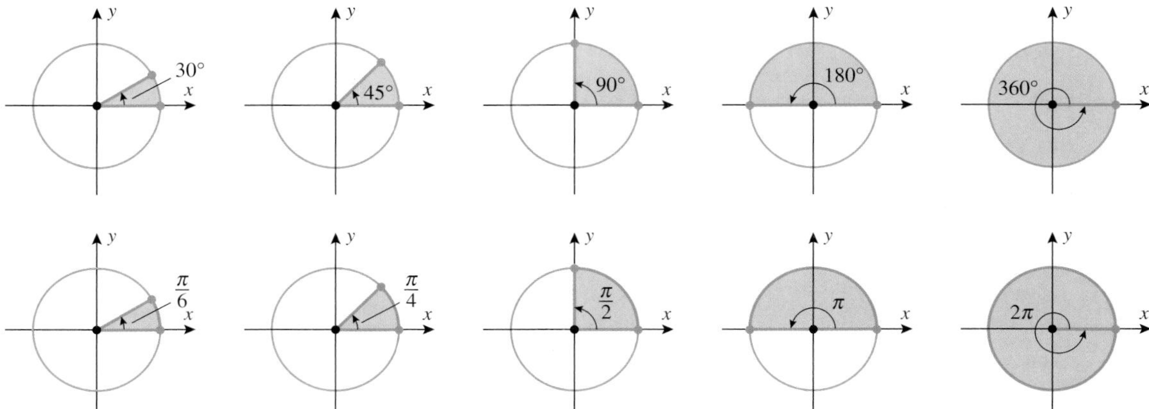

Figure A.3

Observe that in Table 1, angles in degrees are designated by the degree symbol, but angles in radians have no units specified. This is standard practice—when no units are specified for an angle, it is understood that the units are radians.

Table 1

DEGREES	30°	45°	60°	90°	120°	135°	150°	180°	270°	360°
RADIANS	$\frac{\pi}{6}$	$\frac{\pi}{4}$	$\frac{\pi}{3}$	$\frac{\pi}{2}$	$\frac{2\pi}{3}$	$\frac{3\pi}{4}$	$\frac{5\pi}{6}$	π	$\frac{3\pi}{2}$	2π

From the fact that π radians corresponds to $180°$, we obtain the following formulas, which are useful for converting from degrees to radians and conversely.

$$1° = \frac{\pi}{180} \text{rad} \approx 0.01745 \text{ rad} \tag{1}$$

$$1 \text{ rad} = \left(\frac{180}{\pi}\right)^{\circ} \approx 57° 17' 44.8'' \tag{2}$$

► **Example 1**

(a) Express 146° in radians. (b) Express 3 radians in degrees.

Solution (a). From (1), degrees can be converted to radians by multiplying by a conversion factor of $\pi/180$. Thus,

$$146° = \left(\frac{\pi}{180} \cdot 146\right) \text{rad} = \frac{73\pi}{90} \text{ rad} \approx 2.5482 \text{ rad}$$

Solution (b). From (2), radians can be converted to degrees by multiplying by a conversion factor of $180/\pi$. Thus,

$$3 \text{ rad} = \left(3 \cdot \frac{180}{\pi}\right)^{\circ} = \left(\frac{540}{\pi}\right)^{\circ} \approx 171.9° \blacktriangleleft$$

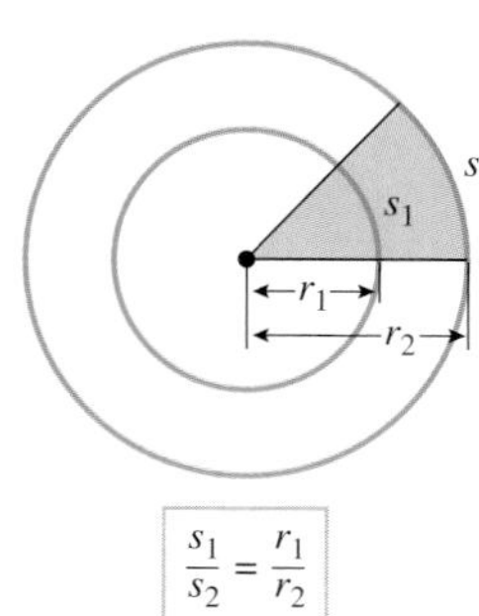

Figure A.4

RELATIONSHIPS BETWEEN ARC LENGTH, ANGLE, RADIUS, AND AREA

There is a theorem from plane geometry which states that for two concentric circles, the ratio of the arc lengths subtended by a central angle is equal to the ratio of the corresponding radii (Figure A.4). In particular, if s is the arc length subtended on a circle of radius r by a central angle of θ radians, then by comparison with the arc length subtended by that angle on a circle of radius 1 we obtain

$$\frac{s}{\theta} = \frac{r}{1}$$

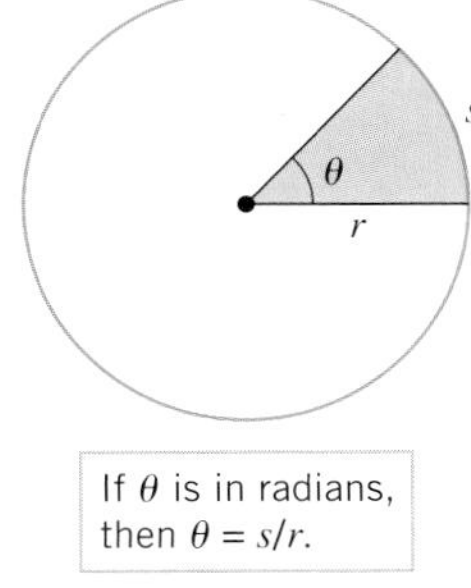

Figure A.5

from which we obtain the following relationships between the central angle θ, the radius r, and the subtended arc length s when θ is in radians (Figure A.5):

$$\theta = s/r \qquad \text{and} \qquad s = r\theta \tag{3–4}$$

The shaded region in Figure A.5 is called a ***sector***. It is a theorem from plane geometry that the ratio of the area A of this sector to the area of the entire circle is the same as the ratio of the central angle of the sector to the central angle of the entire circle; thus, if the angles are in radians, we have

$$\frac{A}{\pi r^2} = \frac{\theta}{2\pi}$$

Solving for A yields the following formula for the area of a sector in terms of the radius r and the angle θ in radians:

$$A = \tfrac{1}{2}r^2\theta \tag{5}$$

TRIGONOMETRIC FUNCTIONS FOR RIGHT TRIANGLES

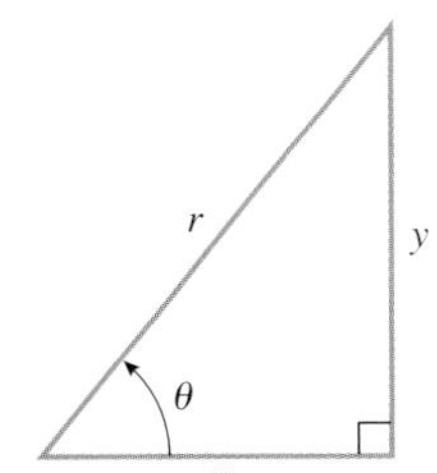

Figure A.6

The ***sine***, ***cosine***, ***tangent***, ***cosecant***, ***secant***, and ***cotangent*** of a positive acute angle θ can be defined as ratios of the sides of a right triangle. Using the notation from Figure A.6, these definitions take the following form:

$$\begin{aligned}
\sin\theta &= \frac{\text{side opposite } \theta}{\text{hypotenuse}} = \frac{y}{r}, & \csc\theta &= \frac{\text{hypotenuse}}{\text{side opposite } \theta} = \frac{r}{y} \\
\cos\theta &= \frac{\text{side adjacent to } \theta}{\text{hypotenuse}} = \frac{x}{r}, & \sec\theta &= \frac{\text{hypotenuse}}{\text{side adjacent to } \theta} = \frac{r}{x} \\
\tan\theta &= \frac{\text{side opposite } \theta}{\text{side adjacent to } \theta} = \frac{y}{x}, & \cot\theta &= \frac{\text{side adjacent to } \theta}{\text{side opposite } \theta} = \frac{x}{y}
\end{aligned} \tag{6}$$

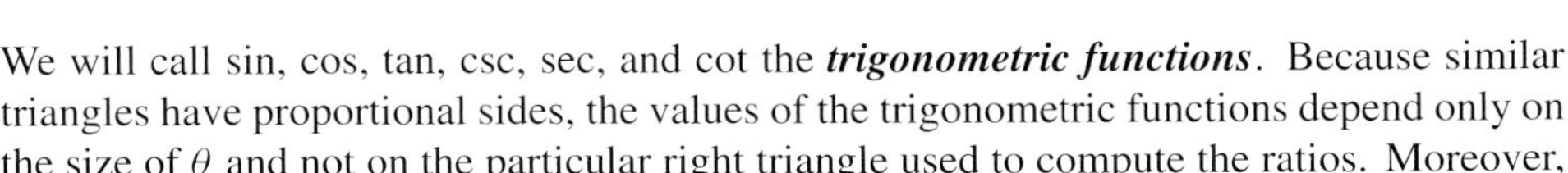
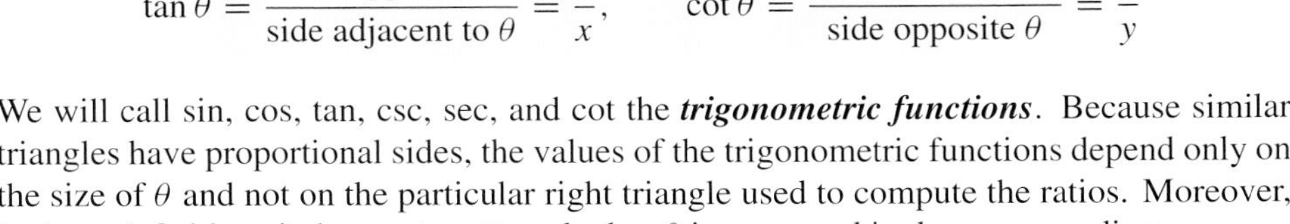

We will call sin, cos, tan, csc, sec, and cot the ***trigonometric functions***. Because similar triangles have proportional sides, the values of the trigonometric functions depend only on the size of θ and not on the particular right triangle used to compute the ratios. Moreover, in these definitions it does not matter whether θ is measured in degrees or radians.

► **Example 2** Recall from geometry that the two legs of a $45°–45°–90°$ triangle are of equal size and that the hypotenuse of a $30°–60°–90°$ triangle is twice the shorter leg, where the shorter leg is opposite the $30°$ angle. These facts and the Theorem of Pythagoras yield Figure A.7. From that figure we obtain the results in Table 2. ◄

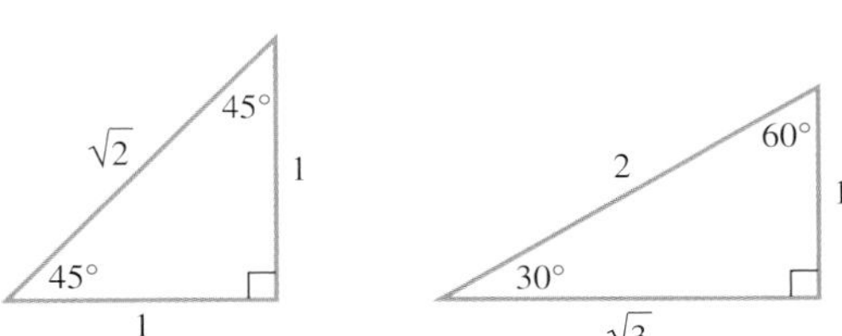

Figure A.7

Table 2

$\sin 45° = 1/\sqrt{2}$,	$\cos 45° = 1/\sqrt{2}$,	$\tan 45° = 1$
$\csc 45° = \sqrt{2}$,	$\sec 45° = \sqrt{2}$,	$\cot 45° = 1$
$\sin 30° = 1/2$,	$\cos 30° = \sqrt{3}/2$,	$\tan 30° = 1/\sqrt{3}$
$\csc 30° = 2$,	$\sec 30° = 2/\sqrt{3}$,	$\cot 30° = \sqrt{3}$
$\sin 60° = \sqrt{3}/2$,	$\cos 60° = 1/2$,	$\tan 60° = \sqrt{3}$
$\csc 60° = 2/\sqrt{3}$,	$\sec 60° = 2$,	$\cot 60° = 1/\sqrt{3}$

ANGLES IN RECTANGULAR COORDINATE SYSTEMS

Because the angles of a right triangle are between 0° and 90°, the formulas in (6) are not directly applicable to negative angles or to angles greater than 90°. To extend the trigonometric functions to include these cases, it will be convenient to consider angles in rectangular coordinate systems. An angle is said to be in ***standard position*** in an xy-coordinate system if its vertex is at the origin and its initial side is on the positive x-axis (Figure A.8).

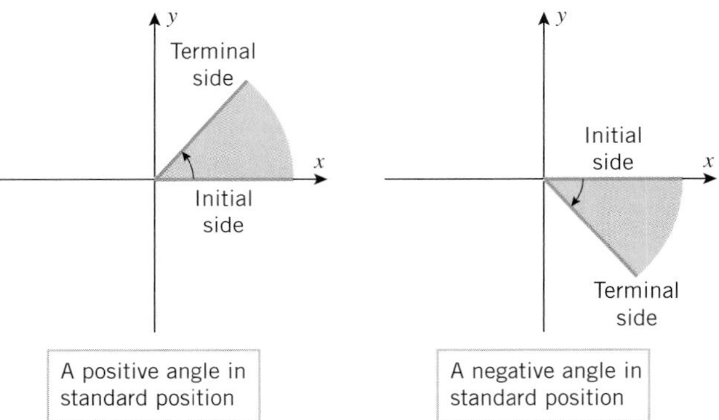

Figure A.8

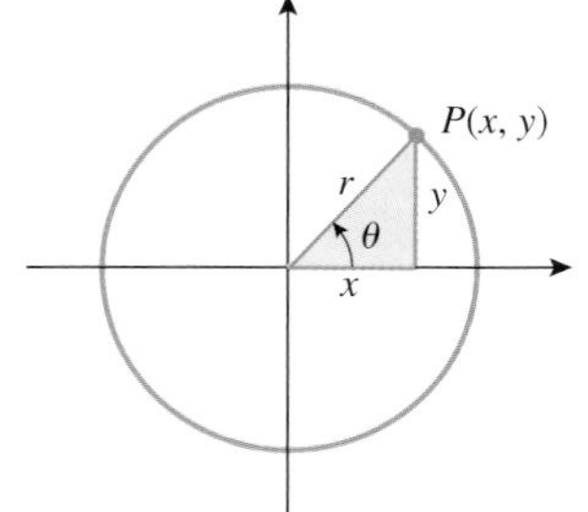

Figure A.9

To define the trigonometric functions of an angle θ in standard position, construct a circle of radius r, centered at the origin, and let $P(x, y)$ be the intersection of the terminal side of θ with this circle (Figure A.9). We make the following definition.

A.1 DEFINITION.

$$\sin\theta = \frac{y}{r}, \quad \cos\theta = \frac{x}{r}, \quad \tan\theta = \frac{y}{x}$$

$$\csc\theta = \frac{r}{y}, \quad \sec\theta = \frac{r}{x}, \quad \cot\theta = \frac{x}{y}$$

Figure A.10

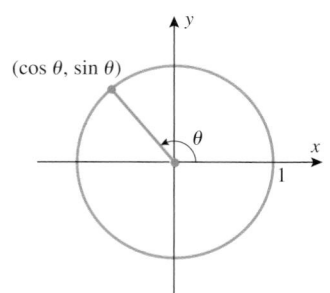

Note that the formulas in this definition agree with those in (6), so there is no conflict with the earlier definition of the trigonometric functions for triangles. However, this definition applies to all angles (except for cases where a zero denominator occurs).

In the special case where $r = 1$, we have $\sin\theta = y$ and $\cos\theta = x$, so the terminal side of the angle θ intersects the unit circle at the point $(\cos\theta, \sin\theta)$ (Figure A.10). It follows from

Definition A.1 that the remaining trigonometric functions of θ are expressible as (verify)

$$\tan\theta = \frac{\sin\theta}{\cos\theta}, \quad \cot\theta = \frac{\cos\theta}{\sin\theta} = \frac{1}{\tan\theta}, \quad \sec\theta = \frac{1}{\cos\theta}, \quad \csc\theta = \frac{1}{\sin\theta} \tag{7–10}$$

These observations suggest the following procedure for evaluating the trigonometric functions of common angles:

- Construct the angle θ in standard position in an xy-coordinate system.
- Find the coordinates of the intersection of the terminal side of the angle and the unit circle; the x- and y-coordinates of this intersection are the values of $\cos\theta$ and $\sin\theta$, respectively.
- Use Formulas (7) through (10) to find the values of the remaining trigonometric functions from the values of $\cos\theta$ and $\sin\theta$.

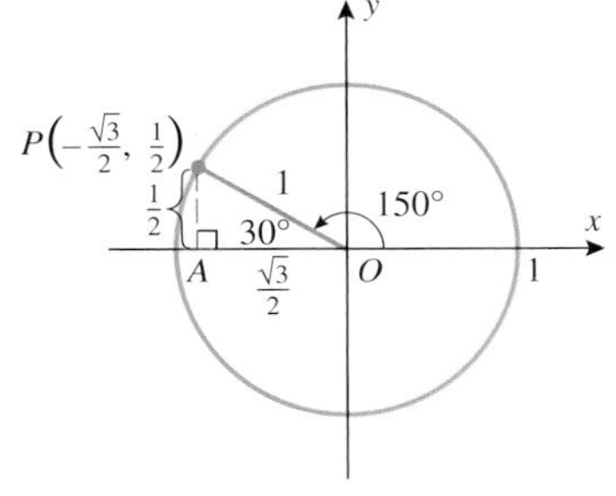

Figure A.11

▶ **Example 3** Evaluate the trigonometric functions of $\theta = 150°$.

Solution. Construct a unit circle and place the angle $\theta = 150°$ in standard position (Figure A.11). Since $\angle AOP$ is 30° and $\triangle OAP$ is a 30°–60°–90° triangle, the leg AP has length $\frac{1}{2}$ (half the hypotenuse) and the leg OA has length $\sqrt{3}/2$ by the Theorem of Pythagoras. Thus, the coordinates of P are $(-\sqrt{3}/2, 1/2)$, from which we obtain

$$\sin 150° = \frac{1}{2}, \quad \cos 150° = -\frac{\sqrt{3}}{2}, \quad \tan 150° = \frac{\sin 150°}{\cos 150°} = \frac{1/2}{-\sqrt{3}/2} = -\frac{1}{\sqrt{3}}$$

$$\csc 150° = \frac{1}{\sin 150°} = 2, \quad \sec 150° = \frac{1}{\cos 150°} = -\frac{2}{\sqrt{3}}$$

$$\cot 150° = \frac{1}{\tan 150°} = -\sqrt{3} \blacktriangleleft$$

▶ **Example 4** Evaluate the trigonometric functions of $\theta = 5\pi/6$.

Solution. Since $5\pi/6 = 150°$, this problem is equivalent to that of Example 3. From that example we obtain

$$\sin\frac{5\pi}{6} = \frac{1}{2}, \quad \cos\frac{5\pi}{6} = -\frac{\sqrt{3}}{2}, \quad \tan\frac{5\pi}{6} = -\frac{1}{\sqrt{3}}$$

$$\csc\frac{5\pi}{6} = 2, \quad \sec\frac{5\pi}{6} = -\frac{2}{\sqrt{3}}, \quad \cot\frac{5\pi}{6} = -\sqrt{3} \blacktriangleleft$$

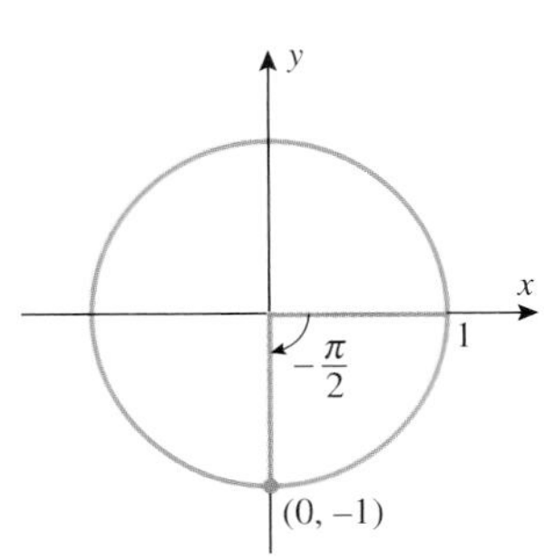

Figure A.12

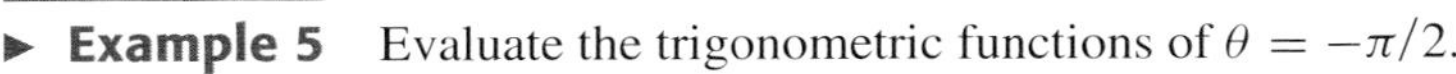

▶ **Example 5** Evaluate the trigonometric functions of $\theta = -\pi/2$.

Solution. As shown in Figure A.12, the terminal side of $\theta = -\pi/2$ intersects the unit circle at the point $(0, -1)$, so

$$\sin(-\pi/2) = -1, \quad \cos(-\pi/2) = 0$$

and from Formulas (7) through (10),

$$\tan(-\pi/2) = \frac{\sin(-\pi/2)}{\cos(-\pi/2)} = \frac{-1}{0} \quad \text{(undefined)}$$

$$\cot(-\pi/2) = \frac{\cos(-\pi/2)}{\sin(-\pi/2)} = \frac{0}{-1} = 0$$

$$\sec(-\pi/2) = \frac{1}{\cos(-\pi/2)} = \frac{1}{0} \quad \text{(undefined)}$$

$$\csc(-\pi/2) = \frac{1}{\sin(-\pi/2)} = \frac{1}{-1} = -1 \blacktriangleleft$$

The reader should be able to obtain all of the results in Table 3 by the methods illustrated in the last three examples. The dashes indicate quantities that are undefined.

Table 3

	$\theta = 0$ (0°)	$\pi/6$ (30°)	$\pi/4$ (45°)	$\pi/3$ (60°)	$\pi/2$ (90°)	$2\pi/3$ (120°)	$3\pi/4$ (135°)	$5\pi/6$ (150°)	π (180°)	$3\pi/2$ (270°)	2π (360°)
$\sin\theta$	0	1/2	$1/\sqrt{2}$	$\sqrt{3}/2$	1	$\sqrt{3}/2$	$1/\sqrt{2}$	1/2	0	−1	0
$\cos\theta$	1	$\sqrt{3}/2$	$1/\sqrt{2}$	1/2	0	−1/2	$-1/\sqrt{2}$	$-\sqrt{3}/2$	−1	0	1
$\tan\theta$	0	$1/\sqrt{3}$	1	$\sqrt{3}$	—	$-\sqrt{3}$	−1	$-1/\sqrt{3}$	0	—	0
$\csc\theta$	—	2	$\sqrt{2}$	$2/\sqrt{3}$	1	$2/\sqrt{3}$	$\sqrt{2}$	2	—	−1	—
$\sec\theta$	1	$2/\sqrt{3}$	$\sqrt{2}$	2	—	−2	$-\sqrt{2}$	$-2/\sqrt{3}$	−1	—	1
$\cot\theta$	—	$\sqrt{3}$	1	$1/\sqrt{3}$	0	$-1/\sqrt{3}$	−1	$-\sqrt{3}$	—	0	—

> It is only in special cases that exact values for trigonometric functions can be obtained; usually, a calculating utility or a computer program will be required.

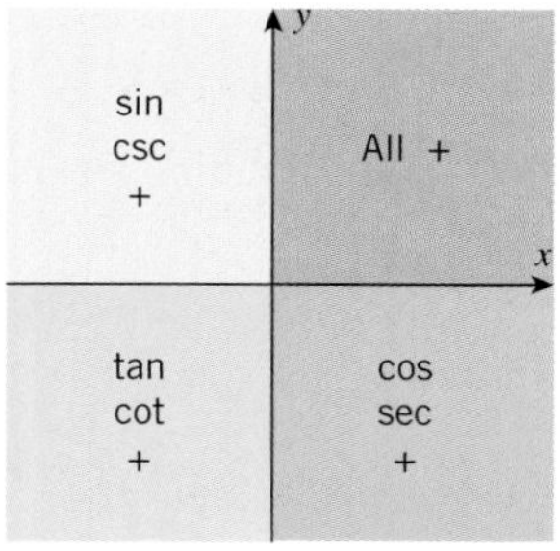

The signs of the trigonometric functions of an angle are determined by the quadrant in which the terminal side of the angle falls. For example, if the terminal side falls in the first quadrant, then x and y are positive in Definition A.1, so all of the trigonometric functions have positive values. If the terminal side falls in the second quadrant, then x is negative and y is positive, so sin and csc are positive, but all other trigonometric functions are negative. The diagram in Figure A.13 shows which trigonometric functions are positive in the various quadrants. The reader will find it instructive to check that the results in Table 3 are consistent with Figure A.13.

Figure A.13

TRIGONOMETRIC IDENTITIES

A ***trigonometric identity*** is an equation involving trigonometric functions that is true for all angles for which both sides of the equation are defined. One of the most important identities in trigonometry can be derived by applying the Theorem of Pythagoras to the triangle in Figure A.9 to obtain

$$x^2 + y^2 = r^2$$

Dividing both sides by r^2 and using the definitions of $\sin\theta$ and $\cos\theta$ (Definition A.1), we obtain the following fundamental result:

$$\sin^2\theta + \cos^2\theta = 1 \qquad (11)$$

The following identities can be obtained from (11) by dividing through by $\cos^2\theta$ and $\sin^2\theta$, respectively, then applying Formulas (7) through (10):

$$\tan^2\theta + 1 = \sec^2\theta \tag{12}$$

$$1 + \cot^2\theta = \csc^2\theta \tag{13}$$

If (x, y) is a point on the unit circle, then the points $(-x, y)$, $(-x, -y)$, and $(x, -y)$ also lie on the unit circle (why?), and the four points form corners of a rectangle with sides parallel to the coordinate axes (Figure A.14*a*). The x- and y-coordinates of each corner represent the cosine and sine of an angle in standard position whose terminal side passes through the corner; hence we obtain the identities in parts (*b*), (*c*), and (*d*) of Figure A.14 for sine and cosine. Dividing those identities leads to identities for the tangent. In summary:

$$\sin(\pi - \theta) = \sin\theta, \quad \sin(\pi + \theta) = -\sin\theta, \quad \sin(-\theta) = -\sin\theta \qquad (14\text{–}16)$$
$$\cos(\pi - \theta) = -\cos\theta, \quad \cos(\pi + \theta) = -\cos\theta, \quad \cos(-\theta) = \cos\theta \qquad (17\text{–}19)$$
$$\tan(\pi - \theta) = -\tan\theta, \quad \tan(\pi + \theta) = \tan\theta, \quad \tan(-\theta) = -\tan\theta \qquad (20\text{–}22)$$

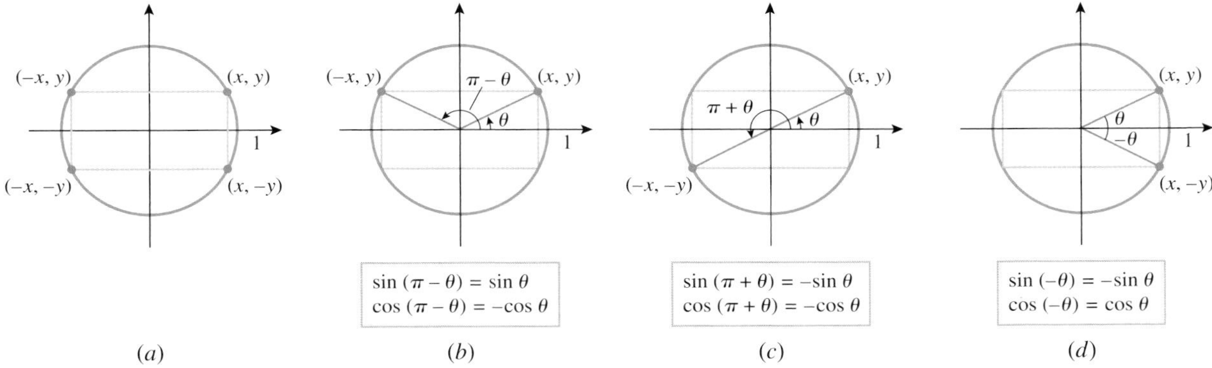

Figure A.14

Two angles in standard position that have the same terminal side must have the same values for their trigonometric functions since their terminal sides intersect the unit circle at the same point. In particular, two angles whose radian measures differ by a multiple of 2π have the same terminal side and hence have the same values for their trigonometric functions. This yields the identities

$$\sin\theta = \sin(\theta + 2\pi) = \sin(\theta - 2\pi) \tag{23}$$

$$\cos\theta = \cos(\theta + 2\pi) = \cos(\theta - 2\pi) \tag{24}$$

and more generally,

$$\sin\theta = \sin(\theta \pm 2n\pi), \quad n = 0, 1, 2, \ldots \tag{25}$$

$$\cos\theta = \cos(\theta \pm 2n\pi), \quad n = 0, 1, 2, \ldots \tag{26}$$

Identity (21) implies that

$$\tan\theta = \tan(\theta + \pi) \qquad \text{and} \qquad \tan\theta = \tan(\theta - \pi) \qquad (27\text{–}28)$$

Identity (27) is just (21) with the terms in the sum reversed, and identity (28) follows from (21) by substituting $\theta - \pi$ for θ. These two identities state that adding or subtracting π

from an angle does not affect the value of the tangent of the angle. It follows that the same is true for any multiple of π; thus,

$$\tan\theta = \tan(\theta \pm n\pi), \quad n = 0, 1, 2, \ldots \tag{29}$$

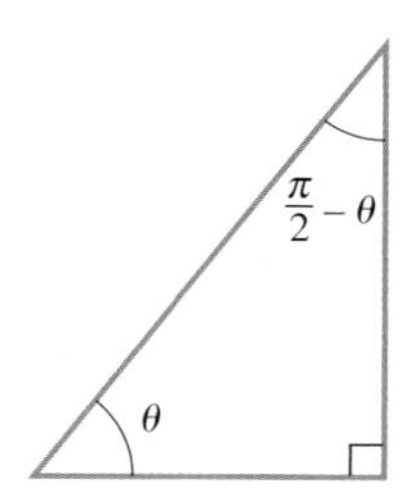

Figure A.15

Figure A.15 shows complementary angles θ and $(\pi/2) - \theta$ of a right triangle. It follows from (6) that

$$\sin\theta = \frac{\text{side opposite }\theta}{\text{hypotenuse}} = \frac{\text{side adjacent to }(\pi/2)-\theta}{\text{hypotenuse}} = \cos\left(\frac{\pi}{2} - \theta\right)$$

$$\cos\theta = \frac{\text{side adjacent to }\theta}{\text{hypotenuse}} = \frac{\text{side opposite }(\pi/2)-\theta}{\text{hypotenuse}} = \sin\left(\frac{\pi}{2} - \theta\right)$$

which yields the identities

$$\sin\left(\frac{\pi}{2} - \theta\right) = \cos\theta, \quad \cos\left(\frac{\pi}{2} - \theta\right) = \sin\theta, \quad \tan\left(\frac{\pi}{2} - \theta\right) = \cot\theta \tag{30–32}$$

where the third identity results from dividing the first two. These identities are also valid for angles that are not acute and for negative angles as well.

THE LAW OF COSINES

The next theorem, called the ***law of cosines***, generalizes the Theorem of Pythagoras. This result is important in its own right and is also the starting point for some important trigonometric identities.

> **A.2 THEOREM (*Law of Cosines*).** *If the sides of a triangle have lengths a, b, and c, and if θ is the angle between the sides with lengths a and b, then*
>
> $$c^2 = a^2 + b^2 - 2ab\cos\theta$$

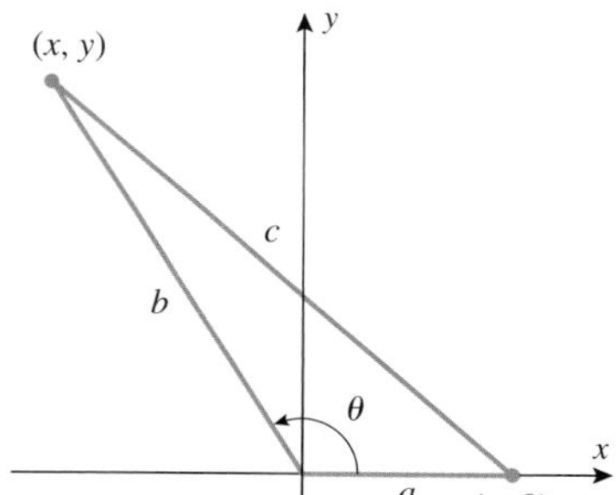

Figure A.16

PROOF. Introduce a coordinate system so that θ is in standard position and the side of length a falls along the positive x-axis. As shown in Figure A.16, the side of length a extends from the origin to $(a, 0)$ and the side of length b extends from the origin to some point (x, y). From the definition of $\sin\theta$ and $\cos\theta$ we have $\sin\theta = y/b$ and $\cos\theta = x/b$, so

$$y = b\sin\theta, \quad x = b\cos\theta \tag{33}$$

From the distance formula in Theorem G.1 of Appendix G, we obtain

$$c^2 = (x - a)^2 + (y - 0)^2$$

so that, from (33),

$$\begin{aligned} c^2 &= (b\cos\theta - a)^2 + b^2\sin^2\theta \\ &= a^2 + b^2(\cos^2\theta + \sin^2\theta) - 2ab\cos\theta \\ &= a^2 + b^2 - 2ab\cos\theta \end{aligned}$$

which completes the proof. ■

We will now show how the law of cosines can be used to obtain the following identities, called the ***addition formulas*** for sine and cosine:

$$\sin(\alpha + \beta) = \sin\alpha\cos\beta + \cos\alpha\sin\beta \tag{34}$$

$$\cos(\alpha + \beta) = \cos\alpha\cos\beta - \sin\alpha\sin\beta \tag{35}$$

$$\sin(\alpha - \beta) = \sin\alpha\cos\beta - \cos\alpha\sin\beta \tag{36}$$

$$\cos(\alpha - \beta) = \cos\alpha\cos\beta + \sin\alpha\sin\beta \tag{37}$$

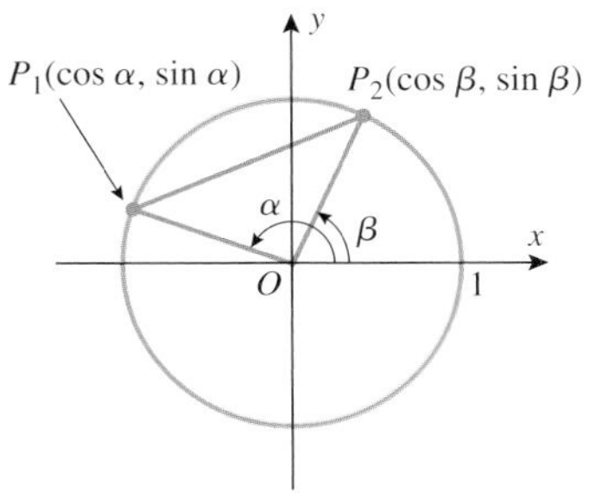

Figure A.17

We will derive (37) first. In our derivation we will assume that $0 \le \beta < \alpha < 2\pi$ (Figure A.17). As shown in the figure, the terminal sides of α and β intersect the unit circle at the points $P_1(\cos\alpha, \sin\alpha)$ and $P_2(\cos\beta, \sin\beta)$. If we denote the lengths of the sides of triangle OP_1P_2 by OP_1, P_1P_2, and OP_2, then $OP_1 = OP_2 = 1$ and, from the distance formula in Theorem G.1 of Appendix G,

$$\begin{aligned}(P_1P_2)^2 &= (\cos\beta - \cos\alpha)^2 + (\sin\beta - \sin\alpha)^2\\ &= (\sin^2\alpha + \cos^2\alpha) + (\sin^2\beta + \cos^2\beta) - 2(\cos\alpha\cos\beta + \sin\alpha\sin\beta)\\ &= 2 - 2(\cos\alpha\cos\beta + \sin\alpha\sin\beta)\end{aligned}$$

But angle $P_2OP_1 = \alpha - \beta$, so that the law of cosines yields

$$\begin{aligned}(P_1P_2)^2 &= (OP_1)^2 + (OP_2)^2 - 2(OP_1)(OP_2)\cos(\alpha - \beta)\\ &= 2 - 2\cos(\alpha - \beta)\end{aligned}$$

Equating the two expressions for $(P_1P_2)^2$ and simplifying, we obtain

$$\cos(\alpha - \beta) = \cos\alpha\cos\beta + \sin\alpha\sin\beta$$

which completes the derivation of (37).

We can use (31) and (37) to derive (36) as follows:

$$\begin{aligned}\sin(\alpha - \beta) &= \cos\left[\frac{\pi}{2} - (\alpha - \beta)\right] = \cos\left[\left(\frac{\pi}{2} - \alpha\right) - (-\beta)\right]\\ &= \cos\left(\frac{\pi}{2} - \alpha\right)\cos(-\beta) + \sin\left(\frac{\pi}{2} - \alpha\right)\sin(-\beta)\\ &= \cos\left(\frac{\pi}{2} - \alpha\right)\cos\beta - \sin\left(\frac{\pi}{2} - \alpha\right)\sin\beta\\ &= \sin\alpha\cos\beta - \cos\alpha\sin\beta\end{aligned}$$

Identities (34) and (35) can be obtained from (36) and (37) by substituting $-\beta$ for β and using the identities

$$\sin(-\beta) = -\sin\beta, \quad \cos(-\beta) = \cos\beta$$

We leave it for the reader to derive the identities

$$\tan(\alpha + \beta) = \frac{\tan\alpha + \tan\beta}{1 - \tan\alpha\tan\beta} \qquad \tan(\alpha - \beta) = \frac{\tan\alpha - \tan\beta}{1 + \tan\alpha\tan\beta} \tag{38–39}$$

Identity (38) can be obtained by dividing (34) by (35) and then simplifying. Identity (39) can be obtained from (38) by substituting $-\beta$ for β and simplifying.

In the special case where $\alpha = \beta$, identities (34), (35), and (38) yield the ***double-angle formulas***

$$\sin 2\alpha = 2\sin\alpha\cos\alpha \tag{40}$$

$$\cos 2\alpha = \cos^2\alpha - \sin^2\alpha \tag{41}$$

$$\tan 2\alpha = \frac{2\tan\alpha}{1 - \tan^2\alpha} \tag{42}$$

By using the identity $\sin^2\alpha + \cos^2\alpha = 1$, (41) can be rewritten in the alternative forms

$$\cos 2\alpha = 2\cos^2\alpha - 1 \qquad \text{and} \qquad \cos 2\alpha = 1 - 2\sin^2\alpha \tag{43–44}$$

If we replace α by $\alpha/2$ in (43) and (44) and use some algebra, we obtain the ***half-angle formulas***

$$\cos^2\frac{\alpha}{2} = \frac{1+\cos\alpha}{2} \quad \text{and} \quad \sin^2\frac{\alpha}{2} = \frac{1-\cos\alpha}{2} \tag{45–46}$$

We leave it for the exercises to derive the following ***product-to-sum formulas*** from (34) through (37):

$$\sin\alpha\cos\beta = \frac{1}{2}[\sin(\alpha-\beta)+\sin(\alpha+\beta)] \tag{47}$$

$$\sin\alpha\sin\beta = \frac{1}{2}[\cos(\alpha-\beta)-\cos(\alpha+\beta)] \tag{48}$$

$$\cos\alpha\cos\beta = \frac{1}{2}[\cos(\alpha-\beta)+\cos(\alpha+\beta)] \tag{49}$$

We also leave it for the exercises to derive the following ***sum-to-product formulas***:

$$\sin\alpha+\sin\beta = 2\sin\frac{\alpha+\beta}{2}\cos\frac{\alpha-\beta}{2} \tag{50}$$

$$\sin\alpha-\sin\beta = 2\cos\frac{\alpha+\beta}{2}\sin\frac{\alpha-\beta}{2} \tag{51}$$

$$\cos\alpha+\cos\beta = 2\cos\frac{\alpha+\beta}{2}\cos\frac{\alpha-\beta}{2} \tag{52}$$

$$\cos\alpha-\cos\beta = -2\sin\frac{\alpha+\beta}{2}\sin\frac{\alpha-\beta}{2} \tag{53}$$

FINDING AN ANGLE FROM THE VALUE OF ITS TRIGONOMETRIC FUNCTIONS

There are numerous situations in which it is necessary to find an unknown angle from a known value of one of its trigonometric functions. The following example illustrates a method for doing this.

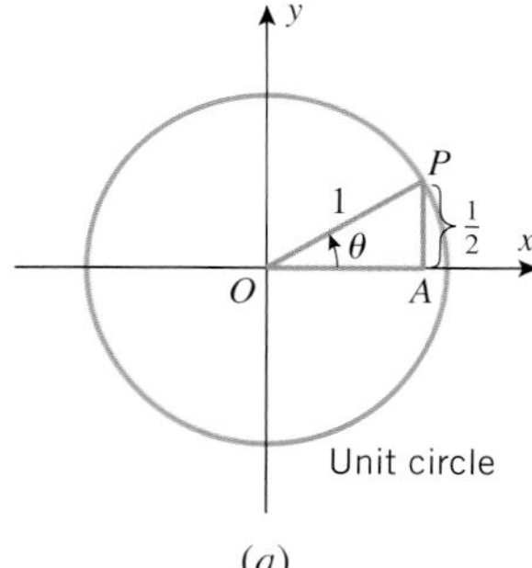

(a)

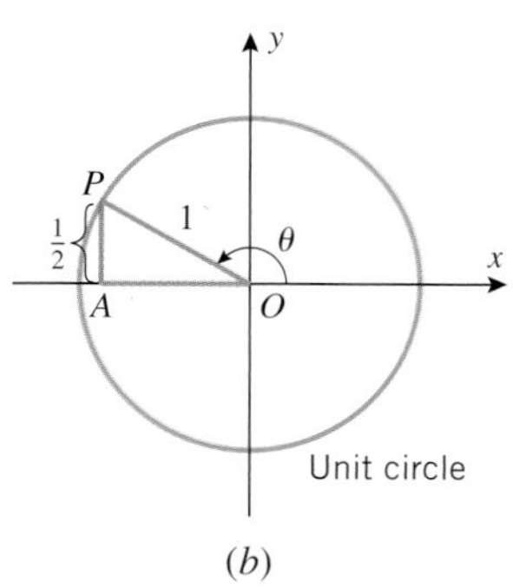

(b)

Figure A.18

► **Example 6** Find θ if $\sin\theta = \frac{1}{2}$.

Solution. We begin by looking for positive angles that satisfy the equation. Because $\sin\theta$ is positive, the angle θ must terminate in the first or second quadrant. If it terminates in the first quadrant, then the hypotenuse of $\triangle OAP$ in Figure A.18a is double the leg AP, so

$$\theta = 30^\circ = \frac{\pi}{6} \text{ radians}$$

If θ terminates in the second quadrant (Figure A.18b), then the hypotenuse of $\triangle OAP$ is double the leg AP, so $\angle AOP = 30^\circ$, which implies that

$$\theta = 180^\circ - 30^\circ = 150^\circ = \frac{5\pi}{6} \text{ radians}$$

Now that we have found these two solutions, all other solutions are obtained by adding or subtracting multiples of 360° (2π radians) to or from them. Thus, the entire set of solutions is given by the formulas

$$\theta = 30^\circ \pm n\cdot 360^\circ, \quad n = 0, 1, 2, \ldots$$

and

$$\theta = 150^\circ \pm n \cdot 360^\circ, \quad n = 0, 1, 2, \ldots$$

or in radian measure,

$$\theta = \frac{\pi}{6} \pm n \cdot 2\pi, \quad n = 0, 1, 2, \ldots$$

and

$$\theta = \frac{5\pi}{6} \pm n \cdot 2\pi, \quad n = 0, 1, 2, \ldots \blacktriangleleft$$

ANGLE OF INCLINATION

The slope of a nonvertical line L is related to the angle that L makes with the positive x-axis. If ϕ is the smallest positive angle measured counterclockwise from the x-axis to L, then the slope of the line can be expressed as

$$m = \tan\phi \tag{54}$$

(Figure A.19a). The angle ϕ, which is called the ***angle of inclination*** of the line, satisfies $0^\circ \le \phi < 180^\circ$ in degree measure (or, equivalently, $0 \le \phi < \pi$ in radian measure). If ϕ is an acute angle, then $m = \tan\phi$ is positive and the line slopes up to the right, and if ϕ is an obtuse angle, then $m = \tan\phi$ is negative and the line slopes down to the right. For example, a line whose angle of inclination is 45° has slope $m = \tan 45^\circ = 1$, and a line whose angle of inclination is 135° has a slope of $m = \tan 135^\circ = -1$ (Figure A.19b). Figure A.20 shows a convenient way of using the line $x = 1$ as a "ruler" for visualizing the relationship between lines of various slopes.

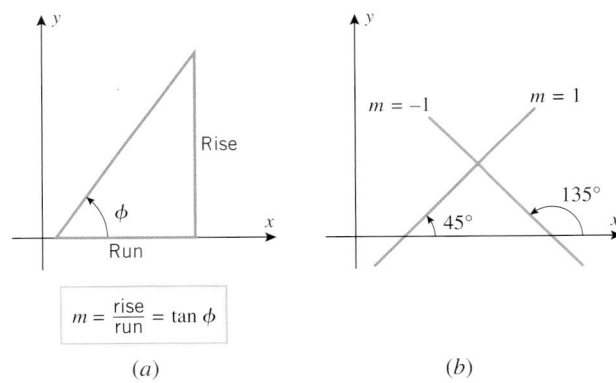

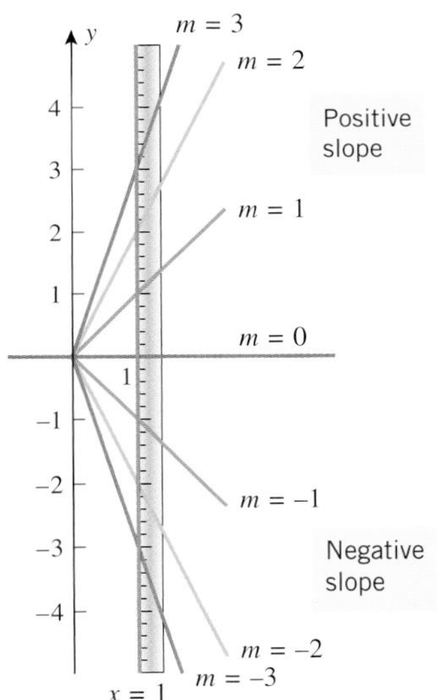

Figure A.20

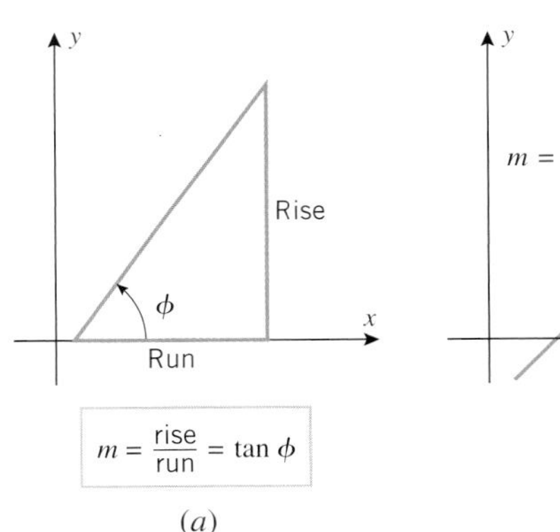

Figure A.19

EXERCISE SET

1–2 Express the angles in radians.

1. (a) 75° (b) 390° (c) 20° (d) 138°

2. (a) 420° (b) 15° (c) 225° (d) 165°

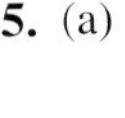

3–4 Express the angles in degrees.

3. (a) $\pi/15$ (b) 1.5 (c) $8\pi/5$ (d) 3π

4. (a) $\pi/10$ (b) 2 (c) $2\pi/5$ (d) $7\pi/6$

5–6 Find the exact values of all six trigonometric functions of θ.

5. (a)

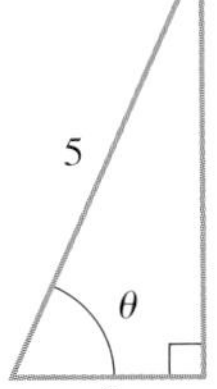

(b)

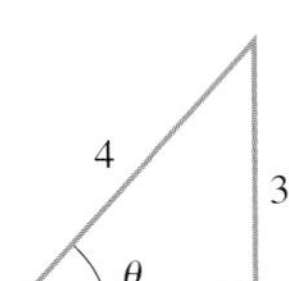

(c)

6. (a)

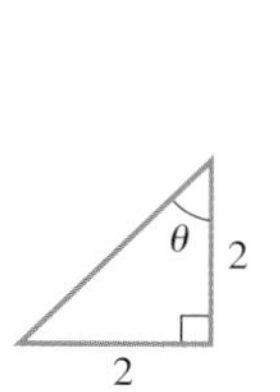

(b)

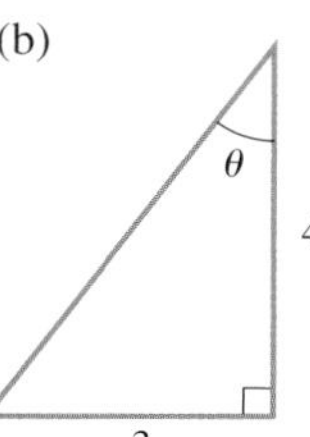

(c)

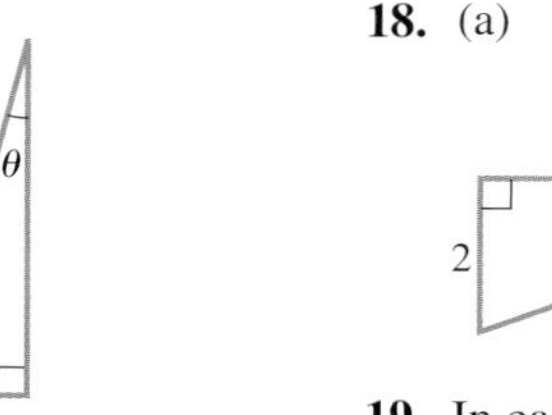

7–12 The angle θ is an acute angle of a right triangle. Solve the problems by drawing an appropriate right triangle. Do *not* use a calculator.

7. Find $\sin\theta$ and $\cos\theta$ given that $\tan\theta = 3$.
8. Find $\sin\theta$ and $\tan\theta$ given that $\cos\theta = \frac{2}{3}$.
9. Find $\tan\theta$ and $\csc\theta$ given that $\sec\theta = \frac{5}{2}$.
10. Find $\cot\theta$ and $\sec\theta$ given that $\csc\theta = 4$.
11. Find the length of the side adjacent to θ given that the hypotenuse has length 6 and $\cos\theta = 0.3$.
12. Find the length of the hypotenuse given that the side opposite θ has length 2.4 and $\sin\theta = 0.8$.

13–14 The value of an angle θ is given. Find the values of all six trigonometric functions of θ without using a calculator.

13. (a) 225° (b) -210° (c) $5\pi/3$ (d) $-3\pi/2$
14. (a) 330° (b) -120° (c) $9\pi/4$ (d) -3π

15–16 Use the information to find the exact values of the remaining five trigonometric functions of θ.

15. (a) $\cos\theta = \frac{3}{5},\ 0 < \theta < \pi/2$
(b) $\cos\theta = \frac{3}{5},\ -\pi/2 < \theta < 0$
(c) $\tan\theta = -1/\sqrt{3},\ \pi/2 < \theta < \pi$
(d) $\tan\theta = -1/\sqrt{3},\ -\pi/2 < \theta < 0$
(e) $\csc\theta = \sqrt{2},\ 0 < \theta < \pi/2$
(f) $\csc\theta = \sqrt{2},\ \pi/2 < \theta < \pi$

16. (a) $\sin\theta = \frac{1}{4},\ 0 < \theta < \pi/2$
(b) $\sin\theta = \frac{1}{4},\ \pi/2 < \theta < \pi$
(c) $\cot\theta = \frac{1}{3},\ 0 < \theta < \pi/2$
(d) $\cot\theta = \frac{1}{3},\ \pi < \theta < 3\pi/2$
(e) $\sec\theta = -\frac{5}{2},\ \pi/2 < \theta < \pi$
(f) $\sec\theta = -\frac{5}{2},\ \pi < \theta < 3\pi/2$

17–18 Use a calculating utility to find x to four decimal places.

17. (a)

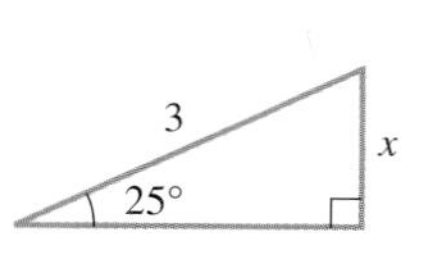

(b)

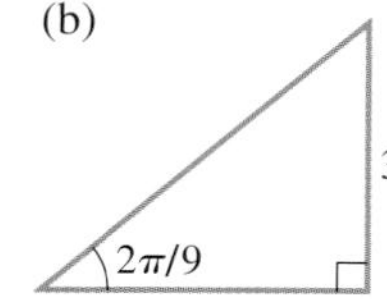

18. (a)

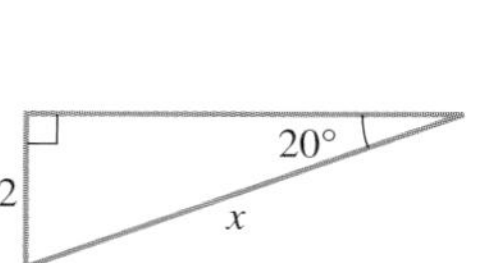

(b)

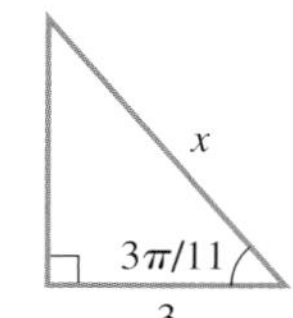

19. In each part, let θ be an acute angle of a right triangle. Express the remaining five trigonometric functions in terms of a.
(a) $\sin\theta = a/3$ (b) $\tan\theta = a/5$ (c) $\sec\theta = a$

20–27 Find all values of θ (in radians) that satisfy the given equation. Do not use a calculator.

20. (a) $\cos\theta = -1/\sqrt{2}$ (b) $\sin\theta = -1/\sqrt{2}$
21. (a) $\tan\theta = -1$ (b) $\cos\theta = \frac{1}{2}$
22. (a) $\sin\theta = -\frac{1}{2}$ (b) $\tan\theta = \sqrt{3}$
23. (a) $\tan\theta = 1/\sqrt{3}$ (b) $\sin\theta = -\sqrt{3}/2$
24. (a) $\sin\theta = -1$ (b) $\cos\theta = -1$
25. (a) $\cot\theta = -1$ (b) $\cot\theta = \sqrt{3}$
26. (a) $\sec\theta = -2$ (b) $\csc\theta = -2$
27. (a) $\csc\theta = 2/\sqrt{3}$ (b) $\sec\theta = 2/\sqrt{3}$

28–29 Find the values of all six trigonometric functions of θ.

28.

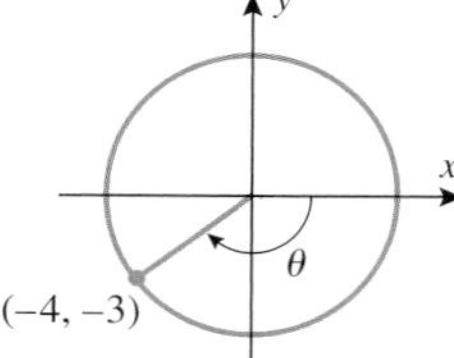

29.

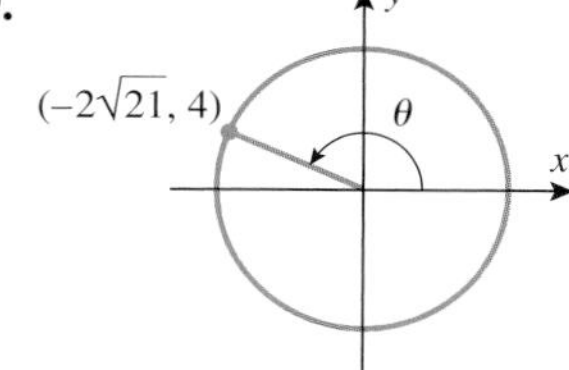

30. Find all values of θ (in radians) such that
(a) $\sin\theta = 1$ (b) $\cos\theta = 1$ (c) $\tan\theta = 1$
(d) $\csc\theta = 1$ (e) $\sec\theta = 1$ (f) $\cot\theta = 1$.

31. Find all values of θ (in radians) such that
(a) $\sin\theta = 0$ (b) $\cos\theta = 0$ (c) $\tan\theta = 0$
(d) $\csc\theta$ is undefined (e) $\sec\theta$ is undefined
(f) $\cot\theta$ is undefined.

32. How could you use a ruler and protractor to approximate $\sin 17^\circ$ and $\cos 17^\circ$?

33. Find the length of the circular arc on a circle of radius 4 cm subtended by an angle of
(a) $\pi/6$ (b) 150°.

34. Find the radius of a circular sector that has an angle of $\pi/3$ and a circular arc length of 7 units.

35. A point P moving counterclockwise on a circle of radius 5 cm traverses an arc length of 2 cm. What is the angle swept out by a radius from the center to P?

36. Find a formula for the area A of a circular sector in terms of its radius r and arc length s.

37. As shown in the accompanying figure, a right circular cone is made from a circular piece of paper of radius R by cutting out a sector of angle θ radians and gluing the cut edges of the remaining piece together. Find

(a) the radius r of the base of the cone in terms of R and θ

(b) the height h of the cone in terms of R and θ.

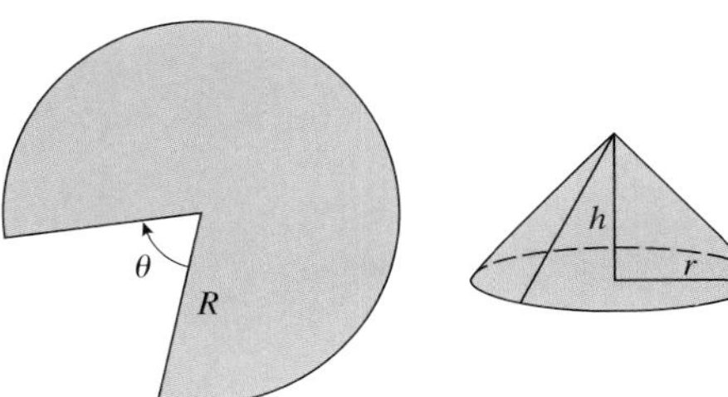

Figure Ex-37

38. As shown in the accompanying figure, let r and L be the radius of the base and the slant height of a right circular cone. Show that the lateral surface area, S, of the cone is $S = \pi r L$. [*Hint:* As shown in the figure in Exercise 37, the lateral surface of the cone becomes a circular sector when cut along a line from the vertex to the base and flattened.]

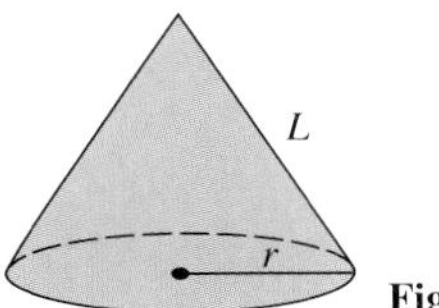

Figure Ex-38

39. Two sides of a triangle have lengths of 3 cm and 7 cm and meet at an angle of 60°. Find the area of the triangle.

40. Let ABC be a triangle whose angles at A and B are 30° and 45°. If the side opposite the angle B has length 9, find the lengths of the remaining sides and the size of the angle C.

41. A 10-foot ladder leans against a house and makes an angle of 67° with level ground. How far is the top of the ladder above the ground? Express your answer to the nearest tenth of a foot.

42. From a point 120 feet on level ground from a building, the angle of elevation to the top of the building is 76°. Find the height of the building. Express your answer to the nearest foot.

43. An observer on level ground is at a distance d from a building. The angles of elevation to the bottoms of the windows on the second and third floors are α and β, respectively. Find the distance h between the bottoms of the windows in terms of α, β, and d.

44. From a point on level ground, the angle of elevation to the top of a tower is α. From a point that is d units closer to the tower, the angle of elevation is β. Find the height h of the tower in terms of α, β, and d.

45–46 Do *not* use a calculator in these exercises.

45. If $\cos\theta = \frac{2}{3}$ and $0 < \theta < \pi/2$, find
(a) $\sin 2\theta$ (b) $\cos 2\theta$.

46. If $\tan\alpha = \frac{3}{4}$ and $\tan\beta = 2$, where $0 < \alpha < \pi/2$ and $0 < \beta < \pi/2$, find
(a) $\sin(\alpha - \beta)$ (b) $\cos(\alpha + \beta)$.

47. Express $\sin 3\theta$ and $\cos 3\theta$ in terms of $\sin\theta$ and $\cos\theta$.

48–58 Derive the given identities.

48. $\dfrac{\cos\theta\sec\theta}{1+\tan^2\theta} = \cos^2\theta$

49. $\dfrac{\cos\theta\tan\theta + \sin\theta}{\tan\theta} = 2\cos\theta$

50. $2\csc 2\theta = \sec\theta\csc\theta$

51. $\tan\theta + \cot\theta = 2\csc 2\theta$

52. $\dfrac{\sin 2\theta}{\sin\theta} - \dfrac{\cos 2\theta}{\cos\theta} = \sec\theta$

53. $\dfrac{\sin\theta + \cos 2\theta - 1}{\cos\theta - \sin 2\theta} = \tan\theta$

54. $\sin 3\theta + \sin\theta = 2\sin 2\theta\cos\theta$

55. $\sin 3\theta - \sin\theta = 2\cos 2\theta\sin\theta$

56. $\tan\dfrac{\theta}{2} = \dfrac{1-\cos\theta}{\sin\theta}$

57. $\tan\dfrac{\theta}{2} = \dfrac{\sin\theta}{1+\cos\theta}$

58. $\cos\left(\dfrac{\pi}{3}+\theta\right) + \cos\left(\dfrac{\pi}{3}-\theta\right) = \cos\theta$

59–60 In these exercises, refer to an arbitrary triangle ABC in which the side of length a is opposite angle A, the side of length b is opposite angle B, and the side of length c is opposite angle C.

59. Prove: The area of a triangle ABC can be written as
$$\text{area} = \tfrac{1}{2}bc\sin A$$
Find two other similar formulas for the area.

60. Prove the ***law of sines***: In any triangle, the ratios of the sides to the sines of the opposite angles are equal; that is,
$$\frac{a}{\sin A} = \frac{b}{\sin B} = \frac{c}{\sin C}$$

61. Use identities (34) through (37) to express each of the following in terms of $\sin\theta$ or $\cos\theta$.
(a) $\sin\left(\dfrac{\pi}{2}+\theta\right)$ (b) $\cos\left(\dfrac{\pi}{2}+\theta\right)$
(c) $\sin\left(\dfrac{3\pi}{2}-\theta\right)$ (d) $\cos\left(\dfrac{3\pi}{2}+\theta\right)$

62. Derive identities (38) and (39).

63. Derive identity
(a) (47) (b) (48) (c) (49).

64. If $A = \alpha + \beta$ and $B = \alpha - \beta$, then $\alpha = \frac{1}{2}(A+B)$ and $\beta = \frac{1}{2}(A-B)$ (verify). Use this result and identities (47) through (49) to derive identity
(a) (50) (b) (52) (c) (53).

65. Substitute $-\beta$ for β in identity (50) to derive identity (51).

66. (a) Express $3 \sin \alpha + 5 \cos \alpha$ in the form

$$C \sin(\alpha + \phi)$$

(b) Show that a sum of the form

$$A \sin \alpha + B \cos \alpha$$

can be rewritten in the form $C \sin(\alpha + \phi)$.

67. Show that the length of the diagonal of the parallelogram in the accompanying figure is

$$d = \sqrt{a^2 + b^2 + 2ab \cos \theta}$$

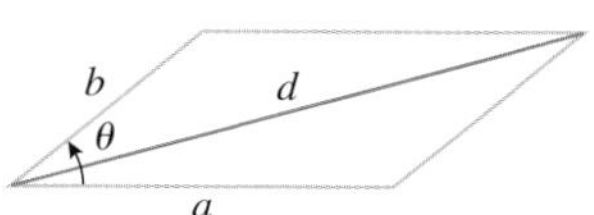

Figure Ex-67

68–69 Find the angle of inclination of the line with slope m to the nearest degree. Use a calculating utility, where needed.

68. (a) $m = \frac{1}{2}$ (b) $m = -1$ (c) $m = 2$ (d) $m = -57$

69. (a) $m = -\frac{1}{2}$ (b) $m = 1$ (c) $m = -2$ (d) $m = 57$

70–71 Find the angle of inclination of the line to the nearest degree. Use a calculating utility, where needed.

70. (a) $3y = 2 - \sqrt{3}x$ (b) $y - 4x + 7 = 0$

71. (a) $y = \sqrt{3}x + 2$ (b) $y + 2x + 5 = 0$

a p p e n d i x b

SOLVING POLYNOMIAL EQUATIONS

We will assume in this appendix that you know how to divide polynomials using long division and synthetic division. If you need to review those techniques, refer to an algebra book.

A BRIEF REVIEW OF POLYNOMIALS

Recall that if n is a nonnegative integer, then a ***polynomial of degree n*** is a function that can be written in the following forms, depending on whether you want the powers of x in ascending or descending order:

$$c_0 + c_1x + c_2x^2 + \cdots + c_nx^n \quad (c_n \neq 0)$$
$$c_nx^n + c_{n-1}x^{n-1} + \cdots + c_1x + c_0 \quad (c_n \neq 0)$$

The numbers $c_0, c_1, \ldots, c_n$ are called the ***coefficients*** of the polynomial. The coefficient c_n (which multiplies the highest power of x) is called the ***leading coefficient***, the term c_nx^n is called the ***leading term***, and the coefficient c_0 is called the ***constant term***. Polynomials of degree 1, 2, 3, 4, and 5 are called ***linear***, ***quadratic***, ***cubic***, ***quartic***, and ***quintic***, respectively. For simplicity, general polynomials of low degree are often written without subscripts on the coefficients:

$$p(x) = a$$ Constant polynomial

$$p(x) = ax + b \quad (a \neq 0)$$ Linear polynomial

$$p(x) = ax^2 + bx + c \quad (a \neq 0)$$ Quadratic polynomial

$$p(x) = ax^3 + bx^2 + cx + d \quad (a \neq 0)$$ Cubic polynomial

When you attempt to factor a polynomial completely, one of three things can happen:

- You may be able to decompose the polynomial into distinct linear factors using only real numbers; for example,

$$x^3 + x^2 - 2x = x(x^2 + x - 2) = x(x-1)(x+2)$$

- You may be able to decompose the polynomial into linear factors using only real numbers, but some of the factors may be repeated; for example,

$$x^6 - 3x^4 + 2x^3 = x^3(x^3 - 3x + 2) = x^3(x-1)^2(x+2) \tag{1}$$

- You may be able to decompose the polynomial into linear and quadratic factors using only real numbers, but you may not be able to decompose the quadratic factors into linear factors using only real numbers (such quadratic factors are said to be ***irreducible*** over the real numbers); for example,

$$\begin{aligned} x^4 - 1 = (x^2-1)(x^2+1) &= (x-1)(x+1)(x^2+1) \\ &= (x-1)(x+1)(x-i)(x+i) \end{aligned}$$

Here, the factor $x^2 + 1$ is irreducible over the real numbers.

In general, if $p(x)$ is a polynomial of degree n with leading coefficient a, and if complex numbers are allowed, then $p(x)$ can be factored as

$$p(x) = a(x - r_1)(x - r_2)\cdots(x - r_n) \tag{2}$$

where $r_1, r_2, \ldots, r_n$ are called the ***zeros*** of $p(x)$ or the ***roots*** of the equation $p(x) = 0$, and (2) is called the ***complete linear factorization*** of $p(x)$. If some of the factors in (2) are repeated, then they can be combined; for example, if the first k factors are distinct and the rest are repetitions of the first k, then (2) can be expressed in the form

$$p(x) = a(x - r_1)^{m_1}(x - r_2)^{m_2}\cdots(x - r_k)^{m_k} \tag{3}$$

where $r_1, r_2, \ldots, r_k$ are the *distinct* roots of $p(x) = 0$. The exponents $m_1, m_2, \ldots, m_k$ tell us how many times the various factors occur in the complete linear factorization; for example, in (3) the factor $(x - r_1)$ occurs m_1 times, the factor $(x - r_2)$ occurs m_2 times, and so forth. Some techniques for factoring polynomials are discussed later in this appendix. In general, if a factor $(x - r)$ occurs m times in the complete linear factorization of a polynomial, then we say that r is a root or zero of ***multiplicity m***, and if $(x - r)$ has no repetitions (i.e., r has multiplicity 1), then we say that r is a ***simple*** root or zero. For example, it follows from (1) that the equation $x^6 - 3x^4 + 2x^3 = 0$ can be expressed as

$$x^3(x - 1)^2(x + 2) = 0 \tag{4}$$

so this equation has three distinct roots—a root $x = 0$ of multiplicity 3, a root $x = 1$ of multiplicity 2, and a simple root $x = -2$.

Note that in (3) the multiplicities of the roots must add up to n, since $p(x)$ has degree n; that is,

$$m_1 + m_2 + \cdots + m_k = n$$

For example, in (4) the multiplicities add up to 6, which is the same as the degree of the polynomial.

It follows from (2) that a polynomial of degree n can have at most n distinct roots; if all of the roots are simple, then there will be *exactly* n, but if some are repeated, then there will be fewer than n. However, when counting the roots of a polynomial, it is standard practice to count multiplicities, since that convention allows us to say that a polynomial of degree n has n roots. For example, from (1) the six roots of the polynomial $p(x) = x^6 - 3x^4 + 2x^3$ are

$$r = 0,\ \ 0,\ \ 0,\ \ 1,\ \ 1,\ \ -2$$

In summary, we have the following important theorem.

B.1 THEOREM. *If complex roots are allowed, and if roots are counted according to their multiplicities, then a polynomial of degree n has exactly n roots.*

THE REMAINDER THEOREM

When two positive integers are divided, the numerator can be expressed as the quotient plus the remainder over the divisor, where the remainder is less than the divisor. For example,

$$\tfrac{17}{5} = 3 + \tfrac{2}{5}$$

If we multiply this equation through by 5, we obtain

$$17 = 5 \cdot 3 + 2$$

which states that the *numerator is the divisor times the quotient plus the remainder.*

The following theorem, which we state without proof, is an analogous result for division of polynomials.

B.2 THEOREM. *If $p(x)$ and $s(x)$ are polynomials, and if $s(x)$ is not the zero polynomial, then $p(x)$ can be expressed as*

$$p(x) = s(x)q(x) + r(x)$$

where $q(x)$ and $r(x)$ are the quotient and remainder that result when $p(x)$ is divided by $s(x)$, and either $r(x)$ is the zero polynomial or the degree of $r(x)$ is less than the degree of $s(x)$.

In the special case where $p(x)$ is divided by a first-degree polynomial of the form $x - c$, the remainder must be some constant r, since it is either zero or has degree less than 1. Thus, Theorem B.2 implies that

$$p(x) = (x - c)q(x) + r$$

and this in turn implies that $p(c) = r$. In summary, we have the following theorem.

B.3 THEOREM (*Remainder Theorem*). *If a polynomial $p(x)$ is divided by $x - c$, then the remainder is $p(c)$.*

▶ **Example 1** According to the Remainder Theorem, the remainder on dividing

$$p(x) = 2x^3 + 3x^2 - 4x - 3$$

by $x + 4$ should be

$$p(-4) = 2(-4)^3 + 3(-4)^2 - 4(-4) - 3 = -67$$

Show that this is so.

Solution. By long division

$$\begin{array}{r@{}l}
 & 2x^2 - 5x + 16 \\
x+4\,\big) & 2x^3 + 3x^2 - 4x - 3 \\
 & \underline{2x^3 + 8x^2} \\
 & \quad -5x^2 - 4x \\
 & \quad \underline{-5x^2 - 20x} \\
 & \qquad\qquad 16x - 3 \\
 & \qquad\qquad \underline{16x + 64} \\
 & \qquad\qquad\qquad -67
\end{array}$$

which shows that the remainder is -67.

Alternative Solution. Because we are dividing by an expression of the form $x - c$ (where $c = -4$), we can use synthetic division rather than long division. The computations are

$$\begin{array}{r|rrrr}
-4 & 2 & 3 & -4 & -3 \\
 & & -8 & 20 & -64 \\
\hline
 & 2 & -5 & 16 & -67
\end{array}$$

which again shows that the remainder is -67. ◀

THE FACTOR THEOREM

To ***factor*** a polynomial $p(x)$ is to write it as a product of lower-degree polynomials, called ***factors*** of $p(x)$. For $s(x)$ to be a factor of $p(x)$ there must be no remainder when $p(x)$ is divided by $s(x)$. For example, if $p(x)$ can be factored as

$$p(x) = s(x)q(x) \tag{5}$$

then

$$\frac{p(x)}{s(x)} = q(x) \tag{6}$$

so dividing $p(x)$ by $s(x)$ produces a quotient $q(x)$ with no remainder. Conversely, (6) implies (5), so $s(x)$ is a factor of $p(x)$ if there is no remainder when $p(x)$ is divided by $s(x)$.

In the special case where $x - c$ is a factor of $p(x)$, the polynomial $p(x)$ can be expressed as

$$p(x) = (x - c)q(x)$$

which implies that $p(c) = 0$. Conversely, if $p(c) = 0$, then the Remainder Theorem implies that $x - c$ is a factor of $p(x)$, since the remainder is 0 when $p(x)$ is divided by $x - c$. These results are summarized in the following theorem.

B.4 THEOREM (*Factor Theorem*). *A polynomial $p(x)$ has a factor $x - c$ if and only if $p(c) = 0$.*

It follows from this theorem that the statements below say the same thing in different ways:

- $x - c$ is a factor of $p(x)$.
- $p(c) = 0$.
- c is a zero of $p(x)$.
- c is a root of the equation $p(x) = 0$.
- c is a solution of the equation $p(x) = 0$.
- c is an x-intercept of $y = p(x)$.

► **Example 2** Confirm that $x - 1$ is a factor of

$$p(x) = x^3 - 3x^2 - 13x + 15$$

by dividing $x - 1$ into $p(x)$ and checking that the remainder is zero.

Solution. By long division

$$\begin{array}{r|l} & x^2 - 2x - 15 \\ \hline x - 1 & x^3 - 3x^2 - 13x + 15 \\ & \underline{x^3 - x^2} \\ & \quad -2x^2 - 13x \\ & \quad \underline{-2x^2 + 2x} \\ & \qquad\quad -15x + 15 \\ & \qquad\quad \underline{-15x + 15} \\ & \qquad\qquad\qquad 0 \end{array}$$

which shows that the remainder is zero.

Alternative Solution. Because we are dividing by an expression of the form $x - c$, we can use synthetic division rather than long division. The computations are

$$\begin{array}{r|rrrr} 1 & 1 & -3 & -13 & 15 \\ & & 1 & -2 & -15 \\ \hline & 1 & -2 & -15 & 0 \end{array}$$

which again confirms that the remainder is zero. ◄

USING ONE FACTOR TO FIND OTHER FACTORS

If $x - c$ is a factor of $p(x)$, and if $q(x) = p(x)/(x - c)$, then

$$p(x) = (x - c)q(x) \tag{7}$$

so that additional linear factors of $p(x)$ can be obtained by factoring the quotient $q(x)$.

▸ **Example 3** Factor

$$p(x) = x^3 - 3x^2 - 13x + 15 \tag{8}$$

completely into linear factors.

Solution. We showed in Example 2 that $x - 1$ is a factor of $p(x)$ and we also showed that $p(x)/(x - 1) = x^2 - 2x - 15$. Thus,

$$x^3 - 3x^2 - 13x + 15 = (x - 1)(x^2 - 2x - 15)$$

Factoring $x^2 - 2x - 15$ by inspection yields

$$x^3 - 3x^2 - 13x + 15 = (x - 1)(x - 5)(x + 3)$$

which is the complete linear factorization of $p(x)$. ◄

METHODS FOR FINDING ROOTS

A general quadratic equation $ax^2 + bx + c = 0$ can be solved by using the quadratic formula to express the solutions of the equation in terms of the coefficients. Versions of this formula were known since Babylonian times, and by the seventeenth century formulas had been obtained for solving general cubic and quartic equations. However, attempts to find formulas for the solutions of general fifth-degree equations and higher proved fruitless. The reason for this became clear in 1829 when the French mathematician Evariste Galois (1811–1832) proved that it is impossible to express the solutions of a general fifth-degree equation or higher in terms of its coefficients using algebraic operations.

Today, we have powerful computer programs for finding the zeros of specific polynomials. For example, it takes only seconds for a computer algebra system, such as *Mathematica*, *Maple*, or *Derive*, to show that the zeros of the polynomial

$$p(x) = 10x^4 - 23x^3 - 10x^2 + 29x + 6 \tag{9}$$

are

$$x = -1, \quad x = -\tfrac{1}{5}, \quad x = \tfrac{3}{2}, \quad \text{and} \quad x = 2 \tag{10}$$

The algorithms that these programs use to find the integer and rational zeros of a polynomial, if any, are based on the following theorem, which is proved in advanced algebra courses.

> **B.5 THEOREM.** *Suppose that*
>
> $$p(x) = c_n x^n + c_{n-1}x^{n-1} + \cdots + c_1 x + c_0$$
>
> *is a polynomial with integer coefficients.*
>
> (a) *If r is an integer zero of $p(x)$, then r must be a divisor of the constant term c_0.*
>
> (b) *If $r = a/b$ is a rational zero of $p(x)$ in which all common factors of a and b have been canceled, then a must be a divisor of the constant term c_0, and b must be a divisor of the leading coefficient c_n.*

For example, in (9) the constant term is 6 (which has divisors ± 1, ± 2, ± 3, and ± 6) and the leading coefficient is 10 (which has divisors ± 1, ± 2, ± 5, and ± 10). Thus, the only possible integer zeros of $p(x)$ are

$$\pm 1,\ \pm 2,\ \pm 3,\ \pm 6$$

and the only possible noninteger rational zeros are

$$\pm\tfrac{1}{2},\ \pm\tfrac{1}{5},\ \pm\tfrac{1}{10},\ \pm\tfrac{2}{5},\ \pm\tfrac{3}{2},\ \pm\tfrac{3}{5},\ \pm\tfrac{3}{10},\ \pm\tfrac{6}{5}$$

Using a computer, it is a simple matter to evaluate $p(x)$ at each of the numbers in these lists to show that its only rational zeros are the numbers in (10).

▶ **Example 4** Solve the equation $x^3 + 3x^2 - 7x - 21 = 0$.

Solution. The solutions of the equation are the zeros of the polynomial

$$p(x) = x^3 + 3x^2 - 7x - 21$$

We will look for integer zeros first. All such zeros must divide the constant term, so the only possibilities are ± 1, ± 3, ± 7, and ± 21. Substituting these values into $p(x)$ (or using the method of Exercise 6) shows that $x = -3$ is an integer zero. This tells us that $x + 3$ is a factor of $p(x)$ and that $p(x)$ can be written as

$$x^3 + 3x^2 - 7x - 21 = (x + 3)q(x)$$

where $q(x)$ is the quotient that results when $x^3 + 3x^2 - 7x - 21$ is divided by $x + 3$. We leave it for you to perform the division and show that $q(x) = x^2 - 7$; hence,

$$x^3 + 3x^2 - 7x - 21 = (x + 3)(x^2 - 7) = (x + 3)(x + \sqrt{7})(x - \sqrt{7})$$

which tells us that the solutions of the given equation are $x = 3$, $x = \sqrt{7} \approx 2.65$, and $x = -\sqrt{7} \approx -2.65$. ◀

EXERCISE SET B [C] CAS

1–2 Find the quotient $q(x)$ and the remainder $r(x)$ that result when $p(x)$ is divided by $s(x)$.

1. (a) $p(x) = x^4 + 3x^3 - 5x + 10$; $s(x) = x^2 - x + 2$
(b) $p(x) = 6x^4 + 10x^2 + 5$; $s(x) = 3x^2 - 1$
(c) $p(x) = x^5 + x^3 + 1$; $s(x) = x^2 + x$

2. (a) $p(x) = 2x^4 - 3x^3 + 5x^2 + 2x + 7$; $s(x) = x^2 - x + 1$
(b) $p(x) = 2x^5 + 5x^4 - 4x^3 + 8x^2 + 1$; $s(x) = 2x^2 - x + 1$
(c) $p(x) = 5x^6 + 4x^2 + 5$; $s(x) = x^3 + 1$

3–4 Use synthetic division to find the quotient $q(x)$ and the remainder r that result when $p(x)$ is divided by $s(x)$.

3. (a) $p(x) = 3x^3 - 4x - 1$; $s(x) = x - 2$
(b) $p(x) = x^4 - 5x^2 + 4$; $s(x) = x + 5$
(c) $p(x) = x^5 - 1$; $s(x) = x - 1$

4. (a) $p(x) = 2x^3 - x^2 - 2x + 1$; $s(x) = x - 1$
(b) $p(x) = 2x^4 + 3x^3 - 17x^2 - 27x - 9$; $s(x) = x + 4$
(c) $p(x) = x^7 + 1$; $s(x) = x - 1$

5. Let $p(x) = 2x^4 + x^3 - 3x^2 + x - 4$. Use synthetic division and the Remainder Theorem to find $p(0)$, $p(1)$, $p(-3)$, and $p(7)$.

6. Let $p(x)$ be the polynomial in Example 4. Use synthetic division and the Remainder Theorem to evaluate $p(x)$ at $x = \pm 1, \pm 3, \pm 7$, and ± 21.

7. Let $p(x) = x^3 + 4x^2 + x - 6$. Find a polynomial $q(x)$ and a constant r such that
(a) $p(x) = (x - 2)q(x) + r$
(b) $p(x) = (x + 1)q(x) + r$.

8. Let $p(x) = x^5 - 1$. Find a polynomial $q(x)$ and a constant r such that
(a) $p(x) = (x + 1)q(x) + r$
(b) $p(x) = (x - 1)q(x) + r$.

9. In each part, make a list of all possible candidates for the rational zeros of $p(x)$.
(a) $p(x) = x^7 + 3x^3 - x + 24$
(b) $p(x) = 3x^4 - 2x^2 + 7x - 10$
(c) $p(x) = x^{35} - 17$

10. Find all integer zeros of
$$p(x) = x^6 + 5x^5 - 16x^4 - 15x^3 - 12x^2 - 38x - 21$$

11–15 Factor the polynomials completely.

11. $p(x) = x^3 - 2x^2 - x + 2$

12. $p(x) = 3x^3 + x^2 - 12x - 4$

13. $p(x) = x^4 + 10x^3 + 36x^2 + 54x + 27$

14. $p(x) = 2x^4 + x^3 + 3x^2 + 3x - 9$

15. $p(x) = x^5 + 4x^4 - 4x^3 - 34x^2 - 45x - 18$

C **16.** For each of the factorizations that you obtained in Exercises 11–15, check your answer using a CAS.

17–21 Find all real solutions of the equations.

17. $x^3 + 3x^2 + 4x + 12 = 0$

18. $2x^3 - 5x^2 - 10x + 3 = 0$

19. $3x^4 + 14x^3 + 14x^2 - 8x - 8 = 0$

20. $2x^4 - x^3 - 14x^2 - 5x + 6 = 0$

21. $x^5 - 2x^4 - 6x^3 + 5x^2 + 8x + 12 = 0$

C **22.** For each of the equations you solved in Exercises 17–21, check your answer using a CAS.

23. Find all values of k for which $x - 1$ is a factor of the polynomial $p(x) = k^2x^3 - 7kx + 10$.

24. Is $x + 3$ a factor of $x^7 + 2187$? Justify your answer.

C **25.** A 3-cm-thick slice is cut from a cube, leaving a volume of 196 cm^3. Use a CAS to find the length of a side of the original cube.

26. (a) Show that there is no positive rational number that exceeds its cube by 1.
(b) Does there exist a real number that exceeds its cube by 1? Justify your answer.

27. Use the Factor Theorem to show each of the following.
(a) $x - y$ is a factor of $x^n - y^n$ for all positive integer values of n.
(b) $x + y$ is a factor of $x^n - y^n$ for all positive even integer values of n.
(c) $x + y$ is a factor of $x^n + y^n$ for all positive odd integer values of n.

appendix C

SELECTED PROOFS

PROOFS OF BASIC LIMIT THEOREMS

An extensive excursion into proofs of limit theorems would be too time consuming to undertake, so we have selected a few proofs of results from Section 2.2 that illustrate some of the basic ideas.

> **C.1 THEOREM.** *Let a be any real number, let k be a constant, and suppose that* $\lim_{x\to a} f(x) = L_1$ *and that* $\lim_{x\to a} g(x) = L_2$. *Then*
>
> (*a*) $\lim_{x\to a} k = k$
>
> (*b*) $\lim_{x\to a} [f(x) + g(x)] = \lim_{x\to a} f(x) + \lim_{x\to a} g(x) = L_1 + L_2$
>
> (*c*) $\lim_{x\to a} [f(x)g(x)] = \left(\lim_{x\to a} f(x)\right)\left(\lim_{x\to a} g(x)\right) = L_1 L_2$

PROOF (*a*). We will apply Definition 2.4.1 with $f(x) = k$ and $L = k$. Thus, given $\epsilon > 0$, we must find a number $\delta > 0$ such that

$$|k - k| < \epsilon \quad \text{if} \quad 0 < |x - a| < \delta$$

or, equivalently,

$$0 < \epsilon \quad \text{if} \quad 0 < |x - a| < \delta$$

But the condition on the left side of this statement is *always* true, no matter how δ is chosen. Thus, any positive value for δ will suffice.

PROOF (*b*). We must show that given $\epsilon > 0$ we can find a number $\delta > 0$ such that

$$|(f(x) + g(x)) - (L_1 + L_2)| < \epsilon \quad \text{if} \quad 0 < |x - a| < \delta \tag{1}$$

However, from the limits of f and g in the hypothesis of the theorem we can find numbers δ_1 and δ_2 such that

$$|f(x) - L_1| < \epsilon/2 \quad \text{if} \quad 0 < |x - a| < \delta_1$$
$$|g(x) - L_2| < \epsilon/2 \quad \text{if} \quad 0 < |x - a| < \delta_2$$

Moreover, the inequalities on the left sides of these statements *both* hold if we replace δ_1 and δ_2 by any positive number δ that is less than both δ_1 and δ_2. Thus, for any such δ it follows that

$$|f(x) - L_1| + |g(x) - L_2| < \epsilon \quad \text{if} \quad 0 < |x - a| < \delta \tag{2}$$

However, it follows from the triangle inequality [Theorem E.5 of Appendix E] that

$$\begin{aligned} |(f(x) + g(x)) - (L_1 + L_2)| &= |(f(x) - L_1) + (g(x) - L_2)| \\ &\le |f(x) - L_1| + |g(x) - L_2| \end{aligned}$$

so that (1) follows from (2).

PROOF (c). We must show that given $\epsilon > 0$ we can find a number $\delta > 0$ such that

$$|f(x)g(x) - L_1L_2| < \epsilon \quad \text{if} \quad 0 < |x - a| < \delta \tag{3}$$

To find δ it will be helpful to express (3) in a different form. If we rewrite $f(x)$ and $g(x)$ as

$$f(x) = L_1 + (f(x) - L_1) \quad \text{and} \quad g(x) = L_2 + (g(x) - L_2)$$

then the inequality on the left side of (3) can be expressed as (verify)

$$|L_1(g(x) - L_2) + L_2(f(x) - L_1) + (f(x) - L_1)(g(x) - L_2)| < \epsilon \tag{4}$$

Since

$$\lim_{x\to a} f(x) = L_1 \quad \text{and} \quad \lim_{x\to a} g(x) = L_2$$

we can find positive numbers δ_1, δ_2, δ_3, and δ_4 such that

$$\begin{aligned}
|f(x) - L_1| &< \sqrt{\epsilon/3} && \text{if} \quad 0 < |x - a| < \delta_1 \\
|f(x) - L_1| &< \frac{\epsilon}{3(1 + |L_2|)} && \text{if} \quad 0 < |x - a| < \delta_2 \\
|g(x) - L_2| &< \sqrt{\epsilon/3} && \text{if} \quad 0 < |x - a| < \delta_3 \\
|g(x) - L_2| &< \frac{\epsilon}{3(1 + |L_1|)} && \text{if} \quad 0 < |x - a| < \delta_4
\end{aligned} \tag{5}$$

Moreover, the inequalities on the left sides of these four statements *all* hold if we replace δ_1, δ_2, δ_3, and δ_4 by any positive number δ that is smaller than δ_1, δ_2, δ_3, and δ_4. Thus, for any such δ it follows with the help of the triangle inequality that

$$\begin{aligned}
&|L_1(g(x) - L_2) + L_2(f(x) - L_1) + (f(x) - L_1)(g(x) - L_2)| \\
&\quad \le |L_1(g(x) - L_2)| + |L_2(f(x) - L_1)| + |(f(x) - L_1)(g(x) - L_2)| \\
&\quad = |L_1||g(x) - L_2| + |L_2||f(x) - L_1| + |f(x) - L_1||g(x) - L_2| \\
&\quad < |L_1|\frac{\epsilon}{3(1 + |L_1|)} + |L_2|\frac{\epsilon}{3(1 + |L_2|)} + \sqrt{\epsilon/3}\sqrt{\epsilon/3} && \text{From (5)} \\
&\quad = \frac{\epsilon}{3}\frac{|L_1|}{1 + |L_1|} + \frac{\epsilon}{3}\frac{|L_2|}{1 + |L_2|} + \frac{\epsilon}{3} \\
&\quad < \frac{\epsilon}{3} + \frac{\epsilon}{3} + \frac{\epsilon}{3} = \epsilon && \text{Since } \frac{|L_1|}{1 + |L_1|} < 1 \text{ and } \frac{|L_2|}{1 + |L_2|} < 1
\end{aligned}$$

which shows that (4) holds for the δ selected. ■

Do not be alarmed if the proof of part (c) seems difficult; it takes some experience with proofs of this type to develop a feel for choosing a valid δ. Your initial goal should be to understand the ideas and the computations.

PROOF OF A BASIC CONTINUITY PROPERTY

Next we will prove Theorem 2.5.5 for two-sided limits.

C.2 THEOREM (*Theorem 2.5.5*). *If $\lim_{x\to c} g(x) = L$ and if the function f is continuous at L, then $\lim_{x\to c} f(g(x)) = f(L)$. That is,*

$$\lim_{x\to c} f(g(x)) = f\Big(\lim_{x\to c} g(x)\Big)$$

PROOF. We must show that given $\epsilon > 0$, we can find a number $\delta > 0$ such that

$$|f(g(x)) - f(L)| < \epsilon \quad \text{if} \quad 0 < |x - c| < \delta \tag{6}$$

Since f is continuous at L, we have

$$\lim_{u\to L} f(u) = f(L)$$

and hence we can find a number $\delta_1 > 0$ such that

$$|f(u) - f(L)| < \epsilon \quad \text{if} \quad |u - L| < \delta_1$$

In particular, if $u = g(x)$, then

$$|f(g(x)) - f(L)| < \epsilon \quad \text{if} \quad |g(x) - L| < \delta_1 \tag{7}$$

But $\lim_{x \to c} g(x) = L$, and hence there is a number $\delta > 0$ such that

$$|g(x) - L| < \delta_1 \quad \text{if} \quad 0 < |x - c| < \delta \tag{8}$$

Thus, if x satisfies the condition on the right side of statement (8), then it follows that $g(x)$ satisfies the condition on the right side of statement (7), and this implies that the condition on the left side of statement (6) is satisfied, completing the proof. ■

PROOF OF THE CHAIN RULE

Next we will prove the chain rule (Theorem 3.6.1), but first we need a preliminary result.

C.3 THEOREM. *If f is differentiable at x and if $y = f(x)$, then*

$$\Delta y = f'(x)\Delta x + \epsilon \Delta x$$

where $\epsilon \to 0$ as $\Delta x \to x$ and $\epsilon = 0$ if $x = 0$.

PROOF. Define

$$\epsilon = \begin{cases} \dfrac{f(x + \Delta x) - f(x)}{\Delta x} - f'(x) & \text{if } \Delta x \neq 0 \\ 0 & \text{if } \Delta x = 0 \end{cases} \tag{9}$$

If $\Delta x \neq 0$, it follows from (9) that

$$\epsilon \Delta x = [f(x + \Delta x) - f(x)] - f'(x)\Delta x \tag{10}$$

But

$$\Delta y = f(x + \Delta x) - f(x) \tag{11}$$

so (10) can be written as

$$\epsilon \Delta x = \Delta y - f'(x)\Delta x$$

or

$$\Delta y = f'(x)\Delta x + \epsilon \Delta x \tag{12}$$

If $\Delta x = 0$, then (12) still holds, (why?), so (12) is valid for all values of Δx. It remains to show that $\epsilon \to 0$ as $\Delta x \to 0$. But this follows from the assumption that f is differentiable at x, since

$$\lim_{\Delta x \to 0} \epsilon = \lim_{\Delta x \to 0} \left[\frac{f(x + \Delta x) - f(x)}{\Delta x} - f'(x) \right] = f'(x) - f'(x) = 0$$ ■

We are now ready to prove the chain rule.

C.4 THEOREM (*Theorem 3.6.1*). *If g is differentiable at the point x and f is differentiable at the point $g(x)$, then the composition $f \circ g$ is differentiable at the point x. Moreover, if $y = f(g(x))$ and $u = g(x)$, then*

$$\frac{dy}{dx} = \frac{dy}{du} \cdot \frac{du}{dx}$$

PROOF. Since g is differentiable at x and $u = g(x)$, it follows from Theorem C.3 that

$$\Delta(u) = g'(x)\Delta x + \epsilon_1 \Delta x \tag{13}$$

where $\epsilon_1 \to 0$ as $\Delta x \to 0$. And since $y = f(u)$ is differentiable at $u = g(x)$, it follows from Theorem C.3 that

$$\Delta y = f'(u)\Delta u + \epsilon_2 \Delta u \tag{14}$$

where $\epsilon_2 \to 0$ as $\Delta u \to 0$.

Factoring out the Δu in (14) and then substituting (13) yields

$$\Delta y = [f'(u) + \epsilon_2][g'(x)\Delta x + \epsilon_1 \Delta x]$$

or

$$\Delta y = [f'(u) + \epsilon_2][g'(x) + \epsilon_1]\Delta x$$

or if $\Delta x \neq 0$,

$$\frac{\Delta y}{\Delta x} = [f'(u) + \epsilon_2][g'(x) + \epsilon_1] \tag{15}$$

But (13) implies that $\Delta u \to 0$ as $\Delta x \to 0$, and hence $\epsilon_1 \to 0$ and $\epsilon_2 \to 0$ as $\Delta x \to 0$. Thus, from (15)

$$\lim_{\Delta x \to 0} \frac{\Delta y}{\Delta x} = f'(u)g'(x)$$

or

$$\frac{\Delta y}{\Delta x} = f'(u)g'(x) = \frac{dy}{du} \cdot \frac{du}{dx}$$ ■

PROOF THAT RELATIVE EXTREMA OCCUR AT CRITICAL POINTS

In this subsection we will prove Theorem 5.2.2, which states that the relative extrema of a function occur at critical points.

> **C.5 THEOREM (*Theorem 5.2.2*).** *Suppose that f is a function defined on an open interval containing the point x_0. If f has a relative extremum at $x = x_0$, then $x = x_0$ is a critical point of f; that is, either $f'(x_0) = 0$ or f is not differentiable at x_0.*

PROOF. Suppose that f has a relative maximum at x_0. There are two possibilities—either f is differentiable at a point x_0 or it is not. If it is not, then x_0 is a critical point for f and we are done. If f is differentiable at x_0, then we must show that $f'(x_0) = 0$. We will do this by showing that $f'(x_0) \geq 0$ and $f'(x_0) \leq 0$, from which it follows that $f'(x_0) = 0$. From the definition of a derivative we have

$$f'(x_0) = \lim_{h \to 0} \frac{f(x_0 + h) - f(x_0)}{h}$$

so that

$$f'(x_0) = \lim_{h \to 0^+} \frac{f(x_0 + h) - f(x_0)}{h} \tag{16}$$

and

$$f'(x_0) = \lim_{h \to 0^-} \frac{f(x_0 + h) - f(x_0)}{h} \tag{17}$$

Because f has a relative maximum at x_0, there is an open interval (a, b) containing x_0 in which $f(x) \leq f(x_0)$ for all x in (a, b).

Assume that h is sufficiently small so that $x_0 + h$ lies in the interval (a, b). Thus,

$$f(x_0 + h) \leq f(x_0) \quad \text{or equivalently} \quad f(x_0 + h) - f(x_0) \leq 0$$

Thus, if h is negative,

$$\frac{f(x_0 + h) - f(x_0)}{h} \geq 0 \tag{18}$$

and if h is positive,

$$\frac{f(x_0+h)-f(x_0)}{h} \leq 0 \tag{19}$$

But an expression that never assumes negative values cannot approach a negative limit and an expression that never assumes positive values cannot approach a positive limit, so that

$$f'(x_0) = \lim_{h\to 0^-} \frac{f(x_0+h)-f(x_0)}{h} \geq 0 \qquad \boxed{\text{From (17) and (18)}}$$

and

$$f'(x_0) = \lim_{h\to 0^+} \frac{f(x_0+h)-f(x_0)}{h} \leq 0 \qquad \boxed{\text{From (16) and (19)}}$$

Since $f'(x_0) \geq 0$ and $f'(x_0) \leq 0$, it must be that $f'(x_0) = 0$.

A similar argument applies if f has a relative minimum at x_0. ■

PROOFS OF TWO SUMMATION FORMULAS

We will prove parts (a) and (b) of Theorem 6.4.2. The proof of part (c) is similar to that of part (b) and is omitted.

C.6 THEOREM (*Theorem 6.4.2*).

(*a*) $\displaystyle\sum_{k=1}^{n} k = 1+2+\cdots+n = \frac{n(n+1)}{2}$

(*b*) $\displaystyle\sum_{k=1}^{n} k^2 = 1^2+2^2+\cdots+n^2 = \frac{n(n+1)(2n+1)}{6}$

(*c*) $\displaystyle\sum_{k=1}^{n} k^3 = 1^3+2^3+\cdots+n^3 = \left[\frac{n(n+1)}{2}\right]^2$

PROOF (*a*). Writing

$$\sum_{k=1}^{n} k$$

two ways, with summands in increasing order and in decreasing order, and then adding, we obtain

$$\begin{aligned}
\sum_{k=1}^{n} k &= 1 + 2 + 3 + \cdots + (n-2) + (n-1) + n \\
\sum_{k=1}^{n} k &= n + (n-1) + (n-2) + \cdots + 3 + 2 + 1 \\
\hline
2\sum_{k=1}^{n} k &= (n+1) + (n+1) + (n+1) + \cdots + (n+1) + (n+1) + (n+1) \\
&= n(n+1)
\end{aligned}$$

Thus,

$$\sum_{k=1}^{n} k = \frac{n(n+1)}{2}$$

PROOF (*b*). Note that

$$(k+1)^3 - k^3 = k^3 + 3k^2 + 3k + 1 - k^3 = 3k^2 + 3k + 1$$

So,

$$\sum_{k=1}^{n}[(k+1)^3 - k^3] = \sum_{k=1}^{n}(3k^2 + 3k + 1) \tag{20}$$

Writing out the left side of (20) with the index running *down* from $k = n$ to $k = 1$, we have

The sum in (21) is an example of a ***telescoping sum***, since the cancellation of each of the two parts of an interior summand with parts of its neighboring summands allows the entire sum to collapse like a telescope.

$$\begin{aligned}\sum_{k=1}^{n}[(k+1)^3 - k^3] &= [(n+1)^3 - n^3] + \cdots + [4^3 - 3^3] + [3^3 - 2^3] + [2^3 - 1^3] \\ &= (n+1)^3 - 1\end{aligned} \tag{21}$$

Combining (21) and (20), and expanding the right side of (20) by using Theorem 6.4.1 and part (*a*) of this theorem yields

$$\begin{aligned}(n+1)^3 - 1 &= 3\sum_{k=1}^{n}k^2 + 3\sum_{k=1}^{n}k + \sum_{k=1}^{n}1 \\ &= 3\sum_{k=1}^{n}k^2 + 3\frac{n(n+1)}{2} + n\end{aligned}$$

So,

$$\begin{aligned}3\sum_{k=1}^{n}k^2 &= [(n+1)^3 - 1] - 3\frac{n(n+1)}{2} - n \\ &= (n+1)^3 - 3(n+1)\left(\frac{n}{2}\right) - (n+1) \\ &= \frac{n+1}{2}[2(n+1)^2 - 3n - 2] \\ &= \frac{n+1}{2}[2n^2 + n] = \frac{n(n+1)(2n+1)}{2}\end{aligned}$$

Thus,

$$\sum_{k=1}^{n}k^2 = \frac{n(n+1)(2n+1)}{6}$$

PROOF OF THE LIMIT COMPARISON TEST

C.7 THEOREM (*Theorem 10.5.4*). *Let $\sum a_k$ and $\sum b_k$ be series with positive terms and suppose that*

$$\rho = \lim_{k\to+\infty}\frac{a_k}{b_k}$$

If ρ is finite and $\rho > 0$, then the series both converge or both diverge.

PROOF. We need only show that $\sum b_k$ converges when $\sum a_k$ converges and that $\sum b_k$ diverges when $\sum a_k$ diverges, since the remaining cases are logical implications of these (why?). The idea of the proof is to apply the comparison test to $\sum a_k$ and suitable multiples of $\sum b_k$. For this purpose let ϵ be any positive number. Since

$$\rho = \lim_{k\to+\infty}\frac{a_k}{b_k}$$

it follows that eventually the terms in the sequence $\{a_k/b_k\}$ must be within ϵ units of ρ; that is, there is a positive integer K such that for $k \geq K$ we have

$$\rho - \epsilon < \frac{a_k}{b_k} < \rho + \epsilon$$

In particular, if we take $\epsilon = \rho/2$, then for $k \geq K$ we have

$$\frac{1}{2}\rho < \frac{a_k}{b_k} < \frac{3}{2}\rho \quad \text{or} \quad \frac{1}{2}\rho b_k < a_k < \frac{3}{2}\rho b_k$$

Thus, by the comparison test we can conclude that

$$\sum_{k=K}^{\infty} \frac{1}{2}\rho b_k \quad \text{converges if} \quad \sum_{k=K}^{\infty} a_k \quad \text{converges} \tag{22}$$

$$\sum_{k=K}^{\infty} \frac{3}{2}\rho b_k \quad \text{diverges if} \quad \sum_{k=K}^{\infty} a_k \quad \text{diverges} \tag{23}$$

But the convergence or divergence of a series is not affected by deleting finitely many terms or by multiplying the general term by a nonzero constant, so (22) and (23) imply that

$$\sum_{k=1}^{\infty} b_k \quad \text{converges if} \quad \sum_{k=1}^{\infty} a_k \quad \text{converges}$$

$$\sum_{k=1}^{\infty} b_k \quad \text{diverges if} \quad \sum_{k=1}^{\infty} a_k \quad \text{diverges}$$

■

PROOF OF THE RATIO TEST

C.8 THEOREM (*Theorem 10.5.5*). *Let $\sum u_k$ be a series with positive terms and suppose that*

$$\rho = \lim_{k \to +\infty} \frac{u_{k+1}}{u_k}$$

(*a*) *If $\rho < 1$, the series converges.*

(*b*) *If $\rho > 1$ or $\rho = +\infty$, the series diverges.*

(*c*) *If $\rho = 1$, the series may converge or diverge, so that another test must be tried.*

PROOF (*a*). The number ρ must be nonnegative since it is the limit of u_{k+1}/u_k, which is positive for all k. In this part of the proof we assume that $\rho < 1$, so that $0 \leq \rho < 1$.

We will prove convergence by showing that the terms of the given series are eventually less than the terms of a convergent geometric series. For this purpose, choose any real number r such that $0 < \rho < r < 1$. Since the limit of u_{k+1}/u_k is ρ, and $\rho < r$, the terms of the sequence $\{u_{k+1}/u_k\}$ must eventually be less than r. Thus, there is a positive integer K such that for $k \geq K$ we have

$$\frac{u_{k+1}}{u_k} < r \quad \text{or} \quad u_{k+1} < ru_k$$

This yields the inequalities

$$\begin{aligned} u_{K+1} &< ru_K \\ u_{K+2} &< ru_{K+1} < r^2 u_K \\ u_{K+3} &< ru_{K+2} < r^3 u_K \\ u_{K+4} &< ru_{K+3} < r^4 u_K \\ &\vdots \end{aligned} \tag{24}$$

But $0 < r < 1$, so

$$ru_K + r^2 u_K + r^3 u_K + \cdots$$

is a convergent geometric series. From the inequalities in (24) and the comparison test it follows that

$$u_{K+1} + u_{K+2} + u_{K+3} + \cdots$$

must also be a convergent series. Thus, $u_1 + u_2 + u_3 + \cdots + u_k + \cdots$ converges by Theorem 10.4.3(c).

PROOF (b). In this part we will prove divergence by showing that the limit of the general term is not zero. Since the limit of u_{k+1}/u_k is ρ and $\rho > 1$, the terms in the sequence $\{u_{k+1}/u_k\}$ must eventually be greater than 1. Thus, there is a positive integer K such that for $k \geq K$ we have

$$\frac{u_{k+1}}{u_k} > 1 \quad \text{or} \quad u_{k+1} > u_k$$

This yields the inequalities

$$\begin{aligned} &u_{K+1} > u_K \\ &u_{K+2} > u_{K+1} > u_K \\ &u_{K+3} > u_{K+2} > u_K \\ &u_{K+4} > u_{K+3} > u_K \\ &\quad\vdots \end{aligned} \tag{25}$$

Since $u_K > 0$, it follows from the inequalities in (25) that $\lim_{k\to+\infty} u_k \neq 0$, and thus the series $u_1 + u_2 + \cdots + u_k + \cdots$ diverges by part (a) of Theorem 10.4.1. The proof in the case where $\rho = +\infty$ is omitted.

PROOF (c). The divergent harmonic series and the convergent p-series with $p = 2$ both have $\rho = 1$ (verify), so the ratio test does not distinguish between convergence and divergence when $\rho = 1$. ■

PROOF OF THE REMAINDER ESTIMATION THEOREM

C.9 THEOREM (*Theorem 10.7.4*). *If the function f can be differentiated $n+1$ times on an interval I containing the number x_0, and if M is an upper bound for $|f^{(n+1)}(x)|$ on I, that is, $|f^{(n+1)}(x)| \leq M$ for all x in I, then*

$$|R_n(x)| \leq \frac{M}{(n+1)!}|x - x_0|^{n+1}$$

for all x in I.

PROOF. We are assuming that f can be differentiated $n+1$ times on an interval I containing the number x_0 and that

$$|f^{(n+1)}(x)| \leq M \tag{26}$$

for all x in I. We want to show that

$$|R_n(x)| \leq \frac{M}{(n+1)!}|x - x_0|^{n+1} \tag{27}$$

for all x in I, where

$$R_n(x) = f(x) - \sum_{k=0}^{n} \frac{f^{(k)}(x_0)}{k!}(x - x_0)^k \tag{28}$$

In our proof we will need the following two properties of $R_n(x)$:

$$R_n(x_0) = R_n'(x_0) = \cdots = R_n^{(n)}(x_0) = 0 \tag{29}$$

$$R_n^{(n+1)}(x) = f^{(n+1)}(x) \quad \text{for all } x \text{ in } I \tag{30}$$

These properties can be obtained by analyzing what happens if the expression for $R_n(x)$ in Formula (28) is differentiated j times and x_0 is then substituted in that derivative. If $j < n$, then the jth derivative of the summation in Formula (28) consists of a constant term $f^{(j)}(x_0)$ plus terms involving powers of $x - x_0$ (verify). Thus, $R_n^{(j)}(x_0) = 0$ for $j < n$, which proves all but the last equation in (29). For the last equation, observe that the nth derivative of the summation in (28) is the constant $f^{(n)}(x_0)$, so $R_n^{(n)}(x_0) = 0$. Formula (30) follows from the observation that the $(n+1)$-st derivative of the summation in (28) is zero (why?).

Now to the main part of the proof. For simplicity we will give the proof for the case where $x \geq x_0$ and leave the case where $x < x_0$ for the reader. It follows from (26) and (30) that $|R_n^{(n+1)}(x)| \leq M$, and hence

$$-M \leq R_n^{(n+1)}(x) \leq M$$

Thus,

$$\int_{x_0}^{x} -M\,dt \leq \int_{x_0}^{x} R_n^{(n+1)}(t)\,dt \leq \int_{x_0}^{x} M\,dt \tag{31}$$

However, it follows from (29) that $R_n^{(n)}(x_0) = 0$, so

$$\int_{x_0}^{x} R_n^{(n+1)}(t)\,dt = R_n^{(n)}(t)\Big]_{x_0}^{x} = R_n^{(n)}(x)$$

Thus, performing the integrations in (31) we obtain the inequalities

$$-M(x - x_0) \leq R_n^{(n)}(x) \leq M(x - x_0)$$

Now we will integrate again. Replacing x by t in these inequalities, integrating from x_0 to x, and using $R_n^{(n-1)}(x_0) = 0$ yields

$$-\frac{M}{2}(x - x_0)^2 \leq R_n^{(n-1)}(x) \leq \frac{M}{2}(x - x_0)^2$$

If we keep repeating this process, then after $n+1$ integrations we will obtain

$$-\frac{M}{(n+1)!}(x - x_0)^{n+1} \leq R_n(x) \leq \frac{M}{(n+1)!}(x - x_0)^{n+1}$$

which we can rewrite as

$$|R_n(x)| \leq \frac{M}{(n+1)!}(x - x_0)^{n+1}$$

This completes the proof of (27), since the absolute value signs can be omitted in that formula when $x \geq x_0$ (which is the case we are considering). ■

PROOF OF THE TWO-VARIABLE CHAIN RULE

C.10 THEOREM (*Theorem 14.5.1*). *If $x = x(t)$ and $y = y(t)$ are differentiable at t, and if $z = f(x, y)$ is differentiable at the point $(x(t), y(t))$, then $z = f(x(t), y(t))$ is differentiable at t and*

$$\frac{dz}{dt} = \frac{\partial z}{\partial x}\frac{dx}{dt} + \frac{\partial z}{\partial y}\frac{dy}{dt}$$

PROOF. Let Δx, Δy, and Δz denote the changes in x, y, and z, respectively, that correspond to a change of Δt in t. Then

$$\frac{dz}{dt} = \lim_{\Delta t \to 0} \frac{\Delta z}{\Delta t}, \qquad \frac{dx}{dt} = \lim_{\Delta t \to 0} \frac{\Delta x}{\Delta t}, \qquad \frac{dy}{dt} = \lim_{\Delta t \to 0} \frac{\Delta y}{\Delta t}$$

Since $f(x, y)$ is differentiable at $(x(t), y(t))$, it follows from (5) in Section 14.4 that

$$\Delta z = \frac{\partial z}{\partial x}\Delta x + \frac{\partial z}{\partial y}\Delta y + \epsilon(\Delta x, \Delta y)\sqrt{\Delta x^2 + \Delta y^2} \tag{32}$$

where the partial derivatives are evaluated at $(x(t), y(t))$ and where $\epsilon(\Delta x, \Delta y)$ satisfies $\epsilon(\Delta x, \Delta y) \to 0$ as $(\Delta x, \Delta y) \to (0, 0)$ and $\epsilon(0, 0) = 0$. Dividing both sides of (32) by Δt yields

$$\frac{\Delta z}{\Delta t} = \frac{\partial z}{\partial x}\frac{\Delta x}{\Delta t} + \frac{\partial z}{\partial y}\frac{\Delta y}{\Delta t} + \frac{\epsilon(\Delta x, \Delta y)\sqrt{\Delta x^2 + \Delta y^2}}{\Delta t} \tag{33}$$

Since

$$\begin{aligned}\lim_{\Delta t \to 0} \frac{\sqrt{\Delta x^2 + \Delta y^2}}{|\Delta t|} &= \lim_{\Delta t \to 0} \sqrt{\left(\frac{\Delta x}{\Delta t}\right)^2 + \left(\frac{\Delta y}{\Delta t}\right)^2} = \sqrt{\left(\lim_{\Delta t \to 0} \frac{\Delta x}{\Delta t}\right)^2 + \left(\lim_{\Delta t \to 0} \frac{\Delta y}{\Delta t}\right)^2} \\ &= \sqrt{\left(\frac{dx}{dt}\right)^2 + \left(\frac{dy}{dt}\right)^2}\end{aligned}$$

we have

$$\begin{aligned}\lim_{\Delta t \to 0} \left| \frac{\epsilon(\Delta x, \Delta y)\sqrt{\Delta x^2 + \Delta y^2}}{\Delta t} \right| &= \lim_{\Delta t \to 0} \frac{|\epsilon(\Delta x, \Delta y)|\sqrt{\Delta x^2 + \Delta y^2}}{|\Delta t|} \\ &= \lim_{\Delta t \to 0} |\epsilon(\Delta x, \Delta y)| \cdot \lim_{\Delta t \to 0} \frac{\sqrt{\Delta x^2 + \Delta y^2}}{|\Delta t|} \\ &= 0 \cdot \sqrt{\left(\frac{dx}{dt}\right)^2 + \left(\frac{dy}{dt}\right)^2} = 0\end{aligned}$$

Therefore,

$$\lim_{\Delta t \to 0} \frac{\epsilon(\Delta x, \Delta y)\sqrt{\Delta x^2 + \Delta y^2}}{\Delta t} = 0$$

Taking the limit as $\Delta t \to 0$ of both sides of (33) then yields the equation

$$\frac{dz}{dt} = \frac{\partial z}{\partial x}\frac{dx}{dt} + \frac{\partial z}{\partial y}\frac{dy}{dt}$$

■

ANSWERS TO ODD-NUMBERED EXERCISES

Exercise Set 1.1 (Page 12)

1. **(a)** $-2.9, -2.0, 2.35, 2.9$ **(b)** none **(c)** 0 **(d)** $-1.75 \le x \le 2.15$ **(e)** $y_{\max} = 2.8$ at $x = -2.6$; $y_{\min} = -2.2$ at $x = 1.2$
3. **(a)** yes **(b)** yes **(c)** no **(d)** no
5. **(a)** 1943 **(b)** 1960; 4200 **(c)** no, you need the year's population **(d)** war, marketing **(e)** news of health risk, social pressure, anti-smoking campaigns, increased taxation
7. **(a)** 1999, about \$43,400 **(b)** 1985, \$37,000 **(c)** second year
9. **(a)** -2; 10; 10; 25; 4; $27t^2 - 2$ **(b)** 0; 4; -4; 6; $2\sqrt{2}$; $f(3t) = 1/3t$ for $t > 1$ and $f(3t) = 6t$ for $t \le 1$
11. **(a)** $x \ne 3$ **(b)** $x \le -\sqrt{3}, x \ge \sqrt{3}$ **(c)** $(-\infty, +\infty)$ **(d)** $x \ne 0$ **(e)** $x \ne (2n + \frac{1}{2})\pi, n = 0, \pm 1, \pm 2, \ldots$
13. **(a)** $x \le 3$ **(b)** $-2 \le x \le 2$ **(c)** $x \ge 0$ **(d)** all x **(e)** all x
15. **(a)** no; war, pestilence, flood, earthquakes **(b)** decreases for 8 hours, takes a jump upward, and repeats
17.

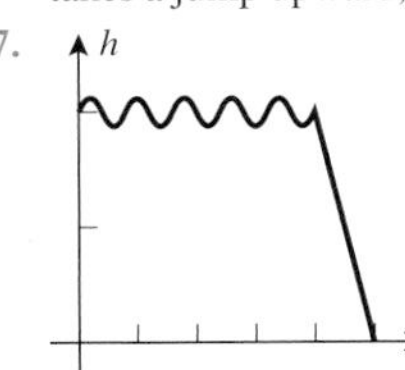

19. **(a)** 2, 4 **(b)** none **(c)** $x \le 2$; $4 \le x$ **(d)** $y_{\min} = -1$; no maximum
21. $h = L(1 - \cos\theta)$
23. **(a)** $f(x) = \begin{cases} 2x + 1, & x < 0 \\ 4x + 1, & x \ge 0 \end{cases}$ **(b)** $g(x) = \begin{cases} 1 - 2x, & x < 0 \\ 1, & 0 \le x \le 1 \\ 2x - 1, & x \ge 1 \end{cases}$
25. **(a)** $V = (8 - 2x)(15 - 2x)x$ **(b)** $0 < x < 4$ **(c)** $0 < V \le 90$, approximately

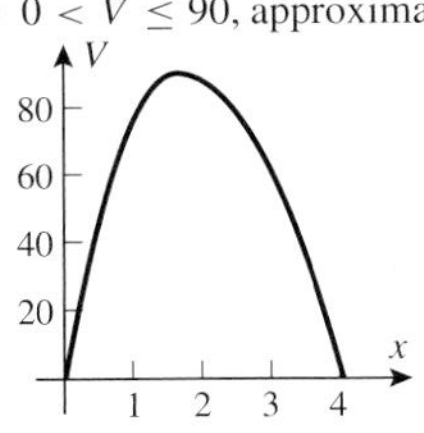

(d) V appears to be maximal for $x \approx 1.7$

27. **(a)** $L = x + 2y$ **(b)** $L = x + 2000/x$ **(c)** $0 < x \le 100$ **(d)** $x \approx 45$ ft, $y \approx 22$ ft
29. **(a)** $r \approx 3.4, h \approx 13.8$ **(b)** taller **(c)** $r \approx 3.1$ cm, $h \approx 16.0$ cm, $C \approx 4.76$ cents
31. **(i)** $x = 1, -2$ **(ii)** $g(x) = x + 1$, all x
33. **(a)** 25°F **(b)** 13°F **(c)** 5°F 35. 15°F

Exercise Set 1.2 (Page 24)

1. (e) 3. (b), (c) 5. $[-3, 3] \times [0, 5]$
9. $[-5, 14] \times [-60, 40]$

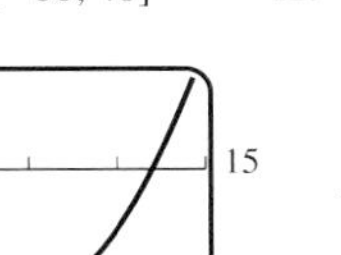

11. $[-0.1, 0.1] \times [-3, 3]$

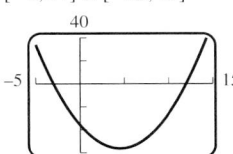

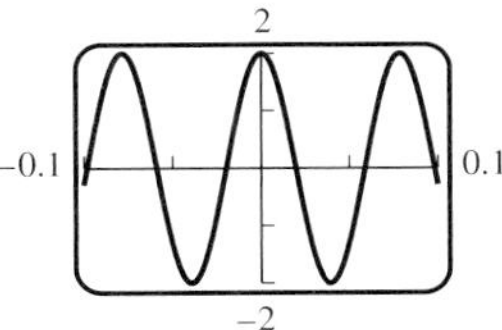

13. $[-400, 1050] \times [-1500000, 10000]$

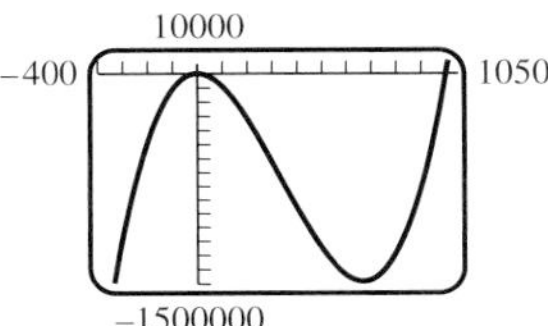

15. $[-2, 2] \times [-20, 20]$

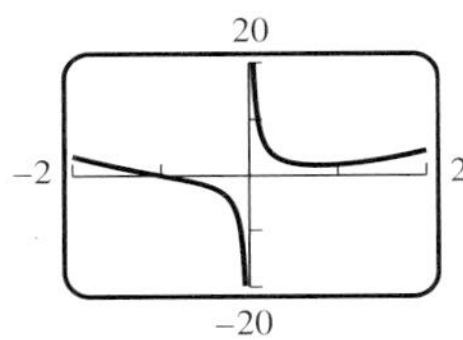

19. **(a)** $f(x) = \sqrt{16 - x^2}$ **(b)** $f(x) = -\sqrt{16 - x^2}$ **(e)** no
21. **(a)**

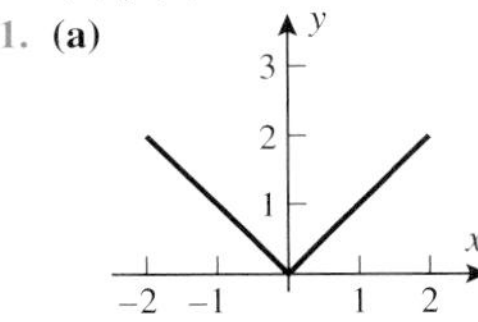

(b)

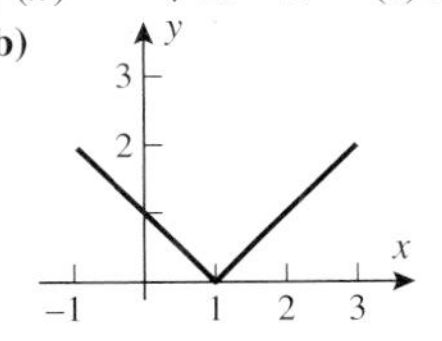

(c)

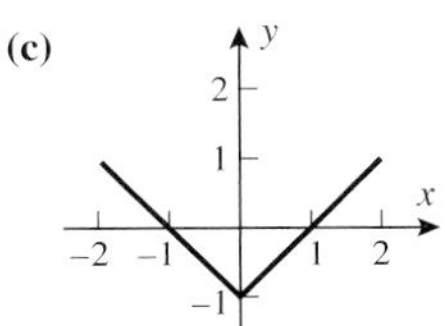

(d)

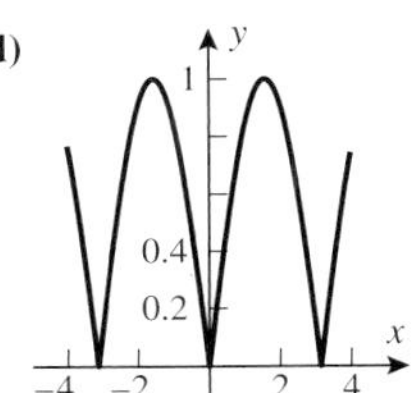

(e)

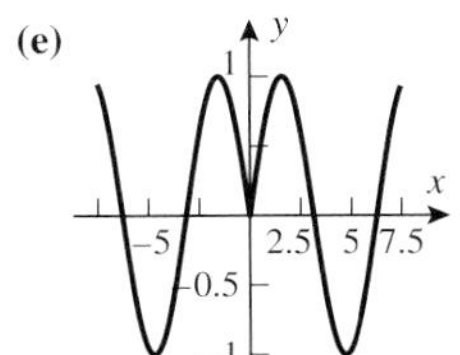

(f)

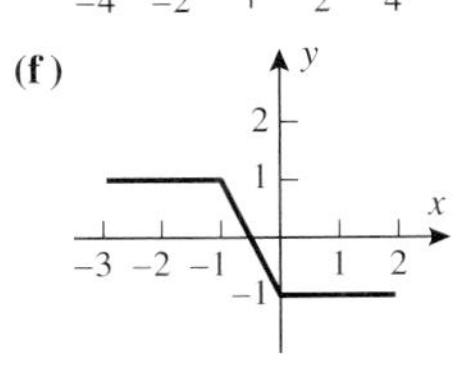

23. The graph of $y = |f(x)|$ consists of those parts of the graph of $y = f(x)$ which lie above the x-axis, together with the reflections across the x-axis of those parts of the graph of $y = f(x)$ which lie below the x-axis.

25. (a)
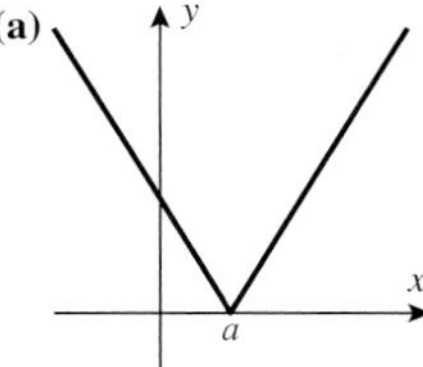

(b)
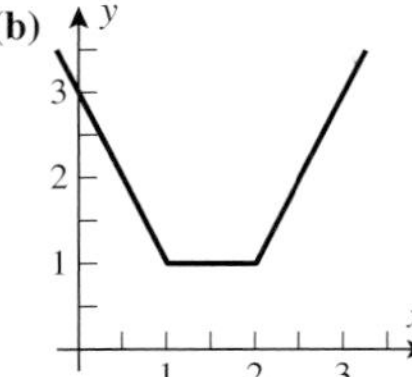

27.
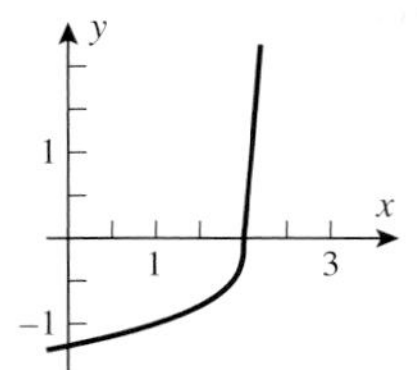

31. (a)
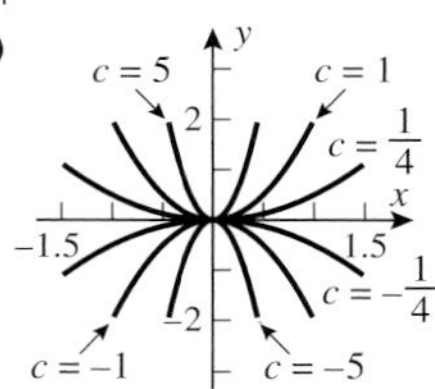

The graph is stretched in the vertical direction, and reflected across the x-axis if $c < 0$.

(b)
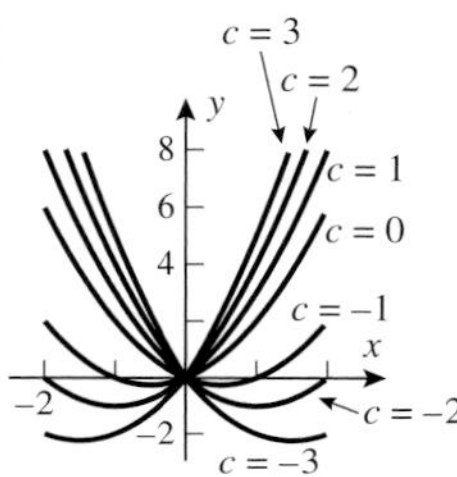

The graph is translated so its vertex is on the parabola $y = -x^2$

(c) The graph is translated vertically.

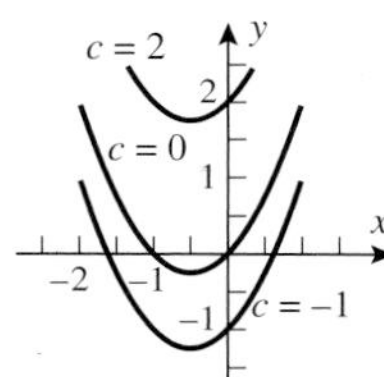

33. $y = x$, 10, -4π, 4π, −10, $y = -x$

▶ Exercise Set 1.3 (Page 36)

1. (a)
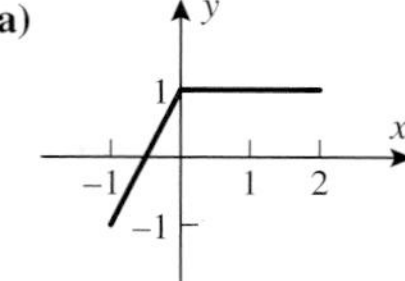

(b)
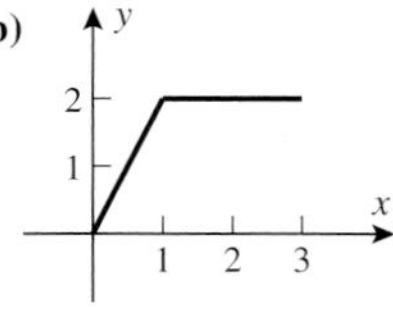

(c)
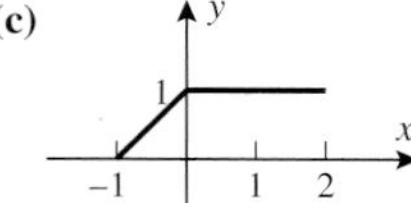

(d)
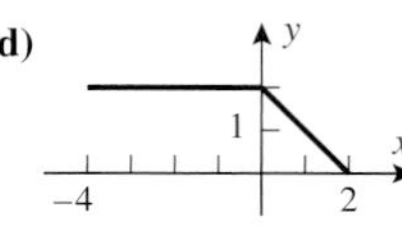

3. (a)
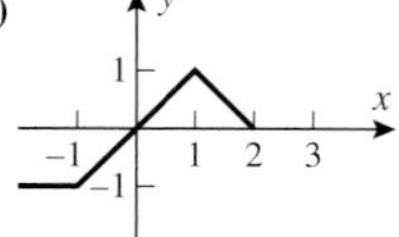

(b)
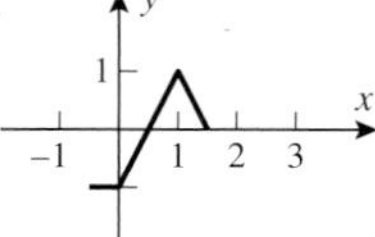

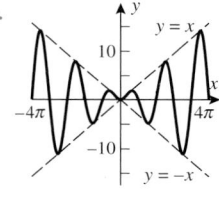

(c)
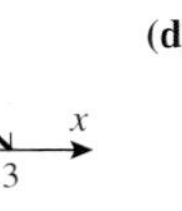

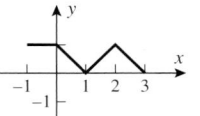

(d)
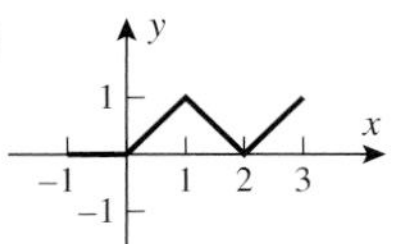

5.
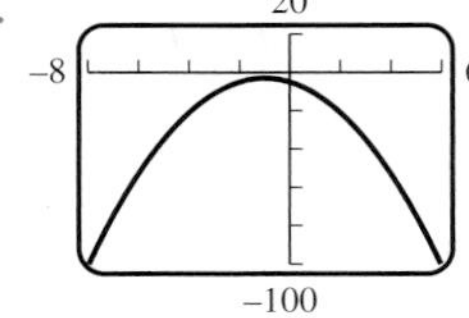

7.
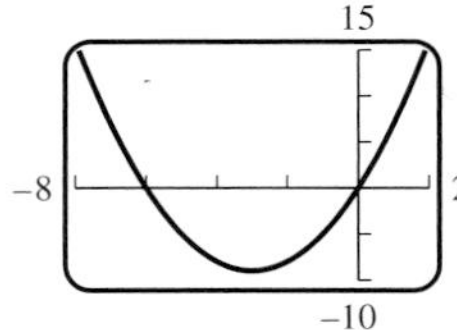

9.
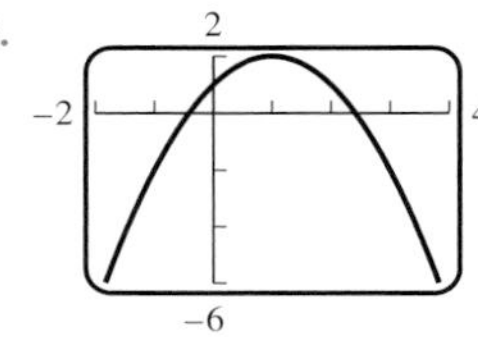

11.
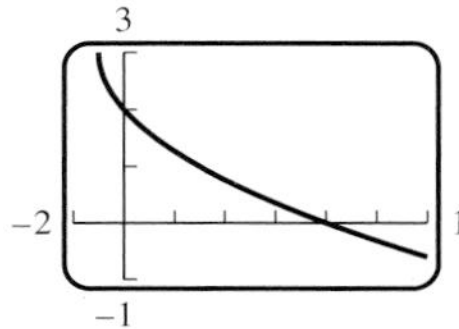

13.
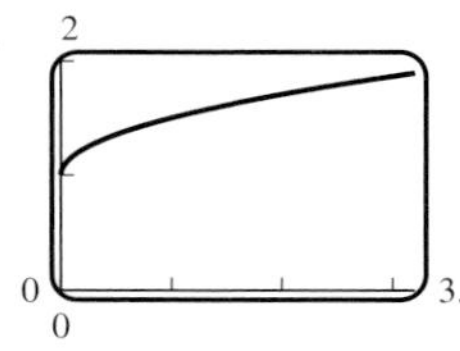

15.
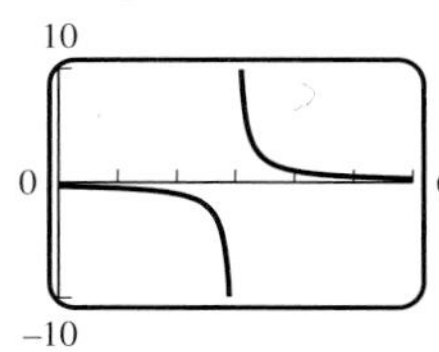

17.
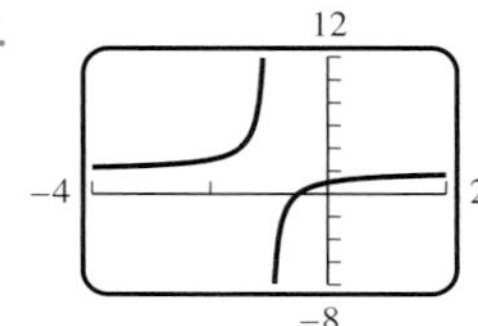

19.
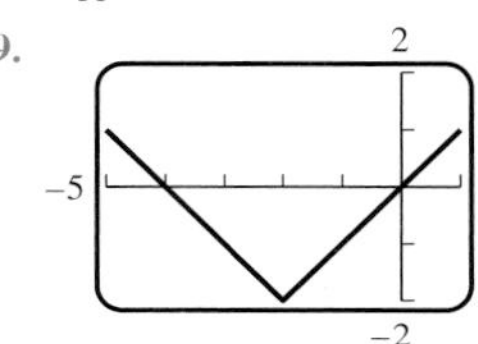

21.
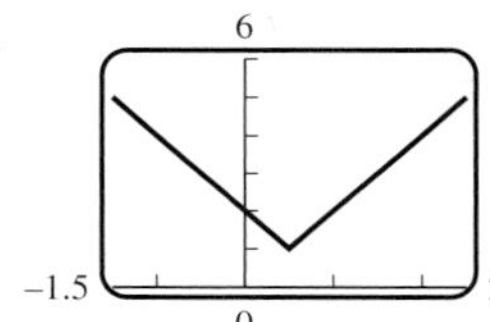

23.
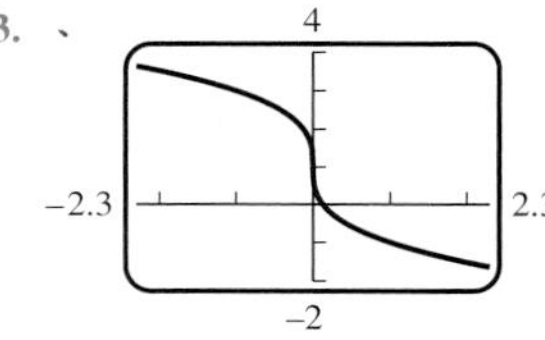

25.
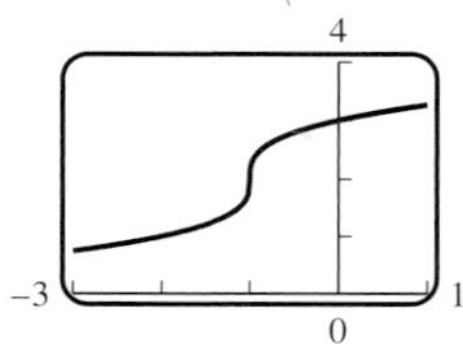

27. (a)
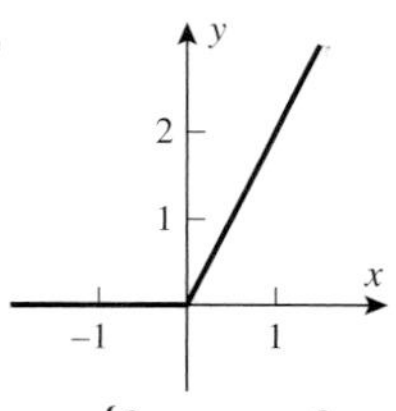

(b) $y = \begin{cases} 0, & x \le 0 \\ 2x, & x > 0 \end{cases}$

29. $3\sqrt{x-1}, x \ge 1$; $\sqrt{x-1}, x \ge 1$; $2x - 2, x \ge 1$; $2, x > 1$

31. (a) 3 (b) 9 (c) 2 (d) 2

33. (a) $t^4 + 1$ (b) $t^2 + 4t + 5$ (c) $x^2 + 4x + 5$ (d) $\dfrac{1}{x^2} + 1$
(e) $x^2 + 2xh + h^2 + 1$ (f) $x^2 + 1$ (g) $x + 1$ (h) $9x^2 + 1$

35. $1 - x, x \le 1$; $\sqrt{1 - x^2}, |x| \le 1$

37. $\dfrac{1}{1-2x}, x \ne \dfrac{1}{2}, 1$; $-\dfrac{1}{2x} - \dfrac{1}{2}, x \ne 0, 1$ 39. $x^{-6} + 1$

41. (a) $g(x) = \sqrt{x}, h(x) = x + 2$ (b) $g(x) = |x|, h(x) = x^2 - 3x + 5$

43. (a) $g(x) = x^2, h(x) = \sin x$ (b) $g(x) = 3/x, h(x) = 5 + \cos x$

45. (a) $f(x) = x^3, g(x) = 1 + \sin x, h(x) = x^2$
(b) $f(x) = \sqrt{x}, g(x) = 1 - x, h(x) = \sqrt[3]{x}$

47.

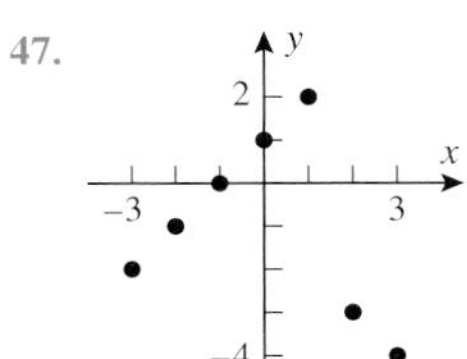

49.

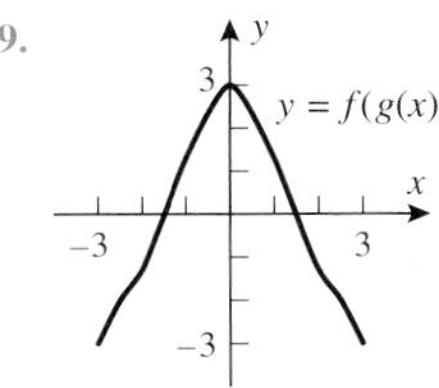

51. $\pm 1.5, \pm 2$ **53.** $3w + 3x, 6x + 3h$ **55.** $-\frac{1}{xw}, -\frac{1}{x(x+h)}$

57. f = neither, g = odd, h = even

59. **(a)**

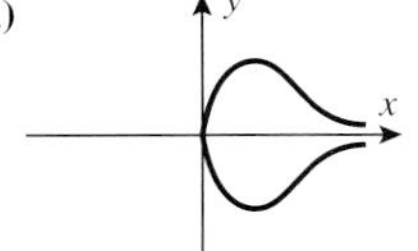

(b)

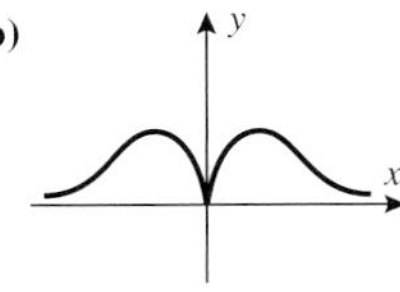

(c)

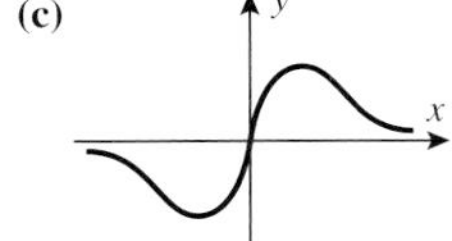

61. **(a)** even **(b)** odd **(c)** odd **(d)** neither

63. **(a)** even **(b)** odd **(c)** even **(d)** neither **(e)** odd **(f)** even

67. **(a)** y-axis
(b) origin
(c) x-axis, y-axis, origin

69.

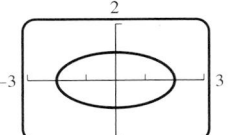

73. **(a)**

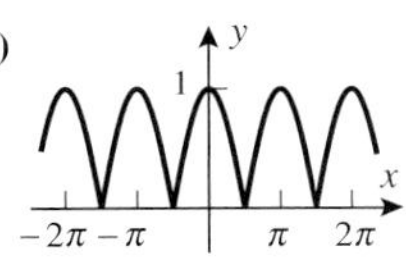

(b)

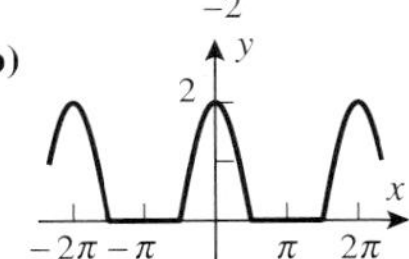

75. yes; $f(x) = x^k, g(x) = x^n$

▶ Exercise Set 1.4 (Page 48)

1. **(a)** $y = 3x + b$ **(c)**
(b) $y = 3x + 6$

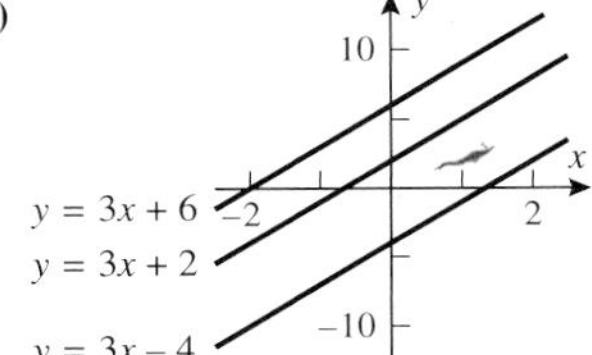

3. **(a)** $y = mx + 2$ **(c)**
(b) $y = -x + 2$

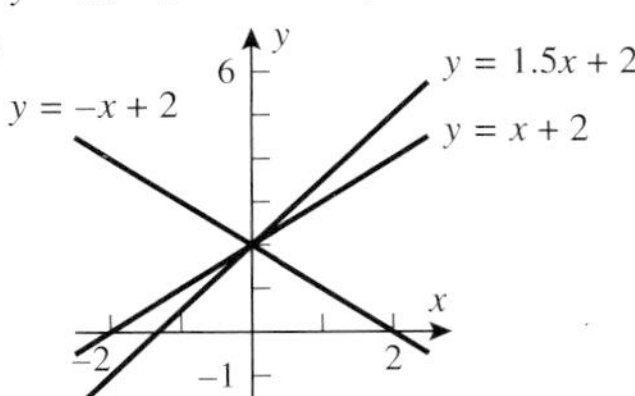

5. $y = \pm\frac{9 - x_0 x}{\sqrt{9 - x_0^2}}$ **7.**

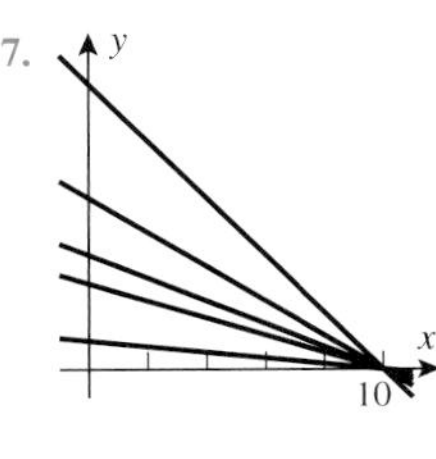

y-intercepts represent current value of item being depreciated.

9. **(a)** slope: -1

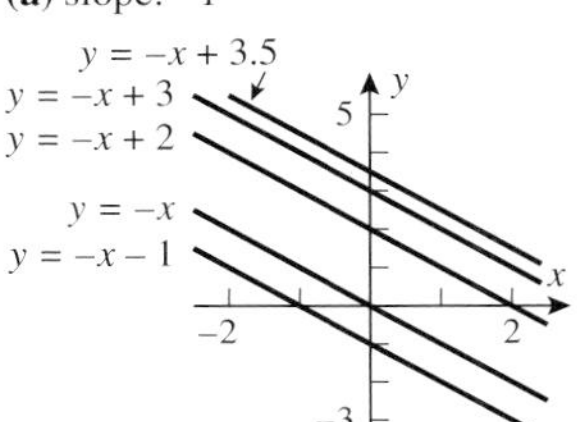

(b) y-intercept: $y = -1$

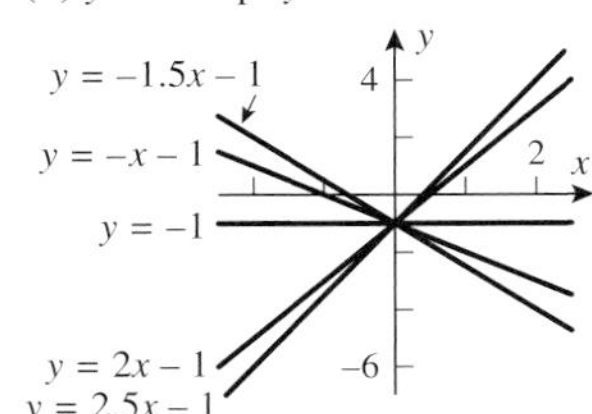

(c) pass through $(-4, 2)$

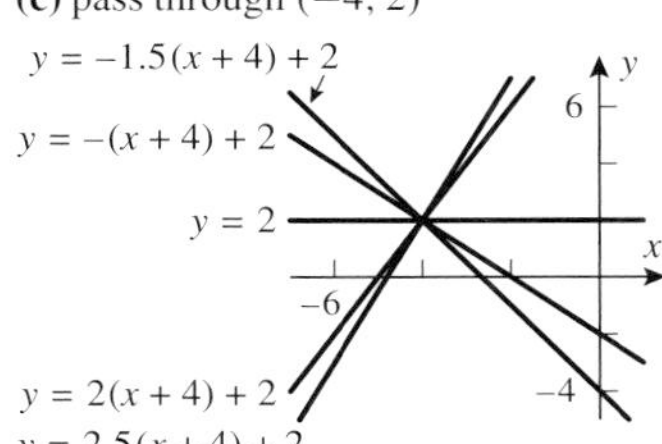

(d) x-intercept: $x = 1$

11. **(a)** VI
(b) IV
(c) III
(d) V
(e) I
(f) II

13. **(a)**

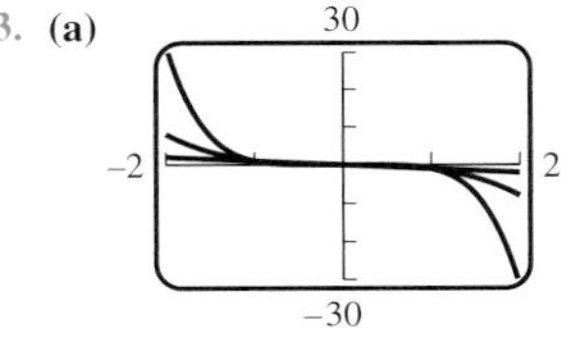

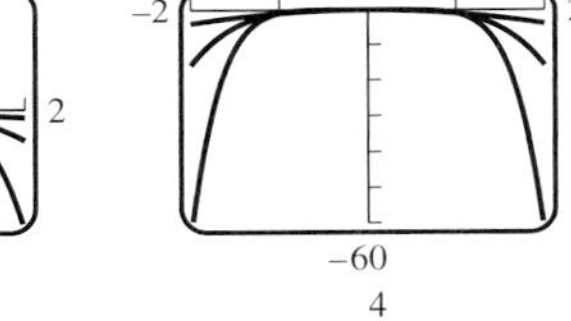

(b)

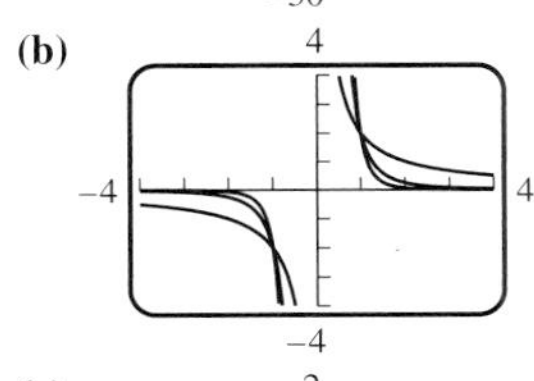

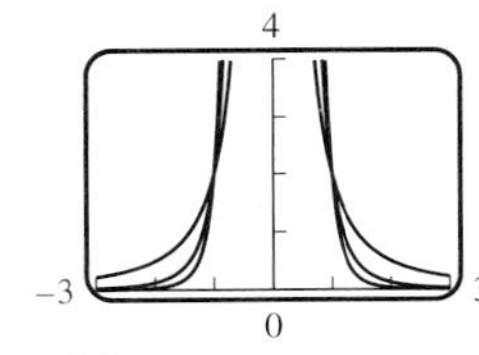

(c)

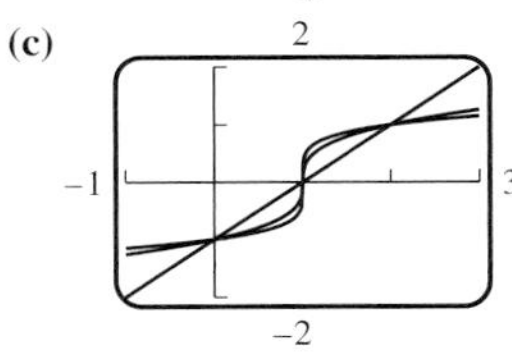

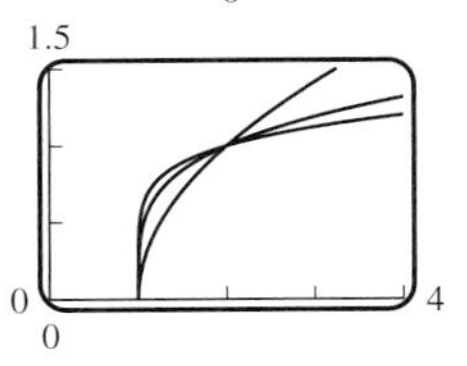

15. **(a)**

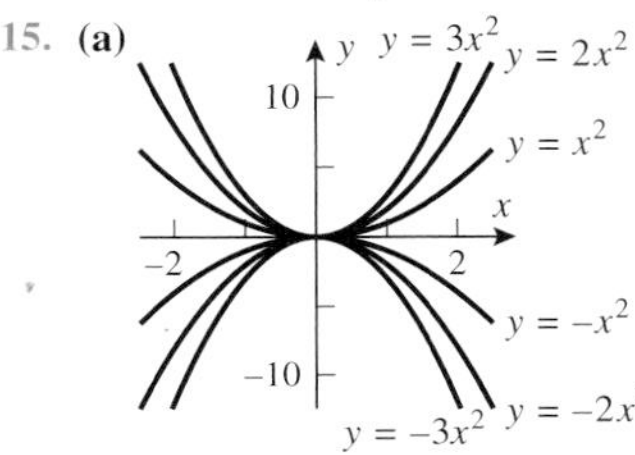

(b)

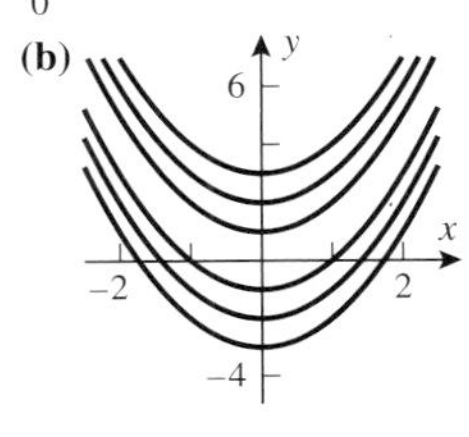

(c)

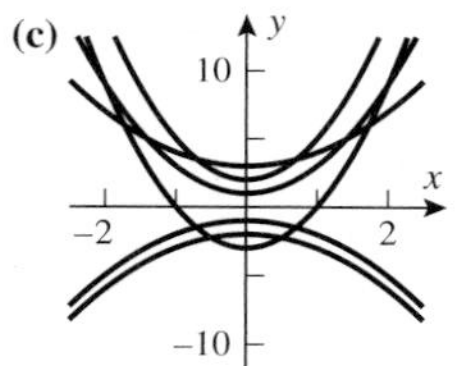

17. (a)

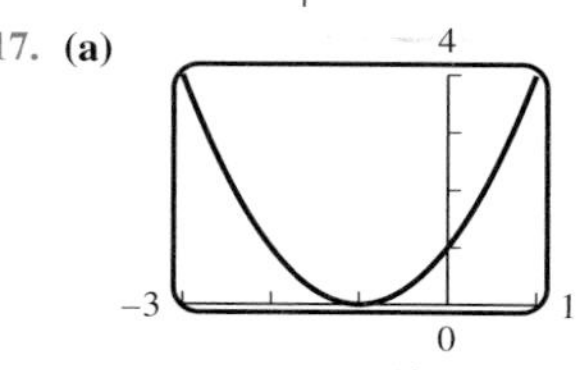

(b)

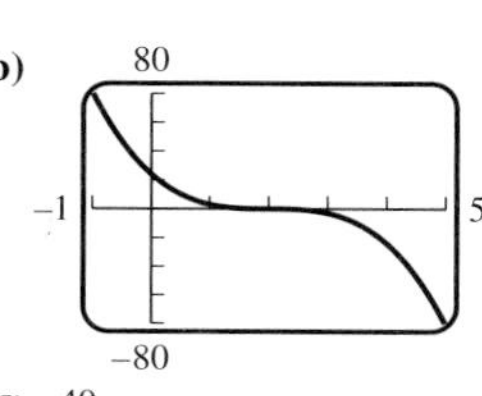

(c)

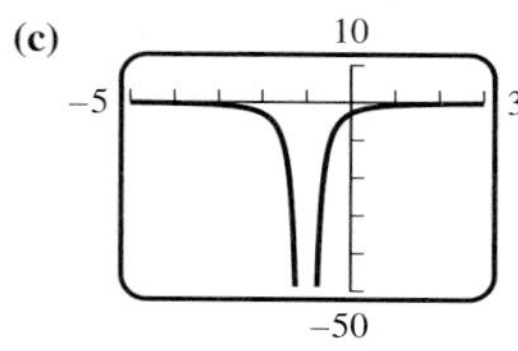

(d)

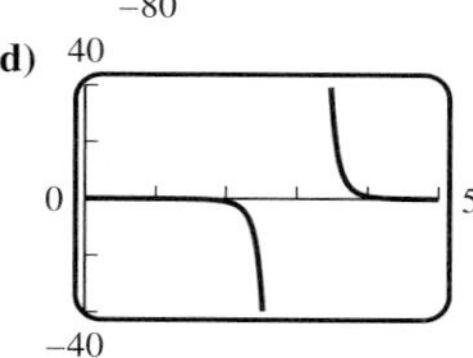

19. (a)

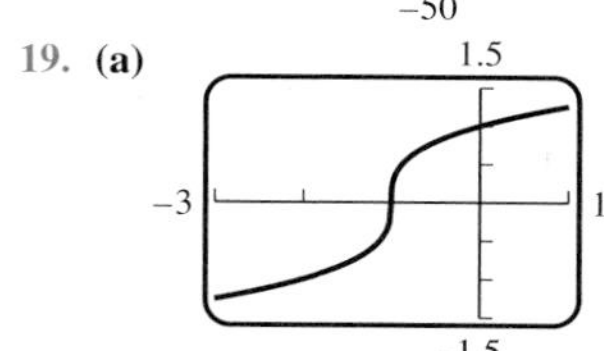

(b)

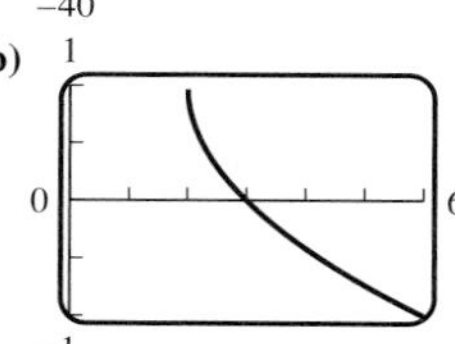

(c)

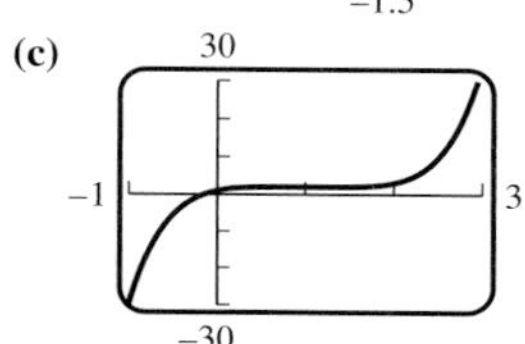

(d)

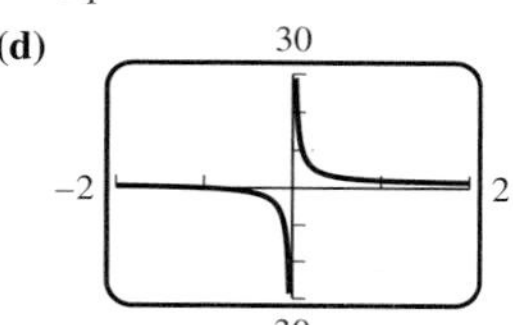

21.

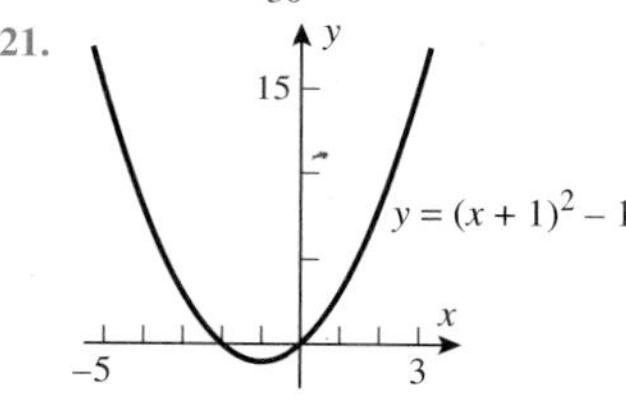

23. **(a)** newton-meters (N·m) **(b)** 20 N·m

(c)

V (L)	0.25	0.5	1.0	1.5	2.0
P (N/m²)	80×10^3	40×10^3	20×10^3	13.3×10^3	10×10^3

(d)

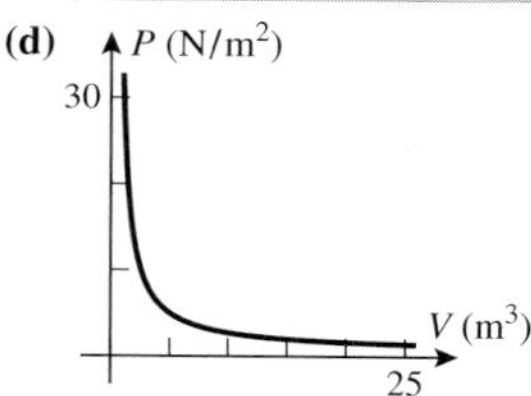

25. **(a)** $k = 0.000045$ N·m² **(b)** 0.000005 N **(d)** The force becomes infinite; the force tends to zero.

(c)

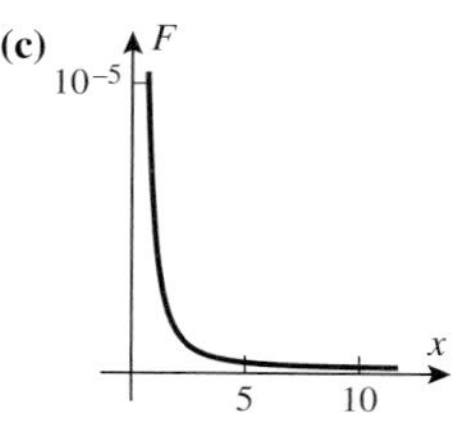

27. **(a)** II; $y = 1, x = -1, 2$ **(b)** I; $y = 0, x = -2, 3$ **(c)** IV; $y = 2$ **(d)** III; $y = 0, x = -2$

29. **(a)** $y = 3\sin(x/2)$ **(b)** $y = 4\cos 2x$ **(c)** $y = -5\sin 4x$

31. **(a)** $y = \sin[x + (\pi/2)]$ **(b)** $y = 3 + 3\sin(2x/9)$ **(c)** $y = 1 + 2\sin[2(x - \frac{\pi}{4})]$

33. **(a)** $3, \pi/2$ **(b)** 2, 2

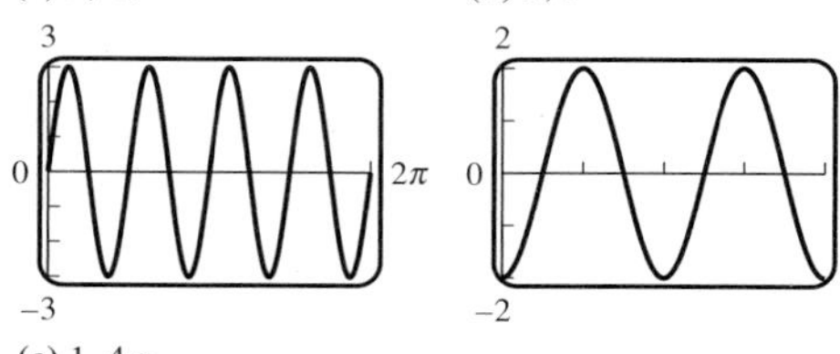

(c) $1, 4\pi$

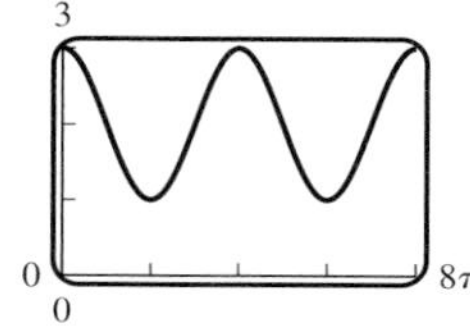

35. **(b)** $x = 2\sqrt{2}\sin\left(2\pi t + \frac{\pi}{3}\right)$

Exercise Set 1.5 (Page 58)

1. **(a)** yes **(b)** no **(c)** yes **(d)** no
3. **(a)** yes **(b)** yes **(c)** no **(d)** yes **(e)** no **(f)** no
5. **(a)** yes **(b)** no
7. **(a)** $8, -1, 0$ **(b)** $[-2, 2], [-8, 8]$ **(c)**
9. $\frac{1}{7}(x + 6)$
11. $\sqrt[3]{(x + 5)/3}$
13. $(x^3 + 1)/2$
15. $-\sqrt{3/x}$
17. $\begin{cases} (5/2) - x, & x > 1/2 \\ 1/x, & 0 < x \le 1/2 \end{cases}$
19. $x^{1/4} - 2$ for $x \ge 16$
21. $\frac{1}{2}(3 - x^2)$ for $x \le 0$
23. $\frac{1}{10}(1 + \sqrt{1 - 20x})$ for $x \le -4$
25. **(a)** $y = (6.214 \times 10^{-4})x$ **(b)** $x = \dfrac{10^4}{6.214}y$ **(c)** how many meters in y miles
27. **(b)** symmetric about the line $y = x$
29. 10

Exercise Set 1.6 (Page 66)

1. II
3. $S = 0.6414t - 1054.52, 0.575435$
5. **(a)** $p = 0.0146T + 3.98, 0.9999$ **(b)** 3.25 atm **(c)** $\approx -272°$C
7. **(a)** $R = 0.00723T + 1.55$ **(b)** $\approx -214°$C
9. **(a)** $S = 0.50179\omega - 0.00643$ **(b)** ≈ 16.0 lb
11. **(a)** $R = 0.2087h - 1.0549, 0.842333$ **(b)**

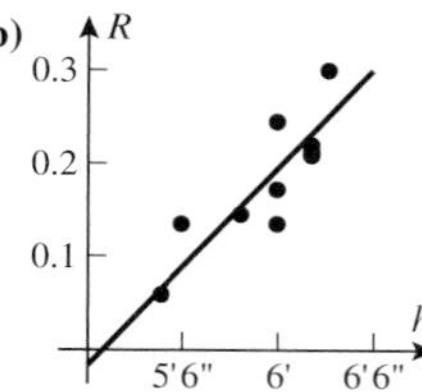

13. **(a)** $3b^2 + 2b + 1$ **(b)** $y = x - 1/3$
15. The linear regression line joins the third point to the midpoint of the vertical line segment between the other two points.
17. **(a)** 181.8 km/s/Mly **(b)** 1.492×10^{10} years **(c)** increase
19. $T = 849.5 + 143.5\sin\left[\frac{\pi}{183}t - \frac{\pi}{2}\right]$
21. $t = 0.445\sqrt{d}$

Exercise Set 1.7 (Page 76)

1. **(a)** $y = x + 2$

(c)

t	0	1	2	3	4	5
x	−1	0	1	2	3	4
y	1	2	3	4	5	6

(d)

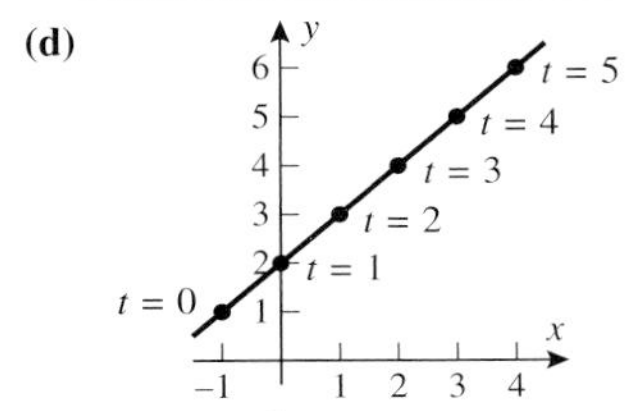

3.

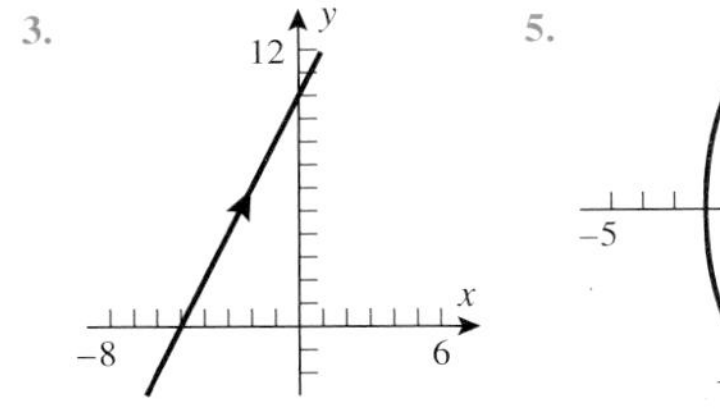

5.

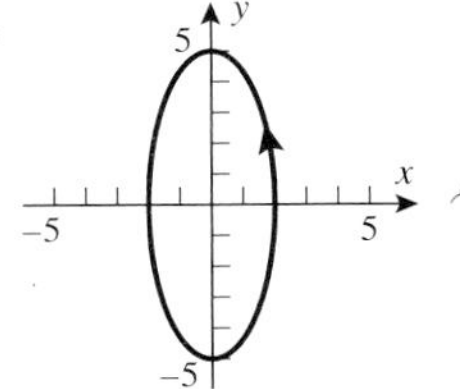

7.

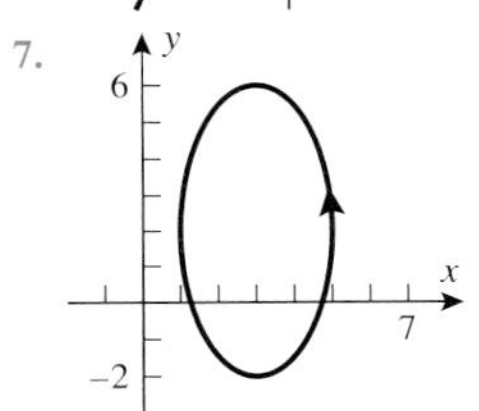

9.

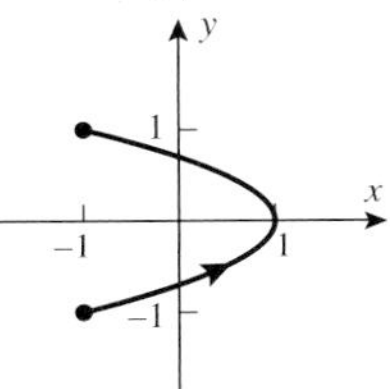

11.

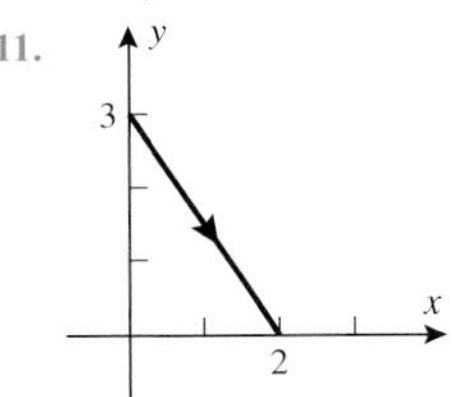

13. $x = 5\cos t,\ y = -5\sin t,\ 0 \le t \le 2\pi$ 15. $x = 2,\ y = t$

17. $x = t^2,\ y = t,\ -1 \le t \le 1$

19. **(a)**

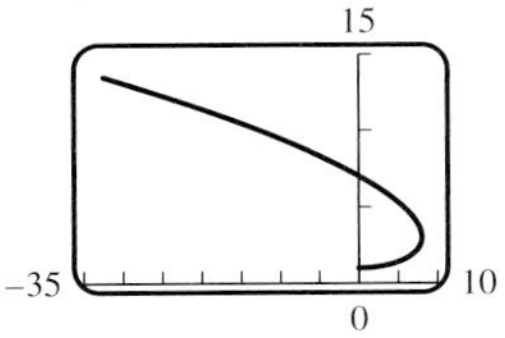

(b)

t	0	1	2	3	4	5
x	0	5.5	8	4.5	−8	−32.5
y	1	1.5	3	5.5	9	13.5

(c) $t = 0, 2\sqrt{3}$ **(d)** $0 < t < 2\sqrt{2}$ **(e)** 2

21. **(a)** **(b)**

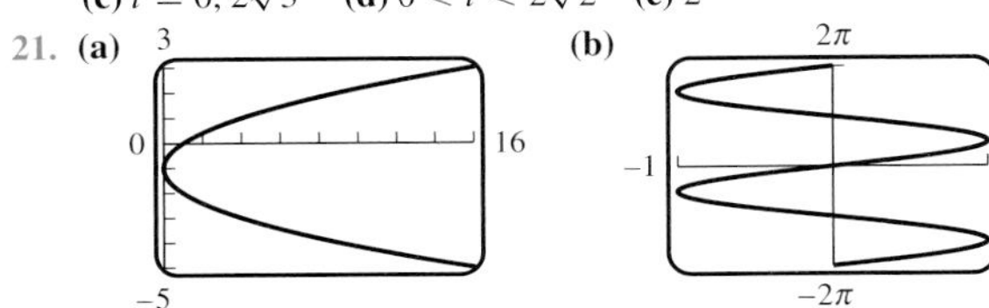

23. **(c)** $x = 1 + t,\ y = -2 + 6t,\ 0 \le t \le 1$
(d) $x = 2 - t,\ y = 4 - 6t,\ 0 \le t \le 1$

25. **(b)** $\frac{1}{2}$ **(c)** $\frac{3}{4}$ 27. **(a)** IV **(b)** II **(c)** V **(d)** VI **(e)** III **(f)** I

29. As t varies, the point moves in both directions along part of the parabola $y = x^2$.

31. **(b)** 33.

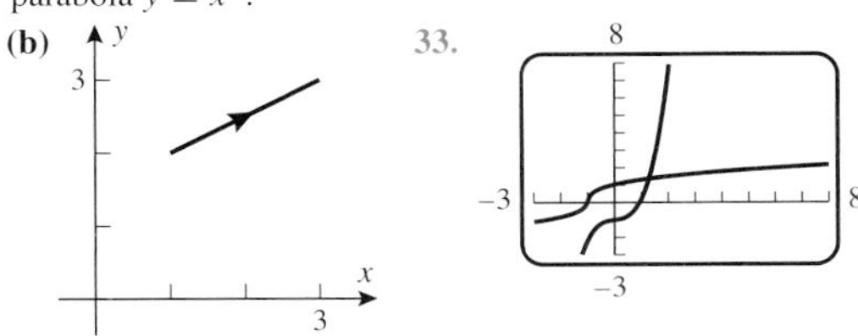

35. 37.

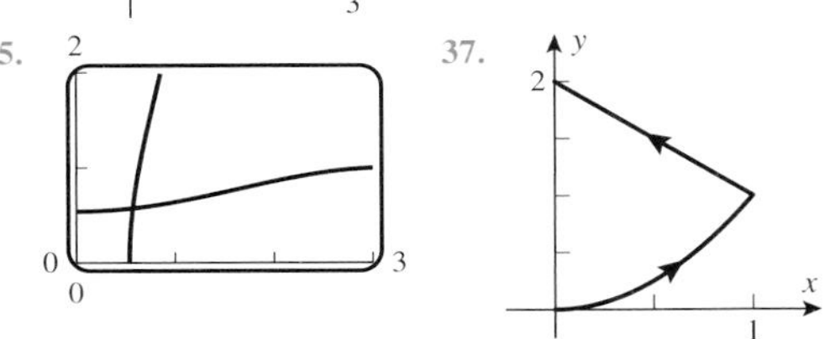

39. **(a)** $x = 4\cos t,\ y = 3\sin t$ **(b)** $x = -1 + 4\cos t,\ y = 2 + 3\sin t$

41. **(a)** $x = 400\sqrt{2}t,\ y = 400\sqrt{2}t - 4.9t^2$ **(b)** 16,326.53 m
(c) 65,306.12 m

43. **(a)** ellipses with fixed center, varying axes of symmetry **(b)** (assume $a \ne 0, b \ne 0$) ellipses with varying center, fixed axes of symmetry
(c) circles of radius 1 with centers on line $y = x - 1$

45. **(a)**

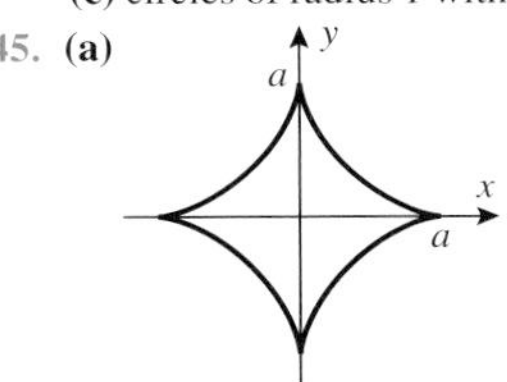

Chapter 1. Review Exercises (Page 80)

1. 3.

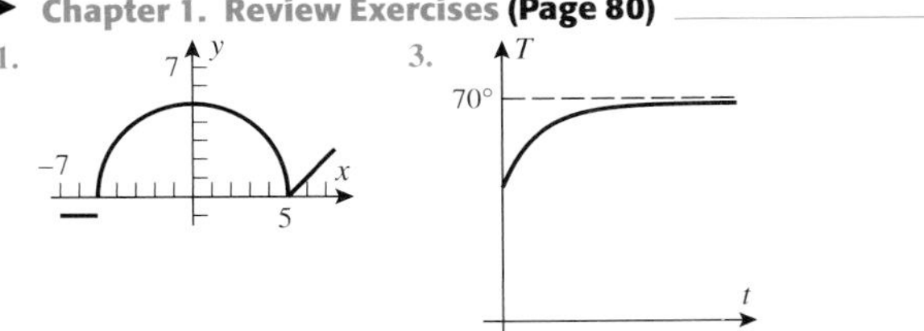

5. **(a)** $C = 5x^2 + (64/x)$ **(b)** $x > 0$ 9.

7. **(a)** $V = (6 - 2x)(5 - x)x$ ft³
(b) $0 < x < 3$
(c) 3.57 ft × 3.79 ft × 1.21 ft

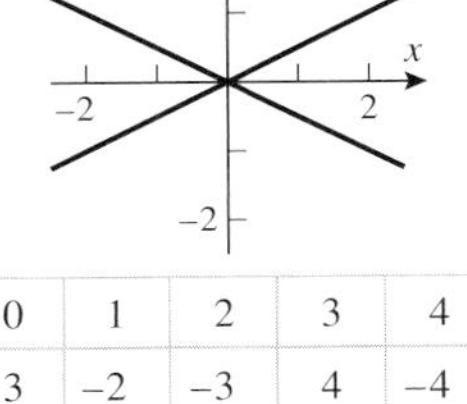

11.

x	−4	−3	−2	−1	0	1	2	3	4
$f(x)$	0	−1	2	1	3	−2	−3	4	−4
$g(x)$	3	2	1	−3	−1	−4	4	−2	0
$(f \circ g)(x)$	4	−3	−2	−1	1	0	−4	2	3
$(g \circ f)(x)$	−1	−3	4	−4	−2	1	2	0	3

13. 0, −2 15. $1/(2 - x^2),\ x \ne \pm 1, \pm\sqrt{2}$

17. **(a)** odd **(b)** even **(c)** neither **(d)** even

19. **(a)** circles of radius 1 centered on the parabola $y = x^2$
(b) parabolas that open up with vertices on the line $y = x/2$

21. **(a)**

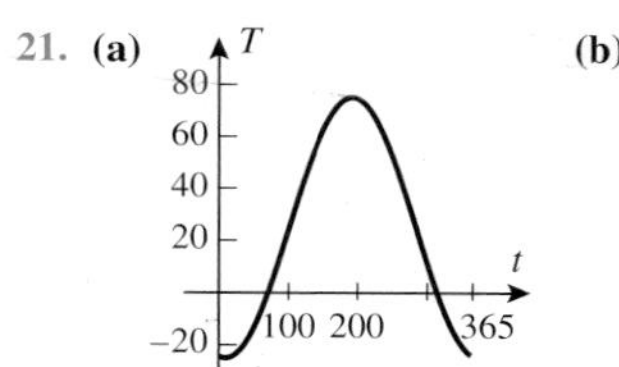

(b) January 11 **(c)** 122

23. $A: (-\frac{2}{3}\pi, 1-\sqrt{3})$; $B: (\frac{1}{3}\pi, 1+\sqrt{3})$; $C: (\frac{2}{3}\pi, 1+\sqrt{3})$; $D: (\frac{5}{3}\pi, 1-\sqrt{3})$

27. **(a)**

1.90	1.92	1.94	1.96	1.98	2.00
3.4161	3.4639	3.5100	3.5543	3.5967	3.6372

2.02	2.04	2.06	2.08	2.10
3.6756	3.7119	3.7459	3.7775	3.8068

(b) $y = 1.9590x - 0.29101$ **(c)** $y = 1.9590x - 0.2808$

(d)

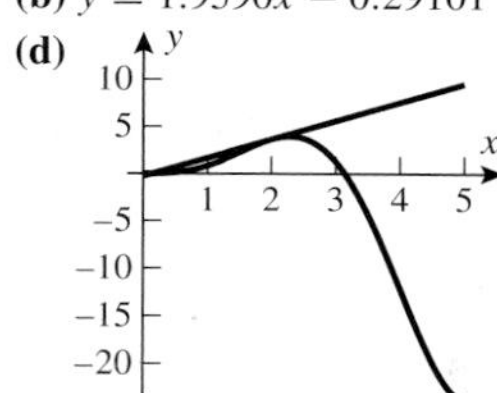

29. $T = 0.537 + 0.495\sin\left[\frac{\pi}{6}(t-6.5)\right]$

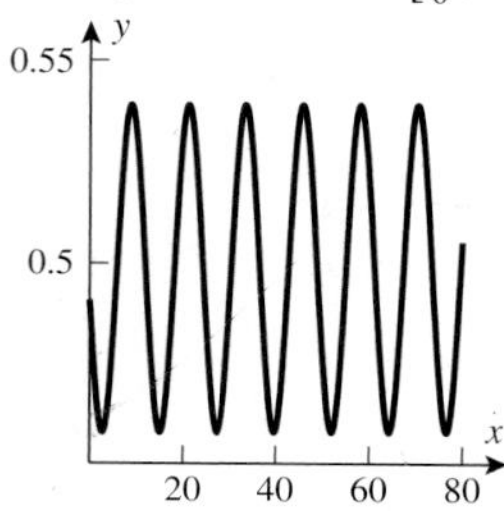

31. $x = \sqrt{2}\cos t$, $y = -\sqrt{2}\sin t$, $0 \le t \le \frac{3\pi}{2}$

33.

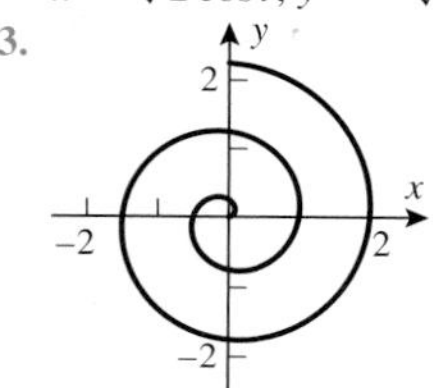

Exercise Set 2.1 (Page 93)

1. **(a)** 0 **(b)** 0 **(c)** 0 **(d)** 3 3. **(a)** $-\infty$ **(b)** $-\infty$ **(c)** $-\infty$ **(d)** 1
5. for all $x_0 \neq -4$.
7.

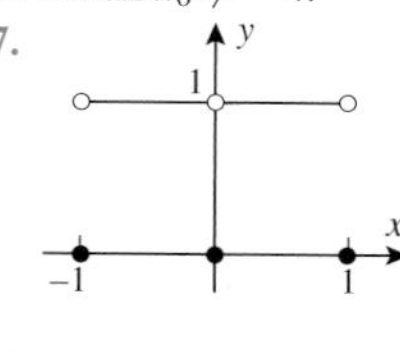

9.

11.

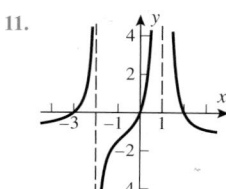

13. **(a)** $\frac{1}{3}$ **(b)** $+\infty$ **(c)** $-\infty$
15. **(a)** 3 **(b)** does not exist
17. $y = -2x - 1$
19. $y = 4x - 3$

23. **(a)** Catastrophic substraction when the x-interval is small (the size depending on the calculating utility) **(c)** no
25. **(a)** rest length **(b)** 0. As speed approaches c, length shrinks to zero.

Exercise Set 2.2 (Page 104)

1. **(a)** -6 **(b)** 13 **(c)** -8 **(d)** 16 **(e)** 2 **(f)** $-\frac{1}{2}$ **(g)** does not exist **(h)** does not exist 3. 6 5. 3/4 7. 4 9. $-\frac{4}{5}$ 11. -3 13. $\frac{3}{2}$
15. $+\infty$ 17. does not exist 19. $-\infty$ 21. $+\infty$ 23. does not exist
25. $+\infty$ 27. $+\infty$ 29. 6 31. **(a)** 2 **(b)** 2 **(c)** 2 33. **(a)** 3
35. **(a)** Theorem 2.2.2(a) does not apply.
(b) $\lim_{x\to 0^+}\left(\frac{1}{x}-\frac{1}{x^2}\right) = \lim_{x\to 0^+}\left(\frac{x-1}{x^2}\right) = -\infty$ 37. $\frac{1}{4}$
39. The left and/or right limits could be $\pm\infty$; or the limit could exist and equal any preassigned real number.

Exercise Set 2.3 (Page 113)

1. **(a)** $-\infty$ **(b)** $+\infty$ 3. **(a)** 0 **(b)** -1
5. **(a)** -12 **(b)** 21 **(c)** -15 **(d)** 25 **(e)** 2 **(f)** $-\frac{3}{5}$ **(g)** 0 **(h)** does not exist
7. $-\infty$ 9. $+\infty$ 11. $\frac{3}{2}$ 13. 0 15. 0 17. $-\frac{\sqrt[3]{5}}{2}$ 19. $-\sqrt{5}$
21. $1/\sqrt{6}$ 23. $\sqrt{3}$ 25. $-\infty$ 27. $-\frac{1}{7}$
29. $\lim_{t\to+\infty} n(t) = +\infty$; $\lim_{t\to+\infty} e(t) = c$ 31. **(a)** $+\infty$ **(b)** -5 33. 0
35. $a/2$ 37. $\lim_{x\to+\infty} p(x) = \begin{cases} -\infty, n \text{ is odd} \\ +\infty, n \text{ is even} \end{cases}$ and $\lim_{x\to-\infty} p(x) = +\infty$
39. For $m > n$, the limits are both zero; for $m = n$, the limits are equal to the leading coefficient of p; for $n > m$, the limits are $\pm\infty$
41. $\lim_{x\to-\infty}\frac{2+3x^n}{1-x^m} = \begin{cases} 0 & \text{if } m > n \\ -3 & \text{if } m = n \\ +\infty & \text{if } m < n \text{ and } n-m \text{ odd} \\ -\infty & \text{if } m < n \text{ and } n-m \text{ even} \end{cases}$
43. They equal L. 45. $x+2$ 47. $1-x^2$ 49. $\sin x$

Exercise Set 2.4 (Page 122)

1. **(a)** $|x| < 0.1$ **(b)** $|x-3| < 0.0025$ **(c)** $|x-4| < 0.000125$
3. **(a)** $x_0 = 3.8025$, $x_1 = 4.2025$ **(b)** $\delta = 0.1975$
5. $\delta = 0.0442$

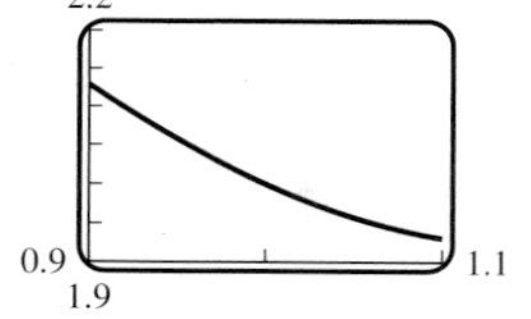

7. $\delta = 0.13$
9. $\delta = 0.05$
11. $\delta = 0.05$

13. $\delta = \sqrt[3]{8.001} - 2 \approx 8.332986\cdot 10^{-5}$ 15. $\delta = 1/505 \approx 0.000198$
17. **(a)** $\lim_{x\to 4} f(x) = 3$ **(b)** $\delta = 0.0001$
19. $\delta = 1/5200 \approx 0.0001923$ 21. $\delta = \frac{1}{3}\epsilon$ 23. $\delta = \epsilon/2$ 25. $\delta = \epsilon$
27. **(b)** 65 **(c)** $\epsilon/65$; 65; 65; $\epsilon/65$ 29. $\delta = \min(1, \frac{1}{6}\epsilon)$
31. $\delta = \min(1, \epsilon/(1+\epsilon))$ 33. $\delta = 2\epsilon$
35. **(a)** $\sqrt{10}$ **(b)** 99 **(c)** -10 **(d)** -101
37. **(a)** $-\sqrt{\frac{1-\epsilon}{\epsilon}}$; $\sqrt{\frac{1-\epsilon}{\epsilon}}$ **(b)** $\sqrt{\frac{1-\epsilon}{\epsilon}}$ **(c)** $-\sqrt{\frac{1-\epsilon}{\epsilon}}$ 39. 10 41. 999
43. -202 45. -57.5 47. $N = \frac{1}{\sqrt{\epsilon}}$ 49. $N = -\frac{5}{2} - \frac{11}{2\epsilon}$
51. $N = (1+2/\epsilon)^2$
53. **(a)** $|x| < \frac{1}{10}$ **(b)** $|x-1| < \frac{1}{1000}$ **(c)** $|x-3| < \frac{1}{10\sqrt{10}}$ **(d)** $|x| < \frac{1}{10}$
55. $\delta = 1/\sqrt{M}$ 57. $\delta = 1/M$ 59. $\delta = 1/(-M)^{1/4}$ 61. $\delta = \epsilon$
63. $\delta = \epsilon^2$ 65. $\delta = \epsilon$ 67. **(a)** $\delta = -1/M$ **(b)** $\delta = 1/M$
69. **(a)** $N = M - 1$ **(b)** $N = M - 1$ 71. $\delta = \min\left(2, \frac{1}{8}\epsilon\right)$
73. **(a)** 0.4 amps **(b)** about 0.39474 to 0.40541 amps **(c)** $3/(7.5+\delta)$ to $3/(7.5-\delta)$ **(d)** $\delta \approx 0.01870$ **(e)** current approaches $+\infty$

▶ Exercise Set 2.5 (Page 134)

1. **(a)** not continuous, $x = 2$ **(b)** not continuous, $x = 2$ **(c)** not continuous, $x = 2$ **(d)** continuous **(e)** continuous **(f)** continuous
3. **(a)** not continuous, $x = 1, 3$ **(b)** continuous **(c)** not continuous, $x = 1$ **(d)** continuous **(e)** not continuous, $x = 3$ **(f)** continuous
5. **(a)** 3 **(b)** 3
7. **(a)**

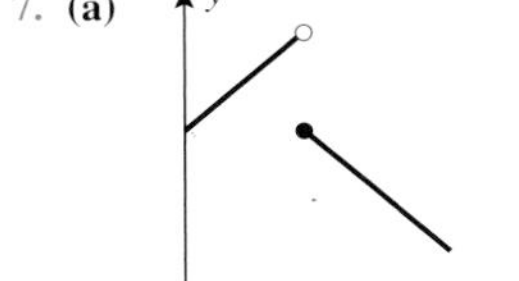

(b)

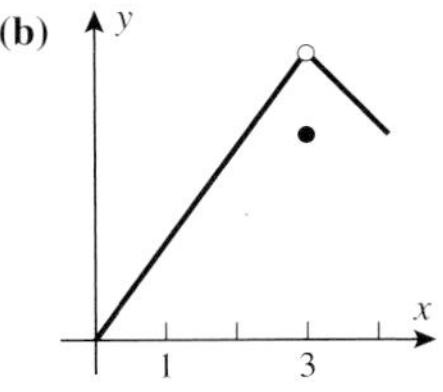

(c)

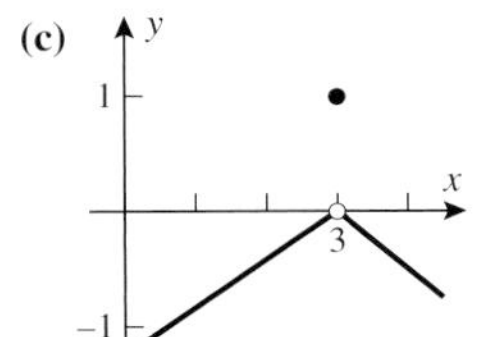

(d)

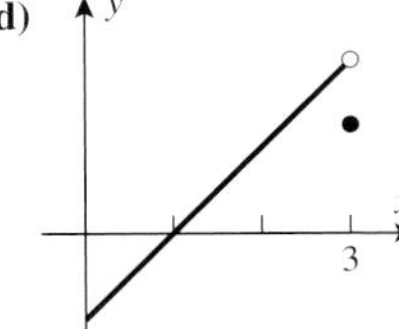

9. **(a)**

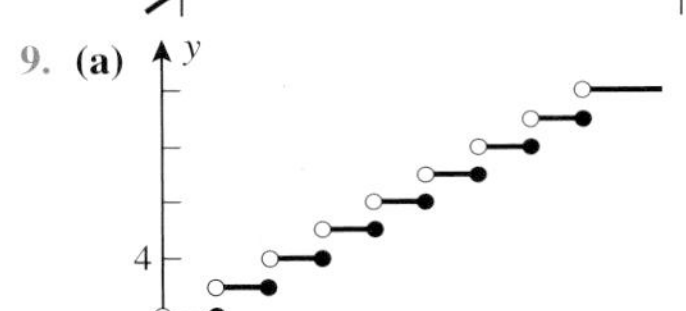

(b) One second could cost you one dollar.

11. none 13. none 15. $-1/2, 0$ 17. $-1, 0, 1$ 19. none
21. none 23. **(a)** $k = 5$ **(b)** $k = \frac{4}{3}$ 25. $k = 4, m = 5/3$
27. **(a)**

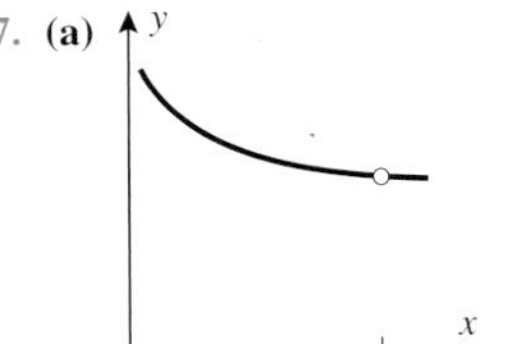

(b)

29. **(a)** $x = 0$, not removable **(b)** $x = -3$, removable **(c)** $x = 2$, removable; $x = -2$, not removable
31. **(a)** $x = \frac{1}{2}$, not removable; at $x = -3$, removable **(b)** $(2x - 1)(x + 3)$

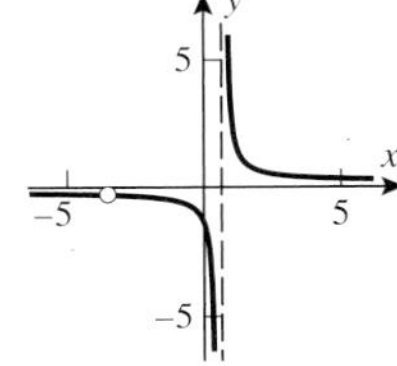

35. **(a)** $f(x) = k$ for $x \neq c$, $f(c) = 0$; $g(x) = \ell$ for $x \neq c$, $g(c) = 0$. If $k = -\ell$, then $f + g$ is continuous; otherwise it is not. **(b)** $f(x) = k$ for $x \neq c$, $f(c) = 1$; $g(x) = \ell \neq 0$ for $x \neq c$, $g(c) = 1$. If $k\ell = 1$, then fg is continuous; otherwise it is not.
39. $f(x) = 1$ for $0 \le x < 1$, $f(x) = -1$ for $1 \le x \le 2$
45. $x = -1.25, x = 0.75$ 47. $x = -1.605, x = 1.375$
49. $x = 2.24$ 51. $x = 4.847$ cm

▶ Exercise Set 2.6 (Page 141)

1. none 3. $x = n\pi, n = 0, \pm 1, \pm 2, \ldots$
5. $x = n\pi, n = 0, \pm 1, \pm 2, \ldots$
7. $2n\pi + (\pi/6), 2n\pi + (5\pi/6), n = 0, \pm 1, \pm 2, \ldots$
9. **(a)** $\sin x, x^3 + 7x + 1$ **(b)** $|x|, \sin x$ **(c)** $x^3, \cos x, x + 1$ **(d)** $\sqrt{x}, 3 + x, \sin x, 2x$ **(e)** $\sin x, \sin x$ **(f)** $x^5 - 2x^3 + 1, \cos x$
11. 1 13. 3 15. $+\infty$ 17. $\frac{7}{3}$ 19. 0 21. 0 23. 1
25. 2 27. does not exist 29. 0 31. a/b 33. **(b)** 1/10 35. **(b)** -1
37. $\lim_{x \to 0} \sin(1/x)$ does not exist 39. $k = \frac{1}{2}$ 41. **(a)** 1 **(b)** 0 **(c)** 1
43. $-\pi$ 45. $-|x| \le x \cos(50\pi/x) \le |x|$
47. $\lim_{x \to 0} f(x) = 1$ by the squeezing Theorem.

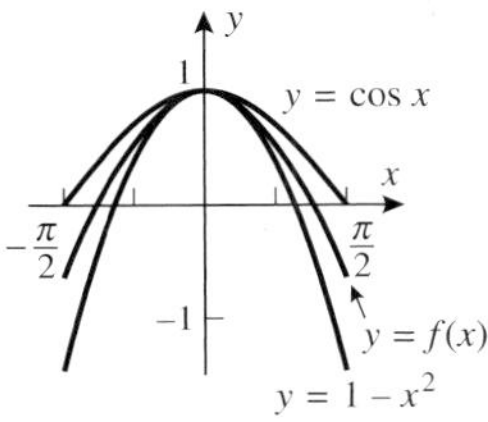

49. $g(x) = -\frac{1}{x}, h(x) = \frac{1}{x}$; $\lim_{x \to +\infty} \frac{\sin x}{x} = 0$ by the Squeezing Theorem.
53. **(a)** 0.17365 **(b)** 0.17453 55. **(a)** 0.08749 **(b)** 0.08727
57. **(b)**

(c) 0.739

59. **(a)** Gravity is strongest at the poles and weakest at the equator.

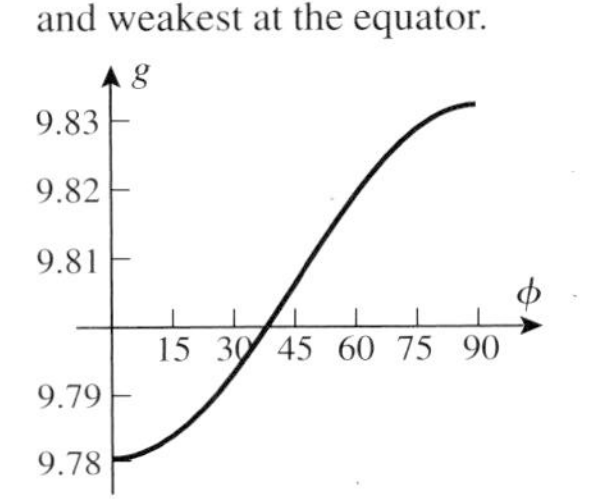

▶ Chapter 2. Review Exercises (Page 143)

1. **(a)** 1 **(b)** does not exist **(c)** does not exist **(d)** 1 **(e)** 3 **(f)** 0 **(g)** 0 **(h)** 2 **(i)** $\frac{1}{2}$
3. **(a)** 0.405 5. 1 7. $-3/2$ 9. $32/3$
11. **(a)** $y = 0$ **(b)** none **(c)** $y = 2$ 13. 1 15. $3 - k$ 17. $-1/2$
19. **(a)** $2x/(x - 1)$ is one example.
21. **(a)** $\lim_{x \to 2} f(x) = 5$ **(b)** $\delta = 0.0045$
23. **(a)** $\delta = 0.0025$ **(b)** $\delta = 0.0025$ **(c)** $\delta = 1/9000$ (Some larger values also work.)
27. **(a)** $-1, 1$ **(b)** none **(c)** $-3, 0$ 29. no; not continuous at $x = 2$
31. Consider $f(x) = x$ for $x \neq 0$, $f(0) = 1$, $a = -1$, $b = 1$, $k = 0$.

▶ Exercise Set 3.1 (Page 157)

1. **(a)** 4 m/s 3. **(a)** 33 m/s **(b)** 55 m/s
5. **(a)** t_0 **(b)** 0 **(c)** speeding up **(d)** slowing down
7. straight line with slope equal to the velocity
9. **(a)** 2 **(b)** 0 **(c)** $4x_0$ **(d)**
11. **(a)** $-\frac{1}{6}$ **(b)** $-\frac{1}{4}$ **(c)** $-1/x_0^2$ **(d)**
13. **(a)** $2x_0$ **(b)** -2 15. **(a)** $1/(2\sqrt{x_0})$ **(b)** $\frac{1}{2}$
17. **(a)** 72° F at about 4:30 P.M. **(b)** 4° F/h **(c)** -7° F/h at about 9 P.M.

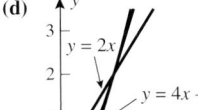

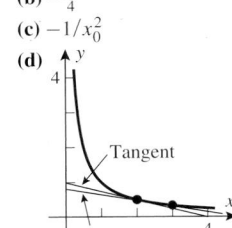

19. **(a)** first year **(b)** 6 cm/year **(c)** 10 cm/year at about age 14 **(d)**

21. **(a)** $6400\sqrt{10}$ ft **(b)** $160\sqrt{10}$ ft/s **(c)** $\frac{9}{2}\sqrt[6]{4000}$ ft/s **(d)** $480\sqrt{10}$ ft/s
23. **(a)** 720 ft/min **(b)** 192 ft/min

▶ Exercise Set 3.2 (Page 168)

1. 2, 0, −2, −1 3. **(b)** 3 **(c)** 3 7. $y = 5x - 16$ 9. $4x$, $y = 4x - 2$
11. $3x^2$; $y = 0$ 13. $\dfrac{1}{2\sqrt{x+1}}$; $y = \frac{1}{6}x + \frac{5}{3}$ 15. $-1/x^2$ 17. $2x - 1$
19. $-1/(2x^{3/2})$ 21. $8t + 1$
23. **(a)** D **(b)** F **(c)** B **(d)** C **(e)** A **(f)** E
25. **(a)** 

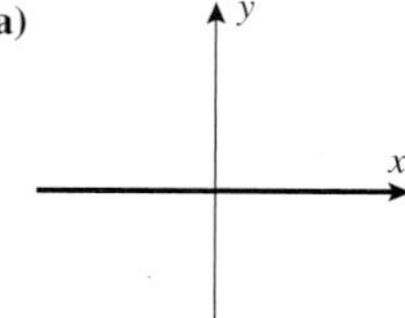

(b) 

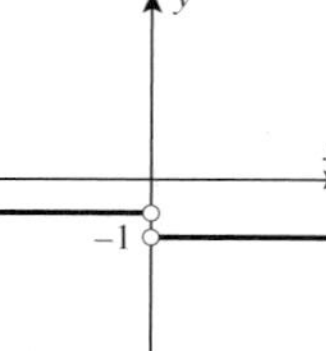

(c)

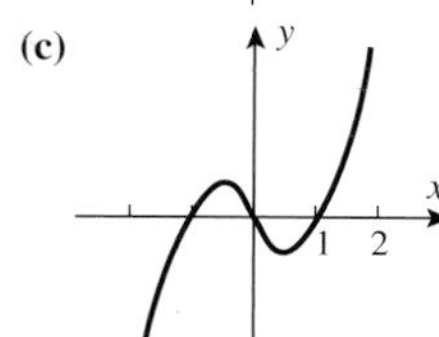

27. **(a)** $\sqrt{x}$, 1 **(b)** x^2, 3 31. $y = -2x + 1$
29. −2

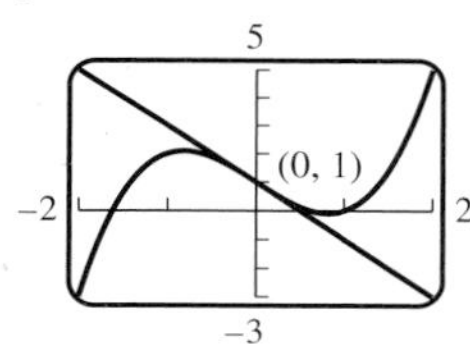

33. **(b)**

h	0.5	0.1	0.01	0.001	0.0001	0.00001
$[f(1+h) - f(1)]/h$	1.6569	1.4355	1.3911	1.3868	1.3863	1.3863

35. **(a)** dollars per foot **(b)** the price per additional foot **(c)** positive **(d)** $1000
37. **(a)** $F \approx 200$ lb, $dF/d\theta \approx 50$ lb/rad **(b)** $\mu = -0.25$
39. **(a)** $T \approx 115°$F, $dT/dt \approx -3.35°$F/min **(b)** $k = -0.084$

▶ Exercise Set 3.3 (Page 177)

1. $28x^6$ 3. $24x^7 + 2$ 5. 0 7. $-\frac{1}{3}(7x^6 + 2)$
9. $-3x^{-4} - 7x^{-8}$ 11. $24x^{-9} + (1/\sqrt{x})$ 13. $12x(3x^2 + 1)$
15. $3ax^2 + 2bx + c$ 17. 7 19. $2t - 1$ 21. 15 23. −8 25. 0
27. 0 29. $32t$ 31. $3\pi r^2$ 33. **(a)** $4\pi r^2$ **(b)** 100π 35. $y = 5x + 17$
37. **(a)** $42x - 10$ **(b)** 24 **(c)** $2/x^3$ **(d)** $700x^3 - 96x$
39. **(a)** $-210x^{-8} + 60x^2$ **(b)** $-6x^{-4}$ **(c)** $6a$
41. **(a)** 0 **(b)** 112 **(c)** 360 45. $(1, \frac{5}{6})(2, \frac{2}{3})$

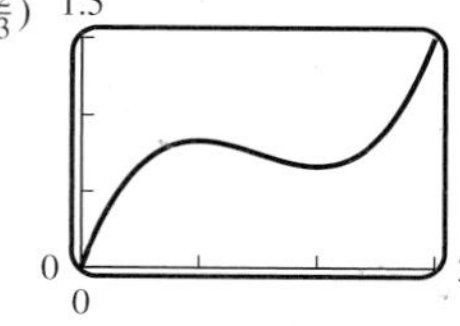

47. $y = 3x^2 - x - 2$ 49. $x = \frac{1}{2}$
51. $(2 + \sqrt{3}, -6 - 4\sqrt{3})$, $(2 - \sqrt{3}, -6 + 4\sqrt{3})$
53. $-2x_0$ 57. $-\dfrac{2GmM}{r^3}$ 59. $f'(x) > 0$ for all $x \neq 0$
61. yes, 3

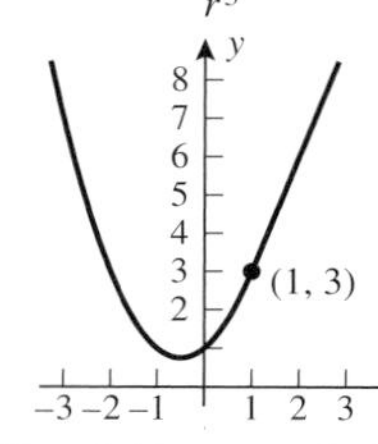

63. not differentiable at $x = 1$ 65. **(a)** $x = \frac{2}{3}$ **(b)** $x = \pm 2$ 67. **(b)** yes
69. **(a)** $n(n-1)(n-2)\cdots 1$ **(b)** 0 **(c)** $a_n n(n-1)(n-2)\cdots 1$
75. $-12/(2x+1)^3$ 77. $-2/(x+1)^3$

▶ Exercise Set 3.4 (Page 184)

1. $4x + 1$ 3. $4x^3$ 5. $18x^2 - \frac{3}{2}x + 12$
7. $-15x^{-2} - 14x^{-3} + 48x^{-4} + 32x^{-5}$ 9. $3x^2$ 11. $-\frac{5}{4}$ 13. $\frac{7}{16}$
15. −29 17. 0 19. **(a)** $-\frac{37}{4}$ **(b)** $-\frac{23}{16}$
21. **(a)** 10 **(b)** 19 **(c)** 9 **(d)** −1 23. $-2 \pm \sqrt{3}$ 25. none 27. −2
31. $F''(x) = xf''(x) + 2f'(x)$
35. **(a)** $2(1 + x^{-1})(x^{-3} + 7) + (2x + 1)(-x^{-2})(x^{-3} + 7) + (2x + 1)(1 + x^{-1})(-3x^{-4})$ **(b)** $3(7x^6 + 2)(x^7 + 2x - 3)^2$
37. $g'(x) = n(f(x))^{n-1}(x)$ 39. $f'(x) = -nx^{-n-1}$

▶ Exercise Set 3.5 (Page 188)

1. $-4\sin x + 2\cos x$ 3. $4x^2 \sin x - 8x \cos x$
5. $(1 + 5\sin x - 5\cos x)/(5 + \sin x)^2$ 7. $\sec x \tan x - \sqrt{2}\sec^2 x$
9. $-4\csc x \cot x + \csc^2 x$ 11. $\sec^3 x + \sec x \tan^2 x$ 13. $-\dfrac{\csc x}{1 + \csc x}$
15. 0 17. $\dfrac{1}{(1 + x\tan x)^2}$ 19. $-x\cos x - 2\sin x$
21. $-x\sin x + 5\cos x$ 23. $-4\sin x \cos x$
25. **(a)** $y = x$ **(b)** $y = 2x - (\pi/2) + 1$ **(c)** $y = 2x + (\pi/2) - 1$
29. **(a)** $x = \pm\pi/2, \pm 3\pi/2$ **(b)** $x = -3\pi/2, \pi/2$ **(c)** no horizontal tangent line **(d)** $x = \pm 2\pi, \pm\pi, 0$
31. 0.087 ft/degree 33. 1.75 m/degree 35. **(a)** $-\cos x$ **(b)** $\cos x$
37. 3, 7, 11, ...
39. **(a)** all x **(b)** all x **(c)** $x \neq (\pi/2) + n\pi, n = 0, \pm 1, \pm 2, \ldots$ **(d)** $x \neq n\pi, n = 0, \pm 1, \pm 2, \ldots$ **(e)** $x \neq (\pi/2) + n\pi, n = 0, \pm 1, \pm 2, \ldots$ **(f)** $x \neq n\pi, n = 0, \pm 1, \pm 2, \ldots$ **(g)** $x \neq (2n + 1)\pi, n = 0, \pm 1, \pm 2, \ldots$ **(h)** $x \neq n\pi/2, n = 0, \pm 1, \pm 2, \ldots$ **(i)** all x

▶ Exercise Set 3.6 (Page 195)

1. 6 3. **(a)** $(2x - 3)^5$, $10(2x - 3)^4$ **(b)** $2x^5 - 3$, $10x^4$
5. **(a)** −7 **(b)** −8 7. $37(x^3 + 2x)^{36}(3x^2 + 2)$
9. $-2\left(x^3 - \dfrac{7}{x}\right)^{-3}\left(3x^2 + \dfrac{7}{x^2}\right)$ 11. $\dfrac{24(1 - 3x)}{(3x^2 - 2x + 1)^4}$
13. $\dfrac{3}{4\sqrt{x}\sqrt{4 + 3\sqrt{x}}}$ 15. $-\dfrac{2}{x^3}\cos\left(\dfrac{1}{x^2}\right)$ 17. $-20\cos^4 x \sin x$
19. $-\dfrac{3}{\sqrt{x}}\cos(3\sqrt{x})\sin(3\sqrt{x})$ 21. $28x^6 \sec^2(x^7)\tan(x^7)$
23. $-\dfrac{5\sin(5x)}{2\sqrt{\cos(5x)}}$
25. $-3[x + \csc(x^3 + 3)]^{-4}[1 - 3x^2\csc(x^3 + 3)\cot(x^3 + 3)]$
27. $10x^3 \sin 5x \cos 5x + 3x^2 \sin^2 5x$
29. $-x^3 \sec\left(\dfrac{1}{x}\right)\tan\left(\dfrac{1}{x}\right) + 5x^4 \sec\left(\dfrac{1}{x}\right)$
31. $\sin(\cos x)\sin x$ 33. $-6\cos^2(\sin 2x)\sin(\sin 2x)\cos 2x$
35. $35(5x + 8)^6(1 - \sqrt{x})^6 - \dfrac{3}{\sqrt{x}}(5x + 8)^7(1 - \sqrt{x})^5$

37. $\dfrac{33(x-5)^2}{(2x+1)^4}$ 39. $-\dfrac{2(2x+3)^2(52x^2+96x+3)}{(4x^2-1)^9}$

41. $5[x\sin 2x+\tan^4(x^7)]^4[2x\cos 2x+\sin 2x+28x^6\tan^3(x^7)\sec^2(x^7)]$

43. $y=-x$ 45. $y=-1$ 47. $y=8\sqrt{\pi}x-8\pi$ 49. $y=\frac{7}{2}x-\frac{3}{2}$

51. $-25x\cos(5x)-10\sin(5x)-2\cos(2x)$ 53. $4(1-x)^{-3}$

55. $3\cot^2\theta\csc^2\theta$ 57. $\pi(b-a)\sin 2\pi\omega$

59. **(a)**

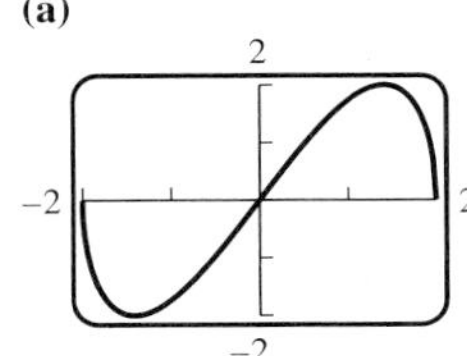

(c) $\dfrac{4-2x^2}{\sqrt{4-x^2}}$

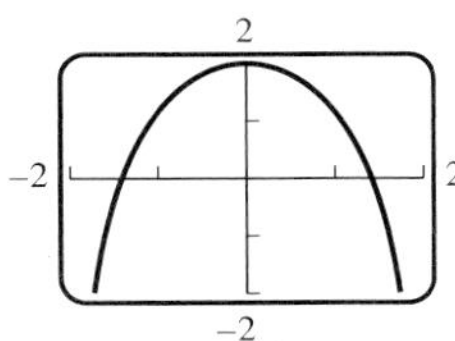

(d) $y-\sqrt{3}=\frac{2}{\sqrt{3}}(x-1)$

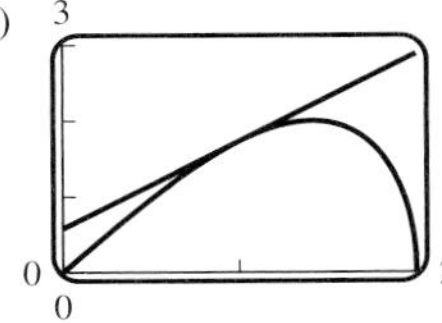

61. **(c)** $f=1/T$ **(d)** amplitude $=0.6$ cm, $T=2\pi/15$ seconds per oscillation, $f=15/(2\pi)$ oscillations per second

63. $\frac{7}{24}\sqrt{6}$ 65. **(a)** 10 lb/in^2, -2 lb/in^2/mi **(b)** -0.6 lb/in^2/s

67. $\begin{cases}\cos x, & x<x<\pi \\ -\cos x, & -\pi<x<0\end{cases}$

69. **(c)** $-\dfrac{1}{x}\cos\dfrac{1}{x}+\sin\dfrac{1}{x}$ **(d)** limit as x goes to 0 does not exist

71. **(a)** 21 **(b)** -36 73. $1/(2x)$ 75. $\frac{2}{3}x$

79. $f'(g(h(x)))g'(h(x))h'(x)$

▶ Exercise Set 3.7 **(Page 204)**

1. $\dfrac{2}{3}(2x-5)^{-2/3}$ 3. $-\dfrac{2}{(x-2)^2}\left[\dfrac{x+1}{x-2}\right]^{-1/3}$

5. $\dfrac{1}{3}x^2(5x^2+1)^{-5/3}(25x^2+9)$ 7. $-\dfrac{15[\sin(3/x)]^{3/2}\cos(3/x)}{2x^2}$

9. **(a)** $(6x^2-y-1)/x$ **(b)** $4x-2/x^2$ 11. $-\dfrac{x}{y}$ 13. $\dfrac{1-2xy-3y^3}{x^2+9xy^2}$

15. $\dfrac{-y^{3/2}}{x^{3/2}}$ 17. $\dfrac{1-2xy^2\cos(x^2y^2)}{2x^2y\cos(x^2y^2)}$

19. $\dfrac{1-3y^2\tan^2(xy^2+y)\sec^2(xy^2+y)}{3(2xy+1)\tan^2(xy^2+y)\sec^2(xy^2+y)}$ 21. $-\dfrac{8}{9y^3}$ 23. $\dfrac{2y}{x^2}$

25. $\dfrac{\sin y}{(1+\cos y)^3}$ 27. $-1/\sqrt{3}, 1/\sqrt{3}$ 29. $-15^{-3/4}\approx-0.1312$

31. $-\dfrac{9}{13}$ 33. $\dfrac{2t^3+3a^2}{2a^3-6at}$ 35. $-\dfrac{b^2\lambda}{a^2\omega}$

37. **(a)**

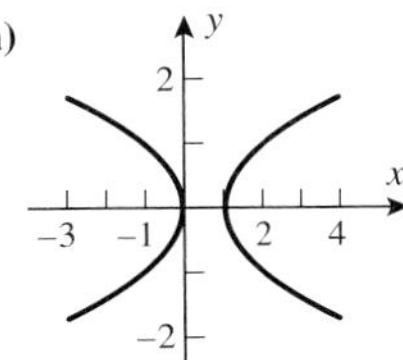

(c) $x=-y^2$ or $x=y^2+1$

39. **(a)**

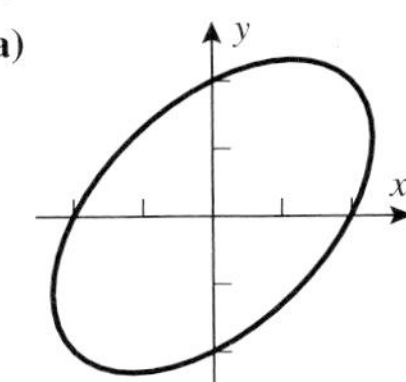

(b) ±1.1547

(c) $x=\pm\dfrac{2}{\sqrt{3}}$

41. points $(2,2), (-2,-2)$; $y'=-1$ at both points

43. $a=\dfrac{1}{4}, b=\dfrac{5}{4}$

45. **(a)**

(c) $2\sqrt[3]{2}/3$

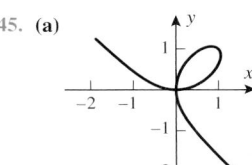

47. $y=(\sqrt{3}/3)x,\ y=-(\sqrt{3}/3)x$ 51. $-\dfrac{2y^3+3t^2y}{(6ty^2+t^3)\cos t}$ 53. $-1, \dfrac{2}{3}$

▶ Exercise Set 3.8 **(Page 210)**

1. **(a)** 6 **(b)** $-\frac{1}{3}$ 3. **(a)** -2 **(b)** $6\sqrt{5}$

5. **(b)** $A=x^2$ **(c)** $\dfrac{dA}{dt}=2x\dfrac{dx}{dt}$ **(d)** 12 ft^2/min

7. **(a)** $\dfrac{dV}{dt}=\pi\left(r^2\dfrac{dh}{dt}+2rh\dfrac{dr}{dt}\right)$ **(b)** -20π in^3/s; decreasing

9. **(a)** $\dfrac{d\theta}{dt}=\dfrac{\cos^2\theta}{x^2}\left(x\dfrac{dy}{dt}-y\dfrac{dx}{dt}\right)$ **(b)** $-\frac{5}{16}$ rad/s; decreasing

11. $\dfrac{4\pi}{15}$ in^2/min 13. $\dfrac{1}{\sqrt{\pi}}$ mi/h 15. 4860π cm^3/min 17. $\frac{5}{6}$ ft/s

19. $\dfrac{125}{\sqrt{61}}$ ft/s 21. 704 ft/s

23. **(a)** 500 mi, 1716 mi **(b)** 1354 mi; 27.7 mi/min

25. $\dfrac{9}{20\pi}$ ft/min 27. 125π ft^3/min 29. 250 mi/h 31. $\dfrac{36\sqrt{69}}{25}$ ft/min

33. $\dfrac{8\pi}{5}$ km/s 35. $600\sqrt{7}$ mi/h

37. **(a)** $-\frac{60}{7}$ units per second **(b)** falling 39. -4 units per second

41. $x=\pm\sqrt{\dfrac{-5+\sqrt{33}}{2}}$ 43. 4.5 cm/s; away 47. $\dfrac{20}{9\pi}$ cm/s

▶ Exercise Set 3.9 **(Page 219)**

1. **(a)** $f(x)\approx1+3(x-1)$ **(b)** $f(1+\Delta x)\approx1+3\Delta x$ **(c)** 1.06

3. **(a)** $1+\frac{1}{2}x$, 0.95, 1.05 13. $|x|<1.692$

(b)

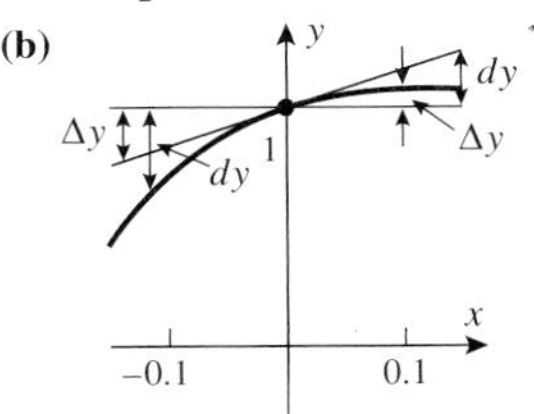

15. $|x|<0.3158$ 17. **(a)** 0.0174533 **(b)** $x_0=45°$ **(c)** 0.694765

19. 83.16 21. 8.0625 23. 8.9944 25. 0.1 27. 0.8573

29. **(a)** 4, 5

(b)

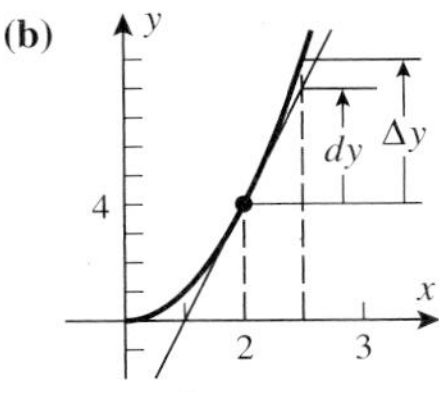

31. **(a)** 0.5, 1

(b)

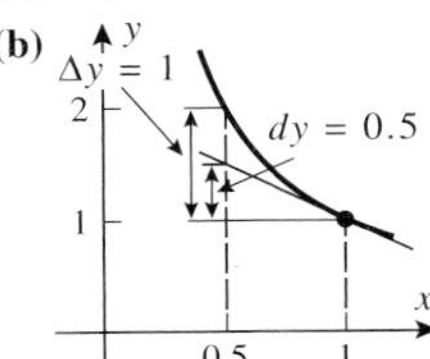

33. $3x^2dx,\ 3x^2\Delta x+3x(\Delta x)^2+(\Delta x)^3$

35. $(2x-2)dx,\ 2x\Delta x+(\Delta x)^2-2\Delta x$

37. **(a)** $(12x^2-14x)dx$ **(b)** $(-x\sin x+\cos x)dx$

39. **(a)** $\dfrac{2-3x}{2\sqrt{1-x}}dx$ **(b)** $-17(1+x)^{-18}\,dx$

41. 0.0225 43. 0.0048 45. **(a)** ±2 ft^2 **(b)** side: $\pm1\%$; area: $\pm2\%$

47. **(a)** opposite: ±0.151 in; adjacent: ±0.087 in **(b)** opposite: $\pm3.0\%$; adjacent: $\pm1.0\%$

49. $\pm10\%$ 51. ±0.017 cm^2 53. $\pm6\%$ 55. $\pm0.5\%$ 57. $15\pi/2$ cm^3

59. **(a)** $\alpha = 1.5 \times 10^{-5}/^\circ\text{C}$ **(b)** 180.1 cm long

Chapter 3. Review Exercises (Page 221)

3. **(a)** $2x$ **(b)** 4 5. 58.75 ft/s 7. **(a)** 13 mi/h **(b)** 7 mi/h
9. **(a)** $-2/\sqrt{9-4x}$ **(b)** $1/(x+1)^2$
11. **(a)** $x = -2, -1, 1, 3$ **(b)** $(-\infty, -2), (-1, 1), (3, +\infty)$ **(c)** $(-2, -1), (1, 3)$ **(d)** 4
13. **(a)** 78 million people per year **(b)** 1.3% per year
15. **(a)** $x^2 \cos x + 2x \sin x$ **(c)** $4x \cos x + (2 - x^2) \sin x$
17. **(a)** $(6x^2 + 8x - 17)/(3x+2)^2$ **(c)** $118/(3x+2)^3$
19. **(a)** 2000 gal/min **(b)** 2500 gal/min 21. **(a)** 3.6 **(b)** -0.777778
23. $f(1) = 0, f'(1) = 5$ 25. $y = -16x, y = -145x/4$
29. **(a)** $8x^7 - \dfrac{3}{2\sqrt{x}} - 15x^{-4}$ **(b)** $(2x+1)^{100}(1030x^2 + 10x - 1414)$
(c) $\dfrac{(x-1)(15x+1)}{2\sqrt{3x+1}}$ **(d)** $-3(3x+1)^2(3x+2)/x^7$
31. $x = \frac{7}{2}, 2, -\frac{1}{2}$ 33. $y = \pm 2x$
35. $x = n\pi \pm (\pi/4), n = 0, \pm 1, \pm 2, \ldots$ 37. $y = -3x + (1 + 9\pi/4)$
39. **(a)** $40\sqrt{3}$ **(b)** 7500 41. $\frac{1}{3}(x^2 + x)^{-2/3}(2x+1)$
43. **(a)** $\dfrac{2 - 3x^2 - y}{x}$ **(b)** $-\dfrac{1}{x^2} - 2x$ 45. $-\dfrac{y^2}{x^2}$
47. $\dfrac{y \sec(xy) \tan(xy)}{1 - x \sec(xy) \tan(xy)}$ 49. $-\dfrac{21}{16y^3}$ 51. 500π m^2/min
53. **(a)** $-0.5, 1, 0.5$ **(b)** $\frac{\pi}{4}, 1, \frac{\pi}{2}$ **(c)** $3, -1, 0$
55. **(a)** between 139.48 m and 144.55 m **(b)** $|d\phi| \le 0.98^\circ$

Exercise Set 4.1 (Page 231)

1. **(a)** $f' > 0, f'' > 0$ **(b)** $f' > 0, f'' < 0$ **(c)** $f' < 0, f'' > 0$ **(d)** $f' < 0, f'' < 0$

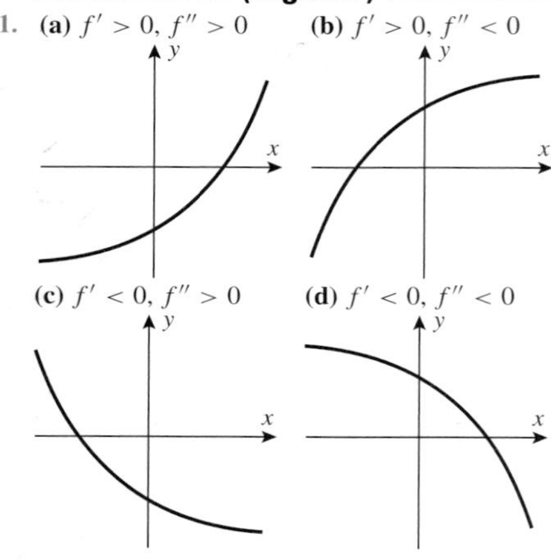

3. $A: dy/dx < 0, d^2y/dx^2 > 0$ $B: dy/dx > 0, d^2y/dx^2 < 0$
$C: dy/dx < 0, d^2y/dx^2 < 0$ 5. $x = -1, 0, 1, 2$
7. **(a)** [4, 6] **(b)** [1, 4], [6, 7] **(c)** (1, 2), (3, 5) **(d)** (2, 3), (5, 7) **(e)** $x = 2, 3, 5$
9. **(a)** [1,3] **(b)** $(-\infty, 1], [3, +\infty)$ **(c)** $(-\infty, 2), (4, +\infty)$ **(d)** (2,4) **(e)** $x = 2, 4$
11. **(a)** $[3/2, +\infty]$ **(b)** $(-\infty, 3/2]$ **(c)** $(-\infty, +\infty)$ **(d)** none **(e)** none
13. **(a)** $(-\infty, +\infty)$ **(b)** none **(c)** $(-1/2, +\infty)$ **(d)** $(-\infty, -1/2)$ **(e)** $-1/2$
15. **(a)** $[1, +\infty)$ **(b)** $(-\infty, 1]$ **(c)** $(-\infty, 0), (\frac{2}{3}, +\infty)$ **(d)** $(0, \frac{2}{3})$ **(e)** $0, \frac{2}{3}$
17. **(a)** $\left[\dfrac{3-\sqrt{5}}{2}, \dfrac{3+\sqrt{5}}{2}\right]$ **(b)** $\left(-\infty, \dfrac{3-\sqrt{5}}{2}\right], \left[\dfrac{3+\sqrt{5}}{2}, +\infty\right)$
(c) $\left(0, \dfrac{4-\sqrt{6}}{2}\right), \left(\dfrac{4+\sqrt{6}}{2}, +\infty\right)$ **(d)** $(-\infty, 0), \left(\dfrac{4-\sqrt{6}}{2}, \dfrac{4+\sqrt{6}}{2}\right)$
(e) $0, \dfrac{4 \pm \sqrt{6}}{2}$
19. **(a)** $[-1/2, +\infty)$ **(b)** $(-\infty, -1/2]$ **(c)** $(-2, 1)$ **(d)** $(-\infty, -2), (1, +\infty)$ **(e)** $-2, 1$
21. **(a)** $[-1, 0], [1, +\infty)$ **(b)** $(-\infty, -1], [0, 1]$ **(c)** $(-\infty, 0), (0, +\infty)$ **(d)** none **(e)** none
23. increasing: $[-\pi/4, 3\pi/4]$; decreasing: $[-\pi, -\pi/4], [3\pi/4, \pi]$; concave up: $(-3\pi/4, \pi/4)$; concave down: $(-\pi, -3\pi/4), (\pi/4, \pi)$; inflection points: $-3\pi/4, \pi/4$
25. increasing: none; decreasing: $(-\pi, \pi)$; concave up: $(-\pi, 0)$; concave down: $(0, \pi)$; inflection point: 0
27. increasing: $[-\pi, -3\pi/4], [-\pi/4, \pi/4], [3\pi/4, \pi]$; decreasing: $[-3\pi/4, -\pi/4], [\pi/4, 3\pi/4]$; concave up: $(-\pi/2, 0), (\pi/2, \pi)$; concave down: $(-\pi, -\pi/2), (0, \pi/2)$; inflection points: $0, \pm\pi/2$
29. **(a)** **(b)** **(c)**

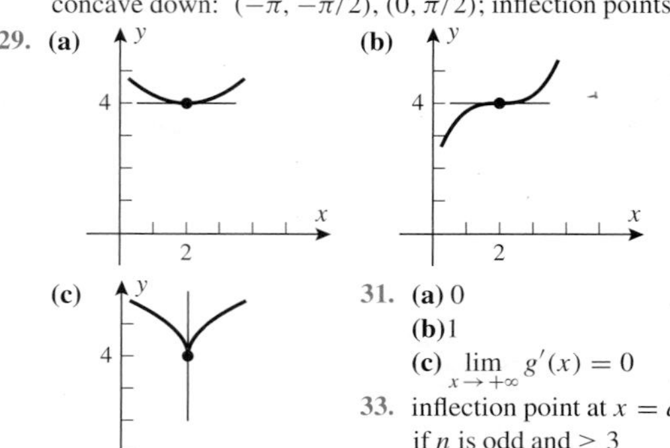

31. **(a)** 0 **(b)** 1 **(c)** $\lim_{x \to +\infty} g'(x) = 0$
33. inflection point at $x = a$ if n is odd and ≥ 3

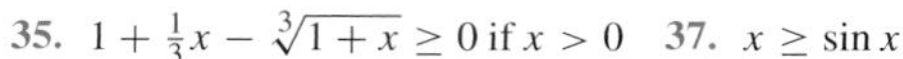

35. $1 + \frac{1}{3}x - \sqrt[3]{1+x} \ge 0$ if $x > 0$ 37. $x \ge \sin x$

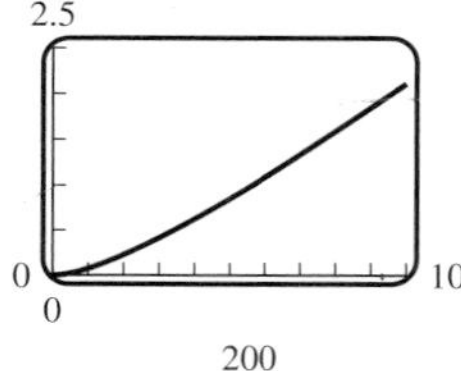

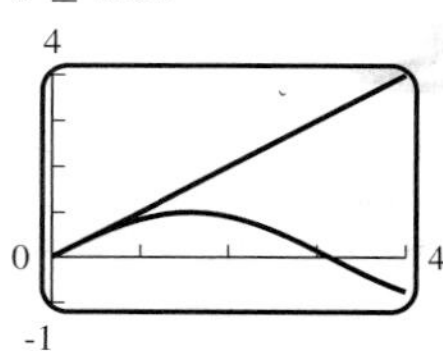

39.

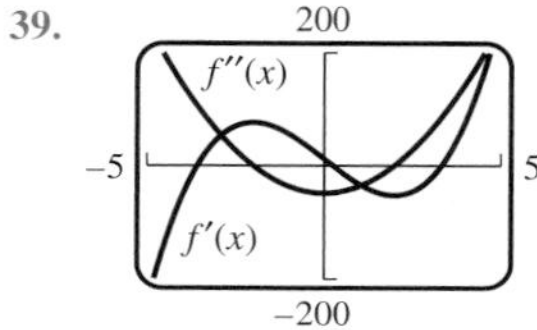

points of inflection at $x = -2, 2$; concave up on $[-5, -2], [2, 5]$; concave down on $[-2, 2]$; increasing on $[-3.5829, 0.2513]$ and $[3.3316, 5]$; decreasing on $[-5, -3.5829]$, $[0.2513, 3.3316]$

41. $-2.464202, 0.662597, 2.701605$ 45. **(a)** true **(b)** false
49. **(c)** inflection point $(1, 0)$; concave up on $(1, +\infty)$; concave down on $(-\infty, 1)$
55. 57.

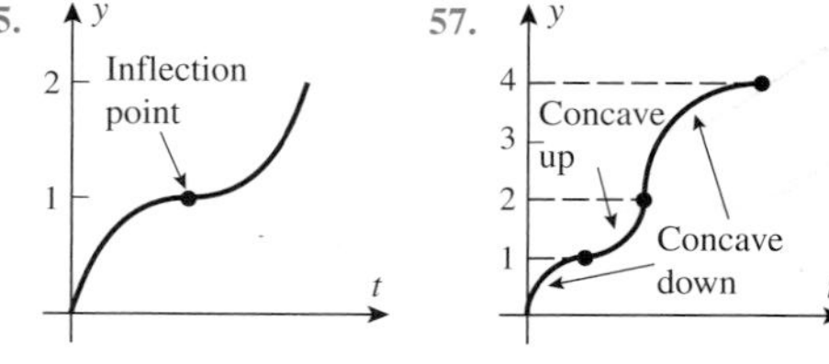

Exercise Set 4.2 (Page 242)

1. **(a)** **(b)**

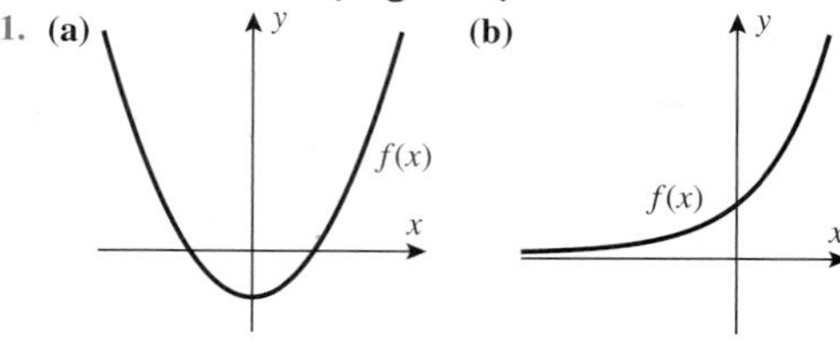

(c) **(d)**

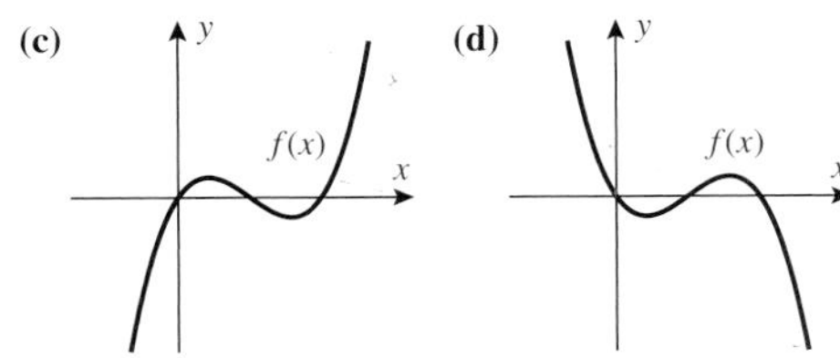

5. **(b)** nothing **(c)** f has a relative minimum at $x = 1$, g has no relative extremum at $x = 1$.
7. Critical: $0, \pm\sqrt{2}$; stationary: $0, \pm\sqrt{2}$
9. Critical: $-3, 1$; stationary: $-3, 1$ 11. Critical: $0, \pm 5$; stationary: 0
13. Critical: $n\pi/2$ for every integer n; stationary: $n\pi + \pi/2$ for every integer n
15. **(a)** none **(b)** $x = 1$ **(c)** none 17. **(a)** 2 **(b)** 0 **(c)** 1, 3
19. 0 (neither), $\sqrt[3]{5}$ (min) 21. -2 (min), $2/3$ (max)
23. relative maximum at $(4/3, 19/3)$
25. relative maximum at $(\pi/4, 1)$; relative minimum at $(3\pi/4, -1)$
27. relative maximum at $(1, 1)$; relative minima at $(0, 0)$, $(2, 0)$
29. relative maximum at $(-1, 0)$; relative minimum at $(-3/5, -108/3125)$
31. relative maximum at $(-1, 1)$; relative minimum at $(0, 0)$
33. no relative extrema
35. relative maximum at $(3/2, 9/4)$; relative minima at $(0, 0)$, $(3, 0)$
37. intercepts: $(0, -4)$, $(-1, 0)$, $(4, 0)$
stationary point: $(3/2, -25/4)$ (min)
inflection points: none

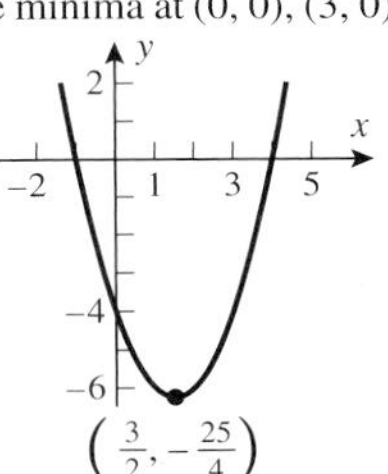

39. intercepts: $(0, 5)$, $\left(\dfrac{-7 \pm \sqrt{57}}{4}, 0\right)$, $(5, 0)$
stationary points: $(-2, 49)$ (max), $(3, -76)$ (min)
inflection point: $(1/2, -27/2)$

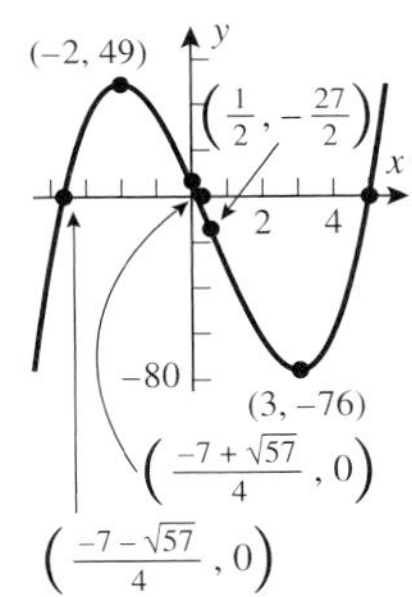

41. intercepts: $(-1, 0)$, $(0, 0)$, $(2, 0)$
stationary points: $(-1, 0)$ (max), $\left(\frac{1-\sqrt{3}}{2}, \frac{9-6\sqrt{3}}{4}\right)$ (min), $\left(\frac{1+\sqrt{3}}{2}, \frac{9+6\sqrt{3}}{4}\right)$ (max)
inflection points: $\left(-\frac{1}{\sqrt{2}}, \frac{5}{4} - \sqrt{2}\right)$, $\left(\frac{1}{\sqrt{2}}, \frac{5}{4} + \sqrt{2}\right)$,

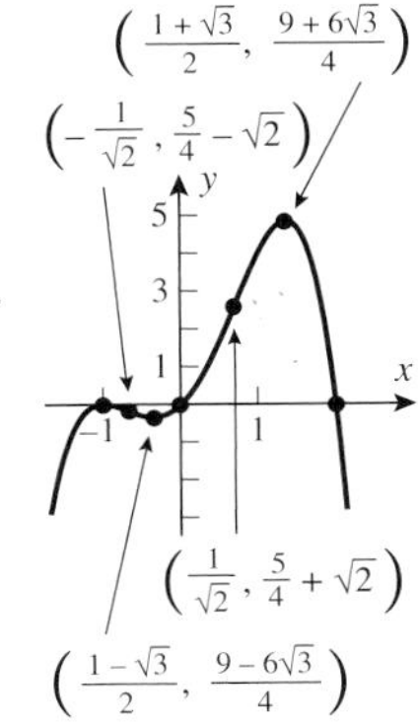

43. intercepts: $(0, -1)$, $(-1, 0)$, $(1, 0)$
stationary points: $(-1/2, -27/16)$ (min), $(1, 0)$ (neither)
inflection points: $(0, -1)$, $(1, 0)$

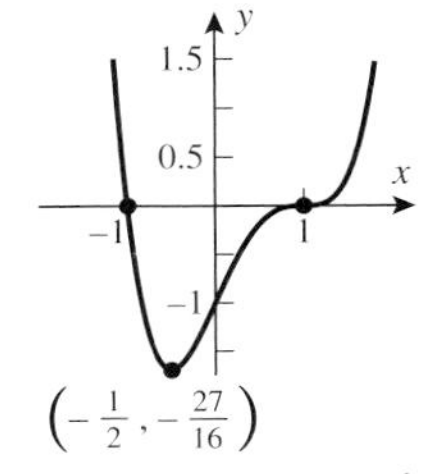

45. intercepts: $(-1, 0)$, $(0, 0)$, $(1, 0)$
stationary points: $(-1, 0)$ (max), $\left(-\frac{1}{\sqrt{5}}, -\frac{16}{25\sqrt{5}}\right)$ (min), $\left(\frac{1}{\sqrt{5}}, \frac{16}{25\sqrt{5}}\right)$ (max), $(1, 0)$ (min)
inflection points: $\left(-\sqrt{\frac{3}{5}}, -\frac{4}{25}\sqrt{\frac{3}{5}}\right)$,

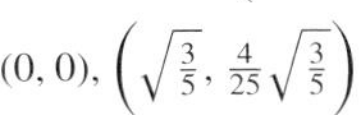
$(0, 0)$, $\left(\sqrt{\frac{3}{5}}, \frac{4}{25}\sqrt{\frac{3}{5}}\right)$

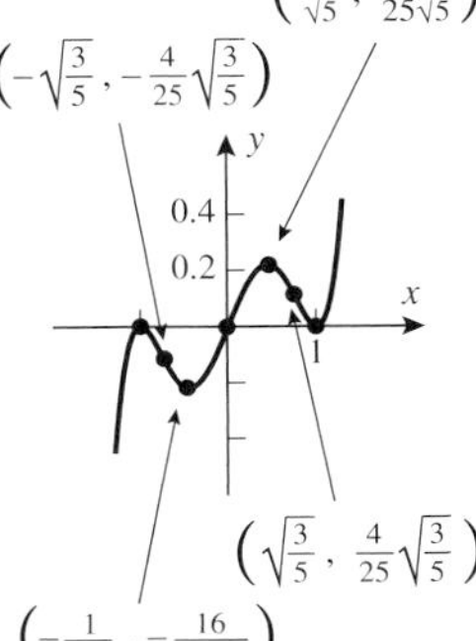

47. **(a)** **(b)** **(c)** **(d)**

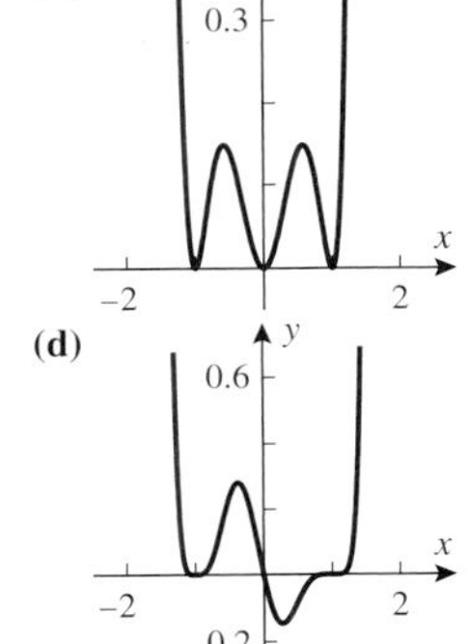

49. relative min of 0 at $x = \pi/2, \pi, 3\pi/2$;
relative max of 1 at $x = \pi/4, 3\pi/4, 5\pi/4, 7\pi/4$

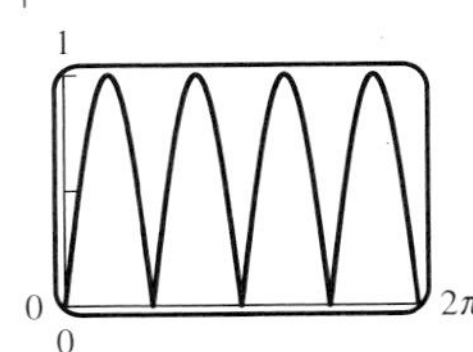

51. relative min of 0 at $x = \pi/2, 3\pi/2$;
relative max of 1 at $x = \pi$

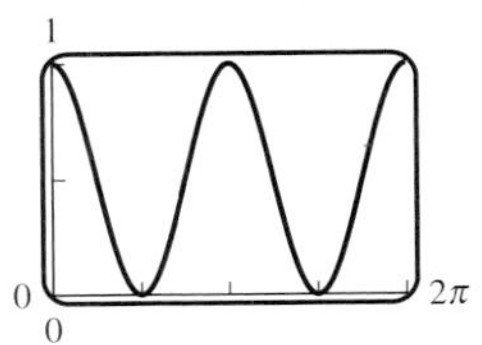

53. relative minima at $x = -3.58, 3.33$;
relative max at $x = 0.25$

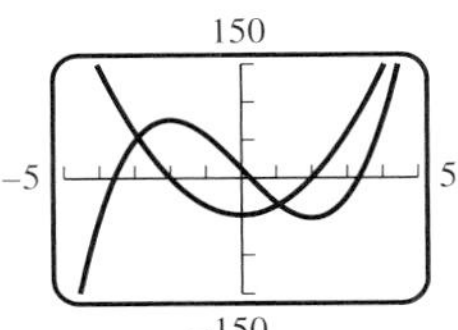

55. relative maximum at $x \approx -0.272$; relative minimum at $x \approx 0.224$
57. relative maximum at $x = 0$; relative minima at $x \approx \pm 0.618$

59. **(a)** 54 **(b)** 9

61. **(a)** $(-2.2, 4)$, $(2, 1.2)$, $(4.2, 3)$
(b) critical numbers at $x = -5.1, -2, 0.2, 2$;
local min at $x = -5.1, 2$; local max at $x = -2$;
no extrema at $x = 0.2$; $f''(1) \approx -1.2$

63. $a = -2, b = 3, c = d = 0$

65. **(a)**

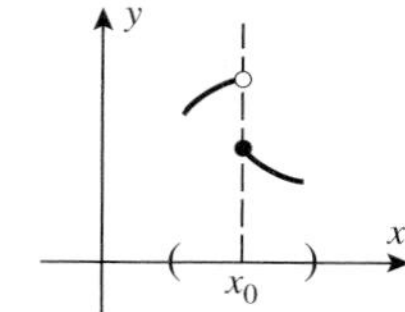

$f(x_0)$ is not an extreme value

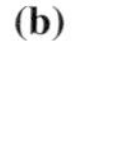

(b)

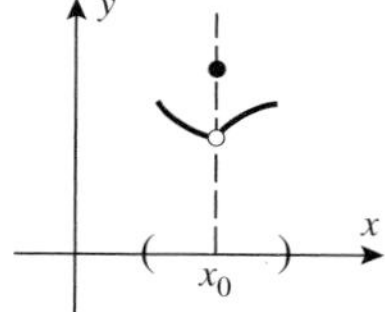

$f(x_0)$ is a relative maximum

(c)

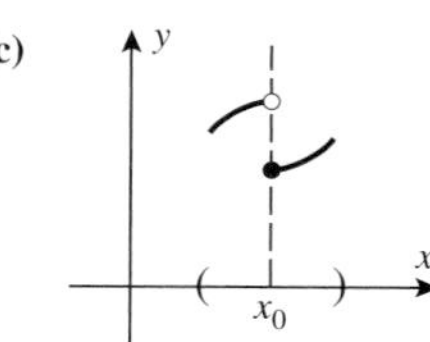

$f(x_0)$ is a relative minimum

▶ Exercise Set 4.3 (Page 252)

1. stationary points: none;
inflection points: none;
asymptotes: $x = 4$, $y = -2$;
asymptote crossings: none

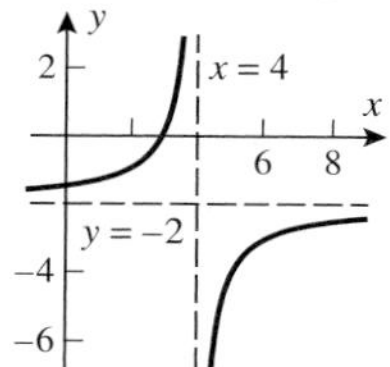

3. stationary points: none;
inflection point: $(0,0)$;
asymptotes: $x = \pm 2$, $y = 0$;
asymptote crossings: $(0,0)$

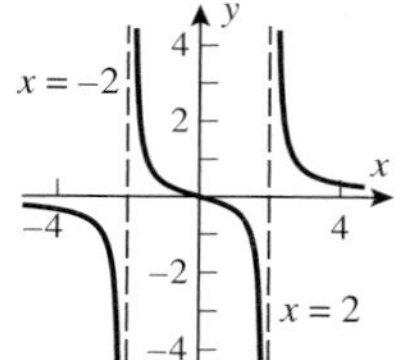

5. stationary point: $(0,0)$;
inflection points: $\left(\pm\frac{2}{\sqrt{3}}, \frac{1}{4}\right)$;
asymptote: $y = 1$;
asymptote crossings: none

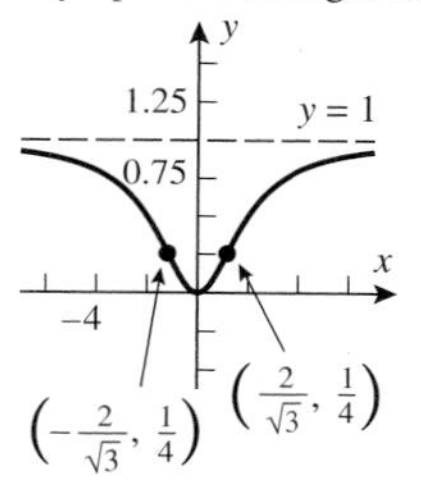

7. stationary point: $(0, -1)$;
inflection points: $(0, -1)$,
$\left(-\frac{1}{\sqrt[3]{2}}, -\frac{1}{3}\right)$;
asymptotes: $x = 1$, $y = 1$;
asymptote crossings: none

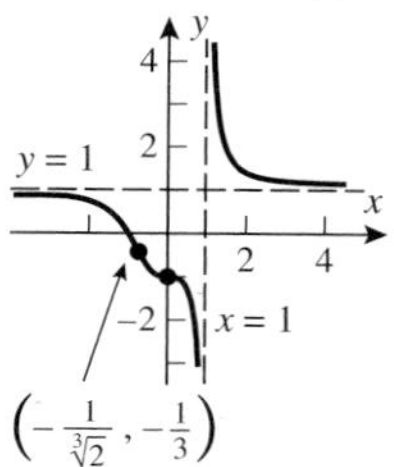

9. stationary point: $(4, 11/4)$;
inflection point: $(6, 25/9)$;
asymptotes: $x = 0$, $y = 3$;
asymptote crossing: $(2,3)$

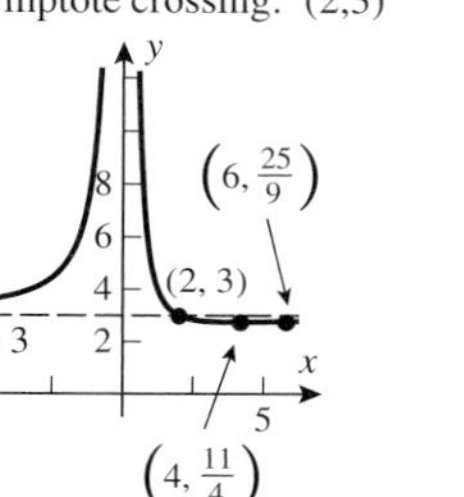

11. stationary point: $(-1/3, 0)$;
inflection point: $(-1, 1)$;
asymptotes: $x = 1$, $y = 9$;
asymptote crossing: $(1/3, 9)$

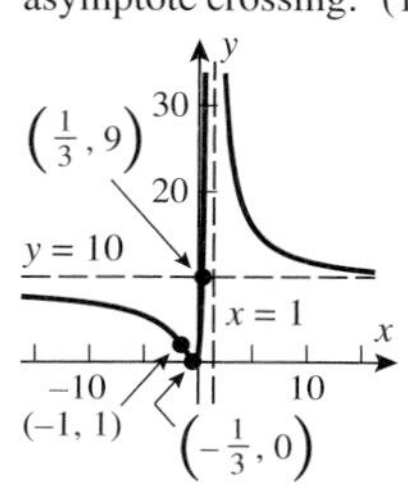

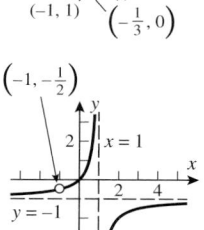

13. stationary points: none;
inflection points: none;
asymptotes: $x = 1$, $y = -1$;
asymptote crossings: none

15. **(a)**

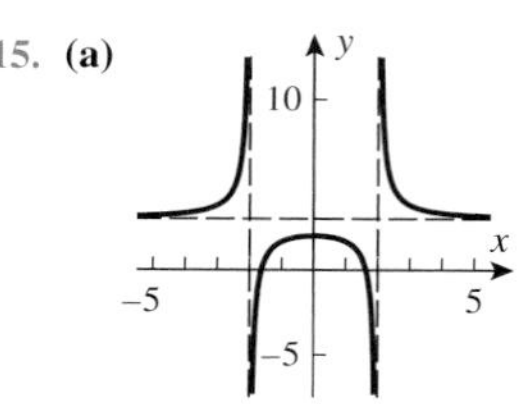

(b)

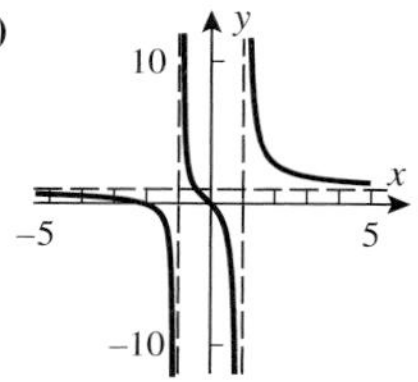

(c)

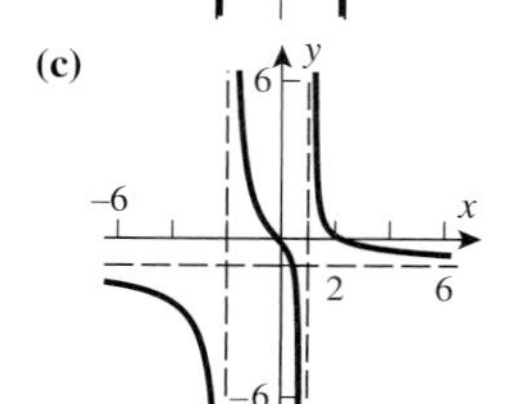

(d)

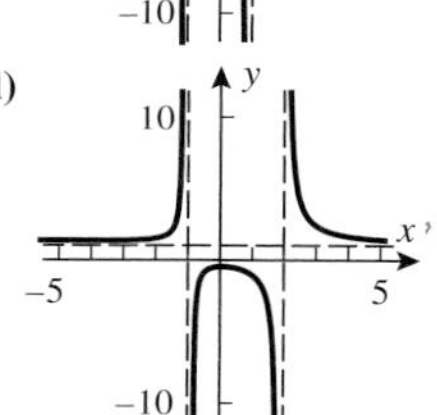

17.

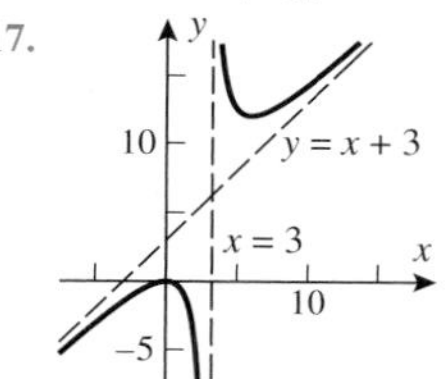

19. stationary point: $\left(-\frac{1}{\sqrt[3]{2}}, \frac{3}{2}\sqrt[3]{2}\right)$;
inflection point: $(1, 0)$;
asymptotes: $y = x^2$, $x = 0$;
asymptote crossings: none

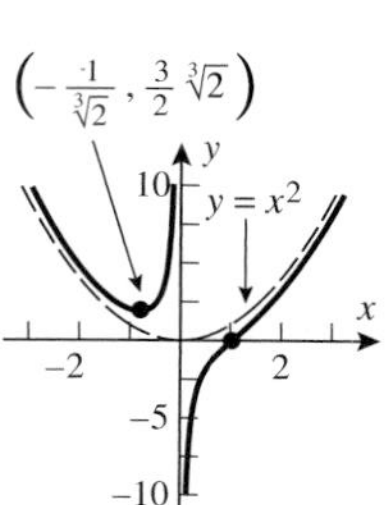

21. stationary points: $(-4, -27/2)$, $(2, 0)$;
inflection point: $(2,0)$;
asymptotes: $x = 0$, $y = x - 6$;
asymptote crossing: $(2/3, -16/3)$

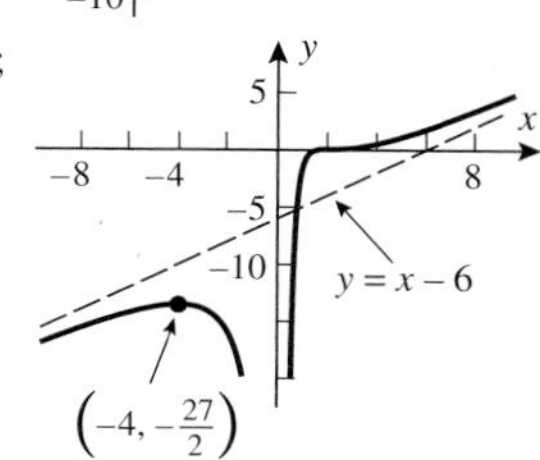

23. stationary points: $(-3, 23)$, $(0, -4)$; inflection point: $(0, -4)$; asymptotes: $x = -2$, $y = x^2 - 2x$; asymptote crossings: none

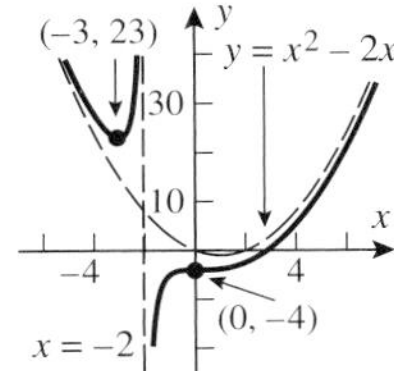

25. (a). VI (b). I (c). III (d). V (e). IV (f). II

27. critical points: $(\pm 1/2, 0)$; inflection points: none

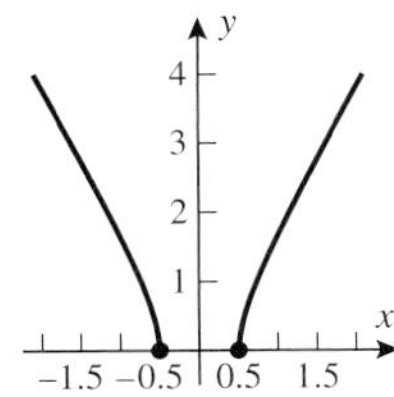

29. critical points: $(-1, 1)$, $(0, 0)$; inflection points: none

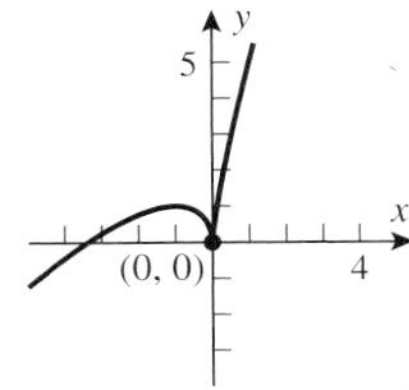

31. critical points: $(0, 0)$, $(1, 3)$; inflection point: $(-2, -6\sqrt[3]{2})$ It's hard to see all the important features in one graph, so two graphs are shown:

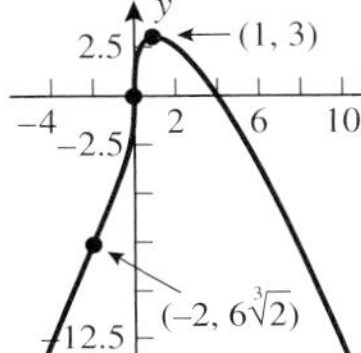

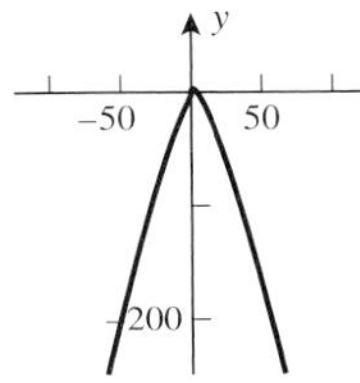

33. critical points: $(0, 4)$, $(1, 3)$; inflection points: $(0, 4)$, $(8, 4)$

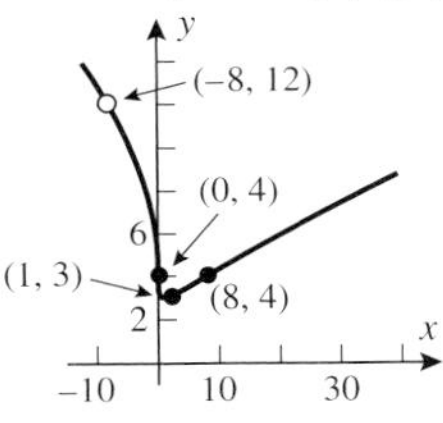

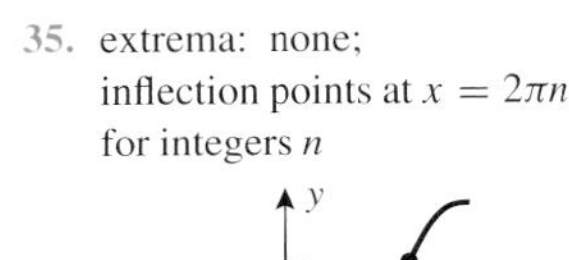

35. extrema: none; inflection points at $x = 2\pi n$ for integers n

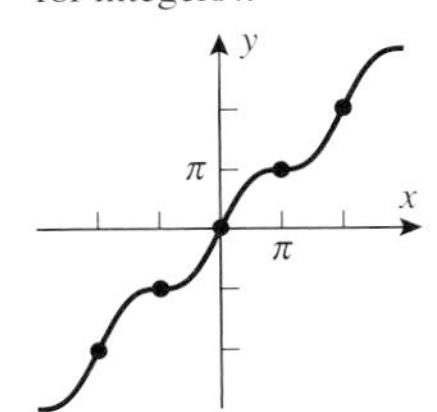

37. minima: $x = 7\pi/6 + 2\pi n$ for integers n; maxima: $x = \pi/6 + 2\pi n$ for integers n; inflection points: $x = 2\pi/3 + \pi n$ for integers n

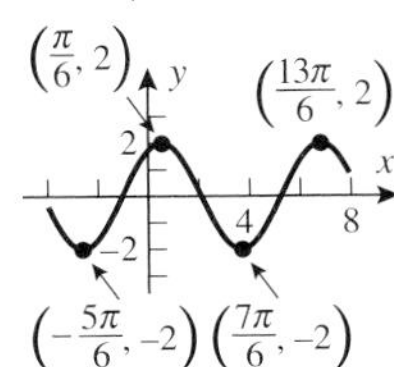

39. relative minima: 1 at $x = -\pi, \pi, 3\pi$, -1 at $x = 0, 2\pi$; relative maxima: $5/4$ at $x = -2\pi/3, 2\pi/3, 4\pi/3, 8\pi/3$; inflection points where $\cos x = \dfrac{-1 \pm \sqrt{33}}{8}$: $(-2.57, 1.13)$, $(-0.94, 0.06)$, $(0.94, 0.06)$, $(2.57, 1.13)$, $(3.71, 1.13)$, $(5.35, 0.06)$, $(7.22, 0.06)$, $(8.86, 1.13)$

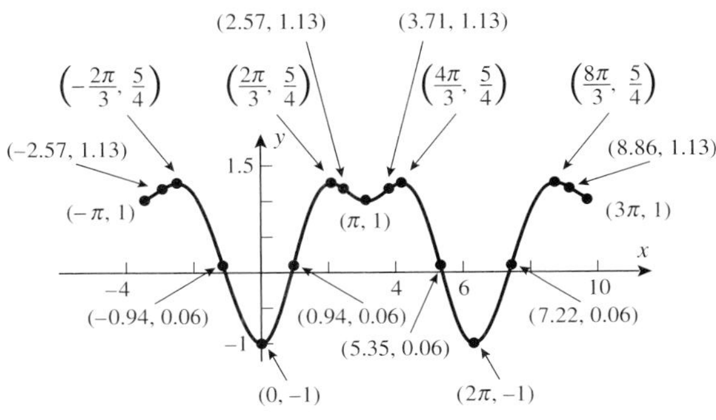

41. **(a)** $x = 1, 2.5, 3, 4$ **(b)** $(-\infty, 1]$, $[2.5, 3]$ **(c)** relative max at $x = 1, 3$; relative min at $x = 2.5$ **(d)** $x \approx 0.6, 1.9, 4$

43.

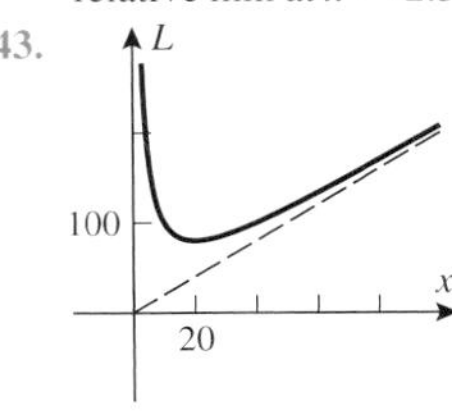

45. Graph misses zeroes at $x = 0, 1$ and min at $x = 5/6$

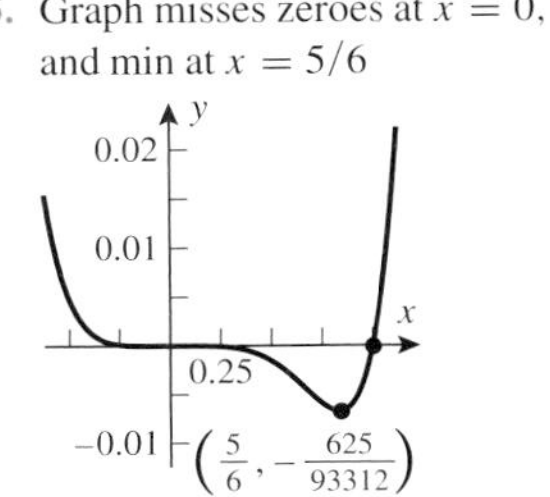

▶ Exercise Set 4.4 (Page 260)

1. relative maxima at $x = 2, 6$; absolute max at $x = 6$; relative and absolute min at $x = 4$

3. **(a)**

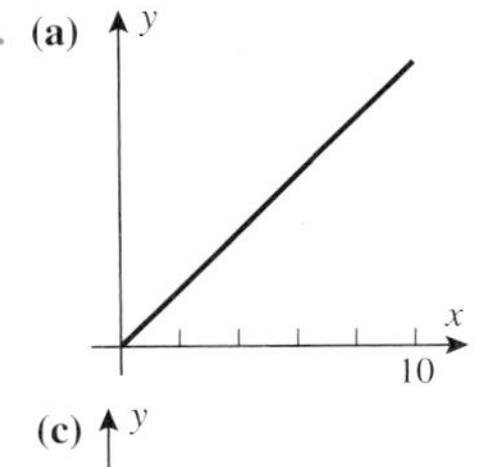

(b)

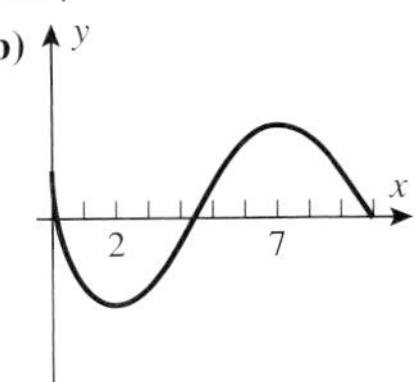

(c)

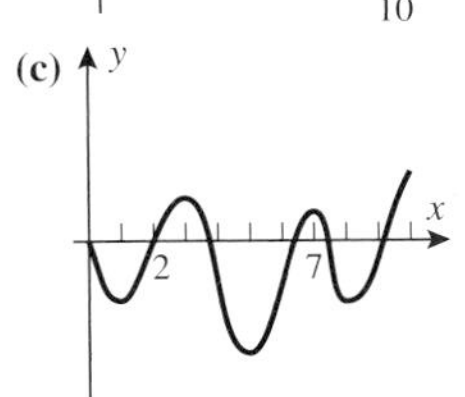

7. max $= 2$ at $x = 1, 2$; min $= 1$ at $x = 3/2$

9. max $= 8$ at $x = 4$; min $= -1$ at $x = 1$

11. maximum value $3/\sqrt{5}$ at $x = 1$, minimum value $-3/\sqrt{5}$ at $x = -1$

13. max $= \sqrt{2} - \pi/4$ at $x = -\pi/4$; min $= \pi/3 - \sqrt{3}$ at $x = \pi/3$

15. maximum value 17 at $x = -5$, minimum value 1 at $x = -3$.

17. no max; min $= -9/4$ at $x = 1/2$

19. maximum value $f(1) = 1$, no minimum

21. no max or min

23. max $= -2 - 2\sqrt{2}$ at $x = -1 - \sqrt{2}$; no min

25. no max; min $= 0$ at $x = 0, 2$

27. maximum value 48 at $x = 8$, minimum value 0 at $x = 0, 20$

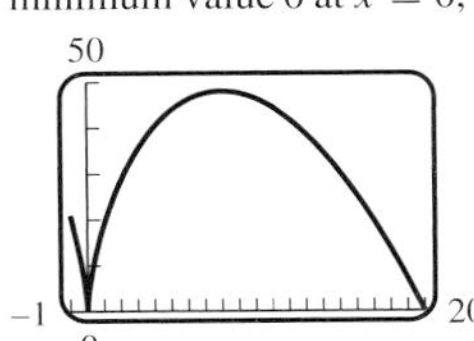

29. no maximum or minimum

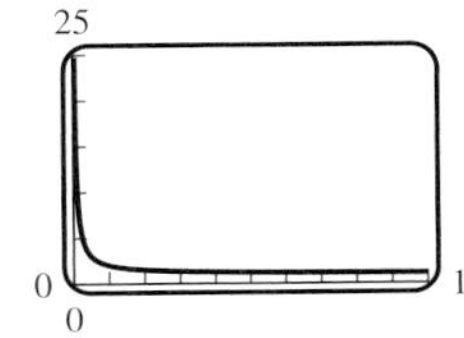

31. max $= 2\sqrt{2} + 1$ at $x = 3\pi/4$, min $= \sqrt{3}$ at $x = \pi/3$

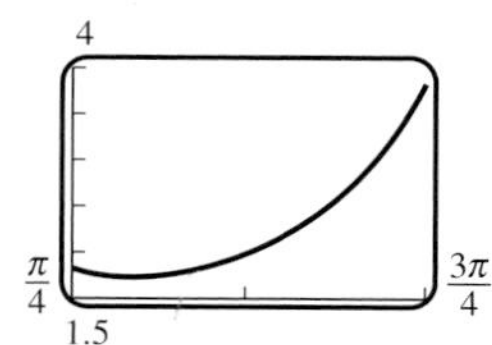

33. maximum value $\sin(1) \approx 0.84147$, minimum value $-\sin(1) \approx -0.84147$

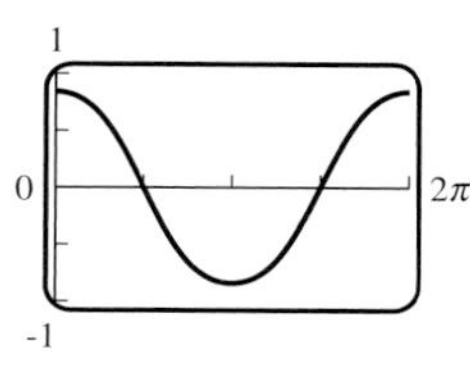

35. maximum value 2, minimum value $-\frac{1}{4}$
37. max $= 3$ at $x = 2n\pi$, min $= -3/2$ at $x = \pm 2\pi/3 + 2n\pi$ for any integer n
41. 2 45. $(\frac{1}{2}, -\frac{1}{4})$ is closest, $(-1, -1)$ is farthest
47. maximum $y = 4$ at $t = \pi, 3\pi$; minimum $y = 0$ at $t = 0, 2\pi$

▶ Exercise Set 4.5 (Page 271)

1. **(a)** 1 **(b)** $\frac{1}{2}$ 3. 500 ft parallel to stream, 250 ft perpendicular
5. 500 ft ($3 fencing) × 750 ft ($2 fencing) 7. 5 in × $\frac{12}{5}$ in
9. $10\sqrt{2} \times 10\sqrt{2}$ 11. 80 ft ($1 fencing), 40 ft ($2 fencing)
15. 150 yd × 150 yd × $150\sqrt{2}$ yd
17. 11664 in^3 19. $\dfrac{200}{27}$ ft^3 21. base 10 cm square, height 20 cm
23. ends $\sqrt[3]{3V/4}$ units square, height $\frac{4}{3}\sqrt[3]{3V/4}$
25. height $= 2\sqrt{(5-\sqrt{5})/10}\,R$, radius $= \sqrt{(5+\sqrt{5})/10}\,R$
29. height = radius = $\sqrt[3]{500/\pi}$ cm 31. $L/12$ by $L/12$ by $L/12$
33. height $= L/\sqrt{3}$, radius $= \sqrt{2/3}L$
35. radius $= \sqrt[6]{450/\pi^2}$ cm, height $= \dfrac{30}{\pi}\sqrt[3]{\pi^2/450}$ cm
37. height $= 4R$, radius $= \sqrt{2}R$ 39. $\pi/3$ 41. $5\sqrt{5}$ ft
43. **(a)** 7000 units **(b)** yes **(c)** $15 45. 13,722 lb 47. $1/\sqrt{5}$
53. **(a)** $(2,2), (-2,-2), (2/\sqrt{3}, -2/\sqrt{3}), (-2/\sqrt{3}, 2/\sqrt{3})$
55. $\left(\sqrt{2}, \frac{1}{2}\right)$ 57. $\left(-1/\sqrt{3}, \frac{3}{4}\right)$ 59. $4(1 + 2^{2/3})^{3/2}$ ft
61. 30 cm from the weaker source 63. $x = 1 + 2\sqrt{2}$
67. **(c)** $\frac{1}{4}$ mile downstream from the house

▶ Exercise Set 4.6 (Page 280)

1. 1.414213562 3. 1.817120593 5. $x \approx 1.76929$
7. $x \approx 1.224439550$ 9. $x \approx -1.24962$ 11. $x \approx 1.02987$
13. $x \approx 4.493409458$ 15. $x \approx 0.68233$
17. −0.474626618, 1.395336994

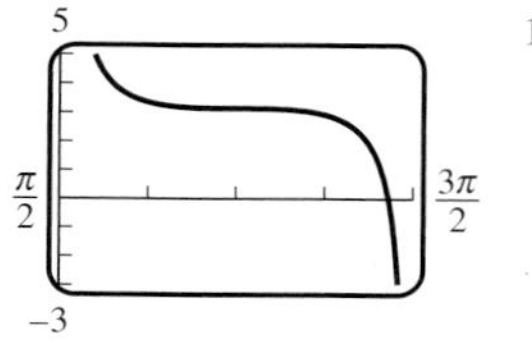

19. **(b)** 3.162277660
21. −4.098859132
23. (0.589754512, 0.347810385)
25. **(b)** $\theta \approx 2.99156$ rad or 171°
27. −1.220744085, 0.724491959
29. $i = 0.053362$ or 5.33%
33. **(d)** $f(c) = 0$

▶ Exercise Set 4.7 (Page 287)

1. $c = 4$ 3. $c = \pi$ 13. **(a)** $[-2, 1]$
5. $c = 1$ 7. $c = 1$ **(b)** $c \approx -1.29$
9. $\frac{5}{4}$ 11. $-\sqrt{5}$ **(c)** −1.2885843

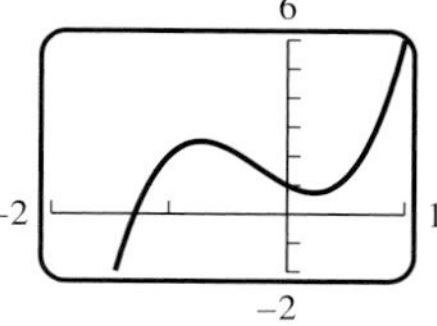

15. **(b)** $\tan x$ is not continuous on $[0, \pi]$.
35.

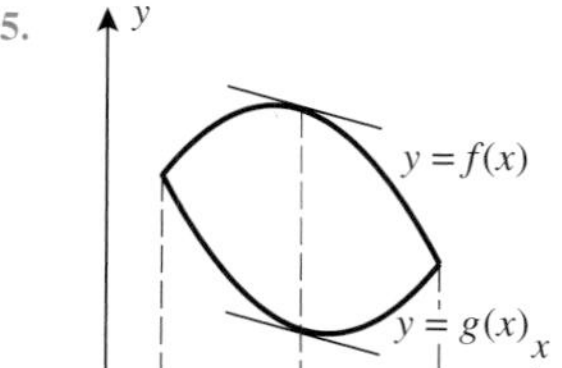

39. $a = 6, b = -3$

▶ Exercise Set 4.8 (Page 294)

1. **(a)** positive, negative, slowing down
(b) positive, positive, speeding up
(c) negative, positive, slowing down
3. **(a)** left
(b) negative
(c) speeding up
(d) slowing down
5.

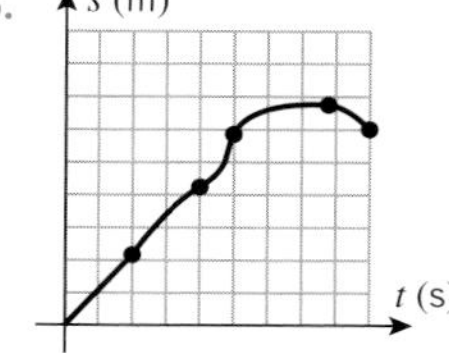

7.

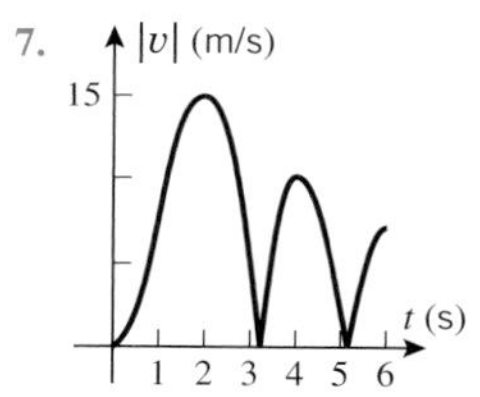

9. **(a)** 7.5 ft/s^2 **(b)** $t = 0$ s
11. **(a)**

t	s	v	a
1	0.71	0.56	−0.44
2	1	0	−0.62
3	0.71	−0.56	−0.44
4	0	−0.79	0
5	−0.71	−0.56	0.44

(b) stopped at $t = 2$; moving right at $t = 1$; moving left at $t = 3, 4, 5$
(c) speeding up at $t = 3$; slowing down at $t = 1, 5$; neither at $t = 2, 4$

13. **(a)** $v(t) = 3t^2 - 12t$, $a(t) = 6t - 12$ **(b)** $s(1) = -5$ ft, $v(1) = -9$ ft/s, $|v(1)| = 9$ ft/s, $a(1) = -6$ ft/s^2 **(c)** 0, 4 **(d)** speeding up for $0 < t < 2$ and $4 < t$, slowing down for $2 < t < 4$ **(e)** 39 ft
15. **(a)** $v(t) = 3\pi\sin(\pi t/3)$; $a(t) = \pi^2\cos(\pi t/3)$ **(b)** $s(1) = 9/2$ ft; $v(1) = \text{speed} = 3\sqrt{3}\pi/2$ ft/s; $a(1) = \pi^2/2$ ft/s^2 **(c)** $t = 0$ s, 3 s **(d)** speeding up: $0 < t < 1.5$, $3 < t < 4.5$; slowing down: $1.5 < t < 3$, $4.5 < t < 5$ **(e)** 31.5 ft
17. **(a)** $\sqrt{5}$ **(b)** $\sqrt{5}/10$

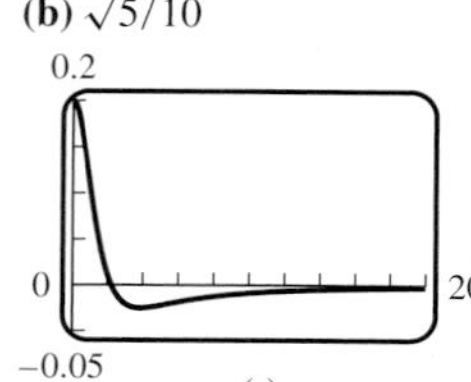

(c) speeding up for $\sqrt{5} < t < \sqrt{15}$, slowing down for $0 < t < \sqrt{5}$ and $\sqrt{15} < t$

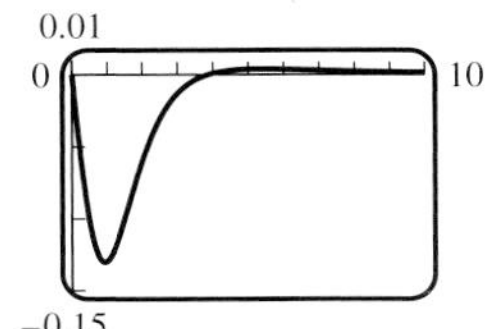

19.

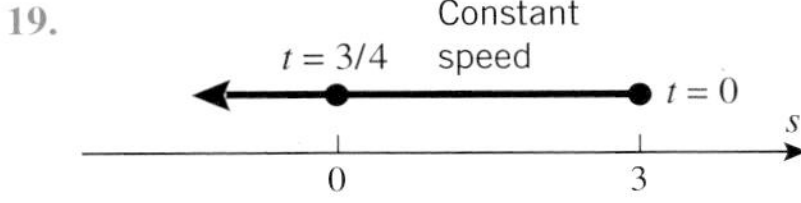

21.

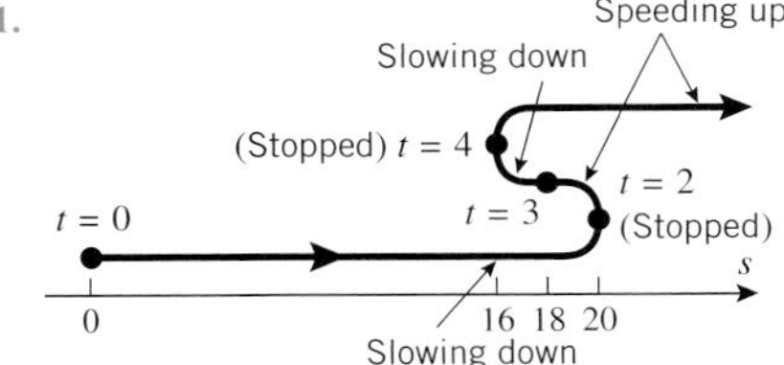

23.

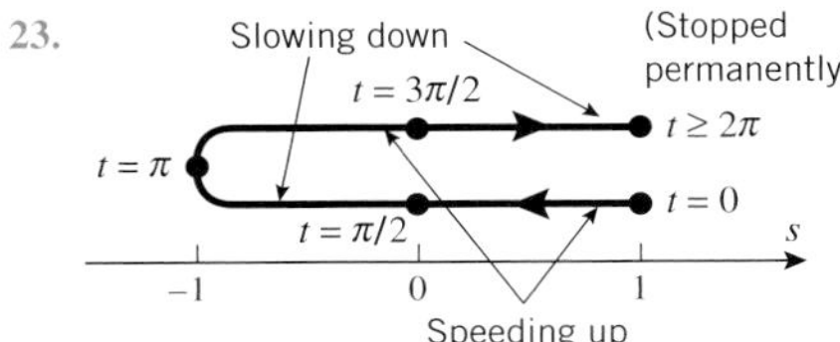

25. **(a)** 12 ft/s **(b)** $t = 2.2$ s, $s = -24.2$ ft

27. **(a)** $s = 0$, $v = 2$ **(b)** $s = 1$, $a = -4$

29. **(a)** 1.5 **(b)** $\sqrt{2}$

1.5
0
5
0

31. **(b)** $\frac{2}{3}$ unit **(c)** $0 \le t < 1$ and $t > 2$

▶ Chapter 4. Review Exercises (Page 297)

1. **(a)** $f(x_1) < f(x_2)$; $f(x_1) > f(x_2)$; $f(x_1) = f(x_2)$
   (b) $f' > 0$; $f' < 0$; $f' = 0$

3. **(a)** $\left[\frac{5}{2}, +\infty\right)$ **(b)** $\left(-\infty, \frac{5}{2}\right]$ **(c)** $(-\infty, +\infty)$ **(d)** none **(e)** none

5. **(a)** $[0, +\infty)$ **(b)** $(-\infty, 0]$ **(c)** $(-\sqrt{2/3}, \sqrt{2/3})$
   (d) $(-\infty, -\sqrt{2/3})$, $(\sqrt{2/3}, +\infty)$ **(e)** $-\sqrt{2/3}$, $\sqrt{2/3}$

7. **(a)** $[-1, +\infty)$ **(b)** $(-\infty, -1]$ **(c)** $(-\infty, 0)$, $(2, +\infty)$
   (d) $(0, 2)$ **(e)** 0, 2

9. increasing; $[\pi, 2\pi]$
   decreasing; $[0, \pi]$
   concave up; $(\pi/2, 3\pi/2)$
   concave down; $(0, \pi/2)$, $(3\pi/2, 2\pi)$
   inflection points; $\pi/2$, $3\pi/2$

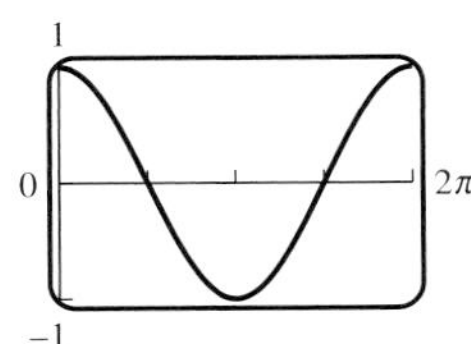

11. increasing; $[0, \pi/4]$, $[3\pi/4, \pi]$
    decreasing; $[\pi/4, 3\pi/4]$
    concave up; $(\pi/2, \pi)$
    concave down; $(0, \pi/2)$
    inflection points; $\pi/2$

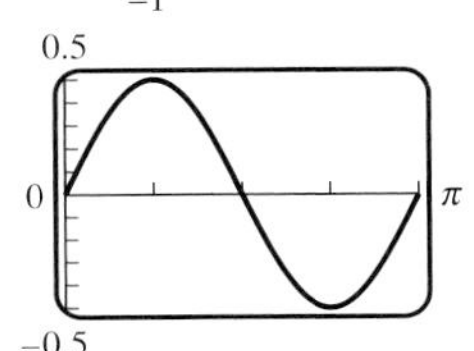

13. **(a)** **(b)** **(c)**

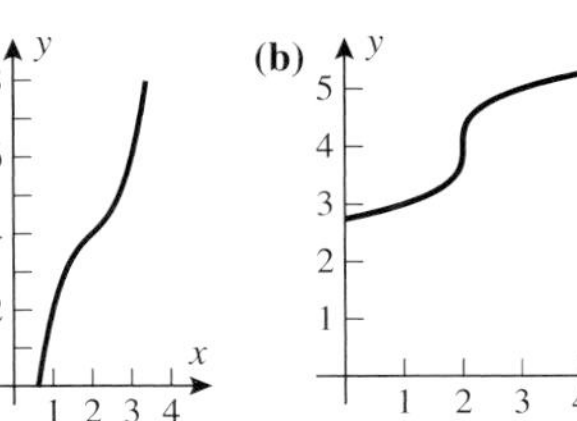

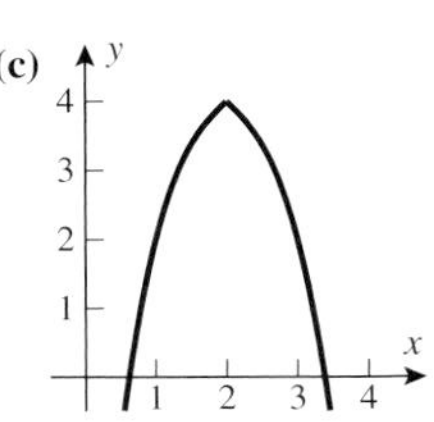

15. $-\dfrac{b}{2a} \le 0$ 17. **(a)** at an inflection point

21. **(a)** $x = \pm\sqrt{2}$ (stationary points) **(b)** $x = 0$ (stationary point)

23. **(a)** relative max at $x = 1$, relative min at $x = 7$, neither at $x = 0$
    (b) relative max at $x = \pi/2, 3\pi/2$; relative min at $x = 7\pi/6, 11\pi/6$
    (c) relative max at $x = 5$

25. $\lim_{x\to-\infty} f(x) = +\infty$, $\lim_{x\to+\infty} f(x) = +\infty$;
    relative min at $x = 0$;
    points of inflection at $x = \frac{1}{2}, 1$;
    no asymptotes

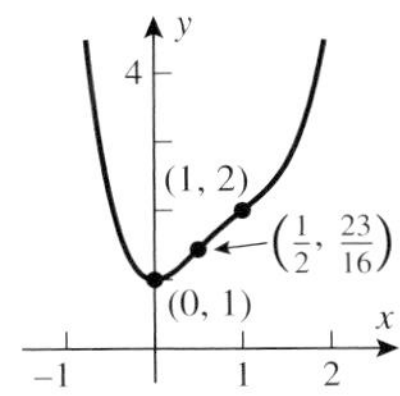

27. $\lim_{x\to\pm\infty} f(x)$ does not exist; critical point at $x = 0$; relative min at $x = 0$; point of inflection when $1 + 4x^2\tan(x^2+1) = 0$; vertical asymptotes at $x = \pm\sqrt{\pi(n+\frac{1}{2}) - 1}$, $n = 0, 1, 2, \ldots$

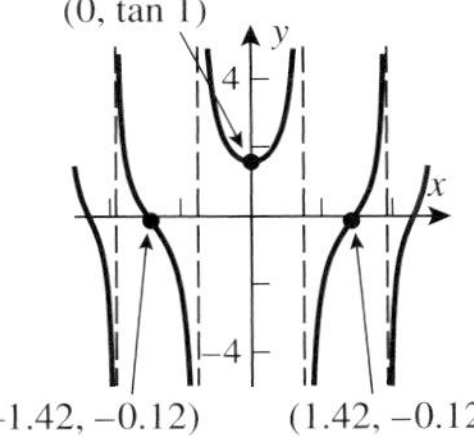

29. critical points at $x = -5, 0$; relative max at $x = -5$, relative min at $x = 0$; points of inflection at $x = -7.26, -1.44, 1.20$; horizontal asymptote $y = 1$ for $x \to \pm\infty$

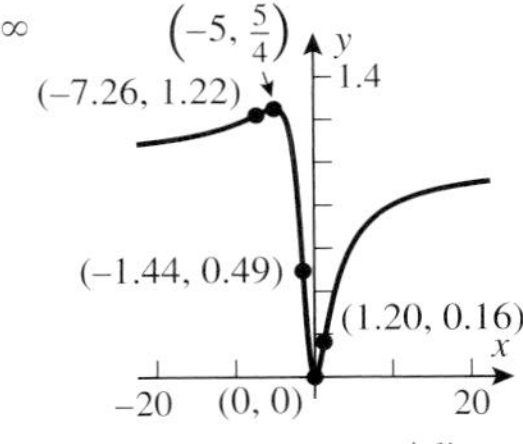

31. $\lim_{x\to-\infty} f(x) = +\infty$, $\lim_{x\to+\infty} f(x) = -\infty$;
    critical point at $x = 0$;
    no extrema;
    inflection point at $x = 0$
    (f changes concavity);
    no asymptotes

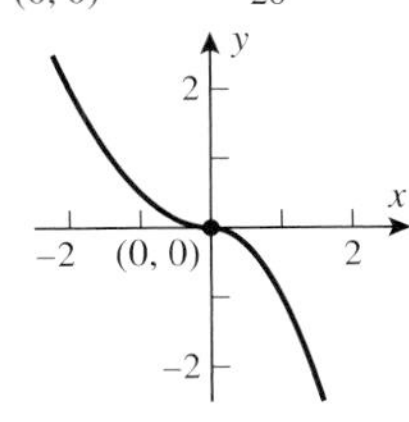

33. no relative extrema 35. relative min of 0 at $x = 0$

37. relative min of 0 at $x = 0$

39. **(a)**

40
−5
5
−40

(b) relative max at $x = -\frac{1}{20}$, relative min at $x = \frac{1}{20}$

(c) The finer details can be seen when graphing over a much smaller x-window.

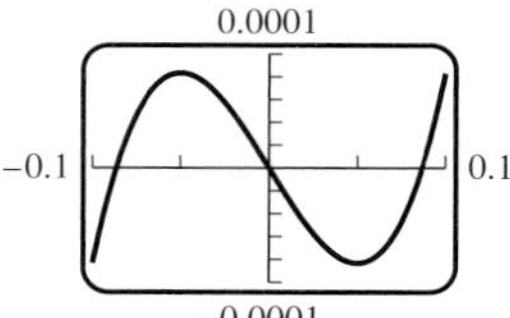

41. **(a)**

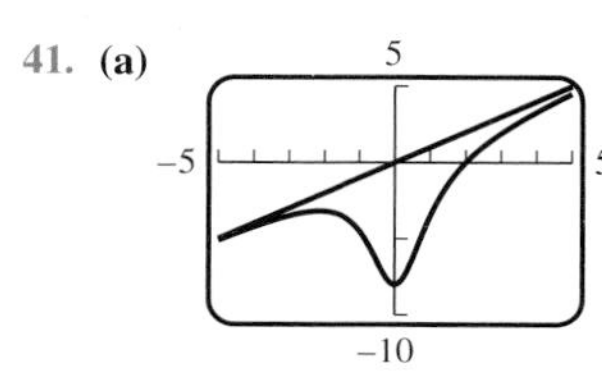

43. $f(x) = \dfrac{x^2 + x - 7}{3x^2 + x - 1}, x \neq \dfrac{1}{2}$

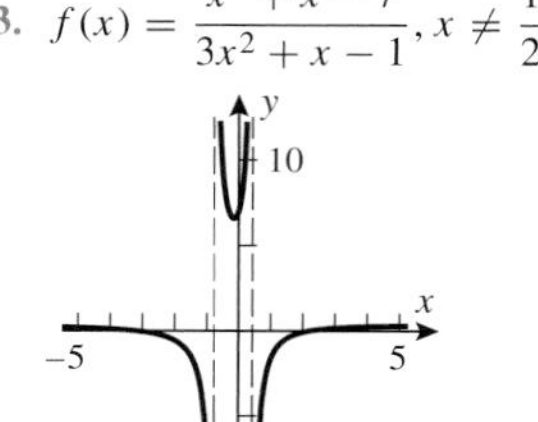

vertical asymptotes
$x = (-1 \pm \sqrt{13})/6$

47. **(a)** true **(b)** false

49. **(a)** no max; min $= -13/4$ at $x = 3/2$ **(b)** no max or min
(c) max $= 0$ at $x = 0$ and $x = 2$; no mim

51. **(a)** minimum value 0 for $x = \pm 1$, no maximum **(b)** max $= 1/2$ at $x = 1$; min $= -1/2$ at $x = -1$

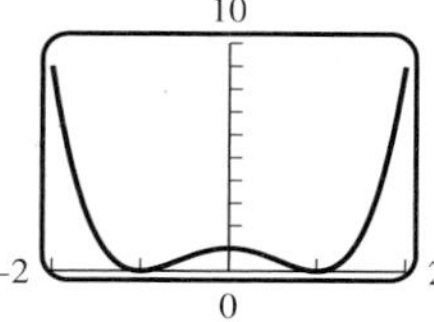

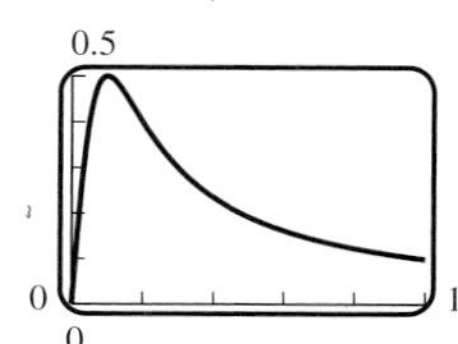

(c) maximum value 2 at $x = 0$, minimum value $\sqrt{3}$ at $x = \pi/6$

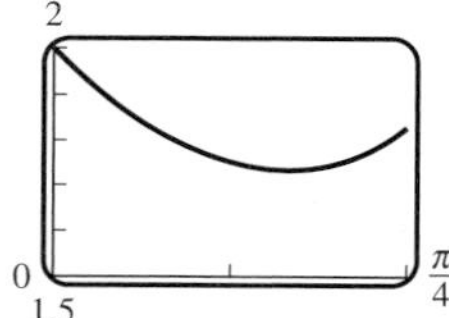

53. **(a)**

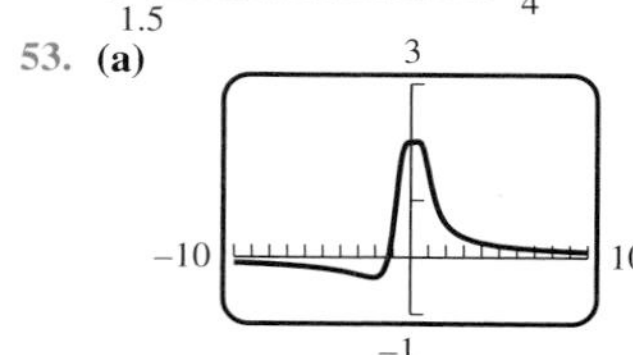

(b) minimum: $(-2.111985, -0.355116)$; maximum: $(0.372591, 2.012931)$

55. width $= 4\sqrt{2}$, height $= 3\sqrt{2}$ 57. 2 in square

59. $x \approx -2.11491, 0.25410, 1.86081$

61. -1.165373043

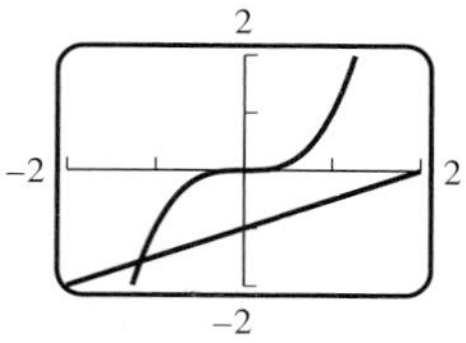

63. 249×10^6 km

65. **(a)** yes, $c = 0$ **(b)** no
(c) yes, $c = \sqrt{\pi/2}$

69. **(a)** yes **(b)** yes

71. **(a)** $v = -2\dfrac{t(t^4 + 2t^2 - 1)}{(t^4 + 1)^2}, a = 2\dfrac{3t^8 + 10t^6 - 12t^4 - 6t^2 + 1}{(t^4 + 1)^3}$
(c) $t = 0.64, s = 1.2$ **(d)** $0 \leq t \leq 0.64$ s
(e) speeding up when $0 \leq t < 0.36$ and $0.64 < t < 1.1$, otherwise slowing down **(f)** maximum speed $= 1.05$ m/s when $t = 1.10$ s

Exercise Set 5.1 (Page 307)

1.

n	2	5	10	50	100
A_n	0.853553	0.749739	0.710509	0.676095	0.671463

3.

n	2	5	10	50	100
A_n	1.57080	1.93376	1.98352	1.99935	1.99984

5.

n	2	5	10	50	100
A_n	0.583333	0.645635	0.668771	0.688172	0.690653

7.

n	2	5	10	50	100
A_n	0.433013	0.659262	0.726130	0.774567	0.780106

9. $3(x - 1)$ 11. $x(x + 2)$ 13. $(x + 3)(x - 1)$

17. area $= A(6) - A(3)$ 19. $f(x) = 2x; a = 2$

Exercise Set 5.2 (Page 316)

1. **(a)** $\displaystyle\int \frac{x}{\sqrt{1 + x^2}}\,dx = \sqrt{1 + x^2} + C$
(b) $\displaystyle\int x^2 \cos(1 + x^3)\,dx = \frac{1}{3}\sin(1 + x^3) + C$

5. $\dfrac{d}{dx}\left[\sqrt{x^3 + 5}\right] = \dfrac{3x^2}{2\sqrt{x^3 + 5}}$, so $\displaystyle\int \frac{3x^2}{2\sqrt{x^3 + 5}}\,dx = \sqrt{x^3 + 5} + C.$

7. $\dfrac{d}{dx}[\sin(2\sqrt{x})] = \dfrac{\cos(2\sqrt{x})}{\sqrt{x}}$, so $\displaystyle\int \frac{\cos(2\sqrt{x})}{\sqrt{x}}\,dx = \sin(2\sqrt{x}) + C.$

9. **(a)** $(x^9/9) + C$ **(b)** $\frac{7}{12}x^{12/7} + C$ **(c)** $\frac{2}{9}x^{9/2} + C$

11. $\dfrac{5}{2}x^2 - \dfrac{1}{6x^4} + C$ 13. $-\dfrac{1}{2}x^{-2} - \dfrac{12}{5}x^{5/4} + \dfrac{8}{3}x^3 + C$

15. $(x^2/2) + (x^5/5) + C$ 17. $3x^{4/3} - \frac{12}{7}x^{7/3} + \frac{3}{10}x^{10/3} + C$

19. $\dfrac{x^2}{2} - \dfrac{2}{x} + \dfrac{1}{3x^3} + C$

21. $-3\cos x - 2\tan x + C$ 23. $\tan x + \sec x + C$

25. $\tan\theta + C$ 27. $\sec x + C$ 29. $\theta - \cos\theta + C$

31. $\tan x - \sec x + C$

33.

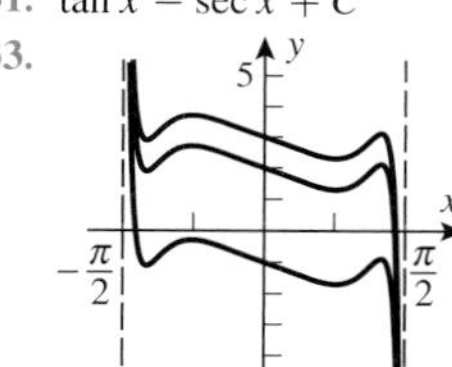

35. $f(x) = \cos x + 1$

37. **(a)** $y(x) = \frac{3}{4}x^{4/3} + \frac{5}{4}$
(b) $y = -\cos t + t + 1 - \pi/3$
(c) $y(x) = \frac{2}{3}x^{3/2} + 2x^{1/2} - \frac{8}{3}$

39. $f(x) = \dfrac{4}{15}x^{5/2} + C_1 x + C_2$

41. $y = x^2 + x - 6$ 43. $y = x^3 - 6x + 7$

45. **(a)** **(b)** **(c)** $f(x) = \dfrac{x^2}{2} - 1$

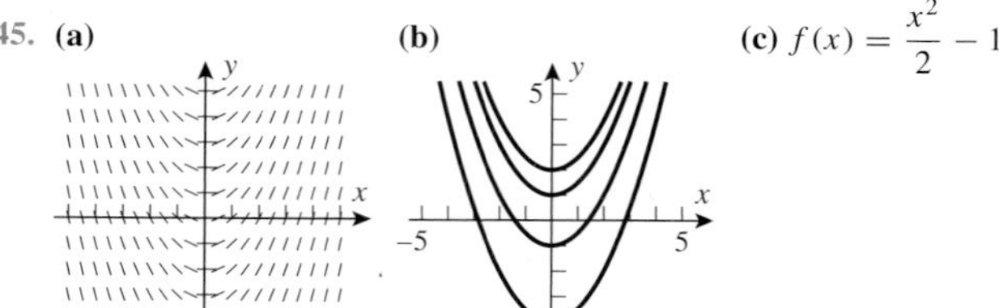

47. (b) $dy/dx = -x$ 49. (c) $dy/dx = x^2 - 4$

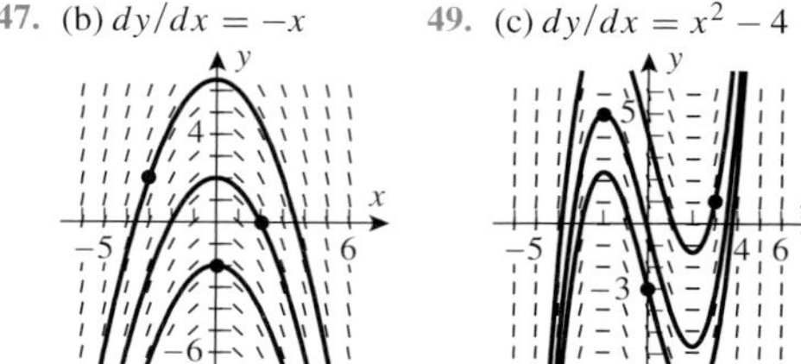

51. **(b)** $F(0) - G(0) = \frac{8}{3}$ 53. $F'(x) = G'(x) = 1,\ x \neq 0$

55. $\tan x - x + C$ 57. $\frac{1}{2}(x - \sin x) + C$ 59. $v = \dfrac{1087}{\sqrt{273}}T^{1/2}$ ft/s

▶ Exercise Set 5.3 (Page 323)

1. (a) $\frac{(x^2+1)^{24}}{24}+C$ (b) $-\frac{\cos^4 x}{4}+C$
(c) $-2\cos\sqrt{x}+C$ (d) $\frac{3}{4}\sqrt{4x^2+5}+C$
3. (a) $-\frac{1}{2}\cot^2 x+C$ (b) $\frac{1}{10}(1+\sin t)^{10}+C$
(c) $\frac{1}{2}\sin 2x+C$ (d) $\frac{1}{2}\tan(x^2)+C$
7. $\frac{1}{40}(4x-3)^{10}+C$ 9. $-\frac{1}{7}\cos 7x+C$ 11. $\frac{1}{4}\sec 4x+C$
13. $\frac{1}{21}(7t^2+12)^{3/2}+C$
15. $\frac{3}{2(1-2x)^2}+C$ 17. $-\frac{1}{40(5x^4+2)^2}+C$
19. $\frac{1}{5}\cos(5/x)+C$
21. $-\frac{1}{15}\cos^5 3t+C$ 23. $\frac{1}{2}\tan(x^2)+C$ 25. $-\frac{1}{6}(2-\sin 4\theta)^{3/2}+C$
27. $\frac{1}{6}\sec^3 2x+C$
29. $\frac{1}{6}(2y+1)^{3/2}-\frac{1}{2}(2y+1)^{1/2}+C$
31. $-\frac{1}{2}\cos 2\theta+\frac{1}{6}\cos^3 2\theta+C$
33. $\frac{1}{b}\frac{(a+bx)^{n+1}}{n+1}+C$ 35. $\frac{1}{b(n+1)}\sin^{n+1}(a+bx)+C$
37. (a) $\frac{1}{2}\sin^2 x+C_1;\ -\frac{1}{2}\cos^2 x+C_2$ (b) They differ by a constant.
39. $\frac{2}{15}(5x+1)^{3/2}-\frac{158}{15}$
41. (a) $\sqrt{x^2+1}+C$ 43. $f(x)=\frac{2}{9}(3x+1)^{3/2}+\frac{7}{9}$
(b)

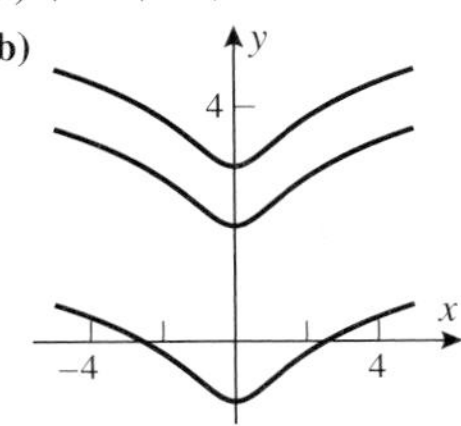

▶ Exercise Set 5.4 (Page 334)

1. (a) 36 (b) 55 (c) 40 (d) 6 (e) 11 (f) 0
3. $\sum_{k=1}^{10} k$ 5. $\sum_{k=1}^{10} 2k$ 7. $\sum_{k=1}^{6}(-1)^{k+1}(2k-1)$
9. (a) $\sum_{k=1}^{50} 2k$ (b) $\sum_{k=1}^{50}(2k-1)$
11. 5050 13. 2870 15. 214,365 17. $\frac{3}{2}(n+1)$
19. $\frac{1}{4}(n-1)^2$ 23. $\frac{n+1}{2n};\ \frac{1}{2}$ 25. $\frac{5(n+1)}{2n};\ \frac{5}{2}$
27. (a) $\sum_{j=0}^{5} 2^j$ (b) $\sum_{j=1}^{6} 2^{j-1}$ (c) $\sum_{j=2}^{7} 2^{j-2}$
29. (a) $\left(2+\frac{3}{n}\right)^4\cdot\frac{3}{n},\ \left(2+\frac{6}{n}\right)^4\cdot\frac{3}{n},\ \left(2+\frac{9}{n}\right)^4\cdot\frac{3}{n},$
$\left(2+\frac{3(n-1)}{n}\right)^4\cdot\frac{3}{n},\ (2+3)^4\cdot\frac{3}{n}$ (b) $\sum_{k=0}^{n-1}\left(2+k\cdot\frac{3}{n}\right)^4\frac{3}{n}$
31. (a) 46 (b) 52 (c) 58 33. (a) $\frac{\pi}{4}$ (b) 0 (c) $-\frac{\pi}{4}$
35. (a) 0.7188, 0.7058, 0.6982 (b) 0.6688, 0.6808, 0.6882
(c) 0.6928, 0.6931, 0.6931
37. (a) 4.8841, 5.1156, 5.2488 (b) 5.6841, 5.5156, 5.4088
(c) 5.3471, 5.3384, 5.3346
39. $\frac{15}{4}$ 41. 18 43. 320 45. $\frac{15}{4}$ 47. 18 49. 16 51. $\frac{1}{3}$ 53. 0
55. $\frac{2}{3}$ 57. $\frac{1}{2}m(b^2-a^2)$ 59. (b) $\frac{1}{4}(b^4-a^4)$
61. $\frac{n^2+2n}{4}$ if n is even; $\frac{(n+1)^2}{4}$ if n is odd 63. $3^{17}-3^4$
65. $-\frac{399}{400}$ 67. (b) $\frac{1}{2}$
71. (a) $\frac{3}{2}(3^{20}-1)$ (b) $2^{31}-2^5$ (c) $-\frac{2}{3}\left(1+\frac{1}{2^{101}}\right)$
73. (a) yes (b) yes

▶ Exercise Set 5.5 (Page 344)

1. (a) $\frac{71}{6}$ (b) 2 3. (a) $-\frac{117}{16}$ (b) 3 5. $\int_{-1}^{2} x^2\,dx$
7. $\int_{-3}^{3} 4x(1-3x)\,dx$ 9. (a) $\lim_{\max \Delta x_k\to 0}\sum_{k=1}^{n} 2x_k^*\Delta x_k;\ a=1, b=2$
(b) $\lim_{\max \Delta x_k\to 0}\sum_{k=1}^{n}\frac{x_k^*}{x_k^*+1}\Delta x_k;\ a=0, b=1$
11. (a) $A=\frac{9}{2}$ (b) $-A=-\frac{3}{2}$

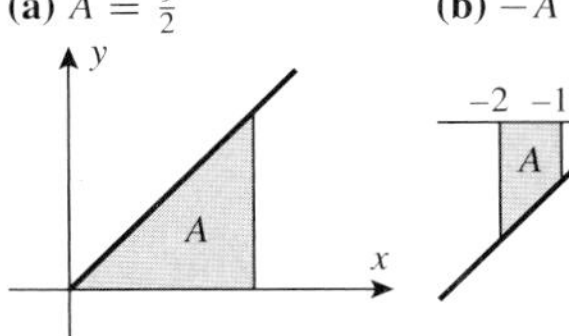

(c) $-A_1+A_2=\frac{15}{2}$ (d) $-A_1+A_2=0$

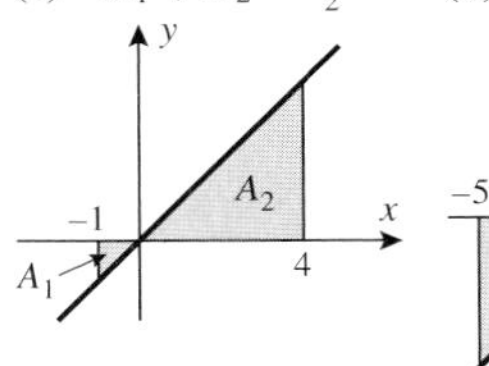

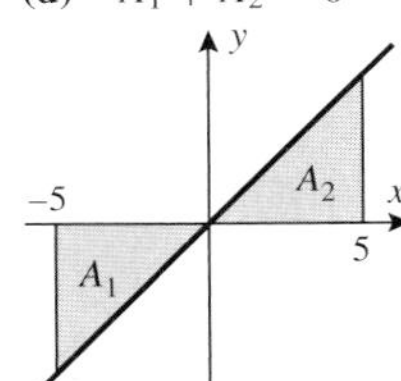

13. (a) $A=10$ (b) $A_1-A_2=0$ by symmetry

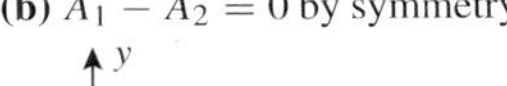

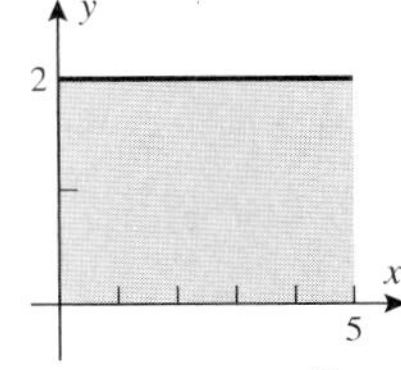

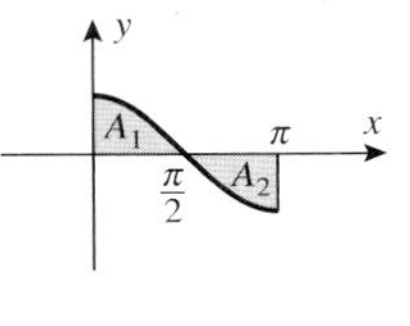

(c) $A_1+A_2=\frac{13}{2}$ (d) $\pi/2$ 15. (a) 2 (b) 4 (c) 10 (d) 10 17. (a) 0.8 (b) −2.6 (c) −1.8 (d) −0.3

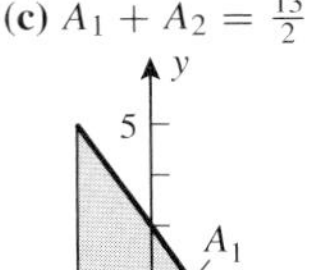

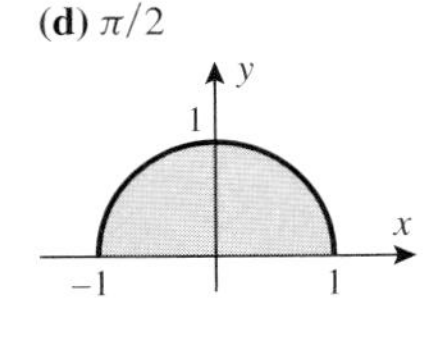

19. −1 21. 3 23. −4 25. $(1+\pi)/2$ 27. (a) negative (b) positive
29. $\frac{25}{2}\pi$ 31. $\frac{5}{2}$ 37. largest: $\frac{\pi}{24}(16+3\sqrt{2})$; smallest: $\frac{7\pi}{24}$ 39. $\frac{16}{3}$

▶ Exercise Set 5.6 (Page 357)

1. (a) $\int_0^2(2-x)\,dx=2$ (b) $\int_{-1}^{1}2\,dx=4$ (c) $\int_1^3(x+1)\,dx=6$
3. $\frac{65}{4}$ 5. 14 7. 48 9. 3 11. $\frac{844}{5}$ 13. 0 15. $\sqrt{2}$
17. −12 19. $\frac{\pi^2}{9}+2\sqrt{3}$ 21. (a) 5/2 (b) $2-\frac{\sqrt{2}}{2}$
23. (a) $\frac{17}{6}$ (b) $F(x)=\begin{cases}\frac{x^2}{2} & x\le 1\\ \frac{x^3}{3}+\frac{1}{6} & x>1\end{cases}$ 25. 0.6659; $\frac{2}{3}$
27. 3.1060; 2 tan 1 29. 12 31. $\frac{9}{2}$
33. area = 1 35. area = 3/2

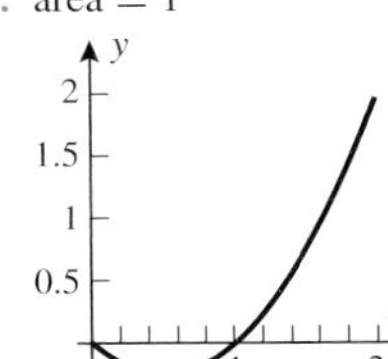

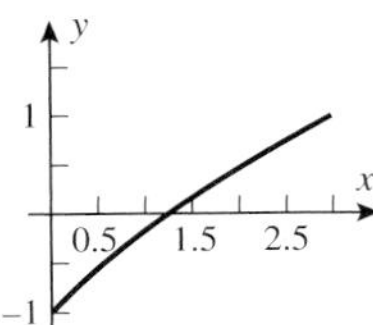

37. **(a)** $\int_0^{0.8} \cos x\,dx = \sin 0.8$ **(b)** degree mode; 0.7174
39. **(a)** The integral is zero. 41. **(a)** $3x^2 - 3$
43. **(a)** $\sin(x^2)$ **(b)** $\sqrt{1 - \cos x}$
45. $-x \sec x$ 47. **(a)** 0 **(b)** 5 **(c)** 4/5
49. **(a)** $x = 3$ **(b)** increasing on $[3, +\infty)$, decreasing on $(-\infty, 3]$ **(c)** concave up on $(-1, 7)$, concave down on $(-\infty, -1)$ and $(7, +\infty)$
51. **(a)** $(0, +\infty)$ **(b)** $x = 1$ 53. **(a)** 4/3 **(b)** -7
55. $3\sqrt{2} \le \int_0^3 \sqrt{x^3 + 2}dx \le 3\sqrt{29}$
59. **(a)** change in height from age 0 to age 10 years; inches **(b)** change in radius from time $t = 1$ sec to time $t = 2$ sec; centimeters **(c)** difference between speed of sound at 100° and at 32°F; feet per second **(d)** change in position from time t_1 to time t_2; centimeters
61. **(a)** 120 gal **(b)** 420 gal **(c)** 2076.36 gal 63. 1

▶ Exercise Set 5.7 (Page 367)

1. **(a)** displacement $= -\frac{1}{2}$; distance $= \frac{3}{2}$ **(b)** displacement $= \frac{3}{2}$; distance $= 2$
3. **(a)** 35.3m/s **(b)** 51.4 m/s 5. **(a)** $t^3 - t^2 + 1$ **(b)** $4t + 3 - \frac{1}{3}\sin 3t$
7. **(a)** $\frac{3}{2}t^2 + t - 4$ **(b)** $t + (1/t)$
9. **(a)** displacement = 1 m; distance = 1 m **(b)** displacement = -1 m; distance = 3 m
11. **(a)** displacement $= \frac{9}{4}$ m; distance $= \frac{11}{4}$ m **(b)** displacement $= 2\sqrt{3} - 6$ m; distance $= 6 - 2\sqrt{3}$ m
13. 4, 13/3 15. 296/27, 296/27
17. **(a)** $s = 2/\pi$, $v = 1$, $|v| = 1$, $a = 0$ **(b)** $s = \frac{1}{2}$, $v = -\frac{3}{2}$, $|v| = \frac{3}{2}$, $a = -3$ 19. $t \approx 1.3$ s
21.

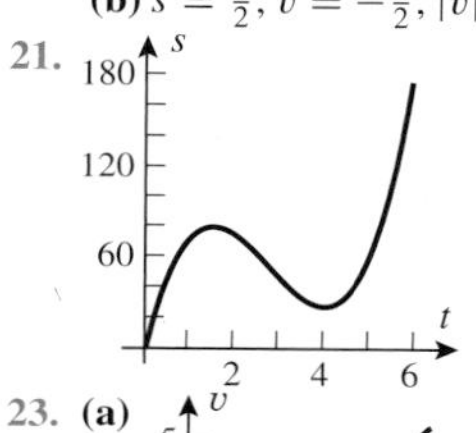

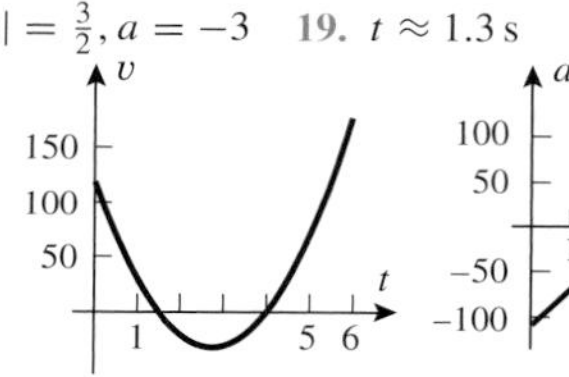

23. **(a)**

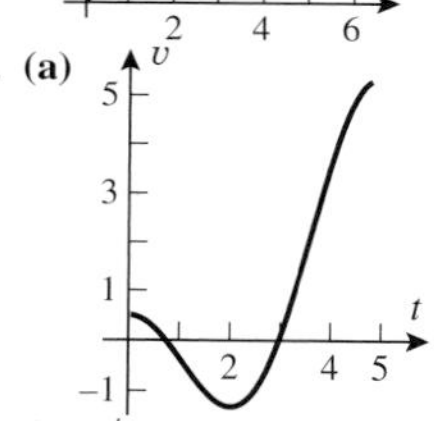

(b) $5/2 - \sin(5) + 5\cos(5)$
25. **(a)**

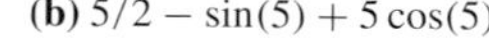

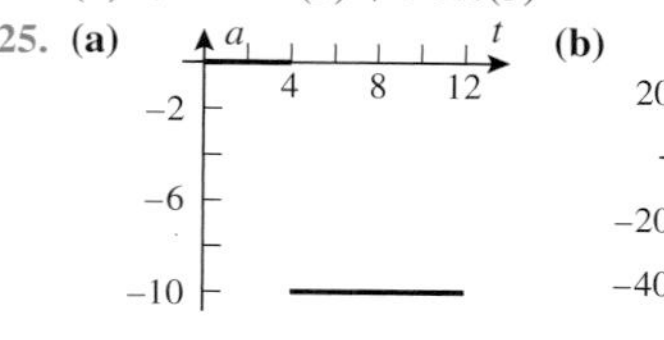

(b)

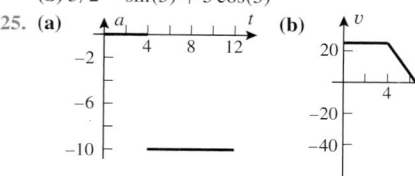

(c) 120 cm, -20 cm **(d)** 131.25 cm at $t = 6.5$ s
27. **(a)** -2.2 ft/s^2 **(b)** $\frac{1}{7200}$ km/s^2
29. **(a)** $-\frac{121}{5}$ ft/s^2 **(b)** $\frac{70}{33}$ s **(c)** $\frac{60}{11}$ s 31. 280 m
33. 50 s, 5000 ft 35. **(a)** 16 ft/s, -48 ft/s **(b)** 196 ft **(c)** 112 ft/s
37. **(a)** 1 s **(b)** $\frac{1}{2}$ s 39. **(a)** 6.122 s **(b)** 183.7 m **(c)** 6.122 s **(d)** 60 m/s
41. **(a)** 5 s **(b)** 272.5 m **(c)** 10 s **(d)** -49 m/s **(e)** 12.46 s **(f)** 73.1 m/s 43. 113.42 ft/s

▶ Exercise Set 5.8 (Page 373)

1. **(a)** $\frac{1}{2}\int_1^5 u^3\,du$ **(b)** $\frac{3}{2}\int_9^{25}\sqrt{u}\,du$ **(c)** $\frac{1}{\pi}\int_{-\pi/2}^{\pi/2}\cos u\,du$ **(d)** $\int_1^2 (u + 1)u^5\,du$
3. 10 5. 0 7. $\frac{1192}{15}$ 9. $8 - (4\sqrt{2})$ 11. $-\frac{1}{48}$ 13. $25\pi/6$
15. $\pi/8$ 17. $2/\pi$ 19. 6 21. 2 23. $\frac{2}{3}(\sqrt{10} - 2\sqrt{2})$
25. $2(\sqrt{7} - \sqrt{3})$ 27. 1 29. 0 31. $(\sqrt{3} - 1)/3$ 33. $\frac{106}{405}$
35. $\frac{23}{4480}$ 37. **(a)** $\frac{5}{3}$ **(b)** $\frac{5}{3}$ **(c)** $-\frac{1}{2}$ 41. **(a)** $2/\pi$
43. **(b)** $\frac{3}{2}$ **(c)** $\pi/4$

▶ Chapter 5. Review Exercises (Page 375)

3. $-\frac{1}{4x^2} + \frac{8}{3}x^{3/2} + C$ 5. $-4\cos x + 2\sin x + C$
7. **(a)** $y(x) = 2\sqrt{x} - \frac{2}{3}x^{3/2} - \frac{4}{3}$ **(b)** $y(x) = \sin x - \frac{5}{2}x^2 + 1$
11. $\frac{1}{10}(x^4 + 2)^{5/2} - \sqrt{x^4 + 2} + C$ 13. $\frac{1}{3}\sqrt{5 + 2\sin 3x} + C$
15. $-\frac{1}{3a}\frac{1}{ax^3 + b} + C$ 21. $A \approx 8$ 23. 32/3
25. 0.718771, 0.668771, 0.692835 27. 1.98352, 1.98352, 2.00825
29. **(a)** $\frac{3}{4}$ **(b)** $-\frac{3}{2}$ **(c)** $-\frac{35}{4}$ **(d)** -2 **(e)** not enough information **(f)** not enough information
31. **(a)** $2 + (\pi/2)$ **(b)** $\frac{1}{3}(10^{3/2} - 1) - \frac{9\pi}{4}$ **(c)** $\pi/8$ 33. $\frac{35\pi}{128}$
35. **(a)** $\int_a^b \left[\sum_{i=1}^n f_i(x)\right] dx = \sum_{i=1}^n \int_a^b f_i(x)dx$
37. $\frac{52}{3}$ 39. 48 41. $\frac{2}{3}$ 43. $3/2 - \sec 1$ 45. $\frac{5}{2}$
47. area = 1/6

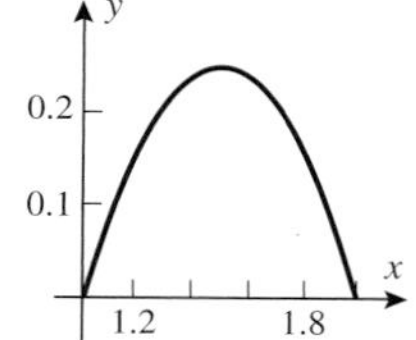

49. **(a)** $x^3 + 1$
51. $1/(x^4 + 5)$
53. $|x - 1|$
59. **(a)** $F(x)$ is 0 if $x = 1$, positive if $x > 1$, and negative if $x < 1$. **(b)** $F(x)$ is 0 if $x = -1$, positive if $-1 < x \le 2$, and negative if $-2 \le x < -1$.
61. **(a)** 4/3 **(b)** $1 \pm \frac{1}{3}\sqrt{3}$ 65. **(a)** $\frac{1}{4}t^4 - \frac{2}{3}t^3 + t + 1$
67. **(a)** $t^2 - 3t + 7$ 69. 12, 20 71. $1/3, 10/3 - 2\sqrt{2}$
73. displacement = -6 m; distance $= \frac{13}{2}$ m 75. $\frac{22}{3}$ 77. 37/12
79. **(a)** 2.2 s **(b)** 387.2 ft 81. $v_0/2$ 83. 121/5 85. $\frac{2}{3}$ 87. 0
89. $f(x) = -8/(x + 3)^2$, $a = 1$

▶ Exercise Set 6.1 (Page 386)

1. 9/2 3. 1 5. **(a)** 32/3 **(b)** 32/3 7. 49/192 9. 1/2 11. $\sqrt{2}$
13. 24 15. 37/12 17. $4\sqrt{2}$ 19. $\frac{1}{2}$
21. 9152/105 23. $9/\sqrt[3]{4}$
25. **(a)** 4/3 **(b)** $m = 2 - \sqrt[3]{4}$ 27. 1.180898334
29. 2.54270
31. Racer 1's lead over racer 2 at time $t = 0$
33. **(a)** (Area above graph of g and below graph of f) minus (area above graph of f and below graph of g) **(b)** Area between graphs of f and g 35. $a^2/6$

▶ Exercise Set 6.2 (Page 394)

1. 8π 3. $13\pi/6$ 5. 32/5 7. $(1 - \sqrt{2}/2)\pi$ 9. $256\pi/3$
11. $2048\pi/15$
13. 3/5 15. 8π 17. 2π
19. $72\pi/5$ 21. $4\pi ab^2/3$ 23. π

25. $\int_a^b \pi[f(x)-k]^2 dx$ 27. **(b)** $40\pi/3$ 29. $648\pi/5$ 31. $\pi/2$
33. $40{,}000\pi$ ft^3 35. $1/30$ 37. **(a)** $2\pi/3$ **(b)** $16/3$ **(c)** $4\sqrt{3}/3$
39. 0.710172176
43. **(b)** left ≈ 11.157; right ≈ 11.771; $V \approx$ average $= 11.464$ cm^3
45. $V = \begin{cases} 3\pi h^2, & 0 \le h < 2 \\ \frac{1}{3}\pi(12h^2 - h^3 - 4), & 2 \le h \le 4 \end{cases}$ 47. $\frac{2}{3}r^3 \tan\theta$ 49. $16r^3/3$

Exercise Set 6.3 (Page 402)

1. $15\pi/2$ 3. $\pi/3$ 5. $2\pi/5$ 7. 4π 9. $20\pi/3$
11. $\pi/2$
13. $\pi/5$ 15. $2\pi^2$ 17. 1.73680 19. **(a)** $7\pi/30$ **(b)** easier
21. $7\pi/4$ 23. $\pi r^2 h/3$ 25. $V = \frac{4}{3}\pi(L/2)^3$ 27. $b = 1$

Exercise Set 6.4 (Page 407)

1. $L = \sqrt{5}$ 3. $(85\sqrt{85} - 8)/243$ 5. $(80\sqrt{10} - 13\sqrt{13})/27$
7. $17/6$ 9. $(2\sqrt{2} - 1)/3$ 11. π
13. **(a)**

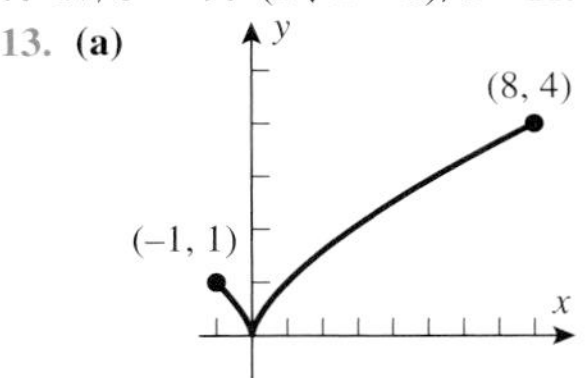

(b) dy/dx does not exist at $x = 0$. **(c)** $L = (13\sqrt{13} + 80\sqrt{10} - 16)/27$

15. **(a)** They are mirror images across the line $y = x$.

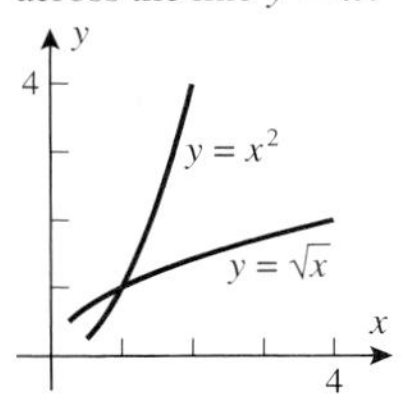

(b) $\int_{1/2}^{2} \sqrt{1+4x^2}\,dx$, $\int_{1/4}^{4} \sqrt{1+\frac{1}{4x}}\,dx$, $x = \sqrt{u}$ transforms the first integral into the second.
(c) $\int_{1/4}^{4} \sqrt{1+\frac{1}{4y}}\,dy$, $\int_{1/2}^{2} \sqrt{1+4y^2}\,dy$
(d) 4.0724, 4.0716
(e) The first: Both are understimates of the arc length, so the larger one is more accurate.
(f) 4.0724, 4.0662 **(g)** 4.0729

17. **(a)** They are mirror images across the line $y = x$.

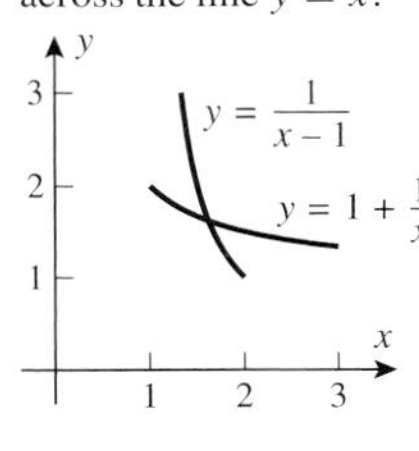

(b) $\int_1^3 \sqrt{1+\frac{1}{x^4}}\,dx$, $\int_{4/3}^{2} \sqrt{1+\frac{1}{(x-1)^4}}\,dx$, $u = 1 + \frac{1}{x}$ transforms the first integral into the second.
(c) $\int_{4/3}^{2} \sqrt{1+\frac{1}{(y-1)^4}}\,dy$, $\int_1^3 \sqrt{1+\frac{1}{y^4}}\,dy$
(d) 2.1459, 2.1463
(e) The second is more accurate since Δx is smaller,
(f) 2.1443, 2.1371 **(g)** 2.1466
21. $k = 1.83$ 23. 196.31 yards
25. **(a)**

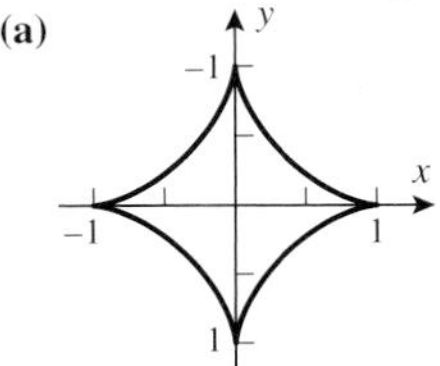

(b) 6 27. **(b)** 9.69 **(c)** 5.16 cm

Exercise Se6 6.5 (Page 412)

1. $35\pi\sqrt{2}$ 3. 8π 5. $40\pi\sqrt{82}$ 7. 24π 9. $16\pi/9$
11. $16{,}911\pi/1024$ 13. 29.9649 15. 14.39

21. $S = \int_a^b 2\pi(f(x)+k)\sqrt{1+[f'(x)]^2}\,dx$ 27. $\frac{8}{3}\pi(17\sqrt{17}-1)$
29. $\frac{\pi}{24}(17\sqrt{17}-1)$

Exercise Set 6.6 (Page 417)

1. **(a)** 4 **(b)** 2 **(c)**

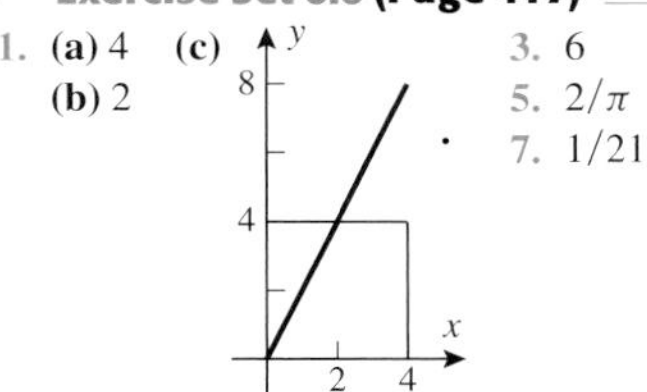

3. 6
5. $2/\pi$
7. $1/21$
9. **(a)** 5.28 **(b)** 4.305 **(c)** 4 11. **(a)** $-1/6$ **(b)** $1/2$
15. **(a)** $263/4$ **(b)** 31 17. 1404π lb 19. 97 cars/min
21. **(a)** 0.451 **(b)** 0.461 23. 27 25. **(b)** 169.7 V

Exercise Set 6.7 (Page 425)

1. **(a)** 210 ft·lb **(b)** 5/6 ft·lb 3. $d = 7/4$ 5. raising the cup of coffee
7. 100 ft·lb 9. 160 J 11. 20 lb/ft 13. $900\pi\rho$ ft·lb 15. 261,600 J
17. **(a)** 926,640 ft·lb **(b)** hp of motor $= 0.468$ 19. 75,000 ft·lb
21. 120,000 ft·tons 23. **(a)** $2{,}400{,}000{,}000/x^2$ lb **(b)** $(9.6 \times 10^{10})/(x+4000)^2$ lb **(c)** $2{,}5344 \times 10^{10}$ ft·lb
25. $v_f = 100$ m/s
27. **(a)** decreases of 4.5×10^{14} J **(b)** ≈ 0.107 **(c)** ≈ 8.24 bombs

Exercise Set 6.8 (Page 432)

1. **(a)** $F = 31{,}200$ lb; $P = 312$ lb/ft^2 **(b)** $F = 2{,}452{,}500$ N; $P = 98.1$ kPa
3. 499.2 lb 5. 8.175×10^5 N 7. 1,098,720 N 9. yes
11. $\rho a^3/\sqrt{2}$ lb 13. 63,648 lb 15. 9.81×10^9 N
17. **(b)** $80\rho_0$ lb/min

Chapter 6. Review Exercises (Page 433)

7. **(a)** $\int_a^b (f(x)-g(x))dx + \int_b^c (g(x)-f(x))dx + \int_c^d (f(x)-g(x))dx$ **(b)** $11/4$
9. $4352\pi/105$ 11. 9 13. $\frac{\pi}{6}\left(65^{3/2} - 37^{3/2}\right)$
15. $3/10$ 17. $3\sqrt{3}$ 19. **(a)** $W = \frac{1}{16}$ J **(b)** 5 m
21. **(a)** $F = \int_0^1 \rho x 3\,dx$ N **(b)** $F = \int_1^4 \rho(1+x)2x\,dx$ lb/ft^2 **(c)** $\int_{-10}^{0} 9810|y|2\sqrt{\frac{125}{8}(y+10)}\,dy$ N

Exercise Set 7.1 (Page 444)

1. **(a)** -4 **(b)** 4 **(c)** $\frac{1}{4}$ 3. **(a)** 2.9690 **(b)** 0.0341
5. **(a)** 4 **(b)** -5 **(c)** 1 **(d)** $\frac{1}{2}$ 7. **(a)** 1.3655 **(b)** -0.3011
9. **(a)** $2r + \frac{1}{2}s + \frac{1}{2}t$ **(b)** $s - 3r - t$
11. **(a)** $1 + \log x + \frac{1}{2}\log(x-3)$ **(b)** $2\ln|x| + 3\ln\sin x - \frac{1}{2}\ln(x^2+1)$
13. $\log\frac{256}{3}$ 15. $\ln\frac{\sqrt[3]{x}(x+1)^2}{\cos x}$ 17. 0.01 19. e^2 21. 4 23. 10^5
25. $\sqrt{3/2}$ 27. $-\frac{\ln 3}{2\ln 5}$ 29. $\frac{1}{3}\ln\frac{7}{2}$ 31. -2 33. $0, -\ln 2$
35. **(a)** **(b)**

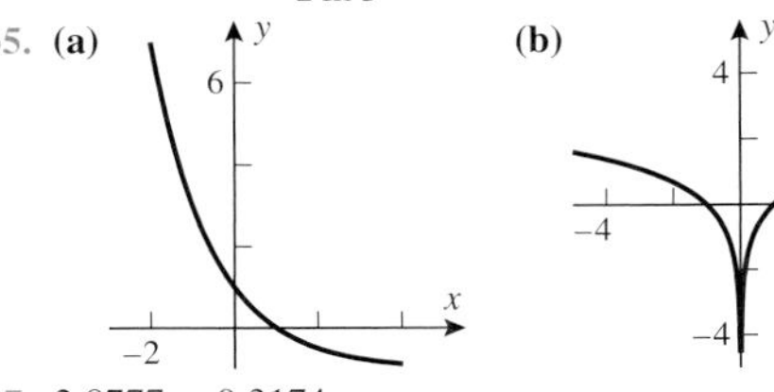

37. 2.8777, -0.3174

39\.

41\. $x \approx 1.471, 7.857$

43\. **(a)** no **(b)** $y = 2^{x/4}$ **(c)** $y = 2^{-x}$ **(d)** $y = (\sqrt{5})^x$

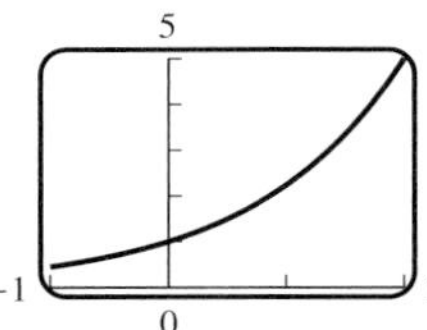

45\. $\log \frac{1}{2} < 0$, so $3 \log \frac{1}{2} < 2 \log \frac{1}{2}$ 47. 201 days

49\. **(a)** 7.4, basic **(b)** 4.2, acidic **(c)** 6.4, acidic **(d)** 5.9, acidic

51\. **(a)** 140 dB, damage **(b)** 120 dB, damage **(c)** 80 dB, no damage **(d)** 75 dB, no damage

53\. ≈ 200 55. **(a)** $\approx 5 \times 10^{16}$J **(b)** ≈ 0.67

Exercise Set 7.2 (Page 451)

1\. $1/x$ 3. $1/(1+x)$ 5. $2x/(x^2-1)$ 7. $\dfrac{1-x^2}{x(1+x^2)}$ 9. $2/x$

11\. $\dfrac{1}{2x\sqrt{\ln x}}$ 13. $1 + \ln x$ 15. $2x \log_2(3-2x) - \dfrac{2x^2}{(\ln 2)(3-2x)}$

17\. $\dfrac{2x(1+\log x) - x/(\ln 10)}{(1+\log x)^2}$ 19. $1/(x \ln x)$ 21. $2 \csc 2x$

23\. $-\dfrac{1}{x}\sin(\ln x)$ 25. $2 \cot x/(\ln 10)$ 27. $\dfrac{3}{x-1} + \dfrac{8x}{x^2+1}$

29\. $-\tan x + \dfrac{3x}{4-3x^2}$ 31. $x\sqrt[3]{1+x^2}\left[\dfrac{1}{x} + \dfrac{2x}{3(1+x^2)}\right]$

33\. $\dfrac{(x^2-8)^{1/3}\sqrt{x^3+1}}{x^6-7x+5}\left[\dfrac{2x}{3(x^2-8)} + \dfrac{3x^2}{2(x^3+1)} - \dfrac{6x^5-7}{x^6-7x+5}\right]$

35\. ex^{e-1} 37. **(a)** $-\dfrac{1}{x(\ln x)^2}$ **(b)** $-\dfrac{\ln 2}{x(\ln x)^2}$ 39. $y = ex - 2$

41\. $y = -x/e$ 43. $y = x/e$ 45. $A(w) = w/2$

49\. $f(x) = \ln(x+1)$ 51. **(a)** $1/e^2$ **(b)** 1 53. $2 \ln x - 3 \cos x + C$

55\. $\ln|\ln x| + C$ 57. $\frac{1}{5}\ln|1+x^5| + C$ 59. $t + \ln|t| + C$

61\. $\frac{3}{2}\ln 5$ 63. 3/2 65. $(\ln 3)/2$ 67. $y = \ln|t| + 5$

Exercise Set 7.3 (Page 458)

1\. **(b)** 1/9 3. $-2/x^2$ 5. **(a)** no **(b)** yes **(c)** yes **(d)** yes

7\. $\dfrac{1}{15y^2+1}$ 9. $\dfrac{1}{10y^4+3y^2}$ 11. $7e^{7x}$ 13. $x^2e^x(x+3)$

15\. $\dfrac{4}{(e^x+e^{-x})^2}$ 17. $(x\sec^2 x + \tan x)e^{x\tan x}$ 19. $(1-3e^{3x})e^{x-e^{3x}}$

21\. $\dfrac{x-1}{e^x - x}$ 23. $2^x \ln 2$ 25. $\pi^{\sin x}(\ln \pi)\cos x$

27\. $(x^3-2x)^{\ln x}\left[\dfrac{3x^2-2}{x^3-2x}\ln x + \dfrac{1}{x}\ln(x^3-2x)\right]$

29\. $(\ln x)^{\tan x}\left[\dfrac{\tan x}{x \ln x} + (\sec^2 x)\ln(\ln x)\right]$ 31. ex^{e-1}

33\. **(b)** $1 - (\sqrt{3}/3)$

35\. **(b)** $y = (88x - 89)/7$ 37. **(a)** $k^n e^{kx}$ **(b)** $(-1)^n k^n e^{-kx}$

39\. $-\dfrac{1}{\sqrt{2\pi}\sigma^3}(x-\mu)\exp\left[-\dfrac{1}{2}\left(\dfrac{x-\mu}{\sigma}\right)^2\right]$ 45. $r = 1, K = 12$

47\. $\ln 10$ 49. $2\ln|x| + 3e^x + C$ 51. $-\frac{1}{5}e^{-5x} + C$ 53. $\frac{1}{2}e^{2x} + C$

55\. $e^{\sin x} + C$ 57. $-\frac{1}{6}e^{-2x^3} + C$ 59. $-e^{-x} + C$

61\. C 63. $5e^3 - 10$ 65. $\frac{1}{2}(e - e^{-1})$ 67. $\ln(21/13)$

69\. $\approx 48{,}233{,}500{,}000$ 71. $\frac{1}{2}\ln 7$

Exercise Set 7.4 (Page 465)

1\. 0 (min) 3. -1 (min), 1 (max)

5\. maximum value $\frac{27}{8}e^{-3}$ at $x = \frac{3}{2}$, minimum value $64/e^8$ at $x = 4$

7\. max $= 5 \ln 10 - 9$ at $x = 3$; min $= 5 \ln(10/9) - 1$ at $x = 1/3$

9\. **(a)** $+\infty, 0$

(b) relative min $= -1/e$ at $x = -1$; no relative max; inflection point $(-2, -2/e^2) \approx (-2, -0.27)$; asymptote: $y = 0$

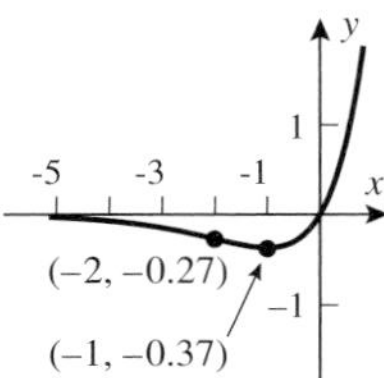

11\. **(a)** $0, +\infty$

(b) relative min $= 0$ at $x = 0$; relative max $= 1/e^2$ at $x = 1$; inflection points: $\left(\frac{1}{2}(2-\sqrt{2}), \frac{1}{4}(2-\sqrt{2})^2e^{-2+\sqrt{2}}\right) \approx (0.29, 0.05)$ and $\left(\frac{1}{2}(2+\sqrt{2}), \frac{1}{4}(2+\sqrt{2})^2e^{-2-\sqrt{2}}\right) \approx (1.71, 0.10)$; asymptote: $y = 0$

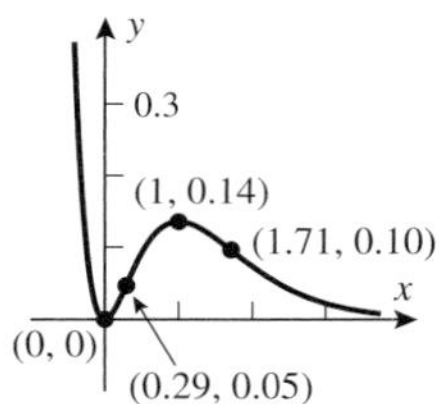

13\. **(a)** 0, 0

(b) relative max $= 1/e$ at $x = \pm 1$; relative min $= 0$ at $x = 0$; inflection points where $x = \pm\sqrt{\dfrac{5 \pm \sqrt{17}}{4}}$; about $(\pm 0.47, 0.18), (\pm 1.510, 0.23)$; asymptote: $y = 0$

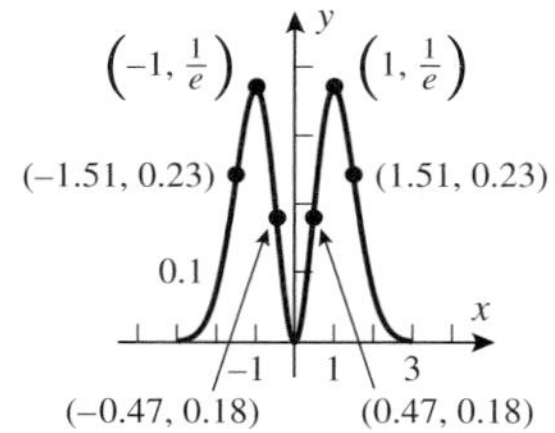

15\. **(a)** $-\infty, 0$

(b) relative max $= -e^2$ at $x = 2$; no relative min; no inflection points; asymptotes: $y = 0, x = 1$

17\. critical points at $x = 0.2$; relative min at $x = 0$, relative max at $x = 2$; points of inflection at $x = 2 \pm \sqrt{2}$; horizontal asymptote $y = 0$ as $x \to +\infty$; $\lim_{x \to -\infty} f(x) = +\infty$

19. **(a)** $+\infty, 0$
(b) relative min $= -1/e$ at $x = 1/e$;
no relative max; no inflection point; no asymptote

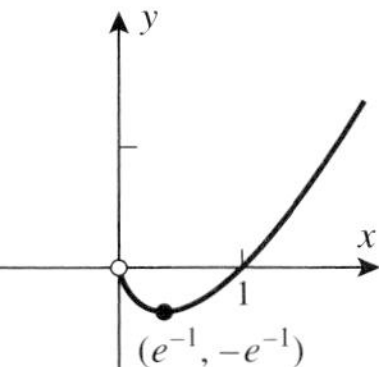

21. **(a)** $+\infty, 0$
(b) relative min $-1/(8e)$ at $x = 1/(2\sqrt{e})$; no relative max;
inflection point $(1/(2e^{3/2})), -3/(8e^3)) \approx (0.11, -0.02)$;
no asymptote

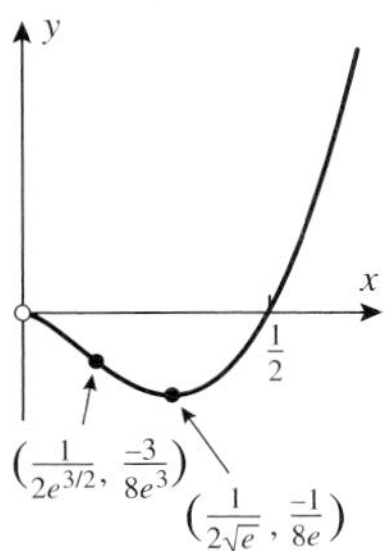

23. **(a)** $+\infty, 0$
(b) no relative max; relative min $= -\frac{3}{2e}$ at $x = e^{-3/2}$;
inflection point: $(e^{3/2}, 3e/2)$;
no asymptotes. It's hard to see all the important features in one graph, so two graphs are shown:

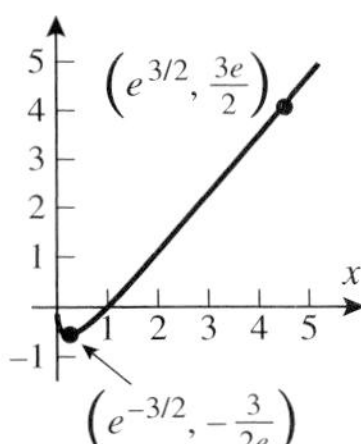

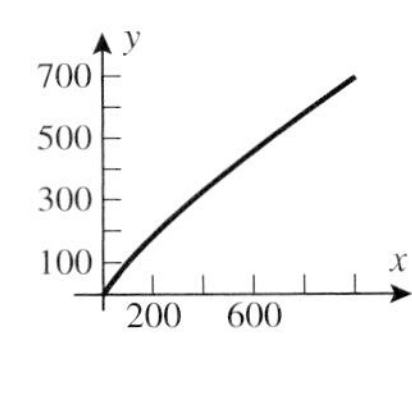

25. **(a)**

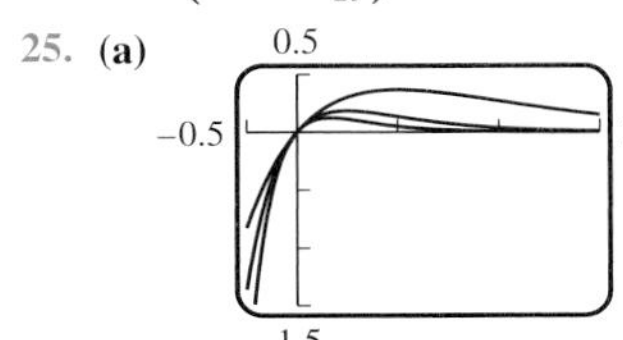

(b) relative max at $x = 1/b$; inflection point at $x = 2/b$

27. **(a)** does not exist, 0
(b) $y = e^x$ and $y = e^x \cos x$ intersect for $x = 2\pi n$, and $y = -e^x$ and $y = e^x \cos x$ intersect for $x = 2\pi n + \pi$, for all integers n.

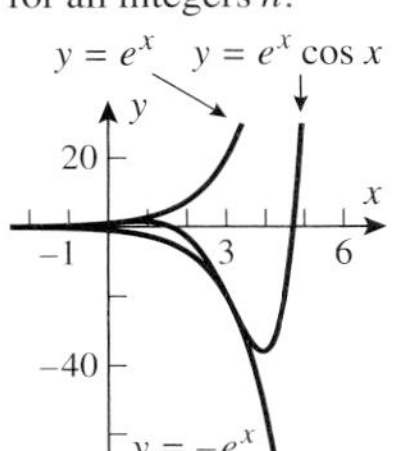

(c)

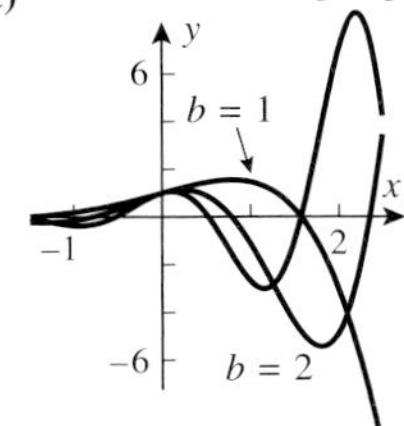

29. **(a)** $\dfrac{LAk}{(1+A)^2}$ **(c)** $\dfrac{1}{k}\ln A$ 31. the eighth day

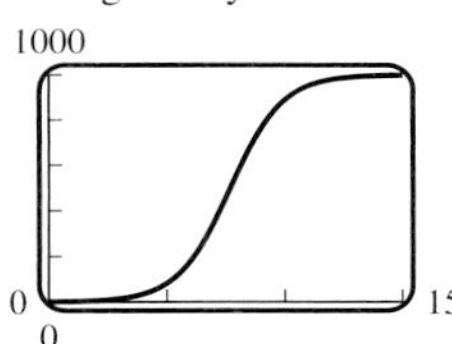

33. **(a)** largest number $\approx 161{,}788$, smallest number $= 125{,}000$
(b) $t = 40$

35. $-\dfrac{k_0 q}{2T^2}\exp\left(-\dfrac{q(T-T_0)}{2T_0 T}\right)$ 37. $3/2$ 39. $\frac{1}{2}$

41. area $= e + e^{-1} - 2$

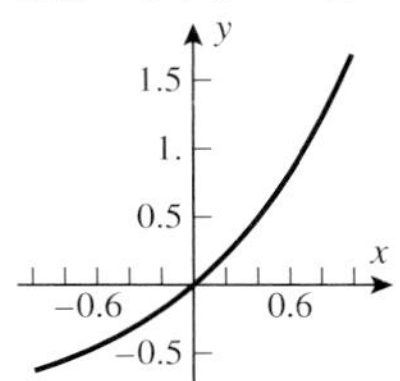

43. $\dfrac{1}{e-1}$ 45. $\dfrac{1-e^{-8}}{8}$

47. **(a)**

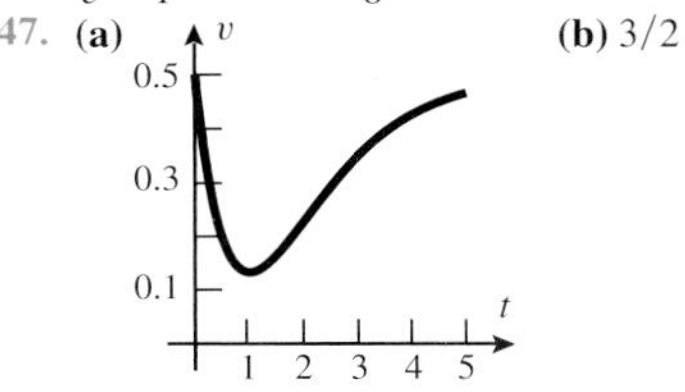

(b) $3/2 + 6e^{-5}$

49. $x \approx 0.58853$ or 3.09636 51. $x = -1$ or $x \approx 0.17951$
53. 4π 55. $\pi \ln 2$ 57. $L = \sqrt{2}(e^{\pi/2} - 1)$ 59. $L = \ln(1 + \sqrt{2})$
61. $S \approx 22.94$ 63. $S = \dfrac{2\sqrt{2}}{5}\pi(2e^{\pi} + 1)$

▶ Exercise Set 7.5 (Page 474)

1. **(a)** $\frac{2}{3}$ **(b)** $\frac{2}{3}$
3. **(b)** $T_f(x) = -2x$, $T_g(x) = -3x$ **(c)** limit $= 2/3$
5. 1 7. 1 9. -1 11. 0 13. $-\infty$ 15. 0 17. 0
19. π 21. $-\frac{5}{3}$ 23. e^{-3} 25. e^2 27. $e^{2/\pi}$ 29. 0 31. $\frac{1}{2}$
33. $+\infty$ 37. **(b)** 2
39. 0 41. e^3

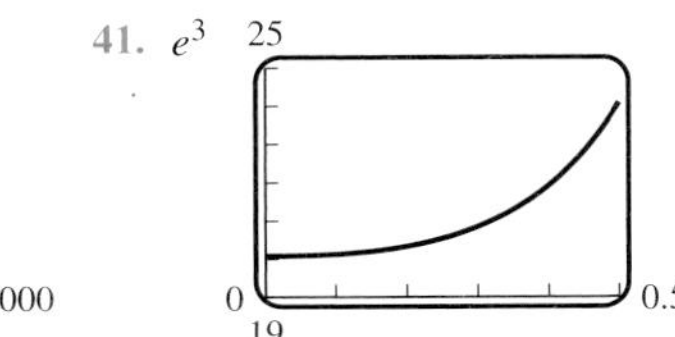

43. no horizontal asymptote 45. $y = 1$

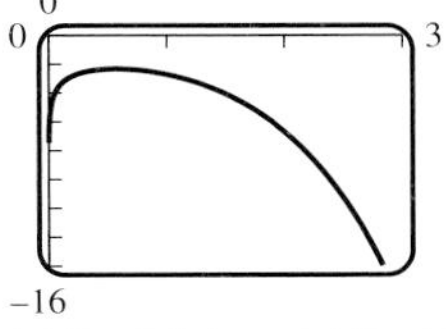

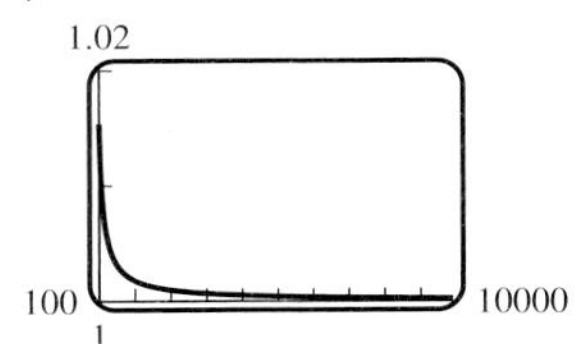

47. **(a)** 0 **(b)** $+\infty$ **(c)** 0 **(d)** $-\infty$ **(e)** $+\infty$ **(f)** $-\infty$ 49. 1
51. does not exist 53. Vt/L 57. **(b)** Both limits equal 0.
59. does not exist

Exercise Set 7.6 (Page 485)

1. **(a)**
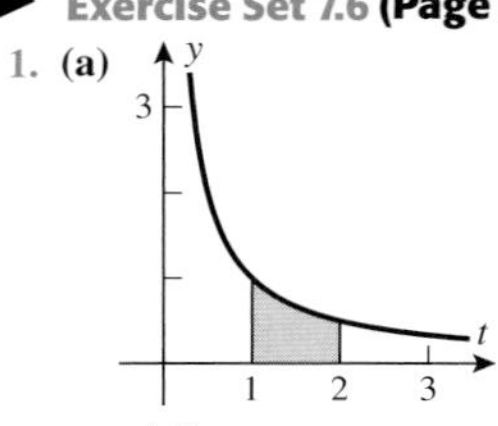
(b)
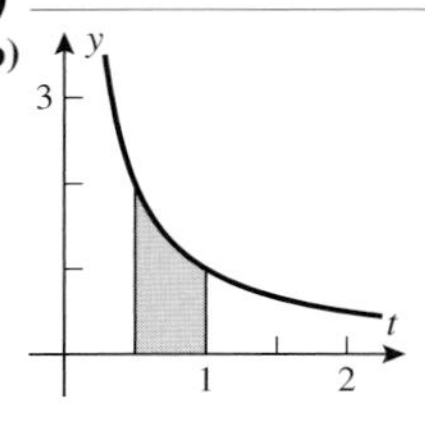
(c)
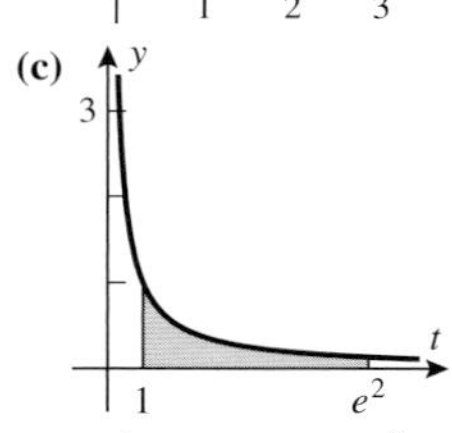

3. **(a)** 7 **(b)** -5 **(c)** -3 **(d)** 6
5. 1.603210678; magnitude of error is < 0.0063

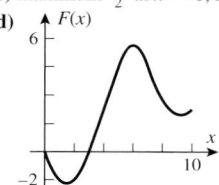

7. **(a)** $x^{-1}, x > 0$ **(b)** $x^2, x \neq 0$ **(c)** $-x^2, -\infty < x < +\infty$ **(d)** $-x, -\infty < x < +\infty$ **(e)** $x^3, x > 0$ **(f)** $\ln x + x, x > 0$ **(g)** $x - \sqrt[3]{x}, -\infty < x < +\infty$ **(h)** $\dfrac{e^x}{x}, x > 0$
9. **(a)** $e^{\pi \ln 3}$ **(b)** $e^{\sqrt{2}\ln 2}$ 11. **(a)** $\sqrt{e}$ **(b)** e^2 13. $x^2 - x$
15. **(a)** $3/x$ **(b)** 1 17. **(a)** 0 **(b)** 0 **(c)** 1
19. **(a)** $2x^3\sqrt{1+x^2}$ **(b)** $-\frac{2}{3}(x^2+1)^{3/2} + \frac{2}{5}(x^2+1)^{5/2} - \frac{4\sqrt{2}}{15}$
21. **(a)** $-\cos(x^3)$ **(b)** $-\tan^2 x$ 23. $-3\dfrac{3x-1}{9x^2+1} + 2x\dfrac{x^2-1}{x^4+1}$
25. **(a)** $3x^2\sin^2(x^3) - 2x\sin^2(x^2)$ **(b)** $\dfrac{2}{1-x^2}$
27. **(a)** $F(0) = 0, F(3) = 0, F(5) = 6, F(7) = 6, F(10) = 3$ **(b)** increasing on $[\frac{3}{2}, 6]$ and $[\frac{37}{4}, 10]$, decreasing on $[0, \frac{3}{2}]$ and $[6, \frac{37}{4}]$ **(c)** maximum $\frac{15}{2}$ at $x = 6$, minimum $-\frac{9}{4}$ at $x = \frac{3}{2}$ **(d)**
29. $F(x) = \begin{cases} (1-x^2)/2, & x < 0 \\ (1+x^2)/2, & x \geq 0 \end{cases}$ 31. $y(x) = x^2 + \ln x + 1$
33. $y(x) = \tan x + \cos x - (\sqrt{2}/2)$
35. $P(x) = P_0 + \int_0^x r(t)dt$ individuals 37. I is the derivative of II.
39. **(a)** $t = 3$ **(b)** $t = 1, 5$ **(c)** $t = 5$ **(d)** $t = 3$ **(e)** F is concave up on $(0, \frac{1}{2})$ and (2,4), concave down on $(\frac{1}{2}, 2)$ and (4,5). **(f)**
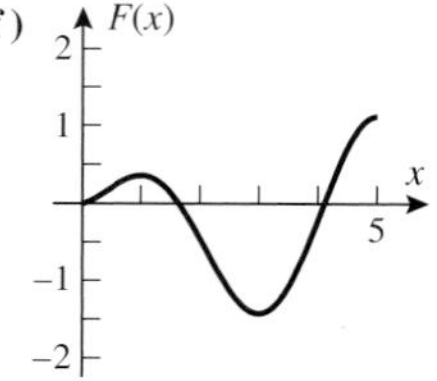
41. **(a)** relative maxima at $x = \pm\sqrt{4k+1}, k = 0, 1, \ldots$; relative minima at $x = \pm\sqrt{4k-1}, k = 1, 2, \ldots$ **(b)** $x = \pm\sqrt{2k}, k = 1, 2, \ldots$, and at $x = 0$
43. $f(x) = 2e^{2x}, a = \ln 2$ 45. 0.06
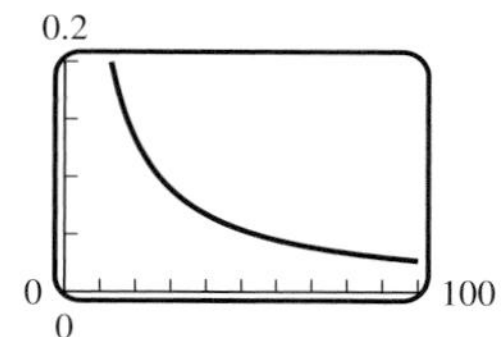

Exercise Set 7.7 (Page 494)

1. $\frac{4}{5}, \frac{3}{5}, \frac{3}{4}, \frac{5}{3}, \frac{5}{4}$
3. **(a)** $0 \leq x \leq \pi$ **(b)** $-1 \leq x \leq 1$ **(c)** $-\pi/2 < x < \pi/2$ **(d)** $-\infty < x < +\infty$ 5. 24/25
7. **(a)** $\dfrac{1}{\sqrt{1+x^2}}$ **(b)** $\dfrac{\sqrt{1-x^2}}{x}$ **(c)** $\dfrac{\sqrt{x^2-1}}{x}$ **(d)** $\dfrac{1}{\sqrt{x^2-1}}$
9. **(a)** **(b)**
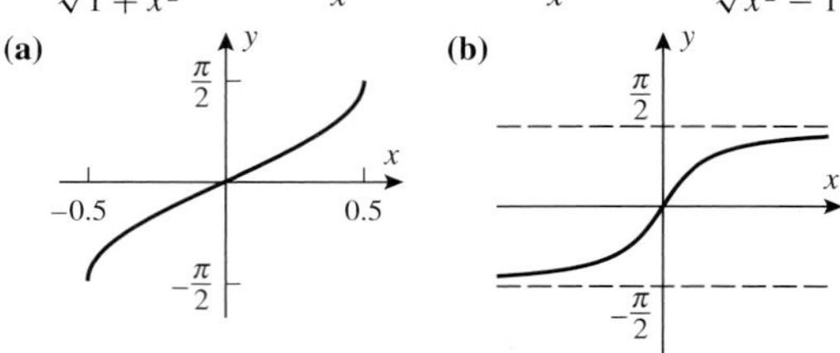
11. **(a)** $x = 3.6964$ rad **(b)** $\theta = -76.7°$
13. **(a)** 0.25545, error **(b)** $|x| \leq \sin 1$
15. **(a)**
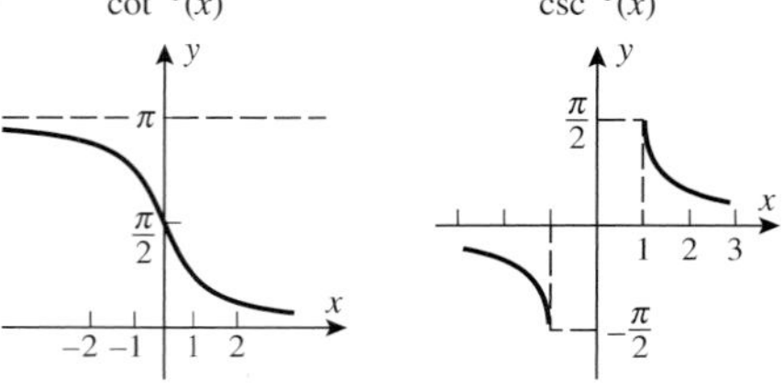

(b) $\cot^{-1} x$: all x, $0 < y < \pi$; $\csc^{-1} x$: $|x| \geq 1$, $0 < |y| < \pi/2$
17. $3/\sqrt{1-9x^2}$
19. $-\dfrac{1}{|x|\sqrt{x^2-1}}$ 21. $3x^2/(1+x^6)$ 23. $-\dfrac{1}{(1+x^2)\tan^2 x}$
25. $\dfrac{e^x}{|x|\sqrt{x^2-1}} + e^x \sec^{-1} x$ 27. 0 29. 0 31. $-\dfrac{1}{2\sqrt{x}(1+x)}$
33. $\dfrac{(3x^2 + \tan^{-1} y)(1+y^2)}{(1+y^2)e^y - x}$ 41. $\frac{1}{2}\sin^{-1} x - 3\tan^{-1} x + C$
43. $\frac{1}{2}\sin^{-1}(2x) + C$ 45. $\tan^{-1} e^x + C$ 47. $\sin^{-1}(\tan x) + C$
49. $\pi/4$ 51. $\pi/12$ 53. $\dfrac{2}{3}\pi^{3/2}\left(\dfrac{1}{3\sqrt{3}} - \dfrac{1}{8}\right)$ 55. $\pi/6$
57. $\dfrac{\pi}{6\sqrt{3}}$ 59. $\pi/9$
61. **(a)** $\sin^{-1}(\frac{1}{3}x) + C$ **(b)** $\dfrac{1}{\sqrt{5}}\tan^{-1}\dfrac{x}{\sqrt{5}} + C$ **(c)** $\dfrac{1}{\sqrt{\pi}}\sec^{-1}\dfrac{x}{\sqrt{\pi}} + C$
63. **(a)** 55.0° **(b)** 33.6° **(c)** 25.8° 65. **(a)** 21.1 hours **(b)** 2.9 hours
67. 29° 69. $\pi - 1$ 71. $k \approx 0.9973$ 73. $\pi^2/4$
75. $\frac{1}{4}\pi(\pi^2 - 8) \approx 1.46838$ 77. $\dfrac{\pi}{12(\sqrt{3}-1)}$
81. $3\sin^{-1} t - \pi$ 83. $y = \dfrac{1}{15}\tan^{-1}\dfrac{3t}{5} + \dfrac{\pi}{20}$

Exercise Set 7.8 (Page 507)

1. **(a)** ≈ 10.0179 **(b)** ≈ 3.7622 **(c)** $\approx 15/17 \approx 0.8824$ **(d)** ≈ -1.4436 **(e)** ≈ 1.7627 **(f)** ≈ 0.9730
3. **(a)** $\dfrac{4}{3}$ **(b)** $\dfrac{5}{4}$ **(c)** $\dfrac{312}{313}$ **(d)** $-\dfrac{63}{16}$
5.

	$\sinh x_0$	$\cosh x_0$	$\tanh x_0$	$\coth x_0$	$\text{sech}\, x_0$	$\text{csch}\, x_0$
(a)	2	$\sqrt{5}$	$2/\sqrt{5}$	$\sqrt{5}/2$	$1/\sqrt{5}$	1/2
(b)	3/4	5/4	3/5	5/3	4/5	4/3
(c)	4/3	5/3	4/5	5/4	3/5	3/4

9. $4\cosh(4x - 8)$ 11. $-\dfrac{1}{x}\text{csch}^2(\ln x)$
13. $\dfrac{1}{x^2}\text{csch}\left(\dfrac{1}{x}\right)\coth\left(\dfrac{1}{x}\right)$ 15. $\dfrac{2 + 5\cosh(5x)\sinh(5x)}{\sqrt{4x + \cosh^2(5x)}}$
17. $x^{5/2}\tanh(\sqrt{x})\,\text{sech}^2(\sqrt{x}) + 3x^2\tanh^2(\sqrt{x})$

19. $\dfrac{1}{\sqrt{9+x^2}}$ 21. $\dfrac{1}{(\cosh^{-1}x)\sqrt{x^2-1}}$ 23. $-\dfrac{(\tanh^{-1}x)^{-2}}{1-x^2}$

25. $\dfrac{|\sinh x|}{|\sinh x|} = \begin{cases} 1, & x > 0 \\ -1, & x < 0 \end{cases}$ 27. $-\dfrac{e^x}{2x\sqrt{1-x}} + e^x \operatorname{sech}^{-1} x$

31. $\frac{1}{7}\sinh^7 x + C$ 33. $\frac{2}{3}(\tanh x)^{3/2} + C$ 35. $\ln(\cosh x) + C$

37. $37/375$ 39. $\frac{1}{3}\sinh^{-1} 3x + C$ 41. $-\operatorname{sech}^{-1}(e^x) + C$

43. $-\operatorname{csch}^{-1}|2x| + C$ 45. $\frac{1}{2}\ln 3$ 49. $16/9$ 51. 5π 53. $\frac{3}{4}$

55. **(a)** $+\infty$ **(b)** $-\infty$ **(c)** 1 **(d)** -1 **(e)** $+\infty$ **(f)** $+\infty$

63. $|u| < 1 : \tanh^{-1} u + C$; $|u| > 1 : \tanh^{-1}(1/u) + C$

65. **(a)** $\ln 2$ **(b)** $1/2$ 71. 405.9 ft

73. **(a)**

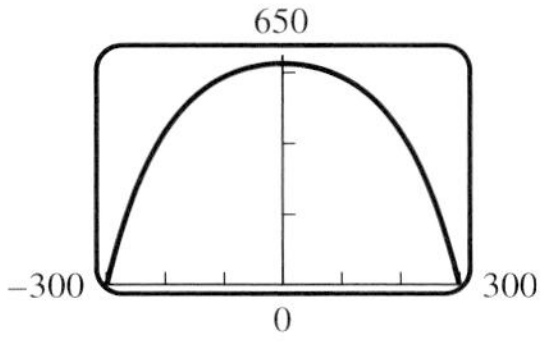

(b) 1480.2798 ft **(c)** ± 283.6249 ft **(d)** $82°$

75. **(b)** 14.44 m **(c)** $15\ln 3 \approx 16.48$ m

▶ Chapter 7. Review Exercises (Page 510)

1. **(a)** $\frac{1}{2}\ln(x-1)$ **(b)** $\dfrac{1}{2+\sin^{-1}x}$

(c) $\tan\left(\dfrac{1}{3x} - \dfrac{1}{3}\right)$, $x \le -\dfrac{2}{3\pi-2}$ or $x \ge \dfrac{2}{3\pi+2}$

3. **(a)** $\dfrac{33}{65}$ **(b)** $\dfrac{56}{65}$ 5. $e^{60}/63360 \approx 1.8\ 10^{21}$ miles 7. $15x+2$

9. **(a)**

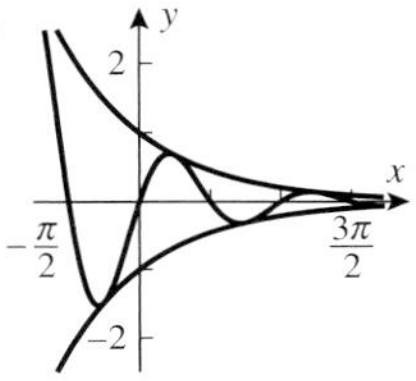

(b) $-\dfrac{\pi}{2}, 0, \dfrac{\pi}{2}, \pi, \dfrac{3\pi}{2}$; $-\dfrac{\pi}{4}, \dfrac{\pi}{4}, \dfrac{3\pi}{4}, \dfrac{5\pi}{4}$

11. **(a)**

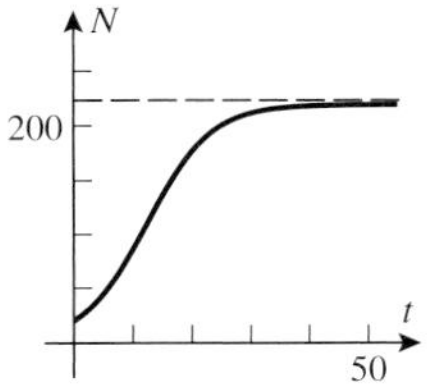

(b) about 10 years **(c)** 220 sheep

13. **(b)** 3.654, 332105.108 15. 0 17. e^{-3}

19. $\dfrac{1}{x+1} + \dfrac{2}{x+2} - \dfrac{3}{x+3} - \dfrac{4}{x+4}$ 21. $\dfrac{1}{x}$ 23. $\dfrac{1}{3x(\ln x+1)^{2/3}}$

25. $\dfrac{1}{(\ln 10)x\ln x}$ 27. $\dfrac{3}{2x} + \dfrac{2x^3}{1+x^4}$ 29. $2x$ 31. $e^{\sqrt{x}}(2+\sqrt{x})$

33. $\dfrac{2}{\pi(1+4x^2)}$ 35. $e^x x^{(e^x)}\left(\ln x + \dfrac{1}{x}\right)$ 37. $\dfrac{1}{|2x+1|\sqrt{x^2+x}}$

39. $\dfrac{x^3}{\sqrt{x^2+1}}\left(\dfrac{3}{x} - \dfrac{x}{x^2+1}\right)$

41. **(b)**

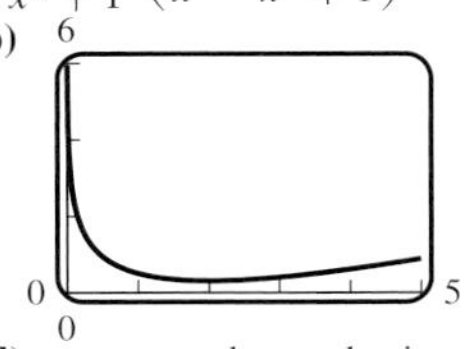

(d) curve must have a horizontal tangent line between $x = 1$ and $x = e$ **(e)** $x = 2$

43. e^2 45. $e^{1/e}$ 47. No; e.g. $f(x) = x^3$. 49. $(1/3, e)$

53. **(a)** 100

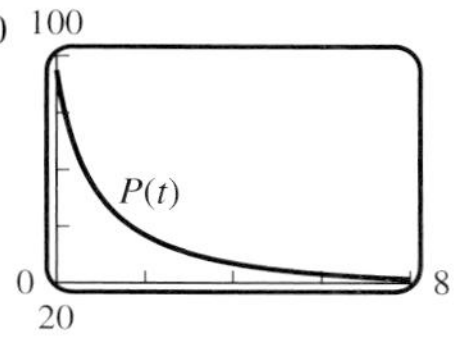

(b) The population tends to 19.
(c) The *rate* tends to zero.

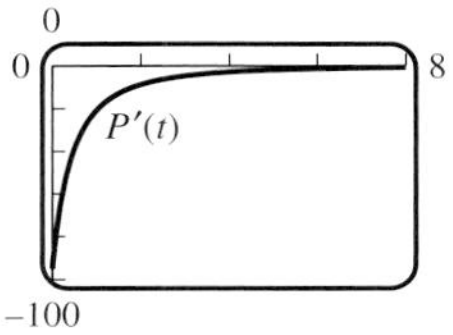

55. $+\infty, +\infty$: yes; $+\infty, -\infty$: no; $-\infty, +\infty$: no; $-\infty, -\infty$: yes

57. $+\infty$ 59. $1/9$

61. **(a)** $(-\infty, 0]$ **(b)** $[0, +\infty)$ **(c)** $(-\infty, -1/\sqrt{2}), (1/\sqrt{2}, +\infty)$
(d) $(-1/\sqrt{2}, 1/\sqrt{2})$ **(e)** $\pm 1/\sqrt{2}$

63. relative min of 0 at $x = 0$ 65. $m = e^2/4$ at $x = 2$

67. maximum value $f(-2-\sqrt{3}) \approx 0.84$,
minimum value $f(-2+\sqrt{3}) \approx -0.06$

69. $3x^{1/3} - 5e^x + C$ 71. $\tan^{-1}x + 2\sin^{-1}x + C$

73. 0.35122, 0.42054, 0.38650 77. $e - 1$ 79. $e^3 - e$

81. **(a)** $y = \sin x - 5e^x + 5$ **(b)** $y = \frac{1}{2}(e^{x^2} - 1)$ 83. $2 - 2/\sqrt{e}$

85. $3/2 + \ln 4$

▶ Exercise Set 8.1 (Page 516)

1. $-2(x-2)^4 + C$ 3. $\frac{1}{2}\tan(x^2) + C$ 5. $-\frac{1}{3}\ln(2+\cos 3x) + C$

7. $\cosh(e^x) + C$ 9. $e^{\tan x} + C$ 11. $-\dfrac{1}{30}\cos^6 5x + C$

13. $\ln(e^x + \sqrt{e^{2x}+4}) + C$ 15. $2e^{\sqrt{x-1}} + C$ 17. $2\sinh\sqrt{x} + C$

19. $-\dfrac{2}{\ln 3}3^{-\sqrt{x}} + C$ 21. $\dfrac{1}{2}\coth\dfrac{2}{x} + C$ 23. $-\dfrac{1}{4}\ln\left|\dfrac{2+e^{-x}}{2-e^{-x}}\right| + C$

25. $\sin^{-1}(e^x) + C$ 27. $-\dfrac{1}{2}\cos(x^2) + C$ 29. $-\dfrac{1}{\ln 16}4^{-x^2} + C$

31. **(a)** $\frac{1}{2}\sin^2 x + C$ **(b)** $-\frac{1}{4}\cos 2x + C$

33. **(b)** $\ln\left|\tan\dfrac{x}{2}\right| + C$ **(c)** $\ln\left|\cot\left(\dfrac{\pi}{4} - \dfrac{x}{2}\right)\right| + C$

▶ Exercise Set 8.2 (Page 524)

1. $-e^{-2x}\left(\dfrac{x}{2} + \dfrac{1}{4}\right) + C$ 3. $x^2e^x - 2xe^x + 2e^x + C$

5. $-\dfrac{1}{3}x\cos 3x + \dfrac{1}{9}\sin 3x + C$ 7. $x^2\sin x + 2x\cos x - 2\sin x + C$

9. $\dfrac{x^2}{2}\ln x - \dfrac{x^2}{4} + C$ 11. $x(\ln x)^2 - 2x\ln x + 2x + C$

13. $x\ln(3x-2) - x - \dfrac{2}{3}\ln(3x-2) + C$ 15. $x\sin^{-1}x + \sqrt{1-x^2} + C$

17. $x\tan^{-1}(3x) - \dfrac{1}{6}\ln(1+9x^2) + C$ 19. $\frac{1}{2}e^x(\sin x - \cos x) + C$

21. $\dfrac{e^{ax}}{a^2+b^2}(a\sin bx - b\cos bx) + C$

23. $(x/2)[\sin(\ln x) - \cos(\ln x)] + C$ 25. $x\tan x + \ln|\cos x| + C$

27. $\frac{1}{2}x^2e^{x^2} - \frac{1}{2}e^{x^2} + C$ 29. $\dfrac{1}{4}(3e^4+1)$ 31. $(2e^3+1)/9$

33. $3\ln 3 - 2$ 35. $\dfrac{5\pi}{6} - \sqrt{3} + 1$ 37. $-\pi/2$

39. $\dfrac{1}{3}\left(2\sqrt{3}\pi - \dfrac{\pi}{2} - 2 + \ln 2\right)$

41. **(a)** $2(\sqrt{x}-1)e^{\sqrt{x}} + C$ **(b)** $2\sqrt{x}\sin\sqrt{x} + 2\cos\sqrt{x} + C$

43. $-(3x^2+5x+7)e^{-x} + C$

45. $(4x^3-6x)\sin 2x - (2x^4-6x^2+3)\cos 2x + C$

47. **(a)** $\frac{1}{2}\sin^2 x + C$ **(b)** $\frac{1}{2}\sin^2 x + C$

49. **(a)** $A = 1$ **(b)** $V = \pi(e-2)$ 51. $V = 2\pi^2$ 53. $\pi^3 - 6\pi$

55. **(a)** $-\frac{1}{4}\sin^3 x\cos x - \dfrac{3}{8}\sin x\cos x + \dfrac{3}{8}x + C$ **(b)** $8/15$

59. **(a)** $\frac{1}{3}\tan^3 x - \tan x + x + C$ **(b)** $\frac{1}{3}\sec^2 x\tan x + \frac{2}{3}\tan x + C$
(c) $x^3e^x - 3x^2e^x + 6xe^x - 6e^x + C$

63. $(x+1)\ln(x+1) - x + C$ 65. $\frac{1}{2}(x^2+1)\tan^{-1}x - \frac{1}{2}x + C$

▶ Exercise Set 8.3 (Page 533)

1. $-\frac{1}{4}\cos^4 x + C$ 3. $\frac{\theta}{2} - \frac{1}{20}\sin 10\theta + C$
5. $\frac{1}{3a}\cos^3 a\theta - \cos a\theta + C$ 7. $\frac{1}{2a}\sin^2 ax + C$
9. $\frac{1}{3}\sin^3 t - \frac{1}{5}\sin^5 t + C$ 11. $\frac{1}{8}x - \frac{1}{32}\sin 4x + C$
13. $-\frac{1}{10}\cos 5x + \frac{1}{2}\cos x + C$ 15. $-\frac{1}{3}\cos(3x/2) - \cos(x/2) + C$
17. $2/3$ 19. 0 21. $7/24$ 23. $\frac{1}{2}\tan(2x-1) + C$
25. $\ln|\cos(e^{-x})| + C$ 27. $\frac{1}{4}\ln|\sec 4x + \tan 4x| + C$
29. $\frac{1}{3}\tan^3 x + C$ 31. $\frac{1}{16}\sec^4 4x + C$ 33. $\frac{1}{7}\sec^7 x - \frac{1}{5}\sec^5 x + C$
35. $\frac{1}{4}\sec^3 x\tan x - \frac{5}{8}\sec x\tan x + \frac{3}{8}\ln|\sec x + \tan x| + C$
37. $\frac{1}{3}\sec^3 t + C$ 39. $\tan x + \frac{1}{3}\tan^3 x + C$
41. $\frac{1}{8}\tan^2 4x + \frac{1}{4}\ln|\cos 4x| + C$ 43. $\frac{2}{3}\tan^{3/2} x + \frac{2}{7}\tan^{7/2} x + C$
45. $\frac{1}{2} - \frac{\pi}{8}$ 47. $-\frac{1}{2} + \ln 2$ 49. $-\frac{1}{5}\csc^5 x + \frac{1}{3}\csc^3 x + C$
51. $-\frac{1}{2}\csc^2 x - \ln|\sin x| + C$ 55. $L = \ln(\sqrt{2}+1)$ 57. $V = \pi/2$
63. $-\frac{1}{\sqrt{a^2+b^2}}\ln\left[\frac{\sqrt{a^2+b^2} + a\cos x - b\sin x}{a\sin x + b\cos x}\right] + C$
65. **(a)** $\frac{2}{3}$ **(b)** $3\pi/16$ **(c)** $\frac{8}{15}$ **(d)** $5\pi/32$

▶ Exercise Set 8.4 (Page 539)

1. $2\sin^{-1}(x/2) + \frac{1}{2}x\sqrt{4-x^2} + C$ 3. $8\sin^{-1}\left(\frac{x}{4}\right) - \frac{x\sqrt{16-x^2}}{2} + C$
5. $\frac{1}{16}\tan^{-1}(x/2) + \frac{x}{8(4+x^2)} + C$ 7. $\sqrt{x^2-9} - 3\sec^{-1}(x/3) + C$
9. $-(x^2+2)\sqrt{1-x^2} + C$ 11. $\frac{\sqrt{9x^2-4}}{4x} + C$ 13. $\frac{x}{\sqrt{1-x^2}} + C$
15. $\ln|\sqrt{x^2-9} + x| + C$ 17. $\frac{-x}{9\sqrt{4x^2-9}} + C$
19. $\frac{1}{2}\sin^{-1}(e^x) + \frac{1}{2}e^x\sqrt{1-e^{2x}} + C$ 21. $2/3$ 23. $(\sqrt{3}-\sqrt{2})/2$
25. $\frac{10\sqrt{3}+18}{243}$ 27. $\frac{1}{2}\ln(x^2+4) + C$
29. $L = \sqrt{5} - \sqrt{2} + \ln\frac{2+2\sqrt{2}}{1+\sqrt{5}}$ 31. $S = \frac{\pi}{32}[18\sqrt{5} - \ln(2+\sqrt{5})]$
33. $\tan^{-1}(x-2) + C$ 35. $\sin^{-1}\left(\frac{x-1}{2}\right) + C$
37. $\ln(x - 3 + \sqrt{(x-3)^2+1}) + C$
39. $2\sin^{-1}\left(\frac{x+1}{2}\right) + \frac{1}{2}(x+1)\sqrt{3-2x-x^2} + C$
41. $\frac{1}{\sqrt{10}}\tan^{-1}\sqrt{\frac{2}{5}}(x+1) + C$ 43. $\pi/6$
45. $u = \sin^2 x,\ \frac{1}{2}\int\sqrt{1-u^2}\,du$
$= \frac{1}{4}[\sin^2 x\sqrt{1-\sin^4 x} + \sin^{-1}(\sin^2 x)] + C$
47. **(a)** $\sinh^{-1}(x/3) + C$ **(b)** $\ln\left(\frac{\sqrt{x^2+9}}{3} + \frac{x}{3}\right) + C$

▶ Exercise Set 8.5 (Page 547)

1. $\frac{A}{x-3} + \frac{B}{x+4}$ 3. $\frac{A}{x} + \frac{B}{x^2} + \frac{C}{x-1}$
5. $\frac{A}{x} + \frac{B}{x^2} + \frac{C}{x^3} + \frac{Dx+E}{x^2+2}$ 7. $\frac{Ax+B}{x^2+5} + \frac{Cx+D}{(x^2+5)^2}$
9. $\frac{1}{5}\ln\left|\frac{x-4}{x+1}\right| + C$ 11. $\frac{5}{2}\ln|2x-1| + 3\ln|x+4| + C$
13. $\ln\left|\frac{x(x+3)^2}{x-3}\right| + C$ 15. $\frac{x^2}{2} - 3x + \ln|x+3| + C$
17. $3x + 12\ln|x-2| - \frac{2}{x-2} + C$
19. $x + \frac{x^3}{3} + \ln\left|\frac{(x-1)^2(x+1)}{x^2}\right| + C$
21. $3\ln|x| - \ln|x-1| - \frac{5}{x-1} + C$
23. $\frac{2}{x-3} + \ln|x-3| + \ln|x+1| + C$
25. $\frac{2}{x+1} - \frac{1}{2(x+1)^2} + \ln|x+1| + C$
27. $-\frac{7}{34}\ln|4x-1| + \frac{6}{17}\ln(x^2+1) + \frac{3}{17}\tan^{-1} x + C$
29. $3\tan^{-1} x + \frac{1}{2}\ln(x^2+3) + C$
31. $\frac{x^2}{2} - 2x + \frac{1}{2}\ln(x^2+1) + C$ 33. $\frac{1}{6}\ln\left(\frac{1-\sin\theta}{5+\sin\theta}\right) + C$
35. $V = \pi\left(\frac{19}{5} - \frac{9}{4}\ln 5\right)$ 37. $\frac{1}{\sqrt{2}}\tan^{-1}\left(\frac{x+1}{\sqrt{2}}\right) + \frac{1}{x^2+2x+3} + C$
39. $\frac{1}{8}\ln|x-1| - \frac{1}{5}\ln|x-2| + \frac{1}{12}\ln|x-3| - \frac{1}{120}\ln|x+3| + C$

▶ Exercise Set 8.6 (Page 557)

1. Formula (60): $\frac{4}{3}x + \frac{4}{9}\ln|3x-1| + C$
3. Formula (65): $\frac{1}{5}\ln\left|\frac{x}{5+2x}\right| + C$
5. Formula (102): $\frac{1}{5}(x-1)(2x+3)^{3/2} + C$
7. Formula (108): $\frac{1}{2}\ln\left|\frac{\sqrt{4-3x}-2}{\sqrt{4-3x}+2}\right| + C$
9. Formula (69): $\frac{1}{8}\ln\left|\frac{x+4}{x-4}\right| + C$
11. Formula (73): $\frac{x}{2}\sqrt{x^2-3} - \frac{3}{2}\ln|x + \sqrt{x^2-3}| + C$
13. Formula (95): $\frac{x}{2}\sqrt{x^2+4} - 2\ln(x + \sqrt{x^2+4}) + C$
15. Formula (74): $\frac{x}{2}\sqrt{9-x^2} + \frac{9}{2}\sin^{-1}\frac{x}{3} + C$
17. Formula (79): $\sqrt{4-x^2} - 2\ln\left|\frac{2+\sqrt{4-x^2}}{x}\right| + C$
19. Formula (38): $-\frac{\sin 7x}{14} + \frac{1}{2}\sin x + C$
21. Formula (50): $\frac{x^4}{16}[4\ln x - 1] + C$
23. Formula (42): $\frac{e^{-2x}}{13}[-2\sin(3x) - 3\cos(3x)] + C$
25. Formula (62): $\frac{1}{2}\int\frac{u\,du}{(4-3u)^2} = \frac{1}{18}\left[\frac{4}{4-3e^{2x}} + \ln\left|4-3e^{2x}\right|\right] + C$
27. Formula (68): $\frac{2}{3}\int\frac{du}{u^2+4} = \frac{1}{3}\tan^{-1}\frac{3\sqrt{x}}{2} + C$
29. Formula (76): $\frac{1}{2}\int\frac{du}{\sqrt{u^2-9}} = \frac{1}{2}\ln|2x + \sqrt{4x^2-9}| + C$
31. Formula (81): $\frac{1}{4}\int\frac{u^2}{\sqrt{2-u^2}}\,du = -\frac{1}{4}x^2\sqrt{2-4x^4}$
$+ \frac{1}{4}\sin^{-1}\left(\sqrt{2}x^2\right) + C$
33. Formula (26): $\int\sin^2 u\,du = \frac{1}{2}\ln x + \frac{1}{4}\sin(2\ln x) + C$
35. Formula (51): $\frac{1}{4}\int ue^u\,du = \frac{1}{4}(-2x-1)e^{-2x} + C$
37. $u = \sin 3x$, Formula (67): $\frac{1}{3}\int\frac{du}{u(u+1)^2}$
$= \frac{1}{3}\left(\frac{1}{\sin 3x + 1} + \left|\frac{\sin 3x}{\sin 3x + 1}\right|\right) + C$
39. $u = 4x^2$, Formula (70): $\frac{1}{8}\int\frac{du}{u^2-1} = \frac{1}{16}\ln\left|\frac{4x^2-1}{4x^2+1}\right| + C$
41. $u = 2e^x$, Formula (74): $\frac{1}{2}\int\sqrt{3-u^2}\,du = \frac{1}{2}e^x\sqrt{3-4e^{2x}}$
$+ \frac{3}{4}\sin^{-1}\left(\frac{2e^x}{\sqrt{3}}\right) + C$
43. $u = 3x$, Formula (112): $\frac{1}{3}\int\sqrt{\frac{5}{3}u - u^2}\,du = \frac{18x-5}{36}\sqrt{5x-9x^2}$
$+ \frac{25}{216}\sin^{-1}\left(\frac{18x-5}{5}\right) + C$

45. $u = 2x$, Formula (44): $\int u \sin u\, du = \sin 2x - 2x\cos 2x + C$

47. $u = -\sqrt{x}$, Formula (51): $2\int ue^u\, du = -2(\sqrt{x}+1)e^{-\sqrt{x}} + C$

49. $x^2 + 6x - 7 = (x+3)^2 - 16$, $u = x + 3$, Formula (70): $\int \frac{du}{u^2-16} = \frac{1}{8}\ln\left|\frac{x-1}{x+7}\right| + C$

51. $x^2 - 4x - 5 = (x-2)^2 - 9$, $u = x - 2$, Formula (77): $\int \frac{u+2}{\sqrt{9-u^2}}\,du = -\sqrt{5+4x-x^2} + 2\sin^{-1}\left(\frac{x-2}{3}\right) + C$

53. $u = \sqrt{x-2}$, $\frac{2}{5}(x-2)^{5/2} + \frac{4}{3}(x-2)^{3/2} + C$

55. $u = \sqrt{x^3+1}$, $\frac{2}{3}\int u^2(u^2-1)\,du = \frac{2}{15}(x^3+1)^{5/2} - \frac{2}{9}(x^3+1)^{3/2} + C$

57. $u = x^{1/3}$, $\int \frac{3u^2}{u^3-u}\,du = \frac{3}{2}\ln|x^{2/3}-1| + C$

59. $u = x^{1/4}$, $4\int \frac{1}{u(1-u)}\,du = 4\ln\frac{x^{1/4}}{|1-x^{1/4}|} + C$

61. $u = x^{1/6}$, $6\int \frac{u^3}{u-1}\,du = 2x^{1/2} + 3x^{1/3} + 6x^{1/6} + 6\ln|x^{1/6}-1| + C$

63. $u = \sqrt{1+x^2}$, $\int (u^2-1)\,du = \frac{1}{3}(1+x^2)^{3/2} - (1+x^2)^{1/2} + C$

65. $\int \frac{1}{1 + \frac{2u}{1+u^2} + \frac{1-u^2}{1+u^2}}\frac{2}{1+u^2}\,du = \int \frac{1}{u+1}\,du = \ln|\tan(x/2)+1| + C$

67. $\int \frac{d\theta}{1-\cos\theta} = \int \frac{1}{u^2}\,du = -\cot(\theta/2) + C$

69. $\int \frac{1}{\frac{2u}{1+u^2} + \frac{2u}{1+u^2}\cdot\frac{1+u^2}{1-u^2}}\cdot\frac{2}{1+u^2}\,du = \int \frac{1-u^2}{2u}\,du = \frac{1}{2}\ln|\tan(x/2)| - \frac{1}{4}\tan^2(x/2) + C$

71. $x = \frac{4e^2}{1+e^2}$ 73. $A = 6 + \frac{25}{2}\sin^{-1}\frac{4}{5}$ 75. $A = \frac{1}{40}\ln 9$

77. $V = \pi(\pi - 2)$ 79. $V = 2\pi(1 - 4e^{-3})$

81. $L = \sqrt{65} + \frac{1}{8}\ln(8+\sqrt{65})$ 83. $S = 2\pi[\sqrt{2} + \ln(1+\sqrt{2})]$

85.

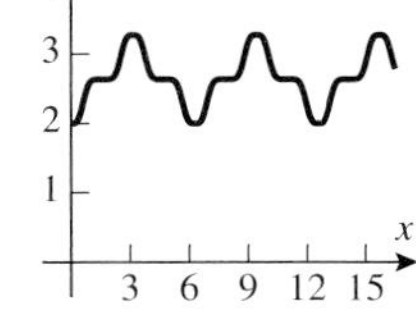

91. $\frac{1}{31}\cos^{31}x\sin^{31}x + C$

93. $-\frac{1}{9}\ln|1+x^{-9}| + C$

▶ Exercise Set 8.7 (Page 570)

1. exact value $= 14/3 \approx 4.666666667$
   (a) 4.667600663, $|E_M| \approx 0.000933996$
   (b) 4.664795679, $|E_T| \approx 0.001870988$
   (c) 4.666651630, $|E_S| \approx 0.000015037$

3. exact value $= 2$
   (a) 2.008248408, $|E_M| \approx 0.008248408$
   (b) 1.983523538, $|E_T| \approx 0.016476462$
   (c) 2.000109517, $|E_S| \approx 0.000109517$

5. exact value $= e^{-1} - e^{-4} \approx 0.349563802$
   (a) 0.348256371, $|E_M| \approx 0.001307431$
   (b) 0.352181607, $|E_T| \approx 0.002617804$
   (c) 0.349579366, $|E_S| \approx 0.000015563$

7. **(a)** $|E_M| \le \frac{27}{2400}(1/4) = 0.002812500$
   (b) $|E_T| \le \frac{27}{1200}(1/4) = 0.005625000$
   (c) $|E_S| \le \frac{243}{180\times 10^4}(15/16) \approx 0.000126563$

9. **(a)** $|E_M| \le \frac{\pi^3}{2400}(1) \approx 0.012919282$
   (b) $|E_T| \le \frac{\pi^3}{1200}(1) \approx 0.025838564$
   (c) $|E_S| \le \frac{\pi^5}{180\times 10^4}(1) \approx 0.000170011$

11. **(a)** $|E_M| \le \frac{9}{800}(e^{-1}) \approx 0.004138644$
   (b) $|E_T| \le \frac{9}{400}(e^{-1}) \approx 0.008277287$
   (c) $|E_S| \le \frac{27}{20000}(e^{-1}) \approx 0.000049664$

13. **(a)** $n = 24$ **(b)** $n = 34$ **(c)** $2n = 8$

15. **(a)** $n = 36$ **(b)** $n = 51$ **(c)** $2n = 8$

17. **(a)** $n = 644$ **(b)** $n = 910$ **(c)** $2n = 28$

19. $g(x) = \frac{1}{24}x^2 - \frac{3}{8}x + \frac{13}{12}$ 21. 0.746824948, 0.746824133

23. 1.511518748, 1.515927142 25. 0.805376152, 0.804776489

27. **(a)** 3.142425985, $|E_M| \approx 0.00083331$
   (b) 3.139925989, $|E_T| \approx 0.001666665$
   (c) 3.141592614, $|E_S| \approx 0.000000040$

29. $S_{14} = 0.693147984$, $|E_S| \approx 0.000000803 = 8.03 \times 10^{-7}$

31. $n = 116$ 35. ≈ 3.820187624 37. 1071 ft. 39. 37.9 mi 41. 9.3L

43. **(a)** $\max|f''(x)| \approx 3.844880$ **(b)** $n = 18$ **(c)** 0.904741

45. **(a)** $\max|f^{(4)}| \approx 42.551816$ **(b)** $2n = 8$ **(c)** 0.904524

▶ Exercise Set 8.8 (Page 580)

1. **(a)** improper; infinite discontinuity at $x = 3$ **(b)** not improper **(c)** improper; infinite discontinuity at $x = 0$ **(d)** improper; infinite interval of integration **(e)** improper; infinite interval of integration and infinite discontinuity at $x = 1$ **(f)** not improper

3. 1/2 5. ln 2 7. $\frac{1}{2}$ 9. $-\frac{1}{4}$ 11. $\frac{1}{3}$ 13. divergent 15. 0

17. divergent 19. divergent 21. $\pi/2$ 23. 1 25. divergent

27. $\frac{9}{2}$ 29. divergent 31. 2 33. 2 35. $\frac{1}{2}$

37. **(a)** 2.726585 **(b)** 2.804364 **(c)** 0.219384 **(d)** 0.504067 39. 12

41. -1 43. $\frac{1}{3}$ 45. **(a)** $V = \pi/2$ **(b)** $S = \pi[\sqrt{2} + \ln(1+\sqrt{2})]$

47. **(b)** $1/e$ **(c)** It is convergent. 53. $\frac{2\pi NI}{kr}\left(1 - \frac{a}{\sqrt{r^2+a^2}}\right)$

55. **(b)** 2.4×10^7 mi.lb 57. **(a)** $\frac{1}{s^2}$ **(b)** $\frac{2}{s^3}$ **(c)** $\frac{e^{-3s}}{s}$ 61. **(a)** 1.047

65. 1.809 67. **(a)** $\Gamma(1) = 1$ **(c)** $\Gamma(2) = 1$, $\Gamma(3) = 2$, $\Gamma(4) = 6$

69. **(b)** 1.37078 seconds

▶ Chapter 8. Review Exercises (Page 584)

1. $\frac{2}{27}(4+9x)^{3/2} + C$ 3. $-\frac{2}{3}\cos^{3/2}\theta + C$ 5. $\frac{1}{6}\tan^3(x^2) + C$

7. **(a)** $2\sin^{-1}(\sqrt{x/2}) + C$; $-2\sin^{-1}(\sqrt{2-x}/\sqrt{2}) + C$; $\sin^{-1}(x-1) + C$

9. $-xe^{-x} - e^{-x} + C$ 11. $x\ln(2x+3) - x + \frac{3}{2}\ln(2x+3) + C$

13. $(4x^4 - 12x^2 + 6)\sin(2x) + (8x^3 - 12x)\cos(2x) + C$

15. $\frac{1}{2}\theta - \frac{1}{20}\sin 10\theta + C$ 17. $-\frac{1}{6}\cos 3x + \frac{1}{2}\cos x + C$

19. $-\frac{1}{8}\sin^3(2x)\cos 2x - \frac{3}{16}\cos 2x\sin 2x + \frac{3}{8}x + C$

21. $\frac{9}{2}\sin^{-1}(x/3) - \frac{1}{2}x\sqrt{9-x^2} + C$ 23. $\ln|x + \sqrt{x^2-1}| + C$

25. $\frac{x\sqrt{x^2+9}}{2} - \frac{9\ln(|\sqrt{x^2+9}+x|)}{2} + C$ 27. $\frac{1}{5}\ln\left|\frac{x-1}{x+4}\right| + C$

29. $\frac{1}{2}x^2 - 2x + 6\ln|x+2| + C$

31. $\ln|x+2| + \frac{4}{x+2} - \frac{2}{(x+2)^2} + C$

35. Formula (40): $-\frac{\cos 16x}{32} + \frac{\cos 2x}{4} + C$

37. Formula (113): $\frac{1}{24}(8x^2 - 2x - 3)\sqrt{x - x^2} + \frac{1}{16}\sin^{-1}(2x-1) + C$

39. Formula (28): $\frac{1}{2}\tan 2x - x + C$

41. exact value $= 14/3 \approx 4.666666667$
   (a) 4.667600663, $|E_M| \approx 0.000933996$

(b) $4.664795679, |E_T| \approx 0.001870988$
(c) $4.666651630, |E_S| \approx 0.000015037$

43. **(a)** $|E_M| \le \frac{27}{2400}(1/4) = 0.002812500$
(b) $|E_T| \le \frac{27}{1200}(1/4) = 0.005625000$
(c) $|E_S| \le \dfrac{243}{180 \times 10^4}(15/16) \approx 0.000126563$

45. **(a)** $n = 24$ **(b)** $n = 34$ **(c)** $n = 8$ 47. 1 49. 6

51. e^{-1} 53. $a = \pi/2$ 55. $\dfrac{x}{3\sqrt{3+x^2}} + C$ 57. $\dfrac{5}{12} - \dfrac{1}{2}\ln 2$

59. $\dfrac{1}{6}\sin^3 2x - \dfrac{1}{10}\sin^5 2x + C$

61. $\dfrac{2}{13}e^{2x}\cos 3x + \dfrac{3}{13}e^{2x}\sin 3x + C$

63. $-\frac{1}{6}\ln|x-1| + \frac{1}{15}\ln|x+2| + \frac{1}{10}\ln|x-3| + C$

65. $4 - \pi$ 67. $\ln\dfrac{\sqrt{e^x+1}-1}{\sqrt{e^x+1}+1} + C$ 69. $\dfrac{\pi}{12} + \dfrac{\sqrt{3}}{2} - 1$

71. $\sqrt{x^2+2x+2} + 2\ln(\sqrt{x^2+2x+2} + x + 1) + C$ 73. $\dfrac{1}{2(a^2+1)}$

▶ Exercise Set 9.1 (Page 596)

3. **(a)** first order **(b)** second order 7. **(a)** $y = Ce^{-3x}$ **(b)** $y = Ce^{2t}$

9. $y = e^{-3x} + Ce^{-4x}$ 11. $y = e^{-x}\sin(e^x) + Ce^{-x}$

13. $y = \dfrac{C}{\sqrt{x^2+1}}$ 15. $y = Cx$ 17. $y = Ce^{-\sqrt{1+x^2}} - 1, C \neq 0$

19. $2\ln|y| + y^2 = e^x + C$ 21. $y = \ln(\sec x + C)$

23. $y = \dfrac{1}{1 - C(\csc x - \cot x)}, C \neq 0$ and $y = 0$

25. **(a)** $y = \dfrac{x}{2} + \dfrac{3}{2x}$ **(b)** $y = \dfrac{x}{2} - \dfrac{5}{2x}$ 27. $y = 4e^{x^2} - 1$

29. $y^2 + \sin y = x^3 + \pi^2$ 31. $y^2 - 2y = t^2 + t + 3$

33. **(a)**

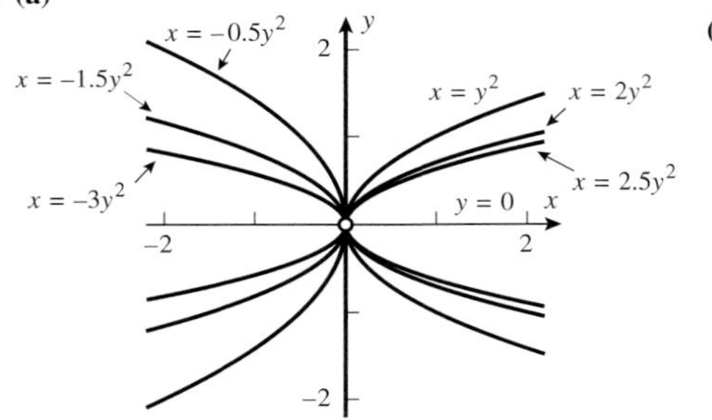

(b) $y^2 = x/4$

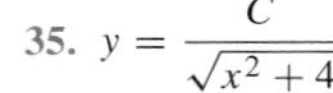

35. $y = \dfrac{C}{\sqrt{x^2+4}}$

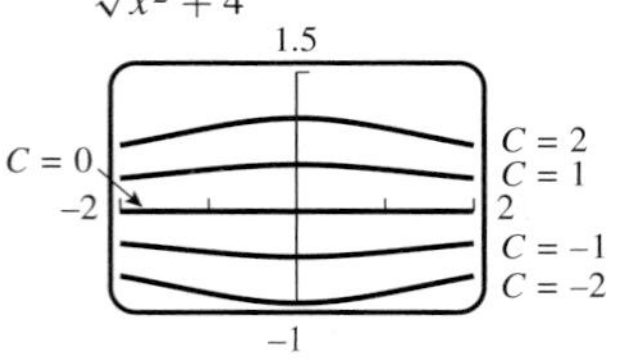

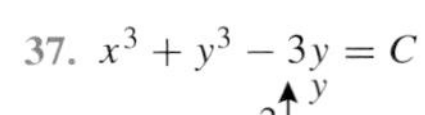

37. $x^3 + y^3 - 3y = C$

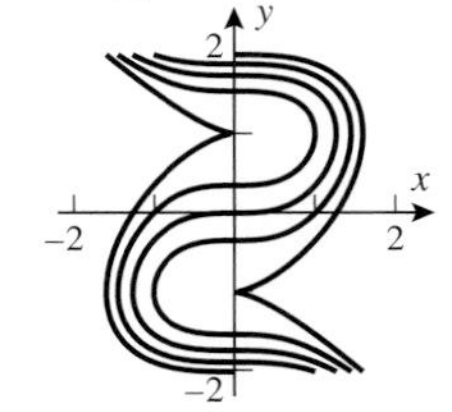

41. $y = \ln\left(\dfrac{x^2}{2} - 1\right)$ 43. **(a)** $200 - 175e^{-t/25}$ oz **(b)** 136 oz

45. 25 lb 49. **(a)** $I(t) = 2 - 2e^{-2t}$ **(b)** $I(t) \to 2$

51. **(a)** $v = c\ln\dfrac{m_0}{m_0 - kt} - gt$ **(b)** 3044 m/s

53. **(a)** $h \approx (2 - 0.003979t)^2$ **(b)** 8.4 min

55. **(a)** $v = 128/(4t+1), x = 32\ln(4t+1)$

57. $\dfrac{dy}{dx} = -\sin x + e^{-x^2}, y(0) = 1$

▶ Exercise Set 9.2 (Page 604)

1.

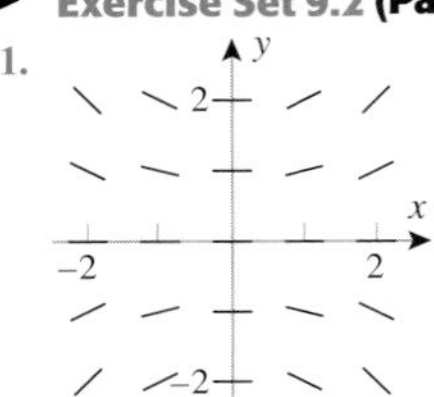

3.

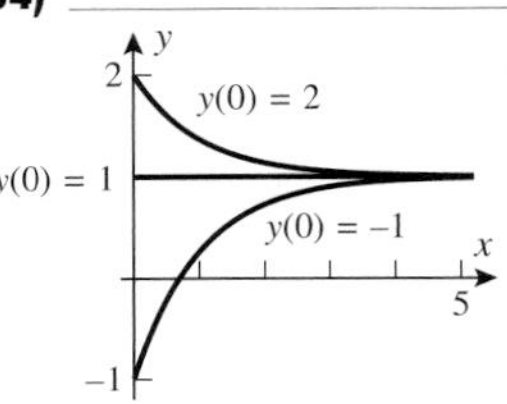

5.

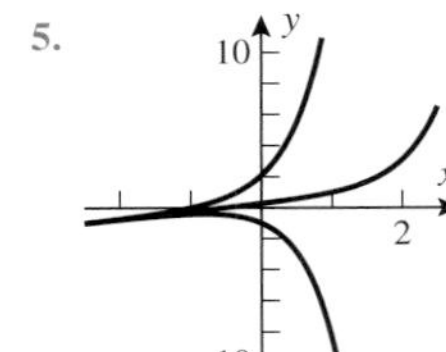

9. **(a)** IV **(b)** VI **(c)** V **(d)** II **(e)** I **(f)** III

11. **(a)**

n	0	1	2	3	4	5
x_n	0	0.2	0.4	0.6	0.8	1.0
y_n	1	1.20	1.48	1.86	2.35	2.98

(b) $y = -(x+1) + 2e^x$

x_n	0	0.2	0.4	0.6	0.8	1.0
$y(x_n)$	1	1.24	1.58	2.04	2.65	3.44
absolute error	0	0.04	0.10	0.19	0.30	0.46
percentage error	0	3	6	9	11	13

(c)

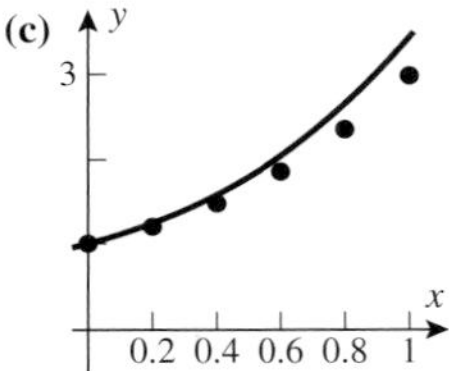

13.

n	0	1	2	3	4	5	6	7	8
x_n	0	0.5	1	1.5	2	2.5	3	3.5	4
y_n	1.00	1.50	2.07	2.71	3.41	4.16	4.96	5.82	6.72

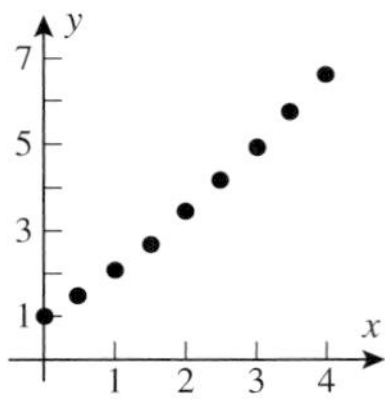

15.

n	0	1	2	3	4
t_n	0.00	0.50	1.00	1.50	2.00
y_n	1.00	1.27	1.42	1.49	1.53

y
2
1
t
0.5 1 1.5 2

17. 0.615537 19. **(b)** $y(1/2) = \sqrt{3}/2$

23. **(a)** $y' = \dfrac{2xy - y^3}{3xy^2 - x^2}$ **(c)** $xy^3 - x^2y = 2$

▶ Exercise Set 9.3 (Page 612)

1. **(a)** $\dfrac{dy}{dt} = ky^2, y(0) = y_0 (k > 0)$ **(b)** $\dfrac{dy}{dt} = -ky^2, y(0) = y_0 (k > 0)$

3. **(a)** $\dfrac{ds}{dt} = \dfrac{1}{2}s$ **(b)** $\dfrac{d^2s}{dt^2} = 2\dfrac{ds}{dt}$

5. **(a)** $y'(t) = y(t)/50$, $y(0) = 10000$ **(b)** $y(t) = 10000e^{t/50}$ **(c)** $50 \ln 2 \approx 34.66$ hr **(d)** $50 \ln(4.5) \approx 75.20$ hr

7. **(a)** $\frac{dy}{dt} = -ky$, $k \approx 0.1810$ **(b)** $y = 5.0 \times 10^7 e^{-0.181t}$ **(c)** $\approx 219,000$ atoms **(d)** 12.72 days

9. $50 \ln(100) \approx 230.26$ days 11. 3.30 days

13. **(a)** $y = 3e^{((\ln 2)/6)t}$ **(b)** $y = 4e^{t/50}$ **(c)** $y = 200^{-1/9}e^{((\ln 200)/9)t}$ **(d)** $y = 2^{5/6}e^{((\ln 2)/6)t}$

17. **(b)** 70 years **(c)** 20 years **(d)** 7% 21. **(a)** no **(b)** same, $r\%$

23. **(c)** $y = 4e^{t \ln 2}$ **(d)** $y = 4e^{-t \ln 2}$ 25. $\ln(2)/\ln(5/4) \approx 3.106$ hr

27. **(a)** \$1491.82 **(b)** \$4493.29 **(c)** 8.7 years

29. **(a)** $\frac{dT}{dt} = -k(T - 21)$, $T(0) = 95$; $T = 21 + 74e^{-kt}$ **(b)** 6.22 min

33. **(d)** $L/2$

35. **(a)** **(b)** **(c)** **(d)**

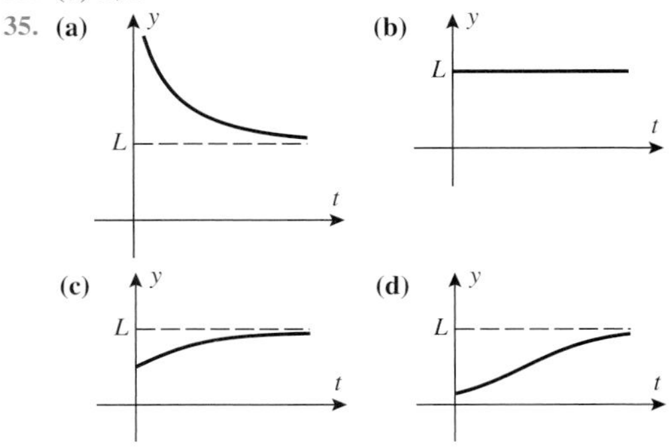

37. $y_0 \approx 2$, $L \approx 8$, $k \approx 0.5493$

39. **(a)** $y_0 = 5$ **(b)** $L = 12$ **(c)** $k = 1$ **(d)** $t = 0.3365$ **(e)** $\frac{dy}{dt} = \frac{1}{2}y(12 - y)$, $y(0) = 5$

41. Assume that $y(t)$ students have had the flu t days after semester break. Then $y(0) = 20$, $y(5) = 35$. **(a)** $\frac{dy}{dt} = ky(1000 - y)$, $y_0 = 20$ **(b)** $y = \frac{1000}{1 + 49e^{-0.115t}}$; $k = 0.115$

(c)

t	0	1	2	3	4	5	6	7
$y(t)$	20	22	25	28	31	35	39	44
t	8	9	10	11	12	13	14	
$y(t)$	49	54	61	67	75	83	93	

(d)

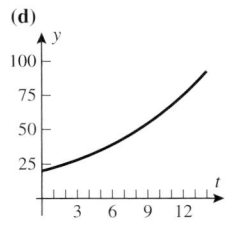

▶ Exercise Set 9.4 (Page 623)

3. $y = c_1e^x + c_2e^{-4x}$ 5. $y = c_1e^x + c_2xe^x$

7. $y = c_1 \cos x + c_2 \sin x$ 9. $y = c_1 + c_2e^x$

11. $y = c_1e^{2t} + c_2te^{2t}$ 13. $y = e^{-2x}(c_1 \cos 3x + c_2 \sin 3x)$

15. $y = c_1e^{-x/4} + c_2e^{x/2}$ 17. $y = 3e^x - 2e^{-3x}$

19. $y = 2e^{-3x} + xe^{-3x}$ 21. $y = -e^{-2x}(3 \cos x + 6 \sin x)$

23. **(a)** $y'' - 3y' - 10y = 0$ **(b)** $y'' - 8y' + 16y = 0$ **(c)** $y'' + 2y' + 17y = 0$

25. **(a)** $k < 0$ or $k > 4$ **(b)** 0, 4 **(c)** $0 < k < 4$

27. **(a)** $y = (1/x)[c_1 \cos(\ln x) + c_2 \sin(\ln x)]$ **(b)** $y = c_1x^{1+\sqrt{3}} + c_2x^{1-\sqrt{3}}$

33. **(a)** $y(t) = 0.4 \cos(t/2)$ m **(b)** period $= 4\pi$ s, frequency $= \frac{1}{4\pi}$ Hz

(c)

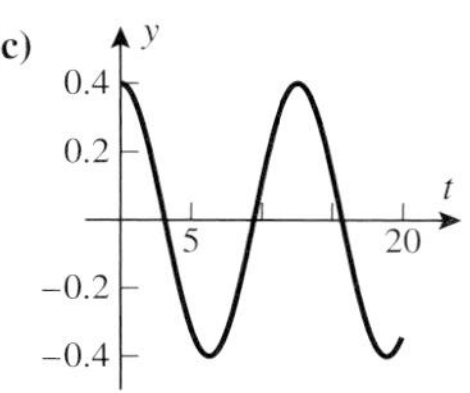

(d) $t = \pi$ s **(e)** $t = 2\pi$ s

35. **(a)** $y = -0.12 \cos 14t$ **(b)** $T = \pi/7$ s, $f = 7/\pi$ Hz

(c)

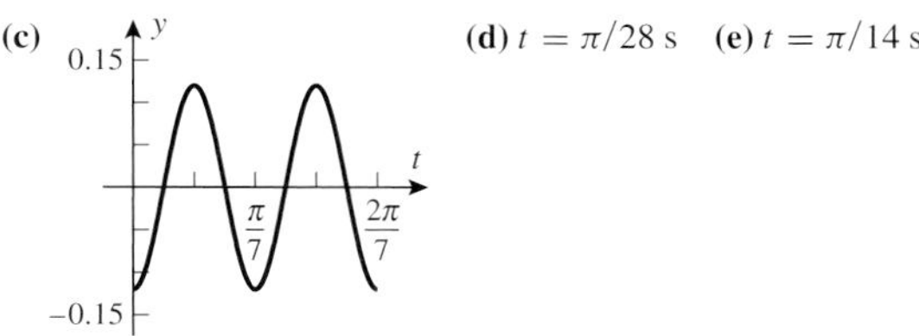

(d) $t = \pi/28$ s **(e)** $t = \pi/14$ s

37. **(a)** Maximum speed occurs when $y = 0$. **(b)** Minimum speed occurs when $y = \pm y_0$.

39. $Mx''(t) + kx(t) = 0$, $x(0) = x_0$, $x'(0) = 0$

43. **(a)** $y = e^{-1.2t} + 3.2te^{-1.2t}$ **(b)** 1.427364 cm

45. **(a)** $y = e^{-t/2} \cos(\sqrt{19}t/2) - \frac{6}{19}\sqrt{19}e^{-t/2} \sin(\sqrt{19}t/2)$

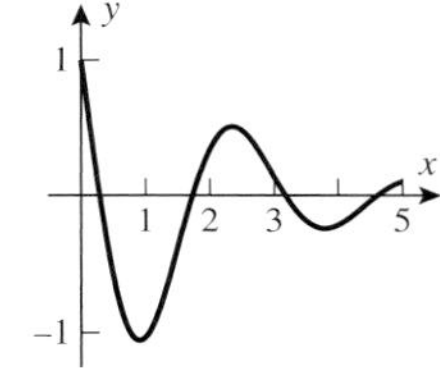

(b) 1.0545 cm **(c)** -3.210357 cm/s **(d)** 3.210357 cm/s^2

47. **(a)** $y = (4 + 2v_0)e^{-3t/2} - (3 + 2v_0)e^{-2t}$ **(b)** $8e^{-3t/2} - 7e^{-2t}$, $2e^{-3t/2} - e^{-2t}$, $-4e^{-3t/2} + 5e^{-2t}$

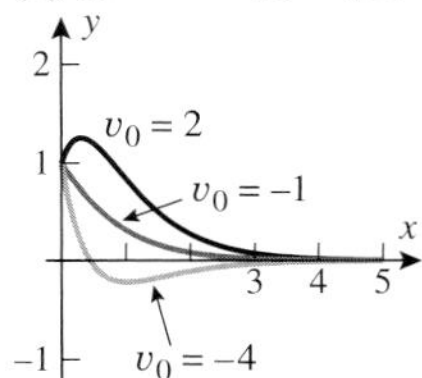

▶ Chapter 9. Review Exercises (Page 626)

3. **(a)** linear **(b)** both **(c)** separable **(d)** neither

5. no 7. $y = \tan(x^3/3 + C)$ 9. $\ln|y| + y^2/2 = e^x + C$ and $y = 0$

11. $y = -1 + 4e^{x^2/2}$ 13. $y = 2 \operatorname{sech} x + \frac{1}{2}(x \operatorname{sech} x + \sinh x)$

15. $y^{-4} + 4 \ln(x/y) = 1$

17. **(a)** $y = \left(-\frac{3}{10}x - \frac{3}{50}\right) \cos 3x + \left(-\frac{1}{10}x + \frac{2}{25}\right) \sin 3x + \frac{53}{50}e^x$

(c)

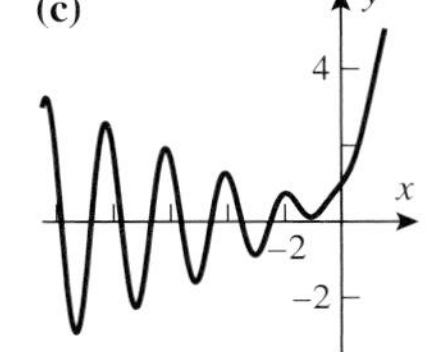

19. about 646 oz 21.

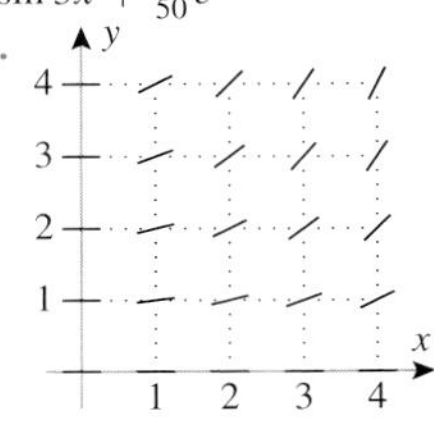

23.

n	0	1	2	3	4	5	6	7	8
x_n	0	0.5	1	1.5	2	2.5	3	3.5	4
y_n	1	1.50	2.11	2.84	3.68	4.64	5.72	6.91	8.23

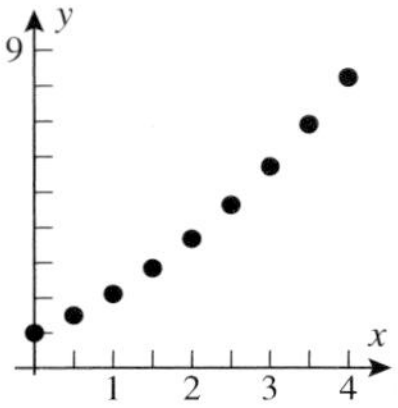

25. $y(1) \approx 1.00$

n	0	1	2	3	4	5
t_n	0	0.2	0.4	0.6	0.8	1.0
y_n	1.00	1.20	1.26	1.10	0.94	1.00

27. **(a)** $y \approx 2e^{0.1386t}$ **(b)** $y = 5e^{0.015t}$ **(c)** $y \approx 0.5995e^{0.5117t}$ **(d)** $y \approx 0.8706e^{0.1386t}$ 29. about 2000.6 years

31. **(a)** $y = C_1e^x + C_2e^{2x}$ **(b)** $y = C_1e^{x/2} + C_2xe^{x/2}$
(c) $y = e^{-x/2}\left[c_1 \cos \frac{\sqrt{7}x}{2} + C \sin \frac{\sqrt{7}x}{2}\right]$

33. **(a)** $y = 0.3\cos(t/2)$
(b) $T = 4\pi$ s, $f = 1/(4\pi)$ Hz
(d) $t = \pi$ s
(e) $t = 2\pi$ s
(c)

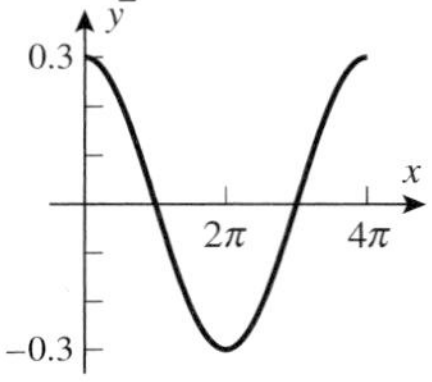

35. $\lim_{n\to+\infty} \sqrt[n]{n^3} = 1$

39. **(a)** $1, \frac{3}{4}, \frac{2}{3}, \frac{5}{8}$ **(c)** $\lim_{n\to+\infty} a_n = \frac{1}{2}$

43. **(a)** $(0.5)^{2n}$ **(c)** $\lim_{n\to+\infty} a_n = 0$ **(d)** $-1 \le a_0 \le 1$

45. **(a)**

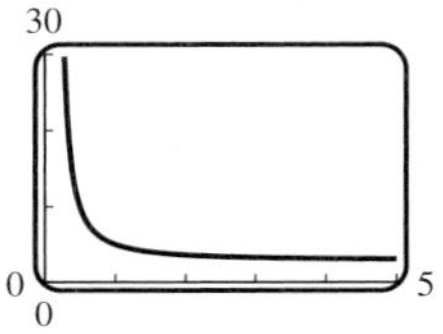

(b) $\lim_{n\to+\infty} (2^n + 3^n)^{1/n} = 3$

47. converges to 0 49. **(a)** $N = 3$ **(b)** $N = 11$ **(c)** $N = 1001$

▶ Exercise Set 10.1 (Page 637)

1. **(a)** $\frac{1}{3^{n-1}}$ **(b)** $\frac{(-1)^{n-1}}{3^{n-1}}$ **(c)** $\frac{2n-1}{2n}$ **(d)** $\frac{n^2}{\pi^{1/(n+1)}}$
3. **(a)** 2, 0, 2, 0 **(b)** 1, −1, 1, −1 **(c)** $2(1 + (-1)^n)$; $2 + 2\cos n\pi$
5. $\frac{1}{3}, \frac{2}{4}, \frac{3}{5}, \frac{4}{6}, \frac{5}{7}$; converges, $\lim_{n\to+\infty} \frac{n}{n+2} = 1$
7. 2, 2, 2, 2, 2; converges, $\lim_{n\to+\infty} 2 = 2$
9. $\frac{\ln 1}{1}, \frac{\ln 2}{2}, \frac{\ln 3}{3}, \frac{\ln 4}{4}, \frac{\ln 5}{5}$; converges, $\lim_{n\to+\infty} \frac{\ln n}{n} = 0$
11. 0, 2, 0, 2, 0; diverges 13. $-1, \frac{16}{9}, -\frac{54}{28}, \frac{128}{65}, -\frac{250}{126}$; diverges
15. $\frac{6}{2}, \frac{12}{8}, \frac{20}{18}, \frac{30}{32}, \frac{42}{50}$; converges, $\lim_{n\to+\infty} \frac{1}{2}\left(1 + \frac{1}{n}\right)\left(1 + \frac{2}{n}\right) = \frac{1}{2}$
17. $\cos 3, \cos \frac{3}{2}, \cos 1, \cos \frac{3}{4}, \cos \frac{3}{5}$; converges, $\lim_{n\to+\infty} \cos(3/n) = 1$
19. $e^{-1}, 4e^{-2}, 9e^{-3}, 16e^{-4}, 25e^{-5}$; converges, $\lim_{n\to+\infty} n^2e^{-n} = 0$
21. $2, \left(\frac{5}{3}\right)^2, \left(\frac{6}{4}\right)^3, \left(\frac{7}{5}\right)^4, \left(\frac{8}{6}\right)^5$; converges, $\lim_{n\to+\infty} \left[\frac{n+3}{n+1}\right]^n = e^2$
23. $\left\{\frac{2n-1}{2n}\right\}_{n=1}^{+\infty}$; converges, $\lim_{n\to+\infty} \frac{2n-1}{2n} = 1$
25. $\left\{(-1)^{n+1}\frac{1}{3^n}\right\}_{n=1}^{+\infty}$; converges, $\lim_{n\to+\infty} (-1)^{n+1}\frac{1}{3^n} = 0$
27. $\left\{(-1)^{n+1}\left(\frac{1}{n} - \frac{1}{n+1}\right)\right\}_{n=1}^{+\infty}$;
converges, $\lim_{n\to+\infty} (-1)^{n+1}\left(\frac{1}{n} - \frac{1}{n+1}\right) = 0$
29. $\{\sqrt{n+1} - \sqrt{n+2}\}_{n=1}^{+\infty}$; converges, $\lim_{n\to+\infty} (\sqrt{n+1} - \sqrt{n+2}) = 0$
31. For example, $\{(-1)^n\}_{n=1}^{+\infty}$ and $\{\sin(\pi n/2) + 1/n\}_{n=1}^{+\infty}$
33. **(a)** 1,2,1,4,1,6 **(b)** $a_n = \begin{cases} n, & n \text{ odd} \\ 1/2^n, & n \text{ even} \end{cases}$
(c) $a_n = \begin{cases} 1/n, & n \text{ odd} \\ 1/(n+1); & n \text{ even} \end{cases}$
(d) (a) diverges; (b) diverges; (c) $\lim_{n\to+\infty} a_n = 0$

▶ Exercise Set 10.2 (Page 645)

1. strictly decreasing 3. strictly increasing 5. strictly decreasing
7. strictly increasing 9. strictly decreasing 11. strictly increasing
13. strictly increasing 15. strictly decreasing 17. strictly decreasing
19. eventually strictly increasing 21. eventually strictly decreasing
23. eventually strictly increasing
25. **(a)** Yes; the limit lies in the interval [1, 2].
(b) No, but if so, then the limit is ≤ 2.

▶ Exercise Set 10.3 (Page 653)

1. **(a)** $2, \frac{12}{5}, \frac{62}{25}, \frac{312}{125}$; $\frac{5}{2}\left(1 - \left(\frac{1}{5}\right)^n\right)$; $\lim_{n\to+\infty} s_n = \frac{5}{2}$ (converges)
(b) $\frac{1}{4}, \frac{3}{4}, \frac{7}{4}, \frac{15}{4}$; $-\frac{1}{4}(1 - 2^n)$; $\lim_{n\to+\infty} s_n = +\infty$ (diverges)
(c) $\frac{1}{6}, \frac{1}{4}, \frac{3}{10}, \frac{1}{3}$; $\frac{1}{2} - \frac{1}{n+2}$; $\lim_{n\to+\infty} s_n = \frac{1}{2}$ (converges)
3. $\frac{4}{7}$ 5. 6 7. $\frac{1}{3}$ 9. $\frac{1}{6}$ 11. diverges 13. $\frac{448}{3}$
15. **(a)** Exercise 5 **(b)** Exercise 3 **(c)** Exercise 7 **(d)** Exercise 9
17. $\frac{4}{9}$ 19. $\frac{532}{99}$ 23. 70 m
25. **(a)** $S_n = -\ln(n+1)$; $\lim_{n\to+\infty} S_n = -\infty$ (diverges)
(b) $S_n = \sum_{k=2}^{n+1}\left[\ln \frac{k-1}{k} - \ln \frac{k}{k+1}\right]$, $\lim_{n\to+\infty} S_n = -\ln 2$
27. **(a)** converges for $|x| < 1$; $S = \frac{x}{1+x^2}$
(b) converges for $|x| > 2$; $S = \frac{1}{x^2 - 2x}$
(c) converges for $x > 0$; $S = \frac{1}{e^x - 1}$
29. $a_n = \frac{1}{2^{n-1}}a_1 + \frac{1}{2^{n-1}} + \frac{1}{2^{n-2}} + \cdots + \frac{1}{2}$, $\lim_{n\to+\infty} a_n = 1$
37. **(b)** $A = 1, B = -2$
(c) $S_n = 2 - \frac{2^{n+1}}{3^{n+1} - 2^{n+1}}$,
$\lim_{n\to+\infty} S_n = \lim_{n\to+\infty}\left[2 - \frac{(2/3)^{n+1}}{1 - (2/3)^{n+1}}\right] = 2$

▶ Exercise Set 10.4 (Page 661)

1. **(a)** $\frac{4}{3}$ **(b)** $-\frac{3}{4}$
3. **(a)** $p = 3$, converges **(b)** $p = \frac{1}{2}$, diverges
(c) $p = 1$, diverges **(d)** $p = \frac{2}{3}$, diverges
5. **(a)** diverges **(b)** diverges **(c)** diverges **(d)** no information
7. **(a)** diverges **(b)** converges
9. diverges 11. diverges 13. diverges 15. diverges 17. diverges
19. converges 21. diverges 23. converges 25. converges for $p > 1$
29. **(a)** diverges **(b)** diverges
31. **(a)** $(\pi^2/2) - (\pi^4/90)$ **(b)** $(\pi^2/6) - (5/4)$ **(c)** $\pi^4/90$
33. **(d)** $\frac{1}{11} < \frac{1}{6}\pi^2 - s_{10} < \frac{1}{10}$
35. **(b)** $n = 5$ **(c)** 1.20
37. **(d)** $n > e^{100} - 1$ 39. converges

▶ Exercise Set 10.5 (Page 668)

1. **(a)** converges **(b)** diverges 5. converges 7. converges 9. diverges 11. converges 13. inconclusive 15. diverges 17. diverges 19. converges 21. converges 23. converges 25. converges 27. converges 29. diverges 31. converges 33. diverges 35. converges 37. converges 39. diverges 41. converges 43. converges

45. $u_k = \dfrac{k!}{1 \cdot 3 \cdot 5 \cdots (2k-1)}$; $\rho = \lim\limits_{k\to+\infty} \dfrac{k+1}{2k+1} = \dfrac{1}{2}$; converges

47. converges 49. diverges 51. **(a)** converges **(b)** diverges

▶ Exercise Set 10.6 (Page 677)

3. diverges 5. converges 7. converges absolutely 9. diverges 11. converges absolutely 13. conditionally convergent 15. divergent 17. conditionally convergent 19. conditionally convergent 21. divergent 23. conditionally convergent 25. converges absolutely 27. conditionally convergent 29. converges absolutely 31. $|\text{error}| < 0.125$ 33. $|\text{error}| < 0.1$ 35. $n = 9999$ 37. $n = 39{,}999$ 39. $|\text{error}| < 0.00074$; $s_{10} \approx 0.4995$; $S = 0.5$ 41. 0.84 43. 0.41 45. **(c)** $n = 50$

47. **(a)** If $a_k = \dfrac{(-1)^k}{\sqrt{k}}$, then $\sum a_k$ converges and $\sum a_k^2$ diverges. If $a_k = \dfrac{(-1)^k}{k}$, then $\sum a_k$ converges and $\sum a_k^2$ also converges.
(b) If $a_k = \dfrac{1}{k}$, then $\sum a_k^2$ converges and $\sum a_k$ diverges. If $a_k = \dfrac{1}{k^2}$, then $\sum a_k^2$ converges and $\sum a_k$ also converges.

55. **(a)** $124.58 < d < 124.77$ **(b)** $1243 < s < 1424$

▶ Exercise Set 10.7 (Page 688)

1. **(a)** $1 - x + \frac{1}{2}x^2$, $1 - x$ **(b)** $1 - \frac{1}{2}x^2$, 1 **(c)** $1 - \frac{1}{2}(x - \pi/2)^2$, 1
(d) $1 + \frac{1}{2}(x-1) - \frac{1}{8}(x-1)^2$, $1 + \frac{1}{2}(x-1)$

3. **(a)** $1 + \frac{1}{2}(x-1) - \frac{1}{8}(x-1)^2$ **(b)** 1.04875 5. 1.80397443

7. $p_0(x) = 1$, $p_1(x) = 1 - x$, $p_2(x) = 1 - x + \frac{1}{2}x^2$, $p_3(x) = 1 - x + \frac{1}{2}x^2 - \frac{1}{3!}x^3$, $p_4(x) = 1 - x + \frac{1}{2}x^2 - \frac{1}{3!}x^3 + \frac{1}{4!}x^4$; $\displaystyle\sum_{k=0}^{n} \frac{(-1)^k}{k!}x^k$

9. $p_0(x) = 1$, $p_1(x) = 1$, $p_2(x) = 1 - \dfrac{\pi^2}{2!}x^2$, $p_3(x) = 1 - \dfrac{\pi^2}{2!}x^2$, $p_4(x) = 1 - \dfrac{\pi^2}{2!}x^2 + \dfrac{\pi^4}{4!}x^4$; $\displaystyle\sum_{k=0}^{\lfloor n/2 \rfloor} \frac{(-1)^k \pi^{2k}}{(2k)!}x^{2k}$ (See Exercise 74 of Section 1.3.)

11. $p_0(x) = 0$, $p_1(x) = x$, $p_2(x) = x - \frac{1}{2}x^2$, $p_3(x) = x - \frac{1}{2}x^2 + \frac{1}{3}x^3$, $p_4(x) = x - \frac{1}{2}x^2 + \frac{1}{3}x^3 - \frac{1}{4}x^4$; $\displaystyle\sum_{k=1}^{n} \frac{(-1)^{k+1}}{k}x^k$

13. $p_0(x) = 1$, $p_1(x) = 1$, $p_2(x) = 1 + \dfrac{x^2}{2}$, $p_3(x) = 1 + \dfrac{x^2}{2}$, $p_4(x) = 1 + \dfrac{x^2}{2} + \dfrac{x^4}{4!}$; $\displaystyle\sum_{k=0}^{\lfloor n/2 \rfloor} \frac{1}{(2k)!}x^{2k}$ (See Exercise 74 of Section 1.3.)

15. $p_0(x) = 0$, $p_1(x) = 0$, $p_2(x) = x^2$, $p_3(x) = x^2$, $p_4(x) = x^2 - \frac{1}{6}x^4$; $\displaystyle\sum_{k=0}^{\lfloor n/2 \rfloor - 1} \frac{(-1)^k}{(2k+1)!}x^{2k+2}$ (See Exercise 74 of Section 1.3.)

17. $p_0(x) = e$, $p_1(x) = e + e(x-1)$, $p_2(x) = e + e(x-1) + \dfrac{e}{2}(x-1)^2$, $p_3(x) = e + e(x-1) + \dfrac{e}{2}(x-1)^2 + \dfrac{e}{3!}(x-1)^3$, $p_4(x) = e + e(x-1) + \dfrac{e}{2}(x-1)^2 + \dfrac{e}{3!}(x-1)^3 + \dfrac{e}{4!}(x-1)^4$; $\displaystyle\sum_{k=0}^{n} \frac{e}{k!}(x-1)^k$

19. $p_0(x) = -1$, $p_1(x) = -1 - (x+1)$, $p_2(x) = -1 - (x+1) - (x+1)^2$, $p_3(x) = -1 - (x+1) - (x+1)^2 - (x+1)^3$, $p_4(x) = -1 - (x+1) - (x+1)^2 - (x+1)^3 - (x+1)^4$; $\displaystyle\sum_{k=0}^{n} (-1)(x+1)^k$

21. $p_0(x) = p_1(x) = 1$, $p_2(x) = p_3(x) = 1 - \dfrac{\pi^2}{2}\left(x - \dfrac{1}{2}\right)^2$, $p_4(x) = 1 - \dfrac{\pi^2}{2}\left(x - \dfrac{1}{2}\right)^2 + \dfrac{\pi^4}{4!}\left(x - \dfrac{1}{2}\right)^4$; $\displaystyle\sum_{k=0}^{\lfloor n/2 \rfloor} \frac{(-1)^k \pi^{2k}}{(2k)!}\left(x - \frac{1}{2}\right)^{2k}$ (See Exercise 74 of Section 1.3.)

23. $p_0(x) = 0$, $p_1(x) = (x-1)$, $p_2(x) = (x-1) - \frac{1}{2}(x-1)^2$, $p_3(x) = (x-1) - \frac{1}{2}(x-1)^2 + \frac{1}{3}(x-1)^3$, $p_4(x) = (x-1) - \frac{1}{2}(x-1)^2 + \frac{1}{3}(x-1)^3 - \frac{1}{4}(x-1)^4$; $\displaystyle\sum_{k=1}^{n} \frac{(-1)^{k-1}}{k}(x-1)^k$

25. **(a)** $p_3(x) = 1 + 2x - x^2 + x^3$
(b) $p_3(x) = 1 + 2(x-1) - (x-1)^2 + (x-1)^3$

27. $p_0(x) = 1$, $p_1(x) = 1 - 2x$, $p_2(x) = 1 - 2x + 2x^2$, $p_3(x) = 1 - 2x + 2x^2 - \frac{4}{3}x^3$

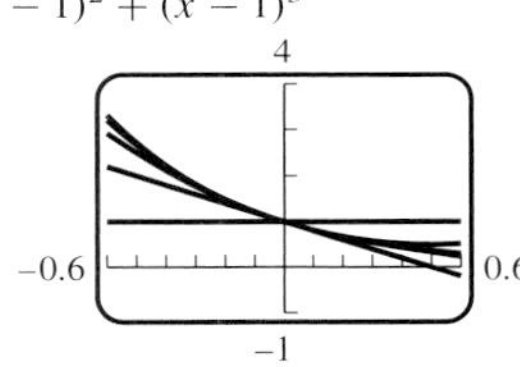

29. $p_0(x) = -1$, $p_2(x) = -1 + \frac{1}{2}(x-\pi)^2$, $p_4(x) = -1 + \frac{1}{2}(x-\pi)^2 - \frac{1}{24}(x-\pi)^4$, $p_6(x) = -1 + \frac{1}{2}(x-\pi)^2 - \frac{1}{24}(x-\pi)^4 + \frac{1}{720}(x-\pi)^6$

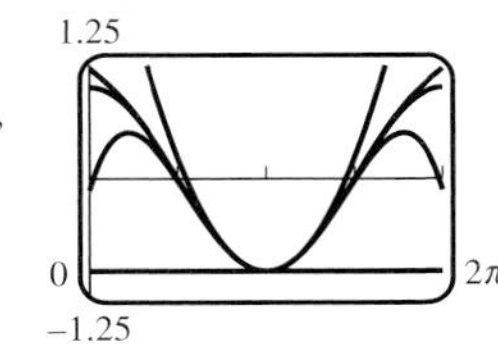

31. 1.64870 35. IV

37. **(a)**

(b)

x	−1.000	−0.750	−0.500	−0.250	0.000	0.250	0.500	0.750	1.000
$f(x)$	0.431	0.506	0.619	0.781	1.000	1.281	1.615	1.977	2.320
$p_1(x)$	0.000	0.250	0.500	0.750	1.000	1.250	1.500	1.750	2.000
$p_2(x)$	0.500	0.531	0.625	0.781	1.000	1.281	1.625	2.031	2.500

(c) $|e^{\sin x} - (1 + x)| < 0.01$ for $-0.14 < x < 0.14$

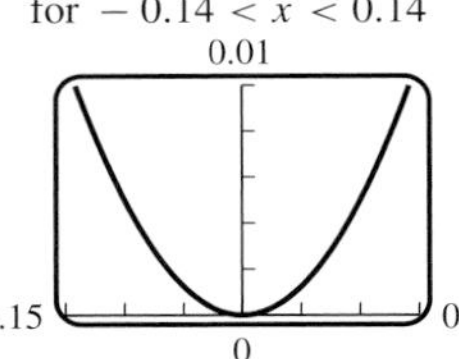

(d) $\left|e^{\sin x} - \left(1 + x + \frac{x^2}{2}\right)\right| < 0.01$ for $-0.50 < x < 0.50$

0.015

-0.6 0.6

0

39. **(a)** $(-0.569, 0.569)$

0.0005

-0.7 0.7

0

41. $(-.311, .311)$

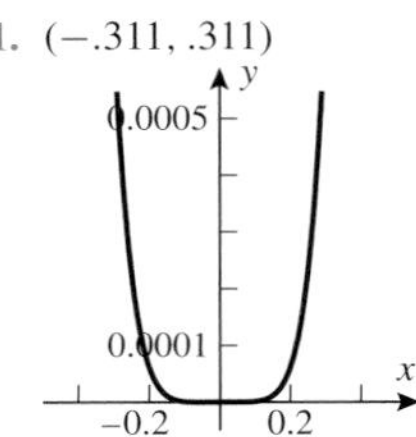

▶ Exercise Set 10.8 (Page 697)

1. $\sum_{k=0}^{\infty} \frac{(-1)^k}{k!} x^k$ **3.** $\sum_{k=0}^{\infty} \frac{(-1)^k \pi^{2k}}{(2k)!} x^{2k}$ **5.** $\sum_{k=1}^{\infty} \frac{(-1)^{k+1}}{k} x^k$

7. $\sum_{k=0}^{\infty} \frac{1}{(2k)!} x^{2k}$ **9.** $\sum_{k=0}^{\infty} \frac{(-1)^k}{(2k+1)!} x^{2k+2}$ **11.** $\sum_{k=0}^{\infty} \frac{e}{k!}(x-1)^k$

13. $\sum_{k=0}^{\infty} (-1)(x+1)^k$ **15.** $\sum_{k=0}^{\infty} \frac{(-1)^k \pi^{2k}}{(2k)!} \left(x - \frac{1}{2}\right)^{2k}$

17. $\sum_{k=1}^{\infty} \frac{(-1)^{k-1}}{k} (x-1)^k$ **19.** $-1 < x < 1$; $\frac{1}{1+x}$

21. $1 < x < 3$, $\frac{1}{3-x}$ **23.** **(a)** $-2 < x < 2$ **(b)** $f(0) = 1$; $f(1) = \frac{2}{3}$

25. $R = 1$; $[-1, 1)$ **27.** $R = +\infty$; $(-\infty, +\infty)$ **29.** $R = \frac{1}{5}$; $[-\frac{1}{5}, \frac{1}{5}]$

31. $R = 1$; $[-1, 1]$ **33.** $R = 1$; $(-1, 1]$ **35.** $R = +\infty$; $(-\infty, +\infty)$

37. $R = +\infty$; $(-\infty, +\infty)$ **39.** $R = 1$; $[-1, 1]$ **41.** $R = 1$; $(-2, 0]$

43. $R = \frac{4}{3}$; $(-\frac{19}{3}, -\frac{11}{3})$ **45.** $R = 1$; $[-2, 0]$

47. $R = +\infty$; $(-\infty, +\infty)$ **49.** $(-\infty, +\infty)$

55. radius $= R$

▶ Exercise Set 10.9 (Page 706)

1. 0.069756 **3.** 0.99500 **5.** 0.99619 **7.** 0.5208

11. **(a)** $\sum_{k=1}^{\infty} 2\frac{(1/9)^{2k-1}}{2k-1}$ **(b)** 0.223

13. **(a)** 0.4635; 0.3218 **(b)** 3.1412 **(c)** no

17. **(a)** $\sum_{k=0}^{\infty} (-1)^k x^k$ **(b)** $1 + \frac{x}{3} + \sum_{k=2}^{\infty} (-1)^{k-1} \frac{2 \cdot 5 \cdots (3k-4)}{3^k k!} x^k$

(c) $\sum_{k=0}^{\infty} (-1)^k \frac{(k+2)(k+1)}{2} x^k$ **23.** 23.406%

25. **(a)** 0.78539816339744830961566608

(b)

n	s_n
0	0.3183098 78 . . .
1	0.3183098 861837906 067 . . .
2	0.3183098 861837906 7153776 695 . . .
3	0.3183098 861837906 7153776 752674502 34 . . .

▶ Exercise Set 10.10 (Page 716)

1. **(a)** $1 - x + x^2 - \cdots + (-1)^k x^k + \cdots$; $R = 1$

(b) $1 + x^2 + x^4 + \cdots + x^{2k} + \cdots$; $R = 1$

(c) $1 + 2x + 4x^2 + \cdots + 2^k x^k + \cdots$; $R = \frac{1}{2}$

(d) $\frac{1}{2} + \frac{1}{2^2}x + \frac{1}{2^3}x^2 + \cdots + \frac{1}{2^{k+1}}x^k + \cdots$; $R = 2$

3. **(a)** $(2 + x)^{-1/2} = \frac{1}{2^{1/2}} - \frac{1}{2^{5/2}}x + \frac{1 \cdot 3}{2^{9/2} \cdot 2!}x^2 - \frac{1 \cdot 3 \cdot 5}{2^{13/2} \cdot 3!}x^3 + \cdots$

(b) $(1 - x^2)^{-2} = 1 + 2x^2 + 3x^4 + 4x^6 + \cdots$

5. **(a)** $2x - \frac{2^3}{3!}x^3 + \frac{2^5}{5!}x^5 - \frac{2^7}{7!}x^7 + \cdots$; $R = +\infty$

(b) $1 - 2x + 2x^2 - \frac{4}{3}x^3 + \cdots$; $R = +\infty$

(c) $1 + x^2 + \frac{1}{2!}x^4 + \frac{1}{3!}x^6 + \cdots$; $R = +\infty$

(d) $x^2 - \frac{\pi^2}{2}x^4 + \frac{\pi^4}{4!}x^6 - \frac{\pi^6}{6!}x^8 + \cdots$; $R = +\infty$

7. **(a)** $x^2 - 3x^3 + 9x^4 - 27x^5 + \cdots$; $R = \frac{1}{3}$

(b) $2x^2 + \frac{2^3}{3!}x^4 + \frac{2^5}{5!}x^6 + \frac{2^7}{7!}x^8 + \cdots$; $R = +\infty$

(c) $x - \frac{3}{2}x^3 + \frac{3}{8}x^5 + \frac{1}{16}x^7 + \cdots$; $R = 1$

9. **(a)** $x^2 - \frac{2^3}{4!}x^4 + \frac{2^5}{6!}x^6 - \frac{2^7}{8!}x^8 + \cdots$

(b) $12x^3 - 6x^6 + 4x^9 - 3x^{12} + \cdots$

11. **(a)** $1 - (x-1) + (x-1)^2 - \cdots + (-1)^k(x-1)^k + \cdots$ **(b)** (0,2)

13. **(a)** $x + x^2 + \frac{x^3}{3} - \frac{x^5}{30} + \cdots$ **(b)** $x - \frac{x^3}{24} + \frac{x^4}{24} - \frac{71}{1920}x^5 + \cdots$

15. **(a)** $1 + \frac{1}{2}x^2 + \frac{5}{24}x^4 + \frac{61}{720}x^6 + \cdots$ **(b)** $x - x^2 + \frac{1}{3}x^3 - \frac{1}{30}x^5 + \cdots$

19. $2 - 4x + 2x^2 - 4x^3 + 2x^4 + \cdots$

25. **(a)** $\sum_{k=0}^{\infty} x^{2k+1}$ **(b)** $f^{(5)}(0) = 5!$, $f^{(6)}(0) = 0$

(c) $f^{(n)}(0) = n!c_n = \begin{cases} n! & \text{if } n \text{ odd} \\ 0 & \text{if } n \text{ even} \end{cases}$

27. **(a)** 1 **(b)** $-\frac{1}{3}$ **29.** 0.3103 **31.** 0.200

33. **(a)** $\sum_{k=0}^{\infty} \frac{x^{4k}}{k!}$; $R = +\infty$ **35.** **(a)** 3/4 **(b)** ln(4/3)

37. **(a)** $x - \frac{1}{6}x^3 + \frac{3}{40}x^5 - \frac{5}{112}x^7 + \cdots$

(b) $x + \sum_{k=1}^{\infty} (-1)^k \frac{1 \cdot 3 \cdot 5 \cdots (2k-1)}{2^k k!(2k+1)} x^{2k+1}$ **(c)** $R = 1$

39. **(a)** $y(t) = y_0 \sum_{k=0}^{\infty} \frac{(-1)^k (0.000121)^k t^k}{k!}$ **(c)** $0.9998790073 y_0$

41. **(a)** $T \approx 2.00709$ **(b)** $T \approx 2.008044621$ **(c)** 2.008045644

43. **(a)** $F = mg\left(1 - \frac{2h}{R} + \frac{3h^2}{R^2} - \frac{4h^3}{R^3} + \cdots\right)$ **(d)** about 0.27% less

▶ Chapter 10. Review Exercises (Page 718)

9. **(a)** true **(b)** sometimes false **(c)** sometimes false **(d)** true **(e)** sometimes false **(f)** sometimes false **(g)** false **(h)** sometimes false **(i)** true **(j)** true **(k)** sometimes false **(l)** sometimes false

11. **(a)** $\left\{\frac{n+2}{(n+1)^2 - n^2}\right\}_{n=1}^{+\infty}$; converges, $\lim_{n\to+\infty} \frac{n+2}{(n+1)^2 - n^2} = \frac{1}{2}$

(b) $\left\{(-1)^{n+1}\frac{n}{2n+1}\right\}_{n=1}^{+\infty}$; diverges

15. **(a)** converges **(b)** converges **17.** **(a)** converges **(b)** diverges

19. **(a)** diverges **(b)** converges **21.** $\frac{1}{4 \cdot 5^{99}}$

23. **(a)** 2 **(b)** diverges **(c)** 3/4 **(d)** $\pi/4$ **25.** $p > 1$

29. **(a)** $p_0(x) = 1$, $p_1(x) = 1 - 7x$, $p_2(x) = 1 - 7x + 5x^2$, $p_3(x) = 1 - 7x + 5x^2 + 4x^3$, $p_4(x) = 1 - 7x + 5x^2 + 4x^3$

33. **(a)** $e^2 - 1$ **(b)** 0 **(c)** $\cos e$ **(d)** $\frac{1}{3}$

37. **(a)** $x - \frac{2}{3}x^3 + \frac{2}{15}x^5 - \frac{4}{315}x^7$ **(b)** $x - \frac{2}{3}x^3 + \frac{2}{15}x^5 - \frac{4}{315}x^7$

▶ Exercise Set 11.1 (Page 732)

1.

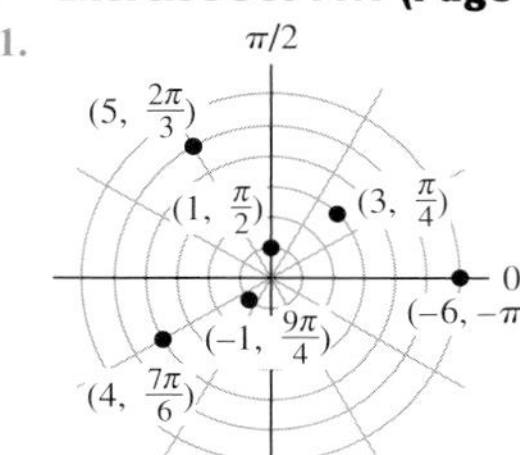

3. **(a)** $(3\sqrt{3}, 3)$
(b) $(-7/2, 7\sqrt{3}/2)$
(c) $(3\sqrt{3}, 3)$
(d) $(0, 0)$
(e) $(-7\sqrt{3}/2, 7/2)$
(f) $(-5, 0)$

5. **(a)** $(5, \pi)$, $(5, -\pi)$ **(b)** $(4, 11\pi/6)$, $(4, -\pi/6)$
(c) $(2, 3\pi/2)$, $(2, -\pi/2)$ **(d)** $(8\sqrt{2}, 5\pi/4)$, $(8\sqrt{2}, -3\pi/4)$
(e) $(6, 2\pi/3)$, $(6, -4\pi/3)$ **(f)** $(\sqrt{2}, \pi/4)$, $(\sqrt{2}, -7\pi/4)$

7. **(a)** $(5, 0.92730)$ **(b)** $(10, -0.92730)$ **(c)** $(1.27155, 2.47582)$

9. **(a)** circle **(b)** line **(c)** circle **(d)** line

11. **(a)** $r = 3\sec\theta$ **(b)** $r = \sqrt{7}$ **(c)** $r = -6\sin\theta$
(d) $r^2\cos\theta\sin\theta = 4/9$

13.

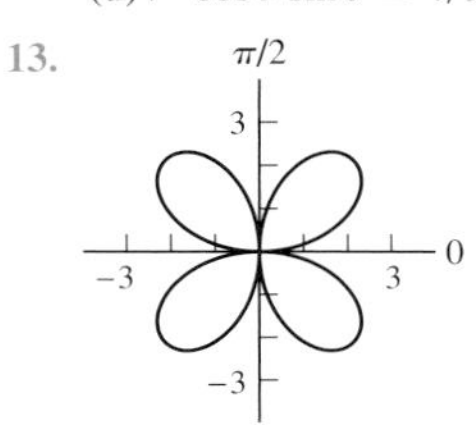

15.

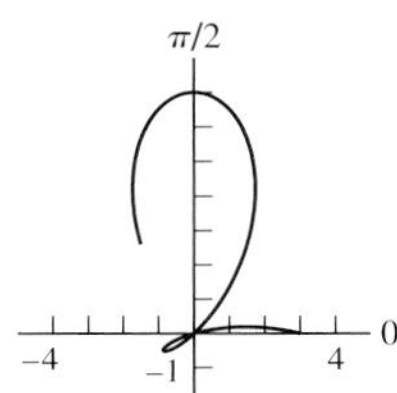

17. **(a)** $r = 5$ **(b)** $r = 6\cos\theta$ **(c)** $r = 1 - \cos\theta$

19. **(a)** $r = 3\sin 2\theta$ **(b)** $r = 3 + 2\sin\theta$ **(c)** $r^2 = 9\cos 2\theta$

21.

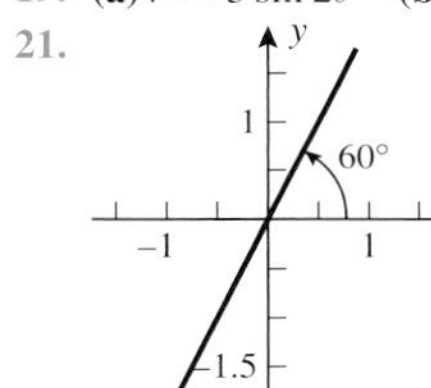

23.

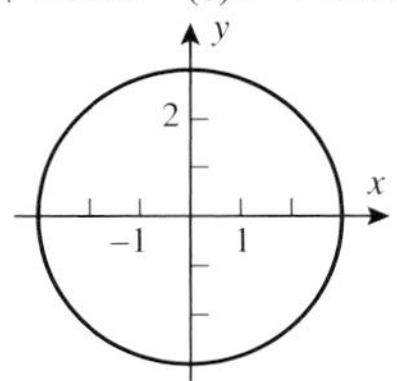

25.

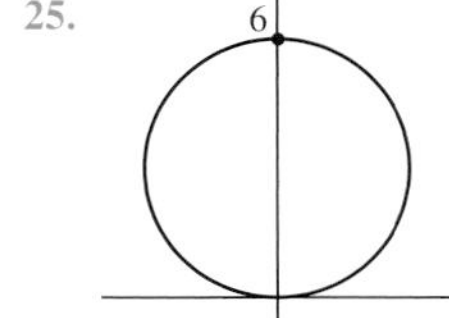

Circle

27.

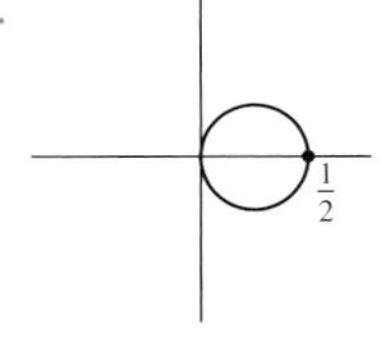

Circle

29.

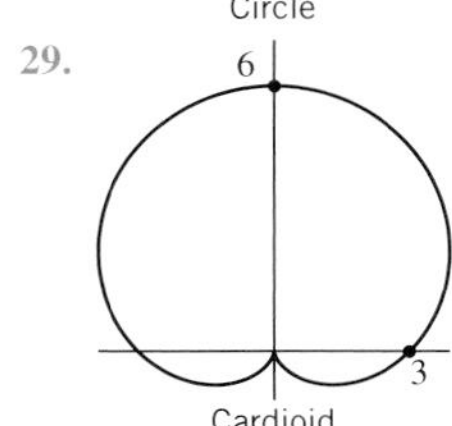

Cardioid

31.

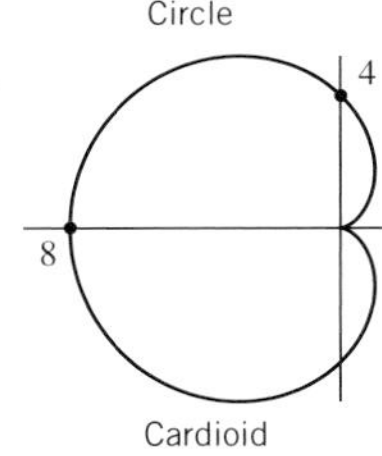

Cardioid

33.

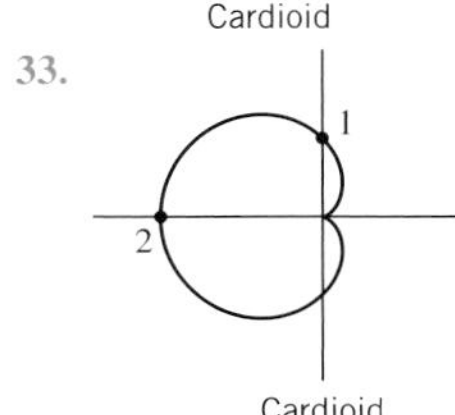

Cardioid

35.

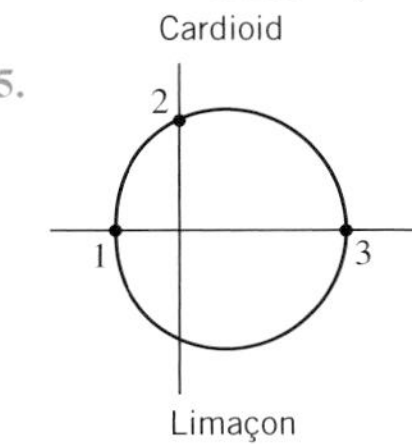

Limaçon

37.

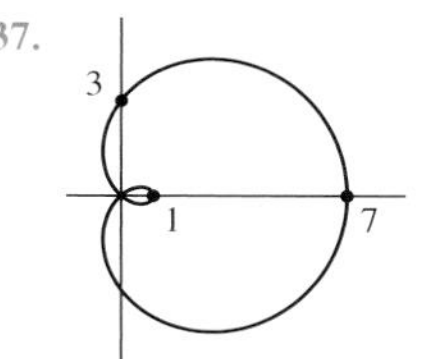

Limaçon

39.

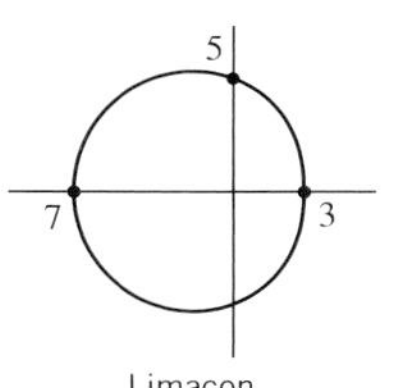

Limaçon

41.

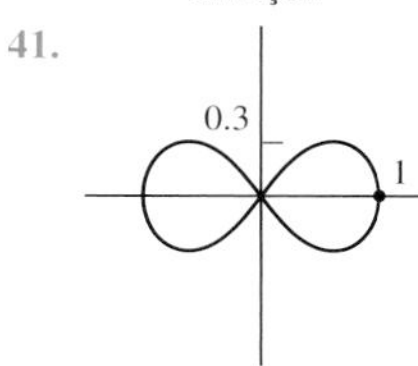

Lemniscate

43.

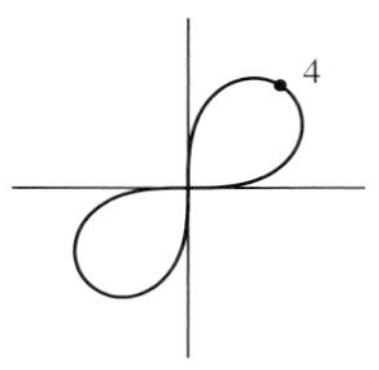

Lemniscate

45.

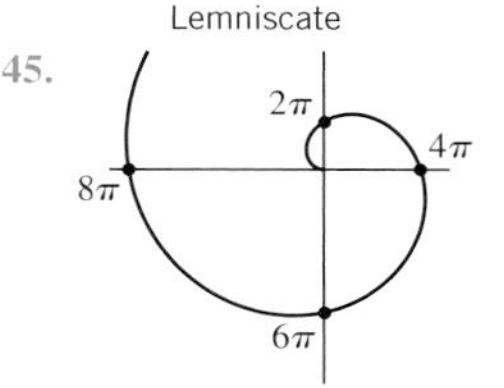

Spiral

47.

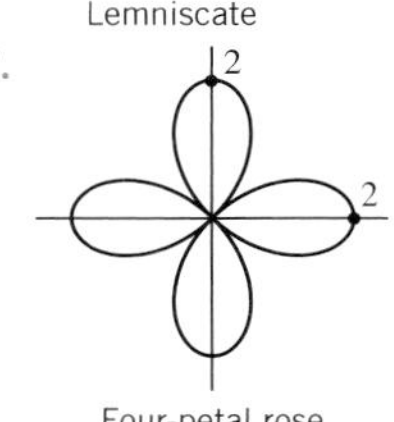

Four-petal rose

49.

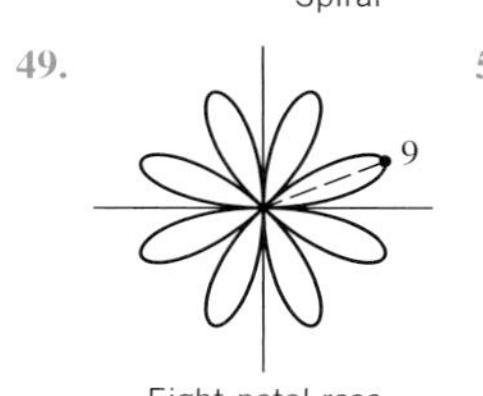

Eight-petal rose

51.

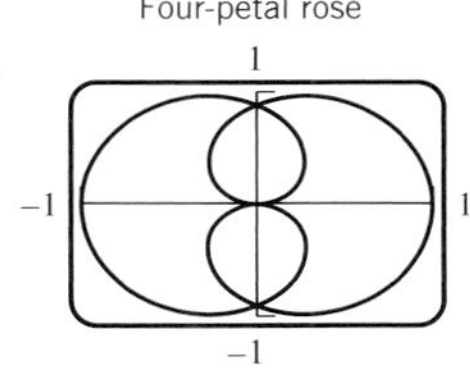

53.

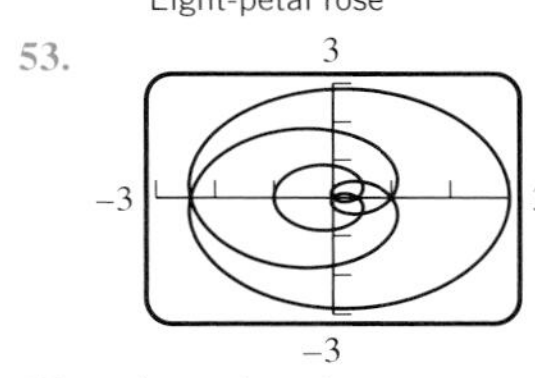

55.

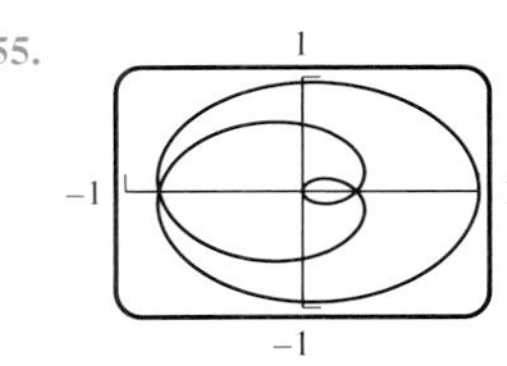

57. $-4\pi < \theta < 4\pi$

61. **(a)**

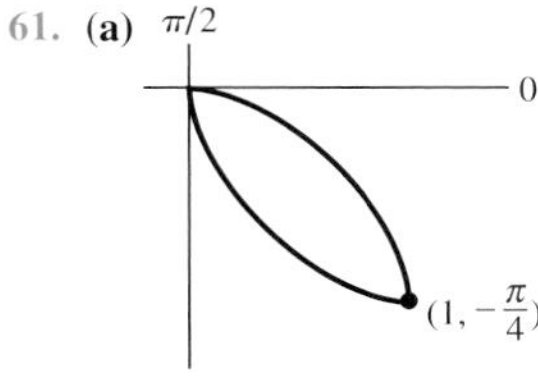

(b)

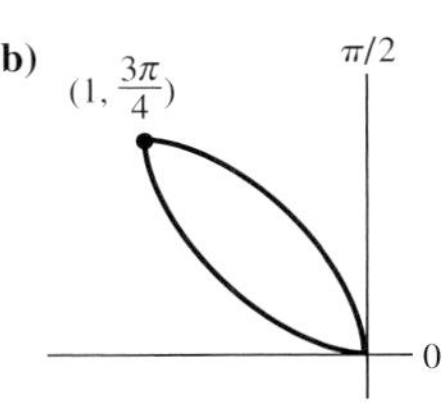

(c)

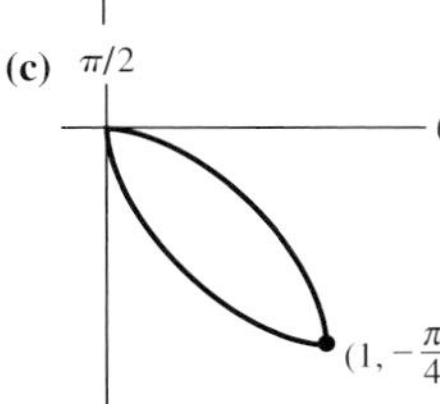

(d)

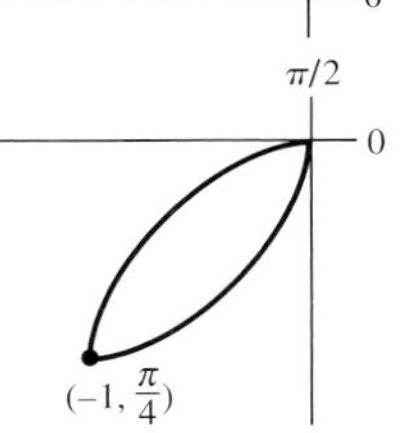

(e)

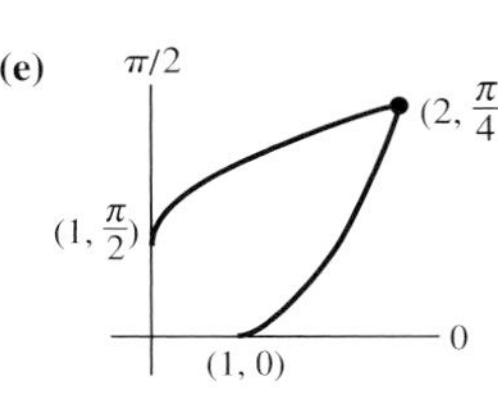

65. **(a)** $r = 1 + \dfrac{\sqrt{2}}{2}(\cos\theta + \sin\theta)$
(b) $r = 1 + \sin\theta$
(c) $r = 1 - \cos\theta$
(d) $r = 1 - \dfrac{\sqrt{2}}{2}(\cos\theta + \sin\theta)$

67.

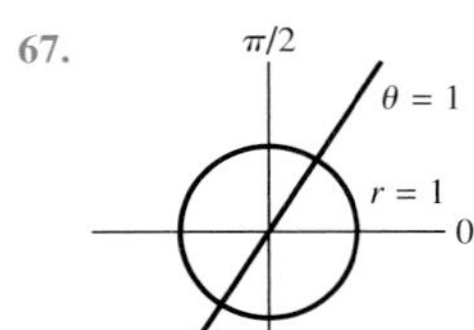

69. $(3/2, \pi/3)$ 73. **(c)** $\sqrt{13 - 6\sqrt{3}} \approx 1.615$ **(d)** $A = 1$
75. $\sqrt{3}$

Exercise Set 11.2 (Page 741)

1. $-4, 4$ 3. both are positive 5. $4, 4$ 7. $2/\sqrt{3}, -1/(3\sqrt{3})$
9. $\sqrt{3}, 4$ 11. **(a)** $y = -e^{-2}x + 2e^{-1}$ 13. **(a)** $0, \pi, 2\pi$ **(b)** $\pi/2, 3\pi/2$
15. $y = -2x, y = 2x$ 17. $y = 2x - 8, y = -2x + 8$
19.

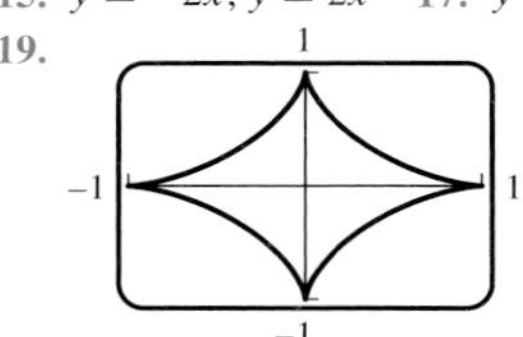

21. $\sqrt{3}$
23. $\dfrac{\tan 2 - 2}{2\tan 2 + 1}$
25. $1/2$
27. $1, 0, -1$

29. horizontal: $(3a/2, \pi/3), (0, \pi), (3a/2, 5\pi/3)$; vertical: $(2a, 0), (a/2, 2\pi/3), (a/2, 4\pi/3)$
31. $(0, 0), (\sqrt{2}/4, \pi/4), (\sqrt{2}/4, 3\pi/4)$
33.

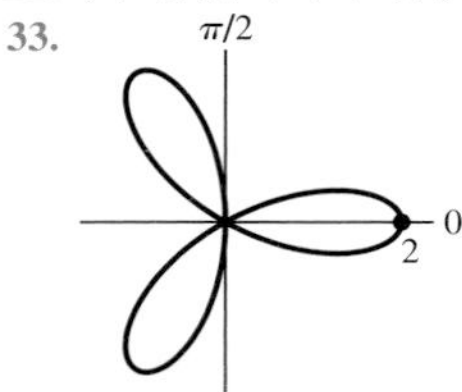

$\theta_0 = \pi/6$

35.

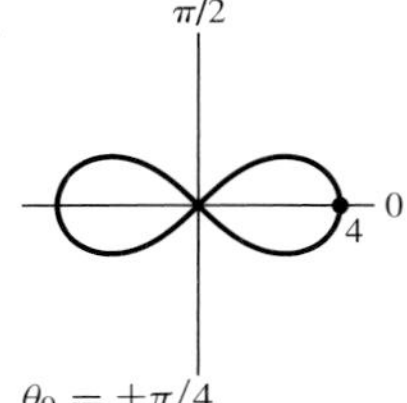

$\theta_0 = \pm\pi/4$

37.

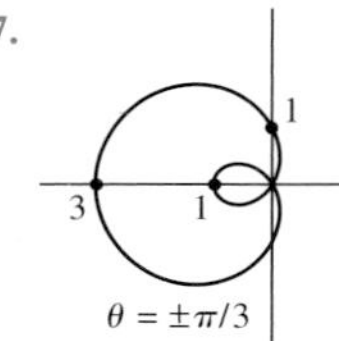

$\theta = \pm\pi/3$

39. $L = 2\pi a$
41. $L = 8a$
43. $L = \sqrt{10}(e^6 - 1)/3$

45. **(a)** $\dfrac{dy}{dx} = \dfrac{3\sin t}{1 - 3\cos t}$ **(b)** $\theta = -0.4345$
47. **(b)** ≈ 2.42
(c)

n	2	3	4	5	6	7
L	2.42211	2.22748	2.14461	2.10100	2.07501	2.05816

n	8	9	10	11	12	13	14
L	2.04656	2.03821	2.03199	2.02721	2.02346	2.02046	2.01802

n	15	16	17	18	19	20
L	2.01600	2.01431	2.01288	2.01167	2.01062	2.00971

49. $S = 49\pi$ 51. $S = \sqrt{2}\pi$ 55. **(a)** $r = 2\theta + 10$ **(b)** 75.7 mm
57. **(b)** $x = \cos t + \cos 2t, y = \sin t + \sin 2t$ **(c)** yes
59. **(b)** North: $\theta = \pi/2$; South: $\theta = 3\pi/2$; East: $\theta = 0, 2\pi$; West: $\theta = \pi$

Exercise Set 11.3 (Page 748)

1. **(a)** πa^2 **(b)** πa^2 3. 6π 5. 4π 7. $\pi - 3\sqrt{3}/2$ 9. $\pi/2 - \frac{1}{4}$
11. $10\pi/3 - 4\sqrt{3}$ 13. π 15. $9\sqrt{3}/2 - \pi$ 17. $(\pi + 3\sqrt{3})/4$
19. $\pi - 2$ 21. **(b)** a^2 **(c)** $2\sqrt{3} - \dfrac{2\pi}{3}$ 23. $8\pi^3 a^2$

25. π^2 27. $32\pi/5$

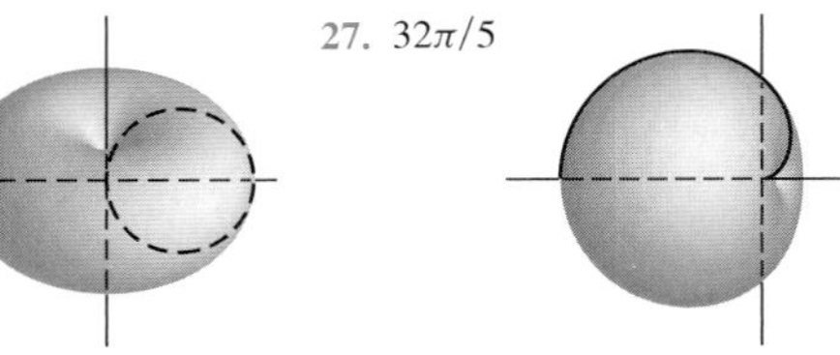

33. $\pi/16$

Exercise Set 11.4 (Page 764)

1. **(a)** $x = y^2$ **(b)** $-3y = x^2$ **(c)** $\dfrac{x^2}{9} + \dfrac{y^2}{4} = 1$ **(d)** $\dfrac{x^2}{4} + \dfrac{y^2}{9} = 1$
(e) $y^2 - x^2 = 1$ **(f)** $\dfrac{x^2}{4} - \dfrac{y^2}{4} = 1$
3. **(a)** focus: $(1, 0)$; vertex: $(0, 0)$; directrix: $x = -1$
(b) focus: $(0, -2)$; vertex: $(0, 0)$; directrix: $y = 2$

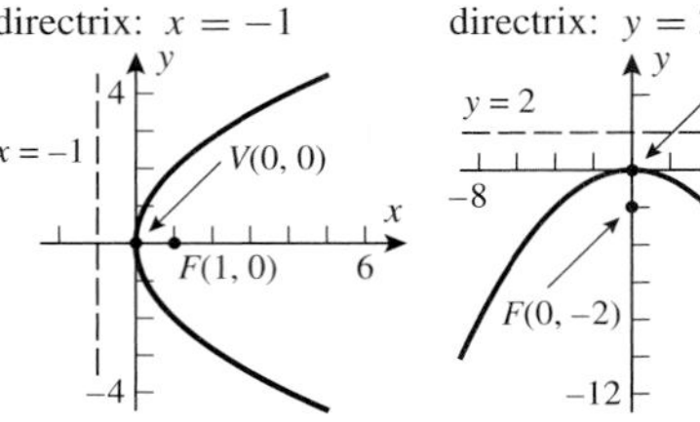

5. **(a)**

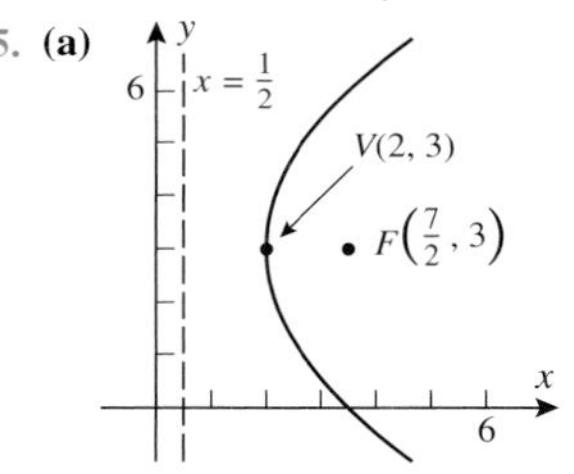

(b)

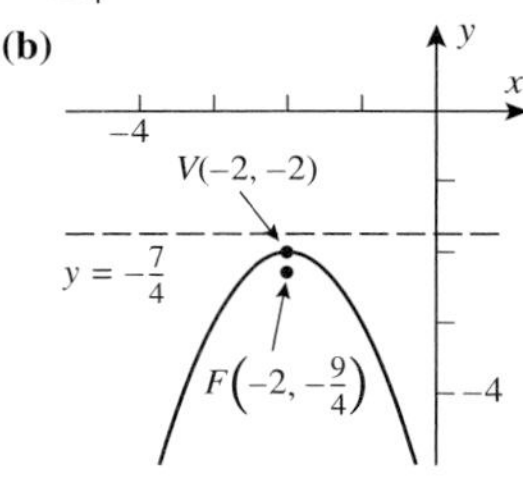

7. **(a)**

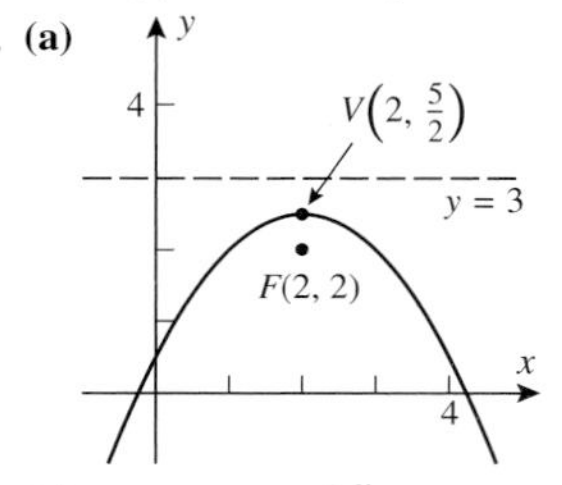

(b)

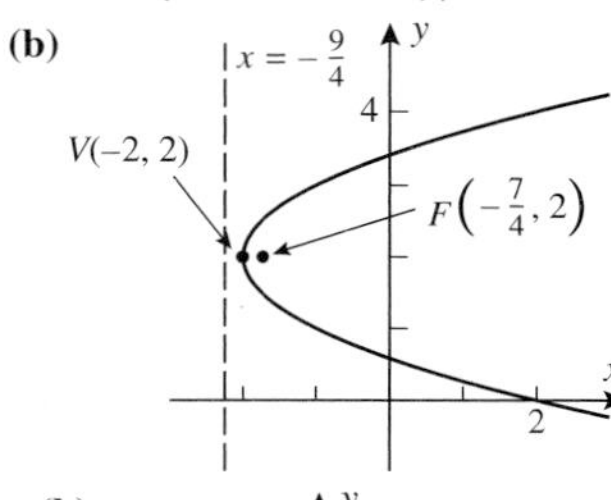

9. **(a)**

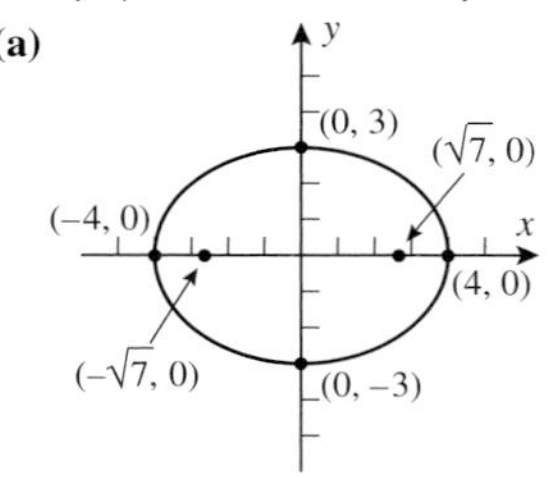

(b)

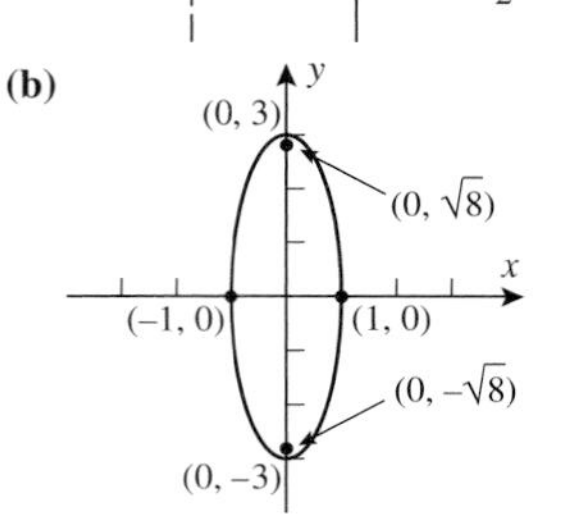

11. **(a)** foci: $(1, 3 \pm \sqrt{7})$; vertices: $(1, -1)$, $(1, 7)$; ends: $(-2, 3)$, $(4, 3)$

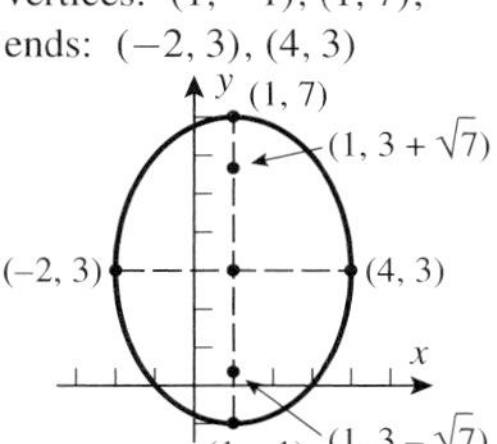

(b) foci: $(-2, -1 \pm \sqrt{5})$; vertices: $(-2, 2)$, $(-2, -4)$; ends: $(-4, -1)$, $(0, -1)$

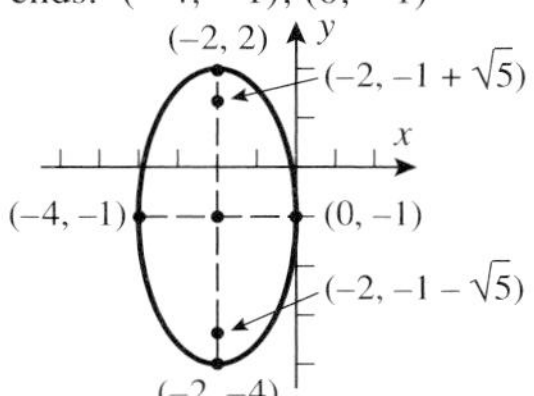

13. **(a)**

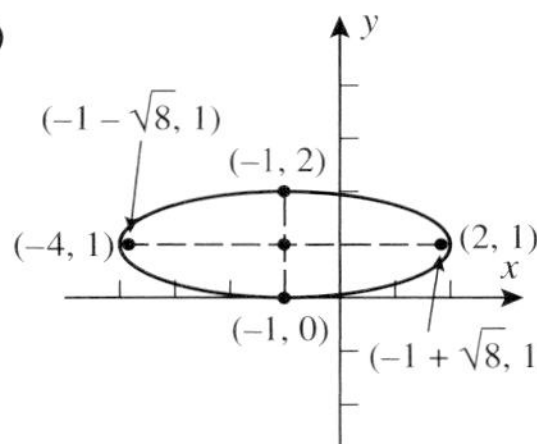

(b)

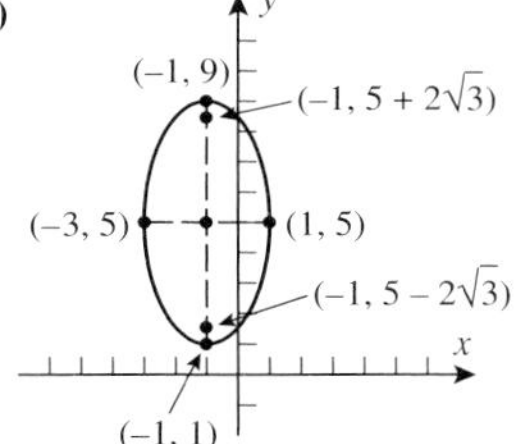

15. **(a)** vertices: $(\pm 4, 0)$; foci: $(\pm 5, 0)$; asymptotes: $y = \pm 3x + 4$

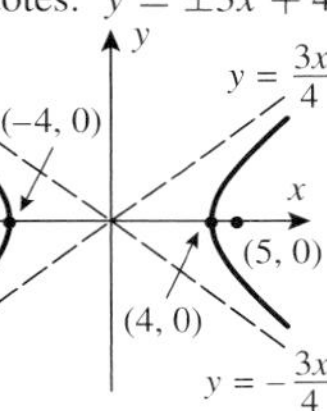

(b) vertices: $(0, \pm 2)$; foci: $(0, \pm 2\sqrt{10})$; asymptotes: $y = \pm x/3$

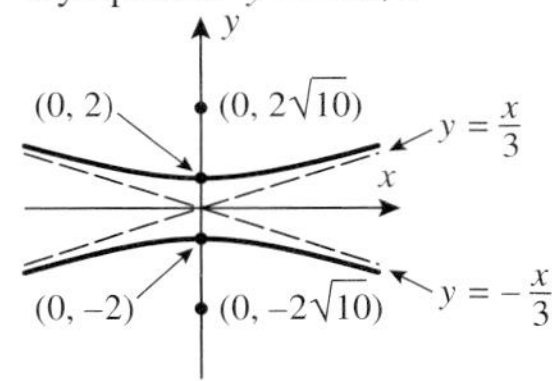

17. **(a)** vertices: $(-2, -2)$, $(4, -2)$; foci: $(1 \pm \sqrt{13}, -2)$; asymptotes: $y + 2 = \pm 2(x - 1)/3$

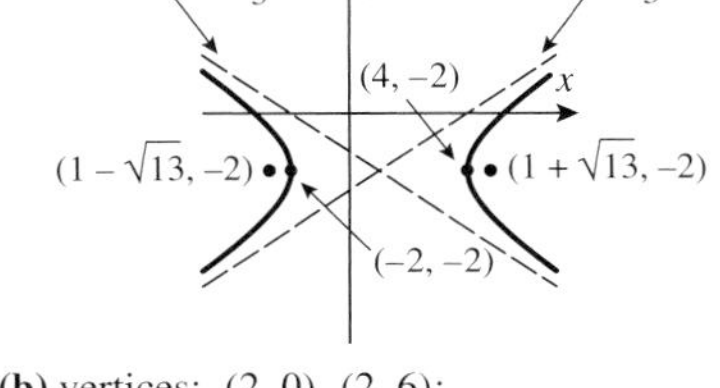

(b) vertices: $(2, 0)$, $(2, 6)$; foci: $(2, 3 \pm \sqrt{13})$; asymptotes: $y = 3x/2$, $y = 6 - 3x/2$

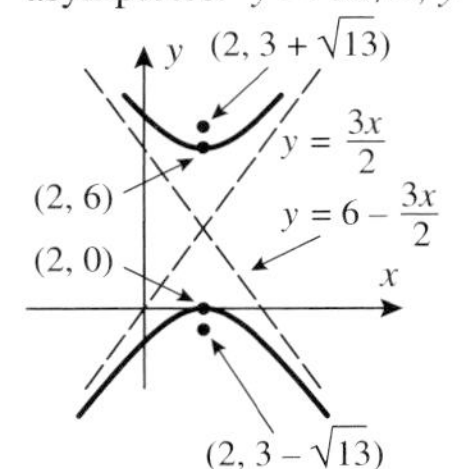

19. **(a)**

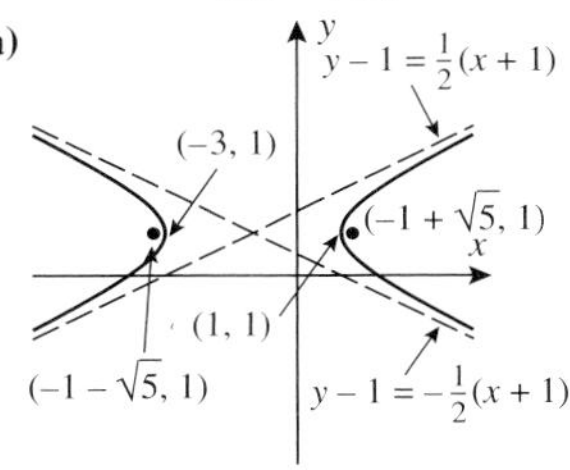

(b)

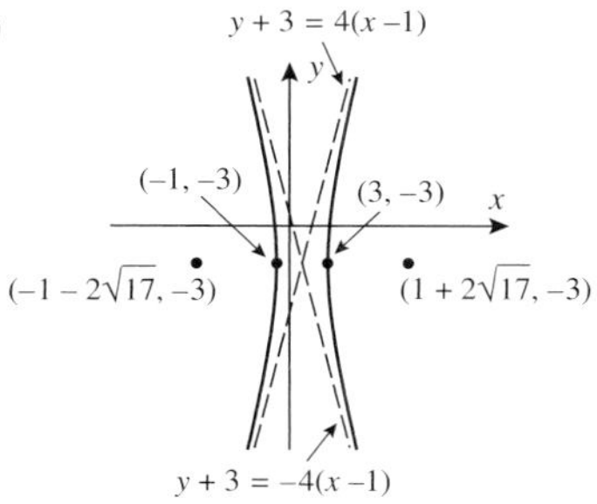

$y + 3 = 4(x - 1)$

$(-1, -3)$ $(3, -3)$ $(-1 - 2\sqrt{17}, -3)$ $(1 + 2\sqrt{17}, -3)$

$y + 3 = -4(x - 1)$

21. **(a)** $y^2 = 12x$ **(b)** $y^2 = -28x$

23. **(a)** $x^2 = -12y$ **(b)** $(x - 1)^2 = 12(y - 1)$ 25. $y^2 = 2(x - 1)$

27. **(a)** $\frac{1}{9}x^2 + \frac{1}{4}y^2 = 1$ **(b)** $\frac{1}{169}x^2 + \frac{1}{144}y^2 = 1$

29. **(a)** $\frac{1}{3}x^2 + \frac{1}{2}y^2 = 1$ **(b)** $\frac{1}{4}x^2 + \frac{1}{16}y^2 = 1$

31. **(a)** $\dfrac{x^2}{81/8} + \dfrac{y^2}{36} = 1$ **(b)** $\dfrac{(x + 1)^2}{4} + \dfrac{(y - 2)^2}{5} = 1$

33. **(a)** $\frac{1}{4}x^2 - \frac{1}{5}y^2 = 1$ **(b)** $x^2 - \frac{1}{4}y^2 = 1$

35. **(a)** $\frac{9}{64}x^2 - \frac{1}{16}y^2 = 1$, $\frac{1}{36}y^2 - \frac{1}{16}x^2 = 1$ **(b)** $\frac{1}{20}y^2 - \frac{1}{5}x^2 = 1$

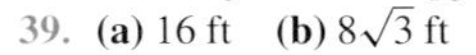

37. **(a)** $\dfrac{(x - 3)^2}{9} - \dfrac{(y - 6)^2}{16} = 1$ **(b)** $(x - 3)^2 - (y - 1)^2 = 1$

39. **(a)** 16 ft **(b)** $8\sqrt{3}$ ft

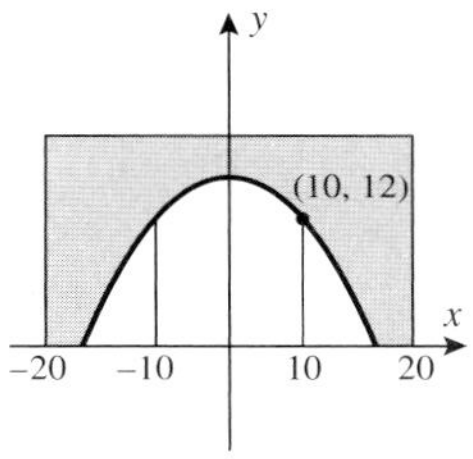

y (10, 12) x –20 –10 10 20

43. $\frac{1}{16}$ ft

45. **(a)** $P : (b \cos t, b \sin t)$; $Q : (a \cos t, a \sin t)$; $R : (a \cos t, b \sin t)$

49. $\frac{1}{32}(x - 4)^2 + \frac{1}{36}(y - 3)^2 = 1$

51. 96

61. $L = D\sqrt{1 + p^2}$, $T = \frac{1}{2}pD$

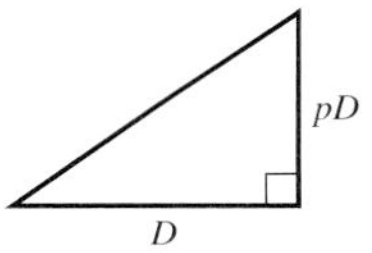

pD D

67. $\left(\pm\dfrac{3}{\sqrt{5}}, \dfrac{4}{\sqrt{5}}\right), \left(\pm\dfrac{3}{\sqrt{5}}, -\dfrac{4}{\sqrt{5}}\right)$

71. **(b)** 14.30465, 24, 33.69535 in

73. **(a)** $(x - 1)^2 - 5(y + 1)^2 = 5$, hyperbola
(b) $x^2 - 3(y + 1)^2 = 0$, $x = \pm\sqrt{3}(y + 1)$, two lines
(c) $4(x + 2)^2 + 8(y + 1)^2 = 4$, ellipse
(d) $3(x + 2)^2 + (y + 1)^2 = 0$, the point $(-2, -1)$ (degenerate case)
(e) $(x + 4)^2 + 2y = 2$, parabola
(f) $5(x + 4)^2 + 2y = -14$, parabola

▶ Exercise Set 11.5 (Page 774)

1. **(a)** $x' = -1 + 3\sqrt{3}$, $y' = 3 + \sqrt{3}$ **(b)** $3x'^2 - y'^2 = 12$
(c)

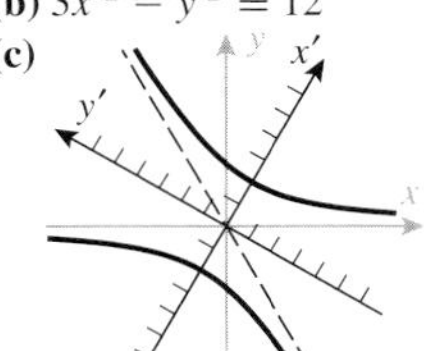

3. $y'^2 - x'^2 = 18$, hyperbola

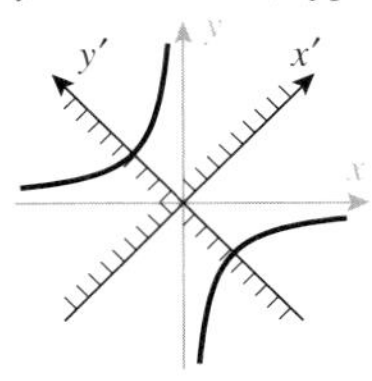

5. $\frac{1}{3}x'^2 - \frac{1}{2}y'^2 = 1$, hyperbola 7. $y' = x'^2$, parabola

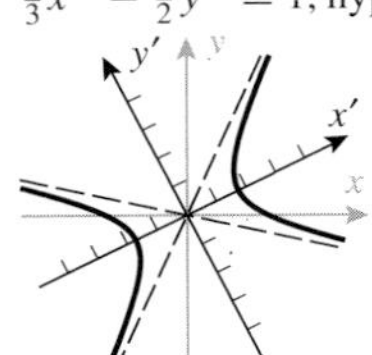

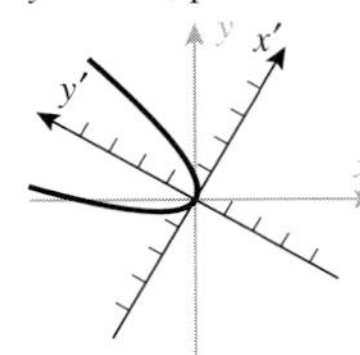

9. $y'^2 = 4(x' - 1)$, parabola 11. $\frac{1}{4}(x' + 1)^2 + y'^2 = 1$, ellipse

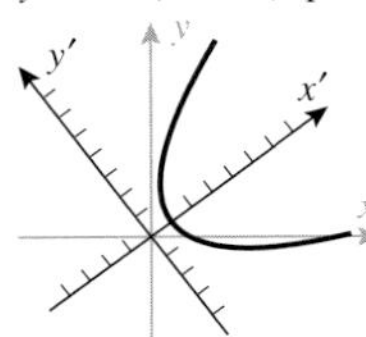

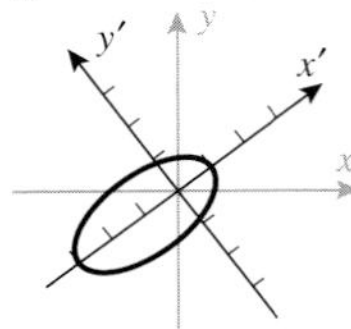

13. $x^2 + xy + y^2 = 3$
19. vertex: $(0, 0)$; focus: $(-1/\sqrt{2}, 1/\sqrt{2})$; directrix: $y = x - \sqrt{2}$
21. vertex: $(4/5, 3/5)$; focus: $(8/5, 6/5)$; directrix: $4x + 3y = 0$
23. foci: $\pm(4\sqrt{7}/5, 3\sqrt{7}/5)$; vertices: $\pm(16/5, 12/5)$;
ends: $\pm(-9/5, 12/5)$
25. foci: $(1 - \sqrt{5}/2, -\sqrt{3} + \sqrt{15}/2)$, $(1 + \sqrt{5}/2, -\sqrt{3} - \sqrt{15}/2)$;
vertices: $(-1/2, \sqrt{3}/2)$, $(5/2, -5\sqrt{3}/2)$;
ends: $(1 + \sqrt{3}, 1 - \sqrt{3})$, $(1 - \sqrt{3}, -1 - \sqrt{3})$
27. foci: $\pm(\sqrt{15}, \sqrt{5})$; vertices: $\pm(2\sqrt{3}, 2)$;
asymptotes: $y = \dfrac{5\sqrt{3} \pm 8}{11}x$
29. foci: $\left(-\dfrac{4}{\sqrt{5}} \pm 2\sqrt{\dfrac{13}{5}}, \dfrac{8}{\sqrt{5}} \pm \sqrt{\dfrac{13}{5}}\right)$;
vertices: $(2/\sqrt{5}, 11/\sqrt{5})$, $(-2\sqrt{5}, \sqrt{5})$;
asymptotes: $y = 7x/4 + 3\sqrt{5}$, $y = -x/8 + 3\sqrt{5}/2$

▶ Exercise Set 11.6 (Page 782)

1. **(a)** $e = 1, d = \frac{3}{2}$ **(b)** $e = \frac{1}{2}, d = 3$

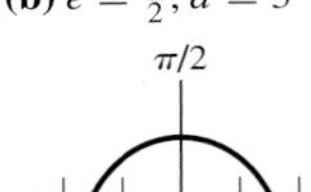
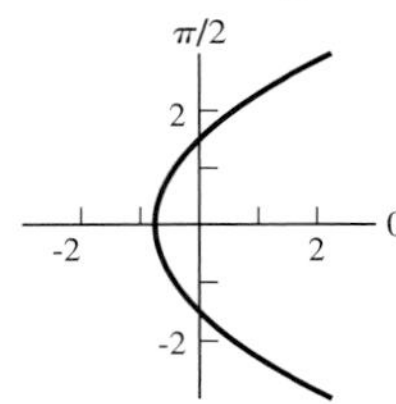

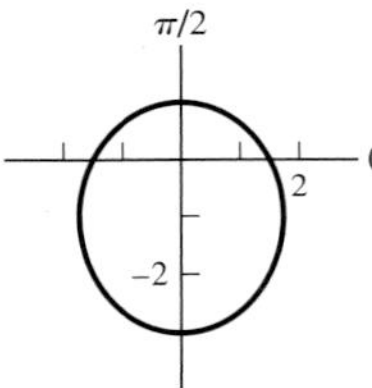

(c) $e = \frac{3}{2}, d = \frac{4}{3}$ **(d)** $e = 1, d = \frac{5}{3}$

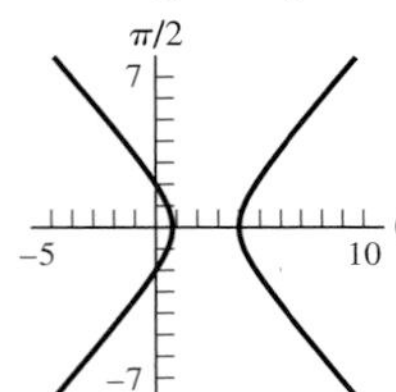

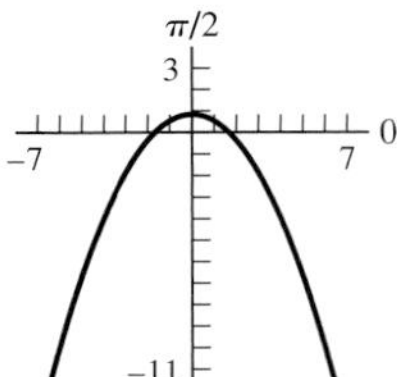

3. **(a)** parabola, opens up **(b)** ellipse, directrix above the pole

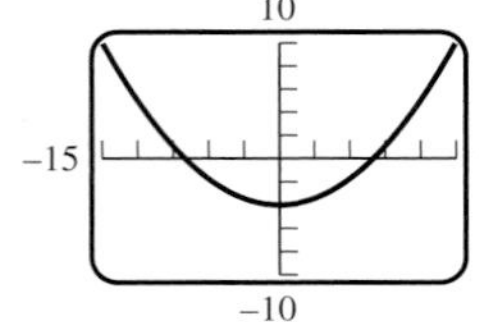

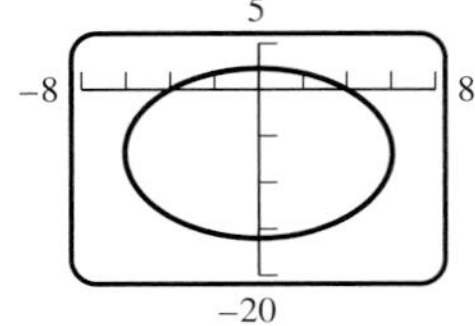

(c) hyperbola, directrix below the pole **(d)** ellipse, directrix to the right of the pole

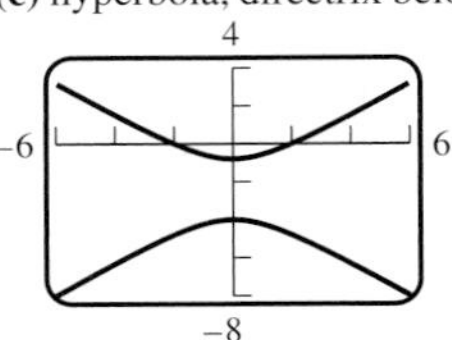

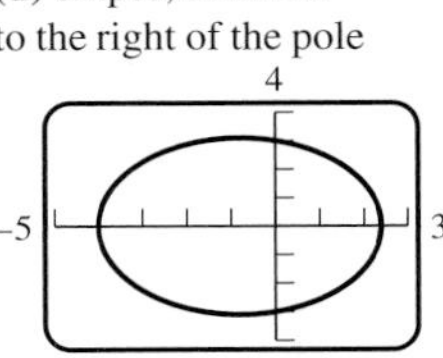

5. **(a)** $r = \dfrac{6}{(4 + 3\cos\theta)}$ **(b)** $r = \dfrac{1}{(1 + \cos\theta)}$ **(c)** $r = \dfrac{12}{(3 + 4\sin\theta)}$
7. **(a)** $r = \dfrac{24}{5 - \cos\theta}$ **(b)** $r = \dfrac{2}{1 - \sin\theta}$ **(c)** $r = \dfrac{21}{2 + 5\sin\theta}$
9. **(a)** $d = 6$; $\frac{1}{12}x^2 + \frac{1}{16}(y + 2)^2 = 1$
(b) $d = 1$; $\frac{9}{4}\left(x - \frac{1}{3}\right)^2 + 3y^2 = 1$
11. **(a)** $d = 1, 3$; $(y - 2)^2 - x^2/3 = 1$
(b) $d = 1, 5$; $(x + 3)^2/4 - y^2/5 = 1$
13. **(a)** $r = \dfrac{12}{2 + \cos\theta}$ **(b)** $r = \dfrac{64}{25 - 15\sin\theta}$
(c) $r = \dfrac{16}{5 - 3\cos\theta}$ **(d)** $r = \dfrac{120}{5 + \sin\theta}$
17. distance approaches $+\infty$
21. $r = \dfrac{(5\sqrt{2} + 5)}{(1 + \sqrt{2}\cos\theta)}$ or $r = \dfrac{(5\sqrt{2} - 5)}{(1 + \sqrt{2}\cos\theta)}$
23. **(a)** $T \approx 248$ yr
(b) $r_0 \approx 4{,}449{,}675{,}000$ km $r_1 \approx 7{,}400{,}325{,}000$ km
(c) $r \approx \dfrac{37.05}{1 + 0.249\cos\theta}$ AU **(d)**

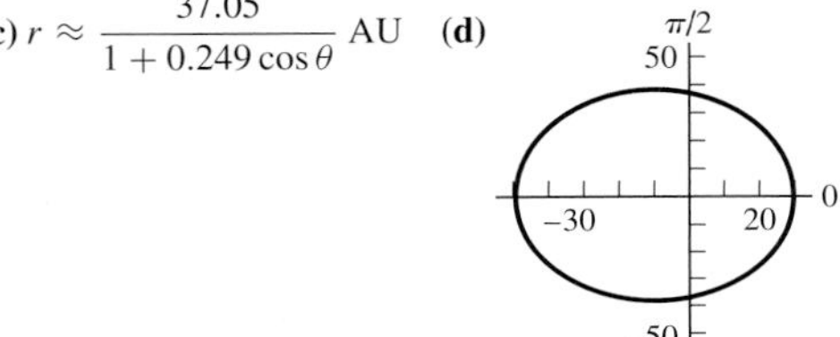

25. **(a)** $a \approx 178.26$ AU
(b) $r_0 \approx 0.8735$ AU,
$r_1 \approx 355.64$ AU
(c) $r \approx \dfrac{1.74}{1 + 0.9951\cos\theta}$ AU
(d) π/2, 4, 0, −180
27. 563 km, 4286 km

▶ Chapter 11. Review Exercises (Page 784)

1. **(a)** $(-4\sqrt{2}, -4\sqrt{2})$ **(b)** $(7/\sqrt{2}, -7/\sqrt{2})$ **(c)** $(4\sqrt{2}, 4\sqrt{2})$
(d) $(5, 0)$ **(e)** $(0, -2)$ **(f)** $(0, 0)$
3. **(a)** $(5, 0.6435)$ **(b)** $(\sqrt{29}, 5.0929)$
5. **(a)** parabola **(b)** hyperbola **(c)** line **(d)** circle
7. 9. 11.

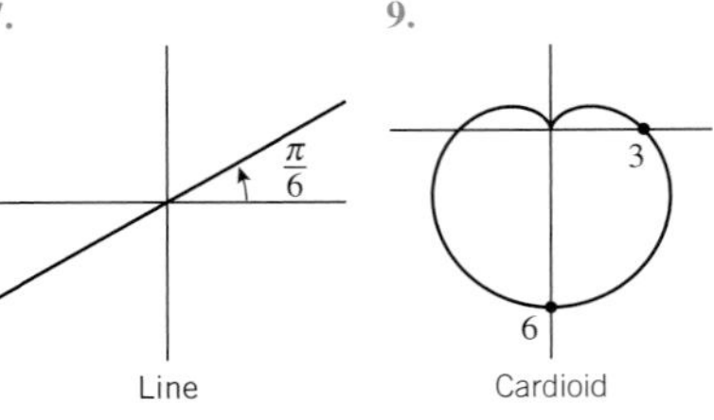

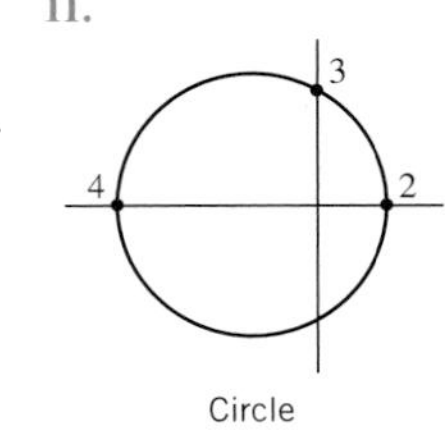

13. **(a)** $-2, 1/4$ **(b)** $-3\sqrt{3}/4, 3\sqrt{3}/4$ 15. **(a)** $-1/4, 1/4$
17. **(a)** $t = \pi/2 + n\pi$ for $n = 0, \pm 1, \ldots$ **(b)** $t = n\pi$ for $n = 0, \pm 1, \ldots$
19. **(a)** The top is traced from right to left as t goes from 0 to π. The bottom is traced from right to left as t goes from π to 2π, except for the loop, which is traced counterclockwise as t goes from $\pi + \sin^{-1}(1/4)$ to $2\pi - \sin^{-1}(1/4)$. **(b)** $y = 1$
(c) horizontal: $t = \pi/2, 3\pi/2$; vertical: $t = \pi + \sin^{-1}(1/\sqrt[3]{4})$, $2\pi - \sin^{-1}(1/\sqrt[3]{4})$

(d) $r = 4 + \csc\theta$, $\theta = \pi + \sin^{-1}(1/4)$, $\theta = 2\pi - \sin^{-1}(1/4)$

21. $A = 6\pi$ 23. $A = \dfrac{5\pi}{12} - \dfrac{\sqrt{3}}{2}$ 25.

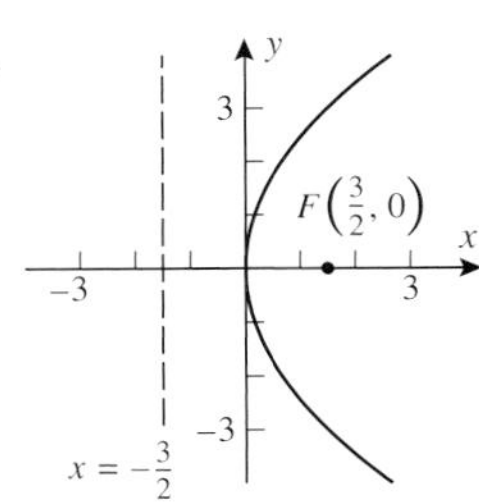

27.

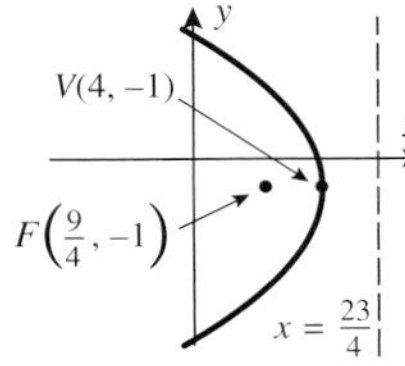

focus: $(9/4, -1)$;
vertex: $(4, -1)$;
directrix: $x = 23/4$

29.

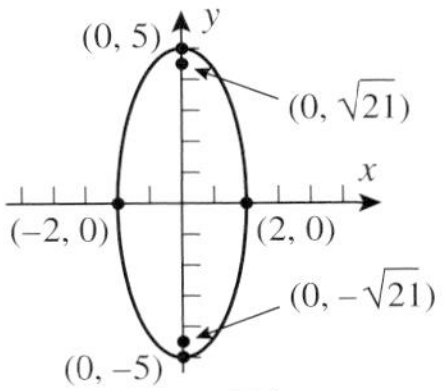

(0, 5), $(0, \sqrt{21})$, (−2, 0), (2, 0), $(0, -\sqrt{21})$, (0, −5)

foci: $(0, \pm\sqrt{21})$;
vertices: $(0, \pm 5)$;
ends: $(\pm 2, 0)$

31.

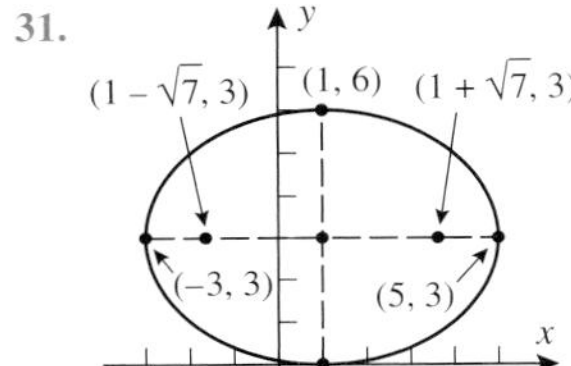

33.

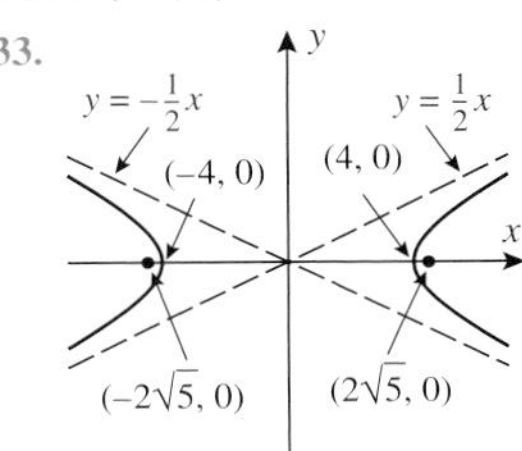

35.

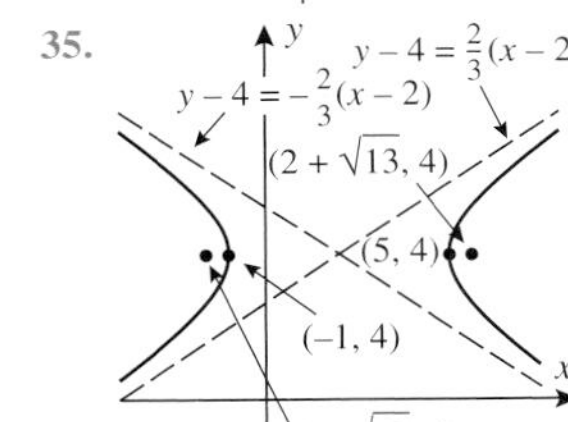

37. **(a)**

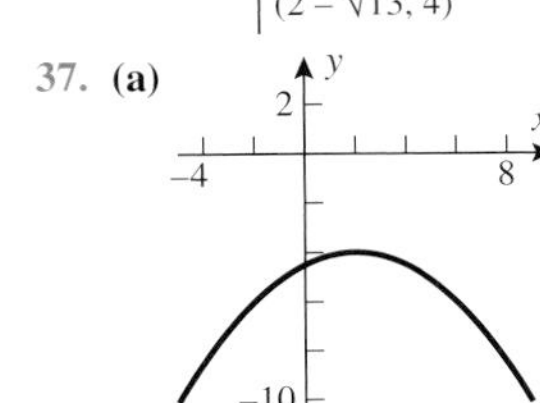

(b)

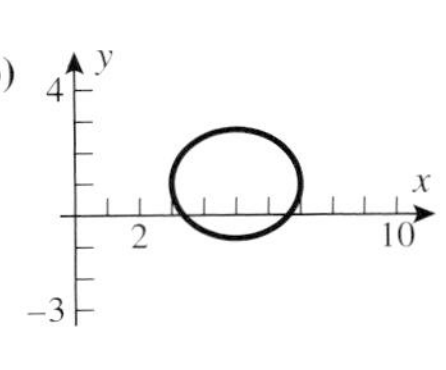

(c)

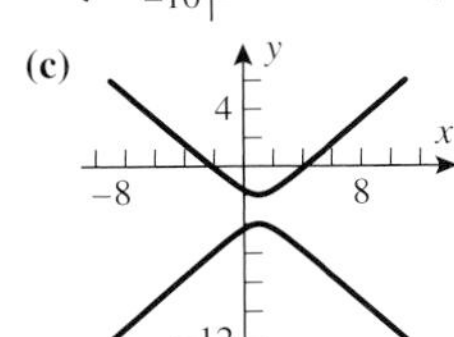

39. $x^2 = -16y$ 41. $y^2 - x^2 = 9$

43. **(b)** $x = \dfrac{v_0^2}{g}\sin\alpha\cos\alpha$; $y = y_0 + \dfrac{v_0^2\sin^2\alpha}{2g}$

45. **(a)** $V = \dfrac{\pi b^2}{3a^2}(b^2 - 2a^2)\sqrt{a^2 + b^2} + \dfrac{2}{3}ab^2\pi$ **(b)** $V = \dfrac{2b^4}{3a}\pi$

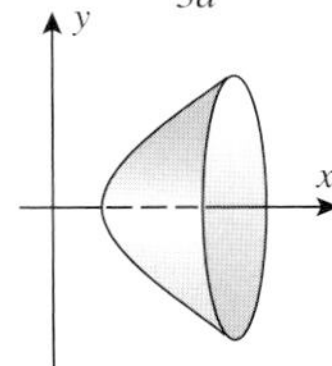

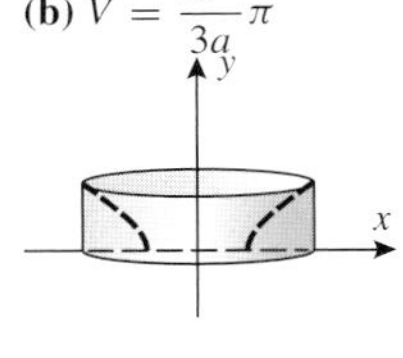

47. $\theta = \pi/4$; $5(y')^2 - (x')^2 = 6$; hyperbola
49. $\theta = \tan^{-1}(1/2)$; $y' = (x')^2$; parabola
51. **(a)** (i) ellipse; (ii) right; (iii) 1 **(b)** (i) hyperbola (ii) left; (iii) 1/3 **(c)** (i) parabola; (ii) above; (iii) 1/3 **(d)** (i) parabola; (ii) below; (iii) 3
53. **(a)** $(x + 3)^2/25 + (y - 2)^2/9 = 1$ **(b)** $(x + 2)^2 = -8y$ **(c)** $(y - 5)^2/4 - 16(x + 1)^2 = 1$
55. 15.86543959

▶ Exercise Set 12.1 (Page 794)

1. **(a)** (0, 0, 0), (3, 0, 0), (3, 5, 0), (0, 5, 0), (0, 0, 4), (3, 0, 4), (3, 5, 4), (0, 5, 4)
(b) (0, 1, 0), (4, 1, 0), (4, 6, 0), (0, 6, 0), (0, 1, −2), (4, 1, −2), (4, 6, −2), (0, 6, −2)
3. (4, 2, −2), (4, 2, 1), (4, 1, 1), (4, 1, −2), (−6, 1, 1), (−6, 2, 1), (−6, 2, −2), (−6, 1, −2)

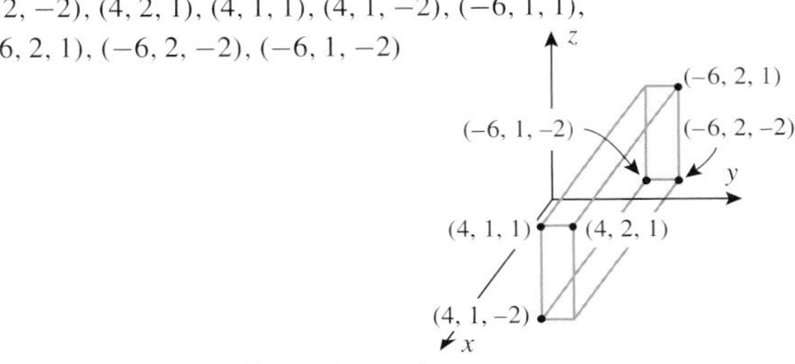

5. **(a)** point **(b)** line parallel to the y-axis
(c) plane parallel to the yz-plane
7. radius $\sqrt{74}$, center $(2, 1, -4)$ 9. **(b)** (2, 1, 6) **(c)** area 49
11. **(a)** $(x - 1)^2 + y^2 + (z + 1)^2 = 16$
(b) $(x + 1)^2 + (y - 3)^2 + (z - 2)^2 = 14$
(c) $\left(x + \frac{1}{2}\right)^2 + (y - 2)^2 + (z - 2)^2 = \frac{5}{4}$
13. $(x - 2)^2 + (y + 1)^2 + (z + 3)^2 = r^2$;
(a) $r^2 = 9$ **(b)** $r^2 = 1$ **(c)** $r^2 = 4$
17. sphere, center $(-5, -2, -1)$, radius 7
19. sphere; center $\left(\frac{1}{2}, \frac{3}{4}, -\frac{5}{4}\right)$, radius $\dfrac{3\sqrt{6}}{4}$
21. no graph
23. **(a)** **(b)**

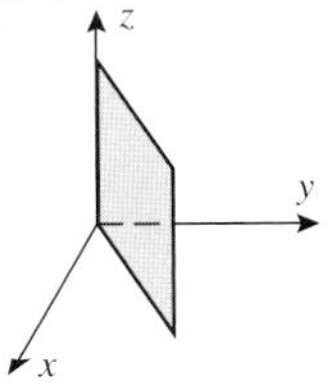

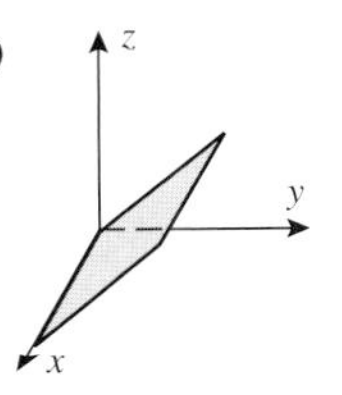

(c)

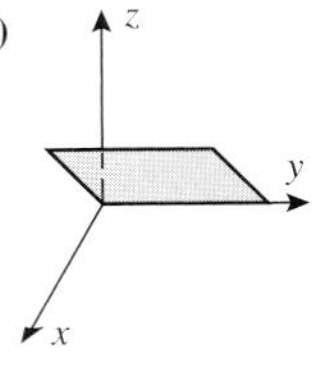

25. **(a)**

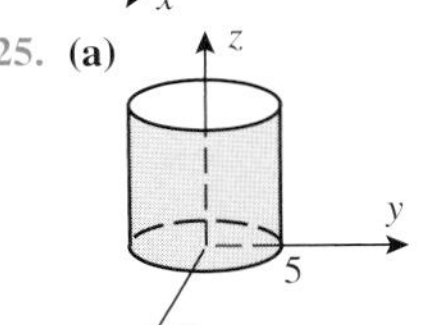

(b)

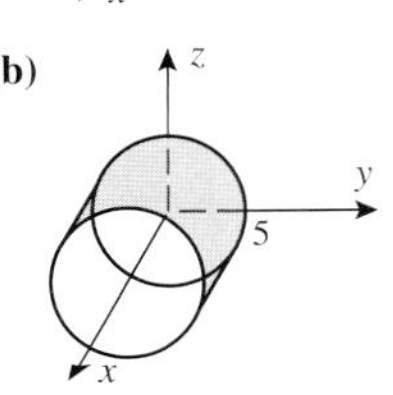

(c)

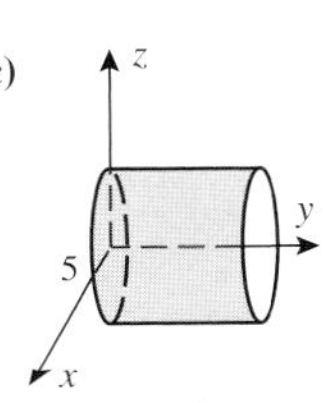

27. **(a)** $-2y + z = 0$ **(b)** $-2x + z = 0$ **(c)** $(x - 1)^2 + (y - 1)^2 = 1$
(d) $(x - 1)^2 + (z - 1)^2 = 1$

29. 31.

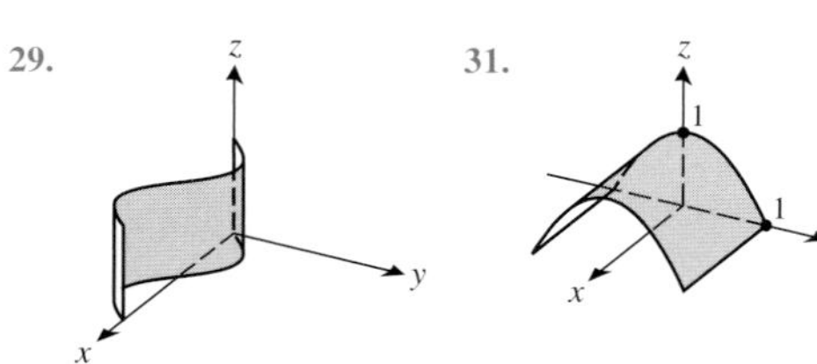

33. 35.

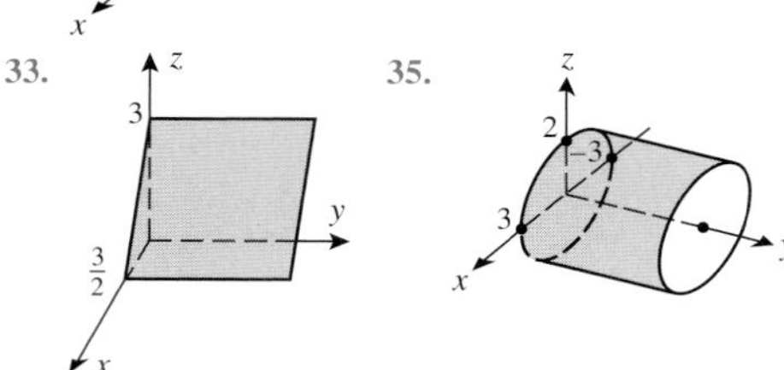

37.

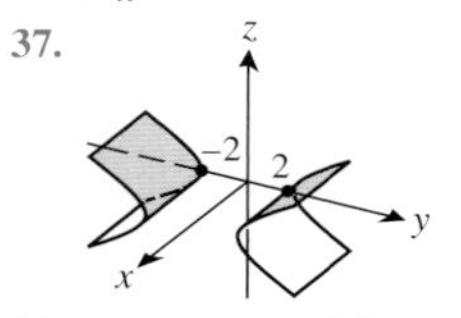

39.

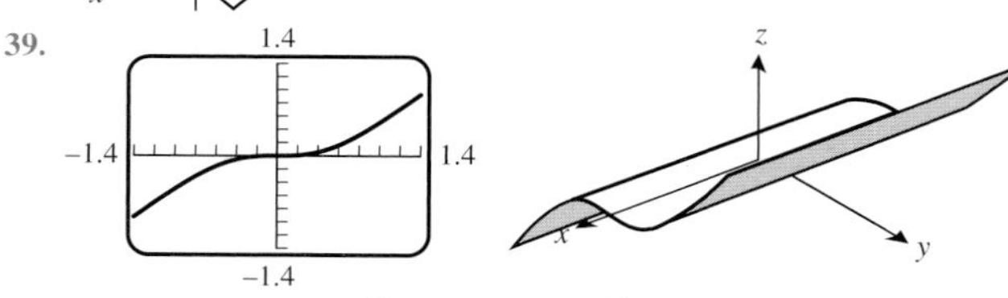

41. largest distance $3 + \sqrt{6}$, smallest $3 - \sqrt{6}$
43. all points outside the circular cylinder $(y + 3)^2 + (z - 2)^2 = 16$
45. $r = (2 - \sqrt{3})R$ 47. **(b)** $y^2 + z^2 = e^{2x}$

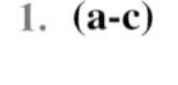

Exercise Set 12.2 (Page 805)

1. **(a-c)**

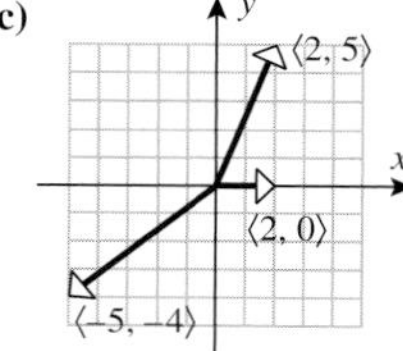

(d-f)

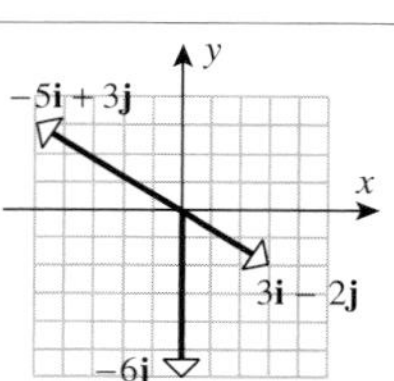

3. **(a,b)**

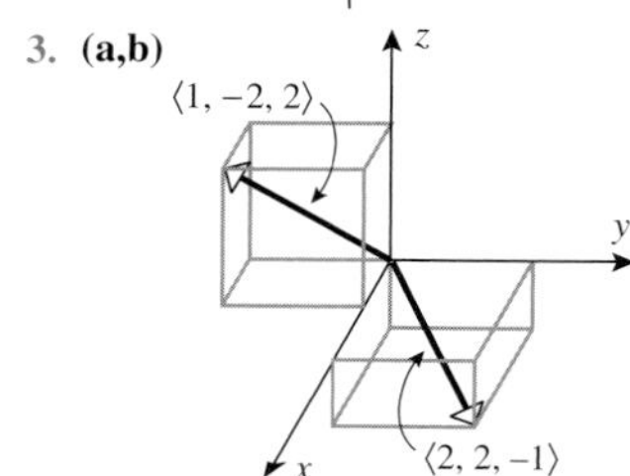

(c,d)

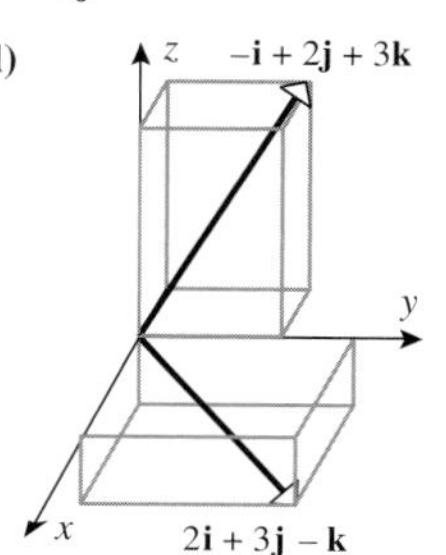

5. **(a)** $\langle 3, -4\rangle$ **(b)** $\langle -2, -3, 4\rangle$

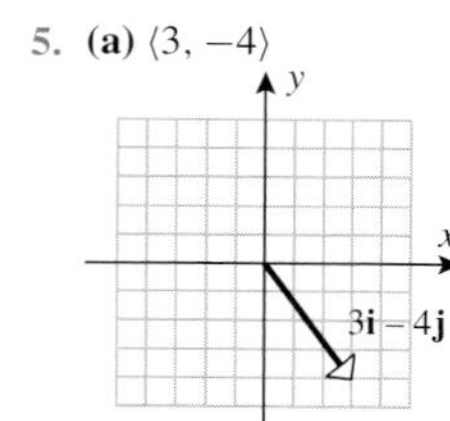

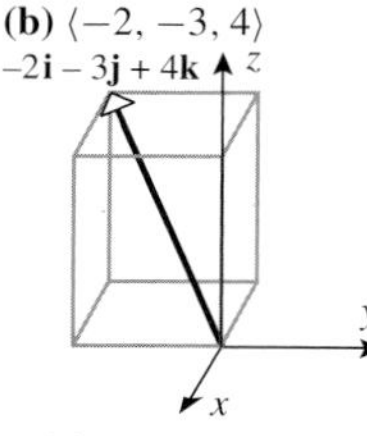

7. **(a)** $\langle -1, 3\rangle$ **(b)** $\langle -7, 2\rangle$ **(c)** $\langle -3, 6, 1\rangle$
9. **(a)** $(4, -4)$ **(b)** $(8, -1, -3)$
11. **(a)** $-\mathbf{i} + 4\mathbf{j} - 2\mathbf{k}$ **(b)** $18\mathbf{i} + 12\mathbf{j} - 6\mathbf{k}$ **(c)** $-\mathbf{i} - 5\mathbf{j} - 2\mathbf{k}$ **(d)** $40\mathbf{i} - 4\mathbf{j} - 4\mathbf{k}$ **(e)** $-2\mathbf{i} - 16\mathbf{j} - 18\mathbf{k}$ **(f)** $-\mathbf{i} + 13\mathbf{j} - 2\mathbf{k}$
13. **(a)** $\sqrt{2}$ **(b)** $5\sqrt{2}$ **(c)** $\sqrt{21}$ **(d)** $\sqrt{14}$
15. **(a)** $2\sqrt{3}$ **(b)** $\sqrt{14} + \sqrt{2}$ **(c)** $2\sqrt{14} + 2\sqrt{2}$ **(d)** $2\sqrt{37}$ **(e)** $(1/\sqrt{6})\mathbf{i} + (1/\sqrt{6})\mathbf{j} - (2/\sqrt{6})\mathbf{k}$ **(f)** 1
17. **(a)** $(-1/\sqrt{17})\mathbf{i} + (4/\sqrt{17})\mathbf{j}$ **(b)** $(-3\mathbf{i} + 2\mathbf{j} - \mathbf{k})/\sqrt{14}$ **(c)** $(4\mathbf{i} + \mathbf{j} - \mathbf{k})/(3\sqrt{2})$
19. **(a)** $\langle -\frac{3}{2}, 2\rangle$ **(b)** $\dfrac{1}{\sqrt{5}}\langle 7, 0, -6\rangle$
21. **(a)** $\langle 3\sqrt{2}/2, 3\sqrt{2}/2\rangle$ **(b)** $\langle 0, 2\rangle$ **(c)** $\langle -5/2, 5\sqrt{3}/2\rangle$ **(d)** $\langle -1, 0\rangle$
23. $\langle(\sqrt{3} - \sqrt{2})/2, (1 + \sqrt{2})/2\rangle$
25. **(a)** $\langle -2, 5\rangle$ −2i + 5j **(b)** $\langle 3, -8\rangle$

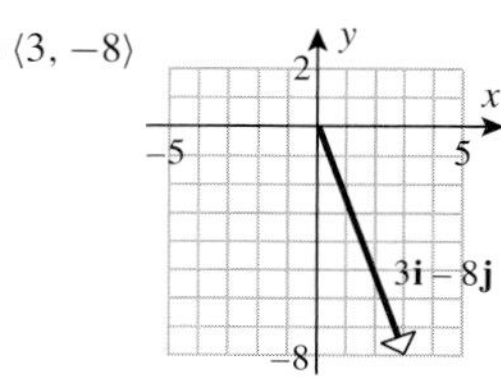

27. $\langle -\frac{2}{3}, 1\rangle$ 29. $\mathbf{u} = \frac{5}{7}\mathbf{i} + \frac{2}{7}\mathbf{j} + \frac{1}{7}\mathbf{k}$, $\mathbf{v} = \frac{8}{7}\mathbf{i} - \frac{1}{7}\mathbf{j} - \frac{4}{7}\mathbf{k}$
31. $\sqrt{5}, 3$ 33. **(a)** $\pm\frac{5}{3}$ **(b)** 3
35. **(a)** $\langle 1/\sqrt{10}, 3/\sqrt{10}\rangle$, $\langle -1/\sqrt{10}, -3/\sqrt{10}\rangle$ **(b)** $\langle 1/\sqrt{2}, -1/\sqrt{2}\rangle$, $\langle -1/\sqrt{2}, 1/\sqrt{2}\rangle$ **(c)** $\pm\dfrac{1}{\sqrt{26}}\langle 5, 1\rangle$
37. **(a)** the circle of radius 1 about the origin **(b)** the closed disk of radius 1 about the origin **(c)** all points outside the closed disk of radius 1 about the origin
39. **(a)** the (hollow) sphere of radius 1 about the origin **(b)** the closed ball of radius 1 about the origin **(c)** all points outside the closed ball of radius 1 about the origin
41. magnitude $= 30\sqrt{5}$ lb, $\theta \approx 26.57°$
43. magnitude ≈ 207.06 N, $\theta = 45°$
45. magnitude ≈ 94.995 N, $\theta \approx 28.28°$
47. magnitude ≈ 9.165 lb, angle $\approx -70.890°$
49. ≈ 183.02 lb, 224.13 lb
51. **(a)** $c_1 = -2, c_2 = 1$

Exercise Set 12.3 (Page 814)

1. **(a)** -10; $\cos\theta = -1/\sqrt{5}$ **(b)** -3; $\cos\theta = -3/\sqrt{58}$ **(c)** 0; $\cos\theta = 0$ **(d)** -20; $\cos\theta = -20/(3\sqrt{70})$
3. **(a)** obtuse **(b)** acute **(c)** obtuse **(d)** orthogonal
5. $\sqrt{2}/2, 0, -\sqrt{2}/2, -1, -\sqrt{2}/2, 0, \sqrt{2}/2$
7. **(a)** vertex B **(b)** $82°, 60°, 38°$ 13. $r = 7/5$
15. **(a)** $\alpha = \beta \approx 55°, \gamma \approx 125°$ **(b)** $\alpha \approx 48°, \beta \approx 132°, \gamma \approx 71°$
19. **(a)** $\approx 35°$ **(b)** $90°$
21. $64°, 41°, 60°$ 23. $71°, 61°, 36°$
25. **(a)** $\left\langle \frac{2}{3}, \frac{4}{3}, \frac{4}{3}\right\rangle, \left\langle \frac{4}{3}, -\frac{7}{3}, \frac{5}{3}\right\rangle$ **(b)** $\left\langle -\frac{74}{49}, -\frac{111}{49}, \frac{222}{49}\right\rangle, \left\langle \frac{270}{49}, \frac{62}{49}, \frac{121}{49}\right\rangle$
27. **(a)** $\langle 1, 1\rangle + \langle -4, 4\rangle$ **(b)** $\left\langle 0, -\frac{8}{5}, \frac{4}{5}\right\rangle + \left\langle -2, \frac{13}{5}, \frac{26}{5}\right\rangle$ **(c)** $\mathbf{v} = \langle 1, 4, 1\rangle$ is orthogonal to $\mathbf{b}$.
29. $\sqrt{564/29}$ 31. 98 N
33. **(a)**

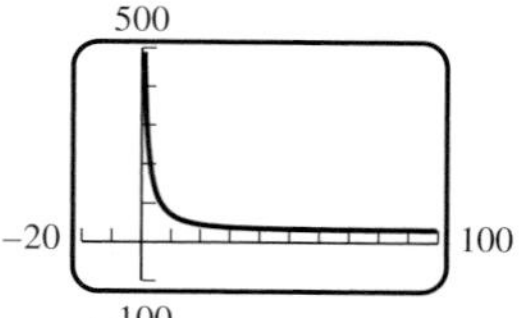

(b) decrease **(c)** $40/\sqrt{65}$ ft
35. $-5\sqrt{3}$ J 37. $W = 375$ ft·lb
45. **(a)** $40°$ **(b)** $x \approx -0.682328$

Exercise Set 12.4 (Page 825)

1. **(a)** $-\mathbf{j}+\mathbf{k}$ 3. $\langle 7, 10, 9\rangle$ 5. $\langle -4, -6, -3\rangle$
7. **(a)** $\langle -20, -67, -9\rangle$ **(b)** $\langle -78, 52, -26\rangle$
(c) $\langle 0, -56, -392\rangle$ **(d)** $\langle 0, 56, 392\rangle$
9. $\frac{1}{\sqrt{2}}, -\frac{1}{\sqrt{2}}, 0$ 11. $\pm\frac{1}{\sqrt{6}}\langle 2, 1, 1\rangle$ 13. $\sqrt{59}$ 15. $\sqrt{374}/2$
17. 80 19. -3 21. 16 23. **(a)** yes **(b)** yes **(c)** no
25. **(a)** 9 **(b)** $\sqrt{122}$ **(c)** $\sin^{-1}\left(\frac{9}{14}\right)$
27. **(a)** $2\sqrt{141/29}$ **(b)** $6/\sqrt{5}$ 29. $\frac{2}{3}$ 33. $\theta = \pi/4$
35. **(a)** $10\sqrt{2}$ lb·ft, direction of rotation about P is counterclockwise looking along $\overrightarrow{PQ} \times \mathbf{F} = -10\mathbf{i} + 10\mathbf{k}$ toward its initial point
(b) 10 lb·ft, direction of rotation about P is counterclockwise looking along $-10\mathbf{i}$ toward its initial point
(c) 0 lb·ft, no rotation about P
37. ≈ 36.19 N·m 41. $-8\mathbf{i} - 20\mathbf{j} + 2\mathbf{k}, -8\mathbf{i} - 8\mathbf{k}$ 45. 1.887850

Exercise Set 12.5 (Page 832)

1. **(a)** $L_1: x = 1, y = t, L_2: x = t, y = 1, L_3: x = t, y = t$
(b) $L_1: x = 1, y = 1, z = t, L_2: x = t, y = 1, z = 1,$
$L_3: x = 1, y = t, z = 1, L_4: x = t, y = t, z = t$
3. **(a)** $x = 3 + 2t, y = -2 + 3t$; line segment: $0 \le t \le 1$
(b) $x = 5 - 3t, y = -2 + 6t, z = 1 + t$; line segment: $0 \le t \le 1$
5. **(a)** $x = 2 + t, y = -3 - 4t$ **(b)** $x = t, y = -t, z = 1 + t$
7. **(a)** $P(2, -1), \mathbf{v} = 4\mathbf{i} - \mathbf{j}$ **(b)** $P(-1, 2, 4), \mathbf{v} = 5\mathbf{i} + 7\mathbf{j} - 8\mathbf{k}$
9. **(a)** $\langle -3, 4\rangle + t\langle 1, 5\rangle$; $-3\mathbf{i} + 4\mathbf{j} + t(\mathbf{i} + 5\mathbf{j})$
(b) $\langle 2, -3, 0\rangle + t\langle -1, 5, 1\rangle$; $2\mathbf{i} - 3\mathbf{j} + t(-\mathbf{i} + 5\mathbf{j} + \mathbf{k})$
11. $x = -5 + 2t, y = 2 - 3t$ 13. $x = 3 + 4t, y = -4 + 3t$
15. $x = -1 + 3t, y = 2 - 4t, z = 4 + t$
17. $x = -2 + 2t, y = -t, z = 5 + 2t$
19. **(a)** $x = 7$ **(b)** $y = \frac{7}{3}$ **(c)** $x = \frac{-1 \pm \sqrt{85}}{6}, y = \frac{43 \mp \sqrt{85}}{18}$
21. $(-2, 10, 0)$; $(-2, 0, -5)$; The line does not intersect the yz-plane.
23. $(0, 4, -2), (4, 0, 6)$ 25. $(1, -1, 2)$ 29. The lines are parallel.
31. The points do not lie on the same line.
35. $\langle x, y\rangle = \langle -1, 2\rangle + t\langle 1, 1\rangle$
37. the point $1/n$ of the way from $(-2, 0)$ to $(1, 3)$
39. the line segment joining the points $(1, 0)$ and $(-3, 6)$
41. $(5, 2)$ 43. $2\sqrt{5}$ 45. distance $= \sqrt{35/6}$
47. **(a)** $x = x_0 + (x_1 - x_0)t, y = y_0 + (y_1 - y_0)t, z = z_0 + (z_1 - z_0)t$
(b) $x = x_1 + at, y = y_1 + bt, z = z_1 + ct$
49. **(b)** $\langle x, y, z\rangle = \langle 1 + 2t, -3 + 4t, 5 + t\rangle$
51. **(b)** 84° **(c)** $x = 7 + t, y = -1, z = -2 + t$
53. $x = t, y = 2 + t, z = 1 - t$
55. **(a)** $\sqrt{17}$ cm **(b)**

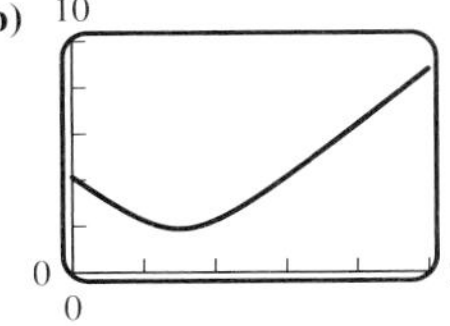

(d) $\sqrt{14}/2$ cm

Exercise Set 12.6 (Page 841)

1. $x = 3, y = 4, z = 5$ 3. $x + 4y + 2z = 28$ 5. $z = 0$
7. $x - y = 0$ 9. $y + z = 1$ 11. $2y - z = 1$
13. **(a)** parallel **(b)** perpendicular **(c)** neither
15. **(a)** parallel **(b)** neither **(c)** perpendicular
17. **(a)** point of intersection is $\left(\frac{5}{2}, \frac{5}{2}, \frac{5}{2}\right)$ **(b)** no intersection
19. 35° 21. $4x - 2y + 7z = 0$ 23. $4x - 13y + 21z = -14$
25. $x + y - 3z = 6$ 27. $x + 5y + 3z = -6$
29. $x + 2y + 4z = \frac{29}{2}$ 31. $x = 5 - 2t, y = 5t, z = -2 + 11t$
33. $7x + y + 9z = 25$ 35. yes 37. yes
39. $x = -\frac{11}{7} - 23t, y = -\frac{12}{7} + t, z = -7t$
41. $\frac{5}{3}$ 43. $5/\sqrt{54}$ 45. $25/\sqrt{126}$
47. $(x - 2)^2 + (y - 1)^2 + (z + 3)^2 = \frac{121}{14}$ 49. $5/\sqrt{12}$

Exercise Set 12.7 (Page 852)

1. **(a)** elliptic paraboloid, $a = 2, b = 3$
(b) hyperbolic paraboloid, $a = 1, b = 5$
(c) hyperboloid of one sheet, $a = b = c = 4$
(d) circular cone, $a = b = 1$ **(e)** elliptic paraboloid, $a = 2, b = 1$
(f) hyperboloid of two sheets, $a = b = c = 1$
3. **(a)** $-z = x^2 + y^2$, circular paraboloid opening down the negative z-axis

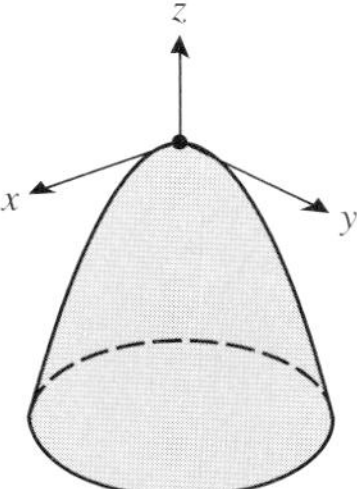

(b) $z = x^2 + y^2$, circular paraboloid, no change
(c) $z = x^2 + y^2$, circular paraboloid, no change
(d) $z = x^2 + y^2$, circular paraboloid, no change

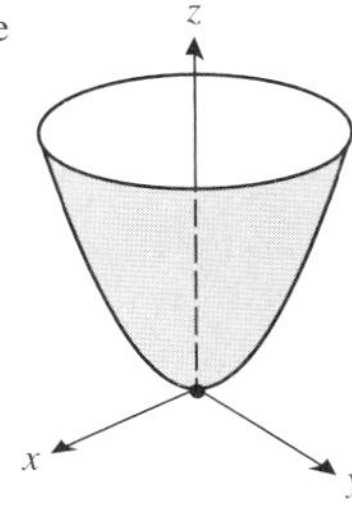

(e) $x = y^2 + z^2$, circular paraboloid opening along the positive x-axis

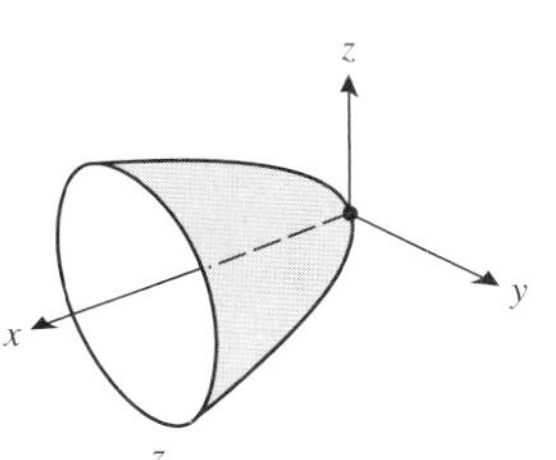

(f) $y = x^2 + z^2$, circular paraboloid opening along the positive y-axis

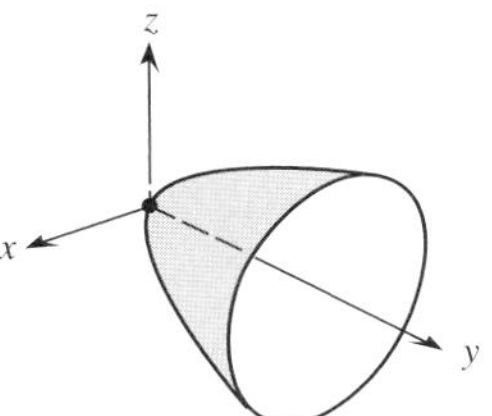

5. **(a)** hyperboloid of one sheet, axis is y-axis
(b) hyperboloid of two sheets separated by yz-plane
(c) elliptic paraboloid opening along the positive x-axis
(d) elliptic cone with x-axis as axis
(e) hyperbolic paraboloid straddling the x-axis
(f) paraboloid opening along the negative y-axis

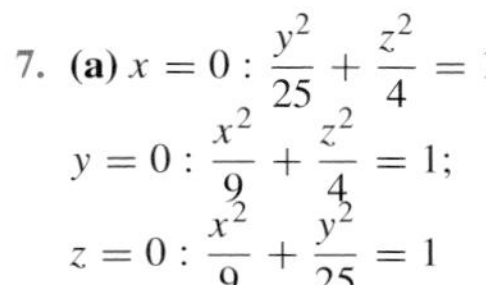

7. **(a)** $x = 0: \dfrac{y^2}{25} + \dfrac{z^2}{4} = 1;$
$y = 0: \dfrac{x^2}{9} + \dfrac{z^2}{4} = 1;$
$z = 0: \dfrac{x^2}{9} + \dfrac{y^2}{25} = 1$

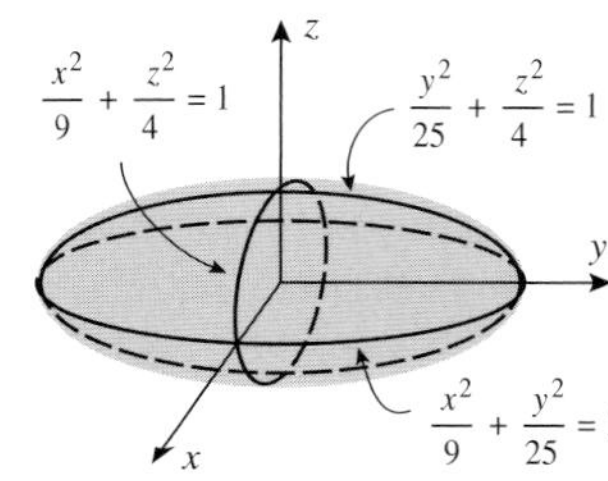

(b) $x = 0: z = 4y^2;$
$y = 0: z = x^2;$
$z = 0: x = y = 0$

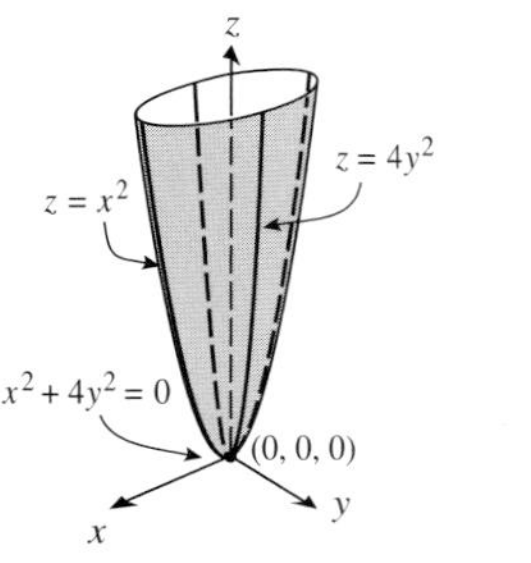

(c) $x = 0: \dfrac{y^2}{16} - \dfrac{z^2}{4} = 1;$
$y = 0: \dfrac{x^2}{9} - \dfrac{z^2}{4} = 1;$
$z = 0: \dfrac{x^2}{9} + \dfrac{y^2}{16} = 1$

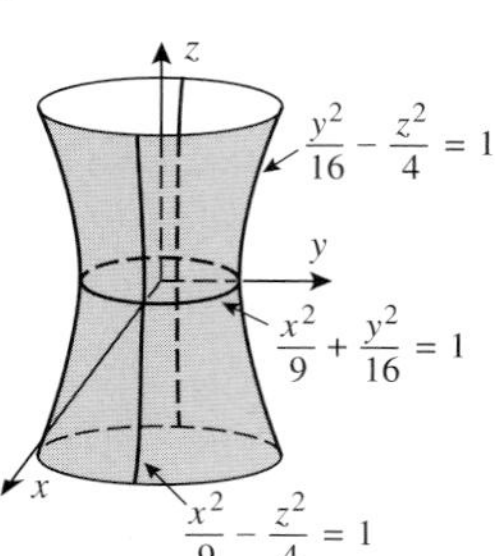

9. **(a)** $4x^2 + z^2 = 3$; ellipse **(b)** $y^2 + z^2 = 3$; circle
(c) $y^2 + z^2 = 20$; circle **(d)** $9x^2 - y^2 = 20$; hyperbola
(e) $z = 9x^2 + 16$; parabola **(f)** $9x^2 + 4y^2 = 4$; ellipse

11.

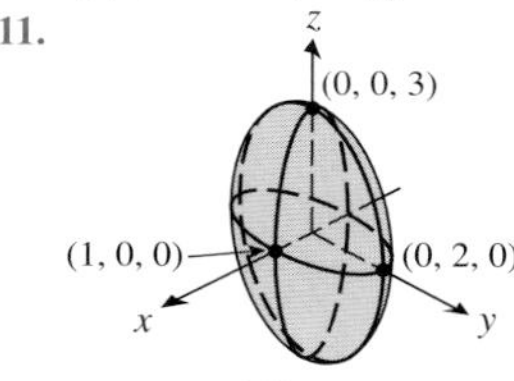

Ellipsoid

13. Hyperboloid of one sheet

(2, 0, 0) (0, 3, 0)

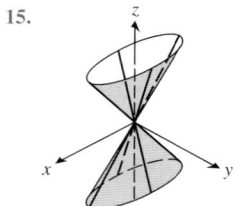

15. Elliptic cone

17.

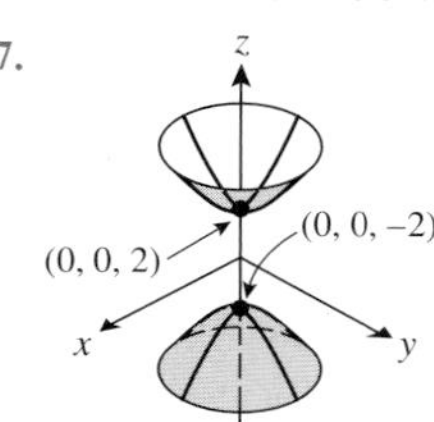

Hyperboloid of two sheets

19. Hyperbolic paraboloid

21. Elliptic paraboloid

23.

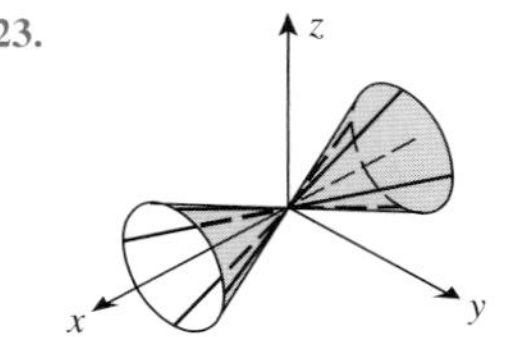

Circular cone

25.

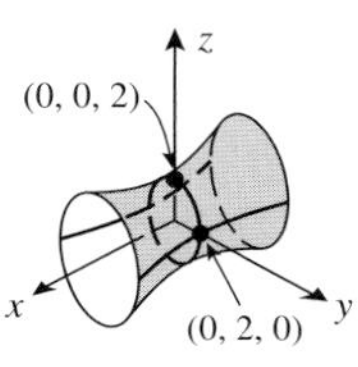

Hyperboloid of one sheet

27.

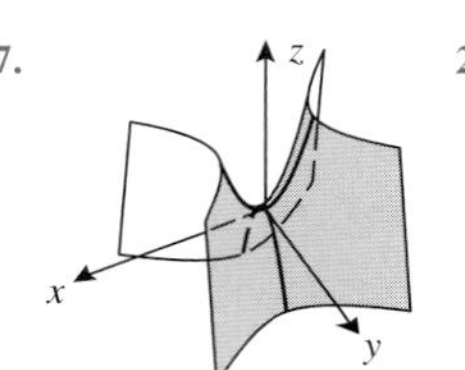

Hyperbolic paraboloid

29.

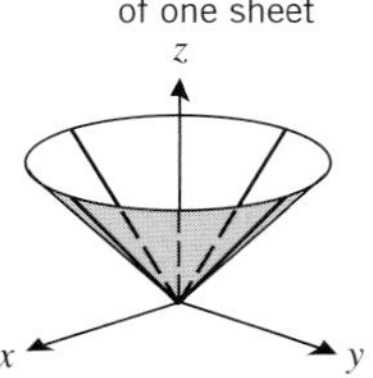

31.

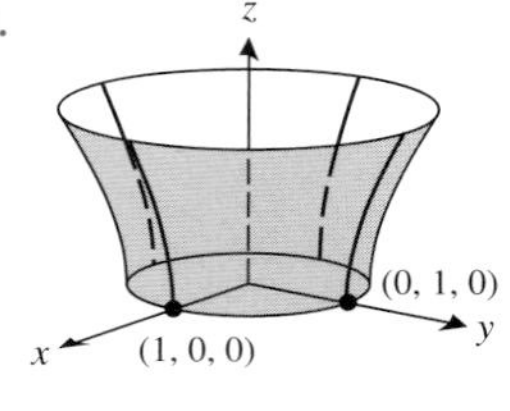

33.

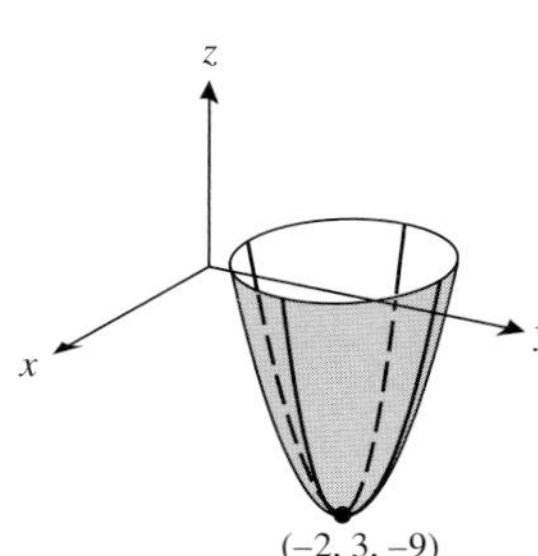

Circular paraboloid

35.

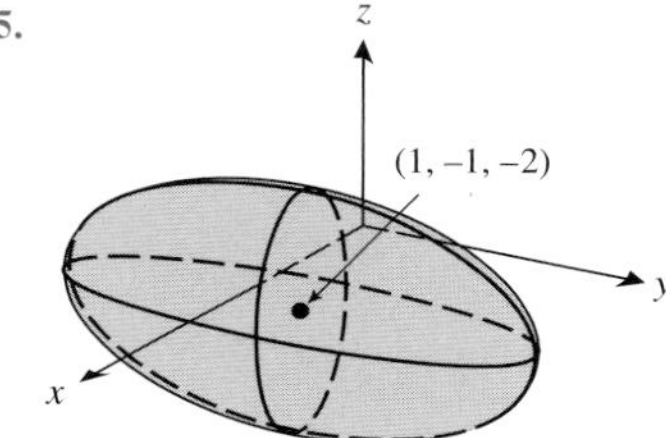

Ellipsoid

37. **(a)** $\dfrac{x^2}{9} + \dfrac{y^2}{4} = 1$ **(b)** 6, 4 **(c)** $(\pm\sqrt{5}, 0, \sqrt{2})$
(d) The focal axis is parallel to the x-axis.

39. **(a)** $\dfrac{y^2}{4} - \dfrac{x^2}{4} = 1$ **(b)** $(0, \pm 2, 4)$ **(c)** $(0, \pm 2\sqrt{2}, 4)$
(d) The focal axis is parallel to the y-axis.

41. **(a)** $z + 4 = y^2$ **(b)** $(2, 0, -4)$ **(c)** $\left(2, 0, -\frac{15}{4}\right)$
(d) The focal axis is parallel to z-axis.

43. circle of radius $\sqrt{2}$ in the plane $z = 2$, centered at $(0, 0, 2)$

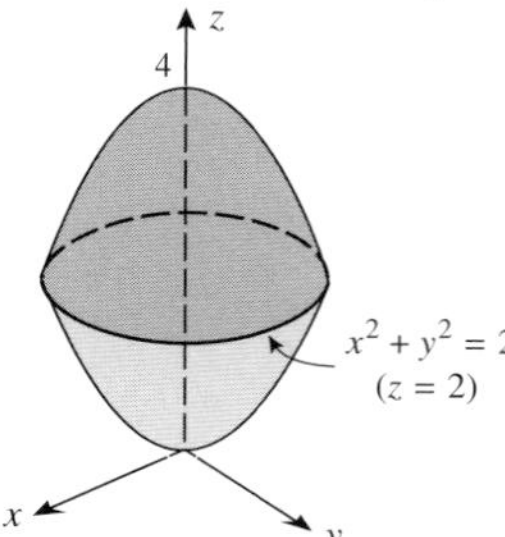

45. $y = 4(x^2 + z^2)$ 47. $z = (x^2 + y^2)/4$ (circular paraboloid)

Exercise Set 12.8 (Page 859)

1. **(a)** $(8, \pi/6, -4)$ **(b)** $(5\sqrt{2}, 3\pi/4, 6)$ **(c)** $(2, \pi/2, 0)$ **(d)** $(8, 5\pi/3, 6)$
3. **(a)** $(2\sqrt{3}, 2, 3)$ **(b)** $(-4\sqrt{2}, 4\sqrt{2}, -2)$ **(c)** $(5, 0, 4)$ **(d)** $(-7, 0, -9)$
5. **(a)** $(2\sqrt{2}, \pi/3, 3\pi/4)$ **(b)** $(2, 7\pi/4, \pi/4)$ **(c)** $(6, \pi/2, \pi/3)$ **(d)** $(10, 5\pi/6, \pi/2)$
7. **(a)** $(5\sqrt{6}/4, 5\sqrt{2}/4, 5\sqrt{2}/2)$ **(b)** $(7, 0, 0)$ **(c)** $(0, 0, 1)$ **(d)** $(0, -2, 0)$
9. **(a)** $(2\sqrt{3}, \pi/6, \pi/6)$ **(b)** $(\sqrt{2}, \pi/4, 3\pi/4)$ **(c)** $(2, 3\pi/4, \pi/2)$ **(d)** $(4\sqrt{3}, 1, 2\pi/3)$
11. **(a)** $(5\sqrt{3}/2, \pi/4, -5/2)$ **(b)** $(0, 7\pi/6, -1)$ **(c)** $(0, 0, 3)$ **(d)** $(4, \pi/6, 0)$

15. 17.

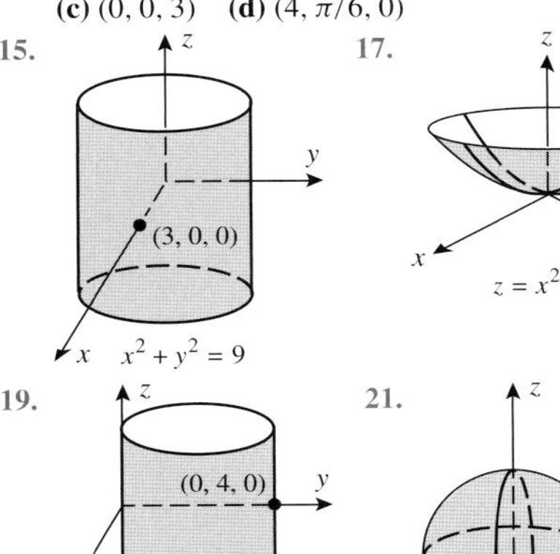

19.

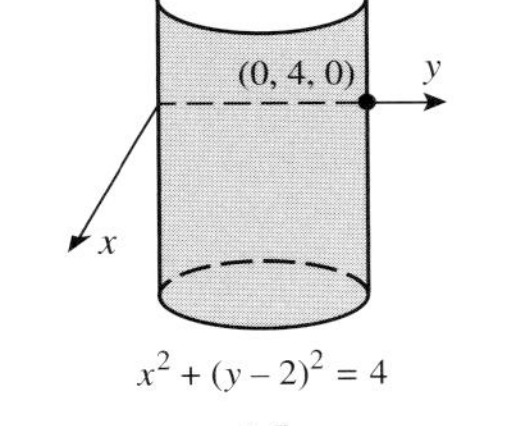

21.

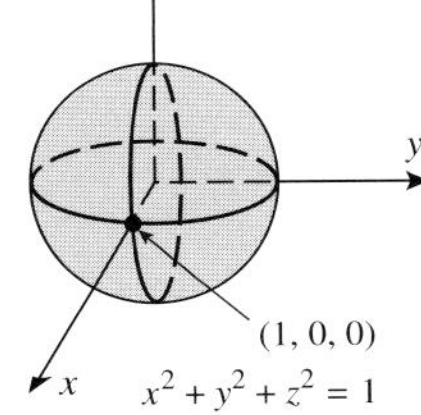

23.

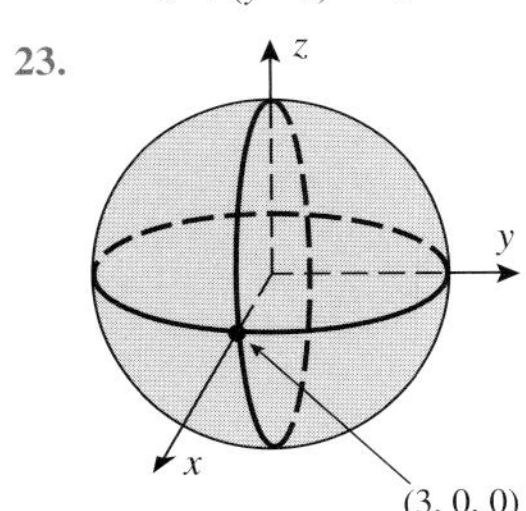

25.

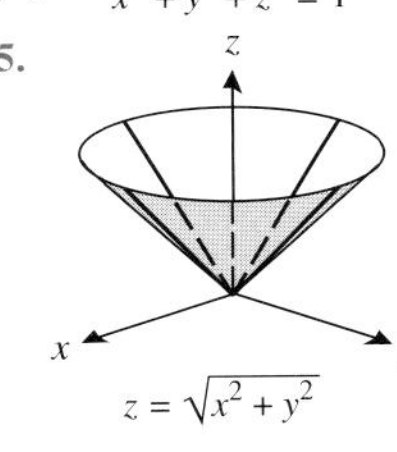

27. 29.

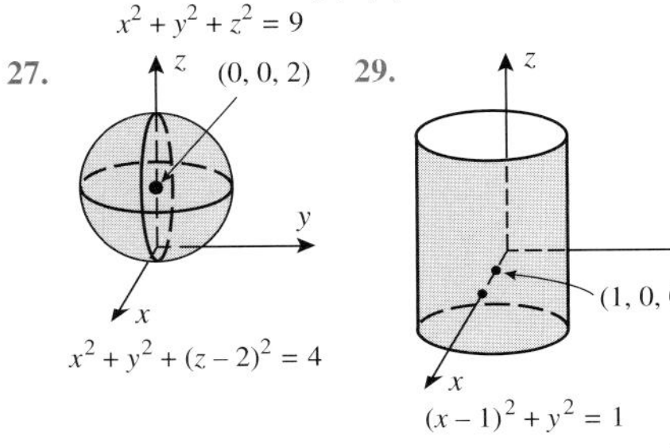

31. **(a)** $z = 3$ **(b)** $\rho = 3 \sec\phi$ 33. **(a)** $z = 3r^2$ **(b)** $\rho = \frac{1}{3}\csc\phi \cot\phi$
35. **(a)** $r = 2$ **(b)** $\rho = 2\csc\phi$ 37. **(a)** $r^2 + z^2 = 9$ **(b)** $\rho = 3$
39. **(a)** $2r\cos\theta + 3r\sin\theta + 4z = 1$ **(b)** $2\rho\sin\phi\cos\theta + 3\rho\sin\phi\sin\theta + 4\rho\cos\phi = 1$
41. **(a)** $r^2\cos^2\theta = 16 - z^2$ **(b)** $\rho^2(1 - \sin^2\phi\sin^2\theta) = 16$
43. all points on or above the paraboloid $z = x^2 + y^2$ that are also on or below the plane $z = 4$
45. all points on or between concentric spheres of radii 1 and 3 centered at the origin
47. spherical $(4000, \pi/6, \pi/6)$, rectangular $(1000\sqrt{3}, 1000, 2000\sqrt{3})$
49. **(a)** $(10, \pi/2, 1)$ **(b)** $(0, 10, 1)$ **(c)** $(\sqrt{101}, \pi/2, \tan^{-1} 10)$
51. ≈ 2927 km

Chapter 12. Review Exercises (Page 860)

3. **(b)** $-1/2, \pm\sqrt{3}/2$ **(d)** true
5. **(a)** $r^2 = 16$ **(b)** $r^2 = 25$ **(c)** $r^2 = 9$
7. $(7, 5)$
9. **(a)** $-\frac{3}{4}$ **(b)** $\frac{1}{7}$ **(c)** $(48 \pm 25\sqrt{3})/11$ **(d)** $c = \frac{4}{3}$
13. 13 ft·lb 15. **(a)** $\sqrt{26}/2$ **(b)** $\sqrt{26}/3$
17. **(a)** 29 **(b)** $\dfrac{29}{\sqrt{65}}$ 19. $x = 4 + t, y = 1 - t, z = 2$
21. $x + 5y - z - 2 = 0$ 23. $a_1a_2 + b_1b_2 + c_1c_2 = 0$
25. **(a)** hyperboloid of one sheet **(b)** sphere **(c)** circular cone
27. **(a)** $z = x^2 - y^2$ **(b)** $xz = 1$
29. **(a)** **(b)**

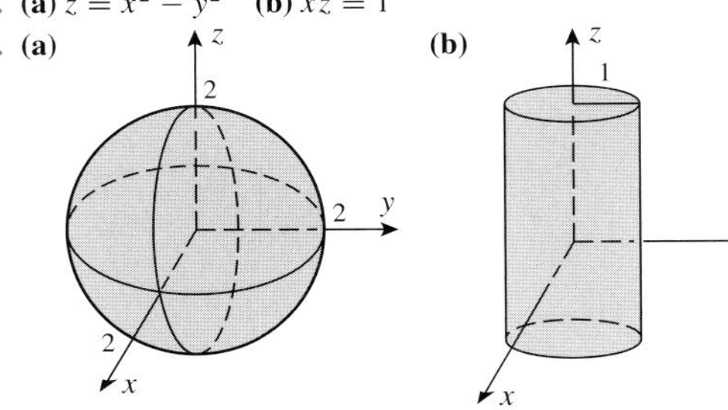

(c)

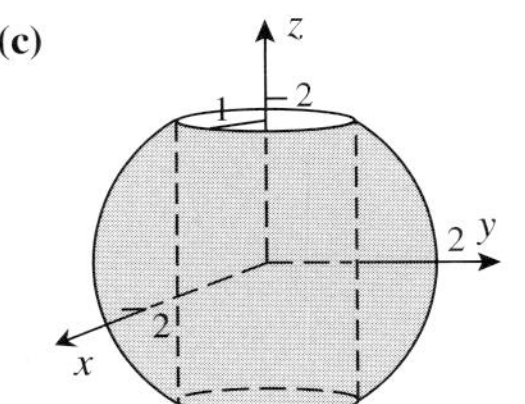

31. **(a)** **(b)**

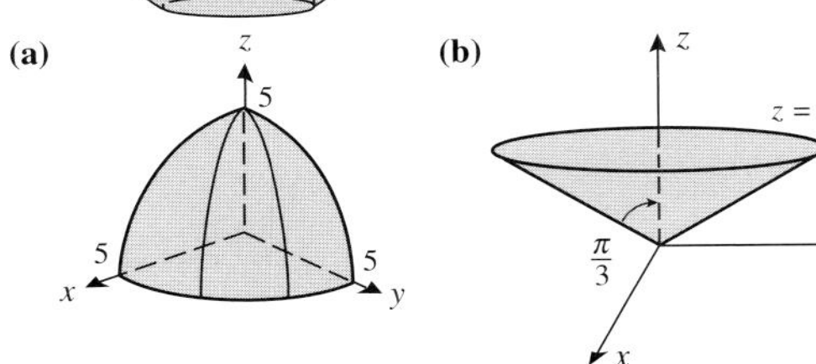

(c)

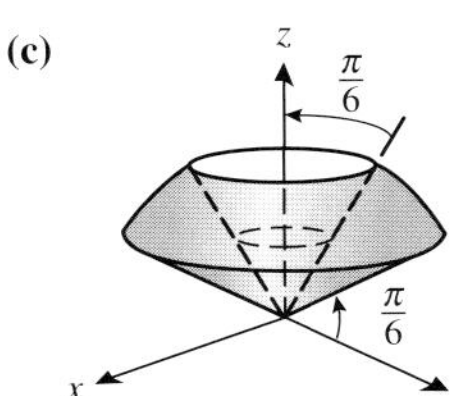

Exercise Set 13.1 (Page 867)

1. $(-\infty, +\infty)$; $\mathbf{r}(\pi) = -\mathbf{i} - 3\pi\mathbf{j}$ 3. $[2, +\infty)$; $\mathbf{r}(3) = -\mathbf{i} - \ln 3\mathbf{j} + \mathbf{k}$
5. $\mathbf{r} = 3\cos t\mathbf{i} + (t + \sin t)\mathbf{j}$ 7. $\mathbf{r} = 2t\mathbf{i} + 2\sin 3t\mathbf{j} + 5\cos 3t\mathbf{k}$
9. $x = 3t^2, y = -2$ 11. $x = 2t - 1, y = -3\sqrt{t}, z = \sin 3t$
13. the line in 2-space through $(3, 0)$ with direction vector $\mathbf{a} = -2\mathbf{i} + 5\mathbf{j}$
15. the line in 3-space through the point $(0, -3, 1)$ and parallel to the vector $2\mathbf{i} + 3\mathbf{k}$
17. an ellipse centered at $(0, 0, 1)$ in the plane $z = 1$
19. **(a)** slope $-\frac{3}{2}$ **(b)** $\left(\frac{5}{2}, 0, \frac{3}{2}\right)$

21. (a)

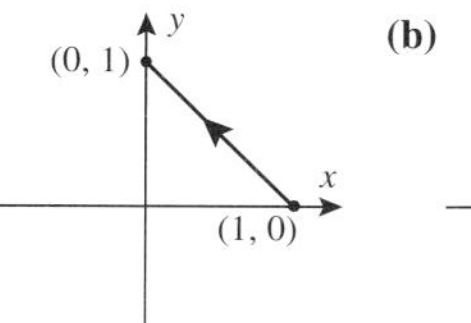

(b)

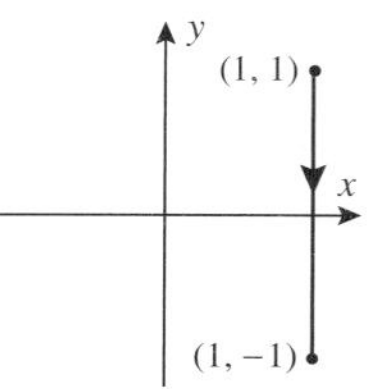

23. $\mathbf{r} = (1-t)(3\mathbf{i}+4\mathbf{j}),\ 0 \le t \le 1$

25. $x = 2$

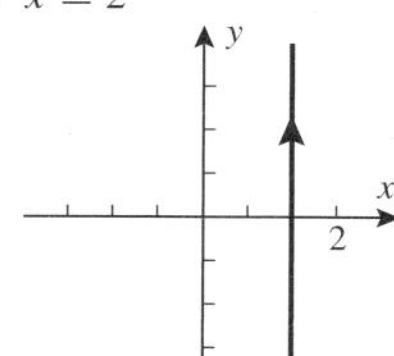

27. $(x-1)^2 + (y-3)^2 = 1$

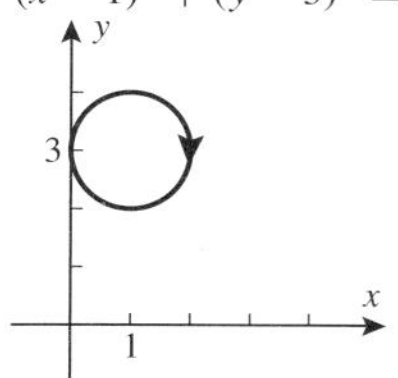

29. $x^2 - y^2 = 1,\ x \ge 1$

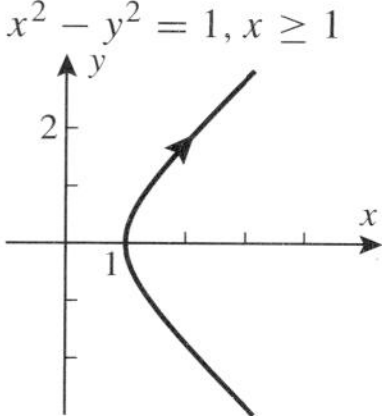

31.

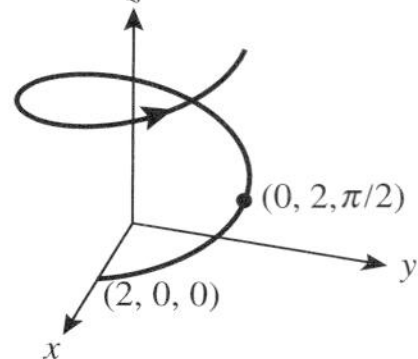

33.

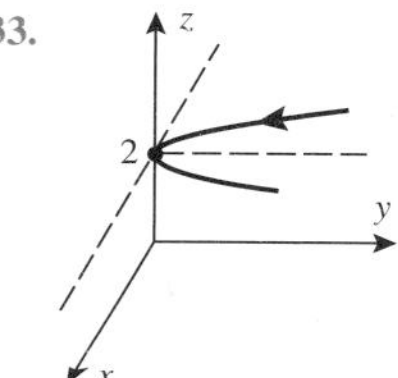

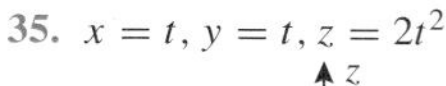
35. $x = t,\ y = t,\ z = 2t^2$

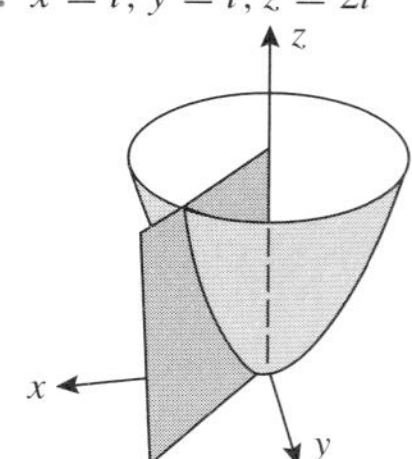

37. $\mathbf{r}(t) = t\mathbf{i} + t^2\mathbf{j} \pm \frac{1}{3}\sqrt{81 - 9t^2 - t^4}\,\mathbf{k}$ 43. $c = 3/(2\pi)$

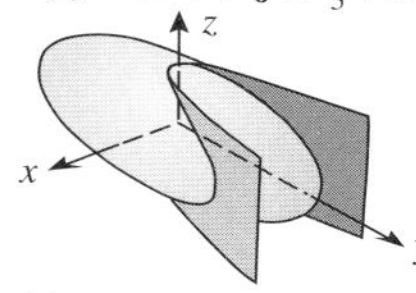

47. (a) III, since the curve is a subset of the plane $y = -x$
(b) IV, since only x is periodic in t and y, z increase without bound
(c) II, since all three components are periodic in t
(d) I, since the projection onto the yz-plane is a circle and the curve increases without bound in the x-direction

49. (a) $x = 3\cos t\ y = 3\sin t,\ z = 9\cos^2 t$ 51.
(b)

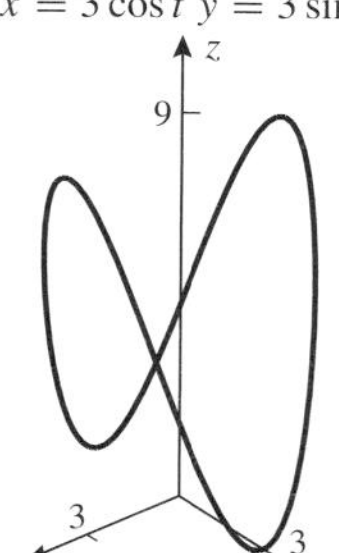

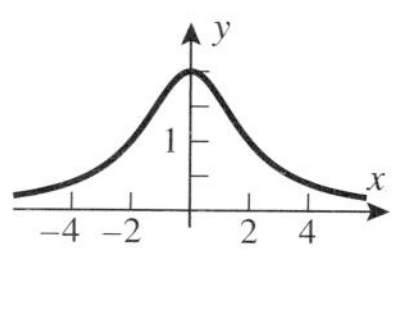

▶ Exercise Set 13.2 (Page 878)

1. $\langle \frac{1}{3}, 0\rangle$ 3. $2\mathbf{i} - 3\mathbf{j} + 4\mathbf{k}$ 5. (a) continuous (b) not continuous

7.

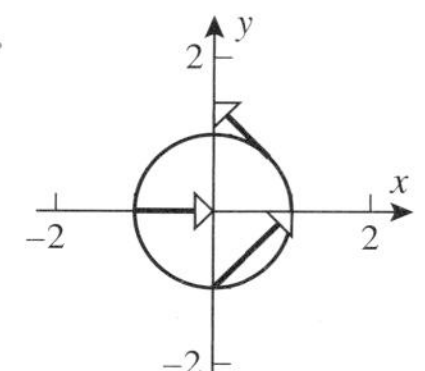

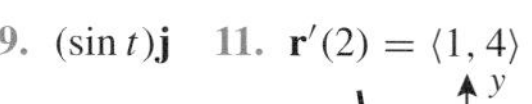
9. $(\sin t)\mathbf{j}$ 11. $\mathbf{r}'(2) = \langle 1, 4\rangle$

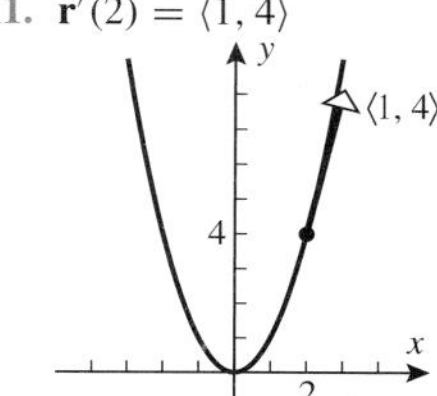

13. $\mathbf{r}'(0) = \mathbf{j}$

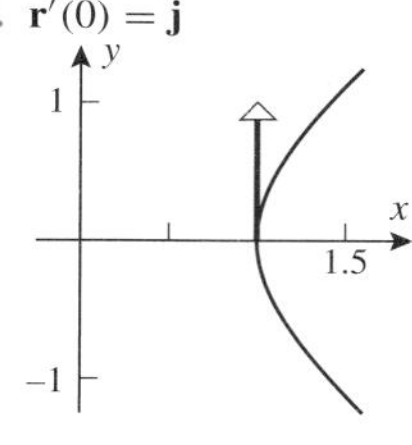

15. $\mathbf{r}'(\pi/2) = -2\mathbf{k}$

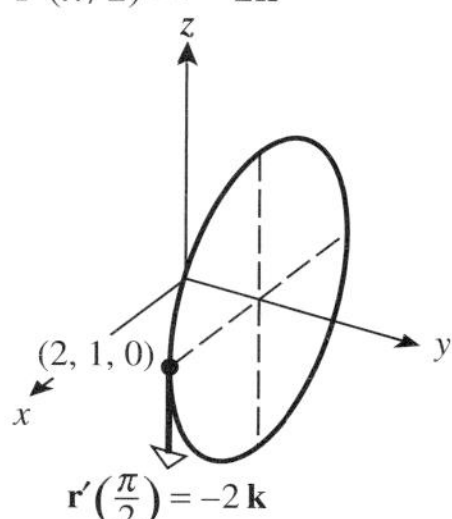

17.

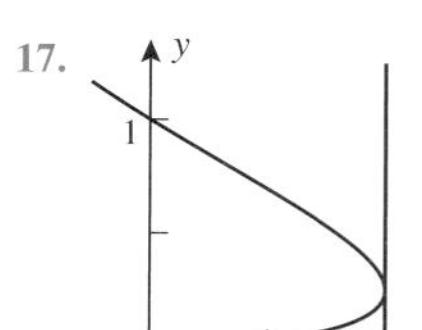

19. $x = 1 + 2t,\ y = 2 - t$
21. $x = 1 - \sqrt{3}\pi t,\ y = \sqrt{3} + \pi t,\ z = 1 + 3t$
23. $\mathbf{r} = (-\mathbf{i} + 2\mathbf{j}) + t\left(2\mathbf{i} + \frac{3}{4}\mathbf{j}\right)$
25. $\mathbf{r} = (4\mathbf{i} + \mathbf{j}) + t(-4\mathbf{i} + \mathbf{j} + 4\mathbf{k})$
27. (a) $\mathbf{i} - \mathbf{j} + \mathbf{k}$ (b) $-\mathbf{i} + \mathbf{k}$ (c) 0
29. $7t^6$; $18t^5\mathbf{i} - 10t^4\mathbf{j}$
31. $3t\mathbf{i} + 2t^2\mathbf{j} + \mathbf{C}$
33. $(-t\cos t + \sin t)\mathbf{i} + t\mathbf{j} + \mathbf{C}$ 35. $(t^3/3)\mathbf{i} - t^2\mathbf{j} + \ln|t|\mathbf{k} + \mathbf{C}$ 37. $\mathbf{j}$
39. $(5\sqrt{5} - 1)/3$ 41. $\frac{52}{3}\mathbf{i} + 4\mathbf{j}$ 43. $(t^2 + 1)\mathbf{i} + (t^3 - 1)\mathbf{j}$
45. $y(t) = \left(\frac{1}{2}t^2 + 2\right)\mathbf{i} + (e^t - 1)\mathbf{j}$
47.

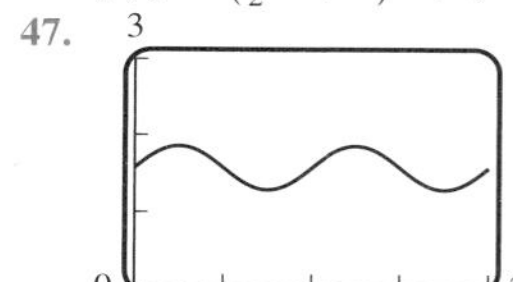

49. (a) $(-2, 4, 6)$ and $(1, 1, -3)$
(b) $76°$, $71°$
51. $68°$

▶ Exercise Set 13.3 (Page 888)

1. smooth 3. not smooth, $\mathbf{r}'(1) = \mathbf{0}$ 5. $L = \frac{3}{2}$ 7. $L = e - e^{-1}$
9. $L = 28$ 11. $L = 2\pi\sqrt{10}$ 13. $\mathbf{r}'(\tau) = 4\mathbf{i} + 8(4\tau + 1)\mathbf{j}$
15. $\mathbf{r}'(\tau) = 2\tau e^{\tau^2}\mathbf{i} - 8\tau e^{-\tau^2}\mathbf{j}$
19. (a) $x = \dfrac{s}{\sqrt{2}},\ y = \dfrac{s}{\sqrt{2}}$ (b) $x = y = z = \dfrac{s}{\sqrt{3}}$
21. (a) $x = 1 + \dfrac{s}{3},\ y = 3 - \dfrac{2s}{3},\ z = 4 + \dfrac{2s}{3}$ (b) $\langle \frac{28}{3}, -\frac{41}{3}, \frac{62}{3}\rangle$
23. $x = 3 + \cos s,\ y = 2 + \sin s,\ 0 \le s \le 2\pi$
25. $x = \frac{1}{3}[(3s+1)^{2/3} - 1]^{3/2},\ y = \frac{1}{2}[(3s+1)^{2/3} - 1],\ s \ge 0$
27. $x = \left(\dfrac{s}{\sqrt{2}} + 1\right)\cos\left[\ln\left(\dfrac{s}{\sqrt{2}} + 1\right)\right],$
$0 \le s \le \sqrt{2}(e^{\pi/2} - 1)$
$y = \left(\dfrac{s}{\sqrt{2}} + 1\right)\sin\left[\ln\left(\dfrac{s}{\sqrt{2}} + 1\right)\right],$
31. $x = 2a\cos^{-1}[1 - s/(4a)]$
$-2a(1 - [1 - s/(4a)]^2)^{1/2}(2[1 - s/(4a)]^2 - 1),$
$y = \dfrac{s(8a - s)}{8a}$ for $0 \le s \le 8a$
33. (a) $9/2$ (b) $9 - 2\sqrt{6}$ 35. (a) $\sqrt{3}(1 - e^{-2})$ (b) $4\sqrt{5}$
37. (a) $g(\tau) = \pi(\tau)$ (b) $g(\tau) = \pi(1 - \tau)$ 39. 44 in.
41. (a) $2t + \dfrac{1}{t}$ (b) $2t + \dfrac{1}{t}$ (c) $8 + \ln 3$

▶ Exercise Set 13.4 (Page 895)

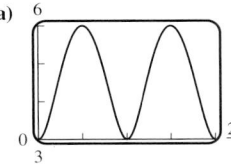

1. (a)

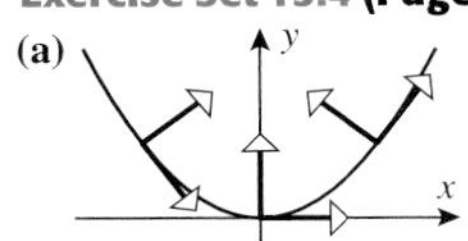

(b)

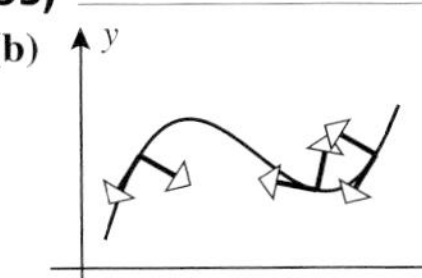

5. $\mathbf{T}(1) = \frac{2}{\sqrt{5}}\mathbf{i} + \frac{1}{\sqrt{5}}\mathbf{j}, \mathbf{N}(1) = \frac{1}{\sqrt{5}}\mathbf{i} - \frac{2}{\sqrt{5}}\mathbf{j}$
7. $\mathbf{T}\left(\frac{\pi}{3}\right) = -\frac{\sqrt{3}}{2}\mathbf{i} + \frac{1}{2}\mathbf{j}, \mathbf{N}\left(\frac{\pi}{3}\right) = -\frac{1}{2}\mathbf{i} - \frac{\sqrt{3}}{2}\mathbf{j}$
9. $\mathbf{T}\left(\frac{\pi}{2}\right) = -\frac{4}{\sqrt{17}}\mathbf{i} + \frac{1}{\sqrt{17}}\mathbf{k}, \mathbf{N}\left(\frac{\pi}{2}\right) = -\mathbf{j}$
11. $\mathbf{T}(0) = \frac{1}{\sqrt{3}}\mathbf{i} + \frac{1}{\sqrt{3}}\mathbf{j} + \frac{1}{\sqrt{3}}\mathbf{k}, \mathbf{N}(0) = -\frac{1}{\sqrt{2}}\mathbf{i} + \frac{1}{\sqrt{2}}\mathbf{j}$
13. $x = s, y = 1$ 15. $\mathbf{B} = \frac{4}{5}\cos t\,\mathbf{i} - \frac{4}{5}\sin t\,\mathbf{j} - \frac{3}{5}\mathbf{k}$ 17. $\mathbf{B} = -\mathbf{k}$
19. $\mathbf{T}\left(\frac{\pi}{4}\right) = \frac{\sqrt{2}}{2}(-\mathbf{i} + \mathbf{j}), \mathbf{N}\left(\frac{\pi}{4}\right) = -\frac{\sqrt{2}}{2}(\mathbf{i} + \mathbf{j}),$
$\mathbf{B}\left(\frac{\pi}{4}\right) = \mathbf{k}$ rectifying: $x + y = \sqrt{2}$; osculating: $z = 1$;
normal: $-x + y = 0$
23. $\mathbf{N} = -\sin t\,\mathbf{i} - \cos t\,\mathbf{j}$

▶ Exercise Set 13.5 (Page 901)

1. $\kappa \approx 2$ 3. (a) I is the curvature of II. (b) I is the curvature of II.
5. $\frac{6}{t(4 + 9t^2)^{3/2}}$ 7. $\frac{12e^{2t}}{(9e^{6t} + e^{-2t})^{3/2}}$ 9. $\frac{4}{17}$ 11. $\frac{1}{2\cosh^2 t}$
13. $\kappa = \frac{2}{5}, \rho = \frac{5}{2}$ 15. $\kappa = \frac{\sqrt{2}}{3}, \rho = \frac{3\sqrt{2}}{2}$ 17. $\kappa = \frac{1}{4}$ 21. 1
23. $\frac{1}{\sqrt{2}}$ 25. $\frac{4}{5\sqrt{5}}$ 27. $\frac{96}{125}$ 29. $\frac{6}{5\sqrt{10}}$ 31. $\frac{1}{\sqrt{2}}$
33. (a)

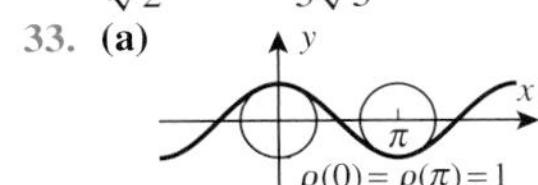

(b)

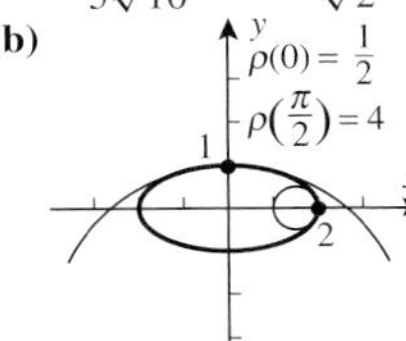

35.

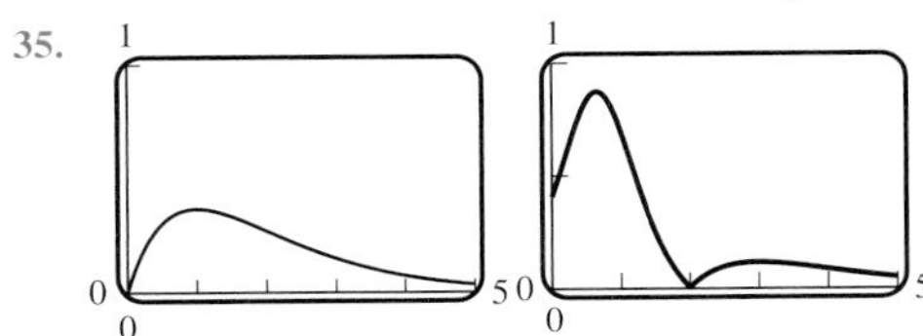

37. (a) $\kappa = \frac{|12x^2 - 4|}{[1 + (4x^3 - 4x)^2]^{3/2}}$ (b)

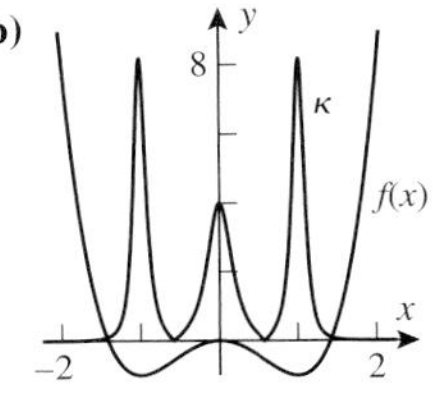

(c) $\rho = \frac{1}{4}$ for $x = 0$ and $\rho = \frac{1}{8}$ when $x = \pm 1$

41. $\frac{3}{2\sqrt{2}}$ 43. $\frac{2}{3}$ 47. $\rho = 2|p|$ 49. $(3, 0), (-3, 0)$

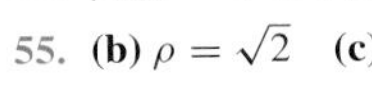

51. $\rho_{\min} = 1/\sqrt{2}$; $\rho_{\max} = 2$
55. (b) $\rho = \sqrt{2}$ (c)

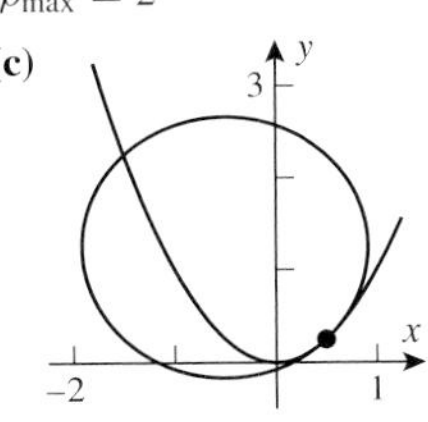

59. $a = \frac{1}{2r}$
69. $\tau = \frac{2}{(t^2 + 2)^2}$
71. $\tau = -\frac{\sqrt{2}}{(e^t + e^{-t})^2}$

▶ Exercise Set 13.6 (Page 914)

1. $\mathbf{v}(t) = -3\sin t\,\mathbf{i} + 3\cos t\,\mathbf{j}$
$\mathbf{a}(t) = -3\cos t\,\mathbf{i} - 3\sin t\,\mathbf{j}$
$\|\mathbf{v}(t)\| = 3$

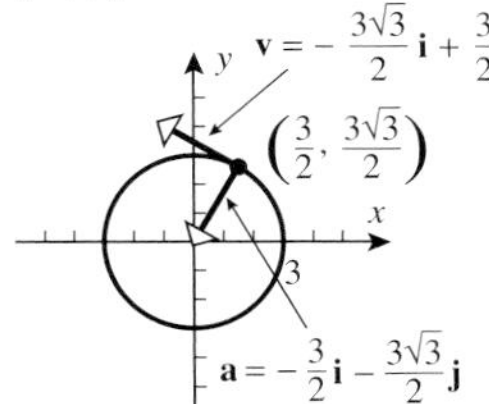

3. $\mathbf{v}(t) = e^t\mathbf{i} - e^{-t}\mathbf{j}$
$\mathbf{a}(t) = e^t\mathbf{i} + e^{-t}\mathbf{j}$
$\|\mathbf{v}(t)\| = \sqrt{e^{2t} + e^{-2t}}$

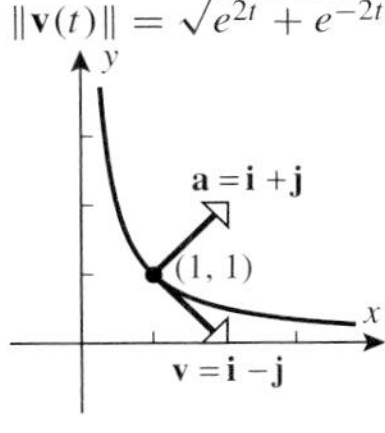

5. $\mathbf{v} = \mathbf{i} + \mathbf{j} + \mathbf{k}, \|\mathbf{v}\| = \sqrt{3}, \mathbf{a} = \mathbf{j} + 2\mathbf{k}$
7. $\mathbf{v} = -\sqrt{2}\mathbf{i} + \sqrt{2}\mathbf{j} + \mathbf{k}, \|\mathbf{v}\| = \sqrt{5}, \mathbf{a} = -\sqrt{2}\mathbf{i} - \sqrt{2}\mathbf{j}$
13. minimum speed $3\sqrt{2}$ when $\mathbf{r} = 24\mathbf{i} + 8\mathbf{j}$
15. (a) 6 … 0, 3, $\frac{2\pi}{3}$ (b) maximum speed = 6, minimum speed = 3
(d) The maximum speed first occurs when $t = \pi/6$.
17. $\mathbf{v}(t) = (1 - \sin t)\mathbf{i} + (\cos t - 1)\mathbf{j}$;
$\mathbf{r}(t) = (t + \cos t - 1)\mathbf{i} + (\sin t - t + 1)\mathbf{j}$
19. $\mathbf{v}(t) = (1 - \cos t)\mathbf{i} + \sin t\,\mathbf{j} + e^t\mathbf{k}$;
$\mathbf{r}(t) = (t - \sin t - 1)\mathbf{i} + (1 - \cos t)\mathbf{j} + e^t\mathbf{k}$
21. 15° 23. (a) $0.7\mathbf{i} + 2.7\mathbf{j} - 3.4\mathbf{k}$ (b) $\mathbf{r}_0 = -0.7\mathbf{i} - 2.9\mathbf{j} + 4.8\mathbf{k}$
25. $\Delta\mathbf{r} = 8\mathbf{i} + \frac{26}{3}\mathbf{j}, s = (13\sqrt{13} - 5\sqrt{5})/3$
27. $\Delta\mathbf{r} = 2\mathbf{i} - \frac{2}{3}\mathbf{j} + \sqrt{2}\ln 3\mathbf{k}$; $s = \frac{8}{3}$
31. (a) $a_T = 0, a_N = \sqrt{2}$ (b) $a_T\mathbf{T} = \mathbf{0}, a_N\mathbf{N} = \mathbf{i} + \mathbf{j}$ (c) $1/\sqrt{2}$
33. (a) $a_T = 2\sqrt{5}, a_N = 2\sqrt{5}$ (b) $a_T\mathbf{T} = 2\mathbf{i} + 4\mathbf{j}, a_N\mathbf{N} = 4\mathbf{i} - 2\mathbf{j}$
(c) $2/\sqrt{5}$
35. (a) $a_T = 20/\sqrt{14}, a_N = 6\sqrt{3}/\sqrt{7}$
(b) $a_T\mathbf{T} = \frac{10}{7}\mathbf{i} + \frac{20}{7}\mathbf{j} + \frac{30}{7}\mathbf{k}, a_N\mathbf{N} = \frac{24}{7}\mathbf{i} - \frac{6}{7}\mathbf{j} + \frac{12}{7}\mathbf{k}$ (c) $\left(\frac{3}{7}\right)^{3/2}$
37. (a) $a_T = 0, a_N = 3$ (b) $a_T\mathbf{T} = \mathbf{0}, a_N\mathbf{N} = -3\mathbf{i}$ (c) $\frac{3}{8}$
39. $a_T = -3, a_N = 2, \mathbf{T} = -\mathbf{j}, \mathbf{N} = \mathbf{i}$
41. $a_T = \frac{4}{3}, a_N = \sqrt{29}/3, \mathbf{T} = \frac{1}{3}(2\mathbf{i} + 2\mathbf{j} + \mathbf{k}),$
$\mathbf{N} = (\mathbf{i} - 8\mathbf{j} + 14\mathbf{k})/(3\sqrt{29})$
43. $\frac{3}{2}$ 45. $-\pi/\sqrt{2}$ 47. $a_N = 8.41 \times 10^{10}$ km/s^2
49. $a_N = 18/(1 + 4x^2)^{3/2}$ 51. $a_N = 0$ 53. ≈ 38.73 m/s
55. (a) $x = 160t, y = 160\sqrt{3}t - 16t^2$ (b) 1200 ft (c) $1600\sqrt{3}$ ft
(d) 320 ft/s
57. $40\sqrt{3}$ ft 59. 800 ft/s 61. 15° or 75° 63. (c) ≈ 14.942 ft
65. (a) $\rho \approx 176.78$ m (b) $\frac{125}{4}$ m
67. (b) R is maximum when $\alpha = 45°$, maximum value v_0^2/g
69. (a) 2.62 s (b) 181.5 ft
71. (a) $v_0 \approx 83$ ft/s, $\alpha \approx 8°$ (b) 268.76 ft

▶ Exercise Set 13.7 (Page 924)

7. 7.75 km/s 9. 10.88 km/s
11. (a) minimum distance = 220, 680 mi,
maximum distance = 246,960 mi (b) 27.5 days
13. (a) 17, 224 mi/h (b) $e \approx 0.071$, apogee altitude = 819 mi

▶ Chapter 13. Review Exercises (Page 925)

3. the circle of radius 3 in the xy-plane, with center at the origin
5. a parabola in the plane $x = -2$, vertex at $(-2, 0, -1)$, opening upward
11. $x = 1 + t, y = -t, z = t$ 13. $(\sin t)\mathbf{i} - (\cos t)\mathbf{j} + \mathbf{C}$
15. $y(t) = \left(\frac{1}{3}t^3 + 1\right)\mathbf{i} + (t^2 + 1)\mathbf{j}$ 17. $15/4$

19. $\mathbf{r}(s) = \frac{s-3}{3}\mathbf{i} + \frac{12-2s}{3}\mathbf{j} + \frac{9+2s}{3}\mathbf{k}$ 25. 3/5 27. 0
29. **(a)** speed **(b)** distance traveled
(c) distance of the particle from the origin
33. **(a)** $\mathbf{r}(t) = \left(\frac{1}{6}t^4 + t\right)\mathbf{i} + \left(\frac{1}{2}t^2 + 2t\right)\mathbf{j} - \left(\frac{1}{4}\cos 2t + t - \frac{1}{4}\right)\mathbf{k}$
(b) 3.475 35. 24.78 ft 37. 36.50 km/s

▶ Exercise Set 14.1 (Page 937)

1. **(a)** 5 **(b)** 3 **(c)** 1 **(d)** −2 **(e)** $9a^3 + 1$ **(f)** $a^3b^2 - a^2b^3 + 1$
3. **(a)** $x^2 - y^2 + 3$ **(b)** $3x^3y^4 + 3$ 5. $x^3e^{x^3(3y+1)}$
7. **(a)** $t^2 + 3t^{10}$ **(b)** 0 **(c)** 3076
9. **(a)** WCI = 17.8°F **(b)** WCI = 22.6°F
11. **(a)** 66% **(b)** 73.5% **(c)** 60.6% 13. **(a)** 19 **(b)** −9 **(c)** 3
(d) $a^6 + 3$ **(e)** $-t^8 + 3$ **(f)** $(a+b)(a-b)^2b^3 + 3$
15. $(y+1)e^{x^2(y+1)z^2}$ 17. **(a)** $80\sqrt{\pi}$ **(b)** $n(n+1)/2$

19.
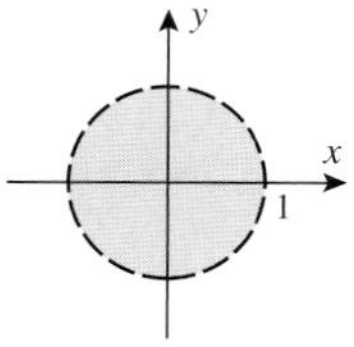

21.
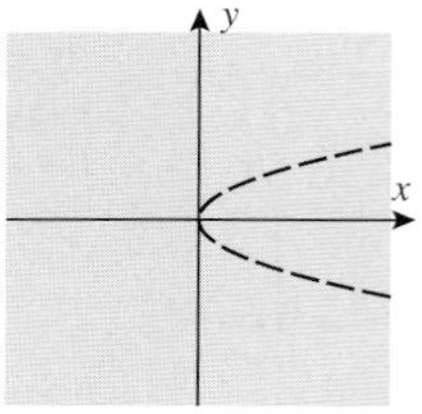

23. **(a)** all points above or on the line $y = -2$ **(b)** all points on or within the sphere $x^2 + y^2 + z^2 = 25$ **(c)** all points in 3-space

25.
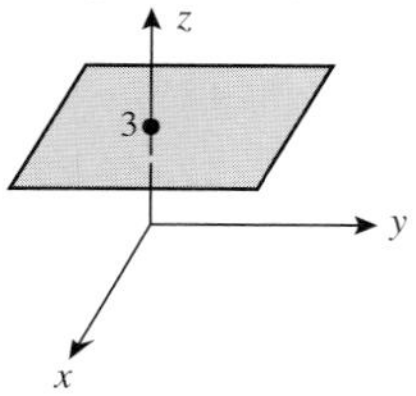

27.
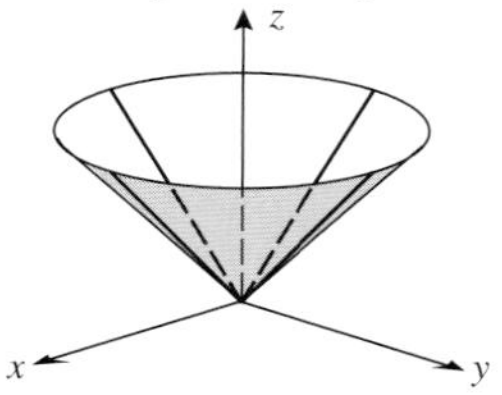

29.
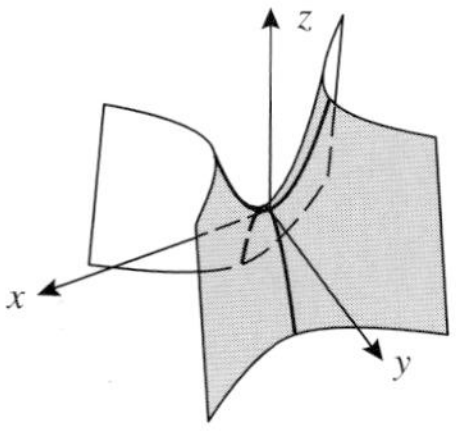

31.
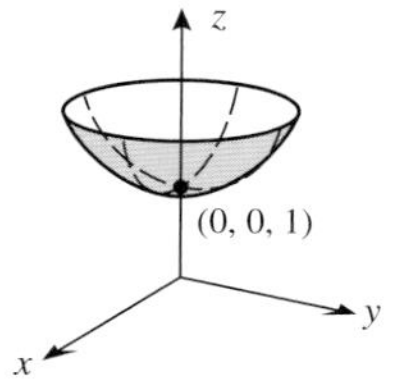

33.
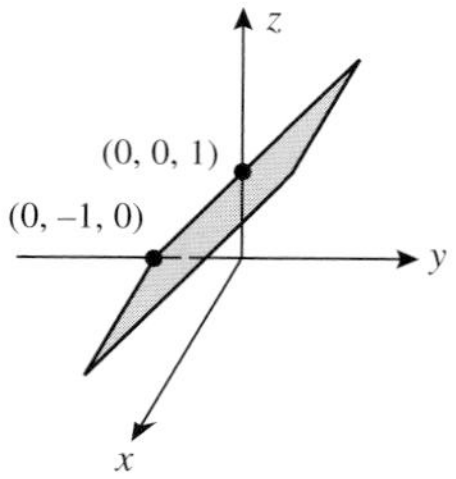

35. **(a)** $1 - x^2 - y^2$
(b) $\sqrt{x^2 + y^2}$
(c) $x^2 + y^2$
37. **(a)** A
(b) B
(c) increase
(d) decrease
(e) increase
(f) decrease

39.
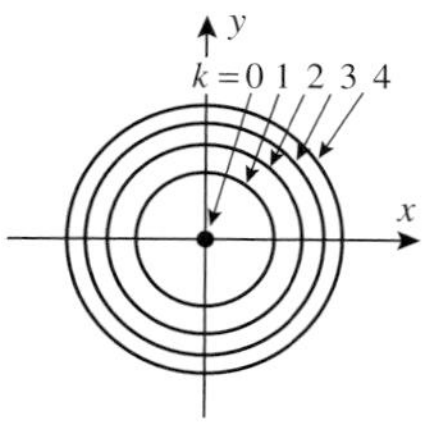

41.
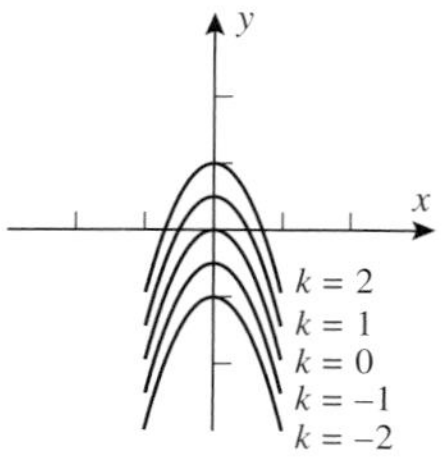

43.
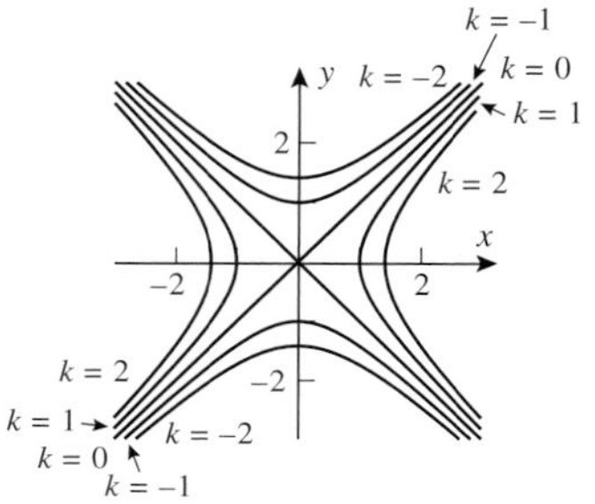

45.
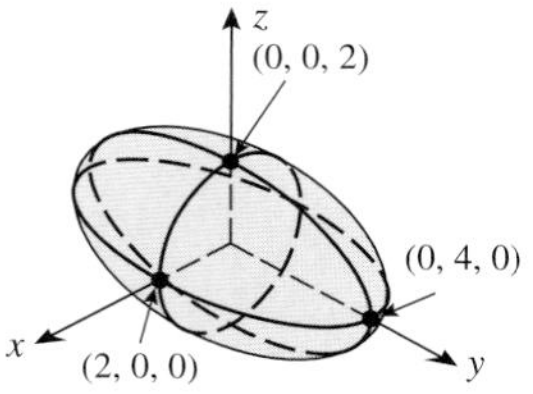

47.
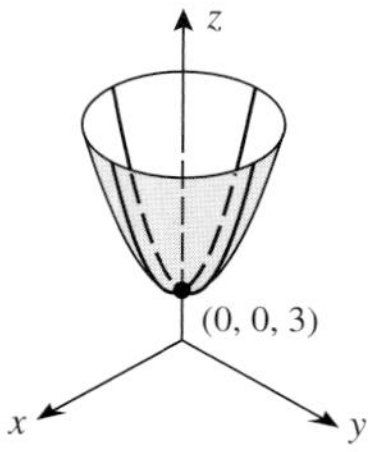

49. concentric spheres, common center at (2, 0, 0)
51. concentric cylinders, common axis the y-axis
53. **(a)** $x^2 - 2x^3 + 3xy = 0$ **(b)** $x^2 - 2x^3 + 3xy = 0$
(c) $x^2 - 2x^3 + 3xy = -18$
55. **(a)** $x^2 + y^2 - z = 5$ **(b)** $x^2 + y^2 - z = -2$ **(c)** $x^2 + y^2 - z = 0$
57. **(a)**
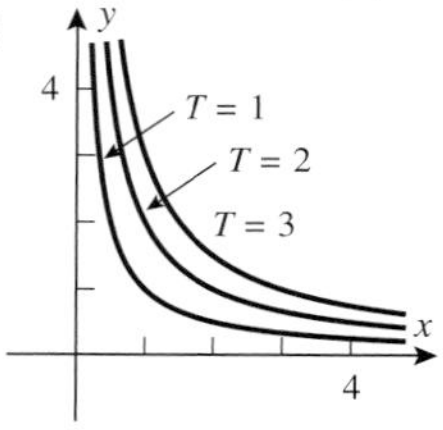

(b) the path $xy = 4$

59. **(a)**
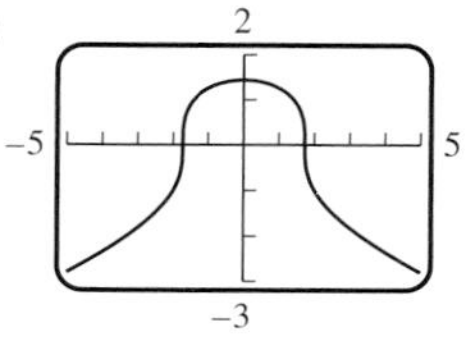

(b)
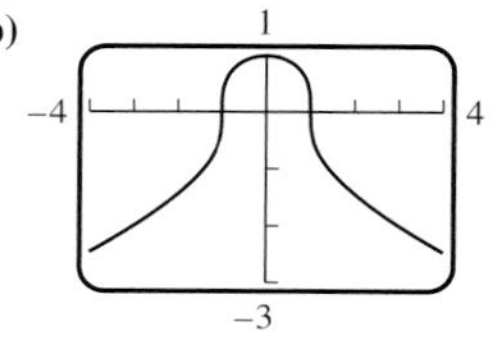

61. **(a)**
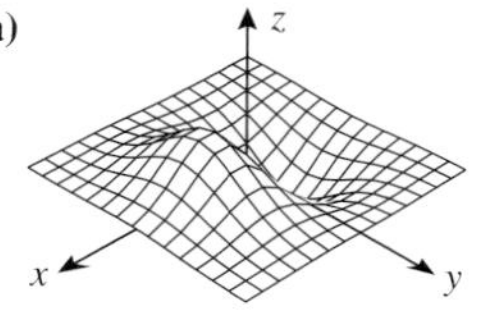

(b)
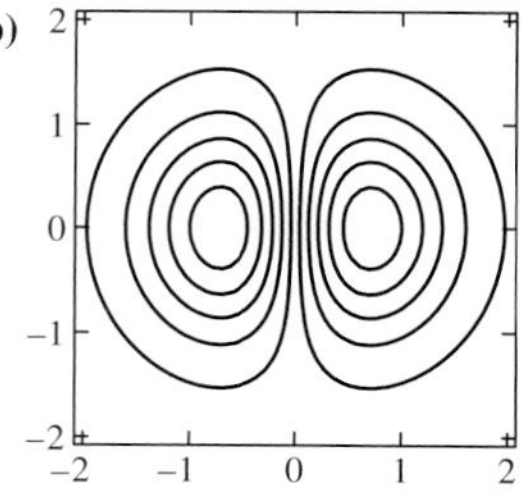

63. **(a)** The graph of g is the graph of f shifted one unit in the positive x-direction.
(b) The graph of g is the graph of f shifted one unit up the z-axis.
(c) The graph of g is the graph of f shifted one unit down the y-axis and then inverted with respect to the plane $z = 0$.

▶ Exercise Set 14.2 (Page 948)

1. 35 3. −8 5. 0
7. **(a)** along $x = 0$ limit does not exist
(b) along $x = 0$ limit does not exist
9. 1 11. 0 13. 0 15. limit does not exist 17. $\frac{8}{3}$ 19. 0 21. 0
23. limit does not exist 25. **(a)** no **(d)** no; yes 29. $-\pi/2$ 31. no

33. 35. 37. 39.

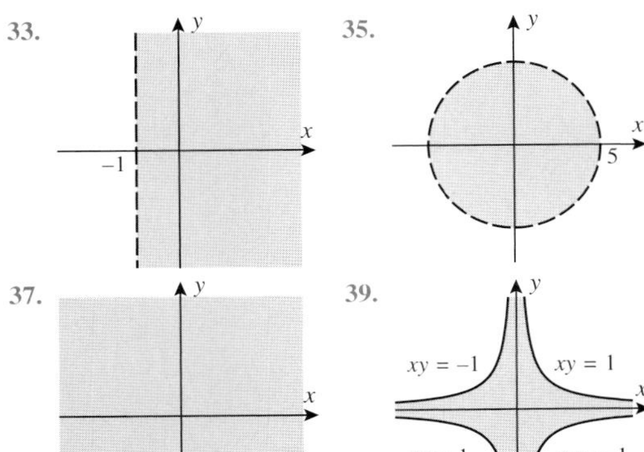

41. all of 3-space
43. all points not on the cylinder $x^2 + z^2 = 1$

▶ Exercise Set 14.3 (Page 959)

1. **(a)** $9x^2y^2$ **(b)** $6x^3y$ **(c)** $9y^2$ **(d)** $9x^2$ **(e)** $6y$ **(f)** $6x^3$ **(g)** 36 **(h)** 12
3. **(a)** $\frac{3}{8}$ **(b)** $\frac{1}{4}$ 5. **(a)** $-4\cos 7$ **(b)** $2\cos 7$
7. $\partial z/\partial x = -4$; $\partial z/\partial y = \frac{1}{2}$ 9. **(a)** 4.9 **(b)** 1.2
11. $z = f(x, y)$ has II as its graph, f_x has I as its graph, and f_y has III as its graph.
13. $8xy^3e^{x^2y^3}$, $12x^2y^2e^{x^2y^3}$
15. $x^3/(y^{3/5} + x) + 3x^2\ln(1 + xy^{-3/5})$, $-\frac{3}{5}x^4/(y^{8/5} + xy)$
17. $-\dfrac{y(x^2 - y^2)}{(x^2 + y^2)^2}, \dfrac{x(x^2 - y^2)}{(x^2 + y^2)^2}$
19. $(3/2)x^2y(5x^2 - 7)(3x^5y - 7x^3y)^{-1/2}$ $(1/2)x^3(3x^2 - 7)(3x^5y - 7x^3y)^{-1/2}$
21. $\dfrac{y^{-1/2}}{y^2 + x^2}, -\dfrac{xy^{-3/2}}{y^2 + x^2} - \dfrac{3}{2}y^{-5/2}\tan^{-1}\left(\dfrac{x}{y}\right)$
23. $-\frac{4}{3}y^2\sec^2 x(y^2\tan x)^{-7/3}$, $-\frac{8}{3}y\tan x(y^2\tan x)^{-7/3}$
25. $-6, -21$ 27. $1/\sqrt{17}, 8/\sqrt{17}$
29. **(a)** $2xy^4z^3 + y$ **(b)** $4x^2y^3z^3 + x$ **(c)** $3x^2y^4z^2 + 2z$ **(d)** $2y^4z^3 + y$ **(e)** $32z^3 + 1$ **(f)** 438
31. $2z/x, z/y, \ln(x^2y\cos z) - z\tan z$
33. $-y^2z^3/(1 + x^2y^4z^6)$, $-2xyz^3/(1 + x^2y^4z^6)$, $-3xy^2z^2/(1 + x^2y^4z^6)$
35. $yze^z\cos(xz)$, $e^z\sin(xz)$, $ye^z(\sin(xz) + x\cos(xz))$
37. $x/\sqrt{x^2 + y^2 + z^2}$, $y/\sqrt{x^2 + y^2 + z^2}$, $z/\sqrt{x^2 + y^2 + z^2}$
39. **(a)** e **(b)** $2e$ **(c)** e
41. **(a)** **(b)**

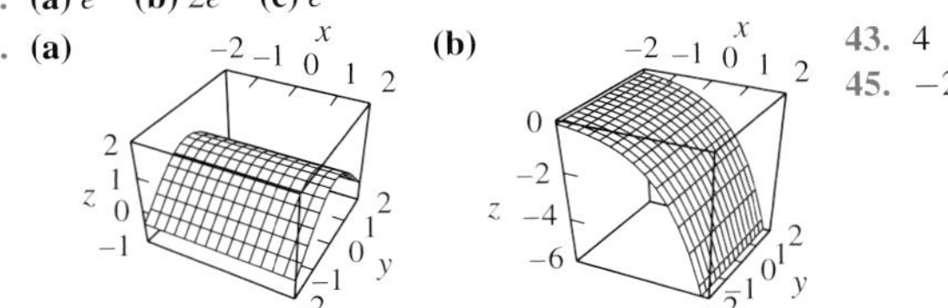

43. 4
45. −2
47. **(a)** $\partial V/\partial r = 2\pi rh$ **(b)** $\partial V/\partial h = \pi r^2$ **(c)** 48π **(d)** 64π
49. **(a)** $\dfrac{1}{5}\dfrac{\text{lb}}{\text{in}^2\cdot\text{K}}$ **(b)** $-\dfrac{25}{8}\dfrac{\text{in}^5}{\text{lb}}$
51. **(a)** $\dfrac{\partial V}{\partial \ell} = 6$ **(b)** $\dfrac{\partial V}{\partial w} = 15$ **(c)** $\dfrac{\partial V}{\partial h} = 10$
55. **(a)** $\pm\sqrt{6}/4$ 57. $-x/z, -y/z$
59. $-\dfrac{2x + yz^2\cos(xyz)}{xyz\cos(xyz) + \sin(xyz)}; -\dfrac{xz^2\cos(xyz)}{xyz\cos(xyz) + \sin(xyz)}$
61. $-x/w, -y/w, -z/w$
63. $-\dfrac{yzw\cos(xyz)}{2w + \sin(xyz)}, -\dfrac{xzw\cos(xyz)}{2w + \sin(xyz)}, -\dfrac{xyw\cos(xyz)}{2w + \cos(xyz)}$
65. $e^{x^2}, -e^{y^2}$
67. **(a)** $-\dfrac{\cos y}{4\sqrt{x^3}}$ **(b)** $-\sqrt{x}\cos y$ **(c)** $-\dfrac{1}{2\sqrt{x}}\sin y$ **(d)** $-\dfrac{1}{2\sqrt{x}}\sin y$
69. $-32y^3$ 71. $-e^x\sin y$ 73. $\dfrac{20}{(4x - 5y)^2}$ 75. $\dfrac{2(x - y)}{(x + y)^3}$
77. **(a)** $\dfrac{\partial^3 f}{\partial x^3}$ **(b)** $\dfrac{\partial^3 f}{\partial y^2\partial x}$ **(c)** $\dfrac{\partial^4 f}{\partial x^2\partial y^2}$ **(d)** $\dfrac{\partial^4 f}{\partial y^3\partial x}$
79. **(a)** $30xy^4 - 4$ **(b)** $60x^2y^3$ **(c)** $60x^3y^2$
81. **(a)** −30 **(b)** −125 **(c)** 150
83. **(a)** $15x^2y^4z^7 + 2y$ **(b)** $35x^3y^4z^6 + 3y^2$ **(c)** $21x^2y^5z^6$ **(d)** $42x^3y^5z^5$ **(e)** $140x^3y^3z^6 + 6y$ **(f)** $30xy^4z^7$ **(g)** $105x^2y^4z^6$ **(h)** $210xy^4z^6$
91. $\dfrac{\partial f}{\partial v} = 8vw^3x^4y^5, \dfrac{\partial f}{\partial w} = 12v^2w^2x^4y^5, \dfrac{\partial f}{\partial x} = 16v^2w^3x^3y^5,$ $\dfrac{\partial f}{\partial y} = 20v^2w^3x^4y^4$
93. $\dfrac{\partial f}{\partial v_1} = \dfrac{2v_1}{v_3^2 + v_4^2}, \dfrac{\partial f}{\partial v_2} = \dfrac{-2v_2}{v_3^2 + v_4^2}, \dfrac{\partial f}{\partial v_3} = \dfrac{-2v_3(v_1^2 - v_2^2)}{(v_3^2 + v_4^2)^2},$ $\dfrac{\partial f}{\partial v_4} = \dfrac{-2v_4(v_1^2 - v_2^2)}{(v_3^2 + v_4^2)^2}$
95. **(a)** 0 **(b)** 0 **(c)** 0 **(d)** 0 **(e)** $2(1 + yw)e^{yw}\sin z\cos z$ **(f)** $2xw(2 + yw)e^{yw}\sin z\cos z$
97. $-i\sin(x_1 + 2x_2 + \cdots + nx_n)$
99. **(a)** xy-plane, $12x^2 + 6x$ **(b)** $y \neq 0, -3x^2/y^2$
101. $f_x(2, -1) = 11, f_y(2, -1) = -8$
103. **(b)** does not exist if $y \neq 0$ and $x = -y$

▶ Exercise Set 14.4 (Page 969)

1. 5.04 3. 4.14 9. $dz = 7\,dx - 2\,dy$ 11. $dz = 3x^2y^2\,dx + 2x^3y\,dy$
13. $dz = \dfrac{y}{1 + x^2y^2}\,dx + \dfrac{x}{1 + x^2y^2}\,dy$ 15. $dw = 8\,dx - 3\,dy + 4\,dz$
17. $dw = 3x^2y^2z\,dx + 2x^3yz\,dy + x^3y^2\,dz$
19. $dw = \dfrac{yz}{1 + x^2y^2z^2}\,dx + \dfrac{xz}{1 + x^2y^2z^2}\,dy + \dfrac{xy}{1 + x^2y^2z^2}\,dz$
21. $df = 0.10, \Delta f = 0.1009$ 23. $df = 0.03, \Delta f \approx 0.029412$
25. $df = 0.96, \Delta f \approx 0.97929$
27. The increase in the area of the rectangle is given by the sum of the areas of the three small rectangles, and the total differential is given by the sum of the areas of the upper left and lower right rectangles.
29. **(a)** $L = \frac{1}{5} - \frac{4}{125}(x - 4) - \frac{3}{125}(y - 3)$ **(b)** 0.000176603
31. **(a)** $L = 0$ **(b)** 0.0024
33. **(a)** $L = 6 + 6(x - 1) + 3(y - 2) + 2(z - 3)$ **(b)** −0.000481
35. **(a)** $L = e + e(x - 1) - e(y + 1) - e(z + 1)$ **(b)** 0.01554
41. 0.5 43. 1, 1, −1, 2 45. (−1, 1) 47. (1, 0, 1) 49. 8%
51. r% 53. 0.3%
55. **(a)** $(r + s)$% **(b)** $(r + s)$% **(c)** $(2r + 3s)$% **(d)** $\left(3r + \dfrac{s}{2}\right)$%
57. ≈ 39 ft²

▶ Exercise Set 14.5 (Page 979)

1. $42t^{13}$ 3. $3t^{-2}\sin(1/t)$ 5. $-\frac{10}{3}t^{7/3}e^{1-t^{10/3}}$ 7. $\dfrac{dw}{dt} = 165t^{32}$
9. $-2t\cos t^2$ 11. 3264 13. 0
17. $24u^2v^2 - 16uv^3 - 2v + 3$, $16u^3v - 24u^2v^2 - 2u - 3$
19. $-\dfrac{2\sin u}{3\sin v}, -\dfrac{2\cos u\cos v}{3\sin^2 v}$ 21. $e^u, 0$
23. $3r^2\sin\theta\cos^2\theta - 4r^3\sin^3\theta\cos\theta$, $-2r^3\sin^2\theta\cos\theta + r^4\sin^4\theta + r^3\cos^3\theta - 3r^4\sin^2\theta\cos^2\theta$
25. $\dfrac{x^2 + y^2}{4x^2y^3}, \dfrac{y^2 - 3x^2}{4xy^4}$ 27. $\dfrac{\partial z}{\partial r} = \dfrac{2r\cos^2\theta}{r^2\cos^2\theta + 1}, \dfrac{\partial z}{\partial \theta} = \dfrac{-2r^2\cos\theta\sin\theta}{r^2\cos^2\theta + 1}$
29. $\dfrac{dw}{d\rho} = 2\rho(4\sin^2\phi + \cos^2\phi), \dfrac{\partial w}{\partial \phi} = 6\rho^2\sin\phi\cos\phi, \dfrac{dw}{d\theta} = 0$

31. $-\pi$ 33. $\sqrt{3}e^{\sqrt{3}}, (2-4\sqrt{3})e^{\sqrt{3}}$ 35. $-\dfrac{2xy^3}{3x^2y^2-\sin y}$

37. $-\dfrac{ye^{xy}}{xe^{xy}+ye^y+e^y}$ 41. $\dfrac{2x+yz}{6yz-xy}, \dfrac{xz-3z^2}{6yz-xy}$

43. $\dfrac{ye^x}{15\cos 3z+3}, \dfrac{e^x}{15\cos 3z+3}$ 45. -39 mi/h 47. $-\frac{7}{36}\sqrt{3}$ rad/s

49. $16{,}200\pi$ in^3/year 51. **(a)** 60 in^3/s **(b)** $\frac{26}{7}$ in/s

53. **(a)** 2 **(b)** 1 **(c)** 3 **(d)** -4

67. $\dfrac{\partial w}{\partial \rho} = (\sin\phi\cos\theta)\dfrac{\partial w}{\partial x} + (\sin\phi\sin\theta)\dfrac{\partial w}{\partial y} + (\cos\phi)\dfrac{\partial w}{\partial z}$,
$\dfrac{\partial w}{\partial \phi} = (\rho\cos\phi\cos\theta)\dfrac{\partial w}{\partial x} + (\rho\cos\phi\sin\theta)\dfrac{\partial w}{\partial y} - (\rho\sin\phi)\dfrac{\partial w}{\partial z}$,
$\dfrac{\partial w}{\partial \theta} = -(\rho\sin\phi\sin\theta)\dfrac{\partial w}{\partial x} + (\rho\sin\phi\cos\theta)\dfrac{\partial w}{\partial y}$

71. **(a)** $\dfrac{dw}{dt} = \sum_{i=1}^{4}\dfrac{\partial w}{\partial x_i}\dfrac{dx_i}{dt}$ **(b)** $\dfrac{\partial w}{\partial v_j} = \sum_{i=1}^{4}\dfrac{\partial w}{\partial x_i}\dfrac{\partial x_i}{\partial v_j}, j = 1, 2, 3$

► Exercise Set 14.6 (Page 990)

1. $6\sqrt{2}$ 3. $-3/\sqrt{10}$ 5. -320 7. $-314/741$ 9. 0 11. $-8\sqrt{2}$
13. $\sqrt{2}/4$ 15. $72/\sqrt{14}$ 17. $-8/63$ 19. $1/2+\sqrt{3}/8$ 21. $2\sqrt{2}$
23. $1/\sqrt{5}$ 25. $-\frac{3}{2}e$ 27. $3/\sqrt{11}$ 29. **(a)** 5 **(b)** 10 **(c)** $-5\sqrt{5}$
31. III 33. $4\mathbf{i}-8\mathbf{j}$
35. $\nabla w = \dfrac{x}{x^2+y^2+z^2}\mathbf{i} + \dfrac{y}{x^2+y^2+z^2}\mathbf{j} + \dfrac{z}{x^2+y^2+z^2}\mathbf{k}$
37. $-36\mathbf{i}-12\mathbf{j}$ 39. $4(\mathbf{i}+\mathbf{j}+\mathbf{k})$
41.

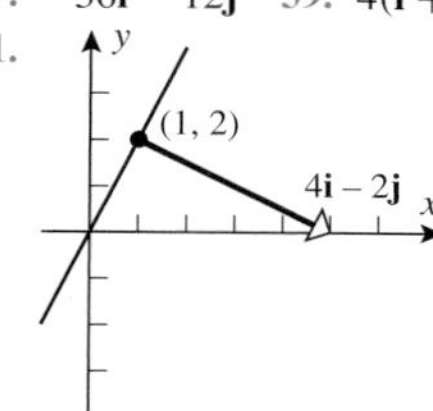

43.

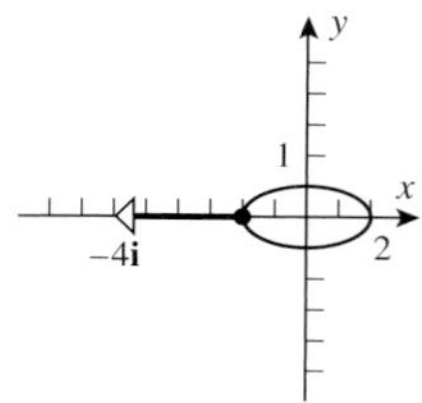

45. $\pm(-4\mathbf{i}+\mathbf{j})/\sqrt{17}$ 47. $\mathbf{u} = (3\mathbf{i}-2\mathbf{j})/\sqrt{13}, \|\nabla f(-1,1)\| = 4\sqrt{13}$
49. $\mathbf{u} = (4\mathbf{i}-3\mathbf{j})/5, \|\nabla f(4,-3)\| = 1$ 51. $\dfrac{1}{\sqrt{2}}(\mathbf{i}-\mathbf{j}), 3\sqrt{2}$
53. $\dfrac{1}{\sqrt{2}}(-\mathbf{i}+\mathbf{j}), \dfrac{1}{\sqrt{2}}$
55. $\mathbf{u} = -(\mathbf{i}+3\mathbf{j})/\sqrt{10}, -\|\nabla f(-1,-3)\| = -2\sqrt{10}$
57. $\mathbf{u} = (3\mathbf{i}-\mathbf{j})/\sqrt{10}, -\|\nabla f(\pi/6,\pi/4)\| = -\sqrt{5}$
59. $(\mathbf{i}-11\mathbf{j}+12\mathbf{k})/\sqrt{266}, -\sqrt{266}$ 61. $8/\sqrt{29}$
63. **(a)** $\approx 1/\sqrt{2}$
(b)

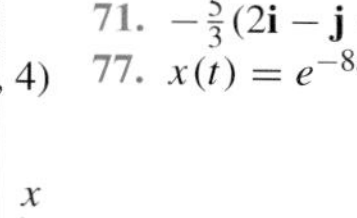

65. $9x^2+y^2=9$
67. $36/\sqrt{17}$
69. **(a)** $2e^{-\pi/2}\mathbf{i}$
71. $-\frac{5}{3}(2\mathbf{i}-\mathbf{j}-2\mathbf{k})$
77. $x(t) = e^{-8t}, y(t) = 4e^{-2t}$

79.

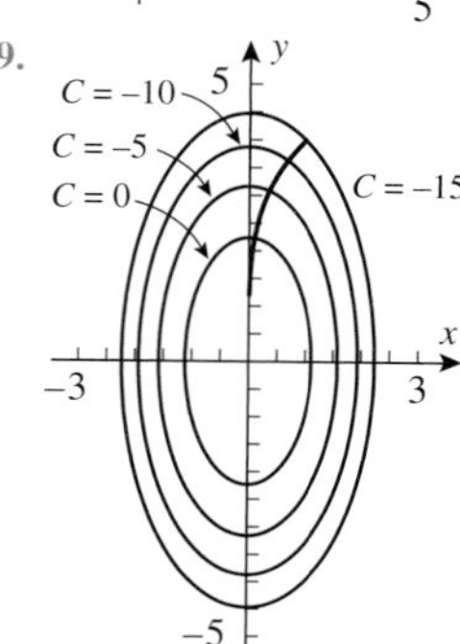

81. **(a)**

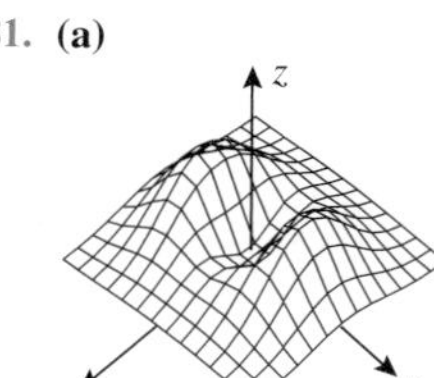

(c) $\nabla f = [2x-2x(x^2+3y^2)]e^{-(x^2+y^2)}\mathbf{i} + [6y-2y(x^2+3y^2)]e^{-(x^2+y^2)}\mathbf{j}$
(d) $x = y = 0$ or $x = 0, y = \pm 1$ or $x = \pm 1, y = 0$

► Exercise Set 14.7 (Page 998)

1. tangent plane: $48x-14y-z=64$;
normal line: $x = 1+48t, y = -2-14t, z = 12-t$
3. tangent plane: $x-y-z=0$;
normal line: $x = 1+t, y = -t, z = 1-t$
5. tangent plane: $3y-z=-1$;
normal line: $x = \pi/6, y = 3t, z = 1-t$
7. tangent plane: $3x-4z=-25$;
normal line: $x = -3+(3t/4), y = 0, z = 4-t$
9. **(a)** all points on the x-axis or y-axis **(b)** $(0,-2,-4)$
11. $\left(\frac{1}{2},-2,-\frac{3}{4}\right)$ 13. **(a)** $(-2,1,5), (0,3,9)$ **(b)** $\dfrac{4}{3\sqrt{14}}, \dfrac{4}{\sqrt{222}}$
15. **(a)** $x+y+2z=6$ **(b)** $x = 2+t, y = 2+t, z = 1+2t$
(c) $35.26°$
17. $\pm\dfrac{1}{\sqrt{365}}(\mathbf{i}-\mathbf{j}-19\mathbf{k})$ 21. $(1, 2/3, 2/3), (-1,-2/3,-2/3)$
23. $x = 1+8t, y = -1+5t, z = 2+6t$
25. $x = 3+4t, y = -3-4t, z = 4-3t$

► Exercise Set 14.8 (Page 1008)

1. **(a)** minimum at $(2,-1)$, no maxima
(b) maximum at $(0,0)$, no minima **(c)** no maxima or minima
3. minimum at $(3,-2)$, no maxima 5. relative minimum at $(0,0)$
7. relative minimum at $(0,0)$; saddle points at $(\pm 2, 1)$
9. saddle point at $(1,-2)$ 11. relative minimum at $(2,-1)$
13. relative minima at $(-1,-1)$ and $(1,1)$ 15. saddle point at $(0,0)$
17. no critical points 19. relative maximum at $(-1,0)$
21. saddle point at $(0,0)$; relative minima at $(1,1)$ and $(-1,-1)$

23. **(b)** relative minimum at $(0,0)$
27. absolute maximum 0, absolute minimum -12
29. absolute maximum 3, absolute minimum -1
31. absolute maximum $\frac{33}{4}$, absolute minimum $-\frac{1}{4}$
33. 16, 16, 16
35. maximum at $(1,2,2)$
37. $2a/\sqrt{3}, 2a/\sqrt{3}, 2a/\sqrt{3}$
39. length and width 2 ft, height 4 ft
41. **(a)** $x = 0$: minimum -3, maximum 0;
$x = 1$: minimum 3, maximum 13/3;
$y = 0$: minimum 0, maximum 4;
$y = 1$: minimum -3, maximum 3
(b) $y = x$: minimum 0, maximum 3;
$y = 1-x$: maximum 4, minimum -3
(c) minimum -3, maximum 13/3
43. length and width $\sqrt[3]{2V}$, height $\sqrt[3]{2V}/2$ 47. $y = \frac{3}{4}x + \frac{19}{12}$
49. $y = 0.5x + 0.8$

51. **(a)** $y = 63.73 + 0.2565t$ **(b)**

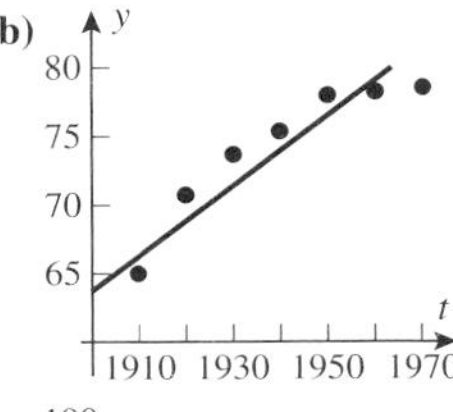

(c) about 84 years

53. **(a)** $P = \dfrac{2798}{21} + \dfrac{171}{350}T$ **(b)**

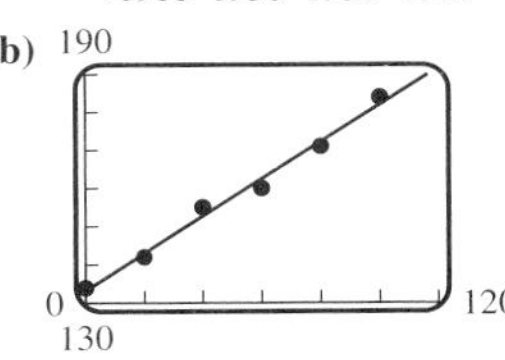

(c) $T \approx -272.7096°\text{C}$

▶ Exercise Set 14.9 (Page 1018)

1. **(a)** 4 3. **(a)**

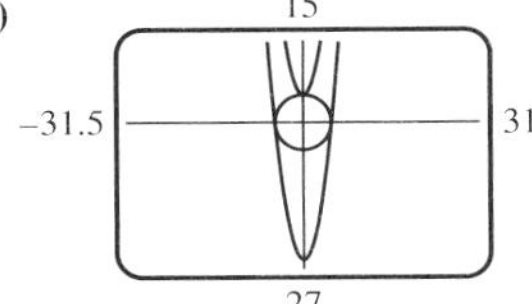

(c) maximum $\frac{101}{4}$, minimum -5

5. maximum $\sqrt{2}$ at $(-\sqrt{2}, -1)$ and $(\sqrt{2}, 1)$, minimum $-\sqrt{2}$ at $(-\sqrt{2}, 1)$ and $(\sqrt{2}, -1)$
7. maximum $\sqrt{2}$ at $(1/\sqrt{2}, 0)$, minimum $-\sqrt{2}$ at $(-1/\sqrt{2}, 0)$
9. maximum 6 at $\left(\frac{4}{3}, \frac{2}{3}, -\frac{4}{3}\right)$, minimum -6 at $\left(-\frac{4}{3}, -\frac{2}{3}, \frac{4}{3}\right)$
11. maximum is $1/(3\sqrt{3})$ at $(1/\sqrt{3}, 1/\sqrt{3}, 1/\sqrt{3})$, $(1/\sqrt{3}, -1/\sqrt{3}, -1/\sqrt{3})$, $(-1/\sqrt{3}, 1/\sqrt{3}, -1/\sqrt{3})$, and $(-1/\sqrt{3}, -1/\sqrt{3}, 1/\sqrt{3})$; minimum is $-1/(3\sqrt{3})$ at $(1/\sqrt{3}, 1/\sqrt{3}, -1/\sqrt{3})$, $(1/\sqrt{3}, -1/\sqrt{3}, 1/\sqrt{3})$, $(-1/\sqrt{3}, 1/\sqrt{3}, 1/\sqrt{3})$, and $(-1/\sqrt{3}, -1/\sqrt{3}, -1/\sqrt{3})$
13. $\left(\frac{3}{10}, -\frac{3}{5}\right)$ 15. $\left(\frac{1}{6}, \frac{1}{3}, \frac{1}{6}\right)$
17. (3, 6) is closest and (−3, −6) is farthest 19. $5(\mathbf{i} + \mathbf{j} + \mathbf{k})/\sqrt{3}$
21. 9, 9, 9 23. $(\pm\sqrt{5}, 0, 0)$ 25. length and width 2 ft, height 4 ft
29. **(a)** $\alpha = \beta = \gamma = \pi/3$, maximum 1/8 **(b)**

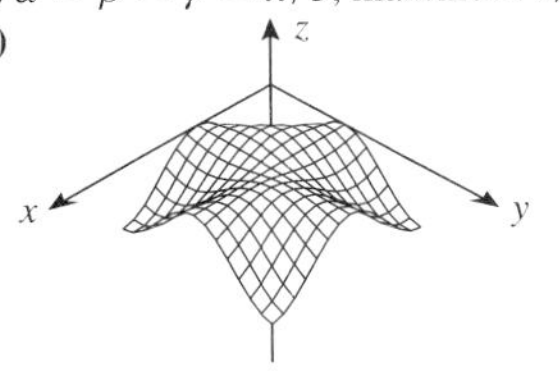

▶ Chapter 14. Review Exercises (Page 1019)

1. **(a)** xy **(b)** $e^{r+s}\ln(rs)$
5. **(a)** not defined on line $y = x$ **(b)** not continuous
9. **(a)** 12 Pa/min **(b)** 240 Pa/min
15. df (the differential of f) is an approximation for Δf (the change in f)
17. $dV = -0.06667\ \text{m}^3$; $\Delta V = -0.07267\ \text{m}^3$ 19. 2
21. $\dfrac{-f_y^2 f_{xx} + 2f_x f_y f_{xy} - f_x^2 f_{yz}}{f_y^3}$ 25. $\dfrac{7}{2} + \dfrac{4}{5}\ln 2$ 27. $-7/\sqrt{5}$
29. (0, 0, 2), (1, 1, 1), (−1, −1, 1) 31. $\left(-\frac{1}{3}, -\frac{1}{2}, 2\right)$
33. relative minimum at (15, −8)
35. saddle point at (0, 0), relative minimum at (3, 9)
37. absolute maximum of 4 at (±1, ±2), absolute minimum of 0 at $(\pm\sqrt{2}, 0)$ and $(0, \pm 2\sqrt{2})$
39. $I_1 : I_2 : I_3 = \dfrac{1}{R_1} : \dfrac{1}{R_2} : \dfrac{1}{R_3}$
41. **(a)** $\partial P/\partial L = c\alpha L^{\alpha-1}K^{\beta}$, $\partial P/\partial K = c\beta L^{\alpha}K^{\beta-1}$

▶ Exercise Set 15.1 (Page 1028)

1. 7 3. 2 5. 2 7. 3 9. $1 - \ln 2$ 11. $\dfrac{1 - \ln 2}{2}$ 13. 0 15. $\frac{1}{3}$
17. **(a)** 37/4 **(b)** exact value = 28/3; differ by 1/12
19. **(a)** **(b)**

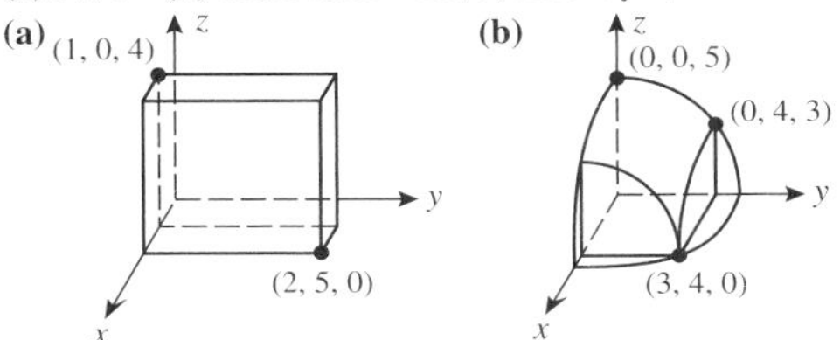

21. 19 23. 8 25. $\dfrac{1}{3\pi}$ 27. $1 - \dfrac{2}{\pi}$ 29. $\frac{14}{3}$ °C 31. 1.381737122
35. first integral equals $\frac{1}{2}$, second equals $-\frac{1}{2}$; no

▶ Exercise Set 15.2 (Page 1037)

1. $\frac{1}{40}$ 3. 9 5. $\dfrac{\pi}{2}$ 7. 1 9. $\frac{1}{12}$
11. **(a)** $\displaystyle\int_0^2\int_0^{x^2} f(x, y)\,dy\,dx$ **(b)** $\displaystyle\int_0^4\int_{\sqrt{y}}^2 f(x, y)\,dx\,dy$
13. **(a)** $\displaystyle\int_1^2\int_{-2x+5}^3 f(x, y)\,dy\,dx + \int_2^4\int_1^3 f(x, y)\,dy\,dx + \int_4^5\int_{2x-7}^3 f(x, y)\,dy\,dx$ **(b)** $\displaystyle\int_1^3\int_{(5-y)/2}^{(y+7)/2} f(x, y)\,dx\,dy$
15. **(a)** 16/3 **(b)** 38 17. 576 19. 0
21. $\dfrac{\sqrt{17} - 1}{2}$ 23. $\frac{50}{3}$ 25. $-\frac{7}{60}$
27. **(a)**

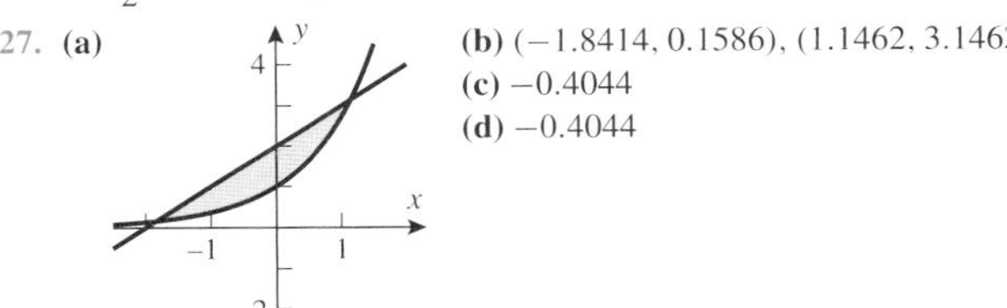

(b) (−1.8414, 0.1586), (1.1462, 3.1462) **(c)** −0.4044 **(d)** −0.4044

29. $\sqrt{2} - 1$ 31. 32 33. 12 35. 27π 37. 170 39. $\dfrac{27\pi}{2}$ 41. $\frac{2000}{3}$
43. $\dfrac{\pi}{2}$ 45. $\displaystyle\int_0^{\sqrt{2}}\int_{y^2}^2 f(x, y)\,dx\,dy$ 47. $\displaystyle\int_1^{e^2}\int_{\ln x}^2 f(x, y)\,dy\,dx$
49. $\displaystyle\int_0^{\pi/2}\int_0^{\sin x} f(x, y)\,dy\,dx$ 51. $\dfrac{1 - e^{-16}}{8}$ 53. $\dfrac{e^8 - 1}{3}$
55. $\dfrac{1 - \cos 8}{3}$ 57. **(a)** 0 **(b)** tan 1 59. 0 61. $\dfrac{\pi}{2} - \ln 2$ 63. $\frac{2}{3}$ °C
65. 0.676089

▶ Exercise Set 15.3 (Page 1045)

1. $\frac{1}{6}$ 3. $\frac{2}{9}a^3$ 5. 0 7. $\dfrac{3\pi}{2}$ 9. $\dfrac{\pi}{16}$ 11. $\displaystyle\int_{\pi/6}^{5\pi/6}\int_2^{4\sin\theta} f(r, \theta)r\,dr\,d\theta$
13. $8\displaystyle\int_0^{\pi/2}\int_1^3 r\sqrt{9 - r^2}\,dr\,d\theta$ 15. $2\displaystyle\int_0^{\pi/2}\int_0^{\cos\theta}(1 - r^2)r\,dr\,d\theta$
17. $\dfrac{64\sqrt{2}}{3}\pi$ 19. $\dfrac{5\pi}{32}$ 21. $\dfrac{27\pi}{16}$ 23. $(1 - e^{-1})\pi$ 25. $\dfrac{\pi}{8}\ln 5$ 27. $\dfrac{\pi}{8}$
29. $\frac{16}{9}$ 31. $\dfrac{\pi}{2}\left(1 - \dfrac{1}{\sqrt{1 + a^2}}\right)$ 33. $\dfrac{\pi}{4}(\sqrt{5} - 1)$ 35. $\pi a^2 h$
37. $\dfrac{(3\pi - 4)a^2 c}{9}$ 39. $\dfrac{4\pi}{3} + 2\sqrt{3} - 2$ 41. **(b)** $\dfrac{\pi}{4}$
43. **(a)** 1.173108605 **(b)** 1.173108605 45. $\dfrac{1}{5} + \dfrac{\pi}{2}$

Exercise Set 15.4 (Page 1057)

1. (a) 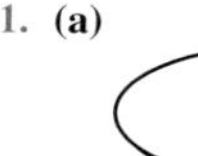 (b) 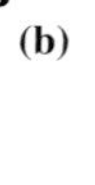 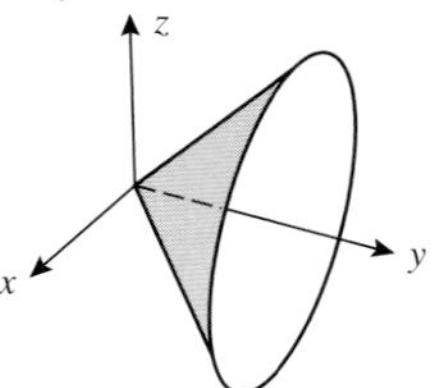

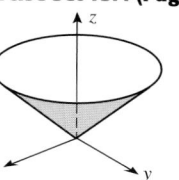

(c) 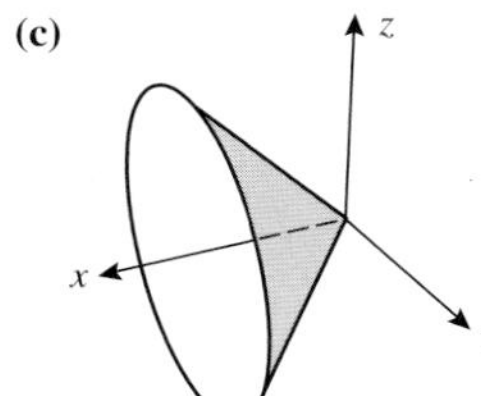

3. (a) $x = u,\ y = v,\ z = \frac{5}{2} + \frac{3}{2}u - 2v$ (b) $x = u,\ y = v,\ z = u^2$
5. (a) $x = \sqrt{5}\cos u,\ y = \sqrt{5}\sin u,\ z = v;\ 0 \le u \le 2\pi,\ 0 \le v \le 1$
   (b) $x = 2\cos u,\ y = v,\ z = 2\sin u;\ 0 \le u \le 2\pi,\ 1 \le v \le 3$
7. $x = u,\ y = \sin u \cos v,\ z = \sin u \sin v$
9. $x = r\cos\theta,\ y = r\sin\theta,\ z = \dfrac{1}{1+r^2}$
11. $x = r\cos\theta,\ y = r\sin\theta,\ z = 2r^2\cos\theta\sin\theta$
13. $x = r\cos\theta,\ y = r\sin\theta,\ z = \sqrt{9-r^2};\ r \le \sqrt{5}$
15. $x = \dfrac{1}{2}\rho\cos\theta,\ y = \dfrac{1}{2}\rho\sin\theta,\ z = \dfrac{\sqrt{3}}{2}\rho$ 17. $z = x - 2y$; a plane
19. $(x/3)^2 + (y/2)^2 = 1;\ 2 \le z \le 4$; part of an elliptic cylinder
21. $(x/3)^2 + (y/4)^2 = z^2;\ 0 \le z \le 1$; part of an elliptic cone
23. (a) $x = r\cos\theta,\ y = r\sin\theta,\ z = r,\ 0 \le r \le 2$;
    $x = u,\ y = v,\ z = \sqrt{u^2+v^2};\ 0 \le u^2 + v^2 \le 4$
25. (a) $0 \le u \le 3,\ 0 \le v \le \pi$ (b) $0 \le u \le 4,\ -\pi/2 \le v \le \pi/2$
27. (a) $0 \le \phi \le \pi/2,\ 0 \le \theta \le 2\pi$ (b) $0 \le \phi \le \pi,\ 0 \le \theta \le \pi$
29. $2x + 4y - z = 5$ 31. $z = 0$ 33. $x - y + \dfrac{\sqrt{2}}{2}z = \dfrac{\pi\sqrt{2}}{8}$
35. 6π 37. $\dfrac{\sqrt{5}}{6}$ 39. $\dfrac{(5\sqrt{5}-1)\pi}{6}$ 41. $\dfrac{(17\sqrt{17}-5\sqrt{5})\pi}{6}$
43. $\dfrac{(10\sqrt{10}-1)\pi}{18}$ 45. 8π 47. $4\pi a^2$
51. $4\pi^2 ab$ 53. 9.099 55. (a) an ellipsoid (b) 111.55
57. $(x/a)^2 + (y/b)^2 + (z/c)^2 = 1$; ellipsoid
59. $(x/a)^2 + (y/b)^2 - (z/c)^2 = -1$; hyperboloid of two sheets

Exercise Set 15.5 (Page 1066)

1. 8 3. $\frac{47}{3}$ 5. $\frac{81}{5}$ 7. $\frac{128}{15}$ 9. $\pi(\pi-3)/2$ 11. $\frac{1}{6}$ 13. 9.425
15. 4 17. $\frac{256}{15}$
19. (a) $\displaystyle\int_{-1}^{1}\int_{-\sqrt{1-x^2}}^{\sqrt{1-x^2}}\int_{4x^2+y^2}^{4-3y^2} f(x,y,z)\,dz\,dy\,dx$
    (b) $\displaystyle\int_{-1}^{1}\int_{-\sqrt{1-y^2}}^{\sqrt{1-y^2}}\int_{4x^2+y^2}^{4-3y^2} f(x,y,z)\,dz\,dy\,dx$
21. $\displaystyle 4\int_{0}^{1}\int_{0}^{\sqrt{1-x^2}}\int_{4x^2+y^2}^{4-3y^2} dz\,dy\,dx$
23. $\displaystyle 2\int_{-3}^{3}\int_{0}^{\frac{1}{3}\sqrt{9-x^2}}\int_{0}^{x+3} dz\,dy\,dx$
25. (a)  (b) (c) 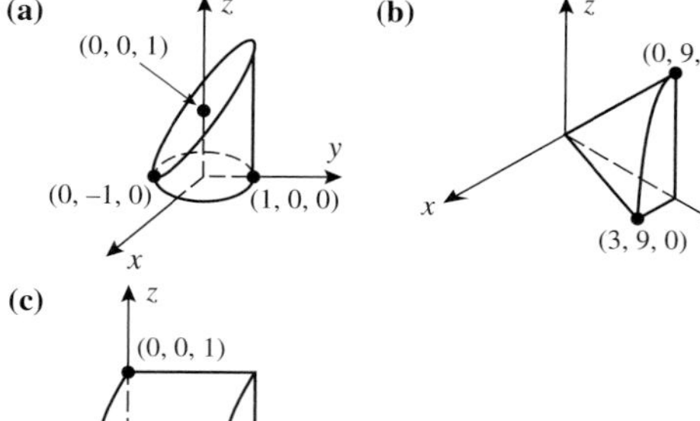

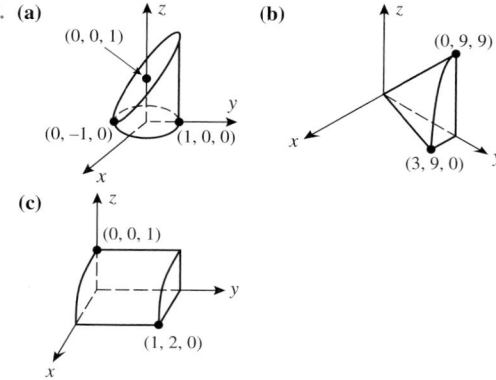

27. $\frac{3}{4}$ 29. 3.291
31. (a) $\displaystyle\int_0^a\int_0^{b(1-x/a)}\int_0^{c(1-x/a-y/z)} dz\,dy\,dx$ is one example
33. (a) $\displaystyle\int_0^2\int_0^{\sqrt{4-x^2}}\int_0^5 f(x,y,z)\,dz\,dy\,dx$
    (b) $\displaystyle\int_0^9\int_0^{3-\sqrt{x}}\int_y^{3-\sqrt{x}} f(x,y,z)\,dz\,dy\,dx$
    (c) $\displaystyle\int_0^2\int_0^{4-x^2}\int_0^{8-y} f(x,y,z)\,dz\,dy\,dx$
35. (a) the sphere $0 \le x^2 + y^2 + z^2 \le 1$ (b) 4.934802202 (c) $\pi^2/2$
37. (a) 0 (b) $\dfrac{e^2-1}{2}$

Exercise Set 15.6 (Page 1077)

1. (a) positive: m_2 is at the fulcrum, so it can be ignored; masses m_1 and m_3 are equidistant from position 5, but $m_1 < m_3$, so the beam will rotate clockwise. (b) The fulcrum should be placed $\frac{50}{7}$ units to the right of m_1
3. $\left(\frac{1}{2}, \frac{1}{2}\right)$ 5. $\left(\frac{2}{3}, \frac{1}{3}\right)$ 7. $\left(\frac{5}{14}, \frac{38}{35}\right)$ 9. $\left(0, \dfrac{4(b^3-a^3)}{3\pi(b^2-a^2)}\right)$ 11. $\left(\frac{1}{2}, \frac{1}{2}\right)$
13. $M = \frac{13}{20}$, center of gravity $\left(\frac{190}{273}, \frac{6}{13}\right)$
15. $M = a^4/8$, center of gravity $(8a/15, 8a/15)$
17. $\left(\frac{1}{2}, \frac{1}{2}, \frac{1}{2}\right)$ 19. $\left(\frac{1}{4}, \frac{1}{4}, \frac{1}{4}\right)$ 21. $\left(\frac{1}{2}, 0, \frac{3}{5}\right)$ 23. $(3a/8, 3a/8, 3a/8)$
25. $M = a^4/2$, center of gravity $(a/3, a/2, a/2)$
27. $M = \frac{1}{6}$, center of gravity $\left(0, \frac{16}{35}, \frac{1}{2}\right)$ 29. (a) $\left(\frac{5}{8}, \frac{5}{8}\right)$ (b) $\left(\frac{2}{3}, \frac{1}{2}\right)$
31. $(1.177406, 0.353554, 0.231557)$ 35. $\left(\dfrac{128}{105\pi}, \dfrac{128}{105\pi}\right)$
39. $2\pi^2 abk$ 41. $(a/3, b/3)$

Exercise Set 15.7 (Page 1088)

1. $\dfrac{\pi}{4}$ 3. $\dfrac{\pi}{16}$
5. the region is bounded by the xy-plane and the upper half of a sphere of radius 1 centered at the origin; $f(r, \theta, z) = z$
7. the region is the portion of the first octant inside a sphere of radius 1 centered at the origin; $f(\rho, \theta, \phi) = \rho\cos\phi$
9. $\dfrac{81\pi}{2}$ 11. $\frac{152}{3}\pi + \frac{80}{3}\pi\sqrt{5}$ 13. $\dfrac{64\pi}{3}$ 15. $\dfrac{11\pi a^3}{3}$ 17. $\dfrac{\pi a^6}{48}$
19. $\dfrac{32(2\sqrt{2}-1)\pi}{15}$
21. (a) $\frac{5}{2}(-8 + 3\ln 3)\ln(\sqrt{5}-2)$ (b) $f(x,y,z) = \dfrac{y^3}{x^3\sqrt{1+z^2}}$;
    G is the cylindrical wedge $1 \le r \le 4,\ \dfrac{\pi}{6} \le \theta \le \dfrac{\pi}{3},\ -2 \le z \le 2$
23. $\dfrac{4\pi a^3}{3}$ 25. $\dfrac{27\pi}{4}$ 27. $\pi k a^4$
29. $\left(0, 0, \dfrac{7}{16\sqrt{2}-14}\right)$ 31. $(3a/8, 3a/8, 3a/8)$

33. $\left(\frac{4}{3}, 0, \frac{10}{9}\right)$ 35. $\dfrac{2(\sqrt{3}-1)\pi}{3}$ 37. $\left(0, 0, \frac{11}{30}\right)$ 39. $(0, 0, 2a/5)$
41. $\frac{4}{3}\pi(1-e^{-1})\delta_0 R^3$ 43. $\frac{1}{2}\delta\pi a^4 h$ 45. $\frac{1}{2}\delta\pi h(a_2^4 - a_1^4)$

▶ Exercise Set 15.8 (Page 1101)

1. -17 3. $\cos(u-v)$ 5. $x = \frac{2}{9}u + \frac{5}{9}v,\ y = -\frac{1}{9}u + \frac{2}{9}v;\ \frac{1}{9}$
7. $x = \dfrac{\sqrt{u+v}}{\sqrt{2}},\ y = \dfrac{\sqrt{v-u}}{\sqrt{2}};\ \dfrac{1}{4\sqrt{v^2-u^2}}$ 9. 5 11. $\dfrac{1}{v}$

13.

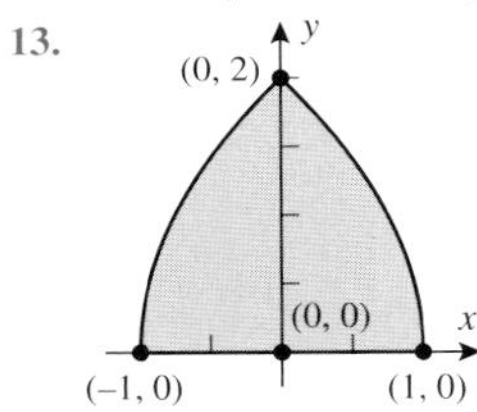

15.

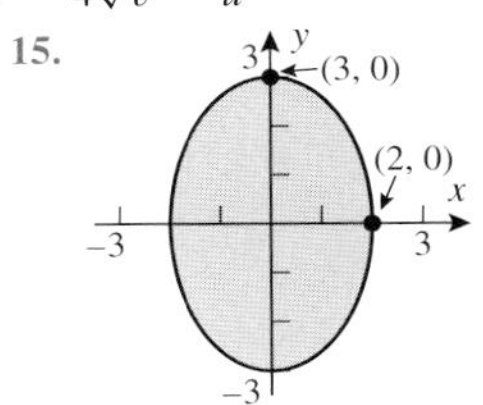

17. $\frac{3}{2}\ln 3$ 19. $1 - \frac{1}{2}\sin 2$ 21. 96π 23. $\dfrac{\pi}{24}(1-\cos 1)$ 25. $\frac{192}{5}\pi$
27. $u = \cot^{-1}(x/y),\ v = \sqrt{x^2+y^2}$
29. $u = (3/7)x - (2/7)y;\ v = (-1/7)x + (3/7)y$ 31. $\frac{1}{4}\ln\frac{5}{2}$
33. $\dfrac{1}{2}\left[\ln(\sqrt{2}+1) - \dfrac{\pi}{4}\right]$ 35. $\frac{35}{256}$ 37. $2\ln 3$ 39. $\dfrac{4}{35}\pi a^3$ 43. $21/8$

▶ Chapter 15. Review Exercises (Page 1104)

3. (a) $\displaystyle\iint_R dA$ (b) $\displaystyle\iiint_G dV$ (c) $\displaystyle\iint_R \sqrt{1 + \left(\frac{\partial z}{\partial x}\right)^2 + \left(\frac{\partial z}{\partial y}\right)^2}\,dA$

7. $\displaystyle\int_0^1 \int_{1-\sqrt{1-y^2}}^{1+\sqrt{1-y^2}} f(x,y)\,dx\,dy$

9. (a) $a = 2, b = 1, c = 1, d = 2$ (b) 3

11. $-\dfrac{1}{\sqrt{2}\pi}$

13. $\displaystyle\int_0^1 \int_{2y}^2 e^x e^y\,dx\,dy$

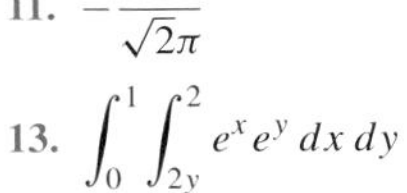

15.

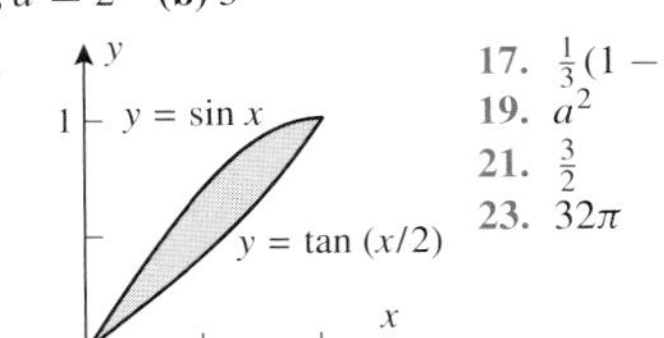

17. $\frac{1}{3}(1-\cos 64)$
19. a^2
21. $\frac{3}{2}$
23. 32π

25. (a) $\displaystyle\int_0^{2\pi}\int_0^{\pi/3}\int_0^a \rho^4 \sin^3\phi\,d\rho\,d\phi\,d\theta$

(b) $\displaystyle\int_0^{2\pi}\int_0^{\sqrt{3}a/2}\int_{r/\sqrt{3}}^{\sqrt{a^2-r^2}} r^3\,dz\,dr\,d\theta$

(c) $\displaystyle\int_{-\sqrt{3}a/2}^{\sqrt{3}a/2}\int_{-\sqrt{(3a^2/4)-x^2}}^{\sqrt{(3a^2/4)-x^2}}\int_{\sqrt{x^2+y^2}/\sqrt{3}}^{\sqrt{a^2-x^2-y^2}} (x^2+y^2)\,dz\,dy\,dx$

27. $\dfrac{\pi a^3}{9}$ 29. $\dfrac{8\pi}{3}(3\sqrt{3}-1)$ 31. $2x + 4y - z = 5$ 33. $\left(\frac{8}{5}, 0\right)$
35. $(0, 0, h/4)$ 37. $\dfrac{3}{4}a$ 39. (a) $\dfrac{1}{2(u+w)}$ (b) $\frac{1}{2}(7\ln 7 - \ln 84, 375)$

▶ Exercise Set 16.1 (Page 1114)

1. (a) III (b) IV 3. (a) true (b) true (c) true

5. 7.

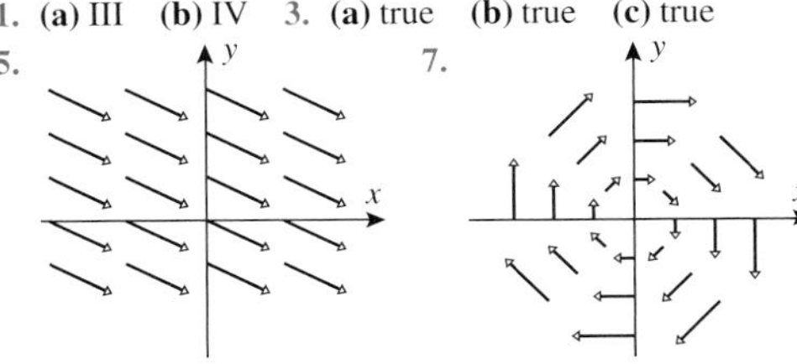

9. (vector field graph with axes x, y)

11. (a) all x, y (b) all x, y
13. div $\mathbf{F} = 2x + y$, curl $\mathbf{F} = z\mathbf{i}$

15. div $\mathbf{F} = 0$, curl $\mathbf{F} = (40x^2z^4 - 12xy^3)\mathbf{i} + (14y^3z + 3y^4)\mathbf{j} - (16xz^5 + 21y^2z^2)\mathbf{k}$
17. div $\mathbf{F} = \dfrac{2}{\sqrt{x^2+y^2+z^2}}$, curl $\mathbf{F} = \mathbf{0}$ 19. $4x$ 21. 0
23. $(1+y)\mathbf{i} + x\mathbf{j}$
35. $\nabla\cdot(k\mathbf{F}) = k\nabla\cdot\mathbf{F},\ \nabla\cdot(\mathbf{F}+\mathbf{G}) = \nabla\cdot\mathbf{F} + \nabla\cdot\mathbf{G},\ \nabla\cdot(\phi\mathbf{F}) = \phi\nabla\cdot\mathbf{F} + \nabla\phi\cdot\mathbf{F},\ \nabla\cdot(\nabla\times\mathbf{F}) = 0$ 43. (b) $x^2 + y^2 = K$
45. $\dfrac{dy}{dx} = \dfrac{1}{x},\ y = \ln x + K$

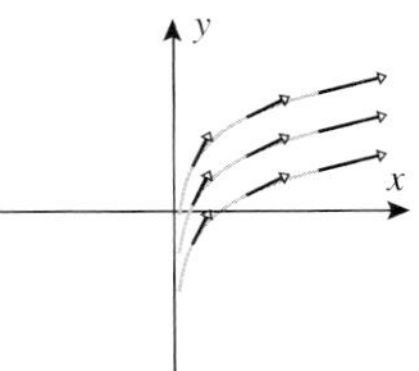

▶ Exercise Set 16.2 (Page 1130)

1. (a) 1 (b) 0 3. 16
7. (a) $-\frac{11}{108}\sqrt{10} - \frac{1}{36}\ln(\sqrt{10}-3) - \frac{4}{27}$ (b) 0 (c) $-\frac{1}{2}$
9. (a) 3 (b) 3 (c) 3 (d) 3 11. 2 13. $\frac{13}{20}$ 15. $1-\pi$ 17. 3
19. $-1-(\pi/4)$ 21. $1-e^3$
23. (a) $63\sqrt{17}/64 + \dfrac{1}{4}\ln(4+\sqrt{17}) - \dfrac{1}{8}\ln\dfrac{\sqrt{17}+1}{\sqrt{17}-1} - \dfrac{1}{4}\ln(\sqrt{2}+1) + \dfrac{1}{8}\ln\dfrac{\sqrt{2}+1}{\sqrt{2}-1}$ (b) $1/2 - \pi/4$
25. (a) -1 (b) -2 27. $\frac{5}{2}$ 29. 0 31. $1-e^{-1}$ 33. $6\sqrt{3}$
35. $5k\tan^{-1}3$ 37. $\frac{3}{5}$ 39. $\frac{27}{28}$ 41. $\frac{3}{4}$ 43. $\dfrac{17\sqrt{17}-1}{4}$
45. (b) $S = \displaystyle\int_C z(t)\,dt$ (c) 4π 47. $\lambda = -12$

▶ Exercise Set 16.3 (Page 1141)

1. conservative, $\phi = \dfrac{x^2}{2} + \dfrac{y^2}{2} + K$ 3. not conservative
5. conservative, $\phi = x\cos y + y\sin x + K$
7. (b) 13 9. -6 11. $9e^2$ 13. 32 15. $W = -\frac{1}{2}$
17. $W = 1 - e^{-1}$ 19. $\ln 2 - 1$ 21. ≈ -0.307 23. no
29. $h(x) = Ce^x$
31. (a) $W = -\dfrac{1}{\sqrt{14}} + \dfrac{1}{\sqrt{16}}$ (b) $W = -\dfrac{1}{\sqrt{14}} + \dfrac{1}{\sqrt{6}}$ (c) $W = 0$

▶ Exercise Set 16.4 (Page 1149)

1. 0 3. 0 5. 0 7. 8π 9. -4 11. -1 13. 0
15. (a) ≈ -3.550999378 (b) ≈ -0.269616482 17. $\frac{3}{8}a^2\pi$ 19. $\frac{1}{2}abt_0$
23. Formula (1) of Section 7.1 25. $\frac{250}{3}$ 27. $-3\pi a^2$ 29. $\left(\frac{8}{15}, \frac{8}{21}\right)$
31. $\left(0, \dfrac{4a}{3\pi}\right)$ 33. the circle $x^2 + y^2 = 1$ 35. 69

▶ Exercise Set 16.5 (Page 1157)

1. $\dfrac{15}{2}\pi\sqrt{2}$ 3. $\dfrac{\pi}{4}$ 5. $-\dfrac{\sqrt{2}}{2}$ 7. 9
9. (b) $2\pi\left[1 - \sqrt{1-r^2} + \dfrac{r^2}{2}\right] \to 3\pi$ as $r \to 1^-$
(c) $\mathbf{r}(\phi,\theta) = \sin\phi\cos\theta\mathbf{i} + \sin\phi\sin\theta\mathbf{j} + \cos\phi\mathbf{k}$, $0 \le \theta \le 2\pi,\ 0 \le \phi \le \pi/2$;

$$\iint (1+z)\,dS = \int_0^{2\pi}\int_0^{\pi/2} (1+\cos\phi)\sin\phi\, d\phi\, d\theta = 3\pi$$

13. **(c)** $4\pi/3$

15. **(a)** $\dfrac{\sqrt{29}}{16}\displaystyle\int_0^6\int_0^{(12-2x)/3} xy(12-2x-3y)\,dy\,dx$

(b) $\dfrac{\sqrt{29}}{4}\displaystyle\int_0^3\int_0^{(12-4z)/3} yz(12-3y-4z)\,dy\,dz$

(c) $\dfrac{\sqrt{29}}{9}\displaystyle\int_0^3\int_0^{6-2z} xz(12-2x-4z)\,dx\,dz$

17. $\dfrac{18\sqrt{29}}{5}$

19. $\displaystyle\int_0^4\int_1^2 y^3 z\sqrt{4y^2+1}\,dy\,dz;\ \frac{1}{2}\int_0^4\int_1^4 xz\sqrt{1+4x}\,dx\,dz$

21. $\dfrac{391\sqrt{17}}{15} - \dfrac{5\sqrt{5}}{3}$ 23. $\frac{4}{3}\pi\delta_0$ 25. $\frac{1}{4}(37\sqrt{37}-1)$ 27. $M = \delta_0 S$

29. $(0, 0, 149/65)$ 31. $\dfrac{93}{\sqrt{10}}$ 33. $\dfrac{\pi}{4}$ 35. 57.895751

▶ Exercise Set 16.6 (Page 1166)

1. **(a)** zero **(b)** zero **(c)** positive **(d)** negative **(e)** zero **(f)** zero
3. **(a)** positive **(b)** zero **(c)** positive **(d)** zero **(e)** positive **(f)** zero
5. **(a)** $\mathbf{n} = -\cos v\mathbf{i} - \sin v\mathbf{j}$ **(b)** inward 7. 2π 9. $\dfrac{14\pi}{3}$ 11. 0
13. 18π 15. $\frac{4}{9}$ 17. **(a)** 8 **(b)** 24 **(c)** 0 19. 3π
21. **(a)** $0\ \text{m}^3/\text{s}$ **(b)** 0 kg/s 23. **(b)** $32/3$
25. **(a)** $4\pi a^{k+3}$ **(b)** $k = -3$

▶ Exercise Set 16.7 (Page 1176)

1. 3 3. $\dfrac{4\pi}{3}$ 5. 12 7. $3\pi a^2$ 9. 180π 11. $\dfrac{192\pi}{5}$ 13. $\dfrac{\pi}{2}$
15. $\dfrac{4608}{35}$ 17. 135π 29. no sources or sinks
31. sources at all points except the origin, no sinks 33. $\dfrac{7\pi}{4}$

▶ Exercise Set 16.8 (Page 1183)

1. $\frac{3}{2}$ 3. 0 5. 2π 7. 16π 9. 0 11. πa^2
13. **(a)** $\frac{3}{2}$ **(b)** -1 **(c)** $-\dfrac{1}{\sqrt{2}}\mathbf{j} - \dfrac{1}{\sqrt{2}}\mathbf{k}$ 19. $-\dfrac{5\pi}{4}$

▶ Chapter 16. Review Exercises (Page 1185)

3. $\dfrac{1-x}{\sqrt{(1-x)^2+(2-y)^2}}\mathbf{i} + \dfrac{2-y}{\sqrt{(1-x)^2+(2-y)^2}}\mathbf{j}$ 5. $\mathbf{i}+\mathbf{j}+\mathbf{k}$

7. **(a)** $\displaystyle\int_a^b \left[f(x(t), y(t))\frac{dx}{dt} + g(x(t), y(t))\frac{dy}{dt}\right] dt$

(b) $\displaystyle\int_a^b f(x(t), y(t))\sqrt{x'(t)^2 + y'(t)^2}\,dt$

11. 0 13. $-7/2$ 17. **(a)** $h(x) = Cx^{-3/2}$ **(b)** $g(y) = C/y^3$

23. $\displaystyle\iint_R f(x(u,v), y(u,v), z(u,v))\|\mathbf{r}_u \times \mathbf{r}_v\|\,du\,dv$ 25. yes 27. 2π

31. -8π 35. **(a)** conservative **(b)** not conservative

▶ Appendix A (Page A1)

1. **(a)** $\frac{5}{12}\pi$ **(b)** $\frac{13}{6}\pi$ **(c)** $\frac{1}{9}\pi$ **(d)** $\frac{23}{30}\pi$
3. **(a)** $12°$ **(b)** $(270/\pi)°$ **(c)** $288°$ **(d)** $540°$
5.

	$\sin\theta$	$\cos\theta$	$\tan\theta$	$\csc\theta$	$\sec\theta$	$\cot\theta$
(a)	$\sqrt{21}/5$	$2/5$	$\sqrt{21}/2$	$5/\sqrt{21}$	$5/2$	$2/\sqrt{21}$
(b)	$3/4$	$\sqrt{7}/4$	$3/\sqrt{7}$	$4/3$	$4/\sqrt{7}$	$\sqrt{7}/3$
(c)	$3/\sqrt{10}$	$1/\sqrt{10}$	3	$\sqrt{10}/3$	$\sqrt{10}$	$1/3$

7. $\sin\theta = 3/\sqrt{10}$, $\cos\theta = 1/\sqrt{10}$ 9. $\tan\theta = \sqrt{21}/2$, $\csc\theta = 5/\sqrt{21}$
11. 1.8

13.

	θ	$\sin\theta$	$\cos\theta$	$\tan\theta$	$\csc\theta$	$\sec\theta$	$\cot\theta$
(a)	$225°$	$-1/\sqrt{2}$	$-1/\sqrt{2}$	1	$-\sqrt{2}$	$-\sqrt{2}$	1
(b)	$-210°$	$1/2$	$-\sqrt{3}/2$	$-1/\sqrt{3}$	2	$-2/\sqrt{3}$	$-\sqrt{3}$
(c)	$5\pi/3$	$-\sqrt{3}/2$	$1/2$	$-\sqrt{3}$	$-2/\sqrt{3}$	2	$-1/\sqrt{3}$
(d)	$-3\pi/2$	1	0	—	1	—	0

15.

	$\sin\theta$	$\cos\theta$	$\tan\theta$	$\csc\theta$	$\sec\theta$	$\cot\theta$
(a)	$4/5$	$3/5$	$4/3$	$5/4$	$5/3$	$3/4$
(b)	$-4/5$	$3/5$	$-4/3$	$-5/4$	$5/3$	$-3/4$
(c)	$1/2$	$-\sqrt{3}/2$	$-1/\sqrt{3}$	2	$-2/\sqrt{3}$	$-\sqrt{3}$
(d)	$-1/2$	$\sqrt{3}/2$	$-1/\sqrt{3}$	-2	$2/\sqrt{3}$	$-\sqrt{3}$
(e)	$1/\sqrt{2}$	$1/\sqrt{2}$	1	$\sqrt{2}$	$\sqrt{2}$	1
(f)	$1/\sqrt{2}$	$-1/\sqrt{2}$	-1	$\sqrt{2}$	$-\sqrt{2}$	-1

17. **(a)** 1.2679 **(b)** 3.5753
19.

	$\sin\theta$	$\cos\theta$	$\tan\theta$	$\csc\theta$	$\sec\theta$	$\cot\theta$
(a)	$a/3$	$\sqrt{9-a^2}/3$	$a/\sqrt{9-a^2}$	$3/a$	$3/\sqrt{9-a^2}$	$\sqrt{9-a^2}/a$
(b)	$a/\sqrt{a^2+25}$	$5/\sqrt{a^2+25}$	$a/5$	$\sqrt{a^2+25}/a$	$\sqrt{a^2+25}/5$	$5/a$
(c)	$\sqrt{a^2-1}/a$	$1/a$	$\sqrt{a^2-1}$	$a/\sqrt{a^2-1}$	a	$1/\sqrt{a^2-1}$

21. **(a)** $3\pi/4 \pm n\pi, n = 0, 1, 2, \ldots$
(b) $\pi/3 \pm 2n\pi$ and $5\pi/3 \pm 2n\pi, n = 0, 1, 2, \ldots$
23. **(a)** $\pi/6 \pm n\pi, n = 0, 1, 2, \ldots$
(b) $4\pi/3 \pm 2n\pi$ and $5\pi/3 \pm 2n\pi, n = 0, 1, 2, \ldots$
25. **(a)** $3\pi/4 \pm n\pi, n = 0, 1, 2, \ldots$
(b) $\pi/6 \pm n\pi, n = 0, 1, 2, \ldots$
27. **(a)** $\pi/3 \pm 2n\pi$ and $2\pi/3 \pm 2n\pi, n = 0, 1, 2, \ldots$
(b) $\pi/6 \pm 2n\pi$ and $11\pi/6 \pm 2n\pi, n = 0, 1, 2, \ldots$
29. $\sin\theta = 2/5$, $\cos\theta = -\sqrt{21}/5$, $\tan\theta = -2/\sqrt{21}$, $\csc\theta = 5/2$, $\sec\theta = -5/\sqrt{21}$, $\cot\theta = -\sqrt{21}/2$
31. **(a)** $\theta = \pm n\pi, n = 0, 1, 2, \ldots$ **(b)** $\theta = \pi/2 \pm n\pi, n = 0, 1, 2, \ldots$ **(c)** $\theta = \pm n\pi, n = 0, 1, 2, \ldots$ **(d)** $\theta = \pm n\pi, n = 0, 1, 2, \ldots$ **(e)** $\theta = \pi/2 \pm n\pi, n = 0, 1, 2, \ldots$
(f) $\theta = \pm n\pi, n = 0, 1, 2, \ldots$
33. **(a)** $2\pi/3$ cm **(b)** $10\pi/3$ cm 35. $\frac{2}{5}$
37. **(a)** $\dfrac{2\pi-\theta}{2\pi}R$ **(b)** $\dfrac{\sqrt{4\pi\theta-\theta^2}}{2\pi}R$ 39. $\frac{21}{4}\sqrt{3}$ 41. 9.2 ft
43. $h = d(\tan\beta - \tan\alpha)$ 45. **(a)** $4\sqrt{5}/9$ **(b)** $-\frac{1}{9}$
47. $\sin 3\theta = 3\sin\theta\cos^2\theta - \sin^3\theta$, $\cos 3\theta = \cos^3\theta - 3\sin^2\theta\cos\theta$
61. **(a)** $\cos\theta$ **(b)** $-\sin\theta$ **(c)** $-\cos\theta$ **(d)** $\sin\theta$
69. **(a)** $153°$ **(b)** $45°$ **(c)** $117°$ **(d)** $89°$ 71. **(a)** $60°$ **(b)** $117°$

▶ Appendix B (Page A20)

1. **(a)** $q(x) = x^2+4x+2, r(x) = -11x+6$
(b) $q(x) = 2x^2+4, r(x) = 9$
(c) $q(x) = x^3-x^2+2x-2, r(x) = 2x+1$
3. **(a)** $q(x) = 3x^2+6x+8, r(x) = 15$
(b) $q(x) = x^3-5x^2+20x-100, r(x) = 504$
(c) $q(x) = x^4+x^3+x^2+x+1, r(x) = 0$
5.

x	0	1	-3	7
$p(x)$	-4	-3	101	5001

7. **(a)** $q(x) = x^2+6x+13, r = 20$ **(b)** $q(x) = x^2+3x-2, r = -4$
9. **(a)** $\pm1, \pm2, \pm3, \pm4, \pm6, \pm8, \pm12, \pm24$
(b) $\pm1, \pm2, \pm5, \pm10, \pm\frac{1}{3}, \pm\frac{2}{3}, \pm\frac{5}{3}, \pm\frac{10}{3}$ **(c)** $\pm1, \pm17$
11. $(x+1)(x-1)(x-2)$ 13. $(x+3)^3(x+1)$
15. $(x+3)(x+2)(x+1)^2(x-3)$ 17. -3 19. $-2, -\frac{2}{3}, -1 \pm \sqrt{3}$
21. $-2, 2, 3$ 23. $2, 5$ 25. 7 cm

PHOTO CREDITS

Chapter 1
Page 1: Courtesy Christopher Evans, The Republican Company.

Chapter 2
Page 84: Joe McBride/Stone/Getty Images.

Chapter 3
Page 146: Roger Ressmeyer/Corbis Images.

Chapter 4
Page 225: Stone/Getty Images.

Chapter 5
Page 302: © Chris Alan Wilton/The Image Bank/Getty Images. Page 309: Reproduced from C.I. Gerhardt, *Briefwechsel von G.W. Leibniz mit Mathematikern* (1899). Page 366: Corbis-Bettmann.

Chapter 6
Page 380: Courtesy NASA. Page 420: Stephen Sutton/Duomo Photography, Inc.

Chapter 7
Page 435: Craig Lovell/Corbis Images. Page 443: Bob Gruen/Star File. Page 499: Glen Allison/Stone/Getty Images.

Chapter 8
Page 514: © AP/Wide World Photos.

Chapter 9
Page 586: Photo by Milton Bell, Texas Archeological Research Laboratory, University of Texas at Austin. Page 611: Patrick Mesner/Liaison Agency, Inc./Getty Images.

Chapter 10
Page 628: Taxi/Getty Images.

Chapter 11
Page 721: Dwight R. Kuhn. Page 730 (left): Thomas Taylor/Photo Researchers. Page 730 (center): Rex Ziak/Stone/Getty Images. Page 730 (right): Courtesy NASA & The Hubble Heritage Team. Page 763: John Mead/Science Photo Library/Photo Researchers. Page 781: Science Photo Library/Photo Researchers.

Chapter 12
Page 790: Craig Aurness/Corbis Images. Page 824: World Perspectives/Stone/Getty Images.

Chapter 13
Page 863: Courtesy Cedar Point. Page 864: Ken Eward/Biografx/Photo Researchers.

Chapter 14
Page 928: Stone/Getty Images. Page 958: Leverett Bradley/Stone/Getty Images.

Chapter 15
Page 1022: Stone/Getty Images.

Chapter 16
Page 1106: Images and animation produced by Hal Pierce, Laboratory for Atmospheres, NASA Goddard Space and Flight Center.

INDEX

RATIONAL FUNCTIONS CONTAINING POWERS OF $a + bu$ IN THE DENOMINATOR

60. $\int \frac{u\,du}{a+bu} = \frac{1}{b^2}[bu - a\ln|a+bu|] + C$

61. $\int \frac{u^2\,du}{a+bu} = \frac{1}{b^3}\left[\frac{1}{2}(a+bu)^2 - 2a(a+bu) + a^2\ln|a+bu|\right] + C$

62. $\int \frac{u\,du}{(a+bu)^2} = \frac{1}{b^2}\left[\frac{a}{a+bu} + \ln|a+bu|\right] + C$

63. $\int \frac{u^2\,du}{(a+bu)^2} = \frac{1}{b^3}\left[bu - \frac{a^2}{a+bu} - 2a\ln|a+bu|\right] + C$

64. $\int \frac{u\,du}{(a+bu)^3} = \frac{1}{b^2}\left[\frac{a}{2(a+bu)^2} - \frac{1}{a+bu}\right] + C$

65. $\int \frac{du}{u(a+bu)} = \frac{1}{a}\ln\left|\frac{u}{a+bu}\right| + C$

66. $\int \frac{du}{u^2(a+bu)} = -\frac{1}{au} + \frac{b}{a^2}\ln\left|\frac{a+bu}{u}\right| + C$

67. $\int \frac{du}{u(a+bu)^2} = \frac{1}{a(a+bu)} + \frac{1}{a^2}\ln\left|\frac{u}{a+bu}\right| + C$

RATIONAL FUNCTIONS CONTAINING $a^2 \pm u^2$ IN THE DENOMINATOR ($a > 0$)

68. $\int \frac{du}{a^2+u^2} = \frac{1}{a}\tan^{-1}\frac{u}{a} + C$

69. $\int \frac{du}{a^2-u^2} = \frac{1}{2a}\ln\left|\frac{u+a}{u-a}\right| + C$

70. $\int \frac{du}{u^2-a^2} = \frac{1}{2a}\ln\left|\frac{u-a}{u+a}\right| + C$

71. $\int \frac{bu+c}{a^2+u^2}\,du = \frac{b}{2}\ln(a^2+u^2) + \frac{c}{a}\tan^{-1}\frac{u}{a} + C$

INTEGRALS OF $\sqrt{a^2+u^2}$, $\sqrt{a^2-u^2}$, $\sqrt{u^2-a^2}$ AND THEIR RECIPROCALS ($a > 0$)

72. $\int \sqrt{u^2+a^2}\,du = \frac{u}{2}\sqrt{u^2+a^2} + \frac{a^2}{2}\ln(u+\sqrt{u^2+a^2}) + C$

73. $\int \sqrt{u^2-a^2}\,du = \frac{u}{2}\sqrt{u^2-a^2} - \frac{a^2}{2}\ln|u+\sqrt{u^2-a^2}| + C$

74. $\int \sqrt{a^2-u^2}\,du = \frac{u}{2}\sqrt{a^2-u^2} + \frac{a^2}{2}\sin^{-1}\frac{u}{a} + C$

75. $\int \frac{du}{\sqrt{u^2+a^2}} = \ln(u+\sqrt{u^2+a^2}) + C$

76. $\int \frac{du}{\sqrt{u^2-a^2}} = \ln|u+\sqrt{u^2-a^2}| + C$

77. $\int \frac{du}{\sqrt{a^2-u^2}} = \sin^{-1}\frac{u}{a} + C$

POWERS OF u MULTIPLYING OR DIVIDING $\sqrt{a^2-u^2}$ OR ITS RECIPROCAL

78. $\int u^2\sqrt{a^2-u^2}\,du = \frac{u}{8}(2u^2-a^2)\sqrt{a^2-u^2} + \frac{a^4}{8}\sin^{-1}\frac{u}{a} + C$

79. $\int \frac{\sqrt{a^2-u^2}\,du}{u} = \sqrt{a^2-u^2} - a\ln\left|\frac{a+\sqrt{a^2-u^2}}{u}\right| + C$

80. $\int \frac{\sqrt{a^2-u^2}\,du}{u^2} = -\frac{\sqrt{a^2-u^2}}{u} - \sin^{-1}\frac{u}{a} + C$

81. $\int \frac{u^2\,du}{\sqrt{a^2-u^2}} = -\frac{u}{2}\sqrt{a^2-u^2} + \frac{a^2}{2}\sin^{-1}\frac{u}{a} + C$

82. $\int \frac{du}{u\sqrt{a^2-u^2}} = -\frac{1}{a}\ln\left|\frac{a+\sqrt{a^2-u^2}}{u}\right| + C$

83. $\int \frac{du}{u^2\sqrt{a^2-u^2}} = -\frac{\sqrt{a^2-u^2}}{a^2u} + C$

POWERS OF u MULTIPLYING OR DIVIDING $\sqrt{u^2 \pm a^2}$ OR THEIR RECIPROCALS

84. $\int u\sqrt{u^2+a^2}\,du = \frac{1}{3}(u^2+a^2)^{3/2} + C$

85. $\int u\sqrt{u^2-a^2}\,du = \frac{1}{3}(u^2-a^2)^{3/2} + C$

86. $\int \frac{du}{u\sqrt{u^2+a^2}} = -\frac{1}{a}\ln\left|\frac{a+\sqrt{u^2+a^2}}{u}\right| + C$

87. $\int \frac{du}{u\sqrt{u^2-a^2}} = \frac{1}{a}\sec^{-1}\left|\frac{u}{a}\right| + C$

88. $\int \frac{\sqrt{u^2-a^2}\,du}{u} = \sqrt{u^2-a^2} - a\sec^{-1}\left|\frac{u}{a}\right| + C$

89. $\int \frac{\sqrt{u^2+a^2}\,du}{u} = \sqrt{u^2+a^2} - a\ln\left|\frac{a+\sqrt{u^2+a^2}}{u}\right| + C$

90. $\int \frac{du}{u^2\sqrt{u^2\pm a^2}} = \mp\frac{\sqrt{u^2\pm a^2}}{a^2u} + C$

91. $\int u^2\sqrt{u^2+a^2}\,du = \frac{u}{8}(2u^2+a^2)\sqrt{u^2+a^2} - \frac{a^4}{8}\ln(u+\sqrt{u^2+a^2}) + C$

92. $\int u^2\sqrt{u^2-a^2}\,du = \frac{u}{8}(2u^2-a^2)\sqrt{u^2-a^2} - \frac{a^4}{8}\ln|u+\sqrt{u^2-a^2}| + C$

93. $\int \frac{\sqrt{u^2+a^2}}{u^2}\,du = -\frac{\sqrt{u^2+a^2}}{u} + \ln(u+\sqrt{u^2+a^2}) + C$

94. $\int \frac{\sqrt{u^2-a^2}}{u^2}\,du = -\frac{\sqrt{u^2-a^2}}{u} + \ln|u+\sqrt{u^2-a^2}| + C$

95. $\int \frac{u^2}{\sqrt{u^2+a^2}}\,du = \frac{u}{2}\sqrt{u^2+a^2} - \frac{a^2}{2}\ln(u+\sqrt{u^2+a^2}) + C$

96. $\int \frac{u^2}{\sqrt{u^2-a^2}}\,du = \frac{u}{2}\sqrt{u^2-a^2} + \frac{a^2}{2}\ln|u+\sqrt{u^2-a^2}| + C$

INTEGRALS CONTAINING $(a^2+u^2)^{3/2}$, $(a^2-u^2)^{3/2}$, $(u^2-a^2)^{3/2}$ ($a > 0$)

97. $\int \frac{du}{(a^2-u^2)^{3/2}} = \frac{u}{a^2\sqrt{a^2-u^2}} + C$

98. $\int \frac{du}{(u^2\pm a^2)^{3/2}} = \pm\frac{u}{a^2\sqrt{u^2\pm a^2}} + C$

99. $\int (a^2-u^2)^{3/2}\,du = -\frac{u}{8}(2u^2-5a^2)\sqrt{a^2-u^2} + \frac{3a^4}{8}\sin^{-1}\frac{u}{a} + C$

100. $\int (u^2+a^2)^{3/2}\,du = \frac{u}{8}(2u^2+5a^2)\sqrt{u^2+a^2} + \frac{3a^4}{8}\ln(u+\sqrt{u^2+a^2}) + C$

101. $\int (u^2-a^2)^{3/2}\,du = \frac{u}{8}(2u^2-5a^2)\sqrt{u^2-a^2} + \frac{3a^4}{8}\ln|u+\sqrt{u^2-a^2}| + C$